Integrated
SCIENCE

Fourth Edition

Bill W. Tillery
Arizona State University

Eldon D. Enger
Delta College

Frederick C. Ross
Delta College

McGraw-Hill
Higher Education

Boston   Burr Ridge, IL   Dubuque, IA   New York   San Francisco   St. Louis
Bangkok   Bogotá   Caracas   Kuala Lumpur   Lisbon   London   Madrid   Mexico City
Milan   Montreal   New Delhi   Santiago   Seoul   Singapore   Sydney   Taipei   Toronto

McGraw-Hill Higher Education

INTEGRATED SCIENCE, FOURTH EDITION

 This book is printed on recycled, acid-free paper containing 10% postconsumer waste.

1 2 3 4 5 6 7 8 9 0 DOW/DOW 0 9 8 7

ISBN 978–0–07–340448–6
MHID 0–07–340448–9

Publisher: *Thomas D. Timp*
Sponsoring Editor: *Debra B. Hash*
Senior Developmental Editor: *Mary E. Hurley*
Senior Marketing Manager: *Lisa Nicks*
Project Manager: *April R. Southwood*
Senior Production Supervisor: *Sherry L. Kane*
Lead Media Project Manager: *Judi David*
Media Producer: *Daniel M. Wallace*
Designer: *Laurie B. Janssen*
(USE) Cover Image: *Masterfile © R. Ian Lloyd*
Senior Photo Research Coordinator: *John C. Leland*
Photo Research: *Editorial Image, LLC*
Supplement Producer: *Melissa M. Leick*
Compositor: *Laserwords Private Limited*
Typeface: 10/12 *Minion*
Printer: *R. R. Donnelley Willard, OH*

The credits section for this book begins on page 733 and is considered an extension of the copyright page.

Library of Congress Cataloging-in-Publication Data

Tillery, Bill W.
   Integrated science / Bill W. Tillery, Eldon D. Enger, Frederick C. Ross. – 4th ed.
      p. cm.
   Includes index.
   ISBN–13 978-0-07-340448-6 — ISBN 0-07-340448-9 (hard copy : alk. paper)
      1. Science—Textbooks. I. Enger, Eldon D. II. Ross, Frederick C. III. Title.
Q161.2.T54 2008
500—dc22
                                                      2007019177

www.mhhe.com

Contents

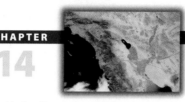

CHAPTER 26

Mendelian and Molecular Genetics 655

Preface

WHAT SETS THIS BOOK APART?

Creating Informed Citizens

Integrated Science is a straightforward, easy-to-read, but substantial introduction to the fundamental behavior of matter and energy in living and nonliving systems. It is intended to serve the needs of nonscience majors who must complete one or more science courses as part of a general or basic studies requirement.

Integrated Science provides an introduction to a scientific way of thinking as it introduces fundamental scientific concepts, often in historical context. Several features of the text provide opportunities for students to experience the methods of science by evaluating situations from a scientific point of view. While technical language and mathematics are important in developing an understanding of science, only the language and mathematics needed to develop central concepts are used. No prior work in science is assumed.

Many features, such as Science and Society readings, as well as basic discussions of the different branches of science help students understand how the branches relate. This allows students to develop an appreciation of the major developments in science and an ability to act as informed citizens on matters that involve science and public policy.

> *"I especially like the application of the concepts and the connections in this text. We try very hard to show that science has connections to the everyday world and why it's important to see those connections. I don't think science can be taught to nonscience majors unless this type of approach is taken."*
>
> —Richard L. Kopec, St. Edward's University

Flexible Organization

The *Integrated Science* sequence of chapters is flexible, and the instructor can determine topic sequence and depth of coverage as needed. The materials are also designed to support a conceptual approach or a combined conceptual and problem-solving approach. The *Integrated Science* ARIS Instructor's Resources offer suggestions for integrating the text's topics around theme options. With laboratory studies, the text contains enough material for the instructor to select a sequence for a one- or two-semester course.

THE GOALS OF *INTEGRATED SCIENCE*

1. **Create an introductory science course aimed at the nonscience major.** The origin of this book is rooted in our concern for the education of introductory-level students in the field of science. Historically, nonscience majors had to

enroll in courses intended for science or science-related majors such as premeds, architects, or engineers. Such courses are important for these majors but are mostly inappropriate for introductory-level nonscience students. To put a nonscience student in such a course is a mistake. Few students will have the time or background to move through the facts, equations, and specialized language to gain any significant insights into the logic or fundamental understandings; instead, they will leave the course with a distaste for science. Today, society has a great need for a few technically trained people but a much larger need for individuals who understand the process of science and its core concepts.

2. **Introduce a course that presents a coherent and clear picture of all science disciplines through an interdisciplinary approach.** Recent studies and position papers have called for an interdisciplinary approach to teaching science to nonmajors. For example, the need is discussed in *Science for All Americans—Project 2061* (American Association for the Advancement of Science), *National Science Education Standards* (National Research Council, 1994), and *Science in the National Interest* (White House, 1994). Interdisciplinary science is an attempt to broaden and humanize science education by reducing and breaking down the barriers that enclose traditional science disciplines as distinct subjects.

> *"The authors obviously feel that emphasizing the interconnectedness of nature should be studied by integrating all of the sciences into a coherent, understandable network of facts, concepts, and interpretations that lead the student to view the universe in a new and enlightened perspective. This philosophy is particularly important in the education of nonscience majors who may never again formally study science."*
>
> —Jay R. Yett, Orange Coast College

3. **Help instructors build their own mix of descriptive and analytical aspects of science, arousing student interest and feelings as they help students reach the educational goals of their particular course.** The spirit of interdisciplinary science is sometimes found in courses called "General Science," "Combined Science," or "Integrated Science." These courses draw concepts from a wide range of the traditional fields of science but are not concentrated around certain problems or questions. For example, rather than just dealing with the physics of energy, an interdisciplinary approach might consider broad aspects of energy—dealing with potential problems of an energy crisis—including social and ethical issues. A number of approaches can be used in interdisciplinary science, including the teaching of science in a *social, historical, philosophical,* or *problem-solving* context, but there is no single best approach. One of the characteristics of interdisciplinary science is that it is not constrained by the necessity of teaching certain facts or by traditions. It likewise cannot be imposed as a formal discipline, with certain facts to be learned. It is justified by its success in attracting and holding the attention and interest of students, making them a little wiser as they make their way toward various careers and callings.

4. **Humanize science for nonscience majors.** Each chapter presents historical background where appropriate, uses everyday examples in developing concepts, and follows a logical flow of presentation. A discussion of the people and events involved in the development of scientific concepts puts a human face on the process of science. The use of everyday examples appeals to the nonscience major, typically accustomed to reading narration, not scientific technical writing, and also tends to bring relevancy to the material being presented. The logical flow of presentation is helpful to students not accustomed to thinking about relationships between what is being read and previous knowledge learned, a useful skill in understanding the sciences.

VALUED INPUT WENT INTO STRIVING TO MEET YOUR NEEDS

Text development today involves a team that includes authors and publishers and valuable input from instructors who share their knowledge and experience with publishers and authors through reviews and focus groups. Such feedback has shaped this edition, resulting in reorganization of existing content and expanded coverage in key areas. This text has continued to evolve as a result of feedback from instructors actually teaching integrated science courses in the classroom. Reviewers point out that current and accurate content, a clear writing style with concise explanations, quality illustrations, and dynamic presentation materials are important factors considered when evaluating textbooks. Those criteria have guided the revision of the *Integrated Science* text and the development of its ancillary resources.

New to This Edition

- A new biographical element, People Behind the Science, has been added to this edition to humanize the sciences and to make them even more relevant to students' lives.
- In addition, a number of organizational changes have been made and new topic areas added to the text:

Chapter 1 What Is Science?: The section on pseudoscience has been revised and expanded, including a new Science and Society boxed reading on herbal medicine, legislation, and pseudoscience. A new biographical People Behind the Science box on Florence Bascom also appears.

Chapter 2 Motion: An overview of the three laws of motion has been added, along with new sections on earth satellites and weightlessness. Two new boxed elements, People Behind the Science: Isaac Newton and A Closer Look: Gravity Problems, also appear.

Chapter 3 Energy: A new section on energy conservation and a biographical People Behind the Science box on James Prescott Joule have been added.

Chapter 4 Heat and Temperature: A biographical People Behind the Science box on Count Rumford (Benjamin Thompson) has been added.

Chapter 5 Wave Motions and Sound: A biographical People Behind the Science box on Johann Christian Doppler is now included.

Chapter 6 Electricity: A biographical People Behind the Science box on Benjamin Franklin has been added.

Chapter 7 Light: All new sections on special relativity and general relativity are now included, along with People Behind the Science: James Clerk Maxwell.

Chapter 8 Atoms and Periodic Properties: People Behind the Science: Dmitri Ivanovich Mendeleyev has been added.

Chapter 9 Chemical Reactions: People Behind the Science: Linus Carl Pauling has been added.

Chapter 10 Water and Solutions: People Behind the Science: Johannes Nicolaus Brönsted is now included.

Chapter 11 Nuclear Reactions: People Behind the Science: Marie Curie has been added.

Chapter 12 The Universe: Information on the big bang theory has been revised and expanded, and this chapter now includes People Behind the Science: Jocelyn Bell Burnell.

Chapter 13 The Solar System: Updated information on Pluto, new information on dwarf planets, and a discussion of new definitions of planets have been added. People Behind the Science: Percival Lowell is also included.

Chapter 14 Earth in Space: People Behind the Science: Carl Edward Sagan has been added.

Chapter 15 Earth: People Behind the Science: Frederick John Vine has been added.

Chapter 16 Earth's Surface: People Behind the Science: James Hutton is now included.

Chapter 17 Earth's Weather: New information on Hurricane Katrina and People Behind the Science: Vilhelm Firman Koren Bjerknes are now included.

Chapter 18 Earth's Waters: This chapter now includes a new Closer Look on rogue waves and People Behind the Science: Rachel Louise Carson.

Chapter 19 Organic and Biochemistry: New Closer Look on enzymes, Connections on waxes, and People Behind the Science: Friedrich Wöhler boxed readings have been added.

Chapter 20 The Nature of Living Things: The introductions to the What Makes Something Alive? and Respiration and Photosynthesis sections have been revised, and People Behind the Science: Matthias Jakob Schleiden and Theodor Schwann has been added.

Chapter 21 The Origin and Evolution of Life: Revised and new information has been added to the Major Events in the Early Evolution of Living Things and Evolution and Natural Selection sections. There is a new subsection on defining evolution. Revised sections on Genetic Diversity Resulting from Sexual Reproduction and Acquired Characteristics Do Not Influence Natural Selection; People Behind the Science: Ernst Mayr; and additional background information in the section on the History of the Development of Evolutionary Thought are also included. New subsections titles and figures have been added to further assist the reader.

Chapter 22 The History of Life on Earth: There are several new subsection headings and six new figures to assist the reader. This chapter also contains a considerable amount of revised and expanded information:

- New Mass Extinctions section
- Revised and new information in the Classifications of Organisms section, with new subsections on the Problem with Common Names, Binomial System of Nomenclature, and Organizing Species into Logical Groups
- Revised and new information on the Phylogeny section, with a new table on the Classification of Humans
- Revised and new information in the Domains Eubacteria and Archaea section
- Revised and new information in the Acellular Infectious Particles section
- Revised new information on the viruses section, with new subsections on How Did Viruses Originate? and How Viruses Cause Disease
- Revised and new information in the Prions: Infectious Proteins section with a new subsection on How Prions Cause Disease (such as mad cow disease)
- New Closer Look on Cladistics—A Tool for Taxonomy and Phylogeny
- People Behind the Science: Lynn (Alexander) Margulis

Chapter 23 Ecology and Environment: Revised and new information has been added in the Community Interactions section. New information and subsections on Mediterranean Shrublands (Chaparral) and Tropical Dry Forest have also been added. This chapter now includes a definition and discussion of *symbiosis* in the Mutualism section. Two new boxes, People Behind the Science: Eugene Odum and A Closer Look: Zebra Mussels: Invaders from Europe, are also included.

Chapter 24 Human Biology: Materials Exchange and Control Mechanisms: There are several new subsection headings and one new figure to assist the reader. Revised and new information has been added to the Exchanging Materials: Basic Principles; Guidelines for Obtaining Adequate Nutrients; and Eating Disorders sections. The section on Obtaining Nutrients: The Digestive System has been reorganized. This chapter also now includes People Behind the Science: William Beaumont.

Chapter 25 Human Biology: Reproduction: The information on cryptorchidism has been placed in A Closer Look box, and People Behind the Science: Robert Geoffrey Edwards and Patrick Christopher Steptoe has been added.

Chapter 26 Mendelian and Molecular Genetics: A Closer Look on Cystic Fibrosis—What Is the Probability? has been revised, and People Behind the Science: Gregor Johann Mendel has been added.

THE LEARNING SYSTEM

To achieve the goals stated, this text includes a variety of features that should make student's study of *Integrated Science* more effective and enjoyable. These aids are included to help you clearly understand the concepts and principles that serve as the foundation of the integrated sciences.

OVERVIEW TO INTEGRATED SCIENCE

Chapter 1 provides an overview or orientation to integrated science in general and this text in particular. It also describes the fundamental methods and techniques used by scientists to study and understand the world around us.

MULTIDISCIPLINARY APPROACH

Chapter Outlines

The chapter outline includes all the major topic headings and subheadings within the body of the chapter. It gives students a quick glimpse of the chapter's contents and helps them locate sections dealing with particular topics.

Core Concept Map

The concept map identifies a core idea for the chapter and shows how the topics in the chapter are related to this core idea. It also outlines that idea's relationship to other science disciplines throughout the text. The core concept map, combined with the chapter outline and overview, help students to see the big picture of the chapter content and the even bigger picture of how that content relates to other science discipline areas.

Chapter Overviews

Each chapter begins with an introductory overview. The overview previews the chapter's contents and what students can

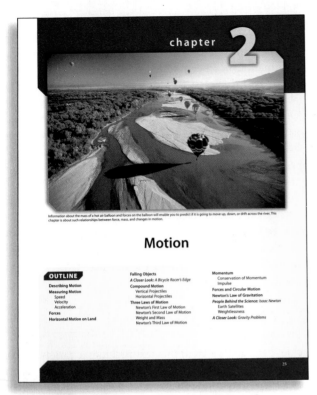

chapter 2

Information about the mass of a hot air balloon and forces on the balloon will enable you to predict if it is going to move up, down, or drift across the river. This chapter is about such relationships between force, mass, and changes in motion.

Motion

25

Physics Connections
▶ Mechanical work is the product of a force and the distance an object moves as a result of the force (Ch. 3).
▶ Resonance occurs when the frequency of an applied force matches a natural frequency (Ch. 5).

Life Science Connections
▶ Biomechanics is the application of principles of motion to living things (Ch. 20).

CORE CONCEPT

A net force is required for any change in a state of motion.

Inertia is the tendency of an object to remain in unchanging motion when the net force is zero (p. 33).

The force of gravity uniformly accelerates falling objects (p. 35).

Every object retains its state of rest or straight-line motion unless acted upon by an unbalanced force (p. 38).

All objects in the universe are attracted to all other objects in the universe (p. 47).

Astronomy Connections
▶ Gravity pulls clouds of gas together in space to form stars (Ch. 12).
▶ The solar system may have formed when gas, dust, and elements from a previously existing star were pulled together by gravity into a large disk (Ch. 12).

Earth Science Connections
▶ Earth's surface is made up of large, rigid plates that move from applied forces (Ch. 15).

OVERVIEW

In chapter 1, you learned some "tools and rules" and some techniques for finding order in your surroundings. Order is often found in the form of patterns, or relationships between quantities that are expressed as equations. Equations can be used to (1) describe properties, (2) define concepts, and (3) describe how quantities change relative to one another. In all three uses, patterns are quantified, conceptualized, and used to gain a general understanding about what is happening in nature.

In the study of science, certain parts of nature are often considered and studied together for convenience. One of the more obvious groupings involves *movement*. Most objects around you appear to spend a great deal of time sitting quietly without

motion. Buildings, rocks, utility poles, and trees rarely, if ever, move from one place to another. Even things that do move from time to time still for a great deal of time. This includes you, automobiles, and bicycles (figure 2.1). On the other hand, the Sun, the Moon, and starry heavens always seem to move, never standing still. Why do things stand still? Why do things move?

Questions about motion have captured the attention of people for thousands of years. But the ancient people answered questions about motion with stories of mysticism and spirits that lived in objects. It was during the classic Greek culture, between 600 B.C. and 300 B.C., that people began to look beyond myths and spirits. One particular Greek philosopher, Aristotle, wrote a theory

about the universe that offered not only explanations about things such as motion but also offered a sense of beauty, order, and perfection. The theory seemed to fit with other ideas that people had and was held to be correct for nearly two thousand years after it was written. It was not until the work of Galileo and Newton during the 1600s that a new, correct understanding about motion was developed. The development of ideas about motion is an amazing and absorbing story. You will learn in this chapter how to describe and use some properties of motion. This will provide some basic understandings about motion and will be very helpful in understanding some important aspects of astronomy and the earth sciences, as well as the movement of living things.

DESCRIBING MOTION

Motion is one of the more common events in your surroundings. You can see motion in natural events such as clouds moving, rain and snow falling, and streams of water, all moving in a never-ending cycle. Motion can also be seen in the activities of people who walk, jog, or drive various machines from place to place. Motion is so common that you would think everyone would intuitively understand the concepts of motion, but history indicates that it was only during the past three hundred years or so that people began to understand motion correctly. Perhaps the correct concepts are subtle and contrary to common sense, requiring a search for simple, clear concepts in an otherwise complex situation. The process of finding such order in a multitude of sensory impressions by taking measurable data and then inventing a concept to

describe what is happening is the activity called *science*. We will now apply this process to motion.

What is motion? Consider a ball that you notice one morning in the middle of a lawn. Later in the afternoon, you notice that the ball is at the edge of the lawn, against a fence, and you wonder if the wind or some person moved the ball. You do not know if the wind blew it at a steady rate, if many gusts of wind moved it, or even if some children kicked it all over the yard. All you know for sure is that the ball has been moved because it is in a different position after some time passed. These are the two important aspects of motion: (1) a change of position and (2) the passage of time.

If you did happen to see the ball rolling across the lawn in the wind, you would see more than the ball at just two locations. You would see the ball moving continuously. You could consider, however, the ball in continuous motion to be a series of individual locations with very small time intervals. Moving

26   Chapter 2   Motion

expect to learn from reading the chapter. It adds to the general outline of the chapter by introducing students to the concepts to be covered. It also expands upon the core concept map, facilitating in the integration of topics. Finally, the overview will help students to stay focused and organized while reading the chapter for the first time. After reading this introduction, students should browse through the chapter, paying particular attention to the topic headings and illustrations so that they get a feel for the kinds of ideas included within the chapter.

APPLYING SCIENCE TO THE REAL WORLD

Concepts Applied

As students look through each chapter, they will find one or more Concepts Applied boxes. These activities are simple exercises that students can perform at home or in the classroom to demonstrate important concepts and reinforce their understanding of them. This feature also describes the application of those concepts to their everyday lives.

Science and Society

These readings relate the chapter's content to current societal issues. Many of these boxes also include Questions to Discuss that provide students an opportunity to discuss issues with their peers.

Myths, Mistakes, and Misunderstandings

These brief boxes provide short, scientific explanations to dispel a societal myth or a home experiment or project that enables students to dispel the myth on their own.

People Behind the Science

NEW! Many chapters also have one or two fascinating biographies that spotlight well-known scientists, past and present. From these People Behind the Science biographies, students learn about the human side of science: science is indeed relevant, and real people do the research and make the discoveries. These readings present the sciences in real-life terms that students can identify with and understand.

Science & Society

Transportation and the Environment

Environmental science is an interdisciplinary study of Earth's environment. The concern of this study is the overall problem of human degradation of the environment and remedies for that damage. As an example of an environmental topic of study, consider the damage that results from current human activities involving the use of transportation. Researchers estimate that overall transportation activities are responsible for about one-third of the total U.S. carbon emissions that are added to the air every day. Carbon emissions are a problem because they are directly harmful in the form of carbon monoxide. They are also indirectly harmful because of the contribution of carbon dioxide to global

warming and the consequences of climate change.

Here is a list of things that people might do to reduce the amount of environmental damage from transportation:

A. Use a bike, carpool, walk, or take public transportation wherever possible.
B. Plan to combine trips to the store, mall, and work, leaving the car parked whenever possible.
C. Purchase hybrid electric or fuel cell–powered cars and vehicles whenever possible.
D. Move to a planned community that makes the use of cars less necessary and less desirable.

Questions to Discuss

Discuss with your group the following questions concerning connections between thought and feeling:

1. What are your positive or negative feelings associated with each item in the list?
2. Would you feel differently if you had a better understanding of the global problem?
3. Do your feelings mean that you have reached a conclusion?
4. What new items could be added to the list?

The relationship between forces and a change of motion is obvious in many everyday situations. When a car, bus, or plane in which you are riding starts moving, you feel a force on your back. Likewise, you feel a force on the bottoms of your feet when an elevator you are in starts moving upward. On the other hand, you seem to be forced toward the dashboard if a car stops quickly, and it feels as if the floor pulls away from your feet when an elevator drops rapidly. These examples all involve patterns between forces and motion, patterns that can be quantified, conceptualized, and used to answer questions about why things move or stand still. These patterns are the subject of Newton's three laws of motion.

Newton's First Law of Motion

Newton's first law of motion is also known as the *law of inertia* and is very similar to one of Galileo's findings about motion. Recall that Galileo used the term *inertia* to describe the tendency of an object to resist changes in motion. Newton's first law describes this tendency more directly. In modern terms (not Newton's words), the **first law of motion** is as follows:

    Every object retains its state of rest or its state of uniform
    straight-line motion unless acted upon by an unbalanced
    force.

This means that an object at rest will remain at rest unless it is put into motion by an unbalanced force; that is, the net force must be greater than zero. Likewise, an object moving with uniform straight-line motion will retain that motion unless a net force causes it to speed up, slow down, or change its direction of travel. Thus, Newton's first law describes the tendency of an object to resist *any* change in its state of motion.

Think of Newton's first law of motion when you ride standing in the aisle of a bus. The bus begins to move, and you, being an independent mass, tend to remain at rest. You take a few steps back as you tend to maintain your position relative to the

FIGURE 2.14   Without a doubt, this baseball player is aware of the relationship between the projection angle and the maximum distance acquired for a given projection velocity.

38   Chapter 2   Motion

Closer Look and Connections

Each chapter of *Integrated Science* also includes one or more **Closer Look** readings that discuss topics of special human or environmental concern, topics concerning interesting technological applications, or topics on the cutting edge of scientific research. These readings enhance the learning experience by taking a more detailed look at related topics and adding concrete examples to help students better appreciate the real-world applications of science.

In addition to the **Closer Look** readings, each chapter contains concrete interdisciplinary **Connections** that are highlighted. **Connections** will help students better appreciate the interdisciplinary nature of the sciences. The **Closer Look** and **Connections** readings are informative materials that are supplementary in nature. These boxed features highlight valuable information beyond the scope of the text and relate intrinsic concepts discussed to real-world issues, underscoring the relevance of integrated science in confronting the many issues we face in our day-to-day lives. They are identified with the following icons:

"A Closer Look: The Compact Disc was, again, an excellent application of optics to everyday life and to something modern students thrive on—CDs and DVDs."
—Treasure Brasher, West Texas A&M University

"Connections—wonderful!!!. . . . . . A Closer Look . . . excellent. Clear, interesting, good figures. You have presented crucial information in a straightforward and uncompromising way."
—Megan M. Hoffman, Berea College

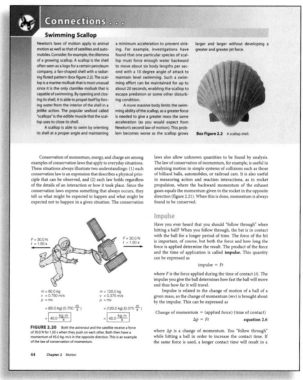

General: This icon identifies interdisciplinary topics that cross over several categories; for example, life sciences and technology.

Life: This icon identifies interdisciplinary life science topics, meaning connections concerning all living organisms collectively: plant life, animal life, marine life, and any other classification of life.

Technology: This icon identifies interdisciplinary technology topics, that is, connections concerned with the application of science for the comfort and well being of people, especially through industrial and commercial means.

 Measurement, Thinking, Scientific Methods: This icon identifies interdisciplinary concepts and understandings concerned with people trying to make sense out of their surroundings by making observations, measuring, thinking, developing explanations for what is observed, and experimenting to test those explanations.

 *Environmental Science:* This icon identifies interdisciplinary concepts and understandings about the problems caused by human use of the natural world and remedies for those problems.

END-OF-CHAPTER FEATURES

At the end of each chapter are the following materials:

- *Summary:* highlights the key elements of the chapter
- *Summary of Equations* (chapters 1–9, 11): highlights the key equations to reinforce retention of them
- *Key Terms:* page-referenced where students will find the terms defined in context
- *Applying the Concepts:* a multiple choice quiz to test students' comprehension of the material covered
- *Questions for Thought:* designed to challenge students to demonstrate their understandings of the topic
- *For Further Analysis:* exercises include analysis or discussion questions, independent investigations, and activities

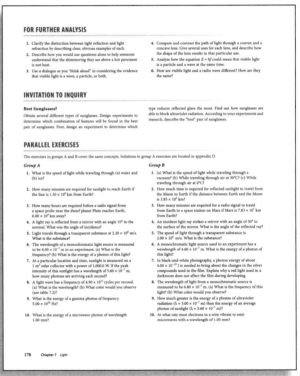

intended to emphasize critical thinking skills and societal issues, and develop a deeper understanding of the chapter content.

- *Invitation to Inquiry:* exercises that consist of short, open-ended activities that allow students to apply investigative skills to the material in the chapter.
- *Parallel Exercises* (chapters 1–9, 11): There are two groups of parallel exercises, Group A and Group B. The Group A parallel exercises have complete solutions worked out, along

with useful comments in appendix D. The Group B parallel exercises are similar to those in Group A but do not contain answers in the text. By working through the Group A parallel exercises and checking the solution in appendix D, students will gain confidence in tackling the parallel exercises in Group B and thus reinforce their problem-solving skills.

"I like the key terms with the page numbers with each one. I always like to see more conceptual- and synthesis-type questions, which is why I like the 'Questions for Thought' and 'For Further Analysis' parts. . . . Exercises such as 'Questions for Thought' number 7, having students think about why oxygen is in Earth's atmosphere but not in Venus or Mars' atmosphere, is a valuable sort of question, because it requires students to know something and apply it."

—Jim Hamm, Big Bend Community College

END-OF-TEXT MATERIAL

At the back of the text are appendices that give additional background details, charts, and answers to chapter exercises. There are also a glossary of all key terms, an index organized alphabetically by subject matter, and special tables printed on the inside covers for reference use.

". . . many books addressing similar disciplines have a tendency to talk over a student's head, making a student frustrated further in a class they do not want to be attending. . . . Personally, I would admit that Integrated Science *has a slight edge. The glossary seems up-to-date and centers in on words many nonscience majors may not understand."*

—David J. DiMattio, St. Bonaventure University

Glossary

Appendix A

Mathematical Review

MULTIMEDIA SUPPLEMENTS

McGraw-Hill's ARIS—Assessment, Review, and Instruction System

McGraw-Hill's ARIS for *Integrated Science* is a complete, online electronic homework and course management system designed for greater ease of use than any other system available. Created specifically for *Integrated Science*, fourth edition text, ARIS allows instructors to create and share course materials and assignments with colleagues with a few clicks of the mouse. For instructors, an instructor's manual, personal response system questions, all PowerPoint lectures, and assignable content are directly tied to text-specific materials in *Integrated Science*. Instructors can also edit questions, import their own content, and create announcements and due dates for assignments. ARIS has automatic grading and reporting of easy-to-assign homework, quizzing, and testing. All student activity within McGraw-Hill's ARIS is automatically recorded and available to the instructor through a fully integrated grade book that can be downloaded to Excel. For students, there are multiple-choice quizzes, author quizzes, crossword puzzles, animations, and even more materials that may be used for self-study or in combination with assigned materials.

ARIS Presentation Center

The ARIS Presentation Center is the ultimate multimedia resource center for the instructor. Located within McGraw-Hill's ARIS—Assessment, Review, and Instruction System, this online site allows instructors to utilize graphics from the textbook and provides the option to search by topic within all other McGraw-Hill ARIS Presentation Center sites. Graphics are available in electronic format to create customized classroom presentations, visually based tests and quizzes, dynamic course website content, or attractive printed support materials.

The following assets are available in digital formats, grouped by chapter:

- **Art and Photo Library:** Full-color digital files of all of the illustrations and many of the photos in the text can be readily incorporated into lecture presentations, exams, or custom-made classroom materials.
- **Worked Example Library and Table Library:** Access the worked examples and tables from the text in electronic format for inclusion in your classroom resources.
- **Animations Library:** Files of animations and videos covering various topics are included so that you can easily make use of these animations in a lecture or classroom setting.
- **Lecture Outlines:** Lecture notes, incorporating illustrations and animated images, have been written to the fourth edition text. They are provided in PowerPoint format so that you may use these lectures as written or customize them to fit your lecture.

Classroom Performance System

The **Classroom Performance System** (CPS) by eInstruction brings interactivity into the classroom or lecture hall. It is a wireless response system that gives the instructor and students immediate feedback from the entire class. The wireless response pads are essentially remotes that are easy to use and engage students. CPS allows instructors to motivate student preparation, interactivity, and active learning. Instructors receive immediate feedback to gauge which concepts students understand. Questions covering the content of the *Integrated Science* text and formatted for the CPS eInstruction software are available on the *Integrated Science* ARIS site.

Instructor's Testing and Resource CD-ROM

The **Instructor's Testing and Resource CD-ROM** contains the *Integrated Science* test bank (over seven hundred test questions in a combination of true/false and multiple choice formats) within McGraw-Hill's EZ Test testing software. EZ Test is a flexible and easy-to-use electronic testing program. The program allows instructors to create tests from book specific items. It accommodates a wide range of question types and instructors may add their own questions. Multiple versions of the test can be created, and any test can be exported for use with course management systems such as WebCT, BlackBoard or PageOut. EZ Test Online is a new service and gives you a place to easily administer your EZ Test created exams and quizzes online. The program is available for Windows and Macintosh environments. Also located on the Instructor's Testing and Resources CD-ROM are Word and PDF files of the test bank, the instructor's manual, the instructor's edition lab manual, quizzes from the ARIS site, and personal response system questions. The Word files for the test bank, instructor's

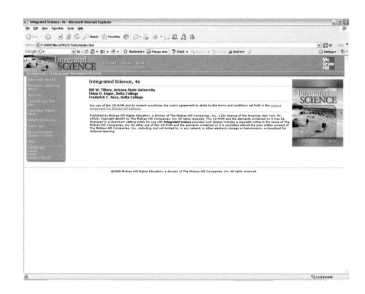

manual, quizzes, and personal response system questions can be used in combination with the testbank software or independently.

PRINTED SUPPLEMENTARY MATERIALS

Laboratory Manual

The laboratory manual, written and classroom-tested by the authors, presents a selection of laboratory exercises specifically written for the interest and abilities of nonscience majors. Each lab begins with an open-ended *Invitations to Inquiry,* designed to pique student interest in the lab concept. This is followed by laboratory exercises that require measurement and data analysis for work in a more structured learning environment. When the laboratory manual is used with *Integrated Science,* students will have an opportunity to master basic scientific principles and concepts, learn new problem-solving and thinking skills, and understand the nature of scientific inquiry from the perspective of hands-on experiences. There is also an **instructor's edition lab manual** available on the *Integrated Science* ARIS site and Instructor's Testing and Resource CD-Rom.

ACKNOWLEDGMENTS

This revision of *Integrated Science* has been made possible by the many users and reviewers of its fourth edition. The authors are indebted to the fourth edition reviewers for their critical reviews, comments, and suggestions. The fourth edition reviewers were:

Barbara Maher Bur, *Arcadia University*
Aaron Fried, *State University of New York, Cortland*
David Garza, *Samford University*
James Hamm, *Big Bend Community College*
Richard L. Kopec, *St. Edward's University*

Stephen Lattanzio, *Orange Coast College*
Kingshuk Majumdar, *Berea College*
Matt Nehring, *Adams State College*
E. Herbert Newman, *Coastal Carolina University*
Siva Paramasivam, *Savannah State University*
Amy Rollins, *Mercer University—Henry County*
Jay R. Yett, *Orange Coast College*
Walter J. Zalot, *Point Park University*
Robert Zdor, *Andrews University*

Many constructive suggestions, new ideas, and invaluable advice were provided by the earlier reviews of the previous three editions. Special thanks and appreciation to those who reviewed in the past:

Douglas Allchin, *University of Texas–El Paso*
B. J. Bateman, *Troy State University*
David J. Bauer, *College of Staten Island, City University of New York*
William Blaker, *Furman University*
Treasure Brasher, *West Texas A&M University*
Katrina Brown, *University of Pittsburgh—Greensburg*
Stephen C. Brown, *University of Texas–San Antonio*
Lauretta Buschar, *Beaver College*
Bruce Callen, *Drury University*
Tracey Cascadden, *University of New Mexico*
Linda Chamberlin, *Valdosta State University*
Timothy D. Champion, *Johnson C. Smith University*
Kailash Chandra, *Savannah State University*
Ken Comer, *Edison College*
Sarah Cooper, *Arcadia University*
Felicia Corsaro-Barbieri, *Gwynedd-Mercy College*
James Courtright, *Marquette University*
David J. DiMattio, *St. Bonaventure University*
David Fawcett, *Columbia State Community College*
Guy E. Ferish, *Adams State College*
Robert Gannon, *Dowling College*
David C. Garza, *Samford University*
Gian S. Ghuman, *Savannah State University*
Michael Giazzoni, *Point Park College*
David Grainger, *Colorado State University*
James J. Grant, *St. Peter's College*
Eamonn F. Healy, *St. Edward's University*
Pam Henderson, *Texas A&M University—Commerce*
Megan M. Hoffman, *Berea College*
Peter M. Jeffers, *State University of New York-Cortland*
Leslie S. Jones, *Valdosta State University*
Phyllis Katz, *Endicott College*
Lance E. Kearns, *James Madison University*
William Keller, *St. Petersburg Junior College*
Tasneem F. Khaleel, *Montana State University–Billings*
Catherine Kleier, *Adams State College*
Mary Hurn Korte, *Concordia University*
Theo Koupelis, *University of Wisconsin-Marathon*
Larry L. Lehr, *Baylor University*
Estela Llinás, *University of Pittsburgh-Greensburg*
Bryan Long, *Columbia State Community College*
Bruce MacLaren, *Eastern Kentucky University*

Kingshuk Majumdar, *Berea College*
Joe Mehaffey, *Francis Marion University*
Kenneth S. Mendelson, *Marquette University*
Donald Miller, *University of Michigan-Dearborn*
Jeffrey J. Miller, *Metropolitan State College of Denver*
Scott Mohr, *Boston University*
Lee Netherton, *John Brown University*
E. Herbert Newman, *Coastal Carolina University*
G. A. Nixon, *Texas A&M University-Commerce*
Robert E. Pyle, *DeVry University*
Dr. William Quintana, *New Mexico State University*
Denice N. Robertson, *Northern Kentucky University*
Joaquin Ruiz, *University of Arizona*
John Oakes, *Marian College*
C. Dianne Phillips, *University of Arkansas*
Richard Sapanaro, *Contra Costa College*
Rhine Singleton, *Franklin Pierce College*
Eric Stavney, *DeVry University*
Herbert Stewart, *Florida Atlantic University*
Ken Sumida, *Chapman University*
Stephen Welton Taber, *Saginaw Valley State University*
Mary Ellen Teasdale, *Texas A&M University-Commerce*
Laura Thurlow, *Jackson Community College*
Deborah Tull, *University of New Mexico*
Andrew Wallace, *Angelo State University*
Jim Westgate, *Lamar University*
Robert Wingfield, Jr., *Fisk University*
Cheryl Lynn Wistrom, *St. Joseph's College*
Jay R. Yett, *Orange Coast College*
Yen-Yuan James Wu, *Texas A&M University-Commerce*
Thad Zaleskiewicz, *University of Pittsburgh-Greensburg*

The authors would also like to thank the theme integration authors for their contributions to the Instructor's Resources on the ARIS site. Those contributors include:

Mary Brown, *Lansing Community College*
David J. DiMattio, *St. Bonaventure University*
Tasneem F. Khaleel, *Montana State University-Billings*
G. A. Nixon and Mary Ellen Teasdale, *Texas A&M-Commerce*
Thad Zaleskiewicz and Jennifer Siegert, *University of Pittsburgh-Greensburg*

The authors would also like to thank the following media ancillary authors for their contributions of the PowerPoint Lecture Outlines and the CPS eInstruction Questions, respectively:

Christine McCreary, *University of Pittsburgh-Greensburg*
Jeffrey J. Miller, *Metropolitan State College of Denver*

MEET THE AUTHORS

Bill W. Tillery

Bill W. Tillery is professor emeritus of Physics at Arizona State University. He earned a bachelor's degree at North-eastern State University (1960) and master's and doctorate degrees from the University of Northern Colorado (1967). Before moving to Arizona State University, he served as director of the Science and Mathematics Teaching Center at the University of Wyoming (1969–73) and as an assistant professor at Florida State University (1967–69). Bill has served on numerous councils, boards, and committees and was honored as the "Outstanding University Educator" at the University of Wyoming in 1972. He was elected the "Outstanding Teacher" in the Department of Physics and Astronomy at Arizona State University in 1995.

During his time at Arizona State, Bill has taught a variety of courses, including general education courses in science and society, physical science, and introduction to physics. He has received more than forty grants from the National Science Foundation, the U. S. Office of Education, private industry (Arizona Public Service), and private foundations (Flinn Foundation) for science curriculum development and science teacher inservice training. In addition to teaching and grant work, Bill has authored or co-authored more than sixty textbooks and many monographs, and has served as editor of three newsletters and journals between 1977 and 1996.

Eldon D. Enger

Eldon D. Enger is professor emeritus of biology at Delta College, a community college near Saginaw, Michigan. He received his B.A. and M.S. degrees from the University of Michigan. Professor Enger has over thirty years of teaching experience, during which he has taught biology, zoology, environmental science, and several other courses. He has been very active in curriculum and course development.

Professor Enger is an advocate for variety in teaching methodology. He feels that if students are provided with varied experiences, they are more likely to learn. In addition to the standard textbook assignments, lectures, and laboratory activities, his classes are likely to include writing assignments, student presentation of lecture material, debates by students on controversial issues, field experiences, individual student projects, and discussions of local examples and relevant current events. Textbooks are very valuable for presenting content, especially if they contain accurate, informative drawings and visual examples. Lectures are best used to help students see themes and make connections, and laboratory activities provide important hands-on activities.

Professor Enger has been a Fulbright Exchange Teacher to Australia and Scotland, received the Bergstein Award for Teaching Excellence and the Scholarly Achievement Award from Delta College, and participated as a volunteer in Earthwatch Research Programs in Costa Rica, the Virgin Islands and Australia. During 2001, he was a member of a People to People delegation to South Africa.

Professor Enger is married, has two adult sons, and enjoys a variety of outdoor pursuits such as cross-country skiing, hiking, hunting, kayaking, fishing, camping, and gardening. Other interests include reading a wide variety of periodicals, beekeeping, singing in a church choir, and preserving garden produce.

Frederick C. Ross

Fred Ross is professor emeritus of biology at Delta College, a community college near Saginaw, Michigan. He received his B.S. and M.S. from Wayne State University, Detroit, Michigan, and has attended several other universities and institutions. Professor Ross has over thirty years' teaching experience, including junior and senior high school, during which he has taught biology, cell biology and biological chemistry, microbiology, environmental science, and zoology. He has been very active in curriculum and course development. These activities included the development of courses in infection control and microbiology, and AIDS and infectious diseases, and a PBS ScienceLine course for elementary and secondary education majors in cooperation with Central Michigan University. In addition, he was involved in the development of the wastewater microbiology technician curriculum offered by Delta College.

He was also actively involved in the National Task Force of Two Year College Biologists (American Institute of Biological Sciences) and in the National Science Foundation College Science Improvement Program, and has been an evaluator for science and engineering fairs, Michigan Community College Biologists, a judge for the Michigan Science Olympiad and the Science Bowl, a member of a committee to develop and update blood-borne pathogen standards protocol, and a member of Topic Outlines in Introductory Microbiology Study Group of the American Society for Microbiology.

Professor Ross involves his students in a variety of learning techniques and has been a prime advocate of the writing-to-learn approach. Besides writing, his students are typically engaged in active learning techniques including use of inquiry-based learning, the Internet, e-mail communications, field experiences, classroom presentation, as well as lab work. The goal of his classroom presentations and teaching is to actively engage the minds of his students in understanding the material, not just memorization of "scientific facts." Professor Ross is married and recently a grandfather. He enjoys sailing, horseback riding, and cross-country skiing.

Science is concerned with your surroundings and your concepts and understanding of these surroundings.

What Is Science?

▶ Energy flows in and out of your surroundings (Ch. 2–7).

Chemistry Connections

▶ Matter is composed of atoms that interact on several different levels (Ch. 8–11).

CORE CONCEPT

Science is a way of thinking about and understanding your surroundings.

Measurement is used to accurately describe properties and events (p. 4).

An equation is a statement of a relationship between variables (p. 10).

Science investigations include collecting observations, developing explanations, and testing explanations (p. 12).

Scientific laws describe relationships between events that happen time after time (p. 15).

Earth Science Connections

▶ Earth is matter and energy that interact through cycles of change (Ch. 14–18).

Astronomy Connections

▶ The stars and solar system are matter and energy that interact through cycles of change (Ch. 12–13).

OVERVIEW

Have you ever thought about your thinking and what you know? On a very simplified level, you could say that everything you know came to you through your senses. You see, hear, and touch things of your choosing, and you can smell and taste things in your surroundings. Information is gathered and sent to your brain by your sense organs. Somehow, your brain processes all this information in an attempt to find order and make sense of it all. Finding order helps you understand the world and what may be happening at a particular place and time. Finding order also helps you predict what may happen next, which can be very important in a lot of situations.

This is a book on thinking about and understanding your surroundings. These surroundings range from the obvious, such as the landscape and the day-to-day weather, to the not so obvious, such as how atoms are put together. Your surroundings include natural things as well as things that people have made and used (figure 1.1). You will learn how to think about your surroundings, whatever your previous experience with thought-demanding situations. This first chapter is about "tools and rules" that you will use in the thinking process.

OBJECTS AND PROPERTIES

Science is concerned with making sense out of the environment. The early stages of this "search for sense" usually involve objects in the environment, things that can be seen or touched. These could be *objects* you see every day, such as a glass of water, a moving automobile, or a running dog. They could be quite large, such as the Sun, the Moon, or even the solar system, or invisible to the unaided human eye. Objects can be any size, but people are usually concerned with objects that are larger than a pinhead and smaller than a house. Outside these limits, the actual size of an object is difficult for most people to comprehend.

As you were growing up, you learned to form a generalized mental image of objects called a *concept*. Your concept of an object is an idea of what it is, in general, or what it should be according to your idea (figure 1.2). You usually have a word stored away in your mind that represents a concept. The word *chair*, for example, probably evokes an idea of "something to sit on." Your generalized mental image for the concept that goes with the word *chair* probably includes a four-legged object with a backrest. Upon close inspection, most of your (and everyone else's) concepts are found to be somewhat vague. For example, if the word *chair* brings forth a mental image of something with four legs and a backrest (the concept), what is the difference between a "high chair" and a "bar stool"? When is a chair a chair and not a stool? These kinds of questions can be troublesome for many people.

Not all of your concepts are about material objects. You also have concepts about intangibles such as time, motion, and relationships between events. As was the case with concepts of material objects, words represent the existence of intangible concepts. For example, the words *second, hour, day,* and *month* represent concepts of time. A concept of the pushes and pulls that come with changes of motion during an airplane flight might be represented with such words as *accelerate* and *falling*. Intangible concepts might seem to be more abstract since they do not represent material objects.

By the time you reach adulthood, you have literally thousands of words to represent thousands of concepts. But most,

FIGURE 1.1 Your surroundings include naturally occurring objects and manufactured objects such as sidewalks and buildings.

you would find on inspection, are somewhat ambiguous and not at all clear-cut. That is why you find it necessary to talk about certain concepts for a minute or two to see if the other person has the same "concept" for words as you do. That is why when one person says, "Boy, was it hot!" the other person may respond, "How hot was it?" The meaning of *hot* can be quite different for two people, especially if one is from Arizona and the other from Alaska!

The problem with words, concepts, and mental images can be illustrated by imagining a situation involving you and another person. Suppose that you have found a rock that you believe would make a great bookend. Suppose further that you are talking to the other person on the telephone, and you want to discuss the suitability of the rock as a bookend, but you do not know the name of the rock. If you knew the name, you would simply state that you found a "_____." Then you would probably discuss the rock for a minute or so to see if the other person really understood what you were talking about. But not knowing the name of the rock and wanting to communicate about the suitability of the object as a bookend, what would you do? You would probably describe the characteristics, or **properties,** of the rock. Properties are the qualities or attributes that, taken together, are usually peculiar to an object. Since you commonly determine properties with your senses (smell, sight, hearing, touch, and taste), you could say that the properties of an object are the effect the object has on your senses. For example, you might say that the rock is a "big, yellow, smooth rock with shiny gold cubes on one side." But consider the mental image that the other person on the telephone forms when you describe these properties. It is entirely possible that the other person is thinking of something very different from what you are describing (figure 1.3)!

FIGURE 1.2 What is your concept of a chair? Are all of these pieces of furniture chairs? Most people have concepts, or ideas of what things in general should be, that are loosely defined. The concept of a chair is one example of a loosely defined concept.

FIGURE 1.3 Could you describe this rock to another person over the telephone so that the other person would know *exactly* what you see? This is not likely with everyday language, which is full of implied comparisons, assumptions, and inaccurate descriptions.

As you can see, the example of describing a proposed bookend by listing its properties in everyday language leaves much to be desired. The description does not really help the other person form an accurate mental image of the rock. One problem with the attempted communication is that the description of any property implies some kind of *referent*. The word **referent** means that you *refer to,* or think of, a given property in terms of another, more familiar object. Colors, for example, are sometimes stated with a referent. Examples are "sky blue," "grass green," or "lemon yellow." The referents for the colors blue, green, and yellow are, respectively, the sky, living grass, and a ripe lemon.

Referents for properties are not always as explicit as they are with colors, but a comparison is always implied. Since the comparison is implied, it often goes unspoken and leads to assumptions in communications. For example, when you stated that the rock was "big," you assumed that the other person knew that you did not mean as big as a house or even as big as a bicycle. You assumed that the other person knew that you meant that the rock was about as large as a book, perhaps a bit larger.

Another problem with the listed properties of the rock is the use of the word *smooth.* The other person would not know if you meant that the rock *looked* smooth or *felt* smooth. After all, some objects can look smooth and feel rough. Other objects can look rough and feel smooth. Thus, here is another assumption, and probably all of the properties lead to implied comparisons, assumptions, and a not very accurate communication. This is the nature of your everyday language and the nature of most attempts at communication.

QUANTIFYING PROPERTIES

Typical day-to-day communications are often vague and leave much to be assumed. A communication between two people, for example, could involve one person describing some person, object, or event to a second person. The description is made by using referents and comparisons that the second person may or may not have in mind. Thus, such attributes as "long"

fingernails or "short" hair may have entirely different meanings to different people involved in a conversation. Assumptions and vagueness can be avoided by using **measurement** in a description. Measurement is a process of comparing a property to a well-defined and agreed-upon referent. The well-defined and agreed-upon referent is used as a standard called a **unit.** The measurement process involves three steps: (1) *comparing* the referent unit to the property being described, (2) following a *procedure,* or operation, which specifies how the comparison is made, and (3) *counting* how many standard units describe the property being considered.

The measurement process thus uses a defined referent unit, which is compared to a property being measured. The *value* of the property is determined by counting the number of referent units. The name of the unit implies the procedure that results in the number. A measurement statement always contains a *number* and *name* for the referent unit. The number answers the question "How much?" and the name answers the question "Of what?" Thus a measurement always tells you "how much of what." You will find that using measurements will sharpen your communications. You will also find that using measurements is one of the first steps in understanding your physical environment.

MEASUREMENT SYSTEMS

Measurement is a process that brings precision to a description by specifying the "how much" and "of what" of a property in a particular situation. A number expresses the value of the property, and the name of a unit tells you what the referent is as well as implying the procedure for obtaining the number. Referent units must be defined and established, however, if others are to understand and reproduce a measurement. It would be meaningless, for example, for you to talk about a length in "clips" if other people did not know what you meant by a "clip" unit. When standards are established, the referent unit is called a **standard unit** (figure 1.4). The use of standard units makes it possible to communicate and duplicate measurements. Standard units are usually defined and established by governments and their agencies that are created for that purpose. In the United States, the agency concerned with measurement standards is the National Institute of Standards and Technology. In Canada, the Standards Council of Canada oversees the National Standard System.

50 leagues
130 nautical miles
150 miles
158 Roman miles
1,200 furlongs
12,000 chains
48,000 rods
452,571 cubits
792,000 feet

FIGURE 1.4 Any of these units and values could have been used at some time or another to describe the same distance between these hypothetical towns. Any unit could be used for this purpose, but when one particular unit is officially adopted, it becomes known as the *standard unit.*

There are two major *systems* of standard units in use today, the English system and the metric system. The metric system is used throughout the world except in the United States, where both systems are in use. The continued use of the English system in the United States presents problems in international trade, so there is pressure for a complete conversion to the metric system. More and more metric units are being used in everyday measurements, but a complete conversion will involve an enormous cost. Appendix A contains a method for converting from one system to the other easily. Consult this section if you need to convert from one metric unit to another metric unit or to convert from English to metric units or vice versa. Conversion factors are listed inside the front cover.

People have used referents to communicate about properties of things throughout human history. The ancient Greek civilization, for example, used units of *stadia* to communicate about distances and elevations. The "stadium" was a unit of length of the racetrack at the local stadium (*stadia* is the plural of stadium), based on a length of 125 paces. Later civilizations, such as the ancient Romans, adopted the stadia and other referent units from the ancient Greeks. Some of these same referent units were later adopted by the early English civilization, which eventually led to the *English system* of measurement. Some adopted units of the English system were originally based on parts of the human body, presumably because you always had these referents with you (figure 1.5). The inch, for example, used the end joint of the thumb for a referent. A foot, naturally, was the length of a foot, and a yard

TABLE 1.1 The SI Standard Units

Property	Unit	Symbol
Length	meter	m
Mass	kilogram	kg
Time	second	s
Electric current	ampere	A
Temperature	kelvin	K
Amount of substance	mole	mol
Luminous intensity	candela	cd

was the distance from the tip of the nose to the end of the fingers on an arm held straight out. A cubit was the distance from the end of an elbow to the fingertip, and a fathom was the distance between the finger-tips of two arms held straight out. As you can imagine, there were problems with these early units because everyone was not the same size. Beginning in the 1300s, the sizes of the units were gradually standardized by various English kings. In 1879, the United States, along with sixteen other countries, signed the *Treaty of the Meter,* defining the English units in terms of the metric system. The United States thus became officially metric but not entirely metric in everyday practice.

The *metric system* was established by the French Academy of Sciences in 1791. The academy created a measurement system that was based on invariable referents in nature, not human body parts. These referents have been redefined over time to make the standard units more reproducible. In 1960, six standard metric units were established by international agreement. The *International System of Units,* abbreviated *SI,* is a modernized version of the metric system. Today, the SI system has seven units that define standards for the properties of length, mass, time, electric current, temperature, amount of substance, and light intensity (table 1.1). The standard units for the properties of length, mass, and time are introduced in this chapter. The remaining units will be introduced in later chapters as the properties they measure are discussed.

STANDARD UNITS FOR THE METRIC SYSTEM

If you consider all the properties of all the objects and events in your surroundings, the number seems overwhelming. Yet, close inspection of how properties are measured reveals that some properties are combinations of other properties (figure 1.6). Volume, for example, is described by the three length measurements of length, width, and height. Area, on the other hand, is described by just the two length measurements of length and width. Length, however, cannot be defined in simpler terms of any other property. There are four properties that cannot be described in simpler terms, and all other properties are combinations of these four. For this reason they are called the *fundamental properties.* A fundamental property cannot be defined in simpler terms other than to describe how it is measured.

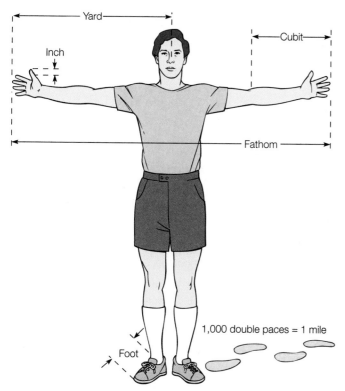

FIGURE 1.5 Many early units for measurement were originally based on the human body. Some of the units were later standardized by governments to become the basis of the English system of measurement.

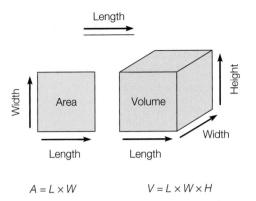

$$A = L \times W \qquad V = L \times W \times H$$

FIGURE 1.6 Area, or the extent of a surface, can be described by two length measurements. Volume, or the space that an object occupies, can be described by three length measurements. Length, however, can be described only in terms of how it is measured, so it is called a *fundamental property*.

These four fundamental properties are (1) *length*, (2) *mass*, (3) *time*, and (4) *charge*. Used individually or in combinations, these four properties will describe or measure what you observe in nature. Metric units for measuring the fundamental properties of length, mass, and time will be described next. The fourth fundamental property, charge, is associated with electricity, and a unit for this property will be discussed in chapter 6.

Length

The standard unit for length in the metric system is the *meter* (the symbol or abbreviation is m) A meter is defined in terms of the distance that light travels in a vacuum during a certain time period, 1/299,792,458 second. The important thing to remember, however, is that the meter is the metric *standard unit* for length. A meter is slightly longer than a yard, 39.3 inches. It is approximately the distance from your left shoulder to the tip of your right hand when your arm is held straight out. Many doorknobs are about 1 meter above the floor. Think about these distances when you are trying to visualize a meter length.

Mass

The standard unit for mass in the metric system is the *kilogram* (kg). The kilogram is defined as the mass of a certain metal cylinder kept by the International Bureau of Weights and Measures in France. This is the only standard unit that is still defined in terms of an object. The property of mass is sometimes confused with the property of weight since they are directly proportional to each other at a given location on the surface of the Earth. They are, however, two completely different properties and are measured with different units. All objects tend to maintain their state of rest or straight-line motion, and this property is called "inertia." The *mass* of an object is a measure of the inertia of an object. The *weight* of the object is a measure of the force of gravity on it. This distinction between weight and mass will be discussed in detail in chapter 2. For now, remember that weight and mass are not the same property.

Time

The standard unit for time is the *second* (s). The second was originally defined as 1/86,400 of a solar day (1/60 × 1/60 × 1/24). Earth's spin was found not to be as constant as thought, so the second was redefined to be the duration required for a certain number of vibrations of a certain cesium atom. A special spectrometer called an "atomic clock" measures these vibrations and keeps time with an accuracy of several millionths of a second per year.

METRIC PREFIXES

The metric system uses prefixes to represent larger or smaller amounts by factors of 10. Some of the more commonly used prefixes, their abbreviations, and their meanings are listed in table 1.2. Suppose you wish to measure something smaller than the standard unit of length, the meter. The meter is subdivided into ten equal-sized subunits called *decimeters*. The prefix *deci-* has a meaning of "one-tenth of," and it takes 10 decimeters to equal the length of 1 meter. For even smaller measurements, each decimeter is divided into ten equal-sized subunits called *centimeters*. It takes 10 centimeters to equal 1 decimeter and 100 to equal 1 meter. In a similar fashion, each prefix up or down the metric ladder represents a simple increase or decrease by a factor of 10 (figure 1.7).

When the metric system was established in 1791, the standard unit of mass was defined in terms of the mass of a certain volume of water. A cubic decimeter (dm^3) of pure water at 4°C was *defined* to have a mass of 1 kilogram (kg). This definition was convenient because it created a relationship between length, mass, and volume. As illustrated in figure 1.8, a cubic decimeter is 10 cm on each side. The volume of this cube is therefore 10 cm × 10 cm × 10 cm, or 1,000 cubic centimeters (abbreviated as cc or cm^3). Thus, a volume of 1,000 cm^3 of water has a mass of 1 kg. Since 1 kg is 1,000 g, 1 cm^3 of water has a mass of 1 g.

TABLE 1.2 Some Metric Prefixes

Prefix	Symbol	Meaning
tera-	T	10^{12} (1,000,000,000,000 times the unit)
giga-	G	10^{9} (1,000,000,000 times the unit)
mega-	M	10^{6} (1,000,000 times the unit)
kilo-	k	10^{3} (1,000 times the unit)
hecto-	h	10^{2} (100 times the unit)
deka-	da	10^{1} (10 times the unit)
Unit		
deci-	d	10^{-1} (0.1 of the unit)
centi-	c	10^{-2} (0.01 of the unit)
milli-	m	10^{-3} (0.001 of the unit)
micro-	μ	10^{-6} (0.000001 of the unit)
nano-	n	10^{-9} (0.000000001 of the unit)
pico-	p	10^{-12} (0.000000000001 of the unit)

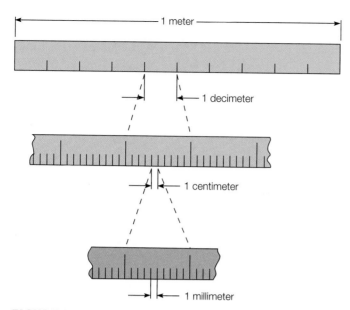

FIGURE 1.7 Compare the units shown above. How many millimeters fit into the space occupied by 1 centimeter? How many millimeters fit into the space of 1 decimeter? Can you express this as multiples of ten?

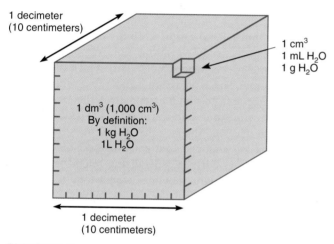

FIGURE 1.8 A cubic decimeter of water (1,000 cm³) has a liquid volume of 1 L (1,000 mL) and a mass of 1 kg (1,000 g). Therefore, 1 cm³ of water has a liquid volume of 1 mL and a mass of 1 g.

The volume of 1,000 cm³ also defines a metric unit that is commonly used to measure liquid volume, the *liter* (L). For smaller amounts of liquid volume, the milliliter (mL) is used. The relationship between liquid volume, volume, and mass of water is therefore

$$1.0 \text{ L} \rightarrow 1.0 \text{ dm}^3 \text{ and has a mass of } 1.0 \text{ kg}$$

or, for smaller amounts,

$$1.0 \text{ mL} \rightarrow 1.0 \text{ cm}^3 \text{ and has a mass of } 1.0 \text{ g}$$

UNDERSTANDINGS FROM MEASUREMENTS

One of the more basic uses of measurement is to *describe* something in an exact way that everyone can understand. For example, if a friend in another city tells you that the weather

Weather Report

Friday (24 hours ended at 5 P.M.)
Highs—airport 73°F, downtown 76°F
Lows—airport 68°F, downtown 70°F
Rainfall .... 0.26 in
Average wind speed .... 5.2 mph
Relative humidity .... High 85%
Low 75%
Rainfall ± normal to date.....+0.94 in

FIGURE 1.9 A weather report gives exact information, data that describe the weather by reporting numerically specified units for each condition.

has been "warm," you might not understand what temperature is being described. A statement that the air temperature is 70°F carries more exact information than a statement about "warm weather." The statement that the air temperature is 70°F contains two important concepts: (1) the numerical value of 70 and (2) the referent unit of degrees Fahrenheit. Note that both a numerical value and a unit are necessary to communicate a measurement correctly. Thus, weather reports describe weather conditions with numerically specified units; for example, 70° Fahrenheit for air temperature, 5 miles per hour for wind speed, and 0.5 inch for rainfall (figure 1.9). When such numerically specified units are used in a description, or a weather report, everyone understands *exactly* the condition being described.

Data

Measurement information used to describe something is called **data.** Data can be used to describe objects, conditions, events, or changes that might be occurring. You really do not know if the weather is changing much from year to year until you compare the yearly weather data. The data will tell you, for example, if the weather is becoming hotter or dryer or is staying about the same from year to year.

Let's see how data can be used to describe something and how the data can be analyzed for further understanding. The cubes illustrated in figure 1.10 will serve as an example. Each cube can be described by measuring the properties of size and surface area.

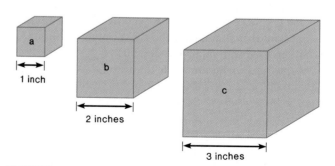

FIGURE 1.10 Cube *a* is 1 inch on each side, cube *b* is 2 inches on each side, and cube *c* is 3 inches on each side. These three cubes can be described and compared with data, or measurement information, but some form of analysis is needed to find patterns or meaning in the data.

First, consider the size of each cube. Size can be described by *volume*, which means *how much space something occupies*. The volume of a cube can be obtained by measuring and multiplying the length, width, and height. The data is

volume of cube *a*	1 in³
volume of cube *b*	8 in³
volume of cube *c*	27 in³

Now consider the surface area of each cube. *Area* means *the extent of a surface*, and each cube has six surfaces, or faces (top, bottom, and four sides). The area of any face can be obtained by measuring and multiplying length and width. The data for the three cubes thus describes them as follows:

	Volume	Surface Area
cube *a*	1 in³	6 in²
cube *b*	8 in³	24 in²
cube *c*	27 in³	54 in²

Ratios and Generalizations

Data on the volume and surface area of the three cubes in figure 1.10 describes the cubes, but whether it says anything about a relationship between the volume and surface area of a cube is difficult to tell. Nature seems to have a tendency to camouflage relationships, making it difficult to extract meaning from raw data. Seeing through the camouflage requires the use of mathematical techniques to expose patterns. Let's see how such techniques can be applied to the data on the three cubes and what the pattern means.

One mathematical technique for reducing data to a more manageable form is to expose patterns through a *ratio*. A ratio is a relationship between two numbers obtained when one number is divided by another number. Suppose, for example, that an instructor has 50 sheets of graph paper for a laboratory group of 25 students. The relationship, or ratio, between the number of sheets and the number of students is 50 papers to 25 students, and this can be written as 50 papers/25 students. This ratio is *simplified* by dividing 25 into 50, and the ratio becomes 2 papers/1 student. The 1 is usually understood (not stated), and the ratio is written as simply 2 papers/student. It is read as 2 papers "for each" student, or 2 papers "per" student. The concept of simplifying with a ratio is an important one, and you will see it time and time again throughout science. It is important that you understand the meaning of *per* and *for each* when used with numbers and units.

Applying the ratio concept to the three cubes in figure 1.10, the ratio of surface area to volume for the smallest cube, cube *a*, is 6 in² to 1 in³, or

$$\frac{6 \text{ in}^2}{1 \text{ in}^3} = 6 \frac{\text{in}^2}{\text{in}^3}$$

meaning there are 6 square inches of area *for each* cubic inch of volume.

The middle-sized cube, cube *b*, had a surface area of 24 in² and a volume of 8 in³. The ratio of surface area to volume for

this cube is therefore

$$\frac{24 \text{ in}^2}{8 \text{ in}^3} = 3 \frac{\text{in}^2}{\text{in}^3}$$

meaning there are 3 square inches of area *for each* cubic inch of volume.

The largest cube, cube *c*, had a surface area of 54 in² and a volume of 27 in³. The ratio is

$$\frac{54 \text{ in}^2}{27 \text{ in}^3} = 2 \frac{\text{in}^2}{\text{in}^3}$$

or 2 square inches of area *for each* cubic inch of volume. Summarizing the ratio of surface area to volume for all three cubes, you have

small cube	*a*–6:1
middle cube	*b*–3:1
large cube	*c*–2:1

Now that you have simplified the data through ratios, you are ready to generalize about what the information means. You can generalize that the surface-area-to-volume ratio of a cube *decreases* as the volume of a cube becomes larger. Reasoning from this generalization will provide an explanation for a number of related observations. For example, why does crushed ice melt faster than a single large block of ice with the same volume? The explanation is that the crushed ice has a larger surface-area-to-volume ratio than the large block, so more surface is exposed to warm air. If the generalization is found to be true for shapes other than cubes, you could explain why a log chopped into small chunks burns faster than the whole log. Further generalizing might enable you to predict if 10 lb of large potatoes would require more or less peeling than 10 lb of small potatoes. When generalized explanations result in predictions that can be verified by experience, you gain confidence in the explanation. Finding patterns of relationships is a satisfying intellectual adventure that leads to understanding and generalizations that are frequently practical.

The Density Ratio

The power of using a ratio to simplify things, making explanations more accessible, is evident when you compare the simplified ratio 6 to 3 to 2 with the hodgepodge of numbers that you would have to consider without using ratios. The power of using the ratio technique is also evident when considering other properties of matter. Volume is a property that is sometimes confused with mass. Larger objects do not necessarily contain more matter than smaller objects. A large balloon, for example, is much larger than this book, but the book is much more massive than the balloon. The simplified way of comparing the mass of a particular volume is to find the ratio of mass to volume. This ratio is called **density,** which is defined as *mass per unit volume*. The *per* means "for each," as previously discussed, and *unit* means "one," or "each." Thus "mass per unit volume" literally means the "mass of one volume" (figure 1.11).

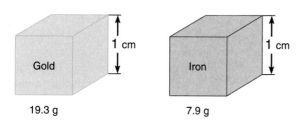

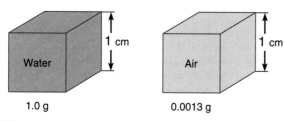

FIGURE 1.11 Equal volumes of different substances do not have the same mass, as these cube units show. Calculate the densities in g/m³. Do equal volumes of different substances have the same density? Explain.

The relationship can be written as

$$\text{density} = \frac{\text{mass}}{\text{volume}}$$

or

$$\rho = \frac{m}{V} \qquad \text{equation 1.1}$$

(ρ is the symbol for the Greek letter *rho*.)

As with other ratios, density is obtained by dividing one number and unit by another number and unit. Thus, the density of an object with a volume of 5 cm³ and a mass of 10 g is

$$\text{density} = \frac{10\ \text{g}}{5\ \text{cm}^3} = 2\ \frac{\text{g}}{\text{cm}^3}$$

The density in this example is the ratio of 10 g to 5 cm³, or 10 g/5 cm³, or 2 g to 1 cm³. Thus, the density of the example object is the mass of *one* volume (a unit volume), or 2 g *for each* cm³.

Any unit of mass and any unit of volume may be used to express density. The densities of solids, liquids, and gases are usually expressed in grams per cubic centimeter (g/cm³), but the densities of liquids are sometimes expressed in grams

TABLE 1.3 Densities of Some Common Substances

Substance	Density (ρ) (g/cm³)
Aluminum	2.70
Copper	8.96
Iron	7.87
Lead	11.4
Water	1.00
Seawater	1.03
Mercury	13.6
Gasoline	0.680

Fish Density

Sharks and rays are marine animals that have an internal skeleton made entirely of cartilage. These animals have no swim bladder to adjust their body density in order to maintain their position in the water; therefore, they must constantly swim or they will sink. The bony fish, on the other hand, have a skeleton composed of bone and most also have a swim bladder. These fish can regulate the amount of gas in the bladder to control their density. Thus, the fish can remain at a given level in the water without expending large amounts of energy.

per milliliter (g/mL). Using SI standard units, densities are expressed as kg/m³. Densities of some common substances are shown in table 1.3.

If matter is distributed the same throughout a volume, the *ratio* of mass to volume will remain the same no matter what mass and volume are being measured. Thus, a teaspoonful, a cup, or a lake full of freshwater at the same temperature will all have a density of about 1 g/cm³ or 1 kg/L.

CONCEPTS APPLIED

Density Examples

1. What is the density of this book? Measure the length, width, and height of this book in cm, then multiply to find the volume in cm³. Use a balance to find the mass of this book in grams. Compute the density of the book by dividing the mass by the volume. Compare the density in g/cm³ with other substances listed in table 1.3.

2. Compare the densities of some common liquids. Pour a cup of vinegar in a large bottle. Carefully add a cup of corn syrup, then a cup of cooking oil. Drop a coin, tightly folded pieces of aluminum foil, and toothpicks into the bottle. Explain what you observe in terms of density. Take care not to confuse the property of *density*, which describes the compactness of matter, with *viscosity*, which describes how much fluid resists flowing under normal conditions. (Corn syrup has a greater viscosity than water—is this true of density, too?)

Myths, Mistakes, and Misunderstandings

Tap a Can?

Some people believe that tapping on the side of a can of carbonated beverage will prevent it from foaming over when the can is opened. Is this true or a myth? Set up a controlled experiment to compare opening cold cans of a carbonated beverage that have been tapped with cans that have not been tapped. Are you sure you have controlled all the other variables?

EXAMPLE 1.1 *(Optional)*

Two blocks are on a table. Block A has a volume of 30.0 cm³ and a mass of 81.0 g. Block B has a volume of 50.0 cm³ and a mass of 135 g. Which block has the greater density? If the two blocks have the same density, what material are they? (See table 1.3.)

SOLUTION

Density is defined as the ratio of the mass of a substance per unit volume. Assuming the mass is distributed equally throughout the volume, you could assume that the ratio of mass to volume is the same no matter what quantity of mass and volume are measured. If you can accept this assumption, you can use equation 1.1 to determine the density:

Block A

$$\begin{aligned}
\text{mass } (m) &= 81.0 \text{ g} \\
\text{volume } (V) &= 30.0 \text{ cm}^3 \\
\text{density } (\rho) &= ?
\end{aligned}$$

$$\begin{aligned}
\rho &= \frac{m}{V} \\
&= \frac{81.0 \text{ g}}{30.0 \text{ cm}^3} \\
&= \frac{81.0}{30.0} \frac{\text{g}}{\text{cm}^3} \\
&= 2.70 \frac{\text{g}}{\text{cm}^3}
\end{aligned}$$

Block B

$$\begin{aligned}
\text{mass } (m) &= 135 \text{ g} \\
\text{volume } (V) &= 50.0 \text{ cm}^3 \\
\text{density } (\rho) &= ?
\end{aligned}$$

$$\begin{aligned}
\rho &= \frac{m}{V} \\
&= \frac{135 \text{ g}}{50.0 \text{ cm}^3} \\
&= \frac{135}{50.0} \frac{\text{g}}{\text{cm}^3} \\
&= 2.70 \frac{\text{g}}{\text{cm}^3}
\end{aligned}$$

As you can see, both blocks have the same density. Inspecting table 1.3, you can see that aluminum has a density of 2.70 g/cm³, so both blocks must be aluminum.

EXAMPLE 1.2 *(Optional)*

A rock with a volume of 4.50 cm³ has a mass of 15.0 g. What is the density of the rock? (Answer: 3.33 g/cm³)

Symbols and Equations

In the previous section, the relationship of density, mass, and volume was written with symbols. Density was represented by ρ, the lowercase letter rho in the Greek alphabet, mass was represented by m, and volume by V. The use of such symbols is established and accepted by convention, and these symbols are like the vocabulary of a foreign language. You learn what the symbols mean by use and practice, with the understanding that each symbol stands for a very specific property or concept. The symbols actually represent **quantities**, or *measured properties.* The symbol m thus represents a quantity of mass that is specified by a number and a unit, for example, 16 g. The symbol V represents a quantity of volume that is specified by a number and a unit, such as 17 cm³.

Symbols

Symbols usually provide a clue about which quantity they represent, such as m for mass and V for volume. However, in some cases, two quantities start with the same letter, such as volume and velocity, so the uppercase letter is used for one (V for volume) and the lowercase letter is used for the other (v for velocity). There are more quantities than upper- and lowercase letters, however, so letters from the Greek alphabet are also used, for example, ρ for mass density. Sometimes a subscript is used to identify a quantity in a particular situation, such as v_i for initial, or beginning, velocity and v_f for final velocity. Some symbols are also used to carry messages; for example, the Greek letter delta (Δ) is a message that means "the change in" a value. Other message symbols are the symbol $\therefore$, which means "therefore," and the symbol $\propto$, which means "is proportional to."

Equations

Symbols are used in an **equation,** a statement that describes a relationship where *the quantities on one side of the equal sign are identical to the quantities on the other side. Identical* refers to both the numbers and the units. Thus, in the equation describing the property of density, $\rho = m/V$, the numbers on both sides of the equal sign are identical (e.g., 5 = 10/2). The units on both sides of the equal sign are also identical (e.g., g/cm³ = g/cm³).

Equations are used to (1) *describe a property,* (2) *define a concept,* or (3) *describe how quantities change relative to each other.* Understanding how equations are used in these three classes is basic to comprehension of physical science. Each class of uses is considered separately in the following discussion.

Describing a property. You have already learned that the compactness of matter is described by the property called density. Density is a ratio of mass to a unit volume, or $\rho = m/V$. The key to understanding this property is to understand the meaning of a ratio and what *per* or *for each* means. Other examples of properties that will be defined by ratios are how fast something is moving (speed) and how rapidly a speed is changing (acceleration).

Defining a concept. A physical science concept is sometimes defined by specifying a measurement procedure. This is called an *operational definition* because a procedure is established that defines a concept as well as telling you how to measure it. Concepts of what is meant by force, mechanical work, and mechanical power and concepts involved in electrical and magnetic interactions can be defined by measurement procedures.

Describing how quantities change relative to each other. Nature is full of situations where one or more quantities change in value, or vary in size, in response to changes in other quantities. Changing quantities are called **variables.** Your weight, for example, is a variable that changes in size in response to changes in another variable; for example, the amount of food you eat. You already know about the pattern, or relationship, between these two variables. With all other factors being equal, an increase in the amount of food you eat results in an increase in your weight. When two variables increase (or decrease) together in the same ratio, they are said to be in *direct proportion.* When two variables are in direct proportion, *an increase or decrease in one variable results in the same relative increase or decrease in a second variable.*

Variables do not always increase or decrease in direct proportion. Sometimes one variable *increases* while a second variable *decreases* in the same ratio. This is an *inverse proportion* relationship. Other common relationships include one variable increasing in proportion to the *square* or to the *inverse square* of a second variable. Here are the forms of these four different types of proportional relationships:

Direct	$a \propto b$
Inverse	$a \propto 1/b$
Square	$a \propto b^2$
Inverse square	$a \propto 1/b^2$

Proportionality statements

Proportionality statements describe in general how two variables change relative to each other, but a proportionality statement is *not* an equation. For example, consider the last time you filled your fuel tank at a service station (figure 1.12). You could say that the volume of gasoline in an empty tank you are filling is directly proportional to the amount of time that the fuel pump was running, or

$$\text{volume} \propto \text{time}$$

FIGURE 1.12 The volume of fuel you have added to the fuel tank is directly proportional to the amount of time that the fuel pump has been running. This relationship can be described with an equation by using a proportionality constant.

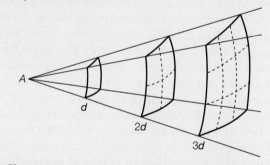
This is not an equation because the numbers and units are not identical on both sides. Considering the units, for example, it should be clear that minutes do not equal gallons; they are two different quantities. To make a statement of proportionality into an equation, you need to apply a *proportionality constant,* which is sometimes given the symbol k. For the fuel pump example, the equation is

$$\text{volume} = (\text{time})(\text{constant})$$

or

$$V = tk$$

In this example, the constant is the flow of gasoline from the pump in gal/min (a ratio). Assume the rate of flow is 10 gal/min. In units, you can see why the statement is now an equality.

$$gal = (min)\left(\frac{gal}{min}\right)$$

$$gal = \frac{\cancel{min} \times gal}{\cancel{min}}$$

$$gal = gal$$

Here are some tips to consider if your instructor uses a problem-solving approach. You could think of an equation as a *set of instructions*. The density equation, for example, is $\rho = m/V$. The equation tells you that mass density is a ratio of mass to volume, and you can find the density by dividing the mass by the volume. If you have difficulty, you either do not know the instructions or do not know how to follow the instructions (algebraic rules). Appendix D has worked examples that can help you with both the instructions and how to follow them. See also appendix E, which deals with problem solving.

THE NATURE OF SCIENCE

Most humans are curious, at least when they are young, and are motivated to understand their surroundings. These traits have existed since antiquity and have proven to be a powerful motivation. In recent times, the need to find out has motivated the launching of space probes to learn what is "out there," and humans have visited the Moon to satisfy their curiosity. Curiosity and the motivation to understand nature were no less powerful in the past than today. Over two thousand years ago, the ancient Greeks lacked the tools and technology of today and could only make conjectures about the workings of nature. These early seekers of understanding are known as *natural philosophers,* and they observed, thought, and wrote about the workings of all of nature. They are called philosophers because their understandings came from reasoning only, without experimental evidence. Nonetheless, some of their ideas were essentially correct and are still in use today. For example, the idea of matter being composed of *atoms* was first reasoned by certain ancient Greeks in the fifth century B.C. The idea of *elements,* basic components that make up matter, was developed much earlier but refined by the ancient Greeks in the fourth century B.C. The concept of what the elements are and the concept of the nature of atoms have changed over time, but the ideas first came from ancient natural philosophers.

The Scientific Method

Some historians identify the time of Galileo and Newton, approximately three hundred years ago, as the beginning of modern science. Like the ancient Greeks, Galileo and Newton were interested in studying all of nature. Since the time of Galileo

and Newton, the content of physical science has increased in scope and specialization, but the basic means of acquiring understanding, the scientific investigation, has changed little. A *scientific investigation* provides understanding through *experimental evidence,* as opposed to the conjectures based on thinking only of the ancient natural philosophers. In chapter 2, for example, you will learn how certain ancient Greeks described how objects fall toward Earth with a thought-out, or reasoned, explanation. Galileo, on the other hand, changed how people thought of falling objects by developing explanations from both creative thinking and precise measurement of physical quantities, providing experimental evidence for his explanations. Experimental evidence provides explanations today, much as it did for Galileo, as relationships are found from precise measurements of physical quantities. Thus, scientific knowledge about nature has grown as measurements and investigations have led to understandings that lead to further measurements and investigations.

What is a scientific investigation and what methods are used to conduct one? Attempts have been made to describe scientific methods in a series of steps (define problem, gather data, make hypothesis, test, make conclusion), but no single description has ever been satisfactory to all concerned. Scientists do similar things in investigations, but there are different approaches and different ways to evaluate what they find. Overall, the similar things might look like this:

1. Observe some aspect of nature.
2. Propose an explanation for something observed.
3. Use the explanation to make predictions.
4. Test predictions by doing an experiment or by making more observations.
5. Modify explanation as needed.
6. Return to step 3.

The exact approach a scientist uses depends on the individual doing the investigation as well as the particular field of science being studied.

Another way to describe what goes on during a scientific investigation is to consider what can be generalized. At least three separate activities seem to be common to scientists in different fields as they conduct scientific investigations, and these generalized activities are:

- Collecting observations.
- Developing explanations.
- Testing explanations.

No particular order or routine can be generalized about these common elements. In fact, individual scientists might not even be involved in all three activities. Some, for example, might spend all of their time out in nature, "in the field" collecting data and generalizing about their findings. This is an acceptable means of scientific investigation in some fields of science. Yet, other scientists might spend all of their time indoors, at computer terminals, developing theoretical equations that offer explanations for generalizations made by others. Again, the work at a computer terminal is an acceptable means of scientific investigation. Thus, there is not an order of five steps that are followed, particularly by today's specialized scientists. This is one reason why many philosophers of science argue there is no such thing as *the* scientific method. There are common activities of observing, explaining, and testing in scientific investigations in different fields, and these activities will be discussed next.

Explanations and Investigations

Explanations in the natural sciences are concerned with things or events observed, and there can be several different means of developing or creating explanations. In general, explanations can come from the results of experiments, from an educated guess, or just from imaginative thinking. In fact, there are several examples in the history of science of valid explanations being developed even from dreams.

Explanations go by various names, each depending on intended use or stage of development. For example, an explanation in an early stage of development is sometimes called a *hypothesis.* A **hypothesis** is a tentative thought- or experiment-derived explanation. It must be compatible with observations and provide understanding of some aspect of nature, but the key word here is *tentative.* A hypothesis is tested by experiment and is rejected, or modified, if a single observation or test does not fit.

The successful testing of a hypothesis may lead to the design of experiments, or it could lead to the development of another hypothesis, which could, in turn, lead to the design of yet more experiments, which could lead to. . . . As you can see, this is a branching, ongoing process that is very difficult to describe in specific terms. In addition, it can be difficult to identify a conclusion, an endpoint in the process. The search for new concepts to explain experimental evidence may lead from a hypothesis to a new theory, which results in more new hypotheses. This is why one of the best ways to understand scientific methods is to study the history of science. Or you can conduct a scientific investigation yourself.

Testing a Hypothesis

In some cases, a hypothesis may be tested by simply making additional observations. For example, if you hypothesize that a certain species of bird uses cavities in trees as places to build nests, you could observe several birds of the species and record the kinds of nests they build and where they are built.

Another common method for testing a hypothesis involves devising an experiment. An experiment is a re-creation of an event or occurrence in a way that enables a scientist to support or disprove a hypothesis. This can be difficult since a particular event may be influenced by a great many separate things. For example, the production of a song by a bird involves many activities of the bird's nervous and muscular systems and is influenced by a wide variety of environmental

factors. It might seem that developing an understanding of the factors involved in birdsong production is an impossible task. To help unclutter such situations, scientists have devised what is known as a controlled experiment. A **controlled experiment** compares two situations that have all the influencing factors identical except one. The situation used as the basis of comparison is called the *control group* and the other is called the *experimental group.* The single influencing factor that is allowed to be different in the experimental group is called the *experimental variable.*

The situation involving birdsong production would have to be broken down into a large number of simple questions, as previously mentioned. Each question would provide the basis on which experimentation would occur. Each experiment would provide information about a small part of the total process of birdsong production. For example, to test the hypothesis that male sex hormones are involved in stimulating male birds to sing, an experiment could be performed in which one group of male birds had their testes removed (the experimental group), while the control group was allowed to develop normally. After the experiment, the new data (facts) gathered would be analyzed. If there were no differences between the two groups, scientists could conclude that the variable evidently did not have a cause-and-effect relationship (i.e., was not responsible for the event). However, if there were a difference, it would be likely that the variable was responsible for the difference between the control and experimental groups. In the case of songbirds, removal of the testes does change their singing behavior.

Accept Results?

Scientists are not likely to accept the results of a single experiment, since a proposed hypothesis has to explain all experimental results. Otherwise, the hypothesis needs revision. For example, the operation necessary to remove the testes of male birds might cause illness or discomfort in some birds, resulting in less singing. A way to overcome this difficulty would be to subject all birds to the same surgery but to remove the testes of only half of them. (The control birds would still have their testes.) The results of the experiment are considered convincing only when there is one variable, many replicates (copies) of the same experiment have been conducted, and the results are consistent.

Furthermore, scientists often apply statistical tests to the results to help decide in an impartial manner if the results obtained are *valid* (meaningful; fit with other knowledge), *reliable* (give the same results repeatedly), and show cause-and-effect, or if they are just the result of random events.

During experimentation, scientists learn new information and formulate new questions that can lead to yet more experiments. One good experiment can result in a hundred new questions and experiments. The discovery of the structure of the DNA molecule by Watson and Crick resulted in thousands of experiments and stimulated the development of the entire field

Basic and Applied Research

Science is the process of understanding your environment. It begins with making observations, creating explanations, and conducting research experiments. New information and conclusions are based on the results of the research.

There are two types of scientific research: basic and applied. *Basic research* is driven by a search for understanding and may or may not have practical applications. Examples of basic research include seeking understandings about how the solar system was created, finding new information about matter by creating a new element in a research lab, or mapping temperature variations on the bottom of the Chesapeake Bay. Such basic research expands our knowledge but will not lead to practical results.

Applied research has a goal of solving some practical problem rather than just looking for answers. Examples of applied research include the creation and testing of a highly efficient fuel cell to run cars on hydrogen fuel, improving the energy efficiency of the refrigerator, or creating a faster computer chip from new materials.

Whether research is basic or applied depends somewhat on the time frame. If a practical use cannot be envisioned in the future, then it is definitely basic research. If a practical use is immediate, then the work is definitely applied research. If a practical use is developed some time in the future, then the research is partly basic and partly practical. For example, when the laser was invented, there was no practical use for it. It was called "an answer waiting for a question." Today, the laser has many, many practical applications.

Knowledge gained by basic research has sometimes resulted in the development of technological breakthroughs. On the other hand, other basic research—such as learning how the solar system formed—has no practical value other than satisfying our curiosity.

Questions to Discuss

1. Should funding priorities go to basic research, applied research, or both?
2. Should universities concentrate on basic research and industries concentrate on applied research, or should both do both types of research?
3. Should research-funding organizations specify which types of research should be funded?

---

of molecular biology (figure 1.13). Similarly, the discovery of molecules that regulate the growth of plants resulted in much research about how the molecules work and which molecules might be used for agricultural purposes.

If the processes of questioning and experimentation continue, and evidence continually and consistently supports the original hypothesis and other closely related hypotheses, the scientific community will begin to see how these hypotheses and facts fit together into a broad pattern.

Scientific Laws

Sometimes you can observe a series of relationships that seem to happen over and over again. There is a popular saying, for example, that "if anything can go wrong, it will." This is called Murphy's law. It is called a *law* because it describes a relationship between events that seems to happen time after time. If you drop a slice of buttered bread, for example, it can land two ways, butter side up or butter side down. According to Murphy's law, it will land butter side down. With this example, you know at least one way of testing the validity of Murphy's law.

Another "popular saying" type of relationship seems to exist between the cost of a houseplant and how long it lives. You could call it the "law of houseplant longevity." The relationship is that the life of a houseplant is inversely proportional to its purchase price. This "law" predicts that a $10 houseplant will wilt and die within a month, but a 50 cent houseplant will live for years. The inverse relationship is between the variables of (1) cost and (2) life span, meaning the more you pay for a plant,

FIGURE 1.13 The structure of the DNA molecule contains the genetic information of a cell. This photograph shows a model of the DNA molecule.

the shorter the time it will live. This would also mean that inexpensive plants will live for a long time. Since the relationship seems to occur time after time, it is called a law.

A **scientific law** describes an important relationship that is observed in nature to occur consistently time after time. Basically, scientific laws describe *what* happens in nature. The law is often identified with the name of a person associated with the formulation of the law. For example, with all other factors being equal, an increase in the temperature of the air in a balloon results in an increase in its volume. Likewise, a decrease in the temperature results in a decrease in the total volume of the balloon.

The volume of the balloon varies directly with the temperature of the air in the balloon, and this can be observed to occur consistently time after time. This relationship was first discovered in the latter part of the eighteenth century by two French scientists, A. C. Charles and Joseph Gay-Lussac. Today, the relationship is sometimes called *Charles' law.* When you read about a scientific *law,* you should remember that a law is a statement that means something about a relationship that you can observe time after time in nature.

Have you ever heard someone state that something behaved a certain way *because* of a scientific law? For example, a big truck accelerated slowly *because* of Newton's laws of motion. Perhaps this person misunderstands the nature of scientific laws. Scientific laws do not dictate the behavior of objects; they simply describe it. They do not say how things ought to act but rather how things *do* act. A scientific law is *descriptive;* it describes how things act.

Models and Theories

Often the part of nature being considered is too small or too large to be visible to the human eye and the use of a *model* is needed. A **model** (figure 1.14) is a description of a theory or idea that accounts for all known properties. The description can come in many different forms, such as an actual physical model, a computer model, a sketch, an analogy, or an equation. No one has ever seen the whole solar system, for example, and all you can see in the real world is the movement of the Sun, Moon, and planets against a background of stars. A physical model or sketch of the solar system, however, will give you a pretty good idea of what the solar system might look like. The physical model and the sketch are both models since they give you a mental picture of the solar system.

At the other end of the size scale, models of atoms and molecules are often used to help us understand what is happening in this otherwise invisible world. Also, a container of small, bouncing rubber balls can be used as a model to explain the relationships of Charles' law. This model helps you see what happens to invisible particles of air as the temperature, volume, and pressure of the gas change. Some models are better than others, and models constantly change along with our understanding about nature. Early twentieth-century models

of atoms, for example, were based on a "planetary model," which had electrons in the role of planets moving around the nucleus, which played the role of the Sun. Today, the model has changed as our understandings about the nature of the atom have changed. Electrons are now pictured as vibrating with certain wavelengths, which can make standing waves only at certain distances from the nucleus. Thus, the model of the atom changed from one with electrons viewed as solid particles to one that views them as vibrations on a string.

The most recently developed scientific theory was refined and expanded during the 1970s. This theory concerns the surface of Earth, and it has changed our model of what the Earth is like. At first, however, the basic idea of today's accepted theory was pure and simple conjecture. The term *conjecture* usually means an explanation or idea based on speculation, or one based on trivial grounds without any real evidence. Scientists would look at a map of Africa and South America, for example, and mull over how the two continents seem to be as pieces of a picture puzzle that had moved apart (figure 1.15). Any talk of moving continents was considered conjecture because it was not based on anything acceptable as real evidence.

Many years after the early musings about moving continents, evidence was collected from deep-sea drilling rigs that the ocean floor becomes progressively older toward the African and South American continents. This was good enough evidence to establish the "seafloor spreading hypothesis" that described the two continents moving apart.

If a hypothesis survives much experimental testing and leads, in turn, to the design of new experiments with the generation of new hypotheses that can be tested, you now have a working *theory.* A **theory** is defined as a broad, working hypothesis that is based on extensive experimental evidence. A scientific theory tells you *why* something happens. For example, the "seafloor spreading hypothesis" did survive requisite experimental testing and, together with other working hypotheses, is today found as part of the *plate tectonic theory.* The plate tectonic theory describes how the continents have moved apart, just like pieces of a picture puzzle. Is this the same idea that was once considered conjecture? Sort of, but this time it is supported by experimental evidence.

The term *scientific theory* is reserved for historic schemes of thought that have survived the test of detailed examination for long periods of time. The *atomic theory,* for example, was developed in the late 1800s and has been the subject of extensive investigation and experimentation over the last century. The atomic theory and other scientific theories form the framework of scientific thought and experimentation today. Scientific theories point to new ideas about the behavior of nature, and these ideas result in more experiments, more data to collect, and more explanations to develop. All of this may lead to a slight modification of an existing theory, a major modification, or perhaps the creation of an entirely new one. These activities continue in an ongoing attempt to satisfy the curiosity of people by understanding nature.

A

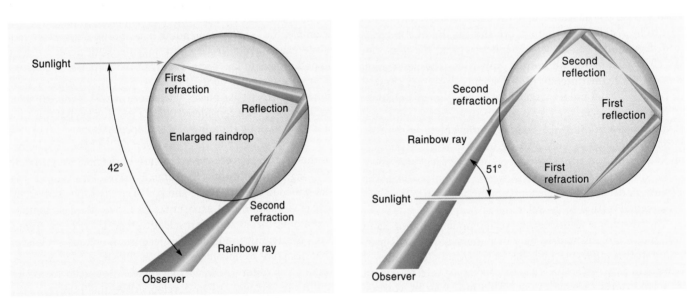

B

FIGURE 1.14 A model helps you visualize something that cannot be observed. You cannot observe what is making a double rainbow, for example, but models of light entering the upper and lower surface of a raindrop help you visualize what is happening. The drawings in (*B*) serve as a model that explains how a double rainbow is produced. (Also, see "The Rainbow" in chapter 7.)

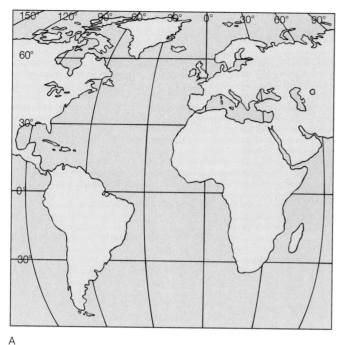

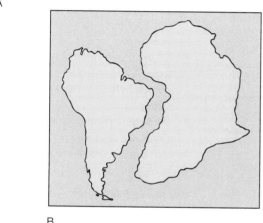

FIGURE 1.15 (*A*) Normal position of the continents on a world map. (*B*) A sketch of South America and Africa, suggesting that they once might have been joined together and subsequently separated by a continental drift.

SCIENCE, NONSCIENCE, AND PSEUDOSCIENCE

As you can see from the discussion of the nature of science, a scientific approach to the world requires a certain way of thinking. There is an insistence on ample supporting evidence by numerous studies rather than easy acceptance of strongly stated opinions. Scientists must separate opinions from statements of fact. A scientist is a healthy skeptic.

Careful attention to detail is also important. Since scientists publish their findings and their colleagues examine their work, there is a strong desire to produce careful work that can be easily defended. This does not mean that scientists do not speculate and state opinions. When they do, however, they should take great care to clearly distinguish fact from opinion.

There is also a strong ethic of honesty. Scientists are not saints, but the fact that science is conducted out in the open in front of one's peers tends to reduce the incidence of dishonesty. In addition, the scientific community strongly condemns and severely penalizes those who steal the ideas of others, perform shoddy science, or falsify data. Any of these infractions could lead to the loss of one's job and reputation.

From Experimentation to Application

The scientific method has helped us to understand and control many aspects of our natural world. Some information is extremely important in understanding the structure and functioning of things in our world but at first glance appears to have little practical value. For example, understanding the life cycle of a star may be important for people who are trying to answer questions about how the universe is changing, but it seems of little practical value to the average citizen. However, as our knowledge has increased, the time between first discovery to practical application has decreased significantly.

For example, scientists known as *genetic engineers* have altered the chemical code system of small organisms (microorganisms) so that they may produce many new drugs such as antibiotics, hormones, and enzymes. The ease with which these complex chemicals are produced would not have been possible had it not been for the information gained from the basic, theoretical sciences of microbiology, molecular biology, and genetics (see figure 1.13). Our understanding of how organisms genetically control the manufacture of proteins has led to the large-scale production of enzymes. Some of these chemicals can remove stains from clothing, deodorize, clean contact lenses, remove damaged skin from burn patients, and "stone wash" denim for clothing.

Another example is Louis Pasteur, a French chemist and microbiologist. Pasteur was interested in the theoretical problem of whether life could be generated from nonliving material. Much of his theoretical work led to practical applications in disease control. His theory that there are microorganisms that cause diseases and decay led to the development of vaccinations against rabies and the development of pasteurization for the preservation of foods (figure 1.16).

FIGURE 1.16 We can enjoy milk and other products because of pasteurization. This food preservation method was a direct result of experiments performed by Louis Pasteur (1822–1895).

Science and Society

Herbal Medicine, Legislation, and Pseudoscience

The Dietary Supplement Health and Education Act of 1994 effectively removed herbal medicines from regulation by the U.S. Food and Drug Administration (FDA). The law, which amended the federal Food, Drug, and Cosmetic Act, defined "dietary supplements" as a separate regulatory category with very few regulations.

Following passage of the Dietary Supplement Health and Education Act,

▶ Manufacturers no longer needed to ask for permission from the FDA to produce and distribute products unless it is a "new product." There is no useful definition of a "new product." Since herbal medicines have been used for centuries, manufacturers could show that most of the things they would market would have been used previously somewhere and would not be new products.

▶ Manufacturers are responsible for determining the safety of the product. The FDA can only act if it finds a product is unsafe.

▶ Manufacturers determine serving size or quantity to be taken.

▶ Manufacturers do not need to prove that the herbal product actually works. The public is not protected

Box Figure 1.2   This is Saint-John's-wort, a herbal medicine claimed to relieve depression.

from the sale of useless but harmless products.

▶ Manufacturers may not make claims about curing particular disease conditions but may make vague claims such as "improves mood," "boosts immune system," or "supports liver function."

▶ If a manufacturer makes a claim about how a product affects the structure or function of the body, it must include the following disclaimer:

"These statements have not been evaluated by the Food and Drug Administration. This product is not intended to diagnose, treat, cure, or prevent any disease."

One outcome of this change in law was a continued growth in the amount of misinformation provided by marketers of herbal medicines under the guise of scientific statements—pseudoscience.

Let's examine an example of such misinformation.

The following is a typical information statement supporting the use of Saint-John's-wort to treat depression.

▶ Provides support for a positive mood balance.*

▶ May help promote general well-being.*

▶ Grows in Europe and the United States.

▶ The botanical species is positively identified by the sophisticated thin layer chromatography (TLC) technology.

▶ TLC verification method is as accurate and reliable for identifying true herbal species as human fingerprinting.

▶ Whole ground herb is minimally processed, dried and pulverized.

▶ Each capsule contains 500 mg of Saint-John's-wort herb.

* These statements have not been evaluated by the Food and Drug Administration. This product is not intended to diagnose, treat, cure, or prevent any disease.

Science and Nonscience

The differences between science and nonscience are often based on the assumptions and methods used to gather and organize information and, most important, the testing of these assumptions. The difference between a scientist and a nonscientist is that a scientist continually challenges and tests principles and assumptions to determine a cause-and-effect relationship, whereas a nonscientist may not feel that this is important.

Once you understand the nature of science, you will not have any trouble identifying astronomy, chemistry, physics, and biology as sciences. But what about economics, sociology, anthropology, history, philosophy, and literature? All of these fields may make use of certain central ideas that are derived in a logical way, but they are also nonscientific in some ways. Some things cannot be approached using the scientific method. Art, literature, theology, and philosophy are rarely thought of as

sciences. They are concerned with beauty, human emotion, and speculative thought rather than with facts and verifiable laws. On the other hand, physics, chemistry, geology, and biology are always considered sciences.

Music is an area of study in a middle ground where scientific approaches may be used to some extent. "Good" music is certainly unrelated to science, but the study of how the human larynx generates the sound of a song is based on scientific principles. Any serious student of music will study the physics of sound and how the vocal cords vibrate to generate sound waves.

Similarly, economists use mathematical models and established economic laws to make predictions about future economic conditions. However, the regular occurrence of unpredicted economic changes indicates that economics is far from scientific, since the reliability of predictions is a central criterion of science. Anthropology and sociology are also scientific in nature in many respects, but they cannot be

Typical Herbal Supplements and Claims

Common Name of Plant	Scientific Name of Plant	Claimed Benefit
St.-John's-wort	*Hypericum perforatum*	Relieves mild or moderate depression
Gingko	*Ginkgo biloba*	Improves memory
Garlic	*Allium sativum*	Improves immune function; prevents atherosclerosis
Ginseng	*Panax quinquefolius* or *P. ginseng*	Elevates energy level
Green tea	*Camellia sinensis*	Prevents certain cancers, atherosclerosis, and tooth decay
Echinacea (purple coneflower)	*Echinacea pupurea*	Improves immune system
Black cohosh	*Cimicifuga racemosa* or *Actaea racemosa*	Relieves menopause symptoms or painful menstruation

How does this information exemplify pseudoscience?

1. The first two statements provide vague statements about mental health. They do not say that it is used to treat depression because that link has not been scientifically established to most scientists' satisfaction. The manufacturer relies on scientifically unverified statements in popular literature and on the Internet, and propagated by word of mouth to provide the demand for the product.

2. The two statements about thin layer chromatography are meant to suggest accuracy and purity. In fact, thin layer chromatography is a poor way to identify plants. It simply looks at the pigments present in plants. The use of this technique is unnecessary, since Saint-John's-wort is a plant that is easily identified by amateurs.

3. The statement about the herb being "minimally processed" is meant to suggest that nothing was lost from the plant during the preparation of the capsule and implies a high level of quality control and purity. Although the statement may be true, it is quite possible that the plant material collected contained dust, insects, and other contaminants. It is also highly likely that individual plants differ greatly in the amount of specific chemicals they contain.

Questions to Discuss

1. Select a herbal medicine and examine the claims made about this herb. Design a means of experimentally testing these claims. Decide with your group what would be acceptable evidence and what would not be acceptable.

2. Discuss with your group why people tend to ignore experimentally testing a claim.

3. Discuss with your group the possibility of a "placebo effect," that is, if someone strongly believes a claim, it might come true.

considered true sciences because many of the generalizations they have developed cannot be tested by repeated experimentation. They also do not show a significantly high degree of cause-and-effect, or they have poor predictive value.

Pseudoscience

Pseudoscience (*pseudo* = false) is a deceptive practice that uses the appearance or language of science to convince, confuse, or mislead people into thinking that something has scientific validity when it does not. When pseudoscientific claims are closely examined, they are not found to be supported by unbiased tests. For example, although nutrition is a respected scientific field, many individuals and organizations make claims about their nutritional products and diets that cannot be supported. Because of nutritional research, we all know that we must obtain certain nutrients such as amino acids, vitamins, and minerals

from the food that we eat or we may become ill. Many scientific experiments reliably demonstrate the validity of this information. However, in most cases, it has not been proven that the nutritional supplements so vigorously promoted are as useful or desirable as advertised. Rather, selected bits of scientific information (amino acids, vitamins, and minerals are essential to good health) have been used to create the feeling that additional amounts of these nutritional supplements are necessary or that they can improve your health. In reality, the average person eating a varied diet will obtain all of these nutrients in adequate amounts, and nutritional supplements are not required.

Another related example involves the labeling of products as organic or natural. Marketers imply that organic or natural products have greater nutritive value because they are organically grown (grown without pesticides or synthetic fertilizers) or because they come from nature. Although there are questions about the health effects of trace amounts of

People Behind the Science

Florence Bascom (1862–1945)

Florence Bascom, a U.S. geologist, was an expert in the study of rocks and minerals and founded the geology department at Bryn Mawr College, Pennsylvania. This department was responsible for training the foremost women geologists of the early twentieth century.

Born in Williamstown, Massachusetts, in 1862, Bascom was the youngest of the six children of suffragist and schoolteacher Emma Curtiss Bascom and William Bascom, professor of philosophy at Williams College. Her father, a supporter of suffrage and the education of women, later became president of the University of Wisconsin, to which women were admitted in 1875. Florence Bascom enrolled there in 1877 and with other women was allowed limited access to the facilities but was denied access to classrooms filled with men. In spite of this, she earned a B.A. in 1882, a B.Sc. in 1884, and an M.S. in 1887. When Johns Hopkins University graduate school opened to women in 1889, Bascom was allowed to enroll to study geology on the condition that she sit behind a screen to avoid distracting the male students. With the support of her advisor, George Huntington Williams, and her father, she managed in 1893 to become the second woman to gain a Ph.D. in geology (the first being Mary Holmes at the University of Michigan in 1888).

Bascom's interest in geology had been sparked by a driving tour she took with her father and his friend Edward Orton, a geology professor at Ohio State University. It was an exciting time for geologists with new areas opening up all the time. Bascom was also inspired by her teachers at Wisconsin and Johns Hopkins, who were experts in the new fields of metamorphism and crystallography. Bascom's Ph.D. thesis was a study of rocks that had previously been thought to be sediments but which she proved to be metamorphosed lava flows.

While studying for her doctorate, Bascom became a popular teacher, passing on her enthusiasm and rigor to her students. She taught at the Hampton Institute for Negroes and American Indians and at Rockford College before becoming an instructor and associate professor of geology at Ohio State University from 1892 to 1895. Moving to Bryn Mawr College, where geology was considered subordinate to the other sciences, she spent two years teaching in a storeroom while building a considerable collection of fossils, rocks, and minerals. While at Bryn Mawr, she took great pride in passing on her knowledge and training to a generation of women who would become successful. At Bryn Mawr, she rose rapidly, becoming reader (1898), associate professor (1903), professor (1906), and finally professor emerita from 1928 till her death in 1945 in Northampton, Massachusetts.

Bascom became, in 1896, the first woman to work as a geologist on the U.S. Geological Survey, spending her summers mapping formations in Pennsylvania, Maryland, and New Jersey and her winters analyzing slides. Her results were published in Geographical Society of America bulletins. In 1924, she became the first woman to be elected a fellow of the Geographical Society and went on, in 1930, to become the first woman vice president. She was associate editor of the *American Geologist* (1896–1905) and achieved a four-star place in the first edition of *American Men and Women of Science* (1906), a sign of how highly regarded she was in her field.

Bascom was the author of over forty research papers. She was an expert on the crystalline rocks of the Appalachian Piedmont, and she published her research on Piedmont geomorphology. Geologists in the Piedmont area still value her contributions, and she is still a powerful model for women seeking status in the field of geology today.

Source: Modified from the *Hutchinson Dictionary of Scientific Biography*. © RM, 2007. Reprinted by permission.

---

pesticides in foods, no scientific study has shown that a diet of natural or organic products has any benefit over other diets. The poisons curare, strychnine, and nicotine are all organic molecules that are produced in nature by plants that could be grown organically, but we would not want to include them in our diet.

Absurd claims that are clearly pseudoscience sometimes appear to gain public acceptance because of promotion in the media. Thus, some people continue to believe stories that psychics can really help solve puzzling crimes, that perpetual energy machines exist, or that sources of water can be found by a person with a forked stick. Such claims could be subjected to scientific testing and disposed of if they fail the test, but this process is generally ignored. In addition to experimentally testing such a claim that appears to be pseudoscience, here are some questions that you should consider when you suspect something is pseudoscience:

1. What is the background and scientific experience of the person promoting the claim?
2. How many articles have been published by the person in peer-reviewed scientific journals?
3. Has the person given invited scientific talks at universities and national professional organization meetings?
4. Has the claim been researched and published by the person in a peer-reviewed scientific journal *and* have other scientists independently validated the claim?
5. Does the person have something to gain by making the claim?

Limitations of Science

By definition, science is a way of thinking and seeking information to solve problems. Therefore, scientific methods can be applied only to questions that have a factual basis. Questions concerning morals, value judgments, social issues, and attitudes cannot be answered using the scientific methods. What makes a painting great? What is the best type of music? Which wine is best? What color should I paint my car? These questions are related to values, beliefs, and tastes; therefore, scientific methods cannot be used to answer them.

Science is also limited by the ability of people to pry understanding from the natural world. People are fallible and do not always come to the right conclusions, because information is lacking or misinterpreted, but science is self-correcting. As new information is gathered, old incorrect ways of thinking must be changed or discarded. For example, at one time, people were sure that the Sun went around Earth. They observed that the Sun rose in the east and traveled across the sky to set in the west. Since they could not feel Earth moving, it seemed perfectly logical that the sun traveled around Earth. Once they understood that Earth rotated on its axis, people began to understand that the rising and setting of the Sun could be explained in other ways. A completely new concept of the relationship between the Sun and Earth developed.

Although this kind of study seems rather primitive to us today, this change in thinking about the Sun and Earth was a very important step in understanding the universe and how the various parts are related to one another. This background information was built upon by many generations of astronomers and space scientists, and finally led to space exploration.

People also need to understand that science cannot answer all the problems of our time. Although science is a powerful tool, there are many questions it cannot answer and many problems it cannot solve. The behavior and desires of people generate most of the problems societies face. Famine, drug abuse, and pollution are human-caused and must be resolved by humans. Science may provide some tools for social planners, politicians, and ethical thinkers, but science does not have, nor does it attempt to provide, all the answers to the problems of the human race. Science is merely one of the tools at our disposal.

SUMMARY

Science is a search for order in our surroundings. People have *concepts,* or mental images, about material *objects* and intangible *events* in their surroundings. Concepts are used for thinking and communicating. Concepts are based on *properties,* or attributes that describe a thing or event. Every property implies a *referent* that describes the property. Referents are not always explicit, and most communications require assumptions.

Measurement is a process that uses a well-defined and agreed upon *referent* to describe a *standard unit.* The unit is compared to the property being defined by an *operation* that determines the *value* of the unit by *counting.* Measurements are always reported with a *number,* or value, and a *name* for the unit.

The two major *systems* of standard units are the *English system* and the *metric system.* The English system uses standard units that were originally based on human body parts, and the metric system uses standard units based on referents found in nature. The metric system also uses a system of prefixes to express larger or smaller amounts of units. The metric standard units for length, mass, and time are the *meter, kilogram,* and *second.*

Measurement information used to describe something is called *data.* One way to extract meanings and generalizations from data is to use a *ratio,* a simplified relationship between two numbers. Density is a ratio of mass to volume, or $\rho = m/V.$

Symbols are used to represent *quantities,* or measured properties. Symbols are used in *equations,* which are shorthand statements that describe a relationship where the quantities (both number values and units) are identical on both sides of the equal sign. Equations are used to (1) *describe* a property, (2) *define* a concept, or (3) *describe* how *quantities change* together.

Quantities that can have different values at different times are called *variables.* Variables that increase or decrease together in the same ratio are said to be in *direct proportion.* If one variable increases while the other decreases in the same ratio, the variables are in *inverse proportion.* Proportionality statements are not necessarily equations. A *proportionality constant* can be used to make such a statement into an equation. Proportionality constants might have numerical value only, without units, or they might have both value and units.

Modern science began about three hundred years ago during the time of Galileo and Newton. Since that time, *scientific investigation* has been used to provide *experimental evidence* about nature. The investigations provide *accurate, specific,* and *reliable* data that are used to develop and test *explanations.* A *hypothesis* is a tentative explanation that is accepted or rejected from experimental data. An accepted hypothesis may result in a *scientific law,* an explanation concerned with important phenomena. Laws are sometimes identified with the name of a scientist and can be expressed verbally, with an equation, or with a graph.

A *model* is used to help understand something that cannot be observed directly, explaining the unknown in terms of things already understood. Physical models, mental models, and equations are all examples of models that explain how nature behaves. A *theory* is a broad, detailed explanation that guides development and interpretations of experiments in a field of study.

Science and *nonscience* can be distinguished by the kinds of laws and rules that are constructed to unify the body of knowledge. Science involves the continuous *testing* of rules and principles by the collection of new facts. If the rules are not testable, or if no rules are used, it is not science. *Pseudoscience* uses scientific appearances to mislead.

Summary of Equations

1.1
$$density = \frac{mass}{volume}$$

$$\rho = \frac{m}{V}$$

KEY TERMS

controlled experiment (p. **13**)
data (p. **7**)
density (p. **8**)
equation (p. **10**)

hypothesis (p. **13**)
measurement (p. **4**)
model (p. **15**)
properties (p. **3**)

pseudoscience (p. **19**)
quantities (p. **10**)
referent (p. **4**)
scientific law (p. **15**)

standard unit (p. **4**)
theory (p. **15**)
unit (p. **4**)
variables (p. **11**) .

APPLYING THE CONCEPTS

1. The process of comparing a property of an object to a well-defined and agreed-upon referent is called the process of
   a. generalizing.
   b. measurement.
   c. graphing.
   d. scientific investigation.

2. The height of an average person is closest to
   a. 1.0 m.
   b. 1.5 m.
   c. 2.5 m.
   d. 3.5 m.

3. Which of the following standard units is defined in terms of an object as opposed to an event?
   a. kilogram
   b. meter
   c. second
   d. none of the above

4. One-half liter of water has a mass of
   a. 0.5 g.
   b. 5 g.
   c. 50 g.
   d. 500 g.

5. A cubic centimeter (cm^3) of water has a mass of about 1
   a. mL.
   b. kg.
   c. g.
   d. dm.

6. Measurement information that is used to describe something is called
   a. referents.
   b. properties.
   c. data.
   d. a scientific investigation.

7. The property of volume is a measure of
   a. how much matter an object contains.
   b. how much space an object occupies.
   c. the compactness of matter in a certain size.
   d. the area on the outside surface.

8. As the volume of a cube becomes larger and larger, the surface area-to-volume ratio
   a. increases.
   b. decreases.
   c. remains the same.
   d. sometimes increases and sometimes decreases.

9. If you consider a very small portion of a material that is the same throughout, the density of the small sample will be
   a. much less.
   b. slightly less.
   c. the same.
   d. greater.

10. A scientific law can be expressed as
    a. a written concept.
    b. an equation.
    c. a graph.
    d. Any of the above is correct.

Answers

1. b 2. b 3. a 4. d 5. c 6. c 7. b 8. b 9. c 10. d

QUESTIONS FOR THOUGHT

1. What is a concept?
2. What are two components of a measurement statement? What does each component tell you?
3. Other than familiarity, what are the advantages of the English system of measurement?
4. Define the metric standard units for length, mass, and time.
5. Does the density of a liquid change with the shape of a container? Explain.
6. Does a flattened pancake of clay have the same density as the same clay rolled into a ball? Explain.
7. Compare and contrast a scientific hypothesis and a scientific law.
8. What is a model? How are models used?
9. Are all theories always completely accepted or completely rejected? Explain.
10. What is pseudoscience and how can you always recognize it?

FOR FURTHER ANALYSIS

1. Select a statement that you feel might represent pseudoscience. Write an essay supporting *and* refuting your selection, noting facts that support one position or the other.
2. Evaluate the statement that science cannot solve human-produced problems such as pollution. What does it mean to say pollution is caused by humans and can only be solved by humans? Provide evidence that supports your position.
3. Make an experimental evaluation of what happens to the density of a substance at larger and larger volumes.
4. If your wage were dependent on your work-time squared, how would it affect your pay if you double your hours?
5. *Merriam-Webster's 11th Collegiate Dictionary* defines science, in part, as "knowledge or a system of knowledge covering general truths or the operation of general laws especially as obtained and tested through scientific method." How would you define science?
6. Are there any ways in which scientific methods differ from common-sense methods of reasoning?
7. The United States is the only country in the world that does not use the metric system of measurement. With this understanding, make a list of advantages and disadvantages for adopting the metric system in the United States.

INVITATION TO INQUIRY

Paper Helicopters

Construct paper helicopters and study the effects that various variables have on their flight. After considering the size you wish to test, copy the pattern shown in figure 1.17A on a sheet of notebook paper. Note that solid lines are to be cut and dashed lines are to be folded. Make three scissor cuts on the solid lines. Fold A toward you and B away from you to form the wings. Then fold C and D inward to overlap, forming the body. Finally, fold up the bottom on the dashed line and hold it together with a paper clip. Your finished product should look like the helicopter in figure 1.17B. Try a preliminary flight test by standing on a chair or stairs and dropping your helicopter.

Decide what variables you would like to study to find out how they influence the total flight time. Consider how you will hold everything else constant while changing one variable at a time. You can change the wing area by making new helicopters with more or less area in the A and B flaps. You can change the weight by adding more paper clips. Study these and other variables to find out who can design a helicopter that will remain in the air the longest. Who can design a helicopter that is most accurate in hitting a target?

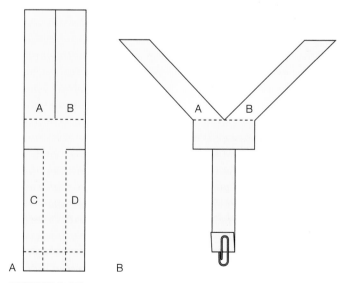

FIGURE 1.17 Pattern for a paper helicopter.

PARALLEL EXERCISES

The exercises in groups A and B cover the same concepts. Solutions to group A exercises are located in appendix D.

Group A

Note: You will need to refer to table 1.3 to complete some of the following exercises.

1. What is your height in meters? In centimeters?
2. What is the mass density of mercury if 20.0 cm³ has a mass of 272 g?
3. What is the mass of a 10.0 cm³ cube of lead?
4. What is the volume of a rock with a mass density of 3.00 g/cm³ and a mass of 600 g?
5. If you have 34.0 g of a 50.0 cm³ volume of one of the substances listed in table 1.3, which one is it?
6. What is the mass of water in a 40 L aquarium?

7. A 2.1 kg pile of aluminum cans is melted, then cooled into a solid cube. What is the volume of the cube?
8. A cubic box contains 1,000 g of water. What is the length of one side of the box in meters? Explain your reasoning.
9. A loaf of bread (volume 3,000 cm³) with a density of 0.2 g/cm³ is crushed in the bottom of a grocery bag into a volume of 1,500 cm³. What is the density of the mashed bread?
10. According to table 1.3, what volume of copper would be needed to balance a 1.00 cm³ sample of lead on a two-pan laboratory balance?

Group B

Note: You will need to refer to table 1.3 to complete some of the following exercises.

1. What is your mass in kilograms? In grams?
2. What is the mass density of iron if 5.0 cm³ has a mass of 39.5 g?
3. What is the mass of a 10.0 cm³ cube of copper?
4. If ice has a mass density of 0.92 g/cm³, what is the volume of 5,000 g of ice?
5. If you have 51.5 g of a 50.0 cm³ volume of one of the substances listed in table 1.3, which one is it?
6. What is the mass of gasoline ($\rho = 0.680$ g/cm³) in a 94.6 L gasoline tank?
7. What is the volume of a 2.00 kg pile of iron cans that are melted, then cooled into a solid cube?
8. A cubic tank holds 1,000.0 kg of water. What are the dimensions of the tank in meters? Explain your reasoning.
9. A hot dog bun (volume 240 cm³) with a density of 0.15 g/cm³ is crushed in a picnic cooler into a volume of 195 cm³. What is the new density of the bun?
10. According to table 1.3, what volume of iron would be needed to balance a 1.00 cm³ sample of lead on a two-pan laboratory balance?

Information about the mass of a hot air balloon and forces on the balloon will enable you to predict if it is going to move up, down, or drift across the river. This chapter is about such relationships between force, mass, and changes in motion.

Motion

OVERVIEW

In chapter 1, you learned some "tools and rules" and some techniques for finding order in your surroundings. Order is often found in the form of patterns, or relationships between quantities that are expressed as equations. Equations can be used to (1) describe properties, (2) define concepts, and (3) describe how quantities change relative to one another. In all three uses, patterns are quantified, conceptualized, and used to gain a general understanding about what is happening in nature.

In the study of science, certain parts of nature are often considered and studied together for convenience. One of the more obvious groupings involves *movement*. Most objects around you appear to spend a great deal of time sitting quietly without

motion. Buildings, rocks, utility poles, and trees rarely, if ever, move from one place to another. Even things that do move from time to time sit still for a great deal of time. This includes you, automobiles, and bicycles (figure 2.1). On the other hand, the Sun, the Moon, and starry heavens always seem to move, never standing still. Why do things stand still? Why do things move?

Questions about motion have captured the attention of people for thousands of years. But the ancient people answered questions about motion with stories of mysticism and spirits that lived in objects. It was during the classic Greek culture, between 600 B.C. and 300 B.C., that people began to look beyond magic and spirits. One particular Greek philosopher, Aristotle, wrote a theory

about the universe that offered not only explanations about things such as motion but also offered a sense of beauty, order, and perfection. The theory seemed to fit with other ideas that people had and was held to be correct for nearly two thousand years after it was written. It was not until the work of Galileo and Newton during the 1600s that a new, correct understanding about motion was developed. The development of ideas about motion is an amazing and absorbing story. You will learn in this chapter how to describe and use some properties of motion. This will provide some basic understandings about motion and will be very helpful in understanding some important aspects of astronomy and the earth sciences, as well as the movement of living things.

DESCRIBING MOTION

Motion is one of the more common events in your surroundings. You can see motion in natural events such as clouds moving, rain and snow falling, and streams of water, all moving in a never-ending cycle. Motion can also be seen in the activities of people who walk, jog, or drive various machines from place to place. Motion is so common that you would think everyone would intuitively understand the concepts of motion, but history indicates that it was only during the past three hundred years or so that people began to understand motion correctly. Perhaps the correct concepts are subtle and contrary to common sense, requiring a search for simple, clear concepts in an otherwise complex situation. The process of finding such order in a multitude of sensory impressions by taking measurable data and then inventing a concept to

describe what is happening is the activity called *science*. We will now apply this process to motion.

What is motion? Consider a ball that you notice one morning in the middle of a lawn. Later in the afternoon, you notice that the ball is at the edge of the lawn, against a fence, and you wonder if the wind or some person moved the ball. You do not know if the wind blew it at a steady rate, if many gusts of wind moved it, or even if some children kicked it all over the yard. All you know for sure is that the ball has been moved because it is in a different position after some time passed. These are the two important aspects of motion: (1) a change of position and (2) the passage of time.

If you did happen to see the ball rolling across the lawn in the wind, you would see more than the ball at just two locations. You would see the ball moving continuously. You could consider, however, the ball in continuous motion to be a series of individual locations with very small time intervals. Moving

FIGURE 2.1 The motion of this windsurfer, and of other moving objects, can be described in terms of the distance covered during a certain time period.

involves a change of position during some time period. Motion is the act or process of something changing position.

The motion of an object is usually described with respect to something else that is considered to be not moving. (Such

a stationary object is said to be "at rest.") Imagine that you are traveling in an automobile with another person. You know that you are moving across the land outside the car since your location on the highway changes from one moment to another. Observing your fellow passenger, however, reveals no change of position. You are in motion relative to the highway outside the car. You are not in motion relative to your fellow passenger. Your motion, and the motion of any other object or body, is the process of a change in position *relative* to some reference object or location. Thus, *motion* can be defined as the act or process of changing position relative to some reference during a period of time.

MEASURING MOTION

You have learned that objects can be described by measuring certain fundamental properties such as mass and length. Since motion involves (1) a change of position and (2) the passage of *time*, the motion of objects can be described by using combinations of the fundamental properties of length and time. Combinations of these measurements describe three properties of motion: *speed, velocity,* and *acceleration.*

Speed

Suppose you are in a car that is moving over a straight road. How could you describe your motion? You need at least two measurements: (1) the distance you have traveled and (2) the time that has elapsed while you covered this distance. Such a distance and time can be expressed as a ratio that describes your motion. This ratio is a property of motion called **speed,** which is a measure of how fast you are moving. Speed is defined as distance per unit of time, or

$$\text{speed} = \frac{\text{distance}}{\text{time}}$$

The units used to describe speed are usually miles/hour (mi/h), kilometers/hour (km/h), or meters/second (m/s).

Let's go back to your car that is moving over a straight highway and imagine you are driving to cover equal distances in equal periods of time (figure 2.2). If you use a stopwatch to measure the time required to cover the distance between

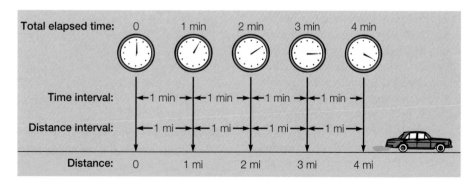

FIGURE 2.2 This car is moving in a straight line over a distance of 1 mi each minute. Therefore, the car moves 60 mi in 60 min and has a speed of 60 mi/h.

highway mile markers (those little signs with numbers along major highways), the time intervals will all be equal. You might find, for example, that one minute elapses between each mile marker. Such a uniform straight-line motion that covers equal distances in equal periods of time is the simplest kind of motion.

If your car were moving over equal distances in equal periods of time, it would have a *constant speed.* This means that the car is neither speeding up nor slowing down. It is usually difficult to maintain a constant speed. Other cars and distractions such as interesting scenery cause you to reduce your speed. At other times you increase your speed. If you calculate your speed over an entire trip, you are considering a large distance between two places and the total time that elapsed. The increases and decreases in speed would be averaged. Therefore, most speed calculations are for an *average speed.* The speed at any specific instant is called the *instantaneous speed.* To calculate the instantaneous speed, you would need to consider a very short time interval—one that approaches zero. An easier way would be to use the speedometer, which shows the speed at any instant.

It is easier to study the relationships between quantities if you use symbols instead of writing out the whole word. The letter v can be used to stand for speed when dealing with straight-line motion, which is the only kind of motion that will be considered in the problems in this text. A bar over the v ($\bar{v}$) is a symbol that means average (it is read "v-bar" or "v-average"). The letter d can be used to stand for distance and the letter t to stand for time. The relationship between average speed, distance, and time is therefore

$$\bar{v} = \frac{d}{t} \qquad \textbf{equation 2.1}$$

This is one of the three types of equations that were discussed earlier, and in this case, the equation defines a motion property.

Constant, instantaneous, or average speeds can be measured with any distance and time units. Common units in the English system are miles/hour and feet/second. Metric units for speed are commonly kilometers/hour and meters/second. The ratio of any distance/time is usually read as distance per time, such as miles per hour.

Velocity

The word *velocity* is sometimes used interchangeably with the word *speed,* but there is a difference. **Velocity** describes the *speed and direction* of a moving object. For example, a speed might be described as 60 km/h. A velocity might be described as 60 km/h to the west. To produce a change in velocity, either the speed or the direction is changed (or both are changed). A satellite moving with a constant speed in a circular orbit around Earth does not have a constant velocity since its direction of movement is constantly changing. Velocities can be represented graphically with arrows. The lengths of the arrows are proportional to the speed, and the arrowheads indicate the direction (figure 2.3).

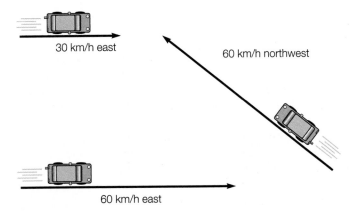

FIGURE 2.3 Velocity can be presented graphically with arrows. Here are three different velocities represented by three different arrows. The length of each arrow is proportional to the speed, and the arrowhead shows the direction of travel.

EXAMPLE 2.1 *(Optional)*

The driver of a car moving at 72.0 km/h drops a road map on the floor. It takes her 3.00 seconds to locate and pick up the map. How far did she travel during this time?

SOLUTION

The car has a speed of 72.0 km/h and the time factor is 3.00 s, so km/h must be converted to m/s. From inside the front cover of this book, the conversion factor is 1 km/h = 0.2778 m/s, so

$$\bar{v} = \frac{0.2778 \frac{m}{s}}{\frac{km}{h}} \times 72.0 \frac{km}{h}$$

$$= (0.2778)(72.0) \frac{m}{s} \times \frac{h}{km} \times \frac{km}{h}$$

$$= 20.0 \frac{m}{s}$$

The relationship between the three variables, $\bar{v}$, t, and d, is found in equation 2.1: $\bar{v} = d/t$

$$\begin{array}{ll} \bar{v} = 20.0 \frac{m}{s} & \bar{v} = \frac{d}{t} \\ t = 3.00 \text{ s} & \bar{v}t = \frac{d\cancel{t}}{\cancel{t}} \\ d = ? & d = \bar{v}t \\ & = \left(20.0 \frac{m}{s}\right)(3.00 \text{ s}) \\ & = (20.0)(3.00) \frac{m}{\cancel{s}} \times \frac{\cancel{s}}{1} \\ & = \boxed{60.0 \text{ m}} \end{array}$$

EXAMPLE 2.2 *(Optional)*

A bicycle has an average speed of 8.00 km/h. How far will it travel in 10.0 seconds? (Answer: 22.2 m)

Acceleration

Motion can be changed in three different ways: (1) by changing the speed, (2) by changing the direction of travel, or (3) by changing both the speed and the direction of travel. Since velocity describes both the speed and the direction of travel, any of these three changes will result in a change of velocity. You need at least one additional measurement to describe a change of motion, which is how much time elapsed while the change was taking place. The change of velocity and time can be combined to define the *rate* at which the motion was changed. This rate is called **acceleration.** Acceleration is defined as a change of velocity per unit time, or

$$\text{acceleration} = \frac{\text{change of velocity}}{\text{time elapsed}}$$

Another way of saying "change in velocity" is the final velocity minus the initial velocity, so the relationship can also be written as

$$\text{acceleration} = \frac{\text{final velocity} - \text{initial velocity}}{\text{time elapsed}}$$

Acceleration due to a change in speed only can be calculated as follows. Consider a car that is moving with a constant, straight-line velocity of 60 km/h when the driver accelerates

---

---

to 80 km/h. Suppose it takes 4 s to increase the velocity of 60 km/h to 80 km/h. The change in velocity is therefore 80 km/h minus 60 km/h, or 20 km/h. The acceleration was

$$\text{acceleration} = \frac{80\,\dfrac{\text{km}}{\text{h}} - 60\,\dfrac{\text{km}}{\text{h}}}{4\text{s}}$$

$$= \frac{20\,\dfrac{\text{km}}{\text{h}}}{4\text{s}}$$

$$= 5\,\frac{\text{km/h}}{\text{s}}\ \text{or,}$$

$$= 5\ \text{km/h/s}$$

The average acceleration of the car was 5 km/h for each ("per") second. This is another way of saying that the velocity increases an average of 5 km/h in each second. The velocity of the car was 60 km/h when the acceleration began (initial velocity). At the end of 1 s, the velocity was 65 km/h. At the end of 2 s, it was 70 km/h; at the end of 3 s, 75 km/h; and at the end of 4 s (total time elapsed), the velocity was 80 km/h (final velocity). Note how fast the velocity is changing with time. In summary,

start (initial velocity)	60 km/h
end of first second	65 km/h
end of second second	70 km/h
end of third second	75 km/h
end of fourth second (final velocity)	80 km/h

As you can see, acceleration is really a description of how fast the speed is changing; in this case, it is increasing 5 km/h each second.

Usually, you would want all the units to be the same, so you would convert km/h to m/s. A change in velocity of 5.0 km/h converts to 1.4 m/s and the acceleration would be 1.4 m/s/s. The units "m/s per s" mean what change of velocity (1.4 m/s) is occurring every second. The combination "m/s/s" is rather cumbersome, so it is typically treated mathematically to simplify the expression (to simplify a fraction, invert the divisor and multiply, or $\text{m/s} \times 1/\text{s} = \text{m/s}^2$). Remember that the expression 1.4 m/s² means the same as 1.4 m/s per s, a change of velocity in a given time period.

The relationship among the quantities involved in acceleration can be represented with the symbols a for average acceleration, v_f for final velocity, v_i for initial velocity, and t for time. The relationship is

$$a = \frac{v_f - v_i}{t} \qquad \textbf{equation 2.2}$$

There are also other changes in the motion of an object that are associated with acceleration. One of the more obvious is a change that results in a decreased velocity. Your car's brakes, for example, can slow your car or bring it to a complete stop. This is *negative acceleration*, which is sometimes called *deceleration*. Another change in the motion of an object is a change of direction. Velocity encompasses both the rate of motion as well as direction, so a change of direction is an acceleration. The satellite

Connections . . .

Travel

The super-speed magnetic levitation (maglev) train is a completely new technology based on magnetically suspending a train 3 to 10 cm (about 1 to 4 in) above a monorail, then moving it along with a magnetic field that travels along the monorail guides. The maglev train does not have friction between wheels and the rails since it does not have wheels. This lack of resistance and the easily manipulated magnetic fields makes very short acceleration distances possible. For example, a German maglev train can accelerate from 0 to 300 km/h (about 185 mi/h) over a distance of just 5 km (about 3 mi). A conventional train with wheels requires about 30 km (about 19 mi) in order to reach the same speed from a standing start. The maglev is attractive for short runs because of its superior acceleration. It is also attractive for longer runs because of its high top speed—up to about 500 km/h (about 310 mi/h). Today, only an aircraft can match such a speed.

moving with a constant speed in a circular orbit around Earth is constantly changing its direction of movement. It is therefore constantly accelerating because of this constant change in its motion. Your automobile has three devices that could change the state of its motion. Your automobile therefore has three accelerators—the gas pedal (which can increase magnitude of

CONCEPTS APPLIED

Acceleration Patterns

Suppose the radiator in your car has a leak and drops of coolant fall constantly, one every second. What pattern would the drops make on the pavement when you accelerate the car from a stoplight? What pattern would they make when you drive a constant speed? What pattern would you observe as the car comes to a stop? Use a marker to make dots on a sheet of paper that illustrate (1) acceleration, (2) constant speed, and (3) negative acceleration. Use words to describe the acceleration in each situation.

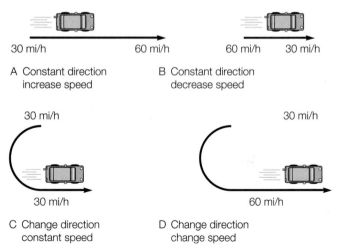

A Constant direction increase speed

B Constant direction decrease speed

C Change direction constant speed

D Change direction change speed

FIGURE 2.4 Four different ways (A–D) to accelerate a car.

velocity), the brakes (which can decrease magnitude of velocity), and the steering wheel (which can change direction of velocity). (See figure 2.4.) The important thing to remember is that acceleration results from any *change* in the motion of an object.

EXAMPLE 2.3 *(Optional)*

A bicycle moves from rest to 5 m/s in 5 s. What was the acceleration?

SOLUTION

$$v_i = 0 \text{ m/s}$$
$$v_f = 5 \text{ m/s}$$
$$t = 5 \text{ s}$$
$$a = ?$$

$$a = \frac{v_f - v_i}{t}$$

$$= \frac{5 \text{ m/s} - 0 \text{ m/s}}{5 \text{ s}}$$

$$= \frac{5}{5} \frac{\text{m/s}}{\text{s}}$$

$$= 1 \frac{\text{m}}{\text{s}} \times \frac{1}{\text{s}}$$

$$= \boxed{1 \frac{\text{m}}{\text{s}^2}}$$

EXAMPLE 2.4 *(Optional)*

An automobile uniformly accelerates from rest at 15 ft/s² for 6 s. What is the final velocity in ft/s? (Answer: 90 ft/s)

FORCES

The Greek philosopher Aristotle considered some of the first ideas about the causes of motion back in the fourth century B.C. However, he had it all wrong when he reportedly stated that a dropped object falls at a constant speed that is determined by its weight. He also incorrectly thought that an object moving across Earth's surface requires a continuously applied force to continue moving. These ideas were based on observing and thinking, not measurement, and no one checked to see if they were correct. It took about two thousand years before people began to correctly understand motion.

Aristotle did recognize an association between force and motion, and this much was acceptable. It is partly correct because a force is closely associated with *any* change of motion, as you will see. This section introduces the concept of a force, which will be developed more fully when the relationship between forces and motion is considered later.

A **force** is a push or a pull that is capable of changing the state of motion of an object. Consider, for example, the movement of a ship from the pushing of two tugboats (figure 2.5). Tugboats can vary the strength of the force exerted on a ship, but they can also push in different directions. What effect does direction have on two forces acting on an object? If the tugboats were side by side, pushing in the same direction, the overall force is the sum of the two forces. If they act in exactly opposite directions, one pushing on each side of the ship, the overall force is the difference between the strength of the two forces. If they have the same strength, the overall effect is to cancel each other without producing any motion. The **net force** is the sum of all the forces acting on an object. Net force means "final," after the forces are added (figure 2.6).

FIGURE 2.5 The rate of movement and the direction of movement of this ship are determined by a combination of direction and size of force from each of the tugboats. Which direction are the two tugboats pushing? What evidence would indicate that one tugboat is pushing with a greater force? If the tugboat by the numbers is pushing with a greater force and the back tugboat is keeping the back of the ship from moving, what will happen?

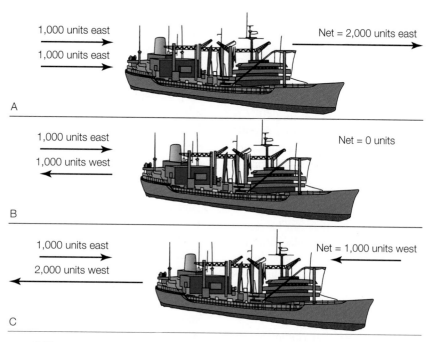

FIGURE 2.6 (A) When two parallel forces are acting on the ship in the same direction, the net force is the two forces added together. (B) When two forces are opposite and of equal size, the net force is zero. (C) When two parallel forces are not of equal size, the net force is the difference in the direction of the larger force.

Weather

Classification schemes are imaginative mental constructions used to show similarities and differences in objects or events. For example, the following describes two schemes used to help us classify storms that are not associated with weather fronts.

What is the difference between a tropical depression, tropical storm, and a hurricane? They are classified according to the *speed* of the maximum sustained surface winds. In the United States, the maximum sustained surface wind is measured by averaging the wind speed over a 1-minute period. Here is the classification scheme:

Tropical Depression. This is a center of low pressure around which the winds are generally moving 55 km/h (about 35 mi/h) or less. The tropical depression might dissolve into nothing, or it might develop into a more intense disturbance.

Tropical Storm. This is a more intense, highly organized center of low pressure with winds between 56 and 120 km/h (about 35 to 75 mi/h).

Hurricane. This is an intense low-pressure center with winds greater than 120 km/h (about 75 mi/h). A strong storm of this type is called a "hurricane" if it occurs over the Atlantic Ocean or the Pacific Ocean east of the international date line. It is called a "typhoon" if it occurs over the North Pacific Ocean west of the international date line. Hurricanes are further classified according to category and damage to be expected. Here is the classification scheme:

Category	Damage	Winds
1	minimal	120–153 km/h (75–95 mi/h)
2	moderate	154–177 km/h (96–110 mi/h)
3	extensive	178–210 km/h (111–130 mi/h)
4	extreme	211–250 km/h (131–155 mi/h)
5	catastrophic	>250 km/h (>155 mi/h)

When two parallel forces act in the same direction, they can be simply added. In this case, there is a net force that is equivalent to the sum of the two forces. When two parallel forces act in opposite directions, the net force is the difference between the two forces and is in the direction of the larger force. When two forces act neither in a way that is exactly together nor exactly opposite each other, the result will be like a new, different force having a new direction and strength.

Forces have a strength and direction that can be represented by force arrows. The tail of the arrow is placed on the object that feels the force, and the arrowhead points in the direction in which the force is exerted. The length of the arrow is proportional to the strength of the force. The use of force arrows helps you visualize and understand all the forces and how they contribute to the net force.

There are four **fundamental forces** that *cannot* be explained in terms of any other force. They are gravitational, electromagnetic, weak, and the strong nuclear force. Gravitational forces act between all objects in the universe—between you and Earth, between Earth and the Sun, between the planets in

the solar systems, and, in fact, hold stars in large groups called galaxies. Switching scales from the very large galaxy to inside an atom, we find electromagnetic forces acting between electrically charged parts of atoms, such as electrons and protons. Electromagnetic forces are responsible for the structure of atoms, chemical change, and electricity and magnetism. Weak and strong forces act inside the nucleus of an atom, so they are not as easily observed at work as are gravitational and electromagnetic forces. The weak force is involved in certain nuclear reactions. The strong nuclear force is involved in close-range holding of the nucleus together. In general, the strong nuclear force between particles inside a nucleus is about 10^2 times stronger than the electromagnetic force and about 10^{39} times stronger than the gravitational force. The fundamental forces are responsible for everything that happens in the universe, and we will learn more about them in chapters on electricity, light, nuclear energy, chemistry, geology, and astronomy.

HORIZONTAL MOTION ON LAND

Everyday experience seems to indicate that Aristotle's idea about horizontal motion on Earth's surface is correct. After all, moving objects that are not pushed or pulled do come to rest in a short period of time. It would seem that an object keeps moving only if a force continues to push it. A moving automobile will slow and come to rest if you turn off the ignition. Likewise, a ball that you roll along the floor will slow until it comes to rest. Is the natural state of an object to be at rest, and is a force necessary to keep an object in motion? This is exactly what people thought until Galileo (figure 2.7) published his book *Two New Sciences* in 1638, which described his findings about motion.

The book had three parts that dealt with uniform motion, accelerated motion, and projectile motion. Galileo described details of simple experiments, measurements, calculations, and thought experiments as he developed definitions and concepts of motion. In one of his thought experiments, Galileo presented an argument against Aristotle's view that a force is needed to keep an object in motion. Galileo imagined an object (such as a ball) moving over a horizontal surface without the force of friction. He concluded that the object would move forever with a constant velocity as long as there was no unbalanced force acting to change the motion.

Why does a rolling ball slow to a stop? You know that a ball will roll farther across a smooth, waxed floor such as a bowling lane than it will across floor covered with carpet. The rough carpet offers more resistance to the rolling ball. The resistance of the floor friction is shown by a force arrow, F_{floor}, in figure 2.8. This force, along with the force arrow for air resistance, F_{air}, oppose the forward movement of the ball. Notice the dashed line arrow in part A of figure 2.8. There is no other force applied to the ball, so the rolling speed decreases until the ball finally comes to a complete stop. Now imagine what force you would need to exert by pushing with your hand, moving along with the ball to keep it rolling at a uniform rate. An examination of the forces in part B of figure 2.8 can help you determine the amount of force. The force you apply, $F_{applied}$, must counteract the resistance forces. It opposes the forces that are slowing down the ball as illustrated

FIGURE 2.7 Galileo (left) challenged the Aristotelian view of motion and focused attention on the concepts of distance, time, velocity, and acceleration.

by the direction of the arrows. To determine how much force you should apply, look at the arrow equation. $F_{applied}$ has the same length as the sum of the two resistance forces, but it is in the opposite direction of the resistance forces. Therefore, the overall force, F_{net}, is zero. The ball continues to roll at a uniform rate when you *balance* the force opposing its motion. It is reasonable, then, that if there were no opposing forces, you would not need to apply a force to keep it rolling. This was the kind of reasoning that Galileo did when he discredited the Aristotelian view that a force was necessary to keep an object moving. Galileo concluded that a moving object would continue moving with a constant velocity if no unbalanced forces were applied; that is, if the net force were zero.

It could be argued that the difference in Aristotle's and Galileo's views of forced motion is really a degree of analysis. After all, moving objects on Earth do come to rest unless continuously pushed or pulled. But Galileo's conclusion describes *why* they must be pushed or pulled and reveals the true nature of the motion of objects. Aristotle argued that the natural state of objects is to be at rest and attempted to explain why objects move. Galileo, on the other hand, argued that it is just as natural for an object to be moving and attempted to explain why they come to rest. The behavior of matter to persist in its state of motion is called **inertia.** Inertia is the *tendency of an object to remain in unchanging motion whether actually moving or at rest, when the net force is zero.* The development of this concept changed the way people viewed the natural state of an object and opened the way for further understandings about motion. Today, it is understood that a satellite moving through free space will continue to do so with no unbalanced forces acting on it (figure 2.9A). An unbalanced force is needed to slow the satellite (figure 2.9B), increase its speed (figure 2.9C), or change its direction of travel (figure 2.9D).

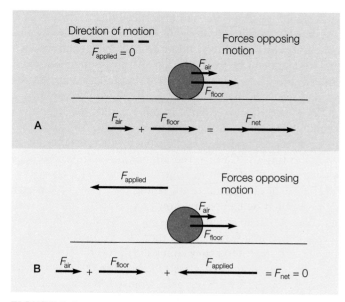

FIGURE 2.8 (*A*) This ball is rolling to your left with no forces in the direction of motion. The sum of the force of floor friction (F_{floor}) and the force of air friction (F_{air}) results in a net force opposing the motion, so the ball slows to a stop. (*B*) A force is applied to the moving ball, perhaps by a hand that moves along with the ball. The force applied ($F_{applied}$) equals the sum of the forces opposing the motion, so the ball continues to move with a constant velocity.

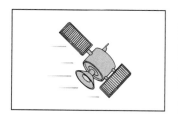
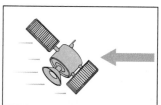

A No force—constant speed in straight line

B Force applied against direction of motion

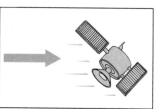
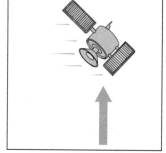

C Force applied in same direction as motion

D Force applied sideways to direction of motion

FIGURE 2.9 Explain how the combination of drawings (*A–D*) illustrates inertia.

A Bicycle Racer's Edge

Galileo was one of the first to recognize the role of friction in opposing motion. As shown in figure 2.8, friction with the surface and air friction combine to produce a net force that works against anything that is moving on the surface. This article is about air friction and some techniques that bike riders use to reduce that opposing force—perhaps giving them an edge in a close race.

The bike riders in box figure 2.1 are forming a single-file line, called a *paceline*, because the slip-stream reduces the air resistance for a closely trailing rider. Cyclists say that riding in the slipstream of another cyclist will save much of their energy. In fact, the cyclists will be able to move 5 mi/h faster than they would expending the same energy while riding alone.

In a sense, riding in a slipstream means that you do not have to push as much air out of your way. It has been estimated that at 20 mi/h a cyclist must move a little less than half a ton of air out of the way every minute. One of the earliest demonstrations of how a slipstream can help a cyclist was done back about the turn of the century. Charles Murphy had a special bicycle trail built down the middle of a railroad track.

Box Figure 2.1 The object of the race is to be in the front, to finish first. If this is true, why are these racers forming a single-file line?

Riding very close behind a special train caboose, Murphy was able to reach a speed of over 60 mi/h for a 1 mi course. More recently, cyclists have reached over 125 mi/h by following close in the slipstream of a race car.

Along with the problem of moving about a half-ton of air out of the way every minute, there are two basic factors related to air resistance. These are (1) a turbulent versus a smooth flow of air and (2) the problem of frictional drag. A turbulent flow of air contributes to air resistance because it causes the air to separate slightly on the back side, which increases the pressure on the front of the moving object. This is

why racing cars, airplanes, boats, and other racing vehicles are streamlined to a tear-droplike shape. This shape is not as likely to have the lower-pressure-producing air turbulence behind (and resulting greater pressure in front) because it smoothes or streamlines the air flow.

The frictional drag of air is similar to the frictional drag that occurs when you push a book across a rough tabletop. You know that smoothing the rough tabletop will reduce the frictional drag on the book. Likewise, the smoothing of a surface exposed to moving air will reduce air friction. Cyclists accomplish this "smoothing" by wearing smooth spandex clothing and by shaving hair from arm and leg surfaces that are exposed to moving air. Each hair contributes to the overall frictional drag, and removal of the arm and leg hair can thus result in seconds saved. This might provide enough of an edge to win a close race. Shaving legs and arms, together with the wearing of spandex or some other tight, smooth-fitting garments, are just a few of the things a cyclist can do to gain an edge. Perhaps you will be able to think of more ways to reduce the forces that oppose motion.

Myths, Mistakes, & Misunderstandings

Walk or Run in Rain?

Is it a mistake to run in rain if you want to stay drier? One idea is that you should run because you spend less time in the rain so you will stay drier. On the other hand, this is true only if the rain lands only on the top of your head and shoulders. If you run, you will end up running into more raindrops on the larger surface area of your face, chest, and front of your legs.

Two North Carolina researchers looked into this question with one walking and the other running over a measured distance while wearing cotton sweatsuits. They then weighed their clothing and found that the walking person's sweatsuit weighed more. This means you should run to stay drier.

FALLING OBJECTS

Did you ever wonder what happens to a falling rock during its fall? Aristotle reportedly thought that a rock falls at a uniform speed that is proportional to its weight. Thus, a heavy

rock would fall at a faster uniform speed than a lighter rock. As stated in a popular story, Galileo discredited Aristotle's conclusion by dropping a solid iron ball and a solid wooden ball simultaneously from the top of the Leaning Tower of Pisa (figure 2.10). Both balls, according to the story, hit the ground nearly at the same time. To do this, they would have to fall with the same velocity. In other words, the velocity of a falling object does not depend on its weight. Any difference in freely falling bodies is explainable by air resistance. Soon after the time of Galileo, the air pump was invented. The air pump could be used to remove the air from a glass tube. The effect of air resistance on falling objects could then be demonstrated by comparing how objects fall in the air with how they fall in an evacuated glass tube. You know that a coin falls faster than a feather when they are dropped together in the air. A feather and heavy coin will fall together in the near vacuum of an evacuated glass tube because the effect of air resistance on the feather has been removed. When objects fall toward Earth without considering air resistance, they are said to be in **free fall**. Free fall considers only gravity and neglects air resistance.

Galileo concluded that light and heavy objects fall together in free fall, but he also wanted to know the details of what was

FIGURE 2.10 According to a widespread story, Galileo dropped two objects with different weights from the Leaning Tower of Pisa. They reportedly hit the ground at about the same time, discrediting Aristotle's view that the speed during the fall is proportional to weight.

going on while they fell. He now knew that the velocity of an object in free fall was *not* proportional to the weight of the object. He observed that the velocity of an object in free fall *increased* as the object fell and reasoned from this that the velocity of the falling object would have to be (1) somehow proportional to the *time* of fall and (2) somehow proportional to the *distance* the object fell. If the time and distance were both related to the velocity of a falling object at a given time and distance, how were they related to one another?

Galileo reasoned that a freely falling object should cover a distance *proportional to the square of the time of the fall* ($d \propto t^2$). In other words, the object should fall 4 times as far in 2 s as in 1 s ($2^2 = 4$), 9 times as far in 3 s ($3^2 = 9$), and so on. Galileo checked this calculation by rolling balls on an inclined board with a smooth groove in it. He used the inclined board to slow

There are two different meanings for the term *free fall*. In physics, *free fall* means the unconstrained motion of a body in a gravitational field, without considering air resistance. Without air resistance, all objects are assumed to accelerate toward the surface at 9.8 m/s².

In the sport of skydiving, *free fall* means falling within the atmosphere without a drag-producing device such as a parachute. Air provides a resisting force that opposes the motion of a falling object, and the net force is the difference between the downward force (weight) and the upward force of air resistance. The weight of the falling object depends on the mass and acceleration from gravity, and this is the force downward. The resisting force is determined by at least two variables: (1) the area of the object exposed to the airstream and (2) the speed of the falling object. Other variables such as streamlining, air temperature, and turbulence play a role, but the greatest effect seems to be from exposed area and the increased resistance as speed increases.

A skydiver's weight is constant, so the downward force is constant. Modern skydivers typically free-fall from about 3,650 m (about 12,000 ft) above the ground until about 750 m (about 2,500 ft), where they open their parachutes. After jumping from the plane, the diver at first accelerates toward the surface, reaching speeds up to about 185–210 km/h (about 115–130 mi/h). The air resistance increases with increased speed and the net force becomes less and less. Eventually, the downward weight force will be balanced by the upward air resistance force, and the net force becomes zero. The person now falls at a constant speed, and we say the terminal velocity has been reached. It is possible to change your body position to vary your rate of fall up or down by 32 km/h (about 20 mi/h). However, by diving or "standing up" in free fall, experienced skydivers can reach speeds of up to 290 km/h (about 180 mi/h). The record free fall speed, done without any special equipment, is 517 km/h (about 321 mi/h). Once the parachute opens, a descent rate of about 16 km/h (about 10 mi/h) is typical.

the motion of descent in order to measure the distance and time relationships, a necessary requirement since he lacked the accurate timing devices that exist today. He found, as predicted, that the falling balls moved through a distance proportional to the square of the time of falling. This also means that the *velocity of the falling object increased at a constant rate.* Recall that a change of velocity during some time period is called *acceleration.* In other words, a falling object *accelerates* toward the surface of Earth.

Since the velocity of a falling object increases at a constant rate, this must mean that falling objects are *uniformly accelerated* by the force of gravity. *All objects in free fall experience a constant acceleration.* During each second of fall, the object on Earth gains 9.8 m/s (32 ft/s) in velocity. This gain is the acceleration of the falling object, 9.8 m/s² (32 ft/s²).

The acceleration of objects falling toward Earth varies slightly from place to place on Earth's surface because of Earth's shape and spin. The acceleration of falling objects decreases from the poles to the equator and also varies from place to place because Earth's mass is not distributed equally. The value of 9.8 m/s² (32 ft/s²) is an approximation that is fairly close to, but not exactly, the acceleration due to gravity in any particular location. The acceleration due to gravity is important in a number of situations, so the acceleration from this force is given a special symbol, **g**.

COMPOUND MOTION

So far we have considered two types of motion: (1) the horizontal, straight-line motion of objects moving on the surface of Earth and (2) the vertical motion of dropped objects that accelerate toward the surface of Earth. A third type of motion occurs when an object is thrown, or projected, into the air. Essentially, such a projectile (rock, football, bullet, golf ball, or whatever) could be directed straight upward as a vertical projection, directed straight out as a horizontal projection, or directed at some angle between the vertical and the horizontal. Basic to understanding such compound motion is the observation that (1) gravity acts on objects *at all times,* no matter where they are, and (2) the acceleration due to gravity (*g*) is *independent of any motion* that an object may have.

Vertical Projectiles

Consider first a ball that you throw straight upward, a vertical projection. The ball has an initial velocity but then reaches a maximum height, stops for an instant, then accelerates back toward Earth. Gravity is acting on the ball throughout its climb, stop, and fall. As it is climbing, the force of gravity is continually reducing its velocity. The overall effect during the climb is deceleration, which continues to slow the ball until the instantaneous stop. The ball then accelerates back to the surface just like a ball that has been dropped. If it were not for air resistance, the ball would return with the same velocity that it had initially. The velocity arrows for a ball thrown straight up are shown in figure 2.11.

Horizontal Projectiles

Horizontal projectiles are easier to understand if you split the complete motion into vertical and horizontal parts. Consider, for example, an arrow shot horizontally from a bow. The force of gravity accelerates the arrow downward, giving it an increasing downward velocity as it moves through the air. This increasing downward velocity is shown in figure 2.12 as increasingly longer velocity arrows (v_v). There are no forces in the horizontal direction if you can ignore air resistance, so the horizontal velocity of the arrow remains the same as shown by the v_h velocity arrows. The combination of the increasing

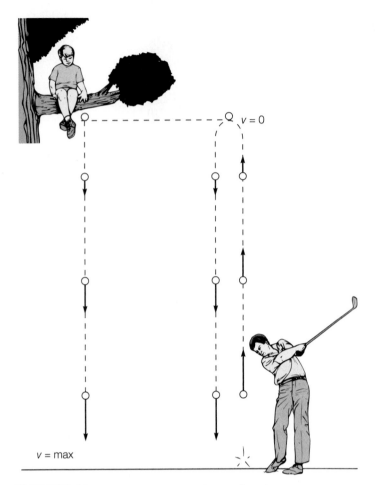

FIGURE 2.11 On its way up, a vertical projectile such as this misdirected golf ball is slowed by the force of gravity until an instantaneous stop; then it accelerates back to the surface, just as another golf ball does when dropped from the same height. The straight up- and down-moving golf ball has been moved to the side in the sketch so we can see more clearly what is happening.

vertical (v_v) motion and the unchanging horizontal (v_h) motion causes the arrow to follow a curved path until it hits the ground.

An interesting prediction that can be made from the shot arrow analysis is that an arrow shot horizontally from a bow will hit the ground at the same time as a second arrow that is simply dropped from the same height (figure 2.12). Would this be true of a bullet dropped at the same time as one fired horizontally from a rifle? The answer is yes; both bullets would hit the ground at the same time. Indeed, without air resistance, all the bullets and arrows should hit the ground at the same time if dropped or shot from the same height.

Golf balls, footballs, and baseballs are usually projected upward at some angle to the horizon. The horizontal motion of these projectiles is constant as before because there are no horizontal forces involved. The vertical motion is the same as that of a ball projected directly upward. The combination of these two motions causes the projectile to follow a curved path called a *parabola,* as shown in figure 2.13. The next time you have the opportunity, observe the path of a ball that has been

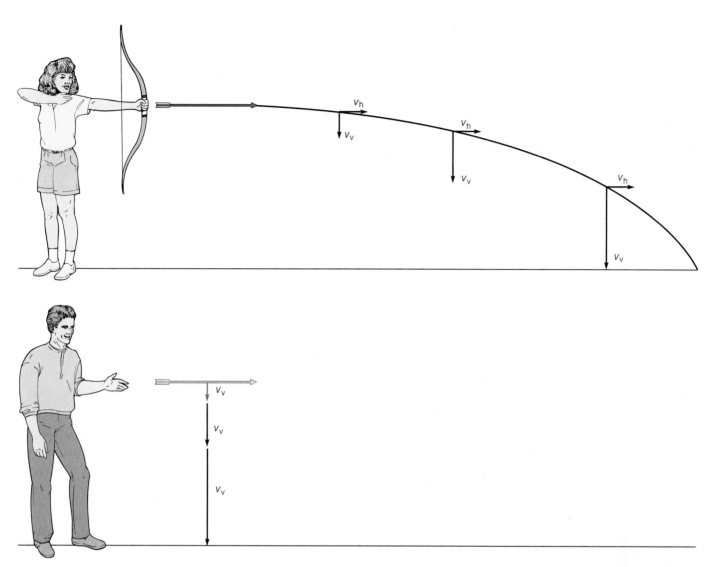

FIGURE 2.12 A horizontal projectile has the same horizontal velocity throughout the fall as it accelerates toward the surface, with the combined effect resulting in a curved path. Neglecting air resistance, an arrow shot horizontally will strike the ground at the same time as one dropped from the same height above the ground, as shown here by the increasing vertical velocity arrows.

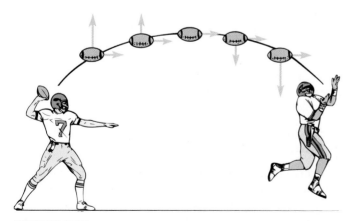

FIGURE 2.13 A football is thrown at some angle to the horizon when it is passed downfield. Neglecting air resistance, the horizontal velocity is a constant, and the vertical velocity decreases, then increases, just as in the case of a vertical projectile. The combined motion produces a parabolic path. Contrary to statements by sportscasters about the abilities of certain professional quarterbacks, it is impossible to throw a football with a "flat trajectory" because it begins to accelerate toward the surface as soon as it leaves the quarterback's hand.

projected at some angle (figure 2.14). Note that the second half of the path is almost a reverse copy of the first half. If it were not for air resistance, the two values of the path would be exactly the same. Also note the distance that the ball travels as compared to the angle of projection. An angle of projection of 45° results in the maximum distance of travel.

THREE LAWS OF MOTION

In the previous sections, you learned how to describe motion in terms of distance, time, velocity, and acceleration. In addition, you learned about different kinds of motion, such as straight-line motion, the motion of falling objects, and the compound motion of objects projected up from the surface of Earth. You were also introduced, in general, to two concepts closely associated with motion: (1) that objects have inertia, a tendency to resist a change in motion, and (2) that forces are involved in a change of motion.

Transportation and the Environment

Environmental science is an interdisciplinary study of Earth's environment. The concern of this study is the overall problem of human degradation of the environment and remedies for that damage. As an example of an environmental topic of study, consider the damage that results from current human activities involving the use of transportation. Researchers estimate that overall transportation activities are responsible for about one-third of the total U.S. carbon emissions that are added to the air every day. Carbon emissions are a problem because they are directly harmful in the form of carbon monoxide. They are also indirectly harmful because of the contribution of carbon dioxide to global warming and the consequences of climate change.

Here is a list of things that people might do to reduce the amount of environmental damage from transportation:

A. Use a bike, carpool, walk, or take public transportation wherever possible.

B. Plan to combine trips to the store, mall, and work, leaving the car parked whenever possible.

C. Purchase hybrid electric or fuel cell-powered cars and vehicles whenever possible.

D. Move to a planned community that makes the use of cars less necessary and less desirable.

Questions to Discuss

Discuss with your group the following questions concerning connections between thought and feeling:

1. What are your positive or negative feelings associated with each item in the list?
2. Would you feel differently if you had a better understanding of the global problem?
3. Do your feelings mean that you have reached a conclusion?
4. What new items could be added to the list?

FIGURE 2.14 Without a doubt, this baseball player is aware of the relationship between the projection angle and the maximum distance acquired for a given projection velocity.

The relationship between forces and a change of motion is obvious in many everyday situations. When a car, bus, or plane in which you are riding starts moving, you feel a force on your back. Likewise, you feel a force on the bottoms of your feet when an elevator you are in starts moving upward. On the other hand, you seem to be forced toward the dashboard if a car stops quickly, and it feels as if the floor pulls away from your feet when an elevator drops rapidly. These examples all involve patterns between forces and motion, patterns that can be quantified, conceptualized, and used to answer questions about why things move or stand still. These patterns are the subject of Newton's three laws of motion.

Newton's First Law of Motion

Newton's first law of motion is also known as the *law of inertia* and is very similar to one of Galileo's findings about motion. Recall that Galileo used the term *inertia* to describe the tendency of an object to resist changes in motion. Newton's first law describes this tendency more directly. In modern terms (not Newton's words), the **first law of motion** is as follows:

> **Every object retains its state of rest or its state of uniform straight-line motion unless acted upon by an unbalanced force.**

This means that an object at rest will remain at rest unless it is put into motion by an unbalanced force; that is, the net force must be greater than zero. Likewise, an object moving with uniform straight-line motion will retain that motion unless a net force causes it to speed up, slow down, or change its direction of travel. Thus, Newton's first law describes the tendency of an object to resist *any* change in its state of motion.

Think of Newton's first law of motion when you ride standing in the aisle of a bus. The bus begins to move, and you, being an independent mass, tend to remain at rest. You take a few steps back as you tend to maintain your position relative to the

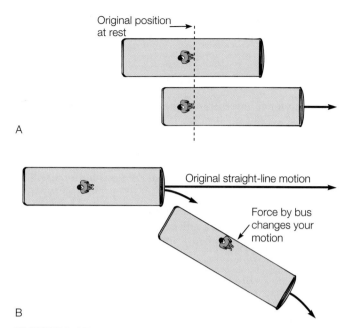

Original position
at rest

A

Original straight-line motion

Force by bus
changes your
motion

B

FIGURE 2.15 Top view of a person standing in the aisle of a bus. (*A*) The bus is at rest and then starts to move forward. Inertia causes the person to remain in the original position, appearing to fall backward. (*B*) The bus turns to the right, but inertia causes the person to retain the original straight-line motion until forced in a new direction by the side of the bus.

ground outside. You reach for a seat back or some part of the bus. Once you have a hold on some part of the bus, it supplies the forces needed to give you the same motion as the bus and you no longer find it necessary to step backward. You now have the same motion as the bus, and no forces are involved, at least until the bus goes around a curve. You now feel a tendency to move to the side of the bus. The bus has changed its straight-line motion, but you, again being an independent mass, tend to move straight ahead. The side of the seat forces you into following the curved motion of the bus. The forces you feel when the bus starts moving or turning are a result of your tendency to remain at rest or follow a straight path until forces correct your motion so that it is the same as that of the bus (figure 2.15).

Newton's Second Law of Motion

Newton successfully used Galileo's ideas to describe the nature of motion. Newton's first law of motion explains that any object, once started in motion, will continue with a constant velocity in a straight line unless a force acts on the moving object. This law not only describes motion but establishes the role of a force as well. A change of motion is therefore *evidence* of the action of a net force. The association of forces and a change of motion is common in your everyday experience. You have felt forces on your back in an accelerating automobile, and you have felt other forces as the automobile turns or stops. You have also learned about gravitational forces that accelerate objects toward the surface of Earth. Unbalanced forces and acceleration are involved in any change of motion. Newton's second law of motion is a relationship between *net force, acceleration,* and *mass* that describes the cause of a change of motion.

Consider the motion of you and a bicycle you are riding. Suppose you are riding your bicycle over level ground in a

straight line at 10 mi/h. Newton's first law tells you that you will continue with a constant velocity in a straight line as long as no external, unbalanced force acts on you and the bicycle. The force that you *are* exerting on the pedals seems to equal some external force that moves you and the bicycle along (more on this later). The force exerted as you move along is needed to *balance* the resisting forces of tire friction and air resistance. If these resisting forces were removed, you would not need to exert any force at all to continue moving at a constant velocity. The net force is thus the force you are applying minus the forces from tire friction and air resistance. The *net force* is therefore zero when you move at a constant speed in a straight line (figure 2.16).

If you now apply a greater force on the pedals, the *extra* force you apply is unbalanced by friction and air resistance. Hence, there will be a net force greater than zero, and you will accelerate. You will accelerate during, and *only* during, the time that the net force is greater than zero. Likewise, you will slow down if you apply a force to the brakes, another kind of resisting friction. A third way to change your velocity is to apply a force on the handlebars, changing the direction of your velocity. Thus, *unbalanced forces* on you and your bicycle produce an *acceleration.*

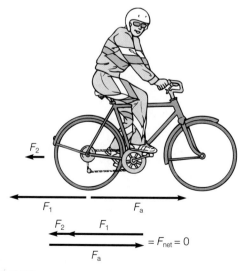

FIGURE 2.16 At a constant velocity, the force of tire friction (F_1) and the force of air resistance (F_2) have a sum that equals the force applied (F_a). The net force is therefore zero.

Starting a bicycle from rest suggests a relationship between force and acceleration. You observe that the harder you push on the pedals, the greater your acceleration. If you double the net force, then you will also double the acceleration, reaching the same velocity in half the time. Likewise, if you triple the net force, you will increase the acceleration threefold. Recall that when quantities increase or decrease together in the same ratio, they are said to be *directly proportional*. The acceleration is therefore directly proportional to the force.

Suppose that your bicycle has two seats, and you have a friend who will ride (but not pedal) with you. Suppose also that the addition of your friend on the bicycle will double the mass of the bike and riders. If you use the same net force as before, the bicycle will undergo a much smaller acceleration. In fact, with all other factors equal, doubling the mass and applying the same extra force will produce an acceleration of only half as much (figure 2.17). An even more massive friend would reduce the acceleration even more. If you triple the mass and apply the same extra force, the acceleration will be one-third as much. Recall that when a relationship between two quantities shows that one quantity increases as another decreases, in the same ratio, the quantities are said to be *inversely proportional*.

The acceleration of an object depends on *both* the *net force applied* and the *mass* of the object. The **second law of motion** is as follows:

The acceleration of an object is directly proportional to the net force acting on it and inversely proportional to the mass of the object.

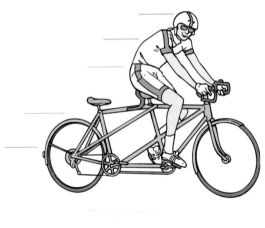

FIGURE 2.17 More mass results in less acceleration when the same force is applied. With the same force applied, the riders and bike with twice the mass will have half the acceleration, with all other factors constant. Note that the second rider is not pedaling.

If we express force in appropriate units, we can write this statement as an equation,

$$a = \frac{F}{m}$$

By solving for F, we rearrange the equation into the form it is most often expressed,

$$F = ma \qquad \qquad \textbf{equation 2.3}$$

In the metric system, you can see that the units for force will be the units for mass (m) times acceleration (a). The unit for mass is kg and the unit for acceleration is m/s². The combination of these units, (kg) (m/s²) is a unit of force called the **newton** (N), in honor of Isaac Newton. So,

$$1 \text{ newton} = 1 \text{ N} = 1 \frac{\text{kg} \cdot \text{m}}{\text{s}^2}$$

Newton's second law of motion is the essential idea of his work on motion. According to this law, there is always a relationship between the acceleration, a net force, and the mass of an object. Implicit in this statement are three understandings: (1) that we are talking about the net force, meaning total external force acting on an object, (2) that the motion statement is concerned with acceleration, not velocity, and (3) that the mass does not change unless specified.

EXAMPLE 2.5 *(Optional)*

A 60 kg bicycle and rider accelerate at 0.5 m/s². How much extra force was applied?

SOLUTION

The mass (m) of 60 kg and the acceleration (a) of 0.5 m/s² are given. The problem asked for the extra force (F) needed to give the mass the acquired acceleration. The relationship is found in equation 2.3, $F = ma$.

$$m = 60 \text{ kg} \qquad F = ma$$
$$a = 0.5\frac{\text{m}}{\text{s}^2} \qquad \quad = (60 \text{ kg})\left(0.5\frac{\text{m}}{\text{s}^2}\right)$$
$$F = ? \qquad \qquad = (60)(0.5)\text{kg} \times \frac{\text{m}}{\text{s}^2}$$
$$\qquad \qquad \quad = 30\frac{\text{kg} \cdot \text{m}}{\text{s}^2}$$
$$\qquad \qquad \quad = \boxed{30 \text{ N}}$$

An extra force of 30 N beyond that required to maintain constant speed must be applied to the pedals for the bike and rider to maintain an acceleration of 0.5 m/s². (Note that the units kg·m/s² form the definition of a newton of force, so the symbol N is used.)

EXAMPLE 2.6 *(Optional)*

What is the acceleration of a 20 kg cart if the net force on it is 40 N? (Answer: 2 m/s²)

Weight and Mass

What is the meaning of weight—is it the same concept as mass? Weight is a familiar concept to most people, and in everyday language, the word is often used as having the same meaning as mass. In physics, however, there is a basic difference between weight and mass, and this difference is very important in Newton's explanation of motion and the causes of motion.

Mass is defined as the property that determines how much an object resists a change in its motion. The greater the mass the greater the *inertia,* or resistance to change in motion. Consider, for example, that it is easier to push a small car into motion than to push a large truck into motion. The truck has more mass and therefore more inertia. Newton originally defined mass as the "quantity of matter" in an object, and this definition is intuitively appealing. However, Newton needed to measure inertia because of its obvious role in motion and redefined mass as a measure of inertia.

You could use Newton's second law to measure a mass by exerting a force on the mass and measuring the resulting acceleration. This is not very convenient, so masses are usually measured on a balance by comparing the force of gravity acting on a standard mass compared to the force of gravity acting on the unknown mass.

The force of gravity acting on a mass is the *weight* of an object. Weight is a force and has different units (N) from those of mass (kg). Since weight is a measure of the force of gravity acting on an object, the force can be calculated from Newton's second law of motion,

$$F = ma$$

or

downward force = (mass)(acceleration due to gravity)

or

weight = (mass)(g)

$$w = mg \qquad \text{equation 2.4}$$

You learned previously that g is the symbol used to represent acceleration due to gravity. Near Earth's surface, g has an approximate value of 9.8 m/s². To understand how g is applied to an object not moving, consider a ball you are holding in your hand. By supporting the weight of the ball, you hold it stationary, so the upward force of your hand and the downward force of the ball (its weight) must add to a net force of zero. When you let go of the ball, the gravitational force is the only force acting on the ball. The ball's weight is then the net force that accelerates it at g, the acceleration due to gravity. Thus, $F_{net} = w = ma = mg$. The weight of the ball never changes in a given location, so its weight is always equal to $w = mg$, even if the ball is not accelerating.

An important thing to remember is that *pounds* and *newtons* are units of *force* (table 2.1). A *kilogram,* on the other hand, is a measure of *mass.* Thus, the English unit of 1.0 lb is comparable to the metric unit of 4.5 N (or 0.22 lb is equivalent to 1.0 N). Conversion tables sometimes show how to convert from pounds (a unit of weight) to kilograms (a unit of mass). This is possible because weight and mass are proportional in a given location on the surface of Earth. Using conversion factors from inside the front cover of this book, see if you can express your weight in pounds and newtons and your mass in kilograms.

EXAMPLE 2.7 (Optional)

What is the weight of a 60.0 kg person on the surface of Earth?

SOLUTION

A mass (m) of 60.0 kg is given, and the acceleration due to gravity (g) 9.8 m/s² is implied. The problem asked for the weight (w). The relationship is found in equation 2.4, $w = mg$, which is a form of $F = ma$.

$$m = 60.0 \text{ kg}$$
$$g = 9.8 \text{ m/s}^2 \quad w = mg$$
$$w = ? \qquad = (60.0 \text{ kg})\left(9.8 \frac{\text{m}}{\text{s}^2}\right)$$
$$= (60.0)(9.8) \text{ kg} \times \frac{\text{m}}{\text{s}^2}$$
$$= 588 \frac{\text{kg} \cdot \text{m}}{\text{s}^2}$$
$$= 588 \text{ N}$$
$$= \boxed{590 \text{ N}}$$

EXAMPLE 2.8 (Optional)

A 60.0 kg person weighs 100.0 N on the Moon. What is the value of g on the Moon? (Answer: 1.67 m/s²)

Newton's Third Law of Motion

Newton's first law of motion states that an object retains its state of motion when the net force is zero. The second law states what happens when the net force is *not* zero, describing how an object with a known mass moves when a given force is applied. The two laws give one aspect of the concept of a force; that is,

Connections . . .

Weight on Different Planets*

Planet	Acceleration of Gravity	Approximate Weight(N)	Approximate Weight (lb)
Mercury	3.72 m/s²	223 N	50 lb
Venus	8.92 m/s²	535 N	120 lb
Earth	9.80 m/s²	588 N	132 lb
Mars	3.72 m/s²	223 N	50 lb
Jupiter	24.89 m/s²	1,493 N	336 lb
Saturn	10.58 m/s²	635 N	143 lb
Uranus	8.92 m/s²	535 N	120 lb
Neptune	11.67 m/s²	700 N	157 lb

*For a 60.0 kg person

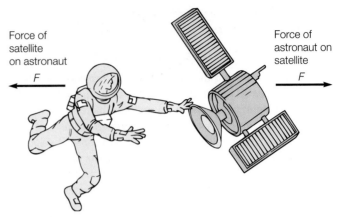

Force of satellite on astronaut F

Force of astronaut on satellite F

FIGURE 2.18 Forces occur in matched pairs that are equal in strength and opposite in direction.

TABLE 2.1 Units of Mass and Weight in the Metric and English Systems of Measurement

	Mass	×	Acceleration	=	Force
Metric system	kg	×	$\frac{m}{s^2}$	=	$\frac{kg \cdot m}{s^2}$ (newton)
English system	$\left(\frac{lb}{ft/s^2}\right)$	×	$\frac{ft}{s^2}$	=	lb (pound)

if you observe that an object starts moving, speeds up, slows down, or changes its direction of travel, you can conclude that an unbalanced force is acting on the object. Thus, any change in the state of motion of an object is *evidence* that an unbalanced force has been applied.

Newton's third law of motion is also concerned with forces. First, consider where a force comes from. A force is always produced by the interaction of two objects. Sometimes we do not know what is producing forces, but we do know that they always come in pairs. Anytime a force is exerted, there is always a matched and opposite force that occurs at the same time. For example, if you push on the wall, the wall pushes back with an equal and opposite force. The two forces are opposite and balanced, and you know this because $F = ma$ and neither you nor the wall accelerated. If the acceleration is zero, then you know from $F = ma$ that the net force is zero (zero equals zero). Note also that the two forces were between two different objects, you and the wall. Newton's third law always describes what happens between two different objects. To simplify the many interactions that occur on Earth, consider a satellite freely floating in space. According to Newton's second law ($F = ma$), a force must be applied to change the state of motion of the satellite. What is a possible source of such a force? Perhaps an astronaut pushes on the satellite for 1 second. The satellite would accelerate *during* the application of the force, then move away from the original position at some constant velocity. The astronaut would also move away from the original position, but in the opposite

direction (figure 2.18). A *single force does not exist* by itself. There is always a matched and opposite force that occurs at the same time. Thus, the astronaut exerted a momentary force on the satellite, but the satellite evidently exerted a momentary force back on the astronaut as well, for the astronaut moved away from the original position in the opposite direction. Newton did not have astronauts and satellites to think about, but this is the kind of reasoning he did when he concluded that forces always occur in matched pairs that are equal and opposite. Thus, the **third law of motion** is as follows:

> Whenever two objects interact, the force exerted on one object is equal in strength and opposite in direction to the force exerted on the other object.

The third law states that forces always occur in matched pairs that act in opposite directions and on two *different* bodies. Sometimes the third law of motion is expressed as follows: "For every action there is an equal and opposite reaction," but this can be misleading. Neither force is the cause of the other. The forces are at every instant the cause of each other, and they appear and disappear at the same time. If you are going to describe the force exerted on a satellite by an astronaut, then you must realize that there is a simultaneous force exerted on the astronaut by the satellite. The forces (astronaut on satellite and satellite on astronaut) are equal in magnitude but opposite in direction.

Perhaps it would be more common to move a satellite with a small rocket. A satellite is maneuvered in space by firing a rocket in the direction opposite to the direction someone wants to move the satellite. Exhaust gases (or compressed gases) are accelerated in one direction and exert an equal but opposite force on the satellite that accelerates it in the opposite direction. This is another example of the third law.

Consider how the pairs of forces work on Earth's surface. You walk by pushing your feet against the ground (figure 2.19). Of course you could not do this if it were not for friction. You would slide as on slippery ice without friction. But since friction does exist, you exert a backward horizontal force on the ground, and, as the third law explains, the ground exerts an equal and opposite force on you. You accelerate forward from the net force as explained by the second law. If Earth had the same mass as you, however, it would accelerate backward at the same rate that you were accelerated forward. Earth is much more massive than you, however, so any acceleration of Earth is a vanishingly small amount. The overall effect is that you are accelerated forward by the force the ground exerts on you.

Return now to the example of riding a bicycle that was discussed previously. What is the source of the *external* force that accelerates you and the bike? Pushing against the pedals is not external to you and the bike, so that force will *not* accelerate you and the bicycle forward. This force is transmitted through the bike mechanism to the rear tire, which pushes against the ground. It is the ground exerting an equal and opposite force against the system of you and the bike that accelerates you forward. You must consider the forces that act on the system of you and the bike before you can apply $F = ma$. The only forces that will affect the forward motion of the bike system are the force of the ground pushing it forward and the frictional forces that oppose the forward motion. This is another example of the third law.

MOMENTUM

Sportscasters often refer to the *momentum* of a team, and newscasters sometimes refer to an election where one of the candidates has *momentum*. Both situations describe a competition where one side is moving toward victory and it is difficult to stop. It seems appropriate to borrow this term from the physical sciences because momentum is a property of movement. It takes a longer time to stop something from moving when it has a lot of momentum. The physical science concept of momentum is closely related to Newton's laws of motion. **Momentum** (p) is defined as the product of the mass (m) of an object and its velocity (v),

$$\text{momentum} = \text{mass} \times \text{velocity}$$

or

$$p = mv \qquad \textbf{equation 2.5}$$

The astronaut in figure 2.20 has a mass of 60.0 kg and a velocity of 0.750 m/s as a result of the interaction with the satellite. The resulting momentum is therefore (60.0 kg) (0.750 m/s), or 45.0 kg·m/s. As you can see, the momentum would be greater if the astronaut had acquired a greater velocity or if the astronaut had a greater mass and acquired the same velocity. Momentum involves both the inertia and the velocity of a moving object.

Conservation of Momentum

Notice that the momentum acquired by the satellite in figure 2.20 is *also* 45.0 kg·m/s. The astronaut gained a certain momentum in one direction, and the satellite gained the *very same momentum in the opposite direction*. Newton originally

FIGURE 2.19 The football player's foot is pushing against the ground, but it is the ground pushing against the foot that accelerates the player forward to catch a pass.

defined the second law in terms of a change of momentum being proportional to the net force acting on an object. Since the third law explains that the forces exerted on both the astronaut and satellite were equal and opposite, you would expect both objects to acquire equal momentum in the opposite direction. This result is observed any time objects in a system interact and the only forces involved are those between the interacting objects (figure 2.20). This statement leads to a particular kind of relationship called a *law of conservation*. In this case, the law applies to momentum and is called the **law of conservation of momentum:**

> **The total momentum of a group of interacting objects remains the same in the absence of external forces.**

Connections . . .

Swimming Scallop

Newton's laws of motion apply to animal motion as well as that of satellites and automobiles. Consider, for example, the dilemma of a growing scallop. A scallop is the shell often seen as a logo for a certain petroleum company, a fan-shaped shell with a radiating fluted pattern (box figure 2.2). The scallop is a marine mollusk that is most unusual since it is the only clamlike mollusk that is capable of swimming. By opening and closing its shell, it is able to propel itself by forcing water from the interior of the shell in a jetlike action. The popular seafood called "scallops" is the edible muscle that the scallop uses to close its shell.

A scallop is able to swim by orienting its shell at a proper angle and maintaining a minimum acceleration to prevent sinking. For example, investigations have found that one particular species of scallop must force enough water backward to move about six body lengths per second with a 10 degree angle of attack to maintain level swimming. Such a swimming effort can be maintained for up to about 20 seconds, enabling the scallop to escape predation or some other disturbing condition.

A more massive body limits the swimming ability of the scallop, as a greater force is needed to give a greater mass the same acceleration (as you would expect from Newton's second law of motion). This problem becomes worse as the scallop grows larger and larger without developing a greater and greater jet force.

Box Figure 2.2   A scallop shell.

Conservation of momentum, energy, and charge are among examples of conservation laws that apply to everyday situations. These situations always illustrate two understandings: (1) each conservation law is an expression that describes a physical principle that can be observed, and (2) each law holds regardless of the details of an interaction or how it took place. Since the conservation laws express something that always occurs, they tell us what might be expected to happen and what might be expected not to happen in a given situation. The conservation laws also allow unknown quantities to be found by analysis. The law of conservation of momentum, for example, is useful in analyzing motion in simple systems of collisions such as those of billiard balls, automobiles, or railroad cars. It is also useful in measuring action and reaction interactions, as in rocket propulsion, where the backward momentum of the exhaust gases equals the momentum given to the rocket in the opposite direction (figure 2.21). When this is done, momentum is always found to be conserved.

Impulse

Have you ever heard that you should "follow through" when hitting a ball? When you follow through, the bat is in contact with the ball for a longer period of time. The force of the hit is important, of course, but both the force and how long the force is applied determine the result. The product of the force and the time of application is called **impulse**. This quantity can be expressed as

$$\text{impulse} = Ft$$

where F is the force applied during the time of contact (t). The impulse you give the ball determines how fast the ball will move and thus how far it will travel.

Impulse is related to the change of motion of a ball of a given mass, so the change of momentum (mv) is brought about by the impulse. This can be expressed as

Change of momentum = (applied force) (time of contact)

$$\Delta p = Ft \qquad\qquad \textbf{equation 2.6}$$

where Δp is a change of momentum. You "follow through" while hitting a ball in order to increase the contact time. If the same force is used, a longer contact time will result in a

$F = 30.0$ N
$t = 1.50$ s

$F = 30.0$ N
$t = 1.50$ s

$m = 60.0$ kg
$v = 0.750$ m/s
$p = mv$

$= (60.0 \text{ kg}) (0.750 \frac{m}{s})$

$= \boxed{45.0 \frac{kg \cdot m}{s}}$

$m = 120.0$ kg
$v = 0.375$ m/s
$p = mv$

$= (120.0 \text{ kg}) (0.375 \frac{m}{s})$

$= \boxed{45.0 \frac{kg \cdot m}{s}}$

FIGURE 2.20   Both the astronaut and the satellite receive a force of 30.0 N for 1.50 s when they push on each other. Both then have a momentum of 45.0 kg·m/s in the opposite direction. This is an example of the law of conservation of momentum.

FIGURE 2.21 According to the law of conservation of momentum, the momentum of the expelled gases in one direction equals the momentum of the rocket in the other direction in the absence of external forces.

greater impulse. A greater impulse means a greater change of momentum, and since the mass of the ball does not change, the overall result is a moving ball with a greater velocity. This means following through will result in more distance from hitting the ball with the same force. That's why it is important to follow through when you hit the ball.

Now consider bringing a moving object to a stop by catching it. In this case, the mass and the velocity of the object are fixed at the time you catch it, and there is nothing you can do about these quantities. The change of momentum is equal to the impulse, and the force and time of force application *can* be manipulated. For example, consider how you would catch a raw egg that is tossed to you. You would probably move your hands with the egg as you catch it, increasing the contact time. Increasing the contact time has the effect of reducing the force, since $\Delta p = Ft$. You changed the force applied by increasing the contact time, and hopefully, you reduced the force sufficiently so the egg does not break.

Contact time is also important in safety. Automobile airbags, the padding in elbow and knee pads, and the plastic barrels off the highway in front of overpass supports are examples of designs intended to increase the contact time. Again, increasing the contact time reduces the force, since $\Delta p = Ft$. The impact force is reduced and so are the injuries. Think about this the next time you see a car that was crumpled and bent by a collision. The driver and passengers were probably saved from injuries that are more serious since more time was involved in stopping the car that crumpled. A car that crumples is a safer car in a collision.

FORCES AND CIRCULAR MOTION

Consider a communications satellite that is moving at a uniform speed around Earth in a circular orbit. According to the first law of motion, there *must be* forces acting on the satellite, since it does *not* move off in a straight line. The second law of motion also indicates forces, since an unbalanced force is required to change the motion of an object.

Recall that acceleration is defined as a rate of change in velocity and that velocity has both strength and direction. Velocity is changed by a change in speed, direction, or both speed and direction. The satellite in a circular orbit is continuously being accelerated. This means that there is a continuously acting unbalanced force on the satellite that pulls it out of a straight-line path.

The force that pulls an object out of its straight-line path and into a circular path is called a **centripetal** (center-seeking) **force.** Perhaps you have swung a ball on the end of a string in a horizontal circle over your head. Once you have the ball moving, the only unbalanced force (other than gravity) acting on the ball is the centripetal force your hand exerts on the ball through the string. This centripetal force pulls the ball from its natural straight-line path into a circular path. There are no outward

Connections . . .

Circular Fun

Amusement park rides are designed to accelerate your body, sometimes producing changes in the acceleration (jerk) as well. This is done by changes in speed, changes in the direction of travel, or changes in both direction and speed. Many rides move in a circular path, since such movement is a constant acceleration.

Why do people enjoy amusement park rides? It is not the high speed, since your body is not very sensitive to moving at a constant speed. Moving at a steady 600 mi/h in an airplane, for example, provides little sensation when you are seated in an aisle seat in the central cabin.

Your body is not sensitive to high-speed traveling, but it is sensitive to acceleration and changes of acceleration. Acceleration affects the fluid in your inner ear, which controls your sense of balance. In most people, acceleration also produces a reaction that results in the release of the hormones epinephrine and norepinephrine from the adrenal medulla, located near the kidney. The heart rate increases, blood pressure rises, blood is shunted to muscles, and the breathing rate increases. You have probably experienced this reaction many times in your life, as when you nearly have an automobile accident or slip and nearly fall. In the case of an amusement park ride, your body adapts and you believe you enjoy the experience.

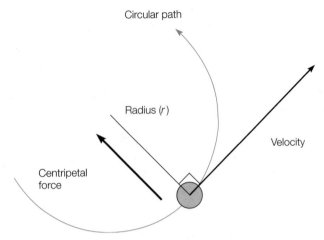

FIGURE 2.22 Centripetal force on the ball causes it to change direction continuously, or accelerate into a circular path. Without the force acting on it, the ball would continue in a straight line.

EXAMPLE 2.9 *(Optional)*

A 0.25 kg ball is attached to the end of a 0.5 m string and moved in a horizontal circle at 2.0 m/s. What net force is needed to keep the ball in its circular path?

SOLUTION

$m = 0.25$ kg

$r = 0.5$ m

$v = 2.0$ m/s

$F = ?$

$$F = \frac{mv^2}{r}$$

$$= \frac{(0.25 \text{ kg})(2.0 \text{ m/s})^2}{0.5 \text{ m}}$$

$$= \frac{(0.25 \text{ kg})(4.0 \text{ m}^2/\text{s}^2)}{0.5 \text{ m}}$$

$$= \frac{(0.25)(4.0)}{0.5} \frac{\text{kg} \cdot \text{m}^2}{\text{s}^2} \times \frac{1}{\text{m}}$$

$$= 2 \frac{\text{kg} \cdot \text{m}^2}{\text{m} \cdot \text{s}^2}$$

$$= 2 \frac{\text{kg} \cdot \text{m}}{\text{s}^2}$$

$$= \boxed{2 \text{ N}}$$

---

forces acting on the ball. The force that you feel on the string is a consequence of the third law; the ball exerts an equal and opposite force on your hand. If you were to release the string, the ball would move away from the circular path in a *straight line* that has a right angle to the radius at the point of release (figure 2.22). When you release the string, the centripetal force ceases, and the ball then follows its natural straight-line motion. If other forces were involved, it would follow some other path. Nonetheless, the apparent outward force has been given a name just as if it were a real force. The outward tug is called a *centrifugal force.*

The magnitude of the centripetal force required to keep an object in a circular path depends on the inertia, or mass, of the object and the acceleration of the object, just as you learned in the second law of motion. The acceleration of an object moving in a circle can be shown to be directly proportional to the square of the speed around the circle (v^2) and inversely proportional to the radius of the circle (r). (A smaller radius requires a greater acceleration.) Therefore, the acceleration of an object moving in uniform circular motion (a_c) is

$$a_c = \frac{v^2}{r} \qquad \textbf{equation 2.7}$$

The magnitude of the centripetal force of an object with a constant mass (m) that is moving with a velocity (v) in a circular orbit of a radius (r) can be found by substituting equation 2.7 in $F = ma$, or

$$F = \frac{mv^2}{r} \qquad \textbf{equation 2.8}$$

EXAMPLE 2.10 *(Optional)*

Suppose you make the string in example 2.9 half as long, 0.25 m. What force is now needed? (Answer: 4.0 N)

NEWTON'S LAW OF GRAVITATION

You know that if you drop an object, it always falls to the floor. You define *down* as the direction of the object's movement and *up* as the opposite direction. Objects fall because of the force of gravity, which accelerates objects on Earth at $g = 9.8$ m/s² (32 ft/s²) and gives them weight, $w = mg$.

Gravity is an attractive force, a pull that exists between all objects in the universe. It is a mutual force that, just like all other forces, comes in matched pairs. Since Earth attracts you with a certain force, you must attract Earth with an exact opposite force. The magnitude of this force of mutual attraction depends on several variables. These variables were first described by Newton in *Principia,* his famous book on motion that was printed in 1687. Newton had, however, worked out his ideas much earlier, by the age of twenty-four, along with ideas about his laws of motion and the formula for centripetal acceleration. In a biography written by a friend in 1752, Newton stated that the notion of gravitation came to mind during a time of thinking that "was occasioned by the fall of an apple." He was thinking about why the Moon stays in orbit around Earth rather than moving off in a straight line as would be predicted by the first law of motion. Perhaps the same force that attracts the Moon toward Earth, he thought, attracts the apple to Earth. Newton developed a theoretical equation for gravitational force that explained not only the motion of the Moon but the motion of the whole solar system. Today, this relationship is known as the **universal law of gravitation:**

> **Every object in the universe is attracted to every other object with a force that is directly proportional to the product of their masses and inversely proportional to the square of the distances between them.**

In symbols, m_1 and m_2 can be used to represent the masses of two objects, d the distance between their centers, and G a constant of proportionality. The equation for the law of universal gravitation is therefore

$$F = G\frac{m_1 m_2}{d^2} \qquad \textbf{equation 2.9}$$

This equation gives the magnitude of the attractive force that each object exerts on the other. The two forces are oppositely directed. The constant G is a universal constant, since the law applies to all objects in the universe. It was first measured experimentally by Henry Cavendish in 1798. The accepted value today is $G = 6.67 \times 10^{-11} \text{N} \cdot \text{m}^2/\text{kg}^2$. Do not confuse G, the universal constant, with g, the acceleration due to gravity on the surface of Earth.

Thus, the magnitude of the force of gravitational attraction is determined by the mass of the two objects and the distance between them (figure 2.23). The law also states that *every* object is attracted to every other object. You are attracted to all the objects around you—chairs, tables, other people, and so forth. Why don't you notice the forces between you and other objects? One or both of the interacting objects must be quite massive before a noticeable force results from the interaction. That is why you do not notice the force of gravitational attraction between you and objects that are not very massive compared to Earth. The attraction between you and Earth overwhelmingly predominates, and that is all you notice.

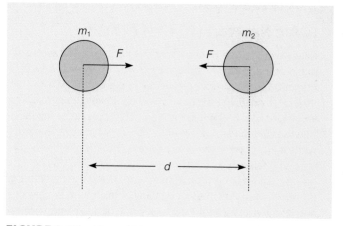

FIGURE 2.23 The variables involved in gravitational attraction. The force of attraction (F) is proportional to the product of the masses (m_1, m_2) and inversely proportional to the square of the distance (d) between the centers of the two masses.

EXAMPLE 2.11 *(Optional)*

What is the force of gravitational attraction between two 60.0 kg (132 lb) students who are standing 1.00 m apart?

SOLUTION

$G = 6.67 \times 10^{-11} \text{ Nm}^2/\text{kg}^2$

$m_1 = 60.0 \text{ kg}$

$m_2 = 60.0 \text{ kg} \qquad F = G\dfrac{m_1 m_2}{d^2}$

$d = 1.00 \text{ m}$

$F = ? \qquad = \dfrac{(6.67 \times 10^{-11}\,\text{Nm}^2/\text{kg}^2)(60.0 \text{ kg})(60.0 \text{ kg})}{(1.00 \text{ m})^2}$

$\qquad = (6.67 \times 10^{-11})(3.60 \times 10^3)\dfrac{\dfrac{\text{Nm}^2\text{kg}^2}{\text{kg}^2}}{m^2}$

$\qquad = 2.40 \times 10^{-7}\,(\text{Nm}^2)\left(\dfrac{1}{\text{m}^2}\right)$

$\qquad = 2.40 \times 10^{-7}\,\dfrac{\text{Nm}^2}{\text{m}^2}$

$\qquad = \boxed{2.40 \times 10^{-7} \text{ N}}$

(Note: A force of 2.40×10^{-7} [0.00000024] N is equivalent to a force of 5.40×10^{-8} [0.00000005] lb, a force that you would not notice. In fact, it would be difficult to measure such a small force.)

EXAMPLE 2.12 *(Optional)*

What would be the value of g if Earth were less dense, with the same mass and double the radius? (Answer: $g = 2.45$ m/s^2)

The acceleration due to gravity, g, is about 9.8 m/s^2 on Earth and is practically a constant for relatively short distances above the surface. Notice, however, that Newton's law of gravitation is an inverse square law. This means if you double the distance,

People Behind the Science

Isaac Newton (1642–1727)

Isaac Newton was a British physicist who is regarded as one of the greatest scientists ever to have lived. He discovered the three laws of motion that bear his name and was the first to explain gravitation, clearly defining the nature of mass, weight, force, inertia, and acceleration. In his honor, the SI unit of force is called the newton. Newton also made fundamental discoveries in light, finding that white light is composed of a spectrum of colors and inventing the reflecting telescope.

Newton was born on January 4, 1643 (by the modern calendar). He was a premature, sickly baby born after his father's death, and his survival was not expected. When he was three, his mother remarried, and the young Newton was left in his grandmother's care. He soon began to take refuge in things mechanical, making water clocks, kites bearing fiery lanterns aloft, and a model mill powered by a mouse, as well as innumerable drawings and diagrams. When Newton was twelve, his mother withdrew him from school with the intention of making him into a farmer. Fortunately, his uncle recognized Newton's ability and managed to get him back into school to prepare for college.

Newton was admitted to Trinity College, Cambridge, and graduated in 1665, the same year that the university was closed because of the plague. Newton returned to his boyhood farm to wait out the plague, making only an occasional visit back to Cambridge. During this period, he performed his first prism experiments and thought about motion and gravitation.

Newton returned to study at Cambridge after the plague had run its course, receiving a master's degree in 1668 and becoming a professor at the age of only twenty-six. Newton remained at Cambridge almost thirty years, studying alone for the most part, though in frequent contact with other leading scientists by letter and through the Royal Society in London. These were Newton's most fertile years. He labored day and night, thinking and testing ideas with calculations.

In Cambridge, he completed what may be described as his greatest single work, *Philosophae Naturalis Principia Mathematica* (*Mathematical Principles of Natural Philosophy*). This was presented to the Royal Society in 1686, which subsequently withdrew from publishing it because of a shortage of funds. The astronomer Edmund Halley (1656–1742), a wealthy man and friend of Newton, paid for the publication of the *Principia* in 1687. In it, Newton revealed his laws of motion and the law of universal gravitation.

Newton's greatest achievement was to demonstrate that scientific principles are of universal application. In *Principia Mathematica,* he built the evidence of experiment and observation to develop a model of the universe that is still of general validity. "If I have seen further than other men," he once said, "it is because I have stood on the shoulders of giants"; and Newton was certainly able to bring together the knowledge of his forebears in a brilliant synthesis.

No knowledge can ever be total, but Newton's example brought about an explosion of investigation and discovery that has never really abated. He perhaps foresaw this when he remarked, "To myself, I seem to have been only like a boy playing on the seashore, and diverting myself in now and then finding a smoother pebble or a prettier shell than ordinary, whilst the great ocean of truth lay all undiscovered before me."

With his extraordinary insight into the workings of nature and rare tenacity in wresting its secrets and revealing them in as fundamental and concise a way as possible, Newton stands as a colossus of science. In physics, only Archimedes (287–212 B.C.) and Albert Einstein (1879–1955), who also possessed these qualities, may be compared to him.

Source: Modified from the *Hutchinson Dictionary of Scientific Biography.* © RM, 2007. Reprinted by permission.

---

the force is $1/(2)^2$ or 1/4 as great. If you triple the distance, the force is $1/(3)^2$ or 1/9 as great. In other words, the force of gravitational attraction and g decrease inversely with the square of the distance from Earth's center.

Newton was able to calculate the acceleration of the Moon toward Earth, about 0.0027 m/s^2. The Moon "falls" toward Earth because it is accelerated by the force of gravitational attraction. This attraction acts as a *centripetal force* that keeps the Moon from following a straight-line path as would be predicted from the first law. Thus, the acceleration of the Moon keeps it in a somewhat circular orbit around Earth. Figure 2.24 shows that the Moon would be in position A if it followed a straight-line path instead of "falling" to position B as it does. The Moon thus "falls" around Earth. Newton was able to analyze the motion of the Moon quantitatively as evidence that it is gravitational force that keeps the Moon in its orbit. The law of gravitation

CONCEPTS APPLIED

Apparent Weightlessness

Use a sharp pencil to make a small hole in the bottom of a Styrofoam cup. The hole should be large enough for a thin stream of water to flow from the cup, but small enough for the flow to continue for 3 or 4 seconds. Test the water flow over a sink.

Hold a finger over the hole on the cup as you fill it with water. Stand on a ladder or outside stairwell as you hold the cup out at arm's length. Move your finger, allowing a stream of water to flow from the cup as you drop it. Observe what happens to the stream of water as the cup is falling. Explain your observations. Also predict what you would see if you were falling with the cup.

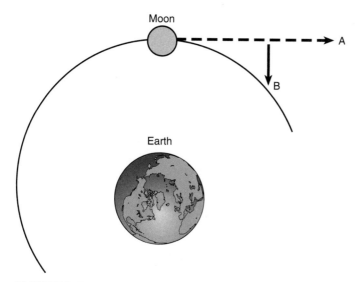

FIGURE 2.24 Gravitational attraction acts as a centripetal force that keeps the Moon from following the straight-line path shown by the dashed line to position A. It was pulled to position B by gravity (0.0027 m/s²) and thus "fell" toward Earth the distance from the dashed line to B, resulting in a somewhat circular path.

FIGURE 2.25 This photograph of Earth from space shows that it is nearly spherical.

was extended to the Sun, other planets, and eventually the universe. The quantitative predictions of observed relationships among the planets were strong evidence that all objects obey the same law of gravitation. In addition, the law provided a means to calculate the mass of Earth, the Moon, the planets, and the Sun. Newton's law of gravitation, laws of motion, and work with mathematics formed the basis of most physics and technology for the next two centuries, as well as accurately describing the world of everyday experience.

Earth Satellites

As you can see in figure 2.25, Earth is round-shaped and nearly spherical. The curvature is obvious in photographs taken from space but not so obvious back on the surface because Earth is so large. However, you can see evidence of the curvature in places on the surface where you have unobstructed vision for long distances. For example, a tall ship appears to "sink" on the horizon as it sails away, following Earth's curvature below your line of sight. The surface of Earth curves away from your line of sight or any other line tangent to the surface, dropping at a rate of about 4.9 m for every 8 km (16 ft in 5 mi). This means that a ship 8 km away will appear to drop about 5 m below the horizon and anything less than about 5 m tall at this distance will be out of sight, below the horizon.

Recall that a falling object accelerates toward Earth's surface at *g*, which has an average value of 9.8 m/s². Ignoring air resistance, a falling object will have a speed of 9.8 m/s at the end of 1 second and will fall a distance of 4.9 m. If you wonder why the object did not fall 9.8 m in 1 s, recall that the object starts with an initial speed of zero and has a speed of 9.8 m/s only during the last instant. The average speed was an average of the initial and final speeds, which is 4.9 m/s. An average

speed of 4.9 m/s over a time interval of 1 second will result in a distance covered of 4.9 m.

Did you know that Newton was the first to describe how to put an artificial satellite into orbit around Earth? He did not discuss rockets, however, but described in *Principia* how to put a cannonball into orbit. He described how a cannonball shot with sufficient speed straight out from a mountaintop would go into orbit around Earth. If it had less than the sufficient speed, it would fall back to Earth following the path of projectile motion, as discussed in the section on "Compound Motion." What speed does it need to go into orbit? Earth curves away from a line tangent to the surface at 4.9 m per 8 km. Any object falling from a resting position will fall a distance of 4.9 m during the first second. Thus, a cannonball shot straight out from a mountaintop with a speed of 8 km/s (nearly 18,000 mi/hr, or 5 mi/s) will fall toward the surface, dropping 4.9 m during the first second. But the surface of Earth drops, too, curving away below the falling cannonball. So the cannonball is still moving horizontally, no closer to the surface than it was a second ago. As it falls 4.9 m during the next second, the surface again curves away 4.9 m over the 8 km distance. This repeats again and again, and the cannonball stays the same distance from the surface, and we say it is now an *artificial satellite* in orbit. The satellite requires no engine or propulsion as it continues to fall toward the surface, with Earth curving away from it continuously. This assumes, of course, no loss of speed from air resistance.

Today, an artificial satellite is lofted by rocket or rockets to an altitude of more than 320 km (about 200 mi), above the air friction of the atmosphere, before being aimed horizontally. The satellite is then "injected" into orbit by giving it the correct tangential speed. This means it has attained an orbital speed of

A Closer Look

Gravity Problems

Gravity does act on astronauts in spacecraft that are in orbit around Earth. Since gravity is acting on the astronaut and spacecraft, the term *zero gravity* is not an accurate description of what is happening. The astronaut, spacecraft, and everything in it are experiencing *apparent weightlessness* because they are continuously falling toward the surface. Everything seems to float because everything is falling together. But, strictly speaking, everything still has weight, because weight is defined as a gravitational force acting on an object ($w = mg$).

Whether weightlessness is apparent or real, however, the effects on people are the same. Long-term orbital flights have provided evidence that the human body changes from the effect of weightlessness. Bones lose calcium and other minerals, the heart shrinks to a much smaller size, and leg muscles shrink so much on prolonged flights that astronauts cannot walk when they return to the surface. These changes occur because on Earth, humans are constantly subjected to the force of gravity. The nature of the skeleton and the strength of the muscles are determined by how the body reacts to this force. Metabolic pathways and physiological processes that maintain strong bones and muscles evolved having to cope with a specific gravitational force. When we are suddenly subjected to a place where gravity is significantly different, these processes result in weakened systems.

If we lived on a planet with a different gravitational force, we would have muscles and bones that were adapted to the gravity on that planet. Many kinds of organisms have been used in experiments in space to try to develop a better understanding of how their systems work without gravity.

The problems related to prolonged weightlessness must be worked out before long-term weightless flights can take place. One solution to these problems might be a large, uniformly spinning spacecraft. The astronauts would tend to move in a straight line, and the side of the turning spacecraft (now the "floor") would exert a force on them to make them go in a curved path. This force would act as an artificial gravity.

---

at least 8 km/s (5 mi/s) but less than 11 km/s (7 mi/s). At a speed less than 8 km/s, the satellite would fall back to the surface in a parabolic path. At a speed more than 11 km/s, it will move faster than the surface curves away and will escape from Earth into space. But with the correct tangential speed, and above the atmosphere and air friction, the satellite will follow a circular orbit for long periods of time without the need for any more propulsion. An orbit injection speed of more than 8 km/s (5 mi/s) would result in an elliptical rather than a circular orbit.

A satellite could be injected into orbit near the outside of the atmosphere, closer to Earth but outside the air friction that might reduce its speed. The satellite could also be injected far away from Earth, where it takes a longer time to complete one orbit. Near the outer limits of the atmosphere—that is, closer to the surface—a satellite might orbit Earth every 90 minutes or so. A satellite as far away as the Moon, on the other hand, orbits Earth in a little less than 28 days. A satellite at an altitude of 36,000 km (a little more than 22,000 mi) has a period of 1 day. In the right spot over the equator, such a satellite is called a **geosynchronous satellite,** since it turns with Earth and does not appear to move. The photographs of the cloud cover you see in weather reports were taken from one or more geosynchronous weather satellites. Communications networks are also built around geosynchronous satellites. One way to locate one of these geosynchronous satellites is to note the aiming direction of backyard satellite dishes that pick up television signals.

Weightlessness

News photos sometimes show astronauts "floating" in the Space Shuttle or next to a satellite (figure 2.26). These astronauts appear to be weightless but technically are no more weightless than a skydiver in free fall or a person in a falling elevator.

Recall that weight is a gravitational force, a measure of the gravitational attraction between Earth and an object (*mg*). The weight of a cup of coffee, for example, can be measured by placing the cup on a scale. The force the cup of coffee exerts against the scale is its weight. You also know that the scale pushes back on the cup of coffee since it is not accelerating, which means the net force is zero.

FIGURE 2.26 Astronauts in an orbiting space station may appear to be weightless. Technically, however, they are no more weightless than a skydiver in free fall or a person in a falling elevator near or on the surface of Earth.

Now consider what happens if a skydiver tries to pour a cup of coffee while in free fall. Even if you ignore air resistance, you can see that the skydiver is going to have a difficult time, at best. The coffee, the cup, and the skydiver will all be falling together. Gravity is acting to pull the coffee downward, but gravity is also acting to pull the cup from under it at the same rate. The coffee, the cup, and the skydiver all fall together, and the skydiver will see the coffee appear to "float" in blobs. If the diver lets go of the cup, it too will appear to float as everything continues to fall together. However, this is only *apparent weightlessness,* since gravity is still acting on everything; the coffee, the cup, and the skydiver only *appear* to be weightless because they are all accelerating at *g.*

Astronauts in orbit are in free fall, falling toward Earth just as the skydiver, so they too are undergoing apparent weightlessness. To experience true weightlessness, the astronauts would have to travel far from Earth and its gravitational field, and far from the gravitational fields of other planets.

SUMMARY

Motion can be measured by speed, velocity, and acceleration. *Speed* is a measure of how fast something is moving. It is a ratio of the distance covered between two locations to the time that elapsed while moving between the two locations. The *average speed* considers the distance covered during some period of time, while the *instantaneous speed* is the speed at some specific instant. *Velocity* is a measure of the speed and direction of a moving object. *Acceleration* is a change of velocity per unit of time.

A *force* is a push or a pull that can change the motion of an object. The *net force* is the sum of all the forces acting on an object.

Galileo determined that a continuously applied force is not necessary for motion and defined the concept of *inertia:* an object remains in unchanging motion in the absence of a net force. Galileo also determined that falling objects accelerate toward Earth's surface independent of the weight of the object. He found the acceleration due to gravity, *g,* to be 9.8 m/s²(32 ft/s²), and the distance an object falls is proportional to the square of the time of free fall ($d \propto t^2$).

Compound motion occurs when an object is projected into the air. Compound motion can be described by splitting the motion into vertical and horizontal parts. The acceleration due to gravity, *g,* is a constant that is acting at all times and acts independently of any motion that an object has. The path of an object that is projected at some angle to the horizon is therefore a parabola.

Newton's *first law of motion* is concerned with the motion of an object and the lack of a net force. Also known as the *law of inertia,* the first law states that an object will retain its state of straight-line motion (or state of rest) unless a net force acts on it.

The *second law of motion* describes a relationship between net force, mass, and acceleration. A *newton* of force is the force needed to give a 1.0 kg mass an acceleration of 1.0 m/s².

Weight is the downward force that results from Earth's gravity acting on the mass of an object. Weight is measured in *newtons* in the metric system and *pounds* in the English system.

Newton's *third law of motion* states that forces are produced by the interaction of *two different* objects. These forces always occur in matched pairs that are equal in size and opposite in direction.

Momentum is the product of the mass of an object and its velocity. In the absence of external forces, the momentum of a group of interacting objects always remains the same. This relationship is the *law of conservation of momentum. Impulse* is a change of momentum equal to a force times the time of application.

An object moving in a circular path must have a force acting on it, since it does not move in a straight line. The force that pulls an object out of its straight-line path is called a *centripetal force.* The centripetal force needed to keep an object in a circular path depends on the mass of the object, its velocity, and the radius of the circle.

The *universal law of gravitation* is a relationship between the masses of two objects, the distance between the objects, and a proportionality constant. Newton was able to use this relationship to show that gravitational attraction provides the centripetal force that keeps the Moon in its orbit.

Summary of Equations

2.1
$$\text{average speed} = \frac{\text{distance}}{\text{time}}$$
$$\bar{v} = \frac{d}{t}$$

2.2
$$\text{acceleration} = \frac{\text{change of velocity}}{\text{time}}$$
$$= \frac{\text{final velocity} - \text{initial velocity}}{\text{time}}$$
$$a = \frac{v_f - v_i}{t}$$

2.3
$$\text{force} = \text{mass} \times \text{acceleration}$$
$$F = ma$$

2.4
$$\text{weight} = \text{mass} \times \text{acceleration due to gravity}$$
$$w = mg$$

2.5
$$\text{momentum} = \text{mass} \times \text{velocity}$$
$$p = mv$$

2.6
$$\text{change of momentum} = \text{force} \times \text{time}$$
$$\Delta p = Ft$$

2.7
$$\text{centripetal acceleration} = \frac{\text{velocity squared}}{\text{radius of circle}}$$
$$a_c = \frac{v^2}{r}$$

2.8 $\text{centripetal force} = \dfrac{\text{mass} \times \text{velocity squared}}{\text{radius of circle}}$

$$F = \frac{mv^2}{r}$$

2.9 $\text{gravitational force} = \text{constant} \times \dfrac{\text{one mass} \times \text{another mass}}{\text{distance squared}}$

$$F = G\frac{m_1 m_2}{d^2}$$

KEY TERMS

acceleration (p. **29**)

centripetal force (p. **45**)

first law of motion (p. **38**)

force (p. **30**)

free fall (p. **34**)

fundamental forces (p. **32**)

g (p. **36**)

geosynchronous satellite (p. **50**)

impulse (p. **44**)

inertia (p. **33**)

law of conservation of
momentum (p. **43**)

mass (p. **41**)

momentum (p. **43**)

net force (p. **30**)

newton (p. **40**)

second law of motion (p. **40**)

speed (p. **27**)

third law of motion (p. **42**)

universal law of gravitation
(p. **47**)

velocity (p. **28**)

APPLYING THE CONCEPTS

1. A quantity of 5 m/s² is a measure of
   a. metric area.
   b. acceleration.
   c. speed.
   d. velocity.

2. An automobile has how many different devices that can cause it to undergo acceleration?
   a. none
   b. one
   c. two
   d. three or more

3. Ignoring air resistance, an object falling toward the surface of Earth has a *velocity* that is
   a. constant.
   b. increasing.
   c. decreasing.
   d. acquired instantaneously but dependent on the weight of the object.

4. Ignoring air resistance, an object falling near the surface of Earth has an *acceleration* that is
   a. constant.
   b. increasing.
   c. decreasing.
   d. dependent on the weight of the object.

5. Two objects are released from the same height at the same time, and one has twice the weight of the other. Ignoring air resistance,
   a. the heavier object hits the ground first.
   b. the lighter object hits the ground first.
   c. they both hit at the same time.
   d. whichever hits first depends on the distance dropped.

6. A ball rolling across the floor slows to a stop because
   a. there is a net force acting on it.
   b. the force that started it moving wears out.
   c. the forces are balanced.
   d. the net force equals zero.

7. Considering the forces on the system of you and a bicycle as you pedal the bike at a constant velocity in a horizontal straight line,
   a. the force you are exerting on the pedal is greater than the resisting forces.
   b. all forces are in balance, with the net force equal to zero.
   c. the resisting forces of air and tire friction are less than the force you are exerting.
   d. the resisting forces are greater than the force you are exerting.

8. If you double the unbalanced force on an object of a given mass, the acceleration will be
   a. doubled.
   b. increased fourfold.
   c. increased by one-half.
   d. increased by one-fourth.

9. If you double the mass of a cart while it is undergoing a constant unbalanced force, the acceleration will be
   a. doubled.
   b. increased fourfold.
   c. half as much.
   d. one-fourth as much.

10. Doubling the distance between the center of an orbiting satellite and the center of Earth will result in what change in the gravitational attraction of Earth for the satellite?
   a. one-half as much
   b. one-fourth as much
   c. twice as much
   d. four times as much

11. If a ball swinging in a circle on a string is moved twice as fast, the force on the string will be
   a. twice as great.
   b. four times as great.
   c. one-half as much.
   d. one-fourth as much.

12. A ball is swinging in a circle on a string when the string length is doubled. At the same velocity, the force on the string will be
    a. twice as great.
    b. four times as great.
    c. one-half as much.
    d. one-fourth as much.

QUESTIONS FOR THOUGHT

1. An insect inside a bus flies from the back toward the front at 5.0 mi/h. The bus is moving in a straight line at 50.0 mi/h. What is the speed of the insect?

2. Disregarding air friction, describe all the forces acting on a bullet shot from a rifle into the air.

3. Can gravity act in a vacuum? Explain.

4. Is it possible for a small car to have the same momentum as a large truck? Explain.

5. What net force is needed to maintain the constant velocity of a car moving in a straight line? Explain.

6. How can an unbalanced force exist if every action has an equal and opposite reaction?

7. Why should you bend your knees as you hit the ground after jumping from a roof?

8. Is it possible for your weight to change as your mass remains constant? Explain.

9. What maintains the speed of Earth as it moves in its orbit around the Sun?

10. Suppose you are standing on the ice of a frozen lake and there is no friction whatsoever. How can you get off the ice? (*Hint:* Friction is necessary to crawl or walk, so that will not get you off the ice.)

11. A rocket blasts off from a platform on a space station. An identical rocket blasts off from free space. Considering everything else to be equal, will the two rockets have the same acceleration? Explain.

12. An astronaut leaves a spaceship that is moving through free space to adjust an antenna. Will the spaceship move off and leave the astronaut behind? Explain.

FOR FURTHER ANALYSIS

1. What are the significant similarities and differences between speed and velocity?

2. What are the significant similarities and differences between velocity and acceleration?

3. Compare your beliefs and your own reasoning about motion before and after learning Newton's three laws of motion.

4. Newton's law of gravitation explains that every object in the universe is attracted to every other object in the universe.

Describe a conversation between yourself and another person who does not believe this law, as you persuade him or her that the law is indeed correct.

5. Why is it that your weight can change by moving from one place to another, but your mass stays the same?

6. Assess the reasoning that Newton's first law of motion tells us that centrifugal force does not exist.

INVITATION TO INQUIRY

The Domino Effect

The *domino effect* is a cumulative effect produced when one event initiates a succession of similar events. In the actual case of dominoes, a row is made by standing dominoes on their end face to face in a line. When one on the end is tipped over toward the others, it will fall into its neighbor, which falls into the next one, and so on until the whole row has fallen. How should the dominoes be spaced so the row falls with maximum speed? Should one domino strike the next one as high as possible, in the center, or as low as possible? If you accept this invitation, you will need to determine how you plan to space the dominoes as well as how you will measure the speed.

PARALLEL EXERCISES

The exercises in groups A and B cover the same concepts. Solutions to group A exercises are located in appendix D.
Note: Neglect all frictional forces in all exercises.

Group A

1. How far away was a lightning strike if thunder is heard 5.00 seconds after seeing the flash? Assume that sound traveled at 350.0 m/s during the storm.

Group B

1. How many meters away is a cliff if an echo is heard one-half second after the original sound? Assume that sound traveled at 343 m/s on that day.

—continued

Group A

2. What is the acceleration of a car that moves from rest to 15.0 m/s in 10.0 s?

3. What is the average speed of a truck that makes a 285-mile trip in 5.0 hours?

4. What force will give a 40.0 kg grocery cart an acceleration of 2.4 m/s²?

5. An unbalanced force of 18 N will give an object an acceleration of 3 m/s². What force will give this very same object an acceleration of 10 m/s²?

6. What is the weight of a 70.0 kg person?

7. What is the momentum of a 100 kg football player who is moving at 6 m/s?

8. A car weighing 13,720 N is speeding down a highway with a velocity of 91 km/h. What is the momentum of this car?

9. A 15 g bullet is fired with a velocity of 200 m/s from a 6 kg rifle. What is the recoil velocity of the rifle?

10. A net force of 5,000.0 N accelerates a car from rest to 90.0 km/h in 5.0 s. (a) What is the mass of the car? (b) What is the weight of the car?

11. How much centripetal force is needed to keep a 0.20 kg ball on a 1.50 m string moving in a circular path with a speed of 3.0 m/s?

12. On Earth, an astronaut and equipment weigh 1,960.0 N. While weightless in space, the astronaut fires a 100 N rocket backpack for 2.0 s. What is the resulting velocity of the astronaut and equipment?

Group B

2. What is the acceleration of a car that moves from a speed of 5.0 m/s to a speed of 15 m/s during a time of 6.0 s?

3. What is the average speed of a car that travels 270.0 miles in 4.50 hours?

4. What force would an asphalt road have to give a 6,000 kg truck in order to accelerate it at 2.2 m/s² over a level road?

5. If a space probe weighs 39,200 N on the surface of Earth, what will be the mass of the probe on the surface of Mars?

6. How much does a 60.0 kg person weigh?

7. What is the momentum of a 30.0 kg shell fired from a cannon with a velocity of 500 m/s?

8. What is the momentum of a 39.2 N bowling ball with a velocity of 7.00 m/s?

9. A 30.0 kg shell fired from a 2,000 kg cannon will have a velocity of 500 m/s. What is the resulting velocity of the cannon?

10. A net force of 3,000.0 N accelerates a car from rest to 36.0 km/h in 5.00 s. (a) What is the mass of the car? (b) What is the weight of the car?

11. What tension must a 50.0 cm length of string support in order to whirl an attached 1,000.0 g stone in a circular path at 5.00 m/s?

12. A 200.0 kg astronaut and equipment move with a velocity of 2.00 m/s toward an orbiting spacecraft. How long will the astronaut need to fire a 100.0 N rocket backpack to stop the motion relative to the spacecraft?

The wind can be used as a source of energy. All you need is a way to capture the energy—such as these wind turbines in California—and to live someplace where the wind blows enough to make it worthwhile.

Energy

Physics Connections

▶ Heat is energy in transit (Ch. 3).
▶ Sound is a mechanical form of energy (Ch. 4).
▶ Electricity is a form of energy (Ch. 6).

Chemistry Connections

▶ Chemical change is a form of energy (Ch. 8–10).
▶ A huge amount of energy exists in the nucleus of an atom (Ch. 11).

CORE CONCEPT

Energy is transformed through working or heating, and the total amount remains constant.

When work is done on an object, it gains energy (p. 59).

Energy comes in various forms (p. 62).

The total energy amount remains the same when one form is converted to another (p. 65).

The main energy sources today are petroleum, coal, nuclear, and moving water (p. 66).

Alternate sources of energy are solar, geothermal, and hydrogen (p. 69).

Earth Science Connections

▶ Energy changes alter Earth and cause materials to be cycled (Ch. 13–18).

Life Science Connections

▶ Living things use energy and materials in complex interactions (Ch. 19–26).

Astronomy Connections

▶ Stars are nuclear reactors (Ch. 12).

OVERVIEW

The term *energy* is closely associated with the concepts of force and motion. Naturally moving matter, such as the wind or moving water, exerts forces. You have felt these forces if you have ever tried to walk against a strong wind or stand in one place in a stream of rapidly moving water. The motion and forces of moving air and moving water are used as *energy sources* (figure 3.1). The wind is an energy source as it moves the blades of a windmill, performing useful work. Moving water is an energy source as it forces the blades of a water turbine to spin, turning an electric generator. Thus, moving matter exerts a force on objects in its path, and objects moved by the force can also be used as an energy source.

Matter does not have to be moving to supply energy; matter *contains* energy. Food supplied the energy for the muscular exertion of the humans and animals that accomplished most of the work before the twentieth century. Today in the developed world, machines do the work that was formerly accomplished by muscular exertion. Machines also use the energy contained in matter. They use gasoline, for example, as they supply the forces and motion to accomplish work.

Moving matter and matter that contains energy can be used as energy sources to perform work. The concepts of work and energy and the relationship to matter are the topics of this chapter. You will learn how energy flows in and out of your surroundings as well as a broad, conceptual view of energy.

WORK

You learned earlier that the term *force* has a special meaning in science that is different from your everyday concept of force. In everyday use, you use the term in a variety of associations such as police force, economic force, or the force of an argument. Earlier, force was discussed in a general way as a push or pull. Then a more precise scientific definition of force was developed from Newton's laws of motion—a force is a result of an interaction that is capable of changing the state of motion of an object.

The word *work* represents another one of those concepts that has a special meaning in science that is different from your everyday concept. In everyday use, work is associated with a task to be accomplished or the time spent in performing the task. You might work at understanding science, for example, or you might tell someone that science is a lot of work. You also probably associate physical work, such as lifting or moving boxes, with how tired you become from the effort. The definition of mechanical work is not concerned with tasks, time, or how tired you become from doing a task. It is concerned with the application of a force to an object and the distance the object moves as a result of the force. The **work** done on the object is defined as *the product of the applied force and the parallel distance through which the force acts*:

$$\text{work} = \text{force} \times \text{distance}$$
$$W = Fd \qquad \text{equation 3.1}$$

FIGURE 3.1 Glen Canyon Dam on the Colorado River between Utah and Arizona is 216 m (710 ft) tall, dropping 940 m³ of water per second (33,000 ft³/s) through eight generating units.

Mechanical work is the product of a force and the distance an object moves as a result of the force. There are two important considerations to remember about this definition: (1) something *must move* whenever work is done, and (2) the movement must be in the *same direction* as the direction of the force. When you move a book to a higher shelf in a bookcase, you are doing work on the book. You apply a vertically upward force equal to the weight of the book as you move it in the same direction as the direction of the applied force. The work done on the book can therefore be calculated by multiplying the weight of the book by the distance it was moved (figure 3.2).

If you simply stand there holding the book, however, you are doing no work on the book. Your arm may become tired from holding the book, since you must apply a vertically upward force equal to the weight of the book. But this force is not acting through a distance, since the book is not moving. According to equation 3.1, a distance of zero results in zero work. Only a force that results in motion in the same direction results in work.

Units of Work

The units of work can be obtained from the definition of work, $W = Fd$. In the metric system, a force is measured in newtons (N), and distance is measured in meters (m), so the unit of work is

$$W = Fd$$
$$W = (\text{newton})(\text{meter})$$
$$W = (\text{N})(\text{m})$$

The newton-meter is therefore the unit of work. This derived unit has a name. The newton-meter is called a **joule** (J) (pronounced "jool").

$$1 \text{ joule} = 1 \text{ newton-meter}$$

The units for a newton are kg·m/s^2, and the unit for a meter is m. It therefore follows that the units for a joule are $\text{kg·m}^2/\text{s}^2$.

In the English system, the force is measured in pounds (lb), and the distance is measured in feet (ft). The unit of work in the English system is therefore the ft·lb. The ft·lb does not have a name of its own as the N·m does (figure 3.3). (Note that although the equation is $W = Fd$, and this means = (pounds)(feet), the unit is called the ft·lb.)

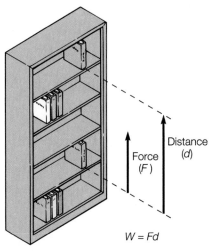

FIGURE 3.2 The force on the book moves it through a vertical distance from the second shelf to the fifth shelf, and work is done, $W = Fd$.

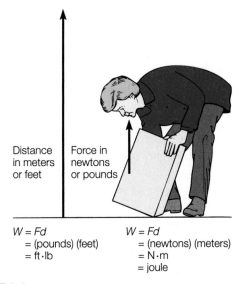

FIGURE 3.3 Work is done against gravity when lifting an object. Work is measured in joules or foot-pounds.

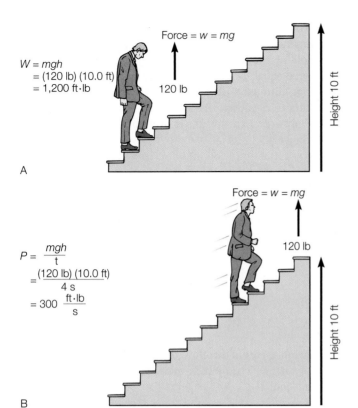

$$W = mgh$$
$$= (120 \text{ lb}) (10.0 \text{ ft})$$
$$= 1{,}200 \text{ ft·lb}$$

Force = $w = mg$

120 lb

Height 10 ft

A

$$P = \frac{mgh}{t}$$
$$= \frac{(120 \text{ lb}) (10.0 \text{ ft})}{4 \text{ s}}$$
$$= 300 \frac{\text{ft·lb}}{\text{s}}$$

Force = $w = mg$

120 lb

Height 10 ft

B

FIGURE 3.4 (A) The work accomplished in climbing a stairway is the person's weight times the vertical distance. (B) The power level is the work accomplished per unit of time.

Power

You are doing work when you walk up a stairway, since you are lifting yourself through a distance. You are lifting your weight (force exerted) the *vertical* height of the stairs (distance through which the force is exerted). Consider a person who weighs 120 lb and climbs a stairway with a vertical distance of 10 ft. This person will do (120 lb)(10 ft) or 1,200 ft·lb of work. Will the amount of work change if the person were to run up the stairs? The answer is no, the same amount of work is accomplished. Running up the stairs, however, is more tiring than walking up the stairs. You use the same amount of energy but at a greater *rate* when running. The rate at which energy is transformed or the rate at which work is done is called **power** (figure 3.4). Power is defined as work per unit of time,

$$\text{power} = \frac{\text{work}}{\text{time}}$$

$$P = \frac{W}{t} \qquad \textbf{equation 3.2}$$

Considering just the work and time factors, the 120 lb person who ran up the 10 ft height of stairs in 4 seconds would have a power rating of

$$P = \frac{W}{t} = \frac{(120 \text{ lb})(10 \text{ ft})}{4 \text{ s}} = 300 \frac{\text{ft·lb}}{\text{s}}$$

If the person had a time of 3 seconds on the same stairs, the power rating would be greater, 400 ft·lb/s. This is a greater *rate* of energy use, or greater power.

EXAMPLE 3.1 (*Optional*)

How much work is needed to lift a 5.0 kg backpack to a shelf 1.0 m above the floor?

SOLUTION

The backpack has a mass (m) of 5.0 kg, and the distance (d) is 1.0 m. To lift the backpack requires a vertically upward force equal to the weight of the backpack. Weight can be calculated from $w = mg$:

$$m = 5.0 \text{ kg} \qquad w = mg$$
$$g = 9.8 \text{ m/s}^2 \qquad\quad = (5.0 \text{ kg})\left(9.8 \frac{\text{m}}{\text{s}^2}\right)$$
$$w = ?$$
$$= (5.0)(9.8) \text{ kg} \times \frac{\text{m}}{\text{s}^2}$$
$$= 49 \frac{\text{kg·m}}{\text{s}^2}$$
$$= 49 \text{ N}$$

The definition of work is found in equation 3.1,

$$F = 49 \text{ N} \qquad W = Fd$$
$$d = 1.0 \text{ m} \qquad\quad = (49 \text{ N})(1.0 \text{ m})$$
$$W = ? \qquad\qquad = (49)(1.0) \text{ N} \times \text{m}$$
$$= 49 \text{ N·m}$$
$$= \boxed{49 \text{ J}}$$

EXAMPLE 3.2 (*Optional*)

How much work is required to lift a 50 lb box vertically a distance of 2 ft? (Answer: 100 ft·lb)

When the steam engine was first invented, there was a need to describe the rate at which the engine could do work. Since people at this time were familiar with using horses to do their

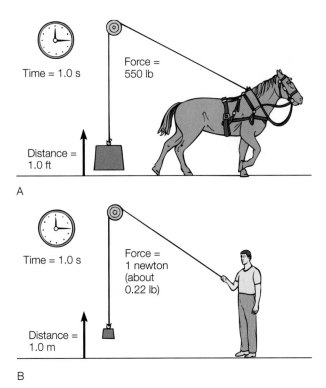

A

B

FIGURE 3.5 (*A*) A horsepower is defined as a power rating of 550 ft·lb/s. (*B*) A watt is defined as a newton-meter per second, or joule per second.

work, the steam engines were compared to horses. James Watt, who designed a workable steam engine, defined **horsepower** as a power rating of 550 ft·lb/s (figure 3.5A). To convert a power rating in the English units of ft·lb/s to horsepower, divide the power rating by 550 ft·lb/s/hp. For example, the 120 lb person who had a power rating of 400 ft·lb/s had a horsepower of 400 ft·lb/s ÷ 550 ft·lb/s/hp, or 0.7 hp.

In the metric system, power is measured in joules per second. The unit J/s, however, has a name. A J/s is called a **watt** (W). The watt (figure 3.5B) is used with metric prefixes for large numbers: 1,000 W = 1 kilowatt (kW) and 1,000,000 W = 1 megawatt (MW). It takes 746 W to equal 1 horsepower. One kilowatt is equal to about 1 1/3 horsepower. The electric utility company charges you for how much electrical energy you have used. Electrical energy is measured by power (kW) times the time of use (h). Thus, electrical energy is measured in kWh. We will return to kilowatts and kilowatt-hours later when we discuss electricity.

EXAMPLE 3.3 (*Optional*)

An electric lift can raise a 500.0 kg mass a distance of 10.0 m in 5.0 s. What is the power of the lift?

SOLUTION

Power is the work per unit time ($P = W/t$), and work is force times distance ($W = Fd$). The vertical force required is the weight lifted, and $w = mg$. Therefore, the work accomplished would be $W = mgh$,

and the power would be $P = mgh/t$. Note that h is for height, a vertical distance (d).

$$
\begin{aligned}
m &= 500.0 \text{ kg} \\
g &= 9.8 \text{ m/s}^2 \\
h &= 10.0 \text{ m} \\
t &= 5.0 \text{ s}
\end{aligned}
\qquad P = \frac{mgh}{t}
$$

$$
= \frac{(500.0 \text{ kg})\left(9.8\,\dfrac{\text{m}}{\text{s}^2}\right)(10.0 \text{ m})}{5.0 \text{ s}}
$$

$$
= \frac{(500.0)(9.8)(10.0)}{5.0}\,\frac{\text{kg} \times \dfrac{\text{m}}{\text{s}^2} \times \text{m}}{\text{s}}
$$

$$
= 9{,}800\,\frac{\text{N}\cdot\text{m}}{\text{s}}
$$

$$
= 9{,}800\,\frac{\text{J}}{\text{s}}
$$

$$
= 9{,}800 \text{ W}
$$

$$
= \boxed{9.8 \text{ kW}}
$$

The power in horsepower (hp) units would be

$$
9{,}800 \text{ W} \times \frac{\text{hp}}{746 \text{ W}} = 13 \text{ hp}
$$

EXAMPLE 3.4 (*Optional*)

A 150 lb person runs up a 15 ft stairway in 10.0 s. What is the horsepower rating of the person? (Answer: 0.41 horsepower)

MOTION, POSITION, AND ENERGY

Closely related to the concept of work is the concept of **energy**. Energy can be defined as the *ability to do work*. This definition of energy seems consistent with everyday ideas about energy and physical work. After all, it takes a lot of energy to do a lot of work. In fact, one way of measuring the energy of something is to see how much work it can do. Likewise, when work is done *on* something, a change occurs in its energy level. The following examples will help clarify this close relationship between work and energy.

Potential Energy

Consider a book on the floor next to a bookcase. You can do work on the book by vertically raising it to a shelf. You can measure this work by multiplying the vertical upward force applied times the distance that the book is moved. You might find, for example, that you did an amount of work equal to 10 J on the book.

Suppose that the book has a string attached to it, as shown in figure 3.6. The string is threaded over a frictionless pulley and attached to an object on the floor. If the book is caused to fall from the shelf, the object on the floor will be vertically lifted through some distance by the string. The falling book exerts a force on the object through the string, and the object is moved

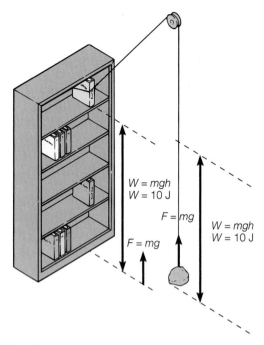

FIGURE 3.6 If moving a book from the floor to a high shelf requires 10 J of work, then the book will do 10 J of work on an object of the same mass when the book falls from the shelf.

through a distance. In other words, the *book* did work on the object through the string, $W = Fd$.

The book can do more work on the object if it falls from a higher shelf, since it will move the object a greater distance. The higher the shelf, the greater the *potential* for the book to do work. The ability to do work is defined as energy. The energy that an object has because of its position is called **potential energy** (PE). Potential energy is defined as *energy due to position*. This type of potential energy is called *gravitational potential energy,* since it is a result of gravitational attraction. There are other types of potential energy, such as that in a compressed or stretched spring.

The gravitational potential energy of an object can be calculated from the work done *on* the object to change its position. You exert a force equal to its weight as you lift it some height above the floor, and the work you do is the product of the weight and height. Likewise, the amount of work the object *could* do because of its position is the product of its weight and height. For the metric unit of mass, weight is the product of the mass of an object times *g*, the acceleration due to gravity, so

$$\text{gravitational potential energy} = \text{weight} \times \text{height}$$
$$PE = mgh \qquad \textbf{equation 3.3}$$

For English units, the pound *is* the gravitational unit of force, or weight, so equation 3.3 becomes $PE = (w)(h)$.

Under what conditions does an object have zero potential energy? Considering the book in the bookcase, you could say that the book has zero potential energy when it is flat on the floor. It can do no work when it is on the floor. But what if that floor happens to be the third floor of a building? You could, after all, drop the book out of a window. The answer is that it makes no difference. The same results would be obtained in either case since it is the *change of position* that is important

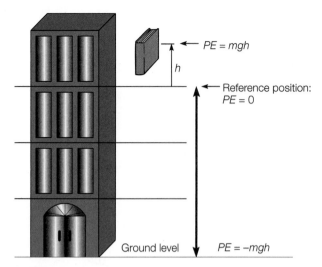

FIGURE 3.7 The zero reference level for potential energy is chosen for convenience. Here the reference position chosen is the third floor, so the book will have a negative potential energy at ground level.

in potential energy. The zero reference position for potential energy is therefore arbitrary. A zero reference point is chosen as a matter of convenience. Note that if the third floor of a building is chosen as the zero reference position, a book on ground level would have negative potential energy. This means that you would have to do work on the book to bring it back to the zero potential energy position (figure 3.7). You will learn more about negative energy levels later in the chapters on chemistry.

EXAMPLE 3.5 *(Optional)*

What is the potential energy of a 2.0 lb book that is on a bookshelf 4.0 ft above the floor?

SOLUTION

Equation 3.3, $PE = mgh$, shows the relationship between potential energy (PE), weight (mg), and height (h).

$$
\begin{aligned}
w &= 2.0 \text{ lb} & PE &= mgh \\
h &= 4.0 \text{ ft} & &= wh \\
PE &= ? & &= (2.0 \text{ lb})(4.0 \text{ ft}) \\
& & &= (2.0)(4.0) \text{ lb} \times \text{ft} \\
& & &= \boxed{8.0 \text{ ft·lb}}
\end{aligned}
$$

EXAMPLE 3.6 *(Optional)*

How much work can a 5.00 kg mass do if it is 5.00 m above the ground? (Answer: 245 J)

Kinetic Energy

Moving objects have the ability to do work on other objects because of their motion. A rolling bowling ball exerts a force on the bowling pins and moves them through a distance, but the ball loses speed as a result of the interaction (figure 3.8). A moving car has the ability to exert a force on a tree and knock it down, again with a corresponding loss of speed. Objects in motion have the ability to do work, so they have energy. The

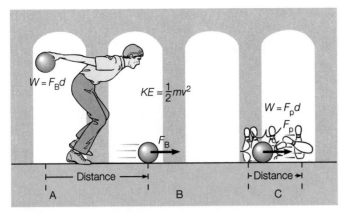

$W = F_B d$

$KE = \frac{1}{2}mv^2$

$W = F_p d$

F_p

F_B

Distance

Distance

A

B

C

FIGURE 3.8 (A) Work is done on the bowling ball as a force (F_B) moves it through a distance. (B) This gives the ball a kinetic energy equal in amount to the work done on it. (C) The ball does work on the pins and has enough remaining energy to crash into the wall behind the pins.

energy of motion is known as **kinetic energy.** Kinetic energy can be measured (1) in terms of the work done to put the object in motion or (2) in terms of the work the moving object will do in coming to rest. Consider objects that you put into motion by throwing. You exert a force on a football as you accelerate it through a distance before it leaves your hand. The kinetic energy that the ball now has is equal to the work (force times distance) that you did on the ball. You exert a force on a baseball through a distance as the ball increases its speed before it leaves your hand. The kinetic energy that the ball now has is equal to the work that you did on the ball. The ball exerts a force on the hand of the person catching the ball and moves it

through a distance. The net work done on the hand is equal to the kinetic energy that the ball had.

A baseball and a bowling ball moving with the same velocity do not have the same kinetic energy. You cannot knock down many bowling pins with a slowly rolling baseball. Obviously, the more-massive bowling ball can do much more work than a less-massive baseball with the same velocity. Is it possible for the bowling ball and the baseball to have the same kinetic energy? The answer is yes, if you can give the baseball sufficient velocity. This might require shooting the baseball from a cannon, however. Kinetic energy is proportional to the mass of a moving object, but velocity has a greater influence. Consider two balls of the same mass, but one is moving twice as fast as the other. The ball with twice the velocity will do *four* times as much work as the slower ball. A ball with three times the velocity will do *nine* times as much work as the slower ball. Kinetic energy is proportional to the square of the velocity ($2^2 = 4$; $3^2 = 9$). The kinetic energy (*KE*) of an object is

$$\text{kinetic energy} = \frac{1}{2}(\text{mass})(\text{velocity})^2$$

$$KE = \frac{1}{2}mv^2 \qquad \textbf{equation 3.4}$$

Kinetic energy is measured in joules, as is work ($F \times d$ or N·m) and potential energy (mgh or N·m).

EXAMPLE 3.7 (Optional)

A 7.00 kg bowling ball is moving in a bowling lane with a velocity of 5.00 m/s. What is the kinetic energy of the ball?

SOLUTION

The relationship between kinetic energy (*KE*), mass (*m*), and velocity (*v*) is found in equation 3.4, $KE = 1/2mv^2$:

$m = 7.00$ kg

$v = 5.00$ m/s

$KE = ?$

$$KE = \frac{1}{2}mv^2$$

$$= \frac{1}{2}(7.00 \text{ kg})\left(5.00\frac{\text{m}}{\text{s}}\right)^2$$

$$= \frac{1}{2}(7.00 \text{ kg})\left(25.0\frac{\text{m}^2}{\text{s}^2}\right)$$

$$= \frac{1}{2}(7.00)(25.0) \text{ kg} \times \frac{\text{m}^2}{\text{s}^2}$$

$$= \frac{1}{2} \times 175\frac{\text{kg}\cdot\text{m}^2}{\text{s}^2}$$

$$= 87.5\frac{\text{kg}\cdot\text{m}}{\text{s}^2} \times \text{m}$$

$$= 87.5 \text{ N}\cdot\text{m}$$

$$= \boxed{87.5 \text{ J}}$$

EXAMPLE 3.8 (Optional)

A 100.0 kg football player moving with a velocity of 6.0 m/s tackles a stationary quarterback. How much work was done on the quarterback? (Answer: 1,800 J)

ENERGY FLOW

The key to understanding the individual concepts of work and energy is to understand the close relationship between the two. When you do work on something, you give it energy of position (potential energy) or you give it energy of motion (kinetic energy). In turn, objects that have kinetic or potential energy can now do work on something else as the transfer of energy continues. Where does all this energy come from and where does it go? The answer to this question is the subject of this section on energy flow.

Energy Forms

Energy comes in various forms, and different terms are used to distinguish one form from another. Although energy comes in various *forms,* this does not mean that there are different *kinds* of energy. The forms are the result of the more common fundamental forces—gravitational, electromagnetic, and nuclear—and objects that are interacting. Energy can be categorized into five forms: (1) *mechanical,* (2) *chemical,* (3) *radiant,* (4) *electrical,* and (5) *nuclear.* The following is a brief discussion of each of the five forms of energy.

Mechanical energy is the form of energy of familiar objects and machines (figure 3.9). A car moving on a highway has kinetic mechanical energy. Water behind a dam has potential mechanical energy. The spinning blades of a steam turbine have kinetic mechanical energy. The form of mechanical energy is usually associated with the kinetic energy of everyday-sized objects and the potential energy that results from gravity. There are other possibilities (e.g., sound), but this description will serve the need for now.

Chemical energy is the form of energy involved in chemical reactions (figure 3.10). Chemical energy is released in the chemical reaction known as *oxidation.* The fire of burning wood is an example of rapid oxidation. A slower oxidation

A

B

FIGURE 3.10 Chemical energy is a form of potential energy that is released during a chemical reaction. Both (*A*) wood and (*B*) coal have chemical energy that has been stored through the process of photosynthesis. The pile of wood might provide fuel for a small fireplace for several days. The pile of coal might provide fuel for a power plant for a hundred days.

releases energy from food units in your body. As you will learn in the chemistry unit, chemical energy involves electromagnetic forces between the parts of atoms. Until then, consider the following comparison. Photosynthesis is carried on in green plants. The plants use the energy of sunlight to rearrange carbon dioxide and water into plant materials and oxygen. Leaving out many steps and generalizing, this could be represented by the following word equation:

energy + carbon dioxide + water = wood + oxygen

The plant took energy and two substances and made two different substances. This is similar to raising a book to a higher shelf in a bookcase. That is, the new substances have more energy than the original ones did. Consider a word equation for the burning of wood:

wood + oxygen = carbon dioxide + water + energy

FIGURE 3.9 Mechanical energy is the energy of motion, or the energy of position, of many familiar objects. This boat has energy of motion.

FIGURE 3.11 This demonstration solar cell array converts radiant energy from the Sun to electrical energy, producing an average of 200,000 watts of electric power (after conversion).

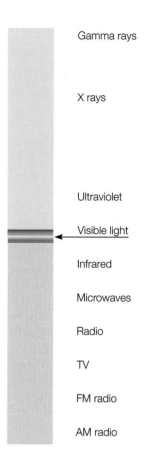

FIGURE 3.12 The electromagnetic spectrum includes many categories of radiant energy. Note that visible light occupies only a tiny part of the entire spectrum.

Notice that this equation is exactly the reverse of photosynthesis. In other words, the energy used in photosynthesis was released during oxidation. Chemical energy is a kind of potential energy that is stored and later released during a chemical reaction.

Radiant energy is energy that travels through space (figure 3.11). Most people think of light or sunlight when considering this form of energy. Visible light, however, occupies only a small part of the complete electromagnetic spectrum, as shown in figure 3.12. Radiant energy includes light and all other parts of the spectrum (see chapter 7). Infrared radiation is sometimes called "heat radiation" because of the association with heating when this type of radiation is absorbed. For example, you feel the interaction of infrared radiation when you hold your hand near a warm range element. However, infrared radiation is another type of radiant energy. In fact, some snakes, such as the sidewinder, have pits between their eyes and nostrils that can detect infrared radiation emitted from warm animals. Microwaves are another type of radiant energy that is used in cooking. As with other forms of energy, light, infrared, and microwaves will be considered in more detail later. For now, consider all types of radiant energy to be forms of energy that travel through space.

Electrical energy is another form of energy from electromagnetic interactions that will be considered in detail later. You are familiar with electrical energy that travels through wires to your home from a power plant (figure 3.13), electrical energy that is generated by chemical cells in a flashlight, and electrical energy that can be "stored" in a car battery.

Nuclear energy is a form of energy often discussed because of its use as an energy source in power plants. Nuclear energy is another form of energy from the atom, but this time the energy involves the nucleus, the innermost part of an atom, and nuclear interactions.

Energy Conversion

Potential energy can be converted to kinetic energy and vice versa. The simple pendulum offers a good example of this

FIGURE 3.13 The blades of a steam turbine. In a power plant, chemical or nuclear energy is used to heat water to steam, which is directed against the turbine blades. The mechanical energy of the turbine turns an electric generator. Thus, a power plant converts chemical or nuclear energy to mechanical energy, which is then converted to electrical energy.

Energy Converter Coaster

A roller coaster is an energy converter that swaps kinetic and potential energy back and forth. An outside energy source is used only to move the cars to the top of the first hill. The first hill is always the highest hill above the ground, and here the cars will have the most potential energy they will have for the entire ride.

When the coaster reaches the top of the first hill, it begins to move down a sloping track with increasing speed as potential energy is converted to kinetic energy. At the bottom of this hill, the track then starts up a second hill, and the cars this time convert kinetic energy back to potential energy. Ideally, all the potential energy is converted to kinetic energy as the cars move down a hill, and all the kinetic energy is converted to potential energy as the cars move up a hill. There is no perfect energy conversion, and some

energy is lost to air resistance and friction. The roller coaster design allows for these losses and leaves room for a slight surplus of kinetic energy. You know this is true because the operator must apply brakes at the end of the ride.

Is the speed of the moving roller coaster at the bottom of a hill directly proportional to the height of the hill? Ignoring friction, the speed of a coaster at the bottom of a hill is proportional to the *square root of the height* of the hill. This means that to double the speed of the coaster at the bottom of the hill, you would need to build the hill more than four times higher. It would need to be more than four times higher because the coaster does not drop straight down. Doubling the height increases the theoretical speed only 40 percent.

What is the relationship between the weight of the people on a roller coaster

and the speed achieved at the bottom of the hill? The answer is that the weight of the people does not matter. Ignoring air resistance and friction, a heavy roller coaster full of people and a lighter one with just a few people will both have the same speed at the bottom of the hill.

Draw a profile of a roller coaster ride to find out about potential and kinetic energy exchanges and the height of each hill. The profile should show the relative differences in height between the crest of each hill and the bottom of each upcoming dip. From such a profile, you could find where in a ride the maximum speed would occur, as well as what speed to expect. As you study a profile keep in mind that the roller coaster is designed to produce many changes of speed—accelerations—rather than high speed alone.

---

conversion. A simple pendulum is an object, called a bob, suspended by a string or wire from a support. If the bob is moved to one side and then released, it will swing back and forth in an arc. At the moment that the bob reaches the top of its swing, it stops for an instant, then begins another swing. At the instant of stopping, the bob has 100 percent potential energy and no kinetic energy. As the bob starts back down through the swing, it is gaining kinetic energy and losing potential energy. At the instant the bob is at the bottom of the swing, it has 100 percent kinetic energy and no potential energy. As the bob now climbs through the other half of the arc, it is gaining potential energy and losing kinetic energy until it again reaches an instantaneous stop at the top, and the process starts over. The kinetic energy of the bob at the bottom of the arc is equal to the potential energy it had at the top of the arc (figure 3.14). Disregarding friction, the sum of the potential energy and the kinetic energy remains constant throughout the swing.

Any *form* of energy can be converted to another form. In fact, most technological devices that you use are nothing more than *energy-form converters* (figure 3.15). A lightbulb, for example, converts electrical energy to radiant energy. A car converts chemical energy to mechanical energy. A solar cell converts radiant energy to electrical energy, and an electric motor converts electrical energy to mechanical energy. Each technological device converts some form of energy (usually chemical or electrical) to another form that you desire (usually mechanical or radiant).

It is interesting to trace the *flow of energy* that takes place in your surroundings. Suppose, for example, that you are riding a bicycle. The bicycle has kinetic mechanical energy as it moves along. Where did the bicycle get this energy? It came from you, as you use the chemical energy of food units to contract your muscles and move the bicycle along. But where did your chemical energy come from? It came from your food, which consists of plants, animals who eat plants, or both plants and animals. In any case, plants are at the bottom of your food chain. Plants convert radiant energy from the Sun into chemical

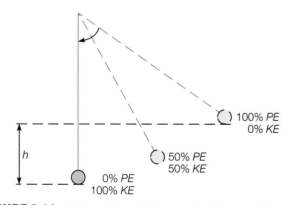

FIGURE 3.14 This pendulum bob loses potential energy (*PE*) and gains an equal amount of kinetic energy (*KE*) as it falls through a distance *h*. The process reverses as the bob moves up the other side of its swing.

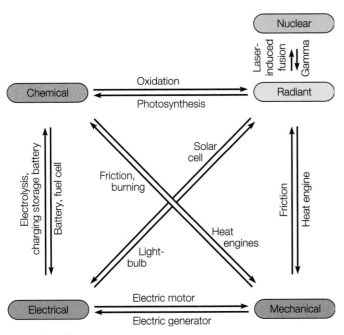

FIGURE 3.15 The energy forms and some conversion pathways.

Grow Your Own Fuel?

Have you heard of biodiesel? Biodiesel is a vegetable-based oil that can be used for fuel in diesel engines. It can be made from soy oils, canola oil, or even recycled deep-fryer oil from a fast-food restaurant. Biodiesel can be blended with regular diesel oil in any amount. Or it can be used in its 100 percent pure form in diesel cars, trucks, buses, or as home heating oil.

Why would we want to use vegetable oil to run diesel engines? First, it is a sustainable (or renewable) resource. It also reduces dependency on foreign oil and lessens the trade deficit. It runs smoother, produces less exhaust smoke, and reduces the health risks associated with petroleum diesel. The only negative aspect seems to occur when recycled oil from fast-food restaurants is used. People behind such a biodiesel-powered school bus complained that it smells like fried potatoes, making them hungry.

A website is maintained by some biodiesel users where you can learn how to produce your own biodiesel from algae. See www.biodieselnow.com and search for algae.

energy. Radiant energy comes to the plants from the Sun because of the nuclear reactions that took place in the core of the Sun. Your bicycle is therefore powered by nuclear energy that has undergone a number of form conversions!

Energy Conservation

Energy can be transferred from one object to another, and it can be converted from one form to another form. If you make a detailed accounting of all forms of energy before and after a transfer or conversion, the total energy will be *constant*. Consider your bicycle coasting along over level ground when you apply the brakes. What happened to the kinetic mechanical energy of the bicycle? It went into heating the rim and brakes of your bicycle, then eventually radiated to space as infrared radiation. All radiant energy that reaches Earth is eventually radiated back to space (figure 3.16). Thus, throughout all the

form conversions and energy transfers that take place, the total sum of energy remains constant.

The total energy is constant in every situation that has been measured. This consistency leads to another one of the conservation laws of science, the **law of conservation of energy:**

Energy is never created or destroyed. Energy can be converted from one form to another but the total energy remains constant.

See also "Thermodynamics" in chapter 4. You may be wondering about the source of nuclear energy. Does a nuclear reaction create energy? Albert Einstein answered this question back in the early 1900s when he formulated his now-famous relationship between mass and energy, $E = mc^2$. This relationship will be discussed in detail in chapter 11. Basically, the relationship states that mass *is* a form of energy, and this has been experimentally verified many times.

Energy Transfer

Earlier it was stated that when you do work on something, you give it energy. The result of work could be increased kinetic mechanical energy, increased gravitational potential energy, or an increase in the temperature of an object. You could summarize this by stating that either *working* or *heating* is always involved any time energy is transformed. This is not unlike your financial situation. To increase or decrease your financial status, you need some mode of transfer, such as cash or checks, as a means of conveying assets. Just as with cash flow from one individual to another, energy flow from one object to another requires a mode of transfer. In energy matters, the mode of transfer is working or heating. Any time you see working

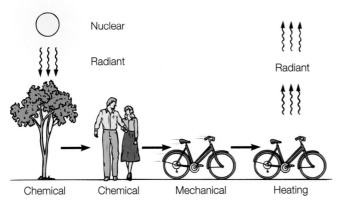

FIGURE 3.16 Energy arrives from the Sun, goes through a number of conversions, then radiates back into space. The total sum leaving eventually equals the original amount that arrived.

or heating occurring, you know that an energy transfer is taking place. The next time you see heating, think about what energy form is being converted to what new energy form. (The final form is usually radiant energy.) Heating is the topic of chapter 4, where you will consider the role of heat in energy matters.

ENERGY SOURCES TODAY

Prometheus, according to ancient Greek mythology, stole fire from heaven and gave it to humankind. Fire has propelled human advancement ever since. All that was needed was something to burn—fuel for Prometheus's fire.

Any substance that burns can be used to fuel a fire, and various fuels have been used over the centuries as humans advanced. First, wood was used as a primary source for heating. Then coal fueled the Industrial Revolution. Eventually, humankind roared into the twentieth century burning petroleum. Today, petroleum is the most widely used source of energy (figure 3.17). It provides about 40 percent of the total energy used by the United States. Natural gas contributes about 23 percent of the total energy used today. The use of coal also provides about 23 percent of the total. Biomass—any material formed by photosynthesis—contributes about 3 percent of the total energy used today. Note that petroleum, coal, biomass, and natural gas are all chemical sources of energy, sources that are mostly burned for their energy. These chemical sources supply about 87 percent of the total energy consumed. About a third of this is burned for heating, and the rest is burned to drive engines or generators.

Nuclear energy and hydropower are the nonchemical sources of energy. These sources are used to generate electrical energy. The alternative sources of energy, such as solar, wind, and geothermal, provide about 1 percent of the total energy consumed in the United States today.

The energy-source mix has changed from past years, and it will change in the future. Wood supplied 90 percent of the energy until the 1850s, when the use of coal increased. Then, by 1910, coal was supplying about 75 percent of the total energy needs. Then petroleum began making increased contributions to the energy supply. Now increased economic and environmental constraints and a decreasing supply of petroleum are

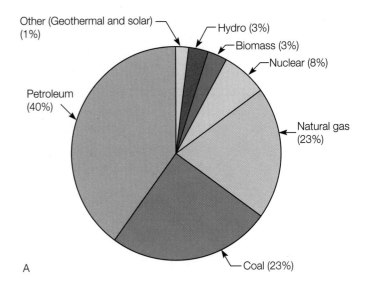

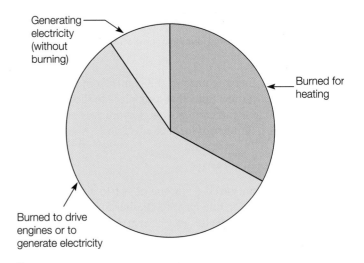

FIGURE 3.17 (*A*) Primary energy consumed in the United States, 2005,* and (*B*) uses of this energy.

*Source: Energy Information Administration (www.eia.doe.gov/emeu/aer/pdf/pages/sec1.pdf).

producing another supply shift. The present petroleum-based energy era is about to shift to a new energy era.

About 92 percent of the total energy consumed today is provided by four sources: (1) petroleum (including natural gas), (2) coal, (3) moving water, and (4) nuclear. The following is a brief introduction to these four sources.

Petroleum

The word *petroleum* is derived from the word *petra,* meaning rock, and the word *oleum,* meaning oil. Petroleum is oil that comes from oil-bearing rock. Natural gas is universally associated with petroleum and has similar origins. Both petroleum and natural gas form from organic sediments, materials that have settled out of bodies of water. Sometimes a local condition permits the accumulation of sediments that are exceptionally rich in organic material. This could occur under special conditions in a freshwater lake, or it could occur on shallow ocean basins. In either case, most of the organic material is from plankton—tiny free-floating animals and plants such as algae. It is from such accumulations of buried organic material that petroleum and natural gas are formed. The exact process by which these materials become petroleum and gas is not understood. It is believed that bacteria, pressure, appropriate temperatures, and time are all important. Natural gas is formed at higher temperatures than is petroleum. Varying temperatures over time may produce a mixture of petroleum and gas or natural gas alone.

Petroleum forms a thin film around the grains of the rock where it formed. Pressure from the overlying rock and water moves the petroleum and gas through the rock until they reach a rock type or structure that stops them. If natural gas is present, it occupies space above the accumulating petroleum. Such accumulations of petroleum and natural gas are the sources of supply for these energy sources.

Discussions about the petroleum supply and the cost of petroleum usually refer to a "barrel of oil." The *barrel* is an accounting device of 42 U.S. gallons. Such a 42-gallon barrel does not exist. When or if oil is shipped in barrels, each drum holds 55 U.S. gallons.

The supply of petroleum and natural gas is limited. Most of the continental drilling prospects appear to be exhausted, and the search for new petroleum supplies is now offshore. In general, over 25 percent of the United States' petroleum is estimated to come from offshore wells. The amount of imported petroleum has ranged from 30 to 50 percent over the years, with most imported oil coming from Canada, Saudi Arabia, Venezuela, Mexico, and Nigeria.

Petroleum is used for gasoline (about 45 percent), diesel (about 40 percent), and heating oil (about 15 percent). Petroleum is also used in making medicine, clothing fabrics, plastics, and ink.

Coal

Petroleum and natural gas formed from the remains of tiny organisms that lived millions of years ago. Coal, on the other hand, formed from an accumulation of plant materials that collected under special conditions millions of years ago. Thus, petroleum, natural gas, and coal are called **fossil fuels.** Fossil fuels contain the stored radiant energy of organisms that lived millions of years ago.

The first thing to happen in the formation of coal was for plants in swamps to die and sink. Stagnant swamp water protected the plants and plant materials from consumption by animals and decomposition by microorganisms. Over time, chemically altered plant materials collected at the bottom of pools of water in the swamp. This carbon-rich material is *peat* (not to be confused with peat moss). Peat is used as a fuel in many places in the world. The flavor of Scotch (whisky) is the result of the peat fires used to brew the liquor. Peat is still being produced naturally in swampy areas today. Under pressure and at high temperatures, peat will eventually be converted to coal.

There are several stages, or *ranks,* in the formation of coal. The lowest rank is lignite (brown coal), then subbituminous, then bituminous (soft coal), and the highest rank is anthracite (hard coal).

Each rank of coal has different burning properties and a different energy content. Coal also contains impurities of clay, silt, iron oxide, and sulfur. The mineral impurities leave an ash when the coal is burned, and the sulfur produces sulfur dioxide, a pollutant.

Most of the coal mined today is burned by utilities to generate electricity (about 80 percent). The coal is ground to a face-powder consistency and blown into furnaces. This greatly increases efficiency but produces *fly ash,* ash that "flies" up the chimney. Industries and utilities are required by the federal Clean Air Act to remove sulfur dioxide and fly ash from plant emissions. About 20 percent of the cost of a new coal-fired power plant goes into air pollution control equipment. Coal is an abundant but dirty energy source.

Moving Water

Moving water has been used as a source of energy for thousands of years. It is considered a renewable energy source, inexhaustible as long as the rain falls. Today, hydroelectric plants generate about 3 percent of the United States' *total* energy consumption at about 2,400 power-generating dams across the nation. Hydropower furnished about 40 percent of the United States' electric power in 1940. Today, dams furnish 9 percent of the electric power. Energy consumption has increased, but hydropower production has not kept pace because geography limits the number of sites that can be built.

Water from a reservoir is conducted through large pipes called penstocks to a powerhouse, where it is directed against turbine blades that turn a shaft on an electric generator.

Nuclear

Nuclear power plants use nuclear energy to produce electricity. Energy is released as the nuclei of uranium and plutonium atoms split, or undergo fission. The fissioning takes place in a large steel vessel called a *reactor.* Water is pumped through the reactor to produce steam, which is used to produce electrical energy, just as in the fossil fuel power plants. The nuclear processes are described in detail in chapter 11, and the process of producing electrical energy is described in detail in chapter 6. Nuclear power plants use nuclear energy to produce electricity, but there are opponents of this process. The electric utility companies view nuclear energy as *one* energy source used to produce electricity. They state that they have no allegiance to any one energy source but are seeking to utilize the most reliable and dependable of several energy sources. Petroleum, coal, and hydropower are also presently utilized as energy sources for electric power production. The electric utility companies are concerned that petroleum and natural gas are becoming increasingly expensive, and there are questions about long-term supplies. Hydropower has limited potential for growth, and solar energy is prohibitively expensive today. Utility companies

James Prescott Joule (1818–1889)

James Joule was a British physicist who helped develop the principle of conservation of energy by experimentally measuring the mechanical equivalent of heat. In recognition of Joule's pioneering work on energy, the SI unit of energy is named the joule.

Joule was born on December 24, 1818, into a wealthy brewing family. He and his brother were educated at home between 1833 and 1837 in elementary math, natural philosophy, and chemistry, partly by the English chemist John Dalton (1766–1844). Joule was a delicate child and very shy, and apart from his early education, he was entirely self-taught in science. He does not seem to have played any part in the family brewing business, although some of his first experiments were done in the laboratory at the brewery.

Joule had great dexterity as an experimenter, and he could measure temperatures very precisely. At first, other scientists could not believe such accuracy and were skeptical about the theories that Joule developed to explain his results. The encouragement of Lord Kelvin from 1847 on changed these attitudes, however, and Kelvin subsequently used Joule's practical ability to great advantage. By 1850, Joule was highly regarded by other scientists and was elected a fellow of the Royal Society. Joule's own wealth funded his scientific career, and he never took an academic post. His funds eventually ran out, however. He was awarded a pension in 1878 by Queen Victoria, but by that time, his mental powers were going. He suffered a long illness and died on October 11, 1889.

Joule realized the importance of accurate measurement very early on, and exact data became his hallmark. His most active research period was between 1837 and 1847. In a long series of experiments, he studied the relationship between electrical, mechanical, and chemical effects and heat, and in 1843, he was able to announce his determination of the amount of work required to produce a unit of heat. This is called the mechanical equivalent of heat (4.184 joules per calorie).

One great value of Joule's work was the variety and completeness of his experimental evidence. He showed that the same relationship could be examined experimentally and that the ratio of equivalence of the different forms of energy did not depend on how one form was converted into another or on the materials involved. The principle that Joule had established is that energy cannot be created or destroyed but only transformed.

Joule lives on in the use of his name to measure energy, supplanting earlier units such as the erg and calorie. It is an appropriate reflection of his great experimental ability and his tenacity in establishing a basic law of science.

Source: Modified from the *Hutchinson Dictionary of Scientific Biography.* © RM, 2007. Reprinted by permission.

---

see two major energy sources that are available for growth: coal and nuclear. There are problems and advantages to each, but the utility companies feel they must use coal and nuclear power until the new technologies, such as solar power, are economically feasible.

Conserving Energy

Conservation is not a way of generating energy, but it is a way of reducing the need for additional energy consumption and it saves money for the consumer. Some conservation technologies are sophisticated, while others are quite simple. For example, if a small, inexpensive wood-burning stove were developed and used to replace open fires in the less-developed world, energy consumption in these regions could be reduced by 50 percent.

Many observers have pointed out that demanding more energy while failing to conserve is like demanding more water to fill a bathtub while leaving the drain open. To be sure, conservation and efficiency strategies by themselves will not eliminate demands for energy, but they can make the demands much easier to meet, regardless of what options are chosen to provide the primary energy. Energy efficiency improvements have significantly reduced the need for additional energy sources. Consider these facts, which are based primarily on data published by the U.S. Energy Information Administration:

- Total primary energy use per capita in the United States in 2003 was almost identical to that in 1973. Over the same thirty-year period, economic output (gross domestic product, or GDP) per capita increased 74 percent.
- National energy intensity (energy use per unit of GDP) fell 43 percent between 1973 and 2002. About 60 percent of this decline is attributable to energy efficiency improvements.
- If the United States had not dramatically reduced its energy intensity over the past thirty years, consumers and businesses would have spent at least $430 billion more on energy purchases in 2003.

Even though the United States is much more energy-efficient today than it was thirty years ago, the potential is still enormous for additional cost-effective energy savings. Some newer energy efficiency measures have barely begun to be adopted. Other efficiency measures could be developed and commercialized in coming years.

Much of the energy we consume is wasted. This statement is not meant as a reminder to simply turn off lights and lower furnace thermostats; it is a technological challenge. Our use of energy is so inefficient that most potential energy in fuel is lost as waste heat, becoming a form of environmental pollution.

The amount of energy wasted through poorly insulated windows and doors alone is about as much energy as the United States receives from the Alaskan pipeline each year. It is estimated that by using inexpensive, energy-efficient measures, the average energy bills of a single home could be reduced by 10 percent to 50 percent and could help reduce the emissions of carbon dioxide into the atmosphere.

Many conservation techniques are relatively simple and highly cost-effective. More efficient and less energy-intensive industry and domestic practices could save large amounts of energy. Improved automobile efficiency, better mass transit, and increased railroad use for passenger and freight traffic are simple and readily available means of conserving transportation energy. In response to the 1970s' oil price shocks, automobile mileage averages in the United States more than doubled, from 5.55 km/L (13 mi/gal) in 1975 to 12.3 km/L (28.8 mi/gal) in 1988. Unfortunately, the oil glut and falling fuel prices of the late 1980s discouraged further conservation. Between 1990 and 1997, the average slipped to only 11.8 km/L (27.6 mi/gal). It remains to be seen if the sharp increase of gasoline prices in the early years of the twenty-first century will translate into increased miles per gallon in new car design.

Several technologies that reduce energy consumption are now available. Highly efficient fluorescent lightbulbs that can be used in regular incandescent fixtures give the same amount of light for 25 percent of the energy and produce less heat. Since lighting and air conditioning (which removes the heat from inefficient incandescent lighting) account for 25 percent of U.S. electricity consumption, widespread use of these lights could significantly reduce energy consumption. Low-emissive glass for windows can reduce the amount of heat entering a building while allowing light to enter. The use of this type of glass in new construction and replacement windows could have a major impact on the energy picture. Many other technologies, such as automatic light-dimming or shutoff devices, are being used in new construction.

The shift to more efficient use of energy needs encouragement. Often, poorly designed, energy-inefficient buildings and machines can be produced inexpensively. The short-term cost is low, but the long-term cost is high. The public needs to be educated to look at the long-term economic and energy costs of purchasing poorly designed buildings and appliances.

Electric utilities have recently become part of the energy conservation picture. In some states, they have been allowed to make money on conservation efforts; previously, they could make money only by building more power plants. This encourages them to become involved in energy conservation education, because teaching their customers how to use energy more efficiently allows them to serve more people without building new power plants.

ENERGY TOMORROW

An *alternative source of energy* is one that is different from the typical sources used today. The sources used today are fossil fuels (coal, petroleum, and natural gas), nuclear, and falling water. Alternative sources could be solar, geothermal, hydrogen gas, fusion, or any other energy source that a new technology could utilize.

Solar Technologies

The term *solar energy* is used to describe a number of technologies that directly or indirectly utilize sunlight as an alternative energy source (figure 3.18). There are eight main categories of these solar technologies:

1. **Solar cells.** A solar cell is a thin crystal of silicon, gallium, or some polycrystalline compound that generates electricity when exposed to light. Also called photovoltaic devices, solar cells have no moving parts and produce electricity directly, without the need for hot fluids or intermediate conversion states. Solar cells have been used extensively in space vehicles and satellites. Here on Earth, however, use has been limited to demonstration projects, remote-site applications, and consumer specialty items such as solar-powered watches and calculators. The problem with solar cells today is that the manufacturing cost is too high (they are essentially handmade). Research is continuing on the development of highly efficient, affordable solar cells that could someday produce electricity for the home. See page 147 to find out how a solar cell is able to create a current.

2. **Power tower.** This is another solar technology designed to generate electricity. One type of planned power tower will have a 171 m (560 ft) tower surrounded by some 9,000 special mirrors called heliostats. The heliostats will focus sunlight

FIGURE 3.18 Wind is another form of solar energy. This wind turbine generates electrical energy for this sailboat, charging batteries for backup power when the wind is not blowing. In case you are wondering, the turbine cannot be used to make a wind to push the boat. In accord with Newton's laws of motion, this would not produce a net force on the boat.

on a boiler at the top of the tower where salt (a mixture of sodium nitrate and potassium nitrate) will be heated to about 566°C (about 1,050°F). This molten salt will be pumped to a steam generator, and the steam will be used to drive a generator, just like other power plants. Water could be heated directly in the power tower boiler. Molten salt is used because it can be stored in an insulated storage tank for use when the Sun is not shining, perhaps for up to 20 hours.

3. **Passive application.** In passive applications, energy flows by natural means, without mechanical devices such as motors, pumps, and so forth. A passive solar house would include considerations such as its orientation to the Sun, the size and positioning of windows, and a roof overhang that lets sunlight in during the winter but keeps it out during the summer. There are different design plans to capture, store, and distribute solar energy throughout a house, and some of these designs are described on pages 88–89.

4. **Active application.** An active solar application requires a solar collector in which sunlight heats air, water, or some liquid. The liquid or air is pumped through pipes in a house to generate electricity, or it is used directly for hot water. Solar water heating makes more economic sense today than the other solar applications.

5. **Wind energy.** The wind has been used for centuries to move ships, grind grain into flour, and pump water. The wind blows, however, because radiant energy from the Sun heats some parts of Earth's surface more than others. This differential heating results in pressure differences and the horizontal movement of air, which is called wind. Thus, wind is another form of solar energy. Wind turbines are used to generate electrical energy or mechanical energy. The biggest problem with wind energy is the inconsistency of the wind. Sometimes the wind speed is too great, and other times it is not great enough. Several methods of solving this problem are being researched. (See "Wind Power" in chapter 17 for more on wind energy.)

6. **Biomass.** Biomass is any material formed by photosynthesis, including small plants, trees, and crops, and any garbage, crop residue, or animal waste. Biomass can be burned directly as a fuel, converted into a gas fuel (methane), or converted into liquid fuels such as alcohol. The problems with using biomass include the energy expended in gathering the biomass, as well as the energy used to convert it to a gaseous or liquid fuel.

7. **Agriculture and industrial heating.** This is a technology that simply uses sunlight to dry grains, cure paint, or do anything that can be done with sunlight rather than using traditional energy sources.

8. **Ocean thermal energy conversion (OTEC).** This is an electric generating plant that would take advantage of the approximately 22°C (about 40°F) temperature difference between the surface and the depths of tropical, subtropical, and equatorial ocean waters. Basically, warm water is drawn into the system to vaporize a fluid, which expands through a turbine generator. Cold water from the depths condenses the vapor back to a liquid form, which is then cycled back to the warm-water side. The concept has been tested and found to be technically successful. The greatest interest seems to be from islands that have warm surface waters and cold depths, such as Hawaii, Puerto Rico, Guam, and the Virgin Islands.

Geothermal Energy

Geothermal energy is energy from beneath Earth's surface. The familiar geysers, hot springs, and venting steam of Yellowstone National Park are clues that this form of energy exists. There is substantially more geothermal energy than is revealed in Yellowstone, however, and geothermal resources are more widespread than once thought. Earth has a high internal temperature, and recoverable geothermal resources may underlie most states. These resources occur in four broad categories of geo-thermal energy: (1) dry steam, (2) hot water, (3) hot, dry rock, and (4) geopressurized resources. Together, the energy contained in these geothermal resources represents about 15,000 times more than is consumed in the United States in a given year. The only problem is getting to the geothermal energy, then using it in a way that is economically attractive.

Most geothermal energy occurs as *hot, dry rock,* which accounts for about 85 percent of the total geothermal resource. Hot, dry rock is usually in or near an area of former volcanic activity. The problem of utilizing this widespread resource is how to get the energy to the surface. Research has been conducted by drilling wells, then injecting water into one well and extracting energy from the heated water pumped from the second well. There is more interest in the less widespread but better understood geothermal systems of hot water and steam.

Geopressurized resources are trapped underground reservoirs of hot water that contain dissolved natural gas. The water temperature is higher than the boiling point, so heat could be used as a source of energy as well as the dissolved natural gas. Such geo-pressurized reservoirs make up about 14 percent of the total accessible geothermal energy found on Earth. They are being studied in some areas to determine whether they are large enough to be economically feasible as an energy source. More is known about recovering energy from other types of hot water and steam resources, so these seem more economically attractive.

Hot water and steam make up the smallest geothermal resource category, together comprising only about 1 percent of the total known resource. However, more is known about the utilization and recovery of these energy sources, which are estimated to contain an amount of energy equivalent to about half of the known reserves of petroleum in the United States. *Steam* is very rare, occurring in only three places in the United States. Two of these places are national parks (Lassen and Yellowstone), so this geothermal steam cannot be used as an energy source. The third place is at the Geysers, an area of fumaroles near San Francisco, California. Steam from the Geysers is used to generate a significant amount of electricity.

Hot water systems make up most of the *recoverable* geothermal resources. Heat from deep volcanic or former volcanic sources create vast, slow-moving convective patterns in groundwater. If the water circulating back near the surface is hot enough, it can be used for generating electricity, heating buildings, or many other possible applications. Worldwide, geothermal energy is used to operate pulp and paper mills, cool hotels, raise fish, heat greenhouses, dry crops, desalt water, and dozens of other things. Thousands of apartments, homes, and businesses are today heated geothermally in Oregon and Idaho in the United States, as well as in Hungary, France, Iceland, and New Zealand. Today, each British thermal unit (Btu) of heat supplied by geothermal energy does not have to be supplied by fossil fuels. Tomorrow, geothermal resources will become more and more attractive as the price and the consequences of using fossil fuels continue to increase.

Hydrogen

Hydrogen is the lightest and simplest of all the elements, occurring as a diatomic gas that can be used for energy directly in a fuel cell or burned to release heat. Hydrogen could be used to replace natural gas with a few modifications of present natural gas burners. A big plus in favor of hydrogen as a fuel is that it produces no pollutants. In addition to the heat produced, the only emission from burning hydrogen is water, as shown in the following equation:

Hydrogen + oxygen → water + 68,300 calories

The primary problem with using hydrogen as an energy source is that *it does not exist* on or under Earth's surface in any but trace amounts! Hydrogen must therefore be obtained by a chemical reaction from such compounds as water. Water is a plentiful substance on Earth, and an electric current will cause decomposition of water into hydrogen and oxygen gas. Measurement of the electric current and voltage will show that:

Water + 68,300 calories → hydrogen + oxygen

Thus, assuming 100 percent efficiency, the energy needed to obtain hydrogen gas from water is exactly equal to the energy released by hydrogen combustion. So hydrogen cannot be used to produce energy, since hydrogen gas is not available, but it can be used as a means of storing energy for later use. Indeed, hydrogen may be destined to become an effective solution to the problems of storing and transporting energy derived from solar energy sources. In addition, hydrogen might serve as the transportable source of energy, such as that needed for cars and trucks, replacing the fossil fuels. In summary, hydrogen has the potential to provide clean, alternative energy for a number of uses, including lighting, heating, cooling, and transportation.

SUMMARY

Work is defined as the product of an applied force and the distance through which the force acts. Work is measured in newton-meters, a metric unit called a *joule*. *Power* is work per unit of time. Power is measured in *watts*. One watt is 1 joule per second. Power is also measured in *horsepower*. One horsepower is 550 ft·lb/s.

Energy is defined as the ability to do work. An object that is elevated against gravity has a potential to do work. The object is said to have *potential energy*, or *energy of position*. Moving objects have the ability to do work on other objects because of their motion. The *energy of motion* is called *kinetic energy*.

Work is usually done *against inertia, fundamental forces, friction, shape,* or *combinations of these*. As a result, there is a gain of *kinetic energy, potential energy, an increased temperature,* or *any combination of these*. Energy comes in the *forms* of *mechanical, chemical, radiant, electrical,* or *nuclear*. Potential energy can be *converted* to kinetic and kinetic can be *converted* to potential. Any form of energy can be *converted* to any other form. Most technological devices are *energy-form converters* that do work for you. Energy flows into and out of the surroundings, but the amount of energy is always constant. The *law of conservation of energy* states that *energy is never created or destroyed.* Energy conversion always takes place through *heating* or *working.*

The basic energy sources today are the chemical *fossil fuels* (petroleum, natural gas, and coal), *nuclear energy,* and *hydropower.* Petroleum and *natural gas* were formed from organic material of plankton, tiny free-floating plants and animals. A barrel of petroleum is 42 U.S. gallons, but such a container does not actually exist. *Coal* formed from plants that were protected from consumption by falling into a swamp.

The decayed plant material, *peat,* was changed into the various *ranks* of coal by pressure and heating over some period of time. Coal is a dirty fuel that contains impurities and sulfur. Controlling air pollution from burning coal is costly. Moving water and nuclear energy are used for the generation of electricity. An *alternative* source of energy is one that is different from the typical sources used today. Alternative sources could be *solar, geothermal,* or *hydrogen.*

Summary of Equations

3.1

$$\text{work} = \text{force} \times \text{distance}$$
$$W = Fd$$

3.2

$$\text{power} = \frac{\text{work}}{\text{time}}$$
$$P = \frac{W}{t}$$

3.3

$$\text{potential energy} = \text{weight} \times \text{height}$$
$$PE = mgh$$

3.4

$$\text{kinetic energy} = \frac{1}{2}(\text{mass})(\text{velocity})^2$$
$$KE = \frac{1}{2}mv^2$$

KEY TERMS

energy (p. **59**)

fossil fuels (p. **67**)

geothermal energy (p. **70**)

horsepower (p. **59**)

joule (p. **57**)

kinetic energy (p. **61**)

law of conservation of
energy (p. **65**)

potential energy (p. **60**)

power (p. **58**)

watt (p. **59**)

work (p. **56**)

APPLYING THE CONCEPTS

1. The metric unit of a joule (J) is a unit of
   a. potential energy.
   b. work.
   c. kinetic energy.
   d. Any of the above is correct.

2. Power is
   a. the rate at which work is done.
   b. the rate at which energy is expended.
   c. work per unit time.
   d. Any of the above is correct.

3. Which of the following is a combination of units called a watt?
   a. N·m/s
   b. kg·m^2/s^2/s
   c. J/s
   d. All of the above are correct.

4. About how many watts are equivalent to 1 horsepower?
   a. 7.5
   b. 75
   c. 750
   d. 7,500

5. A kilowatt-hour is a unit of
   a. power.
   b. work.
   c. time.
   d. electrical charge.

6. The potential energy of a box on a shelf, relative to the floor, is a measure of
   a. the work that was required to put the box on the shelf from the floor.
   b. the weight of the box times the distance above the floor.
   c. the energy the box has because of its position above the floor.
   d. All of the above are correct.

7. A rock on the ground is considered to have zero potential energy. In the bottom of a well, then, the rock would be considered to have
   a. zero potential energy, as before.
   b. negative potential energy.
   c. positive potential energy.
   d. zero potential energy but will require work to bring it back to ground level.

8. Which quantity has the greatest influence on the amount of kinetic energy that a large truck has while moving down the highway?
   a. mass
   b. weight
   c. velocity
   d. size

9. Most energy comes to and leaves Earth in the form of
   a. nuclear energy.
   b. chemical energy.
   c. radiant energy.
   d. kinetic energy.

10. The law of conservation of energy is basically that
   a. energy must not be used up faster than it is created, or the supply will run out.
   b. energy should be saved because it is easily destroyed.
   c. energy is never created or destroyed.
   d. you are breaking a law if you needlessly destroy energy.

11. The most widely used source of energy today is
   a. coal.
   b. petroleum.
   c. nuclear.
   d. moving water.

12. The accounting device of a "barrel of oil" is defined to hold how many U.S. gallons of petroleum?
   a. 24
   b. 42
   c. 55
   d. 100

Answers

1. d **2.** d **3.** d **4.** c **5.** b **6.** d **7.** b **8.** c **9.** c **10.** c **11.** b **12.** b

QUESTIONS FOR THOUGHT

1. How is work related to energy?

2. What is the relationship between the work done while moving a book to a higher bookshelf and the potential energy that the book has on the higher shelf?

3. Does a person standing motionless in the aisle of a moving bus have kinetic energy? Explain.

4. A lamp bulb is rated at 100 W. Why is a time factor not included in the rating?

5. Is a kWh a unit of work, energy, power, or more than one of these? Explain.

6. If energy cannot be destroyed, why do some people worry about the energy supplies?

7. A spring clamp exerts a force on a stack of papers it is holding together. Is the spring clamp doing work on the papers? Explain.

8. Why are petroleum, natural gas, and coal called *fossil fuels?*

9. From time to time, people claim to have invented a machine that will run forever without energy input and develops more energy than it uses (perpetual motion). Why would you have reason to question such a machine?

10. Define a joule. What is the difference between a joule of work and a joule of energy?

11. Compare the energy needed to raise a mass 10 m on Earth to the energy needed to raise the same mass 10 m on the Moon. Explain the difference, if any.

12. What happens to the kinetic energy of a falling book when the book hits the floor?

FOR FURTHER ANALYSIS

1. Evaluate the requirement that something must move whenever work is done. Why is this a requirement?

2. What are the significant similarities and differences between work and power?

3. Whenever you do work on something, you give it energy. Analyze how you would know for sure that this is a true statement.

4. Simple machines are useful because they are able to trade force for distance moved. Describe a conversation between yourself and another person who believes that you do less work when you use a simple machine.

5. Use the equation for kinetic energy to prove that speed is more important than mass in bringing a speeding car to a stop.

6. Describe at least several examples of negative potential energy and how each shows a clear understanding of the concept.

7. The five forms of energy are the result of the fundamental forces—gravitational, electromagnetic, and nuclear—and objects that are interacting. Analyze which force is probably involved with each form of energy.

8. Most technological devices convert one of the five forms of energy into another. Try to think of a technological device that does not convert an energy form to another. Discuss the significance of your finding.

9. Are there any contradictions to the law of conservation of energy in any area of science?

INVITATION TO INQUIRY

New Energy Source?

Is waste paper a good energy source? There are 103 waste-to-energy plants in the United States that burn solid garbage, so waste paper would be a good source, too. The plants burn solid garbage to make steam that is used to heat buildings and generate electricity. Schools might be able to produce a pure waste paper source because waste paper accumulates near computer print stations and in offices. Collecting waste paper from such sources would yield 6,800 Btu/lb, which is about half the heat value of coal.

If you accept this invitation, start by determining how much waste paper is created per month in your school. Would this amount produce enough energy to heat buildings or generate electricity?

PARALLEL EXERCISES

The exercises in groups A and B cover the same concepts. Solutions to group A exercises are located in appendix D.
Note: Neglect all frictional forces in all exercises.

Group A

1. A force of 200 N is needed to push a table across a level classroom floor for a distance of 3 m. How much work was done on the table?

2. An 880 N box is pushed across a level floor for a distance of 5.0 m with a force of 440 N. How much work was done on the box?

3. How much work is done in raising a 10.0 kg backpack from the floor to a shelf 1.5 m above the floor?

Group B

1. Work of 1,200 J is done while pushing a crate across a level floor for a distance of 1.5 m. What force was used to move the crate?

2. How much work is done by a hammer that exerts a 980.0 N force on a nail, driving it 1.50 cm into a board?

3. A 5.0 kg textbook is raised a distance of 30.0 cm as a student prepares to leave for school. How much work did the student do on the book?

—continued

PARALLEL EXERCISES—*Continued*

Group A

4. If 5,000 J of work is used to raise a 102 kg crate to a shelf in a warehouse, how high was the crate raised?

5. A 60.0 kg student runs up a 5.00 m high stairway in a time of 3.92 s. (a) How many watts of power did she develop? (b) How many horsepower is this?

6. What is the kinetic energy of a 2,000 kg car moving at 72 km/h?

7. How much work is needed to stop a 1,000.0 kg car that is moving straight down the highway at 54.0 km/h?

8. A 1,000 kg car stops on top of a 51.02 m hill. (a) How much energy was used in climbing the hill? (b) How much potential energy does the car have?

9. A horizontal force of 10.0 lb is needed to push a bookcase 5 ft across the floor. (a) How much work was done on the bookcase? (b) How much did the gravitational potential energy change as a result?

10. (a) How much work is done in moving a 2.0 kg book to a shelf 2.00 m high? (b) What is the change in potential energy of the book as a result? (c) How much kinetic energy will the book have as it hits the ground as it falls?

11. A 60.0 kg jogger moving at 2.0 m/s decides to double the jogging speed. How did this change in speed change the kinetic energy?

12. A 170.0 lb student runs up a stairway to a classroom 25.0 ft above ground level in 10.0 s. (a) How much work did the student do? (b) What was the average power output in horsepower?

Group B

4. An electric hoist does 196,000 J of work in raising a 250.0 kg load. How high was the load lifted?

5. What is the horsepower of a 1,500.0 kg car that can go to the top of a 360.0 m high hill in exactly one minute?

6. What is the kinetic energy of a 30.0 g bullet that is traveling at 200.0 m/s?

7. How much work will be done by a 30.0 g bullet traveling at 200 m/s?

8. A 10.0 kg box is lifted 15 m above the ground by a construction crane. (a) How much work did the crane do on the box? (b) How much potential energy does the box have relative to the ground?

9. A force of 50.0 lb is used to push a box 10.0 ft across a level floor. (a) How much work was done on the box? (b) What is the change of potential energy as a result of this move?

10. (a) How much work is done in raising a 50.0 kg crate a distance of 1. 5 m above a storeroom floor? (b) What is the change of potential energy as a result of this move? (c) How much kinetic energy will the crate have as it falls and hits the floor?

11. The driver of an 800.0 kg car decides to double the speed from 20.0 m/s to 40.0 m/s. What effect would this have on the amount of work required to stop the car, that is, on the kinetic energy of the car?

12. A 70.0 kg student runs up the stairs of a football stadium to a height of 10.0 m above the ground in 10.0 s. (a) What is the power of the student in kilowatts? (b) in horsepower?

Sparks fly from a plate of steel as it is cut by an infrared laser. Today, lasers are commonly used to cut as well as weld metals, so the cutting and welding is done by light, not by a flame or electric current.

Heat and Temperature

Chemistry Connections

▶ Chemical reactions involve changes in the internal energy of molecules (Ch. 9).

Astronomy Connections

▶ Stars form from compressional heating (Ch. 12).

CORE CONCEPT

A relationship exists between heat, temperature, and the motion and position of molecules.

All matter is made of molecules that move and interact (p. 77).

Temperature is a measure of the average kinetic energy of molecules (p. 78).

Heat is a measure of internal energy that has been transferred or absorbed (p. 82).

The laws of thermodynamics describe a relationship between changes of internal energy, work, and heat (p. 94).

Earth Science Connections

▶ Earth's interior is hot, making possible the movement of plates on the surface (Ch. 15–16).

▶ Movements of air, air masses, and ocean currents are driven in cycles involving temperature differences (Ch. 17–18).

Life Science Connections

▶ The process of aerobic respiration results in a release of heat (Ch. 20).

OVERVIEW

Heat has been closely associated with the comfort and support of people throughout history. You can imagine the appreciation when your earliest ancestors first discovered fire and learned to keep themselves warm and cook their food. You can also imagine the wonder and excitement about 3000 B.C., when people put certain earthlike substances on the hot, glowing coals of a fire and later found metallic copper, lead, or iron. The use of these metals for simple tools followed soon afterwards. Today, metals are used to produce complicated engines that use heat for transportation and do the work of moving soil and rock, construction, and agriculture. Devices made of heat-extracted metals are also used to control the temperature of structures, heating or cooling the air as necessary. Thus, the production and control of heat gradually built the basis of civilization today (figure 4.1).

The sources of heat are the energy forms that you learned about in chapter 3. The fossil fuels are *chemical* sources of heat. Heat is released when oxygen is combined with these fuels. Heat also results when *mechanical* energy does work against friction, such as in the brakes of a car coming to a stop. Heat also appears when *radiant* energy is absorbed. This is apparent when solar energy heats water in a solar collector or when sunlight melts snow. The transformation of *electrical* energy to heat is apparent in toasters, heaters, and ranges. *Nuclear* energy provides the heat to make steam in a nuclear power plant. Thus, all the energy forms can be converted to heat.

The relationship between energy forms and heat appears to give an order to nature, revealing patterns that you will want to understand. All that you need is some kind of explanation for the relationships—a model or theory that helps make sense of it all. This chapter is concerned with heat and temperature and their relationship to energy. It begins with a simple theory about the structure of matter and then uses the theory to explain the concepts of heat, energy, and temperature changes.

THE KINETIC MOLECULAR THEORY

The idea that substances are composed of very small particles can be traced back to certain early Greek philosophers. The earliest record of this idea was written by Democritus during the fifth century B.C. He wrote that matter was empty space filled with tremendous numbers of tiny, indivisible particles called *atoms*. This idea, however, was not acceptable to most of the ancient Greeks, because matter seemed continuous, and empty space was simply not believable. The idea of atoms was rejected by Aristotle as he formalized his belief in continuous matter composed of the earth, air, fire, and water elements.

Aristotle's belief about matter, like his beliefs about motion, predominated through the 1600s. Some people, such as Galileo and Newton, believed the ideas about matter being composed of tiny particles, or atoms, since this theory seemed to explain the behavior of matter. Widespread acceptance of the particle model did not occur, however, until strong evidence was developed through chemistry in the late 1700s and early 1800s. The experiments finally led to a collection of assumptions about the small particles of matter and the space around them. Collectively, the assumptions could be called the **kinetic molecular theory.** The following is a general description of some of these assumptions.

FIGURE 4.1 Heat and modern technology are inseparable. These glowing steel slabs, at over 1,100°C (about 2,000°F), are cut by an automatic flame torch. The slab caster converts 300 tons of molten steel into slabs in about 45 minutes. The slabs are converted to sheet steel for use in the automotive, appliance, and building industries.

Molecules

The basic assumption of the kinetic molecular theory is that all matter is made up of tiny, basic units of structure called *atoms*. Atoms are neither divided, created, nor destroyed during any type of chemical or physical change. There are similar groups of atoms that make up the pure substances known as chemical *elements*. Each element has its own kind of atom, which is different from the atoms of other elements. For example, hydrogen, oxygen, carbon, iron, and gold are chemical elements, and each has its own kind of atom.

In addition to the chemical elements, there are pure substances called *compounds* that have more complex units of structure. Pure substances, such as water, sugar, and alcohol, are composed of atoms of two or more elements that join together in definite proportions. Water, for example, has structural units that are made up of two atoms of hydrogen tightly bound to one atom of oxygen (H_2O). These units are not easily broken apart and stay together as small physical particles of which water is composed. Each is the smallest particle of water that can exist, a molecule of water. A *molecule* is generally defined as a tightly bound group of atoms in which the atoms maintain their identity. How atoms become bound together to form molecules is discussed in chapter 9.

Some elements exist as gases at ordinary temperatures, and all elements are gases at sufficiently high temperatures. At ordinary temperatures, the atoms of oxygen, nitrogen, and other gases are paired in groups of two to form *diatomic molecules*. Other gases, such as helium, exist as single, unpaired atoms at ordinary temperatures. At sufficiently high temperatures, iron, gold, and other metals vaporize to form

gaseous, single, unpaired atoms. In the kinetic molecular theory, the term *molecule* has the additional meaning of the smallest, ultimate particle of matter that can exist. Thus, the ultimate particle of a gas, whether it is made up of two or more atoms bound together or of a single atom, is conceived of as a molecule. A single atom of helium, for example, is known as a *monatomic molecule*. For now, a **molecule** is defined as the smallest particle of a compound, or a gaseous element, that can exist and still retain the characteristic properties of that substance.

Molecules Interact

Some molecules of solids and liquids interact, strongly attracting and clinging to each other. When this attractive force is between the same kind of molecules, it is called *cohesion*. It is a stronger cohesion that makes solids and liquids different from gases, and without cohesion, all matter would be in the form of gases. Sometimes one kind of molecule attracts and clings to a different kind of molecule. The attractive force between unlike molecules is called *adhesion*. Water wets your skin because the adhesion of water molecules and skin is stronger than the cohesion of water molecules. Some substances, such as glue, have a strong force of adhesion when they harden from a liquid state, and they are called adhesives.

Phases of Matter

Three phases of matter are common on Earth under conditions of ordinary temperature and pressure. These phases—or forms of existence—are solid, liquid, and gas. Each of these has a different molecular arrangement (figure 4.2). The different characteristics of each phase can be attributed to the molecular arrangements and the strength of attraction between the molecules (table 4.1).

Solids have definite shapes and volumes because they have molecules that are fixed distances apart and bound by relatively strong cohesive forces. Each molecule is a nearly fixed distance from the next, but it does vibrate and move around an equilibrium position. The masses of these molecules and the spacing between them determine the density of the solid. The hardness of a solid is the resistance of a solid to forces that tend to push its molecules further apart.

Liquids have molecules that are not confined to an equilibrium position as in a solid. The molecules of a liquid are close together and bound by cohesive forces that are not as strong as in a solid. This permits the molecules to move from place to place within the liquid. The molecular forces are strong enough to give the liquid a definite volume but not strong enough to give it a definite shape. Thus, a pint of milk is always a pint of milk (unless it is under tremendous pressure) and takes the shape of the container holding it. Because the forces between the molecules of a liquid are weaker than the forces between the molecules of a solid, a liquid cannot support the stress of a rock placed on it as a solid does. The liquid molecules *flow,*

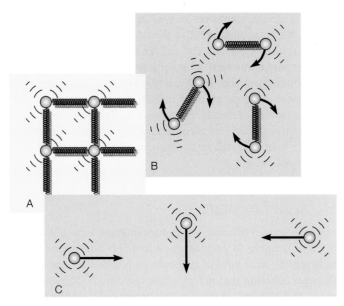

FIGURE 4.2 (*A*) In a solid, molecules vibrate around a fixed equilibrium position and are held in place by strong molecular forces. (*B*) In a liquid, molecules can rotate and roll over each other because the molecular forces are not as strong. (*C*) In a gas, molecules move rapidly in random, free paths.

TABLE 4.1 The Shape and Volume Characteristics of Solids, Liquids, and Gases Are Reflections of Their Molecular Arrangements*

	Solids	Liquids	Gases
Shape	Fixed	Variable	Variable
Volume	Fixed	Fixed	Variable

*These characteristics are what would be expected under ordinary temperature and pressure conditions on the surface of Earth.

rolling over each other as the rock pushes its way between the molecules. Yet, the molecular forces are strong enough to hold the liquid together, so it keeps the same volume.

Gases are composed of molecules with weak cohesive forces acting between them. The gas molecules are relatively far apart and move freely in a constant, random motion that is changed often by collisions with other molecules. Gases therefore have neither fixed shapes nor fixed volumes.

Gases that are made up of positive ions and negative electrons are called *plasmas*. Plasmas have the same properties as gases but also conduct electricity and interact strongly with magnetic fields. Plasmas are found in fluorescent and neon lights on Earth, and in the Sun and other stars. Nuclear fusion occurs in plasmas of stars (see chapter 11), producing starlight as well as sunlight. Plasma physics is studied by scientists in their attempt to produce controlled nuclear fusion.

There are other distinctions between the phases of matter. The term *vapor* is sometimes used to describe a gas that is usually in the liquid phase. Water vapor, for example, is the gaseous form of liquid water. Liquids and gases are collectively

called *fluids* because of their ability to flow, a property that is lacking in most solids.

Molecules Move

Suppose you are in an evenly heated room with no air currents. If you open a bottle of ammonia, the odor of ammonia is soon noticeable everywhere in the room. According to the kinetic molecular theory, molecules of ammonia leave the bottle and bounce around among the other molecules making up the air until they are everywhere in the room, slowly becoming more evenly distributed. The ammonia molecules *diffuse,* or spread, throughout the room. The ammonia odor diffuses throughout the room faster if the air temperature is higher and slower if the air temperature is lower. This would imply a relationship between the temperature and the speed at which molecules move about.

The relationship between the temperature of a gas and the motion of molecules was formulated in 1857 by Rudolf Clausius. He showed that the temperature of a gas is proportional to the average kinetic energy of the gas molecules. This means that ammonia molecules have a greater average velocity at a higher temperature and a slower average velocity at a lower temperature. This explains why gases diffuse at a greater rate at higher temperatures. Recall, however, that kinetic energy involves the mass of the molecules as well as their velocity ($KE = 1/2\, mv^2$). It is the *average kinetic energy* that is proportional to the temperature, which involves the molecular mass as well as the molecular velocity. Whether the kinetic energy is jiggling, vibrating, rotating, or moving from place to place, the **temperature** of a substance is a *measure of the average kinetic energy of the molecules making up the substance* (figure 4.3).

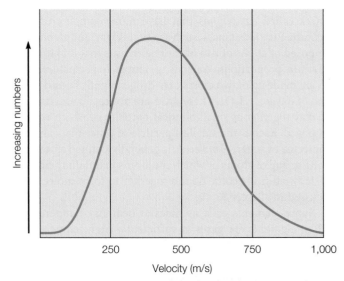

FIGURE 4.3 The number of oxygen molecules with certain velocities that you might find in a sample of air at room temperature. Notice that a few are barely moving and some have velocities over 1,000 m/s at a given time, but the *average* velocity is somewhere around 500 m/s.

TEMPERATURE

If you ask people about the temperature, they usually respond with a referent ("hotter than the summer of '89") or a number ("68°F or 20°C"). Your response, or feeling, about the referent or number depends on a number of factors, including a *relative* comparison. A temperature of 20°C (68°F), for example, might seem cold during the month of July but warm during the month of January. The 20°C temperature is compared to what is expected at the time, even though 20°C is 20°C, no matter what month it is.

When people ask about the temperature, they are really asking *how hot or how cold something is.* Without a thermometer, however, most people can do no better than *hot* or *cold,* or perhaps *warm* or *cool,* in describing a relative temperature. Even then, there are other factors that confuse people about temperature. Your body judges temperature on the basis of the net *direction* of energy flow. You sense situations in which heat is flowing into your body as *warm* and situations in which heat is flowing from your body as *cool.* Perhaps you have experienced having your hands in snow for some time, then washing your hands in cold water. The cold water feels warm. Your hands are colder than the water, energy flows into your hands, and they communicate "warm."

Thermometers

The human body is a poor sensor of temperature, so a device called a *thermometer* is used to measure the hotness or coldness of something. Most thermometers are based on the relationship between some property of matter and changes in temperature. Almost all materials expand with increasing temperatures. A strip of metal is slightly longer when hotter and slightly shorter when cooler, but the change of length is too small to be useful in a thermometer. A more useful, larger change is obtained when two metals that have different expansion rates are bonded together in a strip. The bimetallic (*bi* = two; *metallic* = metal) strip will bend toward the metal with less expansion when the strip is heated (figure 4.4). Such a bimetallic strip is formed into a coil and used in thermostats and dial thermometers (figure 4.5).

The common glass thermometer is a glass tube with a bulb containing a liquid, usually mercury or colored alcohol, that expands up the tube with increases in temperature and contracts back toward the bulb with decreases in temperature. The height of this liquid column is used with a referent scale

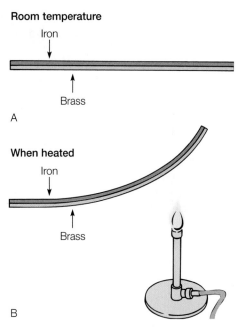

FIGURE 4.4 (*A*) A bimetallic strip is two different metals, such as iron and brass, bonded together as a single unit, shown here at room temperature. (*B*) Since one metal expands more than the other, the strip will bend when it is heated. In this example, the brass expands more than the iron, so the bimetallic strip bends away from the brass.

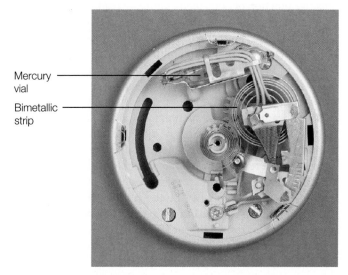

FIGURE 4.5 This thermostat has a coiled bimetallic strip that expands and contracts with changes in the room temperature. The attached vial of mercury is tilted one way or the other, and the mercury completes or breaks an electric circuit that turns the heating or cooling system on or off.

to measure temperature. Some thermometers, such as a fever thermometer, have a small constriction in the tube so the liquid cannot normally return to the bulb. Thus, the thermometer shows the highest reading, even if the temperature it measures has fluctuated up and down during the reading. The liquid must be forced back into the bulb by a small swinging motion, bulb-end down, then sharply stopping the swing with a snap of the wrist. The inertia of the mercury in the bore forces it past the constriction and into the bulb. The fever thermometer is then ready to use again.

Today, scientists have developed a different type of thermometer and a way around the problems of using a glass-mercury fever thermometer. This new approach measures the internal core temperature by quickly reading infrared radiation from the eardrum. All bodies with a temperature above absolute zero emit radiation, including your body. The intensity of the radiation is a sensitive function of body temperature, so reading the radiation emitted will tell you about the temperature of that body.

The human eardrum is close to the hypothalamus, the body's thermostat, so a temperature reading taken here must be close to the temperature of the internal core. You cannot use a mercury thermometer in the ear because of the very real danger of puncturing the eardrum, along with obtaining doubtful readings from a mercury bulb. You can use a pyroelectric material to measure the infrared radiation coming from the entrance to the ear canal, however, to quickly obtain a temperature reading. A pyroelectric material is a polarized crystal that generates an electric charge in proportion to a temperature change. The infrared fever thermometer has a short barrel that is inserted in the ear canal opening. A button opens a shutter inside the battery-powered device, admitting infrared radiation for about 300 milliseconds. Infrared radiation from the ear canal increases the temperature of a thin pyroelectric crystal, which develops an electric charge. A current spike from the pyroelectric sensor now moves through some filters and converters and into a microprocessor chip. The chip is programmed with the relationship between the body temperature and the infrared radiation emitted. Using this information, it calculates the temperature by measuring the current spike produced by the infrared radiation falling on the pyroelectric crystal. The microprocessor now sends the temperature reading to an LCD display on the outside of the device, where it can be read almost immediately.

Thermometer Scales

Several referent scales are used to define numerical values for measuring temperatures (figure 4.6). The **Fahrenheit scale** was developed by the German physicist Gabriel D. Fahrenheit

in about 1715. Fahrenheit invented a mercury-in-glass thermometer with a scale based on two arbitrarily chosen reference points. The original Fahrenheit scale was based on the temperature of an ice and salt mixture for the lower reference point (0°) and the temperature of the human body as the upper reference point (about 100°). Thus, the original Fahrenheit scale was a centigrade scale with 100 divisions between the high and the low reference points. The distance between the two reference points was then divided into equal intervals called *degrees*. There were problems with identifying a "normal" human body temperature as a reference point, since body temperature naturally changes during a given day and from day to day. However, some people "normally" have a higher body temperature than others. Some may have a normal body temperature of 99.1°F, while others have a temperature of 97°F. The average for a large population is 98.6°F. The only consistent thing about the human body temperature is constant change. The standards for the Fahrenheit scale were eventually changed to something more consistent, the freezing point and the boiling point of water at normal atmospheric pressure. The original scale was retained with the new reference points, however, so the "odd" numbers of 32°F (freezing point of water) and 212°F (boiling point of water under normal pressure) came to be the reference points. There are 180 equal intervals, or degrees, between the freezing and boiling points on the Fahrenheit scale.

The **Celsius scale** was invented by Anders C. Celsius, a Swedish astronomer, in about 1735. The Celsius scale uses the freezing point and the boiling point of water at normal atmospheric pressure, but it has different arbitrarily assigned values. The Celsius scale identifies the freezing point of water as 0°C and the boiling point as 100°C. There are 100 equal intervals, or degrees, between these two reference points, so the Celsius scale is sometimes called the *centigrade* scale.

There is nothing special about either the Celsius scale or the Fahrenheit scale. Both have arbitrarily assigned numbers, and one is no more accurate than the other. The Celsius scale is more convenient because it is a decimal scale and because it has a direct relationship with a third scale to be described shortly,

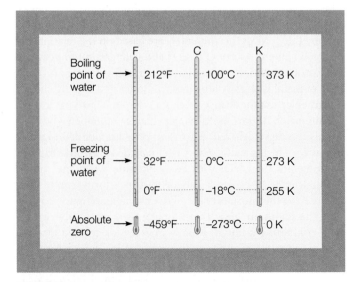

FIGURE 4.6 The Fahrenheit, Celsius, and Kelvin temperature scales.

Goose Bumps and Shivering

For an average age and minimal level of activity, many people feel comfortable when the environmental temperature is about 25°C (77°F). Comfort at this temperature probably comes from the fact that the body does not have to make an effort to conserve or get rid of heat.

Changes in the body that conserve heat occur when the temperature of the air and clothing directly next to a person becomes less than 20°C, or if the body senses rapid heat loss. First, blood vessels in the skin are constricted. This slows the flow of blood near the surface, which reduces heat loss by conduction. Constriction of skin blood vessels reduces body heat loss but may also cause the skin and limbs to become significantly cooler than the body core temperature (producing cold feet, for example).

Sudden heat loss, or a chill, often initiates another heat-saving action by the body. Skin hair is pulled upright, erected to slow heat loss to cold air moving across the skin. Contraction of a tiny muscle attached to the base of the hair shaft makes a tiny knot, or bump, on the skin. These are sometimes called "goose bumps" or "chill bumps." Although "goose bumps" do not significantly increase insulation in humans, the equivalent response in birds and many mammals elevates hairs or feathers and greatly enhances insulation.

Further cooling after the blood vessels in the skin have been constricted results in the body taking yet another action. The body now begins to produce *more* heat, making up for heat loss through involuntary muscle contractions called "shivering." The greater the need for more body heat, the greater the activity of shivering.

If the environmental temperatures rise above about 25°C (77°F), the body triggers responses that cause it to *lose* heat. One response is to make blood vessels in the skin larger, which increases blood flow in the skin. This brings more heat from the core to be conducted through the skin, then radiated away. It also causes some people to have a red blush from the increased blood flow in the skin. This action increases conduction through the skin, but radiation alone provides insufficient cooling at environmental temperatures above about 29°C (84°F). At about this temperature, sweating begins and perspiration pours onto the skin to provide cooling through evaporation. The warmer the environmental temperature, the greater the rate of sweating and cooling through evaporation.

The actual responses to a cool, cold, warm, or hot environment will be influenced by a person's level of activity, age, and gender, and environmental factors such as the relative humidity, air movement, and combinations of these factors. Temperature is the single most important comfort factor. However, when the temperature is high enough to require perspiration for cooling, humidity also becomes an important factor in human comfort.

---

the Kelvin scale. Both scales have arbitrarily assigned reference points and an arbitrary number line that indicates *relative* temperature changes. Zero is simply one of the points on each number line and does *not* mean that there is no temperature. Likewise, since the numbers are relative measures of temperature change, 2° is not twice as hot as a temperature of 1° and 10° is not twice as hot as a temperature of 5°. The numbers simply mean some measure of temperature *relative to* the freezing and boiling points of water under normal conditions.

You can convert from one temperature to the other by considering two differences in the scales: (1) the difference in the degree size between the freezing and boiling points on the two scales, and (2) the difference in the values of the lower reference points.

The Fahrenheit scale has 180° between the boiling and freezing points (212°F − 32°F) and the Celsius scale has 100° between the same two points. Therefore, each Celsius degree is 180/100 or 9/5 as large as a Fahrenheit degree. Each Fahrenheit degree is 100/180 or 5/9 of a Celsius degree. In addition, considering the difference in the values of the lower reference points (0°C and 32°F) gives the equations for temperature conversion.

$$T_F = \frac{9}{5}T_C + 32°$$

equation 4.1

$$T_C = \frac{5}{9}(T_F - 32°)$$

equation 4.2

EXAMPLE 4.1 *(Optional)*

The average human body temperature is 98.6°F. What is the equivalent temperature on the Celsius scale?

SOLUTION

$$
\begin{aligned}
T_C &= \frac{5}{9}(T_F - 32°) \\
&= \frac{5}{9}(98.6° - 32°) \\
&= \frac{5}{9}(66.6°) \\
&= \frac{333°}{9} \\
&= \boxed{37°C}
\end{aligned}
$$

EXAMPLE 4.2 *(Optional)*

A temperature display outside of a bank indicates 20°C (room temperature). What is the equivalent temperature on the Fahrenheit scale? (Answer: 68°F)

There is a temperature scale that does not have arbitrarily assigned reference points and zero *does* mean nothing. This is not a relative scale but an absolute temperature scale called the **absolute scale,** or **Kelvin scale.** The zero point on the absolute scale is thought to be the lowest limit of temperature.

Absolute zero is the *lowest temperature possible,* occurring when all random motion of molecules has ceased. Absolute zero is written as 0 K. A degree symbol is not used, and the K stands for the SI standard scale unit kelvin. The absolute scale uses the same degree size as the Celsius scale and $-273°C = 0$ K. Note in figure 4.6 that 273 K is the freezing point of water and 373 K is the boiling point. You could think of the absolute scale as a Celsius scale with the zero point shifted by 273°. Thus, the relationship between the absolute and Celsius scales is

$$T_K = T_C + 273 \qquad \textbf{equation 4.3}$$

A temperature of absolute zero has never been reached, but scientists have cooled a sample of sodium to 700 nanokelvins, or 700 billionths of a kelvin above absolute zero.

EXAMPLE 4.3 *(Optional)*

A science article refers to a temperature of 300.0 K. (a) What is the equivalent Celsius temperature? (b) What is the equivalent Fahrenheit temperature?

SOLUTION

(a) The relationship between the absolute scale and Celsius scale is found in equation 4.3, $T_K = T_C + 273$. Solving this equation for Celsius yields $T_C = T_K - 273$.

$$\begin{aligned}
T_C &= T_K - 273 \\
&= 300.0 - 273 \\
&= \boxed{27°C}
\end{aligned}$$

(b)
$$\begin{aligned}
T_F &= \frac{9}{5}T_C + 32° \\
&= \frac{9}{5}27° + 32° \\
&= \frac{243°}{5} + 32° \\
&= 48.6° + 32° \\
&= \boxed{81°F}
\end{aligned}$$

HEAT

Suppose you have a bowl of hot soup or a cup of hot coffee that is too hot. What can you do to cool it? You can blow across the surface, which speeds evaporation and therefore results in cooling, but this is a slow process. If you were in a hurry, you would probably add something cooler, such as ice. Adding a cooler substance will cool the hot liquid.

You know what happens when you mix fluids or objects with a higher temperature with fluids or objects with a lower temperature. The warmer temperature object becomes cooler and the cooler temperature object becomes warmer. Eventually, both will have a temperature somewhere between the warmer and the cooler. This might suggest that something is moving between the warmer and cooler objects, changing the temperature. What is doing the moving?

The term **heat** is used to describe the "something" that moves between objects when two objects of different temperature are brought together. As you will learn in the next section, heat flow represents a form of energy transfer that takes place between objects. For now, we will continue to think of heat as "something"—an energy transfer—that moves between objects of different temperatures, such as your bowl of hot soup and a cold ice cube.

The relationship that exists between energy and temperature will help explain the concept of heat, so we will consider it first. If you rub your hands together a few times, they will feel a little warmer. If you rub them together vigorously for a while, they will feel a lot warmer, maybe hot. A temperature increase takes place anytime mechanical energy causes one surface to rub against another (figure 4.7). The two surfaces could be solids, such as the two blocks, but they can also be the surface of a solid and a fluid, such as air. A solid object moving through the air encounters compression, which results in a higher temperature of the surface. A high-velocity meteor enters Earth's atmosphere and is heated so much from the compression that it begins to glow, resulting in the fireball and smoke trail of a "falling star."

To distinguish between the energy of the object and the energy of its molecules, we use the terms *external* and *internal* energy. **External energy** is the total potential and kinetic energy of an everyday-sized object. All the kinetic and potential energy considerations discussed in previous chapters were about the external energy of an object.

Internal energy is the total kinetic and potential energy of the *molecules* of an object. The kinetic energy of a molecule can be much more complicated than straight-line velocity might suggest, however, as a molecule can have many different types of

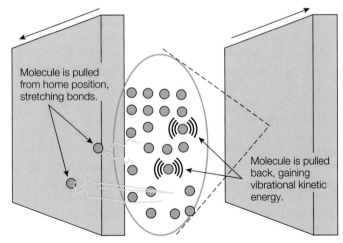

FIGURE 4.7 Here is how friction results in increased temperatures: Molecules on one moving surface will catch on another surface, stretching the molecular forces that are holding it. The molecules are pulled back to their home position with a snap, resulting in a gain of vibrational kinetic energy.

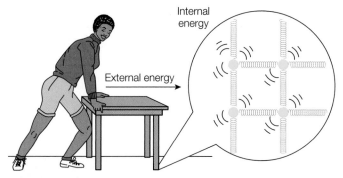

FIGURE 4.8 *External energy* is the kinetic and potential energy that you can see. *Internal energy* is the total kinetic and potential energy of molecules. When you push a table across the floor, you do work against friction. Some of the external mechanical energy goes into internal kinetic and potential energy, and the bottom surface of the legs becomes warmer.

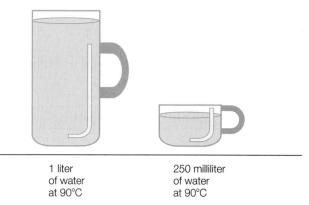

1 liter
of water
at 90°C

250 milliliter
of water
at 90°C

FIGURE 4.9 Heat and temperature are different concepts, as shown by a liter of water (1,000 mL) and a 250 mL cup of water, both at the same temperature. You know the liter of water contains more internal energy since it will require more ice cubes to cool it to, say, 25°C than will be required for the cup of water. In fact, you will have to remove 48,750 *additional* calories to cool the liter of water.

motion at the same time (pulsing, twisting, turning, etc.). Overall, internal energy is characterized by properties such as temperature, density, heat, volume, pressure of a gas, and so forth.

When you push a table across the floor, the observable *external* kinetic energy of the table is transferred to the *internal* kinetic energy of the molecules between the table legs and the floor, resulting in a temperature increase (figure 4.8). The relationship between external and internal kinetic energy explains why the heating is proportional to the amount of mechanical energy used.

Heat as Energy Transfer

Temperature is a measure of the degree of hotness or coldness of a body, a measure that is based on the average molecular kinetic energy. Heat, on the other hand, is based on the *total internal energy* of the molecules of a body. You can see one difference in heat and temperature by considering a cup of water and a pitcher of water. If both the small and the large amount of water have the same temperature, both must have the same average molecular kinetic energy. Now, suppose you wish to cool both by, say, 20°. The pitcher of water would take much longer to cool, so it must be that the large amount of water has more internal energy (figure 4.9). Heat is a measure based on the *total* internal energy of the molecules of a body, and there is more total energy in a pitcher of water than in a cup of water at the same temperature.

How can we measure heat? Since it is difficult to see molecules, internal energy is difficult to measure directly. Thus, heat is nearly always measured during the process of a body gaining or losing energy. This measurement procedure will also give us a working definition of heat:

Heat is a measure of the internal energy that has been absorbed or transferred from one body to another.

The *process* of increasing the internal energy is called "heating" and the *process* of decreasing internal energy is called "cooling." The word *process* is italicized to emphasize that heat is energy in transit, not a material thing you can add or take away. Heat

is understood to be a measure of internal energy that can be measured as energy flows in or out of an object.

There are two general ways that heating can occur. These are (1) from a temperature difference, with energy moving from the region of higher temperature, and (2) from an object gaining energy by way of an energy-form conversion.

When a *temperature difference* occurs, energy is transferred from a region of higher temperature to a region of lower temperature. Energy flows from a hot range element, for example, to a pot of cold water on a range. It is a natural process for energy to flow from a region of a higher temperature to a region of a lower temperature just as it is natural for a ball to roll downhill. The temperature of an object and the temperature of the surroundings determine if heat will be transferred to or from an object. The terms *heating* and *cooling* describe the direction of energy flow, naturally moving from a region of higher energy to one of lower energy.

Measures of Heat

Since heating is a method of energy transfer, a quantity of heat can be measured just like any quantity of energy. The metric unit for measuring work, energy, or heat is the *joule*. However, the separate historical development of the concepts of heat and the concepts of motion resulted in separate units, some based on temperature differences.

The metric unit of heat is called the **calorie** (cal). A calorie is defined as the *amount of energy (or heat) needed to increase the temperature of 1 gram of water 1 degree Celsius*. A more precise definition specifies the degree interval from 14.5°C to 15.5°C because the energy required varies slightly at different temperatures. This precise definition is not needed for a general discussion. A **kilocalorie** (kcal) is the *amount of energy (or heat) needed to increase the temperature of 1 kilogram of water 1 degree Celsius*. The measure of the energy released by the oxidation of food is the kilocalorie, but it is called the Calorie (with a capital C) by nutritionists (figure 4.10). Confusion can be avoided by

Connections . . .

Energy of Some Foods

Note: Typical basal metabolic rate equals 1,200–2,200 Calories per day.

Food	Nutritional Calories	Kilojoules
Candy bar, various	120 to 230	502 to 963
Beer, regular	150	628
Beer, lite	100	419
Carbonated drink	150	628
Arby's roast beef	380	1,590
Whopper, w/cheese	705	2,951
DQ large cone	340	1,423
Domino's deluxe pizza	225 per slice (16")	942
Hardee's big deluxe	500	2,093
Jumbo Jack w/cheese	630	2,638
KFC breast	300	1,256
Big Mac	560	2,344
Pizza Hut supreme	140 per slice (10")	586
Taco Bell taco	160	670
Wendy's big classic	570	2,386

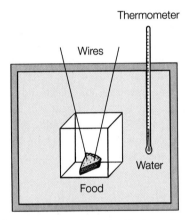

FIGURE 4.10 The Calorie value of food is determined by measuring the heat released from burning the food. If there is 10.0 kg of water and the temperature increases from 10°C to 20°C, the food contains 100 Calories (100,000 calories). The food illustrated here would release much more energy than this.

making sure that the scientific calorie is never capitalized (cal) and the dieter's Calorie is always capitalized. The best solution would be to call the Calorie what it is, a kilocalorie (kcal).

The English system's measure of heating is called the **British thermal unit** (Btu). A Btu is *the amount of energy (or heat) needed to increase the temperature of 1 pound of water 1 degree Fahrenheit.* The Btu is commonly used to measure the heating or cooling rates of furnaces, air conditioners, water heaters, and so forth. The rate is usually expressed or understood to be in Btu per hour. A much larger unit is sometimes mentioned in news reports and articles about the national energy consumption. This unit is the *quad*, which is 1 quadrillion Btu (a million billion or 10^{15} Btu).

In metric units, the mechanical equivalent of heat is

$$4.184 \text{ J} = 1 \text{ cal}$$

or

$$4,184 \text{ J} = 1 \text{ kcal}$$

The establishment of this precise proportionality means that, fundamentally, mechanical energy and heat are different forms of the same thing.

EXAMPLE 4.4 *(Optional)*

A 1,000.0 kg car is moving at 90.0 km/h (25.0 m/s). How many kilocalories are generated when the car brakes to a stop?

SOLUTION

The kinetic energy of the car is

$$
\begin{aligned}
KE &= \frac{1}{2}mv^2 \\
&= \frac{1}{2}(1,000.0 \text{ kg})\left(25.0\frac{\text{m}}{\text{s}}\right)^2 \\
&= \frac{1}{2}(1,000.0 \text{ kg})\left(625\frac{\text{m}^2}{\text{s}^2}\right) \\
&= (500.0)(625)\frac{\text{kg}\cdot\text{m}^2}{\text{s}^2} \\
&= 312,500 \text{ J}
\end{aligned}
$$

You can convert this to kcal by using the relationship between mechanical energy and heat:

$$
(312,500 \text{ J})\frac{1 \text{ kcal}}{4,184 \text{ J}}
$$

$$
\frac{312,500}{4,184} \frac{\cancel{\text{J}}\cdot\text{kcal}}{\cancel{\text{J}}}
$$

$$
\boxed{74.7 \text{ kcal}}
$$

(Note: The temperature increase from this amount of heating could be calculated from equation 4.4.)

Specific Heat

You can observe a relationship between heat and different substances by doing an experiment in "kitchen physics." Imagine that you have a large pot of liquid to boil in preparing a meal. Three variables influence how much heat you need:

1. The initial temperature of the liquid;
2. How much liquid is in the pot; and,
3. The nature of the liquid (water or soup, for example).

What this means specifically, is

1. **Temperature change.** The amount of heat needed is proportional to the temperature change. It takes more heat to raise the temperature of cool water, so this relationship could be written as $Q \propto \Delta T$.
2. **Mass.** The amount of heat needed is also proportional to the amount of the substance being heated. A larger mass requires more heat to go through the same temperature change than a smaller mass. In symbols, $Q \propto m$.
3. **Substance.** Different materials require different amounts of heat to go through the same temperature range when their masses are equal (figure 4.11). This property is called the **specific heat** of a material, which is defined as the amount of heat needed to increase the temperature of 1 gram of a substance 1 degree Celsius.

Considering all the variables involved in our kitchen physics cooking experience, we find the heat (Q) need is described by the relationship of

$$Q = mc\Delta T \qquad \textbf{equation 4.4}$$

where c is the symbol for specific heat.

Specific heat is responsible for the larger variation of temperature observed over land than near a large body of water. Table 4.2 gives the specific heat of soil as 0.200 cal/gC° and the specific heat of water as 1.00 cal/gC°. Since specific

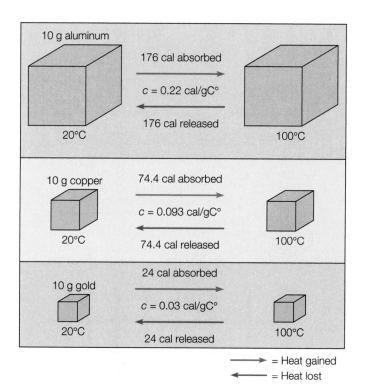

FIGURE 4.11 Of these three metals, aluminum needs the most heat per gram per degree when warmed and releases the most heat when cooled. Why are the cubes different sizes?

TABLE 4.2   The Specific Heat of Selected Substances

Substance	Specific Heat (cal/gC° or kcal/kgC°)
Air	0.17
Aluminum	0.22
Concrete	0.16
Copper	0.093
Glass (average)	0.160
Gold	0.03
Ice	0.500
Iron	0.11
Lead	0.0305
Mercury	0.033
Seawater	0.93
Silver	0.056
Soil (average)	0.200
Steam	0.480
Water	1.00

Note: To convert to specific heat in J/kgC°, multiply each value by 4,184. Also note that 1 cal/gC° = 1 kcal/kgC°.

CONCEPTS APPLIED

More Kitchen Physics

Consider the following information as it relates to the metals of cooking pots and pans:

1. It is easier to change the temperature of metals with low specific heats.
2. It is harder to change the temperature of metals with high specific heats.

Look at the list of metals and specific heats listed in table 4.2 and answer the following questions:

1. Considering specific heat alone, which metal could be used for making practical pots and pans that are the most energy efficient to use?
2. Again considering specific heat alone, would certain combinations of metals provide any advantages for rapid temperature changes?

heat is defined as the amount of heat needed to increase the temperature of 1 gram of a substance 1 degree, this means 1 gram of water exposed to 1 calorie of sunlight will warm 1°C. One gram of soil exposed to 1 calorie of sunlight, on the other hand, will be warmed to 5°C since it only takes 0.2 calories to warm the soil 1°C. Thus, the temperature is more even near large bodies of water since it is harder to change the temperature of the water.

EXAMPLE 4.5 (Optional)

How much heat must be supplied to a 500.0 g pan to raise its temperature from 20.0°C to 100.0°C if the pan is made of (a) iron and (b) aluminum?

SOLUTION

The relationship between the heat supplied (Q), the mass (m), and the temperature change (ΔT) is found in equation 4.4. The specific heats (c) of iron and aluminum can be found in table 4.2.

(a) Iron:

$$m = 500.0 \text{ g}$$
$$c = 0.11 \text{ cal/gC°}$$
$$T_f = 100.0°C$$
$$T_i = 20.0°C$$
$$Q = ?$$

$$Q = mc\Delta T$$
$$= (500.0 \text{ g})\left(0.11\frac{\text{cal}}{\text{gC°}}\right)(80.0°C)$$
$$= (500.0)(0.11)(80.0) \text{ g} \times \frac{\text{cal}}{\text{g} \cdot \text{C°}} \times °C$$
$$= 4,400 \frac{\text{g} \cdot \text{cal} \cdot °C}{\text{g} \cdot \text{C°}}$$
$$= \boxed{4.4 \text{ kcal}}$$

(b) Aluminum:

$$m = 500.0 \text{ g}$$
$$c = 0.22 \text{ cal/gC°}$$
$$T_f = 100.0°C$$
$$T_i = 20.0°C$$
$$Q = ?$$

$$Q = mc\Delta T$$
$$= (500.0 \text{ g})\left(0.22\frac{\text{cal}}{\text{gC°}}\right)(80.0°C)$$
$$= (500.0)(0.22)(80.0) \text{ g} \times \frac{\text{cal}}{\text{g} \cdot \text{C°}} \times °C$$
$$= 8,800 \frac{\text{g} \cdot \text{cal} \cdot °C}{\text{g} \cdot \text{C°}}$$
$$= \boxed{8.8 \text{ kcal}}$$

It takes twice as much heat energy to warm the aluminum pan through the same temperature range as the iron pan. Thus, with equal rates of energy input, the iron pan will warm twice as fast as the aluminum pan.

EXAMPLE 4.6 (Optional)

What is the specific heat of a 2 kg metal sample if 1.2 kcal are needed to increase the temperature from 20.0°C to 40.0°C? (Answer: 0.03 kcal/kgC°)

Heat Flow

In a previous section, you learned that heating is a transfer of energy that involves (1) a temperature difference or (2) energy-form conversions. Heat transfer that takes place because of a temperature difference takes place in three different ways: by conduction, convection, or radiation.

Conduction

Anytime there is a temperature difference, there is a natural transfer of heat from the region of higher temperature to the

region of lower temperature. In solids, this transfer takes place as heat is *conducted* from a warmer place to a cooler one. Recall that the molecules in a solid vibrate in a fixed equilibrium position and that molecules in a higher temperature region have more kinetic energy, on the average, than those in a lower temperature region. When a solid, such as a metal rod, is held in a flame, the molecules in the warmed end vibrate violently. Through molecular interaction, this increased energy of vibration is passed on to the adjacent, slower-moving molecules, which also begin to vibrate more violently. They, in turn, pass on more vibrational energy to the molecules next to them. The increase in activity thus moves from molecule to molecule, causing the region of increased activity to extend along the rod. This is called **conduction**, the transfer of energy from molecule to molecule (figure 4.12).

Most insulating materials are good insulators because they contain many small air spaces (figure 4.13). The small air spaces are poor conductors because the molecules of air are far apart, compared to a solid, making it more difficult to pass the increased vibrating motion from molecule to molecule. Styrofoam, glass wool, and wool cloth are good insulators because they have many small air spaces, not because of the material they are made of. The best insulator is a vacuum, since there are no molecules to pass on the vibrating motion.

Wooden and metal parts of your desk have the same temperature, but the metal parts will feel cooler if you touch them. Metal is a better conductor of heat than wood and feels

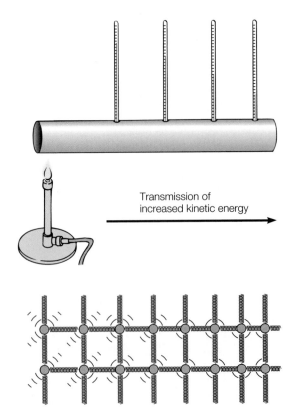

Transmission of increased kinetic energy

FIGURE 4.12   Thermometers placed in holes drilled in a metal rod will show that heat is conducted from a region of higher temperature to a region of lower temperature. The increased molecular activity is passed from molecule to molecule in the process of conduction.

FIGURE 4.13 Fiberglass insulation is rated in terms of R-value, a ratio of the conductivity of the material to its thickness.

cooler because it conducts heat from your finger faster. This is the same reason that a wood or tile floor feels cold to your bare feet. You use an insulating rug to slow the conduction of heat from your feet.

Convection

Convection is the transfer of heat by a large-scale displacement of groups of molecules with relatively higher kinetic energy. In conduction, increased kinetic energy is passed from molecule to molecule. In convection, molecules with higher kinetic energy are moved from one place to another place. Conduction happens primarily in solids, but convection happens only in liquids and gases, where fluid motion can carry molecules with higher kinetic energy over a distance. When molecules gain energy, they move more rapidly and push more vigorously against their surroundings. The result is an expansion as the region of heated molecules pushes outward and increases the volume. Since the same amount of matter now occupies a larger volume, the overall density has been decreased.

In fluids, expansion sets the stage for convection. Warm, less-dense fluid is pushed upward by the cooler, more-dense fluid around it. In general, cooler air is more dense; it sinks and flows downhill. Cold air, being more dense, flows out near the bottom of an open refrigerator. You can feel the cold, dense air pouring from the bottom of an open refrigerator to your toes on the floor. On the other hand, you hold your hands *over* a heater because the warm, less-dense air is pushed upward. In a room, warm air is pushed upward from a heater. The warm air spreads outward along the ceiling and is slowly displaced as newly warmed air is pushed upward to the ceiling. As the air cools it sinks over another part of the room, setting up a circulation pattern known as a *convection current* (figure 4.14). Convection currents can also be observed in a large pot of liquid that is heating on a range. You can see the warmer liquid being forced upward over the warmer parts of the range element, then sink

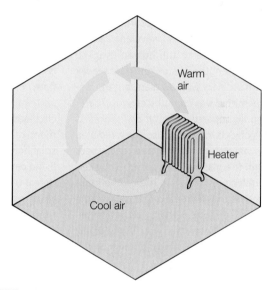

FIGURE 4.14 Convection currents move warm air throughout a room as the air over the heater becomes warmed, expands, and is moved upward by cooler air.

over the cooler parts. Overall, convection currents give the pot of liquid an appearance of turning over as it warms.

Radiation

The third way that heat transfer takes place because of a temperature difference is called **radiation.** Radiation involves the form of energy called *radiant energy,* energy that moves through space. As you will learn in chapter 7, radiant energy includes visible light and many other forms as well. All objects with a temperature above absolute zero give off radiant energy. The absolute temperature of the object determines the rate, intensity, and kinds of radiant energy emitted. You know that visible light is emitted if an object is heated to a certain temperature. A heating element on an electric range, for example, will glow with a reddish-orange light when at the highest setting, but it produces no visible light at lower temperatures, although you feel warmth in your hand when you hold it near the element. Your hand absorbs the nonvisible radiant energy being emitted from the element. The radiant energy does work on the

A Closer Look

Passive Solar Design

Passive solar application is an economically justifiable use of solar energy today. Passive solar design uses a structure's construction to heat a living space with solar energy. Few electric fans, motors, or other energy sources are used. The passive solar design takes advantage of free solar energy; it stores and then distributes this energy through natural conduction, convection, and radiation.

Sunlight that reaches Earth's surface is mostly absorbed. Buildings, the ground, and objects become warmer as the radiant energy is absorbed. Nearly all materials, however, reradiate the absorbed energy at longer wavelengths, wavelengths too long to be visible to the human eye. The short wavelengths of sunlight pass readily through ordinary window glass, but the longer, reemitted wavelengths cannot. Therefore, sunlight passes through a window and warms objects inside a house. The reradiated longer wavelengths cannot pass readily back through the glass but are absorbed by certain molecules in the air. The temperature of the air is thus increased. This is called the "greenhouse effect." Perhaps you have experienced the effect when you left your car windows closed on a sunny, summer day.

In general, a passive solar home makes use of the materials from which it is constructed to capture, store, and distribute solar energy to its occupants. Sunlight enters the house through large windows facing south and warms a thick layer of concrete, brick, or stone. This energy "storage mass" then releases energy during the day and, more important, during the night. This release of energy can be by direct radiation to occupants, by conduction to adjacent air, or by convection of air across the surface of the storage mass. The living space is thus heated without special plumbing or forced air circulation. As you can imagine, the key to a successful passive solar home is to consider every detail of natural energy flow, including the materials of which floors and walls are constructed, convective air circulation patterns, and the size and placement of windows. In addition, a passive solar home requires a different lifestyle and living patterns. Carpets, for example, would defeat the purpose of a storage-mass floor, since they would insulate the storage mass from sunlight. Glass is not a good insulator, so windows must have curtains or movable insulation panels to slow energy loss

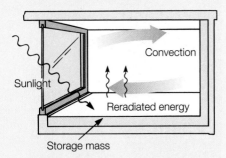

Box Figure 4.1 The direct solar gain design collects and stores solar energy in the living space.

at night. This requires the daily activity of closing curtains or moving insulation panels at night and then opening curtains and moving panels in the morning. Passive solar homes, therefore, require a high level of personal involvement by the occupants.

There are three basic categories of passive solar design: (1) direct solar gain, (2) indirect solar gain, and (3) isolated solar gain.

A *direct solar gain* home is one in which solar energy is collected in the actual living space of the home (box figure 4.1). The advantage of this design is the large, open window space with a calculated overhang

---

molecules of your hand, giving them more kinetic energy. You sense this as an increase in temperature, that is, warmth.

All objects above absolute zero emit radiant energy, but all objects also absorb radiant energy. A hot object, however, emits more radiant energy than a cold object. The hot object will emit more energy than it absorbs from the colder object, and the colder object will absorb more energy from the hot object than it emits. There is, therefore, a net energy transfer that will take place by radiation as long as there is a temperature difference between the two objects.

ENERGY, HEAT, AND MOLECULAR THEORY

The kinetic molecular theory of matter is based on evidence from different fields of science, not just one subject area. Chemists and physicists developed some convincing conclusions about the structure of matter over the past 150 years, using carefully designed experiments and mathematical calculations that explained observable facts about matter. Step by step, the detailed structure of this submicroscopic, invisible world of particles became firmly established. Today, an

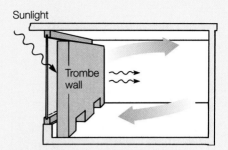

Sunlight

Trombe wall

Box Figure 4.2 The indirect solar gain design uses a Trombe wall to collect, store, and distribute solar energy.

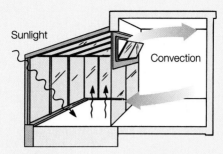

Sunlight

Convection

Box Figure 4.3 The isolated solar gain design uses a separate structure to collect and store solar energy.

that admits maximum solar energy in the winter but prevents solar gain in the summer. The disadvantage is that the occupants are living in the collection and storage components of the design and can place nothing (such as carpets and furniture) that would interfere with warming the storage mass in the floors and walls.

An *indirect solar gain* home uses a massive wall inside a window that serves as a storage mass. Such a wall, called a *Trombe wall,* is shown in box figure 4.2. The Trombe wall collects and stores solar energy, then warms the living space with radiant energy and convection currents. The disadvantage

to the indirect solar gain design is that large windows are blocked by the Trombe wall. The advantage is that the occupants are not in direct contact with the solar collection and storage area, so they can place carpets and furniture as they wish. Controls to prevent energy loss at night are still necessary with this design.

An *isolated solar gain* home uses a structure that is separated from the living space to collect and store solar energy. Examples of an isolated gain design are an attached greenhouse or sun porch (box figure 4.3). Energy flow between the attached structure and the living space can

be by conduction, convection, and radiation, which can be controlled by opening or closing off the attached structure. This design provides the best controls, since it can be completely isolated, opened to the living space as needed, or directly used as living space when the conditions are right. Additional insulation is needed for the glass at night, however, and for sunless winter days.

It has been estimated that building a passive solar home would cost about 10 percent more than building a traditional home of the same size. Considering the possible energy savings, you might believe that most homes would now have a passive solar design. They do not, however, as most new buildings require technology and large amounts of energy to maximize comfort. Yet, it would not require too much effort to consider where to place windows in relation to the directional and seasonal intensity of the Sun and where to plant trees. Perhaps in the future you will have an opportunity to consider using the environment to your benefit through the natural processes of conduction, convection, and radiation.

understanding of this particle structure is basic to physics, chemistry, biology, geology, and practically every other science subject. This understanding has also resulted in present-day technology.

Phase Change

Solids, liquids, and gases are the three common phases of matter, and each phase is characterized by different molecular arrangements. The motion of the molecules in any of the three common phases can be increased through (1) the addition of heat through a temperature difference or (2) the absorption of one of the five forms of energy, which results in heating. In either case, the temperature of the solid, liquid, or gas increases according to the specific heat of the substance, and more heating generally means higher temperatures.

More heating, however, does not always result in increased temperatures. When a solid, liquid, or gas changes from one phase to another, the transition is called a **phase change.** A phase change always absorbs or releases energy, *a quantity of heat that is not associated with a temperature change.* Since the quantity of heat associated with a phase change is not associated

with a temperature change, it is called *latent heat.* Latent heat refers to the "hidden" energy of phase changes, which is energy (heat) that goes into or comes out of *internal potential energy* (figure 4.15).

Three kinds of major phase changes can occur: (1) *solid-liquid,* (2) *liquid-gas,* and (3) *solid-gas.* In each case, the phase change can go in either direction. For example, the solid-liquid phase change occurs when a solid melts to a liquid or when a liquid freezes to a solid. Ice melting to water and water freezing to ice are common examples of this phase change and its two directions. Both occur at a temperature called the *freezing point* or the *melting point,* depending on the direction of the phase change. In either case, however, the freezing and melting points are the same temperature.

The liquid-gas phase change also occurs in two different directions. The temperature at which a liquid boils and changes to a gas (or vapor) is called the *boiling point.* The temperature at which a gas or vapor changes back to a liquid is called the *condensation point.* The boiling and condensation points are the same temperature. There are conditions other than boiling under which liquids may undergo liquid-gas

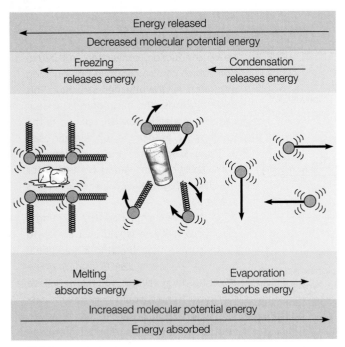

FIGURE 4.15 Each phase change absorbs or releases a quantity of latent heat, which goes into or is released from molecular potential energy.

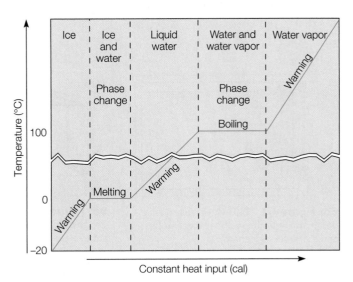

FIGURE 4.16 This graph shows three warming sequences and two phase changes with a constant input of heat. The ice warms to the melting point, then absorbs heat during the phase change as the temperature remains constant. When all the ice has melted, the now-liquid water warms to the boiling point, where the temperature again remains constant as heat is absorbed during this second phase change from liquid to gas. After all the liquid has changed to gas, continued warming increases the temperature of the water vapor.

phase changes, and these conditions are discussed in the next section.

You probably are not as familiar with solid-gas phase changes, but they are common. A phase change that takes a solid directly to a gas or vapor is called *sublimation*. Mothballs and dry ice (solid CO_2) are common examples of materials that undergo sublimation, but frozen water, meaning common ice, also sublimates under certain conditions. Perhaps you have noticed ice cubes in a freezer become smaller with time as a result of sublimation. The frost that forms in a freezer, on the other hand, is an example of a solid-gas phase change that takes place in the other direction. In this case, water vapor forms the frost without going through the liquid state, a solid-gas phase change that takes place in an opposite direction to sublimation.

For a specific example, consider the changes that occur when ice is subjected to a constant source of heat (figure 4.16). Starting at the left side of the graph, you can see that the temperature of the ice increases from the constant input of heat. The ice warms according to $Q = mc\Delta T$, where c is the specific heat of ice. When the temperature reaches the melting point (0°C), it stops increasing as the ice begins to melt. More and more liquid water appears as the ice melts, but the temperature *remains* at 0°C even though heat is still being added at a constant rate. It takes a certain amount of heat to melt all of the ice. Finally, when all the ice is completely melted, the temperature again increases at a constant rate between the melting and boiling points. Then, at constant temperature, the addition of heat produces another phase change, from liquid to gas. The quantity of heat involved in

this phase change is used in doing the work of breaking the molecule-to-molecule bonds in the solid, making a liquid with molecules that are now free to move about and roll over one another. Since the quantity of heat (Q) is absorbed without a temperature change, it is called the **latent heat of fusion** (L_f). The latent heat of fusion is *the heat involved in a solid-liquid phase change in melting or freezing.* For water, the latent heat of fusion is 80.0 cal/g (144.0 Btu/lb). This means that every gram of ice that melts *absorbs* 80.0 cal of heat. Every gram of water that freezes *releases* 80.0 cal. The total heat involved in a solid-liquid phase change depends on the mass of the substance involved, so

$$Q = mL_f \qquad \text{equation 4.5}$$

where L_f is the latent heat of fusion for the substance involved.

Refer again to figure 4.16. After the solid-liquid phase change is complete, the constant supply of heat increases the temperature of the water according to $Q = mc\Delta T$, where c is now the specific heat of liquid water. When the water reaches the boiling point, the temperature again remains constant even though heat is still being supplied at a constant rate. The quantity of heat involved in the liquid-gas phase change again goes into doing the work of overcoming the attractive molecular forces. This time the molecules escape from the liquid state to become single, independent molecules of gas. The quantity of heat (Q) absorbed or released during this phase change is called the **latent heat of vaporization** (L_v). The latent heat of vaporization is *the heat involved in a liquid-gas phase change where there is evaporation or condensation.* For water, the latent heat of vaporization is 540.0 cal/g (970.0 Btu/lb).

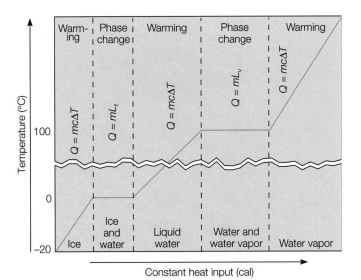

FIGURE 4.17 Compare this graph to the one in figure 4.16. This graph shows the relationships between the quantity of heat absorbed during warming and phase changes as water is warmed from ice at $-20°C$ to water vapor at some temperature above 100°C. Note that the specific heat for ice, liquid water, and water vapor (steam) has different values.

TABLE 4.3   Some Physical Constants for Water and Heat

Specific Heat (c)

Water	$c = 1.00$ cal/gC°
Ice	$c = 0.500$ cal/gC°
Steam	$c = 0.480$ cal/gC°

Latent Heat of Fusion

L_f (water)	$L_f = 80.0$ cal/g

Latent Heat of Vaporization

L_v (water)	$L_v = 540.0$ cal/g

Mechanical Equivalent of Heat

1 kcal	4,184 J

This means that every gram of water vapor that condenses on your bathroom mirror releases 540.0 cal, which warms the bathroom. The total heating depends on how much water vapor condensed, so

$$Q = mL_v \qquad \textbf{equation 4.6}$$

where L_v is the latent heat of vaporization for the substance involved. The relationships between the quantity of heat absorbed during warming and phase changes are shown in figure 4.17. Some physical constants for water and heat are summarized in table 4.3.

EXAMPLE 4.7   *(Optional)*

How much energy does a refrigerator remove from 100.0 g of water at 20.0°C to make ice at −10.0°C?

SOLUTION

This type of problem is best solved by subdividing it into smaller steps that consider (1) the heat added or removed and the resulting temperature changes for each phase of the substance, and (2) the heat flow resulting from any phase change that occurs within the ranges of changes as identified by the problem (see figure 4.17). The heat involved in each phase change and the heat involved in the heating or cooling of each phase are identified as Q_1, Q_2, and so forth. Temperature readings are calculated with absolute values, so you ignore any positive or negative signs.

1. Water in the liquid state cools from 20.0°C to 0°C (the freezing point) according to the relationship $Q = mc\Delta T$, where c is the specific heat of water, and

$$
\begin{aligned}
Q_1 &= mc\Delta T \\
&= (100.0 \text{ g})\left(1.00\frac{\text{cal}}{\text{gC}°}\right)(0°C - 20.0°C) \\
&= (100.0)(1.00)(20.0) \text{ g} \times \frac{\text{cal}}{\text{g} \cdot C°} \times °C \\
&= 2,000 \frac{\text{g} \cdot \text{cal} \cdot °\cancel{C}}{\cancel{\text{g}} \cdot \cancel{C}°} \\
&= 2.00 \text{ kcal}
\end{aligned}
$$

2. The latent heat of fusion must now be removed as water at 0°C becomes ice at 0°C through a phase change, and

$$
\begin{aligned}
Q_2 &= mL_f \\
&= (100.0 \text{ g})\left(80.0\frac{\text{cal}}{\text{g}}\right) \\
&= (100.0)(80.0) \text{ g} \times \frac{\text{cal}}{\text{g}} \\
&= 8,000 \text{ cal} \\
&= 8.00 \text{ kcal}
\end{aligned}
$$

3. The ice is now at 0°C and is cooled to −10°C, as specified in the problem. The ice cools according to $Q = mc\Delta T$, where c is the specific heat of ice. The specific heat of ice is 0.500 cal/gC°, and

$$
\begin{aligned}
Q_3 &= mc\Delta T \\
&= (100.0 \text{ g})\left(0.500\frac{\text{cal}}{\text{gC}°}\right)(10°C - 0°C) \\
&= (100.0)(0.500)(10.0) \text{ g} \times \frac{\text{cal}}{\text{g} \cdot C°} \times °C \\
&= 500 \text{ cal} \\
&= 0.500 \text{ kcal}
\end{aligned}
$$

4. The total energy removed is then

$$
\begin{aligned}
Q_T &= Q_1 + Q_2 + Q_3 \\
&= (2.00 \text{ kcal}) + (8.00 \text{ kcal}) + (0.500 \text{ kcal}) \\
&= \boxed{10.50 \text{ kcal}}
\end{aligned}
$$

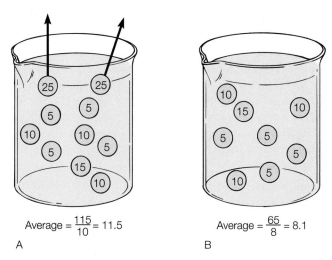

Average $= \dfrac{115}{10} = 11.5$ A

Average $= \dfrac{65}{8} = 8.1$ B

FIGURE 4.18 Temperature is associated with the average energy of the molecules of a substance. These numbered circles represent arbitrary levels of molecular kinetic energy that, in turn, represent temperature. The two molecules with the higher kinetic energy values [25 in (*A*)] escape, which lowers the average values from 11. 5 to 8. 1 (*B*). Thus, evaporation of water molecules with more kinetic energy contributes to the cooling effect of evaporation, in addition to the absorption of latent heat.

Evaporation and Condensation

Liquids do not have to be at the boiling point to change to a gas and, in fact, tend to undergo a phase change at any temperature when left in the open. The phase change occurs at any temperature but does occur more rapidly at higher temperatures. The temperature of the water is associated with the *average* kinetic energy of the water molecules. The word *average* implies that some of the molecules have a greater energy and some have less. If a molecule of water that has an exceptionally high energy is near the surface and is headed in the right direction, it may overcome the attractive forces of the other water molecules and escape the liquid to become a gas. This is the process of *evaporation*. Evaporation reduces a volume of liquid water as water molecules leave the liquid state to become water vapor in the atmosphere (figure 4.18).

Water molecules that evaporate move about in all directions, and some will return, striking the liquid surface. The same forces that they escaped from earlier capture the molecules, returning them to the liquid state. This is called the process of condensation. Condensation is the opposite of evaporation. In *evaporation,* more molecules are leaving the liquid state than are returning. In *condensation,* more molecules are returning to the liquid state than are leaving. This is a dynamic, ongoing process with molecules leaving and returning continuously. The net number leaving or returning determines if evaporation or condensation is taking place (figure 4.19).

When the condensation rate *equals* the evaporation rate, the air above the liquid is said to be *saturated*. The air immediately next to a surface may be saturated, but the condensation of water molecules is easily moved away with air movement. There is no net energy flow when the air is saturated, since the heat carried away by evaporation is returned by condensation.

FIGURE 4.19 The inside of this closed bottle is isolated from the environment, so the space above the liquid becomes saturated. While it is saturated, the evaporation rate equals the condensation rate. When the bottle is cooled, condensation exceeds evaporation and droplets of liquid form on the inside surfaces.

This is why you fan your face when you are hot. The moving air from the fanning action pushes away water molecules from the air near your skin, preventing the adjacent air from becoming saturated, thus increasing the rate of evaporation. Think about this process the next time you see someone fanning his or her face.

There are four ways to increase the rate of evaporation. (1) An increase in the temperature of the liquid will increase the average kinetic energy of the molecules and thus increase the number of high-energy molecules able to escape from the liquid state. (2) Increasing the surface area of the liquid will also increase the likelihood of molecular escape to the air. This is why you spread out wet clothing to dry or spread out a puddle you want to evaporate. (3) Removal of water vapor from near the surface of the liquid will prevent the return of the vapor molecules to the liquid state and thus increase the net rate of evaporation. This is why things dry more rapidly on a windy day. (4) *Pressure* is defined as *force per unit area*, which can be measured in lb/in^2 or N/m^2. Gases exert a pressure, which is interpreted in terms of the kinetic molecular theory.

Evaporative cooling is one of the earliest forms of mechanical air conditioning. An evaporative cooler (sometimes called a "swamp cooler") works by cooling outside air by evaporation, then blowing the cooler but now more humid air through a house. Usually, an evaporative cooler moves a sufficient amount of air through a house to completely change the air every two or three minutes.

An evaporative cooler is a metal or fiberglass box with louvers. Inside the box is a fan and motor, and there might be a small pump to recycle water. Behind the louvers are loose pads made of wood shavings (excelsior) or some other porous material. Water is pumped from the bottom of the cooler and trickles down through the pads, thoroughly wetting them. The fan forces air into the house and dry, warm outside air moves through the wet pads into the house in a steady flow. Windows or a door in the house must be partly open, or the air will not be able to flow through the house.

The water in the pads evaporates, robbing heat from the air moving through them. Each liter of water that evaporates could take over 540 kcal of heat from the air. The actual amount removed depends on the temperature of the water, the efficiency of the cooler, and the relative humidity of the outside air.

An electric fan cools you by evaporation, but an evaporative cooler cools the air that blows around you. Relative humidity is the main variable that determines how much an evaporative cooler will cool the air. The following data show the cooling power of a typical evaporative cooler at various relative humidities when the outside air is 38°C (about 100°F).

The advantage of an evaporative cooler is the low operating cost compared to refrigeration. The disadvantages are that it doesn't cool the air that much when the humidity is high, it adds even more humidity to the air, and outside air with its dust, pollen, and pollution is continually forced through the house. Another disadvantage is that mineral deposits left by evaporating hard water could require a frequent maintenance schedule.

At This Humidity:	Air at 38°C (100°F) Will Be Cooled to:
10%	22°C (72°F)
20%	24°C (76°F)
30%	27°C (80°F)
40%	29°C (84°F)
50%	31°C (88°F)
60%	33°C (91°F)
70%	34°C (93°F)
80%	35°C (95°F)
90%	37°C (98°F)

Atmospheric pressure is discussed in detail in chapter 17. For now, consider that the atmosphere exerts a pressure of about 10.0 N/cm² (14.7 lb/in²) at sea level. The atmospheric pressure, as well as the intermolecular forces, tend to hold water molecules in the liquid state. Thus, reducing the atmospheric pressure will reduce one of the forces holding molecules in a liquid state. Perhaps you have noticed that wet items dry more quickly at higher elevations, where the atmospheric pressure is less.

Relative Humidity

There is a relationship between evaporation-condensation and the air temperature. If the air temperature is decreased, the average kinetic energy of the molecules making up the air is decreased. Water vapor molecules condense from the air when they slow enough that molecular forces can pull them into the liquid state. Fast-moving water vapor molecules are less likely to be captured than slow-moving ones. Thus, as the air temperature increases, there is less tendency for water vapor molecules to return to the liquid state. Warm air can therefore hold more water vapor than cool air. In fact, air at 38°C (100°F) can hold five times as much water vapor as air at 10°C (50°F) (figure 4.20).

The ratio of how much water vapor *is* in the air to how much water vapor *could be* in the air at a certain temperature is called **relative humidity**. This ratio is usually expressed as a percentage, and

$$\text{relative humidity} = \frac{\text{water vapor in air}}{\text{capacity at present temperature}} \times 100\%$$

$$R.H. = \frac{\text{g/m}^3(\text{present})}{\text{g/m}^3(\text{max})} \times 100\% \qquad \textbf{equation 4.7}$$

Figure 4.20 shows the maximum amount of water vapor that can be in the air at various temperatures. Suppose that the air contains 10 g/m³ of water vapor at 10°C (50°F). According to figure 4.20, the maximum amount of water vapor that *can be* in the air when the air temperature is 10°C (50°F) is 10 g/m³. Therefore, the relative humidity is (10 g/m³) ÷ (10 g/m³) × 100%, or 100 percent. This air is therefore saturated. If the air held only 5 g/m³ of water vapor at 10°C, the relative humidity would be 50 percent, and 2 g/m³ of water vapor in the air at 10°C is 20 percent relative humidity.

As the air temperature increases, the capacity of the air to hold water vapor also increases. This means that if the *same amount* of water vapor is in the air during a temperature increase, the relative humidity will decrease. Thus, the relative humidity increases every night because the air temperature decreases, not because water vapor has been added to the air. The relative humidity is important because it is one of the things that controls the rate of evaporation, and the evaporation rate is one of the variables involved in how well you can cool yourself in hot weather.

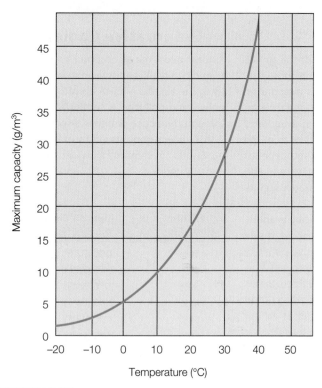

FIGURE 4.20 The curve shows the *maximum* amount of water vapor in g/m³ that can be in the air at various temperatures.

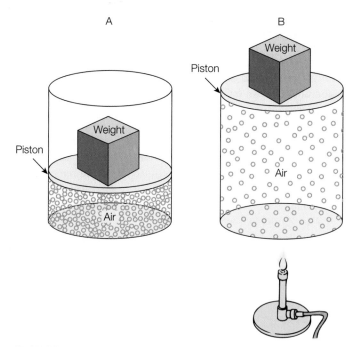

FIGURE 4.21 A very simple heat engine. The air in (*B*) has been heated, increasing the molecular motion and thus the pressure. Some of the heat is transferred to the increased gravitational potential energy of the weight as it is converted to mechanical energy.

THERMODYNAMICS

The branch of physical science called *thermodynamics* is concerned with the study of heat and its relationship to mechanical energy, including the science of heat pumps, heat engines, and the transformation of energy in all its forms. The *laws of* *thermodynamics* describe the relationships concerning what happens as energy is transformed to work and the reverse, also serving as useful intellectual tools in meteorology, chemistry, and biology.

Mechanical energy is easily converted to heat through friction, but a special device is needed to convert heat to mechanical energy. A *heat engine* is a device that converts heat into mechanical energy. The operation of a heat engine can be explained by the kinetic molecular theory, as shown in figure 4.21. This illustration shows a cylinder, much like a big can, with a closely fitting piston that traps a sample of air. The piston is like a slightly smaller cylinder and has a weight resting on it, supported by the trapped air. If the air in the large cylinder is now heated, the gas molecules will acquire more kinetic energy. This results in more gas molecule impacts with the enclosing surfaces, which results in an increased pressure. Increased pressure results in a net force, and the piston and weight move upward, as shown in figure 4.21B. Thus, some of the heat has now been transformed to the increased gravitational potential energy of the weight.

Thermodynamics is concerned with the *internal energy* (U), the total internal potential and kinetic energies of molecules making up a substance, such as the gases in the simple heat engine. The variables of temperature, gas pressure, volume, heat, and so forth characterize the total internal energy, which is called the *state* of the system. Once the system is identified, everything else is called the *surroundings*. A system can exist in a number of states since the variables that characterize a state can have any number of values and combinations of values. Any two systems that have the same values of variables that characterize internal energy are said to be in the same state.

The First Law of Thermodynamics

Any thermodynamic system has a unique set of properties that will identify the internal energy of the system. This state can be changed two ways: (1) by heat flowing into (Q_{in}) or out (Q_{out}) of the system, or (2) by the system doing work (W_{out}) or work being done on the system (W_{in}). Thus, work (W) and heat (Q) can change the internal energy of a thermodynamic system according to

$$JQ - W = U_2 - U_1 \qquad \text{equation 4.8}$$

where J is the mechanical equivalence of heat ($J = 4.184$ joule/calorie) and ($U_2 - U_1$) is the internal energy difference between two states. This equation represents the **first law of thermodynamics,** which states that the energy supplied to a thermodynamic system in the form of heat minus the work done by the system is equal to the change in internal energy. The first law of thermodynamics is an application of the *law of conservation of energy,* which applies to all energy matters. The first law of thermodynamics is concerned specifically with a thermodynamic system. As an example, consider energy conservation that is observed in the thermodynamic system of a heat engine (see figure 4.22). As the engine cycles to the original state of internal energy ($U_2 - U_1 = 0$), all the external work accomplished must be equal to all the heat absorbed in the cycle. The heat supplied to the engine from a high temperature source (Q_H) is partly converted to work (W), and the rest is rejected in the lower-temperature exhaust (Q_L). The work accomplished is therefore the difference in the heat input and the heat output ($Q_H - Q_L$), so the work accomplished represents the heat used,

$$W = J(Q_H - Q_L) \qquad \text{equation 4.9}$$

where J is the mechanical equivalent of heat ($J = 4.184$ joules/calorie). A schematic diagram of this relationship is shown in figure 4.22. You can increase the internal energy (produce heat) as long as you supply mechanical energy (or do work). The first law of thermodynamics states that the conversion of work to heat is reversible, meaning that heat can be changed to work. There are several ways of converting heat to work, for example, the use of a steam turbine or gasoline automobile engine.

The Second Law of Thermodynamics

A heat pump is the opposite of a heat engine, as shown schematically in figure 4.23. The heat pump does work (W) in compressing vapors and moving heat from a region of lower

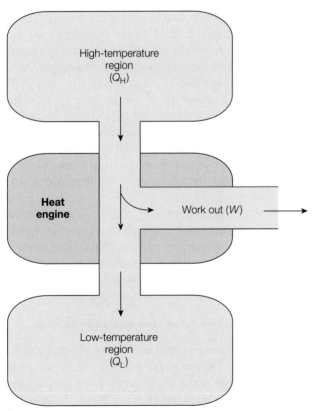

FIGURE 4.22 The heat supplied (Q_H) to a heat engine goes into the mechanical work (W), and the remainder is expelled in the exhaust (Q_L). The work accomplished is therefore the difference in the heat input and output ($Q_H - Q_L$), so the work accomplished represents the heat used, $W = J(Q_H - Q_L)$.

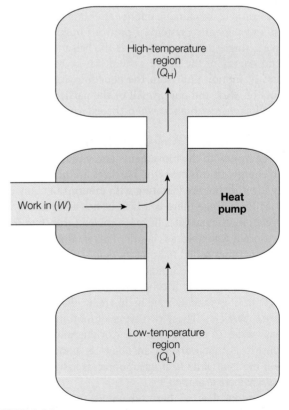

FIGURE 4.23 A heat pump uses work (W) to move heat from a low temperature region (Q_L) to a high temperature region (Q_H). The heat moved (Q_L) requires work (W), so $JQ_L = W$. A heat pump can be used to chill things at the Q_L end or warm things at the Q_H end.

temperature (Q_L) to a region of higher temperature (Q_H). That work is required to move heat this way is in accord with the observation that heat naturally flows from a region of higher temperature to a region of lower temperature. Energy is required for the opposite, moving heat from a cooler region to a warmer region. The natural direction of this process is called the **second law of thermodynamics,** which is that heat flows from objects with a higher temperature to objects with a cooler temperature. In other words, if you want heat to flow from a colder region to a warmer one, you must *cause* it to do so by using energy. And if you do, such as with the use of a heat pump, you necessarily cause changes elsewhere, particularly in the energy sources used in the generation of electricity. Another statement of the second law is that it is impossible to convert heat completely into mechanical energy. This does not say that you cannot convert mechanical energy completely into heat, for example, in the brakes of a car when the brakes bring it to a stop. The law says that the reverse process is not possible, that you cannot convert 100 percent of a heat source into mechanical energy. Both of the preceding statements of the second law are concerned with a *direction* of thermodynamic processes, and the implications of this direction will be discussed next.

The Second Law and Natural Processes

Energy can be viewed from two considerations of scale: (1) the observable *external energy* of an object, and (2) the *internal energy* of the molecules or particles that make up an object. A ball, for example, has kinetic energy after it is thrown through the air, and the entire system of particles making up the ball acts like a single massive particle as the ball moves. The motion and energy of the single system can be calculated from the laws of motion and from the equations representing the concepts of work and energy. All of the particles are moving together in *coherent motion* when the external kinetic energy is considered.

But the particles making up the ball have another kind of kinetic energy, with the movements and vibrations of internal kinetic energy. In this case, the particles are not moving uniformly together but are vibrating with motions in many different directions. Since there is a lack of net motion and a lack of correlation, the particles have a jumbled *incoherent motion*, which is often described as chaotic. This random, chaotic motion is sometimes called *thermal motion.*

Thus, there are two kinds of motion that the particles of an object can have: (1) a coherent motion where they move together, in step, and (2) an incoherent, chaotic motion of individual particles. These two types of motion are related to the two modes of energy transfer, working and heating. The relationship is that *work* on an object is associated with its *coherent motion,* while *heating* an object is associated with its internal *incoherent motion.*

The second law of thermodynamics implies a direction to the relationship between work (coherent motion) and heat

(incoherent motion), and this direction becomes apparent as you analyze what happens to the motions during energy conversions. Some forms of energy, such as electrical and mechanical, have a greater amount of order since they involve particles moving together in a coherent motion. The term *quality of energy* is used to identify the amount of coherent motion. Energy with high order and coherence is called *high-quality energy.* Energy with less order and less coherence, on the other hand, is called *low-quality energy.* In general, high-quality energy can be easily converted to work, but low-quality energy is less able to do work.

High-quality electrical and mechanical energy can be used to do work but then become dispersed as heat through energy-form conversions and friction. The resulting heat can be converted to do more work only if there is a sufficient temperature difference. The temperature differences do not last long, however, as conduction, convection, and radiation quickly disperse the energy even more. Thus, the transformation of high-quality energy into lower-quality energy is a natural process. Energy tends to disperse, both from the conversion of an energy form to heat and from the heat flow processes of conduction, convection, and radiation. Both processes flow in one direction only and cannot be reversed. This is called the *degradation of energy,* which is the transformation of high-quality energy to lower-quality energy. In every known example, it is a natural process of energy to degrade, becoming less and less available to do work. The process is *irreversible,* even though it is possible to temporarily transform heat to mechanical energy through a heat engine or to upgrade the temperature through the use of a heat pump. Eventually, the upgraded mechanical energy will degrade to heat and the increased heat will disperse through the processes of heat flow.

The apparent upgrading of energy by a heat pump or heat engine is always accompanied by a greater degrading of energy someplace else. The electrical energy used to run the heat pump, for example, was produced by the downgrading of chemical or nuclear energy at an electrical power plant. The overall result is that the *total* energy was degraded toward a more disorderly state.

A *thermodynamic measure of disorder* is called **entropy.** Order means patterns and coherent arrangements. Disorder means dispersion, no patterns, and a randomized or spread-out arrangement. Entropy is therefore a measure of chaos, and this leads to another statement about the second law of thermodynamics and the direction of natural change, that

the total entropy of the universe continually increases.

Note the use of the words *total* and *universe* in this statement of the second law. The entropy of a system can decrease (more order); for example, when a heat pump cools and condenses the random, chaotically moving water vapor molecules into the more ordered state of liquid water. When the energy source for the production, transmission, and use of electrical

Thermodynamics in Action

The laws of thermodynamics are concerned with changes in energy and heat. This application explores some of these relationships.

Obtain an electric blender and a thermometer. Fill the blender half way with water, then let it remain undisturbed until the temperature is constant as shown by two consecutive temperature readings.

Run the blender at the highest setting for a short time, then stop and record the temperature of the water. Repeat this procedure several times.

Explain your observations in terms of thermodynamics. See if you can think of other experiments that show relationships between changes in energy and heat.

It Makes Its Own Fuel?

Have you ever heard of a perpetual motion machine? A perpetual motion machine is a hypothetical device that would produce useful energy out of nothing. This is generally accepted as being impossible according to laws of physics. In particular, perpetual motion machines would violate either the first or the second law of thermodynamics.

A perpetual motion machine that violates the first law of thermodynamics is called a "machine of the first kind." In general, the first law says that you can never get something for nothing. This means that without energy input, there can be no change in internal energy, and without a change in internal energy, there can be no work output. Machines of the first kind typically use no fuel or make their own fuel faster than they use it. If this type of machine appears to work, look for some hidden source of energy.

A "machine of the second kind" does not attempt to make energy out of nothing. Instead, it tries to extract random molecular motion into useful work or extract useful energy from some degraded source such as outgoing radiant energy. The second law of thermodynamics says this just cannot happen any more than rocks can roll uphill on their own. This just does not happen.

The American Physical Society states, "The American Physical Society deplores attempts to mislead and defraud the public based on claims of perpetual motion machines or sources of unlimited useful free energy, unsubstantiated by experimentally tested established physical principles."

Visit www.phact.org/e/dennis4.html to see a historical list of perpetual motion and free energy machines.

energy is considered, however, the *total* entropy will be seen as increasing. Likewise, the total entropy increases during the growth of a plant or animal. When all the food, waste products, and products of metabolism are considered, there is again an increase in *total* entropy.

Thus, the *natural process* is for a state of order to degrade into a state of disorder with a corresponding increase in entropy. This means that all the available energy of the universe is gradually diminishing, and over time, the universe should therefore approach a limit of maximum disorder called the *heat death* of the universe. The heat death of the universe is the theoretical limit of disorder, with all molecules spread far, far apart, vibrating slowly with a uniform low temperature.

The heat death of the universe seems to be a logical consequence of the second law of thermodynamics, but scientists are not certain if the second law should apply to the whole universe. What do you think? Will the universe end as matter becomes spread out with slowly vibrating molecules? As has been said, nature is full of symmetry. So why should the universe begin with a bang and end with a whisper?

EXAMPLE 4.8   *(Optional)*

A heat engine operates with 65.0 kcal of heat supplied and exhausts 40.0 kcal of heat. How much work did the engine do?

SOLUTION

Listing the known and unknown quantities:

heat input	$Q_H = 65.0$ kcal
heat rejected	$Q_L = 40.0$ kcal
mechanical equivalent of heat	1 kcal = 4,184 J

The relationship between these quantities is found in equation 4.9 $W = J(Q_H - Q_L)$. This equation states a relationship between the heat supplied to the engine from a high-temperature source (Q_H), which is partly converted to work (W), with the rest rejected in a lower-temperature exhaust (Q_L). The work accomplished is therefore the difference in the heat input and the heat output ($Q_H - Q_L$), so the work accomplished represents the heat used, where J is the mechanical equivalence of heat (1 kcal = 4,184 J). Therefore,

$$W = J(Q_H - Q_L)$$

$$= 4,184 \ \frac{J}{kcal} \ (65.0 \text{ kcal} - 40.0 \text{ kcal})$$

$$= 4,184 \ \frac{J}{kcal} \ (25.0 \text{ kcal})$$

$$= (4,184)(25.0) \ \frac{J \, \cancel{kcal}}{\cancel{kcal}}$$

$$= 104,600 \text{ J}$$

$$= \boxed{105 \text{ kJ}}$$

People Behind the Science

Count Rumford (Benjamin Thompson) (1753–1814)

Count Rumford was a U.S.-born physicist who first demonstrated conclusively that heat is not a fluid but a form of motion. He was born Benjamin Thompson in Woburn, Massachusetts, on March 26, 1753. At the age of nineteen, he became a schoolmaster as a result of much self-instruction and some help from local clergy. He moved to Rumford (now named Concord), New Hampshire, and almost immediately married a wealthy widow many years his senior.

Thompson's first activities seem to have been political. When the War of Independence broke out, he remained loyal to the British crown and acted as some sort of secret agent. Because of these activities, he had to flee to London in 1776 (having separated from his wife the year before). He was rewarded with government work and an appointment as a lieutenant colonel in a British regiment in New York. After the war, he retired from the army and lived permanently in exile in Europe. Thompson moved to Bavaria and spent the next few years with the civil administration there, becoming war and police minister.

In 1791, Thompson was made a count of the Holy Roman Empire in recognition of his work in Bavaria. He took his title from Rumford in his homeland, and it is by this name that we know him today—Count Rumford.

Rumford's early work in Bavaria combined social experiments with his lifelong interests concerning heat in all its aspects. When he employed beggars from the streets to manufacture military uniforms, he faced the problem of feeding them. A study of nutrition led him to recognize the importance of water and vegetables, and Rumford decided that soups would fit his requirements. He devised many recipes and developed cheap food emphasizing the potato. Soldiers were employed in gardening to produce the vegetables. Rumford's enterprise of manufacturing military uniforms led to a study of insulation and to the conclusion that heat was lost mainly through convection. Therefore, he designed clothing to inhibit convection—sort of the first thermal clothing.

No application of heat technology was too humble for Rumford's experiments. He devised the domestic range—the "fire in a box"—and special utensils to go with it. In the interest of fuel efficiency, he devised a calorimeter to compare the heats of combustion of various fuels. Smoky fireplaces also drew his attention, and after a study of the various air movements, he produced designs incorporating all the features now considered essential in open fires and chimneys, such as the smoke shelf and damper. His search for an alternative to alcoholic drinks led to the promotion of coffee and the design of the first percolator.

The work for which Rumford is best remembered took place in 1798. As military commander for the elector of Bavaria, he was concerned with the manufacture of cannons. These were bored from blocks of iron with drills, and it was believed that the cannons became hot because as the drills cut into the iron, heat was escaping in the form of a fluid called caloric. However, Rumford noticed that heat production increased as the drills became blunter and cut less into the metal. If a very blunt drill was used, no metal was removed, yet the heat output appeared to be limitless. Clearly, heat could not be a fluid in the metal but must be related to the work done in turning the drill. Rumford also studied the expansion of liquids of different densities and different specific heats, and showed by careful weighing that the expansion was not due to caloric taking up the extra space.

Rumford's contribution to science in demolishing the caloric theory of heat was very important, because it paved the way to the realization that heat is related to energy and work and that all forms of energy can be converted to heat. However, it took several decades to establish the understanding that caloric does not exist and there was no basis for the caloric theory of heat.

Source: Modified from the *Hutchinson Dictionary of Scientific Biography*. © RM, 2007. Reprinted by permission.

SUMMARY

The kinetic theory of matter assumes that all matter is made up of tiny, ultimate particles of matter called *molecules.* A molecule is defined as the smallest particle of a compound or a gaseous element that can exist and still retain the characteristic properties of that substance. Molecules interact, attracting each other through a force of *cohesion.* Liquids, solids, and gases are the *phases of matter* that are explained by the molecular arrangements and forces of attraction between their molecules. A *solid* has a definite shape and volume because it has molecules that vibrate in a fixed equilibrium position with strong cohesive forces. A *liquid* has molecules that have cohesive forces strong enough to give it a definite volume but not strong enough to give it a definite shape. The molecules of a liquid can flow, rolling over each other. A *gas* is composed of molecules that are far apart, with weak cohesive forces. Gas molecules move freely in a constant, random motion.

The *temperature* of an object is related to the *average kinetic energy* of the molecules making up the object. A measure of temperature tells how hot or cold an object is on two arbitrary scales, the *Fahrenheit scale* and the *Celsius scale.* The *absolute scale,* or *Kelvin scale,* has the coldest temperature possible (−273°C) as zero (0 K).

The observable potential and kinetic energy of an object is the *external energy* of that object, while the potential and kinetic energy of the molecules making up the object is the *internal energy* of the object. Heat refers to the total internal energy and is a transfer of energy that takes place (1) because of a *temperature difference* between two objects or

(2) because of an *energy-form conversion*. An energy-form conversion is actually an energy conversion involving work at the molecular level, so all energy transfers involve *heating* and *working*.

A quantity of heat can be measured in *joules* (a unit of work or energy) or *calories* (a unit of heat). A *kilocalorie* is 1,000 calories, another unit of heat. A *Btu*, or *British thermal unit*, is the English system unit of heat. The *mechanical equivalent of heat* is 4,184 J = 1 kcal.

The *specific heat* of a substance is the amount of energy (or heat) needed to increase the temperature of 1 gram of a substance 1 degree Celsius. The specific heat of various substances is not the same because the molecular structure of each substance is different.

Energy transfer that takes place because of a temperature difference does so through conduction, convection, or radiation. *Conduction* is the transfer of increased kinetic energy from molecule to molecule. Substances vary in their ability to conduct heat, and those that are poor conductors are called *insulators*. Gases, such as air, are good insulators. The best insulator is a vacuum. *Convection* is the transfer of heat by the displacement of large groups of molecules with higher kinetic energy. Convection takes place in fluids, and the fluid movement that takes place because of density differences is called a *convection current. Radiation* is radiant energy that moves through space. All objects with an absolute temperature above zero give off radiant energy, but all objects absorb it as well. Energy is transferred from a hot object to a cold one through radiation.

The transition from one phase of matter to another is called a *phase change*. A phase change always absorbs or releases a quantity of *latent heat* not associated with a temperature change. Latent heat is energy that goes into or comes out of *internal potential energy*. The *latent heat of fusion* is absorbed or released at a solid-liquid phase change. The latent heat of fusion for water is 80.0 cal/g (144.0 Btu/lb). The *latent heat of vaporization* is absorbed or released at a liquid-gas phase change. The latent heat of vaporization for water is 540.0 cal/g (970.0 Btu/lb).

Molecules of liquids sometimes have a high enough velocity to escape the surface through the process called *evaporation*. Evaporation is a cooling process, since the escaping molecules remove the latent heat of vaporization in addition to their high molecular energy. Vapor molecules return to the liquid state through the process called *condensation*. Condensation is the opposite of evaporation and is a warming process. When the condensation rate equals the evaporation rate, the air is said to be *saturated*. The rate of evaporation can be *increased* by (1) increased temperature, (2) increased surface area, (3) removal of evaporated molecules, and (4) reduced atmospheric pressure.

Warm air can hold more water vapor than cold air, and the ratio of how much water vapor is in the air to how much could be in the air at that temperature (saturation) is called *relative humidity*.

Thermodynamics is the study of heat and its relationship to mechanical energy, and the *laws of thermodynamics* describe these relationships: The *first law of thermodynamics* is a thermodynamic statement of the law of conservation of energy. The *second law of thermodynamics* states that heat flows from objects with a higher temperature to objects with a cooler temperature. The second law implies a *degradation of energy* as *high-quality* (more ordered) energy sources undergo *degradation* to *low-quality* (less ordered) sources. *Entropy* is a thermodynamic measure of disorder, and entropy is seen as continually increasing in the universe and may result in the maximum disorder called the *heat death* of the universe.

Summary of Equations

4.1
$$T_F = \frac{9}{5} T_C + 32°$$

4.2
$$T_C = \frac{5}{9}(T_F - 32°)$$

4.3
$$T_K = T_C + 273$$

4.4 Quantity of heat = (mass)(specific heat)(temperature change)
$$Q = mc\Delta T$$

4.5 Heat absorbed or released = (mass)(latent heat of fusion)
$$Q = mL_f$$

4.6 Heat absorbed or released = (mass)(latent heat of vaporization)
$$Q = mL_v$$

4.7 relative humidity = $\dfrac{\text{water vapor in air}}{\text{capacity at present temperature}} \times 100\%$

$$R.H. = \frac{\text{g/m}^3(\text{present})}{\text{g/m}^3(\text{max})} \times 100\%$$

4.8 (mechanical equivalence of heat) − (work) = internal energy difference between two states
$$JQ - W = U_2 - U_1$$

4.9 work = (mechanical equivalence of heat)(difference between heat input and output)
$$W = J(Q_H - Q_L)$$

KEY TERMS

APPLYING THE CONCEPTS

1. The kinetic molecular theory explains the expansion of a solid material with increases of temperature as basically the result of
   a. individual molecules expanding.
   b. increased translational kinetic energy.
   c. molecules moving a little farther apart.
   d. heat taking up the spaces between molecules.

2. Using the absolute temperature scale, the freezing point of water is correctly written as
   a. 0 K.
   b. 0°K.
   c. 273 K.
   d. 273°K.

3. The metric unit of heat called a *calorie* is
   a. the specific heat of water.
   b. the energy needed to increase the temperature of 1 gram of water 1 degree Celsius.
   c. equivalent to a little over 4 joules of mechanical work.
   d. All of the above are correct.

4. Table 4.2 lists the specific heat of soil as 0.20 kcal/kgC° and the specific heat of water as 1.00 kcal/kgC°. This means that if 1 kg of soil and 1 kg of water each receive 1 kcal of energy, ideally
   a. the water will be warmer than the soil by 0.8°C.
   b. the soil will be 4°C warmer than the water.
   c. the water will be 4°C warmer than the soil.
   d. the water will warm by 1°C, and the soil will warm by 0.2°C.

5. The heat transfer that takes place by energy moving directly from molecule to molecule is called
   a. conduction.
   b. convection.
   c. radiation.
   d. None of the above is correct.

6. The heat transfer that does not require matter is
   a. conduction.
   b. convection.
   c. radiation.
   d. impossible, for matter is always required.

7. Styrofoam is a good insulating material because
   a. it is a plastic material that conducts heat poorly.
   b. it contains many tiny pockets of air.
   c. of the structure of the molecules making up the Styrofoam.
   d. it is not very dense.

8. The transfer of heat that takes place because of density difference in fluids is
   a. conduction.
   b. convection.
   c. radiation.
   d. None of the above is correct.

9. As a solid undergoes a phase change to a liquid state, it
   a. releases heat while remaining at a constant temperature.
   b. absorbs heat while remaining at a constant temperature.
   c. releases heat as the temperature decreases.
   d. absorbs heat as the temperature increases.

10. The condensation of water vapor actually
    a. warms the surroundings.
    b. cools the surroundings.
    c. sometimes warms and sometimes cools the surroundings, depending on the relative humidity at the time.
    d. neither warms nor cools the surroundings.

11. No water vapor is added to or removed from a sample of air that is cooling, so the relative humidity of this sample of air will
    a. remain the same.
    b. be lower.
    c. be higher.
    d. be higher or lower, depending on the extent of change.

12. Compared to cooler air, warm air can hold
    a. more water vapor.
    b. less water vapor.
    c. the same amount of water vapor.
    d. less water vapor, the amount depending on the humidity.

Answers

1. c **2.** c **3.** d **4.** b **5.** a **6.** c **7.** b **8.** b **9.** b **10.** a **11.** c **12.** a

QUESTIONS FOR THOUGHT

1. What is temperature? What is heat?

2. Explain why most materials become less dense as their temperature is increased.

3. Would the tight packing of more insulation, such as glass wool, in an enclosed space increase or decrease the insulation value? Explain.

4. A true vacuum bottle has a double-walled, silvered bottle with the air removed from the space between the walls. Describe how this design keeps food hot or cold by dealing with conduction, convection, and radiation.

5. Why is cooler air found in low valleys on calm nights?

6. Why is air a good insulator?

7. A piece of metal feels cooler than a piece of wood at the same temperature. Explain why.

8. What is condensation? Explain, on a molecular level, how the condensation of water vapor on a bathroom mirror warms the bathroom.

9. Which Styrofoam cooler provides more cooling, one with 10 lb of ice at 0°C or one with 10 lb of ice water at 0°C? Explain your reasoning.

10. Explain why a glass filled with a cold beverage seems to sweat. Would you expect more sweating inside a house during the summer or during the winter? Explain.

11. Explain why a burn from 100°C steam is more severe than a burn from water at 100°C.

12. The relative humidity increases almost every evening after sunset. Explain how this is possible if no additional water vapor is added to or removed from the air.

FOR FURTHER ANALYSIS

1. Considering the criteria for determining if something is a solid, liquid, or a gas, what is table salt, which can be poured?

2. What are the significant similarities and differences between heat and temperature?

3. Gas and plasma are phases of matter, yet gas runs a car and plasma is part of your blood. Compare and contrast these terms and offer an explanation for the use of similar names.

4. Analyze the table of specific heats (table 4.2) and determine which metal would make an energy-efficient and practical pan, providing more cooking for less energy.

5. This chapter contains information about three types of passive solar home design. Develop criteria or standards of evaluation that would help someone decide which design is right for their local climate.

6. Could a heat pump move heat without the latent heat of vaporization? Explain.

7. Explore the assumptions on which the "heat death of the universe" idea is based. Propose and evaluate an alternative idea for the future of the universe.

INVITATION TO INQUIRY

Who Can Last Longest?

How can we be more energy efficient? Much of our household energy consumption goes into heating and cooling, and much is lost through walls and ceilings. This invitation is about the insulating properties of various materials and their arrangement.

The challenge of this invitation is to create an insulated container that can keep an ice cube from melting. Decide on a maximum size for the container, and then decide what materials to use. Consider how you will use the materials. For example, if you are using aluminum foil, should it be shiny side in or shiny side out? If you are using newspapers, should they be folded flat or crumpled loosely?

One ice cube should be left outside the container to use as a control. Find out how much longer your insulated ice cube will outlast the control.

PARALLEL EXERCISES

The exercises in groups A and B cover the same concepts. Solutions to group A exercises are located in appendix D.
Note: Neglect all frictional forces in all exercises.

Group A

1. The average human body temperature is 98.6°F. What is the equivalent temperature on the Celsius scale?

2. An electric current heats a 221 g copper wire from 20.0°C to 38.0°C. How much heat was generated by the current? (c_{copper} = 0.093 kcal/kgC°)

3. A bicycle and rider have a combined mass of 100.0 kg. How many calories of heat are generated in the brakes when the bicycle comes to a stop from a speed of 36.0 km/h?

4. A 15.53 kg bag of soil falls 5.50 m at a construction site. If all the energy is retained by the soil in the bag, how much will its temperature increase? (c_{soil} = 0.200 kcal/kgC°)

5. A 75.0 kg person consumes a small order of french fries (250.0 Cal) and wishes to "work off" the energy by climbing a 10.0 m stairway. How many vertical climbs are needed to use all the energy?

6. A 0.5 kg glass bowl (c_{glass} = 0.2 kcal/kgC°) and a 0.5 kg iron pan (c_{iron} = 0.11 kcal/kgC°) have a temperature of 68°F when placed in a freezer. How much heat will the freezer have to remove from each to cool them to 32°F?

7. A sample of silver at 20.0°C is warmed to 100.0°C when 896 cal is added. What is the mass of the silver? (c_{silver} = 0.056 kcal/kgC°)

Group B

1. The Fahrenheit temperature reading is 98° on a hot summer day. What is this reading on the Kelvin scale?

2. A 0.25 kg length of aluminum wire is warmed 10.0°C by an electric current. How much heat was generated by the current? ($c_{aluminum}$ = 0.22 kcal/kgC°)

3. A 1,000.0 kg car with a speed of 90.0 km/h brakes to a stop. How many calories of heat are generated by the brakes as a result?

4. A 1.0 kg metal head of a geology hammer strikes a solid rock with a velocity of 5.0 m/s. Assuming all the energy is retained by the hammer head, how much will its temperature increase? (c_{head} = 0.11 kcal/kgC°)

5. A 60.0 kg person will need to climb a 10.0 m stairway how many times to "work off" each excess Cal (kcal) consumed?

6. A 50.0 g silver spoon at 20.0°C is placed in a cup of coffee at 90.0°C. How much heat does the spoon absorb from the coffee to reach a temperature of 89.0°C?

7. If the silver spoon placed in the coffee in problem 6 causes it to cool 0.75°C, what is the mass of the coffee? (Assume c_{coffee} = 1.0 cal/gC°)

—continued

Group A

8. A 300.0 W immersion heater is used to heat 250.0 g of water from 10.0°C to 70.0°C. About how many minutes did this take?

9. A 100.0 g sample of metal is warmed 20.0°C when 60.0 cal is added. What is the specific heat of this metal?

10. How much heat is needed to change 250.0 g of water at 80.0°C to steam at 100.0°C?

11. A 100.0 g sample of water at 20.0°C is heated to steam at 125.0°C. How much heat was absorbed?

12. In an electric freezer, 400.0 g of water at 18.0°C is cooled, frozen, and the ice is chilled to −5.00°C. How much total heat was removed from the water?

Group B

8. How many minutes would be required for a 300.0 W immersion heater to heat 250.0 g of water from 20.0°C to 100.0°C?

9. A 200.0 g china serving bowl is warmed 65.0°C when it absorbs 2.6 kcal of heat from a serving of hot food. What is the specific heat of the china dish?

10. A 500.0 g pot of water at room temperature (20.0°C) is placed on a stove. How much heat is required to change this water to steam at 100.0°C?

11. Spent steam from an electric generating plant leaves the turbines at 120.0°C and is cooled to 90.0°C liquid water by water from a cooling tower in a heat exchanger. How much heat is removed by the cooling tower water for each kilogram of spent steam?

12. Lead is a soft, dense metal with a specific heat of 0.028 kcal/kgC°, a melting point of 328.0°C, and a heat of fusion of 5.5 kcal/kg. How much heat must be provided to melt a 250.0 kg sample of lead with a temperature of 20.0°C?

Compared to the sounds you hear on a calm day in the woods, the sounds from a waterfall can carry up to a million times more energy.

Wave Motions and Sound

► The brain perceives sounds based on signals from the ear (Ch. 24).

CORE CONCEPT

Sound is transmitted as increased and decreased pressure waves that carry energy.

All sounds originate from vibrating matter (p. 118).

Mechanical waves are longitudinal or transverse; sound waves are longitudinal (p. 107).

An external force that matches the natural frequency results in resonance (p. 116).

The Doppler effect is an apparent change of pitch brought about from motion (p. 120).

OVERVIEW

Sometimes you can feel the floor of a building shake for a moment when something heavy is dropped. You can also feel prolonged vibrations in the ground when a nearby train moves by. The floor of a building and the ground are solids that transmit vibrations from a disturbance. Vibrations are common in most solids because the solids are elastic, having a tendency to rebound, or snap back, after a force or an impact deforms them. Usually you cannot see the vibrations in a floor or the ground, but you sense they are there because you can feel them.

There are many examples of vibrations that you can see. You can see the rapid blur of a vibrating guitar string (figure 5.1). You can see the vibrating up-and-down movement of a bounced-upon diving board. Both the vibrating guitar string and the diving board set up a vibrating motion of air that you identify as a sound. You cannot see the vibrating motion of the air, but you sense it is there because you hear sounds.

There are many kinds of vibrations that you cannot see but can sense. Heat, as you have learned, is associated with molecular vibrations that are too rapid and too tiny for your senses to detect other than as an increase in temperature. Other invisible vibrations include electrons that vibrate, generating spreading electromagnetic radio waves or visible light. Thus, vibrations take place as an observable motion of objects but are also involved in sound, heat, electricity, and light. The vibrations involved in all these phenomena are alike in many ways and all involve energy. Therefore, many topics of science are concerned with vibrational motion. In this chapter, you will learn about the nature of vibrations and how they produce waves in general. These concepts will be applied to sound in this chapter and to electricity and electromagnetic radiation in later chapters.

FORCES AND ELASTIC MATERIALS

If you drop a rubber ball, it bounces because it is capable of recovering its shape when it hits the floor. A ball of clay, on the other hand, does not recover its shape and remains a flattened blob on the floor. An *elastic* material is one that is capable of recovering its shape after a force deforms it. A rubber ball is elastic and a ball of clay is not elastic. You know a metal spring is elastic because you can stretch it or compress it and it always recovers its shape.

A direct relationship exists between the extent of stretching or compression of a spring and the amount of force applied to it. A large force stretches a spring a lot; a small force stretches it a little. As long as the applied force does not exceed the elastic limit of the spring, it will always return to its original shape when you remove the applied force. There are three important considerations about the applied force and the response of the spring:

1. The greater the applied force, the greater the compression or stretch of the spring from its original shape.
2. The spring appears to have an *internal restoring force*, which returns it to its original shape.
3. The farther the spring is pushed or pulled, the *stronger* the restoring force that returns the spring to its original shape.

Forces and Vibrations

A **vibration** is a back-and-forth motion that repeats itself. Such a motion is not restricted to any particular direction and can be in many different directions at the same time. Almost any solid can be made to vibrate if it is elastic. To see how forces are involved in vibrations, consider the spring and mass in figure 5.2. The spring and mass are arranged so that the mass can freely move back and forth on a frictionless surface. When the mass has not been disturbed, it is at rest at an *equilibrium*

FIGURE 5.1 Vibrations are common in many elastic materials, and you can see and hear the results of many in your surroundings. Other vibrations in your surroundings, such as those involved in heat, electricity, and light, are invisible to the senses.

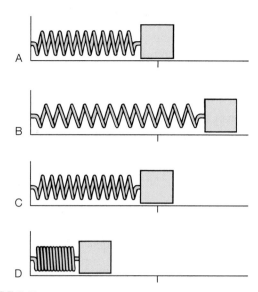

FIGURE 5.2 A mass on a frictionless surface is at rest at an equilibrium position (*A*) when undisturbed. When the spring is stretched (*B*) or compressed (*D*), then released (*C*), the mass vibrates back and forth because restoring forces pull opposite to and proportional to the displacement.

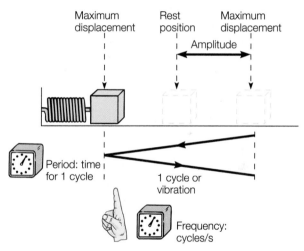

FIGURE 5.3 A vibrating mass attached to a spring is displaced from rest, or equilibrium position, and then released. The maximum displacement is called the *amplitude* of the vibration. A cycle is one complete vibration. The period is the time required for one complete cycle. The frequency is a count of how many cycles it completes in 1 s.

position (figure 5.2A). At the equilibrium position, the spring is not compressed or stretched, so it applies no force on the mass. If, however, the mass is pulled to the right (figure 5.2B), the spring is stretched and applies a restoring force on the mass toward the left. The farther the mass is displaced, the greater the stretch of the spring and thus the greater the restoring force. The restoring force is proportional to the displacement and is in the opposite direction of the applied force.

If the mass is now released, the restoring force is the only force acting (horizontally) on the mass, so it accelerates back toward the equilibrium position. This force will continuously decrease until the moving mass arrives back at the equilibrium position, where the force is zero. The mass will have a maximum velocity when it arrives, however, so it overshoots the equilibrium position and continues moving to the left (figure 5.2C). As it moves to the left of the equilibrium position, it compresses the spring, which exerts an increasing force on the mass. The moving mass comes to a temporary halt (figure 5.2D), but now the restoring force again starts it moving back toward the equilibrium position. The whole process repeats itself again and again as the mass moves back and forth over the same path.

The periodic vibration, or oscillation, of the mass is similar to many vibrational motions found in nature called *simple harmonic motion*. Simple harmonic motion is defined as the vibratory motion that occurs when there is a restoring force opposite to and proportional to a displacement.

The vibrating mass and spring system will continue to vibrate for a while, slowly decreasing with time until the vibrations stop completely. The slowing and stopping is due to air resistance and internal friction. If these could be eliminated or compensated for with additional energy, the mass would continue to vibrate with a repeating, or *periodic,* motion.

Describing Vibrations

A vibrating mass is described by measuring several variables (figure 5.3). The extent of displacement from the equilibrium position is called the **amplitude.** A vibration that has a mass displaced a greater distance from equilibrium thus has a greater amplitude than a vibration with less displacement.

A complete vibration is called a **cycle.** A cycle is the movement from some point, say the far left, all the way to the far right, and back to the same point again, the far left in this example. The **period** (*T*) is simply the time required to complete one cycle. For example, suppose 0.1 s is required for an object to move through one complete cycle, to complete the back-and-forth motion from one point, then back to that point. The period of this vibration is 0.1 s.

Sometimes it is useful to know how frequently a vibration completes a cycle every second. The number of cycles per second is called the **frequency** (f). For example, a vibrating object moves through 10 cycles in 1 s. The frequency of this vibration is 10 cycles per second. Frequency is measured in a unit called a **hertz** (Hz). The unit for a hertz is 1/s because a cycle does not have dimensions. Thus, a frequency of 10 cycles per second is referred to as 10 hertz or 10 1/s.

The period and frequency are two ways of describing the time involved in a vibration. Since the period (T) is the total time involved in one cycle and the frequency (f) is the number of cycles per second, the relationship is

$$T = \frac{1}{f} \qquad \text{equation 5.1}$$

or

$$f = \frac{1}{T} \qquad \text{equation 5.2}$$

EXAMPLE 5.1 *(Optional)*

A vibrating system has a period of 0.5 s. What is the frequency in Hz?

SOLUTION

$$T = 0.5 \text{ s} \qquad f = \frac{1}{T}$$
$$f = ?$$
$$= \frac{1}{0.5 \text{ s}}$$
$$= \frac{1}{0.5} \frac{1}{\text{s}}$$
$$= 2\frac{1}{\text{s}}$$
$$= \boxed{2 \text{ Hz}}$$

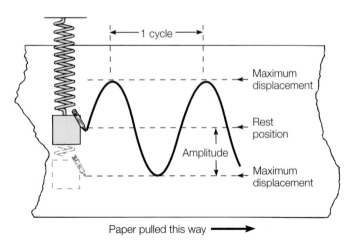

FIGURE 5.4 A graph of simple harmonic motion is described by a sinusoidal curve.

You can obtain a graph of a vibrating object, which makes it easier to measure the amplitude, period, and frequency. If a pen is fixed to a vibrating mass and a paper is moved beneath it at a steady rate, it will draw a curve, as shown in figure 5.4. The greater the amplitude of the vibrating mass, the greater the height of this curve. The greater the frequency, the closer together the peaks and valleys. Note the shape of this curve. This shape is characteristic of simple harmonic motion and is called a *sinusoidal*, or sine, graph. It is so named because it is the same shape as a graph of the sine function in trigonometry.

WAVES

A vibration can be a repeating, or *periodic*, type of motion that disturbs the surroundings. A *pulse* is a disturbance of a single event of short duration. Both pulses and periodic vibrations can create a physical *wave* in the surroundings. A wave is a disturbance that moves through a medium such as a solid or the air. A heavy object dropped on the floor, for example, makes a pulse that sends a mechanical wave that you feel. It might also make a sound wave in the air that you hear. In either case, the medium that transported a wave (solid floor or air) returns to its normal state after the wave has passed. The medium does not travel from place to place, the wave does. Two major considerations about a wave are that (1) a wave is a traveling disturbance and (2) a wave transports energy.

You can observe waves when you drop a rock into a still pool of water. The rock pushes the water into a circular mound as it enters the water. Since it is forcing the water through a distance, it is doing work to make the mound. The mound starts to move out in all directions, in a circle, leaving a depression behind. Water moves into the depression, and a circular wave—mound and depression—moves from the place of disturbance outward (figure 5.5). Any floating object in the path of the wave, such as a leaf, exhibits an up-and-down motion as the mound and

FIGURE 5.5 A water wave moves across the surface. How do you know for sure that it is energy and not water that is moving across the surface?

depression of the wave pass. But the leaf merely bobs up and down and after the wave has passed, it is much in the same place as before the wave. Thus, it was the disturbance that traveled across the water, not the water itself. If the wave reaches a leaf floating near the edge of the water, it may push the leaf up and out of the water, doing work on the leaf. Thus, the wave is a moving disturbance that transfers energy from one place to another.

Kinds of Waves

If you could see the motion of an individual water molecule near the surface as a water wave passed, you would see it trace out a circular path as it moves up and over, down and back. This circular motion is characteristic of the motion of a particle reacting to a water wave disturbance. There are other kinds of waves, and each involves particles in a characteristic motion.

A **longitudinal wave** is a disturbance that causes particles to move closer together or farther apart in the same direction that the wave is moving. If you attach one end of a coiled spring to a wall and pull it tight, you will make longitudinal waves in the spring if you grasp the spring and then move your hand back and forth parallel to the spring. Each time you move your hand toward the length of the spring, a pulse of closer-together coils will move across the spring (figure 5.6A). Each time you pull your hand back, a pulse of farther-apart coils will move across the spring. The coils move back and forth in the same direction that the wave is moving, which is the characteristic movement in reaction to a longitudinal wave.

You will make a different kind of wave in the stretched spring if you now move your hand up and down perpendicular to the length of the spring. This creates a **transverse wave.** A transverse wave is a disturbance that causes motion perpendicular to the direction that the wave is moving. Particles

responding to a transverse wave do not move closer together or farther apart in response to the disturbance; rather, they vibrate back and forth or up and down in a direction perpendicular to the direction of the wave motion (see figure 5.6B).

Whether you make mechanical longitudinal or transverse waves depends not only on the nature of the disturbance creating the waves but also on the nature of the medium. Mechanical transverse waves can move through a material only if there is some interaction, or attachment, between the molecules making up the medium. In a gas, for example, the molecules move about freely without attachments to one another. A pulse can cause these molecules to move closer together or farther apart, so a gas can carry a longitudinal wave. But if a gas molecule is caused to move up and then down, there is no reason for other molecules to do the same, since they are not attached. Thus, a gas will carry mechanical longitudinal waves but not mechanical transverse waves. Likewise a liquid will carry mechanical longitudinal waves but not mechanical transverse waves since the liquid molecules simply slide past one another. The surface of a liquid, however, is another story because of surface tension. A surface water wave is, in fact, a combination of longitudinal and transverse wave patterns that produce the circular motion of a disturbed particle. Solids can and do carry both mechanical longitudinal and transverse waves because of the strong attachments between the molecules.

Waves in Air

Waves that move through the air are longitudinal, so sound waves must be longitudinal waves. A familiar situation will be used to describe the nature of a longitudinal wave moving through air before considering sound specifically. The situation involves a small room with no open windows and two doors that open into the room. When you open one door into the room, the other door closes. Why does this happen? According to the kinetic molecular theory, the room contains many tiny, randomly moving gas molecules that make up the air. As you opened the door, it pushed on these gas molecules, creating a jammed-together zone of molecules immediately adjacent to the door. This jammed-together zone of air now has a greater density and pressure, which immediately spreads outward from

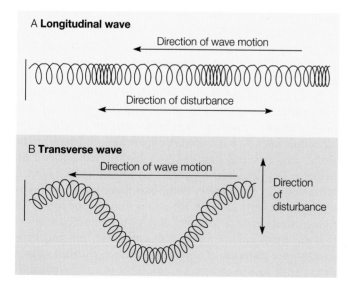

FIGURE 5.6 (A) Longitudinal waves are created in a spring when the free end is moved back and forth parallel to the spring. (B) Transverse waves are created in a spring when the free end is moved up and down.

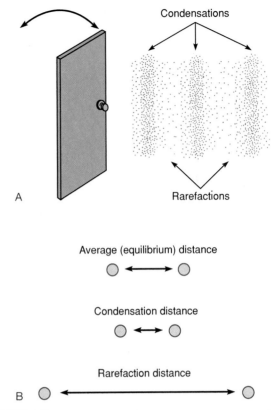

FIGURE 5.7 (A) Swinging the door inward produces pulses of increased density and pressure called *condensations*. Pulling the door outward produces pulses of decreased density and pressure called *rarefactions*. (B) In a condensation, the average distance between gas molecules is momentarily decreased as the pulse passes. In a rarefaction, the average distance is momentarily increased.

the door as a pulse. The disturbance is rapidly passed from molecule to molecule, and the pulse of compression spreads through the room. The pulse of greater density and increased pressure of air reached the door at the other side of the room, and the composite effect of the molecules impacting the door, that is, the increased pressure, caused it to close.

If the door at the other side of the room does not latch, you can probably cause it to open again by pulling on the first door quickly. By so doing, you send a pulse of thinned-out molecules of lowered density and pressure. The door you pulled quickly pushed some of the molecules out of the room. Other molecules quickly move into the region of less pressure, then back to their normal positions. The overall effect is the movement of a thinned-out pulse that travels through the room. When the pulse of slightly reduced pressure reaches the other door, molecules exerting their normal pressure on the other side of the door cause it to move. After a pulse has passed a particular place, the molecules are very soon homogeneously distributed again due to their rapid, random movement.

If you were to swing a door back and forth, it would be a vibrating object. As it vibrates back and forth, it would have a certain frequency in terms of the number of vibrations per second. As the vibrating door moves toward the room, it creates a pulse of jammed-together molecules called a **condensation** (or compression) that quickly moves throughout the room. As the vibrating door moves away from the room, a pulse of thinned-out molecules called a **rarefaction** quickly moves throughout the room. The vibrating door sends repeating pulses of condensation (increased density and pressure) and rarefaction (decreased density and pressure) through the room as it moves back and forth (figure 5.7). You know that the pulses transmit energy because they produce movement, or do work on, the other door. Individual molecules execute a harmonic motion about their equilibrium position and can do work on a movable object. Energy is thus transferred by this example of longitudinal waves.

Hearing Waves in Air

You cannot hear a vibrating door because the human ear normally hears sounds originating from vibrating objects with a frequency between 20 and 20,000 Hz. Longitudinal waves with frequencies less than 20 Hz are called **infrasonic.** You usually *feel* sounds below 20 Hz rather than hear them, particularly if you are listening to a good sound system. Longitudinal waves above 20,000 Hz are called **ultrasonic.** Although 20,000 Hz is usually considered the upper limit of hearing, the actual limit varies from person to person and becomes lower and lower with increasing age. Humans do not hear infrasonic nor ultrasonic sounds, but various animals have different limits. Dogs, cats, rats, and bats can hear higher frequencies than humans. Dogs can hear an ultrasonic whistle when a human hears nothing, for example. Some bats make and hear sounds of frequencies up to 100,000 Hz as they navigate and search for flying insects in total darkness. Scientists discovered recently that elephants communicate with extremely low-frequency sounds over distances of several kilometers. Humans cannot detect such low-frequency sounds. This raises the possibility of infrasonic waves that other animals can detect that we cannot.

Ultrasonic waves are mechanical waves that have frequencies above the normal limit of hearing of the human ear. The arbitrary upper limit is about 20,000 Hz, so an ultrasonic wave has a frequency of 20,000 Hz or greater. Intense ultrasonic waves are used in many ways in industry and medicine.

Industrial and commercial applications of ultrasound utilize lower-frequency ultrasonic waves in the 20,000 to 60,000 Hz range. Commercial devices that send ultrasound through the air include burglar alarms and rodent repellers. An ultrasonic burglar alarm sends ultrasonic spherical waves through the air of a room. The device is adjusted to ignore echoes from the contents of the room. The presence of a person provides a new source of echoes, which activates the alarm. Rodents emit ultrasonic frequencies up to 150,000 Hz that are used in rodent communication when they are disturbed or during aggressive behavior. The ultrasonic rodent repeller generates ultrasonic waves of similar frequency. Other commercial applications of ultrasound include sonar and depth measurements, cleaning and drilling, welding plastics and metals, and material flaw detection. Ultrasonic cleaning baths are used to remove dirt and foreign matter from solid surfaces, usually within a liquid solvent. The

ultrasonic waves create vapor bubbles in the liquid, which vibrate and emit audible and ultrasonic sound waves. The audible frequencies are often heard as a hissing or frying sound.

Medical applications of ultrasound use frequencies in the 1,000,000 to 20,000,000 Hz range. Ultrasound in this frequency range cannot move through the air because the required displacement amplitudes of the gas molecules in the air are less than the average distance between the molecules. Thus, a gas molecule that is set into motion in this frequency range cannot collide with other gas molecules to transmit the energy of the wave. Intense ultrasound is used for cleaning teeth and disrupting kidney stones. Less intense ultrasound is used for therapy (heating and reduction of pain). The least intense ultrasounds are used for diagnostic imaging. The largest source of exposure of humans to ultrasound is for the purpose of ultrasonic diagnostic imaging, particularly in fertility and pregnancy cases (box figure 5.1).

Originally developed in the late 1950s, ultrasonic medical machines have been improved and today make images of outstanding detail and clarity. One type of ultrasonic medical machine uses a transducer probe, which emits an ultrasonic

pulse that passes into the body. Echoes from the internal tissues and organs are reflected back to the transducer, which sends the signals to a computer. Another pulse of ultrasound is then sent out after the echoes from the first pulse have returned. The strength of and number of pulses per second vary with the application, ranging from hundreds to thousands of pulses per second. The computer constructs a picture from the returning echoes, showing an internal view without the use of more dangerous X rays.

Ultrasonic scanners have been refined to the point that the surface of the ovaries can now be viewed, showing the number and placement of developing eggs. The ovary scan is typically used in conjunction with fertility-stimulating drugs, where multiple births are possible, and to identify the exact time of ovulation. As early as four weeks after conception, the ultrasonic scanner is used to identify and monitor the fetus. By the thirteenth week, an ultrasonic scan can show the fetal heart movement, bone and skull size, and internal organs. The ultrasonic scanner is often used to show the position of the fetus and placenta for the purpose of amniocentesis, which involves withdrawing a sample of amniotic fluid from the uterus for testing.

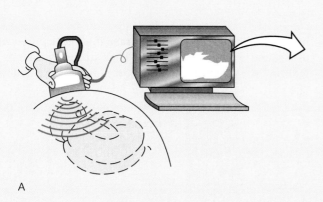

A

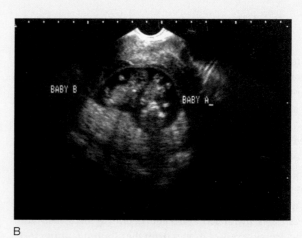

B

Box Figure 5.1 (A) Physicians can "see" a baby inside its mother's body by using the handheld external probe of an ultrasonic imaging system. (B) An ultrasonic image of ten-week-old human twins inside the mother's body.

The Human Ear

Box figure 5.2 shows the anatomy of the ear. The sound that arrives at the ear is first funneled by the external ear to the tympanum, also known as the eardrum. The cone-shaped nature of the external ear focuses sound on the tympanum and causes it to vibrate at the same frequency as the sound waves reaching it. Attached to the tympanum are three tiny bones known as the malleus (hammer), incus (anvil), and stapes (stirrup). The malleus is attached to the tympanum, the incus is attached to the malleus and stapes, and the stapes is attached to a small, membrane-covered opening called the oval window in a snail-shaped structure known as the cochlea. The vibration of the tympanum causes the tiny bones (malleus, incus, and stapes) to vibrate, and they, in turn, cause a corresponding vibration in the membrane of the oval window.

The cochlea of the ear is the structure that detects sound, and it consists of a snail-shaped set of fluid-filled tubes. When the oval window vibrates, the fluid in the cochlea begins to move, causing a membrane in the cochlea, called the basilar membrane, to vibrate. High-pitched, short-wavelength sounds cause the basilar membrane to vibrate at the base of the cochlea near the oval window. Low-pitched, long-wavelength sounds vibrate the basilar membrane far from the oval window. Loud sounds cause the basilar membrane to vibrate more vigorously than do faint sounds. Cells on this membrane depolarize when they are stimulated by its vibrations. Since they synapse with neurons, messages can be sent to the brain.

Because sounds of different wavelengths stimulate different portions of the cochlea, the brain is able to determine the pitch of a sound. Most sounds consist of a mixture of pitches that are heard. Louder sounds stimulate the membrane more forcefully, causing the sensory cells in the cochlea to send more nerve impulses per second. Thus, the brain is able to perceive the loudness of various sounds, as well as the pitch.

Associated with the cochlea are two fluid-filled chambers and a set of fluid-filled tubes called the semicircular canals. These structures are not involved in hearing but

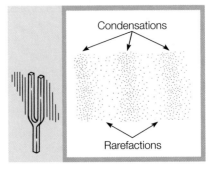

Box Figure 5.2 A schematic sketch of the human ear.

are involved in maintaining balance and posture. In the walls of these canals and chambers are cells similar to those found on the basilar membrane. These cells are stimulated by movements of the head and by the position of the head with respect to the force of gravity. The constantly changing position of the head results in sensory input that is important in maintaining balance.

A tuning fork that vibrates at 260 Hz makes longitudinal waves much like the swinging door, but these longitudinal waves are called *sound waves* because they are within the frequency range of human hearing. The prongs of a struck tuning fork vibrate, moving back and forth. This is more readily observed if the prongs of the fork are struck, then held against a sheet of paper or plunged into a beaker of water. In air, the vibrating prongs first move toward you, pushing the air molecules into a condensation of increased density and pressure. As the prongs then move back, a rarefaction of decreased density and pressure is produced. The alternation of increased and decreased pressure pulses moves from the vibrating tuning fork and spreads outward equally in all directions, much like the surface of a rapidly expanding balloon (figure 5.8). Your eardrum is forced in and out by the pulses it receives. It now vibrates with the same frequency as the tuning fork. The vibrations of the eardrum are transferred by three tiny bones to a fluid in a coiled chamber. Here, tiny hairs respond to the frequency and size of the disturbance, activating nerves that transmit the information to the brain. The brain interprets a frequency as a sound with a certain **pitch.** High-frequency sounds are interpreted as high-pitched musical notes, for example, and

FIGURE 5.8 A vibrating tuning fork produces a series of condensations and rarefactions that move away from the tuning fork. The pulses of increased and decreased pressure reach your ear, vibrating the eardrum. The ear sends nerve signals to the brain about the vibrations, and the brain interprets the signals as sounds.

low-frequency sounds are interpreted as low-pitched musical notes. The brain then selects certain sounds from all you hear, and you "tune" to certain ones, enabling you to listen to whatever sounds you want while ignoring the background noise, which is made up of all the other sounds.

DESCRIBING WAVES

A tuning fork vibrates with a certain frequency and amplitude, producing a longitudinal wave of alternating pulses of increased-pressure condensations and reduced-pressure rarefactions. A graph of the frequency and amplitude of the vibrations is shown in figure 5.9A, and a representation of the condensations and rarefactions is shown in figure 5.9B. The wave pattern can also be represented by a graph of the changing air pressure of the traveling sound wave, as shown in figure 5.9C. This graph can be used to define some interesting concepts associated with sound waves. Note the correspondence between the (1) amplitude, or displacement, of the vibrating prong, (2) the pulses of condensations and rarefactions, and (3) the changing air pressure. Note also the correspondence between the frequency of the vibrating prong and the frequency of the wave cycles.

Figure 5.10 shows the terms commonly associated with waves from a continuously vibrating source. The wave *crest* is the maximum disturbance from the undisturbed (rest) position. For a sound wave, this would represent the maximum increase of air pressure. The wave *trough* is the maximum disturbance in the opposite direction from the rest position. For a sound wave,

this would represent the maximum decrease of air pressure. The *amplitude* of a wave is the displacement from rest to the crest *or* from rest to the trough. The time required for a wave to repeat itself is the *period* (T). To repeat itself means the time required to move through one full wave, such as from the crest of one wave to the crest of the next wave. This length in which the wave repeats itself is called the **wavelength** (the symbol is λ, which is the Greek letter lambda). Wavelength is measured in centimeters or meters just like any other length.

There is a relationship between the wavelength, period, and speed of a wave. The relationship is

$$v = \lambda f \qquad \textbf{equation 5.3}$$

which we will call the *wave equation*. This equation tells you that the velocity of a wave can be obtained from the product of the wavelength and the frequency. Note that it also tells you that the wavelength and frequency are inversely proportional at a given velocity.

EXAMPLE 5.2 *(Optional)*

A sound wave with a frequency of 260 Hz has a wavelength of 1.27 m. With what speed would you expect this sound wave to move?

SOLUTION

$$f = 260 \text{ Hz} \qquad v = \lambda f$$
$$\lambda = 1.27 \text{ m}$$
$$v = ? \qquad\qquad = (1.27 \text{ m})\left(260 \frac{1}{s}\right)$$
$$= (1.27)(260) \text{ m} \times \frac{1}{s}$$
$$= \boxed{330 \frac{m}{s}}$$

EXAMPLE 5.3 *(Optional)*

In general, the human ear is most sensitive to sounds at 2,500 Hz. Assuming that sound moves at 330 m/s, what is the wavelength of sounds to which people are most sensitive? (Answer: 13 cm)

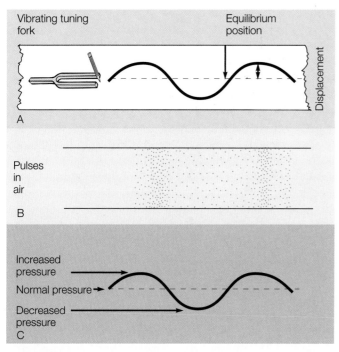

FIGURE 5.9 Compare the (A) back-and-forth vibrations of a tuning fork with (B) the resulting condensations and rarefactions that move through the air and (C) the resulting increases and decreases of air pressure on a surface that intercepts the condensations and rarefactions.

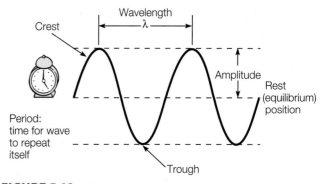

FIGURE 5.10 Here are some terms associated with periodic waves. The *wavelength* is the distance from a part of one wave to the same part in the next wave, such as from one crest to the next. The *amplitude* is the displacement from the rest position. The *period* is the time required for a wave to repeat itself, that is, the time for one complete wavelength to move past a given location.

SOUND WAVES

The transmission of a sound wave requires a medium, that is, a solid, liquid, or gas to carry the disturbance. Therefore, sound does not travel through the vacuum of outer space, since there is nothing to carry the vibrations from a source. The nature of the molecules making up a solid, liquid, or gas determines how well or how rapidly the substance will carry sound waves. The two variables are (1) the inertia of the molecules and (2) the strength of the interaction, if the molecules are attached to one another. Thus, hydrogen gas, with the least massive molecules with no interaction or attachments, will carry a sound wave at 1,284 m/s (4,213 ft/s) when the temperature is 0°C. More massive helium gas molecules have more inertia and carry a sound wave at only 965 m/s (3,166 ft/s) at the same temperature. A solid, however, has molecules that are strongly attached so vibrations are passed rapidly from molecule to molecule. Steel, for example, is highly elastic, and sound will move through a steel rail at 5,940 m/s (19,488 ft/s). Thus, there is a reason for the old saying, "Keep your ear to the ground," because sounds move through solids more rapidly than through a gas (table 5.1).

Velocity of Sound in Air

Most people have observed that sound takes some period of time to move through the air. If you watch a person hammering on a roof a block away, the sounds of the hammering are not in sync with what you see. Light travels so rapidly that you can consider what you see to be simultaneous with what is actually happening for all practical purposes. Sound, however, travels much more slowly and the sounds arrive late in comparison to what you are seeing. This is dramatically illustrated by seeing a flash of lightning, then hearing thunder seconds later. Perhaps you know of a way to estimate the distance to a lightning flash by timing the interval between the flash and boom.

The air temperature influences how rapidly sound moves through the air. The gas molecules in warmer air have a greater kinetic energy than those of cooler air. The molecules of warmer air therefore transmit an impulse from molecule to molecule more rapidly. More precisely, the speed of a sound wave increases 0.60 m/s (2.0 ft/s) for *each* Celsius degree increase in temperature. In *dry* air at sea-level density (normal pressure) and 0°C (32°F), the velocity of sound is about 331 m/s (1,087 ft/s).

Refraction and Reflection

When you drop a rock into a still pool of water, circular patterns of waves move out from the disturbance. These water waves are on a flat, two-dimensional surface. Sound waves, however, move in three-dimensional space like a rapidly expanding balloon. Sound waves are *spherical waves* that move outward from the source. Spherical waves of sound move as condensations and rarefactions from a continuously vibrating source at the center. If you identify the same part of each wave in the spherical waves, you have identified a *wave front*. For example, the crests of each condensation could be considered as a wave front. From one wave front to the next, therefore, identifies one complete wave or wavelength. At some distance from the source, a small part of a spherical wave front can be considered a *linear wave front* (figure 5.11).

Waves move within a homogeneous medium such as a gas or a solid at a fairly constant rate but gradually lose energy to friction. When a wave encounters a different condition (temperature, humidity, or nature of material), however, drastic changes may occur rapidly. The division between two physical conditions is called a *boundary*. Boundaries are usually encountered (1) between different materials or (2) between the

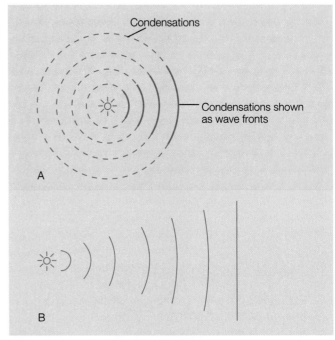

FIGURE 5.11 (*A*) Spherical waves move outward from a sounding source much as a rapidly expanding balloon. This two-dimensional sketch shows the repeating condensations as spherical wave fronts. (*B*) Some distance from the source, a spherical wave front is considered a linear, or plane, wave front.

TABLE 5.1   Speed of Sound in Various Materials

Medium	m/s	ft/s
Carbon dioxide (0°C)	259	850
Dry air (0°C)	331	1,087
Helium (0°C)	965	3,166
Hydrogen (0°C)	1,284	4,213
Water (25°C)	1,497	4,911
Seawater (25°C)	1,530	5,023
Lead	1,960	6,430
Glass	5,100	16,732
Steel	5,940	19,488

same materials with different conditions. An example of a wave moving between different materials is a sound made in the next room that moves through the air to the wall and through the wall to the air in the room where you are. The boundaries are air-wall and wall-air. If you have ever been in a room with "thin walls," it is obvious that sound moved through the wall and air boundaries.

An example of sound waves moving through the same material with different conditions is found when a wave front moves through air of different temperatures. Since sound travels faster in warm air than in cold air, the wave front becomes bent. The bending of a wave front between boundaries is called *refraction*. Refraction changes the direction of travel of a wave front. Consider, for example, that on calm, clear nights, the air near Earth's surface is cooler than air farther above the surface. Air at rooftop height above the surface might be four or five degrees warmer under such ideal conditions. Sound will travel faster in the higher, warmer air than it will in the lower, cooler air close to the surface. A wave front will therefore become bent, or refracted, toward the ground on a cool night and you will be able to hear sounds from farther away than on warm nights (figure 5.12A). The opposite process occurs during the day as Earth's surface becomes warmer from sunlight (figure 5.12B). Wave fronts are refracted upward because part of the wave front travels faster in the warmer air near the surface. Thus, sound does not seem to carry as far in the summer as it does in the winter. What is actually happening is that during the summer, the wave fronts are refracted away from the ground before they travel very far.

When a wave front strikes a boundary that is parallel to the front the wave may be absorbed, transmitted, or undergo *reflection,* depending on the nature of the boundary medium, or the wave may be partly absorbed, partly transmitted, partly reflected, or any combination thereof. Some materials, such as hard, smooth surfaces, reflect sound waves more than they absorb them. Other materials, such as soft, ruffly curtains, absorb sound waves more than they reflect them. If you have ever been in a room with smooth, hard walls and with no curtains, carpets, or furniture, you know that sound waves may be reflected several times before they are finally absorbed.

Do you sing in the shower? Many people do because the tone is more pleasing than singing elsewhere. The walls of a shower are usually hard and smooth, reflecting sounds back and forth several times before they are absorbed. The continuation of many reflections causes a tone to gain in volume. Such mixing of reflected sounds with the original is called **reverberation.** Reverberation adds to the volume of a tone, and it is one of the factors that determines the acoustical qualities of a room, lecture hall, or auditorium. An open-air concert sounds flat without the reverberation of an auditorium and is usually enhanced electronically to make up for the lack of reflected sounds. Too much reverberation in a room or classroom is not good since the spoken word is not as sharp. Sound-absorbing materials are therefore used on the walls and floors where clear, distinct speech is important (figure 5.13). The carpet and drapes you see in a movie theater are not decorator items but are there to absorb sounds.

If a reflected sound arrives after 0.10 s, the human ear can distinguish the reflected sound from the original sound. A reflected sound that can be distinguished from the original is called an **echo.** Thus, a reflected sound that arrives before 0.10 s is perceived as an increase in volume and is called a reverberation, but a sound that arrives after 0.10 s is perceived as an echo.

Sound wave echoes are measured to determine the depth of water or to locate underwater objects by a *sonar* device. The word *sonar* is taken from *so*und *na*vigation *r*anging. The device generates an underwater ultrasonic sound pulse, then measures the elapsed time for the returning echo. Sound waves travel at about 1,531 m/s (5,023 ft/s) in seawater at 25°C (77°F). A 1 s lapse between the ping of the generated sound and the echo return would mean that the sound traveled 5,023 ft for the round trip. The bottom would be half this distance below the surface (figure 5.14).

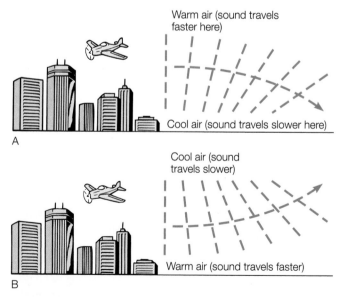

Warm air (sound travels faster here)

Cool air (sound travels slower here)

A

Cool air (sound travels slower)

Warm air (sound travels faster)

B

FIGURE 5.12 (A) Since sound travels faster in warmer air, a wave front becomes bent, or refracted, toward Earth's surface when the air is cooler near the surface. (B) When the air is warmer near the surface, a wave front is refracted upward, away from the surface.

FIGURE 5.13 This closed-circuit TV control room is acoustically treated by covering the walls with sound-absorbing baffles.

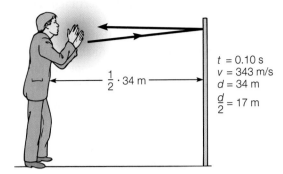

A Echo

$t = 0.10$ s
$v = 343$ m/s
$d = 34$ m
$\frac{d}{2} = 17$ m

$\frac{1}{2} \cdot 34$ m

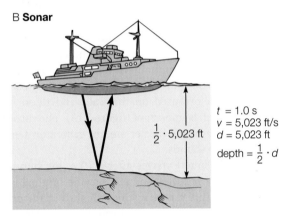

B Sonar

$t = 1.0$ s
$v = 5{,}023$ ft/s
$d = 5{,}023$ ft

depth $= \frac{1}{2} \cdot d$

$\frac{1}{2} \cdot 5{,}023$ ft

FIGURE 5.14 (*A*) At room temperature, sound travels at 343 m/s. In 0.10 s, sound would travel 34 m. Since the sound must travel to a surface and back in order for you to hear an echo, the distance to the surface is one-half the total distance. (*B*) Sonar measures a depth by measuring the elapsed time between an ultrasonic sound pulse and the echo. The depth is one-half the round trip.

Myths, Mistakes, and Misunderstandings

A Duck's Quack Doesn't Echo?

You may have heard the popular myth that "a duck's quack doesn't echo, and no one knows why." An acoustic research experiment was carried out at the University of Salford in Greater Manchester, England, to test this myth. Acoustic experts first recorded a quacking duck in a special chamber that was constructed to produce no sound reflections. Simulations were then done in a reverberation chamber to match the effect of the duck quacking when flying past a cliff face. The tests found that a duck's quack indeed does echo, just like any other sound.

The quack researchers speculated that the myth might have resulted because:

1. The quack does echo, but it is usually too quiet to hear. A duck quacks too quietly, so the echo is too quiet to hear.
2. Ducks quack near water, not near reflecting surfaces such as a mountain or building that would make an echo.
3. It is hard to hear the echo of a sound that fades in and out, as a duck quack does.

You can read the original research report and hear recordings of the tests at the University of Salford website, www.acoustics.salford.ac.uk.

EXAMPLE 5.4 (*Optional*)

The human ear can distinguish a reflected sound pulse from the original sound pulse if 0.10 s or more elapses between the two sounds. What is the minimum distance to a reflecting surface from which we can hear an echo (see figure 5.14A) if the speed of sound is 343 m/s?

SOLUTION

$t = 0.10$ s (minimum)
$v = 343$ m/s
$d = ?$

$$v = \frac{d}{t} \therefore d = vt$$

$$= \left(343 \frac{m}{s}\right)(0.10 \text{ s})$$

$$= (343)(0.10)\frac{m}{s} \times s$$

$$= 34.4 \frac{m \cdot s}{s}$$

$$= 34 \text{ m}$$

Since the sound pulse must travel from the source to the reflecting surface, then back to the source,

$$34 \text{ m} \times 1/2 = \boxed{17 \text{ m}}$$

The minimum distance to a reflecting surface from which we hear an echo when the air is at room temperature is therefore 17 m (about 56 ft).

EXAMPLE 5.5 (*Optional*)

An echo is heard exactly 1.00 s after a sound when the speed of sound is 1,147 ft/s. How many feet away is the reflecting surface? (Answer: 574 ft)

Interference

Waves interact with a boundary much as a particle would, reflecting or refracting because of the boundary. A moving ball, for example, will bounce from a surface at the same angle it strikes the surface, just as a wave does. A particle or a ball, however, can be in only one place at a time, but waves can be spread over a distance at the same time. You know this since many different people in different places can hear the same sound at the same time.

Another difference between waves and particles is that two or more waves can exist in the same place at the same time. When two patterns of waves meet, they pass through each other without refracting or reflecting. However, at the place where they meet, the waves interfere with each other, producing a *new* disturbance. This new disturbance has a different amplitude, which is the algebraic sum of the amplitudes of the two separate wave patterns. If the wave crests or wave troughs arrive at the same place at the same time, the two waves are said to be *in phase*. The result of two waves arriving in phase is a new disturbance with a crest and trough that has greater displacement than either of the two separate waves. This is called *constructive interference* (figure 5.15A). If the trough of one wave arrives at the same place and time as the crest of another wave, the waves are completely *out of phase*. When two waves are completely out of phase, the crest of one wave (positive displacement) will cancel the trough of the other wave (negative displacement), and the result is zero total disturbance, or no wave. This is called *destructive interference* (figure 5.15B). If the two sets of wave patterns do not have the exact same amplitudes or wavelengths, they will be neither completely in phase nor completely out of phase. The result will be partly constructive or destructive interference, depending on the exact nature of the two wave patterns.

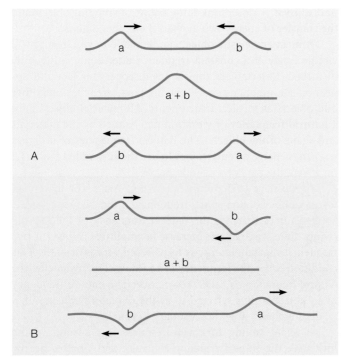

FIGURE 5.15 (*A*) Constructive interference occurs when two equal, in-phase waves meet. (*B*) Destructive interference occurs when two equal, out-of-phase waves meet. In both cases, the wave displacements are superimposed when they meet, but they then pass through one another and return to their original amplitudes.

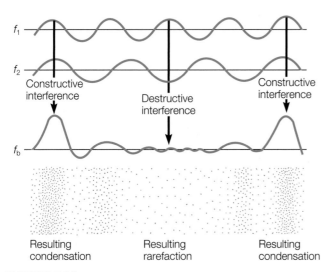

FIGURE 5.16 Two waves of equal amplitude but slightly different frequencies interfere destructively and constructively. The result is an alternation of loudness called a *beat*.

Suppose that two vibrating sources produce sounds that are in phase, equal in amplitude, and equal in frequency. The resulting sound will be increased in volume because of constructive interference. But suppose the two sources are slightly different in frequency, for example, 350 Hz and 352 Hz. You will hear a regularly spaced increase and decrease of sound known as **beats**. Beats occur because the two sound waves experience alternating constructive and destructive interferences (figure5.16). The phase relationship changes because of the difference in frequency, as you can see in the illustration. These alternating constructive and destructive interference zones are moving from the source to the receiver, and the receiver hears the results as a rapidly rising and falling sound level. The beat frequency is the difference between the frequencies of the two sources. A 352 Hz source and 350 Hz source sounded together would result in a beat frequency of 2 Hz. Thus, as two frequencies are closer and closer together, fewer beats will be heard per second. You may be familiar with the phenomenon of beats if you have ever flown in an airplane with two engines. If one engine is running slightly faster than the other, you hear a slow beat. The beat frequency (f_b) is equal to the absolute difference in frequency of two interfering waves with slightly different frequencies, or

$$f_b = f_2 - f_1 \qquad \text{equation 5.4}$$

ENERGY AND SOUND

All waves involve the transportation of energy, including sound waves. The vibrating mass and spring in figure 5.2 vibrate with an amplitude that depends on how much work you did on the mass in moving it from its equilibrium position. More work on the mass results in a greater displacement and a greater amplitude of vibration. A vibrating object that is producing sound waves will produce more intense condensations and rarefactions if it has a greater amplitude. The intensity of a sound wave is a measure of the energy the sound wave is carrying (figure 5.17). **Intensity** is defined as the power (in watts) transmitted by a wave to a unit area (in square meters) that is perpendicular to the waves.

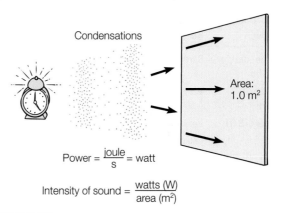

FIGURE 5.17 The intensity of a sound wave is the rate of energy transferred to an area perpendicular to the waves. Intensity is measured in watts per square meter, W/m².

Loudness

The *loudness* of a sound is a subjective interpretation that varies from person to person. Loudness is also related to (1) the energy of a vibrating object, (2) the condition of the air the sound wave travels through, and (3) the distance between you and the vibrating source. Furthermore, doubling the amplitude of the vibrating source will quadruple the *intensity* of the resulting sound wave, but the sound will not be perceived as four times as loud. The relationship between perceived loudness and the intensity of a sound wave is not a linear relationship. In fact, a sound that is perceived as twice as loud requires ten times the intensity, and quadrupling the loudness requires a one-hundredfold increase in intensity.

The human ear is very sensitive, capable of hearing sounds with intensities as low as 10^{-12} W/m^2, and is not made uncomfortable by sound until the intensity reaches about 1 W/m^2. The second intensity is a million million (10^{12}) times greater than the first. Within this range, the subjective interpretation of intensity seems to vary by powers of ten. This observation led to the development of the **decibel scale** to measure the intensity level. The scale is a ratio of the intensity level of a given sound to the threshold of hearing, which is defined as 10^{-12} W/m^2 at 1,000 Hz. In keeping with the power-of-ten subjective interpretations of intensity, a logarithmic scale is used rather than a linear scale. Originally, the scale was the logarithm of the ratio of the intensity level of a sound to the threshold of hearing. This definition set the zero point at the threshold of human hearing. The unit was named the *bel* in honor of Alexander Graham Bell. This unit was too large to be practical, so it was reduced by one-tenth and called a *decibel*. The intensity level of a sound is therefore measured in decibels (table 5.2). Compare the decibel noise level of familiar sounds listed in table 5.2, and note that each increase of 10 on the decibel scale is matched by a *multiple* of 10 on the intensity level. For example, moving from a decibel level of 10 to a decibel level of 20 requires *ten times* more intensity. Likewise, moving from a decibel level of 20 to 40 requires a one-hundredfold increase in the intensity level. As you can see, the decibel scale is not a simple linear scale.

TABLE 5.2 Comparison of Noise Levels in Decibels with Intensity

Example	Response	Decibels	Intensity W/m^2
Least needed for hearing	Barely perceived	0	1×10^{-12}
Calm day in woods	Very, very quiet	10	1×10^{-11}
Whisper (15 ft)	Very quiet	20	1×10^{-10}
Library	Quiet	40	1×10^{-8}
Talking	Easy to hear	65	3×10^{-6}
Heavy street traffic	Conversation difficult	70	1×10^{-5}
Pneumatic drill (50 ft)	Very loud	95	3×10^{-3}
Jet plane (200 ft)	Discomfort	120	1

Resonance

You know that sound waves transmit energy when you hear a thunderclap rattle the windows. In fact, the sharp sounds from an explosion have been known not only to rattle but also break windows. The source of the energy is obvious when thunderclaps or explosions are involved. But sometimes energy transfer occurs through sound waves when it is not clear what is happening. A truck drives down the street, for example, and one window rattles but the others do not. A singer shatters a crystal water glass by singing a single note, but other objects remain undisturbed. A closer look at the nature of vibrating objects and the transfer of energy will explain these phenomena.

Almost any elastic object can be made to vibrate and will vibrate freely at a constant frequency after being sufficiently disturbed. Entertainers sometimes discover this fact and appear on late-night talk shows playing saws, wrenches, and other odd objects as musical instruments. All material objects have a **natural frequency** of vibration determined by the materials and shape of the objects. The natural frequencies of different wrenches enable an entertainer to use the suspended tools as if they were the bars of a xylophone.

If you have ever pumped a swing, you know that small forces can be applied at any frequency. If the frequency of the applied forces matches the natural frequency of the moving swing, there is a dramatic increase in amplitude. When the two frequencies match, energy is transferred very efficiently. This condition, when the frequency of an external force matches the natural frequency, is called **resonance.** The natural frequency of an object is thus referred to as the *resonant frequency,* that is, the frequency at which resonance occurs.

A silent tuning fork will resonate if a second tuning fork with the same frequency is struck and vibrates nearby (figure 5.18). You will hear the previously silent tuning fork sounding if you stop the vibrations of the struck fork by touching it. The waves of condensations and rarefactions produced by the struck tuning fork produce a regular series of impulses that match the natural frequency of the silent tuning fork. This illustrates that at resonance, relatively little energy is required to start vibrations.

A truck causing vibrations as it is driven past a building may cause one window to rattle while others do not. Vibrations caused by the truck have matched the natural frequency

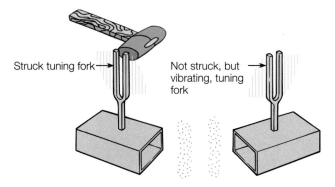

Struck tuning fork →

Not struck, but vibrating, tuning fork →

FIGURE 5.18 When the frequency of an applied force, including the force of a sound wave, matches the natural frequency of an object, energy is transferred very efficiently. The condition is called *resonance.*

Bells, jet planes, sirens, motorcycles, jack-hammers, and construction noises all contribute to the constant racket that has become commonplace in many areas. Such noise pollution is everywhere, and it is a rare place where you can escape the ongoing din. Earplugs help reduce the noise level, but they also block sounds that you want to hear, such as music and human voices. In addition, earplugs are more effective against high-frequency sounds than they are against lower ones such as aircraft engines or wind noise.

A better solution for noise pollution might be the relatively new "antinoise" technology that cancels sound waves before they reach your ears. A microphone detects background noise and transmits information to microprocessors about the noise waveform. The microprocessors then generate an "antinoise" signal that is 180 degrees out of phase with the background noise. When the noise and antinoise meet, they undergo destructive interference and significantly reduce the final sound loudness.

The noise-canceling technology is available today as a consumer portable electronic device with a set of headphones, looking much like a portable tape player. The headphones can be used to combat steady, ongoing background noise such as you might hear from a spinning computer drive, a vacuum cleaner, or while inside the cabin of a jet plane. Some vendors claim their device cancels up to 40 percent of whirring air conditioner noise, 80 percent of ongoing car noise, or up to 95 percent of constant airplane cabin noise. You hear all the other sounds—people talking, warning sounds, and music—as the microprocessors are not yet fast enough to match anything but a constant sound source.

Noise-canceling microphones can also be used to limit background noise that muddles the human voice in teleconferencing or during cellular phone use from a number of high-noise places. With improved microprocessor and new digital applications, noise-canceling technology will help bring higher sound quality to voice-driven applications. It may also find extended practical use in helping to turn down the volume on our noisy world.

A Singing Glass

Did you ever hear a glass "sing" when the rim is rubbed? The trick to making the glass sing is to remove as much oil from your finger as possible. Then you lightly rub around and on the top of the glass rim at the correct speed. Without oil, your finger will imperceptibly catch on the glass as you rub the rim. With the appropriate pressure and speed, your catching finger might match the natural frequency of the glass. The resonate vibration will cause the glass to "sing" with a high-pitched note.

Swinging Resonance

Set up two identical pendulums as shown in figure 5.19A. The bobs should be identical and suspended from identical strings of the same length attached to a tight horizontal string. Start one pendulum vibrating by pulling it back, then releasing it. Observe the vibrations and energy exchange between the two pendulums for the next several minutes. Now change the frequency of vibrations of *one* of the pendulums by shortening the string (figure 5.19B). Again start one pendulum vibrating and observe for several minutes. Compare what you observe when the frequencies are matched and when they are not. Explain what happens in terms of resonance.

of this window but not the others. The window is undergoing resonance from the sound wave impulses that matched its natural frequency. It is also resonance that enables a singer to break a water glass. If the tone is at the resonant frequency of the glass, the resulting vibrations may be large enough to shatter it.

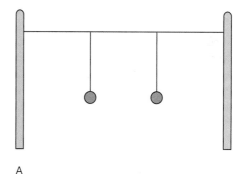

A

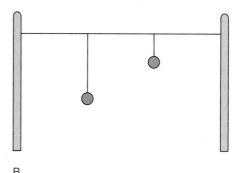

B

FIGURE 5.19 The "Concepts Applied" section demonstrates that one of these pendulum arrangements will show resonance and the other will not. Can you predict which one will show resonance?

Science and Society

Laser Bug

Hold a fully inflated balloon lightly between your fingertips and talk. You will be able to feel the slight vibrations from your voice. Likewise, the sound waves from your voice will cause a nearby window to vibrate slightly. If a laser beam is bounced off the window, the reflection will be changed by the vibrations. The incoming laser beam is coherent; all the light has the same frequency and amplitude (see p. 168). The reflected beam, however, will have different frequencies and amplitudes from the window pane vibrating in and out. The changes can be detected by a receiver and converted into sound in a headphone.

You cannot see an infrared laser beam because infrared is outside the frequencies that humans can see. Any sound-sensitive target can be used by the laser bug, including a windowpane, inflated balloon, hanging picture, or the glass front of a china cabinet.

Questions to Discuss

1. Is it legal for someone to listen in on your private conversations?
2. Should the sale of technology such as the laser bug be permitted? What are the issues?

Resonance considerations are important in engineering. A large water pump, for example, was designed for a nuclear power plant. Vibrations from the electric motor matched the resonant frequency of the impeller blades, and they shattered after a short period of time. The blades were redesigned to have a different natural frequency when the problem was discovered. Resonance vibrations are particularly important in the design of buildings.

SOURCES OF SOUNDS

All sounds have a vibrating object as their source. The vibrations of the object send pulses or waves of condensations and rarefactions through the air. These sound waves have physical properties that can be measured, such as frequency and intensity. Subjectively, your response to frequency is to identify a certain pitch. A high-frequency sound is interpreted as a high-pitched sound, and a low-frequency sound is interpreted as a low-pitched sound. Likewise, a greater intensity is interpreted as increased loudness, but there is not a direct relationship between intensity and loudness as there is between frequency and pitch.

There are other subjective interpretations about sounds. Some sounds are bothersome and irritating to some people but go unnoticed by others. In general, sounds made by brief, irregular vibrations such as those made by a slamming door, dropped book, or sliding chair are called *noise*. Noise is characterized by sound waves with mixed frequencies and jumbled intensities (figure 5.20). On the other hand, there are sounds made by very regular, repeating vibrations such as those made by a tuning fork. A tuning fork produces a *pure tone* with a sinusoidal curved

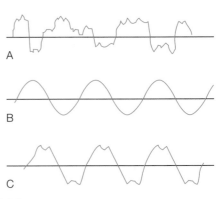

FIGURE 5.20 Different sounds that you hear include (*A*) noise, (*B*) pure tones, and (*C*) musical notes.

pressure variation and regular frequency. Yet a tuning fork produces a tone that most people interpret as bland. You would not call a tuning fork sound a musical note! Musical sounds from instruments have a certain frequency and loudness, as do noise and pure tones, but you can readily identify the source of the very same musical note made by two different instruments. You recognize it as a musical note, not noise and not a pure tone. You also recognize if the note was produced by a violin or a guitar. The difference is in the wave form of the sounds made by the two instruments, and the difference is called the *sound quality*. How does a musical instrument produce a sound of a characteristic quality? The answer may be found by looking at instruments that make use of vibrating strings.

Vibrating Strings

A stringed musical instrument, such as a guitar, has strings that are stretched between two fixed ends. When a string is plucked, waves of many different frequencies travel back and forth on the string, reflecting from the fixed ends. Many of these waves quickly fade away, but certain frequencies

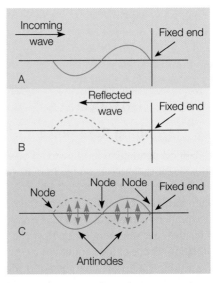

FIGURE 5.21 An incoming wave on a cord with a fixed end (*A*) meets a reflected wave (*B*) with the same amplitude and frequency, producing a standing wave (*C*). Note that a standing wave of one wavelength has three nodes and two antinodes.

Johann Christian Doppler (1803–1853)

Johann Doppler was an Austrian physicist who discovered the Doppler effect, which relates the observed frequency of a wave to the relative motion of the source and the observer. The Doppler effect is readily observed in moving sound sources, producing a fall in pitch as the source passes the observer, but it is of most use in astronomy, where it is used to estimate the velocities and distances of distant bodies.

Doppler was born in Salzburg, Austria, on November 29, 1803, the son of a stonemason. He showed early promise in mathematics and attended the Polytechnic Institute in Vienna from 1822 to 1825. He then returned to Salzburg and continued his studies privately while tutoring in physics and mathematics. From 1829 to 1833, Doppler went back to Vienna to work as a mathematical assistant and produced his first papers on mathematics and electricity. Despairing of ever obtaining an academic post, he decided in 1835 to emigrate to the United States. Then, on the point of departure, he was offered a professorship of mathematics at the State Secondary School in Prague and changed his mind. He subsequently obtained professorships in mathematics at the State Technical Academy in Prague in 1841 and at the Mining Academy in Schemnitz in 1847. Doppler returned to Vienna the following year and, in 1850, became director of the new Physical Institute and professor of experimental physics at the Royal Imperial University of Vienna. He died from a lung disease in Venice on March 17, 1853.

Doppler explained the effect that bears his name by pointing out that sound waves from a source moving toward an observer will reach the observer at a greater frequency than if the source is stationary, thus increasing the observed frequency and raising the pitch of the sound. Similarly, sound waves from a source moving

away from the observer reach the observer more slowly, resulting in a decreased frequency and a lowering of pitch. In 1842, Doppler put forward this explanation and derived the observed frequency mathematically in Doppler's principle.

The first experimental test of Doppler's principle was made in 1845 at Utrecht in Holland. A locomotive was used to carry a group of trumpeters in an open carriage to and fro past some musicians able to sense the pitch of the notes being played. The variation of pitch produced by the motion of the trumpeters verified Doppler's equations.

Doppler correctly suggested that his principle would apply to any wave motion and cited light as an example as well as sound. He believed that all stars emit white light and that differences in color are observed on Earth because the motion of stars affects the observed frequency of the light and hence its color. This idea was not universally true, as stars vary in their basic color. However, Armand Fizeau (1819–1896) pointed out in 1848 that shifts in the spectral lines of stars could be observed and ascribed to the Doppler effect and hence enable their motion to be determined. This idea was first applied in 1868 by William Huggins (1824–1910), who found that Sirius is moving away from the solar system by detecting a small redshift in its spectrum. With the linking of the velocity of a galaxy to its distance by Edwin Hubble (1889–1953) in 1929, it became possible to use the redshift to determine the distances of galaxies. Thus, the principle that Doppler discovered to explain an everyday and inconsequential effect in sound turned out to be of truly cosmological importance.

Source: Modified from the *Hutchinson Dictionary of Scientific Biography.* © RM, 2007. Reprinted by permission.

---

resonate, setting up patterns of waves. Before considering these resonant patterns in detail, keep in mind that (1) two or more waves can be in the same place at the same time, traveling through one another from opposite directions; (2) a confined wave will be reflected at a boundary, and the reflected wave will be inverted (a crest becomes a trough); and (3) reflected waves interfere with incoming waves of the same frequency to produce **standing waves.** Figure 5.21 is a graphic "snapshot" of what happens when reflected wave patterns meet incoming wave patterns. The incoming wave is shown as a solid line, and the reflected wave is shown as a dotted line. The result is (1) places of destructive interference, called *nodes,* which show no disturbance, and (2) loops of constructive interference, called *antinodes,* which take place where the crests and troughs of the two wave patterns produce a disturbance that rapidly alternates upward and downward. This pattern of alternating nodes and antinodes does not move along the string and is thus called a *standing wave.* Note that the standing wave for *one wavelength* will have a node at both ends and in the center,

as well as two antinodes. Standing waves occur at the natural, or resonant, frequencies of the string, which are a consequence of the nature of the string, the string length, and the tension in the string. Since the standing waves are resonant vibrations, they continue as all other waves quickly fade away.

Since the two ends of the string are not free to move, the ends of the string will have nodes. The *longest* wave that can make a standing wave on such a string has a wavelength (λ) that is twice the length (L) of the string. Since frequency (f) is inversely proportional to wavelength ($f = v/\lambda$ from equation 5.3), this longest wavelength has the lowest frequency possible, called the **fundamental frequency.** The fundamental frequency has one antinode, which means that the length of the string has one-half a wavelength. The fundamental frequency (f_1) determines the pitch of the *basic* musical note being sounded and is called the first harmonic. Other resonant frequencies occur at the same time, however, since other standing waves can also fit onto the string. A higher frequency of vibration (f_2) could fit two half-wavelengths between the two fixed nodes. An even

Connections . . .

The Redshift

The Doppler effect occurs for all waves, including light as well as sound. A change in the light occurs anytime there is relative motion between a light source and an observer. If the relative motion is one of moving away from each other, then the observer will notice that the frequency has decreased (and the wavelength has increased). Since the light has moved toward the red end of the spectrum, this is called a *redshift*. Thus, if you see a redshift, you know the light source is moving away from you.

Suppose the relative motion is one of the source and observer moving toward each other. The observer will this time notice that the frequency has increased (and the wavelength has decreased). The light has shifted toward the blue end of the spectrum, and this is called a *blueshift*. If you see a blueshift, you know the light source is moving toward you.

Both red- and blueshifts are seen in light from a spinning star. Light from the side spinning toward you is blueshifted, and light from the side spinning away from you is redshifted.

All faraway galaxies are understood to be moving away from us because light from the galaxies is redshifted. Also, the farther away the galaxy, the greater the amount of redshifting, which means farther-away galaxies are moving away faster than those closer. This means that the universe must be expanding. See chapter 12 for more on the expanding universe.

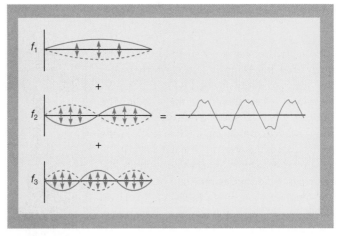

FIGURE 5.23 A combination of the fundamental and overtone frequencies produces a composite waveform with a characteristic sound quality.

higher frequency (f_3) could fit three half-wavelengths between the two fixed nodes (figure 5.22). Any whole number of halves of the wavelength will permit a standing wave to form. The frequencies (f_2, f_3, etc.) of these wavelengths are called the *overtones*, or harmonics. It is the presence and strength of various

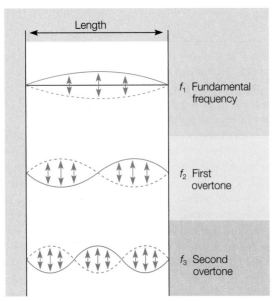

FIGURE 5.22 A stretched string of a given length has a number of possible resonant frequencies. The lowest frequency is the fundamental, f_1; the next higher frequencies, or overtones, shown are f_2 and f_3.

overtones that give a musical note from a certain instrument its characteristic quality. The fundamental and the overtones add together to produce the characteristic *sound quality*, which is different for the same-pitched note produced by a violin and by a guitar (figure 5.23).

The vibrating string produces a waveform with overtones, so instruments that have vibrating strings are called *harmonic instruments*. Instruments that use an air column as a sound maker are also harmonic instruments. These include all the wind instruments such as the clarinet, flute, trombone, trumpet, pipe organ, and many others. The various wind instruments have different ways of making a column of air vibrate. In the flute, air vibrates as it moves over a sharp edge, while in the clarinet, saxophone, and other reed instruments, it vibrates through fluttering thin reeds. The air column in brass instruments, on the other hand, is vibrated by the tightly fluttering lips of the player.

The length of the air column determines the frequency, and woodwind instruments have holes in the side of a tube that are opened or closed to change the length of the air column. The resulting tone depends on the length of the air column and the resonate overtones.

Sounds from Moving Sources

When the source of a sound is stationary, equally spaced sound waves expand from a source in all directions. But if the sounding source starts moving, then successive sound waves become displaced in the direction of movement and this changes the pitch. For example, the siren of an approaching ambulance seems to change pitch when the ambulance passes you. The sound wave is "squashed" as the ambulance approaches you and you hear a higher-frequency siren than the people inside the ambulance. When the ambulance passes you, the sound waves are "stretched" and you hear a lower-frequency siren. The overall effect of a higher pitch as a source approaches and then a lower pitch as it moves away is called the **Doppler effect.** The Doppler effect is evident if you stand by a street and an approaching car sounds its horn as it drives by you. You will

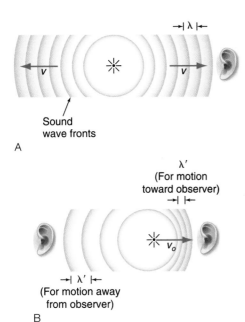

FIGURE 5.24 (*A*) Sound waves emitted by a stationary source and observed by a stationary observer. (*B*) Sound waves emitted by a source in motion toward the right. An observer on the right receives wavelengths that are shortened; an observer on the left receives wavelengths that are lengthened.

hear a higher-pitched horn as the car approaches, which shifts to a lower-pitched horn as the waves go by you. The driver of the car, however, will hear the continual, true pitch of the horn since the driver is moving with the source (figure 5.24).

A Doppler shift is also noted if the observer is moving and the source of sound is stationary. When the observer moves toward the source, the wave fronts are encountered more frequently than if the observer were standing still. As the observer moves away from the source, the wave fronts are encountered less frequently than if the observer were not moving. An observer on a moving train approaching a crossing with a sounding bell thus hears a high-pitched bell that shifts to a lower-pitched bell as the train whizzes by the crossing.

Connections . . .

Doppler Radar

The Doppler effect was named after the Austrian scientist Christian Doppler, who first demonstrated the effect using sound waves back in 1842. The same principle applies to electromagnetic radiation as well as sound, but now the shifts are in frequency of the radiation. A lower frequency is observed when a source of light is moving away, and this is called a "redshift." Also, a "blueshift" toward a higher frequency occurs when a source of light is moving toward an observer. Radio waves will also experience such shifts of frequency, and weather radar that measures frequency changes as a result of motion is called *Doppler radar*.

Weather radar broadcasts short radio waves from an antenna. When directed at a storm, the waves are reflected back to the antenna by rain, snow, and hail. Reflected radar waves are electronically converted and displayed on a monitor, showing the location and intensity of precipitation. A Doppler radar also measures frequency shifts in the reflected radio waves. Waves from objects moving toward the antenna show a higher frequency, and waves from objects moving away from the antenna show a lower frequency. These shifts of frequency are measured, then displayed as the speed and direction of winds moving raindrops and other objects in the storm.

Weather forecasters can direct a Doppler radar machine to measure different elevations of a storm system. This shows highly accurate information that can be used to identify, for example, where and when a tornado might form, as well as the intensity of storm winds in a given area, and even provide an estimate of how much precipitation fell from the storm.

When an object moves through the air at the speed of sound, it keeps up with its own sound waves. All the successive wave fronts pile up on one another, creating a large wave disturbance called a *shock wave*. The shock wave from a supersonic airplane is a cone-shaped wave of intense condensations trailing backward at an angle dependent on the speed of the aircraft. Wherever this cone of superimposed crests passes, a **sonic boom** occurs (figure 5.25). The many crests have been added together, each contributing to the pressure increase. The human ear cannot differentiate between such a pressure wave created by a supersonic aircraft and a pressure wave created by an explosion.

Does a sonic boom occur only when an airplane breaks the sound barrier? The answer is no, an airplane traveling at or faster than the speed of sound produces the shock wave continuously, and a sonic boom will be heard everywhere the plane drags its cone-shaped shock wave. Can you find evidence of shock waves associated with projections on the airplane pictured in figure 5.26?

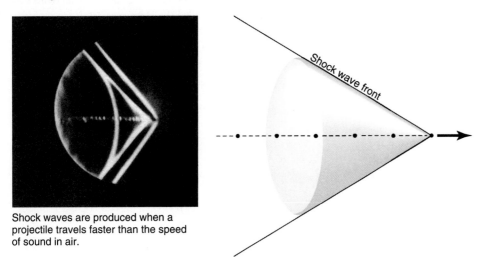

Shock waves are produced when a projectile travels faster than the speed of sound in air.

FIGURE 5.25 A sound source moves with velocity *greater* than the speed of sound in the medium. The envelope of spherical wave front forms the conical shock wave.

FIGURE 5.26 A cloud sometimes forms just as a plane accelerates to break the sound barrier. Moist air is believed to form the cloud water droplets as air pressure drops behind the shock wave.

The Austrian physicist Ernst Mach published a paper in 1877 laying out the principles of supersonics. He also came up with the idea of using a ratio of the velocity of an object to the velocity of sound. Today, this ratio is called the Mach number. A plane traveling at the speed of sound has a Mach number of 1, a plane traveling at twice the speed of sound has a Mach number of 2, and so on. Ernst Mach was also the first to describe what is happening to produce a sonic boom, and he observed the existence of a conical shock wave formed by a projectile as it approached the speed of sound.

SUMMARY

Elastic objects *vibrate,* or move back and forth, in a repeating motion when disturbed by some external force. They are able to do this because they have an *internal restoring force* that returns them to their original positions after being deformed by some external force. If the internal restoring force is opposite to and proportional to the deforming displacement, the vibration is called a *simple harmonic motion.* The extent of displacement is called the *amplitude,* and one complete back-and-forth motion is one *cycle.* The time required for one cycle is a *period.* The *frequency* is the number of cycles per second, and the unit of frequency is the *hertz.* A *graph* of the displacement as a function of time for a simple harmonic motion produces a *sinusoidal* graph.

Periodic, or repeating, vibrations or the *pulse* of a single disturbance can create *waves,* disturbances that carry energy through a medium. A wave that disturbs particles in a back-and-forth motion in the direction of the wave travel is called a *longitudinal wave.* A wave that disturbs particles in a motion perpendicular to the direction of wave travel is called a *transverse wave.* The nature of the medium and the nature of the disturbance determine the type of wave created.

Waves that move through the air are longitudinal and cause a back-and-forth motion of the molecules making up the air. A zone of molecules forced closer together produces a *condensation,* a pulse of increased density and pressure. A zone of reduced density and pressure is a *rarefaction.* A vibrating object produces condensations and rarefactions that expand outward from the source. If the frequency is between 20 Hz and 20,000 Hz, the human ear perceives the waves as *sound* of a certain *pitch.* High frequency is interpreted as high-pitched sound and low frequency as low-pitched sound.

A graph of pressure changes produced by condensations and rarefactions can be used to describe sound waves. The condensations produce *crests,* and the rarefactions produce *troughs.* The *amplitude* is the maximum change of pressure from the normal. The *wavelength* is the distance between any two successive places on a wave train, such as the distance from one crest to the next crest. The *period* is the time required for a wave to repeat itself. The *velocity* of a wave is how quickly a wavelength passes. The *frequency* can be calculated from the *wave equation,* $v = \lambda f$.

Sound waves can move through any medium but not a vacuum. The velocity of sound in a medium depends on the molecular inertia and strength of interactions. Sound, therefore, travels most rapidly through a solid, then a liquid, then a gas. In air, sound has a greater velocity in warmer air than in cooler air because the molecules of air are moving about more rapidly, therefore transmitting a pulse more rapidly.

Sound waves are *reflected* or *refracted* from a *boundary,* which means a change in the transmitting medium. Reflected waves that are *in phase* with incoming waves undergo *constructive interference,* and waves that are *out of phase* undergo *destructive interference.* Two waves that are otherwise alike but with slightly different frequencies produce an alternating increasing and decreasing of loudness called *beats.*

The *energy* of a sound wave is called the wave *intensity,* which is measured in watts per square meter. The intensity of sound is expressed on the *decibel scale,* which relates it to changes in loudness as perceived by the human ear.

All elastic objects have *natural frequencies* of vibration that are determined by the materials they are made of and their shapes. When energy is transferred at the natural frequencies, there is a dramatic increase of amplitude called *resonance.* The natural frequencies are also called *resonant frequencies.*

Sounds are compared by pitch, loudness, and *quality.* The quality is determined by the instrument sounding the note. Each instrument has its own characteristic quality because of the resonant frequencies that it produces. The basic, or *fundamental,* frequency is the longest standing wave that it can make. The fundamental frequency determines the basic note being sounded, and other resonant frequencies, or standing waves called *overtones* or *harmonics,* combine with the fundamental to give the instrument its characteristic quality.

A moving source of sound or a moving observer experiences an apparent shift of frequency called the *Doppler effect.* If the source is moving as fast or faster than the speed of sound, the sound waves pile up into a *shock wave* called a *sonic boom.* A sonic boom sounds very much like the pressure wave from an explosion.

Summary of Equations

5.1
$$\text{period} = \frac{1}{\text{frequency}}$$
$$T = \frac{1}{f}$$

5.2
$$\text{frequency} = \frac{1}{\text{period}}$$
$$f = \frac{1}{T}$$

5.3
$$\text{velocity} = (\text{wavelength})(\text{frequency})$$
$$v = \lambda f$$

5.4
$$\text{beat frequency} = \text{one frequency} - \text{other frequency}$$
$$f_b = f_2 - f_1$$

KEY TERMS

amplitude (p. **105**)
beats (p. **115**)
condensation (p. **108**)
cycle (p. **105**)
decibel scale (p. **116**)
Doppler effect (p. **120**)
echo (p. **113**)

frequency (p. **106**)
fundamental frequency
   (p. **119**)
hertz (p. **106**)
infrasonic (p. **108**)
intensity (p. **115**)
longitudinal wave (p. **107**)

natural frequency (p. **116**)
period (p. **105**)
pitch (p. **110**)
rarefaction (p. **108**)
resonance (p. **116**)
reverberation (p. **113**)
sonic boom (p. **121**)

standing waves (p. **119**)
transverse wave (p. **107**)
ultrasonic (p. **108**)
vibration (p. **104**)
wavelength (p. **111**)

APPLYING THE CONCEPTS

1. The time required for a vibrating object to complete one full cycle is the
   a. frequency.
   b. amplitude.
   c. period.
   d. hertz.

2. The unit of cycles per second is called a
   a. hertz.
   b. lambda.
   c. wave.
   d. watt.

3. The period of a vibrating object is related to the frequency, since they are
   a. directly proportional.
   b. inversely proportional.
   c. frequently proportional.
   d. not proportional.

4. A longitudinal mechanical wave causes particles of a material to move
   a. back and forth in the same direction the wave is moving.
   b. perpendicular to the direction the wave is moving.
   c. in a circular motion in the direction the wave is moving.
   d. in a circular motion opposite the direction the wave is moving.

5. Transverse mechanical waves will move only through
   a. solids.
   b. liquids.
   c. gases.
   d. All of the above are correct.

6. Longitudinal mechanical waves will move only through
   a. solids.
   b. liquids.
   c. gases.
   d. All of the above are correct.

7. The characteristic of a wave that is responsible for what you interpret as pitch is the wave
   a. amplitude.
   b. shape.
   c. frequency.
   d. height.

8. The number of cycles that a vibrating tuning fork experiences each second is related to the resulting sound wave characteristic of
   a. frequency.
   b. amplitude.
   c. wave height.
   d. quality.

9. Sound waves travel faster in
   a. solids as compared to liquids.
   b. liquids as compared to gases.
   c. warm air as compared to cooler air.
   d. All of the above are correct.

10. Sound interference is necessary to produce the phenomenon known as
   a. resonance.
   b. decibels.
   c. beats.
   d. reverberation.

11. The efficient transfer of energy that takes place at a natural frequency is known as
   a. resonance.
   b. beats.
   c. the Doppler effect.
   d. reverberation.

12. An observer on the ground will hear a sonic boom from an airplane traveling faster than the speed of sound
   a. only when the plane breaks the sound barrier.
   b. as the plane is approaching.
   c. when the plane is directly overhead.
   d. after the plane has passed by.

Answers

1. c **2.** a **3.** b **4.** a **5.** a **6.** d **7.** c **8.** a **9.** d **10.** c **11.** a **12.** d

QUESTIONS FOR THOUGHT

1. What is a wave?

2. Is it possible for a transverse wave to move through air? Explain.

3. A piano tuner hears three beats per second when a tuning fork and a note are sounded together and six beats per second after the string is tightened. What should the tuner do next, tighten or loosen the string? Explain.

4. Why do astronauts on the moon have to communicate by radio even when close to one another?

5. What is resonance?

6. Explain why sounds travel faster in warm air than in cool air.

7. Do all frequencies of sound travel with the same velocity? Explain your answer by using the wave equation.

8. What eventually happens to a sound wave traveling through the air?

9. What gives a musical note its characteristic quality?

10. Does a supersonic aircraft make a sonic boom only when it cracks the sound barrier? Explain.

11. What is an echo?

12. Why are fundamental frequencies and overtones also called resonant frequencies?

FOR FURTHER ANALYSIS

1. How would distant music sound if the speed of sound decreased with frequency?

2. What are the significant similarities and differences between longitudinal and transverse waves? Give examples of each.

3. Sometimes it is easier to hear someone speaking in a full room than in an empty room. Explain how this could happen.

4. Describe how you can use beats to tune a musical instrument.

5. Is sound actually destroyed in destructive interference?

6. Are vibrations the source of all sounds? Discuss if this is supported by observations or if it is an inference.

7. How can sound waves be waves of pressure changes if you can hear several people talking at the same time?

8. Why is it not a good idea for a large band to march in unison across a bridge?

INVITATION TO INQUIRY

Does A Noisy Noise Annoy You?

An old question-and-answer game played by children asks, "What annoys an oyster?"

The answer is, "A noisy noise annoys an oyster."

You could do an experiment to find out how much noise it takes to annoy an oyster, but you might have trouble maintaining live oysters, as well as measuring how annoyed they might become. So,

consider using different subjects, including humans. You could modify the question to, "What noise level affects how well we concentrate?"

If you choose to accept this invitation, start by determining how you are going to make the noise, how you can control different noise levels, and how you can measure the concentration level of people. A related question could be, "Does it help or hinder learning to play music while studying?"

PARALLEL EXERCISES

The exercises in groups A and B cover the same concepts. Solutions to group A exercises are located in appendix D.

Group A

1. A vibrating object produces periodic waves with a wavelength of 50 cm and a frequency of 10 Hz. How fast do these waves move away from the object?

2. The distance between the center of a condensation and the center of an adjacent rarefaction is 1.50 m. If the frequency is 112.0 Hz, what is the speed of the wave front?

3. Water waves are observed to pass under a bridge at a rate of one complete wave every 4.0 s. (a) What is the period of these waves? (b) What is the frequency?

4. A sound wave with a frequency of 260 Hz moves with a velocity of 330 m/s. What is the distance from one condensation to the next?

5. The following sound waves have what velocity?
   a. Middle C, or 256 Hz and 1.34 m λ.
   b. Note A, or 440.0 Hz and 78.0 cm λ.
   c. A siren at 750.0 Hz and λ of 45.7 cm.
   d. Note from a stereo at 2,500.0 Hz and λ of 13.7 cm.

6. You hear an echo from a cliff 4.80 s after you shout, "Hello." How many feet away is the cliff if the speed of sound is 1,100 ft/s?

7. During a thunderstorm, thunder was timed 4.63 s after lightning was seen. How many feet away was the lightning strike if the speed of sound is 1,140 ft/s?

8. If the velocity of a 440 Hz sound is 1,125 ft/s in the air and 5,020 ft/s in seawater, find the wavelength of this sound (a) in air, (b) in seawater.

Group B

1. A tuning fork vibrates 440.0 times a second, producing sound waves with a wavelength of 78.0 cm. What is the velocity of these waves?

2. The distance between the center of a condensation and the center of an adjacent rarefaction is 65.23 cm. If the frequency is 256.0 Hz, how fast are these waves moving?

3. A warning buoy is observed to rise every 5.0 s as crests of waves pass by it. (a) What is the period of these waves? (b) What is the frequency?

4. Sound from the siren of an emergency vehicle has a frequency of 750.0 Hz and moves with a velocity of 343.0 m/s. What is the distance from one condensation to the next?

5. The following sound waves have what velocity?
   a. 20.0 Hz, λ of 17.2 m
   b. 200.0 Hz, λ of 1.72 m
   c. 2,000.0 Hz, λ of 17.2 cm
   d. 20,000.0 Hz, λ of 1.72 cm

6. A ship at sea sounds a whistle blast, and an echo returns from the coastal land 10.0 s later. How many kilometers is it to the coastal land if sound travels at 337 m/s?

7. How many seconds will elapse between seeing lightning and hearing the thunder if the lightning strikes 1 mile (5,280 ft) away and sound travels at 1,151 ft/s?

8. A 600.0 Hz sound has a velocity of 1,087.0 ft/s in the air and a velocity of 4,920.0 ft/s in water. Find the wavelength of this sound (a) in the air, (b) in the water.

A thunderstorm produces an interesting display of electrical discharge. Each bolt can carry over 150,000 amperes of current with a voltage of 100 million volts.

Electricity

Physics Connections

▶ A force is a push or pull (Ch. 2).

▶ When you do work on something, you give it energy (Ch. 3).

▶ Power is the rate at which energy is transformed (Ch. 3).

▶ Electrical energy can be changed to other forms (Ch. 3).

▶ Electricity is one form of energy (Ch. 3).

▶ Light is radiation consisting of electrical and magnetic fields (Ch. 7).

CORE CONCEPT

Electric and magnetic fields interact and can produce forces.

Electric forces can be measured (p. 128).

Moving charges produce magnetic fields (p. 140).

Magnetic fields interact (p. 141).

Moving magnets produce electric fields (p. 142).

Chemistry Connections

▶ Matter (atoms) is electrical in nature (Ch. 8).

▶ Chemical reactions involve electric forces (Ch. 9).

Earth Science Connections

▶ Minerals are crystals held together by charge (Ch. 15).

Life Science Connections

▶ Living things modify molecules by chemical change (Ch. 24).

OVERVIEW

The previous chapters have been concerned with *mechanical* concepts, explanations of the motion of objects that exert forces on one another. These concepts were used to explain straight-line motion, the motion of free fall, and the circular motion of objects on Earth as well as the circular motion of planets and satellites. The mechanical concepts were based on Newton's laws of motion and are sometimes referred to as Newtonian physics. The mechanical explanations were then extended into the submicroscopic world of matter through the kinetic molecular theory. The objects of motion were now particles, molecules that exert force on one another, and concepts associated with heat were interpreted as the motion of these particles. In a further extension of Newtonian

concepts, mechanical explanations were given for concepts associated with sound, a mechanical disturbance that follows the laws of motion as it moves through the molecules of matter.

You might wonder, as did the scientists of the 1800s, if mechanical interpretations would also explain other natural phenomena such as electricity, chemical reactions, and light. A mechanical model would be very attractive since it already explained so many other facts of nature, and scientists have always looked for basic, unifying theories. Mechanical interpretations were tried, as electricity was considered as a moving fluid and light was considered as a mechanical wave moving through a material fluid. There were many unsolved puzzles with such a model, and gradually,

it was recognized that electricity, light, and chemical reactions could not be explained by mechanical interpretations. Gradually, the point of view changed from a study of particles to a study of the properties of the *space* around the particles. In this chapter, you will learn about electric charge in terms of the space around particles. This model of electric charge, called the *field model*, will be used to develop concepts about electric current, the electric circuit, and electrical work and power. A relationship between electricity and the fascinating topic of magnetism is discussed next, including what magnetism is and how it is produced. The relationship is then used to explain the mechanical production of electricity (figure 6.1), how electricity is measured, and how electricity is used in everyday technological applications.

ELECTRIC CHARGE

It was a big mystery for thousands of years. No one could figure out why a rubbed piece of amber, which is fossilized tree resin, would attract small pieces of paper, thread, and hair. This unexplained attraction was called the "amber effect." Then about one hundred years ago, Joseph J. Thomson found the answer while experimenting with electric currents. From these experiments, Thomson concluded that negatively charged particles were present in all matter and in fact might be the stuff of which matter is made. The amber effect was traced to the movement of these particles, so they were called *electrons* after the Greek word for amber. The word *electricity* is also based on the Greek word for amber.

Today, we understand that the basic unit of matter is the *atom*, which is made up of electrons and other particles such as

protons and *neutrons.* The atom is considered to have a dense center part called a *nucleus* that contains the closely situated protons and neutrons. The electrons move around the nucleus at some relatively greater distance (figure 6.2). For understanding electricity, you need only to consider the protons in the nucleus, the electrons that move around the nucleus, and the fact that electrons can be moved from an atom and caused to move to or from an object. Details on the nature of protons, neutrons, electrons, and models on how the atom is constructed will be considered in chapter 8.

Electrons and protons have a property called **electric charge.** Electrons have a *negative electric charge* and protons have a *positive electric charge.* The negative and positive description simply means that these two properties are opposite; it does not mean that one charge is better than the other. Charge is as fundamental to these subatomic particles as gravitational

FIGURE 6.1 The importance of electrical power seems obvious in a modern industrial society. What is not so obvious is the role of electricity in magnetism, light, and chemical change, and as the very basis for the structure of matter. All matter, in fact, is electrical in nature, as you will see.

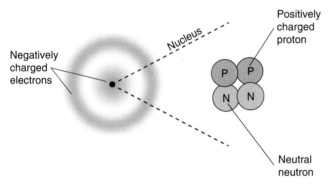

FIGURE 6.2 A very highly simplified model of an atom has most of the mass in a small, dense center called the *nucleus*. The nucleus has positively charged protons and neutral neutrons. Negatively charged electrons move around the nucleus at a much greater distance than is suggested by this simplified model. Ordinary atoms are neutral because there is balance between the number of positively charged protons and negatively charged electrons.

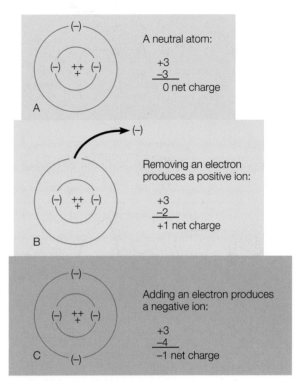

FIGURE 6.3 (A) A neutral atom has no net charge because the numbers of electrons and protons are balanced. (B) Removing an electron produces a net positive charge; the charged atom is called a positive ion. (C) The addition of an electron produces a net negative charge and a negative ion.

attraction is fundamental to masses. This means that you cannot separate gravity from a mass, and you cannot separate charge from an electron or a proton.

Electric charges interact to produce what is called the **electric force.** Like charges produce a repulsive electric force as positive repels positive and negative repels negative. Unlike charges produce an attractive electric force as positive and negative charges attract each other. You can remember how this happens with the simple rule of "*like charges repel and unlike charges attract.*"

Ordinary atoms are usually neutral because there is a balance between the number of positively charged protons and the number of negatively charged electrons. A number of different physical and chemical interactions can result in an atom gaining or losing electrons. In either case, the atom is said to be *ionized* and *ions* are produced as a result. An atom that is ionized by losing electrons results in a *positive ion* since

it has a net positive charge. An atom that is ionized by gaining electrons results in a *negative ion* since it has a net negative charge (figure 6.3).

Electrons can be moved from atom to atom to create ions. They can also be moved from one object to another by friction. Since electrons are negatively charged, an object that acquires an excess of electrons becomes a negatively charged body. The loss of electrons by another body results in a deficiency of electrons, which results in a positively charged object. Thus, *electric charges on objects result from the gain or loss of electrons.* Because the electric charge is confined to an object and is not moving, it is called an **electrostatic charge.** You probably call this charge *static electricity.* Static electricity is an accumulated electric charge at rest, that is, one that is not moving. When you comb your hair with a hard rubber comb, the comb becomes negatively charged because electrons are transferred from your hair to the comb, and your hair acquires a positive charge (figure 6.4). Both the negative charge on the comb from an excess of electrons and the positive charge on your hair from a deficiency of electrons are charges that are momentarily at rest, so they are electrostatic charges.

Once charged by friction, objects such as the rubber comb soon return to a neutral, or balanced, state by the movement of electrons. This happens more quickly on a humid day because water vapor assists with the movement of electrons to or from charged objects. Thus, static electricity is more noticeable on dry days than on humid ones.

A charged object can also exert a force of attraction on a second object that does not have a net charge. For example, you

+3
−3
0 net charge

+3
−3
0 net charge

A

+3
−1
+2 net charge

+3
−5
−2 net charge

B

FIGURE 6.4 Arbitrary numbers of protons (+) and electrons (−) on a comb and in hair (A) before and (B) after combing. Combing transfers electrons from the hair to the comb by friction, resulting in a negative charge on the comb and a positive charge on the hair.

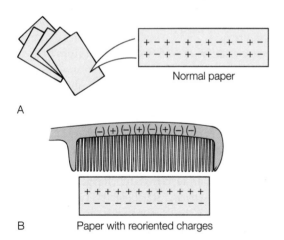

A   Normal paper

B   Paper with reoriented charges

FIGURE 6.5 Charging by induction. The comb has become charged by friction, acquiring an excess of electrons. The paper (A) normally has a random distribution of (+) and (−) charges. (B) When the charged comb is held close to the paper, there is a reorientation of charges because of the repulsion of like charges. This leaves a net positive charge on the side close to the comb, and since unlike charges attract, the paper is attracted to the comb.

can give your hard rubber comb a negative charge by combing your hair. Your charged comb will attract tiny pieces of paper that do not have a charge, pulling them to the comb. This happens because the negative charge on the comb repels electrons on the tiny bits of paper, giving the paper a positive charge on the side nearest the comb (figure 6.5). These unlike charges attract more than the like ones repel since the distance to the like ones is greater, and the paper is pulled to the comb.

Measuring Electric Charge

As you might have experienced, sometimes you receive a slight shock after walking across a carpet, and sometimes you are really zapped. You receive a greater shock when you have accumulated a greater electric charge. Since there is less electric charge at one time and more at another, it should be evident that charge occurs in different amounts, and these amounts can be measured. The magnitude of an electric charge is identified with the number of electrons that have been transferred onto or away from an object. The quantity of such a charge (q) is measured in a unit called a **coulomb** (C). The coulomb is a fundamental metric unit of measure like the meter, kilogram, and second.

Every electron has a charge of -1.60×10^{-19} C and every proton has a charge of $+1.60 \times 10^{-19}$ C. To accumulate a negative charge of 1 C, you would need to accumulate more than 6 billion billion electrons.

The charge on an electron (or proton), 1.60×10^{-19} C, is the smallest common charge known (more exactly, $1.6021892 \times 10^{-19}$ C). It is the **fundamental charge** of the electron ($e^- = 1.60 \times 10^{-19}$ C) and the proton ($p^+ = 1.60 \times 10^{-19}$ C). All charged objects have multiples of this fundamental charge. An object might have a charge on the order of about 10^{-8} to 10^{-6} C.

Measuring Electric Force

Recall that two objects with like charges, (−) and (−) or (+) and (+), produce a repulsive force, and two objects with unlike charges, (−) and (+), produce an attractive force. The size of either force depends on the amount of charge of each object and on the distance between the objects. The relationship is known as **Coulomb's law,** which is

$$F = k\frac{q_1 q_2}{d^2}$$ **equation 6.1**

where k has the value of 9.00×10^9 newton·meters²/coulomb² (9.00×10^9 N·m²/C²).

The force between the two charged objects is repulsive if q_1 and q_2 are the same charge and attractive if they are different (like charges repel, unlike charges attract). Whether the force is attractive or repulsive, you know that both objects feel the same force, as described by Newton's third law of motion. In addition, the strength of this force decreases if the distance between the objects increases. (A doubling of the distance reduces the force to 1/4 of the original value.)

EXAMPLE 6.1 *(Optional)*

Electrons carry a negative electric charge and move about the nucleus of the atom, which carries a positive electric charge from the proton. The electron is held by the force of electrical attraction at a typical distance of 1.00×10^{-10} m. What is the force of electrical attraction between an electron and proton?

SOLUTION

$$q_1 = 1.60 \times 10^{-19}\,\text{C}$$
$$q_2 = 1.60 \times 10^{-19}\,\text{C}$$
$$d = 1.00 \times 10^{-10}\,\text{m}$$
$$k = 9.00 \times 10^9\,\text{N·m}^2/\text{C}^2$$
$$F = ?$$

$$F = k\frac{q_1 q_2}{d^2}$$

$$= \frac{\left(9.00 \times 10^9\,\dfrac{\text{N·m}^2}{\text{C}^2}\right)(1.60 \times 10^{-19}\,\text{C})(1.60 \times 10^{-19}\,\text{C})}{(1.00 \times 10^{-10}\,\text{m})^2}$$

$$= \frac{(9.00 \times 10^9)(1.60 \times 10^{-19})(1.60 \times 10^{-19})}{1.00 \times 10^{-20}}\frac{\left(\dfrac{\text{N·m}^2}{\text{C}^2}\right)(\text{C}^2)}{\text{m}^2}$$

$$= \frac{2.30 \times 10^{-28}}{1.00 \times 10^{-20}}\frac{\text{N·m}^2}{\text{C}^2} \times \frac{\text{C}^2}{1} \times \frac{1}{\text{m}^2}$$

$$= \boxed{2.30 \times 10^{-8}\,\text{N}}$$

The electrical force of attraction between the electron and proton is 2.30×10^{-8} newton.

The model of a *field* is useful in understanding how a charge can attract or repel another charge some distance away. This model does not consider the force that one object exerts on another one through space. Instead, it considers the condition of space around a charge. The condition of space around an electric charge is changed by the presence of the charge. The charge produces a *force field* in the space around it. Since this force field is produced by an electric charge, it is called an **electric field.** Imagine a second electric charge, called a "test charge," that is far enough away from the electric charge that no forces are experienced. As you move the test charge closer and closer, it will experience an increasing force as it enters the electric field.

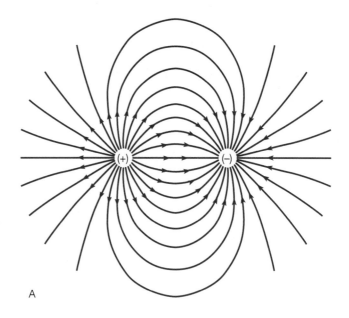

A

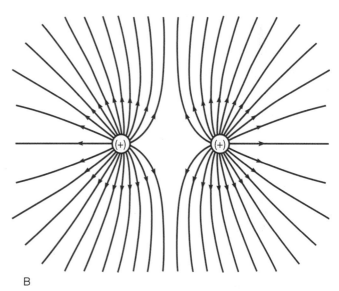

B

FIGURE 6.6 Lines of force diagrams for (*A*) a negative charge and (*B*) a positive charge when the charges have the same strength as the test charge.

The test charge can be used to identify the electric field that spreads out and around the space of an electric charge.

An electric field can be visualized by making a map of the field. The field is represented by electric field lines that indicate the strength and direction of the force the field would exert on the field of another charge. The field lines always point outward around a positively charged particle and point inward around a negatively charged particle. The spacing of the field lines shows the strength of the field. The field is stronger where the lines are closer together and weaker where they are farther apart (figure 6.6).

ELECTRIC CURRENT

Electric current means a flow of charge in the same way that "water current" means a flow of water. An electric current is flow of charge and this flow can be either negative or positive.

Effects of Electric Current on People

Current (A)	Effect (varies with individual)
0.001 to 0.005	Perception threshold
0.005 to 0.01	Mild shock
0.01 to 0.02	Cannot let go of wire
0.02 to 0.05	Breathing difficult
0.05 to 0.1	Breathing stops, heart stops
0.1 and higher	Severe burns, death

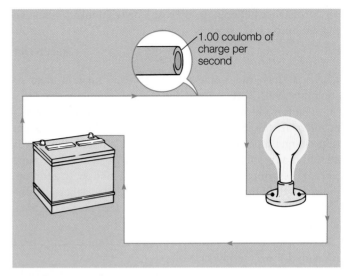

FIGURE 6.7 A simple electric circuit carrying a current of 1.00 coulomb per second through a cross section of a conductor has a current of 1.00 amp.

An electric wire in a car is usually a copper wire covered with a plastic insulation. When the radio is operated by using such a wire, electrons are moved from one terminal of the car battery through the radio, then back toward the battery. In a computer monitor, electrons are accelerated inside the picture tube to the back of the screen, producing an image as directed by the processor. Finally, a computer printer is plugged into a household current, which has an insulated copper wire with electrons moving back and forth. In all of these examples, there was movement of charge, also known as an electric current.

Many examples of different types of current exist, but all can be compared by the amount of electric charge that flows per second. The relationship is

$$\text{electric current} = \frac{\text{quantity of charge}}{\text{time}}$$

$$I = \frac{q}{t} \qquad \textbf{equation 6.2}$$

The unit for current is coulomb/second, which is called an **ampere** (A or **amp** for short). A current of 1 ampere in a wire is 1 coulomb of charge flowing through the wire every second. A 2 amp current is 2 coulombs of charge flowing through the wire every second, and so on (figure 6.7).

Charge can flow easily through some materials, such as metals, because they have many loosely attached electrons that can be easily moved from atom to atom. A substance that allows charges to flow easily is called a *conductor*. Materials such as plastic, wood, and rubber hold tightly to their electrons and do not readily allow for the flow of charge. A substance that does not allow charges to flow is called an *insulator* (table 6.1). Thus, metal wires are used to conduct an electric current from one place to another, and rubber, glass, and plastics are used as insulators to keep the current from going elsewhere.

There is a third class of materials, such as silicon and germanium, that sometimes conduct and sometimes insulate, depending on the conditions and how pure they are. These materials are called *semiconductors,* and their special properties make possible a number of technological devices such as

TABLE 6.1 Electric Conductors and Insulators

Conductors	Insulators
Silver	Rubber
Copper	Glass
Gold	Carbon (diamond)
Aluminum	Plastics
Carbon (graphite)	Wood
Tungsten	
Iron	
Lead	
Nichrome	

computers, electrostatic copying machines, laser printers, solar cells, and so forth.

Resistance

Insulators have a property of limiting a current, and this property is called *electric resistance*. A good electric conductor has a very low electric resistance, and a good electric insulator has a very high electric resistance. The actual magnitude of Electric resistance of a metal wire conductor depends on four variables (figure 6.8):

1. **Material.** Different materials have different resistances, as shown by the list of conductors in table 6.1. Silver, for example, is at the top of the list because it offers the least resistance, followed by copper, gold, then aluminum. Of the materials listed in the table, nichrome is the conductor with the greatest resistance. By definition, conductors have less electric resistance than insulators, which have a very large electric resistance.

Hydrogen and Fuel Cells

There is more than one way to power an electric vehicle, and a rechargeable lead-acid battery is not the best answer. A newly developing technology uses a hydrogen-powered fuel cell to power an electric vehicle. Storage batteries, dry cells, and fuel cells all operate by using a chemical reaction to produce electricity. The difference is that the fuel cell does not need charging like a storage battery or replacing like a dry cell. A fuel cell can use a fuel to run continuously, and that is why it is called a "fuel cell."

Fuel cells are able to generate electricity directly onboard an electric vehicle, so heavy batteries are not needed. A short driving range is not a problem, and time lost charging the batteries is not a problem. Instead, electricity is produced electrochemically, in a device without any moving parts. As you probably know, energy is required to separate water into its component gases of hydrogen and oxygen. Thus, as you might expect, energy is released when hydrogen and oxygen combine to form water. It is this energy that a fuel cell uses to produce an electric current.

One design of a fuel cell has two electrodes, one positive and one negative, where the reactions that produce electricity take place. A proton exchange membrane (PEM) separates the electrodes, which also use catalysts to speed chemical reactions. In general terms, hydrogen atoms enter the fuel cell at the positive terminal where a chemical reaction strips them of their electrons (box figure 6.1). The hydrogen atoms are now positive ions, which pass through the membrane and leave their electrons behind. Once through the membrane, the hydrogen ions combine with oxygen, forming water, which drains from the cell. The electrons on the original side of the membrane increase in number, building up an electrical potential difference. Once a wire connects the two sides of the membrane, the electrons are able to make a current, providing electricity.

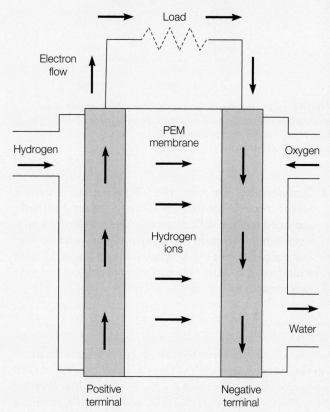

Box Figure 6.1 A schematic of a PEM fuel cell.

As long as such a fuel cell is supplied with hydrogen and oxygen, it will generate electricity to run a car.

The silent-running, nonpolluting fuel cell with no moving parts sounds too good to be true, but the technology works. The technology has been too expensive for everyday use until recently but is now more affordable. Fuel-cell powered vehicles can operate directly on compressed hydrogen gas or liquid hydrogen, and when they do, the only emission is water vapor. Direct use of hydrogen is also very efficient, with a 50 to 60 percent efficiency compared to the typical 15 to 20 percent efficiency for automobiles that run on petroleum in an internal combustion engine. Other fuels can also be used by running them through an onboard reformer, which transforms the fuel to hydrogen. Methanol or natural gas, for example, can be used with significantly less CO_2, CO, HC, and NO_x emissions than produced by an internal combustion engine. An added advantage to the use of methanol is that the existing petroleum fuel distribution system (tanks, pumps, etc.) can be used to distribute this liquid fuel. Liquid or compressed hydrogen, on the other hand, requires a completely new type of distribution system.

A fuel cell–powered electric vehicle gives the emission benefits of a battery-powered vehicle without the problems of constantly recharging the batteries. Before long, you may see a fuel-cell vehicle in your neighborhood. It is the car of the future, which is needed now for the environment.

2. **Length.** The resistance of a conductor varies directly with the length; that is, a longer wire has more resistance and a shorter wire has less resistance. The longer the wire is, the greater the resistance.

3. **Diameter.** The resistance varies inversely with the cross-sectional area of a conductor. A thick wire has a greater cross-sectional area and therefore has less resistance than a thin wire. The thinner the wire is, the greater the resistance.

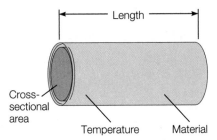

FIGURE 6.8 The four factors that influence the resistance of an electric conductor are the material the conductor is made of, the length of the conductor, the cross-sectional area of the conductor, and the temperature of the conductor.

4. **Temperature.** For most materials, the resistance increases with increases in temperature. This is a consequence of the increased motion of electrons and ions at higher temperatures, which increases the number of collisions. At very low temperatures (100 K or less), the resistance of some materials approaches zero, and the materials are said to be *superconductors.*

AC and DC

Another aspect of the nature of an electric current is the direction the charge is flowing. The wires in a car have currents that always move in one direction, and this is called a **direct current** (DC). Chemical batteries, fuel cells, and solar cells produce a direct current, and direct currents are utilized in electronic devices. Electric utilities and most of the electrical industry, on the other hand, use an **alternating current** (AC). An alternating current, as the name implies, moves the electrons alternately one way, then the other way. Since household electric circuits use alternating current, there is no flow of electrons from the electrical outlets through the wires. Instead an electric *field* moves back and forth through a wire at nearly the speed of light, causing electrons to jiggle back and forth. This constitutes a current that flows one way, then the other with the changing field. The current changes like this 120 times a second in a 60 hertz alternating current.

THE ELECTRIC CIRCUIT

An electric current is established in a conductor when an electric field exerts a force on charges in the conductor. A car battery, for example, is able to light a bulb because the battery produces an electric field that forces electrons to move through the lightbulb filament. An **electric circuit** contains some device, such as a battery or electric generator, that acts as a source of energy as it forces charges to move out one terminal, through the wires of the circuit, and then back in the other terminal (figure 6.9). The charges do work in another part of the circuit as they light bulbs, run motors, or provide heat. The charges flow through connecting wires

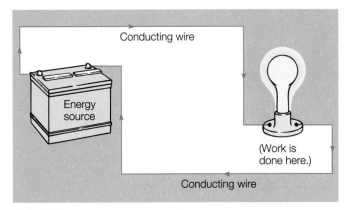

FIGURE 6.9 A simple electric circuit has an energy source (such as a generator or battery), some device (such as a lamp or motor) where work is done, and continuous pathways for the current to follow.

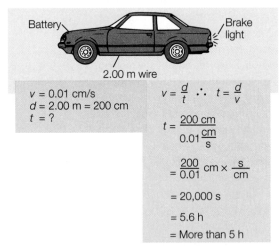

$$v = 0.01 \text{ cm/s}$$
$$d = 2.00 \text{ m} = 200 \text{ cm}$$
$$t = ?$$

$$v = \frac{d}{t} \quad \therefore \quad t = \frac{d}{v}$$

$$t = \frac{200 \text{ cm}}{0.01 \frac{\text{cm}}{\text{s}}}$$

$$= \frac{200}{0.01} \text{ cm} \times \frac{\text{s}}{\text{cm}}$$

$$= 20{,}000 \text{ s}$$

$$= 5.6 \text{ h}$$

$$= \text{More than 5 h}$$

FIGURE 6.10 Electrons move very slowly in a direct current circuit. With a drift velocity of 0.01 cm/s, more than 5 hours would be required for an electron to travel 200 cm from a car battery to the brake light. It is the electric field, not the electrons, that moves at near the speed of light in an electric circuit.

to make a continuous path, and the number of coulombs of charge that leaves one terminal is the same as the number of coulombs of charge that enters the other terminal. The electric field moves through the circuit at nearly the speed of light, forcing the electrons to move along. The electrons, however, actually move through the circuit very slowly (figure 6.10).

As we learned in chapter 3, work is done when a force moves an object over a distance. In a circuit, work is done by the device that creates the electric field (battery, for example) as it exerts a force on the electrons and moves them through a distance in the circuit. Disregarding any losses due to the very small work done in moving electrons through a wire, the work done in some device (lamp, for example) is equal to the work done by the battery. The amount of work can be quantified by considering the work done and the size of the charge moved,

Benjamin Franklin (1706–1790)

Benjamin Franklin was the first great U.S. scientist. He made an important contribution to physics by arriving at an understanding of the nature of electric charge, introducing the terms *positive* and *negative* to describe charges. He also proved in a classic experiment that lightning is electrical in nature and went on to invent the lightning rod. In addition to being a scientist and inventor, Franklin is widely remembered as a statesman. He played a leading role in drafting the Declaration of Independence and the Constitution of the United States.

Franklin was born in Boston, Massachusetts, of British settlers on January 17, 1706. He started life with little formal instruction, and by the age of ten, he was helping his father in the tallow and soap business. Soon, apprenticed to his brother, a printer, he was launched into that trade, leaving home in 1724 to set himself up as a printer in Philadelphia. His business prospered, and he was soon active in journalism and publishing. He started the *Pennsylvania Gazette* but is better remembered for *Poor Richard's Almanack*. The almanac was a collection of articles and advice on a huge range of topics, "conveying instruction among the common people." Published in 1732, it was a great success and brought Franklin a considerable income.

In 1746, his business booming, Franklin turned his thoughts to electricity and spent the next seven years executing a remarkable series of experiments. Although he had little formal education, his voracious reading habits gave him the necessary background, and his practical skills, together with an analytical yet intuitive approach, enabled Franklin to put the whole topic on a very sound basis. It was said that he found electricity a curiosity and left it a science.

In 1752, Franklin carried out his famous experiments with kites. By flying a kite in a thunderstorm, he was able to produce sparks from the end of the wet string, which he held with a piece of insulating silk. The lightning rod used everywhere today owes its origin to these experiments. Furthermore, some of Franklin's last work in this area demonstrated that while most thunderclouds have negative charges, a few are positive—something confirmed in modern times.

Franklin was also influential in areas of physics other than electricity. He rejected the particle theory of light, being unable to account for the vast momentum that the particles should possess if they existed. Always interested in the sea, Franklin produced the first chart of the Gulf Stream following observations made in 1770 that merchantmen crossed the Atlantic Ocean from east to west in two weeks less than the mail ships—the former keeping to the side of the current, not fighting it. In 1775, he used a thermometer to aid navigation by detecting the warm waters of the Gulf Stream. Finally, Franklin also busied himself with such diverse topics as the first public library, bifocal lenses, population control, the rocking chair, and daylight-saving time.

Benjamin Franklin is arguably the most interesting figure in the history of science and not only because of his extraordinary range of interests, his central role in the establishment of the United States, and his amazing willingness to risk his life to perform a crucial experiment—a unique achievement in science. By conceiving of the fundamental nature of electricity, he began the process by which a most detailed understanding of the structure of matter has been achieved.

Source: Modified from the *Hutchinson Dictionary of Scientific Biography*. © RM, 2007. Reprinted by permission.

and this ratio is used to define *voltage*. The voltage is defined by taking the ratio of the work done to the size of the charge that is being moved. So,

$$\text{voltage} = \frac{\text{work}}{\text{charge moved}}$$

$$V = \frac{W}{q} \qquad \textbf{equation 6.3}$$

The unit of voltage is the **volt,** which is the ratio that results when 1 joule of work is used to move 1 coulomb of charge:

$$1 \text{ volt (V)} = \frac{1 \text{ joule (J)}}{1 \text{ coulomb (C)}}$$

Thus, the voltage is the energy transfer per coulomb. The energy transfer can be measured by the work that is done to move the charge or by the work that the charge can do because of its position in the field. This is perfectly analogous to the work that must be done to give an object gravitational potential energy or to the work that the object can potentially do because of its new position. Thus, when a 12-volt battery is charging, 12 joules of work are done to transfer 1 coulomb of charge from an outside source against the electric field of the battery terminal. When the 12-volt battery is used, it does 12 joules of work for each coulomb of charge transferred from one terminal of the battery through the electrical system and back to the other terminal. Household circuits usually have a difference of potential of 120 or 240 volts. A voltage of 120 means that each coulomb of charge that moves through the circuit can do 120 joules of work in some electrical device.

The current in a circuit depends on the resistance as well as the voltage that is causing the current. If a conductor offers a small resistance, less voltage would be required to push an amp of current through the circuit. If a conductor offers more resistance, then more voltage will be required to push the same amp of current through the circuit. Resistance (R) is therefore a ratio between the voltage (V) and the resulting current (I). This ratio is

$$R = \frac{V}{I}$$

In units, this ratio is

$$1 \text{ ohm } (\Omega) = \frac{1 \text{ volt } (V)}{1 \text{ amp } (A)}$$

The ratio of volts/amps is the unit of resistance called an **ohm** (Ω) after the German physicist who discovered the relationship.

Another way to show the relationship between the voltage, current, and resistance is

$$V = IR \qquad \text{equation 6.4}$$

which is known as **Ohm's law.** This is one of three ways to show the relationship, but this way (solved for V) is convenient for easily solving the equation for other unknowns.

EXAMPLE 6.2 *(Optional)*

A lightbulb in a 120 V circuit is switched on, and a current of 0.50 A flows through the filament. What is the resistance of the bulb?

SOLUTION

The current (I) of 0.50 A is given with a potential difference (V) of 120 V. The relationship to resistance (R) is given by Ohm's law (equation 6.4)

$$I = 0.50 \text{ A} \qquad V = IR \therefore R = \frac{V}{I}$$
$$V = 120 \text{ V}$$
$$R = ? \qquad\qquad = \frac{120 \text{ V}}{0.50 \text{ A}}$$
$$= 240 \frac{V}{A}$$
$$= 240 \text{ ohm}$$
$$= \boxed{240 \text{ } \Omega}$$

EXAMPLE 6.3 *(Optional)*

What current would flow through an electrical device in a circuit with a potential difference of 120 V and a resistance of 30 Ω? (Answer: 4 A)

ELECTRIC POWER AND WORK

All electric circuits have three parts in common:

1. A voltage source, such as a battery or electric generator that uses some nonelectric source of energy to do work on electrons.
2. An electric device, such as a lightbulb or electric motor, where work is done by the electric field.
3. Conducting wires that maintain the current between the electric device and voltage source.

The work done by a voltage source (battery, electric generator) is equal to the work done by the electric field in an electric device (lightbulb, electric motor) plus the energy lost to resistance. Resistance is analogous to friction in a mechanical device, so low-resistance conducting wires are used to reduce this loss. Disregarding losses to resistance, electric work can therefore be measured where the voltage source does work in moving charges.

If you include a time factor with the work done in moving charges, you will be considering the *power output* of the voltage source. The power is determined by the voltage and current in the following relationship:

$$\text{power} = \text{voltage} \times \text{current}$$
$$P = VI \qquad \text{equation 6.5}$$

In units, you can see that multiplying the current ($I = C/s$) by the voltage ($V = J/C$) yields

$$\frac{\text{coulomb}}{\text{second}} \times \frac{\text{joule}}{\text{coulomb}} = \frac{\text{joule}}{\text{second}}$$

A joule/second is a unit of power called the **watt.** Therefore, electric power is measured in units of watts. Note that the relationship between power output and the current is directly proportional. Therefore, the greater the current supplied, the greater the power output.

Household electric devices are designed to operate on a particular voltage, usually 120 or 240. They therefore draw a certain current to produce the designed power. Information about these requirements is usually found somewhere on the device. A lightbulb, for example, is usually stamped with the designed power, such as 100 watts. Other electric devices may be stamped with amp and volt requirements. You can determine the power produced in these devices by using equation 6.5; that is, amps $\times$ volts = watts (figure 6.11). Another handy conversion factor to remember is that 746 watts are equivalent to 1 horsepower.

EXAMPLE 6.4 *(Optional)*

A 1,100 W hair dryer is designed to operate on 120 V. How much current does the dryer require?

SOLUTION

The power (P) produced is given in watts with a potential difference of 120 V across the dryer. The relationship between the units of amps, volts, and watts is found in equation 6.5, $P = VI$

$$P = 1,100 \text{ W}$$
$$V = 120 \text{ V}$$
$$I = ?$$

$$P = VI \therefore I = \frac{P}{V}$$

$$= \frac{1,100 \frac{J}{s}}{120 \frac{J}{C}}$$

$$= \frac{1,100}{120} \frac{J}{s} \times \frac{C}{J}$$

$$= 9.2 \frac{J \cdot C}{s \cdot J}$$

$$= 9.2 \frac{C}{s}$$

$$= \boxed{9.2 \text{ A}}$$

EXAMPLE 6.5 *(Optional)*

An electric fan is designed to draw 0.5 A in a 120 V circuit. What is the power rating of the fan? (Answer: 60 W)

A

B

Connections . . .

Inside a Dry Cell

The common dry cell used in a flashlight produces electric energy from a chemical reaction between ammonium chloride and the zinc can (box figure 6.3). The reaction leaves a negative charge on the zinc and a positive charge on the carbon rod. Manganese dioxide takes care of hydrogen gas, which is a by-product of the reaction. Dry cells always produce 1.5 volts, regardless of their size. Larger voltages are produced by combinations of smaller cells (making a true "battery").

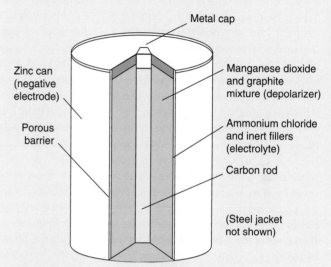

Box Figure 6.3   A schematic sketch of a dry cell.

C

FIGURE 6.11   What do you suppose it would cost to run each of these appliances for one hour? (*A*) This lightbulb is designed to operate on a potential difference of 120 volts and will do work at the rate of 100 W. (*B*) The finishing sander does work at the rate of 1.6 amp × 120 volts, or 192 W. (*C*) The garden shredder does work at the rate of 8 amps × 120 volts, or 960 W.

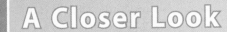
Household Circuits and Safety

The example of a household circuit in box figure 6.4B shows a light and four wall outlets in a parallel circuit. The use of the term *parallel* means that a current can flow through any of the separate branches without having to first go through any of the other devices. It does not imply that the branches are necessarily lined up with each other. Lights and outlets in this circuit all have at least two wires, one that carries the electrical load and one that maintains a potential difference by serving as a system ground. The load-carrying wire is usually black (or red) and the system ground is usually white. A third wire, usually bare or green, is used as a ground for an appliance.

Too many appliances running in a circuit—or a short circuit—can result in a very large current, perhaps great enough to cause strong heating and possibly a fire. A fuse or circuit breaker prevents this by disconnecting the circuit when it reaches a preset value, usually 15 or 20 amps. A fuse contains a short piece of metal that melts when heated by too large a current, creating a gap when the current through the circuit reaches the preset rating. The gap disconnects the circuit, just as cutting the wire or using a switch. The circuit breaker has the same purpose but uses the proportional relationship between the magnitude of a current and the strength

of the magnetic field that forms around the conductor. When the current reaches a preset level, the magnetic field is strong enough to open a spring-loaded switch. The circuit breaker is reset by flipping the switch back to its original position.

Besides a fuse or circuit breaker, a modern household electric circuit has three-pronged plugs, polarized plugs, and ground fault interrupters to help protect people and property from electrical damage. A *three-pronged plug* provides a grounding wire through a (usually round) prong on the plug. The grounding wire connects the housing of an appliance directly to the ground. If there is a short circuit, the current will take the path of least resistance—through the grounding wire—rather than through a person.

A *polarized plug* has one of the two flat prongs larger than the other. An alternating current moves back and forth with a frequency of 60 Hz, and *polarized* in this case has nothing to do with positive or negative. A polarized plug in an AC circuit means that one prong always carries the load. The smaller plug is connected to the load-carrying wire and the larger one is connected to the neutral wire. An ordinary, nonpolarized plug can fit into an outlet either way, which means there is a 50–50 chance that one of the

wires will be the one that carries the load. The polarized plug always has the load-carrying wire on the same side of the circuit, so the switch can be wired in so it is always on the load-carrying side. The switch will function the same on either wire, but when it is on the ground wire, the appliance contains a load-carrying wire, just waiting for a short circuit. When the switch is on the load-carrying side, the appliance does not have this potential safety hazard.

Yet another safety device called a *ground-fault interrupter* (GFI) offers a different kind of protection. Normally, the current in the load-carrying and system ground wire is the same. If a short circuit occurs, some of the current might be diverted directly to the ground or to the appliance ground. A GFI device monitors the load-carrying and system ground wires, and if any difference is detected, it trips, opening the circuit within a fraction of a second. This is much quicker than the regular fuse or general circuit breaker can react, and the difference might be enough to prevent a fatal shock. The GFI device is usually placed in an outside or bathroom circuit, places where currents might be diverted through people with wet feet. Note the GFI can also be tripped by a line surge that might occur during an electrical thunderstorm.

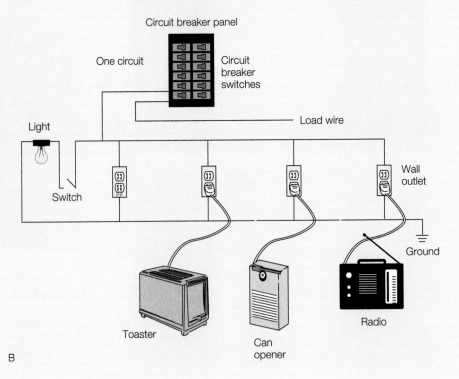

Box Figure 6.4 *(A)* One circuit breaker in this circuit breaker panel has tripped, indicating an overload or short circuit. *(B)* This could be circuit number 3 that is protected by a simple circuit breaker.

The electric utility charge for the electric work done is at a rate of cents per kilowatt-hour. The rate varies from place to place across the country, depending on the cost of producing the power (typically, 5 to 15 cents per kWh). You can predict the cost of running a particular electric appliance with the following equation,

$$cost = \frac{(watts)(time)(rate)}{1,000 \frac{watt}{kilowatt}}$$

equation 6.6

If the watt power rating is not given, it can be obtained by multiplying amps times volts. Also note that since the time unit is in hours, if you want to know the cost of running an appliance for a number of minutes, the time must be converted to the decimal equivalent of an hour (x min ÷ 60 min).

EXAMPLE 6.6 *(Optional)*

What is the cost of operating a 100 W lightbulb for 1.00 h if the utility rate is $0.10 per kWh?

SOLUTION

The power rating is given as 100 W, so the volt and amp units are not needed. Therefore,

$IV = P = 100$ W
$t = 1.00$ h
rate = $0.10/kWh
cost = ?

$$cost = \frac{(watts)(time)(rate)}{1,000 \frac{watts}{kilowatt}}$$

$$= \frac{(100 \text{ W})(1.00 \text{ h})(\$0.10/\text{kWh})}{1,000 \frac{watts}{kilowatt}}$$

$$= \frac{(100)(1.00)(0.10)}{1,000} \frac{W}{1} \times \frac{h}{1} \times \frac{\$}{kWh} \times \frac{kW}{W}$$

$$= \boxed{\$0.01}$$

The cost of operating a 100 W lightbulb at a rate of $0.10/kWh is 1¢/h.

EXAMPLE 6.7 *(Optional)*

An electric fan draws 0.5 A in a 120 V circuit. What is the cost of operating the fan if the rate is $0.10/kWh? (Answer: $0.006, which is 0.6 of a cent per hour)

As you may have noticed, the electric cord to certain appliances such as hair dryers becomes warm when the appliance is used. Heating occurs in any conductor that has any resistance, and it represents a loss of useable energy. The amount of heating depends on the resistance and the size of the current. Increasing the current or increasing any of the four resistance factors discussed earlier will cause an increase in heating. A large current in a wire of high resistance can become very hot, hot enough to melt insulation and ignite materials. This is why

all circuits have fuses or circuit breakers to "break" the circuit if the current exceeds a preset safe limit (see "A Closer Look: Household Circuits and Safety").

MAGNETISM

The ability of a certain naturally occurring rock to attract iron has been known since at least 600 B.C. This rock was called "Magnesian stone" since it was discovered near the ancient city of Magnesia in Turkey. Knowledge about the iron-attracting properties of the Magnesian stone grew slowly. About A.D. 100, the Chinese learned to magnetize a piece of iron with the stone, and sometime before A.D. 1000, they learned to use the magnetized iron or stone as a direction finder (compass). The rock that attracts iron is known today as the mineral named *magnetite*.

Magnetite is a natural magnet that strongly attracts iron and steel but also attracts cobalt and nickel. Such substances that are attracted to magnets are said to have *ferromagnetic* properties, or simply magnetic properties. Iron, cobalt, and nickel are considered to have magnetic properties, and most other common materials are considered not to have magnetic properties. Most of these nonmagnetic materials, however, are slightly attracted or slightly repelled by a strong magnet. In

Magnetic Fields and Instinctive Behavior

Do animals use Earth's magnetic fields to navigate? Since animals move from place to place to meet their needs, it is useful to be able to return to a nest, water hole, den, or favorite feeding spot. This requires some sort of memory of their surroundings (a mental map) and a way of determining direction. Often it is valuable to have information about distance as well. Direction can be determined by such things as magnetic fields, identified landmarks, scent trails, or reference to the Sun or stars. If the Sun or stars are used for navigation, then some sort of time sense is also needed since these bodies move in the sky.

Instinctive behaviors are automatic, preprogrammed, and genetically determined. Such behaviors are found in a wide range of organisms from simple one-celled protozoans to complex vertebrates. These behaviors are performed correctly the first time without previous experience when the proper stimulus is given. A stimulus is some change in the internal or external environment of the organism that causes it to react. The reaction of the organism to the stimulus is called a response.

An organism can respond only to stimuli it can recognize. For example, it is difficult for us as humans to appreciate what the world seems like to a bloodhound. The bloodhound is able to identify individuals by smell, whereas we have great difficulty detecting, let alone distinguishing, many odors. Some animals, such as dogs, deer, and mice, are color-blind and are able to see only shades of gray. Others, such as honeybees, can see ultra-violet light, which is invisible to us. Some birds and other animals are able to detect the magnetic field of Earth.

There is evidence that some birds navigate by compass direction, that is, they fly as if they had a compass in their heads. They seem to be able to sense magnetic north. Their ability to sense magnetic fields has been proven at the U.S. Navy's test facility in Wisconsin. The weak magnetism radiated from this test site has changed the flight pattern of migrating birds, but it is yet to be proven that birds use the magnetism of Earth to guide their migration. Homing pigeons are famous for their ability to find their way home. They make use of a wide variety of clues, but it has been shown that one of the clues they use involves magnetism. Birds with tiny magnets glued to the sides of their heads were very poor navigators, while others with nonmagnetic objects attached to the sides of their heads did not lose their ability to navigate.

---

addition, certain rare earth elements as well as certain metal oxides exhibit strong magnetic properties.

Every magnet has two **magnetic poles,** or ends, about which the force of attraction seems to be concentrated. Iron filings or other small pieces of iron are attracted to the poles of a magnet, for example, revealing their location (figure 6.12). A magnet suspended by a string will turn, aligning itself in a north-south direction. The north-seeking pole is called the *north pole* of the magnet. The south-seeking pole is likewise named the *south pole* of the magnet. All magnets have both a north pole and a south pole, and neither pole can exist by itself. You cannot separate a north pole from a south pole. If a magnet is broken into pieces, each new piece will have its own north and south pole (figure 6.13).

You are probably familiar with the fact that two magnets exert forces on each other. For example, if you move the north pole of one magnet near the north pole of a second magnet, each will experience a repelling force. A repelling force also occurs if two south poles are moved close together. But if the north pole of one magnet is brought near the south pole of a second magnet, an attractive force occurs. The rule is *"like magnetic poles repel and unlike magnetic poles attract."*

A similar rule of "likes charges repel and unlike charges attract" was used for electrostatic charges, so you might wonder if there is some similarity between charges and poles. The answer is they are not related. A magnet has no effect on a charged glass rod and the charged glass rod has no effect on either pole of a magnet.

A magnet moved into the space near a second magnet experiences a magnetic force as it enters the **magnetic field**

FIGURE 6.12 Every magnet has ends, or poles, about which the magnetic properties seem to be concentrated. As this photo shows, more iron filings are attracted to the poles, revealing their location.

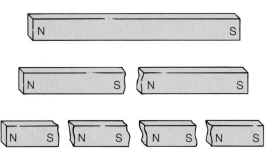

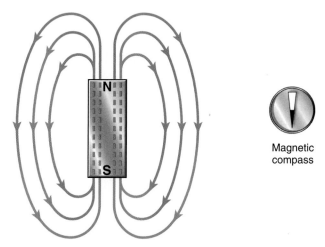

FIGURE 6.13 A bar magnet cut into halves always makes new, complete magnets with both a north and a south pole. The poles always come in pairs, and the separation of a pair into single poles, called monopoles, has never been accomplished.

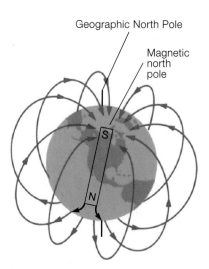

Magnetic compass

FIGURE 6.14 These lines are a map of the magnetic field around a bar magnet. The needle of a magnetic compass will follow the lines, with the north end showing the direction of the field.

The Inventor

Michael Faraday (1791–1867) is often regarded as the greatest experimental scientist of the 1800s. Among other experimental work, he invented the electric motor, electric generator, and transformer, discovered electromagnetic induction, and established the laws of electrolysis. He also coined the terms *anode*, *cathode*, *cation*, *anion*, *electrode*, and *electrolyte* to help explain his new inventions and laws.

Faraday was the son of a poor blacksmith and received a poor education as a child, with little knowledge of mathematics. At fourteen, he was apprenticed to a bookbinder and began to read many books on chemistry and physics. He also began attending public lectures at science societies and in general learned about science and scientific research. At twenty-one, he was appointed an assistant at the Royal Institution in London, where he liquefied gases for the first time and continued to work in electricity and magnetism. To understand electric and magnetic fields that he could not see or represent mathematically (he was not very accomplished at math), Faraday invented field lines—lines that do not exist but help us to picture what is going on.

Michael Faraday was a modest man, content to work in science without undue rewards. He declined both a knighthood and the presidency of the Royal Society. His accomplishments were recognized nonetheless, and the SI unit of capacitance, the farad, was named in his honor, as was the faraday. The faraday is the amount of electricity needed to liberate a standard amount of something through electrolysis.

of the second magnet. A magnetic field can be represented by *magnetic field lines*. By convention, magnetic field lines are drawn to indicate how the *north pole* of a tiny imaginary magnet would point when in various places in the magnetic field. Arrowheads indicate the direction that the north pole would point, thus defining the direction of the magnetic field. The strength of the magnetic field is greater where the lines are closer together and weaker where they are farther apart. Figure 6.14 shows the magnetic field lines around the familiar bar magnet.

The north end of a magnetic compass needle points north because Earth has a magnetic field. Earth's magnetic field is shaped and oriented as if there were a huge bar magnet inside Earth (figure 6.15). The geographic North Pole is the axis of Earth's rotation, and this pole is used to determine the direction of true north on maps. A magnetic compass does not point to true north because the north magnetic pole and the geographic North Pole are in two different places. The difference is called the *magnetic declination*. The map in figure 6.16 shows approximately how many degrees east or west of true north a compass needle will point in different locations. Magnetic declination must be considered when navigating with a compass.

Geographic North Pole

Magnetic north pole

FIGURE 6.15 Earth's magnetic field. Note that the magnetic north pole and the geographic North Pole are not in the same place. Note also that the magnetic north pole acts as if the south pole of a huge bar magnet were inside Earth. You know that it must be a magnetic south pole, since the north end of a magnetic compass is attracted to it, and opposite poles attract.

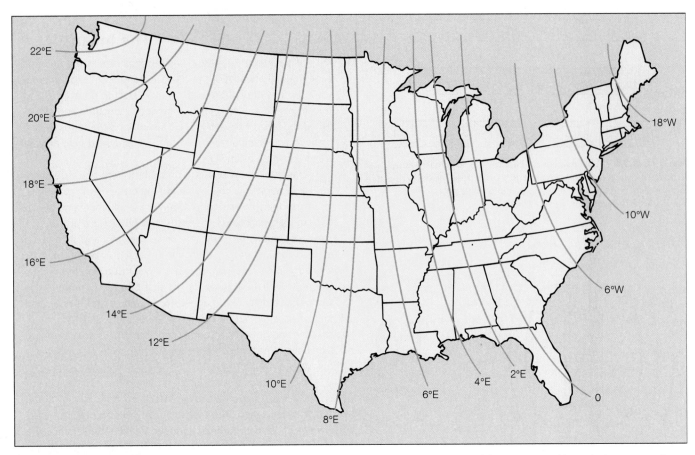

FIGURE 6.16 This magnetic declination map shows the approximate number of degrees east or west of the true geographic north that a magnetic compass will point in various locations.

Note in figure 6.15 that Earth's magnetic field acts as if there is a huge bar magnet inside Earth with a south pole near Earth's geographic North Pole. This is not an error. The north pole of a magnet is attracted to the south pole of a second magnet, and the north pole of a compass needle points to the north. Therefore, the bar magnet must be arranged as shown. This apparent contradiction is a result of naming the magnetic poles after their "seeking" direction.

Moving Charges and Magnetic Fields

Recall that every electric charge is surrounded by an electric field. If the charge is moving, it is surrounded by an electric field *and* a magnetic field. The magnetic field is formed in the shape of circles around the moving charge. A DC current is many charges moving through a conductor in response to an electric field, and these moving charges also produce a magnetic field. The field is in the shape of circles around the length of a current-carrying wire (figure 6.17). This relationship suggests that electricity and magnetism are two different manifestations of charges in motion. The electric field of a charge, for example, is fixed according to the fundamental charge of the particle. The magnetic field, however, changes with the velocity of the moving charge. The magnetic field does not exist at all if the charge is not moving, and the strength of the magnetic

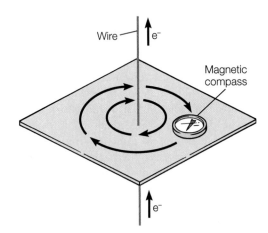

FIGURE 6.17 A magnetic compass shows the presence and direction of the magnetic field around a straight length of current-carrying wire.

field increases with increases in velocity. It seems clear that magnetic fields are produced by the motion of charges, or electric currents. Thus, a magnetic field is a property of the space around a moving charge.

You can see the shape of the magnetic field established by the current by running a straight wire vertically through a sheet of paper. The wire is connected to a battery, and iron filings are sprinkled on the paper. The filings will become aligned as each tiny piece of iron is moved parallel to the field. Overall, filings near the wire form a pattern of circles with the wire in the center.

A current-carrying wire that is formed into a loop has circular magnetic field lines that pass through the inside of the loop in the same direction. This has the effect of concentrating the field lines, which increases the magnetic field intensity. Since the field lines all pass through the loop in the same direction, one side of the loop will have a north pole and the other side a south pole (figure 6.18).

Many loops of wire formed into a cylindrical coil are called a **solenoid.** When a current passes through the loops of wire in a solenoid, each loop contributes field lines along the length of the cylinder (figure 6.19). The overall effect is a magnetic field around the solenoid that acts just like the magnetic field

of a bar magnet. This **electromagnet** can be turned on or off by turning the current on or off. In addition, the strength of the electromagnet depends on the size of the current and the number of loops. The strength of the electromagnet can also be increased by placing a piece of soft iron in the coil.

A solenoid can serve as an electrical switch or valve. It can be used as a water valve by placing a spring-loaded movable piece of iron inside the wire coil. When a current flows in such a coil, the iron is pulled into the coil by the magnetic field, turning the hot or cold water on in a washing machine or dishwasher, for example. Solenoids are also used as mechanical switches on VCRs, automobile starters, and signaling devices such as door bells and buzzers.

Electrons in atoms are moving around the nucleus, so they produce a magnetic field. Electrons also have a magnetic field associated with their spin. In most materials, these magnetic fields cancel one another and neutralize the overall magnetic effect. In other materials, such as an iron magnet, the electrons are arranged and oriented in such a way that individual magnetic fields are aligned. Each field becomes essentially a tiny magnet with a north and south pole. In an unmagnetized piece of iron, the fields are oriented in all possible directions and effectively cancel any overall magnetic effect. The net magnetism is therefore zero or near zero.

Magnetic Fields Interact

An electric charge at rest does not have a magnetic field, so it does not interact with a magnetic field from a magnet. A moving electric charge does produce a magnetic field, and this field will interact with another magnetic field, producing a force. A DC current is many charges moving through a wire, so it follows that a current-carrying wire will also interact with another magnetic field, producing a force. This force varies, depending on the size and direction of the current and the orientation of the external field. With a constant current, the force is at a maximum when the direction of the current is at a right angle to the other magnetic field.

Since you cannot measure electricity directly, it must be measured indirectly through one of the effects that it produces. Recall that the strength of the magnetic field around a current-carrying wire is proportional to the size of the current. Thus, one way to measure a current is to measure the magnetic field that it produces. A device that measures the size of a current from the size of its magnetic field is called a **galvanometer** (figure 6.20). A galvanometer has a coil of wire that can rotate on pivots in the magnetic field of a permanent magnet. When there is a current in the coil, the magnetic field produced is attracted and repelled by the field of the permanent magnet. The larger the current, the greater the force and the more the coil will rotate. The amount of movement of the coil is proportional to the current in the coil, and the pointer shows this on a scale. With certain modifications and applications, the device can be used to measure current (ammeter), voltage (voltmeter), and resistance (ohmmeter).

An electric motor is able to convert electrical energy to mechanical energy by using the force produced by interacting

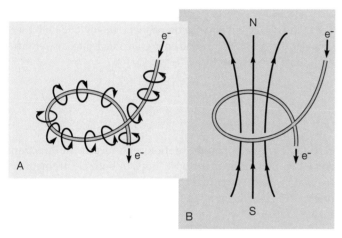

FIGURE 6.18 (A) Forming a wire into a loop causes the magnetic field to pass through the loop in the same direction. (B) This gives one side of the loop a north pole and the other side a south pole.

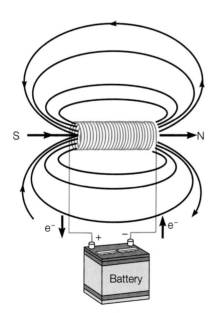

FIGURE 6.19 When a current is run through a cylindrical coil of wire, a solenoid, it produces a magnetic field like the magnetic field of a bar magnet.

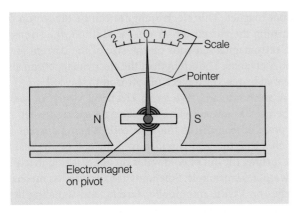

FIGURE 6.20 A galvanometer measures the direction and relative strength of an electric current from the magnetic field it produces. A coil of wire wrapped around an iron core becomes an electromagnet that rotates in the field of a permanent magnet. The rotation moves a pointer on a scale.

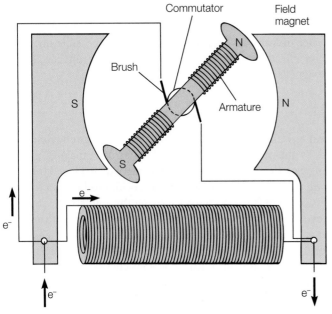

FIGURE 6.21 A schematic of a simple electric motor.

reverses the polarity, and the motion continues in one direction. An actual motor has many wire loops in both electromagnets to obtain a useful force and change the direction of the current often. This gives the motor a smoother operation with a greater turning force.

A Moving Magnet Produces an Electric Field

So far, you have learned what happens when we move a charge: (1) a moving charge and a current-carrying wire produce a magnetic field, and (2) a second magnetic field exerts a force on a moving charge and exerts a force on a current-carrying wire as their magnetic fields interact. Since a moving charge produces a magnetic field, what should we expect if we move a magnet? The answer is that a *moving magnet* produces an *electric field* in the shape of circles around the path of the magnet. Here we see some similarity: A moving charge produces a magnetic field and a moving magnet produces an electric field. Thus, we observe that electricity and magnetism interact, but only when there is motion.

A moving magnet produces an electric field that is circular around the path of the magnet. If you place a coil of wire near the moving magnet, the created electric field will interact with charges in the wire, forcing them to move as a current. The process of creating an electric current with a moving magnetic field is called **electromagnetic induction.** One way to produce electromagnetic induction is to move a bar magnet into or out of a coil of wire (figure 6.22). A galvanometer shows that the induced current flows one way when the bar magnet is moved toward the coil and flows the other way when the bar magnet is

magnetic fields. Basically, a motor has two working parts, a stationary electromagnet and an electromagnet that moves. The movable electromagnet rotates in the magnetic field of the stationary one. It turns fan blades, compressors, drills, pulleys, or other devices that do mechanical work.

Different designs of electric motors are used for various applications, but the simple demonstration motor shown in figure 6.21 can be used as an example of the basic operating principle. Both the stationary and the movable electromagnets are connected to an electric current. When the current is turned on, the unlike poles of the two fields attract, rotating the movable electromagnet for a half turn. The motor has a simple device that now reverses the direction of the current. This switches the magnetic poles on the movable electromagnet, which is now repelled for another half turn. The device again

Lemon Battery

1. You can make a simple compass galvanometer that will detect a small electric current (box figure 6.5). All you need is a magnetic compass and some thin insulated wire (the thinner the better).
2. Wrap the thin insulated wire in parallel windings around the compass. Make as many parallel windings as you can, but leave enough room to see both ends of the compass needle. Leave the wire ends free for connections.
3. To use the galvanometer, first turn the compass so the needle is parallel to the wire windings. When a current passes through the coil of wire, the magnetic field produced will cause the needle to move from its north-south position, showing the presence of a current. The needle will deflect one way or the other, depending on the direction of the current.
4. Test your galvanometer with a "lemon battery." Roll a soft lemon on a table while pressing on it with the palm of your hand. Cut two slits in the lemon about 1 cm apart. Insert a 8 cm (approximate) copper wire in one slit and a same-sized length of a straightened paper clip in the other slit, making sure the metals do not touch inside the lemon. Connect the galvanometer to the two metals. Try the two metals in other fruits, vegetables, and liquids. Can you find a pattern?

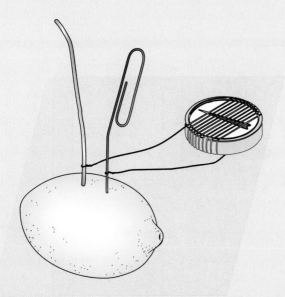

Box Figure 6.5 You can use the materials shown here to create and detect an electric current.

Swinging Coils

The interactions between moving magnets and moving charges can be easily demonstrated with two large magnets, two coils of wire, and a galvanometer.

1. Make a coil of insulated bell wire (#18 copper wire) by wraping fifty windings around a narrow jar. Tape the coil at several places so it does not come apart.
2. Connect the coil to a galvanometer (see "Concepts Applied: Lemon Battery" to make your own).
3. Move a strong bar magnet into and out of the stationary coil of wire and observe the galvanometer. Note the magnetic pole, direction of movement, and direction of current for both in and out movements.
4. Move the coil of wire back and forth over the stationary bar magnet.
5. Now make a second coil of wire from insulated bell wire and tape the coil as before.
6. Suspend the coil of wire on a ring stand or some other support on a table top. The coil should hang so it will swing with the broad circle of the coil moving back and forth. Place a large magnet on supports so it is near the center of the coil.
7. Set up an identical coil of wire, ring stand or support, and magnet on another table. Connect the two coils of wire.
8. Move one of the coils of wire and observe what happens to the second coil. The second coil should move, mirroring the movements of the first coil. (If it does not move, find some stronger magnets.)
9. Explain what happens at the first coil and the second coil.

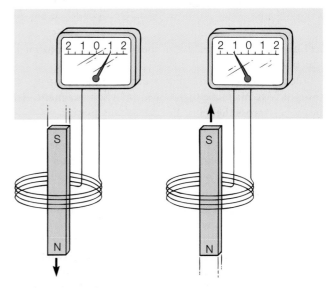

FIGURE 6.22 A current is induced in a coil of wire moved through a magnetic field. The direction of the current depends on the direction of motion.

moved away from the coil. The same effect occurs if you move the coil back and forth over a stationary magnet. Furthermore, no current is detected when the magnetic field and the coil of wire are not moving. Thus, electromagnetic induction depends on the relative motion of the magnetic field and the coil of wire. It does not matter which one moves or changes, but one

Connections . . .

Current War

Thomas Edison built the first electric generator and electrical distribution system to promote his new long-lasting lightbulbs. The DC generator and distribution system was built in lower Manhattan, New York City, and was switched on September 4, 1882. It supplied 110 volts DC to fifty-nine customers. Edison studied both AC and DC systems and chose DC because of the advantages it offered at the time. Direct current was used because batteries are DC, and batteries were used as a system backup. Also, DC worked fine with electric motors, and AC motors were not yet available.

George Westinghouse was in the business of supplying gas for gas lighting, and he could see that electric lighting would soon be replacing all the gaslights. After studying the matter, he decided that Edison's low-voltage system was not efficient enough. In 1885, he began experimenting with AC generators and transformers in Pittsburgh. In 1886, he installed a 500-volt AC system in Great Barrington, Massachusetts. The system stepped up the voltage to 3,000 volts for transmission, then back down to 110 volts, 60 Hz AC to power lightbulbs in homes and businesses.

Westinghouse's promotion of AC led to direct competition with Edison and his DC electrical systems. A "war of currents" resulted, with Edison claiming that transmission of such high voltage was dangerous. He emphasized this point by recommending the use of high-voltage AC in an electric chair as the best way to execute prisoners.

The advantages of AC were greater since you could increase the voltage, transmit for long distances at a lower cost, and then decrease the voltage to a safe level. Eventually, even Edison's own General Electric company switched to producing AC equipment. Westinghouse turned his attention to the production of large steam turbines for producing AC power and was soon setting up AC distribution systems across the nation.

down the AC voltage. It has two basic parts: (1) a *primary* or "input" coil, and (2) a *secondary* or "output" coil, which is close by. Both coils are often wound on a single iron core but are always fully insulated from each other. When an alternating current flows through the primary coil, a magnetic field grows around the coil to a maximum size, collapses to zero, then grows to a maximum size with an opposite polarity. This happens 120 times a second as the alternating current oscillates at 60 hertz. The growing and collapsing magnetic field moves across the wires in the secondary coil, inducing a voltage in the secondary coil.

The size of the induced voltage in the secondary coil is proportional to the number of wire loops in the two coils. Each turn or loop of the output coil has the same voltage induced in it, so more loops in an output coil means higher voltage output. If the output coil has the same number of loops as the input coil, the induced voltage in the secondary coil will be the same as the voltage in the primary coil. If the secondary coil has one-tenth as many loops as the primary coil, then the induced voltage in the secondary coil will be one-tenth the voltage in the primary coil. This is called a **step-down transformer** since the voltage was stepped down in the secondary coil. On the other hand, if the secondary coil has ten times more loops than the primary coil, then the voltage will be *increased* by a factor of 10. This is a **step-up transformer.** How much the voltage is stepped up or stepped down depends on the ratio of wire loops in the primary and secondary coils (figure 6.23).

must move or change relative to the other for electromagnetic induction to occur.

Electromagnetic induction occurs when the loop of wire moves across magnetic field lines or when magnetic field lines move across the loop. The size of the induced voltage is proportional to (1) the number of wire loops passing through the magnetic field lines, (2) the strength of the magnetic field, and (3) the rate at which magnetic field lines pass through the wire.

An **electric generator** is a device that converts mechanical energy into electrical energy. A simple generator is built much like an electric motor, with an axle with many loops in a wire coil that rotates. The coil is turned by some form of mechanical energy, such as a water turbine or a steam turbine, which uses steam generated from fossil fuels or nuclear energy. As the coil rotates in a magnetic field, a current is induced in the coil.

Current from a power plant goes to a transformer to step up the voltage. A **transformer** is a device that steps up or steps

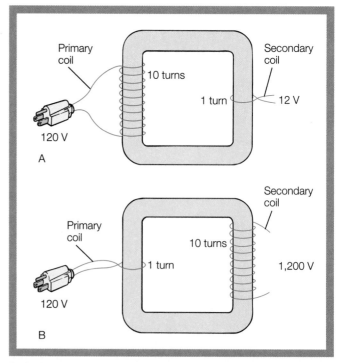

FIGURE 6.23 (*A*) This step-down transformer has ten turns on the primary for each turn on the secondary and reduces the voltage from 120 V to 12 V. (*B*) This step-up transformer increases the voltage from 120 V to 1,200 V, since there are ten turns on the secondary to each turn on the primary.

Note that the volts per wire loop are the same in each coil. The relationship is

$$\frac{\text{volts}_{\text{primary}}}{(\text{number of loops})_{\text{primary}}} = \frac{\text{volts}_{\text{secondary}}}{(\text{number of loops})_{\text{secondary}}}$$

or

$$\frac{V_{\text{p}}}{N_{\text{p}}} = \frac{V_{\text{s}}}{N_{\text{s}}} \qquad \textbf{equation 6.7}$$

EXAMPLE 6.8 (Optional)

A step-up transformer has five loops on its primary coil and twenty loops on its secondary coil. If the primary coil is supplied with an alternating current at 120 V, what is the voltage in the secondary coil?

SOLUTION

$$N_{\text{p}} = 5 \text{ loops}$$
$$N_{\text{s}} = 20 \text{ loops}$$
$$V_{\text{p}} = 120 \text{ V}$$
$$V_{\text{s}} = ?$$

$$\frac{V_{\text{p}}}{N_{\text{p}}} = \frac{V_{\text{s}}}{N_{\text{s}}} \therefore V_{\text{s}} = \frac{V_{\text{p}}N_{\text{s}}}{N_{\text{p}}}$$

$$= \frac{(120 \text{ V})(20 \text{ loops})}{5 \text{ loops}}$$

$$= \frac{(120)(20)}{5} \frac{\text{V} \cdot \text{loops}}{\text{loops}}$$

$$= \boxed{480 \text{ V}}$$

EXAMPLE 6.9 (Optional)

The step-up transformer in example 6.8 is supplied with an alternating current at 120 V and a current of 10.0 A in the primary coil. What current flows in the secondary circuit?

$$V_{\text{p}} = 120 \text{ V}$$
$$I_{\text{p}} = 10.0 \text{ A}$$
$$V_{\text{s}} = 480 \text{ V}$$
$$I_{\text{s}} = ?$$

$$V_{\text{p}}I_{\text{p}} = V_{\text{s}}I_{\text{s}} \therefore I_{\text{s}} = \frac{V_{\text{p}}I_{\text{p}}}{V_{\text{s}}}$$

$$= \frac{(120 \text{ V})(10.0 \text{ A})}{480 \text{ V}}$$

$$= \frac{(120)(10.0)}{480} \frac{\text{V} \cdot \text{A}}{\text{V}}$$

$$= \boxed{2.5 \text{ A}}$$

A step-up or step-down transformer steps up or steps down the voltage of an alternating current according to the ratio of wire loops in the primary and secondary coils. Assuming no losses in the transformer, the *power input* on the primary coil equals the *power output* on the secondary coil. Since $P = IV$, you can see that when the voltage is stepped up the current is correspondingly decreased, as

$$\text{power input} = \text{power output}$$
$$\text{watts input} = \text{watts output}$$
$$(\text{amps} \times \text{volts})_{\text{in}} = (\text{amps} \times \text{volts})_{\text{out}}$$

or

$$V_{\text{p}}I_{\text{p}} = V_{\text{s}}I_{\text{s}} \qquad \textbf{equation 6.8}$$

Energy losses in transmission are reduced by stepping up the voltage. Recall that electrical resistance results in an energy loss and a corresponding absolute temperature increase in the conducting wire. If the current is large, there are many collisions between the moving electrons and positive ions of the wire, resulting in a large energy loss. Each collision takes energy from the electric field, diverting it into increased kinetic energy of the positive ions and thus increased temperature of the conductor. The energy lost to resistance is therefore reduced by lowering the current, which is what a transformer does by increasing the voltage. Hence, electric power companies step up the voltage of generated power for economical transmission. A step-up transformer at a power plant, for example, might step up the voltage from 22,000 volts to 500,000 volts for transmission across the country to a city (figure 6.24A). This step up in voltage correspondingly reduces the current, lowering the resistance losses to a more acceptable 4 or 5 percent over long distances. A step-down transformer at a substation near the city reduces the voltage to several thousand volts for transmission around the city. Additional step-down transformers reduce this voltage to 120 volts for transmission to three or four houses (figure 6.24B).

A

FIGURE 6.24 Energy losses in transmission are reduced by increasing the voltage, so the voltage of generated power is stepped up at the power plant. (*A*) These transformers, for example, might step up the voltage from tens to hundreds of thousands of volts. After a step-down transformer reduces the voltage at a substation, still another transformer (*B*) reduces the voltage to 120 volts for transmission to three or four houses.

B

Solar Cells

You may be familiar with many solid-state devices such as calculators, computers, word processors, digital watches, VCRs, digital stereos, and camcorders. All of these are called solid-state devices because they use a solid material, such as the semiconductor silicon, in an electric circuit in place of vacuum tubes. Solid-state technology developed from breakthroughs in the use of semiconductors during the 1950s, and the use of thin pieces of silicon crystal is common in many electric circuits today.

A related technology also uses thin pieces of a semiconductor such as silicon but not as a replacement for a vacuum tube. This technology is concerned with photovoltaic devices, also called *solar cells*, that generate electricity when exposed to light (box figure 6.6). A solar cell is unique in generating electricity since it produces electricity directly, without moving parts or chemical reactions, and potentially has a very long lifetime. This reading is concerned with how a solar cell generates electricity.

The conducting properties of silicon can be changed by *doping*, that is, artificially forcing atoms of other elements into the crystal lattice. Phosphorus, for example, has five electrons in its outermost shell, compared to the four in a silicon atom. When phosphorus atoms replace silicon atoms in the crystal lattice, there are extra electrons not tied up in the two electron bonds. The extra electrons move easily through the crystal lattice, carrying a charge. Since the phosphorus-doped silicon carries a negative charge, it is called an *n-type* semiconductor. The *n* means negative charge carrier.

A silicon crystal doped with boron will have atoms in the lattice with only three electrons in the outermost shell. This results in a deficiency, that is, electron "holes" that act as positive charges. A hole can move as an electron is attracted to it, but it leaves another hole elsewhere, from where it moved. Thus, a flow of electrons in one direction is equivalent to a flow of holes in the opposite direction. A hole, therefore, behaves as a positive charge. Since

A

B

Box Figure 6.6   Solar cells are economical in remote uses such as (*A*) navigational aids and (*B*) communications. The solar panels in both of these examples are oriented toward the south.

the boron-doped silicon carries a positive charge, it is called a *p-type* semiconductor. The *p* means positive charge carrier.

The basic operating part of a silicon solar cell is typically an 8 cm wide and 3×10^{-1}mm (about one-hundredth of an inch) thick wafer cut from a silicon crystal. One side of the wafer is doped with boron to make p-silicon, and the other side is doped with phosphorus to make n-silicon. The place of contact between the two is called the p-n junction, which creates a *cell barrier*. The cell barrier forms as electrons are attracted from the n-silicon to the holes in the p-silicon. This creates a very thin zone of negatively charged p-silicon and positively charged n-silicon (box figure 6.7). Thus, an internal electric field is established at the p-n junction, and the field is the cell barrier.

The cell is thin, and light can penetrate through the p-n junction. Light impacts the p-silicon, freeing electrons. Low-energy free electrons might combine with a hole, but high-energy electrons cross the cell barrier into the n-silicon. The electron loses

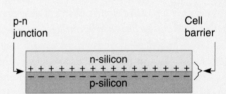

Box Figure 6.7   The cell barrier forms at the p-n junction between the n-silicon and the p-silicon. The barrier creates a "one-way" door that accumulates negative charges in the n-silicon.

some of its energy, and the barrier prevents it from returning, creating an excess negative charge in the n-silicon and a positive charge in the p-silicon. This establishes a potential that will drive a current.

Today, solar cells are essentially handmade and are economical only in remote power uses (navigational aids, communications, or irrigation pumps) and in consumer specialty items (solar-powered watches and calculators). Research continues on finding methods of producing highly efficient, highly reliable solar cells that are affordably priced.

Science and Society

Blackout Reveals Pollution

In August 2003, problems in a huge electrical grid resulted in power plants shutting down and a massive electric power blackout that affected some 50 million people. Scientists from the University of Maryland took advantage of the shutdown, measuring different levels of atmospheric air pollution while the fossil-fueled power plants in the Ohio Valley were shut down. Scooping many air samples with a small airplane 24 hours after the blackout, they found 90 percent less sulfur dioxide, 50 percent less ozone, and 70 percent fewer light-scattering particles from the air in the same area than when the power plants were running. The scientists stated that the result could come from an underestimation of emission from power plants or from unknown chemical reactions in the atmosphere.

Questions to Discuss

1. Are there atmospheric factors that might have contributed to the findings?

2. Are there factors on the ground that might have contributed to the findings?

3. Do the results mean that power plants contribute that much pollution, or what else should be considered?

4. How would you conduct a pollution-measuring experiment that would leave no room for doubt about the results?

SUMMARY

The basic unit of matter is the *atom,* which is made up of *protons, electrons,* and other particles. Protons and electrons have a property called *electric charge;* electrons have a *negative electric charge* and protons have a *positive electric charge.* The charges interact, and *like charges repel* and *unlike charges attract.*

Electrons can be moved and *electrostatic charge,* or *static electricity,* results from a surplus or deficiency of electrons.

A *quantity of charge* (q) is measured in units of *coulombs* (C), the charge equivalent to the transfer of 6.24×10^{18} charged particles such as the electron. The *fundamental charge* of an electron or proton is 1.60×10^{-19} coulomb. The *electric forces* between two charged objects can be calculated from the relationship between the quantity of charge and the distance between two charged objects. The relationship is known as *Coulomb's law.*

A flow of electric charge is called an *electric current* (I). Current (I) is measured as the *rate* of flow of charge, the quantity of charge (q) through a conductor in a period of time (t). The unit of current in coulomb/second is called an *ampere,* or *amp* for short (A).

An *electric circuit* has some device that does work in moving charges through wires to do work in another part of the circuit. The *work done* and the *size of the charge* moved defines *voltage.* A *volt* (V) is the ratio of work to charge moved, $V = W/q$. The ratio of volts/amps in a circuit is the *unit of resistance* called an *ohm. Ohm's law* is $V = IR$.

Disregarding the energy lost to resistance, the *work* done by a voltage source is equal to the work accomplished in electrical devices in a circuit. The *rate* of doing work is *power,* or *work per unit time,* $P = W/t$. Electrical power can be calculated from the relationship of $P = IV$, which gives the power unit of *watts.*

Magnets have two poles about which their attraction is concentrated. When free to turn, one pole moves to the north and the other to the south. The north-seeking pole is called the *north pole* and the south-seeking pole is called the *south pole. Like poles repel* one another and *unlike poles attract.*

A *current-carrying wire* has magnetic field lines of closed, *concentric circles* that are at right angles to the length of wire. The *direction* of the magnetic field depends on the direction of the current. A coil of many loops is called a *solenoid* or *electromagnet.* The electromagnet is the working part in electric meters, electromagnetic switches, and the electric motor.

When a loop of wire is moved in a magnetic field, or if a magnetic field is moved past a wire loop, a voltage is induced in the wire loop. The interaction is called *electromagnetic induction.* An electric generator is a rotating coil of wire in a magnetic field. The coil is rotated by mechanical energy, and electromagnetic induction induces a voltage, thus converting mechanical energy to electrical energy. A *transformer* steps up or steps down the voltage of an alternating current. The ratio of input and output voltage is determined by the number of loops in the primary and secondary coils. Increasing the voltage decreases the current, which makes long-distance transmission of electrical energy economically feasible.

Summary of Equations

6.1 $\text{electric force} = (\text{constant}) \times \dfrac{\text{charge on one object} \times \text{charge on second object}}{\text{distance between objects squared}}$

$$F = k\frac{q_1 q_2}{d^2}$$

where $k = 9.00 \times 10^9$ newton·meters^2/coulomb^2

6.2 $\text{electrical current} = \dfrac{\text{quantity of charge}}{\text{time}}$

$$I = \frac{q}{t}$$

6.3
$$\text{voltage} = \frac{\text{work}}{\text{charge moved}}$$
$$V = \frac{W}{q}$$

6.6
$$\text{cost} = \frac{(\text{work})(\text{time})(\text{rate})}{1,000 \frac{\text{watt}}{\text{kilowatt}}}$$

6.7
$$\frac{\text{volts}_{\text{primary}}}{(\text{number of loops})_{\text{primary}}} = \frac{\text{volts}_{\text{secondary}}}{(\text{number of loops})_{\text{secondary}}}$$
$$\frac{V_{\text{p}}}{N_{\text{p}}} = \frac{V_{\text{s}}}{N_{\text{s}}}$$

6.4
$$\text{voltage} = \text{current} \times \text{resistance}$$
$$V = IR$$

6.5
$$\text{power} = \text{voltage} \times \text{current}$$
$$P = VI$$

6.8
$$(\text{amps} \times \text{volts})_{\text{in}} = (\text{amps} \times \text{volts})_{\text{out}}$$
$$V_{\text{p}}I_{\text{p}} = V_{\text{s}}I_{\text{s}}$$

KEY TERMS

alternating current (p. **132**)
amp (p. **130**)
ampere (p. **130**)
coulomb (p. **128**)
Coulomb's law (p. **129**)
direct current (p. **132**)
electric charge (p. **126**)

electric circuit (p. **132**)
electric current (p. **129**)
electric field (p. **129**)
electric generator (p. **144**)
electric force (p. **127**)
electromagnet (p. **141**)
electromagnetic induction (p. **142**)

electrostatic charge (p. **127**)
fundamental charge (p. **128**)
galvanometer (p. **141**)
magnetic field (p. **138**)
magnetic poles (p. **138**)
ohm (p. **134**)
Ohm's law (p. **134**)

solenoid (p. **141**)
step-down transformer (p. **144**)
step-up transformer (p. **144**)
transformer (p. **144**)
volt (p. **133**)
watt (p. **134**)

APPLYING THE CONCEPTS

1. An object that acquires an excess of electrons becomes a (an)
   a. ion.
   b. negatively charged object.
   c. positively charged object.
   d. electric conductor.

2. Which of the following is most likely to acquire an electrostatic charge?
   a. electric conductor
   b. electric nonconductor
   c. Both are equally likely.
   d. None of the above is correct.

3. A quantity of electric charge is measured in a unit called a (an)
   a. coulomb.
   b. volt.
   c. amp.
   d. watt.

4. The unit that describes the potential difference that occurs when a certain amount of work is used to move a certain quantity of charge is called the
   a. ohm.
   b. volt.
   c. amp.
   d. watt.

5. An electric current is measured in units of
   a. coulomb.
   b. volt.
   c. amp.
   d. watt.

6. In an electric current, the electrons are moving
   a. at a very slow rate.
   b. at the speed of light.
   c. faster than the speed of light.
   d. at a speed described as "Warp 8."

7. If you multiply amps × volts, the answer will be in units of
   a. resistance.
   b. work.
   c. current.
   d. power.

8. The unit of resistance is the
   a. watt.
   b. ohm.
   c. amp.
   d. volt.

9. Compared to a thick wire, a thin wire of the same length, material, and temperature has
   a. less electric resistance.
   b. more electric resistance.
   c. the same electric resistance.
   d. None of the above is correct.

10. A permanent magnet has magnetic properties because
    a. the magnetic fields of its electrons are balanced.
    b. of an accumulation of monopoles in the ends.
    c. the magnetic fields of elections are aligned.
    d. All of the above are correct.

11. A current-carrying wire has a magnetic field around it because
    a. a moving charge produces a magnetic field of its own.
    b. the current aligns the magnetic fields in the metal of the wire.
    c. the metal was magnetic before the current was established and the current enhanced the magnetic effect.
    d. None of the above is correct.

12. A step-up transformer steps up (the)
    a. power.
    b. current.
    c. voltage.
    d. All of the above are correct.

Answers

1. b **2.** b **3.** a **4.** b **5.** c **6.** a **7.** d **8.** b **9.** b **10.** c **11.** a **12.** c

QUESTIONS FOR THOUGHT

1. Explain why a balloon that has been rubbed sticks to a wall for a while.
2. Explain what is happening when you walk across a carpet and receive a shock when you touch a metal object.
3. Why does a positively or negatively charged object have multiples of the fundamental charge?
4. Explain how you know that it is an electric field, not electrons, that moves rapidly through a circuit.
5. Is a kWh a unit of power or a unit of work? Explain.
6. What is the difference between AC and DC?
7. What is a magnetic pole? How are magnetic poles named?
8. How is an unmagnetized piece of iron different from the same piece of iron when it is magnetized?
9. Explain why the electric utility company increases the voltage of electricity for long-distance transmission.
10. Describe how an electric generator is able to generate an electric current.
11. Why does the north pole of a magnet generally point to the geographic North Pole if like poles repel?
12. Explain what causes an electron to move toward one end of a wire when the wire is moved across a magnetic field.

FOR FURTHER ANALYSIS

1. Explain how the model of electricity as electrons moving along a wire is an oversimplification that misrepresents the complex nature of an electric current.
2. What are the significant similarities and differences between AC and DC? What determines which is "better" for a particular application?
3. Transformers usually have signs warning, "Danger—High Voltage." Analyze if this is a contradiction since it is exposure to amps, not volts, that harms people.
4. Will a fuel cell be the automobile engine of the future? Identify the facts, beliefs, and theories that support or refute your answer.
5. Analyze the apparent contradiction in the statement that "solar energy is free" with the fact that solar cells are too expensive to use as an energy source.
6. What are the basic similarities and differences between an electric field and a magnetic field?

INVITATION TO INQUIRY

Earth Power?

Investigate if you can use Earth's magnetic field to induce an electric current in a conductor. Connect the ends of a 10 m wire to a galvanometer. Have a partner hold the ends of the wire on the galvanometer while you hold the end of the wire loop and swing the doubled wire like a jump rope.

If you accept this invitation, try swinging the wire in different directions. Can you figure out a way to measure how much electricity you can generate?

PARALLEL EXERCISES

The exercises in groups A and B cover the same concepts. Solutions to group A exercises are located in appendix D.

Group A

1. A rubber balloon has become negatively charged from being rubbed with a wool cloth, and the charge is measured as 1.00×10^{-14} C. According to this charge, the balloon contains an excess of how many electrons?

2. An electric current through a wire is 6.00 C every 2.00 s. What is the magnitude of this current?

3. A current of 4.00 A flows through a toaster connected to a 120.0 V circuit. What is the resistance of the toaster?

4. What is the current in a 60.0 Ω resistor when the potential difference across it is 120.0 V?

5. A lightbulb with a resistance of 10.0 Ω allows a 1.20 A current to flow when connected to a battery. (a) What is the voltage of the battery? (b) What is the power of the lightbulb?

6. A small radio operates on 3.00 V and has a resistance of 15.0 Ω. At what rate does the radio use electric energy?

7. A 1,200 W hair dryer is operated on a 120 V circuit for 15 min. If electricity costs $0.10/kWh, what was the cost of using the blow dryer?

8. An automobile starter rated at 2.00 hp draws how many amps from a 12.0 V battery?

9. An average-sized home refrigeration unit has a 1/3 hp fan motor for blowing air over the inside cooling coils, a 1/3 hp fan motor for blowing air over the outside condenser coils, and a 3.70 hp compressor motor. (a) All three motors use electric energy at what rate? (b) If electricity costs $0.10/kWh, what is the cost of running the unit per hour? (c) What is the cost for running the unit 12 hours a day for a 30-day month?

10. A 15 ohm toaster is turned on in a circuit that already has a 0.20 hp motor, three 100 W lightbulbs, and a 600 W electric iron that are on. Will this trip a 15 A circuit breaker? Explain.

11. A power plant generator produces a 1,200 V, 40 A alternating current that is fed to a step-up transformer before transmission over the high lines. The transformer has a ratio of 200 to 1 wire loops. (a) What is the voltage of the transmitted power? (b) What is the current?

12. A step-down transformer has an output of 12 V and 0.5 A when connected to a 120 V line. Assuming no losses: (a) What is the ratio of primary to secondary loops? (b) What current does the transformer draw from the line? (c) What is the power output of the transformer?

Group B

1. An inflated rubber balloon is rubbed with a wool cloth until an excess of a billion electrons is on the balloon. What is the magnitude of the charge on the balloon?

2. A wire carries a current of 2.0 A. At what rate is the charge flowing?

3. A current of 0.83 A flows through a lightbulb in a 120 V circuit. What is the resistance of this lightbulb?

4. What is the voltage across a 60.0 Ω resistor with a current of 3⅓ amp?

5. A 10.0 Ω lightbulb is connected to a 12.0 V battery. (a) What current flows through the bulb? (b) What is the power of the bulb?

6. A lightbulb designed to operate in a 120.0 V circuit has a resistance of 192 Ω. At what rate does the bulb use electric energy?

7. What is the monthly energy cost of leaving a 60 W bulb on continuously if electricity costs $0.10 per kWh?

8. An electric motor draws a current of 11.5 A in a 240 V circuit. (a) What is the power of this motor in W? (b) How many horsepower is this?

9. A swimming pool requiring a 2.0 hp motor to filter and circulate the water runs for 18 hours a day. What is the monthly electrical cost for running this pool pump if electricity costs $0.10 per kWh?

10. Is it possible for two people to simultaneously operate 1,300 W hair dryers on the same 120 V circuit without tripping a 15 A circuit breaker? Explain.

11. A step-up transformer has a primary coil with 100 loops and a secondary coil with 1,500 loops. If the primary coil is supplied with a household current of 120 V and 15 A, (a) what voltage is produced in the secondary circuit? (b) What current flows in the secondary circuit?

12. The step-down transformer in a local neighborhood reduces the voltage from a 7,200 V line to 120 V. (a) If there are 125 loops on the secondary, how many are on the primary coil? (b) What current does the transformer draw from the line if the current in the secondary is 36 A? (c) What are the power input and output?

This fiber optics bundle carries pulses of light from an infrared laser to carry much more information than could be carried by electrons moving through wires. This is part of a dramatic change underway, a change that will first find a hybrid "optoelectronics" replacing the more familiar "electronics" of electrons and wires.

Light

OVERVIEW

You use light and your eyes more than any other sense to learn about your surroundings. All of your other senses—touch, taste, sound, and smell—involve matter, but the most information is provided by light. Yet, light seems more mysterious than matter. You can study matter directly, measuring its dimensions, taking it apart, and putting it together to learn about it. Light, on the other hand, can only be studied indirectly in terms of how it behaves (figure 7.1). Once you understand its behavior, you know everything there is to know about light. Anything else is thinking about what the behavior means.

The behavior of light has stimulated thinking, scientific investigations, and debate for hundreds of years. The investigations and debate have occurred because light cannot be directly observed, which makes the exact nature of light very difficult to pin down. For example, you know that light moves energy from one place to another place. You can feel energy from the Sun as sunlight warms you, and you know that light has carried this energy across millions of miles of empty space. The ability of light to move energy like this could be explained (1) as energy transported by waves, just as sound waves carry energy from a source, or (2) as the kinetic energy of a stream of moving particles, which give up their energy when they strike a surface. The movement of energy from place to place could be explained equally well by a wave model of light or by a particle model of light. When two possibilities exist like this in science, experiments are designed and measurements are made to support one model and reject the other. Light, however, presents a baffling dilemma. Some experiments provide evidence that light consists of waves and not a stream of moving particles. Yet other experiments provide evidence of just the opposite, that light is a stream of particles and not a wave. Evidence for accepting a wave or particle model seems to depend on which experiments are considered.

The purpose of using a model is to make new things understandable in terms of what is already known. When these new things concern light, three models are useful in visualizing separate behaviors. Thus, the electromagnetic wave model will be used to describe how light is created at a source. Another model, a model of light as a ray, a small beam of light, will be used to discuss some common properties of light such as reflection and the refraction, or bending, of light. Finally, properties of light that provide evidence for a particle model will be discussed before ending with a discussion of the present understanding of light.

SOURCES OF LIGHT

The Sun, and other stars, lightbulbs, and burning materials all give off light. When something produces light, it is said to be **luminous.** The Sun is a luminous object that provides almost all of the *natural* light on Earth. A small amount of light does reach Earth from the stars but not really enough to see by on a moonless night. The Moon and planets shine by reflected light and do not produce their own light, so they are not luminous.

Burning has been used as a source of *artificial* light for thousands of years. A wood fire and a candle flame are luminous because of their high temperatures. When visible light is given off as a result of high temperatures, the light source is said to be **incandescent.** A flame from any burning source, an ordinary lightbulb, and the Sun are all incandescent sources because of high temperatures.

How do incandescent objects produce light? One explanation is given by the electromagnetic wave model. This model

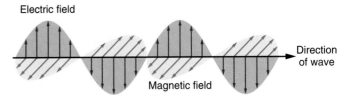

FIGURE 7.2 The electric and magnetic fields in an electromagnetic wave vary together. Here the fields are represented by arrows that indicate the strength and direction of the fields. Note the fields are perpendicular to each other and to the direction of the wave.

FIGURE 7.1 Light, sounds, and odors can identify the pleasing environment of this garden, but light provides the most information. Sounds and odors can be identified and studied directly, but light can only be studied indirectly, that is, in terms of how it behaves. As a result, the behavior of light has stimulated thinking, scientific investigations, and debate for hundreds of years. Perhaps you have wondered about light and its behaviors. What is light?

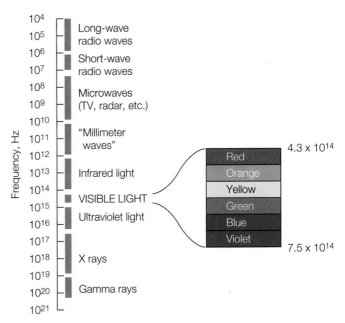

FIGURE 7.3 The electromagnetic spectrum. All electromagnetic waves have the same fundamental character and the same speed in a vacuum, but many aspects of their behavior depend on their frequency.

describes a relationship between electricity, magnetism, and light. The model pictures an electromagnetic wave as forming whenever an electric charge is *accelerated* by some external force. The acceleration produces a wave consisting of electrical and magnetic fields that become separated from the accelerated charge, moving off into space (figure 7.2). As the wave moves through space, the two fields exchange energy back and forth, continuing on until they are absorbed by matter and give up their energy.

The frequency of an electromagnetic wave depends on the acceleration of the charge; the greater the acceleration, the higher the frequency of the wave that is produced. The complete range of frequencies is called the *electromagnetic spectrum* (figure 7.3). The spectrum ranges from radio waves at the low frequency end of the spectrum to gamma rays at the high frequency end. Visible light occupies only a small part of the middle portion of the complete spectrum.

Visible light is emitted from incandescent sources at high temperatures, but actually electromagnetic radiation is given off from matter at *any* temperature. This radiation is called **blackbody radiation,** which refers to an idealized material (the *blackbody*) that perfectly absorbs and perfectly emits electromagnetic radiation. From the electromagnetic wave model, the radiation originates from the acceleration of charged particles

near the surface of an object. The frequency of the black-body radiation is determined by the energy available for accelerating charged particles, that is, the temperature of the object. Near absolute zero, there is little energy available and no radiation is given off. As the temperature of an object is increased, more energy is available, and this energy is distributed over a range of values, so more than one frequency of radiation is emitted. A graph of the frequencies emitted from the range of available energy is thus somewhat bell-shaped. The steepness of the curve and the position of the peak depend on the temperature (figure 7.4). As the temperature of an object increases, there is an increase in the *amount* of radiation given off, and the peak radiation emitted progressively *shifts* toward higher and higher frequencies.

At room temperature, the radiation given off from an object is in the infrared region, invisible to the human eye. When the temperature of the object reaches about 700°C (about 1,300°F), the peak radiation is still in the infrared region, but the peak has shifted enough toward the higher frequencies that a little visible light is emitted as a dull red glow. As the

Connections . . .

Ultraviolet Light and Life

Ultraviolet (UV) light is a specific portion of the electromagnetic spectrum that has a significant impact on living things. UV light has been subdivided into different categories based on wavelength—the shorter the wavelength, the higher the energy level. Ultraviolet light causes mutations when absorbed by DNA.

▶ Ultraviolet light is important in forming vitamin D in the skin. When cholesterol in the skin is struck by ultraviolet light, some of the cholesterol is converted to vitamin D. Therefore, some exposure to ultraviolet light is good. However, since large amounts of ultraviolet light damage the skin, the normal tanning reaction seen in the skin of light-skinned people is a protective reaction to the damaging effects of ultraviolet light. The darker skin protects the deeper layers of the skin from damage.

▶ Tanning booths have mostly UV-A with a small amount of UV-B. Dosages in tanning booths are usually higher than in sunlight.

▶ Since UV-C disrupts DNA, it can be used to sterilize medical instruments, drinking water, and wastewater. A standard mercury vapor light emits much of its light in the UV-C range, but the lamp is coated to prevent light in that range from leaving the lamp. Specially manufactured lamps without the coating and ultraviolet transparent glass are used to provide UV-C for germicidal purposes.

▶ Some animals are able to see light in the ultraviolet spectrum. In particular, bees do not see the color red but are able to see ultraviolet. When flowers are photographed with filters that allow ultraviolet light to be recorded on the film, flowers often have a different appearance from what we see. The pattern of ultraviolet light assists the bees in finding the source of nectar within the flower.

Class	Wavelength (nanometers)	Characteristics	Biological Effects
UV-A	320–400	Most passes through the ozone layer; can pass through window glass	Causes skin damage and promotes skin cancer
UV-B	280–320	90% blocked by ozone layer; amount is highest at midday	Causes sunburn, skin damage, and skin cancer
UV-C	100–280	Nearly all blocked by ozone layer	Kills bacteria and viruses

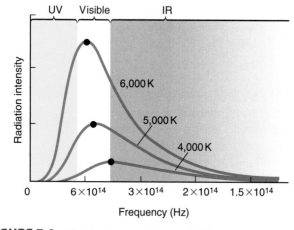

FIGURE 7.4 Blackbody spectra for several different temperatures. The frequency of the peak of the curve (shown by dot) shifts to higher frequency at higher temperatures.

temperature of the object continues to increase, the amount of radiation increases, and the peak continues to shift toward shorter wavelengths. Thus, the object begins to glow brighter, and the color changes from red, to orange, to yellow, and eventually to white. The association of this color change with temperature is noted in the referent description of an object being "red hot," "white hot," and so forth.

The incandescent flame of a candle or fire results from the blackbody radiation of carbon particles in the flame. At a black-body temperature of 1,500°C (about 2,700°F), the carbon particles emit visible light in the red to yellow frequency range. The tungsten filament of an incandescent lightbulb is heated to about 2,200°C (about 4,000°F) by an electric current. At this temperature, the visible light emitted is in the reddish, yellow-white range.

The radiation from the Sun, or sunlight, comes from the Sun's surface, which has a temperature of about 5,700°C (about 10,000°F). As shown in figure 7.5, the Sun's radiation has a broad spectrum centered near the yellow-green frequency. Your eye is most sensitive to this frequency of sunlight. The spectrum of sunlight before it travels through Earth's atmosphere is infrared (about 51 percent), visible light (about 40 percent), and ultraviolet (about 9 percent). Sunlight originated as energy released from nuclear reactions in the Sun's core (see chapter 12). This energy requires about a million years to work its way up to the surface. At the surface, the energy from the core accelerates charged particles, which then emit light like tiny antennas. The sunlight requires about eight minutes to travel the distance from the Sun's surface to Earth.

A cell phone is a small, low-power radio that is used with a system of cells. Cells are about 26 km²(10 mi²) in area, and each has its own cell pod tower and base station. There is also a central switching station that controls all the cell base stations and the phone connections to the regular telephone system. When you make a phone call within a cell, the central switching station connects you to the other phone on a certain electromagnetic (radio) frequency. As you move from one cell to another, the base station notices your changing signal strength. At some point, the central switching station switches you from one cell to a new frequency in the new cell. Thus, your call actually is a two-way radio transmission that is switched one cell to another as you move.

Questions to Discuss

1. Do drivers using a cell phone put themselves and surrounding drivers at risk?

2. Are new laws needed to restrict the use of a cell phone while driving?

3. Comment on this statement: The goal of using a cell phone is to communicate effectively without anybody else noticing or caring.

4. Why do some people seem incapable of talking on a cell phone in a normal tone of voice?

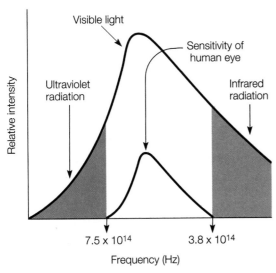

FIGURE 7.5 Sunlight is about 9 percent ultraviolet radiation, 40 percent visible light, and 51 percent infrared radiation before it travels through Earth's atmosphere.

PROPERTIES OF LIGHT

You can see luminous objects from the light they emit, and you can see nonluminous objects from the light they reflect, but you cannot see the path of the light itself. For example, you cannot see a flashlight beam unless you fill the air with chalk dust or smoke. The dust or smoke particles reflect light, revealing the path of the beam. This simple observation must be unknown to the makers of science fiction movies, since they always show visible laser beams zapping through the vacuum of space.

Some way to represent the invisible travels of light is needed in order to discuss some of its properties. Throughout history, a **light ray model** has been used to describe the travels of light. The meaning of this model has changed over time, but it has always been used to suggest that "something" travels in *straight-line paths*. The light ray is a line that is drawn to represent the straight-line travel of light. A line is drawn to represent this imaginary beam to illustrate the law of reflection (as from a mirror) and the law of refraction (as through a lens). There are limits to using a light ray for explaining some properties of light, but it works very well in explaining mirrors, prisms, and lenses.

Light Interacts with Matter

A ray of light travels in a straight line from a source until it encounters some object or particles of matter (figure 7.6). What happens next depends on several factors, including (1) the smoothness of the surface, (2) the nature of the material, and (3) the angle at which the light ray strikes the surface.

The *smoothness* of the surface of an object can range from perfectly smooth to extremely rough. If the surface is perfectly smooth, rays of light undergo *reflection*, leaving the surface parallel to each other. A mirror is a good example of a very smooth surface that reflects light this way (figure 7.7A). If a surface is not smooth, the light rays are reflected in many random directions as *diffuse reflection* takes place (figure 7.7B).

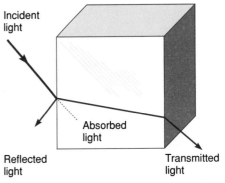

FIGURE 7.6 Light that interacts with matter can be reflected, absorbed, or transmitted through transparent materials. Any combination of these interactions can take place, but a particular substance is usually characterized by what it mostly does to light.

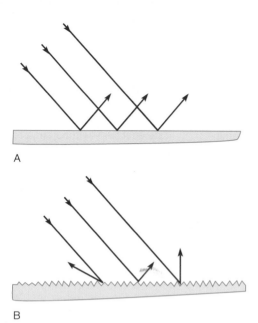

A

B

FIGURE 7.7 (*A*) Rays reflected from a perfectly smooth surface are parallel to each other. (*B*) Diffuse reflection from a rough surface causes rays to travel in many random directions.

FIGURE 7.8 Light travels in a straight line, and the color of an object depends on which wavelengths of light the object reflects. Each of these flowers absorbs the colors of white light and reflects the color that you see.

Rough and irregular surfaces and dust in the air make diffuse reflections. It is diffuse reflection that provides light in places not in direct lighting, such as under a table or a tree. Such shaded areas would be very dark without diffuse reflection of light.

Some materials allow much of the light that falls on them to move through the material without being reflected. Materials that allow transmission of light through them are called *transparent*. Glass and clear water are examples of transparent materials. Many materials do not allow transmission of any light and are called *opaque*. Opaque materials reflect light, absorb light, or some combination of partly absorbing and partly reflecting light (figure 7.8). The light that is reflected varies with wavelength and gives rise to the perception of

color, which will be discussed shortly. Absorbed light gives up its energy to the material and may be reemitted at a different wavelength, or it may simply show up as a temperature increase.

The *angle* of the light ray to the surface and the nature of the material determine if the light is absorbed, transmitted through a transparent material, or reflected. Vertical rays of light, for example, are mostly transmitted through a transparent material with some reflection and some absorption. If the rays strike the surface at some angle, however, much more of the light is reflected, bouncing off the surface. Thus, the glare of reflected sunlight is much greater around a body of water in the late afternoon than when the Sun is directly overhead.

Light that interacts with matter is reflected, transmitted, or absorbed, and all combinations of these interactions are possible. Materials are usually characterized by which of these interactions they *mostly* do, but this does not mean that other interactions are not occurring too. For example, a window glass is usually characterized as a transmitter of light. Yet, the glass always *reflects* about 4 percent of the light that strikes it. The reflected light usually goes unnoticed during the day because of the bright light that is transmitted from the outside. When it is dark outside, you notice the reflected light as the window glass now appears to act much like a mirror. A one-way mirror is another example of both reflection and transmission occurring (figure 7.9). A mirror is usually characterized as a reflector of light. A one-way mirror, however, has a very thin silvering that reflects most of the light but still transmits a little. In a lighted room, a one-way mirror appears to reflect light just as any other mirror does. But a person behind the mirror in a dark room can see into the lighted room by means of the transmitted light. Thus, you know that this mirror transmits as well as reflects light. One-way mirrors are used to unobtrusively watch for shoplifters in many businesses.

Reflection

Most of the objects that you see are visible from diffuse reflection. For example, consider some object such as a tree that you see during a bright day. Each *point* on the tree must reflect light in all directions, since you can see any part of the tree from any angle (figure 7.10). As a model, think of bundles of light rays entering your eye, which enable you to see the tree. This means that you can see any part of the tree from any angle because different bundles of reflected rays will enter your eye from different parts of the tree.

Light rays that are diffusely reflected move in all possible directions, but rays that are reflected from a smooth surface, such as a mirror, leave the mirror in a definite direction. Suppose you look at a tree in a mirror. There is only one place on the mirror where you look to see any one part of the tree. Light is reflecting off the mirror from all parts of the tree, but the only rays that reach your eye are the rays that are reflected at a certain angle from the place where you look. The relationship between the light rays moving from the tree and the direction in which they are reflected from the mirror to reach your eyes can be understood by drawing three lines: (1) a line representing

A

B

FIGURE 7.9 (*A*) The one-way mirror around the top of the wall in this store reflects most of the light that strikes it. (*B*) It also transmits some light to a person behind the mirror in a darkened room.

FIGURE 7.10 Bundles of light rays are reflected diffusely in all directions from every point on an object. Only a few light rays are shown from only one point on the tree in this illustration. The light rays that move to your eyes enable you to see the particular point from which they were reflected.

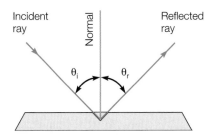

FIGURE 7.11 The law of reflection states that the angle of incidence (θ_i) is equal to the angle of reflection (θ_r). Both angles are measured from the *normal,* a reference line drawn perpendicular to the surface at the point of reflection.

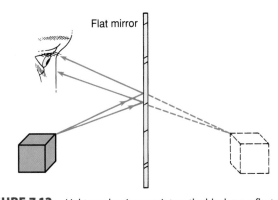

FIGURE 7.12 Light rays leaving a point on the block are reflected according to the law of reflection, and those reaching your eye are seen. After reflecting, the rays continue to spread apart at the same rate. You interpret this to be a block the same distance behind the mirror. You see a virtual image of the block, because light rays do not actually move from the image.

an original ray from the tree, called the *incident ray,* (2) a line representing a reflected ray, called the *reflected ray,* and (3) a reference line that is perpendicular to the reflecting surface and is located at the point where the incident ray struck the surface. This line is called the *normal.* The angle between the incident ray and the normal is called the *angle of incidence,* θ_i, and the angle between the reflected ray and the normal is called the *angle of reflection,* θ_r (figure 7.11). The *law of reflection,* which was known to the ancient Greeks, is that the *angle of incidence equals the angle of reflection,* or

$$\theta_i = \theta_r \qquad \textbf{equation 7.1}$$

Figure 7.12 shows how the law of reflection works when you look at a flat mirror. Light is reflected from all points on the box, and of course only the rays that reach your eyes are detected. These rays are reflected according to the law of reflection, with the angle of reflection equaling the angle of incidence. If you move your head slightly, then a different bundle of rays reaches your eyes. Of all the bundles of rays that reach your eyes, only two rays from a point are shown in the illustration. After these two rays are reflected, they continue to spread apart at the same rate that they were spreading before reflection. Your eyes and brain do not know that the rays have been reflected, and the diverging rays appear to come from behind the mirror, as the dashed lines show. The image, therefore, appears to be the same distance *behind* the mirror as the box is from the front of the mirror. Thus, a mirror image is formed where the rays of light *appear* to originate. This is called a **virtual image.** A virtual image is the result of your eyes' and brain's interpretations of light rays, not actual light rays originating from an image. Light rays that do originate from the other kind of image are called a **real image.** A real image is like the one displayed on a movie screen, with light originating from the image. A virtual image cannot be displayed on a screen, since it results from an interpretation.

Curved mirrors are either *concave,* with the center part curved inward, or *convex,* with the center part bulging outward. A concave mirror can be used to form an enlarged virtual image, such as a shaving or makeup mirror, or it can be used to form a real image, as in a reflecting telescope. Convex mirrors, for example, the mirrors on the sides of trucks and vans, are often used to increase the field of vision. Convex mirrors are also placed above an aisle in a store to show a wide area.

Refraction

You may have observed that an object that is partly in the air and partly in water appears to be broken, or bent, where the air and water meet. When a light ray moves from one transparent material to another, such as from water through air, the ray undergoes a change in the direction of travel at the boundary between the two materials. This change of direction of a light ray at the boundary is called **refraction.** The amount of change can be measured as an angle from the normal, just as it was for the angle of reflection. The incoming ray is called the *incident ray,* as before, and the new direction of travel is called the *refracted ray.* The angles of both rays are measured from the normal (figure 7.13).

Refraction results from a *change in speed* when light passes from one transparent material into another. The speed of light in a vacuum is 3.00×10^8 m/s, but it is slower when moving through a transparent material. In water, for example, the speed of light is reduced to about 2.30×10^8 m/s. The speed of light has a magnitude that is specific for various transparent materials.

When light moves from one transparent material to another transparent material with a *slower* speed of light, the ray is refracted *toward* the normal (figure 7.14A). For example, light travels through air faster than through water. Light traveling from air into water is therefore refracted toward the normal as it enters the water. On the other hand, if light has a *faster* speed in the new material, it is refracted *away* from the normal. Thus, light traveling from water into the air is refracted away from the normal as it enters the air (figure 7.14B).

The magnitude of refraction depends on (1) the angle at which light strikes the surface and (2) the ratio of the speed of light in the two transparent materials. An incident ray that is perpendicular (90°) to the surface is not refracted at all. As the angle of incidence is increased, the angle of refraction is also increased. There is a limit, however, that occurs when the angle of refraction reaches 90°, or along the water surface. Figure 7.15 shows rays of light traveling from water to air at various angles. When the incident ray is about 49°, the angle of refraction that results is 90°, along the water surface. This limit to the angle of incidence that results in an angle of refraction of 90° is called the *critical angle* for a water-to-air surface

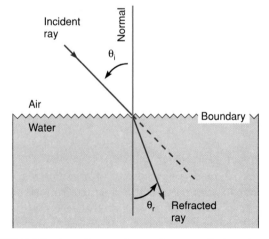

FIGURE 7.13 A ray diagram shows refraction at the boundary as a ray moves from air through water. Note that θ_i does not equal θ_r in refraction.

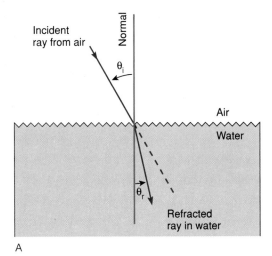

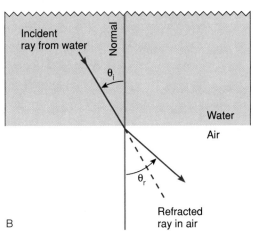

FIGURE 7.14 (A) A light ray moving to a new material with a slower speed of light is refracted toward the normal ($\theta_i > \theta_r$). (B) A light ray moving to a new material with a faster speed is refracted away from the normal ($\theta_i < \theta_r$).

CONCEPTS APPLIED

Seeing Around Corners

Place a coin in an empty cup. Position the cup so the coin appears to be below the rim, just out of your line of sight. Do not move from this position as your helper slowly pours water into the cup. Explain why the coin becomes visible, then appears to rise in the cup. Use a sketch such as the one in figure 7.13 to help with your explanation.

(figure 7.15). At any incident angle greater than the critical angle, the light ray does not move from the water to the air but is *reflected* back from the surface as if it were a mirror. This is called **total internal reflection** and implies that the light is trapped inside if it arrived at the critical angle or beyond. Faceted transparent gemstones such as the diamond are brilliant because they have a small critical angle and thus reflect much light internally. Total internal reflection is also important in fiber optics.

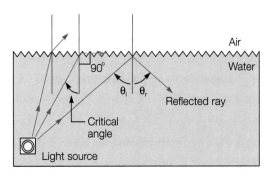

FIGURE 7.15 When the angle of incidence results in an angle of refraction of 90°, the refracted light ray is refracted along the water surface. The angle of incidence for a material that results in an angle of refraction of 90° is called the *critical angle*. When the incident ray is at this critical angle or greater, the ray is reflected internally. The critical angle for water is about 49°, and for a diamond it is about 25°.

CONCEPTS APPLIED

Colors and Refraction

A convex lens is able to magnify by forming an image with refracted light. This application is concerned with magnifying, but it is really more concerned with experimenting to find an explanation.

Here are three pairs of words:

SCIENCE BOOK

RAW HIDE

CARBON DIOXIDE

Hold a cylindrical solid glass rod over the three pairs of words, using it as a magnifying glass. A clear, solid, and transparent plastic rod or handle could also be used as a magnifying glass.

Notice that some words appear inverted but others do not. Does this occur because red letters are refracted differently than blue letters?

Make some words with red and blue letters to test your explanation. What is your explanation for what you observed?

EXAMPLE 7.1 *(Optional)*

What is the speed of light in a diamond?

SOLUTION

The relationship between the speed of light in a material (v), the speed of light in a vacuum ($c = 3.00 \times 10^8$ m/s), and the index of refraction is given in equation 7.2. The index of refraction of a diamond is found in table 7.1 on page 164 ($n = 2.42$).

$$n_{\text{diamond}} = 2.42 \qquad n = \frac{c}{v} \therefore v = \frac{c}{n}$$
$$c = 3.00 \times 10^8 \text{ m/s}$$
$$v = ? \qquad\qquad\qquad = \frac{3.00 \times 10^8 \text{ m/s}}{2.42}$$
$$= \boxed{1.24 \times 10^8 \text{ m/s}}$$

Optics

Historians tell us there are many early stories and legends about the development of ancient optical devices. The first glass vessels were made about 1500 B.C., so it is possible that samples of clear, transparent glass were available soon after. One legend claimed that the ancient Chinese invented eyeglasses as early as 500 B.C. A burning glass (lens) was mentioned in a Greek play written about 424 B.C. Several writers described how Archimedes saved his hometown of Syracuse with a burning glass in about 214 B.C. Syracuse was besieged by Roman ships when Archimedes supposedly used the burning glass to focus sunlight on the ships, setting them on fire. It is not known if this story is true or not, but it is known that the Romans indeed did have burning glasses. Glass spheres, which were probably used to start fires, have been found in Roman ruins, including a convex lens recovered from the ashes of Pompeii.

Today, lenses are no longer needed to start fires, but they are common in cameras, scanners, optical microscopes, eyeglasses, lasers, binoculars, and many other optical devices. Lenses are no longer just made from glass, and today many are made from a transparent, hard plastic that is shaped into a lens.

The primary function of a lens is to form an image of a real object by refracting incoming parallel light rays. Lenses have two basic shapes, with the center of a surface either bulging in or bulging out. The outward bulging shape is thicker at the center than around the outside edge and is called a *convex lens* (box figure 7.1A). The other basic lens shape is just the opposite—thicker around the outside edge than at the center—and is called a *concave lens* (box figure 7.1B).

Convex lenses are used to form images in magnifiers, cameras, eyeglasses, projectors, telescopes, and microscopes (box figure 7.2). Concave lenses are used in some eyeglasses and in combinations with the convex lens to correct for defects. The convex lens is the most commonly used lens shape.

Your eyes are optical devices with convex lenses. Box figure 7.3 shows the basic structure. First, a transparent hole called

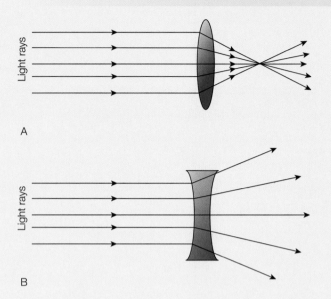

Box Figure 7.1 (*A*) Convex lenses are called converging lenses since they bring together, or converge, parallel rays of light. (*B*) Concave lenses are called diverging lenses since they spread apart, or diverge, parallel rays of light.

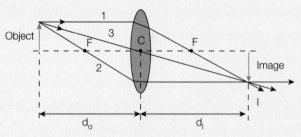

Box Figure 7.2 A convex lens forms an inverted image from refracted light rays of an object outside the focal point. Convex lenses are mostly used to form images in cameras, film or overhead projectors, magnifying glasses, and eyeglasses.

the *pupil* allows light to enter the eye. The size of the pupil is controlled by the *iris*, the colored part that is a muscular diaphragm. The *lens* focuses a sharp image on the back surface of the eye, the *retina*. The retina is made up of millions of light-sensitive structures, and nerves carry signals (impulses) from the retina through the optic nerve to the brain.

The lens is a convex, pliable material held in place and changed in shape by the attached *ciliary muscle*. When the eye is focused on a distant object, the ciliary muscle is completely relaxed. Looking at a closer object requires the contraction of the ciliary muscles to change the curvature of the lens. This adjustment of focus by action of the ciliary muscle is called *accommodation*. The closest distance an object

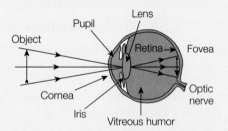

Box Figure 7.3 Light rays from a distant object are focused by the lens onto the retina, a small area on the back of the eye.

can be seen without a blurred image is called the *near point*, and this is the limit to accommodation.

The near point moves outward with age as the lens becomes less pliable. By middle age, the near point may be twice this distance or greater, creating the condition

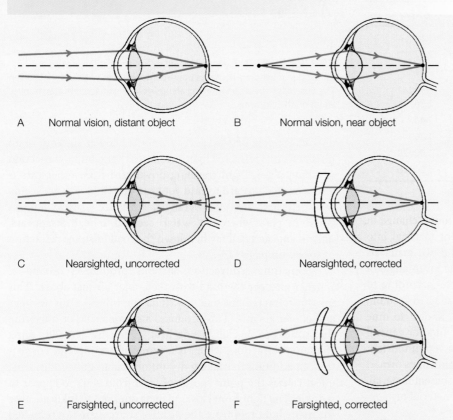

A   Normal vision, distant object

B   Normal vision, near object

C   Nearsighted, uncorrected

D   Nearsighted, corrected

E   Farsighted, uncorrected

F   Farsighted, corrected

Box Figure 7.4   (*A*) The relaxed, normal eye forms an image of distant objects on the retina. (*B*) For close objects, the lens of the normal eye changes shape to focus the image on the retina. (*C*) In a nearsighted eye, the image of a distant object forms in front of the retina. (*D*) A diverging lens corrects for nearsightedness. (*E*) In a farsighted eye, the image of a nearby object forms beyond the retina. (*F*) A converging lens corrects for farsightedness.

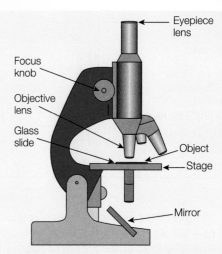

Box Figure 7.5   A simple microscope uses a system of two lenses, which are an objective lens that makes an enlarged image of the specimen and an eyepiece lens that makes an enlarged image of that image.

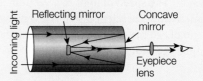

Box Figure 7.6   This illustrates how the path of light moves through a simple reflecting astronomical telescope. Several different designs and placement of mirrors are possible.

known as farsightedness. The condition of farsightedness, or *hyperopia,* is a problem associated with aging (called presbyopia). Hyperopia can be caused at an early age by an eye that is too short or by problems with the cornea or lens that focuses the image behind the retina. Farsightedness can be corrected with a convex lens as shown in box figure 7.4A.

Nearsightedness, or *myopia,* is a problem caused by an eye that is too long or problems with the cornea or lens that focus the image in front of the retina. Nearsightedness can be corrected with a concave lens, as shown in box figure 7.4B.

The microscope is an optical device used to make things look larger. It is essentially a system of two lenses, one to produce an image of the object being studied and the other to act as a magnifying glass and enlarge that image. The power of the microscope is basically determined by the *objective lens,* which is placed close to the

specimen on the stage of the microscope. Light is projected up through the specimen, and the objective lens makes an enlarged image of the specimen inside the tube between the two lenses. The *eyepiece lens* is positioned so that it makes a sharp enlarged image of the image produced by the objective lens (box figure 7.5).

Telescopes are optical instruments used to provide enlarged images of near and distant objects. There are two major types of telescopes, *refracting* telescopes that use two lenses and *reflecting* telescopes that use combinations of mirrors or a mirror and a lens. The refracting telescope has two lenses, with the objective lens forming a reduced image, which is viewed with an eyepiece lens to enlarge that image. In reflecting telescopes, mirrors are used instead of lenses to collect the light (box figure 7.6).

Finally, the *digital camera* is one of the more recently developed light-

gathering and photograph-taking optical instruments. This camera has a group of small photocells, with perhaps thousands lined up on the focal plane behind a converging lens. An image falls on the array, and each photocell stores a charge that is proportional to the amount of light falling on the cell. A microprocessor measures the amount of charge registered by each photocell and considers it as a pixel, a small bit of the overall image. A shade of gray or a color is assigned to each pixel, and the image is ready to be enhanced, transmitted to a screen, printed, or magnetically stored for later use. Modern digital cameras have an advantage since the picture is available instantly, without the use of darkrooms, chemicals, or light-sensitive paper. A photo can be immediately downloaded from the camera into a computer and then perhaps loaded onto your website, ready for the whole world to view.

TABLE 7.1 Index of Refraction

Substance	$n = c/v$
Glass	1.50
Diamond	2.42
Ice	1.31
Water	1.33
Benzene	1.50
Carbon tetrachloride	1.46
Ethyl alcohol	1.36
Air (0°C)	1.00029
Air (30°C)	1.00026

As was stated earlier, refraction results from a change in speed when light passes from one transparent material into another. The ratio of the speeds of light in the two materials determines the magnitude of refraction at any given angle of incidence. The greatest speed of light possible, according to current theory, occurs when light is moving through a vacuum. The speed of light in a vacuum is accurately known to nine decimals but is usually rounded to 3.00×10^8 m/s for general discussion. The speed of light in a vacuum is a very important constant in physical science, so it is given a symbol of its own, c. The ratio of c to the speed of light in some transparent material, v, is called the **index of refraction**, n, of that material or

$$n = \frac{c}{v}$$

equation 7.2

The indexes of refraction for some substances are listed in table 7.1. The values listed are constant physical properties and can be used to identify a specific substance. Note that a larger value means a greater refraction at a given angle. Of the materials listed, diamond refracts light the most and air the least. The index for air is nearly 1, which means that light is slowed only slightly in air.

Note that table 7.1 shows that colder air at 0°C (32°F) has a higher index of refraction than warmer air at 30°C (86°F), which means that light travels faster in warmer air. This difference explains the "wet" highway that you sometimes see at a distance in the summer. The air near the road is hotter on a clear, calm day. Light rays traveling toward you in this hotter

Myths, Mistakes, and Misunderstandings

The Light Saber

The *Star Wars* light saber is an impossibility. Assuming that the light saber is a laser beam, we know that one laser beam will not stop another laser beam. Light beams simply pass through each other. Furthermore, a laser with a fixed length is not possible without a system of lenses that would also scatter the light, in addition to being cumbersome on a saber. Moreover, scattered laser light from reflective surfaces could result in injury to the saber wielder.

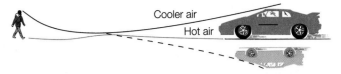

FIGURE 7.16 Mirages are caused by hot air near the ground refracting, or bending, light rays upward into the eyes of a distant observer. The observer believes he is seeing an upside-down image reflected from water on the highway.

air are refracted upward as they enter the cooler air. Your brain interprets this refracted light as *reflected* light, but no reflection is taking place. Light traveling downward from other cars is also refracted upward toward you, and you think you are seeing cars "reflected" from the wet highway (figure 7.16). When you reach the place where the "water" seemed to be, it disappears, only to appear again farther down the road (also see refraction of sound on page 113).

Sometimes convection currents produce a mixing of warmer air near the road with the cooler air just above. This mixing refracts light one way, then the other, as the warmer and cooler air mix. This produces a shimmering or quivering that some people call "seeing heat." They are actually seeing changing refraction, which is a *result* of heating and convection. In addition to causing distant objects to quiver, the same effect causes the point source of light from stars to appear to twinkle. The light from closer planets does not twinkle because the many light rays from the disklike sources are not refracted together as easily as the fewer rays from the point sources of stars. The light from planets will appear to quiver, however, if the atmospheric turbulence is great.

Dispersion and Color

Electromagnetic waves travel with the speed of light with a whole spectrum of waves of various frequencies and wavelengths. The speed of electromagnetic waves (c) is related to the wavelength (λ) and the frequency (f) by a form of the wave equation, or

$$c = \lambda f$$

equation 7.3

Visible light is the part of the electromagnetic spectrum that your eyes can detect, a narrow range of wavelength from about 7.90×10^{-7} m to 3.90×10^{-7} m. In general, this range of visible light can be subdivided into ranges of wavelengths that you perceive as colors (figure 7.17). These are the colors of the rainbow, and there are six distinct colors that blend one into another. These colors are *red, orange, yellow, green, blue,* and *violet.* The corresponding ranges of wavelengths and frequencies of these colors are given in table 7.2.

In general, light is interpreted to be white if it has the same mixture of colors as the solar spectrum. That sunlight is made up of component colors was first investigated in detail by Isaac Newton. While a college student, Newton became interested in grinding lenses, light, and color. At the age of twenty-three, Newton visited a local fair and bought several triangular glass prisms and proceeded to conduct a series of experiments with a beam of sunlight in his room. In 1672, he reported the results

The beauty of a diamond depends largely on optical properties, including degree of refraction, but how well the cutter shapes the diamond determines the brilliance and reflection of light from a given stone. The "finish" describes the precision of facet placement on a cut stone, but overall quality involves much more. Four factors are generally used to determine the quality of a cut diamond. These are sometimes called the four C's of carat, color, clarity, and cut.

Carat is a measure of the weight of a diamond. When measuring diamonds (as opposed to gold), 1 metric carat is equivalent to 0.2 g, so a 5 carat diamond would have a mass of 1 g. Diamonds are weighed to the nearest hundredth of a carat, and 100 points equal 1 carat. Thus, a 3 point diamond is 3/100 of a carat, with a mass of about 0.006 g.

Color is a measure of how much (or how little) color a diamond has. The color grade is based on the color of the body of a diamond, not the color of light dispersion from the stone. The color grades range from colorless, near colorless, faint yellow, very light yellow, and down to light yellow. It is measured by comparing a diamond to a standard of colors, but the color of a diamond weighing one-half carat or less is often difficult to determine.

Clarity is a measure of inclusions (grains or crystals), cleavages, or other blemishes. Such inclusions and flaws may affect transparency, brilliance, and even the durability of a diamond. The clarity grade ranges from flawless down to industrial-grade stones, which may appear milky from too many tiny inclusions.

Cut involves the relationships between the sizes of a stone's major physical features and their various angles. They determine, above all else, the limit to which a diamond will accomplish its optical potential. There are various types of faceted cuts, including the round brilliant cut, with a flat top, a flat bottom, and about fifty-eight facets. Facet shapes of square, triangular, diamond-shaped, and trapezoidal cuts might be used.

The details of a given diamond's four C's determine how expensive the diamond is going to be, and a top grade in all factors would be very expensive. No two stones are alike, and it is possible to find a brilliant diamond with a few lower grades in some of the factors that is a good value. °

FIGURE 7.17 The flowers appear to be red because they reflect light in the 7.9×10^{-7} m to 6.2×10^{-7} m range of wavelengths.

TABLE 7.2 **Range of Wavelengths and Frequencies of the Colors of Visible Light**

Color	Wavelength (in meters)	Frequency (in hertz)
Red	7.9×10^{-7} to 6.2×10^{-7}	3.8×10^{14} to 4.8×10^{14}
Orange	6.2×10^{-7} to 6.0×10^{-7}	4.8×10^{14} to 5.0×10^{14}
Yellow	6.0×10^{-7} to 5.8×10^{-7}	5.0×10^{14} to 5.2×10^{14}
Green	5.8×10^{-7} to 4.9×10^{-7}	5.2×10^{14} to 6.1×10^{14}
Blue	4.9×10^{-7} to 4.6×10^{-7}	6.1×10^{14} to 6.6×10^{14}
Violet	4.6×10^{-7} to 3.9×10^{-7}	6.6×10^{14} to 7.7×10^{14}

EXAMPLE 7.2 *(Optional)*

The colors of the spectrum can be measured in units of wavelength, frequency, or energy, which are alternative ways of describing colors of light waves. The human eye is most sensitive to light with a wavelength of 5.60×10^{-7} m, which is a yellow-green color. What is the frequency of this wavelength?

SOLUTION

The relationship between the wavelength (λ), frequency (f), and speed of light in a vacuum (c) is found in equation 7.3, $c = \lambda f$.

$$\lambda = 5.60 \times 10^{-7} \text{ m} \qquad c = \lambda f \therefore f = \frac{c}{\lambda}$$
$$c = 3.00 \times 10^8 \text{ m/s}$$
$$f = ?$$
$$= \frac{3.00 \times 10^8 \frac{\text{m}}{\text{s}}}{5.60 \times 10^{-7} \text{ m}}$$
$$= \frac{3.00 \times 10^8}{5.60 \times 10^{-7}} \frac{\text{m}}{\text{s}} \times \frac{1}{\text{m}}$$
$$= 5.40 \times 10^{14} \frac{1}{\text{s}}$$
$$= \boxed{5.40 \times 10^{14} \text{ Hz}}$$

of his experiments with prisms and color, concluding that white light is a mixture of all the independent colors. Newton found that a beam of sunlight falling on a glass prism in a darkened room produced a band of colors he called a *spectrum*. Further, he found that a second glass prism would not subdivide each separate color but would combine all the colors back into white sunlight. Newton concluded that sunlight consists of a mixture of the six colors.

A glass prism separates sunlight into a spectrum of colors because the index of refraction is different for different wavelengths of light. The same processes that slow the speed of light in a transparent substance have a greater effect on short wavelengths than they do on longer wavelengths. As a result, violet light is refracted most, red light is refracted least, and the other colors are refracted between these extremes. This results in a beam of white light being separated, or dispersed, into a spectrum when it is refracted. Any transparent material in which the index of refraction varies with wavelength has the property of *dispersion.* The dispersion of light by ice crystals sometimes produces a colored halo around the Sun and the Moon.

EVIDENCE FOR WAVES

The nature of light became a topic of debate toward the end of the 1600s as Isaac Newton published his *particle theory* of light. He believed that the straight-line travel of light could be better explained as small particles of matter that traveled at great speed from a source of light. Particles, reasoned Newton, should follow a straight line according to the laws of motion. Waves, on the other hand, should bend as they move, much as water waves on a pond bend into circular shapes as they move away from a disturbance. About the same time that Newton developed his particle theory of light, Christian Huygens (pronounced *hi-ganz*) was concluding that light is not a stream of particles but rather a longitudinal wave.

Both theories had advocates during the 1700s, but the majority favored Newton's particle theory. By the beginning of the 1800s, new evidence was found that favored the wave theory, evidence that could not be explained in terms of anything but waves.

Interference

In 1801, Thomas Young published evidence of a behavior of light that could only be explained in terms of a wave model of light. Young's experiment is illustrated in figure 7.18A. Light

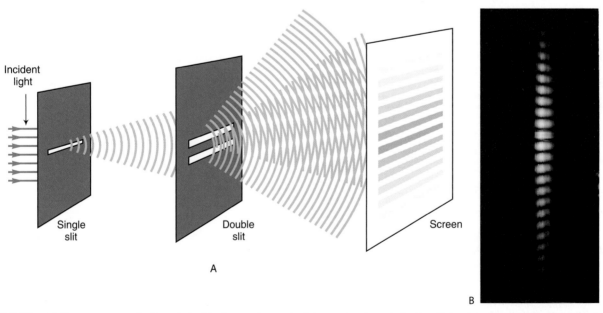

FIGURE 7.18 (*A*) The arrangement for Young's double-slit experiment. Sunlight passing through the first slit is coherent and falls on two slits close to each other. Light passing beyond the two slits produces an interference pattern on a screen. (*B*) The double-slit pattern of a small-diameter beam of light from a helium-neon laser.

A rainbow is a spectacular, natural display of color that is supposed to have a pot of gold under one end. Understanding the why and how of a rainbow requires information about water droplets and knowledge of how light is reflected and refracted. This information will also explain why the rainbow seems to move when you move—making it impossible to reach the end to obtain that mythical pot of gold.

First, note the pattern of conditions that occurs when you see a rainbow. It usually appears when the Sun is shining low in one part of the sky and rain is falling in the opposite part. With your back to the Sun, you are looking at a zone of raindrops that are all showing red light, another zone that are all showing violet light, with zones of the other colors between (ROYGBV). For a rainbow to form like this requires a surface that refracts and reflects the sunlight, a condition met by spherical raindrops.

Water molecules are put together in such a way that they have a positive side and a negative side, and this results in strong molecular attractions. It is the strong attraction of water molecules for one another that results in the phenomenon of surface tension. Surface tension is the name given to the surface of water acting as if it is covered by an ultrathin elastic membrane that is contracting. It is surface tension that pulls raindrops into a spherical shape as they fall through the air.

Box figure 7.7 shows one thing that can happen when a ray of sunlight strikes a single spherical raindrop near the top of the drop. At this point, some of the sunlight is reflected, and some is refracted into the raindrop. The refraction disperses the light into its spectrum colors, with the violet

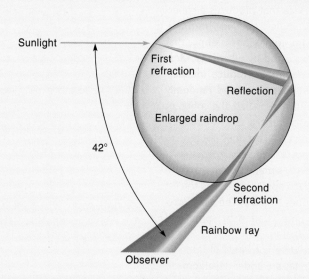

Box Figure 7.7 Light is refracted when it enters a raindrop and when it leaves. The part that leaves the front surface of the raindrop is the source of the light in thousands upon thousands of raindrops from which you see zones of color—a rainbow.

light being refracted most and red the least. The refracted light travels through the drop to the opposite side, where some of it might be reflected back into the drop. The reflected part travels back through the drop again, leaving the front surface of the raindrop. As it leaves, the light is refracted for a second time. The combined refraction, reflection, and second refraction is the source of the zones of colors you see in a rainbow. This also explains why you see a rainbow in the part of the sky opposite from the Sun.

The light from any one raindrop is one color, and that color comes from all drops on the arc of a circle that is a certain angle between the incoming sunlight and the refracted light. Thus, the raindrops in the red region refract red light toward your

eyes at an angle of 42°, and all other colors are refracted over your head by these drops. Raindrops in the violet region refract violet light toward your eyes at an angle of 40° and the red and other colors toward your feet. Thus, the light from any one drop is seen as one color, and all drops showing this color are on the arc of a circle. An arc is formed because the angle between the sunlight and the refracted light of a color is the same for each of the spherical drops.

Sometimes a fainter secondary rainbow, with colors reversed, forms from sunlight entering the bottom of the drop, reflecting twice, and then refracting out the top. The double reflection reverses the colors, and the angles are 50° for the red and 54° for the violet (see figure 1.14 on page 16).

from a single source is used to produce two beams of light that are in phase, that is, having their crests and troughs together as they move away from the source. This light falls on a card with two slits, each less than a millimeter in width. The light moves out from each slit as an expanding arc. Beyond the card, the light from one slit crosses over the light from the other slit to produce a series of bright lines on a screen. Young had produced a phenomenon of light called **interference,** and interference can only be explained by waves!

The pattern of bright lines and dark zones is called an *interference pattern* (figure 7.18B). The light moved from each slit in phase, crest to crest and trough to trough. Light from both slits traveled the same distance directly across to the screen, so they arrived in phase. The crests from the two slits are superimposed here, and constructive interference produces a bright line in the center of the pattern. But for positions above and below the center, the light from the two slits must travel different distances to the screen. At a certain distance above

Connections . . .

Lasers

A laser is a device that produces a coherent beam of single-frequency, in-phase light. The beam comes from atoms that have been stimulated by electricity. Most ordinary light sources produce incoherent light; light that is emitted randomly and at different frequencies. The coherent light from a laser has the same frequency, phase, and direction, so it does not tend to spread out and can be very intense (box figure 7.8). This has made possible a number of specialized applications, and the list of uses continues to grow. The word *laser* is from *l*ight *a*mplification by *s*timulated *e*mission of *r*adiation.

There are different kinds of lasers in use and new ones are under development. One common type of laser is a gas-filled tube with mirrors at both ends. The mirror at one end is only partly silvered, which allows light to escape as the laser beam. The distance between the mirrors matches the resonate frequency of the light produced, so the trapped light will set up an optical standing wave. An electric discharge produces fast electrons that raise the energy level of the electrons of the specific gas atoms in the tube. The electrons of the energized gas atoms emit a particular

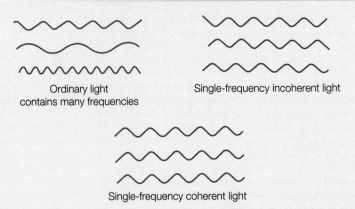

Ordinary light contains many frequencies

Single-frequency incoherent light

Single-frequency coherent light

Box Figure 7.8 A laser produces a beam of light whose waves all have the same frequency and are in step with one another (coherent). The beam is also very narrow and spreads out very little even over long distances.

frequency of light as they drop back to the original level, and this emitted light sets up the standing wave. The standing wave stimulates other atoms of the gas, resulting in the emission of more light at the same frequency and phase.

Lasers are everywhere today and have connections with a wide variety of technologies. At the supermarket, a laser and detector unit reads the bar code on each grocery item. The laser sends the pattern to a computer, which sends a price to the register while also tracking the store inventory. A low-powered laser and detector also reads your CD music disk or MP3 disk and can be used to make a three-dimensional image. Most laser printers use a laser, and a laser makes the operational part of a fiber optics communication system. Stronger lasers are used for cutting, drilling, and welding. Lasers are used extensively in many different medical procedures, from welding a detached retina to bloodless surgery.

and below the bright center line, light from one slit had to travel a greater distance and arrives one-half wavelength after light from the other slit. Destructive interference produces a zone of darkness at these positions. Continuing up and down the screen, a bright line of constructive interference will occur at each position where the distance traveled by light from the two slits differs by any whole number of wavelengths. A dark zone of destructive interference will occur at each position where the distance traveled by light from the two slits differs by any half-wavelength. Thus, bright lines occur above and below the center bright line at positions representing differences in paths of 1, 2, 3, 4, and so on wavelengths. Similarly, zones of darkness occur above and below the center bright line at positions representing differences in paths of 1/2, 1 1/2, 2 1/2, 3 1/2, and so on wavelengths. Young found all of the experimental data such as this in full agreement with predictions from a wave theory of light. About fifteen years later, A. J. Fresnel (pronounced *fray-nel*) demonstrated

mathematically that diffraction as well as other behaviors of light could be fully explained with the wave theory. In 1821, Fresnel determined that the wavelength of red light was about 8×10^{-7} m and of violet light about 4×10^{-7} m, with other colors in between these two extremes. The work of Young and Fresnel seemed to resolve the issue of considering light to be a stream of particles or a wave, and it was generally agreed that light must be waves.

Polarization

Huygens' wave theory and Newton's particle theory could explain some behaviors of light satisfactorily, but there were some behaviors that neither (original) theory could explain. Both theories failed to explain some behaviors of light, such as light moving through certain transparent crystals. For example, a slice of the mineral tourmaline transmits what appears to be a low-intensity greenish light. But if a second slice of tourmaline is placed on the first and rotated, the transmitted light passing through both slices begins to dim. The transmitted light is practically zero when the second slice is rotated 90°. Newton suggested that this behavior had something to do with "sides" or "poles" and introduced the concept of what is now called the *polarization* of light.

The waves of Huygens' wave theory were longitudinal, moving like sound waves, with wave fronts moving in the direction of travel. A longitudinal wave could not explain the polarization behavior of light. In 1817, Young modified Huygens' theory by describing the waves as *transverse,* vibrating at right angles to the direction of travel. This modification helped explain the polarization behavior of light transmitted through the two crystals and provided firm evidence that light is a transverse wave. As shown in figure 7.19A, **unpolarized light** is assumed to consist of transverse waves vibrating in all conceivable random directions. Polarizing materials, such as the tourmaline crystal, transmit light that is vibrating in one direction only, such as the vertical direction in figure 7.19B. Such a wave is said to be **polarized,** or *plane-polarized,* since it vibrates only in one plane. The single crystal polarized light by transmitting only waves that vibrate parallel to a certain direction while selectively absorbing waves that vibrate in all other directions. Your eyes cannot tell the difference between unpolarized and polarized light, so the light transmitted through a single crystal looks just like any other light. When a second crystal is placed on the first, the amount of light transmitted depends on the alignment of the two crystals (figure 7.20). When the two crystals are *aligned,* the polarized light from the first crystal passes through the second with little absorption. When the crystals are *crossed* at 90°, the light transmitted by the first is vibrating in a plane that is absorbed by the second crystal, and practically all the light is absorbed. At some other angle, only a fraction of the polarized light from the first crystal is transmitted by the second.

You can verify whether or not a pair of sunglasses is made of polarizing material by rotating a lens of one pair over a lens

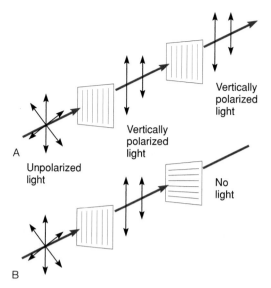

FIGURE 7.20 (*A*) Two crystals that are aligned both transmit vertically polarized light that looks like any other light. (*B*) When the crystals are crossed, no light is transmitted.

of a second pair. Light is transmitted when the lenses are aligned but mostly absorbed at 90° when the lenses are crossed.

Light is completely polarized when all the waves are removed except those vibrating in a single direction. Light is partially polarized when some of the waves are in a particular orientation, and any amount of polarization is possible. There are several means of producing partially or completely polarized light, including (1) selective absorption, (2) reflection, and (3) scattering.

Selective absorption is the process that takes place in certain crystals, such as tourmaline, where light in one plane is transmitted and all the other planes are absorbed. A method of manufacturing a polarizing film was developed in the 1930s by Edwin H. Land. The film is called *Polaroid.* Today, Polaroid is made of long chains of hydrocarbon molecules that are aligned in a film. The long-chain molecules ideally absorb all light waves that are parallel to their lengths and transmit light that is perpendicular to their lengths. The direction that is *perpendicular* to the oriented molecular chains is thus called the polarization direction or the *transmission axis.*

Reflected light with an angle of incidence between 1° and 89° is partially polarized as the waves parallel to the reflecting surface are reflected more than other waves. Complete polarization, with all waves parallel to the surface, occurs at a particular angle of incidence. This angle depends on a number of variables, including the nature of the reflecting material. Figure 7.21 illustrates polarization by reflection. Polarizing sunglasses reduce the glare of reflected light because they have vertically oriented transmission axes. This absorbs the horizontally oriented reflected light. If you turn your head from side to side so as to rotate your sunglasses while looking at a reflected glare, you will see the intensity of the reflected

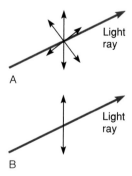

FIGURE 7.19 (*A*) Unpolarized light has transverse waves vibrating in all possible directions perpendicular to the direction of travel. (*B*) Polarized light vibrates only in one plane. In this illustration, the wave is vibrating in a vertical direction only.

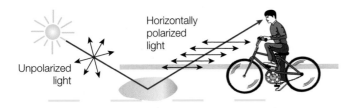

FIGURE 7.21 Light that is reflected becomes partially or fully polarized in a horizontal direction, depending on the incident angle and other variables.

light change. This means that the reflected light is partially polarized.

The phenomenon called *scattering* occurs when light is absorbed and reradiated by particles about the size of gas molecules that make up the air. Sunlight is initially unpolarized. When it strikes a molecule, electrons are accelerated and vibrate horizontally and vertically. The vibrating charges reradiate polarized light. Thus, if you look at the blue sky with a pair of polarizing sunglasses and rotate them, you will observe that light from the sky is polarized. Bees are believed to be able to detect polarized skylight and use it to orient the direction of their flights. Violet and blue light have the shortest wavelengths of visible light, and red and orange light have the longest. The violet and blue rays of sunlight are scattered the most. At sunset, the path of sunlight through the atmosphere is much longer than when the Sun is more directly overhead. Much of the blue and violet have been scattered away as a result of the longer path through the atmosphere at sunset. The remaining light that comes through is mostly red and orange, so these are the colors you see at sunset.

EVIDENCE FOR PARTICLES

In 1850, J. L. Foucault was able to prove that light travels much slower in transparent materials than it does in air. This was in complete agreement with the wave theory and completely opposed to the particle theory. By the end of the 1800s, James Maxwell's theoretical concept of electric and magnetic fields changed the concept of light from mechanical waves to waves of changing electric and magnetic fields. Further evidence removed the necessity for ether, the material supposedly needed for waves to move through. Light was now seen as electromagnetic waves that could move through empty space. By this time, it was possible to explain all behaviors of light moving through empty space or through matter with a wave theory. Yet, there were nagging problems that the wave theory could not explain. In general, these problems concerned light that is absorbed by or emitted from matter.

Photoelectric Effect

Light is a form of energy, and it gives its energy to matter when it is absorbed. Usually, the energy of absorbed light results in a temperature increase, such as the warmth you feel from absorbed sunlight. Sometimes, however, the energy from absorbed light results in other effects. In some materials, the energy is acquired by electrons, and some of the electrons acquire sufficient energy to jump out of the material. The movement of electrons as a result of energy acquired from light is known as the **photoelectric effect.** The photoelectric effect is put to a practical use in a solar cell, which transforms the energy of light into an electric current (figure 7.22).

The energy of light can be measured with great accuracy. The kinetic energy of electrons after they absorb light can also be measured with great accuracy. When these measurements were made of the light and electrons involved in the photoelectric effect, some unexpected results were observed. Monochromatic light, that is, light of a single, fixed frequency, was used to produce the photoelectric effect. First, a low-intensity, or dim, light was used, and the numbers and energy of the ejected electrons were measured. Then a high-intensity light was used, and the numbers and energy of the ejected electrons were again measured. Measurement showed that (1) low-intensity light caused fewer electrons to be ejected, and high-intensity light caused many to be ejected, and (2) all electrons ejected from low- or high-intensity light ideally had the *same* kinetic energy. Surprisingly, the kinetic energy of the ejected electrons was found to be *independent* of the light intensity. This was contrary to what the wave theory of light would predict, since a stronger light should mean that waves with more energy have more energy to give to the electrons. Here is a behavior involving light that the wave theory could not explain.

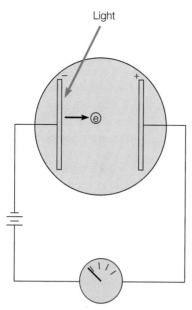

FIGURE 7.22 A setup for observing the photoelectric effect. Light strikes the negatively charged plate, and electrons are ejected. The ejected electrons move to the positively charged plate and can be measured as a current in the circuit.

Quantization of Energy

In addition to the problem of the photoelectric effect, there were problems with blackbody radiation, light emitted from hot objects. The experimental measurements of light emitted through blackbody radiation did not match predictions made from theory. In 1900, Max Planck (pronounced *plonk*), a German physicist, found that he could fit the experimental measurements and theory together by assuming that the vibrating molecules that emitted the light could only have a *discrete amount* of energy. Instead of energy existing through a continuous range of amounts, Planck found that the vibrating molecules could only have energy in multiples of energy in certain amounts, or **quanta** (meaning "discrete amounts"; *quantum* is singular and *quanta*, plural).

Planck's discovery of quantized energy states was a radical, revolutionary development, and most scientists, including Planck, did not believe it at the time. Planck, in fact, spent considerable time and effort trying to disprove his own discovery. It was, however, the beginning of the quantum theory, which was eventually to revolutionize physics.

Five years later, in 1905, Albert Einstein applied Planck's quantum concept to the problem of the photoelectric effect. Einstein described the energy in a light wave as quanta of energy called **photons.** Each photon has an energy E that is related to the frequency f of the light through Planck's constant h, or

$$E = hf \qquad \textbf{equation 7.4}$$

The value of Planck's constant is 6.63×10^{-34} J·s. This relationship says that higher-frequency light (e.g., blue light at 6.50×10^{14} Hz) has more energy than lower-frequency light

(e.g., red light at 4.00×10^{14} Hz). The energy of such high- and low-frequency light can be verified by experiment.

The photon theory also explained the photoelectric effect. According to this theory, light is a stream of moving photons. It is the number of photons in this stream that determines if the light is dim or intense. A high-intensity light has many, many photons, and a low-intensity light has only a few photons. At any particular fixed frequency, all the photons would have the same energy, the product of the frequency and Planck's constant (hf). When a photon interacts with matter, it is absorbed and gives up all of its energy. In the photoelectric effect, this interaction takes place between photons and electrons. When an intense light is used, there are more photons to interact with the electrons, so more electrons are ejected. The energy given up by each photon is a function of the frequency of the light, so at a fixed frequency, the energy of each photon, hf, is the same, and the acquired kinetic energy of each ejected electron is the same. Thus, the photon theory explains the measured experimental results of the photoelectric effect.

EXAMPLE 7.3 *(Optional)*

What is the energy of a photon of red light with a frequency of 4.00×10^{14} Hz?

SOLUTION

The relationship between the energy of a photon (E) and its frequency (f) is found in equation 7.4. Planck's constant is given as 6.63×10^{-34} J·s.

$$
\begin{aligned}
f &= 4.00 \times 10^{14} \text{ m} & E &= hf \\
h &= 6.63 \times 10^{-34} \text{ J·s} \\
E &= ? & &= (6.63 \times 10^{-34} \text{ J·s})\left(4.00 \times 10^{14}\, \frac{1}{\text{s}}\right) \\
& & &= (6.63 \times 10^{-34})(4.00 \times 10^{14}) \text{ J·s} \times \frac{1}{\text{s}} \\
& & &= 2.65 \times 10^{-19}\, \frac{\text{J·s}}{\text{s}} \\
& & &= \boxed{2.65 \times 10^{-19} \text{ J}}
\end{aligned}
$$

EXAMPLE 7.4 *(Optional)*

What is the energy of a photon of violet light with a frequency of 7.00×10^{14} Hz? (Answer: 4.64×10^{-19} J)

The photoelectric effect is explained by considering light to be photons with quanta of energy, not a wave of continuous energy. This is not the only evidence about the quantum nature of light, and more will be presented in chapter 8. But, as you can see, there is a dilemma. The electromagnetic wave theory and the photon theory seem incompatible. Some experiments cannot be explained by the wave theory and seem to support the photon theory. Other experiments are contradictions, providing seemingly equal evidence to reject the photon theory in support of the wave theory.

A Closer Look

The Compact Disc (CD)

A compact disc (CD) is a laser-read (also called *optically read*) data storage device on which music, video, or any type of digital data can be stored. Three types in popular use today are the CD audio, DVD (or MP3), and the CD-ROM used to store any type of digital data. All three kinds of players rotate between 200 to 500 revolutions per minute, but the drive changes speed to move the head at a constant linear velocity over the recording track, faster near the inner hub and slower near the outer edge of the disc. Furthermore, the CD drive reads from the inside out, so the disc will slow as it is played.

The CD itself is a 12 cm diameter, 1.3 mm thick sandwich of a hard plastic core, a mirrorlike layer of metallic aluminum, and a tough, clear plastic over-coating that protects the thin layer of aluminum. The CD records digitized data; music, video, or computer data that have been converted into a string of binary numbers. First, a master disc is made. The binary numbers are translated into a series of pulses that are fed to a laser. The laser is focused onto a photosensitive material on a spinning master disc. Whenever there is a pulse in the signal, the laser burns a small oval pit into the surface, making a pattern of pits and bumps on the track of the master disc. The laser beam is incredibly small, making marks about a micron or so in diameter. A micron is one-millionth of a meter, so you can fit a tremendous number of data tracks onto the disc, which has each track spaced 1.6 microns apart. Next, the CD audio or CD-ROM discs are made by using the master disc as a mold. Soft plastic is pressed against the master disc in a vacuum-forming machine so the small physical marks—the pits and bumps made by the laser—are pressed into the plastic. This makes a record of the strings of binary numbers that were etched into the master disc by the strong

but tiny laser beam. During playback, a low-powered laser beam is reflected off the track to read the binary marks on it. The optical sensor head contains a tiny diode laser, a lens, mirrors, and tracking devices that can move the head in three directions. The head moves side to side to keep the head over a single track (within 1.6 micron), it moves up and down to keep the laser beam in focus, and it moves forward and backward as a fine adjustment to maintain a constant linear velocity.

The advantages of the CD audio and video discs over conventional records or tapes include more uniform and accurate frequency response, a complete absence of background noise, and absence of wear—since nothing mechanical touches the surface of the disc when it is played. The advantages of the CD-ROM over the traditional magnetic floppy disks or magnetic hard disks include storage capacity, long-term reliability, and the impossibility of a head crash with a resulting loss of data.

The disadvantage of the CD audio and CD-ROM discs is the lack of ability to do writing or rewriting. Rewritable optical media are available, and these are called CD-R and CD-RW.

A CD-R records data to a disc by using a laser to burn spots into an organic dye. Such a "burned" spot reflects less light than an area that was not heated by the laser. This is designed to mimic the way light reflects off of pits and bumps of a normal CD, except this time the string of binary numbers are burned (nonreflective) and not burned areas (reflective). Since this is similar to how data on a normal CD are represented, a CD-R disc can generally be used in a CD audio or CD-ROM player as if it were a pressed CD. The dyes in a CD-R disc are photosensitive organic compounds that are similar to those used in making photographs. The color of a CD-R disc is a

result of the kind of dye that was used in the recording layer combined with the type of reflective coating used. Some of these dye and reflective coating combinations appear green, some appear blue, and others appear to be gold. Once a CD-R disc is burned, it cannot be rewritten or changed.

The CD-RW is designed to have the ability to do writing or rewriting. It uses a different technology but again mimics the way light reflects off the pits and bumps of a pressed CD. Instead of a dye-based recording layer, the CD-RW uses a compound made from silver, indium, antimony, and tellurium. This layer has a property that permits rewriting the information on a disc. This property is that when it is heated to a certain temperature and cooled, it becomes crystalline. However, when it is heated to a higher temperature and cooled, it becomes noncrystalline. A crystalline surface reflects a laser beam, while a noncrystalline surface absorbs the laser beam. The CD-RW is again designed to mimic the way light reflects off of pits and bumps of a normal CD, except this time the string of binary numbers are noncrystalline (nonreflective) and crystalline areas (reflective). In order to write, erase, and read, the CD-RW recorder must have three different laser powers. It must have (1) a high power to heat spots to about 600°C, which cool rapidly and make noncrystalline spots that are less reflective. It must have (2) a medium power to erase data by heating the media to about 200°C, which allows the media to crystallize and have a uniform reflectivity. Finally, it must have (3) a low setting that is used for finding and reading nonreflective and the more reflective areas of a disc. The writing and rewriting of a CD-RW can be repeated hundreds of times. A problem is that not all CD players can read the less reflective data spots of a CD-RW unless they have a dual-wavelength head that meets multiread specifications.

THE PRESENT THEORY

Today, light is considered to have a dual nature, sometimes acting like a wave and sometimes acting like a particle. A wave model is useful in explaining how light travels through space and how it exhibits such behaviors as refraction, interference, and diffraction. A particle model is useful in explaining how light is emitted from and absorbed by matter, exhibiting such behaviors as blackbody radiation and the photoelectric effect. Together, both of these models are part of a single theory of light, a theory that pictures light as having both particle and wave properties. Some properties are more useful when explaining some observed behaviors, and other properties are more useful when explaining other behaviors.

Frequency is a property of a wave, and the energy of a photon is a property of a particle. Both frequency and the energy of a photon are related in equation 7.4, $E = hf$. It is thus possible to describe light in terms of a frequency (or wavelength) or in terms of a quantity of energy. Any part of the electromagnetic spectrum can thus be described by units of frequency, wavelength, or energy, which are alternative means of describing light. The radio radiation parts of the spectrum are low-frequency, low-energy, and long-wavelength radiations. Radio radiations have more wave properties and practically no particle properties, since the energy levels are low. Gamma radiation, on the other hand, is high-frequency, high-energy, and short wavelength radiation. Gamma radiation has more particle properties, since the extremely short wavelengths have very high energy levels. The more familiar part of the spectrum, visible light, is between these two extremes and exhibits both wave and particle properties, but it never exhibits both properties at the same time in the same experiment.

Part of the problem in forming a concept or mental image of the exact nature of light is understanding this nature in terms of what is already known. The things you already know about are observed to be particles, or objects, or they are observed to be waves. You can see objects that move through the air, such as baseballs or footballs, and you can see waves on water or in a field of grass. There is nothing that acts like a moving object in some situations but acts like a wave in other situations. Objects are objects, and waves are waves, but objects do not become waves, and waves do not become objects. If this dual nature did exist, it would seem very strange. Imagine, for example, holding a book at a certain height above a lake (figure 7.23). You can make measurements and calculate the kinetic energy the book will have when dropped into the lake. When it hits the water, the book disappears, and water waves move away from the point of impact in a circular pattern that moves across the water. When the waves reach another person across the lake,

a book identical to the one you dropped pops up out of the water as the waves disappear. As it leaves the water across the lake, the book has the same kinetic energy that your book had when it hit the water in front of you. You and the other person could measure things about either book, and you could measure things about the waves, but you could not measure both at the same time. You might say that this behavior is not only strange but impossible. Yet, it is an analogy to the observed behavior of light.

As stated, light has a dual nature, sometimes exhibiting the properties of a wave and sometimes exhibiting the properties of moving particles but never exhibiting both properties at the same time. Both the wave and the particle nature are accepted as being part of one model today, with the understanding that the exact nature of light is not describable in terms of anything that is known to exist in the everyday-sized world. Light is an extremely small-scale phenomenon that must be different, without a sharp distinction between a particle and a wave. Evidence about this strange nature of an extremely small-scale phenomenon will be considered again in chapter 8 as a basis for introducing the quantum theory of matter.

RELATIVITY

The electromagnetic wave model brought together and explained electric and magnetic phenomena, and explained that light can be thought of as an electromagnetic wave (see figure 7.2). There remained questions, however, that would not be answered until Albert Einstein developed a revolutionary new theory. Even at the age of seventeen, Einstein was already thinking about ideas that would eventually lead to his new theory. For example, he wondered about chasing a beam of light if you were also moving at the speed of light. Would you see the light as an oscillating electric and magnetic field at rest? He realized there was no such thing, either on the basis of experience or according to Maxwell's theory of electromagnetic waves.

In 1905, at the age of twenty-six, Albert Einstein published an analysis of how space and time are affected by motion between an observer and what is being measured. This analysis is called the *special theory of relativity*. Eleven years later, Einstein published an interpretation of gravity as distortion of the structure of space and time. This analysis is called the *general theory of relativity*. A number of remarkable predictions have been made based on this theory, and all have been verified by many experiments.

Special Relativity

The **special theory of relativity** is concerned with events as observed from different points of view, or different "reference frames." Here is an example: You are on a bus traveling straight down a highway at a constant 100 km/h. An insect is observed to fly from the back of the bus at 5 km/h. With respect to the bus, the insect is flying at 5 km/h. To someone observing from the ground, however, the speed of the insect is 100 km/h plus

FIGURE 7.23 It would seem very strange if there were not a sharp distinction between objects and waves in our everyday world. Yet this appears to be the nature of light.

5 km/h, or 105 km/h. If the insect is flying toward the back of the bus, its speed is 100 km/h minus 5 km/h, or 95 km/h with respect to Earth. Generally, the reference frame is understood to be Earth, but this is not always stated. Nonetheless, we must specify a reference frame whenever a speed or velocity is measured.

Einstein's special theory is based on two principles. The first concerns frames of reference and the fact that all motion is relative to a chosen frame of reference. This principle could be called the **consistent law principle**:

> **The laws of physics are the same in all reference frames that are moving at a constant velocity with respect to each other.**

Ignoring vibrations, if you are in a windowless bus, you will not be able to tell if the bus is moving uniformly or if it is not moving at all. If you were to drop something—say, your keys—in a moving bus, they would fall straight down, just as they would in a stationary bus. The keys fall straight down with respect to the bus in either case. To an observer outside the bus, in a different frame of reference, the keys would appear to take a curved path because they have an initial velocity. Moving objects follow the same laws in a uniformly moving bus or any other uniformly moving frame of reference (figure 7.24).

The second principle concerns the speed of light and could be called the **constancy of speed principle**:

> **The speed of light in empty space has the same value for all observers regardless of their velocity.**

The speed of light in empty space is 3.00×10^8 m/s (186,000 mi/s). An observer traveling toward a source would measure the speed of light in empty space as 3.00×10^8 m/s.

An observer not moving with respect to the source would measure this very same speed. This is not like the insect moving in a bus—you do not add or subtract the velocity of the source from the velocity of light. The velocity is always 3.00×10^8 m/s for all observers, regardless of the velocity of the observers and regardless of the velocity of the source of light. Light behaves differently than anything in our everyday experience.

The special theory of relativity is based solely on the consistent law principle and the constancy of speed principle. Together, these principles result in some very interesting results if you compare measurements from the ground of the length, time, and mass of a very fast airplane with measurements made by someone moving with the airplane. You, on the ground, would find that

- The length of an object is shorter when it is moving.
- Moving clocks run more slowly.
- Moving objects have increased mass.

The special theory of relativity shows that measurements of length, time, and mass are different in different moving reference frames. Einstein developed equations that describe each of the changes described above. These changes have been verified countless times with elementary particle experiments, and the data always fit Einstein's equations with predicted results.

General Relativity

Einstein's **general theory of relativity** could also be called Einstein's geometric theory of gravity. According to Einstein, a gravitational interaction does not come from some mysterious force called gravity. Instead, the interaction is between a mass and the geometry of space and time where the mass is located. Space and time can be combined into a fourth-dimensional "spacetime" structure. A mass is understood to interact with the spacetime, telling it how to curve. Spacetime also interacts with a mass, telling it how to move. A gravitational interaction is considered to be a local event of movement along a geodesic (shortest distance between two points on a curved surface) in curved spacetime (figure 7.25). This different viewpoint has led to much more accurate measurements and has been tested by many events in astronomy.

FIGURE 7.24 All motion is relative to a chosen frame of reference. Here the photographer has turned the camera to keep pace with one of the cyclists. Relative to him, both the road and the other cyclists are moving. There is no fixed frame of reference in nature and, therefore, no such thing as "absolute motion"; all motion is relative.

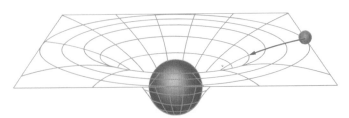

FIGURE 7.25 General relativity pictures gravity as a warping of the structure of space and time due to the presence of a body of matter. An object nearby experiences an attractive force as a result of this distortion in spacetime, much as a marble rolls toward the bottom of a saucer-shaped hole in the ground.

James Clerk Maxwell (1831–1879)

James Maxwell was a British physicist who discovered that light consists of electromagnetic waves and established the kinetic theory of gases. He also proved the nature of Saturn's rings and demonstrated the principles governing color vision.

Maxwell was born at Edinburgh, Scotland, on November 13, 1831. He was educated at Edinburgh Academy from 1841 to 1847, when he entered the University of Edinburgh. He then went on to study at Cambridge University in 1850, graduating in 1854.

He became professor of natural philosophy at Marischal College, Aberdeen, in 1856 and moved to London in 1860 to take up the post of professor of natural philosophy and astronomy at King's College. On the death of his father in 1865, Maxwell returned to his family home in Scotland and devoted himself to research. However, in 1871, he was persuaded to move to Cambridge, where he became the first professor of experimental physics and set up the Cavendish Laboratory, which opened in 1874.

Maxwell continued in this position until 1879, when he contracted cancer. He died at Cambridge on November 5, 1879, at the age of 48.

Maxwell demonstrated his great analytical ability at the age of fifteen, when he discovered an original method for drawing a perfect oval. His first important contributions to science were made from 1849 onward, when Maxwell applied himself to color vision. Maxwell showed how colors could be built up from mixtures of the primary colors red, green, and blue by spinning discs containing sectors of these colors in various sizes. In the 1850s, he refined this approach by inventing a color box in which the three primary colors could be selected from the Sun's spectrum and combined together. He explained fully how the addition and subtraction of primary colors produces all other colors, and he crowned this achievement in 1861 by producing the first color photograph using a three-color process. This picture, the ancestor of all color photography, printing, and television, was taken of a tartan ribbon by using red, green, and blue filters to photograph the tartan and to project a colored image.

Maxwell's development of the electromagnetic theory of light took many years. It began with the paper *On Faraday's Lines of Force*, in which Maxwell built on the views of Michael Faraday (1791–1867) that electric and magnetic effects result from fields of lines of force that surround conductors and magnets. Maxwell drew an analogy between the behavior of the lines of force and the flow of an incompressible liquid, thereby deriving equations that represented known electric and magnetic effects. The next step toward the electromagnetic theory took place with the publication of the paper *On Physical Lines of Force* (1861–1862). In it, Maxwell developed a model for the medium in which electric and magnetic effects could occur. He devised a hypothetical medium consisting of an incompressible fluid containing rotating vortices responding to magnetic intensity separated by cells responding to electric current.

By considering how the motion of the vortices and cells could produce magnetic and electric effects, Maxwell successfully explained all known effects of electromagnetism, showing that the lines of force must behave in a similar way. However, Maxwell went further and considered what effects would be caused if the medium were elastic. It turned out that the movement of a charge would set up a disturbance in the medium, forming transverse waves that would be propagated through the medium. The velocity of these waves would be equal to the ratio of the value for a current when measured in electrostatic units and electromagnetic units. This had been determined earlier, and it was equal to the velocity of light. Maxwell thus inferred that light consists of transverse waves in the same medium that causes electric and magnetic phenomena.

Maxwell was reinforced in this opinion by work undertaken to make basic definitions of electric and magnetic quantities in terms of mass, length, and time. In *On the Elementary Regulations of Electric Quantities* (1863), he found that the ratio of the two definitions of any quantity based on electric and magnetic forces is always equal to the velocity of light. He considered that light must consist of electromagnetic waves, but he first needed to prove this by abandoning the vortex analogy and arriving at an explanation based purely on dynamic principles. This he achieved in *A Dynamical Theory of the Electromagnetic Field* (1864), in which he developed the fundamental equations that describe the electromagnetic field. These showed that light is propagated in two waves, one magnetic and the other electric, which vibrate perpendicular to each other and to the direction of propagation. This was confirmed in Maxwell's *Note on the Electromagnetic Theory of Light* (1868), which used an electrical derivation of the theory instead of the dynamical formulation, and Maxwell's whole work on the subject was summed up in *Treatise on Electricity and Magnetism* in 1873.

The treatise also established that light has a radiation pressure and suggested that a whole family of electromagnetic radiations must exist, of which light was only one. This was confirmed in 1888 with the sensational discovery of radio waves by Heinrich Hertz (1857–1894). Sadly, Maxwell did not live long enough to see this triumphant vindication of his work.

Maxwell is generally considered to be the greatest theoretical physicist of the 1800s, as his forebear Michael Faraday was the greatest experimental physicist. His rigorous mathematical ability was combined with great insight to enable him to achieve brilliant syntheses of knowledge in the two most important areas of physics at that time. In building on Faraday's work to discover the electromagnetic nature of light, Maxwell not only explained electromagnetism but also paved the way for the discovery and application of the whole spectrum of electromagnetic radiation that has characterized modern physics.

Source: Modified from the *Hutchinson Dictionary of Scientific Biography*. © RM, 2007. Reprinted by permission.

SUMMARY

Electromagnetic radiation is emitted from all matter with a temperature above absolute zero, and as the temperature increases, more radiation and shorter wavelengths are emitted. Visible light is emitted from matter hotter than about 700°C, and this matter is said to be *incandescent*. The Sun, a fire, and the ordinary lightbulb are incandescent sources of light.

The behavior of light is shown by a light ray model that uses straight lines to show the straight-line path of light. Light that interacts with matter is *reflected* with parallel rays, moves in random directions by *diffuse reflection* from points, or is *absorbed*, resulting in a temperature increase. Matter is *opaque*, reflecting light, or *transparent*, transmitting light.

In reflection, the incoming light, or *incident ray*, has the same angle as the *reflected ray* when measured from a perpendicular from the point of reflection, called the *normal*. That the two angles are equal is called the *law of reflection*. The law of reflection explains how a flat mirror forms a *virtual image*, one from which light rays do not originate. Light rays do originate from the other kind of image, a *real image*.

Light rays are bent, or *refracted*, at the boundary when passing from one transparent media to another. The amount of refraction depends on the *incident angle* and the *index of refraction*, a ratio of the speed of light in a vacuum to the speed of light in the media. When the refracted angle is 90°, *total internal reflection* takes place. This limit to the angle of incidence is called the *critical angle*, and all light rays with an incident angle at or beyond this angle are reflected internally.

Each color of light has a range of wavelengths that forms the *spectrum* from red to violet. A glass prism has the property of *dispersion*, separating a beam of white light into a spectrum. Dispersion occurs because the index of refraction is different for each range of colors, with short wavelengths refracted more than larger ones.

A wave model of light can be used to explain diffraction, interference, and polarization, all of which provide strong evidence for the wavelike nature of light. *Interference* occurs when light passes through two small slits or holes and produces an *interference pattern* of bright lines and dark zones. *Polarized light* vibrates in one direction only, in a plane. Light can be polarized by certain materials, by reflection, or by scattering. Polarization can only be explained by a transverse wave model.

A wave model fails to explain observations of light behaviors in the *photoelectric effect* and *blackbody radiation*. Max Planck found that he could modify the wave theory to explain blackbody radiation by assuming that vibrating molecules could only have discrete amounts, or *quanta*, of energy and found that the quantized energy is related to the frequency and a constant known today as *Planck's constant*. Albert Einstein applied Planck's quantum concept to the photoelectric effect and described a light wave in terms of quanta of energy called *photons*. Each photon has an energy that is related to the frequency and Planck's constant.

Today, the properties of light are explained by a model that incorporates both the wave and the particle nature of light. Light is considered to have both wave and particle properties and is not describable in terms of anything known in the everyday-sized world.

Summary of Equations

7.1   angle of incidence = angle of reflection
$$\theta_i = \theta_r$$

7.2   index of refraction $= \dfrac{\text{speed of light in vacuum}}{\text{speed of light in material}}$
$$n = \frac{c}{v}$$

7.3   speed of light in vacuum = (wavelength)(frequency)
$$c = \lambda f$$

7.4 $\begin{array}{c}\text{energy of}\\\text{photon}\end{array} = \left(\begin{array}{c}\text{Planck's}\\\text{constant}\end{array}\right)(\text{frequency})$
$$E = hf$$

KEY TERMS

blackbody radiation (p. **155**)
consistent law principle (p. **174**)
constancy of speed principle (p. **174**)
general theory of relativity (p. **174**)

incandescent (p. **154**)
index of refraction (p. **164**)
interference (p. **167**)
light ray model (p. **157**)
luminous (p. **154**)
photoelectric effect (p. **170**)

photons (p. **171**)
polarized (p. **169**)
quanta (p. **171**)
real image (p. **160**)
refraction (p. **160**)
special theory of relativity (p. **173**)

total internal reflection (p. **161**)
unpolarized light (p. **169**)
virtual image (p. **160**)

APPLYING THE CONCEPTS

1. An object is hot enough to emit a dull red glow. When this object is heated even more, it will emit
   a. shorter-wavelength, higher-frequency radiation.
   b. longer-wavelength, lower-frequency radiation.
   c. the same wavelengths as before but with more energy.
   d. more of the same wavelengths with more energy.

2. The difference in the light emitted from a candle, an incandescent lightbulb, and the Sun is basically from differences in
   a. energy sources.   c. temperatures.
   b. materials.   d. phases of matter.

3. Before it travels through Earth's atmosphere, sunlight is mostly
   a. infrared radiation.
   b. visible light.
   c. ultraviolet radiation.
   d. blue light.

4. You are able to see in shaded areas, such as under a tree, because light has undergone
   a. refraction.   c. a change in speed.
   b. incident bending.   d. diffuse reflection.

5. The ratio of the speed of light in a vacuum to the speed of light in some transparent materials is called
   a. the critical angle.
   b. total internal reflection.
   c. the law of reflection.
   d. the index of refraction.

6. Any part of the electromagnetic spectrum, including the colors of visible light, can be measured in units of
   a. wavelength.
   b. frequency.
   c. energy.
   d. Any of the above is correct.

7. A prism separates the colors of sunlight into a spectrum because
   a. each wavelength of light has its own index of refraction.
   b. longer wavelengths are refracted more than shorter wavelengths.
   c. red light is refracted the most, violet the least.
   d. All of the above are correct.

8. Which of the following can only be explained by a wave model of light?
   a. reflection
   b. refraction
   c. interference
   d. photoelectric effect

9. The polarization behavior of light is best explained by considering light to be
   a. longitudinal waves.
   b. transverse waves.
   c. particles.
   d. particles with ends, or poles.

10. Max Planck made the revolutionary discovery that the energy of vibrating molecules involved in blackbody radiation existed only in
    a. multiples of certain fixed amounts.
    b. amounts that smoothly graded one into the next.
    c. the same, constant amount of energy in all situations.
    d. amounts that were never consistent from one experiment to the next.

11. Einstein applied Planck's quantum discovery to light and found
    a. a direct relationship between the energy and frequency of light.
    b. that the energy of a photon divided by the frequency of the photon always equaled a constant known as Planck's constant.
    c. that the energy of a photon divided by Planck's constant always equaled the frequency.
    d. All of the above are correct.

12. Today, light is considered to be
    a. tiny particles of matter that move through space, having no wave properties.
    b. electromagnetic waves only, with no properties of particles.
    c. a small-scale phenomenon without a sharp distinction between particle and wave properties.
    d. something that is completely unknown.

Answers

1. a 2. c 3. a 4. d 5. d 6. d 7. a 8. c 9. b 10. a 11. d 12. c

QUESTIONS FOR THOUGHT

1. What determines if an electromagnetic wave emitted from an object is a visible light wave or a wave of infrared radiation?

2. What model of light does the polarization of light support? Explain.

3. Which carries more energy, red light or blue light? Should this mean anything about the preferred color of warning and stop lights? Explain.

4. What model of light is supported by the photoelectric effect? Explain.

5. What happens to light that is absorbed by matter?

6. One star is reddish, and another is bluish. Do you know anything about the relative temperatures of the two stars? Explain.

7. When does total internal reflection occur? Why does this occur in the diamond more than other gemstones?

8. Why does a highway sometimes appear wet on a hot summer day when it is not wet?

9. How can you tell if a pair of sunglasses is polarizing or not?

10. Explain why the intensity of reflected light appears to change if you tilt your head from side to side while wearing polarizing sunglasses.

11. What was so unusual about Planck's findings about blackbody radiation? Why was this considered revolutionary?

12. Why are both the photon model and the electromagnetic wave model accepted today as a single theory? Why was this so difficult for people to accept at first?

FOR FURTHER ANALYSIS

1. Clarify the distinction between light reflection and light refraction by describing clear, obvious examples of each.

2. Describe how you would use questions alone to help someone understand that the shimmering they see above a hot pavement is not heat.

3. Use a dialogue as you "think aloud" in considering the evidence that visible light is a wave, a particle, or both.

4. Compare and contrast the path of light through a convex and a concave lens. Give several uses for each lens, and describe how the shape of the lens results in that particular use.

5. Analyze how the equation $E = hf$ could mean that visible light is a particle and a wave at the same time.

6. How are visible light and a radio wave different? How are they the same?

INVITATION TO INQUIRY

Best Sunglasses?

Obtain several different types of sunglasses. Design experiments to determine which combination of features will be found in the best pair of sunglasses. First, design an experiment to determine which type reduces reflected glare the most. Find out how sunglasses are able to block ultraviolet radiation. According to your experiments and research, describe the "best" pair of sunglasses.

PARALLEL EXERCISES

The exercises in groups A and B cover the same concepts. Solutions to group A exercises are located in appendix D.

Group A

1. What is the speed of light while traveling through (a) water and (b) ice?

2. How many minutes are required for sunlight to reach Earth if the Sun is 1.50×10^8 km from Earth?

3. How many hours are required before a radio signal from a space probe near the dwarf planet Pluto reaches Earth, 6.00×10^9 km away?

4. A light ray is reflected from a mirror with an angle 10° to the normal. What was the angle of incidence?

5. Light travels through a transparent substance at 2.20×10^8 m/s. What is the substance?

6. The wavelength of a monochromatic light source is measured to be 6.00×10^{-7} m in an experiment. (a) What is the frequency? (b) What is the energy of a photon of this light?

7. At a particular location and time, sunlight is measured on a 1 m^2 solar collector with a power of 1,000.0 W. If the peak intensity of this sunlight has a wavelength of 5.60×10^{-7} m, how many photons are arriving each second?

8. A light wave has a frequency of 4.90×10^{14} cycles per second. (a) What is the wavelength? (b) What color would you observe (see table 7.2)?

9. What is the energy of a gamma photon of frequency 5.00×10^{20} Hz?

10. What is the energy of a microwave photon of wavelength 1.00 mm?

Group B

1. (a) What is the speed of light while traveling through a vacuum? (b) While traveling through air at 30°C? (c) While traveling through air at 0°C?

2. How much time is required for reflected sunlight to travel from the Moon to Earth if the distance between Earth and the Moon is 3.85×10^5 km?

3. How many minutes are required for a radio signal to travel from Earth to a space station on Mars if Mars is 7.83×10^7 km from Earth?

4. An incident light ray strikes a mirror with an angle of 30° to the surface of the mirror. What is the angle of the reflected ray?

5. The speed of light through a transparent substance is 2.00×10^8 m/s. What is the substance?

6. A monochromatic light source used in an experiment has a wavelength of 4.60×10^{-7} m. What is the energy of a photon of this light?

7. In black-and-white photography, a photon energy of about 4.00×10^{-19} J is needed to bring about the changes in the silver compounds used in the film. Explain why a red light used in a darkroom does not affect the film during developing.

8. The wavelength of light from a monochromatic source is measured to be 6.80×10^{-7} m. (a) What is the frequency of this light? (b) What color would you observe?

9. How much greater is the energy of a photon of ultraviolet radiation ($\lambda = 3.00 \times 10^{-7}$ m) than the energy of an average photon of sunlight ($\lambda = 5.60 \times 10^{-7}$ m)?

10. At what rate must electrons in a wire vibrate to emit microwaves with a wavelength of 1.00 mm?

This is a picture of pure zinc, one of the eighty-nine naturally occurring elements found on Earth.

Atoms and Periodic Properties

Chemistry Connections

▶ Electron configuration can be used to explain how atoms join together (Ch. 9).

▶ Water and its properties can be explained by considering the electron structure of hydrogen and oxygen atoms (Ch. 10).

▶ The nature of the atomic nucleus will explain radioactivity and nuclear energy (Ch. 11).

Astronomy Connections

▶ The energy from stars originates in nuclear reactions in their core (Ch. 12).

▶ Stars are nuclear reactors (Ch. 12).

CORE CONCEPT

Different fields of study contributed to the model of the atom.

The electron was discovered from experiments with electricity (p. 181).

The nucleus and proton were discovered from experiments with radioactivity (p. 182).

Experiments with light and line spectra and application of the quantum concept led to the Bohr model of the atom (pp. 184–85).

Application of the wave properties of electrons led to the quantum mechanics model of the atom (pp. 186–87).

Earth Science Connections

▶ Materials that make up Earth can be understood by considering their atomic structures (Ch. 15–16).

▶ Earth cycles materials through change (Ch. 15–18).

Life Science Connections

▶ Living things use energy and materials in complex interactions (Ch. 19–26).

OVERVIEW

The development of the modern atomic model illustrates how modern scientific understanding comes from many different fields of study. For example, you will learn how studies of electricity led to the discovery that atoms have subatomic parts called *electrons.* The discovery of radioactivity led to the discovery of more parts, a central nucleus that contains protons and neutrons. Information from the absorption and emission of light was used to construct a model of how these parts are put together, a model resembling a miniature solar system with electrons circling the nucleus. The solar system model had initial, but limited, success and was inconsistent with other understandings about matter and energy. Modifications of this model were attempted, but none solved the problems. Then the discovery of wave properties of matter led to an entirely new model of the atom (figure 8.1).

The atomic model will be put to use in later chapters to explain the countless varieties of matter and the changes that matter undergoes. In addition, you will learn how these changes can be manipulated to make new materials, from drugs to ceramics. In short, you will learn how understanding the atom and all the changes it undergoes not only touches your life directly but shapes and affects all parts of civilization.

ATOMIC STRUCTURE DISCOVERED

Did you ever wonder how scientists could know about something so tiny that you cannot see it, even with the most powerful optical microscope? The atom is a tiny unit of matter, so small that 1 gram of hydrogen contains about 600,000,000,000,000,000,000,000 (six-hundred-thousand-billion billion or 6×10^{23}) atoms. Even more unbelievable is that atoms are not individual units but are made up of even smaller particles. How is it possible that scientists are able to tell you about the parts of something so small that it cannot be seen? The answer is that these things cannot be observed directly, but their existence can be inferred from experimental evidence. The following story describes the evidence and how scientists learned about the parts—electrons, the nucleus, protons, neutrons, and others— and how they are all arranged in the atom.

The atomic concept is very old, dating back to ancient Greek philosophers some 2,500 years ago. The ancient Greeks also reasoned about the way that pure substances are put together. A glass of water, for example, appears to be the same throughout. Is it the same? Two plausible, but conflicting, ideas were possible as an intellectual exercise. The water could have a *continuous* structure; that is, it could be completely homogeneous throughout. The other idea was that the water only appears to be continuous but is actually *discontinuous.* This means that if you continue to divide the water into smaller and smaller volumes, you would eventually reach a limit to this dividing, a particle that could not be further subdivided. The Greek philosopher Democritus (460–362 B.C.) developed this model in the fourth century B.C., and he called the indivisible particle an *atom,* from a Greek word meaning "uncuttable." However, neither Plato nor Aristotle accepted the atomic theory of matter, and it was not until about two thousand years later that the atomic concept of matter was reintroduced. In the early 1800s, the English chemist John Dalton brought back the ancient Greek idea of hard, indivisible atoms to explain

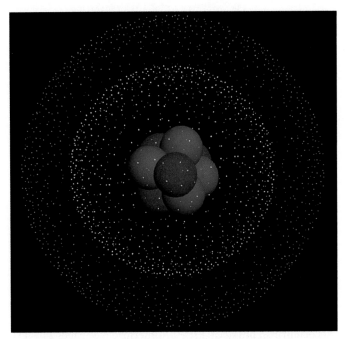

FIGURE 8.1 This is a computer-generated model of a beryllium atom, showing the nucleus and electron spacing relationships. This configuration can also be predicted from information on a periodic table. (Not to scale.)

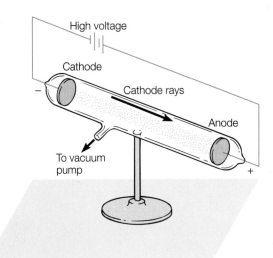

FIGURE 8.2 A vacuum tube with metal plates attached to a high-voltage source produces a greenish beam called *cathode rays*. These rays move from the cathode (negative charge) to the anode (positive charge).

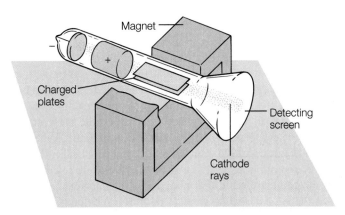

FIGURE 8.3 A cathode ray passed between two charged plates is deflected toward the positively charged plate. The ray is also deflected by a magnetic field. By measuring the deflection by both, J. J. Thomson was able to calculate the ratio of charge to mass. He was able to measure the deflection because the detecting screen was coated with zinc sulfide, a substance that produces a visible light when struck by a charged particle.

chemical reactions. Five statements will summarize his theory. As you will soon see, today we know that statement 2 is not strictly correct:

1. Indivisible minute particles called atoms make up all matter.
2. All the atoms of an element are exactly alike in shape and mass.
3. The atoms of different elements differ from one another in their masses.
4. Atoms chemically combine in definite whole-number ratios to form chemical compounds.
5. Atoms are neither created nor destroyed in chemical reactions.

During the 1800s, Dalton's concept of hard, indivisible atoms was familiar to most scientists. Yet, the existence of atoms was not generally accepted by all scientists. There was skepticism about something that could not be observed directly. Strangely, full acceptance of the atom came in the early 1900s with the discovery that the atom was not indivisible after all. The atom has parts that give it an internal structure. The first part to be discovered was the *electron*, a part that was discovered through studies of electricity.

Discovery of the Electron

Scientists of the late 1800s were interested in understanding the nature of the recently discovered electric current. To observe a current directly, they tried to produce one by itself, away from wires, by removing the air from a tube and then running electricity through the vacuum. When metal plates inside a tube

were connected to the negative and positive terminals of a high-voltage source (figure 8.2), a greenish beam was observed that seemed to move from the cathode (negative terminal) through the empty tube and collect at the anode (positive terminal). Since this mysterious beam seemed to come out of the cathode, it was said to be a *cathode ray*.

The English physicist J. J. Thomson figured out what the cathode ray was in 1897. He placed charged metal plates on each side of the beam (figure 8.3) and found that the beam was deflected away from the negative plate. Since it was known that like charges repel, this meant that the beam was composed of negatively charged particles.

The cathode ray was also deflected when caused to pass between the poles of a magnet. By balancing the deflections made by the magnet with the deflections made by the electric field, Thomson could determine the ratio of the charge to mass for an individual particle. Today, the charge-to-mass ratio is considered to be 1.7584×10^{11} coulomb/kilogram. A significant

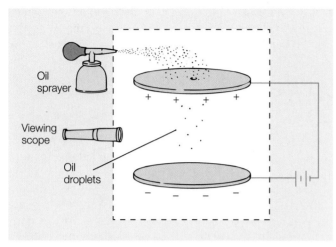

FIGURE 8.4 Millikan measured the charge of an electron by balancing the pull of gravity on oil droplets with an upward electrical force. Knowing the charge-to-mass ratio that Thomson had calculated, Millikan was able to calculate the charge on each droplet. He found that all the droplets had a charge of 1.60×10^{-19} coulomb or multiples of that charge. The conclusion was that this had to be the charge of an electron.

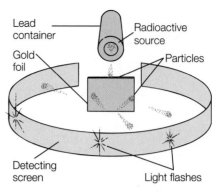

FIGURE 8.5 Rutherford and his co-workers studied particle scattering from a thin metal foil. The particles struck the detecting screen, producing a flash of visible light. Measurements of the angles between the flashes, the metal foil, and the source of the particles showed that they were scattered in all directions, including straight back toward the source.

part of Thomson's experiments was that he found the charge-to-mass ratio was the same no matter what gas was in the tube and of what materials the electrodes were made. Thomson had discovered the **electron,** a fundamental particle of matter.

A method for measuring the charge and mass of the electron was worked out by an American physicist, Robert A. Millikan, around 1906. Millikan used an apparatus like the one illustrated in figure 8.4 to measure the charge on tiny droplets of oil. Millikan found that none of the droplets had a charge less than one particular value (1.60×10^{-19} coulomb) and that larger charges on various droplets were always multiples of this unit of charge. Since all of the droplets carried the single unit of charge or multiples of the single unit, the unit of charge was understood to be the charge of a single electron.

Knowing the charge of a single electron and knowing the charge-to-mass ratio that Thomson had measured now made it possible to calculate the mass of a single electron. The mass of an electron was thus determined to be about 9.11×10^{-31} kg, or about 1/1,840 of the mass of the lightest atom, hydrogen.

Thomson had discovered the negatively charged electron, and Millikan had measured the charge and mass of the electron. But atoms themselves are electrically neutral. If an electron is part of an atom, there must be something else that is positively charged, canceling the negative charge of the electron. The next step in the sequence of understanding atomic structure would be to find what is neutralizing the negative charge and to figure out how all the parts are put together.

Thomson had proposed a model for what was known about the atom at the time. He suggested that an atom could be a blob of massless, positively charged stuff in which electrons were stuck like "raisins in plum pudding." If the mass of a hydrogen atom is due to the electrons embedded in a massless, positively charged matrix, then 1,840 electrons would be needed together with sufficient positive matter to make the atom electrically neutral.

The Nucleus

The nature of radioactivity and matter were the research interests of a British physicist, Ernest Rutherford (see pages 244–45). In 1907, Rutherford was studying the scattering of radioactive particles directed toward a thin sheet of metal. As shown in figure 8.5, the particles from a radioactive source were allowed to move through a small opening in a lead container, so only a narrow beam of the massive, fast-moving particles would penetrate a very thin sheet of gold. The particles were detected by plates, which produced a small flash of light when struck by the particles.

Rutherford found that most of the particles went straight through the foil. However, he was astounded to find that some were deflected at very large angles and some were even reflected backward. He could account for this only by assuming that the massive, positively charged particles were repelled by a massive positive charge concentrated in a small region of the atoms in the gold foil (figure 8.6). He concluded that an atom must have a tiny, massive, and positively charged **nucleus** surrounded by electrons.

From measurements of the scattering, Rutherford estimated electrons must be moving around the nucleus at a distance 100,000 times the radius of the nucleus. This means the volume of an atom is mostly empty space. A few years later, Rutherford was able to identify the discrete unit of positive charge, which we now call a **proton.** Rutherford also speculated about the existence of a neutral particle in the nucleus, a neutron. The **neutron** was eventually identified in 1932 by James Chadwick.

Today, the number of protons in the nucleus of an atom is called the **atomic number.** All the atoms of a particular element have the same number of protons in their nucleus, so all atoms of an element have the same atomic number. Hydrogen has an atomic number of 1, so any atom that has one proton in its nucleus is an atom of the element hydrogen. Today, scientists have identified 117 different kinds of elements, each with a different number of protons.

The neutrons of the nucleus, along with the protons, contribute to the mass of an atom. Although all the atoms of an

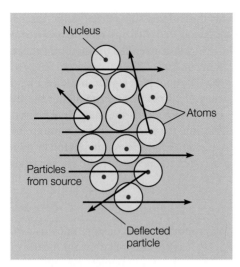

FIGURE 8.6 Rutherford's nuclear model of the atom explained the particle-scattering results as positive particles experiencing a repulsive force from the positive nucleus. Measurements of the percent of particles passing straight through and of the various angles of scattering of those coming close to the nuclei gave Rutherford a means of estimating the size of the nucleus.

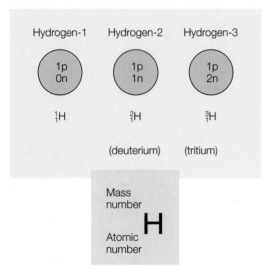

FIGURE 8.7 The three isotopes of hydrogen have the same number of protons but different numbers of neutrons. Hydrogen-1 is the most common isotope. Hydrogen-2, with an additional neutron, is named *deuterium,* and hydrogen-3 is called *tritium.*

element must have the same number of protons in their nuclei, the number of neutrons may vary. Atoms of an element that have different numbers of neutrons are called **isotopes.** There are three isotopes of hydrogen illustrated in figure 8.7. All three isotopes have the same number of protons and electrons, but one isotope has no neutrons, one isotope has 1 neutron (deuterium), and one isotope has 2 neutrons (tritium).

An atom is very tiny, and it is impossible to find the mass of a given atom. It is possible, however, to compare the mass of one atom to another. The mass of any atom is compared to the mass of an atom of a particular isotope of carbon. This particular carbon isotope is assigned a mass of exactly 12.00 . . . units called **atomic mass units** (u). Since this isotope is *defined* to be exactly 12 u, it can have an infinite number of significant figures. This isotope, called *carbon-12,* provides the standard to which the masses of all other isotopes are compared. The relative mass of any isotope is based on the mass of a carbon-12 isotope.

The relative mass of the hydrogen isotope without a neutron is 1.007 when compared to carbon-12. The relative mass of the hydrogen isotope with 1 neutron is 2.0141 when compared to carbon-12. Elements occur in nature as a mixture of isotopes, and the contribution of each is calculated in the atomic weight. **Atomic weight** is a weighted average of the isotopes based on their mass compared to carbon-12 and their relative abundance found on Earth. Of all the hydrogen isotopes, for example, 99.985 percent occurs as the isotope without a neutron and 0.015 percent is the isotope with 1 neutron (the other isotope is not considered because it is radioactive). The fractional part of occurrence is multiplied by the relative atomic mass for each isotope, and the results are summed to obtain the atomic weight. Table 8.1 gives the atomic weight of hydrogen as 1.0079 as a result of this calculation.

The sum of the number of protons and neutrons in a nucleus of an atom is called the **mass number** of that atom. Mass numbers are used to identify isotopes. A hydrogen atom with 1 proton and 1 neutron has a mass number of $1 + 1$, or 2, and is referred to as hydrogen-2. A hydrogen atom with 1 proton and 2 neutrons has a mass number of $1 + 2$, or 3, and is referred to as hydrogen-3. Using symbols, hydrogen-3 is written as

$$^{3}_{1}\text{H}$$

where H is the chemical symbol for hydrogen, the subscript to the bottom left is the atomic number, and the superscript to the top left is the mass number.

How are the electrons moving around the nucleus? It might occur to you, as it did to Rutherford and others, that an atom might be similar to a miniature solar system. In this analogy, the nucleus is in the role of the Sun, electrons in the role of moving planets in their orbits, and electrical attractions between the nucleus and electrons in the role of gravitational attraction. There are, however, big problems with this idea. If electrons were moving in circular orbits, they would continually change

TABLE 8.1 Selected Atomic Weights Calculated from Mass and Abundance of Isotopes

Stable Isotopes	Mass of Isotope Compared to C-12		Abundance		Atomic Weight
$^{1}_{1}\text{H}$	1.007	×	99.985%		
$^{2}_{1}\text{H}$	2.0141	×	0.015%	=	1.0079
$^{9}_{4}\text{Be}$	9.01218	×	100%	=	9.01218
$^{14}_{7}\text{N}$	14.00307	×	99.63%		
$^{15}_{7}\text{N}$	15.00011	×	0.37%	=	14.0067

their direction of travel and would therefore be accelerating. According to the Maxwell model of electromagnetic radiation, an accelerating electric charge emits electromagnetic radiation such as light. If an electron gave off light, it would lose energy. The energy loss would mean that the electron could not maintain its orbit, and it would be pulled into the oppositely charged nucleus. The atom would collapse as electrons spiraled into the nucleus. Since atoms do not collapse like this, there is a significant problem with the solar system model of the atom.

THE BOHR MODEL

Niels Bohr was a young Danish physicist who visited Rutherford's laboratory in 1912 and became very interested in questions about the solar system model of the atom. He wondered what determined the size of the electron orbits and the energies of the electrons. He wanted to know why orbiting electrons did not give off electromagnetic radiation. Seeking answers to questions such as these led Bohr to incorporate the *quantum concept* of Planck and Einstein with Rutherford's model to describe the electrons in the outer part of the atom. This quantum concept will be briefly reviewed before proceeding with the development of Bohr's model of the hydrogen atom.

The Quantum Concept

In 1900, Max Planck introduced the idea that matter emits and absorbs energy in discrete units that he called **quanta.** Planck had been trying to match data from spectroscopy experiments with data that could be predicted from the theory of electromagnetic radiation. In order to match the experimental findings

with the theory, he had to assume that specific, discrete amounts of energy were associated with different frequencies of radiation. In 1905, Albert Einstein extended the quantum concept to light, stating that light consists of discrete units of energy that are now called **photons.** The energy of a photon is directly proportional to the frequency of vibration, and the higher the frequency of light, the greater the energy of the individual photons. In addition, the interaction of a photon with matter is an "all-or-none" affair; that is, matter absorbs an entire photon or none of it. The relationship between frequency (f) and energy (E) is

$$E = hf \qquad \text{equation 8.1}$$

where h is the proportionality constant known as *Planck's constant* (6.63×10^{-34} J·s). This relationship means that higher-frequency light, such as ultraviolet, has more energy than lower-frequency light, such as red light.

EXAMPLE 8.1 *(Optional)*

What is the energy of a photon of red light with a frequency of 4.60×10^{14} Hz?

SOLUTION

$f = 4.60 \times 10^{14}$ Hz
$h = 6.63 \times 10^{-34}$ J·s
$E =$

$E = hf$

$= (6.63 \times 10^{-34} \text{ J·s})\left(4.60 \times 10^{14} \frac{1}{s}\right)$

$= (6.63 \times 10^{-34})(4.60 \times 10^{14}) \text{J·s} \times \frac{1}{s}$

$= \boxed{3.05 \times 10^{-19} \text{ J}}$

EXAMPLE 8.2 *(Optional)*

What is the energy of a photon of violet light with a frequency of 7.30×10^{14} Hz? (Answer: 4.84×10^{-19} J)

Atomic Spectra

Planck was concerned with hot solids that emit electromagnetic radiation. The nature of this radiation, called *blackbody radiation,* depends on the temperature of the source. When this light is passed through a prism, it is dispersed into a *continuous spectrum,* with one color gradually blending into the next as in a rainbow. Today, it is understood that a continuous spectrum comes from solids, liquids, and dense gases because the atoms interact, and all frequencies within a temperature-determined range are emitted. Light from an incandescent gas, on the other hand, is dispersed into a **line spectrum,** narrow lines of colors with no light between the lines (figure 8.8). The atoms in the incandescent gas are able to emit certain characteristic frequencies, and each frequency is a line of color that represents a definite value of energy. The line spectra are specific for a substance, and increased or decreased temperature changes only the intensity of the lines of colors. Thus, hydrogen always produces the same colors of lines in

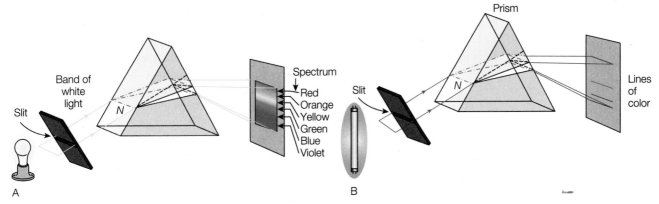

FIGURE 8.8 (*A*) Light from incandescent solids, liquids, or dense gases produces a continuous spectrum as atoms interact to emit all frequencies of visible light. (*B*) Light from an incandescent gas produces a line spectrum as atoms emit certain frequencies that are characteristic of each element.

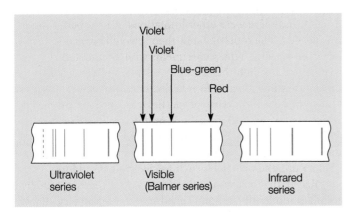

FIGURE 8.9 Atomic hydrogen produces a series of characteristic line spectra in the ultraviolet, visible, and infrared parts of the total spectrum. The visible light spectra always consist of two violet lines, a blue-green line, and a bright red line.

the same position. Helium has its own specific set of lines, as do other substances. Line spectra are a kind of fingerprint that can be used to identify a gas. A line spectrum might also extend beyond visible light into ultraviolet, infrared, and other electromagnetic regions.

In 1885, a Swiss mathematics teacher named J. J. Balmer was studying the regularity of spacing of the hydrogen line spectra. Balmer was able to develop an equation that fit all the visible lines. These four lines became known as the *Balmer series*. Other series were found later, outside the visible part of the spectrum (figure 8.9).

Such regularity of observable spectral lines must reflect some unseen regularity in the atom. At this time it was known that hydrogen had only one electron. How could one electron produce series of spectral lines with such regularity?

Bohr's Theory

An acceptable model of the hydrogen atom would have to explain the characteristic line spectra and their regularity as described by Balmer. In fact, a successful model should be able to predict the occurrence of each color line as well as account for its origin. By 1913, Bohr was able to do this by applying

the quantum concept to a solar system model of the atom. He began by considering the single hydrogen electron to be a single "planet" revolving in a circular orbit around the nucleus. Three sets of rules described this electron:

1. **Allowed orbits.** An electron can revolve around an atom only in specific allowed orbits. Bohr considered the electron to be a particle with a known mass in motion around the nucleus and used Newtonian mechanics to calculate the distances of the allowed orbits. According to the Bohr model, electrons can exist only in one of these allowed orbits and nowhere else.

2. **Radiationless orbits.** An electron in an allowed orbit does not emit radiant energy as long as it remains in the orbit. According to Maxwell's theory of electromagnetic radiation, an accelerating electron should emit an electromagnetic wave, such as light, which would move off into space from the electron. Bohr recognized that electrons moving in a circular orbit are accelerating, since they are changing direction continuously. Yet, hydrogen atoms did not emit light in their normal state. Bohr decided that the situation must be different for orbiting electrons and that electrons could stay in their allowed orbits and not give off light. He postulated this rule as a way to make his theory consistent with other scientific theories.

3. **Quantum leaps.** An electron gains or loses energy only by moving from one allowed orbit to another (figure 8.10). In the Bohr model, the energy an electron has depends on which allowable orbit it occupies. The only way that an electron can change its energy is to jump from one allowed orbit to another in quantum "leaps." An electron must acquire energy to jump from a lower orbit to a higher one. Likewise, an electron gives up energy when jumping from a higher orbit to a lower one. Such jumps must be all at once, not part way and not gradual. An electron acquires energy from high temperatures or from electrical discharges to jump to a higher orbit. An electron jumping from a higher to a lower orbit gives up energy in the form of light. A single photon is emitted when a downward jump occurs, and the energy of the photon is *exactly* equal to the difference in the energy level of the two orbits.

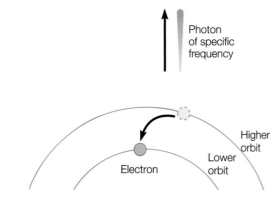

Photon
of specific
frequency

Higher
orbit

Lower
orbit

Electron

FIGURE 8.10 Each time an electron makes a "quantum leap," moving from a higher-energy orbit to a lower-energy orbit, it emits a photon of a specific frequency and energy value.

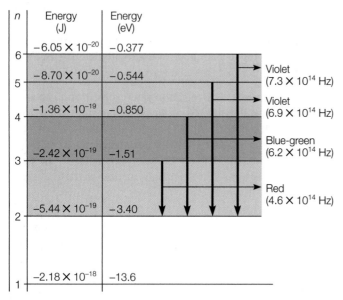

n	Energy (J)	Energy (eV)	
6	-6.05×10^{-20}	-0.377	
5	-8.70×10^{-20}	-0.544	Violet (7.3×10^{14} Hz)
4	-1.36×10^{-19}	-0.850	Violet (6.9×10^{14} Hz)
3	-2.42×10^{-19}	-1.51	Blue-green (6.2×10^{14} Hz)
2	-5.44×10^{-19}	-3.40	Red (4.6×10^{14} Hz)
1	-2.18×10^{-18}	-13.6	

FIGURE 8.11 An energy level diagram for a hydrogen atom, not drawn to scale. The energy levels (n) are listed on the left side, followed by the energies of each level in J and eV. The color and frequency of the visible light photons emitted are listed on the right side, with the arrow showing the orbit moved from and to.

The energy level diagram in figure 8.11 shows the energy states for the orbits of a hydrogen atom. The lowest energy state is the **ground state** (or normal state). The higher states are the **excited states.** The electron in a hydrogen atom would normally occupy the ground state, but high temperatures or electric discharge can give the electron sufficient energy to jump to one of the excited states. Once in an excited state, the electron immediately jumps back to a lower state, as shown by the arrows in the figure. The length of the arrow represents the frequency of the photon that the electron emits in the process. A hydrogen atom can give off only one photon at a time, and the many lines of a hydrogen line spectrum come from many atoms giving off many photons at the same time.

As you can see, the energy level diagram in figure 8.11 shows how the change of known energy levels from known orbits results in the exact frequencies of the color lines in the Balmer series. Bohr's theory did explain the lines in the hydrogen spectrum with a remarkable degree of accuracy. However, the model did not have much success with larger atoms. The Bohr model could not explain the spectra line of larger atoms with its single quantum number. A German physicist, A. Sommerfeld, tried to modify Bohr's model by adding elliptical orbits in addition to Bohr's circular orbits. It soon became apparent that the "patched up" model, too, was not adequate. Bohr had made the rule that there were radiationless orbits without an explanation, and he did not have an explanation for the quantized orbits. There was something fundamentally incomplete about the model.

QUANTUM MECHANICS

The Bohr model of the atom successfully accounted for the line spectrum of hydrogen and provided an understandable mechanism for the emission of photons by atoms. However, the model did not predict the spectra of any atom larger than hydrogen, and there were other limitations. A new, better theory was needed. The roots of a new theory would again come from experiments with light. Experiments with light had established that sometimes light behaves like a stream of particles and at other times like a wave. Eventually, scientists began to accept that light has both wave properties and particle properties, which is now referred to as the *wave-particle duality of light*. In 1923, Louis de Broglie, a French physicist, reasoned that symmetry is usually found in nature, so if a particle of light has a dual nature, then particles such as electrons should too.

Recall that waves confined on a fixed string establish resonant modes of vibration called *standing waves* (see chapter 5). Only certain fundamental frequencies and harmonics can exist on a string, and the combination of the fundamental and overtones gives the stringed instrument its particular quality. The same result of resonant modes of vibrations is observed in *any* situation where waves are confined to a fixed space. Characteristic standing wave patterns depend on the wavelength and wave velocity for waves formed on strings, in enclosed columns of air, or for any kind of wave in a confined space. Electrons are confined to the space near a nucleus, and electrons have wave properties, so an electron in an atom must be a confined wave. Does an electron form a characteristic wave pattern? This was the question being asked in about 1925 when Heisenberg, Schrödinger, Dirac, and others applied the wave nature of the electron to develop a new model of the atom based on the mechanics of electron waves. The new theory is now called *wave mechanics*, or quantum mechanics.

Erwin Schrödinger, an Austrian physicist, treated the atom as a three-dimensional system of waves to derive the *Schrödinger equation*. Instead of the simple circular planetary orbits of the Bohr model, solving the Schrödinger equation results in a description of three-dimensional shapes of the patterns that develop when electron waves are confined by a nucleus. Schrödinger first considered the hydrogen atom, calculating the states of vibration that would be possible for an electron wave confined by a nucleus. He found that the frequency of these vibrations, when multiplied by Planck's constant, matched exactly, to the last decimal point, the observed energies of the quantum states of the hydrogen atom. The conclusion is that the wave nature of the electron is the important property to consider for a successful model of the atom.

The quantum mechanics theory of the atom proved to be very successful; it confirmed all the known experimental facts and predicted new discoveries.

Quantum mechanics does share at least one idea with the Bohr model. This is that an electron emits a photon when jumping from a higher state to a lower one. The Bohr model, however, considered the *particle* nature of an electron moving in a circular orbit with a definitely assigned position at a given time. Quantum mechanics considers the *wave* nature, with the electron as a confined wave with well-defined shapes and frequencies. A wave is not localized like a particle and is spread out in space. The quantum mechanics model is, therefore, a series of orbitlike smears, or fuzzy statistical representations, of where the electron might be found.

The quantum mechanics model is a highly mathematical treatment of the mechanics of matter waves. In addition, the wave properties are considered as three-dimensional problems, and three quantum numbers are needed to describe the fuzzy electron cloud. The mathematical detail will not be presented here. The following is a qualitative description of the main ideas in the quantum mechanics model. It will describe the results of the mathematics and will provide a mental visualization of what it all means.

First, understand that the quantum mechanical theory is not an extension or refinement of the Bohr model. The Bohr model considered electrons as particles in circular orbits that could be only certain distances from the nucleus. The quantum mechanical model, on the other hand, considers the electron as a wave and considers the energy of its harmonics, or modes, of standing waves. In the Bohr model, the location of an electron was certain—in an orbit. In the quantum mechanical model, the electron is a spread-out wave.

Quantum mechanics describes the energy state of an electron wave with four *quantum numbers:*

1. **Distance from the nucleus.** The *principal quantum number* describes the *main energy level* of an electron in terms of its most probable distance from the nucleus. The lowest energy state possible is closest to the nucleus and is assigned the principal quantum number of 1 ($n = 1$). Higher states are assigned progressively higher positive whole numbers of $n = 2$, $n = 3$, $n = 4$, and so on. Electrons with higher principal quantum numbers have higher energies and are located farther from the nucleus.

2. **Energy sublevel.** The *angular momentum quantum number* defines energy sublevels within the main energy levels. Each sublevel is identified with a letter. The first four of these letters, in order of increasing energy, are s, p, d, and f. The letter s represents the lowest sublevel, and the letter f represents the highest sublevel. A principal quantum number and a letter indicating the angular momentum quantum number are combined to identify the main energy state and energy sublevel of an electron. For an electron in the lowest main energy level, $n = 1$, and in the lowest sublevel, s, the number and letter are 1s (read as "one-s"). Thus, 1s indicates an electron that is as close to the nucleus as possible in the lowest energy sublevel possible.

   There are limits to how many sublevels can occupy each of the main energy levels. Basically, the lowest main energy level can have only the lowest sublevel, and another sublevel is added as you move up through the main energy levels. Thus, the lowest main energy level, $n = 1$, can have only the s sublevel. The $n = 2$ can have s and p sublevels. The $n = 3$ main energy level can have the s, p, and the d sublevels. Finally, the $n = 4$ main energy level can have all four sublevels, with s, p, d, and f. Therefore, the number of possible sublevels is the same as the principal quantum number.

   The Bohr model considered the location of an electron as certain, like a tiny shrunken marble in an orbit. The quantum mechanics model considers the electron as a wave, and knowledge of its location is very uncertain. The **Heisenberg uncertainty principle** states that you cannot measure the exact position of a wave because a wave is spread out. One cannot specify the position and the momentum of a spread-out electron. The location of the electron can be described only in terms of *probabilities* of where it might be at a given instant. The probability of location is described by a fuzzy region of space called an **orbital.** An orbital defines the space where an electron is likely to be found. Orbitals have characteristic three-dimensional shapes and sizes and are identified with electrons of characteristic energy levels (figure 8.12). An orbital shape represents where an electron could probably be located at any particular instant. This "probability cloud" could likewise have any particular orientation in space, and the direction of this orientation is uncertain.

3. **Orientation in space.** An external magnetic field applied to an atom produces different energy levels that are related to the orientation of the orbital to the magnetic field. The orientation of an orbital in space is described by the *magnetic quantum number*. This number is related to the energies of orbitals as they are oriented in space relative to an external magnetic field, a kind of energy sub-sublevel. In general, the lowest-energy sublevel (s) has only one orbital orientation (figure 8.12A). The next higher-energy sublevel (p) can have three orbital orientations (figure 8.12B). The d sublevel can have five orbital orientations, and the highest sublevel, f, can have a total of seven different orientations (figure 8.12C) (also see table 8.2).

4. **Direction of spin.** Detailed studies have shown that an electron spinning one way, say clockwise, in an external magnetic field would have a different energy than one spinning the other way, say counterclockwise. The *spin quantum number* describes these two spin orientations (figure 8.13).

Electron spin is an important property that helps determine the electronic structure of an atom. As it turns out, two electrons spinning in opposite directions produce unlike magnetic fields that are attractive, balancing some of the normal repulsion from two like charges. Two electrons of opposite spin, called an electron pair, can thus occupy the same orbital. Wolfgang Pauli, a German physicist, summarized this idea in 1924. His summary, now known as the **Pauli exclusion principle,** states that *no two electrons in an atom can have the same four quantum numbers.* This principle provides the key for understanding the electron structure of atoms.

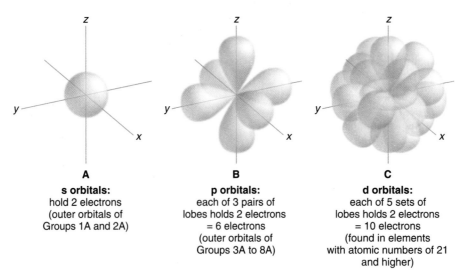

FIGURE 8.12 The general shapes of s, p, and d orbitals, the regions of space around the nuclei of atoms in which electrons are likely to be found. (The f orbital is too difficult to depict.)

A

s orbitals:
hold 2 electrons
(outer orbitals of
Groups 1A and 2A)

B

p orbitals:
each of 3 pairs of
lobes holds 2 electrons
= 6 electrons
(outer orbitals of
Groups 3A to 8A)

C

d orbitals:
each of 5 sets of
lobes holds 2 electrons
= 10 electrons
(found in elements
with atomic numbers of 21
and higher)

TABLE 8.2   Quantum Numbers and Electron Distribution to $n = 4$

Main Energy Level	Energy Sublevels	Maximum Number of Electrons	Maximum Number of Electrons per Main Energy Level
$n = 1$	s	2	2
$n = 2$	s	2	
	p	6	8
$n = 3$	s	2	
	p	6	
	d	10	18
$n = 4$	s	2	
	p	6	
	d	10	
	f	14	32

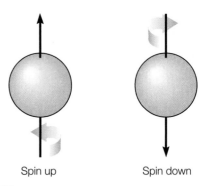

Spin up   Spin down

FIGURE 8.13 Experimental evidence supports the concept that electrons can be considered to spin one way or the other as they move about an orbital under an external magnetic field.

ELECTRON CONFIGURATION

The arrangement of electrons in orbitals is called the *electron configuration*. Before you can describe the electron arrangement, you need to know how many electrons are present in an atom. An atom is electrically neutral, so the number of protons (positive charge) must equal the number of electrons (negative charge). The atomic number therefore identifies the number of electrons as well as the number of protons. Now that you have a means of finding the number of electrons, consider the various energy levels to see how the electron configuration is determined.

According to the Pauli exclusion principle, no two electrons in an atom can have all four quantum numbers the same. As it works out, this means there can only be *a maximum of two electrons in any given orbital*. There are four things to consider: (1) the main energy level, (2) the energy sublevel, (3) the number of orbital orientations, and (4) the electron spin. Recall

that the lowest-energy level is $n = 1$ and successive numbers identify progressively higher-energy levels. Recall also that the energy sublevels, in order of increasing energy, are s, p, d, and f. This electron configuration is written in shorthand, with 1s standing for the lowest-energy sublevel of the first energy level. A superscript gives the number of electrons present in a sublevel. Thus, the electron configuration for a helium atom, which has two electrons, is written as $1s^2$. This combination of symbols has the following meaning: an atom with two electrons in the s sublevel of the first main energy level.

Table 8.3 gives the electron configurations for the first twenty elements. The configurations of the p energy sublevel have been condensed in this table. There are three possible orientations of the p orbital, each with two electrons. This is shown as p^6, which designates the number of electrons in all of the three possible p orientations. Note that the sum of the electrons in all the orbitals equals the atomic number. Note also that as you proceed from a lower atomic number to a higher one, the higher element has the same configuration as the element before it with the addition of one more electron. In general, it is then possible to begin with the simplest atom, hydrogen, and add one electron at a time to the order of energy

CONCEPTS APPLIED

Firework Configuration

Certain strontium (atomic number 38) chemicals are used to add the pure red color to flares and fireworks. Write the electron configuration of strontium and do this before looking at the solution below.

First, note that an atomic number of 38 means a total of thirty-eight electrons. Second, refer to the order of filling the matrix in figure 8.14. Remember that only two electrons can occupy an orbital, but there are three orientations of the p orbital, for a total of six electrons. There are likewise five possible orientations of the d orbital, for a total of ten electrons. Starting at the lowest energy level, two electrons go in 1s, making $1s^2$; then two go in 2s, making $2s^2$. That is a total of four electrons so far. Next, $2p^6$ and $3s^2$ use eight more electrons, for a total of twelve so far. The $3p^6$, $4s^2$, $3d^{10}$, and $4p^6$ use up twenty-four more electrons, for a total of thirty-six. The remaining two go into the next sublevel, $5s^2$, and the complete answer is

Strontium: $1s^2\ 2s^2\ 2p^6\ 3s^2\ 3p^6\ 4s^2\ 3d^{10}\ 4p^6\ 5s^2$

TABLE 8.3 Electron Configuration for the First Twenty Elements

Atomic Number	Element	Electron Configuration
1	Hydrogen	$1s^1$
2	Helium	$1s^2$
3	Lithium	$1s^2 2s^1$
4	Beryllium	$1s^2 2s^2$
5	Boron	$1s^2 2s^2 2p^1$
6	Carbon	$1s^2 2s^2 2p^2$
7	Nitrogen	$1s^2 2s^2 2p^3$
8	Oxygen	$1s^2 2s^2 2p^4$
9	Fluorine	$1s^2 2s^2 2p^5$
10	Neon	$1s^2 2s^2 2p^6$
11	Sodium	$1s^2 2s^2 2p^6 3s^1$
12	Magnesium	$1s^2 2s^2 2p^6 3s^2$
13	Aluminum	$1s^2 2s^2 2p^6 3s^2 3p^1$
14	Silicon	$1s^2 2s^2 2p^6 3s^2 3p^2$
15	Phosphorus	$1s^2 2s^2 2p^6 3s^2 3p^3$
16	Sulfur	$1s^2 2s^2 2p^6 3s^2 3p^4$
17	Chlorine	$1s^2 2s^2 2p^6 3s^2 3p^5$
18	Argon	$1s^2 2s^2 2p^6 3s^2 3p^6$
19	Potassium	$1s^2 2s^2 2p^6 3s^2 3p^6 4s^1$
20	Calcium	$1s^2 2s^2 2p^6 3s^2 3p^6 4s^2$

Atomic Research

There are two types of scientific research: basic and applied. Basic research is driven by a search for understanding and may or may not have practical applications. Applied research has a goal of solving some practical problems rather than just looking for understanding.

Some people feel that all research should result in something practical, so all research should be applied. Hold that thought while considering if the following research discussed in this chapter is basic or applied:

1. J. J. Thomson investigates cathode rays.
2. Robert Millikan measures the charge of an electron.
3. Ernest Rutherford studies radioactive particles striking gold foil.
4. Niels Bohr proposes a solar system model of the atom by applying the quantum concept.
5. Erwin Schrödinger proposes a model of the atom based on the wave nature of the electron.

Questions to Discuss:

1. Would we ever have developed a model of the atom if all research had to be practical?

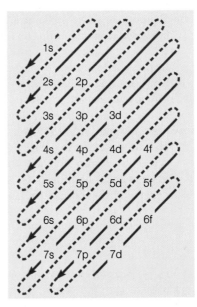

FIGURE 8.14 A matrix showing the order in which the orbitals are filled. Start at the top left, then move from the head of each arrow to the tail of the one immediately below it. This sequence moves from the lowest-energy level to the next higher level for each orbital.

sublevels and obtain the electron configuration for all the elements. The exclusion principle limits the number of electrons in any orbital, and allowances will need to be made for the more complex behavior of atoms with many electrons.

The energies of the orbital are different for each element, and several factors influence their energies. The first orbitals are filled in a straightforward 1s, 2s, 2p, 3s, then 3p order. Then the order becomes contrary to what you might expect. One useful way of figuring out the order in which orbitals are filled is illustrated in figure 8.14. Each row of this matrix represents a principal energy level with possible energy sublevels increasing from left to right. The order of filling is indicated by the diagonal arrows. There are exceptions to the order of filling shown by the matrix, but it works for most of the elements.

THE PERIODIC TABLE

The periodic table is made up of rows and columns of cells, with each element having its own cell in a specific location. The cells are not arranged symmetrically. The arrangement has a meaning, both about atomic structure and about chemical behaviors. It will facilitate your understanding of the code if you refer frequently to a periodic table during the following discussion (figure 8.15).

An element is identified in each cell with its chemical symbol. The number above the symbol is the atomic number of the element, and the number below the symbol is the rounded atomic weight of the element. Horizontal rows of elements run from left to right with increasing atomic numbers. Each row is called a *period.* The periods are numbered from 1 to 7 on the left side. A vertical column of elements is called a *family* (or group) of elements. Elements in families have similar properties, but this is more true of some families than others. The table is subdivided into A and B groups. The members of the A-group families are called the *main group,* or **representative elements.** The members of the B-group are called the **transition elements** (or metals). Some science organizations use a "1 to 18" designation for the A and B groups, as shown in figure 8.15. The A and B designations will be used throughout this text.

As shown in table 8.4, all of the elements in the first column have an outside electron configuration of one electron. With the exception of hydrogen, the elements of the first column are shiny, low-density metals that are so soft you can cut them easily with a knife. These metals are called the *alkali metals* because they react violently with water to form an alkaline solution. The alkali metals do not occur in nature as free elements because they are so reactive. Hydrogen is a unique element in the periodic table. It is not an alkali metal, but it is placed in the group because it seems to fit there because it has one electron in its outer orbital.

The elements in the second column all have an outside configuration of two electrons and are called the *alkaline earth metals.* The alkaline earth metals are soft, reactive metals but not as reactive or soft as the alkali metals. Calcium and magnesium, in the form of numerous compounds, are familiar examples of this group.

The elements in group VIIA all have an outside configuration of seven electrons, needing only one more electron to completely fill the outer orbitals. These elements are called the *halogens.* The halogens are very reactive nonmetals. The halogens fluorine and chlorine are pale yellow and yellowish green gases. Bromine is a reddish brown liquid and iodine is a dark purple solid. Halogens are used as disinfectants, bleaches, and combined with a metal as a source of light in halogen lights. Halogens react with metals to form a group of chemicals called *salts,* such as sodium chloride. In fact, the word *halogen* is Greek, meaning "salt former."

As shown in table 8.5, the elements in group VIIIA have orbitals that are filled to capacity. These elements are colorless, odorless gases that almost never react with other elements to form compounds. Sometimes they are called the *noble gases* because they are chemically inert, perhaps indicating they are above the other elements. They have also been called the *rare*

TABLE 8.4 Electron Structures of the Alkali Metal Family

Element	Electron Configuration	Number of Electrons in Main Energy Level						
		1ST	2ND	3RD	4TH	5TH	6TH	7TH
Lithium (Li)	[He] 2s^1	2	1	—	—	—	—	—
Sodium (Na)	[Ne] 3s^1	2	8	1	—	—	—	—
Potassium (K)	[Ar] 4s^1	2	8	8	1	—	—	—
Rubidium (Rb)	[Kr] 5s^1	2	8	18	8	1	—	—
Cesium (Cs)	[Xe] 6s^1	2	8	18	18	8	1	—
Francium (Fr)	[Rn] 7s^1	2	8	18	32	18	8	1

TABLE 8.5 Electron Structures of the Noble Gas Family

Element	Electron Configuration	Number of Electrons in Main Energy Level						
		1ST	2ND	3RD	4TH	5TH	6TH	7TH
Helium (He)	1s^2	2	—	—	—	—	—	—
Neon (Ne)	[He] 2s^{2}2p^6	2	8	—	—	—	—	—
Argon (Ar)	[Ne] 3s^{2}3p^6	2	8	8	—	—	—	—
Krypton (Kr)	[Ar] 4s^{2}3d^{10}4p^6	2	8	18	8	—	—	—
Xenon (Xe)	[Kr] 5s^{2}4d^{10}5p^6	2	8	18	18	8	—	—
Radon (Rn)	[Xe] 6s^{2}4f^{14}5d^{10}6p^6	2	8	18	32	18	8	—

Periodic Table of the Elements

Alkali Metals: IA (1); Alkaline Earth Metals: IIA (2); Transition Elements: IIIB (3) – IIB (12); Halogens: VIIA (17); Noble gases: VIIIA (18)

Metals | Semiconductors | Nonmetals

Period	IA (1)	IIA (2)	IIIB (3)	IVB (4)	VB (5)	VIB (6)	VIIB (7)	VIIIB (8)	VIIIB (9)	VIIIB (10)	IB (11)	IIB (12)	IIIA (13)	IVA (14)	VA (15)	VIA (16)	VIIA (17)	VIIIA (18)
1	Hydrogen 1 **H** 1.008																	Helium 2 **He** 4.003
2	Lithium 3 **Li** 6.941	Beryllium 4 **Be** 9.012											Boron 5 **B** 10.81	Carbon 6 **C** 12.01	Nitrogen 7 **N** 14.01	Oxygen 8 **O** 16.00	Fluorine 9 **F** 19.00	Neon 10 **Ne** 20.18
3	Sodium 11 **Na** 22.99	Magnesium 12 **Mg** 24.31											Aluminum 13 **Al** 26.98	Silicon 14 **Si** 28.09	Phosphorus 15 **P** 30.97	Sulfur 16 **S** 32.07	Chlorine 17 **Cl** 35.45	Argon 18 **Ar** 39.95
4	Potassium 19 **K** 39.10	Calcium 20 **Ca** 40.08	Scandium 21 **Sc** 44.96	Titanium 22 **Ti** 47.88	Vanadium 23 **V** 50.94	Chromium 24 **Cr** 52.00	Manganese 25 **Mn** 54.94	Iron 26 **Fe** 55.85	Cobalt 27 **Co** 58.93	Nickel 28 **Ni** 58.69	Copper 29 **Cu** 63.55	Zinc 30 **Zn** 65.39	Gallium 31 **Ga** 69.72	Germanium 32 **Ge** 72.61	Arsenic 33 **As** 74.92	Selenium 34 **Se** 78.96	Bromine 35 **Br** 79.90	Krypton 36 **Kr** 83.80
5	Rubidium 37 **Rb** 85.47	Strontium 38 **Sr** 87.62	Yttrium 39 **Y** 88.91	Zirconium 40 **Zr** 91.22	Niobium 41 **Nb** 92.91	Molybdenum 42 **Mo** 95.94	Technetium 43 **Tc** (98)	Ruthenium 44 **Ru** 101.1	Rhodium 45 **Rh** 102.9	Palladium 46 **Pd** 106.4	Silver 47 **Ag** 107.9	Cadmium 48 **Cd** 112.4	Indium 49 **In** 114.8	Tin 50 **Sn** 118.7	Antimony 51 **Sb** 121.8	Tellurium 52 **Te** 127.6	Iodine 53 **I** 126.9	Xenon 54 **Xe** 131.3
6	Cesium 55 **Cs** 132.9	Barium 56 **Ba** 137.3	Lanthanum 57 **La** 138.9 †	Hafnium 72 **Hf** 178.5	Tantalum 73 **Ta** 180.9	Tungsten 74 **W** 183.8	Rhenium 75 **Re** 186.2	Osmium 76 **Os** 190.2	Iridium 77 **Ir** 192.2	Platinum 78 **Pt** 195.1	Gold 79 **Au** 197.0	Mercury 80 **Hg** 200.6	Thallium 81 **Tl** 204.4	Lead 82 **Pb** 207.2	Bismuth 83 **Bi** 209.0	Polonium 84 **Po** (209)	Astatine 85 **At** (210)	Radon 86 **Rn** (222)
7	Francium 87 **Fr** (223)	Radium 88 **Ra** (226)	Actinium 89 **Ac** (227) ‡	Rutherfordium 104 **Rf** (261)	Dubnium 105 **Db** (262)	Seaborgium 106 **Sg** (266)	Bohrium 107 **Bh** (264)	Hassium 108 **Hs** (277)	Meitnerium 109 **Mt** (268)	Ununnilium 110 **Uun** (281)	Unununium 111 **Uuu** (272)	Ununbium 112 **Uub** (285)	Ununtrium 113 **Uut** (284)	Ununquadium 114 **Uuq** (289)	Ununpentium 115 **Uup** (288)	Ununhexium 116 **Uuh** (292)		Ununoxtium 118 **Uuo** (294)

Inner Transition Elements

†Lanthanides 6	Cerium 58 **Ce** 140.1	Praseodymium 59 **Pr** 140.9	Neodymium 60 **Nd** 144.2	Promethium 61 **Pm** (145)	Samarium 62 **Sm** 150.4	Europium 63 **Eu** 152.0	Gadolinium 64 **Gd** 157.3	Terbium 65 **Tb** 158.9	Dysprosium 66 **Dy** 162.5	Holmium 67 **Ho** 164.9	Erbium 68 **Er** 167.3	Thulium 69 **Tm** 168.9	Ytterbium 70 **Yb** 173.0	Lutetium 71 **Lu** 175.0
‡Actinides 7	Thorium 90 **Th** 232.0	Protactinium 91 **Pa** 231.0	Uranium 92 **U** 238.0	Neptunium 93 **Np** (237)	Plutonium 94 **Pu** (244)	Americium 95 **Am** (243)	Curium 96 **Cm** (247)	Berkelium 97 **Bk** (247)	Californium 98 **Cf** (251)	Einsteinium 99 **Es** (252)	Fermium 100 **Fm** (257)	Mendelevium 101 **Md** (258)	Noblelium 102 **No** (259)	Lawrencium 103 **Lr** (262)

Key: element name — Hydrogen; atomic number — 1; symbol of element — **H**; atomic weight — 1.008

Values in parentheses are the mass numbers of the most stable or best-known isotopes.

Names and symbols for elements 110–118 are under review.

FIGURE 8.15 The periodic table of the elements.

The Rare Earths

Compounds of the rare earths were first identified when they were isolated from uncommon minerals in the late 1700s. The elements are very reactive and have similar chemical properties, so they were not recognized as elements until some fifty years later. Thus, they were first recognized as earths, that is, non-metal substances, when in fact they are metallic elements. They were also considered to be rare since, at that time, they were known to occur only in uncommon minerals. Today, these metallic elements are known to be more abundant in the Earth than gold, silver, mercury, or tungsten. The rarest of the rare earths, thulium, is twice as abundant as silver. The rare earth elements are neither rare nor earths, and they are important materials in glass, electronic, and metallurgical industries.

You can identify the rare earths in the two lowermost rows of the periodic table. These rows contain two series of elements that actually belong in periods 6 and 7, but they are moved below so that the entire table is not so wide. Together, the two series are called the inner transition elements. The top series is fourteen elements wide from elements 58 through 71. Since this series belongs next to element 57, lanthanum, it is sometimes called the *lanthanide series*. This series is also known as the rare earths. The second series of fourteen elements is called the *actinide series*. These are mostly the artificially prepared elements that do not occur naturally.

You may never have heard of the rare earth elements, but they are key materials in many advanced or high-technology products. Lanthanum, for example, gives glass special refractive properties and is used in optic fibers and expensive camera lenses. Samarium, neodymium, and dysprosium are used to manufacture crystals used in lasers. Samarium, ytterbium, and terbium have special magnetic properties that have made possible new electric motor designs, magnetic-optical devices in computers, and the creation of a ceramic superconductor. Other rare earth metals are also being researched for use in possible high-temperature superconductivity materials. Many rare earths are also used in metal alloys; for example, an alloy of cerium is used to make heat-resistant jet-engine parts. Erbium is also used in high-performance metal alloys. Dysprosium and holmium have neutron-absorbing properties and are used in control rods to control nuclear fission. Europium should be mentioned because of its role in making the red color of color television screens. The rare earths are relatively abundant metallic elements that play a key role in many common and high-technology applications. They may also play a key role in superconductivity research.

CONCEPTS APPLIED

Outer Orbitals

How many outer orbital electrons are found in an atom of (a) oxygen, (b) calcium, and (c) aluminum? Write your answers before reading the answers in the next paragraph.

(a) According to the list of elements on the inside back cover of this text, oxygen has the symbol O and an atomic number of 8. The square with the symbol O and the atomic number 8 is located in the column identified as VIA. Since the A family number is the same as the number of electrons in the outer orbital, oxygen has six outer orbital electrons. (b) Calcium has the symbol Ca (atomic number 20) and is located in column IIA, so a calcium atom has two outer orbital electrons. (c) Aluminum has the symbol Al (atomic number 13) and is located in column IIIA, so an aluminum atom has three outer orbital electrons.

gases because of their scarcity and *inert gases* because they are mostly chemically inert, not forming compounds. The noble gases are inert because they have filled outer electron configurations, a particularly stable condition.

Each period *begins* with a single electron in a new orbital. Second, each period *ends* with the filling of an orbital, completing the maximum number of electrons that can occupy that main energy level. Since the first A family is identified as IA, this means that all the atoms of elements in this family have one electron in their outer orbitals. All the atoms of elements in family IIA have two electrons in their outer orbitals. This pattern continues on to family VIIIA, in which all the atoms of elements have eight electrons in their outer orbitals except helium. Thus, the number identifying the A families *also identifies the number of electrons in the outer orbitals* (also called *valence electrons*), with the exception of helium. Helium is nonetheless similar to the other elements in this family, since all have filled outer orbitals. The electron theory of chemical bonding, which is discussed in chapter 9, states that only the electrons in the outermost orbitals of an atom are involved in chemical reactions. Thus, *the outer orbital electrons are mostly responsible for the chemical properties of an element.* Since the members of a family all have similar outer configurations, you would expect them to have similar chemical behaviors, and they do.

METALS, NONMETALS, AND SEMICONDUCTORS

As indicated earlier, chemical behavior is mostly concerned with the outer shell electrons. The outer shell electrons, that is, the highest energy level electrons, are conveniently represented with an **electron dot notation,** made by writing the chemical symbol with dots around it indicating the number of outer shell electrons. Electron dot notations are shown for the representative elements in figure 8.16. Again, note the pattern

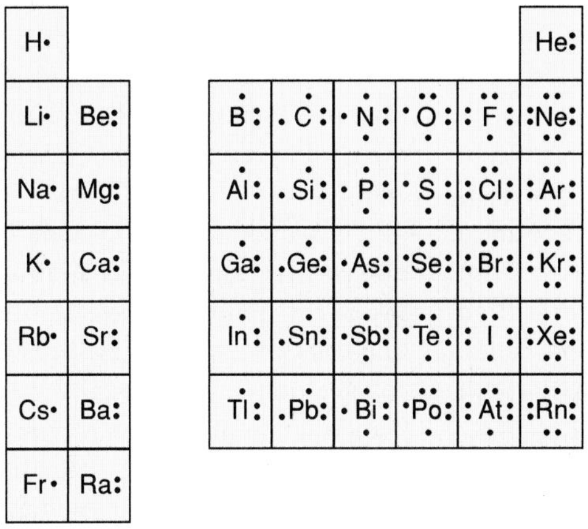

FIGURE 8.16 Electron dot notation for the representative elements.

in figure 8.16—all the noble gases are in group VIIIA, and all (except helium) have eight outer electrons. All the group IA elements (alkali metals) have one dot, all the IIA elements have two dots, and so on. This pattern will explain the difference in metals, nonmetals, and a third group of in-between elements called semiconductors.

There are several ways to classify matter according to properties. One example is to group substances according to the physical properties of metals and nonmetals—luster, conductivity, malleability, and ductility. Metals and nonmetals also have certain chemical properties that are related to their positions in the periodic table. Figure 8.17 shows where the *metals, nonmetals,* and *semiconductors* are located. Note that about 80 percent of all the elements are metals.

The noble gases have completely filled outer orbitals in their highest energy levels, and this is a particularly stable arrangement. Other elements react chemically, either *gaining or losing electrons to attain a filled outermost energy level like the noble gases.* When an atom loses or gains electrons, it acquires

an unbalanced electron charge and is called an **ion.** An atom of lithium, for example, has three protons (plus charges) and three electrons (negative charges). If it loses the outermost electron, it now has an outer filled orbital structure like helium, a noble gas. It is also now an ion, since it has three protons (3+) and two electrons (2−), for a net charge of 1+. A lithium ion thus has a 1+ charge.

Elements with one, two, or three outer electrons tend to lose electrons to form positive ions. The metals lose electrons like this, and the *metals are elements that lose electrons to form positive ions* (figure 8.18). Nonmetals, on the other hand, are elements with five to seven outer electrons that tend to acquire electrons to fill their outer orbitals. *Nonmetals are elements that gain electrons to form negative ions.* In general, elements located in the left two-thirds or so of the periodic table are metals. The nonmetals are on the right side of the table.

The dividing line between the metals and nonmetals is a steplike line from the left top of group IIIA down to the bottom left of group VIIA. This is not a line of sharp separation between the metals and nonmetals, and elements *along* this line sometimes act like metals, sometimes like nonmetals, and sometimes like both. These hard-to-classify elements are called **semiconductors** (or *metalloids*). Silicon, germanium, and arsenic have physical properties of nonmetals; for example, they are brittle materials that cannot be hammered into a new shape. Yet these elements conduct electric currents under certain conditions. The ability to conduct an electric current is a property of a metal, and nonmalleability is a property of nonmetals, so as you can see, these semiconductors have the properties of both metals and nonmetals.

The transition elements, which are all metals, are located in the B-group families. Unlike the representative elements, which form vertical families of similar properties, the transition elements tend to form horizontal groups of elements with similar properties. Iron (Fe), cobalt (Co), and nickel (Ni) in group VIIIB, for example, are three horizontally arranged metallic elements that show magnetic properties.

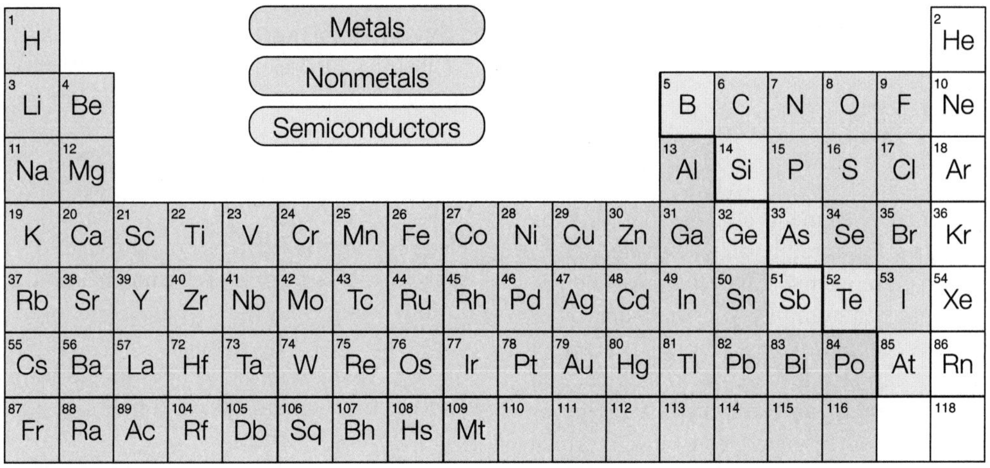

FIGURE 8.17 The location of metals, nonmetals, and semiconductors in the periodic table.

Connections . . .

Nutrient Elements

All molecules required to support living things are called nutrients. Some nutrients are inorganic molecules such as calcium, iron, or potassium, and others are organic molecules such as carbohydrates, proteins, fats, and vitamins. All minerals are elements and cannot be synthesized by the body. Because they are elements, they cannot be broken down or destroyed by metabolism or cooking. They commonly occur in many foods and in water. Minerals retain their characteristics whether they are in foods or in the body, and each plays a different role in metabolism.

Minerals* can function as regulators, activators, transmitters, and controllers of various enzymatic reactions. For example, sodium ions (Na^+) and potassium ions (K^+) are important in the transmission of nerve impulses, while magnesium ions (Mg^{2+}) facilitate energy release during reactions involving ATP. Without iron, not enough hemoglobin would be formed to transport oxygen, a condition called anemia, and a lack of calcium may result in osteoporosis. Osteoporosis is a condition that results from calcium loss leading to painful, weakened bones. There are many minerals that are important in your diet. In addition to those just mentioned, you need chlorine, cobalt, copper, iodine, phosphorus, potassium, sulfur, and zinc to remain healthy. With few exceptions, adequate amounts of minerals are obtained in a normal diet. Calcium and iron supplements may be necessary, particularly in women.

Osteoporosis is a nutritional deficiency disease that results in a change in the density of the bones as a result of the loss of bone mass. Bones that have undergone this change look lacy or like Swiss cheese, with larger than normal holes. A few risk factors found to be associated with this disease are being female and fair skinned; having a sedentary lifestyle; using alcohol, caffeine, and tobacco; being anorexic; and having reached menopause.

*Note that the term *mineral* has a completely different meaning in geology. See page 335 in chapter 15.

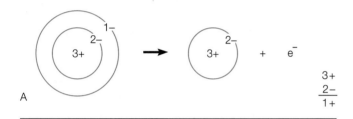

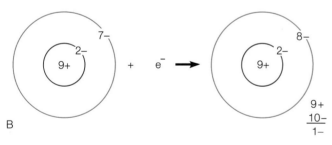

FIGURE 8.18 (A) Metals lose their outer electrons to acquire a noble gas structure and become positive ions. Lithium becomes a 1+ ion as it loses its one outer electron. (B) Nonmetals gain electrons to acquire an outer noble gas structure and become negative ions. Fluorine gains a single electron to become a 1− ion.

CONCEPTS APPLIED

Metals and Charge

Is strontium a metal, nonmetal, or semiconductor? What is the charge on a strontium ion?

The list of elements inside the back cover identifies the symbol for strontium as Sr (atomic number 38). In the periodic table, Sr is located in family IIA, which means that an atom of strontium has two electrons in its outer shell. For several reasons, you know that strontium is a metal: (1) An atom of strontium has two electrons in its outer shell, and atoms with one, two, or three outer electrons are identified as metals; (2) strontium is located in the IIA family, the alkaline earth metals; and (3) strontium is located on the left side of the periodic table, and in general, elements located in the left two-thirds of the table are metals.

Elements with one, two, or three outer electrons tend to lose electrons to form positive ions. Since strontium has an atomic number of 38, you know that it has thirty-eight protons (38+) and thirty-eight electrons (38−). When it loses its two outer shell electrons, it has 38+ and 36− for a charge of 2+ .

A family of representative elements all form ions with the same charge. Alkali metals, for example, all lose an electron to form a 1+ ion. The transition elements have *variable charges*. Some transition elements, for example, lose their one outer electron to form 1+ ions (copper, silver). Copper, because of its special configuration, can also lose an additional electron to form a 2+ ion. Thus, copper can form either a 1+ ion or a 2+ ion. Most transition elements have two outer s orbital electrons and lose them both to form 2+ ions (iron, cobalt, nickel), but some of these elements also have special configurations that permit them to lose more of their electrons. Thus, iron and cobalt, for example, can form either a 2+ ion or a 3+ ion. Much more can be interpreted from the periodic table, and more generalizations will be made as the table is used in the following chapters.

Dmitri Ivanovich Mendeleyev (1834–1907)

Dmitri Mendeleyev was a Russian chemist whose name will always be linked with his outstanding achievement, the development of the periodic table. He was the first chemist to understand that all elements are related members of a single, ordered system. He converted what had been a highly fragmented and speculative branch of chemistry into a true, logical science. The spelling of his name has been a source of confusion for students and frustration for editors for more than a century, and the forms Mendeléeff, Mendeléev, and even Mendelejeff can all be found in print.

Mendeleyev was born in Tobol'sk, Siberia, on February 7, 1834, the youngest of the seventeen children of the head of the local high school. His father went blind when Mendeleyev was a child, and the family had to rely increasingly on their mother for support. He was educated locally but could not gain admission to any Russian university because of prejudice toward the supposedly backward attainments of those educated in the provinces. In 1855, he finally qualified as a teacher at the Pedagogical Institute in St. Petersburg. He took an advanced-degree course in chemistry. In 1859, he was sent by the government for further study at the University of Heidelberg, and in 1861, he returned to St. Petersburg and became professor of general chemistry at the Technical Institute in 1864. He could find no textbook adequate for his students' needs, so he decided to produce his own. The resulting *Principles of Chemistry* (1868–1870) won him international renown; it was translated into English in 1891 and 1897.

Before Mendeleyev produced his periodic law, understanding of the chemical elements had long been an elusive and frustrating task. According to Mendeleyev, the properties of the elements are periodic functions of their atomic weights. In 1869, he stated that "the elements arranged according to the magnitude of atomic weights show a periodic change of properties." Other chemists, notably Lothar Meyer in Germany, had meanwhile come to similar conclusions, with Meyer publishing his findings independently.

Mendeleyev compiled the first true periodic table, listing all the sixty-three elements then known. Not all elements would "fit" properly using the atomic weights of the time, so he altered indium from 76 to 114 (modern value 114.8) and beryllium from 13.8 to 9.2 (modern value 9.013). In 1871, he produced a revisionary paper showing the correct repositioning of seventeen elements.

To make the table work, Mendeleyev also had to leave gaps, and he predicted that further elements would eventually be discovered to fill them. These predictions provided the strongest endorsement of the periodic law. Three were discovered in Mendeleyev's lifetime: gallium (1871), scandium (1879), and germanium (1886), all with properties that tallied closely with those he had assigned to them.

Farsighted though Mendeleyev was, he had no notion that the periodic recurrences of similar properties in the list of elements reflected anything in the structures of their atoms. It was not until the 1920s that it was realized that the key parameter in the periodic system is not the atomic weight but the atomic number of the elements—a measure of the number of protons in the atom. Since then, great progress has been made in explaining the periodic law in terms of the electronic structures of atoms and molecules.

Source: Modified from the *Hutchinson Dictionary of Scientific Biography.* © RM, 2007. Reprinted by permission.

SUMMARY

Attempts at understanding matter date back to ancient Greek philosophers, who viewed matter as being composed of *elements,* or simpler substances. Two models were developed that considered matter to be (1) *continuous,* or infinitely divisible, or (2) *discontinuous,* made up of particles called *atoms.*

In the early 1800s, Dalton published an *atomic theory,* reasoning that matter was composed of hard, indivisible atoms that were joined together or dissociated during chemical change.

When a good air pump to provide a vacuum was invented in 1885, *cathode rays* were observed to move from the negative terminal in an evacuated glass tube. The nature of cathode rays was a mystery. The mystery was solved in 1887 when Thomson discovered they were negatively charged particles now known as *electrons.* Thomson had discovered the first elementary particle of which atoms are made and measured their charge-to-mass ratio.

Rutherford developed a solar system model based on experiments with alpha particles scattered from a thin sheet of metal. This model had a small, massive, and positively charged *nucleus* surrounded by moving electrons. These electrons were calculated to be at a distance from the nucleus of 100,000 times the radius of the nucleus, so the volume of an atom is mostly empty space. Later, Rutherford proposed that the nucleus contained two elementary particles: *protons* with a positive charge and *neutrons* with no charge. The *atomic number* is the number of protons in an atom. Atoms of elements with different numbers of neutrons are called *isotopes.* The mass of each isotope is compared to the mass of carbon-12, which is assigned a mass of exactly 12.00 *atomic mass units.* The mass contribution of the isotopes of an element according to their abundance is called the *atomic weight* of an element. Isotopes are identified by their *mass number,* which is the sum of the number of protons and neutrons in the nucleus. Isotopes are identified by their chemical symbol, with the atomic number as a subscript and the mass number as a superscript.

Bohr developed a model of the hydrogen atom to explain the characteristic *line spectra* emitted by hydrogen. His model specified that (1) electrons can move only in allowed orbits, (2) electrons do not emit radiant energy when they remain in an orbit, and (3) electrons move from one allowed orbit to another when they gain or lose energy. When an electron jumps from a higher orbit to a lower one, it gives up energy in the form of a single photon. The energy of the photon corresponds to the difference in energy between the two levels. The Bohr model worked well for hydrogen but not for other atoms.

Schrödinger and others used the wave nature of the electron to develop a new model of the atom called *wave mechanics,* or *quantum mechanics.* This model was found to confirm exactly all the experimental data as well as predict new data. The quantum mechanical model describes the energy state of the electron in terms of quantum numbers based on the wave nature of the electron. The quantum numbers defined the *probability* of the location of an electron in terms of fuzzy regions of space called *orbitals.*

The *periodic table* has horizontal rows of elements called *periods* and vertical columns of elements called *families.* Members of a given family have the same outer orbital electron configurations, and it is the electron configuration that is mostly responsible for the chemical properties of an element.

Summary of Equations

8.1   energy = (Planck's constant)(frequency)
$$E = hf$$
$$\text{where } h = 6.63 \times 10^{-34} \, \text{J} \cdot \text{s}$$

KEY TERMS

atomic mass units (p. **183**)

atomic number (p. **182**)

atomic weight (p. **183**)

electron (p. **182**)

electron dot notation (p. **192**)

excited states (p. **186**)

ground state (p. **186**)

Heisenberg uncertainty principle (p. **187**)

ion (p. **193**)

isotope (p. **183**)

line spectrum (p. **184**)

mass number (p. **183**)

neutron (p. **182**)

nucleus (p. **182**)

orbital (p. **187**)

Pauli exclusion principle (p. **187**)

photons (p. **184**)

proton (p. **182**)

quanta (p. **184**)

representative elements (p. **190**)

semiconductors (p. **193**)

transition elements (p. **190**)

APPLYING THE CONCEPTS

1. The electron was discovered through experiments with
   a. radioactivity.
   b. light.
   c. matter waves.
   d. electricity.

2. Thomson was convinced that he had discovered a subatomic particle, the electron, from the evidence that
   a. the charge-to-mass ratio was the same for all materials.
   b. cathode rays could move through a vacuum.
   c. electrons were attracted toward a negatively charged plate.
   d. the charge was always 1.60×10^{-19} coulomb.

3. The existence of a tiny, massive, and positively charged nucleus was deduced from the observation that
   a. fast, massive, and positively charged alpha particles all move straight through metal foil.
   b. alpha particles were deflected by a magnetic field.
   c. some alpha particles were deflected by metal foil.
   d. None of the above is correct.

4. According to Rutherford's calculations, the volume of an atom is mostly
   a. occupied by protons and neutrons.
   b. filled with electrons.
   c. occupied by tightly bound protons, electrons, and neutrons.
   d. empty space.

5. Hydrogen, with its one electron, produces a line spectrum in the visible light range with
   a. one color line.
   b. two color lines.
   c. three color lines.
   d. four color lines.

6. According to the Bohr model, an electron gains or loses energy only by
   a. moving faster or slower in an allowed orbit.
   b. jumping from one allowed orbit to another.
   c. being completely removed from an atom.
   d. jumping from one atom to another atom.

7. When an electron in a hydrogen atom jumps from an orbit farther from the nucleus to an orbit closer to the nucleus, it
   a. emits a single photon with an energy equal to the energy difference of the two orbits.
   b. emits four photons, one for each of the color lines observed in the line spectrum of hydrogen.
   c. emits a number of photons dependent on the number of orbit levels jumped over.
   d. None of the above is correct.

8. The quantum mechanics model of the atom is based on
   a. the quanta, or measured amounts of energy of a moving particle.
   b. the energy of a standing electron wave that can fit into an orbit.
   c. calculations of the energy of the three-dimensional shape of a circular orbit of an electron particle.
   d. Newton's laws of motion but scaled down to the size of electron particles.

9. The Bohr model of the atom described the energy state of electrons with one quantum number. The quantum mechanics model uses how many quantum numbers to describe the energy state of an electron?
   a. one
   b. two
   c. four
   d. ten

10. An electron in the second main energy level and the second sublevel is described by the symbols
    a. 1s.
    b. 2s.
    c. 1p.
    d. 2p.

11. The space in which it is probable that an electron will be found is described by a(an)
    a. circular orbit.
    b. elliptical orbit.
    c. orbital.
    d. geocentric orbit.

12. Two electrons can occupy the same orbital because they have different
    a. principal quantum numbers.
    b. angular momentum quantum numbers.
    c. magnetic quantum numbers.
    d. spin quantum numbers.

Answers

1. d 2. a 3. c 4. d 5. d 6. b 7. a 8. b 9. c 10. d 11. c 12. d

QUESTIONS FOR THOUGHT

1. What was the experimental evidence that Thomson had discovered the existence of a subatomic particle when working with cathode rays?

2. Describe the experimental evidence that led Rutherford to the concept of a nucleus in an atom.

3. What is the main problem with a solar system model of the atom?

4. Compare the size of an atom to the size of its nucleus.

5. What does *atomic number* mean? How does the atomic number identify the atoms of a particular element? How is the atomic number related to the number of electrons in an atom?

6. An atom has 11 protons in the nucleus. What is the atomic number? What is the name of this element? What is the electron configuration of this atom?

7. How is the atomic weight of an element determined?

8. Describe the three main points in the Bohr model of the atom.

9. Why do the energies of electrons in an atom have negative values? (*Hint:* It is *not* because of the charge of the electron.)

10. Which has the lowest energy, an electron in the first energy level ($n = 1$) or an electron in the third energy level ($n = 3$)? Explain.

11. What is similar about the Bohr model of the atom and the quantum mechanical model? What are the fundamental differences?

12. What is the difference between a hydrogen atom in the ground state and one in the excited state?

FOR FURTHER ANALYSIS

1. Evaluate Millikan's method for finding the charge of an electron. Are there any doubts about the results of using this technique?

2. What are the significant similarities and differences between the isotopes of a particular element?

3. Thomson's experiments led to the discovery of the electron. Analyze how you know for sure that he discovered the electron.

4. Describe a conversation between yourself and another person as you correct his or her belief that atomic weight has something to do with gravity.

5. Analyze the significance of the observation that matter only emits and absorbs energy in discrete units.

6. Describe at least several basic differences between the Bohr and quantum mechanics models of the atom.

INVITATION TO INQUIRY

Too Small to See?

As Rutherford knew when he conducted his famous experiment with radioactive particles and gold foil, the structure of an atom is too small to see, so it must be inferred from other observations. To illustrate this process, pour 50 mL of 95 percent ethyl alcohol (or some other almost pure alcohol) into a graduated cylinder. In a second graduated cylinder, measure 50 mL of water. Mix the alcohol and water together thoroughly and record the volume of the combined liquids. Assuming no evaporation took place, is the result contrary to your expectation? What question could be asked about the result?

Answers to questions about things that you cannot see often require a model that can be observed. For example, a model might represent water and alcohol molecules by using beans for alcohol molecules and sand for water molecules. Does mixing 50 mL of beans and 50 mL of sand result in 100 mL of mixed sand and beans? Explain the result of mixing alcohol and water based on your observation of mixing beans and sand.

To continue this inquiry, fill a water glass to the brim with water. Add a small amount of salt to the water. You know that two materials cannot take up the same space at the same time, so what happened to the salt? How can you test your ideas about things that are too tiny to see?

PARALLEL EXERCISES

The exercises in groups A and B cover the same concepts. Solutions to group A exercises are located in appendix D.

Group A

1. How much energy is needed to move an electron in a hydrogen atom from $n = 2$ to $n = 6$? (See figure 8.11 for needed values.)

2. What frequency of light is emitted when an electron in a hydrogen atom jumps from $n = 6$ to $n = 2$? What color would you see?

3. How much energy is needed to completely remove the electron from a hydrogen atom in the ground state?

4. Thomson determined the charge-to-mass ratio of the electron to be -1.76×10^{11} coulomb/kilogram. Millikan determined the charge on the electron to be -1.60×10^{-19} coulomb. According to these findings, what is the mass of an electron?

5. Using any reference you wish, write the complete electron configurations for (a) boron, (b) aluminum, and (c) potassium.

6. Explain how you know that you have the correct *total* number of electrons in your answers for 5a, 5b, and 5c.

7. Refer to figure 8.15 *only,* and write the complete electron configurations for (a) argon, (b) zinc, and (c) bromine.

Group B

1. How much energy is needed to move an electron in a hydrogen atom from the ground state ($n = 1$) to $n = 3$?

2. What frequency of light is emitted when an electron in a hydrogen atom jumps from $n = 2$ to the ground state ($n = 1$)?

3. How much energy is needed to completely remove an electron from $n = 2$ in a hydrogen atom?

4. If the charge-to-mass ratio of a proton is 9.58×10^7 coulomb/kilogram and the charge is 1.60×10^{-19} coulomb, what is the mass of the proton?

5. Using any reference you wish, write the complete electron configurations for (a) nitrogen, (b) phosphorus, and (c) chlorine.

6. Explain how you know that you have the correct *total* number of electrons in your answers for 5a, 5b, and 5c.

7. Referring to figure 8.15 *only,* write the complete electron configuration for (a) neon, (b) sulfur, and (c) calcium.

A chemical change occurs when iron rusts, and rust is a different substance with physical and chemical properties different from iron. This rusted anchor makes a colorful display on the bow of a grain ship.

Chemical Reactions

CORE CONCEPT

Electron structure will explain how and why atoms join together in certain numbers.

Chemical reactions are changes in matter in which different substances are created by forming or breaking chemical bonds (p. 203).

A chemical bond is an attractive force that holds atoms together in a compound (p. 205).

An ionic bond is a chemical bond of electrostatic attraction between ions (p. 206).

A covalent bond is a chemical bond formed by the sharing of electrons (p. 207).

Earth Science Connections

► Chemical change will explain one way solid rocks on Earth's surface are weathered (Ch. 16).

Life Science Connections

► Organic chemistry and biochemistry consider chemical changes in the important organic compounds of life (Ch. 19).

OVERVIEW

In chapter 8, you learned how the modern atomic theory is used to describe the structures of atoms of different elements. The electron structures of different atoms successfully account for the position of elements in the periodic table as well as for groups of elements with similar properties. On a large scale, all metals were found to have a similarity in electron structure, as were nonmetals. On a smaller scale, chemical families such as the alkali metals were found to have the same outer electron configurations. Thus, the modern atomic theory accounts for observed similarities between elements in terms of atomic structure.

So far, only individual, isolated atoms have been discussed; we have not considered how atoms of elements join together to produce compounds. There is a relationship between the electron structure of atoms and the reactions they undergo to produce specific compounds. Understanding this relationship will explain the changes that matter itself undergoes. For example, hydrogen is a highly flammable, gaseous element that burns with an explosive reaction. Oxygen, on the other hand, is a gaseous element that supports burning. As you know, hydrogen and oxygen combine to form water. Water is a liquid that neither burns nor supports burning. What happens when atoms of elements such as hydrogen and oxygen join to form molecules such as water? Why do such atoms join and why do they stay together? Why does water have properties different from the elements that

combine to produce it? And finally, why is water H_2O and not H_3O or H_4O?

Answers to questions about why and how atoms join together in certain numbers are provided by considering the electronic structures of the atoms. Chemical substances are formed from the interactions of electrons as their structures merge, forming new patterns that result in molecules with new properties. It is the new electron pattern of the water molecule that gives water properties that are different from the oxygen or hydrogen from which it formed (figure 9.1). Understanding how electron structures of atoms merge to form new patterns is understanding the changes that matter itself undergoes, the topic of this chapter.

COMPOUNDS

Matter that you see around you seems to come in a wide variety of sizes, shapes, forms, and kinds. Are there any patterns in the apparent randomness that will help us comprehend matter? Yes, there are many patterns, and one of the more obvious is that all matter occurs as either a mixture or as a pure substance (figure 9.2). A *mixture* has unlike parts and a composition that varies from sample to sample. For example, sand from a beach is a variable mixture of things such as bits of rocks, minerals, and sea shells.

There are two distinct ways in which mixtures can have unlike parts with a variable composition. A *heterogeneous mixture* has physically distinct parts with different properties. Beach

sand, for example, is usually a heterogeneous mixture of tiny pieces of rocks and tiny pieces of shells that you can see. It is said to be heterogeneous because any two given samples will have a different composition with different kinds of particles.

A solution of salt dissolved in water also meets the definition of a mixture since it has unlike parts and can have a variable composition. A solution, however, is different from a sand mixture since it is a *homogeneous mixture,* meaning it is the same throughout a given sample. A homogeneous mixture, or solution, is the same throughout. The key to understanding that a solution is a mixture is found in its variable composition; that is, a given solution might be homogeneous, but one solution can vary from the next. Thus, you can have a salt solution with a 1 percent concentration, another with a 7 percent

FIGURE 9.1 Water is the most abundant liquid on Earth and is necessary for all life. Because of water's great dissolving properties, any sample is a solution containing solids, other liquids, and gases from the environment. This stream also carries suspended, ground-up rocks, called *rock flour*, from a nearby glacier.

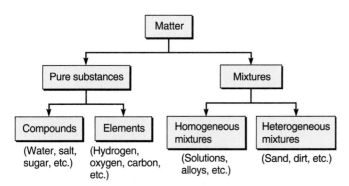

FIGURE 9.2 A classification scheme for matter.

concentration, yet another with a 10 percent concentration, and so on. Solutions, mixed gases, and metal alloys are all homogeneous mixtures since they are made of unlike parts and do not have a fixed, definite composition.

Mixtures can be separated into their component parts by physical means. For example, you can physically separate the parts making up a sand mixture by using a magnifying glass and tweezers to move and isolate each part. A solution of salt in water is a mixture since the amount of salt dissolved in water can have a variable composition. But how do you separate the parts of a solution? One way is to evaporate the water, leaving the salt behind. There are many methods for separating mixtures, but all involve a **physical change.** A physical change does not alter the *identity* of matter. When water is evaporated from the salt solution, for example, it changes to a different state of matter (water vapor) but is still recognized as water. Physical changes involve physical properties only; no new substances are formed. Examples of physical changes include evaporation,

A   B

FIGURE 9.3 Sugar (*A*) is a compound that can be easily decomposed to simpler substances by heating. (*B*) One of the simpler substances is the black element carbon, which cannot be further decomposed by chemical or physical means.

condensation, melting, freezing, and dissolving, as well as reshaping processes such as crushing or bending.

Mixtures can be physically separated into *pure substances,* materials that are the same throughout and have a fixed, definite composition. If you closely examine a sample of table salt, you will see that it is made up of hundreds of tiny cubes. Any one of these cubes will have the same properties as any other cube, including a salty taste. Sugar, like table salt, has all of its parts alike. Unlike salt, sugar grains have no special shape or form, but each grain has the same sweet taste and other properties as any other sugar grain.

If you heat salt and sugar in separate containers, you will find very different results. Salt, like some other pure substances, undergoes a physical change and melts, changing back to a solid upon cooling with the same properties that it had originally. Sugar, however, changes to a black material upon heating, while it gives off water vapor. The black material does not change back to sugar upon cooling. The sugar has *decomposed* to a new substance, while the salt did not. The new substance is carbon, and it has properties completely different from sugar. The original substance (sugar) and its properties have changed to a new substance (carbon) with new properties. The sugar has gone through a **chemical change.** A chemical change alters the identity of matter, producing new substances with different properties. In this case, the chemical change was one of decomposition. Heat produced a chemical change by decomposing the sugar into carbon and water vapor.

The decomposition of sugar always produces the same mass ratio of carbon to water, so sugar has a fixed, definite composition. A **compound** is a pure substance that can be decomposed by a chemical change into simpler substances with a fixed mass ratio. This means that sugar is a compound (figure 9.3).

A pure substance that cannot be broken down into anything simpler by chemical or physical means is an **element.**

Connections . . .

Strange Symbols

Some elements have symbols that are derived from the earlier use of their Latin names. For example, the symbol Au is used for gold because the metal was earlier known by its Latin name of *aurum,* meaning "shining dawn." There are ten elements with symbols derived from Latin names and one with a symbol from a German name. These eleven elements are listed below, together with the sources of their names and their symbols.

Atomic Number	Name	Source of Symbol	Symbol
11	Sodium	Latin: *Natrium*	Na
19	Potassium	Latin: *Kalium*	K
26	Iron	Latin: *Ferrum*	Fe
29	Copper	Latin: *Cuprum*	Cu
47	Silver	Latin: *Argentum*	Ag
50	Tin	Latin: *Stannum*	Sn
51	Antimony	Latin: *Stibium*	Sb
74	Tungsten	German: *Wolfram*	W
79	Gold	Latin: *Aurum*	Au
80	Mercury	Latin: *Hydrargyrum*	Hg
82	Lead	Latin: *Plumbum*	Pb

Sugar is decomposed by heating into carbon and water vapor. Carbon cannot be broken down further, so carbon is an element. It has been known since about 1800 that water is a compound that can be broken down by electrolysis into hydrogen and oxygen, two gases that cannot be broken down to anything simpler. So sugar is a compound made from the elements of carbon, hydrogen, and oxygen.

But what about the table salt? Is table salt a compound? Table salt is a stable compound that is not decomposed by heating. It melts at a temperature of about 800°C and then returns to the solid form with the same salty properties upon cooling. Electrolysis—the splitting of a compound by means of electricity—can be used to decompose table salt into the elements sodium and chlorine, positively proving that it is a compound.

Pure substances are either compounds or elements. Decomposition through heating and decomposition through electrolysis are two means of distinguishing between compounds and elements. If a substance can be decomposed into something simpler, you know for sure that it is a compound. If the substance cannot be decomposed, it might be an element, or it might be a stable compound that resists decomposition. More testing would be necessary before you can be confident that you have identified an element. Most pure substances are compounds. There are millions of different compounds but only 113 known elements at the present time. These elements are the fundamental materials of which all matter is made.

TABLE 9.1   Elements Making up 99 Percent of Earth's Crust

Element (Symbol)	Percent by Weight
Oxygen (O)	46.6
Silicon (Si)	27.7
Aluminum (Al)	8.1
Iron (Fe)	5.0
Calcium (Ca)	3.6
Sodium (Na)	2.8
Potassium (K)	2.6
Magnesium (Mg)	2.1

ELEMENTS

Elements are not equally abundant and only a few are common. In table 9.1, for example, you can see that only eight elements make up about 99 percent of the solid surface of Earth. Oxygen is most abundant, making up about 50 percent of the weight of Earth's crust. Silicon makes up more than 25 percent, so these two nonmetals alone make up about 75 percent of Earth's solid surface. Almost all the rest is made up of just six metals, as shown in the table.

The number of common elements is limited elsewhere, too. Only two elements make up about 99 percent of the atmospheric air around Earth. Air is mostly nitrogen (about 78 percent) and oxygen (about 21 percent), with traces of five other elements and compounds. Water is hydrogen and oxygen, of course, but seawater also contains elements in solution. These elements are chlorine (55 percent), sodium (31 percent), sulfur (8 percent), and magnesium (4 percent). Only three elements make up about 97 percent of your body. These elements are hydrogen (60 percent), oxygen (26 percent), and carbon (11 percent). Generally, all of this means that the elements are not equally distributed or equally abundant in nature (figure 9.4).

CHEMICAL CHANGE

The air you breathe, the liquids you drink, and all the things around you are elements, compounds, or mixtures. Most are compounds, however, and very few are pure elements. Water, sugar, gasoline, and chalk are examples of compounds. Each can be broken down into the elements that make it up. Recall that elements are basic substances that cannot be broken down into simpler substances. Examples of elements are hydrogen, carbon, and calcium. Why and how these elements join together in different ways to form different compounds is the subject of this chapter.

You have already learned that elements are made up of atoms that can be described by the modern atomic theory. You can also consider an **atom** to be *the smallest unit of an element that can exist alone or in combination with other elements.* Compounds are formed when atoms are held together by an

FIGURE 9.4 The elements of aluminum, iron, oxygen, and silicon make up about 88 percent of Earth's solid surface. Water on the surface and in the air as clouds and fog is made up of hydrogen and oxygen. The air is 99 percent nitrogen and oxygen. Hydrogen, oxygen, and carbon make up 97 percent of a person. Thus, almost everything you see in this picture is made up of just six elements.

FIGURE 9.5 Magnesium is an alkaline earth metal that burns brightly in air, releasing heat and light. As chemical energy is released, a new chemical substance is formed. The new chemical material is magnesium oxide, a soft powdery material that forms an alkaline solution in water (called *milk of magnesia*).

attractive force called a *chemical bond.* The chemical bond binds individual atoms together in a compound. A molecule is generally thought of as a tightly bound group of atoms that maintains its identity. More specifically, a **molecule** is defined as *the smallest particle of a compound, or a gaseous element, that can exist and still retain the characteristic chemical properties of a substance.* Compounds with one type of chemical bond, as you will see, have molecules that are electrically neutral groups of atoms held together strongly enough to be considered independent units. For example, water is a compound. The smallest unit of water that can exist alone is an electrically neutral unit made up of two hydrogen atoms and one oxygen atom held together by chemical bonds. The concept of a molecule will be expanded as chemical bonds are discussed.

Compounds occur naturally as gases, liquids, and solids. Many common gases occur naturally as molecules made up of two or more atoms. For example, at ordinary temperatures, hydrogen gas occurs as molecules of two hydrogen atoms bound together. Oxygen gas also usually occurs as molecules of two oxygen atoms bound together. Both hydrogen and oxygen occur naturally as *diatomic molecules* (*di-* means "two"). Oxygen sometimes occurs as molecules of three oxygen atoms bound together. These *triatomic molecules* (*tri-* means "three") are called *ozone.*

When molecules of any size are formed or broken down into simpler substances, new materials with new properties are produced. This kind of a change in matter is called a chemical change, and the process is called a chemical reaction. A **chemical reaction** is defined as

a change in matter in which different chemical substances are created by forming or breaking chemical bonds.

In general, chemical bonds are formed when atoms of elements are bound together to form compounds. Chemical bonds are broken when a compound is decomposed into simpler substances. Chemical bonds are electrical in nature, formed by electrical attractions, as discussed in chapter 6.

Chemical reactions happen all the time, all around you. A growing plant, burning fuels, and your body's utilization of food all involve chemical reactions. These reactions produce different chemical substances with greater or smaller amounts of internal potential energy (see chapter 4 for a discussion of internal potential energy). Energy is *absorbed* to produce new chemical substances with more internal potential energy. Energy is *released* when new chemical substances are produced with less internal potential energy (figure 9.5). In general, changes in internal potential energy are called **chemical energy.** For example, new chemical substances are produced in green plants through the process called *photosynthesis.* A green plant uses radiant energy (sunlight), carbon dioxide, and water to produce new chemical materials and oxygen. These new chemical materials, the stuff that leaves, roots, and wood are made of, contain more chemical energy than the carbon dioxide and water they were made from.

A **chemical equation** is a way of describing what happens in a chemical reaction. The chemical reaction of photosynthesis can be described by using words in an equation:

$$\begin{array}{c}\text{energy}\\\text{(sunlight)}\end{array} + \begin{array}{c}\text{carbon}\\\text{dioxide}\\\text{molecules}\end{array} + \begin{array}{c}\text{water}\\\text{molecules}\end{array} \rightarrow \begin{array}{c}\text{plant}\\\text{material}\\\text{molecules}\end{array} + \begin{array}{c}\text{oxygen}\\\text{molecules}\end{array}$$

The substances that are changed are on the left side of the word equation and are called *reactants*. The reactants are carbon dioxide molecules and water molecules. The equation also indicates that energy is absorbed, since the term *energy* appears on the left side. The arrow means *yields*. The new chemical substances are on the right side of the word equation and are called *products*. Reading the photosynthesis reaction as a sentence, you would say, "Carbon dioxide and water absorb energy to react, yielding plant materials and oxygen."

The plant materials produced by the reaction have more internal potential energy, also known as *chemical energy,* than the reactants. You know this from the equation because the term *energy* appears on the left side but not the right. This means that the energy on the left went into internal potential energy on the right. You also know this because the reaction can be reversed to release the stored energy (figure 9.6). When plant materials (such as wood) are burned, the materials react with oxygen, and chemical energy is released in the form of radiant energy (light) and high kinetic energy of the newly formed gases and vapors. In words,

$$\begin{array}{c}\text{plant}\\\text{material}\\\text{molecules}\end{array} + \begin{array}{c}\text{oxygen}\\\text{molecules}\end{array} \rightarrow \begin{array}{c}\text{carbon}\\\text{dioxide}\\\text{molecules}\end{array} + \begin{array}{c}\text{water}\\\text{molecules}\end{array} + \text{energy}$$

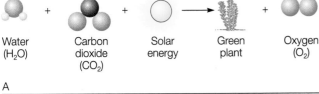

Water (H_2O)   Carbon dioxide (CO_2)   Solar energy   Green plant   Oxygen (O_2)

A

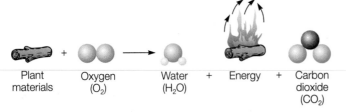

Plant materials   Oxygen (O_2)   Water (H_2O) + Energy + Carbon dioxide (CO_2)

B

FIGURE 9.6 (*A*) New chemical bonds are formed as a green plant makes new materials and stores solar energy through the photosynthesis process. (*B*) The chemical bonds are later broken, and the same amount of energy and the same original materials are released. The same energy and the same materials are released rapidly when the plant materials burn, and they are released slowly when the plant decomposes.

If you compare the two equations, you will see that burning is the opposite of the process of photosynthesis! The energy released in burning is exactly the same amount of solar energy that was stored as internal potential energy by the plant. Such chemical changes, in which chemical energy is stored in one reaction and released by another reaction, are the result of the making, then the breaking, of chemical bonds. Chemical bonds were formed by utilizing energy to produce new chemical substances. Energy was released when these bonds were broken then re-formed to produce the original substances. In this example, chemical reactions and energy flow can be explained by the making and breaking of chemical bonds. Chemical bonds can be explained in terms of changes in the electron structures of atoms. Thus, the place to start in seeking understanding about chemical reactions is the electron structure of the atoms themselves.

VALENCE ELECTRONS AND IONS

As discussed in chapter 8, it is the number of electrons in the outermost orbital that usually determines the chemical properties of an atom. These outer electrons are called **valence electrons,** and it is the valence electrons that participate in chemical bonding. The inner electrons are in stable, fully occupied orbitals and do not participate in chemical bonds. The representative elements (the A-group families) have valence electrons in the outermost orbitals, which contain from one to eight valence electrons. Recall that you can easily find the number of valence electrons by referring to a periodic table. The number at the top of each representative family is the same as the number of outer orbital electrons (with the exception of helium).

The noble gases have filled outer orbitals and do not normally form compounds. Apparently, half-filled and filled orbitals are particularly stable arrangements. Atoms have a tendency to seek such a stable, filled outer orbital arrangement such as the one found in the noble gases. For the representative elements, this tendency is called the **octet rule.** The octet rule states that *atoms attempt to acquire an outer orbital with eight electrons* through chemical reactions. This rule is a generalization, and a few elements do not meet the requirement of eight electrons but do seek the same general trend of stability. There are a few other exceptions, and the octet rule should be considered a generalization that helps keep track of the valence electrons in most representative elements.

The family number of the representative element in the periodic table tells you the number of valence electrons and what the atom must do to reach the stability suggested by the octet rule. For example, consider sodium (Na). Sodium is in family IA, so it has one valence electron. If the sodium atom can get rid of this outer valence electron through a chemical reaction, it will have the same outer electron configuration as an atom of the noble gas neon (Ne) (compare figure 9.7B and 9.7C).

When a sodium atom (Na) loses an electron to form a sodium ion (Na^+), it has the same, stable outer electron configuration as a neon atom (Ne). The sodium ion (Na^+) is still a form of sodium since it still has eleven protons. But it is now

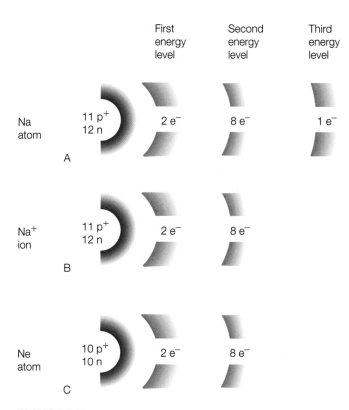

First energy level | Second energy level | Third energy level

Na atom — 11 p⁺ 12 n — $2\,e^-$ — $8\,e^-$ — $1\,e^-$

A

Na⁺ ion — 11 p⁺ 12 n — $2\,e^-$ — $8\,e^-$

B

Ne atom — 10 p⁺ 10 n — $2\,e^-$ — $8\,e^-$

C

FIGURE 9.7 (*A*) A sodium atom has two electrons in the first energy level, eight in the second energy level, and one in the third level. (*B*) When it loses its one outer, or valence, electron, it becomes a positively charged sodium ion with the same electron structure as an atom of neon (*C*).

a sodium *ion,* not a sodium *atom,* since it has eleven protons (eleven positive charges) and now has ten electrons (ten negative charges) for a total of

$$11 + \text{(protons)}$$
$$\underline{10 - \text{(electrons)}}$$
$$1 + \text{(net charge on sodium ion)}$$

CONCEPTS APPLIED

Calcium

What is the symbol and charge for a calcium ion?

From the list of elements on the inside back cover, the symbol for calcium is Ca, and the atomic number is 20. The periodic table tells you that Ca is in family IIA, which means that calcium has two valence electrons. According to the octet rule, the calcium ion must lose two electrons to acquire the stable outer arrangement of the noble gases. Since the atomic number is 20, a calcium atom has twenty protons (20 +) and twenty electrons (20 −). When it is ionized, the calcium ion will lose two electrons for a total charge of (20 +) + (18 −), or 2 +. The calcium ion is represented by the chemical symbol for calcium and the charge shown as a superscript: Ca^{2+}. What is the symbol and charge for an aluminum ion? (Answer: Al^{3+})

This charge is shown on the chemical symbol of Na⁺ for the *sodium ion.* Note that the sodium nucleus and the inner orbitals do not change when the sodium atom is ionized. The sodium ion is formed when a sodium atom loses its valence electron, and the process can be described by

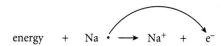

energy + Na· ⟶ Na⁺ + e⁻

where Na is the electron dot symbol for sodium, and the e⁻ is the electron that has been pulled off the sodium atom.

CHEMICAL BONDS

Atoms gain or lose electrons through a chemical reaction to achieve a state of lower energy, the stable electron arrangement of the noble gas atoms. Such a reaction results in a **chemical bond,** an *attractive force that holds atoms together in a compound.* There are three general classes of chemical bonds: (1) ionic bonds, (2) covalent bonds, and (3) metallic bonds.

Ionic bonds are formed when atoms *transfer* electrons to achieve the noble gas electron arrangement. Electrons are given up or acquired in the transfer, forming positive and negative ions. The electrostatic attraction between oppositely charged ions forms ionic bonds, and ionic compounds are the result. In general, ionic compounds are formed when a metal from the left side of the periodic table reacts with a nonmetal from the right side.

Covalent bonds result when atoms achieve the noble gas electron structure by *sharing* electrons. Covalent bonds are generally formed between the nonmetallic elements on the right side of the periodic table.

Metallic bonds are formed in solid metals such as iron, copper, and the other metallic elements that make up about 80 percent of all the elements. The atoms of metals are closely packed and share many electrons in a "sea" that is free to move throughout the metal, from one metal atom to the next. Metallic bonding accounts for metallic properties such as high electrical conductivity.

Ionic, covalent, and metallic bonds are attractive forces that hold atoms or ions together in molecules and crystals. There are two ways to describe what happens to the electrons when one of these bonds is formed: by considering (1) the new patterns formed when atomic orbitals overlap to form a combined orbital called a *molecular orbital* or (2) the atoms in a molecule as *isolated atoms* with changes in their outer shell arrangements. The molecular orbital description considers that the electrons belong to the whole molecule and form a molecular orbital with its own shape, orientation, and energy levels. The isolated atom description considers the electron energy levels as if the atoms in the molecule were alone, isolated from the molecule. The isolated atom description is less accurate than the molecular orbital description, but it is less complex and more easily understood. Thus, the following details about chemical bonding will mostly consider individual atoms and ions in compounds.

Ionic Bonds

An **ionic bond** is defined as the *chemical bond of electrostatic attraction* between negative and positive ions. Ionic bonding occurs when an atom of a metal reacts with an atom of a nonmetal. The reaction results in a transfer of one or more valence electrons from the metal atom to the valence shell of the nonmetal atom. The atom that loses electrons becomes a positive ion, and the atom that gains electrons becomes a negative ion. Oppositely charged ions attract one another, and when pulled together, they form an ionic solid with the ions arranged in an orderly geometric structure (figure 9.8). This results in a crystalline solid that is typical of salts such as sodium chloride (figure 9.9).

As an example of ionic bonding, consider the reaction of sodium (a soft reactive metal) with chlorine (a pale yellow-green gas). When an atom of sodium and an atom of chlorine collide, they react violently as the valence electron is transferred from the sodium to the chlorine atom. This produces a sodium ion and a chlorine ion. The reaction can be illustrated with electron dot symbols as follows:

$$Na \cdot \; + \; \cdot \ddot{\underset{..}{Cl}} : \longrightarrow \; Na^+ \; (: \ddot{\underset{..}{Cl}} :)^-$$

As you can see, the sodium ion transferred its valence electron, and the resulting ion now has a stable electron configuration. The chlorine atom accepted the electron in its outer orbital to acquire a stable electron configuration. Thus, a stable positive ion and a stable negative ion are formed. Because of opposite electrical charges, the ions attract each other to produce an ionic bond. When many ions are involved, each Na^+ ion is surrounded by six Cl^- ions, and each Cl^- ion is surrounded by six Na^+ ions. This gives the resulting solid NaCl its crystalline cubic structure, as shown in figure 9.9. In the solid state, all the sodium ions and all the chlorine ions are bound together in one giant unit. Thus, the term *molecule* is not really appropriate for ionic solids such as sodium chloride. But the term is sometimes used anyway, since any given sample will have the same number of Na^+ ions as Cl^- ions.

The sodium-chlorine reaction can be represented with electron dot notation as occurring in three steps:

1. $\quad energy \; + \; Na \cdot \longrightarrow Na^+ \; + \; e^-$

2. $\quad \cdot \ddot{\underset{..}{Cl}} : \; + \; e^- \longrightarrow \; (: \ddot{\underset{..}{Cl}} :)^- + \; energy$

3. $\quad Na^+ \; + \; (: \ddot{\underset{..}{Cl}} :)^- \longrightarrow Na^+ \; (: \ddot{\underset{..}{Cl}} :)^- + \; energy$

The energy released in steps 2 and 3 is greater than the energy absorbed in step 1, and an ionic bond is formed. The energy released is called the **heat of formation**. It is also the amount of energy required to decompose the compound (sodium chloride) into its elements. The reaction does not take place in steps as described, however, but occurs all at once. Note again, as in the photosynthesis-burning reactions described earlier, that the total amount of chemical energy is conserved. The energy released by the formation of the sodium chloride compound is the *same* amount of energy needed to decompose the compound.

Ionic bonds are formed by electron transfer, and electrons are conserved in the process. This means that electrons are not created or destroyed in a chemical reaction. The same total number of electrons exists after a reaction that existed before the reaction. There are two rules you can use for keeping track of electrons in ionic bonding reactions:

1. Ions are formed as atoms gain or lose valence electrons to achieve the stable noble gas structure.

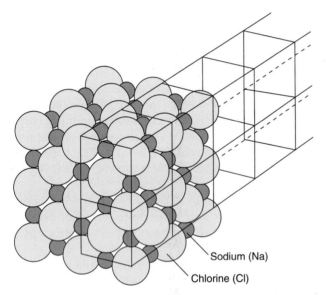

FIGURE 9.8 Sodium chloride crystals are composed of sodium and chlorine ions held together by electrostatic attraction. A crystal builds up, giving the sodium chloride crystal a cubic structure.

FIGURE 9.9 You can clearly see the cubic structure of these ordinary table salt crystals because they have been magnified about ten times.

2. There must be a balance between the number of electrons lost and the number of electrons gained by atoms in the reaction.

The sodium-chlorine reaction follows these two rules. The loss of one valence electron from a sodium atom formed a stable sodium ion. The gain of one valence electron by the chlorine atom formed a stable chlorine ion. Thus, both ions have noble gas configurations (rule 1), and one electron was lost and one was gained, so there is a balance in the number of electrons lost and the number gained (rule 2).

The **formula** of a compound *describes what elements are in the compound and in what proportions.* Sodium chloride contains one positive sodium ion for each negative chlorine ion. The formula of the compound sodium chloride is NaCl. If there are no subscripts at the lower right part of each symbol, it is understood that the symbol has a number "1." Thus, NaCl indicates a compound made up of the elements sodium and chlorine, and there is one sodium atom for each chlorine atom.

Calcium (Ca) is an alkaline metal in family IIA, and fluorine (F) is a halogen in family VIIA. Since calcium is a metal and fluorine is a nonmetal, you would expect calcium and fluorine atoms to react, forming a compound with ionic bonds. Calcium must lose two valence electrons to acquire a noble gas configuration. Fluorine needs one valence electron to acquire a noble gas configuration. So calcium needs to lose two electrons and fluorine needs to gain one electron to achieve a stable configuration (rule 1). Two fluorine atoms, each acquiring one electron, are needed to balance the number of electrons lost and the number of electrons gained. The compound formed from the reaction, calcium fluoride, will therefore have a calcium ion with a charge of plus two for every two fluorine ions with a charge of minus one. Recalling that electron dot symbols show only the outer valence electrons, you can see that the reaction is

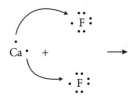

which shows that a calcium atom transfers two electrons, one each to two fluorine atoms. Now showing the results of the reaction, a calcium ion is formed from the loss of two electrons (charge 2+) and two fluorine ions are formed by gaining one electron each (charge 1−):

The formula of the compound is therefore CaF$_2$, with the subscript 2 for fluorine and the understood subscript 1 for calcium. This means that there are two fluorine atoms for each calcium atom in the compound.

Sodium chloride (NaCl) and calcium fluoride (CaF$_2$) are examples of compounds held together by ionic bonds. Such compounds are called **ionic compounds.** Ionic compounds of the representative elements are generally white, crystalline solids that form colorless solutions. Sodium chloride, the most common example, is common table salt. Many of the transition elements form colored compounds that make colored solutions. Ionic compounds dissolve in water, producing a solution of ions that can conduct an electric current.

In general, the elements in families IA and IIA of the periodic table tend to form positive ions by losing electrons. The ion charge for these elements equals the family number of these elements. The elements in families VIA and VIIA tend to form negative ions by gaining electrons. The ion charge for these elements equals their family number minus 8. The elements in families IIIA and VA have less of a tendency to form ionic compounds, except for those in higher periods. Common ions of some representative elements are given in table 9.2. The transition elements form positive ions of several different charges. Common ions of some transition elements are listed in table 9.3.

Covalent Bonds

Most substances do not have the properties of ionic compounds since they are not composed of ions. Most substances are molecular, composed of electrically neutral groups of atoms that are tightly bound together. As noted earlier, many gases are diatomic, occurring naturally as two atoms bound together as an electrically neutral molecule. Hydrogen, for example, occurs as molecules of H$_2$ and no ions are involved. The hydrogen atoms are held together by a covalent bond. A **covalent bond** is a *chemical bond formed by the sharing of at least a pair of electrons.* In the diatomic hydrogen molecule, each hydrogen atom contributes a single electron to the shared pair. Both hydrogen atoms count the shared pair of electrons in achieving

TABLE 9.2 Common Ions of Some Representative Elements

Element	Symbol	Ion
Lithium	Li	1+
Sodium	Na	1+
Potassium	K	1+
Magnesium	Mg	2+
Calcium	Ca	2+
Barium	Ba	2+
Aluminum	Al	3+
Oxygen	O	2−
Sulfur	S	2−
Hydrogen	H	1+, 1−
Fluorine	F	1−
Chlorine	Cl	1−
Bromine	Br	1−
Iodine	I	1−

TABLE 9.3 **Common Ions of Some Transition Elements**

Single-Charge Ions		
Element	**Symbol**	**Charge**
Zinc	Zn	2+
Tungsten	W	6+
Silver	Ag	1+
Cadmium	Cd	2+

Variable-Charge Ions		
Element	**Symbol**	**Charge**
Chromium	Cr	2+, 3+, 6+
Manganese	Mn	2+, 4+, 7+
Iron	Fe	2+, 3+
Cobalt	Co	2+, 3+
Nickel	Ni	2+, 3+
Copper	Cu	1+, 2+
Tin	Sn	2+, 4+
Gold	Au	1+, 3+
Mercury	Hg	1+, 2+
Lead	Pb	2+, 4+

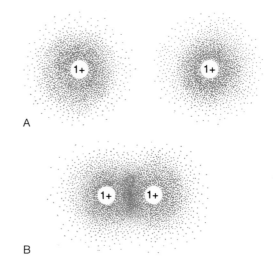

A

B

FIGURE 9.10 (A) Two hydrogen atoms, each with its own probability distribution of electrons about the nucleus. (B) When the hydrogen atoms bond, a new electron distribution pattern forms around the entire molecule, and both atoms share the two electrons.

their noble gas configuration. Hydrogen atoms both share one pair of electrons, but other elements might share more than one pair to achieve a noble gas structure.

Consider how the covalent bond forms between two hydrogen atoms by imagining two hydrogen atoms moving toward one another. Each atom has a single electron. As the atoms move closer and closer together, their orbitals begin to overlap. Each electron is attracted to the oppositely charged nucleus of the other atom and the overlap tightens. Then the repulsive forces from the like-charged nuclei will halt the merger. A state of stability is reached between the two nuclei and two electrons, and an H_2 molecule has been formed. The two electrons are now shared by both atoms, and the attraction of one nucleus for the other electron and vice versa holds the atoms together (figure 9.10).

Electron dot notation can be used to represent the formation of covalent bonds. For example, the joining of two hydrogen atoms to form an H_2 molecule can be represented as

$$H\bullet + H\bullet \longrightarrow H\!:\!H$$

Since an electron pair is *shared* in a covalent bond, the two electrons move throughout the entire molecular orbital. Since each hydrogen atom now has both electrons on an equal basis, each can be considered now to have the noble gas configuration of helium. A dashed circle around each symbol shows that both atoms have two electrons:

$$H\bullet + H\bullet \longrightarrow \left(H\!:\!H\right)$$

Hydrogen and fluorine react to form a covalent molecule (how this is known will be discussed shortly), and this bond can be represented with electron dots. Fluorine is in the VIIA family, so you know an atom of fluorine has seven valence electrons in the outermost energy level. The reaction is

$$H\bullet + \bullet\ddot{\underset{\cdot\cdot}{F}}\!: \longrightarrow \left(H\!:\!\ddot{\underset{\cdot\cdot}{F}}\!:\right)$$

Each atom shares a pair of electrons to achieve a noble gas configuration. Hydrogen achieves the helium configuration, and fluorine achieves the neon configuration. All the halogens have seven valence electrons, and all need to gain one electron (ionic bond) or share an electron pair (covalent bond) to achieve a noble gas configuration. This also explains why the halogen gases occur as diatomic molecules. Two fluorine atoms can achieve a noble gas configuration by sharing a pair of electrons:

$$\bullet\ddot{\underset{\cdot\cdot}{F}}\!: + \bullet\ddot{\underset{\cdot\cdot}{F}}\!: \longrightarrow \left(\ddot{\underset{\cdot\cdot}{F}}\!:\!\ddot{\underset{\cdot\cdot}{F}}\!:\right)$$

The diatomic hydrogen (H_2) and fluorine (F_2), hydrogen fluoride (HF), and water (H_2O) are examples of compounds held together by covalent bonds. A compound held together by covalent bonds is called a **covalent compound.** In general, covalent compounds form from nonmetallic elements on the right side of the periodic table. For elements in families IVA through VIIA, the number of unpaired electrons (and thus the number of covalent bonds formed) is eight minus the family number. You can get a lot of information from the periodic table from generalizations like this one. For another generalization, compare table 9.4 with

Aluminum Fluoride

Use electron dot notation to predict the formula of a compound formed when aluminum (Al) combines with fluorine (F).

Aluminum, atomic number 13, is in family IIIA, so it has three valence electrons and an electron dot notation of

$$\overset{\displaystyle .}{\underset{\displaystyle .}{Al}} \,.$$

According to the octet rule, the aluminum atom would need to lose three electrons to acquire the stable noble gas configuration. Fluorine, atomic number 9, is in family VIIA, so it has seven valence electrons and an electron dot notation of

$$.\,\overset{\displaystyle ..}{\underset{\displaystyle ..}{F}}\,:$$

Fluorine would acquire a noble gas configuration by accepting one electron. Three fluorine atoms, each acquiring one electron, are needed to balance the three electrons lost by aluminum. The reaction can be represented as

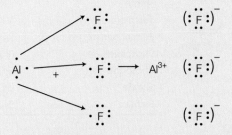

The ratio of aluminum atoms to fluorine atoms in the compound is 1:3. The formula for aluminum fluoride is therefore AlF_3.

Predict the formula of the compound formed between aluminum and oxygen using electron dot notation. (Answer: Al_2O_3)

TABLE 9.4 Structures and Compounds of Nonmetallic Elements Combined with Hydrogen

Nonmetallic Element	Elements (E represents any element of family)	Compound
Family IVA: C, Si, Ge	$\cdot\overset{.}{E}\cdot$	H:E:H with H above and below (H:E:H)
Family VA: N, P, As, Sb	$\cdot\overset{.}{\underset{.}{E}}\cdot$	H:E:H with H below
Family VIA: O, S, Se, Te	$\cdot\overset{.}{\underset{..}{E}}\cdot$	H:E:H
Family VIIA: F, Cl, Br, I	$\cdot\overset{.}{\underset{..}{E}}:$	H:E:

the periodic table. The table gives the structures of nonmetals combined with hydrogen and the resulting compounds.

Two dots can represent a lone pair of valence electrons, or they can represent a bonding pair, a single pair of electrons being shared by two atoms. Bonding pairs of electrons are often represented by a simple line between two atoms. For example,

H : H   is shown as   H — H

and

:O: (with H, H below)   is shown as   O with H, H

Note that the line between the two hydrogen atoms represents an electron pair, so each hydrogen atom has two electrons in the outer shell, as does helium. In the water molecule, each hydrogen atom has two electrons as before. The oxygen atom has two lone pairs (a total of four electrons) and two bonding pairs (a total of four electrons) for a total of eight electrons. Thus, oxygen has acquired a stable octet of electrons.

A covalent bond in which a single pair of electrons is shared by two atoms is called a *single covalent bond,* or simply a **single bond.** Some atoms have two unpaired electrons and can share more than one electron pair. A **double bond** is a covalent bond formed when *two pairs* of electrons are shared by two atoms. This happens mostly in compounds involving atoms of the elements C, N, O, and S. Ethylene, for example, is a gas given off from ripening fruit. The electron dot formula for ethylene is

H, H around :C :: C: with H, H   or   H, H on C = C with H, H

The ethylene molecule has a double bond between two carbon atoms. Since each line represents two electrons, you can simply count the lines around each symbol to see if the octet rule has been satisfied. Each H has one line, so each H atom is sharing two electrons. Each C has four lines so each C atom has eight electrons, satisfying the octet rule.

A **triple bond** is a covalent bond formed when *three pairs* of electrons are shared by two atoms. Triple bonds occur mostly in compounds with atoms of the elements C and N. Acetylene, for example, is a gas often used in welding torches (figure 9.11). The electron dot formula for acetylene is

H:C :::C:H   or   H — C ≡ C — H

Name That Compound

Ionic Compound Names

Ionic compounds formed by metal ions are named by stating the name of the metal (positive ion), then the name of the nonmetal (negative ion). Ionic compounds formed by variable-charge ions have an additional rule to identify which variable-charge ion is involved. There was an old way of identifying the charge on the ion by adding either -ic or -ous to the name of the metal. The suffix -ic meant the higher of two possible charges, and the suffix -ous meant the lower of two possible charges. For example, iron has two possible charges, 2+ or 3+. The old system used the Latin name for the root. The Latin name for iron is *ferrum,* so a higher charged iron ion (3+) was named a ferric ion. The lower charged iron ion (2+) was called a ferrous ion. You still hear the old names sometimes, but chemists now have a better way to identify the variable-charge ion. The newer system uses the English name of the metal with Roman numerals in parentheses to indicate the charge number. Thus, an iron ion with a charge of 2+ is called an iron(II) ion and an iron ion with a charge of 3+ is an iron(III) ion. Box table 9.1 gives some of the modern names for variable-charge ions. These names are used with the name of a nonmetal ending in -ide, just like the single-charge ions in ionic compounds made up of two different elements.

Box Table 9.1

Modern Names of Some Variable-Charge Ions	
Ion	**Name of Ion**
Fe^{2+}	Iron(II) ion
Fe^{3+}	Iron(III) ion
Cu^+	Copper(I) ion
Cu^{2+}	Copper(II) ion
Pb^{2+}	Lead(II) ion
Pb^{4+}	Lead(IV) ion
Sn^{2+}	Tin(II) ion
Sn^{4+}	Tin(IV) ion
Cr^{2+}	Chromium(II) ion
Cr^{3+}	Chromium(III) ion
Cr^{6+}	Chromium(VI) ion

Some ionic compounds contain three or more elements, and so are more complex than a combination of a metal ion and a nonmetal ion. This is possible because they have **polyatomic ions,** groups of two or more atoms that are bound together tightly and behave very much like a single monatomic ion (box table 9.2). For example, the OH^- ion is an oxygen atom bound to a hydrogen atom with a net charge of $1-$. This polyatomic ion is called a *hydroxide ion.* The hydroxide compounds make up one of the main groups of ionic compounds, the *metal hydroxides.* A metal hydroxide is an ionic compound consisting of a metal with the hydroxide ion. Another main group consists of the salts with polyatomic ions.

The metal hydroxides are named by identifying the metal first and the term *hydroxide* second. Thus, NaOH is named

Box Table 9.2

Some Common Polyatomic Ions	
Ion Name	**Formula**
Acetate	$(C_2H_3O_2)^-$
Ammonium	$(NH_4)^+$
Borate	$(BO_3)^{3-}$
Carbonate	$(CO_3)^{2-}$
Chlorate	$(ClO_3)^-$
Chromate	$(CrO_4)^{2-}$
Cyanide	$(CN)^-$
Dichromate	$(Cr_2O_7)^{2-}$
Hydrogen carbonate (or bicarbonate)	$(HCO_3)^-$
Hydrogen sulfate (or bisulfate)	$(HSO_4)^-$
Hydroxide	$(OH)^-$
Hypochlorite	$(ClO)^-$
Nitrate	$(NO_3)^-$
Nitrite	$(NO_2)^-$
Perchlorate	$(ClO_4)^-$
Permanganate	$(MnO_4)^-$
Phosphate	$(PO_4)^{3-}$
Phosphite	$(PO_3)^{3-}$
Sulfate	$(SO_4)^{2-}$
Sulfite	$(SO_3)^{2-}$

FIGURE 9.11 Acetylene is a hydrocarbon consisting of two carbon atoms and two hydrogen atoms held together by a triple covalent bond between the two carbon atoms. When mixed with oxygen gas (the tank to the right), the resulting flame is hot enough to cut through most metals.

CONCEPTS APPLIED

Household Chemicals

Pick a household product that has a list of ingredients with names of covalent compounds or of ions you have met in this chapter. Write the brand name of the product and the type of product (example: Sani-Flush; toilet-bowl cleaner), then list the ingredients as given on the label, writing them one under the other (column 1). Beside each name put the formula, if you can figure out what it should be (column 2). Also, in a third column, put whatever you know or can guess about the function of that substance in the product. (Example: This is an acid; helps dissolve mineral deposits.)

sodium hydroxide and KOH is potassium hydroxide. The salts are similarly named, with the metal (or ammonium ion) identified first, then the name of the polyatomic ion. Thus, $NaNO_3$ is named sodium nitrate and $NaNO_2$ is sodium nitrite. Note that the suffix -*ate* means the polyatomic ion with one more oxygen atom than the -*ite* ion. For example, the chlor*ate* ion is $(ClO_3)^-$ and the chlor*ite* ion is $(ClO_2)^-$. Sometimes more than two possibilities exist, and more oxygen atoms are identified with the prefix *per*- and less with the prefix *hypo*-. Thus, the *per*chlor*ate* ion is $(ClO_4)^-$ and the *hypo*chlor*ite* ion is $(ClO)^-$.

Covalent Compound Names

Covalent compounds are molecular, and the molecules are composed of two *nonmetals,* as opposed to the metal and nonmetal elements that make up ionic compounds. The combinations of nonmetals do not present simple names as the ionic compounds did, so a different set of rules for naming and formula writing is needed.

Ionic compounds were named by stating the name of the positive metal ion, then the name of the negative nonmetal ion with an -*ide* ending. This system is not adequate for naming the covalent compounds. To begin, covalent compounds are composed of two or more nonmetal atoms that form a molecule. It is possible for some atoms to form single, double, or even triple bonds with other atoms, including atoms of the same element. The net result is that the same two elements can form more than one kind of covalent compound. Carbon and oxygen, for example, can combine to form the gas released from burning and respiration, carbon dioxide (CO_2). Under certain conditions, the very same elements combine to produce a different gas, the poisonous carbon monoxide (CO). Similarly, sulfur and oxygen can combine differently to produce two different covalent compounds. A successful system for naming covalent compounds must therefore provide a means of identifying different compounds made of the same elements. This is accomplished by using a system of Greek prefixes (see box table 9.3). The rules are as follows:

1. The first element in the formula is named first with a prefix indicating the number of atoms if the number is greater than one.
2. The stem name of the second element in the formula is next. A prefix is used with the stem if two elements form more than one compound. The suffix -*ide* is again used to indicate a compound of only two elements. For example, CO is carbon monoxide and CO_2 is carbon dioxide. The compound BF_3 is boron trifluoride and N_2O_4 is dinitrogen tetroxide. Knowing the formula and the prefix and stem information in box table 9.3, you can write the name of any covalent compound made up of two elements by ending it with -*ide*. Conversely, the name will tell you the formula. However, there are a few polyatomic ions with -*ide* endings that are compounds made up of more than just two elements (hydroxide and cyanide). Compounds formed with the ammonium ion will also have an -*ide* ending, and these are also made up of more than two elements.

Box Table 9.3

Prefixes and Element Stem Names			
Prefixes		**Stem Names**	
PREFIX	MEANING	ELEMENT	STEM
Mono-	1	Hydrogen	Hydr-
Di-	2	Carbon	Carb-
Tri-	3	Nitrogen	Nitr-
Tetra-	4	Oxygen	Ox-
Penta-	5	Fluorine	Fluor-
Hexa-	6	Phosphorus	Phosph-
Hepta-	7	Sulfur	Sulf-
Octa-	8	Chlorine	Chlor-
Nona-	9	Bromine	Brom-
Deca-	10	Iodine	Iod-

Note: The *a* or *o* ending on the prefix is often dropped if the stem name begins with a vowel, for example, "tetroxide," not "tetraoxide."

Myths, Mistakes, and Misunderstandings

Ban DHMO?

"Dihydrogen monoxide (DHMO) is colorless, odorless, tasteless, and kills uncounted thousands of people every year. Most of these deaths are caused by accidental inhalation of DHMO, but the dangers do not end there. Prolonged exposure to its solid form causes severe tissue damage. Symptoms of DHMO ingestion can include excessive sweating and urination, and possibly a bloated feeling, nausea, vomiting, and body electrolyte imbalance. For those who have become dependent, DHMO withdrawal means certain death."

The above is part of a hoax that was recently circulated on the Internet. The truth is that dihydrogen monoxide is the chemical name of H_2O—water.

The acetylene molecule has a triple bond between two carbon atoms. Again, note that each line represents two electrons. Each C atom has four lines, so the octet rule is satisfied.

COMPOSITION OF COMPOUNDS

As you can imagine, there are literally millions of different chemical compounds from all the possible combinations of over ninety natural elements held together by ionic or covalent bonds. Each of these compounds has its own name, so there are millions of names and formulas for all the compounds. In the early days, compounds were given *common names* according to how they were used, where they came from, or some other means of identifying them. Thus, sodium carbonate was called soda, and closely associated compounds were called baking soda (sodium bicarbonate), washing soda (sodium

How to Write a Chemical Formula

Ionic Compound Formulas

The formulas for ionic compounds are easy to write. There are two rules:

1. **The symbols.** Write the symbol for the positive element first, followed by the symbol for the negative element (same order as in the name).
2. **The subscripts.** Add subscripts to indicate the numbers of ions needed to produce an electrically neutral compound.

As an example, let us write the formula for the compound calcium chloride. The name tells you that this compound consists of positive calcium ions and negative chlorine ions. The suffix -*ide* tells you there are only two elements present. Following rule 1, the symbols would be CaCl.

For rule 2, note the calcium ion is Ca^{2+} and the chlorine ion is Cl^-. You know the calcium is plus two and chlorine is negative one by applying the atomic theory, knowing their positions in the periodic table, or by using a table of ions and their charges. To be electrically neutral, the compound must have an equal number of pluses and minuses. Thus, you will need two negative chlorine ions for every calcium ion with its 2+ charge. Therefore, the formula is $CaCl_2$.

The total charge of two chlorines is thus 2−, which balances the 2+ charge on the calcium ion.

One easy way to write a formula showing that a compound is electrically neutral is to cross over the absolute charge numbers (without plus or minus signs) and use them as subscripts. For example, the symbols for the calcium ion and the chlorine ion are

$$Ca^{2+}Cl^{1-}$$

Crossing the absolute numbers as subscripts, as follows

and then dropping the charge numbers gives

$$Ca_1\ Cl_2$$

No subscript is written for 1; it is understood. The formula for calcium chloride is thus

$$CaCl_2$$

The crossover technique works because ionic bonding results from a transfer of electrons, and the net charge is conserved. A calcium ion has a 2+ charge because the atom lost two electrons and two chlorine atoms gain one electron each, for a total of two electrons gained. Two electrons lost equals two electrons gained, and the net charge on calcium chloride is zero, as it has to be.

When using the crossover technique, it is sometimes necessary to reduce the ratio to the lowest common multiple. Thus, Mg_2O_2 means an equal ratio of magnesium and oxygen ions, so the correct formula is MgO.

The formulas for variable-charge ions are easy to write, since the Roman numeral tells you the charge number. The formula for tin(II) fluoride is written by crossing over the charge numbers (Sn^{2+}, F^{1-}), and the formula is SnF_2.

The formulas for ionic compounds with polyatomic ions are written from combinations of positive metal ions or the ammonium ion with the polyatomic ions, as listed in box table 9.2. Since the polyatomic ion is a group of atoms that has a charge and stays together in a unit, it is sometimes necessary to indicate this with parentheses. For example, magnesium hydroxide is composed of Mg^{2+} ions and $(OH)^{1-}$ ions.

carbonate), and caustic soda (sodium hydroxide), and the bubbly drink made by reacting soda with acid was called soda water, later called soda pop (figure 9.12). Potassium carbonate was extracted from charcoal by soaking in water and came to be called potash. Such common names are colorful, and some are descriptive, but it was impossible to keep up with the names as the number of known compounds grew. So a systematic set of rules was developed to determine the name and formula of each compound. Once you know the rules, you can write the formula when you hear the name. Conversely, seeing the formula will tell you the systematic name of the compound. This can be an interesting intellectual activity and can also be important when reading the list of ingredients to understand the composition of a product.

A different set of systematic rules is used with ionic compounds and covalent compounds, but there are a few rules in common. For example, a compound made of only two different elements always ends with the suffix -*ide*. So when you hear the name of a compound ending with -*ide*, you automatically know that the compound is made up of only two elements. Sodium chlor*ide* is an ionic compound made up of sodium and chlorine

ions. Carbon diox*ide* is a covalent compound with carbon and oxygen atoms. Thus, the systematic name tells you what elements are present in a compound with an -*ide* ending.

FIGURE 9.12 These substances are made up of sodium and some form of a carbonate ion. All have common names with the term *soda* for this reason. Soda water (or "soda pop") was first made by reacting soda (sodium carbonate) with an acid, so it was called "soda water."

Using the crossover technique to write the formula, you get

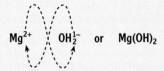

The parentheses are used and the subscript is written *outside* the parenthesis to show that the entire hydroxide unit is taken twice. The formula $Mg(OH)_2$ means

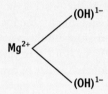

which shows that the pluses equal the minuses. Parentheses are not used, however, when only one polyatomic ion is present. Sodium hydroxide is $NaOH$, not $Na(OH)_1$.

Covalent Compound Formulas

The systematic name tells you the formula for a covalent compound. The gas that dentists use as an anesthetic, for example, is dinitrogen monoxide. This tells you there are two nitrogen atoms and one oxygen atom in the molecule, so the formula is N_2O. A different molecule composed of the very same elements is nitrogen dioxide. Nitrogen dioxide is the pollutant responsible for the brownish haze of smog. The formula for nitrogen dioxide is NO_2. Other examples of formulas from systematic names are carbon dioxide (CO_2) and carbon tetrachloride (CCl_4).

Formulas of covalent compounds indicate a pattern of how many atoms of one element combine with atoms of another. Carbon, for example, combines with no more than two oxygen atoms to form carbon dioxide. Carbon combines with no more than four chlorine atoms to form carbon tetrachloride. Electron dot formulas show these two molecules as

$$\ddot{O} :: C :: \ddot{O} \qquad \begin{matrix} & : \ddot{Cl} : & \\ : \ddot{Cl} : & C : & \ddot{Cl} : \\ & : \ddot{Cl} : & \end{matrix}$$

Using a dash to represent bonding pairs, we have

$$O = C = O \qquad \begin{matrix} & Cl & \\ & | & \\ Cl - & C & - Cl \\ & | & \\ & Cl & \end{matrix}$$

In both of these compounds, the carbon atom forms four covalent bonds with another atom. The number of covalent bonds that an atom can form is called its *valence*. Carbon has a valence of four and can form single, double, or triple bonds. Here are the possibilities for a single carbon atom (combining elements not shown):

$$-\overset{|}{\underset{|}{C}}- \qquad -\overset{|}{C}=$$

$$=C= \qquad -C\equiv$$

Hydrogen has only one unshared electron, so the hydrogen atom has a valence of one. Oxygen has a valence of two, and nitrogen has a valence of three. Here are the possibilities for hydrogen, oxygen, and nitrogen:

$$H- \qquad -\overset{\cdot\cdot}{\underset{\cdot\cdot}{O}}- \qquad :\overset{\cdot\cdot}{O}=$$

$$-\overset{\cdot\cdot}{\underset{|}{N}}- \qquad -\overset{\cdot\cdot}{N}= \qquad :N\equiv$$

CHEMICAL EQUATIONS

Chemical reactions occur when bonds between the outermost parts of atoms are formed or broken. Bonds are formed, for example, when a green plant uses sunlight—a form of energy—to create molecules of sugar, starch, and plant fibers. Bonds are broken and energy is released when you digest the sugars and starches or when plant fibers are burned. Chemical reactions thus involve changes in matter, the creation of new materials with new properties, and energy exchanges. So far, you have considered chemical symbols as a concise way to represent elements and formulas as a concise way to describe what a compound is made of. There is also a concise way to describe a chemical reaction, the *chemical equation*.

Word equations are useful in identifying what has happened before and after a chemical reaction. The substances that existed before a reaction are called *reactants*, and the substances that exist after the reaction are called the *products*. The equation has a general form of

$$\text{reactants} \rightarrow \text{products}$$

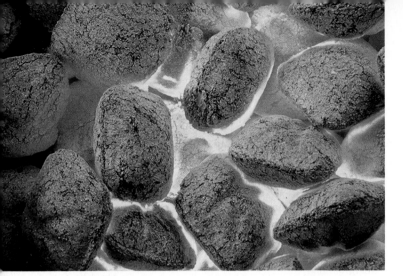

FIGURE 9.13 The charcoal used in a grill is basically carbon. The carbon reacts with oxygen to yield carbon dioxide. The chemical equation for this reaction, $C + O_2 \rightarrow CO_2$, contains the same information as the English sentence but has quantitative meaning as well.

where the arrow signifies a separation in time; that is, it identifies what existed before the reaction and what exists after the reaction. For example, the charcoal used in a barbecue grill is carbon (figure 9.13). The carbon reacts with oxygen while burning, and the reaction (1) releases energy and (2) forms carbon dioxide. The reactants and products for this reaction can be described as

$$\text{carbon} + \text{oxygen} \rightarrow \text{carbon dioxide}$$

The arrow means *yields,* and the word equation is read as, "Carbon reacts with oxygen to yield carbon dioxide." This word equation describes what happens in the reaction but says nothing about the quantities of reactants or products.

Chemical symbols and formulas can be used in the place of words in an equation and the equation will have a whole new meaning. For example, the equation describing carbon reacting with oxygen to yield carbon dioxide becomes

$$C + O_2 \rightarrow CO_2 \qquad \textbf{(balanced)}$$

The new, added meaning is that one atom of carbon (C) reacts with one molecule of oxygen (O_2) to yield one molecule of carbon dioxide (CO_2). Note that the equation also shows one atom of carbon and two atoms of oxygen (recall that oxygen occurs as a diatomic molecule) as reactants on the left side and one atom of carbon and two atoms of oxygen as products on the right side. Since the same number of each kind of atom appears on both sides of the equation, the equation is said to be *balanced.*

You would not want to use a charcoal grill in a closed room because there might not be enough oxygen. An insufficient supply of oxygen produces a completely different product, the poisonous gas carbon monoxide (CO). An equation for this reaction is

$$C + O_2 \rightarrow CO \qquad \textbf{(not balanced)}$$

As it stands, this equation describes a reaction that violates the **law of conservation of mass,** that matter is neither created nor

destroyed in a chemical reaction. From the point of view of an equation, this law states that

$$\text{mass of reactants} = \text{mass of products}$$

Mass of reactants here means all that you start with, including some that might not react. Thus, elements are neither created nor destroyed, and this means the elements present and their mass. In any chemical reaction, the kind and mass of the reactive elements are identical to the kind and mass of the product elements.

From the point of view of atoms, the law of conservation of mass means that *atoms are neither created nor destroyed in the chemical reaction.* A chemical reaction is the making or breaking of chemical bonds between atoms or groups of atoms. Atoms are not lost or destroyed in the process, nor are they changed to a different kind. The equation for the formation of carbon monoxide has two oxygen atoms in the reactants (O_2) but only one in the product (CO). An atom of oxygen has disappeared somewhere, and that violates the law of conservation of mass. You cannot fix the equation by changing the CO to a CO_2, because this would change the identity of the compounds. Carbon monoxide is a poisonous gas that is different from carbon dioxide, a relatively harmless product of burning and respiration. *You cannot change the subscript in a formula* because that would change the formula. A different formula means a different composition and thus a different compound.

You cannot change the subscripts of a formula, but you can place a number called a *coefficient* in *front* of the formula. Changing a coefficient changes the *amount* of a substance, not the identity. Thus 2 CO means two molecules of carbon monoxide and 3 CO means three molecules of carbon monoxide. If there is no coefficient, 1 is understood, as with subscripts. The meaning of coefficients and subscripts is illustrated in figure 9.14.

C	means	●	One atom of carbon
O	means	●	One atom of oxygen
O_2	means	●●	One molecule of oxygen consisting of two atoms of oxygen
CO	means	●●	One molecule of carbon monoxide consisting of one atom of carbon attached to one atom of oxygen
CO_2	means	●●●	One molecule of carbon dioxide consisting of one atom of carbon attached to two atoms of oxygen
3 CO_2	means	●●● ●●● ●●●	Three molecules of carbon dioxide, each consisting of one atom of carbon attached to two atoms of oxygen

FIGURE 9.14 The meaning of subscripts and coefficients used with a chemical formula. The subscripts tell you how many atoms of a particular element are in a compound. The coefficient tells you about the quantity, or number, of molecules of the compound.

On Balancing Equations

Natural gas is mostly methane, CH_4, which burns by reacting with oxygen (O_2) to produce carbon dioxide (CO_2) and water vapor (H_2O). Write a balanced chemical equation for this reaction by following a procedure of four steps.

Step 1. Write the correct formulas for the reactants and products in an unbalanced equation. For the burning of methane, the unbalanced, but otherwise correct, formula equation would be

$$CH_4 + O_2 \rightarrow CO_2 + H_2O \quad \textbf{(not balanced)}$$

Step 2. Inventory the number of each kind of atom on both sides of the unbalanced equation. In the example, there are

Reactants	Products
1 C	1 C
4 H	2 H
2 O	3 O

This step shows that the H and O are unbalanced.

Step 3. Determine where to place coefficients in front of formulas to balance the equation. It is often best to focus on the simplest thing you can do with whole number ratios. The H and the O are unbalanced, for example, and there are 4 H atoms on the left and 2 H atoms on the right. Placing a coefficient 2 in front of H_2O will balance the H atoms:

$$CH_4 + O_2 \rightarrow CO_2 + 2\,H_2O \quad \textbf{(not balanced)}$$

Now take a second inventory:

Reactants	Products
1 C	1 C
4 H	4 H
2 O	4 O (O_2+2 O)

This shows the O atoms are still unbalanced with 2 on the left and 4 on the right. Placing a coefficient of 2 in front of O_2 will balance the O atoms.

$$CH_4 + 2\,O_2 \rightarrow CO_2 + 2\,H_2O \quad \textbf{(balanced)}$$

Step 4. Take another inventory to determine if the numbers of atoms on both sides are now equal. If they are, determine if the coefficients are in the lowest possible whole-number ratio. The inventory is now

Reactants	Products
1 C	1 C
4 H	4 H
4 O	4 O

The number of each kind of atom on each side of the equation is the same, and the ratio of 1:2 → 1:2 is the lowest possible whole-number ratio. The equation is balanced, which is illustrated with sketches of molecules in box figure 9.1.

Balancing chemical equations is mostly a trial-and-error procedure. But with practice, you will find there are a few generalized "role models" that can be useful in balancing equations for many simple reactions. The key to success at balancing equations is to think it out step-by-step while remembering the following:

1. Atoms are neither lost nor gained nor do they change their identity in a chemical reaction. The same kind and number of atoms in the reactants must appear in the products, meaning atoms are conserved.

2. A correct formula of a compound cannot be changed by altering the number or placement of subscripts. Changing subscripts changes the identity of a compound and the meaning of the entire equation.

3. A coefficient in front of a formula multiplies everything in the formula by that number.

Here are a few generalizations that can be helpful in balancing equations:

1. Look first to formulas of compounds with the most atoms and try to balance the atoms or compounds from which they were formed or to which they decomposed.

2. You should treat polyatomic ions that appear on both sides of the equation as independent units with a charge. That is, consider the polyatomic ion as a unit while taking an inventory rather than the individual atoms making up the polyatomic ion. This will save time and simplify the procedure.

3. Both the "crossover technique" and the use of "fractional coefficients" can be useful in finding the least common multiple to balance an equation.

4. The physical state of reactants and products in a reaction is often identified by the symbols (g) for gas, (l) for liquid, (s) for solid, and (aq) for an aqueous solution, which means water. If a gas escapes, this is identified with an arrow pointing up (↑). A solid formed from a solution is identified with an arrow pointing down (↓). The Greek symbol delta (Δ) is often used under or over the yield sign to indicate a change of temperature or other physical values.

Reaction:
Methane reacts with oxygen to yield carbon dioxide and water

Balanced equation:
$$CH_4 + 2\,O_2 \longrightarrow CO_2 + 2\,H_2O$$

Sketches representing molecules:

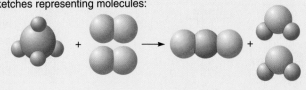

Meaning:

| 1 molecule of methane | + | 2 molecules of oxygen | ⟶ | 1 molecule of carbon dioxide | + | 2 molecules of water |

Box Figure 9.1 Compare the numbers of each kind of atom in the balanced equation with the numbers of each kind of atom in the sketched representation. Both the equation and the sketch have the same number of atoms in the reactants and in the products.

Placing a coefficient of 2 in front of the C and a coefficient of 2 in front of the CO in the equation will result in the same numbers of each kind of atom on both sides:

$$2\,C + O_2 \rightarrow 2\,CO$$

Reactants: 2 C   Products: 2 C
         2 O   2 O

The equation is now balanced.

Generalizing from groups of chemical reactions also makes it possible to predict what will happen in similar reactions. For example, the combustion of methane (CH_4), propane (C_3H_8), and octane (C_8H_{18}) all involve a *hydrocarbon,* a compound of the elements hydrogen and carbon. Each hydrocarbon reacts with O_2, yielding CO_2 and releasing the energy of combustion. Generalizing from these reactions, you could predict that the combustion of *any* hydrocarbon would involve the combination of atoms of the hydrocarbon molecule with O_2 to produce CO_2 and H_2O with the release of energy. Such reactions could be analyzed by chemical experiments, and the products could be identified by their physical and chemical properties. You would find your predictions based on similar reactions would be correct, thus justifying predictions from such generalizations. Butane, for example, is a hydrocarbon with the formula C_4H_{10} The balanced equation for the combustion of butane is

$$2\,C_4H_{10(g)} + 13\,O_{2(g)} \rightarrow 8\,CO_{2(g)} + 10\,H_2O_{(g)}$$

You could extend the generalization further, noting that the combustion of compounds containing oxygen as well as carbon and hydrogen also produces CO_2 and H_2O (figure 9.15). These compounds are *carbohydrates,* composed of carbon and water. Glucose, for example, is a compound with the formula $C_6H_{12}O_6$.

FIGURE 9.15 *Hydrocarbons* are composed of the elements hydrogen and carbon. Propane (C_3H_8) and gasoline, which contains a mixture of molecules such as octane (C_8H_{18}), are examples of hydrocarbons. *Carbohydrates* are composed of the elements of hydrogen, carbon, and oxygen. Table sugar, for example, is the carbohydrate $C_{12}H_{22}O_{11}$. Generalizing, all hydrocarbons and carbohydrates react completely with oxygen to yield CO_2 and H_2O.

Glucose combines with oxygen to produce CO_2 and H_2O, and the balanced equation is

$$C_6H_{12}O_{6(s)} + 6\,O_{2(g)} \rightarrow 6\,CO_{2(g)} + 6\,H_2O_{(g)}$$

Note that three molecules of oxygen were not needed from the O_2 reactant since the other reactant, glucose, contains six oxygen atoms per molecule. An inventory of atoms will show that the equation is thus balanced.

Combustion is a rapid reaction with O_2 that releases energy, usually with a flame. A very similar, although much slower, reaction takes place in plant and animal respiration. In respiration, carbohydrates combine with O_2 and release energy used for biological activities. This reaction is slow compared to combustion and requires enzymes to proceed at body temperature. Nonetheless, CO_2 and H_2O are the products.

TYPES OF CHEMICAL REACTIONS

The reactions involving hydrocarbons and carbohydrates with oxygen are examples of an important group of chemical reactions called *oxidation-reduction* reactions. When the term *oxidation* was first used, it specifically meant reactions involving the combination of oxygen with other atoms. But fluorine, chlorine, and other nonmetals were soon understood to have similar reactions as oxygen, so the definition was changed to one concerning the shifts of electrons in the reaction.

An **oxidation-reduction reaction** (or **redox reaction**) is broadly defined as a reaction in which electrons are transferred from one atom to another. As is implied by the name, such a reaction has two parts and each part tells you what happens to the electrons. *Oxidation* is the part of a redox reaction in which there is a loss of electrons by an atom. *Reduction* is the part of a redox reaction in which there is a gain of electrons by an atom. The name also implies that in any reaction in which oxidation

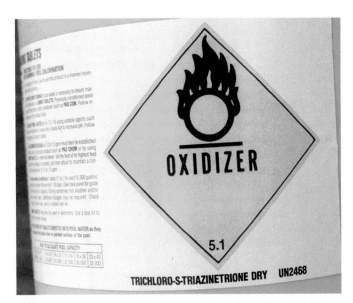

FIGURE 9.16 Oxidizing agents take electrons from other substances that are being oxidized. Oxygen and chlorine are commonly used, strong oxidizing agents.

occurs, reduction must take place, too. One cannot take place without the other.

Substances that take electrons from other substances are called **oxidizing agents.** Oxidizing agents take electrons from the substances being oxidized. Oxygen is the most common oxidizing agent, and several examples have already been given about how it oxidizes foods and fuels. Chlorine is another commonly used oxidizing agent, often for the purposes of bleaching or killing bacteria (figure 9.16).

A **reducing agent** supplies electrons to the substance being reduced. Hydrogen and carbon are commonly used reducing agents. Carbon is commonly used as a reducing agent to extract metals from their ores. For example, carbon (from coke, which is coal that has been baked) reduces Fe_2O_3, an iron ore, in the reaction

$$2\ Fe_2O_{3(s)} + 3\ C_{(s)} \rightarrow 4\ Fe_{(s)} + 3\ CO_2 \uparrow$$

The Fe in the ore gained electrons from the carbon, the reducing agent in this reaction.

Many chemical reactions can be classified as redox or nonredox reactions. Another way to classify chemical reactions is to consider what is happening to the reactants and products. This type of classification scheme leads to four basic categories of chemical reactions, which are (1) *combination,* (2) *decomposition,* (3) *replacement,* and (4) *ion exchange* reactions. The first three categories are subclasses of redox reactions. It is in the ion exchange reactions that you will find the first example of a reaction that is not a redox reaction.

Combination Reactions

A **combination reaction** is a synthesis reaction in which two or more substances combine to form a single compound. The combining substances can be (1) elements, (2) compounds, or (3) combinations of elements and compounds. In generalized form, a combination reaction is

$$X + Y \rightarrow XY$$

Many redox reactions are combination reactions. For example, metals are oxidized when they burn in air, forming a metal oxide. Consider magnesium, which gives off a bright white light as it burns:

$$2\ Mg_{(s)} + O_{2(g)} \rightarrow 2\ MgO_{(s)}$$

The rusting of metals is oxidation that takes place at a slower pace than burning, but metals are nonetheless oxidized in the process. Again noting the generalized form of a combination reaction, consider the rusting of iron:

$$4\ Fe_{(s)} + 3\ O_{2(g)} \rightarrow 2\ Fe_2O_{3(s)}$$

Nonmetals are also oxidized by burning in air, for example, when carbon burns with a sufficient supply of O_2:

$$C_{(s)} + O_{2(g)} \rightarrow CO_{2(g)}$$

Note that all the combination reactions follow the generalized form of $X + Y \rightarrow XY$.

Decomposition Reactions

A **decomposition reaction,** as the term implies, is the opposite of a combination reaction. In decomposition reactions, a compound is broken down (1) into the elements that make up the compound, (2) into simpler compounds, or (3) into elements and simpler compounds. Decomposition reactions have a generalized form of

$$XY \rightarrow X + Y$$

Decomposition reactions generally require some sort of energy, which is usually supplied in the form of heat or electrical energy. An electric current, for example, decomposes water into hydrogen and oxygen:

$$2\ H_2O_{(l)} \xrightarrow{\text{electricity}} 2\ H_{2(g)} + O_{2(g)}$$

Mercury(II) oxide is decomposed by heat, an observation that led to the discovery of oxygen:

$$2\ HgO_{(s)} \xrightarrow{\Delta} 2\ Hg_{(s)} + O_2 \uparrow$$

Note that all the decomposition reactions follow the generalized form of $XY \rightarrow X + Y$.

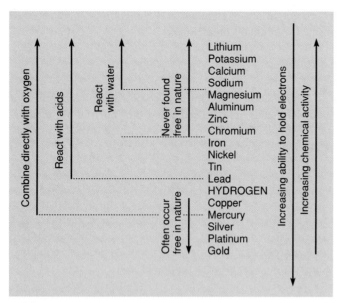

FIGURE 9.17 The activity series for common metals, together with some generalizations about the chemical activities of the metals. The series is used to predict which replacement reactions will take place and which reactions will not occur. (Note that hydrogen is not a metal and is placed in the series for reference to acid reactions.)

Replacement Reactions

In a **replacement reaction,** an atom or polyatomic ion is replaced in a compound by a different atom or polyatomic ion. The replaced part can be either the negative or positive part of the compound. In generalized form, a replacement reaction is

$$XY + Z \rightarrow XZ + Y$$

(negative part replaced)

or

$$XY + A \rightarrow AY + X$$

(positive part replaced)

Replacement reactions occur because some elements have a stronger electron-holding ability than other elements. Elements that have the least ability to hold on to their electrons are the most chemically active. Figure 9.17 shows a list of chemical activity of some metals, with the most chemically active at the top. Hydrogen is included because of its role in acids. Take a few minutes to look over the generalizations listed in figure 9.17. The generalizations apply to combination, decomposition, and replacement reactions.

Replacement reactions take place as more active metals give up electrons to elements lower on the list with a greater electron-holding ability. For example, aluminum is higher on the activity series than copper. When aluminum foil is placed in a solution of copper(II) chloride, aluminum is oxidized, losing electrons to the copper. The loss of electrons from metallic aluminum forms aluminum ions in solution, and the copper comes out of solution as a solid metal (figure 9.18).

$$2\,Al_{(s)} + 3\,CuCl_{2(aq)} \rightarrow 2\,AlCl_{3(aq)} + 3\,Cu_{(s)}$$

FIGURE 9.18 This shows a reaction between metallic aluminum and the blue solution of copper(II) chloride. Aluminum is above copper in the activity series, and aluminum replaces the copper ions from the solution as copper is deposited as a metal. The aluminum loses electrons to the copper and forms aluminum ions in solution.

A metal will replace any metal ion in solution that it is above in the activity series. If the metal is listed below the metal ion in solution, no reaction occurs. For example, $Ag_{(s)} + CuCl_{2(aq)} \rightarrow$ no reaction.

The very active metals (lithium, potassium, calcium, and sodium) react with water to yield metal hydroxides and hydrogen. For example,

$$2\,Na_{(s)} + 2\,H_2O_{(1)} \rightarrow 2\,NaOH_{(aq)} + H_2 \uparrow$$

Acids yield hydrogen ions in solution, and metals above hydrogen in the activity series will replace hydrogen to form a metal salt. For example,

$$Zn_{(s)} + H_2SO_{4(aq)} \rightarrow ZnSO_{4(aq)} + H_2 \uparrow$$

In general, the energy involved in replacement reactions is less than the energy involved in combination or decomposition reactions.

Ion Exchange Reactions

An **ion exchange reaction** is a reaction that takes place when the ions of one compound interact with the ions of another compound, forming (1) a solid that comes out of solution (a precipitate), (2) a gas, or (3) water.

A water solution of dissolved ionic compounds is a solution of ions. For example, solid sodium chloride dissolves in water to become ions in solution,

$$NaCl_{(s)} \rightarrow Na^+_{(aq)} + Cl^-_{(aq)}$$

If a second ionic compound is dissolved with a solution of another, a mixture of ions results. The formation of a precipitate, a gas, or water, however, removes ions from the solution,

The modern automobile produces two troublesome products in the form of (1) nitrogen monoxide and (2) hydrocarbons from the incomplete combustion of gasoline. These products from the exhaust enter the air to react in sunlight, eventually producing an irritating haze known as photochemical smog. To reduce photochemical smog, modern automobiles are fitted with a catalytic converter as part of their exhaust system (box figure 9.2).

Molecules require a certain amount of energy to change chemical bonds. This certain amount of energy is called the *activation energy,* and it represents an energy barrier that must be overcome before a chemical reaction can take place. This explains why chemical reactions proceed at a faster rate at higher temperatures. At higher temperatures, molecules have greater average kinetic energies; thus, they already have part of the minimum energy needed for a reaction to take place.

The rate at which a chemical reaction proceeds is affected by a *catalyst,* a material that speeds up a chemical reaction without being permanently changed by the reaction. A catalyst appears to speed a chemical reaction by lowering the activation energy. Molecules become temporarily attached to the surface of the catalyst, which weakens the chemical bonds holding the molecule together. The weakened molecule is easier to break apart and the activation energy is lowered. Some catalysts do this better with some specific compounds than others, and extensive chemical research programs are devoted to finding new and more effective catalysts.

Automobile catalytic converters use metals such as platinum and transition metal oxides such as copper(II) oxide and chromium(III) oxide. Catalytic reactions that occur in the converter can reduce or oxidize about 90 percent of the hydrocarbons, 85 percent of the carbon monoxide, and 40 percent of the nitrogen monoxide from exhaust gases. Other controls, such as exhaust gas recirculation, are used to reduce further nitrogen monoxide formation.

Box Figure 9.2 This silver-colored canister is the catalytic converter. The catalytic converter is located between the engine and the muffler, which is farther back toward the rear of the car.

and this must occur before you can say that an ionic exchange reaction has taken place. For example, water being treated for domestic use sometimes carries suspended matter that is removed by adding aluminum sulfate and calcium hydroxide to the water. The reaction is

$$3\, Ca(OH)_{2(aq)} + Al_2(SO_4)_{3(aq)} \rightarrow 3\, CaSO_{4(aq)} + 2\, Al(OH)_3 \downarrow$$

The aluminum hydroxide is a jellylike solid that traps the suspended matter for sand filtration. The formation of the insoluble aluminum hydroxide removed the aluminum and hydroxide ions from the solution, so an ion exchange reaction took place.

In general, an ion exchange reaction has the form

$$AX + BY \rightarrow AY + BX$$

where one of the products removes ions from the solution. The calcium hydroxide and aluminum sulfate reaction took place as the aluminum and calcium ions traded places. A solubility table such as the one in appendix B will tell you if an ionic exchange reaction has taken place. Aluminum hydroxide is insoluble, according to the table, so the reaction did take place. No ionic exchange reaction occurred if the new products are both soluble.

Another way for an ion exchange reaction to occur is if a gas or water molecule forms to remove ions from the solution. When an acid reacts with a base (an alkaline compound), a salt and water are formed

$$HCl_{(aq)} + NaOH_{(aq)} \rightarrow NaCl_{(aq)} + H_2O_{(l)}$$

The reactions of acids and bases are discussed in chapter 10.

CONCEPTS APPLIED

Chemical Reactions

Look around your school and home for signs that a chemical reaction has taken place. Can you find evidence that a reaction has taken place with oxygen? Can you find new substances being made or decomposition taking place?

Linus Carl Pauling (1901–1994)

Linus Pauling was a U.S. theoretical chemist and biologist whose achievements ranked among the most important of any in twentieth-century science. His main contribution was to the understanding of molecular structure and chemical bonding. He was one of the very few people to have been awarded two Nobel prizes: he received the 1954 Nobel Prize for chemistry (for his work on intermolecular forces) and the 1962 Peace Prize. Throughout his career, his work was noted for the application of intuition and inspiration, assisted by his phenomenal memory; he often carried over principles from one field of science to another.

Pauling was born in Portland, Oregon, on February 28, 1901, the son of a pharmacist. He began his scientific studies at Oregon State Agricultural College, from which he graduated in chemical engineering in 1922. He then began his research at the California Institute of Technology, Pasadena, gaining his Ph.D. in 1925. He became a full professor at Pasadena in 1931 and left there in 1936 to take up the post of director of the Gates and Crellin Laboratories, which he held for the next twenty-two years. He also held appointments at the University of California, San Diego, and Stanford University. His last appointment was as director of the Linus Pauling Institute of Science and Medicine at Menlo Park, California.

In 1931, Pauling published a classic paper, "The Nature of the Chemical Bond," in which he used quantum mechanics to explain that an electron-pair bond is formed by the interaction of two unpaired electrons, one from each of two atoms, and that once paired, these electrons cannot take part in the formation of other bonds. It was followed by the book *Introduction to Quantum Mechanics* (1935), of which he was coauthor. He was a pioneer in the application of quantum mechanical principles to the structures of molecules.

It was Pauling who introduced the concept of hybrid orbitals in molecules to explain the symmetry exhibited by carbon atoms in most of its compounds. Pauling also investigated electronegativity of atoms and polarization in chemical bonds. He assigned electronegativities on a scale up to 4.0. A pair of electrons in a bond is pulled preferentially toward an atom with a higher electronegativity. In hydrogen chloride (HCl), for example, hydrogen has an electronegativity of 2.1 and chlorine of 3.5. The bonding electrons are pulled toward the chlorine atom, giving it a small excess negative charge (and leaving the hydrogen atom with a small excess positive charge), polarizing the hydrogen-chlorine bond.

Pauling's ideas on chemical bonding are fundamental to modern theories of molecular structure. Much of this work was consolidated in his book *The Nature of the Chemical Bond, The Structure of Molecules and Crystals* (1939). In the 1940s, Pauling turned his attention to the chemistry of living tissues and systems. He applied his knowledge of molecular structure to the complexity of life, principally to proteins in blood. With Robert Corey, he worked on the structures of amino acids and polypeptides. They proposed that many proteins have structures held together with hydrogen bonds, giving them helical shapes. This concept assisted Francis Crick and James Watson in their search for the structure of DNA, which they eventually resolved as a double helix.

In his researches on blood, Pauling investigated immunology and sickle-cell anemia. Later work confirmed his hunch that the disease is genetic and that normal hemoglobin and the hemoglobin in abnormal "sickle" cells differ in electrical charge. Throughout the 1940s, he studied living materials; he also carried out research on anesthesia. At the end of this period, he published two textbooks, *General Chemistry* (1948) and *College Chemistry* (1950), which became best-sellers.

Source: Modified from the *Hutchinson Dictionary of Scientific Biography.* © RM, 2007. Reprinted by permission.

SUMMARY

Mixtures are made up of *unlike parts* with a *variable composition. Pure substances* are the *same throughout* and have a *definite composition.* Mixtures can be separated into their components by *physical changes,* changes that do not alter the identity of matter. Some pure substances can be broken down into simpler substances by a *chemical change,* a change that *alters the identity of matter as it produces new substances with different properties.* A pure substance that can be decomposed by chemical change into simpler substances with a definite composition is a *compound.* A pure substance that cannot be broken down into anything simpler is an *element.*

A *chemical change* produces new substances by making or breaking chemical bonds. The process of chemical change is called a *chemical reaction.* During a chemical reaction, different chemical substances with greater or lesser amounts of internal potential energy are produced. *Chemical energy* is the change of internal potential energy during a chemical reaction. A *chemical equation* is a shorthand way of describing a chemical reaction. An equation shows the substances that are changed, the *reactants,* on the left side and the new substances produced, the *products,* on the right side.

Chemical reactions involve *valence electrons,* the electrons in the outermost energy level of an atom. Atoms tend to lose or acquire electrons to achieve the configuration of the noble gases with stable, filled outer orbitals. This tendency is generalized as the *octet rule,* that atoms lose or gain electrons to acquire the noble gas structure of eight electrons in the outer orbital. Atoms form negative or positive *ions* in the process.

A chemical bond is an attractive force that holds atoms together in a compound. Chemical bonds formed when atoms transfer electrons to become ions are *ionic bonds*. An ionic bond is an electrostatic attraction between oppositely charged ions. Chemical bonds formed when ions share electrons are *covalent bonds.*

Ionic bonds result in *ionic compounds* with a crystalline structure. The energy released when an ionic compound is formed is called the *heat of formation*. It is the same amount of energy that is required to decompose the compound into its elements.

A *formula* of a compound uses symbols to tell what elements are in a compound and in what proportions. Ions of representative elements have a single, fixed charge, but many transition elements have variable charges. Electrons are conserved when ionic compounds are formed, and the ionic compound is electrically neutral. A formula shows the overall balance of charges.

Covalent compounds are molecular, composed of electrically neutral groups of atoms bound together by *covalent bonds*. The sharing of a pair of electrons, with each atom contributing a single electron to the shared pair, forms a *single covalent bond*. Covalent bonds formed when two pairs of electrons are shared are called *double bonds,* and a *triple bond* is the sharing of three pairs of electrons.

Compounds are named with different rules for ionic and covalent compounds. Both ionic and covalent compounds that are made up of only two different elements always end with an *-ide* suffix, but there are a few exceptions.

The rule for naming variable-charge ions states the English name and gives the charge with Roman numerals in parentheses.

Ionic compounds are electrically neutral, and formulas must show a balance of charge. The *crossover technique* is an easy way to write formulas that show a balance of charge.

Covalent compounds are molecules of two or more nonmetal atoms held together by a covalent bond. The system for naming covalent compounds uses Greek prefixes to identify the numbers of atoms, since more than one compound can form from the same two elements (CO and CO_2, for example).

A concise way to describe a chemical reaction is to use formulas in a *chemical equation*. A chemical equation with the same number of each kind of atom on both sides is called a *balanced equation*. A balanced equation is in accord with the *law of conservation of mass,* which states that atoms are neither created nor destroyed in a chemical reaction. To balance a chemical equation, *coefficients* are placed in front of chemical formulas. Subscripts of formulas may not be changed since this would change the formula, meaning a different compound.

One important group of chemical reactions is called *oxidation reduction reactions,* or *redox* reactions for short. Redox reactions are reactions where shifts of electrons occur. The process of losing electrons is called *oxidation,* and the substance doing the losing is said to be *oxidized*. The process of gaining electrons is called *reduction,* and the substance doing the gaining is said to be *reduced*. Substances that take electrons from other substances are called *oxidizing agents*. Substances that supply electrons are called *reducing agents*.

Chemical reactions can also be classified as (1) *combination,* (2) *decomposition,* (3) *replacement,* or (4) *ion exchange*. The first three of these are redox reactions, but ion exchange is not.

KEY TERMS

atom (p. **202**)
chemical bond (p. **205**)
chemical change (p. **201**)
chemical energy (p. **203**)
chemical equation (p. **204**)
chemical reaction (p. **203**)
combination reaction (p. **217**)
compound (p. **201**)
covalent bond (p. **207**)

covalent compound (p. **208**)
decomposition reaction (p. **217**)
double bond (p. **209**)
element (p. **201**)
formula (p. **207**)
heat of formation (p. **206**)
ion exchange reaction (p. **218**)
ionic bond (p. **206**)

ionic compounds (p. **207**)
law of conservation of mass (p. **214**)
molecule (p. **203**)
octet rule (p. **204**)
oxidation-reduction reaction (p. **216**)
oxidizing agents (p. **217**)
physical change (p. **201**)

polyatomic ion (p. **210**)
redox reaction (p. **216**)
reducing agent (p. **217**)
replacement reaction (p. **218**)
single bond (p. **209**)
triple bond (p. **209**)
valence electrons (p. **204**)

APPLYING THE CONCEPTS

1. Which of the following represents a chemical change?
   a. heating a sample of ice until it melts
   b. tearing a sheet of paper into tiny pieces
   c. burning a sheet of paper
   d. All of the above are correct.

2. A pure substance that cannot be decomposed into anything simpler by chemical or physical means is called a (an)
   a. element.
   b. compound.
   c. mixture.
   d. isotope.

3. The electrons that participate in chemical bonding are (the)
   a. valence electrons.
   b. electrons in fully occupied orbitals.
   c. stable inner electrons.
   d. All of the above are correct.

4. Which type of chemical bond is formed between two atoms by the sharing of electrons?
   a. ionic
   b. covalent
   c. metallic

5. Which combination of elements forms crystalline solids that will dissolve in water, producing a solution of ions that conduct an electric current?
   a. metal and metal
   b. metal and nonmetal
   c. nonmetal and nonmetal
   d. All of the above are correct.

6. An inorganic compound made of only two different elements has a name that always ends with the suffix
   a. *-ite.*   c. *-ide.*
   b. *-ate.*   d. *-ous.*

7. Dihydrogen monoxide is a compound with the common name of
   a. laughing gas.   c. smog.
   b. water.   d. rocket fuel.

8. A chemical equation is balanced by changing (the)
   a. subscripts.
   b. superscripts.
   c. coefficients.
   d. any of the above as necessary to achieve a balance.

9. Since wood is composed of carbohydrates, you should expect what gases to exhaust from a fireplace when complete combustion takes place?
   a. carbon dioxide, carbon monoxide, and pollutants
   b. carbon dioxide and water vapor
   c. carbon monoxide and smoke
   d. It depends on the type of wood being burned.

10. When carbon burns with an insufficient supply of oxygen, carbon monoxide is formed according to the following equation: $2 C + O_2 \rightarrow 2 CO$. What category of chemical reaction is this?
    a. combination
    b. ion exchange
    c. replacement
    d. None of the above is correct.

11. Of the elements listed below, the one with the greatest chemical activity is
    a. aluminum.
    b. zinc.
    c. iron.
    d. mercury.

12. A balanced chemical equation has
    a. the same number of molecules on both sides of the equation.
    b. the same kinds of molecules on both sides of the equation.
    c. the same number of each kind of atom on both sides of the equation.
    d. All of the above are correct.

Answers

1. c 2. a 3. a 4. b 5. b 6. c 7. b 8. c 9. b 10. a 11. a 12. c

QUESTIONS FOR THOUGHT

1. What is the difference between a chemical change and a physical change? Give three examples of each.

2. Describe how the following are alike and how they are different: (a) a sodium atom and a sodium ion, and (b) a sodium ion and a neon atom.

3. What is the difference between an ionic and covalent bond? What do atoms forming the two bond types have in common?

4. What is the octet rule?

5. What is a polyatomic ion? Give the names and formulas for several common polyatomic ions.

6. Write the formula for magnesium hydroxide. Explain what the parentheses mean.

7. What is the basic difference between a single bond and a double bond?

8. What is the law of conservation of mass? How do you know if a chemical equation is in accord with this law?

9. Describe in your own words how a chemical equation is balanced.

10. How is the activity series for metals used to predict if a replacement reaction will occur or not?

11. What must occur in order for an ion exchange reaction to take place? What is the result if this does not happen?

12. Predict the products for the following reactions: (a) the combustion of ethyl alcohol (C_2H_5OH), and (b) the rusting of aluminum (Al).

FOR FURTHER ANALYSIS

1. What are the significant similarities and differences between a physical change and a chemical change?

2. Analyze how you would know for sure that a pure substance you have is a compound and not an element.

3. Analyze how you would know for sure that a pure substance you have is an element and not a compound.

4. Make up an explanation for why ionic compounds are formed when a metal from the left side of the periodic table reacts with a nonmetal from the right side, but covalent bonds are formed between nonmetallic elements on the right side of the table.

5. What are the advantages and disadvantages to writing a chemical equation with chemical symbols and formulas rather than just words?

6. Provide several examples of each of the four basic categories of chemical reactions and describe how each illustrates a clear representation of the category.

7. Summarize for another person the steps needed for successfully writing a balanced chemical equation.

INVITATION TO INQUIRY

Rate of Chemical Reactions

Temperature is one of the more important factors that influence the rate of a chemical reaction. You can use a "light stick" or "light tube" to study how temperature can influence a chemical reaction. Light sticks and tubes are devices that glow in the dark and are very popular on July 4 and at other times when people might be outside after sunset. They work from a chemical reaction that is similar to the chemical reaction that produces light in a firefly. Design an experiment that uses light sticks to find out the effect of temperature on the brightness of light and how long the device will continue providing light. Perhaps you will be able to show by experimental evidence that use at a particular temperature produces the most light for the longest period of time.

PARALLEL EXERCISES

The exercises in groups A and B cover the same concepts. Solutions to group A exercises are located in appendix D.

Group A

1. How many outer orbital electrons are found in an atom of:
   a. Li
   b. N
   c. F
   d. Cl
   e. Ra
   f. Be

2. Write electron dot notations for the following elements:
   a. Boron
   b. Bromine
   c. Calcium
   d. Potassium
   e. Oxygen
   f. Sulfur

3. Identify the charge on the following ions:
   a. Boron
   b. Bromine
   c. Calcium
   d. Potassium
   e. Oxygen
   f. Nitrogen

4. Name the following polyatomic ions:
   a. $(OH)^-$
   b. $(SO_3)^{2-}$
   c. $(ClO)^-$
   d. $(NO_3)^-$
   e. $(CO_3)^{2-}$
   f. $(ClO_4)^-$

5. Use the crossover technique to write formulas for the following compounds:
   a. Iron(III) hydroxide
   b. Lead(II) phosphate
   c. Zinc carbonate
   d. Ammonium nitrate
   e. Potassium hydrogen carbonate
   f. Potassium sulfite

6. Write formulas for the following covalent compounds:
   a. Carbon tetrachloride
   b. Dihydrogen monoxide
   c. Manganese dioxide
   d. Sulfur trioxide
   e. Dinitrogen pentoxide
   f. Diarsenic pentasulfide

Group B

1. How many outer orbital electrons are found in an atom of:
   a. Na
   b. P
   c. Br
   d. I
   e. Te
   f. Sr

2. Write electron dot notations for the following elements:
   a. Aluminum
   b. Fluorine
   c. Magnesium
   d. Sodium
   e. Carbon
   f. Chlorine

3. Identify the charge on the following ions:
   a. Aluminum
   b. Chlorine
   c. Magnesium
   d. Sodium
   e. Sulfur
   f. Hydrogen

4. Name the following polyatomic ions:
   a. $(C_2H_3O_2)^-$
   b. $(HCO_3)^-$
   c. $(SO_4)^{2-}$
   d. $(NO_2)^-$
   e. $(MnO_4)^-$
   f. $(CO_3)^{2-}$

5. Use the crossover technique to write formulas for the following compounds:
   a. Aluminum hydroxide
   b. Sodium phosphate
   c. Copper(II) chloride
   d. Ammonium sulfate
   e. Sodium hydrogen carbonate
   f. Cobalt(II) chloride

6. Write formulas for the following covalent compounds:
   a. Silicon dioxide
   b. Dihydrogen sulfide
   c. Boron trifluoride
   d. Dihydrogen dioxide
   e. Carbon tetrafluoride
   f. Nitrogen trihydride

7. Name the following covalent compounds:
   a. CO
   b. CO_2
   c. CS_2
   d. N_2O
   e. P_4S_3
   f. N_2O_3

8. Write balanced chemical equations for each of the following unbalanced reactions:
   a. $SO_2 + O_2 \rightarrow SO_3$
   b. $P + O_2 \rightarrow P_2O_5$
   c. $Al + HCl \rightarrow AlCl_3 + H_2$
   d. $NaOH + H_2SO_4 \rightarrow Na_2SO_4 + H_2O$
   e. $Fe_2O_3 + CO \rightarrow Fe + CO_2$
   f. $Mg(OH)_2 + H_3PO_4 \rightarrow Mg_3(PO_4)_2 + H_2O$

9. Identify the following as combination, decomposition, replacement, or ion exchange reactions:
   a. $NaCl_{(aq)} + AgNO_{3(aq)} \rightarrow NaNO_{3(aq)} + AgCl\downarrow$
   b. $H_2O_{(l)} + CO_{2(g)} \rightarrow H_2CO_{3(l)}$
   c. $2\ NaHCO_{3(s)} \rightarrow Na_2CO_{3(s)} + H_2O_{(g)} + CO_{2(g)}$
   d. $2\ Na_{(s)} + Cl_{2(g)} \rightarrow 2\ NaCl_{(s)}$
   e. $Cu_{(s)} + 2\ AgNO_{3(aq)} \rightarrow Cu(NO_3)_{2(aq)} + 2\ Ag_{(s)}$
   f. $CaO_{(s)} + H_2O_{(l)} \rightarrow Ca(OH)_{2(aq)}$

10. Write complete, balanced equations for each of the following reactions:
   a. $C_5H_{12(g)} + O_{2(g)} \rightarrow$
   b. $HCl_{(aq)} + NaOH_{(aq)} \rightarrow$
   c. $Al_{(s)} + Fe_2O_{3(s)} \rightarrow$
   d. $Fe_{(s)} + CuSO_{4(aq)} \rightarrow$
   e. $MgCl_{(aq)} + Fe(NO_3)_{2(aq)} \rightarrow$
   f. $C_6H_{10}O_{5(s)} + O_{2(g)} \rightarrow$

7. Name the following covalent compounds:
   a. N_2O
   b. SO_2
   c. SiC
   d. PF_5
   e. $SeCl_6$
   f. N_2O_4

8. Write balanced chemical equations for each of the following unbalanced reactions:
   a. $NO + O_2 \rightarrow NO_2$
   b. $KClO_3 \rightarrow KCl + O_2$
   c. $NH_4Cl + Ca(OH)_2 \rightarrow CaCl_2 + NH_3 + H_2O$
   d. $NaNO_3 + H_2SO_4 \rightarrow Na_2SO_4 + HNO_3$
   e. $PbS + H_2O_2 \rightarrow PbSO_4 + H_2O$
   f. $Al_2(SO_4)_3 + BaCl_2 \rightarrow AlCl_3 + BaSO_4$

9. Identify the following as combination, decomposition, replacement, or ion exchange reactions:
   a. $ZnCO_{3(s)} \rightarrow ZnO_{(s)} + CO_2\uparrow$
   b. $2\ NaBr_{(aq)} + Cl_{2(g)} \rightarrow 2\ NaCl_{(aq)} + Br_{2(g)}$
   c. $2\ Al_{(s)} + 3\ Cl_{2(g)} \rightarrow 2\ AlCl_{3(s)}$
   d. $Ca(OH)_{2(aq)} + H_2SO_{4(aq)} \rightarrow CaSO_{4(aq)} + 2\ H_2O_{(l)}$
   e. $Pb(NO_3)_{2(aq)} + H_2S_{(g)} \rightarrow 2\ HNO_{3(aq)} + PbS\downarrow$
   f. $C_{(s)} + ZnO_{(s)} \rightarrow Zn_{(s)} + CO\uparrow$

10. Write complete, balanced equations for each of the following reactions:
   a. $C_3H_{6(g)} + O_{2(g)} \rightarrow$
   b. $H_2SO_{4(aq)} + KOH_{(aq)} \rightarrow$
   c. $C_6H_{12}O_{6(s)} + O_{2(g)} \rightarrow$
   d. $Na_3PO_{4(aq)} + AgNO_{3(aq)} \rightarrow$
   e. $NaOH_{(aq)} + Al(NO_3)_{3(aq)} \rightarrow$
   f. $Mg(OH)_{2(aq)} + H_3PO_{4(aq)} \rightarrow$

Water is often referred to as the *universal solvent* because it makes so many different kinds of solutions. Eventually, moving water can dissolve solid rock, carrying it away in solution.

Water and Solutions

Earth Science Connections

▶ Water cycles in and out of the atmosphere (Ch. 7).

▶ Less than 1 percent of all water on Earth is fit for human consumption or agriculture (Ch. 18).

CORE CONCEPT

Water and solutions of water have unique properties.

Water is a universal solvent and has a high specific heat and a high latent heat of vaporization (p. 227).

A water molecule is polar and able to establish hydrogen bonding (p. 228).

Solutions of acids, bases, and salts are evident in environmental quality, food, and everyday living (pp. 234–40).

OVERVIEW

What do you think about when you see a stream (figure 10.1)? Do you wonder about the water quality and what might be dissolved in the water? Do you wonder where the stream comes from and if it will ever run out of water?

Many people can look at a stream, but they might think about different things. A farmer might think about how the water could be diverted and used for his crops.

A city planner might wonder if the water is safe for domestic use, and if not, what it would cost to treat the water. Others might wonder if the stream has large fish they could catch. Many large streams can provide water for crops, domestic use, and recreation, and still meet the requirements for a number of other uses.

It is the specific properties of water that make it important for agriculture, domestic use, and recreation. Living things evolved in a watery environment, so water and its properties are essential to life on Earth. Some properties of water, such as the ability to dissolve almost anything, also make water very easy to pollute. This chapter is concerned with some of the unique properties of water, water solutions, and household use of water.

HOUSEHOLD WATER

Water is an essential resource, not only because it is required for life processes but also because of its role in a modern society. (See chapter 18.) Water is used in the home for drinking and cooking (2 percent), washing dishes (6 percent), laundry (11 percent), bathing (23 percent), toilets (29 percent), and maintaining lawns and gardens (29 percent).

The water supply is obtained from streams, lakes, and reservoirs on the surface or from groundwater pumped from below the surface. Surface water contains more sediments, bacteria, and possible pollutants than water from a well because it is exposed to the atmosphere and water runs off the land into streams and rivers. Surface water requires filtering to remove suspended particles, treatment to kill bacteria, and sometimes processing to remove pollution. Well water is generally cleaner but still might require treatment to kill bacteria and remove pollution that has seeped through the ground from waste dumps, agricultural activities, or industrial sites.

Most pollutants are usually too dilute to be considered a significant health hazard, but there are exceptions. There are five types of contamination found in U.S. drinking water that are responsible for the most widespread danger, and these are listed in table 10.1. In spite of these general concerns and other occasional local problems, the U.S. water supply is considered to be among the cleanest in the world.

TABLE 10.1 Possible Pollution Problems in the U.S. Water Supply

Pollutant	Source	Risk
Lead	Lead pipes in older homes; solder in copper pipes, brass fixtures	Nerve damage, miscarriage, birth defects, high blood pressure, hearing problems
Chlorinated solvents	Industrial pollution	Cancer
Trihalomethanes	Chlorine disinfectant reacting with other pollutants	Liver damage, kidney damage, possibly cancer
PCBs	Industrial waste, older transformers	Liver damage, possibly cancer
Bacteria and viruses	Septic tanks, outhouses, overflowing sewer lines	Gastrointestinal problems, serious diseases

FIGURE 10.1 A freshwater stream has many potential uses.

Science and Society

Who Has the Right?

As the population grows and new industries develop, more and more demands are placed on the water supply. This raises some issues about how water should be divided among agriculture, industries, and city domestic use. Agricultural interests claim they should have the water because they produce the food and fibers that people must have. Industrial interests claim they should have the water because they create the jobs and the products that people must have. Cities, on the other hand, claim that domestic consumption is the most important because people cannot survive without water. Yet, others claim that no group has a right to use water when it is needed to maintain habitats.

Questions to Discuss:

1. Who should have the first priority for water use?

Demand for domestic water sometimes exceeds the immediate supply in some growing metropolitan areas. This is most common during the summer, when water demand is high and rainfall is often low. Communities in these areas often have public education campaigns designed to help reduce the demand for water. For example, did you know that taking a tub bath can use up to 135 liters (about 36 gal) of water compared to only 95 liters (about 25 gal) for a regular shower? Even more water is saved by a shower that does not run continuously—wetting down, soaping up, and rinsing off uses only 15 liters (about 4 gal) of water. You can also save about 35 liters (about 9 gal) of water by not letting the water run continuously while brushing your teeth.

It is often difficult to convince people to conserve water when it is viewed as an inexpensive, limitless supply. However, efforts to conserve water increase dramatically as the cost to the household consumer increases.

The issues involved in maintaining a safe water supply are better understood by considering some of the properties of water and water solutions. These are the topics of the following sections.

PROPERTIES OF WATER

Water is essential for life since living organisms are made up of cells filled with water and a great variety of dissolved substances. Foods are mostly water, with fruits and vegetables containing up to 95 percent water and meat consisting of about 50 percent water. Your body is over 70 percent water by weight. Since water is such a large component of living things, understanding the properties of water is important to understanding life. One important property is water's unusual ability to act as a solvent. Water is called a "universal solvent" because of its ability to dissolve most molecules. In living things, these dissolved molecules can be transported from one place to another by diffusion or by some kind of a circulatory system.

The usefulness of water does not end with its unique abilities as a solvent and transporter; it has many more properties that are useful, although unusual. For example, unlike other liquids, water in its liquid phase has a greater density than solid water (ice). This important property enables solid ice to float on the surface of liquid water, insulating the water below and permitting fish and other water organisms to survive the winter. If ice were denser than water, it would sink, freezing all lakes and rivers from the bottom up. Fish and most organisms that live in water would not be able to survive in a lake or river of solid ice.

As described in chapter 4, water is also unusual because it has a high specific heat. The same amount of sunlight falling on equal masses of soil and water will warm the soil 5°C for each 1°C increase in water temperature. Thus, it will take five times more sunlight to increase the temperature of the water to as much as the soil. This enables large bodies of water to moderate the temperature, making it more even.

A high latent heat of vaporization is yet another unusual property of water. This particular property enables people to dissipate large amounts of heat by evaporating a small amount of water. Since people carry this evaporative cooling system with them, they can survive some very warm desert temperatures, for example.

Finally, other properties of water are not crucial for life but are interesting nonetheless. For example, why do all snowflakes have six sides? Is it true that no two snowflakes are alike? The unique structure of the water molecule will explain water's unique solvent abilities, why solid water is less dense than liquid water, its high specific heat, its high latent heat of vaporization, and perhaps why no two snowflakes seem to be alike.

Structure of the Water Molecule

In chapter 9, you learned that atoms combine two ways. Atoms from opposite sides of the periodic table form ionic bonds after transferring one or more electrons. Atoms from the right side of the periodic table form covalent bonds by sharing one or more pairs of electrons. This distinction is clear-cut in many compounds but not in water. The way atoms share electrons in a water molecule is not exactly covalent, but it is not ionic either.

In a molecule of water, an oxygen atom shares a pair of electrons with two hydrogen atoms. Oxygen has six outer electrons and needs two more to satisfy the octet rule, achieving the noble gas structure of eight. Each hydrogen atom needs one more electron to fill its outer orbital with two. Therefore, one oxygen atom bonds with two hydrogen atoms, forming H_2O. Both oxygen and hydrogen are more stable with the outer orbital configuration of the noble gases (neon and helium in this case).

Electrons are shared in a water molecule but not equally. Oxygen, with its eight positive protons, has a greater attraction for the shared electrons than do either of the hydrogens with a single proton. Therefore, the shared electrons spend more time around the oxygen part of the molecule than they do around the hydrogen part. This results in the oxygen end of the molecule being more negative than the hydrogen end. When electrons in a covalent bond are not equally shared, the molecule is said to be polar. A **polar molecule** has a *dipole* (*di* = two; *pole* = side or end), meaning it has a positive end and a negative end.

A water molecule has a negative center at the oxygen end and a positive center at the hydrogen end. The positive charges on the hydrogen end are separated, giving the molecule a bent arrangement rather than a straight line. Figure 10.2A shows a model of a water molecule showing its polar nature.

It is the polar structure of the water molecule that is responsible for many of the unique properties of water. Polar molecules of any substances have attractions between the positive end of a molecule and the negative end of another molecule. When the polar molecule has hydrogen at one end and fluorine, oxygen, or nitrogen on the other part, the attractions are strong enough to make a type of bonding called **hydrogen bonding.** Hydrogen bonding is a bond that occurs between the hydrogen end of a molecule and the fluorine, oxygen, or nitrogen end of other similar molecules. A better name for this would be a hydrogen-fluorine bond, a hydrogen-oxygen bond, or a hydrogen-nitrogen bond. However, for brevity, the second part of the bond is not named and all the hydrogen-something bonds are simply known of as "hydrogen" bonds. The dotted line between the hydrogen and oxygen molecules in

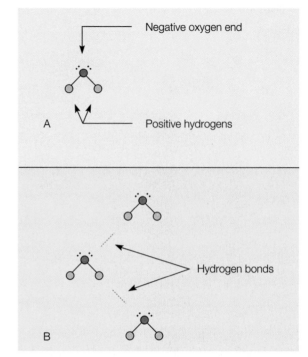

FIGURE 10.2 (*A*) The water molecule is polar, with centers of positive and negative charges. (*B*) Attractions between these positive and negative centers establish hydrogen bonds between adjacent molecules.

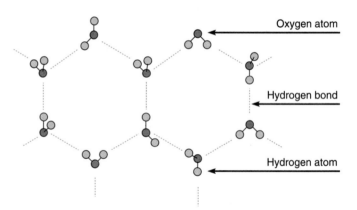

FIGURE 10.3 The hexagonal structure of ice. Hydrogen bonding between the oxygen atom and two hydrogen atoms of other water molecules results in an arrangement, which forms the open, hexagonal structure of ice. Note the angles of the water molecules do not change but have different orientations.

figure 10.2B represents a hydrogen bond. A dotted line is used to represent a bond that is not as strong as the bond represented by the solid line of a covalent compound.

Hydrogen bonding accounts for the physical properties of water, including its unusual density changes with changes in temperature. Figure 10.3 shows the hydrogen-bonded structure of ice. Water molecules form a six-sided hexagonal structure that extends out for billions of molecules. The large channels, or holes, in the structure result in ice being less dense than water. The shape of the hexagonal arrangement also suggests why snowflakes always have six sides. Why does it seem like no two snowflakes are alike? Perhaps the answer can be found in the almost infinite variety of shapes that can be built from billions and billions of tiny hexagons of ice crystals.

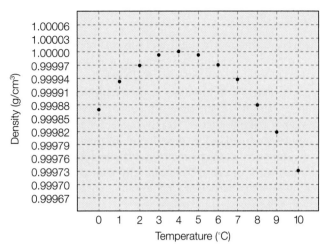

FIGURE 10.4 The density of water from 0°C to 10°C. The density of water is at a maximum at 4°C, becoming less dense as it is cooled or warmed from this temperature. Hydrogen bonding explains this unusual behavior.

Myths, Mistakes, and Misunderstandings

Teardrops Keep Falling?

It is a mistake to represent raindrops or drops of falling water with teardrop shapes. Small raindrops are pulled into a spherical shape by surface tension. Larger raindrops are also pulled into a spherical shape, but the pressure of air on the bottom of the falling drop somewhat flattens the bottom. If the raindrop is too large, the pressure of air on the falling drop forms a concave depression on the bottom, which grows deeper and deeper until the drop breaks up into smaller spherical drops.

When ice is warmed, the increased vibrations of the molecules begin to expand and stretch the hydrogen bond structure. When ice melts, about 15 percent of the hydrogen bonds break and the open structure collapses into the more compact arrangement of liquid water. As the liquid water is warmed from 0°C, still more hydrogen bonds break down, and the density of the water steadily increases. At 4°C, the expansion of water from the increased molecular vibrations begins to predominate, and the density decreases steadily with further warming (figure 10.4). Thus, water has its greatest density at a temperature of 4°C.

The heat of fusion, specific heat, and heat of vaporization of water are unusually high when compared to other chemically similar substances. These high values are accounted for by the additional energy needed to break hydrogen bonds.

The Dissolving Process

A **solution** is a homogeneous mixture of ions or molecules of two or more substances. *Dissolving* is the process of making a solution. During dissolving, the different components that make up the solution become mixed. For example, when sugar dissolves in water, molecules of sugar become uniformly dispersed throughout the molecules of water. The uniform taste of sweetness of any part of the sugar solution is a result of this uniform mixing.

The general terms of *solvent* and *solute* identify the components of a solution. The solvent is the component present in the larger amount. The solute is the component that dissolves in the solvent. Atmospheric air, for example, is about 78 percent nitrogen, so nitrogen is considered the solvent. Oxygen (about 21 percent), argon (about 0.9 percent), and other gases make up the solutes. If one of the components of a solution is a liquid, it is usually identified as the solvent. An *aqueous solution* is a solution of a solid, a liquid, or a gas in water.

A solution is formed when the molecules or ions of two or more substances become homogeneously mixed. But the process of dissolving must be more complicated than the simple mixing together of particles because (1) solutions become saturated, meaning there is a limit on solubility, and (2) some substances are *insoluble*, not dissolving at all or at least not noticeably. In general, the forces of attraction between molecules or ions of the solvent and solute determine if something will dissolve and if there are limits on the solubility. These forces of attraction and their role in the dissolving process will be considered in the following examples.

First, consider the dissolving process in gaseous and liquid solutions. In a gas, the intermolecular forces are small, so gases can mix in any proportion. Fluids that can mix in any proportion like this are called **miscible fluids.** Fluids that do not mix are called *immiscible fluids.* Air is a mixture of gases, so gases (including vapors) are miscible.

Liquid solutions can dissolve a gas, another liquid, or a solid. Gases are miscible in liquids, and a carbonated beverage (your favorite cola) is the common example, consisting of carbon dioxide dissolved in water. Whether or not two given liquids form solutions depends on some similarities in their molecular structures. The water molecule, for example, is a polar molecule with a negative end and a positive end. On the other hand, carbon tetrachloride (CCl_4) is a molecule with polar bonds that are symmetrically arranged. Because of the symmetry, CCl_4 has no negative or positive ends, so it is nonpolar. Thus, some liquids have polar molecules, and some have nonpolar molecules. The general rule for forming solutions is *like dissolves like.* A nonpolar compound, such as carbon tetrachloride, will dissolve oils and greases because they are nonpolar compounds. Water, a polar compound, will not dissolve the nonpolar oils and greases. Carbon tetrachloride was at one time used as a cleaning solvent because of its oil and grease dissolving abilities. Its use is no longer recommended because it causes liver damage.

Some molecules, such as soap, have a part of the molecule that is polar and a part that is nonpolar. Washing with water alone will not dissolve oils because water and oil are immiscible. When soap is added to the water, however, the polar end of the soap molecule is attracted to the polar water molecules, and the nonpolar end is absorbed into the oil. A particle (larger than a molecule) is formed, and the oil is washed away with the water.

The "like dissolves like" rule applies to solids and liquid solvents as well as liquids and liquid solvents. Polar solids, such as salt, will readily dissolve in water, which has polar molecules, but do not dissolve readily in oil, grease, or other nonpolar solvents. Polar water readily dissolves salt because the charged polar water molecules are able to exert an attraction on the ions, pulling them away from the crystal structure. Thus, ionic compounds dissolve in water.

Ionic compounds vary in their solubilities in water. This difference is explained by the existence of two different forces involved in an ongoing "tug of war." One force is the attraction between an ion on the surface of the crystal and a water molecule, an *ion-polar molecule force.* When solid sodium chloride and water are mixed together, the negative ends of the water molecules (the oxygen ends) become oriented toward the positive sodium ions on the crystal. Likewise, the positive ends of water molecules (the hydrogen ends) become oriented toward the negative chlorine ions. The attraction of water molecules for ions is called **hydration.** If the force of hydration is greater than the attraction between the ions in the solid, they are pulled away from the solid, and dissolving occurs (figure 10.5). Considering sodium chloride only, the equation is

$$Na^+Cl^-_{(s)} \rightarrow Na^+_{(aq)} + Cl^-_{(aq)}$$

which shows that the ions were separated from the solid to become a solution of ions. In other compounds, the attraction between the ions in the solid might be greater than the energy of hydration. In this case, the ions of the solid would win the "tug of war," and the ionic solid is insoluble.

The saturation of soluble compounds is explained in terms of hydration eventually occupying a large number of the polar water molecules. Fewer available water molecules means less attraction on the ionic solid, with more solute ions being pulled back to the surface of the solid. The tug of war continues back and forth as an equilibrium condition is established.

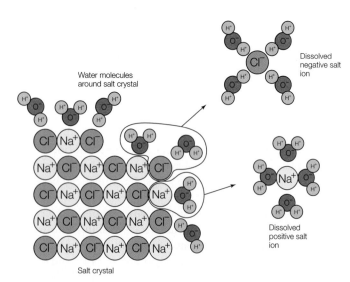

FIGURE 10.5 An ionic solid dissolves in water because the number of water molecules around the surface is greater than the number of other ions of the solid. The attraction between polar water molecules and a charged ion enables the water molecules to pull ions away from the crystal so the salt crystals dissolve in the water.

Concentration of Solutions

The relative amounts of solute and solvent are described by the concentration of a solution. In general, a solution with a large amount of solute is *concentrated,* and a solution with much less solute is *dilute.* The terms *dilute* and *concentrated* are somewhat arbitrary, and it is sometimes difficult to know the difference between a solution that is "weakly concentrated" and one that is "not very diluted." More meaningful information is provided by measurement of the *amount of solute in a solution.* There are different ways to express concentration measurements, each lending itself to a particular kind of solution or to how the information will be used. For example, you read about concentrations of parts per million in an article about pollution, but most of the concentration of solutions sold in stores are reported in percent by volume or percent by weight. Each of these concentrations is concerned with the amount of *solute* in the *solution.*

Concentration ratios that describe small concentrations of solute are sometimes reported as a ratio of *parts per million* (ppm) or *parts per billion* (ppb). This ratio could mean ppm by volume or ppm by weight, depending on whether the solution is a gas or a liquid. For example, a drinking water sample with 1 ppm Na^+ by weight has 1 weight measure of solute, sodium ions, *in* every 1,000,000 weight measures of the total solution. By way of analogy, 1 ppm expressed in money means 1 cent in every $10,000 (which is 1 million cents). A concentration of 1 ppb means 1 cent in $10,000,000. Thus, the concentrations of very dilute solutions, such as certain salts in seawater, minerals in drinking water, and pollutants in water or in the atmosphere, are often reported in ppm or ppb.

The concentration term of *percent by volume* is defined as the *volume of solute in 100 volumes of solution.* This concentration term is just like any other percentage ratio, that is, "part" divided by the "whole" times 100 percent. The distinction is that the part and the whole are concerned with a volume of solute and a volume of solution. Knowing the meaning of percent by volume can be useful in consumer decisions. Rubbing alcohol, for example, can be purchased at a wide range of prices. The various brands range from a concentration, according to the labels, of "12% by volume" to "70% by volume." If the volume unit is mL, a "12% by volume" concentration contains 12 mL of pure isopropyl (rubbing) alcohol in every 100 mL of solution. The "70% by volume" contains 70 mL of isopropyl alcohol in every 100 mL of solution. The relationship for percent by volume is

$$\frac{\text{volume solute}}{\text{volume solution}} \times 100\% \text{ solution} = \% \text{ solute}$$

The concentration term of *percent by weight* is defined as the *weight of solute in 100 weight units of solution.* This concentration term is just like any other percentage composition, the difference being that it is concerned with the weight of solute (the part) in a weight of solution (the whole). Hydrogen peroxide, for example, is usually sold in a concentration of "3% by weight." This means that 3 oz (or other weight units) of pure hydrogen peroxide are in 100 oz of solution. Since weight is proportional to mass in a given location, mass units such as

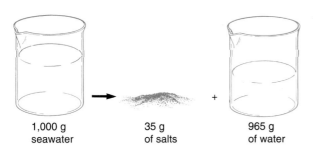

FIGURE 10.6 Salinity is a measure of the amount of salts dissolved in 1 kg of solution. If 1,000 g of seawater were evaporated, 35.0 g of salts would remain as 965.0 g of water leave.

1,000 g
seawater

35 g
of salts

965 g
of water

grams are sometimes used to calculate a percent by weight. The relationship for percent by weight (using mass units) is

$$\frac{\text{mass of solute}}{\text{mass of solution}} \times 100\% \text{ solution} = \% \text{ solute}$$

Both percent by volume and percent by weight are defined as the volume or weight per 100 units of solution because percent *means* parts per hundred. The measure of dissolved salts in seawater is called *salinity*. **Salinity** is defined as the mass of salts dissolved in 1,000 g of solution. As illustrated in figure 10.6, evaporation of 965 g of water from 1,000 g of seawater will leave an average of 35 g salts. Thus, the average salinity of the seawater is 35‰. Note the ‰, which means parts per thousand just as % means parts per hundred. Thus, the average salinity of seawater is 35‰, which means there are 35 g of salts dissolved in every 1,000 g of seawater. The equivalent percent measure for salinity is 3.5%, which equals 350‰.

A **mole** is a measure of amount used in chemistry. One mole is defined as the amount of a substance that contains the same number of elementary units as there are atoms in exactly 12 grams of the carbon-12 isotope. The number of units in this case is called *Avogadro's number,* which is 6.02×10^{23}—a very large number. This measure can be compared with identifying amounts in the grocery store by the dozen. You know that a dozen is 12 of something. Now you know that a *mole* is 6.02×10^{23} of whatever you are measuring.

Chemists use a measure of concentration that is convenient for considering chemical reactions of solutions. The measure is based on moles of solute, since a mole is a known number of particles (atoms, molecules, or ions). The concentration term of **molarity** (M) is defined as the number of moles of solute dissolved in one liter of solution. Thus,

$$\frac{\text{molarity}}{\text{(M)}} = \frac{\text{moles of solute}}{\text{liters of solution}}$$

An aqueous solution of NaCl that has a molarity of 1 contains 1 mole of NaCl per liter of solution. To make such a solution, you would place 58.5 g (1.0 mole) NaCl in a beaker, then add water to make 1 liter of solution.

Solubility

Gases and liquids appear to be soluble in all proportions, but there is a limit to how much solid can be dissolved in a liquid. You may have noticed that a cup of hot tea will dissolve several

Connections . . .

Ppm or Ppb to Percent

Sometimes it is useful to know the conversion factors between ppm or ppb and the more familiar percent concentration by weight. These factors are ppm ÷ (1×10^4) = percent concentration and ppb ÷ (1×10^7) = percent concentration. For example, very hard water (water containing Ca^{2+} or Mg^{2+} ions), by definition, contains more than 300 ppm of the ions. This is a percent concentration of 300 ÷ 1 = 10^4, or 0.03 percent. To be suitable for agricultural purposes, irrigation water must not contain more than 700 ppm of total dissolved salts, which means a concentration no greater than 0.07 percent salts.

teaspoons of sugar, but the limit of solubility is reached quickly in a glass of iced tea. The limit of how much sugar will dissolve seems to depend on the temperature of the tea. More sugar added to the cold tea after the limit is reached will not dissolve, and solid sugar granules begin to accumulate at the bottom of the glass. At this limit, the sugar and tea solution is said to be *saturated*. Dissolving does not actually stop when a solution becomes saturated and undissolved sugar continues to enter the solution. However, dissolved sugar is now returning to the undissolved state at the same rate as it is dissolving. The overall equilibrium condition of sugar dissolving as sugar is coming out of solution is called a **saturated solution.** A saturated solution is a *state of equilibrium that exists between dissolving solute and solute coming out of solution.* You actually cannot see the dissolving and coming out of solution that occurs in a saturated solution because the exchanges are taking place with particles the size of molecules or ions.

Not all compounds dissolve as sugar does, and more or less of a given compound may be required to produce a saturated solution at a particular temperature. In general, the difficulty of dissolving a given compound is referred to as *solubility*. More specifically, the **solubility** of a solute is defined as the *concentration that is reached in a saturated solution at a particular temperature.* Solubility varies with the temperature, as the sodium and potassium salt examples show in figure 10.7. These solubility curves describe the amount of solute required to reach the saturation equilibrium at a particular temperature. In general, the solubilities of most ionic solids increase with temperature, but there are exceptions. In addition, some salts release heat when dissolved in water, and other salts absorb heat when dissolved. The "instant cold pack" used for first aid is a bag of water containing a second bag of ammonium nitrate (NH_4NO_3). When the bag of ammonium nitrate is broken, the compound dissolves and absorbs heat.

You can usually dissolve more of a solid, such as salt or sugar, as the temperature of the water is increased. Contrary to what you might expect, gases usually become *less* soluble in water as the temperature increases. As a glass of water warms, small bubbles collect on the sides of the glass as dissolved air comes out of solution. The first bubbles that appear when warming a pot of water to boiling are also bubbles of dissolved air coming out of solution. This is why water that has been boiled usually tastes "flat." The dissolved air has been removed by the heating. The

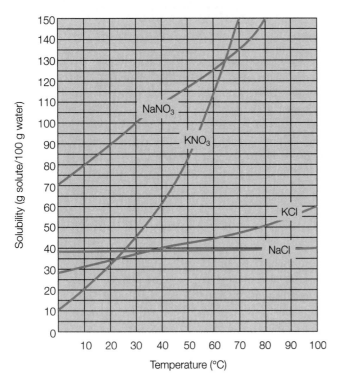

FIGURE 10.7 Approximate solubility curves for sodium nitrate, potassium nitrate, potassium chloride, and sodium chloride.

"normal" taste of water can be restored by pouring the boiled water back and forth between two glasses. The water dissolves more air during this process, restoring the usual taste.

Changes in pressure have no effect on the solubility of solids in liquids but greatly affect the solubility of gases. The release of bubbles (fizzing) when a bottle or can of soda is opened occurs because pressure is reduced on the beverage and dissolved carbon dioxide comes out of solution. In general, *gas solubility decreases with temperature and increases with pressure.* As usual, there are exceptions to this generalization.

PROPERTIES OF WATER SOLUTIONS

Pure solvents have characteristic physical and chemical properties that are changed by the addition of a solute. Following are some of the more interesting changes.

Electrolytes

Water solutions of ionic substances will conduct an electric current, so they are called **electrolytes.** Ions must be present and free to move in a solution to carry the charge, so electrolytes are solutions containing ions. Pure water will not conduct an electric current as it is a covalent compound, which ionizes only very slightly. Water solutions of sugar, alcohol, and most other covalent compounds are nonconductors, so they are called *nonelectrolytes.* Nonelectrolytes are covalent compounds that form molecular solutions, so they cannot conduct an electric current.

Some covalent compounds are nonelectrolytes as pure liquids but become electrolytes when dissolved in water. Pure

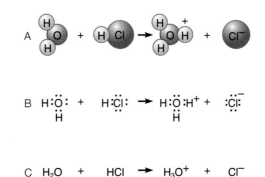

FIGURE 10.8 Three representations of water and hydrogen chloride in an ionizing reaction. (*A*) Sketches of molecules involved in the reaction. (*B*) Electron dot equation of the reaction. (*C*) The chemical equation for the reaction. Each of these representations shows the hydrogen being pulled away from the chlorine atom to form H_3O^+, the hydronium ion.

hydrogen chloride (HCl), for example, does not conduct an electric current, so you can assume that it is a molecular substance. When dissolved in water, hydrogen chloride does conduct a current, so it must now contain ions. Evidently, the hydrogen chloride has become *ionized* by the water. The process of forming ions from molecules is called *ionization.* Hydrogen chloride, just as water, has polar molecules. The positive hydrogen atom on the HCl molecule is attracted to the negative oxygen end of a water molecule, and the force of attraction is strong enough to break the hydrogen-chlorine bond, forming charged particles (figure 10.8). The reaction is

$$H_2O_{(1)} + HCl_{(1)} \rightarrow H_3O^+_{(aq)} + Cl^-_{(aq)}$$

The H_3O^+ ion is called a **hydronium ion.** A hydronium ion is basically a molecule of water with an attached hydrogen ion. The presence of the hydronium ion gives the solution new chemical properties; the solution is no longer hydrogen chloride but is *hydrochloric acid.* Hydrochloric acid, and other acids, will be discussed shortly.

Boiling Point

Boiling occurs when the pressure of the vapor escaping from a liquid (vapor pressure) is equal to the atmospheric pressure on the liquid. The *normal* boiling point is defined as the temperature at which the vapor pressure is equal to the average atmospheric pressure at sea level. For pure water, this temperature is 100°C (212°F). It is important to remember that boiling is a purely physical process. No bonds within water molecules are broken during boiling.

The vapor pressure over a solution is *less* than the vapor pressure over the pure solvent at the same temperature. Molecules of a liquid can escape into the air only at the surface of the liquid, and the presence of molecules of a solute means that fewer solvent molecules can be at the surface to escape. Thus, the vapor pressure over a solution is less than the vapor pressure over a pure solvent (figure 10.9).

Because the vapor pressure over a solution is less than that over the pure solvent, the solution boils at a higher temperature. A higher temperature is required to increase the vapor pressure

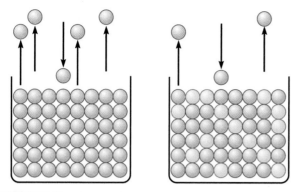

FIGURE 10.9 The rate of evaporation, and thus the vapor pressure, is less for a solution than for a solvent in the pure state. The greater the solute concentration, the less the vapor pressure.

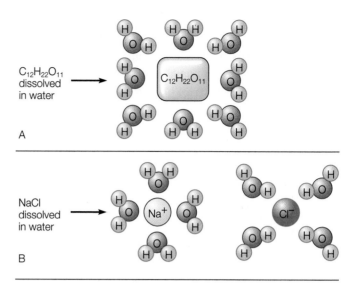

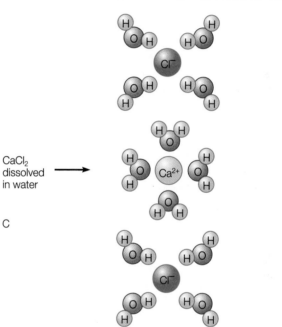

FIGURE 10.10 Since ionic compounds dissolve by the separation of ions, they provide more particles in solution than molecular compounds. (A) A mole of sugar provides Avogadro's number of particles. (B) A mole of NaCl provides two times Avogadro's number of particles. (C) A mole of $CaCl_2$ provides three times Avogadro's number of particles.

to that of the atmospheric pressure. Some cooks have been observed to add a "pinch" of salt to a pot of water before boiling. Is this to increase the boiling point and therefore cook the food more quickly? How much does a pinch of salt increase the boiling temperature? The answers are found in the relationship between the concentration of a solute and the boiling point of the solution.

It is the number of solute particles (ions or molecules) at the surface of a solution that increases the boiling point. A mole is a measure that can be defined as a number of particles called Avogadro's number. Since the number of particles at the surface is proportional to the ratio of particles in the solution, the concentration of the solute will directly influence the increase in the boiling point. In other words, the boiling point of any dilute solution is increased proportional to the concentration of the solute. For water, the boiling point is increased 0.521°C for every mole of solute dissolved in 1,000 g of water. Thus, any water solution will boil at a higher temperature than pure water. Since it boils at a higher temperature, it also takes a longer time to reach the boiling point.

It makes no difference what substance is dissolved in the water; 1 mole of solute in 1,000 g of water will elevate the boiling point by 0.521°C. A mole contains Avogadro's number of particles, so a mole of any solute will lower the vapor pressure by the same amount. Sucrose, or table sugar, for example, is $C_{12}H_{22}O_{11}$ and has a gram-formula weight of 342 g. Thus, 342 g of sugar in 1,000 g of water (about a liter) will increase the boiling point by 0.521°C. Therefore, if you measure the boiling point of a sugar solution, you can determine the concentration of sugar in the solution. For example, pancake syrup that boils at 100.261°C (sea-level pressure) must contain 171 g of sugar dissolved in 1,000 g of water. You know this because the increase of 0.261°C over 100°C is one-half of 0.521°C. If the boiling point were increased by 0.521°C over 100°C, the syrup would have the full gram-formula weight (342 g) dissolved in a kg of water.

Since it is the number of particles of solute in a specific sample of water that elevates the boiling point, different effects are observed in dissolved covalent and dissolved ionic compounds (figure 10.10). Sugar is a covalent compound, and the solute is molecules of sugar moving between the water molecules. Sodium chloride, on the other hand, is an ionic compound and dissolves by the separation of ions, or

$$Na^+Cl^-_{(s)} \rightarrow Na^+_{(aq)} + Cl^-_{(aq)}$$

This equation tells you that 1 mole of NaCl separates into 1 mole of sodium ions and 1 mole of chlorine ions for a total of *2* moles of solute. The boiling point elevation of a solution made from 1 mole of NaCl (58.5 g) is therefore multiplied by two, or $2 \times 0.521°C = 1.04°C$. The boiling point of a solution made by adding 58.5 g of NaCl to 1,000 g of water is therefore 101.04°C at normal sea-level pressure.

Now back to the question of how much a pinch of salt increases the boiling point of a pot of water. Assuming the pot contains about a liter of water (about a quart) and assuming that a "pinch" of salt has a mass of about 0.2 gram, the boiling point will be increased by 0.0037°C. Thus, there must be some reason other than increasing the boiling point that a cook adds a pinch of salt to a pot of boiling water. Perhaps the salt is for seasoning?

Freezing Point

Freezing occurs when the kinetic energy of molecules has been reduced sufficiently so the molecules can come together, forming the crystal structure of the solid. Reduced kinetic energy of the molecules, that is, reduced temperature, results in a specific freezing point for each pure liquid. The *normal* freezing point for pure water, for example, is 0°C (32°F) under normal pressure. The presence of solute particles in a solution interferes with the water molecules as they attempt to form the six-sided hexagonal structure. The water molecules cannot get by the solute particles until the kinetic energy of the solute particles is reduced, that is, until the temperature is below the normal freezing point. Thus, the presence of solute particles lowers the freezing point, and solutions freeze at a lower temperature than the pure solvent.

The freezing-point depression of a solution has a number of interesting implications for solutions such as seawater. When seawater freezes, the water molecules must work their way around the salt particles as was described earlier. Thus, the solute particles are *not* normally included in the hexagonal structure of ice. Ice formed in seawater is practically pure water. Since the solute was *excluded* when the ice formed, the freezing of seawater increases the salinity. Increased salinity means increased concentration, so the freezing point of seawater is further depressed and more ice forms only at a lower temperature. When this additional ice forms, more pure water is removed and the process goes on. Thus, seawater does not have a fixed freezing point but has a lower and lower freezing point as more and more ice freezes.

The depression of the freezing point by a solute has a number of interesting applications in colder climates. Salt, for example, is spread on icy roads to lower the freezing point (and thus the melting point) of the ice. Calcium chloride, $CaCl_2$, is a salt that is often used for this purpose. Water in a car radiator would also freeze in colder climates if a solute, called antifreeze, were not added to the radiator water. Methyl alcohol has been used as an antifreeze because it is soluble in water and does not damage the cooling system. Methyl alcohol, however, has a low boiling point and tends to boil away. Ethylene glycol has a higher boiling point, so it is called a "permanent" antifreeze. Like other solutes, ethylene glycol also raises the boiling point, which is an added benefit for summer driving.

ACIDS, BASES, AND SALTS

The electrolytes known as *acids*, *bases*, and *salts* are evident in environmental quality, foods, and everyday living. Environmental quality includes the hardness of water, which is determined by the presence of certain salts, the acidity of soils, which determines how well plants grow, and acid rain, which is a by-product of industry and automobiles. Many concerns about air and water pollution are often related to the chemistry concepts of acids, bases, and salts.

Properties of Acids and Bases

Acids and bases are classes of chemical compounds that have certain characteristic properties. These properties can be used to identify if a substance is an acid or a base (tables 10.2 and 10.3). The following are the properties of *acids* dissolved in water:

1. Acids have a sour taste, such as the taste of citrus fruits.
2. Acids change the color of certain substances; for example, litmus changes from blue to red when placed in an acid solution (figure 10.11A).

TABLE 10.2   Some Common Acids

Name	Formula	Comment
Acetic acid	CH_3COOH	A weak acid found in vinegar
Boric acid	H_3BO_3	A weak acid used in eyedrops
Carbonic acid	H_2CO_3	The weak acid of carbonated beverages
Formic acid	$HCOOH$	Makes the sting of insects and certain plants
Hydrochloric acid	HCl	Also called muriatic acid; used in swimming pools, soil acidifiers, and stain removers
Lactic acid	$CH_3CHOHCOOH$	Found in sour milk, sauerkraut, and pickles; gives tart taste to yogurt
Nitric acid	HNO_3	A strong acid
Phosphoric acid	H_3PO_4	Used in cleaning solutions; added to carbonated beverages for tartness
Sulfuric acid	H_2SO_4	Also called oil of vitriol; used as battery acid and in swimming pools

TABLE 10.3   Some Common Bases

Name	Formula	Comment
Sodium hydroxide	$NaOH$	Also called lye or caustic soda; a strong base used in oven cleaners and drain cleaners
Potassium hydroxide	KOH	Also called caustic potash; a strong base used in drain cleaners
Ammonia	NH_3	A weak base used in household cleaning solutions
Calcium hydroxide	$Ca(OH)_2$	Also called slaked lime; used to make brick mortar
Magnesium hydroxide	$Mg(OH)_2$	Solution is called milk of magnesia; used as antacid and laxative

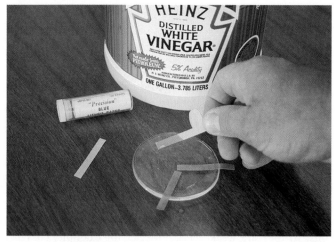

A

B

FIGURE 10.11 (*A*) Acid solutions will change the color of blue litmus to red. (*B*) Solutions of bases will change the color of red litmus to blue.

3. Acids react with active metals, such as magnesium or zinc, releasing hydrogen gas.
4. Acids *neutralize* bases, forming water and salts from the reaction.

Likewise, *bases* have their own characteristic properties. Bases are also called alkaline substances, and the following are the properties of bases dissolved in water:

1. Bases have a bitter taste, for example, the taste of caffeine.
2. Bases reverse the color changes that were caused by acids. Red litmus is changed back to blue when placed in a solution containing a base (figure 10.11B).
3. Basic solutions feel slippery on the skin. They have a *caustic* action on plant and animal tissue, converting tissue into soluble materials. A strong base, for example, reacts with fat to make soap and glycerine. This accounts for the slippery feeling on the skin.
4. Bases *neutralize* acids, forming water and salts from the reaction.

Tasting an acid or base to see if it is sour or bitter can be hazardous, since some are highly corrosive or caustic. Many

organic acids are not as corrosive and occur naturally in foods. Citrus fruit, for example, contains citric acid, vinegar is a solution of acetic acid, and sour milk contains lactic acid. The stings or bites of some insects (bees, wasps, and ants) and some plants (stinging nettles) are painful because an organic acid, formic acid, is injected by the insect or plant. Your stomach contains a solution of hydrochloric acid. In terms of relative strength, the hydrochloric acid in your stomach is about ten times stronger than the carbonic acid (H_2CO_3) of carbonated beverages.

Examples of bases include solutions of sodium hydroxide (NaOH), which has a common name of lye or caustic soda, and potassium hydroxide (KOH), which has a common name of caustic potash. These two bases are used in products known as drain cleaners. They open plugged drains because of their caustic action, turning grease, hair, and other organic "plugs" into soap and other soluble substances that are washed away. A weaker base is a solution of ammonia (NH_3), which is often used as a household cleaner. A solution of magnesium hydroxide, $Mg(OH)_2$, has a common name of milk of magnesia and is sold as an antacid and laxative.

Many natural substances change color when mixed with acids or bases. You may have noticed that tea changes color slightly, becoming lighter, when lemon juice (which contains citric acid) is added. Some plants have flowers of one color when grown in acidic soil and flowers of another color when grown in basic soil. A vegetable dye that changes color in the presence of acids or bases can be used as an *acid-base indicator*. An indicator is simply a vegetable dye that is used to distinguish between acid and base solutions by a color change. Litmus, for example, is an acid-base indicator made from a dye extracted from certain species of lichens. The dye is applied to paper strips, which turn red in acidic solutions and blue in basic solutions.

Explaining Acid-Base Properties

Comparing the lists in tables 10.2 and 10.3, you can see that acids and bases appear to be chemical opposites. Notice in table 10.2 that the acids all have an H, or hydrogen atom, in their formulas. In table 10.3, most of the bases have a hydroxide ion, OH$^-$, in their formulas. Could this be the key to acid-base properties?

The modern concept of an acid considers the properties of acids in terms of the hydronium ion, H_3O^+. As was mentioned earlier, the hydronium ion is a water molecule to which an H^+ ion is attached. Since a hydrogen ion is a hydrogen atom without its single electron, it could be considered as an ion consisting of a single proton. Thus, the H^+ ion can be called a *proton*. An **acid** is defined as any substance that is a *proton donor* when dissolved in water, increasing the hydronium ion concentration. For example, hydrogen chloride dissolved in water has the following reaction:

$$\text{(H)Cl}_{(aq)} \quad + \quad H_2O_{(l)} \quad \longrightarrow \quad H_3O^+_{(aq)} \quad + \quad Cl^-_{(aq)}$$

The dotted circle and arrow were added to show that the hydrogen chloride donated a proton to a water molecule. The resulting solution contains H_3O^+ ions and has acid properties, so the solution is called hydrochloric acid. It is the H_3O^+ ion that is responsible for the properties of an acid.

The bases listed in table 10.3 all appear to have a hydroxide ion, OH^-. Water solutions of these bases do contain OH^- ions, but the definition of a base is much broader. A **base** is defined as any substance that is a *proton acceptor* when dissolved in water, increasing the hydroxide ion concentration. For example, ammonia dissolved in water has the following reaction:

$$NH_{3(g)} \quad + \quad \text{(H}_2\text{)O}_{(l)} \quad \longrightarrow \quad (NH_4)^+ \quad + \quad OH^-$$

The dotted circle and arrow show that the ammonia molecule accepted a proton from a water molecule, providing a hydroxide ion. The resulting solution contains OH^- ions and has basic properties, so a solution of ammonium hydroxide is a base.

Carbonates, such as sodium carbonate (Na_2CO_3), form basic solutions because the carbonate ion reacts with water to produce hydroxide ions.

$$(CO_3)^{2-}_{(aq)} + H_2O_{(l)} \rightarrow (HCO_3)^-_{(aq)} + OH^-_{(aq)}$$

Thus, sodium carbonate produces a basic solution.

Acids could be thought of as simply solutions of hydronium ions in water, and bases could be considered solutions of hydroxide ions in water. The proton donor and proton acceptor definition is much broader, and it does include the definition of acids and bases as hydronium and hydroxide compounds. The broader, more general definition covers a wider variety of reactions and is therefore more useful.

The modern concept of acids and bases explains why the properties of acids and bases are **neutralized,** or lost, when acids and bases are mixed together. For example, consider the hydronium ion produced in the hydrochloric acid solution and the hydroxide ion produced in the ammonia solution. When these solutions are mixed together, the hydronium ion reacts with the hydroxide ion, and

$$H_3O^+_{(aq)} + OH^+_{(aq)} \rightarrow H_2O_{(l)} + H_2O_{(l)}$$

Thus, a proton is transferred from the hydronium ion (an acid), and the proton is accepted by the hydroxide ion (a base). Water

is produced, and both the acid and base properties disappear or are neutralized.

Strong and Weak Acids and Bases

Acids and bases are classified according to their degree of ionization when placed in water. *Strong acids* ionize completely in water, with all molecules dissociating into ions. Nitric acid, for example, reacts completely in the following equation:

$$HNO_{3(aq)} + H_2O_{(l)} \rightarrow H_3O^+_{(aq)} + (NO_3)^-_{(aq)}$$

Nitric acid, hydrochloric acid (figure 10.12), and sulfuric acid are common strong acids.

Acids that ionize only partially produce fewer hydronium ions, so they are *weak acids*. Vinegar, for example, contains acetic acid that reacts with water in the following reaction:

$$HC_2H_3O_2 + H_2O \rightarrow H_3O^+ + (C_2H_3O_2)^-$$

Only about 1 percent or less of the acetic acid molecules ionize, depending on the concentration.

Bases are also classified as strong or weak. A *strong base* is completely ionized in solution. Sodium hydroxide, or lye, is the

CONCEPTS APPLIED

Acid or Base?

Pick some household product that probably has an acid or base character (Example: pH increaser for aquariums). On a separate paper, write the listed ingredients and identify any you believe would be distinctly acidic or basic in a water solution. Tell whether you expect the product to be an acid or a base. Describe your findings of a litmus paper test.

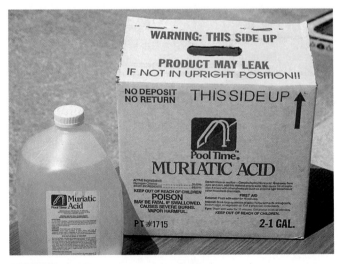

FIGURE 10.12 Hydrochloric acid (HCl) has the common name of *muriatic* acid. Hydrochloric acid is a strong acid used in swimming pools, soil acidifiers, and stain removers.

most common example of a strong base. It dissolves in water to form a solution of sodium and hydroxide ions:

$$Na^+OH^-_{(s)} \rightarrow Na^+_{(aq)} + OH^-_{(aq)}$$

A *weak base* is only partially ionized. Ammonia, magnesium hydroxide, and calcium hydroxide are examples of weak bases. Magnesium and calcium hydroxide are only slightly soluble in water, and this reduces the *concentration* of hydroxide ions in a solution.

The pH Scale

The strength of an acid or a base is usually expressed in terms of a range of values called a **pH scale.** The pH scale is based on the concentration of the hydronium ion (in moles/L) in an acidic or a basic solution. To understand how the scale is able to express both acid and base strength in terms of the hydronium ion, first note that pure water is very slightly ionized in the reaction:

$$H_2O_{(1)} + H_2O_{(1)} \rightarrow H_3O^+_{(aq)}OH^-_{(aq)}$$

The amount of self-ionization by water has been determined through measurements. In pure water at 25°C or any neutral water solution at that temperature, the H_3O^+ concentration is 1×10^{-7} moles/L, and the OH^- concentration is also 1×10^{-7} moles/L. Since both ions are produced in equal numbers, then the H_3O^+ concentration equals the OH^- concentration, and pure water is neutral, neither acidic nor basic.

In general, adding an acid substance to pure water increases the H_3O^+ concentration. Adding a base substance to pure water increases the OH^- concentration. Adding a base also *reduces* the H_3O^+ concentration as the additional OH^- ions are able to combine with more of the hydronium ions to produce unionized water. Thus, at a given temperature, an increase in OH^- concentration is matched by a *decrease* in H_3O^+ concentration. The concentration of the hydronium ion can be used as a measure of acidic, neutral, and basic solutions. In general, (1) acidic solutions have H_3O^+ concentrations above 1×10^{-7} moles/L, (2) neutral solutions have H_3O^+ concentrations equal to 1×10^{-7} moles/L, and (3) basic solutions have H_3O^+ concentrations less than 1×10^{-7} moles/L. These three statements lead directly to the pH scale, which is named from the French *pouvoir hydrogene,* meaning "hydrogen power." Power refers to the exponent of the hydronium ion concentration, and the pH is a *power of ten notation that expresses the H_3O^+ concentration.*

A neutral solution has a pH of 7.0. Acidic solutions have pH values below 7, and smaller numbers mean greater acidic properties. Increasing the OH^- concentration decreases the H_3O^+ concentration, so the strength of a base is indicated on the same scale with values greater than 7. Note that the pH scale is logarithmic, so a pH of 2 is ten times as acidic as a pH of 3. Likewise, a pH of 10 is one hundred times as basic as a pH of 8. Figure 10.13 is a diagram of the pH scale, and table 10.4 compares the pH of some common substances (figure 10.14).

H_3O^+ Concentration (moles/liters)	pH	Meaning
1×10^{-0} (=1)	0	
1×10^{-1}	1	
1×10^{-2}	2	
1×10^{-3}	3	Increasing acidity
1×10^{-4}	4	
1×10^{-5}	5	
1×10^{-6}	6	
1×10^{-7}	7	Neutral
1×10^{-8}	8	
1×10^{-9}	9	
1×10^{-10}	10	Increasing basicity
1×10^{-11}	11	
1×10^{-12}	12	
1×10^{-13}	13	
1×10^{-14}	14	

FIGURE 10.13   The pH scale.

FIGURE 10.14   The pH increases as the acidic strength of these substances decreases from left to right. Did you know that lemon juice is more acidic than vinegar? That a soft drink is more acidic than orange juice or grapefruit juice?

Properties of Salts

Salt is produced by a neutralization reaction between an acid and a base. A **salt** is defined as any ionic compound except those with hydroxide or oxide ions. Table salt, NaCl, is but one example of this large group of ionic compounds. As an example of a salt produced by a neutralization reaction, consider the reaction of HCl (an acid in solution) with $Ca(OH)_2$ (a base in solution). The reaction is

$$2\,HCl_{(aq)} + Ca(OH)_{2(aq)} \rightarrow CaCl_{2(aq)} + 2\,H_2O_{(1)}$$

This is an ionic exchange reaction that forms molecular water, leaving Ca^{2+} and Cl^- in solution. As the water is evaporated, these ions begin forming ionic crystal structures as the solution concentration increases. When the water is all evaporated, the white crystalline salt of $CaCl_2$ remains.

TABLE 10.4 The Approximate pH of Some Common Substances

Substance	pH (or pH Range)
Hydrochloric acid (4%)	0
Gastric (stomach) solution	1.6–1.8
Lemon juice	2.2–2.4
Vinegar	2.4–3.4
Carbonated soft drinks	2.0–4.0
Grapefruit	3.0–3.2
Oranges	3.2–3.6
Acid rain	4.0–5.5
Tomatoes	4.2–4.4
Potatoes	5.7–5.8
Natural rainwater	5.6–6.2
Milk	6.3–6.7
Pure water	7.0
Seawater	7.0–8.3
Blood	7.4
Sodium bicarbonate solution	8.4
Milk of magnesia	10.5
Ammonia cleaning solution	11.9
Sodium hydroxide solution	13.0

If sodium hydroxide had been used as the base instead of calcium hydroxide, a different salt would have been produced:

$$HCl_{(aq)} + NaOH_{(aq)} \rightarrow NaCl_{(aq)} + H_2O_{(l)}$$

Salts are also produced when elements combine directly, when an acid reacts with a metal, and by other reactions.

Salts are essential in the diet both as electrolytes and as a source of certain elements, usually called *minerals* in this context. Plants must have certain elements that are derived from water-soluble salts. Potassium, nitrates, and phosphate salts are often used to supply the needed elements. There is no scientific evidence that plants prefer to obtain these elements from natural sources, as compost, or from chemical fertilizers. After all, a nitrate ion is a nitrate ion, no matter what its source. Table 10.5 lists some common salts and their uses.

Hard and Soft Water

Salts vary in their solubility in water, and a solubility chart appears in appendix B. Table 10.6 lists some generalizations concerning the various common salts. Some of the salts are dissolved by water that will eventually be used for domestic supply. When the salts are soluble calcium or magnesium compounds, the water will contain calcium or magnesium ions in solution. A solution of Ca^{2+} or Mg^{2+} ions is said to be *hard water* because it is hard to make soap lather in the water. "Soft" water, on the other hand, makes a soap lather easily. The difficulty occurs because soap is a sodium or potassium compound that is soluble in water. The calcium or magnesium ions, when present, replace the sodium or potassium ions in the soap compound, forming an insoluble compound. It is this insoluble compound

TABLE 10.5 Some Common Salts and Their Uses

Common Name	Formula	Use
Alum	$KAl(SO_4)_2$	Medicine, canning, baking powder
Baking soda	$NaHCO_3$	Fire extinguisher, antacid, deodorizer, baking powder
Bleaching powder (chlorine tablets)	$CaOCl_2$	Bleaching, deodorizer, disinfectant in swimming pools
Borax	$Na_2B_4O_7$	Water softener
Chalk	$CaCO_3$	Antacid tablets, scouring powder
Chile saltpeter	$NaNO_3$	Fertilizer
Cobalt chloride	$CoCl_2$	Hygrometer (pink in damp weather, blue in dry weather)
Epsom salt	$MgSO_4 \cdot 7\,H_2O$	Laxative
Fluorspar	CaF_2	Metallurgy flux
Gypsum	$CaSO_4 \cdot 2\,H_2O$	Plaster of Paris, soil conditioner
Lunar caustic	$AgNO_3$	Germicide and cauterizing agent
Niter (or saltpeter)	KNO_3	Meat preservative, makes black gunpowder (75 parts KNO_3, 15 of carbon, 10 of sulfur)
Potash	K_2CO_3	Makes soap, glass
Rochelle salt	$KNaC_4H_4O_6$	Baking powder ingredient
TSP	Na_3PO_4	Water softener, fertilizer

TABLE 10.6 Generalizations About Salt Solubilities

Salts	Solubility	Exceptions
Sodium Potassium Ammonium	Soluble	None
Nitrate Acetate Chlorate	Soluble	None
Chlorides	Soluble	Ag and Hg(I) are insoluble
Sulfates	Soluble	Ba, Sr, and Pb are insoluble
Carbonates Phosphates Silicates	Insoluble	Na, K, and NH_4 are soluble
Sulfides	Insoluble	Na, K, and NH_4 are soluble: Mg, Ca, Sr, and Ba decompose

that forms a "bathtub ring" and also collects on clothes being washed, preventing cleansing.

The key to "softening" hard water is to remove the troublesome calcium and magnesium ions (figure 10.15). If the hardness is caused by magnesium or calcium *bicarbonates*, the removal is accomplished by simply heating the water. Upon heating, they decompose, forming an insoluble compound that

Acid Rain

Acid rain is a general term used to describe any acidic substances, wet or dry, that fall from the atmosphere. Wet acidic deposition could be in the form of rain, but snow, sleet, and fog could also be involved. Dry acidic deposition could include gases, dust, or any solid particles that settle out of the atmosphere to produce an acid condition.

Pure, unpolluted rain is naturally acidic. Carbon dioxide in the atmosphere is absorbed by rainfall, forming carbonic acid (H_2CO_3). Carbonic acid lowers the pH of pure rainfall to a range of 5.6 to 6.2. Decaying vegetation in local areas can provide more CO_2, making the pH even lower. A pH range of 4.5 to 5.0, for example, has been measured in remote areas of the Amazon jungle. Human-produced exhaust emissions of sulfur and nitrogen oxides can lower the pH of rainfall even more, to a 4.0 to 5.5 range. This is the pH range of acid rain.

The sulfur and nitrogen oxides that produce acid rain come from exhaust emissions of industries and electric utilities that burn coal and from the exhaust of cars, trucks, and buses (box figure 10.1). The emissions are sometimes called "SO_x" and "NO_x," which is read "socks" and "knox." The x subscript implies the variable presence of any or all of the oxides, for example, nitrogen monoxide (NO), nitrogen dioxide (NO_2), and dinitrogen tetroxide (N_2O_4) for NO_x.

SO_x and NO_x are the raw materials of acid rain and are not themselves acidic. They react with other atmospheric chemicals to form sulfates and nitrates, which combine with water vapor to form sulfuric acid (H_2SO_4) and nitric acid (HNO_3). These are the chemicals of concern in acid rain.

Many variables influence how much and how far SO_x and NO_x are carried in the atmosphere and if they are converted to acid rain or simply return to the surface as a dry gas or particles. During the 1960s and 1970s, concerns about local levels of

Box Figure 10.1 Natural rainwater has a pH of 5.6 to 6.2. Exhaust emissions of sulfur and nitrogen oxides can lower the pH of rainfall to a range of 4.0 to 5.5. The exhaust emissions come from industries, electric utilities, and automobiles. Not all emissions are as visible as those pictured in this illustration.

pollution led to the replacement of short smokestacks of about 60 m (about 200 ft) with taller smokestacks of about 200 m (about 650 ft). This reduced the local levels of pollution by dumping the exhaust higher in the atmosphere where winds could carry it away. It also set the stage for longer-range transport of SO_x and NO_x and their eventual conversion into acids.

There are two main reaction pathways by which SO_x and NO_x are converted to acids: (1) reactions in the gas phase and (2) reactions in the liquid phase, such as in water droplets in clouds and fog. In the gas phase, SO_x and NO_x are oxidized to acids, mainly by hydroxyl ions and ozone, and the acid is absorbed by cloud droplets and precipitated as rain or snow. Most of the nitric acid in acid rain and about one-fourth

of the sulfuric acid is formed in gas-phase reactions. Most of the liquid-phase reactions that produce sulfuric acid involve the absorbed SO_x and hydrogen peroxide (H_2O_2), ozone, oxygen, and particles of carbon, iron oxide, and manganese oxide particles. These particles also come from the exhaust of fossil fuel combustion.

Acid rain falls on land, bodies of water, forests, crops, buildings, and people. Concerns about acid rain center on its environmental impact on lakes, forests, crops, materials, and human health. Lakes in different parts of the world, for example, have been increasing in acidity over the past fifty years. Lakes in northern New England, the Adirondacks, and parts of Canada now have a pH of less than 5.0, and correlations have been established between lake acidity and decreased fish populations. Trees, mostly conifers, are dying at unusually rapid rates in the northeastern United States. Red spruce in Vermont's Green Mountains and the mountains of New York and New Hampshire have been affected by acid rain, as have pines in New Jersey's Pine Barrens. It is believed that acid rain leaches essential nutrients, such as calcium, from the soil and mobilizes aluminum ions. The aluminum ions disrupt the water equilibrium of fine root hairs, and when the root hairs die, so do the trees.

Human-produced emissions of sulfur and nitrogen oxides from burning fossil fuels are the cause of acid rain. The heavily industrialized northeastern part of the United States, from the Midwest through New England, releases sulfur and nitrogen emissions that result in a precipitation pH of 4.0 to 4.5. This region is the geographic center of the nation's acid rain problem. The solution to the problem is found in (1) using fuels other than fossil fuels and (2) reducing the thousands of tons of SO_x and NO_x that are dumped into the atmosphere per day when fossil fuels are used.

effectively removes the ions from solution. The decomposition reaction for calcium bicarbonate is

$$Ca^{2+}(HCO_3)^-_{2(aq)} \rightarrow CaCO_{3(s)} + H_2O_{(l)} + CO_2\uparrow$$

The reaction is the same for magnesium bicarbonate. As the solubility chart in appendix B shows, magnesium and calcium carbonates are insoluble, so the ions are removed from solution in the solid that is formed. Perhaps you have noticed such a white compound forming around faucets if you live where bicarbonates are a problem. Commercial products to remove such deposits usually contain an acid, which reacts with the carbonate to make a new, soluble salt that can be washed away.

People Behind the Science

Johannes Nicolaus Brönsted (1879–1947)

Johannes Brönsted was a Danish physical chemist whose work in solution chemistry, particularly electrolytes, resulted in a new theory of acids and bases.

Brönsted was born on February 22, 1879, in Varde, Jutland, the son of a civil engineer. He was educated at local schools before going to study chemical engineering at the Technical Institute of the University of Copenhagen in 1897. He graduated two years later and then turned to chemistry, in which he qualified in 1902. After a short time in industry, he was appointed an assistant in the university's chemical laboratory in 1905, becoming professor of physical and inorganic chemistry in 1908.

Brönsted's early work was wide-ranging, particularly in the fields of electrochemistry, the measurement of hydrogen ion concentrations, amphoteric electrolytes, and the behavior of indicators. He discovered a method of eliminating potentials in the measurement of hydrogen ion concentrations and devised a simple equation that connects the activity and osmotic coefficients of an electrolyte, as well as another that relates activity coefficients to reaction velocities. From the absorption spectra of chromium(III) salts, he concluded that strong electrolytes are completely dissociated and that the changes of molecular conductivity and freezing point that accompany changes in concentration are caused by the electrical forces between ions in solution.

In 1887, Svante Arrhenius had proposed a theory of acidity that explained its nature on an atomic level. He defined an acid as a compound that could generate hydrogen ions in aqueous solution and an alkali as a compound that could generate hydroxyl ions. A strong acid is completely ionized (dissociated) and produces many hydrogen ions, whereas a weak acid is only partly dissociated and produces few hydrogen ions. Conductivity measurements confirm the theory, as long as the solutions are not too concentrated.

In 1923, Brönsted published (simultaneously with Thomas Lowry in Britain) a new theory of acidity, which has certain important advantages over that of Arrhenius. Brönsted defined an acid as a proton donor and a base as a proton acceptor. The definition applies to all solvents, not just water. It also explains the different behavior of pure acids in solution. Pure liquid sulfuric acid or acetic acid does not change the color of indicators nor does it react with carbonates or metals. But as soon as water is added, all of these reactions occur.

Source: Modified from the *Hutchinson Dictionary of Scientific Biography*. © RM, 2007. Reprinted by permission.

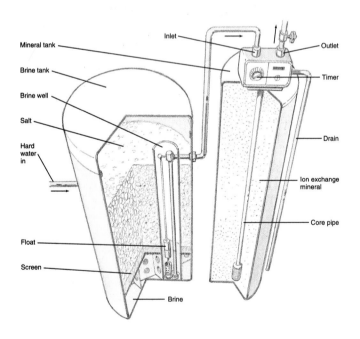

FIGURE 10.15 A water softener exchanges sodium ions for the calcium and magnesium ions of hard water. Thus, the water is now soft, but it contains the same number of ions as before.

Water hardness is also caused by magnesium or calcium *sulfate*, which requires a different removal method. Certain chemicals such as sodium carbonate (washing soda), trisodium phosphate (TSP), and borax will react with the troublesome ions, forming an insoluble solid that removes them from solution. For example, washing soda and calcium sulfate react as follows:

$$Na_2CO_{3(aq)} + CaSO_{4(aq)} \rightarrow Na_2SO_{4(aq)} + CaCO_3\downarrow$$

Calcium carbonate is insoluble; thus, the calcium ions are removed from solution before they can react with the soap. Many laundry detergents have Na_2CO_3, TSP, or borax ($Na_2B_4O_7$) added to soften the water. TSP causes other problems, however, as the additional phosphates in the waste water can act as a fertilizer, stimulating the growth of algae to such an extent that other organisms in the water die.

A water softener unit is an ion exchanger. The unit contains a mineral that exchanges sodium ions for calcium and magnesium ions as water is run through it. The softener is regenerated periodically by flushing with a concentrated sodium chloride solution. The sodium ions replace the calcium and magnesium ions, which are carried away in the rinse water. The softener is then ready for use again. The frequency of renewal cycles depends on the water hardness, and each cycle can consume from 4 to 20 pounds of sodium chloride per renewal cycle. In general, water with less than 75 ppm calcium and magnesium ions is called soft water; with greater concentrations, it is called hard water. The greater the concentration above 75 ppm, the harder the water.

SUMMARY

A water molecule consists of two hydrogen atoms and an oxygen atom with bonding and electron pairs in a tetrahedral arrangement. Electrons spend more time around the oxygen, producing a *polar molecule,* with centers of negative and positive charge. Polar water molecules interact. The force of attraction is called a *hydrogen bond.* The hydrogen bond accounts for the decreased density of ice, the high heat of fusion, and the high heat of vaporization of water. The hydrogen bond is also involved in the *dissolving* process.

A *solution* is a homogeneous mixture of ions or molecules of two or more substances. The substance present in the large amount is the *solvent,* and the *solute* is dissolved in the solvent. If one of the components is a liquid, however, it is called the solvent.

Fluids that mix in any proportion are called *miscible fluids,* and *immiscible fluids* do not mix. Polar substances dissolve in polar solvents but not nonpolar solvents, and the general rule is *like dissolves like.* Thus, oil, a nonpolar substance, is immiscible in water, a polar substance.

The relative amount of solute in a solvent is called the *concentration* of a solution. Concentrations are measured (1) in *parts per million* (ppm) or *parts per billion* (ppb), (2) *percent by volume,* the volume of a solute per 100 volumes of solution, (3) *percent by weight,* the weight of solute per 100 weight units of solution, and (4) *salinity,* the mass of salts in 1 kg of solution.

A limit to dissolving solids in a liquid occurs when the solution is *saturated.* A *saturated solution* is one with equilibrium between solute dissolving and solute coming out of solution. The *solubility* of a solid is the concentration of a saturated solution at a particular temperature.

Water solutions that carry an electric current are called *electrolytes,* and nonconductors are called *nonelectrolytes.* In general, ionic substances make electrolyte solutions, and molecular substances make nonelectrolyte solutions. Polar molecular substances may be *ionized* by polar water molecules, however, making an electrolyte from a molecular solution.

The *boiling point of a solution* is greater than the boiling point of the pure solvent, and the increase depends only on the concentration of the solute (at a constant pressure). For water, the boiling point is increased 0.521°C for each mole of solute in each kg of water. The *freezing point of a solution* is lower than the freezing point of the pure solvent, and the depression also depends on the concentration of the solute.

Acids, bases, and salts are chemicals that form ionic solutions in water, and each can be identified by simple properties. These properties are accounted for by the modern concepts of each. *Acids* are *proton donors* that form *hydronium ions* (H_3O^+) in water solutions. *Bases* are *proton acceptors* that form *hydroxide ions* (OH^-) in water solutions. The strength of an acid or base is measured on the *pH scale,* a power of ten notation of the hydronium ion concentration. On the scale, numbers from 0 up to 7 are acids, 7 is neutral, and numbers above 7 and up to 14 are bases. Each unit represents a tenfold increase or decrease in acid or base properties.

A *salt* is any ionic compound except those with hydroxide or oxide ions. Salts provide plants and animals with essential elements. The solubility of salts varies with the ions that make up the compound. Solutions of magnesium or calcium produce *hard water,* water in which it is hard to make soap lather. Hard water is softened by removing the magnesium and calcium ions.

KEY TERMS

acid (p. **236**)
base (p. **236**)
electrolytes (p. **232**)
hydration (p. **230**)
hydrogen bonding (p. **228**)

hydronium ion (p. **232**)
miscible fluids (p. **229**)
molarity (p. **231**)
mole (p. **231**)
neutralized (p. **236**)

pH scale (p. **237**)
polar molecule (p. **228**)
salinity (p. **231**)
salt (p. **237**)
saturated solution (p. **231**)

solubility (p. **231**)
solution (p. **229**)

APPLYING THE CONCEPTS

1. Which of the following is *not* a solution?
   a. seawater   **c.** sand
   b. carbonated water   **d.** brass

2. Atmospheric air is a homogeneous mixture of gases that is mostly nitrogen gas. The nitrogen is therefore (the)
   a. solvent.
   b. solution.
   c. solute.
   d. None of the above is correct.

3. A concentration of 500 ppm is reported in a news article. This is the same concentration as
   a. 0.005%.   **c.** 5%.
   b. 0.05%.   **d.** 50%.

4. According to the label, a bottle of vodka has a 40% by volume concentration. This means the vodka contains 40 mL of pure alcohol
   a. in each 140 mL of vodka.
   b. to every 100 mL of water.
   c. to every 60 mL of vodka.
   d. mixed with water to make 100 mL vodka.

5. A bottle of vinegar is 4% by weight, so you know that the solution contains 4 weight units of pure vinegar with
   a. 96 weight units of water.
   b. 100 weight units of water.
   c. 104 weight units of water.

6. If a salt solution has a salinity of 40‰, what is the equivalent percentage measure?
   a. 400%
   b. 40%
   c. 4%
   d. 0.4%
7. Water has the greatest density at what temperature?
   a. 100°C
   b. 20°C
   c. 4°C
   d. 0°C
8. A solid salt is insoluble in water, so the strongest force must be the
   a. ion-water molecule force.
   b. ion-ion force.
   c. force of hydration.
   d. polar molecule force.
9. The ice that forms in freezing seawater is
   a. pure water.
   b. the same salinity as liquid seawater.
   c. more salty than liquid seawater.
   d. more dense than liquid seawater.

10. Which of the following would have a pH of *less* than 7?
    a. a solution of ammonia
    b. a solution of sodium chloride
    c. pure water
    d. carbonic acid
11. Which of the following would have a pH of *more* than 7?
    a. a solution of ammonia
    b. a solution of sodium chloride
    c. pure water
    d. carbonic acid
12. Substance A has a pH of 2 and substance B has a pH of 3. This means that
    a. substance A has more basic properties than substance B.
    b. substance B has more acidic properties than substance A.
    c. substance A is ten times more acidic than substance B.
    d. substance B is ten times more acidic than substance A.

Answers

1. c 2. a 3. b 4. d 5. a 6. c 7. c 8. b 9. a 10. d 11. a 12. c

QUESTIONS FOR THOUGHT

1. How is a solution different from other mixtures?
2. Explain why some ionic compounds are soluble while others are insoluble in water.
3. Explain why adding salt to water increases the boiling point.
4. A deep lake in Minnesota is covered with ice. What is the water temperature at the bottom of the lake? Explain your reasoning.
5. Explain why water has a greater density at 4°C than at 0°C.
6. What is hard water? How is it softened?

7. According to the definition of an acid and the definition of a base, would the pH increase, decrease, or remain the same when NaCl is added to pure water? Explain.
8. What is a hydrogen bond? Explain how a hydrogen bond forms.
9. What feature of a soap molecule gives it cleaning ability?
10. What ion is responsible for (a) acidic properties? (b) for basic properties?

FOR FURTHER ANALYSIS

1. What are the basic differences and similarities between the concentration measures of salinity and percent by weight?
2. Compare and contrast the situations where you would express concentration in (1) parts per million, (2) parts per billion, (3) percent (volume or weight), and (4) salinity.
3. Analyze the basic reason that water is a universal solvent, becomes less dense when it freezes, has a high heat of fusion, has a high specific heat, and has a high heat of vaporization.
4. What is the same and what is different between a salt that will dissolve in water and one that is insoluble?

5. There are at least three ways to change the boiling point of water, so describe how you know for sure that 100°C (212°F) is the boiling point.
6. What are the significant similarities and differences between an acid, a base, and a salt?
7. Describe how you would teach someone why the pH of an acid is a low number (less than 7), while the pH of a base is a larger number (greater than 7).
8. Describe at least four different examples of how you could make hard water soft.

INVITATION TO INQUIRY

Which Freezes Faster?

Is it true that hot water that has boiled freezes faster than fresh cold water from the tap? Investigate freezing of hot, boiled water and fresh

cold water. Make sure you control all the variables in your experimental design. Compare your findings to those of classmates.

With the top half of the steel vessel and control rods removed, fuel rod bundles can be replaced in the water-flooded nuclear reactor.

Nuclear Reactions

CORE CONCEPT

Nuclear reactions involve changes in the nucleus of the atom.

Natural radioactivity is the spontaneous emission of particles or energy from a disintegrating nucleus (p. 244).

Nuclear instability results from an imbalance between the attractive nuclear force and the repulsive electromagnetic force (pp. 247–48).

An unstable nucleus becomes more stable by emitting alpha, beta, or gamma radiation from the nucleus (p. 248).

The relationship between energy and mass changes is $E = mc^2$ (p. 254).

Earth Science Connections

► Radiometric dating is a powerful tool in Earth sciences (Ch. 22).

OVERVIEW

The ancient alchemist dreamed of changing one element into another, such as lead into gold. The alchemist was never successful, however, because such changes were attempted with chemical reactions. Chemical reactions are reactions that involve only the electrons of atoms. Electrons are shared or transferred in chemical reactions, and the internal nucleus of the atom is unchanged. Elements thus retain their identity during the sharing or transferring of electrons. This chapter is concerned with a different kind of reaction, one that involves the *nucleus* of the atom.

In nuclear reactions, the nucleus of the atom is often altered, changing the identity of the elements involved. The ancient alchemist's dream of changing one element into another was actually a dream of achieving a nuclear change, that is, a nuclear reaction.

Understanding nuclear reactions is important because although fossil fuels are the major source of energy today, there are growing concerns about (1) air pollution from fossil fuel combustion, (2) increasing levels of CO_2 from fossil fuel combustion, which contributes to the warming of Earth

(the greenhouse effect), and (3) the dwindling fossil fuel supply, which cannot last forever. Energy experts see nuclear energy as a means of meeting rising energy demands in an environmentally acceptable way. However, the topic of nuclear energy is controversial, and discussions of it often result in strong emotional responses. Decisions about the use of nuclear energy require some understandings about nuclear reactions and some facts about radioactivity and radioactive materials (figure 11.1). These understandings and facts are the topics of this chapter.

NATURAL RADIOACTIVITY

Natural **radioactivity** is the spontaneous emission of particles or energy from an atomic nucleus as it disintegrates. It was discovered in 1896 by Henri Becquerel, a French scientist who was very interested in the recent discovery of X rays. Becquerel was experimenting with fluorescent minerals, minerals that give off visible light after being exposed to sunlight. He wondered if fluorescent minerals emitted X rays in addition to visible light. From previous work with X rays, Becquerel knew that they would penetrate a wrapped, light-tight photographic plate, exposing it as visible light exposes an unprotected plate. Thus, Becquerel decided to place a fluorescent uranium mineral on a protected photographic plate while the mineral was exposed to sunlight. Sure enough, he found a silhouette

of the mineral on the plate when it was developed. Believing the uranium mineral emitted X rays, he continued his studies until the weather turned cloudy. Storing a wrapped, protected photographic plate and the uranium mineral together during the cloudy weather, Becquerel returned to the materials later and developed the photographic plate to again find an image of the mineral (figure 11.2). He concluded that the mineral was emitting an "invisible radiation" that was not induced by sunlight. The emission of invisible radiation was later named *radioactivity*. Materials that have the property of radioactivity are called *radioactive* materials.

Becquerel's discovery led to the beginnings of the modern atomic theory and to the discovery of new elements. Ernest Rutherford studied the nature of radioactivity and found that there are three kinds, which are today known by the first three letters of the Greek alphabet—alpha (α), beta (β), and gamma (γ). These Greek letters were used at first before the nature of the radiation was known. Today, an **alpha particle** (sometimes called an alpha ray) is known to be the nucleus of a helium atom, that is, two protons and two neutrons. A **beta particle** (or beta ray) is a high-energy electron. A **gamma ray** is electromagnetic radiation, as is light, but of very short wavelength (figure 11.3).

It was Rutherford's work with alpha particles that resulted in the discovery of the nucleus and the proton (see chapter 8). At Becquerel's suggestion, Marie Curie searched for other radioactive materials and, in the process, discovered two new

FIGURE 11.1 Decisions about nuclear energy require some understanding of nuclear reactions and the nature of radioactivity. This is one of the three units of the Palo Verde Nuclear Generating Station in Arizona. With all three units running, enough power is generated to meet the electrical needs of nearly 4 million people.

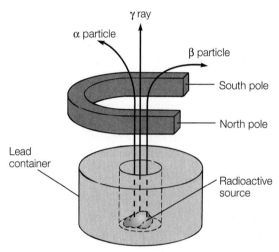

FIGURE 11.3 Radiation passing through a magnetic field shows that massive, positively charged alpha particles are deflected one way and less massive beta particles with their negative charge are greatly deflected in the opposite direction. Gamma rays, like light, are not deflected.

A

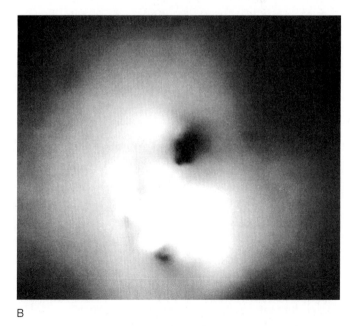

B

FIGURE 11.2 Radioactivity was discovered by Henri Becquerel when he exposed a light-tight photographic plate to a radioactive mineral, then developed the plate. (A) A photographic film is exposed to a uraninite ore sample. (B) The film, developed normally after a four-day exposure to uraninite. Becquerel found an image like this one and deduced that the mineral gave off invisible radiation.

elements, polonium and radium. More radioactive elements have been discovered since that time, and, in fact, all the isotopes of all the elements with an atomic number greater than 83 (bismuth) are radioactive. As a result of radioactive disintegration, the nucleus of an atom often undergoes a change of identity, becoming a simpler nucleus. The spontaneous disintegration of a given nucleus is a purely natural process and cannot be controlled or influenced. The natural spontaneous disintegration or decomposition of a nucleus is also called **radioactive decay.** Although it is impossible to know *when* a given nucleus will undergo radioactive decay, as you will see later, it is possible to deal with the *rate* of decay for a given radioactive material with precision.

Nuclear Equations

There are two main subatomic particles in the nucleus, the proton and the neutron. The proton and neutron are called **nucleons.** Recall that the number of protons, the *atomic number,* determines what element an atom is and that all atoms of a given element have the same number of protons. The number of neutrons varies in *isotopes,* which are atoms with the same atomic number but different numbers of neutrons. The number of protons and neutrons together determines the *mass number,* so different isotopes of the same element are identified with their mass numbers. Thus, the two most common, naturally occurring isotopes of uranium are referred to as uranium-238 and uranium-235, and the 238 and 235 are the mass numbers of these isotopes. Isotopes are also represented by the following symbol:

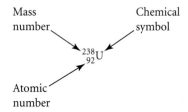

Subatomic particles involved in nuclear reactions are represented by symbols with the following form:

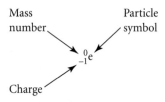

Symbols are used in an equation for a nuclear reaction that is written much like a chemical reaction with reactants and products. When a uranium-238 nucleus emits an alpha particle ($_2^4$He), for example, it loses two protons and two neutrons. The nuclear reaction is written in equation form as

$$_{92}^{238}\text{U} \rightarrow _{90}^{234}\text{Th} + _2^4\text{He}$$

Name	Symbol	Mass Number	Charge
Proton	$_1^1\text{H}$(or $_1^1\text{p}$)	1	1+
Electron	$_{-1}^0\text{e}$ (or $_{-1}^0\beta$)	0	1−
Neutron	$_0^1\text{n}$	1	0
Gamma photon	$_0^0\gamma$	0	0

The *products* of this nuclear reaction from the decay of a uranium-238 nucleus are (1) the alpha particle ($_2^4$He) given off and (2) the nucleus, which remains after the alpha particle leaves the original nucleus. What remains is easily determined since all nuclear equations must show conservation of charge and conservation of the total number of nucleons. Therefore, (1) the number of protons (positive charge) remains the same, and the sum of the subscripts (atomic number, or numbers of protons) in the reactants must equal the sum of the subscripts in the products; and (2) the total number of nucleons remains the same, and the sum of the superscripts (atomic mass, or number of protons plus neutrons) in the reactants must equal the sum of the superscripts in the products. The new nucleus remaining after the emission of an alpha particle, therefore, has an atomic number of $90(92 - 2 = 90)$. According to the table of atomic numbers on the inside back cover of this text, this new nucleus is thorium (Th). The mass of the thorium isotope is 238 minus 4, or 234. The emission of an alpha particle thus decreases the number of protons by 2 and the mass number by 4. From the subscripts, you can see that the total charge is conserved ($92 = 90 + 2$). From the superscripts, you can see that the total number of nucleons is also conserved ($238 = 234 + 4$). The mass numbers (superscripts) and the atomic numbers (subscripts) are *balanced* in a correctly written nuclear equation. Such nuclear equations are considered to be independent of any chemical form or chemical reaction. Nuclear reactions are independent and separate from chemical reactions, whether or not the atom is in the pure element or in a compound. Each particle that is involved in nuclear reactions has its own symbol with a superscript indicating mass number and a subscript indicating the charge. These symbols, names, and numbers are given in table 11.1.

The Nature of the Nucleus

The modern atomic theory does not picture the nucleus as a group of stationary protons and neutrons clumped together by some "nuclear glue." The protons and neutrons are understood to be held together by a *nuclear force,* a strong fundamental force of attraction that is functional only at very short distances, on the order of 10^{-15} m or less. At distances greater than about 10^{-15} m, the nuclear force is negligible, and the weaker *electromagnetic force,* the force of repulsion between like charges, is the operational force. Thus, like-charged protons experience a

Plutonium Decay Equation

A plutonium-242 nucleus undergoes radioactive decay, emitting an alpha particle. Write the nuclear equation for this nuclear reaction.

Step 1: The table of atomic weights on the inside back cover gives the atomic number of plutonium as 94. Plutonium-242 therefore has a symbol of $^{242}_{94}\text{Pu}$. The symbol for an alpha particle is (^4_2He), so the nuclear equation so far is

$$^{242}_{94}\text{Pu} \rightarrow \,^4_2\text{He} + \,?$$

Step 2: From the subscripts, you can see that $94 = 2 + 92$, so the new nucleus has an atomic number of 92. The table of atomic weights identifies element 92 as uranium with a symbol of U.

Step 3: From the superscripts, you can see that the mass number of the uranium isotope formed is $242 - 4 = 238$, so the product nucleus is $^{238}_{92}\text{U}$ and the complete nuclear equation is

$$^{242}_{94}\text{Pu} \rightarrow \,^4_2\text{He} + \,^{238}_{92}\text{U}$$

Step 4: Checking the subscripts ($94 = 2 + 92$) and the superscripts ($242 = 4 + 238$), you can see that the nuclear equation is balanced.

What is the product nucleus formed when radium emits an alpha particle? (Answer: Radon-222, a chemically inert, radioactive gas.)

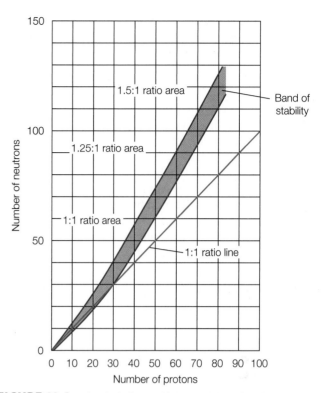

FIGURE 11.4 The shaded area indicates stable nuclei, which group in a band of stability according to their neutron-to-proton ratio. As the size of nuclei increases, so does the neutron-to-proton ratio that represents stability. Nuclei outside this band of stability are radioactive.

repulsive force when they are farther apart than about 10^{-15} m. When closer together than 10^{-15} m, the short-range, stronger nuclear force predominates, and the protons experience a strong attractive force. This explains why the like-charged protons of the nucleus are not repelled by their like electric charges.

Observations of radioactive decay reactants and products and experiments with nuclear stability have led to a *shell model of the nucleus*. This model considers the protons and neutrons moving in energy levels in the nucleus analogous to the orbital structure of electrons in the outermost part of the atom. As in the electron orbitals, there are certain configurations of nuclear shells that have a greater stability than others. Considering electrons, filled and half-filled orbitals are more stable than other arrangements, and maximum stability occurs with the noble gases and their 2, 10, 18, 36, 54, and 86 electrons. Considering the nucleus, atoms with 2, 8, 20, 28, 50, 82, or 126 protons or neutrons have a maximum nuclear stability. The stable numbers are not the same for electrons and nucleons because of differences in nuclear and electromagnetic forces.

Isotopes of uranium, radium, and plutonium, as well as other isotopes, emit an alpha particle during radioactive decay to a simpler nucleus. The alpha particle is a helium nucleus, (^4_2He). The alpha particle contains two protons as well as two neutrons, which is one of the nucleon numbers of stability, so

you would expect the helium nucleus (or alpha particle) to have a stable nucleus, and it does. *Stable* means it does not undergo radioactive decay. Pairs of protons and pairs of neutrons have increased stability, just as pairs of electrons in a molecule do. As a result, nuclei with an *even number* of both protons and neutrons are, in general, more stable than nuclei with odd numbers of protons and neutrons. There are a little more than 150 stable isotopes with an even number of protons and an even number of neutrons, but there are only 5 stable isotopes with odd numbers of each. Just as in the case of electrons, other factors come into play as the nucleus becomes larger and larger with increased numbers of nucleons.

The results of some of these factors are shown in figure 11.4, which is a graph of the number of neutrons versus the number of protons in nuclei. As the number of protons increases, the neutron-to-proton ratio of the *stable nuclei* also increases in a **band of stability.** Within the band, the neutron-to-proton ratio increases from about 1:1 at the bottom left to about 1½:1 at the top right. The increased ratio of neutrons is needed to produce a stable nucleus as the number of protons increases. Neutrons provide additional attractive *nuclear* (not electrical) forces, which counter the increased electrical repulsion from a larger number of positively charged protons. Thus, more neutrons are required in larger nuclei to produce a stable nucleus. However, there is a limit to the additional attractive forces that can be provided by more and more neutrons, and all isotopes of all elements with more than 83 protons are unstable and thus undergo radioactive decay.

The generalizations about nuclear stability provide a means of predicting if a particular nucleus is radioactive. The generalizations are as follows:

1. All isotopes with an atomic number greater than 83 have an unstable nucleus.
2. Isotopes that contain 2, 8, 20, 28, 50, 82, or 126 protons or neutrons in their nucleus occur in more stable isotopes than those with other numbers of protons or neutrons.
3. Pairs of protons and pairs of neutrons have increased stability, so isotopes that have nuclei with even numbers of both protons and neutrons are generally more stable than nuclei with odd numbers of both protons and neutrons.
4. Isotopes with an atomic number less than 83 are stable when the ratio of neutrons to protons in the nucleus is about 1:1 in isotopes with up to 20 protons, but the ratio increases in larger nuclei in a band of stability (see figure 11.4). Isotopes with a ratio to the left or right of this band are unstable and thus will undergo radioactive decay.

Types of Radioactive Decay

Through the process of radioactive decay, an unstable nucleus becomes a more stable one with less energy. The three more familiar types of radiation emitted—alpha, beta, and gamma— were introduced earlier.

1. **Alpha emission.** Alpha (α) emission is the expulsion of an alpha particle ($_2^4$He) from an unstable, disintegrating nucleus. The alpha particle, a helium nucleus, travels from 2 to 12 cm through the air, depending on the energy of emission from the source. An alpha particle is easily stopped by a sheet of paper close to the nucleus. As an example of alpha emission, consider the decay of a radon-222 nucleus,

$$_{86}^{222}\text{Rn} \rightarrow _{84}^{218}\text{Po} + _2^4\text{He}$$

The spent alpha particle eventually acquires two electrons and becomes an ordinary helium atom.

2. **Beta emission.** Beta (β^-) emission is the expulsion of a different particle, a beta particle, from an unstable disintegrating nucleus. A beta particle is simply an electron ($_{-1}^0$e) ejected from the nucleus at a high speed. The emission of a beta particle *increases the number of protons* in a nucleus. It is as if a neutron changed to a proton by emitting an electron, or

$$_0^1\text{n} \rightarrow _1^1\text{p} + _{-1}^0\text{e}$$

Carbon-14 is a carbon isotope that decays by beta emission:

$$_6^{14}\text{C} \rightarrow _7^{14}\text{N} + _{-1}^0\text{e}$$

Note that the number of protons increased from six to seven, but the mass number remained the same. The mass number is unchanged because the mass of the expelled electron (beta particle) is negligible.

Beta particles are more penetrating than alpha particles and may travel several hundred centimeters through the air. They can be stopped by a thin layer of metal close to the emitting nucleus, such as a 1 cm thick piece of aluminum. A spent beta particle may eventually join an ion to become part of an atom, or it may remain a free electron.

3. **Gamma emission.** Gamma (γ) emission is a high-energy burst of electromagnetic radiation from an excited nucleus. It is a burst of light (photon) of a wavelength much too short to be detected by the eye. Other types of radioactive decay, such as alpha or beta emission, sometimes leave the nucleus with an excess of energy, a condition called an *excited state*. As in the case of excited electrons, the nucleus returns to a lower energy state by emitting electromagnetic radiation. From a nucleus, this radiation is in the high-energy portion of the electromagnetic spectrum. Gamma is the most penetrating of the three common types of nuclear radiation. Like X rays, gamma rays can pass completely through a person, but all gamma radiation can be stopped by a 5 cm thick piece of lead close to the source. As with other types of electromagnetic radiation, gamma radiation is absorbed by and gives its energy to materials. Since the product nucleus changed from an excited state to a lower energy state, there is no change in the number of nucleons. For example, radon-222 is an isotope that emits gamma radiation:

$$_{86}^{222}\text{Rn}^* \rightarrow _{86}^{222}\text{Rn} + _0^0\gamma$$

(* denotes excited state)

Radioactive decay by alpha, beta, and gamma emission is summarized in table 11.2, which also lists the unstable nuclear conditions that lead to the particular type of emission. Just as electrons seek a state of greater stability, a nucleus undergoes radioactive decay to achieve a balance between nuclear attractions, electromagnetic repulsions, and a low quantum of nuclear shell energy. The key to understanding the types of reactions that occur is found in the band of stable nuclei illustrated in figure 11.4. The isotopes within this band have achieved the state of stability, and other isotopes above, below, or beyond the band are unstable and thus radioactive.

Nuclei that have a neutron-to-proton ratio beyond the upper right part of the band are unstable because of an imbalance between the proton-proton electromagnetic repulsions and all the combined proton and neutron nuclear attractions.

TABLE 11.2 Radioactive Decay

Unstable Condition	Type of Decay	Emitted	Product Nucleus
More than 83 protons	Alpha emission	$_2^4$He	Lost 2 protons and 2 neutrons
Neutron-to-proton ratio too large	Beta emission	$_{-1}^0$e	Gained 1 proton, no mass change
Excited nucleus	Gamma emission	$_0^0\gamma$	No change
Neutron-to-proton ratio too small	Other emission	$_1^0$e	Lost 1 proton, no mass change

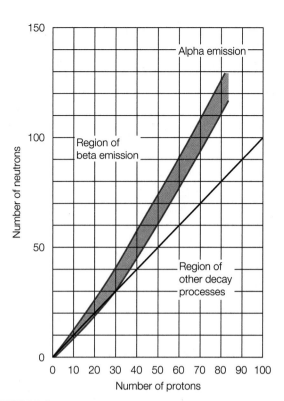

FIGURE 11.5 Unstable nuclei undergo different types of radioactive decay to obtain a more stable nucleus. The type of decay depends, in general, on the neutron-to-proton ratio, as shown.

Recall that the neutron-to-proton ratio increases from about 1:1 to about 1½:1 in the larger nuclei. The additional neutron provided additional nuclear attractions to hold the nucleus together, but atomic number 83 appears to be the upper limit to this additional stabilizing contribution. Thus, all nuclei with an atomic number greater than 83 are outside the upper right limit of the band of stability. Emission of an alpha particle reduces the number of protons by two and the number of neutrons by two, moving the nucleus more toward the band of stability. Thus, you can expect a nucleus that lies beyond the upper right part of the band of stability to be an alpha emitter (figure 11.5).

A nucleus with a neutron-to-proton ratio that is too large will be on the left side of the band of stability. Emission of a beta particle decreases the number of neutrons and increases the number of protons, so a beta emission will lower the neutron-to-proton ratio. Thus, you can expect a nucleus with a large neutron-to-proton ratio, that is, one to the left of the band of stability, to be a beta emitter.

A nucleus that has a neutron-to-proton ratio that is too small will be on the right side of the band of stability. These nuclei can increase the number of neutrons and reduce the number of protons in the nucleus by other types of radioactive decay. As usual when dealing with broad generalizations and trends, there are exceptions to the summarized relationships between neutron-to-proton ratios and radioactive decay.

Radioactive Decay Series

A radioactive decay reaction produces a simpler and eventually more stable nucleus than the reactant nucleus. As discussed in

Radiation and Food Preservation

Radiation can be used to delay food spoilage and preserve foods by killing bacteria and other pathogens, just as heat is used to pasteurize milk. Foods such as wheat, flour, fruits, vegetables, pork, chicken, turkey, ground beef, and other uncooked meats are exposed to gamma radiation from cobalt-60 or cesium-137 isotopes, X rays, and electron beams. This kills insects, parasites such as *Trichinella spiralis* and tapeworms, and bacteria such as *E. coli, Listeria, Salmonellae,* and *Staphylococcus.* The overall effect is that many food-borne causes of human disease are eliminated and it is possible to store foods longer.

Food in the raw state, processed, or frozen is passed through a machine where it is irradiated. This process does not make the food radioactive since the food does not touch any radioactive substance. In addition, the radiation used in the process is not strong enough to disrupt the nucleus of atoms in food molecules, so it does not produce radioactivity, either.

In addition to killing the parasites and bacteria, the process might result in some nutritional loss but no more than that which normally occurs in canning. Some new chemical products may be formed by the exposure to radiation, but studies in several countries have not been able to identify any health problems or ill effects from these compounds.

Treatment with radiation works better for some foods than others. Dairy products undergo some flavor changes that are undesirable, and some fruits such as peaches become soft. Irradiated strawberries, on the other hand, remain firm and last for weeks instead of a few days in the refrigerator. Foods that are sterilized with stronger doses of irradiation can be stored for years without refrigeration just like canned foods that have undergone heat pasteurization.

In the United States, the Food and Drug Administration regulates which products can be treated by radiation and the dosages used in the treatment. The U.S. Department of Agriculture is responsible for the inspection of irradiated meat and poultry products. All foods that have undergone a radiation treatment must show the international logo for this process and a statement. The logo, called a radura, is a stylized flower inside a circle with five openings on the top part (box figure 11.1).

Box Figure 11.1

the previous section, large nuclei with an atomic number greater than 83 decay by alpha emission, giving up two protons and two neutrons with each alpha particle. A nucleus with an atomic number greater than 86, however, will emit an alpha particle and *still* have an atomic number greater than 83, which means the product nucleus will also be radioactive. This nucleus will also undergo radioactive decay, and the process will continue through a series of decay reactions until a stable nucleus is achieved. Such a series of decay reactions that (1) begins with one radioactive nucleus, which (2) decays to a second nucleus, which (3) then decays to a third nucleus, and so on until (4) a stable nucleus is reached is called a *radioactive decay series.* There are three naturally occurring radioactive decay series. One begins with thorium-232 and ends with lead-208, another begins with uranium-235 and ends with lead-207, and the third series begins with uranium-238 and ends with lead-206. Figure 11.6 shows the uranium-238 radioactive decay series.

As figure 11.6 illustrates, the uranium-238 begins with uranium-238 decaying to thorium-234 by alpha emission. Thorium has a new position on the graph because it now has a new atomic number and a new mass number. Thorium-234 is unstable and decays to protactinium-234 by beta emission, which is also unstable and decays by beta emission to uranium-234. The process continues with five sequential alpha emissions, then two beta-beta-alpha decay steps before the series terminates with the stable lead-206 nucleus.

The rate of radioactive decay is usually described in terms of its *half-life.* The **half-life** is the time required for one-half of the unstable nuclei to decay. Since each isotope has a characteristic decay constant, each isotope has its own characteristic half-life. Half-lives of some highly unstable isotopes are measured in fractions of seconds, and other isotopes have half-lives

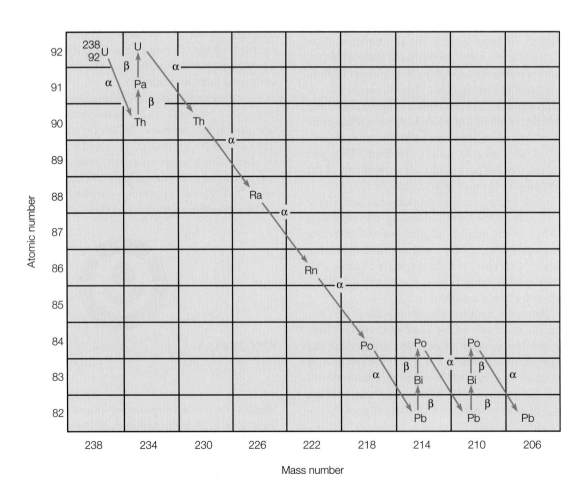

FIGURE 11.6   The radioactive decay series for uranium-238. This is one of three naturally occurring series.

TABLE 11.3   Half-Lives of Some Radioactive Isotopes

Isotope	Half-Life	Mode of Decay
$^{3}_{1}H$ (tritium)	12.26 years	Beta
$^{14}_{6}C$	5,730 years	Beta
$^{90}_{38}Sr$	28 years	Beta
$^{131}_{53}I$	8 days	Beta
$^{133}_{54}Xe$	5.27 days	Beta
$^{238}_{92}U$	4.51×10^{9} years	Alpha
$^{242}_{94}Pu$	3.79×10^{5} years	Alpha
$^{240}_{94}Pu$	6,760 years	Alpha
$^{239}_{94}Pu$	24,360 years	Alpha

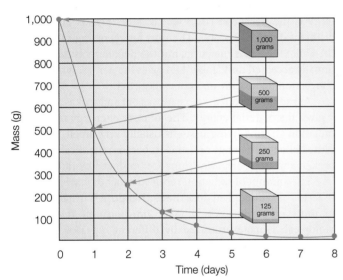

FIGURE 11.7   Radioactive decay of a hypothetical isotope with a half-life of one day. The sample decays each day by one-half to some other element. Actual half-lives may be in seconds, minutes, or any time unit up to billions of years.

measured in seconds, minutes, hours, days, months, years, or billions of years. Table 11.3 lists half-lives of some of the isotopes, and the process is illustrated in figure 11.7.

As an example of the half-life measure, consider a hypothetical isotope that has a half-life of one day. The half-life is independent of the amount of the isotope being considered, but suppose you start with a 1.0 kg sample of this element with a half-life of one day. One day later, you will have half of the original sample, or 500 g. The other half did not disappear, but it is now the decay product, that is, some new element. During the next day, half of the remaining nuclei will disintegrate, and only 250 g of the initial sample is still the original element. One-half of the remaining sample will disintegrate each day until the original sample no longer exists.

A Closer Look

How Is Half-Life Determined?

It is not possible to predict when a radioactive nucleus will decay because it is a random process. It is possible, however, to deal with nuclear disintegration statistically since the rate of decay is not changed by any external conditions of temperature, pressure, or any chemical state. When dealing with a large number of nuclei, the ratio of the rate of nuclear disintegration per unit of time to the total number of radioactive nuclei is a constant, or

$$\text{radioactive decay constant} = \frac{\text{decay rate}}{\text{number of nuclei}}$$

The radioactive decay constant is a specific constant for a particular isotope, and each isotope has its own decay constant that can be measured. For example, a 238 g sample of uranium-238 (1 mole) that has 2.93×10^{6} disintegrations per second would have a decay constant of

$$\text{radioactive decay constant} = \frac{\text{decay rate}}{\text{number of nuclei}}$$
$$= \frac{2.93 \times 10^{6} \text{ nuclei/s}}{6.01 \times 10^{23} \text{ nuclei}}$$
$$= 4.87 \times 10^{-18} \text{ 1/s}$$

The half-life of a radioactive nucleus is related to its radioactive decay constant by

$$\text{half-life} = \frac{\text{a mathematical constant}}{\text{decay constant}}$$

The half-life of uranium-238 is therefore

$$\text{half-life} = \frac{\text{a mathematical constant}}{\text{decay constant}}$$
$$= \frac{0.693}{4.87 \times 10^{-18} \text{ 1/s}}$$
$$= 1.42 \times 10^{17} \text{ s}$$

This is the half-life of uranium-238 in seconds. There are $60 \times 60 \times 24 \times 365$, or 3.15×10^{7} s in a year, so

$$\frac{1.42 \times 10^{17} \text{ s}}{3.15 \times 10^{7} \text{ s/yr}} = 4.5 \times 10^{9} \text{ yr}$$

The half-life of uranium-238 is thus 4.5 billion years.

MEASUREMENT OF RADIATION

The measurement of radiation is important in determining the half-life of radioactive isotopes. Radiation measurement is also important in considering biological effects, which will be discussed in the next section. As is the case with electricity, it is not possible to make direct measurements on things as small as electrons and other parts of atoms. Indirect measurement methods are possible, however, by considering the effects of the radiation.

Measurement Methods

As Becquerel discovered, radiation affects photographic film, exposing it as visible light does. Since the amount of film exposure is proportional to the amount of radiation, photographic film can be used as an indirect measure of radiation. Today, people who work around radioactive materials or X rays carry light-tight film badges. The film is replaced periodically and developed. The optical density of the developed film provides a record of the worker's exposure to radiation since the darkness of the developed film is proportional to the exposure.

There are also devices that indirectly measure radiation by measuring an effect of the radiation. An *ionization counter* is one type of device that measures ions produced by radiation. A second type of device is called a *scintillation counter*. *Scintillate* is a word meaning "sparks or flashes," and a scintillation counter measures the flashes of light produced when radiation strikes a phosphor.

The most common example of an ionization counter is known as a *Geiger counter*. The working components of a Geiger counter are illustrated in figure 11.8. Radiation is received in a metal tube filled with an inert gas, such as argon. An insulated wire inside the tube is connected to the positive terminal of a direct current source. The metal cylinder around the insulated wire is connected to the negative terminal. There is not a current between the center wire and the metal cylinder because the gas acts as an insulator. When radiation passes through the window, however, it ionizes some of the gas atoms, releasing free electrons. These electrons are accelerated by the field between the wire and cylinder, and the accelerated electrons ionize more gas molecules, which results in an *avalanche* of free electrons. The avalanche creates a pulse of current that is amplified and then measured. More radiation means more avalanches, so the pulses are an indirect means of measuring radiation. When connected to a speaker or earphone, each avalanche produces a "pop" or "click."

Some materials are *phosphors*, substances that emit a flash of light when excited by radiation. Zinc sulfide, for example, gives off a tiny flash of light when struck by radiation from a disintegrating radium nucleus. A scintillation counter measures the flashes of light through the photoelectric effect, producing free electrons that are accelerated to produce a pulse of current. Again, the pulses of current are used as an indirect means to measure radiation.

Radiation Units

You have learned that *radioactivity* is a property of isotopes with unstable, disintegrating nuclei and *radiation* is emitted particles (alpha or beta) or energy traveling in the form of photons (gamma). Radiation can be measured (1) at the source of radioactivity or (2) at a place of reception, where the radiation is absorbed.

The *activity* of a radioactive source is a measure of the number of nuclear disintegrations per unit of time. The unit of activity at the source is called a **curie** (Ci), which is defined as 3.70×10^{10} nuclear disintegrations per second. Activities are usually expressed in terms of fractions of curies, for example, a *picocurie* (pCi), which is a millionth of a millionth of a curie. Activities are sometimes expressed in terms of so many picocuries per liter (pCi/L).

The International System of Units (SI) unit for radioactivity is the *Becquerel* (Bq), which is defined as one nuclear disintegration per second. The unit for reporting radiation in the United States is the curie, but the Becquerel is the internationally accepted unit. Table 11.4 gives the names, symbols, and conversion factors for units of radioactivity.

As radiation from a source moves out and strikes a material, it gives the material energy. The amount of energy released by radiation striking living tissue is usually very small, but it can cause biological damage nonetheless because chemical bonds are broken and free polyatomic ions are produced by radiation.

The amount of radiation received by a human is expressed in terms of radiological dose. Radiation dose is usually written in units of a **rem,** which takes into account the possible biological damage produced by different types of radiation. Doses are usually expressed in terms of fractions of the rem, for example, a *millirem* (mrem). A millirem is 1/1,000 of a rem and is the unit of choice when low levels of radiation are discussed. The SI unit for radiation dose is the *millisievert* (mSv). Both the millirem and the millisievert relate ionizing radiation and biological effect to humans. The natural radiation that people receive from nature in one day is less than 1 millirem (0.01 millisievert). A single dose of 100,000 to 200,000 millirem

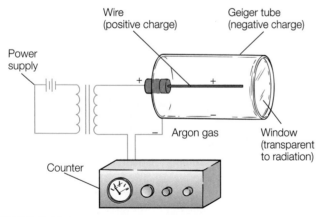

FIGURE 11.8 The working parts of a Geiger counter.

TABLE 11.4 Names, Symbols, and Conversion Factors for Radioactivity

Name	Symbol	To Obtain	Multiply By
becquerel	Bq	Ci	2.7×10^{11}
gray	Gy	rad	100
sievert	Sv	rem	100
curie	Ci	Bq	3.7×10^{10}
rem	rem	Sv	0.01
millirem	millirem	rem	0.001
rem	rem	millirem	1,000

TABLE 11.5 Approximate Single-Dose, Whole-Body Effects of Radiation Exposure

Level	Comment
0.240 rem	Average annual exposure to natural background radiation
0.500 rem	Upper limit of annual exposure to general public
25.0 rem	Threshold for observable effects such as blood count changes
100.0 rem	Fatigue and other symptoms of radiation sickness
200.0 rem	Definite radiation sickness, bone marrow damage, possibility of developing leukemia
500.0 rem	Lethal dose for 50 percent of individuals
1,000.0 rem	Lethal dose for all

(1,000 to 2,000 millisievert) can cause radiation sickness in humans (table 11.5). A single dose of 500,000 millirem (5,000 millisievert) results in death about 50 percent of the time.

Another measure of radiation received by a material is the **rad.** The term *rad* is from *radiation absorbed dose.* The SI unit for radiation received by a material is the *gray* (Gy). One gray is equivalent to an exposure of 100 rad.

Radiation Exposure

Natural radioactivity is a part of your environment, and you receive between 100 and 500 millirems each year from natural sources. This radiation from natural sources is called **background radiation.** Background radiation comes from outer space in the form of cosmic rays and from unstable isotopes in the ground, building materials, and foods. Many activities and situations will increase your yearly exposure to radiation. For example, the atmosphere absorbs some of the cosmic rays from space, so the less atmosphere above you, the more radiation you will receive. You are exposed to one additional millirem per year for each 100 feet you live above sea level. You receive approximately 0.3 millirem for each hour spent on a jet flight. Airline crews receive an additional 300 to 400 millirems per year because they spend so much time high in the atmosphere. Additional radiation exposure comes from medical X rays and television sets. In general, the worldwide average background radiation exposure for the average person is about 240 millirems per year.

What are the consequences of radiation exposure? Radiation can be a hazard to living organisms because it produces ionization along its path of travel. This ionization can (1) disrupt chemical bonds in essential macromolecules such as DNA and (2) produce molecular fragments, which are free polyatomic ions that can interfere with enzyme action and other essential cell functions. Tissues with highly active cells are more vulnerable to radiation damage than others, such as blood-forming tissue. Thus, one of the symptoms of an excessive radiation exposure is an altered blood count. Table 11.5 compares the estimated results of various levels of acute radiation exposure.

Radiation is not a mysterious, unique health hazard. It is a hazard that should be understood and neither ignored nor exaggerated. Excessive radiation exposure should be avoided, just as you avoid excessive exposure to other hazards such as certain chemicals, electricity, or even sunlight. Everyone agrees that *excessive* radiation exposure should be avoided, but there is some controversy about long-term, low-level exposure and its possible role in cancer. Some claim that tolerable low-level exposure does not exist because that is not possible. Others point to many studies comparing high and low background radioactivity with cancer mortality data. For example, no cancer mortality differences could be found between people receiving 500 or more millirems a year and those receiving fewer than 100 millirems a year. The controversy continues, however, because of lack of knowledge about long-term exposure. Two models of long-term, low-level radiation exposure have been proposed: (1) a linear model and (2) a threshold model. The *linear model* proposes that any radiation exposure above zero is damaging and can produce cancer and genetic damage. The *threshold model* proposes that the human body can repair damage and get rid of damaging free polyatomic ions up to a certain exposure level called the threshold (figure 11.9). The controversy over long-term, low-level radiation exposure will probably continue until there is clear evidence about which model is correct. Whichever is correct will not lessen the need for rational risks versus cost-benefit analyses of all energy alternatives.

Myths, Mistakes, and Misunderstandings

Antiradiation Pill?

It is a myth that an antiradiation pill exists that will protect you from ionizing radiation. There is a pill, an iodine supplement, that is meant to saturate your thyroid with a nonradioactive isotope of iodine. Once saturated, your thyroid will not absorb radioactive isotopes for storage in the gland, which could be dangerous.

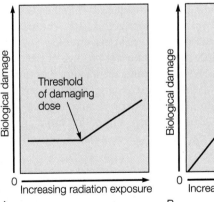
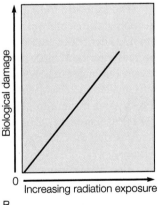

FIGURE 11.9 Graphic representation of the (*A*) threshold model and (*B*) linear model of low-level radiation exposure. The threshold model proposes that the human body can repair damage up to a threshold. The linear model proposes that any radiation exposure is damaging.

NUCLEAR ENERGY

Some nuclei are unstable because they are too large or because they have an unstable neutron-to-proton ratio. These unstable nuclei undergo radioactive decay, eventually forming products of greater stability. An example of this radioactive decay is the alpha emission reaction of uranium-238 to thorium-234,

$$^{238}_{92}\text{U} \rightarrow \,^{234}_{90}\text{Th} + \,^{4}_{2}\text{He}$$

$$238.0003 \text{ u} \rightarrow 233.9942 \text{ u} + 4.00150 \text{ u}$$

The numbers below the nuclear equation are the *nuclear* masses (u) of the reactant and products. As you can see, there seems to be a loss of mass in the reaction,

$$233.9942 + 4.00150 - 238.0003 = -0.0046 \text{ u}$$

This change in mass is related to the energy change according to the relationship that was formulated by Albert Einstein in 1905. The relationship is

$$E = mc^2 \qquad \textbf{equation 11.1}$$

where E is a quantity of energy, m is a quantity of mass, and c is a constant equal to the speed of light in a vacuum, 3.00×10^8 m/s. According to this relationship, matter and energy are the same thing, and energy can be changed to matter and vice versa.

The relationship between mass and energy explains why the mass of a nucleus is always *less* than the sum of the masses of the individual particles of which it is made.

The difference between the mass of the individual nucleons making up a nucleus and the actual mass of the nucleus is called the **mass defect** of the nucleus. The explanation for the mass defect is again found in $E = mc^2$. When nucleons join to make a nucleus, energy is released as the more stable nucleus is formed.

The energy equivalent released when a nucleus is formed is the same as the **binding energy,** the energy required to break the nucleus into individual protons and neutrons. The binding energy of the nucleus of any isotope can be calculated from the mass defect of the nucleus.

The ratio of binding energy to nucleon number is a reflection of the stability of a nucleus (figure 11.10). The greatest binding energy per nucleon occurs near mass number 56, then decreases for both more massive and less massive nuclei. This means that more massive nuclei can gain stability by splitting into smaller nuclei with the release of energy. It also means that less massive nuclei can gain stability by joining together with the release of energy. The slope also shows that more energy is released in the coming-together process than in the splitting process.

The nuclear reaction of splitting a massive nucleus into more stable, less massive nuclei with the release of energy is **nuclear fission** (figure 11.11). Nuclear fission occurs rapidly in an atomic bomb explosion and occurs relatively slowly in a nuclear reactor. The nuclear reaction of less massive nuclei, coming together to form more stable, and more massive, nuclei with the release of energy is **nuclear fusion.** Nuclear fusion

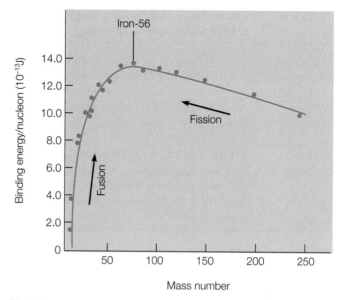

FIGURE 11.10 The maximum binding energy per nucleon occurs around mass number 56, then decreases in both directions. As a result, fission of massive nuclei and fusion of less massive nuclei both release energy.

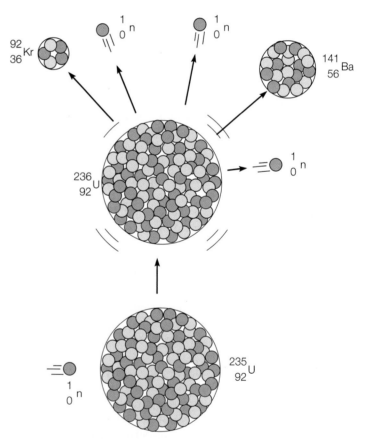

FIGURE 11.11 The fission reaction occurring when a neutron is absorbed by a uranium-235 nucleus. The deformed nucleus splits any number of ways into lighter nuclei, releasing neutrons in the process.

occurs rapidly in a hydrogen bomb explosion and occurs continually in the Sun, releasing the energy essential for the continuation of life on Earth.

Nuclear medicine had its beginnings in 1946 when radioactive iodine was first successfully used to treat thyroid cancer patients. Then physicians learned that radioactive iodine could also be used as a diagnostic tool, providing a way to measure the function of the thyroid and to diagnose thyroid disease. More and more physicians began to use nuclear medicine to diagnose thyroid disease as well as to treat hyperthyroidism and other thyroid problems. Nuclear medicine is a branch of medicine using radiation or radioactive materials to diagnose as well as treat diseases.

The development of new nuclear medicine technologies, such as new cameras, detection instruments, and computers, has led to a remarkable increase in the use of nuclear medicine as a diagnostic tool. Today, there are nearly one hundred different nuclear medicine imaging procedures. These provide unique, detailed information about virtually every major organ system within the body, information that was unknown just years ago. Treatment of disease with radioactive materials continues to be a valuable part of nuclear medicine, too. The material that follows will consider some techniques of using nuclear medicine as a diagnostic tool, followed by a short discussion of the use of radioactive materials in the treatment of disease.

Nuclear medicine provides diagnostic information about organ function, compared to conventional radiology, which provides images about the structure. For example, a conventional X-ray image will show if a bone is broken or not, while a bone imaging nuclear scan will show changes caused by tumors, hair-line fractures, or arthritis. There are procedures for making detailed structural X-ray pictures of internal organs such as the liver, kidney, or heart, but these images often cannot provide diagnostic information, showing only the structure. Nuclear medicine scans, on the other hand, can provide information about how much heart tissue is still alive after a heart attack or if a kidney is working, even when there are no detectable changes in organ appearance.

An X-ray image is produced when X rays pass through the body and expose photographic film on the other side. Some X-ray exams improve photographic contrast by introducing certain substances. A barium sulfate "milk shake," for example, can be swallowed to highlight the esophagus, stomach, and intestine. More information is provided if X rays are used in a CAT scan (CAT stands for computed axial tomography). The CAT scan is a diagnostic test that combines the use of X rays with computer technology. The CAT scan shows organs of interest by making X-ray images from many different angles as the source of the X rays moves around the patient. Contrast-improving substances, such as barium sulfate, might also be used with a CAT scan. In any case, CAT scan images are assembled by a computer into a three-dimensional picture that can show organs, bones, and tissues in great detail.

The gamma camera is a key diagnostic imaging tool used in nuclear medicine. Its use requires a radioactive material, called a radiopharmaceutical, to be injected into or swallowed by the patient. A given radiopharmaceutical tends to go to a specific organ of the body, for example, radioactive iodine tends to go to the thyroid gland, and others go to other organs. Gamma-emitting radiopharmaceuticals are used with the gamma camera, and the gamma camera collects and processes these gamma rays to produce images. These images provide a way of studying the structure as well as measuring the function of the selected organ. Together, the structure and function provide a way of identifying tumors, areas of infection, or other problems. The patient experiences little or no discomfort and the radiation dose is small.

A SPECT (single photon emission computerized tomography) scan is an imaging technique employing a gamma camera that rotates around the patient, measuring gamma rays and computing their point of origin. Cross-sectional images of a three-dimensional internal organ can be obtained from such data, resulting in images that have higher resolution and thus more diagnostic information than a simple gamma camera image. A gallium radiopharmaceutical is often used in a scan to diagnose and follow the progression of tumors or infections. Gallium scans also can be used to evaluate the heart, lungs, or any other organ that may be involved with inflammatory disease.

Use of MRI (magnetic resonance imaging) also produces images as an infinite number of projections through the body. Unlike CAT, gamma, or SPECT scans, MRI does not use any form of ionizing radiation. MRI uses magnetic fields, radio waves, and a computer to produce detailed images. As the patient enters an MRI scanner, his body is surrounded by a large magnet. The technique requires a very strong magnetic field, a field so strong that it aligns the nuclei of the person's atoms. The scanner sends a strong radio signal, temporarily knocking the nuclei out of alignment. When the radio signal stops, the nuclei return to the aligned position, releasing their own faint radio frequencies. These radio signals are read by the scanner, which uses them in a computer program to produce very detailed images of the human anatomy (box figure 11.2).

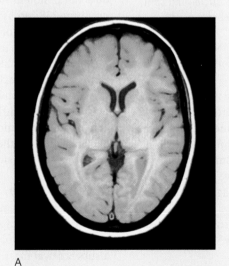

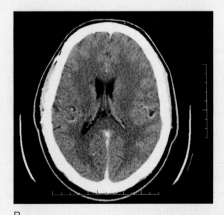

A   B

Box Figure 11.2 (*A*) MRI scan of a brain, (*B*) CAT scan of a brain.

—Continued top of next page

The PET (positron emission tomography) scan is the most recent technological development in nuclear medicine imaging, producing 3-D images superior to gamma camera images. This technique is built around a radiopharmaceutical that emits positrons (like an electron with a positive charge). Positrons collide with electrons, releasing a burst of energy in the form of photons. Detectors track the emissions and feed the information into a computer. The computer has a program to plot the source of radiation and translates the data into an image. Positron-emitting radiopharmaceuticals used in a PET scan can be low atomic weight elements such as carbon, nitrogen, and oxygen. This is important for certain purposes since these are the same elements found in many biological substances such as sugar, urea, or carbon dioxide. Thus, a PET scan can be used to study processes in organs such as the brain and heart where glucose is being broken down or oxygen is being consumed. This diagnostic method can be used to detect epilepsy or brain tumors, among other problems.

Radiopharmaceuticals used for diagnostic examinations are selected for their affinity for certain organs, if they emit sufficient radiation to be easily detectable in the body, and if they have a rather short half-life, preferably no longer than a few hours. Useful radioisotopes that meet these criteria for diagnostic purposes are technetium-99, gallium-67, indium-111, iodine-123, iodine-131, thallium-201, and krypton-81.

The goal of therapy in nuclear medicine is to use radiation to destroy diseased or cancerous tissue while sparing adjacent healthy tissue. Few radioactive therapeutic agents are injected or swallowed, with the exception of radioactive iodine—mentioned earlier as a treatment for cancer of the thyroid. Useful radioisotopes for therapeutical purposes are iodine-131, phosphorus-32, iridium-192, and gold-198. The radioactive source placed in the body for local irradiation of a tumor is normally iridium-192. A nuclear pharmaceutical is a physiologically active carrier to which a radioisotope is attached. Today, it is possible to manufacture chemical or biological carriers that migrate to a particular part of the human body, and this is the subject of much ongoing medical research.

Nuclear Fission

Nuclear fission was first accomplished in the late 1930s when researchers were attempting to produce isotopes by bombarding massive nuclei with neutrons. In 1938, two German scientists, Otto Hahn and Fritz Strassman, identified the element barium in a uranium sample that had been bombarded with neutrons. Where the barium came from was a puzzle at the time, but soon afterward, Lise Meitner, an associate who had moved to Sweden, deduced that the uranium nuclei had split, producing barium. The reaction might have been

$$\ _{0}^{1}\text{n} + \ _{92}^{235}\text{U} \rightarrow \ _{56}^{141}\text{Ba} + \ _{36}^{92}\text{Kr} + 3\ _{0}^{1}\text{n}$$

The phrase "might have been" is used because a massive nucleus can split in many different ways, producing different products. About thirty-five different, less massive elements have been identified among the fission products of uranium-235. Some of these products are fission fragments, and some are produced by unstable fragments that undergo radioactive decay. Selected fission fragments are listed in table 11.6 on page 257, together with their major modes of radioactive decay and half-lives. Some of the isotopes are the focus of concern about nuclear wastes, the topic of the "Science and Society" reading at the end of this chapter.

The fission of a uranium-235 nucleus produces two or three neutrons along with other products. These neutrons can each collide with other uranium-235 nuclei, where they are absorbed, causing fission with the release of more neutrons, which collide with other uranium-235 nuclei to continue the process. A reaction where the products are able to produce more reactions in a self-sustaining series is called a **chain reaction.** A chain reaction is self-sustaining until all the uranium-235 nuclei have fissioned or until the neutrons fail to strike a uranium-235 nucleus (figure 11.12).

You might wonder why all the uranium in the universe does not fission in a chain reaction. Natural uranium is mostly uranium-238, an isotope that does not fission easily. Only about 0.7 percent of natural uranium is the highly fissionable uranium-235. This low ratio of readily fissionable uranium-235 nuclei makes it unlikely that a stray neutron would be able to achieve a chain reaction.

To achieve a chain reaction, there must be (1) a sufficient mass with (2) a sufficient concentration of fissionable nuclei. When the mass and concentration are sufficient to sustain a chain reaction, the amount is called a **critical mass.** Likewise, a mass too small to sustain a chain reaction is called a *subcritical mass.* A mass of sufficiently pure uranium-235 (or plutonium-239) that is large enough to produce a rapidly accelerating chain reaction is called a *supercritical mass.* An atomic bomb is simply a device that uses a small, conventional explosive to push subcritical masses of fissionable material into a supercritical mass. Fission occurs almost instantaneously in the supercritical mass, and tremendous energy is released in a violent explosion.

TABLE 11.6   **Fragments and Products from Nuclear Reactors Using Fission of Uranium-235**

Isotope	Major Mode of Decay	Half-Life	Isotope	Major Mode of Decay	Half-Life
Tritium	Beta	12.26 years	Cerium-144	Beta, gamma	285 days
Carbon-14	Beta	5,730 years	Promethium-147	Beta	2.6 years
Argon-41	Beta, gamma	1.83 hours	Samarium-151	Beta	90 years
Iron-55	Electron capture	2.7 years	Europium-154	Beta, gamma	16 years
Cobalt-58	Beta, gamma	71 days	Lead-210	Beta	22 years
Cobalt-60	Beta, gamma	5.26 years	Radon-222	Alpha	3.8 days
Nickel-63	Beta	92 years	Radium-226	Alpha, gamma	1,620 years
Krypton-85	Beta, gamma	10.76 years	Thorium-229	Alpha	7,300 years
Strontium-89	Beta	5.4 days	Thorium-230	Alpha	26,000 years
Strontium-90	Beta	28 years	Uranium-234	Alpha	2.48×10^5 years
Yttrium-91	Beta	59 days	Uranium-235	Alpha, gamma	7.13×10^8 years
Zirconium-93	Beta	9.5×10^5 years	Uranium-238	Alpha	4.51×10^9 years
Zirconium-95	Beta, gamma	65 days	Neptunium-237	Alpha	2.14×10^6 years
Niobium-95	Beta, gamma	35 days	Plutonium-238	Alpha	89 years
Technetium-99	Beta	2.1×10^5 years	Plutonium-239	Alpha	24,360 years
Ruthenium-106	Beta	1 year	Plutonium-240	Alpha	6,760 years
Iodine-129	Beta	1.6×10^7 years	Plutonium-241	Beta	13 years
Iodine-131	Beta, gamma	8 days	Plutonium-242	Alpha	3.79×10^5 years
Xenon-133	Beta, gamma	5.27 days	Americium-241	Alpha	458 years
Cesium-134	Beta, gamma	2.1 years	Americium-243	Alpha	7,650 years
Cesium-135	Beta	2×10^6 years	Curium-242	Alpha	163 days
Cesium-137	Beta	30 years	Curium-244	Alpha	18 years
Cerium-141	Beta	32.5 days			

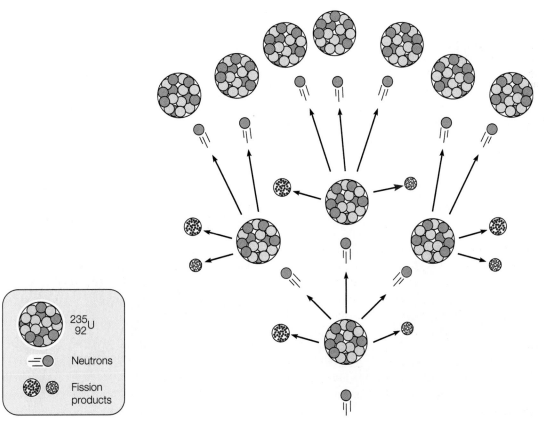

FIGURE 11.12   A schematic representation of a chain reaction. Each fissioned nucleus releases neutrons, which move out to fission other nuclei. The number of neutrons can increase quickly with each series.

Nuclear Power Plants

The nuclear part of a nuclear power plant is the *nuclear reactor,* a steel vessel in which a controlled chain reaction of fissionable material releases energy (figure 11.13). In the most popular design, called a pressurized light-water reactor, the fissionable material is enriched 3 percent uranium-235 and 97 percent uranium-238 that has been fabricated in the form of small ceramic pellets (figure 11.14A). The pellets are encased in a long zirconium alloy tube called a *fuel rod.* The fuel rods are locked into a *fuel rod assembly* by locking collars, arranged to permit pressurized water to flow around each fuel rod (figure 11.14B) and to allow the insertion of *control rods* between the fuel rods. *Control rods* are constructed of

materials, such as cadmium, that absorb neutrons. The lowering or raising of control rods within the fuel rod assemblies slows or increases the chain reaction by varying the amount of neutrons absorbed. When they are lowered completely into the assembly, enough neutrons are absorbed to stop the chain reaction.

It is physically impossible for the low-concentration fuel pellets to form a supercritical mass. A nuclear reactor in a power plant can only release energy at a comparatively slow rate, and it is impossible for a nuclear power plant to produce a nuclear explosion. In a pressurized water reactor, the energy released is carried away from the reactor by pressurized water in a closed pipe called the *primary loop* (figure 11.15). The water is pressurized at about 150 atmospheres (about 2,200 lb/in²) to keep it from boiling, since its temperature may be 350°C (about 660°F).

In the pressurized light-water (ordinary water) reactor, the circulating pressurized water acts as a coolant, carrying heat away from the reactor. The water also acts as a *moderator,* a substance that slows neutrons so they are more readily absorbed by uranium-235 nuclei. Other reactor designs use heavy water (deuterium dioxide) or graphite as a moderator.

Water from the closed primary loop is circulated through a heat exchanger called a *steam generator* (figure 11.15). The pressurized high-temperature water from the reactor moves through hundreds of small tubes inside the generator as *feedwater* from the *secondary loop* flows over the tubes. The water in the primary loop heats feedwater in the steam generator and then returns to the nuclear reactor to become heated again. The feedwater is heated to steam at about 235°C (455°F) with a pressure of about 68 atmospheres (1,000 lb/in²). This steam is piped to the turbines, which turn an electric generator (figure 11.16).

After leaving the turbines, the spent steam is condensed back to liquid water in a second heat exchanger receiving water from the cooling towers. Again, the cooling water does not mix with the closed secondary loop water. The cooling-tower water enters the condensing heat exchanger at about 32°C (90°F) and

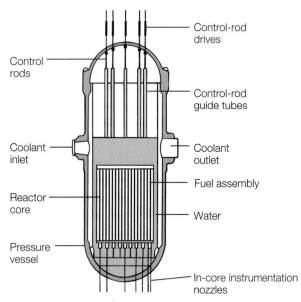

FIGURE 11.13 A schematic representation of the basic parts of a nuclear reactor. The largest commercial nuclear power plant reactors are 23 to 28 cm (9 to 11 in) thick steel vessels with a stainless steel liner, standing about 12 m (40 ft) high with a diameter of about 5 m (16 ft). Such a reactor has four pumps, which move 1,665,581 liters (440,000 gallons) of water per minute through the primary loop.

A

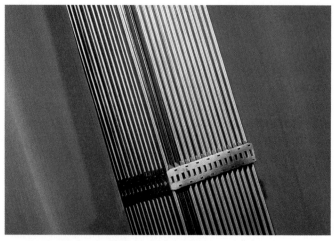

B

FIGURE 11.14 (*A*) These are uranium oxide fuel pellets that are stacked inside fuel rods, which are then locked together in a fuel rod assembly. (*B*) A fuel rod assembly. See also figure 11.17, which shows a fuel rod assembly being loaded into a reactor.

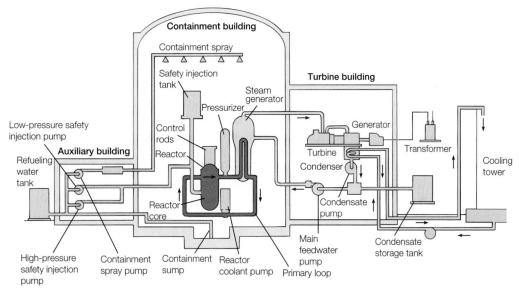

FIGURE 11.15 A schematic general system diagram of a pressurized water nuclear power plant, not to scale. The containment building is designed to withstand an internal temperature of 149°C (300°F) at a pressure of 414 kilopascals (60 lbs/in²) and still maintain its leak-tight integrity.

FIGURE 11.16 The turbine deck of a nuclear generating station. One large generator is in line with four steam turbines in this non-nuclear part of the plant. The large silver tanks are separators that remove water from the steam after it has left the high-pressure turbine and before it is recycled back into the low-pressure turbines.

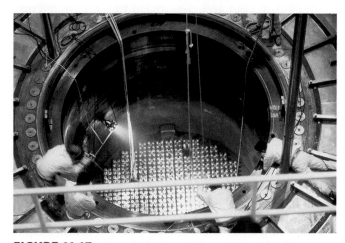

FIGURE 11.17 Spent fuel rod assemblies are removed and new ones are added to a reactor head during refueling. This shows an initial fuel load to a reactor, which has the upper part removed and set aside for the loading.

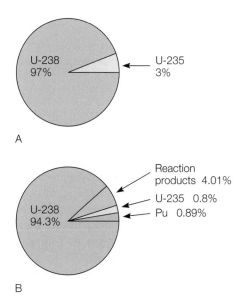

FIGURE 11.18 The composition of the nuclear fuel in a fuel rod (A) before and (B) after use over a three-year period in a nuclear reactor.

leaves at about 50°C (about 120°F) before returning to a cooling tower, where it is cooled by evaporation. The feedwater is preheated, then recirculated to the steam generator to start the cycle over again. The steam is condensed back to liquid water because of the difficulty of pumping and reheating steam.

After a period of time, the production of fission products in the fuel rods begins to interfere with effective neutron transmission, so the reactor is shut down annually for refueling. During refueling, about one-third of the fuel that had the longest exposure in the reactor is removed as "spent" fuel. New fuel rod assemblies are inserted to make up for the part removed (figure 11.17). However, only about 4 percent of the "spent" fuel is unusable waste, about 94 percent is uranium-238, 0.8 percent is uranium-235, and about 0.9 percent is plutonium (figure 11.18). Thus, "spent" fuel rods contain an appreciable amount of usable uranium and plutonium. For

A Closer Look

Three Mile Island and Chernobyl

Three Mile Island (downriver from Harrisburg, Pennsylvania) and Chernobyl (former USSR, now Ukraine) are two nuclear power plants that became famous because of accidents. Here is a brief accounting of what happened.

Three Mile Island

It was March 28, 1979, and the 880-megawatt Three Mile Island Nuclear Plant, operated by Metropolitan Edison Company, was going full blast. At 4:00 A.M. that morning the main feedwater pump that pumps water to the steam generator failed for some unexplained reason (follow this description in figure 11.15). Backup feedwater pumps kicked in, but the valves that should have been open were closed for maintenance, blocking the backup source of water for the steam generator. All of a sudden there was not a source of water for the steam generator that removes heat from the primary loop. Events began to happen quickly at that point.

The computer sensors registered that the steam generator was not receiving water, and the computer began to follow a shutdown procedure. First, the turbine was shut down as steam was vented from the steam line out through the turbine building, sounding much like a large jet plane. Within six seconds, the reactor was shut down (scrammed) with the control rods dropping down between the fuel rods in the reactor vessel. Fissioning began to slow, but the reactor was still hot.

Between three and six seconds, a significant event occurred when a pressure relief valve opened on the primary loop, relieving the excess pressure that was generated because feedwater was not entering the steam generator to remove heat from the primary loop. The valve should have closed when the excess pressure was released, but it did not. It was stuck in an open position. Pressurized water and steam was pouring from the primary loop into the containment building. As water was lost from the primary loop, temperatures inside the reactor began to climb. The loss of pressure resulted in high-pressure water flashing into steam. If an operator had pressed the right button in the control room, it would have closed the open valve, but this did not happen until thirty-two minutes later.

At this point, the reactor could have recovered from the events. Two minutes after the initial shutdown, the computer sensors noted the loss of pressure from the open valve and kicked in the emergency core cooling system, which pumps more water into the reactor vessel. However, for some unknown reason, the control room operators shut down one pump four and a half minutes after the initial event and the second pump six minutes later.

Water continued to move through the open pressure relief valve into the containment building. At seven and a half minutes after the start of the event, the radioactive water on the floor was 2 feet deep and the sump pumps started pumping water into tanks in the auxiliary building. This water would become the source of the radioactivity that did escape the plant. It escaped because the pump seals leaked and spilled radioactive water. The filters had been removed from the auxiliary building air vents, allowing radioactive gases to escape.

Eleven minutes after the start of the event, an operator restarted the emergency core cooling system that had been turned off. With the cooling water flowing again, the pressure in the reactor stopped falling. The fuel rods, some thirty-six thousand in this reactor, had not yet suffered any appreciable damage. This would be taken care of by the next incredible event.

What happened next was that operators began turning off the emergency cooling pumps, perhaps because they were vibrating too much. In any case, with the pumps off, the water level in the reactor fell again, this time uncovering the fuel rods. Within minutes, the temperature was high enough to rupture fuel rods, dropping radioactive oxides into the bottom of the reactor vessel. The operators now had a general emergency.

Eleven hours later, the operators decided to start the main reactor coolant pump. This pump had shut down at the beginning of the series of events. Water again covered the fuel rods, and the pressure and temperature stabilized.

The consequences of this series of events were as follows:

1. Local residences did receive some radiation from the release of gases. They received 10 millirem (0.1 millisievert) in a low-exposure area and up to 25 millirem (0.25 millisievert) in a high-exposure area.

---

now, spent reactor fuel rods are mostly stored in cooling pools at the nuclear plant sites. In the future, a decision will be made either to reprocess the spent fuel, recovering the uranium and plutonium through chemical reprocessing, or put the fuel in terminal storage. Concerns about reprocessing are based on the fact that plutonium-239 and uranium-235 are fissionable and could possibly be used by terrorist groups to construct nuclear explosive devices. Six other countries do have reprocessing plants, however, and the spent fuel rods represent a significant energy source. Some energy experts say that it would be inappropriate to dispose of such an energy source.

The technology to dispose of fuel rods exists if the decision is made to do so. The longer half-life waste products are mostly alpha emitters. These metals could be converted to oxides, mixed with powdered glass (or a ceramic), melted, and then poured into stainless steel containers. The solidified canisters would then be buried in a stable geologic depository. The glass technology is used in France for disposal of high-level wastes. Buried at 610 to 914 m (2,000–3,000 ft) depths in solid granite, the only significant means of the radioactive wastes reaching the surface would be through groundwater dissolving the stainless steel, glass, and waste products and then transporting them

2. Cleaning up the damaged reactor vessel and core required more than ten years to cut it up, pack it into canisters, and ship everything to the Federal Nuclear Reservation at Idaho Falls.
3. The cost of the cleanup was more than $1 billion.
4. Changes were implemented at other nuclear power plants as a consequence: Pressure relief valves have been removed, operators can no longer turn off the emergency cooling system, and operators must now spend about one-fourth of their time in training.

Chernobyl

The Soviet-designed Chernobyl reactor was a pressurized water reactor with individual fuel channels, which is very different from the pressurized water reactors used in the United States. The Chernobyl reactor was constructed with each fuel assembly in an individual pressure tube with circulating pressurized water. This heated water was circulated to a steam separator, and the steam was directed to turbines, which turned a generator to produce electricity. Graphite blocks surrounded the pressure tubes, serving as moderators to slow the neutrons involved in the chain reaction. The graphite was cooled by a mixture of helium and nitrogen. The reactor core was located in a concrete bunker that acted as a radiation shield. The top part was a steel cap and shield that supported the fuel assemblies. There were no containment buildings around the Soviet reactors as there are in the United States.

The Chernobyl accident was the result of a combination of a poorly engineered reactor design, poorly trained reactor operators, and serious mistakes made by the operators on the day of the accident.

The reactor design was flawed because at low power, steam tended to form pockets in the water-filled fuel channels, creating a condition of instability. Instability occurred because (1) steam is not as efficient at cooling as is liquid water, and (2) liquid water acts as a moderator and neutron absorber while steam does not. Excess steam therefore leads to overheating and increased power generation. Increased power can lead to increased steam generation, which leads to further increases in power. This coupled response is very difficult to control because it feeds itself.

On April 25, 1986, the operators of Chernobyl unit 4 undertook a test to find out how long the turbines would spin and supply power following the loss of electrical power. The operators disabled the automatic shutdown mechanisms and then started the test early on April 26. The plan was to stabilize the reactor at 1,000 MW, but an error was made and the power fell to about 30 MW and pockets of steam became a problem. They tried to increase the power by removing all the control rods. At 1:00 A.M., they were able to stabilize the reactor at 200 MW. Then instability returned and the operators were making continuous adjustments to maintain a constant power. They reduced the feedwater to maintain steam pressure, and this created even more steam voids in the fuel channels. Power surged to a very high

level and fuel elements ruptured. A steam explosion moved the reactor cap, exposing individual fuel channels and releasing fission products to the environment. A second explosion knocked a hole in the roof, exposed more of the reactor core, and the reactor graphite burst into flames. The graphite burned for nine days, releasing about 324 million Ci (12×10^{18} Bq) into the environment.

The fiery release of radioactivity was finally stopped by using a helicopter to drop sand, boron, lead, and other materials onto the burning graphite reactor. After the fire was out, the remains of the reactor were covered with a large concrete shelter.

In addition to destroying the reactor, the accident killed 30 people, 28 of whom died from radiation exposure. Another 134 people were treated for acute radiation poisoning and all recovered from the immediate effects.

Cleanup crews over the next year received about 10 rem (100 millisievert) to 25 rem (250 millisievert) and some received as much as 50 rem (500 millisievert). In addition to this direct exposure, large expanses of Belarus, Ukraine, and Russia were contaminated by radioactive fallout from the reactor fire. Hundreds of thousands of people have been resettled into less contaminated areas. The World Health Organization and other international agencies have studied the data to understand the impact of radiation-related disease. These studies do confirm a rising incidence of thyroid cancer but no increases in leukemia so far.

back to the surface. Many experts believe that if such groundwater dissolving were to take place, it would require thousands of years. The radioactive isotopes would thus undergo natural radioactive decay by the time they could reach the surface. Nonetheless, research is continuing on nuclear waste and its disposal. In the meantime, the question of whether it is best to reprocess fuel rods or place them in permanent storage remains unanswered.

What is the volume of nuclear waste under question? If all the spent fuel rods from all the commercial nuclear plants

accumulated since coming online were reprocessed, then mixed with glass, the total amount of glassified waste would make a pile on one football field an estimated 4 m (about 13 ft) high.

Nuclear Fusion

As the graph of nuclear binding energy versus mass numbers shows (see figure 11.10), nuclear energy is released when (1) massive nuclei such as uranium-235 undergo fission and

Nuclear Waste

There are two general categories of nuclear wastes: (1) low-level wastes and (2) high-level wastes. *Low-level wastes* are produced by the normal operation of a nuclear reactor. Radioactive isotopes sometimes escape from fuel rods in the reactor and in the spent fuel storage pools. These isotopes are removed from the water by ion-exchange resins and from the air by filters. The used resins and filters will contain the radioactive isotopes and will become low-level wastes. In addition, any contaminated protective clothing, tools, and discarded equipment also become low-level wastes.

Low-level liquid wastes are evaporated, mixed with cement, then poured into 55 gallon steel drums. Solid wastes are compressed and placed in similar drums. The drums are currently disposed of by burial in government-licensed facilities. In general, low-level waste has an activity of less than 1.0 curie per cubic foot. Contact with the low-level waste could expose a person to up to 20 millirems per hour of contact.

High-level wastes from nuclear power plants are spent nuclear fuel rods. At the present time, most of the commercial nuclear power plants have these rods in temporary storage at the plant site. These rods are "hot" in the radioactive sense, producing about 100,000 curies per cubic foot. They are also hot in the thermal sense, continuing to generate heat for months after removal from the reactor. The rods are cooled by heat exchangers connected to storage pools; they could otherwise achieve an internal temperature as high as 800°C for several decades. In the future, these spent fuel rods will be reprocessed or disposed of through terminal storage.

Agencies of the United States government have also accumulated millions of gallons of high-level wastes from the manufacture of nuclear weapons and nuclear research programs. These liquid wastes are stored in million-gallon stainless steel containers that are surrounded by concrete. The future of this large amount of high-level wastes may be evaporation to a solid form or mixture with a glass or ceramic matrix, which is melted and poured into stainless steel containers. These containers would be buried in solid granite rock in a stable geologic depository. Such high-level wastes must be contained for thousands of years as they undergo natural radioactive decay (box figure 11.3). Burial at a depth of 610 to 914 m (2,000–3,000 ft) in solid

Box Figure 11.3 This is a standard warning sign for a possible radioactive hazard. Such warning signs would have to be maintained around a nuclear waste depository for thousands of years.

granite would provide protection from exposure by explosives, meteorite impact, or erosion. One major concern about this plan is that a hundred generations later, people might lose track of what is buried in the nuclear garbage dump.

---

(2) when less massive nuclei come together to form more massive nuclei through nuclear fusion. Nuclear fusion is responsible for the energy released by the Sun and other stars. At the present halfway point in the Sun's life—with about 5 billion years to go—the core is now 35 percent hydrogen and 65 percent helium. Through fusion, the Sun converts about 650 million tons of hydrogen to 645 million tons of helium every second. The other roughly 5 million tons of matter are converted into energy. Even at this rate, the Sun has enough hydrogen to continue the process for an estimated 5 billion years. Several fusion reactions take place between hydrogen and helium isotopes, including the following:

$$^1_1H + ^1_1H \rightarrow ^2_1H + ^0_1e$$

$$^2_1H + ^2_1H \rightarrow ^3_2He + ^1_0n$$

$$^3_2He + ^3_2He \rightarrow ^4_2He + 2\,^1_1H$$

The fusion process would seem to be a desirable energy source on Earth because (1) two isotopes of hydrogen, deuterium (^2_1H) and tritium (^3_1H), undergo fusion at a relatively low temperature; (2) the supply of deuterium is practically unlimited, with each gallon of seawater containing about a teaspoonful of deuterium dioxide; and (3) enormous amounts of energy are released with no radioactive by-products.

The oceans contain enough deuterium to generate electricity for the entire world for millions of years, and tritium can be constantly produced by a fusion device. Researchers know what needs to be done to tap this tremendous energy source. The problem is *how* to do it in an economical, continuous energy-producing fusion reactor. The problem, one of the most difficult engineering tasks ever attempted, is meeting three basic fusion reaction requirements of (1) temperature, (2) density, and (3) time (figure 11.19):

1. **Temperature.** Nuclei contain protons and are positively charged, so they experience the electromagnetic repulsion of like charges. This force of repulsion can be overcome, moving the nuclei close enough to fuse together, by giving the nuclei sufficient kinetic energy. The fusion reaction of deuterium and tritium, which has the lowest temperature

High-Level Nuclear Waste

In 1982, the U.S. Congress established a national policy to solve the problem of nuclear waste disposal. This policy is a federal law called the Nuclear Waste Policy Act. Congress based this policy on what most scientists worldwide agree is the best way to dispose of nuclear waste.

The Nuclear Waste Policy Act made the U.S. Department of Energy (DOE) responsible for finding a site, building, and operating an underground disposal facility called a geologic repository.

In 1983, the DOE selected nine locations in six states for consideration as potential repository sites. This selection was based on data collected for nearly ten years. The nine sites were studied and results of these preliminary studies were reported in 1985. Based on these reports, President Ronald Reagan approved three sites for intensive scientific study called site characterization. The three sites were Hanford, Washington; Deaf Smith County, Texas; and Yucca Mountain, Nevada.

In 1987, Congress amended the Nuclear Waste Policy Act and directed the Department of Energy to study only Yucca Mountain.

On July 9, 2002, the U.S. Senate cast the final legislative vote approving the development of a repository at Yucca Mountain. On July 23, 2002, President George W. Bush signed House Joint Resolution 87, allowing the DOE to take the next step in establishing a safe repository in which to store our nation's nuclear waste.

In July 2004, the DOE issued a report that describes how commercial spent nuclear fuel and high-level waste will be acquired, transported to federal facilities, and disposed of in the Yucca Mountain Geologic Repository. The plan is to accept scheduled allocations from nuclear power plant owners beginning in 2010. In 2010, 400 metric tons of uranium will be accepted. This number will grow to 3,000 metric tons per year in 2014 and for each of the next five years.

The Yucca Mountain Project is currently focused on preparing an application to obtain a license from the U.S. Nuclear Regulatory Commission to construct a repository.

Source: www.ocrwm.doe.gov/ym_repository/index.shtml

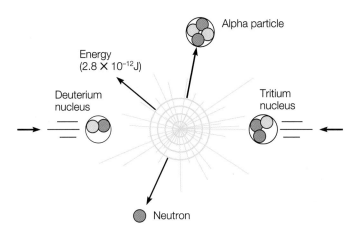

FIGURE 11.19   A fusion reaction between a tritium nucleus and a deuterium nucleus requires a certain temperature, density, and time of containment to take place.

requirements of any fusion reaction known at the present time, requires temperatures on the order of 100 million degrees Celsius.

2. **Density.**   There must be a sufficiently dense concentration of heavy hydrogen nuclei, on the order of $10^{14}/cm^3$, so many reactions occur in a short time.

3. **Time.**   The nuclei must be confined at the appropriate density up to a second or longer at pressures of at least 10 atmospheres to permit a sufficient number of reactions to take place.

The temperature, density, and time requirements of a fusion reaction are interrelated. A short time of confinement, for example, requires an increased density, and a longer confinement time requires less density. The primary problems of fusion research are the high-temperature requirements and confinement. No material in the world can stand up to a temperature of 100 million degrees Celsius, and any material container would be instantly vaporized. Thus, research has centered on meeting the fusion reaction requirements without a material container. Two approaches are being tested, *magnetic confinement* and *inertial confinement.*

Magnetic confinement utilizes a very hot *plasma,* a gas consisting of atoms that have been stripped of their electrons because of the high kinetic energies. The resulting positively and negatively charged particles respond to electrical and magnetic forces, enabling researchers to develop a "magnetic bottle," that is, magnetic fields that confine the plasma and avoid the problems of material containers that would vaporize. A magnetically confined plasma is very unstable, however, and researchers have compared the problem to trying to carry a block of Jell-O on a pair of rubber bands. Different magnetic field geometries and magnetic "mirrors" are the topics of research in attempts to stabilize the hot, wobbly plasma. Electric currents, injection of fast ions, and radio frequency (microwave) heating methods are also being studied.

Inertial confinement is an attempt to heat and compress small frozen pellets of deuterium and tritium with energetic laser beams or particle beams, producing fusion. The focus of this research is new and powerful lasers, light ion and heavy ion beams. If successful, magnetic or inertial confinement will provide a long-term solution for future energy requirements.

People Behind the Science

Marie Curie (1867–1934)

Marie Curie was a Polish-born French scientist who, with her husband, Pierre Curie (1859–1906), was an early investigator of radioactivity. From 1896, the Curies worked together, building on the results of Henri Becquerel, who had discovered radioactivity from uranium salts. Marie Curie discovered that thorium also emits radiation and found that the mineral pitchblende was even more radioactive than could be accounted for by any uranium and thorium content. The Curies then carried out an exhaustive search and in July 1898 announced the discovery of polonium, followed in December of that year with the discovery of radium. They shared the 1903 Nobel Prize for physics with Becquerel for the discovery of radioactivity. The Curies did not participate in Becquerel's discovery but investigated radioactivity and gave the phenomenon its name. Marie Curie went on to study the chemistry and medical applications of radium, and was awarded the 1911 Nobel Prize for chemistry in recognition of her work in isolating the pure metal.

At the outbreak of World War I in 1914, Marie Curie helped to equip ambulances with X-ray equipment and drove the ambulances to the front lines. The International Red Cross made her head of its Radiological Service. She taught medical orderlies and doctors how to use the new technique. By the late 1920s, her health began to deteriorate: continued exposure to high-energy radiation had given her leukemia. She entered a sanatorium and died on July 4, 1934.

Throughout much of her life, Marie Curie was poor, and the painstaking radium extractions were carried out in primitive conditions. The Curies refused to patent any of their discoveries, wanting them to benefit everyone freely. They used the Nobel Prize money and other financial rewards to finance further research. One of the outstanding applications of their work has been the use of radiation to treat cancer, one form of which cost Marie Curie her life.

Source: Modified from the *Hutchinson Dictionary of Scientific Biography*. © RM, 2007. Reprinted by permission.

SUMMARY

Radioactivity is the spontaneous emission of particles or energy from an unstable atomic nucleus. The modern atomic theory pictures the nucleus as protons and neutrons held together by a short-range *nuclear force* that has moving *nucleons* (protons and neutrons) in *energy shells* analogous to the shell structure of electrons. A graph of the number of neutrons to the number of protons in a nucleus reveals that stable nuclei have a certain neutron-to-proton ratio in a *band of stability*. Nuclei that are above or below the band of stability and nuclei that are beyond atomic number 83 are radioactive and undergo *radioactive decay*.

Three common examples of radioactive decay involve the emission of an *alpha particle*, a *beta particle*, and a *gamma ray*. An alpha particle is a helium nucleus, consisting of two protons and two neutrons. A beta particle is a high-speed electron that is ejected from the nucleus. A gamma ray is a short-wavelength electromagnetic radiation from an excited nucleus. In general, nuclei with an atomic number of 83 or larger become more stable by alpha emission. Nuclei with a neutron-to-proton ratio that is too large become more stable by beta emission. Gamma ray emission occurs from a nucleus that was left in a high-energy state by the emission of an alpha or beta particle.

Each radioactive isotope has its own specific *radioactive decay rate*. This rate is usually described in terms of *half-life*, the time required for one-half the unstable nuclei to decay.

Radiation is measured by (1) its effects on photographic film, (2) the number of ions it produces, or (3) the flashes of light produced on a phosphor. It is measured at a source in units of a *curie*, defined as 3.70×10^{10} nuclear disintegrations per second. It is measured where received in units of a *rad*. A *rem* is a measure of radiation that takes into account the biological effectiveness of different types of radiation damage. In general, the natural environment exposes everyone to 100 to 500 millirems per year, an exposure called *background radiation*. Lifestyle and location influence the background radiation received, but the worldwide average is 240 millirems per year.

Energy and mass are related by Einstein's famous equation of $E = mc^2$, which means that *matter can be converted to energy and energy to matter*. The mass of a nucleus is always less than the sum of the masses of the individual particles of which it is made. This *mass defect* of a nucleus is equivalent to the energy released when the nucleus was formed according to $E = mc^2$. It is also the *binding energy*, the energy required to break the nucleus apart into nucleons.

When the binding energy is plotted against the mass number, the greatest binding energy per nucleon is seen to occur for an atomic number near that of iron. More massive nuclei therefore release energy by fission, or splitting to more stable nuclei. Less massive nuclei release energy by fusion, the joining of less massive nuclei to produce a more stable, more massive nucleus. Nuclear fission provides the energy for atomic explosions and nuclear power plants. Nuclear fusion is the energy source of the Sun and other stars and also holds promise as a future energy source for humans.

Summary of Equations

11.1   energy = mass × the speed of light squared

$$E = mc^2$$

KEY TERMS

alpha particle (p. 245)

background radiation (p. 253)

band of stability (p. 247)

beta particle (p. 245)

binding energy (p. 254)

chain reaction (p. 256)

critical mass (p. 256)

curie (p. 252)

gamma ray (p. 245)

half-life (p. 250)

mass defect (p. 254)

nuclear fission (p. 254)

nuclear fusion (p. 254)

nucleons (p. 246)

rad (p. 253)

radioactive decay (p. 246)

radioactivity (p. 244)

rem (p. 252)

APPLYING THE CONCEPTS

1. A high-speed electron ejected from a nucleus during radioactive decay is called a (an)
   a. alpha particle.
   b. beta particle.
   c. gamma ray.
   d. None of the above is correct.

2. The ejection of an alpha particle from a nucleus results in
   a. an increase in the atomic number by one.
   b. an increase in the atomic mass by four.
   c. a decrease in the atomic number by two.
   d. None of the above is correct.

3. An atom of radon-222 loses an alpha particle to become a more stable atom of
   a. radium.
   b. bismuth.
   c. polonium.
   d. radon.

4. A sheet of paper will stop a (an)
   a. alpha particle.
   b. beta particle.
   c. gamma ray.
   d. None of the above is correct.

5. The most penetrating of the three common types of nuclear radiation is the
   a. alpha particle.
   b. beta particle.
   c. gamma ray.
   d. All have equal penetrating ability.

6. An atom of an isotope with an atomic number greater than 83 will probably emit a (an)
   a. alpha particle.
   b. beta particle.
   c. gamma ray.
   d. None of the above is correct.

7. An atom of an isotope with a large neutron-to-proton ratio will probably emit a (an)
   a. alpha particle.
   b. beta particle.
   c. gamma ray.
   d. None of the above is correct.

8. The rate of radioactive decay can be increased by increasing the
   a. temperature.
   b. pressure.
   c. size of the sample.
   d. None of the above is correct.

9. Isotope A has a half-life of seconds, and isotope B has a half-life of millions of years. Which isotope is more radioactive?
   a. It depends on the sample size.
   b. isotope A
   c. isotope B
   d. Unknown from the information given.

10. A measure of radioactivity at the *source* is (the)
    a. curie.
    b. rad.
    c. rem.
    d. Any of the above is correct.

11. A measure of radiation received that considers the biological effect resulting from the radiation is (the)
    a. curie.
    b. rad.
    c. rem.
    d. Any of the above is correct.

12. Used fuel rods from a nuclear reactor contain about
    a. 96% usable uranium and plutonium.
    b. 33% usable uranium and plutonium.
    c. 4% usable uranium and plutonium.
    d. 0% usable uranium and plutonium.

Answers

1. b 2. c 3. c 4. a 5. c 6. a 7. b 8. c 9. b 10. a 11. c 12. a

QUESTIONS FOR THOUGHT

1. How is a radioactive material different from a material that is not radioactive?

2. What is radioactive decay? Describe how the radioactive decay rate can be changed if this is possible.

3. Describe three kinds of radiation emitted by radioactive materials. Describe what eventually happens to each kind of radiation after it is emitted.

4. How are positively charged protons able to stay together in a nucleus since like charges repel?

5. What is half-life? Give an example of the half-life of an isotope, describing the amount remaining and the time elapsed after five half-life periods.

6. Would you expect an isotope with a long half-life to be more, the same, or less radioactive than an isotope with a short half-life? Explain.

7. What is meant by background radiation? What is the normal radiation dose for the average person from background radiation?

8. Why is there controversy about the effects of long-term, low levels of radiation exposure?

9. What is a mass defect? How is it related to the binding energy of a nucleus? How can both be calculated?

10. Compare and contrast nuclear fission and nuclear fusion.

FOR FURTHER ANALYSIS

1. What are the significant differences between a radioactive isotope and an isotope that is not radioactive?

2. Analyze the different types of radioactive decay to explain how each is a hazard to living organisms.

3. Make up a feasible explanation for why some isotopes have half-lives of seconds, yet other kinds of isotopes have half-lives in the billions of years.

4. Suppose you believe the threshold model of radiation exposure is correct. Describe a conversation between yourself and another person who feels strongly that the linear model of radiation exposure is correct.

5. Explain how the fission of heavy elements and the fusion of light elements both release energy.

6. Write a letter to your congressional representative describing why used nuclear fuel rods should be reprocessed rather than buried as nuclear waste.

7. What are the similarities and differences between a nuclear fission power plant and a nuclear fusion power plant?

INVITATION TO INQUIRY

How Much Radiation?

Ionizing radiation is understood to be potentially harmful if certain doses are exceeded. How much ionizing radiation do you acquire from the surroundings where you live, from your lifestyle, and from medical procedures? Investigate radiation from cosmic sources, the Sun, television

sets, time spent in jet airplanes, and dental or other X-ray machines. What are other sources of ionizing radiation in your community? How difficult is it to find relevant information and make recommendations? Does any agency monitor the amount of radiation that people receive? What are the problems and issues with such monitoring?

PARALLEL EXERCISES

The exercises in groups A and B cover the same concepts. Solutions to group A exercises are located in appendix D.

Group A

Note: You will need the table of atomic weights inside the back cover of this text.

1. Give the number of protons and the number of neutrons in the nucleus of each of the following isotopes:
   a. cobalt-60   c. neon-24
   b. potassium-40   d. lead-208

2. Write the nuclear symbols for each of the nuclei in exercise 1.

3. Predict if the nuclei in exercise 1 are radioactive or stable, giving your reasoning behind each prediction.

4. Write a nuclear equation for the decay of the following nuclei as they give off a beta particle:
   a. $^{56}_{26}Fe$ d. $^{24}_{11}Na$
   b. $^{7}_{4}Be$ e. $^{214}_{82}Pb$
   c. $^{64}_{29}Cu$ f. $^{32}_{15}P$

5. Write a nuclear equation for the decay of the following nuclei as they undergo alpha emission:
   a. $^{235}_{92}U$ d. $^{214}_{83}Bi$
   b. $^{226}_{88}Ra$ e. $^{230}_{90}Th$
   c. $^{239}_{94}Pu$ f. $^{210}_{84}Po$

6. The half-life of iodine-131 is 8 days. How much of a 1.0 oz sample of iodine-131 will remain after 32 days?

7. How much energy must be supplied to break a single iron-56 nucleus into separate protons and neutrons? (The mass of an iron-56 nucleus is 55. 9206 u, one proton is 1.00728 u, and one neutron is 1.00867 u.)

Group B

Note: You will need the table of atomic weights inside the back cover of this text.

1. Give the number of protons and the number of neutrons in the nucleus of each of the following isotopes:
   a. aluminum-25   c. tin-120
   b. technetium-95   d. mercury-200

2. Write the nuclear symbols for each of the nuclei in exercise 1.

3. Predict if the nuclei in exercise 1 are radioactive or stable, giving your reasoning behind each prediction.

4. Write a nuclear equation for the beta emission decay of each of the following:
   a. $^{14}_{6}C$ d. $^{241}_{94}Pu$
   b. $^{60}_{27}Co$ e. $^{131}_{53}I$
   c. $^{24}_{11}Na$ f. $^{210}_{82}Pb$

5. Write a nuclear equation for each of the following alpha emission decay reactions:
   a. $^{241}_{95}Am$ d. $^{234}_{92}U$
   b. $^{232}_{90}Th$ e. $^{242}_{96}Cm$
   c. $^{223}_{88}Ra$ f. $^{237}_{93}Np$

6. If the half-life of cesium-137 is 30 years, how much time will be required to reduce a 1.0 kg sample to 1.0 g?

7. How much energy is needed to separate the nucleons in a single lithium-7 nucleus? (The mass of a lithium-7 nucleus is 7.01435 u, one proton is 1.00728 u, and one neutron is 1.00867 u.)

This is a planetary nebula in the constellation Aquarius. Planetary nebulae are clouds of ionized gases with no relationship to any planet. They were named long ago, when they appeared similar to the planets Neptune and Uranus when viewed through early telescopes.

The Universe

OVERVIEW

Astronomy is an exciting field of science that has fascinated people since the beginnings of recorded history. Ancient civilizations searched the heavens in wonder, some recording on clay tablets what they observed. Many religious and philosophical beliefs were originally based on interpretations of these ancient observations. Today, we are still awed by space, but now we are fascinated with ideas of space travel, black holes, and the search for extraterrestrial life. Throughout history, people have speculated about the universe and their place in it and watched the sky and wondered (figure 12.1). What is out there and what does it all mean? Are there other people on other planets, looking at the star in their sky that is our Sun, wondering if we exist?

Until about thirty years ago, progress in astronomy was limited to what could be observed and photographed. Developments in technology then began to provide the details of what is happening in the larger expanses of space away from Earth. This included new data made available from the development of infrared, radio, and X-ray telescopes and the Hubble Space Telescope. New data and the discovery of pulsars, neutron stars, and black holes began to fit together like the pieces of a puzzle. Theoretical models emerged about how stars, galaxies, and the universe have evolved. This chapter is concerned with these topics and how the stars are arranged in space. The chapter concludes with theoretical models of how the universe began and what may happen to it in the future.

THE NIGHT SKY

Early civilizations had a much better view of the night sky before city lights, dust, and pollution obscured much of it. Today, you must travel far from cities, perhaps to a remote mountain-top, to see a clear night sky as early people observed it. Back then, people could clearly see the motion of the moon and stars night after night, observing recurring cycles of motion. These cycles became important as people associated them with certain events. Thus, watching the Sun, Moon, and star movements became a way to identify when to plant crops, when to harvest, and when it was time to plan for other events. Observing the sky was an important activity, and many early civilizations built observatories with sighting devices to track and record astronomical events. Stonehenge, for example, was an ancient observatory built in England around 2600 B.C. by Neolithic people (figure 12.2).

Light from the stars and planets must pass through Earth's atmosphere to reach you and this affects the light. Stars appear as point *sources* of light, and each star generates its own light. The stars seem to twinkle because density differences in the atmosphere refract the point of starlight one way, then the other, as the air moves. The result is the slight dancing about and change in intensity called twinkling. The points of starlight are much steadier when viewed on a calm night or from high in the mountains, where there is less atmosphere for the starlight to pass through. Astronauts outside the atmosphere see no twinkling, and the stars appear as steady point sources of light.

Back at ground level, within the atmosphere, the *reflected* light from a planet does not seem to twinkle. A planet appears as a disk of light rather than a point source, so refraction from moving air of different densities does not affect the image as much. Sufficient air movement can cause planets to appear to

A

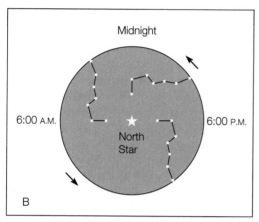

B

FIGURE 12.1 Ancient civilizations used celestial cycles of motion as clocks and calendars. (*A*) This photograph shows the path of stars around the North Star. (*B*) A "snapshot" of the position of the Big Dipper over a period of twenty-four hours as it turns around the North Star. This shows how the Big Dipper can be used to help you keep track of time.

FIGURE 12.2 The stone pillars of Stonehenge were positioned so that they could be used to follow the movement of the Sun and Moon with the seasons of the year.

shimmer, however, just as a road appears to shimmer on a hot summer day.

How far away is a star? When you look at the sky, it appears that all the stars are at the same distance. It seems impossible to know anything about the actual distance to any given star.

Standard referent units of length such as kilometers or miles have little meaning in astronomy since there are no referent points of comparison. Distance without a referent point can be measured in terms of angles or time. The unit of astronomical distance that uses time is the **light-year** (ly). A light-year is the distance that light travels in one year, about 9.5×10^{12} km (about 6×10^{12} mi).

If you could travel by spaceship a few hundred light-years from Earth, you would observe the Sun shrink to a bright point of light among the billions and billions of other stars. The Sun is just an ordinary star with an average brightness. Like the other stars, the Sun is a massive, dense ball of gases with a surface heated to incandescence by energy released from fusion reactions deep within. Since the Sun is an average star, it can be used as a reference for understanding all the other stars.

ORIGIN OF STARS

Theoretically, stars are born from swirling clouds of hydrogen gas in the deep space between other stars. Such interstellar (between stars) clouds are called **nebulae.** These clouds consist of random, swirling atoms of gases that have little gravitational attraction for one another because they have little mass. Complex motion of stars, however, can produce a shock wave that causes particles to move closer together, making local compressions. Their mutual gravitational attraction then begins to pull them together into a cluster. The cluster grows as more atoms are pulled in, which increases the mass and thus the gravitational attraction, and still more atoms are pulled in from farther away. Theoretical calculations indicate that on the order of 1×10^{57} atoms are necessary, all within a distance of 3 trillion km (about 1.9 trillion mi). When these conditions occur, the cloud of gas atoms begins to condense by gravitational attraction to a **protostar,** an accumulation of gases that will become a star.

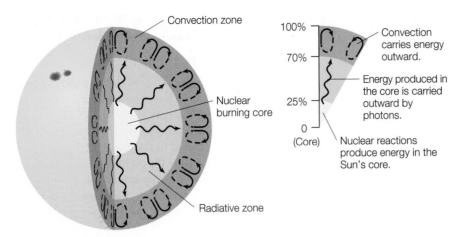

FIGURE 12.3 Energy-producing nuclear reactions occur only within the inner 25 percent of the Sun's radius. The energy produced by these reactions is carried outward by photons to 70 percent of the Sun's radius. From that distance outward, convection carries most of the Sun's energy.

Gravitational attraction pulls the average protostar from a cloud with a diameter of trillions of kilometers (trillions of miles) down to a dense sphere with a diameter of 2.5 million km (1.6 million mi) or so. As gravitational attraction accelerates the atoms toward the center, they gain kinetic energy, and the interior temperature increases. Over a period of some 10 million years of contracting and heating, the temperature and density conditions at the center of the protostar are sufficient to start nuclear fusion reactions. Pressure from hot gases and energy from increasing fusion reactions begin to balance the gravitational attraction over the next 17 million years, and the newborn, average star begins its stable life, which will continue for the next 10 billion years.

The interior of an average star, such as the Sun, is modeled after the theoretical pressure, temperature, and density conditions that would be necessary to produce the observed energy and light from the surface. This model describes the interior as a set of three shells: (1) the core, (2) a radiation zone, and (3) the convection zone (figure 12.3).

Our model describes the **core** as a dense, very hot region where nuclear fusion reactions release gamma and X-ray radiation. The density of the core is about twelve times that of solid lead. Because of the exceedingly hot conditions, however, the core remains in a plasma state even at this density.

The model describes the **radiation zone** as less dense than the core, having a density about the same as the density of water. Energy in the form of gamma and X rays from the core is absorbed and reemitted by collisions with atoms in this zone. The radiation slowly diffuses outward because of the countless collisions over a distance comparable to the distance between Earth and the Moon. It could take millions of years before this radiation finally escapes the radiation zone.

In the model, the **convection zone** begins about seven-tenths of the way to the surface, where the density of the plasma is about 1 percent the density of water. Plasma at the bottom of this zone is heated by radiation from the radiation zone below, expands from the heating, and rises to the surface by convection. At the surface, the plasma emits energy in the form of visible light, ultraviolet radiation, and infrared radiation,

which moves out into space. As it loses energy, the plasma contracts in volume and sinks back to the radiation zone to become heated again, continuously carrying energy from the radiation zone to the surface in convection cells. The surface is continuously heated by the convection cells as it gives off energy to space, maintaining a temperature of about 5,800 K (about 5,500°C).

As an average star, the Sun converts about 1.4×10^{17} kg of matter to energy every year as hydrogen nuclei are fused to produce helium. The Sun was born about 5 billion years ago and has sufficient hydrogen in the core to continue shining for another 4 or 5 billion years. Other stars, however, have masses that are much greater or much less than the mass of the Sun so they have different life spans. More massive stars generate higher temperatures in the core because they have a greater gravitational contraction from their greater masses. Higher temperatures mean increased kinetic energy, which results in increased numbers of collisions between hydrogen nuclei with the end result an increased number of fusion reactions. Thus, a more massive star uses up its hydrogen more rapidly than a less massive star. On the other hand, stars that are less massive than the Sun use their hydrogen at a slower rate so they have longer life spans. The life spans of the stars range from a few million years for large, massive stars, to 10 billion years for average stars such as the Sun, to trillions of years for small, less massive stars.

BRIGHTNESS OF STARS

Stars generate their own light, but some stars appear brighter than others in the night sky. As you can imagine, this difference in brightness could be related to (1) the amount of light produced by the stars, (2) the size of each star, or (3) the distance to a particular star. A combination of these factors is responsible for the brightness of a star as it appears to you in the night sky. A classification scheme for different levels of brightness that you see is called the **apparent magnitude** scale (table 12.1).

CONCEPTS APPLIED

A Near Miss?

An asteroid will come very close to Earth on April 13, 2029, according to radar measurements by astronomers. The asteroid, 2004 MN4, is 300 m (about 1,000 ft) wide and will glow like a third-magnitude star as it passes by Earth at a distance of 30,000 km (about 18,640 mi). Learn to recognize a third-magnitude star and mark it on your calendar!

Source: Space Weather News, 8 February 2005 (http://SpaceWeather.com). Also see www.jpl.nasa.gov/templates/flash/neo/neo.htm.

TABLE 12.1   The Apparent Magnitude Scale Comparing Some Familiar Objects and Some Observable Limits

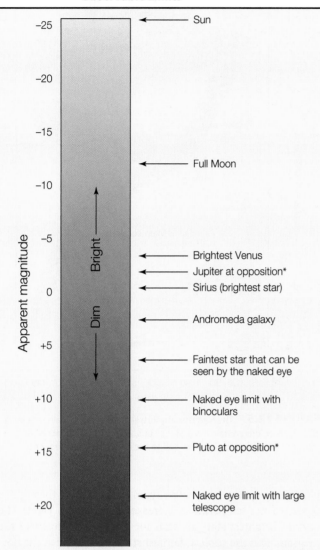

*When the planet is opposite the Sun in the sky.

The apparent magnitude scale is based on a system established by the Greek astronomer Hipparchus over two thousand years ago. Hipparchus made a catalog of stars he could see and assigned a numerical value to each to identify its relative brightness. The brightness values ranged from 1 to 6, with the number 1 assigned to the brightest star and the number 6 assigned to the faintest star that could be seen. Later, some stars were found to be brighter than the apparent magnitude of +1, which extended the scale into negative numbers. The brightest star in the night sky is Sirius, for example, with an apparent magnitude of −1.42 (see table 12.1).

The apparent magnitude of a star depends on how far away stars are in addition to differences in the stars themselves. Stars at a farther distance will appear fainter, and those closer will appear brighter, just as any other source

of light. To compensate for distance differences, astronomers calculate the brightness that stars would appear to have if they were all at a defined, standard distance (32.6 light-years). The brightness of a star at this distance is called the **absolute magnitude.** The Sun, for example, is the closest star and has an apparent magnitude of −26.7 at an average distance from Earth. When viewed from the standard distance, the Sun would have an absolute magnitude of +4.8, which is about the brightness of a faint star.

STAR TEMPERATURE

If you observe the stars on a clear night, you will notice that some are brighter than others, and you will also notice some color differences. Some stars have a reddish color, some have a bluish white color, and others have a yellowish color. This color difference is understood to be a result of the relationship that exists between the color and the temperature of a glowing object. The colors of the various stars are a result of the temperatures of the stars (figure 12.4). You see a cooler star as reddish in color and comparatively hotter stars as bluish white. Stars with in-between temperatures, such as the Sun, appear to have a yellowish color.

Astronomers use information about the star temperature and spectra as the basis for a star classification scheme. Originally, the classification scheme was based on sixteen categories according to the strength of the hydrogen line spectra (see line spectrum in chapter 8). The groups were identified alphabetically with A for the group with the strongest hydrogen line spectrum, B for slightly weaker lines, and on to the last group with the faintest lines. Later, astronomers realized that the star temperature was the important variable, so they rearranged the categories according to decreasing temperatures. The original letter categories were retained, however, resulting in classes of stars with the hottest temperature first and the coolest last with

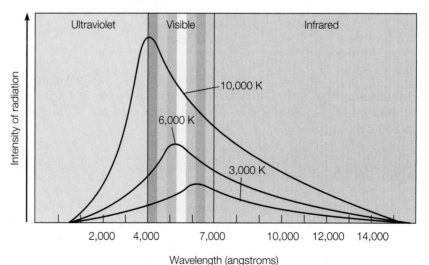

FIGURE 12.4   The distribution of radiant energy emitted is different for stars with different surface temperatures. Note that the peak radiation of a cooler star is more toward the red part of the spectrum and the peak radiation of a hotter star is more toward the blue part of the spectrum.

Seeing Spectra

Inexpensive diffraction grating on plastic film is available from many scientific materials supply houses. If available, view the light from gas discharge tubes to see bright line spectra. Use the grating to examine light from incandescent lightbulbs of different wattages, fluorescent lights, lighted "neon" signs of different colors, and street lights. Describe the type of spectrum each produces. If it has lines, see if you can identify the elements present.

TABLE 12.2   **Major Stellar Spectral Types and Temperatures**

Type	Color	Temperature (K)	Comment
O	Bluish	30,000–80,000	Spectrum with ionized helium and hydrogen but little else; short-lived and rare stars
B	Bluish	10,000–30,000	Spectrum with neutral helium, none ionized
A	Bluish	7,500–10,000	Spectrum with no helium, strongest hydrogen, some magnesium and calcium
F	White	6,000–7,500	Spectrum with ionized calcium, magnesium, neutral atoms of iron
G	Yellow	5,000–6,000	The spectral type of the Sun. Spectrum shows sixty-seven elements in the Sun
K	Orange-red	3,500–5,000	Spectrum packed with lines from neutral metals
M	Reddish	2,000–3,500	Band spectrum of molecules, e.g., titanium oxide; other related spectral types (R, N, and S) are based on other molecules present in each spectral type

the sequence O B A F G K M. Table 12.2 compares the color, temperature ranges, and other features of the stellar spectra classification scheme.

STAR TYPES

In 1910, Henry Russell in the United States and Ejnar Hertzsprung in Denmark independently developed a scheme to classify stars with a temperature-luminosity graph. The graph is called the **Hertzsprung-Russell diagram,** or the *H-R diagram* for short. The diagram is a plot with temperature indicated by spectral types, and the true brightness indicated by absolute magnitude. The diagram, as shown in figure 12.5,

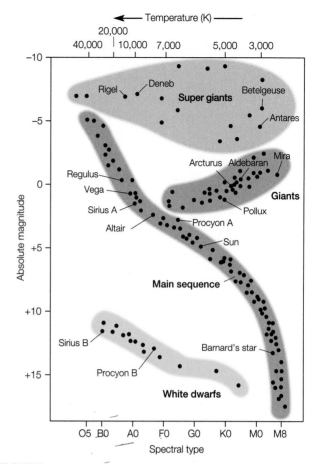

FIGURE 12.5   The Hertzsprung-Russell diagram. The main sequence and giant regions contain most of the stars, whereas hot underluminous stars, the white dwarfs, lie below and to the left of the main sequence.

plots temperature by spectral types sequenced O through M types, so the temperature decreases from left to right. The hottest, brightest stars are thus located at the top left of the diagram, and the coolest, faintest stars are located at the bottom right.

Each dot is a data point representing the surface temperature and brightness of a particular star. The Sun, for example, is a type G star with an absolute magnitude of about +5, which places the data point for the Sun almost in the center of the diagram. This means that the Sun is an ordinary, average star with respect to both surface temperature and true brightness.

Most of the stars plotted on an H-R diagram fall in or close to a narrow band that runs from the top left to the lower right. This band is made up of **main sequence stars.** Stars along the main sequence band are normal, mature stars that are using their nuclear fuel at a steady rate. Those stars on the upper left of the main sequence are the brightest, bluest, and most massive stars on the sequence. Those at the lower right are the faintest, reddest, and least massive of the stars on the main sequence. In general, most of the main sequence stars have masses that fall between a range from ten times greater than the mass of the Sun (upper left) to one-tenth the mass of the Sun (lower right). The extremes, or ends, of the main sequence range from about sixty times more massive than the Sun to one-twenty-fifth of the Sun's mass. It is the *mass* of a main sequence star that determines its brightness, its temperature, and its location on the H-R diagram. High-mass stars on the main sequence are brighter, hotter, and have shorter lives than low-mass stars. These relationships do not apply to the other types of stars in the H-R diagram.

There are groups of stars that have a different set of properties than the main sequence stars. The **red giant stars** are bright but low-temperature stars. These reddish stars are enormously bright for their temperature because they are very large, with an enormous surface area giving off light. A red giant might be one hundred times larger but have the same mass as the Sun. These low-density red giants are located in the upper right part of the H-R diagram. The **white dwarf stars,** on the other hand, are located at the lower left because they are faint, white-hot stars. A white dwarf is faint because it is small, perhaps twice the size of Earth. It is also very dense, with a mass approximately equal to the Sun's. During its lifetime, a star will be found in different places on the H-R diagram as it undergoes changes. Red giants and white dwarfs are believed to be evolutionary stages that aging stars pass through, and the path a star takes across the diagram is called an evolutionary track. During the lifetime of the Sun, it will be a main sequence star, a red giant, and then a white dwarf.

Stars such as the Sun emit a steady light because the force of gravitational contraction is balanced by the outward flow of energy. *Variable stars,* on the other hand, are stars that change in brightness over a period of time. A **Cepheid variable** is a bright variable star that is used to measure distances. There is a general relationship between the period and the brightness: the longer the time needed for one pulse, the greater the apparent brightness of that star. The period-brightness relationship to distance was calibrated by comparing the apparent brightness with the absolute magnitude (true brightness) of a Cepheid at a known distance with a known period. Using the period to predict how bright the star would appear at various distances allowed astronomers to calculate the distance to a Cepheid given its apparent brightness.

Edwin Hubble used the Cepheid period-brightness relationship to find the distances to other galaxies and discovered yet another relationship, that the greater the distance to a galaxy, the greater a shift in spectral lines toward the red end of the spectrum (redshift). This relationship is called Hubble's law, and it forms the foundation for understanding our expanding universe. Measuring redshift provides another means of establishing distances to other, far-out galaxies.

THE LIFE OF A STAR

A star is born in a gigantic cloud of gas and dust in interstellar space, then spends billions of years of calmly shining while it fuses hydrogen nuclei in the core. How long a star shines and what happens to it when it uses up the hydrogen in the core depends on the mass of the star. Of course, no one has

observed a life cycle of over billions of years. The life cycle of a star is a theoretical outcome based on what is known about nuclear reactions. The predicted outcomes seem to agree with observations of stars today, with different groups of stars that can be plotted on the H-R diagram. Thus, the groups of stars on the diagram—main sequence, red giants, and white dwarfs, for example—are understood to be stars in various stages of their lives.

Protostar Stage. The first stage in the theoretical model of the life cycle of a star is the formation of the protostar. As gravity pulls the gas of a protostar together, the density, pressure, and temperature increase from the surface down to the center. Eventually, the conditions are right for nuclear fusion reactions to begin in the core, which requires a temperature of 10 million kelvins. The initial fusion reaction essentially combines four hydrogen nuclei to form a helium nucleus with the release of much energy. This energy heats the core beyond the temperature reached by gravitational contraction, eventually to 16 million kelvins. Since the star is plasma, the increased temperature expands the volume of the star. The outward pressure of expansion balances the inward pressure from gravitational collapse, and the star settles down to a balanced condition of calmly converting hydrogen to helium in the core, radiating the energy released into space (figure 12.6). The theoretical time elapsed from the initial formation and collapse of the protostar to the main sequence is about 50 million years for a star of a solar mass.

Main Sequence Stage. Where the star is located on the main sequence and what happens to it next depend only on how massive it is. The more massive stars have higher core temperatures and use up their hydrogen more rapidly as they shine at higher surface temperatures (O type stars). Less massive stars shine at lower surface temperatures (M type stars) as they use their fuel at a slower rate. The overall life span on the main sequence ranges from millions of years for O type stars to trillions of years for M type stars. An average one-solar-mass star will last about 10 billion years.

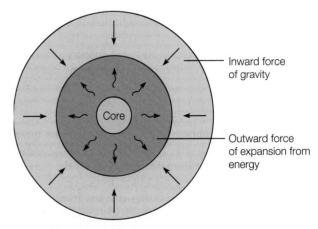

FIGURE 12.6 A star becomes stable when the outward forces of expansion from the energy released in nuclear fusion reactions balance the inward forces of gravity.

Red Giant Stage.

The next stage in the theoretical life of a star begins when much of the hydrogen in the core has been fused into helium. With fewer hydrogen fusion reactions, less energy is released and less outward balancing pressure is produced, so the star begins to collapse. The collapse heats the core, which now is composed primarily of helium, and the surrounding shell where hydrogen still exists. The increased temperature causes the hydrogen in the shell to undergo fusion, and the increased release of energy causes the outer layers of the star to expand. With an increased surface area, the amount of radiation emitted per unit area is less, and the star acquires the properties of a brilliant red giant. Its position on the H-R diagram changes since it now has different brightness and temperature properties. (The star has not physically *moved*. The changing properties move its temperature brightness data point, not the star, to a new position.)

Back Toward Main Sequence.

After about five hundred million years as a red giant, the star now has a surface temperature of about 4,000 kelvins compared to its main sequence surface temperature of 6,000 kelvins. The radius of the red giant is now a thousand times greater, a distance that will engulf Earth when the Sun reaches this stage, assuming Earth is in the same position as it is today. Even though the surface temperature has decreased from the expansion, the helium core is continually heating and eventually reaches a temperature of 100 million kelvins, the critical temperature necessary for the helium nuclei to undergo fusion to produce carbon. The red giant now has helium fusion reactions in the core and hydrogen fusion reactions in a shell around the core. This changes the radius, the surface temperature, and the brightness with the overall result depending on the composition of the star. In general, the radius and brightness decrease when this stage is reached, moving the star back toward the main sequence (figure 12.7).

Beginning of the End for Less Massive Stars.

After millions of years of helium fusion reactions, the core is gradually converted to a carbon core and helium fusion begins in the shell surrounding the core. The core reactions decrease as the star now has a helium-fusing shell surrounded by a second, hydrogen-fusing shell. This releases additional energy, and the star again expands to a red giant for the second time. A star the size of the Sun or less massive may cool enough at this point that nuclei at the surface become neutral atoms rather than a plasma. As neutral atoms, they can absorb radiant energy coming from within the star, heating the outer layers. Changes in temperature produce changes in pressure, which change the balance between the temperature, pressure, and the internal energy generation rate. The star begins to expand outward from heating. The expanded gases are cooled by the expansion process, however, and are pulled back to the star by gravity, only to be heated and expand outward again. In other words, the outer layers of the star begin to pulsate in and out. Finally, a violent expansion blows off the outer layers of the star, leaving the hot core. Such blown-off outer layers of a star form circular nebulae called *planetary nebulae* (figure 12.8). The nebulae

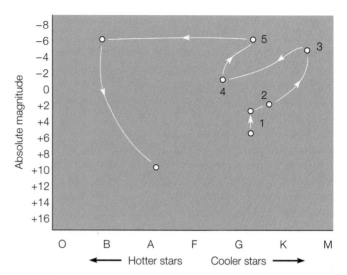

FIGURE 12.7 The evolution of a star of solar mass as it depletes hydrogen in the core (1), fuses hydrogen in the shell to become a red giant (2 to 3), becomes hot enough to produce helium fusion in the core (3 to 4), then expands to a red giant again as helium and hydrogen fusion reactions move out into the shells (4 to 5). It eventually becomes unstable and blows off the outer shells to become a white dwarf star.

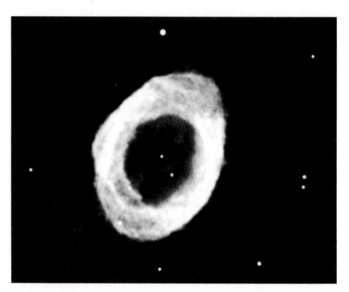

FIGURE 12.8 The blown-off outer layers of stars form ringlike structures called *planetary nebulae.*

continue moving away from the core, eventually adding to the dust and gases between the stars. The remaining carbon core and helium-fusing shell begin gravitationally to contract to a small, dense *white dwarf* star. A star with the original mass of the Sun or less slowly cools from white to red, then to a black lump of carbon in space (figure 12.9).

Beginning of the End for Massive Stars.

A more massive star will have a different theoretical ending than the slow cooling of a white dwarf. A massive star will contract, just as the less massive stars, after blowing off its outer shells. In a more massive star, however, heat from the contraction may reach the

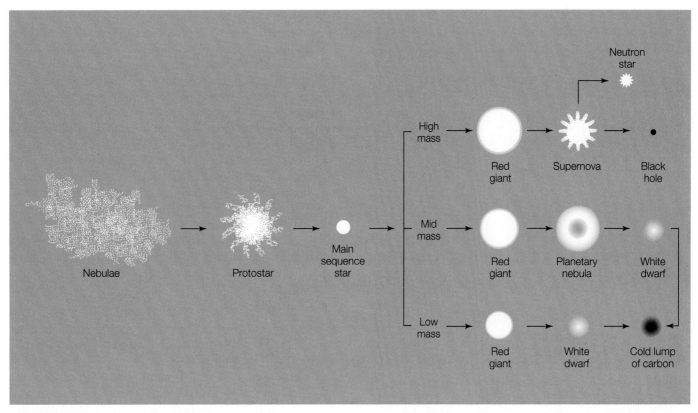

FIGURE 12.9   This flowchart shows some of the possible stages in the birth and aging of a star. The differences are determined by the mass of the star.

critical temperature of 600 million kelvins to begin carbon fusion reactions. Thus, a more massive star may go through a carbon fusing stage and other fusion reaction stages that will continue to produce new elements until the element iron is reached. (See binding energy and figure 11.10 on page 254.) After iron, energy is no longer released by the fusion process, and the star has used up all of its energy sources. Lacking an energy source, the star is no longer able to maintain its internal temperature. The star loses the outward pressure of expansion from the high temperature, which had previously balanced the inward pressure from gravitational attraction. The star thus collapses, then rebounds like a compressed spring into a catastrophic explosion called a **supernova.** A supernova produces a brilliant light in the sky that may last for months before it begins to dim as the new elements that were created during the life of the star diffuse into space. These include all the elements up to iron that were produced by fusion reactions during the life of the star and heavier elements that were created during the instant of the explosion. All the elements heavier than iron were created as some less massive nuclei disintegrated in the explosion, joining with each other and with lighter nuclei to produce the nuclei of the elements from iron to uranium. As you will see in chapter 13, these newly produced, scattered elements will later become the building blocks for new stars and planets such as the Sun and Earth.

The remains of the compressed core after the supernova have yet another fate if the core has a remaining mass greater than 1.4 solar masses. The gravitational forces on the remaining matter, together with the compressional forces of the supernova

explosion, are great enough to collapse nuclei, forcing protons and electrons together into neutrons, forming the core of a **neutron star.** A neutron star is the very small (10 to 20 km diameter), superdense (10^{11} kg/cm^3 or greater) remains of a supernova with a center core of pure neutrons.

Because it is a superdense form of matter, the neutron star also has an extremely powerful magnetic field, capable of becoming a pulsar. A **pulsar** is a very strongly magnetized neutron star that emits a uniform series of equally spaced electromagnetic pulses. Evidently, the magnetic field of a rotating neutron star makes it a powerful electric generator, capable of accelerating charged particles to very high energies. These accelerated charges are responsible for emitting a beam of electromagnetic radiation, which sweeps through space with amazing regularity (figure 12.10). The pulsating radio signals from a pulsar were a big mystery when first discovered. For a time, extraterrestrial life was considered as the source of the signals, so they were jokingly identified as LGM (for "little green men"). Over three hundred pulsars have been identified, and most emit radiation in the form of radio waves. Two, however, emit visible light, two emit beams of gamma radiation, and one emits X-ray pulses.

Another theoretical limit occurs if the remaining core has a mass of about 3 solar masses or more. At this limit, the force of gravity overwhelms *all* nucleon forces, including the repulsive forces between like charged particles. If this theoretical limit is reached, nothing can stop the collapse, and the collapsed star will become so dense that even light cannot escape. The star is now a **black hole** in space. Since nothing can stop the collapsing

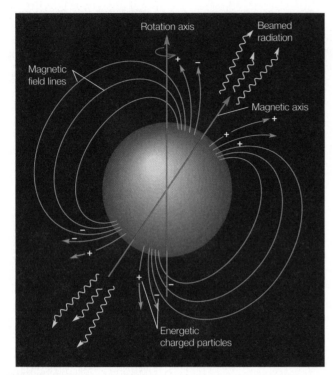

FIGURE 12.10 The magnetic axis of the pulsar is inclined with respect to the rotation axis. Rapidly moving electrons in the regions near the magnetic poles emit radiation in a beam pointed outward. When the beam sweeps past Earth, a pulse is detected.

star, theoretically a black hole would continue to collapse to a pinpoint and then to a zero radius called a *singularity*. This event seems contrary to anything that can be directly observed in the physical universe, but it does agree with the general theory of relativity and concepts about the curvature of space produced by such massively dense objects. Black holes are theoretical and none has been seen, of course, because a black hole theoretically pulls in radiation of all wavelengths and emits nothing. Evidence for the existence of a black hole is sought by studying X rays that would be given off by matter as it is accelerated into a black hole.

Another form of evidence for the existence of a black hole has now been provided by the Hubble Space Telescope. Hubble pictured a disk of gas only about 60 light-years out from the center of a galaxy (M87), moving at more than 1.6 million km/h (about 1 million mi/h). The only known possible explanation for such a massive disk of gas moving with this velocity at the distance observed would require the presence of a 1–2 billion solar-mass black hole. This gas disk could only be resolved by the Hubble Space Telescope, so this telescope has provided the first observational evidence of a black hole.

GALAXIES

Stars are associated with other stars on many different levels, from double stars that orbit a common center of mass, to groups of tens or hundreds of stars that have gravitational links

and a common origin, to the billions and billions of stars that form the basic unit of the universe, a **galaxy.** The Sun is but one of an estimated four hundred billion stars that are held together by gravitational attraction in the Milky Way galaxy (figure 12.11). The numbers of stars and vastness of the Milky Way galaxy alone seem almost beyond comprehension, but there is more to come. The Milky Way is but one of *billions* of galaxies that are associated with other galaxies in clusters, and these clusters are associated with one another in superclusters. Through a large telescope, you can see more galaxies than individual stars in any direction, each galaxy with its own structure of billions of stars. Yet, there are similarities that point to a common origin. Some of the similarities and associations of stars will be introduced in this section along with the Milky Way galaxy, the vast, flat, spiraling arms of stars, gas, and dust where the Sun is located.

The Milky Way Galaxy

Away from city lights, you can clearly see the faint, luminous band of the Milky Way galaxy on a moonless night. Through a telescope or a good pair of binoculars, you can see that the luminous band is made up of countless numbers of stars. You may also be able to see the faint glow of nebulae, concentrations of gas and dust. There are dark regions in the Milky Way that also give an impression of something blocking starlight, such as dust. You can also see small groups of stars called **galactic clusters.** Galactic clusters are gravitationally bound subgroups of as many as one thousand stars that move together within the Milky Way. Other clusters are more symmetrical and tightly packed, containing as many as a million stars, and are known as **globular clusters.**

Viewed from a distance in space, the Milky Way would appear to be a huge, flattened cloud of spiral arms radiating out from the center. There are three distinct parts: (1) the spherical

---

Myths, Mistakes, and Misunderstandings

Hello, Enterprise?

It is a myth that spaceships could have instant two-way communication. Radio waves travel at the speed of light, but distances in space are huge. The table below gives the approximate time required for a radio signal to travel from Earth to a planet or place. Two-way communication would require twice the time listed.

Mars	4.3 min
Jupiter	35 min
Saturn	1.2 h
Uranus	2.6 h
Neptune	4 h
Across Milky Way galaxy	100,000 yr
To Andromeda galaxy	2,000,000 yr

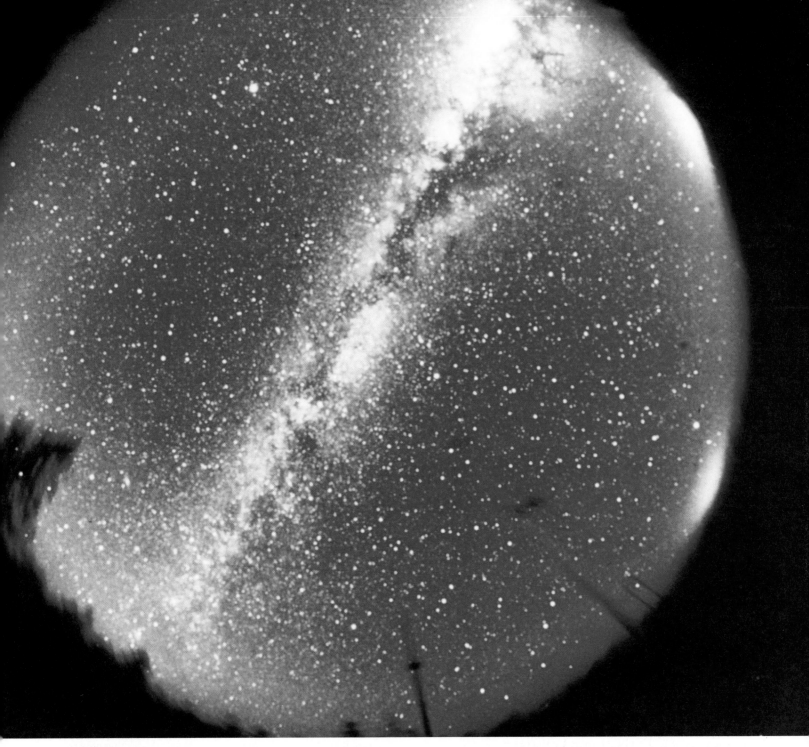

FIGURE 12.11 A wide-angle view toward the center of the Milky Way galaxy. Parts of the white, milky band are obscured from sight by gas and dust clouds in the galaxy.

concentration of stars at the center of the disk called the *galactic nucleus;* (2) the rotating *galactic disk,* which contains most of the bright, blue stars along with much dust and gas; and (3) a spherical *galactic halo,* which contains some 150 globular clusters located outside the galactic disk (figure 12.12). The Sun is located in one of the arms of the galactic disk, some 25,000 to 30,000 light-years from the center. The galactic disk rotates, and the Sun completes one full rotation every 200 million years.

The diameter of the *galactic disk* is about 100,000 light-years. Yet, in spite of the recent estimate of four hundred billion stars in the Milky Way, it is mostly full of emptiness. By way of analogy, imagine reducing the size of the Milky Way disk until stars like the Sun were reduced to the size of tennis balls. The distance between two of these tennis-ball-sized stars would now compare to the distance across the state of Texas. The space between the stars is not actually empty since it contains a thin concentration of gas, dust, and molecules of

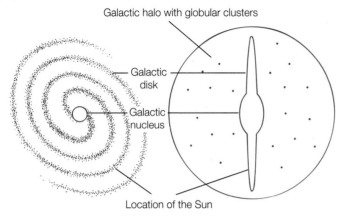

FIGURE 12.12 The structure of the Milky Way galaxy.

chemical compounds. The gas particles outnumber the dust particles about 10^{12} to 1. The gas is mostly hydrogen, and the dust is mostly solid iron, carbon, and silicon compounds. Over forty different chemical molecules have been discovered in the space between the stars, including many organic molecules. Some nebulae consist of clouds of molecules with a maximum density of about 10^6 molecules/cm^3. The gas, dust, and chemical compounds make up part of the mass of the galactic disk, and the stars make up the remainder. The gas plays an important role in the formation of new stars, and the dust and chemical compounds play an important role in the formation of planets.

Other Galaxies

Outside the Milky Way is a vast expanse of emptiness, lacking even the few molecules of gas and dust spread thinly through the galactic nucleus. There is only the light from faraway galaxies and the time that it takes for this light to travel across the vast vacuum of intergalactic space. How far away is the nearest galaxy? Recall that the Milky Way is so large that it takes light 100,000 years to travel the length of its diameter.

The nearest galactic neighbor to the Milky Way is a dwarf spherical galaxy only 80,000 light-years from the solar system. The nearby galaxy is called a dwarf because it has a diameter of only about 1,000 light-years. It is apparently in the process of being pulled apart by the gravitational pull of the Milky Way, which now is known to have eleven satellite galaxies.

The nearest galactic neighbor similar to the Milky Way is Andromeda, about 2 million light-years away. Andromeda is similar to the Milky Way in size and shape, with about four hundred billion stars, gas, and dust turning in a giant spiral pinwheel (figure 12.13). Other galaxies have other shapes and other characteristics. The American astronomer Edwin Hubble developed a classification scheme for the structure of galaxies based on his 1926 study of some six hundred different galaxies. The basic galactic structures were identified as elliptical, spiral, barred, and irregular.

FIGURE 12.13 The Andromeda galaxy, which is believed to be similar in size, shape, and structure to the Milky Way galaxy.

The Life of a Galaxy

Hubble's classification of galaxies into distinctly different categories of shape was an exciting accomplishment because it suggested that some relationship or hidden underlying order might exist in the shapes. Finding underlying order is important because it leads to the discovery of the physical laws that govern the universe. Soon after Hubble published his classification results in 1926, two models of galactic evolution were proposed. One model, which was suggested by Hubble, had extremely slowly spinning spherical galaxies forming first, which gradually flattened out as their rate of spin increased while they condensed. This is a model of spherical galaxies flattening out to increasingly elliptical shapes, eventually spinning off spirals until they finally broke up into irregular shapes over a long period of time.

Among many uses, the Hubble Space Telescope is used to study young galaxies and galaxies on collision courses. Based on these studies, astronomers today recognize that the different shapes of galaxies do not represent an evolutionary sequence.

Extraterrestrial is a descriptive term, meaning a thing or event outside Earth or its atmosphere. The term is also used to describe a being, a life-form that originated away from Earth. This reading is concerned with the search for extraterrestrials, intelligent life that might exist beyond Earth and outside the solar system.

Why do people believe that extraterrestrials might exist? The affirmative answer comes from a mixture of current theories about the origin and development of stars, statistical odds, and faith. Considering the statistical odds, note that our Sun is one of the some 400 billion stars that make up our Milky Way galaxy. The Milky Way galaxy is one of some 10 billion galaxies in the observable universe. Assuming an average of about 400 billion stars per galaxy, this means there are some 400 billion times 10 billion stars, or 4×10^{21} stars in the observable universe. There is nothing special or unusual about our Sun, and astronomers believe it to be quite ordinary among all the other stars (all 4,000,000,000,000,000, 000,000 or so).

So the Sun is an ordinary star, but what about our planet? Not too long ago, most people, including astronomers, thought our solar system with its life-supporting planet (Earth) to be unique. Evidence collected over the past decade or so, however, has strongly suggested that this is not so. Planets are now believed to be formed as a natural part of the star-forming process. Evidence of planets around other stars has also been detected by astronomers. One of the stars with planets is "only" 53 light-years from Earth.

Even with a very low probability of planetary systems forming with the development of stars, a population of 4×10^{21} stars means there are plenty of planetary

Box Figure 12.1 The 300 m (984 ft) diameter radio telescope at Arecibo, Puerto Rico, is the largest fixed-dish radio telescope in the world.

systems in existence, some with the conditions necessary to support life. (Note: If 1 percent have planetary systems, this means 4×10^{19} stars have planets.) Thus, it is a statistical observation that suitable planets for life are very likely to exist. In addition, radio astronomers have found that many organic molecules exist, even in the space between the stars. Based on statistics alone, there should be life on other planets, life that may have achieved intelligence and developed into a technological civilization.

If extraterrestrial life exists, why have we not detected them or why have they not contacted us? The answer to this question is found in the unbelievable distances involved in interstellar space. For example, a logical way to search for extraterrestrial intelligence is to send radio signals to space and analyze those coming from space. Modern radio telescopes can send powerful radio beams and present-day computers and data processing techniques can now search through incoming radio

signals for the patterns of artificially generated radio signals (box figure 12.1). Radio signals, however, travel through space at the speed of light. The diameter of our Milky Way galaxy is about 100,000 light-years, which means 100,000 years would be required for a radio transmission to travel across the galaxy. For this discussion, assume Earth is located on one edge of the Milky Way. If we were to transmit a super, super strong radio beam from Earth, it would travel at the speed of light and cross the distance of our galaxy in 100,000 years. If some extraterrestrials on the other side of the Milky Way galaxy did detect the message and send a reply, it could not arrive at Earth until 200,000 years after the message was sent. Now consider the fact that of all the 10 billion other galaxies in the observable universe, our nearest galactic neighbor similar to the Milky Way is Andromeda. Andromeda is 2 million light-years from the Milky Way galaxy.

In addition to problems with distance and time, there are questions about in which part of the sky you should send and look for radio messages, questions about which radio frequency to use, and problems with the power of present-day radio transmitters and detectors. Realistically, the hope for any exchange of radio-transmitted messages would be restricted to within several hundred light-years of Earth.

Considering all the limitations, what sort of signals should we expect to receive from extraterrestrials? Probably a series of pulses that somehow indicate counting, such as 1, 2, 3, 4, and so on, repeated at regular intervals. This is the most abstract, while at the same time the simplest, concept that an intelligent being anywhere would have. It could provide the foundation for communications between the stars.

Their shapes are understood to be a result of the various conditions under which the galaxies were formed.

The current model of how galaxies form is based on the **big bang theory** of the creation of the universe. The big bang theory considers the universe to have had an explosive beginning. According to this theory, all matter in the universe was located together in an arbitrarily dense state from which it began to expand, an expansion that continues today. Evidence that supports the big bang theory comes from Albert Einstein's theory of general relativity as well as three areas of physical observations:

1. **Expansion of the universe.** The initial evidence for the big bang theory came from Edwin Hubble and his earlier

Redshift and Hubble's Law

As described in chapter 5, the Doppler effect tells us that the frequency of a wave depends on the relative motion of the source and observer. When the source and observer are moving toward each other, the frequency appears to be higher. If the source and observer are moving apart, the frequency appears to be lower.

Light from a star or galaxy is changed by the Doppler effect, and the frequency of the observed spectral lines depends on the relative motion. The Doppler effect changes the frequency from what it would be if the star or galaxy were motionless relative to the observer. If the star or galaxy is moving toward the observer, a shift occurs in the spectral lines toward a higher frequency (blueshift). If the star or galaxy is moving away from the observer, a shift occurs in the spectral lines toward a lower frequency (redshift). Thus, a redshift or blueshift in the spectral lines will tell you if a star or galaxy is moving toward or away from you.

One of the first measurements of the distance to other galaxies was made by Edwin Hubble at Mount Wilson Observatory in California. When Hubble compared the distance figures with the observed redshifts, he found that the recession speeds were proportional to the distance. Farther-away galaxies were moving away from the Milky Way, but galaxies that are more distant are moving away faster than closer galaxies. This proportional relationship between galactic speed and distances was discovered in 1929 by Hubble and today is known as *Hubble's law*. The conclusion was that all the galaxies are moving away from one another and an observer on any given galaxy would have the impression that all galaxies were moving away in all directions. In other words, the universe is expanding with component galaxies moving farther and farther apart.

work with galaxies. Hubble had determined the distances to some of the galaxies that had redshifted spectra. From this expansive redshift, it was known that these galaxies were moving away from the Milky Way. Hubble found a relationship between the distance to a galaxy and the velocity with which it was moving away. He found the velocity to be directly proportional to the distance; that is, the greater the distance to a galaxy, the greater the velocity. This means that a galaxy twice as far from the Milky Way is moving away from the Milky Way at twice the speed. Since this relationship was seen in all directions, it meant that the universe is expanding uniformly. The same effect would be viewed from any particular galaxy; that is, all the other galaxies are moving away with a velocity proportional to the distance to the other galaxies. This points to a common beginning, a time when all matter in the universe was together.

2. **Background radiation.** The big bang occurred as some unstable form of energy expanded and cooled, eventually creating matter and space. The initial temperature was 10 billion kelvins or so, and began to cool as the universe expanded. The afterglow of the big bang is called cosmic background radiation. A measurement of cosmic background radiation today agrees with the radiation that should be present according to an expanding model of the universe.

3. **Abundance of elements.** The proportion of helium in the universe should be about 24 percent, based on an expanding model of the universe. A measurement of the abundance of helium verifies the proportion as predicted by the big bang theory.

Cosmic Background Explorer (COBE) spacecraft studied diffuse cosmic background radiation to help answer such questions as how matter is distributed in the universe, whether the universe is uniformly expanding, and how and when galaxies first formed.

The 2003 results from NASA's orbiting *Wilkinson Microwave Anisotropy Probe (WMAP)* produced a precision map of the remaining cosmic microwave background from the big bang. *WMAP* surveyed the entire sky for a whole year with a resolution some 40 times greater than *COBE*. Analysis of *WMAP* data revealed that the universe is 13.7 billion years old, with a 1 percent margin of error. The *WMAP* data found strong support for the big bang and expanding universe theories. It also revealed that the content of the universe includes 4 percent ordinary matter, 23 percent of an unknown type of dark matter, and 73 percent of a mysterious dark energy.

How old is the universe? As mentioned earlier, astronomical and physical "clocks" indicate that the universe was created in a "big bang" some 13.7 billion years ago, expanding as an intense and brilliant fireball with a temperature of some 10 billion kelvins. This estimate of the age is based on precise measurements of the rate at which galaxies are moving apart, an expansion that started with the big bang. Astronomers used data on the expansion to back-calculate the age, much like running a movie backward, to arrive at the age estimate. At first, this technique found that the universe was about 18 to 20 billion years old. Then astronomers found that the universe is not expanding at a constant rate. Instead, the separation of galaxies is actually accelerating, pushed by a mysterious force known as "**dark energy.**" By adding in calculations for this poorly understood force, the estimate of 13 to 14 billion years was developed.

Other age-estimating techniques agree with this ballpark age. Studies of white dwarfs, for example, established their rate of cooling. By looking at the very faintest and oldest white dwarfs with the Hubble Space Telescope, astronomers were able to use the cooling rate to estimate the age of the universe. The result found the dimmest of the white dwarfs to be about 13 billion years old, plus or minus about half a billion years. More recently, data from the orbiting *WMAP* spacecraft has

Will the universe continue to expand forever, or will it be pulled back together into a really big crunch? Whether the universe will continue expanding or gravity will pull it back together again depends on the *critical density* of the universe. If there is enough matter, the actual density of the universe will be above the critical density, and gravity will stop the current expansion and pull everything back together again. On the other hand, if the actual density of the universe is less than or equal to the critical density, the universe will be able to escape its own gravity and continue expanding. Is the actual density small enough for the universe to escape itself?

All the detailed calculations of astrophysicists point to an actual density that is *less* than the critical density, meaning the universe will continue to expand forever. However, these calculations also show that there is more matter in the universe than can be accounted for in the stars and galaxies. This means there must be matter in the universe that is not visible, or not shining at least, so we cannot see it. Three examples follow that show why astrophysicists believe more matter exists than we can see.

1. There is a relationship between the light emitted from a star and the mass of that star. Thus, you can indirectly measure the amount of matter in a galaxy by measuring the light output of that galaxy. Clusters of galaxies have been observed moving, and the motion of galaxies within a cluster does not agree with matter information from the light. The motion suggests that galaxies are attracted by a gravitational force from about ten times more matter than can be accounted for by the light coming from the galaxies.

2. There are mysterious variations in the movement of stars within an individual galaxy. The rate of rotation of stars about the center of rotation of a galaxy is related to the distribution of matter in that galaxy. The outer stars are observed to rotate too fast for the amount of matter that can be seen in the galaxy. Again, measurements indicate there is about ten times more matter in each galaxy than can be accounted for by the light coming from the galaxies.

3. Finally, there are estimates of the total matter in the universe that can be made by measuring ratios such as deuterium to ordinary hydrogen, lithium−7 to helium−3, and helium to hydrogen. In general, theoretical calculations all seem to account for only about 10 percent of all the matter in the universe.

Calculations from a variety of sources all seem to agree that 90 percent or more of the matter that makes up the universe is missing. The missing matter and the mysterious variations in the orbits of galaxies and stars can be accounted for by the presence of **dark matter,** which is invisible and unseen. What is the nature of this dark matter? You could speculate that dark matter is simply normal matter that has been overlooked, such as dark galaxies, brown dwarfs, or planetary material such as rock, dust, and the like. You could also speculate that dark matter is dark because it is in the form of subatomic particles, too small to be seen. As you can see, scientists have reasons to believe that dark matter exists, but they can only speculate about its nature.

The nature of dark matter represents one of the major unsolved problems in astronomy today. There are really two dark matter problems: (1) the nature of the dark matter and (2) how much dark matter contributes to the actual density of the universe. Not everyone believes that dark matter is simply normal matter that has been overlooked. There are at least two schools of thought on the nature of dark matter and what should be the focus of research. One school of thought focuses on *particle physics* and contends that dark matter consists primarily of exotic undiscovered particles or known particles such as the neutrino. Neutrinos are electrically neutral, stable subatomic particles. This school of thought is concerned with forms of dark matter called WIMPs (after *w*eakly *i*nteracting *m*assive *p*articles). In spite of the name, particles under study are not always massive. It is entirely possible that some low-mass species of neutrino—or some as-of-yet unidentified WIMP—will be experimentally discovered and found to have the mass needed to account for the dark matter and thus close the universe.

The other school of thought about dark matter focuses on *astrophysics* and contends that dark matter is ordinary matter in the form of brown dwarfs, unseen planets outside the solar system, and galactic halos. This school of thought is concerned with forms of dark matter called MACHOs (after *m*assive *a*strophysical *c*ompact *h*alo *o*bjects). Dark matter considerations in this line of reasoning consider massive objects, but ordinary matter (protons and neutrons) is also considered as the material making up galactic halos. Protons and neutrons belong to a group of subatomic particles called the *baryons,* so they are sometimes referred to as *baryonic dark matter.* Some astronomers feel that baryonic dark matter is the most likely candidate for halo dark matter.

In general, astronomers can calculate the probable cosmological abundance of WIMPs, but there is no proof of their existence. By contrast, MACHOs dark matter candidates are known to exist, but there is no way to calculate their abundance. MACHOs astronomers assert that halo dark matter and probably all dark matter may be baryonic. Do you believe WIMPs or MACHOs will provide an answer about the future of the universe? What is the nature of dark matter, how much dark matter exists, and what is the fate of the universe? Answers to these and more questions await further research.

Connections . . .

Dark Energy

The first evidence of dark energy came from astronomers trying to measure how fast the expanding universe is slowing. At the time, the generally accepted model of the universe was that it was still expanding but had been slowing from gravitational attraction since the big bang. The astronomers intended to measure the slowing rate to determine the average density of matter in the universe, providing a clue about the extent of dark matter. Their idea was to compare light traveling from a supernova in the distant universe with light traveling from a much closer supernova. Light from the distant universe is just now reaching us since it was emitted when the universe was very young. The brightness of the older and younger supernova would provide information about distance, and expansive redshift data would provide information about their speed of expansion. By comparing these two light sources, they would be able to calculate the expansion rate for now and in the distant past.

To their surprise, the astronomers did not find a rate of slowing expansion. Rather, they found that the expansion is speeding up. There was no explanation for this speeding up other than to assume that some unknown antigravity force was at work. This unknown force must be pushing the galaxies farther apart at the same time that gravity is trying to pull them together. The unknown repulsive force became known as "dark energy."

The idea of an unknown repulsive force is not new. It was a new idea when Einstein added a "cosmological constant" to the equations in the theory of general relativity. This constant represents a force that opposes gravity, and the force grows as a function of space. This means there was not much—shall we say, dark energy—when the universe was smaller, and gravity slowed the expansion. As the distance between galaxies increased, there was an increase of dark energy and the expansion accelerated. Einstein removed this constant when Edwin Hubble reported evidence that the universe is expanding. Removing the constant may have been a mistake. One of the problems in understanding what was happening to the early universe was a lack of information about the distant past as is represented by the distant universe of the present. Today, this is changing as more new technology is helping astronomers study the distant universe. Will they find the meaning of dark energy? What is dark energy? The answer is that no one knows. Stay tuned . . . this story is to be continued.

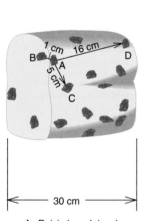

A Raisin bread dough before rising

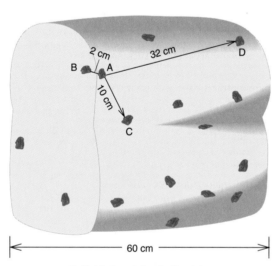

B Raisin bread dough after rising

FIGURE 12.14 As the universe (here represented by a loaf of raisin bread) expands, the expansion carries galaxies (represented by the raisins) away from each other at speeds that are proportional to their distances from each other. It doesn't matter within which galaxy an astronomer resides; the other galaxies all appear to be moving away.

produced a precision map of the cosmic microwave background that shows the universe to be 13.7 billion years old. An age of 13.7 billion years may not be the final answer for the age of the universe, but when three independent measurement techniques agree closely, the answer becomes more believable.

An often-used analogy for the movement of galaxies after the big bang is a loaf of rising raisin bread. Consider galaxies as raisins in a loaf of raisin bread dough as it rises (figure 12.14). As the dough expands in size, it carries along the raisins, which are moved farther and farther apart by the expansion. If you were on a raisin, you would see all the other raisins moving away from you, and the speed of this movement would be directly proportional to their distance away. It would not matter which raisin you were on since all the raisins would appear to be moving away from all the other raisins. In other words, the galaxies are not expanding into space that is already there. It is space itself that is expanding, and the galaxies move with the expanding space (figure 12.15).

The ultimate fate of the universe will strongly depend on the mass of the universe and if the expansion is slowing or continuing. All the evidence tells us it began by expanding with a big bang. Will it continue to expand, becoming increasingly cold and diffuse? Will it slow to a halt, then start to contract to a fiery finale of a big crunch (figure 12.16)? Researchers continue searching for answers with experimental data and theoretical models, looking for clues in matches between the data and models.

People are fascinated with ideas of space travel as well as new discoveries about the universe and how it formed. Unmanned space probes, new kinds of instruments, and new kinds of telescopes have resulted in new information. We are learning what is happening in outer space away from Earth, as well as about the existence of other planets.

Few would deny that the space program has provided valuable information. However, some people wonder if it is worth the cost. They point to many problems here on Earth, such as growing energy and water needs, pollution problems, and ongoing health problems. They say that the money spent on exploring space could be better used for helping resolve problems on Earth.

In addition to new information and understanding coming out of the space program, supporters of space exploration point to new technology that helps people living on Earth. Satellites now provide valuable information for agriculture, including land use and weather monitoring. Untold numbers of lives have been saved thanks to storm warnings provided by weather satellites. There are many other spin-offs from the space program, including improvements in communication systems.

Questions to Discuss

1. What do you think: have the gain of knowledge and the spin-offs merited the expense? Or would the funds be better spent on solving other problems?

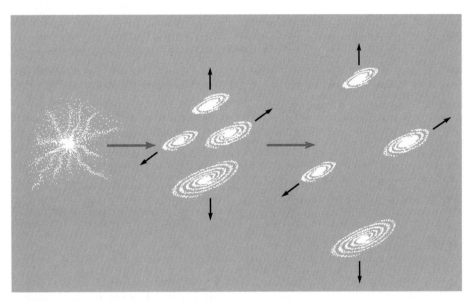

FIGURE 12.15   Will the universe continue expanding as the dust and gas in galaxies become locked up in white dwarf stars, neutron stars, and black holes?

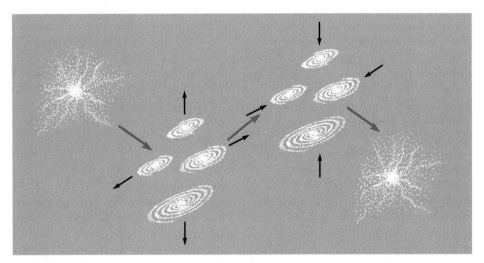

FIGURE 12.16   One theory of the universe assumes that the space between the galaxies is expanding, as does the big bang theory, but the galaxies gradually come back together to begin all over in another big bang.

People Behind the Science

Jocelyn (Susan) Bell Burnell (1943–)

Jocelyn Bell is a British astronomer who discovered pulsating radio stars—pulsars—an important astronomical discovery of the 1960s.

Bell was born in Belfast, Ireland, on July 15, 1943. The Armagh Observatory, of which her father was architect, was sited near her home, and the staff there was particularly helpful and offered encouragement when they learned of her early interest in astronomy. From 1956 to 1961, she attended the Mount School in York, England. She then went to the University of Glasgow, receiving her B.Sc. degree in 1965. In the summer of 1965, she began to work for her Ph.D. under the supervision of Anthony Hewish at the University of Cambridge. It was during the course of this work that she discovered pulsars.

She spent her first two years in Cambridge building a radio telescope that was specially designed to track quasars—her Ph.D. research topic. The telescope that she and her team built had the ability to record rapid variations in signals. It was also nearly 2 hectares (about 5 acres) in area, equivalent to a dish of 150 m (about 500 ft) in diameter, making it an extremely sensitive instrument.

The sky survey began when the telescope was finally completed in 1967, and Bell was given the task of analyzing the signals received. One day, while scanning the charts of recorded signals, she noticed a rather unusual radio source that had occurred during the night and had been picked up in a part of the sky that was opposite in direction to the Sun. This was curious because strong variations in the signals from quasars are caused by solar wind and are usually weak during the night. At first, she thought that the signal might be due to a local interference. After a month of further observations, it became clear that the position of the peculiar signals remained fixed with respect to the stars, indicating that it was neither terrestrial nor solar in origin. A more detailed examination of the signal showed that it was in fact composed of a rapid set of pulses that occurred precisely every 1.337 seconds. The pulsed signal was as regular as the most regular clock on Earth.

One attempted explanation of this curious phenomenon was that it represented an interstellar beacon sent out by extraterrestrial life on another star, so initially it was nicknamed LGM, for "little green men." Within a few months of noticing this signal, however, Bell located three other similar sources. They too pulsed at an extremely regular rate, but their periods varied over a few fractions of a second, and they all originated from widely spaced locations in our galaxy. Thus, it seemed that a more likely explanation of the signals was that they were being emitted by a special kind of star—a pulsar.

Since the astonishing discovery was announced, other observatories have searched the heavens for new pulsars. Some three hundred are now known to exist, their periods ranging from hundredths of a second to four seconds. It is thought that neutron stars are responsible for the signal. These are tiny stars, only about 7 km (about 4.3 mi) in diameter, but they are incredibly massive. The whole star and its associated magnetic field are spinning at a rapid rate, and the rotation produces the pulsed signal.

Source: Modified from the *Hutchinson Dictionary of Scientific Biography.* © RM, 2007. Reprinted by permission.

SUMMARY

Stars are theoretically born in clouds of hydrogen gas and dust in the space between other stars. Gravity pulls huge masses of hydrogen gas together into a *protostar,* a mass of gases that will become a star. The protostar contracts, becoming increasingly hotter at the center, eventually reaching a temperature high enough to start *nuclear fusion* reactions between hydrogen atoms. Pressure from hot gases balances the gravitational contraction, and the average newborn star will shine quietly for billions of years. The average star has a dense, hot *core* where nuclear fusion releases radiation, a less dense *radiation zone* where radiation moves outward, and a thin *convection zone* that is heated by the radiation at the bottom, then moves to the surface to emit light to space.

The brightness of a star is related to the amount of energy and light it is producing, the size of the star, and the distance to the star. The *apparent magnitude* is the brightness of a star as it appears to you. To compensate for differences in brightness due to distance, astronomers calculate the brightness that stars would have at a standard distance. This standard-distance brightness is called the *absolute magnitude.*

Stars appear to have different colors because they have different surface temperatures. A graph of temperature by spectral types and brightness by absolute magnitude is called the *Hertzsprung-Russell diagram,* or H-R diagram for short. Such a graph shows that normal, mature stars fall on a narrow band called the *main sequence* of stars. Where a star falls on the main sequence is determined by its brightness and temperature, which in turn are determined by the mass of the star. Other groups of stars on the H-R diagram have different sets of properties that are determined by where they are in their evolution.

The life of a star consists of several stages, the longest of which is the *main sequence* stage after a relatively short time as a *protostar.* After using up the hydrogen in the core, a star with an average mass expands to a *red giant,* then blows off the outer shell to become a *white dwarf star,* which slowly cools to a black lump of carbon. The blown-off outer shell forms a *planetary nebula,* which disperses over time to become the gas and dust of interstellar space. More massive stars collapse into *neutron stars* or *black holes* after a violent *supernova* explosion.

Galaxies are the basic units of the universe. The Milky Way galaxy has three distinct parts: (1) the *galactic nucleus,* (2) a rotating *galactic disk,* and (3) a *galactic halo.* The galactic disk contains subgroups of stars that move together as *galactic clusters.* The halo contains symmetrical and tightly packed clusters of millions of stars called *globular clusters.*

Evidence from astronomical and physical "clocks" indicates that the galaxies formed some 13.7 billion years ago, expanding ever since from a common origin in a *big bang.* The *big bang theory* describes how the universe began by expanding.

KEY TERMS

absolute magnitude (p. **271**)
apparent magnitude (p. **270**)
big bang theory (p. **279**)
black hole (p. **275**)
Cepheid variable (p. **273**)
convection zone (p. **270**)

core (p. **270**)
dark energy (p. **280**)
dark matter (p. **281**)
galactic clusters (p. **276**)
galaxy (p. **276**)
globular clusters (p. **276**)

Hertzsprung-Russell diagram (p. **272**)
light-year (p. **269**)
main sequence stars (p. **272**)
nebulae (p. **269**)
neutron star (p. **275**)

protostar (p. **269**)
pulsar (p. **275**)
radiation zone (p. **270**)
red giant stars (p. **273**)
supernova (p. **275**)
white dwarf stars (p. **273**)

APPLYING THE CONCEPTS

1. Stars twinkle and planets do not twinkle because
   a. planets shine by reflected light, and stars produce their own light.
   b. all stars are pulsing light sources.
   c. stars appear as point sources of light, and planets are disk sources.
   d. All of the above are correct.

2. Which of the following stars would have the longer life spans?
   a. the less massive
   b. between the more massive and the less massive
   c. the more massive
   d. All have the same life span.

3. A bright blue star on the main sequence is probably
   a. very massive.
   b. less massive.
   c. between the more massive and the less massive.
   d. None of the above is correct.

4. The basic property of a main sequence star that determines most of its other properties, including its location on the H-R diagram, is
   a. brightness.
   b. color.
   c. temperature.
   d. mass.

5. All the elements that are more massive than the element iron were formed in a
   a. nova.
   b. white dwarf.
   c. supernova.
   d. black hole.

6. If the core remaining after a supernova has a mass between 1.5 and 3 solar masses, it collapses to form a
   a. white dwarf.
   b. neutron star.
   c. red giant.
   d. black hole.

7. The basic unit of the universe is a
   a. star.
   b. solar system.
   c. galactic cluster.
   d. galaxy.

8. The greater the distance to a galaxy, the greater a redshift in its spectral lines. This is known as
   a. Doppler's law.
   b. Cepheid's law.
   c. Hubble's law.
   d. Kepler's law.

9. Dark energy calculations and the age of cooling white dwarfs indicate that the universe is about how old?
   a. 6,000 years
   b. 4.5 billion years
   c. 13.7 billion years
   d. 100,000 billion years

10. Whether the universe will continue to expand or will collapse back into another big bang seems to depend on what property of the universe?
    a. the density of matter in the universe
    b. the age of galaxies compared to the age of their stars
    c. the availability of gases and dust between the galaxies
    d. the number of black holes

Answers

1. c 2. a 3. a 4. d 5. c 6. b 7. d 8. c 9. c 10. a

QUESTIONS FOR THOUGHT

1. What is a light-year and how is it defined?
2. Why are astronomical distances not measured with standard referent units of distance such as kilometers or miles?
3. Which size of star has the longest life span, a star sixty times more massive than the Sun, one just as massive as the Sun, or a star that has a mass of one-twenty-fifth that of the Sun? Explain.
4. What is the Hertzsprung-Russell diagram? What is the significance of the diagram?
5. Describe, in general, the life history of a star with an average mass like the Sun.

6. What, if anything, is the meaning of the Hubble classification scheme of the galaxies?
7. What is a nova? What is a supernova?
8. Describe the theoretical physical circumstances that lead to the creation of (a) a white dwarf star, (b) a red giant, (c) a neutron star, (d) a black hole, and (e) a supernova.
9. Describe the two forces that keep a star in a balanced, stable condition while it is on the main sequence. Explain how these forces are able to stay balanced for a period of billions of years or longer.

10. What is the source of all the elements in the universe that are more massive than helium but less massive than iron? What is the source of all the elements in the universe that are more massive than iron?

11. What is a red giant star? Explain the conditions that lead to the formation of a red giant. How can a red giant become brighter than it was as a main sequence star if it now has a lower surface temperature?

12. Describe the structure of the Milky Way galaxy. Where are new stars being formed in the Milky Way? Explain why they are formed in this part of the structure and not elsewhere.

FOR FURTHER ANALYSIS

1. A star is 513 light-years from Earth. During what event in history did the light now arriving at Earth leave the star?

2. What are the significant differences between the life and eventual fate of a massive star and an average-sized star such as the Sun?

3. Analyze when apparent magnitude is a better scale of star brightness and when absolute magnitude is a better scale of star brightness.

4. What is the significance of the Hertzsprung-Russell diagram?

5. The Milky Way galaxy is a huge, flattened cloud of spiral arms radiating out from the center. Describe several ideas that explain why it has this shape. Identify which idea you favor and explain why.

INVITATION TO INQUIRY

It Keeps Going, And Going, And . . .

Pioneer 10 was the first space probe to visit an outer planet of our solar system. It was launched March 2, 1972, and successfully visited Jupiter on June 13, 1983. After transmitting information and relatively close-up pictures of Jupiter, *Pioneer 10* continued on its trajectory, eventually becoming the first space probe to leave the solar system. It continued to move silently into deep space and sent the last signal on January 22, 2003, when it was 12.2 billion km (7.6 billion mi) from Earth. It will now continue to drift for the next 2 million years toward the star Aldebaran in the constellation Taurus.

As the first human-made object out of the solar system, *Pioneer 10* carries a gold-plated plaque with the image shown in box figure 12.2. Perhaps intelligent life will find the plaque and decipher the image to learn about us. What information is in the image? Try to do your own deciphering to reveal the information. When you have exhausted your efforts, see grin.hq.nasa.gov/ABSTRACTS/GPN-2000-001623.html.

For more on the *Pioneer 10* mission, see nssdc.gsfc.nasa.gov/nwc/tmp/1972-012A.html.

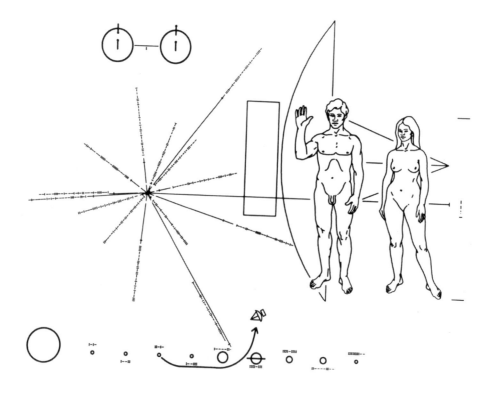

Box Figure 12.2 *Pioneer 10* plaque symbology.

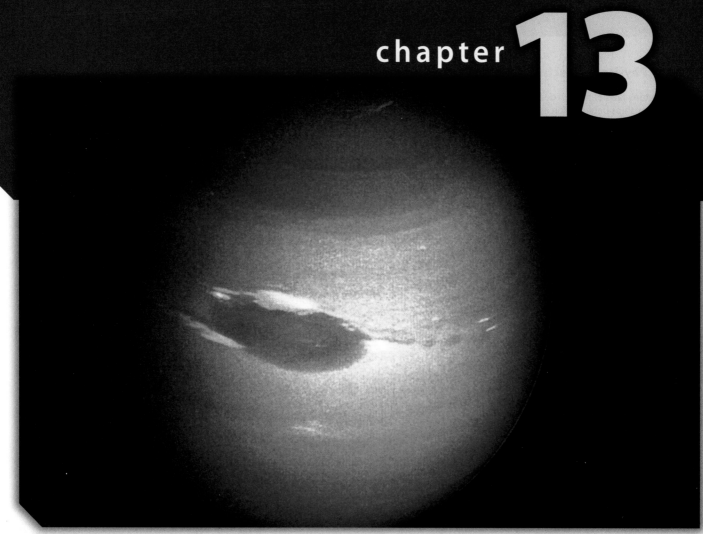

Neptune, the most distant and smallest of the gas giant planets, is a cold and interesting place. It has a Great Dark Spot, as you can see in this photograph made by *Voyager*. This spot is about the size of Earth and is similar to the Great Red Spot on Jupiter. Neptune has the strongest winds of any planet of the solar system—up to 2,000 km/h (1,200 mi/h). Clouds were observed by *Voyager* to be "scooting" around Neptune every sixteen hours or so. *Voyager* scientists called these fast-moving clouds "scooters."

The Solar System

Physics Connections

▶ Newton's laws of motion describe the relationships between forces, mass, and motion (Ch. 2).

▶ Newton's universal law of gravitation describes how every object in the universe is attracted to every other object in the universe (Ch. 2).

CORE CONCEPT

The solar system is composed of the Sun, a system of related planets, moons, comets, and asteroids.

The Sun, planets, moons, asteroids, and comets are believed to have formed from gas, dust, and chemical elements of previously existing stars (p. 306).

The interior planets of Mercury, Venus, Earth, and Mars are composed of rocky materials with a metallic nickel and iron core (p. 289).

The outer planets of Jupiter, Saturn, and Uranus are giant planets mostly composed of hydrogen, helium, and methane (p. 289).

Astronomy Connections

▶ A massive star ages to become a supernova, which spreads elements and gas in space (Ch. 12).

OVERVIEW

For generations, people have observed the sky in awe, wondering about the bright planets moving across the background of stars, but they could do no more than wonder. You are among the first generations on Earth to see close-up photographs of the planets, comets, and asteroids and to see Earth as it appears from space. Spacecrafts have now made thousands of photographs of the other planets and their moons, measured properties of the planets, and in some cases, studied their surfaces with landers. Astronauts have left Earth and visited the Moon, bringing back rock samples, data, and photographs of Earth as seen from the Moon (figure 13.1). All of these photographs and findings have given a new perspective of Earth, the planets, and the moons, comets, and asteroids that make up the solar system.

Viewed from the Moon, Earth is a spectacular blue globe with expanses of land and water covered by huge changing patterns of white clouds. Viewed from a spacecraft, other planets present a very different picture, each unique in its own way. Mercury has a surface that is covered with craters, looking very much like the surface of Earth's Moon. Venus is covered with clouds of sulfuric acid over an atmosphere of mostly carbon dioxide, which is under great pressures with surface temperatures hot enough to melt lead. The surface of Mars has great systems of canyons, inactive volcanoes, dry riverbeds and tributaries, and ice beneath the surface. The giant planets Jupiter and Saturn have orange, red, and white bands of organic and sulfur compounds and storms with gigantic lightning discharges much

larger than anything ever seen on Earth. One moon of Jupiter has active volcanoes spewing out liquid sulfur and gaseous sulfur dioxide. The outer giant planets Uranus and Neptune have moons and particles in rings that appear to be covered with powdery, black carbon.

These and many more findings, some fascinating surprises and some expected, have stimulated the imagination as well as added to our comprehension of the frontiers of space. The new information about the Sun's impressive system of planets, moons, comets, and asteroids has also added to speculations and theories about the planets and how they evolved over time in space. This information, along with the theories and speculations, will be presented in this chapter to give you a picture of the solar system.

PLANETS, MOONS, AND OTHER BODIES

The International Astronomical Union (IAU) is the governing authority over names of celestial bodies. At the August 24, 2006, meeting of the IAU, the definitions of planets, dwarf planets, and small solar system bodies were clarified and approved. To be a classical **planet,** an object must be orbiting the Sun, must be nearly spherical, and must be large enough to clear all matter from its orbital zone. A **dwarf planet** is defined as an object that is orbiting the Sun, is nearly spherical but has

not cleared matter from its orbital zone and is *not* a satellite. All other objects orbiting the Sun are referred to collectively as **small solar system bodies.**

Some astronomers had dismissed Pluto as a true planet for years because it has properties that do not fit with those of the other planets. The old definition of a planet was anything spherical that orbits the Sun, which resulted in nine planets. But in 2003, a new astronomical body was discovered. This body was named "Eris" after the Greek goddess of discord. Eris is larger than Pluto, round, and circles the Sun. Is it a planet? If so, the

FIGURE 13.1 This view of rising Earth was seen by the *Apollo 11* astronauts after they entered the orbit around the Moon. Earth is just above the lunar horizon in this photograph.

asteroid Ceres would also be a planet as would the fifty or so large, icy bodies believed to be orbiting the Sun far beyond Pluto. The idea of fifty or sixty planets in the solar system spurred astronomers to clarify the definition of a planet. However, Pluto does not clear its orbital zone, so it was downgraded to a dwarf planet. Today, there are eight planets, three dwarf planets (Pluto, Eris, and the giant asteroid Ceres), and many, many small solar system bodies. These definitions may change again in the future as more is learned about our solar system.

In this chapter, we will visit each of the planets, Earth's Moon, and other bodies of the solar system. (figure 13.2).

The Sun has seven hundred times the mass of all the planets, moons, and minor members of the solar system together. It is the force of gravitational attraction between the comparatively massive Sun and the rest of our solar system that holds it all together. The distance from Earth to the Sun is known as one **astronomical unit** (AU). One AU is about 1.5×10^8 km (about 9.3×10^7 mi). The astronomical unit is used to describe distances in the solar system, for example, Earth is 1 AU from the Sun.

Table 13.1 compares the basic properties of the eight planets. From this table, you can see that the planets can be classified into two major groups based on size, density, and nature of the atmosphere. The interior planets of Mercury, Venus, and Mars have densities and compositions similar to those of Earth, so these planets, along with Earth, are known as the **terrestrial planets.**

Outside the orbit of Mars are four **giant planets,** which are similar in density and chemical composition. The terrestrial planets are mostly composed of rocky materials and metallic nickel and iron. The giant planets of Jupiter, Saturn, Uranus, and Neptune, on the other hand, are massive giants mostly composed of hydrogen, helium, and methane. The density of the giant planets suggests the presence of rocky materials and iron as a core surrounded by a deep layer of compressed gases beneath a deep atmosphere of vapors and gases. Note that the terrestrial planets are separated from the giant planets by the asteroid belt.

We will start with the planet closest to the Sun and work our way outward, moving farther and farther from the Sun as we learn about our solar system.

Mercury

Mercury is the innermost planet, moving rapidly in a highly elliptical orbit that averages about 0.4 astronomical unit, or about 0.4 of the average distance of Earth from the Sun. Mercury is the smallest planet and is slightly larger than Earth's Moon. Mercury is very bright because it is so close to the Sun, but it is difficult to observe because it only appears briefly for a few

FIGURE 13.2 The order of the planets out from the Sun. The planets are (1) Mercury, (2) Venus, (3) Earth, (4) Mars, (5) Jupiter, (6) Saturn, (7) Uranus, and (8) Neptune. The orbits and the planet sizes are not drawn to scale and not all rings or moons are shown. Also, the planets are not in a line as shown.

TABLE 13.1   Properties of the Planets

	Mercury	Venus	Earth	Mars	Jupiter	Saturn	Uranus	Neptune
Average Distance from the Sun:								
in 10^6 km	58	108	150	228	778	1,400	3,000	4,497
in AU	0.38	0.72	1.0	1.5	5.2	9.5	19.2	30.1
Inclination to Ecliptic	7°	3.4°	0°	1.9°	1.3°	2.5°	0.8°	1.8°
Revolution Period								
(Earth years)	0.24	0.62	1.00	1.88	11.86	29.46	84.01	164.8
Rotation Period*	59 days	−243 days	23 h	24 h	9 h	10 h	−17 h	16 h
(Earth days,			56 min	37 min	50 min	39 min	14 min	6.7 min
min, and s)			4 s	23 s	30 s			
Mass (Earth = 1)	0.05	0.82	1.00	0.11	317.9	95.2	14.6	17.2
Equatorial Dimensions:								
diameter in km	4,880	12,104	12,756	6,787	142,984	120,536	57,118	49,528
in Earth radius = 1	0.38	0.95	1.00	0.53	11	9	4	4
Density (g/cm^3)	5.43	5.25	5.52	3.95	1.33	0.69	1.29	1.64
Atmosphere								
(major compounds)	None	CO_2	N_2, O_2	CO_2	H_2, He	H_2, He	H_2, He, CH_4	H_2, He, CH_4
Solar Energy Received								
(cal/cm^2/s)	13.4	3.8	2.0	0.86	0.08	0.02	0.006	0.002

*A negative value means spin is opposite to motion in orbit.

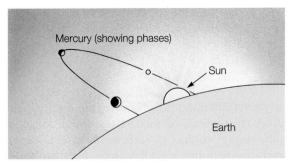

FIGURE 13.3   Mercury is close to the Sun and is visible from Earth only briefly before or after sunrise or sunset, showing phases. Mercury actually appears much smaller in an orbit that is not tilted as much as shown in this figure.

hours immediately after sunset or before sunrise. This appearance, low on the horizon, means that Mercury must be viewed through more of Earth's atmosphere, making the study of such a small object difficult at best (figure 13.3).

Mercury moves around the Sun in about three Earth months, giving Mercury the shortest "year" of all the planets. With the highest orbital velocity of all the planets, Mercury was appropriately named after the mythical Roman messenger of speed. Oddly, however, this speedy planet has a rather long day in spite of its very short year. With respect to the stars, Mercury rotates once every fifty-nine days. This means that Mercury rotates on its axis three times every two orbits.

The long Mercury day with a nearby large, hot Sun means high temperatures on the surface facing the Sun. High temperatures mean higher gas kinetic energies, and with a low

gravity, gases easily escape from Mercury so it has only trace gases for an atmosphere. The lack of an atmosphere to even the heat gains from the long days and heat losses from the long nights results in some very large temperature differences. The temperature of the surface of Mercury ranges from above the melting point of lead on the sunny side to below the temperature of liquid oxygen on the dark side.

Mercury has been visited by *Mariner 10,* which flew by three times in 1973 and 1974. The photographs transmitted by *Mariner 10* revealed that the surface of Mercury is covered with craters and very much resembles the surface of Earth's Moon. There are large craters, small craters, superimposed craters, and craters with lighter colored rays coming from them just like the craters on the Moon. Also as on Earth's Moon, there are hills and smooth areas with light and dark colors that were covered by lava in the past, some time after most of the impact craters were formed (figure 13.4).

A new spacecraft, named *MESSENGER,* is now on its way to investigate the planet Mercury. Its name is an acronym for *ME*rcury *S*urface, *S*pace *EN*vironment, *GE*ochemistry and *R*anging. It was launched August 3, 2004, on a 7.9 billion km (about 4.9 billion mi) trip designed to slow the spacecraft as it falls toward the Sun. Overall, it looped by Earth in August 2005, then twice by Venus in October 2006 and June 2007, and then three times by Mercury in January 2008, October 2008, and September 2009. Eventually, it will be slow enough to be captured by the planet Mercury as it flies by in March 2011. It will then become the first spacecraft to orbit Mercury. Mercury has a high surface temperature, and the spacecraft instruments will be protected from high radiant energy from Mercury and the Sun by a sun-shade of heat-resistant ceramic fabric.

FIGURE 13.4 A photomosaic of Mercury made from pictures taken by the *Mariner 10* spacecraft. The surface of Mercury is heavily cratered, looking much like the surface of Earth's Moon. All the interior planets and the Moon were bombarded early in the life of the solar system.

Mercury has no natural satellites or moons, it has a weak magnetic field, and it has an average density more similar to that of Venus or Earth than to the Moon's. The presence of the magnetic field and the relatively high density for such a small body probably means that Mercury has a relatively large core of iron with at least part of the core molten. Because of its high density, Mercury is believed to have lost much of its less dense, outer layer of rock materials sometime during its formation.

Venus

Venus is the brilliant evening and morning "star" that appears near sunrise or sunset, sometimes shining so brightly that you can see it while it is still daylight. Venus orbits the Sun at an average distance of about 0.7 AU. Venus is sometimes to the left of the Sun, appearing as the evening star, and sometimes to the right of the Sun, appearing as the morning star. Venus also has phases just as the Moon does. When Venus is in the full phase, it is small and farthest away from Earth. A crescent Venus appears much larger and thus the brightest because it is closest to Earth when it is in its crescent phase. You can see the phases of Venus with a good pair of binoculars.

Venus shines brightly because it is covered with clouds that reflect about 80 percent of the sunlight, making it the brightest object in the sky after the Sun and the Moon. These same clouds prevented any observations of the surface of Venus until the early 1960s, when radar astronomers were able to penetrate the clouds and measure the planet's spin rate. Venus was found to spin slowly, so slowly that each day on Venus is longer than a Venus year! Also a surprise, Venus was found to spin in the *opposite* direction to its direction of movement in its orbit. On Venus, you would observe the Sun to rise in the west and set in the east, if you could see it, that is, through all the clouds.

In addition to early studies by radio astronomers, Venus has been the target of many American and Soviet spacecraft probes (Table 13.2).

Venus has long been called Earth's sister planet since its mass, size, and density are very similar. That is where the similarities with Earth end, however, as Venus has been found to have a very hostile environment. Spacecraft probes found a hot, dry surface under tremendous atmospheric pressure (figure 13.5). The atmosphere consists mostly of carbon dioxide, a few percent of nitrogen, and traces of water vapor and other gases. The atmospheric pressure at the surface of Venus is almost one hundred times the pressure at the surface of Earth, a pressure many times beyond what a human could tolerate. The average surface temperature is comparable to the surface temperature on Mercury, which is hot enough to melt lead. The hot temperature on Venus, which is nearly twice the distance from the Sun as Mercury, is a result of the greenhouse effect. Sunlight filters through the atmosphere of Venus, warming the surface. The surface reemits the energy in

TABLE 13.2 Spacecraft Missions to Venus

Date	Name	Owner	Remark
Feb 12, 1961	*Venera 1*	USSR	Flyby
Aug 27, 1962	*Mariner 2*	U.S.	Flyby
Apr 2, 1964	*Zond 1*	USSR	Flyby
Nov 12, 1965	*Venera 2*	USSR	Flyby
Nov 16, 1965	*Venera 3*	USSR	Crashed on Venus
June 12, 1967	*Venera 4*	USSR	Impacted Venus
June 14, 1967	*Mariner 5*	U.S.	Flyby
Jan 5, 1969	*Venera 5*	USSR	Impacted Venus
Jan 10, 1969	*Venera 6*	USSR	Impacted Venus
Aug 17, 1970	*Venera 7*	USSR	Venus landing
Mar 27, 1972	*Venera 8*	USSR	Venus landing
Nov 3, 1973	*Mariner 10*	U.S.	Venus, Mercury flyby photos
June 8, 1975	*Venera 9*	USSR	Lander/orbiter
June 14, 1975	*Venera 10*	USSR	Lander/orbiter
May 20, 1978	*Pioneer 12* (also called *Pioneer Venus 1* or *Pioneer Venus*)	U.S.	Orbital studies of Venus
Aug 8, 1978	*Pioneer 13* (also called *Pioneer Venus 2* or *Pioneer Venus*)	U.S.	Orbital studies of Venus
Sept 9, 1978	*Venera 11*	USSR	Lander; sent photos
Sept 14, 1978	*Venera 12*	USSR	Lander; sent photos
Oct 30, 1981	*Venera 13*	USSR	Lander; sent photos
Nov 4, 1981	*Venera 14*	USSR	Lander; sent photos
June 2, 1983	*Venera 15*	USSR	Radar mapper
June 7, 1983	*Venera 16*	USSR	Radar mapper
Dec 15, 1984	*Vega 1*	USSR	Venus/Comet Halley probe
Dec 21, 1984	*Vega 2*	USSR	Venus/Comet Halley probe
May 4, 1989	*Magellan*	U.S.	Orbital radar mapper
Oct 18, 1989	*Galileo*	U.S.	Flyby measurements and photos

FIGURE 13.5 This is an image of an 8 km (5 mile) high volcano on the surface of Venus. The image was created by a computer using *Magellan* radar data, simulating a viewpoint elevation of 1.7 km (1 mi) above the surface. The lava flows extend for hundreds of kilometers across the fractured plains shown in the foreground. The simulated colors are based on color images recorded by the Soviet *Venera 13* and *14* spacecraft.

Earth but without the erosion caused by ice, rain, and running water.

Venus, like Mercury, has no satellites. Venus also does not have a magnetic field, as might be expected. Two conditions seem to be necessary in order for a planet to generate a magnetic field: a molten center and a relatively rapid rate of rotation. Since Venus takes 243 days to complete one rotation, the slowest of all the planets, it does not have a magnetic field even if some of the interior of Venus is still liquid as in Earth.

Earth's Moon

Next to the Sun, the Moon is the largest, brightest object in the sky. The Moon is Earth's nearest neighbor and surface features can be observed with the naked eye. With the aid of a telescope or a good pair of binoculars, you can see light-colored mountainous regions called the *lunar highlands,* smooth dark areas called *maria,* and many sizes of craters, some with bright streaks extending from them (figure 13.6). The smooth dark areas are called maria after a Latin word meaning "sea." They acquired this name from early observers who thought the dark areas were oceans and the light areas were continents. Today, the maria are understood to have formed from ancient floods of molten lava that poured across the surface and solidified to form the "seas" of today. There is no water or atmosphere on the Moon.

Many facts known about the Moon were established during the *Apollo* missions, the first human exploration of a place away from Earth. A total of twelve *Apollo* astronauts walked on the Moon, taking thousands of photographs, conducting hundreds of experiments, and returning to Earth with over 380 kg (about 840 lb) of moon rocks (Table 13.3). In addition, instruments were left on the Moon that continued to radio data back to

the form of infrared radiation, which is absorbed by the almost pure carbon dioxide atmosphere. Carbon dioxide molecules absorb the infrared radiation, increasing their kinetic energy and the temperature.

The surface of Venus is mostly a flat, rolling plain. There are several raised areas, or "continents," on about 5 percent of the surface. There is also a mountain larger than Mount Everest, a great valley deeper and wider than the Grand Canyon, and many large, old impact craters. In general, the surface of Venus appears to have evolved much as did the surface of

FIGURE 13.6 You can easily see the light-colored lunar highlands, smooth and dark maria, and many craters on the surface of Earth's nearest neighbor in space.

Earth after the *Apollo* program ended in 1972. As a result of the *Apollo* missions, many questions were answered about the Moon, but unanswered questions still remain.

The *Apollo* astronauts found that the surface of the Moon is covered by a 3 m (about 10 ft) layer of fine gray dust that contains microscopic glass beads. The dust and beads were formed from millions of years of continuous bombardment of micrometeorites. These very small meteorites generally burn up in Earth's atmosphere. The Moon does not have an atmosphere, so meteorites have continually fragmented and pulverized the surface in a slow, steady rain. The glass beads are believed to have formed when larger meteorite impacts melted part of the surface, which was immediately forced into a fine spray that cooled rapidly while above the surface.

The rocks on the surface of the Moon were found to be mostly *basalts*, a type of rock formed on Earth from the cooling and solidification of molten lava. The dark-colored rocks from the maria are similar to Earth's basalts (see chapter 15) but contain greater amounts of titanium and iron oxides. The light-colored rocks from the highlands are mostly *breccias*, a

TABLE 13.3 The *Apollo* Missions

Mission	Date	Crew	Comments
Apollo 1	Jan 27, 1967	Gus Grissom, Ed White, Roger Chaffee	The crew died in their spacecraft during a test three weeks before they would have flown in space.
Apollo 7	Oct 11–22, 1968	Wally Schirra, Donn F. Eisele, Walter Cunningham	This was the first *Apollo* mission in space following the *Apollo 1* launchpad fire. It was an eleven-day mission to validate *Apollo* hardware in low Earth orbit.
Apollo 8	Dec 21–27, 1968	Frank Borman, Jim Lovell, Bill Anders	First space mission to orbit the Moon. The first picture of Earth taken from deep space.
Apollo 9	Mar 3–13, 1969	James A. McDivitt, David R. Scott, Russel L. Schweickhart	First test of the lunar module in space. *Apollo 9* was an Earth orbital mission. *Apollo*-type rendezvous and docking was tested after a 6-hour, 113-mile separation.
Apollo 10	May 18–26, 1969	Tom Stafford, John Young, Gene Cernan	Trial rehearsal of Moon landing. The lunar module (LM) was taken to the Moon and separated from the command module (CM), but it did not land on the Moon. The LM was tested in lunar orbit.
Apollo 11	Jul 16–24, 1969	Neil Armstrong, Mike Collins, Buzz Aldrin	The lunar module, *Eagle,* landed the first man, Neil Armstrong, on the Moon on July 20, 1969, and established the first manned moon base.
Apollo 12	Nov 14–24, 1969	Peter Conrad, Dick Gordon, Al Bean	Landing was very accurate (only 535 ft from *Surveyor III*). The crew conducted two moon walks and put up both a geophysical station and a nuclear power station.
Apollo 13	Apr 11–17, 1970	Jim Lovell, Jack Swigert, Fred Haise	This was the first abort in deep space (200,000 mi from Earth). The lunar module was used as a lifeboat to return the crew safely.
Apollo 14	Jan 31–Feb, 9, 1971	Alan Shepard, Stuart A. Roosa, Edgar D. Mitchell	This was the third mission to land on the Moon, landing in the Frai Mauro region. Alan Shepard hit two golf balls on the Moon. Lunar specimens (95 lb) were collected.
Apollo 15	Jul 26–Aug 7, 1971	Dave Scott, Alfred M. Worden, James B. Irwin	During their record time on the Moon (66 h 54 min), the crew placed a subsatellite in lunar orbit and were the first to use the lunar rover.
Apollo 16	Apr 16–27, 1972	John Young, Thomas K. Mattingly II, Charles M. Duke	Highest landing on the Moon (elevation 25, 688 ft); lunar rover land speed record of 11.2 mi/h and distance record of 22.4 miles covered. The crew returned 213 lb of lunar samples.
Apollo 17	Dec 7–19, 1972	Gene Cernan, Ronald E. Evans, Harrison H. Schmitt	This, the last of the *Apollo* flights, was the first time an *Apollo* flight was launched at night. A record of 75 hours was set for time spent on the Moon, and 250 lb of lunar samples were returned.
Apollo-Soyuz	Jul 15–24, 1975	Tom Stafford, Deke Slayton, Vance Brand	First international space rendezvous. This was the first coordinated launch of two spacecraft from different countries.

Source: Data from NASA.

kind of rock made up of rock fragments that have been compacted together. On the Moon, the compacting was done by meteorite impacts. The rocks from the highlands contain more aluminum and less iron than the maria basalts and thus have a lower density than the darker rocks.

All the moon rocks contained a substantial amount of radioactive elements, which made it possible to precisely measure their age. The light-colored rocks from the highlands were formed about 4 billion years ago. The dark-colored rocks from the maria were much younger, with ages ranging from 3.1 to 3.8 billion years. This indicates a period of repeated volcanic eruptions and lava flooding over a seven hundred-million-year period that ended about three billion years ago.

The moon rocks brought back to Earth, the results of the lunar seismographs, and all the other data gathered through the *Apollo* missions have increased our knowledge about the Moon, leading to new understandings of how it formed. This model pictures the present Moon developing through four distinct stages.

The *origin stage* describes how the moon originally formed. The Moon is believed to have originated from the impact of Earth with a very large object, perhaps as large as Mars or larger. The Moon formed from ejected material produced by this collision. The collision is believed to have vaporized the colliding body as well as part of Earth. Some of the debris condensed away from Earth to form the Moon.

The collision that resulted in the Moon took place after Earth's iron core formed, so there is not much iron in moon rocks. The difference in moon and earth rocks can be accounted for by the presence of materials from the impacting body.

The *molten surface stage* occurred during the first two hundred million years after the collision. Heating from a number of sources could have been involved in the melting of the entire lunar surface 100 km (about 60 mi) or so deep. The heating required to melt the surface is believed to have resulted from the impacts of rock fragments, which were leftover debris from the formation of the solar system that intensely bombarded the Moon. After a time, there were fewer fragments left to bombard the Moon, and the molten outer layer cooled and solidified to solid rock. The craters we see on the Moon today are the result of meteorites hitting the Moon between 3.9 and 4.2 billion years ago after the crust formed.

The *molten interior stage* involved the melting of the interior of the Moon. Radioactive decay had been slowly heating the interior, and 3.8 billion years ago, or about a billion years after the Moon formed, sufficient heat accumulated to melt the interior. The light and heavier rock materials separated during this period, perhaps producing a small iron core. Molten lava flowed into basins on the surface during this period, forming the smooth, darker maria seen today. The lava flooding continued for about 700 million years, ending about 3.1 billion years ago.

The *cold and quiet stage* began 3.1 billion years ago as the last lava flow cooled and solidified. Since that time, the surface of the Moon has been continually bombarded by micrometeorites and a few larger meteorites. With the exception of a few new craters, the surface of the Moon has changed little in the last 3 billion years.

Mars

Mars has always attracted attention because of its unique, bright reddish color. The properties and surface characteristics have also attracted attention, particularly since Mars seems to have similarities to Earth. It orbits the Sun at an average distance of about 1.5 AU. It makes a complete orbit every 687 days, about twice the time that Earth takes. Mars rotates on its axis in twenty-four hours, thirty-seven minutes, so the length of a day on Mars is about the same as the length of a day on Earth. The observations that Mars has an atmosphere, light and dark regions that appear to be greenish and change colors with the seasons, and white polar caps that grow and shrink with the seasons led to early speculations (and many fantasies) about the possibilities of life on Mars. These speculations increased dramatically in 1877 when Schiaparelli, an Italian astronomer, reported seeing "channels" on the Martian surface. Other astronomers began interpreting the dark greenish regions as vegetation and the white polar caps as ice caps as Earth has. In the early part of the twentieth century, the American astronomer who founded the Lowell Observatory in Arizona, Percival Lowell, published a series of popular books showing a network of hundreds of canals on Mars. Lowell and other respectable astronomers interpreted what they believed to be canals as evidence of intelligent life on Mars. Other astronomers, however, interpreted the greenish colors and the canals to be illusions, imagined features of astronomers working with the limited telescopes of that time. Since canals never appeared in photographs, said the skeptics, the canals were the result of the human tendency to see patterns in random markings where no patterns actually exist.

This speculation ended in the late 1960s and early 1970s with extensive studies and probes by spacecraft (Table 13.4). Limited photographs by *Mariner* flybys in 1965 and 1969 had provided some evidence that the surface of Mars was much like the Moon, with no canals, vegetation, or much of anything else. Then in 1971, *Mariner 9* became the first spacecraft to orbit Mars, photographing the entire surface as well as making extensive measurements of the Martian atmosphere, temperature ranges, and chemistry. For about a year, *Mariner 9* sent a flood of new and surprising information about Mars back to Earth.

Mariner 9 found the surface of Mars not to be a crater-pitted surface as is found on the Moon. Mars has had a geologically active past and has four provinces, or regions, of related surface features. There are (1) volcanic regions with inactive volcanoes, one larger than any found on Earth, (2) regions with systems of canyons, some larger than any found on Earth, (3) regions of terraced plateaus near the poles, and (4) flat regions pitted with impact craters. Surprisingly, dry channels suggesting former water erosion were discovered near the cratered regions. These are sinuous, dry riverbed features with dry tributaries. Liquid water may have been present on Mars in the past, but none is to be found today.

The atmosphere of Mars is very thin, exerting an average pressure at the surface that is only 0.6 percent of the average atmospheric pressure on Earth's surface. Moreover, this thin Martian atmosphere is about 95 percent carbon dioxide, and

TABLE 13.4   Completed Spacecraft Missions to Mars

Date	Name	Owner	Remark
Nov 5, 1964	*Mariner 3*	U.S.	Flyby
Nov 28, 1964	*Mariner 4*	U.S.	First photos
Feb 24, 1969	*Mariner 6*	U.S.	Flyby
Mar 27, 1969	*Mariner 7*	U.S.	Flyby
May 19, 1971	*Mars 2*	USSR	Lander
May 28, 1971	*Mars 3*	USSR	Orbiter/lander
May 30, 1971	*Mariner 9*	U.S.	Orbiter
Jul 21, 1973	*Mars 4*	USSR	Probe
Jul 25, 1973	*Mars 5*	USSR	Orbiter
Aug 5, 1973	*Mars 6*	USSR	Lander
Aug 9, 1973	*Mars 7*	USSR	Flyby/lander
Aug 20, 1975	*Viking 1*	U.S.	Lander/orbiter
Sept 9, 1975	*Viking 2*	U.S.	Lander/orbiter
July 7, 1988	*Phobos 1*	USSR	Orbiter/Phobos lander
July 12, 1988	*Phobos 2*	USSR	Orbiter/Phobos lander
Nov 7, 1996	*Global Surveyor*	U.S.	Orbiter
Sept 11, 1997	*Mars Global Surveyor*	U.S.	Orbiter
Dec 4, 1997	*Pathfinder*	U.S.	Lander/surface rover
Oct 23, 2001	*2001 Mars Odyssey*	U.S.	Orbiter
Jan 4, 2004	*Spirit*	U.S.	Lander/surface rover
Jan 25, 2004	*Opportunity*	U.S.	Lander/surface rover
Aug 26, 2006	*Mars Reconnaissance Orbiter*	U.S.	Orbiter

FIGURE 13.7   Researchers used the rover *Spirit*'s rock abrasion tool to help them study a rock dubbed "Uchben" in the "Colombia Hills" of Mars. The tool ground into the rock, creating a shallow hole 4.5 cm (1.8 in) in diameter in the central upper portion of this image. It also used wire bristles to brush a portion of the surface below and to the right of the hole. *Spirit* used its panoramic camera during the rover's 293rd martian day (October 29, 2004) to take the frames combined into this approximately true-color image.

Source: NASA/JPL/Cornell.

20 percent of this freezes as dry ice at the Martian South Pole every winter.

Does life exist on Mars? Two *Viking* spacecraft were sent to Mars in 1975 to search for signs of life. The two *Viking* spacecraft were identical, each consisting of an orbiter and a lander. After eleven months of travel time, *Viking 1* entered an orbit around Mars in June 1976 and spent a month sending high-resolution images of the surface back to Earth. From these images, a landing site was selected for the *Viking 1* lander. Using retrorockets, parachutes, and descent rockets, the *Viking 1* lander arrived on a dusty, rocky slope in the southern hemisphere on July 20, 1976. The *Viking 2* lander arrived forty-five days later but farther to the north. The *Viking* landers contained a mechanical soil-retrieving arm and a miniature computerized lab to analyze the soil for evidence of metabolism, respiration, and other life processes. Neither lander detected any evidence of life processes or any organic compounds that would indicate life now or in the past.

The *Viking* spacecraft continued sending images and weather data back to Earth until 1982. During their six-year life, the orbiters sent about fifty-two thousand images and mapped about 97 percent of the Martian surface. The landers sent an additional forty-five hundred images, recorded a major "Marsquake," and recorded data about regular dust storms that occur on Mars with seasonal changes.

The *Mars Exploration Rovers,* named *Spirit* and *Opportunity,* landed on Mars on January 4 and January 25, 2004, respectively, to answer questions about the history of water on Mars. The spacecraft were sent to sites on opposite sides of Mars that appear to have been affected by liquid water in the past. After parachute and airbag landings, the 185 kg (408 lb) rovers charged their solar-powered batteries. They then began driving to different locations to perform on-site scientific investigations over the course of their mission (figure 13.7).

What did the rovers find? They found that Mars is made of basalt rock (see chapter 15) and groundwater that is dilute sulfuric acid. The acid interacts with the rock, dissolving things out of it, and then evaporates and leaves sulfur-rich salts. The *Spirit* and *Opportunity* rover results confirmed that sufficient amounts of water to alter the rocks have been present in the past (figure 13.8). The results also confirm the premission interpretations of remote-sensing data. This provides evidence that other present and future remote-sensing data is accurate.

The presence of past or present life on Mars remains an open question. Scientists already knew there was liquid water in the past, and water and life go together. Beyond that, the *Rover* mission has really not changed the prospect of finding evidence of past or present life, so the search goes on.

Planets and Astrology

Do you read the astrology forecasts in daily newspapers or on Internet sites? Below is a brief background of astrology as developed by the Babylonians, followed by some questions intended for class or small group discussions.

As early as 2000 B.C., the Babylonians began keeping track of time by dividing the year into 12 months, with 7 days to a week and 360 days to a year. They maintained observatories, noting for example that the Sun, the Moon, and five planets known at the time (Mercury, Venus, Mars, Jupiter, and Saturn) moved across the sky only along a certain path, which is today called the ecliptic. This movement was independent of the stars, which followed the motions of the seven celestial bodies but kept the same position relative to each other. The Sun appeared to move completely around the ecliptic each year.

To keep track of the Sun and the time of year, the Babylonians imagined the arrangements of certain stars to be the shapes of gods, objects, or animals. These patterns, today called constellations, were used to identify twelve equal divisions of the path the Sun followed for a year. The twelve constellations are called the zodiac, so there are twelve signs of the zodiac in a year. By 540 B.C., the Babylonians had fully developed the art of studying the zodiac, the Sun, and planets as a guide to human affairs, and this activity today is known as astrology.

First, consider astrology forecasts as they are made today. A daily horoscope might include a forecast for those with a birthday on this day and forecasts for people born during each of the twelve signs of the zodiac. In the horoscope, you can find such predictions as "your computer could crash today," "focus on your ability to tear down before you rebuild," and "take a chance on romance with a Virgo." Discuss such forecasts with your group. Consider, for example, the population of a nation and how many people are forecasted to have their computer crash. How many computers would crash without the forecast?

Next, discuss why your passage through the birth canal (your birthday) is so important. Perhaps the time when your embryo formed might be more important if you were going to be "marked" by the planets for certain things to happen to you.

Finally, discuss the topic of how Earth's axis has a slow wobble, called precession, which causes it to swing in a slow circle like the wobble of a spinning top. The axis takes about 26,000 years to complete one turn, or wobble. The moving pole changes over time in which particular signs of the zodiac appear, for example, with the spring equinox. Because of precession, the occurrence of the spring equinox has been moving backward through the zodiac constellations at about 1 degree every 72 years. So, 3,000 years ago, the Sun entered the constellation of Virgo in August, which is still the basis for horoscopes today. However, the Sun is now in the Leo constellation in August.

FIGURE 13.8   A rock dubbed "Palenque" in the "Colombia Hills" of Mars has contrasting textures in upper and lower portions. This view of the rock combines two frames taken by the panoramic camera on NASA's Mars Exploration Rover *Spirit* during the rover's 278th martian day (October 14, 2004). The layers meet each other at an angular unconformity that may mark a change in environmental conditions between the formation of the two portions of the rock. Scientists would have liked the rover to take a closer look, but Palenque is not on a north-tilted slope, which is the type of terrain needed to keep the rover's solar panels tilted toward the winter sun. The exposed portion of the rock is about 100 cm (39 in) long.
Source: NASA/JPL/Cornell.

Jupiter

Jupiter is the largest of all the planets, with a mass equivalent to some 318 Earths and, in fact, is more than twice as massive as all the other planets combined. This massive planet is located an average 5 AU from the Sun in an orbit that takes about twelve Earth years for one complete trip around the Sun. The internal heating from gravitational contraction was tremendous when this giant formed, and today it still radiates twice the energy that it receives from the Sun. The source of this heat is the slow gravitational compression of the planet, not from nuclear reactions as in the Sun. Jupiter would have to be about eighty times as massive to create the internal temperatures needed to start nuclear fusion reactions, or in other words, to become a star itself. Nonetheless, the giant Jupiter and its system of satellites seem almost like a smaller version of a planetary system within the solar system.

Jupiter has an average density that is about a quarter of the density of Earth. This low density indicates that Jupiter is mostly made of light elements, such as hydrogen and helium, but does contain a percentage of heavier rocky substances. The model of Jupiter's interior (figure 13.9) is derived from this and other information from spectral studies, studies of spin rates, and measurements of heat flow. The model indicates a solid, rocky core that is more than twice the size of Earth. Surrounding this core is a thick layer of liquid hydrogen, compressed so tightly by millions of atmospheres of pressure that it is able to conduct electric currents. Liquid hydrogen with this property is called *metallic hydrogen* because it has the conductive ability of metals. Above the layer of metallic hydrogen is a thick layer of ordinary liquid hydrogen, which is under less pressure. The outer layer, or atmosphere, of Jupiter is a 500 km or so (about 300 mi) zone with hydrogen, helium, ammonia gas, crystalline compounds, and a mixture of ice and water. It is the uppermost ammonia clouds, perhaps mixed with sulphur and organic compounds, that form the bright orange, white, and yellow bands around the planet. The banding is believed to be produced by atmospheric convection, in which bright, hot gases

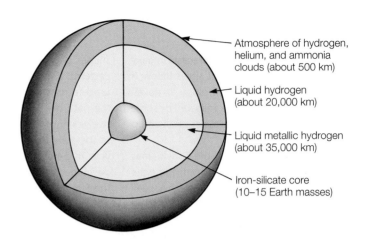

FIGURE 13.9 The interior structure of Jupiter.

are forced to the top where they cool, darken in color, and sink back to the surface.

Jupiter's famous Great Red Spot is located near the equator. This permanent, deep, red oval feature was first observed by Robert Hooke in the 1600s and has generated much speculation over the years. The red oval, some 40,000 km (about 25,000 mi) long, has been identified by infrared observations to be a high-pressure region, with higher and colder clouds, that has lasted for at least three hundred years. The energy source for such a huge, long-lasting feature is unknown (figure 13.10).

A

B

FIGURE 13.10 Photos of Jupiter taken by *Voyager 1*. (*A*) From a distance of about 36 million km (about 22 million mi). (*B*) A closer view, from the Great Red Spot to the South Pole, showing organized cloud patterns. In general, dark features are warmer and light features are colder. The Great Red Spot soars about 25 km (about 15 mi) above the surrounding clouds and is the coldest place on the planet.

UFOs and You

Unidentified flying objects (UFOs) are observed around the world. UFOs are generally sighted near small towns and out in the country, often near a military installation. Statistically, most sightings occur during the month of July at about 9:00 P.M., then at 3:00 A.M. UFOs can be grouped into three categories:

1. **Natural phenomena.** Most sightings of UFOs can be explained as natural phenomena. For example, a bright light that flashes red and green on the horizon could be a star viewed through atmospheric refraction. The vast majority of all UFO sightings are natural phenomena such as atmospheric refraction, ball lightning, swamp gas, or something simple such as a drifting weather balloon or military flares on parachutes, high in the atmosphere and drifting across the sky. This category of UFOs should also include exaggeration and fraud—such as balloons released with burning candles as a student prank.

2. **Aliens.** The idea that UFOs are alien spacecraft from other planets is very popular. However, there is no authentic, unambiguous evidence that would prove the existence of aliens. In fact, most unidentified flying objects are not objects at all. They are lights that can eventually be identified.

3. **Psychological factors.** This group of sightings includes misperceptions of natural phenomena resulting from unusual conditions that influence a person's perception and psychological health. This includes people who claim to receive information from aliens by "channeling" messages.

Questions to Discuss

Divide your group into three subgroups, with each subgroup selecting one of the three categories above. After preparing for a few minutes, have each group present reasons why we should understand UFOs to be natural phenomena, aliens from other planets, or psychological misperceptions. Then have the entire group discuss the three categories and try to come to a consensus.

Jupiter has thirty-nine satellites, and the four brightest and largest can be seen from Earth with a good pair of binoculars. These four are called the *Galilean moons* because they were discovered by Galileo in 1610. The Galilean moons are named Io, Europa, Ganymede, and Callisto (figure 13.11).

Observations by the *Pioneer* and *Voyager* spacecrafts revealed some fascinating and intriguing information about the moons of Jupiter. Io, for example, was discovered to have active volcanoes that eject enormous plumes of molten sulfur and sulfur dioxide gas. Europa is covered with a 19.3 km (about 12 mi) thick layer of smooth water ice, which has a network of long, straight, dark cracks. Ganymede has valleys, ridges, folded mountains, and other evidence of an active geologic history. Callisto, the most distant of the Galilean moons, was found to be the most heavily cratered object in the solar system.

These impact events are still occurring. In 1994, the comet Shoemaker-Levy 9 broke apart into a "string of pearls," (figure 13.12A) and then produced a once-in-a-lifetime spectacle as it proceeded to leave its imprint on Jupiter as well as people of Earth watching from the sidelines. The string of twenty-two comet fragments fell onto Jupiter during July 1994, creating a show eagerly photographed by viewers using telescopes around the world (figure 13.12B). The fragments impacted the upper atmosphere of Jupiter, producing visible, energetic fireballs. The aftereffects of these fireballs were visible for about a year. There are chains of craters on two of the Galilean moons that may have been formed by similar events.

FIGURE 13.11 The four Galilean moons pictured by *Voyager 1*. Clockwise from upper left, Io, Europa, Ganymede, and Callisto. Io and Europa are about the size of Earth's Moon; Ganymede and Callisto are larger than Mercury.

A

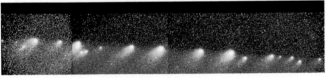

B

FIGURE 13.12 (A) This image, made by the Hubble Space Telescope, clearly shows the large impact site made by fragment G of former comet Shoemaker-Levy 9 when it collided with Jupiter. (B) This is a picture of comet Shoemaker-Levy 9 after it broke into twenty-two pieces, lined up in this row, then proceeded to plummet into Jupiter during July 1994. The picture was made by the Hubble Space Telescope.

Saturn

Saturn is slightly smaller and substantially less massive than Jupiter, and has similar surface features to Jupiter, but it is readily identified by its unique, beautiful system of rings. Saturn's rings consist of thousands of narrow bands of particles. Some rings are composed of particles large enough to be measured in meters and some have particles that are dust-sized (figure 13.13).

Saturn is about 9.5 AU from the Sun, but its system of rings is easily spotted with a good pair of binoculars. Saturn also has the lowest average density of any of the planets, about 0.7 the density of water.

The surface of Saturn, like Jupiter's surface, has bright and dark bands that circle the planet parallel to the equator. Saturn also has a smaller version of Jupiter's Great Red Spot, but in general, the bands and spot are not as highly contrasted or brightly colored as they are on Jupiter. Saturn has thirty satellites, including Titan, the only moon in the solar system with a substantial atmosphere.

Titan is covered with clouds and impossible to observe. It is larger than the planet Mercury and is covered with a deep layer of reddish clouds. Titan's atmosphere is mostly nitrogen, with some hydrocarbons. This could be similar to Earth's atmosphere before life began adding oxygen to the atmosphere. The pressure on the surface is 1.5 atmospheres, but with a surface temperature of about −180°C (about −290°F), it is doubtful that life has developed on Titan.

The international *Cassini-Huygens* mission entered orbit around Saturn on July 1, 2004, after a 3.5 billion km (2.2 billion mi), seven-year voyage from Earth. Establishing orbit was the first step of a four-year study of Saturn and its rings and moons. In December 2004, it delivered a detachable probe, called *Huygens,* to the moon Titan. The *Cassini-Huygens* mission is a cooperative project of the National Aeronautics and Space Administration, the European Space Agency (ESA), and the Italian Space Agency.

Uranus and Neptune

Uranus and Neptune are two more giant planets that are far, far away from Earth. Uranus revolves around the Sun at an average distance of over 19 AU, taking about 84 years to circle the Sun once. Neptune is an average 30 AU from the Sun and takes about 165 years for one complete orbit. Uranus is about twice as far away from the Sun as Saturn, and Neptune is three times as far away. To give you an idea of these tremendous distances, consider that the time required for radio signals to travel from Uranus to Earth is more than 2.5 hours! It would be most difficult to carry on a conversation by radio with someone such a distance away. Even farther away, a radio signal from a transmitter near Neptune would require over 4 hours to reach Earth, which means 8 hours would be required for two people just to say "Hello" to each other!

Uranus and Neptune are more similar to each other than Saturn is to Jupiter (figure 13.14). Both are the smallest of the giant planets, with a diameter of about 50,000 km (about 30,000 mi) and about a third the mass of Jupiter. Both planets are thought to have similar interior structures (figure 13.15), which consist of water and water ice surrounding a rocky core with an atmosphere of hydrogen and helium. Because of their great distances from the Sun, both have very low average surface temperatures.

FIGURE 13.13 A part of Saturn's system of rings, pictured by *Voyager 2* from a distance of about 3 million km (about 2 million mi). More than sixty bright and dark ringlets are seen here; different colors indicate different surface compositions.

FIGURE 13.14 This is a photo image of Neptune taken by *Voyager 2*. Neptune has a turbulent atmosphere over a very cold surface of frozen hydrogen and helium.

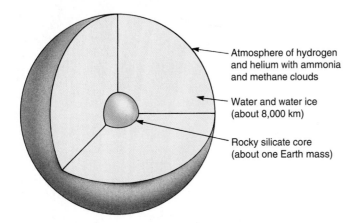

Atmosphere of hydrogen and helium with ammonia and methane clouds

Water and water ice (about 8,000 km)

Rocky silicate core (about one Earth mass)

FIGURE 13.15 The interior structure of Uranus and Neptune.

SMALL BODIES OF THE SOLAR SYSTEM

Comets, asteroids, and meteorites are the leftovers from the formation of the Sun and planets. Presently, the total mass of all these leftovers in and around the solar system may account for a significant fraction of the mass of the solar system, perhaps as much as two-thirds of the total mass. It must have been much greater in the past, however, as evidenced by the intense bombardment that took place on the Moon and other planets up to some four billion years ago.

TABLE 13.5 Completed Spacecraft Missions to Study Comets

Date	Name	Owner	Remark
Sep 11, 1985	*ISSE 3* (or *ICE*)	U.S.	Studies of electric and magnetic fields around Giacobini-Zinner comet from 7,860 km (4,880 mi)
Mar 6, 1986	*Vega 1*	USSR	Photos and studies of nucleus of Halley's comet from 8,892 km (5,525 mi)
Mar 8, 1986	*Suisei*	Japan	Studied hydrogen halo of Halley's comet from 151,000 km (93,800 mi)
Mar 9, 1986	*Vega 2*	USSR	Photos and studies of nucleus of Halley's comet from 8,034 km (4,992 mi)
Mar 11, 1986	*Sakigake*	Japan	Studied solar wind in front of Halley's comet from 7.1 million km (4.4 million mi)
Mar 28, 1986	*Giotto*	ESA	Photos and studies of Halley's comet from 541 km (336 mi)
Mar 28, 1986	*ISSE 3* (or *ICE*)	U.S.	Studies of electric and magnetic fields around Halley's comet from 32 million km (20 million mi)
Jul 4, 2005	*Deep Impact*	U.S.	Projectile shot into comet Tempel 1

Comets

A **comet** is known to be a relatively small, solid body of frozen water, carbon dioxide, ammonia, and methane, along with dusty and rocky bits of materials mixed in. Until the 1950s, most astronomers believed that comet bodies were mixtures of sand and gravel. Fred Whipple proposed what became known as the *dirty-snowball cometary model,* which was recently verified when spacecraft probes observed Halley's comet in 1986 (Table 13.5).

Based on calculation of their observed paths, comets are estimated to originate some 30 AU to a light-year or more from the Sun. Here, according to other calculations and estimates, is a region of space containing billions and billions of objects. There is a spherical "cloud" of the objects beyond the orbit of Pluto from about 30,000 AU out to a light-year or more from the Sun, called the **Oort cloud** (figure 13.16). The icy aggregates of the Oort cloud are understood to be the source of long-period comets, with orbital periods of more than two hundred years.

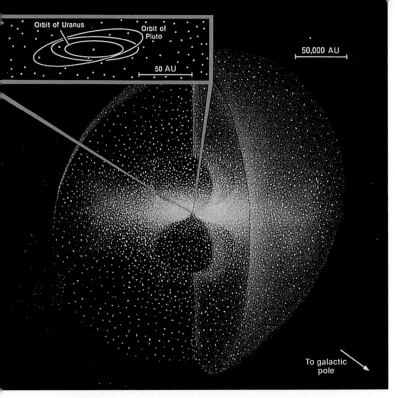

FIGURE 13.16 Although there may be as many as one trillion comets in the Oort cloud, the volume of space that the Oort cloud occupies is so immense that the comets are separated from one another by distances that are typically about 10 AU.

producing its own light, making it visible from Earth. The coma generally appears when a comet is within about 3 AU of the Sun. It reaches its maximum diameter about 1.5 AU from the Sun. The nucleus and coma together are called the *head* of the comet. In addition, a large cloud of invisible hydrogen gas surrounds the head, and this hydrogen *halo* may be hundreds of thousands of kilometers across.

As the comet nears the Sun, the solar wind and solar radiation ionize gases and push particles from the coma, pushing both into the familiar visible *tail* of the comet. Comets may have two types of tails: (1) ionized gases and (2) dust. The dust is pushed from the coma by the pressure from sunlight. It is visible because of reflected sunlight. The ionized gases are pushed into the tail by magnetic fields carried by the solar wind. The ionized gases of the tail are fluorescent, emitting visible light because they are excited by ultraviolet radiation from the Sun. The tail generally points away from the Sun, so it follows the comet as it approaches the Sun but leads the comet as it moves away from the Sun (figure 13.17).

There is also a disk-shaped region of small icy bodies, which ranges from about 30 to 100 AU from the Sun, called the **Kuiper Belt.** The small icy bodies in the Kuiper Belt are understood to be the source of short-period comets, with orbital periods of less than two hundred years. There are thousands of Kuiper Belt objects that are larger than 100 km in diameter, and six are known to be orbiting between Jupiter and Neptune. Called *centaurs,* these objects are believed to have escaped the Kuiper Belt. Centaurs might be small, icy bodies similar to Pluto.

The current theory of the origin of comets was developed by the Dutch astronomer Jan Oort in 1950. According to the theory, the huge cloud and belt of icy, dusty aggregates are leftovers from the formation of the solar system and have been slowly circling the solar system ever since it formed. Something, perhaps a gravitational nudge from a passing star, moves one of the icy bodies enough that it is pulled toward the Sun in what will become an extremely elongated elliptical orbit. The icy, dusty body forms the only substantial part of a comet, and the body is called the comet *nucleus.*

Observations by the *Vega* and *Giotto* spacecrafts found the nucleus of Halley's comet to be an elongated mass of about 8 by 11 km (about 5 by 7 mi) with an overall density less than one-fourth that of solid water ice. As the comet nucleus moves toward the Sun, it warms from the increasing intense solar radiation. Somewhere between Jupiter and Mars, the ice and frozen gases begin to vaporize, releasing both grains of dust and evaporated ices. These materials form a large hazy head around the comet called a *coma.* The coma grows larger with increased vaporization, perhaps several hundred or thousands of kilometers across. The coma reflects sunlight as well as

FIGURE 13.17 As a comet nears the Sun it grows brighter, with the tail always pointing away from the Sun.

Comets are not very massive or solid, and the porous, snow-like mass has a composition more similar to the giant planets than to the terrestrial planets in comparison. Each time a comet passes near the Sun, it loses some of its mass through evaporation of gases and loss of dust to the solar wind. After passing the Sun, the surface forms a thin, fragile crust covered with carbon and other dust particles. Each pass by the Sun means a loss of matter, and the coma and tail are dimmer with each succeeding pass. About 20 percent of the approximately six hundred comets that are known have orbits that return them to the Sun within a two-hundred-year period, some of which return as often as every five or ten years. The other 80 percent have long elliptical orbits that return them at intervals exceeding two hundred years. The famous Halley's comet has a smaller elliptical orbit and returns about every seventy-six years. Halley's comet, like all other comets, may eventually break up into a trail of gas and dust particles that orbit the Sun.

Asteroids

Between the orbits of Mars and Jupiter is a belt, or circular region of thousands of small rocky bodies called **asteroids** (figure 13.18). This belt contains thousands of *asteroids* that range in size from 1 km or less up to the largest asteroid, named Ceres, which has a diameter of about 1,000 km (over 600 mi). The asteroids are thinly distributed in the belt, 1 million km or so apart (about 600,000 mi), but there is evidence of collisions occurring in the past. Most asteroids larger than 50 km (about 30 mi) have been studied by analyzing the sunlight reflected from their surfaces. These spectra provide information about the composition of the asteroids. Asteroids on the inside of the belt are made of stony materials, and those on the outside of the belt are dark with carbon minerals. Still other asteroids are metallic, containing iron and nickel. These spectral composition studies, analyses of the orbits of asteroids, and studies of meteorites that have fallen to Earth all indicate that the asteroids are not the remains of a planet

or planets that were broken up. The asteroids are now believed to have formed some 4.6 billion years ago from the original solar nebula. During their formation, or shortly thereafter, their interiors were partly melted, perhaps from the heat of short-lived radioactive decay reactions. Their location close to Jupiter, with its gigantic gravitational field, prevented the slow gravitational clumping-together process that would have formed a planet.

Jupiter's gigantic gravitational field also captured some of the asteroids, pulling them into its orbit. Today, there are two groups of asteroids, called the *Trojan asteroids,* which lead and follow Jupiter in its orbit. They lead and follow at a distance where the gravitational forces of Jupiter and the Sun balance to keep them in the orbit. A third group of asteroids, called the *Apollo asteroids,* has orbits that cross the orbit of Earth. It is possible that one of the Apollo asteroids could collide with Earth. One theory about what happened to the dinosaurs is based on evidence that such a collision indeed did occur some sixty-five million years ago. The chemical and physical properties of the two satellites of Mars, Phobos and Deimos, are more similar to the asteroids than to Mars. It is probable that the Martian satellites are captured asteroids.

Meteors and Meteorites

Comets leave trails of dust and rock particles after encountering the heat of the Sun, and collisions between asteroids in the past have ejected fragments of rock particles into space. In space, the remnants of comets and asteroids are called **meteoroids.** When a meteoroid encounters Earth moving through space, it accelerates toward the surface with a speed that depends on its direction of travel and the relative direction that Earth is moving. It soon begins to heat from air compression, melting into a visible trail of light and smoke. The streak of light and smoke in the sky is called a **meteor.** The "falling star" or "shooting star" is a meteor. Most meteors burn up or evaporate completely within seconds after reaching an altitude of about 100 km (about 60 mi) because they are nothing more than a speck of dust. A **meteor shower** occurs when Earth passes through a stream of particles left by a comet in its orbit. Earth might meet the stream of particles concentrated in such an orbit on a regular basis as it travels around the Sun, resulting in predictable meteor showers (Table 13.6). In the third week of October, for example, Earth crosses the orbital path of Halley's comet,

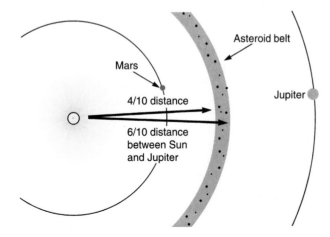

FIGURE 13.18 Most of the asteroids in the asteroid belt are about halfway between the Sun and Jupiter.

TABLE 13.6 Some Annual Meteor Showers

Name	Date of Maximum	Hour Rate
Quadrantid	January 3	30
Aquarid	May 4	5
Perseid	August 12	40
Orionid	October 22	15
Taurids	November 1, 16	5
Leonid	November 17	5
Geminid	December 12	55

A

B

FIGURE 13.19 (*A*) A stony meteorite. The smooth, black surface was melted by friction with the atmosphere. (*B*) An iron meteorite that has been cut, polished, and etched with acid. The pattern indicates that the original material cooled from a molten material over millions of years.

resulting in a shower of some ten to fifteen meteors per hour. Meteor showers are named for the constellation in which they appear to originate. The October meteor shower resulting from an encounter with the orbit of Halley's comet, for example, is called the Orionid shower because it appears to come from the constellation Orion.

Did you know that atom-bomb-sized meteoroid explosions often occur high in Earth's atmosphere? Most smaller meteors melt into the familiar trail of light and smoke. Larger meteors may fragment upon entering the atmosphere, and the smaller fragments will melt into multiple light trails. Still larger meteors may actually explode at altitudes of about 32 km (about 20 mi) or so. Military satellites that watch Earth for signs of rockets blasting off or nuclear explosions record an average of eight meteor explosions a year. These are big explosions, with an energy equivalent estimated to be similar to a small nuclear bomb. Actual explosions, however, may be ten times larger than the estimation. Based on statistical data, scientists have estimated that every ten million years, Earth should be hit by a very, very large meteor. The catastrophic explosion and aftermath would devastate life over much of the planet, much like the theoretical dinosaur-killing impact of sixty-five million years ago.

If a meteoroid survives its fiery trip through the atmosphere to strike the surface of Earth, it is called a **meteorite.** Most meteors are from fragments of comets, but most meteorites generally come from particles that resulted from collisions between asteroids that occurred long ago. Meteorites are classified into three basic groups according to their composition: (1) *iron meteorites,* (2) *stony meteorites,* and (3) *stony-iron meteorites* (figure 13.19). The most common meteorites are stony, composed of the same minerals that make up rocks on Earth. The stony meteorites are further subdivided into two groups according to their structure, the *chondrites* and the *achondrites.* Chondrites have a structure of small spherical lumps of silicate minerals or glass, called *chondrules,* held together by a fine-grained cement. The achondrites do not have the chondrules, as their name implies, but have a homogeneous texture more like volcanic rocks such as basalt that cooled from molten rock.

The iron meteorites are about half as abundant as the stony meteorites. They consist of variable amounts of iron and nickel, with traces of other elements. In general, there is proportionally much more nickel than is found in the rocks of Earth. When cut, polished, and etched, beautiful crystal patterns are observed on the surface of the iron meteorite. The patterns mean that the iron was originally molten, then cooled very slowly over millions of years as the crystal patterns formed.

A meteorite is not, as is commonly believed, a ball of fire that burns up the landscape where it lands. The iron or rock has been in the deep freeze of space for some time, and it travels rapidly through Earth's atmosphere. The outer layers become hot enough to melt, but there is insufficient time for this heat to be conducted to the inside. Thus, a newly fallen iron meteorite will be hot since metals are good heat conductors, but it will not be hot enough to start a fire. A stone meteorite is a poor conductor of heat so it will be merely warm.

ORIGIN OF THE SOLAR SYSTEM

Any model of how the solar system originated presents a problem in testing or verification. This problem is that the solar system originated a long time ago, some five billion years ago, according to a number of different independent sources of evidence. Also, there are no other planetary systems that can be

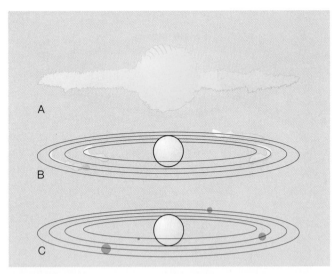

FIGURE 13.20 Formation of the solar system according to the protoplanet nebular model, not drawn to scale. (*A*) The process starts with a nebula of gas, dust, and chemical elements from previously existing stars. (*B*) The nebula is pulled together by gravity, collapsing into the protosun and protoplanets. (*C*) As the planets form, they revolve around the Sun in newly formed orbits.

directly observed for comparison either in existence or in the process of being formed. At the distance they occur, even the Hubble Space Telescope would not be able to directly observe planets around their suns. Astronomers have identified about one hundred extrasolar planets. They identify the presence of planets by measuring the very slight wobble of a central star and then use the magnitude of this motion to determine the presence of orbiting planets, the size and shape of their orbits, and their mass. The technique works only for larger planets and cannot detect those much smaller than about half the mass of Saturn. The technique does not provide a visual image of the planets but only measures the gravitational effect of the planets on the star.

The most widely accepted theory of the origin of the solar system is called the **protoplanet nebular model.** A *protoplanet* is the earliest stage in the formation of a planet. The model can be considered in stages, which are not really a part of the model but are simply a convenient way to organize the total picture (figure 13.20).

Stage A

The first important event in the formation of our solar system involves stars that disappeared billions of years ago, long before the Sun was born. Earth, the other planets, and all the members of the solar system are composed of elements that were manufactured by these former stars. In a sequence of nuclear reactions, hydrogen fusion in the core of large stars results in the formation of the elements up to iron. Elements heavier than iron are formed in rare supernova explosions of dying massive stars. Thus, *stage A* of the formation of the solar

system consisted of the formation of elements heavier than hydrogen in many, many previously existing stars, including the supernovas of more massive stars. Many stars had to live out their life cycles to provide the raw materials of the solar system. The death of each star, including supernovas, added newly formed elements to the accumulating gases and dust in interstellar space. Over a long period of time, these elements began to concentrate in one region of space as dust, gases, and chemical compounds, but hydrogen was still the most abundant element in the nebula that was destined to become the solar system.

Stage B

During *stage B*, the hydrogen gas, dust, elements, and chemical compounds from former stars began to form a large, slowly rotating nebula that was much, much larger than the present solar system. Under the influence of gravity, the large but diffuse, slowly rotating nebula began to contract, increasing its rate of spin. The largest mass pulled together in the center, contracting to the protostar, which eventually would become the Sun. The remaining gases, elements, and dusts formed an enormous, fat, bulging disk called an *accretion disk,* which would eventually form the planets and smaller bodies. The fragments of dust and other solid matter in the disk began to stick together in larger and larger accumulations from numerous collisions over the first million years or so. All of the present-day elements of the planets must have been present in the nebula along with the most abundant elements of hydrogen and helium. The elements and the familiar chemical compounds accumulated into basketball-sized or larger chunks of matter.

Did the planets have an icy slush beginning? Over a period of time, perhaps one-hundred million years or so, huge accumulations of frozen water, frozen ammonia, and frozen crystals of methane began to accumulate, together with silicon, aluminum, and iron oxide plus other metals in the form of rock and mineral grains. Such a slushy mixture would no doubt have been surrounded by an atmosphere of hydrogen, helium, and other vapors thinly interspersed with smaller rocky grains of dust. Local concentrations of certain minerals might have occurred throughout the whole accretion disk, with a greater concentration of iron, for example, in the disk where the protoplanet Mars was forming compared to where the protoplanet Earth was forming. Evidence for this part of the model is found in Mars today, with its greater abundance of iron. The abundant iron oxides are what color Mars the red planet.

All of the protoplanets might have started out somewhat similarly as huge accumulations of a slushy mixture with an atmosphere of hydrogen and helium gases. Gravitational attraction must have compressed the protoplanets as well as the protosun. During this period of contraction and heating, gravitational adjustments continue, and about a fifth of the disk nearest to the protosun must have been pulled into the central

Percival Lowell (1855–1916)

Percival Lowell was an American astronomer, mathematician, and the founder of an important observatory in the United States whose main field of research was the planets of the solar system. Responsible for the popularization in his time of the theory of intelligent life on Mars, he also predicted the existence of a planet beyond Neptune that was later discovered and named Pluto. Today, Pluto is recognized as a dwarf planet.

Lowell was born in Boston, Massachusetts, on March 13, 1855. His interest in astronomy began to develop during his early school years. In 1876, he graduated from Harvard University, where he had concentrated on mathematics, and then traveled for a year before entering his father's cotton business. Six years later, Lowell left the business and went to Japan. He spent most of the next ten years traveling around the Far East, partly for pleasure, partly to serve business interests, but also holding a number of minor diplomatic posts.

Lowell returned to the United States in 1893 and soon afterward decided to concentrate on astronomy. He set up an observatory at Flagstaff, Arizona, at an altitude more than 2,000 m (6,564 ft) above sea level, on a site chosen for the clarity of its air and its favorable atmospheric conditions. He first used borrowed telescopes of 12- and 18-inch (30 and 45 cm) diameters to study Mars, which at that time was in a particularly suitable position. In 1896, he acquired a larger telescope and studied Mars by night and Mercury and Venus during the day. Overwork led to deterioration in Lowell's health, and from 1897 to 1901, he could do little research, although he was able to participate in an expedition to Tripoli in 1900 to study a solar eclipse.

He was made nonresident professor of astronomy at the Massachusetts Institute of Technology in 1902 and gave several lecture series in that capacity. He led an expedition to the Chilean Andes in 1907 that produced the first high-quality photographs of Mars. The author of many books and the holder of several honorary degrees, Lowell died in Flagstaff on November 12, 1916.

The planet Mars was a source of fascination for Lowell. Influenced strongly by the work of Giovanni Schiaparelli (1835–1910)—and possibly misled by the current English translation of "canals" for the Italian *canali* ("channels")—Lowell set up his observatory at Flagstaff originally with the sole intention of confirming the presence of advanced life forms on the planet. Thirteen years later, the expedition to South America was devoted to the study and photography of Mars. Lowell "observed" a complex and regular network of canals and believed that he detected regular seasonal variations that strongly indicated agricultural activity. He found darker waves that seemed to flow from the poles to the equator and suggested that the polar caps were made of frozen water. (The waves were later attributed to dust storms, and the polar caps are now known to consist not of ice but mainly of frozen carbon dioxide. Lowell's canal system also seems to have arisen mostly out of wishful thinking; part of the system does indeed exist, but it is not artificial and is apparent only because of the chance apposition of dark patches on the Martian surface.)

Lowell is remembered as a scientist of great patience and originality. He contributed to the advancement of astronomy through his observations and his establishment of a fine research center, and he did much to bring the excitement of the subject to the general public.

Source: Modified from the *Hutchinson Dictionary of Scientific Biography*. © RM, 2007. Reprinted by permission.

---

body of the protosun, leaving a larger accumulation of matter in the outer part of the accretion disk.

Stage C

During *stage C*, the warming protosun became established as a star, perhaps undergoing an initial flare-up that has been observed today in other newly forming stars. Such a flareup might have been of such a magnitude that it blasted away the hydrogen and helium atmospheres of the interior planets (Mercury, Venus, Earth, and Mars) out past Mars, but it did not reach far enough out to disturb the hydrogen and helium atmospheres of the outer planets. The innermost of the outer planets, Jupiter and Saturn, might have acquired some of the matter blasted away from the inner planets, becoming the giants of the solar system by comparison. This is just speculation, however, and the two giants may have simply formed from greater concentrations of matter in that part of the accretion disk.

The evidence, such as separation of heavy and light mineral matter, shows that the protoplanets underwent heating early in their formation. Much of the heating may have been provided by gravitational contraction, the same process that gave the protosun sufficient heat to begin its internal nuclear fusion reactions. Heat was also provided from radioactive decay processes inside the protoplanets, and the initial greater heating from the Sun may have played a role in the protoplanet heating process. Larger bodies were able to retain this heat better than smaller ones, which radiated it to space more readily. Thus, the larger bodies underwent a more thorough heating and melting, perhaps becoming completely molten early in their history. In the larger bodies, the heavier elements, such as iron, were pulled to the center of the now molten mass, leaving the lighter elements near the surface. The overall heating and cooling process took millions of years as the planets and smaller bodies were formed. Gases from the hot interiors formed secondary atmospheres of water vapor, carbon dioxide, and nitrogen on the larger interior planets.

TABLE 13.7 Distances from the Sun to Planets Known in the 1790s

Planet	n	Distance Predicted by $(n + 4)/10$ (AU)	Actual Distance (AU)
Mercury	0	0.4	0.39
Venus	3	0.7	0.72
Earth	6	1.0	1.0
Mars	12	1.6	1.5
(Asteroid belt)	24	2.8	—
Jupiter	48	5.2	5.2
Saturn	96	10.0	9.5
Uranus	192	19.6	19.2

Interestingly, the asteroid belt was discovered from a prediction made by the German astronomer Bode at the end of the eighteenth century. Bode had noticed a pattern of regularity in the spacing of the planets that were known at the time. He found that by expressing the distances of the planets from the Sun in astronomical units, these distances could be approximated by the relationship $(n + 4)/10$, where n is a number in the sequence 0, 3, 6, 12, and so on where each number (except the first) is doubled in succession. When these calculations were done, the distances turned out to be very close to the distances of all the planets known at that time, but the numbers also predicted a planet between Mars and Jupiter where there was none. Later, a belt of asteroids was found where the Bode numbers predicted there should be a planet. This suggested to some people that a planet had existed between Mars and Jupiter in the past and somehow this planet was broken into pieces, perhaps by a collision with another large body (Table 13.7).

Such patterns of apparent regularity in the spacing of planetary orbits were of great interest because, if a true pattern existed, it could hold meaning about the mechanism that determined the location of planets at various distances from the Sun. Many attempts have been made to explain the mechanism of planetary spacing and why a belt of asteroids exists where the Bode numbers predict there should be a planet. The most successful explanations concern Jupiter and the influence of its gigantic gravitational field on the formation of clumps of matter at certain distances from the Sun. In other words, a planet does not exist today between Mars and Jupiter because there never was a planet there. The gravitational influence of Jupiter prevented the clumps of matter from joining together to form a planet, and a belt of asteroids formed instead.

SUMMARY

The planets can be classified into two major groups: (1) the *terrestrial planets* of Mercury, Venus, Mars, and Earth and (2) the *giant planets* of Jupiter, Saturn, Uranus, and Neptune.

Comets are porous aggregates of water ice, frozen methane, frozen ammonia, dry ice, and dust. The solar system is surrounded by the *Kuiper Belt* and the *Oort cloud* of these objects. Something nudges one of the icy bodies and it falls into a long elliptical orbit around the Sun. As it approaches the Sun, increased radiation evaporates ices and pushes ions and dust into a long visible tail. *Asteroids* are rocky or metallic bodies that are mostly located in a belt between Mars and Jupiter. The remnants of comets, fragments of asteroids, and dust are called *meteoroids*. A meteoroid that falls through Earth's atmosphere and melts to a visible trail of light and smoke is called a *meteor*. A meteoroid that survives the trip through the atmosphere to strike the surface of Earth is called a *meteorite*.

Most meteors are fragments and pieces of dust from comets. Most meteorites are fragments that resulted from collisions between asteroids.

The *protoplanet nebular model* is the most widely accepted theory of the origin of the solar system, and this theory can be considered as a series of events, or stages. *Stage A* is the creation of all the elements heavier than hydrogen in previously existing stars. *Stage B* is the formation of a nebula from the raw materials created in stage A. The nebula contracts from gravitational attraction, forming the *protosun* in the center with a fat, bulging *accretion disk* around it. The Sun will form from the protosun, and the planets will form in the accretion disk. *Stage C* begins as the protosun becomes established as a star. The icy remains of the original nebula are the birthplace of *comets*. *Asteroids* are other remains that did undergo some melting.

KEY TERMS

asteroids (p. **304**)
astronomical unit (p. **289**)
comet (p. **302**)
dwarf planet (p. **288**)

giant planets (p. **289**)
Kuiper Belt (p. **303**)
meteor (p. **304**)
meteorite (p. **305**)

meteoroids (p. **304**)
meteor shower (p. **304**)
Oort cloud (p. **302**)
planet (p. **288**)

protoplanet nebular model (p. **306**)
small solar system bodies
(p. **288**)
terrestrial planets (p. **289**)

APPLYING THE CONCEPTS

1. Earth, other planets, and all the members of the solar system
   a. have always existed.
   b. formed thousands of years ago from elements that have always existed.
   c. formed millions of years ago, when the elements and each body were created at the same time.
   d. formed billions of years ago from elements that were created in many previously existing stars.

2. The belt of asteroids between Mars and Jupiter is probably
   a. the remains of a planet that exploded.
   b. clumps of matter that condensed from the accretion disk but never got together as a planet.
   c. the remains of two planets that collided.
   d. the remains of a planet that collided with an asteroid or comet.

3. Which of the following planets would be mostly composed of hydrogen, helium, and methane and have a density of less than 2 g/cm³?
   a. Uranus
   b. Mercury
   c. Mars
   d. Venus

4. Which of the following planets probably still has its original atmosphere?
   a. Mercury
   b. Venus
   c. Mars
   d. Jupiter

5. Venus appears the brightest when it is in the
   a. full phase.
   b. half phase.
   c. quarter phase.
   d. crescent phase.

6. The largest planet is
   a. Saturn.
   b. Jupiter.
   c. Uranus.
   d. Neptune.

7. The small body with a composition and structure closest to the materials that condensed from the accretion disk is a (an)
   a. asteroid.
   b. meteorite.
   c. comet.
   d. None of the above is correct.

8. A small body from space that falls on the surface of Earth is a
   a. meteoroid.
   b. meteor.
   c. meteor shower.
   d. meteorite.

Answers

1. d 2. b 3. a 4. d 5. d 6. b 7. c 8. d

QUESTIONS FOR THOUGHT

1. Describe the surface and atmospheric conditions on Mars.
2. What evidence exists that Mars at one time had abundant liquid water? If Mars did have liquid water at one time, what happened to it and why?
3. Describe the internal structure of Jupiter and Saturn.
4. Describe some of the unusual features found on the moons of Jupiter, Saturn, and Neptune.
5. What are the similarities and the differences between the Sun and Jupiter?
6. What evidence exists today that the number of rocks and rock particles floating around in the solar system was much greater in the past soon after the planets formed?
7. Explain why oxygen is a major component of Earth's atmosphere but not the atmospheres of Venus or Mars.
8. Using the properties of the planets other than Earth, discuss the possibilities of life on each.
9. What are "shooting stars"? Where do they come from? Where do they go?
10. What is an asteroid? What evidence indicates that asteroids are parts of a broken-up planet? What evidence indicates that asteroids are not parts of a broken-up planet?
11. Where do comets come from? Why are astronomers so interested in studying the physical and chemical structure of a comet?
12. What is a meteorite? What is the most likely source of meteorites?

FOR FURTHER ANALYSIS

1. What are the significant similarities and differences between the terrestrial and giant planets? Speculate why these similarities and differences exist.

2. Draw a sketch showing the positions of the Earth, Sun, and Venus when it appears as the morning star. Draw a second sketch showing the positions when Venus appears as the evening star.

3. Evaluate the statement that Venus is Earth's sister planet.

4. Describe the possibility and probability of life on each of the other planets.

5. Provide arguments with evidence that Pluto should be considered a planet. Counter this argument with evidence that it should not be classified as a planet.

6. Describe and analyze why it would be important to study the nucleus of a comet.

INVITATION TO INQUIRY

What's Your Sign?

Form a team to investigate horoscope forecasts in a newspaper or on an Internet site. Each team member should select one birthday and track what is forecast to happen and what actually happens each day for a week. Analyze the way the forecasts are written that may make them "come true." Compare the prediction, the actual results, and the analysis for each team member.

This shows part of Earth as seen from space, with the Salton Sea in the center of the photo. Could you tell someone where on Earth the Salton Sea is located? One topic of this chapter is identifying places on Earth, which should help you describe the location of any part of Earth's surface.

Earth in Space

Earth moves around the Sun in a yearly revolution (p. 315).

Earth spins around its axis in a daily rotation (p. 315).

Earth's axis is inclined 23.5° and keeps the same orientation all year (p. 315).

Earth's axis serves as a reference point for direction and location on the entire surface (p. 318).

OVERVIEW

Earth is a common object in the solar system, one of eight planets that goes around the Sun once a year in an almost circular orbit. Earth is the third planet out from the Sun, it is fifth in mass and diameter, and it has the greatest density of all the planets (figure 14.1). Earth is unique because of its combination of an abundant supply of liquid water, a strong magnetic field, and a particular atmospheric composition. In addition to these physical properties, Earth has a unique set of natural motions that humans have used for thousands of years as a frame of reference to mark time and to identify the events of their lives. These references to Earth's motions are called the day, the month, and the year.

Eventually, about three hundred years ago, people began to understand that their references for time came from an Earth that spins like a top as it circles the Sun. It was still difficult, however, for them to understand Earth's place in the universe. The problem was not unlike that of a person trying to comprehend the motion of a distant object while riding a moving merry-go-round being pulled by a cart. Actually, the combined motions of Earth are much more complex than a simple moving merry-go-round being pulled by a cart. Imagine trying to comprehend the motion of a distant object while undergoing a combination of Earth's more conspicuous motions, which are as follows:

1. A daily rotation of 1,670 km/h (about 1,040 mi/h) at the equator and less at higher latitudes.
2. A monthly revolution of Earth around the Earth-Moon center of gravity at about 50 km/h (about 30 mi/h).
3. A yearly revolution around the Sun at about an average 106,000 km/h (about 66,000 mi/h).
4. A motion of the solar system around the core of the Milky Way at about 370,000 km/h (about 230,000 mi/h).
5. A motion of the local star group that contains the Sun as compared to other star clusters of about 1,000,000 km/h (about 700,000 mi/h).
6. Movement of the entire Milky Way galaxy relative to other, remote galaxies at about 580,000 km/h (about 360,000 mi/h).
7. Minor motions such as cycles of change in the size and shape of Earth's orbit and the tilt of Earth's axis. In addition to these slow changes, there is a gradual slowing of the rate of Earth's daily rotation.

Basically, Earth is moving through space at fantastic speeds, following the Sun in a spiral path of a giant helix as it spins like a top (figure 14.2). This ceaseless and complex motion in space is relative to various frames of reference, however, and the limited perspective from Earth's surface can result in some very different ideas about Earth and its motions. This chapter is about the more basic, or fundamental, motions of Earth and its Moon. In addition to conceptual understandings and evidences for the motions, some practical human uses of the motions will be discussed.

SHAPE AND SIZE OF EARTH

The most widely accepted theory about how the solar system formed pictures the planets forming in a disk-shaped nebula with a turning, swirling motion. The planets formed from separate accumulations of materials within this disk-shaped, turning nebula, so the orbit of each planet was established along with its rate of rotation as it formed. Thus, all the planets move around the Sun in the same direction in elliptical orbits that are nearly circular. The flatness of the solar system results in the observable planets moving in, or near, the plane of Earth's orbit, which is called the **plane of the ecliptic.**

FIGURE 14.1 Artist's concept of the solar system. Shown are the orbits of the planets, Earth being the third planet from the Sun, and the other planets and their relative sizes and distances from each other and to the Sun. Also shown is the solar system as seen looking toward Earth from the Moon.

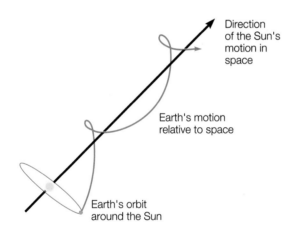

Direction of the Sun's motion in space

Earth's motion relative to space

Earth's orbit around the Sun

FIGURE 14.2 Earth undergoes many different motions as it moves through space. There are seven more conspicuous motions, three of which are more obvious on the surface. Earth follows the path of a gigantic helix, moving at fantastic speeds as it follows the Sun and the galaxy through space.

When viewed from Earth, the planets appear to move only within a narrow band across the sky as they move in the plane of the ecliptic. The Sun also appears to move in the center of this band, which is called the *ecliptic.* As viewed from Earth, the Sun appears to move across the background of stars, completely around the ecliptic each year.

Today, almost everyone has seen pictures of Earth from space, and it is difficult to deny that it has a rounded shape. During the fifth and sixth centuries b.c., the ancient Greeks decided that Earth must be round because (1) philosophically, they considered the sphere to be the perfect shape and they

considered Earth to be perfect, so therefore Earth must be a sphere, (2) Earth was observed to cast a circular shadow on the Moon during a lunar eclipse, and (3) ships were observed to slowly disappear below the horizon as they sailed off into the distance. More abstract evidence of a round Earth was found in the observation that the altitude of the North Star above the horizon appeared to increase as a person traveled northward. This established that Earth's surface was curved, at least, which seemed to fit with other evidence.

The shape and size of Earth have been precisely measured by artificial satellites circling Earth. These measurements have found that Earth is not a perfectly round sphere as believed by the ancient Greeks. It is flattened at the poles and has an equatorial bulge, as do many other planets. In fact, you can observe through a telescope that both Jupiter and Saturn are considerably flattened at the poles. A shape that is flattened at the poles has a greater distance through the equator than through the poles, which is described as an *oblate* shape. Earth, like a water-filled, round balloon resting on a table, has an oblate shape. It is not perfectly symmetrically oblate, however, since the North Pole is slightly higher and the South Pole is slightly lower than the average surface. In addition, it is not perfectly circular around the equator, with a lump in the Pacific and a depression in the Indian Ocean. The shape of the earth is a slightly pear-shaped, slightly lopsided *oblate spheroid.* All the elevations and depressions are less than 85 m (about 280 ft), however, which is practically negligible compared to the size of Earth (figure 14.3). Thus, Earth is very close to, but not exactly, an oblate spheroid. The significance of this shape will become apparent when Earth's motions are discussed next (figure 14.4).

The Celestial Sphere

To locate the ecliptic, planets, or anything else in the sky, you need something to refer to, a referent system. A referent system is easily established by first imagining the sky to be a celestial sphere just as the ancient Greeks did. A coordinate system of lines can be visualized on this sphere. Imagine that you could inflate Earth until its surface touched the celestial sphere. If you now transfer lines to the celestial sphere, you will have a system of sky coordinates. From the surface of Earth, you can see that the *celestial equator* is a line on the celestial sphere directly above Earth's equator and the *north celestial pole* is a point directly above the North Pole of Earth. Likewise, the *south celestial pole* is a point directly above the South Pole of Earth.

You can only see half of the overall celestial sphere from any one place on the surface of Earth. Imagine a point on the celestial sphere directly above where you are located. An imaginary line that passes through this point, then passes north through the north celestial pole, continuing all the way around through the south celestial pole and back to the point directly above you makes a big circle called the *celestial meridian* (box figure 14.1). Note that the celestial meridian location is determined by where you are on Earth.

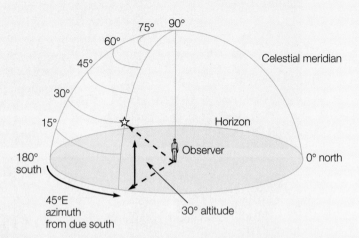

Box Figure 14.1 Once you have established the celestial equator, the celestial poles, and the celestial meridian, you can use a two-coordinate horizon system to locate positions in the sky. One popular method of using this system identifies the altitude angle (in degrees) from the horizon up to an object on the celestial sphere and the azimuth angle (again in degrees) the object on the celestial sphere is east or west of due south, where the celestial meridian meets the horizon. The illustration shows an altitude of 30° and an azimuth of 45° east of due south.

The celestial equator and the celestial poles, on the other hand, are always in the same place no matter where you are.

Overall, the celestial sphere appears to spin, turning on an axis through the celestial poles. A photograph made by pointing a camera at the north celestial pole and leaving the shutter open for several hours will show the apparent motion of the celestial sphere with star trails (see figure 12.1A).

The moderately bright star near the center is the North Star, Polaris. Polaris is almost, but not exactly, at the north celestial pole. If you observe the celestial sphere night after night, you will see that the stars maintain their positions relative to one another as they turn counterclockwise around Polaris. Those near Polaris pivot around it and are called "circumpolar." Those farther out rise in the east, move in an arc, then set in the west.

FIGURE 14.3 Earth as seen from space.

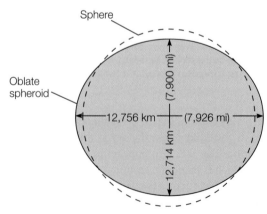

FIGURE 14.4 Earth has an irregular, slightly lopsided, slightly pear-shaped form. In general, it is considered to have the shape of an oblate spheroid, departing from a perfect sphere as shown here.

MOTIONS OF EARTH

Ancient civilizations had a fairly accurate understanding of the size and shape of Earth but had difficulty accepting the idea that Earth moves for at least two reasons: (1) they could not sense any motion of Earth, and (2) they had ideas about being at the center of a universe that was created for them. It was not until the 1700s that the concept of an Earth in motion became generally accepted. Today, Earth is understood to move a number of different ways, seven of which were identified in the introduction to this chapter. Three of these motions are independent of motions of the Sun and the galaxy. These are (1) a yearly revolution around the Sun, (2) a daily rotation on its axis, and (3) a slow clockwise wobble of its axis.

Revolution

Earth moves constantly around the Sun in a slightly elliptical orbit that requires an average of one year for one complete circuit. The movement around the Sun is called a **revolution,** and all points of Earth's orbit lie in the plane of the ecliptic. The average distance between Earth and the Sun is about 150 million km (about 93 million mi).

Earth's orbit is slightly elliptical, so it moves with a speed that varies. It moves fastest when it is closer to the Sun in January and moves slowest when it is farthest away from the Sun in early July. Earth is about 2.5 million km (about 1.5 million mi) closer to the Sun in January and about the same distance farther away in July than it would be if the orbit were a circle. This total difference of about 5 million km (about 3 million mi) results in a January Sun with an apparent diameter that is 3 percent larger than the July Sun, and Earth as a whole receives about 6 percent more solar energy in January. The effect of being closer to the Sun is much less than the effect of some other relationships, and winter occurs in the Northern Hemisphere when Earth is closest to the Sun. Likewise, summer occurs in the Northern Hemisphere when the Sun is at its greatest distance from Earth (figure 14.5).

The important directional relationships that override the effect of Earth's distance from the Sun involve the daily **rotation,** or spinning, of Earth around an imaginary line through the geographic poles called Earth's *axis.* The important directional relationships are a constant inclination of Earth's axis to the plane of the ecliptic and a constant orientation of the axis to the stars. The *inclination of Earth's axis* to the plane of the ecliptic is about 66.5° (or 23.5° from a line perpendicular to the plane). This relationship between the plane of Earth's orbit and the tilt of its axis is considered to be the same day after day throughout the year, even though small changes do occur in the inclination over time. Likewise, the *orientation of Earth's axis* to the stars is considered to be the same throughout the year as Earth moves through its orbit. Again, small changes do occur in the orientation over time. Thus, in general, the axis points in the same direction, remaining essentially parallel to its position during any day of the year. The essentially constant orientation and inclination of the axis result in the axis pointing toward the Sun as Earth moves in one part of its orbit, then pointing away from the Sun six months later. The generally constant inclination and orientation of the axis, together with Earth's rotation and revolution, combine to produce three related effects: (1) days and nights that vary in length, (2) changing seasons, and (3) climates that vary with latitude.

Figure 14.5 shows how the North Pole points toward the Sun on June 21 or 22, then away from the Sun on December 22 or 23 as it maintains its orientation to the stars. When the North Pole is pointed toward the Sun, it receives sunlight for a full twenty-four hours, and the South Pole is in Earth's shadow for a full twenty-four hours. This is summer in the Northern Hemisphere, with the longest daylight periods and the Sun at its maximum noon height in the sky. Six months later, on December 22 or 23, the orientation is reversed with winter in the Northern Hemisphere, the shortest daylight periods, and the Sun at its lowest noon height in the sky.

The beginning of a season can be recognized from any one of the three related observations: (1) the length of the daylight period, (2) the altitude of the Sun in the sky at noon, or (3) the length of a shadow from a vertical stick at noon. All of these observations vary with changes in the direction of Earth's axis of rotation relative to the Sun. On about June 22 and December 22, the Sun reaches its highest and lowest noon altitudes as Earth moves to point the North Pole directly toward the Sun (June 21 or 22) and directly away from the Sun (December 22 or 23). Thus, the Sun appears to stop increasing or decreasing its altitude in the sky, stop, then reverse its movement twice a year. These times are known as **solstices** after the Latin meaning "Sun stand still." The Northern Hemisphere's **summer solstice** occurs on about June 22 and identifies the beginning of the summer season. At the summer solstice, the Sun at noon has the highest altitude, and the shadow from a vertical stick is shorter than on any other day of the year. The Northern Hemisphere's **winter solstice** occurs on about December 22 and identifies the beginning of the winter season. At the winter solstice, the Sun at noon has the lowest altitude, and the shadow from a vertical stick is longer than on any other day of the year (figure 14.6).

As Earth moves in its orbit between pointing its North Pole toward the Sun on about June 22 and pointing it away on about December 22, there are two times when it is halfway

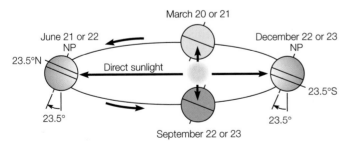

FIGURE 14.5 The consistent tilt and orientation of Earth's axis as it moves around its orbit is the cause of the seasons. The North Pole is pointing toward the Sun during the summer solstice and away from the Sun during the winter solstice.

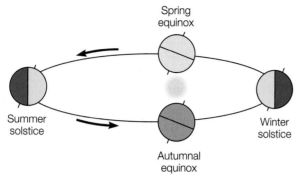

FIGURE 14.6 The length of daylight during each season is determined by the relationship of Earth's shadow to the tilt of the axis. At the equinoxes, the shadow is perpendicular to the latitudes, and day and night are of equal length everywhere. At the summer solstice, the North Pole points toward the Sun and is completely out of the shadow for a twenty-four-hour day. At the winter solstice, the North Pole is in the shadow for a twenty-four-hour night. The situation is reversed for the South Pole.

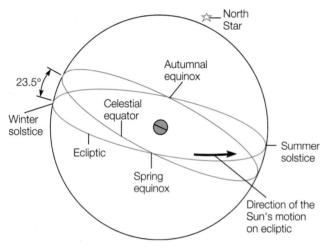

FIGURE 14.7 The position of the Sun on the celestial sphere at the solstices and the equinoxes.

between. At these times, Earth's axis is perpendicular to a line between the center of the Sun and Earth, and daylight and night are of equal length. These are called the **equinoxes** after the Latin meaning "equal nights." The **spring equinox** (also called the **vernal equinox**) occurs on about March 21 and identifies the beginning of the spring season. The **autumnal equinox** occurs on about September 23 and identifies the beginning of the fall season.

The relationship between the apparent path of the Sun on the celestial sphere and the seasons is shown in figure 14.7. The celestial equator is a line on the celestial sphere directly above Earth's equator. The equinoxes are the points on the celestial sphere where the ecliptic, the path of the Sun, crosses the celestial equator. Note also that the summer solstice occurs when the ecliptic is 23.5° north of the celestial equator, and the winter solstice occurs when it is 23.5° south of the celestial equator.

Rotation

Observing the apparent turning of the celestial sphere once a day and seeing the east-to-west movement of the Sun, Moon, and stars, it certainly seems as if it is the heavenly bodies and not Earth doing the moving. You cannot sense any movement, and there is little apparent evidence that Earth indeed moves. Evidence of a moving Earth comes from at least three different observations: (1) the observation that the other planets and the Sun rotate, (2) the observation of the changing plane of a long, heavy pendulum at different latitudes on Earth, and (3) the observation of the direction of travel of something moving across, but above, Earth's surface, such as a rocket.

Other planets, such as Jupiter, and the Sun can be observed to rotate by keeping track of features on the surface such as the Great Red Spot on Jupiter and sunspots on the Sun. While such observations are not direct evidence that Earth also rotates, they do show that other members of the solar system spin on their axes. As described earlier, Jupiter is also observed to be oblate, flattened at its poles with an equatorial bulge. Since Earth is also oblate, this is again indirect evidence that it rotates, too.

The most easily obtained and convincing evidence about Earth's rotation comes from a *Foucault pendulum*, a heavy mass swinging from a long wire. This pendulum is named after the French physicist Jean Foucault, who first used a long pendulum in 1851 to prove that Earth rotates. Foucault started a long, heavy pendulum moving just above the floor, marking the plane of its back-and-forth movement. Over some period of time, the pendulum appeared to slowly change its position, smoothly shifting its plane of rotation. Science museums often show this shifting plane of movement by setting up small objects for the pendulum to knock down. Foucault demonstrated that the pendulum actually maintains its plane of movement in space (inertia) while Earth rotates eastward (counterclockwise)

FIGURE 14.8 As is being demonstrated in this old woodcut, Foucault's insight helped people understand that Earth turns. The pendulum moves back and forth without changing its direction of movement, and we know this is true because no forces are involved. We turn with Earth, and this makes the pendulum appear to change its plane of rotation. Thus, we know Earth rotates.

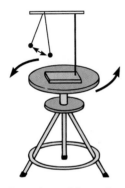

FIGURE 14.9 The Foucault pendulum swings back and forth in the same plane while a stool is turned beneath it. Likewise, a Foucault pendulum on Earth's surface swings back and forth in the same plane while Earth turns beneath it. The amount of turning observed depends on the latitude of the pendulum.

under the pendulum. It is Earth that turns under the pendulum, causing the pendulum to appear to change its plane of rotation. It is difficult to imagine the pendulum continuing to move in a fixed direction in space while Earth, and everyone on it, turns under the swinging pendulum (figure 14.8).

Figure 14.9 illustrates the concept of the Foucault pendulum. A pendulum is attached to a support on a stool that is free to rotate. If the stool is slowly turned while the pendulum is swinging, you will observe that the pendulum maintains its plane of rotation while the stool turns under it. If you were much smaller and looking from below the pendulum, it would appear to turn as you rotate with the turning stool. This is what happens on Earth. Such a pendulum at the North Pole would make a complete turn in about twenty-four hours. Moving south from the North Pole, the change decreases with latitude until, at the equator, the pendulum would not appear to turn

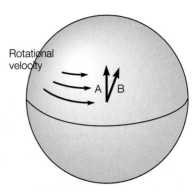

FIGURE 14.10 Earth has a greater rotational velocity at the equator and less toward the poles. As an object moves north or south (A), it passes over land with a different rotational velocity, which produces a deviation to the right in the Northern Hemisphere (B) and to the left in the Southern Hemisphere.

at all. At higher latitudes, the plane of the pendulum appears to move clockwise in the Northern Hemisphere and counterclockwise in the Southern Hemisphere.

More evidence that Earth rotates is provided by objects that move above and across Earth's surface. As shown in figure 14.10, Earth has a greater rotational velocity at the equator than at the poles. As an object leaves the surface and moves north or south, the surface has a different rotational velocity, so it rotates beneath the object as it proceeds in a straight line. This gives the moving object an apparent deflection to the right of the direction of movement in the Northern Hemisphere and to the left in the Southern Hemisphere. The apparent deflection caused by Earth's rotation is called the **Coriolis effect.** The Coriolis effect will explain Earth's prevailing wind systems as well as the characteristic direction of wind in areas of high pressure and areas of low pressure (see chapter 17).

Precession

If Earth were a perfect spherically shaped ball, its axis would always point to the same reference point among the stars. The reaction of Earth to the gravitational pull of the Moon and the Sun on its equatorial bulge, however, results in a slow wobbling of Earth as it turns on its axis. This slow wobble of Earth's axis, called **precession,** causes it to swing in a slow circle like the wobble of a spinning top (figure 14.11). It takes

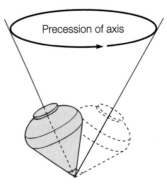

FIGURE 14.11 A spinning top wobbles as it spins, and the axis of the top traces out a small circle. The wobbling of the axis is called *precession*.

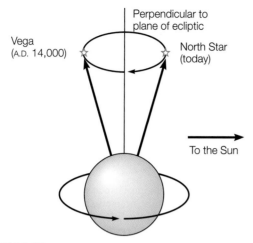

FIGURE 14.12 The slow, continuous precession of Earth's axis results in the North Pole pointing around a small circle over a period of about twenty-six thousand years.

Earth's axis about twenty-six thousand years to complete one turn, or wobble. Today, the axis points very close to the North Star, Polaris, but is slowly moving away to point to another star. In about twelve thousand years, the star Vega will appear to be in the position above the North Pole, and Vega will be the new North Star. The moving pole also causes changes over time in which particular signs of the zodiac appear with the spring equinox. Because of precession, the occurrence of the spring equinox has been moving backward (westward) through the zodiac constellations at about 1 degree every seventy-two years. Thus, after about twenty-six thousand years, the spring equinox will have moved through all the constellations and will again approach the constellation of Aquarius for the next "age of Aquarius" (figure 14.12).

PLACE AND TIME

The continuous rotation and revolution of Earth establishes an objective way to determine direction, location, and time on Earth. If Earth were an unmoving sphere, there would be no side, end, or point to provide a referent for direction and location. Earth's rotation, however, defines an axis of rotation, which serves as a reference point for determination of direction and location on the entire surface. Earth's rotation and revolution together define cycles, which define standards of time. The following describes how Earth's movements are used to identify both place and time.

Identifying Place

A system of two straight lines can be used to identify a point, or position, on a flat, two-dimensional surface. The position of the letter *X* on this page, for example, can be identified by making a line a certain number of measurement units from the top of the page and a second line a certain number of measurement units from the left side of the page. Where the two lines intersect will identify the position of the letter *X,* which can be recorded or communicated to another person (figure 14.13).

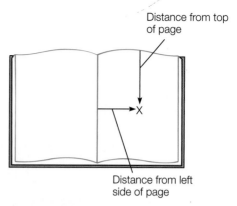

FIGURE 14.13 Any location on a flat, two-dimensional surface is easily identified with two references from two edges. This technique does not work on a motionless sphere because there are no reference points.

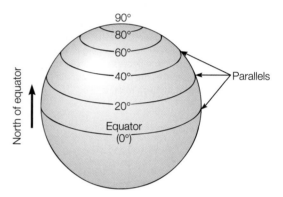

FIGURE 14.14 A circle that is parallel to the equator is used to specify a position north or south of the equator. A few of the possibilities are illustrated here.

A system of two straight lines can also be used to identify a point, or position, on a sphere, except this time the lines are circles. The reference point for a sphere is not as simple as in the flat, two-dimensional case, however, since a sphere does not have a top or side edge. Earth's axis provides the north-south reference point. The equator is a big circle around Earth that is exactly halfway between the two ends, or poles, of the rotational axis. An infinite number of circles are imagined to run around Earth parallel to the equator as shown in figure 14.14. The east- and west-running parallel circles are called **parallels.** Each parallel is the same distance between the equator and one of the poles all the way around Earth. The distance from the equator to a point on a parallel is called the **latitude** of that point. Latitude tells you how far north or south a point is from the equator by telling you the parallel the point is located on. The distance is measured northward from the equator (which is 0°) to the North Pole (90° north) or southward from the equator (0°) to the South Pole (90° south) (figure 14.15). If you are somewhere at a latitude of 35° north, you are somewhere on Earth on the 35° latitude line north of the equator.

Since a parallel is a circle, a location of 40°N latitude could be anyplace on that circle around Earth. To identify a location, you need another line, this time one that runs pole to pole and perpendicular to the parallels. These north-south running arcs

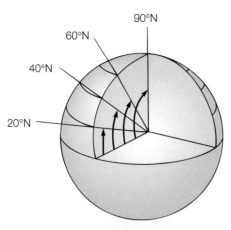

FIGURE 14.15 If you could see to Earth's center, you would see that latitudes run from 0° at the equator north to 90° at the North Pole (or to 90° south at the South Pole).

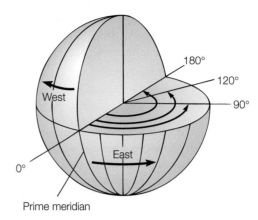

FIGURE 14.17 If you could see inside Earth, you would see 360° around the equator and 180° of longitude east and west of the prime meridian.

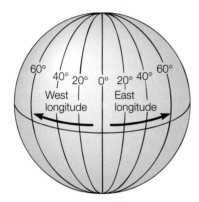

FIGURE 14.16 Meridians run pole to pole perpendicular to the parallels and provide a reference for specifying east and west directions.

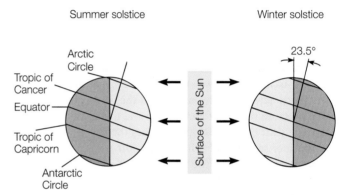

FIGURE 14.18 At the summer solstice, the noon Sun appears directly overhead at the tropic of Cancer (23.5°N) and twenty-four hours of daylight occurs north of the Arctic Circle (66.5°N). At the winter solstice, the noon Sun appears overhead at the tropic of Capricorn (23.5°S) and twenty-four hours of daylight occurs south of the Antarctic Circle (66.5°S).

that intersect at both poles are called **meridians** (figure 14.16). There is no naturally occurring, identifiable meridian that can be used as a point of reference such as the equator serves for parallels, so one is identified as the referent by international agreement. The referent meridian is the one that passes through the Greenwich Observatory near London, England, and this meridian is called the **prime meridian.** The distance from the prime meridian east or west is called the **longitude.** The degrees of longitude of a point on a parallel are measured to the east or to the west from the prime meridian up to 180° (figure 14.17). New Orleans, Louisiana, for example, has a latitude of about 30°N of the equator and a longitude of about 90°W of the prime meridian. The location of New Orleans is therefore described as 30°N, 90°W.

Locations identified with degrees of latitude north or south of the equator and degrees of longitude east or west of the prime meridian are more precisely identified by dividing each degree of latitude into subdivisions of 60 minutes (60') per degree, and each minute into 60 seconds (60"). On the other hand, latitudes near the equator are sometimes referred to in general as the *low latitudes,* and those near the poles are sometimes called the *high latitudes.*

In addition to the equator (0°) and the poles (90°), the parallels of 23.5°N and 23.5°S from the equator are important references for climatic consideration. The parallel of 23.5°N is called the **tropic of Cancer,** and 23.5°S is called the **tropic of Capricorn.** These two parallels identify the limits toward the poles within which the Sun appears directly overhead during the course of a year. The parallel of 66.5°N is called the **Arctic Circle,** and the parallel of 66.5°S is called the **Antarctic Circle.** These two parallels identify the limits toward the equator within which the Sun appears above the horizon all day during the summer (figure 14.18). This starts with six months of daylight every day at the pole, then decreases as you get fewer days of full light until reaching the limit of one day of twenty-four-hour daylight at the 66.5° limit.

Measuring Time

Standards of time are determined by intervals between two successive events that repeat themselves in a regular way. Since ancient civilizations, many of the repeating events used to mark time have been recurring cycles associated with the rotation of Earth on its axis and its revolution around the Sun. Thus, the day, month, season, and year are all measures of time based on recurring natural motions of Earth. All other measures of time are based on other events or definitions of events. There are, however, several different ways to describe the day, month, and year, and each depends on a different set of events. These events are described in the following section.

Daily Time

The technique of using astronomical motions for keeping time originated some four thousand years ago with the Babylonian culture. The Babylonians marked the yearly journey of the Sun against the background of the stars, which was divided into twelve periods, or months, after the signs of the zodiac. Based on this system, the Babylonian year was divided into twelve months with a total of 360 days. In addition, the Babylonians invented the week and divided the day into hours, minutes, and seconds. The week was identified as a group of seven days, each based on one of the seven heavenly bodies that were known at the time. The hours, minutes, and seconds of a day were determined from the movement of the shadow around a straight, vertical rod.

As seen from a place in space above the North Pole, Earth rotates counterclockwise turning toward the east. On Earth, this motion causes the Sun to appear to rise in the east, travel across the sky, and set in the west. The changing angle between the tilt of Earth's axis and the Sun produces an apparent shift of the Sun's path across the sky, northward in the summer season and southward in the winter season. The apparent movement of the Sun across the sky was the basis for the ancient as well as the modern standard of time known as the day.

Today, everyone knows that Earth turns as it moves around the Sun, but it is often convenient to regard space and astronomical motions as the ancient Greeks did, as a celestial sphere that turns around a motionless Earth. Recall that the celestial meridian is a great circle on the celestial sphere that passes directly overhead where you are and continues around Earth through both celestial poles. The movement of the Sun across the celestial meridian identifies an event of time called **noon.** As the Sun appears to travel west, it crosses meridians that are farther and farther west, so the instant identified as noon moves west with the Sun. The instant of noon at any particular longitude is called the **apparent local noon** for that longitude because it identifies noon from the apparent position of the Sun in the sky. The morning hours before the Sun crosses the meridian are identified as *ante meridiem* (A.M.) hours, which is Latin for "before meridian." Afternoon hours are identified as *post meridiem* (P.M.) hours, which is Latin for "after the meridian."

There are several ways to measure the movement of the Sun across the sky. The ancient Babylonians, for example, used a vertical rod called a *gnomon* to make and measure a shadow that moved as a result of the apparent changes of the Sun's position. The gnomon eventually evolved into a *sundial,* a vertical or slanted gnomon with divisions of time marked on a horizontal plate beneath the gnomon. The shadow from the gnomon indicates the **apparent local solar time** at a given place and a given instant from the apparent position of the Sun

FIGURE 14.19 A sundial indicates the apparent local solar time at a given instant in a given location. The time read from a sundial, which is usually different from the time read from a clock, is based on an average solar time.

in the sky. If you have ever read the time from a sundial, you know that it usually does not show the same time as a clock or a watch (figure 14.19). In addition, sundial time is nonuniform, fluctuating throughout the course of a year, sometimes running ahead of clock time and sometimes running behind clock time.

A sundial shows the apparent local solar time, but clocks are set to measure a uniform standard time based on **mean solar time.** Mean solar time is a uniform time averaged from the apparent solar time. The apparent solar time is nonuniform, fluctuating because (1) Earth moves sometimes faster and sometimes slower in its elliptical orbit around the Sun and (2) the equator of Earth is inclined to the ecliptic. The combined consequence of these two effects is a variable, nonuniform sundial time as compared to the uniform mean solar time, otherwise known as clock time.

A day is defined as the length of time required for Earth to rotate once on its axis. There are different ways to measure this rotation, however, which result in different definitions of the day. A **sidereal day** is the interval between two consecutive crossings of the celestial meridian by a particular star (*sidereal* means "star"). This interval of time depends only on the time Earth takes to rotate 360° on its axis. One sidereal day is practically the same length as any other sidereal day because Earth's rate of rotation is constant for all practical purposes.

An **apparent solar day** is the interval between two consecutive crossings of the celestial meridian by the Sun, for example, from one local solar noon to the next solar noon. Since Earth is moving in orbit around the Sun, it must turn a little bit farther to compensate for its orbital movement, bringing the Sun back to local solar noon (figure 14.20). As a consequence, the apparent solar day is about four minutes longer than the sidereal day. This additional time accounts for the observation that the stars and constellations of the zodiac rise about four minutes earlier every night, appearing higher in the sky at the same clock time until they complete a yearly cycle. A sidereal

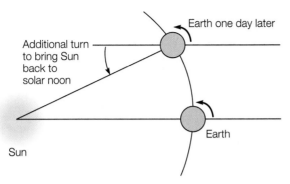

FIGURE 14.20 Because Earth is moving in orbit around the Sun, it must rotate an additional distance each day, requiring about four minutes to bring the Sun back across the celestial meridian (local solar noon). This explains why the stars and constellations rise about four minutes earlier every night.

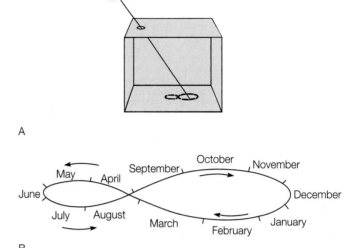

A

B

FIGURE 14.21 (A) During a year, a beam of sunlight traces out a lopsided figure eight on the floor if the position of the light is marked at noon every day. (B) The location of the point of light on the figure eight during each month.

day is twenty-three hours, fifty-six minutes, and four seconds long. A **mean solar day** is twenty-four hours long, averaged from the mean solar time to keep clocks in closer step with the Sun than would be possible using the variable apparent solar day. Just how out of synchronization the apparent solar day can become with a clock can be illustrated with another ancient way of keeping track of the Sun's motions in the sky, the "hole in the wall" sun calendar and clock.

Variations of the "hole in the wall" sun calendar were used all over the world by many different ancient civilizations, including the early Native Americans of the American Southwest. More than one ancient Native American ruin has small holes in the western wall aligned in such a way as to permit sunlight to enter a chamber only on the longest and shortest days of the year. This established a basis for identifying the turning points in the yearly cycle of seasons.

A hole in the roof can be used as a sun clock, but it will require a whole year to establish the meaning of a beam of sunlight shining on the floor. Imagine a beam of sunlight passing through a small hole to make a small spot of light on the floor. For a year, you mark the position of the spot of light on the floor *each day* when your clock tells you the *mean solar time is noon.* You trace out an elongated, lopsided figure eight with the small end pointing south and the larger end pointing north (figure 14.21A). Note by following the monthly markings shown in figure 14.21B that the figure-eight shape is actually traced out by the spot of sunlight making two S shapes as the Sun changes its apparent position in the sky. Together, the two S shapes make the shape of the figure eight.

Why did the sunbeam trace out a figure eight over a year? The two extreme north-south positions of the figure are easy to understand because by December, Earth is in its orbit with the North Pole tilted away from the Sun. At this time, the direct rays of the Sun fall on the tropic of Capricorn (23.5° south of the equator), and the Sun appears low in the sky as seen from the Northern Hemisphere. Thus, on this date, the winter solstice, a beam of sunlight strikes the floor at its northernmost position beneath the hole. By June, Earth has moved halfway around its orbit, and the North Pole is now tilted toward the Sun. The direct rays of the Sun now fall on the tropic of Cancer (23.5° north of the equator), and the Sun appears high in the sky as seen from the Northern Hemisphere (figure 14.22). Thus, on this date, the summer solstice, a beam of sunlight strikes the floor at its southernmost position beneath the hole.

If everything else were constant, the path of the spot would trace out a straight line between the northernmost and southernmost positions beneath the hole. The east and west movements of the point of light as it makes an S shape on the floor must mean, however, that the Sun crosses the celestial meridian (noon) earlier one part of the year and later the other part. This early and late arrival is explained in part by Earth moving at different speeds in its orbit.

If changes in orbital speed were the only reason that the Sun does not cross the sky at the same rate during the year, the spot of sunlight on the floor would trace out an oval rather than a figure eight. The plane of the ecliptic, however, does not coincide with the plane of Earth's equator, so the Sun appears at different angles in the sky, and this makes it appear to change its speed during different times of the year. This effect changes the length of the apparent solar day by making the Sun up to ten minutes later or earlier than the mean solar time four times a year between the solstices and equinoxes.

The two effects add up to a cumulative variation between the apparent local solar time (sundial time) and the mean solar time (clock time) (figure 14.23). This cumulative variation is known as the **equation of time,** which shows how many minutes sundial time is faster or slower than clock time during different days of the year. The equation of time is often shown on globes in the figure-eight shape called an *analemma,* which also can be used to determine the latitude of direct solar radiation for any day of the year.

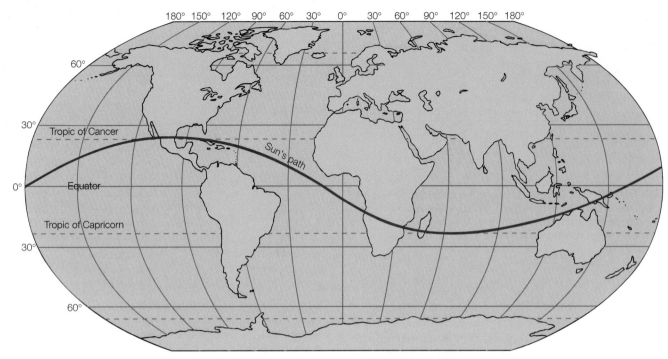

FIGURE 14.22 The path of the Sun's direct rays during a year. The Sun is directly over the tropic of Cancer at the summer solstice and high in the Northern Hemisphere sky. At the winter solstice, the Sun is directly over the tropic of Capricorn and low in the Northern Hemisphere sky.

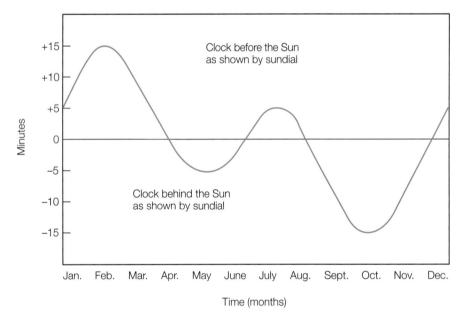

FIGURE 14.23 The equation of time, which shows how many minutes sundial time is faster or slower than clock time during different months of the year.

Since the local mean time varies with longitude, every place on an east-west line around Earth could possibly have clocks that were a few minutes ahead of those to the west and a few minutes behind those to the east. To avoid the confusion that would result from many clocks set to local mean solar time, Earth's surface is arbitrarily divided into one-hour **standard time zones** (figure 14.24). Since there are 360° around Earth and 24 hours in a day, this means that each time zone is 360° divided by 24, or 15° wide. These 15° zones are adjusted so that whole states are in the same time zone or for other political reasons. The time for each zone is defined as the mean solar time at the middle of each zone. When you cross a boundary between two zones, the clock is set ahead one hour if you are traveling east and back one hour if you are traveling west. Most states adopt **daylight saving time** during the summer, setting clocks ahead one hour in the spring and back one hour in the fall ("spring ahead and fall back"). Daylight saving time results in an extra hour of daylight during summer evenings.

Saving Time?

The purpose of daylight saving time is to make better use of daylight during the summer by moving an hour of daylight from the morning to the evening. In the United States, daylight saving time is observed from the second Sunday in March to the first Sunday in November. Clocks are changed on these Sundays according to the saying, "Spring ahead, fall back." Arizona and Hawaii choose not to participate and stay on standard time all year.

Americans who say they like daylight saving time say they like it because it gives them more light in the evenings and it saves energy. Some people do not like daylight saving time because it requires them to reset all their clocks and adjust their sleep schedule twice a year. They also complain that the act of changing the clock is not saving daylight at all but it is sending them to bed an hour earlier. Farmers also complain that plants and animals are regulated by the Sun, not the clock, so they have to plan all their nonfarm interactions on a different schedule.

Questions to Discuss:

Divide your group into two subgroups, one representing those who like daylight saving time and the other representing those who do not. After a few minutes of preparation, have a short debate about the advantages and disadvantages of daylight saving time.

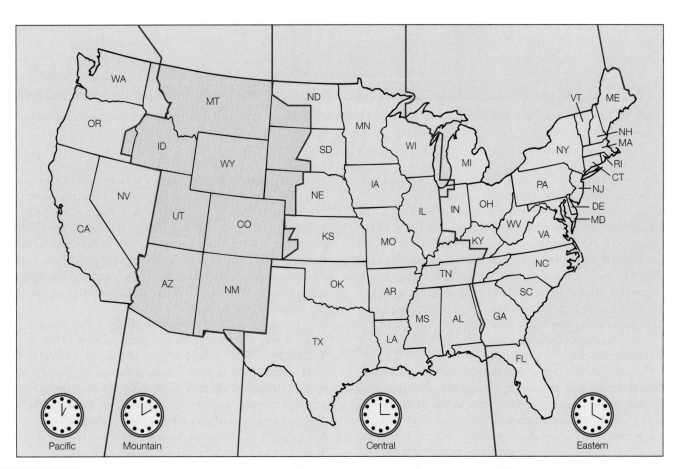

FIGURE 14.24 The standard time zones. Hawaii and most of Alaska are two hours earlier than Pacific Standard Time.

The 180° meridian is arbitrarily called the **international date line,** an imaginary line established to compensate for cumulative time zone changes (figure 14.25). A traveler crossing the date line gains or loses a day just as crossing a time zone boundary results in the gain or loss of an hour. A person moving across the line while traveling westward gains a day; for example, the day after June 2 would be June 4. A person crossing the line while traveling eastward repeats a day; for example, the day after June 6 would be June 6. Note that the date line is curved around land masses to avoid local confusion.

Yearly Time

A *year* is generally defined as the interval of time required for Earth to make one complete revolution in its orbit. As was the case for definitions of a day, there are different definitions of

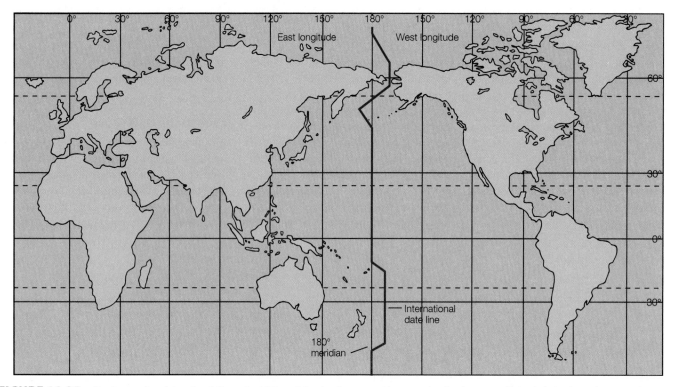

FIGURE 14.25 The international date line follows the 180° meridian but is arranged in a way that land areas and island chains have the same date.

what is meant by a year. The most common definition of a year is the interval between two consecutive spring equinoxes, which is known as the **tropical year** (*trope* is Greek for "turning"). The tropical year is 365 days, 5 hours, 48 minutes, and 46 seconds, or 365.24220 mean solar days.

A **sidereal year** is defined as the interval of time required for Earth to move around its orbit so the Sun is again in the same position relative to the stars. The sidereal year is slightly longer than the tropical year because Earth rotates more than 365.25 times during one revolution. Thus, the sidereal year is 365.25636 mean solar days, which is about 20 minutes longer than the tropical year.

The tropical and sidereal years would be the same interval of time if Earth's axis pointed in a consistent direction. The precession of the axis, however, results in the axis pointing in a slightly different direction with time. This shift of direction over the course of a year moves the position of the spring equinox westward, and the equinox is observed twenty minutes before the orbit has been completely circled. The position of the spring equinox against the background of the stars thus moves westward by some fifty seconds of arc per year.

It is the *tropical year* that is used as a standard time interval to determine the calendar year. Earth does not complete an exact number of turns on its axis while completing one trip around the Sun, so it becomes necessary to periodically adjust the calendar so it stays in step with the seasons. The calendar system that was first designed to stay in step with the seasons was devised by the ancient Romans. Julius Caesar reformed the calendar, beginning in 46 B.C., to have

a 365-day year with a 366-day year (leap year) every fourth year. Since the tropical year of 365.24220 mean solar days is very close to 365¼ days, the system, called a *Julian calendar,* accounted for the ¼ day by adding a full day to the calendar every fourth year. The Julian calendar was very similar to the one now used, except the year began in March, the month of the spring equinox. The month of July was named in honor of Julius Caesar, and the following month was later named after his successor, Augustus.

There was a slight problem with the Julian calendar because it was longer than the tropical year by 365.25 minus 365.24220, or 0.0078 day per year. This small interval (which is 11 minutes, 14 seconds) does not seem significant when compared to the time in a whole year. But over the years, the error of minutes and seconds grew to an error of days. By 1582, when Pope Gregory XIII revised the calendar, the error had grown to 13 days but was corrected for 10 days of error. This revision resulted in the *Gregorian calendar,* which is the system used today. Since the accumulated error of 0.0078 day per year is almost 0.75 day per century, it follows that four centuries will have 0.75 times 4, or 3 days of error. The Gregorian system corrects for the accumulated error by dropping the additional leap year day three centuries out of every four. Thus, the century year of 2000 was a leap year with 366 days, but the century years of 2100, 2200, and 2300 will not be leap years. You will note that this approximation still leaves an error of 0.0003 day per century, so another calendar revision will be necessary in a few thousand years to keep the calendar in step with the seasons.

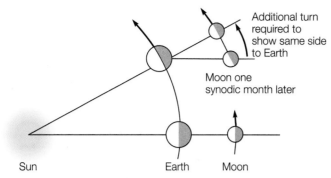

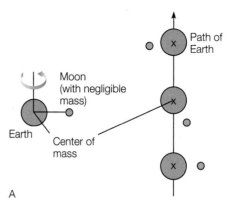

FIGURE 14.26 As the Moon moves in its orbit around Earth, it must revolve a greater distance to bring the same part to face Earth. The additional turning requires about 2.2 days, making the synodic month longer than the sidereal month.

Monthly Time

In ancient times, people often used the Moon to measure time intervals that were longer than a day but shorter than a year. The word *month*, in fact, has its origins in the word *moon* and its period of revolution. The Moon revolves around Earth in an orbit that is inclined to the plane of Earth's orbit, the plane of the ecliptic, by about 5°. The Moon is thus never more than about ten apparent diameters from the ecliptic. It revolves in this orbit in about 27⅓ days as measured by two consecutive crossings of any star. This period is called a **sidereal month.** The Moon rotates in the same period as the time of revolution, so the sidereal month is also the time required for one rotation. Because the rotation and revolution rates are the same, you always see the same side of the Moon from Earth.

The ancient concept of a month was based on the **synodic month,** the interval of time from new moon to new moon (or any two consecutive identical phases). The synodic month is longer than a sidereal month at a little more than 29½ days. The Moon's phases (see "Phases of the Moon" in this chapter) are determined by the relative positions of Earth, Moon, and Sun. As shown in figure 14.26, the Moon moves with Earth in its orbit around the Sun. During one sidereal month, the Moon has to revolve a greater distance before the same phase is observed on Earth, and this greater distance requires 2.2 days. This makes the synodic month about 29½ days long, only a little less than 1/12 of a year, or the period of time the present calendar identifies as a "month."

THE EARTH-MOON SYSTEM

Earth and its Moon are unique in the solar system because of the size of the Moon. It is not the largest satellite, but the ratio of its mass to Earth's mass is greater than the mass ratio of any other moon to its planet. The Moon has a diameter of 3,476 km (about 2,159 mi), which is about one-fourth the diameter of Earth, and a mass of about 1/81 of Earth's mass. This is a small fraction of Earth's mass, but it is enough to affect Earth's motion as it revolves around the Sun.

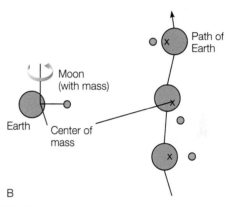

FIGURE 14.27 (*A*) If the Moon had a negligible mass, the center of gravity between the Moon and Earth would be Earth's center, and Earth would follow a smooth orbit around the Sun. (*B*) The actual location of the center of mass between Earth and Moon results in a slightly in and out, or wavy, path around the Sun.

If the Moon had a negligible mass, it would circle Earth with a center of rotation (center of mass) located at the center of Earth. In this situation, the center of Earth would follow a smooth path around the Sun (figure 14.27A). The mass of the Moon, however, is great enough to move the center of rotation away from Earth's center toward the Moon. As a result, both bodies act as a system, moving around a center of mass. The center of mass between Earth and the Moon follows a smooth orbit around the Sun. Earth follows a slightly wavy path around the Sun as it slowly revolves around the common center of mass (figure 14.27B).

Phases of the Moon

The phases of the Moon are a result of the changing relative positions of Earth, the Moon, and the Sun as the Earth-Moon system moves around the Sun. Sunlight always illuminates half of the Moon, and half is always in shadow. As the Moon's path takes it between Earth and the Sun, then to the dark side of Earth, you see different parts of the illuminated half called *phases* (figure 14.28). When the Moon is on the dark side of Earth, you see the entire illuminated half of the Moon called the **full moon** (or the full phase). Halfway around the orbit,

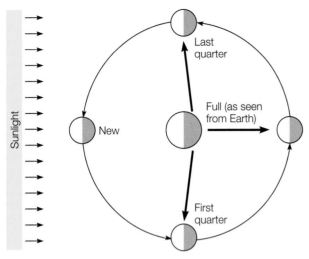

FIGURE 14.28 Half of the Moon is always lighted by the Sun, and half is always in the shadow. The Moon phases result from the view of the lighted and dark parts as the Moon revolves around Earth.

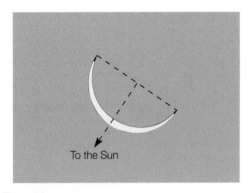

FIGURE 14.29 The cusps, or horns, of the Moon always point away from the Sun. A line drawn from the tip of one cusp to the other is perpendicular to a straight line between the Moon and the Sun.

the lighted side of the Moon now faces away from Earth, and the unlighted side now faces Earth. This dark appearance is called the **new moon** (or the new phase). In the new phase, the Moon is not *directly* between Earth and the Sun, so it does not produce an eclipse (see "Eclipses of the Sun and Moon" in this chapter).

As the Moon moves from the new phase in its orbit around Earth, you will eventually see half the lighted surface, which is known as the **first quarter.** Often the unlighted part of the Moon shines with a dim light of reflected sunlight from Earth called *earthshine.* Note that the division between the lighted and unlighted part of the Moon's surface is curved in an arc. A straight line connecting the ends of the arc is perpendicular to the direction of the Sun (figure 14.29). After the first quarter, the Moon moves to its full phase, then to the **last quarter** (see figure 14.28). The period of time between two consecutive phases, such as new moon to new moon, is the synodic month, or about 29.5 days.

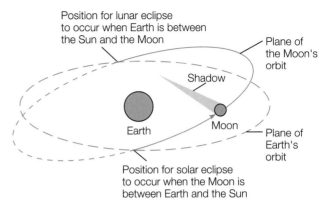

FIGURE 14.30 The plane of the Moon's orbit is inclined to the plane of Earth's orbit by about 5°. An eclipse occurs only where the two planes intersect and Earth, the Moon, and the Sun are in a line.

Eclipses of the Sun and Moon

Sunlight is not visible in the emptiness of space because there is nothing to reflect the light, so the long conical shadow behind each spherical body is not visible, either. One side of Earth and one side of the Moon are always visible because they reflect sunlight. The shadow from Earth or from the Moon becomes noticeable only when it falls on the illuminated surface of the other body. This event of Earth's or the Moon's shadow falling on the other body is called an **eclipse.** Most of the time, eclipses do not occur because the plane of the Moon's orbit is inclined to Earth's orbit about 5° (figure 14.30). As a result, the shadow from the Moon or the shadow from Earth usually falls above or below the other body, too high or too low to produce an eclipse. An eclipse occurs only when the Sun, Moon, and Earth are in a line with each other.

The shadow from Earth and the shadow from the Moon are long cones that point away from the Sun. Both cones have two parts, an inner cone of a complete shadow called the **umbra** and an outer cone of partial shadow called the **penumbra.** When and where the umbra of the Moon's shadow falls on Earth, people see a **total solar eclipse.** During a total solar eclipse, the Moon completely covers the disk of the Sun. The total solar eclipse is preceded and followed by a partial eclipse, which is seen when the observer is in the penumbra. If the observer is in a location where only the penumbra passes, then only a partial eclipse will be observed (figure 14.31). More people see partial than full solar eclipses because the penumbra covers a larger area. The occurrence of a total solar eclipse is a rare event in a given location, occurring once every several hundred years and then lasting for less than seven minutes.

The Moon's cone-shaped shadow averages a length of 375,000 km (about 233,000 mi), which is less than the average distance between Earth and the Moon. The Moon's elliptical orbit brings it sometimes closer to and sometimes farther from Earth. A total solar eclipse occurs only when the Moon is close enough so at least the tip of its umbra reaches the surface of Earth. If the Moon's umbra fails to reach Earth, an **annular**

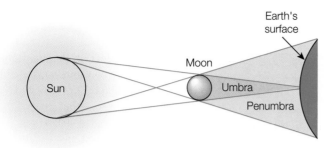

FIGURE 14.31 People in a location where the tip of the umbra falls on the surface of Earth see a total solar eclipse. People in locations where the penumbra falls on Earth's surface see a partial solar eclipse.

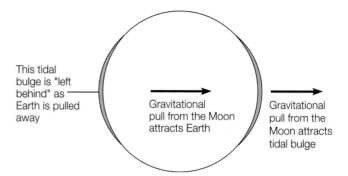

FIGURE 14.32 Gravitational attraction pulls on Earth's waters on the side of Earth facing the Moon, producing a tidal bulge. A second tidal bulge on the side of Earth opposite the Moon is produced when Earth, which is closer to the Moon, is pulled away from the waters.

eclipse occurs. Annular means "ring-shaped," and during this eclipse, the edge of the Sun is seen to form a bright ring around the Moon. As before, people located in the area where the penumbra falls will see a partial eclipse. The annular eclipse occurs more frequently than the total solar eclipse.

When the Moon is full and the Sun, Moon, and Earth are lined up so Earth's shadow falls on the Moon, a **lunar eclipse** occurs. Earth's shadow is much larger than the Moon's diameter, so a lunar eclipse is visible to everyone on the night side of Earth. This larger shadow also means a longer eclipse that may last for hours. As the umbra moves over the Moon, the darkened part takes on a reddish, somewhat copper-colored glow from light refracted and scattered into the umbra by Earth's atmosphere. This light passes through the thickness of Earth's atmosphere on its way to the eclipsed Moon, and it acquires the reddish color for the same reason that a sunset is red: much of the blue light has been removed by scattering in Earth's atmosphere.

Tides

If you live near or have ever visited a coastal area of the ocean, you are familiar with the periodic rise and fall of sea level known as **tides.** The relationship between the motions of the Moon and the magnitude and timing of tides has been known and studied since very early times. These relationships are that (1) the greatest range of tides occurs at full and new moon phases, (2) the least range of tides occurs at quarter moon phases, and (3) in most oceans, the time between two high tides or between two low tides is an average of twelve hours and twenty-five minutes. The period of twelve hours and twenty-five minutes is half the average time interval between consecutive passages of the Moon across the celestial meridian. A location on the surface of Earth is directly under the Moon when it crosses the meridian, and directly opposite it on the far side of Earth an average twelve hours and twenty-five minutes later. There are two *tidal bulges* that follow the Moon as it moves around Earth, one on the side facing the Moon and one on the opposite side. In general, tides are a result of these bulges moving westward around Earth.

A simplified explanation of the two tidal bulges involves two basic factors: the gravitational attraction of the Moon and the motion of the Earth-Moon system (figure 14.32). Water on Earth's surface is free to move, and the Moon's gravitational attraction pulls the water to the tidal bulge on the side of Earth facing the Moon. This tide-raising force directed toward the Moon bulges the water in mid-ocean some .75 m (about 2.5 ft), but it also bulges the land, producing a land tide. Since Earth is much more rigid than water, the land tide is much smaller at about 12 cm (about 4.5 in). Since all parts of the land bulge together, this movement is not evident without measurement by sensitive instruments.

The tidal bulge on the side of Earth opposite the Moon occurs as Earth is pulled away from the ocean by the Earth-Moon gravitational interaction. Between the tidal bulge facing the Moon and the tidal bulge on the opposite side, sea level is depressed across the broad surface. The depression is called a *tidal trough,* even though it does not actually have the shape of a trough. The two tidal bulges, with the trough between, move slowly eastward following the Moon. Earth turns more rapidly on its axis, however, which forces the tidal bulge to stay in front of the Moon moving through its orbit. Thus, the tidal axis is not aligned with the Earth-Moon gravitational axis.

The tides do not actually appear as alternating bulges that move around Earth. There are a number of factors that influence the making and moving of the bulges in complex interactions that determine the timing and size of a tide at a given time in a given location. Some of these factors include (1) the relative positions of Earth, Moon, and Sun, (2) the elliptical orbit of the Moon, which sometimes brings it closer to Earth, and (3) the size, shape, and depth of the basin holding the water.

The relative positions of Earth, Moon, and Sun determine the size of a given tide because the Sun, as well as the Moon, produces a tide-raising force. The Sun is much more massive than the Moon, but it is so far away that its tide-raising force is about half that of the closer Moon. Thus, the Sun basically

Carl Edward Sagan (1934–1996)

Carl Edward Sagan was an American astronomer and popularizer of astronomy whose main research was on planetary atmospheres, including that of the primordial Earth. His most remarkable achievement was to provide valuable insights into the origin of life on our planet.

Sagan was born on November 19, 1934, in New York City. Completing his education at the University of Chicago, he obtained his bachelor's degree in 1955 and his doctorate in 1960. Then, for two years, he was a research fellow at the University of California in Berkeley before he transferred to the Smithsonian Astrophysical Observatory in Cambridge, Massachusetts, lecturing also at Harvard University, where he became assistant professor. Finally, in 1968, Sagan moved to Cornell University in Ithaca (New York) to become director of the Laboratory for Planetary Studies; in 1970, he became professor of astronomy and space science there. He died on December 20, 1996.

The editor of the astronomical journal *Icarus*, Sagan wrote a number of popular books including *Broca's Brain: Reflections on the Romance of Science* (1979); *Cosmos* (1980), based on his television series of that name; and a science-fiction novel, *Contact*.

In the early 1960s, Sagan's first major research was into the planetary surface and atmosphere of Venus. At the time, although intense emission of radiation had shown that the dark-side temperature of Venus was nearly 600K, it was thought that the surface itself remained relatively cool—leaving open the possibility that there was some form of life on the planet. Various hypotheses were put forward to account for the strong emission actually observed: perhaps it was due to interactions between charged particles in Venus' dense upper atmosphere; perhaps it was glow discharge between positive and negative charges in the atmosphere; or perhaps emission was due to a particular radiation from charged particles trapped in the Venusian equivalent of a Van Allen Belt. Sagan showed that each of these hypotheses was incompatible with other observed characteristics or with implications of these characteristics. The positive part of Sagan's proposal was to show that all the observed characteristics were compatible with the straightforward hypothesis that the surface of Venus was very hot. On the basis of radar and optical observations, the distance between surface and clouds was calculated to be between 44 km (27 mi) and 65 km (40 mi); given the cloud-top temperature and Sagan's expectation

of a "greenhouse effect" in the atmosphere, surface temperature on Venus was computed to be between 500K (227°C/440°F) and 800K (527°C/980°F)—the range that would also be expected on the basis of emission rate.

Sagan then turned his attention to the early planetary atmosphere of Earth, with regard to the origins of life. One way of understanding how life began is to try to form the compounds essential to life in conditions analogous to those of the primeval atmosphere. Before Sagan, Stanley Miller and Harold Urey had used a mixture of methane, ammonia, water vapor, and hydrogen, sparked by a corona discharge that simulated the effect of lightning, to produce amino and hydroxy acids of the sort found in life-forms. Later experiments used ultraviolet light or heat as sources of energy, and even these had less energy than would have been available in Earth's primordial state. Sagan followed a similar method and, by irradiating a mixture of methane, ammonia, water, and hydrogen sulfide, was able to produce amino acids—and, in addition, glucose, fructose, and nucleic acids. Sugars can be made from formaldehyde under alkaline conditions and in the presence of inorganic catalysts. These sugars include five-carbon sugars, which are essential to the formation of nucleic acids, glucose, and fructose—all common metabolites found as constituents of present-day life-forms. Sagan's simulated primordial atmosphere not only showed the presence of those metabolites, it also contained traces of adenosine triphosphate (ATP)—the foremost agent used by living cells to store energy.

In 1966, in work done jointly with Pollack and Goldstein, Sagan was able to provide evidence supporting a hypothesis about Mars put forward by Wells, who observed that in regions on Mars where there were both dark and light areas, the clouds formed over the lighter areas aligned with boundaries of adjacent dark areas. Wells suggested that they were lee clouds formed by the Martian wind as it crossed dark areas. The implication, that dark areas mark the presence of ridges, was given support by Sagan's finding that dark areas had a high radar reflectivity that was slightly displaced in longitude. Sagan concluded that these dark areas were elevated areas with ridges of about 10 km (6 mi) and low slopes extending over long distances.

Source: Modified from the *Hutchinson Dictionary of Scientific Biography.* © RM, 2007. Reprinted by permission.

---

modifies lunar tides rather than producing distinct tides of its own. For example, Earth, Moon, and Sun are nearly in line during the full and new moon phases. At these times, the lunar and solar tide-producing forces act together, producing tides that are unusually high and corresponding low tides that are unusually low. The unusually high and low tides are called **spring tides** (figure 14.33A). Spring tides occur every two weeks and

have nothing to do with the spring season. When the Moon is in its quarter phases, the Sun and Moon are at right angles to one another and the solar tides occur between the lunar tides, causing unusually less pronounced high and low tides called **neap tides** (figure 14.33B). Neap tides also occur every two weeks.

The size of the lunar-produced tidal bulge varies as the Moon's distance from Earth changes. The Moon's elliptical

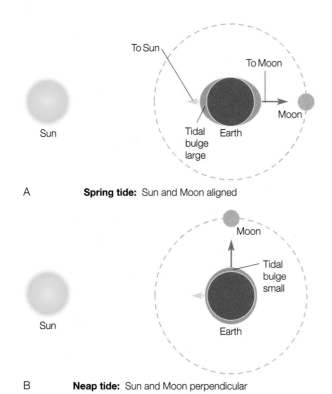

A **Spring tide:** Sun and Moon aligned

B **Neap tide:** Sun and Moon perpendicular

FIGURE 14.33 (*A*) When the Sun and Moon lie in the same or in opposite directions as seen from Earth, the tidal bulges they cause coincide, producing a large *spring* tide. (*B*) When the directions to the Sun and Moon are perpendicular, the tides of the Sun and Moon cancel to some extent, producing a weak *neap* tide.

orbit brings it closest to Earth at a point called **perigee** and farthest from Earth at a point called **apogee**. At perigee, the Moon is about 44,800 km (about 28,000 mi) closer to Earth than at apogee, so its gravitational attraction is much greater. When perigee coincides with a new or full moon, especially high spring tides result.

The open basins of oceans, gulfs, and bays are all connected but have different shapes and sizes and have bordering landmasses in all possible orientations to the westward-moving tidal bulges. Water in each basin responds differently to the tidal forces, responding as periodic resonant oscillations that move back and forth much like the water in a bowl shifts when carried. Thus, coastal regions on open seas may experience tides that range between about 1 and 3 m (about 3 to 10 ft), but mostly enclosed basins such as the Gulf of Mexico have tides less than about 1/3 m (about 1 ft). The Gulf of Mexico, because of its size, depth, and limited connections with the open ocean, responds only to the stronger tidal attractions and has only one high and one low tide per day. Even lakes and ponds respond to tidal attractions, but the result is too small to be noticed. Other basins, such as the Bay of Fundy in Nova Scotia, are funnel-shaped and undergo an unusually high tidal range. The Bay of Fundy has experienced as much as a 15 m (about 50 ft) tidal range.

As the tidal bulges are pulled against a rotating Earth, friction between the moving water and the ocean basin tends to slow Earth's rotation over time. This is a very small slowing effect that is increasing the length of each day by about a second per year.* Evidence for this slowing comes from a number of sources, including records of ancient solar eclipses. The solar eclipses of two thousand years ago occurred three hours earlier than would be expected by using today's time but were on the mark if a lengthening day is considered. Fossils of a certain species of coral still living today provide further evidence of a lengthening day. This particular coral adds daily growth rings, and five-hundred-million-year-old fossils show that the day was about twenty-one hours long at that time. Finally, the Moon is moving away from Earth at a rate of about 4 cm (about 1.5 in) per year. This movement out to a larger orbit is a necessary condition to conserve angular momentum as Earth slows. As the Moon moves away from Earth, the length of the month increases. Some time in the distant future, both the day and the month will be equal, about fifty of the present days long.

SUMMARY

Earth is an *oblate spheroid* that undergoes three basic motions: (1) a yearly *revolution* around the Sun, (2) a daily *rotation* on its axis, and (3) a slow wobble of its axis called *precession*.

As Earth makes its yearly *revolution* around the Sun, it maintains a generally *constant inclination of its axis* to the *plane of the ecliptic* of 66.5°, or 23.5° from a line perpendicular to the plane. In addition, Earth maintains a generally constant *orientation of its axis* to the stars, which always points in the same direction. The constant inclination and orientation of the axis, together with Earth's rotation and revolution, produce three effects: (1) days and nights that vary in length, (2) seasons that change during the course of a year, and (3) climates that vary with latitude. When Earth is at a place in its orbit so the axis points toward the Sun, the Northern Hemisphere experiences the longest days and the summer season. This begins on June 21 or 22, which is called the *summer solstice*. Six months later, the axis points

away from the Sun and the Northern Hemisphere experiences the shortest days and the winter season. This begins on December 22 or 23 and is called the *winter solstice*. On March 20 or 21, Earth is halfway between the solstices and has days and nights of equal length, which is called the *spring* (or *vernal*) *equinox*. On September 22 or 23, the *autumnal equinox*, another period of equal nights and days, identifies the beginning of the fall season.

Precession is a slow wobbling of the axis as Earth spins. Precession is produced by the gravitational tugs of the Sun and Moon on Earth's equatorial bulge.

Lines around Earth that are parallel to the equator are circles called *parallels*. The distance from the equator to a point on a parallel is called the *latitude* of that point. North and south arcs that intersect

*tycho.usno.navy.mil/leapsec.html

at the poles are called *meridians*. The meridian that runs through the Greenwich Observatory is a reference line called the *prime meridian*. The distance of a point east or west of the prime meridian is called the *longitude* of that point.

The event of time called *noon* is the instant the Sun appears to move across the celestial meridian. The instant of noon at a particular location is called the *apparent local noon*. The time at a given place that is determined by a sundial is called the *apparent local solar time*. It is the basis for an averaged, uniform standard time called the *mean solar time*. Mean solar time is the time used to set clocks.

A *sidereal day* is the interval between two consecutive crossings of the celestial meridian by a star. An *apparent solar day* is the interval between two consecutive crossings of the celestial meridian by the Sun, from one apparent solar noon to the next. A *mean solar day* is twenty-four hours as determined from mean solar time. The *equation of time* shows how the local solar time is faster or slower than the clock time during different days of the year.

Earth's surface is divided into one-hour *standard time zones* that are about 15° of meridian wide. The *international date line* is the 180° meridian; you gain a day if you cross this line while traveling westward and repeat a day if you are traveling eastward.

A *tropical year* is the interval between two consecutive spring equinoxes. A *sidereal year* is the interval of time between two consecutive crossings of a star by the Sun. It is the tropical year that is used as a standard time interval for the calendar year. A *sidereal month* is the interval of time between two consecutive crossings of a star by the Moon. The *synodic month* is the interval of time from a new moon to the next new moon. The synodic month is about 29 1/2 days long, which is about 1/12 of a year.

Earth and the Moon act as a system, with both bodies revolving around a common center of mass located under Earth's surface. This combined motion around the Sun produces three phenomena: (1) as the Earth-Moon system revolves around the Sun, different parts of the illuminated lunar surface, called *phases*, are visible from Earth; (2) a *solar eclipse* is observed where the Moon's shadow falls on Earth, and a *lunar eclipse* is observed where Earth's shadow falls on the Moon; and (3) the *tides*, a periodic rising and falling of sea level, are produced by gravitational attractions of the Moon and Sun and by the movement of the Earth-Moon system.

KEY TERMS

annular eclipse (p. **326**)
Antarctic Circle (p. **319**)
apogee (p. **329**)
apparent local noon (p. **320**)
apparent local solar time (p. **320**)
apparent solar day (p. **320**)
Arctic Circle (p. **319**)
autumnal equinox (p. **316**)
Coriolis effect (p. **317**)
daylight saving time (p. **322**)
eclipse (p. **326**)
equation of time (p. **321**)
equinoxes (p. **316**)

first quarter (p. **326**)
full moon (p. **325**)
international date line (p. **323**)
last quarter (p. **326**)
latitude (p. **318**)
longitude (p. **319**)
lunar eclipse (p. **327**)
mean solar day (p. **321**)
mean solar time (p. **320**)
meridians (p. **319**)
neap tides (p. **328**)
new moon (p. **326**)
noon (p. **320**)

parallels (p. **318**)
penumbra (p. **326**)
perigee (p. **329**)
plane of the ecliptic (p. **312**)
precession (p. **317**)
prime meridian (p. **319**)
revolution (p. **315**)
rotation (p. **315**)
sidereal day (p. **320**)
sidereal month (p. **325**)
sidereal year (p. **324**)
solstices (p. **315**)
spring equinox (p. **316**)

spring tides (p. **328**)
standard time zones (p. **322**)
summer solstice (p. **315**)
synodic month (p. **325**)
tides (p. **327**)
total solar eclipse (p. **326**)
tropic of Cancer (p. **319**)
tropic of Capricorn (p. **319**)
tropical year (p. **324**)
umbra (p. **326**)
vernal equinox (p. **316**)
winter solstice (p. **315**)

APPLYING THE CONCEPTS

1. If you are located at 20°N latitude, when will the Sun appear directly overhead?
   a. never
   b. once a year
   c. twice a year
   d. four times a year

2. If you are located on the equator (0° latitude), when will the Sun appear directly overhead?
   a. never
   b. once a year
   c. twice a year
   d. four times a year

3. If you are located at 40°N latitude, when will the Sun appear directly overhead?
   a. never
   b. once a year
   c. twice a year
   d. four times a year

4. In about twelve thousand years, the star Vega will be the North Star, not Polaris, because of Earth's
   a. uneven equinox.
   b. tilted axis.
   c. precession.
   d. recession.

5. The time as read from a sundial is the same as the time read from a clock
   a. all the time.
   b. only once a year.
   c. twice a year.
   d. four times a year.

6. You are traveling west by jet and cross three time zone boundaries. If your watch reads 3:00 P.M. when you arrive, you should reset it to
   a. 12:00 noon.
   b. 6:00 P.M.
   c. 12:00 midnight.
   d. 6:00 A.M.

7. If it is Sunday when you cross the international date line while traveling westward, the next day is
   a. Wednesday.
   b. Sunday.
   c. Tuesday.
   d. Saturday.

8. If you see a full moon, an astronaut on the Moon looking back at Earth at the same time would see a
   a. full Earth.
   b. new Earth.
   c. first quarter Earth.
   d. last quarter Earth.

9. A lunar eclipse can occur only during the moon phase of
   a. full moon.
   b. new moon.
   c. first quarter.
   d. last quarter.

10. A total solar eclipse can occur only during the moon phase of
    a. full moon.
    b. new moon.
    c. first quarter.
    d. last quarter.

11. A lunar eclipse does not occur every month because
    a. the plane of the Moon's orbit is inclined to the ecliptic.
    b. of precession.
    c. Earth moves faster in its orbit when closest to the Sun.
    d. Earth's axis is tilted with respect to the Sun.

12. The smallest range between high and low tides occurs during
    a. full moon.
    b. new moon.
    c. quarter moon phases.
    d. an eclipse.

Answers

1. c 2. c 3. a 4. c 5. d 6. a 7. c 8. b 9. a 10. b 11. a 12. c

QUESTIONS FOR THOUGHT

1. Use sketches with brief explanations to describe how the constant inclination and constant orientation of Earth's axis produce (a) a variation in the number of daylight hours and (b) a variation in seasons throughout a year.

2. Where on Earth are you if you observe the following at the instant of apparent local noon on September 23? (a) The shadow from a vertical stick points northward. (b) There is no shadow on a clear day. (c) The shadow from a vertical stick points southward.

3. Briefly describe how Earth's axis is used as a reference for a system that identifies locations on Earth's surface.

4. Use a map or a globe to identify the latitude and longitude of your present location.

5. The tropic of Cancer, tropic of Capricorn, Arctic Circle, and Antarctic Circle are parallels that are identified with specific names. What parallels do the names represent? What is the significance of each?

6. When it is 12:00 noon in Texas, what time is it (a) in Jacksonville, Florida? (b) in Bakersfield, California? (c) at the North Pole?

7. Explain why a lunar eclipse is not observed once a month.

8. Using sketches, briefly describe the positions of Earth, Moon, and Sun during each of the major moon phases.

9. What phase is the Moon in if it rises at sunset? Explain your reasoning.

10. Does an eclipse of the Sun occur during any particular moon phase? Explain.

FOR FURTHER ANALYSIS

1. What are the significant similarities and differences between a solstice and an equinox?

2. On what date is Earth the closest to the Sun? What season is occurring in the Northern Hemisphere at this time? Explain this apparent contradiction.

3. Explain why an eclipse of the Sun does not occur at each new moon phase when the Moon is between Earth and the Sun.

4. Explain why sundial time is often different than the time as shown by a clock.

5. Analyze why the time between two consecutive tides is twelve hours and twenty-five minutes rather than twelve hours. Explore many different explanations that you imagine, then select the best and explain how it could be tested.

INVITATION TO INQUIRY

Hello, Moon!

Observe where the Moon is located relative to the landscape where you live each day or night that it is visible. Make a sketch of the outline of the landscape. For each date, draw an accurate location of where the Moon is located and note the time. Also, sketch the moon phase for each date.

Continue your observations until you have enough information for analysis. Analyze the data for trends that would enable you to make predictions. For each day, draw some circles to represent the relative positions of Earth, the Moon, and the Sun.

These rose-red rhodochrosite crystals are a naturally occurring form of manganese carbonate. Rhodochrosite is but one of about twenty-five hundred minerals that are known to exist, making up the solid materials of Earth's crust.

Earth

► Changes in Earth's surface create barriers
that lead to the development of new
species (Ch. 21).

CORE CONCEPT

Earth is a dynamic body, which cycles materials on its surface and between its surface and interior.

Most of Earth's surface is composed of minerals made of oxygen and silicon (p. 334).

Igneous, sedimentary, and metamorphic rocks are continuously changed through cycles of one type of rock to another (p. 341).

Earth's surface is broken into a number of rigid plates that move (p. 347).

Earth Science Connections

► Changes in magnetic fields are used to determine the sequence of past geologic events (Ch. 22).

► Organisms trapped in sediments form fossils in sedimentary rock (Ch. 22).

OVERVIEW

The separation of the earth sciences into independent branches—such as geology, oceanography, and meteorology—was traditionally done for convenience. This made it easier to study a large and complex Earth. In the past, scientists in each branch studied their field without considering Earth as an interacting whole. Today, most earth scientists consider changes in Earth as taking place in an overall dynamic system. The parts of Earth's interior, the rocks on the surface, the oceans, the atmosphere, and the environmental conditions are today understood to be parts of a complex, interacting system with a cyclic movement of materials from one part to another.

How can materials cycle through changes from the interior to the surface and to the atmosphere and back? The answer to this question is found in the unique combination of fluids of Earth. No other known planet has Earth's combination of (1) an atmosphere consisting mostly of nitrogen and oxygen, (2) a surface that is mostly covered with liquid water, and (3) an interior that is partly fluid, partly semi-fluid, and partly solid. Earth's atmosphere is unique in terms of both its composition and interactions with the liquid water surface (figure 15.1). These interactions have cycled materials, such as carbon dioxide, from the atmosphere to the land and oceans of Earth. The internal flow of rock materials, on the other hand, produces the large-scale motion of Earth's continents and the associated phenomena of earthquakes and volcanoes. Volcanoes cycle carbon dioxide back into the atmosphere and the movement of land cycles rocks from Earth's interior to the surface and back to the interior again. Altogether, Earth's atmosphere, liquid water, and motion of its landmasses make up a dynamic cycling system that is found only on the planet Earth.

Earth also seems to be unique because there is life on Earth but apparently not on the other planets. The cycling of atmospheric gases and vapors, waters of the surface, and flowing interior rock materials sustain a wide diversity of life on Earth. There are millions of different species of plants, animals, and other kinds of organisms. Yet, there is no evidence of even one species of life existing outside Earth. The existence of life on Earth must be related to Earth's unique, dynamic system of interacting fluids. This chapter is concerned with earth materials, the internal structure of Earth, the movement of landmasses across the surface of Earth, and how earth materials are recycled.

EARTH MATERIALS

Earth, like all other solid matter in the universe, is made up of chemical elements. The different elements are not distributed equally throughout the mass of Earth, nor are they equally abundant. Chemical analysis of thousands of rocks from Earth's surface found that only eight elements make up about 98.6 percent of the crust. All the other elements make up the remaining 1.4 percent of the crust. Oxygen is the most abundant element, making up about 50 percent of the weight of the crust. Silicon makes up over 25 percent, so oxygen and silicon alone make up about 75 percent of Earth's solid surface.

Figure 15.2 shows the eight most abundant elements that occur as elements or combine to form the chemical compounds of Earth's crust. Earth was apparently molten during an early stage in its development, as evidenced by distribution of the elements. During the molten stage, the heavier abundant elements, such as iron and nickel, apparently sank to the deep interior of Earth, leaving a thin layer of lighter elements on the surface. This relatively thin layer is called the *crust*. The rocks and rock materials that you see on the surface and the materials sampled in the deepest mines and well holes are all materials of Earth's crust. The bulk of Earth's mass lies below the crust and has not been directly sampled.

FIGURE 15.1 No other planet in the solar system has the unique combination of fluids of Earth. Earth has a surface that is mostly covered with liquid water, water vapor in the atmosphere, and both frozen and liquid water on the land.

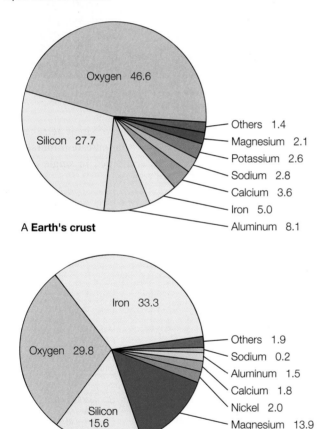

A **Earth's crust**

Oxygen 46.6
Silicon 27.7
Others 1.4
Magnesium 2.1
Potassium 2.6
Sodium 2.8
Calcium 3.6
Iron 5.0
Aluminum 8.1

Iron 33.3
Oxygen 29.8
Silicon 15.6
Others 1.9
Sodium 0.2
Aluminum 1.5
Calcium 1.8
Nickel 2.0
Magnesium 13.9

B **Whole Earth**

FIGURE 15.2 (A) The percentage by weight of the elements that make up Earth's crust. (B) The percentage by weight of the elements that make up the whole Earth.

Minerals

In everyday usage, the word *mineral* can have several different meanings. It can mean something your body should have (vitamins and minerals), something a fertilizer furnishes for a plant (nitrogen, potassium, and phosphorus), or sand, rock,

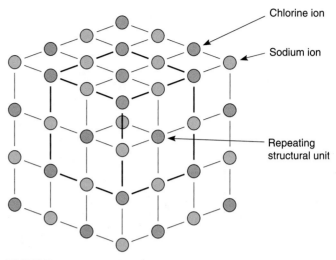

FIGURE 15.3 A crystal is composed of a structural unit that is repeated in three dimensions. This is the basic structural unit of a crystal of sodium chloride, the mineral halite.

and coal taken from Earth for human use (mineral resources). In the earth sciences, a **mineral** is defined as a naturally occurring, inorganic solid element or compound with a constant chemical composition in a crystalline structure (figure 15.3). This definition means that the element or compound cannot be synthetic (must be naturally occurring), cannot be made of organic molecules (see chapter 19), and must have atoms arranged in a regular, repeating pattern (a crystal structure). Note that the crystal structure of a mineral can be present on the microscopic scale and is not necessarily obvious to the unaided eye. Even crystals that could be observed with the unaided eye are sometimes not noticed (figure 15.4).

The crystal structure of a mineral can be made up of atoms of one or more kinds of elements. Diamond, for example, is a mineral with only carbon atoms in a strong crystal structure. Quartz, on the other hand, is a mineral with atoms of silicon and oxygen in a different crystal structure (figure 15.5). No matter how many kinds of atoms are present, each mineral has its own defined chemical composition or range of chemical compositions. A range of chemical composition is possible because the composition of some minerals can vary with the substitution of chemically similar elements. For example, some atoms of magnesium might be substituted for some chemically similar atoms of calcium. Such substitutions might slightly alter some properties but not enough to make a different mineral.

Silicon and oxygen are the most abundant elements in Earth's crust and, as you would expect, the most common minerals contain these two elements. All minerals are classified on the basis of whether the mineral structure contains these two elements or not. The two main groups are thus called the *silicates* and the *nonsilicates* (table 15.1). Note, however, that the silicates can contain some other elements in addition to silicon and oxygen. The silicate minerals are by far the most abundant, making up about 92 percent of Earth's crust. When an ion of silicon (Si^{-4}) combines with four oxygen ions (O^{-2}), a tetrahedral structure of $(SiO_4)^{-4}$ forms (see figure 15.6). All **silicates** have a basic silicon-oxygen tetrahedral unit either isolated or joined together in the crystal structure. The structure has a total

FIGURE 15.4 The structural unit for a crystal of table salt, sodium chloride, is cubic, as you can see in the individual grains.

FIGURE 15.5 These quartz crystals are hexagonal prisms.

of four unattached electrons on the oxygen atoms that can combine with metallic ions such as iron or magnesium. They can also combine with the silicon atoms of *other* tetrahedral units.

Some silicate minerals are thus made up of single tetrahedral units combined with metallic ions. Other silicate minerals are combinations of tetrahedral units combined in single chains, double chains, or sheets (figure 15.7).

TABLE 15.1   Classification Scheme of Some Common Minerals

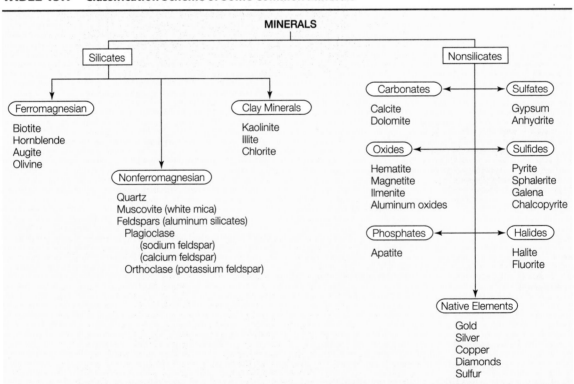

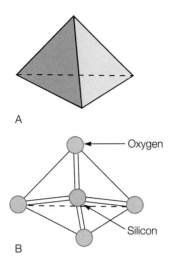

FIGURE 15.6 (A) The geometric shape of a tetrahedron with four equal sides. (B) A silicon and four oxygen atoms are arranged in the shape of a tetrahedron with the silicon in the center. This is the basic building block of all silicate minerals.

CONCEPTS APPLIED

Grow Your Own

Experiment with growing crystals from solutions of alum, copper sulfate, salt, or potassium permanganate. Write a procedure that will tell others what the important variables are for growing large, well-formed crystals.

The silicate minerals can be conveniently subdivided into two groups based on the presence of iron and magnesium. The basic tetrahedral structure joins with ions of iron, magnesium, calcium, and other elements in the *ferromagnesian silicates*. Examples of ferromagnesian silicates are *olivine, augite, hornblende,* and *biotite* (figure 15.8). They have a greater density and

a darker color than the other silicates because of the presence of the metal ions. Augite, hornblende, and biotite are very dark in color, practically black, and olivine is light green.

The *nonferromagnesian silicates* have a light color and a low density compared to the ferromagnesians. This group includes the minerals *muscovite (white mica)*, the *feldspars*, and *quartz* (figure 15.9).

The remaining 8 percent of minerals making up Earth's crust that do not have silicon-oxygen tetrahedrons in their crystal structure are called *nonsilicates*. There are eight subgroups of nonsilicates: (1) carbonates, (2) sulfates, (3) oxides, (4) sulfides, (5) halides, (6) phosphates, (7) hydroxides, and (8) native elements. Some common nonsilicates are identified in table 15.1. The carbonates are the most abundant of the nonsilicates, but others are important as fertilizers, sources of metals, and sources of industrial chemicals.

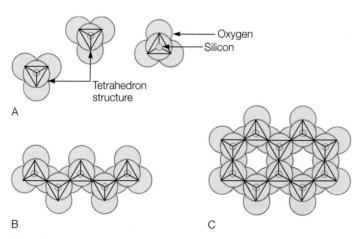
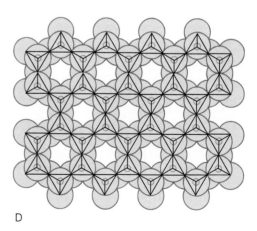

FIGURE 15.7 (A) Isolated silicon-oxygen tetrahedra do not share oxygens. This structure occurs in the mineral olivine. (B) Single chains of tetrahedra are formed by each silicon ion having two oxygens all to itself and sharing two with other silicons at the same time. This structure occurs in augite. (C) Double chains of tetrahedra are formed by silicon ions sharing either two or three oxygens. This structure occurs in hornblende. (D) The sheet structure in which each silicon shares three oxygens occurs in the micas, resulting in layers that pull off easily because of cleavage between the sheets.

Asbestos

Asbestos is a common name for any of several minerals that can be separated into fireproof fibers, fibers that will not melt or ignite. The fibers can be woven into a fireproof cloth, used directly as fireproof insulation material, or mixed with plaster or other materials. People now have a fear of all asbestos because it is presumed to be a health hazard (box figure 15.1). However, there are about six commercial varieties of asbestos. Five of these varieties are made from an amphibole mineral and are commercially called "brown" and "blue" asbestos. The other variety is made from *chrysotile*, a *serpentine* family of minerals, and is commercially called "white" asbestos. White asbestos is the asbestos mined and most commonly used in North America. It is only the amphibole asbestos (brown and blue asbestos) that has been linked to cancer, even for a short exposure time. There is, however, no evidence that exposure to white asbestos results in an increased health hazard. It makes sense to ban the use of and remove all the existing amphibole asbestos from public buildings. It does not make sense to ban or remove the serpentine asbestos since it is not a proven health hazard.

Box Figure 15.1 Did you know there are different kinds of asbestos? Are all kinds of asbestos a health hazard?

FIGURE 15.8 Compare the dark colors of the ferromagnesian silicates augite (right), hornblende (left), and biotite to the light-colored nonferromagnesian silicates in figure 15.9.

FIGURE 15.9 Compare the light colors of the nonferromagnesian silicates mica (front center), white and pink orthoclase (top and center), and quartz (left center) to the dark-colored ferromagnesian silicates in figure 15.8.

Using Mineral Resources

Most people understand the fact that our mineral resources are limited, and when we use them, they are gone. Of course, some mineral resources can be recycled, reducing the need to mine increasingly more minerals. For example, aluminum can be recycled and used over and over again repeatedly. Glass, copper, iron, and other metals can similarly be recycled repeatedly. Other critical resources, however, cannot be recycled and, unfortunately, cannot be replaced. Crude oil, for example, is a dwindling resource that will eventually become depleted. Oil is not recyclable once it is burned, and no new supplies are being created, at least not at a rate that would make them available in the immediate future. Even if Earth were a hollow vessel completely filled with oil, it would eventually become depleted, perhaps sooner than you might think.

There is also another of our mineral resources that is critically needed for our survival but will probably soon be depleted. That resource is phosphorus derived from phosphate rock. Phosphorus is an essential nutrient required for plant growth, and if its concentration in soils is too low, plants grow poorly, if at all. Most agricultural soils are artificially fertilized with phosphate. Without this amendment, their productivity would decline and, in some cases, cease altogether.

Phosphate occurs naturally as the mineral apatite. Deposits of apatite were formed where ocean currents carried cold, deep water rich in dissolved phosphate ions to the upper continental slope and outer continental shelf. Here, phosphate ions replaced the carbonate ions in limestone, forming the mineral apatite. Apatite also occurs as a minor accessory mineral in most igneous, sedimentary, and metamorphic rock types. Some igneous rocks serve as a source of phosphate fertilizer, but most phosphate is mined from formerly submerged coastal areas of limestone, such as those found in Florida.

Trends in phosphate production and use over the past forty years suggest that the proved world reserves of phosphate rock will be exhausted within a few decades. New sources might be discovered, but some time soon, phosphate rock will be mined out and no longer available. When this happens, the food supply will have to be grown on lands that are naturally endowed with an adequate phosphate supply. Estimates are that this worldwide existing land area with adequate phosphate supply will supply food for only two billion people on the entire Earth. Phosphate is an essential element for all life on Earth, and no other element can function in its place.

Questions to Discuss:

Discuss with your group the following questions concerning the use of mineral resources:

1. Should the mining industry be permitted to exhaust an important mineral resource? Provide reasons with your answer.
2. What are the advantages and disadvantages of a controlled mining industry?
3. If phosphate minerals supplies become exhausted, who should be responsible for developing new supplies or substitutes: the mining industry or governments?

Rocks

Elements are *chemically* combined to make minerals. Minerals are *physically* combined to make rocks. A **rock** is defined as an aggregation of one or more minerals and perhaps other materials that have been brought together into a cohesive solid. These materials include volcanic glass, a silicate that is not considered a mineral because it lacks a crystalline structure. Thus, a rock can consist of one or more kinds of minerals that are somewhat "glued" together by other materials such as glass. Most rocks are composed of silicate minerals, as you might expect since most minerals are silicates.

There is a classification scheme that is based on the way the rocks were formed. There are three main groups: (1) *igneous rocks* formed as a hot, molten mass of rock materials cooled and solidified; (2) *sedimentary rocks* formed from particles or dissolved materials from previously existing rocks; and (3) *metamorphic rocks* formed from rocks that were subjected to high temperatures and pressures that deformed or recrystallized the rock without complete melting.

Igneous Rocks

The word *igneous* comes from the Latin *ignis*, which means "fire." This is an appropriate name for **igneous rocks** that are defined as rocks that formed from a hot, molten mass of melted rock materials. The first step in forming igneous rocks is the creation of a very high temperature that is hot enough to melt rocks. A mass of melted rock materials is called a *magma*. A magma may cool and crystallize to solid igneous rock either below or on the surface of Earth. Earth has a history of beginning as a molten material, all rocks of Earth were at one time igneous rocks. Today, about two-thirds of the outer layer, or crust, is made up of igneous rocks. This is not apparent in many locations because the surface is covered by other kinds of rocks and rock materials (sand, soil, etc.). When they are found, igneous rocks are usually granite or basalt. *Granite* is a light-colored igneous rock that is primarily made up of three silicate minerals: quartz, mica, and feldspar. You can see the grains of these three minerals in a freshly broken surface of most samples of granite (figure 15.10). *Basalt* is a dark, more dense igneous rock with mineral grains too tiny to see. Basalt is the most common volcanic rock found on the surface of Earth.

Sedimentary Rocks

Sedimentary rocks are rocks that formed from particles or dissolved materials from previously existing rocks. Chemical reactions with air and water tend to break down and dissolve the less chemically stable parts of existing rocks, freeing more stable

FIGURE 15.10 Granite is a coarse-grained igneous rock composed mostly of light-colored, light-density, nonferromagnesian minerals. Earth's continental areas are dominated by granite and by rocks with the same mineral composition of granite.

TABLE 15.2 Classification Scheme for Sedimentary Rocks

Sediment Type	Particle or Composition	Rock
Fragment	Larger than sand	Conglomerate or breccia
Fragment	Sand	Sandstone
Fragment	Silt and clay	Siltstone, claystone, or shale
Chemical	Calcite	Limestone
Chemical	Dolomite	Dolomite
Chemical	Gypsum	Gypsum
Chemical	Halite (sodium chloride)	Salt

CONCEPTS APPLIED

Collect Your Own

Make a collection of rocks and minerals that can be found in your location, showing the name and location found for each. What will determine the total number of rocks and minerals it is possible to collect in your particular location?

particles and grains in the process. The remaining particles are usually transported by moving water and deposited as sediments. *Sediments* are accumulations of silt, clay, sand, or other weathered materials. Weathered rock fragments and dissolved rock materials both contribute to sediment deposits (table 15.2).

Most sediments are deposited as many separate particles that accumulate as loose materials. Such accumulations of rock fragments, chemical deposits, or animal shells must become consolidated into a solid, coherent mass to become sedimentary rock. There are two main parts to this *lithification*, or rock-forming process: (1) compaction and (2) cementation (figure 15.11).

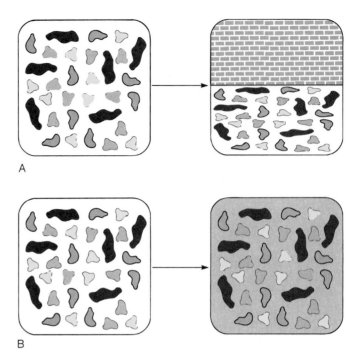

A

B

FIGURE 15.11 (*A*) In compaction, the sediment grains are packed more tightly together, often by overlying sediments, as represented by the bricks. (*B*) In cementation, dissolved minerals from fluids are precipitated in the space between the grains, cementing them together into a rigid, solid mass.

The weight of an increasing depth of overlying sediments causes an increasing pressure on the sediments below. This pressure squeezes the deeper sediments together, gradually reducing the pore space between the individual grains. This *compaction* of the grains reduces the thickness of a sediment deposit, squeezing out water as the grains are packed more tightly together. Compaction alone is usually not enough to make loose sediment into solid rock. Cementation is needed to hold the compacted grains together.

In *cementation,* the spaces between the sediment particles are filled with a chemical deposit. The chemical deposit binds the particles together into the rigid, cohesive mass of a sedimentary rock. Compaction and cementation may occur at the same time, but the cementing agent must have been introduced before compaction restricts the movement of the fluid through the open spaces. Many soluble minerals can serve as cementing agents, and calcite (calcium carbonate) and silica (silicon dioxide) are common.

Sediments accumulate from rocks that are in various stages of being broken down, so there is a wide range of sizes of particles in sediments. The size and shape of the particles are used as criteria to name the sedimentary rocks (table 15.3).

Metamorphic Rocks

The third group of rocks is called metamorphic. **Metamorphic rocks** are previously existing rocks that have been changed by heat, pressure, or hot solutions into a distinctly different rock. The heat, pressure, or hot solutions that produced the changes are associated with geologic events of (1) movement of the crust, which will be discussed in chapter 16, and with (2) heating

TABLE 15.3 Simplified Classification Scheme for Sediment Fragments and Rocks

Sediment Name	Size Range	Rock
Boulder	Over 256 mm (10 in)	
Gravel	2 to 256 mm (0.08–10 in)	Conglomerate or breccia*
Sand	1/16 to 2 mm (0.025–0.08 in)	Sandstone
Silt (or dust)	1/256 to 1/16 mm (0.00015–0.025 in)	Siltstone**
Clay (or dust)	Less than 1/256 mm (less than 0.00015 in)	Claystone**

*Conglomerate has a rounded fragment; breccia has an angular fragment.
**Both are also known as mudstone; called shale if it splits along parallel planes.

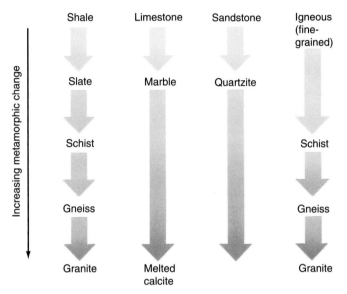

FIGURE 15.12 Increasing metamorphic change occurs with increasing temperatures and pressures. If the melting point is reached, the change is no longer metamorphic, and igneous rocks are formed.

CONCEPTS APPLIED

Sand Sort

Collect dry sand from several different locations. Use a magnifying glass to determine the minerals found in each sample.

---

and hot solutions from magma. Pressures from movement of the crust can change the rock texture by flattening, deforming, or realigning mineral grains. Temperatures from magma must be just right to produce a metamorphic rock. They must be high enough to disrupt the crystal structures to cause them to recrystallize but not high enough to melt the rocks and form igneous rocks (figure 15.12).

The Rock Cycle

Earth is a dynamic planet with a constantly changing surface and interior. As you will see in chapters 16, 17, and 18, internal

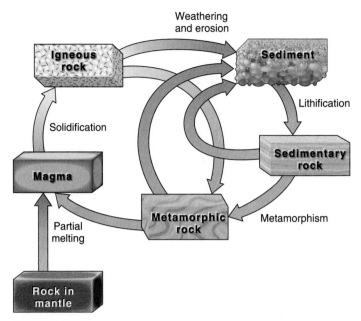

FIGURE 15.13 The rock cycle.

changes alter Earth's surface by moving the continents and, for example, building mountains that are eventually worn away by weathering and erosion. Seas advance and retreat over the continents as materials are cycled from the atmosphere to the land and from the surface to the interior of Earth and then back again. Rocks are transformed from one type to another through this continual change. There is not a single rock on Earth's surface today that has remained unchanged through Earth's long history. The concept of continually changing rocks through time is called the *rock cycle* (figure 15.13). The rock cycle concept views an igneous, a sedimentary, or a metamorphic rock as the present but temporary stage in the ongoing transformation of rocks to new types. Any particular rock sample today has gone through countless transformations in the 4.6-billion-year history of Earth and will continue to do so in the future.

Chapter 22, The History of Life on Earth, is about how rocks are studied to find information about the history of Earth and the organisms on it. This is one of the main reasons that geologists are interested in rocks. For example, sedimentary rocks can be studied to learn about how the environment and the kinds of organisms found on Earth have changed over time.

EARTH'S INTERIOR

Earth is not a solid, stable sphere that is "solid as a rock." All rocks and rock materials can be made to flow, behaving as warm wax or putty that can be molded. Pressure and temperature just a few kilometers below the surface can be high enough so that rock flows very slowly. Rock flowage takes place as hot, buoyant material deep within Earth moves slowly upward toward the cooler surface. Elsewhere, cold, denser material moves downward.

Earth's internal heat and rock movement are related to some things that happen on the surface, such as moving continents, earthquakes, and volcanoes. Before considering how this happens, we will look at Earth's interior and some internal structures. Information about Earth's interior is deducted mostly from a study of earthquake waves as they move through different parts of Earth. Earthquake waves furnish a sort of "X-ray" machine to view Earth's interior.

The sudden movement of very large sections of rock can produce vibrations that move out as waves throughout the entire Earth. These vibrations are called *seismic waves.* Strong seismic waves are what people feel as a shaking, quaking, or vibrating during an earthquake. Seismic waves are generated when a huge mass of rock breaks and slides into a different position.

Seismic waves radiate outward from an earthquake, spreading in all directions through the solid Earth's interior like sound waves from an explosion. There are basically three kinds of waves:

1. A longitudinal (compressional) wave called a *P-wave.* P-waves are the fastest and move through surface rocks and solid and liquid materials below the surface.
2. A transverse (shear) wave called an *S-wave.* S-waves are second fastest after the P-wave. S-waves do not travel through liquids since liquids do not have the cohesion necessary to transmit a shear, or side-to-side, motion.
3. An up-and-down (crest and trough) wave that travels across the surface called a *surface wave* that is much like a wave on water that moves across the solid surface of Earth. Surface waves are the slowest and occur where S- or P-waves reach the surface. There are two important types of surface waves: *Love waves* and *Rayleigh waves.* Love waves are horizontal S-waves that move side to side. This motion knocks buildings off their foundations and can also destroy bridges and overpasses. Rayleigh waves are more like rolling water waves. Rayleigh waves are more destructive because they produce more up, down, and sideways ground movement for a longer time.

Using data from seismic waves that move through Earth, scientists were able to plot three main parts of the interior (figure 15.14). The *crust* is the outer layer of rock that forms a thin shell around Earth. Below the crust is the *mantle,* a much thicker shell than the crust. The mantle separates the crust from the center part, which is called the *core.* The following section starts on Earth's surface, at the crust, and then digs deeper and deeper into Earth's interior.

The Crust

Earth's **crust** is a thin layer of rock that covers the entire Earth, existing below the oceans as well as making up the continents. According to seismic waves, there are differences in the crust making up the continents and the crust beneath the oceans (table 15.4). These differences are (1) the oceanic crust is much thinner than the continental crust and (2) seismic waves move through the oceanic crust faster than they do through the continental crust. The two types of crust vary because they are made up of different kinds of rock.

The boundary between the crust and the mantle is marked by a sharp increase in the velocity of seismic waves as they pass

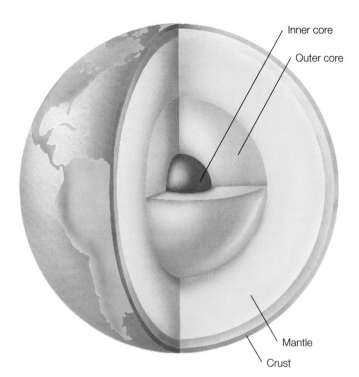

FIGURE 15.14 The structure of Earth's interior.

TABLE 15.4   Comparison of Oceanic Crust and Continental Crust

	Oceanic Crust	**Continental Crust**
Age	Less than 200 million years	Up to 3.8 billion years old
Thickness	5 to 8 km (3 to 5 mi)	10 to 70 km (6 to 45 mi)
Density	3.0 g/cm³	2.7 g/cm³
Composition	Basalt	Granite, schist, gneiss

from the crust to the mantle. Today, this boundary is called the *Mohorovicic discontinuity,* or the "Moho" for short. The boundary is a zone where seismic P-waves increase in velocity because of changes in the composition of the materials. The increase occurs because the composition on both sides of the boundary is different. The mantle is richer in ferromagnesian minerals and poorer in silicon than the crust.

Studies of the Moho show that the crust varies in thickness around Earth's surface. It is thicker under the continents and much thinner under the oceans.

The age of rock samples from the continent has been compared with the age samples of rocks taken from the seafloor by oceanographic ships. This sampling has found the continental crust to be much older, with parts up to 3.8 billion years old. By comparison, the oldest oceanic crust is less than 200 million years old.

Comparative sampling also found that continental crust is a less dense, granite-type rock with a density of about 2.7 g/cm³. Oceanic crust, on the other hand, is made up of basaltic rock with a density of about 3.0 g/cm³. The less dense crust behaves as if it were floating on the mantle, much as less dense ice floats

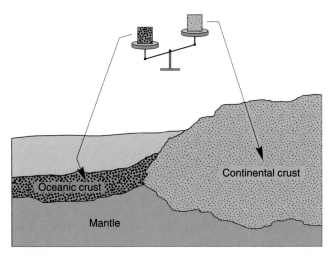

FIGURE 15.15 Continental crust is less dense, granite-type rock, while the oceanic crust is more dense, basaltic rock. Both types of crust behave as if they were floating on the mantle, which is more dense than either type of crust.

on water. There are explainable exceptions, but in general, the thicker, less dense continental crust "floats" in the mantle above sea level and the thin, dense oceanic crust "floats" in the mantle far below sea level (figure 15.15).

The Mantle

The middle part of Earth's interior is called the **mantle.** The mantle is a thick shell between the core and the crust. This shell takes up about 80 percent of the total volume of Earth and accounts for about two-thirds of Earth's total mass. Information about the composition and nature of the mantle comes from (1) studies of seismological data, (2) studies of the nature of meteorites, and (3) studies of materials from the mantle that have been ejected to Earth's surface by volcanoes. The evidence from these separate sources all indicates that the mantle is composed of silicates, predominantly the ferromagnesian silicate *olivine.* Meteorites, as mentioned in chapter 13, are basically either iron meteorites or stony meteorites. Most of the stony meteorites are silicates with a composition that would produce the chemical composition of olivine if they were melted and the heavier elements were separated by gravity. This chemical composition also agrees closely with the composition of basalt, the most common volcanic rock found on the surface of Earth.

The Core

Information about the nature of the **core,** the center part of Earth, comes from several sources of information. Seismological data provide the primary evidence for the structure of the core of Earth. Seismic P-waves spread through Earth from a large earthquake. Figure 15.16 shows how the P-waves spread out, soon arriving at seismic measuring stations all around the world. However, there are places between 103° and 142° of arc from the earthquake that do not receive P-waves. This region is called the *P-wave shadow zone.* The P-wave shadow zone is explained by P-waves being refracted by the core, leaving a shadow. The paths

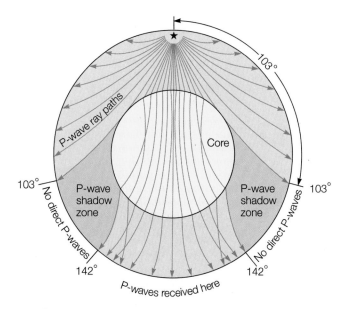

FIGURE 15.16 The P-wave shadow zone, caused by refraction of P-waves within Earth's core.

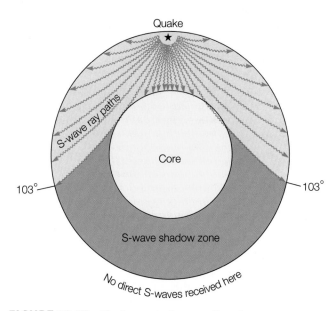

FIGURE 15.17 The S-wave shadow zone. Since S-waves cannot pass through a liquid, at least part of the core is either a liquid or has some of the same physical properties as a liquid.

of P-waves can be accurately calculated, so the size and shape of Earth's core can also be accurately calculated.

Seismic S-waves leave a different pattern at seismic receiving stations around Earth. S- (sideways or transverse) waves can travel only through solid materials. An *S-wave shadow zone* also exists and is larger than the P-wave shadow zone (figure 15.17). S-waves are not recorded in the entire region more than 103° away from the epicenter. The S-wave shadow zone seems to indicate that S-waves do not travel through the core at all. If this is true, it implies that the core of Earth is a liquid, or at least acts like a liquid.

Analysis of P-wave data suggests that the core has two parts: a *liquid outer core* and a *solid inner core.* Both the P-wave

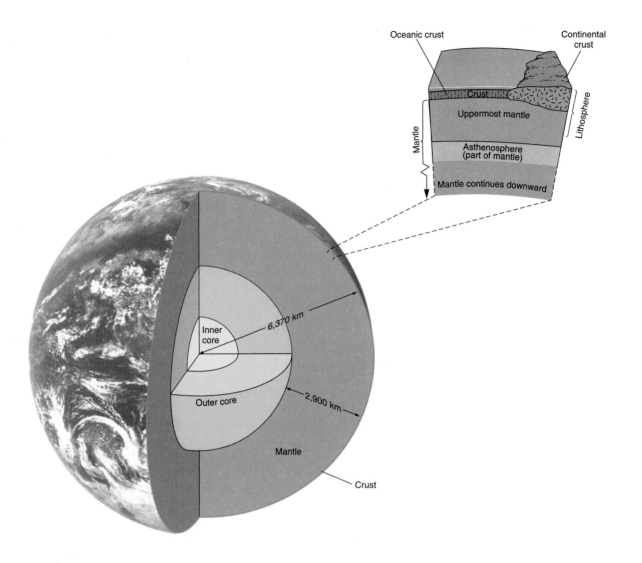

FIGURE 15.18 Earth's interior, showing the weak, plastic layer called the *asthenosphere*. The rigid, solid layer above the asthenosphere is called the *lithosphere*. The lithosphere is broken into plates that move on the upper mantle like giant ice sheets floating on water. This arrangement is the foundation for plate tectonics, which explains many changes that occur on Earth's surface such as earthquakes, volcanoes, and mountain building.

and S-wave data support this conclusion. Overall, the core makes up about 15 percent of Earth's total volume and about one-third of its mass.

Evidence from the nature of meteorites indicates that Earth's core is mostly iron. Earth has a strong magnetic field that has its sources in the turbulent flow of the liquid part of its core. To produce such a field, the material of the core would have to be an electrical conductor, that is, a metal such as iron. There are two general kinds of meteorites that fall to Earth: (1) stony meteorites that are made of silicate minerals and (2) iron meteorites that are made of iron or of a nickel-iron alloy. Since Earth has a silicate-rich crust and mantle, by analogy, Earth's core must consist of iron or a nickel and iron alloy.

A More Detailed Structure

There is strong evidence that Earth has a layered structure with a core, mantle, and crust. This description of the structure is important for historical reasons and for understanding how Earth evolved over time. There is another, more detailed structure that can be described. This structure is far more important in understanding the history and present appearance of Earth's surface, including the phenomena of earthquakes and volcanoes.

The important part of this different detailed description of Earth's interior was first identified from seismic data. There is a thin zone in the mantle where seismic waves undergo a sharp decrease in velocity (figure 15.18). This low-velocity zone is evidently a hot, elastic semiliquid layer that extends around the entire Earth. It is called the **asthenosphere,** after the Greek word for "weak shell." The asthenosphere is weak because it is plastic, mobile, and yields to stresses. In some regions, the asthenosphere contains pockets of magma.

The rocks above and below the asthenosphere are rigid, solid, and brittle. The solid layer above the asthenosphere is called the **lithosphere,** after the Greek word for "stone shell." The lithosphere is also known as the "strong layer" in contrast to the "weak layer" of the asthenosphere. The lithosphere includes the entire crust, the Moho, and the upper part of the mantle. As you will see in the section on plate tectonics, the asthenosphere is one important source of magma that reaches Earth's surface. It is also a necessary part of the mechanism involved in the movement of the crust. The lithosphere is made up of comparatively rigid plates that are moving, floating in the upper mantle like giant ice sheets floating in the ocean.

PLATE TECTONICS

If you observe the shape of the continents on a world map or a globe, you will notice that some of the shapes look as if they would fit together like the pieces of a puzzle. The most obvious is the eastern edge of North and South America, which seems to fit the western edge of Europe and Africa in a slight S-shaped curve. Such patterns between continental shapes seem to suggest that the continents were at one time together, breaking apart and moving to their present positions some time in the past (figure 15.19).

In the early 1900s, a German scientist named Alfred Wegener became enamored with the idea that the continents had shifted positions and published papers on the subject for nearly two decades. Wegener supposed that at one time there was a single large landmass that he called "Pangaea," which is from the Greek word meaning "all lands." He pointed out that similar fossils found in landmasses on both sides of the Atlantic Ocean today must be from animals and plants that lived in Pangaea, which later broke up and split into smaller continents. Wegener's concept came to be known as *continental drift,* the idea that individual continents could shift positions on Earth's surface. Some people found the idea of continental drift plausible, but most had difficulty imagining how a huge and massive continent could "drift" around on a solid Earth. Since Wegener had provided no good explanation of why or how continents might do this, most scientists found the concept unacceptable. The concept of continental drift was dismissed as an interesting but odd idea. Then new evidence indicated that the continents had indeed moved. The first of this evidence came from the bottom of the ocean and led to a new, broader theory about movement of Earth's crust.

Evidence from Earth's Magnetic Field

Earth's magnetic field is probably created by electric currents within the slowly circulating liquid part of the iron core. However, there is nothing static about Earth's magnetic poles. Geophysical studies have found that the magnetic poles are moving slowly around the geographic poles. Studies have also found that Earth's magnetic field occasionally undergoes magnetic reversal. Magnetic reversal is the flipping of polarity of Earth's magnetic field. During a magnetic reversal, the north magnetic pole and the south magnetic pole exchange positions. The present magnetic field orientation has persisted for the past seven-hundred thousand years and, according to the evidence, is now preparing for another reversal. The evidence, such as

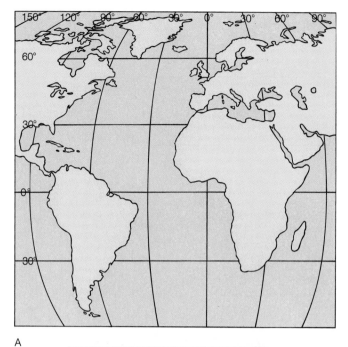

A

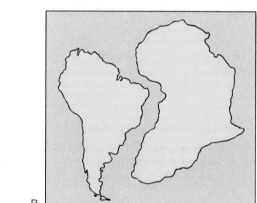

B

FIGURE 15.19 (*A*) Normal position of the continents on a world map. (*B*) A sketch of South America and Africa, showing how they once might have been joined together and subsequently separated by a continental drift.

the magnetized iron particles found in certain Roman ceramic artifacts, shows that the magnetic field was 40 percent stronger two thousand years ago than it is today. If the present rate of weakening were to continue, Earth's magnetic field would be near zero by the end of the next two thousand years—if it weakens that far before reversing orientation and then increases to its usual value.

Many igneous rocks contain a record of the strength and direction of Earth's magnetic field at the time the rocks formed. Iron minerals, such as magnetite (Fe_3O_4), crystallize as particles in a cooling magma and become magnetized and oriented to Earth's magnetic field, like tiny compass needles, at the time they were formed. Such rocks thus provide evidence of the direction and distance to Earth's ancient magnetic poles. The study of ancient magnetism, called *paleomagnetics,* provides the information that Earth's magnetic field has undergone twenty-two magnetic reversals during the past 4.5 million years (figure 15.20).

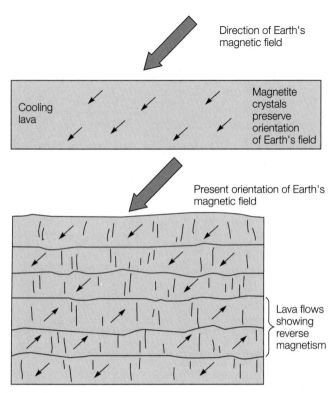

Direction of Earth's magnetic field

Cooling lava

Magnetite crystals preserve orientation of Earth's field

Present orientation of Earth's magnetic field

Lava flows showing reverse magnetism

FIGURE 15.20 Magnetite mineral grains align with Earth's magnetic field and are frozen into position as the magma solidifies. This magnetic record shows Earth's magnetic field has reversed itself in the past.

The record shows the time between pole flips is not consistent, sometimes reversing in as little as ten thousand years and sometimes taking as long as twenty-five million years. Once a reversal starts, however, it takes about five thousand years to complete the process.

Evidence from the Ocean

The first important studies concerning the movement of continents came from studies of the ocean basin, the bottom of the ocean floor. The basins are covered by 4 to 6 km (about 3 to 4 mi) of water and were not easily observed during Wegener's time. It was not until the development and refinement of sonar and other new technologies that scientists began to learn about the nature of the ocean basin. They found that it was not the flat, featureless plain that many had imagined. There are valleys, hills, mountains, and mountain ranges. Long, high, and continuous chains of mountains that seem to run clear around Earth were discovered, and these chains are called **oceanic ridges.** The *Mid-Atlantic Ridge* is one such oceanic ridge, which is located in the center of the Atlantic Ocean basin. The Mid-Atlantic Ridge divides the Atlantic Ocean into two nearly equal parts. Where it is high enough to reach sea level, it makes oceanic islands such as Iceland (figure 15.21). The basins also contain **oceanic trenches.** These trenches are long, narrow, and deep troughs with steep sides. Oceanic trenches always run parallel to the edge of continents.

Studies of the Mid-Atlantic Ridge found at least three related groups of data and observations: (1) submarine earthquakes were discovered and measured, but the earthquakes

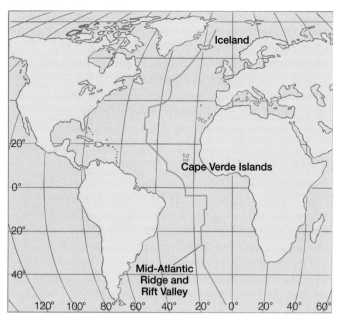

FIGURE 15.21 The Mid-Atlantic Ridge divides the Atlantic Ocean into two nearly equal parts. Where the ridge reaches above sea level, it makes oceanic islands, such as Iceland.

were all observed to occur mostly in a narrow band under the crest of the Mid-Atlantic Ridge; (2) a long, continuous, and deep valley, called a **rift,** was observed to run along the crest of the Mid-Atlantic Ridge for its length; and (3) a large amount of heat was found to be escaping from the rift. One explanation of the related groups of findings is that the rift might be a crack in Earth's crust, a fracture through which lava flowed to build up the ridge. The evidence of excessive heat flow, earthquakes along the crest of the ridge, and the very presence of the ridge all led to a **seafloor spreading** hypothesis. This hypothesis explained that hot, molten rock moved up from the interior of Earth to emerge along the rift, flowing out in both directions to create new rocks along the ridge. The creation of new rock like this would tend to spread the seafloor in both directions, thus the name. The test of this hypothesis would come from further studies on the ages and magnetic properties of the seafloor along the ridge (figure 15.22).

Evidence of the age of sections of the seafloor was obtained by drilling into the ocean floor from a research ship. From these drillings, scientists were able to obtain samples of fossils and sediments at progressive distances outward from the Mid-Atlantic Ridge. They found thin layers of sediments near the ridge that became progressively thicker toward the continents. This is a pattern you would expect if the seafloor were spreading, because older layers would have more time to accumulate greater depths of sediments. The fossils and sediments in the bottom of the layer were also progressively older at increasing distances from the ridge. The oldest, which were about 150 million years old, were near the continents. This would seem to indicate that the Atlantic Ocean did not exist until 150 million years ago. At that time, a fissure formed between Africa and South America, and new materials have been continuously flowing, adding new lithosphere to the edges of the fissure.

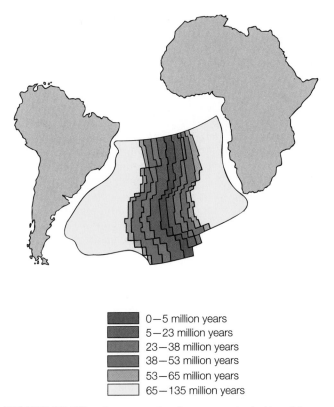

▓	0—5 million years
▓	5—23 million years
▓	23—38 million years
▒	38—53 million years
░	53—65 million years
□	65—135 million years

FIGURE 15.22 The pattern of seafloor ages on both sides of the Mid-Atlantic Ridge reflects seafloor spreading activity. Younger rocks are found closer to the ridge.

More convincing evidence for the support of seafloor spreading came from the paleomagnetic discovery of patterns of magnetic strips in the rocks of the ocean floor. Earth's magnetic field has been reversed many times in the last 150 million years. The periods of time between each reversal were not equal, ranging from thousands to millions of years. Since iron minerals in molten basalt formed, became magnetized, then frozen in the orientation they had when the rock cooled, they made a record of reversals in Earth's ancient magnetic field (figure 15.23). Analysis of the magnetic pattern in the rocks along the Mid-Atlantic Ridge found identical patterns of magnetic bands on both sides of the ridge. This is just what you would expect if molten rock flowed out of the rift, cooled to solid basalt, then moved away from the rift on both sides. The pattern of magnetic bands also matched patterns of reversals measured elsewhere, providing a means of determining the age of the basalt. This showed that the oceanic crust is like a giant conveyer belt that is moving away from the Mid-Atlantic Ridge in both directions. It is moving at an average 5 cm (about 2 in) a year, which is about how fast your fingernails grow. This means that in fifty years, the seafloor will have moved 5 cm/yr × 50 yr, or 2.5 m (about 8 ft). This slow rate is why most people do not recognize that the seafloor—and the continents—move.

Lithosphere Plates and Boundaries

The strong evidence for seafloor spreading soon led to the development of a new theory called **plate tectonics.** According to plate tectonics, the lithosphere is broken into a number of fairly rigid plates that move on the asthenosphere. Some plates, as

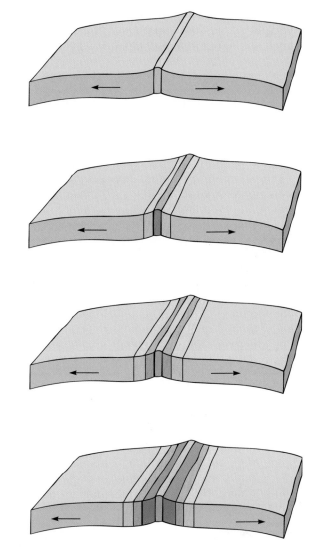

FIGURE 15.23 Formation of magnetic strips on the seafloor. As each new section of seafloor forms at the ridge, iron minerals become magnetized in a direction that depends on the orientation of Earth's field at that time. This makes a permanent record of reversals of Earth's magnetic field.

shown in figure 15.24, contain continents and part of an ocean basin, while other plates contain only ocean basins. The plates move, and the movement is helping to explain why mountains form where they do, the occurrence of earthquakes and volcanoes, and in general, the entire changing surface of Earth.

Earthquakes, volcanoes, and most rapid changes in Earth's crust occur at the edge of a plate, which is called a *plate boundary.* Three general kinds of plate boundaries describe how one plate moves relative to another: divergent, convergent, and transform boundaries.

Divergent boundaries occur between two plates moving away from each other. Magma forms as the plates separate, decreasing pressure on the mantle below. This molten material from the asthenosphere rises, cools, and adds new crust to the edges of the separating plates. The new crust tends to move horizontally from both sides of the divergent boundary, usually known as an oceanic ridge. A divergent boundary is thus a **new crust zone.** Most new crust zones are presently on the seafloor, producing seafloor spreading (figure 15.25).

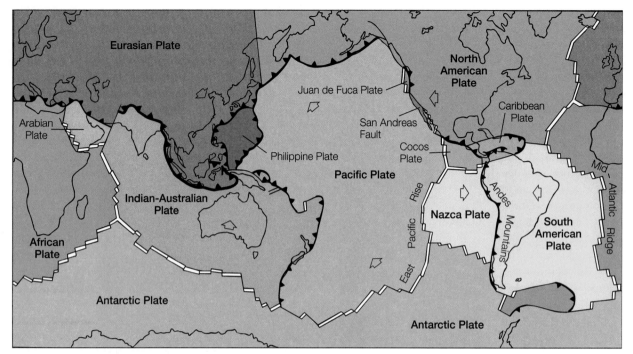

FIGURE 15.24 The major plates of the lithosphere that move on the asthenosphere.

Source: After W. Hamilton, U.S. Geological Survey.

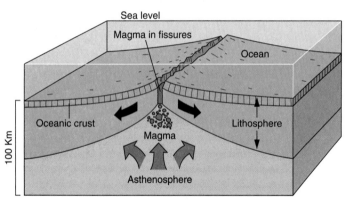

FIGURE 15.25 A diverging boundary at a mid-oceanic ridge. Hot asthenosphere wells upward beneath the ridge crest. Magma forms and squirts into fissures. Solid material that does not melt remains as mantle in the lower part of the lithosphere. As the lithosphere moves away from the spreading axis, it cools, becomes denser, and sinks to a lower level.

The Mid-Atlantic Ridge is a divergent boundary between the South American and African Plates, extending north between the North American and Eurasian Plates (see figure 15.24). This ridge is one segment of the global mid-ocean ridge system that encircles Earth. The results of divergent plate movement can be seen in Iceland, where the Mid-Atlantic Ridge runs as it separates the North American and Eurasian Plates. In the northeastern part of Iceland, ground cracks are widening, often accompanied by volcanic activity. The movement was measured extensively between 1975 and 1984, when displacements caused a total separation of about 7 m (about 23 ft).

The measured rate of spreading along the Mid-Atlantic Ridge ranges from 1 to 6 centimeters per year. This may seem slow, but the process has been going on for millions of years and has caused a tiny inlet of water between the continents of Europe, Africa, and the Americas to grow into the vast Atlantic Ocean that exists today.

Another major ocean may be in the making in East Africa, where a divergent boundary has already moved Saudi Arabia away from the African continent, forming the Red Sea. If this spreading between the African Plate and the Arabian Plate continues, the Indian Ocean will flood the area and the easternmost corner of Africa will become a large island.

Convergent boundaries occur between two plates moving toward each other. The creation of new crust at a divergent boundary means that old crust must be destroyed somewhere else at the same rate, or else Earth would have a continuously expanding diameter. Old crust is destroyed by returning to the asthenosphere at convergent boundaries. The collision produces an elongated belt of down-bending called a **subduction zone.** The lithosphere of one plate, which contains the crust, is subducted beneath the second plate and partially melts, then becoming part of the mantle. The more dense components of this may become igneous materials that remain in the mantle. Some of it may eventually migrate to a spreading ridge to make new crust again. The less dense components may return to the surface as a silicon, potassium, and sodium-rich lava, forming volcanoes on the upper plate, or it may cool below the surface to form a body of granite. Thus, the oceanic lithosphere is being recycled through this process, which explains why ancient seafloor rocks do not exist. Convergent boundaries produce related characteristic geologic features depending on the nature of the materials in the plates, and there are three general possibilities: (1) converging continental and oceanic plates, (2) converging oceanic plates, and (3) converging continental plates.

As an example of *ocean-continent plate convergence*, consider the plate containing the South American continent (the South American Plate) and its convergent boundary with an oceanic plate (the Nazca Plate) along its western edge. Continent-oceanic plate convergence produces a characteristic set of geologic features as the oceanic plate of denser basaltic material is subducted beneath the less dense granite-type continental plate (figure 15.26). The subduction zone is marked by an oceanic trench (the Peru-Chile Trench), deep-seated earthquakes, and volcanic mountains on the continent (the Andes Mountains). The trench is formed from the down-bending associated with subduction and the volcanic mountains from subducted and melted crust that rise up through the overlying plate to the surface. The earthquakes are associated with the movement of the subducted crust under the overlying crust.

Ocean-ocean plate convergence produces another set of characteristics and related geologic features (figure 15.27). The northern boundary of the oceanic Pacific Plate, for example, converges with the oceanic part of the North American Plate near the Bering Sea. The Pacific Plate is subducted, forming the Aleutian oceanic trench with a zone of earthquakes that are shallow near the trench and progressively more deep-seated toward the continent. The deeper earthquakes are associated with the movement of more deeply subducted crust into the mantle. The Aleutian Islands are typical **island arcs,** curving chains of volcanic islands that occur over the belt of deep-seated earthquakes. These islands form where the melted subducted material rises up through the overriding plate above sea level. The Japanese, Marianas, and Indonesians are similar groups of arc islands associated with converging oceanic-oceanic plate boundaries.

During *continent-continent plate convergence*, subduction does not occur as the less dense, granite-type materials tend to resist subduction (figure 15.28). Instead, the colliding plates pile up into a deformed and thicker crust of the lighter material. Such a collision produced the thick, elevated crust known as the Tibetan Plateau and the Himalayan Mountains.

Transform boundaries occur between two plates sliding by each other. Crust is neither created nor destroyed at transform boundaries as one plate slides horizontally past another along a long, vertical fault. The movement is neither smooth nor equal along the length of the fault, however, as short segments move independently with sudden jerks that are separated by periods without motion. The Pacific Plate, for example, is moving slowly to the northwest, sliding past the North American Plate. The San Andreas fault is one boundary along the California coastline. Vibrations from plate movements along this boundary are the famous California earthquakes.

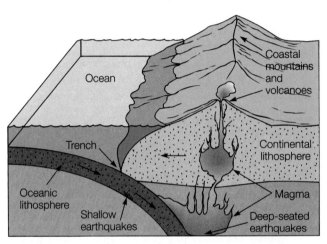

FIGURE 15.26 Ocean-continent plate convergence. This type of plate boundary accounts for shallow and deep-seated earthquakes, an oceanic trench, volcanic activity, and mountains along the coast.

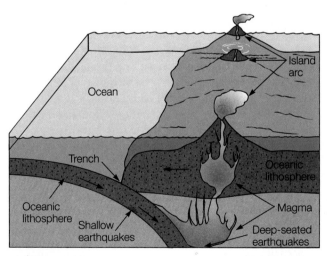

FIGURE 15.27 Ocean-ocean plate convergence. This type of plate convergence accounts for shallow and deep-focused earthquakes, an oceanic trench, and a volcanic arc above the subducted plate.

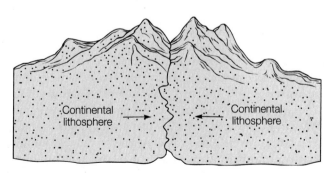

FIGURE 15.28 Continent-continent plate convergence. Rocks are deformed, and some lithosphere thickening occurs, but neither plate is subducted to any great extent.

Seismic Tomography

The CAT scan is a common diagnostic imaging procedure that combines the use of X rays with computer technology. CAT stands for "computed axial tomography" and the word *tomography* means "drawing slices." The medical CAT scan shows organs of interest by making X-ray images from many different angles as the source of the X rays moves around the patient. CAT scan images are assembled by a computer into a three-dimensional picture that can show organs, bones, and tissues in great detail.

The CAT scan is applied to Earth's interior in a *seismic tomography* procedure. It works somewhat like the medical CAT scan, but seismic tomography uses seismic waves instead of X rays. The velocities of S- and P-seismic waves vary with depth, changing with density, temperature, pressure, and the composition of Earth's interior. Interior differences in temperature, pressure, and composition are what cause the motion of tectonic plates on Earth's surface. Thus,

a picture of Earth's interior can be made by mapping variations in seismic wave speeds.

Suppose an earthquake occurs and a number of seismic stations record when the S- and P- waves arrive. From these records, you could compare the seismic velocity data and identify if there were low seismic velocities between the source and some of the receivers. The late arrival of seismic waves could mean some difference in the structure of Earth between the source and the observing station. When an earthquake occurs in a new location, more data will be collected at different observing stations, which provides additional information about the shape and structure of whatever is slowing the seismic waves. Now, imagine repeating these measurements for many new earthquakes until you have enough data to paint a picture of what is beneath the surface.

Huge amounts of earthquake data—perhaps 10 million data points in 5 million

groups—are needed to construct a picture of Earth's interior. A really fast computer may take days to process this much data and construct cross sections through some interesting place, such as a subduction zone where an oceanic plate dives into the mantle.

Seismic tomography has also identified massive plumes of molten rock rising toward the surface from deep within Earth. These *superplumes* originate from the base of the mantle, rising to the lithosphere. The hot material was observed to spread out horizontally under the lithosphere toward midocean ridges. This may contribute to tectonic plate movement. Regions above the superplumes tend to bulge upward, and other indications of superplumes, such as variations in gravity, have been measured. Scientists continue to work with new seismic tomography data with more precise images and higher resolutions to better describe the interior structure of Earth.

Present-Day Understandings

The theory of plate tectonics, developed during the late 1960s and early 1970s, is new compared to most major scientific theories. Measurements are still being made, evidence is being gathered and evaluated, and the exact number of plates and their boundaries are yet to be determined with certainty. The major question that remains to be answered is what drives the plates, moving them apart, together, and past each other? One explanation is that slowly turning *convective cells* in the plastic asthenosphere drive the plates (figure 15.29). According to this hypothesis, hot fluid materials rise at the diverging boundaries. Some of the material escapes to form new crust, but most of it spreads out beneath the lithosphere. As it moves beneath the lithosphere, it drags the overlying plate with it. Eventually, it cools and sinks back inward under a subduction zone.

There is uncertainty about the existence of convective cells in the asthenosphere and their possible role because of a lack of clear evidence. Seismic data is not refined enough to show convective cell movement beneath the lithosphere. In addition, deep-seated earthquakes occur to depths of about 700 km (435 mi), which means that descending materials—parts of a subducted plate—must extend to that depth. This could mean that a convective cell might operate all the way down to the core-mantle boundary some 2,900 km (about 1,800 mi) below the surface. This presents another kind of problem because little is known about the lower mantle and how it interacts with the upper mantle. Theorizing without information is

called speculation, and that is the best that can be done with existing data. The full answer may include the role of heat and the role of gravity.

Heat and gravity are important in a proposed mechanism of plate motion called "*ridge-push*" (figure 15.30). This idea has a plate cooling and thickening as it moves away from a divergent boundary. As it subsides, it cools asthenospheric mantle to lithospheric mantle, forming a sloping boundary between the lithosphere and the asthenosphere. The plate slides down this boundary.

Another proposed mechanism of plate motion is called "*slab-pull*" (figure 15.31). In this mechanism, the subducting plate is colder and therefore denser than the surrounding hot mantle, so it pulls the surface part of the plate downward. Density of the slab may also be increased by loss of water and reforming of minerals into more dense forms.

What is generally accepted about the plate tectonic theory is the understanding that the solid materials of Earth are engaged in a continual cycle of change. Oceanic crust is subducted, melted, then partly returned to the crust as volcanic igneous rocks in island arcs and along continental plate boundaries. Other parts of the subducted crust become mixed with the upper mantle, returning as new crust at diverging boundaries. The materials of the crust and the mantle are thus cycled back and forth in a mixing that may include the deep mantle and the core as well. There is more to this story of a dynamic Earth that undergoes a constant change. The story continues in chapters 16, 17, and 18 with different cycles to consider.

Measuring Plate Movement

According to the theory of plate tectonics, Earth's outer shell is made up of moving plates. The plates making up the continents are about 100 km (62 mi) thick and are gradually drifting at a rate of about 0.5 to 10 cm (0.2 to 4 in) per year. This reading is about one way that scientists know that Earth's plates are moving and how this movement is measured.

The very first human lunar landing mission took place in July 1969. Astronaut Neil Armstrong stepped onto the lunar surface, stating, "That's one small step for a man, one giant leap for mankind." In addition to fulfilling a dream, the *Apollo* project carried out a program of scientific experiments. The *Apollo 11* astronauts placed a number of experiments on the lunar surface in the Sea of Tranquility. Among the experiments was the first Laser Ranging Retro-Reflector

Experiment, which was designed to reflect pulses of laser light from Earth. Later, three more reflectors were placed on the Moon, including two by other *Apollo* astronauts and one by an unmanned Soviet *Lunakhod 2* lander.

The McDonald Observatory in Texas, the Lure Observatory on the island of Maui, Hawaii, and a third observatory in southern France have regularly sent laser beams through optical telescopes to the reflectors. The return signals, which are too weak to be seen with the unaided eye, are detected and measured by sensitive detection equipment at the observatories. The accuracy of these measurements, according to NASA reports, is equivalent to determining the distance between a point on the east and a point on the west coasts of the United States to an accuracy of 0.5 mm, about

the size of the period at the end of this sentence.

Reflected laser light experiments have found that the Moon is pulling away from Earth at about 4 cm/yr (about 1.6 in/yr), that the shape of Earth is slowly changing, undergoing isostatic adjustment from the compression by the glaciers during the last ice age, and that the observatory in Hawaii is slowly moving away from the one in Texas. This provides a direct measurement of the relative drift of two of Earth's tectonic plates. Thus, one way that changes on the surface of Earth are measured is through lunar ranging experiments. Results from lunar ranging, together with laser ranging to the artificial satellites in Earth's orbit, have revealed the small but constant drift rate of the plates making up Earth's dynamic surface.

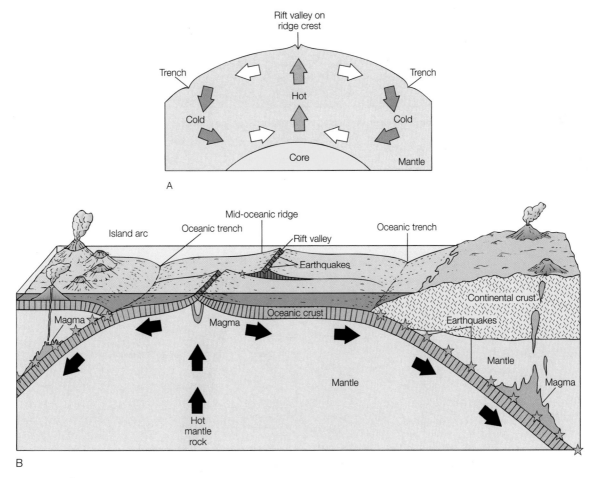

FIGURE 15.29   Not to scale. One idea about convection in the mantle has a convection cell circulating from the core to the lithosphere, dragging the overlying lithosphere laterally away from the oceanic ridge.

Frederick John Vine (1939–1988)

Frederick Vine was an English geophysicist whose work was an important contribution to the development of the theory of plate tectonics.

Vine was born in Brentford, Essex, England, and was educated at Latymer Upper School, London, and at St. John's College, Cambridge. From 1967 to 1970, he worked as assistant professor in the Department of Geological and Geophysical Sciences at Princeton University, before returning to England to become reader (1970) and then professor (1974) in the School of Environmental Sciences at the University of East Anglia.

In Cambridge in 1963, Vine collaborated with his supervisor, Drummond Hoyle Matthews (1931–), and wrote a paper, "Magnetic Anomalies over Ocean Ridges," that provided additional evidence for Harry H. Hess' (1962) seafloor spreading hypothesis. Alfred Wegener's original 1912 theory of continental drift (Wegener's hypothesis) had been met with hostility at the time because he could not explain why the continents had drifted apart, but Hess had continued his

work developing the theory of seafloor spreading to explain the fact that as the oceans grew wider, the continents drifted apart.

Following the work of Brunhes and Matonori Matuyama in the 1920s on magnetic reversals, Vine and Matthews predicted that new rock emerging from the oceanic ridges would intermittently produce material of opposing magnetic polarity to the old rock. They applied paleomagnetic studies to the ocean ridges in the North Atlantic and were able to argue that parallel belts of different magnetic polarities existed on either side of the ridge crests. This evidence was vital proof of Hess' hypothesis. Studies on ridges in other oceans also showed the existence of these magnetic anomalies. Vine and Matthews' hypothesis was widely accepted in 1966 and confirmed Dietz and Hess' earlier work. Their work was crucial to the development of the theory of plate tectonics, and it revolutionized the earth sciences.

Source: Modified from the *Hutchinson Dictionary of Scientific Biography.* © RM, 2007. Reprinted by permission.

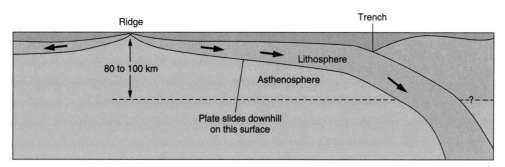

FIGURE 15.30 Ridge-push. A plate may slide downhill on the sloping boundary between the lithosphere and the asthenosphere at the base of the plate.

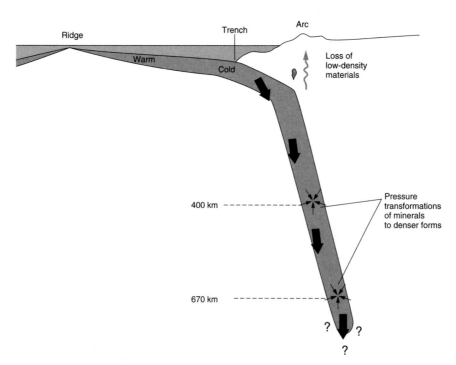

FIGURE 15.31 Slab-pull. The dense, leading edge of a subducting plate pulls the rest of the plate along. Plate density increases due to cooling, loss of low-density material, and pressure transformation of minerals to denser forms.

SUMMARY

The elements silicon and oxygen make up 75 percent of all the elements in the outer layer, or *crust,* of Earth. The elements combine to make crystalline chemical compounds called minerals. A *mineral* is defined as a naturally occurring, inorganic solid element or compound with a crystalline structure.

About 92 percent of the minerals of Earth's crust are composed of silicon and oxygen, the *silicate minerals.* The basic unit of the silicates is a *tetrahedral structure* that combines with positive metallic ions or with other tetrahedral units to form chains, sheets, or an interlocking framework. The *ferromagnesian silicates* are tetrahedral structures combined with ions of iron, magnesium, calcium, and other elements. The ferromagnesian silicates are darker in color and more dense than other silicates. The *nonferromagnesian silicates* do not have irons or magnesium ions, and they are lighter in color and less dense than the ferromagnesians. The *nonsilicate minerals* do not contain silicon and are carbonates, sulfates, oxides, halides, sulfides, and native elements.

A *rock* is defined as an aggregation of one or more minerals that have been brought together into a cohesive solid. *Igneous rocks* formed as hot, molten *magma* cooled and crystallized to firm, hard rocks. *Sedimentary rocks* are formed from *sediments,* accumulations of weathered rock materials that settle out of the atmosphere or out of water. Sediments become sedimentary rocks through a rock-forming process that involves both the *compaction* and *cementation* of the sediments. *Metamorphic rocks* are previously existing rocks that have been changed by heat, pressure, or hot solution into a different kind of rock without melting. The *rock cycle* is a concept that an igneous, a sedimentary, or a metamorphic rock is a temporary stage in the ongoing transformation of rocks to new types.

Earth has a layered interior that formed as Earth's materials underwent *differentiation,* the separation of materials while in the molten state. The center part, or *core,* is predominantly iron with a solid inner part and a liquid outer part. The core makes up about 15 percent of Earth's total volume and about a third of its total mass. The *mantle* is the middle part of Earth's interior that accounts for about two-thirds of Earth's total mass and about 80 percent of its total volume. The *Mohorovicic discontinuity* separates the outer layer, or *crust,* of Earth from the mantle. The crust of the *continents* is composed mostly of less dense granite-type rock. The crust of the *ocean basins* is composed mostly of the more dense basaltic rocks.

Another way to consider Earth's interior structure is to consider the weak layer in the upper mantle, the *asthenosphere* that extends around the entire Earth. The rigid, solid, and brittle layer above the asthenosphere is called the *lithosphere.* The lithosphere includes the entire crust, the Moho, and the upper part of the mantle.

Evidence from the ocean floor revived interest in the idea that continents could move. The evidence for *seafloor spreading* came from related observations concerning oceanic ridge systems, sediment and fossil dating of materials outward from the ridge, and magnetic patterns of seafloor rocks. Confirmation of seafloor spreading led to the *plate tectonic theory.* According to plate tectonics, new basaltic crust is added at *diverging boundaries* of plates, and old crust is *subducted* at *converging boundaries.* Mountain building, volcanoes, and earthquakes are seen as *related geologic features* that are caused by plate movements. The force behind the movement of plates is uncertain, but it may involve *convection* in the deep mantle.

KEY TERMS

asthenosphere (p. **344**)
convergent boundaries (p. **348**)
core (p. **343**)
crust (p. **342**)
divergent boundaries (p. **347**)
igneous rocks (p. **339**)

island arcs (p. **349**)
lithosphere (p. **344**)
mantle (p. **343**)
metamorphic rocks (p. **340**)
mineral (p. **335**)
new crust zone (p. **347**)

oceanic ridges (p. **346**)
oceanic trenches (p. **346**)
plate tectonics (p. **347**)
rift (p. **346**)
rock (p. **339**)
seafloor spreading (p. **346**)

sedimentary rocks (p. **339**)
silicates (p. **335**)
subduction zone (p. **348**)
transform boundaries (p. **349**)

APPLYING THE CONCEPTS

1. Sedimentary rocks are formed by the processes of compaction and
   a. pressurization.
   b. melting.
   c. cementation.
   d. heating but not melting.

2. Which type of rock probably existed first, starting the rock cycle?
   a. metamorphic
   b. igneous
   c. sedimentary
   d. All of the above are correct.

3. From seismological data, Earth's shadow zone indicates that part of Earth's interior must be
   a. liquid.
   b. solid throughout.
   c. plastic.
   d. hollow.

4. The Mohorovicic discontinuity is a change in seismic wave velocity that is believed to take place because of
   a. structural changes in minerals of the same composition.
   b. changes in the composition on both sides of the boundary.
   c. a shift in the density of minerals of the same composition.
   d. changes in the temperature with depth.

5. The oldest rocks are found in (the)
   a. continental crust.
   b. oceanic crust.
   c. neither, since both are the same age.

6. The least dense rocks are found in (the)
   a. continental crust.
   b. oceanic crust.
   c. neither, since both are the same density.

7. The idea of seafloor spreading along the Mid-Atlantic Ridge was supported by evidence from
   a. changes in magnetic patterns and ages of rocks moving away from the ridge.
   b. faulting and volcanoes on the continents.
   c. the observation that there was no relationship between one continent and another.
   d. All of the above are correct.

8. According to the plate tectonics theory, seafloor spreading takes place at a
   a. convergent boundary.
   b. subduction zone.
   c. divergent boundary.
   d. transform boundary.

9. The presence of an oceanic trench, a chain of volcanic mountains along the continental edge, and deep-seated earthquakes is characteristic of a (an)
   a. ocean-ocean plate convergence.
   b. ocean-continent plate convergence.
   c. continent-continent plate convergence.
   d. None of the above is correct.

10. The presence of an oceanic trench with shallow earthquakes and island arcs with deep-seated earthquakes is characteristic of a (an)
    a. ocean-ocean plate convergence.
    b. ocean-continent plate convergence.
    c. continent-continent plate convergence.
    d. None of the above is correct.

11. The ongoing occurrence of earthquakes without seafloor spreading, oceanic trenches, or volcanoes is most characteristic of a
    a. convergent boundary between plates.
    b. subduction zone.
    c. divergent boundary between plates.
    d. transform boundary between plates.

12. The evidence that Earth's core is part liquid or acts like a liquid comes from (the)
    a. P-wave shadow zone.
    b. S-wave shadow zone.
    c. meteorites.
    d. All of the above are correct.

Answers

1. c 2. b 3. a 4. b 5. a 6. a 7. a 8. c 9. b 10. a 11. d 12. b

QUESTIONS FOR THOUGHT

1. Briefly describe the rock-forming process that changes sediments into solid rock.

2. What are metamorphic rocks? What limits the maximum temperatures possible in metamorphism? Explain.

3. What evidence provides information about the nature of Earth's core?

4. What is the asthenosphere? Why is it important in modern understandings of Earth?

5. Describe the origin of the magnetic strip patterns found in the rocks along an oceanic ridge.

6. Explain why ancient rocks are not found on the seafloor.

7. Describe the three major types of plate boundaries and what happens at each.

8. Briefly describe the theory of plate tectonics and how it accounts for the existence of certain geologic features.

9. What is an oceanic trench? What is its relationship to major plate boundaries? Explain this relationship.

10. Describe the probable source of all the earthquakes that occur in southern California.

11. The northwestern coast of the United States has a string of volcanoes running along it. According to plate tectonics, what does this mean about this part of the North American Plate? What geologic feature would you expect to find on the seafloor off the northwestern coast? Explain.

12. Explain how the crust of Earth is involved in a dynamic, ongoing recycling process.

FOR FURTHER ANALYSIS

1. Is ice a mineral? Describe reasons to argue that ice is a mineral. Describe reasons to argue that ice is not a mineral.

2. What are the significant similarities and differences between igneous, sedimentary, and metamorphic rocks?

3. Why are there no active volcanoes in the eastern United States or Canada? Explain why you would or would not expect volcanoes there in the future.

4. Describe cycles that occur on Earth's surface and cycles that occur between the surface and the interior. Explain why these cycles do not exist on other planets of the solar system.

5. The rock cycle describes how igneous, metamorphic, and sedimentary rocks are changed into each other. If this is true, analyze why most of the rocks on Earth's surface are sedimentary.

6. If ice is a mineral, is a glacier a rock? Describe reasons to support or oppose calling a glacier a rock according to the definition of a rock.

7. Discuss evidence that would explain why plate tectonics occurs on Earth but not on other planets.

INVITATION TO INQUIRY

Measuring Plate Motion

Tectonic plate motion can be measured with several relatively new technologies, including satellite laser ranging and use of the Global Positioning System. Start your inquiry by visiting the Tectonic Plate Motion website at cddisa.gsfc.nasa.gov/926/slrtecto.html. Study the regional plate motion of the North American Plate, for example. Note the scale of 50 mm/yr, then measure to find the rate of movement at the different stations shown. Is this recent data on plate motion consistent with other available information on plate motion? How can you account for the different rates of motion at adjacent stations?

Folding, faulting, and lava flows, such as the one you see here, tend to build up, or elevate, Earth's surface.

Earth's Surface

Earth Science Connections

▶ Earth's surface is made up of a number of rigid plates that are moving (Ch. 15).

CORE CONCEPT

The surface of Earth is involved in an ongoing cycle of destruction and renewal.

Rocks are subjected to forces associated with plate tectonics and other forces (p. 358).

Mountain ranges are features of folding and faulting on a very large scale (p. 364).

Weathering and erosion wear down the surface of Earth (p. 368).

Gravity, streams, glaciers, and wind are agents of erosion (p. 372).

Life Science Connections

▶ Erosion of rock releases phosphorus for use by plants (Ch. 23).

▶ Sedimentation traps organisms and forms fossils (Ch. 22)

▶ Layers of sedimentary rock reveal the history of life on Earth (Ch. 22).

OVERVIEW

The central idea of plate tectonics, which was discussed in chapter 15, is that Earth's surface is made up of rigid plates that are slowly moving. Since the plates and the continents riding on them are in constant motion, any given map of the world is only a snapshot that shows the relative positions of the continents at a given time. The continents occupied different positions in the distant past. They will occupy different positions in the distant future. The surface of Earth, which seems so solid and stationary, is in fact mobile.

Plate tectonics has changed the accepted way of thinking about the solid, stationary nature of Earth's surface and ideas about the permanence of the surface as well. The surface of Earth is no longer viewed as having a permanent nature but is understood to be involved in an ongoing cycle of destruction and renewal. Old crust is destroyed as it is plowed back into the mantle through subduction, becoming mixed with the mantle. New crust is created as molten materials move from the mantle through seafloor spreading and volcanoes. Over time, much of the crust must cycle into and out of the mantle.

The movement of plates, the crust-mantle cycle, and the rock cycle all combine to produce a constantly changing surface. There are basically two types of surface changes: (1) changes that originate within Earth, resulting in a building up of the surface (figure 16.1) and (2) changes that occur when rocks are exposed to the atmosphere and water, resulting in a sculpturing and tearing down of the surface. This chapter is about changes in the land. The concepts of this chapter will provide you with something far more interesting about Earth's surface than the scenic aspect. The existence of different features (such as mountains, folded hills, islands) and the occurrence of certain events (such as earthquakes, volcanoes, faulting) are all related. The related features and events also have a story to tell about Earth's past, a story about the here and now, and yet another story about the future.

INTERPRETING EARTH'S SURFACE

Because many geologic changes take place slowly, it is difficult for a person to see significant change occur to mountains, canyons, and shorelines in the brief span of a lifetime. Given a mental framework based on a lack of appreciation of change over geologic time, how do you suppose people interpreted the existence of features such as mountains and canyons? Some believed, as they had observed in their lifetimes, that the mountains and canyons had "always" been there. Statements such as "unchanging as the hills" or "old as the hills" illustrate this lack of appreciation of change over geologic time. Others did not believe the features had always been there but believed they were formed by a sudden, single catastrophic event (figure 16.2).

A catastrophe created a feature of Earth's surface all at once, with little or no change occurring since that time. The Grand Canyon, for example, was not interpreted as the result of incomprehensibly slow river erosion but as the result of a giant crack or rip that appeared in the surface. The canyon that you see today was interpreted as forming when Earth split open and the Colorado River fell into the split. This interpretation was used to explain the formation of major geologic features based on the little change observed during a person's lifetime.

About two hundred years ago, the idea of unchanging, catastrophically formed landscapes was challenged by James Hutton, a Scottish physician. Hutton, who is known today as the founder of modern geology, traveled widely throughout the British Isles. Hutton was a keen observer of rocks, rock

FIGURE 16.1 An aerial view from the south of the eruption of Mount St. Helens volcano on May 18, 1980.

FIGURE 16.2 Would you believe that this rock island has "always" existed where it is? Would you believe it was formed by a sudden, single event? What evidence would it take to convince you that the rock island formed ever so slowly, starting as a part of southern California and moving very slowly, at a rate of centimeters per year, to its present location near the coast of Alaska?

structures, and other features of the landscape. He noted that sandstone, for example, was made up of rock fragments that appeared to be (1) similar to the sand being carried by rivers and (2) similar to the sand making up the beaches next to the sea. He also noted fossil shells of sea animals in sandstone on the land, while the living relatives of these animals were found in the shallow waters of the sea. This and other evidence led Hutton to realize that rocks were being ground into fragments, then carried by rivers to the sea. He surmised that these particles would be reformed into rocks later, then lifted and shaped into the hills and mountains of the land. He saw all this as quiet, orderly change that required only *time* and the ongoing work of the water and some forces to make the sediments back into rocks. With Hutton's logical conclusion came the understanding that Earth's history could be interpreted by tracing it backwards, from the present to the past. This tracing required a frame of reference of slow, uniform change, not the catastrophic frame of reference of previous thinkers. The frame of reference of uniform changes is today called the **principle of uniformity** (also called *uniformitarianism*). The principle of uniformity is often represented by a statement that "the present is the key to the past." This statement means that the same geologic processes you see changing rocks today are the very same processes that changed them in the ancient past, although not necessarily at the same rate. The principle of uniformity does not *exclude* the happening of sudden or catastrophic events on the surface of Earth. A violent volcanic explosion, for example, is a catastrophic event that most certainly modifies the surface of Earth. What the principle of uniformity does state is that the physical and chemical laws we observe today operated exactly the same in the past. The rates of operation may or may not have been the same in the past, but the events you see occurring today are the same events that occurred in the past. Given enough time, you can explain the formation of the structures of Earth's surface with known events and concepts.

The principle of uniformity has been used by geologists since the time of Hutton. The concept of how the constant changes occur has evolved with the development of plate tectonics, but the basic frame of reference is the same. You will see how the principle of uniformity is applied by first considering what can happen to rocks and rock layers that are deeply buried.

PROCESSES THAT BUILD UP THE SURFACE

All the possible movements of Earth's plates, including drift toward or away from other plates, and any process that deforms Earth's surface are included in the term *diastrophism*. Diastrophism is the process of deformation that changes Earth's surface. It produces many of the basic structures you see on Earth's surface, such as plateaus, mountains, and folds in the crust. The movement of magma is called *vulcanism* or *volcanism*. Diastrophism, volcanism, and earthquakes are closely related, and their occurrence can usually be explained by events involving plate tectonics. The results of diastrophism are discussed in the section on stress and strain, which is followed by a discussion of earthquakes, volcanoes, and mountain chains.

Stress and Strain

Any solid material responds to a force in a way that depends on the extent of coverage (force per unit area, or pressure), the nature of the material, and other variables such as the temperature. Consider, for example, what happens if you place the point of a ballpoint pen on the side of an aluminum pop (soda) can and apply an increasing pressure. With increasing pressure, you can observe at least four different and separate responses:

1. At first, the metal successfully resists a slight pressure and *nothing happens.*
2. At a somewhat greater pressure, you will be able to deform, or bend, the metal into a concave surface. The metal will return to its original shape, however, when the pressure is removed. This is called an *elastic deformation* since the metal was able to spring back into its original shape.
3. At a still greater pressure, the metal is deformed to a concave surface, but this time, the metal does not return to its original shape. This means the *elastic limit* of the metal has been exceeded, and it has now undergone a *plastic deformation.* Plastic deformation permanently alters the shape of a material.
4. Finally, at some great pressure, the metal will rupture, resulting in a *break* in the material.

Many materials, including rocks, respond to increasing pressures in this way, showing (1) no change, (2) an elastic change with recovery, (3) a plastic change with no recovery, and (4) finally breaking from the pressure.

A **stress** is a force that tends to compress, pull apart, or deform a rock. Rocks in Earth's solid outer crust are subjected to forces as Earth's plates move into, away from, or alongside each other. However, not all stresses are generated directly by plate interaction. Three types of forces from plate interaction that cause rock stress are:

1. *Compressive stress* is caused by two plates moving together or by one plate pushing against another plate that is not moving.
2. *Tensional stress* is the opposite of compressional stress. It occurs when one part of a plate moves away, for example, and another part does not move.
3. *Shear stress* is produced when two plates slide past one another or by one plate sliding past another plate that is not moving.

Just like the metal in the soda can, a rock is able to withstand stress up to a limit. Then it might undergo elastic deformation, plastic deformation, or breaking with progressively greater pressures. The adjustment to stress is called **strain.** A rock unit might respond to stress by changes in volume, changes in shape, or by breaking. Thus, there are three types of strain: elastic, plastic, and fracture.

1. In *elastic strain,* rock units recover their original shape after the stress is released.
2. In *plastic strain,* rock units are molded or bent under stress and do not return to their original shape after the stress is released.
3. In *fracture strain,* rock units crack or break, as the name suggests.

The relationship between stress and strain, that is, exactly how the rock responds, depends on at least four variables. They are (1) the nature of the rock, (2) the temperature of the rock, (3) how slowly or quickly the stress is applied over time, and (4) the confining pressure on the rock. The temperature and confining pressure are generally a function of how deeply the rock is buried. In general, rocks are better able to withstand compressional rather than pulling-apart stresses. Cold rocks are more likely to break than warm rocks, which tend to undergo plastic deformation. In addition, a stress that is applied quickly tends to break the rock, where stress applied more slowly over time, perhaps thousands of years, tends to result in plastic strain.

In general, rocks at great depths are under great pressure at higher temperatures. These rocks tend to undergo plastic deformation, then plastic flow, so rocks at great depths are bent and deformed extensively. Rocks at less depth can also bend, but they have a lower elastic limit and break more readily (figure 16.3). Rock deformation often results in recognizable surface features called folds and faults, the topics of the next sections.

Folding

Sediments that form most sedimentary rocks are deposited in nearly flat, horizontal layers at the bottom of a body of water. Conditions on the land change over time, and different mixtures of sediments are deposited in distinct layers of varying thickness. Thus, most sedimentary rocks occur naturally as structures of horizontal layers, or beds (figure 16.4).

A sedimentary rock layer that is not horizontal may have been subjected to some kind of compressive stress. The source

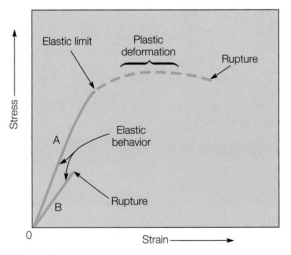

FIGURE 16.3 Stress and deformation relationships for deeply buried, warm rocks under high pressure (*A*) and for cooler rocks near the surface (*B*). Breaking occurs when stress exceeds rupture strength.

A

B

FIGURE 16.4 (*A*) Rock bedding on a grand scale in the Grand Canyon. (*B*) A closer example of rock bedding can be seen in this roadcut.

FIGURE 16.5 These folded rock layers are in the Calico Hills, California. Can you figure out what might have happened here to fold flat rock layers like this?

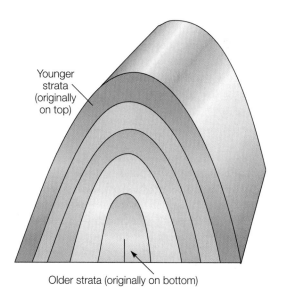

Younger strata (originally on top)

Older strata (originally on bottom)

FIGURE 16.6 An anticline, or arching fold, in layered sediments. Note that the oldest strata are at the center.

of such a stress could be from colliding plates, from the intrusion of magma from below, or from a plate moving over a rising superplume of magma. Seismic tomography has identified massive plumes of molten rock rising toward the surface from deep within Earth. These superplumes originate from the base of the mantle, rising to the lithosphere. The hot material was observed to spread out horizontally, causing bulges as it exerts forces under the lithosphere.

Stress on buried layers of horizontal rocks can result in plastic strain, resulting in a wrinkling of the layers into *folds*. **Folds** are bends in layered bedrock (figure 16.5). They are analogous to layers of rugs or blankets that were stacked horizontally, then pushed into a series of arches and troughs. Folds in layered bedrock of all shapes and sizes can occur from plastic strain, depending generally on the regional or local nature of the stress and other factors. Of course, when the folding

occurred, the rock layers were in a plastic condition and were probably under considerable confining pressure from deep burial. However, you see the results of the folding when the rocks are under very different conditions at the surface.

The most common regional structures from deep plastic deformation are arch-shaped and trough-shaped folds. In general, an arch-shaped fold is called an **anticline** (figure 16.6). The corresponding trough-shaped fold is called a **syncline** (figure 16.7). Anticlines and synclines sometimes alternate across the land like waves on water. You can imagine that a great compressional stress must have been involved over a wide region to wrinkle the land like this.

Anticlines, synclines, and other types of folds are not always visible as such on Earth's surface. The ridges of anticlines are constantly being weathered into sediments. The sediments, in turn, tend to collect in the troughs of synclines, filling them

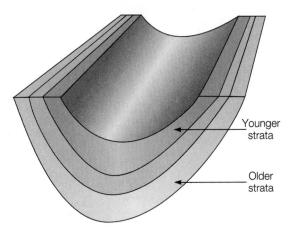

FIGURE 16.7 A syncline, showing the reverse age pattern.

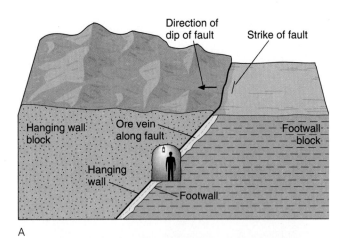

A

B

FIGURE 16.8 (*A*) Relationship between the hanging wall block and footwall block of a fault. (*B*) A photo of a fault near Kingman, Arizona, showing how the hanging wall has moved relative to the footwall.

in. The Appalachian Mountains have ridges of rocks that are more resistant to weathering, forming hills and mountains. The San Joaquin Valley, on the other hand, is a very large syncline in California.

Note that any kind of rock can be folded. Sedimentary rocks are usually the best example of folding, however, since the fold structures of rock layers are easy to see and describe. Folding is much harder to see in igneous or metamorphic rocks that are blends of minerals without a layered structure.

Faulting

Rock layers do not always respond to stress by folding. Rocks near the surface are cooler and under less pressure, so they tend to be more brittle. A sudden stress on these rocks may reach the rupture point, resulting in a cracking and breaking of the rock structure. When there is relative movement between the rocks on either side of a fracture, the crack is called a **fault.** When faulting occurs, the rocks on one side move relative to the rocks on the other side along the surface of the fault, which is called the *fault plane.* Faults are generally described in terms of (1) the steepness of the fault plane, that is, the angle between the plane and imaginary horizontal plane, and (2) the direction of relative movement. There are basically three ways that rocks on one side of a fault can move relative to the rocks on the other side: (1) up and down (called "dip"), (2) horizontally, or sideways (called "strike"), and (3) with elements of both directions of movement (called "oblique").

One classification scheme for faults is based on an orientation referent borrowed from mining (many ore veins are associated with fault planes). Imagine a mine with a fault plane running across a horizontal shaft. Unless the plane is perfectly vertical, a miner would stand on the mass of rock below the fault plane and look up at the mass of rock above. Therefore, the mass of rock below is called the *footwall* and the mass of rock above is called the *hanging wall* (figure 16.8). How the footwall and hanging wall have moved relative to one another describes three basic classes of faults: (1) normal, (2) reverse, and (3) thrust faults. A **normal fault** is one in which the hanging wall has moved downward relative to the footwall. This

seems "normal" in the sense that you would expect an upper block to slide *down* a lower block along a slope (figure 16.9A). Sometimes a huge block of rock bounded by normal faults will drop down, creating a *graben* (figure 16.9B). The opposite of a graben is a *horst,* which is a block bounded by normal faults that is uplifted (figure 16.9C). A very large block lifted sufficiently becomes a faultblock mountain. Many parts of the western United States are characterized by numerous faultblock mountains separated by adjoining valleys.

In a **reverse fault,** the hanging wall block has moved upward relative to the footwall block. As illustrated in

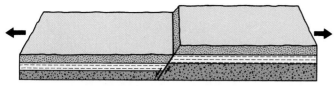

A Normal fault

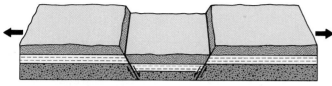

B Graben

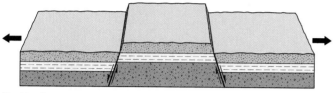

C Horst

FIGURE 16.9 How tensional stress could produce (A) a normal fault, (B) a graben, and (C) a horst.

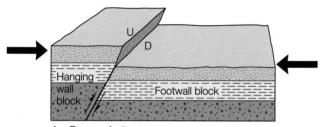

A Reverse fault

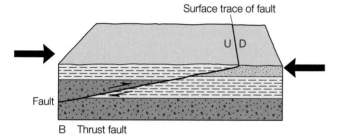

B Thrust fault

FIGURE 16.10 How compressive stress could produce (A) a reverse fault and (B) a thrust fault.

figure 16.10A, a reverse fault is probably the result of horizontal compressive stress.

A reverse fault with a low-angle fault plane is also called a **thrust fault** (figure 16.10B). In some thrust faults, the hanging wall block has completely overridden the lower footwall for 10 to 20 km (6 to 12 mi). This is sometimes referred to as an "over-thrust."

As shown in figures 16.9 and 16.10, the relative movement of blocks of rocks along a fault plane provides information about the stresses that produced the movement. Reverse and thrust faulting result from compressional stress in the direction of the movement. Normal faulting, on the other hand, results from a pulling-apart stress that might be associated with diverging plates. It might also be associated with the stretching and bulging up of the crust over a hot spot.

EARTHQUAKES

What is an earthquake? An **earthquake** is a quaking, shaking, vibrating, or upheaval of the ground. Earthquakes are the result of the sudden release of energy that comes from *stress* on rock beneath Earth's surface. In the section on folding, you learned that rock units can bend and become deformed in response to stress, but there are limits as to how much stress rock can take before it fractures. When it does fracture, the sudden movement of blocks of rock produces vibrations that move out as waves throughout Earth. These vibrations are called **seismic waves.** It is strong seismic waves that people feel as a shaking, quaking, or vibrating during an earthquake.

Seismic waves are generated when a huge mass of rock breaks and slides into a different position. As you learned in the section on folding, the plane between two rock masses that have moved into new relative positions is called a *fault.* Major earthquakes occur along existing fault planes or when a new fault is formed by the fracturing of rock. In either case, most earthquakes occur along a fault plane when there is displacement of one side relative to the other.

Most earthquakes occur along a fault plane and near Earth's surface. You might expect this to happen since the rocks near the surface are brittle and those deeper are more plastic from increased temperature and pressure. Shallow earthquakes are typical of those that occur at the boundary of the North American Plate, which is moving against the Pacific Plate. In California, the boundary between these two plates is known as the *San Andreas fault* (figure 16.11). The San Andreas fault runs north-south through California, with the Pacific Plate moving on one side and the North American Plate moving on the other. The two plates are tightly pressed against each other, and friction between the rocks along the fault prevents them from moving easily. Stress continues to build along the entire fault as one plate attempts to move along the other. Some elastic deformation does occur from the stress, but eventually, the rupture strength of the rock (or the friction) is overcome. The stressed rock, now released of the strain, snaps suddenly into new positions in the phenomenon known as **elastic rebound** (figure 16.12). The rocks are displaced to new positions on either side of the fault,

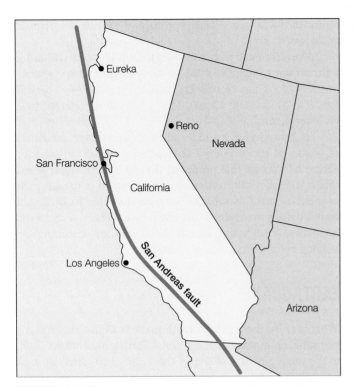

FIGURE 16.11 The San Andreas fault runs north-south for about 1,300 km (800 mi) through California.

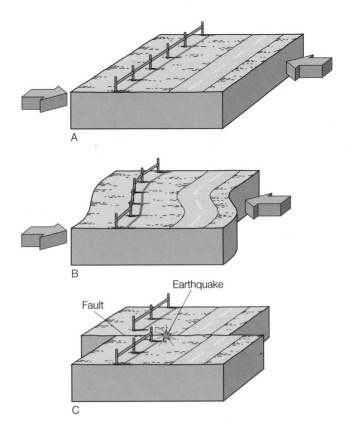

FIGURE 16.12 The elastic rebound theory of the cause of earthquakes. (*A*) Rock with stress acting on it. (*B*) Stress has caused strain in the rock. Strain builds up over a long period of time. (*C*) Rock breaks suddenly, releasing energy, with rock movement along a fault. Horizontal motion is shown; rocks can also move vertically. (*D*) Horizontal offset of rows in a lettuce field, 1979, El Centro, California.

and the vibrations from the sudden movement are felt as an earthquake. The elastic rebound and movement tend to occur along short segments of the fault at different times rather than along long lengths. Thus, the resulting earthquake tends to be a localized phenomenon rather than a regional one.

Most earthquakes occur near plate boundaries. They do happen elsewhere, but those are rare. The actual place where seismic waves originate beneath the surface is called the *focus* of the earthquake. The focus is considered to be the center of the earthquake and the place of initial rock movement on a fault. The point on Earth's surface directly above the focus is called the earthquake *epicenter* (figure 16.13).

Seismic waves radiate outward from an earthquake focus, spreading in all directions through the solid Earth's interior like the sound waves from an explosion. As introduced in chapter 15, there are three kinds of waves: *P-waves*, *S-waves*, and *surface waves*. A *seismometer* is an instrument used to measure seismic waves. Information about S- and P-waves as well as about the size of an earthquake can be read from seismometer recordings. S- and P-waves provide information about the location and magnitude of an earthquake as well as information about Earth's interior.

Seismic S- and P-waves leave the focus of an earthquake at essentially the same time. As they travel away from the focus, they gradually separate because the P-waves travel faster than the S-waves. To locate an epicenter, at least three recording stations measure the time lag between the arrival of the first P-waves and the first slower S-waves. The difference in the speed between the two waves is a constant. Therefore, the farther they travel, the greater the time lag between the arrival

of the faster P-waves and the slower S-waves (figure 16.14A). By measuring the time lag and knowing the speed of the two waves, it is possible to calculate the distance to their source. However, the calculated distance provides no information about the direction or location of the source of the waves. The location is found by first using the calculated distance as the radius of a circle drawn on a map. The place where the circles from the three recording stations intersect is the location of the source of the waves (figure 16.14B).

Earthquakes range from the many that are barely detectable to the few that cause widespread damage. The *intensity* of

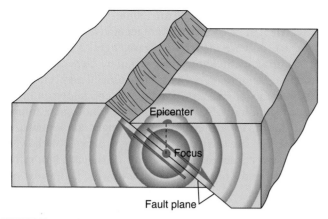

FIGURE 16.13 Simplified diagram of a fault, illustrating component parts and associated earthquake terminology.

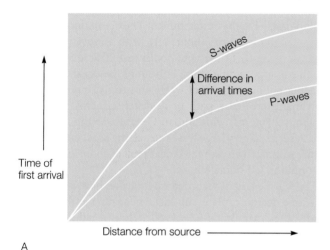

A

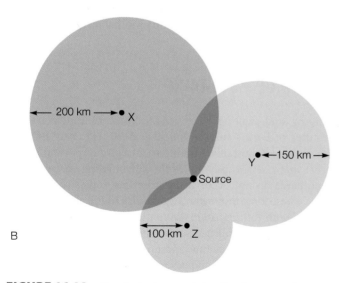

B

FIGURE 16.14 Use of seismic waves in locating the source of a disturbance. (*A*) Difference in times of first arrivals of P-waves and S-waves is a function of the distance from the source. (*B*) Triangulation using data from several seismograph stations allows location of the source.

an earthquake is a measure of the effect on people and buildings. Destruction is caused by the seismic waves, which cause the land and buildings to vibrate or shake. Vibrations during small quakes can crack windows and walls, but vibrations in strong quakes can topple bridges and buildings. Injuries and death are usually the result of falling debris, crumbling buildings, or falling bridges. Fire from broken gas pipes was a problem in the 1906 and 1989 earthquakes in San Francisco and the 1994 earthquake in Los Angeles. Broken water mains made it difficult to fight the 1906 San Francisco fires, but in 1989, fireboats and fire hoses using water pumped from the bay were able to extinguish the fires. Table 16.1 lists some recommendations of things to do during and after an earthquake.

Other effects of earthquakes include landslides, displacement of the land surface, and tsunamis. Vertical, horizontal, or both vertical and horizontal displacement of the land can occur during a quake. People sometimes confuse cause and effect when they see a land displacement saying things like, "Look what the earthquake did!" The fact is that the movement of the land probably produced the seismic waves (the earthquake). The seismic waves did not produce the land displacement. Displacements from a single earthquake can be up to 10 to 15 m (about 30 to 50 ft), but such displacements rarely happen.

TABLE 16.1   Earthquake Safety Rules from the National Oceanic and Atmospheric Administration

During the Shaking

1. *Don't panic.*
2. If you are indoors, stay there. Seek protection under a table or desk or in a doorway. Stay away from glass. Don't use matches, candles, or any open flame; douse all fires.
3. If you are outside, move away from buildings and power lines and stay in the open. Don't run through or near buildings.
4. If you are in a moving car, bring it to a stop as quickly as possible* but stay in it. The car's springs will absorb some of the shaking, and it will offer you protection.

After the Shaking

1. Check but do *not* turn on utilities. If you smell gas, open windows, shut off the main valve, and leave the building. Report the leak to the utility company and don't reenter the building until it has been checked out. If water mains are damaged, shut off the main valve. If electrical wiring is shorting, close the switch at the main meter box.
2. Turn on radio or television (if possible) for emergency bulletins.
3. Stay off the telephone except to report an emergency.
4. Stay out of severely damaged buildings that could collapse in aftershocks.
5. Don't go sightseeing; you will only hamper the efforts of emergency personnel and repair crews.

*Not near buildings, an overpass, or any other structure.
Source: National Oceanic and Atmospheric Administration.

TABLE 16.2   Effects of Earthquakes of Various Magnitudes

Richter Magnitudes	Description
0–2	Smallest detectable earthquake
2–3	Detected and measured but not generally felt
3–4	Felt as small earthquake but no damage occurs
4–5	Minor earthquake with local damage
5–6	Moderate earthquake with structural damage
6–7	Strong earthquake with destruction
7–8	Major earthquake with extensive damage and destruction
8–9	Great earthquake with total destruction

The size of an earthquake (if it was a "big one" or a "little one") can be measured in terms of vibrations, in terms of displacement, or in terms of the amount of energy released at the site of the earthquake. The larger the quake, the larger the waves recorded on a seismometer. From these recorded waves, scientists assign a number called the *magnitude.* Magnitude is a measure of the energy released during an earthquake. Earthquake magnitude is often reported using the *Richter scale* (Table 16.2). The Richter scale was developed by Charles Richter, who was a seismologist at the California Institute of Technology in the early 1930s. The scale was based on the widest swing in the back-and-forth line traces of seismograph recording. This scale assigns a number that increases with the magnitude of an earthquake. The numbers have meaning about (1) the severity of the ground-shaking vibrations and (2) the energy released by the earthquake. Each higher number indicates about ten times more ground movement and about thirty times more energy released than the preceding number. An earthquake measuring below 3 on the scale is usually not felt by people near the epicenter. The largest earthquake measured so far had a magnitude near 9, but there is actually no upper limit to the scale. Today, professional seismologists rate the size of earthquakes in different ways, depending on what they are comparing and why, but each way results in logarithmic scales similar to the Richter scale.

Tsunami is a Japanese term used to describe the very large ocean waves that can be generated by an earthquake, landslide, or volcanic explosion. Such large waves were formerly called "tidal waves." Since the large, fast waves were not associated with tides or tidal forces in any way, the term *tsunami* or *seismic sea wave* is preferred.

A tsunami, like other ocean waves, is produced by a disturbance. Most common ocean waves are produced by winds, travel at speeds of 90 km/h (55 mi/h), and produce a wave height of 0.6 to 3 m (2 to 10 ft) when they break on the shore. A tsunami, on the other hand, is produced by some strong disturbance in the seafloor, travels at speeds of 725 km/h (450 mi/h), and can produce a wave height of over 8 m (about 25 ft) when it breaks on the shore. A tsunami may have a very long wavelength of up to 200 km (about 120 mi) compared to the wavelength of ordinary wind-generated waves of 400 m (1,300 ft). Because of its great wavelength, a tsunami does not just break on the shore, then withdraw. Depending on the seafloor topography, the water from a tsunami may continue to rise for five to ten minutes, flooding the coastal region before the wave withdraws. A gently sloping seafloor and a funnel-shaped bay can force tsunamis to great heights as they break on the shore. The size of a particular tsunami depends on how the seafloor was disturbed. Generally, an earthquake that causes the seafloor suddenly to rise or fall favors the generation of a large tsunami.

On December 26, 2004, an earthquake with a magnitude of 9.0 occurred west of the Indonesian island of Sumatra. This was the largest quake worldwide in four decades. The focus was about 10 km (about 6 mi) beneath the ocean floor at the interface of the India and Burma Plates and was caused by the release of stresses that develop as the India Plate subducts beneath the overriding Burma Plate. A 100 km (62 mi) wide rupture occurred in the ocean floor, about 1,200 km (746 mi) long and parallel to the Sunda trench. The ocean floor was suddenly uplifted over 2 m (about 7 ft), with a movement about 10 m (about 33 ft) to the west-southwest. This displacement acted like a huge paddle at the bottom of the ocean, vertically displacing billions of tons of water and triggering a tsunami. The tsunami created a path of destruction across the 4,500 km (about 2,800 mi) wide Indian Ocean over the next seven hours. A series of very large waves struck the coast of the Indian Ocean, resulting in an estimated death toll of about 295,000 and over a million left homeless.

The world's largest recorded earthquakes have all occurred where one tectonic plate subducts beneath another. These include the magnitude-9.5 1960 Chile earthquake, the magnitude-9.2 1964 Prince William Sound, Alaska, earthquake, the magnitude-9.1 1957 Andreanof Islands, Alaska, earthquake, and the magnitude-9.0 1952 Kamchatka earthquake. Such earthquakes often create large tsunamis that result in death and destruction over a wide area.

ORIGIN OF MOUNTAINS

Folding and faulting have created most of the interesting features of Earth's surface, and the most prominent of these features are **mountains.** Mountains are elevated parts of Earth's crust that rise abruptly above the surrounding surface. Most mountains do not occur in isolation but rather in chains or belts. These long, thin belts are generally found along the edges of continents rather than in the continental interior. There are a number of complex processes involved in the origin of mountains and mountain chains, and no two mountains are exactly alike. For convenience, however, mountains can be classified according to three basic origins: (1) folding, (2) faulting, and (3) volcanic activity.

The major mountain ranges of Earth—the Appalachian, Rocky, and Himalayan Mountains, for example—have a great height that involves complex folding on a very large scale. The

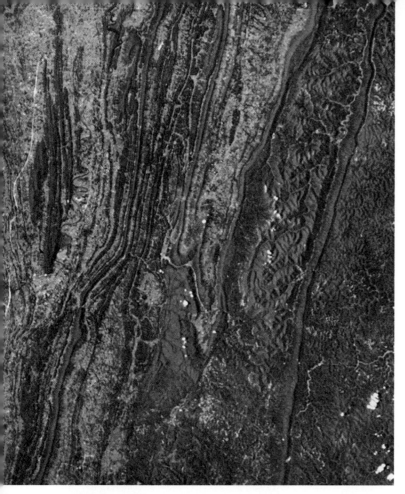

FIGURE 16.15 The folded structure of the Appalachian Mountains, revealed by weathering and erosion, is obvious in this *Skylab* photograph of the Virginia-Tennessee-Kentucky boundary area. The clouds are over the Blue Ridge Mountains.

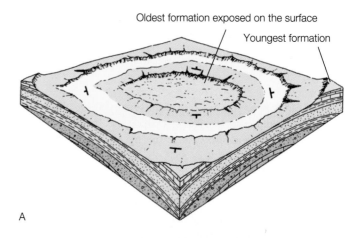

Oldest formation exposed on the surface

Youngest formation

A

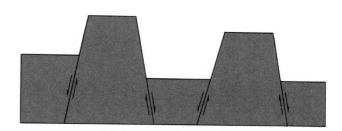

B

FIGURE 16.16 (*A*) A sketch of an eroded structural dome where all the rock layers dip away from the center. (*B*) A photo of a dome named Little Sundance Mountain (Wyoming), showing the more resistant sedimentary layers that dip away from the center.

Original sharply bounded fault blocks softened by erosion and sedimentation

Sediments

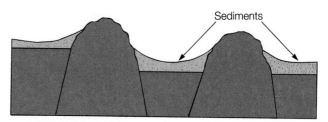

FIGURE 16.17 Faultblock mountains are weathered and eroded as they are elevated, resulting in a rounded shape and sedimentation rather than sharply edged fault blocks.

crust was thickened in these places as compressional forces produced tight, almost vertical folds. Thus, folding is a major feature of the major mountain ranges, but faulting and igneous intrusions are invariably also present. Differential weathering of different rock types produced the parallel features of the Appalachian Mountains that are so prominent in satellite photographs (figure 16.15). The folded sedimentary rocks of the Rockies are evident in the almost upright beds along the flanks of the front range.

A broad arching fold, which is called a dome, produced the Black Hills of South Dakota. The sedimentary rocks from the top of the dome have been weathered away, leaving a somewhat circular area of more resistant granite hills surrounded by upward-tilting sedimentary beds (figure 16.16). The Adirondack Mountains of New York are another example of this type of mountain formed from folding, called domed mountains.

Compression and relaxation of compressional forces on a regional scale can produce large-scale faults, shifting large crustal blocks up or down relative to one another. Huge blocks of rocks can be thrust to mountainous heights, creating a series of faultblock mountains. Faultblock mountains rise sharply from the surrounding land along the steeply inclined fault plane. The mountains are not in the shape of blocks, however, as weathering has carved them into their familiar mountainlike shapes (figure 16.17). The Teton Mountains

Volcanoes Change the World

A volcanic eruption changes the local landscape; that much is obvious. What is not so obvious are the worldwide changes that can happen just because of the eruption of a single volcano. Perhaps the most discussed change brought about by a volcano occurred back in 1815–16 after the eruption of Tambora in Indonesia. The Tambora eruption was massive, blasting huge amounts of volcanic dust, ash, and gas high into the atmosphere. Most of the ash and dust fell back to Earth around the volcano, but some dust particles and sulfur dioxide gas were pushed high into the stratosphere. It is known today that the sulfur dioxide from such explosive volcanic eruptions reacts with water vapor in the stratosphere, forming tiny droplets of diluted sulfuric acid. In the stratosphere, there is no convection, so the droplets of acid and dust from the volcano eventually form a layer around the entire globe. This forms a haze that remains in the stratosphere for years, reflecting and scattering sunlight.

What were the effects of a volcanic haze around the entire world? There were fantastic, brightly colored sunsets from the added haze in the stratosphere. On the other hand, it was also cooler than usual, presumably because of the reflected sunlight that did not reach Earth's surface. It snowed in New England in June 1816, and the cold continued into July. Crops failed, and 1816 became known as the "year without summer."

More information is available about the worldwide effects of present-day volcanic eruptions because there are now instruments to make more observations in more places. However, it is still necessary to do a great deal of estimating because of the relative inaccessibility of the worldwide stratosphere. It was estimated, for example, that the 1982 eruption of El Chichon in Mexico created enough haze in the stratosphere to reflect 5 percent of the solar radiation away from Earth. Researchers also estimated that the effects of the El Chichon eruption cooled the global temperatures by a few tenths of a degree for two or three years. The cooling did take place, but the actual El Chichon contribution to the cooling is not clear because of other interactions. Earth may have been undergoing global warming from the greenhouse effect, for example, so the El Chichon cooling effect could have actually been much greater. Other complicating factors such as the effects of El Niño make changes difficult to predict.

In June 1991, the Philippine volcano Mount Pinatubo erupted, blasting twice as much gas and dust into the stratosphere as El Chichon had about a decade earlier. The haze from such eruptions has the potential to cool the climate about 0.5°C (1°F). The overall result, however, will always depend on a possible greenhouse effect, a possible El Niño effect, and other complications.

---

of Wyoming and the Sierra Nevadas of California are classic examples of faultblock mountains that rise abruptly from the surrounding land. The various mountain ranges of Nevada, Arizona, Utah, and southeastern California have large numbers of faultblock mountains that generally trend north and south.

Lava and other materials from volcanic vents can pile up to mountainous heights on the surface. These accumulations can form local volcano-formed mountains near mountains produced by folding or faulting. Such mixed-origin mountains are common in northern Arizona, New Mexico, and western Texas. The Cascade Mountains of Washington and Oregon are a series of towering volcanic peaks, most of which are not active today. As a source of mountains, volcanic activity has an overall limited impact on the continents. The major mountains built by volcanic activity are the mid-oceanic ridges formed at diverging plate boundaries.

Deep within Earth, previously solid rock melts at high temperatures to become *magma*, a pocket of molten rock. Magma is not just melted rock alone, however, as the melt contains variable mixtures of minerals (resulting in different types of lava flows). It also includes gases such as water vapor, sulfur dioxide, hydrogen sulfide, carbon dioxide, and hydrochloric acid. You can often smell some of these gases around volcanic vents and hot springs. Hydrogen sulfide smells like rotten eggs or sewer gas. The sulfur smells like a wooden match that has just been struck.

The gases dissolved in magma play a major role in forcing magma out of the ground. As magma nears the surface, it comes under less pressure, and this releases some of the dissolved gases from the magma. The gases help push the magma out of the ground. This process is similar to releasing the pressure on a can of warm soda, which releases dissolved carbon dioxide.

Magma works its way upward from its source below to Earth's surface, here to erupt into a lava flow or a volcano. A **volcano** is a hill or mountain formed by the extrusion of lava or rock fragments from magma below. Some lavas have a lower viscosity than others, are more fluid, and flow out over the land rather than forming a volcano. Such *lava flows* can accumulate into a plateau of basalt, the rock that the lava formed as it cooled and solidified. The Columbia Plateau of the states of Washington, Idaho, and Oregon is made up of layer after layer of basalt that accumulated from lava flows. Individual flows of lava formed basalt layers up to 100 m (about 330 ft) thick, covering an area of hundreds of square kilometers. In places, the Columbia Plateau is up to 3 km (about 2 mi) thick from the accumulation of many individual lava flows.

The hill or mountain of a volcano is formed by ejected material that is deposited in a conical shape. The materials are deposited around a central *vent*, an opening through which an eruption takes place. The *crater* of a volcano is a basinlike depression over a vent at the summit of the cone. Figure 16.18 is an aerial view of Mount St. Helens, looking down into the crater at a volcanic dome that formed as magma periodically welled

FIGURE 16.18 This is the top of Mount St. Helens several years after the 1980 explosive eruption.

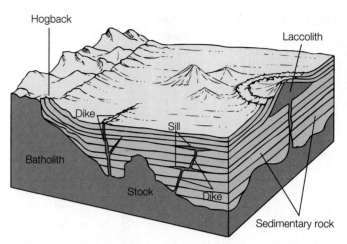

FIGURE 16.19 Here are the basic intrusive igneous bodies that form from volcanic activity.

upward into the floor of the crater. This photo was taken several years after Mount St. Helens erupted in May 1980. Since that time, the volcano had been quiet up until the fall of 2004, when thousands of small earthquakes preceded the renewed growth of the lava dome inside the crater. (Go to *http://vulcan.wr.usgs. gov/Volcanoes/Cascades/CurrentActivity/current_updates.html* for an update on the status of Mount St. Helens.) Geologists can tell when eruptions are near, since they are preceded by thousands of small earthquakes. Mount Garibaldi, Mount Rainier, Mount Hood, Mount Baker, and about ten other volcanic cones of the Cascade Range in Washington and Oregon could erupt next year, in the next decade, or in the next century. There are strong reasons that make it very likely that any one of these volcanic cones of the Cascade Range will indeed erupt again.

A volcano forms when magma breaks out at Earth's surface. Only a small fraction of all the magma generated actually reaches Earth's surface. Most of it remains below the ground, cooling and solidifying to form igneous rocks that were described in chapter 15. A large amount of magma that has crystallized below the surface is known as a *batholith*. A small protrusion from a batholith is called a *stock*. By definition, a stock has less than 100 km^2 (40 mi^2) of exposed surface area, and a batholith is larger. Both batholiths and stocks become exposed at the surface through erosion of the overlying rocks and rock materials, but not much is known about their shape below. The sides seem to angle away with depth, suggesting that they become larger with depth. The intrusion of a batholith sometimes tilts rock layers upward, forming a *hogback*, a ridge with equal slopes on its sides (figure 16.19). Other forms of intruded rock were formed as moving magma took the paths of least resistance, flowing into joints, faults, and planes between sedimentary bodies of rock. An intrusion that has flowed into a joint or fault that cuts across rock bodies is called a *dike*. A dike is usually tabular in shape, sometimes

appearing as a great wall when exposed at the surface. One dike can occur by itself, but frequently dikes occur in great numbers, sometimes radiating out from a batholith like spokes around a wheel. If the intrusion flowed into the plane of contact between sedimentary rock layers, it is called a *sill*. A *laccolith* is similar to a sill but has an arched top where the intrusion has raised the overlying rock into a blisterlike uplift (figure 16.19).

Overall, the origin of mountain systems and belts of mountains such as the Cascades involves a complex mixture of volcanic activity as well as folding and faulting. An individual mountain, such as Mount St. Helens, can be identified as having a volcanic origin. The overall picture is best seen, however, from generalizations about how the mountains have grown along the edge of plates that are converging. Such converging boundaries are the places of folding, faulting, and associated earthquakes. They are also the places of volcanic activities, events that build and thicken Earth's crust. Thus, plate tectonics explains that mountains are built as the crust thickens at a convergent boundary between two plates. These mountains are slowly weathered and worn down as the next belt of mountains begins to build at the new continental edge.

PROCESSES THAT TEAR DOWN THE SURFACE

Sculpturing of Earth's surface takes place through agents and processes acting so gradually that humans are usually not aware that it is happening. Sure, some events such as a landslide or the movement of a big part of a beach by a storm are noticed. But the continual, slow, downhill drift of all the soil on a slope or the constant shift of grains of sand along a beach are outside the awareness of most people. People do notice the muddy water moving rapidly downstream in the swollen river after a storm, but few are conscious of the slow, steady dissolution of limestone by acid rain percolating through it. Yet, it is the processes of slow-moving, shifting grains and bits of rocks, and

FIGURE 16.20 The piles of rocks and rock fragments around a mass of solid rock is evidence that the solid rock is slowly crumbling away. This solid rock that is crumbling to rock fragments is in the Grand Canyon, Arizona.

slow dissolving that will wear down the mountains, removing all the features of the landscape that you can see.

A mountain of solid granite on the surface of Earth might appear to be a very solid, substantial structure, but it is always undergoing slow and gradual changes. Granite on Earth's surface is exposed to and constantly altered by air, water, and other agents of change. It is altered both in appearance and in composition, slowly crumbling, and then dissolving in water. Smaller rocks and rock fragments are moved downhill by gravity or streams, exposing more granite that was previously deeply buried. The process continues, and ultimately—over much time—a mountain of solid granite is reduced to a mass of loose rock fragments and dissolved materials. The photograph in figure 16.20 is a snapshot of a mountain-sized rock mass in a stage somewhere between its formation and its eventual destruction to rock fragments. Can you imagine the length of time that such a process requires?

Weathering

The slow changes that result in the breaking up, crumbling, and destruction of any kind of solid rock are called **weathering.** The term implies changes in rocks from the action of the weather, but it actually includes chemical, physical, and biological processes. These weathering processes are important and necessary in (1) the rock cycle, (2) the formation of soils, and (3) the movement of rock materials over Earth's surface. Weathering is important in the rock cycle because it produces sediment, the raw materials for new rocks. It is important in the formation of soils because soil is an accumulation of rock fragments and organic matter. Because weathering reduces the sizes of rock particles, it is important in preparing the particles to be transported by wind or moving water.

Weathering breaks down rocks physically and chemically, and this breaking down can occur while the rocks are stationary or while they are moving. The process of physically removing weathered materials is called **erosion.** *Weathering* prepares the way for erosion by breaking solid rock into fragments. The fragments are then *eroded*, physically picked up by an agent such as a stream or a glacier. After they are eroded, the materials are then removed by *transportation*. Transportation is the movement of eroded materials by agents such as rivers, glaciers, wind, or waves. The weathering process continues during transportation. A rock being tumbled downstream, for example, is physically worn down as it bounces from rock to rock. It may be chemically altered as well as it is bounced along by the moving water. Overall, the combined action of weathering and erosion wears away and lowers the elevated parts of Earth and sculpts their surfaces.

There are two basic kinds of weathering that act to break down rocks: *mechanical weathering* and *chemical weathering*. *Mechanical weathering* is the physical breaking up of rocks without any changes in their chemical composition. Mechanical weathering results in the breaking up of rocks into smaller and smaller pieces, so it is also called *disintegration*. If you smash a sample of granite into smaller and smaller pieces, you are mechanically weathering the granite. *Chemical weathering* is the alteration of minerals by chemical reactions with water, gases of the atmosphere, or solutions. Chemical weathering results in the dissolving or breaking down of the minerals in rocks, so it is also called *decomposition*. If you dissolve a sample of limestone in a container of acid, you are chemically weathering the limestone.

Mechanical Weathering

Examples of mechanical weathering in nature include the disintegration of rocks caused by (1) *wedging effects* and (2) the *effects of reduced pressure*. Wedging effects are often caused by the repeated freezing and thawing of water in the pores and small cracks of otherwise solid rock. If you have ever seen what happens when water in a container freezes, you know that freezing water expands and exerts a pressure on the sides of its container. As water in a pore or a crack of a rock freezes, it also expands, exerting a pressure on the walls of the pore or crack, making it slightly larger. The ice melts and the enlarged pore or crack again becomes filled with water for another cycle

Water freezing in crack expands, widening crack.

Eventually rock is broken free by progressive fracturing.

A

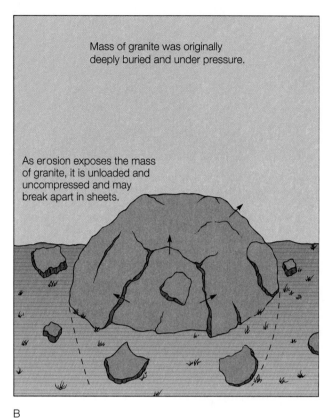

Mass of granite was originally deeply buried and under pressure.

As erosion exposes the mass of granite, it is unloaded and uncompressed and may break apart in sheets.

B

FIGURE 16.21 (*A*) Frost wedging and (*B*) exfoliation are two examples of mechanical weathering, or disintegration, of solid rock.

A

B

FIGURE 16.22 Growing trees can break, separate, and move solid rock. (*A*) Note how this tree has raised the sidewalk. (*B*) This tree is surviving by growing roots into tiny joints and cracks, which become larger as the tree grows.

of freezing and thawing. As the process is repeated many times, small pores and cracks become larger and larger, eventually forcing pieces of rock to break off. This process is called **frost wedging** (figure 16.21A). It is an important cause of mechanical weathering in mountains and other locations where repeated cycles of water freezing and thawing occur. The roots of trees and shrubs can also mechanically wedge rocks apart as they grow into cracks. You may have noticed the work of roots when trees or shrubs have grown next to a sidewalk for some period of time (figure 16.22).

The other example of mechanical weathering is believed to be caused by the reduction of pressure on rocks. As more and more weathered materials are removed from the surface, the downward pressure from the weight of the material on the rock below becomes less and less. The rock below begins to expand upward, fracturing into concentric sheets from the effect of reduced pressure. These curved, sheetlike plates fall away

FIGURE 16.23 Spheroidal weathering of granite. The edges and corners of an angular rock are attacked by weathering from more than one side and retreat faster than flat rock faces. The result is rounded granite boulders, which often shed partially weathered minerals in onionlike layers.

later in the mechanical weathering process called *exfoliation* (figure 16.21B). *Exfoliation* is the term given to the process of spalling off of layers of rock, somewhat analogous to peeling layers from an onion. Granite commonly weathers by exfoliation, producing characteristic dome-shaped hills and rounded boulders. Stone Mountain, Georgia, is a well-known example of an exfoliation-shaped dome. The onionlike structure of exfoliated granite is a common sight in the Sierras, Adirondacks, and any mountain range where older exposed granite is at the surface (figure 16.23).

Chemical Weathering

Examples of chemical weathering include (1) oxidation, (2) carbonation, and (3) hydration. *Oxidation* is a chemical reaction between oxygen and the minerals making up rocks. The ferromagnesian minerals contain iron, magnesium, and other metal ions in a silicate structure. Iron can react with oxygen to produce several different iron oxides, each with its own characteristic color. The most common iron oxide (hematite) has a deep red color. Other oxides of iron are brownish to yellow-brownish. It is the presence of such iron oxides that colors many sedimentary rocks and soils. The red soils of Oklahoma, Georgia, and many other places are colored by the presence of iron oxides produced by chemical weathering.

Carbonation is a chemical reaction between carbonic acid and the minerals making up rocks. Rainwater is naturally somewhat acidic because it dissolves carbon dioxide from the air. This forms a weak acid known as carbonic acid (H_2CO_3), the same acid that is found in your carbonated soda pop. Carbonic acid rain falls on the land, seeping into cracks and crevices where it reacts with minerals. Limestone, for example, is easily weathered to a soluble form by carbonic acid. The limestone caves of Missouri, Kentucky, New Mexico, and elsewhere were produced by the chemical weathering of limestone by carbonation (figure 16.24). Minerals containing calcium, magnesium, sodium, potassium, and iron are chemically weathered by carbonation to produce salts that are soluble in water.

A

B

FIGURE 16.24 Limestone caves develop when slightly acidic groundwater dissolves limestone along joints and bedding planes, carrying away rock components in solution. (*A*) Joints and bedding planes in a limestone bluff. (*B*) This stream has carried away less-resistant rock components, forming a cave under the ledge.

Hydration is a reaction between water and the minerals of rocks. The process of hydration includes (1) the dissolving of a mineral and (2) the combining of water directly with a mineral. Some minerals, for example, halite (which is sodium

Acid rain is a general term used to describe any acidic substances, wet or dry, that fall from the atmosphere. Wet acidic deposition could be in the form of rain, but snow, sleet, and fog could also be involved. Dry acidic deposition could include gases, dust, or any solid particles that settle out of the atmosphere to produce an acid condition.

Pure, unpolluted rain is naturally acidic. Carbon dioxide in the atmosphere is absorbed by rainfall, forming carbonic acid (H_2CO_3). Carbonic acid lowers the pH of pure rainfall to a range of 5.6–6.2. Decaying vegetation in local areas can provide more CO_2, making the pH even lower. A pH range of 4.5–5.0, for example, has been measured in rainfall of the Amazon jungle. Human-produced exhaust emissions of sulfur and nitrogen oxides can lower the pH of rainfall even more, to a 4.0–5.5 range. This is the pH range of acid rain (box table 16.1).

The sulfur and nitrogen oxides that produce acid rain come from exhaust emissions of industries and electric utilities that burn fossil fuels and from the exhaust of cars, trucks, and buses. The oxides are the raw materials of acid rain and are not themselves acidic. They react with other atmospheric chemicals to form sulfates and nitrates, which combine with water vapor to form sulfuric acid and nitric acid. These are the chemicals of concern in acid rain.

Acid rain falls on the land, bodies of water, forests, crops, buildings, and people, so the concerns about acid rain center on its environmental impact on lakes, forest, crops, materials, and human health. Acid rain accelerates chemical weathering, leaches essential nutrients from the soil, and acidifies lakes and streams. All of

Box Table 16.1

Approximate pH of Some Common Acidic Substances

Substance	pH (or pH Range)
Hydrochloric acid (4%)	0
Gastric (stomach) solution	1.6–1.8
Lemon juice	2.2–2.4
Vinegar	2.4–3.4
Carbonated soft drinks	2.0–4.0
Grapefruit	3.0–3.2
Oranges	3.2–3.6
Acid rain	**4.0–5.5**
Tomatoes	4.2–4.4
Potatoes	5.7–5.8
Natural rainwater	**5.6–6.2**
Milk	6.3–6.7
Pure water	7.0

these processes affect plants, animals, and microbes. Chemical weathering by acid rain also can cause deterioration of buildings and other structures.

The type of rocks making up the local landscape can either moderate or aggravate the problems of acid rain. Limestone and the soils of arid climates tend to neutralize acid, *while waters in granite rocks and soils* cannot neutralize acid and tend already to be somewhat acidic.

Although natural phenomena, such as volcanoes, contribute acids to the atmosphere, human-produced emissions of sulfur and nitrogen oxides from burning fossil fuels are the primary cause of acid rain. The heavily industrialized northeastern part of the United States, from the Midwest through New England, releases sulfur and nitrogen emissions that result in a

precipitation pH of 4.0 to 4.5. Unfortunately, the area of New England and adjacent Canada downwind of major acid rain sources has granite bedrock, which means that the effects of acid rain will not be moderated as they would in the West or Midwest. This region is the geographic center of the acid rain problem in North America. The solution to the problem is being sought by (1) using fuels other than fossil fuels when possible and (2) reducing the thousands of tons of sulfur and nitrogen oxides that are dumped into the atmosphere each day when fossil fuels are used.

Questions to Discuss

Discuss with your group the following questions concerning acid rain.

1. Should the use of fossil fuels be taxed to cut the source of acid rain and fund solutions? Provide reasons with your answer.
2. Electric utilities are required to remove sulfur dioxide from power plant exhaust according to the best technology that was available when the plant was constructed. Should they retrofit all plants with more expensive technology to reduce the amount released even more? What if doing so would increase your electric bill by 50 percent? Would you still support further reductions in the amount of sulfur dioxide released?
3. What are the advantages and disadvantages of a total ban on the use of fossil fuels?
4. Brainstorm with your group to see how many solutions you can think of to stop acid rain damage.

---

chloride), dissolve in water to form a solution. The carbonates formed from carbonation are mostly soluble, so they are easily leached from a rock by dissolving. Water also combines directly with some minerals to form new, different minerals. The feldspars, for example, undergo hydration and carbonation to produce (1) water-soluble potassium carbonate, (2) a chemical product that combines with water to produce a clay mineral, and (3) silica. The silica, which is silicon dioxide (SiO_2), may appear as a suspension of finely divided particles or in solution.

Mechanical and chemical weathering are interrelated, working together in breaking up and decomposing solid rocks of Earth's surface. In general, mechanical weathering results in cracks in solid rocks and broken-off coarse fragments. Chemical weathering results in finely pulverized materials and ions in solution, the ultimate decomposition of a solid rock. Consider, for example, a mountain of solid granite, the most common rock found on continents. In general, granite is made up of 65 percent feldspars, 25 percent quartz, and about 10 percent ferromagnesian minerals. Mechanical weathering begins the destruction process as exfoliation and frost wedging create cracks in the solid mass of granite. Rainwater, with dissolved oxygen and carbon dioxide, flows and seeps into the cracks and reacts with ferromagnesian minerals to form soluble carbonates and metal oxides. Feldspars undergo carbonation and hydration, forming clay minerals and soluble salts, which are washed away. Quartz is less susceptible to chemical weathering and remains mostly unchanged to form sand grains. The end products of the complete weathering of granite are quartz sand, clay minerals, metal oxides, and soluble salts.

Erosion

Weathering has prepared the way for erosion to pick up and for some agent of transportation to move or carry away the fragments, clays, and solutions that have been produced from solid rock. The weathered materials can be moved to a lower elevation by the direct result of gravity acting alone. They can also be moved to a lower elevation by gravity acting through some intermediate agent, such as running water, wind, or glaciers. The erosion of weathered materials as a result of gravity alone will be considered first.

Mass Movement

Gravity constantly acts on every mass of materials on the surface of Earth, pulling parts of elevated regions toward lower levels. Rocks in the elevated regions are able to temporarily resist this constant pull through their cohesiveness with a main rock mass or by the friction between the rock and the surface of the slope. Whenever anything happens to reduce the cohesiveness or to reduce the friction, gravity pulls the freed material to a lower elevation. Thus, gravity acts directly on individual rock fragments and on large amounts of surface materials as a mass, pulling all to a lower elevation. Erosion caused by gravity acting directly is called **mass movement** (also called mass wasting). Mass movement can be so slow that it is practically imperceptible. *Creep,* the slow downhill movement of soil down

a steep slope, for example, is detectable only from the peculiar curved growth patterns of trees growing in the slowly moving soil (figure 16.25). At the other extreme, mass movement can be as sudden and swift as a rock bounding and clattering down a slope below a cliff. A *landslide* is a generic term used to describe any slow to rapid movement of any type or mass of materials, from the short slump of a hillside to the slide of a whole mountainside. Either slow or sudden, mass movement is a small victory for gravity in the ongoing process of leveling the landmass of Earth.

Running Water

Running water is the most important of all the erosional agents of gravity that remove rock materials to lower levels. Erosion by running water begins with rainfall. Each raindrop impacting the soil moves small rock fragments about but also begins to dissolve some of the soluble products of weathering. If the rainfall is heavy enough, a shallow layer, or sheet, of water forms on the surface, transporting small fragments and dissolved materials across the surface. This *sheet erosion* picks up fragments and dissolved material, then transports them to small streams at lower levels (figure 16.26). The small streams move to larger channels,

FIGURE 16.25 The slow creep of soil is evidenced by the strange growth pattern of these trees.

FIGURE 16.26 Moving streams of water carry away dissolved materials and sediments as they slowly erode the land.

FIGURE 16.27   A river usually stays in its channel, but during a flood, it spills over and onto the adjacent flat land called the *floodplain*.

and the running water transports materials three different ways: (1) as dissolved rock materials carried in solution, (2) as clay minerals and small grains carried in suspension, and (3) as sand and larger rock fragments that are rolled, bounced, and slid along the bottom of the streambed. Just how much material is eroded and transported by the stream depends on the volume of water, its velocity, and the load that it is already carrying.

Streams and major rivers are at work, for the most part, twenty-four hours a day every day of the year moving rock fragments and dissolved materials from elevated landmasses to the oceans. Any time you see mud, clay, and sand being transported by a river, you know that the river is at work moving mountains, bit by bit, to the ocean. It has been estimated that rivers remove enough dissolved materials and sediments to lower the whole surface of the United States flat in a little over 20 million years, a very short time compared to the 4.6-billion-year age of Earth.

In addition to transporting materials that were weathered and eroded by other agents of erosion, streams do their own erosive work. Streams can dissolve soluble materials directly from rocks and sediments. They also quarry and pluck fragments and pieces of rocks from beds of solid rock by hydraulic action. Most of the erosion accomplished directly by streams, however, is done by the more massive fragments that are rolled, bounced, and slid along the streambed and against each other. This results in a grinding action on the fragments and a wearing away of the streambed.

As a stream cuts downward into its bed, other agents of erosion such as mass movement begin to widen the channel as materials on the stream bank tumble into the moving water. The load that the stream carries is increased by this activity, which slows the stream. As the stream slows, it begins to develop bends, or *meanders* along the channel. Meanders have a dramatic effect on stream erosion because the water moves faster around an outside bank than it does around the inside bank downstream. This difference in stream velocity means that the stream has a greater erosion ability on the outside, downstream side and less on the sheltered area inside of curves. The stream begins to widen the floor of the valley through which it runs by eroding on the outside of the meander, then depositing the eroded material on the inside of another bend downstream. The stream thus begins to erode laterally, slowly working its way across the land. Sometimes two bends in the stream meet, forming a cutoff meander called an *oxbow lake*.

A stream, along with mass movement, develops a valley on a widening floodplain. A **floodplain** is the wide, level

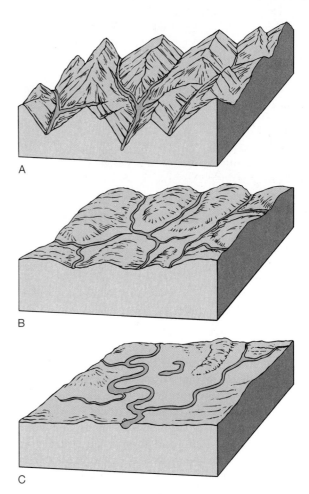

FIGURE 16.28   Three stages in the aging and development of a stream valley: (*A*) youth, (*B*) maturity, and (*C*) old age.

floor of a valley built by a stream (figure 16.27). It is called a floodplain because this is where the stream floods when it spills out of its channel. The development of a stream channel into a widening floodplain seems to follow a general, idealized aging pattern (figure 16.28). When a stream is on a recently uplifted landmass, it has a steep gradient, a vigorous,

FIGURE 16.29 The waterfall and rapids on the Yellowstone River in Wyoming indicate that the river is actively down cutting. Note the V-shaped cross-profile and lack of floodplain, characteristics of a young stream valley.

energetic ability to erode the land, and characteristic features known as the stage of youth. *Youth* is characterized by a steep gradient, a V-shaped valley without a floodplain, and the presence of features that interrupt its smooth flow such as boulders in the streambed, rapids, and waterfalls (figure 16.29). Stream erosion during youth is predominantly downward. The stream eventually erodes its way into *maturity* by eroding away the boulders, rapids, and waterfalls, and in general smoothing and lowering the stream gradient. During maturity, meanders form over a wide floodplain that now occupies the valley floor. The higher elevations are now more sloping hills at the edge of the wide floodplain rather than steep-sided walls close to the river channel. *Old age* is marked by a very low gradient in extremely broad, gently sloping valleys. The stream now flows slowly in broad meanders over the wide floodplain. Floods are more common in old age since the stream is carrying a full load of sediments and flows sluggishly.

Many assumptions are made in any generalized scheme of the erosional aging of a stream. Streams and rivers are dynamic systems that respond to local conditions, so it is possible to find an "old age feature" such as meanders in an otherwise youthful valley. This is not unlike finding a gray hair on an eighteen-year-old youth, and in this case, the presence of the gray hair does not mean old age. In general, old age characteristics are observed near the *mouth* of a stream, where it flows into an ocean, lake, or another stream. Youthful characteristics are observed at the *source*, where the water collects to first form the stream channel. As the stream slowly lowers the land, the old age characteristics will move slowly but surely upstream from the mouth toward the source.

When the stream flows into the ocean or a lake, it loses all of its sediment-carrying ability. It drops the sediments, forming a deposit at the mouth called a **delta** (figure 16.30). Large rivers such as the Mississippi River have large and extensive deltas that actually extend the landmass more and more over time. In a way, you could think of the Mississippi River delta as being formed from pieces and parts of the Rocky Mountains, the Ozark Mountains, and other elevated landmasses that the Mississippi has carried there over time.

Glaciers

Glaciers presently cover only about 10 percent of Earth's continental land area, and since much of this is at higher latitudes, it might seem that glaciers would not have much of an overall effect in eroding the land. However, ice has sculptured much of the present landscape, and features attributed to glacial episodes are found over about three-quarters of the continental surface. Only a few tens of thousands of years ago, sheets of ice covered major portions of North America, Europe, and Asia. Today, the most extensive glaciers in the United States are those of Alaska, which cover about 3 percent of the state's land area. Less extensive glacier ice is found in the mountainous regions of Washington, Montana, California, Colorado, and Wyoming.

A **glacier** is a mass of ice on land that moves under its own weight. Glacier ice forms gradually from snow, but the quantity of snow needed to form a glacier does not fall in a single winter. Glaciers form in cold climates where some snow and ice persist throughout the year. The amount of winter snowfall must exceed the summer melting to accumulate a sufficient

A

FIGURE 16.30 (*A*) Delta of Nooksack River, Washington. Note the sediment-laden water and how the land is being built outward by river sedimentation. (*B*) Cross section showing how a small delta might form. Large deltas are more complicated than this.

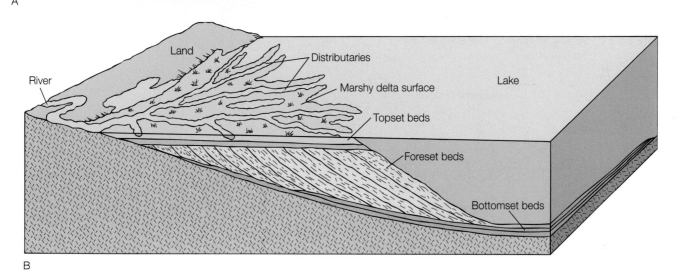

B

mass of snow to form a glacier. As the snow accumulates, it is gradually transformed into ice. The weight of the overlying snow packs it down, driving out much of the air, and causing it to recrystallize into a coarser, denser mass of interlocking ice crystals that appears to have a blue to deep blue color. Complete conversion of snow into glacial ice may take from five to thirty-five hundred years, depending on such factors as climate and rate of snow accumulation at the top of the pile. Eventually, the mass of ice will become large enough that it begins to flow, spreading out from the accumulated mass. Glaciers that form at high elevations in mountainous regions, which are called *alpine glaciers,* tend to flow downhill through a valley so they are also called *valley glaciers* (figure 16.31). Glaciers that cover a large area of a continent are called *continental glaciers.* Continental glaciers can cover whole continents and reach a thickness of 1 km (about 3,280 ft) or more. Today, the remaining continental glaciers are found on Greenland and the Antarctic.

Glaciers move slowly and unpredictably, spreading like a huge blob of putty under the influence of gravity. As an

alpine glacier moves downhill through a V-shaped valley, the sides and bottom of the valley are eroded wider and deeper. When the glacier later melts, the V-shaped valley has been transformed into a U-shaped valley that has been straightened and deepened by the glacial erosion. The glacier does its erosional work using three different techniques: (1) by bulldozing, (2) by plucking, and (3) by abrasion. *Bulldozing,* as the term implies, is the pushing along of rocks, soil, and sediments by the leading edge of an advancing glacier. Deposits of bulldozed rocks and other materials that remain after the ice melts are called *moraines. Plucking* occurs as water seeps into cracked rocks and freezes, becoming a part of the solid glacial mass. As the glacier moves on, it pulls the fractured rock apart and plucks away chunks of it. The process is accelerated by the frost-wedging action of the freezing water. Plucking at the uppermost level of an alpine glacier, combined with weathering of the surrounding rocks, produces a rounded or bowl-like depression known as a *cirque* (figure 16.32). *Abrasion* occurs as the rock fragments frozen into the moving glacial ice scratch, polish, and grind against surrounding rocks at the base and

FIGURE 16.31   Valley glacier on Mount Logan, Yukon Territory.

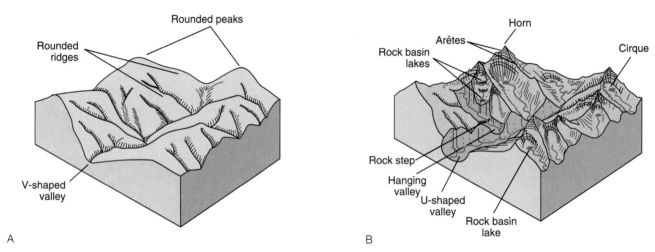

FIGURE 16.32   (*A*) A stream-carved mountainside before glaciation. (*B*) The same area after glaciation, with some of the main features of mountain glaciation labeled.

along the valley walls. The result of this abrasion is the pulverizing of rock into ever finer fragments, eventually producing a powdery, silt-sized sediment called *rock flour*. Suspended rock flour in meltwater from a glacier gives the water a distinctive gray to blue-gray color.

Glaciation is continuously at work eroding the landscape in Alaska and many mountainous regions today. The glaciation that formed the landscape features in the Rockies, the Sierras, and across the northeastern United States took place thousands of years ago.

Wind

Like running water and moving glaciers, wind also acts as an agent shaping the surface of the land. It can erode, transport, and deposit materials. However, wind is considerably less

efficient than ice or water in modifying the surface. Air is much less dense and does not have the eroding or carrying power of water or ice. In addition, a stream generally flows most of the time, but the wind blows only occasionally in most locations. Thus, on a worldwide average, winds move only a few percent as much material as do streams. Wind also lacks the ability to attack rocks chemically as water does through carbonation and other processes, and the wind cannot carry dissolved sediments in solution. Even in many deserts, more sediment is moved during the brief periods of intense surface runoff following the occasional rainstorms than is moved by wind during the prolonged dry periods.

Flowing air and moving water do have much in common as agents of erosion since both are fluids. Both can move larger particles by rolling them along the surface and can move finer particles by carrying them in suspension. Both can move larger

James Hutton was a Scottish natural philosopher who pioneered uniformitarian geology. The son of an Edinburgh merchant, Hutton studied at Edinburgh University, Paris, and Leiden, training first for the law but taking his doctorate in medicine in 1749 (though he never practiced). He spent the next two decades traveling and farming in the southeast of Scotland. During this time, he cultivated a love of science and philosophy, developing a special taste for geology. About 1768, he returned to his native Edinburgh. A friend of Joseph Black, William Cullen, and James Watt, Hutton shone as a leading member of the scientific and literary establishment, playing a large role in the early history of the Royal Society of Edinburgh and in the Scottish Enlightenment.

Hutton wrote widely on many areas of natural science, including chemistry (where he opposed Lavoisier), but he is best known for his geology, set out in his *Theory of the Earth*, of which a short version appeared in 1788, followed by the definitive statement in 1795. In that work, Hutton attempted (on the basis of both theoretical considerations and personal fieldwork) to demonstrate that Earth formed a steady-state system in which terrestrial causes had always been of the same kind as at present, acting with comparable intensity (the principle later known as uniformitarianism). In Earth's economy, in the imperceptible creation and devastation of landforms, there was no vestige of a beginning nor prospect of an end. Continents were continually being gradually eroded by rivers and weather. Denuded debris accumulated on the seabed, to be consolidated into strata and subsequently thrust upward to form new continents, thanks to the action of Earth's central heat. Nonstratified rocks such as granite were of igneous origin. All Earth's processes were exceptionally leisurely, and hence, Earth must be incalculably old.

Though supported by the experimental findings of Sir James Hall, Hutton's theory was vehemently attacked in its day, partly because it appeared to point to an eternal Earth and hence to atheism. It found more favor when popularized by Hutton's friend, John Playfair, and later by Charles Lyell. The notion of uniformitarianism still forms the groundwork for much geological reasoning.

Source: Modified from the *Hutchinson Dictionary of Scientific Biography*. © RM, 2007. Reprinted by permission.

---

and more massive particles with increased velocities. Water is denser and more viscous than air, so it is more efficient at transporting quantities of material than is the wind, but the processes are quite similar.

Two major processes of wind erosion are called (1) abrasion and (2) deflation. *Wind abrasion* is a natural sandblasting process that occurs when the particles carried along by the wind break off small particles and polish what they strike. Generally, the harder mineral grains such as quartz sand accomplish this best near the ground where the wind is bouncing them along. Wind abrasion can strip paint from a car exposed to the moving particles of a dust storm, eroding the paint along with rocks on the surface. Rocks and boulders exposed to repeated wind storms where the wind blows consistently from one or a few directions may be planed off from the repeated action of this natural sandblasting. Rocks sculptured by wind abrasion are called *ventifacts*, after the Latin meaning "wind-made" (figure 16.33).

Deflation, after the Latin meaning "to blow away," is the widespread picking up of loose materials from the surface. Deflation is naturally most active where winds are unobstructed and the materials are exposed and not protected by vegetation. These conditions are often found on deserts, beaches, and unplanted farmland between crops. During the 1930s, several years of drought killed the native vegetation in the Plains states during a period of increased farming activity. Unusually strong winds eroded the unprotected surface, removing and transporting hundreds of millions of tons of soil. This period of prolonged drought, dust storms, and general economic disaster for farmers in the area is known as the Dust Bowl episode.

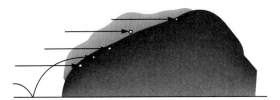

If wind is predominantly from one direction, rocks will be planed off or flattened on the upwind side.

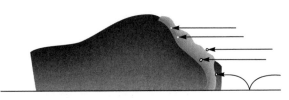

With a persistent shift in wind direction, additional facets are cut in the rock.

FIGURE 16.33 Ventifact formation by abrasion from one or several directions.

The most common wind-blown deposits are (1) dunes and (2) loess. A **dune** is a low mound or ridge of sand or other sediments. Dunes form when sediment-bearing wind encounters an obstacle that reduces the wind velocity. With a slower velocity, the wind cannot carry as large a load, so sediments are deposited on the surface. This creates a windbreak, which results in a growing obstacle, a dune. Once formed, a dune tends to migrate, particularly if the winds blow predominantly from one direction. Dunes are commonly found in semiarid areas or near beaches.

Another common wind deposit is called *loess* (pronounced "luss"). Loess is a very fine dust, or silt, that has been deposited over a large area. One such area is located in the central part of the United States, particularly to the east sides of the major rivers of the Mississippi basin. Apparently, this deposit originated from the rock flour produced during the last great ice age. The rock flour was probably deposited along the major river valleys and later moved eastward by the prevailing westerly winds.

Since rock flour is produced by the mechanical grinding action of glaciers, it has not been chemically broken down. Thus, the loess deposit contains many minerals that were not leached out of the deposit as typically occurs with chemical weathering. It also has an open, porous structure since it does not have as much of the chemically produced clay minerals. The good moisture-holding capacity from this open structure, together with the presence of minerals that serve as plant nutrients, make farming soils formed from loess deposits particularly productive.

SUMMARY

The *principle of uniformity* is the frame of reference that the same geologic processes you see changing rocks today are the same processes that changed them in the past.

Diastrophism is the process of deformation that changes Earth's surface, and the movement of magma is called *vulcanism. Diastrophism, vulcanism,* and *earthquakes* are closely related, and their occurrence can be explained most of the time by events involving *plate tectonics.*

Stress is a force that tends to compress, pull apart, or deform a rock, and the adjustment to stress is called *strain.* Rocks respond to stress by (1) withstanding the stress without change, (2) undergoing *elastic strain,* (3) undergoing *plastic strain,* or (4) by *breaking* in *fracture strain.* Exactly how a particular rock responds to stress depends on (1) the nature of the rock, (2) the temperature, and (3) how quickly the stress is applied over time.

Deeply buried rocks are at a higher temperature and tend to undergo *plastic deformation,* resulting in a wrinkling of the layers into *folds.* The most common are an arch-shaped fold called an *anticline* and a trough-shaped fold called a *syncline.* Anticlines and synclines are most easily observed in sedimentary rocks because they have bedding planes, or layers.

Rocks near the surface tend to break from a sudden stress. A break with movement on one side of the break relative to the other side is called a *fault.* The vibrations that move out as waves from the movement of rocks are called an *earthquake.* The actual place where an earthquake originates is called its *focus.* The place on the surface directly above a focus is called an *epicenter.* There are three kinds of waves that travel from the focus: *S-, P-,* and *surface waves.* The magnitude of earthquake waves is measured on the *Richter scale.*

Folding and faulting produce prominent features on the surface called mountains. *Mountains* can be classified as having an origin of *folding, faulting,* or *volcanic.* In general, mountains that occur in long narrow belts called *ranges* have an origin that can be explained by *plate tectonics.*

Weathering is the breaking up, crumbling, and destruction of any kind of solid rock. The process of physically picking up weathered rock materials is called *erosion.* After the eroded materials are picked up, they are removed by *transportation* agents. The combined action of weathering, erosion, and transportation wears away and lowers the surface of Earth.

The physical breaking up of rocks is called *mechanical weathering.* Mechanical weathering occurs by *wedging effects* and the *effects of reduced pressure. Frost wedging* is a wedging effect that occurs from repeated cycles of water freezing and thawing. The process of spalling off of curved layers of rock from reduced pressure is called *exfoliation.*

The breakdown of minerals by chemical reactions is called *chemical weathering.* Examples include *oxidation,* a reaction between oxygen and the minerals making up rocks; *carbonation,* a reaction between carbonic acid (carbon dioxide dissolved in water) and minerals making up rocks; and *hydration,* the dissolving or combining of a mineral with water. When the end products of complete weathering of rocks are removed directly by gravity, the erosion is called *mass movement.* Erosion and transportation also occur through the agents of *running water, glaciers,* or *wind.* Each creates their own characteristic features of erosion and deposition.

KEY TERMS

anticline (p. **359**)	fault (p. **360**)	mountains (p. **364**)	stress (p. **358**)
delta (p. **374**)	floodplain (p. **373**)	normal fault (p. **360**)	syncline (p. **359**)
dune (p. **377**)	folds (p. **359**)	principle of uniformity (p. **357**)	thrust fault (p. **361**)
earthquake (p. **361**)	frost wedging (p. **369**)	reverse fault (p. **360**)	tsunami (p. **364**)
elastic rebound (p. **361**)	glacier (p. **374**)	seismic waves (p. **361**)	volcano (p. **366**)
erosion (p. **368**)	mass movement (p. **372**)	strain (p. **358**)	weathering (p. **368**)

APPLYING THE CONCEPTS

1. The basic difference in the frame of reference called the principle of uniformity and the catastrophic frame of reference used by previous thinkers is
   a. the energy for catastrophic changes is much less.
   b. the principle of uniformity requires more time.
   c. catastrophic changes have a greater probability of occurring.
   d. None of the above is correct.

2. The difference between elastic deformation and plastic deformation of rocks is that plastic deformation
   a. permanently alters the shape of a rock layer.
   b. always occurs just before a rock layer breaks.
   c. returns to its original shape after the pressure is removed.
   d. All of the above are correct.

3. Whether a rock layer subjected to stress undergoes elastic deformation, plastic deformation, or rupture depends on
   a. the temperature of the rock.
   b. the confining pressure on the rock.
   c. how quickly or slowly the stress is applied over time.
   d. All of the above are correct.

4. When subjected to stress, rocks buried at great depths are under great pressure at high temperatures, so they tend to undergo
   a. no change because of the pressure.
   b. elastic deformation because of the high temperature.
   c. plastic deformation.
   d. breaking or rupture.

5. Earthquakes that occur at the boundary between two tectonic plates moving against each other occur along
   a. the entire length of the boundary at once.
   b. short segments of the boundary at different times.
   c. the entire length of the boundary at different times.
   d. None of the above is correct.

6. Each higher number of the Richter scale
   a. increases with the magnitude of an earthquake.
   b. means ten times more ground movement.
   c. indicates about thirty times more energy released.
   d. means all of the above.

7. Other than igneous activity, all mountain ranges have an origin resulting from
   a. folding.   c. stresses.
   b. faulting.   d. sedimentation.

8. The preferred name for the very large ocean waves that are generated by an earthquake, landslide, or volcanic explosion is
   a. tidal wave.   c. tidal bore.
   b. tsunami.   d. Richter wave.

9. Freezing water exerts pressure on the wall of a crack in a rock mass, making the crack larger. This is an example of
   a. mechanical weathering.
   b. chemical weathering.
   c. exfoliation.
   d. hydration.

10. Which of the following would have the greatest overall effect in lowering the elevation of a continent such as North America?
    a. continental glaciers
    b. alpine glaciers
    c. wind
    d. running water

11. Broad meanders on a very wide, gently sloping floodplain with oxbow lakes are characteristics you would expect to find in a river valley during what stage?
    a. newborn
    b. youth
    c. maturity
    d. old age

12. A glacier forms when
    a. the temperature does not rise above freezing.
    b. snow accumulates to form ice, which begins to flow.
    c. a summer climate does not occur.
    d. a solid mass of snow moves downhill under the influence of gravity.

Answers
1. b **2.** a **3.** d **4.** c **5.** b **6.** d **7.** c **8.** b **9.** a **10.** d **11.** d **12.** b

QUESTIONS FOR THOUGHT

1. What is the principle of uniformity? What are the underlying assumptions of this principle?
2. Describe the responses of rock layers to increasing compressional stress when it increases slowly on deeply buried, warm layers, when it increases slowly on cold rock layers, and when it is applied quickly to rock layers of any temperature.
3. Describe the conditions that would lead to faulting as opposed to folding of rock layers.
4. Where would the theory of plate tectonics predict that earthquakes would occur?
5. Identify some areas of probable active volcanic activity today in the United States. Explain your reasoning for selecting these areas.
6. Granite is the most common rock found on continents. What are the end products after granite has been completely weathered? What happens to these weathering products?
7. What other erosion processes are important as a stream of running water carves a valley in the mountains? Explain.
8. Describe the characteristic features associated with stream erosion as the stream valley passes through the stages of youth, maturity, and old age.
9. Could a stream erode the land lower than sea level? Explain.

10. Explain why glacial erosion produces a U-shaped valley but stream erosion produces a V-shaped valley.

11. Name and describe as many ways as you can think of that mechanical weathering occurs in nature. Do not restrict your thinking to those discussed in this chapter.

12. Compare the features caused by stream erosion, wind erosion, and glacial erosion.

FOR FURTHER ANALYSIS

1. Evaluate the statement "the present is the key to the past" as it represents the principle of uniformity. What evidence supports this principle?

2. Does the theory of plate tectonics support or not support the principle of uniformity? Provide evidence to support your answer.

3. What are the significant similarities and differences between elastic deformation and plastic deformation?

4. Explain the combination of variables that results in solid rock layers folding rather than faulting.

5. Analyze why you would expect most earthquakes to occur as localized, shallow occurrences near a plate boundary.

6. What are the significant similarities and differences between weathering and erosion?

7. Speculate if the continents will ever be weathered and eroded flat at sea level. Provide evidence to support your speculation.

8. Is it possible for any agent of erosion to erode the land to below sea level? Provide evidence or some observation to support your answer.

INVITATION TO INQUIRY

Building Rocks

Survey the use of rocks used in building construction in your community. Compare the type of rocks that are used for building interiors and those that are used for building exteriors. Where were the rocks quarried? Are any trends apparent for buildings constructed in the past and those built more recently? If so, are there reasons (cost, shipping, other limitations) underlying a trend, or is it simply a matter of style?

This cloud forms a thin covering over the mountaintop. Likewise, Earth's atmosphere forms a thin shell around Earth, with 99 percent of the mass within 32 km (about 20 mi) of the surface.

Earth's Weather

CORE CONCEPT

Solar radiation drives cycles in Earth's atmosphere, and some of these cycles determine weather and climate.

Water cycles into Earth's atmosphere as water vapor and out as condensation and precipitation (p. 391).

A uniform body of air is called an air mass (p. 398).

Movement of air masses brings about rapid changes in the weather (p. 398).

Climate is the general pattern of weather (p. 407).

▶ Organisms show a high degree of adaptation to their local climate (Ch. 23).

▶ Through the process of photosynthesis, organisms remove carbon dioxide and add oxygen to the atmosphere (Ch. 23).

▶ Through the process of respiration, organisms remove oxygen and add carbon dioxide to the atmosphere (Ch. 23).

▶ Nitrogen is cycled between the atmosphere and the bodies of organisms (Ch. 23).

OVERVIEW

Earth's atmosphere has a unique composition because of the cyclic flow of materials. Some of these cycles involve the movement of materials in and out of Earth's atmosphere. Carbon dioxide, for example, is a very minor part of Earth's atmosphere. It has been maintained as a minor component in a mostly balanced state for about the past 570 million years, cycling into and out of the atmosphere.

Water is also involved in a global cyclic flow between the atmosphere and the surface. Water on the surface is mostly in the ocean, with lesser amounts in lakes, streams, and underground. Not much water is found in the atmosphere at any one time on a worldwide basis, but billions of tons are constantly evaporating into the atmosphere each year and returning as precipitation in an ongoing cycle.

The cycling of carbon dioxide and water to and from the atmosphere takes place in a dynamic system that is energized by the Sun. Radiant energy from the Sun heats some parts of Earth more than others. Winds redistribute this energy with temperature changes, rain, snow, and other changes that are generally referred to as the *weather*.

Understanding and predicting the weather is the subject of *meteorology*. Meteorology is the science of the atmosphere and weather phenomena, from understanding everyday rain and snow to predicting not-so-common storms and tornadoes (figure 17.1). Understanding weather phenomena depends on a knowledge of the atmosphere and the role of radiant energy on a spinning Earth that is revolving around the Sun. This chapter is concerned with understanding the atmosphere of Earth, its cycles, and the influence of radiant energy on the atmosphere.

THE ATMOSPHERE

The atmosphere is a relatively thin shell of gases that surrounds the solid Earth. If you could see the molecules making up the atmosphere, you would see countless numbers of rapidly moving particles, all undergoing a terrific jostling from the billions of collisions occurring every second. Since this jostling mass of tiny particles is pulled toward Earth by gravity, more are found near the surface than higher up. Thus, the atmosphere thins rapidly with increasing distance above the surface, gradually merging with the very diffuse medium of outer space.

To understand how rapidly the atmosphere thins with altitude, imagine a very tall stack of open boxes. At any given instant, each consecutively higher box would contain fewer of the jostling molecules than the box below it. Molecules in the lowest box on the surface, at sea level, might be able to move a distance of only 1×10^{-8} m (about 3×10^{-6} in) before colliding with another molecule. A box moved to an altitude of 80 km (about 50 mi) above sea level would have molecules that could move perhaps 10^{-2} m (about 1/2 in) before colliding with another molecule. At 160 km (about 100 mi), the distance traveled would be about 2 m (about 7 ft). As you can see, the distance between molecules increases rapidly with increasing altitude. Since air density is defined by the number of molecules in a unit volume, the density of the atmosphere decreases rapidly with increasing altitude (figure 17.2).

It is often difficult to imagine a distance above the surface because there is nothing visible in the atmosphere for comparison. Recall our stack of boxes from the previous example. Imagine that this stack of boxes is so tall that it reaches from the surface past the top of the atmosphere. Now imagine that this tremendously tall stack of boxes is tipped over and carefully laid out horizontally on the surface of Earth. How far would you have to move along these boxes to reach the box that was in

FIGURE 17.1 The probability of a storm can be predicted, but nothing can be done to stop or slow a storm. Understanding the atmosphere may help in predicting weather changes, but it is doubtful that weather will ever be controlled on a large scale.

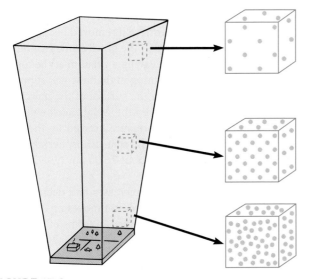

FIGURE 17.2 At greater altitudes, the same volume contains fewer molecules of the gases that make up the air. This means that the density of air decreases with increasing altitude.

outer space, outside of the atmosphere? From the bottom box, you would cover a distance of only 5.6 km (about 3.5 mi) to reach the box that was above 50 percent of the mass of Earth's atmosphere. At 12 km (about 7 mi), you would reach the box that was above 75 percent of Earth's atmosphere. At 16 km (about 10 mi), you would reach the box that was above about 90 percent of the atmosphere. And, after only 32 km (about 20 mi), you would reach the box that was above 99 percent of Earth's atmosphere. The significance of these distances might be better appreciated if you can imagine the distances to some familiar locations; for example, from your campus to a store 16 km (about 10 mi) away would place you above 90 percent of the atmosphere if you were to travel this same distance straight up.

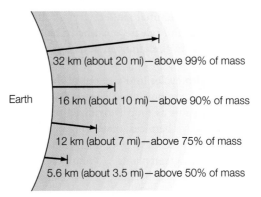

FIGURE 17.3 Earth's atmosphere thins rapidly with increasing altitude and is much closer to Earth's surface than most people realize.

Since the average radius of the solid Earth is about 6,373 km (3,960 mi), you can see that the atmosphere is a very thin shell with 99 percent of the mass within 32 km (about 20 mi) by comparison. The outer edge of the atmosphere is much closer to Earth than most people realize (figure 17.3).

Composition of the Atmosphere

A sample of pure, dry air is colorless, odorless, and composed mostly of the molecules of just three gases, nitrogen (N_2), oxygen (O_2), and argon (Ar). Nitrogen is the most abundant (about 78 percent of the total volume), followed by oxygen (about 21 percent), then argon (about 1 percent). The molecules of these three gases are well mixed, and this composition is nearly constant everywhere near Earth's surface (figure 17.4).

Nitrogen does not readily enter into chemical reactions with rocks, so it has accumulated in the atmosphere. Some nitrogen is involved in a nitrogen cycle, however, as it is removed from the atmosphere by certain bacteria in the soil and by lightning. Both form nitrogen compounds that are essential to the growth of plants. These nitrogen compounds are absorbed by plants and consequently utilized throughout the food chain. Eventually, the nitrogen is returned to the atmosphere through

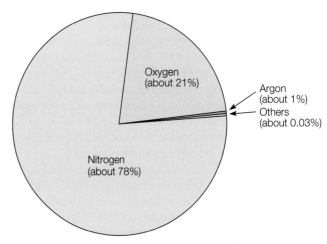

FIGURE 17.4 Earth's atmosphere has a unique composition of gases when compared to that of the other planets in the solar system.

the decay of plant and animal matter. Overall, these processes of nitrogen removal and release must be in balance since the amount of nitrogen in the atmosphere is essentially constant over time.

Oxygen gas also cycles into and out of the atmosphere in balanced processes of removal and release. Oxygen is removed (1) by living organisms as food is oxidized to carbon dioxide and water and (2) by chemical weathering of rocks as metals and other elements combine with oxygen to form oxides. Oxygen is released by green plants as a result of photosynthesis, and the amount released balances the amount removed by organisms and weathering. So oxygen, as well as nitrogen, is maintained in a state of constant composition through balanced chemical reactions.

The third major component of the atmosphere, argon, is inert and does not enter into any chemical reactions or cycles. It is produced as a product of radioactive decay and, once released, remains in the atmosphere as an inactive filler.

In addition to the relatively fixed amounts of nitrogen, oxygen, and argon, the atmosphere contains variable amounts of water vapor. Water vapor is the invisible, molecular form of water in the gaseous state, which should not be confused with fog or clouds. Fog and clouds are tiny droplets of liquid water, not water in the single molecular form of water vapor. The amount of water vapor in the atmosphere can vary from a small fraction of a percent composition by volume in cold, dry air to about 4 percent in warm, humid air. This small, variable percentage of water vapor is essential in maintaining life on Earth. It enters the atmosphere by evaporation, mostly from the ocean, and leaves the atmosphere as rain or snow.

Apart from the variable amounts of water vapor, the relatively fixed amounts of nitrogen, oxygen, and argon make up about 99.97 percent of the volume of a sample of dry air. The remaining gases are mostly carbon dioxide (CO_2) and traces of inert gases. Carbon dioxide makes up only 0.03 percent of the atmosphere, but it is important to life on Earth and it is constantly cycled through the atmosphere by the process of photosynthesis that removes CO_2, and respiration that releases CO_2.

In addition to gases and water vapor, the atmosphere contains particles of dust, smoke, salt crystals, and tiny solid or liquid particles called *aerosols*. These particles become suspended and are dispersed among the molecules of the atmospheric gases. Aerosols are produced by combustion, often resulting in air pollution. Aerosols are also produced by volcanoes and forest fires. Volcanoes, smoke from combustion, and the force of the wind lifting soil and mineral particles into the air all contribute to dust particles larger than aerosols in the atmosphere. These larger particles, which range in size up to 500 micrometers, are not suspended as the aerosols are, and they soon settle out of the atmosphere as dust and soot.

Tiny particles of salt crystals that are suspended in the atmosphere come from the mist created by ocean waves and the surf. This mist forms an atmospheric aerosol of seawater that evaporates, leaving the solid salt crystals suspended in the air. The aerosol of salt crystals and dust becomes well mixed in the lower atmosphere around the globe, playing a large and important role in the formation of clouds.

Atmospheric Pressure

At Earth's surface (sea level), the atmosphere exerts a force of about 10.0 newtons on each square centimeter (14.7 lb/in²). As you go to higher altitudes above sea level the pressure rapidly decreases with increasing altitude. At an altitude of about 5.6 km (about 3.5 mi), the air pressure is about half of what it is at sea level, about 5.0 newtons/cm²(7.4 lb/in²). At 12 km (about 7 mi), the air pressure is about 2.5 newtons/cm²(3.7 lb/in²). Compare this decreasing air pressure at greater elevations to figure 17.3. Again, you can see that most of the atmosphere is very close to Earth, and it thins rapidly with increasing altitude. Even a short elevator ride takes you high enough that the atmospheric pressure on your eardrum is reduced. You equalize the pressure by opening your mouth, allowing the air under greater pressure inside the eardrum to move through the eustachian tube. This makes a "pop" sound that most people associate with changes in air pressure.

Atmospheric pressure is measured by an instrument called a **barometer.** The mercury barometer was invented in 1643 by an Italian named Torricelli. He closed one end of a glass tube and then filled it with mercury. The tube was then placed, open end down, in a bowl of mercury while holding the mercury in the tube with a finger. When Torricelli removed his finger with the open end below the surface in the bowl, a small amount of mercury moved into the bowl leaving a vacuum at the top end of the tube. The mercury remaining in the tube was supported by the atmospheric pressure on the surface of the mercury in the bowl. The pressure exerted by the weight of the mercury in the tube thus balanced the pressure exerted by the atmosphere. At sea level, Torricelli found that atmospheric pressure balanced a column of mercury about 76.00 cm (29.92 in) tall (figure 17.5).

As the atmospheric pressure increases and decreases, the height of the supported mercury column moves up and down. Atmospheric pressure can be expressed in terms of the height

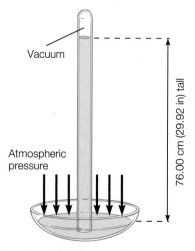

FIGURE 17.5 The mercury barometer measures the atmospheric pressure from the balance between the pressure exerted by the weight of the mercury in a tube and the pressure exerted by the atmosphere. As the atmospheric pressure increases and decreases, the mercury rises and falls. This sketch shows the average height of the column at sea level.

of such a column of mercury. Public weather reports give the pressure by referring to such a mercury column; for example, "The pressure is 30 inches (about 76 cm) and rising." If the atmospheric pressure at sea level is measured many times over long periods of time, an average value of 76.00 cm (29.92 in) of mercury is obtained. This average measurement is called the **standard atmospheric pressure** and is sometimes referred to as the *normal pressure*. It is also called *one atmosphere of pressure*.

Warming the Atmosphere

Radiation from the Sun must pass through the atmosphere before reaching Earth's surface. The atmosphere filters, absorbs, and reflects incoming solar radiation, as shown in figure 17.6. On the average, Earth as a whole reflects about 30 percent of the total radiation back into space, with two-thirds of the reflection occurring from clouds. The amount reflected at any one time depends on the extent of cloud cover, the amount of dust in the atmosphere, and the extent of snow and vegetation on the surface. Substantial changes in any of these influencing variables could increase or decrease the reflectivity, leading to increased heating or cooling of the atmosphere.

As figure 17.6 shows, only about one-half of the incoming solar radiation reaches Earth's surface. The reflection and selective filtering by the atmosphere allow a global average of about 240 watts per square meter to reach the surface. Wide variations from the average occur with latitude as well as with the season.

The incoming solar radiation that does reach Earth's surface is absorbed. Rocks, soil, water, and the ground become warmer as a result. These materials emit the absorbed solar energy as infrared radiation, wavelengths longer than the visible part of the electromagnetic spectrum. This longer-wavelength infrared radiation has a frequency that matches some of the natural frequencies of vibration of carbon dioxide and water molecules. This match means that carbon dioxide and water

molecules readily absorb infrared radiation that is emitted from the surface of Earth. The absorbed infrared energy shows up as an increased kinetic energy of the molecules, which is indicated by an increase in temperature. Carbon dioxide and water vapor molecules in the atmosphere now emit infrared radiation of their own, this time in all directions. Some of this reemitted radiation is again absorbed by other molecules in the atmosphere, some is emitted to space, and significantly, some is absorbed by the surface to start the process all over again. The net result is that less of the energy from the Sun escapes immediately to space after being absorbed and emitted as infrared. It is retained through the process of being redirected to the surface, increasing the surface temperature more than it would have otherwise been. The more carbon dioxide that is present in the atmosphere, the more energy that will be bounced around and redirected back toward the surface, increasing the temperature near the surface. The process of heating the atmosphere in the lower parts by the absorption of solar radiation and reemission of infrared radiation is called the **greenhouse effect.** It is called the greenhouse effect because greenhouse glass allows the short wavelengths of solar radiation to pass into the greenhouse but does not allow all of the longer infrared radiation to leave. This analogy is misleading, however, because carbon dioxide and water vapor molecules do not "trap" infrared radiation, but they are involved in a dynamic absorption and downward reemission process that increases the surface temperature. The more carbon dioxide molecules that are involved in this dynamic process, the more infrared radiation will be directed back to Earth and the more the temperature will increase. More layers of glass on a greenhouse will not increase the temperature significantly. The significant heating factor in a real greenhouse is the blockage of convection by the glass, a process that does not occur from the presence of carbon dioxide and water vapor in the atmosphere.

Structure of the Atmosphere

Convection currents and the repeating absorption and reemission processes of the greenhouse effect tend to heat the atmosphere from the ground up. In addition, the higher-altitude gases lose radiation to space more readily than the lower-altitude gases. Thus, the lowest part of the atmosphere is warmer, and the temperature decreases with increasing altitude. On the average, the temperature decreases about 6.5°C for each kilometer of altitude (3.5°F/1,000 ft). This change of temperature with altitude is called the *observed lapse rate*. The observed lapse rate applies only to air that is not rising or sinking, and the actual change with altitude can be very different from this average value. For example, a stagnant mass of very cold air may settle over an area, producing colder temperatures near the surface than in the air layers above. Such a layer where the temperature increases with height is called an **inversion** (figure 17.7). Inversions often occur on calm winter days after the arrival of a cold front. They also occur on calm, clear, and cool nights ("C" nights), when the surface rapidly loses radiant energy to space. In either case, the situation results in a "cap" of

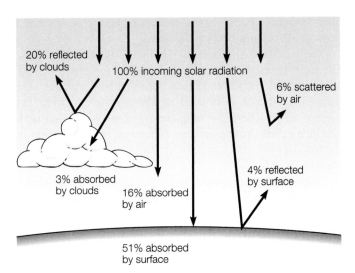

FIGURE 17.6 On the average, Earth's surface absorbs only 51 percent of the incoming solar radiation after it is filtered, absorbed, and reflected.

Hole in the Ozone Layer?

Ozone is triatomic oxygen (O_3) that is concentrated mainly in the upper portions of the stratosphere. Diatomic molecules of oxygen (O_2) are concentrated in the troposphere, and monatomic molecules of oxygen (O) are found in the outer edges of the atmosphere. Although the amount of ozone present in the stratosphere is not great, its presence is vital to life on Earth's surface. Ultraviolet (UV) radiation causes mutations to the DNA of organisms. Without the stratospheric ozone, much more ultraviolet radiation would reach the surface of Earth, causing mutations in all kinds of organisms. In humans, ultraviolet radiation is known to cause skin cancer and damage the cornea of eyes. It is believed that the incidence of skin cancer would rise dramatically without the protection offered by the ozone.

Here is how stratospheric ozone shields Earth from ultraviolet radiation. The ozone concentration is not static because there is an ongoing process of ozone formation and destruction. For ozone to form, diatomic oxygen (O_2) must first be broken down into the monatomic form (O). Shortwave ultraviolet radiation is absorbed by diatomic oxygen, breaking it down into the single-atom form. This reaction is significant because of (1) the high-energy ultraviolet radiation that is removed from the sunlight and (2) the monatomic oxygen that is formed, which will combine with diatomic oxygen to make triatomic oxygen (ozone) that will absorb even more ultraviolet radiation. This initial reaction is

$$O_2 + UV \rightarrow O + O$$

When the O molecule collides with an O_2 molecule and any third, neutral molecule (NM), the following reaction takes place:

$$O_2 + O + NM \rightarrow O_3 + NM$$

When O_3 is exposed to ultraviolet radiation, the ozone absorbs the UV radiation and breaks down to two forms of oxygen in the following reaction:

$$O_3 + UV \rightarrow O_2 + O$$

The monatomic molecule that is produced combines with an ozone molecule to produce two diatomic molecules,

$$O + O_3 \rightarrow 2\,O_2$$

and the process starts all over again.

Much concern exists about Freon (CF_2Cl_2) and other similar chemicals that make their way to the stratosphere. These chemicals are broken down by UV radiation, releasing chlorine (Cl), which reacts with ozone. The reaction might be

$$CF_2Cl_2 + UV \rightarrow CF_2Cl\text{-}^* + Cl\text{-}^*$$
$$Cl\text{-}^* + O_3 \rightarrow\ ^*\text{-}ClO + O_2$$
$$^*\text{-}ClO + O \rightarrow O_2 + Cl\text{-}^*$$

$-^*$ is an unattached bond

Note that ozone is decomposed in the second step, so it is removed from the UV radiation–absorbing process. Furthermore, the second and third steps repeat, so the destruction of one molecule of Freon can result in the decomposition of many molecules of ozone.

A regional zone of decreased ozone availability in the stratosphere is referred to as a "hole in the ozone layer," since much high-energy UV radiation can now reach the surface below. Concerns about the impact chlorine-containing chemicals such as Freon have on the ozone layer led to an international agreement known as the Montreal Protocol. The agreement has resulted in a phasing out of the use of Freon and similar compounds in most of the world. Subsequently, it appears that the sizes of the "holes in the ozone layer" are getting smaller.

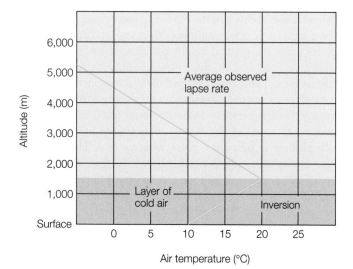

FIGURE 17.7 On the average, the temperature decreases about 6.5°C/km, which is known as the *observed lapse rate*. An inversion is a layer of air in which the temperature increases with height.

cooler, more dense air overlying the warmer air beneath. This often leads to an increase of air pollution because the inversion prevents dispersion of the pollutants.

Temperature decreases with height at the observed lapse rate until an average altitude of about 11 km (about 6.7 mi), where it then begins to remain more or less constant with increasing altitude. The layer of the atmosphere from the surface up to where the temperature stops decreasing with height is called the *troposphere*. Almost all weather occurs in the troposphere, which is derived from Greek words meaning "turning layer."

Above the troposphere is the second layer of the atmosphere called the *stratosphere*. This layer is derived from the Greek terms for "stratified layer." It is stratified, or layered, because the temperature increases with height. Cooler air below means that consecutive layers of air are denser on the bottom, which leads to a stable situation rather than the turning turbulence of the troposphere below. The stratosphere contains little moisture or dust and lacks convective turbulence, making it a

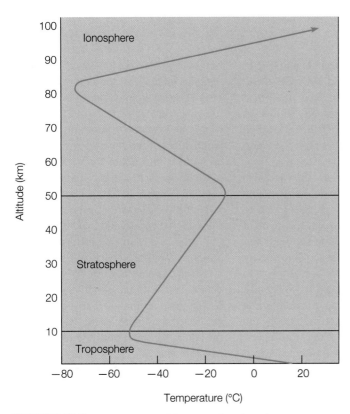

FIGURE 17.8 The structure of the atmosphere based on temperature differences.

desirable altitude for aircraft to fly. Temperature in the lower stratosphere increases gradually with increasing altitude to a height of about 48 km (about 30 mi) (figure 17.8).

The outermost layer is where the molecules merge with the diffuse vacuum of space. Molecules of this layer that have sufficient kinetic energy are able to escape and move off into space. This outer layer is sometimes called the *ionosphere* because of the free electrons and ions at this altitude. The electrons and ions here are responsible for reflecting radio waves around Earth and for the northern lights.

THE WINDS

The troposphere is heated from the bottom up as the surface of Earth absorbs sunlight. Uneven heating of Earth's surface sets the stage for *convection*. As a local region of air becomes heated, the air expands, reducing its density. This less dense air is pushed upward by nearby cooler, more dense air. This results in three general motions of air: (1) the upward movement of air over a region of greater heating, (2) the sinking of air over a cooler region, and (3) a horizontal air movement between the cooler and warmer regions. In general, a horizontal movement of air is called **wind,** and the direction of a wind is defined as the direction from which it blows.

Air in the troposphere rises, moves as the wind, and sinks. All three of these movements are related, and all occur at the same time over regions of the landscape. During a day with gentle breezes on the surface, the individual, fluffy clouds you see are forming over areas where the air is moving upward. The clear air between the clouds is over areas where the air is moving downward. On a smaller scale, air can be observed moving from a field of cool grass toward an adjacent asphalt parking lot on a calm, sunlit day. Soap bubbles or smoke will often reveal the gentle air movement of this localized convection.

Local Wind Patterns

Considering average conditions, there are two factors that are important for a generalized model to help you understand local wind patterns. These factors are (1) the relationship between air temperature and air density, and (2) the relationship between air pressure and the movement of air. The relationship between air temperature and air density is that cool air has a greater density than warm air because a mass of warming air expands and a mass of cooling air contracts. Warm, less dense air is buoyed upward by cooler, more dense air, which results in the upward, downward, and horizontal movement of air called a convection cell.

The upward and downward movement of air leads to the second part of the generalized model, that (1) the upward movement produces a "lifting" effect on the surface that results in an area of lower atmospheric pressure and (2) the downward movement produces a "piling up" effect on the surface that results in an area of higher atmospheric pressure. On the surface, air is seen to move from the "piled up" area of higher pressure horizontally to the "lifted" area of lower pressure (figure 17.9). In other words, air generally moves from an area of higher pressure to an area of lower pressure. The movement of air and the pressure differences occur together, and neither is the cause of the other. This is an important relationship in a working model of air movement that can be observed and measured on a very small scale, such as between an asphalt

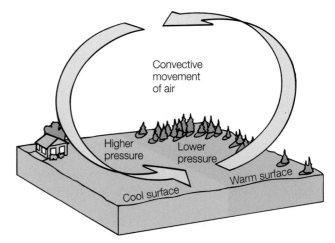

FIGURE 17.9 A model of the relationships between differential heating, the movement of air, and pressure difference in a convective cell. Cool air pushes the less-dense, warm air upward, reducing the surface pressure. As the uplifted air cools and becomes more dense, it sinks, increasing the surface pressure.

A Closer Look

The Wind Chill Factor

The term *wind chill* is attributed to the Antarctic explorer Paul A. Siple. During the 1940s, Siple and Charles F. Passel conducted experiments on how long it took a can of water to freeze at various temperatures and wind speeds at a height of 10 m (33 ft) above the ground (typical height of an anemometer). They found that the time depended on the air temperature and the wind speed. From these data, an equation was developed to calculate the **wind chill factor** for humans.

In 2001, the National Weather Service changed to a new method of computing wind chill temperatures. The new wind chill formula was developed by a year-long cooperative effort between the U.S. and Canadian governments and university scientists. The new standard is based on

wind speeds at an average height of 1.5 m (5 ft) above the ground, which is closer to the height of a human face rather than the height of an anemometer and takes advantage of advances in science, technology, and computers. The new chart also highlights the danger of frostbite.

Here is the reason why the wind chill factor is an important consideration. The human body constantly produces heat to maintain a core temperature, and some of this heat is radiated to the surroundings. When the wind is not blowing (and you are not moving), your body heat is also able to warm some of the air next to your body. This warm blanket of air provides some insulation, protecting your skin from the colder air farther away. If the wind blows, however, it moves this air away from your

body and you feel cooler. How much cooler depends on how fast the air is moving and on the outside temperature—which is what the wind chill factor tells you. Thus, wind chill is an attempt to measure the combined effect of low temperature and wind on humans (box figure 17.1). It is just one of the many factors that can affect winter comfort. Others include the type of clothes, level of physical exertion, amount of sunshine, humidity, age, and body type.

There is a wind chill calculator at the National Weather Service website, *www.nws.noaa.gov/om/windchill/*. All you need to do is enter the air temperature (in degrees Fahrenheit) and the wind speed (in miles per hour), and the calculator will give you the wind chill using both the old and the new formula.

Wind (mph)	Temperature (°F)																	
	40	35	30	25	20	15	10	5	0	−5	−10	−15	−20	−25	−30	−35	−40	−45
5	36	31	25	19	13	7	1	−5	−11	−16	−22	−28	−34	−40	−46	−52	−57	−63
10	34	27	21	15	9	3	−4	−10	−16	−22	−28	−35	−41	−47	−53	−59	−66	−72
15	32	25	19	13	6	0	−7	−13	−19	−26	−32	−39	−45	−51	−58	−64	−71	−77
20	30	24	17	11	4	−2	−9	−15	−22	−29	−35	−42	−48	−55	−61	−68	−74	−81
25	29	23	16	9	3	−4	−11	−17	−24	−31	−37	−44	−51	−58	−64	−71	−78	−84
30	28	22	15	8	1	−5	−12	−19	−26	−33	−39	−46	−53	−60	−67	−73	−80	−87
35	28	21	14	7	0	−7	−14	−21	−27	−34	−41	−48	−55	−62	−69	−76	−82	−89
40	27	20	13	6	−1	−8	−15	−22	−29	−36	−43	−50	−57	−64	−71	−78	−84	−91
45	26	19	12	5	−2	−9	−16	−23	−30	−37	−44	−51	−58	−65	−72	−79	−86	−93
50	26	19	12	4	−3	−10	−17	−24	−31	−38	−45	−52	−60	−67	−74	−81	−88	−95
55	25	18	11	4	−3	−11	−18	−25	−32	−39	−46	−54	−61	−68	−75	−82	−89	−97
60	25	17	10	3	−4	−11	−19	−26	−33	−40	−48	−55	−62	−69	−76	−84	−91	−98

Frostbite times ☐ 30 minutes ☐ 10 minutes ☐ 5 minutes

Box Figure 17.1   Wind chill chart.

parking lot and a grass field. It can also be observed and measured for local, regional wind patterns and for worldwide wind systems.

Adjacent areas of the surface can have different temperatures because of different heating or cooling rates. The difference is very pronounced between adjacent areas of land and water. Under identical conditions of incoming solar radiation, the temperature changes experienced by the water will be much less than the changes experienced by the adjacent land. There

are three principal reasons for this difference: (1) The specific heat of water is about twice the specific heat of soil. This means that it takes more energy to increase the temperature of water than it does for soil. Equal masses of soil and water exposed to sunlight will result in the soil heating about 1°C while the water heats 1/2°C from absorbing the same amount of solar radiation. (2) Water is a transparent fluid that is easily mixed, so the incoming solar radiation warms a body of water throughout, spreading out the heating effect. Incoming solar radiation on

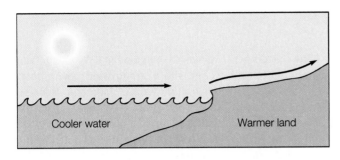

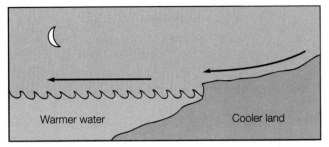

FIGURE 17.10 The land warms and cools more rapidly than an adjacent large body of water. During the day, the land is warmer, and air over the land expands and is buoyed up by cooler, more dense air from over the water. During the night, the land cools more rapidly than the water, and the direction of the breeze is reversed.

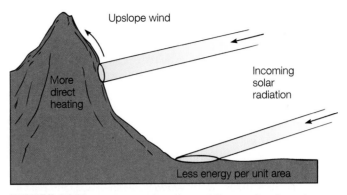

FIGURE 17.11 Incoming solar radiation falls more directly on the side of a mountain, which results in differential heating. The same amount of sunlight falls on the areas shown in this illustration, with the valley floor receiving a more spread-out distribution of energy per unit area. The overall result is an upslope mountain breeze during the day. During the night, dense cool air flows downslope for a reverse wind pattern.

land, on the other hand, warms a relatively thin layer on the top, concentrating the heating effect. (3) The water is cooled by evaporation, which helps keep a body of water at a lower temperature than an adjacent landmass under identical conditions of incoming solar radiation.

A local wind pattern may result from the resulting temperature differences between a body of water and adjacent landmasses. If you have ever spent some time along a coast, you may have observed that a cool, refreshing gentle breeze blows from the water toward the land during the summer. During the day, the temperature of the land increases more rapidly than the water temperature. The air over the land is therefore heated more, expands, and becomes less dense. Cool, dense air from over the water moves inland under the air over the land, buoying it up. The air moving from the sea to the land is called a *sea breeze*. The sea breeze along a coast may extend inland several miles during the hottest part of the day in the summer. The same pattern is sometimes observed around the Great Lakes during the summer, but this breeze usually does not reach more than several city blocks inland. During the night, the land surface cools more rapidly than the water, and a breeze blows from the land to the sea (figure 17.10).

Another pattern of local winds develops in mountainous regions. If you have ever visited a mountain in the summer, you may have noticed that there is usually a breeze or wind blowing up the mountain slope during the afternoon. This wind pattern develops because the air over the mountain slope is heated more than the air in a valley. As shown in figure 17.11, the air over the slope becomes warmer because it receives more direct sunlight than the valley floor. Sometimes this air movement is so gentle that it would be unknown except for the evidence of

clouds that form over the peaks during the day and evaporate at night. During the night, the air on the slope cools as the land loses radiant energy to space. As the air cools, it becomes denser and flows downslope, forming a reverse wind pattern to the one observed during the day.

During cooler seasons, cold, dense air may collect in valleys or over plateaus, forming a layer or "puddle" of cold air. Such an accumulation of cold air often results in some very cold night-time temperatures for cities located in valleys, temperatures that are much colder than anywhere in the surrounding region. Some weather disturbance, such as an approaching front, can disturb such an accumulation of cold air and cause it to pour out of its resting place and through canyons or lower valleys. Air moving from a higher altitude like this becomes compressed as it moves to lower elevations under increasing atmospheric pressure. Compression of air increases the temperature by increasing the kinetic energy of the molecules. This creates a wind called a *Chinook*, which is common to mountainous and adjacent regions. A Chinook is a wind of compressed air with sharp temperature increases that can melt away any existing snow cover in a single day. The *Santa Ana* is a well-known compressional wind that occurs in southern California.

Global Wind Patterns

Local wind patterns tend to mask the existence of the overall global wind pattern that is also present. The global wind pattern is not apparent if the winds are observed and measured for a particular day, week, or month. It does become apparent when the records for a long period of time are analyzed. These records show that Earth has a large-scale pattern of atmospheric circulation that varies with latitude. There are belts in which the winds average an overall circulation in one direction, belts of higher atmospheric pressure averages, and belts of lower atmospheric pressure averages. This has led to a generalized pattern of atmospheric circulation and a global atmospheric model. This model, as you will see, today provides the basis for the daily weather forecast for local and regional areas.

Science and Society

Use Wind Energy?

Millions of windmills were installed in rural areas of the United States between the late 1800s and the late 1940s. These windmills used wind energy to pump water, grind grain, or generate electricity. Some are still in use today, but most were no longer needed after inexpensive electric power became generally available in those areas.

In the 1970s, wind energy began making a comeback as a clean, renewable energy alternative to fossil fuels. The windmills of the past were replaced by wind turbines of today. A wind turbine with blades that are rotated by the wind is usually mounted on a tower. The blades' rotary motion drives a generator that produces electricity. A location should have average yearly wind speeds of at least 19 km/h (12 mi/h) to provide enough wind energy for a turbine, and a greater yearly average means more energy is available. Farms, homes, and businesses in these locations can use smaller turbines, which are generally 50 kilowatts or less. Large turbines of 500 kilowatts or more are used in "wind farms," which are large clusters of interconnected wind turbines connected to a utility power grid.

Many areas of the United States have a high potential for wind-power use. North Dakota, South Dakota, and Texas have enough wind resources to provide electricity for the entire nation. Today, only California has extensively developed wind farms, with more than thirteen thousand wind turbines on three wind farms in the Altamont Pass region (east of San Francisco), Tehachapi (southeast of Bakersfield), and San Gorgonio (near Palm Springs). With a total of 2,361 megawatts of installed capacity in California, wind energy generates enough electricity to more than meet the needs of a city the size of San Francisco. The wind farms in California have a rated capacity that is comparable to two large coal-fired power plants but without the pollution and limits of this nonrenewable energy source. Wind energy makes economic as well as environmental sense, and new wind farms are being developed in Minnesota, Oregon, and Wyoming. Other states with a strong wind power potential include Kansas, Montana, Nebraska, Oklahoma, Iowa, Colorado, Michigan, and New York. All of these states, in fact, have a greater wind-energy potential than California.

Questions to Discuss

Discuss with your group the following questions concerning wind power:

1. Why have electric utilities not used much wind power as an energy source?
2. Should governments provide a tax break to encourage people to use wind power? Why or why not?
3. What are the advantages and disadvantages of using wind power in the place of fossil fuels?
4. What are the advantages and disadvantages of the government building huge wind farms in North Dakota, South Dakota, and Texas to supply electricity for the entire nation?

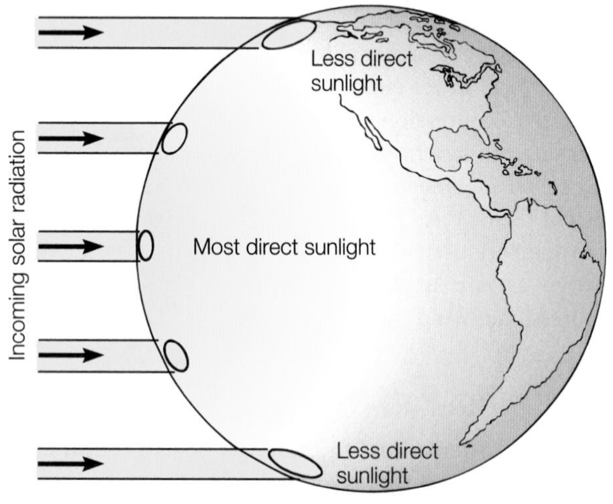

FIGURE 17.12 On a global, yearly basis, the equatorial region of Earth receives more direct incoming solar radiation than the higher latitudes. As a result, average temperatures are higher in the equatorial region and decrease with latitude toward both poles. This sets the stage for worldwide patterns of prevailing winds, high and low areas of atmospheric pressure, and climatic patterns.

As with local wind patterns, it is temperature imbalances that drive the global circulation of the atmosphere. Earth receives more direct solar radiation in the equatorial region than it does at higher latitudes (figure 17.12). As a result, the temperatures of the lower troposphere are generally higher in the equatorial region, decreasing with latitude toward both poles. The lower troposphere from 10°N to 10°S of the equator is heated, expands, and becomes less dense. Hot air rises in this belt around the equator, known as the *intertropical convergence zone.* The rising air cools because it expands as it rises, resulting in heavy average precipitation. The tropical rainforests of Earth occur in this zone of high temperatures and heavy rainfall. As the now dry, rising air reaches the upper parts of the troposphere, it begins to spread toward the north and the south, sinking back toward Earth's surface (figure 17.13). The descending air reaches the surface to form a high-pressure belt that is centered about 30°N and 30°S of the equator. Air moving on the surface away from this high-pressure belt produces the prevailing northeast trade winds and the prevailing westerly winds of the Northern Hemisphere. The great deserts of Earth are also located in this high-pressure belt of descending dry air.

Poleward of the belt of high pressure, the atmospheric circulation is controlled by a powerful belt of wind near the top of the troposphere called a **jet stream.** Jet streams are sinuous, meandering loops of winds that tend to extend all the way around Earth, moving generally from the west in both hemispheres at speeds of 160 km/h (about 100 mi/h) or more. A jet stream may occur as a single belt, or loop, of wind, but sometimes it divides into two or more parts. The jet stream develops north and south loops of waves much like the waves

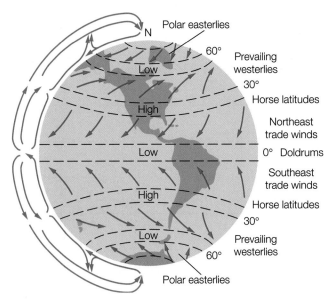

FIGURE 17.13 Simplified pattern of horizontal and vertical circulation in the actual atmosphere. Regions of high and low pressure are indicated.

you might make on a very long rope. These waves vary in size, sometimes beginning as a small ripple but then growing slowly as the wave moves eastward. Waves that form on the jet stream bulge toward the poles (called a crest) or toward the equator (called a trough). Warm air masses move toward the poles ahead of a trough, and cool air masses move toward the equator behind a trough as it moves eastward. The development of a wave in the jet stream is understood to be one of the factors that influences the movement of warm and cool air masses, a movement that results in weather changes on the surface.

The intertropical convergence zone, the 30° belt of high pressure, and the northward and southward migration of a meandering jet stream all shift toward or away from the equator during the different seasons of the year. The troughs of the jet stream influence the movement of alternating cool and warm air masses over the belt of the prevailing westerlies, resulting in frequent shifts of fair weather to stormy weather, then back again. The average shift during the year is about 6° of latitude, which is sufficient to control the overall climate in some locations. The influence of this shift of the global circulation of Earth's atmosphere will be considered as a climatic influence after considering the roles of water and air masses in frequent weather changes.

WATER AND THE ATMOSPHERE

Water exists on Earth in all three states: (1) as a liquid when the temperature is generally above the freezing point of 0°C (32°F), (2) as a solid in the form of ice, snow, or hail when the temperature is generally below the freezing point, and (3) as the invisible, molecular form of water in the gaseous state, which is called *water vapor*.

Over 98 percent of all the water on Earth exists in the liquid state, mostly in the ocean, and only a small, variable amount

of water vapor is in the atmosphere at any given time. Since so much water seems to fall as rain or snow at times, it may be a surprise that the overall atmosphere really does not contain very much water vapor. If the average amount of water vapor in Earth's atmosphere were condensed to liquid form, the vapor *and* all the droplets present in clouds would form a uniform layer around Earth only 3 cm (about 1 in) thick. Nonetheless, it is this small amount of water vapor that is eventually responsible for (1) contributing to the greenhouse effect, which helps make Earth a warmer planet, (2) serving as one of the principal agents in the weathering and erosion of the land, which creates soils and sculptures the landscape, and (3) maintaining life, for all organisms (bacteria, protozoa, algae, fungi, plants, and animals) cannot survive without water. It is the ongoing cycling of water vapor into and out of the atmosphere that makes all this possible. Understanding this cycling process and the energy exchanges involved is also closely related to understanding Earth's weather patterns.

Evaporation and Condensation

Water tends to undergo a liquid-to-gas or a gas-to-liquid phase change at any temperature. The phase change can occur in either direction at any temperature.

In *evaporation,* more molecules are leaving the liquid state than are returning. In *condensation,* more molecules are returning to the liquid state than are leaving. This is a dynamic, ongoing process with molecules leaving and returning continuously (figure 17.14). If the air were perfectly dry and still, more molecules would leave (evaporate) the liquid state than would return (condense). Eventually, however, an equilibrium would be reached with as many molecules returning to the liquid state per unit of time as are leaving. An equilibrium condition between evaporation and condensation occurs in **saturated air.** Saturated air occurs when the processes of evaporation and condensation are in balance.

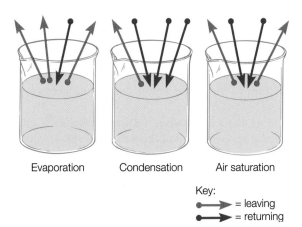

FIGURE 17.14 Evaporation and condensation are occurring all the time. If the number of molecules leaving the liquid state exceeds the number returning, the water is evaporating. If the number of molecules returning to the liquid state exceeds the number leaving, the water vapor is condensing. If both rates are equal, the air is saturated; that is, the relative humidity is 100 percent.

Air will remain saturated as long as (1) the temperature remains constant and (2) the processes of evaporation and condensation remain balanced. Temperature influences the equilibrium condition of saturated air because increases or decreases in the temperature mean increases or decreases in the kinetic energy of water vapor molecules. Water vapor molecules usually undergo condensation when attractive forces between the molecules can pull them together into the liquid state. Lower temperature means lower kinetic energies, and slow-moving water vapor molecules spend more time close to one another and close to the surface of liquid water. Spending more time close together means an increased likelihood of attractive forces pulling the molecules together. On the other hand, higher temperature means higher kinetic energies, and molecules with higher kinetic energy are less likely to be pulled together. As the temperature increases, there is therefore less tendency for water molecules to return to the liquid state. If the temperature is increased in an equilibrium condition, more water vapor must be added to the air to maintain the saturated condition. Warm air can therefore hold more water vapor than cooler air. In fact, warm air on a typical summer day can hold five times as much water vapor as cold air on a cold winter day.

Humidity

The amount of water vapor in the air is referred to generally as **humidity.** Damp, moist air is more likely to have condensation than evaporation, and this air is said to have a *high humidity.* Dry air is more likely to have evaporation than condensation, on the other hand, and this air is said to have a *low humidity.* A measurement of the amount of water vapor in the air at a particular time is called the **absolute humidity** (figure 17.15). At room temperature, for example, humid air might contain 15 grams of water vapor in each cubic meter of air. At the same temperature, air of low humidity might have an absolute humidity of only 2 grams per cubic meter. The absolute humidity can range from near zero up to a maximum that is determined by the temperature at the time. Since the temperature of the water vapor present in the air is the same as the temperature of the air, the maximum absolute humidity is usually said to be determined by the air temperature. What this really means is that the maximum absolute humidity is determined by the temperature of the water vapor, that is, the average kinetic energy of the water vapor.

The relationship between the *actual* absolute humidity at a particular temperature and the *maximum* absolute humidity that can occur at that temperature is called the **relative humidity.** Relative humidity is a ratio between (1) the amount of water vapor in the air and (2) the amount of water vapor needed to saturate the air at that temperature. The relationship is

$$\text{relative humidity} = \frac{\text{absolute humidity at present temperature}}{\text{maximum absolute humidity at present temperature}} \times 100\%$$

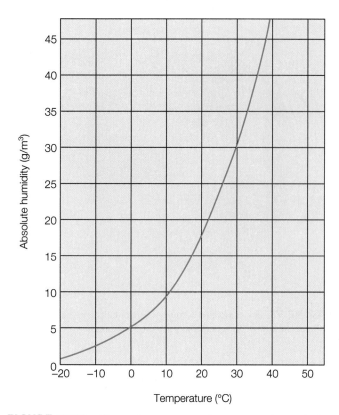

FIGURE 17.15 The maximum amount of water vapor that can be in the air at different temperatures. The amount of water vapor in the air at a particular temperature is called the *absolute humidity.*

For example, suppose a measurement of the water vapor in the air at 10°C (50°F) finds an absolute humidity of 5.0 g/m³. According to figure 17.15, the maximum amount of water vapor that can be in the air when the temperature is 10°C is about 10 g/m³. The relative humidity is then

$$\frac{5.0 \text{ g/m}^3}{10 \text{ g/m}^3} \times 100\% = 50\%$$

If the absolute humidity had been 10 g/m³, then the air would have all the water vapor it could hold, and the relative humidity would be 100 percent. A humidity of 100 percent means that the air is saturated at the present temperature.

The important thing to understand about relative humidity is that the capacity of air to hold water vapor changes with the temperature. Cold air cannot hold as much water vapor, and warming the air will increase its capacity. With the capacity *increased,* the relative humidity *decreases* because you can now add more water vapor to the air than you could before. Just warming the air, for example, can reduce the relative humidity from 50 percent to 3 percent. Lower relative humidity results because warming the air increases the capacity of air to hold water vapor. This explains the need to humidify a home in the winter. Evaporation occurs very rapidly when the humidity is low. Evaporation is a cooling process since the molecules with higher kinetic energy are the ones to escape, lowering the average kinetic energy as they evaporate. Dry air will therefore cause you to feel cool even though the air temperature is fairly high. Adding moisture to the air will enable you to feel warmer at lower air temperatures and thus lower your fuel bill.

The relationship between the capacity of air to hold water vapor and temperature also explains why the relative humidity increases in the evening after the Sun goes down. A cooler temperature means less capacity of air to hold water vapor. With the same amount of vapor in the air, a reduced capacity means a higher relative humidity.

The Condensation Process

Condensation, depends on two factors: (1) the relative humidity and (2) the temperature of the air. During condensation, molecules of water vapor join together to produce liquid water on the surface as dew or in the air as the droplets of water making up fog or clouds. Water molecules may also join together to produce solid water in the form of frost or snow. Before condensation can occur, however, the air must be saturated, which means that the relative humidity must be 100 percent with the air containing all the vapor it can hold at the present temperature. A parcel of air can become saturated as a result of (1) water vapor being added to the air from evaporation, (2) cooling, which reduces the capacity of the air to hold water vapor and therefore increases the relative humidity, or (3) a combination of additional water vapor with cooling.

The process of condensation of water vapor explains a number of common observations. You are able to "see your breath" on a cold day, for example, because the high moisture content of your exhaled breath is condensed into tiny water droplets by cold air. The small fog of water droplets evaporates as it spreads into the surrounding air with a lower moisture content. The white trail behind a high-flying jet aircraft is also a result of condensation of water vapor. Water is one of the products of combustion, and the white trail is condensed water vapor, a trail of tiny droplets of water in the cold upper atmosphere. The trail of water droplets is called a *contrail* after "condensation trail." Back on the surface, a cold glass of beverage seems to "sweat" as water vapor molecules near the outside of the glass are cooled, moving more slowly. Slowly moving water vapor molecules spend more time closer together, and the molecular forces between the molecules pull them together, forming a thin layer of liquid water on the outside of the cold glass. This same condensation process sometimes results in a small stream of water from the cold air conditioning coils of an automobile or home mechanical air conditioner.

As air is cooled, its capacity to hold water vapor is reduced to lower and lower levels. Even without water vapor being added to the air, a temperature will eventually be reached at which saturation, 100 percent humidity, occurs. Further cooling below this temperature will result in condensation. The temperature at which condensation begins is called the **dew point temperature.** If the dew point is above 0°C (32°F), the water vapor will condense on surfaces as a liquid called **dew.** If the temperature is at or below 0°C, the vapor will condense on surfaces as a solid called **frost.** Note that dew and frost form on the tops, sides, and bottoms of objects. Dew and frost condense directly on objects and do not "fall out" of the air. Note also that the temperature that determines if dew or frost forms is the temperature of the object where they condense. This temperature near the open surface can be very different from the reported air temperature, which is measured more at eye level in a sheltered instrument enclosure.

Observations of where and when dew and frost form can lead to some interesting things to think about. Dew and frost, for example, seem to form on "C" nights, nights that can be described by the three "C" words of *clear, calm,* and *cool.* Dew and frost also seem to form more (1) in open areas rather than under trees or other shelters, (2) on objects such as grass rather than on the flat, bare ground, and (3) in low-lying areas before they form on slopes or the sides of hills. What is the meaning of these observations?

Dew and frost are related to clear nights and open areas because these are the conditions best suited for the loss of infrared radiation. Air near the surface becomes cooler as infrared radiation is emitted from the grass, buildings, streets, and everything else that absorbed the shorter-wavelength radiation of incoming solar radiation during the day. Clouds serve as a blanket, keeping the radiation from escaping to space so readily. So a clear night is more conducive to the loss of infrared radiation and therefore to cooling. On a smaller scale, a tree serves the same purpose, holding in radiation and therefore retarding the cooling effect. Thus, an open area on a clear, calm night would have cooler air near the surface than would be the case on a cloudy night or under the shelter of a tree.

The observation that dew and frost form on objects such as grass before forming on flat, bare ground is also related to loss of infrared radiation. Grass has a greater exposed surface area than the flat, bare ground. A greater surface area means a greater area from which infrared radiation can escape, so grass blades cool more rapidly than the flat ground. Other variables, such as specific heat, may be involved, but overall, frost and dew are more likely to form on grass and low-lying shrubs before they form on the flat, bare ground.

Dew and frost form in low-lying areas before forming on slopes and the sides of hills because of the density differences of cool and warm air. Cool air is more dense than warm air and is moved downhill by gravity, pooling in low-lying areas. You may have noticed the different temperatures of low-lying areas if you have ever driven across hills and valleys on a clear, calm, and cool evening. Citrus and other orchards are often located on slopes of hills rather than on valley floors because of the gravity drainage of cold air.

FIGURE 17.16 Fans like this one are used to mix the warmer, upper layers of air with the cooling air in the orchard on nights when frost is likely to form.

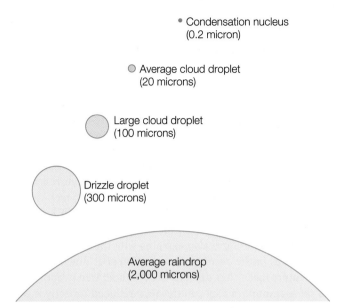

FIGURE 17.17 This figure compares the size of the condensation nuclei to the size of typical condensation droplets. Note that 1 micron is 1/1,000 mm.

It is air near the surface that is cooled first by the loss of radiation from the surface. Calm nights favor dew or frost formation because the wind mixes the air near the surface that is being cooled with warmer air above the surface. If you have ever driven near a citrus orchard, you may have noticed the huge, airplanelike fans situated throughout the orchard on poles. These fans are used on "C" nights when frost is likely to form to mix the warmer, upper layers of air with the cooling air in the orchard (figure 17.16).

Condensation occurs on the surface as frost or dew when the dew point is reached. When does condensation occur in the air? Water vapor molecules in the air are constantly colliding and banging into each other, but they do not just join together to form water droplets, even if the air is saturated. The water molecules need something to condense upon. Condensation of water vapor into fog or cloud droplets takes place on tiny particles present in the air. The particles are called **condensation nuclei.** There are hundreds of tiny dust, smoke, soot, and salt crystals suspended in each cubic centimeter of the air that serve as condensation nuclei. Tiny salt crystals, however, are particularly effective condensation nuclei because salt crystals attract water molecules. You may have noticed that salt in a salt shaker becomes moist on a humid day because of the way it attracts water molecules. Tiny salt crystals suspended in the air act the same way, serving as nuclei that attract water vapor into tiny droplets of liquid water.

After water vapor molecules begin to condense on a condensation nucleus, other water molecules will join the liquid water already formed, and the tiny droplet begins to increase in volume. The water droplets that make up a cloud are about fifteen hundred times larger than a condensation nuclei, and these droplets can condense out of the air in a matter of minutes. As the volume increases, however, the process slows, and

hours and days are required to form the even larger droplets and drops. For comparison to the sizes shown in figure 17.17, consider that the average human hair is about 100 microns in diameter. This is about the same diameter as the large cloud droplet of water. Large raindrops have been observed falling from clouds that formed only a few hours previously, so it must be some process or processes other than the direct condensation of raindrops that form precipitation. These processes are discussed in the section on precipitation.

Fog and Clouds

Fog and clouds are both accumulations of tiny droplets of water that have been condensed from the air. These water droplets are very small, and a very slight upward movement of the air will keep them from falling. If they do fall, they usually evaporate. Fog is sometimes described as a cloud that forms at or near the surface. A fog, as a cloud, forms because air containing water vapor and condensation nuclei has been cooled to the dew point. Some types of fog form under the same "C" night conditions favorable for dew or frost to form, that is, on clear, cool, and calm nights when the relative humidity is high. Sometimes this type of fog forms only in valleys and low-lying areas where cool air accumulates (figure 17.18). This type of fog is typical of inland fogs, those that form away from bodies of water. Other types of fog may form somewhere else, such as in the humid air over an ocean, and then move inland. Many fogs that occur along coastal regions were formed over the ocean and then carried inland by breezes. A third type of fog looks much like a steamy mist rising from melting snow on a street, over a body of water into cold air, or a steamy mist over streets after a summer rain shower. These are examples of a temporary fog that forms as a lot of water vapor is added to cool air. This is a cool fog, like other fogs, and is not hot as the steamlike appearance may lead you to believe.

Sometimes a news report states something about the Sun "burning off" a fog. A fog does not burn, of course, because it is made up of droplets of water. What the reporter really means is that the Sun's radiation will increase the temperature, which increases the air capacity to hold water vapor. With an increased capacity to hold water, the relative humidity drops, and the fog simply evaporates back to the state of invisible water vapor molecules.

Clouds, like fogs, are made up of tiny droplets of water that have been condensed from the air. Luke Howard, an English weather observer, made one of the first cloud classification schemes in 1803. He used the Latin terms *cirrus* (curly), *cumulus* (piled up), and *stratus* (spread out) to identify the basic shapes of clouds (figure 17.19). The clouds usually do not occur just in these basic cloud shapes but in combinations of the different shapes. Later, Howard's system was modified by expanding the different shapes of clouds into ten classes by using the basic cloud shapes and altitude as criteria. Clouds give practical hints about the approaching weather. The relationship between the different cloud shapes and atmospheric conditions and what clouds can mean about the coming weather are discussed in the section on precipitation.

Clouds form when a mass of air above the surface is cooled to its dew point temperature. In general, the mass of air is cooled because something has given it an upward push, moving it to higher levels in the atmosphere. There are three major causes of upward air movement: (1) *convection* resulting from differential heating, (2) mountain ranges that serve as *barriers* to moving air masses, and (3) the meeting of *moving air masses* with different densities, for example, a cold, dense mass of air meeting a warm, less dense mass of air.

The three major causes of uplifted air sometimes result in clouds, but just as often they do not. Whether clouds form or not depends on the condition of the atmosphere at the time. As a parcel of warm air is moved upward, it tends to stay together, mixing very little with the surrounding air. As it is forced upward, it becomes cooler because it is expanding. Similarly, the temperature of a gas increases when it is compressed. So rising air is cooled and descending air is warmed.

What happens to a parcel of air that is pushed upward depends on the difference in density between the parcel and the surrounding air. Air temperature will tell you about air density since the density of air is determined by its temperature. Instruments attached to a weather balloon can measure the change of temperature with altitude. By comparing this change with the rate of cooling by expansion, the state of atmospheric stability can be determined. There are many different states of atmospheric stability, and the following is a simplified description of just a few of the possible states, first considering dry air only.

The atmosphere is in a state of *stability* when a lifted parcel of air is cooler than the surrounding air. Being cooler, the parcel of air will be more dense than the surroundings. If it is moved up to a higher level and released in a stable atmosphere, it will move back to its former level. A lifted parcel of air always returns to its original level when the atmosphere is stable. Any clouds that do develop in a stable atmosphere are usually arranged in the horizontal layers of stratus-type clouds.

A

B

FIGURE 17.18 (*A*) An early morning aerial view of fog that developed close to the ground in cool, moist air on a clear, calm night. (*B*) Air moves from a warm current, then over a cool current, forming fog. The fog often moves inland at night.

The atmosphere is in a state of *instability* when a lifted parcel of air is warmer than the surrounding air. Being warmer, the parcel of air will be less dense than the surroundings. If it is moved up to a higher level, it will continue moving after the uplifting force is removed. Cumulus clouds usually develop in an unstable atmosphere, and the rising parcels of air, called thermals, can result in a very bumpy airplane ride.

So far, only dry air has been considered. As air moves upward and cools from expansion, sooner or later the dew point is reached and the air becomes saturated. As some of the water vapor in the rising parcel condenses to droplets, the latent heat of vaporization is released. The rising parcel of air now

FIGURE 17.19 (*A*) Cumulus clouds. (*B*) Stratus and stratocumulus. Note the small stratocumulus clouds forming from increased convection over each of the three small islands. (*C*) An aerial view between the patchy cumulus clouds below and the cirrus and cirrostratus above (the patches on the ground are clear-cut forests). (*D*) Altocumulus. (*E*) A rain shower at the base of a cumulonimbus. (*F*) Stratocumulus.

cools at a slower rate because of the release of this latent heat of vaporization. The release of latent heat warms the air in the parcel and decreases the density even more, accelerating the ascent. This leads to further condensation and the formation of towering cumulus clouds, often leading to rain.

Precipitation

Water that returns to the surface of Earth, in either the liquid or solid form, is called **precipitation** (figure 17.20). Note that dew and frost are not classified as precipitation because they form directly on the surface and do not fall through the air. Precipitation seems to form in clouds by one of two processes: (1) the *coalescence* of cloud droplets or (2) the *growth of ice crystals*. It would appear difficult for cloud droplets to merge, or coalesce, with one another since any air movement would seem to move them all at the same time, not bring them together. Condensation nuclei come in different sizes, however, and cloud droplets of many different sizes form on these different-sized nuclei. Larger cloud droplets are slowed less by air friction as they drift downward, and they collide and merge with smaller droplets as they fall. They may merge, or coalesce, with a million other droplets before they fall from the cloud as raindrops. This *coalescence process* of forming precipitation is thought to take place in warm cumulus clouds that form near the ocean in the tropics. These clouds contain giant salt condensation nuclei and have been observed to produce rain within about twenty minutes after forming.

Clouds at middle latitudes, away from the ocean, also produce precipitation, so there must be a second way that precipitation forms. The *ice-crystal process* of forming precipitation is important in clouds that extend high enough in the atmosphere

FIGURE 17.20 Precipitation is water in liquid or solid form that returns to the surface of Earth. The precipitation you see here is liquid, and each raindrop is made from billions of the tiny droplets that make up the clouds. The tiny droplets of clouds become precipitation by merging to form larger droplets or by the growth of ice crystals that melt while falling.

to be above the freezing point of water. Water molecules are more strongly bonded to each other in an ice crystal than in liquid water. Thus, an ice crystal can capture water molecules and grow to a larger size while neighboring water droplets are evaporating. As they grow larger and begin to drift toward the surface, they may coalesce with other ice crystals or droplets of water, soon falling from the cloud. During the summer, they fall through warmer air below and reach the ground as raindrops. During the winter, they fall through cooler air below and reach the ground as snow.

Tiny water droplets do not freeze as readily as a larger mass of liquid water, and many droplets do not freeze until the temperature is below about −40°C(−40°F). Water that is still in the liquid state when the temperature is below the freezing temperature is said to be *supercooled*. Supercooled clouds of water droplets are common between the temperatures of −40°C and 0°C (−40°F and 32°F), a range of temperatures that is often found in the upper atmosphere. The liquid droplets at these temperatures need solid particles called **ice-forming nuclei** to freeze upon. Generally, dust from the ground serves as ice-forming nuclei that start the ice-crystal process of forming precipitation. Artificial rainmaking has been successful by (1) dropping crushed dry ice, which is cooler than −40°C, on top of a supercooled cloud and (2) introducing "seeds" of ice-forming nuclei in supercooled clouds. Tiny crystals from the burning of silver iodide are effective ice-forming nuclei, producing ice crystals at temperatures as high as −4.0°C (about 25°F). Attempts at ground-based cloud seeding with silver iodide in the mountains of the western United States have suggested up to 15 percent more snowfall, but it is difficult to know how much snowfall would have resulted without the seeding.

In general, the basic form of a cloud has meaning about the general type of precipitation that can occur as well as the coming weather. Cumulus clouds usually produce showers or thunderstorms that last only brief periods of time. Longer periods of drizzle, rain, or snow usually occur from stratus clouds. Cirrus clouds do not produce precipitation of any kind, but they may have meaning about the coming weather, which is discussed in the section on weather producers.

WEATHER PRODUCERS

The idealized model of the general atmospheric circulation starts with the poleward movement of warm air from the tropics. The region between 10°N and 10°S of the equator receives more direct radiation, on the average, than other regions of Earth's surface. The air over this region is heated more, expands, and becomes less dense as a consequence of the heating. This less dense air is buoyed up by convection to heights up to 20 km (about 12 mi). As it rises, it is cooled by radiation to less than −73°C (about −110°F). This accumulating mass of cool, dry air spreads north and south toward both poles (see figure 17.13), then sinks back toward the surface at about 30°N and 30°S. The descending air is warmed by compression and is warm and dry by the time it reaches the surface. Part of the sinking air then moves back toward the equator across the surface, completing a large convective cell. This giant cell has a low-pressure belt over the equator and high-pressure belts over the subtropics near latitudes of 30°N and 30°S. The other part of the sinking air moves poleward across the surface, producing belts of westerly winds in both hemispheres to latitudes of about 60°.

The overall pattern of pressure belts and belts of prevailing winds is seen to shift north and south with the seasons, resulting in a seasonal shift in the types of weather experienced at a location. This shift of weather is related to three related weather producers: (1) the movement of large bodies of air, called *air masses,* that have acquired the temperature and moisture conditions where they have been located, (2) the leading *fronts* of air masses when they move, and (3) the local *high- and low-pressure* patterns that are associated with air masses and fronts. These are the features shown on almost all daily weather maps, and they are the topics of this section.

Air Masses

An **air mass** is defined as a large, more or less uniform body of air with nearly the same temperature and moisture conditions. An air mass forms when a large body of air, perhaps covering millions of square kilometers, remains over a large area of land or water for an extended period of time. While it is stationary, it acquires the temperature and moisture characteristics of the land or water through the heat transfer processes of conduction, convection, and radiation, and through the moisture transfer processes of evaporation and condensation. For example, a large body of air that remains over the cold, dry, snow-covered surface of Siberia for some time will become cold and dry. A large body of air that remains over a warm tropical ocean, on the other hand, will become warm and moist. Knowledge about the condition of air masses is important because they tend to retain the acquired temperature and moisture characteristics when they finally break away, sometimes moving long distances. An air mass that formed over Siberia can bring cold, dry air to your location, while an air mass that formed over a tropical ocean will bring warm, moist air.

Air masses are classified according to the temperature and moisture conditions where they originate. There are two temperature extreme possibilities, a *polar air mass* from a cold region and a *tropical air mass* from a warm region. There are also two moisture extreme possibilities, a moist *maritime air mass* from over the ocean and a generally dry *continental air mass* from over the land. Thus, there are four main types of air masses that can influence the weather where you live: (1) continental polar, (2) maritime polar, (3) continental tropical, and (4) maritime tropical. Figure 17.21 shows the general direction in which these air masses usually move over the mainland United States.

Once an air mass leaves its source region, it can move at speeds of up to 800 km (about 500 mi) per day while mostly retaining the temperature and moisture characteristics of the source region (figure 17.22). If it slows and stagnates over a new location, however, the air may again begin to acquire a new temperature and moisture equilibrium with the surface. When a location is under the influence of an air mass, the location is having a period of *air mass weather.* This means that the weather conditions will generally remain the same from day to day with

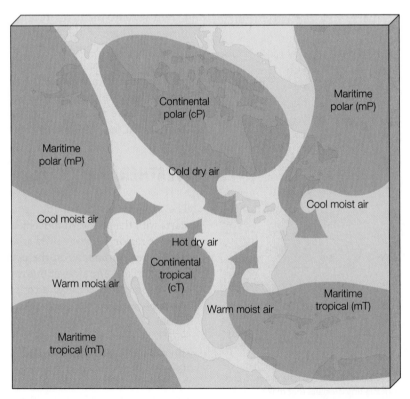

FIGURE 17.21 The air masses that affect weather in North America. The importance of the various air masses depends on the season. In winter, for instance, the continental tropical air mass disappears and the continental polar air mass exerts its greatest influence.

FIGURE 17.22 This satellite photograph shows the result of a polar air mass moving southeast over the southern United States. Clouds form over the warmer waters of the Gulf of Mexico and the Atlantic Ocean, showing the state of atmospheric instability from the temperature differences.

slow, gradual changes. Air mass weather will remain the same until a new air mass moves in or until the air mass acquires the conditions of the new location. This process may take days or several weeks, and the weather conditions during this time depend on the conditions of the air mass and conditions at the new location. For example, a polar continental air mass arriving over a cool, dry land area may produce a temperature inversion with the air colder near the surface than higher up. When the temperature increases with height, the air is stable and cloudless, and cold weather continues with slow, gradual warming. The temperature inversion may also result in hazy periods of air pollution in some locations. A polar continental air mass arriving over a generally warmer land area, on the other hand, results

in a condition of instability. In this situation, each day will start clear and cold, but differential heating during the day develops cumulus clouds in the unstable air. After sunset, the clouds evaporate, and a clear night results because the thermals during the day carried away the dust and air pollution. Thus, a dry, cold air mass can bring different weather conditions, each depending on the properties of the air mass and the land it moves over.

Weather Fronts

The boundary between air masses of different temperatures is called a **front.** A front is actually a thin transition zone between two air masses that ranges from about 5 to 30 km wide

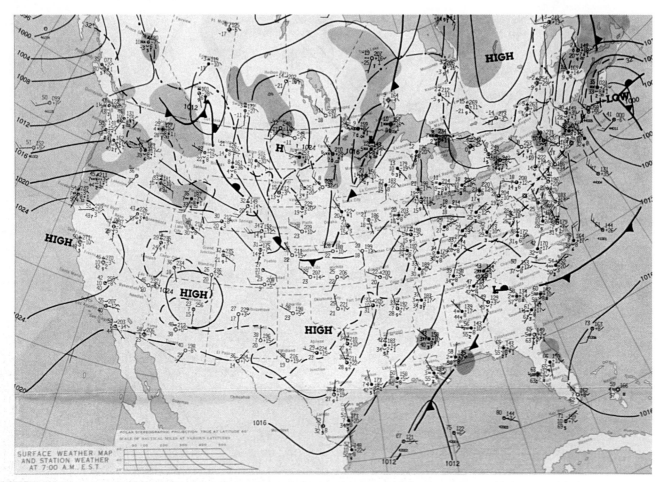

FIGURE 17.23 This weather map of the United States shows two fronts with associated low-pressure areas and five areas with high pressure.

(about 3 to 20 mi) wide, and the air masses do not mix other than in this narrow zone. The density differences between the two air masses prevent any general mixing since the warm, less dense air mass is forced upward by the cooler, more dense air moving under it. You may have noticed on a daily weather map that fronts are usually represented with a line bulging outward in the direction of cold air mass movement (figure 17.23). A cold air mass is much like a huge, flattened bubble of air that moves across the land (figure 17.24). The line on a weather map represents the place where the leading edge of this huge, flattened bubble of air touches the surface of Earth.

A **cold front** is formed when a cold air mass moves into warmer air, displacing it in the process. A cold front is generally steep, and when it runs into the warmer air, it forces it to rise quickly. If the warm air is moist, it is quickly cooled to the dew point temperature, resulting in large, towering cumulus clouds and thunderclouds along the front (figure 17.25). You may have observed that thunderstorms created by an advancing cold front often form in a line along the front. These thunderstorms can be intense but are usually over quickly, soon followed by a rapid drop in temperature from the cold air mass moving past your location. The passage of the cold front is also marked by a rapid shift in the wind direction and a rapid increase in the barometric pressure. Before the cold front arrives, the wind is generally moving toward the front as warm, less dense air is forced upward by the cold, more dense air. The lowest

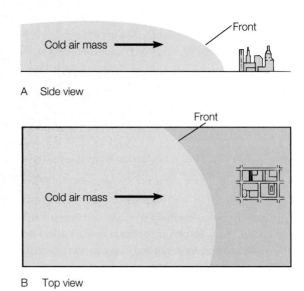

FIGURE 17.24 (*A*) A cold air mass is similar to a huge, flattened bubble of cold air that moves across the land. The front is the boundary between two air masses, a narrow transition zone of mixing. (*B*) A front is represented by a line on a weather map, which shows the location of the front at ground level.

barometric pressure reading is associated with the lifting of the warm air at the front. After the front passes your location, you are in the cooler, more dense air that is settling outward,

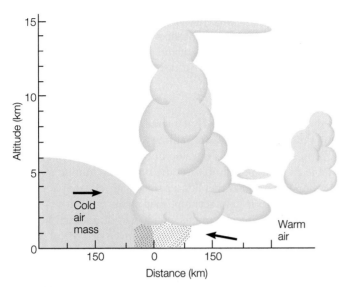

FIGURE 17.25 An idealized cold front, showing the types of clouds that might occur when an unstable cold air mass moves through unstable warm air. Stable air would result in more stratus clouds rather than cumulus clouds.

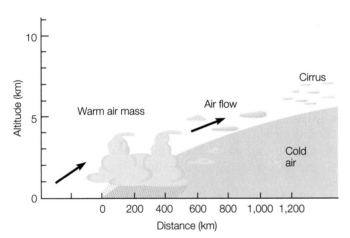

FIGURE 17.26 An idealized warm front, showing a warm air mass overriding and pushing cold air in front of it. Notice that the overriding warm air produces a predictable sequence of clouds far in advance of the moving front.

so the barometric pressure increases and the wind shifts with the movement of the cold air mass.

A **warm front** forms when a warm air mass advances over a mass of cooler air. Since the advancing warm air is less dense than the cooler air it is displacing, it generally overrides the cooler air, forming a long, gently sloping front. Because of this, the overriding warm air may form clouds far in advance of the ground-level base of the front (figure 17.26). This may produce high cirrus clouds a day or more in advance of the front, which are followed by thicker and lower stratus clouds as the front advances. Usually these clouds result in a broad band of drizzle, fog, and the continuous light rain associated with stratus clouds. This light rain (and snow in the winter) may last for days as the warm front passes.

Sometimes the forces influencing the movement of a cold or warm air mass lessen or become balanced, and the front stops

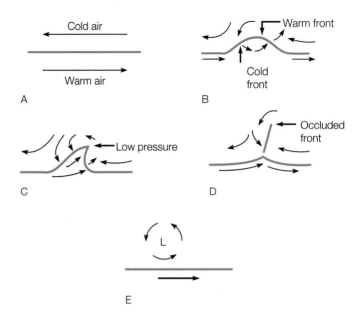

FIGURE 17.27 The development of a low-pressure center, or cyclonic storm, along a stationary front as seen from above. (*A*) A stationary front with cold air on the north side and warm air on the south side. (*B*) A wave develops, producing a warm front moving northward on the rightside and a cold front moving southward on the left side. (*C*) The cold front lifts the warm front off the surface at the apex, forming a low-pressure center. (*D*) When the warm front is completely lifted off the surface, an occluded front is formed. (*E*) The cyclonic storm is now a fully developed low-pressure center.

advancing. When this happens, a stream of cold air moves along the north side of the front, and a stream of warm air moves along the south side in an opposite direction. This is called a **stationary front** because the edge of the front is not advancing. A stationary front may sound as if it is a mild frontal weather maker because it is not moving. Actually, a stationary front represents an unstable situation that can result in a major atmospheric storm. This type of storm is discussed in the section on waves and cyclones.

Waves and Cyclones

A slowly advancing cold front and a stationary front often develop a bulge, or *wave,* in the boundary between cool and warm air moving in opposite directions (figure 17.27B). The wave grows as the moving air is deflected, forming a warm front moving northward on the right side and a cold front moving southward on the left side. Cold air is more dense than warm air, and the cold air moves faster than the slowly moving warm front. As the faster moving cold air catches up with the slower moving warm air, the cold air underrides the warm air, lifting it upward. This lifting action produces a low-pressure area at the point where the two fronts come together (figure 17.27C). The lifted air expands, cools, and reaches the dew point. Clouds form and precipitation begins from the lifting and cooling action. Within days after the wave first appears, the cold front completely overtakes the warm front, forming an occlusion (figure 17.27D). An **occluded front** is one that has been lifted completely off the ground into the atmosphere. The disturbance is now a *cyclonic storm* with a fully developed low-pressure center. Since its formation, this low-pressure cyclonic storm has been moving, taking its associated stormy weather with it in a generally easterly direction. Such cyclonic storms usually follow principal tracks along a front. Since they are observed generally to follow these same tracks, it is possible to predict where the storm might move next.

A *cyclone* is defined as a low-pressure center in which the winds move into the low-pressure center and are forced upward. As air moves in toward the center, the Coriolis effect and friction with the ground cause the moving air to change direction. In the Northern Hemisphere, this produces a counterclockwise circulation pattern of winds around the low-pressure center (figure 17.28). The upward movement associated with the low-pressure center of a cyclone cools the air, resulting in clouds, precipitation, and stormy conditions.

Air is sinking in the center of a region of high pressure, producing winds that move outward. In the Northern Hemisphere, the Coriolis effect and frictional forces deflect this wind to the right, producing a clockwise circulation (figure 17.28). A high-pressure center is called an *anticyclone,* or simply a *high.* Since air in a high-pressure zone sinks, it is warmed, and the relative humidity is lowered. Thus, clear, fair weather is usually associated with a high. By observing the barometric pressure, you can watch for decreasing pressure, which can mean the coming of a cyclone and its associated stormy weather. You can also watch for increasing pressure, which means a high and its associated fair weather are coming. Consulting a daily weather map makes such projections a much easier job, however.

Major Storms

A wide range of weather changes can take place as a front passes, because there is a wide range of possible temperature, moisture, stability, and other conditions between the new air mass and the air mass that it is displacing. The changes that accompany some fronts may be so mild that they go unnoticed. Others are noticed only as a day with breezes or gusty winds. Still other fronts are accompanied by a rapid and violent weather change called a **storm.** A snowstorm, for example, is

a rapid weather change that may happen as a cyclonic storm moves over a location. The most rapid and violent changes occur with three kinds of major storms: (1) thunderstorms, (2) tornadoes, and (3) hurricanes.

Thunderstorms

A **thunderstorm** is a brief but intense storm with rain, lightning and thunder, gusty and often strong winds, and sometimes hail. Thunderstorms usually develop in warm, very moist, and unstable air. These conditions set the stage for a thunderstorm to develop when something lifts a parcel of air, starting it moving upward. This is usually accomplished by the same three general causes that produce cumulus clouds: (1) differential heating, (2) mountain barriers, or (3) along an occluded or cold front. Thunderstorms that occur from differential heating usually occur during warm, humid afternoons after the Sun has had time to establish convective thermals. In the Northern Hemisphere, most of these convective thunderstorms occur during the month of July. Frontal thunderstorms, on the other hand, can occur any month and any time of the day or night that a front moves through warm, moist, and unstable air.

Frontal thunderstorms generally move with the front that produced them. Thunderstorms that developed in mountains or over flat lands from differential heating can move miles after they form, sometimes appearing to wander aimlessly across the land. These storms are not just one big rain cloud but are sometimes made up of cells that are born, grow to maturity, then die out in less than an hour. The thunderstorm, however, may last longer than an hour because new cells are formed as old ones die out. Each cell is about 2 to 8 km (about 1 to 5 mi) in diameter and goes through three main stages in its life: (1) cumulus, (2) mature, and (3) final (figure 17.29).

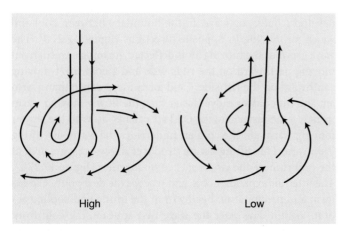

FIGURE 17.28 Air sinks over a high-pressure center and spreads over the surface in a clockwise pattern in the Northern Hemisphere. Air over the surface moves toward a low-pressure center in a counterclockwise pattern in the Northern Hemisphere.

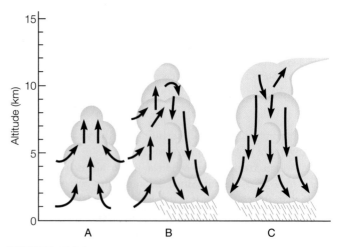

FIGURE 17.29 Three stages in the life of a thunderstorm cell. (*A*) The cumulus stage begins as warm, moist air is lifted in an unstable atmosphere. All the air movement is upward in this stage. (*B*) The mature stage begins when precipitation reaches the ground. This stage has updrafts and downdrafts side by side, which create violent turbulence. (*C*) The final stage begins when all the updrafts have been cut off and only downdrafts exist. This cuts off the supply of moisture, and the rain decreases as the thunderstorm dissipates. The anvil-shaped top is a characteristic sign of this stage.

Damage from a thunderstorm is usually caused by the associated lightning, strong winds, or hail. As illustrated in figure 17.29, the first stage of a thunderstorm begins as convection, mountains, or a dense air mass slightly lifts a mass of warm, moist air in an unstable atmosphere. The lifted air mass expands and cools to the dew point temperature, and a cumulus cloud forms. The latent heat of vaporization released by the condensation process accelerates the upward air motion, called an *updraft,* and the cumulus cloud continues to grow to towering heights. Soon the upward-moving, saturated air reaches the freezing level and ice crystals and snowflakes begin to form. When they become too large to be supported by the updraft, they begin to fall toward the surface, melting into raindrops in the warmer air they fall through. When they reach the surface, this marks the beginning of the mature stage. As the raindrops fall through the air, friction between the falling drops and the cool air produces a downdraft in the region of the precipitation. The cool air accelerates toward the surface at speeds up to 90 km/h (about 55 mi/h), spreading out on the ground when it reaches the surface. In regions where dust is raised by the winds, this spreading mass of cold air from the thunderstorm has the appearance of a small cold front with a steep, bulging leading edge. This miniature cold front may play a role in lifting other masses of warm, moist air in front of the thunderstorm, leading to the development of new cells. This stage in the life of a thunderstorm has the most intense rainfall, winds, and possibly hail. As the downdraft spreads throughout the cloud, the supply of new moisture from the updrafts is cut off and the thunderstorm enters the final, dissipating stage. The entire life cycle, from cumulus cloud to the final stage, lasts for about an hour as the thunderstorm moves across the surface. During the mature stage of powerful updrafts, the top of the thunderstorm may reach all the way to the top of the troposphere, forming a cirrus cloud that is spread into an anvil shape by the strong winds at this high altitude.

The updrafts, downdrafts, and falling precipitation separate tremendous amounts of electric charges that accumulate in different parts of the thundercloud. Large drops of water tend to carry negative charges, and cloud droplets tend to lose them. The upper part of the thunderstorm develops an accumulation of positive charges as cloud droplets are uplifted, and the middle portion develops an accumulation of negative charges from larger drops that fall. The voltage of these charge centers builds to the point that the electrical insulating ability of the air between them is overcome and a giant electrical discharge called *lightning* occurs (figure 17.30). Lightning discharges occur from the cloud to the ground, from the ground to a cloud, from one part of the cloud to another part, or between two different clouds. The discharge takes place in a fraction of a second and may actually consist of a number of strokes rather than one big discharge. The discharge produces an extremely high temperature around the channel, which may be only 6 cm (about 2 in) or so wide. The air it travels through is heated quickly, expanding into a sudden pressure wave that you hear as *thunder.* A nearby lightning strike produces a single, loud crack. Farther away strikes sound more like a rumbling boom as the sound from the separate strokes become separated over

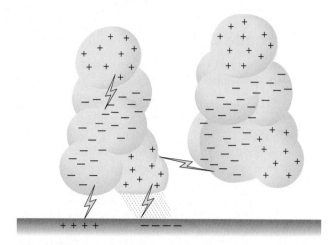

FIGURE 17.30 Different parts of a thunderstorm cloud develop centers of electric charge. Lightning is a giant electric spark that discharges the accumulated charges.

FIGURE 17.31 These hailstones fell from a thunderstorm in Iowa, damaging automobiles, structures, and crops.
© Telegraph Herald/Photo by Patti Carr

distance. Echoing of the thunder produced at farther distances also adds to the rumbling sounds. Lightning can present a risk for people in the open, near bodies of water, or under a single, isolated tree during a thunderstorm. The safest place to be during a thunderstorm is inside a car or a building with a metal frame.

Updrafts are also responsible for **hail,** a frozen form of precipitation that can be very destructive to crops, automobiles, and other property. Hailstones can be irregular, somewhat spherical, or flattened forms of ice that range from the size of a BB to the size of a softball (figure 17.31). Most hailstones, however, are less than 2 cm (about 1 in) in diameter. The larger hailstones have alternating layers of clear and opaque, cloudy ice. These layers are believed to form as the hailstone goes through cycles of falling then being returned to the upper parts of the thundercloud by updrafts. The clear layers are believed to form as

Tornado Damage

Tornadoes are rated on wind speed and damage. Here is the scale with approximate wind speeds:

0: Light damage, winds under 120 km/h (75 mi/h). Damage to chimneys, tree limbs broken, small trees pushed over, signs damaged.

1: Moderate damage, winds 121 to 180 km/h (76 to 112 mi/h). Roofing materials removed, mobile homes and moving autos pushed around or overturned.

2: Considerable damage, winds 181 to 250 km/h (113 to 155 mi/h). Roofs torn off homes, mobile homes demolished, boxcars overturned, large trees snapped or uprooted, light objects become missiles.

3: Severe damage, winds 251 to 330 km/h (156 to 205 mi/h). Roofs and some walls torn off homes, whole trains overturned, most trees uprooted, cars lifted off the ground and thrown.

4: Devastating damage, winds 331 to 418 km/h (206 to 260 mi/h). Homes leveled, cars thrown, large missiles generated.

5: Incredible damage, winds 419 to 512 km/h (261 to 318 mi/h). Homes demolished and swept away, automobile-sized missiles fly through the air more than 90 m (100 yds), trees debarked.

FIGURE 17.32 A tornado might be small, but it is the most violent storm that occurs on Earth. This tornado, moving across an open road, eventually struck Dallas, Texas.

the hailstone moves through heavy layers of supercooled water droplets, which accumulate quickly on the hailstone but freeze slowly because of the release of the latent heat of fusion. The cloudy layers are believed to form as the hailstone accumulates snow crystals or moves through a part of the cloud with less supercooled water droplets. In either case, rapid freezing traps air bubbles, which result in the opaque, cloudy layer. Thunderstorms with hail are most common during the month of May in Colorado, Kansas, and Nebraska.

Tornadoes

A **tornado** is the smallest, most violent weather disturbance that occurs on Earth (figure 17.32). Tornadoes occur with intense thunderstorms and resemble a long, narrow funnel or ropelike structure that drops down from a thundercloud and may or may not touch the ground. This ropelike structure is a rapidly whirling column of air, usually 100 to 400 m (about 330 to 1,300 ft) in diameter. An average tornado will travel 6 to 8 km (about 4 to 5 mi) on the ground, lasting only a few minutes. The bottom of the column moves across the ground, sometimes skipping into the air, then back down again at speeds that average about 50 km/h (about 30 mi/h) across the surface. The speed of the whirling air in the column has been estimated to be up to about 480 km/h (about 300 mi/h), but most tornadoes have winds of less than 180 km/h (112 mi/h). The destruction is produced by the powerful winds, the sudden drop in atmospheric pressure that occurs at the center of the funnel, and the debris that is flung through the air like projectiles. A passing

tornado sounds like very loud, continuous rumbling thunder with cracking and hissing noises that are punctuated by the crashing of debris projectiles.

On the average, several hundred tornadoes are reported in the United States every year. These occur mostly during spring and early summer afternoons over the Great Plains states. Texas, Oklahoma, Kansas, and Iowa have such a high occurrence of tornadoes that the region is called "tornado alley." During the spring and early summer, this region has maritime tropical air from the Gulf of Mexico at the surface. Above this warm, moist layer is a layer of dry, unstable air that has just crossed the Rocky Mountains, moved along rapidly by the jet stream. The stage is now set for some event, such as a cold air mass moving in from the north, to shove the warm, moist air upward, and the result will be violent thunderstorms with tornadoes.

Hurricanes

What is the difference between a tropical depression, tropical storm, and a hurricane? In general, they are all storms with

FIGURE 17.33 This is a satellite photo of hurricane John, showing the eye and counterclockwise motion.

Hurricane Damage

Hurricanes are classified according to category and damage to be expected. Here is the classification scheme:

Category	Damage	Winds	
1	minimal	120–153 km/h	(75–95 mi/h)
2	moderate	154–177 km/h	(96–110 mi/h)
3	extensive	178–210 km/h	(111–130 mi/h)
4	extreme	211–250 km/h	(131–155 mi/h)
5	catastrophic	>250 km/h	(>155 mi/h)

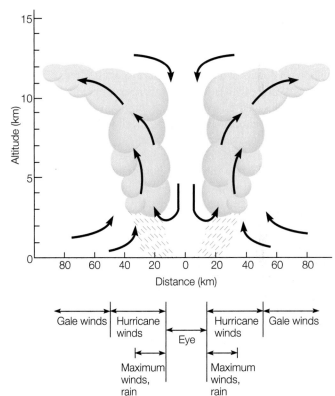

FIGURE 17.34 Cross section of a hurricane.

strong upward atmospheric motion and a cyclonic surface wind circulation (figure 17.33). They are born over tropical or subtropical waters and are not associated with a weather front. The varieties of storm intensities are classified according to the *speed* of the maximum sustained surface winds.

A *tropical depression* is an area of low pressure around which the winds are generally moving 55 km/h (about 35 mi/h) or less. The tropical depression might dissolve into nothing, or it might develop into a more intense disturbance. A *tropical storm* is a more intense low pressure area with winds between 56 and 120 km/h (about 35 to 75 mi/h). A **hurricane** is a very intense low-pressure area with winds greater than 120 km/h (about 75 mi/h). A strong storm of this type is called a hurricane if it occurs over the Atlantic Ocean or the Pacific Ocean east of the international date line. It is called a **typhoon** if it occurs over the North Pacific Ocean west of the international date line.

A tropical cyclone is similar to the wave cyclone of the mid-latitudes because both have low-pressure centers with a counterclockwise circulation in the Northern Hemisphere.

They are different because a wave cyclone is usually about 2,500 km (about 1,500 mi) wide, has moderate winds, and receives its energy from the temperature differences between two air masses. A tropical cyclone, on the other hand, is often less than 200 km (about 125 mi) wide, has very strong winds, and receives its energy from the latent heat of vaporization released during condensation.

A fully developed hurricane has heavy bands of clouds, showers, and thunderstorms that rapidly rotate around a relatively clear, calm eye (figure 17.34). As a hurricane approaches a location, the air seems unusually calm as a few clouds appear, then thicken as the wind begins to gust. Over the next six hours or so, the overall wind speed increases as strong gusts and intense rain showers occur. Thunderstorms, perhaps with tornadoes, and the strongest winds occur just before the winds suddenly die down and the sky clears with the arrival of the eye of the hurricane. The eye is an average of 10 to 15 km (about 6 to 9 mi) across, and it takes about an hour or so to cross a location. When the eye passes, the intense rain showers, thunderstorms, and hurricane-speed winds begin again, this time blowing from the opposite direction. The whole sequence of events may be over in a day or two, but hurricanes are unpredictable and sometimes stall in one location for days. In general, they move at a rate of 15 to 50 km/h (about 10 to 30 mi/h).

Most of the damage from hurricanes results from strong winds, flooding, and the occasional tornado. Flooding occurs from the intense, heavy rainfall but also from the increased sea level that results from the strong, constant winds blowing seawater toward the shore. The sea level can be raised some 5 m (about 16 ft) above normal, with storm waves up to 15 m (about 50 ft) high on top of this elevated sea level. Overall, large inland areas can be flooded with extensive property damage. A single hurricane moving into a populated coastal region has caused billions of dollars of damage and the loss of hundreds of lives in the past. Today, the National Weather Service tracks hurricanes by weather satellites. Warnings of hurricanes, tornadoes, and severe thunderstorms are broadcast locally over special weather alert stations located across the country.

In August 2005, hurricane Katrina initially struck near Miami, Florida, as a category 1 hurricane. It then moved into the Gulf of Mexico and grew to a strong category 3 hurricane that moved up the Gulf to the eastern Louisiana and western Mississippi coast. This massive storm had hurricane-force winds that extended outward 190 km (120 mi) from the center, resulting in severe storm damage over a wide area as it struck the coast on August 29. Damage resulted from a storm surge that exceeded 7 m (25 ft), heavy rainfall, wind damage, and the failure of the levee system in New Orleans. Overall, this resulted in an estimated $81 billion in damages and more than 1,836 fatalities.

Katrina had sustained winds of 200 km/h (125 mi/h) as it stuck the shore, but smaller hurricanes have had stronger sustained winds when they struck the shore. These include

- August 17, 1969: Hurricane Camille hit Mississippi with 306 km/h (190 mi/h) sustained winds.
- August 24, 1992: Hurricane Andrew hit south Florida with 266 km/h (165 mi/h) sustained winds.
- August 13, 2004: Hurricane Charley hit Punta Gorda, Florida, with 240 km/h (150 mi/h) sustained winds.*

WEATHER FORECASTING

Today, weather predictions are based on information about the characteristics, location, and rate of movement of air masses and associated fronts and pressure systems. This information is summarized as average values, then fed into a computer model of the atmosphere. The model is a scaled-down replica of the

*Hurricane data from National Climatic Data Center: www.ncdc.noaa.gov/oa/climate/research/2005/katrina.html.

real atmosphere, and changes in one part of the model result in changes in another part of the model just as they do in the real atmosphere. Underlying the computer model are the basic scientific laws concerning solar radiation, heat, motion, and the gas laws. All these laws are written as a series of mathematical equations, which are applied to thousands of data points in a three-dimensional grid that represents the atmosphere. The computer is given instructions about the starting conditions at each data point, that is, the average values of temperature, atmospheric pressure, humidity, wind speed, and so forth. The computer is then instructed to calculate the changes that will take place at each data point, according to the scientific laws, within a very short period of time. This requires billions of mathematical calculations when the program is run on a worldwide basis. The new calculated values are then used to start the process all over again, and it is repeated some 150 times to obtain a one-day forecast (figure 17.35).

A problem with the computer model of the atmosphere is that small-scale events are inadequately treated, and this introduces errors that grow when predictions are attempted for farther and farther into the future. Small eddies of air, for example, or gusts of wind in a region have an impact on larger-scale atmospheric motions such as those larger than a cumulus cloud. But all of the small eddies and gusts cannot be observed without filling the atmosphere with measuring instruments. This lack of ability to observe small events that can change the large-scale events introduces uncertainties in the data, which, over time, will increasingly affect the validity of a forecast.

To find information about the accuracy of a forecast, the computer model can be run several different times, with each run having slightly different initial conditions. If the results of all the runs are close to each other, the forecasters can feel confident that the atmosphere is in a predictable condition, and this means the forecast is probably accurate. In addition, multiple computer runs can provide forecasts in the form of probabilities.

FIGURE 17.35 Supercomputers make routine weather forecasts possible by solving mathematical equations that describe changes in a mathematical model of the atmosphere. This "fish-eye" view was necessary to show all of this Cray supercomputer.

For example, if eight out of ten forecasts indicate rain, the "averaged" forecast might call for an 80 percent chance of rain.

The use of new computer technology has improved the accuracy of next-day forecasts tremendously, and the forecasts up to three days are fairly accurate, too. For forecasts of more than five days, however, the number of calculations and the effect of uncertainties increase greatly. It has been estimated that the reductions of observational errors could increase the range of accurate forecasting up to two weeks. The ultimate range of accurate forecasting will require a better understanding—and thus, an improved model—of patterns of changes that occur in the ocean as well as in the atmosphere. All of this increased understanding and reduction of errors leads to an estimated ultimate future forecast of three weeks, beyond which any pinpoint forecast would be only slightly better than a wild guess. In the meantime, regional and local daily weather forecasts are fairly accurate, and computer models of the atmosphere now provide the basis for extending the forecasts for up to about a week.

CLIMATE

Changes in the atmospheric condition over a brief period of time, such as a day or a week, are referred to as changes in the *weather*. These changes are part of a composite, larger pattern called **climate**. Climate is the general pattern of the weather that occurs for a region over a number of years. Among other things, the climate determines what types of vegetation grow in a particular region, resulting in characteristic groups of plants associated with the region (figure 17.36). For example, orange, grapefruit, and palm trees grow in a region that has a climate with warm monthly temperatures throughout the year. On the other hand, blueberries, aspen, and birch trees grow in a region that has cool temperature patterns throughout the year. Climate determines what types of plants and animals live in a location, the types of houses that people build, and the lifestyles of people. Climate also influences the processes that shape the

FIGURE 17.36 The climate determines what types of plants and animals live in a location, the types of houses that people build, and the lifestyles of people. This orange tree, for example, requires a climate that is relatively frost-free, yet it requires some cool winter nights to produce a sweet fruit.

landscape, the type of soils that form, the suitability of the region for different types of agriculture, and how productive the agriculture will be in a region. This section is about climate, what determines the climate of a region, and how climate patterns are classified.

Major Climate Groups

Earth's atmosphere is heated directly by incoming solar radiation and by absorption of infrared radiation from the surface. The amount of heating at any particular latitude on the surface depends primarily on two factors: (1) the *intensity* of the incoming solar radiation, which is determined by the angle at which the radiation strikes the surface, and (2) the amount of *time* that the radiation is received at the surface, that is, the number of daylight hours compared to the number of hours of night.

Earth is so far from the Sun that all rays of incoming solar radiation reaching Earth are essentially parallel. Earth, however, has a mostly constant orientation of its axis with respect to the stars as it moves around the Sun in its orbit. Since the inclined axis points toward the Sun part of the year and away from the Sun the other part, radiation reaches different latitudes at different angles during different parts of the year. The orientation of Earth's axis to the Sun during different parts of the year also results in days and nights of nearly equal length in the equatorial region but increasing differences at increasing latitudes to the poles. During the polar winter months, the night is twenty-four hours long, which means no solar radiation is received at all. The equatorial region receives more solar radiation during a year, and the amount received decreases toward the poles as a result of (1) yearly changes in intensity and (2) yearly changes in the number of daylight hours.

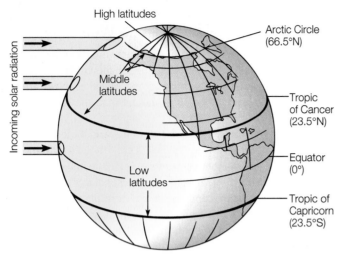

FIGURE 17.37 Latitude groups based on incoming solar radiation. The low latitudes receive vertical solar radiation at noon some time of the year, the high latitudes receive no solar radiation at noon during some time of the year, and the middle latitudes are in between.

To generalize about the amount of radiation received at different latitudes, some means of organizing, or grouping, the latitudes is needed (figure 17.37). For this purpose, the latitudes are organized into three groups:

1. The *low latitudes,* those that some time of the year receive *vertical* solar radiation at noon.
2. The *high latitudes,* those that some time of the year receive *no* solar radiation at noon.
3. The *middle latitudes,* which are between the low and high latitudes.

This definition of low, middle, and high latitudes means that the low latitudes are between the tropics of Cancer and Capricorn (between 23.5°N and 23.5°S latitudes) and that the high latitudes are above the Arctic and Antarctic Circles (above 66.5°N and above 66.5°S latitudes).

In general:

1. The low latitudes receive a high amount of incoming solar radiation that varies little during a year. Temperatures are high throughout the year, varying little from month to month.
2. The middle latitudes receive a higher amount of incoming radiation during one part of the year and a lower amount during the other part. Overall temperatures are cooler than in the low latitudes and have a wide seasonal variation.
3. The high latitudes receive a maximum amount of radiation during one part of the year and none during another part. Overall temperatures are low, with the highest range of annual temperatures.

The low, middle, and high latitudes provide a basic framework for describing Earth's climates. These climates are associated with the low, middle, and high latitudes illustrated in figure 17.37, but they are defined in terms of yearly temperature averages. Temperature and moisture are the two most important climate factors, and temperature will be considered first.

The principal climate zones are defined in terms of yearly temperature averages, which occur in broad regions (figure 17.38). They are

1. The *tropical climate zone* of the low latitudes (figure 17.39).
2. The *polar climate zone* of the high latitudes (figure 17.40).
3. The *temperate climate zone* of the middle latitudes (figure 17.41).

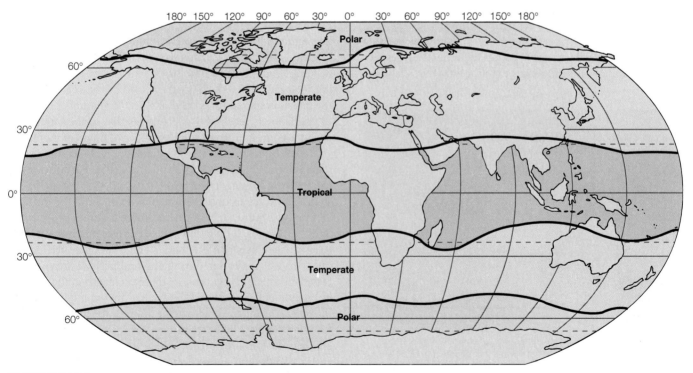

FIGURE 17.38 The principal climate zones are defined in terms of yearly temperature averages, which are determined by the amount of solar radiation received at the different latitude groups.

FIGURE 17.39 A wide variety of plant life can grow in a tropical climate, as you can see here.

FIGURE 17.40 Polar climates occur at high elevations as well as high latitudes. This mountain location has a highland polar climate and tundra vegetation but little else.

FIGURE 17.41 This temperate-climate deciduous forest responds to seasonable changes in autumn with a show of color.

The tropical climate zone is near the equator and receives the greatest amount of sunlight throughout the year. Overall, the tropical climate zone is hot. Average monthly temperatures stay above 18°C (64°F), even during the coldest month of the year.

The other extreme is found in the polar climate zone, where the sun never sets during some summer days and never rises during some winter days. Overall, the polar climate zone is cold. Average monthly temperatures stay below 10°C (50°F), even during the warmest month of the year.

The temperate climate zone is between the polar and tropical zones, with average temperatures that are neither very cold nor very hot. Average monthly temperatures stay between 10°C and 18°C (50°F and 64°F) throughout the year.

General patterns of precipitation and winds are also associated with the low, middle, and high latitudes. An idealized model of the global atmospheric circulation and pressure patterns was described in the section on weather forecasting. Recall that this model described a huge convective movement of air in the low latitudes, with air being forced upward over the equatorial region. This air expands, cools to the dew point, and produces abundant rainfall throughout most of the year. On the other hand, air is slowly sinking over 30°N and 30°S of the equator, becoming warm and dry as it is compressed. Most of the great deserts of the world are near 30°N or 30°S latitude

A Closer Look

El Niño and La Niña

The term *El Niño* was originally used to describe an occurrence of warm, above-normal ocean temperatures off the South American coast. Fishermen along this coast learned long ago to expect associated changes in fishing patterns about every three to seven years, which usually lasted for about eighteen months. They called this event El Niño, which is Spanish for "the boy child" or "Christ child," because it typically began near Christmas. The El Niño event occurs when the trade winds along the equatorial Pacific become reduced or calm, allowing sea surface temperatures to increase much above normal. The warm water drives the fish to deeper waters or farther away from usual fishing locations.

Today, El Niño is understood to be much more involved than just a warm ocean in the Pacific. It is more than a local event, and the bigger picture is sometimes called the "El Niño–Southern Oscillation," or ENSO. In addition to the warmer tropical Pacific water of El Niño, "the boy," the term *La Niña*, "the girl," has been used to refer to the times when the water of the tropical Pacific is *colder* than normal. The "Southern Oscillation" part of the name comes from observations that atmospheric pressure around Australia seems to be inversely linked to the atmospheric pressure in Tahiti. They seem to be linked because when the pressure is low in Australia, it is high in Tahiti. Conversely, when the atmospheric pressure is high in Australia, it is low in Tahiti. The strength of this Southern Oscillation is measured by the Southern Oscillation Index (SOI), which is defined as the pressure at Darwin, Australia, subtracted from that at Tahiti. Negative values of SOI are usually associated with El Niño events, so the Southern Oscillation and the El Niño are obviously linked. How ENSO can impact the weather in other parts of the world has only recently become better understood.

The atmosphere is a system that responds to incoming solar radiation, the spinning Earth, and other factors, such as the amount of water vapor present. The ocean and atmospheric systems undergo changes by interacting with each other,

most visibly in the tropical cyclone. The ocean system supplies water vapor, latent heat, and condensation nuclei, which are the essential elements of a tropical cyclone as well as everyday weather changes and climate. The atmosphere, on the other hand, drives the ocean with prevailing winds, moving warm or cool water to locations where they affect the climate on the land. There is a complex, interdependent relationship between the ocean and the atmosphere, and it is probable that even small changes in one system can lead to bigger changes in the other.

Normally, during non–El Niño times, the Pacific Ocean along the equator has established systems of prevailing wind belts, pressure systems, and ocean currents. In July, these systems push the surface seawater offshore from South America, westward along the equator, and toward Indonesia. During El Niño times, the trade winds weaken and the warm water moves back eastward, across the Pacific to South America, where it then spreads north and south along the coast. Why the trade winds weaken and become calm is unknown and is the subject of ongoing research.

Warmer waters along the coast of South America bring warmer, more humid air and the increased possibility of thunderstorms. Thus, the possibility of towering thunderstorms, tropical storms, or hurricanes increases along the Pacific coast of South America as the warmer waters move north and south. This creates the possibility of weather changes not only along the western coast but elsewhere, too. The towering thunderstorms reach high into the atmosphere, adding tropical moisture and creating changes in prevailing wind belts. These wind belts carry or steer weather systems across the middle latitudes of North America, so typical storm paths are shifted. This shifting can result in

▶ increased precipitation in California during the fall to spring season;

▶ a wet winter with more and stronger storms along regions of the southern United States and Mexico;

▶ a warmer and drier-than-average winter across the northern regions of Canada and the United States;

▶ a variable effect on central regions of the United States, ranging from reduced snowfall to no effect at all; and

▶ other changes in the worldwide global complex of ocean and weather events, such as droughts in normally wet climates and heavy rainfall in normally dry climates.

One major problem in these predictions is a lack of understanding of what causes many of the links and a lack of consistency in the links themselves. For example, southern California did not always have an unusually wet season every time an El Niño occurred and in fact experienced a drought during one event.

Scientists have continued to study the El Niño since the mid-1980s, searching for patterns that will reveal consistent cause-and-effect links. Part of the problem may be that other factors, such as a volcanic eruption, may influence part of the linkage but not another part. Another part of the problem may be the influence of unknown factors, such as the circulation of water deep beneath the ocean surface, the track taken by tropical cyclones, or the energy released by tropical cyclones in one year compared to the next.

The results so far have indicated that atmosphere-ocean interactions are much more complex than early theoretical models had predicted. Sometimes a new model will predict some weather changes that occur with El Niño, but no model is yet consistently correct in predicting the conditions that lead to the event and the weather patterns that result. All this may someday lead to a better understanding of how the ocean and the atmosphere interact on this dynamic planet.

Recent years in which El Niño events have occurred are 1951, 1953, 1957–1958, 1965, 1969, 1972–1973, 1976, 1982–1983, 1986–1987, 1991–1992, 1994, 1997–1998, 2002–2003, and 2006–.

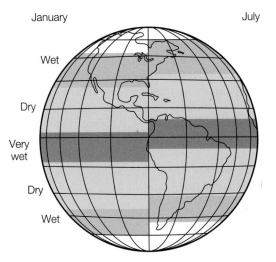

January   July

Wet

Dry

Very wet

Dry

Wet

FIGURE 17.42 The idealized general rainfall patterns over Earth change with seasonal shifts in the wind and pressure areas of the planet's general atmospheric circulation patterns.

for this reason. There is another wet zone near 60° latitudes and another dry zone near the poles. These wet and dry zones are shifted north and south during the year with the changing seasons. This results in different precipitation patterns in each season. Figure 17.42 shows where the wet and dry zones are in winter and in summer seasons.

Regional Climatic Influence

Latitude determines the basic tropical, temperate, and polar climatic zones, and the wet and dry zones move back and forth over the latitudes with the seasons. If these were the only factors influencing the climate, you would expect to find the same climatic conditions at all locations with the same latitude. This is not what is found, however, because there are four major factors that affect a regional climate. These are (1) altitude, (2) mountains, (3) large bodies of water, and (4) ocean currents. The following describes how these four factors modify the climate of a region.

The first of the four regional climate factors is *altitude*. The atmosphere is warmed mostly by the greenhouse effect from the surface upward, and air at higher altitudes increasingly radiates more and more of its energy to space. Average air temperatures therefore decrease with altitude, and locations with higher altitudes will have lower average temperatures. This is why the tops of mountains are often covered with snow when none is found at lower elevations. St. Louis, Missouri, and Denver, Colorado, are located almost at the same latitude (within 1° of 39°N), so you might expect the two cities to have about the same average temperature. Denver, however, has an altitude of 1,609 m (5,280 ft), and the altitude of St. Louis is 141 m (465 ft). The yearly average temperature for Denver is about 10°C (about 50°F) and for St. Louis it is about 14°C (about 57°F). In general, higher altitude means lower average temperature.

The second of the regional climate factors is the influence of *mountains*. In addition to the temperature change caused

by the altitude of the mountain, mountains also affect the conditions of a passing air mass. The western United States has mountainous regions along the coast. When a moist air mass from the Pacific meets these mountains, it is forced upward and cools. Water vapor in the moist air mass condenses, clouds form, and the air mass loses much of its moisture as precipitation falls on the western side of the mountains. Air moving down the eastern slope is compressed and becomes warm and dry. As a result, the western slopes of these mountains are moist and have forests of spruce, redwood, and fir trees. The eastern slopes are dry and have grassland or desert vegetation.

The third of the regional climate factors is the presence of a large body of *water*. Water, as discussed previously, has a higher specific heat than land material, it is transparent, and it loses energy through evaporation. All of these affect the temperature of a landmass located near a large body of water, making the temperatures more even from day to night and from summer to winter. San Diego, California, and Dallas, Texas, for example, are at about the same latitude (both almost 33°N), but San Diego is at a seacoast and Dallas is inland. Because of its nearness to water, San Diego has an average summer temperature about 7°C (about 13°F) cooler and an average winter temperature about 5°C (about 9°F) warmer than the average temperatures in Dallas. Nearness to a large body of water keeps the temperature more even at San Diego.

The fourth of the regional climate factors is *ocean currents*. In addition to the evenness brought about by being near the ocean, currents in the ocean can bring water that has a temperature different from the land. For example, currents can move warm water northward or they can move cool water southward (figure 17.43). This can influence the temperatures of air masses that move from the water to the land and, thus, the temperatures of the land. For example, the North Pacific current brings warm waters to the western coast of North America, which results in warmer temperatures for cities near the coast.

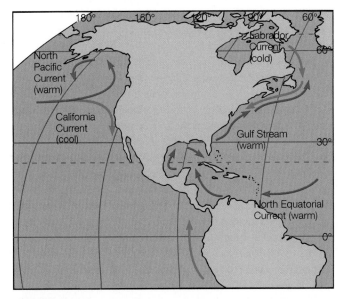

FIGURE 17.43 Ocean currents can move large quantities of warm or cool water to influence the air temperatures of nearby landmasses.

People Behind the Science

Vilhelm Firman Koren Bjerknes (1862–1951)

Vilhelm Bjerknes was the Norwegian scientist who created modern meteorology.

Bjerknes came from a talented family. His father was professor of mathematics at the Christiania (now Oslo) University and a highly influential geophysicist who clearly shaped his son's studies. Bjerknes held chairs at Stockholm and Leipzig before founding the Bergen Geophysical Institute in 1917. Bjerknes made momentous contributions that transformed meteorology into an accepted science. Not least, he showed how weather prediction could be put on a statistical basis, dependent on the use of mathematical models.

During World War I, Bjerknes instituted a network of weather stations throughout Norway; coordination of the findings from such stations led him and his coworkers to develop the highly influential theory of polar fronts, on the basis of the discovery that the atmosphere is made up of discrete air masses displaying dissimilar features. Bjerknes coined the word *front* to delineate the boundaries between such air masses. One of the many contributions of the "Bergen frontal theory" was to explain the generation of cyclones over the Atlantic, at the junction of warm and cold air wedges. Bjerknes's work gave modern meteorology its theoretical tools and methods of investigation.

Source: Modified from the *Hutchinson Dictionary of Scientific Biography*. © RM, 2007. Reprinted by permission.

Describing Earth's climates presents a problem because there are no sharp boundaries that exist naturally between two adjacent regions with different climates. Even if two adjacent climates are very different, one still blends gradually into the other. For example, if you are driving from one climate zone to another, you might drive for miles before becoming aware that the vegetation is now different than it was an hour ago. Since the vegetation is very different from what it was before, you know that you have driven from one regional climate zone to another. Chapter 23, Ecology and Environment, discusses the typical plants and animals associated with the various climate types.

Actually, no two places on Earth have exactly the same climate. Some plants will grow on the north or south side of a building, for example, but not on the other side. The two sides of the building could be considered as small, local climate zones within a larger, major climate zone.

SUMMARY

Earth's *atmosphere* thins rapidly with increasing altitude. Pure, dry air is mostly *nitrogen, oxygen,* and *argon,* with traces of *carbon dioxide* and other gases. Atmospheric air also contains a variable amount of *water vapor.* Water vapor cycles into and out of the atmosphere through evaporation and precipitation.

Atmospheric pressure is measured with a *mercury barometer.* At sea level, the atmospheric pressure will support a column of mercury about 76.00 cm (about 29.92 in) tall. This is the average pressure at sea level, and it is called the *standard atmospheric pressure, normal pressure,* or *one atmosphere of pressure.*

Materials on Earth's surface absorb sunlight, emitting more and more *infrared radiation* as they are warmed. *Carbon dioxide* and *molecules* of water vapor in the atmosphere absorb infrared radiation, which then reemit the energy many times before it reaches outer space again. The overall effect warms the lower atmosphere from the bottom up in a process called the *greenhouse effect.*

The layer of the atmosphere from the surface up to where the temperature stops decreasing with height is called the *troposphere.* The *stratosphere* is the layer above the troposphere. Temperatures in the stratosphere increase because of the interaction between ozone (O_3) and ultraviolet radiation from the Sun.

The surface of Earth is not heated uniformly by sunlight. This results in a *differential heating,* which sets the stage for *convection.* The horizontal movement of air on the surface from convection is called *wind.* A generalized model for understanding why the wind blows involves (1) the relationship between *air temperature and air density* and (2) the relationship between *air pressure and the movement of air.*

This model explains local wind patterns and wind patterns observed on a global scale.

The amount of water vapor in the air at a particular time is called the *absolute humidity.* The *relative humidity* is a ratio between the amount of water vapor that is in the air and the amount needed to saturate the air at the present temperature.

When the air is saturated, condensation can take place. The temperature at which this occurs is called the *dew point temperature.* If the dew point temperature is above freezing, *dew* will form. If the temperature is below freezing, *frost* will form. Both dew and frost form directly on objects and do not fall from the air.

Water vapor condenses in the air on *condensation nuclei.* If this happens near the ground, the accumulation of tiny water droplets is called a *fog. Clouds* are accumulations of tiny water droplets in the air above the ground. In general, there are three basic shapes of clouds: *cirrus, cumulus,* and *stratus.* These basic cloud shapes have meaning about the atmospheric conditions and about the coming weather conditions.

Water that returns to Earth in liquid or solid form falls from the clouds as *precipitation.* Precipitation forms in clouds through two processes: (1) the *coalescence* of cloud droplets or (2) the *growth of ice crystals* at the expense of water droplets.

Weather changes are associated with the movement of large bodies of air called *air masses,* the leading *fronts* of air masses when they move, and local *high-* and *low-pressure* patterns that accompany air masses or fronts. Examples of air masses include (1) *continental polar,* (2) *maritime polar,* (3) *continental tropical,* and (4) *maritime tropical.*

When a location is under the influence of an air mass, the location is having *air mass weather* with slow, gradual changes. More rapid changes take place when the *front,* a thin transition zone between two air masses, passes a location.

A *stationary front* often develops a bulge, or *wave,* that forms into a moving cold front and a moving warm front. The faster-moving cold front overtakes the warm front, lifting it into the air to form an *occluded front.* The lifting process forms a low-pressure center called a *cyclone.* Cyclones are associated with heavy clouds, precipitation, and stormy conditions because of the lifting action.

A *thunderstorm* is a brief, intense storm with rain, lightning and thunder, gusty and strong winds, and sometimes hail. A *tornado* is the smallest, most violent weather disturbance that occurs on Earth. A *hurricane* is a *tropical cyclone,* a large, violent circular storm that is born over warm tropical waters near the equator.

The general pattern of the weather that occurs for a region over a number of years is called *climate.* The three principal climate zones are (1) the *tropical climate zone,* (2) the *polar climate zone,* and (3) the *temperate climate zone.* The climate in these zones is influenced by four factors that determine the local climate: (1) *altitude,* (2) *mountains,* (3) *large bodies of water,* and (4) *ocean currents.* The climate for a given location is described by first considering the principal climate zone, then looking at subdivisions within each that result from local influences.

KEY TERMS

absolute humidity (p. **392**)
air mass (p. **398**)
barometer (p. **384**)
climate (p. **407**)
cold front (p. **400**)
condensation nuclei (p. **394**)
dew (p. **393**)
dew point temperature (p. **393**)

front (p. **399**)
frost (p. **393**)
greenhouse effect (p. **385**)
hail (p. **403**)
humidity (p. **392**)
hurricane (p. **405**)
ice-forming nuclei (p. **397**)
inversion (p. **385**)

jet stream (p. **390**)
occluded front (p. **401**)
precipitation (p. **397**)
relative humidity (p. **392**)
saturated air (p. **391**)
standard atmospheric pressure
 (p. **385**)
stationary front (p. **401**)

storm (p. **402**)
thunderstorm (p. **402**)
tornado (p. **404**)
typhoon (p. **405**)
warm front (p. **401**)
wind (p. **387**)
wind chill factor (p. **388**)

APPLYING THE CONCEPTS

1. Without adding or removing any water vapor, a sample of air experiencing an increase in temperature will have
   a. a higher relative humidity.
   b. a lower relative humidity.
   c. the same relative humidity.
   d. a changed absolute humidity.

2. Cooling a sample of air results in a (an)
   a. increased capacity to hold water vapor.
   b. decreased capacity to hold water vapor.
   c. unchanged capacity to hold water vapor.

3. On a clear, calm, and cool night, dew or frost is most likely to form
   a. under trees or other shelters.
   b. on bare ground on the side of a hill.
   c. under a tree on the side of a hill.
   d. on grass in an open, low-lying area.

4. Longer periods of drizzle, rain, or snow usually occur from which basic form of a cloud?
   a. stratus
   c. cirrus
   b. cumulus
   d. None of the above is correct.

5. Brief periods of showers are usually associated with which type of cloud?
   a. stratus
   c. cirrus
   b. cumulus
   d. None of the above is correct.

6. The type of air mass weather that results from the arrival of polar continental air is
   a. frequent snowstorms with rapid changes.
   b. clear and cold with gradual changes.
   c. unpredictable but with frequent and rapid changes.
   d. much the same from day to day, the conditions depending on the air mass and the local conditions.

7. The appearance of high cirrus clouds followed by thicker, lower stratus clouds and then continuous light rain over several days probably means which of the following air masses has moved to your area?
   a. continental polar
   c. continental tropical
   b. maritime tropical
   d. maritime polar

8. A fully developed cyclonic storm is most likely to form
   a. on a stationary front.
   c. from differential heating.
   b. in a high-pressure center.
   d. over a cool ocean.

9. The basic difference between a tropical storm and a hurricane is
   a. size.
   c. wind speed.
   b. location.
   d. amount of precipitation.

10. Most of the great deserts of the world are located
    a. near the equator.
    b. 30° north or south latitude.
    c. 60° north or south latitude.
    d. anywhere, as there is no pattern to their location.

11. The average temperature of a location is made more even by the influence of
    a. a large body of water.
    b. elevation.
    c. nearby mountains.
    d. dry air.

12. The climate of a specific location is determined by
    a. its latitude.
    b. how much sunlight it receives.
    c. its altitude and nearby mountains and bodies of water.
    d. All of the above are correct.

Answers
1. b **2.** b **3.** d **4.** a **5.** b **6.** d **7.** b **8.** a **9.** c **10.** b **11.** a **12.** d

QUESTIONS FOR THOUGHT

1. Explain why frost is more likely to form on a clear, calm, and cool night than on nights with other conditions.
2. What is a cloud? Describe how a cloud forms.
3. Describe two ways that precipitation may form from the water droplets of a cloud.
4. What is an air mass?
5. What kinds of clouds and weather changes are usually associated with the passing of (a) a warm front? (b) a cold front?
6. Describe the wind direction, pressure, and weather conditions that are usually associated with (a) low-pressure centers and (b) high-pressure centers.
7. In which of the four basic types of air masses would you expect to find afternoon thunderstorms? Explain.
8. Describe the three main stages in the life of a thunderstorm cell, identifying the events that mark the beginning and end of each stage.
9. What is a tornado? When and where do tornadoes usually form?
10. What is a hurricane? Describe how the weather conditions change as a hurricane approaches, passes directly over, then moves away from a location.
11. Describe the average conditions found in the three principal climate zones.
12. Identify the four major factors that influence the climate of a region and explain how each does its influencing.

FOR FURTHER ANALYSIS

1. Describe how you could use a garden hose and a bucket of water to make a barometer. How high a column of water would standard atmospheric pressure balance in a water barometer?
2. If heated air rises, why is there snow on top of a mountain and not at the bottom?
3. According to the U.S. National Oceanic and Atmospheric Administration, the atmospheric concentration of CO_2 has been increasing, global surface temperatures have increased about 0.2°C (0.4°F) over the past twenty-five years, and sea level has been rising 1 to 2 mm/yr since the late nineteenth century. Describe what evidence you would look for to confirm these increases are due to human activity rather than changes in the Sun's output or energy, or changes in Earth's orbit.
4. Evaluate the requirement that differential heating must take place before wind will blow. Do any winds exist without differential heating?
5. Given the current air temperature and relative humidity, explain how you could use the graph in figure 17.15 to find the dew point temperature.
6. Explain why dew is not considered to be a form of precipitation.
7. What are the significant similarities and differences between air mass weather and frontal weather?
8. Analyze and compare the potential damage caused by a hurricane to the potential damage caused by a tornado.
9. Describe several examples of regional climate factors completely overriding the expected weather in a given principal climatic zone. Explain how this happens.

INVITATION TO INQUIRY

Microclimate Experiments

The local climate of a small site or habitat is called a *microclimate*. Certain plants will grow within one microclimate but not another. For example, a north-facing slope of a hill is cooler and loses snow later than a south-facing slope. A different microclimate exists on each side of the hill, and some plants grow better on one side than the other.

Investigate the temperature differences in the microclimate on the north side of a building and on the south side of the same building. Determine the heights, distances, and time of day that you will take temperature readings. Remember to shade the thermometer from direct sunlight for each reading. In your report, discuss the variables that may be influencing the temperature of a microclimate. Design experiments to test your hypotheses about each variable.

Earth's vast oceans cover more than 70 percent of the surface of Earth. Freshwater is generally abundant on the land because the supply is replenished from ocean waters through the hydrologic cycle.

Earth's Waters

CORE CONCEPT

The hydrologic cycle is water evaporating from the ocean, transport by moving air masses, precipitation on the land, and movement of water back to the ocean.

Most water on Earth is stored in Earth's oceans (p. 416).

Water fit for human consumption is replenished by the hydrologic cycle (p. 417).

Precipitation either evaporates, runs across the surface, or soaks in to become groundwater (p. 418).

OVERVIEW

Throughout history, humans have diverted rivers and reshaped the land to ensure a supply of freshwater. There is evidence, for example, that ancient civilizations along the Nile River diverted water for storage and irrigation some five thousand years ago. The ancient Greeks and Romans built systems of aqueducts to divert streams to their cities some two thousand years ago. Some of these aqueducts are still standing today. More recent water diversion activities were responsible for the name of Phoenix, Arizona. Phoenix was named after a mythical bird that arose from its ashes after being consumed by fire. The city was given this name because it is built on a system of canals that were first designed and constructed by ancient Native Americans, then abandoned hundreds of years before settlers reconstructed the ancient canal system (figure 18.1). Water is and always has been an essential resource. Where water is in short supply, humans have historically turned to extensive diversion and supply projects to meet their needs.

Precipitation is the basic source of the water supply found today in streams, lakes, and beneath Earth's surface. Much of the precipitation that falls on the land, however, evaporates back into the atmosphere before it has a chance to become a part of this supply. The water that does not evaporate mostly moves directly to rivers and streams, flowing back to the ocean, but some soaks into the land. The evaporation of water, condensation of water vapor, and the precipitation-making processes were introduced in chapter 17 as important weather elements. They are also part of the generalized *hydrologic cycle* of evaporation from the ocean, transport through the atmosphere by moving air masses, precipitation on the land, and movement of water back to the ocean. Only part of this cycle was considered previously, however, and this was the part from evaporation through precipitation. This chapter is concerned with the other parts of the hydrologic cycle, that is, what happens to the water that falls on the land and makes it back to the ocean. It begins with a discussion of how water is distributed on Earth and a more detailed look at the hydrologic cycle. Then the travels of water across and into the land will be considered as streams, wells, springs, and other sources of usable water are discussed as limited resources. The tracing of the hydrologic cycle will be completed as the water finally makes it back to the ocean. This last part of the cycle will consider the nature of the ocean floor, the properties of seawater, and how waves and currents are generated. The water is now ready to evaporate, starting another one of Earth's never-ending cycles.

WATER ON EARTH

Some water is tied up in chemical bonds deep in Earth's interior, but free water is the most abundant chemical compound near the surface. Water is five or six times more abundant than the most abundant mineral in the outer 6 km (about 4 mi) of Earth, so it should be no surprise that water covers about 70 percent of the surface. On the average, about 98 percent of this water exists in the liquid state in depressions on the surface and in sediments. Of the remainder, about 2 percent exists in the solid state as snow and ice on the surface in colder locations. Only a fraction of a percent exists as a variable amount of water vapor in the atmosphere at a given time. Water is continually moving back and forth between these "reservoirs," but the percentage found in each is assumed to be essentially constant.

As shown in figure 18.2, over 97 percent of Earth's water is stored in Earth's oceans. This water contains a relatively high level of dissolved salts, which will be discussed in the section on

FIGURE 18.1 This is one of the water canals of the present-day system in Phoenix, Arizona. These canals were reconstructed from a system that was built by American Indians, then abandoned. Phoenix is named after a mythical bird that was consumed by fire and then arose from its ashes.

the nature of seawater. These dissolved salts make ocean water unfit for human consumption and for most agricultural purposes. All other water, which is fit for human consumption and agriculture, is called **freshwater.** About two-thirds of Earth's freshwater supply is locked up in the ice caps of Greenland and the Antarctic and in glaciers. This leaves less than 1 percent of all the water found on Earth as available freshwater. There is a generally abundant supply, however, because the freshwater supply is continually replenished by the hydrologic cycle.

Evaporation of water from the ocean is an important process of the hydrologic cycle because (1) water vapor leaves the dissolved salts behind, forming precipitation that is freshwater, and (2) the gaseous water vapor is easily transported in the atmosphere from one part of Earth to another. Over a year, this natural desalination process produces and transports enough freshwater to cover the entire Earth with a layer about 85 cm (about 33 in) deep. Precipitation is not evenly distributed like this, of course, and some places receive much more, while other places receive almost none. Considering global averages, more water is evaporated from the ocean than returns directly to it by precipitation. On the other hand, more water is precipitated over the land than evaporates from the land surface back to the atmosphere. The net amount evaporated and precipitated over the land and over the ocean is balanced by the return of water to the ocean by streams and rivers. This cycle of evaporation, precipitation, and return of water to the oceans is known as the *hydrologic cycle* (figure 18.3). This freshwater returning on and under the land is the source of freshwater. What happens to the freshwater during its return to the ocean is discussed in the section on freshwater.

Freshwater

The basic source of freshwater is precipitation, but not all precipitation ends up as part of the freshwater supply. Liquid water is always evaporating, even as it falls. In arid climates, rain sometimes evaporates completely before reaching the

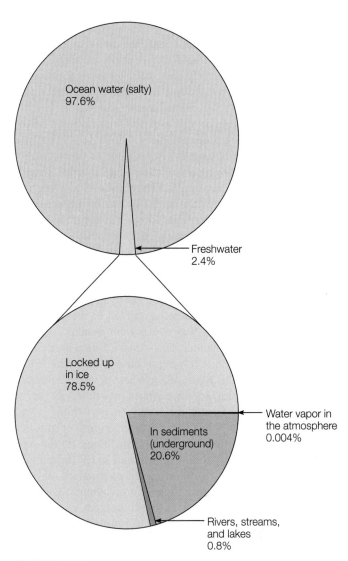

FIGURE 18.2 Estimates of the distribution of all the water found on Earth's surface.

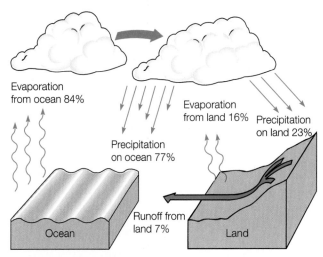

100% is based on a global average of 85 cm/yr precipitation.

FIGURE 18.3 On the average, more water is evaporated from the ocean than is returned by precipitation. More water is precipitated over the land than evaporates. The difference is returned to the ocean by rivers and streams.

surface, even from a fully developed thunderstorm. Evaporation continues from the water that does reach the surface. Puddles and standing water on the hard surface of city parking lots and streets, for example, gradually evaporate back to the atmosphere after a rain and the surface is soon dry. Many factors determine how much of a particular rainfall evaporates, but in general, more than two-thirds of the rain eventually returns to the atmosphere. The remaining amount either (1) flows downhill across the surface of the land toward a lower place or (2) soaks into the ground. Water moving across the surface is called **runoff.** Runoff begins as rain accumulates in thin sheets of water that move across the surface of the land. These sheets collect into a small body of running water called a *stream.* A stream is defined as any body of water that is moving across the land, from one so small that you could step across it to the widest river. Water that soaks into the ground moves downward to a saturated zone and is now called *groundwater.* Groundwater moves through sediments and rocks beneath the surface, slowly moving downhill. Streams carry the runoff of a recent rainfall or melting snow, but otherwise most of the flow comes from groundwater that seeps into the stream channel. This explains how a permanent stream is able to continue flowing when it is not being fed by runoff or melting snow (figure 18.4). Where or when the source of groundwater is in low supply, a stream may flow only part of the time, and it is designated as an *intermittent stream.*

The amount of a rainfall that becomes runoff or groundwater depends on a number of factors, including (1) the type of soil on the surface, (2) how dry the soil is, (3) the amount and type of vegetation, (4) the steepness of the slope, and (5) if the rainfall is a long, gentle one or a cloudburst. Different combinations of these factors can result in from 5 percent to almost 100 percent of a rainfall event running off, with the

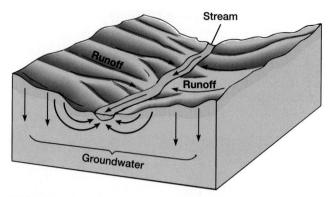

FIGURE 18.4 Some of the precipitation soaks into the ground to become groundwater. Groundwater slowly moves underground, and some of it emerges in streambeds, keeping the streams running during dry spells.

rest evaporating or soaking into the ground. On the average, however, about 70 percent of all precipitation evaporates back into the atmosphere, about 30 percent becomes runoff, and less than 1 percent soaks into the ground.

Surface Water

If you could follow the smallest of streams downhill, you would find that it eventually merges with other streams until they form a major river. The land area drained by a stream is known as the stream's drainage basin, or **watershed.** Each stream has its own watershed, but the watershed of a large river includes all the watersheds of the smaller streams that feed into the larger river. Figure 18.5 shows the watersheds of the Columbia River, the Colorado River, and the Mississippi River. Note that the water from the Columbia River and the Colorado River watersheds empties

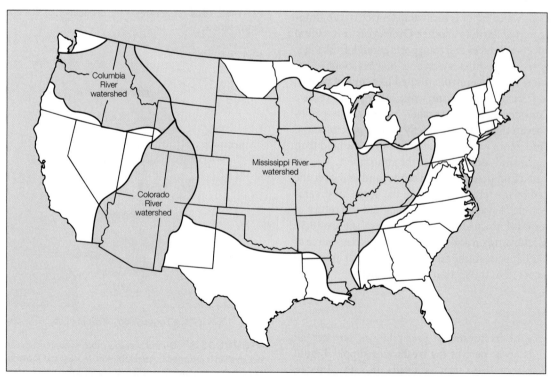

FIGURE 18.5 The approximate watersheds of the Columbia, Colorado, and Mississippi Rivers.

into the Pacific Ocean. The Mississippi River watershed drains into the Gulf of Mexico, which is part of the Atlantic Ocean.

Two adjacent watersheds are separated by a line called a *divide.* Rain that falls on one side of a divide flows into one watershed, and rain that falls on the other side flows into the other watershed. A *continental divide* separates river systems that drain into opposite sides of a continent. The North American continental divide trends northwestward through the Rocky Mountains. Imagine standing over this line with a glass of water in each hand, then pouring the water to the ground. The water from one glass will eventually end up in the Atlantic Ocean, and the water from the other glass will end up in the Pacific Ocean. Sometimes the Appalachian Mountains are considered to be an eastern continental divide, but water from both sides of this divide ends up on the same side of the continent, in the Atlantic Ocean.

Water moving downhill is sometimes stopped by a depression in a watershed, a depression where water temporarily collects as a standing body of freshwater. A smaller body of standing water is usually called a *pond,* and one of much larger size is called a *lake.* A pond or lake can occur naturally in a depression, or it can be created by building a dam on a stream. A natural pond, a natural lake, or a pond or lake created by building a dam is called a *reservoir* if it is used for (1) water storage, (2) flood control, or (3) generating electricity. A reservoir can be used for one or two of these purposes but not generally for all three. A reservoir built for water storage, for example, is kept as full as possible to store water. This use is incompatible with use for flood control, which would require a low water level in the reservoir in order to catch runoff, preventing waters from flooding the land. In addition, extensive use of reservoir water to generate electricity requires the release of water, which could be incompatible with water storage. The water of streams, ponds, lakes, and reservoirs is collectively called *surface water,* and all serve as sources of freshwater. The management of surface water, as you can see, can present some complicated problems.

Groundwater

Precipitation soaks into the ground, or *percolates* slowly downward until it reaches an area, or zone, where the open spaces between rock and soil particles are completely filled with water. Water from such a saturated zone is called **groundwater.** There is a tremendous amount of water stored as groundwater, which makes up a supply about twenty-five times larger than all the surface water on Earth. Groundwater is an important source of freshwater for human consumption and for agriculture. Groundwater is often found within 100 m (about 330 ft) of the surface, even in arid regions where little surface water is found. Groundwater is the source of water for wells in addition to being the source that keeps streams flowing during dry periods.

Water is able to percolate down to a zone of saturation because sediments contain open spaces between the particles called *pore spaces.* The more pore space a sediment has, the more water it will hold. The total amount of pore spaces in a given sample of sediment is a measure of its *porosity.* Sand and gravel sediments, for example, have grains that have large pore spaces between them, so these sediments have a high porosity. In order for water to move through a sediment, however, the

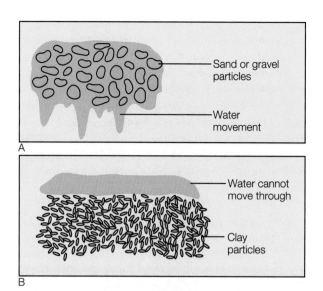

FIGURE 18.6 (*A*) Sand and gravel have large, irregular particles with large pore spaces, so they have a high porosity. Water can move from one pore space to the next, so they also have a high permeability. (*B*) Clay has small, flat particles, so it has a low porosity and is practically impermeable because water cannot move from one pore to the next.

pore spaces must be connected. The ability of a given sample of sediment to transmit water is a measure of its *permeability.* Sand and gravel have a high permeability because the grains do not fit tightly together, allowing water to move from one pore space to the next. Sand and gravel sediments thus have a high porosity as well as a high permeability. Clay sediments, on the other hand, have small, flattened particles that fit tightly together. Clay thus has a low permeability, and when saturated or compressed, clay becomes *impermeable,* meaning water cannot move through it at all (figure 18.6).

The amount of groundwater available in a given location depends on a number of factors, such as the present and past climate, the slope of the land, and the porosity and permeability of the sediments beneath the surface. Generally, sand and gravel sediments, along with solid sandstone, have the best porosity and permeability for transmitting groundwater. Other solid rocks, such as granite, can also transmit groundwater if they are sufficiently fractured by joints and cracks. In any case, groundwater will percolate downward until it reaches an area where pressure and other conditions have eliminated all pores, cracks, and joints. Above this impermeable layer, it collects in all available spaces to form a *zone of saturation.* Water from the zone of saturation is considered to be groundwater. Water from the zone above is not considered to be groundwater. The surface of the boundary between the zone of saturation and the zone above is called the **water table.** The surface of a water table is not necessarily horizontal, but it tends to follow the topography of the surface in a humid climate. A hole that is dug or drilled through the surface to the water table is called a well. The part of the well that is below the water table will fill with groundwater, and the surface of the water in the well is generally at the same level as the water table.

Precipitation falls on the land and percolates down to the zone of saturation, then begins to move laterally, or sideways, to lower and lower elevations until it finds its way back to the

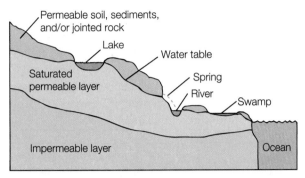

FIGURE 18.7 Groundwater from below the water table seeps into lakes, streams, and swamps and returns to the surface naturally at a spring. Groundwater eventually returns to the ocean, but the trip may take hundreds of years.

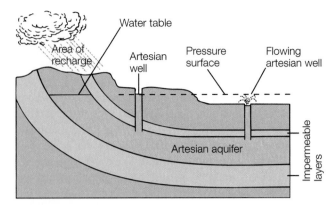

FIGURE 18.8 An artesian aquifer has groundwater that is under pressure because the groundwater is confined between two impermeable layers and has a recharge area at a higher elevation. The pressure will cause the water to rise in a well drilled into the aquifer, becoming a flowing well if the pressure is sufficiently high.

surface. This surface outflowing could take place at a stream, pond, lake, swamp, or spring (figure 18.7). Groundwater flows gradually and very slowly through the tiny pore spaces, moving at a rate that ranges from kilometers (miles) per day to meters (feet) per year. Surface streams, on the other hand, move much faster, at rates up to about 30 km per hour (about 20 mi/h).

An **aquifer** is a layer of sand, gravel, sandstone, or other highly permeable material beneath the surface that is capable of producing water in usable quantities. In some places, an aquifer carries water from a higher elevation, resulting in a pressure on water trapped by impermeable layers at lower elevations. Groundwater that is under such a confining pressure is in an *artesian* aquifer. *Artesian* refers to the pressure, and groundwater from an artesian well rises above the top of the aquifer but not necessarily to the surface. Some artesian wells are under sufficient pressure to produce a fountainlike flow or spring (figure 18.8). Some people call groundwater from any deep well "artesian water," which is technically incorrect.

Freshwater as a Resource

Water is an essential resource, not only because it is required for life processes but also because of its role in a modern industrialized society (see chapter 10). Water is used in the home for drinking, cooking, and cleaning, as a carrier to remove wastes, and for maintaining lawns and gardens. These domestic uses lead to an equivalent consumption of about 570 liters per person each day (about 150 gal/person/day), but this is only about 10 percent of the total consumed. Average daily use of water in the United States amounts to some 5,700 liters per person each day (about 1,500 gal/person/day), or about enough water to fill a small swimming pool once a week. The bulk of the water is used by agriculture (about 40 percent), for the production of electricity (about 40 percent), and

for industrial purposes (about 10 percent). These overall percentages of use vary from one region of the country to another.

Most of the water supply is obtained from the surface water resources of streams, lakes, and reservoirs, and 37 percent of the municipal water supply comes from groundwater. If you then add farms, villages, and many suburban areas, the percentage of groundwater used by humans is well above 40 percent. Surface water contains more sediments, bacteria, and possible pollutants than groundwater. This means that surface water requires filtering to remove suspended particles, treatment to kill bacteria, and sometimes processing to remove dissolved chemicals. Groundwater is naturally filtered as it moves through the pore spaces of an aquifer, so it is usually relatively free of suspended particles and bacteria. Thus, the processing or treatment of groundwater is usually not necessary (figure 18.9). But groundwater, on the other hand, will cost more to use as a resource because it must be pumped to the surface. The energy required for this pumping can be very expensive. In addition, groundwater generally contains more dissolved minerals (hard water), which may require additional processing or chemical treatment to remove the troublesome minerals.

The use of surface water as a source of freshwater means that the supply depends on precipitation. When a drought occurs, low river and lake levels may require curtailing water consumption. In some parts of the western United States, such as the Colorado River watershed, *all* of the surface water is already being used, with certain percentages allotted for domestic, industrial, and irrigation uses. Groundwater is also used in this watershed, and in some locations, it is being pumped from the ground faster than it is being replenished by precipitation (figure 18.10). As the population grows and new industries develop, more and more demands are placed on the surface water supply, which has already been committed to other uses, and on the diminishing supply of groundwater. This raises some very controversial issues about how freshwater should be divided among agriculture, industries, and city domestic use. Agricultural interests claim they should have the water because they produce the food and fibers that people must have. Industrial interests claim they should have the water because they create

What do you think of when you see a stream? Do you wonder how deep it is? Do you think of something to do with the water such as swimming or fishing? As you might imagine, not all people look at a stream and think about the same thing. A city engineer, for example, might wonder if the stream has enough water to serve as a source to supplement the city water supply. A rancher or farmer might wonder how the stream could be easily diverted to serve as a source of water for irrigation. An electric utility planner, on the other hand, might wonder if the stream could serve as a source of power.

Water in a stream is a resource that can be used many different ways, but using it requires knowing about the water quality as well as quantity. We need to know if the quality of the water is good enough for the intended use—and different uses have different requirements. Water fit for use in an electric power plant, for example, might not be suitable for use as a city water supply.

Indeed, water fit for use in a power plant might not be suitable for irrigation. Water quality is determined by the kinds and amounts of substances dissolved and suspended in the water and the consequences for users. Whether a source of water can be used for drinking water or not, for example, is regulated by stringent rules and guidelines about what cannot be in the water. These rules are designed to protect human health but do not call for pure water.

The water of even the healthiest stream is not absolutely pure. All water contains many naturally occurring substances such as ions of bicarbonates, calcium, and magnesium. A pollutant is not naturally occurring; it is usually a waste material that contaminates air, soil, or water. There are basically two types of water pollutants: degradable and persistent. Examples of degradable pollutants include sewage, fertilizers, and some industrial wastes. As the term implies, degradable pollutants can be broken down

into simple, nonpolluting substances such as carbon dioxide and nitrogen. Examples of persistent pollutants include some pesticides, petroleum and petroleum products, plastic materials, leached chemicals from landfill sites, oil-based paints, heavy metals and metal compounds such as lead, mercury, and cadmium, and certain radioactive materials. The damage they cause is either irreversible or reparable only over long periods of time.

Questions to Discuss

Discuss with your group the following questions concerning water quality:

1. Which water use requires the purest water? Which requires the least pure?
2. Create a hierarchy of possible water uses; for example, can water used in a power plant later be used for agriculture? For domestic use?
3. What can an individual do to improve water quality?

FIGURE 18.9 The filtering beds of a city water treatment facility. Surface water contains more sediments, bacteria, and other suspended materials because it is on the surface and exposed to the atmosphere. This means that surface water must be filtered and treated when used as a domestic resource. Such processing is not required when groundwater is used as the resource.

FIGURE 18.10 This is groundwater pumped from the ground for irrigation. In some areas, groundwater is being removed from the ground faster than it is being replaced by precipitation, resulting in a water table that is falling. It is thus possible that the groundwater resource will soon become depleted in some areas.

the jobs and the products that people must have. Cities, on the other hand, claim that domestic consumption is the most important because people cannot survive without water. Yet, others claim that no group has a right to use water when it is needed to maintain habitats. Who should have the first priority for water use in such cases?

Some have suggested that people should not try to live and grow food in areas that have a short water supply, that plenty of freshwater is available elsewhere. Others have suggested

Wastewater Treatment

One of the most common forms of pollution control in the United States is wastewater treatment. The United States has a vast system of sewer pipes, pumping stations, and treatment plants, and about 74 percent of all Americans are served by such wastewater systems. Sewer pipelines collect the wastewater from homes, businesses, and many industries and deliver it to treatment plants. Most of these plants were designed to make wastewater fit for discharge into streams or other receiving waters.

The basic function of a waste treatment plant is to speed up the natural process of purifying the water. There are two basic stages in the treatment, called the *primary stage* and the *secondary stage*. The primary stage physically removes solids from the wastewater. The secondary stage uses biological processes to further purify wastewater. Sometimes, these stages are combined into one operation.

As raw sewage enters a treatment plant, it first flows through a screen to remove large floating objects such as rags and sticks that might cause clogs. After this initial screening, it passes into a grit chamber where cinders, sand, and small stones settle to the bottom (box figure 18.1). A grit chamber is particularly important in communities with combined sewer systems where sand or gravel may wash into the system along with rain, mud, and other stuff, all with the storm water.

After screening and grit removal, the sewage is basically a mixture of organic and

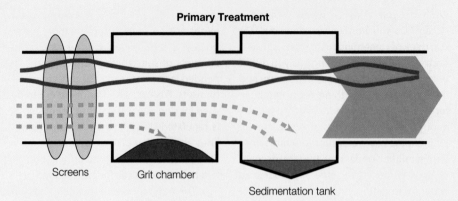

Primary Treatment

Screens   Grit chamber   Sedimentation tank

Box Figure 18.1

inorganic matter along with other suspended solids. The solids are minute particles that can be removed in a sedimentation tank. The speed of the flow through the larger sedimentation tank is slower, and suspended solids gradually sink to the bottom of the tank. They form a mass of solids called *raw primary sludge,* which is usually removed from the tank by pumping. The sludge may be further treated for use as a fertilizer or disposed of through incineration, if necessary.

Once the effluent has passed through the primary treatment process, which is primarily physical filtering and settling, it enters a secondary stage that involves biological activities of microorganisms. The secondary stage of treatment removes about 85 percent of the organic matter in sewage by making use of the bacteria that are naturally a part of the sewage. There are two principal techniques used to provide

secondary treatment: (1) trickling filters or (2) activated sludge. A trickling filter is simply a bed of stones from 1 to 2 m (3 to 6 ft) deep through which the effluent from the sedimentation tank flows. Interlocking pieces of corrugated plastic or other synthetic media have also been used in trickling beds, but the important part is that it provides a place for bacteria to live and grow. Bacteria grow on the stones or synthetic media and consume most of the organic matter flowing by in the effluent. The now cleaner water trickles out through pipes to another sedimentation tank to remove excess bacteria. Disinfection of the effluent with chlorine is generally used to complete this secondary stage of basic treatment.

The trend today is toward the use of an activated sludge process instead of trickling filters. The activated sludge process speeds up the work of the bacteria by bringing

---

that humans have historically moved rivers and reshaped the land to obtain water, so perhaps one answer to the problem is to find new sources of freshwater. Possible sources include the recycling of wastewater and turning to the largest supply of water in the world, the ocean. About 90 percent of the water used by industries is presently dumped as a waste product. In some areas, treated city waste water is already being recycled for use in power plants and for watering parks. A practically limitless supply of freshwater could be available by desalting ocean water, something which occurs naturally in the hydrologic cycle. The treatment of seawater to obtain a new supply of freshwater is presently too expensive because of the cost of energy to accomplish the task. New technologies, perhaps ones that use solar energy, may make this more practical in the future. In the meantime, the best sources of extending the

supply of freshwater appear to be control of pollution, recycling of wastewater, and conservation of the existing supply.

SEAWATER

More than 70 percent of the surface of Earth is covered by seawater, with an average depth of 3,800 m (about 12,500 ft). The land areas cover 30 percent, less than a third of the surface, with an average elevation of only about 830 m (about 2,700 ft). With this comparison, you can see that humans live on and fulfill most of their needs by drawing from a small part of the total Earth. As populations continue to grow and as resources of the land continue to diminish, the ocean will be looked at more as a resource rather than a convenient place for dumping wastes.

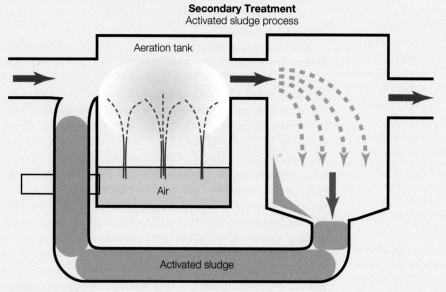

Secondary Treatment
Activated sludge process

Aeration tank

Air

Activated sludge

Box Figure 18.2

air and sludge heavily laden with bacteria into close contact with the effluent (box figure 18.2). After the effluent leaves the sedimentation tank in the primary stage, it is pumped into an aeration tank, where it is mixed with air and sludge loaded with bacteria and allowed to remain for several hours. During this time, the bacteria break down the organic matter into harmless by-products.

The sludge, now activated with additional millions of bacteria, can be used again by returning it to the aeration tank for mixing with new effluent and ample amounts of air. As with trickling, the final step is generally the addition of chlorine to the effluent, which kills more than 99 percent of the harmful bacteria. Some municipalities are now manufacturing chlorine solution on site to avoid the necessity of transporting and storing large amounts of chlorine, sometimes in a gaseous form. Alternatives to chlorine disinfection, such as ultraviolet light or ozone, are also being used in situations where chlorine in sewage effluents can be harmful to fish and other aquatic life.

New pollution problems have placed additional burdens on wastewater treatment systems. Today's pollutants may be more difficult to remove from water. Increased demands on the water supply only aggravate the problem. These challenges are being met through better and more complete methods of removing pollutants at treatment plants or through prevention of pollution at the source. Pretreatment of industrial waste, for example, removes many troublesome pollutants at the beginning, rather than at the end, of the pipeline.

The increasing need to reuse water calls for better and better wastewater treatment. Every use of water—whether at home, in the factory, or on a farm—results in some change in its quality. New methods for removing pollutants are being developed to return water of more usable quality to receiving lakes and streams. Advanced waste treatment techniques in use or under development range from biological treatment capable of removing nitrogen and phosphorus to physical-chemical separation techniques such as filtration, carbon adsorption, distillation, and reverse osmosis. These activities typically follow secondary treatment and are known as tertiary treatment.

These wastewater treatment processes, alone or in combination, can achieve almost any degree of pollution control desired. As waste effluents are purified to higher degrees by such treatment, the effluent water can be used for industrial, agricultural, or recreational purposes, or even drinking water supplies.

Drawings and some text from *How Wastewater Treatment Works . . . The Basics*, U.S. Environmental Protection Agency, Office of Water, www.epa.gov/owm/featinfo.htm.

The ocean already provides some food and is a source of some minerals, but it can possibly provide freshwater, new sources of food, new sources of important minerals, and new energy sources in the future. There are vast deposits of phosphorus and manganese minerals on the ocean bottom, for example, that can provide valuable resources. Phosphorus is an element that can be used to manufacture an important fertilizer needed in agriculture, and the land supplies are becoming depleted. Manganese nodules, which occur in great abundance on the ocean bottom, can be a source of manganese, iron, copper, cobalt, and nickel. Seawater contains enough deuterium to make it a feasible source of energy. One gallon of seawater contains about a spoonful of deuterium, with the energy equivalent of 300 gallons of gasoline. It has been estimated there is sufficient deuterium in the oceans to supply power at one hundred times the present consumption for the next ten billion years. The development of controlled nuclear fusion is needed, however, to utilize this potential energy source. The sea may provide new sources of food through *aquaculture,* the farming of the sea the way that the land is presently farmed. Some aquaculture projects have already started with the farming of oysters, clams, lobsters, shrimp, and certain fishes, but these projects have barely begun to utilize the full resources that are possible.

Part of the problem of utilizing the ocean is that the ocean has remained mostly unexplored and a mystery until recent times. Only now are scientists beginning to understand the complex patterns of the circulation of ocean waters, the nature of the chemical processes at work in the ocean, and the interactions of the ocean and the atmosphere, and to chart the topography of the ocean floor.

Oceans and Seas

The vast body of salt water that covers more than 70 percent of Earth's surface is usually called the *ocean* or the *sea*. In general, the **ocean** is a single, continuous body of salt water on the surface of Earth. Although there is really only one big ocean on Earth, specific regions have been given names for convenience in describing locations. For this purpose, three principal regions are recognized: the (1) Atlantic Ocean, (2) Indian Ocean, and (3) Pacific Ocean, as shown in figure 18.11. Specific regions (Atlantic, Indian, and Pacific) are often subdivided further into North Atlantic Ocean, South Atlantic Ocean, and so on.

A *sea* is usually a smaller part of the ocean, a region with characteristics that distinguish it from the larger ocean of which it is a part. Often the term *sea* is used in the name of certain inland bodies of salty water.

The Pacific Ocean is the largest of the three principal ocean regions. It has the largest surface area, covering 180 million km² (about 70 million mi²) and has the greatest average depth of 3.9 km (about 2.4 mi). The Pacific is circled by active converging plate boundaries, so it is sometimes described as being circled by a "rim of fire." It is called this because of the volcanoes associated with the converging plates. The "rim" also has the other associated features of converging plate boundaries such as oceanic trenches, island arcs, and earthquakes. The Atlantic Ocean is second in size, with a surface area of 107 million km² (about 41 million mi²) and the shallowest average depth of only 3.3 km (about 2.1 mi). The Atlantic Ocean is bounded by nearly parallel continental margins with a diverging plate boundary between. It lacks the trench and island arc features of the Pacific, but it does have islands, such as Iceland, that are a part of the Mid-Atlantic Ridge at the plate boundary. The shallow

seas of the Atlantic, such as the Mediterranean, Caribbean, and Gulf of Mexico, contribute to the shallow average depth of the Atlantic. The Indian Ocean has the smallest surface area, with 74 million km² (about 29 million mi²) and an average depth of 3.8 km (about 2.4 mi).

As mentioned earlier, a sea is usually a part of an ocean that is identified because some characteristic sets it apart. For example, the Mediterranean, Gulf of Mexico, and Caribbean seas are bounded by land, and they are located in a warm, dry climate. Evaporation of seawater is greater than usual at these locations, which results in the seawater being saltier. Being bounded by land and having saltier seawater characterizes these locations as being different from the rest of the Atlantic. The Sargasso Sea, on the other hand, is a part of the Atlantic that is not bounded by land and has a normal concentration of sea salts. This sea is characterized by having an abundance of floating brown seaweeds that accumulate in this region because of the global wind and ocean current patterns. The Arctic Sea, which is also sometimes called the Arctic Ocean, is a part of the North Atlantic Ocean that is less salty. Thus, the terms *ocean* and *sea* are really arbitrary terms that are used to describe different parts of Earth's one continuous ocean.

The Nature of Seawater

According to one theory, the ocean is an ancient feature of Earth's surface, forming at least three billion years ago as Earth cooled from its early molten state. The seawater and much of the dissolved materials are believed to have formed from the degassing of water vapor and other gases from molten rock materials. The degassed water vapor soon condensed, and over a period of time, it began collecting as a liquid in the depression

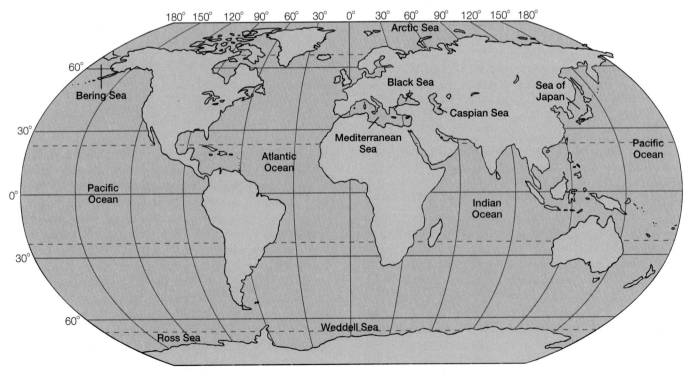

FIGURE 18.11   Distribution of the oceans and major seas on Earth's surface. There is really only one ocean; for example, where is the boundary between the Pacific, Atlantic, and Indian oceans in the Southern Hemisphere?

of the early ocean basin. Ever since, seawater has continuously cycled through the hydrologic cycle, returning water to the ocean through the world's rivers. For millions of years, these rivers have carried large amounts of suspended and dissolved materials to the ocean. These dissolved materials, including salts, stay behind in the seawater as the water again evaporates, condenses, falls on the land, and then brings more dissolved materials much like a continuous conveyor belt.

You might wonder why the ocean basin has not become filled in by the continuous supply of sediments and dissolved materials that would accumulate over millions of years. The basin has not filled in because (1) accumulated sediments have been recycled to Earth's interior through plate tectonics and (2) dissolved materials are removed by natural processes just as fast as they are supplied by the rivers. Some of the dissolved materials, such as calcium and silicon, are removed by organisms to make solid shells, bones, and other hard parts. Other dissolved materials, such as iron, magnesium, and phosphorus, form solid deposits directly and also make sediments that settle to the ocean floor. Hard parts of organisms and solid deposits are cycled to Earth's interior along with suspended sediments that have settled out of the seawater. Studies of fossils and rocks indicate that the composition of seawater has changed little over the past six-hundred million years.

The dissolved materials of seawater are present in the form of ions because of the strong dissolving ability of water molecules. Almost all of the chemical elements are present, but only six ions make up more than 99 percent of any given sample of seawater. As shown in table 18.1, chlorine and sodium are the most abundant ions. These are the elements of sodium chloride, or common table salt. As a sample of seawater evaporates, the positive metal ions join with the different negative ions to form a complex mixture of ionic compounds known as *sea salt*. Sea salt is mostly sodium chloride, but it also contains salts of the four metal ions (sodium, magnesium, calcium, and potassium) combined with the different negative ions of chlorine, sulfate, bicarbonate, and so on.

The amount of dissolved salts in seawater is measured as **salinity.** Salinity is defined as the mass of salts dissolved in 1.0 kg, or 1,000 g, of seawater. Since the salt content is reported in parts per thousand, the symbol ‰ is used (% means parts per hundred). Thus, 35 ‰ means that 1,000 g of seawater contains 35 g of dissolved salts (and 965 g of water). This is the

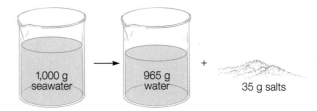

FIGURE 18.12 Salinity is defined as the mass of salts dissolved in 1.0 kg of seawater. Thus, if a sample of seawater has a salinity of 35 ‰, a 1,000 g sample would evaporate 965 g of water and leave 35 g of sea salts behind.

same concentration as a 3.5 percent salt solution (figure 18.12). Oceanographers use the salinity measure because the mass of a sample of seawater does not change with changes in the water temperature. Other measures of concentration are based on the volume of a sample, and the volume of a liquid does vary as it expands and contracts with changes in the temperature. Thus, by using the salinity measure, any corrections due to temperature differences are eliminated.

The average salinity of seawater is about 35 ‰, but the concentration varies from a low of about 32 ‰ in some locations up to a high of about 36 ‰ in other locations. The salinity of seawater in a given location is affected by factors that tend to increase or decrease the concentration. The concentration is increased by two factors, evaporation and the formation of sea ice. Evaporation increases the concentration because it is water vapor only that evaporates, leaving the dissolved salts behind in a greater concentration. Ice that forms from freezing seawater increases the concentration because when ice forms, the salts are excluded from the crystal structure. Thus, sea ice is freshwater, and the removal of this water leaves the dissolved salts behind in a greater concentration. The salinity of seawater is decreased by three factors: heavy precipitation, the melting of ice, and the addition of freshwater by a large river. All three of these factors tend to dilute seawater with freshwater, which lowers the concentration of salts.

Note that increases or decreases in the salinity of seawater are brought about by the addition or removal of freshwater. This changes only the amount of water present in the solution. The *kinds* or *proportions* of the ions present (table 18.1) in seawater do not change with increased or decreased amounts of freshwater. The same proportion, meaning the same chemical composition, is found in seawater of any salinity of any sample taken from any location anywhere in the world, from any depth of the ocean, or taken any time of the year. Seawater has a remarkably uniform composition that varies only in concentration. This means that the ocean is well mixed and thoroughly stirred around the entire Earth. How seawater becomes so well mixed and stirred on a worldwide basis is discussed in the section on movement of seawater.

If you have ever allowed a glass of tap water to stand for a period of time, you may have noticed tiny bubbles collecting as the water warms. These bubbles are atmospheric gases, such as nitrogen and oxygen, that were dissolved in the water (figure 18.13). Seawater contains dissolved gases in addition to the dissolved salts. Near the surface, seawater contains mostly nitrogen and oxygen in similar proportions to the mixture that is found in the atmosphere. There is more carbon dioxide than

TABLE 18.1 Major Dissolved Materials in Seawater

Ion	Percent (by weight)
Chloride (Cl^-)	55.05
Sodium (Na^+)	30.61
Sulfate (SO_4^{-2})	7.68
Magnesium (Mg^{-2})	3.69
Calcium (Ca^{+2})	1.16
Potassium (K^+)	1.10
Bicarbonate (HCO_3^-)	0.41
Bromine (Br^-)	0.19
Total	99.89

FIGURE 18.13 Air will dissolve in water, and cooler water will dissolve more air than warmer water. The bubbles you see here are bubbles of carbon dioxide that came out of solution as the soda became warmer.

you would expect, however, as seawater contains a large amount of this gas. More carbon dioxide can dissolve in seawater because it reacts with water to form carbonic acid, H_2CO_3, the same acid that is found in a bubbly cola. In seawater, carbonic acid breaks down into bicarbonate and carbonate ions, which tend to remain in solution. Water temperature and the salinity have an influence on how much gas can be dissolved in seawater, and increasing either or both will reduce the amount of gases that can be dissolved. Cold, lower salinity seawater in colder regions will dissolve more gases than the warm, higher salinity seawater in tropical locations. Abundant seaweeds (large algae) and phytoplankton (microscopic algae) in the upper, sunlit water tends to reduce the concentration of carbon dioxide and increase the concentration of dissolved oxygen through the process of photosynthesis. With increasing depth, less light penetrates the water, and below about 80 m (about 260 ft), there is insufficient light for photosynthesis. Thus, more algae and phytoplankton and more dissolved oxygen are found above this depth. Below this depth, there is more dissolved carbon dioxide and less dissolved oxygen. The oxygen-poor, deep ocean water does eventually circulate back to the surface, but the complete process may take several thousand years.

Movement of Seawater

Consider the enormity of Earth's ocean, which has a surface area of some 361 million km² (about 139 million mi²) and a volume of 1,370 million km³ (about 328 million mi³) of seawater. There must be a terrific amount of stirring in such an enormous amount of seawater to produce the well-mixed, uniform chemical composition that is found in seawater throughout the world. The amount of mixing required is more easily imagined if you consider the long history of the ocean, the very long period of time over which the mixing has occurred. Based on investigations of the movement of seawater, it has been estimated that there is a complete mixing of all Earth's seawater about every 2,000 years or so. With an assumed age of 3 billion years, this means that

Earth's seawater has been mixed 3,000,000,000 ÷ 2,000, or 1.5 million times. With this much mixing, you would be surprised if seawater were *not* identical all around Earth.

How does seawater move to accomplish such a complete mixing? Seawater is in a constant state of motion, both on the surface and far below the surface. The surface has two types of motion: (1) *waves,* which have been produced by some disturbance, such as the wind, and (2) *currents,* which move water from one place to another. Waves travel across the surface as a series of wrinkles. Waves crash on the shore as booming breakers and make the surf. This produces local currents as water moves along the shore and back out to sea. There are also permanent, worldwide surface currents that move ten thousand times more water across the ocean than all the water moving in all the large rivers on the land. Beneath the surface, there are currents that move water up in some places and move it down in other places. Finally, there are enormous deep ocean currents that move tremendous volumes of seawater. The overall movement of many of the currents on the surface and their relationship to the deep ocean currents are not yet fully mapped or understood. The surface waves are better understood. The general trend and cause of permanent, worldwide currents in the ocean can also be explained. The following is a brief description and explanation of waves, currents, and the deep ocean movements of seawater.

Waves

Any slight disturbance will create ripples that move across a water surface. For example, if you gently blow on the surface of water in a glass, you will see a regular succession of small ripples moving across the surface. These ripples, which look like small, moving wrinkles, are produced by the friction of the air moving across the water surface. The surface of the ocean is much larger, but a gentle wind produces patches of ripples in a similar way. These patches appear, then disappear as the wind begins to blow over calm water. If the wind continues to blow, larger and longer-lasting ripples are made, and the moving air can now push directly on the side of the ripples. A ripple may eventually grow into an **ocean wave,** a moving disturbance that travels across the surface of the ocean. In its simplest form, each wave has a ridge, or mound, of water called a *crest,* which is followed by a depression called a *trough.* Ocean waves are basically repeating series of these crests and troughs that move across the surface like wrinkles (figure 18.14).

The simplest form of an ocean wave can be described by measurements of three distinct characteristics: (1) the *wave height,* which is the vertical distance between the top of a crest and the bottom of the next trough, (2) the *wavelength,* which is the horizontal distance between two successive crests (or other successive parts of the wave), and (3) the *wave period,* which is the time required for two successive crests (or other successive parts) of the wave to pass a given point (figure 18.15).

The characteristics of an ocean wave formed by the wind depend on three factors: (1) the wind speed, (2) the length of time that the wind blows, and (3) the *fetch,* which is the distance the wind blows across the open ocean. As you can imagine, larger waves are produced by strong winds that blow

Estuary Pollution

Pollution is usually understood to mean something that is not naturally occurring and contaminates air, soil, or water to interfere with human health, well-being, or quality of the environment. Important factors in understanding pollution are the size of the human population and the amount of material that might become a pollutant. When the human population was small and produced few biological wastes, there was no pollution problem. The decomposers broke down the material into simpler nonpolluting substances such as water and carbon dioxide, and no harm was done. For example, suppose one person empties the tea leaves remaining from a cup of tea into a nearby river once a week. In this case, decomposer organisms in the water would break down the tea leaves almost as fast as they are added to the river. But imagine one-hundred thousand people doing this every day. In this case, the tea leaves are released faster than they decompose and the leaves become pollutants.

The part of the wide lower course of a river where the freshwater of the river mixes with the saline water from the oceans is called a coastal *estuary*. Estuary waters include bays and tidal rivers that serve as nursery areas for many fish and shellfish populations, including shrimp, oysters, crabs, and scallops. Unfortunately, the rivers carry pollution from their watersheds and adjacent wetlands to the estuary, where it impacts the fish and shellfish industry, swimming, and recreation.

In 1996, the U.S. Environmental Protection Agency asked the coastal states to rate the general water quality in their estuaries. The states reported that pollutants impact aquatic life in 31 percent of the area surveyed, violate shellfish harvesting criteria in 27 percent of the area surveyed, and violate swimming-use criteria in 16 percent of the area surveyed.

The most common pollutants affecting the surveyed estuaries were excessive *nutrients*, which were found in 22 percent of all the estuaries surveyed. Excessive nutrients stimulate population explosions of algae. Fast-growing masses of algae block light from the habitat below, stressing the aquatic life. The algae die and eventually decompose, and this depletes the available oxygen supply, leading to further fish and shellfish kills.

The second most common pollutant was the presence of *bacteria,* which pollute 16 percent of all the estuary waters surveyed. Because *Escherichia coli* is a bacterium commonly found in the intestines of humans and other warm-blooded animals, the presence of *E. coli* is evidence that sewage is polluting the water. Bacteria interfere with recreation activities of people and can contaminate fish and shellfish.

The states also reported that *toxic organic chemicals* pollute 15 percent of the surveyed waters, *oxygen-depleting chemicals* pollute 12 percent, and *petroleum products* pollute another 8 percent. These pollutants impact the fish and shellfish industry, swimming, and recreation activities that require contact with the water.

The leading sources of the pollutants were identified as industrial discharges, urban runoff, municipal wastewater, agricultural runoff, and wastes from landfills. Of these, agricultural runoff is the primary contributor to water pollution, followed by urban runoff. All the other sources together make up a distant third source of pollution.

FIGURE 18.14   The surface of the ocean is rarely, if ever, still. Any disturbance can produce a wave, but most waves on the open ocean are formed by a local wind.

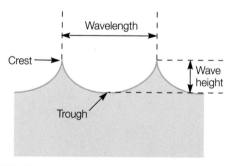

FIGURE 18.15   The simplest form of ocean waves, showing some basic characteristics. Most waves do not look like this representation because most are complicated mixtures of superimposed waves with a wide range of sizes and speeds.

for a longer time over a long fetch. In general, longer-blowing, stronger winds produce waves with greater wave heights, longer wavelengths, and longer periods, but a given wind produces waves with a wide range of sizes and speeds. In addition, the wind does not blow in just one direction, and shifting winds produce a chaotic pattern of waves of many different heights and wavelengths. Thus, the surface of the ocean in the area of a storm or strong wind has a complicated mixture of many sizes and speeds of superimposed waves. The smaller waves soon die out from friction within the water, and the larger ones grow as the wind pushes against their crests. Ocean waves range in height from a few centimeters up to more than 30 m (about 100 ft), but giant waves more than 15 m (about 50 ft) are extremely rare.

The larger waves of the chaotic, superimposed mixture of waves in a storm area last longer than the winds that formed them, and they may travel for hundreds or thousands of kilometers from their place of origin. The longer-wavelength waves travel faster and last longer than the shorter-wavelength waves, so the longer-wavelength waves tend to outrun the shorter-wavelength waves as they die out from energy losses to water friction. Thus, the irregular, superimposed waves created in the area of a storm become transformed as they travel away from the area. They become regular groups of long-wavelength waves with low wave height that are called **swell**. The regular waves of swell that you might observe near a shore may have been produced by a storm that occurred days before thousands of kilometers across the ocean.

The regular crests and troughs of swell carry energy across the ocean, but they do not transport water across the open ocean. If you have ever been in a boat that is floating in swell, you know that you move in a regular pattern of up and forward on each crest, then backward and down on the following trough. The boat does not move along with the waves unless it is moved along by a wind or by some current. Likewise, a particle of water on the surface moves upward and forward with each wave crest, then backward and down on the following trough, tracing out a nearly circular path through this motion. The particle returns to its initial position, without any forward movement while tracing out the small circle. Note that the diameter of the circular path is equal to the wave height (figure 18.16). Water particles farther below the surface also trace out circular paths as a wave passes. The diameters of these circular paths below the surface are progressively smaller with increasing

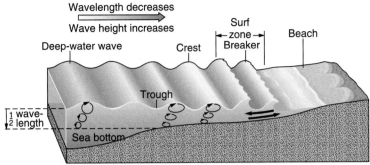

FIGURE 18.17 As a pattern of swell approaches a gently sloping beach, friction between the circular motion of the water particles and the bottom slows the wave and the wave front becomes steeper and steeper. When the depth is about one and one-third times the wave height, the wave breaks forward, moving water toward the beach.

depth. Below a depth equal to about half the wavelength (the wave base), there is no circular movement of the particles. Thus, you can tell how deeply the passage of a wave disturbs the water below if you measure the wavelength.

As swell moves from the deep ocean to the shore, the waves pass over shallower and shallower water depths. When a depth is reached that is equal to about half the wavelength, the circular motion of the water particles begins to reach the ocean bottom. The water particles now move across the ocean bottom, and the friction between the two results in the waves moving slower as the wave height increases. These important modifications result in a change in the direction of travel and in an increasingly unstable situation as the wave height increases.

Most waves move toward the shore at some angle. As the wave crest nearest the shore starts to slow, the part still over deep water continues on at the same velocity. The slowing at the shoreward side *refracts,* or bends, the wave so it is more parallel to the shore. Thus, waves always appear to approach the shore head-on, arriving at the same time on all parts of the shore.

After the waves reach water that is less than one-half the wavelength, friction between the bottom and the circular motion of the water particles progressively slow the bottom part of the wave. The wave front becomes steeper and steeper as the top overruns the bottom part of the wave. When the wave front becomes too steep, the top part breaks forward and the wave is now called a *breaker* (figure 18.17). In general, this occurs where the water depth is about one and one-third times the wave height. The zone where the breakers occur is called **surf** (figure 18.18).

Waves break in the foamy surf, sometimes forming smaller waves that then proceed to break in progressively shallower water. The surf may have several sets of breakers before the water is finally thrown on the shore as a surging sheet of seawater. The turbulence of the breakers in the surf zone and the final surge expend all the energy that the waves may have brought from thousands of kilometers away. Some of the energy does work in eroding the shoreline, breaking up rock masses into the sands that are carried by local currents back to the ocean. The rest of the energy goes into the kinetic energy of water molecules, which appears as a temperature increase.

Swell does not transport water with the waves over a distance, but small volumes of water are moved as a growing wave is pushed

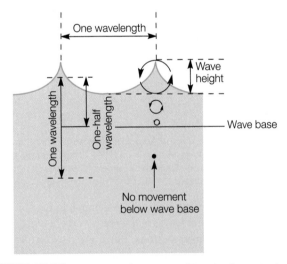

FIGURE 18.16 Water particles are moved in a circular motion by a wave passing in the open ocean. On the surface, a water particle traces out a circle with a diameter that is equal to the wave height. The diameters of the circles traced out by water particles decrease with depth to a depth that is equal to one-half the wavelength of the ocean wave. Bottom sediment cannot be moved by waves when the sediment is below the wave base.

FIGURE 18.18 The white foam is in the surf zone, which is where the waves grow taller and taller, then break forward into a froth of turbulence. Do you see any evidence of rip currents in this picture (see figure 18.19)?

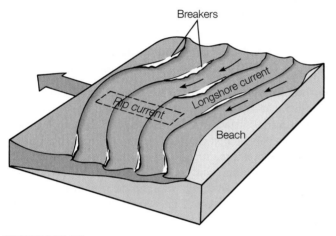

FIGURE 18.19 Breakers result in a buildup of water along the beach that moves as a long-shore current. Where it finds a shore bottom that allows it to return to the sea, it surges out in a strong flow called a *rip current*.

to greater heights by the wind over the open ocean. A strong wind can topple such a wave on the open ocean, producing a foam-topped wave known as a *whitecap*. In general, whitecaps form when the wind is blowing at 30 km/h (about 20 mi/h) or more.

Waves do transport water where breakers occur in the surf zone. When a wave breaks, it tosses water toward the shore, where the water begins to accumulate. This buildup of water tends to move away in currents, or streams, as the water returns to a lower level. Some of the water might return directly to the sea by moving beneath the breakers. This direct return of water forms a weak current known as *undertow*. Other parts of the accumulated water might be pushed along by the waves, producing a *longshore current* that moves parallel to the shore in the surf zone. This current moves parallel to the shore until it finds a lower place or a channel that is deeper than the adjacent bottom. Where the current finds such a channel, it produces a *rip current*, a strong stream of water that bursts out against the waves and returns water through the surf to the sea (figure 18.19). The rip current usually extends beyond the surf zone and then diminishes. A rip current, or where rip currents are occurring, can usually be located by looking for

Rogue Waves

A rogue wave is an unusually large wave that appears with smaller waves. The rogue wave has also been called a "freak" wave. Whatever the name, it is generally one wave or a group of two or three waves that are more than twice the size of the normal surrounding waves. The rogue wave is one or several very large "walls of water" and has unpredictable behavior, not following the wind direction, for example. Large rogue waves have been reported from 21 to 35 m (69 to 114 ft) tall, and have been observed to almost capsize large ships.

It is believed that a rogue wave is an extreme storm wave. It probably forms during a storm from constructive interference (see page 114) between smaller waves when the crest and troughs happen to match, coming together to form a mountainous wave that lasts several minutes before subsiding. Other processes, such as wave focusing by the shape of the coast or movement by currents, may play a part in forming rogue waves. The source of rogue waves continues to be a mystery and an active topic of research. This much is known—rogue waves do exist.

the combination of (1) a lack of surf, (2) darker-looking water, which means a deeper channel, and (3) a turbid, or muddy, streak of water that extends seaward from the channel indicated by the darker water that lacks surf. See chapter 16 for information on how earthquakes can also produce waves.

Ocean Currents

Waves generated by the winds, earthquakes, and tidal forces keep the surface of the ocean in a state of constant motion. Local, temporary currents associated with this motion, such as rip currents or tidal currents, move seawater over a short distance. Seawater also moves in continuous **ocean currents,** streams of water that stay in about the same path as they move through other seawater over large distances. Ocean currents can be difficult to observe directly since they are surrounded by water that looks just like the water in the current. Wind is likewise difficult to observe directly since the moving air looks just like the rest of the atmosphere. Unlike the wind, an ocean current moves *continuously* in about the same path, often carrying water with different chemical and physical properties than the water it is moving through. Thus, an ocean current can be identified and tracked by measuring the physical and chemical characteristics of the current and the surrounding water. This shows where the current is coming from and where in the world it is going. In general, ocean currents are produced by (1) density differences in seawater and (2) winds that blow persistently in the same direction.

Density Currents. The density of seawater is influenced by three factors: (1) the water temperature, (2) salinity, and (3) suspended sediments. Cold water is generally more dense than warm water, thus sinking and displacing warmer water. Seawater of a high salinity has a higher relative density than less salty water, so it sinks and displaces water of less salinity. Likewise, seawater with a larger amount of suspended sediments has a higher relative density than clear water, so it sinks and displaces clear water. The following describes how these three ways of changing the density of seawater result in the ocean current known as a *density current*, which is an ocean current that flows because of density differences.

Earth receives more incoming solar radiation in the tropics than it does at the poles, which establishes a temperature difference between the tropical and polar oceans. The surface water in the polar ocean is often at or below the freezing point of freshwater, while the surface water in the tropical ocean averages about 26°C (about 79°F). Seawater freezes at a temperature below that of freshwater because the salt content lowers the freezing point. Seawater does not have a set freezing point, however, because as it freezes, the salinity is increased as salt is excluded from the ice structure. Increased salinity lowers the freezing point more, so the more ice that freezes from seawater, the lower the freezing point for the remaining seawater. Cold seawater near the poles is therefore the densest, sinking and creeping slowly as a current across the ocean floor toward the equator. Where and how such a cold, dense bottom current moves is influenced by the shape of the ocean floor, the rotation of Earth, and other factors. The size and the distance that cold bottom currents move can be a surprise. Cold, dense water from the Arctic, for example, moves in a 200 m (about 660 ft) diameter current on the ocean bottom between Greenland and Iceland. This current carries an estimated 5 million m³/s (about 177 million ft³/s) of seawater to the 3.5 km (about 2.1 mi) deep water of the North Atlantic Ocean. This is a flow rate about 250 times larger than that of the Mississippi River. At about 30°N, the cold Arctic waters meet even denser water that has moved in currents all the way from the Antarctic to the deepest part of the North Atlantic Basin (figure 18.20).

A second type of density current results because of differences in salinity. The water in the Mediterranean, for example, has a high salinity because it is mostly surrounded by land in a warm, dry climate. The Mediterranean seawater, with its higher salinity, is more dense than the seawater in the open Atlantic Ocean. This density difference results in two separate currents that flow in opposite directions between the Mediterranean and the Atlantic. The greater density seawater flows from the bottom of the Mediterranean into the Atlantic, while the less dense Atlantic water flows into the Mediterranean near the surface. The dense Mediterranean seawater sinks to a depth of about 1,000 m (about 3,300 ft) in the Atlantic, where it spreads over a large part of the North Atlantic Ocean. This increases the salinity of this part of the ocean, making it one of the more saline areas in the world.

The third type of density current occurs when underwater sediments on a slope slide toward the ocean bottom, producing a current of muddy or turbid water called a *turbidity current.* Turbidity currents are believed to be a major mechanism that moves sediments from the continents to the ocean basin. They may also be responsible for some undersea features, such as submarine canyons. Turbidity currents are believed to occur only occasionally, however, and none has ever been directly

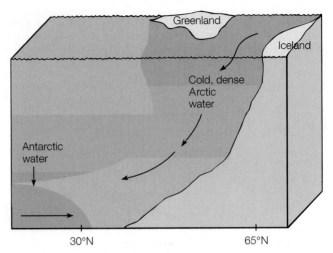

FIGURE 18.20 A cold density current carries about 250 times more water than the Mississippi River from the Arctic and between Greenland and Iceland to the deep Atlantic Ocean. At about 30°N latitude, it meets water that has moved by cold density currents all the way from the Antarctic.

observed or studied. There is thus no data or direct evidence of how they form or what effects they have on the ocean floor.

Surface Currents. There are broad and deep-running ocean currents that slowly move tremendous volumes of water relatively near the surface. As shown in figure 18.21, each current is actually part of a worldwide system, or circuit, of currents. This system of ocean currents is very similar to the worldwide system of prevailing winds. This similarity exists because it is the friction of the prevailing winds on the seawater surface that drives the ocean currents. The currents are modified by other factors, such as the rotation of Earth and the shape of the ocean basins, but they are basically maintained by the wind systems.

Each ocean has a great system of moving water called a **gyre** that is centered in the mid-latitudes. The gyres rotate to the right in the Northern Hemisphere and to the left in the Southern Hemisphere. The movement of water around these systems, or gyres, plus some smaller systems, forms the surface circulation system of the world ocean. Each part of the system has a separate name, usually based on its direction of flow. All are called "currents" except one that is called a "stream" (the Gulf Stream) and those that are called "drifts." Both the Gulf Stream and the drifts are currents that are part of the connected system.

The major surface currents are like giant rivers of seawater that move through the ocean near the surface. You know that all the currents are connected, for a giant river of water cannot just start moving in one place, then stop in another. The Gulf Stream, for example, is a current about 100 km (about 60 mi) wide that may extend to a depth of 1 km (about 0.6 mi) below the surface, moving more than 75 million m³ of water per second (about 2.6 billion ft³/s). The Gulf Stream carries more than 370 times more water than the Mississippi River. The California Current is weaker and broader, carrying cool water southward at a relatively slow rate. The flow rate of all the currents must be equal, however, since all the ocean basins are connected and the sea level is changing very little, if at all, over long periods of time.

Key Forecasting Tool for the Chesapeake Bay

Submerged aquatic vegetation—SAV, for short—is vital to the Chesapeake Bay ecosystem and a key measuring stick for the Chesapeake's overall health. Eel and wigeon grasses grow in the shallows around the bay and across the bottom, providing food and shelter for baby blue crabs and fish, filtering pollutants, and providing life-sustaining oxygen for the water. So the health of the grasses provides an indication of the overall health of the Chesapeake.

The Chesapeake Bay once had an estimated 600,000 acres of grasses and provided an abundance of oysters, blue crab, shad, haddock, sturgeon, rock-fish, and other prized sport fish and seafood. At that time, the bay water was clear, and watermen reported they could clearly see grasses on the bottom some 6 m (about 20 ft) below their boats. Then the water clouded, the grasses began to die, and the ecosystem of the bay began to decline. The low point was reached in 1984, when less than 40,000 acres of grasses could be found in the now murky waters of the Chesapeake.

With the decline of the grasses came the decline of the aquatic species living in the bay. Shad, haddock, and sturgeon once supported large fisheries but are now scarce. The rockfish (striped bass) as well as the prized blue crab were once abundant but seem to decline and recover over the years. The decline of the blue crab population has shocked watermen because the crab is amazingly fertile. The crabs reach sexual maturity in about a year and one female bears millions of eggs. Part of the problem is the loss of the underwater grasses, which shelter the baby crabs from predators. In addition to the loss of habitat, some believe the crab is being overfished.

What happened to the underwater grasses of the Chesapeake? Scientists believe it is a combination of natural erosion and pollutants that wash from farms in the watershed. The pollutants include nutrient-rich fertilizers, chemical residues, and overflow from sewage treatment plants. There are some 6,000 chicken houses around the Chesapeake, raising more than 600 million chickens a year and producing 750,000 tons of manure. The manure is used as fertilizer, and significant amounts of nitrogen and phosphates wash from these fields. These nutrients accelerate algae growth in the bay, which blocks sunlight. The grass dies and the loss results in muddying of the water, making it impossible for new grasses to begin growing.

There is some evidence to support this idea since the 600,000 acres of underwater grasses died back to a low of 40,000 acres in 1984, then began to rebound when sewage treatment plants were modernized and there were fewer pollutants from industry and farms. The grasses recovered to some 63,500 acres by 1999 but are evidently very sensitive to even slight changes in water quality. Thus, they may be expected to die back with unusual weather conditions that might bring more pollutants or muddy conditions but continue to rebound when the conditions are right. The abundance of blue crabs and other aquatic species can be expected to fluctuate with the health of the grasses. The trends of underwater grass growth do indeed provide a key measure for the health of the Chesapeake Bay.

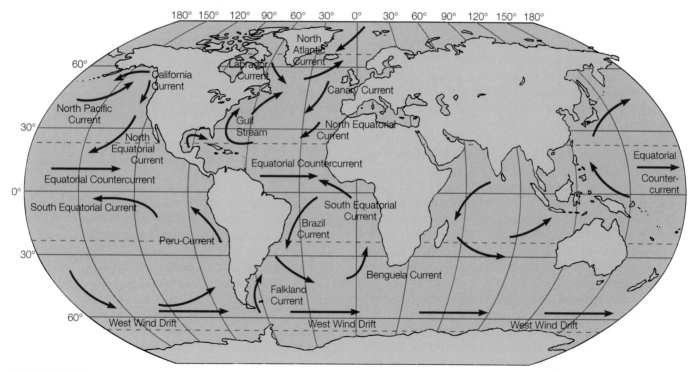

FIGURE 18.21   Earth's system of ocean currents.

Rachel Louise Carson (1907–1964)

Rachel Louise Carson was a U.S. biologist, conservationist, and campaigner. Her writings on conservation and the dangers and hazards that many modern practices imposed on the environment inspired the creation of the modern environmental movement.

Carson was born in Springdale, Pennsylvania, on May 27, 1907, and educated at the Pennsylvania College for Women, studying English to achieve her ambition for a literary career. A stimulating biology teacher diverted her toward the study of science, and she went to Johns Hopkins University, graduating in zoology in 1929. She received her master's degree in zoology in 1932 and was then appointed to the department of zoology at the University of Maryland, spending her summers teaching and researching at the Woods Hole Marine Biological Laboratory in Massachusetts. Family commitments to her widowed mother and orphaned nieces forced her to abandon her academic career, and she worked for the U.S. Bureau of Fisheries, writing in her spare time articles on marine life and fish, and producing her first book on the sea just before the Japanese attack on Pearl Harbor. During World War II, she wrote fisheries information bulletins for the U.S. government and reorganized the publications department of what became known after the war as the U.S. Fish and Wildlife Service. In 1949, she was appointed chief biologist and editor of the service. She also became occupied with fieldwork and regularly wrote freelance articles on the natural world. During this period, she was also working on *The Sea Around Us*. Upon its publication in 1951, this book became an immediate best-seller, was translated into several languages, and won several literary awards. Given a measure of financial independence by this success, Carson resigned from her job in 1952 to become a professional writer. Her second book, *The Edge of the Sea* (1955), an ecological exploration of the seashore, further established her reputation as a writer on biological subjects. Her most famous book, *The Silent Spring* (1962), was a powerful indictment of the effects of the chemical poisons, especially DDT, with which humans were destroying the Earth, sea, and sky. Despite denunciations from the influential agrochemical lobby, one immediate effect of Carson's book was the appointment of a presidential advisory committee on the use of pesticides. By this time, Carson was seriously incapacitated by ill health, and she died in Silver Spring, Maryland, on April 14, 1964. On a larger canvas, *The Silent Spring* alerted and inspired a new worldwide movement of environmental concern. While writing about broad scientific issues of pollution and ecological exploitation, Carson also raised important issues about the reckless squandering of natural resources by an industrial world.

Source: Modified from the Hutchinson Dictionary of Scientific Biography. *© RM, 2007. Reprinted by permission.*

THE OCEAN FLOOR

Some of the features of the ocean floor were discussed in chapter 16 because they were important in developing the theory of plate tectonics. Many features of the present ocean basins were created from the movement of large crustal plates, according to plate tectonics theory, and in fact, some ocean basins are thought to have originated with the movement of these plates. There is also evidence that some features of the ocean floor were modified during the ice ages of the past. During an ice age, much water becomes locked up in glacial ice, which lowers the sea level. The sea level dropped as much as 140 m (about 460 ft) during the most recent major ice age, exposing the margins of the continents to erosion. Today, these continental margins are flooded with seawater, forming a zone of relatively shallow water called the **continental shelf** (figure 18.22). The continental shelf is considered to be a part of the continent and not the ocean, even though it is covered with an average depth of about 130 m (about 425 ft) of seawater. The shelf slopes gently away from the shore for an average of 75 km (about 47 mi), but it is much wider on the edge of some parts of continents than other parts.

The continental shelf is a part of the continent that happens to be flooded by seawater at the present time. It still retains some of the general features of the adjacent land that is above water, such as hills, valleys, and mountains, but these features were smoothed off by the eroding action of waves when the sea level was lower. Today, a thin layer of sediments from the adjacent land covers these smoothed-off features.

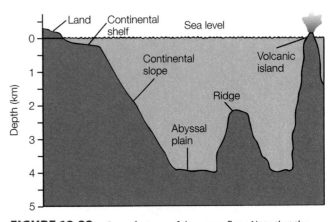

FIGURE 18.22   Some features of the ocean floor. Note that the inclination of the features is not as steep as this vertical exaggeration would suggest.

Beyond the gently sloping continental shelf is a steeper feature called the **continental slope.** The continental slope is the transition between the continent and the deep ocean basin. The water depth at the top of the continental slope is about 120 m (about 390 ft), then plunges to a depth of about 3,000 m (about 10,000 ft) or more. The continental slope is generally 20 to 40 km (about 12 to 25 mi) wide, so the inclination is similar to that encountered driving down a steep mountain road on an interstate highway. At various places around the world, the continental slopes are cut by long, deep, and steep-sided *submarine canyons*. Some of these canyons extend from the top of the slope and down the slope to the ocean basin. Such a submarine canyon can be similar in size and depth to the Grand Canyon on the Colorado River of Arizona. Submarine canyons are believed to have been eroded by turbidity currents, which were discussed in the section on the movement of seawater.

Beyond the continental slope is the bottom of the ocean floor, the **ocean basin.** Ocean basins are the deepest part of the ocean, covered by about 4 to 6 km (about 2 to 4 mi) of seawater. The basin is mostly a practically level plain called the **abyssal plain** and long, rugged mountain chains called *ridges* that rise thousands of meters above the abyssal plain. The Atlantic Ocean and Indian Ocean basins have ridges that trend north and south near the center of the basin. The Pacific Ocean basin has its ridge running north and south near the eastern edge. The Pacific Ocean basin also has more *trenches* than the Atlantic Ocean or the Indian Ocean basins. A trench is a long, relatively narrow, steep-sided trough that occurs along the edges of the ocean basins. Trenches range in depth from about 8 to 11 km (about 5 to 7 mi) deep below sea level.

The ocean basin and ridges of the ocean cover more than half of Earth's surface, accounting for more of the total surface than all the land of the continents. The plain of the ocean basin alone, in fact, covers an area about equal to the area of the land. Scattered over the basin are more than ten thousand steep volcanic peaks called *seamounts.* By definition, seamounts rise more than 1 km (about 0.6 mi) above the ocean floor, sometimes higher than the sea-level surface of the ocean. A seamount that sticks above the water level makes an island. The Hawaiian Islands are examples of such giant volcanoes that have formed islands. Most seamount-formed islands are in the Pacific Ocean. Most islands in the Atlantic, on the other hand, are the tops of volcanoes of the Mid-Atlantic Ridge.

SUMMARY

Precipitation that falls on the land either evaporates, flows across the surface, or soaks into the ground. Water moving across the surface is called *runoff.* Water that moves across the land as a small body of running water is called a *stream.* A stream drains an area of land known as the stream drainage basin or *watershed.* The watershed of one stream is separated from the watershed of another by a line called a *divide.* Water that collects as a small body of standing water is called a *pond,* and a larger body is called a *lake.* A *reservoir* is a natural pond, a natural lake, or a lake or pond created by building a dam for water management or control. The water of streams, ponds, lakes, and reservoirs is collectively called *surface water.*

Precipitation that soaks into the ground *percolates* downward until it reaches a *zone of saturation.* Water from the saturated zone is called *groundwater.* The amount of water that a material will hold depends on its *porosity,* and how well the water can move through the material depends on its *permeability.* The surface of the zone of saturation is called the *water table.*

The *ocean* is the single, continuous body of salt water on the surface of Earth. A *sea* is a smaller part of the ocean with different characteristics. The dissolved materials in seawater are mostly the ions of six substances, but sodium ions and chlorine ions are the most abundant. *Salinity* is a measure of the mass of salts dissolved in 1,000 g of seawater.

An *ocean wave* is a moving disturbance that travels across the surface of the ocean. In its simplest form, a wave has a ridge called a *crest* and a depression called a *trough.* Waves have a characteristic *wave height, wavelength,* and *wave period.* The characteristics of waves made by the wind depend on the wind *speed,* the *time* the wind blows, and the *fetch.* Regular groups of low-profile, long-wavelength waves are called *swell.* When swell approaches a shore, the wave slows and increases in wave height. This slowing *refracts,* or bends, the waves so they approach the shore head-on. When the wave height becomes too steep, the top part breaks forward, forming *breakers* in the *surf zone.* Water accumulates at the shore from the breakers and returns to the sea as *undertow,* as *longshore currents,* or in *rip currents.*

Ocean currents are streams of water that move through other seawater over large distances. Some ocean currents are *density currents,* which are caused by differences in *water temperature, salinity,* or *suspended sediments.* Each ocean has a great system of moving water called a *gyre* that is centered in mid-latitudes. Different parts of a gyre are given different names such as the *Gulf Stream* or the *California Current.*

The ocean floor is made up of the *continental shelf,* the *continental slope,* and the *ocean basin.* The ocean basin has two main parts, the *abyssal plain* and mountain chains called *ridges.*

KEY TERMS

abyssal plain (p. **433**)	groundwater (p. **419**)	ocean wave (p. **426**)	watershed (p. **418**)
aquifer (p. **420**)	gyre (p. **430**)	runoff (p. **418**)	water table (p. **419**)
continental shelf (p. **432**)	ocean (p. **424**)	salinity (p. **425**)	
continental slope (p. **432**)	ocean basin (p. **433**)	surf (p. **428**)	
freshwater (p. **417**)	ocean currents (p. **429**)	swell (p. **428**)	

APPLYING THE CONCEPTS

1. The most abundant chemical compound at the surface of Earth is
   a. silicon dioxide.
   b. nitrogen gas.
   c. water.
   d. minerals of iron, magnesium, and silicon.

2. Of the total supply, the amount of water that is available for human consumption and agriculture is
   a. 97 percent.
   b. about two-thirds.
   c. about 3 percent.
   d. less than 1 percent.

3. In a region of abundant rainfall, a layer of extensively cracked but otherwise solid granite could serve as a limited source of groundwater because it has
   a. limited permeability and no porosity.
   b. average porosity and average permeability.
   c. no permeability and no porosity.
   d. limited porosity and no permeability.

4. How many different oceans are actually on Earth's surface?
   a. 14
   b. 7
   c. 3
   d. 1

5. The largest of the three principal ocean regions of Earth is the
   a. Atlantic Ocean.
   b. Pacific Ocean.
   c. Indian Ocean.
   d. South American Ocean.

6. The Gulf of Mexico is a shallow sea of the
   a. Atlantic Ocean.
   b. Pacific Ocean.
   c. Indian Ocean.
   d. South American Ocean.

7. Measurement of the salts dissolved in seawater taken from various locations throughout the world show that seawater has a
   a. uniform chemical composition and a variable concentration.
   b. variable chemical composition and a variable concentration.
   c. uniform chemical composition and a uniform concentration.
   d. variable chemical composition and a uniform concentration.

8. The salinity of seawater is *increased* locally by
   a. the addition of water from a large river.
   b. heavy precipitation.
   c. the formation of sea ice.
   d. None of the above is correct.

9. Considering only the available light and the dissolving ability of gases in seawater, more abundant life should be found in a
   a. cool, relatively shallow ocean.
   b. warm, very deep ocean.
   c. warm, relatively shallow ocean.
   d. cool, very deep ocean.

10. The regular, low-profile waves called swell are produced from
    a. constant, prevailing winds.
    b. small, irregular waves becoming superimposed.
    c. longer wavelengths outrunning and outlasting shorter wavelengths.
    d. all wavelengths becoming transformed by gravity as they travel any great distance.

11. If the wavelength of swell is 10.0 m, then you know that the fish below the surface feel the waves to a depth of
    a. 5.0 m.
    b. 10.0 m.
    c. 20.0 m.
    d. however deep it is to the bottom.

Answers

1. c 2. d 3. a 4. d 5. b 6. a 7. a 8. c 9. a 10. c 11. a

QUESTIONS FOR THOUGHT

1. Describe in general all the things that happen to the water that falls on the land.

2. Explain how a stream can continue to flow even during a dry spell.

3. What is the water table? What is the relationship between the depth to the water table and the depth that a well must be drilled? Explain.

4. Compare the advantages and disadvantages of using (a) surface water and (b) groundwater as a source of freshwater.

5. Prepare arguments for (a) agriculture, (b) industries, and (c) cities each having first priority in the use of a limited water supply. Identify one of these arguments as being the "best case" for first priority, then justify your choice.

6. Discuss some possible ways of extending the supply of freshwater.

7. The world's rivers and streams carry millions of tons of dissolved materials to the ocean each year. Explain why this does not increase the salinity of the ocean.

8. What is swell and how does it form?

9. Why do waves always seem to approach the shore head-on?

10. What factors determine the size of an ocean wave made by the wind?

11. Describe how a breaker forms from swell. What is surf?

12. Describe what you would look for to avoid where rip current occurs at a beach.

FOR FURTHER ANALYSIS

1. Considering the distribution of all the water on Earth, which presently unavailable category would provide the most freshwater at the least cost for transportation, processing, and storage?

2. Describe a number of ways that you believe would increase the amount of precipitation going into groundwater rather than runoff.

3. Some people believe that constructing a reservoir for water storage is a bad idea because (1) it might change the downstream habitat below the dam and (2) the reservoir will eventually fill with silt and sediments. Write a "letter to the editor" that agrees that a reservoir is a bad idea for the reasons given. Now write a second letter that disagrees with the "bad idea" letter and supports the construction of the reservoir.

4. Explain how the average salinity of seawater has remained relatively constant over the past six-hundred million years in spite of the continuous supply of dissolved salts in the river waters of the world.

5. Can ocean waves or ocean currents be used as an energy source? Explain why or why not.

6. What are the significant similarities and differences between a river and an ocean current?

INVITATION TO INQUIRY

Water Use

Investigate the source of water and the amount used by industrial processes, agriculture, and homes in your area. Make pie graphs to compare your area to national averages and develop explanations for any differences. What could be done to increase the supply of water in your area?

This is a model of the organic molecule deoxyribonucleic acid (DNA). A cell's ability to make a particular protein comes from the genetic information stored in this molecule. It contains the blueprint for making the proteins the cell needs. DNA contains genes, which are specific messages that tell the cell to link particular amino acids in a specific order.

Organic and Biochemistry

Physics Connections

▶ Organic molecules are involved in energy storage and release (Ch. 3).

Chemistry Connections

▶ Octet rule is followed when bonds are formed (Ch. 9).
▶ Chemistry of the carbon atom is responsible for the traits of organic compounds (Ch. 8).

Earth Science Connections

▶ Fossil fuels were originally the organic components of living things (Ch. 3).
▶ Pollution of air, land, and water may be the result of many types of organic molecules (Ch. 15).

CORE CONCEPT

The nature of the carbon atom allows for a great variety of organic compounds, many of which play vital roles in living.

A great variety of organic molecules can be formed (p. 437).

All living things are comprised of organic molecules (p. 448).

Humans have created many organic compounds (p. 460).

Life Science Connections

▶ The structure of living things is comprised of organic molecules (Ch. 20).
▶ Living things are and utilize organic compounds (Ch. 20).

Astronomy Connections

▶ Extraterrestrial organic compounds are known to exist and may have been the source of the original molecules of life on Earth (Ch. 19).
▶ The exploration for extraterrestrial life centers on the identification of organic compounds (Ch. 19).

OVERVIEW

The impact of ancient Aristotelian ideas on the development of understandings of motion, elements, and matter was discussed in earlier chapters. Historians also trace the "vitalist theory" back to Aristotle. According to Aristotle's idea, all living organisms are composed of the four elements (earth, air, fire, and water) and have in addition an *actuating force,* the life or soul that makes the organism different from nonliving things made of the same four elements. Plants, as well as animals, were considered to have this actuating, or vital, force in the Aristotelian scheme of things.

There were strong proponents of the vitalist theory as recently as the early 1800s. Their basic argument was that organic matter, the materials and chemical compounds recognized as being associated with life, could not be produced in the laboratory. Organic matter could only be produced in a living organism, they argued, because the organism had a vital force that

is not present in laboratory chemicals. Then, in 1828, a German chemist named Friedrich Wöhler decomposed a chemical that was *not organic* to produce urea (N_2H_4CO), a known *organic* compound that occurs in urine. Wöhler's production of an organic compound was soon followed by the production of other organic substances by other chemists. The vitalist theory gradually disappeared with each new reaction, and a new field of study, organic chemistry, emerged.

This chapter is an introductory survey of the field of organic chemistry, which is concerned with compounds and reactions of compounds that contain carbon. You will find this an interesting, informative introduction, particularly if you have ever wondered about synthetic materials, natural foods and food products, or any of the thousands of carbon-based chemicals you use every day. The survey begins with the simplest of organic compounds, those consisting of only carbon and hydrogen

atoms, compounds known as hydrocarbons. Hydrocarbons are the compounds of crude oil, which is the source of hundreds of petroleum products (figure 19.1).

Most common organic compounds can be considered derivatives of the hydrocarbons, such as alcohols, ethers, fatty acids, and esters. Some of these are the organic compounds that give flavors to foods, and others are used to make hundreds of commercial products, from face cream to oleo. The main groups, or classes, of derivatives will be briefly introduced, along with some interesting examples of each group. A very strong link exists between organic chemistry and the chemistry of living things, which is called biochemistry, or biological chemistry. Some of the important organic compounds of life, including proteins, carbohydrates, and fats, are discussed next. The chapter concludes with an introduction to synthetic polymers, what they are, and how they are related to the fossil fuel supply.

ORGANIC COMPOUNDS

Today, **organic chemistry** is defined as the study of compounds in which carbon is the principal element, whether the compound was formed by living things or not. The

study of compounds that do not contain carbon as a central element is called **inorganic chemistry.** An *organic compound* is thus a compound in which carbon is the principal element, and an *inorganic compound* is any other compound.

Friedrich Wöhler (1800–1882)

Friedrich Wöhler was a German chemist who generally is credited with having carried out the first laboratory synthesis of an organic compound, although his main interest was inorganic chemistry.

Wöhler was born at Eschershein, near Frankfurt, Germany, on July 31, 1800, the son of a veterinary surgeon. He entered Marburg University in 1820 to study medicine and, after a year, transferred to Heidelberg, where he studied in the laboratory of Leopold Gmelin (1788–1853). He gained his medical degree in 1823, but Gmelin had persuaded Wöhler to study chemistry, and so he spent the following year in Stockholm with Jöns Berzelius, beginning a lifelong association between the two chemists.
From 1825 to 1831, he taught in a technical school in Berlin, and from 1831 to 1836, in another school in Kassel. In 1836, he became professor of chemistry in the medical faculty of Göttingen University and remained there for the rest of his career, making it one of the most prestigious teaching laboratories in Europe. Wöhler died in Göttingen on September 23, 1882.

In Wöhler's first research in 1827, he isolated metallic aluminum; he then prepared many different aluminum salts. In 1828, he used the same procedure to isolate beryllium. Also in 1828, he carried out the reaction for which he is best known. He heated ammonium thiocyanate—a crystalline, inorganic substance—and converted it to urea, an organic substance previously obtained only from natural sources.

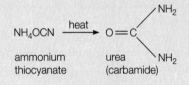

Until that time, there had been a basic misconception in scientific thinking that the chemical changes in living organisms were not governed by the same laws as were those in inanimate substances. It was thought that these "vital" phenomena could not be described in ordinary chemical or physical terms. This theory gave rise to the original division between inorganic (nonvital) and organic (vital) chemistry, and its supporters were known as *vitalists*, who maintained that natural products formed by living organisms could never be synthesized by ordinary chemical means. Wöhler's synthesis of urea was a blow to the vitalists and did much to overthrow their doctrine.

Wöhler worked with Justus von Liebig on a number of important investigations. In the inorganic field, Wöhler isolated boron and silicon. He prepared phosphorus by the modern method and discovered calcium carbide and showed that it can be reacted with water to produce acetylene (ethyne):

$$CaC_2 + 2H_2O \rightarrow Ca(OH)_2 + C_2H_2$$

He demonstrated the analogy between the compounds of carbon and silicon, and just missed being the first to discover vanadium and niobium. He also obtained pure titanium and showed the similarity between this element and carbon and silicon. He published little work after 1845 but concentrated on teaching.

Source: Modified from the *Hutchinson Dictionary of Scientific Biography.* © RM, 2007. Reprinted by permission.

FIGURE 19.1 Refinery and tank storage facilities, such as this one in Texas, are needed to change the hydrocarbons of crude oil into many different petroleum products. The classes and properties of hydrocarbons form one topic of study in organic chemistry.

The majority of known compounds are organic. Several million are known and thousands of new ones are created every year. You use organic compounds every day, including gasoline, plastics, grain alcohol, foods, flavorings, and many others.

It is the unique properties of carbon that allow it to form so many different compounds. The carbon atom has a valence of four and can combine with one, two, three, or four *other carbon atoms,* in addition to a wide range of other kinds of atoms (figure 19.2). The number of possible molecular combinations is almost limitless, which explains why there are so many organic compounds. Fortunately, there are patterns of groups of carbon atoms and groups of other atoms that lead to similar chemical characteristics, making the study of organic chemistry less difficult. The key to success in studying organic chemistry is to recognize patterns and to understand the code and meaning of organic chemical names (see "People Behind the Science in this chapter").

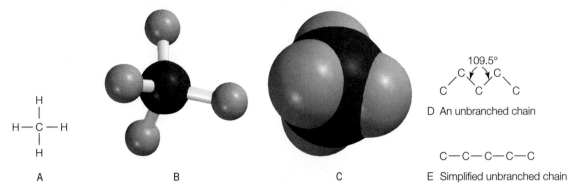

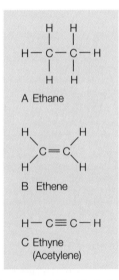

D An unbranched chain

C—C—C—C—C

E Simplified unbranched chain

FIGURE 19.2 Models of a methane molecule. Each of the four bondable electrons in a carbon atom inhabits an area as far away from the other three as possible. Each can share its electron with another. Carbon can share with four other atoms at each of these sites. The structures of molecules can be modeled in many ways. For the sake of simplicity, diagrams of molecules such as the gas methane can be (*A*) two-dimensional drawings, although in reality they are three-dimensional molecules and take up space. Recall from chapter 8 that the higher the atomic number, the larger the space taken up by atoms of that element. The model shown in (*B*) above is called a ball-and-stick model. Part (*C*) is a space-filling model. Each time you see the various ways in which molecules are displayed, try to imagine how much space they actually occupy. (*D*) Carbon-to-carbon bond angles are 109.5°, so a chain of carbon atoms makes a zigzag pattern. (*E*) The unbranched chain of carbon atoms is usually simplified in a way that looks like a straight chain, but it is actually a zigzag, as shown in (*D*).

HYDROCARBONS

A **hydrocarbon** is an organic compound consisting of only two elements. As the name implies, these elements are hydrogen and carbon. The simplest hydrocarbon has one carbon atom and four hydrogen atoms (figure 19.3), but since carbon atoms can combine with one another, there are thousands of possible structures and arrangements. The carbon-to-carbon bonds are covalent and can be single, double, or triple (figure 19.4). Recall that the dash in a structural means one shared electron pair has a covalent bond. To satisfy the octet rule, this means that each satisfied carbon atom must have a total of four dashes around it each attached to another atom, no more and no less. Note that when the carbon atom has double or triple bonds, fewer hydrogen atoms can be attached as the octet rule is satisfied. There are four groups of hydrocarbons that are classified according to how the carbon atoms are put together: the (1) *alkanes*, (2) *alkenes*, (3) *alkynes*, and (4) *aromatic hydrocarbons*.

The **alkanes** are *hydrocarbons with single covalent bonds* between the carbon atoms. Alkanes that are large enough to form chains of carbon atoms occur with a straight structure, a branched structure, or a ring structure, as shown in figure 19.5. (In three dimensions, the "straight" structure is actually a zigzag, as shown in figure 19.2.) You are familiar with many alkanes, for they make up the bulk of petroleum and petroleum products.

The alkanes are also called the *paraffin series*. The alkanes are not as chemically reactive as the other hydrocarbons, and the term *paraffin* means "little affinity." They are called a series because *each higher molecular weight alkane has an additional CH_2*. The simplest alkane is methane, CH_4, and the next highest molecular weight alkane is ethane, C_2H_6. As you can see, C_2H_6 is CH_4 with an additional CH_2.

Note the names of the alkanes listed in table 19.1. The names have a consistent prefix and suffix pattern. The prefix and suffix pattern is a code that provides a clue about the

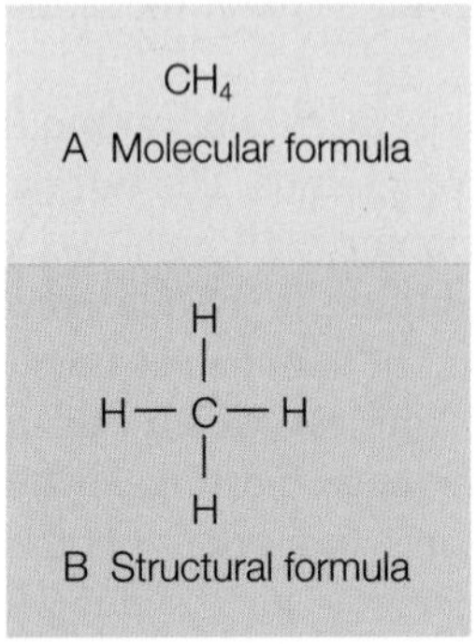

CH_4

A Molecular formula

H—C—H with H above and H below

B Structural formula

FIGURE 19.3 A molecular formula (*A*) describes the numbers of different kinds of atoms in a molecule, and a structural formula (*B*) represents a two-dimensional model of how the atoms are bonded to each other. Each dash represents a bonding pair of electrons.

A Ethane

B Ethene

C Ethyne (Acetylene)

FIGURE 19.4 Carbon-to-carbon bonds can be single (*A*), double (*B*), or triple (*C*). Note that in each example, each carbon atom has four dashes, which represent four bonding pairs of electrons, satisfying the octet rule.

compound. The Greek prefix tells you the *number of carbon atoms* in the molecule; for example, *oct-* means eight, so *oct*ane has eight carbon atoms. The suffix *-ane* tells you this hydrocarbon is a member of the alk*ane* series, so it has single bonds only. With the general alkane formula of C_nH_{2n+2}, you can now write the formula when you hear the name. Octane has eight carbon atoms with single bonds and $n = 8$. Two times 8 plus $2(2n + 2)$ is 18, so the formula for octane is C_8H_{18}. Most organic chemical names provide clues like this.

The alkanes in table 19.1 all have straight chains. A straight, continuous chain is identified with the term *normal*, which is abbreviated *n*. Figure 19.6A shows *n*-butane with a straight chain and a molecular formula of C_4H_{10}. Figure 19.6B shows a different branched structural formula that has the same C_4H_{10} molecular formula. Compounds with the same molecular formulas with different structures are called **isomers**. Since the straight-chained isomer is called *n*-butane, the branched isomer is called *isobutane*. The isomers of a particular alkane, such as butane, have different physical and chemical properties because they have different structures. Isobutane, for example, has a boiling point of $-10°C$. The boiling point of *n*-butane, on the other hand, is $-0.5°C$. In the section on alkenes and alkynes, you will learn that the various isomers

TABLE 19.1   Examples of Straight-Chain Alkanes

Name	Molecular Formula	Structural Formula
Methane	CH_4	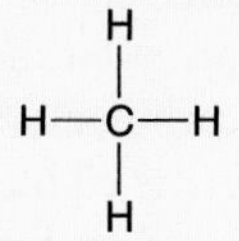
Ethane	C_2H_6	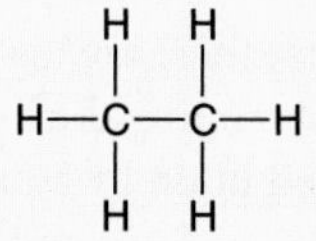
Propane	C_3H_8	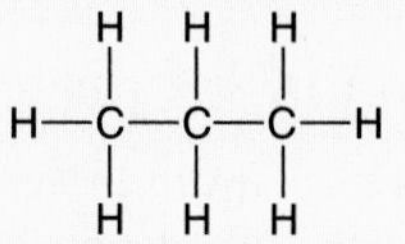
Butane	C_4H_{10}	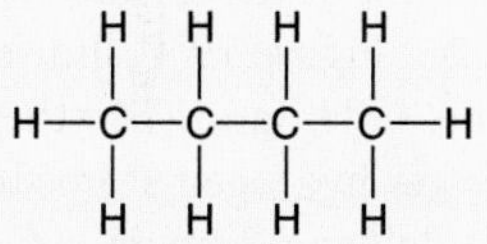
Heptane	C_7H_{16}	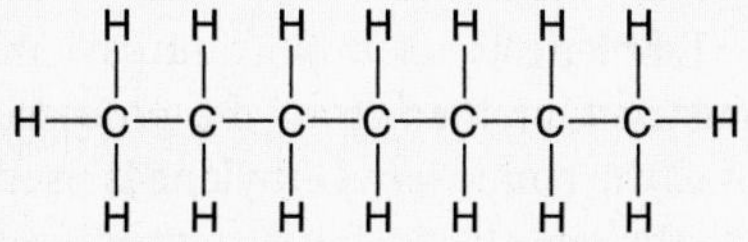
Octane	C_8H_{18}	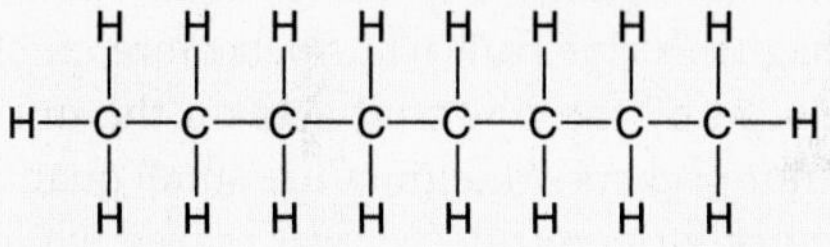

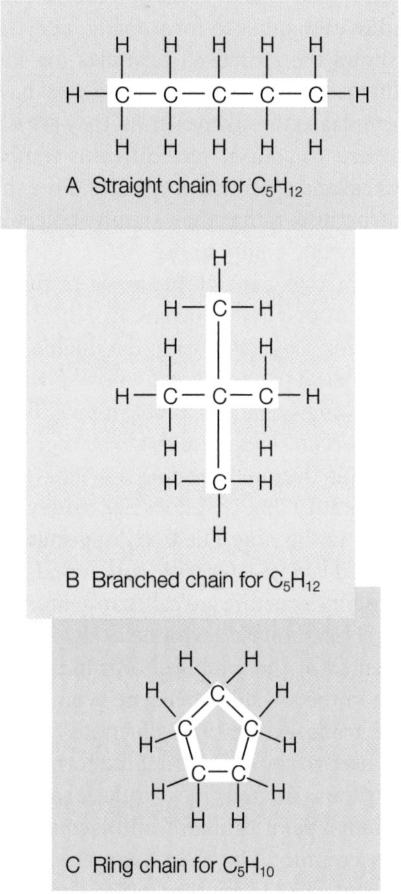

FIGURE 19.5   Carbon-to-carbon chains can be (*A*) straight, (*B*) branched, or (*C*) in a closed ring. (Some carbon bonds are drawn longer but are actually the same length.)

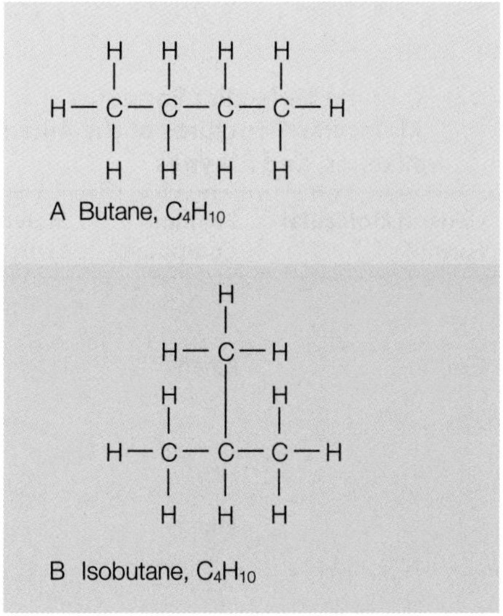

FIGURE 19.6   (*A*) A straight-chain alkane. (*B*) A branched-chain alkane isomer.

of the octane hydrocarbon perform differently in automobile engines, requiring the "re-forming" of *n*-octane to *iso-octane* before it can be used.

Methane, ethane, and propane can have only one structure each, and butane has two isomers. The number of possible isomers for a particular molecular formula increases rapidly as the number of carbon atoms increase. After butane, hexane has five isomers, octane eighteen isomers, and decane seventy-five isomers. Because they have different structures, each isomer has different physical properties. A different naming system is needed because there are just too many isomers to keep track of. The system of naming the branched-chain alkanes is described by rules agreed upon by a professional organization, the International Union of Pure and Applied Chemistry (IUPAC).

Alkenes and Alkynes

The **alkenes** are *hydrocarbons with a double covalent carbon-to-carbon bond.* To denote the presence of a double bond, the *-ane* suffix of the alkanes is changed to *-ene,* as in alk*ene* (table 19.2). Figure 19.4 shows the structural formula for (A) ethane, C_2H_6, and (B) ethene, C_2H_4. Alkenes have room for two fewer hydrogen atoms because of the double bond, so the general alkene formula is C_nH_{2n}. Note the simplest alkene is called ethene but is commonly known as ethylene.

Ethylene is an important raw material in the chemical industry. Obtained from the processing of petroleum, about half of the commercial ethylene is used to produce the familiar polyethylene plastic. It is also produced by plants to ripen fruit, which explains why unripe fruit enclosed in a sealed plastic bag with ripe fruit will ripen more quickly. The ethylene produced by the ripe fruit acts on the unripe fruit. Commercial fruit packers sometimes use small quantities of ethylene gas to quickly ripen fruit that was picked while green.

Perhaps you have heard the terms *saturated* and *unsaturated* in advertisements for cooking oil or margarine. An organic molecule, such as a hydrocarbon, that does not contain the maximum number of hydrogen atoms is an **unsaturated**

hydrocarbon. For example, ethylene can add more hydrogen atoms by reacting with hydrogen gas to form ethane:

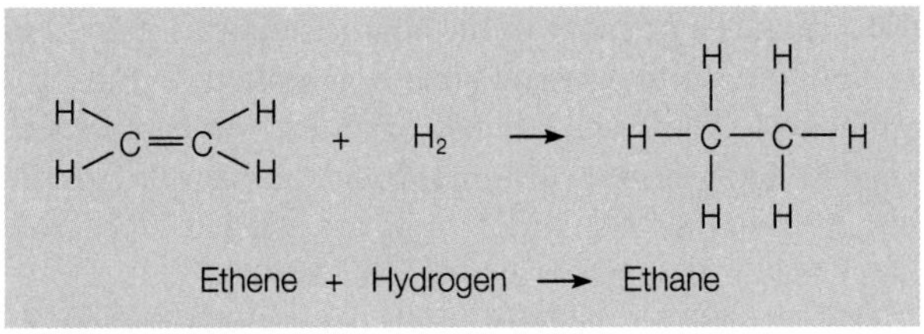

Ethene + Hydrogen ⟶ Ethane

The ethane molecule has all the hydrogen atoms possible, so ethane is a **saturated** hydrocarbon. Unsaturated molecules are less stable, which means that they are more chemically reactive than saturated molecules.

An **alkyne** is a *hydrocarbon with a carbon-to-carbon triple bond* and the general formula of C_nH_{2n-2}. The alkynes are highly reactive, and the simplest one, ethyne, has a common name of acetylene. Acetylene is commonly burned with oxygen gas in a welding torch because the flame reaches a temperature of about 3,000°C. Acetylene is also an important raw material in the production of plastics.

Cycloalkanes and Aromatic Hydrocarbons

The hydrocarbons discussed up until now have been straight or branched open-ended chains of carbon atoms. Carbon atoms can also bond to each other to form a ring, or *cyclic,* structure. Figure 19.7 shows the structural formulas for some of these cyclic structures. Note that the cycloalkanes have the same molecular formulas as the alkenes; thus, they are isomers of the alkenes. They are, of course, very different compounds, with different physical and chemical properties. This shows the importance of structural, rather than simply molecular, formulas in referring to organic compounds.

The six-carbon ring structure shown in figure 19.8A has three double bonds that do not behave like the double bonds in the alkenes. In this six-carbon ring, the double bonds are not localized in one place but are spread over the whole molecule. Instead of alternating single and double bonds, all the bonds are something in between. This gives the C_6H_6 molecule increased stability. As a result, the molecule does not behave like other unsaturated compounds; that is, it does not readily react in order to add hydrogen to the ring. The C_6H_6 molecule is the organic compound named *benzene.* Organic compounds that are based on the benzene ring structure are called *aromatic hydrocarbons.* To denote the six-carbon ring with delocalized electrons, benzene is represented by the symbol shown in figure 19.8B.

The circle in the six-sided benzene symbol represents the delocalized electrons. Figure 19.8B illustrates how this benzene ring symbol is used to show the structural formula of some aromatic hydrocarbons. You may have noticed some of the names on labels of paints, paint thinners, and lacquers. Toluene and the xylenes are commonly used in these products as solvents. A benzene ring attached to another molecule is given the name *phenyl.* Phenyl products are good disinfectants and are found in many pine oil products such as Pine Sol.

TABLE 19.2   General Molecular Formulas and Molecular Structures of the Alkanes, Alkenes, and Alkynes

Group	General Molecular Formula	Example Compound	Molecular Structure
Alkanes	C_nH_{2n+2}	Ethane	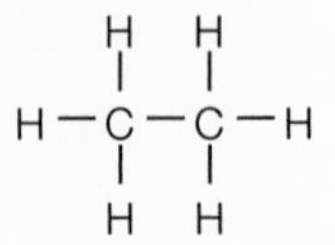
Alkenes	C_nH_{2n}	Ethene	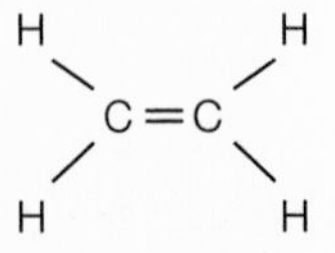
Alkynes	C_nH_{2n-2}	Ethyne	H—C≡C—H

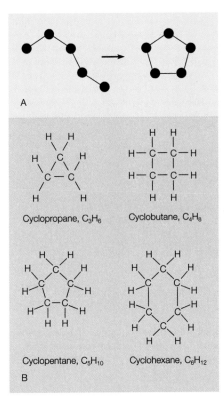

FIGURE 19.7 (A) The "straight" chain has carbon atoms that are able to rotate freely around their single bonds, sometimes linking up in a closed ring. (B) Ring compounds of the first four cycloalkanes.

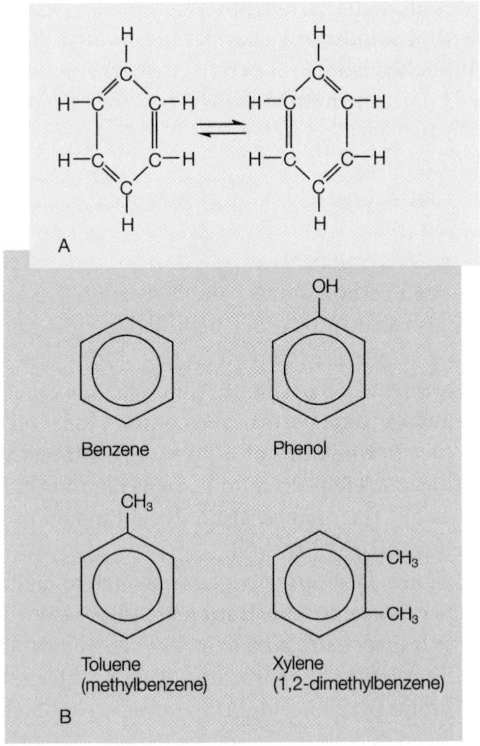

Cyclopropane, C_3H_6

Cyclobutane, C_4H_8

Cyclopentane, C_5H_{10}

Cyclohexane, C_6H_{12}

FIGURE 19.8 (A) The bonds in C_6H_6 are something between single and double, which gives it different chemical properties than double-bonded hydrocarbons. (B) The six-sided symbol with a circle represents the benzene ring. Organic compounds based on the benzene ring are called *aromatic hydrocarbons* because of their aromatic character.

Benzene

Phenol

Toluene (methylbenzene)

Xylene (1,2-dimethylbenzene)

PETROLEUM

Petroleum is a mixture of alkanes, cycloalkanes, and some aromatic hydrocarbons. While they were once the products of photosynthesis, as time passed, these complex carbohydrates were significantly altered. The origin of petroleum is uncertain, but it is believed to have formed from the slow decomposition of buried marine life, primarily plankton and algae in the absence of oxygen (i.e., anaerobic). Time, temperature, pressure, and perhaps bacteria are considered important in the formation of petroleum. As the petroleum formed, it was forced through porous rock until it reached a rock type or rock structure that stopped it. Here, it accumulated to saturate the porous rock, forming an accumulation called an **oil field.** The composition of petroleum varies from one oil field to the next. The oil from a given field might be dark or light in color, and it might have an asphalt base or a paraffin base. Some oil fields contain oil with a high quantity of sulfur, referred to as "sour crude." Because of such variations, some fields have oil with more desirable qualities than oil from other fields.

Early settlers found oil seeps in the eastern United States and collected the oil for medicinal purposes. One enterprising oil peddler tried to improve the taste by running the petroleum through a whiskey still. He obtained a clear liquid by distilling the petroleum and, by accident, found that the liquid made an excellent lamp oil. This was fortunate timing, for the lamp oil used at that time was whale oil, and whale oil production was declining. This clear liquid obtained by distilling petroleum is today known as *kerosene.*

Wells were drilled, and crude oil refineries were built to produce the newly discovered lamp oil. Gasoline was a by-product and was used primarily as a spot remover. With Henry Ford's automobile production and Thomas Edison's electric light invention, the demand for gasoline increased, and the demand for kerosene decreased. The refineries were converted to produce gasoline, and the petroleum industry grew to become one of the world's largest industries. With the expansion of this industry have come downsides. One of great concern has been global warming associated with the emission of carbon dioxide from a worldwide increase in the number of gasoline-burning engines.

Crude oil is petroleum that is pumped from the ground, a complex and variable mixture of hydrocarbons with an upper limit of about fifty carbon atoms. This thick, smelly black mixture is not usable until it is refined, that is, separated into usable groups of hydrocarbons called petroleum products. Petroleum products are separated by distillation. The larger the molecule, the higher the boiling point. Thus, each product has a boiling point range, or "cut," of the distilled vapors. Thus, each product, such as gasoline or heating oil, is made up of hydrocarbons within a range of carbon atoms per molecule (figure 19.9). The products, their boiling ranges, and ranges of carbon atoms per molecule are listed in table 19.3.

The hydrocarbons that have one to four carbon atoms (CH_4 to C_4H_{10}) are gases at room temperature. They can be pumped from certain wells as a gas, but they also occur dissolved in

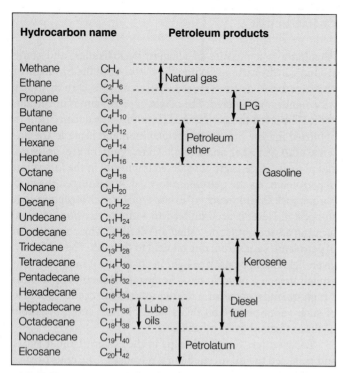

Hydrocarbon name		Petroleum products
Methane	CH₄	}Natural gas
Ethane	C₂H₆	
Propane	C₃H₈	LPG
Butane	C₄H₁₀	
Pentane	C₅H₁₂	Petroleum ether
Hexane	C₆H₁₄	
Heptane	C₇H₁₆	
Octane	C₈H₁₈	Gasoline
Nonane	C₉H₂₀	
Decane	C₁₀H₂₂	
Undecane	C₁₁H₂₄	
Dodecane	C₁₂H₂₆	
Tridecane	C₁₃H₂₈	
Tetradecane	C₁₄H₃₀	Kerosene
Pentadecane	C₁₅H₃₂	
Hexadecane	C₁₆H₃₄	Diesel fuel
Heptadecane	C₁₇H₃₆	Lube oils
Octadecane	C₁₈H₃₈	
Nonadecane	C₁₉H₄₀	Petrolatum
Eicosane	C₂₀H₄₂	

FIGURE 19.9   Petroleum products and the ranges of hydrocarbons in each product.

TABLE 19.3   Petroleum Products

Name	Boiling Range (°C)	Carbon Atoms per Molecule
Natural gas	Less than 0	C_1 to C_4
Petroleum ether	35–100	C_5 to C_7
White "gas"	100–350	C_5 to C_9
Gasoline	35–215	C_5 to C_{12}
Kerosene	35–300	C_{12} to C_{15}
Diesel fuel	300–400	C_{15} to C_{18}
Motor oil, grease	350–400	C_{16} to C_{18}
Paraffin	Solid, melts at about 55	C_{20}
Asphalt	Boiler residue	C_{40} or more

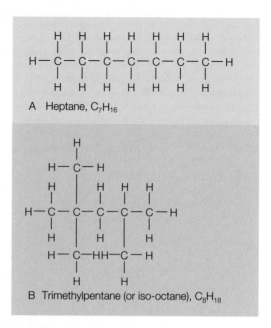

FIGURE 19.10   The octane rating scale is a description of how rapidly gasoline burns. It is based on (*A*) heptane, with an assigned octane number of 0, and (*B*) trimethylpentane, with an assigned number of 100.

crude oil. *Natural gas* is a mixture of hydrocarbon gases, but it is about 95 percent methane (CH_4). Propane (C_3H_8) and butane (C_4H_{10}) are liquified by compression and cooling and are sold as liquified petroleum gas, or *LPG*. LPG is used where natural gas is not available for cooking or heating and is widely used as a fuel in barbecue grills and camp stoves.

Gasoline is a mixture of hydrocarbons that may have five to twelve carbon atoms per molecule. Gasoline distilled from crude oil consists mostly of straight-chain molecules not suitable for use as an automotive fuel. Straight-chain molecules burn too rapidly in an automobile engine, producing more of an explosion than a smooth burn. You hear these explosions as a knocking or pinging in the engine, and they indicate poor efficiency and could damage the engine. On the

other hand, branched chain molecules burn comparatively slower, without the pinging or knocking explosions. The burning rate of gasoline is described by the *octane number* scale. The scale is based on pure *n*-heptane, straight-chain molecules that are assigned an octane number of 0, and a multiple branched isomer of octane, trimethylpentane, which is assigned an octane number of 100 (figure 19.10). Most unleaded gasolines have an octane rating of 87, which could be obtained with a mixture that is 87 percent trimethylpentane and 13 percent heptane. Gasoline, however, is a much more complex mixture.

Kerosene is a mixture of hydrocarbons that have from twelve to fifteen carbon atoms. The petroleum product called kerosene is also known by other names, depending on its use. Some of these names are lamp oil (with coloring and odorants added), jet fuel (with a flash flame retardant added), heating oil, #1 fuel oil, and in some parts of the country, "coal oil."

Diesel fuel is a mixture of a group of hydrocarbons that have from fifteen to eighteen carbon atoms per molecule. Diesel fuel also goes by other names, again depending on its use, that is, distillate fuel oil or #2 fuel oil.

Motor oil and *lubricating oils* have sixteen to eighteen carbon atoms per molecule. Lubricating grease is heavy oil that is thickened with soap. *Petroleum jelly,* also called petrolatum (or Vaseline), is a mixture of hydrocarbons with sixteen to thirty-two carbon atoms per molecule. *Mineral oil* is a light lubricating oil that has been decolorized and purified.

Depending on the source of the crude oil, varying amounts of *paraffin* wax (C_{20} or greater) or *asphalt* (C_{36} or more) may be present. Paraffin is used for candles, waxed paper, and home canning. Asphalt is mixed with gravel and used to surface roads.

HYDROCARBON DERIVATIVES

The hydrocarbons account for only about 5 percent of the known organic compounds, but the other 95 percent can be considered as hydrocarbon derivatives. **Hydrocarbon derivatives** are formed when *one or more hydrogen atoms on a hydrocarbon have been replaced by some element or group of elements other than hydrogen.* For example, the halogens (F_2, Cl_2, Br_2) react with an alkane in sunlight or when heated, replacing a hydrogen:

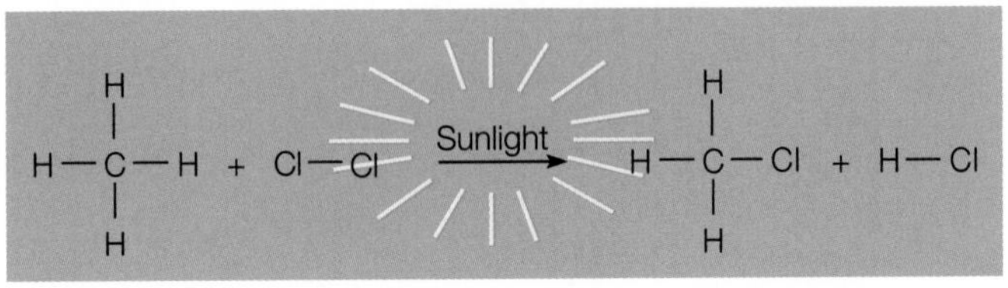

In this particular *substitution reaction,* a hydrogen atom on methane is replaced by a chlorine atom to form methyl chloride. Replacement of any number of hydrogen atoms is possible, and a few *organic halides* are illustrated in figure 19.11. Common organic halides include Freons (chlorofluorocarbons) and Teflon (polytetrafluoroethylene).

If a hydrocarbon molecule is unsaturated (has a double or triple bond), a hydrocarbon derivative can be formed by an *addition reaction:*

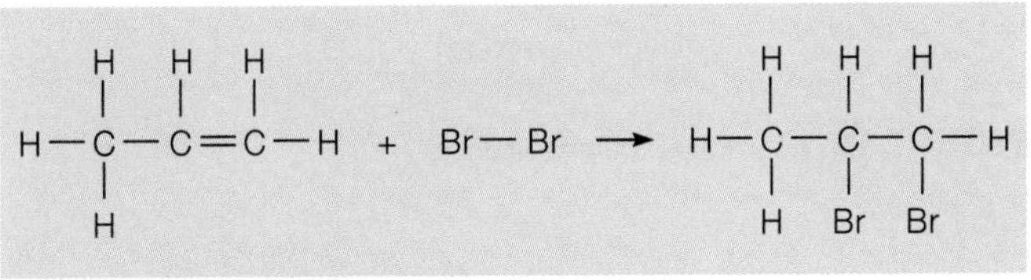

The bromine atoms add to the double bond on propene, forming dibromopropane. This derivative is used in the synthesis of pharmaceuticals and other organic compounds.

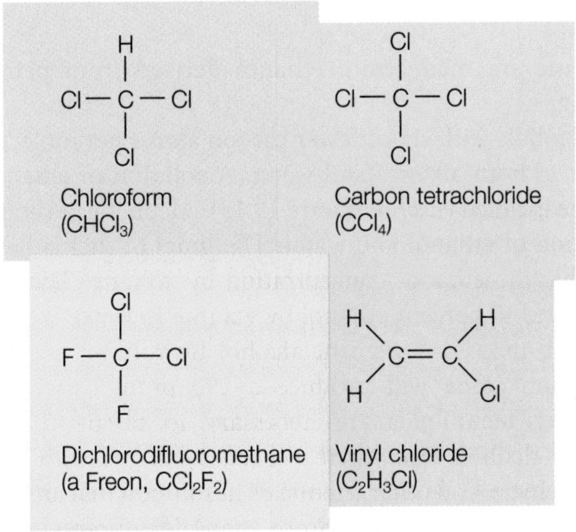

FIGURE 19.11 Common examples of organic halides.

Alkene molecules can also add to each other in an addition reaction to form a very long chain consisting of hundreds of molecules. A long chain of repeating units is called a **polymer** (*poly* = many; *mer* = segment), individual pieces are called **monomers** (*mono* = single; *mer* = segment or piece), and the reaction is called *addition polymerization.* Ethylene, for example, is heated under pressure with a catalyst to form *polyethylene,* the most popular plastic in the world. It is used to make such products as grocery bags, shampoo bottles, children's toys, and bullet-proof vests. Heating breaks the double bond,

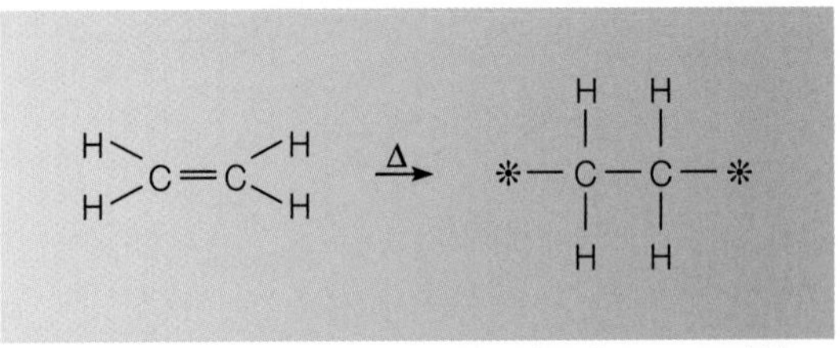

The * indicates that some other atom or series of atoms would be attached at this point.

which provides sites for single covalent bonds to join the ethylene units together,

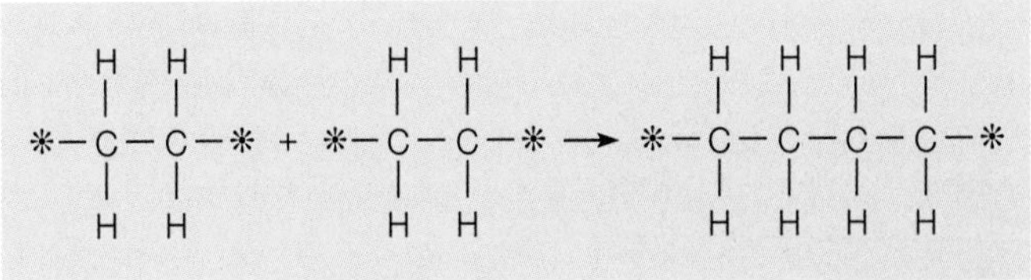

The * indicates that some other atom or series of atoms would be attached at this point.

which continues the addition polymerization until the chain is hundreds of units long.

Functional groups are specific combinations of atoms attached to the carbon skeleton that determine specific chemical properties of the organic compound. Functional groups usually have (1) multiple bonds or (2) lone pairs of electrons that cause them to be sites of reactions. Table 19.4 lists some of the common hydrocarbon functional groups. Look over this list and compare the structure of the functional group with the group name. Some of the more interesting examples from a few of these groups will be considered next. Note that the R and R′ (pronounced, "R prime") stand for one or more hydrocarbon groups. For example, in the reaction between methane and chlorine, the product is methyl chloride. In this case, the R in RCl stands for methyl, but it could represent any hydrocarbon group.

Alcohols

An *alcohol* is an organic compound formed by replacing one or more hydrogens on an alkane with a hydroxyl functional group ($-OH$). The hydroxyl group should not be confused with the hydroxide ion, OH^-. The hydroxyl group is attached to an organic compound and does not form ions in solution, as the hydroxide ion does. It remains attached to a hydrocarbon

TABLE 19.4 Selected Organic Functional Groups

Name of Functional Group	General Formula	General Structure
Organic halide	RCl	R—C̈l:
Alcohol	ROH	R—Ö—H
Ether	ROR'	R—Ö—R'
Aldehyde	RCHO	R—C—H ‖ :O:
Ketone	RCOR'	R—C—R' ‖ :O:
Organic acid	RCOOH	R—C—Ö—H ‖ :O:
Ester	RCOOR'	R—C—Ö—R' ‖ :O:
Amino	RNH₂	R—N̈—H | H
Phosphate	RPO₄	R—O—P=O with O above and O below

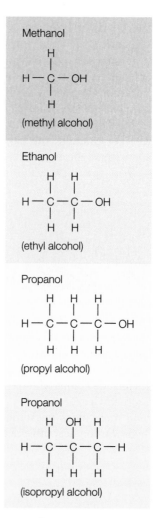

FIGURE 19.12 Four different alcohols.

group (R), giving the compound its set of properties that are associated with alcohols.

The name of the hydrocarbon group determines the name of the alcohol. If the hydrocarbon group in ROH is methyl, for example, the alcohol is called *methyl alcohol,* or *methanol.* Using the IUPAC naming rules, the name of an alcohol has the suffix *-ol* (figure 19.12).

All alcohols have the hydroxyl functional group, and all are chemically similar. Alcohols are toxic to humans except that ethyl alcohol (ethanol) can be consumed in limited quantities. Alcohol has been consumed by people for thousands of years in many cultures. As in the past, there continues to be many health-related problems associated with drinking alcohol, such as fetal alcohol syndrome, liver damage, social disruption, and death. Consumption of small quantities of methanol can result in blindness and death. Ethanol, C_2H_5OH, is produced by the action of yeast or by a chemical reaction of ethylene derived from petroleum refining. Yeast acts on sugars to produce ethanol and CO_2. When beer, wine, and other such beverages are the desired products, the CO_2 escapes during fermentation,

and the alcohol remains in solution. In baking, the same reaction utilizes the CO_2 to make dough rise (leavened), and the alcohol is evaporated during baking. Most alcoholic beverages are produced by the yeast fermentation reaction:

$$\text{sugar} \rightarrow \text{carbon dioxide and ethyl alcohol,}$$

but some are made from ethanol derived from petroleum refining.

Alcohols with six or fewer carbon atoms per molecule are soluble in both alkanes and water. A solution of ethanol and gasoline is called *gasohol* (figure 19.13). Alcoholic beverages are a solution of ethanol and water. The *proof* of such a beverage is double the ethanol concentration by volume. Therefore, a solution of 40 percent ethanol by volume in water is 80 proof, and wine that is 12 percent alcohol by volume is 24 proof. Distillation alone will produce a 190 proof concentration, but other techniques are necessary to obtain 200 proof absolute alcohol. *Denatured alcohol* is ethanol with acetone, formaldehyde, and other chemicals in solution that are difficult to separate by distillation. Since these denaturants are toxic, they make consumption impossible, so denatured alcohol is sold without the consumption tax.

FIGURE 19.13 Gasoline is a mixture of hydrocarbons (C_8H_{18}, for example) that contain no atoms of oxygen. Gasohol contains ethyl alcohol, C_2H_5OH, which does contain oxygen. The addition of alcohol to gasoline, therefore, adds oxygen to the fuel. Since carbon monoxide forms when there is an insufficient supply of oxygen, the addition of alcohol to gasoline helps cut down on carbon monoxide emissions. An atmospheric inversion, with increased air pollution, is likely during the dates shown on the pump, so that is when the ethanol is added.

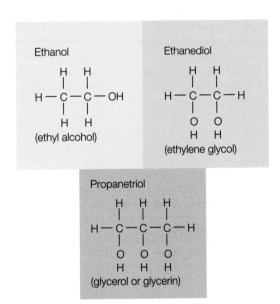

FIGURE 19.14 Common examples of alcohols with one, two, and three hydroxyl groups per molecule.

Methanol, ethanol, and isopropyl alcohol all have one hydroxyl group per molecule. An alcohol with two hydroxyl groups per molecule is called a *glycol.* Ethylene glycol is perhaps the best-known glycol since it is used as an antifreeze. An alcohol with three hydroxyl groups per molecule is called *glycerol* (or *glycerin*). Glycerol is a building block of fat molecules and a by-product in the making of soap. It is added to toothpastes, lotions, cosmetics, and some candies to retain moisture and softness. Ethanol, ethylene glycol, and glycerol are compared in figure 19.14.

Glycerol reacts with nitric acid in the presence of sulfuric acid to produce glyceryl trinitrate, commonly known as *nitroglycerine.* Nitroglycerine is a clear oil that is violently explosive,

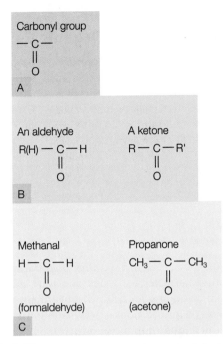

FIGURE 19.15 The carbonyl group (*A*) is present in both aldehydes and ketones, as shown in (*B*). (*C*) The simplest example of each with the more technical name above and the common name below each formula.

and when warmed, it is extremely unstable. In 1867, Alfred Nobel (for whom the Nobel Prizes were named) discovered that a mixture of nitroglycerine and siliceous earth was more stable than pure nitroglycerine but was nonetheless explosive. Siliceous earth is also known as diatomaceous earth. It is composed of the skeletons of algae known as diatoms and provides an enormous surface area upon which the explosive reaction can take place. The cell walls of diatoms are composed of silicon dioxide (SiO_2). The mixture is packed in a tube and is called *dynamite.* The name *dynamite* comes from the Greek word meaning "power."

Ethers, Aldehydes, and Ketones

An *ether* has a general formula of ROR', and the best-known ether is diethylether. In a molecule of diethylether, both the R and the R' are ethyl groups. Diethylether is a volatile, highly flammable liquid that was used as an anesthetic in the past. Today, it is used as an industrial and laboratory solvent.

Aldehydes and *ketones* both have a functional group of a carbon atom doubly bonded to an oxygen atom called a *carbonyl group.* The *aldehyde* has a hydrocarbon group, R (or a hydrogen in one case), and a hydrogen attached to the carbonyl group. A *ketone* has a carbonyl group with two hydrocarbon groups attached (figure 19.15).

The simplest aldehyde is *formaldehyde.* Formaldehyde is soluble in water, and a 40 percent concentration called *formalin* has been used as an embalming agent and to preserve biological specimens. Formaldehyde is also a raw material used to make plastics such as Bakelite. All the aldehydes have odors, and the odors of some aromatic hydrocarbons include the odors of almonds, cinnamon, and vanilla. The simplest ketone is *acetone.* Acetone has a fragrant odor and is used as a solvent in paint removers and nail polish removers.

Chapter 19 *Organic and Biochemistry* **445**

A Very Common Organic Compound—Aspirin

Aspirin went on the market as a prescription drug in 1899 after being discovered only two years earlier! However, even primitive humans were familiar with its value as a pain reliever. The bark of the willow tree was known for its "magical" pain relief power thousands of years ago. People in many cultures stripped and chewed the bark for its medicinal effect. It is estimated that more than 80 million tablets of aspirin are consumed in the United States daily. It is the most widely used drug in the world. Just what does it do? Aspirin is really acetylsalicylic acid and is capable of inhibiting the body's production of compounds known as *prostaglandins,* the cause of the pain.

Questions to Discuss

1. What structural parts of aspirin would make it an organic acid?
2. Go to the Internet and find out just how this organic compound acts as a pain reliever.

Organic Acids and Esters

Mineral acids, such as hydrochloric acid and sulfuric acid, are made of inorganic materials. Acids that were derived from organisms are called **organic acids.** Because many of these organic acids can be formed from fats, they are sometimes called *fatty acids.* Chemically, they are known as the *carboxylic acids* because they contain the carboxyl functional group, $-COOH$, and have a general formula of RCOOH.

The simplest carboxylic acid has been known since the Middle Ages, when it was isolated by the distillation of ants. The Latin word *formica* means "ant," so this acid was given the name *formic acid* (figure 19.16A). Formic acid is

$$H - C - OH$$
$$\overset{\|}{O}$$

It is formic acid, along with other irritating materials, that causes the sting of bees, ants, and certain plants, for example the stinging nettle (*Urtica dioica*) (figure 19.16B).

Acetic acid, the acid of vinegar, has been known since antiquity. Acetic acid forms from the oxidation of ethanol. An oxidized bottle of wine contains acetic acid in place of the alcohol, which gives the wine a vinegar taste. Before wine is served in a restaurant, the person ordering is customarily handed the bottle cork and a glass with a small amount of wine. You first break the cork in half to make sure it is dry, which tells you that the wine has been sealed from oxygen. The small sip is to taste for vinegar before the wine is served. If the wine has been oxidized, the reaction is

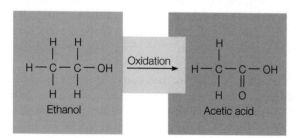

Organic acids are common in many foods. The juice of citrus fruit, for example, contains citric acid, which relieves a thirsty feeling by stimulating the flow of saliva. Lactic acid is found in sour milk, buttermilk, sauerkraut, and pickles. Lactic acid also forms in your muscles as a product of anaerobic carbohydrate metabolism, causing a feeling of fatigue. Anaerobic metabolism of carbohydrates in cells is the breakdown of sugars with the release of usable energy. Citric and lactic acids are small molecules compared to some of the carboxylic acids that are formed from fats. Palmitic acid, for example, is $C_{16}H_{32}O_2$ and comes from palm oil. The structure of palmitic acid is a chain of fourteen CH_2 groups with CH_3- at one end and $-COOH$ at the other. Again, it is the functional carboxyl group, $-COOH$, that gives the molecule its acid properties. Organic acids are also raw materials used in the making of polymers of fabric, film, and paint.

Esters are organic compounds formed from an alcohol and an organic acid by elimination of water. They are common in both plants and animals, giving fruits and flowers their characteristic odor and taste. For example, the ester amyl acetate has the smell of banana. Esters are also used in perfumes and artificial flavorings. A few of the flavors for which particular esters are responsible are listed in table 19.5. These liquid esters

TABLE 19.5 **Flavors and Esters**

Ester Name	Formula	Flavor
Amyl acetate	$CH_3 - \overset{\overset{\displaystyle O}{\|}}{C} - O - C_5H_{11}$	Banana
Octyl acetate	$CH_3 - \overset{\overset{\displaystyle O}{\|}}{C} - O - C_8H_{17}$	Orange
Ethyl butyrate	$C_3H_7 - \overset{\overset{\displaystyle O}{\|}}{C} - O - C_2H_5$	Pineapple
Amyl butyrate	$C_3H_7 - \overset{\overset{\displaystyle O}{\|}}{C} - O - C_5H_{11}$	Apricot
Ethyl formate	$H - \overset{\overset{\displaystyle O}{\|}}{C} - O - C_2H_5$	Rum

Fatlike but Not True Fats—Waxes

Wax is the material secreted by glands on the belly of bees—beeswax (box figure 19.2). The substance is a mixture of alcohols, fatty acids, and other organic compounds. It differs from true fat in that fats contain glycerol in their structure and waxes do not. Waxes are usually harder and less greasy than true fats. Like true fats, they are less dense than water, are soluble in alcohol but water insoluble. Other sources of wax are wool wax, or lanolin, obtained from the surface of sheep wool and used in making cosmetics and soaps. It is also used to softening leather.

Waxes from plants include carnauba wax obtained from the leaves of a Brazilian *palm*, the *carnauba palm* (*Copernicia prunifera*). This wax is used in *automobile waxes*, *shoe polishes*, and floor and furniture

Box Figure 19.2

polishes. The stems of Mexican Candelilla (*Euphorbia antisyphilitica*) are coated with a wax that slows evaporation from this desert

plant. Candelilla plants are boiled in large pots to extract wax and can be used as a polish and water proofer.

A

B

FIGURE 19.16 This red ant (*A*) is classified by biologists as: class: Insecta, order: Hymenoptera, family: Formicidae, genus: *Solenopsis,* species = *Solenopsis invicta.* Like other ants, it makes the simplest of the organic acids, formic acid. The sting of bees, ants, and some plants contains formic acid, along with some other irritating materials. Formic acid is HCOOH. The plant known as the stinging nettle (*B*) (*Urtica dioica*) has small, hollow hairs that contain several irritating substances, including formic acid. These hairs have the ability to scratch the skin and mucous membranes, resulting in almost immediate burning, itching, and irritation.

can be obtained from natural sources, or they can be chemically synthesized. Natural flavors are complex mixtures of these esters along with other organic compounds. Lower molecular-weight esters are fragrant smelling liquids, but higher molecular-weight esters are odorless oils and fats.

ORGANIC COMPOUNDS OF LIFE

Aristotle and the later proponents of the vitalist theory were *partly* correct in their concept that living organisms are different from inorganic substances made of the same elements. Living organisms, for example, have the ability to (1) exchange matter and energy with their surroundings and (2) transform matter and energy into different forms as they (3) respond to changes in their surroundings. In addition, living organisms can use the transformed matter and energy to (4) grow and (5) reproduce. Living organisms are able to do these things through a great variety of organic reactions that are catalyzed by enzymes, however, and not through some mysterious "vital force." These enzyme-regulated organic reactions take place because living organisms are highly organized and have an incredible number of relationships among many different chemical processes.

Organic molecules in living things have molecular weights of thousands or millions of atomic mass units (u's) and are therefore referred to as **macromolecules.** Many types of macromolecules are comprised of similar repeated monomers. Monomers are usually combined to form a polymer by a chemical reaction (*dehydration synthesis*) that results in a water molecule being removed from between two monomers. The reverse of this reaction (*hydrolysis*) is the process of splitting a larger polymer into its parts by the addition of water. Digestion of food molecules in the stomach and small intestine is an important example of this reaction. There are three main types of polymeric macromolecules: (1) carbohydrates, (2) proteins, and (3) nucleic acids. In addition, there is a fourth category, the lipids, that are not true polymers but are macromolecules.

Carbohydrates

One class of organic molecules, **carbohydrates,** is composed of carbon, hydrogen, and oxygen atoms linked together to form monomers called *simple sugars* or *monosaccharides* (*mono* = single; *saccharine* = sweet, sugar) (see table in "Connections . . . How Sweet It Is!"). The name *carbohydrate* literally means "watered carbon," and the empirical formula (CH_2O) for most carbohydrates indicates one carbon (C) atom for each water (H_2O).

The formula for a simple sugar is easy to recognize because there are equal numbers of carbons and oxygens and twice as many hydrogens—for example, $C_3H_6O_3$ or $C_5H_{10}O_5$. We usually describe simple sugars by the number of carbons in the molecule. The ending *-ose* indicates that you are dealing with a carbohydrate. A tri*ose* has three carbons, a pent*ose* has five, and a hex*ose* has six. If you remember that the number of carbons equals the number of oxygen atoms and that the number of hydrogens is double that number, these names tell you the formula for the simple sugar. Carbohydrates play a number of roles in living things. They serve as an immediate source of energy (sugars), provide shape to certain cells (cellulose in plant cell walls), are components of many antibiotics and coenzymes, and are an essential part of the nucleic acids, DNA and RNA.

Simple sugars can be combined with each other to form **complex carbohydrates** (figure 19.18). When two simple sugars bond to each other, a *disaccharide* (*di-* =two) is formed; when three bond together, a *trisaccharide* (*tri-* =three) is formed. Generally, we call a complex carbohydrate that is larger than this a *polysaccharide* (many sugar units). In all cases, the complex carbohydrates are formed by the removal of water from between the sugars. For example, when glucose and fructose are joined together, they form a *disaccharide* (*sucrose*) with the loss of a water molecule,

$$C_6H_{12}O_6 + C_6H_{12}O_6 \rightarrow C_{12}H_{22}O_{11} + H_2O$$
glucose   fructose   sucrose   water

Sucrose, ordinary table sugar, is the most common disaccharide and occurs in high concentrations in sugar cane and sugar beets. It is extracted by crushing the plant materials, then dissolving the sucrose from the materials with water. The water is evaporated and the crystallized sugar is decolorized with charcoal to produce white sugar. Other common

How Sweet It Is!

Simple sugars, such as glucose, fructose, and galactose, provide the chemical energy necessary to keep organisms alive. *Glucose*, $C_6H_{12}O_6$, is the most abundant carbohydrate and serves as a food and a basic building block for other carbohydrates. Glucose (also called dextrose) is found in the sap of plants, and in the human bloodstream, it is called *blood sugar*. Corn syrup, which is often used as a sweetener, is mostly glucose. Fructose, as its name implies, is the sugar that occurs in fruits, and it is sometimes called *fruit sugar*. Both glucose and fructose have the same molecular formula, but glucose is an aldehyde sugar and fructose is a ketone sugar (figure 19.17). A mixture of glucose and fructose is found in honey. This mixture also is formed when table sugar (sucrose) is reacted with water in the presence of an acid, a reaction that takes place in the preparation of canned fruit and candies. The mixture of glucose and fructose is called *invert sugar*. Thanks to fructose, invert sugar is about twice as sweet to the taste as the same amount of sucrose.

Glucose (an aldehyde sugar)

Fructose (a ketone sugar)

FIGURE 19.17 Glucose (blood sugar) is an aldehyde, and fructose (fruit sugar) is a ketone. Both have a molecular formula of $C_6H_{12}O_6$.

Relative Sweetness of Various Sugars and Sugar Substitutes	
Type of Sugar or Artificial Sweetener	**Relative Sweetness**
Lactose (milk sugar)	0.16
Maltose (malt sugar)	0.33
Glucose	0.75
Sucrose (table sugar)	**1.00**
Fructose (fruit sugar)	1.75
Cyclamate	30.00
Aspartame	150.00
Saccharin	350.00
Sucralose	600.00

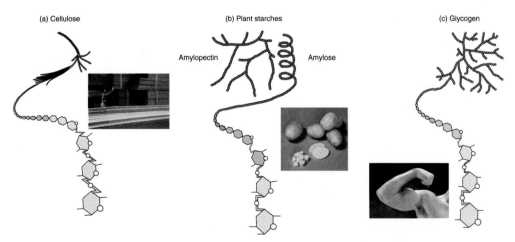

(a) Cellulose (b) Plant starches (c) Glycogen

Amylopectin Amylose

FIGURE 19.18 Simple sugars are attached to each other by the removal of water from between them. Three common complex carbohydrates are (*A*) cellulose (wood fibers), (*B*) plant starch (amylopectin and amylose), and (*C*) glycogen (sometimes called animal starch). Glycogen is found in muscle cells. Notice how each is similar in that they are all polymers of simple sugars, but each differs from one another in how they are joined together. While many organisms are capable of digesting (hydrolyzing) the bonds that are found in glycogen and plant starch molecules, few are able to break those that link together the monosaccharides of cellulose.

Connections . . .

Generic Drugs and Mirror Image Isomers

Isomers that are mirror images of each other are called *mirror image isomers, stereo isomers, enantomers,* or *chiral compounds.* The difference among stereo isomers is demonstrated by shining polarized light through the two types of the same compound. The sugar, glucose, is a good example since there are two stereo isomers, D-glucose and L-glucose. When polarized light is shined on a test tube containing D-glucose, the light coming out the other side of the tube will be turned to the right—that is, *dextro* rotated. When polarized light is shown through a solution of L-glucose, the light coming out the other side will be rotated to the left—that is, *levo* rotated.

The result of this basic research has been utilized in the pharmaceutical and health care industries. When drugs are synthesized in large batches in the lab, many contain 50 percent "D" and 50 percent "L" enantomers. Various so-called "generic drugs" are less expensive because they are a mixture of the two enantomers and have not undergone the more thorough and expensive chemical processes involved in isolating only the "D" or "L" form of the drug. Generic drugs that are mixtures may be (1) just as effective as the pure form,

(2) slower acting, or (3) not effective if prescribed to an individual who is a "nonresponder" to that medication. It is for these last two reasons that physicians must be consulted when there is a choice between a generic and a nongeneric medication.

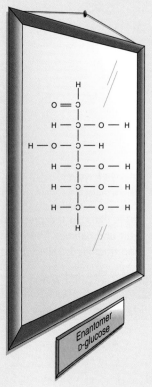

Enantomer
L-glucose

Enantomer
D-glucose

Box Figure 19.3

---

disaccharides include *lactose* (milk sugar) and *maltose* (malt sugar). All three disaccharides have similar properties, but maltose tastes only about one-third as sweet as sucrose. Lactose tastes only about one-sixth as sweet as sucrose. No matter which disaccharide sugar is consumed (sucrose, lactose, or maltose), it is converted into glucose and transported by the bloodstream for use by the body.

Some common examples of polysaccharides are **cellulose, starch,** and **glycogen.** Cellulose is an important polysaccharide used in constructing plant cell walls. Humans cannot digest this complex carbohydrate, so we are not able to use it as an energy source. On the other hand, animals known as ruminants (e.g., cows and sheep) and termites have microorganisms within their digestive tracts that do digest cellulose, making it an energy source for them. Plant cell walls add bulk or fiber to our diet, but no calories. Fiber is an important addition to the diet because it helps control weight and reduces the risks of colon cancer. Its large water-holding molecules also help control constipation and diarrhea. Starch is also a plant product digestible by most other organisms. Once broken down, the

monosaccharide can be used as an energy source or building materials. A close but structurally different polysaccharide is glycogen. This macromolecule is found in the muscle cells of many animals as a storage form of polysaccharide.

Many types of sugars can be used by cells as components in other, more complex molecules. Sugar molecules are a part of molecules such as DNA (deoxyribonucleic acid), RNA (ribonucleic acid), or ATP (adenosine triphosphate).

Proteins

Proteins play many important roles. Enzymes are catalysts that speed the rate of chemical reactions in living things. Proteins such as hemoglobin serve as carriers of other molecules such as oxygen. Others such as collagen provide shape and support, and several kinds of protein in muscle cells are responsible for movement. Proteins also act as chemical messengers and are called hormones. **Hormones** are chemical messengers secreted by endocrine glands to regulate other parts of the body. Certain other proteins called antibodies help defend the body against

Some Interesting Amino Acid Information

Humans require nine amino acids in their diet: threonine, tryptophan, methionine, lysine, phenylalanine, isoleucine, valine, histidine, and leucine. They are called *essential amino acids* because the body is not able to manufacture them. The body uses these essential amino acids in the synthesis of the proteins required for good health. For example, the sulfur-containing amino acid methionine is essential for the absorption and transportation of the elements selenium and potassium. It also prevents excess fat buildup in the liver, and it traps heavy metals, such as lead, cadmium, and mercury, bonding with them so that they can be excreted from the body. Essential amino acids are not readily available in most plant proteins and are most easily acquired through meat, fish, and dairy products.

▶ Lysine is found in foods such as yogurt, fish, chicken, brewer's yeast, cheese, wheat germ, pork, and other meats. It improves calcium uptake; when there is more lysine than the amino acid arginine, it helps control cold sores (herpes virus infection).

▶ Tryptophan is found in turkey, dairy products, eggs, fish, and nuts. It is required for the manufacture of hormones, such as serotonin, prolactin, and growth hormone; it has been shown to be of value in controlling depression, premenstrual syndrome (PMS), insomnia, migraine headaches, and immune function disorders. Other amino acids also play interesting roles in human metabolism:

▶ Glutamic acid is found in animal and vegetable proteins. It is used in monosodium glutamate (MSG), a flavor-enhancing salt. It is required for the synthesis of folic acid. In some people, folic acid can accumulate in brain cells and cause brain damage following a stroke. Folic acid is necessary during pregnancy to decrease the fetus's chance of developing spina bifida.

▶ Asparagine is found in asparagus. Most people have genes that cause asparagine to be converted to very smelly compounds (methyl thioacrylate and methyl 3-(methylthio) thiopropionate), which are excreted in their urine.

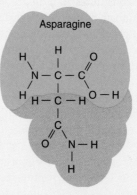

Box Figure 19.4

dangerous microbes and chemicals. An **antibody** is a globular protein made by the body in response to the presence of a foreign or harmful molecule called an antigen. Antigens are in many cases proteins, too. Chemically, proteins are polymers made up of monomers known as *amino acids*. An **amino acid** is a short carbon skeleton that contains an amino group (a nitrogen and two hydrogens) on one end of the skeleton and a carboxylic acid group at the other end (figure 19.19). In addition, the carbon skeleton may have one of *several* different *side chains* on it. These vary in their composition and are generally noted as the amino acid's *R-group*. About twenty common amino acids are important to cells, and each differs from one another in the nature of its R-group (see "A Closer Look: Some Interesting Amino Acid Information").

Any amino acid can form a bond with any other amino acid. They fit together in a specific way, with the amino group of one bonding to the acid group of the next. You can imagine that by using twenty different amino acids as building blocks,

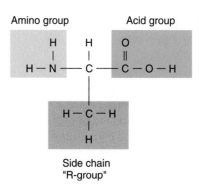

FIGURE 19.19   An amino acid is composed of a short carbon skeleton with three functional groups attached: an amino group, a carboxylic acid group (acid group), and an additional variable group (R-group). It is the variable group that determines which specific amino acid is constructed.

Protein Structure and Sickle-Cell Anemia

Any changes in the arrangement of amino acids within a protein can have far-reaching effects on its function. For example, normal hemoglobin found in red blood cells consists of two kinds of polypeptide chains called the alpha and beta chains. The beta chain is 146 amino acids long. If just one of these amino acids is replaced by a different one, the hemoglobin molecule may not function properly. A classic example of this results in a condition known as *sickle-cell anemia*. In this case, the sixth amino acid in the beta chain, which is normally glutamic acid, is replaced by valine. This minor change causes the hemoglobin to fold differently, and the red blood cells that contain this altered hemoglobin assume a sickle shape when the body is deprived of an adequate supply of oxygen (see figure 26.16).

you can construct millions of different combinations. There are over three million possible combinations for a molecule five amino acids long. Each of these combinations is termed a **polypeptide chain.** A specific polypeptide is composed of a specific sequence of amino acids bonded end to end.

There are four levels or degrees of protein structure. A listing of the amino acids in their proper order within a particular polypeptide constitutes its *primary* structure. The specific sequence of amino acids in a polypeptide is controlled by the genetic information of an organism. Genes are specific messages that tell the cell to link particular amino acids in a specific order; that is, they determine a polypeptide's primary structure.

Some sequences of amino acids in a polypeptide are likely to twist (as a coil or a pleated sheet), whereas other sequences remain straight. These twisted forms are referred to as the *secondary structure* of polypeptides. For example, at this secondary level, some proteins (e.g., hair) take the form of a helix, a shape like that of a coiled spring. Other polypeptides form hydrogen bonds that cause them to make several flat folds that resemble a pleated skirt. This is called a pleated sheet. The way a particular protein folds is important to its function. In bovine spongiform encephalitis (mad cow disease) and Creutzfeldt-Jakob diseases, protein structures are not formed correctly, resulting in characteristic nervous system symptoms.

It is also possible for a single polypeptide to contain one or more coils and pleated sheets along its length. As a result, these different portions of the molecule can interact to form an even more complex globular structure. The complex three-dimensional structure formed in this manner is the polypeptide's *tertiary* (third-degree) *structure.* A good example of a protein with tertiary structure is the myoglobin molecule. This is an oxygen-holding protein found in muscle cells.

Frequently, several different polypeptides, each with its own tertiary structure, twist around each other and chemically combine. The larger, globular structure formed by these interacting polypeptides is referred to as the protein's *quaternary* (fourth-degree) *structure* (figure 19.20). Quaternary structure is displayed by the protein molecules called immunoglobulins or *antibodies,* which are involved in fighting diseases such as mumps and chicken pox. The protein portion of the hemoglobin molecule (globin is globular in shape) also demonstrates quaternary structure.

Energy in the form of heat or light may break the bonds within protein molecules. When this occurs, the chemical and physical properties of the protein are changed and the protein is said to be **denatured.** (Keep in mind, a protein is a molecule, not a living thing, and therefore cannot be "killed.") A common example of this occurs when the gelatinous, clear portion of an egg is cooked and the protein changes to a white solid. Some medications are proteins and must be protected from denaturation so as not to lose their effectiveness. Insulin is an example. For protection, such medications may be stored in brown-colored bottles or kept under refrigeration.

The thousands of kinds of proteins can be placed into three categories. Some proteins are important for maintaining the shape of cells and organisms; they are usually referred to as **structural proteins.** The proteins that make up the cell membrane, muscle cells, tendons, and blood cells are examples of structural proteins. The protein collagen is found throughout the human body and gives tissues shape, support, and strength. The second category of proteins, **regulator proteins,** helps determine what activities will occur in the organism. These regulator proteins include **enzymes** and some hormones. An enzyme is a protein molecule that acts as a catalyst to speed the rate of a reaction. Enzymes can be used over and over again until they are worn out or broken. The production of these protein catalysts is under the direct control of an organism's genetic material (DNA). The instructions for the manufacture of all enzymes are found in the genes of the cell. Organisms make their own enzymes (see "A Closer Look: What Enzymes Are and How They Work"). Enzymes and hormones help control the chemical activities of cells and organisms. Some examples of enzymes are the digestive enzymes in the stomach. Two hormones that are regulator proteins are insulin and oxytocin. Insulin is produced by the pancreas and controls the amount of glucose in the blood. If insulin production is too low or if the molecule is improperly constructed, glucose molecules are not removed from the bloodstream at a fast enough rate. The excess sugar is then eliminated in the urine. Other symptoms of excess sugar in the blood include excessive thirst and even loss of consciousness. The disease caused by improperly functioning insulin is known as *diabetes.* Oxytocin, a second protein hormone, stimulates the contraction of the uterus during childbirth. It is also an example of an organic molecule

What Enzymes Are and How They Work

A *catalyst* is a chemical that speeds a chemical reaction without increasing the temperature and is not used up in the reaction. A protein molecule that acts as a catalyst to speed the rate of a reaction is called an *enzyme*. Enzymes can be used over and over again until they are worn out or broken. The production of these protein catalysts is under the direct control of an organism's genetic material (DNA). Enzymes operate by lowering the amount of energy needed to get a reaction going—the *activation energy*. When this energy is lowered, the nature of the bonds

in the substrate is changed so they are more easily broken. Whereas the cartoon (box figure 19.5) shows the breakdown of a single reactant into many end products (as in a hydrolysis reaction), the lowering of activation energy can also result in bonds being broken so that new bonds may be formed in the construction of a single, larger end product from several reactants (as in a synthesis reaction).

During an enzyme-controlled reaction, the enzyme and substrate come together to form a new molecule—the enzyme-substrate complex molecule. This

molecule exists for only a very short time. During that time, activation energy is lowered and bonds are changed. The result is the formation of a new molecule or molecules called the end products of the reaction. The number of jobs an enzyme can perform during a particular time period is incredibly large—ranging between a thousand (10^3) and ten thousand trillion (10^{16}) times faster per minute than uncatalyzed reactions! Without the enzyme, perhaps only fifty or one-hundred substrate molecules might be altered in the same time.

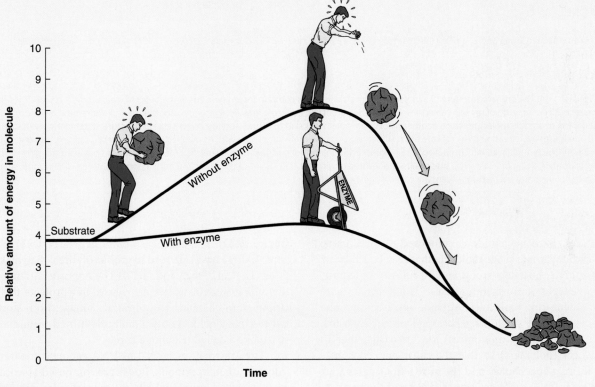

Box Figure 19.5

that has been produced artificially (e.g., pitocin) and is used by physicians to induce labor. The third category of proteins is **carrier proteins.** Proteins in this category pick up and deliver molecules at one place and transport them to another. For example, proteins regularly attach to cholesterol entering the system from the diet, forming molecules called lipoproteins, which are transported through the circulatory system. The cholesterol is released at a distance from the digestive tract, and the proteins return to pick up more dietary cholesterol entering the system.

Nucleic Acids

Nucleic acids are complex polymeric molecules that store and transfer information within a cell. This section is an overview of this important class of organic compounds. More detailed information is presented in chapter 26. There are two types of nucleic acids, DNA and RNA. DNA serves as genetic material determining which proteins will be manufactured, while RNA plays a vital role in the process of protein manufacture. All nucleic acids are constructed of fundamental monomers known

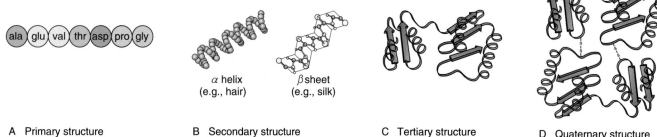

A Primary structure

α helix
(e.g., hair)

β sheet
(e.g., silk)

B Secondary structure

C Tertiary structure
(e.g., myoglobin)

D Quaternary structure
(e.g., antibody)

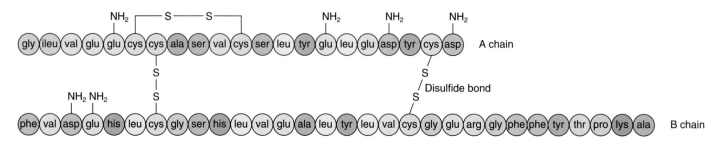

A chain

Disulfide bond

B chain

E Structure of insulin

FIGURE 19.20 (*A*) The primary structure of a molecule is simply a list of its component amino acids in the order in which they occur. (*B*) This figure illustrates the secondary structure of protein molecules, or how one part of the molecule initially attached to another part of the same molecule. (*C*) If already folded parts of a single molecule attach at other places, the molecule is said to display tertiary (third-degree) structure. (*D*) Quaternary (fourth-degree) structure is displayed by molecules that are the result of two separate molecules (each with its own tertiary structure) combining into one large macromolecule. (*E*) The protein insulin is composed of two polypeptide chains bonded together at specific points by reactions between the side chains of particular sulfur-containing amino acids. The side chains of one interact with the side chains of the other and form a particular three-dimensional shape. The bonds that form between the polypeptide chains are called disulfide bonds.

as **nucleotides.** Each nucleotide is composed of three parts: (1) a five-carbon simple sugar molecule that may be ribose or deoxyribose, (2) a phosphate group, and (3) a nitrogenous base. There are five types of nitrogenous bases. Two of the bases are the larger, double-ring molecules, *Adenine* and *Guanine.* The smaller bases are the single-ring bases, *Thymine, Cytosine,* and *Uracil.* Nucleotides (monomers) are linked together in long sequences (polymers) so that the sugar and phosphate sequence forms a "backbone" and the nitrogenous bases stick out to the side. DNA has deoxyribose sugar and the bases A, T, G, and C, while RNA has ribose sugar and the bases A, U, G, and C (figure 19.21).

DNA (*deoxyribo nucleic acid*) is composed of two strands to form a ladderlike structure thousands of nucleotide bases long. The two strands are attached between their bases according to *the base pair rule;* that is, *Adenine* from one strand always pairs with *Thymine* from the other. *Guanine* always pairs with *Cytosine.*

A T (or A U) and G C

One strand of DNA is called the *coding strand* because it has a meaningful genetic message written using the nitrogenous bases as letters (e.g., the base sequence CATTAGACT)

(figure 19.22). This is the basis of the genetic code for all organisms. If these bases are read in groups of three, they make sense to us (i.e., "cat," "tag," and "act"). The opposite strand is called *noncoding* since it makes no "sense" but protects the coding strand from chemical and physical damage. Both strands are twisted into a helix, that is, a molecule turned around a tubular space like a coiled telephone cord.

The information carried by DNA can be compared to the information in a textbook. Books are composed of words (constructed from individual letters) in particular combinations, organized into chapters. In the same way, DNA is composed of tens of thousands of nucleotides in specific sequences (words) organized into genes (a chapter). Each gene carries the information for producing a protein, just as each chapter carries the information relating to one idea. The order of nucleotides in a gene is directly related to the order of amino acids in a protein. Just as chapters in a book are identified by beginning and ending statements, different genes along a DNA strand have beginning and ending signals. They tell when to start and when to stop reading a particular gene. Human body cells contain forty-six strands (books) of helical DNA, each containing thousands of genes (chapters). These strands are called **chromosomes** when they become supercoiled in preparation for cellular reproduction. Before cell reproduction, the DNA makes copies of the coding and noncoding strands, ensuring that the

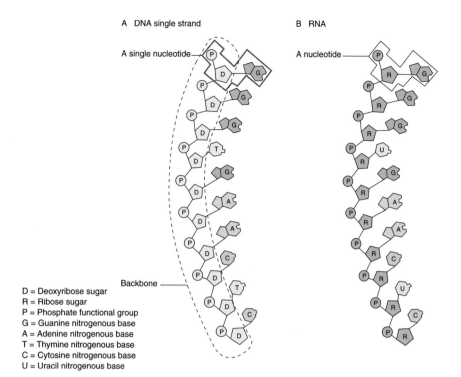

FIGURE 19.21 (A) A single strand of DNA is a polymer composed of nucleotides. Each nucleotide (framed at the top of the molecule) consists of deoxyribose sugar, phosphate, and one of four nitrogenous bases: A, T, G, or C. Notice the backbone of blue-highlighted sugar and phosphate. (B) RNA is also a polymer, but each nucleotide (framed at the top of this RNA molecule) is composed of ribose sugar, phosphate, and one of four nitrogenous bases: A, U, G, or C. The backbone of ribose and phosphate is highlighted in green.

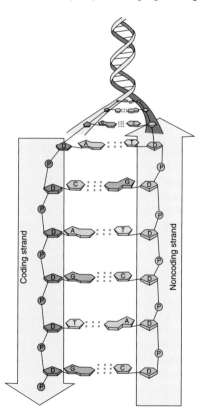

FIGURE 19.22 The genetic material is really double-stranded DNA molecules comprised of sequences of nucleotides that spell out an organism's genetic code. The coding strand of the double molecule is the side that can be translated by the cell into meaningful information. The genetic code has the information for telling the cell what proteins to make, which in turn become the major structural and functional components of the cell. The noncoding strand is unable to code for such proteins.

offspring or *daughter cells* will each receive a full complement of the genes required for their survival. Each chromosome is comprised of a sequence of genes. A **gene** is a segment of DNA that is able to (1) replicate by directing the manufacture of copies of itself; (2) mutate, or chemically change, and transmit these changes to future generations; (3) store information that determines the characteristics of cells and organisms; and (4) use this information to direct the synthesis of structural and regulatory proteins.

Messenger RNA (mRNA) is a single-strand copy of a portion of the coding strand of DNA for a specific gene. When mRNA is formed on the surface of the DNA, the base pair rule applies. However, since **RNA (*ribonucleic acid*)** does not contain thymine, it uses a U-A pairing instead of the T-A pairing between nucleotides. After mRNA is formed and peeled off, it

CONCEPTS APPLIED

Polymers Around the House

The concepts of monomers, polymers, and how they interrelate are extremely important and fundamental to understanding the organic compounds important to life. You may have worked with these concepts as early as preschool when you attached colored paper "links" (monomers) together to form a decorative chain (polymer). Return to that experience and create polymeric molecules that better represent proteins, carbohydrates, and nucleic acids as described in this chapter.

associates with a cellular structure called the *ribosome* where the genetic message can be translated into a protein molecule. Ribosomes contain another type of RNA, **ribosomal RNA (rRNA)**. rRNA is also an RNA copy of DNA, but after being formed, it becomes twisted and covered in protein to form a ribosome. The third form of RNA, **transfer RNA (tRNA),** is also made from copies of different segments of DNA, but when peeled off the surface, each takes the form of a cloverleaf. tRNA molecules are responsible for transferring or carrying specific amino acids to the ribosome where all three forms of RNA come together and cooperate in the manufacture of protein molecules.

Lipids

We generally call molecules in this group the **fats.** They are not polymers, as are the previously discussed carbohydrates, proteins, and nucleic acids. However, there are three different types of **lipids:** *true fats* (pork chop fat or olive oil), *phospholipids* (the primary component of cell membranes), and *steroids* (some hormones). In general, lipids are large, nonpolar (do not have a positive end and a negative end), organic molecules that do not easily dissolve in polar solvents such as water. For example, nonpolar vegetable oil molecules do not dissolve in polar water molecules. Lipids are soluble in nonpolar substances such as ether or acetone. Just like carbohydrates, the lipids are composed of carbon, hydrogen, and oxygen. Lipids generally have very small amounts of oxygen in comparison to the amounts of carbon and hydrogen. Simple lipids cannot be digested into smaller, similar subunits. Complex lipids such as true fats and phospholipids can be digested into smaller, similar units.

True (Neutral) Fats

True (neutral) fats are important, complex organic molecules that are used to provide, among other things, energy. The building blocks of a fat are a glycerol molecule and fatty acids.

Recall that **glycerol** is a carbon skeleton that has three alcohol groups attached to it. Its chemical formula is $C_3H_5(OH)_3$. At room temperature, glycerol looks like clear, lightweight oil. It is used under the name glycerin as an additive to many cosmetics to make them smooth and easy to spread.

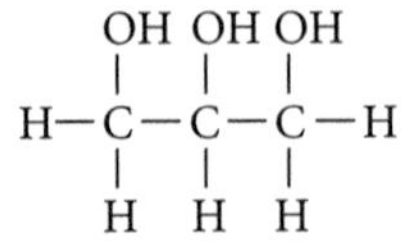

$$
\begin{array}{ccccc}
 & \text{OH} & \text{OH} & \text{OH} & \\
 & | & | & | & \\
\text{H}- & \text{C} & -\text{C}- & \text{C} & -\text{H} \\
 & | & | & | & \\
 & \text{H} & \text{H} & \text{H} &
\end{array}
$$

Glycerol

The **fatty acids** found in fats are a long-chain carbon skeleton that has a carboxylic acid functional group. If the carbon skeleton has as much hydrogen bonded to it as possible, we call it *saturated*. The saturated fatty acid in figure 19.23A is stearic acid, a component of solid meat fats such as mutton tallow. Notice that at every point in this structure, the carbon has as much hydrogen as it can hold. Saturated fats are generally found in animal tissues and tend to be solids at room temperatures. Some examples of saturated fats are butter, whale blubber, suet, lard, and fats associated with such meats as steak or pork chops.

If the carbons in a fatty acid are double-bonded to each other at one or more points, the fatty acid is said to be *unsaturated*. Unsaturated fats are frequently plant fats or oils, and they are usually liquids at room temperature. Peanut, corn, and olive oil are examples of unsaturated fats that are mixtures of different true fats. When glycerol and three fatty acids are combined, a fat is formed. This reaction is almost exactly the same as the reaction that causes simple sugars to bond together.

Fats are important molecules for storing energy. There is more than twice as much energy in a gram of fat as in a gram of sugar, 9 Calories versus 4 Calories. This is important to an organism because fats can be stored in a relatively small space

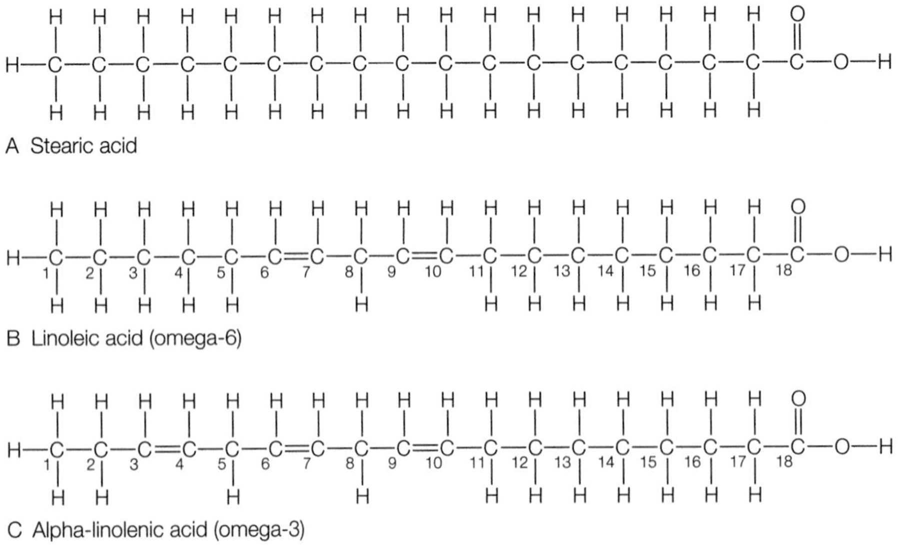

A Stearic acid

B Linoleic acid (omega-6)

C Alpha-linolenic acid (omega-3)

FIGURE 19.23 Structure of saturated and unsaturated fatty acids. (*A*) Stearic acid is an example of a saturated fatty acid. (*B*) Linoleic acid is an example of an unsaturated fatty acid. It is technically an omega-6 fatty acid because the first double bond occurs at carbon number six. (*C*) An omega-3 fatty acid, linolenic acid. Both linoleic and linolenic acids are essential fatty acids for humans.

A Closer Look

Omega Fatty Acids and Your Diet

The occurrence of a double bond in a fatty acid is indicated by the Greek letter ω (omega) followed by a number indicating the location of the first double bond in the molecule. Oleic acid, one of the fatty acids found in olive oil, is comprised of 18 carbons with a single (1) double bond between carbons 9 and 10. Therefore, it is chemically designated C18:1ω9 and is a monounsaturated fatty acid. This fatty acid is commonly referred to as an omega-9 fatty acid. The unsaturated fatty acid in figure 19.23B is linoleic acid, a component of sunflower and safflower oils. Notice that there are two double bonds between

the carbons and fewer hydrogens than in the saturated fatty acid. Linoleic acid is chemically a polyunsaturated fatty acid with two double (2) bonds and is designated C18:2ω6, an omega-6 fatty acid. This indicates that the first double bond of this 18-carbon molecule is between carbons 6 and 7. Since the human body cannot make this fatty acid, it is called an *essential fatty acid* and must be taken in as a part of the diet. The other essential fatty acid, linolenic acid, is C18:3ω3 and has three (3) double bonds. This fatty acid is commonly referred to as an omega-3 fatty acid. One key function of these essential fatty

Sources of Omega-3 Fatty Acids	Sources of Omega-6 Fatty Acids
Certain fish oil (salmon, sardines, herring)	Corn oil
	Peanut oil
Flaxseed oil	Cottonseed oil
Soybeans	Soybean oil
Soybean oil	Sesame oil
Walnuts	Safflower oil
Walnut oil	Sunflower oil

acids is the synthesis of prostaglandin hormones that are necessary in controlling cell growth and specialization.

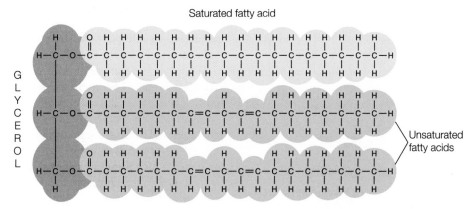

FIGURE 19.24 The arrangement of the three fatty acids attached to a glycerol molecule is typical of the formation of a fat. The structural formula of the fat appears to be very cluttered until you dissect the fatty acids from the glycerol; then it becomes much more manageable. This example of a triglyceride contains a glycerol molecule, two unsaturated fatty acids (linoleic acid), and a third saturated fatty acid (stearic acid).

and still yield a high amount of energy. Fats in animals also provide protection from heat loss. The thick layer of blubber in whales, walruses, and seals prevents the loss of internal body heat to the cold, watery environment in which they live. This same layer of fat, together with the fat deposits around some internal organs—such as the kidneys and heart—serves as a cushion that protects these organs from physical damage. If a fat is formed from a glycerol molecule and three attached fatty acids, it is called a **triglyceride;** if two, a *diglyceride;* and if one, a *monoglyceride* (figure 19.24). Triglycerides account for about 95 percent of the fat stored in human tissue.

Phospholipids

Phospholipids are a class of complex water-insoluble organic molecules that resemble fats but contain a phosphate group (PO_4) in their structure (figure 19.25). One of the reasons phospholipids are important is that they are a major component of membranes in cells. Without these lipids in our membranes, the cell contents

would not be separated from the exterior environment. Some of the phospholipids are better known as the *lecithins*. Lecithins are found in cell membranes and also help in the emulsification of fats. They help separate large portions of fat into smaller units. This allows the fat to mix with other materials. Lecithins are added to many types of food for this purpose (chocolate bars, for example). Some people take lecithin as nutritional supplements because they believe it leads to healthier hair and better reasoning ability; but once inside your intestines, lecithins are destroyed by enzymes, just like any other phospholipid.

Steroids

Steroids, another group of lipid molecules, are characterized by an arrangement of interlocking rings of carbon. They often serve as hormones that aid in regulating body processes. We have already mentioned one familiar steroid molecule: **cholesterol.** Although serum cholesterol (the kind found in your blood associated with lipoproteins) has been implicated

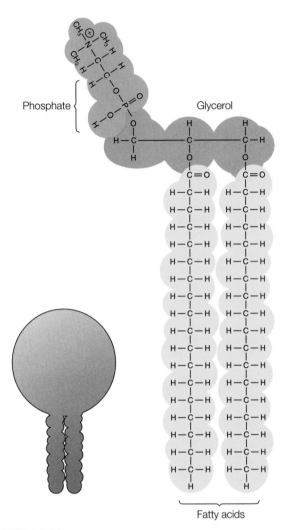

Phosphate

Glycerol

Fatty acids

FIGURE 19.25 This molecule is similar to a fat but has a phosphate group in its structure. The phosphate group and two fatty acids are bonded to the glycerol by a dehydration synthesis reaction. Molecules like these are also known as lecithin. The diagram of phospholipid molecules is shown as a balloon with two strings. The balloon portion is the glycerol and phosphate group, while the strings are the fatty acid segments of the molecule.

in many cases of atherosclerosis, this steroid is made by your body for use as a component of cell membranes. It is also used by your body to make bile acids. These products of your liver are channeled into your intestine to emulsify fats. Cholesterol is also necessary for the manufacture of vitamin D. Cholesterol molecules in the skin react with ultraviolet light to produce vitamin D, which assists in the proper development of bones and teeth. Figure 19.26 illustrates some of the steroid compounds that are typically manufactured by organisms. A large number of steroid molecules are sex hormones. Some of them regulate reproductive processes such as egg and sperm production; others regulate such things as salt concentration in the blood.

As you can see, cholesterol plays both positive and negative roles in metabolism. Regulating the amount of cholesterol in your body to prevent its negative effects can be difficult since your body makes it and it is also acquired in your diet. Recall that saturated fats cause your body to produce more cholesterol, increasing your risk of diseases such as atherosclerosis. By watching your diet, it is possible to reduce the amount of cholesterol in your blood serum by about 20 percent. Therefore, it is best to eat foods that are low in cholesterol and saturated fats. Foods that are touted as having low or no cholesterol often have a high level of saturated fats, so they should also be avoided to control serum cholesterol levels.

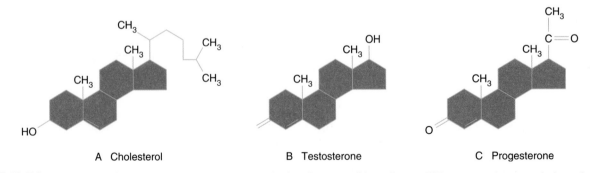

A Cholesterol

B Testosterone

C Progesterone

FIGURE 19.26 (A) Cholesterol is produced by the human body and is found in your cells' membranes. (B) Testosterone increases during puberty, causing the male sex organs to mature. (C) Progesterone is a female sex hormone produced by the ovaries and placenta. Notice the slight structural differences among these molecules.

Fat and Your Diet

When triglycerides are eaten in fat-containing foods, digestive enzymes hydrolyze them into glycerol and fatty acids. These molecules are absorbed by the intestinal tract and coated with protein to form lipoprotein, as shown in box figure 19.6. The combination allows the fat to better dissolve in the blood so that it can move throughout the body in the circulatory system.

Five types of lipoproteins found in the body are:

(1) chylomicrons;
(2) very-low-density lipoproteins (VLDL);
(3) low-density lipoproteins (LDL);
(4) high-density lipoproteins (HDL); and
(5) lipoprotein a—Lp(a).

Chylomicrons are very large particles formed in the intestine and are between 80 percent and 95 percent triglycerides. As the chylomicrons circulate through the body, cells remove the triglycerides in order to make sex hormones, store energy, and build new cell parts. When most of the triglycerides have been removed, the remaining portions of the chylomicrons are harmlessly destroyed.

The VLDLs and LDLs are formed in the liver. VLDLs contain all types of lipid, protein, and 10 percent to 15 percent cholesterol, while the LDLs are about 50 percent cholesterol. As with the chylomicrons, the body uses these molecules for the fats they contain. However, in some people, high levels of LDLs and Lp(a) in the blood are associated with the diseases atherosclerosis, stroke, and heart attack. It appears that saturated fat disrupts the clearance of LDLs from the bloodstream. Thus, while in the blood, LDLs may stick to the insides of the vessels, forming deposits that restrict blood flow and contribute to high blood

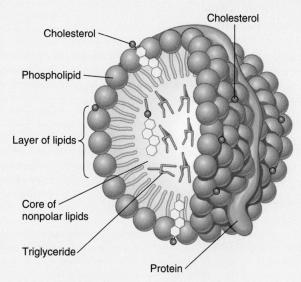

Box Figure 19.6 Diagram of a lipoprotein.

pressure, strokes, and heart attacks. Even though they are 30 percent cholesterol, a high level of HDLs (made in the intestine) in comparison to LDLs and Lp(a) is associated with a lower risk of atherosclerosis. One way to reduce the risk of this disease is to lower your intake of LDLs and Lp(a). This can be done by reducing your consumption of saturated fats. An easy way to remember the association between LDLs and HDLs is "L = lethal" and "H = healthy" or "Low = bad" and "High = good." The federal government's new cholesterol guidelines recommend that all adults get a full lipoprotein profile (total cholesterol, HDL, LDL, and triglycerides) once every five years. They also recommend a sliding scale for desirable LDL levels; however, recent studies suggest your LDL level should be as low as possible.

A way to reduce LDL levels is through a process called *LDL apheresis*. This is similar to kidney dialysis but is designed to remove only particles with LDL, VLDL, and Lp(a). As the blood is removed and passed through the apheresis machine, it is separated into cells and plasma. The plasma is then passed over a column containing a material that grabs onto the LDL particles and removes them, leaving the HDL cholesterol and other important blood components. The plasma is then returned to the patient.

Experimental methods that result in raising the relative amounts of protective HDL levels include

1. use of drugs that stimulate HDL production;
2. injection of an altered form of HDL that is better at helping to protect against heart disease by removing plaque from the bloodstream; and
3. insertion of an omega-3 fatty acid–producing gene into cells that lack the ability to produce such fatty acids.

Normal HDL Values	Normal LDL Values	Normal VLDL Values
Men: 40–70 mg/dL	Men: 91–100 mg/dL	Men: 0–40 mg/dL
Women: 40–85 mg/dL	Women: 69–100 mg/dL	Women: 0–40 mg/dL
Children: 30–65 mg/dL		
Minimum desirable: 40 mg/dL	Heart attack risk level/acceptable LDL levels:	Desirable: below 40 mg/dL
	Low risk: no higher than 70 mg/dL	
	Moderate risk: between 70 and 160 mg/dL	
	High risk: at or above 160 mg/dL	
For Total Cholesterol Levels		
Desirable: Below 200 mg/dL	Borderline: 200–240 mg/dL	Undesirable: Above 240 mg/dL

SYNTHETIC POLYMERS

Natural polymers are huge, chainlike molecules made of hundreds or thousands of smaller, repeating molecular units called monomers. Cellulose and starch are examples of natural polymers made of glucose monomers. *Synthetic polymers* are manufactured from a wide variety of substances, and you are familiar with these polymers as synthetic fibers such as nylon and the inexpensive light plastic used for wrappings and containers (figure 19.27).

The first synthetic polymer was a modification of the naturally existing cellulose polymer. Cellulose was chemically modified in 1862 to produce celluloid, the first *plastic*. The term *plastic* means that celluloid could be molded to any desired shape. Celluloid was produced by first reacting cotton with a mixture of nitric and sulfuric acids. The cellulose in the cell walls of the cotton plant reacts with the acids to produce the ester cellulose nitrate. This ester is an explosive compound known as "guncotton," or smokeless gunpowder. When made with ethanol and camphor, the product is less explosive and can be formed and molded into useful articles. This first plastic, celluloid, was used to make dentures, combs, eyeglass frames, and photographic film. Before the discovery of celluloid, many of these articles, including dentures, were made from wood. Today, only Ping-Pong balls are made from cellulose nitrate.

Cotton reacted with acetic acid and sulfuric acid produces a cellulose acetate ester. This polymer, through a series of chemical reactions, produces rayon filaments when forced through small holes. The filaments are twisted together to

Name	Chemical unit	Uses
Polyethylene	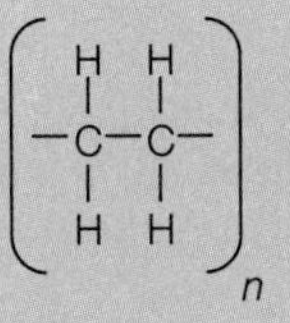	Squeeze bottles, containers, laundry and trash bags, packaging
Polypropylene	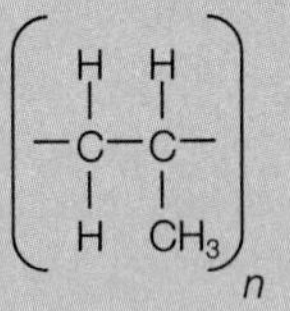	Indoor-outdoor carpet, pipe valves, bottles
Polyvinyl chloride (PVC)	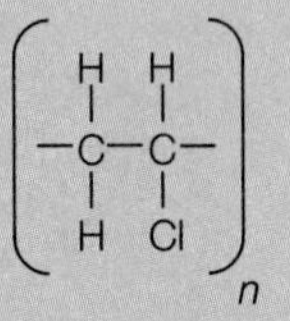	Plumbing pipes, synthetic leather, plastic tablecloths, phonograph records, vinyl tile
Polyvinylidene chloride (Saran)	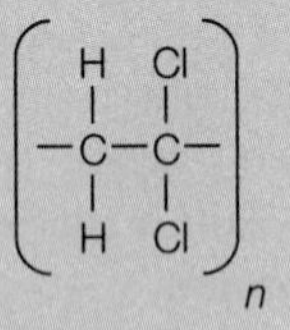	Flexible food wrap
Polystyrene (Styrofoam)	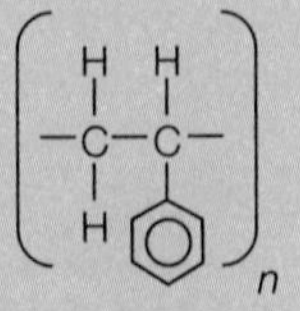	Coolers, cups, insulating foam, shock-resistant packing material, simulated wood furniture
Polytetrafluoroethylene (Teflon)	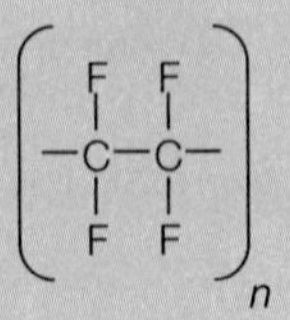	Gears, bearings, coating for nonstick surface of cooking utensils
Polyvinyl acetate	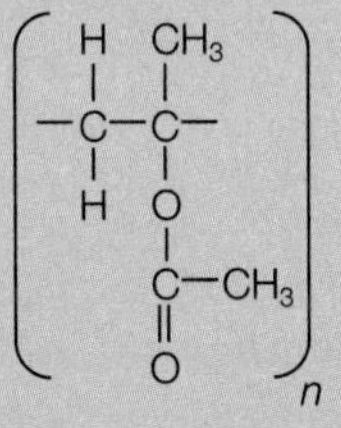	Mixed with vinyl chloride to make vinylite; used as an adhesive and resin in paint
Styrene-butadiene rubber	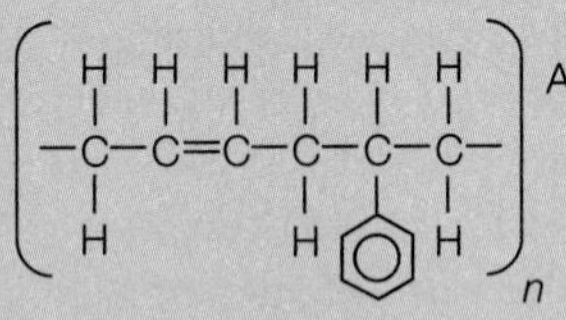	Automobile tires
Polychloroprene (Neoprene)	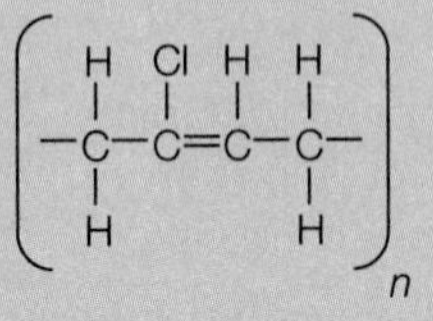	Shoe soles, heels
Polymethyl methacrylate (Plexiglas, Lucite)	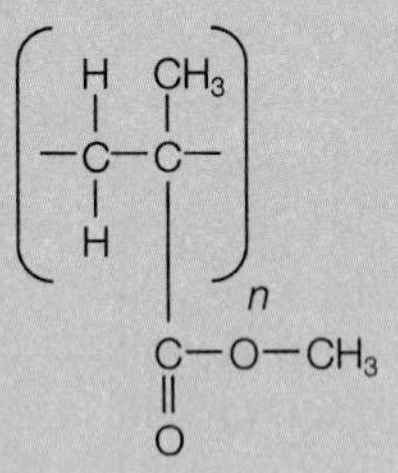	Moldings, transparent surfaces on furniture, lenses, jewelry, transparent plastic "glass"
Polycarbonate (Lexan)	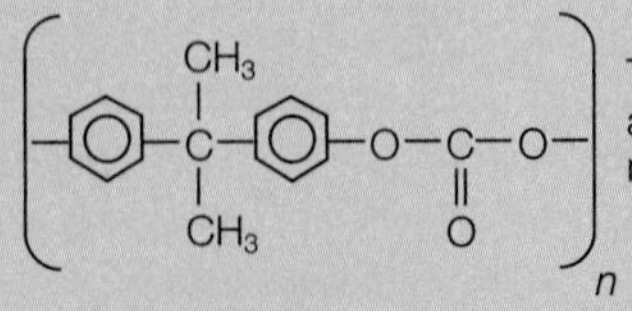	Tough, molded articles such as motorcycle helmets
Polyacrylonitrile (Orlon, Acrilan, Creslan)	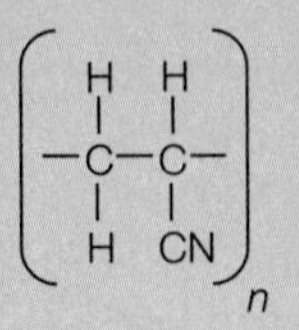	Textile fibers

FIGURE 19.27 Synthetic polymers, the polymer unit, and some uses of each polymer.

form viscose rayon thread. When forced through a thin slit, a sheet is formed rather than filaments, and the transparent sheet is called *cellophane*. Both rayon and cellophane, as celluloid, are manufactured by modifying the natural polymer of cellulose.

The first truly synthetic polymer was produced in the early 1900s by reacting two chemicals with relatively small molecules rather than modifying a natural polymer. Phenol, an aromatic hydrocarbon, was reacted with formaldehyde, the simplest aldehyde, to produce the polymer named *Bakelite*. Bakelite is a *thermosetting* material that forms cross-links between the polymer chains. Once the links are formed during production, the plastic becomes permanently hardened and cannot be softened or made to flow. Some plastics are *thermoplastic* polymers and soften during heating and harden during cooling because they do not have cross-links.

Polyethylene is a familiar thermoplastic polymer produced by a polymerization reaction of ethylene, which is derived from petroleum. This synthetic was invented just before World War II and was used as an electrical insulating material during the war. Today, there are many variations of polyethylene that are produced by different reaction conditions or by substitution of one or more hydrogen atoms in the ethylene molecule. Polyethylene terephthalate (PET) is used extensively for rigid containers, particularly beverage bottles for carbonated drinks and medicine containers. These containers have the recycle code of number 1. When soft polyethylene near the melting point is rolled in alternating perpendicular directions or expanded and compressed as it is cooled, the polyethylene molecules become ordered in a way that improves the rigidity and tensile strength. One form, high-density polyethylene (HDPE), is used in milk and water jugs, as liners in screw-on jar tops and bottle caps, and as a material for toys. HDPE bottles have the recycling code of number 2 and can be reused to make plastic lumber, motor oil containers, playground equipment, and sheeting. The other form, low-density polyethylene (LDPE), has a recycle code of number 4. Recycled LDPE is used for vegetable, dry cleaning, and grocery bags, and plastic squeeze bottles with a recycle code of number 1.

The properties of polyethylene are also changed by replacing one of the hydrogen atoms in a molecule of ethylene. If the hydrogen is replaced by a chlorine atom, the compound is called vinyl chloride, and the polymer formed from vinyl chloride is

polyvinyl chloride (PVC). Polyvinyl chloride is used to make plastic water pipes, synthetic leather, and other vinyl products. It differs from the waxy plastic of polyethylene because of the chlorine atom that replaces hydrogen on each monomer.

Replacement of a hydrogen atom with a benzene ring makes a monomer called *styrene*. Styrene is

and polymerization of styrene produces *polystyrene*. Polystyrene is puffed full of air bubbles to produce the familiar Styrofoam coolers, cups, and insulating materials.

If all hydrogens of an ethylene molecule are replaced with atoms of fluorine, the product is polytetrafluoroethylene, a tough plastic that resists high temperatures and acts more like a metal than a plastic. Since it has a low friction, it is used for bearings, gears, and as a nonstick coating on frying pans. You probably know of this plastic by its trade name of *Teflon*.

There are many different polymers in addition to PVC, Styrofoam, and Teflon, and the monomers of some of these are shown in figure 19.27. There are also polymers of isoprene, or synthetic rubber, in wide use. Fibers and fabrics may be polyamides (such as nylon), polyesters (such as Dacron), or polyacrylonitriles (Orlon, Acrilan, Creslan), which have a CN in place of a hydrogen atom on an ethylene molecule and are called acrylic materials. All of these synthetic polymers have added much to practically every part of your life. It would be impossible to list all of their uses here; however, they present problems since (1) they are manufactured from raw materials obtained from coal and a dwindling petroleum supply and (2) they do not readily decompose when dumped into rivers, oceans, or other parts of the environment. However, research in the polymer sciences is beginning to reflect new understandings learned from research on biological tissues. This could lead to whole new molecular designs for synthetic polymers that will be more compatible with ecosystems.

SUMMARY

Organic chemistry is the study of compounds that have carbon as the principal element. Such compounds are called *organic compounds,* and all the rest are *inorganic compounds*. There are millions of organic compounds because a carbon atom can link with other carbon atoms as well as atoms of other elements.

A *hydrocarbon* is an organic compound consisting of hydrogen and carbon atoms. The simplest hydrocarbon is one carbon atom and four

hydrogen atoms, or CH_4. All hydrocarbons larger than CH_4 have one or more carbon atoms bonded to another carbon atom. The bond can be single, double, or triple, and may be found as chains or rings of carbon. The *alkanes* are hydrocarbons with single carbon-to-carbon bonds, the *alkenes* have a double carbon-to-carbon bond, and the *alkynes* have a triple carbon-to-carbon bond. The alkanes, alkenes, and alkynes can have straight- or branched-chain molecules. When the number of carbon atoms is greater than three, different arrangements can occur for a particular number of carbon atoms. The different arrangements with the same molecular formula are called isomers. *Isomers* have different physical properties, so each isomer is given its own name.

The alkanes have all the hydrogen atoms possible, so they are *saturated* hydrocarbons. The alkenes and the alkynes can add more hydrogens to the molecule, so they are *unsaturated* hydrocarbons. Unsaturated hydrocarbons are more chemically reactive than saturated molecules.

Hydrocarbons that occur in a ring or cycle structure are cyclohydrocarbons. A six-carbon cyclohydrocarbon with three double bonds has different properties than the other cyclohydrocarbons because the double bonds are not localized. This six-carbon molecule is *benzene,* the basic unit of the *aromatic hydrocarbons.*

Petroleum is a mixture of alkanes, cycloalkanes, and a few aromatic hydrocarbons that formed from the slow decomposition of buried marine plankton and algae. Petroleum from the ground, or *crude oil,* is distilled into petroleum products of *natural gas, LPG, petroleum ether, gasoline, kerosene, diesel fuel,* and *motor oils.* Each group contains a range of hydrocarbons and is processed according to use.

In addition to oxidation, hydrocarbons react by *substitution, addition,* and *polymerization* reactions. Reactions take place at sites of multiple bonds or lone pairs of electrons on the *functional groups.* The functional group determines the chemical properties of organic compounds. Changes in functional groups result in the *hydrocarbon derivatives* of *alcohols, ethers, aldehydes, ketones, organic acids, esters,* and *amines.*

Living organisms have an incredible number of highly organized chemical reactions that are catalyzed by *enzymes,* using food and energy to grow and reproduce. These biochemical processes involve building large *macromolecules* such as proteins, carbohydrates, and lipids.

Carbohydrates are a class of organic molecules composed of CHO. The *monosaccharides* are simple sugars such as *glucose* and *fructose.* Glucose is *blood sugar,* a source of energy. The disaccharides are *sucrose* (table sugar), *lactose* (milk sugar), and *maltose* (malt sugar). The polysaccharides are polymers or glucose in straight or branched chains used as a near-term source of stored energy. Plants store the energy in the form of *starch,* and animals store it in the form of *glycogen. Cellulose* is a polymer similar to starch that humans cannot digest.

Proteins are macromolecular polymers of *amino acids.* There are twenty amino acids that are used in various polymer combinations to build structural, carrier, and functional proteins.

Nucleic acids are polymers composed of nucleotide units. Two forms are recognized, DNA and RNA. In most organisms, DNA serves as the genetic material, while RNA plays crucial roles in the synthesis of proteins.

Lipids are a fourth important class of biochemicals. *True or neutral fats* and oils belong in the category known as *lipids* and are esters formed from three fatty acids and glycerol into a *triglyceride.* True fats are usually solid triglycerides associated with animals, and *oils* are liquid triglycerides associated with plant life, but both represent a high-energy storage material. The other two subgroups are the *phospholipids* used in cell membranes, and *steroids,* which primarily serve as hormones.

Polymers occur naturally in plants and animals, and *synthetic polymers* are made today from variations of the ethylene-derived monomers. Among the more widely used synthetic polymers derived from ethylene are polyethylene, polyvinyl chloride, polystyrene, and Teflon. Problems with the synthetic polymers include that (1) they are manufactured from fossil fuels that are also used as the primary energy supply and (2) they do not readily decompose and tend to accumulate in the environment.

KEY TERMS

alkanes (p. **438**)
alkenes (p. **440**)
alkyne (p. **440**)
amino acid (p. **451**)
antibody (p. **451**)
carbohydrates (p. **448**)
carrier proteins (p. **453**)
cellulose (p. **450**)
cholesterol (p. **457**)
chromosomes (p. **454**)
complex carbohydrate (p. **448**)
denatured (p. **452**)
deoxyribonucleic acid (DNA)
  (p. **454**)

enzyme (p. **452**)
fats (p. **456**)
fatty acids (p. **456**)
functional group (p. **443**)
gene (p. **455**)
glycerol (p. **456**)
glycogen (p. **450**)
hormone (p. **450**)
hydrocarbon (p. **438**)
hydrocarbon derivatives
  (p. **443**)
inorganic chemistry (p. **436**)
isomers (p. **439**)
lipid (p. **456**)

macromolecule (p. **448**)
messenger RNA (mRNA)
  (p. **455**)
monomer (p. **443**)
nucleic acid (p. **453**)
nucleotide (p. **454**)
oil field (p. **441**)
organic acids (p. **446**)
organic chemistry (p. **436**)
petroleum (p. **441**)
phospholipid (p. **457**)
polymer (p. **443**)
polypeptide chain (p. **452**)
proteins (p. **450**)

regulator proteins (p. **452**)
ribonucleic acid (RNA)
  (p. **455**)
ribosomal RNA (rRNA)
  (p. **456**)
saturated (p. **440**)
starch (p. **450**)
steroids (p. **457**)
structural proteins (p. **452**)
transfer RNA (tRNA)
  (p. **456**)
triglyceride (p. **457**)
true (neutral) fats (p. **456**)
unsaturated (p. **440**)

APPLYING THE CONCEPTS

1. An organic compound is one that
   a. contains carbon and was formed only by a living organism.
   b. is a natural compound that has not been synthesized.
   c. contains carbon, no matter if it was formed by a living thing or not.
   d. was formed by a plant.

2. There are millions of organic compounds but only thousands of inorganic compounds because
   a. organic compounds were formed by living things.
   b. there is more carbon on Earth's surface than any other element.
   c. atoms of elements other than carbon never combine with themselves.
   d. carbon atoms can combine with up to four other atoms, including other carbon atoms.

3. An alkane with 4 carbon atoms would have how many hydrogen atoms in each molecule?
   a. 4
   b. 8
   c. 10
   d. 16

4. Isomers are compounds with the same
   a. molecular formula with different structures.
   b. molecular formula with different atomic masses.
   c. atoms but different molecular formulas.
   d. structures but different formulas.

5. The hydrocarbons with a double covalent carbon-carbon bond are called
   a. alkanes.
   b. alkenes.
   c. alkynes.
   d. None of the above is correct.

6. Petroleum is believed to have formed mostly from the anaerobic decomposition of buried
   a. dinosaurs.
   b. fish.
   c. pine trees.
   d. plankton and algae.

7. Ethylene molecules can add to each other in a reaction to form a long chain called a
   a. monomer.
   b. dimer.
   c. trimer.
   d. polymer.

8. Chemical reactions usually take place on an organic compound at the site of a
   a. double bond.
   b. lone pair of electrons.
   c. functional group.
   d. Any of the above is correct.

9. The R in ROH represents
   a. a functional group.
   b. a hydrocarbon group with a name ending in "-yl."
   c. an atom of an inorganic element.
   d. a polyatomic ion that does not contain carbon.

10. A protein is a polymer formed from the linking of many
   a. glucose units.
   b. DNA molecules.
   c. amino acid molecules.
   d. monosaccharides.

11. Which of the following is *not* converted to blood sugar by the human body?
   a. lactose
   b. dextrose
   c. cellulose
   d. glycogen

12. Many synthetic polymers become a problem in the environment because they
   a. decompose to nutrients, which accelerates plant growth.
   b. do not readily decompose and tend to accumulate.
   c. do not contain vitamins as natural materials do.
   d. become a source of food for fish, but ruin the flavor of fish meat.

Answers

1. c 2. d 3. c 4. a 5. b 6. d 7. d 8. d 9. b 10. c 11. c 12. b

QUESTIONS FOR THOUGHT

1. What is an organic compound?
2. What features allow organic molecules to form millions of compounds?
3. Is it possible to have an isomer of ethane? Explain.
4. Suggest a reason that ethylene is an important raw material used in the production of plastics but ethane is not.
5. Give an example of each of the following classes of organic molecules: carbohydrates, proteins, nucleic acids, lipids.
6. What are (a) natural gas, (b) LPG, and (c) diesel fuel?
7. What does the octane number of gasoline describe? On what is the number based?
8. What is a functional group? List four examples.
9. Draw a structural formula for the fatty acid linoleic acid. What feature makes this a saturated fat?
10. A soft drink is advertised to "contain no sugar." The label lists ingredients of carbonated water, dextrose, corn syrup, fructose, and flavorings. Evaluate the advertising.
11. What are the three subgroups of lipids? What are the differences between saturated and unsaturated fats?
12. What is a polymer? Give an example of a naturally occurring plant polymer. Give an example of a synthetic polymer.

FOR FURTHER ANALYSIS

1. Many people feel that by using a higher-octane gasoline in their vehicles, they will get better performance. What is the difference between high-octane and regular gasoline? Does scientific evidence bear this out? What octane gasoline is recommended for your vehicle?

2. There have been some health concerns about the additives used in gasoline. What are these additives? What purpose do they serve? What do opponents of using such additives propose are their negative impact?

3. The so-called "birth control pill," or "the pill," has been around since the early 1960s. This medication is composed of a variety of organic molecules. What is the nature of these compounds? How do they work to control conception since they do not control birth? What are some of the negative side effects and what are some of the related benefits of taking "the pill"?

4. Many communities throughout the world are involved in recycling. One of the most important classes of materials recycled is plastic. There has been concern that the cost of recycling plastics is higher than the cost of making new plastic products. Go to the Internet and

   a. Identify the kinds of plastics that are commonly recycled.

   b. Explain how a plastic such as HDPE is recycled.

   c. Present an argument for eliminating the recycling of HDPE and for expanding the recycling of HDPE.

INVITATION TO INQUIRY

Alcohols: What Do You Really Know?

Archaeologists, anthropologists, chemists, biologists, and health care professionals agree that the drinking of alcohol dates back thousands of years. Evidence also exists that this practice has occurred in most cultures around the world. Use the Internet to search out answers to the following questions:

1. What is the earliest date for which there is evidence for the production of ethyl alcohol?

2. In which culture did this occur?

3. What is the molecular formula and structure of ethanol?

4. Do alcohol and water mix?

5. How much ethanol is consumed in the form of beverages in the United States each year?

6. What is the legal limit to be considered intoxicated in your state?

7. How is this level measured?

8. Why is there a tax on alcoholic beverages?

9. How do the negative effects of drinking alcohol compare between men and women?

10. Have researchers demonstrated any beneficial effects of drinking alcohol?

Compare what you thought you knew to what is now supported with scientific evidence.

The cell is the simplest structure capable of existing as an individual living unit. Within this unit, many chemical reactions are required for maintaining life. These reactions do not occur at random but are associated with specific parts of the many kinds of cells.

The Nature of Living Things

> Physics Connections

▶ Energy in a useful form is essential for all life (Ch. 3).

> Chemistry Connections

▶ Molecular movement is required for the transport of atoms and molecules into and out of cells (Ch. 4).

▶ Chemical reactions are the essence of all living things (Ch. 9).

CORE CONCEPT

All living things are composed of cells.

All forms of life demonstrate similar characteristics (p. 466).

All cells come from preexisting cells (p. 470).

The two basic types of cells have many similar structural features (pp. 471–472).

> Life Science Connections

▶ The structure of living things is basically the same—a cell (Ch. 20).

▶ The physiological processes of living things are common to all life forms (Ch. 20).

▶ Cell division is required for life to continue (Ch. 20).

▶ All living things have genetic material (Ch. 20).

▶ Many types of diseases are the result of one life form living on another (Ch. 20).

OVERVIEW

What does it mean to be alive? You would think that a science textbook could answer this question easily. However, this is more than just a theoretical question because in recent years, it has been necessary to construct legal definitions of what life is and especially of when it begins and ends. The legal definition of death is important because it may determine whether a person will receive life insurance benefits or if body parts may be used in transplants. In the case of heart transplants, the person donating the heart may be legally "dead," but the heart certainly isn't since it can be removed while it still has "life." In other words, there are different kinds of death. There is the death of the whole living unit and the death of each cell within the living unit. A person actually "dies" before every cell has died. Death, then, is the absence of life, but that still doesn't tell us what life is. At this point, we won't try to define life, but we will describe some of the basic characteristics of living things.

WHAT MAKES SOMETHING ALIVE?

The science of **biology** is, broadly speaking, the study of living things. It draws on chemistry and physics for its foundation and applies these basic physical laws to life processes. Living things have abilities and structures not typically found in things that were never living. The ability to manipulate energy and matter is unique to living things. Developing an understanding of how they modify matter and use energy will help you appreciate how they differ from nonliving objects. Living things show five characteristics that nonliving things do not: (1) metabolic processes, (2) generative processes, (3) responsive processes, (4) control processes, and (5) a unique structural organization (figure 20.1). For something to be "alive," it must display all these characteristics, that is, they must all work together. It is important to recognize that although these characteristics are typical of all living things, they may not all be present in each organism at every point in time. For example, some individuals may reproduce or grow only at certain times. Should a living organism lose one or more of these features for an extended period, it would "die." *Death,* then, is the absence of life, but this definition still doesn't tell us what life is. Something is living when it displays certain unique chemical processes occurring in association with certain unique structures. This section briefly introduces each of these basic characteristics.

1. **Metabolic processes** are all the chemical reactions and associated energy changes taking place within an organism. Energy is necessary for movement, growth, and many other activities. The energy that organisms use comes from energy stored in the chemical bonds of complex molecules. This energy becomes available through a controlled sequence of chemical reactions.

   There are three essential aspects to metabolic processes: (1) *nutrient uptake,* (2) *nutrient processing,* and (3) *waste elimination.* All living things expend energy to take in nutrients (raw materials) from their environment. Many animals take in these materials by eating or swallowing other organisms. Microorganisms and plants absorb raw materials into their cells to maintain their lives. Once inside, raw materials are used in a series of chemical reactions to manufacture new parts, make repairs, reproduce, and provide energy for essential activities. However, not all the raw materials entering a living thing are valuable to it. There may be portions that are useless or even harmful. Organisms eliminate these portions as waste. Metabolic processes also produce unusable heat energy that may also be considered a waste product.

2. The second group of characteristics of life, **generative processes,** are reactions that result in an increase in the size of an individual organism—*growth*—or an increase in the number of individuals in a population of

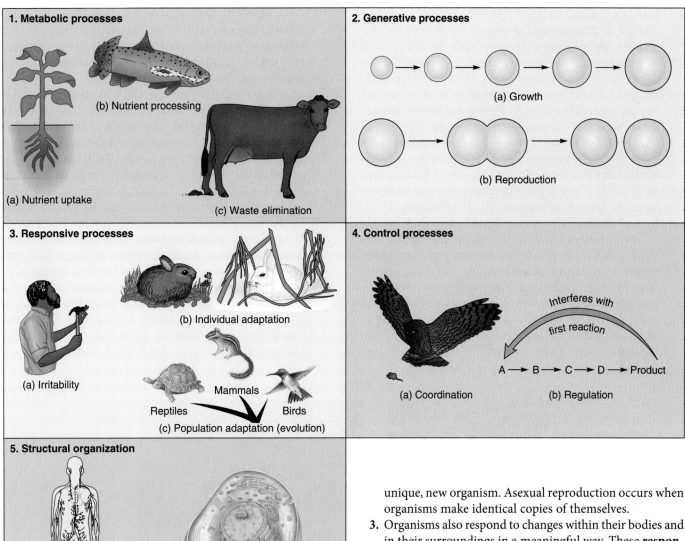

1. Metabolic processes

(a) Nutrient uptake

(b) Nutrient processing

(c) Waste elimination

2. Generative processes

(a) Growth

(b) Reproduction

3. Responsive processes

(a) Irritability

(b) Individual adaptation

Reptiles   Mammals   Birds

(c) Population adaptation (evolution)

4. Control processes

Interferes with first reaction

A → B → C → D → Product

(a) Coordination   (b) Regulation

5. Structural organization

(a) Organismal organization   (b) Cellular organization

FIGURE 20.1   Living things demonstrate many common characteristics.

organisms—*reproduction*. During growth, living things add to their structure, repair parts, and store nutrients for later use. Growth and reproduction are directly related to metabolism, since neither can occur without the acquisition and processing of nutrients.

Reproduction of individuals must occur because each organism eventually dies. Thus, reproduction is the only way that living things can perpetuate themselves. There are a number of different ways that various kinds of organisms reproduce and guarantee their continued existence. Some reproductive processes known as sexual reproduction involve two organisms contributing to the creation of a unique, new organism. Asexual reproduction occurs when organisms make identical copies of themselves.

3. Organisms also respond to changes within their bodies and in their surroundings in a meaningful way. These **responsive processes** have been organized into three categories: *irritability, individual adaptation,* and *population adaptation* or *evolution*. The term *irritability* is not used in the everyday context of an individual being angry. Here, *irritability* means an individual's rapid response to a stimulus, such as your response to a loud noise, beautiful sunset, or noxious odor. This type of response occurs only in the individual receiving the stimulus and is rapid because the mechanism that allows the response to occur (i.e., muscles, bones, and nerves) is already in place. Individual adaptation is also an individual response but is slower since it requires a growth or some other fundamental change in an organism. For example, a weasel's fur color will change from its brown summer coat to its white winter coat when genes responsible for the production of brown pigment are "turned off." Or the response of our body to disease organisms requires a change in the way cells work that eventually gets control of the organism causing the disease. Population adaptation involves changes in the kinds of characteristics displayed by individuals within the population. It is also known as *evolution,* which is a change in the genetic makeup of a *population* of organisms. This process occurs over long periods of time and enables a species (specific

Chapter 20   *The Nature of Living Things*   **467**

kind of organism) to adapt and better survive long-term changes in its environment over many generations. For example, the structures that give birds the ability to fly long distances allow them to respond to a world in which the winter season presents severe conditions that would threaten survival. Similarly, the ability of humans to think and use tools allows them to survive and be successful in a great variety of environmental conditions.

4. The **control processes** of *coordination* and *regulation* constitute the fourth characteristic of life. Control processes are mechanisms that ensure that an organism will carry out all metabolic activities in the proper sequence (coordination) and at the proper rate (regulation). All the chemical reactions of an organism are coordinated and linked together in specific pathways. The orchestration of all the reactions ensures that there will be specific stepwise handling of the nutrients needed to maintain life. The molecules responsible for coordinating these reactions are known as *enzymes*. **Enzymes** are molecules produced by organisms, molecules that are able to increase and control the rate at which life's chemical reactions occur. Enzymes also regulate the amount of nutrients processed into other forms.

Many of the internal activities of organisms are interrelated and coordinated to maintain a constant internal environment. The process of maintaining a constant internal environment is called **homeostasis.** For example, when we begin to exercise, we use up oxygen more rapidly so the amount of oxygen in the blood falls. To maintain a "constant internal environment," the body must obtain more oxygen. This involves more rapid contractions of the muscles that cause breathing and a more rapid and forceful pumping of the heart to get blood to the lungs. These activities must occur together at the right time and at the correct rate, and when they do, the level of oxygen in the blood will remain normal while supporting the additional muscular activity.

5. In addition to these four basic processes that are typical of living things, living things also share some basic **structural similarities.** All living things are organized into complex structural units called *cells.* The cell units have an outer limiting membrane and internal structural units that have specific functions. Some living things, such as you, consist of trillions of cells with specialized abilities that interact to provide the independently functioning unit called an *organism* (figure 20.2). Typically, in such

Yeast

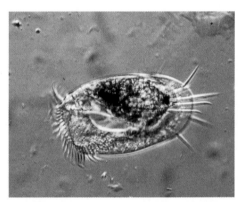
Euplotes

Humans

Orchid

FIGURE 20.2 Each individual organism, whether it is simple or complex, is able to independently carry on metabolic, generative, responsive, and control processes. Some organisms, such as yeast or the protozoan *Euplotes,* consist of single cells, while others, such as humans and orchids, consist of many cells organized into complex structures.

TABLE 20.1 Levels of Organization for Living Things

Category	Characteristics/Explanation	Example/Application
Biosphere	Worldwide ecosystem; human activity affects the climate of Earth.	Global climate change, hole in ozone layer
Ecosystem	Communities (groups of populations) that interact with the physical world in a particular place	The Everglades ecosystem involves many kinds of organisms, the climate, and the flow of water to south Florida.
Community	Populations of different kinds of organisms that interact with one another in a particular place	The populations of trees, insects, birds, mammals, fungi, bacteria, and many other organisms that interact in any location
Population	A group of individual organisms of a particular kind	The human population currently consists of over 6 billion individual organisms. The current population of the California condor is about 220 individuals.
Individual organism	An independent living unit	A single organism Some organisms consist of many cells—you, a morel mushroom, a rose bush. Others are single cells—yeast, pneumonia bacterium, *Amoeba*.
Organ system	Groups of organs that perform particular functions	The circulatory system consists of a heart, arteries, veins, and capillaries, all of which are involved in moving blood from place to place.
Organ	Groups of tissues that perform particular functions	An eye contains nervous tissue, connective tissue, blood vessels, and pigmented tissues, all of which are involved in sight.
Tissue	Groups of cells that perform particular functions	Blood, groups of muscle cells, and the layers of the skin are all groups of cells that perform a particular function.
Cell	The smallest unit that displays the characteristics of life	Some organisms are single cells. Within multicellular organisms are several kinds of cells—heart muscle cells, nerve cells, white blood cells.
Molecules	Specific arrangements of atoms	Living things consist of special kinds of molecules, such as proteins, carbohydrates, and DNA.
Atoms	The fundamental units of matter	Hydrogen, oxygen, nitrogen and about one hundred others

large, multicellular organisms as humans, cells cooperate with one another in units called *tissues* (e.g., muscle, nerves). Groups of tissues are organized into larger units known as *organs* (e.g., heart) and, in turn, into *organ systems* (e.g., circulatory system). Other organisms, such as bacteria or yeast, carry out all four of the life processes within a single cell. Nonliving materials, such as rocks, water, or gases, do not share a structurally complex common subunit.

Biologists and other scientists like to organize vast amounts of information into conceptual chunks that are easier to relate to one another. One important concept in biology is that all living things share the structural and functional characteristics we have just discussed. Another important organizing concept is that organisms are special kinds of matter that interact with their surroundings at several different levels (table 20.1). When biologists seek answers to a particular problem, they may attack it at several different levels simultaneously. They must understand the molecules that make up living things; how the molecules are incorporated into cells; how tissues,

organs, or systems within an organism function; and how populations and ecosystems are affected by changes in individual organisms.

THE CELL THEORY

One of the characteristics of life, the *cell*, is one of the most important ideas in biology because it applies to all living things. It did not emerge all at once but has been developed and modified over many years. It is still being modified today.

Several individuals made key contributions to the cell concept. Anton van Leeuwenhoek (1632–1723) was one of the first to make use of a *microscope* to examine biological specimens. When van Leeuwenhoek discovered that he could see things moving in pond water using his microscope, his curiosity stimulated him to look at a variety of other things. He studied blood, semen, feces, pepper, and tartar, for example. He was the first to see individual cells and recognize them as living units, but he did not call them cells. The name he gave to these "little animals" that he saw moving around in the pond water was *animalcules*.

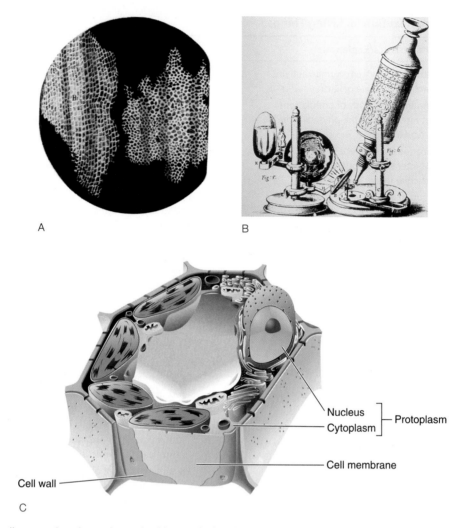

A

B

Nucleus ⎤
Cytoplasm ⎦ Protoplasm

Cell membrane

Cell wall

C

FIGURE 20.3 The cell concept has changed considerably over the last three hundred years. Robert Hooke's idea of a cell (*A*) was based on his observation of slices of cork (cell walls of the bark of the cork oak tree). Hooke invented the compound microscope and illumination system shown in (*B*), one of the best such microscopes of his time. One of the first subcellular differentiations was to divide the protoplasm into cytoplasm and nucleus, as shown in this plant cell (*C*). We now know that cells are much more complex than this; they are composed of many kinds of subcellular structures, some components numbering in the thousands.

The first person to use the term *cell* was Robert Hooke (1635–1703) of England, who was also interested in how things looked when magnified (figure 20.3). He chose to study thin slices of cork from the bark of a cork oak tree. He saw a mass of cubicles fitting neatly together, which reminded him of the barren rooms in a monastery. Hence, he called them *cells.* As it is currently used, the term **cell** refers to the basic functional and structural unit that makes up all living things. When Hooke looked at cork, the tiny boxes he saw were, in fact, only the cell walls that surrounded the living portions of plant cells. We now know that the **cell wall** is composed of the complex carbohydrate cellulose, which provides strength and protection to the living contents of the cell. The cell wall appears to be a rigid, solid layer of material, but in reality, it is composed of many interwoven strands of cellulose molecules. Its structure allows certain very large molecules to pass through it readily, but it acts as a screen to other molecules.

Hooke's use of the term *cell* in 1666 in his publication *Micrographia* was only the beginning, for nearly two hundred years passed before it was generally recognized that all living things are made of cells and that these cells can reproduce themselves; that is, they come from preexisting cells. However, many others were involved in understanding how this concept applied to all living things (see "People Behind the Science" in this chapter).

Soon after the term *cell* caught on, it was recognized that the cell's vitally important portion is inside the cell wall. This living material was termed **protoplasm,** which means *first-formed substance.* The term *protoplasm* allowed scientists to distinguish between the living portion of the cell and the nonliving cell wall. Very soon microscopists were able to distinguish two different regions of protoplasm. One type of protoplasm was more viscous and darker than the other. This core region, called the **cell nucleus**, is a central body within a more fluid material surrounding it. **Cytoplasm** is the

Matthias Jakob Schleiden (1804–1881) and Theodor Schwann (1810–1882)

Matthias Schleiden was a German botanist who, with Theodor Schwann, is best known for establishing the cell theory.

Schleiden was born on April 5, 1804, in Hamburg and studied law at Heidelberg University from 1824 to 1827. After graduating, he practiced as a lawyer in Hamburg but soon returned to university, taking courses in botany and medicine at the universities of Göttingen, Berlin, and Jena. After graduating in 1831, he was appointed professor of botany at Jena, where he remained until he became professor of botany at the University of Dorpat, Estonia (Estonia borders Latvia and Russia), in 1862. He returned to Germany after a short time and died in Frankfurt on June 23, 1881.

The existence of cells had been known since the seventeenth century (Robert Hooke is generally credited with their discovery in 1665). However, Schleiden was the first to recognize their importance as the fundamental units of living organisms. In 1838, he announced that the various parts of plants consist of cells or derivatives of cells. In the following year, Schwann published a paper in which he confirmed for animals Schleiden's idea of the basic importance of cells in the organization of organisms. Thus, Schleiden and Schwann established the cell theory, a concept that is common knowledge today and is as fundamental to biology as atomic theory is to the physical sciences.

Schleiden also researched other aspects of cells. He recognized the importance of the nucleus in cell division. In addition, he noted the active movement of intracellular material in plant tissues.

Theodor Schwann was a German physiologist who did important work on digestion, fermentation, and histology.

Schwann was born on December 7, 1810, in Neuss, Germany. He was educated at the Jesuit College in Cologne, then studied medicine at the universities of Bonn, Würzburg, and Berlin, graduating from the last in 1834. He spent the next four years working as an assistant to the German physiologist Johannes Müller at the Museum of Anatomy in Berlin. In 1839, however, Schwann's work on fermentation attracted so much adverse criticism that he left Germany for Belgium, where he was professor of anatomy at the Roman Catholic University in Louvain from 1839 to 1848, then held the same post at the University of Liège until his death in Cologne on January 11, 1882.

In 1834, Schwann began to investigate digestive processes and two years later isolated a chemical responsible for protein digestion,

Box Figure 20.1 Matthias Schleiden

Box Figure 20.2 Theodor Schwann

which he called pepsin. This was the first *enzyme* to be isolated from animal tissue. Schwann then studied fermentation and showed that the fermentation of sugar is a result of the life processes of living yeast cells. He later coined the term *metabolism* to denote the chemical changes that occur in living tissue. Chemists such as Friedrich Wöhler and Justus von Liebig heavily criticized his work on fermentation. It was not until Louis Pasteur's work on fermentation in the 1850s that Schwann was proved correct. Meanwhile, however, Schwann investigated putrefaction (rotting) in an attempt to disprove the theory of spontaneous generation, repeating, with improved techniques, Lazzaro Spallanzani's earlier experiments. Schwann found no evidence to support the theory, despite which it was still believed by some scientists.

In 1839, Schwann published *Microscopical Researches on the Similarity in the Structure and Growth of Animals and Plants* in which he formulated the cell theory. In the previous year, Matthias Schleiden—whom Schwann knew well—had stated the theory in connection with plants, but Schwann extended the theory to animals. Schwann concluded that all organisms (both animals and plants) consist entirely of cells or of products of cells and that the life of each individual cell is secondary to that of the whole organism.

Source: Modified from the *Hutchinson Dictionary of Scientific Biography.* © RM, 2007. Reprinted by permission.

---

name given to the colloidal fluid portion of the protoplasm (protoplasm = cytoplasm + nucleus). *Colloids* are mixtures that contain suspended particles larger than those in true solutions but smaller than those in a coarse suspension. The solutes in a colloid cannot be filtered out and do not settle out by gravity. While the term *protoplasm* is seldom used today, the term *cytoplasm* is still very common in the vocabulary of cell biologists.

The development of better microscopes and better staining techniques revealed that protoplasm contains many tiny

structures called **organelles.** It has been determined that certain functions characteristic of life are performed in certain organelles. The essential job an organelle does is related to its structure. Each organelle is dynamic in its operation, changing shape and size as it works. Organelles move throughout the cell, and some even self-duplicate.

All living things are cells or composed of cells. To date, most biologists recognize two major cell types, *prokaryotes* and *eukaryotes* (figure 20.4). Whether they are **prokaryotic cells** or **eukaryotic cells,** all have certain things in common: (1) cell

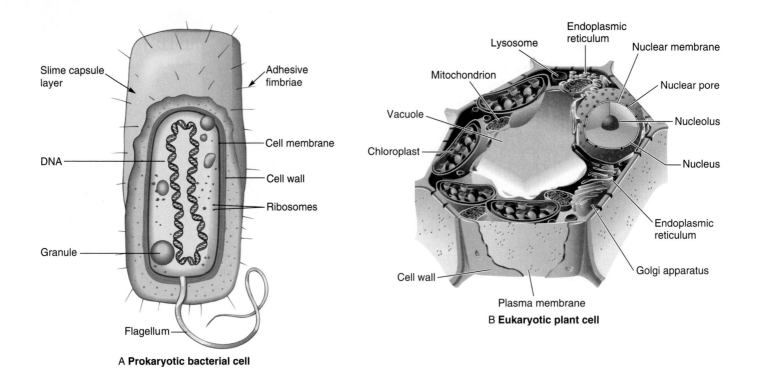

A **Prokaryotic bacterial cell**

B **Eukaryotic plant cell**

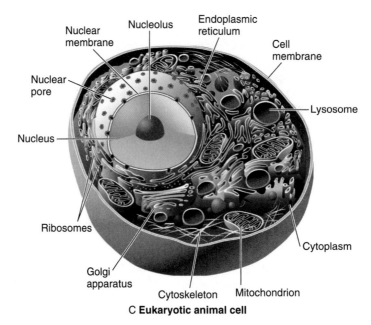

C **Eukaryotic animal cell**

FIGURE 20.4   There are two major types of cells, the prokaryotes and the eukaryotes. Prokaryotic cells are represented by the (*A*) bacteria, and eukaryotic cells by (*B*) plant and (*C*) animal cells.

membranes, (2) cytoplasm, (3) genetic material, (4) energy transfer molecules, (5) enzymes and coenzymes. These are all necessary to carry out life's functions. Should any of these not function properly, a cell would die.

The differences among cell types are found in the details of their structure. While prokaryotic cells lack most of the complex internal organelles typical of eukaryotes, they are cells and can carry out life's functions.

Most single-celled organisms that we commonly refer to as *bacteria* are prokaryotic cells (figure 20.4A). Algae, protozoa, fungi, plants, and animals are all comprised of eukaryotic cells (figure 20.4 B, C).

To view very small objects, we use a magnifying glass as a way of extending our observational powers. A magnifying glass is a lens that bends light in such a way that the object appears larger than it really is. Such a lens might magnify objects ten or even fifty times. Anton van Leeuwenhoek, a Dutch drape and clothing maker, was one of the first individuals to carefully study magnified cells (box figure 20.3A). He made very detailed sketches of the things he viewed with his simple microscopes and communicated his findings to Robert Hooke and the Royal Society of London. His work stimulated further investigation of magnification techniques and descriptions of cell structures. These first microscopes were developed in the early 1600s.

Compound microscopes (box figure 20.3B), developed soon after the simple microscopes, are able to increase magnification by bending light through a series of lenses. One lens, the *objective lens*, magnifies a specimen that is further magnified by the second lens, known as the *ocular lens.*

An *electron microscope* (box figure 20.3C) is able to magnify two-hundred thousand times and still resolve individual structures. The difficulty is, of course, that you are unable to see electrons with your eyes. Therefore, in an electron microscope, the electrons strike a photographic film or video monitor, and this "picture" shows the individual structures. The techniques for preparing the material to be viewed—slicing the specimen very thinly and focusing the electron beam on the specimen—make electron microscopy an art as well as a science.

More recently, the scanning *tunneling microscope* was developed. The developers of this microscope, Gerd Binning and Heinrich Rohrer, were awarded the Nobel Prize in 1985 for their work. The invention of the scanning *tunneling microscope followed by the atomic force microscope* enabled researchers to visualize previously unseen molecules and even the surface of atoms such as chlorine and sodium. A scanning tunneling microscope utilizes a thin platinum and viridium wire to trace the surface of the specimen. Electrons on the surface of the probe interact with those on the specimen's surface to produce a "tunnel" that is seen as a current. The stronger the current, the closer the probe is to the surface. The *atomic force microscope* enables researchers to see three-dimensional objects even as small as molecules and the surface of atoms such as chlorine and sodium.

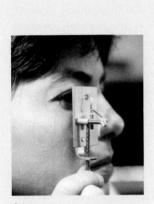
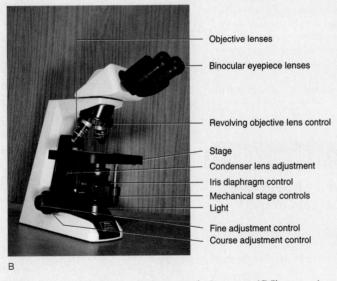

Objective lenses

Binocular eyepiece lenses

Revolving objective lens control

Stage
Condenser lens adjustment
Iris diaphragm control
Mechanical stage controls
Light

Fine adjustment control
Course adjustment control

A   B   C

Box Figure 20.3   (*A*) Replica of van Leeuwenhoek microscope. (*B*) Compound microscope. (*C*) Electron microscope.

CELL MEMBRANES

One feature common to all cells and many of the organelles they contain is a thin layer of material called *membrane.* Membrane can be folded and twisted into many different structures, shapes, and forms. The particular arrangement of membrane of an organelle is related to the functions that it is capable of performing. This is similar to the way a piece of fabric can be fashioned into a pair of pants, a shirt, sheets, pillowcases, or a rag doll. All cellular membranes have a fundamental molecular structure that allows them to be fashioned into a variety of different organelles.

Cell membranes are thin sheets composed primarily of phospholipids (refer to figure 19.25) and proteins. The current hypothesis of how membranes are constructed is known as the *fluid-mosaic model,* which proposes that the various molecules of the membrane are able to flow and move about (figure 20.5). The membrane maintains its form because of the physical interaction of its molecules with its surroundings. The phospholipid molecules of the membrane are polar molecules since they have

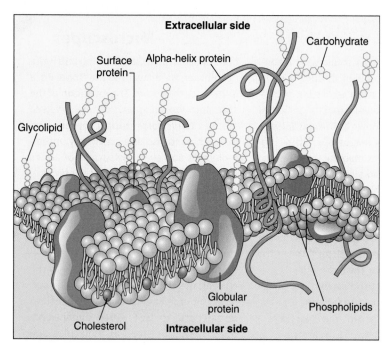

FIGURE 20.5 Notice in this section of a generalized human cell that there is no surrounding cell wall as pictured in Hooke's cell, figure 20.3A. Membranes in all cells are composed of protein and phospholipids. Two layers of phospholipid are oriented so that the hydrophobic fatty ends extend toward each other and the hydrophilic glycerol portions are on the outside. The phosphate-containing chain of the phospholipid is coiled near the glycerol portion. Buried within the phospholipid layer and/or floating on it are the globular proteins. Some of these proteins accumulate materials from outside the cell; others act as sites of chemical activity. Carbohydrates are often attached to one surface of the membrane.

Myths, Mistakes, and Misunderstandings

Why Don't We Have to Worry About HIV Transmission Through Casual Contact?

Misconception: Casual contact (e.g., shaking hands, touching) is likely to be the way people will contract HIV/AIDS.

In fact, cell membranes of all cells have molecules on their surface that make them unique. One such type is known as *clusters of differentiation,* or *CD markers.* These are designated by numbered groups, that is, CD1, CD2, CD3, CD4, etc. Different cell types have different markers on their surfaces. Those with CD4 markers include macrophages, nerve cells, and certain cells of the immune system known as T-helper (T$_H$) and T-delayed-type (T$_D$) hypersensitivity cells. CD4 cells are typically found deep inside the body (e.g., in blood) and not on the skin. The CD4 marker acts as the attachment site for human immunodeficiency virus (HIV), which is responsible for *acquired immunodeficiency syndrome* (AIDS). To cause AIDS, the virus must enter the CD4 cell and take command of its metabolism. HIV cannot attach to cells that do not have CD4 cell surface receptors. Since CD4 cells do not normally occur on the surface of the skin, casual contact is not likely to result in the transmission of HIV/AIDS.

one end (the glycerol portion) that is soluble in water and is therefore called *hydrophilic* (water-loving). The other end that is not water soluble, called *hydrophobic* (water-hating), is comprised of fatty acid. When phospholipid molecules are placed in water, they form a double-layered sheet, with the water-soluble (hydrophilic) portions of the molecules facing away from each other. This is commonly referred to as a *phospholipid bilayer.*

The protein component of cellular membranes can be found on either surface of the membrane or in the membrane, among the phospholipid molecules. Many of the membrane proteins are capable of moving from one side to the other, while others aid in the movement of molecules across the membrane by forming channels through which substances may travel or by acting as transport molecules. Some protein molecules found on the outside surfaces of cellular membranes have carbohydrates or fats attached to them. These combination molecules are important in determining the "sidedness" (inside-outside) of the membrane and also help organisms recognize differences between types of cells. Your body can recognize disease-causing organisms because their surface proteins are different from those of its own cellular membranes. Some of these molecules also serve as attachment sites for specific chemicals, bacteria, protozoa, white blood cells, and viruses. Many dangerous agents cannot stick to the surface of cells and therefore cannot cause harm. For this reason, cell biologists explore the exact structure and function of these molecules. They are also attempting to identify molecules that can interfere with the binding of such agents as viruses and bacteria in the hope of controlling infections.

Other types of molecules found in cell membranes are cholesterol and carbohydrates. Cholesterol appears to play a role in stabilizing the membrane and keeping it flexible. Carbohydrates are usually found on the outside of the membrane, where they are bound to proteins or lipids. These carbohydrates appear to play a role in cell-to-cell interactions and are involved in binding with regulatory molecules.

GETTING THROUGH MEMBRANES

If a cell is to stay alive, it must meet the characteristics of life outlined earlier. This includes taking in nutrients and eliminating wastes and other by-products of metabolism. Several mechanisms allow cells to carry out the processes characteristic of life noted at the start of this chapter. They include diffusion, osmosis, facilitated diffusion, active transport, and phagocytosis.

Diffusion

There is a natural tendency in gases and liquids for molecules of different types to completely mix with each other (refer to chapter 10, Water and Solutions). This is because they are constantly moving about with various levels of kinetic energy. Consider two types of molecules. As the molecules of one type move about, they tend to scatter from the place where they are most concentrated. The other type of molecule also tends to disperse

The Other Outer Layer—The Cell Wall

The *cell walls* of microorganisms, plants, and fungi appear to be rigid, solid layers of material but are really loosely woven layers. And like water pouring through a leaky straw basket, many types of molecules easily pass through a cell wall. The cell wall lends strength and protection to the contents of the cell but hampers flexibility and movement. There are three kinds of materials typically used for cell walls. All the higher plants and most algae have *cellulose* as their wall material. When found in large amounts, it is known as *wood*. Another cell wall material, *chitin*, is found in the fungi. Chitin also constitutes the exoskeleton material in insects, and the outer shell of a beetle or a shrimp is chitin. But in those animals, the chitin surrounds masses of tissue instead of individual cells. The fungi are not crunchy or brittle like shrimp skeletons because the chitin is very thin. Many bacteria have their walls composed of *peptidoglycan*. It is composed of both amino acids and carbohydrates. The lengths of peptidoglycan chains and how they are interlinked may determine the shape of a bacterium.

in the same way. The result of this random motion is that the two types of molecules eventually become mixed throughout.

Remember that the motion of the molecules is completely random. If you follow the paths of molecules from tea leaves placed in a cup of hot water, you will find that some of the colored tea leaf molecules move away from the leaves while others move in the opposite direction. However, more molecules would move away from the leaves because there were more there to start with. If you wait long enough, you will see that the colored molecules become equally distributed throughout the cup of tea.

We generally are not interested in the individual movement but rather in the overall movement, which is called the *net movement*. The direction of greatest movement (net movement) is determined by the relative concentration of the molecules, for example, the amount in one area in comparison to the amount in another area. **Diffusion** is defined as the net movement of a kind of molecule from a place where that molecule is in higher concentration to a place where that molecule is scarcer. When a kind of molecule is completely dispersed and movement is equal in all directions, we say that the system has reached a state of *dynamic equilibrium*. There is no longer a net movement because movement in one direction equals movement in the other. It is dynamic, however, because the system still has energy, and the molecules are still moving.

Because the cell membrane is composed of phospholipid and protein molecules that are in constant motion, temporary openings are formed that allow small molecules to cross from one side of the membrane to the other. Molecules close to the membrane are in constant motion as well. They are able to move into and out of a cell by passing through these openings in the membrane.

The rate of diffusion is related to the kinetic energy and size of the molecules. Since diffusion occurs only when molecules are unevenly distributed, the relative concentration of the molecules is important in determining how fast diffusion occurs. The difference in concentration of the molecules is known as a *concentration gradient* or *diffusion gradient*. When the molecules are equally distributed, no such gradient exists.

Diffusion can take place only as long as there are no barriers to the free movement of molecules. In the case of a cell, the membrane permits some molecules to pass through, while others are not allowed to pass or are allowed to pass more slowly. This permeability is based on size, ionic charge, and solubility of the molecules involved. The membrane does not, however,

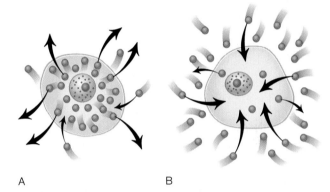

A   B

FIGURE 20.6 As a result of molecular motion, molecules move from areas where they are concentrated to areas where they are less concentrated. (*A*) and (*B*) show molecules leaving and entering animal cells by diffusion. The net direction of movement is controlled by concentration (always high to low concentration), and the energy necessary is supplied by the kinetic energy of the molecules themselves. (*A*) shows net diffusion out of the cell, whereas (*B*) shows net diffusion into the cell.

distinguish direction of movement of molecules; therefore, the membrane does not influence the direction of diffusion. The direction of diffusion is determined by the relative concentration of specific molecules on the two sides of the membrane, and the energy that causes diffusion to occur is supplied by the kinetic energy of the molecules themselves (figure 20.6).

Diffusion is an important means by which materials are exchanged between a cell and its environment. Since the movement of the molecules is random, the cell has little control over the process; thus, diffusion is considered a passive process, that is, chemical bond energy does not have to be expended. For example, animals are constantly using oxygen in various chemical reactions. Consequently, the oxygen concentration in cells always remains low. The cells, then, contain a lower concentration of oxygen than the oxygen level outside of the cells. This creates a diffusion gradient, and the oxygen molecules diffuse from the outside of the cell to the inside of the cell.

Osmosis

An important characteristic of all membranes is that they are selectively permeable. **Selectively permeable** means that a membrane will allow certain molecules to pass across it and

Connections . . .

Kidney Machines in Action

When the kidneys do not function properly, the toxic waste urea builds up in the blood because the ailing kidneys cannot filter it out. This eventually leads to a condition known as *uremia*, "urine in the blood." Unless the urea is removed from the body, the concentration of urea increases and will eventually lead to death. Kidney or dialysis machines use tubular cellulose membranes to separate the urea from the blood.

Cellulose is a polymer of glucose molecules and contains openings through which urea molecules pass. In a dialysis machine, cellulose membrane tubes are immersed in a large volume of "cleansing" or dialysis solution. The blood is pumped from the vein of a person with uremia through this tubing and then returned to the patient. While in the machine, the membrane allows most of the urea in the blood to pass through the tubing walls, but the larger blood proteins and blood cells do not. Therefore, the patient does not lose vital proteins or blood cells. This system works because the dialysis solution in which the tubing is immersed has a lower amount of urea than does the blood. Because of this concentration gradient, urea in the blood passes from the high concentration in the blood through the cellulose membrane tubing pores into the surrounding dialysis solution. This results in a decrease in urea concentration in the blood. As the concentration of the urea increases in the dialysis solution, the solution is flushed out. This enables the dialysis solution to maintain a constant, low concentration of urea and maintain the movement of urea molecules out of the person's body. To maintain the blood's concentration of other vital ions, the solution is made with the same concentration of those ions as found in the blood. This means that these other ions are in dynamic equilibrium, and so their concentration does not change during kidney dialysis (box figure 20.4).

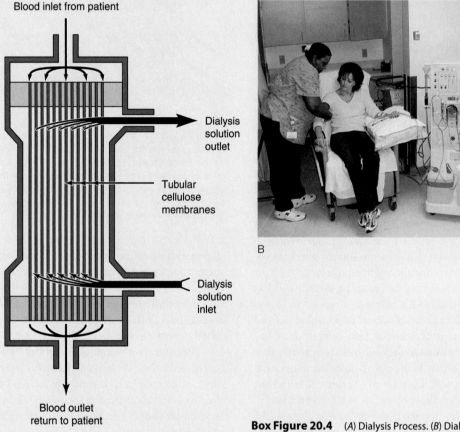

Box Figure 20.4 (*A*) Dialysis Process. (*B*) Dialysis in a hospital.

will prevent others from doing so. Molecules that are able to dissolve in phospholipids, such as vitamins A and D, can pass through the membrane rather easily; however, many molecules cannot pass through at all. In certain cases, the membrane differentiates on the basis of molecular size; that is, the membrane allows small molecules, such as water, to pass through and prevents the passage of larger molecules. The membrane may also regulate the passage of ions. If a particular portion of the membrane has a large number of positive ions on its surface,

positively charged ions in the environment will be repelled and prevented from crossing the membrane.

Water molecules diffuse through cell membranes. The net movement (diffusion) of water molecules through a selectively permeable membrane is known as **osmosis.** In any osmotic situation, there must be a selectively permeable membrane separating two solutions. For example, a solution of 90 percent water and 10 percent sugar separated by a selectively permeable membrane from a different sugar solution, such as one of 80 percent water

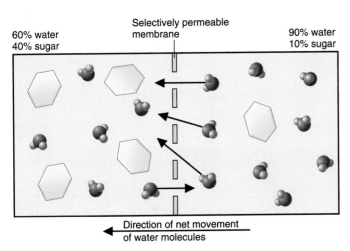

60% water
40% sugar

Selectively permeable
membrane

90% water
10% sugar

Direction of net movement
of water molecules

FIGURE 20.7 When two solutions with different percentages of water are separated by a selectively permeable membrane, there will be a net movement of water from the solution with the highest percentage of water to the one with the lowest percentage of water.

CONCEPTS APPLIED

Beans: Weighing-in

To demonstrate the movement of materials into and out of cells of living things, you do not necessarily need sophisticated equipment. Obtain some dry beans. Count out thirty beans. Set ten dry beans aside for later comparisons and place the remaining twenty beans in tap water overnight. In the morning, compare the size of the dry beans with the size of those that were soaked. (If you have access to a sensitive household scale, you can weigh the beans as well as observe the change in size.) Remove ten of the soaked beans and place them in a strong saltwater solution overnight. Compare the size of the beans after being soaked in salt water to the original dry beans and to those that had been soaked in tap water. Can you explain why these changes occurred?

and 20 percent sugar, demonstrates osmosis (figure 20.7). The membrane allows water molecules to pass freely but prevents the larger sugar molecules from crossing. There is a higher concentration of water molecules in one solution compared to the concentration of water molecules in the other, so more of the water molecules move from the solution with 90 percent water to the other solution with 80 percent water. Be sure that you recognize that osmosis is really diffusion in which the diffusing substance is water and that the regions of different concentrations are separated by a membrane that is more permeable to water.

A proper amount of water is required if a cell is to function efficiently. Too much water in a cell may dilute the cell contents and interfere with the chemical reactions necessary to keep the cell alive. Too little water in the cell may result in a buildup of poisonous waste products. If a cell contains a concentration of water and dissolved materials that is equal to that of its surroundings, the cell is said to be *isotonic* to its surroundings. Cells within the human body must have a concentration of water and dissolved materials within the cells that is equal to the concentration outside the cells. For example, red blood cells are isotonic

Cell Trivia

Here are some tidbits of information about cells:

▶ The smallest cells are the bacteria, some of which are only 0.1 to 0.2 micrometers in diameter. *E. coli*, a very common bacterium in human intestines, is only 0.5 micron long.

▶ The largest cells are the egg cells of birds and mammals. The yolk of a chicken egg is a single cell, which can be 50 mm in diameter. The nerve cell from the human spine to the big toe can be 1 m (3.3 ft) or more in length.

▶ The animal cells that live the longest are the nerve or muscle cells of long-lived reptiles and mammals. The sea tortoise, which lives two hundred years, has the same nerve cells at birth that it does at death.

▶ Cells of certain bacteria can divide every fifteen minutes under optimum conditions. This means that if you start with only one bacterial cell and it divides as often as possible, in only twelve hours, there could be as many as 281 trillion individual cells.

when they have the same percentage of water and dissolved materials as the surrounding plasma. Many products are isotonic. For example, over-the-counter eyewashes are labeled *isotonic*, as are many medically important solutions. *Physiological* or *"normal" saline* solution is isotonic, as are other intravenous (IV) solutions (0.9 percent dissolved salts or 5 percent glucose).

However, if cells or tissues are going to survive in an environment that has a different concentration of water, they must expend energy to maintain this difference. Red blood cells having a lower concentration of water (higher concentration of dissolved materials, solutes) than their surroundings tend to gain water by osmosis very rapidly. They are said to be *hypertonic* to their surroundings and the surroundings are *hypotonic*. These two terms are always used to compare two different solutions. The hypertonic solution is the one with more solutes (dissolved material) and less solvent (water); the hypotonic solution has less solutes (dissolved material) and more solvent (water). It may help to remember that the water goes where the salt is (figure 20.8).

As with the diffusion of other molecules, osmosis is a passive process because the cell has no control over the diffusion of water molecules. This means that the cell can remain in balance with an environment only if that environment does not cause the cell to lose or gain too much water.

The cell walls of bacteria, fungi, and plants provide a way to combat the effects of osmosis. If the water concentration outside the plant cell is higher than the water concentration inside, more water molecules enter the cell than leave. This creates internal pressure within the cell. But bacteria, fungi, and plant cells do not easily burst because they are surrounded by a rigid cell wall. Lettuce cells that are crisp are ones that have gained water so that there is high internal pressure. Wilted lettuce has lost some of its water to its surroundings so that it has only slight internal cellular water pressure. Osmosis occurs when you put salad dressing on a salad. Because the dressing has a very low water concentration, water from the lettuce

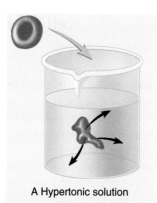

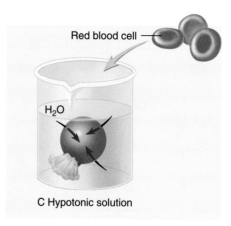

A Hypertonic solution

B Isotonic solution

C Hypotonic solution

FIGURE 20.8   The cells in these three diagrams were subjected to three different environments. (*A*) The cell is in a hypertonic solution. Water has diffused from the cell to the environment, because a high concentration of water was in the cell and the cell has shrunk. (*B*) The cell is isotonic to its surroundings. The water concentration inside the red blood cell and the water concentration in the environment are in balance with each other, so the movement of water into the cell equals the movement of water out of the cell, and the cell has its normal shape. (*C*) The cell is in a hypotonic solution. The cell has accumulated water from the environment, because there is a higher concentration of water outside the cell than inside.

diffuses from the cells into the surroundings. Salad that has been "dressed" too long becomes limp and unappetizing.

So far, we have considered only situations in which cells have no control over the movement of molecules. Cells cannot rely solely on diffusion and osmosis, however, because many of the molecules they require either cannot pass through the cell membranes or occur in relatively low concentrations in the cells' surroundings.

Controlled Methods of Transporting Molecules

Some molecules move across the membrane by combining with specific carrier proteins. When the rate of diffusion of a substance is increased in the presence of a carrier, we call this **facilitated diffusion.** Since this is diffusion, the net direction of movement is in accordance with the concentration gradient. Therefore, this is considered a passive transport method, although it can occur only in living organisms with the necessary carrier proteins. One example of facilitated diffusion is the movement of glucose molecules across the membranes of certain cells. In order for the glucose molecules to pass into these cells, specific proteins are required to carry them across the membrane. The action of the carrier does not require an input of energy other than the kinetic energy of the molecules (figure 20.9).

When molecules are moved across the membrane from an area of *low* concentration to an area of *high* concentration, the cell must expend energy. The process of using a carrier protein to move molecules up a concentration gradient is called **active transport** (figure 20.10). Active transport is very specific: Only certain molecules or ions can be moved in this way, and specific proteins in the membrane must carry them. The action of the carrier requires an input of energy other than the kinetic energy of the molecules; therefore, this process is termed *active* transport. For example, some ions, such as sodium and potassium, are actively pumped across cell membranes. Sodium ions are pumped out of cells up a concentration gradient. Potassium ions are pumped into cells up a concentration gradient.

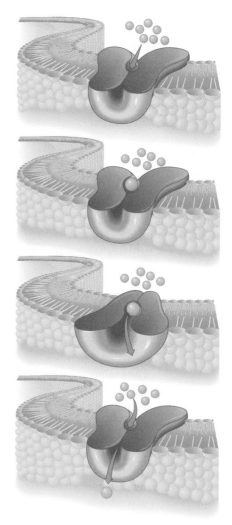

FIGURE 20.9   This method of transporting materials across membranes is a diffusion process (i.e., a movement of molecules from a high to a low concentration). However, the process is helped (facilitated) by a particular membrane protein. No chemical-bond energy is required for this process. The molecules being moved through the membrane attach to a specific transport carrier protein in the membrane. This causes a change in its shape that propels the molecule or ion through to the other side.

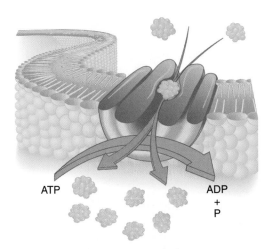

ATP   ADP
+
P

FIGURE 20.10 One possible method whereby active transport could cause materials to accumulate in a cell is illustrated here. Notice that the concentration gradient is such that if simple diffusion were operating, the molecules would leave the cell. The action of the carrier protein requires an active input of energy other than the kinetic energy of the molecules; therefore, this process is termed *active* transport.

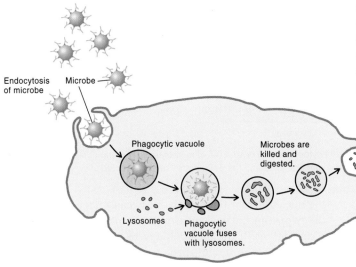

Endocytosis of microbe   Microbe

Phagocytic vacuole

Microbes are killed and digested.

Exocytosis of debris

Lysosomes

Phagocytic vacuole fuses with lysosomes.

FIGURE 20.11 Phagocytosis. The sequence illustrates a cell engulfing a microbe and surrounding it with a membrane. Once encased in a portion of the cell membrane (now called a phagocytic vacuole), a lysosome adds its digestive enzymes to it, which speeds the breakdown of the dangerous microbes. Finally, the hydrolyzed (digested) material moves from the vacuole to the inner surface of the cell membrane, where the contents are discharged by exocytosis.

In addition to active transport, materials can be transported into a cell by *endocytosis* and out by *exocytosis*. **Phagocytosis** is another name for one kind of endocytosis that is the process cells use to wrap membrane around a particle (usually food) and engulf it (figure 20.11). This is the process leukocytes (white blood cells) use to surround invading bacteria, viruses, and other foreign materials. Because of this, these kinds of cells are called *phagocytes*. When phagocytosis occurs, the material to be engulfed touches the surface of the phagocyte and causes a portion of the outer cell membrane to be indented. The indented cell membrane is pinched off inside the cell to form a sac containing the engulfed material. This sac, composed of a single membrane, is called a **vacuole.**

Once inside the cell, the membrane of the vacuole is broken down, releasing its contents inside the cell, or it may combine with another vacuole containing destructive enzymes.

ORGANELLES COMPOSED OF MEMBRANES

Now that you have some background concerning the structure and the function of membranes, let's turn our attention to the way cells use membranes to build the structural components of their protoplasm. The outer boundary of the cell is termed the cell membrane, or **plasma membrane.** It is associated with one of the characteristics of life, metabolism, including taking up and releasing molecules, sensing stimuli in the environment, recognizing other cell types, and attaching to other cells and nonliving objects. In addition to the cell membrane, many other organelles are composed of membranes. Each of these membranous organelles has a unique shape or structure that is associated with particular functions.

One of the most common organelles found in cells is the *endoplasmic reticulum*. The **endoplasmic reticulum** (ER) is a set of folded membranes and tubes throughout the cell. This system of membranes provides a large surface upon which chemical activities take place (figure 20.12). Since the ER has an enormous surface area, many chemical reactions can be carried out in an extremely small space.

The **Golgi apparatus** is another organelle composed of membrane. Even though this organelle is also composed of membrane, the way in which it is structured enables it to perform jobs different from those performed by the ER. The typical Golgi is composed of from five to twenty flattened, smooth, membranous sacs, which resemble a stack of pancakes.

The Golgi apparatus is the site of the synthesis and packaging of certain molecules produced in the cell. It is also the place where particular chemicals are concentrated prior to their release from the cell or distribution within the cell.

An important group of molecules that is necessary to the cell includes the hydrolytic enzymes. This group of enzymes is capable of destroying carbohydrates, nucleic acids, proteins, and lipids. Since cells contain large amounts of these molecules, these enzymes must be controlled in order to prevent the destruction of the cell. The Golgi apparatus is the site where these enzymes are converted from their inactive to their active forms and packaged in membranous sacs. These vesicles are pinched off from the outside surfaces of the Golgi sacs and given the special name **lysosomes,** or "bursting body." Cells use the lysosomes in four major ways:

1. Decompose dead and dying cells.
2. Selectively destroy cells to fashion a developing organism from its immature to mature state (figure 20.13).
3. Digest macromolecules that have been ingested into the cell.
4. Kill dangerous microorganisms that have been taken into the cell by phagocytosis.

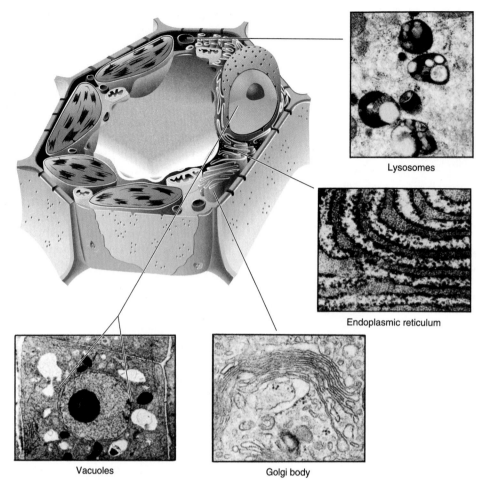

Lysosomes

Endoplasmic reticulum

Vacuoles

Golgi body

FIGURE 20.12 Membranous cytoplasmic organelles. Certain structures in the cytoplasm are constructed of membranes. Membranes are composed of protein and phospholipids. The four structures here—lysosomes, endoplasmic reticulum, vacuoles, and the Golgi body—are constructed of simple membranes.

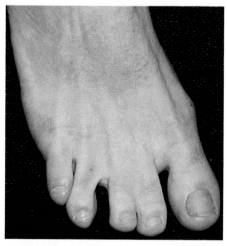

FIGURE 20.13 This person displays the trait known as syndactylism (*syn* = connected; *dactyl* = finger or toe). In most people, lysosomal enzymes break down the cells between the toes, allowing the toes to separate. In this genetic abnormality, these enzymes fail to do their job.

A nucleus is a place in a cell—not a solid mass. Just as a room is a place created by walls, a floor, and a ceiling, the nucleus is a place in the cell created by the **nuclear membrane.** This membrane separates the *nucleoplasm,* liquid material in

the nucleus, from the cytoplasm. Because they are separated, the cytoplasm and nucleoplasm can maintain different chemical compositions. If the membrane were not formed around the genetic material, the organelle we call the nucleus would not exist. The nuclear membrane is formed from many flattened sacs fashioned into a hollow sphere around the genetic material, DNA (*d*eoxyribo*n*ucleic *a*cid). It also has large openings called nuclear pores that allow thousands of relatively large molecules to pass into and out of the nucleus each minute. These pores are held open by donut-shaped molecules that resemble the "eyes" in shoes through which the shoelace is strung.

All of the membranous organelles just described can be converted from one form to another (figure 20.14). For example, phagocytosis results in the formation of vacuolar membrane from cell membrane that fuses with lysosomal membrane, which in turn came from Golgi membrane. Two other organelles composed of membranes are chemically different and are incapable of interconversion. Both types of organelles are associated with energy conversion reactions in the cell. These organelles are the *mitochondrion* and the *chloroplast* (figure 20.15).

The **mitochondrion** is an organelle resembling a small bag with a larger bag inside that is folded back on itself. These inner folded surfaces are known as the *cristae.* Located on the surface of the cristae are particular proteins and enzymes involved in

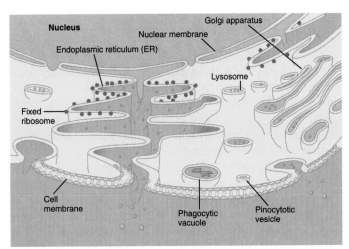

FIGURE 20.14 Eukaryotic cells contain a variety of organelles composed of phospholipids and proteins. Each has a unique shape and function. Many of these organelles are interconverted from one to another as they perform their essential functions. Cell membranes can become vacuolar membrane or endoplasmic reticulum, which can become vesicular membrane, which in turn can become Golgi or nuclear membrane. However, mitochondria cannot exchange membrane parts with other membranous organelles.

aerobic cellular respiration. **Aerobic cellular respiration** is the series of reactions involved in the release of usable energy from food molecules, which requires the participation of oxygen molecules. The average human cell contains upwards of ten thousand mitochondria. Cells that are involved in activities that require large amounts of energy, such as muscle cells, contain many more mitochondria. When properly stained, they can be seen with a compound light microscope. When cells are functioning aerobically, the mitochondria swell with activity. But when this activity diminishes, they shrink and appear as thread-like structures.

A second energy-converting organelle is the **chloroplast.** Some cells contain only one large chloroplast, while others contain hundreds of smaller chloroplasts. A study of the ultrastructure—that is, the structures seen with an electron microscope—of a chloroplast shows that the entire organelle is enclosed by a membrane, while other membranes are folded and interwoven throughout. As shown in figure 20.15A, in some areas, concentrations of these membranes are stacked up or folded back on themselves. Chlorophyll and other photosynthetic molecules are attached to these membranes. These areas of concentrated chlorophyll are called the *grana* of the

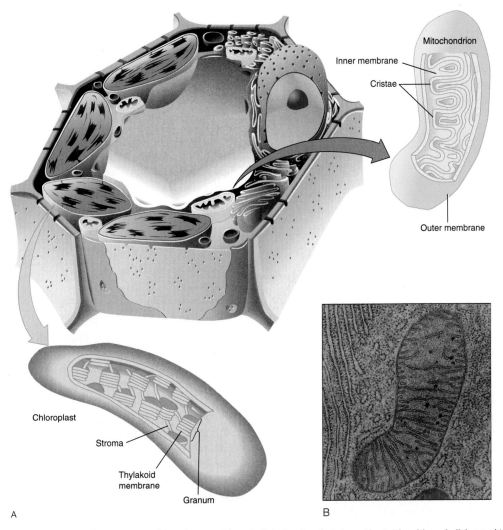

FIGURE 20.15 (*A*) The chloroplast, the container of the pigment chlorophyll, is the site of photosynthesis. The chlorophyll, located in the grana, captures light energy that is used to construct organic molecules in the stroma. (*B*) The mitochondria with their inner folds, called cristae, are the site of aerobic cellular respiration, where food energy is converted to usable cellular energy. Both organelles are composed of phospholipid and protein membranes.

chloroplast. The space between the grana, which has no chlorophyll, is known as the *stroma*.

These membranous, saclike organelles are found only in plants and algae. In this organelle, light energy is converted to chemical-bond energy in a process known as *photosynthesis*. **Photosynthesis** is the *trapping of radiant energy and its conversion into the energy of chemical bonds*. Chloroplasts contain a variety of photosynthetic pigments including green *chlorophyll*.

Mitochondria and chloroplasts are different from other kinds of membranous structures in several ways. First, their membranes are chemically different from those of other membranous organelles. Second, they are composed of double layers of membrane—an inner and an outer membrane. Third, both of these structures have ribosomes and DNA that are similar to those of bacteria. Finally, these two structures have a certain degree of independence from the rest of the cell—they have a limited ability to reproduce themselves but must rely on DNA from the cell's nucleus for assistance.

The biochemical pathways associated with mitochondria and chloroplasts are respiration and photosynthesis, both oxidation-reduction reactions.

RESPIRATION AND PHOTOSYNTHESIS

In living things, oxidation-reduction reactions are metabolic processes that do not take place in one single step but occur in a series of small steps. This allows the potential energy in the oxidized molecule to be released in smaller, more useful amounts. A unique protein catalyst (enzyme) controls each of the steps. Each step begins with a molecule or molecules that serve as the reactant and are called the *substrate*. When the reaction is complete, the substrate is converted to a product, which in turn becomes the new substrate for the next enzyme-controlled reaction. Such a series of enzyme-controlled reactions is often called a *biochemical pathway* or a *metabolic pathway*:

Enzyme 1 Enzyme 2 Enzyme 3 Enzyme 4

Substrate A → Substrate B → Substrate C → Substrate D → End product

Such pathways can be used to create molecules (e.g., photosynthesis), release energy (e.g., aerobic cellular respiration), and perform many other actions. One of the amazing facts of nature is that most organisms use the same basic biochemical or metabolic pathways. So, if you study aerobic cellular respiration in an elephant, it will be essentially the same as that in a petunia, shark, or earthworm. However, since the kinds of enzymes an organism is able to produce depend on the genes that it has, we should expect some variation in the details of biochemical pathways in different organisms. The fact that so many organisms use essentially the same biochemical processes is a strong argument for the idea of evolution from common ancestors. Once a successful biochemical strategy evolved, the genes and the pathway were retained by evolutionary descendants with slight modifications of the scheme. Two such pathways of importance are aerobic cellular respiration and photosynthesis, both of which involve the transfer of chemical bond energy in the form of ATP.

The Energy Transfer Molecules of Living Things—ATP

To transfer the right amount of chemical-bond energy from energy-releasing to energy-requiring reactions, cells use the molecule ATP. **Adenosine *tri*phosphate (ATP)** is a handy source of the right amount of usable chemical-bond energy. Each ATP molecule used in the cell is like a rechargeable AAA battery used to power small toys and electronic equipment. Each contains just the right amount of energy to power the job. When the power has been drained, it can be recharged numerous times before it must be recycled.

Recharging the AAA battery requires getting a small amount of energy from a source of high energy such as a hydroelectric power plant (figure 20.16). Energy from the electric plant is too powerful to directly run a small flashlight or portable recording device. If you plug your recorder directly into the power plant, the recorder would be destroyed. However, the recharged AAA battery delivers just the right amount of energy at the right time and place. ATP functions in much the same manner. After the chemical-bond energy has been drained by breaking one of its bonds:

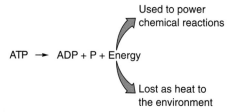

the discharged molecule (ADP) is recharged by "plugging it in" to a high-powered energy source. This source may be (1) sunlight (photosynthesis) or (2) chemical-bond energy (released from cellular respiration):

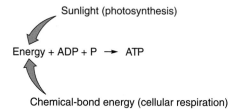

An ATP molecule is formed from adenine (nitrogenous base), ribose (sugar), and phosphates (figure 20.17). These three are chemically bonded to form AMP, *adenosine monophosphate* (one phosphate). When a second phosphate group is added to the AMP, a molecule of ADP (diphosphate) is formed. The ADP, with the addition of more energy, is able to bond to a third phosphate group and form ATP. The covalent bond that attaches the second phosphate to the AMP molecule is easily broken to release energy for energy-requiring cell processes. Because the energy in this bond is so easy for a cell to use, it is called a *high-energy phosphate bond*. ATP has two high-energy phosphate bonds represented by curved solid lines. Both ADP and ATP, because they contain high-energy bonds, are very unstable molecules and readily lose their phosphates. When this occurs, the energy held in the high-energy bonds of the phosphate can be transferred to another molecule or released

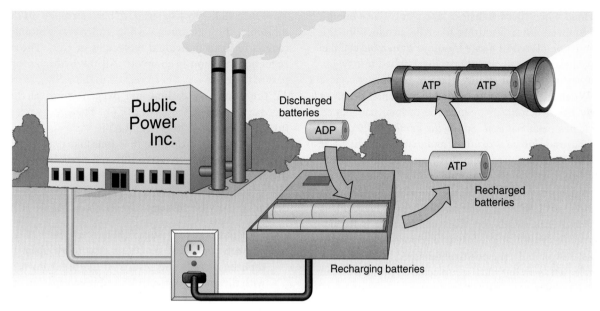

FIGURE 20.16 When rechargeable batteries in a flashlight have been drained of their power, they can be recharged by placing them in a specially designed battery charger. This enables the right amount of power from a power plant to be packed into the batteries for reuse. Cells operate in much the same manner. When the cell's "batteries," ATPs, become drained as a result of powering a job such as muscle contraction, these discharged "batteries," ADPs, can be recharged back to full ATP power.

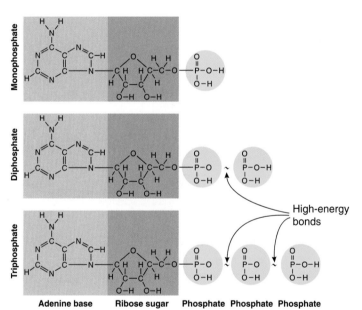

FIGURE 20.17 A macromolecule of ATP consists of a molecule of adenine, a molecule of ribose, and three phosphate groups. The two end phosphate groups are bonded together by high-energy bonds. When these bonds are broken, they release an unusually great amount of energy; therefore, they are known as *high-energy bonds*. These bonds are represented by curved, solid lines. The ATP molecule is considered an energy carrier.

to the environment. It is this ATP energy that is used by cells to perform all the characteristics that keep them alive.

Aerobic Cellular Respiration

Aerobic cellular respiration is a specific series of enzyme-controlled chemical reactions in which oxygen is involved in the breakdown of glucose into carbon dioxide and water, and

the chemical-bond energy from glucose is released to the cell in the form of ATP (adenosine triphosphate). While the actual process of aerobic cellular respiration involves many enzyme-controlled steps, the overall process is a reaction between the substrates sugar and oxygen resulting in the formation of the products carbon dioxide and water with the release of energy. The following equation summarizes this process:

$$\underset{C_6H_{12}O_6}{\text{glucose}} + \underset{O_2}{\text{oxygen}} \rightarrow \underset{CO_2}{\overset{\text{carbon}}{\text{dioxide}}} + \underset{H_2O}{\text{water}} + \underset{ATP + \text{heat}}{\text{energy}}$$

Covalent bonds are formed by atoms sharing pairs of fast-moving, energetic electrons. Therefore, the covalent bonds in the sugar glucose contain chemical potential energy. Of all the covalent bonds in glucose (O—H, C—H, C—C), those easiest to get at are the C—H and O—H bonds on the outside of the molecule. The chemical activities that remove electrons from glucose result in the glucose being oxidized. When these bonds are broken, several things happen:

1. Some ATP is produced;
2. Hydrogen ions (H^+ or protons) are produced; and,
3. Electrons are released and taken up by special carrier molecules.

The ATP is used to power the metabolic activities of the cell. The chemical activities that remove electrons from glucose result in the glucose being oxidized.

The electrons produced during aerobic cellular respiration must be controlled. If they were allowed to fly about at random, they would quickly combine with other molecules, causing cell death. Electron transfer molecules temporarily hold the electrons and transfer them to other electron transfer molecules.

ATP is formed when these transfers take place. Once energy has been removed from electrons for ATP production, the electrons must be placed in a safe location. In *aerobic* cellular respiration, these electrons are ultimately attached to oxygen. Oxygen serves as the final resting place of the less-energetic electrons. When the electrons are added to oxygen, it becomes a negatively charged ion, O=. Since the oxygen has gained electrons, it has been *reduced*. *So, in the aerobic cellular respiration of glucose, glucose is oxidized and oxygen is reduced.* If something is oxidized (loses electrons), something else must be reduced (gains electrons). A simple way to help identify an oxidation-reduction reaction is to use the mnemonic device "LEO the lion says GER." LEO stands for "Loss of Electrons is Oxidation" and GER stands for "Gain of Electrons is Reduction." A molecule cannot simply lose its electrons—they have to go someplace! Eventually, the positively charged hydrogen ions that were released from the glucose molecule combine with the negatively charged oxygen ion to form water.

Once all the hydrogens have been stripped off the glucose molecule, the remaining carbon and oxygen atoms are rearranged to form individual molecules of CO_2. The oxidation-reduction reaction is complete. All the hydrogen originally a part of the glucose has been moved to the oxygen to form water. All the remaining carbon and oxygen atoms of the original glucose are now in the form of CO_2. The energy released from this process is used to generate ATP.

In cells, these oxidation reactions take place in a particular order and in particular places within the cell. In eukaryotic cells, the process of releasing energy from food molecules begins in the cytoplasm and is completed in the mitochondrion. Three distinct enzymatic pathways are involved (figure 20.18):

1. *Glycolysis* (*glyco* = sugar; *lysis* = to split) is a series of enzyme-controlled reactions that takes place in the cytoplasm of cells and results in the breakdown of glucose with the release of electrons and the formation of ATP. The

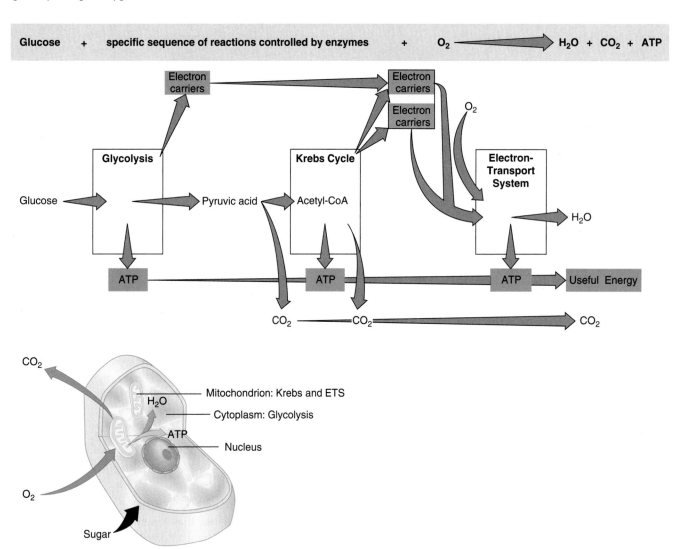

FIGURE 20.18 This sequence of reactions in the aerobic oxidation of glucose is an overview of the energy-yielding reactions of a cell. Glycolysis, the Krebs cycle, and the electron-transport system are each a series of enzyme-controlled reactions that extract energy from the chemical bonds in a glucose molecule. During glycolysis, glucose is split into pyruvic acid and ATP and electrons are released. During the Krebs cycle, pyruvic acid is further broken down to carbon dioxide with the release of ATP and electrons. During the electron-transport system, oxygen is used to accept electrons and water is produced. A very large amount of ATP is also produced. Glycolysis takes place in the cytoplasm of the cell. Pyruvic acid enters mitochondria, where the Krebs cycle and the electron-transport system take place.

electrons are sent to the electron-transport system (ETS) for processing.

2. The *Krebs cycle* is a series of enzyme-controlled reactions that takes place inside the mitochondrion and completes the breakdown of the remaining fragments of the original glucose with the release of carbon dioxide, more ATP, and more electrons are sent to the electron-transport system (ETS) for processing.

3. The *electron-transport system (ETS)* occurs in the mitochondria. It is a series of enzyme-controlled reactions that converts the kinetic energy of electrons it receives from glycolysis and the Krebs cycle to ATP. In fact, the majority of ATP produced during aerobic cellular respiration comes from the ETS. The electrons are transferred through a series of oxidation-reduction reactions involving enzymes until eventually, the electrons are accepted by oxygen atoms to form oxygen ions (O=). The negatively charged oxygen atoms attract two positively charged hydrogen ions to form water (H_2O).

Photosynthesis

Ultimately, the energy to power all organisms comes from the Sun's radiant energy. Chlorophyll is a green pigment that absorbs light energy for the process of photosynthesis. Through the process of photosynthesis, plants, algae, and certain bacteria transform light energy to chemical-bond energy in the form of ATP and then use ATP to produce complex organic molecules such as glucose. In algae and the leaves of green plants, photosynthesis occurs in cells that contain chloroplasts.

The following equation summarizes the chemical reactions green plants and many other photosynthetic organisms use to make ATP and organic molecules:

light energy + carbon dioxide + water → glucose + oxygen

$$\text{light energy} + CO_2 + H_2O \rightarrow C_6H_{12}O_6 + O_2$$

There are three distinct events in the photosynthetic pathway:

1. *Light-capturing events.* In eukaryotic cells, the process of photosynthesis takes place within chloroplasts. Each chloroplast is surrounded by membranes and contains the green pigment, chlorophyll, along with other pigments. Chlorophyll and other *accessory* pigments (yellow, red, and orange) absorb specific wavelengths of light. After specific amounts of light are absorbed by photosynthetic pigments, their electrons become "excited." With this added energy, these excited electrons are capable of entering into the chemical reactions responsible for the production of ATP (figure 20.19).

2. *Light-dependent reactions.* The light-dependent reactions utilize the excited electrons produced by the light-capturing activities. The light-dependent reactions are also known as the *light reactions.* During these reactions, "excited" electrons from the light-capturing reactions are used to do two different things. Some of the "excited" electrons are used to make ATP, while others are used to split water into hydrogen and oxygen. The oxygen from the water

FIGURE 20.19 In the fall, a layer of waterproof tissue forms at the base of each leaf, cutting off the flow of water and other nutrients. The cells of the leaf die and their chlorophyll disintegrates. The color change seen in leaves in the fall in certain parts of the world is the result of the breakdown of the green chlorophyll. Other pigments (red, yellow, orange, brown) are always present but are masked by the green chlorophyll pigments. When the chlorophyll disintegrates, the reds, oranges, yellows, and browns are revealed.

is released to the environment as O_2 molecules, and the hydrogens are transferred to an electron carrier.

3. *Light-independent reactions.* These reactions are also known as the *dark reactions,* since light is not needed for the reactions to take place. During these reactions, ATP and electron-carrying molecules from the light-dependent reactions are used to attach CO_2 to a five-carbon molecule, already present in the cell, to manufacture new, larger organic molecules. Ultimately, glucose ($C_6H_{12}O_6$) is produced. The ADP and the electron carriers produced during the light-independent reactions are recycled back to the light-dependent reactions to be used over again (figure 20.20).

Because prokaryotic cells lack mitochondria and chloroplasts, they carry out photosynthesis and cellular respiration within the cytoplasm, on the inner surfaces of the cell membrane, or on other special membranes. Photosynthesis and cellular respiration both involve a series of chemical reactions that control the flow of energy.

Many people believe that plants only give off oxygen and never require it. This is incorrect! Plants do give off oxygen in the light-dependent reactions of photosynthesis, but in aerobic cellular respiration, they use oxygen as does any other organism. During their life spans, green plants give off more oxygen to the atmosphere than they take in for use in respiration. The surplus oxygen given off is the source of oxygen for aerobic cellular respiration in both plants and animals. Animals are not only dependent on plants for oxygen but are ultimately dependent

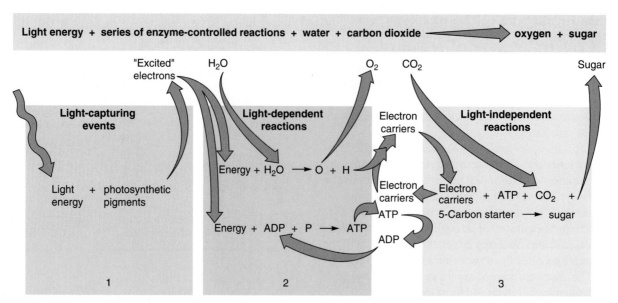

FIGURE 20.20 Photosynthesis is a complex biochemical pathway in plants, algae, and certain bacteria. The upper portion of this figure shows the overall process. Sunlight, along with CO_2 and H_2O, is used to make organic molecules such as sugar. The lower portion illustrates the three parts of the process: (1) the light-capturing events, (2) the light-dependent reactions, and (3) the light-independent reactions. Notice that the end products of the light-dependent reactions, electron carriers and ATP, are necessary to run the light-independent reactions, while the water and carbon dioxide are supplied from the environment.

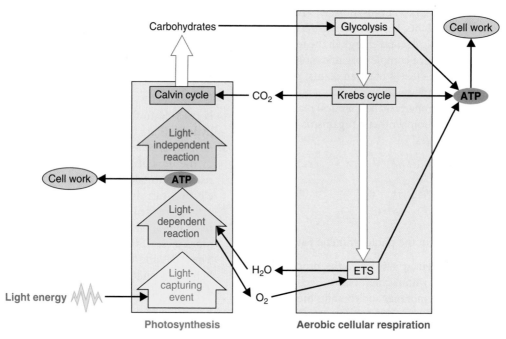

FIGURE 20.21 Although both autotrophs and heterotrophs carry out cellular respiration, the photosynthetic process that is unique to photosynthetic autotrophs provides essential nutrients for both processes. Photosynthesis captures light energy, which is ultimately transferred to heterotrophs in the form of carbohydrates and other organic compounds. Photosynthesis also generates O_2, which is used in aerobic cellular respiration. The ATP generated by cellular respiration in both heterotrophs, (e.g., animals) and autotrophs (e.g., plants) is used to power their many metabolic processes. In return, cellular respiration supplies two of the most important basic ingredients of photosynthesis, CO_2 and H_2O.

on plants for the organic molecules necessary to construct their bodies and maintain their metabolism (figure 20.21).

All of the organelles just described are composed of membranes. Many of these membranes are modified for particular functions. Each membrane is composed of the double phospholipid layer with protein molecules associated with it.

NONMEMBRANOUS ORGANELLES

Suspended in the cytoplasm and associated with the membranous organelles are various kinds of structures that are not composed of phospholipids and proteins arranged in sheets. In the cytoplasm are many very small structures called **ribosomes**

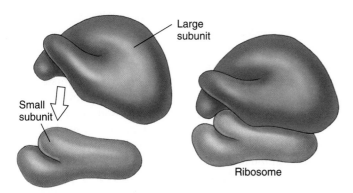

FIGURE 20.22 Each ribosome is constructed of two subunits. Each of the subunits is composed of protein and RNA. These globular organelles are associated with the construction of protein molecules from individual amino acids.

that are composed of ribonucleic acid (RNA) and protein. Ribosomes function in the manufacture of protein. Many ribosomes are found floating freely in the cytoplasm and attached to the endoplasmic reticulum (figure 20.22). Cells that are actively producing protein (e.g., liver cells) have great numbers of free and attached ribosomes.

Among the many types of nonmembranous organelles found there are elongated protein structures that are known as *microtubules, microfilaments,* and *intermediate filaments.* Their various functions are as complex as those provided by the structural framework and cables of a high-rise office building, geodesic dome, or skeletal and muscular systems of a large animal. All three types of organelles interconnect, and some are attached to the inside of the cell membrane, forming what is known as the *cytoskeleton* of the cell (figure 20.23).

These cellular components provide the cell with shape, support, and ability to move about the environment. In certain diseases such as ALS (amyotrophic lateral sclerosis, or Lou Gehrig's disease), Alzheimer's, and Down syndrome, there is a buildup of intermediate-type filaments. As a result, researchers are investigating control methods that prevent such buildup.

An arrangement of two sets of microtubules at right angles to each other makes up the structures known as the *centrioles,* and they operate by organizing microtubules into a complex of strings called *spindle fibers.* The *spindle* is the structure upon which chromosomes are attached in order that they may be properly separated during cell division (refer to figure 20.29). Each set is composed of nine groups of short microtubules arranged in a cylinder (figure 20.24). One curious fact about centrioles is that they are present in most animal cells but not in many types of plant cells.

Many cells have microscopic, hairlike structures projecting from their surfaces; these are *cilia* or *flagella* (figure 20.25). In general, we call them flagella if they are long and few in number, and cilia if they are short and more numerous. The cell has the ability to control the action of these microtubular structures, enabling them to be moved in a variety of different ways. Their coordinated actions either propel the cell through the environment or the environment past the cell surface. The flagella of sperm propel them through the reproductive tract toward the egg. The protozoan *Paramecium* is covered with thousands of cilia that actively beat a rhythmic motion to move the cell through the water. The cilia on the cells that line your trachea move mucus-containing particles from deep within your lungs.

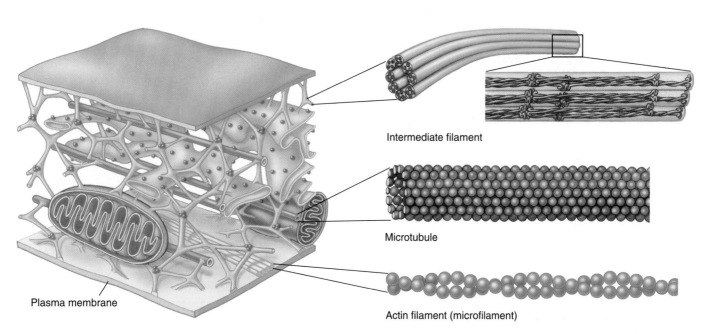

FIGURE 20.23 Microtubules, actin filaments, and intermediate filaments are all interconnected within the cytoplasm of the cell. These structures, along with connections to other cellular organelles, form a cytoskeleton for the cell. The cellular skeleton is not a rigid, fixed-in-place structure but rather changes as the actin and intermediate filaments and microtubule component parts are assembled and disassembled.

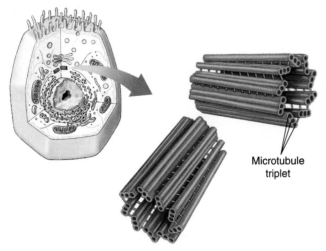

FIGURE 20.24 These two sets of short microtubules are located just outside the nuclear membrane in many types of cells.

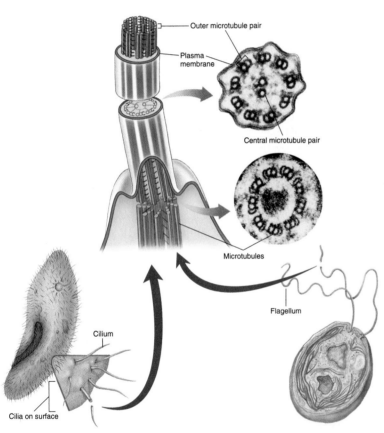

FIGURE 20.25 Cilia and flagella function like oars or propellers that move the cell through its environment or move the environment past the cell. Cilia and flagella are constructed of groups of microtubules, as in the ciliated protozoan shown on the left and the flagellated alga on the right. Flagella are usually less numerous and longer than cilia.

NUCLEAR COMPONENTS

When nuclear structures were first identified, it was noted that certain dyes stained some parts more than others. The parts that stained more heavily were called **chromatin,** which means

"colored material." Chromatin is composed of long molecules of deoxyribonucleic acid (DNA) in association with proteins. Chromatin is loosely organized DNA/protein strands in the nucleus. When the chromatin is tightly coiled into shorter, denser structures, we call them **chromosomes.** Chromatin and chromosomes are really the same molecules but differ in structural arrangement. In addition to chromosomes, the nucleus may also contain one, two, or several *nucleoli.* A **nucleolus** is composed of granules and fibers in association with the cell's DNA used in the manufacture of ribosomes.

The final component of the nucleus is its liquid matrix called the *nucleoplasm.* It is a colloidal mixture composed of water and the molecules used in the construction of ribosomes, nucleic acids, and other nuclear material.

MAJOR CELL TYPES

Not all of the cellular organelles we have just described are located in every cell. Some cells typically have combinations of organelles that differ from others. For example, some cells have a nuclear membrane, mitochondria, chloroplasts, ER, and Golgi; others have mitochondria, centrioles, Golgi, ER, and nuclear membrane. Other cells are even more simple and lack the complex membranous organelles described in this chapter. Because of this fact, biologists have been able to classify cells into two major types: prokaryotic and eukaryotic.

The Prokaryotic Cell Structure

Prokaryotic cells, the *bacteria* and *archaea,* do not have a typical nucleus bound by a nuclear membrane, nor do they contain mitochondria, chloroplasts, Golgi, or extensive networks of ER. However, prokaryotic cells contain DNA and enzymes and are able to reproduce and engage in metabolism. They perform all of the basic functions of living things with fewer and more simple organelles. Members of the Archaea are of little concern to the medical profession because none has been identified as disease-causing. They are typically found growing in extreme environments where the pH, salt concentration, or temperatures make it impossible for most other organisms to survive. The other prokaryotic cells are called bacteria, and about 5 percent cause diseases such as tuberculosis, strep throat, gonorrhea, and acne. Other bacteria are responsible for the decay and decomposition of dead organisms. Although some bacteria have a type of green photosynthetic pigment and carry on photosynthesis, they do so without chloroplasts and use different chemical reactions.

Most prokaryotic cells are surrounded by a *capsule,* slime layer, or spore coat that can be composed of a variety of compounds (figure 20.26). In certain bacteria, this layer is responsible for their ability to stick to surfaces (including host cells) and to resist phagocytosis. Many bacteria also have fimbriae, hairlike protein structures, which help the cell stick to objects. Those with flagella are capable of propelling themselves through the environment. Below the capsule is the rigid cell wall composed of a unique protein-carbohydrate complex called peptidoglycan. This complex provides the cell with the

Antibiotics and Cell Structural Differences

One significant difference between prokaryotic and eukaryotic cells is in the chemical makeup of their ribosomes. The ribosomes of prokaryotic cells contain different proteins from those found in eukaryotic cells. Prokaryotic ribosomes are also smaller. This discovery was important to medicine because many cellular forms of life that cause common diseases are bacterial. As soon as differences in the ribosomes were noted, researchers began to look for ways in which to interfere with the prokaryotic ribosome's function but *not* interfere with the ribosomes of eukaryotic cells. Antibiotics such as streptomycin are the result of this research. This drug combines with prokaryotic ribosomes and causes the death of the prokaryote by preventing the production of proteins essential to its survival. Because eukaryotic ribosomes differ from prokaryotic ribosomes, streptomycin does not interfere with the normal function of ribosomes in human cells.

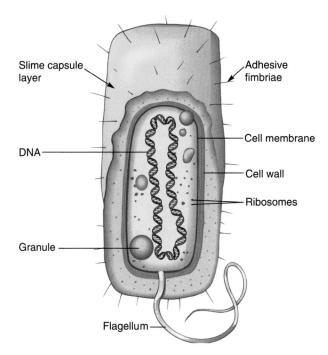

FIGURE 20.26 All bacteria are prokaryotic cells. While smaller and less complex than eukaryotic cells, each is capable of surviving on its own. Most bacteria are involved with decay and decomposition, but a small percentage are pathogens responsible for disease such as strep throat, TB, syphilis, and gas gangrene. The cell illustrated here is a bacillus because it has a rod shape.

TABLE 20.2 Comparison of General Plant and Animal Cell Structure

Plant Cells	Animal Cells
CELL WALL	_____
Cell membrane	Cell membrane
Cytoplasm	Cytoplasm
Nucleus	Nucleus
Mitochondria	Mitochondria
CENTRAL VACUOLE	_____
CHLOROPLASTS	_____
	CENTRIOLE
_____	
Golgi apparatus	Golgi apparatus
Endoplasmic reticulum	Endoplasmic reticulum
Lysosomes	Lysosomes
Vacuoles/vesicles	Vacuoles/vesicles
Ribosomes	Ribosomes
Nucleolus	Nucleolus
Inclusions	Inclusions
Cytoskeleton	Cytoskeleton

strength to resist osmotic pressure changes and gives the cell shape. Just beneath the wall is the cell membrane. Thinner and with a slightly different chemical composition from eukaryotes, it carries out the same functions as the cell membranes in eukaryotes. Most bacteria are either rods (bacilli), spherical (cocci), or curved (spirilla). The genetic material within the cytoplasm is DNA in the form of a loop.

The Eukaryotic Cell Structure

Eukaryotic cells contain a true nucleus and most of the membranous organelles described earlier. Eukaryotic organisms can be further divided into several categories based on the specific combination of organelles they contain. The cells of plants, fungi, protozoa and algae, and animals are all eukaryotic. The most obvious characteristic that sets the plants and algae apart from other organisms is their green color, which indicates that the cells contain chlorophyll. Chlorophyll is necessary for the process of photosynthesis—the conversion of light energy into chemical-bond energy in food molecules. These cells, then, are different from the other cells in that they contain chloroplasts in their cytoplasm. Another distinguishing characteristic of plants and algae is the presence of cellulose in their cell walls (table 20.2).

The group of organisms that has a cell wall but lacks chlorophyll in chloroplasts is collectively known as *fungi*. They were previously thought to be either plants that had lost their ability to make their own food or animals that had developed cell walls. Organisms that belong in this category of eukaryotic cells include yeasts, molds, mushrooms, and the fungi that cause such human diseases as athlete's foot, "jungle rot," and ringworm. Now we have come to recognize this group as different enough from plants and animals to place them in a separate kingdom.

Eukaryotic organisms that lack cell walls and cannot photosynthesize are placed in separate groups. Organisms that consist of only one cell are called protozoans—examples are *Amoeba* and *Paramecium*. They have all the cellular organelles described in this chapter except the chloroplast; therefore, protozoans must consume food as do the fungi and the multicellular animals.

Although the differences in these groups of organisms may seem to set them worlds apart, their similarity in cellular structure is one of the central themes unifying the field of biology. You can obtain a better understanding of how cells operate in general by studying specific examples. Because the organelles have the same general structure and function regardless of the kind of cell in which they are found, you can learn more about how mitochondria function in plants by studying how mitochondria function in animals. There is a commonality among all living things with regard to their cellular structure and function.

THE IMPORTANCE OF CELL DIVISION

The process of cell division is a generative process, one of the characteristics of life. It replaces dead cells with new ones, repairs damaged tissues, and allows living organisms to grow. For example, you began as a single cell that resulted from the union of a sperm and an egg. One of the *first* activities of this single cell was to divide. As this process continued, the number of cells in your body increased, so that as an adult your body consists of several trillion cells.

The *second* function of cell division is to maintain the body. Certain cells in your body, such as red blood cells and cells of the gut lining and skin, wear out. As they do, they must be replaced with new cells. Altogether, you lose about fifty million cells per second; this means that millions of cells are dividing in your body at any given time.

A *third* purpose of cell division is repair. When a bone is broken, the break heals because cells divide, increasing the number of cells available to knit together the broken pieces. If some skin cells are destroyed by a cut or abrasion, cell division produces new cells to repair the damage.

During eukaryotic cell division, two events occur. The replicated genetic information of a cell is equally distributed to two daughter nuclei in a process called **mitosis** (Latin, from Greek *mitos,* meaning "thread"). As the nucleus goes through its division, the cytoplasm also divides into two new cells. This division of the cell's cytoplasm is called **cytokinesis**—cell splitting. Each new cell gets one of the two daughter nuclei so that both have a complete set of genetic information.

THE CELL CYCLE

All eukaryotic cells go through the same basic life cycle, but they vary in the amount of time they spend in the different stages. A generalized picture of a cell's life cycle may help you understand it better (figure 20.27). Once begun, cell division is a continuous process without a beginning or an end. It is a cycle in which cells continue to grow and divide. There are five stages to the life cycle of a eukaryotic cell: (1) G_1, gap (growth)—phase one; (2) S, synthesis; (3) G_2, gap (growth)—phase two; (4) cell division (mitosis and cytokinesis); and (5) G_0, gap (growth)—mitotic dormancy or differentiation.

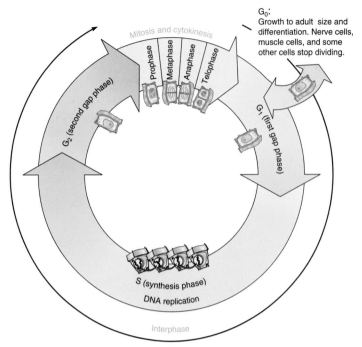

FIGURE 20.27 During the cell cycle, tRNA, mRNA, ribosomes, and enzymes are produced in the G_1 stage. DNA replication occurs in the S stage. Proteins required for the spindles are synthesized in the G_2 stage. The nucleus is replicated in mitosis and two cells are formed by cytokinesis. Once some organs, such as the brain, have completely developed, certain types of cells, such as nerve cells, enter the G_0 stage. The time periods indicated are relative and vary depending on the type of cell and the age of the organism.

During the G_0 phase, cells are not considered to be in the cycle of division but become *differentiated* or specialized in their function. It is at this time that they "mature" to play the role specified by their genetic makeup. Whereas some cells entering the G_0 phase remain there more or less permanently (e.g., nerve cells), others have the ability to move back into the cell cycle of mitosis—G_1, S, and G_2—with ease (e.g., skin cells).

The first three phases of the cell cycle—G_1, S, and G_2—occur during a period of time known as interphase. **Interphase** is the stage between cell divisions. During the G_1 stage, the cell grows in volume as it produces tRNA, mRNA, ribosomes, enzymes, and other cell components. During the S stage, DNA replication occurs in preparation for the distribution of genes to daughter cells. During the G_2 stage that follows, final preparations are made for mitosis with the synthesis of spindle-fiber proteins.

During interphase, the cell is not dividing but is engaged in metabolic activities such as photosynthesis or glandular-cell secretion. During interphase, the nuclear membrane is intact and the individual chromosomes are not visible. The individual loosely coiled *chromatin* strands are too thin and tangled to be seen. Remember that chromosomes are highly coiled strands of DNA and contain coded information that tells the cell such things as (1) how to produce enzymes required for the digestion of nutrients, (2) how to manufacture enzymes that will

Stem cells have been a hot topic in the news recently. Controversy surrounds their use in research because of their sources and how they are viewed in the debate over the definition of life. Just what are stem cells, how are they acquired, and of what value might they be?

Stem cells are primitive, undifferentiated cells that have the ability to become many other types of cells, including liver, skin, and brain cells. Theoretically, they can divide for an indefinite period in culture. They are found in many tissues, such as circulating blood and the red bone marrow of certain bones, including the pelvis, and in embryos, or they can be generated from embryos in the lab. Adult stem cells can differentiate to produce all the specialized cell types of the tissue from which they originated. Embryonic stem cells are undifferentiated cells from the embryo. Their ability to differentiate into specialized cells may make them extremely valuable as replacement cells for individuals who have lost cells as a result of illness or injury.

There are three sources of stem cells: lab cultures, isolation from embryos, and adult donors. One method of obtaining embryonic stem cells is to use donated sperm and an egg and fertilize them in a Petri dish. The fertilized egg or zygote is then allowed to undergo mitosis for several days until it becomes a ball of cells. Some of these cells are then removed and grown in a separate dish, serving as a source of embryonic stem cells.

Since embryonic stem cells can differentiate into many other types of human cells, researchers are hoping that they could potentially be provided to people to resolve many problems. Such problems might include: the repair of spinal injuries to restore function to paralyzed limbs, the replacement of useless scar tissue after a heart attack, or adding cells to have them produce chemicals not normally produced as a result of genetic or other disease (e.g., dopamine in Parkinson's patients, insulin in diabetics, specific receptor sites that help people resist HIV or bubonic plague infections). Embryonic stem cells might also be used to provide sickle-cell disease patients with the healthy cells they need to produce normal red blood cells. This is still in the research stage, and there is still much to be accomplished before such therapies are available to the public.

Adults can also be a source of stem cells. People with certain diseases (e.g., chronic lymphocytic leukemia, CLL) or those receiving chemotherapy or radiation for cancer, lose stem cells and therefore have lower numbers of vital red and white blood cells. However, it is possible to replace these cells with healthy, blood-forming stem cells from an adult donor, provided there is a tissue match. There are two sources of adult stem cells: bone marrow and circulating blood.

Adult stem cells can be collected during an outpatient procedure by using a hypodermic needle and syringe. The sterile needle is inserted into the pelvis and marrow is withdrawn. Stem cells can also be taken in a manner similar to whole blood donation called apheresis. During apheresis, the donor's blood is passed through a machine that separates the cells from their plasma and their stem cells are separated out from other types of blood cells. Once this process is completed, the unneeded cells and plasma are returned to the donor. This may take several hours. To improve the harvest, a stem cell growth factor is given to the donor before apheresis. This stimulates mitosis of the stem cells and increases their numbers. Once harvested, the stem cells are isolated and concentrated for injection into the recipient. If successful, the donated stem cells will repopulate the recipient's marrow, continue to undergo mitosis, and become a source of the red and white blood cells needed by the recovering patient.

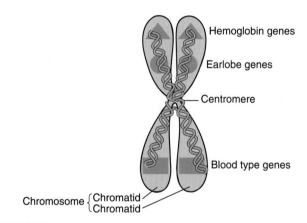

FIGURE 20.28 During interphase, when chromosome replication occurs, the original double-stranded DNA unzips to form two identical double strands that remain attached at the centromere. Each of these double strands is a chromatid. The two identical chromatids of the chromosome are sometimes termed a dyad, to reflect that there are two double-stranded DNA molecules, one in each chromatid. The DNA contains the genetic data. (The examples presented here are for illustrative purposes only. Do not assume that the traits listed are actually located in the positions shown on these hypothetical chromosomes.)

metabolize the nutrients and eliminate harmful wastes, (3) how to repair and assemble cell parts, (4) how to reproduce healthy offspring, (5) when and how to react to favorable and unfavorable changes in the environment, and (6) how to coordinate and regulate all of life's essential functions. The double helix of DNA and the nucleosomes are arranged as a **chromatid,** and there are two attached chromatids for each replicated chromosome after the S stage (figure 20.28). These chromatids (chromosomes) are what will be distributed during mitosis.

THE STAGES OF MITOSIS

All stages in the life cycle of a cell are continuous; there is no precise point when the G_1 stage ends and the S stage begins or when the interphase period ends and mitosis begins. Likewise, in the individual stages of mitosis, there is a gradual transition from one stage to the next. However, for purposes of study and communication, scientists have divided the process of mitosis into four stages based on recognizable events. These four phases are prophase, metaphase, anaphase, and telophase (figure 20.29).

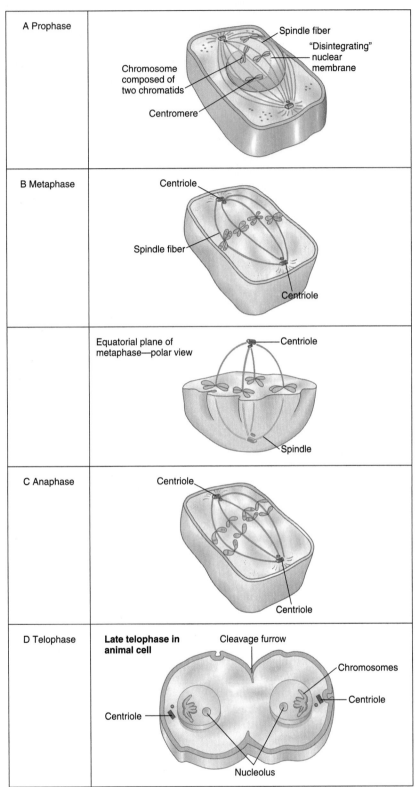

A Prophase	Spindle fiber "Disintegrating" nuclear membrane Chromosome composed of two chromatids Centromere
B Metaphase	Centriole Spindle fiber Centriole
	Equatorial plane of metaphase—polar view Centriole Spindle
C Anaphase	Centriole Centriole
D Telophase	**Late telophase in animal cell** Cleavage furrow Chromosomes Centriole Centriole Nucleolus

FIGURE 20.29 The stages of mitosis.

Prophase

During this phase, the replicated chromatin strands begin to coil into recognizable chromosomes. The nuclear membrane fragments and the pieces become parts of other membranous organelles. The centrioles move away from one another to opposite ends of the cell, forming the cell's poles. Spindle fibers form and grow from one pole to another. As prophase proceeds and as the chromosomes become more visible, we recognize that each chromosome is made of two parallel, threadlike parts lying side by side, the chromatids.

Metaphase

This is the phase when the condensed chromosomes move to the equator of the cell. The phrase "equator of the cell" is used in the same manner as "equator of Earth." The equator is the imaginary circle around Earth's (cell's) surface. When they arrive, the chromosomes attach to the spindle fibers at a point on the chromosomes known as the centromeres. At this stage in mitosis, each chromosome still consists of two chromatids attached at a centromere. In a human cell, there are forty-six chromosomes, or ninety-two chromatids, aligned at the cell's equatorial plane during metaphase. In late metaphase, each chromosome splits as the DNA of each centromere replicates, and the cell enters the next phase, anaphase.

Anaphase

Centromeres complete DNA replication during anaphase, allowing the chromatids to separate and move toward the poles. As this separation of chromatids occurs, the chromatids are now called daughter chromosomes. Daughter chromosomes contain identical genetic information. It is later during anaphase that a second important event occurs, cytokinesis. When cytokinesis (cytoplasm splitting) is completed during the last phase, telophase, the cytoplasm of the original cell is divided so that two smaller, separate daughter cells result.

Telophase

Two daughter cells are formed from the divided cells as cytokinesis (started during anaphase) is completed. The nuclear membranes and nucleoli re-form. Spindle fibers fragment, and the chromosomes unwind and change from chromosomes to chromatin. Depending on the cell type, these newly formed daughter cells can differentiate and become specialized in their functions, or they can reenter the mitotic portion of the cell cycle.

Abnormal Cell Division: Cancer

As we have seen, cells become specialized for a particular function. Each cell type has its cell-division process regulated so that it does not interfere with the activities of other cells or the whole organism. Some cells, however, may begin to divide as if they were "newborn" or undifferentiated cells. Sometimes this division occurs in an uncontrolled fashion. When cells lose their ability to control mitotic divisions, a group of cells forms a *tumor*. A **tumor** is a mass of undifferentiated cells not normally found in a certain portion of the body. A *benign tumor* is a cell mass that does not fragment and spread beyond its original area of growth. A benign tumor can become harmful by growing large enough to interfere with normal body functions. Some tumors are malignant. *Malignant tumors* are nonencapsulated growths of tumor cells that are harmful; they may spread or invade other parts of the body. Cells of

Box Figure 20.5 Malignant melanoma is a type of skin cancer. It forms as a result of a mutation in pigmented skin cells. These cells divide repeatedly, giving rise to an abnormal mass of pigmented skin cells. Only the dark area in the photograph is the cancer; the surrounding cells have the genetic information to develop into normal, healthy skin. This kind of cancer is particularly dangerous because the cells break off and spread to other parts of the body (metastasize).

these tumors move from the original site (**metastasize**) and establish new colonies in other regions of the body. Cells break off from the original tumor and enter the bloodstream. When they get stuck to the inside of a capillary, these cells move through the wall of the blood vessel and invade the tissue, where they begin to reproduce by mitosis. This tumor causes new blood vessels to grow into this new site, which will carry nutrients to this growing mass. These vessels can also bring even more spreading cells to the new tumor site. **Cancer** is the term used to refer to any abnormal growth of cells that has a malignant potential. Agents responsible for causing cancer are called **carcinogens.**

Questions to Discuss

1. What is the difference between squamous cell carcinoma and a melanoma?
2. Do an Internet search and list the five most common types of cancer in the United States.
3. What new kinds of treatments are being developed to control cancer?

CONCEPTS APPLIED

What Is Cancer?

Cancer is a group of many diseases characterized by the uncontrolled growth and spread of abnormal cells. If the spread is not controlled or checked, it results in death. However, many cancers can be cured if detected and treated promptly.

Cancer's Seven Warning Signals

1. A persistent change in bowel or bladder habits
2. A sore that does not heal
3. Unusual bleeding or discharge
4. A lump or thickening in the breast or elsewhere
5. Persistent indigestion or difficulty in swallowing
6. An obvious change in a wart or mole
7. A nagging cough or hoarseness

Some Warning Signs of Some Types of Cancers

1. Bladder and kidney: blood in urine; pain or burning; increased urination
2. Breast: lump or thickening of lumps; itching, redness, or soreness of the nipples that is not caused by pregnancy, breast feeding, or menstruation
3. Cervical, endometrial, and uterine: bleeding between menstrual cycles; any unusual discharge; painful menstruation; heavy periods
4. Colon: rectal bleeding; blood in your stool; changes in bowel habits such as persistent diarrhea or constipation
5. Laryngeal: a persistent cough or hoarse throat
6. Leukemia: paleness; fatigue; weight loss; repeated infections; nosebleeds; bone or joint pain; easy bruising
7. Lung: persistent cough; sputum with blood; heavy chest; chest pain
8. Lymphoma: enlarged, rubbery lymph nodes; itchy skin; night sweats; unexplained fever and weight loss
9. Mouth and throat: any chronic ulcer (sore) of the mouth, tongue or throat that does not heal; white areas in the mouth
10. Ovarian: Unfortunately, there are often no symptoms until it is in the later stages of development.
11. Pancreas: There usually are no symptoms until it has progressed to the later stages, when you may notice jaundiced skin and there may be pain deep in the stomach or back.
12. Skin: moles that change color, size, or appearance; flat sores (lesions that look like moles); a tumor or lump under the skin that resembles a wart; an ulceration that never heals
13. Stomach: vomiting blood; frequent indigestion and pain after eating; weight loss

What percent of Americans will eventually have cancer? Cancer will strike three out of four families. If you notice one of the preceding symptoms, see your doctor.

To become better informed about cancer, contact your local Cancer Society office or write the American Cancer Society, Inc., 90 Park Ave., New York, NY 10016. Request a copy of the latest publication, *Cancer Facts and Figures*. Go to the Internet and search some of the boldface terms used in this chapter to learn more about this topic. Be certain that the sources you access are "good science."

SUMMARY

Living things show the *characteristics* of (1) *metabolic processes*, (2) *generative processes*, (3) *responsive processes*, (4) *control processes*, and (5) *a unique structural organization*. The concept of the *cell* has developed over a number of years. Initially, only two regions, the *cytoplasm* and the *nucleus*, could be identified. At present, numerous *organelles* are recognized as essential components of both major cell types, *prokaryotic* and *eukaryotic*. The structure and function of some of these organelles are compared in table 20.3. This table also indicates whether the organelle is unique to prokaryotic or eukaryotic cells or found in both.

The *cell* is the common unit of life. We study individual cells and their structures to understand how they function as *individual living organisms* and as parts of *many-celled beings*. Knowing how *prokaryotic* and *eukaryotic cell* types resemble or differ from each other helps physicians control some organisms dangerous to humans.

TABLE 20.3 **Summary of the Structure and Function of the Cellular Organelles**

Organelle	Type of Cell in Which Located	Structure	Function
Plasma membrane	Prokaryotic and eukaryotic	Membranous; typical membrane structure; phospholipid and protein present	Controls passage of some materials to and from the environment of the cell
Inclusions (granules)	Prokaryotic and eukaryotic	Nonmembranous; variable	May have a variety of functions
Chromatin material	Prokaryotic and eukaryotic	Nonmembranous; composed of DNA and proteins (histones in eukaryotes and HU proteins in prokaryotes)	Contains the hereditary information that the cell uses in its day-to-day life and passes it on to the next generation of cells
Ribosomes	Prokaryotic and eukaryotic	Nonmembranous; protein and RNA structure	Site of protein synthesis
Microtubules, microfilaments, and intermediate filaments	Eukaryotic	Nonmembranous; strands composed of protein	Provide structural support and allow for movement
Nuclear membrane	Eukaryotic	Membranous; double membrane formed into a single container of nucleoplasm and nucleic acids	Separates the nucleus from the cytoplasm
Nucleolus	Eukaryotic	Nonmembranous; group of RNA molecules and DNA located in the nucleus	Site of ribosome manufacture and storage
Endoplasmic reticulum	Eukaryotic	Membranous; folds of membrane forming sheets and canals	Surface for chemical reactions and intracellular transport system
Golgi apparatus	Eukaryotic	Membranous; stack of single-membrane sacs	Associated with the production of secretions and enzyme activation
Vacuoles and vesicles	Eukaryotic	Membranous; microscopic single-membranous sacs	Containers of materials
Peroxisomes	Eukaryotic	Membranous; submicroscopic membrane-enclosed vesicle	Release enzymes to break down hydrogen peroxide
Lysosomes	Eukaryotic	Membranous; submicroscopic membrane-enclosed vesicle	Isolate very strong enzymes from the rest of the cell
Mitochondria	Eukaryotic	Membranous; double-membranous organelle: large membrane folded inside a smaller membrane	Associated with the release of energy from food; site of aerobic cellular respiration
Chloroplasts	Eukaryotic	Membranous; double-membranous organelle: large membrane folded inside a smaller membrane (grana)	Associated with the capture of light of energy and synthesis of carbohydrate molecules: site of photosynthesis
Centriole	Eukaryotic	Two clusters of 9 microtubules	Associated with cell division
Contractile vacuole	Eukaryotic	Membranous; single-membrane container	Expels excess water
Cilia and flagella	Eukaryotic and prokaryotic	Nonmembranous; prokaryotes composed of single type of protein arranged in a fiber that is anchored into the cell wall and membrane; 9 + 2 tubulin protein in eukaryotes	Flagellar movement in prokaryotic type rotate; ciliary and flagellar movement in eukaryotic type seen as waving or twisting

In the process of respiration, organisms convert foods into energy (ATP) and waste materials (carbon dioxide and water). Aerobic cellular respiration uses oxygen (O_2) in this biochemical pathway. This energy-releasing process is composed of three stages: (1) *glycolysis*, (2) *Krebs cycle*, and (3) the *electron-transport system*. Plants use the products of respiration in the *photosynthesis* pathway. Photosynthesis is comprised of three stages: (1) *light-capturing events*, (2) *light-dependent reactions*, and (3) *light-independent reactions*. Photosynthetic organisms carry out both biochemical pathways. There is also a constant cycling of materials between plants and animals. Sunlight supplies the essential initial energy for making the large organic molecules necessary to maintain the forms of life we know.

All cells come from preexisting cells as a result of cell division. This process is necessary for *growth, repair,* and *reproduction. Eukaryotic cells* go through a *cell cycle* that includes *cell division* (*mitosis* and *cytokinesis*) and *interphase. Interphase* is the period of growth and preparation for division. Mitosis is divided into four stages: *prophase, metaphase, anaphase,* and *telophase.* During mitosis, *two daughter nuclei* are formed from *one parent nucleus.* These nuclei have *identical* sets of *chromosomes* and *genes* that are *exact copies* of those of the parent. Although the process of *mitosis* has been presented as a series of phases, you should realize that it is a *continuous, flowing process* from *prophase* through *telophase.* Following mitosis, *cytokinesis* divides the *cytoplasm,* and the cell returns to *interphase.*

KEY TERMS

active transport (p. **478**)
adenosine triphosphate (ATP)
   (p. **482**)
aerobic cellular respiration
   (p. **481**)
biology (p. **466**)
cancer (p. **493**)
carcinogens (p. **493**)
cell (p. **470**)
cell membrane (p. **473**)
cell nucleus (p. **470**)
cell wall (p. **470**)
chloroplast (p. **481**)

chromatid (p. **491**)
chromatin (p. **488**)
chromosomes (p. **488**)
control processes (p. **468**)
cytokinesis (p. **490**)
cytoplasm (p. **470**)
diffusion (p. **475**)
endoplasmic reticulum
   (p. **479**)
enzymes (p. **468**)
eukaryotic cells (p. **471**)
facilitated diffusion (p. **478**)
generative processes (p. **466**)

Golgi apparatus (p. **479**)
homeostasis (p. **468**)
interphase (p. **490**)
lysosomes (p. **479**)
metabolic processes (p. **466**)
metastasize (p. **493**)
mitochondrion (p. **480**)
mitosis (p. **490**)
nuclear membrane (p. **480**)
nucleolus (p. **488**)
organelles (p. **471**)
osmosis (p. **476**)
phagocytosis (p. **479**)

photosynthesis (p. **482**)
plasma membrane (p. **479**)
prokaryotic cells (p. **471**)
protoplasm (p. **470**)
responsive processes (p. **467**)
ribosomes (p. **486**)
selectively permeable
   (p. **475**)
structural similarities
   (p. **468**)
tumor (p. **493**)
vacuole (p. **479**)

APPLYING THE CONCEPTS

1. Which one of the following represents a generative process?
   a. enzymes
   b. individual adaptation
   c. nutrient uptake
   d. cell division

2. The movement of perfume molecules from one person to another is an example of
   a. osmosis.
   b. diffusion.
   c. phagocytosis.
   d. active transport.

3. Metabolic processes include
   a. nutrient processing.
   b. waste elimination.
   c. aerobic cellular respiration.
   d. All of these are correct.

4. The centromeres split during
   a. anaphase.
   b. prophase.
   c. interphase.
   d. metaphase.

5. A useful chemical bond form of energy used in all cells is
   a. DNA.
   b. ATP.
   c. protein.
   d. centrioles.

6. A series of enzyme-controlled reactions operating in a cell is
   a. photosynthesis.
   b. ATP formation.
   c. biochemical pathway.
   d. All the above are correct.

7. During which stage of the cell cycle does DNA replication occur?
   a. the S stage of interphase
   b. anaphase of mitosis
   c. G_2 stage of metaphase
   d. prophase

8. The *ultimate* energy source for all life is
   a. ATP.
   b. radiant energy from the Sun.
   c. biochemicals.
   d. DNA.

9. The passage of some materials to and from the environment of the cell is controlled by
   a. ribosomes.
   b. microtubules.
   c. cell membrane.
   d. cilia.
10. Associated with the release of energy from food, these are the site of aerobic cellular respiration.
   a. chloroplasts
   b. mitochondria
   c. ribosomes
   d. centrioles
11. Organelles that are membranous and found in plant cells but not animal cells are
   a. nucleoli.
   b. chloroplasts.
   c. nuclear membranes.
   d. mitochondria.

12. Chromosomes move to the equator of the cell and attach to the spindle fibers at a point on the chromosomes known as the centromeres at the _____ stage of mitosis.
   a. interphase
   b. prophase
   c. telophase
   d. metaphase

Answers

1. d **2.** b **3.** d **4.** a **5.** b **6.** d **7.** a **8.** b **9.** c **10.** b **11.** b **12.** d

QUESTIONS FOR THOUGHT

1. Make a list of the membranous organelles of a eukaryotic cell and describe the function of each.
2. How do diffusion, facilitated diffusion, osmosis, and active transport differ?
3. What are the differences between the cell wall and the cell membrane?
4. Diagram a cell and show where proteins, nucleic acids, carbohydrates, and lipids are located.
5. There are several over-the-counter eyewash products on the market. Describe what would happen if they were not isotonic solutions.
6. Why does putting salt on meat preserve it from spoilage by bacteria?
7. In what ways do mitochondria and chloroplasts resemble one another?
8. Name the four stages of mitosis and describe what occurs in each stage.
9. What is meant by cell cycle?
10. During which stage of a cell's cycle does DNA replication occur?
11. At what phase of mitosis does the DNA become most visible?
12. How might spindle fibers aid in the separation of chromatids during mitosis?

FOR FURTHER ANALYSIS

1. Some people believe that the common cold can be controlled by using the same antibiotics that are used to control infections such as strep throat. What is the true story?
2. Local community blood programs are always seeking donors. They want to collect whole blood, red blood cells, and platelets from donors. How do these differ from one another? What is the value and application for each in a medical situation?
3. The sources of red and white blood cells are undifferentiated cells called stem cells. Currently, a great controversy exists over the source and use of these cells for research and medical purposes. What is the nature of the controversy? What kinds of abnormalities are associated with stem cells? To what medical purposes might these cells be used?

INVITATION TO INQUIRY

How Do The Cells Really Work In Different Environments?

A primitive type of cell consists of a membrane and a few other cell organelles. This first kind of cell lives in a sea that contains three major kinds of molecules with the following characteristics:

X
  Inorganic
  High concentration outside of the cell
  Essential to the life of the cell
  Small and can pass through the membrane

Y
  Organic
  High concentration inside of the cell

  Essential to the life of the cell
  Large and cannot pass through the membrane

Z
  Organic
  High concentration inside of the cell
  Poisonous to the cell
  Small and can pass through the membrane

With this information and your background in cell structure and function, osmosis, diffusion, and active transport, decide whether this kind of cell will continue to live in this sea and explain why or why not.

Earth is the only planet with liquid water and the only planet with living things. When seen from space, Earth is blue because of its vast oceans. Early in Earth's history, inorganic materials were converted to organic materials that are thought to have accumulated in the oceans. These organic materials have become combined into living units called cells. Cells are life. While we know a great deal about cells and the organisms they make up, scientists still ask these two questions: What is the nature of life? And how did life originate?

The Origin and Evolution of Life

Physics Connections

▶ Newton's laws of motion (Ch. 2).
▶ Measuring certain radioactive isotopes provides a clock to measure the age of Earth. (Ch. 11 and 22).

Chemistry Connections

▶ Biochemical reactions evolved with living things (Ch. 19).
▶ Carbon chemistry is the chemistry of living things (Ch. 19).

Astronomy Connections

▶ The origin of the universe (Ch. 12).
▶ Conditions on other planets help us understand what the ancient Earth may have been like (Ch. 13).

CORE CONCEPT

Earth and the life on it have changed over billions of years.

There are several different concepts about how life originated on Earth (p. 502).

Most of the species that have ever existed are extinct (p. 515).

Evolution occurs as a result of natural selection (p. 507).

Earth's surface and atmosphere have been modified greatly by the action of organisms (p. 502).

Earth Science Connections

▶ Atmospheric gases have changed over time (Ch. 17).

Life Science Connections

▶ Photosynthesis traps sunlight to make organic matter and release oxygen (Ch. 20).
▶ Respiration releases usable energy from organic molecules (Ch. 20).
▶ DNA stores genetic information that determines the kinds of biochemical reactions an organism can perform. (Ch. 19 and 26).

OVERVIEW

Understanding how Earth came to have such a complex and diverse combination of living things has challenged thinkers for thousands of years. Has Earth always been filled with living things? Were there kinds of living things that have ceased to exist? Did living things originate on Earth? Do new living things originate today? As our understanding of the laws of nature and the nature of living things has developed, the way we approach these questions has changed.

Although at one time these questions were just interesting intellectual exercises, today they have many practical applications and affect the decision making of governments, corporations, and individual citizens. For example, we are concerned about the extinction of various species of organisms, and governments spend considerable time and money to prevent extinctions. Understanding how evolutionary processes occur in populations is important for controlling populations of disease

organisms that continually develop strains resistant to commonly used antibiotics. People are concerned about the manipulation of the DNA of organisms, which can cause major changes in the nature of living things. And governments are spending vast amounts of money to determine if there was life on Mars. However, even from a modern scientific perspective, we are still asking the same basic questions. How did life originate? And how and why does it change?

EARLY ATTEMPTS TO UNDERSTAND THE ORIGIN OF LIFE

In earlier times, no one ever doubted that life originated from nonliving things. The Greeks, Romans, Chinese, and many other ancient peoples believed that maggots arose from decaying meat; mice developed from wheat stored in dark, damp places; lice formed from sweat; and frogs originated from damp mud. This theory was widely believed until the seventeenth century; however, there were some who doubted it. These people subscribed to an opposing concept that life originates only from preexisting life. The concept that life could be generated from nonliving matter is known as **spontaneous generation.** The concept that life can come only from other living things is known as **biogenesis.**

One of the earliest challenges to the idea that life can be generated from nonliving matter came in 1668. Francesco Redi, an Italian physician, set up a controlled experiment (figure 21.1). He used two sets of jars that were identical except for one aspect. Both sets of jars contained decaying meat,

and both were exposed to the atmosphere; however, one set of jars was covered by gauze, and the other was uncovered. Redi observed that flies settled on the meat in the open jar, but the gauze blocked their access to the covered jars. When maggots appeared on the meat in the uncovered jars but not on the meat in the covered ones, Redi concluded that the maggots arose from the eggs of the flies and were not produced from the decaying meat.

In 1861, the French chemist Louis Pasteur convinced most scientists that spontaneous generation could not occur. He placed a fermentable sugar solution in a flask that had a long swan neck. The mixture and the flask were boiled for a long time, which killed any organisms already in the solution. The long swan neck of the flask was left open to allow oxygen to enter. This was an important experiment because at the time, many thought that oxygen was the "vital element" necessary for the spontaneous generation of living things. Pasteur postulated that it was not the oxygen that caused life to "originate" in such mixtures but that there were organisms in the air that caused

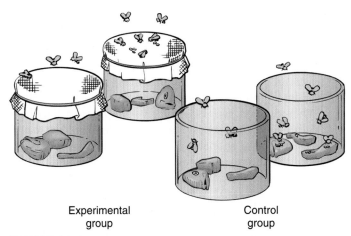

Experimental
group

Control
group

FIGURE 21.1 Francesco Redi performed an experiment in which he prepared two sets of jars that were identical in every way except one. One set of jars had a gauze covering. The uncovered set was the control group; the covered set was the experimental group. Any differences seen between the control and the experimental groups were the result of a single variable—being covered by gauze. In this manner, Redi concluded that the presence of maggots in meat was due to flies laying their eggs on the meat and not spontaneous generation.

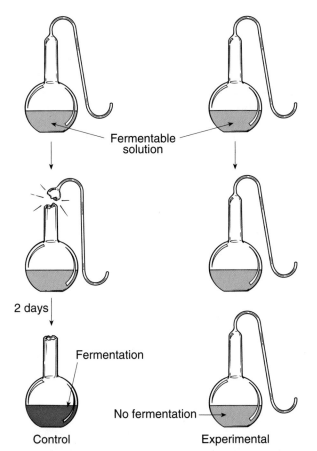

Fermentable
solution

2 days

Fermentation

No fermentation

Control

Experimental

FIGURE 21.2 Louis Pasteur conducted an experiment to test the idea that oxygen was necessary for spontaneous generation to take place. He used swan-neck flasks that allowed oxygen but not airborne organisms to enter the flask. The flasks contained a fermentable solution that had been boiled. One flask was kept intact, but he broke the neck off another flask. The intact flask was the experimental flask and the flask with the broken neck was the control. Within two days, there was growth in the flask with the broken neck but none in the intact flask. Thus, Pasteur demonstrated that it was not oxygen in the air that caused growth in the flasks but living things, which were prevented from entering the flask with the unbroken swan neck. This provided additional evidence against the theory of spontaneous generation.

fermentation. He further postulated that airborne organisms would settle on the bottom of the curved portion of the neck and be unable to reach the sugar-water mixture and thus the sugar solution should not ferment. When he performed the experiment, the solution in the swan-neck flask did not ferment. As a control he cut off the swan neck (figure 21.2). This allowed microorganisms from the air to fall into the flask, and within two days, the fermentable solution was supporting a population of microorganisms. In his address to the French Academy, Pasteur stated, "Never will the doctrine of spontaneous generation recover from the mortal blow of this simple experiment. No, there is now no circumstance known in which it can be affirmed that microscopic beings came into the world without germs, without parents similar to themselves."

CURRENT THINKING ABOUT THE ORIGIN OF LIFE

Today, when we look at the question of how life on Earth originated, we still have basically the same two theories. One holds that life arrived on Earth from some extraterrestrial source. This is essentially a variation of the biogenesis argument. The other maintains that life was created on Earth from nonliving material through a process of chemical evolution, which is a variation of the spontaneous generation argument. It is important to recognize that we will probably never know for sure how life on Earth came to be, but it is interesting to speculate and examine the evidence related to this fundamental question.

Extraterrestrial Origin for Life on Earth

The concept that life arrived from extraterrestrial sources received some support in 1996, when a meteorite from Antarctica was analyzed. It has been known for many years that meteorites often contain organic molecules, and this suggested that life might have existed elsewhere in the solar system. The chemical makeup of the Antarctic meteorite suggests that it was a portion of the planet Mars that was ejected from Mars as a result of an asteroid colliding with the planet. Analysis of the meteorite shows the presence of complex organic molecules and small structures that resemble those found on Earth that are thought to be the result of the activity of ancient microorganisms. While scientists no longer think these structures are from microorganisms, many still think it is quite possible Mars may have had living things in the past. Since Mars currently has some water as ice and shows features that resemble dried-up river systems and seas, it is highly likely that Mars had much more water in the past. Since water is the most common compound in living things, many consider the presence of water to be a prerequisite for life. For these reasons, many feel it is reasonable to consider that life of a nature similar to that presently found on Earth could have existed on Mars. While previous exploration of Mars' surface has failed to find evidence for life, many researchers are not satisfied with the results. Thus, current explorations of the Martian surface by the Mars robots—*Spirit* and *Opportunity*—continue to look for evidence of current or past life.

Earth Origin for Life on Earth

The alternative view, that life originated on Earth, has also received support. There has been much research and speculation about the conditions that would have been necessary for life to originate on Earth and the kinds of problems that early living things would have had to solve. Several different kinds of information are important to this discussion.

1. **Water is necessary for life.**   Earth is currently the only planet in our solar system that has a temperature range that allows for water to exist as a liquid on its surface, and water is the most common compound in most kinds of living things.

2. **Oxygen is produced by living things.**   Analysis of the atmospheres of other planets shows that they all lack oxygen. The oxygen in Earth's current atmosphere is the result of photosynthetic activity. Therefore, before there was life on Earth, the atmosphere probably lacked oxygen.

3. **Organic molecules can form spontaneously.**   Experiments demonstrate that organic molecules can be generated in an atmosphere that lacks oxygen. Furthermore, these organic molecules could have accumulated in the oceans, since there would have been no oxygen to encourage their breakdown. (See figure 21.3.)

4. **The early Earth was hot.**   Since it is assumed that all of the planets have been cooling off as they age, it is very likely that Earth was much hotter in the past. The large portions of Earth's surface that are of volcanic origin strongly support this idea.

5. **Simple organisms today often live in conditions similar to those thought to have existed on a primitive Earth.**   Prokaryotic organisms (commonly called bacteria) are relatively simple and extremely common. Today, certain prokaryotic organisms live in extreme environments of high temperature, high salinity, low pH, or the absence of oxygen. This suggests that they may have been adapted to life in a world that is very different from today's Earth. These specialized organisms are found today in unusual locations such as hot springs and around thermal vents in the ocean floor and may be descendants of the first organisms formed on the primitive, more hostile Earth.

If we put all of these ideas together, we can create the following scenario. A hot planet Earth with an atmosphere that lacked oxygen could have allowed organic molecules to form from inorganic molecules. These organic molecules could have

- ● Carbon
- ● Nitrogen
- ● Oxygen
- ● Hydrogen

FIGURE 21.3   The environment of the primitive Earth was harsh and lifeless. But many scientists believe that it contained the necessary molecules and sources of energy to fashion the first living cell. The energy furnished by volcanoes, lightning, and ultraviolet light could have broken the bonds in the simple inorganic molecules such as carbon dioxide (CO_2), ammonia (NH_3), and methane (CH_4) in the atmosphere. New bonds could have formed as the atoms from the smaller molecules were rearranged and bonded to form simple organic compounds in the atmosphere. Rain and runoff from the land would have carried these chemicals into the oceans. Here, they could have reacted with each other to form more complex organic molecules.

collected in the oceans and produced a dilute organic "soup." These organic molecules could have served as "building blocks" for simple organisms as well as a source of energy. The earliest forms of life would have been adapted to conditions that are very different from those that exist on Earth today.

To "invent" life, there are several problems that would have to have been solved. By focusing on alternative ways of solving these problems, scientists have developed several different ideas about what the first living thing might have been like, regardless of whether life originated on Earth or arrived from an extraterrestrial source.

MEETING METABOLIC NEEDS— HETEROTROPHS OR AUTOTROPHS

Fossil evidence indicates that there were primitive, bacterialike forms of life on Earth at least 3.5 billion years ago. Regardless of how they developed, these first primitive cells would have needed a way to add new organic molecules to their structure as the organisms grew or as previously existing molecules were lost or destroyed. There are two ways to accomplish this.

Heterotrophs consume organic molecules from their surroundings, which they use to make new molecules and to provide themselves with a source of energy. Today, we recognize all animals, fungi, and most protozoa and bacteria as being heterotrophs.

Autotrophs use an external energy source such as the energy from inorganic chemical reactions (chemoautotrophs) or sunlight (photoautotrophs) to combine simple inorganic molecules such as water and carbon dioxide to make new organic molecules. These new organic molecules can then be used as building materials for new cells or can be broken down at a later date to provide a source of energy. Today, we recognize eukaryotic plants and algae as being photoautotrophs. Among the prokaryotic organisms, many are photoautotrophs and some are chemoautotrophs.

The Heterotroph Hypothesis

Many scientists support the idea that the first living things produced on Earth were primitive, bacterialike heterotrophs that lived off the organic molecules that would have accumulated in the oceans. Since the early heterotrophs are thought to have developed in an atmosphere that lacked oxygen, they would have been, of necessity, anaerobic organisms. Therefore, they did not obtain the maximum amount of energy from the organic molecules they obtained from their environment. At first, this would not have been a problem. The organic molecules that had been accumulating in the ocean for hundreds of millions of years served as an ample source of organic material for the heterotrophs.

Evidence exists to suggest that a wide variety of compounds were present in the early oceans, some of which could have been used unchanged by the heterotrophs. The heterotrophs did not have to modify the compounds to meet their needs, and they probably carried out a minimum of enzyme-controlled, metabolic reactions. Those compounds that could be used easily by heterotrophs would have been the first to become depleted from the early environment.

As easily used organic molecules became scarce, it is possible that some of the primitive cells may have had altered (mutated) genetic material that allowed them to convert material that was not directly usable into a compound that could be used. Genetic mutations may have been common because the amount of ultraviolet light, which is known to cause mutations, would have been high. Today, ozone is formed from oxygen in the atmosphere, and ozone screens out much of the ultraviolet light. Before there was oxygen in the atmosphere, there would have been no ozone and much more ultraviolet light would have reached Earth. Heterotrophs with such mutations could have survived, while those without it would have become extinct as the compounds they used for food became scarce. It has been suggested that through a series of mutations in the early heterotrophs, a more complex series of biochemical reactions originated within some of the cells. Such cells could use controlled reactions to modify chemicals from their surroundings and convert them into usable organic compounds.

The Autotroph Hypothesis

As with many areas of science, there are often differences of opinion. Although the heterotroph hypothesis for the origin of living things was the prevailing theory for many years, recent discoveries have caused many scientists to consider an alternative. They propose that the first organism was an autotroph that used the energy released from inorganic chemical reactions to synthesize organic molecules. Such organisms are known as chemoautotrophs. Several kinds of information support this theory. Many kinds of prokaryotic organisms are chemoautotrophs and live in extremely hostile environments around the world. These organisms are found in hot springs such as those in Yellowstone National Park or near hot thermal vents—areas where hot, mineral-rich water enters seawater from the ocean floor (figure 21.4). They use inorganic chemical reactions as a

FIGURE 21.4 Thermal vents are places where hot mineral-rich water enters the ocean water from holes in the ocean bottom. Surrounding these vents are collections of many kinds of animals. Chemoautotrophic members of the domain Archaea use the hot mineral-rich water as a source of energy with which they synthesize organic molecules. These organisms and the organic molecules they produce are the base of the food chain for a variety of kinds of animals found only in these locations.

source of energy to allow them to synthesize organic molecules from inorganic components. The fact that many of these organisms live in very hot environments suggests that they may have originated on an Earth that was much hotter than it is currently. If the first organism was an autotroph, there could have been subsequent evolution of a variety of kinds of cells, both autotrophic and heterotrophic, that could have led to the variety of different prokaryotic cells seen today.

Summary of Ideas About the Origin of Life

As a result of this discussion, you should understand that we do not know how life on Earth originated. Scientists look at many kinds of evidence and continue to explore new avenues of research. So we currently have three competing scientific theories for the origin of life on Earth:

1. Life originated elsewhere in the universe and arrived from some extraterrestrial source.
2. Life originated on Earth as a heterotroph.
3. Life originated on Earth as an autotroph.

From our modern perspective, we can see that today, all life comes into being as a result of reproduction. Life is generated from other living things, the process of biogenesis. However, reproduction does not answer the question: Where did life come from in the first place? We can speculate, test hypotheses, and discuss various possibilities, but we will probably never know for sure. Life either always was, or it started at some point in the past. If it started, then spontaneous generation of some type had to occur at least once, but it is not happening today.

MAJOR EVENTS IN THE EARLY EVOLUTION OF LIVING THINGS

Regardless of what the first living things were like, once they existed, a process began that resulted in changes in their abilities and structures. Furthermore, the nature of the planet Earth also changed. When we compare the primitive nature of the first living things with the natures of current organisms, we recognize that there were several major steps necessary to go from those first organisms to the great diversity and complexity we see today.

Reproduction and the Origin of Genetic Material

In the previous discussion about possible metabolic activities of early life-forms, we made the assumption that there was some kind of genetic material to direct their metabolic activities. There has been considerable thought about the nature of this genetic material. Today, we know that two important *nucleic acid* molecules, DNA and RNA, store and transfer information within a cell. RNA is a simpler molecule than DNA. In most current cells, DNA stores genetic information and RNA assists the DNA in carrying out its instructions. The details of how DNA and RNA function are discussed in chapter 26.

However, it has been discovered that in some viruses, RNA rather than DNA can serve as the genetic material. Other research about the nature of RNA provides interesting information. RNA can be assembled from simpler subunits that could have been present on the early Earth. Scientists have also shown that RNA molecules are able to make copies of themselves without the need for enzymes, and they can do so without being inside cells.

These pieces of evidence suggest that RNA may have been the first genetic material, which helps to solve one of the problems associated with the origin of life: how genetic information was stored in these primitive life-forms. Once a primitive life-form had the ability to copy its genetic material, it would be able to reproduce. Reproduction is one of the most fundamental characteristics of living things.

Once living things existed and had a molecule that stored information, the stage was set for the evolution of new forms of life. Our current knowledge about how DNA and RNA store and use genetic information includes the idea that these molecules are changeable. In other words, *mutations* can occur. Thus, the stage was set for living things to proliferate into a variety of kinds that were adapted to specific environmental conditions.

The Development of an Oxidizing Atmosphere

Ever since its formation, Earth has undergone constant change. In the beginning, it was too hot to support an atmosphere. Later, as it cooled and as gases escaped from volcanoes, a reducing atmosphere (lacking oxygen) was likely to have been formed. The early life-forms would have lived in this reducing atmosphere. However, today we have an oxidizing atmosphere and most organisms use this oxygen as a way to extract energy from organic molecules through a process of aerobic respiration. But what caused the atmosphere to change?

Atmospheric Oxygen Produced by Photosynthesis

Today, it is clear that the oxygen in our atmosphere is the result of the process of photosynthesis. Prokaryotic cyanobacteria are the simplest organisms that are able to photosynthesize. Because the fossil record shows that living things existed before there was oxygen in the atmosphere and that cyanobacteria were present early in the evolution of life, it seemed logical that the first organisms could have accumulated many mutations over time that could have resulted in photosynthetic autotrophs (photoautotrophs). This would have been a significant change because the release of oxygen from photosynthesis would have resulted in an accumulation of oxygen in the atmosphere—an **oxidizing atmosphere.** The development of an oxidizing atmosphere created an environment unsuitable for the spontaneous formation of organic molecules in the atmosphere, because organic molecules tend to break down (oxidize) when oxygen is present. The presence of oxygen in the atmosphere would make it impossible for life to spontaneously originate in the manner described earlier in this chapter, because an oxidizing atmosphere would not allow for the accumulation of organic molecules in the seas. However, it is clear that living things existed before oxygen was present in the atmosphere. The first fossils of primitive living things show up at about 3.5 billion years ago, and it appears that an oxidizing atmosphere began to develop about 2 billion years ago.

The Significance of Ozone

The presence of oxygen in the atmosphere had an additional significance. Oxygen molecules also reacted with one another to form ozone (O_3). Ozone collected in the upper atmosphere and acted as a screen to prevent most of the ultraviolet light from reaching Earth's surface. The reduction of ultraviolet light diminished the spontaneous formation of complex organic molecules. It also reduced the number of mutations in cells. In an oxidizing atmosphere, it was no longer possible for organic molecules to accumulate over millions of years to be later incorporated into living material.

Evolution of Aerobic Respiration Possible

The presence of oxygen in the atmosphere also allowed for the evolution of aerobic respiration. Since the first heterotrophs were, of necessity, anaerobic organisms, they did not derive large amounts of energy (ATP) from the organic materials available as food. With the evolution of aerobic heterotrophs, there could be a much more efficient conversion of food into usable energy. They could use the oxygen for aerobic respiration and, therefore, generate many more energy-rich ATP molecules from the food molecules they consumed. Because of this, aerobic organisms would have a significant advantage over anaerobic organisms.

The Establishment of Three Major Domains of Life

Although biologists have traditionally divided organisms into kingdoms based on their structure and function, it was very difficult to do this with microscopic organisms. In 1977, Carl Woese published the idea that the "bacteria" (organisms that lack a nucleus), which had been considered a group of similar organisms, were really made up of two very different kinds of organisms, the Eubacteria and Archaea. With the newly developed ability to decode the sequence of nucleic acids, it became possible to look at the genetic nature of organisms without being confused by their external structures. Woese studied the sequences of ribosomal RNA and compared similarities and differences. As a result of his studies and those of many others, a new concept of the relationships between various kinds of organisms has emerged. There are three main kinds of living things, Eubacteria, Archaea, and Eucarya, that have been labeled **"domains."**

This new picture of living things requires us to reorganize our thinking. It appears that the oldest organisms may have been prokaryotic organisms that were able to live in hot situations and that they gave rise to the Archaea, many of which still require extreme environments. Perhaps most interesting is the idea that the Archaea and Eucarya share many characteristics, suggesting that they are more closely related to each other than either is to the Eubacteria. It appears that each domain

developed specific abilities. The Archaea are primarily chemoautotrophs that use inorganic chemical reactions to generate the energy they need to make organic matter. The reactions that provide them with energy often produce methane (CH_4) or hydrogen sulfide (H_2S). In addition, most of these organisms are found in extreme environments such as hot springs or in extremely salty or acidic environments.

The Eubacteria developed many different metabolic abilities. Today, many are able to use organic molecules as a source of energy (heterotrophs), some are able to carry on photosynthesis (photoautotrophs), and still others are able to get energy from inorganic chemical reactions (chemoautotrophs) similar to the Archaea.

The Eucarya (animals, plants, fungi, protozoa, and algae) are familiar to most people. The cells of eukaryotic organisms are much larger than those of prokaryotic organisms and appear to have incorporated cells of other organisms within their own cellular structure. Chloroplasts and mitochondria are both bacterialike structures found inside eukaryotic cells. The section entitled "The Endosymbiotic Theory and the Origin of Eukaryotic Cells" discusses current thinking about how eukaryotic cells developed.

The Endosymbiotic Theory and the Origin of Eukaryotic Cells

The earliest fossils occur at about 3.5 billion years ago and appear to be similar in structure to that of present-day Eubacteria and Archaea. Therefore, it is likely that the early heterotrophs and autotrophs were probably simple one-celled prokaryotic organisms. The earliest fossils of eukaryotic cells appear at about 1.8 billion years ago.

Since the fossil record shows prokaryotic organisms existed for about 1.7 billion years before eukaryotic organisms came on the scene, biologists generally believe that the eukaryotes evolved from the prokaryotes. The **endosymbiotic theory** attempts to explain this evolution. This theory states that eukaryotic cells came into being as a result of the combining of several different types of primitive prokaryotic cells. It is thought that some organelles found in eukaryotic cells may have originated as free-living prokaryotes. For example, since mitochondria and chloroplasts contain bacterialike DNA and ribosomes, control their own reproduction, and synthesize their own enzymes, it has been suggested that they originated as free-living prokaryotic bacteria. These bacterial cells could have established a symbiotic relationship with another primitive nuclear-membrane-containing cell type (figure 21.5). When this theory was first suggested, it met with a great deal of criticism and even ridicule. However, continuing research has uncovered several other instances of the probable joining of two different prokaryotic cells to form one. If these cells adapted to one another and were able to survive and reproduce better as a team, it is possible that this relationship may have evolved into present-day eukaryotic cells.

If this relationship had included only a nuclear-membrane-containing cell and aerobic bacteria, the newly evolved cell would have been similar to present-day heterotrophic protozoa,

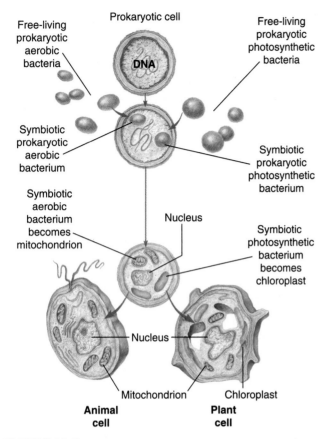

FIGURE 21.5 The endosymbiotic theory proposes that some free-living prokaryotic bacteria developed symbiotic relationships with a host cell. When some aerobic bacteria developed into mitochondria and photosynthetic bacteria developed into chloroplasts, a eukaryotic cell evolved. These cells evolved into eukaryotic plant and animal cells.

fungi, and animal cells. If this relationship had included both aerobic bacteria and photosynthetic bacteria, the newly formed cell would have been similar to present-day autotrophic algae and plant cells. In addition, it is likely that endosymbiosis occurred among eukaryotic organisms as well. Several kinds of eukaryotic red and brown algae contain chloroplastlike structures that appear to have originated as free-living eukaryotic cells.

A Summary of the Early Evolution of Life

Let us return for a moment to the question that perplexed early scientists and caused the controversies surrounding the opposing theories of spontaneous generation and biogenesis. If we incorporate this information along with information about the nature of genetic material, we can develop a diagram that helps us see how these various kinds of organisms are related to one another.

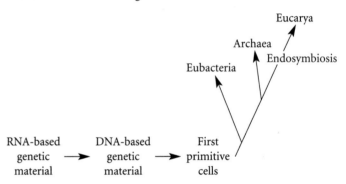

Table 21.1 summarizes the relationships among these three major kinds of organisms. Figure 21.6 summarizes several of the major events in the formation of the planet Earth and the development of life on it.

TABLE 21.1   Major Domains of Life

Eubacteria	Archaea	Eucarya
No nuclear membrane	No nuclear membrane	Nuclear membrane present
No membranous structures	No membranous structures	Many kinds of membranous organelles present in cells
Probably ancestral to both Archaea and Eucarya	Probably have a common ancestor with Eucarya	Probably have a common ancestor with Archaea
Chlorophyll-based, oxygen-generating photosynthesis was an invention of the cyanobacteria.	None is photosynthetic.	Chloroplasts are probably derived from cyanobacteria.
Large number of heterotrophs oxidize organic molecules for energy.	A few are heterotrophs.	Photosynthetic organisms are autotrophs; all others are heterotrophs.
Many use aerobic respiration to obtain energy from organic molecules.		Mitochondria probably derived from certain aerobic bacteria.
Some use energy from inorganic chemical reactions to produce organic molecules (chemoautotrophs).	Many are chemoautotrophs.	None is a chemoautotroph.
Some live at high temperatures and may be ancestral to Archaea.	Typically found in extreme environments	
Much metabolic diversity among closely related organisms	None has been identified as pathogenic.	Common evolutionary theme is the development of complex cells through endosymbiosis with other organisms.
Example: *Streptococcus pneumoniae,* one cause of pneumonia	Example: *Pyrolobus fumarii,* deep-sea hydrothermal vents, hot springs, volcanic areas, growth to 113°C (235°F)	Example: *Spirogyra,* a filamentous alga found in freshwater

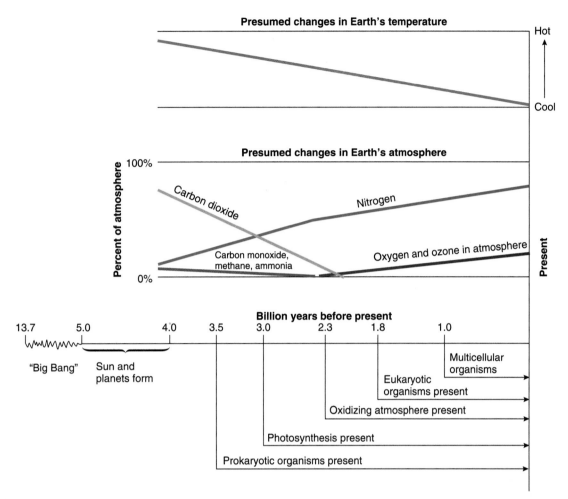

FIGURE 21.6 Current thinking suggests that several factors shaped the nature of Earth and its atmosphere. Conceivably, the initial atmosphere was primarily carbon dioxide (CO_2) with smaller amounts of nitrogen (N_2), ammonia (NH_3), methane (CH_4), and carbon monoxide (CO). Carbon dioxide decreased as a result of chemical reactions at the surface of Earth. Living organisms have had a significant effect on the atmosphere because once photosynthesis began, carbon dioxide was further reduced and oxygen (O_2) began to increase. The presence of oxygen allowed for the increase in ozone (O_3), which reduces the amount of ultraviolet light penetrating the atmosphere to reach the surface of Earth. Because ultraviolet light causes mutations, a protective layer of ozone probably has reduced the frequency of mutations from that which was present on primitive Earth.

EVOLUTION AND NATURAL SELECTION

In many cultural contexts, the word *evolution* means progressive change. We talk about the evolution of economies, fashion, or musical tastes. From a biological perspective, the word has a more specific meaning. One of the key characteristics of life is that organisms are able to respond to their surroundings. Evolution is a responsive process that takes place at the population level.

Defining Evolution

Evolution is the change in frequency of genetically determined characteristics within a population over time. There are three key points to this definition. First of all, evolution only occurs in populations. *Populations* are groups of organisms of the same species that are able to interbreed and are thus genetically similar. The second point is that *genes* (specific pieces of DNA) determine the characteristics displayed by organisms. The third

point is that the mix of genes (DNA) within populations can change. Thus, evolution involves changes in the genes that are present in a population. By definition, individual organisms are not able to evolve—only populations can.

The Role of the Environment in Evolution

The organism's surroundings determine which characteristics favor survival and reproduction (i.e., which characteristics best fit the organism to its environment). The mechanism by which evolution occurs involves the selective passage of genes from one generation to the next through sexual reproduction. This adaptation can take place at different levels and over different time periods. At the smallest level, populations show changes in the genes within a local population. For example, some populations of potato beetles became resistant to certain insecticides and some populations of crabgrass became resistant to specific herbicides when these chemicals became a constant part of their environment. The susceptible members died and

A Closer Look

The Voyage of HMS *Beagle*, 1831–1836

Probably the most significant event in Charles Darwin's life was his opportunity to sail on the British survey ship HMS *Beagle*. Surveys were common at this time; they helped refine maps and chart hazards to shipping. Darwin was twenty-two years old and probably would not have had the opportunity had his uncle not persuaded Darwin's father to allow him to take the voyage. Darwin was to be a gentleman naturalist and companion to the ship's captain, Robert Fitzroy. When the official naturalist left the ship and returned to England, Darwin became the official naturalist for the voyage. The appointment was not a paid position.

The voyage of the *Beagle* lasted nearly five years. During the trip, the ship visited South America, the Galápagos Islands, Australia, and many Pacific Islands. The *Beagle's* entire route is shown on the accompanying map. Darwin suffered greatly from seasickness, and perhaps because of it, he made extensive journeys by mule and on foot some distance inland from wherever the *Beagle* happened to be at anchor. These inland trips provided Darwin the opportunity to make many of his observations. His experience was unique for a man so young

and very difficult to duplicate because of the slow methods of travel used at that time.

Although many people had seen the places that Darwin visited, never before had a student of nature collected volumes of information on them. Also, most other people who had visited these faraway places were military men or adventurers who did not recognize the significance of what they saw. Darwin's notebooks included information on plants, animals, rocks, geography, climate, and the native peoples he encountered. The natural history notes he took during the voyage served as a vast storehouse of information that he used in his writings for the rest of his life. Because Darwin was wealthy, he did not need to work to earn a living and could devote a good deal of his time to the further study of natural history and the analysis of his notes. He was a semi-invalid during much of his later life. Many people think his ill health was caused by a tropical disease he contracted during the voyage of the *Beagle*. As a result of his experiences, he wrote several volumes detailing the events of the voyage, which were first published in 1839 in conjunction with other information related to the voyage of the *Beagle*. His volumes were

revised several times and eventually were entitled *The Voyage of the Beagle*. He also wrote books on barnacles, the formation of coral reefs, how volcanos might have been involved in reef formation, and, finally, *On the Origin of Species by Means of Natural Selection, or the Preservation of Favoured Races in the Struggle for Life*. This last book, written twenty-three years after his return from the voyage, changed biological thinking for all time.

Box Figure 21.1 Young Charles Darwin examining specimens on the Galápagos Islands

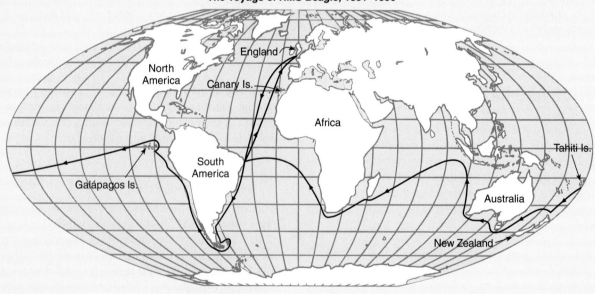

The Voyage of HMS *Beagle*, 1831–1836

England
North America
Canary Is.
Africa
Tahiti Is.
South America
Galápagos Is.
Australia
New Zealand

0 1000 2000 3000
Equatorial scale of miles

Box Figure 21.2

the resistant ones lived, leading to populations that consisted primarily of resistant individuals. These populations became genetically adapted to an altered environment. The genes that allowed the organisms to resist the effects of the poisons became more common in these populations. When we look at evolutionary change over millions or billions of years, we recognize that Earth has gone through major changes of climate, sea level, and other conditions. These changes have resulted in changes in the kinds of organisms present as well. Some species went extinct as new species came into being.

Natural Selection Leads to Evolution

The various processes that encourage the passage of beneficial genes to future generations and discourage the passage of harmful or less valuable genes are collectively known as **natural selection.** The idea that some individuals whose gene combinations favor life in their surroundings will be most likely to survive, reproduce, and pass their genes on to the next generation is known as the **theory of natural selection.** The *theory of evolution,* however, states that populations of organisms become genetically adapted to their surroundings over time. Thus, natural selection is the process that brings about evolution by "selecting" which genes will be passed to the next generation.

A theory is a well-established generalization supported by many different kinds of evidence. At one time, the idea that populations of species changed over time was revolutionary. Today, the concepts of evolution and natural selection are central to the study of all of the biological sciences.

The theory of natural selection was first proposed by Charles Darwin and Alfred Wallace and was clearly set forth in 1859 by Darwin in his book *On the Origin of Species by Means of Natural Selection, or the Preservation of Favoured Races in the Struggle for Life.* Since the time it was first proposed, the theory of natural selection has been subjected to countless tests and remains the core concept for explaining how evolution occurs.

There are two common misinterpretations associated with the process of natural selection. The first involves the phrase "survival of the fittest." Individual survival is certainly important because those that do not survive will not reproduce. But the more important factor is the number of descendants an organism leaves. An organism that has survived for hundreds of years but has not reproduced has not contributed any of its genes to the next generation and so has been selected against. The key, therefore, is not survival alone but survival and reproduction of the more fit organisms.

Second, the phrase "struggle for life" does not necessarily refer to open conflict and fighting. It is usually much more subtle than that. When a resource such as nesting material, water, sunlight, or food is in short supply, some individuals survive and reproduce more effectively than others. For example, many kinds of birds require holes in trees as nesting places (see figure 21.7). If these are in short supply, some birds will be fortunate and find a top-quality nesting site, others will occupy less suitable holes, and some may not find any. There may or may not be fighting for possession of a site. If a site is already occupied, a bird may not necessarily try to dislodge its

FIGURE 21.7 Many kinds of birds, such as this red-bellied woodpecker, nest in holes in trees. If old and dead trees are not available, they may not be able to breed. Many people build birdhouses that provide artificial "tree holes" to encourage such birds to nest near their homes.

occupant but may just continue to search for suitable but less valuable sites. Those that successfully occupy good nesting sites will be much more successful in raising young than will those that must occupy poor sites or those that do not find any.

Similarly, on a forest floor where there is little sunlight, some small plants may grow fast and obtain light while shading out plants that grow more slowly. The struggle for life in this instance involves a subtle difference in the rate at which the plants grow. But the plants are indeed engaged in a struggle, and a superior growth rate is the weapon for survival.

GENETIC DIVERSITY IS IMPORTANT FOR NATURAL SELECTION

Now that we have a basic understanding of how natural selection works, we can look in more detail at factors that influence it. For natural selection to occur there must be genetic differences among the many individuals of an interbreeding population of organisms. If all individuals are identical genetically, it does not matter which ones reproduce—the same genes will be passed to the next generation and natural selection cannot occur. There are two primary ways in which genetic diversity is generated within the organisms of a species: mutation and genetic recombination during sexual reproduction.

Genetic Diversity Resulting from Mutation

A **mutation** is any change in the genetic information (DNA) of an organism. **Spontaneous mutations** are changes in DNA that cannot be tied to a particular causative agent. It is suspected that cosmic radiation or naturally occurring mutagenic chemicals might be the cause of many of these mutations. It is known that subjecting organisms to high levels of radiation or to certain chemicals increases the rate at which mutations occur. It is for this reason that people who work with radioactive materials or other mutagenic agents take special safety precautions (figure 21.8). It is also known that when cells make copies of DNA, the DNA may not be copied perfectly and small errors can be introduced.

Naturally occurring mutation rates are low (perhaps one chance in one hundred thousand that a gene will be altered). There are three possible outcomes when a gene is altered. The mutation may be so minor that it has no effect, the mutation may be harmful, or the mutation may be beneficial. Although it is likely that most mutations are neutral or harmful, beneficial mutations do occur. In populations of millions of individuals, each of whom has thousands of genes, over thousands of

FIGURE 21.8 Because radiation and certain chemicals increase the likelihood of mutations, people who work in hazardous environments receive special training and use protective measures to reduce their exposure to mutagenic agents. It is particularly important to protect the ovaries and testes, since mutations that alter the DNA of eggs and sperm can be passed on to their children.

generations it is highly likely that a new beneficial piece of genetic information could come about as a result of mutation. When we look at the various genes that exist in humans or in any other organism, we should remember that every gene originated as a modification of a previously existing gene. For example, the gene for blue eyes may be a mutated brown-eye gene, or blond hair may have originated as a mutated brown-hair gene. Thus, mutations have been very important for introducing new genetic material into species over time.

In order for mutations to be important in the evolution of organisms, they must be in cells that will become sex cells. Mutations to the cells of the skin or liver will affect only those specific cells and will not be passed to the next generation.

Genetic Diversity Resulting from Sexual Reproduction

A second very important process involved in generating genetic diversity is sexual reproduction. While sexual reproduction does not generate new genetic information the way mutation does, it allows for the recombination of genes into mixtures that did not occur previously. Each individual entering a population by sexual reproduction carries a unique combination of genes; half are donated by the mother and half are donated by the father. During the formation of sex cells, the manipulation of chromosomes results in new combinations of genes. This means that there are millions of possible combinations of genes in the sex cells of any individual. When fertilization occurs, one of the millions of possible sperm unites with one of the millions of possible eggs, resulting in a genetically unique individual. The gene mixing that occurs during sexual reproduction is known as **genetic recombination.** The new individual has a complete set of genes that is different from that of any other organism that ever existed.

The power of genetic recombination is easily seen when you look at members of your family. Compare yourself with your siblings, parents, and grandparents. Although your genes are traceable back to your grandparents, your specific combination of genes is unique, and this uniqueness shows up in differences in physiology, physical structure, and behavior.

Acquired Characteristics Do Not Influence Natural Selection

Many individual organisms acquire characteristics during their lifetime that are not the result of the genes that they have. For example, a squirrel may learn where a bird feeder is and visit it frequently. It may even learn to defeat the mechanics of special bird feeders designed to prevent it from getting to the food. Such an ability can be very important for survival if food becomes scarce. Such **acquired characteristics** are gained during the life of the organism; they are not genetically determined and, therefore, cannot be passed on to future generations through sexual reproduction.

For example, we often desire a specific set of characteristics in our domesticated animals. The breed of dog known as boxers, for example, is "supposed" to have short tails. However, the genes for short tails are rare in this breed. Consequently, the tails of these dogs are amputated—a procedure called docking. Similarly, the tails of lambs are also usually amputated. These acquired characteristics are not passed on to the next generation. Removing the tails of these animals does not remove the genes for tail production from their genetic information, and each generation of puppies or sheep is born with tails.

PROCESSES THAT DRIVE NATURAL SELECTION

Several mechanisms allow for selection of certain individuals for successful reproduction. The specific environmental factors that favor certain characteristics are called **selecting agents.** If predators must pursue swift prey organisms, then the faster predators will be selected for, and the selecting agent is the swiftness of available prey. If predators must find prey that are slow but hard to see, then the selecting agent is the camouflage coloration of the prey, and keen eyesight is selected for. If plants are eaten by insects, then the production of toxic materials in the leaves is selected for. All selecting agents influence the likelihood that certain characteristics will be passed to subsequent generations. Three kinds of mechanisms influence which genes are passed to future generations: differential survival, differential reproductive rates, and differential mate selection.

CONCEPTS APPLIED

Simulating Natural Selection

You can simulate the effects of natural selection with a deck of playing cards. Assume the black cards are recessive alleles that make an insect vulnerable to a specific insecticide. Assume that the red cards are dominant alleles that allow the insect to survive exposure to the insecticide. Initially, the dominant resistance allele will be rare in the population. To simulate this situation, keep all the black cards and six of the red cards. The frequency of the red cards is 6/32 = 18.75 percent.

If this insect population were sprayed with the insecticide, what would happen? Use the cards to simulate the effect. Shuffle the cards and draw two cards at random. You are drawing two cards because every individual has two genes (alleles) for each characteristic. If you draw two black cards, discard them. This individual did not have any alleles that allowed it to survive exposure to the insecticide. If you draw two red cards, or a red and a black card, save them. These individuals had a dominant allele that allowed them to resist the insecticide.

After you have drawn all the cards, examine those you did not discard. Calculate the frequency of the red cards (number of red cards/total red and black cards). This change in the frequency of the resistance gene is the result of the different rates of mortality and an example of evolution. This simulation should help you understand why populations of insects can become resistant to insecticides in a few generations.

Differential Survival

As stated previously, the phrase "survival of the fittest" is often associated with the theory of natural selection. Although this is recognized as an oversimplification of the concept, survival is an important factor in influencing the flow of genes to subsequent generations. If a population consists of a large number of genetically and structurally different individuals, it is likely that some of them will possess characteristics that make their survival difficult. Therefore, they are likely to die early in life and not have an opportunity to pass their genes on to the next generation.

Charles Darwin described several species of finches on the Galápagos Islands, and scientists have often used these birds in scientific studies of evolution. On one of the islands, scientists studied one of the species of seed-eating ground finches, *Geospiza fortis*. They measured the size of the animals and the size of their bills and related these characteristics to their survival. They found the following: During a drought, the birds ate the smaller, softer seeds more readily than the larger, harder seeds. As the birds consumed the more easily eaten seeds, only the larger, harder seeds remained. During the drought, mortality was extremely high. When scientists looked at ground finch mortality, they found that the larger birds that had stronger, deeper bills survived better than smaller birds with weaker, narrower bills. They also showed that the offspring of the survivors tended to show larger body and bill size as well. The lack of small, easily eaten seeds due to the drought resulted in selection for larger birds with stronger bills that could crack open larger, tougher seeds. Table 21.2 shows data on two of the parameters measured in this study.

TABLE 21.2 Changes in the Body Structure of *Geospiza fortis*

	Before Drought	After Drought
Average Body Weight	16.06 g	17.13 g
Average Bill Depth	9.21 mm	9.70 mm

As another example of how differential survival can lead to changes in the genes within a population, consider what has happened to many insect populations as we have subjected them to a variety of insecticides. Since there is genetic diversity within all species of insects, an insecticide that is used for the first time on a particular species kills all those that are genetically susceptible. However, the insecticide may not kill individuals with slightly different genetic compositions that give them resistance to the insecticide.

Suppose that in a population of a particular species of insect, 5 percent of the individuals have genes that make them resistant to a specific insecticide. The first application of the insecticide could, therefore, kill 95 percent of the population. However, tolerant individuals would then constitute the majority of the breeding population that survived. This would mean that a larger proportion of the insects in the second generation would be tolerant. The second use of the insecticide on this population would not be as effective as the first. With continued use of the same insecticide, each generation would become more tolerant. Please note in this example that the spraying did not cause mutations that made the population resistant. The use of insecticide killed susceptible individuals, thus leaving resistant individuals to reproduce and pass their resistance genes to the next generation.

Many species of insects produce a new generation each month. In organisms with a short generation time, 99 percent of the population could become resistant to the insecticide in just five years. As a result, the insecticide would no longer be useful in controlling the species. As a new factor (the insecticide) was introduced into the environment of the insect, natural selection resulted in a population that was tolerant of the insecticide.

A recent study of insecticide resistance of houseflies in poultry facilities used in egg production illustrates this point. Poultry facilities are an excellent situation for developing resistance since the flies spend their entire lives within the facility and are subjected to repeated applications of insecticide. Table 21.3 shows the effect of use of the insecticide Cyfluthrin on resistance in a particular housefly population. The standard dose that killed susceptible houseflies did not kill the resistant population. Even at a dosage of one hundred times the standard dose, nearly 40 percent of the houseflies in the resistant population survived.

TABLE 21.3 **Insecticide Resistance of Houseflies to the Insecticide Cyfluthrin**

	Houseflies from a Susceptible Population	Houseflies from a Poultry Facility Population		
Pesticide Concentration (ng/cm^2)	8.3 (standard dose)	8.3 (standard dose)	83 (10 × standard dose)	830 (100 × standard dose)
Percent Surviving	0	100	90	38

Source: Data from Scott, Jeffrey G., et al. "Insecticide resistance in houseflies from caged-layer poultry facilities," *Pest Management Science* 56 (2000):147–53.

Differential Reproductive Rates

Survival alone does not always ensure reproductive success. For a variety of reasons, some organisms may be better able to utilize available resources to produce offspring. If one individual leaves 100 offspring and another leaves only 2, the first organism has passed more copies of its genetic information to the next generation than has the second. If we assume that all 102 individual offspring have similar survival rates, the first organism has been selected for and its genes have become more common in the subsequent population.

Scientists have conducted studies of the frequencies of genes for the height of clover plants (figure 21.9). Two identical fields of clover were planted and cows were allowed to graze in one of them. Cows acted as a selecting agent by eating the taller plants first. These tall plants rarely got a chance to reproduce. Only the shorter plants flowered and produced seeds. After some time, seeds were collected from both the grazed and ungrazed fields and grown in a greenhouse under identical conditions. The average height of the plants from the ungrazed field was compared to that of the plants from the grazed field. The seeds from the ungrazed field produced some tall, some short, but mostly medium-sized plants. However, the seeds from the grazed field produced many more shorter plants than medium or tall ones. The cows had selectively eaten the plants that had the genes for tallness. Since the flowers are at the tip of the plant, tall plants were less likely to successfully reproduce, even though they might have been able to survive grazing by cows.

Differential Mate Selection

Within animal populations, some individuals may be chosen as mates more frequently than others. Obviously, those that are frequently chosen have an opportunity to pass on more copies of their genes than those that are rarely chosen. Characteristics of the more frequently chosen individuals may involve general characteristics, such as body size or aggressiveness, or specific conspicuous characteristics attractive to the opposite sex.

For example, male red-winged blackbirds establish territories in cattail marshes where females build their nests. A male will chase out all other males but not females. Some males have large territories, some have small territories, and some are unable to establish territories. Although it is possible for any male to mate, it has been demonstrated that those who have no territory are least likely to mate. Those who defend large territories may have two or more females nesting in their territories and are very likely to mate with those females. It is unclear exactly why females choose one male's territory over another, but the fact is that some males are chosen as mates and others are not.

In other cases, it appears that the females select males that display conspicuous characteristics. Male peacocks have very conspicuous tail feathers. Those with spectacular tails are more likely to mate and have offspring (figure 21.10). Darwin was puzzled by such cases as the peacock, in which the large and conspicuous tail should have been a disadvantage to the bird. Long tails require energy to produce, make it more difficult to fly, and make it more likely that predators will capture the individual. The current theory that seeks to explain this paradox involves female choice. If the females have an innate (genetic) tendency to choose the most elaborately decorated males, genes that favor such plumage will be regularly passed to the next generation because females choose the gaudy males as mates. Such special cases in which females choose males with specific characteristics have been called sexual selection.

FIGURE 21.9 The clover field to the left of the fence is undergoing natural selection as a result of cattle eating the tall plants and causing them to reproduce less than do the short plants. The other field is not subjected to this selection pressure, so its clover population has more genes for tallness.

FIGURE 21.10 In many animal species, the males display very conspicuous characteristics that are attractive to females. Because the females choose the males they will mate with, those males with the most attractive characteristics will have more offspring, and in future generations, there will be a tendency to enhance the characteristic. With peacocks, those individuals with large colorful displays are more likely to mate. The male is displaying to a female. Note that the female is not as highly colored and has a short tail.

THE HARDY-WEINBERG CONCEPT

In the section on evolution and natural selection earlier in this chapter, evolution was described as the change in the genetic makeup of a population over time. We can think of all the genes of all the individuals in a population as a **gene pool** and that a change in the gene pool indicates that evolution is taking place.

In the early 1900s, an English mathematician, G. H. Hardy, and a German physician, Wilhelm Weinberg, recognized that it was possible to apply a simple mathematical relationship to the study of genes in populations if certain conditions were met. An unchanging gene pool over several generations would imply that evolution is *not* taking place. A changing gene pool would indicate that evolution is taking place.

The conditions they felt were necessary for the genetic makeup to *remain constant* are:

1. Mating must be completely random.
2. Mutations must not occur.
3. Migration of individual organisms into and out of the population must not occur.
4. The population must be very large.
5. All genes must have an equal chance of being passed on to the next generation. (Natural selection is not occurring.)

A Closer Look

The Development of New Viral Diseases

Mutations are important in the process of evolution. Viruses are particles that are parasites within the cells of organisms. They have either DNA or RNA as their genetic material, and this genetic material only expresses itself, and the virus only displays characteristics of life, when the virus is inside a cell. As parasites and their hosts interact, they are constantly reacting to each other in an evolutionary fashion. Hosts develop new mechanisms to combat parasites, and parasites develop new mechanisms to overcome the defenses of hosts. One of the mechanisms employed by viruses involves high rates of mutation. This ability to mutate has resulted in many new, serious human diseases. In addition, many new diseases arise when viruses that cause disease in another animal are able to establish themselves in humans. Many kinds of influenza originated in pigs, ducks, or chickens and were passed to humans through close contact with or eating infected animals. Since in many parts of the world these domesticated animals live in close contact with humans (often in the same building), conditions are favorable for the transmission of animal viruses to humans.

Each year, mutations result in new varieties of influenza and colds that pass through the human population. Occasionally, the new varieties are particularly deadly. In 1918, a new variety of influenza virus originated in pigs in the United States and spread throughout the world. During the 1918–1919 influenza pandemic that followed, twenty million to forty million people died. In 1997 in Hong Kong, a new kind of influenza was identified that killed six of eighteen people infected. Public health officials discovered the virus had come from chickens and ordered the slaughter of all live chickens in Hong Kong. This stopped the spread of the disease. Since 1997 bird flu has been a constant concern and outbreaks have occurred in Vietnam, Indonesia, and Thailand. Isolated cases have occurred in other places. However, most cases appear to be related to human contact with infected birds. In a few cases it is suspected that the disease has been passed from one human to another but so far these cases are rare. If the virus mutates so that human-to-human transmission becomes easy it would become a major world health problem.

In early 2003, an outbreak of a new viral disease known as *severe acute respiratory syndrome* (SARS) originated in China. It is a variation of a coronavirus, a class of virus commonly associated with the common cold, but it caused severe symptoms and resulted in many deaths. In June 2003, the SARS virus was isolated from an animal known as the masked palm civet (*Paguma larvata*). This animal is used for food in China and is the likely origin of this new human disease. Isolation of infected persons prevented the spread, and this new disease was brought under control.

In addition to colds and influenza, other kinds of diseases often make the leap from nonhuman to human hosts. AIDS is a new disease caused by the human immunodeficiency virus (HIV), which has been traced to a similar virus in monkeys. In June 2003, public health officials announced that monkeypox virus was causing disease in people in Wisconsin, Illinois, and Indiana. The monkeypox virus was traced to a shipment of prairie dogs that were sold as pets. It is likely that the prairie dogs became infected with the virus through contact with an infected Gambian giant rat, which was part of a shipment of animals from Ghana in April 2003. Thus, it appears that the disease entered North America from Africa by way of the Gambian giant rat and was transferred to prairie dogs and subsequently from prairie dogs to people.

The concept that the genetic makeup will remain constant if these five conditions are met has become known as the **Hardy-Weinberg concept.** The Hardy-Weinberg concept is important because it allows a simple comparison of genes in populations to indicate if genetic changes are occurring within the population. Two different populations of the same species can be compared to see if they have the same genetic makeup, or populations can be examined at different times to see if the genetic makeup is changing.

Although Hardy and Weinberg described the conditions necessary to prevent evolution, an examination of these conditions becomes a powerful argument for why evolution must occur.

Random mating rarely occurs. For example, mating between individuals is more likely between those that are nearby than those that are distant. In addition, in animal populations, the possibility exists that females may choose specific males with which to mate.

Mutations occur and alter genetic diversity by either changing one genetic message into another or introducing an entirely new piece of genetic information into the population.

Migration alters the gene pool, since organisms carry their genes with them when they migrate from one place to another. Immigration introduces new genetic information, and emigration removes genes from the gene pool.

Population size influences the gene pool. Small populations typically have less genetic diversity than large populations.

Finally, *natural selection* systematically filters some genes from the population, allowing other genes to remain and become more common. The primary mechanisms involved in natural selection are differences in death rates, reproductive rates, and the rate at which individuals are selected as mates. The diagram in figure 20.11 summarizes these ideas.

Antibiotic Resistance and Human Behavior

Antibiotics have been a major factor in improving the health of people and other animals throughout the world. We have become very dependent on these molecules to combat various kinds of disease-causing organisms—particularly bacteria. However, when a specific antibiotic is used widely, it often becomes ineffective against many kinds of bacteria because exposure of the bacteria to the antibiotic selects for those individual bacteria that have mechanisms for resisting its effects. When the surviving bacteria reproduce, they pass their antibiotic resistance genes on to their offspring, and eventually, the majority of that particular species of bacteria become resistant. The gene pool of the population has changed; thus, we can say that evolution has occurred. Some species of disease-causing bacteria are resistant to several antibiotics.

Recognition that antibiotic resistance is a problem has led to requests to change the way we use antibiotics. Widespread use of antibiotics in animal feed has the high probability of causing the development of resistant strains of bacteria such as tuberculosis that occur in both cattle and humans. Indiscriminate prescription of antibiotics for healthy patients with minor bacterial infections has led to resistant strains. Thus, public health officials are calling on veterinarians and physicians to change their behavior and use antibiotics only when needed to protect the health of the patient.

Even the behavior of the patient can encourage the development of resistant strains. Often people stop taking an antibiotic when they start to feel better. They may still have the disease organisms present, however, and taking a lower dose of an antibiotic or stopping the medication before the disease is under control can result in the development of resistant strains. Tuberculosis is a particularly difficult disease, since it is a chronic disease that requires long periods of antibiotic drug therapy and is likely to be caused by organisms that are already resistant to several antibiotics. Resistant strains of tuberculosis have become so common that one of the standard methods of administering the drug is to have a health care professional directly observe the patient taking the medication as prescribed. Some particularly difficult or stubborn patients have been institutionalized until their course of treatment is complete.

Questions to Discuss

1. Should persons be deprived of their liberty so that a disease can be controlled?
2. Should there be laws about when physicians and veterinarians can prescribe antibiotics?
3. Should physicians withhold antibiotics from patients who are very likely to recover without the antibiotic?

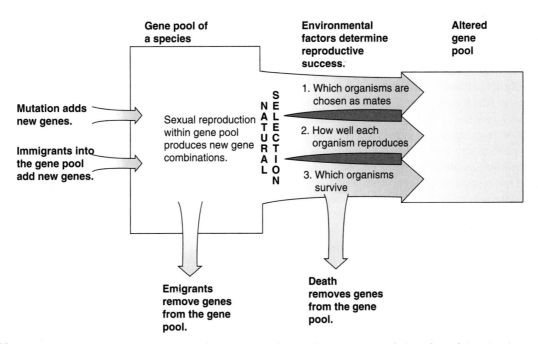

FIGURE 21.11 Several different processes cause gene frequencies to change. Genes enter populations through immigration and mutation. Genes leave populations through death and emigration. Natural selection operates within populations through death and rates of reproduction.

A Closer Look

Human-Designed Organisms

Humans have designed several kinds of plants and animals for their own purposes through the process of domestication. Most modern cereal grains are special plants that rely on human activity for their survival; most would not live without fertilizer, cultivation, and other helps. These grains are the descendants of wild plants. Initially, the process of domestication of plants by humans probably was not a conscious effort to develop crops but the unconscious selection of individual plants that had particularly useful characteristics from among all the plants of that species. For example, the seeds may have been larger than those of other plants of the same species or they may have tasted better or they may have been easier to chew. The unconscious selection of these seeds could have led to the dispersal of these particular seeds along traditional travel routes and eventually through trading with other groups of people. Wheat, barley, rice, and corn are all plants that have been domesticated and currently supply a major part of the food for the world's population. However, modern genetic techniques have resulted in domesticated plants that only superficially resemble their wild ancestors.

Most domesticated animals are mammals or birds that are herbivores that grow rapidly, are reasonably docile, and will reproduce in captivity. Examples are cattle, sheep, pigs, goats, horses, camels, chickens, turkeys, ducks, and geese. (Carnivorous cats and dogs are obvious exceptions to the general herbivore rule.) Obviously, animals that do not reproduce in captivity would not be good candidates for domestication, and those that had behaviors that made them dangerous to be around would not be chosen to be domesticated. Again, the choices by humans would have been "unconscious." They would not have planned to domesticate an animal for a particular purpose (milk, eggs, power, meat), and modern science has greatly modified domesticated animals to something that is quite different from their wild ancestors.

In fact, many of the wild ancestors of modern domesticated plants and animals are extinct. Thousands of generations of selection have, in effect, caused the development of species that are known only by their domesticated remnants.

In recent years, the genetic manipulation of organisms has expanded into a new arena. With the development of biotechnology techniques, genes can be moved from one species to another. The transfer of genes can be between plants and animals, bacteria and plants, or bacteria and animals. Genes from certain kinds of bacteria produce a compound that kills insects. This gene has been inserted into several kinds of plants so that the plants produce the natural insecticide and are protected from some of their insect pests. In addition, animal genes have been inserted into plants and human genes have been inserted into bacteria. It appears that gene transfer between species occurs naturally as well. This knowledge requires us to rethink our concept of what a species is and how evolution occurs.

ACCUMULATING EVIDENCE OF EVOLUTION

The theory of evolution has become the major unifying theory of the biological sciences. Medicine recognizes the dangers of mutations, the similarity in function of the same organ in related species, and the way in which the environment can interfere with the preprogrammed process of embryological development. Agricultural science recognizes the importance of selecting specific genes for passage into new varieties of crop plants and animals. The concepts of mutation, selection, and evolution are so fundamental to understanding what happens in biology that we often forget to take note of the many kinds of observations (facts) that support the theory of evolution. The following list describes some of the more important pieces of evidence that support the idea that evolution has been and continues to be a major force in shaping the nature of living things.

1. Species and populations are not genetically fixed. Change occurs in individuals and populations.
   a. Mutations cause slight changes in the genetic makeup of an individual organism.
   b. Different populations of the same species show adaptations suitable for their local conditions and have different combinations of genes.
   c. Changes in the characteristics displayed by species can be linked to environmental changes.
   d. Selective breeding of domesticated plants and animals indicates that the shape, color, behavior, metabolism, and many other characteristics of organisms can be selected for.
   e. Extinction of poorly adapted species is common.
2. Evolution occurs by small, incremental steps, not by major changes. All evidence suggests that, once embarked on a particular evolutionary road, the system is not abandoned, only modified. The following list supports the concept that evolution proceeds by modification of previously existing structures and processes rather than by catastrophic change.
   a. All species use the same DNA code.
   b. All species use the same left-handed, amino acid building blocks in their proteins.
   c. It is difficult to eliminate a structure when it is part of a developmental process controlled by genes. Vestigial structures such as the appendix and tailbone in humans are evidence of genetic material from previous stages in evolution.
   d. Embryological development of related animals is similar, regardless of the peculiarities of adult anatomy. The embryos of all vertebrates have an early stage that contains structures that resemble gill slits.
   e. Species of organisms that are known to be closely related show greater similarity in their DNA than those that are distantly related.

3. The fossil record supports the concept of evolution.
   a. The nature of Earth has changed significantly over time.
   b. The fossil record shows vast changes in the kinds of organisms present on Earth. New species appear and most go extinct. This is evidence that living things change in response to changes in their environment.
   c. The fossils found in old rocks do not reappear in younger rocks. Once an organism goes extinct, it does not reappear, but new organisms arise that are modifications of previous organisms.

Myths, Mistakes, and Misunderstandings

Common Misconceptions About the Theory of Evolution

1. *Evolution happened only in the past and is not occurring today.* In fact, we see lots of evidence that genetic changes are occurring in the populations of current species (antibiotic resistance, pesticide resistance, and domestication).
2. *Evolution has a specific goal.* Evolution does not move toward a specific goal. Natural selection selects those organisms that best fit the current environment. As the environment changes, so do the characteristics that have value. Random events such as changes in sea level, major changes in climate such as ice ages, or collisions with asteroids have had major influences on subsequent natural selection and evolution. Evolution results in organisms that "fit" the current environment.
3. *Changes in the environment cause mutations that are needed by an organism to survive under the new environmental conditions.* Mutations are random events and are not necessarily adaptive. However, when the environment changes, mutations that were originally detrimental or neutral may have greater value. The gene did not change but the environmental conditions did. In some cases, the mutation rate may increase or there may be more frequent exchanges of genes between individuals when the environment changes, but the mutations are still random. They are not directed to a particular goal.
4. *Individual organisms evolve.* Individuals are stuck with the genes they inherited from their parents. Although individuals may adapt by changing their behavior or physiology, they cannot evolve; only populations can show changes in the gene pool.
5. *Today's species frequently can be shown to be derived from other present-day species. For example, it is commonly stated that apes gave rise to humans.* There are few examples in which it can be demonstrated that one current species gave rise to another. Apes did not become humans, but apes and humans had a common ancestor several million years ago.

4. New techniques and discoveries invariably support the theory of evolution.
   a. The recognition that Earth was formed billions of years ago supports the slow development of new kinds of organisms.
   b. The recognition that the continents of Earth have separated and drifted apart helps explain why organisms on Australia are so different from those found elsewhere.
   c. The discovery of DNA and how it works helps explain mutation and allows us to demonstrate the genetic similarity of closely related species.

SPECIES: A WORKING DEFINITION

The smallest irreversible step in the evolutionary process is the development of a new species. Before we consider how new species are produced, let's establish how one species is distinguished from another. A **species** is commonly defined as a population of organisms whose members have the potential to interbreed naturally to produce fertile offspring but do not interbreed with other groups. This is a working definition since it applies in most cases but must be interpreted to encompass some exceptions.

There are two key ideas within this definition. First, a species is a population of organisms. An individual—you, for example—is not a species. You can only be a member of a group that is recognized as a species. The human species, *Homo sapiens,* consists of more than six billion individuals, whereas the endangered California condor species, *Gymnogyps californianus,* consists of about one hundred individuals. Second, the definition involves the ability of individuals within the group to produce fertile offspring. Obviously, we cannot check every individual to see if it is capable of mating with any other individual that is similar to it, so we must make some judgment calls. Do most individuals within the group potentially have the capability of interbreeding to produce fertile offspring? In the case of humans, we know that some individuals are sterile and cannot reproduce, but we don't exclude them from the human species because of this. If they were not sterile, they would have the potential to interbreed. We recognize that humans from all parts of the world are potentially capable of interbreeding. We know this to be true because of the large number of instances of reproduction involving people of different ethnic and racial backgrounds. The same is true for many other species that have local subpopulations but have a wide geographic distribution.

Another way to look at this question is to think about **gene flow.** Gene flow is the movement of genes from one generation to the next or from one region to another. Two or more populations that demonstrate gene flow between them constitute a single species. Conversely, two or more populations that do not demonstrate gene flow between them are generally considered to be different species. Some examples will clarify this working definition.

The mating of a male donkey and a female horse produces young that grow to be adult mules, incapable of reproduction

People Behind the Science

Ernst Mayr (1904–2005)

Ernst Mayr was born in 1904 at Kempten in southern Germany near the Austrian border to Otto and Helene Mayr. He was the second of three sons. His father was a judge and his mother was from a banking family. His parents were interested in plant and animal life, and the family took nature walks on weekends. By the age of ten, he could recognize all of the local bird species by call as well as sight. Following his father's death when Ernst was twelve, the family moved to Dresden, where he completed high school.

In March 1923, just after completing high school, Mayr visited a lake about 15 km (9 mi) north of Dresden and observed a pair of red-crested pochard ducks, which are rare in Germany. He was encouraged to stop in Berlin on his way to the University of Griefswald, where he was to study medicine, to tell noted ornithologist Edwin Stresemann about his discovery. This was to change the direction of his life. After Mayr had completed his preclinical studies, Stresemann invited him to complete a Ph.D. with him at the University of Berlin. He completed his Ph.D. at age twenty-one. Stresemann introduced Mayr to Lord Walter Rothschild of England, an avid collector of birds. Rothschild sponsored Mayr to visit the island of New Guinea to study and collect birds.

While in New Guinea, Mayr met researchers from the American Museum of Natural History who invited him to accompany them on a trip to the Solomon Islands. Shortly after returning to Germany, Mayr was invited to join the staff of the American Museum of

Natural History (1931), and he joined the faculty of Harvard University in 1953.

He married Margarete Simon in 1935 and they had two daughters.

Mayr was one of the twentieth century's leading evolutionary biologists. In 1942, he published *Systematics and the Origin of Species* in which he introduced the biological species concept—the idea that a species is a population of organisms potentially capable of interbreeding, and, furthermore, if individuals of two populations are not able to interbreed, they are of different species. Previously, scientists had sought to identify individuals as belonging to distinct species by the physical characteristics they possessed.

He also described the process of allopatric speciation, which is the idea that an isolated population is subjected to unique environmental conditions and may have different mutations from the parent population from which it was split. This can ultimately lead to the isolated population becoming a distinct species. These two concepts are central ideas in modern evolutionary thought.

Mayr received numerous awards and honors throughout his life. He retired in 1975 but continued to write about evolution and the philosophy of biology until his death at age one hundred in 2005.

Source: Modified from the *Hutchinson Dictionary of Scientific Biography*. © RM, 2007. Reprinted by permission.

A Horse

B Donkey

C Mule

FIGURE 21.12 Even though they don't do so in nature, (*A*) horses and (*B*) donkeys can be mated. The offspring is called a (*C*) mule and is sterile. Because the mule is sterile, the horse and the donkey are considered to be of different species.

(figure 21.12). Since mules are nearly always sterile, there can be no gene flow between horses and donkeys and they are considered to be separate species. Similarly, lions and tigers can be mated in zoos to produce offspring. However, this does not happen in nature, so gene flow does not occur naturally; thus, they are considered to be two separate species.

Still another way to try to determine if two organisms belong to different species is to determine their genetic similarity. The recent advances in molecular genetics allow scientists to examine the structure of the genes present in individuals from a variety of different populations. Those that have a great deal of similarity are assumed to have resulted from populations

that have exchanged genes through sexual reproduction in the recent past. If there are significant differences in the genes present in individuals from two populations, they have not exchanged genes recently and are more likely to be members of separate species. Interpretation of the results obtained by examining genetic differences still requires the judgment of experts. It will not unequivocally settle every dispute related to the identification of species, but it is another tool that helps to clarify troublesome situations.

This concept that species can be distinguished from one another by their inability to interbreed is often called the **biological species concept.** Although this concept of a species is

Is the Red Wolf a Species?

The red wolf (*Canis rufus*) is listed as an endangered species, so the U.S. Fish and Wildlife Service has instituted a captive breeding program to preserve the animal and reintroduce it to a suitable habitat in the southeastern United States, where it was common into the 1800s. Biologists have long known that red wolves will hybridize with both the coyote, *Canis latrans,* and the gray wolf, *Canis lupus,* and many suspect that the red wolf is really a hybrid between the gray wolf and the coyote. Gray wolf–coyote hybrids are common in nature where one or the other species is rare. Some have argued that if it is a hybrid, the red wolf does not meet the definition of a species and should not be protected under the Endangered Species Act.

Museums have helped shed light on this situation by providing skulls of all three kinds of animals preserved in the early 1900s. It is known that during the early 1900s as the number of red wolves in the southeastern United States declined, they readily interbred with coyotes, which were very common. The gray wolf had been exterminated by the early 1900s. Some scientists believe that the skulls of the few remaining "red wolves" might not represent the true red wolf but a "red wolf" with many coyote characteristics. Studies of the structure of the skulls of red wolves, coyotes, and gray wolves show that the red wolves were recognizably different and intermediate in structure between coyotes and gray wolves. This supports the hypothesis that the red wolf is a distinct species.

DNA studies were performed using DNA from preserved red wolf pelts. The red wolf DNA was compared to coyote and gray wolf DNA. These studies show that red wolves contain DNA sequences typical of both gray wolves and coyotes but do not appear to have distinct base sequences found only in the red wolf. These studies support the hypothesis that the red wolf is not a species but a population that resulted from hybridization between gray wolves and coyotes.

Thus, there is still no consensus on the status of the red wolf. Independent researchers disagree with one another and with Fish and Wildlife Service scientists, who have been responsible for developing and administering a captive breeding program and planning reintroductions of the red wolf.

---

useful from a theoretical point of view, it is often not a practical way to distinguish species. Thus, biologists often use specific observable physical, chemical, or behavioral characteristics as guides to distinguishing species. This method of using physical characteristics to identify species is called the **morphological species concept.** Structural differences are useful but not foolproof ways to distinguish species. However, we must rely on such indirect ways to identify species because we cannot possibly test every individual by breeding it with another to see if they will have fertile offspring. Furthermore, many kinds of organisms reproduce primarily by asexual means. Because organisms that reproduce exclusively by asexual methods do not exchange genes with any other individuals, they do not fit our *biological species* definition very well. In addition, the study of fossil species must rely on structural characteristics to make species distinctions since it is impossible to breed extinct organisms.

HOW NEW SPECIES ORIGINATE

The geographic area over which a species can be found is known as its **range.** The range of the human species is the entire world, while that of a bird known as a snail kite is a small region of southern Florida. As a species expands its range or environmental conditions change in some parts of the range, portions of the population can become separated from the rest. Thus, many species consist of partially isolated populations that display characteristics that differ significantly from other local populations. Many of the differences observed may be directly related to adaptations to local environmental conditions. This means that new colonies or isolated populations may have infrequent gene exchange with their geographically distant relatives. These genetically distinct populations are known as subspecies.

Speciation is the process of generating new species. The process can occur in several ways. One common way that speciation occurs is through geographic isolation.

The Role of Geographic Isolation in Speciation

A portion of a species can become totally isolated from the rest of the gene pool by some geographic change, such as the formation of a mountain range, river valley, desert, or ocean. When this happens, the portion of the species is said to be in **geographic isolation** from the rest of the species. If two populations of a species are geographically isolated, they are also reproductively isolated, and gene exchange is not occurring between them. The geographic features that keep the different portions of the species from exchanging genes are called **geographic barriers.** The uplifting of mountains, the rerouting of rivers, and the formation of deserts all may separate one portion of a gene pool from another. For example, two kinds of squirrels are found on opposite sides of the Grand Canyon. Some people consider them to be separate species, while others consider them to be different isolated subpopulations of the same species (figure 21.13). Even small changes may cause geographic isolation in species that have little ability to move. A fallen tree, a plowed field, or even a new freeway may effectively isolate populations within such species. Snails in two valleys separated by a high ridge have been found to be closely related but different species. The snails cannot get from one valley to the next because of the height and climatic differences presented by the ridge (figure 21.14).

Dogs and Wolves Are the Same Species

Scientists compared the mitochondrial DNA sequences from 67 breeds of dogs and 162 wolves and coyotes and jackals. Mitochondrial DNA passes only from the mother to the offspring. All the wolf and dog sequences were similar, and both differed significantly from coyotes and jackals. Careful analysis of the differences in mitochondrial DNA suggests that there were two major domestication events from different parent wolf populations and that there have been incidences of interbreeding of wolves and dogs several times. Furthermore, the various "breeds" of dogs could not be distinguished by the sequence of DNA. Dogs that superficially resemble one another may have more differences in mitochondrial DNA than those that appear quite different. The breeds of dogs are really rather superficial differences in the basic animal. The unmistakable conclusion is that dogs are simply domesticated wolves.

Kaibab squirrel

Aberts squirrel

FIGURE 21.13 These two squirrels are found on opposite sides of the Grand Canyon. Some people consider them to be different species; others consider them to be distinct populations of the same species.

The Role of Natural Selection in Speciation

The separation of a species into two or more isolated sub-populations is not enough to generate new species. Even after many generations of geographic isolation, these separate groups may still be able to exchange genes (mate and produce fertile offspring) if they overcome the geographic barrier, because they have not accumulated enough genetic differences to prevent reproductive success. Differences in environments and natural selection play very important roles in the process of forming new species. Following separation from the main portion of the gene pool by geographic isolation, the organisms within the small, local population are likely to experience different environmental conditions. If, for example, a mountain range has separated a species into two populations, one of them may receive more rain or more sunlight than the other (figure 21.15). These environmental differences act as natural selecting agents on the two gene pools and, acting over a long period of time, account for different genetic combinations in the two places. Furthermore, different mutations may occur in the two isolated populations, and each may generate different random combinations of genes as a result of sexual reproduction. This would be particularly true if one of the populations was very small. As a result, the two populations may show differences in color, height, enzyme production, time of seed germination, or many other characteristics.

Reproductive Isolation

Over a long period of time, the genetic differences that accumulate may result in regional populations called **subspecies** that are significantly modified structurally, physiologically, or behaviorally. The differences among some subspecies may be so great that they reduce reproductive success when the subspecies mate. If enough genetic differences accumulate that two populations are not able to interbreed, then speciation has occurred. Speciation has occurred only if gene flow between isolated populations

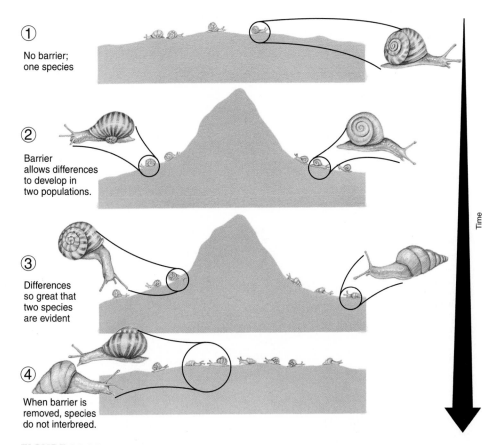

① No barrier; one species

② Barrier allows differences to develop in two populations.

③ Differences so great that two species are evident

④ When barrier is removed, species do not interbreed.

Time

FIGURE 21.14 If a single species of snail were to be divided into two different populations by the development of a ridge between them, the two populations could be subjected to different environmental conditions. This could result in a slow accumulation of changes that could ultimately result in two populations that would not be able to interbreed even if the ridge between them were to erode. They would be different species.

breeding populations. In plants, genetically determined incompatibility of the pollen of one population of flowering plants with the flowers of other populations of the same species could lead to separate species.

THE HISTORY OF THE DEVELOPMENT OF EVOLUTIONARY THOUGHT

Today, most scientists consider speciation an important first step in the process of evolution. However, this was not always the case. For centuries, people believed that the various species of plants and animals were fixed and unchanging; that is, they were thought to have remained unchanged from the time of their creation. This was a reasonable assumption because people knew nothing about DNA, how sperm and eggs were formed, or population genetics. Furthermore, the process of evolution is so slow that the results of evolution were usually not evident during a human lifetime. It is even difficult for modern scientists to recognize this slow change in many kinds of organisms.

does not occur even after barriers are removed. In other words, the process of speciation can begin with the geographic isolation of a portion of the species, but new species are generated only if isolated populations become separate from one another genetically. Speciation by this method is really a three-step process. It begins with geographic isolation, is followed by the action of selective agents that choose specific genetic combinations as being valuable, and ends with the genetic differences becoming so great that reproduction between the two groups is impossible.

Speciation Without Isolation

It is also possible to envision ways in which speciation could occur without geographic isolation being necessary. Any process that could result in the reproductive isolation of a portion of a species could lead to the possibility of speciation. For example, within populations, some individuals may breed or flower at a somewhat different time of the year. If the difference in reproductive time is genetically based, different breeding populations could be established, which could eventually lead to speciation. Among animals, variations in the genetically determined behaviors related to courtship and mating could effectively separate one species into two or more separate

A Historical Perspective

Charles Darwin (**1809–1882**) is famous for his contribution of the idea of natural selection as the process that directs evolution. However, he probably would not have recognized the importance of his observations of nature if he had not had the advantage of reading about the ideas of or engaging in discussions with others who had considered the possibility that organisms change over time. The following discussion lists some of the people known to have influenced Darwin's thinking.

Georges-Louis Leclerc, Comte de Buffon (1707–1788)— In his writings, he questioned the established church doctrines of the day that all organisms were created as we currently see them and that they are unchanging. In addition, he questioned the doctrine that Earth was only about six thousand years old. However, he did not suggest any mechanism for evolution. Buffon produced a thirty-six–volume set of books that covered essentially everything known about the natural world. Darwin referred to Buffon as the first person to suggest the idea of natural selection.

Erasmus Darwin (1731–1802)—Erasmus Darwin was Charles Darwin's grandfather. He was a major thinker of his day and had many thoughts about the processes by

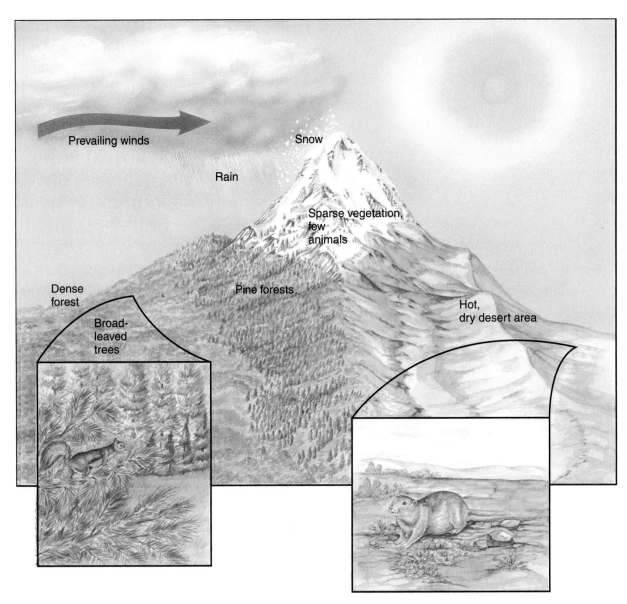

FIGURE 21.15 Most mountain ranges affect the local environment. Because of the prevailing winds, most rain falls on the windward side of the mountain. This supports abundant vegetation. The other side of the mountain receives much less rain and is drier. Often a desert may exist. Both plants and animals must be adapted to the kind of climate typical for their regions. Cactus and ground squirrels would be typical of the desert, and pine trees and tree squirrels would be typical of the windward side of the mountain.

which organisms could change over time. Although he died before Charles was born, in many ways his thinking anticipated what his grandson would make a basic tenet of biological thought. Many of these ideas are in evidence in the following poem.

Organic life beneath the shoreless waves
Was born and nurs'd in ocean's pearly caves;
First forms minute, unseen by spheric glass,
Move on the mud, or pierce the watery mass;
These, as successive generations bloom,
New powers acquire and larger limbs assume;
Whence countless groups of vegetation spring,

And breathing realms of fin and feet and wing.

Erasmus Darwin, *The Temple of Nature* (1802)

Jean-Baptiste de Lamarck (1744–1829)—In 1809, Lamarck, a student of Buffon's, suggested a process by which evolution could occur. He proposed that acquired characteristics could be transmitted to offspring. For example, he postulated that giraffes originally had short necks. Since giraffes constantly stretched their necks to obtain food, their necks got slightly longer. This slightly longer neck acquired through stretching could be passed to the offspring, who were themselves stretching their necks, and over time, the necks of giraffes would get longer and longer. Although we now know Lamarck's theory was wrong (because acquired characteristics are not inherited), it stimulated further thought as to how evolution could occur. All during this period, from the mid-1700s to the mid-1800s, lively arguments continued about the possibility of evolutionary change. Darwin was aware of Lamarck's writings and considered him to be an important thinker of the day.

Thomas Malthus (1766–1834)—Malthus' *Essay on the Principle of Population* put forward the concept that organisms reproduce faster than do resources needed to sustain the population. Therefore, there would be a constant struggle among the members of the population for the limited resources available. Thus, the human population could look forward to poverty and famine unless human reproduction was controlled. Darwin's reading of this essay and its implication of a struggle for limited resources was instrumental in his formulating the theory of natural selection.

Charles Lyell (1797–1875)—Lyell was an eminent geologist of his day. He is most famous for his elucidation of the idea of uniformitarianism. The central idea of uniformitarianism is that we can interpret evidence of past geologic events by understanding how things are working today. He felt that processes such as sedimentation, volcanic activity, and erosion did not operate differently today from the past. One of the obvious outcomes of this thinking is that the age of Earth must be much greater than a few thousand years. Darwin knew Lyell and Lyell's ideas greatly influenced Darwin's thinking about the time available for the mechanisms of evolution to operate.

Alfred Russel Wallace (1823–1913)—Wallace was the co-discoverer of the theory of natural selection. Like Darwin, Wallace traveled extensively and, as a result of his travels, came to the same conclusions about natural selection as did Darwin. He was also influenced by Malthus' ideas, as was Darwin. Although today Wallace is often mentioned as an afterthought, it was his essay concerning his ideas about natural selection that prompted Darwin to publish *On the Origin of Species by Means of Natural Selection, or the Preservation of Favoured Races in the Struggle for Life.*

A Modern View of Natural Selection

In 1858, Charles Darwin and Alfred Wallace suggested the theory of natural selection as a mechanism for evolution. They based their theory on the following assumptions about the nature of living things:

1. All organisms produce more offspring than can survive.
2. No two organisms are exactly alike.
3. Among organisms, there is a constant struggle for survival.
4. Individuals that possess favorable characteristics for their environment have a higher rate of survival and produce more offspring.
5. Favorable characteristics become more common in the species, and unfavorable characteristics are lost.

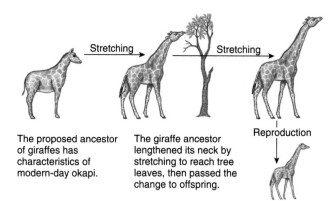

The proposed ancestor of giraffes has characteristics of modern-day okapi.

The giraffe ancestor lengthened its neck by stretching to reach tree leaves, then passed the change to offspring.

Reproduction

A Lamarck's theory: variation is acquired.

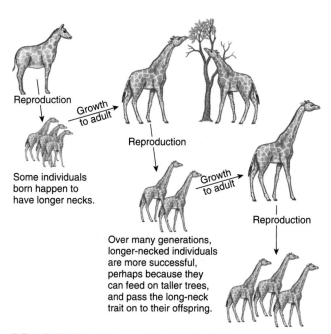

Reproduction

Growth to adult

Reproduction

Some individuals born happen to have longer necks.

Growth to adult

Reproduction

Over many generations, longer-necked individuals are more successful, perhaps because they can feed on taller trees, and pass the long-neck trait on to their offspring.

B Darwin-Wallace theory: variation is inherited.

FIGURE 21.16 (*A*) Lamarck thought that acquired characteristics could be passed on to the next generation. Therefore, he postulated that as giraffes stretched their necks to get food, their necks got slightly longer. This characteristic was passed on to the next generation, which would have longer necks. (*B*) The Darwin-Wallace theory states that there is variation within the population and that those with longer necks would be more likely to survive, reproduce, and pass on their genes for long necks to the next generation.

Using these assumptions, the Darwin-Wallace theory of evolution by natural selection offers a different explanation for the development of long necks in giraffes (figure 21.16):

1. In each generation, more giraffes would be born than the food supply could support.
2. In each generation, some giraffes would inherit longer necks, and some would inherit shorter necks.
3. All giraffes would compete for the same food sources.
4. Giraffes with longer necks would obtain more food, have a higher survival rate, and produce more offspring.
5. As a result, succeeding generations would show an increase in the neck length of the giraffe species.

This logic seems simple and obvious today, but remember that at the time Darwin and Wallace proposed their theory, the processes of sperm and egg formation and fertilization were poorly understood, and the concept of the gene was only beginning to be discussed. Nearly fifty years after Darwin and Wallace suggested their theory, the rediscovery of the work of Gregor Mendel provided an explanation for how characteristics could be transmitted from one generation to the next. Not only did Mendel's idea of the gene provide a means of passing traits from one generation to the next, it also provided the first step in understanding mutations, gene flow, and the significance of reproductive isolation. All of these ideas are interwoven into the modern concept of evolution. If we look at the same five ideas from the thinking of Darwin and Wallace and update them with modern information, they might look something like this:

1. An organism's capacity to overproduce results in surplus organisms.
2. Because of mutation, new genes enter the gene pool. Because of sexual reproduction, involving sex-cell formation and fertilization, new combinations of genes are present in every generation. These processes are so powerful that each individual in a sexually reproducing population is genetically unique. The genes present are expressed in the structure and function of the organism.
3. Resources such as food, soil nutrients, water, mates, and nest materials are in short supply, so some individuals will do without. Other environmental factors such as disease organisms, predators, or helpful partnerships with other species also affect survival. All of these factors that affect survival are called selecting agents.
4. Selecting agents favor individuals with the best combination of genes. They will be more likely to survive and reproduce, passing more of their genes on to the next generation. An organism is selected against if it has fewer offspring than other individuals that have a more favorable combination of genes. It does not need to die to be selected against.
5. Therefore, genes or gene combinations that produce characteristics favorable to survival will become more common, and the species will become better adapted to its environment.

THE TENTATIVE NATURE OF THE EVOLUTIONARY HISTORY OF ORGANISMS

It is important to understand that thinking about the concept of evolution can take us in several different directions. First, it is clear that genetic changes do occur. Mutations introduce new genes into a species. This has been demonstrated repeatedly with chemicals and radiation. Our recognition of this danger is evident by the ways we protect ourselves against excessive exposure to mutagenic agents. We also recognize that species can change. We purposely manipulate the genetic constitution of our domesticated plants and animals and change their characteristics to suit our needs. We also recognize that different populations of the same species show genetic differences. Examination of fossils shows that species of organisms that once existed are no longer in existence. We even have historical examples of plants and animals that are now extinct. We can also demonstrate that new species come into existence, and this is easily done with plants. It is clear from this evidence that species are not fixed, unchanging entities.

However, when we try to piece together the evolutionary history of organisms over long periods of time, we must use much indirect evidence, and it becomes difficult to state definitively the specific sequence of steps that the evolution of a species followed. Although it is clear that evolution occurs, it is not possible to state unconditionally that evolution of a particular group of organisms has followed a specific path. There will always be new information that will require changes in thinking, and equally reputable scientists will disagree on the evolutionary processes or the sequence of events that led to a specific group of organisms. But there can be no question that evolution occurred in the past and continues to occur today.

SUMMARY

Current theories about the *origin of life* include the two competing ideas (1) that the primitive Earth environment led to the *spontaneous organization of organic chemicals* into primitive cells and (2) that primitive forms of life *arrived on Earth from space*. Basic units of life were probably similar to present-day *prokaryotes*. These primitive cells could have *changed through time* as a result of *mutation* and *in response to a changing environment*. The presence of living things has affected the nature of Earth's atmosphere. The first organisms would have been *anaerobic* since there was no oxygen in the atmosphere. *Photosynthesis* by organisms such as cyanobacteria would have resulted in oxygen being added to the atmosphere. The presence of oxygen allowed for the development of *aerobic respiration*. Three major kinds of organisms are referred to as *domains: Eubacteria, Archaea,* and *Eucarya*. It appears that Eubacteria are the oldest group, followed by Archaea and Eucarya. The Eucarya appear to have arisen as a result of the process of *endosymbiosis*.

All sexually reproducing organisms naturally exhibit genetic diversity among the individuals in the population as a result of *mutations* and the *genetic recombination* resulting from sexual reproduction. The *genetic differences* are reflected in physical differences among individuals. These genetic differences are important for the survival of the species because *natural selection* must have *genetic diversity* to select from. Natural selection by the *environment* results in better-suited individual organisms that have greater numbers of offspring than those that are less well-off genetically.

Organisms with wide *geographic distribution* often show genetic differences in separate parts of their *range*. A *species* is a group of *organisms* that can interbreed to produce fertile offspring. This definition of a species is also known as the *biological species concept*. People often distinguish species from one another based on structural differences. This is known as a *morphological species concept*. The process of *speciation* usually involves the *geographic separation* of the species into two or more isolated *populations*. While they are separated, natural selection operates to *adapt each population to its environment*.

At one time, people thought that all organisms had remained *unchanged* from the time of their creation. Lamarck suggested that *change did occur* and thought that *acquired characteristics could be passed from generation to generation*. Darwin and Wallace proposed the *theory of natural selection* as the mechanism that drives *evolution*.

KEY TERMS

acquired characteristics (p. **508**)
autotroph (p. **501**)
biogenesis (p. **498**)
biological species concept (p. **516**)
domains (p. **503**)
endosymbiotic theory (p. **503**)
evolution (p. **505**)

gene flow (p. **515**)
gene pool (p. **511**)
genetic recombination (p. **508**)
geographic barriers (p. **517**)
geographic isolation (p. **517**)
Hardy-Weinberg concept
    (p. **512**)

heterotroph (p. **501**)
morphological species concept
    (p. **517**)
mutation (p. **508**)
natural selection (p. **507**)
oxidizing atmosphere (p. **502**)
range (p. **517**)

selecting agents (p. **509**)
speciation (p. **517**)
species (p. **515**)
spontaneous generation (p. **498**)
spontaneous mutation (p. **508**)
subspecies (p. **518**)
theory of natural selection (p. **507**)

APPLYING THE CONCEPTS

1. Which was *not* a major component of the early Earth's reducing atmosphere?
   a. H_2
   b. CO_2
   c. NH_3
   d. O_2

2. Evidence from fossils shows that prokaryotic cells came into existence approximately _____ years ago.
   a. 20 billion
   b. 4–5 billion
   c. 3.5 billion
   d. 1.5 billion

3. Which one of the following researchers found evidence not to support spontaneous generation?
   a. Redi
   b. Darwin
   c. Lamarck
   d. Woese

4. The three main kinds of living things, Eubacteria, Archaea, and Eucarya, have been labeled as
   a. species.
   b. kingdoms.
   c. domains.
   d. families.

5. Hybrid animals such as mules are not considered to be a species because they
   a. do not reproduce through many generations.
   b. are not common enough.
   c. can be maintained only by humans.
   d. look different from both parents.

6. Two closely related organisms are not considered to be separate species unless they
   a. look different.
   b. are reproductively isolated.
   c. are able to interbreed.
   d. are in different geographic parts of the world.

7. The Darwin-Wallace theory of natural selection differs from Lamarck's ideas in that
   a. Lamarck understood the role of genes, and Darwin and Wallace did not.
   b. Lamarck assumed that characteristics obtained during an organism's lifetime could be passed to the next generation; Darwin and Wallace did not.
   c. Lamarck did not think that evolution took place; Darwin and Wallace did.
   d. Lamarck developed the basic ideas of speciation, which Darwin and Wallace refined.

8. Oxygen present in our atmosphere is the result of
   a. organisms carrying on respiration.
   b. gases from volcanoes.
   c. the breakdown of carbon dioxide into carbon and oxygen.
   d. organisms carrying on photosynthesis.

9. Which of the following is *not* necessary for speciation?
   a. genetic isolation from other species
   b. genetic diversity within a species
   c. hundreds of millions of years
   d. reproduction

10. It is assumed that the original Earth had an atmosphere that was different from today. One way it is different today is that it has
    a. less ozone.
    b. less oxygen.
    c. less nitrogen.
    d. less carbon dioxide.

11. Two groups of organisms belong to different species if
    a. gene flow between the two groups is not possible, even in the absence of physical barriers.
    b. physical barriers separate the two groups, thereby preventing cross matings.
    c. the two groups of organisms have a different physical appearance.
    d. individuals from the two groups, when mated, produce fertile offspring.

12. An organism carries on photosynthesis, lacks a nucleus, and has DNA. It is a member of the Domain
    a. Eucarya.
    b. Archaea.
    c. Eubacteria.
    d. Protista.

Answers
1. d 2. c 3. a 4. c 5. a 6. b 7. b 8. d 9. c 10. d 11. a 12. c

QUESTIONS FOR THOUGHT

1. In what sequence did the following things happen: living cell, oxidizing atmosphere, eukaryotes developed, reducing atmosphere, first organic molecule?

2. What is meant by *spontaneous generation?* What is meant by *biogenesis?*

3. List two important effects caused by the increase of oxygen in the atmosphere.

4. What evidence supports the theory that eukaryotic cells arose from the development of a symbiotic relationship between primitive prokaryotic cells?

5. Why are acquired characteristics of little interest to evolutionary biologists?

6. What is natural selection? How does it work?

7. Give two examples of selecting agents and explain how they operate.

8. Why is geographic isolation important in the process of speciation?

9. Why aren't mules considered a species?

10. Why has Lamarck's theory been rejected?

11. List the series of events necessary for speciation to occur.

12. "Evolution is a fact." "Evolution is a theory." Explain how both of these statements can be true.

FOR FURTHER ANALYSIS

1. People who are hospitalized often develop infections after they have entered the hospital. Often these infections are very difficult to treat with conventional antibiotics. Why do you think this is so?

2. In earlier times, people were certain that horse hairs could turn into worms. There is even a group of simple kinds of worms known as horse-hair worms. Consider the following evidence:

   - Watering troughs were provided for horses to drink from.
   - Hairs were observed to fall from the horse's mane into the water of the watering trough.
   - Long, thin worms were seen to be swimming in the watering troughs.

   Since hair is not alive, if this were to occur, it would be a case of spontaneous generation. Devise an experiment that would determine if nonliving horse hairs become living worms.

3. How much diversity is there in hair color in the students of your class? Assume all colors seen are natural. Have the students stand in groups based on distinguishable differences in hair color. How many different categories do you have? What does this tell you about genetic diversity within our species?

4. During the early part of the twentieth century, a major social movement known as the eugenics movement occurred. The purpose of eugenics was to improve the quality of the human gene pool by eliminating "bad genes" from the population. Universities taught courses in eugenics, states passed laws that allowed certain persons to be sterilized against their will, and judging contests were held at state fairs to determine the family with the best characteristics. Hitler's concept of developing a "master race" was an extension of this thinking. Scientists who study evolutionary processes have thoroughly discredited eugenics. However, there are still vestiges of this thinking in society. Describe several examples.

INVITATION TO INQUIRY

The Evolution of Technology

A circular computer disc stores information on its surface in the form of magnetic "spots." This disc spins and is read by a device that detects the different magnetic spots. Assume you are trying to determine the evolutionary process that led to this current level of technology. How would you fit the following "fossils" into your evolutionary scheme: Edison's cylinder phonograph, a magnetic tape machine, an LP record, and a CD.

Reread "A Closer Look: The Compact Disc" on page 172 to review how a CD stores information. This is similar to how Edison's cylinder and an LP record store information. Determine the approximate time each technology was developed and describe a logical evolutionary relationship among these technologies. Can you see a direct development from one technology to another? Were there instances of "endosymbiosis" in which one technology was applied within a new system?

These are modern-day stromatolites from Hamlin Pool in Western Australia. The dome-shaped structures shown in the photograph are composed of layers of cyanobacteria and materials they secrete. They grow up to 60 cm (about 2 ft) tall. Some of the oldest fossils are of ancient stromatolites that developed in shallow marine environments about 3.5 billion years ago. When samples from fossil stromatolites are cut into slices, microscopic images can be produced that show the fossil remains of some of the world's oldest cells.

The History of Life on Earth

Physics Connections

▶ The nature of radioactivity (Ch. 11).

Chemistry Connections

▶ Fossil fuels are products of past living things (Ch. 20).

Earth Science Connections

▶ Different types of rocks form under specific conditions (Ch. 15).

▶ Plate tectonics results in the movement of continents (Ch. 15).

▶ Weathering processes are important geologic forces (Ch. 16).

CORE CONCEPT

Earth and its kinds of living things changed greatly over billions of years.

The same geological processes operate today as in the past (p. 528).

Fossils provide evidence for past life and show that the kinds of organisms on Earth have changed (p. 533).

In the past, the positions of the continents changed, the climate changed, and sea level changed (p. 538).

Humans are a recent addition to Earth (p. 539).

Living things can be organized into logical evolutionary sequences (p. 543).

Life Science Connections

▶ Evolution of organisms occurs (Ch. 21).

▶ Ecological principles can be applied to past biological communities (Ch. 23).

▶ Endosymbiosis probably led to eukaryotic organisms (Ch. 21).

Astronomy Connections

▶ Meteorite impacts occurred (Ch. 13).

OVERVIEW

At one time, people assumed that Earth and its inhabitants were fixed and unchanging. However, as they began to understand the nature of events such as volcanic eruptions, earthquakes, and erosion, it became obvious that Earth was changeable.

Furthermore, the presence of the fossilized remains of extinct organisms showed that the nature of the kinds of things that inhabited Earth had changed as well. These ideas challenged core beliefs and began a long process of modifying the way we look at the history of Earth and life on it. In particular, people needed to develop an understanding that the history of Earth and its life was very long and should be measured in billions of years rather than hundreds or thousands of years.

GEOLOGIC TIME

Reading history from the rocks of Earth's crust requires both a feel for the immensity of geologic time and an understanding of geologic processes. By "geologic time," we mean the age of Earth; the very long span of Earth's history. This span of time is difficult for most of us to comprehend. Human history is measured in units of hundreds and thousands of years, and even the events that can take place in a thousand years are hard to imagine. Geologic time, on the other hand, is measured in units of millions and billions of years.

The understanding of geologic processes has been made possible through the development of various means of measuring ages and time spans in geologic systems. An understanding of geologic time leads to an understanding of geologic processes, which then leads to an understanding of the environmental conditions that must have existed in the past. Thus, the mineral composition, texture, and sedimentary structure of rocks are clues about past events, events that make up the geological and biological history of Earth.

Early Attempts at Earth Dating

Most human activities are organized by time intervals of minutes, hours, days, months, and years. These time intervals are based on the movements of Earth and the Moon. Short time intervals are measured in minutes and hours, which are typically tracked by watches and clocks. Longer time intervals are measured in days, months, and years, which are typically tracked by calendars. How do you measure and track time intervals for something as old as Earth? First, you would need to know the age of Earth; then you would need some consistent, measurable events to divide the overall age into intervals. Questions about the age of Earth have puzzled people for thousands of years, dating back at least to the time of the ancient Greek philosophers. Many people have attempted to answer this question and understand geologic time but with little success until the last few decades.

Biblical Calculations

One early estimate of the age of Earth was attempted by Archbishop James Ussher of Ireland in the seventeenth century. He painstakingly counted up the generations of people mentioned in biblical history, added some numerological considerations, and arrived at the conclusion that Earth was created on October 23, in the year 4004 B.C. On the authority of biblical scholars, this date was generally accepted for the next century or so, even though some people thought that the geology of Earth seemed to require far longer to develop. The date of 4004 B.C.

meant that Earth and all of the surface features on Earth had formed over a period of about six thousand years. This required a model of great cataclysmic catastrophes to explain how all Earth's features could possibly have formed over a span of six thousand years.

Early Scientific Approaches

Near the end of the eighteenth century, James Hutton reasoned out the concept that the geologic processes we see today have operated since Earth began. This is known as the *principle of uniformity,* and since many geologic processes such as erosion and sedimentation are slow, people began to assume a much older Earth. (Many of Hutton's ideas were restated and popularized by Charles Lyell, whose writings greatly influenced Charles Darwin.) The problem then became one of finding some uniform change or process that could serve as a geologic clock to measure the age of Earth. To serve as a geologic clock, a process or change would need to meet three criteria: (1) the process must have been operating since Earth began, (2) the process must be uniform or at least subject to averaging, and (3) the process must be measurable.

During the nineteenth century, many attempts were made to find earth processes that would meet the criteria to serve as a geologic clock. Among others, the processes explored were (1) the rate that salt is being added to the ocean, (2) the rate that sediments are being deposited, and (3) the rate that Earth is cooling. Comparing the load of salts being delivered to the ocean by all the rivers, and assuming the ocean was initially pure water, it was calculated that about one-hundred million years would be required for the present salinity to be reached. The calculations did not consider the amount of materials being removed from the ocean by organisms and by chemical sedimentation, however, so this technique was considered to be unacceptable. Even if the amount of materials removed were known, it would actually result in the age of the ocean, not the age of Earth.

A number of separate and independent attempts were made to measure the rate of sediment deposition, then compare that rate to the thickness of sedimentary rocks found on Earth. Dividing the total thickness by the rate of deposition resulted in estimates of an Earth age that ranged from about twenty million to fifteen-hundred million years. The wide differences occurred because there are gaps in many sedimentary rock sequences, periods when sedimentary rocks were being eroded away to be deposited again elsewhere as sediments. There were just too many unknowns for this technique to be considered as acceptable.

The idea of measuring the rate that Earth is cooling for use as a geologic clock assumed that Earth was initially a molten mass that has been cooling ever since. Calculations estimating the temperature that Earth must have been to be molten were compared to Earth's present rate of cooling. This resulted in an estimated age of twenty to forty million years. These calculations were made back in the nineteenth century before it was understood that natural radioactivity is adding heat to Earth's interior, so it has required much longer to cool down to its present temperature.

Modern Techniques for Determining the Age of Earth

Soon after the beginning of the twentieth century, the discovery of the radioactive decay process in the elements of minerals and rocks led to the development of a new, accurate geologic clock. This clock finds the *radiometric age* of rocks in years by measuring how much of an unstable radioactive isotope within the crystals of certain minerals has decayed. Since radioactive decay for each kind of radioactive isotope occurs at a constant, known rate, the ratio of the remaining amount of a radioactive element to the amount of decay products present can be used to calculate the time that the radioactive element has been a part of that crystal (see chapter 11). Certain radioactive isotopes of potassium, uranium, and thorium are often included in the minerals of rocks, so they are often used as "radioactive clocks." By using radiometric aging techniques along with other information, we arrive at a generally accepted age for Earth of about 4.5 billion years. It should be noted that radiometric aging is only useful in aging igneous rocks, since sedimentary rocks are the result of weathering and deposition of other rocky materials.

Table 22.1 lists several radioactive isotopes, their decay products, and half-lives. Often two or more isotopes are used to determine an age for a rock. Agreement between them increases the scientist's confidence in the estimates of the age of the rock. Because there are great differences in the half-lives, some are useful for dating things back to several billion years, while others, such as carbon-14, are only useful for dating things to perhaps fifty thousand years. Carbon-14 is not used to age rock but is very useful in aging materials that are of relatively recent biological origin, since carbon is an important part of all living things. Also, a slightly different method is used to determine carbon-14 dating. (See "A Closer Look: What Is Carbon-14 Dating?" on p. 533.)

A recently developed geologic clock is based on the magnetic orientation of magnetic minerals. These minerals become aligned with Earth's magnetic field when the igneous rock crystallizes, making a record of the magnetic field at that time. Earth's magnetic field is global and has undergone a number of reversals in the past. A *geomagnetic time scale* has been established from the number and duration of magnetic field reversals occurring during the past six million years. Combined with radiometric age dating, the geomagnetic time scale is making possible a worldwide geologic clock that can be used to determine local chronologies.

TABLE 22.1   Radioactive Isotopes and Half-Lives

Radioactive Isotope	Stable Daughter Product	Half-Life
Samarium-147	Neodymium-143	106 billion years
Rubidium-87	Strontium-87	48.8 billion years
Thorium-232	Lead-208	14.0 billion years
Uranium-238	Lead-206	4.5 billion years
Potassium-40	Argon-40	1.25 billion years
Uranium-235	Lead-207	704 million years
Carbon-14	Nitrogen-14	5,730 years

Arranging Events in Order

The clues provided by thinking about geologic processes that must have occurred in the past are interpreted within a logical frame of reference that can be described by several basic principles. The following is a summary of these basic guiding principles that are used to read a story of geologic events from the rocks.

The *principle of uniformity* is the cornerstone of the logic used to guide thinking about geologic time. As described in chapter 16, this principle is sometimes stated as "the present is the key to the past." This means that the geologic features that you see today were formed in the past by the same processes of crustal movement, erosion, and deposition that are observed today. By studying the processes now shaping Earth, you can understand how it has evolved through time. This principle establishes the understanding that the surface of Earth has been continuously and gradually modified over the immense span of geologic time.

The *principle of original horizontality* is a basic principle that is applied to sedimentary rocks. It is based on the observation that, on a large scale, sediments are commonly deposited in flat-lying layers. Old rocks are continually being changed to new ones in the continuous processes of crustal movement, erosion, and deposition. As sediments are deposited in a basin of deposition, such as a lake or ocean, they accumulate in essentially flat-lying, approximately horizontal layers (figure 22.1). Thus, any layer of sedimentary rocks that is not horizontal has been subjected to forces that have deformed Earth's surface.

The *principle of superposition* is another logical and obvious principle that is applied to sedimentary rocks. Layers of sediments are usually deposited in succession in horizontal layers, which later are compacted and cemented into layers of sedimentary rock. (See chapter 15 for a discussion of kinds of rocks.) An undisturbed sequence of horizontal layers is thus arranged in chronological order with the oldest layers at the bottom. Each consecutive layer will be younger than the one below it (figure 22.2). This is true, of course, only if the layers have not been turned over by deforming forces.

FIGURE 22.2 The Grand Canyon, Arizona, provides a majestic cross section of horizontal sedimentary rocks. According to the principle of superposition, traveling deeper and deeper into the Grand Canyon means that you are moving into older and older rocks.

The *principle of crosscutting relationships* is concerned with igneous and metamorphic rock, in addition to sedimentary rock layers. Any geologic feature that cuts across or is intruded (forced) into a rock mass must be younger than the rock mass. Thus, if a fault cuts across a layer of sedimentary rock, the fault is the youngest feature. Faults, folds, and igneous intrusions are always younger than the rocks in which they occur. Often the presence of metamorphic rock provides a further clue to the correct sequence. When hot igneous rock is intruded into preexisting rock, the rock surrounding the hot igneous rock can be "baked," or metamorphosed, into a different form. Thus, the rock from which the metamorphic rock formed must have preceded the igneous rock (figure 22.3).

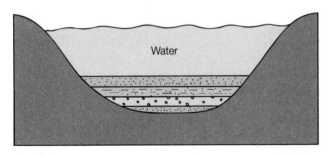

A

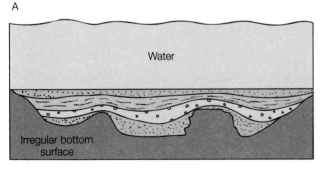

B

FIGURE 22.1 (*A*) The principle of original horizontality states that sediments tend to be deposited in horizontal layers. (*B*) Even where the sediments are draped over an irregular surface, they tend toward the horizontal.

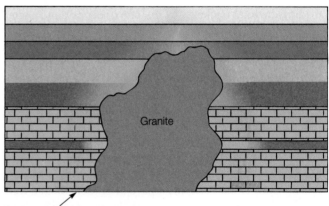

Rocks adjacent to intruding magma may also be metamorphosed by its heat.

FIGURE 22.3 According to the principle of crosscutting relationships, since the granite intrusion is cutting across the sedimentary rock, the granite intrusion is younger than the surrounding sedimentary rock.

Shifting sites of erosion and deposition: The principle of uniformity states that the earth processes going on today have always been occurring. This does not mean, however, that they always occur in the same place. As erosion wears away the rock layers at a site, the sediments produced are deposited someplace else. Later, the sites of erosion and deposition may shift, and the sediments are deposited on top of the eroded area. When the new sediments later are formed into new sedimentary rocks, there will be a time lapse between the top of the eroded layer and the new layers. A time break in the rock record is called an **unconformity.** The unconformity is usually shown by a surface within a sedimentary sequence on which there was a lack of sediment deposition or where active erosion may even have occurred for some period of time. When the rocks are later examined, that time span will not be represented in the record, and if the unconformity is erosional, some of the record once present will have been lost. An unconformity may occur within a sedimentary sequence of the same kind or between different kinds of rocks. The most obvious kind of unconformity to spot is an *angular unconformity.* An angular unconformity, as illustrated in figure 22.4, is one in which the bedding planes above and below the unconformity are not parallel. An angular unconformity usually implies some kind of tilting or folding, followed by a significant period of erosion, which in turn was followed by a period of deposition (figure 22.5).

The principle of superposition, the principle of crosscutting relationships, and the presence of an unconformity help us interpret the order in which geologic events occurred. This order can be used to unravel a complex sequence of events, such as the one shown in figure 22.6.

FIGURE 22.5 This photograph shows an angular unconformity in the rock record in the Grand Canyon, Arizona. The horizontal sedimentary rock layers overlie almost vertically oriented metamorphic rocks. Metamorphic rocks form deep in Earth, so they must have been uplifted, and the overlying material eroded away before being buried again.

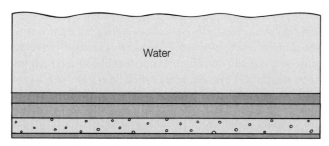

1. Deposition

2. Rocks tilted, eroded

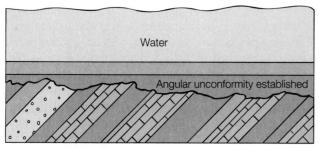

3. Subsequent deposition

FIGURE 22.4 An unconformity is a gap in the geologic record caused when the original deeper rocks were subject to erosion and then a subsequent layer of sediments was deposited on top of them. An angular unconformity is easy to identify because the layers of the deeper rocks are not parallel to the horizontal surface rocks.

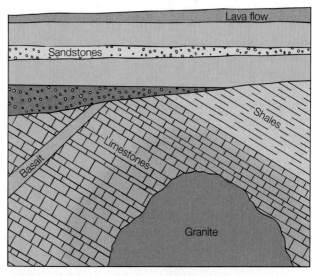

FIGURE 22.6 A complex rock sequence can be deciphered by using geologic principles. The limestones must be oldest (law of superposition), followed by the shales. The granite and basalt must both be younger than the limestone they crosscut (note the metamorphosed zone around the granite). It is not possible to tell whether the igneous rocks (granite and basalt) predate or postdate the shales or to determine whether the sedimentary rocks were tilted before or after the igneous rocks were emplaced. After the limestones and shales were tilted, they were eroded, and then the sandstones were deposited on top. Finally, the lava flow covered the entire sequence.

GEOLOGIC TIME AND THE FOSSIL RECORD

A **fossil** is *any* evidence of former life, so the term means more than fossilized remains such as those pictured in figure 22.7. Evidence can include actual or altered remains of plants, animals, fungi, algae, or even bacteria. It could also be just simple evidence of former life such as the imprint of a leaf, the footprint of a dinosaur, or droppings from bats in a cave.

What would you think if you were on the top of a mountain, broke open a rock, and discovered the fossil fish pictured in figure 22.7? How could you explain what you found? There are several ways that a fish could end up on a mountaintop as a fossil. For example, perhaps the ocean was once much deeper and covered the mountaintop. On the other hand, maybe the mountaintop was once below sea level and pushed its way up to its present high elevation. Another explanation might be that someone left the fossil on the mountain as a practical joke. What would you look for to help you figure out what actually happened? In every rock and fossil there are fascinating clues that help you read what happened in the past, including clues that tell you if an ocean had covered the area or if a mountain pushed its way up from lower levels. There are even clues that tell you if a rock has been brought in from another place. This section is about some of the clues found in fossils and rocks, and what the clues mean.

Early Ideas About Fossils

Our current understanding of fossils requires a great deal of information about geologic time and geologic processes. Without this information, it was difficult for people to appreciate how fossils formed and how to interpret them. The ancient Greek historian Herodotus was among the first to realize that fossil shells found in rocks far from any ocean were remnants of organisms left by a bygone sea. However, one hundred years later, Aristotle still believed that fossils had no relation to living things and were formed inside the rocks. Note that he also believed that living organisms could arise by spontaneous generation from mud. A belief that the fossils must have "grown" in place in rocks would be consistent with a belief in spontaneous generation.

Even when people recognized some similarities between living organisms and certain fossils, they did not make a connection between the fossil and previously living things. Fossils were considered to be the same as quartz crystals or any other mineral crystals that either formed with Earth or grew there later (depending on the philosophical view of the interpreter). With the development of Judeo-Christian beliefs, many people came to regard well-preserved fossils that were very similar to living organisms as remains of once-living organisms—that were buried in Noah's flood. Leonardo da Vinci, like other Renaissance scholars, argued that fossils were the remains of organisms that had lived in the past.

By the early 1800s, the true nature of fossils was becoming widely accepted. William "Strata" Smith, an English surveyor, discovered at this time that sedimentary rock strata could be identified by the fossils they contain. Smith grew up in a region of England where fossils were particularly plentiful. He became a collector, keeping careful notes on where he found each fossil and in which type of sedimentary rock layers. During his travels, he discovered that the succession of rock layers on the south coast of England was the same as the succession of rock layers on the east coast. Through his keen observations, Smith found that each kind of sedimentary rock had a distinctive group of fossils that was unlike the group of fossils in other rock layers. Smith amazed his friends by telling them where and in what type of rock they had found their fossils.

Today, the science of discovering fossils, studying the fossil record, and deciphering the history of life from fossils is known as **paleontology.** The word *paleontology* was invented in 1838 by the British geologist Charles Lyell to describe this newly established branch of geology. It is derived from classic Greek roots and means "study of ancient life," and this means a study of fossils.

People sometimes blur the distinction between *paleontology* and *archaeology.* **Archaeology** is the study of past human life and culture from material evidence of artifacts, such as graves, buildings, tools, pottery, landfills, and so on (*artifact* literally means "something made"). The artifacts studied in archaeology can be of any age, from the garbage added to the city landfill yesterday to the fragments of pottery of an ancient tribe that disappeared hundreds of years ago. The word *fossil* originally meant "anything dug up" but today carries the meaning of any *evidence of ancient organisms in the history of life.* Artifacts are therefore not fossils, as you can see from the definitions.

Types of Fossils

Considering all the different things that can happen to the remains of an organism, and considering the conditions needed

FIGURE 22.7 The hard parts of this fish fossil are beautifully preserved, along with a carbon film, showing a detailed outline of the fish and some of its internal structure.

to form a fossil, it seems amazing that any fossils are formed and then found. For example, the animals you see killed beside the road or the dead trees that fall over in a forest rarely become fossils because scavengers eat the remains of dead animals, and decay organisms break down the organic remains of plants and animals. As a result, very little digestible organic matter escapes destruction.

Thus, a fossil is not likely to form unless there is rapid burial of a recently deceased organism. The presence of hard parts, such as a shell or a skeleton, will also favor the formation of a fossil if there is rapid burial. Several different kinds of fossils can be differentiated based on the kind of material that is preserved and the processes of preservation.

Preserved Whole Organisms

Occasionally, entire organisms are preserved, and in rare cases, even the unaltered remains of an organism's *soft parts* are found. For this to occur, the organism must be quickly protected from scavengers and decomposers following death. Several conditions allow this to occur. The best examples of this uncommon method of fossilization include protection by freezing or entombing in tree resin. Mammoths, for example, have been found frozen and preserved by natural refrigeration in the ices of Alaska and Siberia. The bodies of ice age humans have also been discovered in glaciers, and early inhabitants of South America have been found in ice caves high in mountains. Insects and spiders, complete with delicate appendages, have been found preserved in amber, which is fossilized tree resin (figure 22.8). In each case—ice or resin—the soft parts were protected from scavengers, insects, and bacteria. Additional examples of organisms that have been preserved in a relatively unaltered state include organisms buried in acid bogs or those quickly desiccated following death.

FIGURE 22.8 This fly was stuck in and covered over with plant pitch. When the pitch fossilized to form amber, the fly was preserved as well. The entire fly is preserved in the amber.

Preserved Hard Parts

Although fossils of soft parts are occasionally formed, the *remains of hard parts* such as shells, bones, and teeth of animals or the pollen and spores of plants are much more common. The mammoth bones shown in figure 22.9 are a good example of fossils of hard parts of organisms. The bones were covered with sediment and preserved.

Because hard parts are more easily preserved, organisms that have such parts are more often found as fossils than those that lack hard parts. Worms of various kinds are hard to find as fossils for this reason. The hard parts of various organisms are composed of different kinds of materials. Calcium carbonate is present in the shells of most mollusks and certain kinds of protozoans, and the skeletal structures of corals and some sponges. Some thick deposits of limestone are composed of the calcium carbonate shells of tiny marine protozoans known as foraminiferans. Calcium phosphate was found in the shells of some kinds of extinct brachiopods. Silica is found in the skeletal structures of certain protozoans and sponges. The silica shells of the marine microorganisms known as diatoms have accumulated to such a degree that ancient deposits of the shells are mined as diatomaceous earth and used for filters and certain buffing compounds. Chitin is the tough organic material that makes up the exoskeletons of insects, spiders, crabs, lobsters, and similar organisms. Since chitin resists decomposition, organisms that have exoskeletons of chitin are common in the fossil record.

FIGURE 22.9 Skeletons of recent organisms are often preserved as fossils in sediments. These are the bones of a mammoth.

Traces, Molds, and Casts

Many kinds of fossils do not preserve the organisms but give a general indication of shape and size. A carbon trace is an example of another kind of fossil, one that does not preserve the detail of internal structures but gives a good idea of the general shape of the organism. Plant fossils are often found as carbon traces, sometimes looking like a photograph of a leaf on a slab of shale or limestone. Other fossils that give a good idea of the general shape of an organism are molds and casts. Calcium carbonate shell material may be dissolved by groundwater in

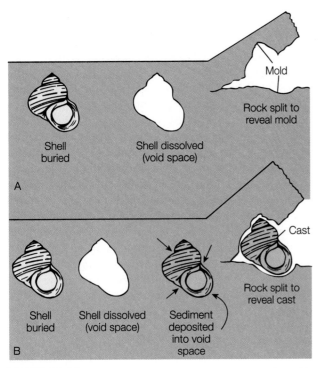

FIGURE 22.10 Origin of molds and casts. (*A*) Formation of a mold. (*B*) Formation of a cast.

certain buried environments, leaving an empty *mold* in the rock. Sediment, or groundwater deposits may fill the mold and make a *cast* of the organism (figure 22.10).

Petrified Fossils

Some fossils are formed by the modification of the original chemical nature of the organism. Figure 22.11 is a photograph of part of the Petrified Forest National Park in Arizona. Petrified trees are not trees that have been "turned to stone." They are stony *replicas* that were formed as the wood was altered by circulating groundwater carrying elements in solution. There are two processes involved in the making of petrified fossils, and they are not restricted to wood. The processes involve (1) *mineralization,* which is the filling of pore spaces within the structure of the buried organism with deposits of calcium carbonate, silica, or pyrite, and/or (2) *replacement,* which is the dissolving of the original material and depositing of new material an ion at a time. Petrified wood is formed by both processes over a long period of time. As it decayed, the original wood was replaced by mineral matter an ion at a time. Over time, the "mix" of minerals being deposited changed, resulting in various colors. Since the mineralization or replacement processes take place an ion at a time, a great deal of detail can be preserved in such fossils. The size and shape of the cells and growth rings in petrified wood are preserved well enough that they can be compared to modern plants. The skeletons or shells of many extinct organisms are also typically preserved in this way. The "fossilized bones" of dinosaurs or the "fossilized shells" of many invertebrates are examples.

FIGURE 22.11 Petrified fossils are the result of an ion-for-ion replacement of the buried organisms with mineral material. These logs of petrified wood are in the Petrified Forest National Park, Arizona.

Fossils That Are Not Remains

Finally, there are many kinds of fossils that are not the preserved remains of an organism but are preserved indications of the activities of organisms. Some of the most interesting of such fossils are the footprints of various kinds of animals. There are examples of footprints of dinosaurs and various kinds of extinct mammals, including ancestors of humans. If an animal walked through mud or other soft substrates and the substrates were covered with silt or volcanic ash, the pattern of the footprints can be preserved (figure 22.12). The tunnels of burrowing animals, the indentations left by crawling worms, and the nests of dinosaurs have been preserved in a similar manner. Even the dung of dinosaurs and other animals has been preserved in a mineralized form.

As you can see, there are many different ways in which a fossil can be formed, but it must be *found and studied* if it is to reveal its part in the history of life. This means the rocks in which the fossil formed must somehow make it back to the surface of Earth. This usually involves movement and uplift of the rock and weathering and erosion of the surrounding

FIGURE 22.12 This dinosaur footprint is in shale near Tuba City, Arizona. It tells you something about the relative age of the shale, since it must have been soft mud when the dinosaur stepped here.

What Is Carbon-14 Dating?

Carbon is an element that occurs naturally as several isotopes. The most common isotope is carbon-12. A second, heavier radioactive isotope, carbon-14, is constantly being produced in the atmosphere by cosmic rays. Radioactive elements are unstable and break down into other forms of matter. Hence, radioactive carbon-14 naturally decays. The rate at which carbon-14 is formed and the rate at which it decays are about the same; therefore, the concentration of carbon-14 on Earth stays relatively constant. All living things contain large quantities of the element carbon. Plants take in carbon in the form of carbon dioxide from the atmosphere, and animals obtain carbon from the

food they eat. While an organism is alive, the proportion of carbon-14 to carbon-12 within its body is equal to its surroundings. When an organism dies, it is no longer adding carbon-12 or carbon-14 to its tissues. As the carbon-14 within its tissues undergoes nuclear decay, it is not replaced, and the relative percentages of carbon-12 and carbon-14 change. Therefore, the age of plant and animal remains can be determined by the ratio of carbon-14 to carbon-12 in the tissues. The older the specimen, the less carbon-14 present. Radioactive decay rates are measured in half-life. One half-life is the amount of time it takes for one-half of a radioactive sample to decay.

The half-life of carbon-14 is 5,730 years. Therefore, a bone containing one-half the normal proportion of carbon-14 is 5,730 years old. If the bone contains one-quarter of the normal proportion of carbon-14, it is $2 \times 5{,}730 = 11{,}460$ years old, and if it contains one-eighth of the naturally occurring proportion of carbon-14, it is $3 \times 5{,}730 = 17{,}190$ years old. As the amount of carbon-14 in a sample becomes smaller, it becomes more difficult to measure the amount remaining. Therefore, carbon-14 dating is generally only useful for dating things that are less than 50,000 years old.

rock to release or reveal the fossil. Most fossils are found in recently eroded sedimentary rocks—sometimes atop mountains that were under the ocean a long, long time ago. The complete record of what has happened in the past is not found in the fossil alone but requires an understanding of the layers of rocks present, the relationship of the layers to each other, and their age.

Using Fossils to Determine the Order of Geologic Events

The *principle of faunal succession* recognizes that life-forms have changed through time. Old life-forms disappear from the fossil record and new ones appear, but the same form is never exactly duplicated independently at two different times in history. This principle implies that the same type of fossil organisms that lived only a brief geologic time should occur only in rocks that are the same age. According to the principle of faunal succession, then, once the basic sequence of fossil forms in the rock record is determined, rocks can be placed

in their correct relative chronological position on the basis of the fossils contained in them. The principle also means that if the same type of fossil organism is preserved in two different rocks, the rocks should be the same age. This is logical even if the two rocks have very different compositions and are from places far, far apart (figure 22.13).

Distinctive fossils of plant or animal species that were distributed widely over Earth but lived only a brief time are called **index fossils.** Index fossils, together with the other principles used in reading rocks, make it possible to compare the ages of rocks exposed in two different locations. This is called *age correlation* between rock units. Correlations of exposed rock units separated by a few kilometers are easier to do, but correlations have been done with exposed rock units that are separated by an ocean. Correlation allows the ordering of geologic events

CONCEPTS APPLIED

Examining Sedimentary Rock

Visit a place where layers of sedimentary rock are exposed. In many parts of the country, this may require you to travel to a place where a hill has been cut through to allow for the building of a road. Make a drawing of the layers and measure their thickness. If it took a hundred years to form a millimeter of sediment, how many years are represented by each different layer? Look for fossils in each of the layers.

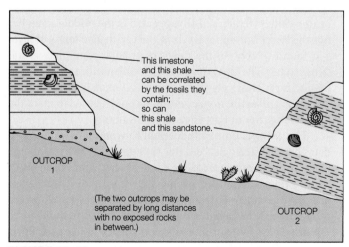

This limestone and this shale can be correlated by the fossils they contain; so can this shale and this sandstone.

OUTCROP 1

(The two outcrops may be separated by long distances with no exposed rocks in between.)

OUTCROP 2

FIGURE 22.13 Index fossils can be used to help to compare the ages of rocks in different parts of the world. Similarity of fossils suggests similarity of ages, even in different rocks widely separated in space.

according to age. Since this process is only able to determine the age of a rock unit or geologic event relative to some other unit or event, it is called *relative dating.*

The usefulness of correlation and relative dating through the concept of faunal succession is limited because the principles can be applied only to rocks in which fossils are well preserved, which are almost exclusively sedimentary rocks. Correlation also can be based on the occurrence of unusual rock types, distinctive rock sequences, or other geologic similarities. All this is useful in clarifying relative age relationships among rock units. It is not useful in answering questions about the age of rocks or the time required for certain events, such as the eruption of a volcano, to occur. Questions requiring numerical answers went unanswered until the twentieth century, when such tools as radiometric and geomagnetic dating techniques became available.

The Geologic Time Scale

A yearly calendar helps you keep track of events over periods of time by dividing the year into months, weeks, and days. In a similar way, the **geologic time scale** helps you keep track of events that have occurred in Earth's geologic history. The first development of this scale came from the work of William "Strata" Smith, the English surveyor described in the section on fossils earlier in this chapter. Recall that Smith discovered that certain sedimentary rock layers in England occurred in the same order, top to bottom, wherever they were located. He also found that he could correlate and identify each layer by the kinds of fossils in the rocks of the layers. In 1815, he published a geologic map of England, identifying the rock layers in a sequence from oldest to youngest. Smith's work was followed by extensive geological studies of the rock layers in other countries. Soon it was realized that similar, distinctive index fossils appeared in rocks of the same age when the principle of superposition was applied. For example, the layers at the bottom contained fossils of trilobites (figure 22.14A), but trilobites were not found in the upper levels. (Trilobites are extinct marine arthropods that may be closely related to living crustaceans, spiders, and horseshoe crabs.) On the other hand, fossil shells of ammonites (figure 22.14B) appeared in the middle levels but not the lower nor upper levels of the rocks. The topmost layer was found to contain the fossils of animals identified as still living today. The early appearance and later disappearance of fossils in progressively younger rocks is explained by organic evolution and extinction, events that could be used to mark the time boundaries of Earth's geologic history.

The geologic time scale developed in the 1800s was without dates because geologists did not yet have a way to measure the length of the eras, periods, or epochs. Their time scale was a *relative time scale,* based on the superposition of rock layers and fossil records of organic evolution. With the development of radiometric dating, it became possible to attach numbers to the time scale. When this was accomplished, it became apparent that geologic history spanned very long periods of time. In addition, the part of the time scale with evidence of life makes up about 75 percent of all of Earth's history.

A Trilobites

B Ammonites

FIGURE 22.14   Since both trilobites (*A*) and ammonites (*B*) were common in the oceans throughout the world, they have been used as index fossils. Trilobites are always found in older rocks than are ammonites. Trilobites were common in the Paleozoic era, and ammonites were common in the earlier parts of the Mesozoic era.

Names for Units of Geologic Time

The major blocks of time in Earth's geologic history are divided into several different categories. The largest are called **eons.** The *Phanerozoic eon* spans the period of time from about 540 million years ago, when the various groups of animals we currently see on Earth developed, until the present. The *Proterozoic eon* encompasses the period of time (2,500 million to 540 million years ago) when most living things were simple, marine organisms. The first eukaryotic organisms developed during this time. The *Archean eon* (3,800 million to 2,500 million years ago) constitutes a time when all organisms were prokaryotic and marine. The *Hadean eon* is the earlier period before life and before Earth solidified.

The Phanerozoic eon is divided into periods of time called **eras,** and each era is identified by the appearance and disappearance of particular fossils in the sedimentary rock record. The eras are: (1) *Cenozoic,* which refers to the time of recent life. Recent life means that the fossils for this time period are similar to the life found on Earth today. (2) *Mesozoic,* which refers to the time of middle life. Middle life means that some of the fossils for this time period are similar to the life found on Earth today, but many are different from anything living today. (3) *Paleozoic,* which refers to the time of ancient life. Ancient life means that the fossils for this time period are very different from anything living on Earth today. (4) Sometimes the Proterozoic, Archean and Hadean, eons are referred to as the *Precambrian,* which refers to the time before the time of ancient life. This means that the rocks for this time period contain very few fossils.

The eras were divided into blocks of time called **periods,** and the periods were further subdivided into smaller blocks of time called **epochs** (table 22.2).

Geologic Periods and Typical Fossils

The Precambrian period of time contains the earliest fossils. The Precambrian fossils that have been found are chiefly those of deposits from bacteria, algae, a few fungi, unusual soft-bodied animals, and the burrow holes of worms. It appears that there were no animals with hard parts; thus, the fossil record is incomplete since it is the hard parts of animals or plants that form fossils, usually after rapid burial. Another problem in finding fossils of soft-bodied or extremely small life-forms in these extremely old rocks is that heat and pressure have altered many of the ancient rocks over time, destroying any fossil evidence that may have been present.

The Paleozoic era was a time when there was great change in the kinds of plants and animals present. In general, the earliest abundant fossils are found in rocks from the Cambrian period at the beginning of the Paleozoic era. They show an abundance of oceanic life that represents all the major groups of marine animals found today. There is no fossil evidence of life of any kind living on the land during the Cambrian. The dominant life-forms of the Cambrian ocean were echinoderms, mollusks, trilobites, and brachiopods. The trilobites, now extinct, made up more than half of the kinds of living things during the Cambrian.

During the Ordovician and Silurian periods, most living things were still marine organisms with various kinds of jawless fish becoming prominent. Also, by the end of the Silurian, some primitive plants were found on land. The Devonian period saw the further development of different kinds of fish, including those that had jaws, and many kinds of land plants and animals. Coral reefs were also common in the Devonian. The Carboniferous period was a time of vast swamps of ferns, horsetails, and other primitive nonseed plants that would form great coal deposits. Fossils of the first reptiles and the first winged insects are found in rocks from this age. The Paleozoic era closed with the extinction at the end of the Permian period of about 90 percent of plant and animal life of that time.

The Mesozoic era was a time when the development of life on land flourished. The dinosaurs first appeared in the Triassic period, outnumbering all the other reptiles until the close of the Mesozoic. Fossils of the first mammals and modern forms of

gymnosperms (cone-bearing plants) developed in the Triassic. The first flowering plants, the first deciduous trees, and the first birds appeared in the Cretaceous period. The Cretaceous is the final period of the Mesozoic era and is characterized by the dominance of dinosaurs and extensive evolution of flowering plants and the insects that pollinate them. Birds and mammals also increased in variety. Like the close of the Paleozoic, the Mesozoic era ended with a great dying of land and marine life that resulted in the extinction of many species, including the dinosaurs.

As the Cenozoic era opened, the dinosaurs were extinct and the mammals became the dominant vertebrate life-form. The Cenozoic is thus called the Age of the Mammals. However, there were also major increases in the kinds of insects, flowering plants (particularly grasses), and birds. Finally, toward the end of this period of time, humans arrived on the scene, and many other kinds of large mammals such as mammoths, mastodons, giant ground sloths, and saber-toothed tigers went extinct.

Mass Extinctions

When we look at the fossil record, there is evidence of several mass extinctions. Five are recognized for causing the extinction of 50 percent or more of the species present. It is important to understand that although these extinctions were "sudden" in geologic time, they occurred over millions of years. Each resulted in a change in the kinds of organisms present with major groups going extinct and the evolution of new kinds of organisms. The boundaries between many of the geologic periods are defined by major extinction events. Geologists have developed theories about the causes of each of these mass extinctions. Many of these theories involve changes in the size and location of continents as a result of plate tectonics.

TABLE 22.2 Geologic Time and the History of Life (To trace the history of life, start with the oldest time period, the Hadean.)

Eon	Era	Period	Epoch	Million Years Ago*	Major Physical Events	Major Biological Events
Phanerozoic	Cenozoic	Quaternary	Holocene (Recent)	0.01 to present	—Much of Earth has been modified by human activity	—Dominance of modern humans
			Pleistocene	0.01 ↑ 1.8	—Extinction of many kinds of large mammals in late Pleistocene, perhaps due to human activity —Most recent periods of glaciation	—Origin of modern humans —Many kinds of large mammals present in early Pleistocene
		Tertiary	Pliocene	1.8 ↑ 5.3	—Cool, dry period —North and South America join —The Indian Plate collides with Asia to form the Himalaya Mountains	—Origin of ancestors of humans —Grasslands and grazing mammals widespread —All modern groups of mammals present
			Miocene	5.3 ↑ 23.8	—Warm, moist period —Current arrangement of continents present	—Grasslands and grazers common —Mammals widespread —Marine kelp forests common
			Oligocene	23.8 ↑ 33.7	—Cool period —India, South America, and Australia are separate continents	—Tropics diminish —Woodlands and grasslands expand —Many new grazing mammals
			Eocene	33.7 ↑ 54.8	—Australia and South America separate from Antarctica —North America separates from Europe	—All modern forms of flowering plants present —Forests common —All major groups of mammals present —Grasses found for the first time —First primates
			Paleocene	54.8 ↑ 65.0	—Warm period	—Many new kinds of mammals —Many new kinds of birds
	Mesozoic	Cretaceous		65 ↑ 144	—Major extinction at the end of Cretaceous affected both land and ocean organisms, probably because of meteorite impact —Warm, moist period —Continents continue to separate	—Dinosaurs dominant —Evolution of birds —Coevolution of flowering plants and their insect pollinators —The three modern mammal groups found —First flowering plants
		Jurassic		144 ↑ 206	—Pangaea begins to split up —North and South America split —South America and Africa split —Climate cooler and wetter as Pangaea splits	—Dinosaurs dominant —First birds —Many small primitive mammals
		Triassic		206 ↑ 248	—Major extinction of about 50% of species —No clear reason for extinction —Supercontinent, Pangaea, near equator —Warm, dry climate	—Many modern insects —Explosion in reptile diversity —Evolution of modern gymnosperms —Evolution of modern corals —First dinosaurs —First mammals —Mollusks are dominant aquatic invertebrates
	Paleozoic	Permian		248 ↑ 290	—Largest of all extinctions occurred at the end of the Permian (90% of all species) but no clear cause for this extinction —New giant continent of Pangaea forms —Much of Pangaea was probably desert —Modern levels of oxygen in atmosphere	—Gymnosperms dominant —Many new species of insects —Amphibians and reptiles dominant

Eon	Era	Period	Epoch	Million Years Ago*	Major Physical Events	Major Biological Events
Phanerozoic (continued)	Paleozoic (continued)	Carboniferous		290 ↑ 354	—Major coal deposits formed —Two major continents moving toward each other to form Pangaea —Warm climate	—Gymnosperms (cone-bearing plants) present by the end of the period —Reptiles present by the end of the period —Vast swamps of club mosses, horsetails, and ferns common —Many kinds of sharks —First winged insects, some very large —Amphibians common —Armored fish extinct
		Devonian		354 ↑ 417	—Glaciation probable cause of mass extinction at end of Devonian —About 60% of genera extinct —Extinction primarily affected warm-water, marine life —Warm, shallow seas common for much of this period —Two large continents present	—First seed plants (seed ferns) —Both jawless and jawed fish abundant —Coral reefs abundant —Land plants abundant (first forests) —Many new kinds of wingless insects —Amphibians on land
		Silurian		417 ↑ 443	—Melting of glaciers led to higher sea level —A large continent, Gondwana, is present at South Pole, but there are many smaller equatorial continents	—Numerous coral reefs —Fungi present —First vascular plants —Widespread nonvascular plants —First land animals—Arthropods (spiders and centipedes) —First jawed fish —Jawless armored fish abundant —Some jawless fish found in freshwater
		Ordovician		443 ↑ 490	—Giant continent, Gondwana, located at South Pole resulted in formation of glaciers —Formation of glaciers led to drop in sea level, which is the probable cause of mass extinction at the end of Ordovician (60% of genera go extinct)	—Primitive, nonvascular land plants present —Jawless, armored fish common —Diverse marine invertebrates
		Cambrian		490 ↑ 543	—Rodinia breaks up into pieces —Glaciation probable cause of mass extinction at the end of Cambrian	—All major groups of animals present —Mollusks, brachiopods, echinoderms, trilobites, and nautiloids common
Proterozoic		This period of time is also known as the Precambrian		543 ↑ 2,500	—Single large supercontinent (Rodinia) existed about 1.1 billion years ago —Oxidation of terrestrial iron deposits forms red beds that indicate presence of atmospheric oxygen (oxidized iron is red) at about 2.5 billion years ago —About 1.8 billion years ago, the atmospheric oxygen concentration was probably about 15% of current level	—Primitive multicellular, soft-bodied marine animals present about 1 billion years ago —Algae present about 1 billion years ago —Fossil record poorly preserved —First eukaryotic cells present about 1.8 billion years ago
Archean				2,500 ↑ 3,800	—Formation of continents —Atmosphere lacked oxygen —Oldest rocks 3.8 billion years ago	—Fossil stromatolites (layers of cyanobacteria) common —Cyanobacteria carried on photosynthesis, releasing oxygen —Fossil cyanobacteria present 3.5 billion years ago —Origin of life about 3.8 billion years ago
Hadean				3,800 ↑ 4,500	—Crust of Earth in process of solidifying —No rocks of this age —Origin of Earth	

* Time scale is based on *1999 Geologic Time Scale* of the Geological Society of America.

The mass extinction at the end of the Ordovician period resulted in the extinction of 60 percent of genera of organisms. At that time, most organisms lived in the oceans. It is thought that the large continent of Gondwana migrated to the South Pole and this resulted in the development of large glaciers and a drastic drop in sea level along with a cooling of the waters.

At the end of the Devonian period, there was a mass extinction that affected primarily marine organisms. Approximately 60 percent of genera went extinct. Since many of the organisms that went extinct were warm-water, marine organisms, glaciation along with a cooling of the oceans is a widely held theory for the cause of this extinction.

The mass extinction at the end of the Permian period is unusual in several ways. It resulted in the extinction of about 90 percent of organisms and took place over a very short time—less than a million years. Because of this, it is often referred to as the "Great Dying." Both marine and terrestrial organisms were affected. Because this extinction event occurred over a short time and affected all species, it is assumed that a major, worldwide event was responsible. However, at this time, there is no clearly identifiable cause. Suggestions include a meteorite impact, a supernova, massive volcanic activity, or a combination of factors.

The extinction at the end of the Triassic period was relatively mild compared to others. About 50 percent of species appear to have gone extinct. There is no clear cause for this extinction.

The mass extinction at the end of the Cretaceous period resulted in the extinction of about 60 percent of species. Based on evidence of a thin clay layer marking the boundary between the Cretaceous and Tertiary periods, one theory proposes that a huge (16 km diameter and 10^{15} kg mass) meteorite struck Earth (figure 22.15). The impact would have thrown a tremendous amount of dust into the atmosphere, obscuring the Sun and significantly changing the climate and thus the conditions of life on Earth. The resulting colder climate may have led to the extinction of many plant and animal species, including the dinosaurs. This theory is based on the clay layer, which theoretically formed as the dust settled, and its location in the rock record at the time of the extinctions. The layer is enriched with a rare metal, iridium, which is not found on Earth in abundance but occurs in certain meteorites in greater abundance.

As we approach current times, the extinction of the many species of large mammals during the Quaternary period is thought to be due to either a major change in climate at the end of the last ice age or hunting by humans as they expanded their range from Africa to Europe, Asia, and the Americas. Many people are convinced that we are currently experiencing a mass extinction because of our ability to alter the face of Earth and destroy the habitats needed by plants and animals.

Interpreting Geologic History—A Summary

When interpreting the fossil record in any part of the world, there are several things to keep in mind:

1. *We are dealing with long periods of time.* The history of life on Earth goes back to 3.5 billion years ago, and the evolution of humans took place over a period of several million years. It is important to understand that many of the processes of sedimentation, continental drift, and climate change took place slowly over many millions of years.

2. *Earth has changed greatly over its history.* There have been repeated periods of warming and cooling. During periods of cooling, glaciers formed, which resulted in lower sea level, which in turn exposed more land and changed the climate of any particular continent. In addition, the continents were not fixed in position. Changes in position affected the climate that the continents experienced. For example, at one time, what is now North America was attached to Antarctica near the South Pole.

3. *There have been many periods in the history of Earth when most of the organisms went extinct.* Cooling climates, changes in sea level, and meteorite impacts are all suspected of causing mass extinctions. However, it should not be thought that these were sudden extinctions. Most took place over millions of years. Some of the extinctions affected as much as 90 percent of the things living at the time.

4. *New forms of life evolved that replaced those that went extinct.* The earliest organisms we see in the fossil record were prokaryotic marine organisms similar to present-day bacteria. The oldest fossils of these organisms date to about 3.5 billion years ago. The next major step was the development of eukaryotic marine organisms about 1.8 billion years ago. The first eukaryotic multicellular organisms were present by about 1 billion years ago. The development of multicellular organisms ultimately led to the colonization of land by plants and animals, with plants colonizing about 500 million years ago and animals at about 450 million years ago.

FIGURE 22.15 A widely held theory is that a large meteorite was responsible for the mass extinction at the end of the Cretaceous period. Several of the other mass extinctions may have also been caused by meteorite impacts. This is an artist's depiction of the kinds of impact a large meteorite could have had on Earth.

5. *Although there were massive extinctions, there are many examples of the descendants of early life-forms present today.* Bacteria and many kinds of simple eukaryotic organisms are extremely common today, as are various kinds of algae and primitive forms of plants. In the oceans, many kinds of marine animals such as starfish, jelly fish, and clams are descendants of earlier forms.

6. *The kinds of organisms present have changed the nature of Earth.* The oxygen in the atmosphere is the result of the process of photosynthesis. Its presence has altered the amount of ultraviolet radiation reaching Earth. Plants tend to reduce the erosive effects of running water and humans have significantly changed the surface of Earth.

Table 22.2 lists the various geologic time periods and major events that occurred during those time periods.

PALEONTOLOGY, ARCHAEOLOGY, AND HUMAN EVOLUTION

There is intense curiosity about how our species (*Homo sapiens*) came to be, and we recognize that as animals, humans are subject to the same evolutionary forces as other organisms. Since all of our close evolutionary relatives are extinct, however, it is difficult for us to visualize our evolutionary development, and we tend to think we are unique and not subject to the laws of nature. Thus, the evolution of the human species remains an interesting and controversial topic.

We use several kinds of evidence to try to sort out our evolutionary history. Fossils of various kinds of prehuman and ancient human ancestors have been found, but these are often only fragments of skeletons that are often hard to date. Stone tools of various kinds have also been found associated with prehuman and early human sites. Finally, other aspects of the culture of our human ancestors have been found in burial sites, cave paintings, and the creation of ceremonial objects. Various methods have been used to age these findings. Some can be dated quite accurately, while others are more difficult to pinpoint.

When fossils are examined, anthropologists can identify differences in the structures of bones that are consistent with changes in species. Based on the amount of change they see and the ages of the fossils, these scientists make judgments about the species to which the fossil belongs. As new discoveries are made, opinions of experts will change and our evolutionary history may become clearer as old ideas are replaced. It is also clear from the fossil record that humans are relatively recent additions to the forms of life.

As evidence of how new discoveries change the way we think, the term *hominin* is a collective term currently used to refer to humans and their humanlike ancestors. Previously, the term *hominid* was used to refer to this group but is now used to refer to the broader group that includes all humanlike organisms plus the great apes. Consequently, when you read material about this topic, you will need to determine how the terms are being used. Although we do not have a clear picture of how humans evolved, several points are well accepted.

1. There is a great deal of fossil evidence that several bipedal species of the genera *Australopithecus* and *Paranthropus* were common in Africa beginning about four million years ago. These organisms are often referred to collectively as australopiths.

2. Based on fossil evidence, it appears that the climate of Africa was becoming drier during much of the time when the evolution of humans was occurring.

3. The earliest *Australopithecus* fossils are from about 4.2 million years ago. Earlier fossils such as *Ardipithecus* and the recently discovered *Orrorin* and *Sahelanthropus* may be ancestral to *Australopithecus*. *Australopithecus* and *Paranthropus* were herbivores and walked upright. Their fossils and the fossils of earlier organisms such as *Ardipithecus, Orrorin,* and *Sahelanthropus* are found only in Africa.

4. The australopiths were sexually dimorphic with the males much larger than the females, and they had relatively small brains (cranial capacity 530 cm³ or less—about the size of a standard softball).

5. Several species of the genus *Homo* became prominent in Africa beginning about 2.2 million years ago and appear to have made a change from a primarily herbivorous diet to a carnivorous or omnivorous diet.

6. All members of the genus *Homo* have relatively large brains compared to australopiths. The cranial capacity ranges from about 650 cm³ for early fossils of *Homo* to about 1,450 cm³ for modern humans. These organisms are also associated with various degrees of stone tool construction and use. It is possible that some of the australopiths also may have constructed stone tools.

7. Fossils of several later species of the genus *Homo* have larger brains and are found in Africa, Europe, and Asia but not in Australia or the Americas. Only *Homo sapiens* is found in Australia and the Americas.

8. Since the fossils of *Homo* species found in Asia and Europe are generally younger than the early *Homo* species found in Africa, it is assumed that they moved from Africa to Europe and Asia.

9. Differences in size between males and females are less prominent in members of the genus *Homo*, so perhaps there were fewer sexual differences in activities than with the australopiths.

When we try to put all of these bits of information together, we can construct the following plausible scenario for the evolution of our species. Early primates, like modern monkeys, chimpanzees, and apes, were adapted to living in forested areas where their grasping hands, opposable thumbs and big toes, and wide range of movement of the shoulders allowed them to move freely in the trees. As the climate became drier, grasslands replaced the forests. Early hominins were adapted to drier conditions. Walking upright was probably an adaptation to these conditions. Later hominins had larger brains and used tools. The sole surviving member of this evolutionary process is *Homo sapiens*. The following sections expand on this general scenario.

Neandertals Were Probably a Different Species

An ongoing controversy surrounds the relationship between Neandertals and other forms of prehistoric humans. One position is that Neandertals were a small separate race or subspecies of human that could have interbred with other humans and may have disappeared because their subspecies was eliminated by interbreeding with more populous, successful groups. (Many small, remote tribes have been eliminated in the same way in recent history.) Others maintain that the Neandertals show such great difference from other early humans that they must be a different species and became extinct because they could not compete with more successful *Homo sapiens* immigrants from Africa. (The names of these ancient people typically are derived from the place where the fossils were first discovered. For example, Neandertals were first found in the Neander Valley of Germany and Cro-Magnons [considered to be *Homo sapiens*] were initially found in the Cro-Magnon caves in France.)

The use of molecular genetic technology has shed some light on the relationship of Neandertals to other kinds of humans. Examination of the mitochondrial DNA obtained from the bones of a Neandertal individual reveals significant differences between Neandertals and other kinds of early humans. This greatly strengthens the argument that Neandertals were a separate species, *Homo neanderthalensis*.

The Australopiths

Various species of *Australopithecus* and *Paranthropus* were present in Africa from over four million years ago until about one million years ago. It is important to recognize that there are few fossils of these early humanlike organisms and that often they are fragments of the whole organism. This has led to much speculation and argument among experts about the specific position each fossil has in the evolutionary history of humans. However, from examining the fossil bones of the leg, pelvis, and foot, it is apparent that the australopiths were relatively short. Males were 1.5 m (5 ft) or less and females about 1.1 m (3.5 ft). They were stocky and walked upright like humans. Anthropologists recognize several different species of *Australopithecus* and *Paranthropus*.

An upright posture had several advantages in a world that was becoming drier. It allowed for more rapid movement over long distances, the ability to see longer distances, and reduced the amount of heat gained from the Sun. In addition, upright posture freed the arms for other uses such as carrying and manipulating objects, and using tools. The various species of *Australopithecus* and *Paranthropus* shared these characteristics and, based on the structure of their skulls, jaws, and teeth, appeared to have been herbivores with relatively small brains.

The Genus *Homo*

About 2.5 million years ago, the first members of the genus *Homo* appeared on the scene. There is considerable disagreement about how many species there were, but *Homo habilis* is one of the earliest. *H. habilis* had a larger brain (650 cm³) and smaller teeth than australopiths and made much more use of stone tools. Some people feel that it was a direct descendant of *Australopithecus africanus*. Many experts feel that *H. habilis* was a scavenger that made use of group activities, tools, and higher intelligence to commandeer the kills made by other carnivores. The higher-quality diet would have supported the metabolic needs of the larger brain.

About 1.8 million years ago, *Homo ergaster* appeared on the scene. It was much larger, up to 1.6 m (6 ft) tall, than *H. habilis*, who was about 1.3 m (4 ft) tall, and also had a much larger brain (cranial capacity of 850 cm³). A little later, a similar species (*Homo erectus*) appears in the fossil record. Some people consider *H. ergaster* and *H. erectus* to be variations of the same species. Along with their larger brain and body size, *H. ergaster* and *H. erectus* are distinguished from earlier species by their extensive use of stone tools. Hand axes were manufactured and used to cut the flesh of prey and crush the bones for marrow. These organisms appear to have been predators, whereas *H. habilis* was a scavenger. The use of meat as food allows animals to move about more freely, because appropriate food is available almost everywhere. By contrast, herbivores are often confined to places that have foods appropriate to their use: fruits for fruit eaters, grass for grazers, forests for browsers, etc. In fact, fossils of *H. erectus* have been found in the Middle East and Asia as well as Africa. Most experts feel that *H. erectus* originated in Africa and migrated through the Middle East to Asia.

At about eight hundred thousand years ago, another hominin classified as *Homo heidelbergensis* appears in the fossil record. Since fossils of this species are found in Africa, Europe, and Asia, it appears that they constitute a second wave of migration of early *Homo* from Africa to other parts of the world. Both *H. erectus* and *H. heidelbergensis* disappear from the fossil record as two new species (*Homo neanderthalensis* and *Homo sapiens*) become common.

The Neandertals were primarily found in Europe and adjoining parts of Asia and were not found in Africa. Therefore, many people feel Neandertals are descendants of *H. heidelbergensis*, which was common in Europe.

The Origin of *Homo sapiens*

Homo sapiens is found throughout the world and is now the only hominin species remaining of a long line of ancestors. There are two different theories that seek to explain the origin of *Homo sapiens*. One theory, known as the **out-of-Africa hypothesis,** states that modern humans, *H. sapiens*, originated in Africa as

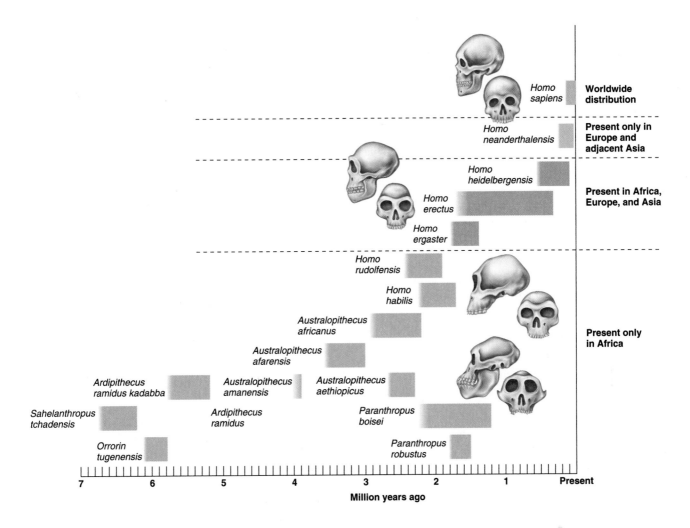

FIGURE 22.16 This diagram shows the various organisms thought to be relatives of humans. The bars represent approximate times the species are thought to have existed. Notice that (1) all species are extinct today except for modern humans; (2) several different species of organisms coexisted for extensive periods; (3) all the older species are found only in Africa; and (4) more recent species of *Homo* are found in Europe and Asia as well as Africa.

had several other hominin species and migrated from Africa to Asia and Europe and displaced species such as *H. erectus* and *H. heidelbergensis* that had migrated into these areas previously. The other theory, known as the **multiregional hypothesis,** states that *H. erectus* evolved into *H. sapiens*. During a period of about 1.7 million years, fossils of *H. erectus* showed a progressive increase in the size of the cranial capacity and reduction in the size of the jaw, so that it becomes difficult to distinguish *H. erectus* from *H. heidelbergensis* and *H. heidelbergensis* from *H. sapiens*. Proponents of this hypothesis believe that *H. heidelbergensis* is not a distinct species but an intermediate between the earlier *H. erectus* and *H. sapiens*. According to this theory, various subgroups of *H. erectus* existed throughout Africa, Asia, and Europe, and interbreeding among the various groups gave rise to the various races of humans we see today.

Another continuing puzzle is the relationship of humans that clearly belong to the species *Homo sapiens* with a contemporary group known as Neandertals. Some people consider Neandertals to be a race of *H. sapiens* specially adapted to life in the harsh conditions found in postglacial Europe. Others

consider them to be a separate species (*Homo neanderthalensis*). The Neandertals were muscular, had a larger brain capacity than modern humans, and had many elements of culture, including burials. The cause of their disappearance from the fossil record at about twenty-five thousand years ago remains a mystery. Perhaps climate change to a warmer climate was responsible. Perhaps contact with *H. sapiens* resulted in their elimination either through hostile interactions or, if they were able to interbreed with *H. sapiens,* they could have been absorbed into the larger *H. sapiens* population.

Large numbers of fossils of prehistoric humans have been found in all parts of the world. Many of these show evidence of a collective group memory we call *culture*. Cave paintings, carvings in wood and bone, tools of various kinds, and burials are examples. These are also evidence of a capacity to think and invent, and "free time" to devote to things other than gathering food and other necessities of life. We may never know how we came to be, but we will always be curious and will continue to search and speculate about our beginnings. Figure 22.16 summarizes the current knowledge of the historical record of humans and their relatives.

THE CLASSIFICATION OF ORGANISMS

To talk about items in our world, we must have names for them. As new items come into being or are discovered, new words are devised to describe them. For example, the words *laptop, palm pilot,* and *Internet* describe new technology that did not exist thirty years ago. Similarly, in the biological world, people have identified different kinds of organisms by giving them names.

The Problem with Common Names

The common names used by people of distinct cultures are usually different. A *dog* in English is *chien* in French, *perro* in Spanish, and *cane* in Italian. It is even possible that different names can be used to identify the same organism in different regions within a country. For example, the common garter snake may be called a garden snake or gardner snake depending on where you live (figure 22.17). Actually, there are several different species of "garter snakes" that have been identified as distinct from one another. Thus, common names can be confusing, and scientists sought a more acceptable way to give organisms names that would eliminate confusion and would be used by all scientists.

The naming of organisms is a technical process, but it is extremely important. When biologists are describing their research, common names such as robin and blackbird or garter snake are not good enough. They must be able to accurately identify the organisms involved so that everyone who reads the report, wherever they live in the world, will be able to know what organism is being discussed. The scientific identification of organisms really involves two different but related activities. One, *taxonomy,* involves the naming of organisms, and the other, *phylogeny,* involves showing how organisms are related evolutionarily.

Taxonomy

Taxonomy is the science of naming organisms and grouping them into logical categories. Various approaches have been used to classify organisms. The Greek philosopher Aristotle (384–322 B.C.) had an interest in nature and was the first person to attempt a logical classification system. The root word for *taxonomy* is the Greek word *taxis,* which means "arrangement." Aristotle used the size of plants to divide them into the categories of trees, shrubs, and herbs.

During the Middle Ages, Latin was widely used as the scientific language. As new species were identified, they were given Latin names, often using as many as fifteen words to describe a single organism. Although using Latin meant that most biologists, regardless of their native language, could understand a species name, it did not completely do away with duplicate names. Because many of the organisms could be found over wide geographic areas and communication was slow, there could still be two or more Latin names for a species. To make the situation even more confusing, ordinary people still used common local names.

Binomial System of Nomenclature

The modern system of classification began in 1758, when Carolus Linnaeus (1707–1778), a Swedish doctor and botanist, published his tenth edition of *Systema Naturae* (figure 22.18). (Linnaeus's original name was Karl von Linne, which he "latinized" to Carolus Linnaeus.) In the previous editions, Linnaeus had used a polynomial (many-names) Latin system. However, in the tenth edition, he introduced the *binomial system of nomenclature.* The **binomial system of nomenclature** uses only

FIGURE 22.17 Depending on where you live you may call this organism a garter snake, a garden snake, or a gardner snake. These common names can lead to confusion about what kind of snake a person might be talking about.

FIGURE 22.18 Carolus Linnaeus (1707–1778), a Swedish doctor and botanist, originated the modern system of taxonomy.

two Latin names—the genus name and the specific epithet (*epithet* = descriptive word)—for each species of organism. Recall that a species is a population of organisms capable of interbreeding and producing fertile offspring. Individual organisms are members of a species. A **genus** (plural, *genera*) is a group of closely related species of organisms; the specific epithet is a word added to the genus name to identify which one of several species within the genus we are discussing. This is similar to the naming system we use with people. When you look in the phone book, you look for the last name (surname), which gets you in the correct general category. Then you look for the first name (given name) to identify the individual you wish to call. The unique name given to a particular type of organism is called its species name or scientific name. To clearly identify the scientific name from other words, binomial names are either *italicized* or underlined. The first letter of the genus name is capitalized. The specific epithet is always written in lowercase. For example, *Thamnophis sirtalis* is the binomial name for the common garter snake.

In addition to assigning a specific name to each species, Linnaeus recognized a need for placing organisms into meaningful groups. In his system, he divided all forms of life into two *kingdoms,* Plantae and Animalia, and then further subdivided the kingdoms into other smaller units.

Organizing Species into Logical Groups

Since Linnaeus's initial attempts to place all organisms into categories, there have been many changes. One of the most fundamental is the recent recognition that there are three major categories of organisms that have been called *domains.*

A **domain** is the largest category into which organisms are classified. There are three domains: Eubacteria, Archaea, and Eucarya (figure 22.19.). Organisms are separated into these three domains based on specific structural and biochemical features of their cells. The Eubacteria and Archaea are prokaryotic, and the Eucarya are eukaryotic.

A **kingdom** is a subdivision of a domain. There are several kingdoms within the Eubacteria and Archaea that are based primarily on differences in the metabolism and genetic composition of the organisms. Within the domain Eucarya, there are four kingdoms: Plantae, Animalia, Fungi, and Protista (protozoa and algae). Figure 22.20 provides examples of the three domains. However, no clear consensus exists at this time about how many kingdoms there are in the Eubacteria and Archaea.

A **phylum** is a subdivision of a kingdom. However, microbiologists and botanists often use the term *division* in place of the term *phylum.*

All kingdoms have more than one phylum. For example, the kingdom Plantae contains several phyla, including flowering plants, conifer trees, mosses, ferns, and several other groups. Organisms are placed in phyla based on careful investigation of the specific nature of their structure, metabolism, and biochemistry. An attempt is made to identify natural groups rather than artificial or haphazard arrangements. For example, while nearly all plants are green and carry on photosynthesis, only flowering plants have flowers and produce seeds; conifers lack flowers but have seeds in cones; ferns lack flowers, cones, and seeds; and mosses lack tissues for transporting water.

A **class** is a subdivision within a phylum. For example, within the phylum Chordata, there are seven classes: mammals,

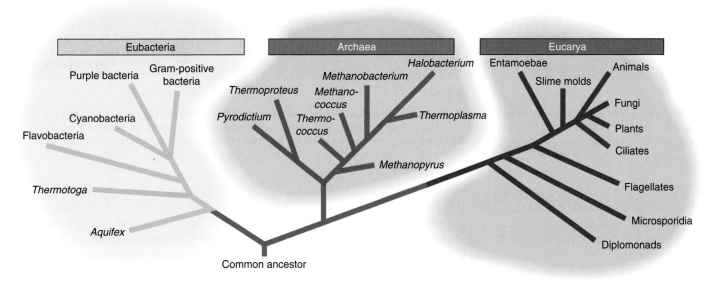

FIGURE 22.19 This diagram shows the three domains of living things and the way they are related to one another evolutionarily. The domain Eubacteria is the oldest group. The domains Archaea and Eucarya are derived from the Eubacteria.

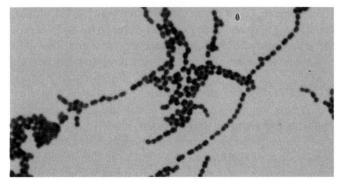

A *Streptococcus pyogenes*

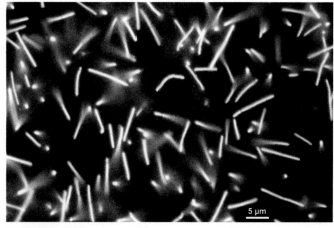

5 µm

B *Methanopyrus*

C *Morchella esculenta*

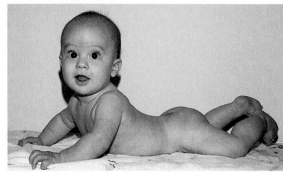

D *Amoeba proteus*

E *Homo sapiens*

F *Acer saccharum*

FIGURE 22.20 (*A*) The domain Eubacteria is represented by the bacterium *Streptococcus pyogenes*. (*B*) The domain Archaea is represented by the methane-producing *Methanopyrus*. The domain Eucarya is represented by members of the following kingdoms: (*C*) Fungi, by the ascomycete *Morchella esculenta*; (*D*) Protista, by the one-celled *Amoeba proteus*; (*E*) Animalia, by the animal *Homo sapiens*; and (*F*) Plantae, by the tree *Acer saccharum*.

birds, reptiles, amphibians, and three classes of fishes. An **order** is a category within a class. Carnivora is an order of meat-eating animals within the class Mammalia. There are several other orders of mammals, including horses and their relatives, cattle and their relatives, rodents, rabbits, bats, seals, whales, humans, and many others. A **family** is a subdivision of an order and consists of a group of closely related genera, which in turn are composed of groups of closely related species. The cat family, Felidae, is a subgrouping of the order Carnivora and includes many species in several genera, including the Canada lynx and bobcat (genus *Lynx*), the cougar (genus *Puma*), the leopard, tiger, jaguar, and lion (genus *Panthera*), the house cat (genus *Felis*), and several others. Thus, in the present-day science of taxonomy, each organism that has been classified has its own unique binomial name. In turn, it is assigned to larger groupings that are thought to have a common evolutionary history. Table 22.3 uses the classification of humans to show how the various categories are used.

Phylogeny

Phylogeny is the science that explores the evolutionary relationships among organisms and seeks to reconstruct evolutionary history. Taxonomists and phylogenists work together so that the products of their work are compatible. A taxonomic ranking should reflect the phylogenetic (evolutionary) relationships among the organisms being classified. Although taxonomy and phylogeny are sciences, there is no complete agreement as to how organisms are classified or how they are related. New organisms and new information about organisms are discovered constantly. Therefore, taxonomic and phylogenetic relationships are constantly being revised. During this revision process, scientists often have differences of opinion about the significance of new information.

Evidence Used to Establish Phylogenetic Relationships

Phylogenists use several lines of evidence to develop evolutionary histories: fossils, comparative anatomy, life-cycle information, and biochemical/molecular evidence.

Fossils are evidence of previously existing life. Evidence obtained from the discovery and study of fossils allows biologists to place organisms in a time sequence. It is also possible to compare subtle changes in particular kinds of fossils over time. For example, the size of the leaf of a specific fossil plant has been found to change extensively through long geological periods. A comparison of the extremes, the oldest with the newest, would lead to their classification into different categories. However, the fossil links between the extremes clearly show that the younger plant is a descendant of the older.

Comparative anatomy studies of fossils or currently living organisms can be very useful in developing a phylogeny. Since the structures of an organism are determined by its genes and developmental processes, those organisms having similar structures are thought to be related. For example, plants can be divided into several categories: All plants that have flowers are thought to be more closely related to one another than to plants like ferns, which do not have flowers. In the animal kingdom, all organisms that have hair and mammary glands are grouped together, and all animals in the bird category have feathers, wings, and beaks.

Life-cycle information is another line of evidence useful to phylogenists and taxonomists. Many organisms have complex life cycles that include many completely different stages. After fertilization, some organisms grow into free-living developmental stages that do not resemble the adults of their species. These are called *larvae* (singular, *larva*). Larval stages often provide clues to the relatedness of organisms. For example, barnacles live attached to rocks and other solid marine objects and look like small, hard cones. Their outward appearance does not suggest that they are related to shrimp; however, the larval stages of barnacles and shrimp are very similar. Detailed anatomical studies of barnacles confirm that they share many

TABLE 22.3 Classification of Humans

Taxonomic Category	Human Classification	Characteristics	Other Representatives
Domain	Eucarya	Cells contain a nucleus and many other kinds of organelles.	Plants, animals, fungi, protozoa, algae
Kingdom	Animalia	Eukaryotic, heterotrophs that are usually motile and have specialized tissues	Sponges, jellyfish, worms, clams, insects, snakes, cats
Phylum	Chordata	Animals that have a stiffening rod down their back	Fish, amphibians, reptiles, birds, mammals
Class	Mammalia	Animals with hair and mammary glands	Kangaroos, mice, whales, skunks, monkeys
Order	Primates	Mammals with relatively large brains and opposable thumbs	Monkeys, gorillas, chimpanzees, baboons
Family	Hominidae	Primates that lack a tail and have upright posture	Humans and extinct relatives (*Australopithecus, Paranthropus* and *Homo*)
Genus	*Homo*	Hominids with large brains	Humans are the only surviving member of the genus. Other species existed in the past (*Homo erectus, Homo neanderthalensis*).
Species	*Homo sapiens*	Humans	

A Barnacle

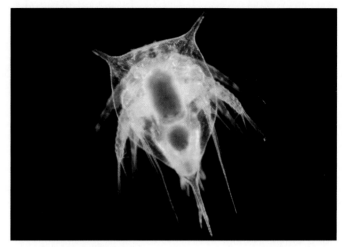

C Nauplius larva of barnacle

B Shrimp

FIGURE 22.21 The adult barnacle (*A*) and shrimp (*B*) are very different from each other, but their early larval stages (*C*) look very much alike.

structures with shrimp; their outward appearance tends to be misleading (figure 22.21).

Both birds and reptiles lay eggs with shells. However, reptiles lack feathers and have scales covering their bodies. The fact that these two groups share this fundamental eggshell characteristic implies that they are more closely related to each other than they are to other groups, but they can be divided into two groups based on their anatomical differences.

This same kind of evidence is available in the plant kingdom. Many kinds of plants, such as peas, peanuts, and lima beans, produce large, two-parted seeds in pods (you can easily split the seeds into two parts). Even though peas grow as vines, lima beans grow as bushes, and peanuts have their seeds underground, all these plants are considered to be related.

Biochemical and molecular studies are recent additions to the toolbox of phylogenists. Like all aspects of biology, the science of phylogeny is constantly changing as new techniques develop. Recent advances in DNA analysis are being used to determine genetic similarities among species. In the field of ornithology, which is the study of birds, there are those who believe that storks and flamingos are closely related; others believe that flamingos are more closely related to geese. An analysis of the DNA points to a higher degree of affinity between flamingos and storks than between flamingos and geese. This is interpreted to mean that the closest relationship is between flamingos and storks.

Algae and plants have several different kinds of chlorophyll: chlorophyll *a, b, c, d,* and *e.* Most photosynthetic organisms contain a combination of two of these chlorophyll molecules. Members of the kingdom Plantae have chlorophyll *a* and *b.* The large seaweeds, such as kelp, superficially resemble terrestrial plants like trees and shrubs. However, a comparison of the chlorophylls present shows that kelp has chlorophyll *a* and *d.* When another group of algae, called the *green algae,* are examined, they are found to have chlorophyll *a* and *b,* as do plants. Along with other anatomical and developmental evidence, this biochemical information has helped to establish an evolutionary link between the green algae and plants. All of these kinds of evidence (fossils, comparative anatomy, developmental stages, and biochemical evidence) have been used to develop the various taxonomic categories, including kingdoms.

Given all these sources of evidence, biologists have developed an idea of how all organisms are related evolutionarily (figure 22.22). At the base of this evolutionary scheme is the biochemical evolution of cells. These first cells are thought to be the origin of all organisms. While these first cells no longer exist, their descendants have diversified over millions of years. Of these groups, the members of the domains Eubacteria and Archaea have the simplest structure and are probably most similar to some of the first cellular organisms on Earth. Members of the domain Eucarya evolved later and have greater structural and functional complexity.

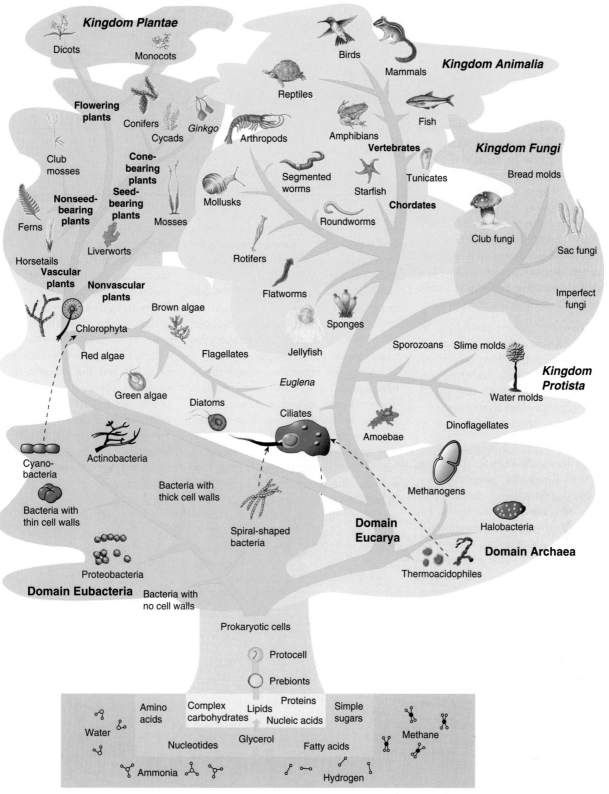

FIGURE 22.22 The theory of chemical evolution proposes that the molecules in the early atmosphere and early oceans accumulated to form prebionts—nonliving structures composed of carbohydrates, proteins, lipids, and nucleic acids. The prebionts are believed to have been the forerunners of the protocells—the first living cells. These protocells probably evolved into prokaryotic cells, on which the domains Eubacteria and Archaea are based. Some prokaryotic cells probably gave rise to eukaryotic cells. The organisms formed from these early eukaryotic cells were probably similar to members of the kingdom Protista. Members of this kingdom are thought to have given rise to the kingdoms Animalia, Plantae, and Fungi. Thus, all present-day organisms evolved from the protocells.

A Closer Look

Cladistics—A Tool for Taxonomy and Phylogeny

Classification, or taxonomy, is one part of the much larger field of phylogenetic systematics. Classification involves placing organisms into logical categories and assigning names to those categories. Phylogeny, or *systematics,* is an effort to understand the evolutionary relationships of living things in order to interpret the way in which life has diversified and changed over billions of years of biological history. Phylogeny attempts to understand how organisms have changed over time. *Cladistics (klados = branch)* is a method biologists use to evaluate the degree of relatedness among organisms within a group, based on how similar they are genetically. The basic assumptions behind cladistics are that

1. Groups of organisms are related by descent from a common ancestor.
2. The relationships among groups can be represented by a branching pattern, with new evolutionary groups arising from a common ancestor.
3. Changes in characteristics occur in organisms over time.

Several steps are involved in applying cladistics to a particular group of organisms. First, you must select characteristics that vary and collect information on the characteristics displayed by the group of organisms you are studying. The second step is to determine which expression of a characteristic is ancestral and which is more recently derived. Usually, this involves comparing the group in which you are interested with an *outgroup* that is related to but not a part of the group you are studying. The characteristics of the outgroup are then considered to be ancestral. Finally, you must compare the characteristics displayed by the group you are studying and construct a diagram known as a *cladogram.* For

Box Table 22.1

Characteristic Organism	Lungs Present	Skin Dry	Warm-Blooded	Hair Present
Shark	0	0	0	0
Frog	+	0	0	0
Lizard	+	+	0	0
Crow	+	+	+	0
Bat	+	+	+	+

example, if you were interested in studying how various kinds of terrestrial vertebrates are related, you could look at the characteristics shown in box table 22.1.

In this example, the shark is the outgroup, and the ancestral conditions are lungs absent, skin not dry, cold-blooded, and hair absent. Using this information, you could construct the cladogram below.

All of the organisms, except sharks, have lungs. Lizards, crows, and bats have dry skin, as well as lungs and so on. Crows and bats share the following characteristics; they have lungs, they have dry skin, and they are warm-blooded. Because they share more characteristics with each other than with the other groups, they are considered to be more closely related.

The choice of characteristics with which to make comparisons is important. Two organisms may share many characteristics but not be members of the same evolutionary group if the characteristics being compared are not from the same genetic background. For example, if you were to compare butterflies, birds, and squirrels using the presence or absence of wings and the presence or absence of bright colors as your characteristics, you would conclude that butterflies and birds are more closely related than birds and squirrels. However, this is not a valid comparison, because the wings of birds and butterflies are not of the same evolutionary origin.

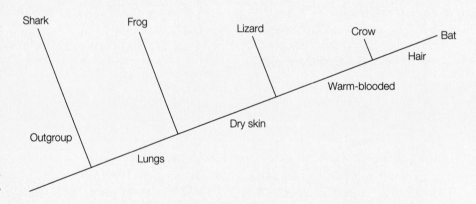

A BRIEF SURVEY OF BIODIVERSITY

There is great variety in the kinds of living things on Earth. Some groups such as the Eubacteria and Archaea have been in existence for over three billion years. Other groups such as plants and animals have been in existence for less than a billion years. In this section, we will look at the three domains of living things and briefly describe some of their more distinctive characteristics.

Domains Eubacteria and Archaea

Because both the Eubacteria and Archaea are prokaryotic, they were previously lumped together as one taxonomic unit and are still commonly referred to as bacteria. However, members of the two domains are very different from one another and now are assigned separate positions on the evolutionary tree. Evidence gained from studying DNA and RNA nucleotide sequences and a comparison of the amino acid sequences of proteins indicates that the Eubacteria are older evolutionarily than the Archaea.

Eubacteria

The "true bacteria" (*eu* = true) are small, prokaryotic, single-celled organisms. Their cell walls contain a unique, complex organic molecule known as peptidoglycan. Peptidoglycan is only found in the Eubacteria and is composed of two kinds of sugars linked together by amino acids. One of these sugars, muramic acid, is only found in the Eubacteria. This characteristic is used to distinguish the Eubacteria from the Archaea and Eucarya.

There are thousands of species of Eubacteria that can be divided into groups based on their metabolic abilities. Most Eubacteria are heterotrophs that require organic molecules as a source of food. Many are important decomposers of dead organic matter. Sewage treatment plants rely on Eubacteria along with other organisms to break down organic wastes. A few kinds are **parasites** that live on the tissues of other organisms and cause disease. Because parasitic Eubacteria cause us problems, they have been intensively studied. Pneumonia, tuberculosis, syphilis, gonorrhea, strep throat, and staphylococcus infections are just a few Eubacteria we know all too well. In addition, heterotrophic bacteria can be differentiated based on their ability to use oxygen. Some are aerobic (use oxygen to break down organic food molecules), while others are anaerobic (do not use oxygen). Since Earth's early atmosphere is thought to have been a reducing atmosphere, the first Eubacteria were probably anaerobic organisms.

Several kinds of autotrophic Eubacteria exist. The cyanobacteria (blue-green bacteria) were probably the first organisms to carry on photosynthesis. They contain a blue-green pigment that allows them to capture sunlight and carry on a kind of photosynthesis. Cyanobacteria are extremely common in marine and freshwater, and contribute significantly to the production of oxygen in the atmosphere. Other Eubacteria are able to use inorganic chemical reactions to provide the energy needed to build new organic molecules. One that has important environmental implications is *Thiobacillus ferrooxidans,* which derives energy from the oxidation of iron and sulfur, and is at least partly responsible for the development of acid mine drainage.

Archaea

The term *Archaea* comes from the Greek term *archaios,* meaning "ancient." This is a little misleading since the Eubacteria preceded them and the Archaea are thought to have branched off from the Eubacteria somewhere between two to three billion years ago. The Archaea differ from Eubacteria in several fundamental ways. The Archaea do not have peptidoglycan in their cell walls but do have a unique chemical structure for their cell membranes that is not found in either the Eubacteria or Eucarya. The DNA of Archaea appears to have a large proportion of genes that are different from those of either the Eubacteria or Eucarya. However, the structure of the DNA is similar in many ways to that found in the Eucarya.

Members of this group show a wide variety of metabolic activities and typically live in very hostile environments. Because they are found in many kinds of extreme environments, they have become known as extremophiles. The Archaea are divided into three groups based on their metabolic abilities and the kinds of environments in which they live.

1. *Methanogens* are heterotrophic methane-producing bacteria that are anaerobic. They can be found in the intestinal tracts of animals (including humans), sewage, and swamps. Much of the methane gas released into the atmosphere is from these organisms.
2. *Halobacteria* (*halo* = salt) are found growing in very salty environments such as the Great Salt Lake (Utah), salt ponds, and brine solutions. Some contain a molecule called bacteriorhodopsin (similar to rhodopsin found in the rods of the eye), which allows them to use light to generate ATP in a unique form of photosynthesis.
3. The *thermophilic* Archaea live in environments that normally have very high temperatures and high concentrations of sulfur (e.g., hot sulfur springs or around deep-sea hydrothermal vents). Over five hundred species of these thermophiles have been identified at the openings of hydrothermal vents in the ocean. Some of these organisms are heterotrophs, while others are chemoautotrophs and use reactions with sulfur to provide energy to synthesize organic molecules. One such thermophile, *Pyrolobus fumarii,* grows in a hot spring in Yellowstone National Park. Its maximum growth temperature is 113°C (235°F), its optimum is 106°C (223°F), and its minimum is 90°C (194°F).

Domain Eucarya

Most biologists feel that eukaryotic cells evolved from prokaryotic cells by a process of endosymbiosis. (See "The Endosymbiotic Theory and the Origin of Eukaryotic Cells" section in chapter 21.) This hypothesis proposes that structures such as mitochondria, chloroplasts, and other membranous organelles originated from separate cells that were ingested by larger, more primitive cells. Once inside, these structures and their functions became integrated with the host cell and ultimately became essential to its survival. This new type of cell was the forerunner of present-day eukaryotic cells, which are usually much larger than the prokaryotes, typically having more than a thousand times the volume of prokaryotic cells. Their larger size was made possible by the presence of specialized membranous organelles, such as mitochondria, the endoplasmic reticulum, chloroplasts, and nuclei. Members of the kingdoms Protista, Fungi, Plantae, and Animalia are eukaryotic. Single-celled eukaryotic organisms are members of the kingdom Protista.

Kingdom Protista

The changes in cell structure that led to eukaryotic organisms most probably gave rise to single-celled organisms similar to those currently grouped in the kingdom Protista. Most members of this kingdom are one-celled organisms, although there are some colonial forms.

There is a great deal of diversity among the sixty-thousand known species of Protista. Many species live in freshwater; others are found in marine or terrestrial habitats. Some are *parasitic* and live on the tissues of other living things. Some are *commensal* organisms that live in another organism without causing harm, and some are *mutualistic* organisms that live in partnership with another organism in a relationship in which both organisms benefit. All species can undergo mitosis, resulting in asexual reproduction. Most species can also reproduce sexually. Many contain chlorophyll in chloroplasts and are autotrophic; others require organic molecules as a source of energy and are heterotrophic. Both autotrophs and heterotrophs have mitochondria and respire aerobically. Some specialized parasitic types lack mitochondria and other cellular structures.

Because members of this kingdom are so diverse with respect to form, metabolism, and reproductive methods, most biologists do not feel that the Protista form a valid phylogenetic unit. However, it is still a convenient taxonomic grouping. By placing these organisms together in this group, it is possible to gain a useful perspective on how they relate to other kinds of organisms. After the origin of eukaryotic organisms, evolution proceeded along several different pathways. Three major lines of evolution can be seen today in the plantlike autotrophs (algae), animal-like heterotrophs (protozoa), and the fungus-like heterotrophs (slime molds). *Amoeba* and *Paramecium* are commonly encountered examples of protozoa. Many seaweeds and pond scums are collections of large numbers of algal cells. Slime molds are less frequently seen because they live in and on the soil in moist habitats; they are most often encountered as slimy masses on decaying logs. Figure 22.23 shows some examples of this diverse group of organisms.

Through the process of evolution, the plantlike autotrophs probably gave rise to the kingdom Plantae, the animal-like heterotrophs probably gave rise to the kingdom Animalia, and the funguslike heterotrophs were probably the forerunners of the kingdom Fungi.

Kingdom Fungi

Fungus is the common name for members of the kingdom Fungi. The majority of fungi are not able to move about. They have a rigid, thin cell wall, which in most species is composed of chitin, a complex carbohydrate containing nitrogen. This is an important diagnostic feature, since plants have cell walls of cellulose. Members of the kingdom Fungi are nonphotosynthetic, eukaryotic organisms. The majority (mushrooms and molds) are multicellular, but a few, such as yeasts, are single-celled. In the multicellular fungi, the basic structural unit is a network of multicellular filaments.

Because all of these organisms are heterotrophs, they must obtain nutrients from organic sources. Most secrete enzymes outside their cells that digest large molecules into smaller units that can be absorbed. They are very important as decomposers in all ecosystems. They feed on a variety of nutrients ranging

Green Algae

Kelp

Slime mold

Protozoan

FIGURE 22.23 The kingdom Protista includes a wide variety of organisms that are simple in structure. They are not a phylogenetic group. These are some examples of this diverse group.

from dead organisms to such products as shoes, foodstuffs, and clothing. Most synthetic organic molecules are not attacked as readily by fungi; this is one reason plastic bags, foam cups, and organic pesticides are slow to decompose.

Some fungi are parasitic and others are mutualistic. Many of the parasitic fungi are important plant pests. Some attack and kill plants (chestnut blight, Dutch elm disease); others injure the fruit, leaves, roots, or stems and reduce yields. The fungi that are human parasites are responsible for athlete's foot, vaginal yeast infections, valley fever, "ringworm," and other diseases. Many kinds of fungi form mutualistic relationships with other kinds of organisms. Mutualistic fungi are important in lichens and in combination with the roots of certain kinds of plants, where they assist the plant in obtaining nutrients from the soil. Figure 22.24 shows some examples of this group of organisms.

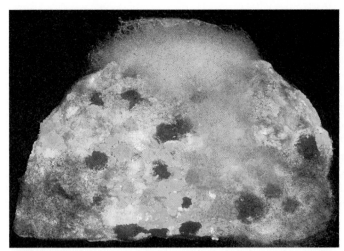
Several kinds of Mold

Puffball

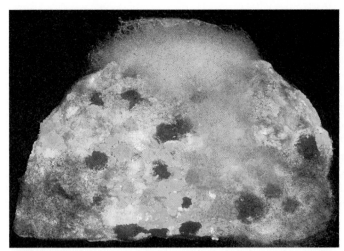
Mushroom

FIGURE 22.24   Molds, mushrooms, and puffballs are commonly seen examples of the kingdom Fungi.

Kingdom Plantae

Another major group thought to be derived from the kingdom Protista is the green, photosynthetic plant kingdom. The ancestors of plants were most likely specific kinds of algae commonly called *green algae*. Members of the kingdom Plantae are nonmotile, terrestrial, multicellular organisms that contain chlorophyll and produce their own organic compounds. All plant cells have a cellulose cell wall. Over three-hundred thousand species of plants have been classified; about 85 percent are flowering plants, 14 percent are mosses and ferns, and the remaining 1 percent are cone-bearers and several other small groups within the kingdom.

A wide variety of plants exist on Earth today. Members of the kingdom Plantae range from simple mosses to vascular plants with stems, roots, leaves, and flowers. Most biologists feel that the evolution of this kingdom began nearly five-hundred million years ago when the green algae of the kingdom Protista gave rise to two lines: The nonvascular plants such as the mosses evolved as one type of plant, and the vascular plants such as the ferns evolved as a second type (figure 22.25). Some of the vascular plants evolved into seed-producing plants, which today are the cone-bearing and flowering plants, while the ferns lack seeds. The development of vascular plants was a major step in the evolution of plants from an aquatic to a terrestrial environment.

The World's Oldest and Largest Living Organisms

Several organisms have been suggested as record holders for the title of oldest and largest organisms. Several of these are plants. Bristlecone pines (*Pinus longaeva*) in the White Mountains of California have been determined to have an age of over five thousand years. Creosote bush (*Larrea divaricata*) forms clones that grow out from the center as the central portion of the plant dies. Several clones of creosote bush in the Mojave Desert have been estimated to have an age of twelve thousand years. The title of the largest organism can be determined in several ways. The Giant Redwood (*Sequoiadendron giganteum*) is the tree with the single largest stem. The General Sherman tree probably weighs about 1,400 tons. However, a clone of trembling aspen (*Populus tremuloides*) consists of many individual stems that are probably all joined together by roots. One such clone in the Rocky Mountains covers about 0.4 km² and probably weighs about 6,000 tons. However, it may be a clone of a fungus (*Armillaria*) that lives in the soil that holds the record for the largest organism. A clone in the state of Washington is estimated to cover about 3 km².

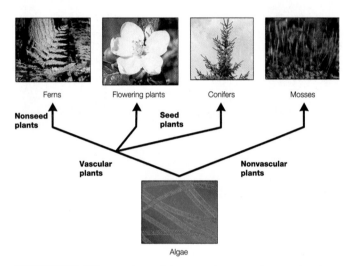

FIGURE 22.25 Two lines of plants are thought to have evolved from green algae in the kingdom Protista. The nonvascular mosses evolved as one type of plant. The second type, the vascular plants, evolved into the seed and nonseed plants.

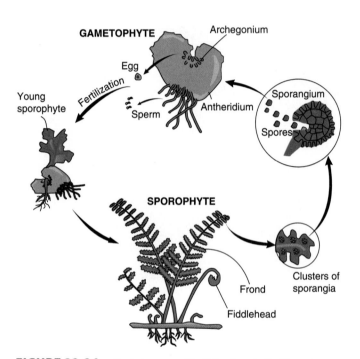

FIGURE 22.26 Plants have a multicellular stage in the life cycle (sporophyte generation) that undergoes meiosis to form spores that are haploid. These spores give rise to haploid organisms (gametophyte generation) that form the gametes—sperm cells and egg cells. The union of these gametes forms the diploid stage, which develops into the sporophyte. In this fern life cycle, the large leafy part of the life cycle is the sporophyte generation and the gametophyte is a tiny structure. All plants show this alternation of sporophyte and gametophyte generations, but the details are different for each major group of plants.

Plants have a unique life cycle that has two distinctly different stages. This is known as alternation of generations. The presence of this kind of life cycle is a unifying theme that ties together all members of this kingdom. There is a gametophyte stage that produces sex cells by mitosis. The sex cells unite and give rise to a sporophyte stage that produces spores by meiosis. The spores give rise to a new gametophyte stage. Mosses and ferns have life cycles with relatively clearly defined gametophyte and sporophyte generations. Figure 22.26 shows a fern life cycle. However, cone-bearing plants and flowering plants have a somewhat more complicated life cycle. Figure 22.27 illustrates the life cycle of a flowering plant. In addition to sexual reproduction, plants are able to reproduce asexually.

Kingdom Animalia

Like fungi and plants, animals are thought to have evolved from the Protista. Over a million species of animals have been classified. These range from microscopic types, such as mites or aquatic larvae of marine animals, to huge animals such as elephants or whales. Regardless of their type, all animals have some common traits. They all are composed of eukaryotic cells, and all species are heterotrophic and multicellular. Most animals are motile; however, some, such as the sponges, barnacles, mussels, and corals are sessile (not able to move). All animals are capable of sexual reproduction, but many of the less complex animals are also able to reproduce asexually.

It is thought that animals originated from certain kinds of Protista that had flagella. This idea proposes that colonies of flagellated Protista gave rise to simple multicellular forms of animals

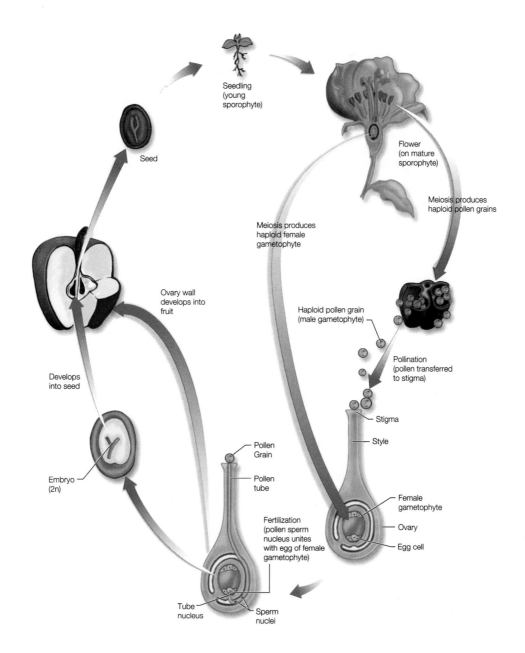

Seedling
(young
sporophyte)

Flower
(on mature
sporophyte)

Meiosis produces
haploid pollen grains

Meiosis produces
haploid female
gametophyte

Seed

Ovary wall
develops into
fruit

Haploid pollen grain
(male gametophyte)

Pollination
(pollen transferred
to stigma)

Stigma

Develops
into seed

Style

Pollen
Grain

Pollen
tube

Female
gametophyte

Embryo
(2n)

Ovary

Egg cell

Fertilization
(pollen sperm
nucleus unites
with egg of female
gametophyte)

Tube
nucleus

Sperm
nuclei

FIGURE 22.27 In flowering plants, the sporophyte generation consists of roots, stems, leaves, and flowers. The flower produces two kinds of gametophyte plants. The pollen grain is the male gametophyte, and the female gametophyte is produced within the ovary of the flower. The male gametophyte produces the equivalent of sperm and the female gametophyte produces an egg. Pollination involves the transfer of pollen to the stigma of a flower. Fertilization occurs following pollination, when the pollen tube releases a sperm nucleus to fertilize the egg of the female gametophyte inside the ovary of the flower. The fertilized egg develops into the embryo plant, and other cells of the female gametophyte produce stored food. A tough coat develops around the embryo and food. This package is known as a seed. The seed germinates to produce a new sporophyte plant.

such as the ancestors of present-day sponges. These first animals lacked specialized tissues and organs. As cells became more specialized, organisms developed special organs and systems of organs and the variety of kinds of animals increased.

Animals originated in the ancient sea, and the majority of kinds of animals are still found there. Most of the major groups of animals can be identified from fossils of the Cambrian period over five-hundred million years ago. Sponges, jellyfish, worms, crustaceans, mollusks, starfish, sharks, and bony fishes are examples of groups that are primarily marine,

although some may be freshwater or terrestrial. Four major groups have become predominantly air-breathing, terrestrial organisms: insects, reptiles, birds, and mammals. In addition to being artificially divided into groups based on where they live, organisms are often categorized by major structural differences. One such division is the difference between vertebrates that have backbones (fish, amphibians, reptiles, birds, mammals) and invertebrates, which include all the other animal groups. Figure 22.28 shows the major groups of animals and their evolutionary relationships.

Lynn (Alexander) Margulis (1938–)

Lynn (Alexander) Margulis is the eldest of four daughters, born in Chicago to Morris and Leone Alexander. Her father was a lawyer and her mother operated a travel agency. She entered the University of Chicago at age sixteen and received an A.B. degree in 1957 at age nineteen. That same year, she married the astronomer Carl Sagan; with whom she had two children. They divorced in 1964. She received a Ph.D. from the University of California, Berkeley, in 1963. In 1967, she married crystallographer Thomas N. Margulis, with whom she had two children. They divorced in 1980.

She taught at Brandeis University and Boston College, and is currently a professor in the geology department at the University of Massachusetts Amherst.

Margulis has had a major impact in two areas related to taxonomy and phylogeny. She championed the idea that eukaryotic cells came about as a result of the combining of prokaryotic cells (the endosymbiotic theory) and argued for changes in taxonomy that better reflect the true nature of living things. Her study of the organisms in the kingdom Protista (algae, protozoa, related organisms) led her to recognize that they were distinct from plants and animals. Thus, she argued that the traditional way of dividing living things into two kingdoms (plants and animals) was inadequate. At least in part due to her efforts, today the former plant and animal kingdoms have been divided into Protists, Fungi, Plants, and Animals.

Her support for the Gaia hypothesis, which suggests that Earth and all the living things on it are in a symbiotic relationship, has led her to be a champion for various environmental concerns.

Source: Modified from the *Hutchinson Dictionary of Scientific Biography.* © RM, 2007. Reprinted by permission.

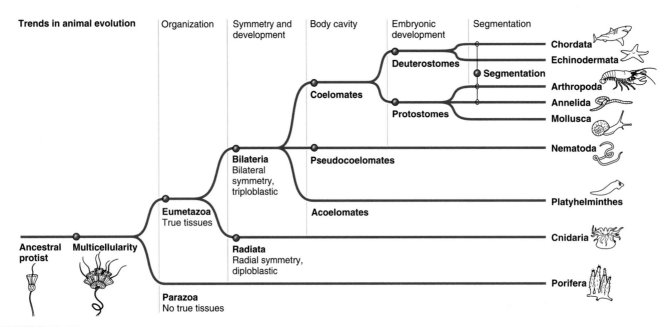

FIGURE 22.28 The classification of animals reflects complexity of body form and developmental features.

Although all animals are heterotrophs, they use various methods for obtaining nutrients. Most collect and consume food. However, some are parasites and live within or upon other organisms and use them as food. Many kinds of insects, worms, and ticks, and even some mollusks and fish are involved in parasitic relationships. Other animals are mutualistic and cooperate with other organisms to obtain food. For example, coral organisms have a mutualistic relationship with algae, which allows the coral to obtain some of its nutrients from the photosynthesis of the algae. (See chapter 23 for a more complete discussion of how animals interact.)

ACELLULAR INFECTIOUS PARTICLES

All of the groups discussed so far fall under the category of cellular forms of life. They all have at least the following features in common: They have (a) cell membranes, (b) nucleic acids as their genetic material, (c) cytoplasm, (d) enzymes and coenzymes, (e) ribosomes, and (f) use ATP as their source of chemical-bond energy. However, there are some particles that show some characteristics of life and cause disease but do not have a cell structure. Because they lack a cell structure, they are referred to as *acellular* (the prefix *a* = lacking). Because they enter cells and cause disease and can be passed from one organism to another, they are often called infectious particles. In the process of causing disease, they make copies of themselves. There is no clear explanation for how these particles came to be. Therefore, they are not included in the classification system used for cellular organisms. There are three kinds of these acellular infectious particles: *viruses, viroids,* and *prions.*

Viruses

A **virus** is an acellular infectious particle that consists of a nucleic acid core surrounded by a coat of protein (figure 22.29). Viruses

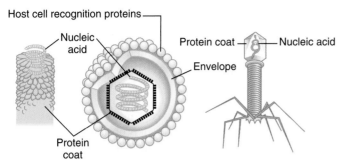

FIGURE 22.29 Viruses consist of a core of nucleic acid, either DNA or RNA, depending on the kind of virus. Some are surrounded by an envelope.

are often called **obligate intracellular parasites,** which means they are infectious particles that can function only when inside a living cell. Viruses are not considered to be living because they are not capable of living and reproducing by themselves and show some characteristics of life only when inside living cells.

How Did Viruses Originate?

Soon after viruses were discovered in the late part of the nineteenth century, biologists began to speculate on how they originated. One early hypothesis was that they were descendents of the first precells that did not evolve into cells. This idea was discarded as biologists learned more about the complex relationship between viruses and host cells. A second hypothesis was that viruses developed from intracellular parasites that became so specialized that they needed only the nucleic acid to continue their existence. Once inside a cell, this nucleic acid can take over and direct the host cell to provide for all of the virus' needs. A third hypothesis is that viruses are runaway genes that have escaped from cells and must return to a host cell to replicate. Regardless of how the viruses came into being, they are important today as parasites in all forms of life.

How Viruses Cause Disease

Viruses are typically host-specific, which means that they usually attack only one kind of host cell that provides what the virus needs to function. To enter cells, viruses must attach to a cell surface. Viruses can infect only those cells that have the proper receptor sites to which the virus can attach. This site is usually a glycoprotein molecule on the surface of the cell membrane. For example, the virus responsible for measles attaches to the membranes of skin cells, hepatitis viruses attach to liver cells, and mumps viruses attach to cells in the salivary glands. Host cells for the human immunodeficiency virus (HIV) include some types of human brain cells and several types belonging to the immune system.

Once it has attached to the host cell, the virus either enters the cell or injects its nucleic acid into the cell. If it enters the cell, the virus loses its protein coat, releasing the nucleic acid. Once released into the cell, the nucleic acid of the virus may remain free in the cell or it may link with the host's genetic material. Some viruses contain as few as three genes, others contain as many as five hundred. A typical eukaryotic cell contains tens of thousands of genes. Most viruses need only a small number of genes since they rely on the host to perform most of the activities necessary for viral reproduction.

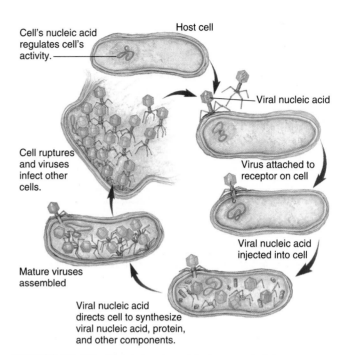

FIGURE 22.30 The viral nucleic acid takes control of the activities of the host cell. Because the virus has no functional organelles of its own, it can become metabolically active only while it is within a host cell.

Some viruses have DNA as their genetic material but others have RNA. Viruses do not divide as do true cells; they are *replicated,* which means copies are made, but the one virus does not become two. Viral genes are able to take command of the host's metabolic pathways and direct it to carry out the work of making new copies of the original virus. The virus makes use of the host's available enzymes and ATP for this purpose. When enough new viral nucleic acid and protein coat are produced, complete virus particles are assembled and released from the host (figure 22.30). In many cases, this process results in the death of the host cell. When the virus particles are released, they can infect adjacent cells and the infection spreads. The number of viruses released ranges from ten to thousands. The virus that causes polio affects nerve cells and releases about ten thousand new virus particles from each infected human host cell. Some viruses remain in cells and are only occasionally triggered to replicate, causing symptoms of disease. Herpes viruses, which cause cold sores, genital herpes, and shingles, are such viruses that reside in nerve cells and occasionally become active.

We know much about the viruses that cause human disease and disease of other organisms we value, but we know very little about viruses that may infect other kinds of organisms. It is likely that most species serve as hosts to some form of virus (table 22.4).

Viroids: Infectious RNA

The term **viroid** refers to infectious particles that are composed solely of a small, single strand of RNA in the form of a loop. To date, no viroids have been found that parasitize animals. The hosts in which they have been found are cultivated crop plants such as potatoes, tomatoes, and cucumbers. Viroid infections result in stunted or distorted growth and may sometimes cause

TABLE 22.4 Viral Diseases

Type of Virus	Disease
Papovaviruses	Warts in humans
Paramyxoviruses	Mumps and measles in humans; distemper in dogs
Adenoviruses	Respiratory infections in most mammals
Poxviruses	Smallpox
Wound-tumor viruses	Diseases in corn and rice
Potexviruses	Potato diseases
Bacteriophage	Infections in many types of bacteria

the plant to die. Pollen, seeds, or farm machinery can transmit viroids from one plant to another. Some scientists believe that viroids may be parts of normal RNA that have gone wrong.

Prions: Infectious Proteins

Prions are proteins that can be passed from one organism to another and cause disease. All the diseases of this type currently known cause changes in the brain that result in a spongy appearance called spongiform encephalopathies. Because the disease can be transmitted from one animal to another, the disease is often called a transmissible spongiform encephalopathy. The symptoms typically involve abnormal behavior and eventually death. In animals, the most common examples are scrapie in sheep and goats and mad cow disease in cattle. Scrapie got its name because one of the symptoms of the disease is an itching of the skin associated with nerve damage that causes the animals to rub against objects and scrape off their hair.

The occurrence of mad cow disease (bovine spongiform encephalopathy—BSE) in Great Britain was apparently caused by the spread of prions from sheep to cattle. This occurred because of the practice of processing unusable parts of sheep carcasses into a protein supplement that was fed to cattle. A human form of this disease is called Creutzfeldt-Jakob disease (CJD). It now appears that the original form of BSE has changed to a variety that is able to infect humans. This new form is called vCJD, which makes scientists believe that BSE and CJD (Creutzfeldt-Jakob disease) are in fact the same prion.

How Prions Cause Disease

It now appears to be well-established that these proteins can be spread from one animal to another and they do cause disease, but how are they formed and how do they multiply? The multiplication of the prion appears to result from the disease-causing prion protein coming in contact with a normal body protein and converting it into the disease-causing form, a process called *conversion*. Since this normal protein is produced as a result of translating a DNA message, scientists looked for the genes that make the protein and have found it in a wide variety of mammals. The normal gene produces a protein that does not cause disease but is able to be changed by the invading prion protein into the prion form. Prions do not "reproduce" or "replicate," as do viruses or viroids. A prion protein (pathogen) presses up against a normal (not harmful) body protein and may cause it to change shape to that of the dangerous protein. When this conversion happens to a number of proteins, they stack up and interlock, as do the individual pieces of a Lego toy. When enough link together, they have a damaging effect—they form plaques (patches) of protein on the surface of nerve cells that disrupt the flow of the nerve impulses and eventually cause nerve cell death. Brain tissues taken from animals that have died of such diseases appear to be full of holes, thus the name "spongiform encephalopathy."

A person's susceptibility to acquiring a prion disease such as CJD depends on many factors, such as their genetic makeup. If a person produces a functional protein with a particular amino acid sequence, the prion may not be able to convert it to its dangerous form. Other people may produce a protein with a slightly different amino acid sequence that can be converted to the prion form. Once formed, these abnormal proteins resist being destroyed by enzymes and most other agents used to control infectious diseases. Therefore, individuals with the disease-causing form of the protein can serve as the source of the infectious prions. There is still much to learn about the function of the prion protein and how the abnormal, infectious protein can cause copies of itself to be made. A better understanding of the genes that produce proteins that can be transformed by prions will eventually lead to effective treatment and prevention of these serious diseases in humans and other animals.

SUMMARY

Geologic time is measured through the radioactive decay process, determining the *radiometric age* of rocks in years. A *geomagnetic time scale* has been established from the number and duration of reversals in the magnetic field of Earth's past.

Correlation and the determination of the numerical ages of rocks and events have led to the development of a *geologic time scale*. The major blocks of time on this calendar are called eons. The Hadean eon, 3.8 to 4.5 billion years ago, is the period of time before Earth's surface solidified. The Archean eon, 2.5 to 3.8 billion years ago, is a time when the atmosphere lacked oxygen and only certain kinds of prokaryotic organisms existed. The Proterozoic eon, 2.5 billion to 540 million years ago, saw the development of an oxidizing atmosphere, the first eukaryotic cells, and the first multicellular organisms. The time encompassed by the Hadean, Archaean, and Proterozoic is also referred to as the Precambrian. The Phanerozoic eon, 540 million years

ago to the present, is a period of time that has seen the elaboration of the various kinds of living things we see today as well as many groups that have gone extinct. The Phanerozoic eon is divided into smaller units of time known as *eras*. The eras are the (1) *Cenozoic,* the time of recent life; (2) *Mesozoic,* the time of middle life; and (3) *Paleozoic,* the time of ancient life. The eras are divided into smaller blocks of time called *periods,* and the periods are further subdivided into *epochs*.

There are many different kinds of *fossils*—evidence of former living things. They are formed when organisms are covered over and prevented from being destroyed. Most are modified chemically but still give information about past living things. The fossil record is seen to change over geologic time, with certain kinds of fossils being associated with certain periods of time. There have been several *great extinctions* of living things. Some of the extinctions eliminated more than 80 percent of the organisms alive at the time. These extinctions appear to be related to changes in climate that may have been initiated by changes in the location and arrangement of continents, volcanic activity, or meteorite impact.

The early evolution of humans has been difficult to piece together because of the fragmentary evidence. Beginning about 4.4 million years ago, the earliest forms of *Australopithecus* and *Paranthropus* showed upright posture and other humanlike characteristics. The structure of the jaw and teeth indicates that the various kinds of australopiths were herbivores. *Homo habilis* had a larger brain and appears to have been a scavenger. Several other species of the genus *Homo* arose in Africa. These forms appear to have been carnivores. Some of these migrated to Europe and Asia. The origin of *Homo sapiens* is in dispute. It may have arisen in Africa and migrated throughout the world or evolved from earlier ancestors found throughout Africa, Asia, and Europe.

To facilitate accurate *communication,* biologists assign a *specific name* to each *species* that is cataloged. The various species are cataloged into larger groups on the basis of similar traits. *Taxonomy* is the science of classifying and naming organisms. *Phylogeny* is the science of trying to figure out the evolutionary history of a particular organism. The *taxonomic ranking of organisms* reflects their *evolutionary relationships. Fossil evidence, comparative anatomy, developmental stages,* and *biochemical evidence* are employed in the sciences of taxonomy and phylogeny.

The first organisms thought to have evolved were single-celled, prokaryotic organisms. These simple organisms gave rise to two different prokaryotic domains, Eubacteria and Archaea. The Eubacteria are a very diverse group of organisms. Some are autotrophs, while others are heterotrophs. The cyanobacteria were probably the first organisms to carry on photosynthesis. Many kinds of heterotrophic Eubacteria are decomposers, while some are parasites, mutualistic, or commensal organisms. The Archaea are distinguished by their ability to live in very hostile environments. Eukaryotic organisms were probably derived from the Eubacteria and Archaea and constitute the domain Eucarya. There are four kingdoms in the Eucarya: *Protista, Fungi, Plantae,* and *Animalia.* The Protista includes organisms such as protozoa and algae that consist of single cells or small colonies of cells. The Fungi kingdom includes multicellular, heterotrophic organisms with cell walls made of chitin. The Plantae are complex, multicellular, autotrophic organisms that have cell walls made of cellulose and are primarily terrestrial. The Animalia are complex, multicellular, heterotrophic organisms that lack cell walls and typically have the ability to move.

KEY TERMS

archaeology (p. **530**)
binomial system of nomenclature (p. **542**)
class (p. **543**)
domain (p. **543**)
eon (p. **534**)
epoch (p. **535**)
era (p. **534**)

family (p. **544**)
fossil (p. **530**)
fungus (p. **550**)
genus (p. **543**)
geologic time scale (p. **534**)
index fossils (p. **533**)
kingdom (p. **543**)
multiregional hypothesis (p. **541**)

obligate intracellular parasite (p. **555**)
order (p. **544**)
out-of-Africa hypothesis (p. **540**)
paleontology (p. **530**)
parasite (p. **549**)
period (p. **535**)
phylogeny (p. **545**)

phylum (p. **543**)
prions (p. **556**)
taxonomy (p. **542**)
unconformity (p. **529**)
viroid (p. **555**)
virus (p. **554**)

APPLYING THE CONCEPTS

1. Some of the oldest fossils are about how many years old?
   a. 4.55 billion
   b. 3.5 billion
   c. 250 million
   d. 10,000
2. A fossil of a jellyfish, if found, would most likely be
   a. a preserved fossil.
   b. one formed by mineralization.
   c. a carbon film.
   d. a cast or a mold.
3. In any sequence of sedimentary rock layers that has not been subjected to stresses, you would expect to find
   a. essentially horizontal stratified layers.
   b. the oldest layers at the bottom and the youngest at the top.
   c. younger faults, folds, and intrusions in the rock layers.
   d. All of the above are correct.
4. You would expect to find the least number of fossils in rocks from which geologic time?
   a. Cenozoic
   b. Mesozoic
   c. Paleozoic
   d. Precambrian
5. The numerical dates associated with events on the geologic time scale were determined by
   a. relative dating of the rate of sediment deposition.
   b. radiometric dating using radioactive decay.
   c. the temperature of Earth.
   d. the rate that salt is being added to the ocean.
6. The three main kinds of living things, Eubacteria, Archaea, and Eucarya, have been labeled as
   a. species.
   b. kingdoms.
   c. domains.
   d. families.

7. Which is commonly used as sources of information when developing a phylogeny?
   a. fossil evidence
   b. biochemical information
   c. embryological development
   d. All of the above are correct.
8. Which of the following kingdoms contain members that are autotrophs?
   a. Protista
   b. Plantae
   c. Eubacteria
   d. All of the above are correct.
9. Plants and fungi differ in that
   a. plants are autotrophic and fungi are heterotrophic.
   b. plants have cell walls of cellulose and fungi have cell walls of chitin.
   c. plants carry on photosynthesis and fungi do not.
   d. All of the above are correct.
10. Viruses
    a. can be free-living or parasitic.
    b. belong to the domain Archaea.
    c. can function and reproduce only inside a host cell.
    d. All of the above are correct.

11. In the evolution of humans, the fossil evidence suggests that
   a. the earliest humanlike primates were present in Africa, Europe, and North America.
   b. *Homo sapiens* has been around for about two million years.
   c. there were several different species of human ancestors in Africa.
   d. large numbers of fossils clearly show the way humans evolved.

12. *Homo sapiens* is the only member of its genus today,
   a. however, at one time there were several members of the genus.
   b. and the only other member of the genus is *Homo neanderthalensis*.
   c. because *Homo neanderthalensis* evolved into *Homo sapiens*.
   d. None of the above is true.

Answers

1. b 2. c 3. d 4. d 5. b 6. c 7. d 8. d 9. d 10. c 11. c 12. a

QUESTIONS FOR THOUGHT

1. Why does the rock record go back only 3.8 billion years? If this missing record were available, what do you think it would show? Explain.
2. What major event marked the end of the Paleozoic and Mesozoic eras according to the fossil record? Describe one theory that proposes to account for this.
3. Describe how the principles of superposition, horizontality, and faunal succession are used in the relative dating of sedimentary rock layers.
4. Describe four different kinds of fossils and how they were formed.
5. Describe four things that fossils can tell you about Earth's history.
6. What are some of the major steps thought to have been involved in the evolution of humans?
7. How do the domains Eubacteria and Archaea differ?
8. What is the value of taxonomy?
9. How do viruses reproduce?
10. Why are Latin names used for genus and species?

FOR FURTHER ANALYSIS

1. Take a long narrow strip of paper. Tear it in half crossways. Tear one of the remaining pieces in half. Continue until you have made five tears. Measure the length of the piece you have remaining. Use this information and the number of times you "halved" the paper to determine the length of the original strip of paper. This is analogous to the way in which the measurement of radioactive isotopes can be used to determine the age of rocks.
2. Develop a "phylogeny" for motor vehicles and draw a diagram of your phylogeny. Consider the following questions among those you include in determining your phylogeny:

   How many "kinds" currently exist?

   Are there "fossils" that may have been precursors to motor vehicles?

   What was the ancestor of motor vehicles?

   What was the first motor vehicle?

   What motor vehicles have gone extinct?

   What major evolutionary changes have occurred?

   What "environmental factors" shaped the evolution of motor vehicles?

   Present your phylogeny to your class and have them critique your effort.

3. Consider the following questions:

   Could our early hominin ancestors have seen living dinosaurs?

   What did the animals in the Cambrian period eat?

   What organisms have dominated most of the history of Earth?

   How might viruses and viroids be related?

INVITATION TO INQUIRY

Understanding Geological Time

It is often difficult to appreciate the extremely long period of time involved in the history of Earth and life. It is also often difficult to appreciate that the kinds of living things on Earth have changed significantly. From table 22.2, obtain the approximate dates for the following significant events and place them on the time line.

1. First fossil evidence of life
2. First evidence of eukaryotic cells
3. First animals
4. First terrestrial plants
5. First dinosaurs
6. First mammals
7. Dinosaurs go extinct
8. First humans

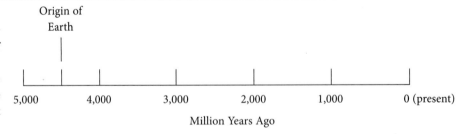

Million Years Ago

Now do the following:

1. Circle the period of time in which only prokaryotic organisms were present.
2. Circle the period of time during which dinosaurs were present.
3. Circle the period of time in which humans have been present.
4. Circle the period of time in which only marine organisms were present.

Organisms interact in many ways.

Ecology and Environment

OVERVIEW

Today, we recognize that environmental problems are a worldwide concern. Poor agricultural practices in Africa and China result in dust storms that affect the local people and are also carried to the rest of the world. Concern about global warming has led to the convening of international conferences and the drafting of treaties whose goal is to reduce the impact of energy use. Many are concerned about protecting endangered species and the forests, grasslands, and oceans that provide habitats for them. In many ways, these problems are the result of an incredible increase in the size of the human population. As the human population has increased, humans have sought to use land for agriculture and other purposes. Converting land to agricultural use results in its loss for other purposes such as habitat and watershed protection. To understand the nature of environmental problems and steps we can take to solve them, we need to be familiar with some of the central ideas of the science of ecology.

A DEFINITION OF ENVIRONMENT

Ecology is the branch of biology that studies the relationships between organisms and their environments. This is a very simple definition for a very complex branch of science. Most ecologists define the word **environment** very broadly as anything that affects an organism during its lifetime. Environmental influences can be divided into two categories: **biotic factors** are other living things that affect an organism, and **abiotic factors** are nonliving influences that affect an organism (figure 23.1). If we consider a fish in a stream, we can identify many environmental factors that are important to its life. The temperature of the water is extremely important as an abiotic factor that may be influenced by the presence of trees (biotic factor) along the stream bank that shade the stream and prevent the Sun from heating it. Obviously, the kind and number of food organisms in the stream are important biotic factors as well. The type of material that makes up the stream bottom and the amount of oxygen dissolved in the water are other important abiotic factors, both of which are related to how rapidly the water is flowing.

Characterizing the environment of an organism is a complex and challenging process; everything seems to be influenced or modified by other factors. A plant is influenced by many

A   B

FIGURE 23.1 (A) The woodpecker feeding its young in the hole in this tree is influenced by several biotic factors. The tree itself is a biotic factor as is the disease that weakened it, causing conditions that allowed the woodpecker to make a hole in the rotting wood. (B) The irregular shape of the trees is the result of wind and snow, both abiotic factors. Snow driven by these prevailing winds tends to "sandblast" one side of the tree and prevent limb growth.

different factors during its lifetime: the types and amounts of minerals in the soil, the amount of sunlight hitting the plant, the animals that eat the plant, and the wind, water, and temperature. Each item on this list can be further subdivided into other areas of study. For instance, water is important in the life of plants, so rainfall is studied in plant ecology. But even the study of rainfall is not simple. In some areas of the world, it rains only during one part of the year, while in others, rain is evenly distributed throughout the year. The rainfall could be hard and driving, or it could come as gentle, misty showers of long duration. The water could soak into the soil for later use, or it could run off into streams and be carried away.

The animals in an area are influenced as much by environmental factors as are the plants. If environmental factors do not favor the growth of plants, there will be little food and few hiding places for animal life. Two types of areas that support only small numbers of living animals are deserts and polar regions. Near the poles, the low temperature and short growing season inhibit growth; therefore, there are relatively few species of animals with relatively small numbers of individuals. Deserts receive little rainfall and therefore have poor plant growth and low concentrations of animals. On the other hand, tropical rainforests have high rates of plant growth and large numbers of animals of many kinds.

THE ORGANIZATION OF ECOLOGICAL SYSTEMS

Ecologists can study ecological relationships at several different levels. The smallest living unit is the individual **organism,** which is composed of atoms and molecules arranged in a highly organized manner. Each kind of organism has a particular functional role in the community, which is known as its **niche.** Groups of organisms of the same species are called **populations.** Interacting groups of populations of different species are called **communities.** An **ecosystem** consists of all the interacting organisms in an area and their interactions with their abiotic surroundings. Figure 23.2 shows how these different levels of organization are related to one another.

All living things require a continuous supply of energy to maintain life. Therefore, many people like to organize living systems by the energy relationships that exist among the different kinds of organisms present. An ecosystem contains several different kinds of organisms. Those that trap sunlight for photosynthesis, resulting in the production of organic material from inorganic material, are called **producers.** Green plants and other photosynthetic organisms such as algae and cyanobacteria are, in effect, converting sunlight energy into the energy contained within the chemical bonds of organic compounds. There is a flow of energy from the Sun into the living matter of plants.

The energy that plants trap can be transferred through a number of other organisms in the ecosystem. Since all organisms, other than producers, must obtain energy in the form of organic matter, they are called **consumers.** Consumers cannot capture energy from the Sun as plants do. All animals are consumers. They either eat plants directly or eat other sources of organic matter derived from plants. Each time the energy enters a different organism, it is said to enter a different **trophic level,** which is a step, or stage, in the flow of energy through an ecosystem (figure 23.3). The plants (producers) receive their energy directly from the Sun and are said to occupy the *first trophic level.*

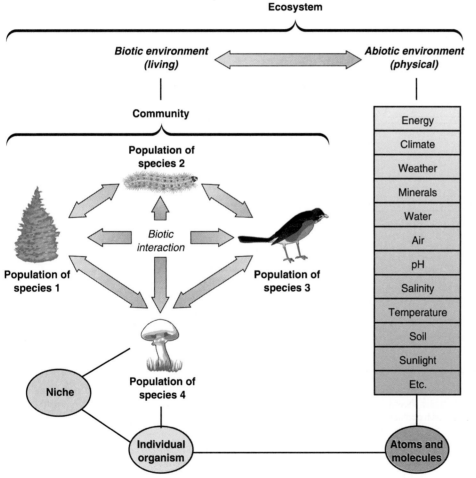

FIGURE 23.2 Ecology is the science that deals with the interactions between organisms and their environment. This study can take place at several different levels, from the broad ecosystem level through community interactions and population studies to the study of the niche of individual organisms. Ecology also involves study of the physical environment and the atoms and molecules that make up both the living and nonliving parts of an ecosystem. The same organism can be viewed in many ways. We can study it as an individual, as a member of a population, or as a participant in a community or ecosystem.

Eugene Odum (1913–2002)

Eugene Odum was born in 1913 to Howard W. and Anna K. Odum, the first of three children. His father was a professor of sociology at the University of North Carolina. His mother was an urban planner. As a teenager, he wrote about bird life for a local newspaper. He continued to be interested in birds for the rest of his life. He received his bachelor's and master's degrees from the University of North Carolina and his doctorate from the University of Illinois.

In 1939, he married Martha Ann Huff. They had two sons. He joined the faculty of the University of Georgia in 1940 and retired in 1984.

One of his major contributions to science was the popularization of the ecosystem concept. Earlier ecologists had looked primarily at individual organisms and how they reacted with their surroundings or at small systems such as ponds that were easy to quantify.

In 1953, Eugene Odum and his brother, Howard T. Odum, published *Fundamentals of Ecology*, which clearly described ecology as an integrated discipline that could be approached from a systems point of view. Furthermore, it was possible to look at ecology from a regional point of view. For ten years, this was the only book of its type available.

Eugene Odum's interest in ecosystem-level processes quite naturally led to concerns about how humans interact with their surroundings. Thus, his thinking was an important catalyst of the environmental movement of the 1970s.

Source: Modified from the *Hutchinson Dictionary of Scientific Biography.* © RM, 2007. Reprinted by permission.

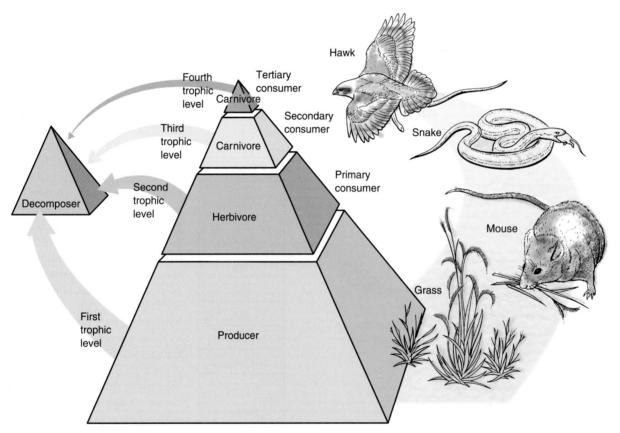

FIGURE 23.3 Organisms within ecosystems can be divided into several different trophic levels on the basis of how they obtain energy. Several different sets of terminology are used to identify these different roles. This illustration shows how the different sets of terminology are related to one another. Each colored block in this diagram represents the amount of energy or living material at each trophic level. The amount of energy or matter decreases sharply as it is transferred to higher trophic levels.

Various kinds of consumers can be divided into several categories, depending on how they fit into the flow of energy through an ecosystem. Animals that feed directly on plants are called **herbivores,** or **primary consumers,** and occupy the *second trophic level.* Animals that eat other animals are called **carnivores,** or **secondary consumers,** and can be subdivided into different trophic levels depending on what animals they eat. Animals that feed on herbivores occupy the *third trophic level* and are known as **primary carnivores.** Animals that feed on the primary carnivores are known as **secondary carnivores** and occupy the *fourth trophic level.*

This sequence of organisms feeding on one another is known as a **food chain.** Often food chains are very complex and can involve many trophic levels. For example, a human may eat a fish that ate a frog that ate a spider that ate an insect that consumed plants for food. Figure 23.4 shows the six different trophic levels in this food chain. Obviously, there can be higher categories, and some organisms don't fit neatly into

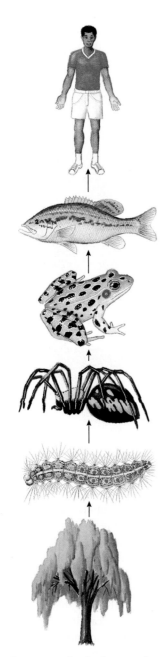

FIGURE 23.4 As one organism feeds on another organism, there is a flow of energy from one trophic level to the next. This illustration shows six trophic levels.

this theoretical scheme. Some animals are carnivores at some times and herbivores at others; they are called **omnivores.** They are classified into different trophic levels depending on what they happen to be eating at the moment.

If an organism dies, the energy contained within the organic compounds of its body is finally released to the environment as heat by organisms that decompose the dead body into carbon dioxide, water, ammonia, and other simple inorganic molecules. Organisms of decay, called **decomposers,** are things such as bacteria, fungi, and other organisms that use dead organisms as sources of energy. This group of organisms efficiently converts nonliving organic matter into simple inorganic molecules that can be used by producers in the process of trapping energy. Decomposers are thus very important components of ecosystems that cause materials to be recycled. As long as the Sun supplies the energy, elements are cycled through ecosystems repeatedly. Table 23.1 summarizes the various categories of organisms within an ecosystem. Now that we have a better idea of how ecosystems are organized, we can look more closely at energy flow through ecosystems.

ENERGY FLOW IN ECOSYSTEMS

All living things need a constant source of energy. Two fundamental physical laws of energy are important when looking at ecological systems from an energy point of view. The *first law of thermodynamics*—also known as the *law of conservation of energy*—states that energy is neither created nor destroyed. This means that we should be able to describe the amounts of energy in each trophic level and follow energy as it flows through successive trophic levels. The *second law of thermodynamics* states

TABLE 23.1   Roles in an Ecosystem

Classification	Description	Examples
Producers	Organisms that convert simple inorganic compounds into complex organic compounds by photosynthesis	Trees, flowers, grasses, ferns, mosses, algae, cyanobacteria
Consumers	Organisms that rely on other organisms as food; animals that eat plants or other animals	
Herbivore	Eats plants directly	Deer, goose, cricket, vegetarian human, many snails
Carnivore	Eats meat	Wolf, pike, dragonfly
Omnivore	Eats plants and meat	Rat, most humans
Scavenger	Eats food left by others	Coyote, skunk, vulture, crayfish
Parasite	Lives in or on another organism, using it for food	Tick, tapeworm, many insects
Decomposers	Organisms that return organic compounds to inorganic compounds; important components in recycling elements	Bacteria, fungi

that natural processes proceed toward a state of greater disorder. This disordered condition is known as entropy. Another way to state the second law of thermodynamics is to say that when energy is converted from one form to another, useful energy is lost, which is to say there is an increase in entropy. This means that as energy passes from one trophic level to the next, there will be a reduction in the amount of available energy in living things and an increase in the amount of heat in their surroundings.

The total energy of an ecosystem can be measured in several ways. An indirect way to get an idea of the amount of energy at a trophic level is to determine its **biomass,** which is the weight of the organisms at that trophic level. Since organic matter will burn, it is possible to convert the energy in biomass into heat and light. For example, the total producer trophic level can be harvested and burned. The number of kilocalories of heat energy produced by burning is equivalent to the energy content of the organic material of the plants. This is what you do when you burn wood at a campfire. Another way of determining the energy present is to measure the rate of photosynthesis and respiration, and calculate the amount of energy being trapped in the living material of the plants. However, this is technically difficult. It requires careful measurements of tiny changes in the amount of carbon dioxide and oxygen in the atmosphere as a result of photosynthesis and respiration or the use of techniques to follow traceable atoms in the bodies of the organisms.

Since only the plants, algae, and cyanobacteria in the producer trophic level are capable of capturing energy from the Sun, all other organisms are directly or indirectly dependent on the producer trophic level. The second trophic level consists of herbivores that eat the producers. This trophic level has significantly less energy and biomass in it for several reasons. *In general, there is about a 90 percent loss of available energy as we proceed from one trophic level to the next higher level.* Actual measurements will vary from one ecosystem to another, but 90 percent is a good rule of thumb. This loss in energy content at the second and subsequent trophic levels is predicted by the second law of thermodynamics. When the energy in producers is converted to the energy of herbivores, much of the energy is lost as heat to the surroundings.

In addition to this loss of available energy, there is an additional loss involved in the capture and processing of food material by herbivores. Although herbivores don't need to chase their food, they do need to travel to where food is available, then gather, chew, digest, and metabolize it. All these processes require energy.

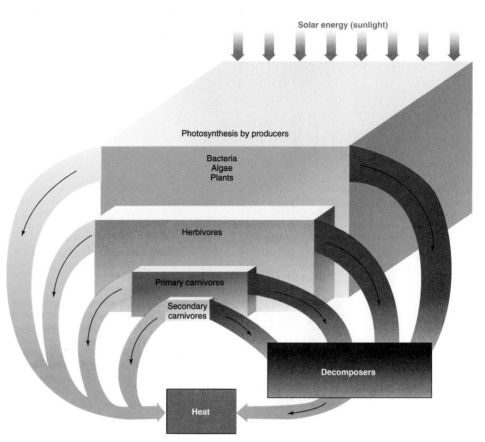

FIGURE 23.5   As energy flows from one trophic level to the next, approximately 90 percent of it is lost. This means that the amount of energy at the producer level must be ten times larger than the amount of energy at the herbivore level.

Just as the herbivore trophic level experiences a 90 percent loss in energy content, the higher trophic levels of primary carnivores, secondary carnivores, and tertiary carnivores also experience a reduction in the energy available to them. This energy loss is reflected in the number of organisms at each trophic level. A field may have millions of producers (grass and other plants), thousands of herbivores that eat grass (grasshoppers, prairie dogs, etc.), and a few carnivores (birds, weasels, etc.) that eat the herbivores. Finally, decomposers act on dead organisms with this same kind of available energy loss. Figure 23.5 shows a diagram in which the energy content decreases by 90 percent as we pass from one trophic level to the next.

COMMUNITY INTERACTIONS

In the section on energy flow in ecosystems, we looked at ecological relationships from the point of view of ecosystems and the way energy flows through them. But we can also study relationships at the community level and focus on the kinds of interactions that take place among organisms.

As you know, one of the ways that organisms interact is by feeding on one another. A community includes many different food chains, and many organisms may be involved in several of the food chains at the same time, so the food chains become interwoven into a *food web* (figure 23.6). In general, communities are relatively stable in terms of the kinds of organisms present and how they interact. Some communities, such as tropical rainforests, have large numbers of different kinds of organisms present. Such communities have high **biodiversity**. Others, such as tundra communities, have low biodiversity. People also talk about a loss of biodiversity when a specific kind of organism is eliminated from a region. For example, when wolves were eliminated from much of North America, there was a loss of biodiversity. Although communities are relatively stable, we need to recognize that they are also dynamic collections of organisms. Although the kinds of organisms present may not change, the numbers of each kind may change significantly throughout the year or over several years. For example, a drought will reduce the survival of plants, and an epidemic of disease will reduce the survival of many birds.

If numbers of a particular kind of organism in a community increase or decrease significantly, there will be a ripple effect through the community. For example, the populations of

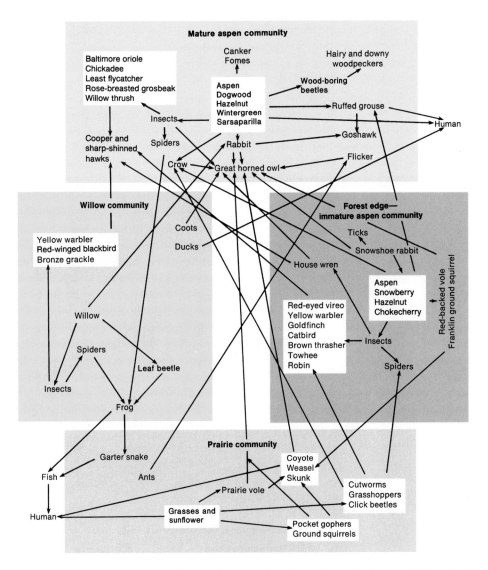

FIGURE 23.6 When many different food chains are interlocked with one another, a food web results. The arrows indicate the direction of energy flow. Notice that some organisms are a part of several food chains—the great horned owl in particular. Because of the interlocking nature of the food web, changing conditions may shift the way in which food flows through this system.

many kinds of small mammals fluctuate from year to year. This results in changes in the numbers of their predators, and often the predators switch to other prey species, which impacts other parts of the community.

Not all organisms within a community have the same level of importance. Obviously, plants are key organisms that supply the energy for all other organisms. However, even certain animal species can have a very high impact on the nature of a community. For example, sea otters feed on sea urchins that feed on a seaweed called kelp. When sea otter populations were reduced significantly, sea urchin populations increased, they ate much of the kelp, and other species of animals had fewer places to hide. Thus, the sea otters played a key role in shaping the nature of the community. Such organisms are often referred to as **keystone species.**

TYPES OF TERRESTRIAL COMMUNITIES

Although each community of organisms is unique, similar communities can be combined into broad categories based on specific characteristics of the physical environment and the kinds of organisms that live in the area. The kind of terrestrial community that develops in an area is determined primarily by climatic factors of precipitation patterns and temperature ranges. These large regional communities are known as **biomes.** The map in figure 23.7 shows the distribution of the major terrestrial biomes of the world.

Temperate Deciduous Forest

The *temperate deciduous forest* biome is found in parts of the world (primarily eastern North America, parts of Europe, Japan, Korea, and parts of China) that have 75 to 100 cm (30 to 40 in) of rainfall and cold weather for a significant part of the year. Precipitation is spread throughout the year. The predominant plants are large trees that lose their leaves more or less completely during the fall of the year and are therefore called *deciduous.* Aspen, birch, cottonwood, oak, hickory, beech, and maple are typical trees found in this geographic region.

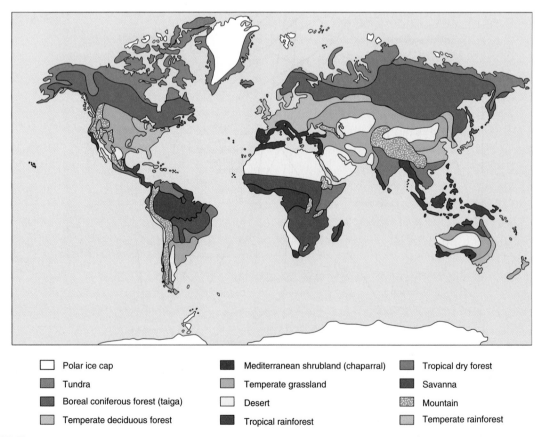

☐ Polar ice cap	◼ Mediterranean shrubland (chaparral)	◼ Tropical dry forest
◼ Tundra	◼ Temperate grassland	◼ Savanna
◼ Boreal coniferous forest (taiga)	☐ Desert	▨ Mountain
◼ Temperate deciduous forest	◼ Tropical rainforest	◻ Temperate rainforest

FIGURE 23.7 Major climatic differences determine the kind of vegetation that can live in a region of the world. Associated with specialized groups of plants are particular kinds of animals. These regional ecosystems are called biomes.

Typical animals of this biome are skunks, porcupines, deer, frogs, opossums, owls, mosquitoes, and beetles.

In much of this region, the natural vegetation has been removed to allow for agriculture, so the original character of the biome is gone except where farming is not practical or the original forest has been preserved.

Temperate Grassland or Prairie

Temperate grasslands are common in temperate regions of western North America and parts of Eurasia, Africa, Australia, and South America. The dominant vegetation in this region is made up of various species of grasses.

The annual rainfall is 25 to 75 cm (10 to 30 in), an amount that is not adequate to support the growth of dense forests. Trees in grasslands grow primarily along streams where they can obtain sufficient water. Animals found in this area include the prairie dog, pronghorn antelope, prairie chicken, grasshopper, rattlesnake, and meadowlark. Most of the original grasslands, like the temperate deciduous forest, have been converted to agricultural uses.

Savanna

Savannas are found in tropical regions of central Africa, northern Australia, and parts of South America that have pronounced rainy and dry seasons. Although these regions may receive 100 cm (40 in) of rain per year, there is an extended dry season of three months or more.

Therefore, savannas consist of grasses with scattered trees. Grazing animals such as antelope, zebra, and buffalo are common, as are termites and many other insects, which serve as food for many birds and reptiles.

Desert

Very dry areas known as *deserts* are found throughout the world wherever rainfall is low and irregular. They receive less than 25 cm (10 in) of rain per year. Some deserts are extremely hot, while others can be quite cold during much of the year. The distinguishing characteristic of desert biomes is low rainfall, not high temperature.

Deserts are characterized by scattered, thorny plants that lack leaves or have reduced leaves. Since leaves tend to lose water rapidly, the lack of leaves is an adaptation to dry conditions. Although a desert is a very harsh environment, many kinds of insects, reptiles, birds, and mammals can live in this biome. The animals usually avoid the hottest part of the day by staying in burrows or other shaded, cooler areas. Staying underground or in the shade also allows the animals to conserve water.

Boreal Coniferous Forest

The *boreal coniferous forest* (also known as the *northern coniferous forest,* or *taiga*) is found through parts of southern Canada, extending southward along the mountains of the United States, and in much of northern Europe and Asia. This region has short, cool summers and long, cold winters with abundant snowfall. Precipitation ranges from 25 to 100 cm (10 to 40 in) per year. However, even if precipitation is low, the extensive snowmelt in spring and low evaporation rate result in humid summers. The dominant vegetation consists of evergreen trees that are especially adapted to withstand long, cold winters with abundant snowfall. Spruces and firs are common tree species that are intermingled with many other kinds of vegetation. Small lakes and bogs are common. Characteristic animals in this biome include mice, bears, wolves, squirrels, moose, midges, and flies.

Mediterranean Shrublands (Chaparral)

Mediterranean shrublands are located near an ocean and are dominated by shrubby plants. As the name implies, this biome is typical of the Mediterranean coast and is also found in coastal southern California, the southern tip of Africa, a portion of the west coast of Chile, and southern Australia. The climate has wet, cool winters and hot, dry summers. Rainfall is 40 to 100 cm (15 to 40 in) per year. The vegetation is dominated by woody shrubs that are adapted to withstand the hot, dry summer. Often the plants are dormant during the summer. Fire is a common feature of this biome, and the shrubs are adapted to withstand occasional fires. The kinds of animals vary widely in the different regions of the world with this biome. Many kinds of insects, reptiles, birds, and mammals are found in these areas. In the chaparral of California, rattlesnakes, spiders, coyotes, lizards, and rodents are typical inhabitants.

Temperate Rainforest

The coastal areas of northern California, Oregon, Washington, British Columbia, and southern Alaska contain an unusual set of environmental conditions that supports a *temperate rainforest*. There are also small areas of temperate rainforest in southern Chile and on the west coast of New Zealand. The temperate rainforests occur along the coast where the prevailing winds from the west bring moisture-laden air. As the air meets the coastal mountains and is forced to rise, it cools and the moisture falls as rain or snow. Most of these areas receive 200 cm (80 in) or more of precipitation per year. This abundance of water, along with fertile soil and mild temperatures, results in a lush growth of plants.

Sitka spruce, Douglas fir, and western hemlock are typical evergreen coniferous trees in the temperate rainforest of North America. Undisturbed (old growth) forests of this region have trees as old as eight hundred years that are nearly 100 m (about 300 ft) tall. Deciduous trees of various kinds (red alder, big leaf maple, black cottonwood) also exist in open areas where they can get enough light. All trees are covered with mosses, ferns, and other plants that grow on the surface of the trees. The kinds of animals present are similar to those of a temperate deciduous forest.

Tundra

North of the coniferous forest biome in North America and northern Europe and Asia is an area known as the *tundra*. It is characterized by extremely long, severe winters and short, cool summers. The deeper layers of the soil remain permanently frozen, forming a layer called the *permafrost*. Although precipitation is typically less than 25 cm (10 in) per year, the melting of snow during the brief summer (about one hundred days) results in moist conditions. Summer temperatures may be near freezing at night and rise to about 10°C (50°F) during the day. Because the deeper layers of the soil are frozen, when the surface thaws, the water forms puddles.

Under these conditions of low temperature and short growing season, very few kinds of animals and plants can survive. No trees can live in this region. Typical plants and animals of the area are dwarf willow and some other shrubs, reindeer moss (actually a lichen), some flowering plants, caribou, wolves, musk oxen, fox, snowy owls, mice, and many kinds of insects. Many kinds of birds are summer residents only.

Tropical Rainforest

Tropical rainforests are found primarily near the equator in Central and South America, Africa, parts of southern Asia, and some Pacific Islands. The temperature is quite warm and constant, and rain falls nearly every day. Most areas receive over 200 cm (80 in) of rain per year and some exceed 500 cm (200 in).

These areas have high biodiversity. Balsa (a very light wood), teak (used in furniture), and ferns the size of trees are examples of plants from the tropical rainforest. Typically, every plant has other plants growing on it. Tree trunks are likely to be covered with orchids, many kinds of vines, and mosses. Tree frogs, bats, lizards, birds, monkeys, and an almost infinite variety of insects inhabit the rainforest. These forests are very dense and little sunlight reaches the forest floor. When the forest is opened up (by a hurricane or the death of a large tree) and sunlight reaches the forest floor, the opened area is rapidly overgrown with vegetation.

Tropical Dry Forest

Tropical dry forests are found in parts of Central and South America, Australia, Africa, and Asia (particularly India and Myanmar). A major characteristic of this biome is seasonal rainfall. Many of the tropical dry forests have a monsoon climate in which several months of heavy rainfall are followed by extensive periods without rain. Some tropical dry forests have as much as eight months without rain. The rainfall may be as low as 50 cm (20 in) or as high as 200 cm (80 in). Since the rainfall is highly seasonal, many of the plants have special adaptations for enduring drought. In many of the regions that have extensive dry periods, many of the trees drop their leaves during the dry period. Many of the species of animals found here are also found in more moist tropical forests of the region. However, there are fewer kinds in dry forests than in rainforests.

TYPES OF AQUATIC COMMUNITIES

Terrestrial biomes are determined by the amount and kind of precipitation and by temperature ranges. Other factors, such as soil type and wind, also play a part. Aquatic ecosystems also are shaped by key environmental factors. Several important factors are the ability of the Sun's rays to penetrate the water, the depth of the water, the nature of the bottom substrate, the water temperature, and the amount of dissolved salts. Aquatic systems are often divided into **marine communities,** which have a high salt content, and **freshwater communities,** which have low salt concentrations.

Marine Communities

In the open ocean, many kinds of organisms float or swim actively and are called **pelagic.** Among the pelagic organisms are two types: larger organisms that actively swim and smaller, weakly swimming organisms. The term **plankton** is used to describe aquatic organisms that are so small and weakly swimming that they are simply carried by currents. The planktonic organisms that carry on photosynthesis are called *phytoplankton.* In the open ocean, a majority of these organisms are microscopic floating algae and bacteria. The upper layer of the ocean, where the Sun's rays penetrate, is known as the *euphotic zone.* It is in this euphotic zone where phytoplankton are most common. Small, weakly swimming animals of many kinds, known as *zooplankton,* feed on the phytoplankton. The zooplankton are in turn eaten by larger animals such as fish and larger shrimp, which are eaten by larger fish such as salmon, tuna, sharks, and mackerel (figure 23.8).

Those marine organisms that live on the bottom are called **benthic.** Many seaweeds, some fish, clams, oysters, various crustaceans, sponges, sea anemones, and many other kinds of organisms live on the bottom. In shallow water, sunlight can penetrate to the bottom, and a variety of attached seaweeds trap this energy as they carry on photosynthesis. Many of the benthic animals graze on the seaweeds and are in turn eaten by other larger animals. Some benthic animals, such as clams, filter water to obtain plankton for food.

As with terrestrial communities, environmental conditions influence the kinds of marine communities that exist. Coral reefs are found in warm, tropical, shallow seas, and "forests" of

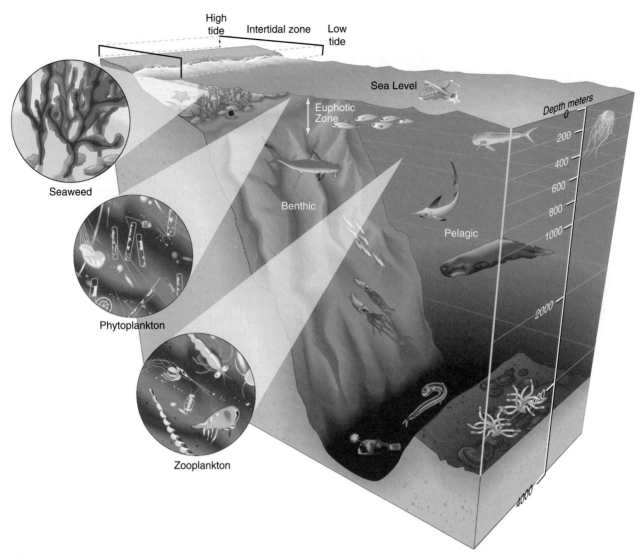

High tide
Intertidal zone
Low tide

Seaweed

Phytoplankton

Zooplankton

Sea Level

Euphotic Zone

Benthic

Pelagic

Depth meters
0
200
400
600
800
1000
2000
4000

FIGURE 23.8 All of the photosynthetic activity of the ocean occurs in shallow water called the euphotic zone, either by attached algae near the shore or by minute phytoplankton in the upper levels of the open ocean. Consumers are either free-swimming pelagic organisms or benthic organisms that live on the bottom. Small animals that feed on phytoplankton are known as zooplankton.

large seaweeds are found along cool, rocky shores. Sandy shores with lots of wave action typically have few plants, while shallow, protected, muddy areas may support mangrove swamps or salt marshes.

Freshwater Communities

Freshwater ecosystems differ from marine ecosystems in several ways. In comparison to marine communities, the amount of salt present is much less, the temperature of the water can change greatly, the water is in the process of moving to the ocean, oxygen can often be in short supply, and the organisms that inhabit freshwater systems are different. Freshwater ecosystems can be divided into two categories: those in which the water is relatively stationary, such as lakes, ponds, and reservoirs, and those in which the water is running downhill, such as streams and rivers.

Large lakes have many of the same characteristics as the ocean. If the lake is deep, there is a euphotic zone at the top,

with many kinds of phytoplankton, and zooplankton that feed on the phytoplankton. Small fish feed on the zooplankton, which are in turn eaten by larger fish. The species of organisms found in freshwater lakes are different from those found in the ocean, but the roles played are similar, so the same terminology is used.

Along the shore and in the shallower parts of lakes, many kinds of flowering plants are rooted in the bottom. Some are completely submerged, while others have leaves that float on the surface or protrude above the water. Cattails, bulrushes, arrowhead plants, and water lilies are examples. In addition, many kinds of freshwater algae are present.

Associated with the plants and algae are a large number of different kinds of animals. Fish, crayfish, clams, and many kinds of aquatic insects are common inhabitants of this mixture of plants and algae. This region, with rooted vegetation, is known as the *littoral zone,* and the portion of the lake that does not have rooted vegetation is called the *limnetic zone* (figure 23.9).

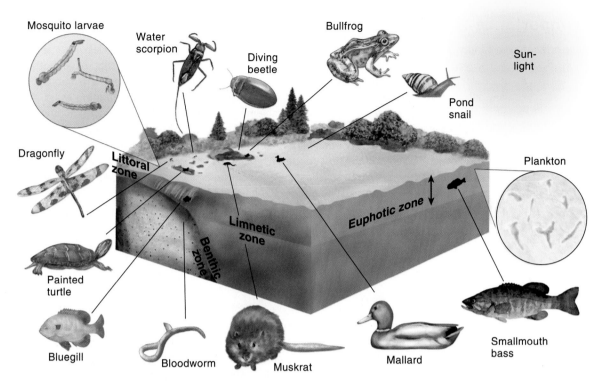

Mosquito larvae

Water scorpion

Diving beetle

Bullfrog

Sun-light

Dragonfly

Littoral zone

Pond snail

Plankton

Limnetic zone

Euphotic zone

Benthic zone

Painted turtle

Bluegill

Bloodworm

Muskrat

Mallard

Smallmouth bass

FIGURE 23.9 Lakes are similar in structure to oceans except that the species are different because most marine organisms cannot live in freshwater. Insects are common organisms in freshwater lakes, as are many kinds of fish, zooplankton, and phytoplankton.

Streams and rivers present a different set of conditions. Since the water is moving, planktonic organisms are less important than are attached organisms. Most algae grow attached to rocks and other objects on the bottom. Since the water is shallow, light can penetrate easily to the bottom (except for large or extremely muddy rivers). Even so, it is difficult for photosynthetic organisms to accumulate the nutrients necessary for growth, and most streams are not very productive. As a matter of fact, the major input of nutrients is from organic matter that falls into the stream from terrestrial sources. These are primarily the leaves from trees and other vegetation, as well as the bodies of living and dead insects.

Within the stream is a community of organisms specifically adapted to use the debris as a source of food. Bacteria and fungi colonize the organic matter, and many kinds of insects shred and eat this organic matter as well as the fungi and bacteria living on it. The feces (intestinal wastes) of these insects and the tiny particles produced during the eating process become food for other insects that build nets to capture the tiny bits of organic matter that drift their way. These herbivorous insects are in turn eaten by carnivorous insects and fish.

Estuaries

An **estuary** is a special category of aquatic ecosystem that consists of shallow, partially enclosed areas where freshwater enters the ocean. The saltiness of the water in the estuary changes with tides and the flow of water from rivers. The organisms that live here are specially adapted to this set of physical conditions, and the number of species is less than in the ocean or in freshwater. Estuaries are especially productive ecosystems because of the large quantity of nutrients introduced into the basin from the rivers that run into them. This productivity is further enhanced by the fact that the shallow water allows light to penetrate to most of the water in the basin. Phytoplankton and attached algae and plants are able to use the sunlight and the nutrients for rapid growth. This photosynthetic activity supports many kinds of organisms in the estuary. Estuaries are especially important as nursery sites for fish and crustaceans such as flounder and shrimp. The adults enter these productive, sheltered areas to reproduce and then return to the ocean. The young spend their early life in the estuary and eventually leave as they get larger and are more able to survive in the ocean.

INDIVIDUAL SPECIES REQUIREMENTS: HABITAT AND NICHE

People approach the study of organism interactions in two major ways. Many people look at interrelationships from the broad ecosystem point of view, while others focus on individual organisms and the specific things that affect them in their daily lives. The first approach involves the study of all the organisms that interact with one another—the community—and usually looks at general relationships among them. For example, organisms are lumped into categories such as producers, consumers, and decomposers because they perform different functions in a community.

A Closer Look

The Importance of Habitat Size

Many people interested in songbird populations have documented a significant decrease in the numbers of certain songbird species. Species particularly affected are those that migrate between North America and South America and require relatively large areas of undisturbed forest in both their northern and southern homes. Many of these species are being hurt by human activities that fragment large patches of forest into many smaller patches, creating more edges between different habitat types. Bird and other animal species that thrive in edge habitats replace the songbirds, which require large patches of undisturbed forest. Cowbirds that normally live in open areas are a particular problem. Cowbirds do not build nests but lay their eggs in the nests of other birds after removing the eggs of the host species. When forests are broken into small patches, cowbirds reach a larger percentage of forest-nesting birds because the forest birds must nest closer to the edge.

The species most severely affected are those that have both their northern and southern habitats disturbed. A study of migrating sharp-shinned hawks indicated that their numbers have also been greatly reduced in recent years. It is thought that since sharp-shinned hawks use small songbirds as their primary source of food, the reduction in hawks is directly related to the reduction in migratory songbirds.

Habitat

A second way of looking at interrelationships is to study in detail the ecological relationships of a certain species of organism. Each organism has particular requirements for life and lives where the environment provides what it needs. The environmental requirements of a whale include large expanses of ocean but with seasonally important feeding areas and protected locations used for giving birth. The kind of place, or part of an ecosystem, occupied by an organism is known as its **habitat.** Habitats are usually described in terms of conspicuous or particularly significant features in the area where the organism lives. For example, the habitat of a prairie dog is usually described as a grassland, while the habitat of a tuna is described as the open ocean. The key thing to keep in mind when you think of habitat is the *place* in which a particular kind of organism lives. When describing the habitats of organisms, we sometimes use the terminology of the major biomes of the world, such as desert, grassland, or savanna. It is also possible to describe the habitat of the bacterium *Escherichia coli* as the gut of humans and other mammals or the habitat of a fungus as a rotting log.

Niche

Each species has particular requirements for life and places specific demands on the habitat in which it lives. The specific functional role of an organism is its **niche.** Its niche is the way it goes about living its life. Just as the word *place* is the key to understanding the concept of habitat, the word *function* is the key to understanding the concept of a niche. Understanding the niche of an organism involves a detailed understanding of the impacts an organism has on its biotic and abiotic surroundings as well as all of the factors that affect the organism. For example, the niche of an earthworm includes abiotic items such as soil particle size, soil texture, and the moisture, pH, and temperature of the soil. The earthworm's niche also includes biotic impacts such as serving as food for birds, moles, and shrews; as bait for anglers; or as a consumer of dead plant organic matter (figure 23.10). In addition, an earthworm serves as a host for a variety of parasites, transports minerals and nutrients from deeper soil layers to the surface, incorporates organic matter into the soil, and creates burrows that allow air and water to penetrate the soil more easily. And this is only a limited sample of all of the aspects of its niche.

Some organisms have rather broad niches; others, with very specialized requirements and limited roles to play, have niches that are quite narrow. The opossum (figure 23.11A) is an animal with a very broad niche. It eats a wide variety of plant and animal foods, can adjust to a wide variety of climates, is used as food by many kinds of carnivores (including humans), and produces large numbers of offspring. By contrast, the koala of Australia (figure 23.11B) has a very narrow niche. It can live only in areas of Australia with specific species of *Eucalyptus* trees, because it eats the leaves of only a few kinds of these trees. Furthermore, it cannot tolerate low temperatures and does not produce large numbers of offspring. As you might guess, the opossum is expanding its range and the koala is endangered in much of its range.

It is often easy to overlook important roles played by some organisms. For example, when Europeans introduced cattle into Australia—a continent where there had previously been no large, hoofed mammals—they did not think about the impact of cow manure or the significance of the niche of a group of beetles called *dung beetles*. These beetles rapidly colonize fresh dung and cause it to be broken down. No such beetles existed in Australia; therefore, in areas where cattle were raised, a significant amount of land became covered with accumulated cow dung. This reduced the area where grass could grow and reduced productivity. The problem was eventually solved by the importation of several species of dung beetles from Africa, where large, hoofed mammals are common. The dung beetles made use of what the cattle did not digest, returning it to a form that plants could more easily recycle into plant biomass.

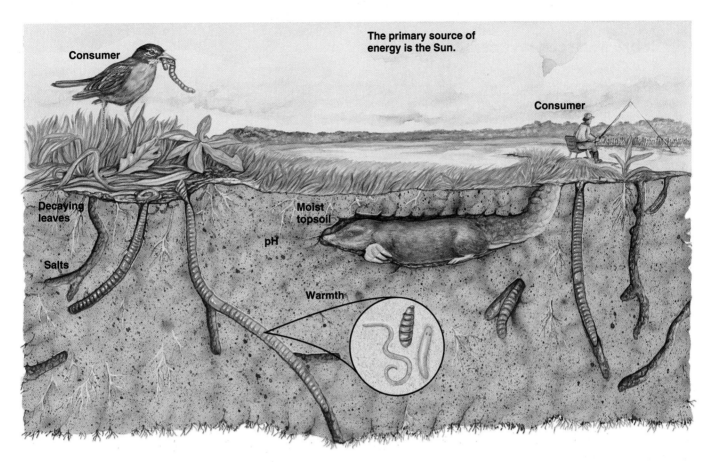

The primary source of energy is the Sun.

Consumer

Consumer

Decaying leaves

Moist topsoil

pH

Salts

Warmth

FIGURE 23.10 The niche of an earthworm involves a great many factors. It includes the fact that the earthworm is a consumer of dead organic matter, a source of food for other animals, host to parasites, and bait for an angler. Furthermore, the earthworm loosens the soil by its burrowing and "plows" the soil when it deposits materials on the surface. In addition, the pH, texture, and moisture content of the soil have an impact on the earthworm. Keep in mind that this is but a small part of what the niche of the earthworm includes.

A Opossum

FIGURE 23.11 (*A*) The opossum has a very broad niche. It eats a variety of foods, is able to live in a variety of habitats, and has a large reproductive capacity. It is generally extending its range in the United States. (*B*) The koala has a narrow niche. It feeds on the leaves of only a few species of *Eucalyptus* trees, is restricted to relatively warm, forested areas of Australia, and is generally endangered in much of its habitat.

B Koala

Zebra Mussels: Invaders from Europe

In the mid-1980s, a clamlike organism called the zebra mussel, *Dressenia polymorpha,* was introduced into the waters of the Great Lakes. It probably arrived in the ballast water of a ship from Europe. Ballast water is pumped into empty ships to make them more stable when crossing the ocean. Immature stages of the zebra mussel were probably emptied into Lake St. Clair, near Detroit, Michigan, when a ship discharged its ballast water to take on cargo. This organism has since spread to many areas of the Great Lakes and smaller inland lakes. It has also been discovered in other parts of the United States, including the mouth of the Mississippi River. Zebra mussels attach to any hard surface and reproduce rapidly. Densities of more than twenty thousand individuals per square meter have been documented in Lake Erie.

These invaders are of concern for several reasons. First, they coat the intake pipes of municipal water plants and other facilities, requiring expensive measures to clean the pipes. Second, they coat any suitable surface, preventing native organisms from using the space. Third, they introduce a new organism into the food chain. Zebra mussels filter small aquatic organisms from the water very efficiently and may remove the food organisms required by native species. Their filtering activity has significantly increased the clarity of the water in several areas in the Great Lakes. This can affect the kinds of fish present, because greater clarity allows predator fish to find prey more easily. In addition, the clarity of the water has allowed for greater production of aquatic vegetation and algae, because light penetrates the water better and to a greater depth. There is concern that zebra mussels have significantly changed the ecological organization of the Great Lakes.

Box Figure 23.1

KINDS OF ORGANISM INTERACTIONS

When organisms encounter one another, they can influence one another in numerous ways. Some interactions are harmful to one or both of the organisms. Others are beneficial. Ecologists have classified kinds of interactions between organisms into several broad categories.

Predation

Predation occurs when one animal captures, kills, and eats another animal. The organism that is killed is called the *prey,* and the one that does the killing is called the *predator.* The predator obviously benefits from the relationship, while the prey organism is harmed. Most predators are relatively large compared to their prey and have specific adaptations that aid them in catching prey. Many spiders build webs that serve as nets to catch flying insects. The prey are quickly paralyzed by the spider's bite and wrapped in a tangle of silk threads. Other rapidly moving spiders, such as wolf spiders and jumping spiders, have large eyes that help them find prey without using webs. Dragonflies patrol areas where they can capture flying insects. Hawks and owls have excellent eyesight that allows them to find their prey. Many predators, such as leopards, lions, and cheetahs, use speed to run down their prey, while others such as frogs, toads, and many kinds of lizards blend in with their surroundings and strike quickly when a prey organism happens by (figure 23.12).

Parasitism

Another kind of interaction in which one organism is harmed and the other aided is the relationship of parasitism. In fact, there are more species of parasites in the world than there are nonparasites, making this a very common kind of relationship. **Parasitism** involves one organism living in or on another living organism from which it derives nourishment. The *parasite* derives the benefit and harms the *host,* the organism it lives in or on (figure 23.13). Many kinds of fungi live on trees and other kinds of plants, including those that are commercially valuable. Dutch elm disease is caused by a fungus that infects the living, sap-carrying parts of the tree. Mistletoe is a common plant that is a parasite on other plants. The mistletoe plant invades the tissues of the tree it is living on and derives nourishment from the tree.

Many kinds of worms, protozoa, bacteria, and viruses are important parasites. Parasites that live on the outside of their hosts are called *external parasites.* For example, fleas live on the outside of the bodies of mammals such as rats, dogs, cats, and humans, where they suck blood and do harm to their host. At the same time, the host could also have a tapeworm in its intestine. Since the tapeworm lives inside the host, it is called an *internal parasite.* Bacterial and viral parasites are common causes of disease in humans and other organisms.

Commensalism

Predation and parasitism both involve one organism benefiting while the other is harmed. Another common relationship is one

A

B

FIGURE 23.12 (A) Many predators capture prey by making use of speed. The cheetah can reach estimated speeds of 100 km/h (about 70 mi/h) during sprints to capture its prey. (B) Other predators, such as this veiled chameleon, blend in with their surroundings, lie in wait, and ambush their prey. Because strength is needed to kill the prey, the predator is generally larger than the prey. Obviously, predators benefit from the food they obtain to the detriment of the prey organism.

A Tapeworm

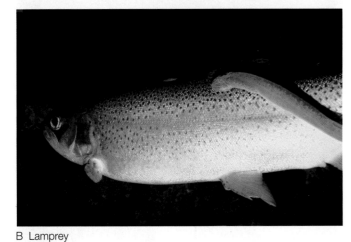
B Lamprey

C Mistletoe

FIGURE 23.13 Parasites benefit from the relationship because they obtain nourishment from the host. Tapeworms (A) are internal parasites in the guts of their host, where they absorb food from the host's gut. The lamprey (B) is an external parasite that sucks body fluids from its host. Mistletoe (C) is a photosynthesizing plant that also absorbs nutrients from the tissues of its host tree. The host in any of these three situations may not be killed directly by the relationship, but it is often weakened, thus becoming more vulnerable to predators or diseases. There are more species of parasites in the world than species of organisms that are not parasites.

A Remora

B Epiphytes

FIGURE 23.14 In the relationship called commensalism, one organism benefits and the other is not affected. (*A*) The remora fish shown here hitchhike a ride on the shark. They eat scraps of food left over by the messy eating habits of the shark. The shark does not seem to be hindered in any way. (*B*) The epiphytic plants growing on this tree do not harm the tree but are aided by using the tree surface as a place to grow.

in which one organism benefits while the other is not affected. This is known as **commensalism.** For example, sharks often have another fish, the remora, attached to them. The remora has a sucker on the top side of its head that allows it to attach to the shark and get a free ride (figure 23.14A). While the remora benefits from the free ride and by eating leftovers from the shark's meals, the shark does not appear to be troubled by this uninvited guest, nor does it benefit from the presence of the remora.

Another example of commensalism is the relationship between trees and epiphytic plants. Epiphytes are plants that live on the surface of other plants but do not derive nourishment from them (figure 23.14B). Many kinds of plants (e.g., orchids, ferns, and mosses) use the surface of trees as a place to live. These kinds of organisms are particularly common in tropical rainforests. Many epiphytes derive benefit from the relationship because they are able to be located in the top of the tree, where they receive more sunlight and moisture. The trees derive no benefit from the relationship, nor are they harmed; they simply serve as a support surface for epiphytes.

Mutualism

So far in our examples, only one species has benefited from the association of two species. There are also many situations in which two species live in close association with one another, and both benefit. This is called **mutualism.** One interesting example of mutualism involves digestion in animals such as rabbits, cows, and termites. These animals eat plant material that is high in cellulose, even though they do not produce the enzymes capable of breaking down cellulose molecules into simple sugars. They manage to get energy out of these cellulose molecules with the help of special microorganisms living in their digestive tracts. The microorganisms produce cellulose-digesting enzymes, called *cellulases*, which break down cellulose into smaller carbohydrate molecules that the host's digestive enzymes can break down into smaller glucose molecules. The microorganisms benefit because

the host's gut provides them with a moist, warm, nourishing environment in which to live. The hosts benefit because the microorganisms provide them with a source of food.

Another kind of mutualistic relationship exists between flowering plants and bees. Undoubtedly, you have observed bees and other insects visiting flowers to obtain nectar from the blossoms. Usually the flowers are constructed in such a manner that the bees pick up pollen (sperm-containing packages) on their hairy bodies and transfer it to the female part of the next flower they visit (figure 23.15). Because bees normally visit

FIGURE 23.15 Mutualism is an interaction between two organisms in which both benefit. The plant benefits because cross-fertilization (exchange of gametes from a different plant) is more probable; the bee benefits by acquiring nectar for food.

many individual flowers of the same species for several minutes and ignore other species of flowers, they can serve as pollen carriers between two flowers of the same species. Plants pollinated in this manner produce less pollen than do plants that rely on the wind to transfer pollen. This saves the plant energy because it doesn't need to produce huge quantities of pollen. It does, however, need to transfer some of its energy savings into the production of showy flowers and nectar to attract the bees. The bees benefit from both the nectar and pollen; they use both for food.

Symbiosis is a close physical relationship between two kinds of organisms. This term is sometimes used to mean the same thing as mutualism. At other times, it is used to include mutualism, parasitism, and commensalism as different kinds of symbiotic relationships. So you will need to determine the meaning from the context in which the word is used.

Competition

So far in our discussion of organism interactions we have left out the most common one. It is reasonable to envision every organism on the face of Earth being involved in *competitive* interactions. **Competition** is a kind of interaction between organisms in which both organisms are harmed to some extent. Competition occurs whenever two organisms both need a vital resource that is in short supply (figure 23.16). The vital resource could be food, shelter, nesting sites, water, mates, or space. It may involve a snarling tug-of-war between two dogs over a scrap of food, or it can be a silent struggle between plants for access to available light. If you have ever started tomato seeds (or other garden plants) in a garden and failed to eliminate the weeds, you have witnessed competition. If the weeds are not removed, they compete with the garden plants for available

FIGURE 23.16 Whenever a needed resource is in limited supply, organisms compete for it. This competition may be between members of the same species (*intraspecific*), illustrated in the photograph, or may involve different species (*interspecific*).

sunlight, water, and nutrients, resulting in poor growth of both the garden plants and the weeds.

Competition and Natural Selection

It is important to recognize that although competition results in harm to both organisms, there can still be winners and losers. The two organisms may not be harmed to the same extent, which results in one having greater access to the limited resource. Furthermore, even the loser can continue to survive if it migrates to an area where competition is less intense. Over many generations, when individuals of two different species compete, it is possible that the species may evolve through natural selection to exploit different niches. Thus, competition provides a major mechanism for natural selection. With the development of slight differences between niches, the intensity of competition is reduced. For example, many birds catch flying insects as food. However, they do not compete directly with each other because some feed at night, some feed high in the air, some feed only near the ground, and still others perch on branches and wait for insects to fly past. The insect-eating niche can be further subdivided by specializing on particular sizes or kinds of insects.

Many of the relationships just described involve the transfer of nutrients from one organism to another (predation, parasitism, mutualism). Another important way scientists look at ecosystems is to look at how materials are cycled from organism to organism.

THE CYCLING OF MATERIALS IN ECOSYSTEMS

Although some new atoms are being added to Earth from cosmic dust and meteorites, this amount is not significant in relation to the entire mass of Earth. Therefore, Earth can be considered to be a closed ecosystem as far as matter is concerned. Only sunlight energy comes to Earth in a continuous stream, and even this is ultimately returned to space as radiant energy. However, it is this flow of energy that drives all biological processes. Living systems have evolved ways of using this energy to continue life through growth and reproduction and the continual reuse of existing atoms. In this recycling process, inorganic molecules are combined to form the organic compounds of living things. If there were no way of recycling this organic matter back into its inorganic forms, organic material would build up as the bodies of dead organisms. This is thought to have occurred millions of years ago when the present deposits of coal, oil, and natural gas were formed. However, under most conditions, decomposers are available to break down organic material to inorganic material that can then be reused by other organisms to rebuild organic material.

Carbon, hydrogen, phosphorus, potassium, nitrogen, sulfur, oxygen, calcium, and many other kinds of atoms are involved in the structure of living things. These atoms are constantly recycled. One way to get an appreciation of how various kinds of organisms interact to cycle materials is to look at a specific kind of atom and follow its progress through an ecosystem.

THE CARBON CYCLE

The **carbon cycle** is the series of activities of organisms that result in the cycling of carbon atoms between the atmosphere and living things. All living things contain organic molecules that are composed of long chains of carbon atoms. Carbon and oxygen atoms combine to form the molecule carbon dioxide (CO_2), which is a gas found in small quantities (about 0.04 percent) in the atmosphere. Producers combine carbon dioxide (CO_2) from the atmosphere and water (H_2O) to form complex organic molecules such as sugar ($C_6H_{12}O_6$) during photosynthesis. At the same time, oxygen molecules (O_2) are released into the atmosphere.

The organic matter in the bodies of plants may be used by herbivores as food. When an herbivore eats a plant, it breaks down the complex organic molecules into simpler organic molecules, such as simple sugars, amino acids, glycerol, and fatty acids. These can be used as building blocks in the construction of its own body. Thus, the atoms in the body of the herbivore can be traced back to the plants that were eaten. Similarly, when carnivores eat herbivores, these same atoms are transferred to them. Finally, the waste products of plants and animals and the remains of dead organisms are used by decomposer organisms as sources of carbon and other atoms they need for survival.

In addition, all the organisms in this cycle—plants, herbivores, carnivores, and decomposers—obtain energy (ATP) from the process of respiration, in which oxygen (O_2) is used to break down organic compounds into carbon dioxide (CO_2) and water (H_2O). Thus, the carbon atoms that started out as components of carbon dioxide (CO_2) molecules in the atmosphere have passed through the bodies of living organisms as parts of their organic molecules and returned to the atmosphere as carbon dioxide, ready to be cycled again. Similarly, the oxygen atoms (O) released as oxygen molecules (O_2) during photosynthesis have been used during the process of respiration (figure 23.17).

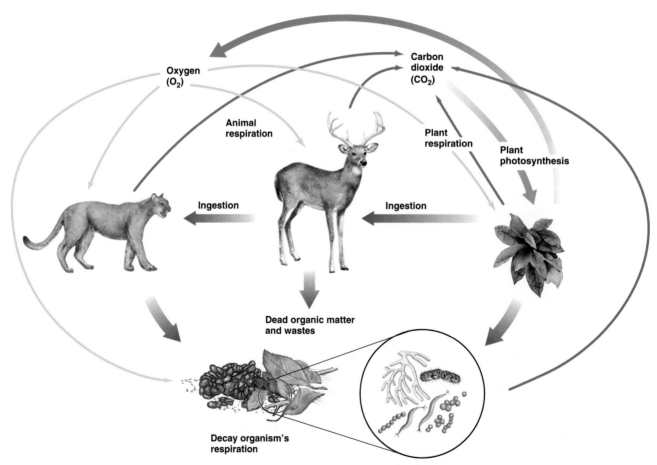

FIGURE 23.17 Carbon atoms are cycled through ecosystems. Carbon dioxide (green arrows) produced by respiration is the source of carbon that plants incorporate into organic molecules when they carry on photosynthesis. These carbon-containing organic molecules (black arrows) are passed to animals when they eat plants and other animals. Organic molecules in waste or dead organisms are consumed by decomposer organisms in the soil when they break down organic molecules into carbon dioxide and water. All organisms (plants, animals, and decomposers) return carbon atoms to the atmosphere as carbon dioxide when they carry on cellular respiration. Oxygen (blue arrows) is being cycled at the same time that carbon is. The oxygen is released into the atmosphere and into the water during photosynthesis and taken up during cellular respiration.

Carbon Dioxide and Global Warming

Humans have significantly altered the carbon cycle. As we burn fossil fuels, the amount of carbon dioxide in the atmosphere continually increases. It has risen nearly 21 percent, from about 317 parts per million (0.0317 percent) in 1960 to about 383 parts per million (0.0383 percent) in 2007. Although present in tiny amounts, it has a significant effect. Carbon dioxide, methane, and chlorofluorocarbons are often referred to as greenhouse gases because they have the effect of warming the Earth. When solar radiation hits the Earth, the Earth is warmed and infrared radiation is radiated away from the Earth. Molecules of carbon dioxide and other greenhouse gases absorb some of the infrared radiation. These molecules, in turn, radiate energy in all directions, some of it back to the Earth. Although the mechanisms are different, many people have compared the warming of the Earth by these gases to the warming that occurs in a greenhouse and the warming of the Earth is often referred to as the greenhouse effect. (See box figure 23.2.)

Many are concerned that increased carbon dioxide levels are causing a warming of the planet that will cause major changes in weather and climate, leading to the flooding of coastal cities and major changes in agricultural production. The Intergovernmental Panel on Climate Change (IPCC) established by the United Nations concluded that there has been an increase in average temperature of Earth. The panel further concluded that the human activities of burning fossil fuels and destroying forests are the cause. There is no doubt that the amount of carbon dioxide in the atmosphere has been increasing.

Despite the fact that carbon dioxide levels have increased and the conclusions reached by the IPCC, controversy still exists about this topic, and some people doubt that warming is occurring. At a meeting in Kyoto, Japan, in 1998, many countries agreed to reduce the amount of carbon dioxide and other greenhouse gases they release into the atmosphere. This formal international agreement is known as the Kyoto Protocol. In 2005, Russia ratified the treaty. Russia's action provided the minimum number of countries needed to put the treaty into effect. However, emissions of carbon dioxide are directly related to economic activity and the energy usage that fuels it. Therefore, it remains to be seen if countries will meet their goals or will succumb to economic pressures to allow continued use of large amounts of fossil fuels. Significantly, the United States, the largest producer of carbon dioxide, has withdrawn from the treaty.

Some countries, however, have sought to control the change in the amount of carbon dioxide in the atmosphere by planting millions of trees or by preventing the destruction of forests. The thought is that the trees carry on photosynthesis, grow, and store carbon in their bodies, leading to reduced carbon dioxide levels. At the same time, people in other parts of the world continue to destroy forests at a rapid rate. When the two activities are compared, tree planting does not offset deforestation. In addition, the trees that have been planted will ultimately die and decompose, releasing carbon dioxide back into the atmosphere, so it is not clear that this is an effective means of reducing atmospheric carbon dioxide over the long term.

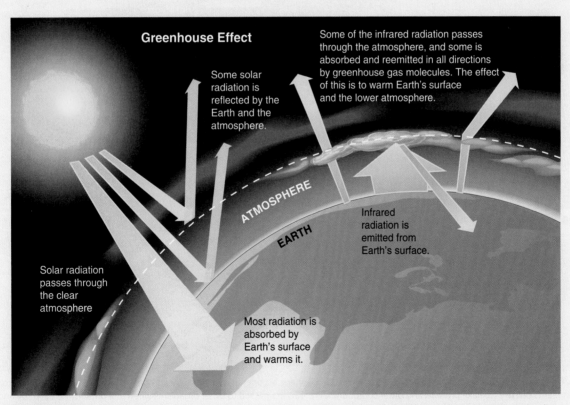

Greenhouse Effect

Some solar radiation is reflected by the Earth and the atmosphere.

Some of the infrared radiation passes through the atmosphere, and some is absorbed and reemitted in all directions by greenhouse gas molecules. The effect of this is to warm Earth's surface and the lower atmosphere.

ATMOSPHERE

EARTH

Infrared radiation is emitted from Earth's surface.

Solar radiation passes through the clear atmosphere

Most radiation is absorbed by Earth's surface and warms it.

Box Figure 23.2 The greenhouse effect naturally warms Earth's surface. Without it, Earth would be 33°C (60°F) cooler than it is today— uninhabitable for life as we know it.

Source: Data from *Climate Change—State of Knowledge.* October 1997, Office of Science and Technology Policy, Washington, D.C.

The Nitrogen Cycle

Another important element for living things is nitrogen (N). Nitrogen is essential in the formation of amino acids, which are needed to form proteins, and in the formation of nitrogenous bases, which are a part of ATP and the nucleic acids DNA and RNA. Nitrogen (N) is found as molecules of nitrogen gas (N_2) in the atmosphere. Although nitrogen gas (N_2) makes up approximately 80 percent of Earth's atmosphere, only a few kinds of bacteria are able to convert it into nitrogen compounds that other organisms can use. Therefore, in most terrestrial ecosystems, the amount of nitrogen available limits the amount of plant biomass that can be produced. (Most aquatic ecosystems are limited by the amount of phosphorus rather than the amount of nitrogen.) Plants utilize several different nitrogen-containing compounds to obtain the nitrogen atoms they need to make amino acids and other compounds (figure 23.18).

Symbiotic nitrogen-fixing bacteria live in the roots of certain kinds of plants, where they convert nitrogen gas molecules into compounds that the plants can use to make amino acids and nucleic acids. The most common plants that enter into this mutualistic relationship with bacteria are the legumes, such as beans, clover, peas, alfalfa, and locust trees. Some other organisms, such as alder trees and even a kind of aquatic fern, can also participate in this relationship. There are also **free-living nitrogen-fixing bacteria** in the soil that provide nitrogen compounds that can be taken up through the roots, but the bacteria do not live in a close physical union with plants.

Another way plants get usable nitrogen compounds involves a series of different bacteria. *Decomposer bacteria* convert organic nitrogen-containing compounds into ammonia

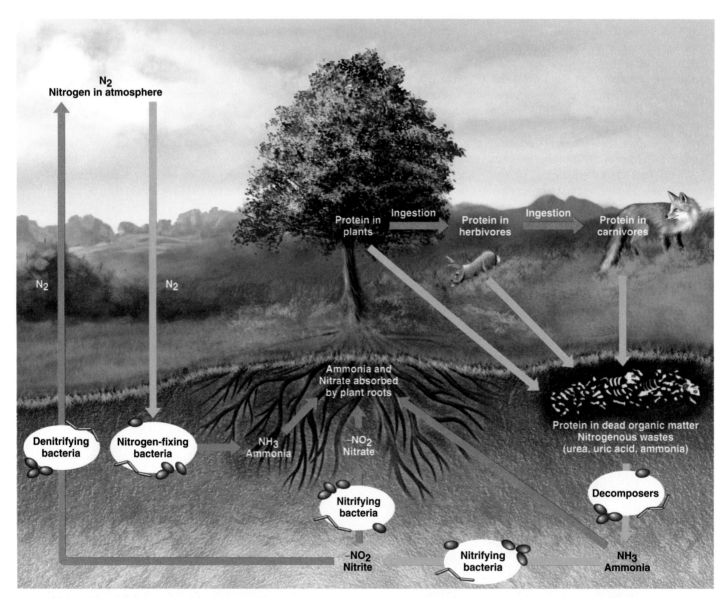

FIGURE 23.18 Nitrogen atoms are cycled through ecosystems. Atmospheric nitrogen is converted by nitrogen-fixing bacteria to nitrogen-containing compounds that plants can use to make proteins and other compounds. Proteins are passed to other organisms when one organism is eaten by another. Dead organisms and their waste products are acted upon by decomposer organisms to form ammonia, which may be reused by plants and converted to other nitrogen compounds by nitrifying bacteria. Denitrifying bacteria return nitrogen as a gas to the atmosphere.

Chemical Fertilizers and the Nitrogen Cycle

Modern agriculture depends on the use of fertilizers to replace the elements removed from the soil by the plants that are harvested. The three most important elements in fertilizers are nitrogen, phosphorus, and potassium. The numbers on bags of fertilizer tell you how much of each of these three components is present. For example, a 6-12-12 fertilizer has 6 percent nitrogen, 12 percent phosphorus, and 12 percent potassium compounds. Although molecular nitrogen makes up 80 percent of the atmosphere, it is not available to plants. From a chemical point of view, it is very difficult to get molecular nitrogen, which consists of two nitrogen atoms bonded to each other, to react with other atoms. Although some bacteria are able to attach hydrogen to nitrogen to form ammonia, a chemical method for producing ammonia was not determined until 1909, when Fritz Haber determined the conditions under which nitrogen and hydrogen would combine to form ammonia. The process of attaching hydrogen to ammonia requires a great deal of energy. This was an important discovery because it led to a practical way to produce a variety of nitrogen-containing compounds that can be used for fertilizer. Farmers apply ammonia directly into the soil to provide nitrogen for many kinds of plants. Ammonia can be reacted with other materials to produce a variety of compounds that are used as nitrogen fertilizers, including urea, ammonium nitrate, and ammonium sulfate.

(NH_3). *Nitrifying bacteria* can convert ammonia (NH_3) into nitrite-containing (NO_2^-) compounds, which in turn can be converted into nitrate-containing (NO_3^-) compounds. Many kinds of plants can use either ammonia (NH_3) or nitrate (NO_3^-) from the soil as building blocks for amino acids and nucleic acids.

All animals obtain their nitrogen from the food they eat. The ingested proteins are broken down into their component amino acids during digestion. These amino acids can then be reassembled into new proteins characteristic of the animal. All dead organic matter and waste products of plants and animals are acted upon by decomposer organisms, and the nitrogen is released as ammonia (NH_3), which can be taken up by plants or acted upon by nitrifying bacteria to make nitrate (NO_3^-).

Finally, other kinds of bacteria called *denitrifying bacteria* are capable of converting nitrite (NO_2^-) to nitrogen gas (N_2), which is released into the atmosphere. Thus, in the **nitrogen cycle,** nitrogen from the atmosphere is passed through a series of organisms, many of which are bacteria, and ultimately returns to the atmosphere to be cycled again. However, the nitrogen cycle differs from the carbon cycle in a fundamental way. Inorganic nitrogen compounds are not necessarily returned to the atmosphere but are returned to plants by way of decomposer, nitrate, and nitrite bacteria.

The Phosphorus Cycle

Phosphorus is another kind of atom common in the structure of living things. It is present in many important biological molecules, such as DNA, and in the membrane structure of cells. In addition, the bones and teeth of animals contain significant quantities of phosphorus. The ultimate source of phosphorus atoms is rock. In nature, new phosphorus compounds are released by the erosion of rock and dissolved in water. Plants use the dissolved phosphorus compounds to construct the molecules they need. Animals obtain the phosphorus they need when they consume plants or other animals.

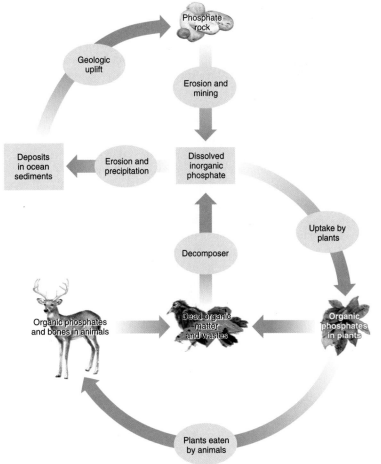

FIGURE 23.19 The source of phosphorus is rock that, when dissolved, provides phosphorus used by plants and animals.

When an organism dies or excretes waste products, decomposer organisms recycle the phosphorus compounds back into the soil. Phosphorus compounds that are dissolved in water are ultimately precipitated as deposits. Geologic processes elevate these deposits and expose them to erosion, thus making these deposits available to organisms (figure 23.19). Waste products of animals often have significant amounts of phosphorus. In places where large numbers of seabirds or bats congregate for hundreds of years, the thickness of their droppings (called *guano*) can be a significant source of phosphorus for fertilizer. In many soils, phosphorus is in short supply and must be provided to crop plants to get maximum yields. Phosphorus is also in short supply in aquatic ecosystems, and the addition of phosphorus from detergents in sewage and runoff from lawns and farmland can cause undesirable growth of algae and plants in lakes, rivers, estuaries, and ocean waters near land.

POPULATION CHARACTERISTICS

A **population** is a group of organisms of the same species located in the same place at the same time. Examples are the number of dandelions in your front yard, the rat population in the sewers of your city, or the number of people in New York City. On a larger scale, all the people of the world constitute the human population. The terms *species* and *population* are interrelated because a species is a population—the largest possible population of a particular kind of organism. The term *population,* however, is often used to refer to portions of a species by specifying a space and time. For example, the size of the human population in a city changes from hour to hour during the day and varies according to where you set the boundaries of the city.

Since each local population is a small portion of a species, we should expect distinct populations of the same species to show differences. Some of these differences will be genetic. *Gene frequency* is a measure of how often a specific gene for a characteristic shows up in the individuals of a population. Two populations of the same species often have quite different gene frequencies. For example, many populations of mosquitoes have high frequencies of insecticide-resistant genes, whereas others do not. The frequency of the genes for tallness in humans is greater in certain African tribes than in any other

human population. Figure 23.20 shows that the frequency of the genetic information for type B blood differs significantly from one human population to another.

Another feature of a population is its *age distribution,* which is the number of organisms of each age in the population. In addition, organisms are often grouped into the following categories:

- prereproductive juveniles—insect larvae, plant seedlings, or babies;
- reproductive adults—mature insects, plants producing seeds, or humans in early adulthood; or
- postreproductive adults no longer capable of reproduction—annual plants that have shed their seeds, salmon that have spawned, and many elderly humans.

A population is not necessarily divided into equal thirds (figure 23.21). In some situations, a population may be made up of a majority of one age group. If the majority of the population is prereproductive, then a "baby boom" should be anticipated in the future. If a majority of the population is reproductive, the population should be growing rapidly. If the majority of the population is postreproductive, a population decline should be anticipated. Many organisms that only live a short time and have high reproductive rates can have age distributions that change significantly in a matter of weeks or months. For example, many birds have a flurry of reproductive activity during the summer months. Therefore, if you sample the population of a specific species of bird at different times during the summer, you would have widely different proportions of reproductive and prereproductive individuals.

Populations can also differ in their sex ratios. The *sex ratio* is the number of males in a population compared to the number of females. In bird and mammal species where strong pair-bonding occurs, the sex ratio may be nearly one to one (1:1). Among mammals and birds that do not have strong pair-bonding, sex ratios may show a larger number of females than

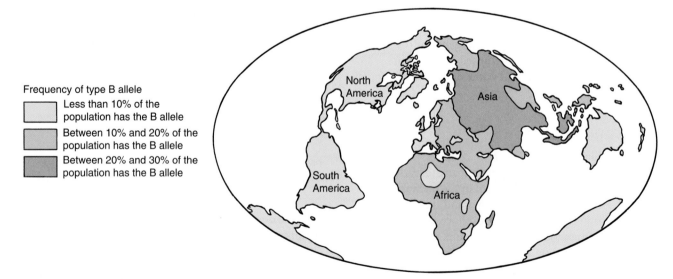

Frequency of type B allele

Less than 10% of the population has the B allele

Between 10% and 20% of the population has the B allele

Between 20% and 30% of the population has the B allele

FIGURE 23.20 The allele for type B blood is not evenly distributed in the world. This map shows that the type B allele is most common in parts of Asia and has been dispersed to the Middle East and parts of Europe and Africa. There has been very little flow of the allele to the Americas.

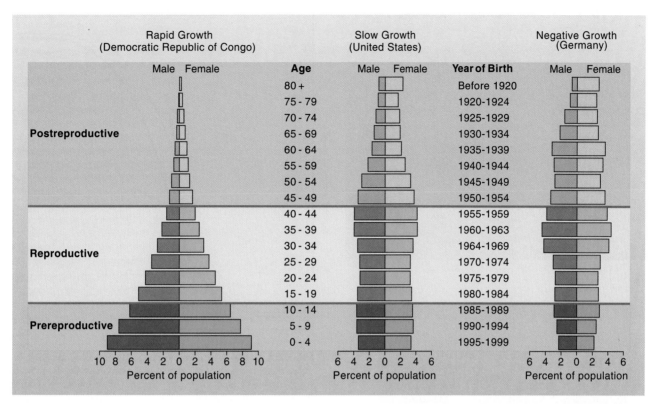

FIGURE 23.21 The relative numbers of individuals in each of the three categories (prereproductive, reproductive, and postreproductive) are good clues to the future growth of a population. The Democratic Republic of Congo has a large number of young individuals who will become reproducing adults. Therefore, this population is likely to grow rapidly. The United States has a large proportion of reproductive individuals and a moderate number of prereproductive individuals. Therefore, this population is likely to grow slowly. Germany has a declining number of reproductive individuals and a very small number of prereprodutive individuals. Therefore, its population has begun to decline.

Source: Data from Population Reference Bureau, Inc.

males. This is particularly true among species that are hunted, because more males than females are shot, resulting in a higher proportion of females surviving. Since one male can fertilize several females, the population can remain large even though the females outnumber the males. Many species of animals such as bison, horses, and elk have mating systems in which one male maintains a harem of females. The sex ratio in these small groups is quite different from a 1:1 ratio (figure 23.22). Some male elk can maintain a harem of up to twenty-five females. There are very few situations in which the number of males exceeds the number of females.

Regardless of the sex ratio of a population, most species can generate large numbers of offspring, producing a concentration of organisms in an area. *Population density* is the number of organisms of a species per unit area. Some populations are extremely concentrated into a limited space, while others are well dispersed. As the population density increases, competition among members of the population for the necessities of life increases. This increases the likelihood that some individuals will explore new habitats and migrate to new areas. Increases in the intensity of competition that cause changes in the environment and lead to dispersal are often referred to as *population pressure*. The dispersal of individuals to new areas can relieve the pressure on the home area and lead to the establishment of new populations. Among animals, it is often the juveniles who participate in this dispersal process.

FIGURE 23.22 Some male animals defend a harem of females; therefore, the sex ratio in these groups is several females per male.

In plant populations, dispersal is not very useful for relieving population density; instead, the death of weaker individuals usually results in reduced population density. In the lodgepole pine, seedlings become established in areas following fire and dense thickets of young trees are established. As the stand ages, many small trees die. The remaining trees grow larger as the population density drops (figure 23.23).

A

B

FIGURE 23.23 This population of lodgepole pine seedlings consists of a large number of individuals very close to one another (*A*). As the trees grow, many of the weaker trees will die, the distance between individuals will increase, and the population density will be reduced (*B*).

THE POPULATION GROWTH CURVE

Because most species of organisms have a high reproductive capacity, there is a tendency for populations to grow if environmental conditions permit. For example, if the usual litter size for a pair of mice is four, the four would produce eight, which in turn would produce sixteen, and so forth. Figure 23.24 shows a graph of change in population size over time known as a **population growth curve.** This kind of curve is typical for situations where a species is introduced into a previously unutilized area.

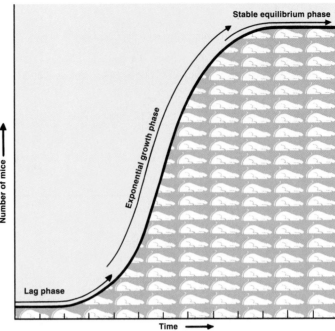

FIGURE 23.24 In this mouse population, the period of time in which there is little growth is known as the lag phase. This is followed by a rapid increase in population as the offspring of the originating population begin to reproduce themselves; this is known as the exponential growth phase. Eventually, the population reaches a stable equilibrium phase, during which the birthrate equals the death rate.

The change in the size of a population depends on the rate at which new organisms enter the population compared to the rate at which they leave. The number of new individuals added to the population by reproduction per thousand individuals is called **natality.** The number of individuals leaving the population by death per thousand individuals is called **mortality.** When a small number of organisms (two mice) first invades an area, there is a period of time before reproduction takes place when the population remains small and relatively constant. This part of the population growth curve is known as the **lag phase.** During the lag phase, both natality and mortality are low. The lag phase occurs because reproduction is not an instantaneous event. Even after animals enter an area, they must mate and produce young. This may take days or years, depending on the animal.

In organisms that take a long time to mature and produce young, such as elephants, deer, and many kinds of plants, the lag phase may be measured in years. With the mice in our example, it will be measured in weeks. The first litter of young will be able to reproduce in a matter of weeks. Furthermore, the original parents will probably produce an additional litter or two during this time period. Now we have several pairs of mice reproducing more than just once. With several pairs of mice reproducing, natality increases while mortality remains low; therefore, the population begins to grow at an ever-increasing (accelerating) rate. This portion of the population growth curve is known as the **exponential growth phase.**

The number of mice (or any other organism) cannot continue to increase at a faster and faster rate because, eventually, something in the environment will become limiting and cause an increase in the number of deaths. For animals, food, water, or nesting sites may be in short supply, or predators or disease may kill many individuals. Plants may lack water, soil nutrients, or sunlight. Eventually, the number of individuals entering the population by reproduction or immigration will come to equal the number of individuals leaving it by death or migration, and the population size becomes stable. Often there is both a decrease in natality and an increase in mortality at this point. This portion of the population growth curve is known as the **stable equilibrium phase.** It is important to recognize that this is still a population with births, deaths, migration, and a changing mix of individuals; however, the size of the population is stable.

POPULATION-SIZE LIMITATIONS

Populations cannot continue to increase indefinitely; eventually, some factor or set of factors acts to limit the size of a population, leading to the development of a stable equilibrium phase or even to a reduction in population size. The identifiable factors that prevent unlimited population growth are known as **limiting factors.** All of the different limiting factors that act on a population are collectively known as **environmental resistance,** and the maximum population that an area can support is known as the **carrying capacity** of the area. In general, organisms that are small and have short life spans tend to have fluctuating populations and do not reach a carrying capacity, whereas large organisms that live a long time tend to reach an optimum population size that can be sustained over an extended period (figure 23.25). A forest ecosystem contains populations of many insect species that fluctuate widely and rarely reach a carrying capacity, but the number of specific tree species or large animals such as owls or deer is relatively constant.

However, the carrying capacity is not an inflexible number. Often environmental changes such as climate variations, disease epidemics, forest fires, or floods can change the carrying capacity of an area for specific species. In aquatic ecosystems, one of the major factors that determines the carrying capacity is the amount of nutrients in the water. In areas where nutrients are abundant, the numbers of various kinds of organisms are high. Often nutrient levels fluctuate with changes in current or runoff from the land, and plant and animal populations fluctuate as well. In addition, a change that negatively affects the carrying capacity for one species may increase the carrying capacity for another. For example, the cutting down of a mature forest followed by the growth of young trees increases the carrying capacity for deer and rabbits, which use the new growth for food, but decreases the carrying capacity for squirrels, which need mature, fruit-producing trees as a source of food and old, hollow trees for shelter.

LIMITING FACTORS TO HUMAN POPULATION GROWTH

Today, we hear differing opinions about the state of the world's human population. On one hand, we hear that the population is growing rapidly. In contrast, we hear that some European countries have shrinking populations. Other countries are concerned about the aging of their populations because birthrates and death rates are low. In magazines and on television, we see that

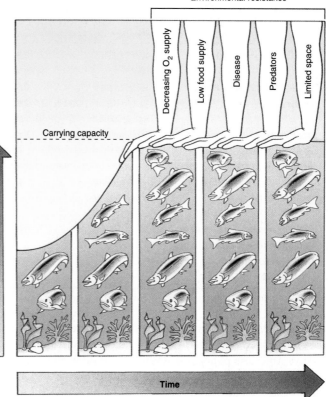

FIGURE 23.25   A number of factors in the environment, such as food, oxygen supply, diseases, predators, and space, determine the number of organisms that can survive in a given area—the carrying capacity of that area. The environmental factors that limit populations are collectively known as environmental resistance.

there are starving people in the world. At the same time, we hear discussions about the problem of food surpluses and obesity in many countries. Some have even said that the most important problem in the world today is the rate at which the human population is growing; others maintain that the growing population will provide markets for goods and be an economic boon. How do we reconcile this mass of conflicting information?

The world population is about 6.6 billion people and is expected to grow to nearly 8 billion by 2025. This is a 20 percent increase in about eighteen years. The distribution of this growth, however, will not be even. The human population is divided into two rather distinct segments: the more economically developed countries of the world (Europe, North America, Australia, New Zealand, and Japan) and the less developed nations. Sharp contrasts exist between the populations of these two groups of nations. Table 23.2 highlights some of these differences.

Regardless of whether we are considering the developed world, the less developed world, or the total world population, it is important to realize that human populations follow the same patterns of growth and are acted upon by the same kinds of limiting factors as are populations of other organisms. When we look at the curve of population growth over the past several thousand years, estimates are that the human population remained low and constant for thousands of years but has increased explosively (from 1 billion to 6.6 billion) in

TABLE 23.2   Comparison of Population Characteristics of More and Less Developed Nations (2006)

Population Characteristic	More Developed Nations	Less Developed Nations	Significance
Total population (2006)	1.216 billion	5.339 billion	Less developed nations constitute 81 percent of world's population.
Expected 2025 population	1.255 billion	6.685 billion	Less developed nations are growing much faster.
Birthrate	11 births per 1,000 in population	23 births per 1,000 in population	Less developed nations have high birthrates.
Death rate	10 deaths per 1,000 in population	8 deaths per 1,000 in population	More developed nations have higher death rates because their populations are older.
Total fertility rate	1.6 births per woman per lifetime	2.9 births per woman per lifetime	Women in less developed nations have nearly twice as many children as those in more developed nations.
Infant mortality rate	6 deaths per 1,000 births	57 deaths per 1,000 births	Less developed nations have much higher infant mortality rates.
Percent under 15 years of age	17	32	Less developed nations have young populations.
Percent over 65 years of age	15	5	More developed nations have aging populations.
Gross national income per person	US$ 27,790	US$ 4,950	Incomes in less developed nations are about 18 percent of those in developed nations.

Source: Data from the Population Reference Bureau, *2006 World Population Data Sheet.*

the past two hundred years (figure 23.26). For example, it has been estimated that when Columbus discovered America, the Native American population was about 1 million and was at or near the carrying capacity. Today, the population of the United States and Canada is about 332 million people. Does this mean that humans are different from other animal species? Can the human population continue to grow forever?

The human species is no different from other animals. It has an upper limit set by the carrying capacity of the environment, but the human population has been able to increase astronomically because technological changes and displacement of other species have allowed us to shift the carrying capacity upward. Much of the exponential growth phase of the human population can be attributed to improvements in sanitation, control of diseases, improvement in agricultural methods, and replacement of natural ecosystems with artificial agricultural ecosystems. But even these conditions have their limits. There will be some limiting factors that eventually cause a leveling off of our population growth curve. We cannot increase beyond our ability to get raw materials and energy, nor can we ignore the waste products we produce or the other organisms with which we interact. Probably the ultimate limiting factor will be the ability to produce food. Improvements in agricultural production have not kept pace with population growth. Indeed, in many parts of the world, food is in short supply despite surpluses in much of the developed world.

Humans are different from most other organisms in a fundamental way: We are able to predict the outcome of a specific course of action. Current technology and medical knowledge are available to control human population and improve the health and well-being of the people of the world. Why then does the human population continue to grow, resulting in human suffering and stressing the environment in which we live? Because we are social animals that have freedom of choice, we frequently do not do what is considered "best" from an unemotional, unselfish, biological point of view. People make decisions based on

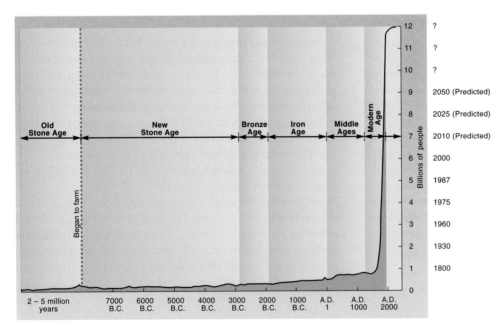

FIGURE 23.26   The number of humans doubled from A.D. 1800 to 1930 (from 1 billion to 2 billion), doubled again by 1975 (4 billion), and is projected to double again by the year 2025. How long can the human population continue to double before Earth's ultimate carrying capacity is reached?

Source: Data from Jean Van Der Tak, et al., "Our Population Predicament: A New Look," in *Population Bulletin,* vol. 34, no. 5, December 1979, Population Reference Bureau, Washington, D.C., and more recent data from the Population Reference Bureau.

Government Policy and Population Control

The actions of government can have a significant impact on the population growth patterns of nations. Some countries have policies that encourage couples to have children. The U.S. tax code indirectly encourages births by providing tax advantages to the parents of children. Some countries in Europe are concerned about the lack of working-age people in the future and are considering ways to encourage births.

China and India are the two most populous countries in the world. Both have over 1 billion people. However, China has taken steps to control its population and now has a total fertility rate of 1.6 children per woman, while India has a total fertility rate of 2.9. The total fertility rate is the average number of children born to a woman during her lifetime. The difference in the total fertility rate between these two countries is the result of different policy decisions over the last fifty years. The history of China's population policy is an interesting study of how government policy affects reproductive activity among its citizens. When the People's Republic of China was established in 1949, the official policy of the government was to encourage births because more Chinese would be able to produce more goods and services, and production was the key to economic prosperity. The population grew from 540 million to 614 million between 1949 and 1955 while economic progress was slow. Consequently, the government changed its policy and began to promote population control.

The first family-planning program began in 1955, as a means of improving maternal and child health. Birthrates fell, but other social changes resulted in widespread famine and increased death rates and low birthrates in the late 1950s and early 1960s.

The present family-planning policy began in 1971 with the launching of the *wan xi shao* campaign. Translated, this phrase means "later" (marriages), "longer" (intervals between births), and "fewer" (children). As part of this program, the legal ages for marriage were raised. For women and men in rural areas, the ages were raised to twenty-three and twenty-five, respectively;

for women and men in urban areas, the ages were raised to twenty-five and twenty-eight, respectively. These policies resulted in a reduction of birthrates by nearly 50 percent between 1970 and 1979.

An even more restrictive "one child" campaign was begun in 1978–1979. The program offered incentives for couples to restrict their family size to one child. Couples enrolled in the program would receive free medical care, cash bonuses for their work, special housing treatment, and extra old-age benefits. Those who broke their pledge were penalized by the loss of these benefits as well as other economic penalties. By the mid-1980s, less than 20 percent of the eligible couples were signing up for the program. Rural couples, particularly, desired more than one child. In fact, in a country where about 60 percent of the population is rural, the rural total fertility rate was 2.5 children per woman. In 1988, a second child was sanctioned for rural couples if their first child was a girl, which legalized what had been happening anyway.

The programs appear to have had an effect because the current total fertility rate has fallen to 1.6 children per woman. Replacement fertility, the total fertility rate at which the population would eventually stabilize, is 2.1 children per woman per lifetime. Furthermore, over 85 percent of couples use contraception. Abortion is also an important aspect of this program, with a ratio of more than six hundred abortions per thousand live births.

Although the program has been effective in reducing the birthrate, there has been one unexpected outcome: the sex ratio shifting in favor of males. It is not clear what is causing the change; however, as in many societies, there is a strong social desire for sons. Because of this desire for sons, many speculate that one or more of the following may be the cause: female fetuses may be aborted more frequently than males; there may be a higher infant mortality rate among female children due to neglect; or female off-spring may be given up for adoption more frequently than males.

By contrast, during the same fifty years, India has had little success in controlling its population. In 2000, a new plan was unveiled that includes the goal of bringing the total fertility rate from the current 3.2 children per woman to 2 (replacement rate) by 2010. In the past, the emphasis of government programs was on meeting goals of sterilization and contraceptive use, but this has not been successful. Today, less than 50 percent of couples use contraceptives. This new plan emphasizes improvements in the quality of life of the people. The major thrusts are to reduce infant and maternal death, immunize children against preventable disease, and encourage girls to attend school. It is hoped that improved health will remove the perceived need for large numbers of births. Currently, less than 50 percent of the women in India can read and write. The emphasis on improving the educational status of women is related to the experiences of other developing countries. In many other countries, it has been shown that an increase in the education level of women has been linked to lower fertility rates.

It seems overly optimistic to think that the total fertility rate can be reduced from 3.2 to 2 in just ten years, but programs that emphasize improvements in maternal and child health and increasing the educational level of women have been very effective in reducing total fertility in other countries.

Questions to Discuss

1. Under what conditions should governments interfere with the reproductive desires of their people?
2. Compare the population control plans of China and India. List the advantages and disadvantages of each.
3. Assume the U.S. government needs to develop specific population control measures. Develop a list of at least three population control measures you would support.

Source: Data from H. Yuan Tien, "China's Demographic Dilemmas," in *Population Bulletin, 1992,* Population Reference Bureau, Washington, D.C., and Natural Family Planning Commission of China; and more recent data taken from Population Reference Bureau.

historical, social, cultural, ethical, and personal considerations. What is best for the population as a whole may be bad for you as an individual. The biggest problems associated with control of the human population are not biological problems but rather require the efforts of philosophers, theologians, politicians, and sociologists. If our population increases, so will political, social, and biological problems; there will be less individual freedom, there will be intense competition for resources, and famine and starvation will become even more common. The knowledge and technology necessary to control the human population are available, but the will is not. What will eventually limit the size of our population? Will it be lack of resources, lack of energy, accumulated waste products, competition among ourselves, or rational planning of family size?

HUMAN POPULATION GROWTH AND THE GLOBAL ECOSYSTEM

The growth of the human population has had major consequences for the global ecosystem. As the size of the human population has increased, more and more resources have been diverted to human use and fewer are available for other organisms. Nearly Earth's entire surface has been affected by human activity. Over 80 percent of the land surface has been modified for human uses such as agricultural and grazing land, managed forests, urban space, and transportation corridors. Essentially all of the land that can support agriculture has been converted from forests or grasslands to raising crops. Oceans have been significantly impacted as well. Most fish populations are either overfished or are being fished at capacity.

When humans convert terrestrial or aquatic ecosystems to their use, these ecosystems are unavailable for their original inhabitants. This habitat destruction has resulted in a loss of biodiversity due to extinction and endangerment of species that compete with humans for resources. For example, the bison population of North America is restricted to a few protected areas, since nearly all the grasslands have been converted to

agriculture and grazing. Habitat destruction in the ocean also occurs. The use of trawls—weighted nets that are pulled along the ocean bottom—to catch fish has greatly modified the ocean bottom in areas of high fishing activity.

Other human activities affect Earth globally. Global climate change is clearly the result of increased amounts of carbon dioxide, methane, and other greenhouse gases in the atmosphere. Increased greenhouse gases lead to a general warming of earth that results in the melting of glaciers, affects precipitation patterns, and causes a rise in sea level. These changes have the potential to cause significant change in biomes. The increase in atmospheric carbon dioxide is a direct result of the use of fossil fuels and the destruction of forests. The increases in methane can be traced to, in large part, domesticated livestock that release methane from their guts and decomposition that occurs in flooded rice paddies.

In addition, persistent organic pollutants such as DDT, polychlorinated biphenyls (PCBs), and many other compounds from industry and mercury from the burning of coal in electric power plants have a global circulation and are found in humans and other animals throughout the world. Many kinds of ocean fish have elevated levels of mercury.

Lifestyles affect the degree of impact. Consumption of energy and other resources is much higher in the nations of the developed world than in those of the less developed world. The developed world, with less than 20 percent of the world's human population, consumes about 60 percent of the world's energy. The United States, with less than 5 percent of the world's population, consumes about 25 percent of the world's energy. The consumption of goods and services by the people of the developed world has a global impact. Even the food consumption habits of people have wide-ranging effects. Rainforests in Brazil are burned to create grazing land to raise beef to sell to developed nations.

As the human population continues to increase, so will the pressure to raise more food, produce more consumer products, and supply more energy. These needs will place increased pressure on the natural resources of Earth.

SUMMARY

Ecology is the study of how organisms interact with their environment. The *environment* consists of *biotic* and *abiotic* components that are interrelated in an *ecosystem*. All *ecosystems* must have a constant *input of energy* from the Sun. *Producer organisms* are capable of trapping the Sun's energy and converting it into *biomass*. *Herbivores* feed on producers and are in turn eaten by *carnivores*, which may be eaten by *other carnivores*. Each level in the *food chain* is known as a *trophic level*. Other kinds of organisms involved in food chains are *omnivores*, which eat both plant and animal food, and *decomposers*, which break down dead organic matter and waste products. All *ecosystems* have a large *producer* base with successively smaller amounts of energy at the *herbivore, primary carnivore,* and *secondary carnivore trophic levels*. This is because each time *energy* passes from *one trophic level* to the next, about 90 percent of the energy is lost from the *ecosystem*. A *community* consists of the interacting *populations of organisms* in an area. The organisms are interrelated in many ways in food chains that

interlock to create *food webs*. Because of this interlocking, changes in one part of the *community* can have effects elsewhere.

Major land-based regional *ecosystems* are known as *biomes*. The *temperate deciduous forest, boreal coniferous forest, temperate rainforest, tropical rainforest, tropical dry forest, Mediterranean shrublands, temperate grasslands, desert, savanna,* and *tundra* are examples of biomes. Aquatic communities can be divided into marine, freshwater, and estuarine communities.

Each organism in a community occupies a specific space known as its *habitat* and has a specific functional role to play known as its *niche*. An *organism's habitat* is usually described in terms of some conspicuous element of its surroundings. The *niche* is difficult to describe because it involves so many interactions with the physical environment and other living things.

Interactions between organisms fit into several categories. *Predation* is one organism benefiting (*predator*) at the expense of the organism killed and eaten (*prey*). *Parasitism* is one organism benefiting (*parasite*) by living in or on another organism (*host*) and deriving nourishment from it. *Commensal* relationships exist when one organism is helped but the other is not affected. *Mutualistic* relationships benefit both organisms. *Competition* causes harm to both of the organisms involved, although one may be harmed more than the other and may become extinct, evolve into a different niche, or be forced to migrate.

Many atoms are cycled through ecosystems. The *carbon atoms* of living things are trapped by *photosynthesis,* passed from organism to organism as food, and released to the atmosphere by *respiration. Nitrogen* originates in the atmosphere, is trapped by *nitrogen-fixing bacteria,* passes through a series of organisms, and is ultimately released to the atmosphere by *denitrifying bacteria. Phosphorus* originates in rock and is used by organisms and eventually deposited as sediments.

A *population* is a group of organisms of the same species in a particular place at a particular time. *Populations* differ from one another in *gene frequency, age distribution, sex ratio,* and *population density.* A typical population growth curve consists of a *lag phase* in which population rises very slowly, followed by an *exponential growth phase* in which the population increases at an accelerating rate, followed by a leveling off of the population in a *stable equilibrium phase* as the *carrying capacity* of the environment is reached.

Humans as a species have the same limits and influences that other organisms do. However, humans can reason and predict, thus offering the possibility of population control through *conscious population limitation.* Human activities have had many impacts on the global ecosystem. These include: habitat destruction, release of pollutants, climate change, and reduced biodiversity.

KEY TERMS

abiotic factors (p. **560**)
benthic (p. **569**)
biodiversity (p. **565**)
biomass (p. **564**)
biomes (p. **566**)
biotic factors (p. **560**)
carbon cycle (p. **578**)
carnivore (p. **562**)
carrying capacity (p. **585**)
commensalism (p. **576**)
community (p. **561**)
competition (p. **577**)
consumer (p. **561**)
decomposers (p. **563**)

ecology (p. **560**)
ecosystem (p. **561**)
environment (p. **560**)
environmental resistance (p. **585**)
estuary (p. **571**)
exponential growth phase (p. **584**)
food chain (p. **562**)
free-living nitrogen-fixing
    bacteria (p. **580**)
freshwater community (p. **569**)
habitat (p. **572**)
herbivore (p. **562**)
keystone species (p. **566**)
lag phase (p. **584**)

limiting factors (p. **585**)
marine community (p. **569**)
mortality (p. **584**)
mutualism (p. **576**)
natality (p. **584**)
niche (p. **572**)
nitrogen cycle (p. **581**)
omnivore (p. **563**)
organism (p. **561**)
parasitism (p. **574**)
pelagic (p. **569**)
plankton (p. **569**)
population (p. **582**)
population growth curve (p. **584**)

predation (p. **574**)
primary carnivore (p. **562**)
primary consumer (p. **562**)
producer (p. **561**)
secondary carnivore (p. **562**)
secondary consumer (p. **562**)
stable equilibrium phase
    (p. **585**)
symbiosis (p. **577**)
symbiotic nitrogen-fixing
    bacteria (p. **580**)
trophic level (p. **561**)

APPLYING THE CONCEPTS

1. As energy is passed from one trophic level to the next in a food chain, about _____ percent is lost at each transfer.
   a. 10
   b. 50
   c. 75
   d. 90

2. The primary factor that determines whether a geographic area will support temperate deciduous forest or prairie is
   a. the amount of rainfall.
   b. the severity of the winters.
   c. the depth of the soil.
   d. the kinds of animals present.

3. Which one of the following organisms is most likely to be at the second trophic level?
   a. a maple tree
   b. a snake
   c. a fungus
   d. an elephant

4. Which one of the following processes is necessary for the production of organic molecules from inorganic molecules?
   a. respiration by animals
   b. decomposer organisms releasing carbon dioxide
   c. carbon dioxide uptake by plants
   d. herbivores consuming plants

5. If two species of organisms occupy the same niche,
   a. mutualism will result.
   b. competition will be very intense.
   c. both organisms will become extinct.
   d. both will need to enlarge their habitat.

6. If nitrogen-fixing bacteria were to become extinct
   a. life on Earth would stop immediately since there would be no source of nitrogen.
   b. life on Earth would continue indefinitely.
   c. life on Earth would slowly dwindle as useful nitrogen became less available.
   d. life on Earth would be unchanged except that proteins would be less common.

7. You obtain nitrogen for the organic molecules in your body from
   a. the air you breathe.
   b. the water you drink.
   c. carbohydrates in the food you eat.
   d. proteins in the food you eat.

8. Many plants (flowers) provide nectar for insects. The insects, in turn, pollinate the flower. This relationship between the insect and plant represents
   a. parasitism.
   b. commensalism.
   c. mutualism.
   d. predation.

9. As population density increases, which one of the following is likely to occur?
   a. Natality will increase.
   b. Mortality will decrease.
   c. The population will experience exponential growth.
   d. Individuals will migrate from the area of highest density.
10. A population made up primarily of prereproductive individuals will
    a. increase rapidly in the future.
    b. become extinct.
    c. rarely occur.
    d. remain stable for several generations.

11. Listed below are the sex ratios for four populations. All other things being equal (including current population size), which population should experience the greatest future growth?
    a. 1 male:1 female   c. 1 male:2 female
    b. 2 male:1 female   d. 3 male:1 female
12. The current human population of the world is experiencing
    a. a population decline.   c. stable equilibrium.
    b. slow steady growth.   d. exponential growth.

Answers

1. d **2.** a **3.** d **4.** c **5.** b **6.** c **7.** d **8.** c **9.** d **10.** a **11.** c **12.** d

QUESTIONS FOR THOUGHT

1. Describe the flow of energy through an ecosystem.
2. What role does each of the following play in an ecosystem: sunlight, plants, the second law of thermodynamics, consumers, decomposers, herbivores, carnivores, and omnivores?
3. Why is there usually a larger herbivore biomass than a carnivore biomass?
4. List a predominant abiotic factor in each of the following biomes: temperate deciduous forest, boreal coniferous forest, desert, tundra, tropical rainforest, and savanna.
5. Can energy be recycled through an ecosystem? Explain why or why not.
6. Describe your niche.

7. What do parasites, commensal organisms, and mutualistic organisms have in common? How are they different?
8. Trace the flow of carbon atoms through a community that contains plants, herbivores, decomposers, and parasites.
9. Describe four different roles played by bacteria in the nitrogen cycle.
10. List four kinds of limiting factors that help to set the carrying capacity for a species.
11. If competition is intense, how can both kinds of competing organisms continue to exist?
12. How does the population growth curve of humans compare with that of other kinds of animals?

FOR FURTHER ANALYSIS

1. Many cities have serious problems with large populations of deer. They eat the plants in people's yards and cause serious traffic accidents. You are a consultant hired to help a city solve its deer population problem. Using what you have learned about the principles of population growth, describe two alternative plans that would work.
2. Sulfur is present in all organisms, particularly in proteins. Based on what you know about the carbon and nitrogen cycles, develop an outline of what you think the sulfur cycle would look like.
3. Many people advocate the use of biofuels (biodiesel, alcohol, etc.) as a substitute for fossil fuels. They suggest that these fuels cause less environmental damage and are renewable. Based on what you know about energy flow through ecosystems and the

effect of agriculture on biodiversity, what kinds of questions would you like to ask these advocates of biofuels production?
4. Biomes are regional ecological systems. What is the predominant biome where you live? Develop a list of ten characteristics you consider important in describing your biome.
5. Compare your social interactions with other people to the kinds of ecological interactions that occur among organisms. List examples of social:
   mutualism
   parasitism
   commensalism
   competition

INVITATION TO INQUIRY

World Population Characteristics

The Population Reference Bureau maintains a website that contains the World Population Data Sheet. It produces a new data sheet each year. Go to that website and access the World Population Data Sheet. Determine which countries have each of the following:

1. The highest infant mortality rate
2. The highest total fertility rate
3. The lowest life expectancy

Compare the countries you have listed. In what ways are they similar?

Now determine which countries have each of the following:

1. The lowest infant mortality rate
2. The lowest total fertility rate
3. The highest life expectancy

Compare the countries you have listed. In what ways are they similar?

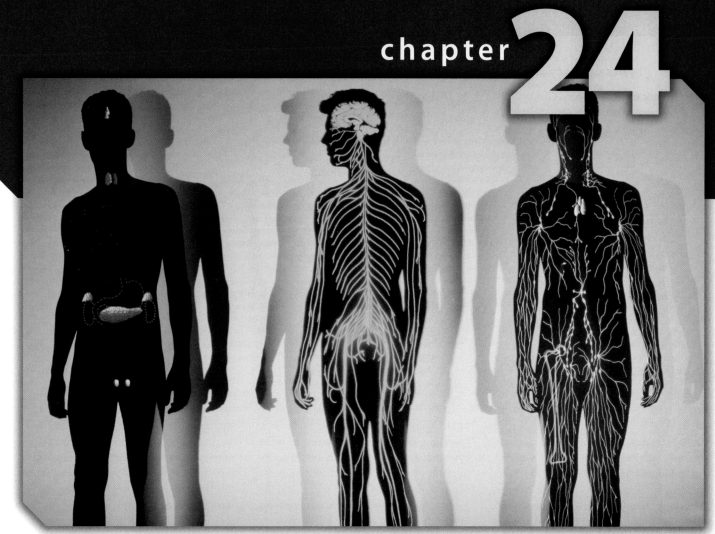

Because many different systems each perform a specific set of tasks in the human body, it is necessary to coordinate these various activities with one another.

Human Biology: Materials Exchange and Control Mechanisms

Physics Connections

▶ Kinetic molecular theory (Ch. 4).
▶ Pressure is a force (Ch. 4).
▶ Gas laws (Ch. 4).

Chemistry Connections

▶ pH is a measure of hydrogen ion concentration (Ch. 10).
▶ Many large organic molecules are polymers (Ch. 19).
▶ Ions are positively or negatively charged (Ch. 10).

CORE CONCEPT

The human body consists of many interacting systems.

The circulatory system is a distribution system (p. 593).

The respiratory system exchanges oxygen and carbon dioxide (p. 602).

The digestive system breaks down complex molecules to simpler molecules, (p. 602).

The excretory system rids the body of metabolic wastes (p. 612).

The nervous and endocrine systems regulate bodily processes (p. 613).

Earth Science Connections

▶ Minerals are inorganic materials (Ch. 15).

Life Science Connections

▶ Enzymes control biological reactions (Ch. 20).
▶ Aerobic cellular respiration requires oxygen (Ch. 20).
▶ Cell membranes are selectively permeable (Ch. 20).

OVERVIEW

All cells must obtain nutrients and get rid of waste products through their cell membranes. In single-celled organisms, the cell membrane is in contact with the cell's surroundings so diffusion and other processes accomplish this exchange. However, large, multicellular organisms consist of trillions of individual cells, most of which are not located near the surface of the organism yet must exchange materials in the same manner as in single-celled creatures. Therefore, large organisms consist of many different systems that assist in the exchange of materials such as food, wastes, and gases between the cells of the organisms and the external environment. The primary systems involved are the circulatory, respiratory, digestive, and excretory systems.

These systems must be coordinated and regulated in such a way that they supply the needs of the individual cells. Two other systems, the nervous and endocrine systems, are involved in managing these functions. This chapter will look at the systems involved in the exchange of materials and coordination of activities within the human body.

EXCHANGING MATERIALS: BASIC PRINCIPLES

Living things are complex machines with many parts that must function in a coordinated fashion. All systems are integrated and affect one another in many ways. For example, when you run up a hill, your leg and arm muscles move in a coordinated way to provide power. They burn fuel (glucose) for energy and produce carbon dioxide and lactic acid as waste products, which tend to lower the pH of the blood. Your heart beats faster to provide oxygen and nutrients to the muscles, you breathe faster to supply the muscles with oxygen and get rid of carbon dioxide, and the blood vessels in the muscles dilate to allow more blood to flow to them. As you run, you generate excess heat. As a result, more blood flows to the skin to get rid of the heat and sweat glands begin to secrete, thus cooling the skin. All of these automatic internal adjustments help the body maintain a constant level of oxygen, carbon dioxide, and glucose in the blood; constant pH; and constant body temperature. These processes can be summed up in the concept of *homeostasis*. **Homeostasis** is the process of maintaining a constant internal environment as a result of monitoring and modifying the functioning of various systems.

The cells of all organisms are highly organized units that require a constant flow of energy and materials to maintain themselves. The energy they require is provided in the form of nutrient molecules that enter the cell. Oxygen is required for the efficient release of energy from the large organic molecules that serve as fuel. Inevitably, as oxidation takes place, waste products form that are useless or toxic. These must be removed from the cell. All of these processes and exchanges must be regulated by control processes in order to maintain homeostasis. Furthermore, all these exchanges of food, oxygen, and waste products must take place through the cell surface.

The ability to transport materials into or out of a cell is determined by its surface area, while its metabolic demands are determined by its volume. As you increase the size of an object, the volume increases faster than the surface area. (See chapter 1, page 8, for a discussion of surface-area-to-volume ratio.) So, the larger a cell becomes, the more difficult it is to satisfy its needs. Some cells overcome this handicap by having highly folded cell membranes that substantially increase their surface area. This is particularly true of cells that line the intestine or are involved in the transport of large numbers of nutrient molecules. These

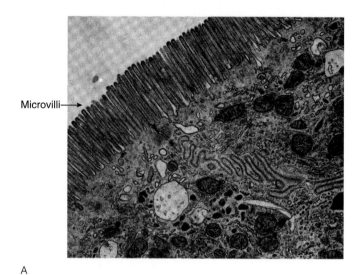

A

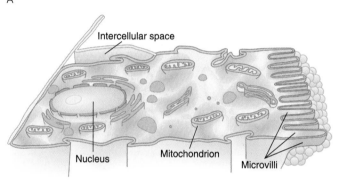

B

FIGURE 24.1 Intestinal cells that are in contact with the food in the gut have highly folded surfaces. The tiny projections of these cells are called microvilli. These can be clearly seen in the photomicrograph in (*A*). The drawing in (*B*) shows that only one surface has these projections.

cells have many tiny, folded extensions of the cell membrane called *microvilli* (figure 24.1).

In addition to the limitation that surface area presents to the transport of materials, large cells also have a problem with diffusion. The molecular process of diffusion is quite rapid over short distances but becomes very slow over longer distances. Diffusion is generally insufficient to handle the needs of cells if it must take place over a distance of more than 1 mm. The center of the cell would die before it received the molecules it needed if the distance were greater. Because of this, it is understandable that the basic unit of life, the cell, must remain small.

When we look at the evolution of organisms, we find that the first organisms were small, single-celled, and marine. Under these conditions, most exchanges could simply occur through their cell membranes. More complex, multicellular organisms have many cells that are not in contact with their surroundings and developed other ways to allow for efficient exchange of materials. The evolution of organs and organ systems allowed them to exchange and distribute materials to all cells even if many cells were not exposed to the sea. Often it is possible to see a pattern of evolutionary development of the structures

involved in the exchange of materials. For example, many tiny multicellular marine animals still can get sufficient oxygen through their surfaces without special structures. Larger marine animals have some sort of gill mechanism (a folding of the body surface) to allow for gas exchange, and terrestrial vertebrates have internal lungs thought to be derived from the swim bladder of fish to accomplish this function.

TRANSPORTING MATERIALS: THE CIRCULATORY SYSTEM

Large, multicellular organisms such as humans consist of trillions of cells. Since many of these cells are buried within the organism far from the body surface, there must be some sort of distribution system to assist them in solving their materials exchange problems. The primary mechanism used is the circulatory system.

The circulatory system consists of several fundamental parts. **Blood** is the fluid medium in animals that assists in the transport of materials and heat. The **heart** is a pump that forces the fluid blood from one part of the body to another. The heart pumps blood into **arteries,** which distribute blood to organs. It flows into successively smaller arteries until it reaches tiny vessels called **capillaries**, where materials are exchanged between the blood and tissues through the walls of the capillaries. The blood flows from the capillaries into **veins** that combine into larger veins that ultimately return the blood to the heart from the tissues.

The Nature of Blood

Blood is a fluid that consists of several kinds of cells suspended in a watery matrix called **plasma.** This fluid plasma also contains many kinds of dissolved molecules. The primary function of the blood is to transport molecules, cells, and heat from one part of the body to another. The major kinds of molecules that are distributed by the blood are respiratory gases (oxygen and carbon dioxide), nutrients of various kinds, waste products, disease-fighting antibodies, and chemical messengers (hormones). Blood has special characteristics that allow it to distribute respiratory gases very efficiently. Although little oxygen is carried as free, dissolved oxygen molecules in the plasma, *red blood cells* (*RBCs*)—also known as *erythrocytes*—contain **hemoglobin,** an iron-containing molecule, to which oxygen molecules readily bind. This allows for much more oxygen to be carried than could be possible if it were simply dissolved in the blood. Thus, the circulating blood is constantly picking up and delivering molecules to where they are needed. This is important for maintaining homeostasis.

Table 24.1 lists the variety of cells found in blood. Another important regulatory function of the blood involves the various kinds of cells involved in immunity. The *white blood cells* (*WBCs*) carried in the blood are involved in defending against harmful agents. These cells help the body resist many diseases. They constitute the core of the *immune system.* The various

TABLE 24.1   The Composition of Blood

Component	Quantity Present	Function
Plasma	55%	Maintain fluid nature of blood
Water	91.5%	
Protein	7.0%	
Other materials	1.5%	
Cellular material	45%	
Red blood cells (erythrocytes)	4.3–5.8 million/mm^3	Carry oxygen and carbon dioxide
White blood cells (leukocytes)	5–9 thousand/mm^3	Immunity
Lymphocytes	25%–30% of white cells present	
Monocytes	3%–7% of white cells present	
Neutrophils	57%–67% of white cells present	
Eosinophils	1%–3% of white cells present	
Basophils	Less than 1% of white cells present	
Platelets	250–400 thousand/mm^3	Clotting

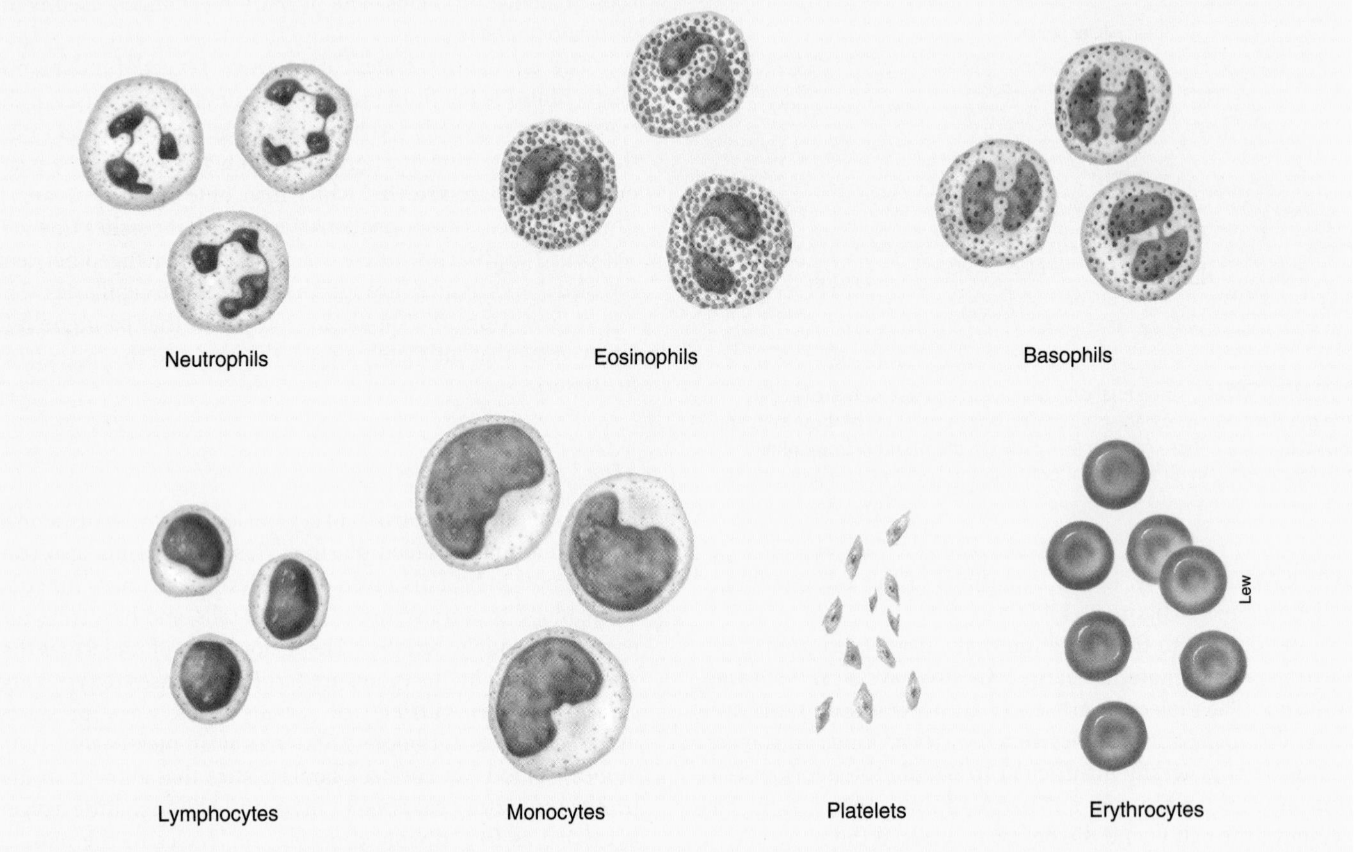

Neutrophils

Eosinophils

Basophils

Lymphocytes

Monocytes

Platelets

Erythrocytes

WBCs participate in providing immunity in several ways. First of all, immune system cells are able to recognize cells and molecules that are harmful to the body. If a molecule is recognized as harmful, certain white blood cells called *lymphocytes* produce *antibodies* (*immunoglobulins*) that attach to the dangerous materials. The dangerous molecules that stimulate the production of antibodies are called *antigens* (*immunogens*). When harmful microorganisms (e.g., bacteria, viruses, fungi) or toxic molecules enter the body or cancer cells are produced, other WBCs (1) recognize, (2) boost their abilities to respond to,

(3) move toward, and (4) destroy the problem causers. *Neutrophils* and *monocytes* are specific kinds of WBCs capable of engulfing harmful material, a process called phagocytosis. Thus, they are often called *phagocytes*. While most can move from the bloodstream into the surrounding tissue, monocytes undergo such a striking increase in size that they are given a different name—*macrophages*. Macrophages can be found throughout the body and are the most active of the phagocytes. *Eosinophils* are primarily involved in destroying multicellular parasites such as worms, and *basophils* appear to be primarily involved in the

inflammation response, which results in increased blood flow to an injured area, causing swelling and reddening.

Another kind of cellular particle in the blood is the *platelet*. These are fragments of specific kinds of white blood cells and are important in blood clotting. They collect at the site of a wound, where they break down, releasing specific molecules, which begin a series of reactions that results in the formation of fibers that trap blood cells and form a plug in the opening of the wound. This mixture of fibers and trapped blood cells dries and hardens to form a scab.

The Heart

Blood can perform its transportation function only if it moves. The organ responsible for providing the energy to pump the blood is the heart. The heart is a muscular pump that provides the pressure necessary to propel the blood throughout the body. It must continue its cycle of contraction and relaxation, or blood stops flowing and body cells are unable to obtain nutrients or get rid of wastes. Some cells, such as brain cells, are extremely sensitive to having their flow of blood interrupted because they require a constant supply of glucose and oxygen. Others, such as muscle cells or skin cells, are much better able to withstand temporary interruptions of blood flow.

The hearts of humans, other mammals, and birds consist of four chambers and four sets of valves that work together to ensure that blood flows in one direction only. Two of these chambers, the right and left *atria* (singular, *atrium*), are relatively thin-walled structures that collect blood from the major veins and empty it into the larger, more muscular ventricles (figure 24.2).

The right and left *ventricles* are chambers that have powerful muscular walls whose contraction forces blood to flow through the arteries to all parts of the body. The valves between the atria and ventricles, known as *atrioventricular valves,* are important one-way valves that allow the blood to flow from the atria to the ventricles but prevent flow in the opposite direction. Similarly, there are valves in the aorta and pulmonary artery, known as *semilunar valves.* The **aorta** is the large artery that carries blood from the left ventricle to the body, and the **pulmonary artery** carries blood from the right ventricle to the lungs. The semilunar valves prevent blood from flowing back into the ventricles. If the atrioventricular or semilunar valves are damaged or function improperly, the efficiency of the heart as a pump is diminished, and the person may develop an enlarged heart or other symptoms.

The right and left sides of the heart have slightly different jobs, because they pump blood to different parts of the body. The right side of the heart receives blood from the general body and pumps it through the pulmonary arteries to the lungs, where exchange of oxygen and carbon dioxide takes place, and the blood returns from the lungs to the left atrium. This is called **pulmonary circulation.** The larger, more powerful left side of the heart receives blood from the lungs, delivers it through the aorta to all parts of the body, and returns it to the right atrium by way of large veins known as the *superior vena*

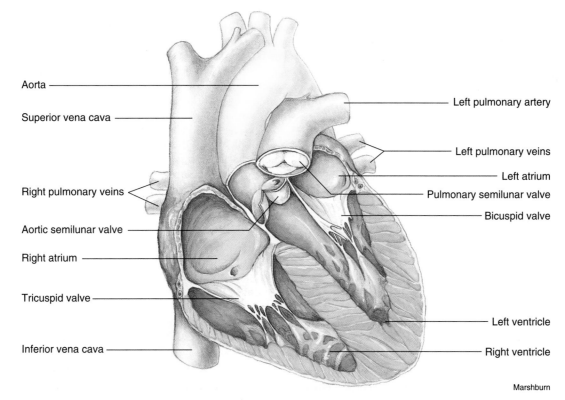

Aorta
Superior vena cava
Right pulmonary veins
Aortic semilunar valve
Right atrium
Tricuspid valve
Inferior vena cava

Left pulmonary artery
Left pulmonary veins
Left atrium
Pulmonary semilunar valve
Bicuspid valve
Left ventricle
Right ventricle

Marshburn

FIGURE 24.2 The heart consists of two thin-walled chambers called atria that contract to force blood into the two ventricles. When the ventricles contract, the atrioventricular valves (bicuspid and tricuspid) close, and blood is forced into the aorta and pulmonary artery. Semilunar valves in the aorta and pulmonary artery prevent the blood from flowing back into the ventricles when they relax.

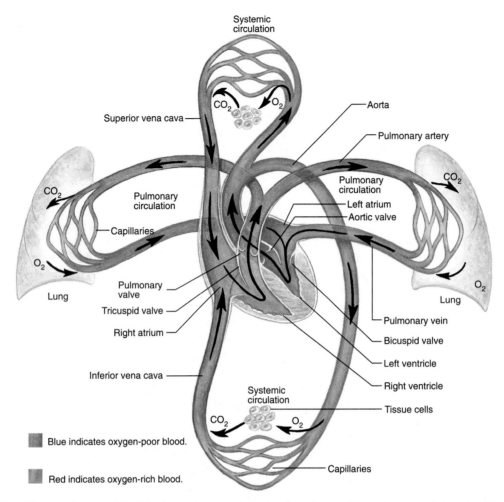

Systemic circulation

Superior vena cava — CO_2 ← → O_2

Aorta

Pulmonary artery

Pulmonary circulation

Pulmonary circulation

CO_2

Capillaries

Left atrium

Aortic valve

O_2

Lung

Pulmonary valve

Tricuspid valve

Right atrium

CO_2

O_2

Lung

Pulmonary vein

Bicuspid valve

Inferior vena cava

Left ventricle

Systemic circulation

Right ventricle

CO_2 O_2

Tissue cells

Blue indicates oxygen-poor blood.

Red indicates oxygen-rich blood.

Capillaries

FIGURE 24.3 The right ventricle pumps blood that is poor in oxygen to the two lungs by way of the pulmonary arteries, where it receives oxygen and turns bright red. The blood is then returned to the left atrium by way of four pulmonary veins. This part of the circulatory system is known as pulmonary circulation. The left ventricle pumps oxygen-rich blood by way of the aorta to all parts of the body except the lungs. This blood returns to the right atrium, depleted of its oxygen, by way of the superior vena cava from the head region and the inferior vena cava from the rest of the body. This portion of the circulatory system is known as systemic circulation.

cava and *inferior vena cava*. This is known as **systemic circulation.** Both circulatory pathways are shown in figure 24.3. The systemic circulation is responsible for gas, nutrient, and waste exchange in all parts of the body except the lungs.

Arteries, Veins, and Capillaries

Arteries and veins are the tubes that transport blood from one place to another within the body. Figure 24.4 compares the structure and function of arteries and veins.

Blood Pressure

Arteries carry blood away from the heart because the contraction of the walls of the ventricles increases the pressure in the arteries compared to that in the capillaries and veins. A typical pressure recorded in a large artery while the heart is contracting is about 120 millimeters of mercury. This is known as the **systolic blood pressure.** The pressure recorded while the heart is relaxing is about 80 millimeters of mercury. This is known as the **diastolic blood pressure.** A blood pressure

reading includes both of these numbers and is recorded as 120/80. Most physicians consider a systolic pressure above 140 or a diastolic pressure above 90 to be cause for concern and will recommend treatment for high blood pressure.

Structure and Function of Arteries and Veins

The walls of arteries are relatively thick and muscular yet elastic. Healthy arteries have the ability to expand as blood is pumped into them and return to normal as the pressure drops. This ability to expand absorbs some of the pressure and reduces the peak pressure within the arteries, thus reducing

CONCEPTS APPLIED

Go with the Flow

Look at figure 24.3. Start with blood in the left lung and trace the flow of blood until it arrives back at the left lung. Make a list of the structures encountered.

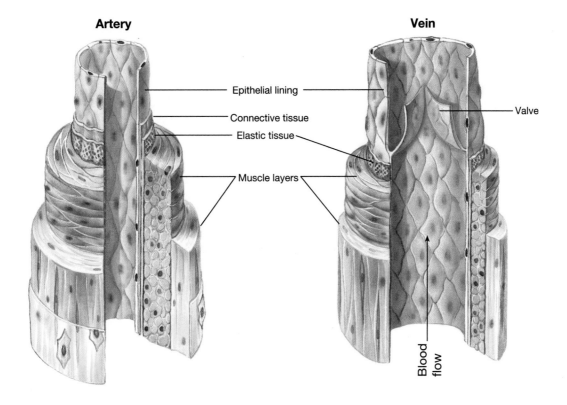

Artery

Vein

Epithelial lining

Connective tissue

Elastic tissue

Muscle layers

Valve

Blood flow

FIGURE 24.4 The walls of arteries are much thicker than the walls of veins. (The pressure in arteries is much higher than the pressure in veins.) The pressure generated by the ventricles of the heart forces blood through the arteries. Veins often have very low pressure. The valves in the veins prevent the blood from flowing backward, away from the heart.

the likelihood that they will burst. If arteries become hardened and less resilient, the peak blood pressure rises and they are more likely to rupture. The elastic nature of the arteries is also responsible for assisting the flow of blood. When they return to normal from their stretched condition, they give a little push to the blood that is flowing through them.

Blood is distributed from the large aorta through smaller and smaller blood vessels to millions of tiny capillaries. Some of the smaller arteries, called **arterioles,** may contract or relax to regulate the flow of blood to specific parts of the body. Major parts of the body that receive differing amounts of blood, depending on need, are the digestive system, muscles, and skin.

Veins collect blood from the capillaries and return it to the heart. The pressure in these blood vessels is very low. Some of the largest veins may have a blood pressure of 0.0 mmHg for brief periods. Since pressure in veins is so low, muscular movements of the body are important in helping to return blood to the heart. When muscles of the body contract, they compress nearby veins, and this pressure pushes blood along in the veins. Because valves in the veins allow blood to flow only toward the heart, this activity acts as an additional pump to help return blood to the heart.

Structure and Function of Capillaries

Although the arteries are responsible for distributing blood to various parts of the body and arterioles regulate where blood goes, it is the function of capillaries to assist in the exchange of materials between the blood and cells. Capillaries are tiny,

thin-walled tubes that receive blood from arterioles. They are so small that red blood cells must go through them in single file. They are so numerous that each cell in the body has a capillary located near it. It is estimated that there are about 1,000 m^2 of surface area represented by the capillary surface in a typical human. Each capillary wall consists of a single layer of cells and therefore presents only a thin barrier to the diffusion of materials between the blood and cells. It is also possible for liquid to flow through tiny spaces between the individual cells of most capillaries (figure 24.5). The flow of blood through these smallest blood vessels is relatively slow. This allows time for the diffusion of such materials as oxygen, glucose, and water from the blood to surrounding cells and for the movement of such materials as carbon dioxide, lactic acid, and ammonia from the cells into the blood.

In addition to molecular exchange by diffusion, considerable amounts of water and dissolved materials leak through the small holes in the capillaries. This liquid is known as **lymph.** Lymph is produced when the blood pressure forces water and some small dissolved molecules through the walls of the capillaries. Lymph bathes the cells but must eventually be returned to the circulatory system by lymph vessels or swelling will occur. Return is accomplished by the *lymphatic system,* a collection of thin-walled tubes that branch throughout the body. These tubes collect lymph that left the circulatory system and ultimately empty it into major blood vessels near the heart. Figure 24.6 shows the structure of the lymphatic system.

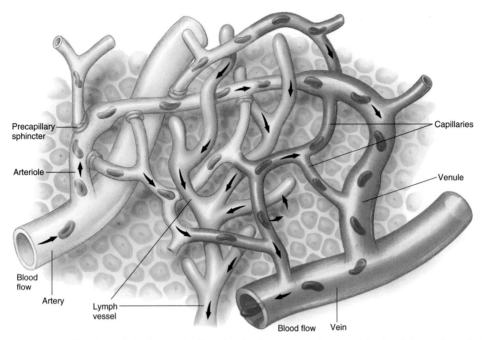

FIGURE 24.5 Capillaries are tiny blood vessels. Exchange of cells and molecules can occur between blood and tissues through their thin walls. Molecules diffuse in and out of the blood, and cells such as monocytes can move from the blood through the thin walls into the surrounding tissue. There is also a flow of liquid through holes in the capillary walls. This liquid, called lymph, bathes the cells and eventually enters small lymph vessels that return lymph to the circulatory system near the heart.

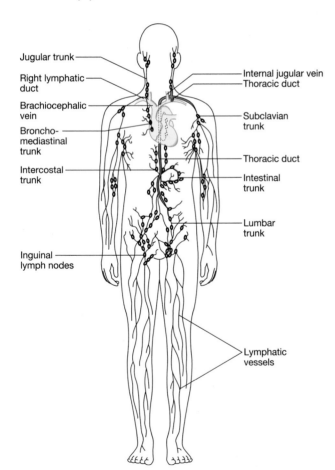

FIGURE 24.6 The lymphatic system consists of many ducts that transport lymph fluid back toward the heart. Along the way the lymph is filtered in the lymph nodes, and bacteria and other foreign materials are removed before the lymph is returned to the circulatory system near the heart.

SKIN: THE BODY'S CONTAINER

The importance of the skin is often overlooked; however, it serves several important functions. One important function is to be an impermeable barrier to the loss of water from the body and the entry of foreign microorganisms. While aquatic animals such as fish and amphibians can get by with a water-permeable skin, terrestrial animals such as reptiles, birds, and mammals must prohibit the loss of water from their body surface. The skin is actually the largest organ of the body. For an average human, it constitutes about 7,600 cm^2 (3,000 in^2). This is about the surface area of a 4-by-5 ft rug.

The large surface area of the skin is important in getting rid of heat. Shunting blood to this large surface area of the skin results in heat loss (as long as the environmental temperature is less than the body temperature), and the secretions of sweat glands spread over this large surface provide evaporative cooling.

The Structure of the Skin

The skin is divided into three distinct layers: the epidermis, the dermis, and the subcutaneous layers. The epidermis is the surface of the skin that you can see. It consists of a layer of cells at its base, next to the dermis, that undergoes mitosis to produce new cells. This layer of cells is called the germinating layer. The cells produced by the germinating layer migrate toward the surface and undergo major changes as they do so; they produce a protein called keratin, lose their nuclei and other structures, and become flattened disks. This outer layer of the epidermis is called the cornified layer and consists of several layers of these flattened, scalelike, keratin-filled cell remnants. The cornified layer provides toughness to the epidermis and is impermeable to most chemicals and microorganisms. Thus, the cornified layer prevents the loss of water from and the entry of microorganisms to the underlying tissues of the body. Since this layer is in direct contact with the environment, it is constantly worn away by friction with surroundings. The constant replenishment of this layer from below, as a result of the germinating layer, is essential to maintain this function of the skin. If this layer is destroyed, people lose fluids very rapidly and are subject to invasion by various kinds of disease microorganisms. This is probably most dramatically illustrated by the problems experienced by victims of heat or chemical burns in which the cornified layer is destroyed. They must receive intravenous fluids to replace the liquids lost and often succumb to generalized infections because they have no barrier to microorganisms.

Immediately below the epidermis is the dermis, which contains a great many connective tissue fibers that provide the flexible but tough texture of the skin. There are many blood vessels and nerve endings in the dermis. The overlying epidermis does not have blood vessels or nerve endings. The cells of the epidermis receive nutrition from the blood vessels in the dermis, and pain, touch, and temperature are sensed by the nerve endings in the dermis.

Beneath the dermis is the subcutaneous layer, which contains a great many connective tissue fibers and fat cells. The fibers in this layer bind the skin to underlying tissues, primarily muscle. There are also blood vessels and nerves in this layer.

Other Features of the Skin

Several other features of the skin deserve comment. Melanin is a brown-black pigment produced by special cells in the germinating layer. Dark-skinned people have melanin-producing cells that are more active than those of light-skinned people. Melanin is a protection against the damaging effects of ultraviolet light. When people are exposed to ultraviolet light, the melanin-producing cells are stimulated to produce more melanin and tanning occurs. Though tanning is most noticeable in light-skinned people, the melanin-producing cells of all people increase activity in response to ultraviolet light.

Hair and nails (and hooves, claws, and horns in other mammals) are composed of keratin. Therefore, they are specialized accumulations produced by the epidermis of the skin.

Several kinds of glands are derived from the epidermis but project down into the underlying dermis. Oil glands associated with hair follicles secrete oily material that is released onto the skin surface where it lubricates the skin and hair. Scent glands are also associated with hair follicles and produce a watery secretion that contains particular odor molecules. In humans, scent glands are found primarily in the armpit and pubic area. Other mammals may have scent glands near their eyes, around the anus, on their feet, or on their belly. Humans and a few other mammals have sweat glands over much of the skin surface. As mentioned earlier, evaporation of sweat from the skin surface cools the body. Finally, mammary glands, which are responsible for milk production, are specialized sweat glands. Figure 24.7 illustrates the various parts of the skin.

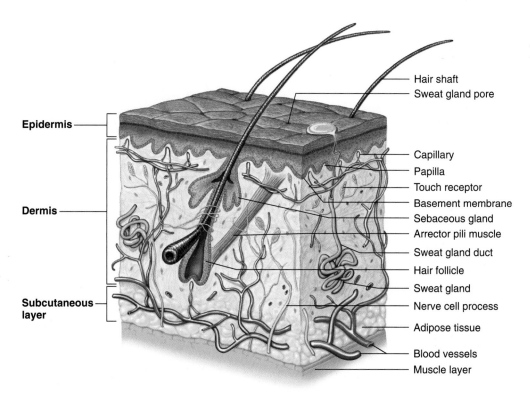

FIGURE 24.7 The skin is the largest organ of the body and consists of several layers and structures that prevent water loss, protect against infection, provide for cooling, and serve as sense organs.

EXCHANGING GASES: THE RESPIRATORY SYSTEM

In terrestrial vertebrates, the primary structures involved in gas exchange are the lungs. Since the lungs are internal structures, the amount of water lost is much less than if these moist surfaces were on the outer surface of the body, as are the gills of fish.

Structure and Function of Lungs

The **lungs** are organs of the body that allow gas exchange to take place between the air and blood. Associated with the lungs is a set of tubes that conducts air from outside the body to the lungs. The single large-diameter **trachea** is supported by rings of cartilage that prevent its collapse. It branches into two major **bronchi** (singular, *bronchus*) that deliver air to smaller and smaller branches. Bronchi are also supported by cartilage. The smallest tubes, known as **bronchioles,** contain smooth muscle and are therefore capable of constricting. Finally, the bronchioles deliver air to clusters of tiny sacs, known as **alveoli** (singular, *alveolus*), where the exchange of gases takes place between the air and blood. The alveoli are very thin-walled and surrounded by capillaries. This close association makes gas exchange by diffusion relatively easy. Figure 24.8 illustrates the close relationship between the alveoli of the lungs and the capillaries of the circulatory system.

The nose, mouth, and throat are also important parts of the air-transport pathway because they modify the humidity and temperature of the air and clean the air as it passes. The lining of the trachea contains cells that have cilia that beat in a direction that moves mucus and foreign materials from the lungs. The foreign matter may then be expelled by swallowing, coughing, or other means, thus keeping the air passage clear.

The Mechanism of Breathing

Breathing is the process of moving air in and out of the lungs. It is accomplished by the movement of a muscular organ known as the **diaphragm,** which separates the chest cavity and the lungs from the abdominal cavity. In addition, muscles located between the ribs (*intercostal* muscles) are attached to the ribs in such a way that their contraction causes the chest wall to move outward and upward, which increases the size of the chest cavity. During inhalation, the diaphragm moves downward and the external intercostal muscles of the chest wall contract, causing the volume of the chest cavity to increase. This results in a lower pressure in the chest cavity compared to the outside air pressure. Consequently, air flows from the outside, higher-pressure area through the trachea, bronchi, and bronchioles to the alveoli. During normal relaxed breathing, exhalation is accomplished by the chest wall and diaphragm simply returning to their normal, relaxed position. Muscular contraction is not involved. During vigorous exhalation, internal intercostal

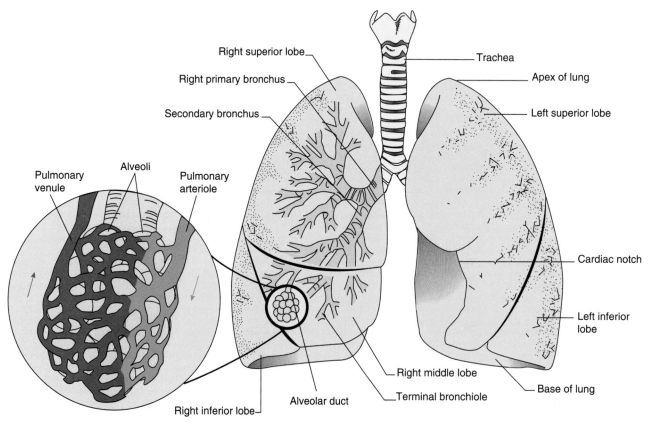

FIGURE 24.8 The exchange of gases takes place between the air-filled alveolus and the blood-filled capillary. The capillaries form a network around the saclike alveoli. The thin walls of the alveolus and capillary are in direct contact with one another; their combined thickness is usually less than 1 micron (a thousandth of a millimeter). This close relationship and the thinness of the walls allow for the easy diffusion of gases.

Cigarette Smoking and Your Health

Cigarette smoking is becoming less and less acceptable in our society. The banning of smoking in public buildings and on domestic air flights attests to this fact. Yet in spite of social pressure to quit smoking, research linking smoking with lung and heart disease, and evidence that even secondhand smoke can be harmful, over one-fifth of American adults are smokers (a smoker in this case is defined as someone who has smoked one hundred or more cigarettes in his or her lifetime). Among Americans with a high school education or less, smoking is even more prevalent.

Hazards of Cigarette Smoking

Bronchitis. Cigarette smoking is the leading cause of chronic bronchitis, which involves the inflammation of the bronchi. A common symptom of bronchitis is a harsh cough that expels a greenish-yellow mucus.

Emphysema. Emphysema is a progressive disease in which some of the alveoli are lost. People afflicted with this disease have less and less respiratory surface area and experience greater difficulty getting adequate oxygen, even though they may be breathing more rapidly. It may be caused by

cigarette smoke and other air pollutants that damage alveoli. This damage reduces the capacity of the lungs to exchange gases with the bloodstream. A common symptom of emphysema is difficulty exhaling. Several years of an emphysema sufferer's forced breathing can increase the size of the chest and give it a barrel appearance.

Asthma. Cigarette smoke is one of many environmental factors that may trigger an asthma attack. Asthma is an allergic reaction that results in the narrowing of the bronchioles of the lungs and the excess production of fluids that limit the amount of air that can enter the lungs. Symptoms of asthma include coughing, wheezing, and difficulty breathing.

Lung cancer. Lung cancer develops twenty times more frequently in heavy smokers than in nonsmokers. Typically, lung cancer starts in the bronchi. Cigarette smoke and other pollutants cause cells below the surface of the bronchi to divide at an abnormally high rate. This malignant growth may spread through the lung and move into other parts of the body. Occurrence of cancers of the mouth, larynx,

esophagus, pancreas, and bladder is also significantly greater in smokers than in nonsmokers.

Pneumonia. Cigarette smokers have an increased risk of developing pneumonia. Pneumonia involves the infection or inflammation of alveoli, which leads to fluid filling the alveolar sacs. Pneumonia is typically caused by the bacterium *Streptococcus pneumoniae* but in some cases is caused by other bacteria, fungi, protists, or viruses.

Smoking during pregnancy. Cigarette smoking during pregnancy has been linked to low birth weight and higher rates of fetal and infant death. Children of mothers who smoke during pregnancy have a higher incidence of heart abnormalities, cleft palate, and sudden infant death syndrome (SIDS). Nursing infants of smoking mothers have higher than normal rates of intestinal problems, and infants exposed to secondhand smoke have an increased incidence of respiratory disorders.

Heart disease. Smoking is a major contributor to heart disease. The action of nicotine from cigarette smoke results in constriction of blood vessels and the reduction of blood flow.

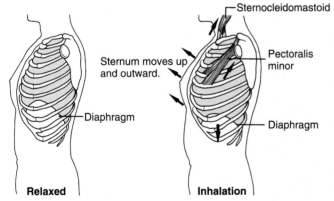

FIGURE 24.9 During inhalation, the diaphragm and external intercostal muscles between the ribs contract, causing the volume of the chest cavity to increase. During a normal exhalation, these muscles relax, and the chest volume returns to normal.

muscles of the chest wall and muscles of the abdominal wall contract to compress the chest cavity and more completely empty the lungs of air. Also during vigorous inhalation, additional muscles of the chest wall are involved, making the chest cavity larger (figure 24.9).

CONCEPTS APPLIED

Heart Rate, Breathing Rate, and Exercise

While sitting quietly, measure you heart rate and breathing rate for one minute. Then run in place for two minutes. Immediately sit down and measure your heart rate. Do a second bout of running in place for two minutes and immediately determine your breathing rate. Based on these changes in heart rate and breathing rate, how much faster were you using oxygen when running?

Homeostasis and Breathing

Exercising causes an increase in the amount of carbon dioxide in the blood because muscles are oxidizing glucose more rapidly. When carbon dioxide dissolves in water, it forms a weak acid (carbonic acid). This lowers the pH of the blood. Certain brain cells and specialized cells in the aortic arch and carotid arteries are sensitive to changes in blood pH. When they sense a lower pH, nerve impulses are sent more frequently to the diaphragm and intercostal muscles. These muscles contract more rapidly and more forcefully, resulting in more rapid, deeper breathing. Because more air is being exchanged per minute, carbon dioxide is lost from the lungs more rapidly. When exercise stops, the carbon dioxide level drops, blood pH rises, and breathing eventually returns to normal. Bear in mind, however, that moving air in and out of the lungs is of no value unless oxygen is diffusing into the blood and carbon dioxide is diffusing out.

OBTAINING NUTRIENTS: THE DIGESTIVE SYSTEM

All cells must have a continuous supply of nutrients that provides the energy they require and the building blocks needed to construct the macromolecules typical of living things. This section will deal with the processing and distribution of different kinds of nutrients.

The digestive system consists of a muscular tube with several specialized segments. In addition, there are glands that secrete digestive juices into the tube. Several different kinds of activities are involved in getting nutrients to the cells that need them.

Processing Food

The digestive system is designed as a disassembly system. Its purpose is to take large chunks of food and break them down to smaller molecules that can be taken up by the circulatory system and distributed to cells. The first steps in this process take place in the mouth.

Digestive Activities in the Mouth

It is important to grind large particles into small pieces by chewing in order to increase their surface area and allow for more efficient chemical reactions. It is also important to add water to the food, which further disperses the particles and provides the watery environment needed for these chemical reactions. Materials must also be mixed, so that all the molecules that need to interact with one another have a good chance of doing so. The mouth and the stomach are the major body regions involved in reducing the size of food particles. The teeth are involved in cutting and grinding food to increase its surface area. The watery mixture that is added to the food in the mouth is known as *saliva,* and the three pairs of glands that produce saliva are known as *salivary glands.* Saliva contains the enzyme *salivary amylase,* which initiates the chemical breakdown of starch. Saliva also lubricates the mouth and helps to bind food before swallowing. Figure 24.10 shows the structures of the digestive system.

Swallowing

Once the food has been chewed, it is swallowed and passes down the esophagus to the stomach. The process of swallowing

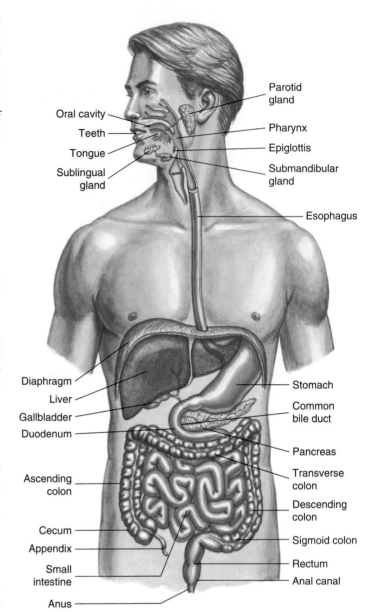

FIGURE 24.10   In the digestive system, the teeth, tongue, and enzymes from the salivary glands modify the food before it is swallowed. The stomach adds acid and enzymes and further changes the texture of the food. The food is eventually emptied into the duodenum, where the liver and pancreas add their secretions. The small intestine also adds enzymes and is involved in absorbing nutrients. The large intestine is primarily involved in removing water.

involves a complex series of events. First, a ball of food, known as a *bolus,* is formed by the tongue and moved to the back of the mouth cavity. Here it stimulates the walls of the throat, also known as the *pharynx.* Nerve endings in the lining of the pharynx are stimulated, causing a reflex contraction of the walls of the esophagus, which transports the bolus to the stomach. Since both food and air pass through the pharynx, it is important to prevent food from getting into the lungs. During swallowing, the larynx is pulled upward. This causes a flap of tissue called the *epiglottis* to cover the opening to the trachea and prevent food from entering the trachea.

William Beaumont (1785–1853)

William Beaumont was born in 1785 in Lebanon, Connecticut, to Samuel and Lucretia Beaumont. He was the third of nine children. William attended school in Lebanon and, at the age of twenty-two, went to live in Champlain, New York, near the border with Vermont, just south of the Canadian border. He had relatives living in the area. In 1807, he became the schoolmaster of Champlain. In 1809, he began to study medicine. As was the custom of the time, he began reading about medical subjects and served an apprenticeship under the guidance of two local doctors in St. Albans, Vermont, about 55 km (35 mi) from Champlain. In June 1812, he was licensed to practice medicine. In September 1812, he joined the U.S. Army as a surgeon's mate and was assigned to the Sixth Infantry Regiment in Plattsburgh, New York. During the War of 1812, he treated the various illnesses and wounds related to the poor living conditions and battle injuries of the soldiers.

In 1815, after the war, he entered private practice in Plattsburgh. But in 1819, he reenlisted in the Army and was sent to Fort Mackinac on Mackinac Island in Lake Huron near the Canadian border. While living in Plattsburgh, he met Deborah Green, who was divorced from Nathanial Platt. In 1821, Beaumont returned to Plattsburgh from Fort Mackinac and married Green, who returned with him to Fort Mackinac. They had four children, two of whom were born while the Beaumonts were at Fort Mackinac.

At Fort Mackinac, Dr. Beaumont had an experience that would make him an expert on the physiology of digestion.

On the morning of June 6, 1822, a nineteen-year-old French-Canadian fur trapper named Alexis St. Martin was shot in the stomach by an accidental discharge from a shotgun. Dr. Beaumont was called to attend the wounded man. Part of the stomach and body wall had been shot away and parts of St. Martin's clothing were imbedded in the wound. Although Dr. Beaumont did not expect St. Martin to live, he quickly cleaned the wound, pushed portions of the lung and stomach that were protruding back into the cavity, and dressed the wound. The next day, Beaumont was surprised to find St. Martin alive and was encouraged to do what he could to extend the young man's life. In fact, Beaumont cared for St. Martin for two years. When the wound was completely healed, the stomach had fused to the body wall and a hole allowed Beaumont to look into the stomach. Fortunately for St. Martin, a flap of tissue from the lining of the stomach closed off the opening so that what he ate did not leak out.

Beaumont found that he could look through the opening and observe the activities in the stomach. He recognized that this presented an opportunity to study the function of the stomach in a way that had not been done before. He gathered gastric juice, had its components identified, introduced food into the hole with a string attached so that he could retrieve the food particles that were partially digested for examination, and observed the effect of emotion on digestion. He discovered many things that were new to science and contrary to the teachings of the time. He recounted many of his observations and experiments in his journal.

I consider myself but a humble inquirer after truth—a simple experimenter. And if I have been led to conclusions opposite to the opinions of many who have been considered luminaries of physiology, and, in some instances, from all the professors of this science, I hope the claim of sincerity will be conceded to me, when I say that such difference of opinion has been forced upon me by the convictions of experiment, and the fair deductions of reasoning.

In 1833, Beaumont published *Experiments and Observations on the Gastric Juice and the Physiology of Digestion,* which contained descriptions of some 240 experiments. The following are some of his important discoveries.

1. He measured the temperature of the stomach and found that it does not heat up when food is introduced, as was thought at the time. *"But from the result of a great number of experiments and examinations, made with a view to asserting the truth of this opinion, in the empty and full state of the organ, ... I am convinced that there is no alteration of temperature. ..."*

2. He found that pure gastric juice contains large amounts of hydrochloric acid. This was contrary to the prevailing opinion that gastric juice was simply water. *"I think I am warranted, from the result of all the experiments, in saying, that the gastric juice, so far from being 'inert as water,' as some authors assert, is the most general solvent in nature, of alimentary matter—even the hardest bone cannot withstand its action."*

3. He observed that gastric juice is not stored in the stomach but is secreted when food is eaten. *"The gastric juice does not accumulate in the cavity of the stomach, until alimentary matter is received, and excite its vessels to discharge their contents, for the immediate purpose of digestion."*

4. He realized that digestion begins immediately when food enters the stomach. The prevailing opinion of the day was that nothing happened for an hour or more. *"At 2 o'clock P.M.— twenty minutes after having eaten an ordinary dinner of boiled, salted beef, bread, potatoes, and turnips, and drank [sic] a gill of water, I took from his stomach, through the artificial opening, a gill of the contents. ... Digestion had evidently commenced, and was perceptually progressing, at the time."*

5. He discovered that food in the stomach satisfies hunger even though it is not eaten. *"To ascertain whether the sense of hunger would be allayed without food being passed through the oesophagus [sic], he fasted from breakfast time, 'til 4 o'clock P.M., and became quite hungry. I then put in at the aperture, three and a half drachms of lean, boiled beef. The sense of hunger immediately subsided, and stopped the borborygmus, or croaking noise, caused by the motion of the air in the stomach and intestines, peculiar to him since the wound, and almost always observed when the stomach is empty."*

Dr. Beaumont was eventually posted to St. Louis, Missouri, where he left the military in 1839 and went into private practice. In 1853, he slipped coming out of a patient's home, fell, and hit his head. The resulting injury became infected and resulted in his death at the age of sixty-eight. Today, his role as an army surgeon and his major contribution to an understanding of gastric physiology are recognized in the naming of the William Beaumont Army Medical Center in El Paso, Texas.

St. Martin did not take kindly to the probings and experiments of Beaumont and twice ran away from Beaumont's care back to Canada, where he married, had two children, and resumed his former life as a voyageur, fur trapper, and wood chopper. He did not die until 1880 at the age of eighty-six, having lived over sixty years with a hole in his stomach.

Source: Modified from the *Hutchinson Dictionary of Scientific Biography.* © RM, 2007. Reprinted by permission.

Digestive Activities in the Stomach

In the stomach, additional liquid, called *gastric juice,* is added to the food. Gastric juice contains enzymes and hydrochloric acid. The major enzyme of the stomach is *pepsin,* which initiates the chemical breakdown of protein. The pH of gastric juice is very low, generally around pH 2. The entire mixture is churned by the contractions of the three layers of muscle in the stomach wall. The combined activities of enzymatic breakdown, chemical breakdown by hydrochloric acid, and mechanical processing by muscular movement result in a thoroughly mixed liquid called *chyme.* Chyme eventually leaves the stomach through a valve known as the pyloric sphincter and enters the small intestine.

Digestive Activities in the Small Intestine

The first part of the **small intestine** is known as the **duodenum.** In addition to producing enzymes, the duodenum secretes several kinds of hormones that regulate the release of food from the stomach and the release of secretions from the pancreas and liver. The **pancreas** produces a number of different digestive enzymes and also secretes large amounts of bicarbonate ions, which neutralize stomach acid so that the pH of the duodenum is about pH 8. The **liver** is a large organ in the upper abdomen that performs several functions. One of its functions is the secretion of *bile.* When bile leaves the liver, it is stored in the *gallbladder* prior to being released into the duodenum. When bile is released from the gallbladder, it assists mechanical mixing by breaking large fat globules into smaller particles. This process is called *emulsification.*

Digestive Activities in the Large Intestine

Along the length of the small intestine, additional watery juices are added until the mixture reaches the **large intestine.** The large intestine is primarily involved in reabsorbing the water that has been added to the food tube along with saliva, gastric juice, bile, pancreatic secretions, and intestinal juices. The large intestine is also home to a variety of different kinds of bacteria. Most live on the undigested food that makes it through the small intestine. Some provide additional benefit by producing vitamins that can be absorbed from the large intestine. A few kinds may cause disease.

Nutrient Uptake

The process of digestion has resulted in a variety of simple organic molecules that are available for absorption from the tube of the gut into the circulatory system. Nearly all nutrients are absorbed in the small intestine. Many features of the small intestine provide a large surface area for the absorption of nutrients. First of all, the small intestine is a very long tube; the longer the tube, the greater the internal surface area. In a typical adult human, it is about 6 m (about 20 ft) long. In addition to length, the lining of the intestine consists of millions of tiny projections called *villi,* which increase the surface area. When we examine the cells that make up the villi, we find

that they also have folds in their surface membranes. All of these characteristics increase the surface area available for the transport of materials from the gut into the circulatory system (figure 24.11).

Scientists estimate that the cumulative effect of all of these features produces a total intestinal surface area of about 250 m^2 (about 2,600 ft^2). That is equivalent to about the area of a basketball court.

The surface area by itself would be of little value if it were not for the intimate contact of the circulatory system with this lining. Each villus contains several capillaries and a branch of the lymphatic system called a *lacteal.* The close association between the intestinal surface and the circulatory and lymphatic systems allows for the efficient uptake of nutrients from the cavity of the gut into the circulatory system.

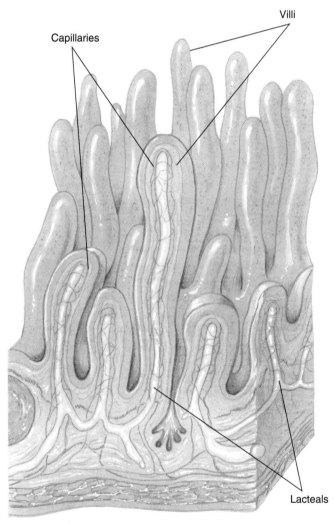

FIGURE 24.11 The surface area of the intestinal lining is increased by the many fingerlike projections known as villi. Within each villus are capillaries and lacteals. Most kinds of materials enter the capillaries, but most fat-soluble substances enter the lacteals, giving them a milky appearance. Lacteals are part of the lymphatic system. Because the lymphatic system empties into the circulatory system, fat-soluble materials also eventually enter the circulatory system. The close relationship between these vessels and the epithelial lining of the villus allows for efficient exchange of materials from the intestinal cavity to the circulatory system.

Several different kinds of processes are involved in the transport of materials from the intestine to the circulatory system. Some molecules, such as water and many ions, simply diffuse through the wall of the intestine into the circulatory system. Other materials, such as amino acids and simple sugars, are assisted across the membrane by carrier molecules. Fatty acids and glycerol are absorbed into the intestinal lining cells where they are resynthesized into fats and enter lacteals in the villi. Since the lacteals are part of the lymphatic system, which eventually empties its contents into the circulatory system, fats are also transported by the blood. They just reach the blood by a different route.

NUTRITION

Organisms maintain themselves by constantly processing molecules to provide building blocks for new living material and energy to sustain themselves. The word **nutrition** is used in two related contexts. First of all, nutrition is a branch of science that seeks to understand food, its nutrients, how the nutrients are used by the body, and how inappropriate combinations or quantities of nutrients lead to ill health. The word *nutrition* is also used in a slightly different context to refer to all the processes by which we take in food and utilize it, including *ingestion, digestion, absorption,* and *assimilation. Ingestion* involves the process of taking food into the body through eating. *Digestion* involves the breakdown of complex food molecules to simpler molecules. *Absorption* involves the movement of simple molecules from the digestive system to the circulatory system for dispersal throughout the body. *Assimilation* involves the modification and incorporation of absorbed molecules into the structure of the organism.

Kinds of Nutrients

Nutrients are the kinds of molecules that must be consumed to maintain the body. They can be organized into several categories based on their chemical structure and the functions they serve in the body. Commonly accepted categories are: carbohydrates; fats; proteins; vitamins, which are organic molecules; and minerals and water, which are inorganic. The chemical nature of organic molecules is discussed in chapter 19.

Carbohydrates are organic molecules that contain carbon, hydrogen, and oxygen atoms in the following ratio: CH_2O. Common carbohydrates that serve as nutrients are various kinds of sugars and starches. The primary function served by these molecules is to provide energy. Each gram of carbohydrate provides about 4 kilocalories of energy. The sugar glucose undergoes aerobic cellular respiration to provide carbon dioxide, water, and ATP. See chapter 20 for a discussion of the process of aerobic cellular respiration. Starch and some other kinds of sugars can be converted to glucose and thus are able to undergo aerobic cellular respiration.

Fats are also organic molecules that consist of carbon, hydrogen, and oxygen, but fats have fewer oxygen atoms than carbohydrates. Fats are excellent sources of energy since they can be broken down into smaller units that enter the Krebs cycle to provide carbon dioxide, water, and ATP. Fats provide about 9 kilocalories of energy per gram—more than twice the energy of carbohydrates.

Proteins are polymers of amino acids that contain carbon, hydrogen, oxygen, and nitrogen, and some contain sulfur. Proteins are important in the structure of cells and the body. All cells and parts of the body contain protein. Thus, protein is needed as a nutrient to provide for growth and repair of cells, tissues, and organs. Proteins also contain energy (about 4 kilocalories per gram) but yield less energy than fats or carbohydrates because they are more difficult to break down in aerobic cellular respiration.

Vitamins are organic molecules that cannot be manufactured by the body but are needed in small amounts to assist enzymes in their functions.

Minerals are inorganic molecules needed for a variety of purposes. Some such as calcium are needed as a component of bone. Others are needed as components of enzymes or other important molecules. Most minerals are needed in relatively small amounts.

Water is often not even classified as a nutrient but is vitally important to sustain life. Water is the most common molecule in the body. Since it is constantly lost in urine, feces, and by evaporation, it must be replaced.

All of these nutrients must be obtained in adequate amounts on a daily basis or ill health results.

Guidelines for Obtaining Adequate Nutrients

In an attempt to give people some guidelines for planning a diet that provides adequate amounts of the six classes of nutrients, nutritional scientists and government officials in the United States and many other countries developed nutrient standards.

Dietary Reference Intakes

The U.S. Department of Agriculture has over the years published various guidelines as aids to maintaining good nutritional health. The current guidelines are called the **Dietary Reference Intakes** and provide information on the amounts of certain nutrients various members of the public should receive. These guidelines are very detailed and include different guidelines for males and females of various ages and for pregnant and nursing mothers. They also provide information about the maximum amount of certain nutrients that people should get. In other words, you can have too much of a good thing.

Dietary Reference Intakes are used when preparing product labels. Federal law requires that labels list ingredients from the greatest to the least in quantity. These labels typically provide information about the amounts of various nutrients present and the number of kilocalories present. In addition to carbohydrates, fats, proteins, and fiber, there are about

TABLE 24.2   Dietary Reference Intakes for Some Common Nutrients

Nutrient	Women 19–30 Years Old	Men 19–30 Years Old	Maximum for Persons 19–30 Years Old	Value of Nutrient
Carbohydrates	130 g/day (45–65% of kilocalories)	130 g/day (45–65% of kilocalories)	No maximum set, but refined sugars should not exceed 25% of total kilocalories.	A source of energy
Proteins	46 g/day (10–35% of kilocalories)	56 g/day (10–35% of kilocalories)	No maximum set, but high protein diets stress kidneys.	Proteins are structural components of all cells. There are ten essential amino acids that must be obtained in the diet.
Fats	20–35% of kilocalories	20–35% of kilocalories	Up to 35% of total kilocalories	Energy source and building blocks needed for many molecules
Saturated and trans fatty acids	No reference amount determined	No reference amount determined	As low as possible	
Linoleic acid	12 g/day	17 g/day	No maximum set	Essential fatty acid needed for enzyme function and to maintain epithelial cells
Linolenic acid	1.1 g/day	1.6 g/day	No maximum set	Essential fatty acid needed to reduce risk of coronary heart disease
Cholesterol	No reference amount determined	No reference amount determined	As low as possible	None needed, since the liver makes cholesterol
Total fiber	25 g/day	38 g/day	No maximum set	Improve gut function
Calcium	1 g/day	1 g/day	2.5 g/day	Needed for the structure of bones and many other functions
Iron	18 mg/day	8 mg/day	45 mg/day	Needed to build hemoglobin of red blood cells
Sodium	1.5 g/day	1.5 g/day	5.8 g/day (most people exceed this limit)	Needed for normal cell function
Vitamin A	700 ug/day	900 ug/day	3,000 ug/day	Maintains skin and intestinal lining
Vitamin C	75 mg/day	90 mg/day	2,000 mg/day	Maintains connective tissue and skin
Vitamin D	5 ug/day	5 ug/day	50 ug/day	Needed to absorb calcium for bones

twenty-five vitamins and minerals for which Dietary Reference Intakes standards are set. Table 24.2 give examples of some of the more common nutrients and their reference amounts for young adults. The food and drink consumed from day to day constitutes a person's **diet.** A good understanding of nutrition can promote good health and involves an understanding of the energy and nutrient content in various foods.

Counting Calories

One of the primary pieces of information on the nutritional facts label on foods you purchase is the energy content of the food item. The unit used to measure the amount of energy in foods is the **kilocalorie (kcal).** One kilocalorie is the amount of heat needed to increase the temperature of one *kilogram* of water one degree Celsius. Remember that the prefix *kilo-*means "1,000 times" the value listed. Therefore, a kilocalorie is 1,000 times more heat energy than a **calorie,** which is the amount of heat needed to raise the temperature of one *gram* of water one degree Celsius. However, the amount of energy contained in food is usually called a Calorie with a capital C. This is unfortunate because it is easy to confuse a Calorie, which is really a kilocalorie, with a calorie. Most books on nutrition and dieting use the term *Calorie* to refer to *food Calories.* A commonly used standard for a daily diet is 2,000 Calories. Thus, the information on a nutrition facts label can be used to help a person determine the number of Calories they are consuming per day.

The Food Guide Pyramid

Using Dietary Reference Intakes and product labels or counting kilocalories is a pretty complicated way for a person to plan a diet. Planning a diet around basic food groups is generally easier. The federal government adopted the **Food Guide Pyramid** of the Department of Agriculture as one of its primary tools to help the general public plan for good nutrition. Over the years, as new nutritional information became available, nutritional suggestions have been modified. The current Food Guide Pyramid, released in 2005, is a modification of earlier versions (see figure 24.12). It contains six basic groups of foods

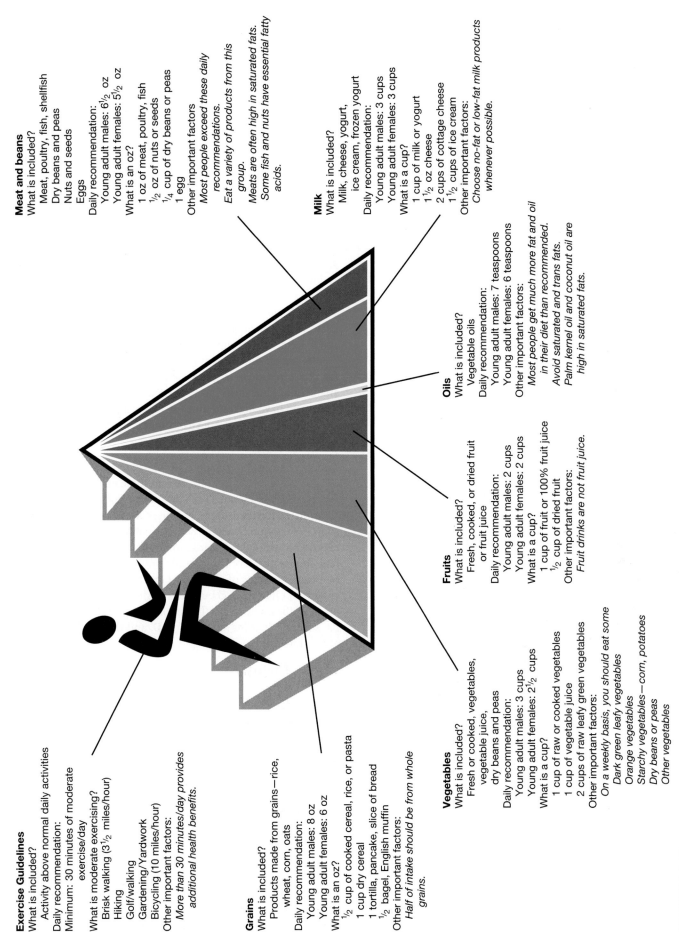

Exercise Guidelines

What is included?
  Activity above normal daily activities
Daily recommendation:
  Minimum: 30 minutes of moderate
    exercise/day
What is moderate exercising?
  Brisk walking (3½ miles/hour)
  Hiking
  Golf/walking
  Gardening/Yardwork
  Bicycling (10 miles/hour)
Other important factors:
  More than 30 minutes/day provides
    additional health benefits.

Grains

What is included?
  Products made from grains—rice,
    wheat, corn, oats
Daily recommendation:
  Young adult males: 8 oz
  Young adult females: 6 oz
What is an oz?
  ½ cup of cooked cereal, rice, or pasta
  1 cup dry cereal
  1 tortilla, pancake, slice of bread
  ½ bagel, English muffin
Other important factors:
  Half of intake should be from whole
    grains.

Vegetables

What is included?
  Fresh or cooked, vegetables,
    vegetable juice,
    dry beans and peas
Daily recommendation:
  Young adult males: 3 cups
  Young adult females: 2½ cups
What is a cup?
  1 cup of raw or cooked vegetables
  1 cup of vegetable juice
  2 cups of raw leafy green vegetables
Other important factors:
  On a weekly basis, you should eat some
    Dark green leafy vegetables
    Orange vegetables
    Starchy vegetables—corn, potatoes
    Dry beans or peas
    Other vegetables

Meat and beans

What is included?
  Meat, poultry, fish, shellfish
  Dry beans and peas
  Nuts and seeds
  Eggs
Daily recommendation:
  Young adult males: 6½ oz
  Young adult females: 5½ oz
What is an oz?
  1 oz of meat, poultry, fish
  ½ oz of nuts or seeds
  ¼ cup of dry beans or peas
  1 egg
Other important factors:
  Most people exceed these daily
    recommendations.
  Eat a variety of products from this
    group.
  Meats are often high in saturated fats.
  Some fish and nuts have essential fatty
    acids.

Milk

What is included?
  Milk, cheese, yogurt,
    ice cream, frozen yogurt
Daily recommendation:
  Young adult males: 3 cups
  Young adult females: 3 cups
What is a cup?
  1 cup of milk or yogurt
  1½ oz cheese
  2 cups of cottage cheese
  1½ cups of ice cream
Other important factors:
  Choose no-fat or low-fat milk products
    whenever possible.

Oils

What is included?
  Vegetable oils
Daily recommendation:
  Young adult males: 7 teaspoons
  Young adult females: 6 teaspoons
Other important factors:
  Most people get much more fat and oil
    in their diet than recommended.
  Avoid saturated and trans fats.
  Palm kernel oil and coconut oil are
    high in saturated fats.

Fruits

What is included?
  Fresh, cooked, or dried fruit
    or fruit juice
Daily recommendation:
  Young adult males: 2 cups
  Young adult females: 2 cups
What is a cup?
  1 cup of fruit or 100% fruit juice
  ½ cup of dried fruit
Other important factors:
  Fruit drinks are not fruit juice.

FIGURE 24.12 In 2005, the U.S. Department of Agriculture published a new food guide pyramid. It includes guidelines for people of different ages. Only the guidelines for young adults nineteen to thirty years of age are included here. See the website http://www.mypyramid.gov/ for more information.

Source: U.S. Department of Agriculture, 2005.

607

Measuring the Caloric Value of Foods

A *bomb calorimeter* is an instrument that is used to determine the heat of combustion or caloric value of foods. To operate the instrument, a small food sample is formed into a pellet and sealed inside a strong container called the bomb. The bomb is filled with 30 atmospheres of oxygen and then placed in a surrounding jacket that is filled with water. The sample is electrically ignited, and as it aerobically reacts with the oxygen, the bomb releases heat to the surrounding water jacket. As the heat is released, the temperature of the water rises. Recall that a calorie is the amount of energy (or heat) needed to increase the temperature of one gram of water one degree Celsius. Therefore, if one gram of food results in a water temperature increase of 13°C for each gram of water, that food has the equivalent of 13 calories.

with guidelines for the amounts one needs daily from each group for ideal nutritional planning. The Food Guide Pyramid is designed to encourage people to reduce the amount of fat, sugar, and salt in the diet, while increasing daily servings of fruits and vegetables. It also includes guidelines on the amount of daily exercise people should get.

1. Grain Products Group   Grains include vitamin-enriched or whole-grain cereals and products such as breads, bagels, buns, crackers, dry and cooked cereals, pancakes, pasta, and tortillas. Because whole grain foods contain more nutrients than those that are not whole grain, it is recommended that half of all grains consumed be whole grain products. The Food Guide Pyramid recommends that people consume 5–8 ounces of grain products per day depending on their age and sex. This is relatively easy to meet, since a slice of bread, tortilla, half a bagel, or a cup of dry cereal is equivalent to an ounce. Grains should provide most of your kilocalorie requirements in the form of complex carbohydrates such as starch, which is the main ingredient in most grain products.

Significant nutritional components: carbohydrate, fiber, several B vitamins, vitamin E, and the minerals iron, selenium, and magnesium.

2. Vegetables Group   Vegetables include either raw or cooked nonsweet plant materials, such as broccoli, carrots, cabbage, corn, green beans, tomatoes, potatoes, lettuce, and spin-

ach. The Food Guide Pyramid recommends 1–3 cups of items from this food group each day depending on a person's age and sex. Since different vegetables contain different amounts of vitamins, it is wise to include as much variety as possible in this group. Therefore, vegetables are divided into five subgroups: dark green vegetables, orange vegetables, dry beans and peas, starchy vegetables, and other vegetables. On a weekly basis, people should be sure to consume some vegetables from each of the following three groups: dark green, orange, and dry beans and peas. The amount of starchy vegetables can vary considerably depending on a person's activity level and caloric needs.

Significant nutritional components: carbohydrate, fiber, several B vitamins, vitamins A, C, E, and K, and the minerals iron, potassium, and magnesium.

3. Fruits Group   Fruits include such sweet plant products as melons, berries, apples, oranges, and bananas. The Food Guide Pyramid recommends 1–2 cups of fruit per day depending on a person's age and sex. Fruit may be fresh or preserved or fruit juice. Many individual pieces of fruit (apple, orange, banana) are about 1 cup. A half cup of dried fruit is equivalent to 1 cup.

However, since these foods tend to be high in natural sugars, consumption of large amounts of fruits can add significant amounts of kilocalories to your diet. Fruit drinks are not fruit juices and have high amounts of sugars added to them.

Significant nutritional components: carbohydrate, fiber, water, and vitamin C.

4. Dairy Products Group   Dairy products include all cheeses, ice cream, yogurt, and milk. Two to three cups from this food group are recommended each day depending on a person's age. Vitamin D fortified dairy products are the primary dietary source of vitamin D. Vitamin D is also manufactured in the skin exposed to sunlight. Because milk products contain saturated fats, choose no-fat or low-fat milk products. Many cheeses contain large amounts of cholesterol and fat for each serving. The cholesterol in low fat cheeses is significantly reduced.

Significant nutritional components: protein, carbohydrate, fat, some B vitamins and vitamin D, and the minerals calcium and potassium.

5. Meat, Poultry, Fish, and Dry Beans Group   This group contains most of the things we eat as a source of protein. In addition to meat, poultry, fish and beans, nuts, peas, tofu, and eggs are considered a part of this group. The Food Guide Pyramid recommends that adults receive 5–6½ ounces of protein per day depending on a person's age and sex. Since the amount needed per day is so small, most people eat many times what is needed.

Daily protein intake is essential since protein is not stored in the body like fats or carbohydrates. Furthermore, of the amino acids that combine to form proteins, there are about ten that the body cannot manufacture. These are called essential amino acids since it is essential that we receive them in the diet. Animal sources of protein have all the amino acids present and are called complete proteins. Many plant proteins may lack specific amino acids and are called incomplete proteins. Vegetarians must pay particular attention to acquiring adequate sources of protein because they have eliminated animal flesh

The Dynamic Skeleton

Most people think of the skeleton as a nonliving framework that supports the more "soft" body parts. However, the skeleton is anything but inert and serves functions in addition to that of support. Box figure 24.2 shows the primary parts of the human skeleton.

A primary function of the skeleton is to provide support while allowing motion. At joints, tough, fibrous cords called ligaments hold bones together, and muscles are attached to the bones by tough, fibrous cords called *tendons*. Muscles are typically attached to two different bones, and muscular contraction results in one bone moving with respect to another, creating a bending or rotation at the joint. The actions of muscles are also important in providing support because they generate tension that provides rigidity to the skeletal framework, like the cables that hold up the mast of a ship.

Protection is another major function of the skeleton. The brain, spinal cord, and organs of the chest are enclosed by and protected by the bony structures of the skull, vertebrae, and ribs and associated structures, respectively.

Chemically, the skeleton is a combination of an inorganic mineral, hydroxyapatite, and organic fibers of various kinds of protein. Hydroxyapatite contains calcium and phosphorus and has the chemical formula $Ca_{10}(PO_4)_6(OH)_2$. There is constant turnover of both hydroxyapatite and protein in the bones of the skeleton. For example, calcium and phosphorus have several important functions in the body and are distributed throughout the body by the circulatory system. If the blood does not contain adequate calcium and phosphorus from the diet, these minerals are taken from the skeleton to supply immediate needs. These substances can be replaced later, when they are present in the diet. This function is particularly important in pregnant and nursing mothers, who must provide calcium and phosphorus for bone development of embryos and as a source of calcium in the mother's milk for nursing infants.

Without adequate calcium and phosphorus in the diet, the mother can experience significant bone loss and skeletal development in the infant is impaired.

Nowhere is the dynamic nature of the skeleton more evident than in the process of growth. Because long bones are typically connected to other bones at a joint, it is important that the joint continue to function while the bone is growing. Long bones of the body have special growth areas near their ends but not on the very end. Thus, new bone is added near the ends so that the bones become longer without interfering with the function of the joint. Bones of the skull and certain other bones grow directly from their edges, eventually growing together and fusing into a solid structure. The bones of the skull of newborn babies are not joined and are capable of being moved around during the birth process. As the individual bones of the skull grow, the bones eventually knit together to form a solid skull, but a "soft spot" persists on the top of the head for a time after birth.

One of the activities that stimulates change in the nature of the skeleton is the stress and strain placed on it. When the skeleton is "challenged," it gets stronger by adding material. When it is not challenged, material is lost and the bones get weaker. For example, astronauts in weightless conditions lose bone mass. Therefore, exercise programs are very important parts of the schedule of astronauts. Since exercise strengthens

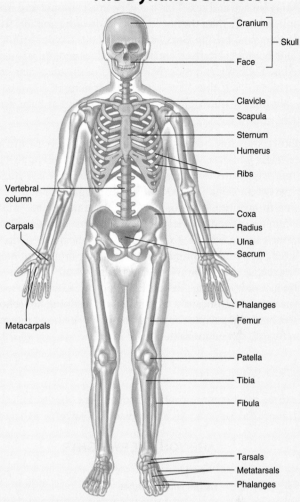

Box Figure 24.2   Major bones of the skeleton.

bones, it is particularly important in elderly persons who tend to have reduced bone mass for age-related reasons.

The skeleton is even involved in the production of blood. The red bone marrow, found in the core of certain bones of the skeleton (vertebrae, ribs, breast bone, and upper arm and leg bones), is the site of production of red and white blood cells and platelets. The lymph nodes, tonsils, spleen, and thymus also produce some kinds of white blood cells. Bone marrow transplants are often provided for people whose blood cell production capacity has been destroyed by chemotherapy or radiation.

from their diet. They can get all the essential amino acids if they eat proper combinations of plant materials.

Since many sources of protein (particularly meat and nuts) also include significant fat and health recommendations suggest reduced fat, choosing low-fat sources of protein is important.

Selecting foods that have less fat, broiling rather than frying, or removing the fat before cooking reduces fat.

Significant nutritional components: protein, fat, several B vitamins and vitamin E from seeds and nuts, and the minerals iron, zinc, and magnesium.

6. Oils Oils are fats that are liquids at room temperature. The Food Guide Pyramid recommends 3–7 teaspoons per day depending on age and sex. This is a very small amount and most people get more than they need in salad dressing and oils used in cooking and baking. Small amounts of oils are important in the diet because certain essential fatty acids cannot be manufactured by the body and must be obtained in the diet. However, since oils are fats, they have a high caloric content.

Significant nutritional components: Essential fatty acids and vitamin E.

7. Exercise Since over 65 percent of adult Americans are overweight and over 30 percent are obese, exercise is important to improving nutritional health. Therefore, exercise is included as a part of the Food Guide Pyramid. Although it is not directly related to nutrition, the amount of exercise people get affects the number of food calories they can consume on a daily basis without gaining weight. Exercise has other health benefits as well. The pyramid recommends at least thirty minutes of moderate exercise per day, and longer periods or more vigorous exercise has additional health benefits. Moderate exercise elevates the heart rate significantly. Activities such as doing household chores or walking while shopping do not elevate heart rate and therefore do not count as moderate exercise.

Eating Disorders

The three most common eating disorders are obesity, bulimia, and anorexia nervosa. All three disorders involve behaviors that lead to ill health. The causes of the behaviors are complex and involve a strong psychological component. Metabolic imbalances may also contribute to the development of these disorders, particularly obesity. These disorders are also related to the prevailing perceptions and values of the culture in which we live. Some studies have shown that there is also a shared familial predisposition to these disorders. It has been suggested that genetic factors probably influence the risk for eating disorders. Eating disorders are often associated with psychological depression.

Obesity

Obesity is the condition of being overweight to the extent that a person's health and life span are adversely affected. Obesity occurs when people consistently take in more food energy than is necessary to meet their daily requirements. About 30 percent of the U.S. population is considered to be obese based on their having a *body mass index* (BMI = weight in kilograms/height in meters^2) above 30. Figure 24.13 shows the body mass index for various heights and weights.

At first glance, it would appear that obesity is a simple problem to solve. To lose weight, all that people must do is consume fewer kilocalories, exercise more, or do both. While all obese people have an imbalance between their need for kilocalories and the amount of food they eat, the reasons for this imbalance are complex and varied.

People differ metabolically. Even at rest, energy is required to maintain breathing, heart rate, and other normal body

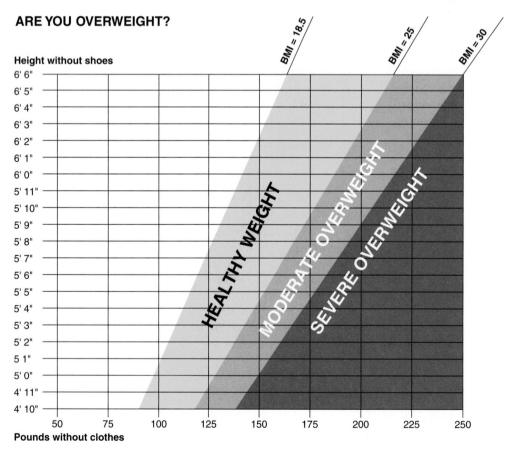

FIGURE 24.13 Find your height and weight and determine your approximate BMI.

TABLE 24.3 Typical Energy Requirements for Common Activities

Light Activities (120–150 kcal/h)	Light to Moderate Activities (150–300 kcal/h)	Moderate Activities (300–400 kcal/h)	Heavy Activities (420–600 kcal/h)
Dressing	Sweeping floors	Pulling weeds	Chopping wood
Typing	Painting	Walking behind a lawnmower	Shoveling snow
Slow walking	Walking 2–3 mi/h	Walking 3.5–4 mi/h on level surface	Walking or jogging 5 mi/h
Standing	Bowling	Golf (no cart)	Walking up hills
Studying	Store clerking	Doubles tennis	Cross-country skiing
Sitting activities	Canoeing 2.5–3 mi/h	Canoeing 4 mi/h	Swimming
	Bicycling on level surface at 5.5 mi/h	Volleyball	Bicycling 11–12 mi/h or up and down hills

functions. The **basal metabolic rate** is the rate at which a person uses energy when at rest. Some people have a low basal metabolic rate and, therefore, use fewer kilocalories for maintenance than other people.

A primary cause of obesity is inactivity. Often exercise is all that is needed to control a weight problem. In addition to increasing metabolism during the exercise itself, exercise tends to raise the basal metabolic rate for a period of time following exercise. Table 24.3 lists several common activities and the amount of energy needed to sustain the activity.

Finally, our culture encourages food consumption. Social occasions and business meetings frequently involve eating, and restaurants and other food preparers have increased the size of portions significantly over the past fifty years. For example, in the 1950s, a fast-food serving of French fries was 2.4 ounces. Today, that size is still available, but it is the small size, and medium and large sizes contain two to three times the quantity of the small size.

So, in the final analysis, maintaining a desirable weight boils down to two things: proper choice of the kinds and amounts of food and an activity level that consumes the caloric content of the food we eat.

Bulimia

Bulimia ("hunger of an ox" in Greek) is a disease condition in which the person engages in a cycle of eating binges followed by purging the body of the food by inducing vomiting or using

---

Myths, Mistakes, and Misunderstandings

Diet and Nutrition

There is probably no area of modern life in which there is more misinformation, pseudoscience, and myth than nutrition. Diet books abound, each purporting to provide the answer to nutritional and weight control problems. It is helpful to look at the science behind these various statements. The list below provides several examples of the mismatch between claims and scientific reality.

Myth or Misunderstanding	Scientific Basis
1. Exercise burns calories.	Calories are not molecules, so they cannot be burned in the physical sense, but you do oxidize (burn) the fuels to provide yourself with the energy needed to perform various activities.
2. Active people who are increasing their fitness need more protein.	The amount of protein needed is very small—about 50 g. Most people get many times the amount of protein required.
3. Vitamins give you energy.	Most vitamins assist enzymes in bringing about chemical reactions, some of which may be energy-yielding, but they are not sources of energy.
4. Large amounts of protein are needed to build muscle.	A person can build only a few grams of new muscle per day. Therefore, consuming large amounts of protein will not increase the rate of muscle growth.
5. Large quantities (mega doses) of vitamins will fight disease, build strength, and increase length of life.	Quantities of vitamins that greatly exceed recommendations have not been shown to be beneficial. Large doses of some vitamins (vitamins A, D, B_3) are toxic.
6. Special protein supplements are more quickly absorbed and can build muscle faster.	Adequate protein is obtained in nearly all diets. The supplements may be absorbed faster, but that does not mean that they will be incorporated into muscle mass faster.
7. Vitamins prevent cancer, heart disease, and other health problems.	Vitamins are important to health. However, it is a gross oversimplification to suggest that consumption of excess amounts of specific vitamins will prevent certain diseases. Many factors contribute to the cause of disease.

Nutritional Needs Associated with Pregnancy and Lactation

Risk-management practices that help in avoiding chronic adult diseases become even more important when planning pregnancy. Studies have shown that an inadequate supply of the essential nutrients can result in infertility, spontaneous abortion, and abnormal fetal development. The period of pregnancy and milk production (lactation) requires that special attention be paid to the diet to ensure proper fetal development, a safe delivery, and a healthy milk supply.

The daily amount of essential nutrients must be increased, as should the kilocaloric intake. Kilocalories must be increased by 300 per day to meet the needs of an increased basal metabolic rate; the development of the uterus, breasts, and placenta; and the work required for fetal growth. Some of these kilocalories can be obtained by drinking milk, which simultaneously supplies calcium needed for fetal bone development. Individuals who cannot tolerate milk should use supplementary sources of calcium. In addition, the daily intake of protein should be at least 65 g per day. As was mentioned earlier, most people consume much more than this amount of protein per day. Two essential nutrients, folic acid and iron, should be obtained through prenatal supplements because they are so essential to cell division and development of the fetal blood supply.

The mother's nutritional status affects the developing baby in several ways. If she is under fifteen years of age or has had two or more pregnancies in a two-year period, her nutritional stores often are inadequate to support a successful pregnancy. The use of drugs such as alcohol, caffeine, nicotine, and "hard" drugs (e.g., heroin) can result in decreased nutrient exchange between the mother and fetus. In particular, heavy smoking can result in low birth weights, and alcohol abuse is responsible for fetal alcohol syndrome.

---

laxatives. Many bulimics also use diuretics that cause the body to lose water and therefore reduce weight. It is often called the silent killer because it is difficult to detect. Bulimics are usually of normal body weight or overweight. The behavior has a strong psychological component and is often associated with depression or other psychological problems.

Vomiting may be induced physically or by the use of some nonprescription drugs. Case studies have shown that bulimics may take forty to sixty laxatives a day to rid themselves of food. For some, the laxative becomes addictive. The binge-purge cycle and associated use of diuretics results in a variety of symptoms that can be deadly.

Anorexia Nervosa

Anorexia nervosa is a nutritional deficiency disease characterized by severe, prolonged weight loss as a result of a voluntary severe restriction in food intake. It is most common among adolescent and preadolescent females. An anorexic person's fear of becoming overweight is so intense that even though weight loss occurs, it does not lessen the fear of obesity, and the person continues to diet, often even refusing to maintain the optimum body weight for his or her age, sex, and height. Persons who have this disorder have a distorted perception of their bodies. They see themselves as fat when in fact they are starving to death. Society's preoccupation with body image, particularly among young people, may contribute to the incidence of this disease.

WASTE DISPOSAL: THE EXCRETORY SYSTEM

Because cells are modifying molecules during metabolic processes, harmful waste products are constantly being formed. Urea is a common waste; many other toxic materials must be eliminated as well. Among these are large numbers of hydrogen

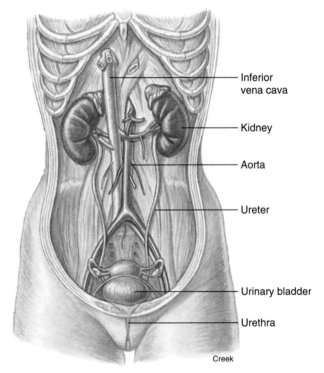

FIGURE 24.14 The primary organs involved in removing materials from the blood are the kidneys. The urine produced by the kidneys is transported by the ureters to the urinary bladder. From the bladder, the urine is emptied to the outside of the body by way of the urethra.

ions produced by metabolism. This excess of hydrogen ions must be removed from the bloodstream. Other molecules, such as water and salts, may be consumed in excessive amounts and must be removed. The primary organs involved in regulating the level of toxic or unnecessary molecules are the **kidneys.** The urine they produce flows to the urinary bladder through tubes known as *ureters*. The urine leaves the bladder through a tube known as the urethra (figure 24.14).

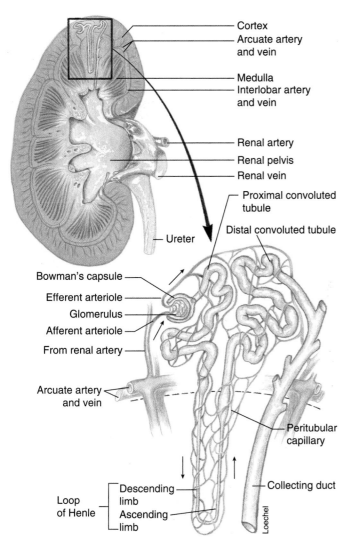

Cortex
Arcuate artery and vein
Medulla
Interlobar artery and vein
Renal artery
Renal pelvis
Renal vein
Proximal convoluted tubule
Distal convoluted tubule
Ureter
Bowman's capsule
Efferent arteriole
Glomerulus
Afferent arteriole
From renal artery
Arcuate artery and vein
Peritubular capillary
Collecting duct
Loop of Henle
Descending limb
Ascending limb
Loechel

FIGURE 24.15 The nephron and the closely associated blood vessels create a system that allows for the passage of materials from the circulatory system to the nephron by way of the glomerulus and Bowman's capsule. Materials are added to and removed from the fluid in the nephron via the tubular portions of the nephron.

The kidneys consist of about 2.4 million tiny units called **nephrons.** At one end of a nephron is a cup-shaped structure called *Bowman's capsule,* which surrounds a knot of capillaries known as a *glomerulus.* In addition to Bowman's capsule, a nephron consists of three distinctly different regions: the *proximal convoluted tubule,* the *loop of Henle,* and the *distal convoluted tubule.* The distal convoluted tubule of a nephron is connected to a collecting duct that transports fluid to the ureters and ultimately to the urinary bladder, where it is stored until it can be eliminated (figure 24.15).

As in the other systems, the excretory system involves a close connection between the circulatory system and a surface. In this case, the large surface is provided by the walls of the millions of nephrons, which are surrounded by capillaries. Three major activities occur at these surfaces: filtration, reabsorption, and secretion. The glomerulus presents a large surface for the filtering of material from the blood to Bowman's

capsule. Blood that enters the glomerulus is under pressure from the muscular contraction of the heart. The capillaries of the glomerulus are quite porous and provide a large surface area for the movement of water and small dissolved molecules from the blood into Bowman's capsule. Normally, only the smaller molecules, such as glucose, amino acids, and ions, are able to pass through the glomerulus into Bowman's capsule at the end of the nephron. The various kinds of blood cells and larger molecules such as proteins do not pass out of the blood into the nephron. This physical filtration process allows many kinds of molecules to leave the blood and enter the nephron. The volume of material filtered in this way is about 7.5 liters per hour. Since your entire blood supply is about 5 to 6 liters, there must be some method of recovering much of this fluid. The proximal convoluted tubule reabsorbs most of the useful molecules as the material passes through the nephron. The loop of Henle is involved in the reabsorption of water, and the distal convoluted tubule regulates ions, such as hydrogen ions and potassium ions.

Some molecules that pass through the nephron are relatively unaffected by the various activities going on in the kidney. One of these is urea, which is filtered through the glomerulus into Bowman's capsule. As it passes through the nephron, much of it stays in the tubule and is eliminated in the urine. Many other kinds of molecules, such as minor metabolic waste products and some drugs, are also lost in the urine.

CONTROL MECHANISMS

The nervous and endocrine systems are the major systems of the body that integrate stimuli and generate appropriate responses necessary to maintain homeostasis. There are many kinds of sense organs located within organs and on the surface of the body that respond to specific kinds of stimuli. A **stimulus** is any change in the environment that the organism can detect. Some stimuli, such as light or sound, are typically external to the organism; others, such as the pain generated by an infection, are internal. The reaction of the organism to a stimulus is known as a **response** (figure 24.16).

The *nervous system* consists of a network of cells with fibrous extensions that carry information throughout the body. The *endocrine system* consists of a number of glands that communicate with one another and with other tissues through chemicals distributed throughout the organism. **Glands** are organs that manufacture molecules that are secreted either through ducts or into surrounding tissue, where they are picked up by the circulatory system. *Endocrine glands* have no ducts and secrete their products, called **hormones,** into the circulatory system; other glands, such as the digestive glands and sweat glands, empty their contents through ducts. These kinds of glands are called *exocrine glands.*

Although the functions of the nervous and endocrine systems can overlap and be interrelated, these two systems have quite different methods of action. The nervous system functions very much like a computer. A message is sent along established pathways from a specific initiating point to a specific end point,

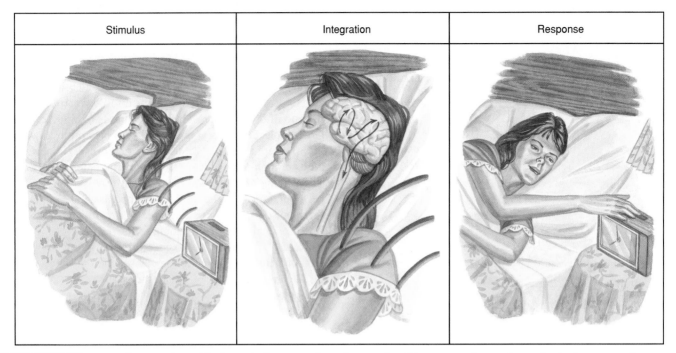

Stimulus	Integration	Response

FIGURE 24.16 A stimulus is any detectable change in the surroundings of an organism. When an organism receives a stimulus, it processes the information and may ignore the stimulus or generate a response to it.

and the transmission is very rapid. The endocrine system functions in a manner analogous to a radio broadcast system. Radio stations send their signals in all directions, but only those radio receivers that are tuned to the correct frequency can receive the message. Messenger molecules (hormones) are typically distributed throughout the body by the circulatory system, but only those cells that have the proper receptor sites can receive and respond to the molecules.

The Structure of the Nervous System

The basic unit of the nervous system is a specialized cell called a *neuron,* or *nerve cell.* A typical neuron consists of a central body called the *soma,* or *cell body,* that contains the nucleus, and several long, protoplasmic extensions called nerve fibers. There are two kinds of fibers: *axons,* which carry information away from the cell body, and *dendrites,* which carry information toward the cell body. Most nerve cells have one axon and several dendrites.

Neurons are arranged into two major systems. The *central nervous system,* which consists of the brain and spinal cord, is surrounded by the skull and the vertebrae of the spinal column. It receives input from sense organs, interprets information, and generates responses. The *peripheral nervous system* is located outside the skull and spinal column and consists of bundles of long axons and dendrites called nerves. There are two different sets of neurons in the peripheral nervous system. *Motor neurons* carry messages from the central nervous system to muscles and glands, and *sensory neurons* carry input from sense organs to the central nervous system. Motor neurons typically have one long axon that runs from the spinal cord to

a muscle or gland, while sensory neurons have long dendrites that carry input from the sense organs to the central nervous system (figure 24.17).

The Nature of the Nerve Impulse

Since most nerve cells have long fibrous extensions, it is possible for information to be passed along the nerve cell from one end to the other. The message that travels along a neuron is known as a **nerve impulse.** A nerve impulse is not a simple electric current but involves a specific sequence of chemical events involving activities at the cell membrane.

Since all cell membranes are differentially permeable, it is difficult for some ions to pass through the membrane, and the combination of ions inside the membrane is different from that on the outside. Cell membranes also contain proteins that actively transport specific ions from one side of the membrane to the other. Active transport involves the cell's use of ATP to move materials from one side of the cell membrane to the other. One of the ions that is actively transported from cells is the sodium ion (Na^+). At the same time sodium ions (Na^+) are being transported out of cells, potassium ions (K^+) are being transported into the normal resting cells. However, there are more sodium ions (Na^+) transported out than potassium ions (K^+) transported in.

Because a normal resting cell has more positively charged Na^+ ions on the outside of the cell than on the inside, a small but measurable voltage exists across the membrane of the cell. The voltage difference between the inside and outside of a cell membrane is about 70 millivolts (0.07 volt). The two sides of the cell membrane are, therefore, polarized in the same sense that a

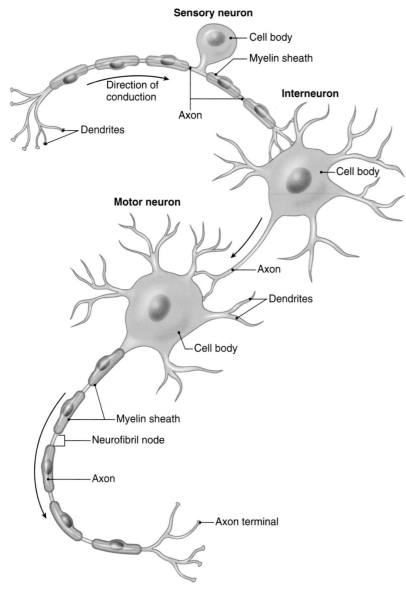

Sensory neuron
- Cell body
- Myelin sheath
- Direction of conduction
- Axon
- Dendrites

Interneuron
- Cell body
- Axon
- Dendrites
- Cell body

Motor neuron
- Myelin sheath
- Neurofibril node
- Axon
- Axon terminal

FIGURE 24.17 Nerve cells consist of a nerve cell body that contains the nucleus and several fibrous extensions. The fibers that carry impulses to the nerve cell body are dendrites. The long fiber that carries the impulse away from the cell body is the axon. Sensory neurons typically have long fibers running to and from the cell body. Motor neurons have short dendrites and long axons leaving the cell body.

the adjacent portion of the cell membrane to change its permeability as well, and it also depolarizes. Thus, a wave of depolarization passes along the length of the neuron from one end to the other (figure 24.18). The depolarization and passage of an impulse along any portion of the neuron is a momentary event. As soon as a section of the membrane has been depolarized, a series of events begins to reestablish the original polarized state, and the membrane is said to be *repolarized.* When the nerve impulse reaches the end of the axon, it stimulates the release of a molecule that stimulates depolarization of the next neuron in the chain.

Activities at the Synapse

Neurons are arranged end-to-end to form long chains of cells. Between the fibers of adjacent neurons is a space called the **synapse.** Many chemical events occur in the synapse that are important in the function of the nervous system. When a neuron is stimulated, an impulse passes along its length from one end to the other. When the impulse reaches a synapse, a molecule called a **neurotransmitter** is released into the synapse from the axon. It diffuses across the synapse and binds to specific receptor sites on the dendrite of the next neuron. When enough neurotransmitter molecules have bound to the second neuron, an impulse is initiated in it as well. Several kinds of neurotransmitters are produced by specific neurons. These include dopamine, epinephrine, acetylcholine, and several other molecules. The first neurotransmitter identified was *acetylcholine.*

As long as a neurotransmitter is bound to its receptor, it continues to stimulate the nerve cell. Thus, if acetylcholine continues to occupy receptors, the neuron continues to be stimulated again and again. An enzyme called *acetylcholinesterase* destroys acetylcholine and prevents this from happening. The breakdown products of the acetylcholine can be used to remanufacture new acetylcholine molecules at the end of the axon. The destruction of acetylcholine allows the second neuron in the chain to return to normal. Thus, it will be ready to accept another burst of acetylcholine from the first neuron a short time later. Neurons must also constantly manufacture new acetylcholine molecules, or they will exhaust their supply and be unable to conduct an impulse across a synapse (figure 24.19).

Caffeine and nicotine are considered stimulants because they make it easier for impulses to pass through the synapse. The nerve cells are sensitized and will respond to smaller amounts of acetylcholine than normal. Drinking coffee, taking caffeine pills, or smoking tobacco will increase the sensitivity of the nervous system to stimulation. However, as with most kinds of drugs, continual use or abuse tends to lead to a loss of the effect as the nervous system adapts to the constant presence of the drugs.

Because of the way the synapse works, impulses can go in only one direction: Only axons secrete acetylcholine, and only

battery is polarized, with a positive and negative pole. A resting neuron has its positive pole on the outside of the cell membrane and its negative pole on the inside of the membrane.

When a cell is stimulated at a specific point on the cell membrane, the cell membrane changes its permeability and lets sodium ions (Na^+) pass through it from the outside to the inside and potassium ions (K^-) diffuse from the inside to the outside. The membrane is thus *depolarized;* it loses its difference in charge as sodium ions (Na^+) diffuse into the cell from the outside. Sodium ions diffuse into the cell because initially they are in greater concentration outside the cell than inside. When the membrane becomes more permeable, they are able to diffuse into the cell, toward the area of lower concentration. The depolarization of one point on the cell membrane causes

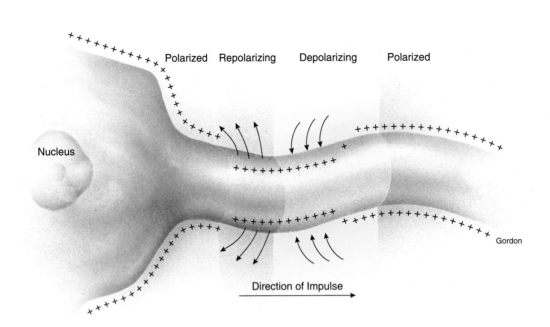

Polarized   Repolarizing   Depolarizing   Polarized

Nucleus

Gordon

Direction of Impulse

FIGURE 24.18   When a nerve cell is stimulated, a small portion of the cell membrane depolarizes as Na⁺ flows into the cell through the membrane. This encourages the depolarization of an adjacent portion of the membrane, and it depolarizes a short time later. In this way a wave of depolarization passes down the length of the nerve cell. Shortly after a portion of the membrane is depolarized, the ionic balance is reestablished. It is repolarized and ready to be stimulated again.

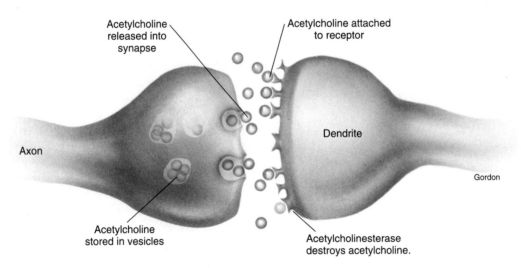

Acetylcholine released into synapse

Acetylcholine attached to receptor

Dendrite

Axon

Gordon

Acetylcholine stored in vesicles

Acetylcholinesterase destroys acetylcholine.

FIGURE 24.19   When a nerve impulse reaches the end of an axon, it releases a neurotransmitter into the synapse. In this illustration, the neurotransmitter is acetylcholine. When acetylcholine is released into the synapse, acetylcholine molecules diffuse across the synapse and bind to receptors on the dendrite, initiating an impulse in the next neuron. Acetylcholinesterase is an enzyme that destroys acetylcholine, preventing continuous stimulation of the dendrite.

dendrites have receptors. This explains why there are sensory and motor neurons to carry messages to and from the central nervous system.

Endocrine System Function

As mentioned previously, the endocrine system is basically a broadcasting system in which glands secrete messenger molecules, called hormones, which are distributed throughout the body by the circulatory system (figure 24.20). However, each kind of hormone attaches only to appropriate receptor molecules on the surfaces of certain cells.

How Cells Respond to Hormones

The cells that receive the message typically respond in one of three ways: (1) Some cells release products that have been previously manufactured, (2) other cells are stimulated to synthesize molecules or to begin metabolic activities, and (3) some are stimulated to divide and grow.

These different kinds of responses mean that some endocrine responses are relatively rapid, while others are very slow. For example, the release of the hormones *epinephrine* and *norepinephrine* from the adrenal medulla, located near the kidney, causes a rapid change in the behavior of an organism.

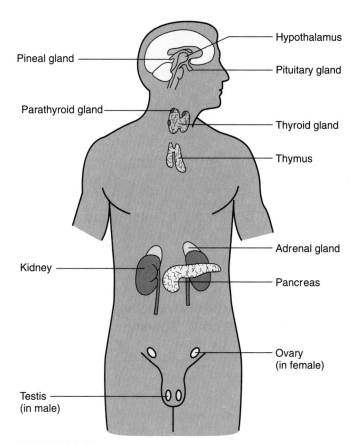

Pineal gland

Parathyroid gland

Kidney

Testis
(in male)

Hypothalamus

Pituitary gland

Thyroid gland

Thymus

Adrenal gland

Pancreas

Ovary
(in female)

FIGURE 24.20 The endocrine glands are located at various places within the body and cause their effects by secreting hormones.

These hormones also often are referred to as adrenalin and noradrenalin. (Epinephrine and norepinephrine are also released from certain nerve cells.) The heart rate increases, blood pressure rises, blood is shunted to muscles, and the breathing rate increases. You have certainly experienced this reaction many times in your lifetime as when you nearly have an automobile accident or slip and nearly fall.

Another hormone, called *antidiuretic hormone (ADH),* acts more slowly. It is released from the posterior pituitary gland at the base of the brain and regulates the rate at which the body loses water through the kidneys. It does this by allowing the reabsorption of water from the kidney back into the bloodstream. The effects of this hormone can be noticed in a matter of minutes to hours. Insulin is another hormone whose effects are quite rapid. Insulin is produced by the pancreas, located near the stomach, and stimulates cells—particularly muscle, liver, and fat cells—to take up glucose from the blood. After a meal that is high in carbohydrates, the level of glucose in the blood begins to rise, stimulating the pancreas to release insulin. The increased insulin causes glucose levels to fall as the sugar is taken up by cells. People with diabetes have insufficient or improperly acting insulin or lack the receptors to respond to the insulin and therefore have difficulty regulating glucose levels in their blood.

The responses that result from the growth of cells may take weeks or years to occur. For example, *growth-stimulating hormone (GSH)* is produced by the anterior pituitary gland over a period of years and results in typical human growth. After sexual maturity, the amount of this hormone generally drops to very low levels and body growth stops. Sexual development is also largely the result of the growth of specific tissues and organs. The male sex hormone *testosterone,* produced by the testes, causes the growth of male sex organs and a change to the adult body form. The female counterpart, *estrogen,* results in the development of female sex organs and the adult body form. In all of these cases, it is the release of hormones over long periods, continually stimulating the growth of sensitive tissues, that results in the normal developmental pattern. The absence or inhibition of any of these hormones early in life changes the normal growth process.

Interactions of the Nervous and Endocrine Systems

Because the pituitary is constantly receiving information from the hypothalamus of the brain, many kinds of sensory stimuli to the body can affect the functioning of the endocrine system. One example is the way in which the nervous system and endocrine system interact to influence the menstrual cycle. At least three different hormones are involved in the cycle of changes that affect the ovary and the lining of the uterus. It is well documented that stress caused by tension or worry can interfere with the normal cycle of hormones and delay or stop menstrual cycles. In addition, young women living in groups, such as in college dormitories, often find that their menstrual cycles become synchronized. Although the exact mechanism involved in this phenomenon is unknown, it is suspected that input from the nervous system causes this synchronization. (Odors and sympathetic feelings have been suggested as causes.)

Although we still tend to think of the nervous and endocrine systems as being separate and different, it is becoming clear that they are interconnected. These two systems cooperate to bring about appropriate responses to environmental challenges. The nervous system is specialized for receiving and sending short-term messages, whereas activities that require long-term, growth-related actions are handled by the endocrine system.

SENSORY INPUT

The activities of the nervous and endocrine systems are often responses to some kind of input received from the sense organs. Sense organs of various types are located throughout the body. Many of them are located on the surface, where environmental changes can be easily detected. Hearing, sight, and touch are good examples of such senses. Other sense organs are located within the body and indicate to the organism how its various parts are changing. For example, pain and pressure are often used to monitor internal conditions. The sense organs detect changes, but the brain is responsible for perception—the recognition that a stimulus has been received. Sensory abilities involve many different kinds of mechanisms, including chemical recognition, the detection of energy changes, and the monitoring of forces.

Chemical Detection

All cells have receptors on their surfaces that can bind selectively to molecules they encounter. This binding process can cause changes in the cell in several ways. In some cells, it causes depolarization. When this happens, the cells can stimulate neurons and cause messages to be sent to the central nervous system, informing it of some change in the surroundings. In other cases, a molecule binding to the cell surface may cause certain genes to be expressed, and the cell responds by changing the molecules it produces. This is typical of the way the endocrine system receives and delivers messages.

Taste

Most cells have specific binding sites for particular molecules. Others, such as the taste buds on the tongue, appear to respond to classes of molecules. Traditionally we have distinguished four kinds of tastes: sweet, sour, salt, and bitter. However, recently, a fifth kind of taste, *umami* (meaty), has been identified that responds to the amino acid glutamate, which is present in many kinds of foods and can be added as a flavor enhancer (monosodium glutamate). The taste buds that give us the sour sensation appear to respond to the presence of hydrogen ions (H^+), because acids taste sour. Sodium chloride stimulates the taste buds that give us the sensation of a salty taste. However, the sensations of sweetness, bitterness, and umami occur when molecules bind to specific surface receptors on the cell. Sweetness can be stimulated by many kinds of organic molecules, including sugars and artificial sweeteners, and also by inorganic lead compounds. When a molecule binds to a sweetness receptor, a molecule is split, and its splitting stimulates an enzyme that leads to the depolarization of the cell. The sweet taste of lead salts in old paints partly explains why children sometimes eat paint chips. Because the lead interferes with normal brain development, this behavior can have disastrous results. Many other kinds of compounds of diverse structures give the bitter sensation. The cells that respond to bitter sensations have a variety of receptor molecules on their surface. When a substance binds to one of the receptors, the cell depolarizes. In the case of umami, it is the glutamate molecule that binds to receptors on the cells of the taste buds.

It is also important to understand that much of what we often refer to as taste involves such inputs as temperature, texture, and smell. Cold coffee has a different taste than hot coffee even though they are chemically the same. Lumpy, cooked cereal and smooth cereal have different tastes. If you are unable to smell food, it doesn't taste as it should, which is why you sometimes lose your appetite when you have a stuffy nose. We still have much to learn about how the tongue detects chemicals and the role that other associated senses play in modifying taste.

Smell

The other major chemical sense, the sense of smell, is much more versatile; it can detect thousands of different molecules at very low concentrations. The cells that make up the *olfactory epithelium,* the cells that line the nasal cavity and respond to smells, apparently bind molecules to receptors on their surface. Exactly how this can account for the large number of recognizably different odors is unknown, but the receptor cells are extremely sensitive. In some cases, a single molecule of a substance is sufficient to cause a receptor cell to send a message to the brain, where the sensation of odor is perceived. These sensory cells for smell also fatigue rapidly. You have probably noticed that when you first walk into a room, specific odors are readily detected, but after a few minutes, you are unable to detect them. Most perfumes and aftershaves are undetectable after fifteen minutes of continuous stimulation.

Other Chemical Senses

Many internal sense organs also respond to specific molecules. For example, the brain and aorta contain cells that respond to concentrations of hydrogen ions, carbon dioxide, and oxygen in the blood. Remember, too, that the endocrine system relies on the detection of specific messenger molecules to trigger its activities.

Light Detection

The eyes primarily respond to changes in the flow of light energy. The structure of the eye is designed to focus light on a light-sensitive layer of the back of the eye known as the **retina** (figure 24.21). There are two kinds of receptors in the retina of the eye. The cells called *rods* respond to a broad range of wavelengths of light and are responsible for black-and-white vision. Since rods are very sensitive to light, they are particularly useful in dim light. Rods are located over most of the retinal surface except for the area of most acute vision known as the *fovea centralis.* The other receptor cells, called *cones,* are found throughout the retina but are particularly concentrated in the fovea centralis. Cones are not as sensitive to light, but they can detect light of different wavelengths (color). This combination of receptors gives us the ability to detect color when light levels are high, but we rely on black-and-white vision at night. There are three different varieties of cones: One type responds best to red light, another responds best to green light, and the third responds best to blue light. Stimulation of various combinations of these three kinds of cones allows us to detect different shades of color.

Rods and the three different kinds of cones each contain a pigment that decomposes when struck by light of the proper wavelength and sufficient strength. The pigment found in rods is called *rhodopsin.* This change in the structure of rhodopsin causes the rod to stimulate the nerve cell connected to it and send messages to the brain. Cone cells have a similar mechanism of action, and each of the three kinds of cones has a different pigment. Thus, the pattern of color and light intensity recorded on the retina is detected by rods and cones and converted into a series of nerve impulses that are received and interpreted by the brain.

Sound Detection

The ears respond to changes in sound waves. Sound is produced by the vibration of molecules. Consequently, the ears are detecting changes in the quantity of energy and the quality of sound waves. Sound has several characteristics. Loudness, or volume, is a measure of the intensity of sound energy that arrives at the ear. Very loud sounds will literally vibrate your body and can cause hearing loss if they are too intense. Pitch is a quality of sound that is determined by the frequency of the sound vibrations. High-pitched sounds have short wavelengths; low-pitched sounds have long wavelengths.

Figure 24.22 shows the anatomy of the ear. The sound that arrives at the ear is first funneled by the external ear to the **tympanum,** also known as the *eardrum.* The cone-shaped nature of the external ear focuses sound on the tympanum and causes it to vibrate at the same frequency as the sound waves reaching it. Attached to the tympanum are three tiny bones known as the *malleus* (hammer), *incus* (anvil), and *stapes* (stirrup). The malleus is attached to the tympanum, the incus is attached to the malleus and stapes, and the stapes is attached to a small, membrane-covered opening called the oval window in a snail-shaped structure known as the cochlea. The vibration of the tympanum causes the tiny bones (malleus, incus, and stapes) to vibrate, and they in turn cause a corresponding vibration in the membrane of the *oval window.*

The *cochlea* of the ear is the structure that detects sound and consists of a snail-shaped set of fluid-filled tubes. When the oval window vibrates, the fluid in the cochlea begins to move, causing a membrane in the cochlea, called the *basilar membrane,* to vibrate. High-pitched, short-wavelength sounds cause the basilar membrane to vibrate at the base of the cochlea near the oval window. Low-pitched, long-wavelength sounds vibrate the basilar membrane far from the oval window. Loud

Connections . . .

Negative Feedback Control

Many common mechanical and electronic systems are controlled by negative feedback. In negative feedback control, some output of a process influences the process so that it produces less of the product. For example, a thermostat is used to control the temperature in a building or room. When the temperature becomes too low, an electronic circuit is completed and a heating device is turned on, which raises the temperature. When the temperature reaches the set point, the circuit is disconnected and the heating device is turned off.

A large number of physiological processes are controlled in the same way. For example, for most people the amount of the sugar, glucose, is regulated within very narrow limits. However, when we eat a source of glucose, the amount in the blood rises. The presence of elevated glucose stimulates the pancreas to release insulin. Insulin stimulates muscle and liver cells to take glucose out of the blood and store it. When the glucose level falls, the pancreas is no longer stimulated and the amount of insulin declines.

sounds cause the basilar membrane to vibrate more vigorously than do faint sounds. Cells on this membrane depolarize when they are stimulated by its vibrations. Because they synapse with neurons, messages can be sent to the brain.

Because sounds of different wavelengths stimulate different portions of the cochlea, the brain is able to determine the pitch of a sound. Most sounds consist of a mixture of pitches that are heard. Louder sounds stimulate the membrane more forcefully, causing the sensory cells in the cochlea to send more nerve impulses per second. Thus, the brain is able to perceive the loudness of various sounds, as well as the pitch.

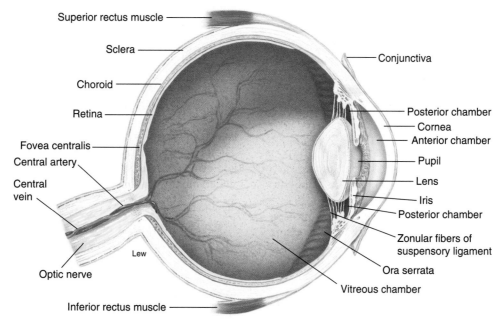

FIGURE 24.21   The eye contains a cornea and lens that focus the light on the retina of the eye. The light causes pigments in the rods and cones of the retina to decompose. This leads to the stimulation of neurons that send messages to the brain.

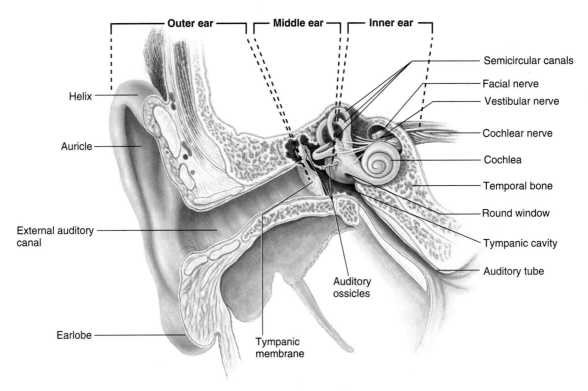

FIGURE 24.22 The ear consists of an external cone that directs sound waves to the tympanum. Vibrations of the tympanum move the ear bones and vibrate the oval window of the cochlea, where the sound is detected. The semicircular canals monitor changes in the position of the head, helping us maintain balance.

Associated with the cochlea are two fluid-filled chambers and a set of fluid-filled tubes called the *semicircular canals.* These structures are not involved in hearing but are involved in maintaining balance and posture. In the walls of these canals and chambers are cells similar to those found on the basilar membrane. These cells are stimulated by movements of the head and by the position of the head with respect to the force of gravity. The constantly changing position of the head results in sensory input that is important in maintaining balance.

Touch

What we normally call the sense of *touch* consists of a variety of different kinds of input. Some receptors respond to pressure, others to temperature, and others, which we call *pain receptors,* usually respond to cell damage. When these receptors are appropriately stimulated, they send a message to the brain. Because receptors are stimulated in particular parts of the body, the brain is able to localize the sensation. However, not all parts of the body are equally supplied with these receptors. The tips of the fingers, lips, and external genitals have the highest density of these nerve endings, while the back, legs, and arms have far fewer receptors.

Some internal receptors, such as pain and pressure receptors, are important in allowing us to monitor our internal activities. Many pains generated by the internal organs are often perceived as if they were somewhere else. For example, the pain associated with heart attack is often perceived to be in the left arm. Pressure receptors in joints and muscles are important in providing information about the degree of stress being placed

on a portion of the body. This is also important information to send back to the brain so that adjustments can be made in movements to maintain posture. If you have ever had your foot "go to sleep" because the nerve stopped functioning, you have experienced what it is like to lose this constant input of nerve messages from the pressure sensors to assist in guiding the movements you make. Your movements become uncoordinated until the nerve function returns to normal.

OUTPUT MECHANISMS

The nervous system and endocrine system cause changes in several ways. Both systems can stimulate muscles to contract and glands to secrete. The endocrine system is also able to change metabolism of cells and regulate the growth of tissues. The nervous system acts upon two kinds of organs: muscles and glands. The actions of muscles and glands are simple and direct: muscles contract and glands secrete.

Muscles

The ability to move is one of the fundamental characteristics of animals. Through the coordinated contraction of many muscles, the intricate, precise movements of a dancer, basketball player, or writer are accomplished. It is important to recognize that muscles can pull only by contracting; they are unable to push by lengthening. The work of any muscle is done during its contraction. Relaxation is the passive state of the muscle. There

Which Type of Exercise Do You Do?

Aerobic exercise occurs when the muscles being contracted are supplied with sufficient oxygen to continue aerobic cellular respiration:

$$C_6H_{12}O_6 + 6O_2 \rightarrow 6CO_2 + 6H_2O + energy$$

This type of exercise involves long periods of activity with elevated breathing and heart rate. It results in strengthened chest muscles, which enable more complete exchange of air during breathing. It also improves the strength of the heart, enabling it to pump more efficiently. Flow of blood to the muscles also improves. All these changes increase endurance.

Anaerobic exercise takes place when insufficient amounts of oxygen reach the contracting muscle cells and they shift to anaerobic cellular respiration:

$$C_6H_{12}O_6 \rightarrow lactic\ acid + less\ energy$$

The buildup of lactic acid results in muscle pain and eventually prevents further contraction. Anaerobic exercise involves explosive bouts of activity, as in sprints or jumping. This kind of exercise increases muscle strength but does little to improve endurance.

Resistance exercise occurs when muscles contract against an object that does not allow the muscles to move that object. This type of exercise does not improve the ability of your body to deliver oxygen to your muscles, nor does it increase your endurance. However, it does stimulate your muscle cells to manufacture more contractile protein fibers.

Make a list of your activities that would be considered aerobic, anaerobic, or resistance exercise. Which activities would result in weight reduction? Weight gain? Improved cardiovascular fitness?

---

must always be some force available that will stretch a muscle after it has stopped contracting and relaxes. Therefore, the muscles that control the movements of the skeleton are present in antagonistic sets—for every muscle's action, there is another muscle that has the opposite action. For example, the biceps muscle causes the arm to flex (bend) as the muscle shortens. The contraction of its antagonist, the triceps muscle, causes the arm to extend (straighten) and at the same time stretches the relaxed biceps muscle (figure 24.23).

There are three major types of muscle: *skeletal, smooth,* and *cardiac.* These differ from one another in several ways. *Skeletal muscle* is voluntary muscle; it is under the control of the nervous system. The brain or spinal cord sends a message to skeletal muscles, and they contract to move the legs, fingers, and other parts of the body. This does not mean that you must make a conscious decision every time you want to move a muscle. Many of the movements we make are learned initially but become automatic as a result of practice. For example, walking, swimming, or riding a bicycle required a great amount of practice originally, but now you probably perform these movements without thinking about them. They are, however, still considered voluntary actions.

Smooth muscles make up the walls of muscular internal organs, such as the gut, blood vessels, and reproductive organs. They have the property of contracting as a response to being stretched. Since much of the digestive system is being stretched constantly, the responsive contractions contribute to the normal rhythmic movements associated with the digestive system. These are involuntary muscles; they can contract on their own without receiving direct messages from the nervous system. Although they are involuntary, they can be stimulated to contract by the presence of certain hormones (the hormone oxytocin causes the smooth muscle of the uterus to contract), or their degree of contraction can be modified by nervous stimulation.

Cardiac muscle is the muscle that makes up the heart. It has the ability to contract rapidly like skeletal muscle but does not require

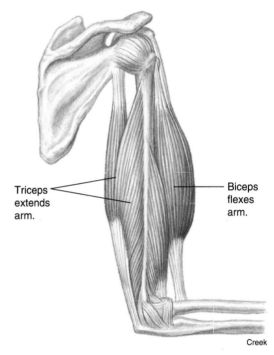

Triceps extends arm.

Biceps flexes arm.

Creek

FIGURE 24.23   Because muscles cannot actively lengthen, it is necessary to have sets of muscles that oppose one another. The contraction and shortening of one muscle cause the stretching of a relaxed muscle.

CONCEPTS APPLIED

Antagonistic Muscles

Close and open your hand. Determine the location of the muscles that flex the hand and the location of the muscles that extend the hand.

nervous stimulation to do so. Nervous stimulation can, however, cause the heart to speed or slow its rate of contraction. Hormones, such as epinephrine and norepinephrine, also influence the heart by increasing its rate and strength of contraction. Cardiac muscle also has the characteristic of being unable to stay contracted. It will contract quickly but must have a short period of relaxation before it will be able to contract a second time. This makes sense in light of its continuous, rhythmic, pumping function.

Glands

The glands of the body are of two different kinds. Those that secrete into the bloodstream are called **endocrine glands.** We have already talked about several of these: the pituitary, thyroid, ovary, and testis are examples. The **exocrine glands** are those that secrete to the surface of the body or into one of the tubular organs of the body, such as the gut or reproductive tract. Examples are the salivary glands, intestinal mucous glands, and sweat glands. Some of these glands, such as salivary glands and sweat glands, are under nervous control. When stimulated by the nervous system, they secrete their contents.

Many other exocrine glands are under hormonal control. Many of the digestive enzymes of the stomach and intestine are secreted in response to local hormones produced in the gut. These are circulated through the blood to the digestive glands, which respond by secreting the appropriate digestive enzymes and other molecules. Endocrine glands also respond to nervous stimulation or stimulation by hormones.

Growth Responses

The hormones produced by the endocrine system can have a variety of effects. As mentioned earlier, hormones can stimulate smooth muscle to contract and can influence the contraction of cardiac muscle as well. Many kinds of glands, both endocrine and exocrine, are caused to secrete as a result of a hormonal stimulus. However, the endocrine system has one major effect that is not equaled by the nervous system: Hormones regulate growth. *Growth-stimulating hormone (GSH)* is produced over a period of years to bring about the increase in size of most of the structures of the body. A low level of this hormone results in a person with small body size. It is important to recognize that the amount of growth-stimulating hormone (GSH) present varies from time to time. It is present in fairly high amounts throughout childhood and results in steady growth. It also appears to be present at higher levels at certain times, resulting in growth spurts. Finally, as adulthood is reached, the level of this hormone falls, and growth stops.

Similarly, testosterone produced during adolescence influences the growth of bone and muscle to provide men with larger, more muscular bodies than those of women. In addition, there is growth of the penis, growth of the larynx, and increased growth of hair on the face and body. The primary female hormone, estrogen, causes growth of reproductive organs and development of breast tissue. It is also involved, along with other hormones, in the cyclic growth and sloughing of the wall of the uterus.

SUMMARY

The *circulatory, respiratory, digestive,* and *excretory systems* are involved in the exchange of materials across cell membranes. All of these systems have special features that provide *large surface* areas to allow for necessary *exchanges.*

The *circulatory system* consists of a pump, the *heart,* and *blood vessels* that distribute the blood to all parts of the body. The *blood* is a carrier fluid that transports molecules and heat. The exchange of materials between the *blood* and body *cells* takes place through the walls of the *capillaries. Hemoglobin* in *red blood cells* is very important in the transport of *oxygen.*

The *skin* is the outer covering of the body and performs several functions, including protecting underlying tissues, preventing water loss, and regulating temperature.

The *respiratory system* consists of the *lungs* and associated tubes that allow air to enter and leave the lungs. The *diaphragm* and *muscles* of the chest wall are important in the process of *breathing.* In the lungs, tiny sacs called *alveoli* provide a large surface area in association with capillaries, which allows for rapid exchange of oxygen and carbon dioxide.

The *digestive system* is involved in disassembling food molecules. This involves several processes: *grinding* by the teeth and stomach, *emulsification* of fats by bile from the liver, *addition of water* to dissolve molecules, and *enzymatic action* to break complex molecules into simpler molecules for *absorption.* The intestine provides a *large surface area* for the *absorption of nutrients* because it is long and its wall contains many tiny projections that increase surface area.

To maintain good health, people must receive *nutrient molecules* that can enter the cells and function in the metabolic processes. The *proper quantity and quality of nutrients* are essential to good health. An important measure of the amount of energy required to sustain a human at rest is the *basal metabolic rate.* To meet this and all additional requirements, the United States has established recommendations for each nutrient. Should there be metabolic or psychological problems associated with a person's *normal metabolism,* a variety of disorders may occur, including *obesity, anorexia nervosa,* and *bulimia.*

The *excretory system* is a filtering system of the body. The *kidneys* consist of *nephrons* into which the *circulatory system* filters fluid. Most of this fluid is useful and is reclaimed by the cells that make up the walls of these *tubules.* Materials that are present in excess or those that are harmful are allowed to escape.

A *nerve impulse* is caused by sodium ions entering the cell as a result of a change in the permeability of the cell membrane. Thus, a *wave of depolarization* passes down the length of a *neuron* to the *synapse.* The *axon* of a neuron secretes a *neurotransmitter,* such as *acetylcholine,* into the *synapse,* where these molecules bind to the dendrite of the next cell in the chain, resulting in an impulse in it as well. The *acetylcholinesterase* present in the synapse destroys acetylcholine so that it does not repeatedly stimulate the dendrite.

The body's various *systems* must be integrated in such a way that the *internal environment* stays relatively constant. This concept is called *homeostasis.*

Several kinds of *sensory inputs* are possible. Many kinds of chemicals can bind to cell surfaces and be recognized. This is probably how the *sense of taste* and the *sense of smell* function. Light energy can be detected because light causes certain molecules in the *retina* of the eye to decompose and stimulate neurons. *Sound* can be detected because fluid in the *cochlea* of the *ear* is caused to vibrate, and special cells detect this movement and stimulate neurons. The sense of *touch* consists of a variety of receptors that respond to *pressure, cell damage,* and *temperature*.

There are two major kinds of organs that respond to nervous and hormonal stimulation: *muscles* and *glands.* There are three kinds of muscle: *skeletal muscle,* which moves parts of the skeleton; *cardiac muscle*—the heart; and *smooth muscle,* which makes up the muscular walls of internal organs.

Glands are of two types: *exocrine glands,* which secrete through ducts into the cavity of an organ or to the surface of the skin, and *endocrine glands,* which release their secretions into the circulatory system. It is becoming clear that the endocrine system and the nervous system are interrelated. Actions of the endocrine system can change how the nervous system functions, and the reverse is also true.

KEY TERMS

alveoli (p. **600**)
anorexia nervosa (p. **612**)
aorta (p. **595**)
arteries (p. 593)
arterioles (p. **597**)
basal metabolic rate (p. **611**)
blood (p. **593**)
bronchi (p. **600**)
bronchioles (p. **600**)
bulimia (p. **611**)
calorie (p. **606**)
capillaries (p. **593**)
diaphragm (p. **600**)

diastolic blood pressure (p. **596**)
diet (p. **606**)
Dietary Reference Intakes (p. **605**)
duodenum (p. **604**)
endocrine glands (p. **622**)
exocrine glands (p. **622**)
Food Guide Pyramid (p. **606**)
glands (p. **613**)
heart (p. **593**)
hemoglobin (p. **593**)
homeostasis (p. **592**)
hormones (p. **613**)

kidney (p. **612**)
kilocalorie (kcal) (p. **606**)
large intestine (p. **604**)
liver (p. **604**)
lung (p. **600**)
lymph (p. **597**)
nephrons (p. **613**)
nerve impulse (p. **614**)
neurotransmitter (p. **615**)
nutrients (p. **605**)
nutrition (p. **605**)
obesity (p. **610**)
pancreas (p. **604**)

plasma (p. **593**)
pulmonary artery (p. **595**)
pulmonary circulation (p. **595**)
response (p. **613**)
retina (p. **618**)
small intestine (p. **604**)
stimulus (p. **613**)
synapse (p. **615**)
systemic circulation (p. **596**)
systolic blood pressure (p. **596**)
trachea (p. **600**)
tympanum (p. **619**)
veins (p. **593**)

APPLYING THE CONCEPTS

1. The primary structures involved in pumping blood are the
   a. veins.
   b. atria.
   c. capillaries.
   d. ventricles.

2. The fluid portion of the blood that leaves the capillaries and surrounds the cells is
   a. hemoglobin.
   b. edema.
   c. lymph.
   d. lacteal.

3. Blood is carried through vessels to all parts of the body except the lungs by
   a. pulmonary circulation.
   b. the pulmonary artery.
   c. systemic circulation.
   d. the lymphatic system.

4. As air passes through the lungs, it follows the path:
   a. trachea → bronchi → bronchioles → alveoli
   b. trachea → bronchioles → bronchi → alveoli
   c. bronchi → trachea → alveoli → bronchioles
   d. bronchioles → alveoli → bronchi → trachea

5. The levels of water, hydrogen ions, salts, and urea in the blood are regulated by the
   a. liver.
   b. kidneys.
   c. bladder.
   d. rectum.

6. An organism's reaction to a change in the environment is a(n)
   a. stimulus.
   b. impulse.
   c. response.
   d. perception.

7. A light stimulus is received by the nervous system, which results in growth. The growth is the direct result of
   a. hormones stimulating cells.
   b. activating muscles.
   c. the endocrine system stimulating the nervous system.
   d. increasing nervous activity.

8. When a nerve cell is stimulated,
   a. acetylcholine is destroyed.
   b. potassium ions enter the neuron.
   c. sodium ions enter the neuron.
   d. calcium attaches to the dendrites.

9. The central nervous system consists of the
   a. brain only.
   b. brain and spinal cord.
   c. brain, spinal cord, and nerves.
   d. motor neurons and sensory neurons.

10. Olfactory senses detect
    a. light.
    b. sound.
    c. chemicals.
    d. pain.

11. Strict vegetarians do not eat animals or their products (milk, eggs). Which one of the following nutritional needs would they have the greatest difficulty meeting from food alone?
    a. carbohydrates
    b. essential amino acids
    c. essential fatty acids
    d. vitamin A

12. Large food molecules are chemically broken down to smaller molecules by
    a. bile salts.
    b. enzymes.
    c. dissolving.
    d. chewing.

Answers
1. d **2.** c **3.** c **4.** a **5.** b **6.** c **7.** a **8.** c **9.** b **10.** c **11.** b **12.** b

QUESTIONS FOR THOUGHT

1. What are the functions of the heart, arteries, veins, arterioles, the blood, and capillaries?
2. Describe three ways in which the digestive system increases its ability to absorb nutrients.
3. Describe the mechanics of breathing.
4. Describe how changing permeability of the cell membrane and the movement of sodium ions cause a nerve impulse.
5. What is the role of acetylcholine in a synapse?
6. What is actually detected by the nasal epithelium, taste buds, cochlea of the ear, and retina of the eye?
7. List the differences between the central and peripheral nervous systems.
8. List two differences in the way the endocrine and nervous systems function.
9. Why do large, multicellular organisms need a circulatory system?
10. List the six groups of the Food Guide Pyramid, along with a major contribution of each to the diet.
11. Describe how villi and the length of the small intestine are related to the ability to absorb nutrients.
12. Describe two ways in which smooth muscle and skeletal muscle differ in the way they function.

FOR FURTHER ANALYSIS

1. The heart can be replaced with a mechanical pump. Major blood vessels can be replaced with plastic tubing. Can you envision artificial capillaries? What technical problems would need to be overcome before these would be possible?
2. The skin is an excellent organ for getting rid of heat. What role does surface area play in this function?
3. If you were allergic to dairy products, what are the particular nutrients for which you would need to find alternative sources?
4. Describe the ways in which the functions of the nervous and endocrine systems are similar. In what ways are they different?
5. How is the structure of each of the following organs able to provide a large surface area for the exchange of molecules: lungs, small intestine, and kidney?

INVITATION TO INQUIRY

Monitor Your Diet

Many people have a poor idea of their normal dietary intake. For twenty-four hours, keep track of what you eat. Record the food and quantity eaten. Include all food and drink (including alcoholic beverages) consumed, not just those foods eaten at mealtime. If you are eating packaged foods, save the package because it has the quantity contained and the nutritional facts about the food. For other items, estimate the size. You can use the following chart to record your food intake.

Meal	Item	Quantity	Nutritional Information
Breakfast	1. 2. 3. 4.		
Snacks	1. 2. 3. 4.		
Lunch	1. 2. 3. 4.		
Snacks	1. 2. 3. 4.		
Dinner	1. 2. 3. 4.		
Snacks	1. 2. 3. 4.		

Now compare your daily diet with the Food Guide Pyramid. What categories were consumed in excess? What categories do you need to consume more of? Did you get a good mixture of vitamins and minerals?

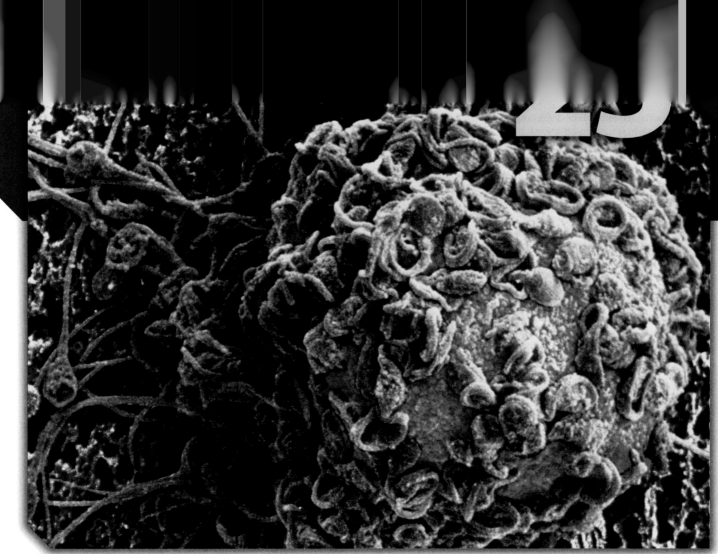

This is a photograph of human sperm cells on the surface of an egg cell. In humans and many other organisms that reproduce sexually, sperm cells are the result of a complex cell division process known as meiosis.

Human Biology: Reproduction

▶ A specialized form of cell division plays a vital role in sexual reproduction (Ch. 25).

▶ One's sexuality is a complex of anatomy, physiology, behavior, and culture (Ch. 25).

▶ Males and females have specialized anatomical features for reproduction (Ch. 25).

▶ Gamete formation in males and females differs (Ch. 25).

▶ Reproduction is under the control of many different hormones (Ch. 25).

▶ Understanding sexuality requires knowledge of the process from gamete production to old age (Ch. 25).

CORE CONCEPT

Humans are sexual beings and possess the structural, functional, and behavioral capabilities of transmitting necessary genetic material to the next generation.

The process of meiosis reduces the number of chromosomes during sex cell formation and introduces genetic variation to the offspring (p. 627).

Sexuality involves distinct hereditary, anatomical, and behavioral aspects (p. 630).

Hormones play important roles in sexual development and function (p. 636).

OVERVIEW

If both parents contribute equally to the genetic information of the child, how can the chromosome number in humans remain constant for generation after generation? There is a specialized kind of cell division that results in the formation of sex cells. In this chapter, we will discuss the mechanics of this process. Knowing the mechanics is essential to understanding how genetic variety can occur in sex cells. This variety ultimately shows up as differences in offspring as a result of sexual reproduction. Sex and your sexuality influence you in many ways throughout life. Before birth, sex-determining chromosomes direct the formation of hormones that control the development of sex organs, after which the effects of these hormones diminish. With the start of puberty, renewed hormonal activity causes major structural and behavioral changes that influence you for the remainder of your life.

SEXUAL REPRODUCTION

Nearly all organisms have a method of shuffling and exchanging genetic information. This involves organisms that have two sets of genetic data, one inherited from each of their parents. **Sexual reproduction** is the formation of a new individual by the union of two sex cells. Before sexual reproduction can occur, each parent must form sex cells. The sex cells produced by male organisms are called **sperm,** and those produced by females are called **eggs.** A general term sometimes used to refer to either eggs or sperm is **gamete** (sex cell). During this process, the two sets of genetic information found in each body cell must be reduced to one set, that is, one of each kind of chromosome. The cellular process that is responsible for forming gametes is called **gametogenesis.** This is somewhat similar to shuffling a deck of cards and dealing out hands; the shuffling and dealing ensure that each hand will be different. An organism with two sets of chromosomes can produce many combinations of chromosomes when it produces sex cells, just as many different hands can be dealt from one pack of cards. When a sperm unites with an egg (*fertilization*), a new organism containing two sets of genetic information is formed. This new organism's genetic information might have survival advantages over the information found in either parent; this is the value of sexual reproduction.

The life cycle of a sexually reproducing organism begins with the fertilization of an egg by a sperm. This new cell, called a **zygote,** divides by mitosis, as do the resulting cells, and growth occurs. When the organism reaches sexual maturity, it is able to produce gametes by another form of cell division, *meiosis* and the life cycle is complete. Notice in figure 25.1 that the zygote and its descendants have two sets of chromosomes. However, male gamete and female gamete each contain only one set of chromosomes. These sex cells are said to be **haploid.** The haploid number of chromosomes is noted as n. A zygote contains two sets and is said to be **diploid.**

Meiosis

Sex cell formation

Single adult cell → Single gamete (sperm or egg)

Contains 2 sets of chromosomes → Contains 1 set of chromosomes

$$2n \rightarrow n$$

The diploid number of chromosomes is noted as $2n$ ($n + n = 2n$). Diploid cells have two sets of chromosomes, one

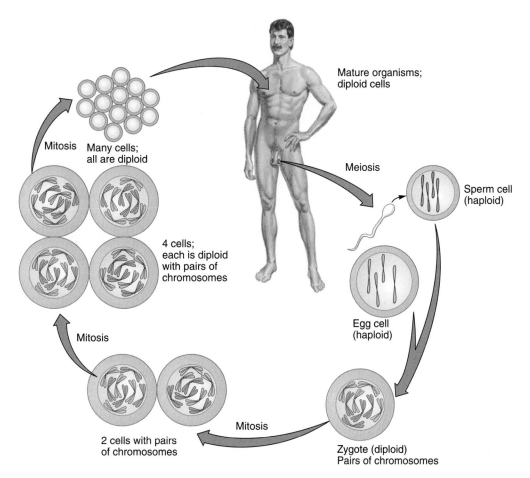

FIGURE 25.1 The cells of this adult human have forty-six chromosomes in their nuclei. In preparation for sexual reproduction, the number of chromosomes must be reduced by half so that fertilization will result in the original number of forty-six chromosomes in the new individual. The offspring will grow and produce new cells by mitosis, completing the life cycle.

set from each parent. Remember, a chromosome is composed of two chromatids, each containing double-stranded DNA. These two chromatids are attached to each other at a point called the *centromere.* In a diploid nucleus, the chromosomes occur as *homologous chromosomes*—a pair of chromosomes in a diploid cell that contain similar (not necessarily identical) genes throughout their length. One of the chromosomes of a homologous pair was donated by the father, the other by the mother (figure 25.2).

It is necessary for organisms that reproduce sexually to form gametes having only one set of chromosomes. If gametes contained two sets of chromosomes, the zygote resulting from their union would have four sets of chromosomes. The number of chromosomes would continue to double with each new generation. However, this does not usually happen; the number of chromosomes remains constant generation after generation. How are sperm and egg cells formed so that they get only half the chromosomes of the diploid cell? The answer lies in the process of **meiosis,** the specialized pair of cell divisions that reduce the chromosome number from diploid ($2n$) to haploid (n). One of the major functions of meiosis is to produce cells that have one set of genetic information. Therefore, when fertilization occurs between a haploid sperm (n) and a haploid egg (n), the resulting zygote will have two sets of chromosomes ($n + n = 2n$), as did each parent.

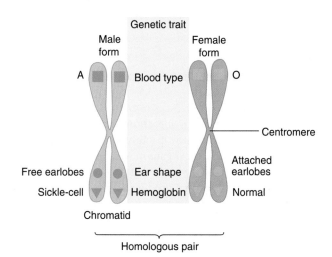

FIGURE 25.2 Pairs of chromosomes that are of similar size and shape that have genes for the same traits are said to be homologous. Notice that members of a gene pair may not be identical but code for the same general type of information. Homologous chromosomes are of the same length, have the same types of genes in the same sequence, and have their centromeres in the same location—one came from the male parent and the other was contributed by the female parent.

Connections . . .

Why Am I Different?

With the possible exception of identical twins, every human who has ever been born is genetically unique. The formation of haploid cells by meiosis and the combination of sex cells from two parents to form a diploid offspring by sexual reproduction result in genetic diversity in the offspring. The five factors that influence genetic diversity in offspring are mutations, crossing-over, segregation and independent assortment, and fertilization.

Mutations

Two types of mutations will be discussed in chapter 26: point mutations and chromosomal aberrations. In point mutations, a change in a DNA nucleotide results in the production of a different protein. In chromosomal aberrations, genes are rearranged when parts of chromosomes are duplicated, deleted, or inverted. By causing the production of different proteins, both types of mutations increase diversity.

Crossing-Over

When crossing-over occurs during gametogenesis in your ovaries or testes, pieces of genetic material are exchanged between the chromosomes you received from your mother and father. This exchange results in a chromosome with a new combination of genes, combinations of traits that were probably never before found in the family.

Segregation and Independent Assortment

The process of segregation relates to how the chromosomes donated by your father line up or assort themselves in pairs with those donated by your mother. How one pair of chromosomes lines up during metaphase I does not affect how the other pair lines up (i.e., they assort themselves independently of the others). If all those donated by your father line up on the right side of the cell during metaphase I and all those donated by your mother line up on the left, when the chromosomes separate or *segregate*, the resulting gametes will be of two kinds: half of your gametes will contain only your father's chromosomes and the other half of the gametes you form will contain only your mother's chromosomes. If however, some of the chromosomes from your father line up on the right side and some of your mother's line up on the left, the gametes you form will contain some of your mother's and some of your father's chromosomes. There are millions of ways (2^{23}) in which these chromosomes can line up and divide.

Fertilization

Because of the large number of possible gametes resulting from independent assortment, segregation, mutations, and crossing-over, an incredibly large number of types of offspring can result. Because human males and females can produce millions of genetically unique sperm and eggs, the number of kinds of offspring possible is infinite.

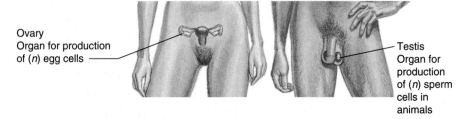

Ovary Organ for production of (*n*) egg cells

Testis Organ for production of (*n*) sperm cells in animals

FIGURE 25.3 Humans produce cells with a haploid number of chromosomes in organs called gonads, the ovaries and the testes.

Not every cell in the body goes through the process of meiosis. Only specialized organs are capable of producing haploid cells (figure 25.3). In humans, the organs in which meiosis occurs are called **gonads.** The female gonads that produce eggs are called **ovaries.** The male gonads that produce sperm are called **testes.**

THE MECHANICS OF MEIOSIS

The whole point of meiosis is to distribute the chromosomes and the genes they carry so that each daughter cell gets one member of each homologous pair. In this way, each daughter cell gets one complete set of genetic information. Meiosis consists of two divisions known as meiosis I and meiosis II. During meiosis I, the number of chromosomes is reduced from the diploid number to the haploid number. Therefore, it is often referred to as a *reduction division.* The sequence

CONCEPTS APPLIED

Meiosis Models

Models can be very useful in helping us to understand complex biological events such as meiosis. You can create model chromosomes very easily by using various lengths of colored strings, threads, or yarns to simulate the twenty-three pairs of homologous chromosomes in a human cell. Each homologous pair should be different from the other pairs, either in color, length, or both. Begin your modeling with each chromosome in its replicated form (i.e., two chromatids per chromosome). Attach the two chromatids with a loose twist. Manipulate your twenty-three pairs of model chromosomes through the stages of meiosis I and II. If you have performed the actions properly, you should end up with four cells, each haploid (*n* = 23).

TABLE 25.1   A Review of the Stages of Meiosis

Prophase I	During prophase I, the cell is preparing itself for division. The chromosomes are composed of two chromatids and coil so that they become shorter and thicker. Homologous chromosomes pair with each other. This pairing is called *synapsis*. While homologous chromosomes are paired, they can exchange equivalent sections of DNA. This process is called *crossing-over* and is important in mixing up the genes that will be divided into separate gametes. Long fibers known as *spindle fibers* become attached to the chromosomes.
Metaphase I	The synapsed homologous chromosomes move to the equator of the cell as single units. How they are arranged on the equator (which one is on the left and which one is on the right) is determined by chance. In the drawing, three green chromosomes from the father and one purple chromosome from the mother are lined up on the left. Similarly, one green chromosome from the father and three purple chromosomes from the mother are on the right. They could have aligned themselves in several other ways.
Anaphase I	During this stage, homologous chromosomes separate and their number is reduced from diploid to haploid. The two members of each pair of homologous chromosomes move away from each other toward opposite poles. The direction each takes is determined by how each pair was originally arranged on the spindle. Each chromosome is independently attached to a spindle fiber at its centromere. Each chromosome still consists of two chromatids. Because the homologous chromosomes and the genes they carry are being separated from one another, this process is called *segregation*. The way in which a single pair of homologous chromosomes segregates does not influence how other pairs of homologous chromosomes segregate; that is, each pair segregates independently of other pairs. This is known as *independent assortment* of chromosomes.
Telophase I	The two newly forming daughter cells are now haploid (n) since each contains only one of each pair of homologous chromosomes.
Prophase II	The chromosomes of each of the two haploid daughter cells shorten and become attached to spindle fibers by their centromeres.
Metaphase II	Chromosomes move to the equator of the cell.
Anaphase II	Centromeres divide, allowing the chromatids to separate toward the poles.
Telophase II	Four haploid (n) cells are formed from the division of the two meiosis I cells. Each of these sex cells is genetically different from all the others. These cells become the sex cells (egg or sperm) of a higher organism.

of events in meiosis I is artificially divided into four phases: prophase I, metaphase I, anaphase I, and telophase I. Meiosis II includes four phases: prophase II, metaphase II, anaphase II, and telophase II. The two daughter cells formed during meiosis I continue through meiosis II, so that four cells usually result from the two divisions. Table 25.1 is a review of the stages of meiosis as they would occur in a cell with diploid number of chromosomes equal to 8, (i.e., $n = 4$).

HUMAN SEXUALITY FROM DIFFERENT POINTS OF VIEW

Probably nothing interests people more than sex and sexuality. **Sex** is the nature of the biological differences between males and females. By **sexuality,** we mean all the factors that contribute to one's female or male nature. These include the

Speculation on the Evolution of Human Sexual Behavior

Much speculation has occurred about how human sexual behavior evolved. It is important to recognize that this speculation is not fact but an attempt to evaluate human sexual behaviors from an evolutionary perspective.

When we compare human sexuality with that of other mammals, human sexuality differs in several ways. Whereas most mammals are sexually active during specific periods of the year, humans may engage in sexual intercourse at any time throughout the year. The sex act appears to be important as a pleasurable activity rather than a purely reproductive act. Associated with this difference is the fact that human females do not display changes that indicate they are releasing eggs (ovulating).

All other female mammals display changes in odor, appearance, or behavior that clearly indicate to the males of the species that the female is ovulating and sexually responsive. This is referred to as "being in heat." This is not true for humans. Human males are unable to differentiate ovulating females from those who are not ovulating.

In other mammals with few exceptions, infants grow to sexual maturity in a year or less. Although extremely long-lived mammals (elephants or whales) do not reach sexual maturity in a year, their young have well-developed muscles that allow them to move about with a high degree of independence. While the young of these species still rely on their mothers for milk and protection, they are capable of obtaining other food for themselves as well. This is not true for human infants, who are extremely immature when born and develop walking skills slowly. They also require several years of training before they are able to function independently (box figure 25.1).

Perhaps the extremely immature condition in which human infants are born is related to human brain size. The size of the head is very large and just fits through the birth canal in the mother's pelvis. One way to accommodate a large brain size and not need to redesign the basic anatomy of the female pelvis would be to have the young be born in a very immature condition while the brain is still small and growing. Having the young born in an immature condition can solve one problem but creates another. The immature condition of human infants is associated with a need to provide extensive care for them.

Box Figure 25.1

Raising young requires a considerable investment of time and resources. Females invest significant resources in the pregnancy itself. Fat stores provide energy necessary to a successful pregnancy. Female mammals, including humans, that have little stored fat often have difficulty becoming pregnant in the first place and are more likely to die of complications resulting from the pregnancy. Nutritional counseling is an important part of modern prenatal care because it protects the health of both mother and developing fetus. The long duration of pregnancy in humans requires good nutrition over an extended period. Once the child is born, the mother continues to require good nutrition because she provides the majority of food for the infant through her breast milk. As the child grows, other food items are added to its diet. Since the young child is unable to find and prepare its own food, the mother or father or both must expend energy to feed the child.

With all these ideas in mind, we can speculate about how human sexual behavior may have evolved. Imagine a primitive Stone Age human culture. Females have a great deal invested in each child produced. They will only be able to produce a few children during their lifetimes, and many children will die because of malnutrition, disease, and accidents. Those females who have genes that will allow more of their offspring to survive will be selected for. Human males, on the other hand, have very little invested in each child and can impregnate many different females. Males who have many children who survive are selected for. How might these different male and female goals fit together to provide insight into the sexual behaviors we see in humans today?

The males of most mammals contribute little toward the raising of young. In many species, males meet with females only for mating (deer, cats, rabbits, mice). In some species, the male and female form short-term pair-bonds for one season and the males share the burden of raising the young (foxes). Only a few (wolves) form pair-bonds lasting for years in which males and females cooperate in the raising of young. However, pair bonding in humans is usually a long-term relationship. The significance of this relationship can be evaluated from an evolutionary perspective. This long-term pair-bond can serve the interests of both males and females. When males form long-term relationships with females, the females gain an additional source of nutrition for their offspring, who will be completely dependent on their parents for food and protection for several years, thus increasing the likelihood that the young will survive. Human males benefit from the long-term pair-bond as well. Because human females do not display the fact that they are ovulating, the only way a male can be assured that the child he is raising is his is to have exclusive mating rights with a specific female. The establishment of bonding involves a great deal of sexual activity, much more than is necessary to just bring about reproduction. It is interesting to speculate that sexual behavior in humans is as much involved in maintaining pair-bonds as it is in creating new humans.

Much has been written about the differences in sexual behavior between men and women and that men and women look for different things when assessing individuals as potential mates. It is very difficult to distinguish behaviors that are truly biologically determined and those that are culturally determined. However, some differences may have biological roots. Females benefit from bonding with males who have access to resources that are shared in the raising of young. Do women look for financial security and a willingness to share in a mate? Because pregnancy and nursing young require a great deal of nutrition, it is in the male's interest to choose a mate who is healthy, young, and in good nutritional condition. Since the breasts and buttocks are places of fat storage in women, do men look for youth and appropriate amounts of nutritionally important fat stored in the breasts and buttocks? If these differences between men and women really exist, are they purely cultural, or is there an evolutionary input from our primitive ancestors?

structure and function of the sex organs, the behaviors that involve these structures, psychological components, and the role culture plays in manipulating our sexual behavior. Males and females have different behavior patterns for a variety of reasons. Some behavioral differences are learned (patterns of dress, use of facial makeup), whereas others appear to be less dependent on culture (degree of aggressiveness, frequency of sexual thoughts). We have an intense interest in the facts about our own sexual nature and the sexual behavior of members of the opposite sex and that of peoples of other cultures.

There are several different ways to look at human sexuality. The behavioral sciences tend to focus on the behaviors associated with being male and female and what is considered appropriate and inappropriate sexual behavior. Sex is considered a strong drive, appetite, or urge by psychologists. They describe the sex drive as a basic impulse to satisfy a biological, social, or psychological need. Other social scientists (sociologists, cultural anthropologists) are interested in sexual behavior as it occurs in different cultures and subcultures. When a variety of cultures is examined, it becomes very difficult to classify various kinds of sexual behavior as normal or abnormal. What is considered abnormal in one culture may be normal in another. For example, public nudity is considered abnormal in many cultures but not in others.

Biologists have studied the sexual behavior of nonhuman animals for centuries. They have long considered the function of sex and sexuality in light of its value to the population or species. Sexual reproduction results in new combinations of genes that are important in the process of natural selection. Many biologists today are attempting to look at human sexual behavior from an evolutionary perspective and speculate on why certain sexual behaviors are common in humans (see "A Closer Look: Speculation on the Evolution of Human Sexual Behavior" in this chapter). The behaviors of courtship, mating, rearing of the young, and the division of labor between the sexes are complex in all social animals, including humans. These are demonstrated in the elaborate social behaviors surrounding

mate selection and the establishment of families. It is difficult to draw the line between the biological development of sexuality and the social establishment of customs related to the sexual aspects of human life. However, the biological mechanism that determines whether an individual will develop into a female or male has been well documented.

Sexuality ranges from strongly heterosexual to strongly homosexual. Human sexual orientation is a complex trait, and evidence suggests that there is no one gene that determines where a person falls on the sexuality spectrum. It is most likely a combination of various genes acting together and interacting with environmental factors (figure 25.4) The primary biological goal of sexual intercourse (coitus, mating) is the union of sperm and egg to form offspring. However, this interaction can also improve the health of and bring pleasure to willing partners.

CHROMOSOMAL DETERMINATION OF SEX AND EARLY DEVELOPMENT

Two of the forty-six chromosomes in an adult are the **sex-determining chromosomes.** The other twenty-two pairs that do not determine the sex of an individual are called **autosomes.** There are two kinds of sex-determining chromosomes: the **X chromosome** and the **Y chromosome** (see box figure 25.2 in "A Closer Look: Karyotyping and Down Syndrome"). The two sex-determining chromosomes, X and Y, do not carry equivalent amounts of information, nor do they have equal functions. The X chromosome carries typical genetic information about the production of specific proteins in addition to its function in determining sex. For example, the X chromosome carries information on blood clotting, color vision, and many other characteristics. The Y chromosome, however, appears to be primarily concerned with determining male sexual differentiation and has few other genes on it.

When a human sperm cell is produced, it carries twenty-two autosomes and a sex-determining chromosome. Unlike eggs, which always carry an X chromosome, half the sperm cells carry an X chromosome and the other half carry a Y chromosome. If an X-carrying sperm cell fertilizes an X-containing egg cell, the resultant embryo will develop into a female. If a Y-carrying sperm cell fertilizes the egg, a male embryo develops. It is the presence or absence of the *SRY* (*s*ex-determining *r*egion *Y*) gene typically located on the short arm of the Y chromosome that determines the sex of the developing individual. The SRY gene produces a chemical called the testes determining factor (TDF) and acts as a master switch that triggers the events that converts the embryo into a male. Without this gene, the embryo would become female.

Chromosomal Abnormalities and Sex

Evidence that the Y chromosome and SRY gene control male development comes as a result of many kinds of studies including individuals who have an abnormal number of chromosomes. An abnormal meiotic division that results in sex cells with too many or too few chromosomes is called *nondisjunction.* If

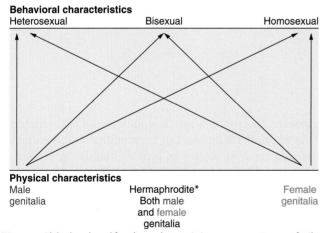

Behavioral characteristics

Heterosexual   Bisexual   Homosexual

Physical characteristics

Male genitalia   Hermaphrodite* Both male and female genitalia   Female genitalia

*Humans with both male and female sex characteristics are rare; most are not fertile, but that does not rule out sexual interest or potential for enjoyment of sexual activity.

FIGURE 25.4 This figure shows how behavioral and physical sexual characteristics interrelate. At the ends of the behavioral spectrum, individuals can be strongly heterosexual and homosexual. They might also find that their physical traits do not match this behavior, but they might be attracted to both sexes or to a person of the opposite sex.

A Closer Look

Karyotyping and Down Syndrome

It is possible to examine cells and count chromosomes. Among the easiest cells to examine in this way are white blood cells. They are dropped onto a microscope slide so that the cells are broken open and the chromosomes are separated. Photographs are taken of chromosomes from cells in the metaphase stage of mitosis. The chromosomes in the photographs can then be cut and arranged for comparison to known samples (box figure 25.2). This picture of an individual's chromosomal makeup is referred to as that person's *karyotype.*

In the normal process of meiosis, diploid cells have their number of chromosomes reduced to haploid. This involves segregating homologous chromosomes into separate cells during the first meiotic division. Occasionally, a pair of homologous chromosomes does not segregate properly during gametogenesis and both chromosomes of a pair end up in the same gamete. This kind of division is known as *nondisjunction.* One example of the effects of nondisjunction is the condition known as Down syndrome. If an abnormal gamete with two number 21 chromosomes has been fertilized by another containing the typical one copy of chromosome number 21, the resulting zygote would have forty-seven chromosomes (e.g., twenty-four from the female plus twenty-three from the male parent) (box figure 25.3). Since the zygote divides by mitosis, the child who develops from this fertilization would have forty-seven chromosomes in

every cell of his or her body and would have the symptoms characteristic of Down syndrome. These may include thickened eyelids, some mental impairment, and faulty speech (box figure 25.4). Premature aging is probably the most significant impact of this genetic disease. On the other hand, a child born with only one chromosome 21 rarely survives.

It is known that the mother's age at childbirth plays an important part in the occurrence of *trisomies* (three of one of the kinds of chromosomes instead of the normal two), such as Down syndrome. In women, gametogenesis begins early in life, but cells destined to become eggs are put on hold during meiosis I. Beginning at puberty and ending at menopause, one of these cells completes meiosis I monthly. This means that cells released for fertilization later in life are older than those released earlier in life. Therefore, it was believed that the chances of abnormalities such as nondisjunction increase as the age of the egg increases. However, the evidence no longer supports this age-egg link. Currently, the increase in frequency of trisomies with age has been correlated

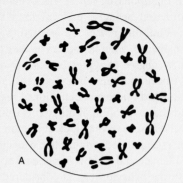

A

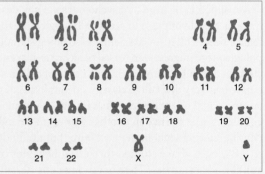

B

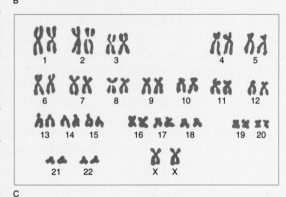

C

Box Figure 25.2

nondisjunction affects the X and Y chromosomes, a gamete might be produced that has only twenty-two chromosomes and lacks a sex-determining chromosome, or it might have twenty-four, with two sex-determining chromosomes. If a cell with too few or too many sex chromosomes is fertilized, sexual development is usually affected. If a normal egg cell is fertilized by a sperm cell with no sex chromosome, the offspring will have only one X chromosome. These people are designated as XO. They develop a collection of characteristics known as *Turner's syndrome* (figure 25.5). An individual with this condition is female, is short for her age, and fails to mature sexually, resulting in sterility. In addition, she may have a thickened neck (termed *webbing*), hearing impairment, and some abnormalities in the

cardiovascular system. When the condition is diagnosed, some of the physical conditions can be modified with treatment. Treatment involves the use of growth-stimulating hormones to increase growth rate and female sex hormones to stimulate sexual development, although sterility is not corrected.

An individual who has XXY chromosomes is basically male (figure 25.6). This genetic condition is termed *Klinefelter's syndrome* and is probably the most common chromosomal variation found in humans. The largest percentage of these men lead healthy, normal lives and are indistinguishable from normal males. However, those with Klinefelter's syndrome may be sterile and show breast enlargement, incomplete masculine body form, lack of facial hair, and have some minor learning

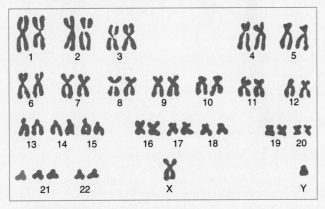

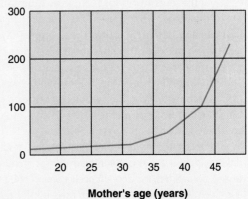

Box Figure 25.3

Box Figure 25.5

Number of
births with
Down syndrome
per 100,000

Mother's age (years)

Box Figure 25.4

with a decrease in the activity of a woman's immune system. As she ages, her immune system is less likely to recognize the difference between an abnormal and a normal embryo. This means that she is more likely to carry an abnormal fetus to full term.

Box figure 25.5 illustrates the frequency of occurrence of Down syndrome at different ages in women. Notice that the frequency increases very rapidly after age thirty-seven. For this reason, many physicians encourage couples to have their children in their early to mid-twenties and not in their late thirties or early forties. Physicians normally encourage older women who are pregnant to have the cells of their fetus checked to see if they have the normal chromosome number. It is important to know that the male parent can also contribute the extra chromosome 21. However, it appears that this occurs less than 30 percent of the time.

Sometimes a portion of chromosome 14 may be cut out and joined to chromosome 21. A person with this 14/21 transfer is monosomic and has only forty-five chromosomes; one 14 and one 21 are missing and replaced by the translocated 14/21. Statistically, about 15 percent of the children of carrier mothers inherit the 14/21 chromosome and have Down syndrome. Fewer of the children born to fathers with the 14/21 translocation inherit the abnormal chromosome and are Downic.

Whenever an individual is born with a chromosomal abnormality, it is recommended that both parents have a karyotype evaluation in an attempt to identify the possible source of the problem. This is not to fix blame but to provide information on the likelihood that another pregnancy would also result in a child with a chromosomal abnormality.

problems. These characteristics vary greatly in degree, and many men are diagnosed with Klinefelter's syndrome only after they undergo testing to determine why they are infertile. This condition is present in about 1 in 500 to 1,000 men. Treatment may involve testosterone therapy and breast-reduction surgery in males who have significant breast development.

Because both conditions involve abnormal numbers of X or Y chromosomes, they provide strong evidence that these chromosomes are involved in determining sexual development. The early embryo resulting from fertilization and cell division is neither male nor female but becomes female or male later in development—based on the sex-determining chromosomes that control the specialization of the cells of the undeveloped, embryonic gonads into female ovaries or male testes. This specialization of embryonic cells is termed **differentiation.** The embryonic gonads begin to differentiate into testes about five to seven weeks after **conception** (fertilization) if the SRY gene is present and functioning. The Y chromosome seems to control this differentiation process in males because the gonads do not differentiate into female sex organs until later and then only if two X chromosomes are present. It is the absence of the Y chromosome that determines female sexual differentiation.

Researchers were interested in how females, with two X chromosomes, handle the double dose of genetic material in comparison to males, who have only one X chromosome. In 1949, M. L. Barr discovered that a darkly staining body was

A Closer Look

Cryptorchidism—Hidden Testes

At about the seventh month of pregnancy (gestation), in normal male fetuses each testis moves from a position in the abdominal cavity to an external sac, called the scrotum. They pass through an opening called the **inguinal canal** (box figure 25.6). This canal closes off but continues to be a weak area in the abdominal wall, and it may rupture later in life. This can happen when strain (e.g., from improperly lifting heavy objects) causes a portion of the intestine to push through the inguinal canal into the scrotum, a condition known as an inguinal hernia.

Occasionally, the testes do not descend, resulting in a condition known as **cryptorchidism** (*crypt*=hidden; *orchidos*= testes). Sometimes, the descent occurs during puberty; if not, there is a twenty-five to fifty times increased risk for testicular cancer. Because of this increased risk, surgery can be done to allow the undescended testes to be moved into the scrotum. Sterility will result if the testes remain in the abdomen. This happens because normal sperm cell development cannot occur in a very warm environment. The temperature in the abdomen is higher than the temperature in the scrotum. Normally, the temperature of the testes is very carefully regulated by muscles that control their distance from the body. Physicians have even diagnosed cases of male infertility as being caused by tight-fitting pants that hold the testes so close to the body that the temperature increase interferes with normal sperm development. Recent evidence has also suggested that teenage boys and young men working with computers in the laptop position for extended periods may also be at risk for lowered sperm counts.

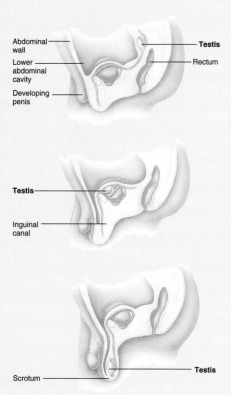

Abdominal wall — Testis
Lower abdominal cavity — Rectum
Developing penis

Testis

Inguinal canal

Scrotum — Testis

Box Figure 25.6

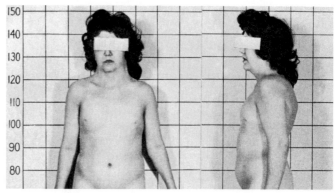

FIGURE 25.5 Turner's syndrome individuals have forty-five chromosomes. They have only one of the sex chromosomes and it is an X chromosome. Individuals with this condition are females, have delayed growth, and fail to develop sexually. This woman is less than 150 cm (5 ft) tall and lacks typical secondary sexual development for her age. She also has a "webbed neck" that is common among Turner's syndrome individuals.

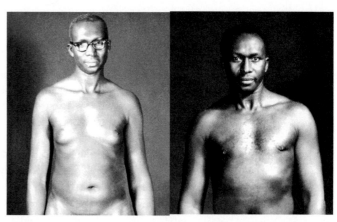

FIGURE 25.6 Klinefelter's syndrome individuals have two X chromosomes and a Y chromosome; they are male, are sterile, and often show some degree of breast development and female body form. They are typically tall. The two photos show a Klinefelter's individual before and after receiving testosterone hormone therapy.

cells have only one dose of X-chromosome genetic information that is functional. The single X chromosome of the male functions as expected, and the Y chromosome directs only male-determining activities. One can identify the blood at a crime scene as male or female based on Barr bodies.

Fetal Development and Sex

Development of embryonic gonads begins very early during fetal growth. First, a group of cells begins to differentiate into primitive gonads at about week 5 (figure 25.7). By week 5 or 7 if a Y chromosome is present, a gene product (TDF) from the chromosome will begin the differentiation of these gonads into testes; they will develop into ovaries beginning about week 12 if two X chromosomes are present (Y chromosome is absent).

As soon as the gonad has differentiated into an embryonic testis at about week 8, it begins to produce testosterone. The presence of testosterone results in the differentiation of male sexual anatomy, and the absence of testosterone results in the differentiation into female sexual anatomy.

generally present in female cells but was not present in male cells. The tightly coiled structure in the cells of female mammals is called a **Barr body** after its discoverer. It was postulated, and has since been confirmed, that this structure is an X chromosome that is largely nonfunctional. Therefore, female

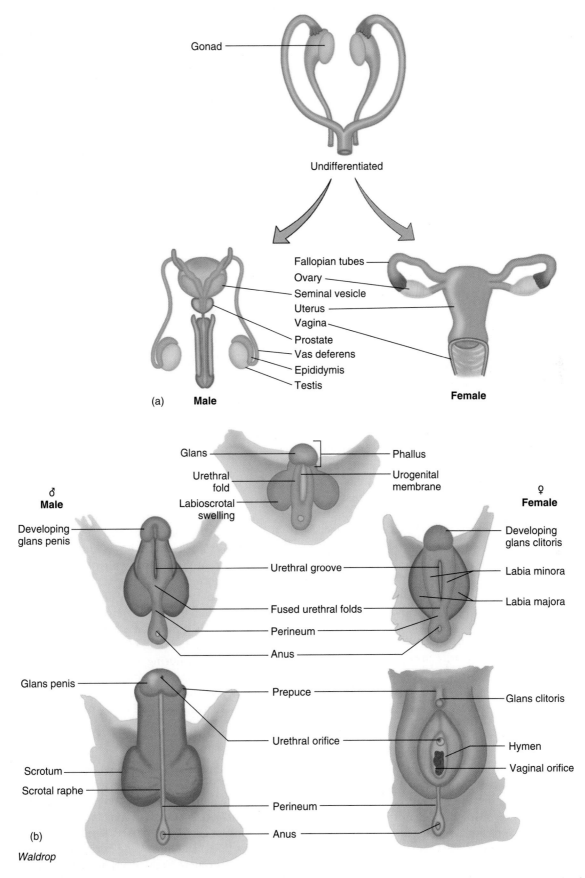

Gonad

Undifferentiated

Fallopian tubes
Ovary
Seminal vesicle
Uterus
Vagina
Prostate
Vas deferens
Epididymis
Testis

(a) **Male** **Female**

♂
Male

♀
Female

Glans
Urethral fold
Labioscrotal swelling

Phallus
Urogenital membrane

Developing glans penis
Developing glans clitoris

Urethral groove
Labia minora

Fused urethral folds
Labia majora

Perineum
Anus

Glans penis
Prepuce
Glans clitoris

Urethral orifice
Hymen
Vaginal orifice

Scrotum
Scrotal raphe

Perineum
Anus

(b)

Waldrop

FIGURE 25.7 The early embryo grows without showing any sexual characteristics. The male and female sexual organs eventually develop from common basic structures. (*A*) shows the development of the internal anatomy. (*B*) shows the development of external anatomy.

SEXUAL MATURATION OF YOUNG ADULTS

Following birth, sexuality plays only a small part in physical development for several years. Culture and environment shape the responses that the individual will come to recognize as normal behavior. During **puberty** (the developmental period when the body changes and becomes able to reproduce), normally between eleven and fourteen years of age, increased production of sex hormones causes major changes as the individual reaches sexual maturity. Generally, females reach puberty six months to two years before males. After puberty, humans are sexually mature and have the capacity to produce offspring.

The Maturation of Females

Female children typically begin to produce quantities of sex hormones from the hypothalamus portion of the brain and the pituitary gland, ovaries, and adrenal glands at nine to twelve years of age. This marks the onset of puberty. The **hypothalamus** controls the functioning of many other glands throughout the body, including the **pituitary gland.** At puberty, the hypothalamus begins to release a hormone known as **gonadotropin-releasing hormone (GnRH)**, which stimulates the pituitary to release luteinizing hormone (LH) and **follicle-stimulating hormone (FSH).** Increased levels of FSH stimulate the development of **follicles,** saclike structures that produce oocytes in the ovary, and the increased luteinizing hormone stimulates the ovary to produce larger quantities of **estrogens.** The increasing supply of estrogen is responsible for the many changes in sexual development that can be noted at this time. These changes include breast growth, changes in the walls of the uterus and vagina, increased blood supply to the clitoris, and changes in the pelvic bone structure. The **clitoris** is a small, elongated erectile structure located between and at the head of the labia; it is equivalent to the penis.

Estrogen also stimulates the female adrenal gland to produce **androgens,** male sex hormones. The androgens are responsible for the production of pubic hair, and they seem to have an influence on the female sex drive. The adrenal gland secretions may also be involved in the development of acne. Those features that are not primarily involved in sexual reproduction but are characteristic of a sex are called **secondary sexual characteristics.** In women, the distribution of body hair, patterns of fat deposits, and a higher voice are examples.

A major development during this time is the establishment of the **menstrual cycle.** This involves the periodic growth and shedding of the lining of the uterus. These changes are under the control of a number of hormones produced by the pituitary and ovaries. The ovaries are stimulated to release their hormones by the pituitary gland, which is in turn influenced by the ovarian hormones. Both follicle-stimulating hormone (FSH) and luteinizing hormone (LH) are produced by the pituitary gland. FSH causes the maturation and development of the ovaries. LH is important in causing ovulation and converting the ruptured follicle into a structure known as the *corpus luteum.* The corpus luteum produces the

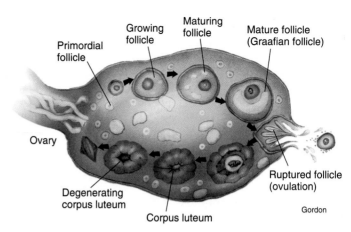

FIGURE 25.8   In the ovary, the egg begins development inside a sac of cells known as a follicle. Each month, one of these follicles develops and releases its product. This release through the wall of the ovary is known as ovulation.

hormone **progesterone,** which is important in maintaining the lining of the uterus. Changes in the levels of progesterone result in a periodic buildup and shedding of the lining of the uterus known as the menstrual cycle. Table 25.2 summarizes the activities of these various hormones. Associated with the menstrual cycle is the periodic release of sex cells from the surface of the ovary, called **ovulation** (figure 25.8). Initially, these two cycles, menstruation and ovulation, may be irregular, which is normal during puberty. Eventually, hormone production becomes regulated so that ovulation and menstruation take place on a regular monthly basis in most women, although normal cycles may vary from twenty-one to forty-five days.

As girls progress through puberty, curiosity about the changing female body form and new feelings lead to self-investigation. Studies have shown that sexual activity such as manipulation of the clitoris, which causes pleasurable sensations, is performed by a large percentage of young women. Self-stimulation, frequently to orgasm, is a common result. This stimulation is termed *masturbation,* and it should be stressed that it is considered a normal part of sexual development. **Orgasm** is the peak of sexual arousal. During orgasm, whether occurring during masturbation or sexual intercourse, the heart rate increases, breathing quickens, and blood pressure rises; muscles throughout the body spasm but mainly those in the vagina, uterus, anus, and pelvic floor.

The Maturation of Males

Males typically reach puberty about two years later (ages eleven to fourteen) than females. Puberty in males also begins with a change in hormone levels. At puberty, the hypothalamus releases increased amounts of gonadotropin-releasing hormone (GnRH), resulting in increased levels of follicle-stimulating hormone (FSH) and luteinizing hormone. These are the same changes that occur in female development. Luteinizing hormone is often called interstitial cell-stimulating hormone (ICSH) in males. LH stimulates the testes to produce **testosterone,** the

TABLE 25.2   Human Reproductive Hormones

Hormone	Production Site	Target Organ	Function
Gonadotropin-releasing hormone (GnRH)	Hypothalamus	Anterior pituitary	Stimulates the release of FSH and LH from anterior pituitary
Luteinizing hormone (LH) or interstitial cell-stimulating hormone (ICSH)	Anterior pituitary	Ovary, testes	Stimulates ovulation in females and sex-hormone (estrogens and testosterone) production in both males and females
Follicle-stimulating hormone (FSH)	Anterior pituitary	Ovary, testes	Stimulates ovary and testis development; stimulates egg production in females and sperm production in males
Estrogens	Ovary	Entire body	Stimulates development of female reproductive tract and secondary sexual characteristics
Testosterone	Testes	Entire body	Stimulates development of male reproductive tract and secondary sexual characteristics
Progesterone	Corpus luteum of ovary	Uterus, breasts	Causes uterine thickening and maturation; maintains pregnancy; contributes to milk production
Androgens	Testes, adrenal glands	Entire body	Stimulates development of male reproductive tract and secondary sexual characteristics in males and females
Oxytocin	Posterior pituitary	Uterus, breasts	Causes uterus to contract and breasts to release milk
Prolactin, lactogenic, or luteotropic hormone	Anterior pituitary	Breast, ovary	Stimulates milk production; also helps maintain normal ovarian cycle
Human chorionic gonadotropin	Placenta	Corpus luteum	Maintains corpus luteum so that it continues to secrete progesterone and maintain pregnancy

primary sex hormone in males. The testosterone produced by the embryonic testes caused the differentiation of internal and external genital anatomy in the male embryo. At puberty, the increased amount of testosterone is responsible for sexual maturation and the development of male secondary sexual characteristics, and is important in the maturation and production of sperm.

The major changes during puberty include growth of the testes and scrotum, pubic-hair development, and increased size of the penis. Secondary sex characteristics also begin to become apparent. Facial hair, underarm hair, and chest hair are some of the most obvious. The male voice changes as the larynx (voice box) begins to change shape. Body contours also change, and a growth spurt increases height. In addition, the proportion of the body that is muscle increases and the proportion of body fat decreases. At this time, a boy's body begins to take on the characteristic adult male shape, with broader shoulders and heavier muscles.

In addition to these external changes, increased testosterone causes the production of seminal fluid by the accessory glands. FSH stimulates the production of sperm cells. The release of sperm cells and seminal fluid begins during puberty and is termed *ejaculation*. This release is generally accompanied by the pleasurable sensations of orgasm. The sensations associated with ejaculation may lead to self-stimulation, or masturbation. Masturbation is a common and normal activity as a boy goes through puberty. Studies of adult sexual behavior have shown that nearly all men masturbate at some time during their lives.

SPERMATOGENESIS

The biological reason for sexual activity is the production of offspring. The process of producing gametes includes meiosis and is called gametogenesis (gamete formation). The term **spermatogenesis** is used to describe gametogenesis that takes place in the testes of males. The two bean-shaped testes are composed of many small sperm-producing tubes, or **seminiferous tubules,** and collecting ducts that store sperm. These are held together by a thin covering membrane. The seminiferous tubules join together and eventually become the epididymis, a long, narrow convoluted tube in which sperm cells are stored and mature before ejaculation (figure 25.9).

Leading from the epididymis is the vas deferens, or sperm duct; this empties into the urethra, which conducts the sperm out of the body through the **penis** (figure 25.10). Before puberty, the seminiferous tubules are packed solid with diploid cells called spermatogonia. These cells, which are found just inside the tubule wall, undergo *mitosis* and produce more spermatogonia. Beginning about age eleven, some of the spermatogonia specialize and begin the process of *meiosis*, whereas others continue to divide by mitosis, ensuring a constant and continuous supply of spermatogonia. Spermatogenesis needs to occur below temperature, which is why the testicles are in a sack, the scrotum, outside the body. Once spermatogenesis begins, the seminiferous tubules become hollow and can transport the mature sperm.

Spermatogenesis involves several steps. It begins when some of the spermatogonia in the walls of the seminiferous tubules differentiate and enlarge to become *primary spermatocytes.*

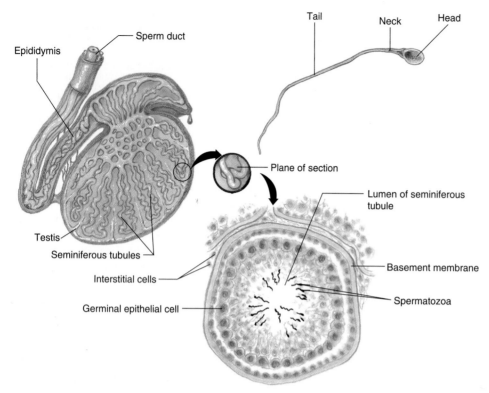

FIGURE 25.9 The testis consists of many tiny tubes called seminiferous tubules. The walls of the tubes consist of cells that continually divide, producing large numbers of sperm. The sperm leave the seminiferous tubules and enter the epididymis, where they are stored prior to ejaculation through the sperm duct. Sperm cells have a head region, with an enzyme-containing cap and the DNA. They also have a neck region, with ATP-generating mitochondria, and a tail flagellum, which propels the sperm.

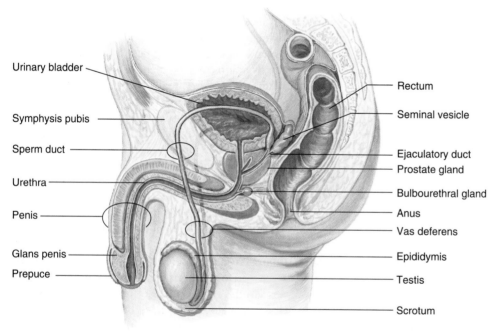

FIGURE 25.10 The male reproductive system consists of two testes that produce sperm, ducts that carry the sperm, and various glands. Muscular contractions propel the sperm through the vas deferens past the seminal vesicles, prostate gland, and bulbourethral gland, where most of the liquid of the semen is added. The semen passes through the urethra of the penis to the outside of the body.

These diploid cells undergo the first meiotic division, which produces two haploid *secondary spermatocytes*. The secondary spermatocytes go through the second meiotic division, resulting in four haploid **spermatids,** which lose much of their cytoplasm and develop long tails to mature into sperm. The sperm have only a small amount of food reserves. Therefore, once they are released and become active swimmers, they live no more than seventy-two hours. However, if the sperm are placed in a special protective solution, the temperature can be lowered drastically to −196°C. Under these conditions, the sperm become deactivated, freeze, and can survive for years outside the testes. This has led to the development of sperm banks. Artificial insemination (placing stored sperm into the reproductive tract of a female) or *in vivo* fertilization of cattle, horses, and other domesticated animals with sperm from sperm banks is common. The same techniques can be used in humans but are most often used when men produce insufficient sperm so that the sperm must be collected and stored to increase the number of sperm to enhance the likelihood of fertilization.

Semen, also known as *seminal fluid,* is a mixture of sperm and secretions from three *accessory glands.* These are the *seminal vesicles, prostate, and bulbourethral glands.* They produce secretions that nourish and activate the spermatozoa, and clear the urethral tract before ejaculation. They also serve to lubricate the tract and act as the vehicle to help propel the spermatozoa.

Seminal vesicles secrete an alkaline fluid that contains fructose and hormones. The alkaline nature of the fluid helps to neutralize the acid environment in the female reproductive tract. This improves the chances that sperm will make their way through the tract to the egg. The fructose provides energy for the sperm. The hormones stimulate contractions in the male's system that help to propel sperm. Seminal vesicle secretions make up about 60 percent of the total

Radioactive Seeds and Prostate Cancer Therapy

Radioactive isotopes of iodine-125 (I^{125}) are placed within thin metal capsules only 4.5 mm long and 0.8 mm in diameter or a rigid, absorbable suture. It only takes about forty-five minutes using a hypodermic needlelike apparatus to insert them through the skin between the scrotum and rectum into a prostate tumor. No incisions are necessary. Radiation released from these "seeds" is absorbed within the tumor and the patient is not considered to be radioactive. Depending on the size of the tumor, twenty, thirty, forty, or more seeds might be inserted. The energy released by the small amount of I^{125} acts locally and lasts only long enough to destroy the tumor cells. Patients are usually up and about in a matter of days and return to their normal activities after two to four weeks.

Myths, Mistakes, and Misunderstandings

Mumps and Sterility

Misunderstanding: Both boys and girls might become sterile if they get the mumps.

In fact, mumps is a viral infection primarily of the salivary glands. It causes them to swell, and the swelling reaches its peak in about forty-eight hours and may last two weeks. Inflammation and swelling of the testes (called orchitis) can also occur in about 15 to 25 percent of the male cases. In rare cases, the swollen testes rupture their thin covering membrane, leading to sterility. In females, inflammation of the ovaries (called oophoritis) occurs in only 5 percent of the mumps cases and does not cause sterility, since the ovaries are not bound by a membrane and can swell without rupturing.

---

seminal fluid. Bulbourethral gland secretions are also alkaline and help neutralize the vaginal tract making fertilization more likely. The prostate gland (note spelling) produces a thin milky fluid with a characteristic odor that makes up about 25 percent of semen and contains sperm-activating enzymes.

Prostate cancer can be a significant problem for men. Age appears to be the greatest risk factor of this disease. It has been estimated that 75 percent of men over sixty-five years of age are diagnosed with this form of cancer. This may be because prostate cancer is a slow, progressive disease. Prostate cancer is more common in African-American men in the United States than in any other group. Treatment of the disease may be watchful waiting, changes in diet, surgery, radiation therapy, hormonal therapy, or a combination of these. Surgery, if not done very carefully, can lead to problems with urination or an inability to obtain an erection. A recent study found that 30 percent of men who had their prostate removed and experienced erectile dysfunction regained their ability to have an erection after taking erectile dysfunction medications for nine months. The prostate also uses zinc when functioning properly. Low levels of dietary zinc can lead to enlargement, which may pinch the urethra, interfering with urination. Dietary supplements of zinc, selenium, and vitamin E are currently being researched as ways of combating this swelling.

Spermatogenesis in human males takes place continuously throughout a male's reproductive life, although the number of sperm produced decreases as a man ages. Sperm counts can be taken and used to determine the probability of successful fertilization. A healthy male probably releases about 150 million sperm with each ejaculation. A man must be able to release at least 100 million sperm per milliliter to be fertile. Men with sperm counts under 50 million/ml are often infertile, and those with sperm counts below 20 million/ml are clinically infertile. These vast numbers of sperm are necessary because so many die during their journey that large numbers are needed for the few survivors to reach the egg. In addition, each sperm contains enzymes in the head of sperm that are needed to digest through mucus and protein found in the female reproductive tract. Millions of sperm contribute in this way to the process of fertilization, but only one is involved in fertilizing the egg.

OOGENESIS

The term **oogenesis** refers to the production of egg cells. This process starts before a girl is born during prenatal development of the ovary. It occurs when diploid oogonia cease dividing by *mitosis* and enlarge to become *primary oocytes.* All of the primary oocytes that a woman will ever have are already formed before her birth. At this time, they number approximately two million, but that number is reduced by cell death to between three-hundred and four-hundred thousand cells by the time of puberty. Primary oocytes begin to undergo meiosis and pause in prophase I. Oogenesis pauses at this point, and all the primary oocytes remain just under the surface of the ovary.

At puberty and on a regular basis thereafter, the sex hormones stimulate a primary oocyte to continue its maturation process. It completes the first meiotic division of a single primary oocyte, which began before the woman's birth, about once a month. But in telophase I, the two cells that form receive unequal portions of cytoplasm. You might think of it as a lopsided division. The smaller of the two cells is called a *polar body,* and the larger haploid cell is the *secondary oocyte* (people commonly refer to this cell as an *egg,* or *ovum,* although technically it is not). The other primary oocytes remain in the ovary. Ovulation begins when the soon-to-be-released secondary oocyte, encased in a saclike structure known as a follicle, grows and moves near the surface of the ovary. When this maturation is complete, the follicle ruptures and the secondary oocyte is released. It is swept into the **oviduct** (fallopian tube) by ciliated cells and travels toward the **uterus** (figure 25.11). Because of the action of the luteinizing hormone, the follicle from which

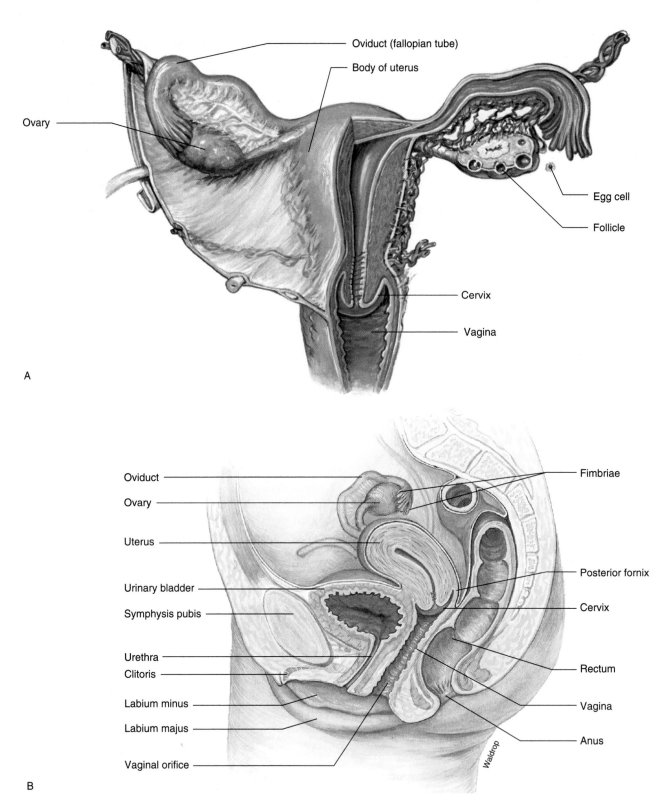

A

B

FIGURE 25.11 In the human female reproductive system (*A*), after ovulation the cell travels down the oviduct to the uterus. If it is not fertilized, it is shed when the uterine lining is lost during menstruation. (*B*) The human female reproductive system, side view.

the oocyte ovulated develops into a glandlike structure, the corpus luteum, which produces hormones (progesterone and estrogen) that prevent the release of other secondary oocytes. If the secondary oocyte is fertilized, it completes meiosis with the sperm DNA inside, and the haploid egg nucleus and sperm nucleus unite to form the zygote. If the secondary oocyte is not fertilized, it passes through the **vagina** to the outside during menstruation. During her lifetime, a female releases about three hundred to five hundred secondary oocytes. Obviously, few of these cells are fertilized.

The Centers for Disease Control and Prevention defines *assisted reproductive technology* (ART) as "all fertility treatments in which both eggs and sperm are manipulated." In general, ART involves surgically removing eggs from a woman's ovaries, combining them with sperm in the laboratory, and returning them to the woman's body or donating them to another woman. It does not include procedures in which only sperm are manipulated (i.e., artificial insemination or intrauterine insemination) or procedures in which a woman takes drugs only to stimulate egg production, without the intention of having eggs retrieved. There are three types of ART: IVF (*in vitro* fertilization), GIFT (gamete intra-fallopian transfer) and ZIFT (zygote intra-fallopian transfer).

In vitro fertilization is a method that uses hormones to stimulate egg production; removing the oocyte from the ovary and fertilizing it with donated sperm. The fertilized egg is incubated to stimulate cell division in a laboratory dish and then placed in the uterus.

GIFT relies on the same hormonal treatment as IVF to stimulate ovulation. Ultrasound is used by the physician to observe the transfer of unfertilized eggs and sperm into the woman's oviduct through small incisions in her abdomen. Once fertilized, the zygote moves down into the uterus and implants. GIFT is only an option for women with open fallopian tubes. The disadvantage to GIFT is a risk of tubal pregnancy and the inability to document actual fertilization.

In the ZIFT procedure, a woman's mature eggs are collected and fertilized in the laboratory. Then, a scope is inserted into the vagina to find a place for the implantation. Another instrument is used to guide the fertilized egg into the identified position.

One distinguishing characteristics is the relative age of male and female sex cells. In males, sperm production is continuous throughout life. Sperm do not remain in the tubes of the male reproductive system for very long. They are either released shortly after they form or die and are harmlessly absorbed.

In females, meiosis begins before birth, but the oogenesis process is not completed, and the cell is not released for many years. A secondary oocyte released when a woman is thirty-seven years old began meiosis thirty-seven years before! During that time, the cell was exposed to many influences, a number of which may have damaged the DNA or interfered with the meiotic process. The increased risk of abnormal births in older mothers may be related to the age of their eggs. Such alterations are less likely to occur in males because new gametes are being produced continuously. Also, defective sperm appear to be much less likely to be involved in fertilization.

Hormonal Control of Female Sexual Cycles

Hormones control the cycle of changes in breast tissue, in the ovaries, and in the uterus. In particular, estrogen and progesterone stimulate milk production by the breasts and cause the lining of the uterus to become thicker and filled with blood vessels before ovulation. This ensures that if fertilization occurs, the resultant embryo will be able to attach itself to the wall of the uterus and receive nourishment. If the cell is not fertilized, the lining of the uterus, *endometrium,* is shed. This is known as *menstruation, menstrual flow,* the *menses,* or a *period.*

The activities of the ovulatory cycle and the menstrual cycle are coordinated. During the first part of the menstrual cycle, increased amounts of FSH cause the follicle to increase in size. Simultaneously, the follicle secretes increased amounts of estrogen that cause the lining of the uterus to increase in thickness. When ovulation occurs, the remains of the follicle are converted into a corpus luteum by the action of LH. The corpus luteum begins to secrete progesterone, and the nature of the uterine lining changes by becoming more vascularized. This is choreographed so that if an embryo arrives in the uterus shortly after ovulation, it meets with a uterine lining prepared to accept it. If pregnancy does not occur, the corpus luteum degenerates, resulting in a reduction in the amount of progesterone needed to maintain the lining of the uterus, and the lining is shed. At the same time that hormones are regulating ovulation and the menstrual cycle, changes are taking place in the breasts. The same hormones that prepare the uterus to receive the embryo also prepare the breasts to produce milk. These changes in the breasts, however, are relatively minor unless pregnancy occurs.

HORMONAL CONTROL OF FERTILITY

An understanding of how various hormones manipulate the menstrual cycle, ovulation, milk production, and sexual behavior has led to the medical use of certain hormones. Some women are unable to have children because they do not release oocytes from their ovaries or they release them at the wrong time. Physicians can now regulate the release of oocytes from the ovary using certain hormones, commonly called *fertility drugs.* These hormones can be used to stimulate the release of oocytes for capture and use in what is called *in vitro* fertilization (*IVF* or *test-tube* fertilization) or to increase the probability of natural conception; that is, *in vivo* fertilization (*in-life* fertilization).

Unfortunately, the use of these techniques often results in multiple embryos being implanted in the uterus. This is likely to occur because the drugs may cause too many secondary oocytes to be released at one time. In the case of *in vitro* fertilization, because there is a high rate of failure and the process is expensive, typically several early-stage embryos are inserted into the uterus to increase the likelihood that one will implant. If several are successful, multiple embryos implant. The implantation of multiple embryos makes it difficult for one embryo to develop

Can Humans Be Cloned?

Recent advances in the understanding of the reproductive biology of humans and other mammals have led to the possibility of cloning in humans. An understanding of the function of hormones allows manipulation of the reproductive cycles of women so that developing oocytes can be harvested from the ovary. These oocytes can be fertilized in dishes, and the early development of the embryos can be observed. These embryos can be implanted into the uterus of a woman who does not need to be the woman who donated the oocyte. In nonhuman mammals, oocytes have been manipulated such that the nucleus of the oocyte was removed and the nucleus from a mature cell was inserted. The manipulated cell begins embryological development and can be implanted in the uterus of a female. In several species (sheep, mouse, monkey), the process has successfully resulted in a cloned offspring that is genetically identical to the individual that donated the nucleus. The technology for cloning humans exists. The oocytes can be harvested. The nuclei can be transferred, and the techniques for introducing them into a uterus are well-known.

Can humans be cloned? In 2001, an Italian embryologist, Dr. Severino Antinori, claimed to have cloned a human embryo. In 2002, the U.S.-based company Clonaid claimed that it had created the first-ever cloned human baby, a girl nicknamed Eve. However, since no evidence—including DNA testing—confirming the event has been provided, skepticism surrounds this claim within the scientific community. Chinese scientists have also claimed to have produced early human clones. Woo Suk Hwany of Seoul National University in Korea announced in 2004 that he had also successfully cloned healthy human embryos. His assertions have since been discredited.

Questions to Discuss

1. What scientific evidence would be needed to confirm such claimed "cloning" births?
2. Should humans be cloned? This is also a question for ethicists and social thinkers to answer.

TABLE 25.3 Common Causes of Infertility

Females

Endometriosis	A condition in which endometrial tissue grows outside the uterus, forming adhesions between organs
Ovulation problems	Usually a hormonal condition that prevents the release of a mature egg from an ovary
Poor egg quality	Eggs that are damaged or develop chromosomal abnormalities; usually age-related
Polycystic ovary syndrome	Ovaries that contain many small cysts causing hormonal imbalances and therefore do not ovulate regularly
Female tube blockages	Causes can include pelvic inflammatory disease (PID), sexually transmitted diseases and previous sterilization surgery.

Males

Male tube blockages	Varicose veins in the testicles are the most common cause of male tube blockages. Sexually transmitted diseases also cause blockage.
Sperm problems	Low or no sperm counts, poor sperm motility (the ability to move), and abnormally shaped sperm
Sperm allergy	An immune reaction to sperm causing them to be killed by antibodies; occurs in fewer than 10 percent of infertility cases
Chronic stress	Decreased LH levels, which cause lower testosterone levels and decreased spermatogenesis

A second medical use of hormones is in the control of conception by the use of birth-control pills—oral contraceptives. Birth-control pills have the opposite effect of fertility drugs. They raise the levels of estrogen and progesterone, which suppresses the production of FSH and LH, preventing the release of secondary oocytes from the ovary. Hormonal control of fertility is not as easy to achieve in men because there is no comparable cycle of gamete release. However, a new, reversible conception control method for males has been developed. It relies on using an implant containing the male sex hormone testosterone and a three-monthly injection of a progestin, a hormone used in female contraceptive pills. The combination of the two hormones temporarily turns off the normal signals from the brain that stimulate sperm production. An extra dose of testosterone may be given to help maintain sex drive. The use of drugs and laboratory procedures to help infertile couples have children has also raised the technical possibility of cloning in humans (Table 25.3).

FERTILIZATION, PREGNANCY, AND BIRTH

In most women, a secondary oocyte is released from the ovary about fourteen days before the next menstrual period. The menstrual cycle is usually said to begin on the first day of menstruation. Therefore, if a woman has a regular twenty-eight-day cycle, the cell is released approximately on day 14 (figure 25.12). If a woman normally has a regular twenty-one-day menstrual cycle, ovulation would occur about day 7 in the cycle. If a woman has a regular forty-day cycle, ovulation would occur about day 26 of her menstrual cycle. Some women, however, have very irregular menstrual cycles, and it is difficult to determine just when the oocyte will be released to become available for fertilization. Once the cell is released, it is swept into the oviduct and moved toward the uterus. If sperm are present, they swarm around the secondary oocyte as it passes down the oviduct, but only one sperm penetrates

properly and be carried through the entire nine-month gestation period. When we understand the action of hormones better, we may be able to control the effects of fertility drugs and eliminate the problem of multiple implantations.

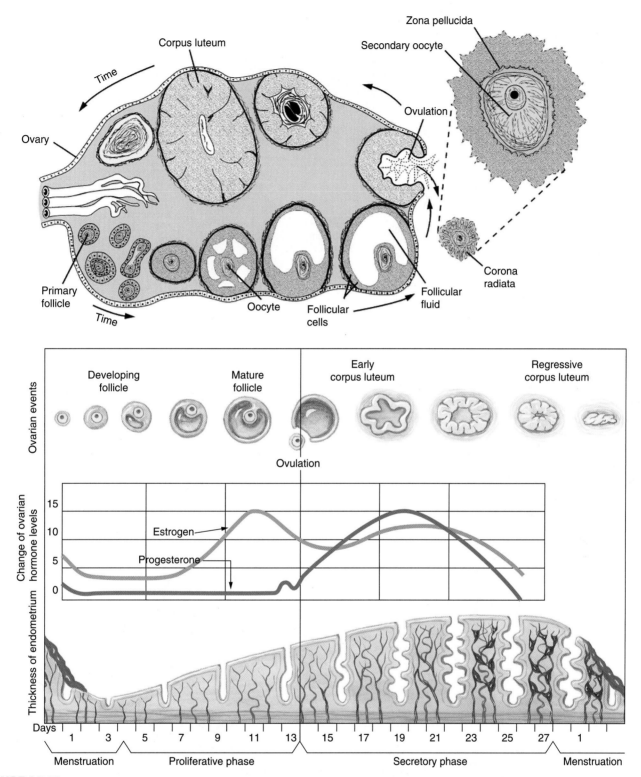

FIGURE 25.12 The release of a secondary oocyte (ovulation) is timed to coincide with the thickening of the lining of the uterus. The uterine cycle in humans involves the preparation of the uterine wall to receive the embryo if fertilization occurs. Knowing how these two cycles compare, it is possible to determine when pregnancy is most likely to occur.

the outer layer to fertilize it and cause it to complete meiosis II. The other sperm contribute enzymes that digest away the protein and mucus barrier between the egg and the successful sperm.

During this second meiotic division, the second polar body is pinched off and the *ovum* (egg) is formed. Because chromosomes from the sperm are already inside, they simply intermingle with those of the ovum, forming a diploid zygote or fertilized egg. As the zygote continues to travel down the oviduct, it begins to divide by mitosis into smaller and smaller cells without having the mass of cells increase in size (figure 25.13). This division process is called *cleavage*. Eventually, a solid ball

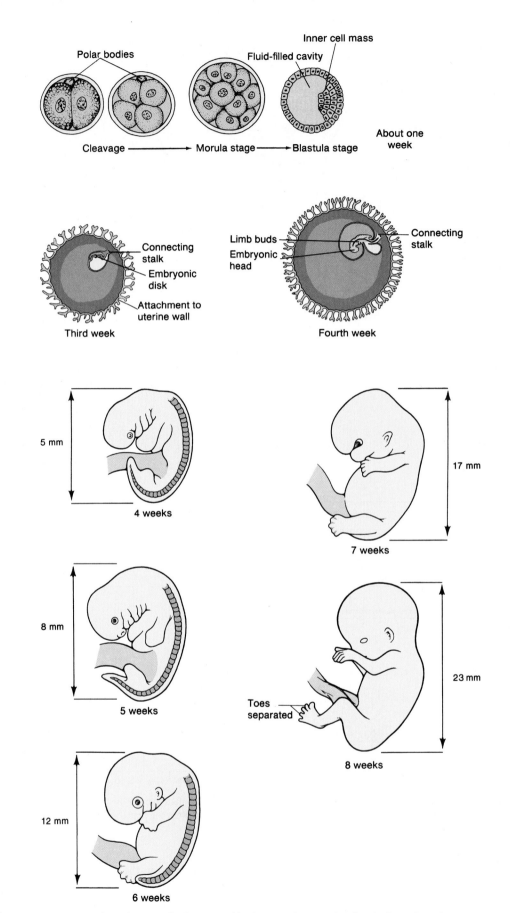

FIGURE 25.13 During the period of time between fertilization and birth, many changes take place in the embryo. Here we see some of the changes that take place during the first eight weeks.

Robert Geoffrey Edwards (1925–) and Patrick Christopher Steptoe (1913–1988)

Robert Geoffrey Edwards and Patrick Christopher Steptoe were British researchers—a physiologist and an obstetric surgeon, respectively—who devised a technique for fertilizing a human egg outside the body and transferring the fertilized embryo to the uterus of a woman. A child born following the use of this technique is popularly known as a "test-tube baby."

Robert Edwards was educated at the University of Wales and the University College of North Wales, Bangor, and from 1951 to 1957 at the University of Edinburgh. In 1957, Edwards went to the California Institute of Technology but returned to England the following year to the National Institute of Medical Research at Mill Hill. He remained there until 1962, when he took an appointment at Glasgow University. A year later, he moved again to the Department of Physiology at Cambridge University. From 1985 to 1989, he was professor of human reproduction at Cambridge.

During his research in Edinburgh, Edwards successfully replanted mouse embryos into the uterus of a mouse and wondered if the same process could be used to replant a human embryo into the uterus of a woman. Edwards was able to obtain human eggs from pieces of ovarian tissue removed during surgery. He found that the ripening process was very slow, the first division beginning only after twenty-four hours.

He studied the maturation of eggs of different species of mammals and in 1965 attempted the fertilization of human eggs. He left mature eggs with sperm overnight and found just one where a sperm had passed through the outer membrane, but it had failed to fertilize the egg. In 1967, Edwards read a paper by Patrick Steptoe describing the use of a new instrument, known as the *laparoscope,* to view the internal organs, which he saw had a possible application to his own research.

Patrick Christopher Steptoe was educated at Kings's College and St. George's Hospital Medical School, London. He was appointed chief assistant obstetrician and gynecologist at St. George's Hospital, London, in 1947, and senior registrar at the Whittington Hospital, London, in 1949. From 1951 to 1978, he was senior obstetrician and gynecologist at Oldham General Hospital, and from 1969, director of the Centre for Human Reproduction.

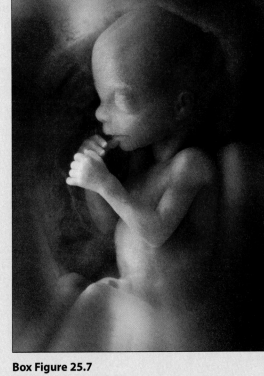

Box Figure 25.7

The paper describing laparoscopy that interested Edwards dealt with Steptoe's method of exploring the interior of the abdomen without a major operation. Steptoe inserted the laparoscope through a small incision near the navel, and by means of this telescopelike instrument, with its object lens inside the body and its eyepiece outside, he was able to examine the ovaries and other internal organs.

Early in 1968, Edwards and Steptoe repeated experiments on the fertilization of human eggs. Steptoe treated volunteer patients with fertility drug to stimulate maturation of the eggs in the ovary. Edwards devised a simple piece of apparatus to be used with the laparoscope for collecting mature eggs from human ovaries. The mature eggs were removed and Edwards then prepared them for fertilization using sperm provided by the patient's husband. For a year, they continued experiments of this kind until they were sure that the fertilized eggs were developing normally. The next step was to see if an eight-celled embryo would develop.

In 1971, Edwards and Steptoe were ready to introduce an eight-celled embryo into the uterus of a volunteer patient. In 1975, an embryo did implant but in the stump of a fallopian tube, where it could not develop properly and was a danger to the mother. It was removed, but it did demonstrate that the basic technique was sound. In 1977, it was decided to abandon the use of the fertility drug and remove the egg at precisely the right natural stage of maturity, an egg was fertilized and then reimplanted (a process called *in vitro fertilization*) in the mother two days later. The patient became pregnant; twelve weeks later, the position of the baby was found to be satisfactory, and its heartbeat could be heard. During the last eight weeks of pregnancy, the mother was kept under close medical supervision, and a healthy girl—Louise Brown—was delivered by cesarean section on July 25, 1978.

In vitro fertilization has also been used to overcome the infertility in men that is due to a low sperm count. Edwards' research has further added to knowledge of the development of the human egg and young embryo, and Steptoe's laparoscope is a valuable instrument capable of wider application.

Source: Modified from the *Hutchinson Dictionary of Scientific Biography.* © RM, 2007. Reprinted by permission.

of cells is produced, known as the morula stage of embryological development. The morula continues down the oviduct and continues to divide by mitosis. The result is called a *blastocyst*. The blastocyst becomes embedded, or implanted, when it reaches the lining of the uterus.

The next stage in the development is known as the gastrula stage because the gut is formed during this time (*gastro* = stomach). In many kinds of animals, the gastrula is formed by a complex folding of the blastocyst walls. The embryo eventually develops a tube that becomes the gut. The formation of the primitive gut is just one of a series of changes that result in an embryo that is recognizable as a miniature human being.

Most of the time during its development, the embryo is enclosed in a water-filled membrane, the amnion, which protects it from blows and keeps it moist. Two other membranes, the chorion and allantois, fuse with the lining of the uterus to form the **placenta** (figure 25.14). A fourth sac, the yolk sac, is well developed in birds, fish, amphibians, and reptiles. The yolk sac in these animals contains a large amount of food used by the developing embryo. Although a yolk sac is present in mammals, it is small and does not contain yolk. The nutritional needs of the embryo are met through the placenta. The placenta also produces the hormone chorionic gonadotropin, which stimulates the corpus luteum to continue producing progesterone and thus prevents menstruation and ovulation during gestation.

As the embryo's cells divide and grow, some of them become differentiated into nerve cells, bone cells, blood cells, or other specialized cells. To divide, grow, and differentiate, cells must receive nourishment. This is provided by the mother through the placenta, in which both fetal and maternal blood vessels are abundant, allowing for the exchange of substances between the mother and embryo. The materials diffusing across the placenta

CONCEPTS APPLIED

Life Without Men—Fiction or Fact?

A great world adventurer discovered a tribe of women in the jungles of Brazil. After many years of very close study and experimentation, he found that sexual reproduction was not possible, yet women in the tribe were getting pregnant and having children. He also noticed that the female children resembled their mothers to a great degree and found that all the women had a gene that prevented meiosis. Ovulation occurred as usual, and pregnancy lasted nine months. The mothers nursed their children for three months after birth and became pregnant the next month. This cycle was repeated in all the women of the tribe.

Consider the topics of meiosis, mitosis, sexual reproduction, and regular hormonal cycles in women, and explain what may be happening in this tribe.

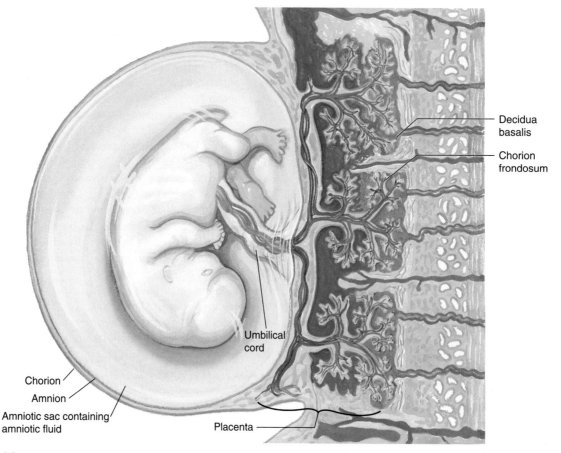

Decidua basalis

Chorion frondosum

Umbilical cord

Chorion
Amnion
Amniotic sac containing amniotic fluid

Placenta

FIGURE 25.14 The embryonic blood vessels that supply the developing embryo with nutrients and remove the metabolic wastes are separate from the blood vessels of the mother. Because of this separation, the placenta can selectively filter many types of incoming materials and microorganisms.

include oxygen, carbon dioxide, nutrients, and a variety of waste products. The materials entering the embryo travel through blood vessels in the umbilical cord. The diet and behavior of the mother are extremely important. Any molecules consumed by the mother can affect the embryo. Cocaine, alcohol, heroin, and chemicals in cigarette smoke can all cross the placenta and affect the development of the embryo. The growth of the embryo results in the development of major parts of the body by the tenth week of pregnancy. After this time, the embryo continues to increase in size, and the structure of the body is refined.

Twins

In the United States, women giving birth have a 1 in 40 chance of delivering twins and a 1 in 650 chance of triplets or other multiple births. Twins happens in two ways. In the case of identical twins (approximately one-third of twins), during cleavage the embryo splits into two separate groups of cells. Each develops into an independent embryo. Because they come from the same single fertilized ovum, they have the same genes and are of the same sex. Should separation be incomplete, the twins would be born attached to one another, a condition referred to as conjoined twins. Conjoined twins occur once in every seventy thousand to one-hundred thousand live births.

Fraternal twins do not contain the same genetic information and may be of different sexes. They result from the fertilization of two separate oocytes by different sperm. Therefore, they no more resemble each other than do ordinary brothers and sisters.

Birth

The process of giving birth is also known as *parturition* or birthing. At the end of about nine months, hormone changes in the mother's body stimulate contractions of the muscles of the uterus during a period before birth called *labor*. These contractions are stimulated by the hormone oxytocin, which is released from the posterior pituitary. The contractions normally move the baby headfirst through the vagina, or birth canal. One of the first effects of these contractions may be bursting of the amnion (bag of water) surrounding the baby. Following this, the uterine contractions become stronger, and shortly thereafter, the baby is born. In some cases, the baby becomes turned in the uterus before labor. If this occurs, the feet or buttocks appear first. Such a birth is called a *breech birth*. This can be a dangerous situation because the baby's source of oxygen is being cut off as the placenta begins to separate from the mother's body.

If for any reason the baby does not begin to breathe on its own, it will not receive enough oxygen to prevent the death of nerve cells; thus, brain damage or death can result.

Occasionally, a baby may not be able to be born normally because of its position in the uterus, the location of the placenta on the uterine wall, the size of the birth canal, the number of babies in the uterus, or many other reasons. A common procedure to resolve this problem is the surgical removal of the baby through the mother's abdomen. This procedure is known as a cesarean, or C-section. The procedure was apparently named after the Roman emperor Julius Caesar, who was said to have been the first child to be delivered by this method. Currently, over 20 percent of births in the United States are by cesarean section. While C-sections are known to have been performed before Caesar, the name stuck. This rate reflects the fact that many women who are prone to problem pregnancies are having children rather than forgoing pregnancy. In addition, changes in surgical techniques have made the procedure much safer. Finally, many physicians who are faced with liability issues related to problem pregnancy may encourage cesarean section rather than normal birth.

Following the birth of the baby, the placenta, also called the *afterbirth,* is expelled. Once born, the baby begins to function on its own. The umbilical cord collapses, and the baby's lungs, kidneys, and digestive system must now support all its bodily needs. This change is quite a shock, but the baby's loud protests fill the lungs with air and stimulate breathing.

Over the next few weeks, the mother's body returns to normal, with one major exception. The breasts, which have undergone changes during the period of pregnancy, are ready to produce milk to feed the baby. Following birth, prolactin, a hormone from the pituitary gland, stimulates the production of milk, and oxytocin stimulates its release. If the baby is breastfed, the stimulus of the baby's sucking will prolong the time during which milk is produced. This response involves both the nervous and endocrine systems. The sucking stimulates nerves in the nipple and breast, which results in the release of prolactin and oxytocin from the pituitary.

In some cultures, breast-feeding continues for two to three years, and the continued production of milk often delays the reestablishment of the normal cycles of ovulation and menstruation. Many people believe that a woman cannot become pregnant while she is nursing a baby. However, because there is so much variation among women, relying on this as a natural conception-control method is not a good choice. Many women have been surprised to find themselves pregnant again a few months after delivery.

CONTRACEPTION

Throughout history, people have tried various methods of conception control (figure 25.15). In ancient times, conception control was encouraged during times of food shortage or when tribes were on the move from one area to another in search of a new home. Writings as early as 1500 B.C. indicate that the Egyptians used a form of tampon medicated with the ground powder of a shrub to prevent fertilization. This may sound primitive, but we use the same basic principle today to destroy sperm in the vagina. As you read about the various methods of contraception, remember that no method described is 100 percent effective for avoiding pregnancy and preventing sexually transmitted diseases (STDs) except abstinence.

Chemical Methods

Contraceptive jellies and foams make the environment of the vagina more acidic, which diminishes the sperm's chances of survival. The spermicidal (sperm-killing) foam or jelly is placed

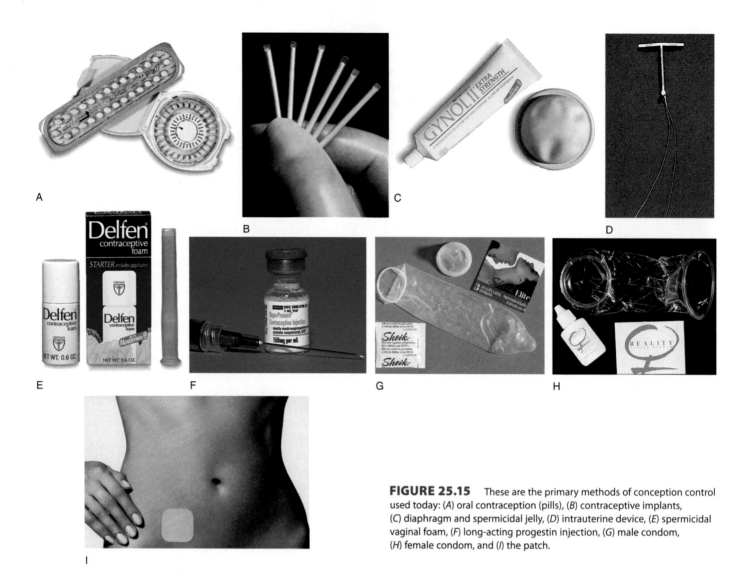

FIGURE 25.15 These are the primary methods of conception control used today: (*A*) oral contraception (pills), (*B*) contraceptive implants, (*C*) diaphragm and spermicidal jelly, (*D*) intrauterine device, (*E*) spermicidal vaginal foam, (*F*) long-acting progestin injection, (*G*) male condom, (*H*) female condom, and (*I*) the patch.

in the vagina before intercourse. When the sperm make contact with the acidic environment, they stop swimming and soon die. Aerosol foams are an effective method of conception control, but interfering with the hormonal regulation of ovulation is more effective. Contraceptive foams and jellies provide very little protection from sexually transmitted diseases.

Hormonal Control Methods

The first successful method of hormonal control was "the pill," or "birth-control pill." However, the way the pill works is by preventing ovulation and secondarily by interfering with implantation. The quantity and balance of hormones (estrogen and progesterone) in the pill serve to fool the ovary into functioning as if the woman was already pregnant. Therefore, conception is highly unlikely. The emergency contraceptive pill (ECP), or "morning-after pill," uses high doses of the same hormones found in oral contraceptives, which prevents the woman from becoming pregnant in the first place. In fact, "the pill" in higher dosages can be used as an ECP. ECPs prevent conception in several ways: (1) delay or inhibit ovulation; (2) inhibit transport of the egg or sperm; (3) prevent

fertilization; (4) inhibit implantation of a fertilized egg; or (5) stimulate an autoimmune response that kills the sperm. This method is effective if used within seventy-two hours after unprotected intercourse. The failure rate for hormonal methods of control is between 1 and 2 per 100 women per year; the procedure provides no protection from sexually transmitted disease.

Another method of conception control that may again become available in the United States also involves hormones. The hormones are contained within small rods or capsules known as implants that are placed under a woman's skin. These rods, when properly implanted, slowly release hormones and prevent the maturation and release of oocytes from the follicle. The major advantage of the implant is its convenience. Once the implant has been inserted, the woman can forget about contraceptive protection for several years. If she wants to become pregnant, the implants are removed, and her normal menstrual and ovulation cycles return over a period of weeks.

The vaginal ring is another method that utilizes hormones. The ring is inserted into the vagina and releases a continuous low dose of estrogen and progestin for twenty-one days. At

Sexually Transmitted Diseases

Diseases currently referred to as *sexually transmitted diseases (STDs)* were formerly called *venereal diseases (VDs)*. The term *venereal* is derived from the name of the Roman goddess for love, Venus. Although these kinds of illnesses are most frequently transmitted by sexual activity, many can also be spread by other methods of direct contact, such as hypodermic needles, blood transfusions, and blood-contaminated materials. Currently, the Centers for Disease Control and Prevention (CDC) in Atlanta, Georgia, recognizes over twenty diseases as being sexually transmitted (Table 25.A).

The United States has the highest rate of sexually transmitted disease among industrially developed countries—65 million people (nearly one-fifth of the population) have incurable sexually transmitted diseases. The CDC estimates there are 15 million new cases of sexually transmitted diseases each year, nearly 4 million among teenagers. Table 25.B lists the most common STDs and estimates of the number of new cases each year. The portions of the public that are most at risk are teenagers, minorities, and women. Some of the most important STDs are described here because of their high incidence in the population and our inability to bring some of them under control. For example, there is no known cure for the HIV virus that is responsible for AIDS. There has also been a sharp rise in the number of gonorrhea cases in the United States caused by a form of the bacterium *Neisseria gonorrhoeae* that has become resistant to the drug penicillin by producing an enzyme that actually destroys the antibiotic. However, most of the infectious agents can be controlled if diagnosis occurs early and treatment programs are carefully followed by the patient.

The spread of STDs during sexual intercourse is significantly diminished by the use of condoms. Other types of sexual contact (i.e., hand, oral, anal) and congenital transmission (i.e., from the mother to the fetus during pregnancy) help maintain some of these diseases in the population at high enough levels to warrant attention by public health officials, the U.S. Public Health Service, the CDC, and state and local public health agencies. All of these agencies are involved in attempts to raise the general public health to a higher level. Their investigations have resulted in the successful control of many diseases and the identification of special problems, such as those associated with the STDs. Because the United States has an incidence rate of STDs that is fifty to one-hundred times higher than other industrially developed countries, there is still much that needs to be done.

High-risk behaviors associated with contracting STDs include sex with multiple partners and failing to use condoms. While some STDs are simply inconvenient or annoying, others severely compromise health and can result in death. As one health official stated, "We should be knowledgeable enough about our own sexuality and the STDs to answer the question, Is what I'm about to do worth dying for?"

Table 25.A

Sexually Transmitted Diseases	
Disease	**Agent**
Genital herpes	Virus
Gonorrhea	Bacterium
Syphilis	Bacterium
Acquired immunodeficiency syndrome (AIDS)	Virus
Candidiasis	Yeast
Chancroid	Bacterium
Genital warts	Virus
Gardnerella vaginalis	Bacterium
Genital *Chlamydia* infection	Bacterium
Genital cytomegalovirus infection	Virus
Genital *Mycoplasma* infection	Bacterium
Group B *Streptococcus* infection	Bacterium
Nongonococcal urethritis	Bacterium
Pelvic inflammatory disease (PID)	Bacterium
Molluscum contagiosum	Virus
Crabs	Body lice
Scabies	Mite
Trichomoniasis	Protozoan
Hepatitis B	Virus
Gay bowel syndrome	Variety of agents

Table 25.B

Cases of Sexually Transmitted Diseases United States, 2003	
Reported cases by state health departments	
	Cases
Syphilis	34,270
Chlamydia	877,478
Gonorrhea	335,104
Chancroid	54
HIV/AIDS	43,171
Reported by initial visits to physician's office	
	Visits
Genital herpes	203,000
Genital warts	264,000
Hepatitis B	73,000
Vaginal trichomoniasis	179,000

the end of the twenty-one days, the ring is removed for seven days to allow the menstrual period to occur. A new ring is inserted monthly.

Contraceptive hormones can also be used to prevent menstrual periods, thereby preventing the symptoms of premenstrual syndrome and improving a woman's sex life. The birth-control pill, Seasonale, limits periods to four a year. The first continuous-use birth-control pill, Lybrel, may soon be available in the United States.

Still another method is the hormone-infused contraceptive patch. It is an adhesive square that can be attached to the abdomen, buttocks, upper arm, or upper torso. The patch works by slowly releasing through the skin a combination of estrogen and progestin that prevents ovulation. The hormones also cause the cervical mucus to thicken, creating a barrier to prevent sperm from entering the uterus. When used correctly, it is about 99 percent effective. It does not protect against reproductive tract infections such as HIV/AIDS.

Timing Method

Killing sperm and preventing ovulation or sperm production are not the only methods of preventing conception. Any method that prevents the sperm from reaching the oocyte prevents conception. One method is to avoid intercourse during those times of the month when a secondary oocyte may be present. This is known as the *rhythm method* of conception control. Although at first glance it appears to be the simplest and least expensive, determining just when a secondary oocyte is likely to be present can be very difficult. A woman with a regular twenty-eight-day menstrual cycle will typically ovulate about fourteen days before the onset of the next menstrual flow. To avoid pregnancy, couples need to abstain from intercourse a few days before and after this date. However, if a woman has an irregular menstrual cycle, there may be only a few days each month for intercourse without the possibility of pregnancy. In addition to calculating safe days based on the length of the menstrual cycle, a woman can better estimate the time of ovulation by keeping a record of changes in her body temperature and vaginal pH. Both changes are tied to the menstrual cycle and can therefore help a woman predict ovulation. In particular, at about the time of ovulation, a woman has a slight rise in body temperature—less than 1°C. Thus, one should use an extremely sensitive thermometer. A digital-readout thermometer on the market spells out the word "yes" or "no." The failure rate for the rhythm method is less than 9 per 100 women per year; the procedure provides no protection from sexually transmitted disease.

Barrier Methods

Other methods of conception control that prevent the sperm from reaching the secondary oocyte include the diaphragm, cap, sponge, and condom. The diaphragm is a specially fitted membranous shield that is inserted into the vagina before intercourse and positioned so that it covers the cervix, which contains the opening of the uterus. Because of anatomical differences among females, diaphragms must be fitted by a physician. The effectiveness of the diaphragm is increased if spermicidal foam or jelly is also used. The vaginal cap functions in a similar way. The contraceptive sponge, as the name indicates, is a small amount of absorbent material that is soaked in a spermicide. The sponge is placed within the vagina and chemically and physically prevents the sperm cells from reaching the oocyte. The contraceptive sponge is no longer available for use in the United States but is still available in many other parts of the world. The failure rate for diaphragms, cervical caps, and sponges including the use of spermicides is between 14 and 50 per 100 women per year; the methods are not an effective protection against sexually transmitted disease.

The male condom is probably the most popular contraceptive device. It is a thin sheath that is placed over the erect penis before intercourse. In addition to preventing sperm from reaching the secondary oocyte, this physical barrier also helps prevent the transmission of the microbes that cause STDs, such as syphilis, gonorrhea, and AIDS, from being passed from one person to another during sexual intercourse. The most desirable condoms are made of a thin layer of latex that does not reduce the sensitivity of the penis. Latex condoms have also been determined to be the most effective in preventing transmission of the AIDS virus. The condom is most effective if it is prelubricated with a spermicidal material such as nonoxynol-9. This lubricant also has the advantage of providing some protection against the spread of the HIV virus. The failure rate for latex condoms is 11 per 100 women per year. Except for abstinence, latex condoms are the best protection against sexually transmitted diseases.

Recently developed condoms for women are now available. One called the Femidom is a polyurethane sheath that, once inserted, lines the contours of the woman's vagina. It has an inner ring that sits over the cervix and an outer ring that lies flat against the labia. Research shows that this device protects against STDs and is as effective a contraceptive as the condom used by men. The failure rate for such barrier protection is 21 per 100 women per year; the method may give some protection against sexually transmitted diseases but is not as effective as latex condoms.

The intrauterine device (IUD) is not a physical barrier that prevents the gametes from uniting. How this mechanical device works is not completely known. It may in some way interfere with the implantation of the embryo. The IUD must be fitted and inserted into the uterus by a physician, who can also remove it if pregnancy is desired. IUDs are used successfully in many countries. Current research with new and different intrauterine implants indicates that they are able to prevent pregnancy, and one is currently available in the United States. The IUD can also be used for "emergency contraception"—in cases of unprotected sex (forced sex) or failure of a conception control method (a condom slips or breaks). The IUD must be inserted within seven days of unprotected sex. The failure rate for the IUD is less than 1 per 100 women per year; the device provides no protection from sexually transmitted disease.

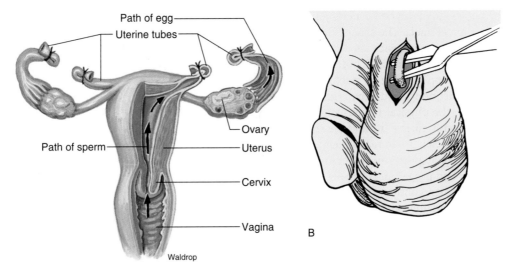

Path of egg
Uterine tubes
Ovary
Path of sperm
Uterus
Cervix
Vagina
Waldrop

A

B

FIGURE 25.16 Two very effective contraceptive methods require surgery. Tubal ligation (*A*) involves severing the oviducts and suturing or sealing the cut ends. This prevents the sperm cell and the secondary oocyte from meeting. This procedure is generally considered ambulatory surgery, or at most requires a short hospitalization period. Vasectomy (*B*) requires minor surgery, usually in a clinic under local anesthesia. Following the procedure, minor discomfort may be experienced for several days. The severing and sealing of the vas deferens prevents the release of sperm cells from the body by ejaculation.

Surgical Methods

Two contraceptive methods that require surgery are tubal ligation and vasectomy (figure 25.16). Tubal ligation is the cutting and tying off of the oviducts and can be done on an outpatient basis in most cases. An alternative to sterilization by tubal ligation involves the insertion of small, flexible devices called micro-inserts into each fallopian tube. Once inserted, tissue grows into the inserts, blocking the tubes. Ovulation continues as usual, but the sperm and egg cannot unite. Vasectomy is not the same as castration. Castration is the surgical removal of testes. Vasectomy can be performed in a physician's office and does not require hospitalization. A small opening is made above the scrotum, and the spermatic cord (vas deferens) is cut and tied. This prevents sperm from moving through the ducts to the outside. Because most of the sperm-carrying fluid (semen) is produced by the seminal vesicles, prostate gland, and bulbourethral glands, a vasectomy does not interfere with normal ejaculation. The sperm that are still being produced die and are reabsorbed in the testes. Neither tubal ligation nor vasectomy interferes with normal sex drives. However, these medical procedures are generally not reversible and should not be considered by those who may want to have children at a future date. The failure rate for sterilization is less than 1 per 100 women per year; the procedure provides no protection from sexually transmitted disease.

TERMINATION OF PREGNANCY—ABORTION

Another medical procedure often associated with birth control is abortion, which has been used throughout history. Abortion involves various medical procedures that cause the death and removal of the developing embryo. Abortion is obviously not a method of conception control; rather, it prevents the normal development of the embryo and causes its death. Abortion is a highly charged subject. Some people believe that abortion should be prohibited by law in all cases. Others think that abortion should be allowed in certain situations, such as in pregnancies that endanger the mother's life or that are the result of rape or incest. Still others think that abortion should be available to any woman under any circumstances. Regardless of the moral and ethical issues that surround abortion, it is still a common method of terminating unwanted pregnancies.

The abortion techniques used in the United States today all involve the possibility of infections, particularly if done by poorly trained personnel. The three most common techniques are scraping the inside of the uterus with special instruments (called a *D and C* or *dilation and curettage*), injecting a saline solution into the uterine cavity, or using a suction device to remove the embryo from the uterus. In the future, abortion may be accomplished by a medication prescribed by a physician. One drug, RU-486, is currently used in about 15 percent or more of the elective abortions in France. It has received approval for use in the United States. RU-486 is an antiprogestin and works by blocking progesterone receptors on cells. The medication is administered orally under the direction of a physician, and several days later, a hormone is administered. This usually results in the onset of contractions that expel the fetus. A follow-up examination of the woman is made after several weeks to ensure that there are no serious side effects of the medication.

CHANGES IN SEXUAL FUNCTION WITH AGE

Although there is a great deal of variation, somewhere around the age of fifty, a woman's hormonal balance begins to change because of changes in the production of hormones by the ovaries. At this time, the menstrual cycle becomes less regular and ovulation is often unpredictable. The changes in hormone levels cause many women to experience mood swings and physical symptoms, including cramps and hot flashes. This period when the ovaries stop producing viable secondary oocytes and the body becomes nonreproductive is known as the **menopause.** Occasionally, the physical impairment becomes so severe that it interferes with normal life and the enjoyment of sexual activity, and a physician might recommend *hormone replacement therapy (HRT)* to augment the natural production of the hormones estrogen and progestin. Normally, the sexual enjoyment of a healthy woman continues during the time of menopause and for many years thereafter.

Hormone Replacement Therapy

The U.S. Preventive Services Task Force (USPSTF) was set up by the U.S. Public Health Service to determine the merits of hormone replacement therapy. In 1996, the USPSTF recommended that physicians advise all women about the potential benefits and problems of HRT, while stating that the evidence was insufficient to recommend for or against HRT for all women after menopause. The task force concluded in 1996 that HRT might have long-term benefits for women with osteoporosis and heart disease but that it might increase the risk for breast cancer.

Although recent evidence confirmed that HRT reduces the risk for bone fractures and may reduce the likelihood of developing colorectal cancer, it has also confirmed that HRT increases the risk for blood clots, breast cancer, and stroke. Most important, new evidence indicates that HRT does not reduce the risk for coronary heart disease and that combined estrogen and progestin may actually increase the risk. In 2003, the USPSTF recommended against the routine use of combined estrogen and progestin for the prevention of chronic conditions in postmenopausal women because the problems are likely to outweigh the benefits for most women. So far, the evidence isn't strong enough to recommend for or against the use of estrogen alone for the prevention of chronic conditions in postmenopausal women who have had a hysterectomy.

Impotence

Human males do not experience a relatively abrupt change in their reproductive or sexual lives. The word *impotence* is used to describe problems that interfere with sexual intercourse and reproduction. These may include a lack of sexual desire, problems with ejaculation or orgasm, and erectile dysfunction. Erectile dysfunction (ED) is the recurring inability to get or keep an erection firm enough for sexual intercourse. Most incidences of ED at any age are physical, not psychological. In older men, this is usually the result of disease, injury, or a side effect of medication. Blood pressure drugs, antihistamines, antidepressants, tranquilizers, appetite suppressants, and certain ulcer drugs have been associated with ED. Damage to nerves, arteries, smooth muscles, and other tissues is the most common cause of ED. Diseases linked with ED include: diabetes, kidney disease, chronic alcoholism, multiple sclerosis, arteriosclerosis, vascular disease, and neurologic disease. Other possible causes are smoking, which reduces blood flow in veins and arteries, and lowered amounts of testosterone.

ED is frequently treated with psychotherapy, behavior modifications, oral or locally injected drugs, vacuum devices, and surgically implanted devices. Oral medications (e.g., Viagra and Levitra) work by enhancing the effects of nitric oxide (NO). NO relaxes smooth muscles in the penis during sexual stimulation and allows increased blood flow. It is increased blood in the penis that causes an erection.

ED and loss of sexual desire are not an inevitable part of aging. Rather, sexual desires tend to wane slowly as a man ages. Men also produce fewer sperm cells and less seminal fluid with age. Nevertheless, healthy individuals can experience a satisfying sex life during aging. Recent evidence indicates that men also experience hormonal and emotional changes similar to those seen as women go through menopause. Human sexual behavior is quite variable. The same is true of older persons. The whole range of responses to sexual partners continues but generally in a diminished form. People who were very active sexually when young continue to be active but less so as they reach middle age. Those who were less active tend to decrease their sexual activity also. It is reasonable to state that one's sexuality continues from before birth until death.

SUMMARY

The human sex drive is a powerful motivator for many activities in our lives. Although it provides for reproduction and improvement of the gene pool, it also has a nonbiological, sociocultural dimension. Sexuality begins before birth, as sexual anatomy is determined by the sex-determining chromosome complement that we receive at fertilization. Females receive two X chromosomes. Males receive one X and one Y *sex-determining chromosome*. It is the presence and activity of the *SRY gene* that causes male development and its absence or inactivity that allows female development.

At *puberty,* hormones influence the development of secondary sex characteristics and the functioning of gonads. As the *ovaries* and

testes begin to produce gametes, fertilization becomes possible. Sexual reproduction involves the production of gametes by meiosis in the ovaries and testes. The production and release of these gametes is controlled by the interaction of hormones. In males, each cell that undergoes *spermatogenesis* results in four sperm; in females, each cell that undergoes *oogenesis* results in one oocyte and two polar bodies. Successful sexual reproduction depends on proper hormone balance, proper meiotic division, fertilization, placenta formation, proper diet of the mother, and birth. *Hormones* regulate ovulation and menstruation and may also be used to encourage or discourage ovulation. *Fertility drugs, the patch* and *birth-control pills,* for example, involve hormonal control. A number of contraceptive methods have been developed, including the diaphragm, condom, IUD, spermicidal jellies and foams, contraceptive implants, the sponge, tubal ligation, and vasectomy.

Hormones continue to direct our sexuality throughout our lives. Even after menopause, when fertilization and pregnancy are no longer possible for a female, normal sexual activity can continue in both women and men.

KEY TERMS

androgens (p. **636**)
autosomes (p. **631**)
Barr body (p. **634**)
clitoris (p. **636**)
conception (p. **633**)
cryptorchidism (p. **634**)
differentiation (p. **633**)
diploid (p. **626**)
egg (p. **626**)
estrogens (p. **636**)
follicle (p. **636**)
follicle-stimulating hormone
    (FSH) (p. **636**)
gamete (p. **626**)

gametogenesis (p. **626**)
gonad (p. **628**)
gonadotropin-releasing hormone
    (GnRH) (p. **636**)
haploid (p. **626**)
hypothalamus (p. **636**)
inguinal canal (p. **634**)
meiosis (p. **627**)
menopause (p. **652**)
menstrual cycle (p. **636**)
oogenesis (p. **639**)
orgasm (p. **636**)
ovaries (p. **628**)
oviduct (p. **639**)

ovulation (p. **636**)
penis (p. **637**)
pituitary gland (p. **636**)
placenta (p. **646**)
progesterone (p. **636**)
puberty (p. **636**)
secondary sexual characteristics
    (p. **636**)
semen (p. **638**)
seminiferous tubules (p. **637**)
sex (p. **629**)
sex-determining chromosome
    (p. **631**)
sexual reproduction (p. **626**)

sexuality (p. **629**)
sperm (p. **626**)
spermatids (p. **638**)
spermatogenesis (p. **637**)
testes (p. **628**)
testosterone (p. **636**)
uterus (p. **639**)
vagina (p. **640**)
X chromosome (p. **631**)
Y chromosome (p. **631**)
zygote (p. **626**)

APPLYING THE CONCEPTS

1. Crossing-over between segments of homologous chromosomes results in
   a. new gene combinations.
   b. zygotes.
   c. diploid cells.
   d. segregation of genes.

2. An event unique to prophase I is
   a. segregation.
   b. synapsis.
   c. reduction division.
   d. independent assortment.

3. The fact that each homologous pair of chromosomes in humans separates and moves to the poles without being influenced by the other pairs is
   a. segregation.   c. independent assortment.
   b. disintegration.   d. fertilization.

4. An organism having a diploid number of 12 forms gametes having
   a. 6 chromosomes.
   b. 12 chromosomes.
   c. 18 chromosomes.
   d. 24 chromosomes.

5. Gametogenesis produces
   a. sex cells.   c. zygotes.
   b. gonads.   d. testes.

6. Human females typically are fertile during which point of their ovarian cycle?
   a. only after sexual intercourse
   b. immediately following their period
   c. about day 13
   d. continuously

7. Which of the following sexually transmitted diseases is caused by a virus?
   a. gonorrhea
   b. scabies
   c. candidiasis
   d. genital herpes

8. Which pituitary hormone stimulates the release of LH and FSH, stimulating the development of a follicle?
   a. estrogen
   b. gonadotropin-releasing hormone (GnRH)
   c. testosterone
   d. androgen

9. Without this, the embryo would become female.
   a. FSH gene   c. SRY gene
   b. estrogen   d. Barr bodies

10. Which of these is not an accessory gland of males?
   a. follicle
   b. seminal vesicle
   c. bulbourethral gland
   d. prostate gland

11. Fertilization normally occurs in the
   a. uterus.   c. oviduct.
   b. ovary.   d. vagina.

12. Which is not a hormonal conception control method?
   a. the pill
   b. condom
   c. the patch
   d. morning-after pill

Answers

1. a **2.** b **3.** a **4.** a **5.** a **6.** c **7.** d **8.** b **9.** c **10.** a **11.** c **12.** b

QUESTIONS FOR THOUGHT

1. How do haploid cells differ from diploid cells?
2. What is unique about metaphase I?
3. How does mitosis differ from meiosis?
4. What are advantages and disadvantages of the rhythm method of conception control?
5. Diagram fertilization as it would occur between a sperm and an egg with the haploid number of 3.
6. What is HRT?
7. What structures are associated with the human female reproductive system? What are their functions?
8. What structures are associated with the human male reproductive system? What are their functions?
9. What are the differences between oogenesis and spermatogenesis in humans?
10. List the hormones associated with the functioning of the male and female reproductive systems. Describe their function.
11. What is the difference between the origin of fraternal and identical twins?
12. List three likely causes of infertility in males and three in females.

FOR FURTHER ANALYSIS

Sexual orientation is a complex concept because there is so much confusion and a lack of information among a large part of society. Some very knowledgeable professionals lead into an explanation of this topic by describing intersex conditions and the complex intersections of what many refer to as "nature and nurture." The intersex concept is one that describes one's sexual orientation as being at some point on a continuum from strongly heterosexual at one end to strongly homosexual on the other. Each person's orientation is somewhere on this continuum. Research this issue by exploring the following:

1. The biological bases of sexual orientation
2. How various cultural biases influence sexual behavior
3. The influence various religious organizations have on an individual's thinking

Lately, there has been news on circumstances surrounding surgically changing one's sex. Thinking critically about this subject, answer the following questions:

1. What genetic reasons are associated with such a decision?
2. What cultural reasons support or rebuff such a decision?
3. What would be your reaction if a relative or close personal friend were to make the decision to change their sex?

INVITATION TO INQUIRY

Sex and Society: How Do Cultures Differ?

Archaeologists, anthropologists, and biologists have studied sexuality in many cultures. They have discovered that there are many differences in attitude and behaviors. Use the Internet to search out answers to the following questions:

1. Is the maturation of males and females the same in all cultures?
2. What differences in attitude are there among cultures with regard to public nudity?
3. What differences exist among cultures with regard to homosexual behavior?
4. Do all cultures show sexual dominance of the male over the female?
5. What methods of conception control are favored in other cultures?
6. What do various religions have to say about sexual behaviors? Which behaviors are acceptable and which are not?

The expression of many genes is influenced by the environment. The gene for dark hair in the cat is sensitive to temperature and expresses itself only in the parts of the body that stay cool.

Mendelian and Molecular Genetics

▶ Hydrogen bonds help hold the genetic material together (Ch. 9).

▶ Covalent bonds are needed to stabilize nucleic acids and proteins (Ch. 9).

▶ Chemical reactions are responsible for DNA replication and protein synthesis (Ch. 9).

▶ Enzymes control all the reaction involved in genetic mechanisms (Ch. 20).

CORE CONCEPT

All living things survive and reproduce only as a result of maintaining molecular genetic information that is passed from one generation to the next.

Sexually reproducing organisms pass genes to offspring (p. 656).

Mutations in the genetic material allow for new combinations in offspring (p. 676).

DNA is the genetic material in eukaryotic organisms and is self-replicating (p. 669).

DNA controls the synthesis of proteins, which in turn control traits (p. 674).

▶ Mendel's laws still explain the transmission of genetic traits from one generation to another (Ch. 25).

▶ There are some simple but many more complex mechanisms used to express a genetic trait (Ch. 26).

▶ Mutations are not all bad, but many result in genetic abnormalities (Ch. 26).

▶ The central dogma explains how genetic information is used to generate physical traits (Ch. 26).

OVERVIEW

Why do you have a particular blood type or hair color? Why do some people have the same skin color as their parents, while others have a skin color different from that of their parents? Why do flowers show such a wide variety of colors? Why is it that generation after generation of plants, animals, and microbes look so much like members of their own kind? These questions and many others can be better answered if you have an understanding of Mendelian and molecular genetics.

GENETICS, MEIOSIS, AND CELLS

A **gene** is a portion of DNA that determines a characteristic. Through meiosis and reproduction, genes can be transmitted from one generation to another. The study of genes, how genes produce characteristics, and how the characteristics are inherited is the field of biology called **genetics.** The first person to systematically study inheritance and formulate laws about how characteristics are passed from one generation to the next was an Augustinian monk named Gregor Mendel (1822–1884). Mendel's work was not generally accepted until 1900, when three men, working independently, rediscovered some of the ideas that Mendel had formulated more than thirty years earlier. Because of his early work, the study of the pattern of inheritance that follows the laws formulated by Gregor Mendel is often called **Mendelian genetics.**

To understand this chapter, you need to know some basic terminology. One term that you have already encountered is *gene.* Mendel thought of a gene as a particle that could be passed from the parents to the *offspring* (children, descendants, or progeny). Today, we know that genes are actually composed of specific sequences of DNA nucleotides. The particle concept is not entirely inaccurate, because a particular gene is located at a specific place on a chromosome called its *locus* (*loci* = pl.; location). When we study genetics, we study how these particles or genes are passed from parents to their offspring or children.

Another important idea to remember is that all sexually reproducing organisms have a diploid ($2n$) stage. Since gametes are haploid (n) and most organisms are diploid, the conversion of diploid to haploid cells during meiosis is an important process.

$$2(n) \rightarrow \text{meiosis } (n) \rightarrow \text{gametes}$$

The diploid cells have two sets of chromosomes—one set inherited from each parent.

$$n + n \text{ gametes} \rightarrow \text{fertilization} \rightarrow 2n$$

Therefore, each individual has two chromosomes of each kind and two genes for each characteristic (refer to figure 25.1), one from the mother and one from the father. When sex cells are produced by meiosis, reduction division occurs, and the diploid number is reduced to haploid. Therefore, the sex cells produced by meiosis have only one chromosome of each of the homologous pairs that were in the diploid cell that began meiosis. Diploid organisms usually result from the fertilization of a haploid egg by a haploid sperm. Therefore, they inherit one gene of each type from each parent. For example, each of us has two genes for earlobe shape: one came with our father's sperm, the other with our mother's egg (figure 26.1).

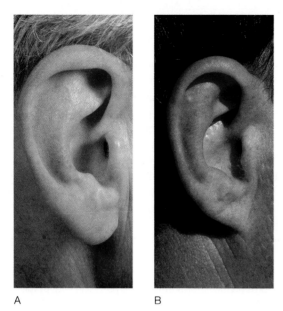

A   B

FIGURE 26.1 Whether your earlobe is free (*A*) or attached (*B*) depends on the genes you have inherited. As genes express themselves, their actions affect the development of various tissues and organs. Some people's earlobes do not separate from the sides of their heads in the same manner as do those of others. How genes control this complex growth pattern and why certain genes function differently from others is yet to be clarified. How much variation do you see among fellow class members?

SINGLE-GENE INHERITANCE PATTERNS

In diploid organisms, there may be two different forms of the gene. In fact, there may be *several* alternative forms or **alleles** of each gene within a population. The word *gene* is used to refer to the genetic material in general, whereas the term *allele* more specifically refers to the alternative forms of the genetic material for a certain characteristic. In people, for example, there are two alleles for earlobe shape. One allele produces an earlobe that is fleshy and hangs free, while the other allele produces a lobe that is attached to the side of the face and does not hang free. The type of earlobe that is present is determined by the type of allele (gene) received from each parent and the way in which these alleles interact with one another. Alleles are located on the same pair of homologous chromosomes—one allele on each chromosome. These alleles are also at the same specific location, or locus (figure 26.2).

The **genome** is a set of all the genes necessary to specify an organism's complete list of characteristics. The term *genome* is used in two ways. It may refer to the diploid (*2n*) or haploid (*n*) number of chromosomes in a cell. Be sure to clarify how your instructor uses this term. The **genotype** of an organism is a listing of the genes present in that organism. It consists of the cell's DNA code; therefore, you cannot see the genotype of an organism. It is not yet possible to know the complete genotype of most organisms, but it is often possible to figure out the genes present that determine a particular characteristic. For example, there are three possible genotypic combinations of the two alleles for earlobe shape. Genotypes are typically represented by upper- and lowercase letters. In the case of the

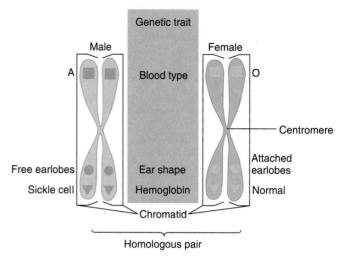

FIGURE 26.2 Homologous chromosomes contain genes for the same characteristics at the same place. Note that the attached-earlobe allele is located at the ear-shape locus on one chromosome and the free-earlobe allele is located at the ear-shape locus on the other member of the homologous pair of chromosomes. The other two genes are for hemoglobin structure (alleles for normal and sickled cells) and blood type (alleles for blood types A and O).The examples presented here are for illustrative purposes only. We do not really know if these particular genes are on these chromosomes.

earlobe trait, the allele for free earlobes is designated "*E*"while that for attached earlobes is "*e*." A person's genotype could be (1) two alleles for attached earlobes (*ee*), (2) one allele for attached earlobes and one allele for free earlobes (*Ee*), or (3) two alleles for free earlobes (*EE*).

How would individuals with each of these three genotypes look? The observable characteristics of an organism are known as its **phenotype** and are determined by the ways in which each combination of alleles expresses (shows) itself. The way combinations of alleles express themselves is also influenced by the environment of the organism. The phrase *gene expression* refers to the degree to which a gene goes through protein synthesis to show itself as a physical feature of the individual.

A person with two alleles for attached earlobes will have earlobes that do not hang free. A person with one allele for attached earlobes and one allele for free earlobes will have a phenotype that exhibits free earlobes. An individual with two alleles for free earlobes will also have free earlobes. Notice that there are three genotypes but only two phenotypes. The individuals with the free-earlobe phenotype can have different genotypes.

Alleles	Genotypes	Phenotypes
E = free earlobes	*EE*	free earlobes
e = attached earlobes	*Ee*	free earlobes
	ee	attached earlobes

The expression of some alleles is directly influenced by the presence of other alleles of the same gene. For any particular pair of alleles in an individual, the two alleles from the two parents are either identical or not identical. A person is **homozygous** for a trait when they have the combination of two identical alleles for that particular characteristic, for example, *EE* and *ee*. A person with two alleles for freckles is said to be homozygous

A Closer Look

Cystic Fibrosis—What Is the Probability?

Cystic fibrosis is among the most common lethal genetic disorders that affect Caucasians in the United States. Cystic fibrosis affects nearly thirty thousand people in North America. One in every twenty persons has a defective recessive allele that causes cystic fibrosis, but most of these individuals display no cystic fibrosis symptoms, because the recessive allele is masked by a normal dominant allele. Only those with two copies of the defective recessive gene develop symptoms. About one thousand new cystic fibrosis cases are identified in the United States each year. The gene for cystic fibrosis occurs on chromosome 7; it is responsible for the manufacture of **c**ystic **f**ibrosis **t**ransmembrane **r**egulator (CFTR) protein. The CFTR protein controls the movement of chloride ions across the cell membrane.

There are many possible types of mutations in the CFTR gene. The most common mutation results in a CFTR protein with a deletion of a single amino acid. As a result, CFTR protein is unable to control the movement of chloride ions across the cell membrane. The major result is mucus filling the bronchioles, resulting in blocked breathing and frequent respiratory infections. It is also responsible for other symptoms:

1. A malfunction of sweat glands in the skin and the secretion of excess chloride ions
2. Bile duct clogging, which interferes with digestion and liver function
3. Mucus clogging the pancreas ducts, preventing the flow of digestive enzymes into the intestinal tract
4. Bowel obstructions caused by thickened stools
5. Sterility in males due to the absence of vas deferens development and, on occasion, female sterility due to dense mucus blocking sperm from reaching eggs

Consider the facts about the frequency of the cystic fibrosis gene in the population. What is the probability that any set of parents will have a child with cystic fibrosis? What is the probability that a person who carries the cystic fibrosis allele will marry someone who also has the allele?

---

for that trait. A person with two alleles for no freckles is also homozygous. If an organism is homozygous, the characteristic expresses itself in a specific manner. A person homozygous for free earlobes has free earlobes, and a person homozygous for attached earlobes has attached earlobes.

An individual is designated as **heterozygous** when they have two different allelic forms of a particular gene, for example, *Ee*. The heterozygous individual received one allelic form of the gene from one parent and a different allele from the other parent. For instance, a person with one allele for freckles and one allele for no freckles is heterozygous. If an organism is heterozygous, these two different alleles interact to determine a characteristic. The phenotype produced by the alleles can vary, depending on how the two different alleles interact.

A SIMPLE MODEL OF INHERITANCE— DOMINANT AND RECESSIVE ALLELES

Phenotypes are determined by the type of allele (gene) received from each parent and the way in which these alleles interact with one another. The simplest rule to use to predict phenotype from genotype describes when one allele in the pair expresses itself and hides the other. This is an inheritance pattern called *dominance* and *recessiveness*. A **dominant allele** (denoted by a capital letter, such as *E* in the earlobe example) masks the effect of other alleles for the trait (table 26.1). For example, if a person has one allele for free earlobes and one allele for attached earlobes, that person has a phenotype of free earlobes. We say the allele for free earlobes is dominant. A **recessive allele** (denoted by a lowercase letter, here *e*) is one that, when present with another allele, has its actions overshadowed by the other; it is

TABLE 26.1

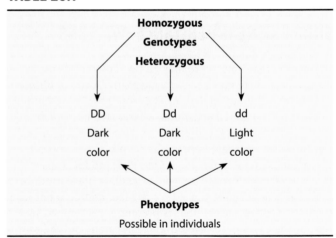

masked by the effect of the other allele. For example, if a person is heterozygous (has one allele for free earlobes and one allele for attached earlobes, *Ee*), the person will have a phenotype of free earlobes. The allele for free earlobes is dominant. While *EE* or *Ee* may show the free earlobe phenotype, recessive phenotypes are only observable when two recessive alleles (*ee*) are present.

Many people have some common misconceptions about recessive alleles. One is that recessive alleles are all bad or harmful. Don't think that recessive genes are necessarily bad. *The term recessive has nothing to do with the significance or value of the gene—it simply describes how it can be expressed. Recessive alleles are not less likely to be inherited but must be present in a homozygous condition to express themselves. This may make recessive alleles less likely to be observed. A second misconception is that recessive alleles occur less frequently in a population.*

This is not always the case. The number of fingers a person has is determined by a gene. The recessive allele for this gene is five fingers. The dominant allele is six. Clearly, the five-finger allele is more frequent in our population because we see many more five-fingered individuals than individuals with six fingers.

MENDEL'S LAWS OF HEREDITY

Heredity problems are concerned with determining which alleles are passed from the parents to the offspring and how likely it is that various types of offspring will be produced.

The Augustinian monk Gregor Mendel developed a method of predicting the outcome of inheritance. He performed experiments that showed the inheritance patterns of certain characteristics in common garden pea plants (*Pisum sativum*). From his work, Mendel concluded which traits were dominant and which were recessive. Some of his results are shown in table 26.2

What made Mendel's work unique was that he studied only one trait at a time. Previous investigations had tried to follow numerous traits at the same time. When this was attempted, the total set of characteristics was so cumbersome to work with that no clear idea could be formed of how the offspring inherited traits. Mendel used traits with clear-cut alternatives, such as purple or white flower color, yellow or green seed pods, and tall or dwarf pea plants. He was very lucky to have chosen pea plants in his study because they naturally self-pollinate. When self-pollination occurs in pea plants over many generations, it is possible to develop a population of plants that is homozygous for a number of characteristics. Such a population is known as a *pure line*.

Mendel took a pure line of pea plants having purple flower color, removed the male parts (anthers), and discarded them so that the plants could not self-pollinate. He then took anthers from a pure-breeding white-flowered plant and pollinated the antherless purple flowers.

When the pollinated flowers produced seeds, Mendel collected, labeled, and planted them.

When these seeds germinated and grew, they eventually produced flowers.

You might be surprised to learn that all of the plants resulting from this cross had purple flowers. One of the prevailing hypotheses of Mendel's day would have predicted that the purple and white colors would have blended, resulting in flowers that were lighter than the parental purple flowers. Another hypothesis would have predicted that the offspring would have

had a mixture of white and purple flowers. The unexpected result—all of the offspring produced flowers like those of one parent and no flowers like those of the other—caused Mendel to examine other traits as well and formed the basis for much of the rest of his work. He repeated his experiments using pure strains for other traits. Pure-breeding tall plants were crossed with pure-breeding dwarf plants. Pure-breeding plants with yellow pods were crossed with pure-breeding plants with green pods. The results were all the same: the offspring showed the characteristic of one parent and not the other.

Next, Mendel crossed the offspring of the white-purple cross (all of which had purple flowers) with each other to see what the third generation would be like. Had the characteristic of the original white-flowered parent been lost completely? This second-generation cross was made by pollinating these purple flowers that had one white parent among themselves. The seeds produced from this cross were collected and grown.

When these plants flowered, three-fourths of them produced purple flowers and one-fourth produced white flowers.

After analyzing his data, Mendel formulated several genetic laws to describe how characteristics are passed from one generation to the next and how they are expressed in an individual.

Mendel's **law of dominance**—When an organism has two different alleles for a given trait, the allele that is expressed, overshadowing the expression of the other allele, is said to be dominant. The gene whose expression is overshadowed is said to be recessive.

Mendel's **law of segregation**—When gametes are formed by a diploid organism, the alleles that control a trait separate from one another into different gametes, retaining their individuality. Parental alleles of a gene separate at meiosis into gametes:

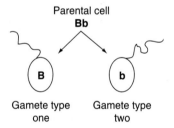

Parental cell
Bb

B — Gamete type one

b — Gamete type two

Mendel's **law of independent assortment**—Members of one *pair of alleles* separate from each other independently of the members of the other pair of alleles. In other words, if two different genes are separate chromosomes, where the one allele goes is not influenced by where the other goes.

At the time of Mendel's research, biologists knew nothing of chromosomes or DNA or of the processes of mitosis and meiosis. Mendel assumed that each gene was separate from other genes. It was fortunate for him that most of the characteristics he picked to study were found on separate chromosomes. If two or more of these genes had been located on the same chromosome (*linked genes*), he probably would not have been able to formulate his laws. The discovery of chromosomes and DNA has led to modifications in Mendel's laws, but it was Mendel's work that formed the foundation for the science of genetics.

TABLE 26.2 **Dominant and Recessive Traits in Pea Plants**

Characteristic	Dominant Allele	Recessive Allele
Plant height	Tall	Dwarf
Pod shape	Full	Constricted
Pod color	Green	Yellow
Seed surface	Round	Wrinkled
Seed color	Yellow	Green
Flower color	Purple	White

People Behind the Science

Gregor Johann Mendel (1822–1884)

Gregor Johann Mendel was an Austrian monk who discovered the basic laws of heredity, thereby laying the foundation of modern genetics—although the importance of his work was not recognized until after his death.

Mendel was born Johann Mendel on July 22, 1822, in Heinzendorf, Austria, the son of a peasant farmer. He studied for two years at the Philosophical Institute in Olomouc, after which, in 1843, he entered the Augustinian monastery in Brünn, Moravia, taking the name Gregor. In 1847, he was ordained a priest. During his religious training, Mendel taught himself a certain amount of science. In 1850, he tried to pass an examination to obtain a teaching license but failed, and in 1851, he was sent by his abbot to the University of Vienna to study physics, chemistry, mathematics, zoology, and botany. Mendel left the university in 1853 and returned to the monastery in Brünn in 1854. He then taught natural science in the local Technical High School until 1868, during which period he again tried, and failed, to gain a teaching certificate that would have enabled him to teach in more advanced institutions. It was also in the period 1854 to 1868 that Mendel performed most of his scientific work on heredity. He was elected abbot of his monastery at Brünn in 1868, and the administrative duties involved left him little time for further scientific investigations. Mendel remained abbot at Brünn until his death on January 6, 1884.

Mendel began the experiments that led to his discovery of the basic laws of heredity in 1856. Much of his work was performed on the edible pea (*Pisum* sp.), which he grew in the monastery garden. He carefully self-pollinated and wrapped (to prevent accidental pollination by insects) each individual plant, collected the seeds produced by the plants, and studied the offspring of these seeds. He found that dwarf plants produced only dwarf offspring and that the seeds produced by this second

Box Figure 26.1

generation also produced only dwarf offspring. With tall plants, however, he found that both tall and dwarf offspring were produced and that only about one-third of the tall plants bred true, from which he concluded that there were two types of tall plants, those that bred true and those that did not. Next, he cross-bred dwarf plants with true-breeding tall plants, planted the resulting seeds, and then self-pollinated each plant from this second generation. He found that all the offspring in the first generation were tall but that the offspring from the self-pollination of this first generation were a mixture of about one-quarter true-breeding dwarf plants, one-quarter true-breeding tall plants, and one-half nontrue-breeding tall plants. Mendel also studied other characteristics in pea plants, such as flower color, seed shape, and flower position, finding that, as with height, simple laws governed the inheritance of these traits. From his findings, Mendel concluded that each parent plant contributes a factor that determines a particular trait and that the pairs of factors in the offspring do not give rise to a merger of traits. These conclusions, in turn, led him to formulate his famous law of segregation and law of independent assortment of characters, which are now recognized as two of the fundamental laws of heredity.

Mendel reported his findings to the Brünn Society for the Study of Natural Science in 1865, and in the following year, he published *Experiments with Plant Hybrids,* a paper that summarized his results. But the importance of his work was not recognized at the time. It was not until 1900, when his work was rediscovered by Hugo De Vries, Carl Erich Correns, and Erich Tschermak von Seysenegg, that Mendel achieved fame—sixteen years after his death.

Source: Modified from the *Hutchinson Dictionary of Scientific Biography.* © RM, 2007. Reprinted by permission.

STEPS IN SOLVING HEREDITY PROBLEMS: SINGLE-FACTOR CROSSES

People have long been interested in understanding why the offspring of plants and animals resemble or don't resemble their parents. Mendel's work helped formulate a way to answer such questions. The first type of problem we will consider is the easiest type, a single-factor cross. A **single-factor cross** (sometimes called a monohybrid cross: *mono* = one; *hybrid* = combination) is a genetic cross or mating in which a single characteristic is followed from one generation to the next.

For example, in humans, the allele for Tourette syndrome (TS) is inherited as a dominant allele. For centuries, people displaying this genetic disorder were thought to be possessed by the devil since they displayed such unusual behaviors. These motor and verbal behaviors or tics are involuntary and range from mild (e.g., leg tapping, eye blinking, face twitching) to the more violent forms such as the shouting of profanities, head

jerking, spitting, compulsive repetition of words, or even barking like a dog. The symptoms result from an excess production of the brain messenger, dopamine.

A Single-Factor Problem: If both parents are heterozygous (have one allele for Tourette syndrome and one allele for no Tourette syndrome), what is the probability that they can have a child without Tourette syndrome? With Tourette syndrome?

Step 1: Assign a symbol for each allele.

Usually a capital letter is used for a dominant allele and a small letter for a recessive allele. Use the symbol *T* for Tourette and *t* for no Tourette.

Allele	Genotype	Phenotype
T = Tourette	*TT*	Tourette syndrome
t = normal	*Tt*	Tourette syndrome
	tt	Normal

Myths, Mistakes, and Misunderstandings

You Don't Always See It Even If You Have It

Misconception: If you've got a dominant gene, you'll show the trait.

In fact, just because you have the dominant allele for a trait doesn't mean you will show it. *Penetrance* is the frequency with which the allele expresses itself when present. Understanding penetrance helps answer the question, "If all these people have this gene, why don't I see more people with the condition?" For example, heterozygous males show a 90 percent penetrance of Tourette syndrome; that is, 90 percent of men with the dominant allele for TS express the trait enough to be diagnosed with TS. There is only 60 to 70 percent penetrance in the female population.

Step 2: Determine the genotype of each parent and indicate a mating.

Because both parents are heterozygous, the male genotype is *Tt*. The female genotype is also *Tt*. The × between them is used to indicate a mating.

$$Tt \times Tt$$

Step 3: Determine all the possible kinds of gametes each parent can produce.

Remember that gametes are haploid; therefore, they can have only one allele instead of the two present in the diploid cell. Because the male has both the Tourette syndrome allele and the normal allele, half of his gametes will contain the Tourette syndrome allele and the other half will contain the normal allele. Because the female has the same genotype, her gametes will be the same as his.

For genetic problems, a *Punnett square* is used. A **Punnett square** is a box figure that allows you to determine the probability of genotypes and phenotypes of the progeny of a particular cross. Remember, because of the process of meiosis, each gamete receives only one allele for each characteristic listed. Therefore, the male will produce sperm with either a *T* or a *t*; the female will produce ova with either a *T* or a *t*. The possible gametes produced by the male parent are listed on the left side of the square and the female gametes are listed on the top. In our example, the Punnett square would show a single dominant allele and a single recessive allele from the male on the left side. The alleles from the female would appear on the top.

Female genotype
Tt

Male genotype
Tt

Possible female gametes
T and *t*

Possible male gametes
T and *t*

	T	*t*
T		
t		

Muscular Dystrophy and Genetics

Because most genes are comprised of thousands of nucleotide base pairs, there are many different changes in these sequences that can result in the formation of multiple "bad" gene products and, therefore, an abnormal phenotype. The same gene can have many different "bad" forms. The fact that a phenotypic characteristic can be determined by many different alleles for a particular characteristic is called *genetic heterogeneity*. For example, two of the best-known forms of *muscular dystrophy* (*MD*) are Duchenne's and Becker's. Duchenne's (DMD) is characterized by a severe, progressive weakening of the muscles, the appearance of a false muscle atrophy (degeneration) in the calves of the legs, onset in early childhood, and a high likelihood of death in the thirties. DMD is caused by a mutation in the dystrophin gene located on the X chromosome and is a recessive trait. Becker's (BMD) is caused by a mutation in the same dystrophin gene but at a different location. BMD is a milder form of MD.

Myths, Mistakes, and Misunderstandings

Another Fatal Mistake?

Misconception: All mutations are fatal.

In fact, this is not true! (1) A mutation may have no effect; that is, it may replace the intended amino acid with the same one. (2) The mutation may result in a new protein that does not interfere with metabolism enough to be lethal. (3) The mutation may be in a gene that is not used.

Step 4: Determine all the gene combinations that can result when these gametes unite.

To determine the possible combinations of alleles that could occur as a result of this mating, simply fill in each of the empty squares with the alleles that can be donated from each parent. Determine all the gene combinations that can result when these gametes unite.

	T	*t*
T	*TT*	*Tt*
t	*Tt*	*tt*

Step 5: Determine the phenotype of each possible gene combination.

In this problem, three of the offspring, *TT*, *Tt*, and *Tt*, have Tourette syndrome. One progeny, *tt*, is normal. Therefore, the answer to the problem is that the probability of having offspring with Tourette syndrome is 3/4; for no Tourette syndrome, it is 1/4.

Take the time to learn these five steps. All single-factor problems can be solved using this method; the only variation in the problems will be the types of alleles and the number of possible types of gametes the parents can produce.

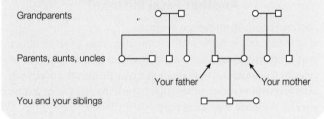
MORE COMPLEX MODELS OF INHERITANCE

So far, we have considered phenotypes that are determined by dominance and recessiveness between two alleles. In fact, dominant/recessive inheritance is probably the least common inheritance model since most inheritance patterns do not fit this simple pattern. In most situations, phenotypes are not only determined by the type of allele (gene) received but also by the alternative ways in which they interact with each other and how the environment influences their expression. There are six generally recognized patterns: X-linked characteristics, codominance, incomplete dominance, multiple alleles, polygenic inheritance, and pleiotropy.

X-Linked Genes

Alleles located on the same chromosome tend to be inherited together. They are said to be linked. The process of crossing-over, which occurs during prophase I of meiosis I, may split up these linkage groups. Crossing-over happens between homologous chromosomes donated by the mother and the father, and results in a mixing of genes. The closer two genes are to each other on a chromosome, the more probable it is that they will be inherited together.

People and many other organisms have two types of chromosomes. Autosomes (22 pairs) are not involved in sex determination and have alleles for the same traits on both members of the homologous pair of chromosomes. *Sex chromosomes* are a pair of chromosomes that control the sex of an organism. In humans and some other animals, there are two types of sex chromosomes—the X chromosome and the Y chromosome. The Y chromosome is much shorter than the X chromosome and has few genes for traits found on the X chromosome. One genetic trait that is located on the Y chromosome contains the testis-determining gene—SRY. Females are normally produced when two X chromosomes are present because they do not have the SRY gene. Males are usually produced when one X chromosome and one Y chromosome are present.

Genes found together on the X chromosomes are said to be **X-linked.** Because the Y chromosome is shorter than the X chromosome, it does not have many of the alleles that are found on the comparable portion of the X chromosome. Therefore, in a man, the presence of a single allele on his only X chromosome will be expressed, regardless of whether it is dominant or recessive because there is no corresponding allele on the Y chromosome. A Y-linked trait in humans is the SRY gene. This gene controls the differentiation of the embryonic gonad to a male testis. By contrast, more than one hundred genes are on the X chromosome. Some of these X-linked genes can result in abnormal traits such as *color deficiency* (formerly referred to as color blindness), *hemophilia, brown teeth,* and at least two forms of *muscular dystrophy* (Becker's and Duchenne's).

Codominance

In some inheritance situations, alleles are neither dominant nor recessive to one another, and both express themselves in the heterozygous individual. This inheritance pattern is called codominance. In a case of codominance, the phenotype of each allele is expressed in the heterozygous condition. Consequently, a person with the heterozygous genotype can have a phenotype very different from either of their homozygous parents. For codominant alleles, all uppercase symbols are used and superscripts are added to represent the different alleles. The uppercase letters call attention to the fact that each allele can be detected phenotypically to some degree even when in the presence of its alternative allele. For example, the coat colors (C) of horses, cattle, and other animals may be phenotypically roan. The roan gene adds white hairs among the other body hairs, whatever their color. The roan gene can be applied to any color. The most common base roan color is red. The color is the result of individual hairs that are two different colors. A red roan would have both red and white hairs intermingled. Animals that are true red would have the genotype $C^R C^R$, red roan would have $C^R C^W$, and true white would have $C^W C^W$ (figure 26.3).

Incomplete Dominance

In the inheritance pattern known as partial or **incomplete dominance**, alleles lack dominant relationships. The results of a cross are heterozygous offspring that have an intermediate

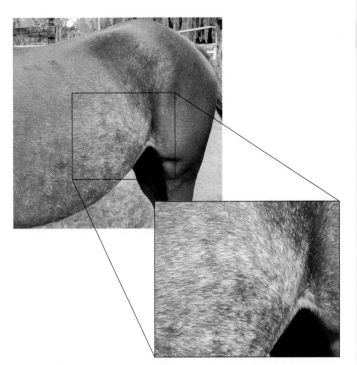

FIGURE 26.3 The color of this horse, an Arab, displays the codominant color pattern called roan. Notice that there are places on the body where both white and red hairs are found side by side.

A Closer Look

Blame That Trait on Your Mother!

Within a eukaryotic cell, the bulk of DNA is located in the nucleus. It is this genetic material that controls the majority of biochemical processes of the cell. It has long been recognized that the mitochondria also contain DNA, mtDNA. This genetic material works in conjunction with that located in the nucleus. However, an interesting thing happens when a sperm is formed as a result of meiosis. All the mitochondria are packed into the "necks" and not the "heads" of the sperm. When a sperm penetrates a secondary oocyte, the head enters but the majority of the mitochondria in the neck remain outside. Therefore, fathers rarely, if ever, contribute mitochondrial genetic information to their children. Should a mutation occur in the mother's mitochondrial DNA, she will pass the abnormality on to her children. Should a mutation occur in the father's mitochondrial DNA, he will not transmit the abnormality. This unusual method of transmission is called *mitochondrial* (or *maternal*) *inheritance*. Most abnormalities transmitted through mitochondrial genes are associated with muscular weakness because the mitochondria are the major source of ATP in eukaryotic cells. People with one form of this rare type of genetic abnormality, *Leber's hereditary optic neuropathy,* show symptoms of sudden loss of central vision due to optic nerve death in young adults with onset at about age twenty. Both males and females may be affected.

phenotype that is not distinct from either homozygous parent.

Another example of incomplete dominance occurs in certain colors of horses. Palomino and buckskin horses show a cream-color allele for the basic color gene (C^{cr}). In these horses, the C^{cr} allele dilutes the hair pigment to yellow. In the homozygous condition ($C^{cr}C^{cr}$), the horse is called cremello; both black and red pigments are intermingled in a single hair to produce yellow.

A classic example of incomplete dominance in plants involves the color of the petals of snapdragons. The gene products interact to produce a blended result. There are two alleles for the color of these flowers. Because neither allele is recessive, we cannot use the traditional capital and small letters as symbols for these alleles. Instead, the allele for white petals is given the symbol F^W, and the one for red petals is given the symbol F^R (figure 26.4).

There are three possible combinations of these two alleles:

Genotype	Phenotype
$F^W F^W$	White flower
$F^R F^R$	Red flower
$F^R F^W$	Pink flower

Notice that there are only two different alleles, red and white, but there are three phenotypes, red, white, and pink. Both the red-flower allele and the white-flower allele partially express themselves when both are present, and this results in pink.

FIGURE 26.4 The colors of these snapdragons are determined by two alleles for petal color, F^W and F^R. There are three different phenotypes because of the way in which the alleles interact with one another. In the heterozygous condition, neither of the alleles dominates the other.

Multiple Alleles

So far, we have discussed only traits that are determined by two alleles, for example, *A, a*. However, there can be more than two different alleles for a single trait. All the various forms of the same gene (alleles) that control a particular trait are referred to as **multiple alleles.** However, one person can have only a maximum of two of the alleles for the characteristic. A good example of a characteristic that is determined by multiple alleles is the ABO blood type. There are three alleles for blood type:

Allele*

I^A = blood has type A antigens on red blood cell surface

I^B = blood has type B antigens on red blood cell surface

i = blood type O has neither type A nor type B antigens on red blood cell surface

*The symbols I^A and I^B stand for the technical term for the antigenic carbohydrates attached to red blood cells, the *Immunogens*. These alleles are located on human chromosome 9. The ABO system is not the only one used to type blood. Others include the Rh, MNS, and Xg systems.

Note in the midst of this example of multiple alleles that in the ABO system, A and B show *codominance* when they are together in the same individual. However, both A and B are dominant over the O allele. These three alleles can be combined as pairs in six different ways, resulting in four different phenotypes:

Genotype		Phenotype
$I^A I^A$	=	Blood type A
$I^A i$	=	Blood type A
$I^B I^B$	=	Blood type B
$I^B i$	=	Blood type B
$I^A I^B$	=	Blood type AB
ii	=	Blood type O

Multiple-allele problems are worked as single-factor problems. Some examples are in the practice problems at the end of this chapter.

Polygenic Inheritance

Thus far, we have considered phenotypic characteristics that are determined by alleles at a single place and chromosome. However, some characteristics are determined by the interaction of genes at several different loci (on different chromosomes or at different places on a single chromosome). This is called **polygenic inheritance.** A number of different pairs of alleles may combine their efforts to determine a characteristic. Skin color in humans is a good example of this inheritance pattern. According to some experts, genes for skin color are located at a minimum of three loci. At each of these loci, the allele for dark skin is dominant over the allele for light skin. Therefore, a wide variety of skin colors is possible depending on how many dark-skin alleles are present (figure 26.5).

Polygenic inheritance is very common in determining characteristics that appear on a gradient in nature. In the skin-color example, and in many other traits as well, the characteristics cannot be categorized in terms of *either/or*, but the variation in phenotypes can be classified as *how much* or *by what amount.*

For instance, people show great variations in height. There are not just tall and short people—there is a wide range. Some people are as short as 1 m, and others are taller than 2 m. This quantitative trait is probably determined by a number of different genes. Intelligence also varies significantly, from those who are severely retarded to those who are geniuses. Many of these traits may be influenced by outside environmental factors such as diet, disease, accidents, and social factors. These are just a few examples of polygenic inheritance patterns.

Pleiotropy

Even though a single gene produces only one product, it often has a variety of effects on the phenotype of the person. This is called *pleiotropy.* **Pleiotropy** (*pleio-* =changeable) is a term used to describe the multiple effects that a gene may have on the phenotype. A good example of pleiotropy is *Marfan syndrome* (figure 26.6), a disease suspected to have occurred in U.S. President Abraham Lincoln. Marfan is a disorder of the body's connective tissue but can also have effects in many other organs, including the eyes, heart, blood, skeleton, and lungs. Symptoms generally appear as a tall, lanky body with long arms and

Locus 1	$d^1 d^1$	$d^1 D^1$	$d^1 D^1$	$D^1 D^1$	$D^1 d^1$	$D^1 d^1$	$D^1 D^1$
Locus 2	$d^2 d^2$	$d^2 d^2$	$d^2 D^2$	$D^2 d^2$	$D^2 d^2$	$D^2 D^2$	$D^2 D^2$
Locus 3	$d^3 d^3$	$d^3 d^3$	$d^3 d^3$	$d^3 d^3$	$D^3 D^3$	$D^3 D^3$	$D^3 D^3$

| Total number of dark-skin genes | 0 | 1 | 2 | 3 | 4 | 5 | 6 |

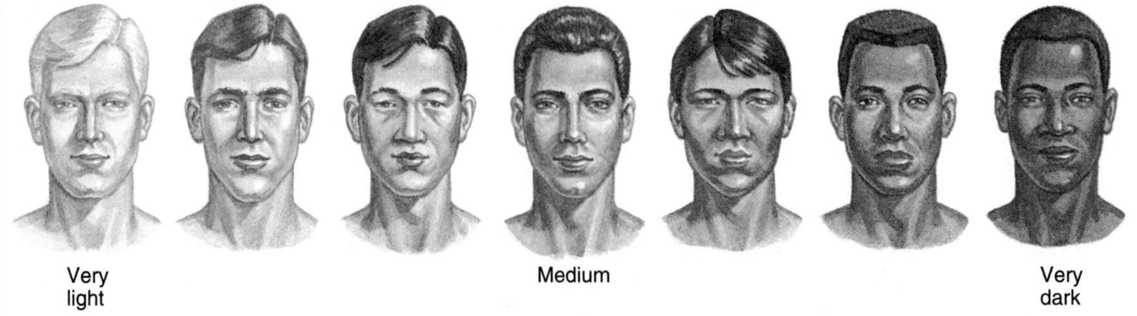

Very light   Medium   Very dark

FIGURE 26.5 Skin color in humans is an example of polygenic inheritance. The darkness of the skin is determined by the number of dark-skin genes a person inherits from his or her parents.

The Inheritance of Eye Color

It is commonly thought that eye color is inherited in a simple dominant/recessive manner. Brown eyes are considered to be dominant over blue eyes. The real pattern of inheritance, however, is considerably more complicated than this. Eye color is determined by the amount of a brown pigment, known as melanin, present in the iris of the eye. If there is a large quantity of melanin present on the anterior surface of the iris, the eyes are dark. Black eyes have a greater quantity of melanin than brown eyes.

If there is not a large amount of melanin present on the anterior surface of the iris, the eyes will appear blue, not because of a blue pigment but because blue light is returned from the iris (see box figure 26.2). The iris appears blue for the same reason that deep bodies of water tend to appear blue. There is no blue pigment in the water, but blue wavelengths of light are returned to the eye from the water. People appear to have blue eyes because the blue wavelengths of light are reflected from the iris.

Just as black and brown eyes are determined by the amount of pigment present, colors such as green, gray, and hazel are produced by the various amounts of melanin in the iris. If a very small amount of brown melanin is present in the iris, the eye tends to appear green, whereas relatively large amounts of melanin produce hazel eyes.

Several different genes are probably involved in determining the quantity and placement of the melanin and, therefore, in determining eye color. These genes interact in such a way that a wide range of eye color is possible. Eye color is probably determined by polygenic inheritance, just as skin color and height are. Some newborn babies have blue eyes that later become brown. This is because they have not yet begun to produce melanin in their irises at the time of birth.

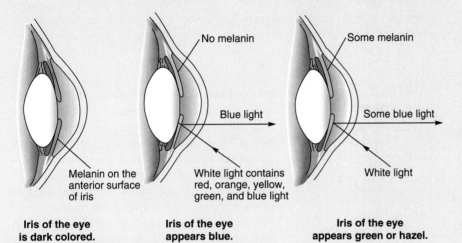

Box Figure 26.2

Iris of the eye is dark colored. — Melanin on the anterior surface of iris

Iris of the eye appears blue. — No melanin; Blue light; White light contains red, orange, yellow, green, and blue light

Iris of the eye appears green or hazel. — Some melanin; Some blue light; White light

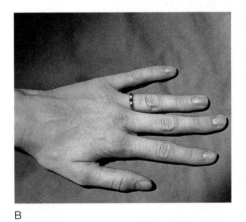

B

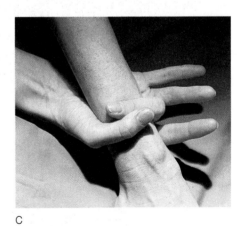

C

FIGURE 26.6 It is estimated that about forty thousand (the incidence is one out of ten thousand) people in the United States have Marfan syndrome, an autosomal dominant abnormality. Notice the lanky appearance to the body and face of this person with Marfan syndrome (*A*). Photos (*B*) and (*C*) illustrate their unusually long fingers.

A

spiderfingers, scoliosis (curvature of the spine), and depression of the chest wall. Many people with Marfan syndrome are nearsighted because they have a dislocation of the lens of the eye. The white of the eye (sclera) may appear bluish. Heart problems include dilation of the aorta and abnormalities of the heart valves. Death may be caused by a dissection (tear) in the aorta from the rupture in a weakened and dilated area of the aorta called an aortic *aneurysm*. Thus, the alleles that produce Marfan syndrome have multiple effects on height, eyesight, and other traits.

Another example of pleiotropy is the genetic abnormality PKU (phenylketonuria). In this example, a single gene affects many different chemical reactions that depend on the way a cell metabolizes the amino acid phenylalanine, commonly found in many foods (refer to figure 26.7).

In this genetic abnormality, people are unable to convert the amino acid phenylalanine into the amino acid tyrosine. The buildup of phenylalanine in the body prevents the normal development of the nervous system. Such individuals suffer from

Connections . . .

Inheritance Pattern of Spina Bifida

Many polygenic traits are influenced by outside environmental factors, such as diet, drugs, disease, accidents, and social factors. The degree to which the environment impacts the genes can determine the nature of phenotype. Patterns of inheritance that are the result of polygenic traits influenced by environmental factors are called multifactorial inheritance. In fact, most polygenic traits are multifactorial. *Multifactorial inheritance* is not yet completely understood and is hard to trace since it does not follow the typical single-gene inheritance pattern. The first generation (children) to follow individuals with such diseases (parents) shows a markedly high risk of acquiring the gene, whereas the second generation (grandchildren) has a much lower risk. Examples of multifactorial traits include neural tube defects (e.g., spina bifida), insulin-dependent diabetes mellitus, cleft palate, different fingerprints in identical twins, and congenital heart disease. In the case of spina bifida, the arches of the vertebrae fail to enclose the spinal cord (typically in the lower back region), and in some cases, the cord and nerves may be exposed at birth. This abnormality is a leading cause of stillbirths and death during early infancy and, if unable to be surgically corrected, is a handicap in survivors. One factor that appears to influence this disease is the amount of the vitamin folic acid the mother has in her system prior to conception. Other factors include chromosome disorders, single-gene mutations, and teratogens (chemicals that cause developmental abnormalities).

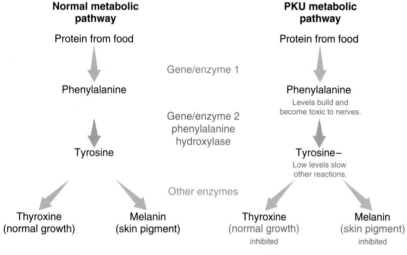

FIGURE 26.7 PKU is an recessive disorder located on chromosome 12. The diagram on the left shows how the normal pathway works. The diagram on the right shows an abnormal pathway. If the enzyme phenylalanine hydroxylase is not produced because of a mutated gene, the amino acid phenylalanine cannot be converted to tyrosine and is converted into phenylpyruvic acid, which accumulates in body fluids. The buildup of phenylpyruvic acid causes the death of nerve cells and ultimately results in mental retardation. Because phenylalanine is not converted to tyrosine, subsequent reactions in the pathway are also affected.

PKU and may become mentally retarded. There is a concern in the medical profession that there may be many adults with PKU who do not know they have the condition. Children diagnosed with PKU are not tracked throughout their life, and many parents do not tell their children they have the condition. After these children become adults, they could easily be unaware of their medical history. The problem comes if a female PKU patient becomes pregnant and does not have proper prenatal care and neither she nor her physician is aware of her medical history.

Environmental Influences on Gene Expression

One of the complicating aspects of genetics is that expression of an allele's phenotype depends on so much more than just the genotype of the organism. Sometimes the physical environment determines whether or not dominant or recessive alleles are expressed. For example, in humans, genes for freckles do not express themselves fully unless the person's skin is exposed to sunlight (figure 26.8). Another example is the allele for six fingers (*polydactylism*), which is dominant over the allele for five fingers in humans. Some people who have received the allele for

A   B   C

FIGURE 26.8 The expression of many genes is influenced by the environment. The allele for dark hair in the cat (A) is sensitive to temperature and expresses itself only in the parts of the body that stay cool. (B and C) The allele for freckles expresses itself more fully when a person is exposed to sunlight.

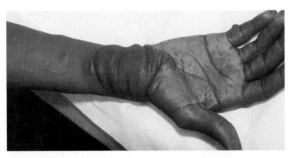

FIGURE 26.9 Neurofibromatosis I is seen in many forms, including benign fibromatous skin tumors, "café-au-lait" spots, nodules in the iris, and possible malignant tumors. It is extremely variable in its expressivity; that is, the traits may be almost unnoticeable or extensive. An autosomal dominant trait, it is the result of a mutation and the production of a protein (neurofibromin) that normally would suppress the activity of a gene that causes tumor formation.

six fingers have a fairly complete sixth finger; in others, it may appear as a little stub. In another case, a dominant allele causes the formation of a little finger that cannot be bent as a normal little finger. However, not all people who are believed to have inherited that allele will have a stiff little finger. In some cases, this dominant characteristic is not expressed or perhaps only shows on one hand. Thus, there may be variation in the degree to which an allele expresses itself in an *individual*. Geneticists refer to this as *variable expressivity*. A good example of this occurs in the genetic abnormality *neurofibromatosis type I* (NF1) (figure 26.9). In some cases, it may not be expressed in the *population* at all. This is referred to as a *lack of penetrance*. Other genes may be interacting with these dominant alleles, causing the variation in expression.

Both internal and external environmental factors can influence the expression of genes. For example, at conception, a male receives genes that will eventually determine the pitch of his voice. However, these genes are expressed differently after puberty. At puberty, male sex hormones are released. This internal environmental change results in the deeper male voice. A male who does not produce these hormones retains a higher-pitched voice in later life. Comparable changes can occur in females when an abnormally functioning adrenal gland causes the release of large amounts of male hormones.

Diet is an external environmental factor that can influence the phenotype of an individual. *Diabetes mellitus*, a metabolic disorder in which glucose is not properly metabolized and is passed out of the body in the urine, has a genetic basis. Some people who have a family history of diabetes are thought to have inherited the trait for this disease. Evidence indicates that they can delay the onset of the disease by reducing the amount of sugar in their diet.

WHAT IS A GENE AND HOW DOES IT WORK?

As scientists began to understand the chemical makeup of the **nucleic acids,** an attempt was made to understand how DNA and RNA relate to inheritance, cell structure, and cell activities. The concept that resulted is known as the *central dogma*. The belief can be written in this form:

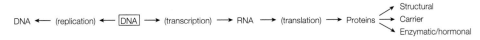

What this concept map says is that at the center of it all is DNA, the genetic material of the cell, and (going to the left) it is capable of reproducing itself, a process called **DNA replication**. Going to the right, DNA is capable of supervising the manufacture of RNA (a process known as **transcription),** which in turn is involved in the production of protein molecules, a process known as **translation.**

DNA replication occurs in cells in preparation for the cell division processes of mitosis and meiosis. Without replication, daughter cells would not receive the library of information required to sustain life. The transcription process results in the formation of a strand of RNA that is a copy of a segment of the DNA on which it is formed. Some of the RNA molecules become involved in various biochemical processes, while others are used in the translation of the RNA information into proteins. Structural proteins are used by the cell as building materials (feathers, collagen, hair), while others are used to direct and control chemical reactions (enzymes or hormones) or carry molecules from place to place (hemoglobin).

It is the processes of transcription and translation that result in the manufacture of all enzymes. Each unique enzyme molecule is made from a blueprint in the form of a DNA nucleotide sequence, or *gene*. Some of the thousands of enzymes manufactured in the cell are the tools required for transcription and translation to take place. *The enzymes made by the process carry out the process of making more enzymes!*

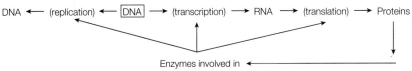

Tools are made to make more tools! The same is true for DNA replication. Enzymes made from the DNA blueprints by transcription and translation are used as tools to make exact copies of the genetic material! More blueprints are made so that future generations of cells will have the genetic materials necessary for them to manufacture their own regulatory and structural proteins. Without DNA, RNA, and enzymes functioning in the proper manner, life as we know it would not occur.

DNA has four properties that enable it to function as genetic material. It is able to (1) *replicate* by directing the manufacture of copies of itself; (2) *mutate,* or chemically change, and transmit these changes to future generations; (3) *store* information that determines the characteristics of cells and organisms; and (4) use this information to *direct* the synthesis of structural and regulatory proteins essential to the operation of the cell or organism.

The Structure of DNA and RNA

Nucleic acid molecules are enormous and complex polymers made up of monomers called **nucleotides.** Each nucleotide is composed of a sugar molecule (S) containing five carbon atoms, a phosphate group (P), and a molecule containing nitrogen that

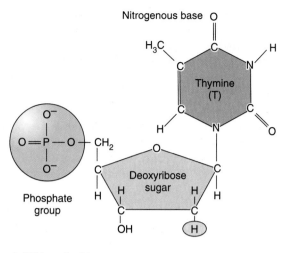

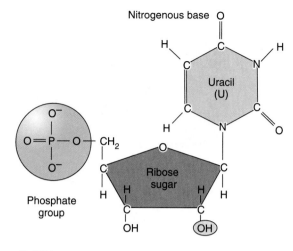

A DNA nucleotide

B RNA nucleotide

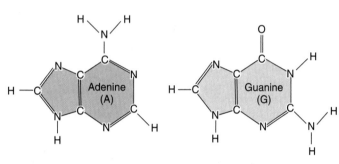

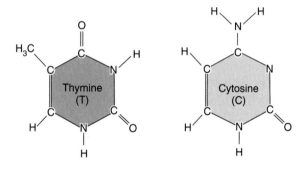

C The four nitrogenous bases that occur in DNA

FIGURE 26.10 (*A*) The nucleotide is the basic structural unit of all nucleic acid molecules. A thymine nucleotide of DNA is comprised of phosphate, deoxyribose sugar, and the nitrogenous base, thymine (T). Notice in the nucleotides that the phosphate group is written in "shorthand" form as a P inside a circle. (*B*) The RNA uracil nucleotide is comprised of a phosphate, ribose sugar, and the nitrogenous base, uracil (U). Notice the difference between the sugars and how the bases differ from one another. (*C*) Using these basic components (phosphate, sugars, and bases), the cell can construct eight common types of nucleotides. Can you describe all eight?

will be referred to as a *nitrogenous base* (B) (figure 26.10). It is possible to classify nucleic acids into two main groups based on the kinds of sugars and nitrogenous bases used in the nucleotides (i.e., DNA and RNA).

In cells, DNA is the nucleic acid that functions as the original blueprint for the synthesis of proteins. It contains the sugar *deoxyribose;* phosphates; and the nitrogenous bases adenine, guanine, cytosine, and thymine (A, G, C, T). RNA is a type of nucleic acid that is directly involved in the synthesis of protein. It contains the sugar *ribose;* phosphates; and adenine, guanine, cytosine, and uracil (A, G, C, U). There is no thymine in RNA and no uracil in DNA.

DNA and RNA differ in one other respect. DNA is actually a double molecule. It consists of two flexible strands held together between their protruding bases. The two strands are twisted about each other in a coil or double helix that resembles a twisted ladder (figure 26.11). The two strands of the molecule are held together because they "fit" each other like two jigsaw puzzle pieces that interlock with one another and are stabilized by weak chemical forces, hydrogen bonds. The four kinds of bases always pair in a definite way: adenine (A) with thymine (T), and guanine (G) with cytosine (C). Notice that the large molecules (A and G) pair with the small ones (T and C),

thus keeping the two complementary (matched) strands parallel. The bases that pair are said to be **complementary bases.**

$$A : T$$

and

$$G : C$$

You can "write" a message in the form of a stable DNA molecule by combining the four different DNA nucleotides (A, T, G, C) in particular sequences. The four DNA nucleotides are being used as an alphabet to construct three-letter words. To make sense out of such a code, it is necessary to read in one direction. Reading the sequence in reverse does not always make sense, just as reading this paragraph in reverse would not make sense.

The genetic material of humans and other eukaryotic organisms is *strands* of coiled *double-stranded DNA,* which has histone proteins attached along its length. When eukaryotic chromatin fibers coil into condensed, highly knotted bodies, they are seen easily through a microscope after staining with dye. Condensed like this, a chromatin fiber is referred to as a **chromosome** (figure 26.12B). The genetic material in bacteria is also double-stranded DNA, but the ends of the molecule are connected to form a *loop* and they do not form condensed chromosomes (figure 26.13).

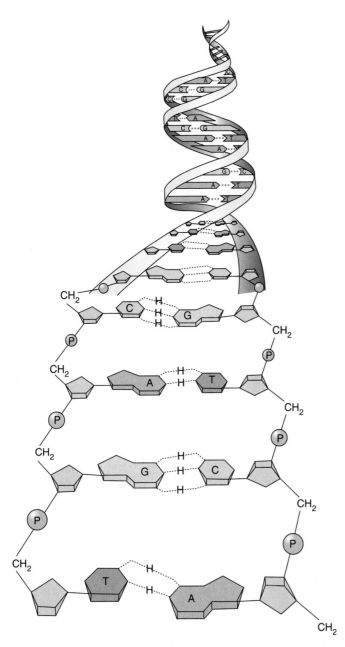

FIGURE 26.11 Polymerized deoxyribonucleic acid (DNA) is a helical molecule. While the units of each strand are held together by covalent bonds, the two parallel strands are interlinked by hydrogen bonds between the paired nitrogenous bases. Four nucleotides are highlighted in the bottom of the structure.

Each chromatin DNA strand is different because each strand has a different chemical code. Coded DNA serves as a central cell library. Tens of thousands of messages are in this storehouse of information. This information tells the cell such things as (1) how to produce enzymes required for the digestion of nutrients, (2) how to manufacture enzymes that will metabolize the nutrients and eliminate harmful wastes, (3) how to repair and assemble cell parts, (4) how to reproduce healthy offspring, (5) when and how to react to favorable and unfavorable changes in the environment, and (6) how to coordinate and regulate all of life's essential functions. If any of these functions is not performed properly, the cell may die. The importance of maintaining essential DNA in a cell becomes clear when we

consider cells that have lost their DNA. For example, human red blood cells lose their nuclei as they mature and become specialized for carrying oxygen and carbon dioxide throughout the body. Without DNA, they are unable to manufacture the essential cell components needed to sustain themselves. They continue to exist for about 120 days, functioning only on enzymes manufactured earlier in their lives. When these enzymes are gone, the cells die. Because these specialized cells begin to die the moment they lose their DNA, they are more accurately called *red blood corpuscles (RBCs)*: "little dying red bodies."

DNA Replication

Because all reproducing cells must maintain a complete set of genetic material, there must be a doubling of DNA in order to have enough to pass on to the offspring (refer to chapter 20 and "The Importance of Cell Division"). DNA replication is the process of duplicating the genetic material prior to its distribution to daughter cells. When a cell divides into two daughter cells, each new cell must receive a complete copy of the parent cell's genetic information, or it will not be able to manufacture all the proteins vital to its existence. Accuracy of duplication is essential to guarantee the continued existence of that type of cell. Should the daughters not receive exact copies, they would most likely die.

1. The DNA replication process begins as an enzyme breaks the attachments between the two strands of DNA. In eukaryotic cells, this occurs in hundreds of different spots along the length of the DNA (figure 26.14).
2. Moving along the DNA, the enzyme "unzips" the halves of the DNA (figure 26.14A). Hydrogen bonds hold each in position (AT, GC), while the new nucleotide is covalently bonded between the sugar and phosphate of the new backbone.
3. Proceeding in opposite directions on each side, the enzyme *DNA polymerase* moves down the length of the DNA, attaching new DNA nucleotides into position. Relatively weak hydrogen bonds hold each in position (AT, GC), while the new nucleotide is bonded to its neighboring nucleotide more strongly with covalent bonds between the sugar and phosphate of the new backbone (figure 26.14B).
4. The enzyme that speeds the addition of new nucleotides to the growing chain works along with another enzyme to make sure that no mistakes are made. If the wrong nucleotide appears to be headed for a match, the enzyme will reject it in favor of the correct nucleotide. If a mistake is made and a wrong nucleotide is paired into position, specific enzymes have the ability to replace it with the correct one (figure 26.14C).
5. Replication proceeds in both directions and at multiple sites along the chromatin, appearing as "bubbles" (figure 26.14D).
6. The complementary molecules (AT, GC) pair with the exposed nitrogenous bases of both DNA strands by forming new hydrogen bonds (figure 26.14C).
7. Once properly aligned, a bond is formed between the sugars and phosphates of the newly positioned nucleotides. A strong sugar and phosphate backbone is formed in the process (figure 26.14C).
8. This process continues until all the replication "bubbles" join (figure 26.14D).

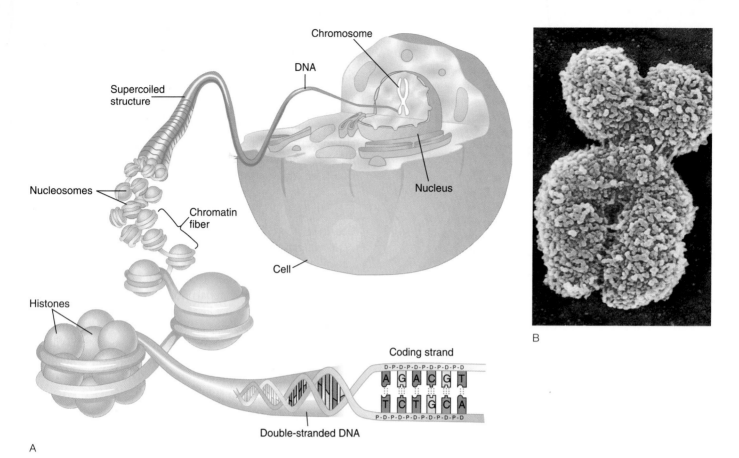

B

FIGURE 26.12 (*A*) Eukaryotic cells contain double-stranded DNA in their nuclei, which takes the form of a three-dimensional helix. One strand is a chemical code (the coding strand) that contains the information necessary to control and coordinate the activities of the cell. The two strands fit together and are bonded by weak hydrogen bonds formed between the complementary, protruding nitrogenous bases according to the base-pairing rule. The length of a DNA molecule is measured in numbers of "base pairs"—the number of rungs on the ladder. (*B*) During certain stages in the reproduction of a eukaryotic cell, the nucleoprotein coils and "supercoils," forming tightly bound masses. When stained, these are easily seen through the microscope. In their supercoiled form, they are called chromosomes, meaning colored bodies.

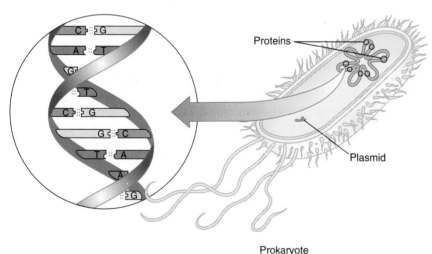

FIGURE 26.13 The nucleic acid of prokaryotic cells (the bacteria) has a different kind of protein compared to eukaryotic cells. In addition, the ends of the giant nucleoprotein molecule overlap and bind with one another to form a loop. The additional small loop of DNA is the plasmid, which contains genes that are not essential to the daily life of the cell.

A new complementary strand of DNA forms on each of the old DNA strands, resulting in the formation of two double-stranded DNA molecules. In this way, the exposed nitrogenous bases of the original DNA serve as a *template,* or pattern, for the formation of the new DNA. As the new DNA is completed, it twists into its double-helix shape.

The completion of the DNA replication process yields two double helices that are identical in their nucleotide sequences. Half of each is new, half is the original parent DNA molecule. The DNA replication process is highly accurate. It has been estimated that there is only one error made for every 2×10^9 nucleotides. A human cell contains forty-six chromosomes consisting of about 3,000,000,000 (3 billion) base pairs. This averages to about 1.5 errors per cell! Don't forget that this figure is an estimate. While some cells may have five errors per replication, others may have more, and some may have no errors at all. It is also important to note that some errors may be major and deadly, while others are insignificant. Because this error rate is so small, DNA replication is considered by most to be essentially error-free. Following DNA replication, the cell now contains twice the amount of genetic information and is ready to begin the process of distributing one set of genetic information to each of its two daughter cells.

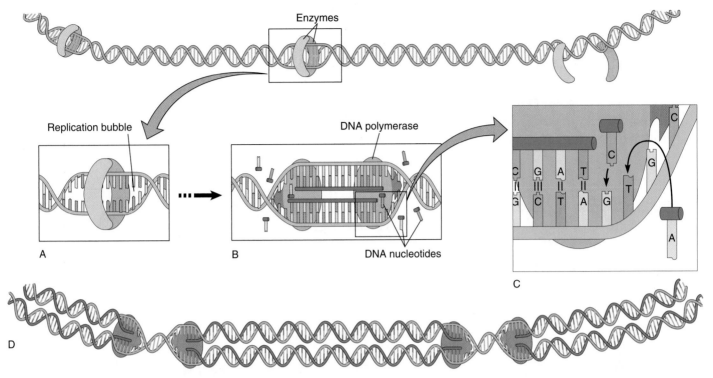

FIGURE 26.14 These illustrations summarize the basic events that occur during DNA replication. (*A*) Enzymes break apart the two strands of DNA. (*B* and *C*) As the DNA strands are separated, new DNA nucleotides are added to the new strands by DNA polymerase. The new DNA strands are synthesized according to base-pairing rules for nucleic acids. (*D*) By working in two directions at once along the DNA strand, the cell is able to more quickly replicate the DNA. Each new daughter cell will receive one of these copies.

The distribution of DNA involves splitting the cell and distributing a set of genetic information to the two new daughter cells. In this way, each new cell has the necessary information to control its activities. The mother cell ceases to exist when it divides its contents between the two smaller daughter cells. A cell does not really die when it reproduces itself; it merely starts over again.

DNA Transcription

DNA functions in the manner of a reference library that does not allow its books to circulate. Information from the originals must be copied for use outside the library. The second major function of DNA is the making of these single-stranded, complementary RNA copies of DNA. This operation is called transcription (*scribe* = to write), which means to transfer data from one form to another. In this case, the data are copied from DNA language to RNA language. The same base-pairing rules that control the accuracy of DNA replication apply to the process of transcription. Using this process, the genetic information stored as a DNA chemical code is carried in the form of an RNA copy to other parts of the cell. It is RNA that is used to guide the assembly of amino acids into structural and regulatory proteins. Without the process of transcription, genetic information would be useless in directing cell functions. Although many types of RNA are synthesized from the genes, the three most important are *messenger RNA (mRNA), transfer RNA (tRNA),* and *ribosomal RNA (rRNA).*

Transcription begins in a way that is similar to DNA replication. The DNA strands are separated by an enzyme, exposing the nitrogenous-base sequences of the two strands. However, unlike DNA replication, transcription occurs only on one of the two DNA strands, which serves as a template, or pattern, for the synthesis of RNA (see figure 26.15). This side is also referred to as the genetic *coding strand* of the DNA. But which strand is copied? Where does it start and when does it stop? Where along the sequence of thousands of nitrogenous bases does the chemical code for the manufacture of a particular enzyme begin and where does it end? If transcription begins randomly, the resulting RNA may not be an accurate copy of the code, and the enzyme product may be useless or deadly to the cell. To answer these questions, it is necessary to explore the nature of the genetic code itself.

We know that genetic information is in chemical-code form in the DNA molecule. When the coded information is used or *expressed,* it guides the assembly of particular amino acids into structural and regulatory polypeptides and proteins. If DNA is molecular language, then each nucleotide in this language can be thought of as a letter within a four-letter alphabet. Each word, or code, is always three letters (nucleotides) long, and only three-letter words can be written. A **DNA code** is a triplet nucleotide sequence that codes for one of the twenty common amino acids. The number of codes in this language is limited because there are only four different nucleotides, which are used only in groups of three. The order of these three letters is just as important in DNA language as it is in our language. We recognize that

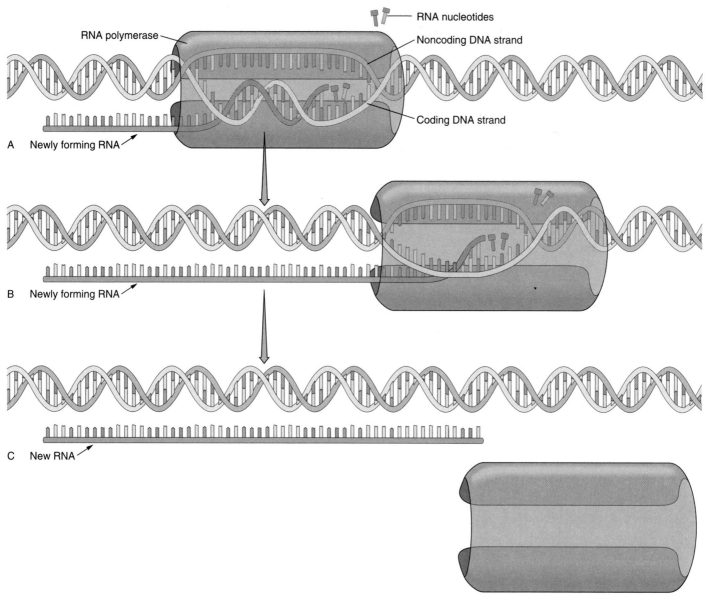

FIGURE 26.15 This summary illustrates the basic events that occur during the transcription of an RNA molecule. (*A*) An enzyme attaches to the DNA at a point that allows it to separate the complementary strands. (*B*) As RNA polymerase moves down the DNA strand, new complementary RNA nucleotides are base-paired to one of the exposed DNA strands. These base-paired RNA nucleotides are linked together by RNA polymerase to form a new RNA molecule that is complementary to the nucleotide sequence of the DNA. (*C*) The newly formed (transcribed) RNA is then separated from the DNA molecule and used by the cell.

CAT is not the same as TAC. If all the possible three-letter codes were written using only the four DNA nucleotides for letters, there would be a total of sixty-four combinations.

$$4^3 = 4 \times 4 \times 4 = 64$$

When codes are found at a particular place along a coding strand of DNA and the sequence has meaning, the sequence is a gene. "Meaning" in this case refers to the fact that the gene can be transcribed into an RNA molecule, which in turn may control the assembly of individual amino acids into a polypeptide.

When a gene is transcribed into RNA, the double-stranded DNA is "unzipped," and an enzyme known as *RNA polymerase* attaches to the DNA at the promoter region. It is from this

region that the enzymes will begin to assemble RNA nucleotides into a complete, single-stranded copy of the gene, including initiation and termination codes. Triplet RNA nucleotide sequences complementary to DNA codes are called **codons**. Remember that there is no thymine in RNA molecules; it is replaced with uracil. Therefore, the initiation code in DNA (TAC) would be base-paired by RNA polymerase to form the RNA codon AUG. When transcription is complete, the newly assembled RNA is separated from its DNA template and made available for use in the cell; the DNA recoils into its original double-helix form.

In summary:

1. The process begins as one portion of the enzyme RNA polymerase breaks the attachments between the two

Basic Steps of Translation

1. An mRNA molecule is placed in the smaller of the two portions of a ribosome so that six nucleotides (two codons) are locked into position.

2. The larger ribosomal unit is added to the ribosome/mRNA combination.

3. A tRNA with bases that match the second mRNA codon attaches to the mRNA. The tRNA is carrying a specific amino acid. Once attached, a second tRNA carrying another specific amino acid moves in and attaches to its complementary mRNA codon right next to the first tRNA/amino acid complex.

4. The two tRNAs properly align their two amino acids so that the amino acids may be chemically attached to one another (box figure 26.3).

5. Once the two amino acids are connected to one another by a covalent peptide bond, the first tRNA detaches from its amino acid and mRNA codon.

6. The ribosome moves along the mRNA to the next codon (the first tRNA is set free to move through the cytoplasm to attach to and transfer another amino acid).

7. The next tRNA/amino acid unit enters the ribosome and attaches to its codon next to the first set of amino acids.

8. The tRNAs properly align their amino acids so that they may be chemically attached to one another, forming a chain of three amino acids.

9. Once three amino acids are connected to one another, the second tRNA is released from its amino acid and

mRNA (this tRNA is set free to move through the cytoplasm to attach to and transfer another amino acid).

10. The ribosome moves along the mRNA to the next codon, and the fourth tRNA arrives (box figure 26.4).

11. This process repeats until all the amino acids needed to form the protein have attached to one another in the proper sequence. This amino acid sequence was encoded by the DNA gene.

12. Once the final amino acid is attached to the growing chain of amino acids, all the molecules (mRNA, tRNA, and newly formed protein) are released from the ribosome. The "stop" mRNA codon signals this action.

13. The ribosome is again free to become involved in another protein-synthesis operation.

14. The newly synthesized chain of amino acids (the new protein) leaves the ribosome to begin its work. However, the protein may need to be altered by the cell before it will be ready for use (box figure 26.5).

Box Figure 26.3

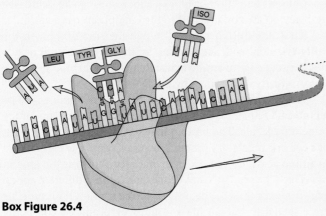

Box Figure 26.4

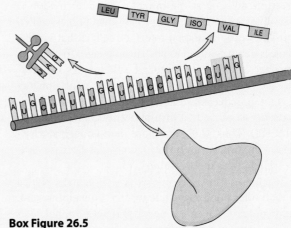

Box Figure 26.5

strands of DNA; the enzyme "unzips" the two strands of the DNA.

2. A second portion of the enzyme RNA polymerase attaches at a particular spot on the DNA. It proceeds in one direction along one of the two DNA strands, attaching new RNA nucleotides into position until it reaches the end of the gene. The enzymes then assemble RNA nucleotides into

a complete, singlestranded RNA copy of the gene. There is no thymine in RNA molecules; it is replaced by uracil. Therefore, the start codon in DNA (TAC) would be paired by RNA polymerase to form the RNA codon AUG.

3. The enzyme that speeds the addition of new nucleotides to the growing chain works along with another enzyme to make sure that no mistakes are made.

4. When transcription is complete, the newly assembled RNA is separated from its DNA template and made available for use in the cell; the DNA recoils into its original double-helix form.

As previously mentioned, three general types of RNA are produced by transcription: messenger RNA, transfer RNA, and ribosomal RNA. Each kind of RNA is made from a specific gene and performs a specific function in the synthesis of polypeptides from individual amino acids at ribosomes. **Messenger RNA (mRNA)** is a mature, straight-chain copy of a gene that describes the exact sequence in which amino acids should be bonded together to form a polypeptide.

Transfer RNA (tRNA) molecules are responsible for picking up particular amino acids and transferring them to the ribosome for assembly into the polypeptide (refer to "A Closer Look: Basic Steps of Translation"). All tRNA molecules are shaped like a cloverleaf. This shape is formed when they fold, and some of the bases form hydrogen bonds that hold the molecule together. One end of the tRNA is able to attach to a specific amino acid. Toward the midsection of the molecule, a triplet nucleotide sequence can base-pair with a codon on mRNA. This triplet nucleotide sequence on tRNA that is complementary to a codon of mRNA is called an **anticodon.** **Ribosomal RNA (rRNA)** is a highly coiled molecule and is used, along with protein molecules, in the manufacture of all ribosomes, the cytoplasmic organelles where tRNA, mRNA, and rRNA come together to help in the synthesis of proteins.

Translation or Protein Synthesis

The mRNA molecule is a coded message written in the biological world's universal nucleic acid language. The code is read in one direction, starting at the initiator. The information is used to assemble amino acids into protein by a process called translation. The word *translation* refers to the fact that nucleic acid language is being changed to protein language. To translate mRNA language into protein language, a dictionary is necessary. Remember, the four letters in the nucleic acid alphabet yield sixty-four possible three-letter words. The protein language has twenty words in the form of twenty common amino acids (table 26.3). Thus, there are more than enough nucleotide words for the twenty amino acid molecules because each nucleotide triplet codes for an amino acid.

Table 26.4 is an amino acid–nucleic acid dictionary. Notice that more than one codon may code for the same amino acid. Some would contend that this is needless repetition, but such "synonyms" can have survival value. If, for example, the gene or the mRNA becomes damaged in a way that causes a particular nucleotide base to change to another type, the chances are still good that the proper amino acid will be read into its proper position. But not all such changes can be compensated for by the codon system, and an altered protein may be produced. Changes can occur that cause great harm. Some damage is so extensive that the entire strand of DNA is broken, resulting in improper protein synthesis or a total lack of synthesis. Any change in DNA is called a **mutation.** Recall that some mutations may be so minor that they have no effect, while others

TABLE 26.3 Twenty Common Amino Acids and Their Codes

Amino Acid	Three-Letter Code
glycine	Gly
alanine	Ala
valine	Val
leucine	Leu
isoleucine	Ile
methionine	Met
phenylalanine	Phe
tryptophan	Trp
proline	Pro
serine	Ser
threonine	Thr
cysteine	Cys
tyrosine	Tyr
asparagine	Asn
glutamine	Gln
aspartic acid	Asp
glutamic acid	Glu
lysine	Lys
arginine	Arg
histidine	His

These are the twenty common amino acids used in the protein synthesis operation of a cell. Each has a known chemical structure.

may be harmful or beneficial. Although most mutations are likely neutral or harmful, beneficial mutations do occur.

The construction site of the protein molecules (i.e., the translation site) is on the ribosome, a cellular organelle that serves as the meeting place for mRNA and the tRNA that is carrying amino acid building blocks. Ribosomes can be found free in the cytoplasm or attached to the endoplasmic reticulum.

Thus, the mRNA moves through the ribosomes with the mRNA's codon sequence allowing for the chemical bonding of a specific sequence of amino acids. Remember that the sequence was originally determined by the DNA.

A protein's three-dimensional shape is determined by its amino acid sequence. This shape determines the activity of the protein molecule. The protein may be a structural component of a cell or a regulatory protein, such as an enzyme. Any change in amino acids or their order changes the action of the protein molecule. The protein insulin, for example, has a different amino acid sequence than the digestive enzyme trypsin. Both proteins are essential to human life and must be produced constantly and accurately. The amino acid sequence of each is determined by a different gene. Each gene is a particular sequence of DNA nucleotides. Any alteration of that sequence can directly alter the protein structure and, therefore, the survival of the organism.

Alterations of DNA

Several kinds of changes to DNA may result in mutations. Phenomena that are either known or suspected causes of DNA damage are called **mutagenic agents.** Agents known to cause

TABLE 26.4 Amino Acid–mRNA Nucleic Acid Dictionary

Second letter

First letter		U	C	A	G	Third letter
U		UUU UUC } Phe UUA UUG } Leu	UCU UCC UCA UCG } Ser	UAU UAC } Tyr UAA Stop UAG Stop	UGU UGC } Cys UGA Stop UGG Try	U C A G
C		CUU CUC CUA CUG } Leu	CCU CCC CCA CCG } Pro	CAU CAC } His CAA CAG } Gln	CGU CGC CGA CGG } Arg	U C A G
A		AUU AUC } Ile AUA AUG Met or start	ACU ACC ACA ACG } Thr	AAU AAC } ASN AAA AAG } Lys	AGU AGC } Ser AGA AGG } Arg	U C A G
G		GUU GUC GUA GUG } Val	GCU GCC GCA GCG } Ala	GAU GAC } Asp GAA GAG } Glu	GGU GGC GGA GGG } Gly	U C A G

damage to DNA are certain viruses (e.g., papillomavirus), weak or "fragile" spots in the DNA, X radiation (X rays), and chemicals found in foods and other products, such as chemicals from burning tobacco. All have been studied extensively, and there is little doubt that they cause mutations. *Chromosomal aberrations* is the term used to describe major changes in DNA. Four types of aberrations are inversions, translocations, duplications, and deletions. An *inversion* occurs when a chromosome is broken and this piece becomes reattached to its original chromosome but in reverse order. It has been cut out and flipped around. A *translocation* occurs when one broken segment of DNA becomes integrated into a different chromosome. *Duplication* occurs when a portion of a chromosome is replicated and attached to the original section in sequence. *Deletion* aberrations result when the broken piece becomes lost or is destroyed before it can be reattached.

In some individuals, a single nucleotide of the gene may be changed. This type of mutation is called a *point mutation.* An example of the effects of altered DNA may be seen in human red blood cells. Red blood cells contain the oxygen-transport molecule, hemoglobin. Normal hemoglobin molecules are composed of 150 amino acids in four chains—two alpha and two beta. The nucleotide sequence of the gene for the beta chain is known, as is the amino acid sequence for this chain. In normal individuals, the sequence begins like this:

Val-His-Leu-Thr-Pro-Glu-Glu-Lys . . .

A single nucleotide change in the DNA sequence results in a mutation that produces a new amino acid sequence in all the red blood cells:

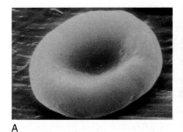

A   B

FIGURE 26.16 (*A*) A normal red blood cell is shown in comparison with (*B*) a cell having the sickle shape. This sickling is the result of a single amino acid change in the hemoglobin molecule.

Val-His-Leu-Thr-Pro-Val-Glu-Lys . . .

This single nucleotide change (known as a *missense point mutation*), which causes a single amino acid to change, may seem minor. However, it is the cause of **sickle-cell anemia,** a disease that affects the red blood cells by changing them from a circular to a sickle shape when oxygen levels are low (figure 26.16). When this sickling occurs, the red blood cells do not flow smoothly through capillaries. Their irregular shapes cause them to clump, clogging the blood vessels. This prevents them from delivering their oxygen load to the oxygen-demanding tissues. A number of physical disabilities may result, including physical weakness, brain damage, pain and stiffness of the joints, kidney damage, rheumatism, and, in severe cases, death.

Changes in the structure of DNA may have harmful effects on the next generation if they occur in the sex cells. Some damage to DNA is so extensive that the entire strand of DNA is

TABLE 26.5 Types of Chromosomal Aberrations

Normal Sequence	THE ONE BIG FLY HAD ONE RED EYE*
KIND OF MUTATION	**SEQUENCE CHANGE**
Missense	THQ ONE BIG FLY HAD ONE RED EYE
Nonsense	THE ONE BIG
Frameshift	THE ONE QBI GFL YHA DON ERE DEY
Deletion	THE ONE BIG HAD ONE RED EYE
Duplication	THE ONE BIG FLY FLY HAD ONE RED EYE
Insertion	THE ONE BIG WET FLY HAD ONE RED EYE
Expanding mutation:	
Parents	THE ONE BIG FLY HAD ONE RED EYE
Children	THE ONE BIG FLY FLY FLY HAD ONE RED EYE
Grandchildren	THE ONE BIG FLY FLY FLY FLY FLY FLY HAD ONE RED EYE

Based on R. Lewis, *Human Genetics, Concepts and Applications,* 2d ed., Dubuque, IA: Wm. C. Brown Publishers, 1997.

*A sentence composed of three-letter words can provide an analogy to the effect of mutations on a gene's nucleotide sequence.

broken, resulting in the synthesis of abnormal proteins or a total lack of protein synthesis. A number of experiments indicate that many street drugs, such as LSD (lysergic acid diethylamide), are mutagenic agents and cause DNA to break. Abnormalities have also been identified that are the result of changes in the number or sequence of bases. One way to illustrate these various kinds of mutations is seen in table 26.5.

USING DNA TO OUR ADVANTAGE

Advances in the field of biotechnology have allowed scientists to do many exciting things. To understand how these accomplishments are achieved by the scientists, we need to explore two key strategies: (1) genetic modification of an organism and (2) DNA sequencing.

Genetic Modification of Organisms

Since biblical times, civilizations have attempted to improve the quality of their livestock and agricultural stock. Cows that produced more milk or more tender meat were valued over those that were dry or tough. Initial attempts to develop improved agricultural stock were limited to selective breeding programs where organisms with the desired characteristics were allowed to breed. With a greater understanding of the genetics, scientists irradiated cells to produce mutations that were then screened to determine if they were desirable. While this approach is a very informative way to learn about the genetics of an organism, it lacked the ability to create a specific change. Creating mutations is a very haphazard process. However, today the results are achieved in a much more directed manner that uses biotechnology's ability to transfer DNA from one organism to another. Techniques such as gene cloning and transformation (introducing new DNA sequences into an organism to alter their genetic makeup) allow scientists to introduce very specific

characteristics to an organism. These techniques use enzymes that manipulate DNAs to form complementary base pairs. Once the new DNA sequences are stable and transferred into the host cell, the cell is genetically altered and begins to read the new DNA and produce new cell products such as enzymes and other important products. The resulting new form of DNA is called **recombinant DNA.** Organisms that contain these genetic changes are referred to as **genetically modified (GM)**. Genetic engineers identify and isolate sequences of nucleotides from a living or dead cell. Organisms such as viruses, bacteria, fungi, plants, and animals or their offspring have been engineered so that they contain genes from at least one unrelated organism.

As this highly sophisticated procedure has been refined, it has become possible to quickly and accurately splice genes from a variety of species into host bacteria or other host cells by a process called *gene cloning.* This makes possible the synthesis of large quantities of important products. For example, recombinant DNA procedures are responsible for the production of human insulin, used in the control of diabetes; nutritionally enriched "golden rice," capable of supplying underdeveloped nations with missing essential amino acids; interferon, used as an antiviral agent; human growth hormone, used to stimulate growth in children lacking this hormone; and somatostatin, a brain hormone also implicated in growth. Over two hundred such products have been manufactured using these methods.

Although some of these chemicals have been produced in small amounts from genetically engineered microorganisms, crops such as turnips, rice, soybeans, potatoes, cotton, corn, and tobacco can generate tens or hundreds of kilograms of specialty chemicals per year. Many of these GM crops also have increased nutritional value and yet can be cultivated using traditional methods. Such crops have the potential of supplying the essential amino acids, fatty acids, and other nutrients now lacking in the diets of people in underdeveloped or developing nations. Researchers have also shown, for example, that turnips can produce interferon (an antiviral agent), tobacco can create antibodies to fight human disease, oilseed rape plants can serve as a source of human brain hormones, and potatoes can synthesize human serum albumin that is indistinguishable from the genuine human blood protein.

Some of the likely rewards of biotechnology are (1) production of additional, medically useful proteins; (2) mapping of the locations of genes on human chromosomes; (3) a more complete understanding of how genes are regulated; (4) production of crop plants with increased yields; and (5) development of new species of garden plants (figure 26.17).

Gene Therapy

The field of biotechnology allows scientists and medical doctors to work together and potentially cure genetic disorders. Unlike contagious diseases, genetic diseases cannot be caught or transmitted because they are caused by a genetic predisposition for a particular disorder—not separate, disease-causing organisms such as a bacterium or virus. Gene therapy involves inserting genes, deleting genes, or manipulating the action of genes in order to cure or lessen the effect of genetic diseases. These

FIGURE 26.17 Using our understanding of how DNA codes for proteins allows scientists to develop treatments for disease as well as develop crops with greater yield and nutritional value.

activities are very new and experimental. While these lines of investigation create hope for many, there are many problems that must be addressed before gene therapy becomes a realistic treatment for many disorders.

The strategy for treating someone with gene therapy varies depending on the individual's disorder. When designing a gene therapy treatment, scientists have to ask themselves, "Exactly what is the problem?" Is the mutant gene not working at all? Is it working normally, but is there too little activity? Is there too much protein being made? Or possibly, is the problem that the gene is acting in a unique and new manner? If there is no gene activity or too little gene activity, the scientists need to somehow introduce a more active version of the gene. If there is too much activity or if the problem is caused by the gene having a new activity, this excess activity must first be stopped and then the normal activity restored.

To stop a mutant gene from working, scientists must change it. This typically involves inserting a mutation into the protein-coding region of the gene or the region that is necessary to activate the gene. Scientists have used some types of viruses to do this in organisms other than humans for some time now. The difficulty in this technique is to mutate only that one gene without disturbing the other genes of the cells and creating more mutations in other genes. Developing reliable methods to accomplish this is a major focus of gene therapy. Once the mutant gene is silenced, the scientists begin the work of introducing a "good" copy of the gene. Again, there are many difficulties in this process:

- Scientists must find a way of returning the corrected DNA to the cell.
- The corrected DNA must be made a part of the cell's DNA so that it is passed on with each cell division, it doesn't interfere with other genes, and it can be transcribed by the cell as needed.
- Finally, cells containing the corrected DNA must be reintroduced to the patient.

Many of these techniques are experimental, and the medical community is still evaluating their usefulness in treating many disorders, as well as the risks that these techniques pose to the patient.

Recently, the first efforts to determine the human genome, the entire human DNA sequence, were completed. Many scientists feel that advances in medical treatments will occur more quickly by having this information available. Three new fields of biology have grown out of these efforts: *genomics, transcriptomics,* and *proteomics.* Genomics is the study of the DNA sequence and looks at the significance of how different genes and DNA sequences are related to each other. When genes are

A Closer Look

The Human Genome Project

The Human Genome Project (HGP) was first proposed in 1986 by the U.S. Department of Energy (DOE) and was co-sponsored soon after by National Institutes of Health (NIH). These agencies were the main research agencies within the U.S. government responsible for developing and planning the project. Later, a private U.S. corporation, Celera Genomics, joined the effort as a competitor. It is one of the most ambitious projects ever undertaken in the biological sciences. The goal was nothing less than the complete characterization of the genetic makeup of humans. The project was completed early in 2003, when the complete nucleotide sequence of all twenty-three pairs of human chromosomes was determined. It is as if we have now identified all the words in the human gene dictionary, but we only have definitions for about half. With this knowledge in hand, scientists will define the rest of the words (learn what the gene does) and produce a map of each of the chromosomes that will show the names and places of all our genes. This international project involving about one hundred laboratories worldwide took only sixteen years to complete. Work began in many of these labs in 1990. Powerful computers are used to store and share the enormous amount of information derived from the analyses of human DNA. To get an idea of the size of this project, consider this: A human Y chromosome (one of the smallest of the human chromosomes) is estimated to be composed of 59 million nitrogenous bases. The larger X chromosomes may be composed of 164 million nitrogenous base pairs! The entire human genome consists of 3.12 gigabases!

Two kinds of work progressed simultaneously. First, *physical maps* were constructed by determining the location of specific "markers" (known sequences of bases) and

their closeness to genes (see box figure 26.6, chromosome 21). A kind of chromosome map already exists that shows patterns of colored bands on chromosomes, a result of chromosome-staining procedures. Using these banded chromosomes, scientists then related the markers to these colored bands on a specific region of a chromosome. Work

is continuing on the HGP to identify the location of specific genes. Each year, a more complete picture is revealed.

The second kind of work was for labs to determine the exact order of nitrogenous bases of the DNA for each chromosome. Techniques exist for determining base sequences, but it is a time-consuming job

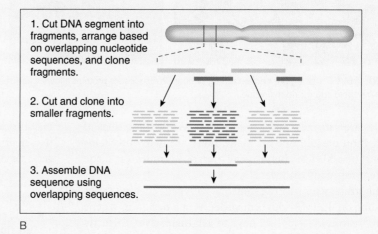

Box Figure 26.6 *(A)* The genetic map shows the approximate positions of disease-associated genes known to be on human chromosome 21. *(B)* To determine the nucleotide sequence of this chromosome, scientists first created a physical map, which consisted of many small pieces. The DNA sequence of these pieces was determined. After this, the DNA sequences of these many small pieces are "knit" together to generate the nucleotide sequence for the entire chromosome. The nucleotide sequence of each of the genes in *A* can be identified within the assembled sequence.

identified from the DNA sequence, transcriptomics looks at when, where, and how much a gene is expressed. Finally, proteomics examines the proteins that are predicted from the DNA sequence. From these types of studies, we are able to identify gene families that can be used to determine how humans have evolved on a molecular level, how genes are used in an organism throughout its body and over its life span, and how to identify common themes from one protein to the next.

Strategy Two: Sequencing

The second strategy of biotechnology works with comparing of nucleotide sequences from different cells. The closer organisms are genetically, the more closely their DNA sequence will resemble one another. Base sequence analysis can be done two ways. Researchers can look at the DNA sequences directly for comparison. However, this is a very tedious and to sort out the several million bases that can be found in any one chromosome. Coming from behind with new, speedier techniques, Celera Genomics was able to catch up to NIH labs and completed its sequencing at almost the same time. The benefit of having these two organizations as competitors is that when they finished their research, they could compare and contrast results. Amazingly, the discrepancies between their findings were declared insignificant. It was originally estimated, for example, that there were between 100,000 and 140,000 genes in the human genome. However, when the results were compared, the evidence from both organizations indicated that there are only about 30,000 protein-coding genes—only about twice as many as in a worm or a fly. However, the genes are more complex, with more alternative splicing, generating a larger number of protein products for each gene. In humans, it appears that each gene is capable of coding for between three and four different proteins. Knowing this information provides insights into the evolution of humans and the mutation rates of males versus females. This will make future efforts to work with the genome through bioengineering much easier.

When the physical maps are finally completed for all of the human genes, it will be possible to examine a person's DNA and identify genetic abnormalities. This could be extremely useful in diagnosing diseases and providing genetic counseling to those considering having children. This kind of information would also identify human genes and proteins as drug targets and create possibilities for new gene therapies. Once it is known where an abnormal gene is located and how it differs in base sequence from the normal DNA sequence, steps could be taken to correct the abnormality. Completing the

HGP will also result in the discovery of new families of proteins and will help explain basic physiological and cell biological processes common to many organisms, thus increasing the breadth and depth of our understanding of basic biology.

However, there is also a concern that as knowledge of our genetic makeup becomes easier to determine, some people may attempt to use this information for profit or political power. This is a real concern because some health insurance companies refuse to insure people with "preexisting conditions" or those at "genetic risk" for certain abnormalities. They fail to realize that in people between the ages of five and fifty, such "conditions" or mutations are normal. Refusing to provide coverage would save these companies the expense of future medical bills incurred by "less-than-perfect" people. Another fear is that attempts may be made to "breed out" certain genes and people from the human population to create a "perfect race."

Here are some other intriguing findings from the human genome and the genome identification projects of other organisms:

- Hundreds of human genes appear likely to have resulted from transfer of genes from bacteria to humans at some point in the vertebrate lineage.
- Human genes are not scattered at random among the human chromosomes. "Forests" or "clusters" of genes are found on certain chromosomes, separated by "deserts" of genes. For example, chromosomes 17 and 19 are forested with thousands of genes, while chromosome 18 has many fewer genes.
- Humans are about 99.9 percent identical at the DNA level! Scientists believe that there is virtually no scientific

reason for the concept of "race" since there is much greater variation within a so-called race than there is between the so-called races.

- Rice appears to have about 50,000 genes.
- Chimpanzees have 98 to 99 percent of the same genes as humans.
- Roundworms have about 26,000 genes.
- Fruits flies contain an estimated 13,600 genes.
- Yeast cells have about 6,241 genes.
- There are numerous and virtually identical genes found in many organisms that appear to be very distantly related—for example, mice, humans, and yeasts.
- Genes jump around (transposons) within the chromosomes more than scientists ever thought.
- The mutation rate of male humans is about twice that of females.

In the last twenty-five years, the genome sequences of hundreds of viruses and viroids, naturally occurring plasmids, organelles, Eubacteria, Archaea, fungi, animals, and plants have been deciphered. This is only the beginning. With the completing of the HGP, scientists look forward to:

- Determining what makes one person more or less susceptible to disease than another;
- Identifying exactly what makes one person different from another;
- Developing tests that can reveal genetic disease or increased susceptibility to disease;
- Developing drugs to treat such diseases; and
- Developing gene therapies to replace the causative gene in an individual.

time-consuming process. A frequently used alternative is to use enzymes to cut the DNA in specific places to create DNA fragments of different lengths, separate out DNA fragments by size, and analyze these for similarities and differences. This second approach scans bigger stretches of DNA more quickly but not as closely. This method can be used by biologists to better understand evolutionary relationships between what at first glance might be unrelated species. This second category

of research involves directly manipulating DNA using the more sophisticated techniques such as the polymerase chain reaction (PCR), genetic fingerprinting, and cloning.

The PCR Reaction

Both the cloning and sequencing strategies of biotechnology have been greatly aided by the development of a technique

Science and Society

Cloning—What This Term Really Means

The term **clone** (Greek word *klön* for "twig") refers to exact copies of biological entities such as genes, cells, or organisms. The term refers to the outcome, not the way that the results are achieved. Many whole organisms "clone" themselves (e.g., bacteria undergo asexual binary fission, and many plants reproduce "clonally" through asexual means, sometimes referred to as *vegetative regeneration*). One process of making exact copies of the cells of a cat, sheep, or other organism in the laboratory is technically known as "*somatic cell nuclear transfer.*" The purpose of somatic cell nuclear transfer is not the "cloning" of a human or other adult but the establishment of a clonal cell line. In this technique, the nucleus of a somatic (body) cell of an adult donor is transferred into an oocyte, or "egg cell," whose own nucleus has been removed. Should such hybrid cells be cultured *in vitro* (*vitro* = glass; e.g., in test tubes) for a few weeks, the result is the production of a line of cloned cells (e.g., human embryonic stem cells). These can be used for a variety of medically valuable therapies or research (e.g., investigating certain genetic or other diseases). Using the other technique, *reproductive cloning,* the hybrid cell can be implanted into the uterus of a host or surrogate mother, where it may develop into a genetic copy or clone of the whole individual who donated the nucleus. A great variety of animals has already been cloned in this fashion, including cats, sheep, mice, monkeys, pigs, cattle, rabbits, mules, deer, horses, fish, and frogs.

In the film *Godsend,* starring Robert De Niro, Greg Kinnear, and Rebecca Romijn-Stamos, parents agree to clone their accidentally killed, eight-year-old son but wish they hadn't when the clone starts behaving erratically.

Questions to Discuss

1. Based on what you have learned, do you think cloning as depicted in *Godsend* would be possible?
2. What are the ethical and moral implications of taking such action?
3. Does it make a difference if the organism cloned is a human or a cat?

A   B

Box Figure 26.7 "CC," the cloned kitty, with surrogate mother and her genetic mother, "Rainbow." Out of 87 implanted cloned embryos, CC (*Copy Cat!*) is the only one to survive—comparable to the success rate in sheep, mice, cows, goats, and pigs. (*A*) Notice that she is completely unlike her tabby surrogate mother. (*B*) "Rainbow" is her genetic donor, and both are female calico domestic shorthair cats.

Source: Nature, AOP, published online: 14 February 2002; DOI: 10.1038/nature723.

---

called PCR. In 1989, the American Association for the Advancement of Science named DNA polymerase "Molecule of the Year." The value of this enzyme in the *polymerase chain reaction (PCR)* is so great that it could not be ignored. Just what is the PCR, how does it work, and what can you do with it?

PCR is a laboratory procedure for copying selected segments of DNA. A single cell can provide enough DNA for analysis and identification! Having a large number of copies of a "target sequence" of nucleotides enables biochemists to more easily work with DNA. This is like increasing the one "needle in the haystack" to such large numbers (one-hundred billion in only a matter of hours) that they are not hard to find, recognize, and work with. The types of specimens that can be used include semen, hair, blood, bacteria, protozoa, viruses, mummified tissue, and frozen cells. The process requires the DNA specimen, free DNA nucleotides, synthetic "primer" DNA, DNA polymerase, and simple lab equipment such as a test tube and a source of heat.

Having decided which target sequence of nucleotides (which "needle") is to be replicated, scientists heat the specimen of DNA to separate the coding and noncoding strands. Molecules of synthetic "primer" DNA are added to the specimen. These primer molecules are specifically designed to attach to the ends of the target sequence. Next, a mixture of nucleotides is added so that they can become the newly replicated DNA. The presence of the primer, attached to the DNA and added nucleotides, serves as the substrate for the DNA polymerase. Once added, the polymerase begins making its way down the length of the DNA from one attached primer end to the other. The enzyme bonds the new DNA nucleotides to the strand, replicating the molecule as it goes. It stops when it reaches the other end, having produced a new copy of the target sequence.

Because the DNA polymerase will continue to operate as long as enzymes and substrates are available, the process continues, and in a short time, there are billions of small pieces of DNA, all replicas of the target sequence.

So what, you say? Well, consider the following. Using the PCR, scientists have been able to:

1. More accurately diagnose such diseases as sickle-cell anemia, cancer, Lyme disease, AIDS, and Legionnaires' disease;
2. Perform highly accurate tissue typing for matching organ-transplant donors and recipients;
3. Help resolve criminal cases of rape, murder, assault, and robbery by matching a suspect's DNA to that found at the crime scene;
4. Detect specific bacteria in environmental samples;
5. Monitor the spread of genetically engineered microorganisms in the environment;
6. Check water quality by detecting bacterial contamination from feces;
7. Identify viruses in water samples;
8. Identify disease-causing protozoa in water;
9. Determine specific metabolic pathways and activities occurring in microorganisms;
10. Determine races, distribution patterns, kinships, migration patterns, evolutionary relationships, and rates of evolution of long-extinct species;
11. Accurately settle paternity suits;
12. Confirm identity in amnesia cases;
13. Identify a person as a relative for immigration purposes;
14. Provide the basis for making human antibodies in specific bacteria;
15. Possibly provide the basis for replicating genes that could be transplanted into individuals suffering from genetic diseases; and
16. Identify nucleotide sequences peculiar to the human genome (an application currently underway as part of the Human Genome Project).

Genetic Fingerprinting

With another genetic engineering accomplishment, *genetic fingerprinting*, it is possible to show the nucleotide sequence differences between individuals since no two people have the same nucleotide sequences. While this sounds like an easy task, the presence of many millions of base pairs in a person's chromosomes makes this a lengthy and sometimes impractical process. Therefore, scientists don't really do a complete fingerprint but focus only on certain shorter, repeating patterns in the DNA. By focusing on these shorter, repeating nucleotide sequences, it is possible to determine whether samples from two individuals have these same repeating segments.

Scientists use a small number of sequences that are known to vary a great deal among individuals and compare those to get a certain probability of a match. The more similar the sequences, the more likely the two samples are from the same person. The less similar the sequences, the less likely the two samples are from the same person. In criminal cases, DNA samples from the crime site can be compared to those taken from suspects. If 100 percent of the short, repeating sequence matches, it is highly probable that the suspect was at the scene of the crime and may be the guilty party. This same procedure can also be used to confirm the identity of a person, as in cases of amnesia, murder, or accidental death (figure 26.18).

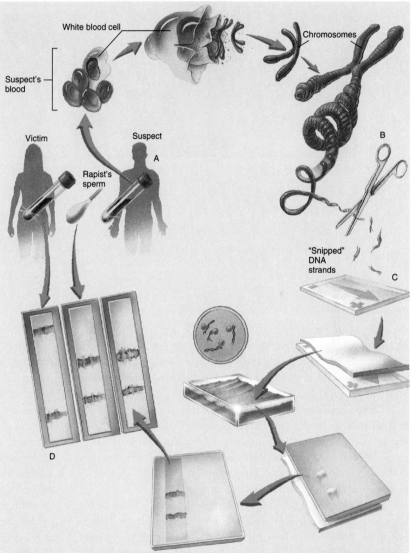

FIGURE 26.18 Because every person's DNA is unique (*A*), when samples of an individual's DNA are subjected to restriction enzymes, the cuts will occur in different places and DNA chunks of different sizes will result. (*B*) Restriction enzymes have the ability to cut DNA at places where specific sequences of nucleotides occur. When the chunks are caused to migrate across an electrophoresis gel (*C*), the smaller fragments migrate farther than the larger fragments, producing a pattern known as a "DNA fingerprint" (*D*). Because of individual differences in DNA sequences, these sites vary from person to person. As a result, the DNA fingerprint that separates DNA fragments on the basis of size can appear different from one person to another. Several controls are done in this type of experiment. One of the controls is the presence of the victim's DNA. Why is this important for the interpretation of this test?

Agarose Gel Electrophoresis

Agarose gel electrophoresis is a process used to separate molecules. Agarose is a complex polysaccharide derived from red algae. When mixed in solution, it forms a gel similar to "finger Jell-O." In the case of DNA, molecules of different size can be separated based on the fact that DNA carries an overall negative electric charge. It is possible to separate a mixture of various lengths of DNA using this method because the heavier segments will move slower than the lighter segments. The various segments will move or migrate through a sheet of gelatin from where the DNA is concentrated toward a positive pole. The DNA sample is placed at one end of an agarose gel tray, and a weak voltage is generated in the gel, usually no more than 5 volts per centimeter to the gel. The negatively charged DNA will then migrate slowly to the positive end. It becomes a race from one end of the tray to the other, with the smaller segments always winning. When the power source is shut off, the location of the DNA segments is determined by viewing the gel under ultraviolet light. The gel is then stained with ethidium bromide and the dark "blotches" in the gel can be cut out for purification and analysis.

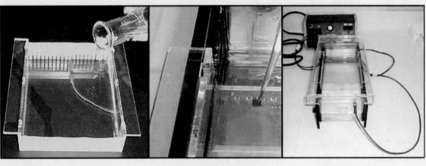

Pouring agarose   Loading a lane   Ready to run

Box Figure 26.8

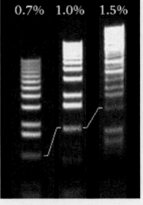

Box Figure 26.9

SUMMARY

Genes are units of heredity composed of specific lengths of *DNA* that determine the characteristics an organism displays. Specific genes are at specific loci on specific chromosomes. The *phenotype* displayed by an organism is the result of the effect of the environment on the ability of the genes to *express* themselves.

Diploid organisms have two genes for each characteristic. The alternative forms of genes for a characteristic are called *alleles*. There may be many different alleles for a particular characteristic. Organisms with two identical alleles are *homozygous* for a characteristic; those with different alleles are *heterozygous*. Some alleles are *dominant* over other alleles that are said to be *recessive*.

Sometimes two alleles express themselves, and often a gene has more than one recognizable effect on the phenotype of the organism. Some characteristics may be determined by several different pairs of alleles. In humans and some other organisms, phenotypes are determined not only by the type of allele (gene) received but also by the alternative ways in which they interact with each other and how the environment influences their expression. There are six generally recognized patterns: *multiple allelic, polygenic, pleiotropic, codominant, incomplete dominant,* and *X-linked characteristics.*

The successful operation of a living cell depends on its ability to accurately use the genetic information found in its DNA. The *enzymes* that can be synthesized using the information in DNA are responsible for the efficient control of a cell's metabolism. However, the production of protein molecules is under the control of the *nucleic acids,* the primary control molecules of the cell. The structure of the nucleic acids, *DNA* and *RNA,* determines the structure of the proteins, whereas the structure of the proteins determines their function in the cell's life cycle. *Protein synthesis* involves the decoding of the DNA into specific protein molecules and the use of the intermediate molecules, mRNA and tRNA, at the ribosome. Errors in any of the codons of these molecules may produce observable changes in the cell's functioning and can lead to cell death.

DNA replication results in an exact doubling of the genetic material. The process virtually guarantees that identical strands of DNA will be passed on to the next generation of cells.

Methods of manipulating DNA have led to the controlled transfer of genes from one kind of organism to another. This has made it possible for bacteria to produce a number of human gene products.

KEY TERMS

alleles (p. **657**)
anticodon (p. **674**)
chromosome (p. **668**)
clone (p. **680**)
codon (p. **672**)
complementary base (p. **668**)
DNA code (p. **671**)
DNA replication (p. **667**)
dominant allele (p. **658**)
gene (p. **656**)
genetically modified (GM)
   organisms (p. **676**)

genetics (p. **656**)
genome (p. **657**)
genotype (p. **657**)
heterozygous (p. **658**)
homozygous (p. **657**)
incomplete dominance
   (p. **662**)
law of dominance (p. **659**)
law of independent assortment
   (p. **659**)
law of segregation (p. **659**)
Mendelian genetics (p. **656**)

messenger RNA (mRNA)
   (p. **674**)
multiple alleles (p. **664**)
mutagenic agent (p. **674**)
mutation (p. **674**)
nucleic acids (p. **667**)
nucleotides (p. **667**)
phenotype (p. **657**)
pleiotropy (p. **664**)
polygenic inheritance
   (p. **664**)
Punnett square (p. **661**)

recessive allele (p. **658**)
recombinant DNA (p. **676**)
ribosomal RNA (rRNA)
   (p. **674**)
sickle-cell anemia (p. **675**)
single-factor cross (p. **660**)
transcription (p. **667**)
transfer RNA (tRNA)
   (p. **674**)
translation (p. **667**)
X-linked gene (p. **662**)

APPLYING THE CONCEPTS

1. An example of a phenotype is
   a. AB type blood.
   b. an allele for type A and B blood.
   c. a sperm with an allele for A blood.
   d. lack of iron in the diet causing anemia.

2. A person's height is determined by the interaction of numerous alleles. This is an example of
   a. autosomes.
   b. pleiotropy.
   c. single-factor crosses.
   d. polygenic inheritance.

3. In order for a recessive X-linked trait to appear in a female, she must inherit a recessive allele from
   a. neither parent.
   b. both parents.
   c. her father only.
   d. her mother only.

4. It was noticed that in certain flowers, if a flower had red petals, they were usually small petals. If the petals were white, pink, or orange, they were usually large petals. This could be the result of
   a. codominance.
   b. polygenic inheritance.
   c. incomplete dominance.
   d. None of these is correct.

5. A woman with blood type O and a man with blood type AB have a child together. What are the possible blood types of this child?
   a. AB or O
   b. A, B, or O
   c. A or B
   d. AB, A, B, or O

6. "She has really long fingers and toes and is exceptionally tall." This statement expresses the _____ of an individual.
   a. genotype
   b. phenotype
   c. monohybridization
   d. locus placement

7. One difference between mRNA and mature tRNA is the:
   a. kinds of polypeptide components.
   b. mRNA is straight while tRNA is cloverleaf-shaped.
   c. part of RNA from which they were coded.
   d. age of the molecules only.

8. While one strand of double-stranded DNA is being transcribed to mRNA:
   a. the complementary strand makes tRNA.
   b. the complementary strand is inactive.
   c. the complementary strand at this point is replicating.
   d. mutations are impossible during this short period.

9. A major difference between the genetic data of prokaryotic and eukaryotic cells is that in prokaryotes, the
   a. genes are RNA not DNA.
   b. histones are arranged differently.
   c. double-stranded DNA is circular.
   d. double-stranded DNA is absent in bacteria.

10. If the DNA gene strand has the base sequence CCA, which of the following DNA nitrogenous bases would pair with them during DNA replication?
    a. CCA   c. GGU
    b. GGU   d. GGT

11. Using the Amino Acid–mRNA Nucleic Acid Dictionary (table 26.4), the codon ACG would control the placement of which amino acid in the synthesis of a new protein?
    a. proline
    b. isoleucine
    c. glycine
    d. threonine

12. This new field of biotechnology examines the proteins that are predicted from the DNA sequence.
    a. human genome
    b. codon technology
    c. proteomics
    d. hormone therapy

Answers

1. a **2.** d **3.** b **4.** c **5.** c **6.** b **7.** b **8.** b **9.** c **10.** d **11.** d **12.** c

QUESTIONS FOR THOUGHT

1. What are the differences between DNA and RNA?
2. List the sequence of events that takes place when a DNA message is translated into protein.
3. Chromosomal and point mutations both occur in DNA. In what ways do they differ? How is this related to recombinant DNA?
4. How does DNA replication differ from the manufacture of an RNA molecule?
5. If a DNA nucleotide sequence is CATAAAGCA, what is the mRNA nucleotide sequence that would base-pair with it?
6. What amino acids would occur in the protein chemically coded by the sequence of nucleotides in question 5?
7. What is the probability of each of the following sets of parents producing the given genotypes in their offspring?

Parents		Offspring	Genotype
(a) AA	× aa		Aa
(b) Aa	× Aa		Aa
(c) Aa	× Aa		aa

8. If an offspring has the genotype Aa, what possible combinations of parental genotypes can exist?
9. In humans, the allele for albinism is recessive to the allele for normal skin pigmentation.
   a. What is the probability that a child of a heterozygous mother and father will be an albino?
   b. If a child is normal, what is the probability that it is a carrier of the recessive albino allele?
10. Dwarfism (achondroplasia) is the result of a mutation in the gene for the production of fibroblast growth factor. It is inherited as an autosomal dominant disorder and often occurs as a new mutation. What might be the explanation if two people of normal height and body stature have five children, one of whom displays achondroplasia?
11. A woman with color-deficient vision marries a man with normal vision. They have ten children—six boys and four girls.
   a. How many are expected to have normal vision?
   b. How many are expected to have color-deficient vision?
12. Hemophilia is a disease that prevents the blood from clotting normally. It is caused by a recessive allele located on the X chromosome. A boy has the disease; neither his parents nor his grandparents have the disease. What are the genotypes of his parents and grandparents?

FOR FURTHER ANALYSIS

1. There are two approaches to gene therapy: somatic cell and germ cell line therapy. Compare and contrast these two approaches. Give examples of genetic abnormalities that might be controlled with each method.
2. It is important to remember that Mendel's work was introduced only a little over a hundred years ago. But in that time, science has progressed from his most basic hypothesis to a level of understanding that has already enabled the control of some of our most notorious genetic abnormalities. List five such diseases. List the kinds of genetic medicine that are proposed to help with each.
3. What is the genetic basis of the following diseases and what are their symptoms?
   a. Kearns-Sayre syndrome
   b. Leber's hereditary optic neuropathy (LHON)
   c. Myoclonic epilepsy and red ragged fibers (MERRF)
4. Polydactyly is the most common genetic abnormality of the human hand and is classified into three types.
   a. What are these types?
   b. Diagram how such phenotypes would arise beginning with a mutation in the DNA, following through the central dogma and ending with how this phenotype would be produced.
5. It is known that the difference between the genomes of humans and chimpanzees is about 1.5 percent. If there is only this small difference between what appears to be two strikingly different organisms, what percentage difference do you think exists among humans? Based on this information, comment on the validity of separating humans of the world into separate races.

INVITATION TO INQUIRY

Smoking and Mutations: What Do You Really Know?

For decades, cancer has been empirically (based on commonplace experiences) linked to the use of tobacco, but only recently have researchers showed a cause-and-effect relationship between the two. Studies of patients with lung cancer revealed that 60 percent exhibited mutations in a gene known as p53. Use the Internet to answer the following questions:

1. What is the p53 gene?
2. What is apoptosis or "programmed cell death"?
3. What does the compound in tobacco (benzopyrene) have to do with this gene?
4. What is known about the p53 gene and lung cancer?

Mathematical Review

WORKING WITH EQUATIONS

Many of the problems of science involve an equation, a shorthand way of describing patterns and relationships that are observed in nature. Equations are also used to identify properties and to define certain concepts, but all uses have well-established meanings, symbols that are used by convention, and allowed mathematical operations. This appendix will assist you in better understanding equations and the reasoning that goes with the manipulation of equations in problem-solving activities.

Background

In addition to a knowledge of rules for carrying out mathematical operations, an understanding of certain quantitative ideas and concepts can be very helpful when working with equations. Among these helpful concepts are (1) the meaning of inverse and reciprocal, (2) the concept of a ratio, and (3) fractions.

The term *inverse* means the opposite, or reverse, of something. For example, addition is the opposite, or inverse, of subtraction, and division is the inverse of multiplication. A *reciprocal* is defined as an inverse multiplication relationship between two numbers. For example, if the symbol n represents any number (except zero), then the reciprocal of n is $1/n$. The reciprocal of a number $(1/n)$ multiplied by that number (n) always gives a product of 1. Thus, the number multiplied by 5 to give 1 is $1/5 (5 \times 1/5 = 5/5 = 1)$. So $1/5$ is the reciprocal of 5, and 5 is the reciprocal of $1/5$. Each number is the *inverse* of the other.

The fraction $1/5$ means 1 divided by 5, and if you carry out the division, it gives the decimal 0.2. Calculators that have a $1/x$ key will do the operation automatically. If you enter 5, then press the $1/x$ key, the answer of 0.2 is given. If you press the $1/x$ key again, the answer of 5 is given. Each of these numbers is a reciprocal of the other.

A *ratio* is a comparison between two numbers. If the symbols m and n are used to represent any two numbers, then the ratio of the number m to the number n is the fraction m/n. This expression means to divide m by n. For example, if m is 10 and n is 5, the ratio of 10 to 5 is 10/5, or 2:1.

Working with *fractions* is sometimes necessary in problem-solving exercises, and an understanding of these operations is needed to carry out unit calculations. It is helpful in many of these operations to remember that a number (or a unit) divided by itself is equal to 1; for example,

$$\frac{5}{5} = 1 \qquad \frac{\text{inch}}{\text{inch}} = 1 \qquad \frac{5 \text{ inches}}{5 \text{ inches}} = 1$$

When one fraction is divided by another fraction, the operation commonly applied is to "invert the denominator and multiply." For example, 2/5 divided by 1/2 is

$$\frac{\dfrac{2}{5}}{\dfrac{1}{2}} = \frac{2}{5} \times \frac{2}{1} = \frac{4}{5}$$

What you are really doing when you invert the denominator of the larger fraction and multiply is making the denominator (1/2) equal to 1. Both the numerator (2/5) and the denominator (1/2) are multiplied by 2/1, which does not change the value of the overall expression. The complete operation is

$$\frac{\dfrac{2}{5}}{\dfrac{1}{2}} \times \frac{\dfrac{2}{1}}{\dfrac{2}{1}} = \frac{\dfrac{2}{5} \times \dfrac{2}{1}}{\dfrac{1}{2} \times \dfrac{2}{1}} = \frac{\dfrac{4}{5}}{\dfrac{2}{2}} = \frac{\dfrac{4}{5}}{1} = \frac{4}{5}$$

Symbols and Operations

The use of symbols seems to cause confusion for some students because it seems different from their ordinary experiences with arithmetic. The rules are the same for symbols as they are for numbers, but you cannot do the operations with the symbols until you know what values they represent. The operation signs, such as $+$, $\div$, $\times$, and $-$, are used with symbols to indicate the operation that you *would* do if you knew the values. Some of the mathematical operations are indicated several ways. For example, $a \times b$, $a \cdot b$, and ab all indicate the same thing, that a is to be multiplied by b. Likewise, $a \div b$, a/b, and $a \times 1/b$ all indicate that a is to be divided by b. Since it is not possible to carry out the operations on symbols alone, they are called *indicated operations*.

Operations in Equations

An equation is a shorthand way of expressing a simple sentence with symbols. The equation has three parts: (1) a left side, (2) an equal sign $(=)$, which indicates the equivalence of the two sides, and (3) a right side. The left side has the same value and units as the right side, but the two sides may have a very different appearance. The two sides may also have the symbols that indicate mathematical operations ($+$, $-$, $\times$, and so forth) and may be in certain forms that indicate operations (a/b, ab, and so forth). In any case, the equation is a complete expression that states the left side has the same value and units as the right side.

Equations may contain different symbols, each representing some unknown quantity. In science, the expression "solve the equation" means to perform certain operations with one symbol (which represents some variable) by itself on one side of the equation. This single symbol is usually, but not necessarily, on the left side and is not present on the other side. For example, the equation $F = ma$ has the symbol F on the left side. In science, you would say that this equation is solved for F. It could also be solved for m or for a, which will be considered shortly. The equation $F = ma$ is solved for F, and the *indicated operation* is to multiply m by a because they are in the form ma, which means the same thing as $m \times a$. This is the only indicated operation in this equation.

A solved equation is a set of instructions that has an order of indicated operations. For example, the equation for the relationship between a Fahrenheit and Celsius temperature, solved for °C, is $C = 5/9(F - 32)$. A list of indicated operations in this equation is as follows:

1. Subtract 32° from the given Fahrenheit temperature.
2. Multiply the result of (1) by 5.
3. Divide the result of (2) by 9.

Why are the operations indicated in this order? Because the bracket means 5/9 of the *quantity* $(F - 32°)$. In its expanded form, you can see that $5/9(F - 32°)$ actually means $5/9(F) - 5/9(32°)$. Thus, you cannot multiply by 5 or divide by 9 until you have found the quantity of $(F - 32°)$. Once you have figured out the order of operations, finding the answer to a problem becomes almost routine as you complete the needed operations on both the numbers and the units.

Solving Equations

Sometimes it is necessary to rearrange an equation to move a different symbol to one side by itself. This is known as solving an equation for an unknown quantity. But you cannot simply move a symbol to one side of an equation. Since an equation is a statement of equivalence, the right side has the same value as the left side. If you move a symbol, you must perform the operation in a way that the two sides remain equivalent. This is accomplished by "canceling out" symbols until you have the unknown on one side by itself. One key to understanding the canceling operation is to remember that a fraction with the same number (or unit) over itself is equal to 1. For example, consider the equation $F = ma$, which is solved for F. Suppose you are considering a problem in which F and m are given, and the unknown is a. You need to solve the equation for a so it is on one side by itself. To eliminate the m, you do the *inverse* of the indicated operation on m, dividing both sides by m. Thus,

Solve an Equation

The equation for finding the kinetic energy of a moving body is $KE = 1/2mv^2$. You need to solve this equation for the velocity, v.

Answer

The order of indicated operations in the equation is as follows:

1. Square v.
2. Multiply v^2 by m.
3. Divide the result of (2) by 2.

To solve for v, this order is *reversed* as the "canceling operations" are used:

Step 1: Multiply both sides by 2

$$KE = \frac{1}{2}mv^2$$

$$2\,KE = \frac{2}{2}mv^2$$

$$2\,KE = mv^2$$

Step 2: Divide both sides by m

$$\frac{2\,KE}{m} = \frac{mv^2}{m}$$

$$\frac{2\,KE}{m} = v^2$$

Step 3: Take the square root of both sides

$$\sqrt{\frac{2\,KE}{m}} = \sqrt{v^2}$$

$$\sqrt{\frac{2\,KE}{m}} = v$$

or

$$v = \sqrt{\frac{2\,KE}{m}}$$

The equation has been solved for v, and you are now ready to substitute quantities and perform the needed operations.

$$F = ma$$

$$\frac{F}{m} = \frac{ma}{m}$$

$$\frac{F}{m} = a$$

Since m/m is equal to 1, the a remains by itself on the right side. For convenience, the whole equation may be flipped to move the unknown to the left side,

$$a = \frac{F}{m}$$

Thus, a quantity that indicated a multiplication (ma) was removed from one side by an inverse operation of dividing by m.

Consider the following inverse operations to "cancel" a quantity from one side of an equation, moving it to the other side:

If the Indicated Operation of the Symbol You Wish to Remove Is:	Perform This Inverse Operation on Both Sides of the Equation
multiplication	division
division	multiplication
addition	subtraction
subtraction	addition
squared	square root
square root	square

SIGNIFICANT FIGURES

The numerical value of any measurement will always contain some uncertainty. Suppose, for example, that you are measuring one side of a square piece of paper as shown in figure A.1. You could say that the paper is *about* 3.5 cm wide and you would be correct. This measurement, however, would be unsatisfactory for many purposes. It does not approach the true value of the length and contains too much uncertainty. It seems clear that the paper width is larger than 3.4 cm but shorter than 3.5 cm. But how much larger than 3.4 cm? You cannot be certain if the paper is 3.44, 3.45, or 3.46 cm wide. As your best estimate, you might say that the paper is 3.45 cm wide. Everyone would agree that you can be certain about the first two numbers (3.4) and they should be recorded. The last number (0.05) has been estimated and is not certain. The two certain numbers, together with one uncertain number, represent the greatest accuracy possible with the ruler being used. The paper is said to be 3.45 cm wide.

A *significant figure* is a number that is believed to be correct with some uncertainty only in the last digit. The value of the width of the paper, 3.45 cm, represents three significant figures. As you can see, the number of significant figures can be determined by the degree of accuracy of the measuring instrument being used. But suppose you need to calculate the area of the paper. You would multiply 3.45 cm × 3.45 cm and the product for the area

would be 11.9025 cm². This is a greater accuracy than you were able to obtain with your measuring instrument. The result of a calculation can be no more accurate than the values being treated. Because the measurement had only three significant figures (two certain, one uncertain), then the answer can have only three significant figures. The area is correctly expressed as 11.9 cm².

There are a few simple rules that will help you determine how many significant figures are contained in a reported measurement:

1. All digits reported as a direct result of a measurement are significant.
2. Zero is significant when it occurs between nonzero digits. For example, 607 has three significant figures, and the zero is one of the significant figures.
3. In figures reported as *larger than the digit one,* the digit zero is not significant when it follows a nonzero digit to indicate place. For example, in a report that "23,000 people attended the rock concert," the digits 2 and 3 are significant but the zeros are not significant. In this situation, the 23 is the measured part of the figure, and the three zeros tell you an estimate of how many attended the concert, that is, 23 thousand. If the figure is a measurement rather than an estimate, then it is written *with a decimal point after the last zero* to indicate that the zeros *are* significant. Thus, 23,000 has *two* significant figures (2 and 3), but 23,000. has *five* significant figures. The figure 23,000 means "about 23 thousand," but 23,000. means 23,000. and not 22,999 or 23,001.
4. In figures reported as *smaller than the digit 1,* zeros after a decimal point that come before nonzero digits *are not* significant and serve only as place holders. For example, 0.0023 has two significant figures, 2 and 3. Zeros alone after a decimal point or zeros after a nonzero digit indicate a measurement, however, so these zeros *are* significant. The figure 0.00230, for example, has three significant figures since the 230 means 230 and not 229 or 231. Likewise, the figure 3.000 cm has four significant figures because the presence of the three zeros means that the measurement was actually 3.000 and not 2.999 or 3.001.

Multiplication and Division

When multiplying or dividing measurement figures, the answer may have no more significant figures than the *least* number of significant figures in the figures being multiplied or divided. This simply means that an answer can be no more accurate than the least accurate measurement entering into the calculation and that you cannot improve the accuracy of a measurement by doing a calculation. For example, in multiplying 54.2 mi/h × 4.0 h to find out the total distance traveled, the first figure (54.2) has three significant figures but the second (4.0) has only two significant figures. The answer can contain only two significant figures since this is the weakest number of those involved in the calculation. The correct answer is therefore 220 mi, not 216.8 mi. This may seem strange since multiplying the two numbers together gives the answer of 216.8 mi. This answer, however, means a greater accuracy than is possible, and the accuracy cannot be improved over the weakest number involved in the calculation. Since the

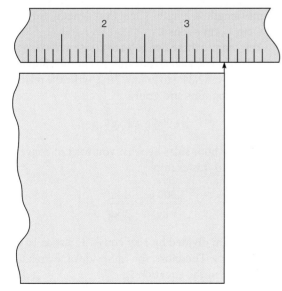

FIGURE A.1

weakest number (4.0) has only two significant figures the answer must also have only two significant figures, which is 220 mi.

The result of a calculation is *rounded* to have the same least number of significant figures as the least number of a measurement involved in the calculation. When rounding numbers, the last significant figure is increased by 1 if the number after it is 5 or larger. If the number after the last significant figure is 4 or less, the nonsignificant figures are simply dropped. Thus, if two significant figures are called for in the answer of the previous example, 216.8 is rounded up to 220 because the last number after the two significant figures is 6 (a number larger than 5). If the calculation result had been 214.8, the rounded number would be 210 miles.

Note that *measurement figures* are the only figures involved in the number of significant figures in the answer. Numbers that are counted or **defined** are not included in the determination of significant figures in an answer. For example, when dividing by 2 to find an average, the 2 is ignored when considering the number of significant figures. Defined numbers are defined exactly and are not used in significant figures. Since 1 kilogram is *defined* to be exactly 1,000 grams, such a conversion is not a measurement.

Addition and Subtraction

Addition and subtraction operations involving measurements, as with multiplication and division, cannot result in an answer that implies greater accuracy than the measurements had before the calculation. Recall that the last digit to the right in a measurement is uncertain; that is, it is the result of an estimate. The answer to an addition or subtraction calculation can have this uncertain number *no farther from the decimal place than it was in the weakest number involved in the calculation.* Thus, when 8.4 is added to 4.926, the weakest number is 8.4, and the uncertain number is .4, one place to the right of the decimal. The sum of 13.326 is therefore rounded to 13.3, reflecting the placement of this weakest doubtful figure.

The rules for counting zeros tell us that the numbers 203 and 0.200 both have three significant figures. Likewise, the numbers 230 and 0.23 only have two significant figures. Once you remember the rules, the counting of significant figures is straightforward. On the other hand, sometimes you find a

---

CONCEPTS APPLIED

Significant Figures

Multiplication. In a problem it is necessary to multiply 0.0039 km by 15.0 km. The result from a calculator is 0.0585 km^2. The least number of significant figures involved in this calculation is two (0.0039 is two significant figures; 15.0 is three—read the rules again to see why). The calculator result is therefore rounded off to have only two significant figures, and the answer is recorded as 0.059 km^2.

Addition. The quantities of 10.3 calories, 10.15 calories, and 16.234 calories are added. The result from a calculator is 36.684 calories. The smallest number of decimal points is one digit to the right of the decimal, so the answer is rounded to 36.7 calories.

---

number that seems to make it impossible to follow the rules. For example, how would you write 3,000 with two significant figures? There are several special systems in use for taking care of problems such as this, including the placing of a little bar over the last significant digit. One of the convenient ways of showing significant figures for difficult numbers is to use scientific notation, which is also discussed in the "Scientific Notation" section in this appendix. The convention for writing significant figures is to display one digit to the left of the decimal. The exponents are not considered when showing the number of significant figures in scientific notation. Thus, if you want to write three thousand showing one significant figure, you would write 3×10^3. To show two significant figures, it is 3.0×10^3, and for three significant figures it becomes 3.00×10^3. As you can see, the correct use of scientific notation leaves little room for doubt about how many significant figures are intended.

CONVERSION OF UNITS

The measurement of most properties results in both a numerical value and a unit. The statement that a glass contains 50 cm^3 of a liquid conveys two important concepts: the numerical value of 50 and the referent unit of cubic centimeters. Both the numerical value and the unit are necessary to communicate correctly the volume of the liquid.

When working with calculations involving measurement units, *both* the numerical value and the units are treated mathematically. As in other mathematical operations, there are general rules to follow.

1. Only properties with *like units* may be added or subtracted. It should be obvious that adding quantities such as 5 dollars and 10 dimes is meaningless. You must first convert to like units before adding or subtracting.
2. Like or unlike units may be multiplied or divided and treated in the same manner as numbers. You have used this rule when dealing with area (length $\times$ length = length^2, for example, cm $\times$ cm = cm^2) and when dealing with volume (length $\times$ length $\times$ length = length^3, for example, cm $\times$ cm $\times$ cm = cm^3).

You can use these two rules to create a *conversion ratio* that will help you change one unit to another. Suppose you need to convert 2.3 kg to grams. First, write the relationship between kilograms and grams:

$$1,000 \text{ g} = 1 \text{ kg}$$

Next, divide both sides by what you wish to convert *from* (kilograms in this example):

$$\frac{1,000 \text{ g}}{1 \text{ kg}} = \frac{1 \text{ kg}}{1 \text{ kg}}$$

One kilogram divided by 1 kg equals 1, just as 10 divided by 10 equals 1. Therefore, the right side of the relationship becomes 1 and the equation is:

$$\frac{1,000 \text{ g}}{1 \text{ kg}} = 1$$

A distance is reported as 100.0 km, and you want to know how far this is in miles.

Answer

First, you need to obtain a *conversion factor* from a textbook or reference book, which usually lists the conversion factors by properties in a table. Such a table will show two conversion factors for kilometers and miles: (1) 1 km = 0.621 mi and (2) 1 mi = 1.609 km. You select the factor that is in the same form as your problem; for example, your problem is 100.0 km = ? mi. The conversion factor in this form is 1 km = 0.621 mi.

Second, you convert this conversion factor into a *conversion ratio* by dividing the factor by what you wish to convert *from*:

conversion factor : $\qquad$ 1 km = 0.621 mi

divide factor by what you
want to convert from : $\qquad \dfrac{1 \text{ km}}{1 \text{ km}} = \dfrac{0.621 \text{ mi}}{1 \text{ km}}$

resulting conversion rate : $\qquad \dfrac{0.621 \text{ mi}}{\text{km}}$

Note that if you had used the 1 mi = 1.609 km factor, the resulting units would be meaningless. The conversion ratio is now multiplied by the numerical value *and unit* you wish to convert:

$$100.0 \text{ km} \times \frac{0.621 \text{ mi}}{\text{km}}$$

$$(100.0)(0.621) \frac{\text{km} \cdot \text{mi}}{\text{km}}$$

$$0.621 \text{ mi}$$

The 1 is usually understood, that is, not stated, and the operation is called *canceling*. Canceling leaves you with the fraction 1,000 g/1 kg, which is a conversion ratio that can be used to convert from kilograms to grams. You simply multiply the conversion ratio by the numerical value and unit you wish to convert:

$$= 2.3 \text{ kg} \times \frac{1,000 \text{ g}}{1 \text{ kg}}$$

$$= \frac{2.3 \times 1,000}{1} \frac{\text{kg} \times \text{g}}{\text{kg}}$$

$$= \boxed{2,300 \text{ g}}$$

The kilogram units cancel. Showing the whole operation with units only, you can see how you end up with the correct unit of grams:

$$\text{kg} \times \frac{\text{g}}{\text{kg}} = \frac{\text{kg} \cdot \text{g}}{\text{kg}} = \text{g}$$

Since you did obtain the correct unit, you know that you used the correct conversion ratio. If you had blundered and used an inverted conversion ratio, you would obtain

$$2.3 \text{ kg} \times \frac{1 \text{ kg}}{1,000 \text{ g}} = .0023 \frac{\text{kg}^2}{\text{g}}$$

which yields the meaningless, incorrect units of kg²/g. Carrying out the mathematical operations on the numbers and the units will always tell you whether or not you used the correct conversion ratio.

SCIENTIFIC NOTATION

Most of the properties of things that you might measure in your everyday world can be expressed with a small range of numerical values together with some standard unit of measure. The range of numerical values for most everyday things can be dealt with by using units (1s), tens (10s), hundreds (100s), or perhaps thousands (1,000s). But the actual universe contains some objects of incredibly large size that require some very big numbers to describe. The Sun, for example, has a mass of about 1,970,000,000,000,000,000,000,000,000,000 kg. On the other hand, very small numbers are needed to measure the size and parts of an atom. The radius of a hydrogen atom, for example, is about 0.00000000005 m. Such extremely large and small numbers are cumbersome and awkward since there are so many zeros to keep track of, even if you are successful in carefully counting all the zeros. A method does exist to deal with extremely large or small numbers in a more condensed form. The method is called *scientific notation*, but it is also sometimes called *powers of ten* or *exponential notation*, since it is based on exponents of 10. Whatever it is called, the method is a compact way of dealing with numbers that not only helps you keep track of zeros but provides a simplified way to make calculations as well.

In algebra you save a lot of time (as well as paper) by writing $(a \times a \times a \times a \times a)$ as a^5. The small number written to the right and above a letter or number is a superscript called an *exponent*. The exponent means that the letter or number is to be multiplied by itself that many times; for example, a^5 means a multiplied by itself five times; or $a \times a \times a \times a \times a$. As you can see, it is much easier to write the exponential form of this operation than it is to write it out in the long form. Scientific notation uses an exponent to indicate the power of the base 10. The exponent tells how many times the base, 10, is multiplied by itself. For example,

$$10,000 = 10^4$$
$$1,000 = 10^3$$
$$100 = 10^2$$
$$10 = 10^1$$
$$1 = 10^0$$
$$0.1 = 10^{-1}$$
$$0.01 = 10^{-2}$$
$$0.001 = 10^{-3}$$
$$0.0001 = 10^{-4}$$

This table could be extended indefinitely, but this somewhat shorter version will give you an idea of how the method works. The symbol 10^4 is read as "ten to the fourth power" and means $10 \times 10 \times 10 \times 10$. Ten times itself four times is 10,000, so 10^4 is the scientific notation for 10,000. It is also equal to the number of zeros between the 1 and the decimal point; that is, to write the longer form of 10^4, you simply write 1, then move the decimal point four places to the *right*—10 to the fourth power is 10,000.

Scientific Notation Expression

What is 26,000,000 in scientific notation?

Answer

Count how many times you must shift the decimal point until one digit remains to the left of the decimal point. For numbers larger than the digit 1, the number of shifts tells you how much the exponent is increased, so the answer is

$$2.6 \times 10^7$$

which means the coefficient 2.6 is multiplied by 10 seven times.

The power of ten table also shows that numbers smaller than 1 have negative exponents. A negative exponent means a reciprocal:

$$10^{-1} = \frac{1}{10} = 0.1$$

$$10^{-2} = \frac{1}{100} = 0.01$$

$$10^{-3} = \frac{1}{1000} = 0.001$$

To write the longer form of 10^{-4}, you simply write 1, then move the decimal point four places to the *left;* 10 to the negative fourth power is 0.0001.

Scientific notation usually, but not always, is expressed as the product of two numbers: (1) a number between 1 and 10 that is called the *coefficient* and (2) a power of ten that is called the *exponent*. For example, the mass of the Sun that was given in long form earlier is expressed in scientific notation as

$$1.97 \times 10^{30} \text{ kg}$$

and the radius of a hydrogen atom is

$$5.0 \times 10^{-11} \text{ m}$$

In these expressions, the coefficients are 1.97 and 5.0, and the power of ten notations are the exponents. Note that in both of these examples, the exponent tells you where to place the decimal point if you wish to write the number all the way out in the long form. Sometimes scientific notation is written without a coefficient, showing only the exponent. In these cases, the coefficient of 1.0 is understood, that is, not stated. If you try to enter a scientific notation in your calculator, however, you will need to enter the understood 1.0, or the calculator will not be able to function correctly. Note also that 1.97×10^{30} kg and the expressions 0.197×10^{31} kg and 19.7×10^{29} kg are all correct expressions of the mass of the Sun. By convention, however, you will use the form that has one digit to the left of the decimal.

It was stated earlier that scientific notation provides a compact way of dealing with very large or very small numbers, but it provides a simplified way to make calculations as well. There are a few mathematical rules that will describe how the use of scientific notation simplifies these calculations.

To *multiply* two scientific notation numbers, the coefficients are multiplied as usual, and the exponents are *added*

Scientific Notation Calculation

Solve the following problem concerning scientific notation:

$$\frac{(2 \times 10^4) \times (8 \times 10^{-6})}{8 \times 10^4}$$

Answer

First, separate the coefficients from the exponents,

$$\frac{2 \times 8}{8} \times \frac{10^4 \times 10^{-6}}{10^4}$$

then multiply and divide the coefficients and add and subtract the exponents as the problem requires,

$$2 \times 10^{\{[(4)+(-6)]=[4]\}}$$

Solving the remaining additions and subtractions of the coefficients gives

$$2 \times 10^{-6}$$

algebraically. For example, to multiply (2×10^2) by (3×10^3), first separate the coefficients from the exponents,

$$(2 \times 3) \times (10^2 \times 10^3),$$

then multiply the coefficients and add the exponents,

$$6 \times 10^{(2+3)} = 6 \times 10^5$$

Adding the exponents is possible because $10^2 \times 10^3$ means the same thing as $(10 \times 10) \times (10 \times 10 \times 10)$, which equals $(100) \times (1,000)$, or 100,000, which is expressed as 10^5 in scientific notation. Note that two negative exponents add algebraically, for example $10^{-2} \times 10^{-3} = 10^{[(-2)+(-3)]} = 10^{-5}$. A negative and a positive exponent also add algebraically, as in $10^5 \times 10^{-3} = 10^{[(+5)+(-3)]} = 10^2$.

If the result of a calculation involving two scientific notation numbers does not have the conventional one digit to the left of the decimal, move the decimal point so it does, changing the exponent according to which way and how much the decimal point is moved. Note that the exponent increases by one number for each decimal point moved to the left. Likewise, the exponent decreases by one number for each decimal point moved to the right. For example, $938. \times 10^3$ becomes 9.38×10^5 when the decimal point is moved two places to the left.

To *divide* two scientific notation numbers, the coefficients are divided as usual and the exponents are *subtracted*. For example, to divide (6×10^6) by (3×10^2), first separate the coefficients from the exponents,

$$(6 \div 3) \times (10^6 \div 10^2)$$

then divide the coefficients and subtract the exponents,

$$2 \times 10^{(6-2)} = 2 \times 10^4$$

Note that when you subtract a negative exponent, for example, $10^{[(3)-(-2)]}$, you change the sign and add, $10^{(3+2)} = 10^5$.

Appendix B

Solubilities Chart

	Acetate	Bromide	Carbonate	Chloride	Fluoride	Hydroxide	Iodide	Nitrate	Oxide	Phosphate	Sulfate	Sulfide
Aluminum	S	S	—	S	s	i	S	S	i	i	S	d
Ammonium	S	S	S	S	S	S	S	S	—	S	S	S
Barium	S	S	i	S	s	S	S	S	S	i	i	d
Calcium	S	S	i	S	i	s	S	S	s	i	s	d
Copper(I)	—	s	i	s	i	—	i	—	i	—	d	i
Copper(II)	S	S	i	S	S	i	S	S	i	i	S	i
Iron(II)	S	S	i	S	s	i	S	S	i	i	S	i
Iron(III)	S	S	i	S	s	i	S	S	i	i	S	d
Lead	S	s	i	s	i	i	s	S	i	i	i	i
Magnesium	S	S	i	S	i	i	S	S	i	i	S	d
Mercury(I)	s	i	i	i	d	d	i	S	i	i	i	i
Mercury(II)	S	s	i	S	d	i	i	S	i	i	i	i
Potassium	S	S	S	S	S	S	S	S	S	S	S	i
Silver	s	i	i	i	S	—	i	S	i	i	i	i
Sodium	S	S	S	S	S	S	S	S	d	S	S	S
Strontium	S	S	s	S	i	s	S	S	—	i	i	i
Zinc	S	S	i	S	S	i	S	S	i	i	S	i

S = soluble
i = insoluble
s = slightly soluble
d = decomposes

Relative Humidity Table

Dry-Bulb Temperature (°C)	Difference Between Wet-Bulb and Dry-Bulb Temperatures (° C)																			
	1	2	3	4	5	6	7	8	9	10	11	12	13	14	15	16	17	18	19	20
0	81	64	46	29	13															
1	83	66	49	33	17															
2	84	68	52	37	22	7														
3	84	70	55	40	26	12														
4	86	71	57	43	29	16														
5	86	72	58	45	33	20	7													
6	86	73	60	48	35	24	11													
7	87	74	62	50	38	26	15													
8	87	75	63	51	40	29	19	8												
9	88	76	64	53	42	32	22	12												
10	88	77	66	55	44	34	24	15	6											
11	89	78	67	56	46	36	27	18	9											
12	89	78	68	58	48	39	29	21	12											
13	89	79	69	59	50	41	32	23	15	7										
14	90	79	70	60	51	42	34	26	18	10										
15	90	80	71	61	53	44	36	27	20	13	6									
16	90	81	71	63	54	46	38	30	23	15	8									
17	90	81	72	64	55	47	40	32	25	18	11									
18	91	82	73	65	57	49	41	34	27	20	14	7								
19	91	82	74	65	58	50	43	36	29	22	16	10								
20	91	83	74	66	59	51	44	37	31	24	18	12	6							
21	91	83	75	67	60	53	46	39	32	26	20	14	9							
22	92	83	76	68	61	54	47	40	34	28	22	17	11	6						
23	92	84	76	69	62	55	48	42	36	30	24	19	13	8						
24	92	84	77	69	62	56	49	43	37	31	26	20	15	10	5					
25	92	84	77	70	63	57	50	44	39	33	28	22	17	12	8					
26	92	85	78	71	64	58	51	46	40	34	29	24	19	14	10	5				
27	92	85	78	71	65	58	52	47	41	36	31	26	21	16	12	7				
28	93	85	78	72	65	59	53	48	42	37	32	27	22	18	13	9	5			
29	93	86	79	72	66	60	54	49	43	38	33	28	24	19	15	11	7			
30	93	86	79	73	67	61	55	50	44	39	35	30	25	21	17	13	9	5		
31	93	86	80	73	67	61	56	51	45	40	36	31	27	22	18	14	11	7		
32	93	86	80	74	68	62	57	51	46	41	37	32	28	24	20	16	12	9	5	
33	93	87	80	74	68	63	57	52	47	42	38	33	29	25	21	17	14	10	7	
34	93	87	81	75	69	63	58	53	48	43	39	35	30	28	23	19	15	12	8	5
35	94	87	81	75	69	64	59	54	49	44	40	36	32	28	24	20	17	13	10	7

Appendix D

Solutions for Group A Parallel Exercises

Note: Solutions that involve calculations of measurements are rounded up or down to conform to the rules for significant figures described in appendix A.

Chapter 1

1.1 Answers will vary but should have the relationship of 100 cm in 1 m, for example, 178 cm = 1.78 m.

1.2 Since density is given by the relationship $\rho = m/V$, then

$$\rho = \frac{m}{V} = \frac{272\,g}{20.0\,cm^3}$$

$$= \frac{272}{20.0}\,\frac{g}{cm^3}$$

$$= \boxed{13.6\,\frac{g}{cm^3}}$$

1.3 The volume of a sample of lead is given and the problem asks for the mass. From the relationship of $\rho = m/V$, solving for the mass (m) tells you that the density (ρ) times the volume (V) equals the mass, or $m = \rho V$. The density of lead, 11.4 g/cm^3, can be obtained from table 1.3, so

$$\rho = \frac{m}{V}$$

$$V\rho = \frac{m\cancel{V}}{\cancel{V}}$$

$$m = \rho V$$

$$m = \left(11.4\,\frac{g}{cm^3}\right)(10.0\,cm^3)$$

$$11.4 \times 10.0\,\frac{g}{cm^3} \times cm^3$$

$$114\,\frac{g \cdot \cancel{cm^3}}{\cancel{cm^3}}$$

$$= \boxed{114\,g}$$

1.4 Solving the relationship $\rho = m/V$ for volume gives $V = m/\rho$, and

$$\rho = \frac{m}{V}$$

$$V\rho = \frac{m\cancel{V}}{\cancel{V}}$$

$$\frac{V\cancel{\rho}}{\cancel{\rho}} = \frac{m}{\rho}$$

$$V = \frac{m}{\rho}$$

$$V = \frac{600\,g}{300\,\dfrac{g}{cm^3}}$$

$$= \frac{600}{3.00}\,\frac{g}{1} \times \frac{cm^3}{g}$$

$$= 200\,\frac{\cancel{g} \cdot cm^3}{\cancel{g}}$$

$$= \boxed{200\,cm^3}$$

1.5 A 50.0 cm^3 sample with a mass of 34.0 grams has a density of

$$\rho = \frac{m}{V} = \frac{34.0\,g}{50.0\,cm^3}$$

$$= \frac{34.0}{50.0}\,\frac{g}{cm^3}$$

$$= \boxed{0.680\,\frac{g}{cm^3}}$$

According to table 1.3, 0.680 g/cm^3 is the density of gasoline, so the substance must be gasoline.

1.6 The problem asks for a mass and gives a volume, so you need a relationship between mass and volume. Table 1.3 gives the density of water as 1.00 g/cm^3, which is a density that is easily remembered. The volume is given in liters (L), which should first be converted to cm^3 because this is the unit in which density is expressed. The relationship of $\rho = m/V$ solved for mass is ρV, so the solution is

$$\rho = \frac{m}{V} \therefore m = \rho V$$

$$m = \left(1.00 \frac{g}{cm^3}\right)(40,000 \text{ cm}^3)$$

$$= 1.00 \times 40,000 \frac{g}{cm^3} \times cm^3$$

$$= 40,000 \frac{g \cdot \cancel{cm^3}}{\cancel{cm^3}}$$

$$= 40,000 \text{ g}$$

$$= \boxed{40 \text{ kg}}$$

1.7 From table 1.3, the density of aluminum is given as 2.70 g/cm³. Converting 2.1 kg to the same units as the density gives 2,100 g. Solving $\rho = m/V$ for the volume gives

$$V = \frac{m}{\rho} = \frac{2,100 \text{ g}}{2.70 \dfrac{g}{cm^3}}$$

$$= \frac{2,100}{2.70} \frac{g}{1} \times \frac{cm^3}{g}$$

$$= 777.78 \frac{\cancel{g} \cdot cm^3}{\cancel{g}}$$

$$= \boxed{780 \text{ cm}^3}$$

1.8 The length of one side of the box is 0.1 m. Reasoning: Since the density of water is 1.00 g/cm³, then the volume of 1,000 g of water is 1,000 cm³. A cubic box with a volume of 1,000 cm³ is 10 cm (since 10 × 10 × 10 = 1,000). Converting 10 cm to m units, the cube is 0.1 m on each edge.

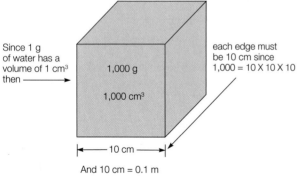

Since 1 g of water has a volume of 1 cm³ then ⟶ 1,000 g 1,000 cm³

each edge must be 10 cm since 1,000 = 10 X 10 X 10

⊢— 10 cm —⊣

And 10 cm = 0.1 m

FIGURE A.4

1.9 The relationship between mass, volume, and density is $\rho = m/V$. The problem gives a volume but not a mass. The mass, however, can be assumed to remain constant during the compression of the bread, so the mass can be obtained from the original volume and density, or

$$\rho = \frac{m}{V} \therefore m = \rho V$$

$$m = \left(0.2 \frac{g}{cm^3}\right)(3,000 \text{ cm}^3)$$

$$= 0.2 \times 3,000 \frac{g}{cm^3} \times cm^3$$

$$= 600 \frac{g \cdot \cancel{cm^3}}{\cancel{cm^3}}$$

$$= 600 \text{ g}$$

A mass of 600 g and the new volume of 1,500 cm³ means that the new density of the crushed bread is

$$\rho = \frac{m}{V}$$

$$= \frac{600 \text{ g}}{1,500 \text{ cm}^3}$$

$$= \frac{600}{1,500} \frac{g}{cm^3}$$

$$= \boxed{0.4 \frac{g}{cm^3}}$$

1.10 According to table 1.3, lead has a density of 11.4 g/cm³. Therefore, a 1.00 cm³ sample of lead would have a mass of

$$\rho = \frac{m}{V} \therefore m = \rho V$$

$$m = \left(11.4 \frac{g}{cm^3}\right)(1.00 \text{ cm}^3)$$

$$= 11.4 \times 1.00 \frac{g}{cm^3} \times cm^3$$

$$= 11.4 \frac{g \cdot \cancel{cm^3}}{\cancel{cm^3}}$$

$$= 11.4 \text{ g}$$

Also according to table 1.3, copper has a density of 8.96 g/cm³. To balance a mass of 11.4 g of lead, a volume of this much copper would be required:

$$\rho = \frac{m}{V} \therefore V = \frac{m}{\rho}$$

$$V = \frac{11.4 \text{ g}}{8.96 \dfrac{g}{cm^3}}$$

$$= \frac{11.4}{8.96} \frac{g}{1} \times \frac{cm^3}{g}$$

$$= 1.27232 \frac{\cancel{g} \cdot cm^3}{\cancel{g}}$$

$$= \boxed{1.27 \text{ cm}^3}$$

Chapter 2

2.1 Listing the quantities with their symbols, we can see the problem involves the quantities found in the definition of average speed:

$$\bar{v} = 350.0 \text{ m/s}$$
$$t = 5.00 \text{ s}$$
$$d = ?$$

$$\bar{v} = \frac{d}{t} \therefore d = \bar{v}t$$

$$d = \left(350.0 \frac{m}{s}\right)(5.00 \text{ s})$$

$$= (350.0)(5.00) \frac{m}{s} \times s$$

$$= \boxed{1,750 \text{ m}}$$

2.2 The initial velocity, final velocity, and time are known and the problem asked for the acceleration. Listing these quantities with their symbols, we have

$$v_i = 0 \text{ m/s}$$
$$v_f = 15.0 \text{ m/s}$$
$$t = 10.0 \text{ s}$$
$$a = ?$$

These are the quantities involved in the acceleration equation, which is already solved for the unknown:

$$a = \frac{v_f - v_i}{t}$$

$$a = \frac{15.0 \text{ m/s} - 0 \text{ m/s}}{10.0 \text{ s}}$$

$$= \frac{15.0}{10.0} \frac{\text{m}}{\text{s}} \times \frac{1}{\text{s}}$$

$$= \boxed{1.50 \frac{\text{m}}{\text{s}^2}}$$

2.3 The distance (d) and the time (t) quantities are given in the problem, and

$$\bar{v} = \frac{d}{t}$$

$$= \frac{285 \text{ mi}}{5.0 \text{ h}}$$

$$= \frac{285}{5.0} \frac{\text{mi}}{\text{h}}$$

$$= \boxed{57 \text{ mi/h}}$$

The units cannot be simplified further. Note two significant figures in the answer, which is the least number of significant figures involved in the division operation.

2.4 Listing the known and unknown quantities:

$$m = 40.0 \text{ kg}$$
$$a = 2.4 \text{ m/s}^2$$
$$F = ?$$

These are the quantities found in Newton's second law of motion, $F = ma$, which is already solved for force (F). Thus,

$$F = (40.0 \text{ kg})\left(2.4 \frac{\text{m}}{\text{s}^2}\right)$$

$$= 40.0 \times 2.4 \frac{\text{kg} \cdot \text{m}}{\text{s}^2}$$

$$= \boxed{96 \text{ N}}$$

2.5 List the known and unknown quantities for the first situation, using an unbalanced force of 18.0 N to give the object an acceleration of 3 m/s²:

$$F_1 = 18 \text{ N}$$
$$a_1 = 3 \text{ m/s}^2$$

For the second situation, we are asked to find the force needed for an acceleration of 10 m/s²:

$$a_2 = 10 \text{ m/s}^2$$
$$F = ?$$

These are the quantities of Newton's second law of motion, $F = ma$, but the mass appears to be missing. The mass can be found from

$$F = ma \therefore m_1 = \frac{F_1}{a_1}$$

$$= \frac{18 \dfrac{\text{kg} \cdot \text{m}}{\text{s}^2}}{3 \dfrac{\text{m}}{\text{s}^2}}$$

$$= \frac{18}{3} \frac{\text{kg} \cdot \text{m}}{\text{s}^2} \times \frac{\text{s}^2}{\text{m}}$$

$$= 6 \text{ kg}$$

Now that we have the mass, we can easily find the force needed for an acceleration of 10 m/s²:

$$F_2 = m_2 a_2$$

$$= (6 \text{ kg})\left(10 \frac{\text{m}}{\text{s}^2}\right)$$

$$= 6 \times 10 \frac{\text{kg} \cdot \text{m}}{\text{s}^2}$$

$$= \boxed{60 \text{ N}}$$

2.6 Listing the known and unknown quantities:

$$m = 70.0 \text{ kg}$$
$$g = 9.8 \text{ m/s}^2$$
$$w = ?$$
$$w = mg$$

$$= (70.0 \text{ kg})\left(9.8 \frac{\text{m}}{\text{s}^2}\right)$$

$$= 70.0 \times 9.8 \text{ kg} \times \frac{\text{m}}{\text{s}^2}$$

$$= 686 \frac{\text{kg} \cdot \text{m}}{\text{s}^2}$$

$$= \boxed{690 \text{ N}}$$

2.7 Listing the known and unknown quantities:

$$m = 100 \text{ kg}$$
$$v = 6 \text{ m/s}$$
$$p = ?$$

These are the quantities found in the equation for momentum, $p = mv$, which is already solved for momentum (p). Thus,

$$p = mv$$

$$= (100 \text{ kg})\left(6 \frac{\text{m}}{\text{s}}\right)$$

$$= \boxed{600 \frac{\text{kg} \cdot \text{m}}{\text{s}}}$$

(Note the lowercase p is the symbol used for momentum. This is one of the few cases where the English letter does not provide a clue about what it stands for. The units for momentum are also somewhat unusual for metric units since they do not have a name or single symbol to represent them.)

2.8 Listing the known and unknown quantities:

$$w = 13{,}720 \text{ N}$$
$$y = 91 \text{ km/h}$$
$$p = ?$$

The equation for momentum is $p = mv$, which is already solved for momentum (p). The weight unit must be first converted to a mass unit:

$$w = mg \therefore m = \frac{w}{g}$$

$$= \frac{13{,}720 \dfrac{\text{kg} \cdot \text{m}}{\text{s}^2}}{9.8 \dfrac{\text{m}}{\text{s}^2}}$$

$$= \frac{13{,}720}{9.8} \frac{\text{kg} \cdot \text{m}}{\text{s}^2} \times \frac{\text{s}^2}{\text{m}}$$

$$= 1{,}400 \text{ kg}$$

The km/h unit should next be converted to m/s. Using the conversion factor from inside the front cover:

$$\frac{0.2778 \dfrac{\text{m}}{\text{s}}}{1 \dfrac{\text{km}}{\text{h}}} \times 91 \frac{\text{km}}{\text{h}}$$

$$0.2778 \times 91 \frac{\text{m}}{\text{s}} \times \frac{\text{h}}{\text{km}} \times \frac{\text{km}}{\text{h}}$$

$$25.2798 \frac{\text{m}}{\text{s}}$$

$$25 \frac{\text{m}}{\text{s}}$$

Now, listing the converted known and unknown quantities:

$$m = 1{,}400 \text{ kg}$$
$$v = 25 \text{ m/s}$$
$$p = ?$$

and solving for momentum (p),

$$p = mv$$

$$= (1{,}400 \text{ kg})\left(25 \frac{\text{m}}{\text{s}}\right)$$

$$= \boxed{35{,}000 \ \frac{\text{kg} \cdot \text{m}}{\text{s}}}$$

2.9 Listing the known and unknown quantities:

Bullet $\rightarrow m = 0.015 \text{ kg}$　Rifle $\rightarrow m = 6 \text{ kg}$
Bullet $\rightarrow v = 200 \text{ m/s}$　Rifle $\rightarrow v = ? \text{ m/s}$

Note the mass of the bullet was converted to kilograms. This is a conservation of momentum question, where the bullet and rifle can be considered as a system of interacting objects:

$$\text{Bullet momentum} = -\text{rifle momentum}$$
$$(mv)_\text{b} = -(mv)_\text{r}$$
$$(mv)_\text{b} - (mv)_\text{r} = 0$$
$$(0.015 \text{ kg})\left(200 \frac{\text{m}}{\text{s}}\right) - (6 \text{ kg})v_\text{r} = 0$$
$$\left(3 \text{ kg} \cdot \frac{\text{m}}{\text{s}}\right) - (6 \text{ kg} \cdot v_\text{r}) = 0$$
$$\left(3 \text{ kg} \cdot \frac{\text{m}}{\text{s}}\right) = (6 \text{ kg} \cdot v_\text{r})$$

$$v_\text{r} = \frac{3 \text{ kg} \cdot \dfrac{\text{m}}{\text{s}}}{6 \text{ kg}}$$

$$= \frac{3}{6} \frac{\text{kg}}{1} \times \frac{1}{\text{kg}} \times \frac{\text{m}}{\text{s}}$$

$$= \boxed{0.5 \ \frac{\text{m}}{\text{s}}}$$

The rifle recoils with a velocity of 0.5 m/s.

2.10 A unit conversion is needed:

$$\left(90.0 \frac{\text{km}}{\text{h}}\right)\left(0.2778 \frac{\dfrac{\text{m}}{\text{s}}}{\dfrac{\text{km}}{\text{h}}}\right) = 25.0 \text{ m/s}$$

a. $F = ma \therefore m = \dfrac{F}{a}$ and $a = \dfrac{v_\text{f} - v_\text{i}}{t}$, so

$$m = \frac{F}{\dfrac{v_\text{f} - v_\text{i}}{t}} = \frac{5{,}000.0 \dfrac{\text{kg} \cdot \text{m}}{\text{s}^2}}{\dfrac{25.0 \text{ m/s} - 0}{5.0 \text{ s}}}$$

$$= \frac{5{,}000.0 \dfrac{\text{kg} \cdot \text{m}}{\text{s}^2}}{5.0 \dfrac{\text{m}}{\text{s}^2}}$$

$$= \frac{5{,}000.0}{5.0} \frac{\text{kg} \cdot \text{m}}{\text{s}^2} \times \frac{\text{s}^2}{\text{m}}$$

$$= 1{,}000 \frac{\text{kg} \cdot \text{m} \cdot \text{s}^2}{\text{m} \cdot \text{s}^2}$$

$$= \boxed{1.0 \times 10^3 \text{ kg}}$$

b. $w = mg$

$$= (1.0 \times 10^3 \text{ kg})\left(9.8 \frac{\text{m}}{\text{s}^2}\right)$$

$$= (1.0 \times 10^3)(9.8) \text{ kg} \times \frac{\text{m}}{\text{s}^2}$$

$$= 9.8 \times 10^3 \frac{\text{kg} \cdot \text{m}}{\text{s}^2}$$

$$= \boxed{9.8 \times 10^3 \text{ N}}$$

2.11

$$F = \frac{mv^2}{r}$$

$$= \frac{(0.20\,\text{kg})\left(3.0\,\dfrac{\text{m}}{\text{s}}\right)^2}{1.5\,\text{m}}$$

$$= \frac{(0.20\,\text{kg})\left(9.0\,\dfrac{\text{m}^2}{\text{s}^2}\right)}{1.5\,\text{m}}$$

$$= \frac{0.20 \times 9.0}{1.5}\,\frac{\text{kg}\cdot\text{m}^2}{\text{s}^2} \times \frac{1}{\text{m}}$$

$$= 1.2\,\frac{\text{kg}\cdot\text{m}\cdot\cancel{\text{m}}}{\text{s}^2\cdot\cancel{\text{m}}}$$

$$= \boxed{1.2\,\text{N}}$$

2.12 Newton's laws of motion consider the resistance to a change of motion, or mass, and not weight. The astronaut's mass is

$$w = mg \therefore m = \frac{w}{g}$$

$$= \frac{1{,}960.0\,\dfrac{\text{kg}\cdot\text{m}}{\text{s}^2}}{9.8\,\dfrac{\text{m}}{\text{s}^2}}$$

$$= \frac{1{,}960.0}{9.8}\,\frac{\text{kg}\cdot\text{m}}{\text{s}^2} \times \frac{\text{s}^2}{\text{m}}$$

$$= 200\,\text{kg}$$

b. From Newton's second law of motion, you can see that the 100 N rocket gives the 200 kg astronaut an acceleration of:

$$F = ma \therefore a = \frac{F}{m}$$

$$= \frac{100\,\dfrac{\text{kg}\cdot\text{m}}{\text{s}^2}}{200\,\text{kg}}$$

$$= \frac{100\,\text{kg}\cdot\text{m}}{200\,\text{s}^2} \times \frac{1}{\text{kg}}$$

$$= 0.5\,\text{m/s}^2$$

c. An acceleration of 0.5 m/s² for 2.0 s will result in a final velocity of

$$a = \frac{v_f - v_i}{t} \therefore v_f = at + v_i$$

$$= (0.5\,\text{m/s}^2)(2.0\,\text{s}) + 0\,\text{m/s}$$

$$= \boxed{1\,\text{m/s}}$$

Chapter 3

3.1 Listing the known and unknown quantities:

$$F = 200\,\text{N}$$
$$d = 3\,\text{m}$$
$$W = ?$$

These are the quantities found in the equation for work, $W = Fd$, which is already solved for work (W). Thus,

$$W = Fd$$

$$= \left(200\,\frac{\text{kg}\cdot\text{m}}{\text{s}^2}\right)(3\,\text{m})$$

$$= (200)(3)\,\text{N}\cdot\text{m}$$

$$= \boxed{600\,\text{J}}$$

3.2 Listing the known and unknown quantities:

$$F = 440\,\text{N}$$
$$d = 5.0\,\text{m}$$
$$w = 880\,\text{N}$$
$$W = ?$$

These are the quantities found in the equation for work, $W = Fd$, which is already solved for work (W). As you can see in the equation, the force exerted and the distance the box was moved are the quantities used in determining the work accomplished. The weight of the box is a different variable and one that is not used in this equation. Thus,

$$W = Fd$$

$$= \left(440\,\frac{\text{kg}\cdot\text{m}}{\text{s}^2}\right)(5.0\,\text{m})$$

$$= 2{,}200\,\text{N}\cdot\text{m}$$

$$= \boxed{2{,}200\,\text{J}}$$

3.3 Note that 10.0 kg is a mass quantity and not a weight quantity. Weight is found from $w = mg$, a form of Newton's second law of motion. Thus, the force that must be exerted to lift the backpack is its weight, or $(10.0\,\text{kg}) \times (9.8\,\text{m/s}^2)$, which is 98 N. Therefore, a force of 98 N was exerted on the backpack through a distance of 1.5 m, and

$$W = Fd$$

$$= \left(98\,\frac{\text{kg}\cdot\text{m}}{\text{s}^2}\right)(1.5\,\text{m})$$

$$= 147\,\text{N}\cdot\text{m}$$

$$= \boxed{150\,\text{J}}$$

3.4 Weight is defined as the force of gravity acting on an object, and the greater the force of gravity, the harder it is to lift the object. The force is proportional to the mass of the object, as the equation $w = mg$ tells you. Thus, the force you exert when lifting is $F = w = mg$, so the work you do on an object you lift must be $W = mgh$.

You know the mass of the box and you know the work accomplished. You also know the value of the acceleration due to gravity, g, so the list of known and unknown quantities is:

$$m = 102\,\text{kg}$$
$$g = 9.8\,\text{m/s}^2$$
$$W = 5{,}000\,\text{J}$$
$$h = ?$$

The equation $W = mgh$ is solved for work, so the first thing to do is to solve it for h, the unknown height in this problem (note that height is also a distance):

$$W = mgh \therefore h = \frac{W}{mg}$$

$$= \frac{5,000 \frac{kg \cdot m}{s^2} \times m}{(102\,kg)\left(9.8\frac{m}{s^2}\right)}$$

$$= \frac{5,000.0}{102 \times 9.8}\frac{kg \cdot m}{s^2} \times \frac{m}{1} \times \frac{1}{kg} \times \frac{s^2}{m}$$

$$= \frac{5,000}{999.6}\,m$$

$$= \boxed{5\,m}$$

3.5 A student running up the stairs has to lift herself, so her weight is the required force needed. Thus, the force exerted is $F = w = mg$, and the work done is $W = mgh$. You know the mass of the student, the height, and the time. You also know the value of the acceleration due to gravity, g, so the list of known and unknown quantities is:

$$m = 60.0\,kg$$
$$g = 9.8\,m/s^2$$
$$h = 5.00\,m$$
$$t = 3.92\,s$$
$$P = ?$$

The equation $P = \dfrac{mgh}{t}$ is already solved for power, so:

a.
$$P = \frac{mgh}{t}$$

$$= \frac{(60.0\,kg)\left(9.8\frac{m}{s^2}\right)(5.00\,m)}{3.92\,s}$$

$$= \frac{(60.0)(9.8)(5.00)}{(3.92)}\frac{\left(\frac{kg \cdot m}{s^2}\right) \times m}{s}$$

$$= \frac{2940}{3.92}\frac{N \cdot m}{s}$$

$$= 750\,\frac{J}{s}$$

$$= \boxed{750\,W}$$

b. A power of 750 watts is almost one horsepower.

3.6 Listing the known and unknown quantities:

$$m = 2,000\,kg$$
$$v = 72\,km/h$$
$$KE = ?$$

These are the quantities found in the equation for kinetic energy, $KE = 1/2mv^2$, which is already solved. However, note that the velocity is in units of km/h, which must be changed to m/s before doing anything else (it must be m/s because all energy and work units are in units of the joule [J]. A joule is a newton-meter, and a newton is a kg·m/s²).

Using the conversion factor from inside the front cover of your text,

$$\frac{0.2778\frac{m}{s}}{1.0\frac{km}{h}} \times 72\frac{km}{h}$$

$$(0.2778)(72)\frac{m}{s} \times \frac{h}{km} \times \frac{km}{h}$$

$$20\frac{m}{s}$$

and

$$KE = \frac{1}{2}mv^2$$

$$= \frac{1}{2}(2,000\,kg)\left(20\frac{m}{s}\right)^2$$

$$= \frac{1}{2}(2,000\,kg)\left(400\frac{m^2}{s^2}\right)$$

$$= \frac{1}{2} \times 2,000 \times 400\frac{kg \cdot m^2}{s^2}$$

$$= 400,000\frac{kg \cdot m}{s^2} \times m$$

$$= 400,000\,N \cdot m$$

$$= \boxed{4 \times 10^5\,J}$$

Scientific notation is used here to simplify a large number and to show one significant figure.

3.7 Recall the relationship between work and energy—that you do work on an object when you throw it, giving it kinetic energy, and the kinetic energy it has will do work on something else when stopping. Because of the relationship between work and energy, you can calculate (1) the work you do, (2) the kinetic energy a moving object has as a result of your work, and (3) the work it will do when coming to a stop, and all three answers should be the same. Thus, you do not have a force or a distance to calculate the work needed to stop a moving car, but you can simply calculate the kinetic energy of the car. Both answers should be the same.

Before you start, note that the velocity is in units of km/h, which must be changed to m/s before doing anything else (it must be m/s because all energy and work units are in units of the joule [J]. A joule is a newton-meter, and a newton is a kg·m/s²). Using the conversion factor from inside the front cover,

$$\frac{0.2778\frac{m}{s}}{1.0\frac{km}{h}} \times 54.0\frac{km}{h}$$

$$0.2778 \times 54.0\frac{m}{s} \times \frac{h}{km} \times \frac{km}{h}$$

$$15.0\frac{m}{s}$$

and

$$KE = \frac{1}{2}mv^2$$

$$= \frac{1}{2}(1,000.0\,\text{kg})\left(15.0\,\frac{\text{m}}{\text{s}}\right)^2$$

$$= \frac{1}{2}(1,000.0\,\text{kg})\left(225\,\frac{\text{m}^2}{\text{s}^2}\right)$$

$$= \frac{1}{2}\times 1,000.0 \times 225\,\frac{\text{kg}\cdot\text{m}^2}{\text{s}^2}$$

$$= 112,500\,\frac{\text{kg}\cdot\text{m}}{\text{s}^2}\times\text{m}$$

$$= 112,500\,\text{N}\cdot\text{m}$$

$$= \boxed{1.13\times 10^5\,\text{J}}$$

Scientific notation is used here to simplify a large number and to easily show three significant figures. The answer could likewise be expressed as 113 kJ.

3.8 a. How much energy was used by a 1,000 kg car climbing a hill 51.02 m high is answered by how much work the car did. In this case, $W = Fd$ and the force exerted is the weight of the car, $w = mgh$. Thus,

$$w = mgh$$

$$= (1,000\,\text{kg})\left(9.8\,\frac{\text{m}}{\text{s}^2}\right)(51.02\,\text{m})$$

$$= 1,000\times 9.8\times 51.02\,\text{kg}\times\frac{\text{m}}{\text{s}^2}\times\text{m}$$

$$= 499,996\,\frac{\text{kg}\cdot\text{m}}{\text{s}^2}\times\text{m}$$

$$= 500,000\,\text{N}\cdot\text{m}$$

$$= 5\times 10^5\,\text{J (or 500 kJ)}$$

b. How much potential energy the car has is found in the potential energy equation, $PE = mgh$. Note the potential energy is path independent, that is, depends only on the vertical height of the hill. Thus,

$$PE = mgh$$

$$= (1,000\,\text{kg})\left(9.8\,\frac{\text{m}}{\text{s}^2}\right)(51.02\,\text{m})$$

$$= 1,000\times 9.8\times 51.02\,\text{kg}\times\frac{\text{m}}{\text{s}^2}\times\text{m}$$

$$= 499,996\,\frac{\text{kg}\cdot\text{m}}{\text{s}^2}\times\text{m}$$

$$= 500,000\,\text{N}\cdot\text{m}$$

$$= \boxed{5\times 10^5\,\text{J (or 500 kJ)}}$$

As you can see, the potential energy of the car is exactly the same as the amount of energy used to climb the hill.

3.9 a.

$$W = Fd$$

$$= (10\,\text{lb})(5\,\text{ft})$$

$$= (10)(5)\,\text{ft}\times\text{lb}$$

$$= 50\,\text{ft}\cdot\text{lb}$$

b. The distance of the bookcase from some horizontal reference level did not change, so the gravitational potential energy does not change.

3.10 The force (F) needed to lift the book is equal to the weight (w) of the book, or $F = w$. Since $w = mg$, then $F = mg$. Work is defined as the product of a force moved through a distance, or $W = Fd$. The work done in lifting the book is therefore $W = mgh$, and:

a.

$$W = mgh$$

$$= (2.0\,\text{kg})(9.8\,\text{m/s}^2)(2.00\,\text{m})$$

$$= (2.0)(9.8)(2.00)\,\frac{\text{kg}\cdot\text{m}}{\text{s}^2}\times\text{m}$$

$$= 39.2\,\frac{\text{kg}\cdot\text{m}^2}{\text{s}^2}$$

$$= 39.2\,\text{J} = \boxed{39\,\text{J}}$$

b.

$$PE = mgh = \boxed{39\,\text{J}}$$

c.

$$PE_{\text{lost}} = KE_{\text{gained}} = mgh = \boxed{39\,\text{J}}$$

(or)

$$v = \sqrt{2gh} = \sqrt{(2)(9.8\,\text{m/s}^2)(2.00\,\text{m})}$$

$$= \sqrt{39.2\,\text{m}^2/\text{s}^2}$$

$$= 6.26\,\text{m/s}$$

$$KE = \frac{1}{2}mv^2 = \left(\frac{1}{2}\right)(2.0\,\text{kg})(6.26\,\text{m/s})^2$$

$$= \left(\frac{1}{2}\right)(2.0\,\text{kg})(39.2\,\text{m}^2/\text{s}^2)$$

$$= (1.0)(39.2)\,\frac{\text{kg}\cdot\text{m}^2}{\text{s}^2}$$

$$= \boxed{39\,\text{J}}$$

3.11

$$KE = \frac{1}{2}mv^2$$

$$= \frac{1}{2}(60.0\,\text{kg})\left(2.0\,\frac{\text{m}}{\text{s}}\right)^2$$

$$= \frac{1}{2}(60.0\,\text{kg})\left(4.0\,\frac{\text{m}^2}{\text{s}^2}\right)$$

$$= 30.0\times 4.0\,\text{kg}\times\left(\frac{\text{m}^2}{\text{s}^2}\right)$$

$$= \boxed{120\,\text{J}}$$

$$KE = \frac{1}{2}mv^2$$

$$= \frac{1}{2}(60.0\,\text{kg})\left(4.0\,\frac{\text{m}}{\text{s}}\right)^2$$

$$= \frac{1}{2}(60.0\,\text{kg})\left(16\,\frac{\text{m}^2}{\text{s}^2}\right)$$

$$= 30.0\times 16\,\text{kg}\times\left(\frac{\text{m}^2}{\text{s}^2}\right)$$

$$= \boxed{480\,\text{J}}$$

Thus, doubling the speed results in a four-fold increase in kinetic energy.

3.12 a. The force needed is equal to the weight of the student. The English unit of a pound is a force unit, so

$$W = Fd$$

$$= (170.0 \text{ lb})(25.0 \text{ ft})$$

$$= \boxed{4,250 \text{ ft·lb}}$$

b. Work (W) is defined as a force (F) moved through a distance (d), or $W = Fd$. Power (P) is defined as work (W) per unit of time (t), or $P = W/t$. Therefore,

$$P = \frac{Fd}{t}$$

$$= \frac{(170.0 \text{ lb})(25.0 \text{ ft})}{10.0 \text{ s}}$$

$$= \frac{(170.0)(25.0)}{10.0} \frac{\text{ft·lb}}{\text{s}}$$

$$= 425 \frac{\text{ft·lb}}{\text{s}}$$

One hp is defined as $550 \dfrac{\text{ft·lb}}{\text{s}}$ and

$$\frac{425 \text{ ft·lb/s}}{550 \dfrac{\text{ft·lb/s}}{\text{hp}}} = \boxed{0.77 \text{ hp}}$$

(Note that the student's power rating [425 ft·lb/s] is less than the power rating defined as 1 horsepower [550 ft·lb/s]. Thus, the student's horsepower must be *less* than 1 horsepower. A simple analysis such as this will let you know if you inverted the ratio or not.)

Chapter 4

4.1 Listing the known and unknown quantities:
body temperature $T_F = 98.6°$

$$T_C = ?$$

These are the quantities found in the equation for conversion of Fahrenheit to Celsius, $T_C = \dfrac{5}{9}(T_F - 32°)$, where T_F is the temperature in Fahrenheit and T_C is the temperature in Celsius. This equation describes a relationship between the two temperature scales and is used to convert a Fahrenheit temperature to Celsius. The equation is already solved for the Celsius temperature, T_C. Thus,

$$T_C = \frac{5}{9}(T_F - 32°)$$

$$= \frac{5}{9}(98.6° - 32°)$$

$$= \frac{333°}{9}$$

$$= \boxed{37°C}$$

4.2 $Q = mc\,\Delta T$

$$= (221 \text{ g})\left(0.093 \frac{\text{cal}}{\text{gC°}}\right)(38.0°C - 20.0°C)$$

$$= (221)(0.093)(18.0)\,\text{g} \times \frac{\text{cal}}{\text{gC°}} \times °C$$

$$= 370 \frac{\text{g·cal·°C}}{\text{gC°}}$$

$$= \boxed{370 \text{ cal}}$$

4.3 First, you need to know the energy of the moving bike and rider. Since the speed is given as 36.0 km/h, convert to m/s by multiplying times 0.2778 m/s per km/h:

$$\left(36.0 \frac{\text{km}}{\text{h}}\right)\left(0.2778 \frac{\text{m/s}}{\text{km/h}}\right)$$

$$= (36.0)(0.2778) \frac{\text{km}}{\text{h}} \times \frac{\text{h}}{\text{km}} \times \frac{\text{m}}{\text{s}}$$

$$= 10.0 \text{ m/s}$$

Then,

$$KE = \frac{1}{2}mv^2$$

$$= \frac{1}{2}(100.0 \text{ kg})(10.0 \text{ m/s})^2$$

$$= \frac{1}{2}(100.0 \text{ kg})(100 \text{ m}^2/\text{s}^2)$$

$$= \frac{1}{2}(100.0)(100) \frac{\text{kg·m}^2}{\text{s}^2}$$

$$= 5,000 \text{ J}$$

Second, this energy is converted to the calorie heat unit through the mechanical equivalent of heat relationship, that 1.0 kcal = 4,184 J, or that 1.0 cal = 4.184 J. Thus,

$$\frac{5,000 \text{ J}}{4,184 \dfrac{\text{J}}{\text{kcal}}}$$

$$1.195 \frac{\text{J}}{1} \times \frac{\text{kcal}}{\text{J}}$$

$$\boxed{1.20 \text{ kcal}}$$

4.4 First, you need to find the energy of the falling bag. Since the potential energy lost equals the kinetic energy gained, the energy of the bag just as it hits the ground can be found from

$$PE = mgh$$

$$= (15.53 \text{ kg})(9.8 \text{ m/s}^2)(5.50 \text{ m})$$

$$= (15.53)(9.8)(5.50) \frac{\text{kg·m}}{\text{s}^2} \times \text{m}$$

$$= 837 \text{ J}$$

In calories, this energy is equivalent to

$$\frac{837 \text{ J}}{4,184 \dfrac{\text{J}}{\text{kcal}}} = 0.200 \text{ kcal}$$

Second, the temperature change can be calculated from the equation giving the relationship between a quantity of heat (Q), mass (m), specific heat of the substance (c), and the change of temperature:

$$Q = mc\,\Delta T \therefore \Delta T = \frac{Q}{mc}$$

$$= \frac{0.200 \text{ kcal}}{(15.53 \text{ kg})\left(0.200 \dfrac{\text{kcal}}{\text{kgC°}}\right)}$$

$$= \frac{0.200}{(15.53)(0.200)} \frac{\text{kcal}}{1} \times \frac{1}{\text{kg}} \times \frac{\text{kgC°}}{\text{kcal}}$$

$$= 0.064 \frac{\text{kcal·kgC°}}{\text{kcal·kg}}$$

$$= \boxed{6.4 \times 10^{-2}°C}$$

4.5 The Calorie used by dietitians is a kilocalorie; thus, 250.0 Cal is 250.0 kcal. The mechanical energy equivalent is 1 kcal = 4,184 J, so (250.0 kcal)(4,184 J/kcal) = 1,046,250 J.

Since $W = Fd$ and the force needed is equal to the weight (mg) of the person, $W = mgh = (75.0 \text{ kg})(9.8 \text{ m/s}^2)(10.0 \text{ m}) = 7,350 \text{ J}$ for each stairway climbed.

A total of 1,046,250 J of energy from the french fries would require 1,046,250 J/7,350 J per climb, or 142.3 trips up the stairs.

4.6 For unit consistency,

$$T_C = \frac{5}{9}(T_F - 32°) = \frac{5}{9}(68° - 32°) = \frac{5}{9}(36°) = 20°C$$

$$= \frac{5}{9}(32° - 32°) = \frac{5}{9}(0°) = 0°C$$

Glass bowl:

$$Q = mc\,\Delta T$$

$$= (0.5 \text{ kg})\left(0.2 \frac{\text{kcal}}{\text{kg°C}}\right)(20°C)$$

$$= (0.5)(0.2)(20)\frac{\text{kg}}{1} \times \frac{\text{kcal}}{\text{kgC°}} \times \frac{°C}{1}$$

$$= \boxed{2 \text{ kcal}}$$

Iron pan:

$$Q = mc\,\Delta T$$

$$= (0.5 \text{ kg})\left(0.11 \frac{\text{kcal}}{\text{kgC°}}\right)(20°C)$$

$$= (0.5)(0.11)(20)\, \text{kg} \times \frac{\text{kcal}}{\text{kgC°}} \times °C$$

$$= \boxed{1 \text{ kcal}}$$

4.7 Note that a specific heat expressed in cal/gC° has the same numerical value as a specific heat expressed in kcal/kgC° because you can cancel the k units. You could convert 896 cal to 0.896 kcal, but one of the two conversion methods is needed for consistency with other units in the problem.

$$Q = mc\,\Delta T \therefore m = \frac{Q}{c\Delta T}$$

$$= \frac{896 \text{ cal}}{\left(0.056 \frac{\text{cal}}{\text{gC°}}\right)(80.0°C)}$$

$$= \frac{896}{(0.056)(80.0)}\frac{\text{cal}}{1} \times \frac{\text{gC°}}{\text{cal}} \times \frac{1}{C°}$$

$$= 200 \text{ g}$$

$$= \boxed{0.20 \text{ kg}}$$

4.8 Since a watt is defined as a joule/s, finding the total energy in joules will tell the time:

$$Q = mc\,\Delta T$$

$$= (250.0 \text{ g})\left(1.00 \frac{\text{cal}}{\text{gC°}}\right)(60.0°C)$$

$$= (250.0)(1.00)(60.0)\, \text{g} \times \frac{\text{cal}}{\text{gC°}} \times °C$$

$$= 1.50 \times 10^4 \text{ cal}$$

This energy in joules is $(1.50 \times 10^4 \text{ cal})\left(4.184 \frac{\text{J}}{\text{cal}}\right) = 62,800 \text{ J}$

A 300 watt heater uses energy at a rate of $300 \frac{\text{J}}{\text{s}}$, so

$$\frac{62,800 \text{ J}}{300 \text{ J/s}} = 209 \text{ s} \text{ is required, which is } \frac{209 \text{ s}}{60 \frac{\text{s}}{\text{min}}} = 3.48 \text{ min,}$$

or

$$\boxed{\text{about } 3\frac{1}{2} \text{ min}}$$

4.9

$$Q = mc\,\Delta \therefore c = \frac{Q}{m\,\Delta T}$$

$$= \frac{60.0 \text{ cal}}{(100.0 \text{ g})(20.0°C)}$$

$$= \frac{60.0}{(100.0)(20.0)}\frac{\text{cal}}{\text{gC°}}$$

$$= \boxed{0.0300 \frac{\text{cal}}{\text{gC°}}}$$

4.10 To change water at 80.0°C to steam at 100.0°C requires two separate quantities of heat that can be called Q_1 and Q_2. The quantity Q_1 is the amount of heat needed to warm the water from 80.0°C to the boiling point, which is 100.0°C at sea level pressure ($\Delta T = 20.0°C$). The relationship between the variables involved is $Q = mc\Delta T$. The quantity Q_2 is the amount of heat needed to take 100.0°C water through the phase change to steam (water vapor) at 100.0°C. The phase change from a liquid to a gas (or gas to liquid) is concerned with the latent heat of vaporization. For water, the latent heat of vaporization is given as 540.0 cal/g.

$m = 250.0 \text{ g}$

$L_{v(\text{water})} = 540.0 \text{ cal/g}$

$Q = ?$

$$Q_1 = mc\Delta T$$

$$= (250.0 \text{ g})\left(1.00 \frac{\text{cal}}{\text{gC°}}\right)(20.0°C)$$

$$= (250.0)(1.00)(20.0)\, \text{g} \times \frac{\text{cal}}{\text{gC°}} \times °C$$

$$= 5,000 \frac{\text{g·cal·C°}}{\text{gC°}}$$

$$= 5,000 \text{ cal}$$

$$= 5.00 \text{ kcal}$$

$$Q_2 = mL_v$$

$$= (250.0 \text{ g})\left(540.0 \frac{\text{cal}}{\text{g}}\right)$$

$$= 250.0 \times 540.0 \frac{\text{g·cal}}{\text{g}}$$

$$= 135,000 \text{ cal}$$

$$= 135.0 \text{ kcal}$$

$$Q_{\text{Total}} = Q_1 + Q_2$$

$$= 5.00 \text{ kcal} + 135.0 \text{ kcal}$$

$$= \boxed{140.0 \text{ kcal}}$$

4.11 To change 20.0°C water to steam at 125.0°C requires three separate quantities of heat. First, the quantity Q_1 is the amount of heat needed to warm the water from 20.0°C to 100.0°C ($\Delta T = 80.0°C$). The quantity Q_2 is the amount of heat needed to take 100.0°C water to steam at 100.0°C. Finally, the quantity Q_3 is the amount of heat needed to warm the steam from 100.0° to 125.0°C. According to table 4.3, the c for steam is 0.480 cal/g°C.

$m = 100.0 \text{ g}$
$\Delta T_{water} = 80.0°C$
$\Delta T_{steam} = 25.0°C$
$L_{v(water)} = 540.0 \text{ cal/g}$
$c_{steam} = 0.480 \text{ cal/gC°}$

$Q_1 = mc\,\Delta T$

$\quad = (100.0 \text{ g})\left(1.00 \dfrac{\text{cal}}{\text{gC°}}\right)(80.0°C)$

$\quad = (100.0)(1.00)(80.0) \text{ g} \times \dfrac{\text{cal}}{\text{gC°}} \times °C$

$\quad = 8,000 \dfrac{\text{g}\cdot\text{cal}\cdot°C}{\text{g}\cdot C°}$

$\quad = 8,000 \text{ cal}$

$\quad = 8.00 \text{ kcal}$

$Q_2 = mL_v$

$\quad = (100.0 \text{ g})\left(540.0 \dfrac{\text{cal}}{\text{g}}\right)$

$\quad = 100.0 \times 540.0 \dfrac{\text{g}\cdot\text{cal}}{\text{g}}$

$\quad = 54,000 \text{ cal}$

$\quad = 54.00 \text{ kcal}$

$Q_3 = mc\,\Delta T$

$\quad = (100.0 \text{ g})\left(0.480 \dfrac{\text{cal}}{\text{gC°}}\right)(25.0°C)$

$\quad = (100.0)(0.480)(25.0) \text{ g} \times \dfrac{\text{cal}}{\text{gC°}} \times °C$

$\quad = 1,200 \dfrac{\text{g}\cdot\text{cal}\cdot°C}{\text{g}\cdot C°}$

$\quad = 1,200 \text{ cal}$

$\quad = 1.20 \text{ kcal}$

$Q_{total} = Q_1 + Q_2 + Q_3$

$\quad = 8.00 \text{ kcal} + 54.00 \text{ kcal} + 1.20 \text{ kcal}$

$\quad = \boxed{63.20 \text{ kcal}}$

4.12 a. Step 1: Cool the water from 18.0°C to 0°C.

$Q_1 = mc\,\Delta T$

$\quad = (400.0 \text{ g})\left(1.00 \dfrac{\text{cal}}{\text{gC°}}\right)(18.0°C)$

$\quad = (400.0)(1.00)(18.0) \text{ g} \times \dfrac{\text{cal}}{\text{gC°}} \times °C$

$\quad = 7,200 \dfrac{\text{g}\cdot\text{cal}\cdot°C}{\text{g}\cdot C°}$

$\quad = 7,200 \text{ cal}$

$\quad = 7.20 \text{ kcal}$

Step 2: Find the energy needed for the phase change of water at 0°C to ice at 0°C.

$Q_2 = mL_f$

$\quad = (400.0 \text{ g})\left(80.0 \dfrac{\text{cal}}{\text{g}}\right)$

$\quad = 400.0 \times 80.0 \dfrac{\text{g}\cdot\text{cal}}{\text{g}}$

$\quad = 32,000 \text{ cal}$

$\quad = 32.0 \text{ kcal}$

Step 3: Cool the ice from 0°C to ice at $-5.00°C$.

$Q_3 = mc\,\Delta T$

$\quad = (400.0 \text{ g})\left(0.50 \dfrac{\text{cal}}{\text{gC°}}\right)(5.00°C)$

$\quad = 400.0 \times 0.50 \times 5.00 \text{ g} \times \dfrac{\text{cal}}{\text{gC°}} \times °C$

$\quad = 1,000 \dfrac{\text{g}\cdot\text{cal}\cdot°C}{\text{g}\cdot C°}$

$\quad = 1,000 \text{ cal}$

$\quad = 1.0 \text{ kcal}$

$Q_{total} = Q_1 + Q_2 + Q_3$

$\quad = 7.20 \text{ kcal} + 32.0 \text{ kcal} + 1.0 \text{ kcal}$

$\quad = \boxed{40.2 \text{ kcal}}$

Chapter 5

5.1

$v = f\lambda$

$\quad = \left(10 \dfrac{1}{s}\right)(0.50 \text{ m})$

$\quad = 5 \dfrac{\text{m}}{\text{s}}$

5.2 The distance between two *consecutive* condensations (or rarefactions) is one wavelength, so $\lambda = 3.00$ m and

$v = f\lambda$

$\quad = \left(112.0 \dfrac{1}{s}\right)(3.00 \text{ m})$

$\quad = 336 \dfrac{\text{m}}{\text{s}}$

5.3 a. One complete wave every 4 s means that $T = 4.00$ s. (Note that the symbol for the *time of a cycle* is T. Do not confuse this symbol with the symbol for temperature.)

b.

$f = \dfrac{1}{T}$

$\quad = \dfrac{1}{4.0 \text{ s}}$

$\quad = \dfrac{1}{4.0} \dfrac{1}{s}$

$\quad = 0.25 \dfrac{1}{s}$

$\quad = \boxed{0.25 \text{ Hz}}$

5.4 The distance from one condensation to the next is one wavelength, so

$$v = f\lambda \therefore \lambda = \frac{v}{f}$$

$$= \frac{330 \frac{m}{s}}{260 \frac{1}{s}}$$

$$= \frac{330}{260} \frac{m}{s} \times \frac{s}{1}$$

$$= \boxed{1.3 \text{ m}}$$

5.5 a. $v = f\lambda = \left(256 \frac{1}{s}\right)(1.34 \text{ m}) = \boxed{343 \text{ m/s}}$

b. $= \left(440.0 \frac{1}{s}\right)(0.780 \text{ m}) = \boxed{343 \text{ m/s}}$

c. $= \left(750.0 \frac{1}{s}\right)(0.457 \text{ m}) = \boxed{343 \text{ m/s}}$

d. $= \left(2,500.0 \frac{1}{s}\right)(0.137 \text{ m}) = \boxed{343 \text{ m/s}}$

5.6 The speed of sound and time are given and you are looking for a distance.

$$v = 1,100 \text{ ft/s}$$
$$t = 4.80 \text{ s}$$
$$d = ?$$

Calculating the total distance the sound traveled,

$$v = \frac{d}{t} \therefore d = vt$$

$$= \left(1,100 \frac{ft}{s}\right)(4.80 \text{ s})$$

$$= (1,100)(4.80)\frac{ft}{s} \times s$$

$$= 4,800 \text{ ft}$$

Since the sound travels from you to the cliff, then back to you, the cliff must be half the total distance:

$$\frac{4,800 \text{ ft}}{2} = \boxed{2,400 \text{ ft}}$$

5.7 The speed of the sound and the time between the lightning and thunder are given:

$$v = 1,140 \text{ ft/s}$$
$$t = 4.63 \text{ s}$$
$$d = ?$$

The distance that a sound with this velocity travels in the given time is

$$v = \frac{d}{t} \therefore d = vt$$

$$= \left(1,140 \frac{ft}{s}\right)(4.63 \text{ s})$$

$$= (1,140)(4.63)\frac{ft}{s} \times s$$

$$= \boxed{5,280 \text{ ft or 1 mile}}$$

5.8 a. $v = f\lambda \therefore \lambda = \frac{v}{f}$

$$= \frac{1,125 \frac{ft}{s}}{440 \frac{1}{s}}$$

$$= \frac{1,125}{440} \frac{ft}{s} \times \frac{s}{1}$$

$$= 2.56 \frac{ft \cdot s}{s}$$

$$= \boxed{2.6 \text{ ft}}$$

b. $v = f\lambda \therefore \lambda = \frac{v}{f}$

$$= \frac{5,020}{440} \frac{ft}{s} \times \frac{s}{1}$$

$$= 11.4 \text{ ft} = \boxed{11 \text{ ft}}$$

Chapter 6

6.1 First, recall that a negative charge means an excess of electrons. Second, the relationship between the total charge (q), the number of electrons (n), and the charge of a single electron (e) is $q = ne$. The fundamental charge of a single ($n = 1$) electron (e) is 1.60×10^{-19} C. Thus

$$q = ne \therefore n = \frac{q}{e}$$

$$= \frac{100 \times 10^{-14} \text{ C}}{1.60 \times 10^{-19} \frac{C}{electron}}$$

$$= \frac{1.00 \times 10^{-14}}{1.60 \times 10^{-19}} \frac{C}{1} \times \frac{electron}{C}$$

$$= 6.25 \times 10^4 \frac{C \cdot electron}{C}$$

$$= \boxed{6.25 \times 10^4 \text{ electron}}$$

6.2 $\frac{electric}{current} = \frac{charge}{time}$

or

$$I = \frac{q}{t}$$

$$= \frac{6.00 \text{ C}}{2.00 \text{ s}}$$

$$= 3.00 \frac{C}{s}$$

$$= \boxed{3.00 \text{ A}}$$

6.3 $R = \frac{V}{I}$

$$= \frac{120.0 \text{ V}}{4.00 \text{ A}}$$

$$= 30.0 \frac{V}{A}$$

$$= \boxed{30.0 \text{ }\Omega}$$

6.4

$$R = \frac{V}{I} \therefore I = \frac{V}{R}$$

$$= \frac{120.0\ \text{V}}{60.0\ \dfrac{\text{V}}{\text{A}}}$$

$$= \frac{120.0}{60.0}\ \text{V} \times \frac{\text{A}}{\text{V}}$$

$$= \boxed{2.00\ \text{A}}$$

6.5 a.

$$R = \frac{V}{I} \therefore V = IR$$

$$= (1.20\ \text{A})\left(10.0\ \frac{\text{V}}{\text{A}}\right)$$

$$= \boxed{12.0\ \text{V}}$$

b. Power = (current)(potential difference)

or

$$P = IV$$

$$= \left(1.20\ \frac{\text{C}}{\text{s}}\right)\left(12.0\ \frac{\text{J}}{\text{C}}\right)$$

$$= (1.20)(12.0)\frac{\text{C}}{\text{s}} \times \frac{\text{J}}{\text{C}}$$

$$= 14.4\ \frac{\text{J}}{\text{s}}$$

$$= \boxed{14.4\ \text{W}}$$

6.6 Note that there are two separate electrical units that are rates: (1) the amp (coulomb/s) and (2) the watt (joule/s). The question asked for a rate of using energy. Energy is measured in joules, so you are looking for the power of the radio in watts. To find watts ($P = IV$), you will need to calculate the current (I) since it is not given. The current can be obtained from the relationship of Ohm's law:

$$I = \frac{V}{R}$$

$$= \frac{3.00\ \text{V}}{15.0\ \dfrac{\text{V}}{\text{A}}}$$

$$= 0.200\ \text{A}$$

$$P = IV$$

$$= (0.200\ \text{C/s})(3.00\ \text{J/C})$$

$$= \boxed{0.600\ \text{W}}$$

6.7

$$\frac{(1{,}200\ \text{W})(0.25\ \text{h})(\$0.10/\text{kWh})}{1{,}000\ \dfrac{\text{W}}{\text{kW}}}$$

$$\frac{(1{,}200)(0.25)(0.10)}{1{,}000}\ \frac{\text{W}}{1} \times \frac{\text{h}}{1} \times \frac{\$}{\text{kWh}} \times \frac{\text{kW}}{\text{W}}$$

$$\boxed{\$0.03}\ (3\ \text{cents})$$

6.8 The relationship between power (P), current (I), and voltage (V) will provide a solution. Since the relationship considers

power in watts, the first step is to convert horsepower to watts. One horsepower is equivalent to 746 watts, so:

$$(746\ \text{W/hp})(2.00\ \text{hp}) = 1{,}492\ \text{W}$$

$$P = IV \therefore I = \frac{P}{V}$$

$$= \frac{1{,}492\ \dfrac{\text{J}}{\text{s}}}{12.0\ \dfrac{\text{J}}{\text{C}}}$$

$$= \frac{1{,}492}{12.0}\ \frac{\text{J}}{\text{s}} \times \frac{\text{C}}{\text{J}}$$

$$= 124.3\ \frac{\text{C}}{\text{s}}$$

$$= \boxed{124\ \text{A}}$$

6.9 a. The rate of using energy is joule/s, or the watt. Since 1.00 hp = 746 W,
inside motor: (746 W/hp)(1/3 hp) = 249 W
outside motor: (746 W/hp)(1/3 hp) = 249 W
compressor motor: (746 W/hp)(3.70 hp) = 2,760 W

$$249\ \text{W} + 249\ \text{W} + 2{,}760\ \text{W} = \boxed{3{,}258\ \text{W}}$$

b. $\dfrac{(3{,}258\ \text{W})(1.00\ \text{h})(\$0.10/\text{kWh})}{1{,}000\ \text{W/kW}} = \$0.33\ \text{per hour}$

c. $(\$0.33/\text{h})(12\ \text{h/day})(30\ \text{day/mo}) = \boxed{\$118.80}$

6.10 The solution is to find how much current each device draws and then to see if the total current is less or greater than the breaker rating:

Toaster: $I = \dfrac{V}{R} = \dfrac{120\ \text{V}}{15\ \text{V/A}} = 8.0\ \text{A}$

Motor: $(0.20\ \text{hp})(746\ \text{W/hp}) = 150\ \text{W}$

$$I = \frac{P}{V} = \frac{15\ \text{J/s}}{120\ \text{J/C}} = 1.3\ \text{A}$$

Three 100 W bulbs: $3 \times 100\ \text{W} = 300\ \text{W}$

$$I = \frac{P}{V} = \frac{300\ \text{J/s}}{120\ \text{J/C}} = 2.5\ \text{A}$$

Iron: $I = \dfrac{P}{V} = \dfrac{600\ \text{J/s}}{120\ \text{J/C}} = 5.0\ \text{A}$

The sum of the currents is 8.0A + 1.3A + 2.5A + 5.0A = 16.8A, so the total current is greater than 15.0 amp and the circuit breaker will trip.

6.11 a.

$$V_p = 1{,}200\ \text{V}$$
$$N_p = 1\ \text{loop}$$
$$N_s = 200\ \text{loops}$$
$$V_s = ?$$

$$\frac{V_p}{N_p} = \frac{V_s}{N_s} \therefore V_s = \frac{V_p N_s}{N_p}$$

$$V_s = \frac{(1{,}200\ \text{V})(200\ \text{loops})}{1\ \text{loop}}$$

$$= \boxed{240{,}000\ \text{V}}$$

b.

$$I_p = 40 \text{ A} \quad V_p I_p = V_s I_s \therefore I_s = \frac{V_p I_p}{V_s}$$

$$I_s = ? \qquad I_s = \frac{1,200 \text{ V} \times 40 \text{ A}}{240,000 \text{ V}}$$

$$= \frac{1,200 \times 40}{240,000} \frac{\cancel{V} \cdot A}{\cancel{V}}$$

$$= \boxed{0.2 \text{ A}}$$

6.12 a.

$$V_s = 12 \text{ V}$$
$$I_s = 0.5 \text{ A} \qquad \frac{V_p}{N_p} = \frac{V_s}{N_s} \therefore \frac{N_p}{N_s} = \frac{V_p}{V_s}$$
$$V_p = 120 \text{ V} \qquad \frac{N_p}{N_s} = \frac{120 \cancel{V}}{12 \cancel{V}} = \frac{10}{1}$$
$$\frac{N_p}{N_s} = ?$$

or

$$\boxed{10 \text{ primary to 1 secondary}}$$

b.

$$I_p = ? \quad V_p I_p = V_s I_s \therefore I_p = \frac{V_s I_s}{V_p}$$

$$I_p = \frac{(12 \text{ V})(0.5 \text{ A})}{120 \text{ V}}$$

$$= \frac{12 \times 0.5}{120} \frac{\cancel{V} \cdot A}{\cancel{V}}$$

$$= \boxed{0.05 \text{ A}}$$

c.

$$P_s = ? \quad P_s = I_s V_s$$
$$= (0.5 \text{ A})(12 \text{ V})$$
$$= 0.5 \times 12 \frac{\cancel{C}}{s} \times \frac{J}{\cancel{C}}$$
$$= 6 \frac{J}{s}$$
$$= \boxed{6 \text{ W}}$$

Chapter 7

7.1 The relationship between the speed of light in a transparent material (v), the speed of light in a vacuum ($c = 3.00 \times 10^8$ m/s), and the index of refraction (n) is $n = c/v$. According to table 7.1, the index of refraction for water is $n = 1.33$ and for ice is $n = 1.31$.

a.

$$c = 3.00 \times 10^8 \text{ m/s}$$
$$n = 1.33$$
$$v = ?$$
$$n = \frac{c}{v} \therefore v = \frac{c}{n}$$
$$v = \frac{3.00 \times 10^8 \text{ m/s}}{1.33}$$
$$= \boxed{2.26 \times 10^8 \text{ m/s}}$$

b.

$$c = 3.00 \times 10^8 \text{ m/s}$$
$$n = 1.31$$
$$v = ?$$
$$v = \frac{3.00 \times 10^8 \text{ m/s}}{1.31}$$
$$= \boxed{2.29 \times 10^8 \text{ m/s}}$$

7.2

$$d = 1.50 \times 10^8 \text{ km}$$
$$= 1.50 \times 10^{11} \text{ m}$$
$$c = 3.00 \times 10^8 \text{ m/s}$$
$$t = ?$$
$$v = \frac{d}{t} \therefore t = \frac{d}{v}$$
$$t = \frac{1.50 \times 10^{11} \text{ m}}{3.00 \times 10^8 \frac{\text{m}}{\text{s}}}$$
$$= \frac{1.50 \times 10^{11}}{3.00 \times 10^8} \text{ m} \times \frac{\text{s}}{\text{m}}$$
$$= 5.00 \times 10^2 \frac{\text{m} \cdot \text{s}}{\text{m}}$$
$$= \frac{5.00 \times 10^2 \text{ s}}{60.0 \frac{\text{s}}{\text{min}}}$$
$$= \frac{5.00 \times 10^2}{60.0} \cancel{s} \times \frac{\text{min}}{\cancel{s}}$$
$$= \boxed{8.33 \text{ min}}$$

7.3

$$d = 6.00 \times 10^9 \text{ km}$$
$$= 6.00 \times 10^{12} \text{ m}$$
$$c = 3.00 \times 10^8 \text{ m/s}$$
$$t = ?$$
$$v = \frac{d}{t} \therefore t = \frac{d}{v}$$
$$t = \frac{6.00 \times 10^{12} \text{ m}}{3.00 \times 10^8 \frac{\text{m}}{\text{s}}}$$
$$= \frac{6.00 \times 10^{12}}{3.00 \times 10^8} \text{ m} \times \frac{\text{s}}{\text{m}}$$
$$= 2.00 \times 10^4 \text{ s}$$
$$= \frac{2.00 \times 10^4 \text{ s}}{3,600 \frac{\text{s}}{\text{h}}}$$
$$= \frac{2.00 \times 10^4}{3.600 \times 10^3} \cancel{s} \times \frac{\text{h}}{\cancel{s}}$$
$$= \boxed{5.56 \text{ h}}$$

7.4 From equation 7.1, note that both angles are measured from the normal and that the angle of incidence (θ_i) equals the angle of reflection (θ_r), or

$$\theta_i = \theta_r \therefore \boxed{\theta_i = 10°}$$

7.5

$$v = 2.20 \times 10^8 \text{ m/s}$$
$$c = 3.00 \times 10^8 \text{ m/s}$$
$$n = ?$$
$$n = \frac{c}{v}$$
$$= \frac{3.00 \times 10^8 \frac{\text{m}}{\text{s}}}{2.20 \times 10^8 \frac{\text{m}}{\text{s}}}$$
$$= 1.36$$

According to table 7.1, the substance with an index of refraction of 1.36 is ethyl alcohol.

7.6 a. From equation 7.3:

$$\lambda = 6.00 \times 10^{-7} \text{ m} \qquad c = \lambda f \ \therefore \ f = \frac{c}{\lambda}$$

$$c = 3.00 \times 10^8 \text{ m/s}$$

$$f = ?$$

$$f = \frac{3.00 \times 10^8 \ \frac{\text{m}}{\text{s}}}{6.00 \times 10^{-7} \text{ m}}$$

$$= \frac{3.00 \times 10^8}{6.00 \times 10^{-7}} \ \frac{\text{m}}{\text{s}} \times \frac{1}{\text{m}}$$

$$= 5.00 \times 10^{14} \ \frac{1}{\text{s}}$$

$$= \boxed{5.00 \times 10^{14} \text{ Hz}}$$

b. From equation 7.4:

$$f = 5.00 \times 10^{14} \text{ Hz}$$

$$h = 6.63 \times 10^{-34} \text{ J} \cdot \text{s}$$

$$E = ?$$

$$E = hf$$

$$= (6.63 \times 10^{-34} \text{ J} \cdot \text{s}) \left(5.00 \times 10^{14} \ \frac{1}{\text{s}} \right)$$

$$= (6.63 \times 10^{-34})(5.00 \times 10^{14}) \text{ J} \cdot \text{s} \times \frac{1}{\text{s}}$$

$$= \boxed{3.32 \times 10^{-19} \text{ J}}$$

7.7 First, you can find the energy of one photon of the peak intensity wavelength (5.60×10^{-7} m) by using equation 7.3 to find the frequency, then equation 7.4 to find the energy:

Step 1:

$$c = \lambda f \ \therefore \ f = \frac{c}{\lambda}$$

$$= \frac{3.00 \times 10^8 \ \frac{\text{m}}{\text{s}}}{5.60 \times 10^{-7} \text{ m}}$$

$$= 5.36 \times 10^{14} \text{ Hz}$$

Step 2: $E = hf$

$$= (6.63 \times 10^{-34} \text{ J} \cdot \text{s})(5.36 \times 10^{14} \text{ Hz})$$

$$= 3.55 \times 10^{-19} \text{ J}$$

Step 3: Since one photon carries an energy of 3.55×10^{-19} J and the overall intensity is 1,000.0 W, each square meter must receive an average of

$$\frac{1,000.0 \ \frac{\text{J}}{\text{s}}}{3.55 \times 10^{-19} \ \frac{\text{J}}{\text{photon}}}$$

$$\frac{1.000 \times 10^3}{3.55 \times 10^{-19}} \ \frac{\text{J}}{\text{s}} \times \frac{\text{photon}}{\text{J}}$$

$$\boxed{2.82 \times 10^{21} \ \frac{\text{photon}}{\text{s}}}$$

7.8 a. $f = 4.90 \times 10^{14}$ Hz $\qquad c = \lambda f \ \therefore \ \lambda = \frac{c}{f}$

$$c = 3.00 \times 10^8 \text{ m/s}$$

$$\lambda = ?$$

$$\lambda = \frac{3.00 \times 10^8 \ \frac{\text{m}}{\text{s}}}{4.90 \times 10^{14} \ \frac{1}{\text{s}}}$$

$$= \frac{3.00 \times 10^8}{4.90 \times 10^{14}} \ \frac{\text{m}}{\text{s}} \times \frac{\text{s}}{1}$$

$$= \boxed{6.12 \times 10^{-7} \text{ m}}$$

b. According to table 7.2, this is the frequency and wavelength of orange light.

7.9

$$f = 5.00 \times 10^{20} \text{ Hz}$$

$$h = 6.63 \times 10^{-34} \text{ J} \cdot \text{s}$$

$$E = ?$$

$$E = hf$$

$$= (6.63 \times 10^{-34} \text{ J} \cdot \text{s}) \left(5.00 \times 10^{20} \ \frac{1}{\text{s}} \right)$$

$$= (6.63 \times 10^{-34})(5.00 \times 10^{20}) \text{ J} \cdot \text{s} \times \frac{1}{\text{s}}$$

$$= \boxed{3.32 \times 10^{-13} \text{ J}}$$

7.10

$$\lambda = 1.00 \text{ mm}$$

$$= 0.001 \text{ m}$$

$$f = ?$$

$$c = 3.00 \times 10^8 \text{ m/s}$$

$$h = 6.63 \times 10^{-34} \text{ J} \cdot \text{s}$$

$$E = ?$$

Step 1: $c = \lambda f \ \therefore \ f = \frac{c}{\lambda}$

$$f = \frac{3.00 \times 10^8 \ \frac{\text{m}}{\text{s}}}{1.00 \times 10^{-3} \text{ m}}$$

$$= \frac{3.00 \times 10^8}{1.00 \times 10^{-3}} \ \frac{\text{m}}{\text{s}} \times \frac{1}{\text{m}}$$

$$= 3.00 \times 10^{11} \text{ Hz}$$

Step 2: $E = hf$

$$= (6.63 \times 10^{-34} \text{ J} \cdot \text{s}) \left(3.00 \times 10^{11} \ \frac{1}{\text{s}} \right)$$

$$= (6.63 \times 10^{-34})(3.00 \times 10^{11}) \text{ J} \cdot \text{s} \times \frac{1}{\text{s}}$$

$$= \boxed{1.99 \times 10^{-22} \text{ J}}$$

Chapter 8

8.1 Energy is related to the frequency and Planck's constant in equation 8.1, $E = hf$.

$$\text{For } n = 6, E_H = 6.05 \times 10^{-20} \text{ J}$$
$$\text{For } n = 2, E_L = 5.44 \times 10^{-19} \text{ J}$$
$$E = ? \text{ J}$$
$$E = E_H - E_L$$
$$= (-6.05 \times 10^{-20} \text{ J}) - (-5.44 \times 10^{-19} \text{ J})$$
$$= \boxed{4.84 \times 10^{-19} \text{ J}}$$

8.2
$$\text{For } n = 6, E_H = -6.05 \times 10^{-20} \text{ J}$$
$$\text{For } n = 2, E_L = -5.44 \times 10^{-19} \text{ J}$$
$$h = 6.63 \times 10^{-34} \text{ J·s}$$
$$f = ?$$
$$hf = E_H - E_L \therefore f = \frac{E_H - E_L}{h}$$
$$f = \frac{(-6.05 \times 10^{-20} \text{ J}) - (-5.44 \times 10^{-19} \text{ J})}{6.63 \times 10^{-34} \text{ J·s}}$$
$$= \frac{4.84 \times 10^{-19} \text{ J}}{6.63 \times 10^{-34} \text{ J·s}}$$
$$= 7.29 \times 10^{14} \frac{1}{s}$$
$$= \boxed{7.29 \times 10^{14} \text{ Hz (violet)}}$$

8.3 $(n = 1) = -13.6 \text{ eV}$

Since the energy of the electron is -13.6 eV, it will require 13.6 eV (or 2.17×10^{-18} J) to remove the electron.

8.4
$$q/m = -1.76 \times 10^{11} \text{ C/kg}$$
$$q = -1.60 \times 10^{-19} \text{ C}$$
$$m = ?$$
$$\text{mass} = \frac{\text{charge}}{\text{charge/mass}}$$
$$= \frac{-1.60 \times 10^{-19} \text{ C}}{-1.76 \times 10^{11} \dfrac{\text{C}}{\text{kg}}}$$
$$= \frac{-1.60 \times 10^{-19}}{-1.76 \times 10^{11}} \text{ C} \times \frac{\text{kg}}{\text{C}}$$
$$= \boxed{9.09 \times 10^{-31} \text{ kg}}$$

8.5
- **a.** Boron: $1s^2 2s^2 2p^1$
- **b.** Aluminum: $1s^2 2s^2 2p^6 3s^2 3p^1$
- **c.** Potassium: $1s^2 2s^2 2p^6 3s^2 3p^6 4s^1$

8.6
- **a.** Boron is atomic number 5 and there are 5 electrons.
- **b.** Aluminum is atomic number 13 and there are 13 electrons.
- **c.** Potassium is atomic number 19 and there are 19 electrons.

8.7
- **a.** Argon: $1s^2 2s^2 2p^6 3s^2 3p^6$
- **b.** Zinc: $1s^2 2s^2 2p^6 3s^2 3p^6 4s^2 3d^{10}$
- **c.** Bromine: $1s^2 2s^2 2p^6 3s^2 3p^6 4s^2 3d^{10} 4p^5$

Chapter 9

9.1 Recall that the number of outer energy level electrons is the same as the family number for the representative elements:
- **a.** Li: 1
- **b.** N: 5
- **c.** F: 7
- **d.** Cl: 7
- **e.** Ra: 2
- **f.** Be: 2

9.2 The same information that was used in question 1 can be used to draw the dot notation:

(a) $\dot{\underset{\cdot}{\text{B}}}\cdot$ (c) Ca: (e) $\cdot\ddot{\underset{\cdot\cdot}{\text{O}}}\cdot$

(b) $\cdot\ddot{\text{Br}}\cdot$ (d) K·   (f) $\ddot{\underset{\cdot\cdot}{\text{S}}}\cdot$

9.3 The charge is found by identifying how many electrons are lost or gained in achieving the noble gas structure:
- **a.** Boron 3+
- **b.** Bromine 1−
- **c.** Calcium 2+
- **d.** Potassium 1+
- **e.** Oxygen 2−
- **f.** Nitrogen 3−

9.4 The name of some common polyatomic ions are in box table 9.2. Using this table as a reference, the names are
- **a.** hydroxide
- **b.** sulfite
- **c.** hypochlorite
- **d.** nitrate
- **e.** carbonate
- **f.** perchlorate

9.5 The Roman numeral tells you the charge on the variable charge elements. The charges for the polyatomic ions are found in box table 9.2. The charges for metallic elements can be found in tables 9.2 and 9.3. Using these resources and the crossover technique, the formulas are as follows:
- **a.** $Fe(OH)_3$
- **b.** $Pb_3(PO_4)_2$
- **c.** $ZnCO_3$
- **d.** NH_4NO_3
- **e.** $KHCO_3$
- **f.** K_2SO_3

9.6 Box table 9.3 has information about the meaning of prefixes and stem names used in naming covalent compounds. (a), for example, asks for the formula of carbon tetrachloride. Carbon has no prefixes, so there is one carbon atom, and it comes first in the formula because it comes first in the name. The *tetra*-prefix means four, so there are four chlorine atoms. The name ends in -*ide*, so you know there are only two elements in the compound. The symbols can be obtained from the list of elements on the inside back cover of this text. Using all this information from the name, you can think out the formula for carbon tetrachloride. The same process is used for the other compounds and formulas:
- **a.** CCl_4
- **b.** H_2O
- **c.** MnO_2
- **d.** SO_3
- **e.** N_2O_5
- **f.** As_2S_5

9.7 Again using information from box table 9.3, this question requires you to reverse the thinking procedure you learned in question 6.

    a. carbon monoxide

    b. carbon dioxide

    c. carbon disulfide

    d. dinitrogen monoxide

    e. tetraphosphorus trisulfide

    f. dinitrogen trioxide

9.8 **a.** $2\,SO_2 + O_2 \rightarrow 2\,SO_3$

    b. $4\,P + 5\,O_2 \rightarrow 2\,P_2O_5$

    c. $2\,Al + 6\,HCl \rightarrow 2\,AlCl_3 + 3\,H_2$

    d. $2\,NaOH + H_2SO_4 \rightarrow Na_2SO_4 + 2\,H_2O$

    e. $Fe_2O_3 + 3\,CO \rightarrow 2\,Fe + 3\,CO_2$

    f. $3\,Mg(OH)_2 + 2\,H_3PO_4 \rightarrow Mg_3(PO_4)_2 + 6\,H_2O$

9.9 **a.** General form of $XY + AZ \rightarrow XZ + AY$ with precipitate formed: Ion exchange reaction.

    b. General form of $X + Y \rightarrow XY$: Combination reaction.

    c. General form of $XY \rightarrow X + Y + \ldots$: Decomposition reaction.

    d. General form of $X + Y \rightarrow XY$: Combination reaction.

    e. General form of $XY + A \rightarrow AY + X$: Replacement reaction.

    f. General form of $X + Y \rightarrow XY$: Combination reaction.

9.10 **a.** $C_5H_{12(g)} + 8\,O_{2(g)} \rightarrow 5\,CO_{2(g)} + 6\,H_2O_{(g)}$

    b. $HCl_{(aq)} + NaOH_{(aq)} \rightarrow NaCl_{(aq)} + H_2O_{(l)}$

    c. $2\,Al_{(s)} + Fe_2O_{3(s)} \rightarrow Al_2O_{3(s)} + 2\,Fe_{(l)}$

    d. $Fe_{(s)} + CuSO_{4(aq)} \rightarrow FeSO_{4(aq)} + Cu_{(s)}$

    e. $MgCl_{(aq)} + Fe(NO_3)_{2(aq)} \rightarrow$ No reaction (all possible compounds are soluble and no gas or water was formed)

    f. $C_6H_{10}O_{5(s)} + 6\,O_{2(g)} \rightarrow 6\,CO_{2(g)} + 5\,H_2O_{(g)}$

Chapter 11

11.1 **a.** cobalt-60: 27 protons, 33 neutrons

    b. potassium-40: 19 protons, 21 neutrons

    c. neon-24: 10 protons, 14 neutrons

    d. lead-208: 82 protons, 126 neutrons

11.2 **a.** $^{60}_{27}CO$

    b. $^{40}_{19}K$

    c. $^{24}_{10}Ne$

    d. $^{204}_{82}Pb$

11.3 **a.** cobalt-60: Radioactive because odd numbers of protons (27) and odd numbers of neutrons (33) are usually unstable.

    b. potassium-40: Radioactive, again having an odd number of protons (19) and an odd number of neutrons (21).

    c. neon-24: Stable, because even numbers of protons and neutrons are usually stable.

    d. lead-208: Stable, because even numbers of protons and neutrons, *and* because 82 is a particularly stable number of nucleons.

11.4 **a.** $^{56}_{26}Fe \rightarrow {}^{0}_{-1}e + {}^{56}_{27}Co$

    b. $^{7}_{4}Be \rightarrow {}^{0}_{-1}e + {}^{7}_{5}B$

    c. $^{64}_{29}Cu \rightarrow {}^{0}_{-1}e + {}^{64}_{30}Zn$

    d. $^{24}_{11}Na \rightarrow {}^{0}_{-1}e + {}^{24}_{12}Mg$

    e. $^{214}_{82}Pb \rightarrow {}^{0}_{-1}e + {}^{214}_{83}Bi$

    f. $^{32}_{15}P \rightarrow {}^{0}_{-1}e + {}^{32}_{16}S$

11.5 **a.** $^{235}_{92}U \rightarrow {}^{4}_{2}He + {}^{231}_{90}Th$

    b. $^{226}_{88}Ra \rightarrow {}^{4}_{2}He + {}^{222}_{86}Rn$

    c. $^{239}_{94}Pu \rightarrow {}^{4}_{2}He + {}^{235}_{92}U$

    d. $^{214}_{83}Bi \rightarrow {}^{4}_{2}He + {}^{210}_{81}Tl$

    e. $^{230}_{90}Th \rightarrow {}^{4}_{2}He + {}^{226}_{88}Ra$

    f. $^{210}_{84}Po \rightarrow {}^{4}_{2}He + {}^{206}_{82}Pb$

11.6 Thirty-two days is four half-lives. After the first half-life (8 days), 1/2 oz will remain. After the second half-life (8 + 8, or 16 days), 1/4 oz will remain. After the third half-life (8 + 8 + 8, or 24 days), 1/8 oz will remain. After the fourth half-life (8 + 8 + 8 + 8, or 32 days), 1/16 oz will remain, or 6.3×10^{-2} oz.

11.7 The Fe-56 nucleus has a mass of 55.9206 u, but the individual masses of the nucleons are

$$26 \text{ protons} \times 1.00728 \text{ u} = 26.1893 \text{ u}$$

$$30 \text{ neutrons} \times 1.00867 \text{ u} = \frac{30.2601 \text{ u}}{56.4494 \text{ u}}$$

The mass defect is thus

$$
\begin{array}{r}
56.4494 \text{ u} \\
-55.9206 \text{ u} \\
\hline
0.5288 \text{ u}
\end{array}
$$

The atomic mass unit (u) is equal to the mass of a mole (g), therefore 0.5288 u = 0.5288 g. The mass defect is equivalent to the binding energy according to $E = mc^2$. For a molar mass of Fe-56, the mass defect is

$$E = (5.29 \times 10^{-4} \text{ kg})\left(3.00 \times 10^8\,\frac{m}{s}\right)^2$$

$$= (5.29 \times 10^{-4} \text{ kg})\left(9.00 \times 10^{16}\,\frac{m^2}{s^2}\right)$$

$$= 4.76 \times 10^{13}\,\frac{kg \cdot m^2}{s^2}$$

$$= 4.76 \times 10^{13} \text{ J}$$

For a single nucleus,

$$\frac{4.76 \times 10^{13} \text{ J}}{6.02 \times 10^{23} \text{ nuclei}} = 7.90 \times 10^{-11} \text{ J/nuclei}$$

Appendix E

Problem Solving

Students are sometimes apprehensive when assigned problem exercises. This apprehension comes from a lack of experience in problem solving and not knowing how to proceed. This reading is concerned with a basic approach and procedures that you can follow to simplify problem-solving exercises. Thinking in terms of quantitative ideas is a skill that you can learn. Actually, no mathematics beyond addition, subtraction, multiplication, and division is needed. What is needed is knowledge of certain techniques. If you follow the suggested formatting procedures and seek help from the mathematical review appendix as needed, you will find that problem solving is a simple, fun activity that helps you to learn to think in a new way. Here are some more considerations that will prove helpful.

1. Read the problem carefully, perhaps several times, to understand the problem situation. If possible, make a sketch to help you visualize and understand the problem in terms of the real world.

2. Be alert for information that is not stated directly. For example, if a moving object "comes to a stop," you know that the final velocity is zero, even though this was not stated outright. Likewise, questions about "how far?" are usually asking about distance, and questions about "how long?" are usually asking about time. Such information can be very important in procedure step 1 (see procedure steps in right column), the listing of quantities and their symbols. Overlooked or missing quantities and symbols can make it difficult to identify the appropriate equation.

3. Understand the meaning and concepts that an equation represents. An equation represents a relationship that exists between variables. Understanding the relationship helps you to identify the appropriate equation or equations by inspection of the list of known and unknown quantities (see procedure step 2). You will find a list of the equations being considered at the end of each chapter. Information about the meaning and the concepts that an equation represents is found within each chapter.

4. Solve the equation *before* substituting numbers and units for symbols (see procedure step 3). A helpful discussion of the mathematical procedures required, with examples, is in appendix A.

5. Note if the quantities are in the same units. A mathematical operation requires the units to be the same; for example, you cannot add nickels, dimes, and quarters until you first convert them all to the same unit of money. Likewise, you cannot correctly solve a problem if one time quantity is in seconds and another time quantity is in hours. The quantities must be converted to the same units before anything else is done (see procedure step 4). There is a helpful section on how to use conversion ratios in appendix A.

6. Perform the required mathematical operations on the numbers and the units as if they were two separate problems (see procedure step 6). You will find that following this step will facilitate problem-solving activities because the units you obtain will tell you if you have worked the problem correctly. If you just write the units that you *think* should appear in the answer, you have missed this valuable self-check.

7. Be aware that not all learning takes place in a given time frame and that solutions to problems are not necessarily arrived at "by the clock." If you have spent a half an hour or so unsuccessfully trying to solve a particular problem, move on to another problem or do something entirely different for a while. Problem solving often requires time for something to happen in your brain. If you move on to some other activity, you might find that the answer to a problem that you have been stuck on will come to you "out of the blue" when you are not even thinking about the problem. This unexpected revelation of solutions is common to many real-world professions and activities that involve thinking.

Until you are comfortable in a problem-solving situation, you should follow a formatting procedure that will help you organize your thinking. Here is an example of such a formatting procedure:

Step 1: List the quantities involved together with their symbols on the left side of the page, including the unknown quantity with a question mark.

Step 2: Inspect the given quantities and the unknown quantity as listed, and *identify* the equation that expresses a relationship between these quantities. A list of the equations that express the relationships discussed in each chapter is found at the end of that chapter. *Write* the identified equation on the right side of your paper, opposite the list of symbols and quantities.

Step 3: If necessary, *solve* the equation for the variable in question. This step must be done before substituting any numbers or units in the equation. This simplifies things and keeps down the confusion that may otherwise result. If you need help solving an equation, see the section on this topic in appendix A.

Step 4: If necessary, *convert* any unlike units so they are all the same. If the equation involves a time in seconds, for example, and a speed in kilometers per hour, you should convert the km/h to m/s. Again, this step should be done at this point to avoid confusion and incorrect operations in a later step. If you need help converting units, see the section on this topic in appendix A.

Step 5: *Substitute* the known quantities in the equation, replacing each symbol with both the number value and units represented by the symbol.

Step 6: *Perform* the required mathematical operations on the numbers and on the units. This performance is less confusing if you first separate the numbers and units, as shown in the following example and in the examples throughout this text, and then perform the mathematical operations on the numbers and units as separate steps, showing all work.

Step 7: *Draw a box* around your answer (numbers and units) to communicate that you have found what you were looking for. The box is a signal that you have finished your work on this problem.

EXAMPLE PROBLEM

Mercury is a liquid metal with a mass density of 13.6 g/cm³. What is the mass of 10.0 cm³ of mercury?

SOLUTION

The problem gives two known quantities, the mass density (ρ) of mercury and a known volume (V), and identifies an unknown quantity, the mass (m) of that volume. Make a list of these quantities:

$$\rho = 13.6 \, \text{g/cm}^3$$
$$V = 10.0 \, \text{cm}^3$$
$$m = \, ?$$

The appropriate equation for this problem is the relationship between mass density (ρ), mass (m), and volume (V):

$$\rho = \frac{m}{V}$$

The unknown in this case is the mass, m. Solving the equation for m, by multiplying both sides by V, gives:

$$V\rho = \frac{m\cancel{V}}{\cancel{V}}$$
$$V\rho = m, \text{ or}$$
$$m = V\rho$$

Now you are ready to substitute the known quantities in the equation and perform the mathematical operations on the numbers and on the units:

$$m = \left(13.6 \, \frac{\text{g}}{\text{cm}^3}\right)(10.0 \, \text{cm}^3)$$
$$= (13.6)(10.0)\left(\frac{\text{g}}{\text{cm}^3}\right)(\text{cm}^3)$$
$$= 136 \, \frac{\text{g}\cdot\cancel{\text{cm}^3}}{\cancel{\text{cm}^3}}$$
$$= \boxed{136 \, \text{g}}$$

Glossary

A

abiotic factors nonliving parts of an organism's environment

absolute humidity a measure of the actual amount of water vapor in the air at a given time—for example, in grams per cubic meter

absolute magnitude a classification scheme to compensate for the distance differences to stars; calculations of the brightness that stars would appear to have if they were all at a defined, standard distance

absolute scale temperature scale set so that zero is at the theoretical lowest temperature possible, which would occur when all random motion of molecules has ceased

absolute zero the theoretical lowest temperature possible, which occurs when all random motion of molecules has ceased

abyssal plain the practically level plain of the ocean floor

acceleration a change in velocity per change in time; by definition, this change in velocity can result from a change in speed, a change in direction, or a combination of changes in speed and direction

accretion disk fat bulging disk of gas and dust from the remains of the gas cloud that forms around a protostar

acetylcholine a neurotransmitter secreted into the synapse by many axons and received by dendrites

acetylcholinesterase an enzyme present in the synapse that destroys acetylcholine

achondrites homogeneously textured stony meteorites

acid any substance that is a proton donor when dissolved in water; generally considered a solution of hydronium ions in water that can neutralize a base, forming a salt and water

acid-base indicator a vegetable dye used to distinguish acid and base solutions by a color change

acquired characteristics characteristics an organism gains during its lifetime that are not genetically determined and therefore cannot be passed on to future generations

active transport use of a carrier molecule to move molecules through a cell membrane in a direction opposite that of the concentration gradient; the carrier requires an input of energy other than the kinetic energy of the molecules

adenine a double-ring nitrogenous-base molecule in DNA and RNA; the complementary base of thymine or uracil

adenosine triphosphate (ATP) a molecule formed from the building blocks of adenine, ribose, and phosphates; it functions as the primary energy carrier in the cell

aerobic cellular respiration the biochemical pathway that requires oxygen and converts food, such as carbohydrates, to carbon dioxide and water; during this conversion, it releases the chemical-bond energy as ATP molecules

air mass a large, more or less uniform body of air with nearly the same temperature and moisture conditions throughout

air mass weather the weather experienced within a given air mass; characterized by slow, gradual changes from day to day

alcohol an organic compound with a general formula of ROH, where R is one of the hydrocarbon groups; for example, methyl or ethyl

aldehyde an organic molecule with the general formula RCHO, where R is one of the hydrocarbon groups; for example, methyl or ethyl

alkali metals members of family IA of the periodic table, having common properties of shiny, low-density metals that can be cut with a knife and that react violently with water to form an alkaline solution

alkaline earth metals members of family IIA of the periodic table, having common properties of soft, reactive metals that are less reactive than alkali metals

alkanes hydrocarbons with single covalent bonds between the carbon atoms

alkenes hydrocarbons with a double covalent carbon-carbon bond

alkyne hydrocarbon with a carbon-carbon triple bond

alleles alternative forms of a gene for a particular characteristic (e.g., attached-earlobe and free-earlobe are alternative alleles for ear shape)

alpha particle the nucleus of a helium atom (two protons and two neutrons) emitted as radiation from a decaying heavy nucleus; also known as an alpha ray

alpine glaciers glaciers that form at high elevations in mountainous regions

alternating current an electric current that first moves one direction, then the opposite direction with a regular frequency

alternation of generations a term used to describe that aspect of the life cycle in which there are two distinctly different forms of an organism; each form is involved in the production of the other, and only one form is involved in producing gametes

alveoli tiny sacs that are part of the structure of the lungs where gas exchange takes place

amino acids organic molecules that join to form polypeptides and proteins

amp unit of electric current; equivalent to C/s

ampere full name of the unit amp

amplitude the extent of displacement from the equilibrium condition; the size of a wave from the rest (equilibrium) position

anaphase the third stage of mitosis, characterized by dividing of the centromeres and movement of the chromosomes to the poles

androgens male sex hormones produced by the testes that cause the differentiation of typical internal and external genital male anatomy

angle of incidence angle of an incident (arriving) ray or particle to a surface; measured from a line perpendicular to the surface (the normal)

angle of reflection angle of a reflected ray or particle from a surface; measured from a line perpendicular to the surface (the normal)

angular momentum quantum number in the quantum mechanics model of the atom, one of four descriptions of the energy state of an electron wave; this quantum number describes the energy sublevels of electrons within the main energy levels of an atom

annular eclipse occurs when the penumbra reaches the surface of Earth; as seen from Earth, the Sun forms a bright ring around the disk of the Moon

anorexia nervosa a nutritional deficiency disease characterized by severe, prolonged weight loss for fear of becoming obese

Antarctic Circle parallel identifying the limit toward the equator where the Sun appears above the horizon all day for six months during the summer; located at 66.5°S latitude

anther the sex organ in plants that produces the pollen that contains the sperm

antibody a globular protein molecule made by the body in response to the presence of a foreign or harmful molecule called an antigen; these molecules are capable of combining with the foreign molecules or microbes to inactivate or kill them

anticline an arch-shaped fold in layered bed rock

anticodon a sequence of three nitrogenous bases on a tRNA molecule capable of forming hydrogen bonds with three complementary bases on an mRNA codon during translation

anticyclone a high-pressure center with winds flowing away from the center; associated with clear, fair weather

antinode region of maximum amplitude between adjacent nodes in a standing wave

aorta the large blood vessel that carries blood from the left ventricle to the majority of the body

apogee the point at which the Moon's elliptical orbit takes the Moon farthest from Earth

apoptosis also known as "programmed cell death"; death of specific cells that has a genetic basis and is not the result of injury

apparent local noon the instant when the Sun crosses the celestial meridian at any particular longitude

apparent local solar time the time found from the position of the Sun in the sky; the shadow of the gnomon on a sundial

apparent magnitude a classification scheme for different levels of brightness of stars that you see; brightness values range from one to six with the number one (first magnitude) assigned to the brightest star and the number six (sixth magnitude) assigned to the faintest star that can be seen

apparent solar day the interval between two consecutive crossings of the celestial meridian by the Sun

aquifer a layer of sand, gravel, or other highly permeable material beneath the surface that is saturated with water and is capable of producing water in a well or spring

Archaea the domain in which are found prokaryotic organisms that live in extreme habitats

archaeology the study of past human life and culture from material evidence

Arctic Circle parallel identifying the limit toward the equator where the Sun appears above the horizon all day for one day up to six months during the summer; located at 66.5°N latitude

area the extent of a surface; the surface bounded by three or more lines

arid dry climate classification; receives less than 25 cm (10 in) precipitation per year

aromatic hydrocarbon organic compound with at least one benzene ring structure; cyclic hydrocarbons and their derivatives

arteries the blood vessels that carry blood away from the heart

arterioles small arteries located just before capillaries that can expand and contract to regulate the flow of blood to parts of the body

artesian term describing the condition where confining pressure forces groundwater from a well to rise above the aquifer

asbestos the common name for any one of several incombustible fibrous minerals that will not melt or ignite and can be woven into a fireproof cloth or used directly in fireproof insulation; about six different commercial varieties of asbestos are used, one of which has been linked to cancer under heavy exposure

assimilation the physiological process that takes place in a living cell as it converts nutrients in food into specific molecules required by the organism

asteroids small rocky bodies left over from the formation of the solar system; most are accumulated in a zone between the orbits of Mars and Jupiter

asthenosphere a plastic, mobile layer of Earth's structure that extends around Earth below the lithosphere

astronomical unit the radius of Earth's orbit is defined as one astronomical unit (AU)

atom the smallest unit of an element that can exist alone or in combination with other elements

atomic mass unit relative mass unit (u) of an isotope based on the standard of the carbon-12 isotope, which is defined as a mass of exactly 12.00 u; one atomic mass unit (1 u) is 1/12 the mass of a carbon-12 atom

atomic number the number of protons in the nucleus of an atom

atomic weight weighted average of the masses of stable isotopes of an element as they occur in nature, based on the abundance of each isotope of the element and the atomic mass of the isotope compared to carbon-12

atria thin-walled sacs of the heart that receive blood from the veins of the body and empty into the ventricles; singular, *atrium*

atrioventricular valves located between the atria and ventricles of the heart preventing blood from flowing backward from the ventricles into the atria

autosomes chromosomes that typically carry genetic information used by the organism for characteristics other than the primary determination of sex

autotroph an organism that is able to make its food molecules from inorganic raw materials by using basic energy sources such as sunlight

autumnal equinox one of two times a year that daylight and night are of equal length; occurs on or about September 23 and identifies the beginning of the fall season

axis the imaginary line about which a planet or other object rotates

axon a neuronal fiber that carries information away from the nerve cell body

B

background radiation ionizing radiation (alpha, beta, gamma, etc.) from natural sources; between 100 and 500 millirems/yr of exposure to natural radioactivity from the environment

Balmer series a set of four line spectra, narrow lines of color emitted by hydrogen atom electrons as they drop from excited states to the ground state

band of stability a region of a graph of the number of neutrons versus the number of protons in nuclei; nuclei that have the neutron to proton ratios located in this band do not undergo radioactive decay

barometer an instrument that measures atmospheric pressure, used in weather forecasting and in determining elevation above sea level

Barr body the tightly coiled structure in the cells of female mammals

basal metabolic rate (BMR) the amount of energy required to maintain normal body activity while at rest

base any substance that is a proton acceptor when dissolved in water; generally considered a solution that forms hydroxide ions in water that can neutralize an acid, forming a salt and water

basilar membrane a membrane in the cochlea containing sensory cells that are stimulated by the vibrations caused by sound waves

basin a large, bowl-shaped fold in the land into which streams drain; also a small enclosed or partly enclosed body of water

beat rhythmic increases and decreases of volume from constructive and destructive interference between two sound waves of slightly different frequencies

benthic aquatic organisms that live on the bottom

beta particle high-energy electron emitted as ionizing radiation from a decaying nucleus; also known as a beta ray

big bang theory current model of galactic evolution in which the universe was created from an intense and brilliant explosion from a primeval fireball

bile a product of the liver, stored in the gallbladder, which is responsible for the emulsification of fats

binding energy the energy required to break a nucleus into its constituent protons and neutrons; also the energy equivalent released when a nucleus is formed

binomial system of nomenclature a naming system that uses two Latin names, genus and specific epithet, for each type of organism

biodiversity the number of different kinds of organisms found in an area

biogenesis the concept that life originates only from preexisting life

biological species concept the concept that species can be distinguished from one another by their inability to interbreed

biology the science that deals with the study of living things and how living things interact with things around them

biomass (energy) any material formed by photosynthesis, including plants, trees, and crops, and any garbage, crop residue, or animal waste; the mass of living material in an area

biomes large regional communities primarily determined by climate

biotic factors living parts of an organism's environment

black hole the theoretical remaining core of a supernova with an intense gravitational field

blackbody radiation electromagnetic radiation emitted by an ideal material (the blackbody) that perfectly absorbs and perfectly emits radiation

blood the fluid medium consisting of cells and plasma that assists in the transport of materials and heat

Bohr model model of the structure of the atom that attempted to correct the deficiencies of the solar system model and account for the Balmer series

boiling point the temperature at which a phase change of liquid to gas takes place through boiling; the same temperature as the condensation point

boundary the division between two regions of differing physical properties

breaker a wave whose front has become so steep that the top part has broken forward of the wave, breaking into foam, especially against a shoreline

British thermal unit the amount of energy or heat needed to increase the temperature of 1 pound of water 1 degree Fahrenheit (abbreviated Btu)

bronchi major branches of the trachea that ultimately deliver air to bronchioles in the lungs

bronchioles small tubes that deliver air to the alveoli in the lung; they are capable of contracting

bulimia a nutritional deficiency disease characterized by a binge-and-purge cycle of eating

C

calorie the amount of energy (or heat) needed to increase the temperature of 1 gram of water 1 degree Celsius

Calorie the nutritionist's "calorie"; equivalent to 1 kilocalorie

cancer any abnormal growth of cells that has a malignant potential

capillaries tiny blood vessels through the walls of which exchange between cells and the blood takes place

carbohydrates organic compounds that include sugars, starches, and cellulose; carbohydrates are used by plants and animals for structure, protection, and food

carbon cycle the series of processes and organisms through which carbon atoms pass in ecological systems

carbon film a type of fossil formed when the volatile and gaseous constituents of a buried organic structure are distilled away, leaving a carbon film as a record

carbonation in chemical weathering, a reaction that occurs naturally between carbonic acid (H_2CO_3) and rock minerals

carcinogens agents responsible for causing cancer

carnivores animals that eat other animals

carrier protein proteins that pick up molecules from one place in a cell or multicellular organism and transport them to another; for example, certain blood proteins

carrying capacity (ecosystem) the *optimum* population size an area can support over an extended period of time

cast sediments deposited by groundwater in a mold, taking the shape and external features of the organism that was removed to form the mold, then gradually changing to sedimentary rock

cathode rays negatively charged particles (electrons) that are emitted from a negative terminal in an evacuated glass tube

cell the basic structural unit that makes up all living things

cell membrane the outer-boundary membrane of the cell; also known as the *plasma membrane*

cell nucleus the central part of a cell that contains the genetic material

cell plate a plant-cell structure that begins to form in the center of the cell and proceeds to the cell membrane, resulting in cytokinesis

cell wall an outer covering on some cells; may be composed of cellulose, chitin, or peptidoglycan depending on the kind of organism

cellulose a polysaccharide abundant in plants that forms the fibers in cell walls that provide structure for plant materials

Celsius scale referent scale that defines numerical values for measuring hotness or coldness, defined as degrees of temperature; based on the reference points of the freezing point of water and the boiling point of water at sea-level pressure, with 100 degrees between the two points

cementation process by which spaces between buried sediment particles under compaction are filled with binding chemical deposits, binding the particles into a rigid, cohesive mass of a sedimentary rock

Cenozoic the most recent geologic era; the time of recent life, meaning the fossils of this era are identical to the life found on Earth today

central nervous system the portion of the nervous system consisting of the brain and spinal cord

centrifugal force an apparent outward force on an object following a circular path that is a consequence of the third law of motion

centrioles organelles composed of microtubules located just outside the nucleus

centripetal force the force required to pull an object out of its natural straight-line path and into a circular path; *centripetal* means "center seeking"

centromere the unreplicated region where two chromatids are joined

Cepheid variables a bright variable star that can be used to measure distance

cerebellum a region of the brain connected to the medulla oblongata that receives many kinds of sensory stimuli and coordinates muscle movement

cerebrum a region of the brain that surrounds most of the other parts of the brain and is involved in consciousness and thought

chain reaction a self-sustaining reaction where some of the products are able to produce more reactions of the same kind; in a nuclear chain reaction, neutrons are the products that produce more nuclear reactions in a self-sustaining series

chemical bond an attractive force that holds atoms together in a compound

chemical change a change in which the identity of matter is altered and new substances are formed

chemical energy a form of energy involved in chemical reactions associated with changes in internal potential energy; a kind of potential energy that is stored and later released during a chemical reaction

chemical equation concise way of describing what happens in a chemical reaction

chemical reaction a change in matter where different chemical substances are created by forming or breaking chemical bonds

chemical weathering the breakdown of minerals in rocks by chemical reactions with water, gases of the atmosphere, or solutions

chemistry the science concerned with the study of the composition, structure, and properties of substances and the transformations they undergo

Chinook a warm wind that has been warmed by compression; also called Santa Ana

chloroplast an energy-converting, membranous, saclike organelle in plant cells containing the green pigment chlorophyll

cholesterol a lipid of the steroid group produced by the body and found in certain foods; a component of lipoproteins; necessary for the manufacture of cell membranes and vitamin D

chondrites subdivision of stony meteorites containing small spherical lumps of silicate minerals or glass

chondrules small spherical lumps of silicate minerals or glass found in some meteorites

chromatid one of two component parts of a chromosome formed by replication and attached at the centromere

chromatin areas or structures within the nucleus of a cell composed of long, loosely arranged molecules of deoxyribonucleic acid (DNA) in association with proteins

chromatin fibers see *nucleoproteins*

chromosomal aberrations changes in the gene arrangement in chromosomes; for example, translocation and duplication mutations

chromosomes complex tightly coiled structures within the nucleus composed of various kinds of histone proteins and DNA that contains the cell's genetic information

class a group of closely related orders found within a phylum

cleavage furrow an indentation of the cell membrane of an animal cell that pinches the cytoplasm into two parts

climate the general pattern of weather that occurs in a region over a number of years

clitoris a small, elongated erectile structure located between and at the head of the labia

clones genetically identical individuals that were reproduced asexually

coalescence process (meteorology) the process by which large raindrops form from the merging and uniting of millions of tiny water droplets

cochlea the part of the ear that converts sound into nerve impulses

coding strand one of two DNA strands that serves as a template, or pattern, for the synthesis of RNA

codon a sequence of three nucleotides of an mRNA molecule that directs the placement of a particular amino acid during translation

cold front the front that is formed as a cold air mass moves into warmer air

combination chemical reaction a synthesis reaction in which two or more substances combine to form a single compound

comets celestial objects originating from the outer edges of the solar system that move about the Sun in highly elliptical orbits; solar heating and pressure from the solar wind form a tail on the comet that points away from the Sun

commensalism a relationship between two organisms in which one organism is helped and the other is not affected

community all of the kinds of interacting organisms within an ecosystem

compaction the process of pressure from a depth of overlying sediments squeezing together the deeper sediments and squeezing out water

competition a kind of interaction between organisms in which both organisms are harmed to some extent

complementary base a nitrogenous base in nucleic acids that can form hydrogen bonds with another base of a specific nucleotide

complex carbohydrates macromolecules formed as a result of the chemical combining of simple sugars (monomers) to form polysaccharides; for example, starch

compound a pure chemical substance that can be decomposed by a chemical change into simpler substances with a fixed mass ratio

compressive stress a force that tends to compress the surface as Earth's plates move into each other

concentration an arbitrary description of the relative amounts of solute and solvent in a solution; a larger amount of solute makes a concentrated solution, and a small amount of solute makes a dilute concentration

conception fertilization

condensation (sound) a compression of gas molecules; a pulse of increased density and pressure that moves through the air at the speed of sound

condensation (water vapor) when more vapor or gas molecules are returning to the liquid state than are evaporating

condensation nuclei tiny particles such as tiny dust, smoke, soot, and salt crystals that are suspended in the air on which water condenses

condensation point the temperature at which a gas or vapor changes back to a liquid

conduction the transfer of heat from a region of higher temperature to a region of lower temperature by increased kinetic energy moving from molecule to molecule

cones light-sensitive cells in the retina of the eye that respond to different colors of light

consistent law principle the laws of physics are the same in all reference frames that are moving at a constant velocity with respect to each other

constancy of speed principle the speed of light in empty space has the same value for all observers regardless of their velocity

constructive interference the condition in which two waves arriving at the same place, at the same time and in phase, add amplitudes to create a new wave

consumers organisms that must obtain energy in the form of organic matter

continental air mass dry air masses that form over large land areas

continental climate a climate influenced by air masses from large land areas; hot summers and cold winters

continental drift a concept that continents shift positions on Earth's surface, moving across the surface rather than being fixed, stationary landmasses

continental glaciers glaciers that cover a large area of a continent; for example, Greenland and the Antarctic

continental shelf a feature of the ocean floor; the flooded margins of the continents that form a zone of relatively shallow water adjacent to the continents

continental slope a feature of the ocean floor; a steep slope forming the transition between the continental shelf and the deep ocean basin

control group the situation used as the basis for comparison in a controlled experiment

control processes mechanisms that ensure that an organism will carry out all metabolic activities in the proper sequence (coordination) and at the proper rate (regulation)

control rods rods inserted between fuel rods in a nuclear reactor to absorb neutrons and thus control the rate of the nuclear chain reaction

controlled experiment an experiment that allows for a comparison of two events that are identical in all but one respect

convection transfer of heat from a region of higher temperature to a region of lower temperature by the displacement of high-energy molecules; for example, the displacement of warmer, less dense air (higher kinetic energy) by cooler, more dense air (lower kinetic energy)

convection cell complete convective circulation pattern; also, slowly turning regions in the plastic asthenosphere that might drive the motion of plate tectonics

convection zone (of a star) part of the interior of a star according to a model; the region directly above the radiation zone where gases are heated by the radiation zone below and move upward by convection to the surface, where they emit energy in the form of visible light, ultraviolet radiation, and infrared radiation

convergent boundaries boundaries that occur between two plates moving toward each other

core (of Earth) the center part of Earth, which consists of a solid inner part and liquid outer part; makes up about 15 percent of Earth's total volume and about one-third of its mass

core (of a star) dense, very hot region of a star where nuclear fusion reactions release gamma and X-ray radiation

Coriolis effect the apparent deflection due to the rotation of Earth; it is to the right in the Northern Hemisphere

corpus luteum a glandlike structure produced in the ovary that produces hormones (progesterone and estrogen) that prevent the release of other eggs

correlation the determination of the equivalence in geologic age by comparing the rocks in two separate locations

coulomb unit used to measure quantity of electric charge; equivalent to the charge resulting from the transfer of 6.24 billion particles such as the electron

Coulomb's law relationship between charge, distance, and magnitude of the electrical force between two bodies

covalent bond a chemical bond formed by the sharing of a pair of electrons

covalent compound chemical compound held together by a covalent bond or bonds

creep the slow downhill movement of soil down a steep slope

crest the high mound of water that is part of a wave; also refers to the condensation, or high-pressure part, of a sound wave

critical angle limit to the angle of incidence when all light rays are reflected internally

critical mass mass of fissionable material needed to sustain a chain reaction

crossing-over the process that occurs when homologous chromosomes exchange equivalent sections of DNA during meiosis

crude oil petroleum pumped from the ground that has not yet been refined into usable products

crust the outermost part of Earth's interior structure; the thin, solid layer of rock that rests on top of the Mohorovicic discontinuity

cryptorchidism a developmental condition in which the testes do not migrate from the abdomen through the inguinal canal to the scrotum

curie unit of nuclear activity defined as 3.70×10^{10} nuclear disintegrations per second

cycle a complete vibration

cyclone a low-pressure center where the winds move into the low-pressure center and are forced upward; a low-pressure center with clouds, precipitation, and stormy conditions

cytokinesis division of the cytoplasm of one cell into two new cells

cytoplasm the more fluid portion of the protoplasm that surrounds the nucleus

cytosine a single-ring nitrogenous-base molecule in DNA and RNA; complementary to guanine

D

dark energy a recently discovered mystery force that is apparently accelerating the expanding universe

dark matter missing matter of the universe that is believed to exist but is invisible and unseen

data measurement information used to describe something

data points points that may be plotted on a graph to represent simultaneous measurements of two related variables

daughter cells two cells formed by cell division

daughter chromosomes chromosomes produced by DNA replication that contain identical genetic information; formed after chromosome division in anaphase

daughter nuclei two nuclei formed by mitosis

daylight saving time setting clocks ahead one hour during the summer to more effectively utilize the longer days of summer, then setting the clocks back in the fall

decibel scale a nonlinear scale of loudness based on the ratio of the intensity level of a sound to the intensity at the threshold of hearing

decomposers organisms that use dead organic matter as a source of energy

decomposition chemical reaction a chemical reaction in which a compound is broken down into the elements that make up the compound into simpler compounds or into elements and simpler compounds

deflation the widespread removal of base materials from the surface by the wind

delta a somewhat triangular deposit at the mouth of a river formed where a stream flowing into a body of water slowed and lost its sediment-carrying ability

denature a change in the chemical and physical properties of a molecule as a result of the breaking of chemical bonds within the molecule; for example, the change in egg white as a result of cooking

dendrites neuronal fibers that receive information from axons and carry it toward the nerve-cell body

denitrifying bacteria bacteria capable of converting nitrite (NO_2^-) to nitrogen gas (N_2), which is released into the atmosphere

density the compactness of matter described by a ratio of mass per unit volume

deoxyribonucleic acid (DNA) a polymer of nucleotides that serves as genetic information; in prokaryotic cells, it is a double-stranded DNA loop and contains attached HU proteins, and in eukaryotic cells, it is found in strands with attached histone proteins; when tightly coiled, it is known as a chromosome

deoxyribose a 5-carbon sugar molecule; a component of DNA

depolarized having lost the electrical difference existing between two points or objects

destructive interference the condition in which two waves arriving at the same point at the same time out of phase add amplitudes to create zero total disturbance

dew condensation of water vapor into droplets of liquid on surfaces

dew point temperature the temperature at which condensation begins

diaphragm a muscle separating the lung cavity from the abdominal cavity that is involved in exchanging the air in the lungs

diastolic blood pressure the blood pressure recorded while the heart is relaxing

diastrophism all-inclusive term that means any and all possible movements of Earth's plates, including drift, isostatic adjustment, and any other process that deforms or changes Earth's surface by movement

diet the food and drink consumed by a person from day to day

Dietary Reference Intakes current U.S. Department of Agriculture guidelines that provide information on the amounts of certain nutrients members of the public should receive

differentiation the process of forming specialized cells within a multicellular organism

diffuse reflection light rays reflected in many random directions, as opposed to the parallel rays reflected from a perfectly smooth surface such as a mirror

diffusion net movement of a kind of molecule from an area of higher concentration to an area of lesser concentration

diploid having two sets of chromosomes: one set from the maternal parent and one set from the paternal parent

direct current an electrical current that always moves in one direction

direct proportion when two variables increase or decrease together in the same ratio (at the same rate)

disaccharides two monosaccharides joined together with the loss of a water molecule; examples are sucrose (table sugar), lactose, and maltose

dispersion the effect of spreading colors of light into a spectrum with a material that has an index of refraction that varies with wavelength

divergent boundaries boundaries that occur between two plates moving away from each other

divide line separating two adjacent watersheds

DNA code a sequence of three nucleotides of a DNA molecule

DNA polymerase an enzyme that bonds DNA nucleotides together when they base pair with an existing DNA strand

DNA replication the process by which the genetic material (DNA) of the cell reproduces itself prior to its distribution to the next generation of cells

domain a classification group above that of kingdoms

dome a large, upwardly bulging, symmetrical fold that resembles a dome

dominant allele an allele that expresses itself and masks the effects of other alleles for the trait

Doppler effect an apparent shift in the frequency of sound or light due to relative motion between the source of the sound or light and the observer

double bond covalent bond formed when two pairs of electrons are shared by two atoms

dune a hill, low mound, or ridge of windblown sand or other sediments

duodenum the first part of the small intestine, which receives food from the stomach and secretions from the liver and pancreas

dwarf planet an object that is orbiting the Sun, is nearly spherical, but has not cleared matter from its orbital zone and is not a satellite

E

earthquake a quaking, shaking, vibrating, or upheaval of Earth's surface

earthquake epicenter point on Earth's surface directly above an earthquake focus

earthquake focus place where seismic waves originate beneath the surface of Earth

echo a reflected sound that can be distinguished from the original sound, which usually arrives 0.1 s or more after the original sound

eclipse when the shadow of a celestial body falls on the surface of another celestial body

ecology the branch of biology that studies the relationships between organisms and their environment

ecosystem an interacting unit consisting of a collection of organisms and abiotic factors

egg the haploid sex cell produced by sexually mature females

ejaculation the release of sperm cells and seminal fluid through the penis

El Niño changes in the atmospheric pressure systems, ocean currents, water temperatures, and wind patterns that seem to be linked to worldwide changes in the weather

elastic rebound the sudden snap of stressed rock into new positions; the recovery from elastic strain that results in an earthquake

elastic strain an adjustment to stress in which materials recover their original shape after a stress is released

electric charge a fundamental property of electrons and protons; electrons have a negative electric charge and protons have a positive electric charge

electric circuit consists of a voltage source that maintains an electrical potential, a continuous conducting path for a current to follow, and a device where work is done by the electrical potential; a switch in the circuit is used to complete or interrupt the conducting path

electric current the flow of electric charge

electric field force field produced by an electric charge

electric field lines a map of an electric field representing the direction of the force that a test charge would experience; the direction of an electric field shown by lines of force

electric generator a mechanical device that uses wire loops rotating in a magnetic field to produce electromagnetic induction in order to generate electricity

electrical energy a form of energy from electromagnetic interactions; one of five forms of energy—mechanical, chemical, radiant, electrical, and nuclear

electrical force a fundamental force that results from the interaction of electrical charge and is billions and billions of times stronger than the gravitational force; sometimes called the "electromagnetic force" because of the strong association between electricity and magnetism

electrical insulators electrical nonconductors, or materials that obstruct the flow of electric current

electrical nonconductors materials that have electrons that are not moved easily within the material—for example, rubber; electrical nonconductors are also called electrical insulators

electrolyte water solution of ionic substances that conducts an electric current

electromagnet a magnet formed by a solenoid that can be turned on and off by turning the current on and off

electromagnetic force one of four fundamental forces; the force of attraction or repulsion between two charged particles

electromagnetic induction process in which current is induced by moving a loop of wire in a magnetic field or by changing the magnetic field

electron subatomic particle that has the smallest negative charge possible and usually found in an orbital of an atom, but gained or lost when atoms become ions

electron configuration the arrangement of electrons in orbitals and suborbitals about the nucleus of an atom

electron dot notation notation made by writing the chemical symbol of an element with dots around the symbol to indicate the number of outer orbital electrons

electron pair a pair of electrons with different spin quantum numbers that may occupy an orbital

electrostatic charge an accumulated electric charge on an object from a surplus or deficiency of electrons; also called "static electricity"

element a pure chemical substance that cannot be broken down into anything simpler by chemical or physical means; there are over one hundred known elements, the fundamental materials of which all matter is made

endocrine glands secrete chemical messengers into the circulatory system

endocrine system a number of glands that communicate with one another and other tissues through chemical messengers transported throughout the body by the circulatory system

endoplasmic reticulum (ER) folded membranes and tubes throughout the eukaryotic cell that provide a large surface upon which chemical activities take place

endosymbiotic theory a theory suggesting that some organelles found in eukaryotic cells may have originated as free-living prokaryotes

energy the ability to do work

English system a system of measurement that originally used sizes of parts of the human body as referents

entropy the measure of disorder in thermodynamics

environment the surroundings; anything that affects an organism during its lifetime

environmental resistance the collective set of factors that limit population growth

enzyme a protein molecule that acts as a catalyst to speed the rate of a reaction

enzymes protein molecules, produced by organisms, that are able to control the rate at which chemical reactions occur

eon the largest blocks of time in Earth's geologic history: the Hadean, Archean, Proterozoic, and Phanerozoic eons

epinephrine a hormone produced by the adrenal medulla that increases heart rate, blood pressure, and breathing rate

epiphyte a plant that lives on the surface of another plant without doing harm

epochs subdivisions of geologic periods

equation a statement that describes a relationship in which quantities on one side of the equal sign are identical to quantities on the other side

equation of time the cumulative variation between the apparent local solar time and the mean solar time

equinoxes Latin meaning "equal nights"; time when daylight and night are of equal length, which occurs during the spring equinox and the autumnal equinox

eras the major blocks of time in Earth's geologic history: the Cenozoic, Mesozoic, Paleozoic, and Precambrian eras

erosion the process of physically removing weathered materials; for example, rock fragments are physically picked up by an erosion agent such as a stream or a glacier

essential amino acids those amino acids that cannot be synthesized by the human body and must be part of the diet (e.g., lysine, tryptophan, and valine)

esters class of organic compounds with the general structure of RCOOR′, where R is one of the hydrocarbon groups—for example, methyl or ethyl; esters make up fats, oils, and waxes, and some give fruit and flowers their taste and odor

estrogen one of the female sex hormones that cause the differentiation of typical female internal and external genital anatomy; responsible for the changes in breasts, vagina, uterus, clitoris, and pelvic bone structure at puberty

estuary a special category of aquatic ecosystem that consists of shallow, partially enclosed areas where freshwater enters the ocean

ether class of organic compounds with the general formula ROR′, where R is one of the hydrocarbon groups—for example, methyl or ethyl; mostly used as industrial and laboratory solvents

Eubacteria the domain in which are found organisms commonly known as bacteria

Eucarya the domain in which are found all eukaryotic organisms

eukaryotic cells one of the two major types of cells; characterized by cells that have a true nucleus, as in plants, fungi, protists, and animals

euphotic zone the upper layer of the ocean, where the Sun's rays penetrate

evaporation process of more molecules leaving a liquid for the gaseous state than returning from the gas to the liquid; can occur at any given temperature from the surface of a liquid

evolution the continuous genetic adaptation of a population of organisms to its environment over time

excited state as applied to an atom, describes the energy state of an atom that has electrons in a state above the minimum energy state for that atom; as applied to a nucleus, describes the energy state of a nucleus that has particles in a state above the minimum energy state for that nuclear configuration

exfoliation the fracturing and breaking away of curved, sheetlike plates from bare rock surfaces via physical or chemical weathering, resulting in dome-shaped hills and rounded boulders

exocrine glands use ducts to secrete to the surface of the body or into hollow organs of the body; for example, sweat glands or pancreas

exosphere the outermost layer of the atmosphere, where gas molecules merge with the diffuse vacuum of space

experiment a re-creation of an event in a way that enables a scientist to gain valid and reliable empirical evidence

experimental group the group in a controlled experiment that is identical to the control group in all respects but one

exponential growth phase a period of time during population growth when the population increases at an accelerating rate

external energy the total potential and kinetic energy of an everyday-sized object

external parasite a parasite that lives on the outside of its host

F

facilitated diffusion diffusion assisted by carrier molecules

Fahrenheit scale referent scale that defines numerical values for measuring hotness or coldness, defined as degrees of temperature; based on the reference points of the freezing point of water and the boiling point of water at sea-level pressure, with 180 degrees between the two points

family (elements) vertical columns of the periodic table consisting of elements that have similar properties

family (organisms) a group of closely related genera within an order

fats organic compounds of esters formed from glycerol and three long-chain carboxylic acids that are also called triglycerides; called fats in animals and oils in plants

fatty acid one of the building blocks of true fats and phospholipids; composed of a longchain carbon skeleton with a carboxylic acid functional group at one end; for example, linoleic acid

fault a break in the continuity of a rock formation along which relative movement has occurred between the rocks on either side

fault plane the surface along which relative movement has occurred between the rocks on either side; the surface of the break in continuity of a rock formation

ferromagnesian silicates silicates that contain iron and magnesium; examples include the dark-colored minerals olivine, augite, hornblende, and biotite

fertilization the joining of haploid nuclei, usually from an egg and a sperm cell, resulting in a diploid cell called the zygote

fiber natural (plant) or industrially produced polysaccharides that are resistant to hydrolysis by human digestive enzymes

first law of motion every object remains at rest or in a state of uniform straight-line motion unless acted on by an unbalanced force

first law of thermodynamics a statement of the law of conservation of energy in the relationship between internal energy, work, and heat

first quarter the moon phase between the new phase and the full phase when the Moon is perpendicular to a line drawn through Earth and the Sun; one-half of the lighted Moon can be seen from Earth, so this phase is called the first quarter

floodplain the wide, level floor of a valley built by a stream; the river valley where a stream floods when it spills out of its channel

fluids matter that has the ability to flow or be poured; the individual molecules of a fluid are able to move, rolling over or by one another

folds bends in layered bedrock as a result of stress or stresses that occurred when the rock layers were in a ductile condition, probably under considerable confining pressure from deep burial

foliation the alignment of flat crystal flakes of a rock into parallel sheets

follicle the saclike structure near the surface of the ovary that encases the soon-to-be released secondary oocyte

follicle-stimulating hormone (FSH) the pituitary secretion that causes the ovaries to begin to produce larger quantities of estrogen and to develop the follicle and prepare the egg for ovulation

food chain a sequence of organisms that feed on one another, resulting in a flow of energy from a producer through a series of consumers

Food Guide Pyramid a tool developed by the U.S. Department of Agriculture to help the general public plan for good nutrition; guidelines for required daily intake from each of five food groups

food web a system of interlocking food chains

force a push or pull capable of changing the state of motion of an object; since a force has magnitude (strength) as well as direction, it is a vector quantity

formula describes what elements are in a compound and in what proportions

formula weight the sum of the atomic weights of all the atoms in a chemical formula

fossil any evidence of former life

fossil fuels organic fuels that contain the stored radiant energy of the Sun converted to chemical energy by plants or animals that lived millions of years ago; coal, petroleum, and natural gas are the common fossil fuels

Foucault pendulum a heavy mass swinging from a long wire that can be used to provide evidence about the rotation of Earth

fovea centralis the area of sharpest vision on the retina, containing only cones, where light is sharply focused

fracture strain an adjustment to stress in which materials crack or break as a result of the stress

free fall when objects fall toward Earth with no forces acting upward; air resistance is neglected when considering an object to be in free fall

free-living nitrogen-fixing bacteria soil bacteria that provide nitrogen compounds that can be taken up by plants through their roots

freezing point the temperature at which a phase change of liquid to solid takes place; the same temperature as the melting point for a given substance

frequency the number of cycles of a vibration or of a wave occurring in one second, measured in units of cycles per second (hertz)

freshwater water that is not saline and is fit for human consumption

freshwater communities aquatic communities that have low salt concentrations

front the boundary, or thin transition zone, between air masses of different temperatures

frost ice crystals formed by water vapor condensing directly from the vapor phase; frozen water vapor that forms on objects

frost wedging the process of freezing and thawing water in small rock pores and cracks that become larger and larger, eventually forcing pieces of rock to break off

fuel rod long zirconium alloy tubes containing fissionable material for use in a nuclear reactor

full moon the moon phase when Earth is between the Sun and the Moon and the entire side of the Moon facing Earth is illuminated by sunlight

functional group the atom or group of atoms in an organic molecule that is responsible for the chemical properties of a particular class or group of organic chemicals

fundamental charge smallest common charge known; the magnitude of the charge of an electron and a proton

fundamental forces forces that cannot be explained in terms of any other force; gravitational, electromagnetic, weak, and strong nuclear force

fundamental frequency the lowest frequency (longest wavelength) that can set up standing waves in an air column or on a string

fundamental properties a property that cannot be defined in simpler terms other than to describe how it is measured; the fundamental properties are length, mass, time, and charge

fungus the common name for the kingdom Fungi; examples, yeast and mold

G

g symbol representing the acceleration of an object in free fall due to the force of gravity

galactic clusters gravitationally bound subgroups of as many as 1,000 stars that move together within the Milky Way galaxy

galaxy group of billions and billions of stars that form the basic unit of the universe; for example, Earth is part of the solar system, which is located in the Milky Way galaxy

gallbladder an organ attached to the liver that stores bile

galvanometer a device that measures the size of an electric current from the size of the magnetic field produced by the current

gamete a haploid sex cell

gametogenesis the generating of gametes; the meiotic cell-division process that produces sex cells; oogenesis and spermatogenesis

gametophyte stage a life-cycle stage in plants in which a haploid sex cell is produced by mitosis

gamma ray very short wavelength electromagnetic radiation emitted by decaying nuclei

gases a phase of matter composed of molecules that are relatively far apart moving freely in a constant, random motion and have weak cohesive forces acting between them, resulting in the characteristic indefinite shape and indefinite volume of a gas

gasohol solution of ethanol and gasoline

gastric juice the secretions of the stomach that contain enzymes and hydrochloric acid

Geiger counter a device that indirectly measures ionizing radiation (beta and/or gamma) by detecting "avalanches" of electrons that are able to move because of the ions produced by the passage of ionizing radiation

gene a unit of heredity located on a chromosome and composed of a sequence of DNA nucleotides

gene flow the movement of genes from one generation to another or from one place to another

gene frequency a measure of how often a specific gene for a characteristic shows up in the individuals of a population

gene pool all of the genes of a population of organisms

genetically modified (GM) organisms that contain genetic changes

general theory of relativity geometric theory of gravity from an analysis of space and time and how it interacts with a mass

generative processes actions that increase the size of an individual organism (growth) or increase the number of individuals in a population (reproduction)

genetic recombination the gene mixing that occurs as a result of sexual reproduction

genetics the study of genes, how genes produce characteristics, and how the characteristics are inherited

genome a set of all the genes necessary to specify an organism's complete list of characteristics

genotype the catalog of genes of an organism, whether or not these genes are expressed

genus (plural, *genera*) a group of closely related species within a family

geosynchronous satellite a satellite with a period of one day, turning with Earth and appearing to move around a fixed point in the sky

geographic barriers geographic features that keep different portions of a species from exchanging genes

geographic isolation a condition in which part of the gene pool is separated by geographic barriers from the rest of the population

geologic time scale a "calendar" of geologic history based on the appearance and disappearance of particular fossils in the sedimentary rock record

geomagnetic time scale time scale established from the number and duration of magnetic field reversals during the past 6 million years

geothermal energy energy from beneath Earth's surface

giant planets the large outer planets of Jupiter, Saturn, Uranus, and Neptune, which all have similar densities and compositions

glacier a large mass of ice on land that formed from compacted snow and slowly moves under its own weight

gland an organ that manufactures and secretes a material either through ducts or directly into the circulatory system

globular clusters symmetrical and tightly packed clusters of as many as a million stars that move together as subgroups within the Milky Way galaxy

glycerol an alcohol with three hydroxyl groups per molecule; for example, glycerin (1, 2, 3-propanetriol)

glycogen a highly branched polysaccharide synthesized by the human body and stored in the muscles and liver; serves as a direct reserve source of energy

glycol an alcohol with two hydroxyl groups per molecule; for example, ethylene glycol, which is used as an antifreeze

Golgi apparatus a stack of flattened, smooth, membranous sacs; the site of synthesis and packaging of certain molecules in eukaryotic cells

gonad a generalized term for organs in which meiosis occurs to produce gametes; ovary or testis

gonadotropin-releasing hormone (GnRH) a hormone released at puberty by the hypothalamus that stimulates the pituitary

gland to release luteinizing hormone (LH) and follicle-stimulating hormone (FSH)

granite light-colored, coarse-grained igneous rock common on continents; igneous rocks formed by blends of quartz and feldspars, with small amounts of micas, hornblende, and other minerals

greenhouse effect the process of increasing the temperature of the lower parts of the atmosphere through redirecting energy back toward the surface; the absorption and reemission of infrared radiation by carbon dioxide, water vapor, and a few other gases in the atmosphere

ground state energy state of an atom with electrons at the lowest energy state possible for that atom

groundwater water from a saturated zone beneath the surface; water from beneath the surface that supplies wells and springs

growth-stimulating hormone (GSH) a hormone produced by the anterior pituitary gland that stimulates tissues to grow

guanine a double-ring nitrogenous-base molecule in DNA and RNA; the complementary base of cytosine

gyre the great circular systems of moving water in each ocean

H

habitat the place or part of an ecosystem occupied by an organism

hail a frozen form of precipitation, sometimes with alternating layers of clear and opaque, cloudy ice

hair hygrometer a device that measures relative humidity from changes in the length of hair

half-life the time required for one-half of the unstable nuclei in a radioactive substance to decay into a new element

halogen member of family VIIA of the periodic table, having common properties of very reactive nonmetallic elements common in salt compounds

haploid having a single set of chromosomes resulting from the reduction division of meiosis

hard water water that contains relatively high concentrations of dissolved salts of calcium and magnesium

Hardy-Weinberg concept the concept that population must be infinitely large, have random mating, no mutations, no migration, and no selection for specific characteristics in order to prevent evolution from taking place

heart the muscular pump that forces the blood through the blood vessels of the body

heat total internal energy of molecules, which is increased by gaining energy from a temperature difference (conduction, convection, radiation) or by gaining energy from

a form conversion (mechanical, chemical, radiant, electrical, nuclear)

heat of formation energy released in a chemical reaction

Heisenberg uncertainty principle you cannot measure both the exact momentum and the exact position of a subatomic particle at the same time—when the more exact of the two is known, the less certain you are of the value of the other

hemoglobin an iron-containing molecule found in red blood cells, to which oxygen molecules bind

herbivores animals that feed directly on plants

hertz unit of frequency; equivalent to one cycle per second

Hertzsprung-Russell diagram diagram to classify stars with a temperature-luminosity graph

heterotroph an organism that requires a source of organic material from its environment; it cannot produce its own food

heterozygous describes a diploid organism that has two different alleles for a particular characteristic

high short for high-pressure center (anticyclone), which is associated with clear, fair weather

high latitudes latitudes close to the poles; those that for a period of time during the winter months receive no solar radiation at noon

high-pressure center another term for *anticyclone*

homeostasis the process of maintaining a constant internal environment as a result of constant monitoring and modification of the functioning of various systems

homologous chromosomes a pair of chromosomes in a diploid cell that contain similar genes at corresponding loci throughout their length

homozygous describes a diploid organism that has two identical alleles for a particular characteristic

hormones chemical messengers secreted by endocrine glands to regulate other parts of the body

horsepower measurement of power defined as a power rating of 550 ft·lb/s

host an organism that a parasite lives in or on

hot spots sites on Earth's surface where plumes of hot rock materials rise from deep within the mantle

humid moist climate classification; receives more than 50 cm (20 in) precipitation per year

humidity the amount of water vapor in the air; see *relative humidity*

hurricane a tropical cyclone with heavy rains and winds exceeding 120 km/h

hydration the attraction of water molecules for ions; a reaction that occurs between water and minerals that make up rocks

hydrocarbon an organic compound consisting of only the two elements hydrogen and carbon

hydrocarbon derivatives organic compounds that can be thought of as forming when one or more hydrogen atoms on a hydrocarbon have been replaced by an element or a group of elements other than hydrogen

hydrogen bonding a strong bond that occurs between the hydrogen end of one molecule and the fluorine, oxygen, or nitrogen end of another molecule

hydrologic cycle water vapor cycling into and out of the atmosphere through continuous evaporation of liquid water from the surface and precipitation of water back to the surface

hydronium ion a molecule of water with an attached hydrogen ion, H_3O

hypothalamus a region of the brain located in the floor of the thalamus and connected to the pituitary gland that is involved in sleep and arousal; emotions such as anger, fear, pleasure, hunger, sexual response, and pain; and automatic functions such as temperature, blood pressure, and water balance

hypothesis a tentative explanation of a phenomenon that is compatible with the data and provides a framework for understanding and describing that phenomenon

I

ice-crystal process a precipitation-forming process that brings water droplets of a cloud together through the formation of ice crystals

ice-forming nuclei small, solid particles suspended in air; ice can form on the suspended particles

igneous rocks rocks that formed from magma, which is a hot, molten mass of melted rock materials

immune system a system of white blood cells specialized to provide the body with resistance to disease; there are two types, antibody-mediated immunity and cell-mediated immunity

impulse a change of motion is brought about by an impulse; the product of the size of an applied force and the time the force is applied

incandescent matter emitting visible light as a result of high temperature; for example, a lightbulb, a flame from any burning source, and the Sun are all incandescent sources because of high temperature

incident ray line representing the direction of motion of incoming light approaching a boundary

incomplete dominance describes a situation in which the phenotype of a heterozygote is intermediate between the two homozygotes on a phenotypic gradient; that is, the phenotypes appear to be "blended" in heterozygotes

incus the ear bone that is located between the malleus and the stapes

independent assortment the segregation, or assortment, of one pair of homologous chromosomes independently of the segregation, or assortment, of any other pair of chromosomes

index fossils distinctive fossils of organisms that lived only a brief time; used to compare the age of rocks exposed in two different locations

index of refraction the ratio of the speed of light in a vacuum to the speed of light in a material

inertia a property of matter describing the tendency of an object to resist a change in its state of motion; an object will remain in unchanging motion or at rest in the absence of an unbalanced force

inferior vena cava a major vein that returns blood to the heart from the lower body

infrasonic sound waves having too low a frequency to be heard by the human ear; sound having a frequency of less than 20 Hz

inguinal canal opening in the floor of the abdominal cavity through which the testes in a human male fetus descend into the scrotum

initiation code the code on DNA with the base sequence TAC that begins the process of transcription

inorganic chemistry the study of all compounds and elements in which carbon is not the principal element

insulators materials that are poor conductors of heat—for example, heat flows slowly through materials with air pockets because the molecules making up air are far apart; also, materials that are poor conductors of electricity—for example, glass or wood

intensity a measure of the energy carried by a wave

interference phenomenon of light where the relative phase difference between two light waves produces light or dark spots, a result of light's wavelike nature

intermediate-focus earthquakes earthquakes that occur in the upper part of the mantle, between 70 to 350 km below the surface of Earth

intermolecular forces forces of interaction between molecules

internal energy sum of all the potential energy and all the kinetic energy of all the molecules of an object

internal parasite a parasite that lives inside its host

international date line the 180° meridian is arbitrarily called the international date line; used to compensate for cumulative time zone changes by adding or subtracting a day when the line is crossed

interphase the stage between cell divisions in which the cell is engaged in metabolic activities

interstitial cell-stimulating hormone (ICSH) the chemical messenger molecule released from the pituitary gland that causes the testes to produce testosterone, the primary male sex hormone; same as follicle-stimulating hormone in females

intertropical convergence zone a part of the lower troposphere in a belt from 10°N to 10°S of the equator where air is heated, expands, and becomes less dense and rises around the belt

intrusive igneous rocks coarse-grained igneous rocks formed as magma cools slowly deep below the surface

inverse proportion the relationship in which the value of one variable increases while the value of the second variable decreases at the same rate (in the same ratio)

inversion a condition of the troposphere when temperature increases with height rather than decreasing with height; a cap of cold air over warmer air that results in trapped air pollution

ion an atom or a particle that has a net charge because of the gain or loss of electrons; polyatomic ions are groups of bonded atoms that have a net charge

ion exchange reaction a reaction that takes place when the ions of one compound interact with the ions of another, forming a solid that comes out of solution, a gas, or water

ionic bond chemical bond of electrostatic attraction between negative and positive ions

ionic compounds chemical compounds that are held together by ionic bonds; that is, bonds of electrostatic attraction between negative and positive ions

ionization process of forming ions from molecules

ionization counter a device that measures ionizing radiation (alpha, beta, gamma, etc.) by indirectly counting the ions produced by the radiation

ionized an atom or a particle that has a net charge because it has gained or lost electrons

ionosphere refers to that part of the atmosphere—parts of the thermosphere and upper mesosphere—where free electrons and ions reflect radio waves around Earth and where the northern lights occur

iron meteorites meteorite classification group whose members are composed mainly of iron

island arcs curving chains of volcanic islands that occur over belts of deep-seated earthquakes; for example, the Japanese and Indonesian islands

isomers chemical compounds with the same molecular formula but different molecular structure; compounds that are made from the same numbers of the same elements but have different molecular arrangements

isotope atoms of an element with identical chemical properties but with different masses; isotopes are atoms of the same element with different numbers of neutrons

J

jet stream a powerful, winding belt of wind near the top of the troposphere that tends to extend all the way around Earth, moving generally from the west in both hemispheres at speeds of 160 km/h or more

joint a break in the continuity of a rock formation without a relative movement of the rock on either side of the break

joule metric unit used to measure work and energy; can also be used to measure heat; equivalent to newton-meter

K

Kelvin scale a temperature scale that does not have arbitrarily assigned referent points, and zero means nothing; the zero point on the Kelvin scale (also called absolute scale) is the lowest limit of temperature, where all random kinetic energy of molecules ceases

ketone an organic compound with the general formula RCOR′, where R is one of the hydrocarbon groups; for example, methyl or ethyl

keystone species species in an ecosystem that affects many aspects of the ecosystem and whose removal causes significant alteration of the ecosystem

kidneys the primary organs involved in regulating blood levels of water, hydrogen ions, salts, and urea

kilocalorie the amount of energy required to increase the temperature of 1 kilogram of water 1 degree Celsius; equivalent to 1,000 calories

kilogram the fundamental unit of mass in the metric system of measurement

kinetic energy the energy of motion; can be measured from the work done to put an object in motion, from the mass and velocity of the object while in motion, or from the amount of work the object can do because of its motion

kinetic molecular theory the collection of assumptions that all matter is made up of tiny atoms and molecules that interact physically, that explain the various states of matter, and that have an average kinetic energy that defines the temperature of a substance

kingdom a category within a domain used in the classification of organisms

Kuiper Belt a disk-shaped region of small icy bodies some 30 to 100 AU from the Sun; the source of short-period comets

L

lack of dominance the condition of two unlike alleles both expressing themselves, neither being dominant

lag phase a stage of a population growth curve during which both natality and mortality are low

lake a large inland body of standing water

large intestine the last portion of the food tube; primarily involved in reabsorbing water

last quarter the moon phase between the full phase and the new phase when the Moon is perpendicular to a line drawn through Earth and the Sun; one half of the lighted Moon can be seen from Earth, so this phase is called the last quarter

latent heat refers to the heat "hidden" in phase changes

latent heat of fusion the heat absorbed when 1 gram of a substance changes from the solid to the liquid phase, or the heat released by 1 gram of a substance when changing from the liquid phase to the solid phase

latent heat of vaporization the heat absorbed when 1 gram of a substance changes from the liquid phase to the gaseous phase, or the heat released when 1 gram of gas changes from the gaseous phase to the liquid phase

latitude the angular distance from the equator to a point on a parallel that tells you how far north or south of the equator the point is located

lava magma, or molten rock, that is forced to the surface from a volcano or a crack in Earth's surface

law of conservation of energy energy is never created or destroyed; it can only be converted from one form to another as the total energy remains constant

law of conservation of mass same as law of conservation of matter; mass, including single atoms, is neither created nor destroyed in a chemical reaction

law of conservation of momentum the total momentum of a group of interacting objects remains constant in the absence of external forces

law of dominance when an organism has two different alleles for a trait, the allele that is expressed and overshadows the expression of the other allele is said to be dominant;

the allele whose expression is overshadowed is said to be recessive

law of independent assortment members of one allelic pair will separate from each other independently of the members of other allele pairs

law of segregation when gametes are formed by a diploid organism, the alleles that control a trait separate from one another into different gametes, retaining their individuality

light ray model model using lines to show the direction of motion of light to describe the travels of light

light-year the distance that light travels through empty space in one year, approximately 9.5×10^{12} km

limiting factors the identifiable factors that prevent unlimited population growth

limnetic zone the region in a lake that does not have rooted vegetation

line spectrum narrow lines of color in an otherwise dark spectrum; these lines can be used as "fingerprints" to identify gases

linear scale a scale, generally on a graph, where equal intervals represent equal changes in the value of a variable

lines of force lines drawn to make an electric field strength map, with each line originating on a positive charge and ending on a negative charge; each line represents a path on which a charge would experience a constant force and lines closer together mean a stronger electric field

linkage group genes located on the same chromosome that tend to be inherited together

lipid large, nonpolar, organic molecules that do not easily dissolve in polar solvents such as water; there are three different types of lipids: *true fats* (pork chop fat or olive oil), *phospholipids* (the primary component of cell membranes), and *steroids* (most hormones)

liquids a phase of matter composed of molecules that have interactions stronger than those found in a gas but not strong enough to keep the molecules near the equilibrium positions of a solid, resulting in the characteristic definite volume but indefinite shape of a liquid

liter a metric system unit of volume usually used for liquids

lithosphere solid layer of Earth's structure that is above the asthenosphere and includes the entire crust, the Moho, and the upper part of the mantle

littoral zone the region in a lake with rooted vegetation

liver an organ of the body responsible for secreting bile, filtering the blood, detoxifying molecules, and modifying molecules absorbed from the gut

locus the spot on a chromosome where an allele is located

loess very fine dust or silt that has been deposited by the wind over a large area

longitude angular distance of a point east or west from the prime meridian on a parallel

longitudinal wave a mechanical disturbance that causes particles to move closer together and farther apart in the same direction that the wave is traveling

longshore current a current that moves parallel to the shore, pushed along by waves that move accumulated water from breakers

loudness a subjective interpretation of a sound that is related to the energy of the vibrating source, related to the condition of the transmitting medium, and related to the distance involved

low latitudes latitudes close to the equator; those that receive vertical solar radiation at noon during some part of the year

luminous an object or objects that produce visible light; for example, the Sun, stars, lightbulbs, and burning materials are all luminous

lunar eclipse occurs when the Moon is full and the Sun, Moon, and Earth are lined up so the shadow of Earth falls on the Moon

lung a respiratory organ in which air and blood are brought close to one another and gas exchange occurs

L-wave seismic waves that move on the solid surface of Earth much as water waves move across the surface of a body of water

lymph liquid material that leaves the circulatory system to surround cells

lymphatic system a collection of thin-walled tubes that collects, filters, and returns lymph from the body to the circulatory system

lysosome a specialized, submicroscopic organelle that holds a mixture of hydrolytic enzymes

M

macromolecule very large molecule, with a molecular weight of thousands or millions of atomic mass units, that is made up of a combination of many smaller, similar molecules

magma a mass of molten rock material either below or on Earth's crust from which igneous rock is formed by cooling and hardening

magnetic field model used to describe how magnetic forces on moving charges act at a distance

magnetic poles the ends, or sides, of a magnet about which the force of magnetic attraction seems to be concentrated

magnetic quantum number from the quantum mechanics model of the atom, one of four descriptions of the energy state of an electron wave; this quantum number describes the energy of an electron orbital as the orbital is oriented in space by an external magnetic field, a kind of energy sub-sublevel

magnetic reversal the flipping of polarity of Earth's magnetic field as the north magnetic pole and the south magnetic pole exchange positions

main sequence stars normal, mature stars that use their nuclear fuel at a steady rate; stars on the Hertzsprung-Russell diagram in a narrow band that runs from the top left to the lower right

malleus the ear bone that is attached to the tympanum

manipulated variable in an experiment, a quantity that can be controlled or manipulated; also known as the independent variable

mantle middle part of Earth's interior; a 2,870 km (about 1,780 mi) thick shell between the core and the crust

marine climate a climate influenced by air masses from over an ocean, with mild winters and cool summers compared to areas farther inland

marine communities aquatic communities that have a high salt content

maritime air mass a moist air mass that forms over the ocean

mass a measure of inertia, which means a resistance to a change of motion

mass defect the difference between the sum of the masses of the individual nucleons forming a nucleus and the actual mass of that nucleus

mass movement erosion caused by the direct action of gravity

mass number the sum of the number of protons and neutrons in a nucleus defines the mass number of an atom; used to identify isotopes; for example, uranium 238

masturbation stimulation of one's own sex organs

matter anything that occupies space and has mass

meanders winding, circuitous turns or bends of a stream

mean solar day is 24 hours long and is averaged from the mean solar time

mean solar time a uniform time averaged from the apparent solar time

measurement the process of comparing a property of an object to a well-defined and agreed-upon referent

mechanical energy the form of energy associated with machines, objects in motion, and objects having potential energy that results from gravity

mechanical weathering the physical breaking up of rocks without any changes in their chemical composition

medulla oblongata a region of the more primitive portion of the brain connected to the spinal cord that controls such automatic functions as blood pressure, breathing, and heart rate

meiosis the specialized pair of cell divisions that reduces the chromosome number from diploid ($2n$) to haploid (n)

melting point the temperature at which a phase change of solid to liquid takes place; the same temperature as the freezing point for a given substance

Mendelian genetics the pattern of inheriting characteristics that follows the laws formulated by Gregor Mendel

menopause the period when the ovaries stop producing viable secondary oocytes and the body becomes nonreproductive

menstrual cycle (menses, menstrual flow, period) the repeated building up and shedding of the lining of the uterus

meridians north-south running arcs that intersect at both poles and are perpendicular to the parallels

Mesozoic a geologic era; the time of middle life, meaning some of the fossils for this time period are similar to the life found on Earth today, but many are different from anything living today

messenger RNA (mRNA) a molecule composed of ribonucleotides that functions as a copy of the gene and is used in the cytoplasm of the cell during protein synthesis

metabolic processes the total of all chemical reactions within an organism; for example, nutrient uptake and processing, and waste elimination

metal matter having the physical properties of conductivity, malleability, ductility, and luster

metamorphic rocks previously existing rocks that have been changed into a distinctly different rock by heat, pressure, or hot solutions

metaphase the second stage in mitosis, characterized by alignment of the chromosomes at the equatorial plane

metastasize the process of cells of tumors moving from the original site and establishing new colonies in other regions of the body

meteor the streak of light and smoke that appears in the sky when a meteoroid is made incandescent by compression of Earth's atmosphere

meteor shower event when many meteorites fall in a short period of time

meteorite the solid iron or stony material of a meteoroid that survives passage through Earth's atmosphere and reaches the surface

meteoroids remnants of comets and asteroids in space

meteorology the science of understanding and predicting weather

meter the fundamental metric unit of length

metric system a system of referent units based on invariable referents of nature that have been defined as standards

microclimate a local, small-scale pattern of climate; for example, the north side of a house has a different microclimate than the south side

microvilli tiny projections from the surfaces of cells that line the intestine

middle latitudes latitudes equally far from the poles and equator; between the high and low latitudes

mineral (geology) a naturally occurring, inorganic solid element or chemical compound with a crystalline structure

minerals (biology) inorganic elements that cannot be manufactured by the body but are required in low concentrations; essential to metabolism

miscible fluids two or more kinds of fluids that can mix in any proportion

mitochondrion a membranous organelle resembling a small bag with a larger bag inside that is folded back on itself; serves as the site of aerobic cellular respiration

mitosis a process that results in equal and identical distribution of replicated chromosomes into two newly formed nuclei

mixture matter composed of two or more kinds of matter that has a variable composition and can be separated into its component parts by physical means

model a mental or physical representation of something that cannot be observed directly that is usually used as an aid to understanding

moderator a substance in a nuclear reactor that slows fast neutrons so the neutrons can participate in nuclear reactions

molarity (M) a measure of concentration based on the number of moles of solute dissolved in 1 liter of solution

mold the preservation of the shape of an organism by the dissolution of the remains of a buried organism, leaving an empty space where the remains were

mole an amount of a substance that contains Avogadro's number of atoms, ions, molecules, or any other chemical unit; a mole is thus 6.02×10^{23} atoms, ions, or other chemical units

molecule from the chemical point of view, a particle composed of two or more atoms held together by an attractive force called a chemical bond; from the kinetic theory point of view, smallest particle of a compound or gaseous element that can exist and still retain the characteristic properties of a substance

momentum the product of the mass of an object times its velocity

monomer individual, repeating units or segments of complex molecules that chemically combine to form long chainlike molecules called polymers; for example, monosaccharides link to form polysaccharides

monosaccharides simple sugars containing 3 to 6 carbon atoms; the most common kinds are 6-carbon molecules such as glucose and fructose

moraines deposits of rocks and other mounded materials bulldozed into position by a glacier and left behind when the glacier melted

morphological species concept the concept that species can be distinguished from one another by structural characteristics

mortality the number of individuals leaving the population by death per thousand individuals in the population

motor neurons those neurons that carry information from the central nervous system to muscles or glands

motor unit all of the muscle cells stimulated by a single neuron

mountain a natural elevation of Earth's crust that rises above the surrounding surface

multiple alleles a term used to refer to conditions in which there are several different alleles for a particular characteristic, not just two

multiregional hypothesis states that *Homo erectus* evolved into *Homo sapiens*

mutagenic agent anything that causes permanent change in DNA

mutation any change in the genetic information of a cell

mutualism a relationship between two organisms in which both organisms benefit

N

natality the number of individuals entering the population by reproduction per thousand individuals in the population

natural frequency the frequency of vibration of an elastic object that depends on the size, composition, and shape of the object

natural selection the processes that encourage the passage of beneficial genes and discourage the passage of harmful or unfavorable genes from one generation to the next

neap tide period of less-pronounced high and low tides; occurs when the Sun and Moon are at right angles to one another

nebulae a diffuse mass of interstellar clouds of hydrogen gas or dust that may develop into a star

negative electric charge one of the two types of electric charge; repels other negative charges and attracts positive charges

negative ion atom or particle that has a surplus, or imbalance, of electrons and, thus, a negative charge

nephrons millions of tiny tubular units that make up the kidneys, which are responsible for filtering the blood

nerve cell see *neuron*

nerve impulse a series of changes that take place in the neuron, resulting in a wave of depolarization that passes from one end of the neuron to the other

nerves bundles of neuronal fibers

nervous system a network of neurons that carry information from sense organs to the central nervous system and from the central nervous system to muscles and glands

net force the resulting force after all forces have been added; if a net force is zero, all the forces have canceled each other and there is not an unbalanced force

neuron the cellular unit consisting of a cell body and fibers that makes up the nervous system; also called nerve cell

neurotransmitter a molecule released by the axons of neurons that stimulates other cells

neutralized acid or base properties have been lost through a chemical reaction

neutron neutral subatomic particle usually found in the nucleus of an atom

neutron star very small superdense remains of a supernova with a center core of pure neutrons

new crust zone zone of a divergent boundary where new crust is formed by magma upwelling at the boundary

new moon the moon phase when the Moon is between Earth and the Sun and the entire side of the Moon facing Earth is dark

newton a unit of force defined as $kg \cdot m/s^2$; that is, a 1 newton force is needed to accelerate a 1 kg mass 1 m/s^2

niche the functional role of an organism

nitrifying bacteria bacteria that can convert ammonia (NH_3) into nitrite-containing (NO_2^-) compounds, which in turn can be converted into nitrate-containing (NO_3^-) compounds

nitrogen cycle nitrogen in the atmosphere is acted on and used by many bacteria and other organisms and ultimately returned to the atmosphere

nitrogenous base a category of organic molecules found as components of the nucleic acids; there are five common types: thymine, guanine, cytosine, adenine, and uracil

noble gas members of family VIII of the periodic table, having common properties of colorless, odorless, chemically inert gases; also known as rare gases or inert gases

node regions on a standing wave that do not oscillate

noise sounds made up of groups of waves of random frequency and intensity

nonelectrolytes water solutions that do not conduct an electric current; covalent compounds that form molecular solutions and cannot conduct an electric current

nonferromagnesian silicates silicates that do not contain iron or magnesium ions; examples include the minerals of muscovite (white mica), the feldspars, and quartz

nonmetal an element that is brittle (when a solid), does not have a metallic luster, is a poor conductor of heat and electricity, and is not malleable or ductile

nonsilicates minerals that do not have the silicon-oxygen tetrahedra in their crystal structure

noon the event of time when the Sun moves across the celestial meridian

norepinephrine a hormone produced by the adrenal medulla that increases heart rate, blood pressure, and breathing rate

normal a line perpendicular to the surface of a boundary

normal fault a fault where the hanging wall has moved downward with respect to the foot wall

north pole (of a magnet) short for "north seeking"; the pole of a magnet that points northward when it is free to turn

nova a star that explodes or suddenly erupts and increases in brightness

nuclear energy the form of energy from reactions involving the nucleus, the innermost part of an atom

nuclear fission nuclear reaction of splitting a massive nucleus into more stable, less massive nuclei with an accompanying release of energy

nuclear force one of four fundamental forces, a strong force of attraction that operates over very short distances between subatomic particles; this force overcomes the electric repulsion of protons in a nucleus and binds the nucleus together

nuclear fusion nuclear reaction of low-mass nuclei fusing together to form more stable and more massive nuclei with an accompanying release of energy

nuclear membrane the structure surrounding the nucleus that separates the nucleoplasm from the cytoplasm

nuclear reactor steel vessel in which a controlled chain reaction of fissionable materials releases energy

nucleic acids complex molecules that store and transfer genetic information within a cell; constructed of fundamental monomers known as nucleotides; the two common forms are DNA and RNA

nucleoli (singular, *nucleolus*) nuclear structures composed of completed or partially completed ribosomes and the specific parts of chromosomes that contain the information for their construction

nucleons name used to refer to both the protons and neutrons in the nucleus of an atom

nucleoproteins the double-stranded DNA with attached proteins; also called chromatin fibers

nucleosomes histone clusters with their encircling DNA

nucleotide the building block of the nucleic acids, composed of a 5-carbon sugar, a phosphate, and a nitrogenous base

nucleus (atom) tiny, relatively massive and positively charged center of an atom containing protons and neutrons; the small, dense center of an atom

numerical constant a constant without units; a number

nutrients molecules required by the body for growth, reproduction, or repair

nutrition collectively, the processes involved in taking in, assimilating, and utilizing nutrients

O

obesity the condition of being 15 percent to 20 percent above the individual's ideal weight

obligate intracellular parasites infectious particles (viruses) that can function only when inside a living cell

observed lapse rate the rate of change in temperature compared to change in altitude

occluded front a front that has been lifted completely off the ground into the atmosphere, forming a cyclonic storm

ocean the single, continuous body of salt water on the surface of Earth

ocean basin the deep bottom of the ocean floor, which starts beyond the continental slope

ocean currents streams of water within the ocean that stay in about the same path as they move over large distances; steady and continuous onward movement of a channel of water in the ocean

ocean wave a moving disturbance that travels across the surface of the ocean

oceanic ridges long, high, continuous, suboceanic mountain chains; for example, the Mid-Atlantic Ridge in the center of the Atlantic Ocean Basin

oceanic trenches long, narrow, deep troughs with steep sides that run parallel to the edge of continents

octet rule a generalization that helps keep track of the valence electrons in most representative elements; atoms of the representative elements (A families) attempt to acquire an outer orbital with eight electrons through chemical reactions

offspring descendants of a set of parents

ohm unit of resistance; equivalent to volts/amps

Ohm's law the electric potential difference is directly proportional to the product of the current times the resistance

oil field petroleum accumulated and trapped in extensive porous rock structure or structures

oils organic compounds of esters formed from glycerol and three long-chain carboxylic acids that are also called triglycerides; called fats in animals and oils in plants

olfactory epithelium the cells of the nasal cavity that respond to chemicals

omnivores animals that are carnivores at some times and herbivores at others

oogenesis the specific name given to the gametogenesis process that leads to the formation of eggs

Oort cloud a spherical "cloud" of small, icy bodies from 30,000 AU out to a light-year from the Sun; the source of long-period comets

opaque materials that do not allow the transmission of any light

orbital the region of space around the nucleus of an atom where an electron is likely to be found

order a group of closely related families within a class

ore mineral mineral deposits with an economic value

organ a structure composed of two or more kinds of tissues that perform a particular function

organ system a group of organs that performs a particular function

organelles cellular structures that perform specific functions in the cell; the function of an organelle is directly related to its structure

organic acids organic compounds with a general formula of RCOOH, where R is one of the hydrocarbon groups; for example, methyl or ethyl

organic chemistry the study of compounds in which carbon is the principal element

organism an independent living unit

orgasm a complex series of responses to sexual stimulation that results in intense frenzy of sexual excitement

osmosis the net movement of water molecules through a selectively permeable membrane

osteoporosis a disease condition resulting from the demineralization of the bone, resulting in pain, deformities, and fractures; related to a loss of calcium

out-of-Africa hypothesis states that modern humans (*Homo sapiens*) originated in Africa, as had several other hominid species, and migrated from Africa to Asia and Europe and displaced species such as *H. erectus* and *H. heidelbergensis* that had migrated into these areas previously

oval window the membrane-covered opening of the cochlea, to which the stapes is attached

ovaries the female sex organs that produce haploid sex cells—the eggs or ova

overtones higher resonant frequencies that occur at the same time as the fundamental frequency, giving a musical instrument its characteristic sound quality

oviduct the tube (fallopian tube) that carries the oocyte to the uterus

ovulation the release of a secondary oocyte from the surface of the ovary

oxbow lake a small body of water, or lake, that formed when two bends of a stream came together and cut off a meander

oxidation the process of a substance losing electrons during a chemical reaction; a reaction between oxygen and the minerals making up rocks

oxidation-reduction reaction a chemical reaction in which electrons are transferred from one atom to another; sometimes called "redox" for short

oxidizing agents substances that take electrons from other substances

oxidizing atmosphere an atmosphere that contains molecular oxygen

oxytocin a hormone released from the posterior pituitary that causes contraction of the uterus

P

paleontology the science of discovering fossils, studying the fossil record, and deciphering the history of life

Paleozoic a geologic era; time of ancient life, meaning the fossils from this time period are very different from anything living on Earth today

pancreas an organ of the body that secretes many kinds of digestive enzymes into the duodenum

parallels reference lines on Earth used to identify where in the world you are northward or southward from the equator; east and west running circles that are parallel to the equator on a globe with the distance from the equator called the latitude

parasite an organism that lives in or on another organism and derives nourishment from it

parasitism a relationship which involves one organism living in or on another living organism from which it derives nourishment

parts per billion concentration ratio of parts of solute in every 1 billion parts of solution (ppb); could be expressed as ppb by volume or as ppb by weight

parts per million concentration ratio of parts of solute in every 1 million parts of solution (ppm); could be expressed as ppm by volume or as ppm by weight

Pauli exclusion principle no two electrons in an atom can have the same four quantum numbers; thus, a maximum of two electrons can occupy a given orbital

pelagic describes aquatic organisms that float or swim actively

penis the portion of the male reproductive system that deposits sperm in the female reproductive tract

penumbra the zone of partial darkness in a shadow

pepsin an enzyme produced by the stomach that is responsible for beginning the digestion of proteins

percent by volume the volume of solute in 100 volumes of solution

percent by weight the weight of solute in 100 weight units of solution

perception recognition by the brain that a stimulus has been received

perigee when the Moon's elliptical orbit brings the Moon closest to Earth

perihelion the point at which an orbit comes closest to the Sun

period (geologic time) subdivisions of geologic eras

period (periodic table) horizontal rows of elements with increasing atomic numbers; runs from left to right on the element table

period (wave) the time required for one complete cycle of a wave

periodic law similar physical and chemical properties recur periodically when the elements are listed in order of increasing atomic number

peripheral nervous system the fibers that communicate between the central nervous system and other parts of the body

permeability the ability to transmit fluids through openings, small passageways, or gaps

permineralization the process that forms a fossil by alteration of an organism's buried remains by circulating groundwater depositing calcium carbonate, silica, or pyrite

petroleum oil that comes from oil-bearing rock, a mixture of hydrocarbons that is believed to have formed from ancient accumulations of buried organic materials such as remains of algae

pH scale scale that measures the acidity of a solution with numbers below 7 representing acids, 7 representing neutral, and numbers above 7 representing bases

phagocytosis the process by which the cell wraps around a particle and engulfs it

pharynx the region at the back of the mouth cavity; the throat

phase change the action of a substance changing from one state of matter to another; a phase change always absorbs or releases internal potential energy without a temperature change

phases of matter the different physical forms that matter can take as a result of different molecular arrangements, resulting in characteristics of the common phases of a solid, liquid, or gas

phenotype the physical, chemical, and behavioral expression of the genes possessed by an organism

phospholipids a class of complex water-insoluble organic molecules that resemble fats but contain a phosphate group (PO_4) in their structure; a component of cellular membranes

photoelectric effect the movement of electrons in some materials as a result of energy acquired from absorbed light

photon a quantum of energy in a light wave; the particle associated with light

photosynthesis a series of reactions that takes place in chloroplasts and results in the storage of sunlight energy in the form of chemical-bond energy

phylogeny the science that explores the evolutionary relationships among organisms and seeks to reconstruct evolutionary history

phylum a subdivision of a kingdom

physical change a change of the state of a substance but not the identity of the substance

phytoplankton planktonic organisms that carry on photosynthesis

pistil the sex organ in plants that produces eggs or ova

pitch the frequency of a sound wave

pituitary gland the gland at the base of the brain that controls the functioning of other glands throughout the organism

placenta an organ made up of tissues from the embryo and the uterus of the mother that allows for the exchange of materials between the mother's bloodstream and the embryo's bloodstream; it also produces hormones

Planck's constant proportionality constant in the relationship between the energy of vibrating molecules and their frequency of vibration; a value of 6.63×10^{-34} Js

plane of the ecliptic the plane of Earth's orbit

planet an object that is orbiting the Sun, is nearly spherical, and is large enough to clear all matter from its orbital zone

plankton aquatic organisms that are so small and weakly swimming that they are simply carried by currents.

plasma (biology) the watery matrix that contains the molecules and cells of the blood

plasma (physics) a phase of matter; a very hot gas consisting of electrons and atoms that have been stripped of their electrons because of high kinetic energies

plasma membrane the outer-boundary membrane of the cell; see *cell membrane*

plastic strain an adjustment to stress in which materials become molded or bent out of shape under stress and do not return to their original shape after the stress is released

plate tectonics the theory that Earth's crust is made of rigid plates that float on the asthenosphere

pleiotropy the multiple effects that a gene may have on the phenotype of an organism

point mutation a change in the DNA of a cell as a result of a loss or change in a single nitrogenous-base

polar air mass cold air mass that forms in cold regions

polar body the smaller of two cells formed by unequal meiotic division during oogenesis

polar climate zone climate zone of the high latitudes; average monthly temperatures stay below 10°C (50°F), even during the warmest month of the year

polar molecule a molecule that has a negative charge on one part and a positive charge on another part

polarized light whose constituent transverse waves are all vibrating in the same plane; also known as plane-polarized light

Polaroid a film that transmits only polarized light

polyatomic ion ion made up of many atoms

polygenic inheritance the concept that a number of different pairs of alleles may combine their efforts to determine a characteristic

polymers huge, chainlike molecules made of hundreds or thousands of smaller repeating molecular units called monomers

polypeptide chain polymers of amino acids; sometimes called proteins

polysaccharides polymers consisting of monosaccharide units joined together in straight or branched chains; starches, glycogen, or cellulose

pond a small body of standing water, smaller than a lake

pons a region of the brain immediately anterior to the medulla oblongata that connects to the cerebellum and higher regions of the brain and controls several sensory and motor functions of the head and face

population a group of organisms of the same species located in an area

population growth curve a graph of the change in population size over time

porosity the ratio of pore space to the total volume of a rock or soil sample, expressed as a percentage; freely admitting the passage of fluids through pores or small spaces between parts of the rock or soil

positive electric charge one of the two types of electric charge; repels other positive charges and attracts negative charges

positive ion atom or particle that has a net positive charge due to an electron or electrons being torn away

potential energy energy due to position; energy associated with changes in position (e.g., gravitational potential energy) or changes in shape (e.g., compressed or stretched spring)

power the rate at which energy is transferred or the rate at which work is performed; defined as work per unit of time

Precambrian the time before the time of ancient life, meaning the rocks for this time period contain very few fossils

precession the slow wobble of the axis of Earth similar to the wobble of a spinning top

precipitation water that falls to the surface of Earth in the solid or liquid form

predation a relationship between two organisms that involves the capturing, killing, and eating of one by the other

predator an organism that captures, kills, and eats another animal

pressure defined as force per unit area; for example, pounds per square inch (lb/in^2)

prey an organism captured, killed, and eaten by a predator

primary carnivores carnivores that eat herbivores and are therefore on the third trophic level

primary coil part of a transformer; a coil of wire that is connected to a source of alternating current

primary consumers organisms that feed directly on plants—herbivores

primary loop part of the energy-converting system of a nuclear power plant; the closed pipe system that carries heated water from the nuclear reactor to a steam generator

primary oocyte the diploid cell of the ovary that begins to undergo the first meiotic division in the process of oogenesis

primary spermatocyte the diploid cell in the testes that undergoes the first meiotic division in the process of spermatogenesis

prime meridian the referent meridian (0°) that passes through the Greenwich Observatory in England

principal quantum number from the quantum mechanics model of the atom, one of four descriptions of the energy state of an electron wave; this quantum number describes the main energy level of an electron in terms of its most probable distance from the nucleus

principle of uniformity a frame of reference that the same processes that changed the landscape in the past are the same processes you see changing the landscape today

prions proteins that can be passed from one individual to another and cause disease

probability the chance that an event will happen, expressed as a percent or fraction

producers organisms that produce new organic material from inorganic material with the aid of sunlight

progesterone a hormone released by the corpus luteum that is important in maintaining the lining of the uterus

prokaryotic cells one of the two major types of cells; they do not have a typical nucleus bound by a nuclear membrane and lack many of the other membranous cellular organelles; for example, members of the Eubacteria and Archaea

promoter a region of DNA at the beginning of each gene, just ahead of an initiator code

proof a measure of ethanol concentration of an alcoholic beverage; proof is double the concentration by volume; for example, 50 percent by volume is 100 proof

properties qualities or attributes that, taken together, are usually unique to an object; for example, color, texture, and size

prophase the first phase of mitosis during which individual chromosomes become visible

proportionality constant a constant applied to a proportionality statement that transforms the statement into an equation

protein synthesis the process whereby the tRNA utilizes the mRNA as a guide to arrange the amino acids in their proper sequence according to the genetic information in the chemical code of DNA

proteins macromolecular polymers made of smaller molecules of amino acids, with molecular weight from about six thousand to 50 million; proteins are amino acid polymers with roles in biological structures or functions; without such a function, they are known as polypeptides

proton subatomic particle that has the smallest possible positive charge, usually found in the nucleus of an atom

protoplanet nebular model a model of the formation of the solar system that states that the planets formed from gas and dust left over from the formation of the Sun

protoplasm the living portion of a cell as distinguished from the nonliving cell wall

protostar an accumulation of gases that will become a star

pseudoscience use of the appearance of science to mislead; the assertions made are not valid or reliable

psychrometer a two-thermometer device used to measure the relative humidity

puberty a time in the life of a developing individual characterized by the increasing production of sex hormones, which cause it to reach sexual maturity

pulmonary artery the major blood vessel that carries blood from the right ventricle to the lungs

pulmonary circulation the flow of blood through certain chambers of the heart and blood vessels to the lungs and back to the heart

pulsars the source of regular, equally spaced pulsating radio signals believed to be the result of the magnetic field of a rotating neutron star

Punnett square a method used to determine the probabilities of allele combinations in an offspring

pure substance materials that are the same throughout and have a fixed definite composition

pure tone sound made by very regular intensities and very regular frequencies from regular repeating vibrations

P-wave a pressure, or compressional, wave in which a disturbance vibrates materials back and forth in the same direction as the direction of wave movement

P-wave shadow zone a region on Earth between 103° and 142° of arc from an earthquake where no P-waves are received; believed to be explained by P-waves being refracted by the core

pyloric sphincter a valve located at the end of the stomach that regulates the flow of food from the stomach to the duodenum

Q

quad 1 quadrillion Btu (10^{15} Btu); used to describe very large amounts of energy

quanta fixed amounts; usually referring to fixed amounts of energy absorbed or emitted by matter (singular, *quantum*)

quantities measured properties; includes the numerical value of the measurement and the unit used in the measurement

quantum mechanics model of the atom based on the wave nature of subatomic particles, the mechanics of electron waves; also called wave mechanics

quantum numbers numbers that describe energy states of an electron; in the Bohr model of the atom, the orbit quantum numbers could be any whole number 1, 2, 3, and so on out from the nucleus; in the quantum mechanics model of the atom, four quantum numbers are used to describe the energy state of an electron wave

R

rad a measure of radiation received by a material (radiation absorbed dose)

radiant energy the form of energy that can travel through space; for example, visible light and other parts of the electromagnetic spectrum

radiation the transfer of heat from a region of higher temperature to a region of lower temperature by greater emission of radiant energy from the region of higher temperature

radiation zone part of the interior of a star according to a model; the region directly above the core where gamma and X rays from the core are absorbed and reemitted, with the radiation slowly working its way outward

radioactive decay the natural spontaneous disintegration or decomposition of a nucleus

radioactive decay constant a specific constant for a particular isotope that is the ratio of the rate of nuclear disintegration per unit of time to the total number of radioactive nuclei

radioactive decay series series of decay reactions that begins with one radioactive nucleus that decays to a second nucleus that decays to a third nucleus and so on until a stable nucleus is reached

radioactivity spontaneous emission of particles or energy from an atomic nucleus as it disintegrates

radiometric age age of rocks determined by measuring the radioactive decay of unstable elements within the crystals of certain minerals in the rocks

range the geographical distribution of a species

rarefaction a thinning or pulse of decreased density and pressure of gas molecules

ratio a relationship between two numbers, one divided by the other; the ratio of distance per time is speed

real image an image generated by a lens or mirror that can be projected onto a screen

recessive allele an allele that, when present with a dominant allele, does not express itself and is masked by the effect of the dominant allele

recombinant DNA DNA that has been constructed by inserting new pieces of DNA into the DNA of another organism, such as a bacterium

red giant stars one of two groups of stars on the Hertzsprung-Russell diagram that have a different set of properties than the main sequence stars; bright, low-temperature giant stars that are enormously bright for their temperature

redox (chemical) reaction short name for oxidation-reduction reaction

reducing agent supplies electrons to the substance being reduced in a chemical reaction

reduction division (also **meiosis**) a type of cell division in which daughter cells get only half the chromosomes from the parent cell

referent referring to or thinking of a property in terms of another, more familiar object

reflected ray a line representing direction of motion of light reflected from a boundary

reflection the change when light, sound, or other waves bounce backward off a boundary

refraction a change in the direction of travel of light, sound, or other waves crossing a boundary

regulator protein proteins that help determine the activities that will occur in a cell or multicellular organism; for example, enzymes and some hormones

relative dating dating the age of a rock unit or geological event relative to some other unit or event

relative humidity ratio (times 100 percent) of how much water vapor is in the air to the maximum amount of water vapor that could be in the air at a given temperature

reliable a term used to describe results that remain consistent over successive trials

rem measure of radiation that considers the biological effects of different kinds of ionizing radiation

replacement chemical reaction reaction in which an atom or polyatomic ion is replaced in a compound by a different atom or polyatomic ion

replacement (fossil formation) process in which an organism's buried remains are altered by circulating groundwaters carrying elements in solution; the removal of original materials by dissolutions and the replacement of new materials an atom or molecule at a time

representative elements name given to the members of the A-group families of the periodic table; also called the main-group elements

reservoir a natural or artificial pond or lake used to store water, control floods, or generate electricity; a body of water stored for public use

resonance when the frequency of an external force matches the natural frequency of a material and standing waves are set up

response the reaction of an organism to a stimulus

responsive processes those abilities to react to external and internal changes in the environment; for example, irritability, individual adaptation, and evolution

retina the light-sensitive region of the eye

reverberation apparent increase in volume caused by reflections, usually arriving within 0.1 second after the original sound

reverse fault a fault where the hanging wall has moved upward with respect to the foot wall

revolution the motion of a planet as it orbits the Sun

rhodopsin a light-sensitive pigment found in the rods of the retina

ribonucleic acid (RNA) a polymer of nucleotides formed on the template surface of DNA by transcription; three forms that have been identified are mRNA, rRNA, and tRNA

ribose a 5-carbon sugar molecule that is a component of RNA

ribosomal RNA (rRNA) a globular form of RNA; a part of ribosomes

ribosomes small structures composed of two protein and ribonucleic acid subunits; involved in the assembly of proteins from amino acids

Richter scale expresses the intensity of an earthquake in terms of a scale with each higher number indicating ten times more ground movement and about thirty times more energy released than the preceding number

ridges long, rugged mountain chains rising thousands of meters above the abyssal plains of the ocean basin

rift a split or fracture in a rock formation, land formation, or in the crust of Earth

rip current strong, brief current that runs against the surf and out to sea

RNA polymerase an enzyme that attaches to the DNA at the promoter region of a gene and assists in combining RNA nucleotides when the genetic information is transcribed into RNA

rock a solid aggregation of minerals or mineral materials that have been brought together into a cohesive solid

rock cycle understanding of igneous, sedimentary, or metamorphic rock as a temporary state in an ongoing transformation of rocks into new types; the process of rocks continually changing from one type to another

rock flour rock pulverized by a glacier into powdery, silt-sized sediment

rods light-sensitive cells in the retina of the eye that respond to low-intensity light but do not respond to different colors of light

rotation the spinning of a planet on its axis

runoff water moving across the surface of Earth as opposed to soaking into the ground

S

salinity a measure of dissolved salts in seawater, defined as the mass of salts dissolved in 1,000 g of solution

salivary amylase an enzyme present in saliva that breaks starch molecules into smaller molecules

salivary glands glands that produce saliva

salt any ionic compound except one with hydroxide or oxide ions

San Andreas fault in California, the boundary between the North American Plate and the Pacific Plate that runs north-south for some 1,300 km (800 miles) with the Pacific Plate moving northwest and the North American Plate moving southeast

saturated air air in which an equilibrium exists between evaporation and condensation; the relative humidity is 100 percent

saturated molecule an organic molecule that has the maximum number of hydrogen atoms possible

saturated solution the apparent limit to dissolving a given solid in a specified amount of water at a given temperature; a state of equilibrium that exists between dissolving solute and solute coming out of solution

scientific law a relationship between quantities, usually described by an equation in the physical sciences; is more important and describes a wider range of phenomena than a scientific principle

scintillation counter a device that indirectly measures ionizing radiation (alpha, beta, gamma, etc.) by measuring the flashes of light produced when the radiation strikes a phosphor

sea a smaller part of the ocean with characteristics that distinguish it from the larger ocean

sea breeze cool, dense air from over water moving over land as part of convective circulation

seafloor spreading the process by which hot, molten rock moves up from the interior of Earth to emerge along mid-oceanic rifts, flowing out in both directions to create new rocks and spread apart the seafloor

seamounts steep, submerged volcanic peaks on the abyssal plain

second the standard unit of time in both the metric and English systems of measurement

second law of motion the acceleration of an object is directly proportional to the net force acting on that object and inversely proportional to the mass of the object

second law of thermodynamics a statement that the natural process proceeds from a state of higher order to a state of greater disorder

secondary carnivores carnivores that feed on primary carnivores and are therefore at the fourth trophic level

secondary coil part of a transformer, a coil of wire in which the voltage of the original alternating current in the primary coil is stepped up or down by way of electromagnetic induction

secondary consumers animals that eat other animals—carnivores

secondary loop part of a nuclear power plant; the closed pipe system that carries steam from a steam generator to the turbines, then back to the steam generator as feedwater

secondary oocyte the larger of the two cells resulting from the unequal cytoplasmic division of a primary oocyte in meiosis I of oogenesis

secondary sexual characteristics characteristics of the adult male or female, including the typical shape that develops at puberty: broader shoulders, heavier long-bone muscles, development of facial hair, axillary hair, and chest hair, and changes in the shape of the larynx in the male; rounding of the pelvis and breasts and changes in deposition of fat in the female

secondary spermatocyte cells in the seminiferous tubules that go through the second meiotic division, resulting in four haploid spermatids

sedimentary rocks rocks formed from particles or dissolved minerals from previously existing rocks that were deposited from air or water

sediments accumulations of silt, sand, or gravel that settled out of the atmosphere or out of water

segregation the separation and movement of homologous chromosomes to the opposite poles of the cell

seismic waves vibrations that move as waves through any part of Earth, usually associated with earthquakes, volcanoes, or large explosions

seismograph an instrument that measures and records seismic wave data

selecting agents specific environmental factors that favor the passage of certain characteristics from one generation to the next and discourage others

selectively permeable the property of a membrane that allows certain molecules to pass through it, but interferes with the passage of others

semen the sperm-carrying fluid produced by the seminal vesicles, prostate glands, and bulbourethral glands of males

semiarid climate classification between arid and humid; receives between 25 and 50 cm (10 and 20 in) precipitation per year

semicircular canals a set of tubular organs associated with the cochlea that sense changes in the movement or position of the head

semiconductors elements that have properties between those of a metal and those of a nonmetal, sometimes conducting an electric current and sometimes acting like an electrical insulator depending on the conditions and their purity; also called metalloids

semilunar valves pulmonary artery and aorta valves that prevent the flow of blood backward into the ventricles

seminal vesicle a part of the male reproductive system that produces a portion of the semen

seminiferous tubules sperm-producing tubes in the testes

sensory neurons those neurons that send information from sense organs to the central nervous system

sex the nature of biological differences between males and females

sex chromosomes a pair of chromosomes that determines the sex of an organism; X and Y chromosomes

sex-determining chromosome the chromosomes X and Y that are primarily responsible for determining if an individual will develop as a male or female

sexual reproduction the propagation of organisms involving the union of gametes from two parents

sexuality a term used in reference to the totality of the aspects—physical, psychological, and cultural—of our sexual nature

shallow-focus earthquakes earthquakes that occur from the surface down to 70 km deep

shear stress produced when two plates slide past one another or by one plate sliding past another plate that is not moving

shell model of the nucleus model of the nucleus that has protons and neutrons moving in energy levels or shells in the nucleus (similar to the shell structure of electrons in an atom)

shells the layers that electrons occupy around the nucleus

shield volcano a broad, gently sloping volcanic cone constructed of solidified lava flows

shock wave a large, intense wave disturbance of very high pressure; the pressure wave created by an explosion, for example

sickle-cell anemia a disease caused by a point mutation; this malfunction produces sickle-shaped red blood cells

sidereal day the interval between two consecutive crossings of the celestial meridian by a particular star

sidereal month the time interval between two consecutive crossings of the Moon across any star

sidereal year the time interval required for Earth to move around its orbit so that the Sun is again in the same position against the stars

silicates minerals that contain silicon-oxygen tetrahedra either isolated or joined together in a crystal structure

sill a tabular-shaped intrusive rock that formed when magma moved into the plane of contact between sedimentary rock layers

simple harmonic motion the vibratory motion that occurs when there is a restoring force opposite to and proportional to a displacement

single bond covalent bond in which a single pair of electrons is shared by two atoms

single-factor cross a genetic study in which one characteristic is followed from the parental generation to the offspring

small intestine the portion of the digestive system immediately following the stomach; responsible for digestion and absorption

small solar system bodies all objects orbiting the Sun that are not planets or dwarf planets

soil a mixture of unconsolidated weathered earth materials and humus, which is altered, decay-resistant organic matter

solenoid a cylindrical coil of wire that becomes electromagnetic when a current runs through it

solids a phase of matter with molecules that remain close to fixed equilibrium positions due to strong interactions between the molecules, resulting in the characteristic definite shape and definite volume of a solid

solstices times when the Sun is at its maximum or minimum altitude in the sky, known as the summer solstice and the winter solstice

solubility dissolving ability of a given solute in a specified amount of solvent, the concentration that is reached as a saturated solution is achieved at a particular temperature

solute the component of a solution that dissolves in the other component; the solvent

solution a homogeneous mixture of ions or molecules of two or more substances

solvent the component of a solution present in the larger amount; the solute dissolves in the solvent to make a solution

soma the cell body of a neuron, which contains the nucleus

sonic boom sound waves that pile up into a shock wave when a source is traveling at or faster than the speed of sound

sound quality characteristic of the sound produced by a musical instrument; determined by the presence and relative strengths of the overtones produced by the instrument

south pole (of a magnet) short for "south seeking"; the pole of a magnet that points southward when it is free to turn

special theory of relativity analysis of how space and time are affected by motion between an observer and what is being measured

speciation the process of generating new species

species a group of organisms that can interbreed naturally to produce fertile offspring

specific heat each substance has its own specific heat, which is defined as the amount of energy (or heat) needed to increase the temperature of 1 gram of a substance 1 degree Celsius

speed a measure of how fast an object is moving—the rate of change of position per change in time; speed has magnitude only and does not include the direction of change

sperm the haploid sex cells produced by sexually mature males

spermatids haploid cells produced by spermatogenesis that change into sperm

spermatogenesis the specific name given to the gametogenesis process that leads to the formation of sperm

spin quantum number from the quantum mechanics model of the atom, one of four descriptions of the energy state of an electron wave; this quantum number describes the spin orientation of an electron relative to an external magnetic field

spinal cord the portion of the central nervous system located within the vertebral column that carries both sensory and motor information between the brain and the periphery of the body

spindle fibers an array of microtubules extending from pole to pole; used in the movement of chromosomes

spontaneous generation the theory that living organisms arose from nonliving material

spontaneous mutation a change in the DNA of an organism for which there is no known cause

spring equinox one of two times a year that daylight and night are of equal length; occurs on or about March 21 and identifies the beginning of the spring season

spring tides unusually high and low tides that occur every two weeks because of the relative positions of Earth, Moon, and Sun

stable equilibrium phase a stage in a population growth curve following rapid growth in which there is both a decrease in natality and an increase in mortality so that the size of the population is stable

standard atmospheric pressure the average atmospheric pressure at sea level, which is also known as normal pressure; the standard pressure is 29.92 in or 760.0 mm of mercury (1,013.25 millibar)

standard time zones 15° wide zones defined to have the same time throughout the zone, with the time defined as the mean solar time at the middle of each zone

standard unit a measurement unit established as the standard upon which the value of the other referent units of the same type are based

standing waves condition where two waves of equal frequency traveling in opposite directions meet and form stationary regions of maximum displacement due to constructive interference and stationary regions of zero displacement due to destructive interference

stapes the ear bone that is attached to the oval window

starch a group of complex carbohydrates (polysaccharides) that plants use as a stored food source and that serves as an important source of food for animals

stationary front occurs when the edge of a weather front is not advancing

steam generator part of a nuclear power plant; the heat exchanger that heats feedwater from the secondary loop to steam with the very hot water from the primary loop

step-down transformer a transformer that decreases the voltage

step-up transformer a transformer that increases the voltage

steroids a group of lipid molecules characterized by an arrangement of interlocking rings of carbon; they serve as hormones that aid in regulating body processes

stimulus any change in the internal or external environment of an organism that it can detect

stony meteorites meteorites composed mostly of silicate minerals that usually make up rocks on Earth

stony-iron meteorites meteorites composed of silicate minerals and metallic iron

storm a rapid and violent weather change with strong winds, heavy rain, snow, or hail

strain adjustment to stress; a rock unit might respond to stress by changes in volume or shape, or by breaking

stratosphere the layer of the atmosphere above the troposphere where temperature increases with height

stream body of running water

stress a force that tends to compress, pull apart, or deform rock; stress on rocks in Earth's solid outer crust results as Earth's plates move into, away from, or alongside each other

structural protein protein molecules whose function is to provide support and shape to a cell or multicellular organism; for example, muscle protein fibers

structural similarities one of the characteristics of living things; describes the fact that all living things are composed of cells, either prokaryotic or eukaryotic

subduction zone the region of a convergent boundary where the crust of one plate is forced under the crust of another plate into the interior of Earth

sublimation the phase change of a solid directly into a vapor or gas

submarine canyons a feature of the ocean basin; deep, steep-sided canyons that cut through the continental slopes

subspecies regional groups within a species that are significantly different structurally, physiologically, or behaviorally yet are capable of exchanging genes by interbreeding with other members of the species

summer solstice in the Northern Hemisphere, the time when the Sun reaches its maximum altitude in the sky, which occurs on or about June 22 and identifies the beginning of the summer season

superconductors some materials in which, under certain conditions, the electrical resistance approaches zero

supercooled water in the liquid phase when the temperature is below the freezing point

superior vena cava a major vein that returns blood to the heart from the head and upper body

supernova a rare catastrophic explosion of a star into an extremely bright but short-lived phenomenon

supersaturated containing more than the normal saturation amount of a solute at a given temperature

surf the zone where breakers occur; the water zone between the shoreline and the outermost boundary of the breakers

surface wave a seismic wave that moves across Earth's surface, spreading across the surface like water waves spread on the surface of a pond from a disturbance

S-wave a sideways or shear wave in which a disturbance vibrates materials from side to side, perpendicular to the direction of wave movement

S-wave shadow zone a region of Earth more than 103° of arc away from the epicenter of an earthquake where S-waves are not recorded; believed to be the result of the core of Earth that is a liquid, or at least acts like a liquid

swell regular groups of low-profile, long wavelength waves that move continuously

symbiosis a close physical relationship between two kinds of organisms

symbiotic nitrogen-fixing bacteria bacteria that live in the roots of certain kinds of plants, where they convert nitrogen gas molecules into compounds that the plants can use to make amino acids and nucleic acids

synapse the space between the axon of one neuron and the dendrite of the next, where chemicals are secreted to cause an impulse to be initiated in the second neuron

synapsis the condition in which the two members of a pair of homologous chromosomes come to lie close to one another

syncline a trough-shaped fold in layered bedrock

synodic month the interval of time from new moon to new moon (or any two consecutive identical phases)

systemic circulation the flow of blood through certain chambers of the heart and blood vessels to the general body and back to the heart

systolic blood pressure the blood pressure recorded in a large artery while the heart is contracting

T

talus steep, conical, or apronlike accumulations of rock fragments at the base of a slope

taxonomy the science of classifying and naming organisms

telophase the last phase in mitosis, characterized by the formation of daughter nuclei

temperate climate zone climate zone of the middle latitudes; average monthly temperatures stay between 10°C and 18°C (50°F and 64°F) throughout the year

temperature how hot or how cold something is; a measure of the average kinetic energy of the molecules making up a substance

tensional stress the opposite of compressional stress; occurs when one part of a plate moves away from another part that does not move

termination code the DNA nucleotide sequence at the end of a gene with the code ATT, ATC, or ACT that signals "stop here"

terrestrial planets planets Mercury, Venus, Earth, and Mars that have similar densities and compositions as compared to the outer giant planets

testes the male sex organs that produce haploid cells called sperm

testosterone the male sex hormone produced in the testes that controls male sexual development

thalamus a region of the brain that relays information between the cerebrum and lower portions of the brain, providing some level of awareness in that it determines pleasant and unpleasant stimuli and is involved in sleep and arousal

theory a broad, detailed explanation that guides the development of hypotheses and interpretations of experiments in a field of study

theory of natural selection the idea that some individuals within a population will have favorable combinations of genes that make it very likely that those individuals will survive, reproduce, and pass their genes to the next generation

thermometer a device used to measure the hotness or coldness of a substance

third law of motion whenever two objects interact, the force exerted on one object is equal in size and opposite in direction to the force exerted on the other object; forces always occur in matched pairs that are equal and opposite

thrust fault a reverse fault with a low-angle fault plane

thunderstorm a brief, intense electrical storm with rain, lightning, thunder, strong winds, and sometimes hail

thymine a single-ring nitrogenous-base molecule in DNA but not in RNA; it is complementary to adenine

thyroid-stimulating hormone (TSH) a hormone secreted by the pituitary gland that stimulates the thyroid to secrete thyroxine

thyroxine a hormone produced by the thyroid gland that speeds up the metabolic rate

tidal bore a strong tidal current, sometimes resembling a wave, produced in very long, very narrow bays as the tide rises

tidal currents a steady and continuous onward movement of water produced in narrow bays by the tides

tides periodic rise and fall of the level of the sea from the gravitational attraction of the Moon and Sun

tissue a group of specialized cells that work together to perform a specific function

tornado a long, narrow, funnel-shaped column of violently whirling air from a thundercloud that moves destructively over a narrow path when it touches the ground

total internal reflection condition where all light is reflected back from a boundary between materials; occurs when light arrives at a boundary at the critical angle or beyond

total solar eclipse eclipse that occurs when Earth, the Moon, and the Sun are lined up so the Moon completely covers the disk of the Sun; the umbra of the Moon's shadow falls on the surface of Earth

trachea a major tube supported by cartilage that carries air to the bronchi; also known as the windpipe

transcription the process of manufacturing RNA from the template surface of DNA; three forms of RNA that may be produced are mRNA, rRNA, and tRNA

transfer RNA (tRNA) a molecule composed of ribonucleic acid; it is responsible for transporting a specific amino acid into a ribosome for assembly into a protein

transform boundaries in plate tectonics, boundaries that occur between two plates sliding horizontally by each other along a long, vertical fault; sudden jerks along the boundary result in the vibrations of earthquakes

transformer a device consisting of a primary coil of wire connected to a source of alternating current and a secondary coil of wire in which electromagnetic induction increases or decreases the voltage of the source

transition elements members of the B-group families of the periodic table

translation the assembly of individual amino acids into a polypeptide

transparent term describing materials that allow the transmission of light; for example, glass and clear water are transparent materials

transportation the movement of eroded materials by agents such as rivers, glaciers, wind, or waves

transverse wave a mechanical disturbance that causes particles to move perpendicular to the direction that the wave is traveling

triglyceride organic compound of esters formed from glycerol and three long-chain carboxylic acids; also called fats in animals and oils in plants

triple bond covalent bond formed when three pairs of electrons are shared by two atoms

trophic level a step in the flow of energy through an ecosystem

tropic of Cancer parallel identifying the northern limit where the Sun appears directly overhead; located at 23.5°N latitude

tropic of Capricorn parallel identifying the southern limit where the Sun appears directly overhead; located at 23.5°S latitutde

tropical air mass a warm air mass from warm regions

tropical climate zone climate zone of the low latitudes; average monthly temperatures stay above 18°C (64°F), even during the coldest month of the year

tropical cyclone a large, violent circular storm that is born over the warm, tropical ocean near the equator; also called hurricane (Atlantic and eastern Pacific) and typhoon (in western Pacific)

tropical year the time interval between two consecutive spring equinoxes; used as standard for the common calendar year

troposphere layer of the atmosphere from the surface to where the temperature stops decreasing with height

trough the low mound of water that is part of a wave; also refers to the rarefaction, or low-pressure part of a sound wave

true fats also known as *neutral fats;* a category of lipids composed of glycerol and fatty acids; for example, pork chop fat or olive oil

tsunami very large, fast, and destructive ocean wave created by an undersea earthquake, landslide, or volcanic explosion; a seismic sea wave

tumor a mass of undifferentiated cells not normally found in a certain portion of the body

turbidity current a muddy current produced by underwater landslides

tympanum the eardrum

typhoon the name for hurricanes in the western Pacific

U

ultrasonic sound waves too high in frequency to be heard by the human ear; frequencies above 20,000 Hz

umbra the inner core of a complete shadow

unconformity a time break in the rock record

undertow a current beneath the surface of the water produced by the return of water from the shore to the sea

unit in measurement, a well-defined and agreed-upon referent

universal law of gravitation every object in the universe is attracted to every other object with a force directly proportional to the product of their masses and inversely proportional to the square of the distance between the centers of the two masses

unpolarized light light consisting of transverse waves vibrating in all conceivable random directions

unsaturated molecule an organic molecule that does not contain the maximum number of hydrogen atoms; a molecule that can add more hydrogen atoms because of the presence of double or triple bonds

uracil a single-ring nitrogenous-base molecule in RNA but not in DNA; it is complementary to adenine

uterus the organ in female mammals in which the embryo develops

V

vacuole a large sac within the cytoplasm of a cell, composed of a single membrane

vagina the passageway between the uterus and outside of the body; the birth canal

valence the number of covalent bonds an atom can form

valence electrons electrons of the outermost shell; the electrons that determine the chemical properties of an atom and the electrons that participate in chemical bonding

valid a term used to describe meaningful data that fit into the framework of scientific knowledge

Van Allen belts belts of radiation caused by cosmic-ray particles becoming trapped and following Earth's magnetic field lines between the poles

vapor the gaseous state of a substance that is normally in the liquid state

variables changing quantities usually represented by a letter or symbol

veins the blood vessels that return blood to the heart

velocity describes both the speed and direction of a moving object; a change in velocity is a change in speed, in direction of travel, or both

ventifacts rocks sculpted by wind abrasion

ventricles the powerful muscular chambers of the heart whose contractions force blood to flow through the arteries to all parts of the body

vernal equinox another name for the spring equinox, which occurs on or about March 21 and marks the beginning of the spring season

vibration a back-and-forth motion that repeats itself

villi tiny fingerlike projections in the lining of the intestine that increase the surface area for absorption

viroids infectious particles that are composed solely of a small, single strand of RNA

virtual image an image where light rays appear to originate from a mirror or lens; this image cannot be projected on a screen

virus a nucleic acid particle coated with protein that functions as an obligate intracellular parasite

vitamins organic molecules that cannot be manufactured by the body but are required in very low concentrations for good health

volcanism volcanic activity; the movement of magma

volcano a hill or mountain formed by the extrusion of lava or rock fragments from a mass of magma below

volt the ratio of work done to move a quantity of charge

voltage source a device that does work in moving a quantity of charge

volume how much space something occupies

vulcanism volcanic activity; the movement of magma

W

warm front the front that forms when a warm air mass advances against a cool air mass

water table the boundary below which the ground is saturated with water

watershed the region or land area drained by a stream; a stream drainage basin

watt metric unit for power; equivalent to J/s

wave a disturbance or oscillation that moves through a medium

wave equation the relationship of the velocity of a wave to the product of the wavelength and frequency of the wave

wave front a region of maximum displacement in a wave; a condensation in a sound wave

wave height the vertical distance of an ocean wave between the top of the wave crest and the bottom of the next trough

wave mechanics alternate name for quantum mechanics derived from the wavelike properties of subatomic particles

wave period the time required for two successive crests or other successive parts of the wave to pass a given point

wavelength the horizontal distance between successive wave crests or other successive parts of the wave

weak acid acids that only partially ionize because of an equilibrium reaction with water

weak base a base only partially ionized because of an equilibrium reaction with water

weathering slow changes that result in the breaking up, crumbling, and destruction of any kind of solid rock

white dwarf stars one of two groups of stars on the Hertzsprung-Russell diagram that have a different set of properties than the main sequence stars; faint, white-hot stars that are very small and dense

wind a horizontal movement of air that moves along or parallel to the ground, sometimes in currents or streams

wind abrasion the natural sand-blasting process that occurs when wind-driven particles break off small particles of rock and polish the rock they strike

wind chill factor a factor that compares heat loss from bodies in still air with those in moving air; moving air removes heat more rapidly and causes a person to feel that the air is colder than its actual temperature; the cooling power of wind

winter solstice in the Northern Hemisphere, the time when the Sun reaches its minimum altitude, which occurs on or about December 22 and identifies the beginning of the winter season

work the magnitude of applied force times the distance through which the force acts; can be thought of as the process by which one form of energy is transformed to another

X

X chromosome the chromosome in a human female egg (and in one-half of sperm cells) that is associated with the determination of sexual characteristics

X-linked gene a gene located on one of the sex-determining X chromosomes

Y

Y chromosome the sex-determining chromosome in one-half of the sperm cells of human males responsible for determining maleness

Z

zone of saturation zone of sediments beneath the surface in which water has collected in all available spaces

zooplankton small, weakly swimming animals of many kinds

zygote a diploid cell that results from the union of an egg and a sperm; a fertilized egg

Credits

Photographs

Unlimted; Box 20.4b/p. 476: Courtesy Fred Ross; 20.12: © K.G. Murti/Visuals Unlimited; 20.12 top: © K.G. Murti/Visuals Unlimited; 20.12 middle: © Warren Rosenberg/Biological Photo Service; 20.12 bottom right: © Richard Rodewald/Biological Photo Service; 20.12 bottom left: © David M. Phillips/Visuals Unlimited; 20.13: Fred Ross; 20.15b: © Dr. Keith Porter; 20.19: Fred Ross; Box 20.5/p. 493: © James Stevenson/Photo Researchers

Chapter 21 Opener: © Brand X Pictures/Punchstock; 21.4: © Ralph White; Box 21.1/p. 506: © Walt Anderson/Visuals Unlimited, Inc.; 21.7: © Ted Levin/ Animals Animals/Earth Scenes; 21.8: © D.O.E./Science Source/Photo Researchers, Inc.; TA 21.3 left, p. 510: © Dynamic Graphics Group/IT/Stock Free/Alamy; TA 21.3 right, p. 510: © Dynamic Graphics Group/IT/Stock Free/Alamy; 21.10: © Carl & Ann Purcell/Corbis; 21.12a: © Fred Ross; 21.12b: Eldon Enger; 21.12c: © Fred Ross; 21.13a top left, 21.13b top right: Tom & Pat Leeson

Chapter 22 Opener: Courtesy Eldon Enger; 22.2: © McGraw-Hill Higher Education, Inc./Dr. Parvender Sethi, photographer; 22.7: © Bill W. Tillery; 22.8: © W.B. Saunders/Biological Photo Service; 22.9: © PhotoLink/Getty Images; 22.11: © B.A.E. Inc./Alamy; 22.12: U.S. Geological Survey; 22.14a: © Getty Images; 22.14b: © James L. Amos/Corbis; 22.15: © Brand X Pictures/Punchstock; 22.17: © PhotoLink/Getty Images; 22.18: © Stock Montage; 22.20a: © A.M. Siegelman/Visuals Unlimited; 22.20b: Dr. R. Rachel, Dr. H. Huber, and Prof. Dr. K.O. Stetter, University of Regensburg, Lehrstuhl fuer Mikrobiologue, Regensburg, Germany; 22.20c: © John Gerlach/Tom Stack & Associates; 22.20d: © Paul W. Johnson/Biological Photo Service; 22.20e: Eldon Enger; 22.20f: © Glenn Oliver/Visuals Unlimited; 22.21a: © David Denning Photography/BioMEDIA ASSOCIATES; 22.21b: © Don & Pat Valenti/Tom Stacks & Associates; 22.21c: © David Denning Photography/BioMEDIA ASSOCIATES; 22.23 top left: © McGraw-Hill Higher Education, Inc./Steven P. Lynch, photographer; 22.23 top right: © Corbis; 22.23 bottom left: © Bill Beatty/Visuals Unlimited; 22.23 bottom right: © M.I. Walker/Photo Researchers, Inc.; 22.24 top left: © The McGraw-Hill Companies, Inc./Steven P. Lynch, photographer; 22.24 bottom left: © Judy Enger; 22.24 right: © Corbis/Royalty Free; 22.25 left, 22.25 middle left, 22.25 middle right: Eldon Enger; 22.25 right: © Claudia E. Mills/Biological Photo Service; 22.25 bottom: © P. Nuridsany/Photo Researchers, Inc.

Chapter 23 Opener: © Getty Images/Fran Joyce Burek/MHHE DIL; 23.1a: © Gregory K. Scott/Photo Researchers, Inc.; 23.1b: Eldon Enger; TA 23.1/pg. 566: Eldon Enger; pg. 567 top right: © Pat & Tom Leeson/Photo Researchers, Inc.; pg. 567 bottom left: © C. P. Hickman/Visuals Unlimited; pg. 567 top right: Eldon Enger; pg. 567 bottom right: Eldon Enger; pg. 568 top: © The McGraw-Hill Companies, Inc./Steven P. Lynch, photographer; pg. 568 bottom: © Peter K. Ziminski/Visuals Unlimited, Inc.; pg. 568 middle: © Michael Giannenchini/Photo Researchers, Inc.; pg. 569 top: Eldon Enger; pg. 569 bottom: © Doug Wechsler/Animals Animals/Earth Scenes; 23.11a: © John D. Cunningham/Visuals Unlimited; 23.11b: Eldon Enger; Box 23.1/pg. 574: © Judy Enger; 23.12a: © David Waters/Envision; 23.12b: © Stephen Dalton/Photo Researchers, Inc.; 23.13a: © J.H. Robinson/Photo Researchers, Inc.; 23.13b: © Gary Milburn/Tom Stack & Associates; 23.13c: © Frank Blackburn; Ecoscene/Corbis; 23.14a: © Douglas Faulkner/Sally Faulkner Collection; 23.14b: © Kjell Sanved/Visuals Unlimited; 23.15: Alan and Sandy Carey/Getty Images; 23.16: © Kennan Ward/Corbis; 23.22: © G.R. Higbee/Photo Researchers, Inc.; 23.23a: © Stephen J. Krasemann/Photo Researchers, Inc.; 23.23b: © Walt Anderson/Visuals Unlimited

Chapter 24 Opener: © MHHE/Getty Images/M. Freeman/Photolink; 24.1a: © Dr. Keith Porter; Box 24.1/pg. 608: LECO Corporation

Chapter 25 Opener: © David Scharf/Peter Arnold, Inc.; Box 25.1/pg. 630: Courtesy Fred Ross; Box 25.4/pg. 633: Courtesy Darlene Schueller; 25.5: From M. Bartalos and T.A. Baramski, Williams & Wilkins; 25.6 both: From Dr. Kenneth L. Becker, "Fertility and Sterility", 23: 5668-78, 1972 Williams & Wilkins; Box 25.7/pg. 645: © Brand X Pictures/Punchstock; 25.15a-i: © Mcgraw-Hill Education/Bob Coyle, photographer

Chapter 26 Opener: © Renee Lynn/Photo Researchers, Inc.; 26.1a, 26.1b: © McGraw-Hill Higher Education, Inc./Bob Coyle, photographer; Box 26.1/pg. 660: © Hulton Archive/Getty Images; figure 26.3 both: Courtesy Fred Ross; 26.4: © John D. Cunningham/Visuals Unlimited; 26.6a, 26.6b, 26.6c: Courtesy Jeanette Navia; 26.8a: © Renee Lynn/Photo Researchers, Inc.; 26.8b, 26 8c: Courtesy Mary Drapeau; 26.9: Neurofibromatosis Inc., Lanham, MD; 26.12b: © BioPhoto Associates/Photo Researchers, Inc.; 26.16a, 26.16b: © Stanley Flegler/Visuals Unlimited; 26.17 top left: Courtesy USDA Agriculture Research Service; 26.17 top middle: © Vol. 275/Corbis; 26.17 top right: © Vol. 40/Corbis; 26.17 bottom left: Swiss Federal Institute of Technology; 26.17 bottom middle: PhotoDisc Website; 26.17 bottom right: © Vol. 16/PhotoDisc; Box 26.7a-b/pg. 680: Photos Courtesy: College of Verterinary Medicine, Texas A&M University; Box 26.8 left/pg. 682, Box 26.8 middle/pg. 682, Box 26.8 right/pg. 682, Box 26.9/pg. 682: Photo courtesy R.A. Bowen

Illustrations

Chapter 3 Figure 3.17: Source: www.eia.doc.gov/emeu/aer/pdf/pages.secl.pdf

Chapter 11 Chapter 11, Science and Society: Source: www.ocrwm.doe.gov/Ymp/about/history.shtml

Chapter 12 Chapter 12, Concepts Applied: Source: Space Weather News, 8 February 2005 (http://SpaceWeather.com). Also see www.jpl.nasa.gov/tempaltes/flash/neo/neo.htm

Chapter 13 Table 13.3: Source: Data from NASA

Chapter 15 Figure 15.22: From W.C. Pitman, III; R.L. Larson; & E.M. Herron; with permission of the publisher, the Geological Society of America, Boulder, Colorado, USA. Copyright c 1974 Geological Society of America. Figure 15.24: Source: After W. Hamilton, U.S. Geological Survey

Chapter 16 Table 16.1: Source: National Oceanic and Atmospheric Administration

Chapter 17 Box Figure 17.1: From www.nws.noaa.gov/om/windfall/html

Chapter 18 Pages 422–423, A Closer Look: Wastewater Treatment; Box Figure 18.1; and Box Figure 18.2: Drawings, and some text, from HOW WASTEWATER TREATMENT WORKS . . . THE BASICS, The Environmental Protection Agency, Office of Water, http://www.epa.gov/owm/featinfo.htm

Chapter 21 Table 21.3: Data from: Scott, Jeffrey G., et al. "Insecticide resistance in house flies from caged-layer poultry facilities", PEST MANAGEMENT SCIENCE 56:147–153 (2000)

Chapter 22 Table 22.2: Time scale based on 1999 GEOLOGIC TIME SCALE of The Geological Society of America

Chapter 23 Table 23.2: Data from the Population Reference Bureau, 2004 WORLD POPULATION DATA SHEET. Box Figure 23.2: Source: Data from CLIMATE CHANGE-STATE OF KNOWLEDGE, October 1997, Office of Science and Technology Policy, Washington, DC. Figure 23.6: Figure from "Biotic Communities of the Aspan Packland of Central Canada" by Ralph D. Bird from ECOLOGY 11 (2): 410. Reprinted by permission of the Ecological Society of America

Chapter 24 Figure 24.12: Source: U.S. Department of Agriculture, 2005

Chapter 25 Table 25.A: Data from the Centers for Disease Control and Prevention publication, TRACKING THE HIDDEN EPIDEMICS, TRENDS IN STDS IN THE UNITED STATES 2000. Table 25.B: Data from the Centers for Disease Control and Prevention publication, TRACKING THE HIDDEN EPIDEMICS, TRENDS IN STDS IN THE UNITED STATES 2000. Figure 25.11a: Reprinted with permission from CONCEPTS OF HUMAN ANATOMY AND PHYSIOLOGY, 5th Edition by Kent M. Van De Graaff and Stuart Ira Fox

Chapter 26 Table 26.5: Based on R. Lewis, HUMAN GENETICS, CONCEPTS AND APPLICATIONS, 2nd ed., Dubuque, IA: Wm. C. Brown Publishers, 1997. Box Figure 26.7: Source: NATURE, AOP, Published online: 14 February 2002; DOI: 10.1038/nature723

Index

Table of Atomic Weights (Based on Carbon-12)

Name	Symbol	Atomic Number	Atomic Weight	Name	Symbol	Atomic Number	Atomic Weight
Actinium	Ac	89	(227)	Mendelevium	Md	101	258.10
Aluminum	Al	13	26.9815	Mercury	Hg	80	200.59
Americium	Am	95	(243)	Molybdenum	Mo	42	95.94
Antimony	Sb	51	121.75	Neodymium	Nd	60	144.24
Argon	Ar	18	39.948	Neon	Ne	10	20.179
Arsenic	As	33	74.922	Neptunium	Np	93	(237)
Astatine	At	85	(210)	Nickel	Ni	28	58.71
Barium	Ba	56	137.34	Niobium	Nb	41	92.906
Berkelium	Bk	97	(247)	Nitrogen	N	7	14.0067
Beryllium	Be	4	9.0122	Nobelium	No	102	259.101
Bismuth	Bi	83	208.980	Osmium	Os	76	190.2
Bohrium	Bh	107	264	Oxygen	O	8	15.9994
Boron	B	5	10.811	Palladium	Pd	46	106.4
Bromine	Br	35	79.904	Phosphorus	P	15	30.9738
Cadmium	Cd	48	112.40	Platinum	Pt	78	195.09
Calcium	Ca	20	40.08	Plutonium	Pu	94	244.064
Californium	Cf	98	242.058	Polonium	Po	84	(209)
Carbon	C	6	12.0112	Potassium	K	19	39.098
Cerium	Ce	58	140.12	Praseodymium	Pr	59	140.907
Cesium	Cs	55	132.905	Promethium	Pm	61	144.913
Chlorine	Cl	17	35.453	Protactinium	Pa	91	(231)
Chromium	Cr	24	51.996	Radium	Ra	88	(226)
Cobalt	Co	27	58.933	Radon	Rn	86	(222)
Copper	Cu	29	63.546	Rhenium	Re	75	186.2
Curium	Cm	96	(247)	Rhodium	Rh	45	102.905
Dubnium	Db	105	(262)	Rubidium	Rb	37	85.468
Dysprosium	Dy	66	162.50	Ruthenium	Ru	44	101.07
Einsteinium	Es	99	(254)	Rutherfordium	Rf	104	(261)
Erbium	Er	68	167.26	Samarium	Sm	62	150.35
Europium	Eu	63	151.96	Scandium	Sc	21	44.956
Fermium	Fm	100	257.095	Seaborgium	Sg	106	(266)
Fluorine	F	9	18.9984	Selenium	Se	34	78.96
Francium	Fr	87	(223)	Silicon	Si	14	28.086
Gadolinium	Gd	64	157.25	Silver	Ag	47	107.868
Gallium	Ga	31	69.723	Sodium	Na	11	22.989
Germanium	Ge	32	72.59	Strontium	Sr	38	87.62
Gold	Au	79	196.967	Sulfur	S	16	32.064
Hafnium	Hf	72	178.49	Tantalum	Ta	73	180.948
Hassium	Hs	108	(269)	Technetium	Tc	43	(99)
Helium	He	2	4.0026	Tellurium	Te	52	127.60
Holmium	Ho	67	164.930	Terbium	Tb	65	158.925
Hydrogen	H	1	1.0079	Thallium	Tl	81	204.37
Indium	In	49	114.82	Thorium	Th	90	232.038
Iodine	I	53	126.904	Thulium	Tm	69	168.934
Iridium	Ir	77	192.2	Tin	Sn	50	118.69
Iron	Fe	26	55.847	Titanium	Ti	22	47.90
Krypton	Kr	36	83.80	Tungsten	W	74	183.85
Lanthanum	La	57	138.91	Uranium	U	92	238.03
Lawrencium	Lr	103	260.105	Vanadium	V	23	50.942
Lead	Pb	82	207.19	Xenon	Xe	54	131.30
Lithium	Li	3	6.941	Ytterbium	Yb	70	173.04
Lutetium	Lu	71	174.97	Yttrium	Y	39	88.905
Magnesium	Mg	12	24.305	Zinc	Zn	30	65.38
Manganese	Mn	25	54.938	Zirconium	Zr	40	91.22
Meitnerium	Mt	109	(268)				

Note: A value given in parentheses denotes the number of the longest-lived or best-known isotope.

digital communication easily converts all kinds of information into the same format—binary code—people can do all those things (and more) on their laptop or smartphone. Furthermore, computer programs and the accessibility of the Internet make it possible to manipulate and circulate that digital content more easily and in ways people couldn't in the past. Thus, digital communication brought about an information revolution.

By the early 2000s, media critics and analysts used the term **convergence** to describe the changes brought by the digital transition. The term has been used in two interconnected ways. First, as just described, it refers to the technological merging of once distinct and incompatible formats into a single format, which can then be accessed through one device. Second, it refers to the trend of media companies merging together in order to better position themselves for a world in which all media can be digital. We'll explore both of these concepts in more depth in the chapters that follow.

Media Culture in the Digital Era

If our media experience for much of the twentieth century helped form a mass nation, the development of digital communication and the Internet along with industry strategies that accompanied those technologies seems to be creating a **niche nation**—a society in which people navigate a more varied and complex media landscape. Rather than providing widely shared touchstones and consensus narratives, the media technologies we use and the media products we consume now sort us into narrow niches or subcultures, connecting us to some people but disconnecting us from others. Media culture in our niche nation is characterized by three key developments.

New Viewing Practices

One factor enabling the shift to a niche nation is the availability of new technologies that allow people to watch shows on their own timetable. In the pre-digital era of the 1970s, most people watched the evening news or a popular TV show like *Happy Days* at a scheduled time; shows were "pushed" by broadcasters to audiences, who watched them when they aired. Such viewing practices provided common media experiences, connecting people across the country via these weekly rituals. When we watch TV shows today, however, we are more likely to use various "pull" technologies, like a DVR or streaming services such as Hulu, Disney+, or Netflix, which allow us to watch programs at our convenience, in the same way we would pull books off a shelf whenever we like.

Faced with a vastly greater number of channels and choices than were available in the 1970s, we often decide what to watch based on social media recommendations from friends or a streaming service's algorithm, which identifies programs it thinks we will enjoy based on what we've watched in the past. Members of a household can even create separate streaming service profiles to ensure individualized recommendations. These developments have changed television from a technology that brought people together through shared viewing rituals to one that enables, and even encourages, far more individualized media experiences.

Participatory Culture

Digital technologies have helped revive what media scholar Henry Jenkins calls a **participatory culture**—a culture in which it is relatively easy for people to create and share their own content and build connections with others that often reflect and deepen the dynamics of a niche nation.[7] We can participate with and through digital media in many different ways—posting pictures from last night's party to Instagram; live-tweeting during our favorite shows; contributing fan fiction to a community website; or uploading video of ourselves playing an original song to YouTube. In the age of social media posts and video blogging, the once clear line between media producers and consumers has become blurred.

Media industries, however, are also working to tap into—some critics would say redirect or exploit—others' creative energy for their own profit. For example, they encourage people to vote for their favorite contestant on *The Voice* or to create user-generated content by posting to TikTok or adding to the billions of hours of video content on YouTube. The science fiction anthology

CONTENT PRODUCER
Digital media like YouTube enable participatory culture—a way for people to make themselves producers of media content.

series *Black Mirror* even allowed Netflix viewers to help determine how the story would end on the interactive episode "Bandersnatch." Regardless of where the fruits of such labor go, digital-era cultures allow for more people to become producers, creating a more varied media landscape for consumers.

Fragmentation

Partly because of new technologies and partly because of industry strategies, we now have far more media content to choose from than ever before. Much of that content is tailored and marketed to very specific tastes. Netflix subscribers, for example, can look for programming in categories like "Faith & Spirituality Movies," "Emotional LGBTQ+ TV Shows," and "Celebrating Asian Americans and Pacific Islanders," to list just a few. If we turn to YouTube and Spotify, the types of content we can access appear limitless. This wealth of content doesn't mean everything that could or should be produced is, nor does it mean that everything that is produced should be. But it is undeniable that we have access to media products that suit narrow tastes and interests much more than people in earlier eras did.

What are the implications of this change? If culture is a process by which a society's values are established and delivered, and if media products play a role in that process, what happens when people in a society consume very different media from one another—when we trust different news sources, watch different TV shows, follow different Instagram accounts, and have different Facebook feeds? There have always been deep social and political differences in the United States, including at the height of the mass nation. In a highly fragmented media landscape, however, we have fewer media experiences that integrate us into a mainstream culture and more that reinforce certain differences by sorting us into niche cultures or subcultures.

As with most complex social transformations, evaluating the impact of this fragmentation is not easy. On the one hand, we seem to have fewer—or at least less powerful—consensus narratives. In this way, our media culture could be considered more democratic. On the other hand, if the media we consume offer each of us different representations of reality, and if we approach a situation without shared facts or values, solving the kinds of problems democratic societies face is likely to get more difficult as the electorate and our political processes become more polarized. The challenge for all of us will be to find ways of maintaining a healthy democratic system, even in our niche nation.

The Media Environment in Our Digital Era

The word *media* is a Latin plural form of the singular noun *medium*, a term that has evolved to have subtly different usages. When thinking about the mass media, a medium is often understood to be the intervening channel or conduit through which messages are conveyed, usually from a powerful company or organization to a large audience. In this approach, a medium operates in the middle—between a speaker and the audience.

A slightly different usage of the term comes from microbiology, where a medium is seen as a substance in which an organism lives, or its habitat; bacteria in a petri dish, for example, are grown in a medium. Applying this sense of the word, we can think about how media form "a general environment for living—for thinking, perceiving, sensing, [and] feeling"—not just a channel for communicating.[8] To consider today's **media environment** is to think about media as the habitat in which we conduct almost every aspect of our daily lives. Doing so pushes us to define a wider range of experiences as media experiences—not just the things that occur when mass media companies communicate with us, but also the many and varied ways we use technologies to manage and navigate our relationship to the world around us. While some scholars argue that media technologies have always functioned in this way, certain aspects of the digital era make it particularly useful to think of media as an environment in which we live, work, and play.

Online Media as an Environment

To help us see how media function as an environment in the digital era, we can look to online experiences like Facebook. Facebook is certainly a channel for communication, though like many new communication media in the digital age, it complicates the distinction between mass communication and interpersonal communication. Advertisers, magazines, and newspapers can push stories to our feeds (mass communication), but we can also broadcast our own posts, which show up on the feeds of our friends, and send private messages or create group chats (interpersonal communication). But these opportunities for masspersonal communication don't capture all our experiences on Facebook. More than just a channel, Facebook creates an entire ecosystem for its users. It reminds us of friends' birthdays, suggests new people to connect with, and encourages us to manage friendships in specific ways. The profile page even prompts us to define ourselves through limited preset options, such as gender, occupation, education, and relationship status. And for Facebook, Inc., the platform isn't a way to communicate with us but to gather data about us, to surveil us. In the current digital moment, a major function of the media environment—at least from the perspective of corporations—is data gathering. Thus, to think about Facebook solely as a communication channel makes it harder for us to understand all the ways we use, and are used by, the platform.

Thinking about media in this way expands our object of study to include mediated experiences not traditionally associated with mass media. For example, students in both K–12 and college classes increasingly use course management sites like Canvas or Blackboard to access coursework and connect with instructors and classmates. Like Facebook, these sites create an entire habitat. They enable teachers and students to communicate interpersonally in private, one-on-one messages, as well as in group chats and public announcements; they also organize students' schedules with calendars and deadlines, and motivate students with rubrics and grades. In addition, they enable faculty to gather data about students' performance—far more data than most students probably realize. In other words, students and teachers don't just communicate via these platforms—they also work inside them.

Smartphones

Smartphones have become the central hub through which we navigate an increasingly complex media environment. We can use our phones to access traditional mass media products, like news stories, video, games, and songs, and to communicate with other people via voice calls, text messages, or FaceTime. But smartphones also include tools like calendars, calculators, alarm clocks, and timers, and we can link to surveillance cameras to keep an eye on our baby, pet, or front porch. There are apps that play ambient noise to help us sleep, apps that track our daily steps, and apps that ping us to stand up when we've been sitting too long. Google Maps gives us directions, helps us avoid traffic jams, and locates the best tire store or coffeehouse. Our phone also sends out information to telecommunication networks about where we are and what we are doing, integrating us into the digital ecosystem. Many of these media experiences aren't about sending or receiving messages or being

THE MEDIA ENVIRONMENT In the Amazon sci-fi dramedy *Upload* (2020–), a young man's consciousness is uploaded to a luxury virtual resort after a car accident leaves him near death, and he continues to live in this fully digital environment complete with irritating pop-up ads. Set in the near future, the series comments on modern society's increasingly media-infused environment.

entertained or informed in the manner of traditional mass media and communication devices; rather, they are tools to navigate through and manage our relationship to the world around us. The fact that such varied apps sit side by side on the defining media and communication device of our time requires us to reflect on how broad the term *media* can be in this digital era.

The Media Environment and Multitasking

The environment we live in today is increasingly filled with media. Video screens can appear on our appliances and car dashboards, where we pump gas, and as miniature updatable price tags in grocery stores. Machines listen for our commands, and algorithms suggest movies for us to watch. Virtual reality underscores what is already true: The world is increasingly mediated. It isn't surprising that many popular programs, including *Black Mirror*, *Upload*, and *Westworld*, engage with these developments.

Living in this media- and technology-rich environment makes it easier to multitask, and media multitasking has led to growing media consumption—particularly for young people. A Common Sense study found that on average, today's teens report spending nearly ten hours (9:49) a day consuming entertainment media.[9] And roughly 75 percent of teens say they multitask while doing homework by watching TV, texting, or listening to music. If similar studies were to count casual encounters with screens in public or the use of media technologies for homework and other non-entertainment purposes, these averages would be even higher.

Some critics warn about the problems of media multitasking—that we end up more distracted, that we engage less deeply with any one media product, and that we often pay closer attention to the media device in our hand than to people standing next to us. On the other hand, digital-era multitasking can create new kinds of interactions with both media and people. Live-tweeting during a show, for example, can connect us with a wide community of people and enable new forms of engagement with a media product.

Thinking Differently about Media and Culture

In today's digital era, where the media environment constantly surrounds us, our familiarity with media—the fact that we spend more time using media than we do sleeping—can sometimes be an obstacle when we try to analyze media texts. Using and consuming media in our daily lives isn't the same thing as understanding media's various offerings. In this section, we provide a variety of tools for examining the media and its relationship to our culture and ourselves.

Three Roles: Media Consumer, Media Producer, and Media Citizen

Each of us plays three roles in relationship to media (see Figure 1.1). Our most familiar role is that of *media consumer*. Every time we watch a TV show, play a video game, visit a website, scroll through Instagram, or use Google Maps, we consume media. We often have to turn over something in exchange—our money, our attention, or perhaps our data.

We can also play the role of *media producer*. When we post a picture to Snapchat, create a video on TikTok, or stream live on Twitch, we're producing media and often sharing it with large audiences. By pursuing a career in media—as a journalist, for example, or a screenwriter, videographer, game designer, publicist, or PR practitioner—we can also play the role of media producer. Professional media producers like these operate within specific business models, market conditions, and professional norms, though this can also be true for amateur producers; posting to sites like Twitch or YouTube means getting entangled in those companies' regulations and business goals. And, of course, the boundary between amateur and professional producers can be porous; many social media influencers hope to get sponsors and make money.

Finally, we are always *media citizens*, meaning we are members of societies that are saturated with media. As citizens, we have specific rights and responsibilities. In democratic societies, for example, citizenship comes with expectations of being informed, participating in civic life, and caring about the common good. To be a citizen is to act in the ways you think are right and just, and to participate in conversations with your fellow citizens about what society's values should be and how those values should shape our political and legal systems, as well as our culture.

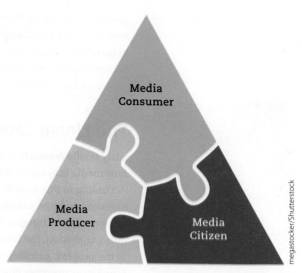

FIGURE 1.1

THE THREE MEDIA ROLES YOU PLAY

It is useful to consider the three different roles you play in relationship to media: the roles of consumer, producer, and citizen. It is also important to reflect on how these roles interact. Sometimes the roles complement one another; other times, they conflict.

Sometimes these three roles dovetail smoothly. As producers, we can draw on our experiences as consumers to make better media products; likewise, our experiences as producers can enrich our media consumption. As citizens, we might seek out media products that reflect our values and boycott those that don't; and as media producers, we may turn down jobs we don't think support our values or the common good. But tensions can also arise between these three roles. Our favorite YouTube channels may include content we feel guilty watching. We may continue to share photos with our friends on Instagram while knowing full well that social media deepens political polarization and hurts our democratic system. Professional media producers often work on projects that conflict with the values they hold as media citizens because they need the job. As you read this book, keep these three roles in mind and reflect on the possible tensions that may exist for you.

Historical Perspective: Connecting the Past to the Present

Another useful tool for analyzing the media is found in every chapter of this book: historical background. Earlier in this chapter, for example, you read about the development of media and communication over five different eras. Historical information like this strengthens your ability to analyze media in two important ways. First, it helps you understand that today's media environment is the result of choices made in the past. For example, your smartphone, the infrastructure that makes it work, the government bodies that regulate it, and much of the content you use on it exist in the form they do because of decisions politicians, corporate leaders, media producers, and activists made decades ago. Recognizing this fact—that the media of today is shaped by the decisions of yesterday—helps you realize that decisions are being made *right now* that will ultimately shape our media future. This in turn might encourage you, in your role as a media citizen, to try to influence these decisions and make your voice heard.

Second, looking to the past can help you see the present with fresh eyes. It can be difficult to grasp the nature of the moment you live in. The ways things are today can seem natural and inevitable, while the past often seems strange, with people living in different situations with different technologies and value systems. Interestingly, though, that strangeness can help us recognize that the ways things "used to be" weren't natural or inevitable. For example, it's easy to recognize the rigid gender norms of the 1950s as products of the era rather than undisputed truths; seeing how gender norms have changed since then makes it clear they were not inevitable to begin with. That insight, in turn, helps you recognize that things in the present that seem natural or unchangeable are not necessarily so—and, again, empowers you to take steps as a citizen to influence the direction of our media and our culture.

The Linear Model of Mass Communication

Early media researchers developed the linear model of communication as a tool to explain how media messages and meanings are constructed and communicated in everyday life. According to this model, mass communication is a linear process of producing and delivering messages to large audiences. *Senders* (authors, producers, and organizations) transmit *messages* (programs, ads, images, sounds) through mass media *channels* (newspapers, books, magazines, radio, television, the Internet) to large groups of *receivers* (readers, viewers, and consumers). In the process, *gatekeepers* (news editors, executive producers, and other media managers) function as message filters, making decisions about what messages actually get produced for particular receivers. The model also allows for *feedback*, in which citizens and consumers can return messages to senders or gatekeepers through phone calls, e-mail, web postings, talk shows, or letters to the editor.

The linear model, with its easy-to-follow flowchart highlighting the function of gatekeepers, channels, and feedback, has shaped how many people think about mass communication for decades. However, like all models, the linear model has limitations. Critics argue that in reality, media messages do not move smoothly from a sender at point A to a receiver at point Z—especially in today's complex digital era, in which TV shows, ads, news reports, social media, smartphone apps, and everyday conversation are layered upon each other. The linear model may also lead us to overestimate the sender's control; a sender may try to craft a message carefully, so that the receiver interprets it in the way the sender wants, but there are no guarantees this will happen.

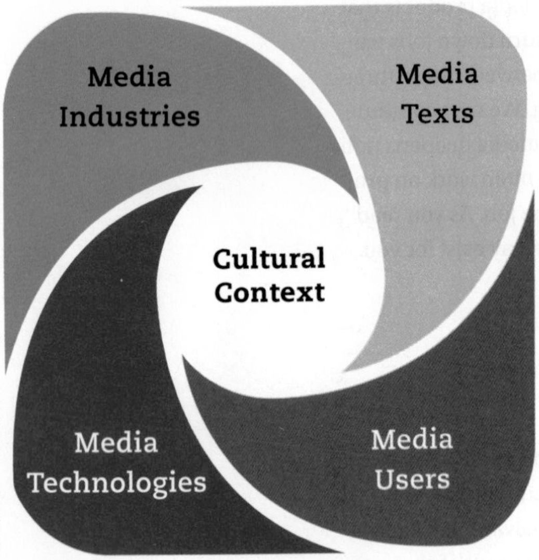

A Cultural Approach to Media and Communication

The cultural approach is another method for analyzing the media—one that doesn't necessarily contradict the linear model but identifies a more intricate fabric. In the cultural approach, every element of the linear model—senders, gatekeepers, messages, channels, and receivers—is an active part of complex cultural conditions and processes, with far more going on than a message moving from point A to point B. The cultural approach has five elements: media texts, technologies, industries, users, and the cultural context within which the other four are embedded. Each individual element focuses our attention on a different aspect of media and its relationship to culture, though ultimately, the model encourages us to analyze the interconnections *among* these five elements (see Figure 1.2).[10] To think through these interconnections, let's use the example of the *Black Panther* movie.

Media Texts

Scholars use the term **text** to refer to anything that conveys meaning or communicates information; a text is anything people "read" or interpret. A book is a text, but so too are all the different media products that we

ananaline/Shutterstock

FIGURE 1.2

A CULTURAL APPROACH TO MEDIA AND COMMUNICATION

The relationship between media and culture is complex, and this figure provides a map for thinking through that relationship. Dividing "media" into four elements—texts, technologies, industries, and users—and locating them within a cultural context encourages us to consider a wide range of issues when analyzing a media-related topic. The graphic's dynamic design reminds us to look at how each element influences—and is influenced by—the others.

encounter every day, including TV series, films, video games, online pop-up ads, Instagram posts, and Spotify playlists. The *Black Panther* film (2018) is certainly a text.

Analyzing media texts leads us to carefully study a text's *formal features*—the various elements that play a role in shaping how we experience and interpret it. Different kinds of texts have different formal elements; for a film, we'd pay attention to narrative (story structure, characters, themes) and genre as well as production design, lighting, sound, editing, camera angles, and framing. For *Black Panther*, we'd consider the fact that the movie is an action film based on the Marvel superhero Black Panther, the king and defender of Wakanda, a resource-rich and technologically superior African country that has isolated itself to keep colonizers at bay. The narrative pits the Black Panther against the villainous Killmonger (who tries to steal Wakanda's advanced weapons to foment a global war against a white world order) and explores a central theme: What responsibility do we have to challenge injustice, and how should we go about doing that? The director, the writer, and a majority of the film's cast are Black—a first for big-budget Hollywood films. The film features fast-paced editing sequences and the elaborate special effects common in today's superhero action films. Although Wakanda is fictional, the sets, costumes, music, and visual effects were created with care to reflect a multiplicity of African cultural styles.

Media Technologies

Media technologies refer to the tools and devices we use to create, send, and receive messages; store and retrieve information; and manage our relationship to the world. It matters which technologies we use, because different technologies enable, encourage, or even force us to communicate in particular ways and ultimately affect us and our culture.

The cultural approach helps us understand that the tools and devices we use shape us. However, it also helps us recognize that we can shape and control these tools and devices, too. Thus, it complicates **technological determinism**—a common but sometimes simplistic way of thinking that sees technology as an independent force that appears out of nowhere and changes everything. Media innovations don't appear fully formed and ready to serve a mass audience; rather, technologies and the ways we use them are embedded in culture, shaped by industry priorities, and sometimes put to unexpected uses by people. For example, although smartphones made it easier to shoot and share video, that affordance was only one factor in the rise of TikTok videos or a political movement for social justice sparked by videos documenting abuse by police. How a device ends up being used, and by whom, is determined not only by the technology itself but also by factors such as investors' profit plans, consumer preferences, and cultural norms.

In the case of *Black Panther*, we could explore various technological issues. We could consider how digital special-effects technologies help immerse viewers into the narrative by making the futuristic world of Wakanda believable. We could consider how the different devices viewers might use to watch the film can influence their experience of it. Did they watch it digitally projected in a crowded ultra-screen theater, on a laptop while sitting alone on their bed, or on their smartphone while riding a bus? We could also look at the ways fans use technologies like Photoshop, social media platforms, or GIF generators to communicate with others about the film by crafting memes and posting fan fiction.

Media Industries

Media industries refer to the organizations that develop and manufacture media technologies; produce, distribute, and sell media texts; or play a role in creating and spreading information through media. This category includes everything from enormous media and technology corporations like Apple and Google to local newspapers, independent bookstores, and YouTube influencers. Analyzing how an industry works—its fundamental business models, workplace cultures, and more—helps us more fully understand why certain media products are made and others aren't, and how specific texts end up looking or sounding the way they do.

In the case of *Black Panther*, we could examine how the film's development was shaped by the business priorities and corporate culture of Disney, the enormous global media company that

launchpadworks.com

***Black Panther* on Film**
The citizens of Wakanda decide whether to challenge Prince T'Challa for the throne in this clip from the Ryan Coogler film.

Discussion: What does this clip tell us about formal elements of the film *Black Panther*, including its narrative (story structure, characters, themes), genre, and production design?

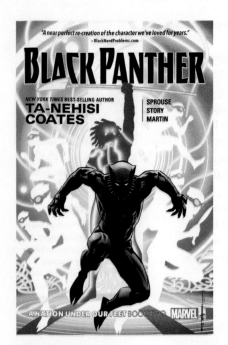

"A near perfect re-creation of the character we've loved for years."
– BlackNerdProblems.com

BLACK PANTHER

NEW YORK TIMES BEST-SELLING AUTHOR
TA-NEHISI COATES

SPROUSE
STORY
MARTIN

A NATION UNDER OUR FEET BOOK TWO · MARVEL

BLACK PANTHER as a movie would probably not have existed if the character hadn't first appeared in comic books. According to acclaimed author Ta-Nehisi Coates—who later wrote a series of *Black Panther* and *Captain America* comics for Marvel—in comic books "there's still more room for a transgressive diversity" compared to big-budget Hollywood products.[11] Marvel comic books give Disney a way of gauging the appeal of characters and story lines long before a movie is proposed.

released the film. Disney had the financial and technological resources to make and market the big-budget film, and having purchased Marvel in 2009, it owned the copyright of the character (which first appeared in Marvel comic books in the 1960s). But producing a film around a Black superhero with a majority Black cast was considered extremely risky; while a Black superhero might have successfully helmed comic books (which are relatively cheap to produce and can be profitable with a smaller fan base), conventional Hollywood logic at the time assumed that white viewers wouldn't show up in the numbers needed to turn a profit on an expensive feature film.

For Disney, such risks were offset by the fact that Black Panther could be introduced to moviegoers through the successful *Avengers* franchise. Disney's decision may also have been affected by the cultural values or maverick sensibilities of key decision makers at the company; by the rising importance of global audiences; or by Disney executives' belief that cultural trends had shifted and conventional logic was out of date. In the end, *Black Panther* was a smash hit, coming in at No. 4 on the list of the all-time biggest blockbusters. Such success will undoubtedly shift industry thinking about films with Black leads.

Media Users

Media users—their values, identities, and beliefs—are influenced by the texts they consume, but they do not passively receive messages, as the linear model suggests. Rather, they actively interpret them. Those interpretations can vary because people bring different experiences to their engagement with a media text—experiences that are shaped by factors like gender, race, age, class, sexuality, region, and religion. In the case of *Black Panther*, for example, Black viewers and non-Black viewers are likely to be moved in different ways by seeing a Black superhero. Similarly, white viewers and nonwhite viewers are likely to bring different frames of reference to their interpretations of Killmonger's call for overturning the white world order.

The cultural model intentionally refers to *users*, which is a broader term than media *receivers* or *audiences*. That's because the decentralized and interactive nature of digital communication means that we don't have to work within an established media industry to produce or distribute media messages widely. Anyone with certain digital tools can upload a video to YouTube or share a meme. Thus, the term *media users* recognizes that people *use* media to both consume and produce media content. In fact, *Black Panther* became the most tweeted-about movie of 2018, generating more than thirty-five million tweets. Clearly, media users were actively interpreting their filmgoing experiences.[12]

Cultural Context

The first four elements of this approach are thoroughly embedded in the fifth: the cultural context, which includes all the interconnected economic, political, and historically specific conditions and practices that characterize our social world. Whether we are media users or media producers working in Hollywood, we are products of our culture. Our social experiences and identities shape how we produce and interpret media texts. Media industries are also influenced by the social world, constantly responding to shifting economic and cultural conditions. Finally, as we saw with the printing press, the Internet, and mobile phones, media technologies alter and are altered by culture.

All of this means that the creation and consumption of *Black Panther* cannot be understood without also considering the cultural context of the era in which it first appeared—a period marked by the legacy of Barack Obama's presidency and the rise of Trumpism; by the Black Lives Matter movement and the white nationalist Unite the Right rally in Charlottesville, Virginia; and by debates over climate change and globalization. Ultimately, the cultural approach pushes us to trace the interactions among its five elements in order to more fully understand the complex forces that shape us, the media we consume, the technologies we use, and the world we live in.

Media and the Politics of Culture

Both the cultural approach and our role as media citizens push us to analyze the ways that media and culture serve as sites of political struggle. In this context, **politics** is broader than electoral politics or political parties; rather, it refers to the process by which power, resources, status, and visibility get distributed in a society—usually unequally. Culture is part of that political process. It is the place where a society's values get established and delivered, where certain ways of life and certain kinds of people get defined as normal and valuable, while others are labeled as abnormal or irrelevant. As a major element of culture, the mass media play an important role in these broader political dynamics. We can better understand that role by examining how media represent the world and how our culture encourages us to feel about the media texts we consume.

Media and Representation

Media texts offer us representations of ourselves, others, and the world around us. Activities like producing a TV show, writing a news article, creating a video game, or crafting a status update all involve representing (or *re*-presenting) a version of reality—repackaging it, picking a frame for it, and molding it into a format that producers hope their audience will consume. For most media companies, these activities also involve representing this reality in a format that will make the producer money.

The Politics of Media Representations

Media representations both reflect and contribute to the distribution of power, status, resources, and visibility in a culture. Media portrayals of people, communities, events, and institutions reflect existing attitudes in a culture and also shape the attitudes of those who consume them. Because they have political consequences, it is important to interrogate the nature of those representations in terms of fairness and equity:

- Who and what is excluded from a culture's media representations? What happens if the experiences of some people remain invisible while the experiences of others are repeatedly made central?
- If you look across a variety of media texts, are there any patterns in the roles different kinds of people are given? For example, what happens when the heroes who save people are generally straight white men, and the victims who need saving are generally straight white women? What *stereotypes* get created and repeated?
- Who controls the production of these media representations? Who gets to choose how to represent our world? Who gets to tell their own story—and who doesn't?

Media critics and activists challenge media producers to consider the power and social responsibility that comes with their jobs. As media citizens, we should also challenge ourselves to understand the ethical implications of the media

MEDIA REPRESENTATION
The Netflix dramedy *Never Have I Ever* (2020–) tells the story of a rebellious South Asian American teenager following the death of her father. The series offers a more culturally and racially diverse representation of teenage life than most of the teen-oriented dramas produced for American viewers that came before it.

texts we consume and produce in our daily lives. The cultural approach pushes us to examine how those texts and their production are entangled in cultural processes linked to race, class, gender, sexuality, disability, religion, and many other aspects of social life. We will explore such issues further in later chapters.

Stories and Media Representation

A large part of representing reality involves turning events and situations into stories, and the stories that circulate in a society can reflect and reinforce dominant values and attitudes. In fact, our media institutions and outlets are all in the **narrative**—or storytelling—business. Media stories put events in context and shape how we understand our daily lives and the larger world. They often frame events within well-established narrative themes and structures—good versus evil, David versus Goliath, coming of age, success against great odds, race to victory. They encourage us to root for some characters and against others and to expect narrative closure; after all, most stories have endings.

As psychologist Jerome Bruner argues, we are storytelling creatures, and as children, we acquire language to tell the stories we have inside us.[13] The common denominator, in fact, between our entertainment and information cultures is narrative. It is the media's main cultural currency—whether in a Super Bowl commercial, an episode of *Keeping Up with the Kardashians*, a Fox News "exclusive," Instagram posts, or a Yelp restaurant review. Despite what we might sometimes think, such popular narratives are complex and varied and have political and ideological implications. In the end, narratives are the dominant way we make sense and meaning of our experiences.

As an example of the power of narratives, when the coronavirus emerged in late 2019 and early 2020, journalists reported statistics about confirmed cases and deaths, but they also told stories to help the public comprehend the magnitude and tragedy of the crisis—stories about the heroism of health-care workers, the fate of victims, the race to develop a vaccine, and battles over governments' responses. Meanwhile, President Trump used Twitter to establish his administration's actions as a success story, and people around the world went online to post stories and videos about life in isolation. Told through a variety of media outlets, such diverse stories shaped the public's understanding and experience of the pandemic. As a result, it is important to not take media stories for granted but to examine them with a careful and critical eye. (For further discussion, see "Examining Ethics: Telling Stories about 'Voices We Seldom Hear.'")

Agenda-Setting and Gatekeeping
Experts discuss how the media exert influence over public discourse.

Discussion: How might the Internet cancel out or reduce the agenda-setting effect in media?

The Politics of Evaluating the Media

Given the role they play in culture, it isn't surprising that media products don't just reflect competing values but are often themselves flash points in political struggles. As we move through our everyday lives, we judge media products at every turn, deciding which ones we think are worth our attention and which ones aren't. Many of those assessments may be automatic and reflect personal tastes, but they also reflect social norms and are tangled up in wider political dynamics.

We can better understand these complex connections by looking to the past, where we see that evaluations of the mass media have always been tied to historical conditions. Historians and cultural critics divide the recent history of Western societies into a *modern era* (roughly 1800–1950s) and a *postmodern era* (roughly 1950s–present). Each period is characterized by large-scale economic, political, and social developments, as well as the cultural values that became dominant during that time. These developments include the emergence of mass media and the cultural hierarchies that shaped how people thought about mass media.

Cultural Values of the Modern Era

The **modern era** coincided with the rise of mass communication industries, which were bound up with the era's faith in expertise, rationalism, and progress. Modernization involved experts—scientists, inventors, and captains of industry—using new technologies to make life better for millions of people.

Medical breakthroughs cured diseases, new appliances transformed people's homes, and efficiencies of industrial production made an array of consumer products affordable. While the period before the modern era was characterized by a strong belief in a divine order, to be modern meant valuing the ability of logical and scientific minds to solve problems, which fostered confidence in experts and an assumption that the future would be better than the present.

In the modern period, the emergence of mass media was seen as both an opportunity and a possible problem. On the one hand, mass media were considered potential tools for modernization. Under the guiding hand of media experts—trained journalists, scholars, and literary and cultural critics—mass media could, people assumed, join other forms of culture to educate citizens, promote social and moral uplift, and nurture an informed democratic society.[14] On the other hand, people feared that if media producers pandered to the tastes of the uneducated masses, the powerful reach of mass media could have a harmful effect. Such assumptions led to a reliance on experts whose judgments differentiated "good" cultural products from "bad" ones and cultivated the idea that to consume "good" culture was to participate in social progress.

High and Low Culture in the Modern Era

Throughout much of the twentieth century, then, critics and audiences generally perceived culture as a hierarchy, with supposedly superior products at the top and inferior ones at the bottom. This can be imagined as a skyscraper. In this model, the top floors of the building house **high culture**, such as ballet, art museums, and classic literature. The bottom floors house popular or **low culture**, including such media as reality television, teen pop music, and violent video games. High culture—identified with "good taste" and higher education and supported by wealthy patrons—is associated with fine art, which is available primarily in museums, theaters, and concert halls. In contrast, low or popular culture is aligned with the "questionable" tastes of the masses, who enjoy the commercial "junk" circulated by the mass media.[15] This hierarchy was based on—and helped to reinforce—certain assumptions about what makes media products better or worse. (See some of these assumptions in Table 1.1, as well as examples of different media products and where they might fit into this hierarchy in Figure 1.3.)

The cultural hierarchy and its assumptions about quality may seem self-evident or objective, but both are products of a specific political context. By the early twentieth century, America's social elites faced a threat to their control over American society from growing working-class and immigrant populations. One way elites responded to this threat was to advance the idea that the culture they controlled and consumed was inherently better. It was at this time that elites (through their financial gifts and connections) established most major museums, symphonies, and operas, giving themselves substantial forums through which they could define "high" culture. Meanwhile, the emerging mass culture of movies and recorded music was often associated with working-class audiences.

The modern era's emphasis on expert opinions and the promise of progress helped elites convince most people that Shakespeare was "better" culture, which in turn allowed elites to retain some of their cultural authority and privilege. The culture they controlled might not have been the most

TABLE 1.1   QUALITIES OF HIGH AND LOW CULTURE

"Superior" qualities of high culture	"Inferior" qualities of low culture
Timeless and enduring	Ephemeral and trendy
Original	Formulaic and derivative
Informative/educational	Sensationalistic/mindless
Promotes moral uplift	Panders to base instincts
Rare	Easily accessible
Free from commercialism	Tainted by commercial forces
Created by an individual genius/artist	Produced by an industrial system

Telling Stories about "Voices We Seldom Hear"

If you live in Iowa, you get to see a little bit of how the sausage is made, so to speak, especially if it's pork. It's a common occurrence to see long semitrailers on the highways, with round pink hog noses poking out through the metal slots, three levels of about 175 hogs. You can smell the trucks before you see them.

In northeast Iowa, the semis are all en route to the Tyson plant in Waterloo. Tyson, a meatpacking giant, has six pork plants (in addition to beef and chicken facilities) in the United States. Collectively, they process 461,000 hogs per week. That's nearly twenty-four million hogs a year.[1]

In Black Hawk County, in the northeast part of Iowa, Tyson is the second-leading employer (after John Deere), generating about twenty-eight hundred jobs. The plant has long drawn Latino, Bosnian, Congolese, Burmese, and Pacific Islander immigrant workers, making Waterloo (along with other meatpacking towns) one of the most diverse cities in the Midwest. Students in Waterloo schools speak forty-five different languages.

The work is hard. Meat cutters often work shoulder to shoulder cutting and packaging meat, on their feet the entire time. Workers need to be able to carry heavy loads, all while working in rooms that can be wet and slippery, with temperatures as low as 35–40 degrees Fahrenheit in some rooms and over 90 degrees in others. The average hourly rate is $14.96. Shifts can run between ten and twelve hours a day, six days a week.

Large institutions with thousands of people in close quarters are perfect places for the coronavirus to spread. When COVID-19 hit Iowa in mid-March of 2020, faculty from the University of Northern Iowa were sent home to teach classes online (like thousands of other professors around the country). On the other side of Black Hawk County, just a few miles away, Tyson workers carried on, in person at the factory. The situation mirrored the COVID-19 class divisions across the country: Many middle-class people were able to work from the safety of their own homes, while working-class "essential" workers risked coronavirus exposure daily as a condition of their labor. The authors of this textbook had many students who worked in essential positions in grocery stores, department stores, and gas stations.

The work of meatpacking is rarely in the news nationally, or even locally in the Midwest. The top four meatpacking companies employ nearly a half million people, in an industry that generates over $231 billion a year, but their difficult work is nearly invisible, off the radar of the middle-class-targeted news audience.

The coronavirus changed that.

A quick review illustrates how the number of stories about meatpacking skyrocketed. A search of the Nexis news database for the first four months of 2020, using the search terms "meatpacker" and "worker," yielded 718 news stories. The monthly story count grew exponentially, almost like the virus itself:

Jan. 2020: 14 stories

Feb. 2020: 33 stories

Mar. 2020: 59 stories

April 2020: 612 stories

The framing of the stories changed, too, as a deep concern about workers and communities spread. All across the Midwest, meatpacking workers were beginning to fall ill with COVID-19 in April 2020. Tony Thompson, Black Hawk County sheriff and chair of the local Emergency Management Commission, inspected the Waterloo plant on April 10. "We didn't see what we would have liked to have seen," he said. He quickly discovered there was no social distancing on the work line, in locker rooms, or in the lunchroom. Only one-third of the workers had any PPE (personal protective equipment), and it consisted mainly of homemade bandana masks.[2] In Waterloo, citizens pleaded with Tyson to close the local plant, Tyson's largest, until all coronavirus mitigation efforts could be completed. Tyson did not respond, and Iowa's governor refused to close the plant, lamenting that shutting down the factory might result in hogs being euthanized. Employees began to call in sick, refusing to work in an environment they called dangerous.

Tyson Plant Shutting Down
Waterloo

THESE IMAGES show the Tyson pork plant in Waterloo, Iowa, where nearly 40 percent of the so-called essential workers had COVID-19, and at least five died from it.

Finally, on April 22, after worker sick-outs, community criticism, and more than 180 Waterloo worker COVID-19 cases and one worker death, Tyson closed the plant. That week, Iowa won the disgraceful honor of having the fastest virus spread of any state in the United States.

The work of meatpacking became a huge news story. Local news media across the state brought workers' voices into the mix. *Waterloo Courier* multimedia reporter Amie Rivers listened to what people were saying in the community and broke the story of the sick-out at Tyson in Waterloo. Iowa AP correspondent Ryan Foley wrote moving stories about people who worked at Tyson plants and died from COVID-19. Iowa Starting Line, an online news outlet, interviewed

young Latinos and Latinas deeply worried about their parents who worked in meatpacking plants. Local TV stations and Iowa Public Radio featured stories of workers and meatpacking, as did CBS, CNN, NBC, and MSNBC. The *New York Times* podcast *The Daily* featured a heartbreaking half-hour interview with Achut Deng, a single mother, United Food and Commercial Workers International Union member, and Sudanese refugee who contracted COVID-19 at the Smithfield factory in Sioux Falls, South Dakota.

In the second week of May, the Tyson pork plant in Waterloo reopened with plastic shields spacing workers six feet apart. The day after reopening, Black Hawk County health officials reported that 1,031 Waterloo Tyson plant workers had COVID-19, nearly 40 percent of the 2,800 workers at the plant.[3] At least five workers died of COVID-19 at the Waterloo plant. Across the United States, more than fourteen thousand unionized meatpacking employees had been infected by COVID-19, and sixty-five had died from it by June 2020.[4]

Meatpacking plants were only part of the story. Mike Elk, a labor reporter and founder of *Payday Report*, literally connected the dots by mapping all the instances of worker resistance. On his COVID-19 Strike Wave Interactive Map, he recorded over two hundred wildcat strikes (actions in which employees decide to withhold their labor, without their labor union formally calling for a strike) in the first ten weeks of the U.S. pandemic.[5] The actions, all pandemic related, include Amazon fulfillment workers walking out in Tracy, California; fast-food workers striking at fifty stores in central Florida; and bus drivers in Richmond, Virginia, conducting a sick-out.

It shouldn't take a pandemic for the news media to feature the stories of meatpackers and other essential workers. The Society of Professional Journalists' Code of Ethics makes a recommendation that applies to all media content producers: "Boldly tell the story of the diversity and magnitude of the human experience. Seek sources whose voices we seldom hear."[6] Media stories help to shape our understanding of our communities and civic life. If we frame stories only around the interests of certain parts of society, we symbolically split our society apart and render other citizens unimportant and invisible. ●

FIGURE 1.3

CULTURE AS A SKYSCRAPER

Culture is diverse and difficult to categorize. Yet throughout the twentieth century, people tended to think of culture not as a social process but as a set of products sorted into high, low, or middle positions on a cultural skyscraper. Reliance on this model has weakened, but it can still shape how we think about the relative quality of culture. Look at the highly arbitrary arrangement shown here, and decide whether you agree or disagree. Write in some of your own examples.

Why do we categorize or classify culture in this way? Who controls this process? Is control of making cultural categories important? Why or why not?

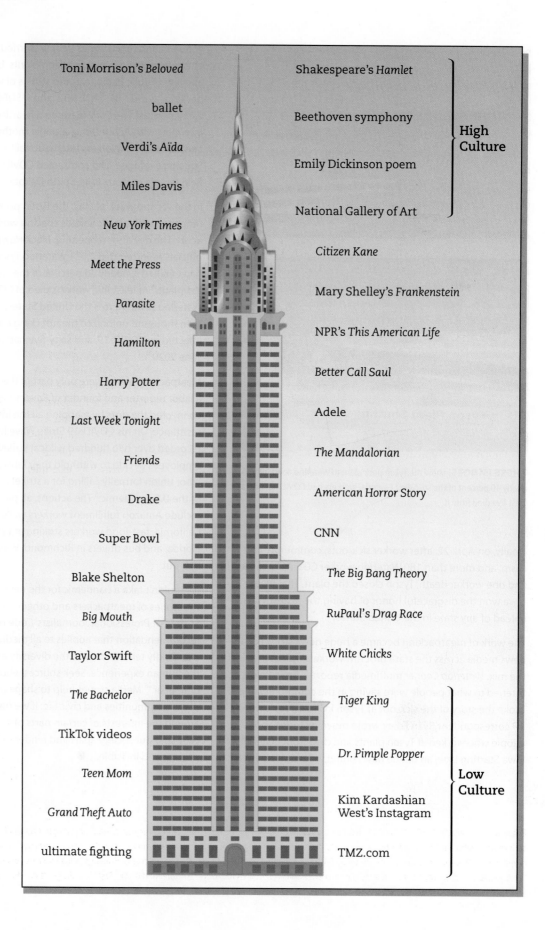

Toni Morrison's *Beloved* — Shakespeare's *Hamlet*

ballet — Beethoven symphony

Verdi's *Aïda* — Emily Dickinson poem

Miles Davis — National Gallery of Art

High Culture

New York Times — Citizen Kane

Meet the Press — Mary Shelley's *Frankenstein*

Parasite — NPR's *This American Life*

Hamilton — *Better Call Saul*

Harry Potter — Adele

Last Week Tonight — *The Mandalorian*

Friends — *American Horror Story*

Drake — CNN

Super Bowl — *The Big Bang Theory*

Blake Shelton — *RuPaul's Drag Race*

Big Mouth — White Chicks

Taylor Swift — Tiger King

The Bachelor — *Dr. Pimple Popper*

TikTok videos — Kim Kardashian West's Instagram

Teen Mom — TMZ.com

Grand Theft Auto

ultimate fighting

Low Culture

popular or profitable, but at least it was the most respected, positioning them as having the best judgment. In this way, history helps us recognize that ideas about "good" and "bad" culture are not simply timeless truths but elements of class politics.

Shifting Values in the Postmodern Era

While the high–low hierarchy still circulates in contemporary culture, shaping the way we discuss "trash" TV, couch potatoes, and video game addiction, its power to affect people's feelings about the media they consume has weakened. This development is symptomatic of the anti-elitism valued in our current **postmodern era**—a period marked by a growing skepticism about expertise and the idea of progress. By the end of the twentieth century, scientific advances were no longer new and miraculous but the "new normal" and expected. In addition, modernization's technological progress had led to unforeseen consequences and disappointments—the threat of nuclear annihilation, global warming, and continued economic and social inequality. As a result, an unquestioning faith in science and a deference to experts has eroded, fueling a growing political and cultural **populism**—a political approach that pits ordinary people against educated elites and nurtures a variety of movements on both the right and the left, such as anti-vaccine activists, the Tea Party, Occupy Wall Street, and the Fight for $15 movement to secure a higher minimum wage.

The media has played a significant role in the shift to the postmodern era. The proliferation of media products and delivery channels in the digital era has transformed the cultural landscape, altering how information is produced, circulated, and consumed. The decline of traditional authority figures' control over the flow of information allows unvetted messages to spread. Whether you believe that such information reflects legitimate alternative views and experiences or illegitimate conspiracy theories and propaganda for immoral lifestyles increasingly depends on where you live, who you socialize with, and what media you consume. In a media landscape with so many choices, the authority of previous cultural experts and hierarchies that drew distinctions between high and low culture has weakened. Those distinctions have also been blurred by postmodern media creators; for example,

Chaplin/Zuma Press

Liam Daniels/©Channel 4/Everett Collection, Inc.

FILMS OFTEN REFLECT THE KEY SOCIAL VALUES of the cultural context in which they are made. For example, Charlie Chaplin's *Modern Times* (1936, *left*) satirized the dehumanizing impact the modern era's factories could have on people who worked there. Similarly, the science fiction TV series *Black Mirror* (2011–present, *right*) takes a dark and satirical look at technology's impact on today's postmodern society. Series creator Charlie Brooker explains how the series relates to our world: "Like an addict, I check my Twitter timeline the moment I wake up. . . . If technology is a drug—and it does feel like a drug—then what, precisely, are the side-effects? This area—between delight and discomfort—is where *Black Mirror* . . . is set."[16]

artist Andy Warhol playfully mixed elements from high and low culture, Queen's song "Bohemian Rhapsody" mixes rock and opera, and classic literature figure Emily Dickinson's life and work is reimagined in the Apple TV+ series, *Dickinson*.

These days, while many of us may still think *Hamlet* is inherently better—and better for us—than *The Bachelor*, we are not likely to feel overly guilty about watching the latter, and are even less likely to be interested in literary critics' pronouncements about what we should watch. That doesn't mean our tastes in media are utterly individualistic or that hierarchies of good and bad don't continue to influence us. In the postmodern era of the niche nation, powerful hierarchies continue to tell us what media products we should and shouldn't consume, but they are more likely to reflect the norms and expectations of narrower social groups we identify with—groups whose contours are often laid out in relation to aspects of social and political life like gender, race, class, religion, political affiliation, education level, and sexual identity.

Media Literacy in Action: The Critical Process

We live at a time when communication technologies and the mass media play an ever more significant role in our daily lives, our political and economic systems, and our culture. For some of us, they function like wallpaper—they are so commonplace we don't spend much time thinking about them. For others, their presence and influence generate anxiety—perhaps for good reason. The quality of the news and entertainment we create and consume is important, yet much of it doesn't seem to serve us or our society well. We are surrounded by complaints about biased journalism, trashy reality television, vapid YouTube influencers, irresponsible corporations, conspiracy theories, cyber-trolls, violence, and other ills. It isn't surprising that many of us are either complacent or cynical about the media (or sometimes both at the same time). Neither attitude, however, is productive. Simply taking the media for granted or cynically dismissing all media as terrible lets us off the hook from making nuanced assessments of the media that surround us.

It's more useful to replace complacency and cynicism with genuine criticism based on a rigorous understanding of the complex dynamics at play in any issue related to the media. In contemporary societies, gaining the ability to carefully analyze and participate in debates over the media, their role in constructing meaning, and their impact on culture—in other words, becoming media literate—should be considered a fundamental skill. Recognizing the different roles you play as a media consumer, producer, and citizen; appreciating the value of historical knowledge; and considering the media through the lens of the cultural approach can help you build media literacy. Using those tools in conjunction with the critical-thinking method laid out in the following section will enable you to not only think about media with precision and nuance but also use those new insights to affect our culture.

The Five Steps of the Critical Process

Developing media literacy requires following a **critical process** that takes us through the steps of **description**, **analysis**, **interpretation**, **evaluation**, and **engagement** (see "Media Literacy and the Critical Process"). We will be aided in our critical process by keeping an open mind, trying to understand the specific cultural forms we are critiquing, and acknowledging the complexity of contemporary culture and the media environment.

To begin this process of critical assessment, we must assume a stance that enables us to get outside our own preferences. We may like or dislike hip-hop, R&B, EDM, or country, but if we want to critique these musical genres intelligently, we should understand what the various types of music have to say and why their messages appeal to particular audiences. If we examine a corporate public relations campaign, we should analyze the persuasive strategies the campaign employs, the technologies it uses, and the relationship the PR agency builds with news outlets. If we analyze a film or TV show, we need to slow down the images in order to understand how they create meaning; consider the economic, technological, and cultural factors that shaped its creation; and consider how different viewers may interpret it.

Benefits of a Critical Perspective

By reading this book, we hope you come to appreciate the important work you are preparing for. Whether you end up in a media-related career or some other profession, you will be responsible for building our culture and safeguarding our democracy. People who work in marketing, public relations, sports media, news, television and film production, journalism, video game design or any other corner of the media industry play a significant role in creating the culture that affects us all. In the digital era, however, when we can post, share, and comment so easily, everyone participates in that work. Developing an informed critical perspective empowers you to help build our culture more thoughtfully and keep our democracy healthy.

It can, for example, help you explore key tensions that emerge in our media culture between democracy and commercial interests. On the one hand, the media can be a catalyst for democracy and social change. Consider the role of smartphone video in documenting police brutality, of websites like FactCheck.org in debunking false information, or of television in shifting attitudes about LGBTQ+ rights. On the other hand, the media can advance powerful commercial interests that sometimes operate in opposition to democratic values. Large parts of the media environment are controlled by relatively few multinational corporations that can greatly limit the ideas that circulate; in addition, much of our media environment is subsidized by advertising, which addresses people as consumers with unequal spending power rather than as citizens with equal rights under the law.

Being an informed and responsible media citizen in the digital era is challenging. In a niche nation, the media environment we navigate is complex and fractured. Working with others to establish a sense of the common good can be difficult when it feels as though we live in a different world than the people down the street—when we watch different TV shows, trust different news sources, and receive different facts. But the fact remains: We still live in the same communities and on the same planet. A healthy democracy requires the active involvement of everyone. Part of this involvement means understanding the nature and consequences of the complex media environment we experience each and every day.

ANTHONY WALLACE/AFP/Getty Images

JOURNALISTS PLAY an important role in democratic societies. Their reporting is often the only way people can learn what is going on. Sometimes they also become part of the story themselves, as was the case with this journalist, who was photographed being pepper sprayed as he covered pro-democracy protests in Hong Kong. But journalists aren't the only ones with an important role to play. As media citizens, each of us can use the critical process to evaluate media information about our world, and in the process become more responsible participants in our democracy.

MEDIA LITERACY & THE CRITICAL PROCESS

It is easy to form a complacent or cynical view about the stream of TV advertising, reality programs, video games, celebrity gossip blogs, political memes, and YouTube videos that floods the cultural landscape. But complacency and cynicism are not substitutes for criticism. To become literate about media involves striking a balance between taking a critical position (developing knowledgeable interpretations and judgments) and becoming tolerant of diverse forms of expression (appreciating the distinctive variety of cultural products and processes).

A cynical view usually involves some form of intolerance and either too little or too much information. For example, after enduring the glut of news coverage and political advertising devoted to the 2020 presidential election, we might have easily become cynical about our political system. However, information in the form of "factual" news bits and knowledge about a complex social process such as a national election are not the same thing. The critical process stresses the subtle distinctions between amassing information and becoming media literate.

Developing a media-literate critical perspective involves mastering five overlapping stages that build on one another:

- **DESCRIPTION:** paying close attention, taking notes, and researching the subject under study

- **ANALYSIS:** discovering and focusing on significant patterns that emerge from the description stage

- **INTERPRETATION:** asking and answering "What does that mean?" and "So what?" questions about one's findings

- **EVALUATION:** arriving at a judgment about whether something is good, bad, or mediocre, which involves subordinating one's personal taste to the critical "bigger picture" resulting from the first three stages

- **ENGAGEMENT:** taking some action that connects our critical perspective with our role as citizens and watchdogs who question our media institutions, adding our voice to the process of shaping the cultural environment

Let's look at each of these stages in greater detail.

1 DESCRIPTION

Let's say we have decided to investigate how well the news media serve democracy. To do so, we might critique the fairness of several segments or individual stories from, say, *60 Minutes* or the *New York Times*. We would start by describing the segments or articles, identifying their reporting strategies, and noting who is featured as interview subjects. We might further identify central characters, conflicts, topics, and themes. From the notes taken at this stage, we can begin comparing what we have found to other stories on similar topics. We can also document what we think is missing from these news narratives—the questions, viewpoints, and persons that were not included—and other ways to tell the story.

2 ANALYSIS

In the second stage of the critical process, we isolate patterns that call for closer attention. At this point, we decide how to focus the critique. Because *60 Minutes* has produced thousands of hours of programming in its more than fifty-year history, our critique might spotlight just a few key patterns. For example, many of the program's reports are organized like detective stories, reporters are almost always visually represented at a medium distance, and interview subjects are generally shot in tight close-ups. In studying the *New York Times*, we might limit our analysis to social or political events in certain countries that get covered more often than events in other areas of the world. Or we could focus on recurring topics chosen for front-page treatment, or what stories are selected to be featured in the *Times'* popular *The Daily* podcast.

❸ INTERPRETATION

In the interpretation stage, we try to determine the meanings of the patterns we have analyzed. The most difficult stage in criticism, interpretation demands an answer to the "So what?" question. For instance, the greater visual space granted to *60 Minutes* reporters—compared with the close-up shots used for interview subjects—might mean that the reporters appear to be in control. They are given more visual space in which to operate, whereas interview subjects have little room to maneuver within the visual frame. As a result, the subjects often look guilty and the reporters look heroic—or at least in charge. Likewise, if we look at the *New York Times*, its attention to particular countries could mean that the paper tends to cover nations in which the United States

has more vital political or economic interests, even though the *Times* might claim to be neutral and evenhanded in its reporting of news from around the world.

❹ EVALUATION

The fourth stage of the critical process focuses on making an informed judgment. Building on description, analysis, and interpretation, we are better able to evaluate the fairness of a group of *60 Minutes* or *New York Times* reports. At this stage, we can grasp the strengths and weaknesses of the news media under study and make critical judgments measured against our own frames of reference—what we like and dislike, as well as what seems to be good or bad or missing, in the stories and coverage we analyzed. This fourth stage differentiates between cynical pundits, who may say that something is bad without knowing anything about it, and informed critics, who make an evaluation based on the first three steps of the critical process.

❺ ENGAGEMENT

To be fully media literate, we must actively work to create a media world that helps serve democracy. Thus, we have added a fifth stage in the critical process—engagement. In our *60 Minutes*

and *New York Times* examples, engagement might involve something as simple as writing a formal letter or an e-mail to these media outlets to offer a critical take on the news narratives we are studying.

But engagement can also mean participating in online discussions; contacting various media producers or governmental bodies, such as the Federal Communications Commission, with critiques and ideas; organizing or participating in public media literacy forums; or learning to construct different types of media narratives ourselves—whether print, audio, video, or online—in order to participate directly in the creation of mainstream or alternative media. Producing actual work for media outlets might involve writing news stories for a local newspaper (and its website), producing a radio program or podcast on a controversial or significant community issue, or constructing a website that critiques various news media. The key to this stage is to challenge our civic imaginations by refusing to sit back and cynically complain about the media without taking some action that lends our own voices and critiques to the process.

Chapter **1**
REVIEW

LAUNCHPAD FOR *MEDIA & CULTURE*

LaunchPad
macmillan learning

Visit LaunchPad for *Media & Culture* at **launchpadworks.com** for additional learning tools:

➔ QUIZ YOURSELF WITH LEARNINGCURVE

LearningCurve uses gamelike quizzing to help you review important concepts from this chapter. LearningCurve includes multiple-choice questions at different levels of difficulty, hints, feedback for both right and wrong answers, and links to e-book material for easy reference. Which answer choice would you select for the sample question below?

> Martin is planning to research the HuffPost news outlet for his communication class. Which action demonstrates Martin's activities in the first step of the critical process?
>
> ○ Martin looks at his research notes and determines what, if any, patterns appear.
>
> ○ Based on his notes and analysis, Martin makes some judgments about the work being done at the HuffPost.
>
> ○ Martin spends time on the HuffPost website, making notes about the stories covered, the writing style, and the reporters working at the site.
>
> ○ Martin has found that a number of stories are reported several times in different ways on the site, and he looks for reasons behind this.

➔ WATCH VIDEO ON KEY CONCEPTS

LaunchPad includes clips from movies, TV shows, online sources, and other media texts, in addition to interviews with media experts and newsmakers. In the videos for this chapter, you'll consider topics like agenda-setting and gatekeeping and the formal features of *Black Panther*.

➔ PRACTICE THE CRITICAL PROCESS

Use the Media Literacy Activity to walk through the critical process step-by-step and develop a critical perspective on what media means to you. This chapter's activity challenges you to go *without* media for a period of time to better appreciate the role that media plays in your life.

➔ LEARN KEY TERMS

Review the definitions for this chapter's key terms with LaunchPad's easy-to-use online flash cards. Try the cards in practice mode, or quiz yourself as a way to focus your study efforts. (Page numbers indicate where each term is highlighted within the chapter.)

media literacy, 5
mass media, 5
communication, 5
culture, 6
affordances, 6
mass communication, 6
masspersonal
    communication, 6
mass nation, 9
consensus narratives, 9

digital communication, 10
convergence, 11
niche nation, 11
participatory culture, 11
media environment, 13
text, 16
technological determinism, 17
politics, 19
narrative, 20
modern era, 20

high culture, 21
low culture, 21
postmodern era, 25
populism, 25
critical process, 26
description, 26
analysis, 26
interpretation, 26
evaluation, 26
engagement, 26

REVIEW QUESTIONS

Use the questions below to revisit key themes and concepts from this chapter.

 Need more practice? Access the LearningCurve multiple-choice questions for this chapter, which are available on LaunchPad. All questions include feedback for correct and incorrect answers, hints, and links to e-book material.

Understanding the Media Today

1. What is *media literacy*, and why is it important?

How We Got Here: Culture, Technology, and the Evolution of Media Communication

2. Define *mass media*, *communication*, and *culture*, and explain their interrelationships.
3. What are *affordances* of technology? Give an example from two different media technologies.
4. Explain what *masspersonal communication* is and how it incorporates elements of both mass and interpersonal communication.
5. What key technological breakthroughs accompanied the transition to the print and electronic eras? Why were these changes significant?

Media in Our Digital Era

6. What are the two ways in which *convergence* has occurred in the digital era?
7. What has changed us from a *mass nation* to a *niche nation*? What is the impact on *consensus narratives*?
8. What are two different ways of thinking of the word *medium* and how it applies to thinking about media in our time?

Thinking Differently about Media and Culture

9. What are the three roles we each play in relation to media? How do these roles relate to one another?
10. What are shortcomings of the linear model of mass communication? How does the cultural approach address those shortcomings?
11. What are the dangers of an overly *technologically deterministic* approach to thinking about our relationship with media?

Media and the Politics of Culture

12. When we talk about the *politics* of culture, what does that mean?
13. Why are *narratives* important to media representation?
14. Describe the skyscraper model of culture and the historical context that existed when it emerged.
15. What are the chief differences between modern and post-modern values?

Media Literacy in Action: The Critical Process

16. What are the five steps in the critical process? Which of these is the most difficult, and why?
17. What is the difference between cynicism and criticism?

QUESTIONING THE MEDIA

Use these critical-thinking questions to reflect on your own experiences with media and apply your understanding of media concepts.

 For an interactive experience, questions on topics like these are available for the iClicker student response system. Instructors can download the questions from LaunchPad and use them to poll the class.

1. Keep track of every type of media technology you use and every media message or product you encounter in a day. What did you learn about your media environment? What surprised you about your relationship to media?

2. Drawing on your personal experience, list the kinds of media stories you like and dislike, and explain why. You might think mostly of movies and TV shows, but remember that news, sports, political ads, and product ads are typically structured as stories as well. Conversations on social media can also be considered narratives.

3. Identify any real-world elements that have been excluded from your favorite media product's *re*-presentation of reality. Does this media product rely on any stereotypes in its version of reality? How might the basic story be told differently?

4. Do you have a guilty media pleasure? Is there a media text you enjoy but are embarrassed to admit it to certain people—your friends, your parents, a professor? What does your embarrassment tell you about how a specific community or subculture that you are a part of evaluates media?

5. Make a critical case either defending or condemning a TV show, radio talk show, music recording, movie, or digital game. Use the five-step critical process to develop your position.

6. The rise of the electronic era transformed how people understood themselves, others, and their relationship to the world around them. How do your own digital-era photos, audio recordings, or videos affect the way you understand and remember your personal history?

The Internet and Digital Media

2

For at least some of us, the socially mediated version of ourselves has become the predominant way we experience the world. As *Time* magazine has noted, "Experiences don't feel fully real" until we have "tweeted them or tumbled them or YouTubed them—and the world has congratulated [us] for doing so."[1] Social media is everywhere, and it is all about us—we are simultaneously creators and subjects. But the flip side of promoting our own experiences as the *most awesome happenings ever (and too bad you aren't here)* is the social anxiety associated with reading about other people's experiences and realizing you're not part of them.

The problem of Fear of Missing Out (FOMO) has been defined as "the uneasy and sometimes all-consuming feeling that you're missing out—that your peers are doing, in the know about or in possession of more or something better than you [are]."[2] There are plenty of platforms for posting about ourselves and anxiously creeping on others; Facebook, Twitter, Snapchat, TikTok, LinkedIn, and Instagram are just a few of the sites that can feed our FOMO problem. The fear of missing out has been around long before social media was invented. Party chatter, photos, postcards, and holiday letters have long put the most positive spin on people's lives. But social media and smartphones have made learning about missed interactions a 24/7 phenomenon. There is potentially *always* something better you could have/should have been doing, right?

With FOMO, there is a "desire to stay continually connected with what others are doing." Therefore, the person suffering from the anxiety continues to be tethered to social media, tracking "friends" and sacrificing time that might be spent having in-person, unmediated experiences.[3] And though spending all this time on social media is a personal choice, it may not make us happy. In fact, a study by University of Pennsylvania researchers found that using Facebook, Instagram, and Snapchat is linked to depression and loneliness. Students who limited their use of those social media platforms to roughly ten minutes per platform per day for a three-week period showed significant decreases in anxiety and FOMO.

◀ FOMO, or Fear of Missing Out, is the anxiety that something exciting may be happening while you're off doing something else. The uninterrupted Internet connection we get from smartphones allows us to be in constant contact with friends through social media. But at what cost?

"Not comparing my life to the lives of others had a much stronger impact than I expected," one of the participants reported, "and I felt a lot more positive about myself during those weeks."[4]

Studies about happiness routinely conclude that the best path to subjective well-being and life satisfaction is having a community of close personal relationships. Social psychologists Ed Diener and Robert Biswas-Diener acknowledge that the high use of mobile phones, text messaging, and social media is evidence that people want to connect. But they also explain that "we don't just need relationships: we need close ones." They conclude, "The close relationships that produce the most happiness are those characterized by mutual understanding, caring, and validation of the other person as worthwhile."[5] Thus, frequent contact via text or chat isn't enough to form the kinds of relationships that produce the most happiness.

Ironically, there has never been a medium better than the Internet and its social media platforms to bring people together. How many people do you know who met online and went on to have successful friendships or romantic relationships? How often have social media connections enhanced close relationships for you? Still, according to Diener and Biswas-Diener, maintaining close relationships may require a "vacation" from social media from time to time, experiencing something together with a friend or friends. Of course (and we hate to say it), you will still need to text, message, e-mail, or call to arrange that date.

→ **How much time do you spend per day on social media apps? Do you or your friends experience FOMO? If so, how big a problem do you feel it is, and what might you do to limit its impact on you?**

The Internet Today

As this exploration of FOMO and the impact of social media illustrates, we live in a world thoroughly shaped by the **Internet**—the vast network of fiber-optic lines, wireless connections, and satellite systems that links laptops, mobile phones, tablets, game consoles, smart TVs, and a growing array of smart devices to enormous data centers around the world. Those data centers house an astonishing number of computers that store, process, and relay a striking amount of digital information.

In 2020, over 4.5 billion people—roughly 60 percent of the world's population—were active Internet users. Rates of Internet access are highest in places like Europe, North America, and South Korea, but rates in Africa, other parts of Asia, and the Middle East are rising more rapidly, and China has more Internet users than any other country.[6] All those users generate, send, and receive immense amounts of digitized data—images, texts, and sounds broken up into discrete pieces of electronic information represented for computers as ones and zeros. Each one or zero is called a *bit* (from *BI*nary digi*T*), and eight bits are called a **byte**, which is the basic unit used to measure digital information. On average, each Internet user currently generates 1.7 million bytes of data every second.[7] If we define data as information that can be stored—whether via a stone tablet, a scroll, a book, a newspaper, celluloid frames, or some other format—reports claim that more data was created in 2015 and 2016 than had been created during the previous five thousand years of human history.[8]

The rise of the Internet and the digital revolution has had a profound impact on traditional media industries. When new technologies like radio, film, and television emerged, they competed alongside existing technologies and the industries built around them. The Internet and digital communication, by contrast, not only competed with but also absorbed previous technologies and industries. New kinds of digital media products—like websites, podcasts, video games, and social media platforms—vied with traditional movies, TV shows, and radio for people's attention. Those traditional industries, however, also went digital. As we will see in later chapters, as older technologies like celluloid film and vinyl records were supplanted by digital formats, legacy (pre-digital) industries had to develop new business strategies to respond to changing consumer practices. As a result, today's media industries have been redefined through technological and industry convergence.

In this chapter, we will examine specific dimensions of the Internet, digital communication, and our digital media environment. We will:

- Review the birth of the Internet, and trace the development of the web over time
- Examine key characteristics of the Internet environment, including decentralization, online communities (both meaningful and malicious), and media manipulation
- Discuss how leading technology companies and governments have sought to control our digital environments (and us) for economic and political advantage
- Investigate the complex relationship between the Internet and democracy

How We Got Here: The Development of the Internet

The evolution of the Internet has been dramatic. From its origins as a U.S. military project in the 1960s, it has grown into a powerful force in the daily lives of billions of people around the world. A variety of factors have shaped the Internet's development, including corporate priorities, government regulations, cultural values, user desires and behaviors, and continual technological innovation. To track these changes over time, we'll divide the Internet's history into four phases: the pre-web Internet, Web 1.0, Web 2.0, and Web 3.0.

The Pre-Web Internet

The U.S. Department of Defense, funded by U.S. taxpayers, developed the Internet. In the 1950s and 1960s, military leaders preparing America for the Cold War looked to address two concerns. First, existing communication systems were highly centralized, which also made them vulnerable: If a Soviet nuclear bomb destroyed a central communication hub, the United States's ability to respond would be compromised. Researchers at the Defense Department's Advanced Research Project Agency (ARPA) envisioned a distributed network system where messages could be rerouted, meaning the loss of any single node would not disrupt communication (see Figure 2.1). Second, anxious to ensure U.S. technological superiority, ARPA worked to strengthen the nation's research capacity, which included finding ways to better utilize expensive supercomputers, which were becoming vital research tools.

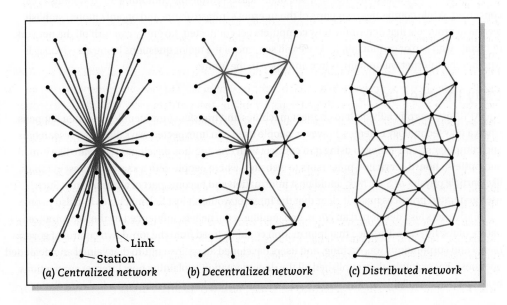

(a) Centralized network   (b) Decentralized network   (c) Distributed network

FIGURE 2.1

DISTRIBUTED NETWORKS

In a centralized network **(a)**, all the paths lead to a single nerve center. Decentralized networks **(b)** contain several main nerve centers. In a distributed network **(c)**, which resembles a net, there are no nerve centers; if any connection is severed, information can be immediately rerouted and delivered to its destination. But is there a downside to distributed networks when it comes to the circulation of network viruses?

Information from: Katie Hafner and Matthew Lyon, *Where Wizards Stay Up Late* (New York: Simon & Schuster, 1996).

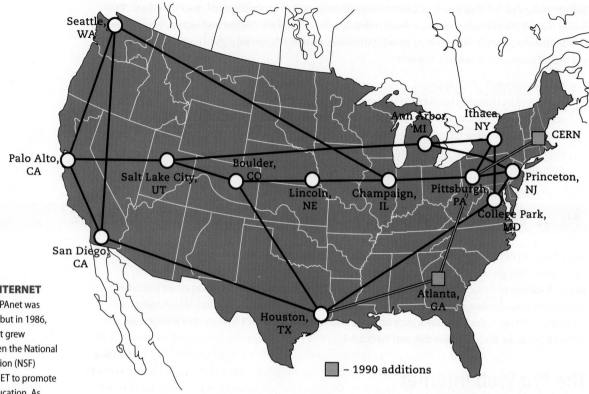

THE EARLY INTERNET EXPANDS ARPAnet was created in 1969, but in 1986, the early Internet grew significantly when the National Science Foundation (NSF) developed NSFNET to promote research and education. As part of this effort, the NSF funded several university supercomputing centers and linked them with a high-speed network, which became the basis for the commercial Internet of the 1990s.

In 1969, ARPA addressed both issues by linking supercomputers at Stanford University, the University of California–Santa Barbara, the University of California–Berkley, and the University of Utah via telephone lines, then establishing the **protocols** (rules) that allowed the supercomputers to join the network and speak to one another. The **ARPAnet**, as the network was called, enabled researchers to pool computing power and established the cornerstone of what would become the Internet's infrastructure. The Internet grew during the 1970s and 1980s as other universities, government research labs, and technology corporations joined the network, eventually shifting it from a military project into a publicly funded and government-managed utility akin to the interstate highway system.

Ironically, one of the most hierarchically structured institutions in our culture—the national defense industry—created the Internet, one of the least controllable communication systems ever conceived. While companies own parts of the Internet's infrastructure and organizations establish technical protocols that determine how computers get connected, no entity can turn off the network or dictate who can or cannot join it. As we will see, one of the major questions being asked today is whether the Internet will remain as free and open in the future.

Early Internet Users

The APRAnet was officially built to connect machines and transfer computer files. As often happens when a new technology emerges, however, people used it in unexpected ways. Early on, researchers and computer scientists began using it to connect with one another, developing software like e-mail and **bulletin board services**—precursors to websites, where people with a shared interest in topics, like particle physics or *Star Trek*, could post information and become part of a community. Such user-generated practices made it clear that the Internet wouldn't just link machines but also people.

Early users also developed an *ethos*, or a guiding set of beliefs, influenced by the decentralized nature of the Internet. The idea that information could be freed from the top-down control of governments and corporations was exciting, and users developed an online culture that valued and defended openness. For example, they published all the documentation explaining how Internet connections work, making it available for anyone to read and implement. In a 1996 manifesto declaring the independence of cyberspace from government interference, poet and Internet activist John Perry Barlow declared, "We are creating a world where anyone, anywhere may express his or her beliefs, no matter how singular, without fear of being coerced into silence or conformity."[9]

Foundations for a New Phase

For its first two decades, the Internet was mainly the domain of research institutions and computer science experts. A number of developments in this period, however, set the stage for the Internet's future as a mass medium. In the 1970s, the introduction of **microprocessors** (miniature circuits that process and store digital electronic signals) helped make computers smaller and more affordable. Together with user-friendly graphical interfaces like Microsoft's Windows, they helped fuel dramatic growth in personal home computer ownership during the 1980s. At the same time, highly efficient **fiber-optic cables** made of glass replaced metal wires as the standard for network connections, dramatically increasing the amount of data networks could handle and the speed by which data could be transmitted. Finally, as the Cold War ended in the late 1980s, responsibility for most aspects of the growing Internet infrastructure shifted from government agencies to private industry. The privatization of the Internet laid the foundation for its commercialization in the 1990s.

Web 1.0: The Internet Becomes a Mass Medium

In the early 1990s, a handful of developments—the creation of the World Wide Web, the first web browsers, and the growth of Internet service providers and search engines—finally pushed the Internet into its Web 1.0 phase, in which it became a mass medium used by many more people. During this time, the Internet quickly became commercialized, as companies rushed to create businesses offering services to the growing ranks of Internet users—a major change from the pre-web era, when commercial activity on the Internet was banned. As a result, the 1990s saw a struggle between competing cultures—one reflecting a commercial ethos that embraced corporate control, advertising, and profitability, and another intent on preserving the original public and noncommercial nature of an open Internet.

The Web Is Invented

Before the 1990s, the Internet was difficult to navigate; users had to have inside knowledge about where things were located in order to find them. The invention of the **World Wide Web** ("the web") changed that. In 1990, Tim Berners-Lee, a computer scientist working at Switzerland's CERN particle physics lab, developed an easier way for physicists around the world to create, post, and locate documents on the Internet. He invented **HTML (hypertext markup language)**, a language for displaying text, images, and other multimedia that allowed users to link files to one another. The links between files formed a virtual web, and the text files created using HTML became web pages. Multiple web pages could be stored together as websites. Importantly, the web was created to be an open environment rather than a proprietary system, meaning that anyone could use HTML to connect without having to pay a licensing fee or get permission. As a result, the number of web pages and websites grew at a fantastic rate during the 1990s.

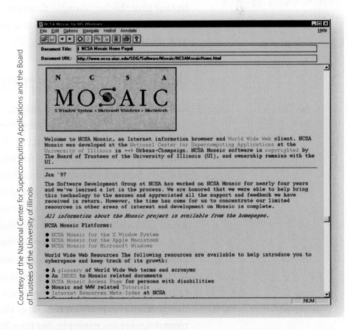

WEB BROWSERS The GUI (graphical user interface) of the World Wide Web changed overnight with the release of Mosaic in 1993. As the first popular web browser, Mosaic unleashed the multimedia potential of the Internet. Mosaic was the inspiration for the commercial browser Netscape Navigator, which was released a year later.

Browsers

The release of web **browsers**—software applications that help users navigate the web—gave mass audiences greater access to online information. In 1993, computer programmers led by Marc Andreessen at the University of Illinois created Mosaic, the first popular browser with an intuitive graphical interface and attractive (for its time) fonts and navigation buttons. A year later, in a move that reflected the emerging role venture capitalism played in the Internet's development, Andreessen joined investors in California's Silicon Valley to produce a commercial browser: Netscape Navigator. Today's most popular browsers—Chrome, Safari, and Firefox—are used by billions of people as their entry point to the web.

Getting Online and Starting to Search

In order for the Internet to become a mass medium, millions of people needed to find a way to connect to it quickly and easily. By the 1990s, new companies called **Internet service providers (ISPs)** emerged to offer Internet access to home users. While these connections initially occurred via a modem and telephone dial-up, the development of higher-capacity **broadband** connections soon led to much faster download speeds. AOL (formerly America Online) became the dominant ISP by the mid-1990s, and it was one of the first companies to offer customers e-mail accounts and other Internet-related services. Today, major ISPs in the United States, like AT&T, Comcast, Verizon, and Cox, compete with hundreds of local services—many offered by regional telephone and cable companies—to provide consumers with Internet access.

As the Internet became more popular and the number of websites grew, search directory and search engine companies emerged to help users navigate the increasingly vast amount of online information. In 1994, two Stanford University graduate students created one of the first directories, Yahoo!, by organizing their favorite websites into various categories and subcategories. Creating a manual directory and using it to navigate the Internet was still viable at that point, when there were only about twenty-two thousand websites.

Eventually, however, the growth of the web made directories impractical. (A billion web pages existed by 2000.)[10] **Search engines**, relying on mathematical algorithms, offered a more automated way to find content by allowing users to enter key words or queries to locate related web pages. The Google search engine, released in 1998, became a major success because it introduced a new algorithm that ranked a page's popularity based on how many other pages linked to it. Today, Google accounts for an astonishingly large percentage of global online searches (some estimates place it as high as 92%), dwarfing competitors like Bing, Yahoo!, and Baidu.[11]

Web 2.0: The Internet Gets Interactive

In the early 2000s, there was a major shift in what users could do online. Up to that point, the web had primarily been a **read-only** system; websites were places people went to view information. In the next phase, which some called "Web 2.0," the web became **read-write**, or interactive—a place where users could read information, contribute their own digital content, and directly engage with other users. In addition, websites became multimedia friendly, integrating text, graphics, pictures, audio, and video, which made the experience of being online more dynamic. Such changes coincided with the development of mobile technologies—most notably the smartphone in the mid-2000s—which transformed where and when people could access these online experiences.

During this time, a number of factors helped make the web more interactive and participatory. E-commerce sites like Amazon created new protocols and software, allowing customers to comfortably search and pay for things online. Broadband connections became cheaper, so more users could both download and upload multimedia content via their home computers; the development of 3G technology did the same for mobile phones. Government regulation also played a role. For example, **Section 230** of the Communications Decency Act, passed by Congress in 1996, protected any company operating an "interactive computer service" from liability for anything published on their service by a third party.[12] In essence, the statute created the legal ground from which iconic Web 2.0 services like YouTube and Twitter sprouted by protecting them from being sued over material posted by users. Such developments helped build a web that fueled collaboration and spawned new websites built on social interactions.

The Collaborative Web

Some observers suggested that the web's new level of interactivity would revolutionize problem solving and content creation by harnessing the collective intelligence of a vast population. Indeed, the development of **wikis**—open and collaborative websites where people work together to edit and create content—has supported such collective engagement. Wikipedia, the online encyclopedia with over fifty-four million articles that is continually updated by millions of editors, is the best-known example.[13] Many other parts of the web also leverage crowdsourcing potential to create user-generated storehouses of knowledge about everything from the quality of local restaurants to video game cheat codes to the difficulty of professors' exams. In certain ways, such efforts follow the ethos of the Internet's earliest users, who celebrated the open exchange of information.

Social Media Platforms

Social media platforms like Facebook, Twitter, Snapchat, Instagram, TikTok, and Pinterest have exploited the interactive, multimedia, and mobile capabilities of Web 2.0 as much as any "interactive computer service." For example, Facebook, founded in 2004, encourages us to create detailed profiles and post our own content as text, image, or video; the site feeds us so much information tailored to our interests that surfing the larger web feels less necessary. The more people and pages we "like," the richer our feeds become; thus, Facebook's very design encourages us to build up a large network of connections. Other features—tagging, commenting, sharing—strengthen those connections, and the program observes our behavior and suggests new people to "friend." Finally, we can access Facebook via a website or mobile app, making it convenient to remain constantly connected to the platform. As we will explore later, features like these are key to Facebook's business model.

TIKTOK is a Chinese video-sharing social networking app with over 800 million users worldwide. The app—which enables people to create and edit short videos, incorporate music, and attach hashtags—makes content creation and sharing easy, fueling viral memes and launching careers for singers, comedians, and models.

Web 3.0: The Internet Starts to Think

Certain changes have pushed the Internet into a new phase—what some have called Web 3.0—that is characterized by two interrelated developments: the Semantic Web and the Internet of Things.

As early as 1999, Tim Berners-Lee, the web's original creator, anticipated the emergence of a **Semantic Web** in which web pages and databases would be created in such a way that a computer—functioning as something akin to artificial intelligence—could examine the web's vast quantities of data and automatically provide useful solutions to people's needs. By the 2010s, that vision had started to become a reality. As one enthusiastic technology solutions company proclaimed, "Semantic experiences make it easier for humans to perform creatively and encounter frictionless content experiences, while a machine does the finding, sifting, sorting, combining, organizing, and presenting of real answers."[14]

Recommendation algorithms woven into services like Amazon and Netflix use these semantic systems. For example, because Netflix employees tag each film and TV series with information about the cast, genre, date of release, and tone, Netflix's algorithm can review the programs you've watched, determine that you like "feel-good coming-of-age movies with a strong female lead," and suggest other similar programs for you. Such systems are likely to continue to evolve, becoming smarter and even more independent data analysts. (For more about algorithms and the challenges inherent in their use, see "Examining Ethics: Algorithmic Bias.")

SMART FRIDGE

With integrated video screens, sensors, Internet connections, and software designed to anticipate our needs, smart appliances like this refrigerator are reshaping our environments and expanding what we consider media devices. While the Internet of Things promises greater convenience, it also raises concerns about privacy and security as we share more information about ourselves.

Algorithmic Bias

I n 2016, Joy Buolamwini, a computer scientist at the MIT Media Lab, founded the Algorithmic Justice League (AJL). The AJL is a nonprofit organization devoted to understanding the social and political consequences of a Web 3.0 world, where the artificial intelligence (AI) of computer algorithms guides the decisions we make. Why is such an organization needed? Because algorithms are powerful, they are everywhere, and, as Buolamwini explains, "Machines can discriminate in harmful ways."[1]

An *algorithm* is a set of instructions, designed and written by computer programmers, that tells a computer how to analyze vast amounts of data. In today's world, the algorithms operating behind our apps, search engines, and websites help us decide what route to take to work, whom to go on a date with, and what songs to listen to on Spotify. They are also used by colleges, banks, and police departments to help determine whether we get accepted to our dream school, receive the loan we need, or become a suspect in a criminal investigation.

In many ways, algorithms enhance human decision making. They can handle more information and juggle more variables than any person ever could. Also, running as they do on machines, algorithms don't get tired or let personal emotions sway their analysis. As a result, it is easy to assume that algorithms take human bias out of decision-making processes. That isn't accurate, however: Algorithms themselves can be deeply biased.

After all, algorithms are created by humans, whose biases can be written into the way information is analyzed. In some cases, such bias is likely planned. For example, some critics believe that Amazon's search engine algorithm is written to rank products that make the company the most money above those that are more popular with consumers or those that are more relevant to a user's query. In other cases, however, the bias can be unintentional. If a human resources algorithm designed to identify the best-qualified applicants for a job awards points for "participating in team sports" because its designers believe that it will help identify people who work well with others, the algorithm will be biased in favor of men (who are more likely to participate in team sports).

Algorithms are also dependent on the data sets they analyze—data that can be deeply biased. Law enforcement agencies, for example, have started to use facial recognition algorithms to help identify suspects. Studies show, however, that those systems falsely identify Black and Asian faces up to one hundred times more often than white faces. Why? Because up to 80 percent of the photos used to develop the algorithm were of white men, making the system much better at correctly identifying them.[2]

The impacts of such biased AI systems can be serious. In January 2020, Robert Julian-Borchak Williams, a Black man, was arrested by Detroit police and held in jail for thirty hours when a facial recognition algorithm incorrectly identified him as a suspect in a theft. Even when the charges were dismissed, he had to fight to get his information removed from the city's criminal databases.[3]

Advocates in the emerging algorithmic accountability movement, like Joy Buolamwini, argue that recognizing and addressing algorithmic bias is an urgent issue and will only grow more important as we and the institutions we interact with become more dependent on them. They advocate for hiring more women and people of color in high-tech industries; fewer than 2 percent of technical employees at Facebook and Google are Black.[4] They also call for regulations designed to make algorithms more transparent so that biases can be identified and corrected. And finally, they urge us to never buy into the false idea that computers are neutral. Like any technology, these machines and their algorithms reflect the priorities of the industries that develop them and the values of the people who build and use them. "What algorithms are doing is giving you a look in the mirror," argues Sandra Wachter, an associate professor of law and ethics at Oxford University. "They reflect the inequalities of our society."[5] ●

The second component of Web 3.0 is technology that allows a growing array of devices—TVs, tablets, and smartphones, but also cars, refrigerators, thermostats, traffic lights, and more—to communicate with each other and with the Internet. The **Internet of Things**, as it has been dubbed, is integrating the Internet into almost every part of our environment, including hospitals, urban infrastructure, factories, financial systems, and our homes. An Internet-connected refrigerator, for example, can take a photo of the interior every time the door closes, making it possible for someone at the supermarket to call up the last photo when deciding whether or not to buy milk.[15]

The Internet of Things and the Semantic Web are expected to work in tandem to alter how we relate to our media environment. Going forward, for example, smart refrigerators might talk to a computer at a local grocery store and automatically order milk for same-day home delivery. Voice recognition assistants like Amazon's Alexa and Apple's Siri already rely on the two Web 3.0 developments. Siri, for example, uses voice recognition to answer questions and interact with various iPhone functions, such as the calendar, reminders, and the music player. And if you are concerned about what might happen when you are stopped by a police officer while driving, you can simply state, "Siri, I'm getting pulled over." Your phone will automatically send a message to a predetermined contact number and start recording using your phone's camera.

Our Complex Digital Environment

The companies that create emerging Web 3.0 communication technologies market them as welcome conveniences or necessary solutions to all sorts of problems. But however useful these technologies may be, they can also be deeply problematic. In this section, we examine the complex dynamics that arise from three key digital-era developments: the decentralization of digital technologies, the creation of online communities, and the manipulation of digital content.

Decentralizing the Creation and Spread of Information

Affordable digital production tools, easy access to a global network, and interactive social media features like tagging, sharing, and liking have democratized the production and distribution of information. Anyone with access to the Internet can share opinions in a tweet, start a petition, organize a fundraising campaign, or livestream a video and instantly reach thousands or even millions of people. Centralized **gatekeepers**—newspaper editors, network TV managers, record executives, and others who have traditionally decided which messages get circulated—have been dramatically reduced, and ideas and content can go viral because millions of users share them. Our open and unregulated communication environment, however, is a double-edged sword.

The Benefits of Unrestricted Communication

An open Internet environment seems to fulfill the dream many people have for unrestricted communication—a dream fueled by the belief that everyone has the right to speak without government or elite gatekeepers interfering, and that this freedom makes our society stronger by allowing diverse voices to be heard. This ethos was picked up by early Internet users and by Silicon Valley companies like Facebook, whose original mission statement declared, "Facebook's mission is to give people the power to share and make the world more open and connected."[16]

It isn't surprising, then, that people have effectively used Twitter, Facebook, and other social media platforms for grassroots activism. Through their decentralized social media accounts, they have challenged repressive regimes, usurped their governments' centralized control of information and propaganda, and exposed government atrocities. One of the earliest instances of democratic action was the wave of protests in numerous Arab nations in North Africa and the Middle East that began in late 2010 and resulted in four rulers' being forced from power by mid-2012. The period, known as the Arab Spring, began in Tunisia. Young activists, using mobile phones and social media, organized

THE BLACK LIVES MATTER MOVEMENT, which began as a response to George Zimmerman's acquittal after he shot and killed Trayvon Martin, has grown to include chapters across the United States, Canada, and the United Kingdom. Per the organization, the Black Lives Matter Global Network "is adaptive and decentralized, with a set of guiding principles. Our goal is to support the development of new Black leaders, as well as create a network where Black people feel empowered to determine our destinies in our communities."[17]

marches and protests across the country. As satellite news networks spread the story and shared protesters' videos with the rest of the world, Tunisia's dictator of nearly twenty-four years fled the country.

The use of digital communication technologies to address injustice and organize grassroots movements has also played out in the United States, where movements on social media like #BlackLivesMatter and #MeToo have helped connect and mobilize millions. Hashtag activism—so called because of the use of the symbol # before a word or phrase that quickly communicates a larger idea, event, or objective—offers a compelling illustration of just how powerful social media can be when it is channeled toward a cause.

The Complications of Unrestricted Communication

While the use of unrestricted communication can be powerful, it can also get complicated quickly. The digital technologies that allow citizens to challenge injustice can also be used by cyberbullies, trolls, criminals, authoritarian governments, and terrorists. For example, the terrorist organization ISIS, one of the warring parties in Syria and Iraq, has successfully used the Internet and social media to recruit young men and women from other countries and to inspire others to commit terrorism around the world.[18] And in 2019, the murder of fifty people by a white supremacist at two mosques in New Zealand was livestreamed on Facebook. Although Facebook removed the video as soon as it was reported, copies were immediately posted on websites across the Internet.[19] The Internet's decentralized structure made the horrific video impossible to contain.

That same structure also makes it difficult to contain **misinformation** (false or misleading information spread by people who assume it's true) and **disinformation** (false or misleading information spread knowingly by people with malicious intent). In 2021, for example, even though Joe Biden beat President Trump by more than 7 million ballots in the popular vote and won the Electoral College vote by a wide margin, strategic disinformation about "a stolen election" (encouraged by Trump) fomented an insurrection. Believing the disinformation, thousands of people mobilized to take over the United States Capitol Building on January 6, 2021, coordinating through social media sites such as Facebook and Parler.[20] Right-wing extremists then continued to plot a government overthrow through a myriad of other Internet communication channels such as Signal and Gab.

False information also played a role in the 2016 election. Disinformation was traced to fake social media accounts linked to the Russian government, which reminds us that the decentralized nature of the Internet can be exploited by highly centralized institutions—governments with malicious intent—that have vast resources. (See "Global Village: Social Media Fraud and Elections.")

Building Online Communities

Humans tend to form connections and interact with people who are similar to themselves. That similarity may be based on racial identity, age, gender, occupation, sexual identity, religion, political beliefs, or other social variables. As the old adage says, birds of a feather flock together. This tendency can produce tightly knit communities that provide meaningful identities and support for their members.

Social Media Fraud and Elections

In the early years of social media, it seemed as if democracy had a new friend. Ideally, social media sites like Facebook, Twitter, and YouTube would open up political conversation, enabling democratic discourse to flourish. In 2010, the London newspaper the *Guardian* optimistically called Facebook "the election's town square" and hoped "that people [were] prepared to let their politics show online."[1] The *Guardian* was right: People *did* let their politics show online. Many of them posted relevant news stories, offered thoughtful discussion, and organized friends to become more politically involved. It turned out, however, that there were a lot of jerks hanging out in the election's town square: people posting nasty political memes, heated arguments, and hate speech. It all came to a head with the 2016 presidential election. "What had been simmering all year suddenly boiled over as [the] presidential election cycle made online friends hostile and prompted many to mute or unfriend those whose political rantings were creating stress," the *Dayton Daily News* wrote.[2]

Unfortunately, that wasn't the worst of it. There were also criminals in the election's town square, as foreign countries infiltrated social media to spread political disinformation and disrupt America's 2016 election. When the assistant director of the FBI's Counterintelligence Division testified about election interference before a U.S. Senate committee in June 2017, he stated:

> Moscow employed a multi-faceted approach intended to undermine confidence in our democratic process. . . . This Russian effort included the weaponization of stolen cyber information, the use of Russia's English-language state media as a strategic messaging platform, and the mobilization of social media bots and trolls to spread disinformation and amplify Russian messaging.[3]

The extent of disinformation was shocking. On Facebook, an estimated 126 million users might have been exposed to the fake ads and event posts produced by a Russian troll farm between 2015 and 2017. According to a Facebook official, "Many of the ads and posts we've seen so far are deeply disturbing—seemingly intended to amplify societal divisions and pit groups of people against each other."[4] There were also Russian-linked accounts spreading disinformation on Twitter, Google, and YouTube.

Election interference didn't stop in 2016. In the lead up to the 2020 election and in its tumultuous aftermath, social media companies faced mounting pressure to protect the democratic processes that many critics felt were being undermined by their platforms and policies. Facebook, Twitter, and YouTube stepped up efforts to identify Russian-linked troll accounts and remove fake content. They also began to place warning labels on misleading social media posts, including those of President Trump (who was ultimately banned by several sites for violating their policies). In addition, Facebook announced that it wouldn't run any new political ads seven days before the election and would step up its efforts to remove content aimed at suppressing voter turnout.[5]

The United States is not alone in battling disinformation, including campaigns by global actors like Russia. Russian trolls have also attempted to disrupt elections in European countries. Because Europe has been dealing with Russian propaganda for decades, most of its nations have a head start on strategies to combat such disinformation:

- In Sweden, there is a school literacy program to teach young people to identify Russian propaganda.
- Lithuania has citizen volunteers engaged in an "elves versus trolls" battle, with the good citizen investigators (elves) researching and revealing Russian trolls.
- Britain and France monitor Facebook closely and pressure the social media company to close down fake accounts.
- In Germany, all political candidates have agreed not to use bots—fake accounts (mostly on Twitter) that make automated posts in an effort to boost a topic's profile—in their political social media campaigns.
- In Slovakia, fourteen hundred advertisers have pulled their business from websites identified by researchers as the work of trolls.
- In Brussels, a European Union task force has published thousands of phony stories to reveal their deception.

In nearly all European countries affected, fact-checking and investigative journalism are additional countermeasures to the Russian troll offensive.[6] What strategies do you think would work best in America to combat political social media fraud? ●

ONLINE CONNECTIONS By enabling us to connect in new ways with others who share our experiences and interests, websites and social media apps can help offset feelings of isolation and even inspire political movements. #MeToo, for example, has given women who have been victims of sexual abuse a sense that they are not alone by building an online community—one that has produced significant changes in numerous industries.

The Internet has transformed the ways we find and interact with others. Its global reach makes it possible to overcome geographic isolation, connect with far more people, and build communities around a seemingly limitless array of interests. The open and interactive nature of the web also makes it possible for online communities to emerge and operate organically, and the anonymity of cyberspace opens up new conditions for participation. As these aspects of digital networks intersect with our tendency to seek out people like ourselves, however, it can lead to complicated outcomes.

The Benefits of Online Communities

The communities people engage with through the Internet can enhance their lives in meaningful ways. Fan communities are among the best-known examples. While fan clubs operated throughout the twentieth century via local chapters and newsletters, the Internet has dramatically changed their scale and accessibility. Fans from around the world who love the *Harry Potter* book and film franchises, for example, can visit websites to chat with other fans, create or listen to podcasts, share and read fan fiction, or participate in online role-play experiences. Online fan communities can also interact offline. *Harry Potter* fans hold conventions and organize live-action role-play events like Quidditch tournaments (a soccer-like team sport featured prominently in the books). Fans inspired by the books' social justice theme can also participate in the Harry Potter Alliance, a nonprofit organization devoted to making the world better through activism. As the popularity of such communities suggests, they play a role in helping people find and build bonds with others who share their interests and values.

The stakes involved in some online communities can be very high. For example, a study of LGBTQ+ teenagers living in rural Kentucky describes the opportunities for communicating with other LGBTQ+ people provided by the Internet. Often isolated and closeted in small towns with few, if any, out peers, these teens were able to connect with others via online message boards, including an Australian-based social network site for LGBTQ+ youth called Mogenic. These online connections helped them feel less alone; they could share their feelings; listen to their peers' coming-out experiences; and get advice about how they might come out to their family, deal with rejection, and cope with depression. Being able to connect with other people experiencing the same things mattered deeply to them, and online communities enabled that connection.[21]

The Complications of Online Communities

The very features of the Internet and social media that help us build meaningful bonds with some people can make it harder to build bonds with others, contributing to issues of fragmentation and polarization. Humans have always socialized with people like themselves, creating some degree of social fragmentation, or division, in the process. But by making it possible to connect with people differently and to form communities around a greater array of interests, the Internet can help increase this social fragmentation, leading some to worry about the consequences for democratic societies.

Concerns about the health of our current political culture have led people to examine social media's role in creating online **filter bubbles**—spaces where we are exposed only to ideas and opinions that match our existing beliefs. Using computer models to analyze how information spreads on Twitter and Facebook, researchers have demonstrated how features of online social networks help separate users into polarized groups.[22] With the mere click of a button, we can unfollow someone whose opinions we dislike. Meanwhile, recommendation algorithms track the people we follow and the posts we "like" or retweet, then encourage us to follow users with similar opinions and push content to our feed that we are likely to enjoy.

Models show that over time, users find themselves in increasingly homogeneous networks; conservative and liberal users, for example, become disconnected from each other, and posts that are shared among one group rarely circulate among the other group. Models also show that surrounding ourselves with networks of like-minded people can make us more vulnerable to false information. Information that upholds a community's preexisting beliefs—whether that information is accurate or not—tends to spread quickly across a homogeneous network, while an article debunking false information is unlikely to be shared widely in the same space.[23] This can deepen differences between groups, fueling an "us-versus-them" mentality and opinion polarization.

Social media's interactive capabilities and dynamic design exploit and amplify our tendency to socialize with people similar to ourselves as well as to favor information that conforms to our preexisting beliefs over information that challenges them (what is known as **confirmation bias**). In doing so, social media operates differently from mass media like broadcast television, which, as past research has shown, had a mainstreaming effect on viewers: People who watched a lot of television tended to share a similar worldview, even if they came from diverse backgrounds. In contrast, the Internet and social media contribute to the complex dynamics that characterize the niche nation (see Chapter 1).

The proliferation of online social networks enables people to build communities that speak to their needs and experiences more directly than a broad mainstream culture usually can; in this way, it is more democratic. At the same time, the resulting fragmentation can separate members of a society into groups with distinct and perhaps conflicting information sources, value systems, and identities—a development that can make it harder for a democratic society to function effectively. It is important to recognize that a diversity of opinions and identities doesn't have to lead to polarization; the task for us as media citizens is to build a society where political systems function well because of the diversity of identities people hold.

Manipulating Media

Digital media products can be manipulated far more easily than can most **analog** media products—a category that refers to various ways of encoding information that existed before binary code. When songs are carved as grooves into vinyl to make records, images are converted by chemical processes onto celluloid to make photographs, and voices are converted into electrical patterns for radio, the media that result are difficult to alter. But when content is digitized—encoded as ones and zeros—it can be processed and changed on a computer far more easily. Software tools like Photoshop, Auto-Tune, and iMovie have transformed our relationship to media products in ways that foster creativity but also raise concerns.

The Benefits of Remix Culture

Digital technologies make it possible for people to do more with media products than simply consume them. This development has fueled a revival of **remix culture**—a term used to describe a society in which people are able to create and communicate by mixing, editing, combining, manipulating, or repurposing existing texts.[24] Understood broadly, the act of remixing has always been essential to how humans express themselves. Artists often remix other works of art (e.g., allusions in poetry, sampling in rap music, and homages in film). But people also remix as part of daily life. If you comment on a situation by saying "methinks thou doth protest too much," you are remixing Shakespeare's *Hamlet*. If you tell a friend to "Let it go," you are remixing the film *Frozen*. We constantly use the cultural texts around us to communicate and participate in our culture.

Because mass media products were delivered in read-only formats for much of the twentieth century, options for remixing them were limited. The digital revolution changed that and helped make remixing an important feature of contemporary culture, influencing how professional producers and everyday users make media products and communicate with other people. **GIFs** allow us to participate in conversations over text or social media using clips culled from popular culture—the visual equivalent of saying "Let it go." Music mash-ups weave two, five, or even fifty songs together. Recut movie trailers posted on YouTube present classic films reimagined in a new genre (e.g., *Mary Poppins* as a horror film or *The Shining* as a romantic comedy). Instagram tools allow users to take a video, insert text, adjust the colors, add music, and post it to their story. While critics debate the artistic value of TikTok videos or Facetuned selfies, it is clear that these digital technologies empower users to do things with—and to—the products media industries deliver, thus expanding the user's role in mass culture beyond that of just consumer.

The Complications of Remix Culture

The fact that digital content can be manipulated so easily has consequences. Media companies, for example, have fought to maintain control of their copyrighted brand assets. Companies like Disney worry that users can easily transform beloved characters, like Mickey Mouse or Cinderella, into GIFs or memes that undermine their corporate goals. For this reason, the digital revolution has led to legal battles, as corporations lobby to strengthen copyright and trademark laws while remix advocates fight to protect the **right of fair use**—a legal doctrine that permits people to use copyrighted material without permission as long as the use does not compromise its value.

Digital content manipulation also raises concerns about the impact modified representations might have on cultural norms. Today, Photoshop artists routinely manipulate fashion spreads and Instagram photos to make men more buff or give women unnaturally thin thighs, smooth skin, long necks, and bright eyes, establishing unrealistic and possibly harmful beauty standards. With the help of apps like Facetune2, anyone with a smartphone can digitally manipulate their own selfies to represent themselves on social media not as they actually are but as they wish to be seen, contributing to a visual environment filled with aspirational—and false—images.

Image manipulation has also become a concern for journalists. In its June 2020 coverage of protests in Seattle over the police killing of George Floyd, Fox News posted digitally altered photos on its website that remixed elements from other photographs, seemingly to intensify the sense of social menace in the resulting image. When reporters challenged the veracity of the photos, a Fox News editor apologized for failing to identify the images as collages. According to the *Washington Post*, "The use of the deceptive material marked a new chapter in an ongoing debate over synthetic media, which includes both sophisticated, computer-generated 'deepfakes,' as well as more rudimentary mash-ups that still may mislead the public."[25] **Deepfakes** are images or videos that use advanced digital editing technology to create fraudulent but convincing content. Deepfake tools can superimpose the face of a politician or celebrity onto the face of someone else in a video, making it look as if the well-known person was somewhere doing something they never actually did.

Such manipulated digital content raises obvious political and social concerns—concerns that are all the more serious because the Internet's open structure makes it easy for disinformation to enter the web, often through social media networks. As we saw earlier, false and manipulated information that confirms a group's preexisting beliefs can spread quickly in these online communities, whereas stories that contradict or debunk those beliefs are less likely to be shared. As Internet disinformation campaigns have weaponized social media, major challenges exist for governments, companies like Facebook, and all of us as media citizens. We need to find a way to maximize the benefits of communicating via digital networks while minimizing the negative consequences of these interactions.

DIGITAL MASHUP Fox News posted this image on its website on June 12, 2020, as part of its coverage of protests in Seattle following the police killing of George Floyd. The image combined elements from three different photographs taken several days apart and in different locations in Seattle. Critics argued that the misleading photo (which appeared with no disclaimer) violated journalistic ethics. Digital mash-ups like these raise deep concerns about the power of disinformation.

The Business of Controlling the Internet

No one actually owns the Internet. The unruly nature of the web, where traditional gatekeepers are relatively powerless, can make it feel like an untamed wild west. But in fact, certain aspects of the Internet are tightly controlled—and powerfully controlling. Corporations and government entities rule key corners of cyberspace, often serving as keepers of new kinds of gates that control our personal data and restrict our freedom across the web.

In the United States, five major tech companies have gained tremendous market dominance in their respective sectors. Apple controls roughly 40 percent of smartphone sales; Amazon, 44 percent of online retail; Facebook, 75 percent of social media; Microsoft, 77 percent of desktop operating systems; and Google, an astonishing 92 percent of online searches.[26] (For more on these companies and their many products and services, see Table 2.1.) Meanwhile, as we'll see in the following section, governments across the globe are using the Internet for both economic and political control.

Controlling Data

When we watch television, listen to the radio, read a book, or go to the movies, we don't usually provide personal information to others. However, when we use the Internet, we give away personal information all the time—often without knowing it. Collecting and leveraging that information has created a new economy and shifted our relationship with both corporations and governments. Such developments have also spawned intense debates over who should control our personal data.

Data Mining for Profit

Google, Amazon, and Facebook are among the most powerful Internet-era companies because of their role in the rise of **surveillance capitalism**—an increasingly important business model that involves making money by controlling the personal data of millions of users. Through its search engine and ad-placement service Google Ads, Google, in particular, has radically changed advertising by using the interactive nature of the web to micro-target consumers in a way that traditional

TABLE 2.1 THE BROAD REACH OF BIG TECH

Amazon, Google, Apple, Facebook, and Microsoft constitute today's leading digital media companies. These companies have realized the potential of dominating the Internet first by selling their gadgets and personal assistants, then by steering their customers to their media services, and finally by turning their customers into huge audiences for advertising and data collection. In the words of tech writer Liz Pelly, we should really think of Big Tech's proprietary hardware as "overpriced, fun-sized plastic and metal surveillance machines."[27]

Big Tech	Gadgets	Personal Assistants	Media Services	Cloud Services
amazon	Kindle, Echo, Fire TV, Fire Phone	Alexa	Amazon (shopping), Amazon Studios, Amazon Prime, Twitch	Amazon Web Services, Amazon Drive
Google	Nexus, Android OS (phone), Pixel, Nest, Nexus (tablet), Chromebook, Pixelbook, Chromecast, Daydream View, Google Home	Google Assistant	Google, YouTube, Google Play, Gmail, Google+, Google Maps	Google Drive
Apple	iPod, iPhone, iPad, iWatch, MacBook, Apple TV, HomePod, AirPlay	Siri	Safari, iTunes, Apple TV, Mail	iCloud
facebook	Oculus Rift, Portal	M (discontinued)	Facebook, Instagram, WhatsApp, Messenger, Facebook Gaming	Dropbox (relationship)
Microsoft	Xbox, Surface (tablet), Windows (phone)	Cortana	Bing, Xbox games, Hotmail	Microsoft Cloud

MEDIA LITERACY & THE CRITICAL PROCESS

Harnessing Creativity by Stepping Away from Your Phone

Manoush Zomorodi is a technology journalist who, through her writing and podcast (*Note to Self*), questions common assumptions about tech culture and offers a set of media literacy challenges to help us use technology more mindfully. How can we make information overload disappear and enjoy healthier digital consumption? To answer these questions, we have adapted a series of challenges based on Zomorodi's book *Bored and Brilliant: How Spacing Out Can Unlock Your Most Productive and Creative Self* (2017). Each of the seven challenges covers the *description* and *analysis* steps of the critical process, but for each, *you* become the subject you are researching.

CHALLENGE 1:
Observe Yourself.

Most smartphones come with a tool to track how much time you spend on your phone; use this tool, or download one of the free tracking apps available for both iOS and Android phones. Then, for the next two to three days, or even for a full week, track your phone use. How much time do you spend on it? When do you most use your phone, and what applications do you spend the most time on? Are your results surprising?

CHALLENGE 2:
Keep Your Devices out of Reach While in Motion.

For the next two days, put your phone in your pocket (or, better yet, your bag) while you walk, bike, or drive from one place to another. As you travel, look around and identify five things you haven't noticed before; as Zomorodi says, it could be ornamental cornices on a house you pass every day, or the patterns made by the clouds in the sky. Write those five things down.

CHALLENGE 3:
Photo-Free Day.

Take no pictures for one entire day and "see the world through your eyes, not your screen."[1] This means getting through one whole day without taking pictures of your food, your friends, or yourself. It means one day without cute animal photos or "wishing you were here" pics. If you don't want to break a Snapchat streak, get a friend to take over for that day.

CHALLENGE 4:
Delete That App.

Select an app on your phone that takes up a lot of your time, then delete it. How do you know which app to choose? According to Zomorodi, "You know which one is your albatross—the one you use too much. The one you use to escape—too often, at the expense of other things (including sleep). The one that makes you feel bad about yourself. Delete said time-wasting, bad-habit app. Uninstall it." Or consider taking this experiment one

magazine or TV ads never could. (See more in Chapter 11.) For example, Google builds unique profiles of each of us that include assumptions about our age, race, location, income, education level, political sensibilities, restaurant preferences, and many other data points. It compiles these profiles by constantly tracking—or **data mining**—our search histories, locations, browser settings, and even the videos we watch on YouTube. If we have a Gmail account, it scans our e-mails. If we use Google Maps, it knows where we are. If we visit one of the millions of websites or apps Google "partners" with, it knows everything we do there. If we use an Android phone, Google monitors how long it takes us to open a particular app, and determines whether we are walking, biking, or driving to certain locations.

This type of surveillance is facilitated by cookies and IP addresses—elements of the web's interactive systems. **Cookies** are computer files that automatically collect and transfer information between a website and a user's browser, which makes revisiting a website easier or more personalized. Cookies also make it possible for companies to track a user's browsing history. An **IP (Internet protocol) address**—the unique number that every device uses when communicating on the Internet—can be used by a data miner like Google to make assumptions about such details as which user is online and where that user is located.

Google uses this information to deliver tailored content and hyper-targeted advertisements to each of us. Every time we type a query into Google Search, an algorithm instantaneously auctions access to us off to advertisers. The winning bidder shows up at the top of the results list as a "sponsored link," and if we click on that link, the advertiser pays Google. Advertisers struggled for years to measure people's attention to ads, but with this system, advertising's effectiveness is easy to track. In addition, each ad is more likely to reach a desired consumer and is therefore relatively inexpensive, since the ad is wasted less often on uninterested parties.

step further and deleting your whole account. That's another way to take back control.

CHALLENGE 5:
Go Phone-Free.

Zomorodi argues that creative problem solving depends on solitude and even boredom. When our phone supplants this kind of quiet time, our brain never gets to relax and think. She recommends breaking the cycle by taking a vacation "from the digital onslaught that exhausts, distracts, and keeps us from thinking beyond the everyday."[2] You can stay phone-free for whatever amount of time works for you. Also, when it's time to step away, it can help to be creative with your away message; Zomorodi suggests going with something fun, like: "Consciously uncoupling from my phone for the day" or "Teched out, checked out. Taking a little break from the devices. Thanks for your patience."[3]

CHALLENGE 6:
Observe Something Else.

Check out what is going on around you. Go spend some time in a public place, like a restaurant, a park, a laundromat, or even a street corner. You can spend five minutes, or even an hour, just looking at your surroundings. Feel free to walk around if it seems awkward to stand in one place the whole time. While you are in this space, "pause and imagine what a single person is thinking, or zoom in on an un-inventable detail. Just make one small observation you might have missed if your nose were glued to a screen . . . noticing is the first step in creating."[4]

CHALLENGE 7:
Be Bored (and Brilliant).

Take thirty minutes out of your day to become completely, "mind-numbingly" bored. Before you do this, spend five minutes thinking about a life issue that you have to resolve—it could be big picture, like your career path, or small, like a disorganized closet. Write this issue down, then get bored. To do this, some people stare at a ceiling or write the numbers "01,01,01" over and over again. Others watch a pot of boiling water. The goal of this exercise is to consciously discover what happens when you let your mind wander. Interestingly, your brain never stops thinking, and you may discover that your creativity gets unleashed in surprising ways. After your thirty minutes of boredom, write down your thoughts (or make a sketch) about the unresolved issue you identified earlier. As Zomorodi writes, "The goal is to use your boredom, and the peace you have away from your phone, to unlock a brilliant solution to your problem."[5]

Now that you've completed all seven exercises, finish up your reflections with the *interpretation*, *evaluation*, and *engagement* phases of the critical process. How was this overall experience for you? What does it mean to you? Was being device-free and bored for a period of time a good thing or not? For the engagement phase, as Zomorodi suggests, try to come up with a broader goal for tackling information overload. This can even be a mantra of sorts—something new to live by—that you can put on a sticky note and affix to your laptop. It might say something like "connect with yourself" as a reminder to stay off your phone, or "walk more, text less." Whatever you decide to do personally, bring your collective experience to the entire class and have a class-wide discussion about digital technology and making information overload a thing of the past.

The information companies gather from us, and about us, every time we use their services has become a highly valuable product. What oil was to the Industrial Age, data is to the Information Age—a development that has helped make Google one of the richest companies in the world. Its website, the most visited spot on the Internet, processes over forty thousand queries every second, with each search adding to its data stockpile, and each click of a sponsored page adding to its coffers.

The Attention Economy and Addictive Design

Controlling our attention helps companies like Google and Facebook collect more data and create opportunities to sell more ads—all essential tactics in the emerging attention economy. For example, Facebook and Instagram (both owned by Facebook Inc.) are designed to mine enormous amounts of data from users through their profiles, likes, and posts. Of course, as noted earlier, these apps also encourage users to build an ever-growing network of friends. This expanding network gives Facebook access to even more layers of data, since the company can look beyond individual users to find patterns in behaviors and preferences across an entire community.

Facebook and Instagram are also created to maximize user engagement. The social nature of these platforms exploits our innate desire to connect with people we know, but the companies behind these platforms use **addictive design** principles developed from research into human behavior modification. Key features—the infinite scroll, notifications, "like" buttons, and continually updated content—exploit aspects of human psychology to increase the time users spend on the platforms and to make checking for updates and messages a habit. (To get a handle on your own social media and smartphone use, see "Media Literacy and the Critical Process: Harnessing Creativity by Stepping Away from Your Phone.")

GOVERNMENTS around the world employ an array of digital surveillance technologies, usually in the name of efficiency and public safety. Such technologies, however, can also be used to maintain political control. In China, tech companies demonstrate the latest tools that the government can use to keep track of its citizens. As one engineer put it, these technologies "apply the ideas of military cyber systems to civilian public security."[30]

Government Surveillance

While companies like Google and Facebook gather our personal data for profit, governments usually do so in order to maintain national security, social order, or political control. Since the inception of the Internet, government agencies worldwide have obtained communication logs, web browser histories, and the online records of individual users who thought their online activities were private. In the United States, for example, the USA PATRIOT Act (which became law shortly after the September 11 attacks in 2001) grants sweeping powers to law enforcement agencies to intercept individuals' online communications, including e-mail and browsing records.

Emerging technologies are extending government surveillance powers even further, leading some to fear the rise of powerful digital-era **surveillance states**—societies in which governments conduct systematic mass surveillance on their populations. The media environment in which we live provides governments with powerful new surveillance tools. The Chinese government, for example, works closely with Chinese technology companies like Tencent to conduct mass surveillance of messages sent on apps like WeChat, the country's most popular messaging service. It is also in the process of building a vast network of urban video cameras, which—in combination with cell phone tracking, facial recognition technologies, and centralized databases—will give the government an unprecedented degree of surveillance power.[28] The system is already being deployed in China's westernmost province as part of the government's efforts to control the ethnic Uyghur people.[29] While China may be ahead of the curve, countries all around the world are developing their own mass surveillance systems.

Controlling Access

The World Wide Web evokes a vast open space that anyone can roam around freely to access seemingly limitless information. Over time, however, a few Internet companies and governments—driven by different motivations and using different strategies—have tried to hem in both users and information. The result has been the emergence of a different kind of online experience: a closed Internet. Such developments challenge the visions many Internet advocates had for the medium as a unifying and neutral global network free of barriers and gatekeepers.

Walled Gardens

When we use our smartphones to find information, it can feel like we have the entire Internet at our fingertips. Instead, we often end up in highly managed environments brought to us by apps—what some have called **walled gardens**.[31] Instagram, Facebook, and Pinterest are good examples of these types of environments. By bringing continuous streams of content directly to our feeds, each of these apps is designed to discourage us from leaving to surf elsewhere. These apps have clean, orderly, and easy-to-use interfaces, but we can wind up trapped in their closed gardens without exactly choosing to be.[32]

Under the direction of cofounder and longtime CEO Steve Jobs, Apple Inc. famously built its own walled garden by developing hardware, software, and retail services like iTunes and its app store, which offer consumers a seamless, integrated experience—as long as they stay loyal to Apple. Although MacBooks can work with Android phones, and iPhones can work with non-Apple laptops, using all Apple products makes it possible to do more, like receive texts and phone calls on our laptops. Apple's iCloud storage and syncing service enables users to instantly access media content purchased from Apple stores on any Apple mobile device. With each new product and service, Apple tempts users to enter and stay inside its insulated ecosystem with the promise of a more secure and efficient experience. It also comes at a price: Not only do Apple products generally cost more, but Apple also profiles its customers across all of their Apple devices and applications.

The Battle for Net Neutrality

Beneath the apps and websites that capture our attention lies the Internet's infrastructure—the fiber-optic lines, servers, and routers that deliver digital traffic. As we saw earlier in this chapter, that infrastructure was initially built and managed by various government agencies and developed within a public service ethos rooted firmly in the assumption that all data being sent across the Internet would be treated the same—that is, it would have the same access to the network and travel across it at the same speed. This vision—what would become known as the principle of **net neutrality**—went unchallenged for decades. By the early 2000s, however, private corporations had taken over responsibility for the Internet's infrastructure in the United States and begun to assert a new gatekeeping power.[33]

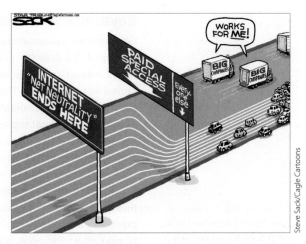

Steve Sack/Cagle Cartoons

Telephone and cable companies like Verizon, Comcast, AT&T, and CenturyLink made money by providing broadband access to their nationwide networks, which are interlaced with one another (and with many other networks around the world) to form the Internet. Those companies increasingly wanted to treat the data that traveled on their networks differently, delivering content faster to clients willing to pay higher rates and providing preferential service for their own content or for content providers who made special deals with them. In 2014, for example, many people who received Internet access from Comcast complained that Netflix videos loaded so slowly they were unwatchable. Comcast, it turns out, had lowered download speeds for Netflix programming to its customers—a move that violated the principle of net neutrality. Comcast claimed that Netflix's large video files were taking up too much of its delivery system's capacity. Netflix, faced with angry subscribers, was forced to pay Comcast to get normal download rates restored.[34]

Companies like Comcast argue that asking content providers to pay different fees will give them resources needed to improve the network's infrastructure and allow them to lower Internet access costs for customers. Their critics, however, argue that they are simply looking to maximize profits and, by doing so, are undermining a fundamental principle of the Internet: that the network should treat all information—whether created by a powerful multinational corporation, a nonprofit organization, or any of us—the same. Supporters of net neutrality—such as bloggers, video gamers, educators, religious groups, unions, nonprofit organizations, and small businesses—claim that the cable and telephone giants have powerful incentives to rig their services and cause network congestion in order to force customers to pay a premium for higher-speed connections. They also argue that an Internet without net neutrality would put small independent websites in the "slow lane," while heavily funded corporate websites would be in the "fast lane." Interestingly, large Internet corporations like Google, Yahoo!, Amazon, eBay, Microsoft, Skype, and Facebook also support net neutrality because their businesses depend on their millions of customers' having equal access to the web.

Although the idea of an open and neutral network existed since the origins of the Internet, there had never been a legal formal policy until 2015, when the Federal Communications Commission (FCC) approved net neutrality rules.[35] In 2017, however, a newly Republican-controlled FCC, which sided with big telecommunications corporations, overturned those rules, allowing broadband companies to slow down or block traffic as they wished. Efforts to defend net neutrality have continued in courts and state legislatures. Attorneys general from twenty-one states and the District of Columbia filed lawsuits to challenge the decision, stating that the FCC vote to repeal net neutrality was "arbitrary, capricious, and an abuse of discretion."[36] While the FCC's repeal was upheld by a federal appeals court in 2019, the ruling did suggest that states and local governments could create rules that establish net neutrality for their citizens. The governor of Montana, for example, ordered broadband providers that have contracts with the state to follow net neutrality principles.[37]

NET NEUTRALITY For decades, the Internet has been a place where all traffic is treated the same. The future of the Internet, however, may be very different. Without net neutrality policies, critics warn that Internet service providers can slow down the flow of some content while allowing other content to flow freely. As ISPs like AT&T and Spectrum look to maximize their profits, how we experience the web could change in profound ways.

LaunchPad
macmillan learning
launchpadworks.com

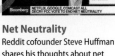

Bloomberg

Net Neutrality
Reddit cofounder Steve Huffman shares his thoughts about net neutrality.

Discussion: Do you support net neutrality? Why or why not?

Government Barriers

Governments can sometimes control user access and mobility as well, though they are often motivated by a desire to maintain moral order or political control in addition to profit. In the United States, strong First Amendment protections limit the government's ability to restrict people's movements on the web. That said, the Children's Internet Protection Act of 2000 requires schools and libraries that receive federal funding for Internet access to use software to filter out any visual content deemed obscene, pornographic, or harmful to minors. Critics like the American Association of School Librarians argue that some schools interpret the guidelines far too broadly, keeping students from accessing websites they would benefit from.[38] Conservative school leaders, for example, can use the law to block most LGBTQ+ websites.

In China, the government has dramatically restricted its citizens' access to the global Internet through a combination of laws and blocking technologies dubbed the Great Firewall. Chinese citizens experience a far different web than do citizens of the United States or Europe. The Chinese government blocks Google, Facebook, WhatsApp, and Twitter, as well as many foreign news and information websites, like Wikipedia. Such restrictions work alongside China's massive surveillance practices to help maintain political control—or, as it is seen by many within China, social stability. The Great Firewall has also protected China's technology start-ups from foreign competition. Today, Chinese companies like Alibaba, Tencent, and Baidu rival—and sometimes surpass—Silicon Valley giants in terms of the number of users served and the sophistication of their products and services. The Chinese government's political and economic agendas are interconnected: Monitoring citizens' activities on apps controlled by Chinese companies is much easier than dealing with U.S.-based companies.

China's Great Firewall may presage a future in which users encounter increasingly controlled national borders on the Internet. In June 2020, for example, the Indian government blocked its citizens from accessing fifty-nine apps created by Chinese companies, including TikTok, which had been installed on over 600 million devices in India. The digital wall was erected after a skirmish with China over a disputed border in the Himalayas left twenty Indian soldiers dead. The move was only the latest in a worldwide trend as governments look to create more walled gardens throughout the Internet landscape, leading two *New York Times* reporters to note that "the digital world, once thought of as a unifying space that transcended old divisions, is being carved up along the same national lines that split the physical one."[39]

Regaining Control

Many people are worried about the security of their data. According to a 2019 Pew Research Center survey, a sizable majority of Americans (79%) were at least somewhat concerned about how much personal information corporations collect, while more than half (64%) were concerned about government surveillance. Over 80 percent said they felt they had little control over the data collected about them.[40]

Why do people feel this way? These attitudes are likely the result of high-profile privacy-related scandals from big tech companies. In 2018, for example, Facebook revealed that a political consulting firm, Cambridge Analytica, gained access to eighty-seven million Facebook user accounts, leading to a congressional hearing in which Facebook CEO Mark Zuckerberg was grilled about the company's data-mining and data-sharing practices.

Scandals like this can be disheartening, but individuals and organizations around the world are finding ways to push back against the powerful tech companies and governments that are tracking our every move, and they are also using high-impact approaches—like open-source software development and digital archiving—to preserve the independent vision of early Internet users.

Pushing Back against Tracking and Data Collection

People are using a number of strategies to regain control over their information, including regulation, protective software, and campaigns for change. Consumer and privacy advocates, for example, have called for stronger regulations related to data collection, such as **opt-in policies**, which would require websites to obtain explicit permission from consumers before the sites can collect browsing history data. In 1998, the Federal Trade Commission (FTC) developed fair information practice principles to address the unauthorized collection of personal data; however, the FTC has no power to enforce these principles, and most websites do not follow the guidelines. Instead, many automatically collect browsing history data unless individual users, as consumers, specifically request to "opt out" of the process.[41] (Europe, by contrast, passed stronger regulations, requiring informed consent before data is collected on any user, with large penalties for violations.)

In the United States, some activists and politicians have also called for the use of antitrust laws to break up tech giants like Facebook, Google, and Amazon, arguing that their business tactics limit innovation and their market domination make them less responsive to consumer demands.[42] This approach gained traction in 2020, as the Department of Justice filed a landmark antitrust lawsuit against Google to stop it from "unlawfully maintaining monopolies through anticompetitive and exclusionary practices in search and search advertising markets."[43]

Nongovernment organizations can also work to protect privacy, as can private citizens. The Electronic Frontier Foundation—a nonprofit group that defines its mission as championing "user privacy, free expression, and innovation through impact litigation, policy analysis, grassroots activism, and technology development"—seeks to protect civil liberties from government overreach.[44] Meanwhile, millions of people take action by signing petitions to protect net neutrality, joining "boycott-Facebook" campaigns, and installing browser plugs-ins to block ads and social media tracking.

Developing Open-Source Software

In the early days of computer code writing, amateur programmers developed **open-source software** on the principle that it was a collective effort. They openly shared program source codes along with their ideas for improving programs. Beginning in the 1970s, Microsoft put an end to much of this activity by transforming software development into a business in which programs were developed privately and users were required to pay for both the software and its periodic upgrades.

However, programmers today are still developing noncommercial open-source software, if on a more limited scale. One open-source operating system, Linux, was established in 1991 by Linus Torvalds, a twenty-one-year-old student at the University of Helsinki in Finland. Since the establishment of Linux, professional computer programmers and hobbyists around the world have participated in improving it, creating a sophisticated software system that even Microsoft has acknowledged is a credible alternative to expensive commercial programs.

Digital Archiving

Librarians have worked tirelessly to build nonprofit digital archives that exist outside any commercial system in order to preserve libraries' tradition of open access to information. One of the biggest and most impressive digital preservation initiatives is the Internet Archive, established in 1996. The Internet Archive aims to ensure that all citizens have universal access to human knowledge—that is, everything that's digital: text, moving images, audio, software, and billions of archived web pages reaching back to the earliest days of the Internet. The archive is growing at a staggering rate, as the general public and partners such as the Smithsonian Institution and the Library of Congress upload cultural artifacts. For example, the Internet Archive stores more than 225,000 live music concerts, including performances by Jack Johnson, the Grateful Dead, and the Smashing Pumpkins.

Media activist David Bollier has likened open-access initiatives to an information "commons," underscoring the idea that the public collectively owns (or should own) certain public resources, like the airwaves, the Internet, and public spaces (such as parks). Says Bollier, "Libraries are one of the few, if not the key, public institutions defending popular access and sharing of information as a right of all citizens, not just those who can afford access."[45]

The Internet, Digital Communication, and Democracy

The democratic potential of communication systems has often been measured against the ideal of **universal access**—the notion that every citizen, regardless of income or location, should have the opportunity to use and benefit from a technology. This principle has long guided the development of the U.S. postal system, as well as broadcast radio and television. As going online evolved from a novelty to an essential tool of daily life—a development only amplified during the 2020 coronavirus pandemic, when many people relied on the Internet for school, work, food, and health-care services—the consequences of unequal access to the Internet grew more serious. Coined to echo the term *economic divide* (the disparity of wealth between the rich and the poor), the term **digital divide** refers to the growing contrast between the "information haves"—those who can afford to purchase a computer and pay for Internet services—and the "information have-nots"—those who may not be able to afford a computer or pay for Internet services.

While roughly 90 percent of Americans respond that they use the Internet at least occasionally or own a smartphone, there are big gaps in access to home broadband service when we factor in demographic differences, such as income and location.[46] For example, about 94 percent of adults with household incomes of $100,000 or more have home broadband, whereas only 56 percent of adults with household incomes less than $30,000 have it.[47] Meanwhile, 79 percent of suburban households have broadband, but only 63 percent of rural households do (see Figure 2.2).[48]

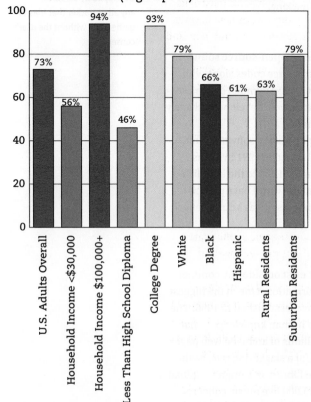

Percentage of U.S. Adults with Home Broadband (High-Speed) Internet Access

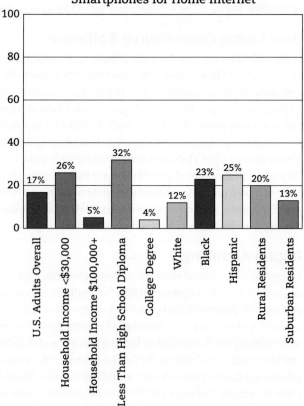

Percentage of U.S. Adults Who Rely on Smartphones for Home Internet

FIGURE 2.2

RATES OF BROADBAND AND SMARTPHONE INTERNET ACCESS

Data from: "Internet/Broadband Fact Sheet," Pew Research Center, June 12, 2019; Monica Anderson and Madhumitha Kumar, "Digital Divide Persists Even as Lower-Income Americans Make Gains in Tech Adoption," Pew Research Center, May 7, 2019.

Unequal access to the Internet has real consequences for people, and the hurdles to getting online can involve more than just access to technology. The study about the importance of an online community for rural LGBTQ+ teens discussed earlier, for example, also revealed the challenges teens faced accessing that community. Some didn't have Internet access at home, but even those who did were nervous about using a family computer to visit LGBTQ+ sites. Some tried to get online at school or libraries, but the sites were often blocked. It is tempting to get caught up in celebrating the Internet revolution and the freedom it offers, but as we saw in Chapter 1, a technology's affordances are never the whole story. Many factors shape how we use the Internet. Where is broadband access available, and at what price? What obstacles do individual users have to deal with to get online? How do cultural norms about "appropriate" content affect access to information? The seemingly limitless possibilities of the Internet are never limitless in practice.

Although not a perfect substitute for home broadband connection, smartphones have helped narrow the digital divide. Rural and lower-income Americans, for example, are more likely to own a smartphone than a computer, and the percentage of lower-income adults who said they have a smartphone but not home broadband access rose from 12 to 26 percent between 2013 and 2019.[49] The current roll out of **5G**—the next generation of mobile networking technology that promises download speeds on par with home broadband—could narrow the divide further. Of course, that would also depend on how much 5G service costs and where it will be available. As is the case with broadband, telecommunications companies are unlikely to provide 5G everywhere unless they are required to by government policy.

Globally, the have-nots face even greater obstacles in crossing the digital divide. Although the web claims to be worldwide, the most economically powerful countries—the United States, Sweden, Japan, South Korea, Australia, and the United Kingdom—have historically accounted for most of its international flavor. In nations such as China, Russia, Turkey, and Pakistan, the government permits limited access or no access to the web. In many underdeveloped countries, the lack of computers and widespread phone networks has hampered Internet access for decades.

In recent years, however, powerful companies, anxious to find new markets, have rushed to provide Internet access via mobile phone networks to underserved parts of the world, most notably sub-Saharan Africa. As a result, Africa has become a front line in the battle between U.S. and Chinese tech companies for global dominance, leading critics to raise concerns about the political ramifications. Facebook's Free Basics project, for example, allows users in thirty African countries to visit Facebook's platform without using any of their plan's data.[50] Facebook, in turn, makes money by selling ads and gathering information about the users' behaviors. As in the United States, critics have raised concerns about privacy rights violations, as well as the effect that disinformation campaigns circulating on Facebook have had on political elections.[51] Meanwhile, Chinese tech giant Huawei has been winning contracts to build much of Africa's 5G infrastructure. Concerned citizens warn that Huawei is equipping some authoritarian-minded African leaders with the same sophisticated surveillance tools the Chinese government has used to maintain political control—a move that they fear could destroy civil liberties in African democracies.

As we consider issues of access related to the digital divide—both globally and in the United States—it helps to reflect on the freedoms the Internet affords us and the inequities it introduces. On the one hand, inexpensive digital production and social media distribution enable greater participation online than did many of our earlier media technologies. On the other hand, even as the Internet matures and becomes more accessible, wealthy users are still able to buy higher levels of privacy and faster speeds of Internet access than are other users. Whereas traditional media made the same information available to everyone who owned a radio or a TV set, the Internet creates economic tiers and classes of service. Policy groups, media critics, and concerned citizens will continue to debate the implications of the digital divide, including the value of all people having an equal opportunity to acquire knowledge.

Chapter REVIEW 2

LAUNCHPAD FOR *MEDIA & CULTURE*

 **LaunchPad**
macmillan learning

Visit LaunchPad for *Media & Culture* at **launchpadworks.com** for additional learning tools:

→ QUIZ YOURSELF WITH LEARNINGCURVE

LearningCurve uses gamelike quizzing to help you review important concepts from this chapter. LearningCurve includes multiple-choice questions at different levels of difficulty, hints, feedback for both right and wrong answers, and links to e-book material for easy reference. Which answer choice would you select for the sample question below?

Which of these people would have had access to ARPAnet in the 1960s?

- ○ a mechanical engineer working in private industry
- ○ the manager of a store that sold computer components
- ○ a scientific researcher working for a major university
- ○ a major-party candidate for political office

→ WATCH VIDEO ON KEY CONCEPTS

LaunchPad includes clips from movies, TV shows, online sources, and other media texts, in addition to interviews with media experts and newsmakers. In the videos for this chapter, you'll consider topics like net neutrality and user-generated content.

→ PRACTICE THE CRITICAL PROCESS

Use the Media Literacy Activity to walk through the critical process step-by-step and develop a critical perspective on what media means to you. This chapter's activity examines issues of online safety and privacy.

→ LEARN KEY TERMS

Review the definitions for this chapter's key terms with LaunchPad's easy-to-use online flash cards. Try the cards in practice mode, or quiz yourself as a way to focus your study efforts. (Page numbers indicate where each term is highlighted within the chapter.)

Internet, 34
byte, 34
protocols, 36
ARPAnet, 36
bulletin board services, 36
microprocessors, 37
fiber-optic cables, 37
World Wide Web, 37
HTML (hypertext markup language), 37
browsers, 37
Internet service providers (ISPs), 38
broadband, 38
search engines, 38

read-only, 38
read-write, 38
Section 230, 38
wikis, 39
Sematic Web, 39
Internet of Things, 41
gatekeepers, 41
misinformation, 42
disinformation, 42
filter bubbles, 44
confirmation bias, 45
analog, 45
remix culture, 45
GIFs, 46
right of fair use, 46

deepfakes, 46
surveillance capitalism, 47
data mining, 48
cookies, 48
IP (Internet protocol) address, 48
addictive design, 49
surveillance states, 50
walled gardens, 50
net neutrality, 51
opt-in policies, 52
open-source software, 53
universal access, 54
digital divide, 54
5G, 55

REVIEW QUESTIONS

Use the questions below to revisit key themes and concepts from this chapter.

 Need more practice? Access the LearningCurve multiple-choice questions for this chapter, which are available on LaunchPad. All questions include feedback for correct and incorrect answers, hints, and links to e-book material.

The Internet Today

1. What is the Internet, and how is Internet use evolving around the world?

How We Got Here: The Development of the Internet

2. How did the Internet originate? What role did the government play?
3. What ethos guided early Internet users?
4. What factors laid the groundwork for the commercialization of the Internet in the 1990s?
5. What are the key distinctions between the pre-web Internet, Web 1.0, Web 2.0, and Web 3.0?

Our Complex Digital Environment

6. What are the benefits and complications of unrestricted communication?
7. What is the relationship between online communities and political polarization?
8. What are the benefits and complications of remixing?

The Business of Controlling the Internet

9. Why and how do companies like Facebook, Amazon, and Google mine our data? How does it shape their business practices?
10. Why and how do governments use digital surveillance technologies?
11. Why and how has the Internet become increasingly closed?
12. What is net neutrality, and why is it such an important issue?
13. How have people begun to push back against corporate and government control of the Internet?

The Internet, Digital Communication, and Democracy

14. What is universal access, and what role has it played in people's visions for democratic media systems?
15. What is the digital divide, and what threat does it pose to the Internet's democratic potential?

QUESTIONING THE MEDIA

Use these critical-thinking questions to reflect on your own experiences with media and apply your understanding of media concepts.

 For an interactive experience, questions on topics like these are available for the iClicker student response system. Instructors can download the questions from LaunchPad and use them to poll the class.

1. What possibilities for the Internet's future are you most excited about? Why? What possibilities are most troubling? Why?
2. In what ways do you participate in remix culture through the media you produce or consume?
3. How are you affected by the rise of surveillance capitalism and surveillance states? What concerns do you have about their growth?
4. In the move from a print-oriented Industrial Age to a digitally based Information Age, how do you think individuals, communities, and nations have been affected positively? How have they been affected negatively?
5. What do you think about the role of universal access in a democratic society?

Digital Gaming and the Media Playground

3

College scholarships, competitions to make the team, grueling practice sessions, matches in arenas drawing thirty thousand people or more, huge television audiences for championships, profiles in *Sports Illustrated*, and hopes to one day compete in the Olympics.

This is not swimming, track and field, soccer, hockey, or skiing. These "athletes" have no incentive to cheat with steroids, and cardio training is optional. These are eSports athletes: athletes who competitively play video games at the highest levels.

In the past two decades, eSports have followed the trajectory of traditional sports, with an increasing number of colleges and universities recruiting scholarship gamers, major sponsors underwriting teams and tournaments, big media corporations offering contracts, and, perhaps in the near future, the International Olympic Committee including them in the Olympic Games.

The Stanford Artificial Intelligence Laboratory hosted the first video game tournament in October 1972. About two dozen people competed playing *Spacewar!*[1] (The Stanford AI Lab was funded by ARPA, the same Defense Department research agency that funded the original Internet, ARPAnet.) Video games were still in their infancy, though; the breakthrough arcade game *Pong* wouldn't be released until the following month.

More organized eSports developed with leagues and tournaments in the late 1990s, as the Internet became a mass medium, connecting digital gamers globally. *Quake* and *Counter-Strike* were among the leading games for league play.[2] By the early 2000s, *StarCraft* and *Warcraft III* became the dominant eSports games, and South Korea fostered eSports with the first twenty-four-hour cable gaming channels, such as OGN.

Dozens of colleges and universities have had club eSports teams for years. In 2014, Robert Morris University, a small private school in Chicago, became the first university in the country to offer college scholarships to eSports athletes. In 2016, the University of California–Irvine became the first public research university to do so, and in 2017, the University of Utah established the first eSports program

◀ Video game players take part in a competitive video gaming event.

at a university belonging to one of the power five athletic conferences. The Utah team, consisting of thirty-three scholarship-supported students (who won positions against approximately two hundred players in tryouts), was launched to compete in four online digital games: *Overwatch*, *League of Legends*, *Rocket League*, and *Hearthstone*.[3]

Just as in other professional sports, college eSports serves as the minor leagues for professional eSports. Growing audience size for professional competitions has drawn media coverage, which has also drawn big money. According to a report by technology consulting firm Activate, the eSports TV/streaming audience had already surpassed that of Major League Soccer (MLS) and the National Hockey League (NHL) by 2018. By 2021, eSports will have exceeded the National Basketball Association (NBA) and Major League Baseball (MLB) in viewership, trailing only the National Football League (NFL), and will have more than $5 billion in annual revenue.[4] Not surprisingly, a number of media corporations have invested in platforms to stream game competitions and content, including Amazon (with Twitch) and Google (YouTube Gaming). In early 2018, Facebook became the latest entry, signing a contract with ESL (the largest eSports company) to carry tournaments and gameplay for *Dota 2* and *Counter-Strike: Global Offensive* on its Facebook Watch platform.[5] Riot Games (maker of *League of Legends*) and Activision Blizzard (maker of *Overwatch*) are two of the biggest forces behind professional eSports leagues. The leagues have team franchise fees of between $10 million and $20 million, and interestingly, some of the leading investors in American eSports teams are owners of NFL, NBA, MLB, and NHL teams.[6]

And the Olympics? First, eSports will be a medal event in Hangzhou, China, at the 2022 Asian Games, the world's largest multisport event besides the Olympics. After that comes the 2024 Summer Olympics in Paris. The International Olympic Committee co-chair held open the possibility: "The youth, yes they are interested in eSport[s]. . . . Let's look at it. Let's meet them. Let's try if we can find some bridges."[7]

 Do you think eSports should count as a sporting competition (like traditional athletic sports) in colleges, universities, and multinational sporting events? Beyond the games listed above, what digital games would you want to watch in eSports competitions?

Digital Gaming Today

It is clear from the rise of eSports that gaming is a popular pastime. But if anyone had lingering doubts about both the enormity and the growth potential of the digital gaming industry, those doubts would have been dispelled in 2020, in the midst of the global coronavirus pandemic. Even before the pandemic became widespread, gaming was a $160 billion a year global industry, with predictions of nearly $200 billion in revenue by 2022. And before the pandemic hit, there were already two billion gamers worldwide, playing on smartphones, consoles, and personal computers (PCs).[8] While Microsoft, maker of the Xbox, had long been involved in digital gaming, other leading digital media companies were also starting to take notice of the booming industry. For example, in fall 2019, Google launched Stadia, its cloud gaming service that lets users play games streamed from its data servers without having to invest in a console or gaming PC. At the same time, Apple launched Apple Arcade, a subscription service to download and play games.

Then, when the coronavirus spread worldwide in early 2020, millions of people were quarantined at home with plenty of extra time. That's when gaming sales really boomed, with digital game purchases 73 percent higher in April 2020 compared to the previous year. Gamers found themselves returning to the comfort of old favorites like *League of Legends* and *Candy Crush Saga* and making new hits of titles like *Animal Crossing: New Horizons* and *Among Us*. At the same time, shipments of smartphones and computer hardware fell sharply, as the virus disrupted global manufacturing.[9] Gaming has always been a popular way to connect with others when in-person contact is restricted, and it became one of the few "safe" environments for play and socializing during the pandemic, when parks, sporting venues, restaurants, and bars were shut down. *Fortnite* demonstrated the sprawling influence of digital game environments during this time, too, as it hosted a multimedia global music release by rapper Travis Scott when concert tours were impossible. And Facebook, which had been planning a June release of a Twitch-like service that allows gamers to livestream their gameplay and watch others play, moved up the launch of Facebook

FORTNITE With the coronavirus pandemic shutting down concert tours, Travis Scott released a brand-new track on *Fortnite* as part of a multi-date event called Astronomical, available for players around the world at five showtimes, April 23–25, 2020.

Gaming to April 2020. As the head of the gaming app said, in somewhat of an understatement, "We're seeing a big rise in gaming during quarantine."[10]

With its increased profile and flexibility across platforms, the gaming industry has achieved a mass medium status on par with film or television. This rise in status has come with stiffer and more complex competition—not just within the gaming industry but across media. Rather than Sony competing with Nintendo, TV networks competing among themselves for viewers, or new movies facing off at the box office, media must now compete against different media for an audience's attention. The cutthroat nature of this media environment makes the global popularity of digital gaming all the more significant.

In this chapter, we will look at the evolving mass medium of digital gaming and:

- Discover how today's digital gaming industry came to be by examining the early history of electronic gaming, including its roots in penny arcades
- Trace the evolving gaming environment, from arcades and bars to living rooms and mobile devices
- Discuss gaming as a masspersonal social medium that forms communities of play
- Analyze the economics of gaming, including the industry's major players and various revenue streams
- Raise questions about the role of digital gaming in our democratic society

How We Got Here: The Development of Digital Gaming

When the Industrial Revolution swept Western civilization two centuries ago, the technological advances involved weren't simply about mass production. They also promoted mass consumption and the emergence of *leisure time*—both of which created moneymaking opportunities for media makers. By the late nineteenth century, the availability of leisure time had sparked the creation of mechanical games. Technology continued to grow, and by the 1950s, computer science students in the United States had developed early versions of the video games we know today.

In their most basic form, digital games immerse users in an interactive computerized environment where they strive to achieve a desired outcome. These days, most digital games go beyond the simple competition of the 1972 table-tennis-style arcade game *Pong*: They often entail sweeping narratives and offer imaginative and exciting adventures, sophisticated problem-solving opportunities, a variety of gameplay, and multiple possible outcomes.

But the boundaries were not always so varied. Over time, digital games evolved from their simplest forms into the four major platforms used for gameplay today: consoles, handheld devices, computers, and the Internet. As these platforms evolved and graphics advanced, distinctive gaming genres emerged and became popular. Together, these varied formats constitute a multibillion dollar media industry that has become a socially driven mass medium.[11]

Mechanical Gaming

In the 1880s, the seeds of the modern entertainment industry were planted by a series of coin-operated contraptions devoted to cashing in on idleness. First appearing in train depots, hotel lobbies, bars, and restaurants, these leisure machines (also called "counter machines") would find a permanent home in the first thoroughly modern indoor playground: the **penny arcade**.[12]

Arcades were like nurseries for fledgling forms of amusement that would mature into mass entertainment industries during the twentieth century. For example, automated phonographs used in arcade machines evolved into the jukebox, while kinetoscopes in arcades (see Chapter 7) set the stage for the cinema. But the machines most relevant to today's digital games were more interactive and primitive than the phonograph and the kinetoscope. Some were strength testers that dared young men to show off their muscles by punching a boxing bag or arm wrestling a robotlike Uncle Sam. Others required more refined skills and sustained play, such as those that simulated bowling, horse racing, and football.[13]

Another arcade game, the bagatelle, spawned the **pinball machine**, the most prominent of the mechanical games. In pinball, players score points by manipulating the path of a metal ball on a slanted table sealed within a glass-covered case. In the 1930s and 1940s, players could control only the launch of the ball. For this reason, pinball was considered a sinister game of chance that, like the slot machine, fed the coffers of the gambling underworld.

As a result, pinball was banned in most American cities, including New York, Chicago, and Los Angeles.[14] However, pinball gained mainstream acceptance and popularity after World War II with the addition of the flipper bumper, which enables players to careen the ball back up the play table, transforming pinball into a challenging game of skill, touch, and timing. With the addition of microprocessors to pinball machines in the mid-1970s, games could post high scores next to players' initials, anticipating key elements of masspersonal communication: Individual players with the highest scores held *personal* bragging rights and set the competitive goal for the *mass* of subsequent players on that machine.

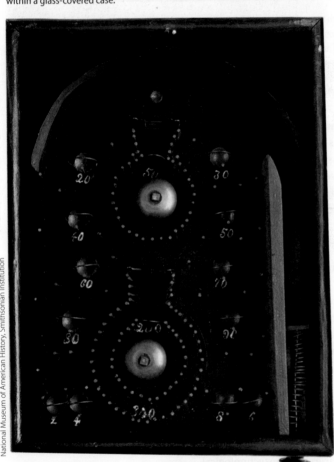

THE EARLIEST PINBALL MACHINES, based on the billiards-like game of bagatelle, were games of chance. Players could control only the launch of the metal ball and watch its path on a slanted table sealed within a glass-covered case.

National Museum of American History, Smithsonian Institution

The First Video Games

Not long after the growth of pinball, the first video game patent was issued on December 14, 1948, to Thomas T. Goldsmith and Estle Ray Mann for what they described as a "Cathode Ray Tube Amusement Device."

The invention would not make much of a splash in the history of digital gaming, but it did feature the key component of the first video games: the cathode ray tube (CRT). CRT-powered screens provided the images for analog television and for early computer displays, where the first video games appeared a few years later. One such game was *Spacewar!*, a two-person game released by computer science students at

MIT in 1962.[15] But because computers consisted of massive mainframes at the time, the games couldn't be easily distributed.

However, more and more people owned televisions, and TVs provided a platform for video games long before people had personal computers at home. *Odyssey*, the first home video game system, was developed by German immigrant and television engineer Ralph Baer. Released by Magnavox in 1972 and sold for a whopping $100 (the equivalent of over $625 today), *Odyssey* used player controllers that moved dots of light around a screen in each of its twelve simple games.

In the next decade, a version of one of the *Odyssey* games brought the delights of video gaming into modern **arcades**, establishments that gathered together multiple coin-operated games. The same year that Magnavox released *Odyssey*, a young American computer engineer named Nolan Bushnell formed a video game development company, Atari, with a friend. The enterprise's first creation was *Pong*, a simple two-dimensional tennis-style game, with two vertical paddles that bounced a white dot back and forth, making a blip sound with each bounce. *Pong* quickly became the first video game to become popular in arcades. In 1974, the *New York Times* took notice of the emerging culture of these new coin-operated video games as they "swept the country . . . popping up alongside pinball machines and jukeboxes in airports and bus stations but especially showing up in the more sedate watering holes which scorn pinballs."[16]

In 1975, Atari began successfully marketing a home version of *Pong* through an exclusive deal with Sears, expanding the home video game market. Just two years later, Bushnell started the Chuck E. Cheese pizza–arcade restaurant chain and sold Atari to Warner Communications for an astounding $28 million. Although Atari folded in 1984, plenty of companies—including Nintendo, Sony, and Microsoft—followed its early lead, transforming the video game business into a full-fledged industry.

ODYSSEY

Elektronisches Fernsehspiel

Bedienungsanleitung

Bitte lesen Sie erst diese Bedienungsanleitung, dann die Spielregeln!

MAGNAVOX'S *ODYSSEY* This image of *Odyssey*, the first home video game system, was featured in the original German manual.

THE ATARI 2600 was followed by the Atari 400, Atari 800, and Atari 5200, but none matched the earlier success of the 2600 model.

Consoles and Advancing Graphics

Home **consoles**—devices people use specifically to play video games—put digital games on a path to becoming a mass medium, bringing gaming into households and sparking unprecedented growth in game development. Until recently, consoles generated more revenue than any other gaming platform, and they have become increasingly more powerful since the appearance of the early Atari consoles in the 1970s. One way of charting the evolution of consoles is to track the number of bits (binary digits) that they can process at one time. The bit rating of a console is a measure of its power at rendering computer graphics. The higher the bit rating, the more detailed and sophisticated the graphics. Bit ratings evolved according to this timeline:

- 1977: Atari released the 2600 with an 8-bit processor
- 1983: the wildly popular Nintendo Entertainment System also used an 8-bit processor
- 1989: Sega Genesis launched the first 16-bit console
- 1992: 32-bit computers appeared on the market
- 1993: 64 bits became the new standard
- 1999: Sega Dreamcast began the 128-bit era
- Today: 256-bit processors are standard

But more detailed graphics have not always replaced simpler games. Nintendo's NES Classic Edition, with thirty built-in classic games mostly from the 1980s, and the Atari Flashback series, with built-in Atari games from the same era, have experienced a surge of nostalgic interest in recent years. Perhaps the best example of enduring games is the *Super Mario Bros.* series. Created by Nintendo, the original *Mario Bros.* game began in arcades. The 1985 sequel *Super Mario Bros.*, developed for the 8-bit Nintendo Entertainment System, became the best-selling video game of all time (holding this title until 2009, when it was unseated by Nintendo's *Wii Sports*). Today, nostalgic references to "8-bit" games are shorthand for old-school, pixelated 2-D games with tinny audio.

Through decades of ups and downs in the digital gaming industry (Atari folded in 1984, and Sega no longer makes video consoles), three major home console makers now compete for gamers: Nintendo, Sony, and Microsoft. Nintendo has been making consoles since the 1980s; Sony and Microsoft came later, but both companies were already major media conglomerates and thus well positioned to support and promote their interests in the video game market.

POPULAR VIDEO GAMES in the 1970s and 1980s were simple and two-dimensional games, with straightforward goals like saving the princess, destroying asteroids, or gobbling up tiny dots. Today, most video games have more complex story lines based in fully fleshed-out worlds.

Nintendo

Headquartered in Kyoto, Nintendo began manufacturing Japanese playing cards in 1889. Decades later, the playing card business was becoming less profitable, so the company ventured into toy production in the 1960s and went on to release its first video game console in 1977. In the early 1980s, Nintendo had two major marketing successes: In 1981, *Donkey Kong*—the platform game (see Table 3.1) in which players help Jumpman rescue Lady from a giant ape—was the company's breakthrough into the U.S. console market. Then, in 1985, the Nintendo Entertainment System console reached U.S. markets bundled with *Super Mario Bros.* With this package, Nintendo set the standard for video game consoles, and Mario and Luigi became household names.

In 2006, Nintendo released a new kind of console, the Wii, which supported traditional video games like *New Super Mario Bros.* but was first to add a wireless motion-sensing controller that took the often-sedentary nature out of gameplay. Games like *Wii Sports* require the user to mimic the full-body motion of bowling or playing tennis, while *Wii Fit* uses a wireless balance board for interactive yoga, strength, aerobic, and balance games.

Nintendo's Switch (2017) is a significant departure from previous Nintendo consoles, as it easily moves from a console to a portable system. When the tablet-size touchscreen is in a dock, it's a console for playing games on a TV monitor; when the touchscreen is removed, it "switches" into a portable system, with detachable joysticks/controllers.

ANIMAL CROSSING: NEW HORIZONS is the fifth in the *Animal Crossing* series of life simulation video games for the Nintendo Switch. *New Horizons* was released in March 2020 and quickly became a hit as people played games to pass the time during coronavirus quarantine.

Sony

Sony emerged after World War II as a manufacturer of tape recorders and radios. (The name Sony is rooted in the Latin word *sonus*, meaning "sound.") Also headquartered in Japan, the company was soon manufacturing televisions, VCRs, computers, cameras, and, beginning in the 1994, the PlayStation. Its venture into video games came about because of a deal gone bad with Nintendo; when a proposed partnership fell through, Sony became Nintendo's direct competition. Sony's initial PlayStation console doubled the microprocessor size introduced by Sega (from 16 bits to 32 bits) and played both full-motion and three-dimensional (3-D) video.[17]

The PlayStation has a reputation as a high-performance gaming console, and the PlayStation 4 (PS4, launched in 2013) has been the best-selling brand of its generation of consoles, topping the Nintendo Switch and Xbox One. By 2020, in its final full year of release, the PS4 console had sold more than 106 million units, and 1.15 billion copies of PS4 games had been sold globally.[18] Sony launched its next-generation PlayStation 5 in late 2020.

YOUNG MAN PLAYS *Ultimate Ninja Storm Revolution.*

Microsoft

In 2001, computer software goliath Microsoft entered into the world of serious gaming. At that time, Microsoft, which had dominated the computer software market with its Windows brand for years, hoped to reinvigorate corporate profits by building a home entertainment business based on a powerful game console with new features.[19] The Xbox, which represented a $500 million commitment from Microsoft, had a number of firsts:

- the first console to feature a built-in hard disk drive
- the first to be connected to an online service (Xbox Live)
- the first to have Dolby Digital sound, for a cinematic sound experience

While Xbox could not offer the arsenal of games that PlayStation gamers had access to, the console did launch with one particular game: *Halo*. Game critics and players immediately recognized this sci-fi first-person shooter game—now a multibillion-dollar franchise—as Microsoft's "killer app."[20] Microsoft also correctly anticipated that Internet-connected computer consoles could be used not only for networked gameplay but also for playing DVDs and, later, streaming video like Netflix.

In 2013, Microsoft released Xbox One, the third-generation Xbox console. Microsoft replaced it with its faster fourth-generation Xbox Series X console in late 2020.

Gaming on PCs

Like the early console games, very early home computer games often mimicked (and sometimes ripped off) popular arcade games like *Frogger, Centipede, Pac-Man,* and *Space Invaders*. They also featured certain genres not often seen on consoles, like digitized card and board games. The early days of the personal computer saw the creation of electronic versions of games like Solitaire, Hearts, and Chess, simple games still popular today.

For a time in the late 1980s and much of the 1990s, personal computers held some clear advantages over console gaming. The versatility of keyboards, compared with the relatively simple early console controllers, allowed for ambitious puzzle-solving games like *Myst*. Moreover, faster processing speeds gave some computer games richer, more detailed 3-D graphics. Many of the most popular early first-person shooter games, like *Doom* and *Quake*, were developed for home computers rather than consoles.

As consoles caught up with greater processing speeds and disc-based games in the late 1990s, elaborate personal computer games attracted less attention. But due to the advent of Internet-based free-to-play games (like *Spelunky* and *League of Legends*), subscription games (such as *World of Warcraft*), social media games (such as *Candy Crush Saga* and *Words with Friends* on Facebook), and the Steam PC game platform, PC gaming has experienced a resurgence. Many PC gamers play on specialized gaming computers with powerful processors, lots of memory, and special graphic cards to handle the rich multimedia of advanced video games. The PC platform constitutes about 20 percent of the game market.

Mobile Gaming

Consoles have historically occupied a significant share of the global gaming market, attracting attention with high-profile game releases. Yet with the release of the first iPhone in 2007, smartphones and tablets have grown to become the biggest gaming platform (surpassing consoles in 2016) and now generate 49 percent of global revenue.[21] (See Figure 3.1.) It turns out that the most popular gaming device is the mobile screen in one's pocket or bag, because it's always available for play.

Portability has long been part of digital games. Nintendo popularized handheld digital games with the release of its Game Boy line of devices and sold nearly 120 million of them from 1989 to 2003, with games like *Tetris, Metroid,* and *Pokémon Red/Blue*.[22] The early handhelds gave way to later generations of devices, and the multipurpose Nintendo Switch, released in 2017, tries to remain true to that heritage. But as smartphone and tablet apps began offering familiar games like *Tetris* and *Solitaire,* and new ones like *Angry Birds, Fruit Ninja,* and *Clash of Clans,* sales of specialized handheld devices fizzled and a whole new wave of casual mobile game players emerged. More recently, mobile games have offered a greater range of game genres, from made-for-mobile augmented reality, geo-based games like *Pokémon Go,* to versions of well-known console and PC platform games like *Call of Duty, Fortnite,* and *Madden NFL*.

This portable and mobile gaming convergence is changing the way people look at digital games and their systems. With games no longer confined to arcades or home consoles, the mobile media have gained power as entertainment tools, reaching a wider and more diverse audience.

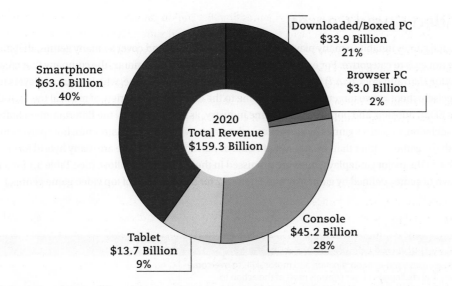

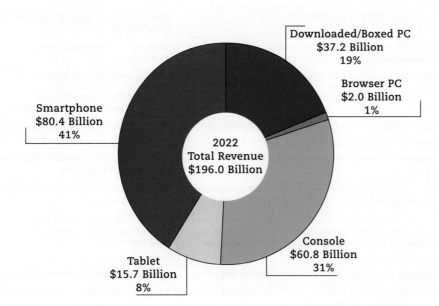

FIGURE 3.1

**GLOBAL GAME
REVENUE BY
PLATFORM,
2020 AND 2022**

Data from: Newzoo, *Global Games
Market Report 2019*, p. 15, and Newzoo,
Global Games Market Report 2020,
p. 14, https://newzoo.com/products/
reports/global-games-market-report.

Note: Data for 2022 are estimates.
Tablet and smartphone combined
represent the mobile market. Revenue
and percentages were rounded.

In fact, the Entertainment Software Association reports that 31 percent of gamers play while waiting
for an appointment, 27 percent play during a break at work or school, and 16 percent play during
a commute.[23] Thus, gaming has become an everyday form of entertainment rather than the niche
pursuit of hard-core enthusiasts.

The Gaming Environment

Digital games occupy an enormous range of styles, from casual games like *Tetris*, *Subway Surfers*,
and *Sonic Dash* to full-blown Hollywood-esque immersive adventure games like *Final Fantasy*.[24] No
matter what the style, digital games are compelling entertainment and mass media because they pose
challenges (mental and physical), allow us to engage in situations both realistic and fantastical, and
allow us to socialize with others as we play.

Video Game Genres

Digital games inhabit so many playing platforms and devices, and cover so many genres, that they are not easy to categorize. For example, *Audica* (developed by Harmonix, the company that created *Guitar Hero* and *Rock Band*) is a virtual reality "rhythm shooter" game, which requires players to precisely shoot at two targets in a cosmic scene to the beat of a wide variety of musical tracks, including EDM, dubstep, and pop. The video game industry, as represented by the Entertainment Software Association, organizes games by **gameplay**—the way in which the rules structure how players interact with the game—rather than by any sort of visual or narrative style. There are many hybrid forms, but a few of the major gameplay genres are discussed in the sections that follow. (See Table 3.1 for a rundown of genres defined by gameplay and Figure 3.2 for a breakdown of top video game genres.)

TABLE 3.1   VIDEO GAME GENRES

Genre		Gameplay	Examples
Action		Player uses hand-eye coordination and motor skills to overcome physical challenges. Player controls most of the action to:	
	Platform games	- move character(s) between various platform levels in order to avoid or chase adversaries	*Super Mario Bros., Never Alone, Canabalt, Super Mario Odyssey, Celeste*
	First-person shooter	- use a range of weapons to obliterate the enemy	*Overwatch, Call of Duty: Advanced Warfare, Halo, Half-Life*
	Fighting	- work in close-range combat against a small number of equally powerful opponents	*Street Fighter, Dragon Ball FighterZ, Soulcalibur*
	Stealth	- engage in subterfuge and precision strikes to beat the enemy	*Dishonored, Mark of the Ninja*
	Survival	- learn to survive in a hostile environment	*Paladins: Battlegrounds, Fortnite, PlayerUnknown's Battlegrounds (PUBG)*
	Rhythm	- demonstrate sense of rhythm, coordination, and musical precision	*Dance Dance Revolution, Rock Band, Audica*
Adventure		Player solves puzzles by interacting with people or the environment	*Myst, Tomb Raider, Lumino City, Monument Valley, Limbo, Hidden Folks*
Action-Adventure		Player navigates horror-fiction elements or constant obstacles (e.g., doors) and acquires special tools or abilities to open them	*Zelda, Metroid, Castlevania, Resident Evil, Grand Theft Auto*
Role-Playing		Player takes on specific characteristics and skill sets, goes on "adventures," and often amasses treasure; the most popular setting is a fantasy world	*Final Fantasy, Fallout, Grand Theft Auto, Minecraft*
MMPORG		Similar to role-playing games but distinguished by the high number of players interacting together	*World of Warcraft, Guild Wars 2, EVE Online*
Simulation		Player simulates a real or fictional reality to:	
	Construction/ Management	- expand or manage fictional communities or projects with limited resources	*SimCity, Minecraft* (in creative mode)
	Life	- "realistically" live the life of a person or being (e.g., a wolf), possibly in a strange world	*SimLife, Spore, Animal Crossing: New Horizons*
	Vehicle	- experience flight, race-car driving, train travel, combat vehicles, and so on	*FlightGear, Microsoft Flight Simulator, Nascar Racing*
Strategy		Player carefully plots out tactics to achieve a goal, usually military or world domination	*Master of Orion, Hogs of War, StarCraft*
	MOBA (Multiplayer Online Battle Arena)	Player on a team competes against another team to destroy the opposing team's main structure; similar to role-playing games but distinguished by the high number of players	*League of Legends, Dota 2, Heroes of the Storm*
Sports		Player takes either a player's or management's perspective in simulating a sport, like soccer, Nascar racing, football, or fighting	*Pong, FIFA, Fight Night, Championship Manager, Madden NFL*
Casual Games		Player makes progress toward a simple reward, increasing the challenge if desired; the rules are simple, and there is no long-term commitment	*Tetris, Candy Crush Saga, Run Sausage Run*

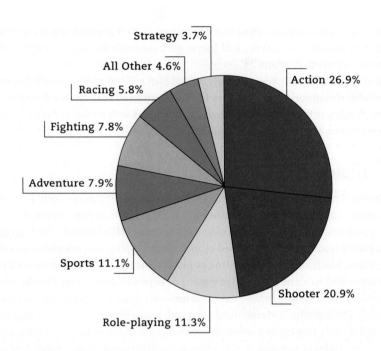

Strategy 3.7%

All Other 4.6%

Racing 5.8%

Fighting 7.8%

Adventure 7.9%

Sports 11.1%

Role-playing 11.3%

Action 26.9%

Shooter 20.9%

FIGURE 3.2

TOP VIDEO GAME GENRES BY UNITS SOLD, 2019

Data from: Entertainment Software Association, *2019 Essential Facts about the Computer and Video Game Industry*, May 2019.

Note: Percentages were rounded.

Social Gaming

The connectivity of the Internet has opened the door to *social gaming*—millions of people playing the same online games in virtual spaces. Types of social gaming include massively multiplayer online role-playing games, multiplayer online battle arenas, and other virtual worlds.

MMORPGS

One of the longest acronyms in the world of gaming is **massively multiplayer online role-playing games (MMORPGs)**. These games are set in virtual worlds that require users to play through an **avatar**—a graphic interactive "character" situated within the world of the game—of their own design. The "massively multiplayer" aspect of MMORPGs indicates that digital games—once designed for solo or small-group play—now engage huge numbers of player-participants. The fantasy adventure game *World of Warcraft* still remains one of the most popular MMORPGs. In the game, users can select from twelve different "races" of avatars, including dwarves, gnomes, night elves, orcs, trolls, goblins, and humans, and form groups with other players.

MOBAs

Multiplayer online battle arenas (MOBAs) can also have hundreds of thousands of active players worldwide. At the game level, a team of up to five players aims to destroy an opposing team's structure. MOBAs like *League of Legends* and *Dota 2* are popular eSports games. **Battle royale games**, which are a subset of MOBAs, are also multiplayer, but the number of players is limited (usually to a hundred players), since the objective is to be the last player (or team) surviving. *Fortnite* and *PlayerUnknown's Battleground* (*PUBG*) are favorite battle royale games and are among the most popular games played and watched on interactive livestreaming platforms like Twitch (discussed later in the chapter).

Virtual Worlds for Children

One of the most overlooked areas in online gaming (at least by adults) is the children's market, which spotlights different types of virtual worlds. For example, toy maker Ganz developed the online *Webkinz* game in 2005 to revive its stuffed-animal sales. Each Webkinz stuffed animal comes with a code that lets players access the online game and care for the virtual version of their plush pets. *Wizard101* is a free-to-play MMORPG for children set in the Ravenwood School of Magical Arts. Players take the role of Ravenwood students who join one of five schools of magic (the Harry Potter influences are quite apparent). *Woozworld* is a fashion virtual world launched in 2009 in which tweens are encouraged to decorate rooms, chat with other players, and "never get bored with thousands of stylish clothes that bring you one step closer to stardom!"[25]

All these virtual worlds offer younger players their own age-appropriate environment in which to experiment with virtual socializing, but they have also attracted criticism for their messages of consumerism. In many of these games, children can buy items with virtual currency or, through a premium membership, acquire bling more quickly.

Players: Inside the Game

Virtual communities often crop up around online video games. Players can perform the role of one of a game's many avatars, and then they can organize into a group, all without ever meeting the other players in person. Players interact in two basic types of in-game groups: **PUGs** (short for "pick-up groups"), which are temporary teams usually assembled by matchmaking programs integrated into the game, and **guilds** or **clans**, which are groups organized by players themselves. Most experienced players join guilds or clans, which can be small and easygoing or large and demanding. Players in both types of groups use two forms of in-game chat—voice and text—to socialize and strategize, even as they are playing the game. These methods of communicating with fellow players who may or may not know one another outside the game create a sense of community around gameplay, and some players have formed lasting friendships or romantic relationships through their video game habit. Avid gamers have even held in-game ceremonies, such as weddings, funerals, and graduations—sometimes for game-only characters but sometimes for real-life events.

In role-playing games, players take on the digital form of avatars graphically situated in the game. The idea of role-playing in games goes back at least as far as the tabletop card game *Dungeons & Dragons* (*D&D*), first published in 1974. Like *D&D*, digital role-playing games require players to choose a character and embark on a series of adventures. Some games offer a single-player experience, but MMORPGs and MOBAs include millions of players. Avatars might be preset and part of the narrative, or they might be invented from various options. Either way, gamers often try on various characters that are wholly different from their earthly identity in terms of gender, race, sexuality, and human versus nonhuman/animal status. According to one study of representation and media, gamers may choose avatars that look like themselves (which raises the question of whether the game offers a diverse range of avatar possibilities), or they may choose characters based solely on what powers those particular avatars have in gameplay (because players want to win games).[26]

A *MINECRAFT* GRADUATION Computer gamers at a number of campuses across the United States—including the University of California–Berkeley, Georgia Southern University, the University of Georgia, Boston University, the University of Connecticut, and Rochester Institute of Technology—created virtual graduations for the class of 2020 in *Minecraft*. Here is a still from the University of Minnesota's *Minecraft* graduation during the COVID-19 pandemic.

Christopher Martin

Communities of Play

The ubiquity of digital games has fostered the development of communities outside games dedicated to gaming in its many forms. This phenomenon is similar to the formation of online and in-person groups devoted to discussing other forms of mass media, such as movies, TV shows, and books. These communities enhance the social experience gained through the games themselves.

Collective Intelligence

Mass media productions are almost always collaborative efforts, as is evident in the credits for movies, television shows, and music recordings. The same goes for digital games. But what is unusual about game developers and the gaming industry is their interest in listening to gamers and their communities in order to gather new ideas and constructive criticism and to gauge popularity. Gamers, too, collaborate with one another to share shortcuts and "cheats" to solve tasks and quests, and to create their own modifications to games. This sharing of knowledge and ideas is an excellent example of **collective intelligence**. French professor Pierre Lévy coined the term *collective intelligence* in 1997 to describe the Internet—"this new dimension of communication"—and its ability to "enable us to share our knowledge and acknowledge it to others."[27] In the world of gaming, where users are active participants (more than in any other medium), the collective intelligence of players informs the entire game environment.

For example, collective intelligence (and action) is necessary to work through levels of many games. In *World of Warcraft*, gamers share ideas, tips, and cheats through chats, wikis, and online posts. The largest of the sites devoted to sharing collective intelligence is the *World of Warcraft* wiki. Similar user-generated sites are dedicated to a range of digital games, including *Assassin's Creed*, *Grand Theft Auto*, and *Pokémon Go*. (See "Global Village: Phones in Hand, the World Finds Pokémon" for more on *Pokémon Go*, collective intelligence, and global gaming.)

The most advanced form of collective intelligence in gaming is **modding**, slang for "modifying game software or hardware." In many media industries, modifying hardware or content would land someone in a copyright lawsuit. In gaming, modding is often encouraged as yet another way players become more deeply invested in a game and improve it for others. For example, *Counter-Strike*, a popular first-person shooter game, originated as a mod of Valve Corporation's first-person shooter game *Half-Life*. The developers of *Half-Life* encouraged mods by including software development tools with it. *Counter-Strike* was released to retailers, eventually selling more copies than *Half-Life*. Although modding engages game fans as they create elements like new environments, characters, or gear, it is worth noting that they rarely get any compensation for their labor, which ultimately benefits the game developers.[28]

Interactive Livestreaming Platforms: Gaming Becomes Television

One of the most notable developments in gaming is the extraordinary popularity of interactive livestreaming platforms. If you had asked people years ago whether they would spend time watching other people play digital games, they might have thought it was a ridiculous question. Yet in the professional sports world, millions of fans watch others play competitive games (e.g., baseball or basketball), so the experience is not all that different. In fact, professional players have emerged on the leading platforms, which include Twitch (owned by Amazon), YouTube Gaming (owned by Google/Alphabet), and Facebook Gaming. Twitch is the leading streaming platform, garnering 73 percent of the total stream hours watched by 2020, but the other two companies have slowly been chipping away at its lead. Worldwide, interactive livestreaming platforms have nearly one billion viewers.[29]

The livestreaming platforms had their big coming-out moment on March 14, 2018, when recording superstar Drake joined game-playing superstar Ninja (a.k.a. Tyler Blevins) on Twitch to play *Fortnite*.

Phones in Hand, the World Finds Pokémon

In a matter of hours, *Pokémon Go*—developed by Nintendo, the Pokémon Company, and Niantic Labs (a private company spun off from Google)—became the most popular mobile game app ever in America. Within a month, it expanded into a global sensation, with over one hundred million downloads and players in more than seventy countries. By 2019, it had been downloaded over one billion times and had earned more than $4.7 billion.[1] Although the game is free to play, in-game purchases, sponsorships, and events help it earn its revenue.

Pokémon Go is an augmented reality (AR), geo-based game. Players (who are called Trainers and who adopt a unique nickname to play) create their own avatars and use their phone cameras as they walk through neighborhoods. The AR technology overlays the game's map on a player's phone, and the map follows the real-world map of where the player is actually walking. Niantic Labs, the game's developer, uses computer servers around the world to operate the game and locate Pokémon, PokéStops, and Pokémon Gyms, where

The unique thing about the release of the digital game *Pokémon Go* on July 6, 2016, is that everyone played in open view. Millions of people were trying to catch wild Pokémon, not on consoles or computers in their living rooms or bedrooms but outdoors and in public, where individuals and groups of people scoured neighborhoods to locate PokéStops and Pokémon Gyms.

Christopher Martin

teams can battle for control of the site. Niantic received complaints about some of its initial Gym sites, which sent people into some inappropriate gaming spaces, including graveyards, airports, a hospital delivery room, the 9/11 memorial in New York, and the demilitarized zone in Korea, where "walking to the location would likely put them in imminent danger of being killed."[2] These days, *Pokémon Go* Gyms are mostly located in public parks and sponsored locations, like Starbucks cafés.[3]

Although play of *Pokémon Go* peaked in summer 2016, it still remains popular, especially with worldwide *Pokémon Go* promotional events. For example, on Valentine's Day 2018, *Pokémon Go* tweeted, "Until February 15, schools of Luvdisc will be out swimming and awarding 3x Stardust for each one caught" (translation: a Luvdisc is a heart-shaped fish; Stardust is valuable because it increases Pokémon combat power). By spring 2020, as outdoor *Pokémon Go* play was set to gear up in the Northern Hemisphere, the global coronavirus pandemic limited gameplay and caused the cancellation of a number of scheduled play gatherings. In response, the *Pokémon Go* makers hosted a virtual global *Pokémon Go* Fest in the summer.

The success of *Pokémon Go* is its unique, interactive AR play, which, as few games do, actually gets players out of the house and interacting in person (pandemic notwithstanding). *Pokémon Go* also has the built-in bonus of nostalgia. Pokémon (short for Pocket Monsters) were invented in Japan and made their debut on the Nintendo Game Boy in 1996. Since that time, the franchise has expanded to trading card games, television shows, films, comic books, and toys.

Niantic continues to expand and refine *Pokémon Go*, and in 2019 it remained the leading mobile augmented reality game, earning $1.4 billion—81 percent of the AR market. The company's launch of another mobile AR game in 2019, *Harry Potter: Wizards Unite*, didn't go as well, and the game hasn't received the same energetic reception that *Pokémon Go* has. Microsoft released the mobile AR game *Minecraft Earth* in late 2019, and *Time* magazine included it as one of the Best Inventions of 2019.[4] Given the popularity of Microsoft's original *Minecraft* and its entirely different gameplay compared to *Pokémon Go*, it may have a chance to become a popular mobile AR game on its own terms. ●

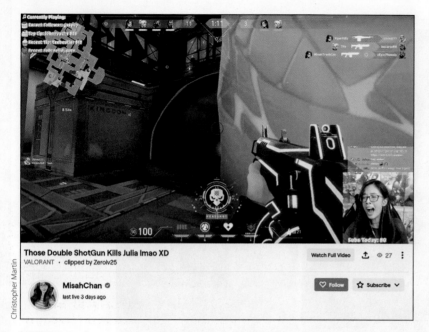

TWITCH A college student from a midwestern university earns extra income through her Twitch channel by creating gameplay content, increasing her following, and endorsing products.

Rapper Travis Scott and NFL player JuJu Smith-Schuster joined in as well. Word traveled quickly on social media, and more than 630,000 viewers tuned in, setting an audience record for a single live gaming stream, and also introducing platforms like Twitch to traditional mass media audiences.[30]

As live game streaming has intensified, competition has also increased for the big-name professional gamers who draw viewers in with their high level of play and entertaining live dialogue. Game streaming platforms and their star players have helped to popularize games like *Fortnite* and *Apex Legends* and have created a new generation of social media personalities. One of YouTube's most popular figures is PewDiePie (Felix Kjellberg, a native of Sweden who has more than 107 million subscribers), who got his start in 2010 by posting his gameplay videos.

Live gaming platforms have also extended beyond gameplay, offering livestreamed talk shows for teens, comedy from performers like Cody Ko and Noel Miller, and livestreamed news events, including localized streams from cities across the United States during the 2020 uprisings that followed a Minneapolis police officer's use of lethal force against George Floyd, an unarmed Black man.

Fantasy Sports

Online fantasy sports games also create communities of play. For decades before the advent of digital games, people used to play in friendly (and technically illegal) "office pool" competitions, wherein workers would pool their small bets on football games or college basketball tournaments, with money going to the winners. The 1980s saw the development of fantasy baseball leagues in which players would act as owners, drafting their own teams with other "owners" and using the performance statistics of real baseball players—published weekly in newspapers—as the measure of their team's performance. By the mid-1990s, the growth of the web enabled fantasy sports to go online, eliminating the need to calculate league statistics by hand, and expanded to cover all major sports. Fantasy sports players can choose to form leagues with friends or be part of a league of other players assigned through sites like Espn.com.

Rather than experiencing the visceral thrills of, say, the most recent release of *Madden NFL*, fantasy football participants take a more detached, managerial perspective on the game—a departure from the classic video game experience. Fantasy sports owners become statistically savvy aficionados of the game overall, rather than rabid fans of a particular team. According to the Fantasy Sports & Gaming Association, 19 percent of Americans age eighteen and older play fantasy sports.[31]

Conventions

In addition to online gaming communities, there are conventions and expos where video game enthusiasts can come together in person to test out new games and other new products, play classic games in competition, and meet video game developers. One of the most significant is the Electronic Entertainment Expo (E3), which draws about sixty-eight thousand industry professionals, investors, developers, and retailers to its annual meeting.

The Penny Arcade Expo (PAX) is a convention created by gamers for gamers, held each year in Seattle, Boston, San Antonio, and Melbourne, Australia. One of its main attractions is the Omegathon, a three-day elimination tournament in which twenty randomly selected attendees compete in games across several genres, culminating in the championship match at the convention's closing. The group also holds an annual PAX Unplugged for tabletop games and an annual PAX Dev for game developers. In addition, the world's largest conventions include the Tokyo Game Show and Gamescom in Cologne, Germany.

Linked to gaming conventions are **cosplay** events. Cosplay, a word coined by a Japanese writer that combines *costume* and *play*, first became popular in the United States at science fiction conferences. Fans of certain texts—comic books, novels, television shows, movies, and games—become performance artists by dressing in detailed costumes designed to accurately represent fictional characters. Gamer conventions like E3, PAX, the Tokyo Game Show, and Gamescom host hundreds of cosplayers. Other fan events associated with cosplay, such as MegaCon (Orlando) and Dragon Con (Atlanta), combine interest in gaming with comic books, anime, fantasy, and sci-fi. Digital game studios also hold their own cosplay events.

A COSPLAYER at New York Comic Con dresses up as the character Mercy from the digital game *Overwatch*.

Digital Gaming and Society

The ever-growing relationship between video games and other media, such as books, movies, and television, leaves no doubt that digital gaming has a permanent place in our culture. But like other media, games are a subject of social concern, too. The potential for playing addiction and violent, misogynistic, and racist content has from time to time spurred calls for more regulation of digital games. But as games permeate more aspects of culture and become increasingly available in nonstandard formats, like completely immersive environments, they may also become harder to define and therefore regulate.

Digital Gaming and Media Culture

Beyond the immediate industry, digital games have had a pronounced effect on media culture. Like television shows, books, and comics before them, digital games have inspired movies, such as *Warcraft* (2016), the *Resident Evil* series (2001–2017), *Rampage* (2018), *Detective Pikachu* (2019), and *Sonic the Hedgehog* (2020). For many Hollywood blockbusters today, a video game spin-off is a must-have. For example, the box-office hit *Guardians of the Galaxy* (2014) inspired the episodic digital game *Guardians of the Galaxy: The Telltale Series* (2017), which is available on console, PC, and mobile platforms.

FILM ADAPTATIONS OF VIDEO GAMES, such as *Resident Evil: The Final Chapter*—the sixth and final film of the series—give game makers another platform for promoting their games while giving gamers an opportunity to become even more immersed in the worlds of the games they love.

Video Games at the Movies
Alice, the hero of the *Resident Evil* film series, fights zombies in this clip.

Discussion: In what ways does this clip replicate the experience of gameplay? In what ways is a film inherently different from a game?

Books and digital games have also had a long history of influencing each other. Japanese manga and anime (comic books and animation) have inspired video games, such as *Akira*, *Astro Boy*, and *Naruto*. *Batman: Arkham Asylum*, a top video game title introduced in 2009, is based closely on the *Batman* comic-book stories, while *The Witcher* series of action role-playing games is based on Polish fantasy writer Andrzej Sapkowski's saga. Perhaps the most unusual link between books and digital games is the *Marvel vs. Capcom* game series. In this series, characters from Marvel comic books (Captain America, Hulk, Spider-Man, Wolverine) battle characters from Capcom games, such as *Street Fighter* and *Resident Evil* (Akuma, Chun-Li, Ryu, Albert Wesker). The influence goes the opposite way, too, with hundreds of books based on digital games, including Karen Traviss's acclaimed trilogy of novels expanding the universe of *Halo*.

Addiction and Other Concerns

Though many people view gaming as a simple leisure activity, the digital gaming industry has sparked controversy. Parents, politicians, the medical establishment, and media scholars have expressed concern about the addictive quality of video games and have raised the alarm about violent and misogynistic game content—standard fare for many of the most heavily played games—and issues of representation.

Addiction

No serious—and honest—gamer can deny the addictive qualities of digital gaming. In 2019, the World Health Organization (WHO) included "gaming disorder" for the first time in its *International Classification of Diseases*. According to the WHO, gaming disorder is characterized "by impaired control over gaming, increasing priority given to gaming over other activities to the extent that gaming takes precedence over other interests and daily activities, and continuation or escalation of gaming despite the occurrence of negative consequences."[32] Studies suggest that a small proportion of gamers—less than 3 percent—may have enough symptoms to warrant the diagnosis.[33]

Still, just as habit formation is a primary goal of virtually every commercial form of media, cultivating compulsiveness is the aim of most game designs. From recognizing high scores to incorporating various difficulty settings (encouraging players to try easy, medium, and hard versions) and levels that gradually increase in complexity, designers provide constant in-game incentives for obsessive play. This is especially true of multiplayer online games—like *League of Legends*, *Call of Duty*, and *World of Warcraft*—which make money from long-term engagement by selling expansion packs; selling avatars, skins, and booster packs; and charging monthly subscription fees. These games have elaborate achievement systems with hard-to-resist rewards that include military ranks like "General" or fanciful titles like "King Slayer," as well as special armor, weapons, and mounts (creatures your avatar can ride, including bears, wolves, or even dragons), all aimed at turning casual players into habitual ones.

This strategy of promoting habit formation may not differ from the cultivation of other media obsessions, like swiping up to the next TikTok video, or live-tweeting the latest episode of *The Bachelor*. Even so, real-life stories—such as that of the South Korean couple whose three-month-old daughter died of malnutrition while the negligent parents spent ten-hour overnight sessions in an Internet café raising a virtual daughter—bring up serious questions about video games and addiction.[34] Meanwhile, industry executives and others cite the positive impact of digital games, such as the mental stimulation and educational benefits of games like *SimCity*, the health benefits of *Wii Fit*, and the socially rewarding benefits of playing games together as a family or with friends.

Violence, Misogyny, and Representation

The Entertainment Software Association (ESA)—the main trade association of the gaming industry—likes to counter the myth that gamers are all male teens playing aggressive games by pointing out that nearly half of game players are women, that more than 90 percent of console games are rated in the family- and teen-friendly categories (the game ratings system of the Entertainment Software Rating Board includes Everyone, Everyone 10+, Teen, Mature 17+, and Adults Only 18+), and that the average age of a game player is thirty-three. While these statements are true, they also mask a troubling aspect of some of game culture's most popular games: their violent and sexist imagery and their emphasis on heterosexual white masculinity.

By the ESA's own data, there is reason for some concern. Although it notes that most console games on the market are family or teen rated, nine of the top twenty best-selling games in 2018 were rated M (Mature 17+), meaning that they "may contain intense violence, blood and gore, sexual content and/or strong language."[35] (The top two best-selling games were *Call of Duty: Black Ops 4* and *Red Dead Redemption 2*, both rated M.)

Most games involving combat, guns, and other weapons are intentionally violent, with representations of violence becoming all the more graphic as game visuals reach cinematic hyperrealism. In fact, an analysis of the 126 new games presented at the E3 conference in 2019 found that 85 percent used combat in gameplay.[36] The possible effects of such bloodthirsty and misanthropic games have been debated for years, and video games have been accused of being a factor in violent episodes, such as the Columbine High School shootings in 1999. Earlier research linked playing violent video games to aggressive thoughts or hostility, but those effects don't necessarily transfer to real-world environments. Instead, more recent studies suggest that the personality traits of certain types of players should be of greater concern than the violence of video games. For example, a study in the *Review of General Psychology* noted that individuals with a combination of "high neuroticism (e.g., easily upset, angry, depressed, emotional, etc.), low agreeableness (e.g., little concern for others, indifferent to others' feelings, cold, etc.), and low conscientiousness (e.g., break rules, don't keep promises, act without thinking, etc.)" are more susceptible to the negative outcomes measured in studies of violent video games.[37] For the vast majority of players, the study concluded, violent video games have no adverse effects.

There is less research on misogyny (hatred of women) in video games. One of the most extreme game narratives is from the hugely successful *Grand Theft Auto* series, in which male characters earn points by picking up female prostitutes, paying money for sex, and then beating or killing them to get their money back. Although women are close to half of the digital game audience in the United States, it's likely that many aren't engaged by this story. There are significant differences in the kinds of game genres that women and men like, according to the ESA. Male millennials (ages eighteen to thirty-four) like action, shooter, and sports games, while female millennials prefer casual and action games. For Generation X (ages thirty-five to fifty-four) women, the favorite genre is casual games, including puzzle and classic arcades. For Gen X men, it's sports, racing, and shooter games. The gender differences dissolve for male and female boomers (ages fifty-five to sixty-four), who both prefer card, puzzle, and virtual board games.[38]

The source of the problem may be the male insularity of the game development industry; few women are on the career path to be involved in game development. According to the National Center for Women & Information Technology, "Women hold 57 percent of all professional occupations in the U.S. workforce, but only 26 percent of computing occupations." And even as the digital game industry gets bigger, a small proportion of women are trained for such jobs. Only 18 percent of the recipients of computer or information science undergraduate degrees are women.[39] More specifically, according to a survey by the International Game Developers Association (IGDA), 71 percent of game developers are male.[40] (See "Examining Ethics: The Gender Problem in Digital Games" and "Media Literacy and the Critical Process: First-Person Shooter Games: Misogyny as Entertainment?" for more on violence and misogyny in video games.)

Representation of diverse people in games is also a concern. One of the liberating elements of many games is that players can choose the identity of their character. A Feminist Frequency analysis of new games presented at the 2019 E3 gamer conference found that most new games that year (65%) offered multiple options to pick from when selecting a protagonist—a positive development for diversity in games. But only 5 percent of the games had an established female protagonist

launchpadworks.com

Portrayals of Women in Video Games
In this clip, Anita Sarkeesian analyzes how women are portrayed in a variety of video games.

Discussion: Do you agree with Sarkeesian's analysis? Why or why not?

The Gender Problem in Digital Games

Anita Sarkeesian has a well-documented love of playing video games, from *Mario Kart* and *Rock Band* to *Plants vs. Zombies* and *Half-Life 2*. But that hasn't stopped her from becoming one of the most outspoken—and targeted—critics of how video games depict and treat women. In 2012, a successful Kickstarter campaign helped her launch her *Tropes vs. Women in Video Games* video series on her nonprofit Feminist Frequency website. As Sarkeesian explained, she was moved to examine video games because as a girl growing up and playing the games, she saw that so many of the troubling stereotypes about women were enmeshed in games and gaming culture. "The games often reinforce a similar message, overwhelmingly casting men as heroes and relegating women to the roles of damsels, victims or hypersexualized playthings," Sarkeesian said. "The notion that gaming was not for women rippled out into society, until we heard it . . . from our families, teachers and friends."[1]

Sarkeesian has gained critical acclaim and visibility for her videos and writing, appearing in the *New York Times*, *Businessweek*, and *Rolling Stone*. However, since she began releasing her videos on digital games, she has also been the target of a global campaign of incredibly graphic and violent threats of rape, torture, and murder on social media. This ongoing online harassment reached a new low in 2014, when another of her Feminist Frequency video releases coincided with what has become known as the #GamerGate controversy.

The story surrounding the event that ostensibly touched off #GamerGate began when a computer programmer, Eron Gjoni, had a bad breakup with game designer Zoe Quinn. Gjoni then went online with their breakup in a 9,425-word blog post, claiming that Quinn had had an affair with a writer at Kotaku, an influential gamers' website that features information about a variety of games. The post went viral (as Gjoni intended). Male gamers who believed Gjoni assumed that the affair had led to a favorable review of Quinn's most recent game on Kotaku, pointing to this as indicative of a larger trend of shady journalistic ethics in the gaming press. They organized their criticisms under the hashtag #GamerGate. Very quickly, however, any supposed concerns over journalistic ethics were overshadowed by those focused on "slut-shaming" Quinn. As *Boston* magazine wrote, "Zoe Quinn's ex-boyfriend was obsessed with destroying her reputation—and thousands of online strangers were eager to help."[2] The misogynistic attacks by supporters of #GamerGate exploded into a global barrage of anonymous

threats and attacks on a number of high-profile women in the gaming industry, including Sarkeesian, game developer Brianna Wu, and journalists Katherine Cross and Maddy Myers. A Reddit discussion board identified #GamerGate supporters in nearly every country around the world.[3]

It was at this point that Sarkeesian (and other critics) spoke up and pointed out that the deeply disturbing threats that female gamers and critics were experiencing proved her point about a deeper problem in gaming culture, which in turn reflected broader cultural misogyny. In response to this criticism, many supporters of #GamerGate started behaving even worse.

Soon Sarkeesian and others weren't just receiving anonymous and graphic threats in places like Twitter; rather, they found themselves victims of doxing and swatting. To "dox" someone means to steal private or personal information (from addresses and personal phone numbers to social security and credit card information in some cases) and make it public. To "swat" someone means to call in an anonymous tip to a police department using the victim's address, in an attempt to provoke a raid—particularly by an armed SWAT team—on the person's home. In one such incident, approximately twenty Portland police officers were dispatched to the scene of a supposed armed hostage situation, when the target of the hoax saw someone bragging about it on a message board and called the police before the situation could escalate.[4] Quinn and Wu both had to flee their homes after being doxed and receiving threats that identified where they lived.

In another case, before a scheduled speech by Sarkeesian at Utah State University, an anonymous person threatened to carry out the biggest school shooting ever if the video game critic spoke. Sarkeesian canceled her speech after campus police said Utah's gun laws prohibited them from turning away any audience member who showed up with a gun. Sarkeesian went into hiding for a time, afraid to return to her home because of the various threats. Her Wikipedia page has been vandalized several times with pornographic pictures, and her Feminist Frequency website has been the target of denial-of-service attacks, which seek to make sites unavailable to users by maliciously overwhelming them with automated requests.

But Sarkeesian is far from giving up. In an ironic twist, the hatred leveled at the critic has brought many supporters her way. For example, donors have sent almost $400,000 to her crowdfunded Feminist Frequency website (which is now officially a nonprofit organization dedicated to providing commercial-free videos critiquing the portrayal of women in video games and mass media). Sarkeesian's Twitter feed (@femfreq) reaches over 700,000 followers, and her YouTube video commentary has drawn millions of views.[5]

Because the gaming news media was largely ignoring the misogyny of #GamerGate, Sarkeesian, Wu, and others began to speak out to mainstream news organizations, and coverage spread to Canada, the United Kingdom, Sweden, and elsewhere. With the #GamerGate controversy subject to greater international media scrutiny, the discussion began to change. "We finally shamed [the gaming news media] into . . . addressing #GamerGate," Wu said in a university speech.[6] After being subjected to more than fifty death threats and constant bullying and still feeling "damaged from this experience," Wu perseveres. "We are making this better. We took #GamerGate and we turned it around in its tracks."[7]

Yet years after #GamerGate, there is a deep scar remaining on our culture. *New York Times* opinion writer Charlie Warzel argues that the ultimate lesson of #GamerGate infects our current politics: "that there's a sinister power afforded to those brazen enough to construct their own false realities and foist them on others."[8] ●

MEDIA LITERACY & THE CRITICAL PROCESS

First-Person Shooter Games: Misogyny as Entertainment?

Historical first-person shooter games are a significant subgenre of action games, the biggest-selling genre of the digital game industry. *Call of Duty: Modern Warfare 3* (set in a fictional World War III) made $775 million in its first five days. And with thirteen million units sold by 2012, Rockstar Games' critically acclaimed *Red Dead Redemption* (*RDR*, set in the Wild West) was applauded for its realism and called a "tour de force" by the *New York Times*.[1] But as these games proliferate through our culture, what are we learning as we are launched back and forth in time and into these virtual worlds? Select your own first-person shooter game for critical analysis, and follow the process modeled here with *RDR*.

1 DESCRIPTION

Describe the characters, setting, and various story lines available in the gameplay of your first-person shooter game. *Red Dead Redemption*, for example, features John Madsen, a white outlaw turned federal agent who journeys to the "uncivilized" West to capture or kill his old gang members. Within this game, which takes place in 1911, gamers encounter breathtaking vistas and ghost towns with saloons, prostitutes, and gunslingers; large herds of cattle; and scenes of the Mexican Rebellion. Shoot-outs are common in towns and on the plains, and gamers earn points for killing animals and people.

2 ANALYSIS

In your game, look for patterns that you identified during the description step. What types of things commonly happen in your game, and what types of

story lines never seem to be included? Also, who are the major and minor characters, and which types of characters are completely absent? In *RDR*, certain patterns emerge. For example, Black Americans and Native Americans are absent from the story line (although they were clearly present in the West of 1911). The roles of women are limited: They are portrayed as untrustworthy and chronically nagging wives, prostitutes, or nuns—and they can be blithely killed in front of sheriffs and husbands without ramifications. One special mission is to hog-tie a nun or prostitute and drop her onto railroad tracks in front of an oncoming train.

3 INTERPRETATION

Think about the patterns you discovered in the analysis step and ways of explaining their meaning. Does the game tap into stereotypes or transcend

them? Is the game bound by current norms of society, or does it imagine a better (or more dystopian) version of society? Why? *RDR* may give us a technologically rich immersion into the Wild West of 1911, but it relies on clichés to do so (macho white gunslinger as leading man, weak or contemptible women, vigilante justice). If the macho-misogynistic narrative possibilities and value system of *RDR* seem familiar, it's because the game is based on Rockstar's other video game hit, *Grand Theft Auto* (*GTA*), which lets players have sex with and then graphically kill prostitutes.

4 EVALUATION

What is your verdict for the first-person shooter game you are analyzing? Is it good, bad, mediocre, or other, and why did you come to this judgment? For example, we might conclude that the problem with *Red Dead Redemption* is its limited view of history, lack of imagination, and reliance on misogyny as entertainment.

5 ENGAGEMENT

Talk to your friends about your evaluation of this game and how you reached your conclusions. Comment on blog sites about the game, and give the game publisher your praise, your criticism, or both in an e-mail or a hashtagged social media post.

(see Figure 3.3), and as the study's authors noted, having female protagonists "is no guarantee whatsoever that those representations will be good ones. Games can and often do center women while also reinforcing harmful stereotypes or turning those women into sexual fantasies for the benefit of straight male players."[41]

Racial representation is another weak point in the industry, as the IGDA found that 81 percent of developers are white, whereas 10 percent are East/South/Southeast Asian, 7 percent are Hispanic, 5 percent are Indigenous or Pacific Islanders, and just 2 percent are Black/African/Afro-Caribbean.[42]

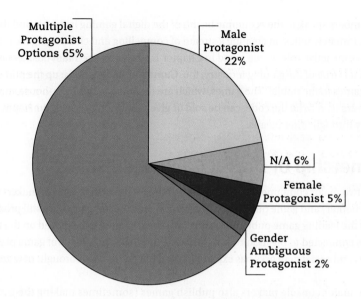

FIGURE 3.3

GENDER REPRESENTATION IN DIGITAL GAMES AT E3

Data from: Anita Sarkeesian and Carolyn Petit, "Female Representation in Videogames Isn't Getting Any Better," *Wired*, June 14, 2019, www.wired.com/story/e3-2019-female-representation-videogames.

Again, the structure of the gaming industry can affect the diversity of the product. *Newsweek* magazine noted that Black protagonist characters in games are rare: "When they do show up, they often appear as sidekicks, overly muscular antagonists or comic relief. Or they're narrowly typecast as athletes, rappers and gangsters."[43] According to Jo Twist, CEO of Ukie, a nonprofit trade group for the United Kingdom's video game industry, this has to change: "Diversity isn't a nicety; it's a necessity if the industry is going to grow, thrive and truly reflect the tens of millions of people who play games every day."[44]

The Future of Gaming and Interactive Environments

Gaming technology of the future promises a more immersive and portable experience that will touch even more aspects of our lives. Virtual reality (VR) headsets, which were first widely released in 2016, offer deep engagement in games, though they can't override the limitations of the actual space the player is in. (As you move in a VR headset, you will eventually encounter furniture or a wall.) To remedy this, a company called the VOID offers experiences in which players wear VR headsets and gear and walk through a set that provides a completely immersive environment. The VOID has about fifteen locations in the United States and Canada, and the experiences the company offers typically extend Hollywood movie narratives, including *Jumanji, Avengers, Star Wars, Wreck-It Ralph*, and *Ghostbusters*. Players walk through the environment in groups of four, and each experience comes with its own story. For example, *Avengers: Damage Control* inserts players into this story: "Shuri has recruited your team to test a powerful new prototype design combining Wakandan and Stark technologies. When an enemy from the past seeks to steal the technology, you'll fight alongside some of your favorite Avengers . . . to stop the attack before they unleash an oppressive new age upon the world."[45] Experiences at the VOID cost about $30 for thirty minutes and demonstrate that premium immersive experiences that combine elements of gaming, amusement parks, and movies are one of the frontiers for digital games.

The Business of Digital Gaming

Today, 65 percent of American households have someone at home who regularly plays video games, with 11 percent of households playing with a VR headset. The entire global digital game market, including portable and console hardware and accessories, is a $160 billion a year industry, with nearly $200 billion in revenue expected by 2022.[46]

These numbers speak to the economic health of the digital gaming industry, and the success of that industry ultimately relies on creating a stream of compelling and challenging narratives in which the players have an active role. As we noted in Chapter 1, stories are the main business of media companies, and that is true of the gaming industry, too. **Game publishers** make up the part of the industry that releases games to the public. The games, which are designed either by in-house studios or hired **game developers** who write the code, can be sold or given away for free, with the hopes of eventually making money through other means.

The Ownership of Digital Game Publishing

For years, the two major components of the gaming industry have been console makers (discussed earlier in the chapter) and game publishers. The biggest blockbuster games are still produced and distributed by the leading game publishers, and many are designed to be played on the three leading game consoles connected to big monitors. At the same time, the emergence of game platforms on mobile devices and social networks has expanded the game market and brought new game publishers into the field.

The three major console makers also publish games (sometimes making the games *proprietary*, meaning they will only play on that company's system). For example, Microsoft famously began publishing its *Halo* game series to drive sales of the Xbox. Similarly, Sony publishes the *Uncharted* game series just for PlayStation, and Nintendo publishes *The Legend of Zelda* series solely for its gaming platforms. More often, however, game publishers are independent companies, distributing games that play across multiple platforms. Sometimes these publishers are also the developers—the people who write the actual code for the games. Other times these publishers serve as distributors for the game developers, in the same way that film studios may distribute the work of independent filmmakers. Activision Blizzard and Electronic Arts—two leading independent game publishers—have been particularly good at adaptation and innovation, producing the most imaginative and ambitious titles and selling the most games across multiple platforms.

Activision Blizzard was created through the merging of Activision and Vivendi's Blizzard Entertainment division in 2008. The Activision half of the company got its start in the 1970s as the first independent game developer and distributor, initially providing games for the Atari platform. Activision was unique in that it rewarded its developers with royalty payments and name credits on game box covers, something that hadn't yet been considered by other game publishers, which kept their developers anonymous. As a result, top game designers and programmers migrated to Activision, and Activision began to produce a number of top-selling games, including the *X-Men* series (2000–), the *Call of Duty* series (2003–), and *Guitar Hero* (2006–2011).

Meanwhile, Blizzard Entertainment, established in 1991 as an independent game publisher, has three famous franchises in game publishing: *Diablo* (1996–), *StarCraft* (1998–), and *World of Warcraft* (2004–). Its widely praised *Overwatch* (2016) reached thirty-five million players in a little over a year and has become Blizzard's fourth major franchise, particularly as a result of its becoming a widely used team-based first-person shooter game for eSports.[47] As one merged company, Activision Blizzard has become a publishing giant in the industry, expanding its offerings even further by purchasing mobile and social gamemaker King (*Candy Crush Saga*) in 2016.

Electronic Arts (EA) got its name by recognizing that the video game is an art form and that game developers are indeed artists; the name Electronic Arts is also a tribute to the United Artists film studio, established in 1919 by three actors and one director—Charlie Chaplin, Mary Pickford, Douglas Fairbanks, and D. W. Griffith—who broke away from the studio-dominated film industry (see Chapter 7). Operating under the same principle that Activision pioneered, EA was able to secure a stable of top talent and begin producing a promising lineup of titles: *Archon*, *Pinball Construction Set*, *M.U.L.E.*, *Seven Cities of Gold*, *The Bard's Tale*, *Starflight*, *Wasteland*, and, perhaps most notably, *Madden NFL*—first released in 1988 and then updated annually beginning in 1990, a practice that has become the modus operandi for EA.

Like Activision Blizzard, EA has moved toward mobile and social gaming platforms. The company acquired PopCap Games, the studio responsible for *Bejeweled* and *Plants vs. Zombies*. EA also sought to compete directly with Activision Blizzard's *World of Warcraft* series by developing (through its Canadian subsidiary, BioWare) the lavish MMORPG game *Star Wars: The Old Republic* (2012), with a price tag approaching $200 million.[48]

One of the newest major game publishers, Zynga, was established in 2007 and specializes in casual games. *FarmVille*, *Draw Something*, *Zynga Poker*, and *Looney Tunes Dash* are among its Facebook hits, though it is losing players to competing developers like Activision Blizzard subsidiary King (*Candy Crush Saga*, *Bubble Witch Saga*) and Wooga (*Diamond Dash*, *Bubble Island*). Rovio, founded in Finland in 2003 and maker of the international phenomenon *Angry Birds* series, is another major casual games publisher.

Other top game publishers around the world include Square Enix (*Deus Ex*, *Final Fantasy*), Ubisoft (*Assassin's Creed*, *Rayman*), Sega (*Sonic the Hedgehog*, *Super Monkey Ball*), and Bandai Namco (*Dark Souls: Remastered*, *Tekken 7*).

MARK RALSTON/Getty Images

DIGITAL GAME DEVELOPERS discuss YouTube Gaming at the E3 at the Los Angeles Convention Center.

The Structure of Digital Game Publishing

AAA game titles (games that represent the current standard for technical excellence) can cost as much as a blockbuster film to make and promote. For example, *The Witcher 3: Wild Hunt* (2015), *Mass Effect: Andromeda* (2017), and *Shadow of the Tomb Raider* (2018) all ranged between $75 million and $125 million in total costs to produce, while *Cyberpunk 2077* (2020) reportedly cost over $300 million. Development, licensing, and marketing constitute major expenditures in game publishing.

Development

The largest part of the development budget—the money spent designing, coding, scoring, and testing a game—goes to paying talent, digital artists, and game testers. Each new generation of gaming platforms doubles the number of people involved in designing, programming, and mixing digitized images and sounds.

That being said, the advent of mobile gaming has provided a new entry point for independent game developers. Whereas the cost of making a major console game is in the millions of dollars, the price is substantially lower for mobile game creation. Independent developers need pay only $25 for a Google Play developer's account or $99 for an Apple developer's account to get started. Of course, there is still the cost and time needed to create the game app for Android or iOS mobile platforms, but doing so can be in the thousands, not the millions, of dollars.

Licensing

Independent gamemakers must deal with two types of licensing. First, they must pay royalties to console manufacturers (Microsoft, Sony, or Nintendo) for the right to distribute a game using their system. These royalties vary from $3 to $10 per unit sold. (Of course, if a console manufacturer such as Nintendo makes games exclusively for the Wii, then it doesn't have to pay a console royalty to itself.) The other form of licensing involves **intellectual property**—stories, characters, personalities, and music that require licensing agreements. In 2005, for instance, John Madden reportedly signed a $150 million deal with EA Sports that allowed the company to use his name and likeness for the next ten years.[49] Madden has continued to license his name and image under new financial terms, while the NFL has also licensed its league for exclusive use in the *Madden NFL* series, signing a $1.5 billion extension with EA from 2021 to 2026. About $1 billion will go to the NFL, and $500 million to the players' union.[50]

Marketing

The marketing costs of launching a digital game often equal or exceed the development costs. The successful launch of a game involves online promotions, banner ads, magazine print ads, in-store displays, and the most expensive element of all: television advertising. In many ways, the marketing blitz associated with introducing a major new franchise title—including cinematic trailers—resembles the promotional campaign surrounding the debut of a blockbuster movie. For example, Rockstar Games reportedly spent $150 million for the marketing of its 2013 blockbuster release *Grand Theft Auto V*, eclipsing its $115 million development budget. *Grand Theft Auto V* remains one of the most expensive digital games made to date.[51]

Selling Digital Games

Just as digital distribution has altered the relationship between other mass media and their audiences, it has transformed the way digital games are sold. Although the selling of $60 AAA console games at retail stores is an enduring model, many games are now free (with opportunities for hooked players to pay for additional play features), and digital stores and downloadable apps make access to games almost immediate.

There are three main pay models in the digital game industry: the boxed game/retail model, the free-to-play model, and the subscription model.

- The *boxed game/retail model* is the most traditional and dates back to the days of cartridges on Atari, Sega, and Nintendo console systems. By the 1990s, games were

VIDEO GAME STORES
Apart from buying boxed game titles at stores like Walmart, Best Buy, and Target, or online stores like Amazon, there is really only one major chain of video game stores devoted entirely to new and used video games: GameStop. The chain, which started in Dallas, Texas, in 1984, operates in the United States and a number of other countries.[53] GameStop is currently trying to negotiate the shifting ground of the digital era and the COVID-19 pandemic by selling customers access codes to digital game downloads in the stores, selling Android tablets and refurbished iPads, and investing in other digital gaming companies.

Bettina Fabos

being released on CD-ROMs (and later DVDs) to better handle the richer game files. On September 17, 2013, the biggest game launch ever—still the biggest launch of *any* media product in history—was the release of *Grand Theft Auto V*. The game, published by Rockstar Games, generated more than $1 billion in sales in just three days, more than any other previous game or movie release.[52] In recent years, though, in-store sales of boxed games are going the way of the video store, as games can be downloaded to a player's computer much more quickly. All three major console companies sell digital downloads through their devices, and several companies—including Steam, Amazon, and Origin Access—compete for the download market in PC games.

- The *free-to-play model* (sometimes called *freemium*) is common with casual and online games like *Super Mario Run*. Free-to-play games are offered online or as downloads at no charge as a way to gain or retain a large audience. These games make money by selling extras, like power boosters (to aid in gameplay), or in-game subscriptions for upgraded play. In addition to free casual games (like *Alto's Odyssey*, *Color Switch*, and *Two Dots*), popular MOBA games like *League of Legends* are free-to-play and generate revenue through in-game purchases. The most ubiquitous casual game distributors are Apple's App Store and Google Play, where users can access games on mobile devices.
- Some games are also sold via a *subscription model*, in which gamers pay a monthly fee to play. In recent years, however, the subscription model for specific games has faltered, and now games that used to be subscription-only (such as *World of Warcraft* and *Star Wars: The Old Republic*) offer the free-to-play model, with premium play as an option. Increasingly, the subscription model has been adapted by major digital corporations and digital game publishers to offer Netflix-style subscriptions that range from about $5 to $20 a month for access to a catalog of online games. Subscription services include Apple Arcade, Google Play Pass, Google Stadia Pro, PlayStation Now, Xbox Game Pass (+Ultimate), EA Play and EA Play Pro, Amazon Luna, and Ubisoft's UPlay+.

Digital Gaming and Advertising

Commercialism is as prevalent in video games as it is in most entertainment media. Advergames—like television's infomercials—are video games created for purely promotional purposes. The first notable advergame debuted in 1992, when Chester Cheetah, the official mascot for Cheetos snacks, starred in two video games for the Sega Genesis and Super Nintendo systems. **In-game advertisements** are more subtle; ads are integrated into the game as billboards, logos, or storefronts (e.g., Dove soap spas appearing in *The Sims Social*), or products appear as components of the game itself (e.g., a classic Porsche 911 automobile in *Cyberpunk 2077*) or as display ads in-between gameplay.[54]

Some in-game advertisements are static, which means the ads are permanently placed in the game. Other in-game ads are dynamic, which means the ads are digitally networked and can be altered remotely, so agencies can tailor them according to release time, geographical location, or user preferences. In-game ads are now an essential part of the revenue mix for major gaming corporations. For example, Activision Blizzard, the largest game company in the United States, reported to its stockholders in 2020, "We continue to focus on in-game advertising as a growing source of additional revenue"—especially in mobile games like *Candy Crush Saga*.[55]

Digital Gaming, Free Speech, and Democracy

As people around the world were urged to quarantine themselves at the spread of the COVID-19 pandemic in spring 2020, it's not surprising that many turned to digital games for entertainment and as a way to step into a completely different environment. The breakout hit of the era emerged for Nintendo Switch gamers: *Animal Crossing: New Horizons*, just about the gentlest game one could imagine

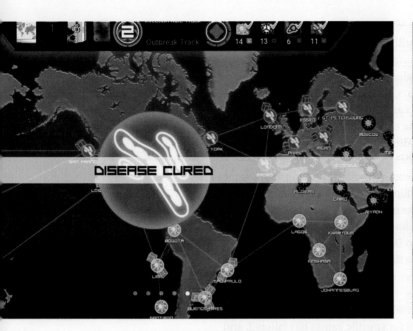

AS COVID-19 spread through the United States, people in quarantine turned to games such as *Animal Crossing: New Horizons* to escape the stress of everyday life and, conversely, games like *Pandemic* to confront the particular problems of a fast-spreading global disease.

(rated E for Everyone, of course). Filled with tiny, wide-eyed human and animal beings in colorful environments, the game invites players "to create [their] personal island paradise on a deserted island brimming with possibility."[56]

Of course, gentle escapism isn't for everyone. For those who really wished to confront the pandemic head-on in gameplay, there already existed a board game (and digital versions) called *Pandemic* (2018). The game's story seemed too familiar: "Humanity is on the brink of extinction. As members of an elite disease control team, you're the only thing standing in the way of the four deadly diseases spreading across the world."[57] Interestingly, game developer Matt Leacock described *Pandemic* as a cooperative, not competitive, game: "Board games have comparatively low stakes, but I've learned they have much to teach us: We all need to play to our strengths, balance short-term threats against long-term goals and make sacrifices for the common good."[58]

Animal Crossing: New Horizons and *Pandemic* match just two of the many reactions we may have had to the pandemic. And there may yet be games in the works or in the imaginations of developers whose stories will resonate with our experiences in still other ways. Because games like these are such a significant element of media culture, it is almost shocking to be reminded that digital games were recognized as a legitimate form of speech only about a decade ago. In a landmark decision handed down in 2011 over a California law enforcing fines for renting or selling M-rated games to minors, the Supreme Court granted digital games speech protections afforded by the First Amendment. According to the opinion written by the late Justice Antonin Scalia, video games communicate ideas worthy of such protection:

Like the protected books, plays, and movies that preceded them, video games communicate ideas—and even social messages—through many familiar literary devices (such as characters, dialogue, plot, and music) and through features distinctive to the medium (such as the player's interaction with the virtual world).[59]

Scalia even mentioned *Mortal Kombat* (a fighting game) in footnote 4 of the decision:

Reading Dante is unquestionably more cultured and intellectually edifying than playing Mortal Kombat. *But these cultural and intellectual differences are not constitutional ones. Crudely violent video games, tawdry TV shows, and cheap novels and magazines are no less forms of speech than* The Divine Comedy. . . . *Even if we can see in them "nothing of any possible value to society . . . they are as much entitled to the protection of free speech as the best of literature."*[60]

Digital games—speech in an interactive medium—provide us entertainment; connect us to communities; and help us express and play through our fears, our fantasies, our hopes, and our dreams.

Chapter
REVIEW 3

LAUNCHPAD FOR *MEDIA & CULTURE*

 **LaunchPad**
macmillan learning

Visit LaunchPad for *Media & Culture* at **launchpadworks.com** for additional learning tools:

➔ QUIZ YOURSELF WITH LEARNINGCURVE

LearningCurve uses gamelike quizzing to help you review important concepts from this chapter. LearningCurve includes multiple-choice questions at different levels of difficulty, hints, feedback for both right and wrong answers, and links to e-book material for easy reference. Which answer choice would you select for the sample question below?

> Whenever she has time to kill, Veata loves playing the game Flappy Bird on her phone because the rules are simple, there is no story, and the game is quick to play. What type of game is Flappy Bird?
>
> ◯ a casual game
>
> ◯ an adventure game
>
> ◯ an action game
>
> ◯ a strategy game

➔ WATCH VIDEO ON KEY CONCEPTS

LaunchPad includes clips from movies, TV shows, online sources, and other media texts, in addition to interviews with media experts and newsmakers. In the videos in this chapter, you'll consider topics like movies based on video games and portrayals of women in digital games.

➔ PRACTICE THE CRITICAL PROCESS

Use the Media Literacy Activity to walk through the critical process step-by-step and develop a critical perspective on what media means to you. This chapter's activity asks you to assess how digital gaming has changed with the development of new media technologies.

➔ LEARN KEY TERMS

Review the definitions for this chapter's key terms with LaunchPad's easy-to-use online flash cards. Try the cards in practice mode, or quiz yourself as a way to focus your study efforts. (Page numbers indicate where each term is highlighted within the chapter.)

penny arcade, 62
pinball machine, 62
arcades, 63
consoles, 64
gameplay, 68
massively multiplayer
  online role-playing games
  (MMORPGs), 69

avatar, 69
multiplayer online battle
  arenas (MOBAs), 69
battle royale games, 69
PUGs, 70
guilds *or* clans, 70
collective intelligence, 71

modding, 71
online fantasy sports, 74
cosplay, 75
game publishers, 82
game developers, 82
intellectual property, 84
in-game advertisements, 85

REVIEW QUESTIONS

Use the questions below to revisit key themes and concepts from this chapter.

✱ Need more practice? Access the LearningCurve multiple-choice questions for this chapter, which are available on LaunchPad. All questions include feedback for correct and incorrect answers, hints, and links to e-book material.

Chapter Opener and Digital Gaming Today

1. What are the main business forces behind the rise of professional eSports?

How We Got Here: The Development of Digital Gaming

2. What sparked the creation of mechanical games in both the nineteenth and the twentieth centuries?

3. How are classic arcade games and the culture of the arcade similar to today's popular console games and gaming culture?

4. What advantages did personal computers have over video game consoles in the late 1980s and much of the 1990s?

The Gaming Environment

5. What are some of the main genres within digital gaming?

6. How are MMORPGs, MOBAs, and virtual worlds built around online social interaction?

Communities of Play

7. How do communities of play—including collective intelligence, interactive livestreaming platforms, online fantasy sports, and conventions—enhance the social experience of gaming and make games different from other mass media?

Digital Gaming and Society

8. How have digital games influenced media culture, and vice versa?

9. To what extent are video game addiction and violent, misogynistic, and racist representations in gaming problems for the gaming industry?

The Business of Digital Gaming

10. What are the roles of two major components of the gaming industry: console makers and game publishers?

11. How do game publishers develop, license, and market new titles?

12. What are the three main pay models for selling video games today?

13. In what ways has advertising become incorporated into digital games?

Digital Gaming, Free Speech, and Democracy

14. Why did the U.S. Supreme Court rule that games count as speech?

QUESTIONING THE MEDIA

Use these critical-thinking questions to reflect on your own experiences with media and apply your understanding of media concepts.

✱ For an interactive experience, questions on topics like these are available for the iClicker student response system. Instructors can download the questions from LaunchPad and use them to poll the class.

---

1. Do you have any strong early memories from playing video games? To what extent did these games define your childhood?

2. Have you ever been upset by the level of violence, misogyny, or racism in a video game you played (or watched being played)? Discuss the game narrative and what made it problematic.

3. Most digital games have a heterosexual white male point of view. Why is that? If you were a game developer, what kinds of game narratives would you like to see developed?

Sound Recording and Popular Music

4

At age nineteen, Montero Lamar Hill of Atlanta, Georgia, dropped out of the University of West Georgia to start a music career. His parents were not pleased. "My dad initially was like, 'There's a million rappers in this industry,'" he said.[1]

But Hill created something about a million other rappers wouldn't: a country rap song. Music fans now know him as Lil Nas X, the artist behind the 2019 hit single "Old Town Road."

Hill took a new path to making his song No. 1: He used TikTok, a social media app for creating and sharing short viral music videos. With more than 500 million global users at the time (and more than 800 million by 2020), the app has quickly amassed an immense following. Videos on TikTok are usually shared with hashtags so users can participate in viral challenges on a certain theme. "Old Town Road" became the music of choice for people participating in the #yeehaw challenge, which helped the song gain wider notice with tens of millions of views.[2] The viral strategy was part of Lil Nas X's plan: "I promoted the song as a meme for months until it caught on to TikTok and it became way bigger."[3]

From TikTok, the song moved up the charts on Spotify and Apple Music. It then landed simultaneously on *Billboard* magazine's Hot 100 chart, Hot Country Songs chart, and Hot R&B/Hip-Hop Songs chart. But then something happened to the country rap song: *Billboard* dropped "Old Town Road" from its Hot Country Songs chart. In a statement released to *Rolling Stone* magazine, *Billboard* said that "upon further review, it was determined that 'Old Town Road' by Lil Nas X . . . does not embrace enough elements of today's country music to chart in its current version."[4]

Some critics cried foul, as the act seemed to suggest that a Black artist doesn't belong on the Hot Country chart. As *Rolling Stone* noted, "At this point, rap's country infiltration is ancient news. . . . But no genre wrestles with its identity as openly as country."[5] Critics also recalled that in the 1940s,

◀ Lil Nas X, whose career launched in 2019 when his song "Old Town Road" went viral on TikTok, became one of the most acclaimed performers of 2020.

Billboard used to have a chart for "Race Records" (later renamed R&B). Was *Billboard* now defending an updated form of racial segregation in music genres?

Lil Nas X ended up having the last laugh, though. He released a remix version with country legend Billy Ray Cyrus, reasserting his country credibility. And regardless of how the industry wanted to categorize the song, fans didn't seem to care. With a record-setting 2.5 million on-demand streams, "Old Town Road" was the most streamed song of 2019.[6] In 2020, it won Grammy Awards for best pop duo/group performance and best music video of the year.

Of course, it was TikTok that helped Lil Nas X upend the music industry. The global app, from the Chinese company ByteDance, was launched in the United States in 2018. Music sound bites often become the basis for TikTok videos, showcasing dance challenges (as in the case of "Old Town Road") and comedy videos featuring people, dogs, and cats. The app makes editing and uploading easy, and the quick production of short videos leads to a never-ending stream of fresh content. By 2020, TikTok rivaled YouTube for the amount of time children ages four to fifteen spent there.[7]

In the time since its introduction to the U.S. market, TikTok has launched hit singles like "Roxanne" by Arizona Zervas, "Say So" by Doja Cat, and "GoodMorningTokyo!" by Tokyo's Revenge. It has also given new life to old songs with distinctive hooks, like Hot Chelle Rae's "Tonight Tonight" (2011), War's "Low Rider" (1975), and Harry Belafonte's calypso classic "Jump in the Line (Shake Senora)" (1961).[8] For earlier generations, radio, MTV, or Myspace might have been where artists first found their audiences. Right now, TikTok is the place where a short sample of a song, incorporated into millions of videos, can help an artist launch a career.

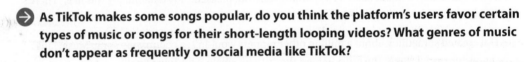 **As TikTok makes some songs popular, do you think the platform's users favor certain types of music or songs for their short-length looping videos? What genres of music don't appear as frequently on social media like TikTok?**

Sound Recording Today

Back in 1999, things looked rosy for the sound recording industry's future. The major record companies had become flush with profits from two strong decades of CD sales, and the industry was set to roll out advanced CD and DVD audio formats. And then a nineteen-year-old at Northeastern University in Boston disrupted the music industry when he created a free file-sharing service called Napster, enabling people to (illegally) share their music. Music lovers quickly moved to MP3 files, abandoning CDs. Since that time, we've gone from illegal file-sharing to a *legal* online space to download music (the Apple iTunes Store) and its subsequent displacement by music streaming services like Spotify and Apple Music.

After some dark years, the music industry enters the 2020s with two important changes making an impact. First, the music industry is growing again on the strength of streaming. Instead of buying physical copies of music (a small resurgence in vinyl recordings notwithstanding), most fans subscribe to massive online streaming libraries, using mobile phones as their main playback device. With streaming, there is more diversity of available musical acts and styles (literally millions of songs) than at any other time in the history of the industry. With social media like TikTok, there are also new opportunities and roles for fans, who can—through just one short music video—become (perhaps unintentionally) super-promoters of recording artists.

The second big change is that musicians are a lot more independent. For most musicians, the big goal had always been a signed recording contract with a major label. Although those big recording labels still exist, about one-third of the music business is now done outside that system, with independent labels, the "artists direct" approach of selling music directly to fans, and some artists who choose not to sell any recorded music at all. (Chance the Rapper is the most famous proponent of this last approach; he makes money through concert touring, sponsorships, and selling his own merchandise.) For music consumers, this means they have the opportunity to hear a wider range of musical artists—not just the relative few who get a contract with a major music label.

Although this extraordinary access to so much music is unique to the current generation, music's influence is nothing new. In modern times, sound recording has helped shape people's identities during the transition from childhood to adulthood and resonated throughout their lives. But music has also caused controversy. In the course of its history, popular music has been banned by parents, school officials, and even governments under the guise of protecting young people from corrupting influences. As far back as the late eighteenth century, authorities in Europe outlawed waltz music, thinking that it was immoral for men and women to dance close together and rotate equally around a common axis.[9] From the jazz age to today, popular music has added chapters to the age-old musical battle between generations.

In this chapter, we will place the impact of popular music in context, examine its influence on our culture, and look at how current developments in the music industry are shaping sound recording's future. We will:

Jeff Kravitz/Getty Images

CHANCE THE RAPPER
Chicago-based artist Chance the Rapper is the first major recording artist to build his career on *not* selling his music.

- Investigate the origins of recording's technological "hardware," from the early phonograph to the flat disk record and the development of audiotape, compact discs, and MP3s
- Study radio's early threat to sound recording and the subsequent alliance between the two media when television arrived in the 1950s
- Explore the impact of the Internet on music, especially how streaming has moved us from a model of music ownership to one of music access
- Examine the content and culture of the music industry, focusing on the predominant role of rock music and its extraordinary impact on mass media forms and a diverse array of cultures, both American and international
- Explore the economic and democratic issues facing the recording industry

How We Got Here: The Development of Sound Recording

New mass media have often been defined in terms of the communication technologies that preceded them. For example, movies were initially called *motion pictures*, a term that derived from photography; radio was known as *wireless telegraphy*, referring to telegraphs; and television was often called *picture radio*. Likewise, sound recording instruments were initially described as talking machines and later as phonographs, drawing on names of existing inventions: the tele*phone* and the tele*graph*.

This early blending of technology foreshadowed our contemporary era, in which media as diverse as newspapers and movies converge on the Internet. Long before the Internet and streaming music, though, the first major media convergence involved the relationship between two industries: sound recording and radio.

From Cylinders to Disks: Sound Recording Becomes a Mass Medium

Between the 1850s and 1880s, men like French printer Édouard-Léon Scott de Martinville and American inventor Thomas Edison conducted experiments in first capturing and then playing back sound, particularly human speech. De Martinville was the first to record sound, speaking into a funnel that had a hog's hair bristle as a needle at the opposite end, which would vibrate and etch grooves into a cylinder coated with black pigment. At that time (1857), no one was thinking about the possibility of playing back sound.

In 1877, Edison succeeded in both recording and playing back sound. He recorded his voice by using a needle to press the sound waves onto tinfoil, which was wrapped around a metal cylinder about the size of a cardboard toilet-paper roll. After recording his voice, Edison played it back by repositioning the needle to retrace the grooves in the foil. The machine that played these foil cylinders became known as the *phonograph*, derived from the Greek terms for "sound" and "writing."

THOMAS EDISON

In addition to inventing the phonograph, Edison (1847–1931) ran an industrial research lab that is credited with inventing the motion-picture camera, the first commercially successful lightbulb, and a system for distributing electricity.

Edison was more than an inventor, however. He wanted to find practical and profitable uses for his inventions. When he patented his phonograph in 1878, for example, Edison initially envisioned it as a device for office work—an answering machine of sorts that would help businesses use their telephones to keep valuable records, rather than having telephones function solely as a medium for "momentary and fleeting communication."[10] At the time, Edison's phonograph and the devices produced by his competitors—like the *graphophone*, which used more durable wax cylinders—were only marginally successful as office machines, but they laid the foundation for numerous voice-recording devices that allow us to archive and time-shift human speech, such as answering machines, audio surveillance equipment, voice memo apps, and voice recognition software. Eventually, these inventors also began to produce cylinders containing prerecorded generic music, which proved to be more popular but were difficult to mass-produce and not very durable for repeated plays.

It was Emile Berliner, a German engineer who had immigrated to America, who developed a better machine that played round, flat disks, or records. Made of zinc and coated with beeswax, these records played on a hand-powered turntable, which Berliner called a *gramophone* and patented in 1887. Berliner's flat disks also enabled the mass production of records. Previously, when using Edison's cylinder, performers had to play or sing

Library of Congress, Prints & Photographs Division, [LC-US262-111421]

Library of Congress, Prints & Photographs Division, [LC-US262-29368]

SOUND IN AND SOUND OUT In the first photo, Red Point (also known as Miguelito), a Navajo artist and storyteller, speaks into the sound cone of an early recording device (ca. 1914) in order to capture his voice on a disk. In the second photo, a young girl (ca. 1901) listens to an early Victor gramophone, on which the sound horn was connected directly to the needle on the disk for playback.

into the speaker for each separate recording. Berliner's technique featured a master recording from which copies could be easily duplicated in mass quantities. In addition, Berliner's records could be stamped with labels, allowing the music to be differentiated by title, performer, and songwriter. This led to the development of a "star system," wherein fans could identify and choose their favorite artists across many records. For example, the acclaimed Italian tenor Enrico Caruso became famous worldwide as one of the first recording stars, making more than 250 recordings between 1902 and 1920. This ability to mass-produce disk recordings and label them with the identity of the performers set the stage for sound recording to become a mass medium.

Between 1900 and 1950, record players became an essential appliance in most American homes, and the emerging recording industry adapted to evolving technological changes. Mechanical, hand-cranked devices were replaced by electric record players in the 1920s. Fragile records made of shellac (which had replaced wax records) were themselves replaced by polyvinyl records, leading to a more durable product with better sound quality. In 1948, CBS Records introduced the 33⅓-rpm (revolutions per minute) *long-playing* record (LP), with about twenty minutes of music on each side, creating a market for multisong albums and classical music. This was an improvement over the three to four minutes of music contained on the existing 78-rpm records. The next year, RCA developed a competing smaller 45-rpm record that featured a quarter-size hole (best for jukeboxes) and one song on each side, invigorating the sales of songs heard on jukeboxes throughout the country.

launchpadworks.com

Shawn Harmsen

Sound Recordings from a Century Ago
A museum curator demonstrates how to operate two hand-cranked working phonographs.

Discussion: If the phonograph was still the main method of music delivery available to listeners, how might this change the way we consume music today?

From Audiotape to CDs: Analog Goes Digital

The 1940s saw key advances in a new form of sound recording—magnetic **audiotape** and tape players. Audiotape significantly altered recording production practices. Its lightweight magnetized strands finally made possible sound editing and multiple-track mixing, in which instrumentals and vocals can be recorded at one location and later mixed onto a master recording in another studio. It also propelled **stereo**—the recording of two separate channels, or tracks, of sound that enabled more natural sound distribution—into commercial use in 1958. By the mid-1960s, engineers had placed miniaturized reel-to-reel audiotape inside small plastic *cassettes* and developed portable cassette players, permitting listeners to bring recorded music anywhere and creating a market for prerecorded cassettes. Audiotape also enabled "home dubbing"; consumers could copy their favorite records onto tape or record songs from the radio, leading to new cultural practices like making and trading mixtapes, precursors to today's playlists.

The biggest recording advancement came in the 1970s, when electrical engineer Thomas Stockham made the first digital audio recordings on standard computer equipment. Although the digital recorder was invented in 1967, Stockham was the first to put it to practical use. In contrast to **analog recording**, which captures the fluctuations of sound waves and stores those signals in a record's grooves or a tape's continuous stream of magnetized particles, **digital recording** translates sound waves into binary on-off pulses and stores that information as numerical code. When a digital recording is played back, a microprocessor translates those numerical codes back into sounds and sends them to loudspeakers. By the late 1970s, Sony and Philips were jointly working on a way to design a digitally recorded disc and player to take advantage of this new technology, which could be produced at a lower cost than either vinyl records or audiocassettes. As a result of their efforts, digitally recorded **compact discs (CDs)** hit the market in 1983.

By 1987, CD sales were double the amount of LP sales. By 2000, CDs had rendered records and audiocassettes nearly obsolete, except for deejays and record enthusiasts who continued to play and collect vinyl LPs (see Figure 4.1). As vinyl devotees point out, the sound produced when

FIGURE 4.1

THE EVOLUTION OF DIGITAL SOUND RECORDING SALES (REVENUE IN BILLIONS)

Data from: Recording Industry Association of America, Annual Year-End Statistics. Figures are rounded.

Note: The year 1999 is the year Napster arrived, and the peak year of industry revenue. In 2011, digital product revenue surpassed physical product revenue for the first time. Synchronization royalties are royalties generated by music that is licensed for use in television, movies, and advertisements.

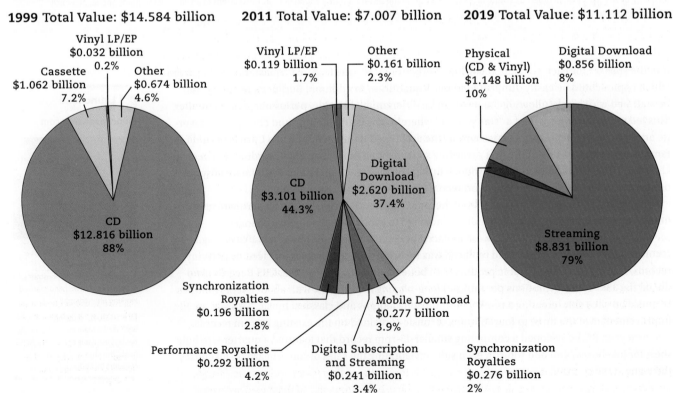

1999 Total Value: $14.584 billion

Vinyl LP/EP
$0.032 billion
0.2%

Cassette
$1.062 billion
7.2%

Other
$0.674 billion
4.6%

CD
$12.816 billion
88%

2011 Total Value: $7.007 billion

Vinyl LP/EP
$0.119 billion
1.7%

Other
$0.161 billion
2.3%

CD
$3.101 billion
44.3%

Digital Download
$2.620 billion
37.4%

Synchronization Royalties
$0.196 billion
2.8%

Performance Royalties
$0.292 billion
4.2%

Digital Subscription and Streaming
$0.241 billion
3.4%

Mobile Download
$0.277 billion
3.9%

2019 Total Value: $11.112 billion

Physical (CD & Vinyl)
$1.148 billion
10%

Digital Download
$0.856 billion
8%

Streaming
$8.831 billion
79%

Synchronization Royalties
$0.276 billion
2%

playing vinyl records is different—some would say richer and more complex—than that produced by compressed digital files. Such appreciation of and nostalgia for vinyl records has created a subculture, or *taste culture*, organized around the analog recording format. Today's vinyl recordings also enable people to own a physical copy of a recording at a time when most music is experienced through online streaming subscriptions.

Sound Recording in the Internet Age

Music recordings, perhaps more so than any other mass medium, are bound up in the social fabric of our lives. People shared and swapped recorded disks from the very first record pressings and created homemade mixtapes for friends on cassette tapes and CDs.

It is not surprising, then, that the Internet, a mass medium that links individuals and communities together like no other medium, became a hub for sharing music. In fact, the reason Napster inventor Shawn Fanning said he developed the groundbreaking file-sharing site in 1999 was "to build communities around different types of music."[11] But the development of a new digital file format and its convergence with the Internet began to unravel the music industry in the 2000s.

MP3s and File-Sharing

The **MP3** file format, developed in 1992, enables digital recordings to be compressed into smaller, more manageable files. With the increasing popularity of the Internet in the mid-1990s, computer users began swapping MP3 music files online because they could be uploaded or downloaded in a fraction of the time it took to exchange noncompressed music files.

By 1999, the year Napster's infamous free file-sharing service brought the MP3 format to popular attention, music files were widely available on the Internet—some for sale, some legally available for free downloading, and many available for trading, in possible violation of copyright laws. Despite the higher quality of industry-manufactured CDs, music fans enjoyed the convenience of downloading MP3 files on Napster, which was the first platform to make music file-sharing so simple. Losing countless music sales to illegal downloading, the music industry fought the proliferation of the MP3 format with an array of lawsuits, but the popularity of MP3s continued to increase.

In 2001, the U.S. Supreme Court ruled in favor of the music industry and against Napster, declaring free music file-swapping illegal and in violation of music copyrights held by recording labels and artists. It was relatively easy for the music industry to shut down Napster (which later relaunched as a legal service) because it required users to log into a centralized system. However, the music industry's elimination of file-sharing was not complete, as decentralized *peer-to-peer* (P2P) systems, such as Grokster, LimeWire, Morpheus, Kazaa, eDonkey, eMule, and BitTorrent, enabled people to continue illegal music file-sharing.

The recording industry fought back with thousands of lawsuits and fines, and by 2010, most P2P systems had been shut down.[12] By 2011, several major Internet service providers, including AT&T, Comcast, and Verizon, had agreed to help the music industry identify customers who were illegally downloading music and try to prevent them from doing so by sending them "copyright alert" warning letters, redirecting them to web pages about digital piracy, and ultimately slowing download speeds or closing their broadband accounts.

As it cracked down on digital theft, the music industry—realizing that it would have to somehow adapt its business to the digital format—embraced services like the iTunes store (launched by Apple in 2003 to accompany the iPod), which had become the model for legal online distribution. In 2008, iTunes became the top music retailer in the United States. But by the time iTunes surpassed twenty-five billion song downloads in 2013, global digital download sales had fallen for the first time.[13] What happened? The next big digital format had arrived.

Streaming Music

If the history of recorded music tells us anything, it is that both musical tastes and formats change over time. Today, streaming music is the format of choice. In the language of the music industry,

LaunchPad
macmillan learning
launchpadworks.com

Recording Music Today
Composer Scott Dugdale discusses technological innovations in music recording.

Discussion: What surprised you the most about the way the video showed a song being produced, and why?

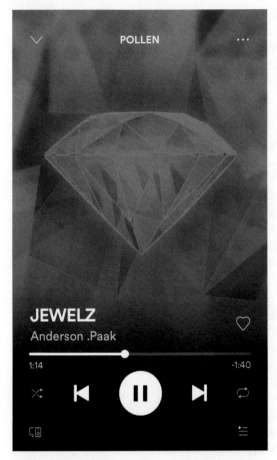

STREAMING MUSIC
services like Spotify (seen here) and Apple Music provide users with more opportunity to fully customize their listening experience. Streaming has transformed the sound recording business, as listeners no longer need to purchase individual recordings. A subscription gives them access to millions of songs, and the ability to more easily discover new artists or go deep into the catalog of established artists. Spotify and Apple Music each carry more than fifty million songs.

we are shifting from *ownership* of music to *access* to music.[14] The access model has been driven by the availability of streaming services such as the Sweden-based Spotify, which made its debut in the United States in 2011 and hit 124 million worldwide subscribers by 2020.[15] Other services include Apple Music, Google Play Music, Amazon Music, Pandora, Tidal, Sound-Cloud, Deezer, and Tencent Music (in China). With these services, listeners can pay a subscription fee (typically $5 to $10 per month) and instantly play millions of songs on demand via the Internet. YouTube and Vevo also supply ad-supported music streaming and have wide international use.

The Ongoing Battles between Records and Radio

Each successful new recording technology, from flat disks played on a gramophone to music streamed on a mobile phone, creates new cultural and business practices but also threatens other traditions and businesses. There is no better (and longer-running) example of media industry rivalry than the relationship between the recording industry and radio, which has flipped from friends to foes and back again multiple times. Radio's very existence sparked the first battle. By 1915, the phonograph had become a popular form of entertainment. The recording industry sold thirty million records that year, and by the end of the decade, sales had more than tripled each year. In 1924, however, record sales dropped to only half of what they had been the previous year. Why? Because radio had arrived as a competing mass medium, providing free entertainment over the airwaves, independent of the recording industry.

The battle heated up when, to the alarm of the recording industry, radio stations began broadcasting recorded music without compensating the industry. The American Society of Composers, Authors and Publishers (ASCAP), founded in 1914 to collect copyright fees for music publishers and writers, charged that radio was contributing to plummeting sales of records and sheet music. By 1925, ASCAP had established fees for radio, charging stations between $250 and $2,500 a week for the right to play recorded music—and causing many stations to leave the air.

But other stations countered by establishing their own live, in-house orchestras, disseminating "free" (ad-supported) music to listeners. This time the recording industry could do nothing, as original radio music did not infringe on any copyrights. Throughout the late 1920s and 1930s, record and phonograph sales continued to fall. The two industries only began to cooperate with each other after television became popular in the early 1950s. Television pilfered radio's variety shows, crime dramas, and comedy programs, along with much of its advertising revenue and audience. Seeking to reinvent itself, radio turned to the recording industry, and this time both industries greatly benefited from radio's new "hit-song" format. The alliance between the recording industry and radio was aided enormously by rock-and-roll music, which was just emerging in the 1950s. Rock created an enduring consumer youth market for sound recordings and provided much-needed content for radio precisely when television made it seem like an obsolete medium.

After the digital revolution, however, the mutually beneficial arrangement between the recording and radio industries began to fray. While Internet streaming radio stations were required to pay royalties to music companies when they played their songs, radio stations still got to play music royalty-free over the air, and the music industry decided that it was time for radio's free ride to end. In 2012, Clear Channel (now iHeartMedia)—the largest radio station chain in the United States and one of the largest music streaming companies, with iHeartRadio—was the first company to strike a new deal with the recording industry and pay royalties for music played over the air. Since that first deal, other radio groups have begun to forge agreements with music labels, paying royalties for on-air play while getting reduced rates for streaming music. As streaming has become more lucrative for the recording industry, leading streamers like Spotify and Apple Music are often competing directly against broadcast radio and broadcast radio's own streaming sites, once again straining the relationship.

The Culture of Popular Music

Popular music, or **pop music**, is music that appeals either to a wide cross section of the public or to sizable subcultures within the larger public. The word *pop* has also been used to distinguish popular music from classical music, which is written primarily for ballet, opera, ensemble, or symphony. Pop music in the United States has its origins in the sheet music that began to be produced in New York in the late nineteenth century (see the next section). As the recording industry developed, pop music in the United States was often, in practice, music by white people for white people; recordings by most Black artists were separately categorized as "race records," before being renamed rhythm and blues (R&B) by the 1950s, soul by the 1970s, and urban by the 1990s. In 2020, as a new civil rights movement coursed through the United States, the urban category was rejected by some in the music industry and replaced with the more accurate genre names of R&B and hip-hop.[16]

The sociological categories of the music industry haven't kept pace with how U.S. popular music has developed organically, with intersections of subcultures constantly creating new forms and reinvigorating older musical styles. (The story about Lil Nas X and the country music charts at the beginning of this chapter is just one instance of the business defending outmoded classifications.) "Popular" used to be determined more by radio listening and music sales, elements that the industry could direct to some degree. These days popularity is determined largely by streaming, where listeners have much more control in actively choosing what they want to hear. Across the landscape of music today, pop music is now a catch-all category for the rich mix of genres that constitute what is in demand, including music by artists such as Rihanna, Lady Gaga, Billie Eilish, J. Cole, Cardi B, Diplo, Cage the Elephant, Anderson .Paak, Rosalía, Kacey Musgraves, and Jason Isbell.

The Rise of Pop Music

Although it is often assumed that pop music developed simultaneously with the phonograph and radio, it actually existed before these media. In the late nineteenth century, the sale of sheet music for piano and other instruments sprang from a section of Broadway in Manhattan known as Tin Pan Alley, a derisive term used to describe the sound of these quickly produced tunes, which supposedly resembled cheap pans clanging together. Tin Pan Alley's tradition of song publishing began in the late 1880s with such music as the marches of John Philip Sousa and the ragtime piano pieces of Scott Joplin. It continued through the first half of the twentieth century with the show tunes and vocal ballads of Irving Berlin, George Gershwin, and Cole Porter, and into the 1950s and 1960s with such rock-and-roll writing teams as Jerry Leiber–Mike Stoller and Carole King–Gerry Goffin.

SCOTT JOPLIN (1868–1917) published more than fifty compositions during his life, including "Maple Leaf Rag"—arguably his most famous piece.

At the turn of the twentieth century, with the newfound ability of song publishers to mass-produce sheet music for a growing middle class, popular songs became a major business enterprise. With the emergence of the phonograph, song publishers also discovered that recorded tunes boosted interest in and sales of sheet music. Thus, songwriting and Tin Pan Alley played a key role in the rise of popular music.

As sheet music grew in popularity, **jazz** developed in New Orleans. An improvisational and mostly instrumental musical form, jazz absorbed and integrated a diverse body of musical styles, including African rhythms, **blues** (a music emerging from Black spirituals, ballads, and work songs from the rural South), and gospel (especially by electric-guitar-wielding gospel singer Sister Rosetta Tharpe). Jazz influenced many bandleaders throughout the 1930s and 1940s. Groups led by Louis Armstrong, Count Basie, Tommy Dorsey, Duke Ellington, Benny Goodman, and Glenn Miller were among the most popular of the swing jazz bands, whose rhythmic music also dominated radio, recordings, and dance halls in their day. The instruments of jazz—piano, saxophone, clarinet, horn, guitar, drum, and bass—as well as its improvisational style, were present in early rock and roll.

The first pop vocalists of the twentieth century were products of the vaudeville circuit, which radio, movies, and the Depression would bring to an end in the 1930s. In the 1920s, Eddie Cantor, Belle Baker, Sophie Tucker, and Al Jolson were extremely popular. By the 1930s, Rudy Vallée and Bing Crosby had established themselves as the first crooners, or singers of pop standards. Bing Crosby also popularized Irving Berlin's "White Christmas," one of the most covered songs in recording history. In one of the first mutually beneficial alliances between sound recording and radio, many early pop vocalists had their own network of regional radio programs, which vastly increased their exposure.

Frank Sinatra arrived in the 1940s, and his romantic ballads foreshadowed the teen love songs of rock and roll's early years. Nicknamed "the Voice" early in his career, Sinatra, like Crosby, parlayed his music and radio exposure into movie stardom. Helped by radio, pop vocalists like Sinatra were among the first vocalists to be followed by a large national teen audience.

Rock and Roll Is Here to Stay

The cultural storm called **rock and roll** hit in the mid-1950s. *Rock and roll*, as with the term *jazz*, was a blues slang term for "sex," lending it instant controversy. Early rock and roll was considered the first "integrationist music," merging the Black sounds of rhythm and blues, gospel, and Robert Johnson's screeching blues guitar with the white influences of country, folk, and pop vocals.[17] From a cultural perspective, only a few musical forms have ever sprung from such a diverse set of influences, and no new style of music has ever had such a widespread impact on so many different cultures as rock and roll. From an economic perspective, rock and roll was the musical form that simultaneously transformed the structure of sound recording and radio. Rock's development set the stage for how music is produced, distributed, and performed today. Many social, cultural, economic, and political factors leading up to the 1950s contributed to the growth of rock and roll, including Black migration, the growth of youth culture, and the beginnings of racial integration.

The migration of Black southerners to northern cities in search of better jobs during the first half of the twentieth century helped spread different popular music styles. In particular, blues—the foundation of rock and roll—came to the North, including places like Chicago. Emerging from the rural South, blues was exemplified in the work of Robert Johnson, Ma Rainey, Son House, Bessie Smith, Charley Patton, and others. The introduction in the 1930s of the electric guitar—a major contribution to rock music—gave southern blues its urban

BESSIE SMITH (1895–1937) is considered the best female blues singer of the 1920s and 1930s. Mentored by the famous Ma Rainey, Smith had many hits, including "Down Hearted Blues" and "Gulf Coast Blues." She also appeared in the 1929 film *St. Louis Blues*.

Library of Congress, Prints & Photographs Division, [LC-DIG-ppmsca-09571]

style, popularized in the work of Muddy Waters, Howlin' Wolf, Sonny Boy Williamson, B.B. King, Buddy Guy, and Bo Diddley.[18]

During this time, blues-based urban Black music began to be marketed under the name **rhythm and blues**, or **R&B**. Featuring "huge rhythm units smashing away behind screaming blues singers," R&B appealed to young listeners fascinated by the explicit (and forbidden) sexual lyrics in songs like "Annie Had a Baby," "Sexy Ways," and "Wild Wild Young Men."[19] Although it was banned on some stations, R&B was continuing to gain airtime by 1953. As previously noted, in those days, Black and white musical forms were segregated: Trade magazines tracked R&B record sales on the race charts, which were kept separate from white record sales, which were tracked on the pop charts.

Yet the segregation of musical forms could not be sustained. Early white rockers such as Buddy Holly and Carl Perkins combined country (or hillbilly) music, southern gospel, and Mississippi delta blues to create a sound called **rockabilly**. At the same time, the R&B influence on early rock came from Black artists like Fats Domino, Willie Mae "Big Mama" Thornton, and Big Joe Turner. The beginning of the integration of white and Black cultures aided the growth of rock and roll, which in turn broke down barriers of segregation. In addition to increased exposure to Black literature, art, and music, several key historical events in the 1950s chipped away at the borders between Black and white cultures. For example, with the Supreme Court's *Brown v. Board of Education* decision in 1954, "separate but equal" laws—which had kept white and Black schools, hotels, restaurants, restrooms, and drinking fountains segregated for decades—were declared unconstitutional. A cultural reflection of the times, rock and roll would burst forth from the midst of the decade's social and political tensions.

Rock Muddies the Waters

In the 1950s, legal integration accompanied a cultural shift, and the music industry's race and pop charts blurred. White deejay Alan Freed had been playing Black music for his young audiences in Cleveland and New York since the early 1950s, and such white performers as Johnnie Ray and Bill Haley had crossed over to the race charts to score R&B hits. Meanwhile, Black artists like Chuck Berry were performing country songs, and for a time, Ray Charles even played in an otherwise all-white country band.

Soaring record sales and the crossover appeal of the music represented an enormous threat to long-standing racial and class boundaries. In 1956, the secretary of the North Alabama White Citizens' Council bluntly spelled out the racism and white fear concerning the new blending of Black and white culture: "Rock and roll is a means of pulling the white man down to the level of the Negro. It is part of a plot to undermine the morals of the youth of our nation."[20]

Although continuing the work of breaking down racial borders was one of rock and roll's most important contributions, the genre also blurred other long-standing distinctions between high and low culture, masculinity and femininity, the North and the South, and the sacred and the secular.

High and Low Culture

In 1956, Chuck Berry's "Roll Over Beethoven" merged rock and roll—considered low culture by many—with the high culture of classical music composer names, forever blurring the traditional boundary between these cultural forms with lyrics like "You know my temperature's risin' / the jukebox is blowin' a fuse . . . / Roll over Beethoven / and tell Tchaikovsky the news." Although such early rock-and-roll lyrics seem tame by today's standards, at the time they seemed like sacrilege. Rock and

ROCK-AND-ROLL PIONEER A major influence on early rock and roll, Chuck Berry, born in 1926, scored major hits between 1955 and 1958, writing "Maybellene," "Roll Over Beethoven," "School Day," "Sweet Little Sixteen," and "Johnny B. Goode." At the time, he was criticized by some Black artists for sounding white, and his popularity among white teenagers was bemoaned by conservative critics. Today, young guitar players routinely imitate his style.

Michael Ochs Archives/Getty Images

rollers also challenged musical decorum and the rules governing how musicians should behave or misbehave: Berry's "duck walk" across the stage, Elvis Presley's pegged pants and gyrating hips, and Bo Diddley's use of the guitar as a phallic symbol were an affront to the norms of well-behaved, culturally elite audiences.

Masculinity and Femininity

Rock and roll was the first popular music genre to overtly confuse issues of sexual identity and orientation. Although early rock and roll largely attracted males as performers, the most fascinating feature of Elvis Presley, according to the Rolling Stones' Mick Jagger, was his androgynous appearance.[21] During this early period, though, the most sexually outrageous rock-and-roll performer was Little Richard (Penniman). Little Richard has said that given the reality of American racism, he blurred the lines between masculinity and femininity because he feared the consequences of becoming a Black sex symbol for white girls: "I decided that my image should be crazy and way out so that adults would think I was harmless. I'd appear in one show dressed as the Queen of England and in the next as the pope."[22] Little Richard's playful blurring of gender identity and sexual orientation paved the way for performers like David Bowie, Elton John, Boy George, Annie Lennox, Prince, Grace Jones, Marilyn Manson, Lady Gaga, and Harry Styles.

The North and the South

Rock and roll also combined northern and southern influences. In the 1950s, less than a century after the Civil War, the North and South were still regions with very distinct cultures. Things that make, for example, Atlanta feel similar to Philadelphia today—such as the interstate highway system, familiar restaurant and hotel chains, and national consumer product brands—hardly existed then. The spread of rock and roll began to break down those cultural boundaries, not only saturating the North with

ELVIS PRESLEY AND HIS LEGACY Elvis Presley remains one of the most popular solo artists of all time. From 1956 to 1962, he recorded seventeen No. 1 hits, from "Heartbreak Hotel" to "Good Luck Charm." According to Little Richard, Presley's main legacy was that he opened doors for many young performers and made Black music popular in mainstream America.

Bettmann/Getty Images

southern blues and R&B but introducing northern styles to southern artists along the way. Musicians such as Carl Perkins and Buddy Holly—both from the South—toured the North, bringing their distinctive sounds to new audiences. Meanwhile, musicians and audiences in the North absorbed blues music as their own, making it not just a specifically southern style. Both the North and the South contributed to the development of rock and roll in the 1950s. For example, Chess Records and Vee-Jay Records in Chicago launched blues and R&B crossover artists, while Sun Records and Stax Records recorded rockabilly and soul crossovers in Memphis.

But the key to record sales and the spread of rock and roll across both the North and the South, according to famed record producer Sam Phillips of Sun Records, was to negotiate the boundaries by finding a white man who sounded Black. Phillips found that man in Mississippi-born and Memphis-bred Elvis Presley. Commenting on Presley's cultural importance, one critic wrote, "White rockabillies like Elvis took poor white southern mannerisms of speech and behavior deeper into mainstream culture than they had ever been taken."[23]

The Sacred and the Secular

Although many mainstream adults in the 1950s complained that rock and roll's sexuality and questioning of moral norms constituted an offense against God, many early rock figures actually had close ties to religion, often transforming gospel tunes into rock and roll. Still, many people did not appreciate crossing the boundaries between the sacred and the secular. In the late 1950s, public outrage over rock and roll was so great that even Little Richard and Jerry Lee Lewis, both sons of southern preachers, became convinced that they were playing the "devil's music." By 1959, Little Richard had left rock and roll to become a minister. Lewis had to be coerced into recording "Great Balls of Fire," a song by Otis Blackwell that turned an apocalyptic biblical phrase into a sexually charged teen love song. The boundaries between sacred and secular music have blurred in the years since, with some churches using rock and roll to appeal to youth, and some Christian-themed rock groups recording music as seemingly incongruous as heavy metal.

Pushback against Early Rock and Roll

The long-standing racial divisions and other conventional boundaries meant that performers and producers were forced to play a tricky game to get rock and roll accepted by the masses. Two prominent white disc jockeys had different methods for achieving this end. Cleveland deejay Alan Freed, credited with popularizing the term *rock and roll*, played original R&B recordings from the race charts and Black versions of early rock and roll on his program. In contrast, Philadelphia deejay Dick Clark believed that making Black music acceptable to white audiences required cover versions by white artists. By the mid-1950s, rock and roll was gaining acceptance among the masses, though Black artists were finding that their music was often undermined by the white cover versions. At the same time, rock-and-roll artists and promoters began to face pushback against the cultural changes brought by the new music, as fears of rock and roll as a contributing factor in juvenile delinquency resulted in censorship.

White Cover Music Undermines Black Artists

A song recorded or performed by someone other than the original writer or artist is called **cover music**, and by the mid-1960s, Black and white artists routinely covered each other's original tunes. For example, established Black R&B artist Otis Redding covered the Rolling Stones' "Satisfaction," Jimi Hendrix covered Bob Dylan's "All Along the Watchtower," and just about every white rock-and-roll band—including the Beatles and the Rolling Stones—established its career by covering R&B classics.

Although commercial cover recordings are common now, back in the 1950s the covering of Black artists' songs by white musicians was almost always an attempt to capitalize on popular songs from the R&B race charts by transforming them into hits on the white pop charts. Often, not only would

Michael Ochs Archives/Getty Images

Bettmann/Getty Images

IN 1955, LITTLE RICHARD *(LEFT)* wrote and recorded his first major hit record, "Tutti Frutti." The next year, Pat Boone *(right)* recorded a cover of "Tutti Frutti" that surpassed the original in popularity, reaching No. 12 on the *Billboard* Top 40 (Little Richard's original peaked at No. 17). In a 1984 interview with the *Washington Post*, Little Richard argued that this difference reflected the racial attitudes of the time, despite his attempts to be a nonthreatening Black man: "The white kids would have Pat Boone up on the dresser and me in the drawer 'cause they liked my version better, but the families didn't want me because of the image that I was projecting."[24]

white producers give cowriting credit to white performers for the tunes they merely covered, but the producers would also buy the rights to potential hits from Black songwriters, who seldom saw a penny in royalties or received songwriting credit.

During this period, Black R&B artists, working for small record labels, saw many of their popular songs being covered by white artists working for major labels. These cover records, boosted by better marketing and ties to white deejays, usually outsold the original Black versions. For instance, the 1954 R&B song "Sh-Boom," written and performed by the Chords on Atlantic's Cat label, was immediately covered by a white group, the Crew Cuts, for the major Mercury label. Record sales declined for the Chords, although jukebox and R&B radio play remained strong for the original version. By 1955, R&B hits regularly crossed over to the pop charts, but inevitably the white cover versions were more successful. Pat Boone's cover of Fats Domino's "Ain't That a Shame" went to No. 1 and stayed on the Top 40 pop chart for twenty weeks, whereas Domino's original made it only to No. 10. Slowly, however, the cover situation changed. After watching Boone outsell his song "Tutti Frutti" in 1956, Little Richard wrote "Long Tall Sally," which included lyrics written and delivered in such a way that he believed Boone would not be able to adequately replicate them. "Long Tall Sally" went to No. 6 for Little Richard and charted for twelve weeks; Boone's version got to No. 8 and stayed there for nine weeks.

Overt racism lingered in the music business well into the 1960s. A turning point, however, came in 1962, the last year that Pat Boone, then aged twenty-eight, ever had a Top 40 rock-and-roll hit. That year, Ray Charles covered "I Can't Stop Loving You," a 1958 country song by the Grand Ole Opry's Don Gibson. This marked the first time that a Black artist, covering a white artist's song, had notched a No. 1 pop hit. In fact, the relative acceptance of Black crossover music provided a more favorable cultural context for the political activism that spurred important Civil Rights legislation in the mid-1960s.

Fears of Rock and Roll as a Corrupting Influence Lead to Censorship

Since rock and roll's inception, one of the uphill battles the genre faced was the perception that it was a cause of juvenile delinquency, which was statistically on the rise in the 1950s. Looking for an easy culprit rather than considering contributing factors such as neglect, the rising consumer culture, or the growing youth population, many people assigned blame to rock and roll. The view that rock and roll corrupted youth was widely accepted by social authorities, and rock-and-roll music was often censored, eventually even by the industry itself.

By late 1959, many key figures in rock and roll had been subdued. Jerry Lee Lewis was exiled from the industry, labeled southern "white trash" for marrying his thirteen-year-old third cousin; Elvis Presley, having already been censored on television, was drafted into the army; Chuck Berry was run out of Mississippi and eventually jailed for gun possession and transporting a minor across state lines; and Little Richard felt forced to tone down his image and left rock and roll to become a minister and sing gospel music. A tragic accident dealt an even more devastating blow to rock and roll's first front line. In February 1959, Buddy Holly ("Peggy Sue"), Ritchie Valens ("La Bamba"), and the Big Bopper ("Chantilly Lace") all died in an Iowa plane crash—a tragedy mourned in Don McLean's 1971 hit "American Pie" as "the day the music died."

Although rock and roll did not die in the late 1950s, the U.S. recording industry decided that it needed a makeover. To protect the enormous profits the new music had been generating, record companies responded to the pushback by disciplining some of rock and roll's rebellious impulses. In the early 1960s, the industry introduced a new generation of clean-cut white singers championed by deejay Dick Clark, like Frankie Avalon, Connie Francis, Ricky Nelson, Lesley Gore, and Fabian. Rock and roll's explosive violations of racial, class, and other boundaries were transformed into simpler generation-gap problems, and the music developed a milder reputation.

Popular Music's Continuing Reinvention

As the 1960s began, rock and roll was tamer and "safer," as reflected in the surf and road music of the Beach Boys and Jan & Dean, but it was also beginning to branch out. For instance, the success of all-female groups, such as the Shangri-Las ("Leader of the Pack") and the Angels ("My Boyfriend's Back"), challenged the male-dominated world of early rock and roll. In the 1960s and the following decades, popular music went through cultural reformations that significantly changed the industry, including the international appeal of the British Invasion, the development of soul and Motown, the political impact of folk-rock, the experimentalism of psychedelic music, the rejection of music's mainstream by the punk and indie rock movements, and the reassertion of Black style in hip-hop.

The British Are Coming!

The global trade of pop music is evident in the exchanges and melding of rhythms, beats, vocal styles, and musical instruments across cultures. The origin of this global impact can be traced to England in the late 1950s, when the young Rolling Stones listened to the blues of Robert Johnson and Muddy Waters, and the young Beatles tried to imitate Chuck Berry and Little Richard.

Until 1964, rock-and-roll recordings had traveled on a one-way ticket to Europe. Even though American artists regularly reached the top of the charts overseas, no British performers

THE BEATLES Ed Sullivan, who booked the Beatles several times on his TV variety show in 1964, helped promote the band's early success. On their first appearance on the show, on Sunday night, February 9, 1964 (depicted in this photo), the Beatles performed five songs to a screaming live audience and a home audience of more than seventy-three million viewers—about 38 percent of the nation's population. The Beatles would appear on the Sullivan show live two more times in their career, and several more times in prerecorded performances.

had yet achieved the same feat in the States. This changed almost overnight. Following the Beatles' journey to America in 1964, British bands as diverse as the Kinks, the Rolling Stones, the Animals, and the Who hit the American Top 40 charts.

With the British Invasion, "rock and roll" unofficially became "rock," sending popular music and the industry in two directions. On the one hand, the Rolling Stones would influence generations of musicians emphasizing gritty, chord-driven, high-volume rock, including those in the glam rock, hard rock, punk, heavy-metal, and grunge genres. On the other hand, the Beatles would influence countless artists interested in a more accessible, melodic, and softer sound, in such genres as pop rock, power pop, new wave, and indie rock. The success of British groups helped change an industry arrangement in which most pop music was produced by songwriting teams hired by major labels and matched with selected performers. Even more important, the British Invasion showed the recording industry how older American musical forms, especially blues and R&B, could be repackaged as rock and exported around the world.

Motor City Music: Detroit Gives America Soul

Ironically, the British Invasion, which took much of its inspiration from Black influences, drew many white listeners away from a new generation of Black performers. Gradually, however, throughout the 1960s, Black singers like James Brown, Aretha Franklin, Otis Redding, Ike and Tina Turner, and Wilson Pickett found large and diverse audiences. Transforming the rhythms and melodies of older R&B and early rock and roll into what became known as **soul**, these artists countered the British invaders with powerful vocal performances. Mixing gospel and blues with emotion and lyrics drawn from the American Black experience, soul contrasted sharply with the emphasis on loud, fast instrumentals and lighter lyrical concerns that characterized much of rock music.[25]

The most prominent independent label that nourished soul and Black popular music was Motown, established in 1959 by former Detroit autoworker and songwriter Berry Gordy with a $700 investment and named after Detroit's "Motor City" nickname. Beginning with Smokey Robinson and the Miracles' 1960 hit "Shop Around," Motown enjoyed a long string of hit records that rivaled the pop success of British bands throughout the decade. Motown's many successful artists included the Temptations ("My Girl"), Martha and the Vandellas ("Heat Wave"), Marvin Gaye ("I Heard It through the Grapevine"), and, in the early 1970s, the Jackson 5 ("ABC"). But the label's most successful group was the Supremes, featuring Diana Ross, which scored twelve No. 1 singles between 1964 and 1969 ("Where Did Our Love Go," "Stop! In the Name of Love"). The Motown groups had a more stylized, softer sound than the grittier southern soul (later known as funk) of Brown and Pickett.

THE SUPREMES One of the most successful groups in rock-and-roll history, the Supremes started out as the Primettes in Detroit in 1959. The group signed with Motown's Tamla label in 1960 and became the Supremes in 1961. Between 1964 and 1969, the group recorded twelve No. 1 hits, including "Where Did Our Love Go," "Baby Love," "Come See about Me," "Stop! In the Name of Love," "I Hear a Symphony," "You Can't Hurry Love," and "Someday We'll Be Together." Lead singer Diana Ross *(center)* left the group in 1969 for a solo career. The group was inducted into the Rock and Roll Hall of Fame in 1988.

Folk and Psychedelic Music Reflect the Times

Popular music has always been a product of its time, so the social upheavals of the Civil Rights movement, the women's movement, the environmental movement, and the Vietnam War naturally brought social concerns into the music of the 1960s and early 1970s. By the late 1960s, the Beatles had transformed from a relatively lightweight pop band to one that spoke for the social and political concerns of its generation, and many other groups followed the same trajectory. (To explore how the times and personal taste influence music choices, see "Media Literacy and the Critical Process: Music Preferences across Generations".)

Folk Inspires Protest

The musical genre that most clearly responded to the political happenings of the time was folk music, which had long been the sound of social activism. In its broadest sense, **folk music** in any culture refers to songs performed by untrained musicians and passed down mainly through oral traditions, from the banjo and fiddle tunes of Appalachia to the accordion-led zydeco of Louisiana and the folk-blues of the legendary Lead Belly (Huddie Ledbetter). During the 1930s, folk was defined by the music of Woody Guthrie ("This Land Is Your Land"), who not only brought folk to the city but also was extremely active in social reform. Groups such as the Weavers, featuring labor activist and songwriter Pete Seeger, carried on Guthrie's legacy and inspired a new generation of singer-songwriters, including Joan Baez; Arlo Guthrie; Peter, Paul, and Mary; and, perhaps the most influential, Bob Dylan.

BOB DYLAN Born Robert Allen Zimmerman in Minnesota, Bob Dylan took his stage name from Welsh poet Dylan Thomas. He led a folk music movement in the early 1960s with engaging, socially provocative lyrics. He also served as an astute media critic, as is evident in the seminal documentary *Don't Look Back* (1967).

MEDIA LITERACY & THE CRITICAL PROCESS

Music Preferences across Generations

We make judgments about music all the time. Older generations do not like some of the music younger people prefer, and young people often dismiss some of the music of previous generations. Even among our peers, we have different musical tastes and often reject certain kinds of music that have become too popular or that do not conform to our preferences. The following exercise aims to understand musical tastes beyond our individual choices. Be sure to include yourself in this project.

1 DESCRIPTION

Arrange to interview four to eight friends or relatives of different ages about their musical tastes and influences. Devise questions about what music they listen to now and have listened to at different stages of their lives. What stories or vivid memories do they relate to particular songs or artists? What music or song was the most meaningful to them as they came of age? What kinds of music, if any, did they reject? Have they ever bought or collected music formats, like records or CDs? Collect demographic and consumer information: age, gender, occupation, educational background, place of birth, and current place of residence.

2 ANALYSIS

Chart and organize your results. Do you recognize any patterns emerging from the data or stories? What kinds of music did your interview subjects listen to when they were younger? What kinds of music do they listen to now? What formed/influenced their musical interests? If their musical interests changed, what happened? (If they stopped listening to music, note that and find out why.) Do they have any associations between music and their everyday lives? Are these music associations and lifetime interactions with songs and artists important to them?

3 INTERPRETATION

Based on what you have discovered and the patterns you have charted, determine what the patterns mean. Does age, gender, geographic location, or education matter when it comes to musical tastes? Are the changes in musical tastes over time significant? Why or why not? What kind of music is most important to your subjects? Finally, and most important, why do you think your subjects' music preferences developed as they did?

4 EVALUATION

Determine how your interview subjects came to like particular kinds of music. What constitutes "good" and "bad" music for them? Did their ideas change over time? How? Are they open- or closed-minded about music? How do they form judgments about music? What criteria did your interview subjects offer for making judgments about music? Do you think their criteria are a valid way to judge music?

5 ENGAGEMENT

To expand on your findings, consider the connections of music across generations, geography, and genres. Search online for *music mapper* sites; on one of these sites, input an artist you like, then use the output of related artists to discover new bands. Search for favorite artists of the people you interviewed in Step 1, and share the results with them. Expand your musical tastes.

Significantly influenced by the blues, Dylan identified folk as "finger-pointing" music that addressed the social circumstances of the day. At a key moment in popular music's history, Dylan walked onstage at the 1965 Newport Folk Festival fronting a full electric rock band. He was booed and cursed by traditional "folkies," who saw amplified music as a sellout to the commercial recording industry. However, Dylan's change inspired the formation of amplified folk music, or **folk-rock**, and led millions to protest during the turbulent 1960s.

Rock Turns Psychedelic

Alcohol and drugs have long been associated with the private lives of musicians, but these links became much more public in the late 1960s and early 1970s, when authorities busted members of the Rolling Stones and the Beatles. With the increasing role of drugs in youth culture and the availability of LSD (not illegal until the mid-1960s), an increasing number of rock musicians experimented with and sang about drugs in what was frequently labeled rock's *psychedelic* era, named for the mind-altering effects of LSD and other drugs. Many groups and performers, including Jefferson Airplane, the Jimi Hendrix Experience, the Doors, and the Grateful Dead, believed that artistic expression and responses to social problems could be enhanced through mind-altering drugs.

But following the surge of optimism that culminated in the historic Woodstock concert in August 1969, the psychedelic movement was quickly overshadowed. In 1969, a similar concert at the Altamont racetrack in California started in chaos and ended in tragedy when one of the Hell's Angels hired as a bodyguard for the show murdered a concertgoer in front of the Rolling Stones' stage. Around the same time, the shocking multiple murders committed by the Charles Manson "family" cast a negative light on hippies, drug use, and psychedelic culture. Then, in quick succession, a number of the psychedelic movement's greatest stars died from drug overdoses, including Janis Joplin, Jimi Hendrix, and Jim Morrison of The Doors.

Punk and Indie Respond to Mainstream Rock

By the 1970s, rock music was increasingly viewed as just another part of mainstream consumer culture. With major music acts earning huge profits, rock soon became another product line for manufacturers and retailers to promote, package, and sell—primarily to middle-class white male teens. Rock musicians like Bruce Springsteen and Elton John; glam artists like David Bowie, Lou Reed, and Iggy Pop; and soul artists like Curtis Mayfield and Marvin Gaye continued to explore the social possibilities of rock or at least keep its legacy of outrageousness alive. But they had, for the most part, been replaced by "faceless" supergroups, like REO Speedwagon, Styx, Boston, and Kansas. By the late 1970s, rock could only seem to define itself by saying what it was not; "Disco Sucks" became a standard rock slogan against the popular dance music of the era.

Punk Revives Rock's Rebelliousness

Punk rock rose in the late 1970s to challenge the orthodoxy and commercialism of the record business. By this time, the glory days of rock's competitive independent labels had ended, and rock music was controlled by just a half dozen major companies. By avoiding rock's consumer popularity, punk attempted to return to the basics of rock and roll: simple chord structures, catchy melodies, and politically or socially challenging lyrics.

The punk movement took root at a small dive bar in New York City, CBGB, around such bands as the Ramones, Blondie, and Talking Heads. (The roots of punk essentially lay in four pre-punk groups from the late 1960s and early 1970s—the Velvet Underground, the Stooges, the New York Dolls, and MC5—none of which experienced commercial success in their day.) Punk quickly spread to England, where a soaring unemployment rate and growing class inequality ensured the success of socially critical rock. Groups like the Sex Pistols, the Clash, the Buzzcocks, and Siouxsie and the Banshees sprang up and even scored Top 40 hits on the U.K. charts.

Punk was not a commercial success in the United States, where it was (not surprisingly) shunned by radio. However, punk's contributions continue to be felt. Punk broke down the "boys' club" mentality of rock, launching unapologetic and unadorned front women like Patti Smith, Joan Jett, Debbie Harry, and Chrissie Hynde, and it introduced all-women bands (writing and performing their own music) like the Go-Go's into the mainstream. It also reopened the door to rock experimentation at a time when the industry had turned music into a purely commercial enterprise. The influence of experimental, or post-punk, music is still felt today in alternative and indie bands such as the Yeah Yeah Yeahs, Joyce Manor, and Guerilla Toss.

Indie Groups Reinterpret Rock

Taking the spirit of punk and updating it, indie groups emerged from the do-it-yourself approach of independent labels and created music that found its audience in live shows and on alternative-format college radio stations beginning in the 1980s. Groups often associated with early indie rock include R.E.M., the Cure, Sonic Youth, the Pixies, the Minutemen, and Hüsker Dü.

Scott Dudelson/Getty Images

KISHI BASHI is the stage name of singer, songwriter, and multi-instrumentalist Kaoru Ishibashi. A member of several indie bands—including Jupiter One and Of Montreal—Ishibashi has performed at major festivals, including SXSW and Austin City Limits. His original songs have been licensed in major commercials for Microsoft, Sony, and Smart USA.

In the Pacific Northwest, a subgenre called **grunge** emerged in the 1980s. After years of limited commercial success, grunge broke into the American mainstream with Nirvana's "Smells Like Teen Spirit" on the 1991 album *Nevermind*. Nirvana opened the floodgates to other "alternative" bands, such as Green Day, Pearl Jam, Sound-garden, Stone Temple Pilots, and Sleater-Kinney.

Mainstream attention illustrates a key dilemma for successful indie acts: that their popularity results in commercial success, a situation that their music often criticizes. Still, independent acts like Arcade Fire, Vampire Weekend, and Belle and Sebastian are among many that have launched and sustained successful recording careers built on independent labels, playing concerts and using the Internet and social media to promote their music and sell merchandise.

Hip-Hop Redraws Musical Lines

With the dominance of mainstream rock by white male performers, the place of Black artists in the rock world diminished from the late 1970s onward. These trends, combined with the rise of "safe" disco dance music by white bands (the Bee Gees), Black artists (Donna Summer), and integrated groups (the Village People), created a space for a new sound to emerge: **hip-hop**, a term for the Black urban culture that includes *rapping*, *cutting* (or *sampling*) by deejays, break dancing, street clothing, poetry slams, and graffiti art.

In the same way that punk opposed commercial rock, hip-hop stood in direct opposition to the polished, professional, and often less political world of dance music. Its combination of social politics, swagger, and confrontational lyrics carried forward long-standing traditions in blues, R&B, soul, and rock and roll. Like punk and early rock and roll, hip-hop was driven by a democratic, nonprofessional spirit and was cheap to produce, requiring only a few mikes, speakers, amps, turntables, and vinyl records. Deejays, like the pioneering Jamaican émigré Clive Campbell (a.k.a. DJ Kool Herc), emerged first in New York, scratching and re-cueing old reggae, disco, soul, and rock albums. These deejays, or MCs (masters of ceremony), used humor, boasts, and trash talking to entertain and keep the peace at parties.

The music industry initially saw hip-hop as a novelty, despite the enormous success of the Sugarhill Gang's "Rapper's Delight" in 1979 (which sampled the bass beat of a disco hit from the same year, Chic's "Good Times"). Then, in 1982, Grandmaster Flash and the Furious Five released "The Message" and infused hip-hop with a political take on the stress of poverty, a tradition continued by artists like Public Enemy and Ice-T. By 1985, hip-hop had exploded as a popular genre with the commercial successes of groups like Run-DMC, the Fat Boys, and LL Cool J. That year, Run-DMC's album *Raising Hell* became a major crossover hit, the first No. 1 hip-hop album on the popular charts (thanks in part to a collaboration with Aerosmith on a rap version of the group's 1976 hit "Walk This Way"). But because most major labels and many Black radio stations rejected the rawness of hip-hop, the music spawned hundreds of new independent labels. Although initially dominated by male performers, hip-hop was open to women, and some—Salt-N-Pepa, Queen Latifah, and Missy Elliott among them—quickly became major players. Soon, white artists like the Beastie Boys, Limp Bizkit, and Kid Rock were combining hip-hop and rock in a commercially successful way, while Eminem found enormous success emulating Black rap artists.

Hip-hop encompasses many different styles, including various Latin and Asian offshoots, and today often takes the form of more danceable music that combines singing and rapping with musical

KENDRICK LAMAR made history in 2018 when he became the first hip-hop musician—and, in fact, the first nonclassical or non-jazz musician—to win a Pulitzer Prize when his album *Damn* took home the award in April of that year.

CARDI B, the New York City native born Belcalis Marlenis Almánzar, first got noticed on social media and then on reality TV. Atlantic Records released her first commercial single, "Bodak Yellow," in 2017, which became a No. 1 hit. Since that time, she has earned several more chart-topping hits and become the top female rap artist.

elements of rock, soul, R&B, and even gospel. By 2017, R&B/hip-hop had surpassed rock to become the top music genre in the United States, with the majority of the top recordings and songs coming from that genre, and continued to be the top genre with music consumers in the following years.[26] Contemporary hip-hop stars include artists like Vince Staples, Kendrick Lamar, Chance the Rapper, and Lizzo, as well as megastars such as Drake, Kanye West, Daddy Yankee, Nicki Minaj, and Cardi B.

Pop in the Age of Music Fragmentation

In the past decade, popular music has been influenced by the rise of electronic dance music (EDM) and hip-hop, as DJs-remixers-producers like David Guetta, Skrillex, Calvin Harris, Zedd, and WondaGurl have collaborated with a number of popular vocalists, using an experimental approach that makes beats and mixes across all genres. Similarly, streaming services such as Spotify, Apple Music, and Deezer have greatly expanded accessibility to music and new remixes. The digital formats in music have resulted in a leap in viability and market share for independent labels and have changed the cultural landscape of the music industry in the twenty-first century. They have also encouraged the development of strong international music artists outside the U.S.-British scene, which dominated the production of much of the popular music of the twentieth century (see "Global Village: Aya Nakamura: France's Global Pop Star").

Aya Nakamura: France's Global Pop Star

BY LIV MARTIN

S tep into any nightclub in Paris, France, and there's a high chance you'll hear the energetic dance song "Djadja." The first major hit from French Malian singer Aya Nakamura, "Djadja" has become wildly popular since its release in 2018, cementing Nakamura's status as a global Afropop sensation.

According to the French newspaper *Le Figaro*, Nakamura's "Djadja" is the first song by a French female artist to take European music charts by storm since the work of Édith Piaf (singer of the famous "La vie en rose"), who was active from the 1930s to the 1960s.[1] The song's YouTube video has also logged more than half a billion views. With her success, Nakamura has become what the French music industry has historically failed to produce: a Black female pop icon, well on her way to becoming the French equivalent of influential stars like Rihanna or Cardi B.

Born Aya Danioko in Mali and raised in the outskirts of Paris, Nakamura grew up surrounded by music. Her mother is a griot—part of a West African community of musicians, story-tellers, and poets who pass down oral history. But Nakamura has always made it clear that she is set on carving out a new genre and persona that is authentically her own.

One way that Nakamura has forged her unique identity as an artist is through her evocative lyrics. "Djadja"—often interpreted as a female empowerment anthem—tracks her experience with a friend who falsely claims the two had a romantic relationship. Her second big hit, "Pookie," also showcases her strong and powerful attitude, making it clear that she will not debase herself for a man.

Nakamura's approach to the French language is wholly orig-inal, too. Though mainly written in French, Nakamura infuses her lyrics with phrases from English, Spanish, Arabic, and the Malian language of Bambara. "I sing exactly how I would talk with my friends," she told *Fader* magazine.[2]

Nakamura's rise to fame, like that of many other artists of this generation, began online. She first gained traction in 2014 on social media after posting early songs like "Karma" and "J'ai mal" (I'm hurting). Her career began to take off with the release of her single "Brisé" (Broken)—a track that draws from 2000s-era R&B and showcases Nakamura's vocal range—the following year. In 2017, she released her first album, *Journal intime* (Personal diary), which went to No. 6 on the French charts and was certified gold.

Her second album, 2018's *Nakamura*, marked a new level of success. The album scored seven Top 10 hits on the French singles chart—a feat never before accomplished by a female artist. "Djadja," first released as a single in April 2018, was an overnight success, reaching No. 1 in its first week on the French charts. It was followed by "Copines" (Girlfriends), released in August 2018, which also climbed the French single charts, eventually reaching No. 1.

In *Nakamura*, her label, Warner Music France, found an artist who appeals not only to French speakers but to listeners throughout Europe and West Africa. "[Nakamura's music] resonated outside of France very quickly. Our colleagues in neighbouring countries were asking us about her story; they could feel there was something happening," said Alain Veille, managing director of digital at Warner Music France.[3]

In the history of French pop, which has overwhelmingly favored white musicians, Aya Nakamura's success as a Black woman marks a potentially groundbreaking new phase for the industry. "It's hard when you're a black woman in this industry, *period*," Nakamura said. "People would ask me to bleach my skin or wear lighter founda-tion to appeal to a broader audience, but I didn't let that stop me at all."[4]

Liv Martin is an arts journalist based in Minneapolis. ●

The Business of Sound Recording

For many in the recording industry, the relationship between music's business and its artistic elements is an uneasy one. The lyrics of hip-hop or punk rock, for example, often question the commercial value of popular music. But the line between commercial success and artistic expression is hazier than simply arguing that the business side is driven by commercialism and the artistic side is free of commercial concerns. The truth, in most cases, is that the business needs artists who are provocative, original, and appealing to the public, and the artists need the expertise of the industry's marketers, promoters, and producers to hone their sound and reach the public. Both sides stand to make a lot of money from this relationship. But such factors as the enormity of the major labels; the power of the major streaming platforms; and the complexities of making, selling, and profiting from music in an industry that has been shaken up by the digital revolution affect the economics of sound recording.

Global Music Corporations Influence the Industry

After several years of steady growth, revenues for the recording industry experienced significant losses beginning in 2000 as file-sharing began to undercut CD sales (see Figure 4.1). In 2019, U.S. music sales were about $11.1 billion, down from a peak of $14.6 billion in 1999 but growing again as subscription streaming music revenue continued to increase. The global music business also increased to a value of about $21.5 billion in 2019.[27] This growth suggests that the music industry has overcome the damage it sustained from MP3 piracy at the turn of the twenty-first century.

Most of the revenue generated in sound recording in the United States and globally is absorbed by three large music corporations, constituting an **oligopoly**: a business situation in which a few firms control most of an industry's production and distribution resources. This global reach gives these firms influence over what types of music gain worldwide distribution and popular acceptance. However, the rise of independent-label market share has challenged the dominance of the big music corporations.

Fewer Major Corporations and Falling Market Share

From the 1950s through the 1980s, the music industry, though powerful, consisted of a large number of competing music corporations (which can own dozens of label subsidiaries), along with many smaller independent labels, which aren't aligned with the big music corporations. (Music businesses are often called *labels* because the company brand name was typically included on the center label of a vinyl recording.)

Over time, the major music corporations began swallowing up the independents and then buying one another. By 1998, only six major corporations remained—Universal, Warner, Sony, BMG, EMI, and Polygram. After a series of acquisitions and mergers, by 2012 only Universal Music Group, Warner Music Group, and Sony Music Entertainment remained. Together, these three companies now control about 67.5 percent of the global recording industry market share (see Figure 4.2). Although their revenue has eroded over the past decade, the major music corporations still wield great power, with a number of major artists under contract and enormous back catalogs of recordings that continue to sell. Despite the oligopoly in the music industry, however, the biggest change has been the growth in market share for independent music labels.

The Indies and Artists Direct Grow with Digital Music

The rise of rock and roll in the 1950s and early 1960s showcased a rich diversity of independent labels—including Sun, Stax, Chess, and Motown—all vying for a share of the new music. That tradition lives on today. In contrast to the three global

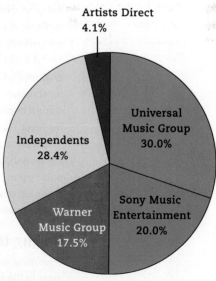

FIGURE 4.2

GLOBAL MARKET SHARE OF THE MAJOR LABELS IN THE RECORDING INDUSTRY, 2019

Data from: Mark Mulligan, "Recorded Music Revenues Hit $21.5 Billion in 2019," MIDiA, March 5, 2020, www.midiaresearch.com /blog/recorded-music-revenues-hit-215-billion-in-2019. Figures are rounded.

players, some five thousand large and small independent production houses—or **indies**—record music that appears to be less commercial. Indies require only a handful of people to operate them.

For years, indies accounted for just 10 to 15 percent of all music releases. But with the advent of downloads, streaming, and social media, independent-label music was able to circumvent the old system of radio airplay and retail store sales and reach listeners more directly, resulting in the market share of indies growing to nearly one-third of the U.S. recording industry. Indies often still have business relationships with major labels; most independent labels use major labels to distribute their music (not unlike how independent film companies rely on major studios for distribution). Independent labels have produced some of the best-selling artists of recent years; these labels include Big Machine Records (Sugarland, Rascal Flatts), Broken Bow Records (Jason Aldean, Zac Brown Band), Dualtone Records (the Lumineers), XL Recordings (Adele, Radiohead), and Cash Money Records (Drake, Nicki Minaj).

Interestingly, the fastest-growing segment of the sound recording industry is categorized as **artists direct**—artists who bypass all label representation, including indies, and sell their music directly to their audience, usually through streaming services. For example, Michael Brun, the Haitian artist and producer based in New York, releases music on his own Kid Coconut label and thus receives 100 percent of his royalties. Artists direct now constitutes more than 4 percent of the market.

Making, Selling, and Profiting from Music

Like most mass media, the music business is divided into several areas, each working in a different capacity. In the music industry, those areas are making the music (signing, developing, and recording the artist), selling the music (selling, distributing, advertising, and promoting the music), and dividing the profits.

Making the Music

Labels are driven by **A&R (artist & repertoire) agents**, the talent scouts of the music business, who discover, develop, and sometimes manage artists. A&R agents seek out and listen to demonstration tapes, or *demos*, from new artists and decide whom to sign and which songs to record. (Today, demos are typically in digital form.) Naturally, these agents look for artists with commercial potential.

A typical recording session is a complex process that involves the artist, the producer, the session engineer, and audio technicians. In charge of the overall recording process, the producer handles most nontechnical elements of the session, including reserving studio space, hiring session musicians if necessary, and making final decisions about the sound of the recording. (As noted earlier, however, top DJs can serve as producers, remixers, and even songwriters for other recording artists.) The session engineer oversees the technical aspects of the recording session, everything from choosing recording equipment to managing the audio technicians. Most popular records are recorded part by part. Using separate microphones, the vocalists, guitarists, drummers, and so on, are digitally recorded onto separate audio tracks, which are edited and remixed during postproduction and ultimately mixed down to a two-track stereo master copy for digital reproduction.

Selling the Music

Streaming services like Spotify, Apple Music, Google Play, and Amazon are the leading revenue generators in the music industry today and account for 79 percent of the U.S. music business.[28] As recently as 2011, physical recordings (CDs and some vinyl) accounted for nearly 50 percent of U.S. music sales (see Figure 4.1), but CD sales continue to decline and now constitute about 5.5 percent of the U.S. market. In some countries, however—such as Japan, Poland, and Germany—CDs retain a much larger market share. Vinyl album sales have carved out a successful and growing niche as a classic format in the United States, accounting for about 4.5 percent of industry revenues. Digital downloads of singles and albums are about 8 percent of the market, down from the days in the early years of the century, when the iTunes store was the dominant seller of music.[29]

Dividing the Profits

When music is sold, money gets divided among the music label, the artists, the songwriters (who may also be the performing artists), and the song publisher (the publisher helps manage the song's copyright). Although streaming is the main format for music today, the system of royalties was built on the traditional notion of selling physical recordings—albums, tapes, or CDs. There are four different types of royalties that are paid, which help determine how the profits are divided:

- **Master recording royalties**. The music label retains the bulk of the profits from a recording but pays the artist/band royalties (a percentage of the profits) as determined by the artist's/band's contract with the label. With traditional formats like CDs, new artists would typically negotiate a master recording royalty rate of between 8 and 12 percent on the retail price of a CD, while more established performers might negotiate for 15 percent or higher. An artist who has negotiated an 11 percent royalty rate would earn about $1.31 for an $11.99 CD sold in a store. (Streaming situations are described later in the section.) Artists who sign with a label may get an *advance*, which is an up-front lump-sum payment that later gets deducted from what they earn in royalties. If artists are operating as artists direct (that is, they are not under contract with a music label), they own the recording copyright themselves and get the full share of sales profits.
- **Public performance royalties**. These are royalties paid out to songwriters and their publishers when music (live or recorded) is played publicly on radio; on television (for example, a performance on *Saturday Night Live*); streaming; or in stores, bars, restaurants, and even stadiums and arenas. Performance rights organizations like ASCAP, BMI, and SESAC collect the royalties and pay the songwriters and publishers. (Your college or university likely has contracts with these agencies to cover the public performance of all music on campus.)
- **Mechanical royalties**. These are royalties paid automatically to songwriters and publishers when music is sold or streamed, and the royalty rate is always 9.1 cents per song. The term *mechanical* originated in a 1909 copyright law requiring songwriters to be paid when their music was mechanically reproduced on music rolls for player pianos. As the music industry progressed, the requirement extended to all recordings and streams, so songwriters and their publishing partners would be paid. In the United States, the Harry Fox Agency based in New York handles most of the mechanical rights permissions and royalty distributions.
- **Synchronization royalties**. These royalties are for the use of music in a film, TV show, or commercial. Film directors or TV producers must get permission (a "sync" license) and pay a negotiated rate before they can use the music. This royalty is split between the artist and the label (if there is one) and songwriters.

With streaming as the leading music distribution format, figuring out what counts as a sale of a song or an album has become harder to gauge than it was when physical recordings were more common. For this reason, the music industry has developed an equivalency standard, with 1,500 song streams from an album equal to the sale of one album, and 150 song streams equal to the sale of a single.[30] This equivalency comes in handy for the industry in deciding which songs or albums go gold (which means 500,000 units sold) or platinum (one million units sold).

The way in which songs get monetized in music streaming is built on a similar system. First, the streaming services generally make money by selling subscriptions to their services. For example, most of the top streaming services—including Spotify, Apple Music, Amazon Music Unlimited, and Deezer—charge $9.99 a month, or $4.99 a month for college students. Other streaming services offer music for free (the freemium model) but earn money by playing ads (like YouTube and Spotify's free tier). Next, the services pay part of their revenue to the labels, artists, and songwriters whose music has been streamed (see Figure 4.3 for a look at how much of the overall revenue paid to the music industry comes from each major streaming company). Generally, the payout to the music industry is in the 70 percent range. That is, the streaming companies, such as Apple Music, keep roughly 30 percent of their revenue and pay the rest to license the music they stream. Of the 70 percent payout, about 57 percent goes to the record label and performing

FIGURE 4.3

PERCENT OF OVERALL STREAMING REVENUE PAID TO THE MUSIC INDUSTRY BY COMPANY, 2020

Data from: "2019–2020 Streaming Price Bible: YouTube Is STILL the #1 Problem to Solve," *The Trichordist* (blog), March 5, 2020, https://thetrichordist.com/2020/03/05/2019-2020-streaming-price-bible-youtube-is-still-the-1-problem-to-solve.

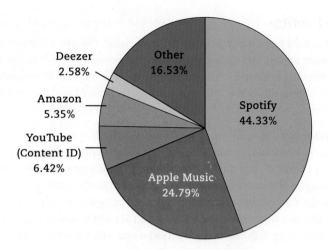

artist (who own the recording), and 13 percent goes to the songwriters and publishers (who own the music) through mechanical royalties.[31] But while streaming services generally pay a certain amount of money per 1,000 streams, that rate actually varies widely depending on the streaming service. For example, Amazon Music Unlimited pays labels $11.23 per 1,000 streams, whereas Apple Music pays $6.75 and Spotify pays $3.48 for the same number of streams. YouTube, which streams more music than any other service, pays only $0.22 per 1,000 streams.[32] This does not sit well with recording artists, who accuse YouTube of "exploiting legal loopholes and shortchanging artists of their fair share."[33] (For more on YouTube, see "Examining Ethics: YouTube: How One of the Richest Companies Shortchanges Music Artists".)

The profits are divided somewhat differently in digital download sales. A $1.29 iTunes download generates about $0.40 for Apple (the company gets 30 percent of every song sale) and the standard $0.09 mechanical royalty for the song publisher and writer, leaving about $0.60 for the record label. Artists at a royalty rate of about 15 percent would get $0.20 from the song download.

Songs streamed on Internet radio (like iHeartRadio) and SiriusXM satellite radio follow yet another formula for determining royalties. In 2003, the nonprofit group SoundExchange was established to collect royalties for Internet radio, charging a fraction of a penny per stream. Large Internet radio stations can pay up to 25 percent of their gross revenue for these fees—less for smaller Internet radio stations and a small flat fee for streaming nonprofit stations. About 50 percent of the fees go to the music label, and the other 50 percent goes to artists. (Internet radio royalties for songwriters and publishers are collected separately by performance rights organizations like ASCAP, BMI, and SESAC.)

Sound Recording, Breaking Barriers, and Democracy

From sound recording's earliest years as a mass medium, when the music industry began stamping out flat records, to the breakthrough of MP3s and Internet-based music services, fans have been sharing music and pushing culture in unpredictable directions. The battles over pop music's controversial aspects speak to the heart of democratic expression. Nevertheless, pop—like other art forms—also has a history of reproducing old stereotypes: limiting women's access as performers and fostering racist and homophobic attitudes.

Popular musical forms that test cultural boundaries face a dilemma: how to uphold a legacy of free expression while resisting giant companies bent on consolidating independents and maximizing profits. Since the 1950s, forms of rock music have been in a recurring pattern of breaking boundaries, becoming commercial, then reemerging as rebellious.

In the past two decades, however, popular music has gone through seismic changes that have taken a great deal of power away from a recording industry that, since the second half of the twentieth century, has attempted to control music by signing only certain artists and promoting those preferred artists through heavy radio airplay. The emergence of the MP3 and file-sharing shook up the industry, and although today there are just three major music corporations, a contract with one of their labels is not the only available path for a successful musical career.

Nevertheless, a powerful music industry is again ascendant, this time on the strength of streaming. And quickly, three of the five major digital media corporations—Apple, Google (YouTube), and Amazon—have added streaming to their already enormous digital ecosystems and are among the top four streaming companies along with Spotify, which leads the pack but lacks the massive holdings that cushion the others. (In fact, Spotify, which was established in 2006, took thirteen years to finally turn a profit.)[34] We can expect that musical artists, some of the media's most independent figures, will continue to call out overcommercialization and exploitation in the music industry. Ironically, a successful music industry is dependent on resistance and innovation, as well as periodic infusions of the diverse sounds that come from ethnic communities, backyard garages, dance parties, and neighborhood clubs.

No matter how it is produced and distributed, popular music endures because it speaks to both individual and universal themes, from a teenager's first romantic adventure to a nation's outrage over social injustice. Music often reflects the personal or political anxieties of a society. It can break down artificial or hurtful barriers and champion a democratic spirit. Writer and free-speech advocate Nat Hentoff addressed this issue in the 1970s when he wrote, "Popular music always speaks, among other things, of dreams—which change with the times."[35] The recording industry continues to capitalize on and spread those dreams globally, but in each generation, musicians and their fans keep imagining new ones.

TAYLOR SWIFT left her first label, independent Big Machine Records, in 2018. But in that contract, she didn't retain ownership of her master recordings. So, in 2019, she revealed plans to re-record her first six albums, which would give her new master versions that she could control. In the first of the re-recordings, she released her 2008 album *Fearless* as the new *Fearless (Taylor's Version)* in 2021.

YouTube: How One of the Richest Companies Shortchanges Music Artists

Streaming music has been the savior of the music industry, rescuing it from a precipitous decline in revenue that began when MP3 file-sharing was released to the world in 1999. In the 2010s, music customers shifted to streaming for their music needs (a monthly fee for access to sixty million tracks sounded like a great deal to most fans), and the billions in music industry revenue began increasing year after year. In fact, by 2020, streaming was earning $1 million every hour for the three major sound recording corporations—Universal, Sony, and Warner.[1]

But for musical artists, transparency in the streaming business model leaves much to be desired. In the pre-streaming world, one could easily count the number of CD units shipped or songs downloaded, and royalties were a calculated rate of these sales. From one music label to the next, the process of paying these royalties was quite similar. But as noted earlier, this is no longer the case: the per-stream rates paid by one company versus another seem to follow no clear rules. The main reason for this variation is that not all streaming environments generate the same amount of revenue. For example, streaming companies generally make more money on paid subscriptions than on free, ad-supported plans. This means that streams in which listeners choose to hear a song (on demand) are more valuable to labels and artists than streams in which listeners of a freemium channel happen to hear a song in a music flow. Also, customers in developed countries pay more for subscriptions than do customers in developing countries—for example, Spotify costs $9.99 a month in the United States but $1.70 a month in India, generating a higher per-stream revenue from U.S. listeners.

Yet even given these differences among streaming company payments, no digital company irritates music labels, artists, and publishers more than YouTube. YouTube has three separate platforms for music payouts (see table). The first platform is YouTube Premium (formerly YouTube Red), a subscription plan for $11.99 a month. For its second platform, the company monetizes per-video streams on artist's YouTube channels. For example, Rihanna has 34.7 million subscribers to her YouTube channel, and views of her videos are collectively in the billions. The third platform, called YouTube Content ID, is the site's most popular. This platform covers user-generated content that includes copyrighted music. (Think of a wedding video that has popular songs in it, for example.)

Before YouTube invented Content ID in 2007, copyright holders would typically ask the site to take down videos that violated their music copyrights. With Content ID, however, copyright holders can have YouTube scan and identify their music on the platform, then have the option of monetizing those uses of their music.[2] Given that five hundred hours of video are uploaded to YouTube *every minute*, YouTube Content ID is its most streamed music platform.[3] It also captures the biggest market share of all the streaming sites, with 51 percent of all music streams (see row 3 of the table). The problem is that the most popular platform for streaming music is also the one that pays the least to the creators of that music—even though it is owned by Google, a company that had $161.9 billion in revenue in 2019, and even though the YouTube subsidiary alone generated $15.1 billion in ad revenue.[4]

In the words of the pro-artist blog *The Trichordist*, "The biggest takeaway by far is that YouTube's Content ID shows a whopping 51% of all streams generate only 6.4% of revenue. Read that again. This is your value gap. Over 50% of all music streams generate less than 7% of revenue."[5]

	Streaming Service	Per-Stream Rate	Share of Total Streams	Share of Total Revenue Paid
1	Spotify	$0.00348	22.09%	44.33%
2	Apple Music	$0.00675	6.36%	24.79%
3	YouTube Content ID	$0.00022	51.00%	6.42%
4	Amazon Unlimited	$0.01123	0.83%	5.35%
5	Deezer	$0.00562	0.80%	2.58%
6	Google Play	$0.00554	0.79%	2.54%
7	Pandora	$0.00203	1.91%	2.24%
8	YouTube	$0.00154	1.90%	1.70%
9	Amazon Music	$0.00426	0.65%	1.60%
10	Facebook	$0.05705	0.05%	1.56%
11	YouTube Premium	$0.01009	0.23%	1.37%
12	Peloton	$0.03107	0.07%	1.28%

Data from: "2019–2020 Streaming Price Bible: YouTube Is STILL the #1 Problem to Solve," *The Trichordist* (blog), March 5, 2020, https://thetrichordist.com/2020/03/05/2019-2020-streaming-price-bible-youtube-is-still-the-1-problem-to-solve.

Conversely, Apple Music accounts for nearly 25 percent of streaming revenue, but only a little over 6 percent of streaming music consumption. The concern with YouTube for the music industry and for artists is fair compensation and the potential for growth. Although YouTube Content ID isn't a typical streaming service like Spotify or Apple Music, there are other atypical music streaming services, like TikTok and Facebook (which license music so users can put the tracks in their personal videos) and Peloton (which licenses music to accompany its video content for its exercise bike network), all of which pay much higher rates than YouTube Content ID.

In 2020, Google reported that "over the past five years, we've paid out over $2 billion to partners who have chosen to monetize their claims using Content ID."[6] This includes film partners as well (if they monetize use of their video clips), so music payouts are only one part of this $2 billion. Given that streaming is worth a total of $8.8 billion *a year* and YouTube Content ID is responsible for over half of the market share of streams, one might expect the site to pay a sizable chunk of the total revenue that comes from streaming sites. By paying less than 7 percent of this revenue instead, YouTube Content ID is clearly shortchanging music artists. ●

Chapter REVIEW 4

LaunchPad
macmillan learning

Visit LaunchPad for *Media & Culture* at **launchpadworks.com** for additional learning tools:

→ QUIZ YOURSELF WITH LEARNINGCURVE

LearningCurve uses gamelike quizzing to help you review important concepts from this chapter. LearningCurve includes multiple-choice questions at different levels of difficulty, hints, feedback for both right and wrong answers, and links to e-book material for easy reference. Which answer choice would you select for the sample question below?

> In 2003, the nonprofit group SoundExchange was established to
>
> ○ mediate royalty disputes for music artists.
>
> ○ provide the public with information about the music industry.
>
> ○ regulate the music streaming industry.
>
> ○ collect royalties for Internet radio.

→ WATCH VIDEO ON KEY CONCEPTS

LaunchPad includes clips from movies, TV shows, online sources, and other media texts, in addition to interviews with media experts and newsmakers. In the videos for this chapter, you'll view working phonographs from years past, take a look at how artists record music today, and consider the latest in streaming.

→ PRACTICE THE CRITICAL PROCESS

Use the Media Literacy Activity to walk through the critical process step-by-step and develop a critical perspective on what media means to you. In this chapter's activity, you'll explore the role of music genres in American society and culture.

→ LEARN KEY TERMS

Review the definitions for this chapter's key terms with LaunchPad's easy-to-use online flash cards. Try the cards in practice mode, or quiz yourself as a way to focus your study efforts. (Page numbers indicate where each term is highlighted within the chapter.)

audiotape, 96
stereo, 96
analog recording, 96
digital recording, 96
compact discs (CDs), 96
MP3, 97
pop music, 99
jazz, 100
blues, 100
rock and roll, 100

rhythm and blues (R&B), 101
rockabilly, 101
cover music, 103
soul, 106
folk music, 107
folk-rock, 108
punk rock, 109
grunge, 110
hip-hop, 110
oligopoly, 113

indies, 114
artists direct, 114
A&R (artist & repertoire) agents, 114
master recording royalties, 115
public performance royalties, 115
mechanical royalties, 115
synchronization royalties, 115

REVIEW QUESTIONS

Use the questions below to revisit key themes and concepts from this chapter.

✶ Need more practice? Access the LearningCurve multiple-choice questions for this chapter, which are available on LaunchPad. All questions include feedback for correct and incorrect answers, hints, and links to e-book material.

Sound Recording Today

1. What two changes have been making an impact on the music industry as it enters the 2020s?

How We Got Here: The Development of Sound Recording

2. The technological configuration of a particular medium sometimes elevates it to mass market status. Why did Emile Berliner's flat disk matter in the history of sound recording as a mass medium? Can you think of other examples in mass media in which the size and shape of the technology have made a difference?

3. How did the music industry attempt to curb illegal downloading and file-sharing?

4. How did sound recording survive the advent of radio?

The Culture of Popular Music

5. How did rock and roll significantly influence the radio and sound recording industries?

6. Although many rock-and-roll lyrics from the 1950s are tame by today's standards, this new musical development represented a threat to many parents and adults at the time. Why?

7. What moral and cultural boundaries were blurred or broken by rock and roll in the 1950s?

8. Why did cover music figure so prominently in the development of rock and roll and the recording industry in the 1950s?

Popular Music's Continuing Reinvention

9. Explain the British Invasion. What was its impact on the recording industry?

10. What domestic music genre brought Black performers back into popular music in the 1960s? How did the music, and its origins, contrast with the British Invasion of the same era?

11. Why did punk rock and hip-hop emerge as significant musical forms in the late 1970s and 1980s? What do their developments have in common, and how are they different?

12. What is often the role of DJs in the post-CD world of popular music?

The Business of Sound Recording

13. What companies control the bulk of worldwide music production and distribution?

14. Why have independent labels and artists direct distribution grown to have a significantly larger market share in the past decade?

15. How do master recording royalties, public performance royalties, mechanical royalties, and synchronization royalties differ from one another?

16. Why are standards that equate a number of streams to the sale of an album or a song important for the music industry?

Sound Recording, Breaking Barriers, and Democracy

17. Why is it ironic that the big digital companies that seek to control the music industry need resistance from musical artists?

QUESTIONING THE MEDIA

Use these critical-thinking questions to reflect on your own experiences with media and apply your understanding of media concepts.

✶ For an interactive experience, questions on topics like these are available for the iClicker student response system. Instructors can download the questions from LaunchPad and use them to poll the class.

1. Music fans used to find out about new music by listening to the radio. Where do you find out about new music today?

2. Think about the role of the 1960s drug culture in rock's history. How are drugs and alcohol treated in contemporary music today?

3. Is it healthy for or detrimental to the music business that streaming has quickly become dominated by three or four big companies? Explain.

4. Do you think signing with a music label helps or hurts musical artists?

5. Consider the platforms through which you most often listen to recorded music (and live music, too). Which of these modes delivers the most money to the artists you enjoy? Which of these modes results in little or no compensation for the artists? How should these modes be fixed to support musical artists?

Popular Radio and the Origins of Broadcasting

5

The iPod was released in 2001, but it's taken about two decades for **podcasts**—spoken-word digital audio programs that can be downloaded or streamed—to become widespread in popularity. More than one million podcasts are now available, and 104 million Americans (37%) listen to podcasts monthly. About sixteen million people in the United States call themselves "avid podcast fans."[1] Just as listeners tuned in during the golden age of radio in the 1930s, when creative storytelling and sound effects transfixed American audiences, podcast listeners today appreciate the intimate and authentic connection they have with their favorite hosts and the simple pleasure they get from listening to stories. These listeners have embraced the fresh perspectives that made sensations out of journalism podcasts *Serial* (2014), *S-Town* (2017), and *1619* (2019), and have made household names out of continuing podcasts like *Pod Save America*, *Reply All*, *My Dad Wrote a Porno*, and NPR's *Alt.Latino*.

Forty-five percent of all podcasts currently address society and culture, with true-crime podcasts (*Criminal, Crime Junkie*) tending to be the most popular. News and politics make up 34 percent of all podcast genres (*Up First, Code Switch*), followed by comedy at 32 percent (*Office Ladies, Call Her Daddy*) and sports at 23 percent (*Spittin Chiclets, The Bill Simmons Podcast*).[2] Listeners have seamlessly integrated podcasts of all types into their lives: while driving, washing dishes, exercising, working, or doing chores at home in the evenings or on weekends. And, not surprisingly, the success of this popular audio format has both challenged and invigorated radio broadcasting.

Just as streaming music services like Spotify have attracted listeners from broadcast radio music formats, podcasts are battling news/talk-radio formats for listeners. In some cases, however, the podcaster and broadcaster are the same, and podcasts actually help to expand the audience for the radio program. National Public Radio (NPR), for example, has been the most successful at making serious inroads in the podcasting space, embracing podcasting as an extension of what radio has been doing since the beginning—creating intimate connections through sound—and making recorded broadcasts (now podcasts) available whenever people have time to listen.

◀ Jasmine Garsd of NPR's podcast *Alt.Latino* interviews musician Ana Tijoux at the SXSW Music + Film + Interactive.

One of the earliest and most popular podcasts is also one of public radio's most popular shows: the hour-long weekly program *This American Life*, hosted by Ira Glass. The program, which features engaging storytelling through original monologues, dramas, and award-winning investigative documentaries, debuted on NPR stations across the country in 1995 and helped launch the careers of essayists-humorists like Sarah Vowell and David Sedaris. In 2006, the show added free podcasts; by 2020, more people—a total of 3.1 million—were listening to each weekly episode via podcast, compared to the 2.2 million listening on the radio.[3]

The conversational style of *This American Life* has influenced a number of other podcasts, including the *New York Times*'s *The Daily* (a program in which the newspaper's reporters focus mainly on one big news story). *The Daily* started in 2017 and is now the No. 1 ranked podcast, with more than two million listeners a day.[4] Interestingly, whereas *This American Life* started as a radio show and later added a podcast, *The Daily* was a podcast first and added a radio program version of the show in 2018.[5] By the end of its first year on radio, more than 150 public radio stations carried *The Daily*.[6]

In a sure sign that podcasting is becoming a big deal, for-profit companies have been muscling in to capture key talent and gain a larger share of the audience. In 2019, Spotify, which is mostly known for music streaming, began to make deep investments in podcasting, buying Gimlet Media (maker of podcasts like *Start Up* and *Reply All*), Anchor (a free app for podcast production and distribution), and Parcast (maker of true-crime podcasts) for a total of nearly $400 million.[7] In 2020, Spotify bought The Ringer (a sports and entertainment podcast company) for close to $200 million, and then reportedly paid $100 million for a multiyear contract to bring *The Joe Rogan Experience*, one of the top podcasts, to the service.[8] Spotify CEO Daniel Ek explained his strategy: "What I didn't know when we launched to consumers in 2008 was that audio—not just music—would be the future of Spotify."[9] It is now clear to the radio and streaming audio industries that podcasts are an essential component of what the listening audience wants.

 Have you listened to podcasts? If so, what is the appeal of podcasts for you, and what kind do you listen to? If not, why haven't you listened to podcasts?

Radio Today

As we will explore in this chapter, the story of radio is the story of three things: wireless technology, a national-international network, and the spoken words or music (both forms of human engagement) that make it all worthwhile. Interestingly, this description could also apply to the Internet and mobile phones, which make use of the same elements—wireless technology (via cell phones and Wi-Fi), a national-international network (via cell towers and the Internet network), and spoken words or music (as content we experience online and on our phones).

Radio and these newer technologies have something else in common: they are constantly evolving. In fact, because the Internet and mobile phones have created new opportunities for mass and masspersonal communication through spoken words and music, the radio industry has been scrambling to expand and migrate audio programming from its old wireless network (which dates back more than a century) to the newer wireless environment of today. Podcasts are just one of those projects. The radio industry is also deeply interested in being on your home digital assistant. (For example, saying "Alexa, play KISS 102.7" would give you L.A.'s top hit music station on an Amazon Echo, while "Alexa, play Top Hip-Hop" would start up one of Amazon Music's automated in-house stations.) Part of the radio industry even built an additional network (now satellite radio's

SiriusXM) to give listeners another way to wirelessly connect to its programming. And SiriusXM itself recently bought streaming music service Pandora in an "if you can't beat them, join them" approach to the newer audio environment.

All of this is having some success. Radio still reaches 92 percent of Americans eighteen and older each week.[10] As the oldest electronic medium—a medium that has given us national networks, sitcoms, the growth of rock and roll, and the fundamental policy of broadcasting in the public interest—radio's continuing prominence has depended on its adaptability. Before the arrival of the Internet and mobile phones, radio had already transformed itself in response to television, which in the 1950s displaced it as the central electronic medium in the home.

In this chapter, we will examine the scientific, cultural, political, and economic factors surrounding radio's development and perseverance. We will:

- Explore the origins of broadcasting, from the early theories about radio waves to the critical formation of RCA as a national radio monopoly
- Probe the evolution of commercial radio, including the rise of NBC as the first network, the development of CBS, and the establishment of the first federal radio legislation
- Review the fascinating ways in which radio reinvented itself in the 1950s
- Examine television's impact on radio programming, the invention of FM radio, radio's convergence with sound recording, and the influence of various formats
- Investigate newer developments, including Internet radio, satellite radio, and podcasting
- Survey the economic health, increasing conglomeration, and cultural impact of commercial and noncommercial radio today, including the emergence of noncommercial low-power FM.

How We Got Here: The Early Development of Radio

Radio did not emerge as a full-blown mass medium until the 1920s, though the technology that made radio possible had been evolving for years. The **telegraph**—the precursor of radio technology—was invented in the 1840s by American inventor Samuel Morse. By interrupting the electrical impulses sent from a transmitter through a cable to a reception point, telegraph operators transmitted a series of dots and dashes that stood for letters in the alphabet—what became known as **Morse code**. By 1861, telegraph lines ran coast to coast, and by 1866, the first transatlantic cable, capable of transmitting about six words per minute, ran between Newfoundland and Ireland along the ocean floor.

Although it was a revolutionary technology, the telegraph had its limitations. It was unable to transmit the human voice, and ships at sea could not use it to communicate with the rest of the world. What was needed was a telegraph without wires.

The Discovery of Radio Waves

The key development in wireless transmissions came from James Maxwell, a Scottish physicist who in the mid-1860s theorized the existence of **electromagnetic waves**: invisible electronic impulses similar to visible light. Maxwell's equations showed that electricity, magnetism, light, and heat are part of the same electromagnetic spectrum and that they radiate in space at the speed of light,

Bettina Fabos/Fortepan Iowa

A TELEGRAPHY CLASS at the Iowa State University, circa 1910. Sending messages using Morse code across telegraph wires was the precursor to radio, which did not fully become a mass medium until the 1920s. Like the Internet, radio was popularized through universities and high schools. Education-related clubs met to build radio transmitters and receivers, and teachers were among the first to develop radio content during the 1910s.

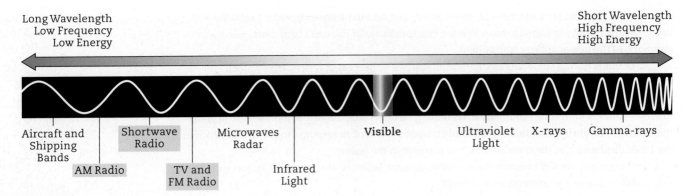

Long Wavelength
Low Frequency
Low Energy

Short Wavelength
High Frequency
High Energy

Aircraft and Shipping Bands

Shortwave Radio

AM Radio

TV and FM Radio

Microwaves Radar

Infrared Light

Visible

Ultraviolet Light

X-rays

Gamma-rays

FIGURE 5.1

THE ELECTROMAGNETIC SPECTRUM

Data from: NASA, "Imagine the Universe!," http://imagine.gsfc
.nasa.gov/docs/science/know_l1
/emspectrum.html.

about 186,000 miles per second (see Figure 5.1). Maxwell further theorized that a portion of these phenomena, later known as **radio waves**, could be harnessed so that signals could be sent from a transmission point to a reception point.

It was German physicist Heinrich Hertz in the 1880s who proved Maxwell's theories. Hertz created a crude device that permitted an electrical spark to leap across a small gap between two steel balls. As the electricity jumped the gap, it emitted waves; this was the first recorded transmission and reception of an electromagnetic wave. Today we might listen to a radio station at 92.3 MHz, or work on a computer with a 3.1 GHz processor; every reference to a *megahertz* or a *gigahertz* refers to the number of cycles per second in electromagnetic waves and memorializes Hertz's famous discovery.

The Inventors of Wireless Telegraphy

A number of inventors around the world worked to invent a way to transmit messages wirelessly by harnessing electromagnetic waves. In 1894, Guglielmo Marconi—a self-educated twenty-year-old Italian engineer—read Hertz's work and set about trying to make wireless technology practical. He attached Hertz's spark-gap transmitter to a telegraph machine, then successfully sent and received dot-dash signals in 1895. In 1896, Marconi traveled to England, where he received a patent for **wireless telegraphy**, a form of voiceless point-to-point communication.

Marconi was not only an inventor but an entrepreneur as well. In London in 1897, he formed the Wireless Telegraph and Signal Company, later known as British Marconi, and began installing wireless technology on British naval and private commercial ships. In 1899, he opened a branch of the company in the United States, nicknamed American Marconi. That same year, he sent the first wireless Morse-code signal across the English Channel to France; and in 1901, he relayed the first wireless signal across the Atlantic Ocean.

History often cites Marconi as the father of radio, but another inventor had been making parallel discoveries about wireless telegraphy in Russia. Unknown to Marconi, Alexander Popov, a professor of physics in Saint Petersburg, successfully sent and received wireless messages in May 1895.[11] Yet Popov was an academic, not an entrepreneur, and after Marconi accomplished a similar feat that same summer, Marconi was the first to apply for and receive a patent.

It is important to note that the work of Popov and Marconi was preceded by that of Nikola Tesla, a Serbian Croatian inventor who immigrated to New York in 1884. Tesla, who also conceived the high-capacity alternating current systems that made worldwide electrification possible, invented a wireless system in 1892 and successfully demonstrated his device a year later.[12] However, Tesla's work was overshadowed by Marconi's. In fact, Marconi used much of Tesla's work in his own developments, and for years Tesla was not associated with the invention of radio. In 1943, a few months after he died

penniless in New York, the U.S. Supreme Court overturned Marconi's wireless patent and deemed Tesla the inventor of radio.[13] Ironically, Tesla's name is probably better known in the United States today than any of the other radio inventors, since entrepreneur Elon Musk used it for Tesla Motors, a company producing electric cars.

NIKOLA TESLA A double-exposed photograph combines the image of inventor Nikola Tesla reading a book in his Colorado Springs, Colorado, laboratory in 1899 with the image of his Tesla coil discharging several million volts.

Wireless Telephony: Radio Gets a Voice

In 1899, inventor Lee De Forest (who, in defiance of other inventors, liked to call himself the father of radio) wrote the first Ph.D. dissertation on wireless technology, building on others' innovations. In 1901, De Forest challenged Marconi, who was covering New York's international yacht races for the Associated Press, by signing up to report on the races for a rival news service. As it turned out, the competing transmitters jammed each other's signals so badly that officials ended up relaying information on the races in the traditional way—with flags and hand signals. The event exemplified a problem that would persist throughout radio's early development: noise and interference from competition for the finite supply of radio frequencies.

In 1902, De Forest set up the Wireless Telephone Company to compete head-on with American Marconi, by then the leader in wireless communication. A major difference between Marconi and De Forest was the latter's interest in wireless voice and music transmissions, later known as **wireless telephony** and, eventually, radio. Although sometimes an unscrupulous competitor (inventor

Reginald Fessenden won a lawsuit against De Forest for using one of his patents without permission), De Forest went on to patent more than three hundred inventions.

De Forest's biggest breakthrough was the development of the Audion, or triode, vacuum tube, which detected radio signals and then amplified them. De Forest's improvements greatly increased listeners' ability to hear dots and dashes—and later speech and music—on a receiver set. His modifications were essential to the development of voice transmission, long-distance radio, and television. In fact, the Audion vacuum tube, which powered radios until the arrival of transistors and solid-state circuits in the 1950s, is considered by many historians to be the beginning of modern electronics. But again, bitter competition taints De Forest's legacy; although De Forest won a twenty-year court battle for the rights to the Audion patent, most engineers at the time agreed that Edwin Armstrong (who later developed FM radio) was the true inventor and disagreed with the U.S. Supreme Court's 1934 decision on the case, which favored De Forest.[14]

The credit for the first voice broadcast belongs to Canadian engineer Reginald Fessenden, formerly a chief chemist for Thomas Edison. Fessenden went to work for the U.S. Navy and eventually for General Electric (GE), where he played a central role in improving wireless signals. On Christmas Eve in 1906, after GE built him a powerful transmitter, Fessenden gave his first public demonstration, sending a voice through the airwaves from his station at Brant Rock, Massachusetts. A radio historian describes what happened:

> *That night, ship operators and amateurs around Brant Rock heard the results: "someone speaking! . . . a woman's voice rose in song. . . . Next someone was heard reading a poem." Fessenden himself played "O Holy Night" on his violin. Though the fidelity was not all that it might be, listeners were captivated by the voices and notes they heard. No more would sounds be restricted to mere dots and dashes of the Morse code.*[15]

Ship operators were astonished to hear voices rather than the familiar Morse code. (Some operators actually thought they were having a supernatural encounter.) This event showed that the wireless medium was moving from a point-to-point communication tool (wireless operator to wireless operator, as with the telegraph and telephone) toward a one-to-many communication tool. **Broadcasting**, once an agricultural term that referred to the process of casting seeds over a large area, would come to mean the transmission of radio waves (and, later, TV signals) to a broad public audience.

Regulating and Controlling a New Medium

In the first decade of the twentieth century, the two most important international issues affecting radio were ship radio requirements and signal interference. Ship safety was a global concern, and regulators soon realized that distressed ships at sea could use wireless equipment to radio for help—provided they could get a clear signal. Signal interference was common at that time, which was a problem not only for ships but also for land transmissions, with amateurs across the United States transmitting their own radio signals and hobbyists on the receiving end curiously listening with headphones to voices from miles away. For example, one of the earliest stations, operated by Charles "Doc" Herrold in San Jose, California, began in 1909 and decades later became KCBS. Additional experimental stations—in places like New York; Detroit; Medford, Massachusetts; Madison, Wisconsin; and Pierre, South Dakota—broadcast voices and music, entertaining listeners but also adding to the cacophony of the unregulated airwaves.

Government officials increasingly recognized that radio airwaves were a valuable and limited resource; if there was no regulation of the airwaves, they could become too chaotic and nonfunctional. In addition, the government began to recognize the significance of the technology that powered radio. Before the end of the second decade of the twentieth century, the U.S. effort to capture control of this technology had become the most important international issue affecting radio.

Radio Waves as a Safety Measure—and a Natural Resource

Recognizing the great utility of a wireless communication medium, Congress passed the **Wireless Ship Act of 1910**, which required that all major U.S. seagoing ships carrying more than fifty passengers and traveling more than two hundred miles off the coast be equipped with wireless equipment with a one-hundred-mile range. The importance of this act would be underscored when Britain's brand-new luxury steamship, the *Titanic*, sank in 1912. Although more than fifteen hundred people died in the tragedy, wireless reports played a critical role in pinpointing the *Titanic*'s location, enabling rescue ships to save over seven hundred lives.

In the wake of the *Titanic* tragedy, Congress passed the **Radio Act of 1912**, which attempted to address the problem of amateur radio operators increasingly cramming the airwaves. Because radio waves crossed state and national borders, legislators determined that broadcasting constituted a "natural resource"—a kind of interstate commerce. This meant that radio waves could not be owned; they were the collective property of all Americans, just like national parks. Therefore, transmitting on radio waves would require licensing from the Commerce Department, which also assigned call letters to stations. Further, the "natural resource" mandate led to the idea that radio, and eventually television, should provide a benefit to society in the form of education and public service. The eventual establishment of public radio stations was one outcome of this idea. (Regulating station power and assigning stations to specific frequencies would, however, continue to be a vexing problem until the next major radio legislation in 1927.)

The Impact of World War I

By 1915, more than twenty American companies sold wireless point-to-point communication systems, primarily for use in ship-to-shore communication. Having established a reputation for efficiency and honesty, American Marconi was the biggest and best of these companies. But in 1914, with the outbreak of World War I in Europe, and America warily watching the conflict, the U.S. Navy questioned the wisdom of allowing a foreign-controlled company to wield so much power. American corporations, especially GE and AT&T, capitalized on the navy's xenophobia and succeeded in undercutting Marconi's influence.

When the United States entered the war in 1917, the navy closed down all radio operations and took control of key radio transmitters to ensure military security. In 1919, after the war, they also took steps to ensure that powerful new radio technology being developed by American companies like GE would not be sold to foreign companies like British Marconi. At this time, President Woodrow Wilson and the navy saw an opportunity to slow Britain's influence over communication and to promote a U.S. plan for the control of emerging wireless operations as part of the larger goal of developing the United States as an international power. Thus, corporate heads and government leaders conspired to make sure radio communication would serve American interests.

The Formation of RCA

Some members of Congress and the corporate community opposed federal legislation that would have granted the U.S. Navy a radio monopoly. Consequently, GE developed a compromise plan that would create a *private sector monopoly*—that is, a private company that would have the government's approval to dominate the radio industry. In 1919, with the government's blessing, GE took the lead in founding a new company, **Radio Corporation of America (RCA)**. RCA soon acquired American Marconi and the radio patents of other U.S. companies, including GE, AT&T, and United Fruit (today's Chiquita Brands, which had an enormous fleet of banana steamships), as well as the navy. Thus, upon its founding, RCA

NEWS OF THE *TITANIC*
Despite the headline in the *St. Louis Post-Dispatch*, 1,523 people died and 705 were rescued when the *Titanic* hit an iceberg on April 14, 1912 (the ship technically sank at 2:20 A.M. on April 15). The crew of the *Titanic* used the Marconi wireless equipment on board to send distress signals to other ships. Of the eight ships nearby, the *Carpathia* was the first to respond with lifeboats.

had *pooled*, or gathered control of, the necessary technology and patents to monopolize the wireless industry and take the lead in expanding American communication technology throughout the world.[16]

Under RCA's patent-pool arrangement, businesses contributing patents to RCA got a percentage of RCA stock, and some also received a role in the newly structured U.S. radio business: AT&T manufactured most transmitters while GE (and later Westinghouse) made radio receivers. RCA was in charge of the arrangement, administering the patent pool, and collecting patent royalties and distributing them to pool members. To protect its partners' profits, the government did not permit RCA to manufacture equipment or to operate radio stations under its own name for several years, although it later made RCA receivers and other audio equipment.

A government restriction at the time mandated that no more than 20 percent of RCA could be owned by foreigners. This restriction, later raised to 25 percent, became law in 1927 and applied to all U.S. broadcasting businesses. (Because of this rule, Rupert Murdoch—the head of Australia's News Corp.—became a U.S. citizen in 1985, so he could buy a number of TV stations and form the Fox television network.) In 2013, the Federal Communications Commission ruled that it would allow exemptions to the 25 percent foreign ownership limit on a case-by-case basis.

RCA's most significant impact was that it gave the United States almost total control over the emerging mass medium of broadcasting. At the time, the United States was the only country that placed broadcasting under the care of commercial, rather than military or government, interests. By pooling more than two thousand patents and sharing research developments, RCA ensured the global dominance of the United States in mass communication, a position it maintained in electronic hardware into the 1960s and maintains in program content today.

The U.K.'s Public-Service Model: The Path Not Taken by the United States

In 1922, upon witnessing the rise of commercial radio in the United States, Great Britain went a different route, beginning what later became its noncommercial state-chartered public broadcasting system, the British Broadcasting Corporation (BBC). The BBC's operations and programming are funded by an annual licensing fee paid by each household; residents today pay about $200 a year for the license, which funds commercial-free TV programming, national and local radio stations, and streaming audio and video on the BBC website. About two-thirds of European countries fund public service broadcasting with TV licenses (or other taxes), as do about half of Asian and African countries, and about 10 percent of nations in the Americas and Caribbean.[17] As historian Paul Starr explained, "The two systems set off on paths of development that in time would take them further apart as American radio turned to advertising and British public-service broadcasting remained more or less insulated from market pressures and kept to its mission of cultural uplift."[18]

The Evolution of U.S. Radio

When Westinghouse engineer Frank Conrad set up a crude radio studio above his Pittsburgh garage in 1916, placing a microphone in front of a phonograph to broadcast music and news to his friends (whom Conrad supplied with receivers) two evenings a week on experimental station 8XK, he unofficially became one of the medium's first disc jockeys. In 1920, a Westinghouse executive, intrigued by Conrad's curious hobby, realized the potential of radio as a mass medium, and understood that compelling programming could sell more of the company's radio sets. Westinghouse then established station KDKA, which is generally regarded as the first commercial broadcast station. (Although other stations claim to be the first, KDKA's success, made possible through the financial backing of Westinghouse, signaled the true start of broadcast radio.) The establishment of the station marked the shift from considering radio a medium used for point-to-point communication to considering it a medium used to broadcast widely to a listening audience. KDKA is most noted for airing national returns from the Cox–Harding presidential election on November 2, 1920, an event most historians consider the first professional broadcast.

Radio grew quickly. In 1921, the U.S. Commerce Department licensed five radio stations for operation; by early 1923, more than six hundred commercial and noncommercial stations were operating. Some of these stations were owned by AT&T, GE, and Westinghouse, but many were run by licensed amateurs or were independently owned by universities or businesses. By the end of 1923, as many as 550,000 radio receivers, most manufactured by GE and Westinghouse, had been sold for about $55 each (over $800 in today's dollars). Just as the guts of the phonograph had been placed inside a piece of furniture to create a consumer product, the vacuum tubes, electrical posts, and bulky batteries that made up the radio receiver were placed inside stylish furniture and marketed to households. By 1925, 5.5 million radio sets were in use across America, and radio was officially a mass medium.

Library of Congress/Getty Images

RUFUS P. TURNER, who attended the Armstrong Technical High School in Washington, D.C., was the first Black American to operate a broadcast radio station, transmitting from Washington's St. Augustine Catholic Church in 1926. Turner became a professional engineer and wrote prolifically about transistors and other technology before becoming an English professor in California later in life.

AT&T's Power Grab: The First Radio Ads and the First Radio Network

In a major power grab in 1922, AT&T, which already had a government-sanctioned monopoly in the telephone business and permission to make transmitters for the radio industry, attempted to monopolize radio as well. Identifying the new medium as the "wireless telephone," AT&T argued that broadcasting was merely an extension of its control over the telephone. Thus, in violation of its early agreements with RCA, AT&T began making and selling its own radio receivers. But that wasn't enough for AT&T, one of the most powerful companies in the United States at the time. In an attempt to dominate the industry, it also brought two more innovations to radio: advertising and a network of stations.

The Rise of the Radio Ad

In 1922, AT&T started WEAF (now WFAN) in New York, the first radio station to regularly sell commercial time to advertisers. AT&T claimed that under the RCA agreements, it had the exclusive right to sell ads, which AT&T called *toll broadcasting*—selling time at a price (or toll) to anyone who wished to buy it. Most people in radio at the time recoiled at the idea of using the medium for advertising, viewing it as crass to commercialize what they considered a public information service. (KDKA, for example, was marketed as a professional public radio station, and it did not sell ads.) In fact, stations that had earlier tried to sell ads received "cease and desist" letters from the Department of Commerce. Yet by August 1922, AT&T had sold its first ad to a New York real estate developer for $50, who received ten minutes to promote the sale of new suburban apartments. For big business, the idea of promoting the new medium as a public service ended when executives realized that radio ads offered an opportunity to earn profits long after radio-set sales had saturated the consumer market.

The Rise and Decline of AT&T's Radio Network

AT&T's final tactic to conquer radio was the invention of a network. Through its agreements with RCA, AT&T retained the rights to interconnect the signals between two or more radio stations via telephone wires. In 1923, when AT&T aired a program simultaneously on its flagship WEAF station and on WNAC in Boston, the phone company created the first **network**: a linked group of broadcast stations that share programming produced at a central location. Networks were potentially lucrative, since they could sell broadcast advertisements that would reach most of the country's population.

A WOMAN LISTENING

to a religious service on the radio in 1923. As radio networks began to expand, so did increasingly polished program offerings. For example, the *Philadelphia Inquirer's* radio listing for October 12, 1922 shows a weekday evening lineup that starts with bedtime stories, switches to "closing prices on stocks, bonds, grain, coffee and sugar," moves to a short blitz on economizing space in men's closets, plays a Schubert quartet concert, shifts to a dramatic reading of Shakespeare's *The Merchant of Venice*, and ends with the weather forecast. Sign off was at 10 P.M.

By the end of 1924, AT&T had interconnected twenty-two stations in order to air a talk by President Calvin Coolidge. Some of these stations were owned by AT&T, but most simply consented to become AT&T "affiliates," agreeing to air the phone company's programs. These network stations informally became known as the *telephone group*, or the WEAF Chain, which extended as far west as Minneapolis–Saint Paul and Kansas City.

In response, GE, Westinghouse, and RCA interconnected a smaller set of competing stations, known as the *radio group*, or the WJZ Chain. Initially, this network linked WGY in Schenectady, New York (then GE's national headquarters), and WJZ (now WABC) in Manhattan. The radio group had to use inferior Western Union telegraph lines when AT&T denied the group access to its telephone wires. By this time, AT&T had sold its stock in RCA and refused to lease its lines to competing radio networks. The telephone monopoly was now enmeshed in a battle to defeat RCA for control of radio.

This clash, among other problems, eventually led to a government investigation and an arbitration settlement in 1925. In the agreement, the Justice Department, irritated by AT&T's power grab, redefined patent agreements. AT&T received a monopoly on providing the wires, known as *long lines*, to interconnect stations nationwide. In exchange, AT&T sold its network to RCA for $1 million and agreed not to reenter broadcasting for eight years.

RCA's Era of Dominating Radio

The rise of commercial radio in the United States was intentionally centered around RCA. The mastermind behind RCA was David Sarnoff, a Russian immigrant who got a job as a personal messenger at American Marconi at age fifteen to support his family after his father became ill. Sarnoff rose rapidly at American Marconi, and as a wireless operator he helped relay information about *Titanic* survivors in 1912. Promoted to a series of management positions, Sarnoff was closely involved in RCA's creation and later became its general manager. When most radio executives saw wireless merely as point-to-point communication, Sarnoff envisioned a national system of broadcasting. He also took steps to expand the reach of radio.

NBC: Building the "Red" and "Blue" Networks

In September 1926, soon after RCA bought AT&T's telephone group network, Sarnoff created a subsidiary called the National Broadcasting Company (NBC). Its ownership was shared by RCA (50 percent), General Electric (30 percent), and Westinghouse (20 percent). The loose network of stations that made up the new subsidiary would be connected by AT&T long lines. Shortly thereafter, the original telephone group became known as the *NBC-Red* network, and the radio group (the network established by RCA, GE, and Westinghouse) became the *NBC-Blue* network.

Although NBC owned a number of stations by the late 1920s, many independent stations began affiliating with the NBC networks to receive programming. NBC affiliates, though independently owned, signed contracts to be part of the network and paid NBC to carry their programs. In exchange, NBC reserved time slots in the programming, which it sold to national advertisers. NBC centralized costs and programming by bringing the best musical, dramatic, and comedic talent to one place, from which programs could be produced and then distributed all over the country. By 1933, NBC-Red had twenty-eight affiliates, and NBC-Blue had twenty-four. One result of network radio is that it linked America culturally, de-emphasizing the local and the regional in favor of national programs broadcast to nearly everyone. Across the country, listeners shared national news events, sports programming, and newly invented forms of radio storytelling.

THE RADIO GAME was released by Milton Bradley soon after David Sarnoff launched the National Broadcasting Company (NBC). This family-friendly board game is played on a stylized map of the United States, illustrated in four colors that represent the four U.S. time zones. Some circles on the map are linked by red lines, while others are linked by blue lines—representing the NBC-Red and NBC-Blue networks.

Government Scrutiny Ends RCA-NBC Monopoly

As early as 1923, the Federal Trade Commission (FTC) had charged RCA with violations of antitrust laws, but it allowed RCA's monopoly to continue. But by the late 1920s, the government, concerned about NBC's growing control over radio content, intensified its scrutiny. Then, in 1930, federal marshals charged RCA-NBC with a number of violations, including exercising too much control over manufacturing and programming. The government had originally sanctioned a closely supervised monopoly for wireless communication, but after the collapse of the stock market in 1929, the public became increasingly distrustful of big business.

RCA acted quickly. To eliminate its monopolizing partnerships, Sarnoff's company proposed buying out GE's and Westinghouse's remaining shares in RCA's manufacturing business. Now RCA would compete directly against GE, Westinghouse, and other radio manufacturers, encouraging more competition in the radio manufacturing industry. In 1932, just days before the antitrust case against RCA was to go to trial, the government accepted RCA's proposal for breaking up its monopoly.

Partnerships and Mergers Expand RCA's Influence

As NBC's broadcasting monopoly was being threatened, Sarnoff worked to extend RCA-NBC's position as a leading media corporation by engineering two business deals in 1929: forging a pact with General Motors for the manufacture of car radios (under the brand name Motorola) and merging RCA with the Victor Talking Machine Company. From 1930 through the mid-1960s, the

CBS HELPED ESTABLISH ITSELF as a premier radio network by attracting top talent from NBC, like Eddie Cantor. *The Eddie Cantor Show* featured lighthearted comedy (Cantor would often tell stories about his wife and four daughters) and singers such as Ethel Merman (shown here), Deanna Durbin, and Dinah Shore. In 1934, Cantor introduced the song "Santa Claus Is Comin' to Town" (other singers had rejected it as being too silly), and by the next day, the song was a hit, with 100,000 copies of its sheet music sold.

record and phonograph company would be known as RCA Victor, adopting as its corporate symbol the famous terrier sitting alertly next to a Victrola radio-phonograph. The merger gave RCA control over Victor's records and recording equipment, making the radio company a major player in the sound recording industry.

CBS: NBC Gets a Strong Rival

Even with RCA's head start and its favored status, the two NBC networks faced competitors in the late 1920s. These competitors, however, found it tough going. One group, a collaboration between the United Independent Broadcasters (UIB) and the Columbia Phonograph Company, launched the Columbia Phonograph Broadcasting System (CPBS), a wobbly sixteen-affiliate network, in 1927. After losing $100,000 the first month, the record company pulled out, and CPBS later dropped "Phonograph" from its title, creating the Columbia Broadcasting System (CBS).

In 1928, William Paley, the twenty-seven-year-old son of Sam Paley, owner of a Philadelphia cigar company, bought a controlling interest in CBS to sponsor the cigar brand La Palina. One of Paley's first moves was to hire public relations pioneer Edward Bernays to polish the new network's image. (Bernays played a significant role in the development of the public relations industry; see Chapter 12.) Paley and Bernays modified a concept called **option time**, in which CBS paid affiliate stations $50 per hour for an option on a portion of their time. The network provided programs to the affiliates and sold ad space or sponsorships to various product companies. Some affiliates received thousands of dollars per week merely to serve as conduits for CBS programs and ads. Because NBC was still charging some of its affiliates as much as $96 a week to carry its network programs, the CBS offer was extremely appealing.

By 1933, Paley's efforts had netted CBS more than ninety affiliates, many of them defecting from NBC. Paley also concentrated on developing news programs and entertainment shows, particularly soap operas and comedy-variety series. In the process, CBS successfully raided NBC, not just for affiliates but for top talent as well. Throughout the 1930s and 1940s, CBS lured a number of radio stars from NBC, including Jack Benny, Frank Sinatra, George Burns and Gracie Allen, and Groucho Marx. During World War II, Edward R. Murrow's powerful firsthand news reports from bomb-riddled London established CBS as the premier radio news network, a reputation it carried forward to television. In 1949, near the end of big-time network radio, CBS finally surpassed NBC as the highest-rated network.

Bringing Order to Chaos with Additional Regulation

In the 1920s, as radio moved from point-to-point transmission to broadcasting, the battle for more frequency space and less channel interference intensified. Manufacturers, engineers, station operators, network executives, and the listening public demanded action. Many wanted more sweeping regulation than the simple licensing function granted under the Radio Act of 1912, which gave the Commerce Department little power to revoke a license or to unclog the airwaves.

In 1924, Herbert Hoover, the secretary of commerce, ordered radio stations to share time by setting aside certain frequencies for entertainment and news, and others for farm and weather reports. To challenge Hoover, a station in Chicago jammed the airwaves, intentionally moving its signal onto an unauthorized frequency. In 1926, the courts decided that based on the existing Radio Act, Hoover had the power only to grant licenses, not to restrict stations from operating. Within the year, two hundred new stations clogged the airwaves, creating a chaotic period in which nearly all radios had poor reception. By early 1927, sales of radio sets had declined sharply.

The Radio Act of 1912 proved to be unsuccessful in controlling the rapid expansion of broadcasters. To restore order to the airwaves, Congress passed the **Radio Act of 1927**, which stated an extremely important principle: Licensees could operate on their assigned channels only as long as

Archive Photos/Getty Images

TABLE 5.1   MAJOR ACTS IN THE HISTORY OF U.S. RADIO

Act	Provisions	Effects
Wireless Ship Act of 1910	Required large U.S. seagoing passenger ships traveling more than two hundred miles off the coast to be equipped with wireless equipment.	Saved lives at sea, including more than seven hundred people rescued by ships responding to the *Titanic*'s distress signals two years later.
Radio Act of 1912	Required radio operators to obtain a license and began a uniform system of assigning call letters to identify stations.	The federal government began to assert control over radio. Penalties were established for stations that interfered with other stations' signals, but enforcement was lax.
Radio Act of 1927	Established the Federal Radio Commission (FRC) as a temporary agency to oversee licenses and negotiate channel assignments.	First expressed the now-fundamental principle that operators could hold a license only if they operated to serve the "public interest, convenience, or necessity."
Communications Act of 1934	Established the Federal Communications Commission (FCC) to replace the FRC. The FCC regulated radio; the telephone; the telegraph; and later television, cable, and the Internet.	Congress tacitly agreed to a system of advertising-supported commercial broadcasting despite concerns among the public.
Telecommunications Act of 1996	Eliminated most radio and television station ownership rules, some dating back more than fifty years.	Enormous national and regional station groups formed, dramatically changing the sound and localism of radio in the United States.

they operated to serve the "public interest, convenience, or necessity" (see Table 5.1). To oversee licenses and negotiate channel problems, the 1927 act created the **Federal Radio Commission (FRC)**, whose members were appointed by the president. Although the FRC was intended as a temporary committee, it grew into a powerful regulatory agency. Most significantly, the "public interest, convenience, or necessity" clause was never clearly defined by Congress—what, for example, would be considered in the public's interest?—but in the years following the 1927 act, the FRC's actions clearly favored commercial broadcasting.

With the **Communications Act of 1934**, the FRC became the **Federal Communications Commission (FCC)**. Its jurisdiction covered not only radio but also the telephone and the telegraph (and later television, cable, and the Internet). By the time the Communications Act of 1934 passed, Congress and the president had sided with the already-powerful radio networks and acceded to a system of advertising-supported commercial broadcasting as best serving the "public interest, convenience, or necessity," overriding the concerns of educational, labor, and citizen broadcasting advocates.[19] An amendment to the 1934 act that would have set aside 25 percent of radio channels for nonprofit broadcasters ended up failing, with strong lobbying opposition from commercial broadcasters. As a result, the United States would be without a nonprofit public broadcasting system until the late 1960s.

In 1941, a more activist FCC went after the networks. Declaring that NBC and CBS could no longer force affiliates to carry programs they did not want, the government outlawed the practice of option time, which Paley had used to build CBS into a major network. The FCC also demanded that RCA sell one of its two NBC networks. RCA and NBC claimed that the rulings would bankrupt them. The Supreme Court sided with the FCC, however, and RCA eventually sold NBC-Blue for $8 million. Soon after, in mid-1944, the network became the American Broadcasting Company (ABC), while NBC-Red became the NBC we know today. These government crackdowns brought long-overdue reform to the radio industry and reined in some of the enormous power held by the networks.

The Golden Age of Radio

Many programs on television today were initially formulated for radio. The first weather forecasts and farm reports on radio began in the 1920s. Regularly scheduled radio news analysis began in 1927, with H. V. Kaltenborn, a reporter for the *Brooklyn Eagle*, providing commentary on AT&T's WEAF. The first regular network news analysis began on CBS in 1930 and featured Lowell Thomas, who would remain on radio for forty-four years.

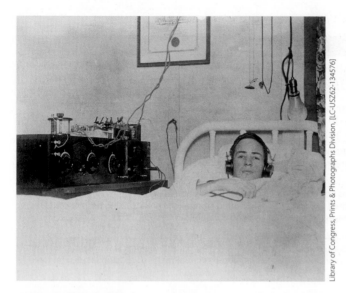

RADIO IN DAILY LIFE

A young man, who had broken his back while building a radio tower, lies in a hospital bed convalescing and listening to the radio *(left)*. By the 1930s, radio had become a main form of home entertainment *(right)*.

Early Radio Programming

Early on, only a handful of stations operated in most large radio markets, and popular stations were affiliated with CBS, NBC-Red, or NBC-Blue. Many large stations employed their own in-house orchestras and aired live music daily. Listeners had favorite evening programs, usually fifteen minutes long, to which they would tune in each night. Families gathered around the radio to hear such shows as *Amos 'n' Andy*, *The Shadow*, *The Lone Ranger*, *The Green Hornet*, and *Fibber McGee and Molly*, or one of President Franklin Roosevelt's fireside chats.

Among the most popular early programs on radio, the variety show was the forerunner to popular TV shows like *America's Got Talent*. The variety show, developed from stage acts and vaudeville, began with the *Eveready Hour* in 1923 on WEAF. Considered experimental, the program presented classical music, minstrel shows, comedy sketches, and dramatic readings. Stars from vaudeville, musical comedy, and New York theater and opera would occasionally make guest appearances.

FIRESIDE CHATS

This bank of radio network microphones makes us wonder today how President Franklin D. Roosevelt managed to project such an intimate and reassuring tone in his famous fireside chats. Conceived originally to promote FDR's New Deal policies amid the Great Depression, these chats were delivered between 1933 and 1944 and touched on topics of national interest. Roosevelt was the first president to effectively use broadcasting to communicate with citizens; he also gave nearly a thousand press conferences during his twelve-plus years as president.

By the 1930s, studio-audience quiz shows—*Professor Quiz* and the *Old Time Spelling Bee*—had emerged. The quiz format was later copied by television, particularly in the 1950s. *Truth or Consequences*, based on a nineteenth-century parlor game, first aired on radio in 1940 and featured guests performing goofy stunts. It ran for seventeen years on radio and another twenty-seven years on television, influencing TV stunt shows like CBS's *Beat the Clock* in the 1950s and NBC's *Fear Factor* in the early 2000s.

Dramatic programs, mostly radio plays that were broadcast live from theaters, developed as early as 1922. Historians mark the appearance of *Clara, Lu 'n' Em* on WGN in 1931 as the first soap opera. One year later, Colgate-Palmolive bought the program, put it on NBC, and began selling the soap products that gave this dramatic genre its distinctive nickname. Early "soaps" were fifteen minutes in length and ran five or six days a week. By 1940, sixty different soap operas occupied nearly eighty hours of network radio time each week.

Most radio programs had a single sponsor that created and produced each show. The networks distributed these programs live around the country, charging each sponsor advertising fees. Many shows—the *Palmolive Hour, General Motors Family Party, Lucky Strike Orchestra*, and the *Eveready Hour* among them—were named after the sole sponsor's product.

Radio Programming as a Cultural Mirror

The situation comedy, a major staple of TV programming today, began on radio in the mid-1920s. By the early 1930s, the most popular comedy was *Amos 'n' Andy*, which started on Chicago radio in 1925 before moving to NBC-Blue in 1929. *Amos 'n' Andy* was based on the conventions of the nineteenth-century minstrel show and featured Black characters stereotyped as shiftless and unintelligent. Created as a blackface stage act by two white comedians, Charles Correll and Freeman Gosden, the program was criticized as racist. But NBC and the program's producers claimed that *Amos 'n' Andy* was as popular among Black audiences as it was among white audiences.[20]

Amos 'n' Andy also launched the idea of the serial show: a program that featured continuing story lines from one day to the next. The serial format would soon be copied by soap operas and other radio dramas. *Amos 'n' Andy* aired six nights a week from 7:00 to 7:15 P.M. During the show's first year on the network, radio-set sales rose nearly 25 percent nationally. To keep people coming to restaurants and movie theaters, owners broadcast *Amos 'n' Andy* in lobbies, restrooms, and entryways. Early radio research estimated that the program aired in more than half of all radio homes in the nation during the 1930–31 season, making it the most popular radio series in history. In 1951, it made a brief transition to television (Correll and Gosden sold the rights to CBS for $1 million), becoming the first TV series to have an entirely Black cast. But amid a strengthening Civil Rights movement and a formal protest by the NAACP (which argued that "every character is either a clown or a crook"), CBS canceled the program in 1953.[21]

Amos 'n' Andy was clearly racist. But there were pioneers of Black radio pushing the boundaries of mostly white radio broadcasts. In Chicago, WSBC began *The All-Negro Hour* in 1929, "the first weekly variety show featuring African American entertainers."[22] Then in 1947, WJJD, a station with a powerful AM signal, aired *Here Comes Tomorrow*, the first-ever soap opera with an all-Black cast. Unlike *Amos 'n' Andy*, the program featured stories about a typical Black Chicago family. The point of the daily fifteen-minute drama was to emphasize messages of equality and push back on racial intolerance.[23]

The Authority of Radio

The most famous single radio broadcast of all time was an adaptation of H. G. Wells's *War of the Worlds* on the radio series *Mercury Theatre on the Air*. Orson Welles produced, hosted, and acted in this popular series, which adapted science fiction, mystery, and historical adventure

JACK GIBSON was a voice actor on the pioneering Black radio soap opera *Here Comes Tomorrow* on WJJD in Chicago in 1947. He later said the show "probably raised the feathers of a lot of people that were in Iowa and downstate Illinois that didn't understand the race relationship that we [were] trying to build."[24]

dramas for radio. On Halloween eve in 1938, the twenty-three-year-old Welles aired the 1898 Martian invasion novel in the style of a radio news program. For people who missed the opening disclaimer, the program sounded like a real news report, with eyewitness accounts of battles between Martian invaders and the U.S. Army.

As the story goes, the program created a panic that lasted several hours. In New Jersey, some people walked through the streets with wet towels around their heads for protection from deadly Martian heat rays; in New York, young men reported to their National Guard headquarters to prepare for battle; and across the nation, calls jammed police switchboards. Yet according to contemporary studies, the story of widespread panic is mostly myth: Newspapers exaggerated the hysteria, which was actually much more limited. Media researcher W. Joseph Campbell further argues that newspapers used the opportunity to "rebuke" radio (increasingly a rival for audience attention and advertising dollars) as "unreliable and untrustworthy."[25] The only reform to emerge after the broadcast was that the FCC called for stricter warnings both before and during programs that imitate the style of radio news.

Radio Reinvents Itself

ADVERTISEMENTS for pocket transistor radios, which became popular in the 1950s, emphasized their portability.

New from Motorola
ALL-TRANSISTOR
Shirt-pocket radio
with powerful built-in speaker

Slips into your pocket...plays with the sound of a set twice its size.
New miniaturized transistor radio lets you enjoy favorite programs wherever you go. Powerful chassis pulls in stations loud and clean . . . plays them clear and mellow through a newly designed built-in speaker. A *single* low-cost battery gives many hours of listening pleasure. Motorola extras include magnifying lens for easy-to-read dial . . . durable case in black, blue, red, or green. 90-day warranty on all parts and labor at no additional cost.

Perfect gift for you or someone extra special

ONLY $29.95 MODEL X11

Built-in Easel Stand in back for stable "stand-up" position.

Power-Packed Chassis with 6 transistors provides plenty of volume.

Built-in Antenna pulls in stations like a magnet for finest reception.

Perfect Companion for sports events. Earphone jack for private listening.

More to enjoy **MOTOROLA**

Older media forms do not generally disappear when confronted by newer forms. Instead, they adapt. For example, although radio threatened sound recording in the 1920s, the recording industry adjusted to the economic and social challenges posed by radio's arrival. Remarkably, the arrival of television in the 1950s marks the only time in media history when a new medium stole virtually every national programming and advertising strategy from an older medium. Television snatched radio's advertisers, program genres, major celebrities, and large evening audiences. The TV set even physically displaced the radio as the living room centerpiece of choice across America.

Nevertheless, radio adapted and survived, a story that is especially important today as newspapers, magazines, books, and other media appear in digital formats. In contemporary culture, we have grown accustomed to such media convergence, but to better understand this blurring of the boundaries between media forms, it is useful to look at the 1950s and the ways in which radio responded to the advent of television with adaptive innovations in technology and program content.

Transistors Make Radio Portable

A key development in radio's adaptation to television occurred with the invention of the transistor by Bell Laboratories in 1947. **Transistors** were small electrical devices that, like vacuum tubes, could receive and amplify radio signals. However, they used less power and produced less heat than vacuum tubes, and they were more durable and less expensive. Best of all, they were tiny. Transistors, which also revolutionized hearing aids, constituted the first step in replacing bulky and delicate tubes, eventually leading to the circuitry of early personal computers.

Texas Instruments marketed the first transistor radio in 1953 for about $40. Using even smaller transistors, Sony introduced the pocket radio in 1957. But it wasn't until the 1960s that transistor radios became cheaper than conventional tube-and-battery radios. For a while, the term *transistor* became a synonym for a small portable radio.

The development of transistors let radio go where television could not—to the beach, to the office, into bedrooms and bathrooms, and into nearly all new cars. (Before the transistor, car radios were a luxury item.) By the 1960s, most radio listening took place outside the home.

The FM Revolution and Edwin Armstrong

By the time the broadcast industry launched commercial television in the 1950s, many people, including David Sarnoff of RCA, were predicting radio's demise. To fund television's development and to protect his existing AM radio holdings, Sarnoff delayed a dramatic breakthrough in broadcast sound that even he called a "revolution"—FM radio.

Edwin Armstrong, who first discovered and developed FM radio in the 1920s and early 1930s, is often considered the most prolific and influential inventor in radio history. His invention of an improved amplification system for radio made him a millionaire as well as RCA's largest private stockholder by the early 1920s. Armstrong also worked on the biggest problem affecting radio reception: electrical interference. Between 1930 and 1933, the inventor filed five patents on **FM**, or frequency modulation. Offering static-free radio reception, FM supplied greater fidelity and clarity than AM did, making FM ideal for music. **AM**, or amplitude modulation (modulation refers to the variation in wave-forms), stressed the volume, or height, of radio waves; FM accentuated the pitch, or distance, between radio waves (see Figure 5.2).

Although David Sarnoff, RCA's president, thought that television would replace radio, he helped Armstrong set up the first experimental FM station atop the Empire State Building in New York City. Eventually, though, Sarnoff thwarted FM's development (which he was able to do because RCA had an option on Armstrong's new patents). Instead, in 1935, Sarnoff threw RCA's considerable weight behind the development of television. With the FCC allocating and reassigning scarce frequency spaces, RCA wanted to ensure that channels went to television before they went to FM. Most of all, however, Sarnoff wanted to protect RCA's existing AM empire. Thus, Sarnoff decided to close down Armstrong's station. (Sarnoff's only use for FM was to use its superior audio technology for the transmission of television sound.)

Armstrong forged ahead without RCA. He founded a new FM station and trained other FM engineers, who established more than twenty experimental stations between 1935 and the early 1940s. In 1941, the FCC approved limited space allocations for commercial FM licenses. During the next few years, FM grew in fits and starts. Between 1946 and early 1949, the number of commercial FM stations expanded from forty-eight to seven hundred. But then the FCC (at RCA's urging) moved FM's frequency space to a new band on the electromagnetic spectrum, rendering some 400,000 prewar FM receiver sets useless. FM's future became uncertain, and by 1954, the number of FM stations had fallen to 560. On January 31, 1954, Edwin Armstrong—weary from years of legal skirmishes over patents with RCA, Lee De Forest, and others—wrote a note apologizing to his wife, removed the air conditioner from the thirteenth-story window of his New York apartment, and jumped to his death.

Although AM stations had greater reach, they could not match the crisp fidelity of FM, which made it preferable for music. In the early 1960s, the FCC opened

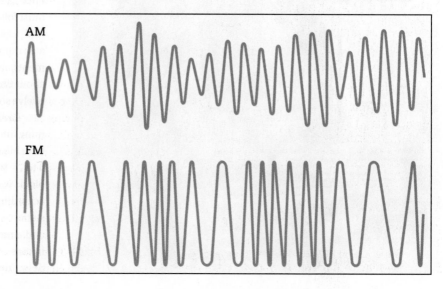

FIGURE 5.2

AM AND FM WAVES

up more spectrum space for the superior sound of FM, infusing new life into radio. In the early 1970s, about 70 percent of listeners tuned almost exclusively to AM radio, but by the 1980s, FM had surpassed AM in profitability. This expansion of FM represented one of the chief ways radio survived the advent of television.

The Rise of Format and Top 40 Radio

Live and recorded music had long been radio's single biggest staple, accounting for 48 percent of all programming in 1938. Although live music on the radio was generally considered superior to recorded music, early disc jockeys (or deejays) made a significant contribution to the latter, demonstrating that recorded music alone could drive radio. Early deejays were curators, using their musical knowledge and taste to introduce listeners to new artists or go deep into certain types of musical genres. By the time television came along and snatched radio's program ideas and national sponsors, radio's dependence on recorded music was well established—and it helped the medium survive the 1950s.

As early as 1949, station owner Todd Storz in Omaha, Nebraska, experimented with formula-driven radio, or **format radio**. Under this system, management controlled programming each hour, curbing the independence of deejays. When Storz and his program manager noticed that bar patrons and waitstaff repeatedly played certain favorite songs from the records available in a jukebox, they began researching record sales to identify the most popular tunes. From observing jukebox culture, Storz hit on the idea of **rotation**: playing the top songs many times during the day. By the mid-1950s, the management-control idea combined with the rock-and-roll explosion, and the **Top 40 format** was born. The term *Top 40* came to refer to the forty most popular hits in a given week as measured by record sales.

As format radio grew, program managers (who are in charge of the overall sound of the station through its programming) combined rapid deejay chatter with the best-selling songs of the day and occasional oldies—popular songs from a few months earlier. By the early 1960s, to avoid "dead air," managers asked deejays to talk over the beginning and the end of a song so that listeners would feel less compelled to switch stations. Ads, news, weather forecasts, and station identifications were all designed to fit a consistent station environment. Listeners, tuning in at any moment, would recognize the station by its distinctive sound.

In format radio, management carefully coordinates, or programs, each hour, dictating what the deejay will do at various intervals throughout each hour of the day. Management creates a program log—once called a *hot clock* in radio jargon—that deejays must follow. In a typical hour on Top 40 radio in the mid-1960s, listeners could expect to hear about twenty ads; numerous weather, time, and contest announcements; multiple recitations of the station's call letters; about three minutes of news; and approximately twelve songs.

Radio managers further sectioned off programming into day parts, which typically consisted of time blocks covering 6 to 10 A.M., 10 A.M. to 3 P.M., 3 to 7 P.M., and 7 P.M. to midnight. Each day part, or block, was programmed through ratings research according to who was listening. For instance, a Top 40 station would feature its top deejays in the morning and afternoon blocks, when audiences—many riding in cars—were largest. From 10 A.M. to 3 P.M., research determined that women at home and secretaries at

BLACK RADIO began to appear in the late 1940s as advertisers and station owners noticed the opportunities available in this market. In Nashville, WLAC began to spin R&B records, and WDIA in Memphis switched to an all-Black format in 1948. In 1949, WERD in Atlanta became the first Black-owned and Black-formatted radio station in the country. These pioneers inspired stations to bring Black deejays and music to cities like Louisville, Pittsburgh, Columbus, Detroit, Chicago, Austin, Houston, and Miami.

Indiana University Archives of African American Music and Culture, Kathy Lewis Collection

work usually controlled the dial, so program managers, capitalizing on the gender stereotypes of the day, played more romantic ballads and less hard rock. Teenagers tended to be heavy evening listeners, so program managers often discarded news breaks at this time, since research showed that teens turned the dial when news came on.

Critics of format radio argued that only the top songs received play and that lesser-known songs deserving airtime received meager attention. Although a few popular star deejays continued to play a role in programming, many others quit when managers introduced formats. Program directors approached programming as a science, whereas deejays considered it an art form. Ultimately, the program directors' position, which generated more revenue, triumphed.

Resisting the Top 40

The expansion of FM in the mid-1960s created room for experimenting, particularly with classical music, jazz, blues, and non–Top 40 rock songs. **Progressive rock** emerged as an alternative to conventional formats. Many noncommercial stations broadcast from college campuses, where student deejays and managers rejected the commercialism associated with Top 40 radio and began playing lesser-known alternative music and longer album cuts (such as Bob Dylan's "Desolation Row" and the Doors' "The End"). Until that time, most rock on the radio had been consigned almost exclusively to Top 40 formats, with song length averaging about three minutes.

Experimental FM stations, both commercial and noncommercial, offered a cultural space for hard-edged political folk music and for rock music that commented on the Civil Rights movement and protested America's involvement in the Vietnam War. By the 1970s, however, progressive rock had been copied, tamed, and absorbed by mainstream radio under a format labeled **album-oriented rock (AOR)**. By 1972, AOR-driven album sales accounted for more than 85 percent of the retail record business. By the 1980s, as first-generation rock and rollers aged and became more affluent, AOR stations became less political and played mostly white post-Beatles music, such as Pink Floyd, Genesis, AC/DC, and Queen.

The Sounds of the Contemporary Radio Environment

Contemporary radio sounds very different from its predecessor. In contrast to the few stations per market in the 1930s, most large markets today include more than forty stations that vie for listener loyalty. Although a few radio personalities—such as Glenn Beck, Ryan Seacrest, and Jim Rome—are nationally prominent, and some shows are syndicated nationally, local deejays and their music are the stars at most radio stations.

Listeners today are also unlike radio's first audiences in several ways. First, listeners in the 1930s tuned in to their favorite shows at set times. But since the 1950s, radio has become a secondary, or background, medium that follows the rhythms of daily life. In addition, people can now listen to radio programs as podcasts any time of the day or night (see the chapter-opening discussion). Second, in the 1930s, peak listening time occurred during the evening hours—dubbed *prime time* in the TV era—while today's heaviest radio listening occurs by commuters during **drive time**, between 6 and 9 A.M. and between 4 and 7 P.M. Third, stations today are more specialized. Listeners are loyal to favorite stations, music formats, and even radio personalities, rather than to specific shows. Although more than fifteen thousand radio stations operate in the United States, people generally listen to only four or five stations.

Format Specialization

Stations today use a variety of formats based on managed program logs and day parts. All told, more than forty different radio formats, plus variations, serve diverse groups of listeners (see Figure 5.3). To please advertisers, who want to know exactly who is listening, formats usually target audiences according to their age, income, gender, or race/ethnicity. Radio's specialization enables advertisers to reach smaller target audiences at much lower costs than those for television.

FIGURE 5.3

THE MOST POPULAR RADIO FORMATS IN THE UNITED STATES AMONG PERSONS AGE SIX AND OLDER

Data from: Nielsen, "Tops of 2019: Radio," December 18, 2019, www .nielsen.com/us/en/insights/article /2019/tops-of-2019-radio.

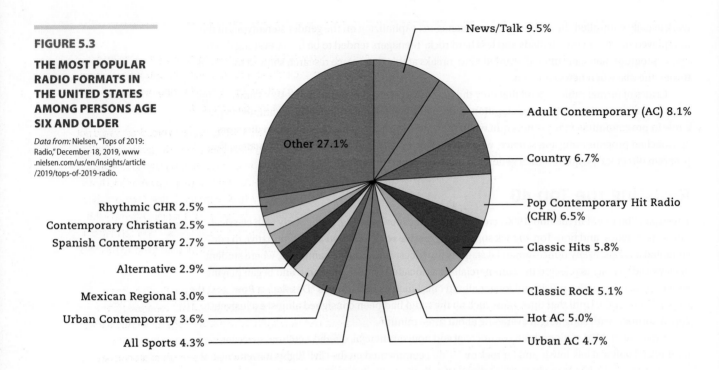

News/Talk 9.5%

Adult Contemporary (AC) 8.1%

Country 6.7%

Pop Contemporary Hit Radio (CHR) 6.5%

Classic Hits 5.8%

Classic Rock 5.1%

Hot AC 5.0%

Urban AC 4.7%

Other 27.1%

Rhythmic CHR 2.5%

Contemporary Christian 2.5%

Spanish Contemporary 2.7%

Alternative 2.9%

Mexican Regional 3.0%

Urban Contemporary 3.6%

All Sports 4.3%

Targeting listeners has become extremely competitive, because forty or fifty stations may be available in a large radio market. About 10 percent of all stations across the country switch formats each year in an effort to find a formula that generates more advertising money. Some stations, particularly those in large cities, even rent blocks of time to various local ethnic or civic groups; this enables the groups to dictate their own formats and sell ads.

News/Talk Radio

The nation's fastest-growing format throughout much of the 1990s was the **news/talk** format (see "Examining Ethics: How Did Talk Radio Become So One-Sided?"). In 1987, only 170 radio stations operated formats dominated by either news programs or talk shows, which tend to appeal to adults over age thirty-five (except for sports talk programs, which draw mostly male sports fans of all ages). Today, more than two thousand stations carry the format—more stations than carry any other format. It is the most dominant format on AM radio and the most popular format (by number of listeners) in the nation (see Figure 5.3 and Table 5.2). A news/talk format, though more expensive to produce than a music format, appeals to advertisers looking to target working- and middle-class adult consumers. Nevertheless, most radio stations continue to be driven by a variety of less expensive music formats.

TABLE 5.2   TOP TALK RADIO WEEKLY AUDIENCE (IN MILLIONS)

Talk-Show Host	2008	2016	2018	2020
Rush Limbaugh (Conservative)	14.25	13.25	14.00	15.50
Sean Hannity (Conservative)	13.25	12.50	13.50	15.00
Dave Ramsey (Financial Advice)	4.50	8.25	13.00	14.00
Mark Levin (Conservative)	5.50	7.00	10.00	11.00
Glenn Beck (Conservative)	6.75	7.00	10.00	10.50
George Noory (Conservative)	*	*	9.00	10.50

Data from: "The Top Talk Radio Audiences," *Talkers*, June 2020.

Notes: * = information unavailable. Limbaugh died in 2021.

How Did Talk Radio Become So One-Sided?

For young adults, news/talk radio is in another universe. Radio listeners ages eighteen to thirty-four are all about music, with pop contemporary hit radio, country, adult contemporary, and urban contemporary as their top radio formats.[1]

It turns out that radio listeners ages fifty-five to sixty-four live in that other universe. Their top radio format is news/talk.[2] According to Nielsen research, the news/talk universe is also mostly white and male. The rulers of that universe in 2020—the top personalities—were Rush Limbaugh (who died in 2021), Sean Hannity, Dave Ramsey, Mark Levin, Glenn Beck, George Noory, Mike Gallagher, Hugh Hewitt, Michael Savage, and Dana Loesch (see Table 5.2). Although one moderate (Jim Bohannon) and one progressive/liberal (Thom Hartmann) were among the top fifteen, their rarity suggests another thing about this universe: it's predominantly politically conservative.[3]

How did one point of view come to dominate news/talk in the United States? Its roots were in the repeal of a little-known regulation that had been on the books of the Federal Communications Commission for decades. Since 1949, the FCC enforced what was called the *Fairness Doctrine*. This allowed "a station to editorialize, provided it made airtime available for 'balanced presentation of all responsible viewpoints on particular issues.'"[4]

The rationale behind the doctrine was that there were a limited number of stations on the public airwaves, and that broadcast stations should serve the public interest of the communities where they were located. Consumer advocate Ralph Nader argued that "such issues as women's rights, the health effects of smoking, and the safety of nuclear power plants would have come to far less public prominence had the fairness doctrine not been in effect."[5]

But by 1987, President Ronald Reagan was pushing business-friendly deregulation. A chief target for the broadcasting industry was the Fairness Doctrine, since broadcasters did not like the additional demands of reporting contrasting positions on controversial public issues. The National Association of Broadcasters (NAB) stated that broadcasters could report evenhandedly without regulation. "Broadcasters believe in fairness and generally report both sides of controversial issues, but want to do so without Uncle Sam looking over our shoulders," the NAB's executive director said.[6]

After the Fairness Doctrine's 1987 repeal, radio broadcasters did not continue to "report both sides of controversial issues," as the NAB had promised, but instead rolled out new programs that *created* controversy with very one-sided political opinions. As media historians Robert Hilliard and Michael Keith explain, the demise of the Fairness Doctrine changed the tenor of talk radio: "Ostensibly this put no limits on any ideas, philosophies, or other political matter a station might wish to advocate. In reality, it swung the tide of radio and television political advocacy to the right."[7]

Limbaugh, long talk radio's biggest star, was the first conservative talk-show host to go national in 1988. He moved from Sacramento to New York City and was a huge hit for ABC Radio, after which a host of conservative personalities flooded the airwaves. "My success has spawned dozens of imitators. It has touched off a frantic scramble to cash in," Limbaugh wrote.[8] Limbaugh's support of conservative causes in his national radio program led to the Republicans gaining control of the House of Representatives in 1994. Republicans acknowledged their debt to Limbaugh and named him an "honorary member of their class."[9]

There was an attempt at a liberal talk-radio network with Air America from 2004 to 2010. It featured personalities like Thom Hartmann, Al Franken, Montel Williams, Janeane Garofalo, and Marc Maron and helped introduce America to current MSNBC host Rachel Maddow. Yet Air America could never build the critical mass that conservative talk had. Rush Limbaugh had about 590 affiliate stations for his three-hour weekday show. Air America could never get more than 70 affiliate stations to run its programs.

But one alternative to commercial news/talk has thrived: noncommercial news/talk. While Limbaugh's rise was meteoric, NPR has slowly built a broad base of listeners for its nonprofit, nonpartisan public radio network since the early 1970s.

Until his death, Rush Limbaugh's audience was commercial talk radio's largest, at 15.5 million weekly listeners. NPR's two-hour flagship *Morning Edition* program has a weekly audience of 14.9 million listeners at more than eight hundred affiliate stations.[10] As a noncommercial alternative, NPR does not focus on personalities (*Morning Edition*'s four hosts are largely unrecognizable compared to Limbaugh or Hannity) and is based in reporting, not commentary, with reporters in seventeen U.S. locations as well as seventeen countries.

Still, commercial news/talk holds a significant edge on the airwaves, in politics, and in overall radio ratings: It has 8.3 percent of the U.S. listening audience, compared to 3.7 percent of the national listening audience for noncommercial news/talk.[11] ●

ANGIE MARTINEZ, a radio deejay specializing in interviews with musicians and actors, has one of the nation's top shows with young listeners in the country. *The Angie Martinez Show* is carried weekdays on The Beat 103.5 in Miami, 10–2 P.M., and on Power 105.1 in New York, 2–6 P.M. Both Hip-Hop/R&B stations are owned by iHeartmedia.

Music Formats

The **adult contemporary (AC)** format is among radio's oldest and most popular formats, reaching about 8.1 percent of all listeners, most of them over age forty, with an eclectic mix of news, talk, oldies, and soft rock music—what *Broadcasting* magazine describes as "not too soft, not too loud, not too fast, not too slow, not too hard, not too lush, not too old, not too new."[26] Now encompassing everything from rap to pop-punk songs, Top 40 radio—also called **contemporary hit radio (CHR)**—still appeals to many teens and young adults. A renewed focus on producing pop singles in the sound recording industry has recently boosted listenership of this format.

Country remains one of the strongest formats in the nation. Many country stations are in tiny markets, where country is traditionally the default format for communities with only one radio station. Country music has old roots in radio, beginning in 1925 with the influential *Grand Ole Opry* program on WSM in Nashville. Although Top 40 drove country music out of many radio markets in the 1950s, the growth of FM in the 1960s brought it back, as station managers looked for market niches not served by rock music.

Many formats appeal to particular ethnic or racial groups. In 1948, WDIA in Memphis was the first station to program exclusively for Black listeners. Now called **urban contemporary**, this format targets a wide variety of Black listeners, primarily in large cities. In 2020, in response to the Black Lives Matter protests across the United States, some companies in the recording and radio industry dropped the *urban* music title from their format names. The largest radio chain, iHeartMedia, renamed those formats hip-hop or R&B. Others criticized the renaming for not directly addressing the problem of a media infrastructure that has far too few Black executives.[27]

Spanish-language radio is concentrated mostly in large Hispanic markets, such as Miami, New York, Chicago, Las Vegas, California, Arizona, New Mexico, and Texas (where KCOR, the first all-Spanish-language station, originated in San Antonio in 1947). Besides talk shows and news segments in Spanish, this format features a variety of Spanish, Caribbean, and Latin American musical styles, including calypso, flamenco, mariachi, merengue, reggae, samba, salsa, and Tejano.

Today there are also formats that are spin-offs from album-oriented rock. Classic rock serves up rock favorites from the mid-1960s through the 1980s to the baby-boom generation and other listeners who have outgrown Top 40. The oldies format originally served adults who grew up on 1950s and early-1960s rock and roll. As that audience has aged, oldies formats now target younger audiences with the classic hits format, featuring songs from the 1970s, 1980s, and 1990s. The alternative format recaptures some of the experimental approach of the FM stations of the 1960s, although with much more controlled playlists, and has helped introduce such artists as Awolnation and Cage the Elephant.

Research indicates that most people identify closely with the music they listened to as adolescents and young adults. This tendency partially explains why classic hits and classic rock stations combined have surpassed CHR stations today. It also helps explain the recent nostalgia for music from the 1980s and 1990s.

Nonprofit Radio and NPR

Nonprofit radio maintains a voice in a landscape dominated by commercial radio conglomerates. But the road to viability for nonprofit radio in the United States has not been easy. In the 1930s, the Wagner-Hatfield amendment to the 1934 Communications Act intended to set aside 25 percent of radio for a wide variety of nonprofit stations. When the amendment was defeated, the future of educational and noncommercial radio looked bleak. Many nonprofits had sold out to for-profit owners during the Great Depression of the 1930s. The stations that remained were often banished from the air during the evening hours or assigned weak signals by federal regulators who favored commercial owners and their lobbying agents. Still, nonprofit public radio survived. Today, more than six thousand full-power and low-power nonprofit stations operate, most of them on the FM band.

The Early Years of Nonprofit Radio

Two government rulings, both in 1948, aided nonprofit radio. First, the government began authorizing noncommercial licenses to stations not affiliated with labor, religious, education, or civic groups. The first license went to Lewis Kimball Hill, a radio reporter and pacifist during World War II who started the **Pacifica Foundation** to run experimental public stations. Like Hill himself, Pacifica stations have often challenged the status quo in both radio and government. Most notably, in the 1950s they aired the poetry, prose, and music of performers who were considered radical, left-wing, or communist and were thus banned by television and seldom acknowledged by AM stations. Over the years, Pacifica has been fined and reprimanded by the FCC and Congress for airing programs that critics considered inappropriate for public airwaves. Today, Pacifica has more than two hundred affiliate stations.

Second, the FCC approved 10-watt FM stations. A 10-watt station with a broadcast range of only about seven miles took very little capital to operate, allowing more people to participate (before this, radio stations had to have at least 250 watts to get licensed). Such stations became training sites for students interested in broadcasting. Although the FCC stopped licensing new 10-watt stations in 1978, about one hundred longtime 10-watters are still in operation.

Creation of the First Noncommercial Networks

During the 1960s, nonprofit broadcasting found a Congress sympathetic to an old idea: using radio and television as educational tools. As a result, **National Public Radio (NPR)** and the **Public Broadcasting Service (PBS)** were created as the first noncommercial networks. Under the provisions of the **Public Broadcasting Act of 1967** and the **Corporation for Public Broadcasting (CPB)**, NPR and PBS were mandated to provide alternatives to commercial broadcasting in radio and on television, respectively. Today, NPR's popular news and interview programs, such as *Morning Edition* and *All Things Considered*, are thriving, contributing to the network's weekly audience of thirty-seven million listeners.

Over the years, however, public radio has faced waning government support and the threat of losing its federal funding. As a result, stations have become more reliant on private donations and corporate sponsorship, which could cause some public broadcasters to steer clear of controversial subjects, especially those that critically examine corporations (see "Media Literacy and the Critical Process: Comparing Commercial and Noncommercial Radio").

Like commercial stations, nonprofit radio has adopted the format style. However, the dominant style in public radio is a loose variety format whereby a station may actually switch from jazz, classical music, and alternative rock to news and talk during different parts of the day. Noncommercial radio remains the place for both tradition and experimentation, as well as for programs that would not draw enough listeners for commercial success.

Radio beyond Broadcasting

Over the past two decades, three alternative radio technologies have helped expand radio beyond its traditional AM and FM bands and bring more diverse sounds to listeners: Internet radio, satellite radio, and podcasting.

MEDIA LITERACY & THE CRITICAL PROCESS

Comparing Commercial and Noncommercial Radio

After the arrival and growth of commercial TV, the Corporation for Public Broadcasting (CPB) was created in 1967 as the funding agent for public broadcasting—an alternative to commercial TV and radio featuring educational and cultural programming that could not be easily sustained by commercial broadcasters in search of large general audiences. As a result, National Public Radio (NPR) developed to provide national programming to public stations to supplement local programming efforts. Today, NPR affiliates get just 2 percent of their funding from the federal government. Most money for public radio comes from corporate sponsorships, individual grants, and private donations.

① DESCRIPTION

Listen to a typical morning or late-afternoon hour of a local commercial news/talk radio station and a typical hour of your local NPR station from the same general time period for two to three days. Keep a log of what topics are covered and what news stories are reported. For the commercial station, log what commercials are carried and how much time in an hour is devoted to ads. For the noncommercial station, note both how much time is devoted to recognizing the station's sources of funding and who the supporters are.

② ANALYSIS

Look for patterns. What kinds of stories are covered? What kinds of topics are discussed? Create a chart to categorize the stories. To cover events and issues, do the stations use actual reporters at the scene? How much time is devoted to reporting compared to opinion? How many sources are cited in each story? What kinds of interview sources are used? Are they expert sources or person-on-the-street interviews? Are the sources people of different genders?

③ INTERPRETATION

What do these patterns mean? Is there a balance between reporting and

opinion? Do you detect any bias, and if so, what is it? Are the stations serving as watchdogs to ensure that democracy's best interests are being served? What effect, if any, do you think the advertisers/supporters have on the programming? What arguments might you make about commercial and noncommercial radio based on your findings?

④ EVALUATION

Which station seems to be doing a better job of serving its local audience? Why? Do you buy the 1930s argument that noncommercial stations serve narrow special interests while commercial stations serve capitalism and the public interest? Why or why not? From which station did you learn the most? Explain. Which station did you find most entertaining? Why? What did you like and dislike about each station?

⑤ ENGAGEMENT

Join your college radio station. Talk to the station manager about the goals for a typical hour of programming and the particular audience the station is trying to reach. Finally, pitch program, podcast, or topic ideas that would improve your college station's programming.

Internet Radio

Internet radio emerged in the 1990s with the popularity of the web. Internet radio stations have two different origins. In some cases, existing AM or FM (or satellite) stations "stream" a simulcast version of their on-air signal over the web, which listeners often access via an app. iHeartRadio is one of the major streaming sites for broadcast and custom digital stations. In other cases, stations are created exclusively for the Internet. Spotify, Apple Music, Pandora, 8tracks, and Last.fm are some of the leading Internet-only streaming audio services that offer personally curated stations, often based on listeners' favorite artists. Apple's Music 1 station is actually staffed by real deejays and operates round-the-clock from Los Angeles, New York, Nashville, and London. Most streaming stations don't have local hosts, choosing instead to focus on delivering a customized flow of music.

When it comes to royalty payments to artists, not all radio is alike. For decades, radio broadcasters have paid royalties to songwriters and music publishers but not to performing artists or record companies. Broadcasters have argued that the promotional value of getting songs played is sufficient compensation. But beginning in 2002, a Copyright Royalty Board established by the Library of Congress began to assess royalty fees on behalf of artists and record companies for streaming copyrighted songs over the Internet, based on a percentage of each station's revenue. Webcasters complained that royalty rates set by the board were too high and threatened their financial viability—particularly compared to satellite radio (see the next section), which pays a lower royalty rate to artists and record companies, and broadcasters, who pay no royalty rates to artists and record companies.

In 2009, Congress passed the Webcaster Settlement Act, which was considered a lifeline for Internet radio. The act enabled Internet stations to negotiate royalty fees directly with the music industry, at rates presumably more reasonable than what the Copyright Royalty Board had proposed. In 2012, Clear Channel (now iHeartMedia) became the first radio company to strike a deal directly with the recording industry. The company pledged to pay royalties to Big Machine Label Group for broadcasting its artists' songs in exchange for a limit on the royalties it had to pay for streaming those artists' music on its iHeartRadio.com site.

Clear Channel's deal with the music industry opened up a new dialogue about equalizing the royalty rates paid by broadcast radio, satellite radio, and Internet radio. Pandora founder Tim Westergren argued before Congress in 2012 that the rates were most unfair to companies like his. In the previous year, Westergren said, Pandora had paid 50 percent of its revenue to artists and record companies, whereas satellite radio had paid 7.5 percent, and broadcast radio had paid nothing. He noted that a car equipped with an AM/FM radio, satellite radio, and streaming Internet radio could deliver the same song to a listener through all three technologies, but the various radio services would pay markedly different levels of royalties to the artist and record company.[28]

LaunchPad
macmillan learning
launchpadworks.com

Radio: Yesterday, Today, and Tomorrow
Scholars and radio producers explain how radio adapts to and influences other media.

Discussion: Do you expect that the Internet will ultimately mean the end of broadcast radio, or will radio stations still be on the air decades from now?

Satellite Radio

A series of satellites launched to cover the continental United States created a subscription-based national **satellite radio** service. Two companies, XM and Sirius, completed their national introduction by 2002 and merged into a single provider (SiriusXM) in 2008. The merger was precipitated by the companies' struggles to make a profit after building competing satellite systems and battling for listeners. SiriusXM offers more than 175 digital music, news, and talk channels to the continental United States for a monthly subscription fee. The satellite radio service requires a special radio tuner, though access is available on mobile devices via an app.

Programming includes a range of music channels, from rock and reggae to Spanish Top 40 and opera, as well as channels dedicated to NASCAR, NPR, cooking, and comedy. Another feature of SiriusXM programming is popular personalities who host their own shows or have their own channels, including Howard Stern, Kenny Chesney, and Bruce Springsteen. U.S. automakers (investors in satellite radio) now equip most new cars with a satellite band, in addition to AM and FM, in order to promote further adoption of satellite radio. SiriusXM had about 34.9 million subscribers by 2020. To extend its audience beyond satellite listeners, SiriusXM bought Pandora in 2019 for $3.5 billion and now cross-markets the two businesses.

Podcasting

Developed in 2004, podcasting (the term marries *iPod* and *broadcasting*) refers to the practice of making audio files available on the Internet so that listeners can download or stream them on their phones or computers (see the chapter-opening discussion). This distribution method quickly became mainstream, as mass media companies created commercial podcasts to promote and extend existing content, such as news and reality TV, while independent producers developed new programs, such as *Crime Junkie*—a popular true-crime audio narrative. As noted earlier, more than one-third of Americans listen to podcasts monthly.[29]

Radio Stories from around the World

R adio can make us laugh, keep us connected, save our lives, and weaken our democratic self-governance with its absence. Here are four stories of radio from four countries.

England

What could the radio station Mansfield 103.2, do? In July 2017, a radio hacker (also called a pirate) with a high-quality transmitter was hijacking its signal and playing a deliberately offensive song about masturbation over the airwaves. It was happening over and over, and children were heard humming the ditty, called "The Winker's Song," in their parents' cars. Written and performed by Doc Cox (a.k.a. Ivor Biggun) in the 1970s, "The Winker's Song"—also called "The Wanker's Song" or "I'm a Wanker"—refers to a "wanker" thirty-six times; the song was never broadcast during the 1970s because of its obscene content.[1]

While the masturbation song was causing some listeners to laugh, others were aghast and incredulous that Britain's communications regulator could not catch the pirate(s). Hacking a radio signal is surprisingly easy; as long as people have portable transmitters and the know-how to lock into the same frequency and modulation as a radio station's receiving equipment, they can override the signal at the receiver. The tactic saw significant use during World War II and the Cold War. Today, signal jamming happens throughout China, Russia, the Middle East, Africa, and Pakistan. However, it is still uncommon in Britain and the

United States. According to Mansfield 103.2's managing director, "There [was] absolutely nothing we could do about it."

South Sudan

In South Sudan, the world's newest country (founded in 2011 in East Central Africa), the threat of COVID-19 in 2020 extended beyond the health problems directly associated with the coronavirus; there were fears that the virus would further burden a fragile health-care system already beset with treatable diseases like malaria, pneumonia, measles, and cholera.[2] With about eleven million people and a land mass just slightly smaller than the states of Washington, Oregon, and Idaho combined, South Sudan, which has been afflicted with political violence since 2013, has a weak mass communication network as well.

Radio has proven to be the best way to connect the country's citizens. The UN Mission in South Sudan built Radio Miraya, which broadcasts from the capital of Juba and is equipped with a relay system that carries the signal to most of the country. "For the great majority of people across South Sudan there is no internet, television or newspapers available. So how do people get their news? Some by word of mouth, but mostly from Radio Miraya," said David Shearer, the Head of UN Mission for South Sudan.[3] Shearer reports that the station reaches more than two-thirds of the country, and 80 percent of people in those areas tune in on a daily basis. The station broadcasts messages calling for a cease-fire, so people can respond to the pandemic, and information on hand washing and social distancing.

India

Radio saves lives in India. The fishermen of Kerala, a state in southern India, have begun to venture farther and farther away from the coast to catch their fish. The reason is overfishing (fish stocks has been depleted) and climate change (fish are migrating to different waters). Because their boats are small and flimsy, Kerala fishermen are at risk: one bad monsoon storm and they might not come home. Fortunately, a radio station called Radio Monsoon has begun providing daily weather forecasts in a number of local languages along the coast. These radio weather broadcasts have made fishing much safer for about thirty thousand families in the area, who depend on fishing for their livelihood.[4]

Hungary

As a post-communist country entering the twenty-first century, Hungary looked as if it would become a leading democracy in eastern Europe. Instead, the country has more recently slid backward, with an authoritarian leader, Viktor Orbán, taking control of nearly every institution (political, legal, cultural) in Hungary. After gaining a political majority in 2010, Orbán's next step was to take over the news media. Radio was an immediate target.

Orbán and his Fidesz Party appointed several new managers to head up Hungarian public radio, who then pushed out about one thousand employees (one-third of the entire radio staff) and enforced a propaganda mentality. The leader also created a National Media and Communications Authority, which set out to impose heavy fines for any news (both print and broadcast) that was critical of the government. Because the government of Hungary is a major advertiser in the national media, Orbán's government regularly withheld advertising from media organizations that it disfavored, which often brought financial ruin to the organizations. The surviving privately owned media are held by Orbán's allies.[5]

In April 2018, the latest casualty in Hungarian independent media was the popular commercial radio station Lánchíd Rádió. "The last independent commercial radio station to broadcast countrywide," it closed at the same time as *Magyar Nemzet*, the last major independent opposition newspaper in Hungary.[6] Both were owned by a political opponent of Orbán and closed within days of Orbán's party retaining power after the election. A Hungarian analyst explained that with the win by Orbán, the newspaper and radio station owner decided that "it was no longer worth his while" to maintain these businesses: "He [saw] the results [of the election], anticipate[d] government revenge, and is shutting down unprofitable media organisations."[7] ●

The Economics of Radio

Radio continues to be one of the most used mass media, reaching about 92 percent of American adults every week.[30] Because of radio's broad reach, the airwaves are very desirable real estate for advertisers, who want to reach people in and out of their homes; for record labels, who want their songs played; and for radio station owners, who want to create large radio groups to dominate multiple markets.

Local and National Advertising

About 10 percent of all U.S. spending on media advertising goes to broadcast radio stations. Like newspapers, radio generates its largest profits by selling local and regional ads. Thirty-second radio spot ads range from $1,500 in large markets to just a few dollars in the smallest markets. In 2019, radio earned $14.1 billion in gross advertising revenue.[31] About 75 percent of radio revenues are from local ad sales, with the remainder in national spot, network, and digital radio sales. Digital sales—advertisements on web pages and apps, for example—are an area of advertising growth for radio and currently account for about 10 percent of revenues.[32] Although industry revenue has dropped from a peak of $21.7 billion in 2006, the number of stations remains relatively stable, totaling 15,460 stations (4,560 AM stations, 6,704 FM commercial stations, and 4,196 FM educational stations).[33]

Unlike television, in which nearly 40 percent of a station's expenses goes toward buying syndicated programs, local broadcast radio stations get much of their content free from the recording industry. Therefore, only about 20 percent of a typical radio station's budget goes toward covering programming costs. But as noted earlier, that free music content is in doubt, as the music industry—which already charges satellite and Internet radio stations royalties for artists and record companies—moves toward charging radio broadcasters similar royalties for playing music on the air.

When radio stations want to purchase programming, they often turn to national network radio, which generates more than $1 billion in ad sales annually by offering dozens of specialized services. For example, Westwood One—the nation's largest radio network service, managed by Cumulus Media—reaches more than 245 million consumers a week with a range of programming, including regular network radio news (CNBC, CNN), entertainment programs (*The Bob & Tom Show*), talk shows (*The Mark Levin Show*), and complete twenty-four-hour formats (*Hot Country, Hits Now!, Classic Hip-Hop*). Dozens of companies offer national program and format services, typically providing local stations with programming in exchange for time slots for national ads. The most successful radio network programs are the shows broadcast by affiliates in the Top 20 markets, which offer advertisers half of the country's radio audience.

Manipulating Playlists with Payola

Because radio airplay can increase music industry profits, there has long been the temptation to illegally influence radio playlists. **Payola**, the practice by which record promoters pay deejays to play particular records, was rampant during the 1950s as record companies sought to guarantee record sales (the more airplay a recording got, the more popular it was). In response, management increasingly took control of programming, arguing that if individual deejays had less say in which records would be played, they would be less susceptible to bribery. But despite congressional hearings and new regulations, payola persisted. In 2010, Univision Radio paid $1 million to settle allegations of payola and end an FCC investigation.

More recently, the practice of payola has resurfaced as streaming radio services have grown. But because streaming services are not broadcasting, they fall outside the FCC's oversight. *Billboard* magazine reports that music promoters have been paying to influence playlists at services like Spotify, Deezer, and Apple Music. These playlists, used by hundreds of thousands of subscribers as a way to discover music, are created by the streaming services, influential individuals, or the music labels themselves. Spotify announced in 2015 that it would prohibit any playlists that had been influenced by money or other compensation. Yet the three major music corporations are still invested in influencing streaming music: Universal Music Group features its music playlists on Digster, Sony showcases its music on Filtr, and Warner Music Group promotes its playlists on Topsify.[34]

In 2020, payola was back as a concern at the FCC, which requested that the three major music companies report on how they were complying with federal anti-payola policies. The request came after *Rolling Stone* magazine reported that pay-for-play still exists in broadcast radio.[35]

Broadcast Radio Ownership: From Diversity to Consolidation

Public airwaves may be public, but the infrastructure behind radio is run by corporate interests. The **Telecommunications Act of 1996** substantially changed the rules concerning access to public airwaves because the FCC eliminated most ownership restrictions on radio. As a result, twenty-one hundred stations and $15 billion changed hands that year alone. From 1995 to 2005, the number of radio station owners declined by one-third, from sixty-six hundred to about forty-four hundred.[36]

Once upon a time, the FCC tried to encourage diversity in broadcast ownership. From the 1950s through the 1980s, a media company could not own more than seven AM, seven FM, and seven TV stations nationally and could own only one radio station per market. Just prior to the 1996 act, the ownership rules were relaxed to allow any single person or company to own up to twenty AM, twenty FM, and twelve TV stations nationwide, but only two in the same market.

The 1996 act allows individuals and companies to acquire as many radio stations as they want, with relaxed restrictions on the number of stations a single broadcaster may own in the same city: The larger the market or area, the more stations a company may own within that market. For example, in areas where forty-five or more stations are available to listeners, a broadcaster may own up to eight stations, but not more than five of one type (AM or FM). In areas with fourteen or fewer stations, a broadcaster may own up to five stations (three of any one type). In very small markets with a handful of stations, a broadcast company may not own more than half the stations.

After passage of the 1996 Telecommunications Act, several enormous radio corporations came to dominate American radio. Consider the cases of Clear Channel Communications (now iHeartMedia) and Cumulus, two of the largest radio chain owners in terms of number of stations owned (see Table 5.3). Clear Channel Communications was formed in 1972 with one San Antonio station. Eventually, it gobbled up enough conglomerates to become the largest radio chain owner in the country, peaking in a pre-recession 2005 with 1,205 stations. Today, as iHeartMedia, it owns 856 radio stations, with 150 of them concentrated in the top twenty-five markets. iHeartMedia also distributes many of the leading syndicated programs—including *The Rush Limbaugh Show*, *The Glenn Beck Program*, *On Air with Ryan Seacrest*, and *Delilah*—through its Premiere Networks business, and sponsors more than twenty thousand local events and eight national events a year, including the iHeartRadio Music Festival. As noted earlier, iHeartMedia is also an Internet radio source with iHeartRadio, which has more than 138 million registered users. Cumulus became the second-largest commercial radio conglomerate when it merged with Citadel in 2011 in a $2.5 billion deal and bought radio network service Dial Global (now Westwood One) in 2013. In 2018, both iHeartMedia and Cumulus declared bankruptcy, burdened with billions of dollars of debt from acquiring so many stations, and laden with the problem of declining radio advertising revenue.[37] Both companies maintained their station broadcasts while restructuring their debt payback plans to creditors.

TABLE 5.3   TOP 10 RADIO COMPANIES (BY NUMBER OF STATIONS), 2020

Rank	Company	Number of Stations
1	iHeartMedia (Top property: WLTW-FM, New York)	856
2	Educational Media Foundation (KLVB, Citrus Heights, CA)	838
3	Cumulus Media (KNBR-AM, San Francisco)	424
4	Townsquare Media (KSAS-FM, Boise)	321
5	Entercom (WEEI-AM, Boston)	235
6	Alpha Media (KINK-FM, Portland, OR)	198
7	American Family Radio (WAFR, Tupelo, MS)	193
8	Saga Communications (WSNY, Columbus, OH)	113
9	Salem Media Group (KLTY, Dallas–Fort Worth)	99
10	Midwest Communications (WTAQ-FM, Green Bay)	82

Data from: The 10-K annual reports and business profiles for each radio company.

Townsquare Media, the fourth largest radio company, launched in 2010 with the buyout of a sixty-two-station group and continued to grow by focusing on acquiring stations in midsize markets. Entercom, the Pennsylvania-based radio chain, nearly doubled its number of stations after its 2017 acquisition of CBS Radio. Two other major radio groups, the Educational Media Foundation and American Family Radio, are nonprofit religious broadcasters. The Educational Media Foundation, which operates the K-Love and Air1 contemporary Christian music formats on its hundreds of stations, has continued to expand. American Family Radio is a conservative Christian activist organization that was originally established by Rev. Donald Wildmon in 1977 as the National Federation for Decency. Salem Media Group is a commercial Christian-format radio chain.

A smaller radio conglomerate, but one that is perhaps the most dominant in a single-format area, is Uforia Audio Network, part of Univision Communications. Through Uforia, Univision is the top Spanish-language radio broadcaster in the United States (with about sixty-one broadcast and digital stations). Univision is also the largest Spanish-language television broadcaster in the United States, owner of the top Spanish-language broadcast and cable networks (including Univision, UniMás, Galavisión, and TUDN, a sports network), and owner of Univision Online, one of the most popular Spanish-language websites in the United States.

Radio and the Democracy of the Airwaves

As radio was the first national electronic mass medium, its influence in the formation of American culture cannot be overestimated. Radio has given us soap operas, situation comedies, and broadcast news; it helped popularize rock and roll, car culture, and the politics of talk radio. Yet for all its national influence, radio's strength in the United States has been its appeal as a local medium (see "Global Village: Radio Stories from Around the World" on pages 148–149). For decades, listeners have tuned in to hear the familiar voices of their community's deejays and talk-show hosts and hear the regional flavor of popular music over publicly owned airwaves.

The early debates over radio gave us one of the most important and enduring ideas in communication policy: a requirement to operate in the "public interest, convenience, or necessity." But the broadcasting industry has long been at odds with this policy, arguing that radio corporations invest heavily in technology and should be able not only to have more control over the radio frequencies on which they operate but also to own as many stations as they want. Deregulation in the past few decades has moved the industry closer to that corporate vision, as nearly every radio market in the nation is dominated by a few owners, whose broadcasting licenses need only be renewed every eight years.

This trend in ownership has moved radio away from its localism, as radio groups often manage hundreds of stations from afar. Large radio groups frequently create audio broadcasts for multiple stations with a smaller staff than they had decades before the wave of consolidation. At many stations, this has undermined the personal connection with listeners that radio has long enjoyed. For decades, the cost-cutting practice of **voice-tracking**—in which deejays digitally record their "breaks" between songs in advance so that they don't have to appear live for their shift—has challenged the concept of "live" radio at stations across the country. At many times during the day, radio station programs run off a computer, with a virtual, voice-tracked deejay and no one actually present in the studio.

In the 1990s, as large corporations gained control of America's radio airwaves, activists in hundreds of communities across the United States protested by starting up their own noncommercial "pirate" radio stations, capable of broadcasting over a few miles with low-power FM signals of 1 to 10 watts, arguing that they were the defenders of the local public interest as radio became more corporatized. The National Association of Broadcasters and other industry groups pressed to stop the pirate broadcasters, citing their illegality and their potential to create interference with existing stations. Between 1995 and 2000, more than five hundred illegal micropower radio stations were shut down. Still, an estimated one hundred to one thousand pirate stations are in operation in the United States, in both large urban areas and small rural towns.

In 2000, the FCC, responding to tens of thousands of inquiries about the development of a local radio broadcasting service, approved a new noncommercial **low-power FM (LPFM)** class of 100-watt stations (with a broadcast reach of about five miles) to give voice to local groups lacking access to the public airwaves. LPFM station licensees included mostly religious groups but also high schools, colleges and universities, Native American nations, labor groups, and museums. By 2020, more than 2,342 LPFM stations were licensed to broadcast. A major advocate of LPFM stations is the Prometheus Radio Project, a nonprofit formed by radio activists in 1998. Prometheus has helped educate community organizations about low-power radio and has sponsored at least a dozen "barn raisings" to build community stations in places like Hudson, New York; Opelousas, Louisiana; and Woodburn, Oregon.

Of course, independent streaming radio stations have offered an alternative to corporate radio as well. The Internet offers an easier route for start-up stations to reach an audience; but because streaming stations do not operate on the public airwaves of the broadcast spectrum, they are not legally obligated to operate in the public interest, convenience, or necessity.

Still, the largest radio broadcasting chains remain dominant in the industry. Given broadcasters' reluctance to publicly raise questions about their own economic arrangements, public debate regarding radio as a natural resource has remained minuscule. As citizens contemplate the future, two big questions remain: With a few large broadcast companies (both commercial and nonprofit religious groups) now permitted to dominate radio ownership nationwide, how much is consolidation of power restricting the number and kinds of voices permitted to speak over the limited spectrum of public airwaves? How important is a diversity of ownership and voices on those public airwaves? To ensure that mass media industries continue to serve democracy and local communities, the public needs to play a role in developing the answers to these questions.

Thomas Fricke/Getty Images

ALTERNATIVE RADIO VOICES can be found on college stations, typically started by students and community members. There are approximately six hundred such stations currently active in the United States, broadcasting in an eclectic variety of formats.[38] Throughout the years, college radio has become a major outlet for new indie bands.

Chapter 5 REVIEW

LAUNCHPAD FOR *MEDIA & CULTURE*

 **LaunchPad**
macmillan learning

Visit LaunchPad for *Media & Culture* at **launchpadworks.com** for additional learning tools:

→ QUIZ YOURSELF WITH LEARNINGCURVE

LearningCurve uses gamelike quizzing to help you review important concepts from this chapter. LearningCurve includes multiple-choice questions at different levels of difficulty, hints, feedback for both right and wrong answers, and links to e-book material for easy reference. Which answer choice would you select for the sample question below?

> By August 1922, AT&T had sold its first radio ad to a New York real estate developer for $50. What effect did this have?
>
> ◯ Cease and desist letters were sent.
>
> ◯ AT&T began making and selling its own radio receivers.
>
> ◯ Radio was no longer seen as a public service.
>
> ◯ AT&T created a monopoly in radio.

→ WATCH VIDEO ON KEY CONCEPTS

LaunchPad includes clips from movies, TV shows, online sources, and other media texts, in addition to interviews with media experts and newsmakers. The videos for this chapter address topics such as radio's past, present, and future and the latest in podcasting.

→ PRACTICE THE CRITICAL PROCESS

Use the Media Literacy Activity to walk through the critical process step-by-step and develop a critical perspective on what media means to you. In this chapter's activity, you'll consider the influence of radio in your life and the lives of your friends.

→ LEARN KEY TERMS

Review the definitions for this chapter's key terms with LaunchPad's easy-to-use online flash cards. Try the cards in practice mode, or quiz yourself as a way to focus your study efforts. (Page numbers indicate where each term is highlighted within the chapter.)

podcasts, 123
telegraph, 125
Morse code, 125
electromagnetic waves, 125
radio waves, 126
wireless telegraphy, 126
wireless telephony, 127
broadcasting, 128
Wireless Ship Act of 1910, 129
Radio Act of 1912, 129
Radio Corporation of America (RCA), 129
network, 131
option time, 134
Radio Act of 1927, 134
Federal Radio Commission (FRC), 135

Communications Act of 1934, 135
Federal Communications Commission (FCC), 135
transistors, 138
FM, 139
AM, 139
format radio, 140
rotation, 140
Top 40 format, 140
progressive rock, 141
album-oriented rock (AOR), 141
drive time, 141
news/talk, 142
adult contemporary (AC), 144
contemporary hit radio (CHR), 144

country, 144
urban contemporary, 144
Pacifica Foundation, 145
National Public Radio (NPR), 145
Public Broadcasting Service (PBS), 145
Public Broadcasting Act of 1967, 145
Corporation for Public Broadcasting (CPB), 145
Internet radio, 146
satellite radio, 147
payola, 150
Telecommunications Act of 1996, 151
voice-tracking, 153
low-power FM (LPFM), 153

REVIEW QUESTIONS

Use the questions below to revisit key themes and concepts from this chapter.

✱ Need more practice? Access the LearningCurve multiple-choice questions for this chapter, which are available on LaunchPad. All questions include feedback for correct and incorrect answers, hints, and links to e-book material.

Radio Today

1. How does radio today involve both mass communication and masspersonal communication?

How We Got Here: The Early Development of Radio

2. Why was the development of the telegraph important in media history? What were some of the disadvantages of telegraph technology?

3. How is the concept of wireless telegraphy different from that of radio?

4. What was Guglielmo Marconi's role in the development of wireless telegraphy?

5. How did broadcasting, unlike print media, come to be federally regulated?

6. Why was the RCA monopoly formed?

The Evolution of U.S. Radio

7. What was AT&T's role in the early days of radio?

8. Why did the government-sanctioned RCA monopoly end?

9. What is the significance of the Radio Act of 1927 and the Communications Act of 1934?

Radio Reinvents Itself

10. How did radio adapt to the arrival of television?

11. What was Edwin Armstrong's role in the advancement of radio technology? Why did RCA hamper Armstrong's work?

12. How did the arrival of the Top 40 radio format change the role of radio deejays?

The Sounds of the Contemporary Radio Environment

13. Why are there so many radio formats today?

14. How did talk radio come to be dominated by conservative voices?

15. What is the state of nonprofit radio today?

16. Why are royalties a topic of debate between broadcast radio, satellite radio, Internet radio, and the recording industry?

17. Why are podcasts so popular now?

The Economics of Radio

18. What are the current ownership rules governing American radio?

19. What has been the main effect of the Telecommunications Act of 1996 on radio station ownership?

Radio and the Democracy of the Airwaves

20. Throughout the history of radio, why did the government encourage monopoly or chain ownership of radio broadcasting?

21. Why did the FCC create a new class of low-power FM stations?

22. What is the relevance of localism to debates about ownership in radio?

QUESTIONING THE MEDIA

Use these critical-thinking questions to reflect on your own experiences with media and apply your understanding of media concepts.

✱ For an interactive experience, questions on topics like these are available for the iClicker student response system. Instructors can download the questions from LaunchPad and use them to poll the class.

1. Count the number and types of radio stations in your area today. What formats do they use? Do a little research, and find out who owns the stations in your market. How much diversity in both ownership and content is there among the highest-rated stations?

2. If you could own and manage a commercial radio station, what format would you choose, and why?

3. How might radio be used to improve social and political discussions in the United States?

4. If you were a broadcast radio executive, what arguments would you make in favor of broadcast radio over streaming radio?

Television: From Broadcasting to Streaming

6

On a Friday night in March 2020, just as much of the world went into coronavirus lockdown, Netflix started streaming *Tiger King: Murder, Mayhem and Madness*, a seven-episode true-crime docuseries about an eccentric big-cat breeder in Oklahoma. By the end of March, the series was seen by over 34 million people in the United States, according to Nielsen rating estimates, making it one of Netflix's biggest hits to date and millions of people's latest guilty pleasure.[1] By contrast, the most watched series on broadcast or cable TV that spring was CBS's *NCIS*, which drew in 15.3 million viewers per episode.[2]

As this example illustrates, Netflix has proven itself to be a major player in television, delivering a diverse array of programming to over 190 million subscribers around the world.[3] In addition to airing talked-about true-crime stories like *Tiger King* and established fan favorites like *Stranger Things* and *The Crown*, the company has been on an acquisition frenzy to continue building its content library, creating licensing deals to air recent episodes of broadcast series like the CBS sitcom *The Unicorn*, signing up late-night talk-show host David Letterman to do a series of high-profile interviews, and building a relationship with Michelle and Barack Obama to stream socially engaged documentaries produced by their new production company.[4] In 2020 alone, Netflix was expected to spend $17.3 billion on original TV shows and movies.[5]

Netflix also looks internationally to build out its programming. In fact, one key to Netflix's success in the United States has been recognizing that viewers are interested in programming from other nations and willing to watch movies and TV shows that are subtitled, a taboo in the old network era of television. In 2018, Netflix aired its first series in Arabic (a supernatural thriller titled *Jinn*) and in Danish (a postapocalyptic series titled *The Rain*), and by 2020, the company was streaming scores of foreign-language TV series and movies from every corner of the globe.

Historically, U.S. film studios, networks, and cable services have created their own versions of programs from other countries, like *The Office*, *Big Brother*, and *The Voice*, but streaming a subtitled original program is a lot cheaper than producing a new U.S. version. As a platform for the global

◀ Original content from streaming providers, like Netflix's *Tiger King*, has shifted the way we consume TV.

delivery of content, then, Netflix is changing both the economics of making and the experience of watching TV. In discussing Netflix's German-language series *Dark*, media analyst Rich Greenfield noted, "[With the Netflix model,] you're not just making German content for Germany, you're making German content for all subscribers all around the world. It makes it much more economical to produce content."[6]

For this and a number of other reasons, many media analysts and traditional network executives view Netflix as a disruptive force, turning the TV business on its head. One of these other reasons is that streaming services like Amazon Prime and Netflix are mostly financed by inexpensive monthly subscription packages and are not subject to the pressures of TV ratings or advertising deals. In addition, Netflix has a stable of executives authorized to find good stories from all over the world. Netflix's chief content officer, Ted Sarandos, has said, "Every one of our creative executives has buying power. . . . They can greenlight a big-budget show in the room without me."[7] Third, in doing original programming, Netflix is not locked into the old network dictate of making twenty-two to twenty-four shows per season, often shooting a single episode in less than two weeks. Instead, the company has adopted the premium cable model of HBO and Showtime, producing series with as few as six and usually no more than twelve episodes per season. This kind of production schedule promotes quality, with fewer episodes and more time to do each one. Finally, Netflix is not beholden to a traditional fall launch schedule, when the most promising network shows debut; in fact, Netflix promotes and releases new programs pretty much every week.

Netflix's enormous subscriber base demonstrates its appeal to viewers who appreciate the wide international choice of stories. As the company operates in almost two hundred countries, Netflix—probably more so than any other entertainment company—understands the cross-cultural power of storytelling and thus is gambling on a model for story distribution that seems to be a safe bet. What is still unclear is the future of network television. One clue to the networks' next steps may be that they, too, are staking part of their future on streaming. By 2021, ABC had Disney+, Hulu, and ESPN+; NBC had Peacock; and CBS had Paramount+, all competing with Netflix, Amazon Prime, HBO Max, and a number of other rival streaming services.

> → **How much of your TV viewing takes place through Netflix or other streaming services versus cable or broadcast? What new content have you discovered through Netflix? Are any of these shows international series?**

Television Today

A decade ago, as people became tethered to their smartphones and laptops, some observers predicted the death of television. That hasn't happened. While viewers—especially young viewers—may not be sitting in front of a traditional TV set watching broadcast programs, they are engaging with TV content through their smartphones and computers at record levels.[8] Instead of getting replaced by digital media, TV has evolved—through sometimes radical and still ongoing changes—into a vital hub in our digital media environment.

TV today is vast and varied. Think of the array of programming TV delivers to us all day, every day—everything from sitcoms, dramas, game shows, and talk shows to cartoons, films, sports, and news. This content is produced and delivered to us by a wide range of industry players: local broadcast stations, networks, cable channels, streaming services, social media platforms, Hollywood studios, and international production companies. It's also supported by a range of funding models, such as advertising, subscription, pay-per-view, and public support. We engage with TV content on a range of devices, from flat-screen TVs and laptops to smartphones and jumbotrons. We can use our TV sets for a range of activities: downloading films, streaming music, playing video games, and surfing YouTube, as well as consuming traditional TV content. And finally, we can experience that content in different ways: live as it is broadcast, time-shifted via a digital recorder, binged on Netflix, or discovered via a YouTube rabbit hole. In today's complex digital environment, saying "I'm watching TV" can mean an increasingly diverse number of things.

This chapter helps us better understand today's television environment by tracing its evolution through four stages: the birth of commercial broadcasting, the network era, the post-network era, and the current digital era (see Figure 6.1). One reason television is so complex is that many of the programs, programming trends, industry practices, government regulations, and cultural attitudes about television that emerged during these different periods still remain today. Rather than being replaced by new developments, they exist alongside them.

In this chapter, we examine television's cultural, social, and economic impact. We will:

- Trace television's early technological development and regulation
- Discuss key features of the broadcast TV model
- Learn about the different ways TV shows have been produced over the years
- Examine when and how the Big Three broadcast networks (NBC, CBS, and ABC) came to dominate the industry
- Learn how the TV industry and viewers' experiences are being transformed by new technologies and increased competition
- Explore how TV's economics have shaped programming
- Investigate the impact of television on democracy and culture

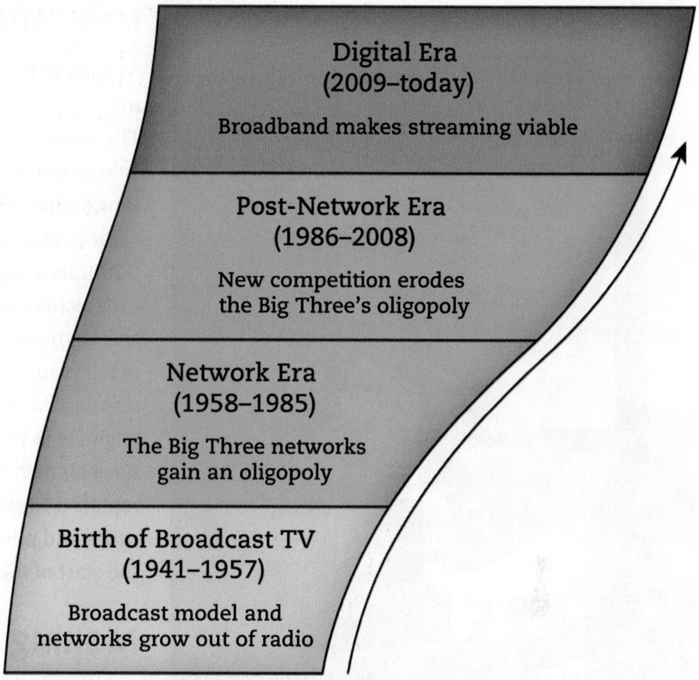

FIGURE 6.1

FOUR ERAS IN THE EVOLUTION OF TELEVISION

The TV industry has experienced dramatic changes in recent years as streaming services challenge old ways of making, delivering, and watching television. Even so, numerous programs, programming trends, viewing habits, institutions, business practices, regulations, and cultural attitudes about television from the medium's past remain firmly in place, coexisting alongside emerging developments. To help you navigate that complex landscape, we have divided the medium's history into four stages, identifying elements from those earlier eras that continue to shape the industry today.

Ollie The Designer/Shutterstock

How We Got Here: The Birth of Broadcast TV

In 1948, only 1 percent of America's households had a TV set, and early television delivered only a few hours of programming on weekday evenings to those viewers. By the early 1960s, over 90 percent of households had televisions, and programming had expanded to morning, afternoon, and even late night, seven days a week. TV's initial, formative phase was a time of some experimentation as producers and regulators figured out the new medium. But it was also a period during which fundamental aspects of the TV industry were established, including many that remain in place today.

CATHODE RAY TUBE A ghostly image of Hollywood actress Joan Crawford appears on the surface of a cathode ray tube in this 1934 demonstration of Philo Farnsworth's electronic TV system.

Inventing Visual Radio

As early as the 1870s, inventors around the world began to ask whether transmitting pictures from one place to another via electrical signals was possible. The question led to decades of experimentation in what was variously called "visual radio," "electric vision," and, of course, "television." In the 1880s, for example, German inventor Paul Nipkow developed a rudimentary TV camera that connected electric sensors to a mechanical spinning disk with perforated holes. Then, in the 1890s, physicists began developing *cathode ray tube* technology that made it possible to project electronic signals onto a glass screen inside a vacuum tube. These tubes would serve as the inner workings of television sets and computer monitors until flat-screen technology became standard in the 2000s. And it was during the 1920s that Vladimir Zworkyin (an engineer at Westinghouse) and Philo Farnsworth (a teenage prodigy who ended up at his own tech start-up) independently invented fully electronic systems that replaced Nipkow's mechanical spinning disk. Their cameras used vacuum tubes and photosensitive plates to convert light rays into electrical signals. By the start of the 1930s, TV's basic technological foundations were in place.

Making TV Ready for the Masses

During the 1930s, powerful companies in the radio industry, such as RCA, CBS, and Westinghouse, saw television as an extension of their existing businesses and as a way to expand their profits. As a result, they battled to control its future. They developed large in-house research labs to advance TV technology, purchased patents to control the manufacture of TV equipment, and established experimental broadcast stations to work out the kinks of producing and transmitting TV content. By the end of the 1930s, industry forces, especially RCA, began pushing for TV's official debut as a commercial mass medium. For that to happen, however, the industry's regulating body, the Federal Communications Commission (FCC), had to resolve technical standards and settle on who was able to broadcast over the airwaves.

TV's First Technical Standards

Figuring out how to push TV as a business and elevate it to a mass medium meant establishing a single technical standard for television in the United States, so consumers could purchase a TV set and have confidence that it wouldn't quickly become obsolete as manufacturers made upgrades. RCA wanted its technology to become the standard and lobbied hard for FCC approval as early as 1938, though most of RCA's competitors argued that adopting RCA's inferior system would be a disservice to the public and give the company an unfair advantage in TV manufacturing.

After a period of intense struggle, a compromise was reached in 1941, when the FCC adopted an analog standard and approved the start of commercial TV (although U.S. entry into World War II later that year effectively put commercial TV on hold until the war ended in 1945). In the United States, TV would be black and white and have a slightly higher picture quality than the system RCA had pushed for: 525-line resolution compared to 441. Although some countries (e.g., Canada and much of Latin America) adopted the U.S. standard, most of Europe and Asia adopted a superior standard with 625-line resolution, creating international incompatibilities.

Only a few years later, another battle erupted between major manufacturers, as RCA and CBS developed competing technical systems for color TV—a battle ultimately won by RCA in large part because existing sets could receive its color images in black and white, meaning customers wouldn't have to replace their televisions. The FCC approved RCA's color TV standard in 1954, but it wasn't until 1966, when the consumer market for color sets really took off, that most evening lineups were broadcast in color.

As TV's early development makes clear, media technologies rarely appear from a lab fully formed and driven by a desire to provide consumers with the best possible product. Instead, they are shaped by corporate priorities and depend on government action. In 2009, for example, after another long process involving competing corporate lobbying efforts, the existing analog standard was replaced by the digital standard we have today (based on binary code), which provides high-definition broadcast TV images and sound quality and requires less spectrum space. With that change, broadcast TV fully entered the digital era.

Who Gets TV? Allocating Spectrum Space and Broadcast Licenses

The second issue that the FCC had to resolve involved licensing space in the electromagnetic spectrum to TV stations. As the birth of radio decades earlier made clear, the spectrum is a limited resource; a country's government can allocate only so many frequencies before stations interfere with one another. Competition for existing frequencies was fierce in the 1930s and 1940s, when both TV and the new FM-radio format needed access to the spectrum in order to grow. TV won this original battle, giving TV stations access to more frequencies and delaying FM radio's development until the 1960s. But equally intense battles over spectrum space are being waged again today, as the rapid growth of mobile devices creates ever more demand for frequencies. In a decision not without controversy, TV stations have been moved to different parts of the spectrum in recent years to open up space for telecommunications companies like AT&T and Verizon.

TV STANDARDS Although color TV sets were on the market as early as 1954, consumers were slow to bring them into their homes, as they were expensive and most programs were still broadcast in black and white. Color TV ownership rates wouldn't increase until the mid-1960s.

Despite its early victory over radio, TV's rollout as a mass medium proved complicated. In 1948, after granting only 108 broadcast licenses, the FCC realized that it needed to devote even more spectrum space to TV stations than originally planned. The commission issued a licensing freeze, meaning that it stopped granting new broadcast licenses until it could work out a solution to the spectrum problem. Meanwhile, existing stations continued to operate, which meant that cities like New York and Chicago had several TV stations, while other cities—even major hubs like Austin and Portland—had none. Accidental though this may have been, creating a divide between cities with and without TV ended up revealing the new medium's enormous impact. In non-TV cities, movie attendance continued to increase during the freeze, while cities with TV stations saw a 20 to 40 percent drop in movie attendance. Taxi receipts and nightclub attendance also fell in cities with TV, as did library book circulation and radio listening. TV was clearly popular.

When the freeze ended in 1952, the number of stations rose quickly, and by the mid-1950s, there were more than four hundred television stations in operation. At last, TV had become a truly national mass medium.

The Broadcast Model Defines Early TV

The TV stations established in the 1940s and 1950s followed the broadcast model popularized by radio: Content was delivered by local, over-the-air stations. Although a few nonprofit educational stations existed, the vast majority of these early TV stations were commercial stations supported by advertisements. As noted in Chapter 5, the United States' reliance on a commercial system differed from noncommercial state-chartered broadcasting systems like the BBC, which evolved in Great Britain, as well as similar systems established in much of Europe, Asia, and Africa. The broadcast model (with and without commercials) was essentially the only way television programming was delivered in the United States for several decades, and it remains the only way viewers can access local and free TV content today. To understand the broadcast model, we need to examine the nature of local TV stations and the broadcast networks to which many of them are connected.

Local Stations

In a TV landscape filled with cable channels and streaming services, it can be easy to overlook broadcast TV stations, but to do so would be a mistake. To understand U.S. television—its past, present, and

BROADCAST TV TECHNOLOGY Until cable penetration rates rose in the 1980s, TV antennas could be seen rising above the rooftops of most American homes, capturing the broadcast signals being sent from TV transmission towers in the area.

H. Armstrong Roberts/ClassicStock/Getty Images

future—we need to understand the role local stations have played in the TV industry. Here are several of local broadcast television's defining features:

- **Free of charge.** Local broadcast television is free TV; viewers can watch any broadcast TV station in their area as long as they have a TV set and an antenna, which might be built into the set or installed outside the house. This is how 90 percent of people in the United States watched TV until the 1980s. Commercial stations cover their operating costs by selling advertising, while noncommercial (public) stations rely on viewer and corporate donations and government funding.

- **Licensed.** Local TV stations, like local radio stations, must get a license from the FCC to broadcast on their specific channel in a specific market, and they are given a four-letter call sign, like WCPO (a station in Cincinnati) or KLTA (in Los Angeles). A license is temporary and must be renewed, usually every five years. In exchange for the exclusive rights to use part of the airwaves (which are understood to be a public resource), license holders agree to operate their station in the public interest, and failure to do so can, theoretically, result in nonrenewal of the broadcast license (for more, see Chapter 5).

- **Geographically divided.** TV stations are assigned to **designated market areas**, or DMAs, and there are 210 TV markets in the United States, ranging in population from the top three—New York, Los Angeles, and Chicago—to No. 210, Glendive, Montana. There are about 1,750 local TV stations in the United States in total, with most markets having at least six stations, and many having over ten.

- **Light on local programming.** In the early decades of television—from the 1950s through the 1970s—local stations produced a wide range of content for their communities, including children's programs, teen dance shows, and locally hosted movie series. Today, broadcast stations produce a limited amount of their own programming, mostly local news and public affairs shows. Filling the bulk of local TV station schedules is network programming and syndicated programming, which they acquire to fill other time slots (more on networks and syndicated programming in later sections).

Today, most viewers access their local broadcast stations via a cable provider or streaming service, like Hulu + Live TV or YouTube TV, rather than by an antenna. As a result, some viewers might struggle to differentiate broadcast stations (which serve a local community, require an FCC license, and are expected to serve the public interest) from cable channels like ESPN, Bravo, or MTV (which are national and not regulated in the same way as broadcast stations).

Broadcast Networks

Networks (e.g., ABC, CBS, the CW, Fox, and NBC) are also key elements in television's broadcast model, playing a twofold role. First, each broadcast network *owns and operates* a limited number of broadcast stations. These **O&Os**, as they are called, are usually located in the largest TV markets. For example, Disney (the parent company of ABC) owns eight local TV stations, including WABC in New York.

Second, and more significant, each network puts together a lineup of TV programs, which it delivers to a nationwide group of stations, which includes the O&Os and almost two hundred **affiliate** stations. In each TV market where a network doesn't have an O&O, the network will usually enter into an exclusive contractual relationship with one local station. In doing so, the network promises to provide programming to the affiliated stations, which in turn promise to air the network's lineup. For example, in Nashville, WRKN-TV, Channel 2, is owned by the Nexstar Media Group but is an ABC affiliate that airs ABC programming. (Nexstar is the largest local TV station owner in the United States, with 198 stations; each of its local stations is affiliated with one of the major networks.)

TV networking in the 1950s grew directly from radio. CBS and NBC, the most powerful players in radio, quickly became powerful players in TV. They did so by channeling their radio profits into building expensive TV production facilities, creating affiliate relationships with the strongest TV stations across the country, and exploiting strong brand recognition. ABC, which was the weakest of the three networks in radio, initially struggled to compete in TV and remained the third place TV network until the 1970s. A fourth early TV network, DuMont, was established in 1942. But DuMont didn't grow out of a radio network, and without those resources and adequate financing, it had to shut down in 1956. As we will discuss later, a new generation of networks appeared decades after (Fox in 1985; UPN and the WB in 1995), but rather than growing out of radio, these new networks were created by Hollywood production companies.

Independent Stations

Some broadcast stations are neither owned by nor affiliated with a network. These *independent stations* must either produce or buy programming themselves. Beginning in the 1950s, some independent stations were set aside for educational purposes. Others were commercial stations that were typically the fourth or fifth station licensed to operate in a TV market; as a result, they appeared after all available network affiliations had already been taken by other stations. In TV's early years, such independent stations were most often located in bigger TV markets, where audiences and potential ad revenue were large enough to keep an independent station financially viable (WTBS in Atlanta, WGN in Chicago). Over time, however, independent stations have appeared in smaller TV markets across the country, usually as religious or low-power stations.

How Should It Look? Making TV Programs

Like the networks themselves, television programming in the 1950s drew heavily on radio. TV imported many of radio's most popular stars and genres, including variety shows, situation comedies, quiz/game shows, crime/detective shows, and soap operas. In fact, several successful radio series moved to television, including situation comedies like *Amos 'n' Andy*, variety shows like *Texaco Star Theater*, and sports shows like *Friday Night Fights*. As a visual medium, however, producers also looked to movies, theater, and vaudeville for inspiration and talent. During the 1950s, as television producers experimented with the new medium, two ways of making TV shows emerged: three-camera live and one-camera film.

Live TV

In the early 1950s, up to 80 percent of TV shows—including dramas, sitcoms, and variety shows—were broadcast live, a fact that was an important selling point for the industry as it tried to convince consumers to buy TV sets. Unlike records or movies, radio and TV technologies made it possible for people to listen to or watch an event as it was happening. Ads often touted that TV could turn a sofa into what felt like the best seat in a Broadway theater.

THE GOLDBERGS (1949–1956), a domestic situation comedy about an immigrant Jewish family assimilating to life in America, was adapted from a long-running successful radio series that debuted in 1929. The TV show was produced with a three-camera system in front of a live audience.

Everett Collection, Inc

Producers developed a *three-camera live* system for producing live television. As the name implies, the system used three electronic TV cameras, which were connected to a control booth where the feeds from all three cameras could be monitored. A device called a *switcher* allowed the director to change which camera's feed left the control booth to go out over the airwaves, meaning the director could instantaneously cut between different angles, characters, and sets.

While some live broadcasts could be done on location, most shows in the 1950s were produced in a studio. For a scripted program like a drama, the actors, director, camera operators, and stage crew spent time rehearsing—that is, blocking scenes, figuring out camera angles, and planning edits. Sets were built strategically, so that actors and cameras could move efficiently between locations. When the show's scheduled time arrived, the performance played out like a theatrical performance; in fact, programs were often produced in front of a live audience.

Live programming offered both advantages and disadvantages. On the upside, producing a show in front of a studio audience created a spontaneous energy, and three-camera live shows could be relatively cheap to make. The downside of early live TV was that it couldn't be recorded—videotape technology capable of recording electronic cameras' signals did not arrive commercially until the late 1950s. Before then, a device called a **kinescope** did exist, which worked by placing a film camera in front of a TV screen to capture a program as it aired, but the image quality of the resulting copy was poor, limiting its usefulness for later broadcasts.

Filmed TV

In the 1950s, the alternative mode of production to three-camera live was *one-camera film*, which relied on the use of a single 35 mm or 16 mm film camera and largely followed the production practices of Hollywood filmmaking. Scenes were often shot out of order over the course of days or weeks, with multiple takes to fine-tune line delivery or to create different camera angles. Filming could take place on a soundstage or on location. Once principle shooting wrapped, the film stock would be developed and sent to postproduction, where it was edited. In the 1950s, TV shows made this way were called *telefilms*.

While one-camera film production could be more expensive, due to the cost of film stock, reshoots, and postproduction, it could also give producers more artistic control. Tight close-ups, precise edits, and intricate lighting and sound designs were easier to do. The biggest advantage of telefilm production, however, was that it created a high-quality copy of the show, which could be broadcast or rebroadcast at any time. To make the best use of this advantage, savvy producers decided to adapt the original three-camera setup by replacing TV cameras with film cameras. This hybrid approach allowed shows like *I Love Lucy* (one of the first to use this system) to retain key aspects of three-camera live, such as the studio audience and spontaneity, while giving producers the option of editing out mistakes and creating a high-quality copy. The technical quality of the original telefilm production is one of the reasons *I Love Lucy* (1951–1957) episodes are still in wide distribution today.

A Lasting Influence: TV Production from Then to Now

In today's digital TV landscape, the original modes of production are still used—with some adaptations. Digital cameras have replaced the use of film and videotape, and *multi-cam production*—an approach developed by the 1980s that employs more than three cameras—has become popular in both live and non-live settings. One-camera film production is still

I LOVE LUCY was produced using an innovative *three-camera film* mode of production. Doing so allowed producers to fully exploit Lucille Ball's extraordinary gift for physical comedy, as seen in this iconic scene. It also gave the production company that owned the series high-quality copies of episodes for use in the rerun market—an outcome that proved to be enormously lucrative.

CBS Photo Archive/Getty Images

the go-to method in certain genres. Here's how the different modes of production commonly break down in television today:

- Multi-cam live is often used for time-sensitive events, like sports, news, and award shows, and by programs that want to stand out. For example, when the sketch comedy-variety show *Saturday Night Live* debuted in 1975, its live broadcasts were marketed as part of its edginess, playing off the fact that anything could happen. And in recent years, broadcast networks have produced multi-cam live broadcasts of musicals like *Grease* and *The Wiz* as ratings stunts.
- Non-live multi-cam production is used widely for afternoon and late-night talk shows, reality TV series, game shows, and soap operas.
- Sitcoms are one of the only genres in which both modes of production have been used consistently throughout TV history; non-live multi-cam sitcoms include *Friends* and *The Big Bang Theory*, while one-camera sitcoms include *Young Sheldon*, *Community*, and *Modern Family*.
- One-camera production is the standard mode for all dramas except daytime soap operas.

The Network Era: The Big Three Dominate TV

By 1958, television had moved from its development phase into a mature industry, becoming the dominant mass medium in the United States. More than four out of five U.S. households owned at least one TV set, and there was more TV programming than ever before. This moment in TV history saw the establishment of economic strategies and cultural norms that would define the medium for years, many of which are still relevant today. Between 1958 and the 1980s, the TV industry was characterized by the power of the **Big Three** broadcast networks—NBC, CBS, and ABC. The three companies established an *oligopoly* in TV—a situation in which a few companies dominate an entire industry—by tightly controlling the production and distribution of prime-time TV programming.

The Networks Take Control of Programming

In the late 1950s, the Big Three made a key move that helped establish their domination: They wrestled control of programming away from advertisers. Like in radio during the 1930s and 1940s, early TV programs were often developed, produced, and supported by a single sponsor. Many top-rated programs even included the sponsor's name in the title (*Buick Circus Hour, Colgate Comedy Hour*). Toward the end of the 1950s, however, the networks became increasingly unhappy with this **single-sponsorship** model of TV advertising. Instead, they looked to replace that model with a new **magazine-sponsorship** model, in which the networks managed the development of programs and sold thirty- or sixty-second ad spots to various sponsors (which appeared throughout the show, like ads in a magazine). The networks preferred that approach because it gave them more creative control over the programs that appeared on their lineups and enabled them to dramatically increase advertising revenue: Selling ten minutes of ad spots in a half-hour show created more profit potential than selling the sponsorship of that show to one advertiser.

A national scandal in the late 1950s involving single-sponsored quiz shows (a type of game show

TWENTY-ONE In 1957, the most popular contestant on the quiz show *Twenty-One* was college professor Charles Van Doren *(left)*. Congressional hearings on rigged quiz shows revealed that Van Doren had been given some of the answers. The networks used the scandal as an excuse to wrestle control of program development away from sponsors.

Everett Collection, Inc

that originated on radio) helped hasten the shift away from the single sponsorship model. The quiz-show genre was a staple on early TV, with twenty-two different quiz shows airing on the networks by 1958. That same year, however, revelations that many quiz shows were rigged at the urging of their sponsors shocked the nation, leading to congressional hearings on the matter. Public outcry over irresponsible sponsors and calls for greater accountability opened the door for the networks to take control of program development.

As the Big Three gained control over development, they also expanded their control over distribution. In the early 1950s, the networks provided their affiliate stations with only a couple of hours of programming each weekday evening, leaving stations to look elsewhere for additional content. Eventually, however, the networks gave their affiliates all the programming they needed during **prime time**, the critical evening hours when TV viewing (and thus advertising revenue) was highest. During the 1960s, the networks delivered four hours of prime-time programming to their affiliates every evening (from 7:00 to 11:00 on the East Coast). As a result, the Big Three were *the* gatekeepers of prime-time television—a role that helped establish their domination of the industry.

Programming for Profits

Firmly in control of a maturing industry and anxious to maximize their profit margins, the Big Three developed efficient ways of producing and distributing television content. These strategies reflect the impact the TV industry's goals and practices can have on our culture—both on the rhythms of our daily lives and on the kind of content we consume.

TV Inserts Itself into Daily Routines

Following the strategy used by radio in the 1930s, the networks created a structured TV schedule for the full day that made the delivery of content routine and efficient for them and predictable for audiences. Schedules were designed to make watching television a daily and weekly habit, with different genres scheduled at different times of the day. Light news shows aired on weekday mornings (when men would be getting ready for work), soap operas on weekday afternoons (when women would be alone at home), children's programs in the late afternoon (when they would return from school), news in the early evening (aimed at parents after dinner), and sports on weekend afternoons (when men would be off work).

Of course, like many industry practices, these schedules reflected sweeping generalizations based on biased assumptions about gender and class. They ignored women who worked during the day, for example, and men who liked soap operas. Nevertheless, as TV became the dominant mass medium, its schedule established the rhythm for millions of people's daily lives. While TV schedules are far more varied in today's TV environment, elements of these early schedules and "appointment viewing" are still present, especially in broadcast TV.

The Networks Abandon Live Anthology Dramas

The fate of live **anthology dramas** illustrates how demands for efficiency shaped programming. During the 1950s, anthology drama series offered new, artistically significant plays written for television, and they were both popular and critically acclaimed. They featured entirely different characters and settings each week, aired live as three-camera productions, and were usually produced in a network's New York studio, allowing them to draw on the city's pool of actors, playwrights, and directors, many of them from Broadway. In the 1952–53 season alone, there were eighteen anthology dramas, including *The Philco Television Playhouse* (1948–1955) and *Kraft Television Theater* (1947–1958).

By the end of the 1950s, however, the genre had virtually disappeared as the TV industry changed its priorities. For one thing, drama anthologies often presented stories about complex and serious human problems that weren't easily resolved—a feature that helped establish the genre as high culture. In the era of single sponsorship, corporations such as U.S. Steel and Ford liked associating their brands with such "quality" content, but as the networks shifted to magazine sponsorship, they found it easier to sell ad spots on programs with upbeat stories about solvable problems. Second, live anthology dramas were expensive to produce—double the cost of most other TV genres in the 1950s—because each week meant a new story line, along with new writers, casts, and sets. Changing casts every week also meant that the series couldn't build actors into stars, who could then grow a loyal audience. While the cost of live drama programs seemed worthwhile in the early 1950s, when the innovative genre helped convince many Americans to buy their first TV set, the cost-benefit analysis changed once most households already owned a set.

Interestingly, the anthology drama format has reappeared in recent years in shows such as *American Horror Story, What/If,* and *Black Mirror.* For example, Netflix's *Black Mirror* (2011–present)—a British sci-fi program about technological dystopias—tells unique stories in each new episode. *Black Mirror* was inspired by an earlier American sci-fi anthology, *The Twilight Zone* (1959–1964), which was shot live. As we will explore in later sections, the format's revival reflects the new economic dynamics of a television industry changed by increased competition and subscription-based services.

Episodic Series Dominate TV

As they moved away from live anthology dramas, the networks turned their focus to filmed **episodic series**—a format in which the main characters remain the same from week to week. Episodic series had a number of built-in economic advantages for producers: Sets could be reused, actors signed long-term contracts, and script development was streamlined by placing a recurring cast of characters in new situations, rather than starting with a blank page every week. The efficiencies of episodic series have made the format a successful staple throughout TV history, and they remain popular today.

Episodic series fit into one of three types: chapter, serial, or hybrid. In **chapter shows**, all story lines wrap up each week, creating self-contained episodes. Examples include most situation comedies and crime-of-the-week police/detective shows. In **serial programs**, story lines continue across episodes. Soap operas, the most obvious example, have influenced other genres, leading to serialized crime, thriller, and fantasy/science-fiction programs. Over time, producers also combined the two show types, creating *hybrid* series in which one plotline may wrap up (e.g., the monster or crime of the week), but other story arcs play out over several episodes or seasons (see Table 6.1).

The episodic format has shaped how television represents our lives in specific ways. Take one of the most common episodic genres, the situation comedy, or *sitcom.* Centered on a limited cast of recurring characters, sitcoms tend to depict relatively insulated worlds in which the main characters are presented as a family unit—either literally, as in *The Brady Bunch* or *Black-ish,* or by analogy, as in *The Office* and *Parks and Recreation.* Each week, something happens to disrupt the characters' normal lives, and their response to it is usually the source of humor. Typically, however, the disruption gets resolved by the end of the episode so a new "situation" can arise the following week. That structure, some critics argue, can make the sitcom a somewhat conservative genre, since the pressure to return to the status quo often makes it difficult for characters and their attitudes to evolve beyond the series' founding premise.

TABLE 6.1   EXAMPLES OF EPISODIC SERIES

Chapter	Serial	Hybrid
The Beverly Hillbillies (1962–1971)	*General Hospital* (1963–)	*Hill Street Blues* (1981–1987)
Columbo (1971–1978)	*Dynasty* (1981–1989)	*Buffy the Vampire Slayer* (1997–2003)
Criminal Minds (2005–2020)	*Melrose Place* (1992–1999)	*Project Runway* (2004–)
The Big Bang Theory (2007–2019)	*Game of Thrones* (2005–2019)	*Grey's Anatomy* (2005–)
Bob's Burgers (2011–)	*Riverdale* (2017–)	*Supernatural* (2005–2020)
Brooklyn Nine-Nine (2013–)	*You* (2018–)	*Schitt's Creek* (2015–2020)

Hollywood Productions Take Center Stage

By the early 1960s, most entertainment TV production had moved from New York to Los Angeles—a change reflected in the decline of live anthology dramas and the rise of filmed episodic series. Although network news, sports, and soap operas continued to be produced in New York (and often still are), the Big Three relied almost exclusively on Hollywood production companies for prime-time sitcoms, dramas, and made-for-TV movies.

That relationship proved advantageous for the Big Three. In this period, many Hollywood production companies were trying to "sell" prime-time TV programs, but the Big Three's oligopoly meant that the only viable buyers were ABC, NBC, and CBS. The networks exploited that market advantage, working together to push down program prices and placing the risk of developing new content on Hollywood production companies. In this environment, production companies cut corners to reduce their costs. They often relied on formulaic genres, used small casts, kept sets and location shoots to a minimum, reduced the complexity of scenes, and followed simple editing rules.

These time and cost-saving strategies were driven by the networks' pricing pressures, but they were also an understandable response to the enormous challenges of series production. In this period, producers were expected to deliver thirty-four episodes of a sitcom or drama each broadcast season. Churning out that much content demanded efficient production practices. Today, most prime-time broadcast series deliver only twenty-two episodes per season, and limited series produced for cable and streaming deliver just eight or ten. Still, many corners of television today (like soap operas, news, and games shows) do require a tremendously fast rate of production—one factor that has contributed to the low-culture status the medium sometimes holds.

Producers Create the Innovation-Imitation-Saturation Cycle

Because it was difficult to predict TV audiences' tastes, shows often failed. In their search for consistent programming success, producers and network executives settled on a clear, risk-averse strategy: whenever a breakout hit emerged, they would copy it. Eventually, the market would get saturated

CLONING was common during the classic network era. For example, *I Dream of Jeannie* (1965–1970), a fantasy sitcom that used special effects to depict a beautiful genie whose "master" was a NASA astronaut, was a clone of *Bewitched* (1964–1972), a fantasy sitcom that used special effects to depict a beautiful suburban housewife/witch. Pictured here are similar scenes from each show's animated opening credits.

with similar shows, viewership would decline, the shows would die out, and producers and executives would hunt for a new programming hit to copy.

Over time, three imitative strategies emerged in this vein: cloning, spin-offs, and franchises.

- **Cloning** involves creating a new series by copying the key features of an innovative and popular program. For example, after *77 Sunset Strip*—a drama featuring swinging-single bachelor detectives solving cases in a glamorized version of Hollywood—became a hit in 1958, a slew of series popped up featuring hip detectives working in exotic locations, including *Hawaiian Eye* set in Honolulu (1959–1963) and *Surfside 6* set in Miami Beach (1960–1962).
- In a **spin-off**, a character from a hit series becomes the lead in a new one. The successful sitcom *All in the Family* (1971–1979), for example, spawned *Maude* (1972–1978) and *The Jeffersons* (1975–1985).
- In a **franchise**, producers leverage the name recognition of a popular show to brand other series. The longest lasting example is the *Star Trek* franchise.

Although the business of TV has changed, these strategies continue to shape program development today. The success of *The Bachelor* (2002–), for example, created a franchise with several related shows (e.g., *The Bachelorette, Bachelor Pad,* and *Bachelor in Paradise*) as well as clones (like *Boy Meets Boy* and *Love Island*). And although less common, producers today still create spin-offs, such as *Young Sheldon* (from *The Big Bang Theory*) and *Better Call Saul* (from *Breaking Bad*). Meanwhile, franchises like *Real Housewives* and *House Hunters* help new programs catch viewers' attention amid an avalanche of TV content.

"A Vast Wasteland"

In 1961, Newton Minnow, the new head of the FCC, spoke at the annual meeting of broadcasters. He offered a withering assessment of television, characterizing it as "a vast wasteland" filled with vapid game shows, formulaic sitcoms, violent westerns, and endlessly "cajoling" commercials. Minnow's speech reflected growing concerns about the path the young medium was taking. These concerns can be categorized into four interconnected areas, each linked to network control:

- **Content quality.** Critics like Minnow were deeply disappointed that commercialism and the networks' and affiliate stations' willingness to place their profit margins above the public interest had significantly undermined TV's potential to educate and uplift viewers. For some, the decision to replace serious-minded drama anthologies with cheaply produced episodic series illustrated the industry's misplaced priorities.
- **Concentrated cultural power.** Complaints about the quality of TV programming amplified concerns about the Big Three's power as cultural gatekeepers. As television became the most important mass medium of the day, just three companies dictated what most Americans watched, which some people worried wasn't good for a democratic society.
- **Unfair competition.** Critics were also concerned that the networks' oligopolistic power in the TV industry gave them an unfair market advantage in their relationship with Hollywood studios, which in turn limited innovation in program development.
- **Limited local control.** A broadcast TV station was awarded a license in exchange for agreeing to operate in the public interest of the community where it was located. However, network-affiliated stations (which were the vast majority of stations in the 1960s) gave much of their airtime over to national programming delivered by the network, leading critics to argue that they were not serving their local communities well.

In the early 1960s, mounting criticism pushed the networks to try to mend their public image by foregrounding their public service contributions. Further changes to television would come from the growth of noncommercial TV.

Bettmann/Getty Images

CIVIL RIGHTS During the 1960s, network news divisions began to produce in-depth documentaries on such topics as the plight of migrant farmworkers, American involvement in Vietnam, and the emerging civil rights movement (shown here). This programming raised awareness about a variety of social issues and made events more "real" to a nation of viewers.

News and the Public Service Mandate

In the early 1960s, the networks used news programming to improve the TV industry's reputation, to help their affiliate stations satisfy their public service obligations, and to keep FCC regulations at bay. All three networks dramatically increased the budgets for their news operations, expanding their evening news programs from fifteen to thirty minutes and establishing teams of experienced reporters and new bureaus around the world. Evening news anchors like CBS's Walter Cronkite became deeply trusted news sources for millions of viewers, illustrating the medium's ability to create a personal connection between viewers and the news anchors they invited into their living room every evening. The networks also funded impressive and costly prime-time documentaries about important social and political topics.

The most visible public service the networks and their affiliates provided, however, was (and still is) interrupting their regular schedules of sitcoms, game shows, and cop dramas to air important political events and breaking news stories—and to do so commercial-free. For presidential debates and Oval Office addresses, as well as coverage of Martin Luther King's assassination and the moon landing, commercial TV temporarily became a nationwide town hall. At the local level, the most common example of this public service—and of the local nature of broadcast stations—has been emergency updates in which a local affiliate station interrupts its network's programming to give information about a tornado, a wildfire, a train derailment, or another dangerous situation in the area.

Over time, the news eventually became a lucrative part of broadcast TV. The profitability of relatively inexpensive news programming at particular parts of the day—such as popular morning shows like ABC's *Good Morning America* and Sunday morning public affairs shows like NBC's *Meet the Press* (television's longest-running show, since 1947)—led to the launch of prime-time newsmagazines like CBS's *60 Minutes* (1968–). The financial success of network news programming also inspired the rise of twenty-four-hour cable news channels like CNN (1980–), MSNBC (1996–), and Fox News Channel (1996–).

LaunchPad
macmillan learning
launchpadworks.com

What Makes Public Television Public?
TV executives and critics explain how public television is different from broadcast and cable networks.

Discussion: In a commercial media system that includes hundreds of commercial channels, what do you see as the role of public systems like PBS and NPR?

Public Television

During the 1960s, critics interested in alternatives to the advertiser-supported commercial TV model increasingly called for publicly funded television. The FCC had set frequencies aside for noncommercial educational TV in the 1950s, and by 1966 there were 114 of them across the country. But because those stations had very little money with which to produce content, they had a very small impact on the TV landscape. In 1967, Congress established the Corporation for Public Broadcasting (CPB) to underwrite programs and provide government grants to these stations, which were becoming known as public television stations. In 1969, the Public Broadcasting Service (PBS) was created with funding from the Ford Foundation to make programs available to local public TV.

In general, public television has sought to serve viewers who are "less attractive" to advertisers and therefore to commercial networks. One such audience is children, and public television has figured prominently in programming for audiences under the age of twelve, with series like *Mister Rogers' Neighborhood* (1968–2001), *Sesame Street* (1969–), and *Barney & Friends* (1991–2009). PBS has also appealed to the tastes of some adult audiences by importing British shows like *Masterpiece Theatre* and *Fawlty Towers* and developing edutainment programs like *This Old House*.

Although public TV advocates have long recommended that Congress create a plan to provide long-term financial support to public television, thus far Congress has declined to do so. As a result, government funding of the CPB has been consistently vulnerable to political interference, rising and falling with the changing priorities of White House administrations and Congress. In 2018 and 2020, for example, the Trump administration proposed eliminating all funding for CPB, forcing public broadcast advocates to lobby Congress to reject the plan. To survive, public broadcasting has increasingly come to depend on support from local viewers and corporate underwriting. In recent years, only about 14 percent of funding for public television has come from the federal government.[9] Public tax support for the CPB is about $445 million a year, which amounts to about $1.35 a year per American.[10]

PUBLIC TELEVISION The most influential children's show in TV history, *Sesame Street* (1969–) has been teaching children their letters and numbers for more than fifty years. The program has also helped break down ethnic, racial, and class barriers by introducing TV audiences to a rich and diverse cast of puppets and people. In 2015, a controversy erupted over the production company's decision to move the program from PBS to the subscription network HBO.

Photofest

The Post-Network Era: Competition Heats Up

By the 1980s, the Big Three networks weren't the only players delivering programming for a national audience. Cable channels like HBO, CNN, and BET began to pull viewers away from broadcast TV, while new broadcast networks like Fox were being established, challenging the Big Three at their own game. In addition, VCRs and home video game consoles became more common, giving people new ways of using their TV sets. The networks were still operating, and in fact continued to be powerful players, but the system of oligopoly that defined the classic network era had eroded; competition had arrived.

Pay for TV? Cable Brings the Subscription Model

In the 1940s, not all communities could receive broadcast TV signals because of mountains or tall buildings that blocked the airwaves. To bring TV programming content to these communities, the solution was to use wires, or coaxial cable. The first cable systems were called **CATV** (community antenna television). Consumers connected their homes to these systems, paid a monthly subscription fee, and received up to twelve channels (all of them broadcast stations).

From the start, cable technology offered two clear advantages over the broadcast system. First, it improved reception quality by eliminating over-the-air interference. Second, cables could carry more TV channels than the airwaves could. On the other hand, cable technology had two main disadvantages: Wiring an entire city for cable was expensive, and people had to pay to receive cable TV, which violated the long-standing idea that access to radio and television should be free to the public. These downsides were significant enough that cable got off to a slow start. By the 1980s, however, cable would transform the TV industry and lay the groundwork for the infrastructure and business model shaping today's digital TV and Internet environment.

Balancing Cable's Growth against Broadcasters' Interests

For decades, the FCC saw cable as nothing more than a support system for delivering over-the-air TV signals, and the commission created regulations designed to keep it that way. First established in 1965 and reaffirmed in 1972, the FCC's **must-carry rules** required cable operators to include all local TV broadcasts on their systems. This rule ensured that local network affiliates, independent stations not carrying network programs, and public television channels would benefit from cable's clearer reception and not be excluded from the local cable system's lineup.

The FCC also pushed cable providers to operate with the public's interest (and not just their own profits) in mind. In 1972, the FCC mandated **access channels** for the nation's top one hundred TV markets, requiring cable systems to provide and fund nonbroadcast channels dedicated to local education, government, and the public. Convinced that the FCC was improperly infringing on their operations, cable companies sued, arguing that they should have the freedom to decide what channels to carry, just as magazines could decide what content to publish. In the 1979 *Midwest Video* case, the U.S. Supreme court agreed, upholding the rights of cable companies to determine their channel lineups and defining cable as a form of "**electronic publishing**," which meant that the industry would have the same publishing freedoms and legal protections that broadcast and print media enjoyed.[11] As a result, the FCC could no longer mandate the inclusion of access channels. The court, however, said that communities could "request" them as part of contract negotiations. Most cable companies continued to offer access channels in some form to remain on good terms with their local communities.

After the *Midwest Video* decision, the cable industry's future was more secure, and competition to obtain franchises to supply local cable service became intense. In 1975, fewer than 15 percent of U.S. households had cable. But with the regulatory path of cable clear, new national cable networks popped up quickly, and the cable industry took off. Penetration rates surged to over 44 percent in 1985 and 63 percent in 1995, peaking at 87 percent in 2010.[12]

Cable Expands with Satellite Distribution

Changes in media regulations helped spur the growth of the cable industry but so, too, did the development of a certain technology: communication satellites. In 1962, AT&T inaugurated the satellite age of telecommunications when it launched *Telstar*: the first satellite used to relay telephone and television signals through space. From then on, satellites helped make live television coverage of events around the world commonplace. They also offered a new way for the cable TV industry to distribute content to cable operators and customers across the United States.

In 1975, Home Box Office (HBO) began uploading a schedule of uncut, commercial-free movies and live sporting events to the recently launched *Westar* satellite. Using a satellite receiving dish, cable operators around the country could download HBO's signal and deliver it via cables to subscribers' homes (see Figure 6.2). Others quickly followed HBO's lead, including media owner Ted Turner, who in 1976 distributed his Atlanta-based independent broadcast station, WTBS, to cable operators via satellite, creating what is now TBS and a cable business that would grow to include CNN, TNT, and Cartoon Network. By 1985, well over fifty cable channels—including ESPN, MTV, and Bravo—were using satellites to nationally distribute content to cable operators, dramatically expanding what cable could offer customers.

Cable Operations

By the 1980s, running a cable system had become highly profitable, and competition to obtain franchises to supply local cable service intensified. A **cable franchise** is a mini-monopoly awarded by a city or town, usually for a fifteen-year period. Most communities grant franchises to only one company, so that there won't be multiple operators tramping over private property to string wire from utility poles.

During the franchising process, a city (or state) outlines its cable system needs and requests bids from various cable companies. In its bid, a company makes a list of promises to the city about subscription rates, channel capacity, types of programming, construction specifications, access channels, and *franchise fees* (the money the cable company pays the city annually for the right to operate the local cable system). Municipalities often use those fees to fund the operation of their public access channels.

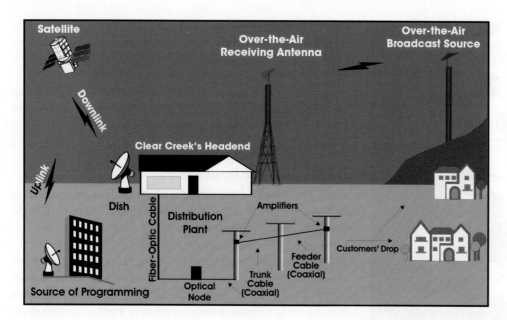

FIGURE 6.2

A BASIC CABLE TELEVISION SYSTEM

Information from: Clear Creek Communications, www.ccmtc.com.

Cable operators sell access to cable TV through an array of tiered packages. Less expensive packages usually include local broadcast stations, access channels, and scores of **basic cable channels**, such as BET, CNN, USA, and Comedy Central. Typically, a cable operator pays each cable channel somewhere between a few cents per month per subscriber (for low-cost, low-demand channels like C-SPAN) and over $8 per month per subscriber (for high-cost, high-demand channels like ESPN). That fee is passed along to consumers as part of their basic monthly cable rate. More expensive packages usually include **premium channels** like HBO, Showtime, or Starz, which provide access to recent Hollywood movies and original series, often without advertising.

While they often call themselves "networks" (e.g., Cable News Network), cable channels are distributed by cable system operators, who include them as part of a package that viewers pay a monthly subscription fee to access. That means they operate very differently from broadcast networks like, say, CBS, which provide programming to a network of local broadcast stations, whose transmissions viewers can access for free via an antenna.

Competition beyond Cable

By offering alternative ways to distribute TV programs, cable complicated the broadcast model and challenged the Big Three's domination—by 1997, the basic cable channels taken together were attracting more prime-time viewers than were the broadcast networks. It was soon clear, however, that cable wasn't the only force upending the status quo.

New Broadcast Networks

Another development of the post-network era was, ironically, the creation of new broadcast networks. By the 1980s, so many independent stations had popped up nationwide—most of them hungry for programming—that a door opened for the creation of a fourth broadcast network. In 1986, Australian-born newspaper mogul Rupert Murdoch created the Fox broadcast network. With the help of shows like *The Simpsons* (1989–) and *In Living Color* (1990–1994) and by outbidding CBS for NFL broadcast rights, Fox gained a sizable market share. Then, in 1995, two more networks debuted: UPN and the WB. Despite success with shows like *Buffy the Vampire Slayer* (1997–2003) and *America's Next Top Model* (2003–2015), both networks struggled financially, eventually shutting down independent operations and combining forces to launch the CW network in 2006.

Despite cable's growth, it made sense for Hollywood studios like Twentieth Century Fox (Fox), Paramount (UPN), and Warner Brothers (WB) to create broadcast networks during this period.

By controlling its own broadcast network, a studio could distribute the TV shows it produced itself, rather than competing with other studios to get its shows onto a Big Three network schedule or relying on cable. Even in the late 1990s, over 20 percent of households in the United States didn't have cable, meaning that millions of people still watched over-the-air television. Twenty-five years later, a similar motivation to control their own programming distribution led Hollywood studios like Disney to launch their own streaming services.

DBS: Cable without Wires

By the 1990s, both cable and the broadcast model faced competition from yet another delivery option: **direct broadcast satellite (DBS)** providers. Instead of using wires, DBS systems transmit signals directly to satellite dishes near or on customers' homes. As early as the 1970s, some people, especially those living in rural areas not yet serviced by cable systems, realized they could purchase a large satellite dish and receive cable channels like MTV and CNN—the same channels that cable companies were providing to wired communities—off of a satellite for free. In response, cable channels began scrambling their signals, forcing satellite owners to subscribe to a provider just as cable customers did. By 1996, full-scale DBS service with smaller receiving dishes was available from DirecTV and Dish, with channels and tiers of service similar to those offered by cable companies.

Direct Broadcast Satellite While the growth of companies like DirecTV during the 1990s gave consumers in the United States an alternative to cable, DBS technology played an even more significant role in other parts of the world, including countries like India and Egypt (shown here).

Frédéric Soltan/Corbis/Getty Images

VHS, DVD, and Video Game Systems

For decades, most people could use their TVs only to tune into a flow of programming controlled entirely by broadcasters. The emergence of several new technologies gave viewers more options for what they could do with their TV sets and more control over what they watched and when they watched it.

- By the 1990s, 90 percent of U.S. households owned a VCR, enabling viewers to **time-shift**, or record programs to watch at their convenience. The simultaneous rise of prerecorded video tapes and rental stores like Blockbuster gave viewers even more control and turned living rooms into home theaters.
- Prerecorded DVDs, which supplanted VHS technology in the early 2000s, disrupted TV even further. Cumbersome VHS tapes, which stored only two hours of content per tape, were poorly suited to the sale of TV series. In contrast, an entire season of a series like *Friends* could be sold in a single compact DVD box—a development that gave consumers another way to access TV content and producers a new distribution option.
- Finally, the emergence of home video game consoles—from Atari's debut in 1977 to the launch of Nintendo, Sega, and PlayStation during the 1980s and 1990s—began the transformation of TV sets into flexible hubs for digital home entertainment. TV sets weren't just for watching TV anymore.

Programming Strategies in the Post-Network Era

Increased competition in the industry changed TV programming strategies significantly during the post-network era. In the network era, executives had followed the **least objectionable programming (LOP)** principle—a strategy aimed at attracting as big an audience as possible by not turning off any viewers. The result was network lineups filled with shows that had broad appeal but politically and aesthetically bland content. At the time, broadcast TV was the preeminent mass marketing medium; networks and stations made billions of dollars selling advertisers access to huge audiences. If such shows didn't actually inspire intense passion in many viewers, broadcasters weren't overly concerned; viewers didn't have a lot of choices at the time, and it was easier to sell commercial spots on "safe" programs to advertisers, who were fearful of being associated with controversial topics. In the post-network era, however, the LOP principle gave way to strategies designed for an environment in which audiences had more choices.

Narrowcasting

When cable channels emerged, they chose not to compete on broadcast TV's terms: Instead of trying to appeal to all viewers, they worked to lure away slices of the networks' huge audience with specialized programming. Some channels focused on a specific genre (CNN on news, ESPN on sports, MTV on music), while others targeted a specific audience segment (Nickelodeon for children, BET for Black viewers, Telemundo for Spanish-language speakers). Still other channels specialized in both ways; when Fox News launched in 1996, for example, it was a news channel that targeted politically conservative white viewers. In the end, this strategy of specializing, called **narrowcasting**, is what convinced customers to consent to paying for television; for many people, this was television that finally spoke to them.

When the Fox (1986), UPN (1993), and WB (1995) broadcast networks launched, they also adopted narrowcasting strategies to differentiate themselves from the Big Three. In the early 1990s, for example, as Fox struggled to establish itself, it targeted young Black viewers (an audience segment that had always been ignored by the Big Three networks) with Black-cast series like *Martin* (1992–1997) and *Living Single* (1993–1998). According to television scholar Kristal Brent Zook, Fox's narrowcasting approach "inadvertently fostered a space for black authorship in television."[13] When Black producers were able to develop series for Black audiences, they created representations of Black life that differed from series like the hit NBC sitcom *The Cosby Show* (1984–1992), which was expected to make its depiction of a Black family appealing to both Black and white viewers. Fox's programs were highly successful with Black viewers, but once Fox executives felt the network could compete with the Big Three for the more lucrative young white male demographic, they canceled several of the series, leaving Black audiences once again underserved by network TV programming.

The Big Three and the Quality Audience

Facing new competition for viewers and advertisers, the Big Three networks began to revise their programming strategies, especially the LOP principle. As they started to lose control of the mass audience, they increasingly focused on what became known as the **quality audience**—the segment of viewers advertisers pay the most money to reach. The quality audience consisted largely of well-educated (code for middle- and upper-class) eighteen- to forty-nine-year-olds—consumers who had money to spend and were at an age where they could buy products for the kids and homes advertisers assumed they had. Efforts to keep the quality audience from moving to cable or home video would, over time, change prime-time network television.

Hill Street Blues, an innovative police drama that aired on NBC from 1981 to 1987, reflected the shifting dynamics of the emerging post-network era. Its producer, Steven Bochco, wanted the series to have a sense of gritty realism, so he used handheld cameras, intricately choreographed long takes, and a soundscape filled with overlapping dialogue and ambient noise to give viewers the sense of being a fly on the wall of a city police precinct. The series broke two key Network-Era programming rules: It was expensive to make and had the potential to turn off some viewers to achieve its artistic goals. Episode production required a large cast and intense rehearsals, and the show's cinema verité style didn't suit everyone's taste. As a result, the show never ranked high in the overall ratings. The viewers it did attract, however, were younger (eighteen to forty-nine) and more upscale than the average TV viewer, a fact NBC used to attract advertisers willing to pay to reach that key demographic.

During the post-network era, then, the Big Three realized that to attract the quality audience, they would have to take risks—both financially and aesthetically. As the *Washington Post* put it in 1989, producers "demanded more

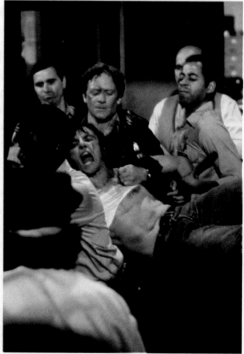

HILL STREET BLUES was an innovative police drama that rewrote the rules for how to produce television. It included a large cast of extras and often featured scenes like this one, in which key action takes place in the background, positioning the viewer as a fly on the wall of a gritty police precinct. Such relatively complex and costly staging, unusual for prime-time television at the time, gave the series a sense of realism and reflected the changing dynamics of an industry that needed to work harder to attract the quality audience.

© NBC/Everett Collection, Inc

creative latitude in the 1980s and got it from networks anxious to halt the defection of habitual viewers."[14] They would repeatedly do that, with innovative series like *Moonlighting* (1985–1989), *Seinfeld* (1989–1998), *Twin Peaks* (1990–1991), and *Lost* (2004–2010). Cable channels like HBO and AMC, which began developing original series in the 1990s, followed a similar strategy with series like *The Sopranos* (1999–2007), *Mad Men* (2007–2015), and *Breaking Bad* (2008–2013).

Gay Material and the Quality Audience

Post-network era dynamics also shifted the Big Three networks' tolerance for controversial content—a change well illustrated by the rise of gay material on TV during the 1990s. In the 1980s and early 1990s, a conservative political climate defined by the Reagan presidency and the moral panic surrounding the AIDS crisis made homosexuality a taboo topic for advertisers and network executives. As a result, viewers could watch TV for months without seeing one gay or lesbian character or even a reference to homosexuality. By the mid-1990s, however, hit shows like *Friends* (1994–2004) and *ER* (1994–2009) featured recurring gay and lesbian characters, and many more had special gay-themed episodes. Gay material became a trend.

The Big Three networks' desire to attract the quality audience helped make that shift possible, as network executives realized that the composition of the well-educated eighteen-to-forty-nine audience segment was changing. They believed that well-educated young adults were more likely to be socially liberal and wanted programs that were "hip" and "edgy"—vague but widely used terms referring to content that pushed boundaries. Executives realized that gay material could give a program the kind of edge their target audience found appealing. If losing some viewers was the price required to attract the quality audience, it was an acceptable trade-off.

Although ABC, CBS, and NBC had given up hope of attracting mass audiences like those they once had, it's important to note that they didn't fully embrace narrowcasting. The Big Three's market share dropped steadily during the post-network era, but each network's audience was still much larger than that of any single cable channel. In 2006, for example, a mildly successful show on CBS could draw eight million viewers, while top cable channels averaged only two million viewers. The fact that they could still attract relatively large audiences made the Big Three skittish about going *too* narrow. Gay material, for example, was rarely used to attract an LGBTQ+ audience, which was considered too small for broadcast networks. Instead, gay content was used mostly to appeal to the broader quality audience of socially liberal, upscale, and presumably straight eighteen-to-forty-nine-year-olds—a fact that shaped how gay characters and themes were represented. Even amid the shifting landscape of the post-network era, the broadcast networks held on to elements of their old business model.

The Rise of Reality Television

Targeting viewers with expensive, artistically innovative, or politically controversial content wasn't the only way programming executives responded to increased competition. Reality shows, which became a staple on network and cable channels beginning in the late 1990s, offered an alternative strategy: lower-cost, formulaic, and sensationalistic but addictive programs that often became people's guilty pleasures. Reality-based programs include everything from *The Voice* and *Deadliest Catch* to *Property Brothers* and *Below Deck*, with subgenres such as structured reality shows, documentary-style shows, game/competition shows, makeover shows, and how-to programs.

Reality shows are popular with many viewers, at least in part because they use comforting formulas and introduce characters and people who seem more like us—or, conversely, not at all like us. For producers, these programs are appealing because they typically cost much less to make than sitcoms and dramas, achieving savings by hiring non-actors, using minimal sets, and avoiding extensive script development. While not scripted in a traditional way, narratives remain central to most reality-based series. In programs like *Survivor*, *RuPaul's Drag Race*, and *Keeping Up with the Kardashians*, heroes and villains, story arcs, and narrative twists are constructed through strategic casting and careful editing. Reality shows have become the main genre on cable channels like Bravo, TLC, and HGTV. (For more, see "Examining Ethics: Is This Entertainment? TV Industry Reconsiders Police Shows" and "Media Literacy and the Critical Process: TV and the State of Storytelling.")

Is This Entertainment? TV Industry Reconsiders Police Shows

Police shows have been a staple from the beginning of television. One of the highest rated programs of the 1950s was *Dragnet*, starring Jack Webb as Sergeant Joe Friday of the Los Angeles Police Department.

Each episode of *Dragnet* began with the first four notes of its familiar theme song, over a close-up of Friday's LAPD badge. The male announcer, in a serious voice, intoned, "Ladies and gentlemen, the story you are about to see is true. The names have been changed to protect the innocent." (Webb, as the show's producer, consulted with LAPD for the show, which depicted dramatizations of real cases.) Webb's narration set up the case, and the show walked the viewer through the tough, exacting police work involved in solving each crime. In all 276 shows of *Dragnet*'s original run (1951–1959), never once did Sgt. Friday and his detective partner ever investigate and arrest someone who was later found not guilty.

That seeming infallibility of police onscreen set the pattern for the thousands of episodes of police and detective shows to follow. That includes *Cops*, the long-running Fox and Spike/Paramount Network show that created more than eleven hundred episodes from 1989 to 2020, and *Live PD*, the A&E live police show that produced nearly three hundred episodes from 2016 to 2020.

Both programs were produced in contractual arrangements with city or county police forces (over 180 different police agencies were involved over the years in *Cops* alone). Because the production crews accompanied officers responding to calls, they literally took the point of view of police in telling these stories for TV. And as was revealed in an investigation of *Cops* detailed in the *New York Times*, the police departments also had final say in the narratives. For example, "In exchange for giving 'Cops' camera access, the Police Department is provided episodes ahead of time to make 'any changes they deem necessary.' . . . In other words, the police can pick and choose what gets aired."[1]

Live PD, which became A&E's top-rated program, ran live for three hours on Friday and Saturday nights, switching between film crews in different cities to find the police department with the most action at the time. Host Dan Abrams analyzed the action with his sidekicks: a TV crime reporter and a Tulsa police sergeant. There was a broadcast delay, which enabled them to eliminate some live content.

Both *Cops* and *Live PD* came to a sudden end in summer 2020 because of a police video they couldn't control: the viral video of George Floyd dying under the knee of a Minneapolis police officer. The subsequent rise of #BlackLivesMatter protests swept the nation, and the networks broadcasting *Cops* and *Live PD* suddenly had to rethink a seventy-year-old tradition of television police shows.

The moment was captured by *Washington Post* columnist Alyssa Rosenberg, who—in a piece titled "Shut Down All Police Movies and TV Shows. Now"—reached back even further into our cultural history: "For a century, Hollywood has been collaborating with police departments, telling stories that whitewash police shootings and valorizing an action-hero style of policing over the harder, less dramatic work of building relationships with the communities cops are meant to serve and protect."[2]

In light of George Floyd's death at the hands of police and nationwide racial justice protests, the Paramount Network pulled *Cops* from its lineup, then announced its cancellation. A&E similarly pulled and canceled *Live PD*. In a statement, A&E said, "Going forward, we will determine if there is a clear pathway to tell the stories of both the community and the police officers whose role it is to serve them. And with that, we will be meeting with community and civil rights leaders as well as police departments."[3]

Given A&E's comment, what should police shows look like in the future? Should "reality" police shows like *Cops*, or live "documentary" shows like *Live PD*, air on TV ever again? ●

MEDIA LITERACY & THE CRITICAL PROCESS

TV and the State of Storytelling

The rise of reality programs over the past two decades has been fueled as much by the cheap cost of producing the genre as by the wild popularity of these programs. In fact, in the history of television and viewer numbers, traditional sitcoms and dramas—and even prime-time news programs like *60 Minutes* and *20/20*—have been far more popular than successful reality programs like *The Bachelor*. But when TV executives cut costs by reducing writing and production staffs and hiring "regular people" instead of trained actors, does the craft of storytelling suffer at the expense of commercial savings? Can good stories be told in a reality program? In this exercise, let's compare the storytelling competence of a reality program with that of a more traditional sitcom or drama.

1 DESCRIPTION

Pick a current reality program and a current sitcom or drama. Choose programs that either started in the last year or two or have been on television for roughly the same period of time. Now develop a "viewing sheet" that will allow you to take notes as you watch the two programs. Keep track of main characters, plotlines, settings, conflicts, and resolutions. Also track the main problems that are posed in the programs and how they are portrayed or worked out in each episode. Find out and compare the basic production costs of each program.

2 ANALYSIS

Look for patterns and differences in the ways stories are told in the two programs. At a general level, what are the conflicts about? (For example, are they about men versus women, managers versus employees, tradition versus change, individuals versus institutions, honesty versus dishonesty, authenticity versus artificiality?) How complicated or simple are the tensions in the two programs, and how are problems resolved? Are there some conflicts that you feel should not be permitted—like pitting older contestants against younger contestants or white contestants against Black contestants? Are there noticeable differences between "the look" of one program and that of the other?

3 INTERPRETATION

What do some of the patterns mean? What seems to be the point of each program? What do the programs say about relationships, values, masculinity or femininity, power, social class, and so on? What is the value of each program for its viewers?

4 EVALUATION

What are the strengths and weaknesses of each program? Which program would you judge as being better at telling a compelling story that you would want to watch? How could each program improve its storytelling?

5 ENGAGEMENT

Through either online forums or personal contacts, find other viewers of these programs. Ask these viewers questions about what they like or do not like about these shows, and what they would change if they could. Then report your findings to the programs' producers through a phone call, an e-mail, or a tagged social media post. Try to elicit responses from the producers about the status of their programs. How did they respond to your findings?

Our Digital Era and the Business of TV

Like every other media industry, television has been transformed by the Internet and the digital revolution. Digital technologies have changed how TV content is produced, stored, delivered, and viewed, and the affordances of the Internet have made television a more interactive and customizable experience. New companies like Netflix, and companies new to television like Apple, Amazon, Google, and Facebook, have entered the picture, and the old TV industry has scrambled to respond. As a result, the last fifteen years have seen a stream of corporate mergers and acquisitions as companies try to position themselves for TV's uncertain future.

The growth of streaming—the hallmark of our digital era—has been a relatively recent development. Although YouTube has been in existence since 2005, the breakthrough streaming moment for the entertainment industry wasn't until 2007, when Netflix decided to make streaming video available to its customers. Netflix began as a service that shipped subscribers hard-copy DVDs by mail.

With streaming, customers could instantly watch films and television shows on the Netflix site—a convenience that made the service wildly popular and changed the direction of television. Although Netflix's streaming success seemed almost immediate, it couldn't have happened without a series of necessary changes to the media infrastructure that had rolled out over the previous decade. In this section, we'll look at how these and other changes have come together to create the television environment we know today, as well as the business model companies use when making television in (and for) a digital world.

The Infrastructure of Digital TV

Infrastructure refers to the technological systems through which viewers access TV content—systems that are shaped by government policies and the strategies of the corporations that typically own them. Although these systems are often less visible to consumers than the programming delivered on them, controlling this technology has always been an important part of the television industry. The digital revolution has been meaningful for infrastructure because it transformed existing systems and opened the door for entirely new players to get involved.

Digital Broadcasting

For decades, television's primary infrastructure system consisted of broadcast stations whose towers transmitted analog signals via the public airwaves. But with the passage of the **Telecommunications Act of 1996**—a massive overhaul of communications law that affected almost every aspect of the U.S. television industry—the government established a plan to switch to a new digital standard. This digital transition, completed in 2009, required every broadcaster to replace its analog transmitting equipment with digital equipment (which ultimately rendered the old big box TVs useless, unless they were fitted with an adapter). Digital transmissions enabled the delivery of higher definition video and better sound quality, as well as the availability of more over-the-air programming. Because digital signals can be compressed, broadcast stations can use a process called **multiplexing** to transmit multiple digital channels via the single channel or frequency allocated to them by the FCC. In Cincinnati, Ohio, for example, WCPO—an ABC affiliate—transmits its main programming on channel 9.1 but also transmits on five subchannels (9.2–9.6). While the digital transition has come with distinct advantages, it has also brought new challenges. For example, the fact that digital transmissions tend to weaken more quickly than analog transmissions has led to weaker reception in rural areas.

Digital Cable, DBS, and Wireless Broadband

Television's other important infrastructure system—the network of coaxial cables that delivered content directly to subscribers' homes—was also transformed during the digital era. In the Telecommunications Act of 1996, Congress dismantled regulatory barriers that had prohibited regional phone companies, long-distance phone carriers, and cable companies from entering one another's markets. That meant that cable companies could begin to offer telephone services, and phone companies could buy or construct cable systems. Even more importantly, as it turned out, it allowed both types of companies to use their networks to provide access to the Internet. As millions of people looked to "get online," providing Internet service became an important and lucrative business.

As a result, cable systems became coveted properties to own, not so much for their ability to carry TV programming as for their wired connections to homes through which an array of digital services could be offered. By the 2000s, thousands of independent cable systems had been snapped up by large **multiple-system operators (MSOs)** like Comcast and Charter Communications—corporations that own many cable systems around the country. Competition with phone-based Internet service providers pushed cable companies to invest $290 billion in technological infrastructure between 1999 and 2018, with most of the funds going to install the high-speed fiber-optic lines that have enabled cable companies to bundle digital cable television, broadband Internet, and telephone service—and created the pathways that have allowed stand-alone streaming services like Netflix and Hulu to emerge and grow.[15]

By the 2010s, the infrastructure of coaxial cables, fiber lines, and copper wires faced growing competition from wireless infrastructures, like the satellite-based systems from DirecTV and Dish. In addition, an even more disruptive development has been the increased capacity of mobile broadband technologies like 5G, which promise swift upload and download speeds through cellular phone service that can compete with wired systems. As a result, the cell phone infrastructure controlled by companies like AT&T, Verizon, and T-Mobile has emerged as a future pathway for the delivery of television content.

Watching TV in the Digital Era

Watching TV today is a different experience than it was in earlier eras. One of the most obvious shifts has been in TV sets themselves, which have transformed from the boxy cabinets that made their way into most U.S. living rooms in the 1950s into today's high-definition flat-screen devices—many with built-in Internet capabilities that integrate television sets with the wider digital environment. But it's not just devices like these that have had an impact; a number of other digital viewing technologies have changed how, when, and where we watch television in various ways.

Time Shifting 2.0

VHS technology changed viewers' relationship to TV in the 1980s, before giving way to DVD players in the 1990s. But DVDs have also begun to disappear—between 2008 and 2019, DVD sales fell 86 percent[16]—and they have largely been replaced by DVRs (digital video recorders), which use computer processing capabilities to record programs to a drive. The large storage capacities and user-friendly interfaces of DVRs allow viewers to pause live transmissions and record vast amounts of content, including every episode of a series. DVRs have taken the time-shifting experiences of VCRs to the next level, helping to transform the experience of **linear TV**—broadcast television's and cable's traditional approach to content delivery, in which a show airs at a specific time—into **on-demand TV**, in which programs are made available to viewers to watch whenever they want. DVR users can ignore live TV and go directly to the menu of programs that they have recorded; viewers of any streaming video service—called **video-on-demand (VOD)** in the industry—can select from that service's many offerings whenever they choose. This availability of large amounts of content at any time has led to a new form of TV watching called *binge watching*, in which viewers watch several episodes (or even a whole season) of a TV show in a day or weekend.

Mobility and Place Shifting

Until the digital era, television was primarily a domestic technology. Although TV sets could be found in bars, schools, airports, and other public places, most people's interactions with television took place in their homes. Today, the Internet and digital TV technologies have made TV mobile and public in new ways by enabling **place shifting**—the practice of accessing stored media from different locations.

The emergence of laptops, tablets, and smartphones has made it possible to access TV content wherever there is adequate wireless broadband connectivity—from one's bedroom or backyard to a park or bus stop. Such mobile technologies can have paradoxical impacts. On the one hand, they make it much easier to move our TV experience out of the privacy of our homes and into public spaces. On the other hand, the nature of such small handheld devices (especially when used with headphones) encourages more individual or privatized viewing experiences, as opposed to the communal experiences that are (at least theoretically) easier to have when watching a traditional TV set.

Today's TV content providers offer a variety of services that make place shifting possible. Subscribers to streaming services like Hulu and Netflix can access content by signing into their account on any Internet-connected device, freeing them from the coaxial lines and cable boxes that once anchored them at home. In response, many cable providers, cable channels, and broadcast networks

TABLE 6.2 TOP 10 VIDEO SUBSCRIPTION SERVICES GLOBALLY, 2020

Rank	Video Subscription Service	Subscribers
1	Netflix (streaming)	195.2 million
2	Amazon (streaming)	150 million
3	Disney+ (streaming)	73.7 million
4	HBO/HBO Max (cable/streaming)	57.0 million
5	Hulu (streaming)	36.6 million
6	Apple TV+ (streaming)	33.6 million
7	Pluto TV (streaming)	26.5 million
8	Peacock (streaming)	22 million
9	Comcast (cable)	20.1 million
10	AT&T (DirecTV)*	17.1 million

Data from: Financial reports and announcements from listed companies

Note: Data are for total global subscribers. Some of the streaming services listed here are currently United States only but with plans to distribute globally. Cable providers are United States only.

*includes DirecTV, U-verse, AT&T TV, and AT&T TV Now

have developed applications that enable similar freedoms—including services like Peacock, ESPN+, and HBO Max. (See Table 6.2 for a list of the top video subscription services globally, which include cable, satellite, and streaming options.)

The Economics of Delivering Content to Viewers

Developing strategies for gathering and delivering TV content to viewers in ways that generate revenue lies at the heart of the TV business. While companies do try to understand audience desires, such delivery strategies are never simply a matter of giving viewers what they want. They are also (and perhaps more) about molding viewers' habits and desires so that they want, or are at least satisfied with, the content that works within a company's broader business model.

In the digital era, there is an ever increasing list of **content delivery services (CDSs)**—companies whose business it is to gather and distribute TV content. Our earlier overview of TV history is helpful here, because the broadcast and cable models form the backdrop against which new players appear and old players evolve—all in response to technological changes and shifting viewer expectations. Silicon Valley tech companies like Netflix, YouTube, Amazon, and Apple have profoundly disrupted the industry and begun a period of experimentation. Although it failed just months after its 2020 launch, Quibi—an app-based CDS designed to deliver ten-minute episodes for people to watch on their mobile phones—hints at the paths TV content delivery might take. Meanwhile, legacy players are scrambling to adapt. Broadcast networks, which once relied on affiliated stations to deliver content for them, can now stream directly to viewers, leaving affiliated stations nervous about their future. All this change means that the business of gathering and delivering TV content has become increasingly complex, especially since content delivery services can differ in how they generate revenue and make content available to viewers.

Revenue Models: Advertisers and Subscribers

Content delivery services can be divided into two categories: free TV and pay TV. Those two models intersect (in sometimes complicated ways) with TV's two primary revenue streams: advertisers and subscribers.

Television content that is "free" to viewers is usually subsidized by advertising. From the beginning, broadcast TV has followed this model (with local stations and networks), which has made advertisers broadcast TV's core customers and shaped programming decisions. In ad-supported free TV, content is fundamentally a tool for attracting the kinds of viewers advertisers will pay money to reach.

FIGURE 6.3

PRIME-TIME NETWORK TV PRICING, 2020–2021

The average costs are shown for a thirty-second commercial during prime-time programs on Monday and Thursday nights during the 2020–2021 season.

Data from: Jeanine Poggi, "What It Costs to Advertise in TV's Biggest Shows in 2020–2021 Season," AdAge.com, October 30, 2020.

Monday

	8:00 P.M.	8:30 P.M.	9:00 P.M.	9:30 P.M.	10:00 P.M.
ABC	Dancing with the Stars ($102,365)				The Good Doctor ($126,055)
CBS	The Neighborhood ($107,785)	Bob Hearts Abishola ($90,701)	All Rise ($75,510)		Bull ($92,557)
NBC	The Voice ($254,224)				Manifest ($113,843)
FOX	L.A.'s Finest ($71,289)		Filthy Rich ($70,506)		no network programming

Thursday

	8:00 P.M.	8:30 P.M.	9:00 P.M.	9:30 P.M.	10:00 P.M.
ABC	Station 19 ($110,588)		Grey's Anatomy ($213,829)		A Million Little Things ($126,783)
CBS	Young Sheldon ($157,213)	B Positive ($97,717)	Mom ($125,187)	The Unicorn ($97,786)	Evil ($91,172)
NBC	Superstore ($111,281)	Brooklyn Nine-Nine ($104,126)	Law & Order: SVU ($149,776)		Law & Order: Organized Crime ($109,352)
FOX	Thursday Night Football ($624,626)				no network programming

Ad-supported TV uses measurements of the size and demographic profile of every program's audience as the base for negotiating ad prices (see Figure 6.3). Nielsen, the company that has tracked TV viewership since 1950, uses a representative national sample to provide various estimates about the TV audience. Each year, Nielsen calculates the number of television households in the United States (there are currently more than 120 million), and for every TV episode, Nielsen reports its rating and its share.

- An episode's Nielsen **rating** answers the question: What percentage of TV households watched that episode? Ratings are provided for various demographic categories, like adults eighteen to forty-nine, men twenty-five to fifty-four, or overall households.
- An episode's **share** answers the question: What percentage of TV households who had their TV set turned on at that time were watching that episode?

The series finale of *The Big Bang Theory*, for example, had a 12.2 rating (meaning 12.2% of TV households watched it) and a 22 share (meaning 22% of TV households that were actually watching TV at the time tuned in to the episode).[17] As digital technologies have changed viewing habits, Nielsen has started reporting other ratings data as well, such as live-plus-3 (the percentage of people who watched the show live or via a DVR or an on-demand streaming service within three days), live-plus-7 (what percentage watched it live or within a week), and social ratings (measurements of social media activity about a show).

Other content delivery services, most notably premium cable channels like HBO and online streaming services like Netflix, rely primarily on fees from subscribers (see Table 6.2) rather than the ad-based model (though they may have product placement in their programs). This economic model leads to different marketing and programming strategies. HBO's famous slogan "It's Not TV, Its HBO," for example, was a dig against ad-supported content. Unlike traditional TV, which is constrained by skittish advertisers, HBO programs like *Game of Thrones*, *Euphoria*, and *Westworld* are free to include nudity, sex, violence, and adult language. Instead of being beholden to advertisers, pay-TV services are dependent on subscribers who are willing and able to pay a monthly fee—a fact that shapes the content they offer.

While they don't use ratings data to negotiate ad sales, subscription sites do carefully track their subscribers' viewing habits. Streaming services like Netflix, for example, gather enormous amounts of data on their users. This data enables recommendation algorithms designed to create personalized user experiences, which, Netflix hopes, will improve subscriber retention rates. The data also informs decisions about which series to cancel and is used in negotiating licensing deals with production companies (more on this later).

Finally, many (perhaps most) content delivery services get revenue from both advertising *and* subscription fees. For example, online sites like YouTube, Hulu, and Peacock offer tiered services, with free or lower-cost subscriptions for programming that includes ads, or premium subscriptions for ad-free viewing. Cable and DBS services offer a different hybrid approach that draws revenue from both sources: Subscribers pay a monthly fee for access to channels, most of which also include ads.

Modes of Content Delivery: Linear versus On Demand

As mentioned earlier, television content can be delivered in two distinctly different ways. First, it can be delivered as a continuous or linear sequence of programs and commercials, pushed out from a CDS to viewers in a live content flow. Sitting in front of one's TV set at a specific time to watch a show, knowing millions of others are doing the same thing, defined the experience of broadcast and cable television for decades. The linear mode was a valuable delivery strategy; it created routinized viewing habits and encouraged viewers to watch whatever was "on" at the time.

In the second delivery mode, a CDS accumulates a library of content, which viewers then access like books on a shelf and watch at their convenience. On-demand TV first emerged when the growth of higher-capacity interactive broadband connections and computer servers capable of storing huge amounts of digital content made video streaming over the Internet practical. By the 2000s, a new category of CDS had emerged: **over-the-top (OTT)** media services. The term refers to streaming sites like Hulu and Netflix, which allow viewers to access TV content through an Internet connection without having to go through broadcast, cable, or satellite providers. The availability of these OTT streaming services led to the first instances of **cord cutting**: consumers cutting the cord to expensive cable subscriptions in favor of a streaming subscription (and maybe free local television) in order to fill their viewing needs at a much lower cost.

Predictions about the imminent death of TV are, in fact, usually predictions that the days of linear TV are numbered. As viewers—especially younger viewers—are being lured away from broadcast TV's and cable's linear flows by on-demand options, traditional linear CDSs are developing their own on-demand services. In 2020, for example, NBCUniversal launched Peacock, an OTT service that would give subscribers access to the vast library of content it owns. Current programs that are part of NBC's linear broadcast schedule are available on demand for Peacock subscribers the next day. Paramount+, ESPN+, and HBO Max are similar ventures.

Importantly, such ventures complement rather than replace linear service. No one is ending linear distribution; in fact, OTT services like YouTube TV and Hulu + Live TV essentially provide the same service as cable providers. Meanwhile, a slew of new **digital subchannel broadcast networks** have emerged to provide content for local broadcast stations' multiplex channels.

Mini-networks—including MeTV, MyNetworkTV, Laff, and Comet—create some original programming but mostly deliver inexpensive reruns and old movies in the linear model. These developments are evidence that, at least for now, the death of linear TV has been greatly exaggerated.

Making Digital TV Content

Digital technologies have transformed both the production practices and aesthetics of TV programs—the industry's core product. Most obviously, the dimensions of TV screens have changed from the original analog 4:3 aspect ratio of old cathode ray tube TV sets to widescreen ratios like 16:9 on flat screens. In addition, digital editing tools make special effects easier and more affordable, fueling everything from CGI in fantasy series like *Game of Thrones* to elaborate overlay graphics in sports coverage. On the other hand, smartphone cameras and sites like YouTube and TikTok have made the stripped-down visual style of amateur do-it-yourself (DIY) video productions a significant part of our TV environment. In the digital era, there is more television content being produced in more places, by more companies and individuals, than ever before.

Original Series and Peak TV

Much of the original programming on broadcast networks, cable channels, and streaming services is produced by major Hollywood studios and independent production companies. That programming can include everything from formulaic design shows like *Flip or Flop* (produced by Pie Town Productions for HGTV) and network dramas like *Grey's Anatomy* (produced by Shondaland for ABC) to big-budget series like *Stranger Things* (produced by 21 Laps Entertainment for Netflix). Development and licensing deals between production companies and content delivery services are complicated. Some CDSs get original content through deals made with the outside production companies, which give a CDS the right to broadcast or stream that content a certain number of times. In other cases, CDSs partner with production companies to create a show. The FX drama anthology *American Horror Story*, for example, is coproduced by Ryan Murphy Television and 20th Television—FX's sister company within the Disney corporate family.

As competition for viewers' attention intensified among broadcast networks, cable channels, and streaming services in the late 2010s, they all raced to lock up production deals for original content, leading to an unprecedented surge in the production of original dramas, comedies, and limited series. Netflix signed producer Shonda Rhimes (*Scandal*) to a multiyear $150 million contract to produce new shows like *Bridgerton* (2020–), as well as signed multimillion dollar contracts with Ryan Murphy (*American Horror Story*), Kenya Barris (*Black-ish*), comedian Chris Rock, and Jenji Kohan (*Orange Is the New Black*). (For more on Netflix's acquisitions, see the chapter-opening discussion.) Meanwhile, competitor Amazon signed Jordan Peele (*Get Out*) and Barry Jenkins (*Moonlight*), and Apple signed Justin Lin (*Fast and the Furious*).

In 2009, 210 series were produced; that number rose to 349 in 2013 and to 532 in 2019.[18] Industry observers have argued that this amount of production—what has been dubbed **peak TV**—is not sustainable and anticipate a collapse in original content production once there are clear winners and losers in the content delivery battles.

PEAK TV *Killing Eve*, a dark British spy drama produced by Sid Gentle Films and aired in the United States on BBC America, is just one of hundreds of original series being produced for a television industry in which scores of content delivery services compete to attract viewers with the promise of must-see original content.

Laura Radford/©BBC-America/Everett Collection, Inc

Syndication Mixes the Old with the New

Although some content delivery services produce a portion of the content they need in-house or develop ongoing relationships with independent production companies, many turn to **syndication**—the process by which producers lease the right to air a program to local TV stations, cable channels, and streaming services. Syndication, a key part of the TV industry, involves many companies that serve as brokers, helping connect program producers or owners with content delivery services.

There are two main types of syndication. **First-run syndication** relates to any program originally produced for sale into syndication markets. Game shows such as *Wheel of Fortune* and daytime talk or advice shows like *The Ellen DeGeneres Show* or *Judge Judy* are made for first-run syndication. The producers of these programs usually license them directly to local markets around the country as well as internationally. (See "Global Village: Telling and Selling Stories around the World.") For local network-affiliated stations, syndicated programs are often used during **fringe time**—immediately before the evening's prime-time schedule (*early fringe*) and after either the local evening news or a network late-night talk show (*late fringe*).

In **second-run syndication** (commonly called "reruns"), older programs that were originally developed for and run on a specific broadcast network, cable channel, or streaming service are later made available to rerun on other content delivery services in the United States and overseas. We can watch reruns of *I Love Lucy* on Hulu, catch a *Law & Order* marathon on USA, or binge-watch *The West Wing* on Netflix because the content delivery services negotiated a deal with the companies that own those series. A show can also be put into rerun syndication even if new episodes are airing on the original CDS. The terms of syndication deals depend on many variables: the popularity of the content, the relationship of the two parties, the size of the potential audience, and, perhaps most importantly, whether the deal will give the CDS exclusive access—that is, will it be the only place viewers can access reruns of that TV show.

Programs from TV's past eras are essential to television in the digital era. Streaming services, for example, work to gain access to the exclusive rights to the complete archives of old shows like *The Office* and *Seinfeld* as a way to lure in new subscribers and expand their content libraries. A cable company like TNT may offer a few original series, like *Snowpiercer* and *Animal Kingdom*, but use reruns of older series, like *Bones* and *Supernatural*, to fill out its schedule.

The demand for second-run series means old TV shows can be an important source of revenue for producers. For hit shows, profits can be astronomical. For instance, early reruns of *Friends* earned an average of $4 million an episode from broadcast and cable syndication deals, totaling over $1 billion. In 2019, *Friends* made even more money, as WarnerMedia outbid Netflix for streaming rights to the program, paying about $425 million for five-year exclusive rights to include it on HBO Max.

TV Production beyond the Industry

Widespread access to digital cameras; digital editing tools; and websites and apps like YouTube, Vimeo, and TikTok have democratized the creation and spread of video content. In the digital era, aspiring writers, directors, and actors can produce scripted content with little money, post it online, and use social media to build a following. Free of many industry constraints, DIY producers often explore characters and story lines or develop visual styles and narrative structures that differ dramatically from those of mainstream TV production. In 2011, for example, inspired by an article about the lack of images of Black female nerds, Issa Rae wrote and starred in a four-minute-long webisode titled *The Mis-Adventures of Awkward Black Girl & Friends*. After it was posted on YouTube, the episode went viral, and a Kickstarter campaign helped fund production of a twelve-episode season. *Awkward Black Girl*'s success helped Rae get a deal to produce *Insecure* (2016–) for HBO. DIY producers like Rae have begun to influence mainstream TV production practices.

Meanwhile, other types of user-generated video content not only blur the boundaries between DIY production and professional production but also draw into question what exactly counts as

Telling and Selling Stories around the World

There is much political talk about how the United States has huge trade deficits—that is, we buy more from other countries than we sell to them. But in one of the country's largest businesses—film and television—the United States always runs a surplus, selling more to other countries than we buy from them and dominating production and distribution around the world. In 2019, for example, the U.S. film and television industry generated $17.2 billion in exports to other countries, which resulted in a trade surplus of $10.3 billion.[1]

As one industry observer puts it, "[American television is] bigger, more profitable, flashier, and at times, *at times*, better than almost any form of television in existence on the planet. America is to television what Switzerland is to clocks and banking, what Holland is to tulips, and what France is to wine."[2]

So what's the deal? Why is our popular culture so popular?

There are at least four reasons.

First, there is the fact that the United States got a historical boost more than one hundred years ago. In the first decade of the motion picture industry, the country had several European rivals. But World War I disrupted the European film industry, and the United States—which never had to fight the war on its soil—was able to create a market for its films in Europe to fill the void. Also one hundred years ago: Whereas much of the world built nationally supported public broadcasting (like Great Britain's BBC), the United States went the commercial route for radio, which then extended to television. Because American television was always treated as a business, not a domestic service, there was great incentive to export it to new markets.

Another reason for the United States' industry dominance is its large affluent domestic market of over 330 million people.

The scale of the U.S. domestic market means that American media companies have a tremendous advantage; the U.S. television industry earns significant profits from telling and selling its stories in the United States and then makes these programs available throughout the world at a discount, collecting billions more in licensing fees from hundreds of countries. Contributing to these additional profits are the more than two billion people in the world who speak English. For decades, no other country's domestic media market has offered media producers the same economic benefits.

A third reason often cited for U.S. pop culture supremacy is innovation.[3] Throughout the twentieth century, American cultural creators established new art forms and genres suited to the emerging mass media of the electronic age. Hollywood filmmakers pioneered westerns, gangster films, screwball comedies, and musicals, while early radio producers invented situation comedies and soap operas. American television fed off of, and added to, this rich pool of cultural creativity. Furthermore, as the American media industry became a global capital for entertainment, talent from around the world gravitated to Hollywood and New York, adding to the U.S. media industry's creative output. Most recently, of course, the United States' development of the web in the early 1990s enabled the rise of Netflix and Amazon, the world's top pop cultural powerhouses in creating and streaming movies and TV programs that dominate the world market.

As a fourth and final reason, Americans are good at telling stories. As one European critic observes, U.S.-made sitcoms and dramas are popular "because they stick to the basics—they portray regular people everyone can recognize and identify with, however dramatic or fanciful the situation they may find themselves in. American culture celebrates the commonplace, the average, the universal; as a result, it has gained a universal audience."[4]

While American television programs remain popular around the world, they have begun to face increasing competition. As discussed at the beginning of the chapter, the growth of streaming services like Netflix and the demand for content they generate is giving TV producers around the world access to global audiences (including the vast U.S. market). With that access comes increased revenue, which might help (at least somewhat) to level the playing field in the global TV marketplace. ●

television. Using only a camera and good lighting, vloggers can attract a few dozen or many millions of subscribers to YouTube channels dedicated to beauty tutorials, comedy sketches, pet videos, political activism, cooking tips, and countless other forms of infotainment. Video streaming services like Twitch allow gamers to produce their own live eSports broadcasts and talk shows. And TikTok encourages users to film, frame, and edit themselves and the world around them for comic, dramatic, endearing, or political effects. In some of these cases, users make a living through the content they create; in most cases, they don't, nor do they want to. But the fact that almost anyone can produce video content, and that audiences in the millions are watching that content, is transforming what counts as TV.

ISSA RAE The success of Issa Rae's self-produced web series *Awkward Black Girl* drew the attention of HBO, which signed her to a development deal that led to the critically acclaimed comedy-drama *Insecure* (shown here).

Television Ownership

Because television in the digital era is vast and complex, with scores of production companies and numerous content delivery services organized around a variety of revenue and distribution models, it can be difficult to recognize the outsized role a few major corporations play in the TV industry. Table 6.3 highlights the major corporations involved in television today and their investments (or lack thereof) in the different categories of television content delivery.

As the table indicates, television has become an arena in which large corporations from diverse corners of the media landscape vie for a place in the medium's still unfolding digital future, each trying to strategically position itself for success (often through mergers or acquisitions). Some of these corporations grew out of traditional broadcast and cable TV companies (Disney/ABC, Comcast, and ViacomCBS), while others—particularly those driving the development of streaming—came from the cash-rich world of Silicon Valley (Netflix, Amazon, Apple, Google, and

TABLE 6.3 MAJOR CORPORATIONS' CONTENT DELIVERY SERVICES IN THE UNITED STATES

Company	Broadcast	Cable/Satellite	Streaming
Amazon			Prime Video, Twitch
Apple			Apple TV+
AT&T	the CW (50%)	AT&T, Fios (MSO), DirecTV, Turner Broadcasting	HBO Max
Comcast	NBC, Telemundo	Xfinity Cable (MSO), MSNBC, CNBC, USA Network, Bravo, and more	Peacock
Disney	ABC	ESPN, Freeform, FX, National Geographic (73%), A&E (50%), and more	Disney+, Hulu, ESPN+
Facebook			Facebook Watch
Fox Broadcasting	Fox, MyNetworkTV	Fox News, Fox Business, FS1, FS2, Big Ten Network (51%), and more	Tubi
Google			YouTube
Netflix			Netflix
ViacomCBS	CBS, the CW (50%)	MTV, BET, Nickelodeon, and more	Paramount+, Showtime, Pluto TV

Facebook). There is also AT&T, a telecommunications company with roots early in the electronic era that bought its way into television with the purchase of the entertainment giant Time Warner (later renamed WarnerMedia) in 2018, uniting its high-capacity broadband network with a major Hollywood studio (Warner Brothers) and content delivery services like HBO Max, the CW, and CNN. As that merger illustrates, corporations have strong financial incentives to gain control over every aspect of how and where TV is produced, distributed, and watched. (See Chapter 13 for more on this aspect of media economics). In 2021, AT&T surprised many with a proposal to spin off WarnerMedia, illustrating that such mergers don't always end happily.

Television and Democracy

As much as in any other corner of the media environment, television has reflected and contributed to America's cultural shift from a mass nation to a niche nation—a shift that has altered the medium's role in the democratic process. For decades, television was the dominant form of mass media in the United States, as its three-network broadcast system created shared experiences for millions of people. In the 1960s and 1970s, television covered events like the presidential debates and the moon landing and, in doing so, interrupted daily life, addressed viewers as citizens, and urged them to identify with and participate in a shared democratic society. Meanwhile, even when there wasn't a breaking news story, limited programming choices meant that millions of people tuned in together to watch the same news broadcasts, sitcoms, and dramas—a dynamic that established a mainstream culture and circulated consensus narratives. By creating such shared cultural touchstones, television during the mass nation greased the wheels of democratic processes in certain ways, but it did so at the expense of letting diverse ideas and values circulate. Many communities' lived experiences were excluded from the vision of American life that television offered up.

TV AND DEMOCRACY
The first televised presidential debates took place in 1960, pitting Massachusetts senator John F. Kennedy against Vice President Richard Nixon. With all three networks preempting regular programming to broadcast such debates without commercials, television helped establish political debates as shared rituals of democracy. While several content delivery services still cover presidential debates, most do not, making it easy for citizens to opt out of the political ritual.

Series: Photographic Materials, ca. 1926–ca. 1994 Collection: Richard Nixon Foundation Collection of Audiovisual Materials, ca. 1926–ca. 1994/National Archives and Records Administration

In the digital era, in which television viewing is highly fragmented, on-demand delivery allows us to disconnect from the unifying flow of broadcast schedules, and content delivery services are organized around people's narrow interests, television adds to the dynamics of the niche nation. A much wider range of experiences and perspectives are included in TV's varied content streams, a fact that makes TV more democratic. Yet the loss of shared experiences and consensus narratives likely contributes to the kind of political polarization that can make certain democratic processes more challenging.

Equally important to note is the fact that the variety of content offered in the digital era is increasingly available only to those who can afford it. The idea of free and universal access to television was a guiding principle in the United States for much of the medium's history, as was the assumption that an industry using the public airwaves should serve the public interest. Yet both visions of the medium have all but disappeared. The buzz that circulates around "high-quality" TV series in the age of peak TV, and the greater control viewers have over when and how to watch TV, obscures the fact that millions of people still rely on free, over-the-air broadcast television for information and entertainment. These viewers and their needs are rarely included in discussions about television's future.

Current developments suggest that access to TV content is likely to get even more expensive in the coming years. For decades, the old model of providing programming for free (the broadcast network model) has been slowly replaced by increasingly costly cable subscriptions and, more recently, the need for increasingly powerful broadband connections in order to access subscriber-based streaming video services. Television may have originated on the public airwaves, and in its latest form it may be delivered via the publicly built Internet, but accessing television today is almost never free.

Chapter REVIEW 6

LAUNCHPAD FOR *MEDIA & CULTURE*

 **LaunchPad**
macmillan learning

Visit LaunchPad for *Media & Culture* at launchpadworks.com for additional learning tools:

→ QUIZ YOURSELF WITH LEARNINGCURVE

LearningCurve uses gamelike quizzing to help you review important concepts from this chapter. LearningCurve includes multiple-choice questions at different levels of difficulty, hints, feedback for both right and wrong answers, and links to e-book material for easy reference. Which answer choice would you select for the sample question below?

> Imagine that you owned a CATV service in the early days of cable television and were speaking to a prospective customer. Which argument could you make to convince the person to sign up for your cable service?
>
> ◯ "Our service offers better reception than your standard TV antenna."
>
> ◯ "Our service is less expensive than over-the-air signals."
>
> ◯ "Our service is completely wireless."
>
> ◯ "Our service offers thirty-five channels."

→ WATCH VIDEO ON KEY CONCEPTS

LaunchPad includes clips from movies, TV shows, online sources, and other media texts, in addition to interviews with media experts and newsmakers. In the videos for this chapter, you'll consider topics like public television and original streaming series.

→ PRACTICE THE CRITICAL PROCESS

Use the Media Literacy Activity to walk through the critical process step-by-step and develop a critical perspective on what media means to you. In this chapter's activity, you'll reflect on the evolving meaning of television in our digital era.

→ LEARN KEY TERMS

Review the definitions for this chapter's key terms with LaunchPad's easy-to-use online flash cards. Try the cards in practice mode, or quiz yourself as a way to focus your study efforts. (Page numbers indicate where each term is highlighted within the chapter.)

designated market areas, 162
O&Os, 162
affiliate, 163
kinescope, 164
Big Three, 165
single sponsorship, 165
magazine sponsorship, 165
prime time, 166
anthology dramas, 166
episodic series, 167
chapter shows, 167
serial programs, 167

cloning, 169
spin-off, 169
franchise, 169
CATV, 171
must-carry rules, 172
access channels, 172
electronic publishing, 172
cable franchise, 172
basic cable channels, 173
premium channels, 173
direct broadcast satellite (DBS), 174
time-shift, 174

least objectionable programming (LOP), 174
narrowcasting, 175
quality audience, 175
Telecommunications Act of 1996, 179
multiplexing, 179
multiple-system operators (MSOs), 179
linear TV, 180
on-demand TV, 180
video-on-demand (VOD), 180
place shifting, 180

content delivery services (CDSs), 181
rating, 182
share, 182
over-the-top (OTT), 183
cord cutting, 183
digital subchannel broadcast networks, 183
peak TV, 184
syndication, 185
first-run syndication, 185
fringe time, 185
second-run syndication, 185

REVIEW QUESTIONS

Use the questions below to revisit key themes and concepts from this chapter.

⊛ Need more practice? Access the LearningCurve multiple-choice questions for this chapter, which are available on LaunchPad. All questions include feedback for correct and incorrect answers, hints, and links to e-book material.

Television Today

1. In what ways and why is TV today so diverse?

How We Got Here: The Birth of Broadcast TV

2. What steps had to be taken before television could become a commercial mass medium?

3. What are the major elements of the broadcast model?

4. What are the key differences between three-camera live and one-camera film systems of TV production?

The Network Era: The Big Three Dominate TV

5. How did the Big Three networks take control of programming?

6. How did the networks' pursuit of efficiency and predictable profits influence the development of TV programming?

7. What concerns did critics have about the path TV was taking in the early 1960s, and how did the networks respond?

The Post-Network Era: Competition Heats Up

8. How and why did the FCC try to protect broadcast TV from cable?

9. How did satellite distribution change the cable industry?

10. Why did Hollywood studios like Warner Brothers launch their own television networks in the 1980s and 1990s?

11. What is the quality audience, and what role did it play in the increase of gay content on network TV in the 1990s?

12. What factors helped fuel the rise of reality TV series?

Our Digital Era and the Business of TV

13. What role did the Telecommunications Act of 1996 play in the shift to the digital era?

14. How does TV viewing in the digital era differ from previous periods?

15. What are the primary revenue models used by content delivery services, and how do they influence program development?

16. Why has there been a record level of original TV content production?

17. Why are programs from TV's past eras important in today's industry?

18. How are the emergence of digital technologies and websites like YouTube and TikTok changing television?

Television and Democracy

19. How has television's cultural role changed over the years?

QUESTIONING THE MEDIA

Use these critical-thinking questions to reflect on your own experiences with media and apply your understanding of media concepts.

⊛ For an interactive experience, questions on topics like these are available for the iClicker student response system. Instructors can download the questions from LaunchPad and use them to poll the class.

1. How much television do you watch today? How has technology influenced your current viewing habits?

2. If you were an executive at a content delivery service, what changes would you make to today's programs and how they are delivered in order to respond to young viewers' habits?

3. Do you think must-carry rules violate a cable company's First Amendment rights? Why or why not?

4. How do you think new technologies will further change TV viewing habits?

5. How could television be used to improve our social and political life?

Movies and the Power of Images

7

"Mickey, meet Yoda."[1] So began the headline of a 2012 *Forbes* article announcing one of the biggest movie developments that year: Megamedia corporation Disney had bought Lucasfilm Ltd. for just over $4 billion. With the purchase, Disney gained control of and responsibility for the *Star Wars* universe—one of Hollywood's most beloved, and most valuable, intellectual property assets.

Since its launch in 1977, the *Star Wars* franchise has continually changed the culture of the movie industry. The first movie, *Star Wars*—produced, written, and directed by George Lucas—departed from the personal filmmaking of the early 1970s and helped establish a blockbuster mentality that identified a new primary audience for Hollywood: teenagers. It had all of the now-typical blockbuster characteristics, including massive promotion and lucrative merchandising tie-ins. Repeat attendance and positive buzz among young people made *Star Wars* the most successful movie of its generation.

The franchise as a whole has also influenced the technical side of moviemaking. To achieve the look and feel he wanted in the original trilogy, Lucas founded the legendary visual effects company Industrial Light and Magic (ILM) in 1975. Through its work on *Star Wars* (1977), *The Empire Strikes Back* (1980), and *Return of the Jedi* (1983), ILM pioneered technologies that are essential elements of moviemaking today: digital animation, motion control cameras, and computer-based film editing. With the second trilogy's *The Phantom Menace* (1999), Lucas and ILM again broke new ground, developing high-definition digital cameras that would eventually become standard in film production.

The success of the *Star Wars* films has spawned an entire ecosystem of other media products, establishing what came to be known as the *Star Wars* Expanded Universe, later changed to *Star Wars* Legends. Video games, comic-book series, and novels began to appear in the late 1970s and have introduced new characters and story arcs built on the world created in the films. The franchise has also expanded into television, with animated series like *Star Wars: The Clone Wars*, which began its highly successful run on Cartoon Network in 2008.

◀ Fans of the classic 1970s *Star Wars* films were skeptical when George Lucas sold Lucasfilm to the Walt Disney Company in 2012 for just over $4 billion, but the partnership brought new life to the *Star Wars* franchise, resulting in a number of new films and TV series.

In purchasing Lucasfilm in 2012, Disney wasn't simply buying a successful film franchise; it was gaining control of a fictional universe from which an array of content could be developed. In announcing the deal, Disney CEO Robert Iger spoke about the business possibilities: "This transaction combines a world-class portfolio of content including *Star Wars*, one of the greatest family entertainment franchises of all time, with Disney's unique and unparalleled creativity across multiple platforms, businesses, and markets to generate sustained growth and drive significant long-term value."[2] In plain language, Disney was promising more *Star Wars* in theaters, digital games, theme parks, and consumer products, as well as on television.

Disney has leveraged the *Star Wars* universe extensively since 2012. In addition to the final trilogy in the Skywalker saga—*The Force Awakens* (2015), *The Last Jedi* (2017), and *The Rise of Skywalker* (2019)—Disney released stand-alone films *Rogue One* (2016) and *Solo* (2018) and created immersive *Star Wars*-themed environments at its various theme parks. The franchise was especially useful for the company when it launched its Disney+ streaming service in 2019. The large archive of past *Star Wars* content was a valuable draw for the new site, as was *Star Wars: The Mandalorian*, the franchise's first live-action television series—in part because it features the highly meme-able character of Grogu (otherwise known as Baby Yoda).

LaunchPad
macmillan learning
launchpadworks.com

Storytelling in *Star Wars*
Poe Dameron and BB-8 battle enemy fighters in this clip from *Star Wars: The Last Jedi*.

Discussion: From what you've seen in this clip, how does the movie use advanced technical tools to tell its story?

While Disney has profited greatly from its control of the *Star Wars* empire, it has also faced challenges. Some die-hard purists have complained about the direction of the franchise under Disney control, and observers (including Lucasfilm executives) wonder if the market has become oversaturated with *Star Wars* content. In addition, although the recent films have all been blockbusters, questions linger about the forty-five-year-old franchise's continued drawing power. Compared to Disney's Marvel Universe, for example, the *Star Wars* brand has underperformed in China. As international ticket sales become increasingly important for media giants like Disney, *Star Wars*' U.S.-centric appeal could become a limitation.

The end of the original Skywalker space epic saga that began with Luke, Leia, Han Solo, and Darth Vader also presents a challenge; as *Fortune* magazine explains, "The company . . . has sold a lot of lightsabers. But without the familiar Jedi clan and their friends, Disney must spend the next several years introducing new characters and story lines."[3] Maintaining franchise viability becomes a constant preoccupation for a company like Disney, which must answer an important question: How can the brand stay relevant to a new generation of moviegoers who didn't grow up surrounded by the *Star Wars* mythology and buzz?

What memories do you have of the *Star Wars* franchise? What other franchise "universes" play a role in your media consumption?

Movies Today

A media powerhouse, Disney is one of a handful of studios that dominate today's truly global movie business. In 2019 alone, the company's big-budget superhero films *Captain Marvel* and *Spider-Man: Far from Home*, as well as family-friendly movies *The Lion King* and *Frozen II*, each earned over a billion dollars from ticket sales around the world. The year's biggest hit was *Avengers: Endgame*—another Disney offering—which earned almost $2.8 billion in global ticket sales. And for movie studios today, the traditional box office represents just one revenue stream; the film industry earns even more money through digital sales and rentals and licensing fees paid by content delivery services like Netflix, HBO, and NBC.

There's more to the movie industry than Hollywood blockbusters, of course. Signs indicate that new competitors, especially the Chinese film industry, are gaining ground in the international market. Although it made only $3.7 million in the United States and Canada, the Chinese animated fantasy *Ne Zha* took in over $720 million internationally—more than *Spider-Man* or *Captain Marvel*'s global earnings. And in the United States, over 800 full-length feature films are produced each year. Only about one-quarter of them are bigger budget movies (over $15 million). Most are fiction, but some are documentaries; some make their way to multiplex screens, others to art-house theaters, and still others go to television or Internet websites. Movies of all types are an integral part of our media environment.

For over one hundred years, movies have had a substantial cultural impact on society, providing communal stories that evoke and shape our desires and form the fabric of our social lives. Watching the annual broadcast of *The Wizard of Oz* was a childhood ritual for generations, and going to the theater with friends has long been a rite of passage for teenagers. Even in the fragmented media environment of today's niche nation, blockbusters like *Avengers: Endgame* remain shared cultural touchstones. Films also help moviegoers sort through experiences that either affirm or deviate from their own values. Some movies have even occasionally been banned from public viewing, including *Scarface* (1983), *The Last Temptation of Christ* (1988), *Brokeback Mountain* (2005), and *Three Billboards Outside Ebbing, Missouri* (2017), all of which have been criticized for appearing to glorify crime and immorality, trample on sacred beliefs, or promote "unpatriotic" viewpoints.

Finally, movies distract us from our daily struggles. They symbolize universal themes of human experience—of childhood, coming of age, family relations, growing older, and coping with death. They can help us understand and respond to major historical events, like the Vietnam War and 9/11, and they encourage us to reexamine contemporary ideas, particularly in terms of how we think about race, class, spirituality, gender, and sexuality.

In this chapter, we examine the rich legacy and current standing of movies. We will:

- Consider film's early technology and the evolution of film as a mass medium
- Understand how the production, distribution, and exhibition of films were organized during Hollywood's studio system era
- Explore the impact of sound technologies, the role of genres and movie stars, and the emergence of a specific film style
- Examine the factors that brought about the end of the studio system era
- Analyze the changing role of film directors and alternatives to Hollywood's style, including foreign films, documentaries, and independent films
- Survey the movie business today—its major players, economic clout, technological advances, and implications for democracy

BLOCKBUSTER STREAMING Streaming services got a big boost in 2020 when the coronavirus took hold and theaters closed for months at a time, prompting studios to debut some movies online—or at least run them concurrently online and in theaters. Viewers could watch Warner Brothers' superhero blockbuster *Wonder Woman 1984* on HBO Max, while movies like *Soul* and *Mulan* went to Disney+.

How We Got Here: The Early Development of Movies

History often credits a handful of enterprising individuals with developing the technologies that lead to new categories of mass media. As the early development of film reveals, however, such innovations are usually the result of simultaneous investigations by numerous people. In addition, the innovations of both known and unknown inventors are propelled by economic and social forces as well as by individual abilities.[4]

The Development of Film Technology

The concept of film goes back as early as Leonardo da Vinci, who theorized in the late fifteenth century about creating a device that would reproduce reality. Other early precursors to film included the magic lantern, developed in the seventeenth century, which projected images painted on glass plates using an oil lamp as a light source, and the *zoetrope*, introduced in 1834, a cylindrical device that rapidly twirled images inside a cylinder, which appeared to make the images move.

Muybridge and Goodwin Make Pictures Move

Moving image technology advanced when inventors started manipulating photographs to make them appear to move while simultaneously projecting them onto a screen. Eadweard Muybridge, an English photographer living in America, is credited with being the first to do both. He studied motion by using multiple cameras to take successive photographs of humans and animals in motion. One of Muybridge's first projects involved using photography to determine if a racehorse actually lifts all four feet from the ground at full gallop (it does). By 1880, Muybridge had developed a method for projecting the photographic images onto a wall for public viewing. These early image sequences were extremely brief, showing a horse jumping over a fence or a man running a few feet, because only so many photographs could be mounted inside the spinning cylinder that projected the images.

Meanwhile, other inventors were also working on capturing and projecting moving images. In 1884, George Eastman developed the first roll film—a huge improvement over the heavy metal and glass plates used to make individual photos. The film had a stiff paper backing that had to be stripped off during the film developing stage. Louis Aimé Augustin Le Prince then invented the first motion picture camera using roll film and is credited with filming the first motion picture, *Roundhay Garden Scene*, in 1888. About two seconds' worth of the film survives today.

In 1889, Hannibal Goodwin improved Eastman's roll film by using thin strips of more pliable material called **celluloid**, which could hold a coating of chemicals sensitive to light. Goodwin's breakthrough solved a major problem: It enabled a strip of film to move through a camera and be

EADWEARD MUYBRIDGE'S study of horses in motion proved that a racehorse gets all four feet off the ground during a full gallop. In his various studies of motion, Muybridge would use up to twelve cameras at a time.

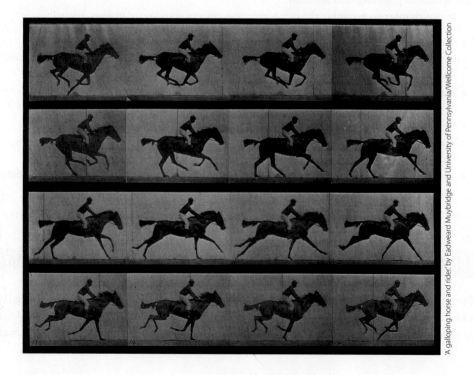

'A galloping horse and rider,' by Eadweard Muybridge and University of Pennsylvania/Wellcome Collection

photographed in rapid succession, producing a series of pictures. Because celluloid was transparent (except for the images made on it during filming), it was ideal for projection, as light could easily shine through it. Eastman, who also announced the development of celluloid film, legally battled Goodwin for years over the patent rights. The courts eventually awarded Goodwin the invention, but Eastman's company (Eastman Kodak) became the major manufacturer of film stock for motion pictures after buying Goodwin's patents.

Edison and the Lumières Create Motion Pictures

As with the development of sound recording, Thomas Edison takes center stage in most accounts of the invention of motion pictures. In the late 1800s, Edison planned to merge phonograph technology and moving images to create talking pictures (which would not happen in feature films until 1927). With no breakthrough, however, Edison lost interest. He directed an assistant, William Kennedy Dickson, to combine his incandescent lightbulb, Goodwin's celluloid, and Le Prince's camera to create another early movie camera, the **kinetograph**, and a single-person viewing system, the **kinetoscope**. This small projection system housed fifty feet of film that revolved on spools. Viewers looked through a hole and saw images moving on a tiny plate. In 1894, the first kinetoscope parlor, featuring two rows of coin-operated machines, opened on Broadway in New York.

Meanwhile, in France, brothers Louis and Auguste Lumière developed the *cinematograph*, a combined camera, film developer, and projection system. The projection system was particularly important, as it allowed more than one person at a time to see the moving images on a large screen. In a Paris café on December 28, 1895, the Lumières projected ten short movies for viewers who paid one franc each, on such subjects as a man falling off a horse and a child trying to grab a fish from a bowl. Within three weeks, twenty-five hundred people were coming each night to see how, according to one Paris paper, film "perpetuates the image of movement."

With innovators around the world now dabbling in moving pictures, Edison's lab renewed its interest in film. Edison patented several inventions and manufactured a new large-screen system called the **vitascope**, which enabled filmstrips of longer lengths to be projected without interruption and hinted at the potential of movies as a future mass medium. Staged at a music hall in New York in April 1896, Edison's first public showing of the vitascope featured shots from a boxing match and

KINETOSCOPES allowed individuals to view motion pictures through a window in a cabinet that held the film. The first kinetoscope parlor opened in 1894 and was such a hit that many others quickly followed.

Everett Collection, Inc

waves rolling onto a beach. Some members of the audience were so taken with the realism of the film images that they stepped back from the screen's crashing waves. By 1900, short movies had become part of the entertainment industry and were showing up in amusement arcades, traveling carnivals, wax museums, and vaudeville theaters.

The Introduction of Narrative in the Silent Era

Movies started out as a novelty, but they became an established mass medium and a more complex art form with the introduction of **narrative films**: movies that tell stories. Audiences quickly tired of films of waves or vaudeville acts. To become a mass medium, films had to offer what books achieved: the ability to create narrative worlds that engage an audience's imagination. Early narrative filmmakers faced unique challenges in the silent film era, which lasted until the end of the 1920s. Movies had no sound except for a musical score, which theater owners provided by either playing a phonograph record or hiring musicians to accompany a film as it was projected; without sound, filmmakers were forced to engage audiences through dramatic acting, clear storytelling, and engaging special effects.

Some of the earliest narrative films were produced and directed by French magician and inventor Georges Méliès, who opened the first public movie theater in France in 1896. Méliès may have been the first director to realize that a movie was not simply a means of recording reality, that it could be artificially planned and controlled like a staged play. Méliès began producing short fantasy films—including *The Vanishing Lady* (1896) and *A Trip to the Moon* (1902)—employing editing and camera tricks, such as slow motion and cartoon animation, which became key ingredients in future narrative filmmaking.

The first American filmmaker to adapt Méliès's innovations to narrative film was Edwin S. Porter. A camera operator who had studied Méliès's work in an Edison lab, Porter edited diverse shots together to tell a coherent story. Porter shot narrative scenes out of order (for instance, some in a studio and some outdoors) and reassembled, or edited, them to make a story. In 1902, he made what is regarded as America's first narrative film, *The Life of an American Fireman*. The film also contained the first close-up shot in U.S. narrative film history—a ringing fire alarm. Until then, moviemakers thought close-ups cheated the audience of the opportunity to see an entire scene. Porter's most important film, *The Great Train Robbery* (1903), introduced the western genre as well as chase scenes. In this popular and influential eleven-minute movie, Porter demonstrated the art of film suspense by alternating shots of the robbers with those of a posse in hot pursuit.

A decade later, film narratives took another leap forward with the work of D. W. Griffith, who was the single most important director in Hollywood's early days. Griffith paved the way for future filmmakers by refining many of the narrative techniques introduced by Méliès and Porter—including varied camera distances, close-up shots, multiple story lines, fast-paced editing, and symbolic imagery—and using nearly all of them in one film for the first time. Griffith's *The Birth of a Nation* (1915) was the first *feature-length film* (a film that is more than an hour long) produced in America. The three-hour epic cost moviegoers a record $2 admission and became a spectacular success. Although considered a technical masterpiece, the film reflected and advanced America's deeply rooted white supremacy. Its racist narrative glorified the Ku Klux Klan and stereotyped Black southerners, leading to a campaign

THE GREAT TRAIN ROBBERY (1903) includes this now iconic close up in which one of the film's outlaws aims and shoots directly at the camera. The film was among the most popular movies of its era.

Library of Congress

producers and distributors had not yet recognized that fans would be drawn not only to particular genres—like dramas, westerns, and romances—but also to particular actors. It was Paramount founder Adolph Zukor who recognized the drawing power of movie stars, and by 1916, he had hired a number of popular actors, including Mary Pickford, Douglas Fairbanks, and Rudolph Valentino. His idea was to dominate the movie business not through patents but through exclusive contracts with actors. His idea caught on with the other major studios, and by the 1930s, each of the Big Five studios had a large stable of movie stars under contract whose careers they worked hard to control.

The introduction of sound technologies into movie production further consolidated the Hollywood studios' dominance. After several decades of false starts, key technological advances made during the 1920s finally meant that **talkies** were ready for their debut. In 1927, Warner Brothers released *The Jazz Singer*, which was essentially a silent film interspersed with musical numbers and brief dialogue. At first there was only modest interest in the movie, which featured just 354 spoken words. But the film grew in popularity as it toured the Midwest, where audiences stood and cheered the short bursts of dialogue. The breakthrough film, however, was Warner Brothers' 1928 release *The Singing Fool*. Costing $200,000 to make, the film took in $5 million and "proved to all doubters that talkies were here to stay."[6] Boosted by the innovation of sound, annual movie attendance in the United States rose from sixty million a week in 1927 to ninety million a week in 1929. By 1935, the world had adopted talking films as the commercial standard.

The conversion to talkies was expensive, however, requiring tremendous investment in new sound stages and equipment. It also required developing new talent—actors, directors, screenwriters—who had the necessary skills for sound production. Many of the smaller film companies that had been trying to compete with the major studios didn't have the deep pockets required to make the transition and either struggled or went out of business. This included production companies that made movies by Black filmmakers, featuring all-Black casts, for Black audiences (see "Examining Ethics: Breaking through Hollywood's Race Barrier" on page 210). In this way, talkies helped centralize control over the movie business in the hands of the major Hollywood studios.

INDEPENDENT BLACK FILMMAKING During the 1920s, Black entrepreneurs established a number of production companies that made movies for Black audiences, featuring all-Black casts in genres ranging from westerns and gangster films to melodramas and war films. But without the economic advantages enjoyed by the Hollywood studios, Black independent producers continually struggled to stay afloat; most went out of business by the 1930s.

Distribution

The major studios reinvented the distribution business as well. **Block booking** was key to their dominance. Under this system, which was eventually outlawed as monopolistic, exhibitors had to agree to rent several **B movies** (marginal films made with small budgets and unknown actors) in order to gain access to any **A movies** (highly anticipated films made with big budgets and famous stars). This system enabled the studios to test-market new actors without taking much financial risk. It also enabled them to flood theaters with inexpensive B movies, which helped keep upstart production companies from getting into the industry—with the numerous B movies foisted on them, theater owners didn't have the money or the need to run non-studio movies.

The studios also employed an elaborate promotional machine to generate interest in their movies. The star system was central to such efforts. Since actors were under long-term contracts, studios worked hard to develop and protect their stars' images—both on- and offscreen. Studio publicists cultivated relationships with the press to keep their stars and films in the public eye. They also established the Academy Awards as a way to generate attention (the first ceremony was held in 1929).

Another distribution strategy involved the marketing of American films in Europe. When World War I disrupted the once-powerful European film production industry, only U.S. studios were able to meet the demand for films in Europe. The war marked a turning point and made the United States the leader in the commercial movie business worldwide. After the war, no other nation's film industry could compete economically with Hollywood. By the mid-1920s, foreign revenue from U.S. films totaled $100 million. Today, Hollywood continues to dominate the world market.

Exhibition

When industrious theater owners across the country came together and tried to push back against the rising power of studios like Paramount, the studio heads conspired to dominate exhibition by buying up theaters. By 1921, Zukor's Paramount owned three hundred theaters, solidifying its ability to show the movies it produced. In 1925, a business merger between Paramount and Publix (then the country's largest theater chain, with more than five hundred screens) gave Zukor enormous influence over movie exhibition.

Zukor and the heads of the other Big Five studios understood that they did not have to own all the theaters to ensure that their movies would be shown. Instead, they needed only to own the first-run theaters (about 15% of the nation's theaters), which premiered new films in major downtown areas and generated 85 to 95 percent of all film revenue.

The studios also quickly realized that to earn revenue from these first-run theaters, they would have to draw the middle and upper-middle classes to the movies. To do so, they built **movie palaces**—full-time single-screen movie theaters that offered a more hospitable moviegoing environment than nickelodeons, providing elegant décor usually reserved for high-society opera, ballet, symphony, and live theater. Another major innovation in exhibition was the development of *mid-city movie theaters*, built in convenient locations near urban mass-transit stations to attract the business of the urban and suburban middle class. This idea continues today, as megaplexes featuring multiple screens lure middle-class crowds to interstate highway crossroads.

Classical Hollywood Cinema Style

During Hollywood's Golden Age, the studio system dictated not only the business but also the style of moviemaking. That style, which scholars call **classical Hollywood cinema**, solidified during the 1930s and 1940s and was firmly rooted in the specific business practices of the studio system—though many of its aspects continue to dominate American filmmaking today.[7]

Classical Hollywood cinema serves up three ingredients that have given Hollywood movies their flavor: the approach to narrative, the reliance on genre, and the adherence to specific production conventions. The right blend of these ingredients—combined with timing, marketing, and luck—has led to many movie hits, from 1930s and 1940s classics like *The Wizard of Oz* and *Casablanca* to recent successes like *Once Upon a Time . . . in Hollywood* (2019) and *Us* (2019).

Hollywood Narratives

American filmmakers from D. W. Griffith and Frank Capra to Steven Spielberg and Ava DuVernay have understood the allure of narrative. As scholars point out, all narratives include two basic components: the story (what happens to whom) and the discourse (how the story is told). During the studio system era, Hollywood codified a familiar narrative structure that remains popular today: Most movies, like most TV shows and novels, feature recognizable character types (protagonist, antagonist, romantic interest, sidekick); a clear beginning, middle, and end (even with flashbacks and flash-forwards, the sequence of events is usually clear to the viewer); and a plot propelled by the main character's experiencing and resolving a conflict by the end of the movie.

Within classical Hollywood narratives, filmgoers find an amazing array of intriguing variations. Familiar narrative conventions of heroes, villains, conflicts, and resolutions, for example, can shift over time in response to historical changes or be transformed by new technologies like computer-generated imagery (CGI). This combination of convention and invention—of standardized Hollywood stories interacting with social and technical changes—provides a powerful economic package that has long satisfied most audiences' appetites for both the familiar and the distinctive.

Hollywood Genres

In general, classical Hollywood narratives fit a **genre**, or category, in which conventions regarding similar characters, scenes, structures, and themes recur in combination. Grouping films by category has been a way for the industry to achieve two related economic goals: **product standardization** (offering generally uniform products as a way to make production efficient and reduce unpredictability for consumers) and **product differentiation** (offering products that are distinct from those of competitors and that target specific consumers). By making films that fall into popular genres, the movie industry provided familiar models that could be imitated. It was also easier for a studio to promote a film to certain audiences when it fit into a preexisting category with which viewers were familiar. There was room for creative adaptation within a genre's boundaries, however, and breaking generic conventions could be a way for a film to stand out. The most popular films during the silent era were historical and religious epics like *The Ten Commandments* (1923) and *Ben-Hur* (1925). Other popular genres established during the studio system era included social drama, slapstick comedy, horror, war, the western, the musical, thriller/suspense, romantic comedy, gangster, fantasy/science fiction, and film noir. Although rooted in the business practices of the studio system, genres remain an integral part of the film business today.

One example of a genre is the horror film, which claims none of the top fifty highest-grossing films of all time. In fact, despite classics like *Psycho* (1960), this often lightly regarded genre has earned only one Oscar for best picture—*Silence of the Lambs* (1991)—and one for original screenplay, *Get Out* (2017). The genre's low cultural status has been linked to its primary audience. For decades, especially during the heyday of the slasher subgenre in the 1970s and 1980s, horror movies were extremely popular with teenagers—a demographic often drawn to media products distinct from those enjoyed by their parents. At the time, critics suggested that the teen appeal of horror movies was similar to the rebellious allure of gangster rap or heavy-metal music. The genre's status and appeal have been evolving, however. Movies like *Pan's Labyrinth* (2006) and *Us* (2019), which add art-house elements or explicitly address political themes, have attracted more adult audiences.

FILM GENRES *Psycho* (1960), a classic horror film, tells the story of Marion Crane (played by Janet Leigh), who flees to a motel after embezzling $40,000 from her employer. There, she meets the motel owner, Norman Bates (played by Anthony Perkins), and her untimely death. The infamous shower scene, pictured here, is widely considered one of the most iconic horror film sequences.

Hollywood Production Conventions

By the 1930s, Hollywood had developed a set of conventions, or norms, for film production that included where to place the camera, how to light a scene, what angles to shoot action from, how to edit shots together, and when to add music. These conventions helped give Hollywood's movies a relatively consistent look and feel.

Many of the rules fell under the umbrella term of **continuity editing**: producing and combining shots so that viewers experience a coherent flow of action as it plays out within and across scenes. The *180-degree rule*, for example, taught filmmakers that if they keep their camera on one side of an imaginary line when filming, it is easier to edit various angles and close-ups together without confusing viewers. Other conventions include opening a scene with an *establishing shot* (a wide angle shot that tells viewers where the upcoming scene takes place) and capturing dialogue between two characters using the *shot/reverse shot* sequence, in which a wider angle shot of the characters is combined with alternating close-ups of each, with their eyelines carefully matched.

Given the studio system's careful division of labor, these conventions helped coordinate the work of various units. Production designers could build sets suitable for standard camera angles, which helped cinematographers shoot the kind of footage editors expected to work with. They also

SHOT/REVERSE SHOT SEQUENCE This four-shot sequence from *Avengers: Endgame* (2019) illustrates one of the most common editing conventions in the classical Hollywood cinema style: the shot/reverse shot sequence used to capture a conversation. The first shot, which establishes the spatial relationship of the two actors, is followed by various close-ups with careful eyeline matching. These types of sequences are so common that they become almost invisible to us. When you watch any edited visual text, try to spot examples of this sequence. You can also observe ways that filmmakers vary camera angles and framing.

established a visual language that became familiar to viewers. Filmmakers would periodically break the rules for dramatic effect, but such moments were usually exceptions. Although these conventions have evolved over time—often in response to new technologies—filmmakers today still rely on many of the basic rules. As a result, they continue to shape how our visual media culture looks and sounds, not just in movies but also in television and video games.

The Decline of the Studio System and Hollywood's Golden Age

After years of thriving, the Hollywood movie industry began to falter after 1946. Weekly movie attendance in the United States peaked at ninety million in 1946, then fell to under twenty-five million by 1963. Critics and observers began talking about the death of Hollywood. Among the changing conditions facing the film industry were communist witch-hunts in Hollywood, the end of the industry's vertical integration, suburbanization, and the arrival of television.

The Hollywood Ten

In 1947, in the wake of the unfolding Cold War with the Soviet Union, conservative members of Congress began investigating Hollywood for alleged anti-government and communist ties. That year, aggressive witch-hunts for political radicals in the film industry by the House Un-American Activities Committee (HUAC) led to hearings and a subsequent trial.

During the investigations, HUAC coerced prominent people from the film industry to declare their patriotism and give up the names of colleagues suspected of having politically questionable tendencies. Upset over labor union strikes and outspoken writers, many film executives were eager to

THE HOLLYWOOD TEN
While many studio heads, producers, and actors named names to HUAC, others, such as the group shown here, held protests to demand the release of the Hollywood Ten.

testify and provide names. For instance, Jack L. Warner of Warner Brothers suggested that whenever film writers made fun of the wealthy or America's political system in their work, or if their movies were sympathetic to people of color, they were engaging in communist propaganda.[8] Other friendly HUAC witnesses included actor Ronald Reagan, director Elia Kazan, and producer Walt Disney. Whether they believed it was their patriotic duty or they feared losing their jobs, many prominent actors, directors, and other film executives also named names.

Eventually, HUAC subpoenaed ten unwilling witnesses who were questioned about their memberships in various organizations. The so-called **Hollywood Ten**—nine screenwriters and one director—refused to discuss their memberships or to identify communist sympathizers. Charged with contempt of Congress in November 1947, they were eventually sent to prison for a maximum of one year. Although jailing the Hollywood Ten clearly violated their free-speech rights, in the atmosphere of the Cold War many people worried that "the American way" could be sabotaged via unpatriotic messages planted in films. Upon release from jail, the Hollywood Ten were denied work by the major studios, and their careers in the film industry were all but ruined. The scandal tarnished Hollywood's reputation, undermining the high esteem the industry had built during the studio system's Golden Age.

The Paramount Decision

Coinciding with the HUAC investigations, the government increased its scrutiny of the movie industry's aggressive business practices. By the mid-1940s, the Justice Department had demanded that the Big Five—Paramount, Warner Brothers, Twentieth Century Fox, MGM, and RKO—end their vertically integrated control over production, distribution, and exhibition. In 1948, after a series of court appeals, the Supreme Court ruled against the film industry in what is commonly known as the **Paramount decision**, forcing the studios to gradually divest themselves of their theaters.

Although the government had hoped that the ruling would increase competition, the Paramount decision never really changed the oligopoly structure of the Hollywood film industry because it failed to challenge the industry's control over distribution. However, the 1948 decision did create opportunities in the exhibition part of the industry for those outside Hollywood, leading to more art-house theaters making documentaries and foreign films available to audiences.

DRIVE-IN THEATERS became increasingly common during the 1950s. As millions of Americans moved to the rapidly expanding suburbs and away from the downtown movie palaces, the film industry scrambled to find ways to make moviegoing more convenient. By the 1980s, most drive-in theaters had gone out of business, replaced by the slew of multiplex theaters that began to appear in suburban communities.

Moving to the Suburbs

While popular memory often assumes that competition from television alone precipitated the decline in post–World War II movie attendance, the most dramatic drop actually occurred in the late 1940s—before most Americans even owned TV sets.[9]

The transformation from a wartime economy and a surge in consumer production had a significant impact on moviegoing. Discretionary income that formerly went to buying movie tickets now went to acquiring consumer products, and the biggest product of all was a new house in the suburbs, far from the downtown movie theaters. Home ownership in the United States doubled between 1945 and 1950, while the moviegoing public decreased just as quickly. In response, thousands of drive-in theaters sprang up in farmers' fields, welcoming new suburbanites who embraced the automobile.

Additionally, after the war, the average age for couples entering marriage dropped from twenty-four to nineteen. Unlike their parents, many postwar couples had their first child before they turned twenty-one. The combination of social and economic changes meant there were significantly fewer couples going on dates at the movies. Thus, when television exploded in the late 1950s, there was already less discretionary income and less reason to go to the movies.

Television Changes Hollywood

In the late 1940s, radio's popularity had a strong impact on film. The years 1948 and 1949 were high points in radio listenership, and with the mass migration to the suburbs, radio offered Americans a convenient and inexpensive entertainment alternative to the movies. As a result, many people stayed home and listened to radio programs—that is, until the mid-1950s, when both radio and movies were displaced by television as the medium of national entertainment. The movie industry responded to this change in a variety of ways.

First, movie content slowly shifted toward more serious subjects. Initially, this shift was a response to the war and the recognition of life's complexity, but eventually movies focused on topics that television avoided. The change began in the 1940s with the rise of *film noir*—a term used to describe a type of Hollywood thriller or crime drama that features fatalistic characters, sexual overtones, and dark, moody lighting. It continued into the 1950s, as commercial movies tackled larger social problems in new ways, such as alcoholism (*The Lost Weekend*, 1945), anti-Semitism (*Gentleman's Agreement*, 1947), mental illness (*The Snake Pit*, 1948), racism (*Pinky*, 1949), adult–teen relationships (*Rebel without a Cause*, 1955), drug abuse (*The Man with the Golden Arm*, 1955), and—perhaps most controversial—sexuality (*Peyton Place*, 1957; *Lolita*, 1962).

These and other films challenged the authority of the industry's own prohibitive Motion Picture Production Code. Hollywood had adopted the Code in the early 1930s to restrict film depictions of violence, crime, drug use, and sexual behavior and to quiet public and political concerns that the movie business was lowering the moral standards of America. (For more on the Code, see Chapter 16.) In 1967, after the Code had been ignored by producers for several years, the Motion Picture Association of America initiated the current ratings system, which rates films for age appropriateness rather than censoring all adult content.

Second, just as radio worked to improve sound in order to compete with television in the 1950s, the film industry introduced a host of technological improvements to lure Americans to movie theaters. Technicolor was used in more movies to draw people away from their black-and-white TV sets. In addition, Cinerama, CinemaScope, and VistaVision arrived in theaters, featuring striking wide-screen images, multiple synchronized projectors, and stereophonic sound. Three-dimensional (3-D) movies also appeared, although the novelty of these films wore off quickly. Finally, Panavision, which used special Eastman color film and camera lenses that decreased the fuzziness of images, became the wide-screen standard throughout the industry. These developments, however, generally failed to address the movies' primary problem: the middle-class flight to the suburbs, away from downtown theaters.

Hollywood and Movies after the Studio System Era

The decline of the studio system transformed Hollywood and movie culture in many ways. Faced with new competition and dwindling audiences, the movie business struggled financially through the 1960s and 1970s, before rebounding in the 1980s with the success of a new strategy focused on big-budget blockbusters (see The Economics of the Movie Business later in the chapter). In particular, several key developments characterized the film industry after the studio system era's decline, developments that continue to influence movie culture today: the changing function of Hollywood studios, the growing cultural status of film directors, and the wider circulation of movies made outside Hollywood.

Major Studios Evolve

Although the studio system ended, most of the Hollywood studios survived (and still do today in one form or another) by evolving. Studios, for example, began to focus much of their energy on film distribution rather than production. Work that had once been done entirely in-house at a studio (such as developing a script; getting financing; and hiring actors, a director, and camera crews) was increasingly outsourced to independent producers and talent agencies. In many ways, Hollywood became a town of free agents and gig workers. Talent agencies like William Morris Endeavor and Creative Artists Agency became powerful players in the movie business by representing creative talent and putting deals together, while producers were increasingly responsible for landing a distribution deal with a studio. As the distributor, a studio would market the film and oversee its delivery to theaters. In addition, the studio often provided the producer with access to its sound stages, film equipment, and post-production facilities. Many films today get produced via some variation of this process.

Hollywood's relationship with television also evolved dramatically. While Hollywood initially saw the new medium as its enemy, television eventually became an integral element of the movie business. First, television became central to Hollywood's distribution strategies; studios developed multiphase release plans that depended on TV, which supplemented theaters as a key site for film exhibition. Second, television redefined the content Hollywood studios produced. By the 1960s, most movie studios created divisions devoted to producing content for television. As a result, the TV and film industries became tightly interwoven—a relationship that is even tighter today.

Directors as Movie "Authors"

Commercial filmmaking is a collaborative project that involves the creative input of scores of professionals, and in the studio system era, directors were often considered just another worker in the factory. After the studio system ended, however, the director came to be seen as a film's main author—a development that has made some directors powerful figures in the movie industry. At the same time, the increased prominence of directors has raised questions about who has and hasn't been given the opportunity to assume that role.

The Auteur and Film Style

In the 1950s and 1960s, some film scholars began to reassess the role of the director, arguing that certain directors, even when working within the constraints of classical Hollywood cinema, were able to develop their own distinctive cinematic style. They called these directors **auteurs** and pointed to Alfred Hitchcock as an example of someone who worked within the studio system but whose unique style still showed through, helping to redefine the suspense drama. In films like *Rear Window* (1954) and *Psycho* (1960), Hitchcock used particular editing and camera techniques designed to heighten tension. These scholars also drew attention to international directors, as working outside Hollywood's studio system could make it easier for a director to develop a distinct authorial voice. Japan's Kenji Mizoguchi, for example, was known for using long takes and emotionally restrained camera framing in films like *Ugetsu* (1953) and *Street of Shame* (1955). The idea of the film auteur remains a powerful concept today. With his unique use of voice-over narration and distinctive production design in films like *The Grand Budapest Hotel* (2014) and *The French Dispatch* (2021), Wes Anderson is often held up as a contemporary auteur.

Celebrity Directors

Beyond critical recognition of auteurs, directors also gained a higher profile in popular culture and more clout within the film business after the decline of the studio system. This new status was influenced by two breakthrough films: Dennis Hopper's *Easy Rider* (1969) and George Lucas's *American Graffiti* (1973), which became surprise box-office hits. Their inexpensive budgets, rock-and-roll soundtracks, and big payoffs created opportunities for a new generation of directors. The success of these films exposed cracks in Hollywood's traditional approach, and studio executives at the time seemed at a loss to predict the tastes of a new generation of moviegoers. Hopper and Lucas, in contrast, had tapped into the anxieties of the postwar baby-boom generation in its search for self-realization, its longing for an innocent past, and its efforts to cope with the turbulence of the 1960s.

Their success opened the door for a new wave of directors who were trained in California or New York film schools in the 1960s, such as Francis Ford Coppola (*The Godfather,* 1972), Steven Spielberg (*Jaws,* 1975), and Martin Scorsese (*Taxi Driver,* 1976). These films signaled the start of a period that saw directors elevated to a higher and higher status. Since the 1970s, a handful of successful directors have gained the kind of economic clout and celebrity standing that had previously belonged almost exclusively to top movie stars.

Diversifying Hollywood's Directors

Although the status of directors grew in the 1970s, opportunities for directing feature films in Hollywood were mostly reserved for white men. Over time, a few women (mostly white) got opportunities to direct because of their prominent standing as popular actors, such as Barbra Streisand, Jodie Foster, and Penny Marshall. More recently, women have come to direct films via their scriptwriting achievements. For example, Jennifer Lee, who wrote *Wreck-It Ralph* (2012), followed up by writing and codirecting *Frozen* (2013) and *Frozen II* (2019). Other women directors—like Kimberly Peirce (*Carrie,* 2013), Greta Gerwig (*Little Women,* 2019), and Patty Jenkins (*Wonder Woman 1984,* 2020)—have moved past debut films and proven themselves as experienced studio auteurs.

In 2016, women were hired to direct only 3.4 percent of major Hollywood releases. Amid mounting criticism, that number rose to 10.6 percent in 2019.[10] While the trend is improving, a clear gender gap in Hollywood remains. Critics argue that providing more opportunities for women to direct (as well as produce) movies has bigger implications than equal access to jobs or the ability to shape how a story is told through film. A competitive and unstable employment environment like the film industry, in which actors, writers, and other talent depend so heavily on the whims of studio executives, producers, and directors for their next gig, can enable an array of discriminatory and abusive behaviors. Fueled by women's historically low status in Hollywood and the dozens of charges of sexual harassment, assault, and rape against producer Harvey Weinstein, a number of powerful women in the entertainment industry formed Time's Up in 2018. Founded by Hollywood celebrities including

Atsushi Nishijima/©Paramount Pictures/Everett Collection, Inc

Oprah Winfrey, Reese Witherspoon, and Ava DuVernay (and inspired by the wider #MeToo movement), the advocacy group established a legal defense fund for those who have experienced sexual harassment or retaliation in the workplace and drew attention to the need for women, including those working in Hollywood, to rise to positions of power.[11]

Both men and women of color have also struggled for opportunities and recognition in Hollywood. Still, some have succeeded as directors, gaining notoriety through independent filmmaking or crossing over from careers as actors. Among the most successful contemporary Black directors are Spike Lee (*Da 5 Bloods*, 2020), Ryan Coogler (*Black Panther*, 2018), Ava DuVernay (*13th*, 2016; *When They See Us* 2019), Jordan Peele (*Get Out*, 2017; *Us*, 2019), Barry Jenkins (*Moonlight*, 2017), and Lee Daniels (*The Butler*, 2013). In addition, besides directing highly successful films like *A Madea Family Funeral* (2018), Tyler Perry established his own studio in Atlanta and has emerged as a powerful figure in both film and TV production. (See "Examining Ethics: Breaking through Hollywood's Race Barrier.")

Some Asian and Asian American directors—such as M. Night Shyamalan (*Glass*, 2019) and Ang Lee (*Life of Pi*, 2012)—have helmed Hollywood movies. Others, like Lulu Wang (*The Farewell*, 2019), have focused on independent productions. At the 2021 Academy Awards, Chloé Zhao (*Nomadland*, 2020; *Eternals*, 2021), who got her start in independent movies, became the first woman of color to be nominated for best director and the first woman of color to win the award.

Chris Eyre (*Hide Away*, 2011) remains the most noted Native American director, working mainly as an independent filmmaker. However, the Sundance Institute Indigenous Program works to support Native American artists, and the 2021 Sundance Film Festival featured four films made by Indigenous filmmakers, with five additional films featured in 2020 and eight in 2019.[12]

Movies Outside Hollywood

The major Hollywood studios have typically focused on feature-length movies that can earn the most money. Hollywood, however, has never been the only source of movies. In the following sections, we look at three alternatives to Hollywood studio output: international films, documentaries, and independent films. The changing dynamics of the film industry have made these types of movies more accessible to current audiences than they were during the studio system era. Today, in fact, the clear distinction that once existed between Hollywood and these alternative traditions has blurred as Hollywood increasingly draws inspiration and talent from the outside.

Breaking through Hollywood's Race Barrier

The problem with the term *Black cinema* is that such a term needs to exist. (Do we, for example, talk about a *white cinema* in the United States?) But there is a long history of Black Amercians' exclusion from the industry as writers, directors, and actors—and even as audience members at theaters—so when a film like *Dope* (2015) by director Rick Famuyiwa gets praised as "revolutionary" and "subversive," it's because this coming-of-age story dares to feature a cast that, for the most part, isn't white.

Despite Black Americans' long support of the film industry, their moviegoing experience has not been the same as that of whites. Until Civil Rights legislation was enacted in the mid-1960s, many theater owners discriminated against Black patrons. In large cities, Black filmgoers often had to attend separate theaters, where new movies might not appear until a year or more after white theaters had shown them. In smaller towns, Black patrons were often required to visit theaters only after midnight or sit in the back.[1]

Confronted with Hollywood films like *The Birth of a Nation*, which either ignored Black lives or reduced them to racist stereotypes, Black entrepreneurs established their own production companies in the 1920s and 1930s. Directors like Oscar Micheaux made what were called "race films"—films made by Black filmmakers for Black audiences with all-Black casts. Produced outside Hollywood, race films offered more complex representations of Black people. Without the resources of Hollywood, however, Black filmmakers continually struggled financially, especially after the conversion to talkies.

Changes began taking place after World War II. In response to the "white flight" from central cities to the suburbs that began in the 1950s, many downtown and neighborhood theaters began catering to Black customers in order to keep from going out of business. By the late 1960s, Hollywood studios began to make films with more complex treatments of race relations, such as *In the Heat of the Night* (1967), *The Learning Tree* (1969), and *Sounder* (1972). Meanwhile, Melvin Van Peebles's controversial *Sweet Sweetback's Baadasssss Song* (1971) helped spark a string of other blacksploitation films like *Shaft* (1971) and *Foxy Brown* (1974), which featured Black lead characters. These movies were popular, but they also drew criticism for depicting Black characters through themes of violence, sex, drugs, and criminality. During the 1980s and 1990s, the growth of independent filmmaking saw the emergence of a new generation of Black filmmakers, including Spike Lee (*Do the Right Thing*, 1989), Julie Dash (*Daughters of the Dust*, 1991), and Marlon Riggs (*Tongues Untied*, 1989).

In recent years, Black filmmakers have achieved greater mainstream success and recognition than ever before. Lee Daniels received only the second Academy Award nomination for a Black director for 2009's *Precious* (the first was John Singleton for *Boyz N the Hood* in 1991). *12 Years a Slave* (2013), by Black British director Steve McQueen, won three Academy Awards, including best picture. And *Moonlight* (2016), a coming-of-age drama by Black director Barry Jenkins, also won best picture. Even so, the lack of regular recognition of actors, writers, and directors of color by the academy led to the trending of the the #OscarsSoWhite hashtag in 2016; in response, the academy has tried to diversify its membership. In 2020, it established a new policy: To be eligible for best picture, a film must meet standards aimed at making films more inclusive, both behind and in front of the camera.

But in Hollywood, nothing speaks more loudly than a blockbuster. Thus, *Black Panther* could have a big impact on the future of Black artists in American cinema. In less than a month after its February 2018 release, *Black Panther* generated more than $1.1 billion in global box-office revenue.[2] It soon became the third highest-grossing movie in the United States. The film is particularly notable for its director, cast, and story. As Jamil Smith of *Time* magazine noted, the film "may be the first mega-budget movie . . . to have an African-American director and a predominantly black cast. Hollywood has never produced a blockbuster this splendidly black." Smith related the movie's cultural significance to his own experience: "Relating to characters onscreen is necessary not merely for [people of color] to feel seen and understood, but also for others who need to see and understand us."[3] As Vox's Sean Rameswaram observes, "Representation matters, and it's good for business."[4] ●

Global Cinema

For generations, Hollywood has dominated the global movie scene. In many countries, American films capture up to 90 percent of the market. In striking contrast, foreign films constitute only a tiny fraction—less than 2 percent—of motion pictures seen in the United States today. Despite Hollywood's domination of global film distribution, other countries have a rich history of producing both successful and provocative short-subject and feature films. For example, cinematic movements of the twentieth century—such as German expressionism (capturing psychological moods), Soviet social realism (presenting a positive view of Soviet life), Italian neorealism (focusing on the everyday lives of Italians), and European new-wave cinema (experimenting with the language of film)—and post–World War II Japanese, Hong Kong, Korean, Australian, Indian, Canadian, and British cinema have all been extremely influential, demonstrating alternatives to classical Hollywood cinema.

Early on, Americans showed interest in British and French short films and in experimental films, such as Germany's *The Cabinet of Dr. Caligari* (1919). Foreign-language movies did reasonably well during the silent era, especially in ethnic neighborhood theaters in large American cities. By the 1930s, however, the daughters and sons of turn-of-the-century immigrants—many of whom were trying to assimilate into mainstream American culture—preferred their Hollywood movies in English.[13]

Postwar prosperity, rising globalism, and the gradual decline of the studios' hold over theater exhibition in the 1950s and 1960s stimulated the rise of art-house theaters, and these decades saw a rebirth of interest in foreign-language films by prominent directors like Sweden's Ingmar Bergman (*The Seventh Seal*, 1957), Italy's Federico Fellini (*La Dolce Vita*, 1960), France's François Truffaut (*Jules and Jim*, 1961), Japan's Akira Kurosawa (*Seven Samurai*, 1954), and India's Satyajit Ray (*Apu Trilogy*, 1955–1959). Catering to academic audiences, art houses made a statement against Hollywood commercialism as they sought to show alternative movies.

The role of foreign films in U.S. film culture has waxed and waned since this period. By the 1970s, new multiplex theater owners rejected the smaller profit margins of most foreign titles, which lacked the promotional hype of U.S. films. As a result, between 1966 and 1990 the number of foreign films released annually in the United States dropped by two-thirds, from nearly three hundred to about one hundred titles per year. However, the growth of home video in the 1980s, followed by the emergence of streaming services like Netflix in the 2000s, has provided new distribution channels for foreign films. As a result, viewers have gained access to a large selection of foreign-language titles. The success of *Amélie* (France, 2001), *Roma* (Mexico, 2018), and *Parasite* (South Korea, 2019) illustrate that a sizable U.S. audience is willing to watch subtitled films with non-Hollywood perspectives.

Today, the largest film industry is in India. Bollywood (a play on words combining cities Bombay—now Mumbai—and Hollywood) produces nearly two thousand films a year—mostly romance or adventure musicals in a distinct style.[14] In comparison, U.S. moviemakers release fewer than eight hundred films a year, and the growing film industry in China produces just under seven hundred movies annually. (The United States and China are first and second in terms of film revenue.) Japan, France, and Nigeria are also major film-producing nations. (For a broader perspective, see "Global Village: Beyond Hollywood: Asian Cinema".)

THE SEVENTH SEAL
Swedish director Ingmar Bergman's *The Seventh Seal* explores existential themes of faith and mortality against a backdrop of a plague-ravaged medieval community. This scene, in which a knight encounters Death and challenges him to a game of chess, has become an icon of European art-house cinema and has been referenced in movies like *Bill and Ted's Bogus Journey*.

Beyond Hollywood: Asian Cinema

Asian nations easily outstrip Hollywood in the quantity of films produced. India alone produces almost two thousand movies a year. But from India to South Korea, Asian films are increasingly challenging Hollywood in terms of quality, and they have become more influential as Asian directors, actors, and film styles are exported to Hollywood and the world.

India

Part musical, part action, part romance, and part suspense, the epic films of Bollywood typically have fantastic sets, hordes of extras, plenty of wet saris, and symbolic fountain bursts (as a substitute for kissing and sex, which are prohibited from being shown). Indian movie fans pay from $.75 to $5 to see these films, and they feel shortchanged if the movies are shorter than three hours. With many films made in less than a week, however, most of the Bollywood fare is cheaply produced and badly acted. Yet these production aesthetics are changing, as bigger-budget releases target the middle and upper classes in India, the twenty-five million Indians living abroad, and Western audiences. *Baahubali 2: The Conclusion* (2017), an action-fantasy film set in medieval India, stars Prabhas—one of India's most famous leading men. Appearing on seven hundred standard screens and another fifty-eight IMAX screens across the United States and Canada, and playing in three different languages (Telugu, Tamil, and Hindi), the film had the most successful U.S. box-office opening of any Bollywood film, grossing $155 million in just ten days.[1]

China

Since the late 1980s, Chinese cinema has developed an international reputation. Leading this generation of directors are Yimou Zhang (*House of Flying Daggers,* 2004; *Coming Home,* 2014) and Kaige Chen (*Farewell My Concubine,* 1993;

Monk Comes Down the Mountain, 2015), whose work has spanned such genres as historical epics, love stories, contemporary tales of city life, and action-fantasy. These directors have also helped make international stars out of Li Gong (*Memoirs of a Geisha*, 2005; *Coming Home*, 2014) and Ziyi Zhang (*Memoirs of a Geisha*, 2005; *Dangerous Liaisons*, 2012). In addition, other films have been very successful at the box office. *The Eight Hundred* (2020), pictured here—a historical war drama set in the content of the Sino-Japanese War—took in over $400 million worldwide in its first month in theaters. An "epic-effects-enhanced patriotic action movie" comparable to U.S. and British films like *Pearl Harbor* and *Dunkirk*, the $80 million film reflects the Chinese film industry's ability to produce its own blockbusters.[2]

THE EIGHT HUNDRED (2020), a war movie that celebrates the heroic sacrifice of a small band of Chinese soldiers who fought for four days against an invading Japanese army, was a huge hit with Chinese moviegoers. The film's release had been delayed for a year by government censors.

Hong Kong

Hong Kong films were the most talked about—and the most influential—film genre in cinema during the late 1980s and 1990s. In those years, Hong Kong studios exported more films than did any other country except the United States. The style of highly choreographed action with often breathtaking, ballet-like violence became hugely popular around the world, reaching American audiences and in some cases even outselling Hollywood blockbusters. Hong Kong directors like John Woo, Ringo Lam, and Jackie Chan (who also acts in his movies) have also directed Hollywood action films, and stars like Jet Li (*Lethal Weapon 4*, 1998; *The Expendables 3*, 2014), Chow Yun-Fat (*Pirates of the Caribbean: At World's End*, 2007; *The Monkey King: The Legend Begins*, 2022), and Malaysia's Michelle Yeoh (*Crazy Rich Asians*, 2018) have landed leading roles in American movies. Since the late 1990s, however, Hong Kong's film industry has been hit hard by a series of challenges, including changing media consumption patterns across Asia; rising Hong Kong rents, which have inflated production costs; aging superstars; increased competition from mainland China and South Korean film industries; and a drain of its talent pool to Hollywood. Some observers believe its day as a global capital for film production may be over.[3]

Japan

Americans may be most familiar with low-budget monster movies like *Godzilla*, but the widely heralded films of the late director Akira Kurosawa have had an even greater impact: His *Seven Samurai* (1954) was remade by Hollywood as the film masterpiece *The Magnificent Seven* (1960), and *The Hidden Fortress* (1958) was George Lucas's inspiration for *Star Wars*. Hayao Miyazaki (*Ponyo*, 2009; *The Wind Rises*, 2013) was the country's top director of anime movies, and with his retirement, a new anime successor has emerged: Makoto Shinkai, with his cultural sensation *Your Name* (2016). Japanese thrillers like *Ringu* (1998), *Ringu 2* (1999), and *Ju-on: The Grudge* (2003) were remade into successful American horror films. Hirokazu Kore-eda's drama *Like Father, Like Son* (2013) won the Jury Prize at the Cannes Film Festival and caught the attention of Steven Spielberg, who acquired the remake rights for his company DreamWorks. *Shoplifters* (2018), Kore-eda's much-praised family drama about a family of rogues living precariously on the edge of poverty, won the Palme d'Or at Cannes.

South Korea

The end of military regimes in the late 1980s and corporate investment in the film business in the 1990s created a new era in Korean moviemaking. Leading directors include Kim Jee-woon (*Illang: The Wolf Brigade*, 2018); Lee Chang-dong (nominated for the Palme d'Or at Cannes for *Poetry*, 2010); and Park Chan-wook, whose Vengeance Trilogy films (*Sympathy for Mr. Vengeance*, 2002; *Oldboy*, 2003; and *Lady Vengeance*, 2005) have won international acclaim, including the Grand Prix at Cannes for *Oldboy*, which was remade in the United States in 2013 by director Spike Lee. Bong Joon-ho's science fiction film *Snowpiercer* (2013)—based on a French graphic novel, filmed in the Czech Republic, and starring a mostly English-speaking cast—epitomizes the international outlook of Korean cinema. South Korean film's profile in the United States reached a new high when Bong Joon-ho's *Parasite* (2019)—a black comedy thriller about class conflict—won best picture at the 2020 Academy Awards. ●

DOCUMENTARY FILMS
American Factory (2019) is a documentary that explores the complex cultural issues that emerge when a Chinese company moves into a former General Motors plant outside Dayton, Ohio. It was produced by Barack and Michelle Obama's Higher Ground Productions and distributed by Netflix—a fact that signals the growing role TV streaming services play in documentary films. The film won the 2020 Academy Award for best documentary feature.

The Documentary Tradition

Both TV news and nonfiction films trace their roots to the *newsreels* of the late 1890s. Pioneered in France and England, newsreels consisted of weekly ten-minute magazine-style compilations of filmed news events from around the world. International news services began supplying theaters and movie studios with newsreels, and by 1911, they had become a regular part of the moviegoing menu.

Over time, the *documentary* film genre developed as a style that interprets reality by recording actual people and settings. As an educational, noncommercial form, the documentary often required the backing of industry, government, or philanthropy to cover costs. In support of a clear alternative to Hollywood cinema, some nations began creating special units, such as Canada's National Film Board, to sponsor documentaries. In the United States, art and film received considerable support from the Roosevelt administration during the Depression.

By the late 1950s and early 1960s, the development of portable cameras led to **cinema verité** (a French term for "truth film"). This documentary style allowed filmmakers to go where cameras could not go before and record fragments of everyday life more unobtrusively. Directly opposed to packaged, high-gloss Hollywood features, cinema verité aimed to track reality, employing a rough, grainy look and shaky handheld-camera work. Among the key innovators in cinema verité was Drew Associates, led by Robert Drew: a former *Life* magazine photographer who shot the groundbreaking documentary *Primary*, which followed the 1960 Democratic presidential primary race between Hubert Humphrey and John F. Kennedy.

Perhaps the major contribution of documentaries has been their willingness to tackle controversial or unpopular subject matter. For example, American documentary filmmakers Kirby Dick and Amy Ziering address complex topics about sex and power. Their films include *The Invisible War* (about sexual assault in the military) and *The Hunting Ground* (about campus assault at Harvard University). Dick and Ziering's films are part of a resurgence in high-profile documentary filmmaking in the United States, which includes *The Cove* (2009), *Hell and Back Again* (2011), *The Queen of Versailles* (2012), *Icarus* (2017), *Step* (2017), and *Won't You Be My Neighbor?* (2018).

Intense competition for programming among content delivery services like Netflix, Amazon, and HBO has opened up new distribution and revenue opportunities for documentary filmmakers. As a result, nonfiction films are finding much wider audiences. Such developments are also blurring the line between film and television production. Netflix series like *Making a Murderer* (2015) and *Tiger King* (2020), for example, combine elements from documentary film, broadcast journalism, and episodic TV series production.

The Rise of Independent Films

The success of documentary films dovetails with the rise of **indies**, or films produced outside the traditional Hollywood system. Filmmakers have produced films without the financial backing and distribution networks of major Hollywood studios since the 1920s, operating on shoestring budgets and struggling to find places to show their movies. The decline of the studio system and the rise of art-house theaters, college campus film screenings, and small film festivals in the 1960s opened up new venues, helping cult filmmakers like John Waters (*Pink Flamingos*, 1972; *Polyester*, 1981) reach audiences. Free from the corporate pressures of Hollywood studios, these cult films usually tackled topics considered too controversial or marginal by Hollywood and experimented with narrative and style, leading to films that looked and felt different.

Ultimately, three interconnected developments helped transform this underground world of cult filmmaking into a wider-reaching indie film movement during the 1980s and 1990s:

- The advent of home video rentals gave independent films that couldn't get into mainstream theaters a way to reach bigger audiences.

- New studios—most notably Miramax—were founded in order to distribute movies made outside the Hollywood establishment. These studios marketed indie films, pushed for access to multiplex theaters, and campaigned for recognition at industry award shows.
- The rise of major film festivals like the Sundance Film Festival (held every January in Park City, Utah); New York's Tribeca Film Festival; and festivals in Toronto, Austin, and Cannes gave independent movies a platform from which to gain publicity—and possibly even a distribution deal with a studio like Miramax.

PINK FLAMINGOS was produced, written, directed, and edited by John Waters and starred Divine, a drag performer and character actor who became a counterculture icon through his partnership with Waters. In the film, Divine's character defends her title as "the filthiest person alive" in a series of increasingly disgusting scenes meant to shock, including this one that echoes a famous shot form Porter's *The Great Train Robbery*. Cult filmmakers like Waters paved the way for the next generation of filmmakers who worked outside Hollywood.

Over time, developments like these helped fuel the mainstream success of indie films like *Sex, Lies and Videotape* (1989), *Pulp Fiction* (1994), and *Shakespeare in Love* (1995). These successes drew the attention of major Hollywood studios looking for films with lower production costs to counterbalance their usual blockbuster offerings, which often had *Titanic*-size budgets. Beginning in 1993, Hollywood and the indie world became increasingly interconnected; Disney bought Miramax, and other major studios created divisions devoted to indie films (Fox Searchlight Pictures, Sony Pictures Classics, Paramount Vantage).

In addition to developing and distributing indie movies, major studios began tapping indie directors like Steven Soderbergh (*Ocean's Eleven*, 2001), Christopher Nolan (*The Dark Knight*, 2008), Alfonso Cuarón (*Harry Potter and the Prisoner of Azkaban*, 2004), and Catherine Hardwicke (*Twilight*, 2008) to direct big-budget projects. In each case, these directors brought some of their quirky indie style and sensibility with them, helping to transform the look and feel of mainstream Hollywood blockbusters. Meanwhile, independent-minded filmmakers like Darren Aronofsky (*Noah*, 2014; *Mother!*, 2017) and David O. Russell (*American Hustle*, 2013; *Joy*, 2015) have established careers somewhere between fully independent and studio backed, often with smaller production companies financing their films before they're picked up for distribution by a major studio or one of the new generation of independent film distributors, like A24.

While Hollywood studios work with some independent production companies, many independent filmmakers remain entirely outside Hollywood's orbit. With digital video camera equipment and computer-based desktop editors, movies can be made for a fraction of what the cost had been on celluloid film. In 2003, Jonathan Couette used free iMovie software on his Mac to edit old photos, VHS tape, and 8 mm footage, along with recorded phone messages, to make a documentary about growing up with his mother, who had a mental illness, on an initial budget of $218 (*Tarnation*).[15] More recently, feature films like *Tangerine* (2015) have been shot entirely on iPhones. Independent filmmakers also have new distribution venues beyond film festivals or studios. For example, Vimeo, YouTube, Netflix, and Amazon have grown into leading Internet sites for the screening of short films, and social media platforms like Instagram and TikTok provide content producers (perhaps the ultimate indie filmmakers) with their most valuable asset—an audience. While the shadow of Hollywood's business model has long defined movies in terms of profitable feature films, the digital era pushes us to reconsider what counts as a movie.

The Economics of the Movie Business

Despite an increasingly competitive media environment, in which network and cable television, social media, streaming services, and video games vie for people's attention, the movie business has largely remained strong. Since 1963, Americans have purchased roughly 1 billion movie tickets each year; in 2019, 1.24 billion tickets were sold in the United States and Canada.[16] With movie tickets in some areas

FIGURE 7.1

NORTH AMERICAN AND GLOBAL BOX-OFFICE REVENUE, 2019 (IN $ BILLIONS)

Data from: Motion Picture Association of America, *Theme Report 2019*, March 11, 2020, 11, www.motionpictures .org/wp-content/uploads/2020/03 /MPA-THEME-2019.pdf.

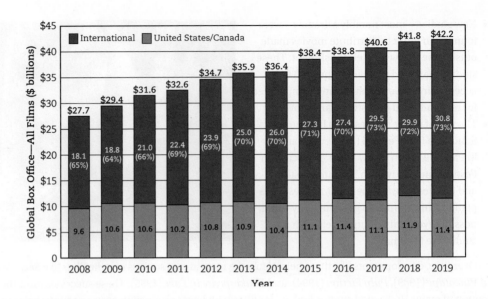

rising to $15, gross revenues from North American box-office sales have climbed above the $11 billion mark, up from $9.6 billion annually in 2008 (see Figure 7.1). The bigger news for Hollywood studios is that global box-office revenues have grown at an even faster rate: In 2019, the global box office reached an all-time high of $42.2 billion.[17]

These positive results wavered in 2020, as the coronavirus pandemic profoundly affected the film industry by closing theaters, halting production, and delaying film releases. While the pandemic's long-term impact on the movie business will take time to play out, for the moment, a small number of powerful Hollywood studios still dominate the global industry, taking in money through a variety of revenue streams.

The Major Studios

The current Hollywood commercial film business is ruled primarily by five studios: Disney (which now owns the former 21st Century Fox), Warner Brothers, Universal, Sony Pictures, and Paramount—the **new Big Five** (see Figure 7.2). The five major studios account for about 83 percent of the revenue

FIGURE 7.2

MARKET SHARE OF U.S. FILM STUDIOS AND DISTRIBUTORS, 2019 (IN $ MILLIONS)

Notes: Based on gross box-office revenue from the domestic (North American) market. Disney acquired 21st Century Fox, including the studios Twentieth Century Fox and Fox Searchlight, in March 2019. Viacom became ViacomCBS in December 2019.

Data from: "Market Share for Each Distributor in 2019," The Numbers, accessed November 30, 2020, www .the-numbers.com/market/2019 /distributors.

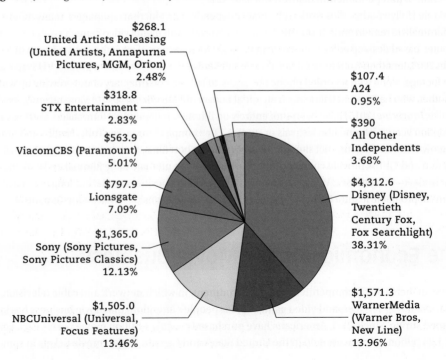

SYNERGIES in feature films can be easy for Disney, a multinational corporation with a vast array of entertainment-related subsidiaries. *Frozen* (2013) and its sequel, *Frozen 2* (2019), are among Disney's biggest animated hits, and movie merchandise was in short supply in North America for fans wanting to celebrate the story of Anna and Elsa, two princess sisters who also became attractions at Disney resort parks. *Frozen's* soundtrack hit No. 1 in sales, and Disney Cruise Line and the Adventures by Disney tour company experienced a huge increase in holiday business to Geirangerfjord, Norway, the fjord that inspired the film's fantasy kingdom of Arendelle.

generated by commercial films domestically. They also control more than half the movie market in Europe and Asia. In addition, independent studios that have modest market share, like Lionsgate (*The Hunger Games*, the *Twilight* series) and STX Entertainment (*Bad Moms*, *Hustlers*), do still exist, and are sometimes called **mini-majors**.

All the major Hollywood studios have recently been part of vast and highly complex conglomerates. For example, Universal is part of NBCUniversal (owned by Comcast), Warner Brothers is part of WarnerMedia (owned by AT&T), and Paramount is part of ViacomCBS (owned by National Amusements). As a result, the business of making movies is fully integrated with other entertainment industries (broadcast television, cable, print, sound recordings, video games, and theme parks) as well as retail stores and telecommunications giants.

Being embedded in large conglomerates allows movie studios to take advantage of **synergy**—opportunities to generate profits that come from interaction and cooperation among a conglomerate's different subsidiaries. In today's movie industry, where success often depends on building a film around a proven commodity and intense advanced promotion, synergy helps on both fronts. For example, Disney's purchase of Pixar in 2006, Marvel in 2009, and Lucasfilm in 2012 has given the studio control of a vast library of popular characters to use in its films, from Buzz Lightyear to Captain America to Luke Skywalker. As the studio makes films with these characters, it can rely on its corporate "partners" for promotional help—such as getting a magazine cover story, booking actors on a talk show, running a "making-of" story on a streaming platform, or creating a theme-park ride based on the film. In return, a studio's movies can generate revenue opportunities for its partners in other divisions, as when Disney's theme-park division created a *Star Wars* land called Galaxy's Edge to draw in visitors who are fans of the films. If managed well, assets like the *Star Wars* franchise and the Marvel universe can lift all boats in the Walt Disney Company.

Making Money on Movies Today

The 1970s marked the rise of **blockbuster** movies, which revealed a new winning formula for Hollywood: Find a proven commodity to base your film on, include special effects, target a youth audience, create anticipation through a massive nationwide ad campaign via press junkets and TV ads (which have been joined by social media blasts today), and win the box-office race on your opening weekend. This formula helped films like *Jaws* (1975) and *Star Wars* (1977) become the first movies to gross more than $100 million in a single year. Since then, the major Hollywood studios have worked to re-create that success year after year (see Table 7.1 for a list of the highest-grossing films of all time and "Media Literacy and the Critical Process: The Blockbuster Mentality").

In the blockbuster playbook, it takes money to make money. Today a major studio film costs, on average, about $66 million to produce and about $37 million for domestic marketing, advertising, and print costs.[18] With such steep budgets, 80 to 90 percent of newly released movies fail to turn a profit at the domestic box office. To remain profitable, studios rely on **tentpoles**—films that studios bank on being hits in order to offset losses on other riskier films. The importance of tentpole films to studios' economic stability has led to Hollywood's reliance on sequels and franchises like *The Avengers*, *Star Wars*, and *Star Trek*. The studios earn their revenue on these movies from a number of different sources, which we will highlight in the next sections.

Domestic Theatrical Exhibition

Theaters are iconic elements of the movie industry and were the primary source of revenue for Hollywood studios for decades. In recent years, studios have been earning less than half of their total revenue from theatrical ticket sales—a downward trend that some predict could accelerate in a post-pandemic era. Nevertheless, a theatrical run has long been considered a vital part of building awareness for a movie, and Hollywood studios and theater owners have traditionally enjoyed a mutually dependent alliance. At the core of that alliance is **theatrical exclusivity**—the roughly three-month window when studios agree to make their films available only in theaters in order to drive ticket sales. For each ticket sold, roughly 55 percent of the purchase price goes back to the studio; the rest is kept by the theater. When Hollywood rebounded in the blockbuster era of the 1980s, theaters did as well. By 2020, there were 40,449 indoor screens in the United States, most of them in **megaplexes** (facilities with fourteen or more screens).[19]

Film exhibition is controlled by a handful of theater chains; the leading five companies operate more than 50 percent of U.S. screens. The major chains—AMC Entertainment, Regal Cinemas, Cinemark USA, Cineplex Entertainment, and Marcus Theatres—own thousands of screens in suburban malls and at highway crossroads, and most have expanded into international markets.[20] Megaplexes offer exhibitors some flexibility when dealing with market unpredictability. For example, they can project a hit on two or three screens at the same time; films that do not debut well are relegated to the smallest theaters or bumped quickly for a new release.

TABLE 7.1　THE TOP 10 ALL-TIME WORLDWIDE BOX-OFFICE CHAMPIONS*

Rank	Title/Date	Worldwide Gross ($ millions)	Domestic Gross ($ millions)
1	*Avengers: Endgame* (2019)	$2,797.8	$858.4
2	*Avatar* (2009)	2,788.7	760.5
3	*Titanic* (1997)	2,208.0	659.4
4	*Star Wars: Episode VII: The Force Awakens* (2015)	2,065.5	936.7**
5	*Avengers: Infinity War* (2018)	2,044.5	678.8
6	*Jurassic World* (2015)	1,669.9	652.3
7	*The Lion King* (2019)	1,654.4	543.6
8	*Furious 7* (2015)	1,517.2	353.0
9	*The Avengers* (2012)	1,515.1	623.4
10	*Frozen II* (2019)	1,447.3	477.4

Data from: "All Time Worldwide Box Office," The Numbers, accessed November 30, 2020, www.the-numbers.com/box-office-records/worldwide/all-movies/cumulative/all-time.

* Worldwide sales include domestic and international box-office numbers. Numbers are rounded. Domestic (North American) gross is provided for comparison.

** *Star Wars: Episode VII: The Force Awakens* is the box-office champion for domestic (North American) earnings.

MEDIA LITERACY & THE CRITICAL PROCESS

The Blockbuster Mentality

After the success of films like *Star Wars* in the 1970s, Hollywood shifted toward a blockbuster mentality. How pervasive is this blockbuster mentality, which targets an audience of young adults, releases action-packed big-budget films featuring heavy merchandising tie-ins, and produces sequels?

❶ DESCRIPTION

Conduct an Internet search to find a list of the highest-grossing movies in the United States.

❷ ANALYSIS

Note patterns in the list. For example, what percentage of the films target young audiences? What percentage feature animated or digitally composited characters (e.g., *Frozen 2*, *Shrek 2*, *Jurassic Park*) or extensive special effects (e.g., *Transformers: Revenge of the Fallen*, *Avengers: Endgame*)? How many either spawned or are part of a series, like *Transformers*, *The Dark Knight*, and *Harry Potter*? How many fit into the action-movie genre? How many had intense merchandising campaigns that featured action figures, fast-food tie-ins, and an incredible variety of products for sale? How many were "surprise" hits?

❸ INTERPRETATION

What do the patterns mean? It's clear, economically, why Hollywood likes to have successful blockbuster movie franchises. But what kinds of films get left out of the mix? Hits like *Forrest Gump* (now bumped way out of the Top 30), which may have had big-budget releases but lack some of the other attributes of blockbusters, are clearly anomalies of the blockbuster mentality, although they illustrate that strong characters and compelling stories can carry a film to great commercial success.

❹ EVALUATION

It is likely that we will continue to see an increase in youth-oriented animated/action-movie franchises that are heavily merchandised and intended for wide international distribution. Indeed, Hollywood does not have a lot of motivation to put out the kinds of movies that don't fit these categories. Is this a good thing? Can you think of a film that you thought was excellent and that would have probably been a bigger hit with better promotion and wider distribution?

❺ ENGAGEMENT

Watch independent and foreign films to see what you're missing. Visit the Sundance Film Festival site and browse through the many films listed. Find these films on Netflix, Amazon, Hulu, or Apple TV (and if the films are unavailable, let these services know), or write your cable company and request to have the Sundance Channel on your cable lineup. Organize an independent film night on your college campus so you can bring these films to a crowd.

Since the advent of TV in the 1950s, movie producers and theater owners have been forced to innovate in order to lure in audiences. In the 1990s, leading theater chains expanded upscale concession services and luxurious screening rooms equipped with stadium-style seating and digital sound to make moviegoing a special event. In the 2010s, studios revived 3-D films, which they hoped would increase ticket sales (by luring audiences away from home theater systems) and ticket revenue (by charging higher ticket prices for 3-D screenings). By 2018, however, audiences' willingness to pay a premium for 3-D had waned, leaving Hollywood to look for new strategies—including bigger screens (like the ultrawide 270-degree views of ScreenX), responsive 4-D seating (with "sway" and "twist" motions to further immerse viewers of action movies), and virtual reality experiences. In 2020, theater owners faced a new question: how to adapt to a post-pandemic world.

Global Theatrical Exhibition

In addition to domestic ticket sales, studios earn revenue from distributing films in foreign markets. In fact, international box-office gross revenues, which can be nearly triple the domestic box-office receipts, have been climbing in recent years. As a result, the international box office (which includes everywhere but the United States and Canada) is becoming more important for Hollywood's bottom line. In 2019, global markets generated 73 percent of the movie industry's revenue—an all-time high.[21] As Hollywood studios seek to remain dominant in these markets, however, they are facing tougher competition from local TV and film producers, especially in China. As international distributers know well, producing and marketing U.S. movies in other countries can be challenging, and always requires an awareness of cultural and political differences.

Home/Mobile Market

Once considered secondary to theatrical exhibition, the various revenue streams linked to home and mobile viewing are now the biggest source of income for most studios. In 2019, the global home/mobile market, which includes digital and physical copies of movies, generated $58.8 billion—a 14 percent increase over 2018. Revenue from digital alone increased 18 percent in the United States and 29 percent internationally.[22] As these numbers indicate, the home/mobile market is being transformed by broadband services like Netflix, which allow viewers to download and stream movies via the Internet.

Internet distribution began in 2006, when Apple's iTunes store began selling digital downloads, and it took another leap forward in 2008, when online DVD rental service Netflix began streaming some movies and television shows. By 2012, movie fans were accessing more movies through digital online media than through DVDs for the first time.[23] The popularity of Netflix's streaming service has opened the door to similar services, like Hulu and Disney+. Today, consumers can purchase and rent movies via numerous other digital stores and streaming platforms, as well as through the video-on-demand services of cable companies and satellite providers.

Typically, studios have followed a relatively standard approach to rolling out a film to the home/mobile market. First, roughly three months after its theatrical release, they make a movie available for purchase via digital download or DVD. Roughly a month later—after a delay meant to encourage more people to purchase the movie—the studio makes it available for rent or for streaming on subscription platforms like Netflix. Eventually a film is licensed for premium cable (such as HBO) and then network and basic cable. Terms and prices for each stage are determined through complex negotiation and can vary depending on the power and priorities of the studio and the outlet involved.

The growing popularity of streaming services, however, has created opportunities for studios to experiment with new models for home and mobile distribution. By launching its own streaming service (Disney+) in 2019, for example, Disney established a digital-era version of vertical integration, giving it control of the production, distribution, and exhibition of its movies. As more studios follow this model, a shake-up in Hollywood's approach to delivering its products will likely result.

The possible consequences of these types of developments have raised concerns for theatrical exhibitors who fear for the future of theatrical exclusivity—and with it their business model. Independent films and documentaries already frequently bypass the theatrical box-office release window entirely, opting to go straight to home-streaming services to avoid steep marketing expenses. Meanwhile, as Netflix and other services become more influential, major Hollywood studios face more pressure to release films to the home/mobile market early and even end theatrical exclusivity altogether. Part of that pressure comes from the fact that Netflix and Amazon have begun to produce their own films. Netflix, for example, released *Bird Box* (2018) and *The Irishman* (2019) in theaters (in part for award consideration) and then streamed them on its platform simultaneously within a few weeks.

As audiences get used to such experiments, major Hollywood studios may consider following suit, as some did during the 2020 global pandemic. When theaters across the world shut down, Disney made the much-anticipated live-action version of *Mulan* available to subscribers of Disney+ who were willing to pay an

THEATRICAL EXCLUSIVITY
The 2020 global pandemic had an enormous impact on the film industry, especially theater owners. When public health orders closed down theaters around the world, Hollywood studios responded by releasing films directly to streaming services like Disney+ and HBO Max. These moves drew into question the future of theatrical exhibition.

Noam Galai/Getty Images

additional $29.99. Similarly, Universal released *Trolls World Tour* via premium video-on-demand. When WarnerMedia announced that *Wonder Woman 1984* (2020) and its entire slate of 2021 releases would stream on HBO Max the same day they were released in theaters, it became especially clear that digital home distribution is shifting the balance of power in the alliance between Hollywood and theater owners.

Digital home distribution is also shifting Hollywood studios' production and distribution strategies. As studios and theaters lean heavily on big-budget blockbuster film franchises with the kind of special effects that play best on big screens, low- and mid-budget films, including the kind that get Oscar nominations, probably won't get wide theatrical distribution like they used to. Even as they are released in theaters in select cities, viewers everywhere will likely be able to stream them in their living rooms at the very same time.

Indie Distribution and Product Placement

Beyond their earnings from the box office and the home/mobile market, studios also make money by distributing the work of independent producers and filmmakers, who hire the studios to gain wider circulation. In exchange, independents give the studios between 30 and 50 percent of the box-office, home video, and streaming revenue that their movies bring in.

In addition, studios earn revenue from merchandise licensing and product placements in movies. In the early days of television and film, characters generally used generic products, or product labels weren't highlighted in shots. But with soaring film production costs, product placements are adding extra revenue while lending an element of authenticity to the staging. Famous product placements in movies include Reese's Pieces in *E.T.* (1982), Pepsi-Cola in *Back to the Future II* (1989), and an entire line of toy products in *The Lego Movie* (2014).

Popular Movies and Democracy

At the cultural level, movies often function as *consensus narratives*, a term that describes cultural products that become popular and provide shared cultural experiences. These consensus narratives operate across different times and cultures. In this sense, movies are part of a long narrative tradition, encompassing "the oral formulaic of Homer's day, the theater of Sophocles, the Elizabethan theater, the English novel from Defoe to Dickens, . . . the silent film, the sound film, and television during the Network Era."[24] Consensus narratives—whether they are dramas, romances, westerns, or mysteries—speak to central myths and values in an accessible language that often bridges cultural and global boundaries.

At the international level, countries struggle with questions about the influence of American films on local customs and culture. As with other American mass media industries, the long reach of Hollywood movies is a key paradox of contemporary life: Do such films contribute to a global village in which people throughout the world share a universal culture that breaks down barriers? Or does an American-based common culture stifle the development of local cultures worldwide and diversity in moviemaking? Clearly, the steady production of profitable action-adventure movies—whether they originate in the United States, Africa, France, or China—continues, not only because these movies appeal to mass audiences but also because they translate easily into other languages.

While massive blockbusters like *Avengers: Endgame* reveal the continued role movies play as a mass medium and the power of the international media conglomerates that produce and distribute them, the digital environment in which movies are made, distributed, and viewed today underscores movies' more complex cultural impacts. Netflix and other streaming services provide easier access to a wider range of movies than in the past and often sort them into niche categories through recommendation algorithms. As such, movies contribute to the emerging dynamics of life in a niche nation. As critical consumers, those of us who enjoy movies and recognize their cultural significance must carefully reflect on the varied ways they can bring us together through shared consensus narratives, while also recognizing that they can sort us into distinct subcultures.

LaunchPad
macmillan learning
launchpadworks.com

More Than a Movie: Social Issues and Film
Independent filmmakers are using social media to get moviegoers involved.

Discussion: Do you think digital media helps social-issue movies make a larger impact? Why or why not?

Chapter 7 REVIEW

LAUNCHPAD FOR *MEDIA & CULTURE*

LaunchPad
macmillan learning

Visit LaunchPad for *Media & Culture* at **launchpadworks.com** for additional learning tools:

➔ QUIZ YOURSELF WITH LEARNINGCURVE

LearningCurve uses gamelike quizzing to help you review important concepts from this chapter. LearningCurve includes multiple-choice questions at different levels of difficulty, hints, feedback for both right and wrong answers, and links to e-book material for easy reference. Which answer choice would you select for the sample question below?

> Katia is interested in the films of Georges Méliès, who opened the first public movie theater in France in 1896. Which film should she see?
>
> ◯ The Birth of a Nation
>
> ◯ A Trip to the Moon
>
> ◯ The Great Train Robbery
>
> ◯ The Life of an American Fireman

➔ WATCH VIDEO ON KEY CONCEPTS

LaunchPad includes clips from movies, TV shows, online sources, and other media texts, in addition to interviews with media experts and newsmakers. In the videos for this chapter, you'll consider topics like storytelling in *Star Wars* and social issues on film.

➔ PRACTICE THE CRITICAL PROCESS

Use the Media Literacy Activity to walk through the critical process step-by-step and develop a critical perspective on what media means to you. In this chapter's activity, you'll examine how the movie-watching experience differs at home and in a theater.

➔ LEARN KEY TERMS

Review the definitions for this chapter's key terms with LaunchPad's easy-to-use online flash cards. Try the cards in practice mode, or quiz yourself as a way to focus your study efforts. (Page numbers indicate where each term is highlighted within the chapter.)

celluloid, 196
kinetograph, 197
kinetoscope, 197
vitascope, 197
narrative films, 198
nickelodeons, 199
studio system, 200
vertical integration, 200
Big Five, 200
Little Three, 200
division of labor, 200
star system, 200

talkies, 201
block booking, 201
B movies, 201
A movies, 201
movie palaces, 202
classical Hollywood cinema, 202
genre, 203
product standardization, 203
product differentiation, 203
continuity editing, 203
Hollywood Ten, 205

Paramount decision, 205
auteurs, 208
cinema verité, 214
indies, 214
new Big Five, 216
mini-majors, 217
synergy, 217
blockbuster, 217
tentpoles, 218
theatrical exclusivity, 218
megaplexes, 218

REVIEW QUESTIONS

Use the questions below to revisit key themes and concepts from this chapter.

 Need more practice? Access the LearningCurve multiple-choice questions for this chapter, which are available on LaunchPad. All questions include feedback for correct and incorrect answers, hints, and links to e-book material.

Movies Today

1. How have movies played an important cultural role?

How We Got Here: The Early Development of Movies

2. What were the key steps in the development of film technologies?
3. What contributions did Méliès, Porter, and Griffith make to the art of film?
4. How did Edison's Trust try to control the movie business?
5. Why did Hollywood end up as the center of film production?

The Rise and Decline of Hollywood's Studio System Era

6. How did the rise of talkies help establish the major studios' control over the movie industry?
7. What strategies and practices did the major Hollywood studios use to dominate the industry during the studio system era?
8. What features are part of the classical Hollywood cinema style?
9. What political and cultural forces changed the Hollywood system in the late 1940s and 1950s?

Hollywood and Movies after the Studio System Era

10. How did the major studios evolve after the end of the studio system era?
11. How has the role of directors evolved in Hollywood?
12. What factors have affected the availability of foreign films in the United States at different times in history?
13. Why are documentaries an important alternative to traditional Hollywood filmmaking? What contributions have they made to the film industry?
14. What factors helped fuel the growth and success of indie films?

The Economics of the Movie Business

15. How does being part of larger conglomerates shape today's major film studios?
16. What are the various ways major movie studios make money from the film business?
17. What changing market forces have affected movie theater owners?

Popular Movies and Democracy

18. Do films contribute to a global village in which people throughout the world share a universal culture? Or do U.S.-based films overwhelm the development of other cultures worldwide? Discuss.

QUESTIONING THE MEDIA

Use these critical-thinking questions to reflect on your own experiences with media and apply your understanding of media concepts.

 For an interactive experience, questions on topics like these are available for the iClicker student response system. Instructors can download the questions from LaunchPad and use them to poll the class.

1. Do some research among family members, then compare your earliest memory of going to a movie with a parent's or grandparent's earliest memory.
2. Do you remember seeing a movie you were not allowed to see? Discuss the experience.
3. Do you prefer viewing films at a movie theater or at home? How might your viewing preferences connect to the way in which the film industry is evolving?
4. If you were a Hollywood film producer or executive, what kinds of films would you like to see made? What changes would you make in what we see in movies?

WE NEED A
RESPONSIBLE
OWNER

JOURNALIST
HOLD THE
POWERFUL
ACCOUNTABL

Denver Newspaper Guild—CWA
dfmworkers.org   #Newsmatters

Newspapers

8

THE RISE AND DECLINE OF MODERN JOURNALISM

Although the U.S. newspaper business has suffered from shaky profits and newspaper closings over the past two decades across the country, the newspaper industry in Iowa likely entered 2020 with a certain amount of pride in its persistence. Every one of Iowa's ninety-nine counties had at least one newspaper, a sign of the industry's health and the presence of civic activity in each of those places.

But just a few months later, things had changed: the COVID-19 pandemic had enveloped the United States, and the economy was suffering its worst downturn since the Great Depression. Interest in the news was up, but the businesses whose advertisements support the news were mostly in a downward spiral. Newspapers already near the breaking point after years of low ad revenue (due to social media competitors), fewer subscribers, and staff cuts started to break.

In Iowa, five newspapers closed within the first few months of the pandemic. The closings left Marion County, which has about thirty-three thousand people, as the first of Iowa's ninety-nine counties without a newspaper. In the terminology of newspaper researchers, Marion County became a **news desert**, joining about two hundred other counties across the United States that have no newspaper. Scholars at the University of North Carolina at Chapel Hill's Center for Innovation and Sustainability in Local Media define a news desert as "a community, either rural or urban, where residents have very limited access to the sort of credible and comprehensive news and information that feed democracy at the grassroots level."[1] Newspapers are the "most significant providers of journalism in their communities," so when a newspaper dies, the community suffers an enormous deficit in original reporting.[2] It's original reporting that gets repeated on social media (with no compensation for the newspaper) and often inspires television and radio news stories. Original reporting is also the main component of journalism's *watchdog* role—closely scrutinizing government and public figures to make sure they are accountable to citizens.

◀ Workers from the *Denver Post*, a newspaper established in 1892, protest against continued staff reductions by Digital First Media—a subsidiary of Alden Global Capital—and demand fair contracts. The editorial board of the newspaper denounced the owners as "vulture capitalists."

The pandemic hastened the decline of many newspapers throughout the United States. From March 2020 through early February 2021, at least 1,450 newspapers and print news organizations experienced layoffs, furloughs, building sales, and closures due to the pandemic, according to the Poynter Institute.[3] The losses hit both sprawling newspaper chains and small independent newspapers. For example, Gannett, the largest publisher of newspapers in the United States, responded to COVID-19 with up to $125 million in expense reductions and with "furloughs, significant pay reductions, reductions in force, and cancellation of non-essential travel and spending."[4]

The coronavirus has been an important factor leading to many newspaper closings, but it hasn't been the only factor. The nature and intent of the companies buying newspapers has further disrupted the stability of an already shaky industry. The fifth largest chain in the United States, CNHI (once known as Community Newspaper Holdings Inc.), made the biggest impact in Iowa with its newspaper closings, bringing the end to the *Knoxville Journal-Express* and the *Pella Chronicle*, both in Marion County, and the *Centerville Daily Iowegian* in Appanoose County. CNHI is a national chain of small newspapers that is owned by Retirement Systems of Alabama, an investment company whose primary holdings are in real estate—from golf courses and resorts to movie theaters and outlet stores. The problem for newspapers owned by CNHI is that the company is actively trying to extract itself from the journalism business due to low profits. In 2018, CNHI announced "plans to sell or close its 114 papers."[5] With the closings of three papers in southern Iowa and elsewhere in the United States, the company was down to ninety-three newspapers by fall 2020.[6]

It is the misfortune of the newspaper industry that at the moment when its historic, advertising-based business model is failing, its main owners are, like CNHI, "investment companies—private equity, hedge funds, financial companies whose interest is maximized profit" rather than a long-term obligation to journalism.[7] Nearly 45 percent of daily circulation is now controlled by these types of owners, who trim down budgets and staff and often shrink the geographic region of news coverage and circulation.[8]

With distant owners focused only on profits, the sense of history and the civic commitment that newspapers bring to their communities are lost. For example, in the first half of the twentieth century, one small-town Iowa newspaper and its editor fought valiantly against white supremacists. "The *Iowegian* waged a fierce—and ultimately successful—fight against the Ku Klux Klan in southern Iowa, where the KKK had gained a foothold that modern Iowans would find astonishing," one long-time reporter recalled.[9] The *Iowegian*, founded in 1883 and bought by CNHI in 1999, is now gone, and the people in that county might well wonder: Who will report on daily events, offer debate on public issues, and be their watchdog now?

 Where do you get your news? If you get news via social media, what is the original source of the reporting that appears in your feed? Do you read just the headlines, or do you follow the story back to the source?

Newspapers Today

Despite their current predicaments, newspapers and their online offspring play many roles in contemporary culture. As chroniclers of daily life, newspapers both inform and entertain. By reporting on scientific, technological, and medical issues, newspapers disseminate specialized knowledge to the public. In reviews of films, concerts, and plays, they shape cultural trends. Investigative teams can expose corruption, and crusading local editors can fight for what they think is best for their communities. Opinion pages trigger public debates and offer differing points of view. Columnists provide everything from advice on raising children to opinions on the United States' role as an economic and military superpower. Newspapers help readers make choices about everything from what kind of food to eat to what kinds of leaders to elect.

Although newspapers have long played a central role in daily life, in today's digital age the industry is losing both papers and readers. In the fifteen-year period from 2005 to 2020, over one-fourth of the nation's newspapers ceased to exist. Correspondingly, "half of newspaper readers and journalists have also vanished over the past 15 years," while many surviving newspapers—just 6,700 in 2020, down from almost 9,000 fifteen years earlier—are **ghost newspapers**, devoid of much actual local news and often carrying only syndicated content from other newspapers.[10] At the same time, a few elite newspapers in the United States, like the *Wall Street Journal* and the *New York Times*, have greatly expanded their audiences with digital subscriptions, charting at least one path for the future of newspapers.[11]

In this chapter, we examine the cultural, social, and economic impact of newspapers. We will:

- Trace the history of newspapers through a number of influential periods and styles
- Explore the early political-commercial press, the penny press, and yellow journalism
- Examine the modern era through the influence of the *New York Times* and journalism's embrace of objectivity
- Look at interpretive journalism in the 1920s and 1930s and the revival of literary journalism in the 1960s
- Review issues of newspaper ownership, new technologies, citizen journalism, declining revenue, and the crucial role of newspapers in our democracy

Chris Martin

THE *NEW YORK TIMES* is one of the few newspapers in the United States to substantially expand its audience with digital subscriptions.

How We Got Here: The Early Development of American Newspapers

The idea of news is as old as language itself. The earliest news was passed along orally from family to family, from tribe to tribe, by community leaders and oral historians. The earliest known written news sheet, *Acta Diurna* (Latin for "daily events"), was developed by Julius Caesar and posted in public spaces and on buildings in Rome in 59 BCE. Even in its oral and early written stages, news informed people on the state of their relations with neighboring tribes and towns. With the development of the printing press in the fifteenth century, a society's ability to send and receive information was greatly accelerated. Throughout history, news has satisfied our need to know things we cannot experience personally. Newspapers today continue to document daily life and bear witness to both ordinary and extraordinary events.

Colonial Newspapers and the Partisan Press

The development of newspapers emerged in Europe with the rise of the printing press. In North America, the first newspaper, *Publick Occurrences, Both Foreign and Domestick*, was published on September 25, 1690, by Boston printer Benjamin Harris. (People who owned printing presses often wore the hats of both printer and editor.) The colonial government objected to Harris's negative tone regarding British rule, and local ministers were offended by his published report

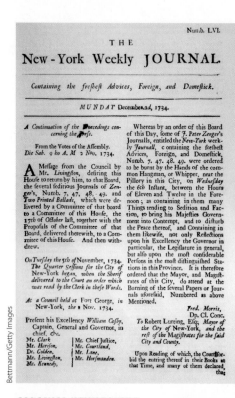

COLONIAL NEWSPAPERS

During the colonial period, New York printer John Peter Zenger was arrested for seditious libel. He eventually won his case, which established the precedent that today allows U.S. journalists and citizens to criticize public officials.

that the king of France had once had an affair with his son's wife. The newspaper was banned after one issue.

Because European news took weeks to travel by ship, early colonial papers were not very timely. In their colony-based reporting, however, the papers did cover local illnesses, public floggings, and even suicides. In 1704, the first regularly published newspaper appeared in the American colonies—the *Boston News-Letter*, published by John Campbell. In 1721, also in Boston, James Franklin, the older brother of Benjamin Franklin, started the *New England Courant*. The *Courant* established a tradition of running stories that interested ordinary readers rather than articles that appealed primarily to business and colonial leaders. In 1729, Benjamin Franklin, at age twenty-four, took over the *Pennsylvania Gazette* and created what historians consider the best of the colonial papers.

Another important colonial paper, the *New-York Weekly Journal*, appeared in 1733. John Peter Zenger had been installed as the printer of the *Journal* by the Popular Party, a political group that opposed British rule and ran articles that criticized the royal governor of New York. Following the dismissal of a Popular Party judge from office, the *Journal* escalated its attack on the governor. Zenger, who shielded the writers of the critical articles, was arrested in 1734 for *seditious libel*—defaming a public official's character in print. Championed by famed Philadelphia lawyer Andrew Hamilton, Zenger ultimately won his case in 1735. A sympathetic jury, in revolt against the colonial government, decided that newspapers had the right to criticize government leaders as long as the reports were true. After the Zenger case, the British never prosecuted another colonial printer. The Zenger decision would later provide a key foundation—the right of a democratic press to criticize public officials—for the First Amendment to the Constitution, adopted as part of the Bill of Rights in 1791. (See Chapter 16 for more on the First Amendment.)

By 1765, about thirty newspapers operated in the American colonies, with the first daily paper beginning in 1784. Newspapers were of two general types: political or commercial. Their development was shaped in large part by social, cultural, and political responses to British rule and its eventual overthrow. Although the political and commercial papers carried both party news and business news, they had different agendas. Political papers, known as the **partisan press**, generally pushed the strategy of the particular political group that subsidized the paper. The *commercial press*, by contrast, served business leaders, who were interested in economic issues. Both types of journalism left a legacy. The partisan press gave us the editorial pages, while the early commercial press was the forerunner of the business section.

In the eighteenth and early nineteenth centuries, even the largest of these papers rarely reached a circulation of fifteen hundred. Readership was primarily confined to educated or wealthy men who controlled local politics and commerce, and the printers-editors who published those papers were almost exclusively white men.

The Penny Press Era: Newspapers Become Mass Media

By the late 1820s, the average newspaper cost six cents a copy and was sold through yearly subscriptions priced at ten to twelve dollars. Because that price was more than a week's salary for most skilled workers, newspaper readers were mostly affluent. By the 1830s, however, the Industrial Revolution made possible the replacement of expensive handmade paper with cheaper machine-made paper. During this time, the rise of the middle class spurred the growth of literacy, setting the stage for a more popular and inclusive press. In addition, breakthroughs in technology, particularly the replacement of mechanical presses with steam-powered presses, permitted publishers to produce as many as four thousand newspapers an hour, which lowered the cost of newspapers. **Penny papers** soon began competing with six-cent papers. Though subscriptions remained the preferred sales tool of many penny papers, they increasingly began relying on daily street sales of individual copies.

NEWSIES sold papers on the streets of New York in the 1890s, a practice that started with the emergence of penny press newspapers in the mid-1800s. With more than a dozen dailies competing, street tactics were ferocious, and publishers often made young "newsies"—newsboys and newsgirls—buy the papers they could not sell.

Day and the *New York Sun*

In 1833, printer Benjamin Day founded the *New York Sun* with no subscriptions and the price set at one penny. The *Sun*—whose slogan was "It shines for all"—highlighted local events, scandals, police reports, and serialized stories. Like today's supermarket tabloids, the *Sun* sometimes fabricated stories, including the infamous Great Moon Hoax of 1835 that reported "scientific" evidence of "man-bats" and other life on the moon.[12] Within six months, the *Sun*'s lower price had generated a circulation of eight thousand, twice that of its nearest New York competitor.

The *Sun*'s success initiated a wave of penny papers that favored **human-interest stories**: news accounts that focus on the daily trials and triumphs of the human condition, often featuring ordinary individuals facing extraordinary challenges. These kinds of stories reveal journalism's ties to literary traditions, such as the archetypal conflicts between good and evil, normal and deviant, or individuals and institutions.

Bennett and the *New York Herald*

The penny press era also featured James Gordon Bennett's *New York Herald*, founded in 1835. Bennett, considered the first U.S. press baron, freed his newspaper from political influence. He established an independent paper serving middle- and working-class readers as well as his own business ambitions. The *Herald* carried political essays, news about scandals, business stories, a letters section, fashion notes, moral reflections, religious news, society gossip, colloquial tales and jokes, sports stories, and eventually reports from the Civil War. In addition, Bennett's paper sponsored balloon races, financed safaris, and overplayed crime stories. By 1860, the *Herald* was reaching nearly eighty thousand readers, making it the world's largest daily paper at the time.

Changing Economics and the Founding of the Associated Press

The penny papers were innovative. For example, they were the first to assign reporters to cover crime, which readers enthusiastically embraced, along with the reporting of local news. By gradually separating daily front-page reporting from overt political viewpoints on an editorial page, penny papers shifted their economic base from political parties to the market—that is, to advertising revenue, classified ads, and street sales. Many partisan papers had taken a moral stand against advertising certain controversial products and "services," such as medical "miracle" cures, abortionists, and especially the slave trade. Most of the penny press papers, in contrast, became more neutral toward advertisers and printed virtually any ad. In fact, many penny papers regarded advertising as consumer news. The rise in ad revenues and circulation accelerated the growth of the newspaper industry. In 1830, 650 weekly and 65 daily papers operated in the United States, reaching a circulation of 80,000. By 1840, a total of 1,140 weeklies and 140 dailies attracted more than 300,000 readers.

In 1848, six New York newspapers formed a cooperative arrangement and founded the Associated Press (AP), the first major news wire service. **Wire services** began as commercial organizations that relayed news stories and information around the country and the world using telegraph lines and, later, radio waves and digital transmissions. In the case of the AP, the New York papers provided access to both their own stories and those from other newspapers. In the 1850s, papers started sending reporters to cover Washington, D.C., and in the early 1860s, more than a hundred reporters from northern papers went south to cover the Civil War, relaying their reports back to their home papers via telegraph and wire services. The news wire companies enabled news to travel rapidly from coast to coast, setting the stage for modern journalism.

The marketing of news as a product and the use of modern technology to dramatically cut costs gradually elevated newspapers to the status of a mass medium. By adapting news content, penny papers captured the middle- and working-class readers of the emerging industrial revolution, who could now afford the paper and also had more leisure time to read it. As newspapers sought to sustain their mass appeal, news and "factual" reports about crimes and other items of human interest eventually became more important than partisan articles about politics and commerce.

The Age of Yellow Journalism: Sensationalism and Investigation

The rise of competitive dailies and the penny press triggered the next significant period in American journalism. In the late 1800s, a new class of extraordinarily profitable papers emerged that carried exciting human-interest stories, crime news, large headlines, and more readable copy. Generally regarded as sensationalistic and the direct forerunner of today's tabloid papers, reality TV, and celebrity-centered shows like *Access Hollywood*, the newspapers of this **yellow journalism** era shared two major characteristics: an emphasis on sensational or overly dramatic stories and a focus on early in-depth "detective" stories—the legacy for twentieth-century **investigative journalism** (news reports that hunt out and expose corruption, particularly in business and government). Reporting during this yellow journalism period increasingly became a crusading force for common people, with the press assuming a watchdog role on their behalf.

During this time, the two largest of these papers were engaged in a high-stakes newspaper circulation war: Joseph Pulitzer's *New York World* and William Randolph Hearst's *New York Journal*. One skirmish in that war saw the two papers fighting for control of the Yellow Kid, the main character in the first popular cartoon strip, *Hogan's Alley*, created in 1895 by artist R. F. Outcault. The phrase *yellow journalism* was inspired by the strip, versions of which appeared in both newspapers during their furious battle for readers in the mid-1890s.

Pulitzer and the *New York World*

Joseph Pulitzer, a Jewish-Hungarian immigrant, began his career in newspaper publishing in the early 1870s as part owner of the *St. Louis Post*. In 1878, he

YELLOW JOURNALISM
Generally considered America's first comic-strip character, the Yellow Kid was created in the mid-1890s by cartoonist Richard (R. F.) Outcault. The cartoon was so popular that newspaper barons Joseph Pulitzer and William Randolph Hearst fought over Outcault's services, giving yellow journalism its name.

GEE I HATE TO KNOCK A FELLER OUT IT MUST HURT HIM

Street & Smith Records, Special Collections Research Center, Syracuse University Libraries.

bought the bankrupt *St. Louis Dispatch* and merged it with the *Post*. The *Post-Dispatch* became known for stories that highlighted "sex and sin" ("A Denver Maiden Taken from Disreputable House") and satires of the upper class ("St. Louis Swells"). Pulitzer also viewed the *Post-Dispatch* as a "national conscience" that promoted the public good. He carried on the legacies of James Gordon Bennett and the *Herald*: making money and developing a "free and impartial" paper that would "serve no party but the people." Within five years, the *Post-Dispatch* had become one of the most influential newspapers in the Midwest.

In 1883, Pulitzer bought the *New York World* for $346,000. He encouraged plain writing and the inclusion of maps and illustrations to help immigrant and working-class readers understand the written text. In addition to running sensational stories on crime and sex, Pulitzer instituted advice columns and women's pages. Like the editors of many of the penny papers before him, Pulitzer treated advertising as a kind of news that displayed consumer products for readers. In fact, department stores became major advertisers during this period. This development contributed directly to the expansion of consumer culture and indirectly to the acknowledgment of women as newspaper readers. Eventually, because of pioneers like Nellie Bly—the *New York World* reporter who got herself committed to the Women's Lunatic Asylum on Blackwell's Island in New York in 1887 in order to investigate deplorable conditions there—newspapers began employing women as reporters.

The *New York World*, soon shortened to *The World*, reflected the contradictory spirit of the yellow press. It crusaded for improved urban housing, better conditions for women, and equal labor laws. It campaigned against monopoly practices by AT&T, Standard Oil, and Equitable Insurance. Such popular crusades helped lay the groundwork for the tightening of federal antitrust laws in the early 1910s. At the same time, the paper manufactured news events and staged stunts, such as sending star reporter Nellie Bly around the world in seventy-two days to beat the fictional "record" in the popular 1873 Jules Verne novel *Around the World in Eighty Days*. By 1887, *The World*'s Sunday circulation had soared to more than 250,000, the largest anywhere.

Pulitzer created a lasting legacy by leaving $2 million to start the graduate school of journalism at Columbia University in 1912. In 1917, part of Pulitzer's Columbia endowment established the Pulitzer Prizes, the prestigious awards given each year for achievements in journalism, literature, drama, and music.

Hearst and the *New York Journal*

In 1895, *The World* faced its fiercest competition when William Randolph Hearst bought the *New York Journal*. Before moving to New York, the twenty-four-year-old Hearst had taken control of the *San Francisco Examiner* when his father, George Hearst, was elected to the U.S. Senate in 1887. Hearst then used an inheritance from his father to buy the ailing *Journal* and raided Joseph Pulitzer's newspaper for editors, writers, and cartoonists.

Taking his cue from Bennett and Pulitzer, Hearst focused on lurid, sensational stories and appealed to immigrant readers by using large headlines and bold layout designs. To boost circulation, the *Journal* invented interviews, faked pictures, and encouraged conflicts that might result in a story. In promoting journalism as mere dramatic storytelling, Hearst reportedly said, "The modern editor of the popular journal does not care for facts. The editor wants novelty. The editor has no objection to facts if they are also novel. But he would prefer a novelty that is not a fact to a fact that is not a novelty."[13]

Hearst is remembered as an unscrupulous publisher who once hired gangsters to distribute his newspapers. He was also, however, considered a champion of the underdog, and his paper's readership soared among the working and middle classes. In 1896, the *Journal*'s daily

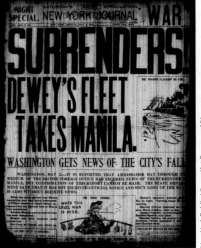

THE PENNY PRESS
The World (top) and the *New York Journal* (bottom) cover the same story in May 1898, during the Spanish-American War.

circulation reached 450,000, and by 1897, the Sunday edition of the paper rivaled the 600,000 circulation of *The World*. By the 1930s, Hearst's holdings included more than forty daily and Sunday papers, thirteen magazines (including *Good Housekeeping* and *Cosmopolitan*), eight radio stations, and two film companies. In addition, he controlled King Features Syndicate, which sold and distributed articles, comics, and features to many of the nation's dailies. Hearst—the model for Charles Foster Kane, the ruthless publisher in Orson Welles's classic 1940 film *Citizen Kane*—operated the largest media business in the world, comparable to today's Disney or Google.

Competing Models of Modern Newspaper Journalism

To some extent, the early commercial and partisan presses had covered important events impartially, often carrying verbatim reports of presidential addresses and murder trials or annual statements of the U.S. Treasury. It was in the late nineteenth century, as newspapers pushed for greater circulation, that newspaper reporting changed. Two distinct types of journalism emerged: the *story-driven model*, which dramatized important events and was favored by the penny papers and the yellow press, and the *"just the facts" objective model*, which appeared to package information more impartially.[14] Implicit in the latter's efforts was a question that is still debated today: Is there some ideal objective news model out there, or does the quest for objectivity actually conflict with journalists' traditional roles of telling news stories, revealing corruption, and questioning abuses of power, especially in government and business? The struggle between the story-driven model and the objective model of newspaper journalism has animated debates about journalism since that time, as would the introduction of other models of journalism—interpretive, literary, and online—that emerged in the twentieth century.

The Rise of "Objectivity" in Modern Journalism

As the consumer marketplace expanded during the Industrial Revolution, facts and news had become marketable products. Throughout the mid-nineteenth century, the more a newspaper appeared not to take sides on its front pages, the more its readership base grew (although, as they are today, editorial pages were still often partisan). To satisfy all clients, readers, and the wide range of political views, newspapers tried to appear more impartial. In addition, wire service organizations were serving a variety of newspaper clients in different regions of the country, which brought a more moderate tone to those news stories.

Ochs and the *New York Times*

The ideal of an impartial, or purely informational, news model was championed by Adolph Ochs, who bought the *New York Times* in 1896. The son of German Jewish immigrants, Ochs grew up in Ohio and Tennessee, where he took over the *Chattanooga Times* in 1878, at the age of twenty-one. Known more for his business and organizational ability than for his writing and editing skills, he transformed the Tennessee paper. Seeking a national stage and business expansion, Ochs moved to New York and invested in the struggling *Times*. Through strategic hiring, Ochs and his editors rebuilt the paper around substantial news coverage and well-argued editorial pages. To distance his New York paper from the yellow press, the editors downplayed sensational stories, favoring the documentation of major events and issues.

Partly as a marketing strategy, Ochs offered a distinct contrast to the more sensational Hearst and Pulitzer newspapers. The *Times* was an informational paper that provided stock and real estate reports to businesses, court reports to legal professionals, treaty summaries to political leaders, and theater and book reviews to educated general readers and intellectuals. Ochs's promotional gimmicks took direct aim at yellow journalism, advertising the *Times* under the motto "It does not soil the breakfast cloth." Ochs's strategy is similar to today's advertising tactic of targeting upscale readers and viewers, who control a disproportionate share of consumer dollars.

With the Hearst and Pulitzer papers capturing the bulk of working- and middle-class readers, managers at the *Times* first focused on using straightforward "no frills" reporting to appeal to more affluent and educated readers. In 1898, however, Ochs lowered the paper's price to a penny. He believed that people bought the *Journal* and *The World* primarily because they were cheap, not because of their stories. The *Times* began attracting middle-class readers who gravitated to the now-affordable paper as a status marker for the educated and well informed. Between 1898 and 1899, its circulation rose from 25,000 to 75,000. In 1911, Joseph Pulitzer died, but the energetic battle between the *Journal* and *The World* had already dissipated as the *Times* ascended. By 1921, the *Times* had a daily circulation of 330,000, and 500,000 on Sunday.

Permafrost, seen at the top of the cliff, melts into the Kolyma River outside of Zyryanka, Russia. This photo and accompanying story were part of The Washington Post's Pulitzer Prize-winning collection of stories about environmental devastation. (Michael Robinson Chavez/The Washington Post)

The Washington Post/Getty Images

"Just the Facts, Please"

The ideal of objectivity began to anchor U.S. journalism in the early twentieth century, and the *New York Times* was the first major newspaper to apply this informational approach to journalism, in which reporters adopt a more "scientific" attitude to news- and fact-gathering. In **objective journalism**, which distinguishes factual reports from opinion columns, modern reporters strive to maintain a neutral attitude toward the issue or event they cover; they also search out competing points of view among the sources for a story.

The story form for packaging and presenting this kind of reporting has been traditionally labeled the **inverted-pyramid style**. According to some historians, Civil War correspondents developed this style by imitating the terse, clinical dispatches that came from President Abraham Lincoln's secretary of war, Edwin M. Stanton.[16] Often stripped of adverbs and adjectives, inverted-pyramid reports began—as they do today—with the most dramatic or newsworthy information. They answered who, what, where, when (and, less frequently, why or how) questions at the top of the story and then narrowed down the information to presumably less significant details. This ensured that if wars or natural disasters disrupted the telegraph transmission of these dispatches, the most important information would have the best chance of getting through.

For much of the twentieth century, the inverted-pyramid style served as an efficient way to arrange a timely story (and it remains so today). It also had the advantage of appearing to present news as straightforward factual information, thereby not offending readers of differing political affiliations. Among other things, the importance of seeming objective and the reliance on the inverted pyramid signaled journalism's break from the partisan tradition. Although impossible to achieve, objectivity nonetheless became a guiding ideal of the modern press.

Interpretive Journalism

By the 1920s, there was a sense, especially after the trauma of World War I, that the objective-style approach to reporting that was common at that time was insufficient for explaining complex national and global conditions. It was partly as a result of "drab, factual, objective reporting," one news scholar contended, that "the American people were utterly amazed when war broke out in August 1914, as they had no understanding of the foreign scene to prepare them for it."[17]

WASHINGTON POST staff won the 2020 Pulitzer Prize for Explanatory Reporting for what the prize committee called "a groundbreaking series that showed with scientific clarity the dire effects of extreme temperatures on the planet."[15] The series, "2°C: Beyond the Limit," consisted of ten richly detailed stories from August through December 2019 that charted "dangerous new hot zones" in the United States and around the world and their disastrous effects. The Pulitzer Prize was established in 1917 by a bequest from Joseph Pulitzer.

The Promise of Interpretive Journalism

Under the sway of objectivity, modern journalism had downplayed an early role of the partisan press: to offer analysis and opinion. But with the world becoming more complex, some papers began to revisit the analytical function of news. The result was the rise of **interpretive journalism**, which aims to explain key issues or events and place them in a broader historical or social context. This shift allowed journalism to take an analytic turn in a world grown more interconnected and complicated.

In 1920, editor and columnist Walter Lippmann insisted that the press should do more to enlighten the public with its reporting. Noting that objectivity and factuality should serve as the foundation for journalism, Lippmann ranked three press responsibilities: (1) "to make a current record"; (2) "to make a running analysis of it"; and (3) "on the basis of both, to suggest plans."[18] Indeed, newspapers had historically distinguished between informational reports and editorials, the latter of which reflected the opinions of the publisher or the editorial board—opinions that were developed independently of reporters in the newsroom. But interpretive journalism offered opinion beyond the editorial pages, often by making a deeper analysis of a situation or suggesting solutions on the news pages. These pieces were identified as analysis or columns so that readers did not confuse them with straight news. It was during this time that political columns developed to evaluate and provide context for news.

Of course, in today's mass and masspersonal media world, analysis and opinion are everywhere. In the nineteenth century, the problem was too little information; today, the problem is too much information. Thus, in this digital age, how do we find reliable sources and the best resources? As always, the best opinions and interpretations are those based on evidence and documentation. What is true of good scholarship is also true of the best journalism: It relies on investigating what happened, verifying information, and documenting multiple sides of the story—as there are usually more than two.

Broadcast News Embraces Interpretive Journalism

In a surprising twist, the rise of broadcast radio in the 1930s forced newspapers to become more analytical in their approach to news. At the time, the newspaper industry was upset that broadcasters took their news directly from papers and wire services. As a result, a battle developed between radio journalism and print news. Although mainstream newspapers tried to copyright the facts they reported and sued radio stations for routinely using newspapers as their main news sources, the papers lost many of these court battles. Editors and newspaper lobbyists argued that radio personalities should be permitted to broadcast only commentary. By conceding this interpretive role to radio, the print press tried to protect its dominion over the facts. It was in this environment that radio analysis began to flourish as a form of interpretive news. Lowell Thomas delivered the first daily network analysis for CBS on September 29, 1930, attacking Hitler's rise to power in Germany. By 1941, twenty regular commentators—the forerunners of today's radio talk-show hosts, cable television talking heads, and political bloggers—were explaining their version of the world to millions of listeners.

In this environment, some print journalists and editors came to believe that interpretive stories, rather than objective reports, could better compete with radio. They realized that interpretation was a way to counter radio's (and later television's) superior ability to report breaking news quickly—or even live. In 1933, the American Society of Newspaper Editors supported the idea of interpretive journalism, though most newspapers did not embrace probing analysis during the 1930s. In most U.S. dailies, then, interpretation remained relegated to a few editorial and opinion pages. In the 1950s—a decade that saw the Korean War, the development of atomic power, tensions with the Soviet Union, and the anticommunist movement—news analysis debuted on the newest medium: television. Interpretive journalism in newspapers grew at the same time, competing with TV and responding to important ethical and policy questions in such areas as the environment, science, agriculture, sports, health, politics, and business. Following the lead of the *New York Times* in 1970, many papers developed their own regular *op-ed page*—an opinion page opposite the editorial page that allowed for a greater variety of columnists, news analyses, and letters to the editor.

Literary Forms of Journalism

By the late 1960s, many people were criticizing America's major social institutions. Political assassinations, Civil Rights protests, the Vietnam War, the drug culture, and the women's movement were not easily explained. Faced with so much change and turmoil, many individuals began to lose faith in the ability of institutions to oversee and ensure the social order. Members of protest movements as well as many middle- and working-class Americans began to suspect the privileges and power of traditional authority. As a result, key institutions—including journalism—lost some of their credibility. A number of reporters responded to the criticism by rethinking the framework of conventional journalism and adopting a variety of alternative techniques.

Journalism as an Art Form

Throughout the first part of the twentieth century—journalism's modern era—journalistic storytelling was downplayed in favor of the inverted-pyramid style and the separation of fact from opinion. Dissatisfied with these limitations, some reporters began exploring a new model of reporting. **Literary journalism**, sometimes dubbed "new journalism," adapted fictional techniques—such as descriptive details and settings and extensive character dialogue—to nonfiction material and in-depth reporting. In the United States, literary journalism's roots are in the work of nineteenth-century novelists like Mark Twain, Stephen Crane, and Theodore Dreiser, all of whom started out as reporters. In the late 1930s and 1940s, literary journalism surfaced: Journalists, such as James Agee and John Hersey, began to demonstrate how writing about real events could achieve an artistry often associated only with fiction.

In the 1960s, Tom Wolfe, a leading practitioner of literary journalism, argued for mixing the *content* of reporting with the *form* of fiction to create a style with both the "objective reality of journalism" and "the subjective reality" of the novel.[19] Writers such as Wolfe (*The Electric Kool-Aid Acid Test*), Truman Capote (*In Cold Blood*), Joan Didion (*The White Album*), James Baldwin (*The Fire Next Time*), and Hunter S. Thompson (*Hell's Angels*) turned to literary journalism to overcome the flaws they perceived in routine reporting. Their often self-conscious treatment of social problems gave their writing a perspective that conventional journalism did not offer. After the 1960s' tide of intense social upheaval ebbed, literary journalism subsided as well. However, literary journalism not only influenced magazines like *Mother Jones* and *Rolling Stone* but also affected daily newspapers by emphasizing longer feature stories on cultural trends and social issues, with detailed description or dialogue. Today, writers such as Dexter Filkins (*The Forever War*), Åsne Seierstad (*The Bookseller of Kabul*), and Ta-Nehisi Coates (*Between the World and Me*) keep this tradition alive.

The Attack on Journalistic Objectivity

Former *New York Times* columnist Tom Wicker argued that in the early 1960s, an objective approach to news remained the dominant model. According to Wicker, the "press had so wrapped itself in the paper chains of 'objective journalism' that it had little ability to report anything beyond the bare and undeniable facts."[20] Eventually, the ideal of objectivity became suspect, along with the authority of experts and professionals in various fields.

In addition to literary journalism's response to the perceived shortcomings of objective-style journalism, there were other new approaches. One of these was *advocacy journalism*, in which the reporter actively promotes a particular cause or viewpoint. (See the later section "Newspapers beyond the Mainstream" for examples of advocacy journalism.) *Precision journalism*, another

JOAN DIDION'S essay collections *Slouching Towards Bethlehem* (1968) and *The White Album* (1979) are considered iconic pieces from the literary journalism movement. Both books detail and analyze Didion's life in California, where she experienced everything from the counterculture movement in San Francisco to encounters with members of the Black Panther Party, the Doors, and even followers of Charles Manson. The first line of the *White Album* reads, "We tell ourselves stories in order to live."

Everett Collection Historical/Alamy

technique, attempts to make the news more scientifically accurate by using poll surveys and question-naires. Today we call this "data journalism," and it has increased in importance as newspapers and other news organizations take advantage of the availability of Internet data and the lack of space and time constraints in online journalism.

Newspapers in the TV and Internet Age

In the early 1980s, a postmodern brand of journalism arose from two important developments: In 1982, the colorful *USA Today*, started by Gannett, radically changed the look of most major U.S. dailies, and in 1980, the *Columbus Dispatch* became the first paper to go online. Later, with the advent of mobile phones and social media, citizens began to disseminate news and information themselves, creating citizen journalism, which again changed the scope of the news.

USA Today Colors the Print Landscape

USA Today made its mark by incorporating features closely associated with postmodern forms, including an emphasis on visual style over substantive news or analysis and the use of brief news items to appeal to readers' busy schedules and shortened attention spans.

Now one of the most widely circulated papers in the nation, *USA Today* represents the only successful launch of a new major U.S. daily newspaper in the last several decades. Showing its marketing savvy, *USA Today* was the first newspaper to acknowledge television's central role in mass culture: The paper used TV-inspired color and designed its first vending boxes to look like color TVs. Even the writing style of *USA Today* mimics TV news by casting many reports in the present tense rather than the past tense in feature articles. The *USA Today* style inspired other major newspapers to add color to their pages. Its style is also present in the eighty daily newspapers that are part of the Gannett chain, of which *USA Today* is the flagship newspaper.

Digital Journalism Redefines News

What started out in the 1980s as simple text-only experiments for newspapers developed into more robust websites in the 1990s, allowing newspapers to develop an online presence. Today, digital journalism has completely changed the news industry in two key ways. First, rather than subscribing to a traditional print paper, most readers now begin their day on their smartphones, tablets, or computers, scanning a wide variety of news websites, or using Facebook friends and Twitter posts to direct them to key news stories on any particular day. Such resources and digital sites are taking over the role of more traditional forms of news, helping to set the nation's cultural, social, and political agendas. Second, without the constraints of print publication or regular broadcast programming, digital news has sped up the news cycle to a constant stream of information, with breaking news on Twitter and Facebook and push notifications available via e-mail and apps.

Initially, there was resistance to the digital revolution, with executive editors trying to get older reporters and editors to embrace the possibilities of telling stories with more than the traditional text and photos. Now, not only are many newspapers using the full range of digital tools to create stories, they are also using various content types—including animation, short documentaries, digital games, interactive data visualizations, and podcasts—that do more than just dump print stories online. These new formats enable readers to experience stories in entirely different ways than they can with a print newspaper. (See "Examining Ethics: The 1619 Project—Journalism Takes the Long View to Rethink History.")

Citizens Expand the Reach of Journalism

The combination of the online news surge and traditional newsroom cutbacks has led to a phenomenon known as **citizen journalism**, or *citizen media*. As a grassroots movement, citizen journalism refers to people—activist amateurs and concerned citizens, not professional journalists—who use the Internet and social media to disseminate news and information. With steep declines in newsroom staffs, many newspapers rely on reports, images, and video from citizen journalists to cover events they might have missed or to reveal breaking news. These reports can take the form of social media posts, livestreams, or blog entries.

In fact, much of the widespread outrage over police violence against Black people has been due to videos from citizen journalists. Videos of the deaths of several Black men at the hands of police—including Eric Garner (New York, 2014), Walter Scott (South Carolina, 2015), Samuel DuBose (Ohio, 2015), Philando Castile (Minnesota, 2016), George Floyd (Minnesota, 2020), and Rayshard Brooks (Georgia, 2020)—and the near-fatal police shooting of Jacob Blake (Wisconsin, 2020) brought renewed attention to the #BlackLivesMatter movement and racial inequality in the criminal justice system. The viral citizen videos became the foundational reports of one of the biggest news stories of the generation: the racial inequalities in policing and the criminal justice system. According to Paul Butler, Georgetown University law professor and a former prosecutor, the citizen videos are "the C-Span of the streets" and supply "corroboration of what African-Americans have been saying for years."[21]

CITIZEN JOURNALISTS capture photos and videos of news events, which can often be used to supplement mainstream news reports or provide news coverage when there might be no professional reporters on the scene.

The Business, Organization, and Ownership of Newspapers

In 2020 in the United States, there were roughly 6,730 newspapers of various sizes. *National newspapers* (such as the *Wall Street Journal*, the *New York Times*, and *USA Today*) serve a broad readership across the country and have circulations (including paid print and online readership) in the millions. *Metropolitan dailies* (like the *Minneapolis Star Tribune*, the *Houston Chronicle*, and the *Boston Globe*) serve specific metro areas or larger geographic regions. Fewer than forty of those have a paid circulation of approximately 100,000. Midsize and small dailies, ethnic newspapers, and nondaily newspapers, such as the *Berkshire Eagle* (a daily in Pittsfield, Massachusetts), *Nguoi Viet Daily News* (a Vietnamese-language daily in Westminster, California), and the *Quoddy Tides* (twice monthly, in Eastport, Maine), may be less well known beyond their communities than larger papers, but they make up the bulk of the newspaper industry—6,576 newspapers fit into this category. No matter the size of the paper, each must determine its approach, target readers, establish social media strategies, and deal with ownership issues in a time of continuing technological transition and declining revenue.

Consensus versus Conflict: Newspapers Play Different Roles

Smaller nondaily papers tend to promote social and economic harmony in their communities. Besides providing community calendars and meeting notices, nondaily papers focus on **consensus-oriented journalism**, carrying articles on local schools, social events, town government, property crimes, and zoning issues. Recalling the partisan spirit of an earlier era, small newspapers are often owned by business leaders who may also serve in local politics. Because consensus-oriented papers have a small advertising base, they are generally careful not to offend local advertisers, who provide the financial underpinnings for many of these papers. At their best, these small-town papers foster a sense of community; at their worst, they overlook or downplay discord and problems.

The 1619 Project— Journalism Takes the Long View to Rethink History

Over a photo of vast dark ocean water separated from gray sky by a long, flat horizon line, the following text was printed on the cover of the *New York Times Magazine*'s August 18, 2019, issue:

> In August of 1619, a ship appeared on this horizon, near Point Comfort, a coastal port in the English colony of Virginia. It carried more than 20 enslaved Africans, who were sold to the colonists.

The cover text concluded that, on the four hundredth anniversary of slavery's introduction to the shores of what would become the United States of America, "it is finally time to tell our story truthfully."

So began the 1619 Project, an ambitious ongoing series of stories, podcasts, and other media from the *New York Times* designed to reframe the history of the United States as one that is undeniably intertwined with the horrors of slavery and the struggle of Black Americans to make this country live up to its promise of democracy for all.

For a newspaper to take on such a project is unexpected. Reporting on history is not a strength of journalism, which has an entrenched bias toward the new. As noted in Chapter 14, reporters seldom link stories to the ebb and flow of history.

Yet for Nikole Hannah-Jones, a *New York Times Magazine* writer who focuses on racial justice and is the force behind the project, the year 1619 is as important to the founding of the United States as the year 1776. Growing up in Waterloo, Iowa, Hannah-Jones became attuned to racial disparities when she rode a bus from her predominantly Black east side neighborhood to the higher-income white west side neighborhood to attend school. "You saw a distinct color line. I was very curious and observant of these things at a young age," she said.[1] After earning degrees at the University of Notre Dame and the University of North Carolina at Chapel Hill, she reported for the Raleigh *News & Observer*, the *Oregonian* in Portland, and *ProPublica* before taking a position at the *New York Times* in 2015. In 2017, she won a MacArthur Foundation Fellowship for her reporting on racial segregation in the United States.

Her work leading the 1619 Project is journalism at its most transformational. The August 18, 2019, edition of the magazine included ten essays on contemporary topics and issues and their historical foundations in the legacy of slavery and Jim Crow, including Wall Street–style capitalism and the plantation, the lack of universal health care and the legacy of segregated health-care systems, traffic jams in Atlanta, the color lines of neighborhoods and civic infrastructure, and the sounds of Black music in the United States and how they have always been appropriated by the white music industry. (In fact, the essay on music reminds readers that the term "Jim Crow," which refers to laws and customs that systematically disenfranchised Black Americans in the late nineteenth and twentieth centuries, came from a white minstrel singer who performed in blackface as Jim Crow.)

Hannah-Jones won a Pulitzer Prize for Commentary for her lead essay for the 1619 Project, and the project's significance continues to reverberate through the country. The *New York Times* has distributed hundreds of thousands of additional copies of that issue of the magazine, a five-part podcast of the project appeared on *The Daily*, Random House announced several related book projects, and the Pulitzer Center (which funds reporting projects but is not affiliated with the Pulitzer Prize) has developed extensive K–12 and college curriculum materials to accompany the

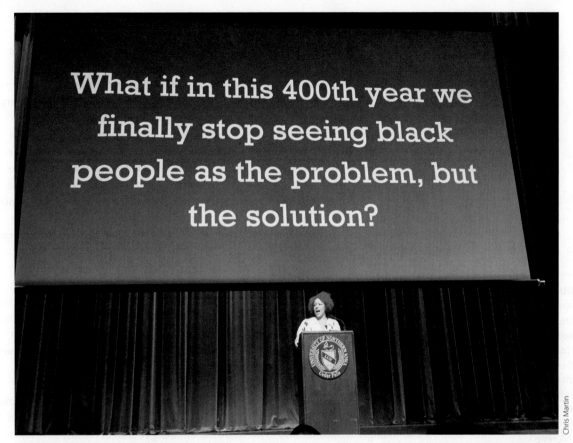

NIKOLE HANNAH-JONES, the *New York Times* writer who developed the 1619 Project, speaks on the four hundredth anniversary of slavery's introduction to North American shores.

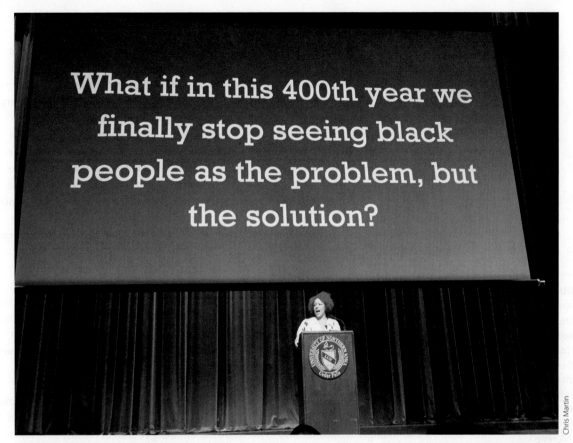
Chris Martin

1619 Project. In 2020, the *New York Times*, Lionsgate, and Oprah Winfrey announced a partnership for films and TV shows based on the project.

It was perhaps inevitable that there would be some pushback against a journalism project that advocates a rethinking of American history. Historians have criticized a few of the project's conclusions. Moreover, Sen. Tom Cotton (R-Arkansas) proposed to withhold federal funds from schools teaching the 1619 Project as part of their curriculum. Then, in September 2020, President Trump lashed out at the 1619 Project, calling it "toxic propaganda, ideological poison that, if not removed, will dissolve the

civic bonds that tie us together."[2] Trump then proposed a 1776 Commission to promote "patriotic education."

Hannah-Jones—whose Twitter account bears the name "Ida Bae Wells," a fond reference to Ida B. Wells, a pioneering Black investigative journalist and one of Hannah-Jones's heroes—replied to the political attacks in a tweet: "When I . . . pitched this project, my editor asked me what was my ultimate goal. I told her I wanted Americans to know the date 1619, to force that foundational date into the national lexicon. The extent to which the project continues to be attacked shows the success of the goal." ●

In contrast, national and metro dailies practice **conflict-oriented journalism**, in which front-page news is often defined primarily as events, issues, or experiences that deviate from social norms. Under this news orientation, journalists see their role not merely as neutral fact-gatherers but as observers who monitor their city's institutions and problems. As such, they often maintain an adversarial relationship with local politicians and public officials. Newspapers practicing this form of journalism offer competing perspectives on such issues as education, government, poverty, crime, and the economy; and their publishers, editors, and reporters avoid playing overt roles in community politics. In theory, modern newspapers believe their role in large cities is to point out problems. However, some newsrooms are asking how they can be more solution oriented in their news coverage. For example, an organization called Your Voice Ohio, staffed by former journalists and activists, is training journalists and citizens to confront problems and consider solutions. The nonprofit organization is responding to the idea that people often "don't see themselves represented in their local news, especially in political news," and envisions "a world where journalists and the communities they serve see each other not only as real people, but partners working together to create a more vibrant Ohio."[22]

Newspapers beyond the Mainstream

Historically, small-town weeklies and daily metro newspapers have served predominantly white, middle-class readers. Outside the mainstream, however, a number of newspapers have taken a different path in terms of organization and ownership, prioritizing the interests of ethnic and racial groups or political ideas that were not (and often still are not) part of mainstream journalism.

Ever since Benjamin Franklin launched the short-lived German-language *Philadelphische Zeitung* in 1732, newspapers aimed at ethnic groups have played a major role in initiating immigrants into American society. During the nineteenth century, Swedish- and Norwegian-language papers informed various immigrant communities in the Midwest. The early twentieth century gave rise to papers written in German, Yiddish, Russian, and Polish, assisting the massive influx of European immigrants.

Today, at least five hundred ethnic or non-English-language newspapers are published in print or online in the United States. Most serve some of the same functions for their constituencies as do mainstream English-language newspapers. These papers, however, are often published outside the social mainstream. Consequently, they provide viewpoints that are different from the mostly white, middle- and upper-class establishment attitudes that have shaped the media throughout much of America's history. The existence of ethnic newspapers is a reminder both of the country's great diversity and of the shortcomings of mainstream newspapers in serving diverse readers.

The Center for Innovation and Sustainability in Local Media reports that ethnic newspapers have been reinvigorated by the growth of new immigrant communities. "While independent ownership has been decreasing among community newspapers, a significant demographic shift in this country—propelled by the growth of Hispanic residents—is nurturing exactly the opposite among ethnic news organizations. Independent ownership of minority and ethnic newspapers and magazines, especially, has increased over the past two decades."[23]

Black Newspapers

Between 1827 and the end of the Civil War in 1865, forty newspapers directed at Black readers and opposed to slavery struggled for survival. These papers not only faced higher rates of illiteracy among potential readers but also encountered hostility from white society and the majority press of the day. The first Black newspaper, *Freedom's Journal*, operated from 1827 to 1829 and opposed the racism of many New York newspapers. In addition, it offered a public voice for antislavery societies. Other notable papers

FREDERICK DOUGLASS helped found the *North Star* in 1847. It was printed in the basement of the Memorial African Methodist Episcopal Zion Church, a gathering spot for abolitionists and "underground" activities in Rochester, New York. At the time, the white-owned *New York Herald* urged Rochester's citizens to throw the *North Star*'s printing press into Lake Ontario. Under Douglass's leadership, the paper came out weekly until 1860, addressing problems facing Black people around the country and offering a forum for Douglass to debate his fellow Black activists.

Library of Congress, Prints & Photographs Division, LC-DIG-ds-07422

included the *Alienated American* (1852–1856) and the *New Orleans Daily Creole*, which began its short life in 1856 as the first Black-owned daily in the South. The most influential oppositional newspaper was Frederick Douglass's *North Star*, a weekly antislavery newspaper in Rochester, New York, which was published from 1847 to 1860 and reached a circulation of three thousand. Douglass, a formerly enslaved person, wrote essays on a variety of national and international topics, including slavery.

Since 1827, fifty-five hundred newspapers have been edited or started by Black Americans.[24] These papers, with an average life span of nine years, have taken stands against race baiting, lynching, and the Ku Klux Klan. They also promoted racial pride long before the Civil Rights movement. The most widely circulated Black-owned paper was Robert C. Vann's weekly *Pittsburgh Courier*, founded in 1910. Its circulation peaked at 350,000 in 1947—the year

BLACK NEWSPAPERS
This 1936 scene reveals the newsroom of Harlem's *Amsterdam News*, one of the nation's leading Black newspapers. Ironically, the Civil Rights movement and the implementation of affirmative action policies in the 1960s served to drain talented reporters from the Black press by encouraging them to work for larger, mainstream newspapers.

professional baseball was integrated by Jackie Robinson, thanks in part to relentless editorials in the *Courier* that denounced the color barrier in pro sports. As they have throughout their history, these papers offer oppositional viewpoints to the mainstream press and record the daily activities of Black communities by listing weddings, births, deaths, graduations, meetings, and church functions.

The circulation rates of most Black papers dropped sharply after the 1960s. The combined circulation of the local and national editions of the *Pittsburgh Courier*, for instance, dropped from 202,080 in 1944 to 20,000 in 1966, when it was reorganized as the *New Pittsburgh Courier*. Several factors contributed to these declines. First, television and Black radio stations tapped into the limited pool of money that businesses allocated for advertising. Second, some advertisers, to avoid controversy, withdrew their support from the Black press when it started giving favorable coverage to the Civil Rights movement in the 1960s. Third, the loss of industrial urban jobs in the 1970s and 1980s not only diminished readership but also hurt small neighborhood businesses, which could no longer afford to advertise in both the mainstream and the Black press. Finally, after the enactment of Civil Rights and affirmative action laws, mainstream papers raided Black papers, seeking to integrate their newsrooms with Black journalists. Black papers could seldom match the offers from large white-owned dailies.

Today, there are roughly 170 Black newspapers, many of them more than a century old, and most now publish as weeklies or online. These papers include Baltimore's *Afro-American*, New York's *Amsterdam News*, the *Chicago Defender*, the *Los Angeles Wave*, and the *Michigan Chronicle*.[25]

Spanish-Language Newspapers

Bilingual and Spanish-language newspapers have served a variety of Mexican, Puerto Rican, Cuban, and other Hispanic readerships since 1808, when *El Misisipi* was founded in New Orleans. Throughout the 1800s, Texas had more than 150 Spanish-language papers.[26] Los Angeles' *La Opinión*, founded in 1926, is now the nation's largest Spanish-language daily. Other prominent publications are in Miami (*Diario Las Américas*), Houston (*La Información*), Chicago (*La Raza*), and New York (*El Diario NY*). A report from the Craig Newmark Graduate School of Journalism at CUNY counted more than 240 Latino-Hispanic newspapers in the United States.[27]

Until the late 1960s, mainstream newspapers virtually ignored Hispanic issues and culture. But with the influx of Mexican, Puerto Rican, and Cuban immigrants throughout the 1980s and 1990s, many mainstream papers began to feature weekly Spanish-language supplements or editions. The first was the *Miami Herald*'s "El Nuevo Herald," introduced in 1977 as a supplement and broken out as a stand-alone publication in 1988. Other mainstream papers also joined in, but some later folded their Spanish-language supplements and editions. Most recently, the *Austin American-Statesman* shut down its weekly *¡Ahora Sí!* in 2018, and the *Chicago Tribune* closed its weekly *Hoy* in 2019. Spanish-language radio, television, and Internet sites triumphed over newspapers in reaching potential customers and advertisers. Nevertheless, in 2020, *USA Today* began a new English- and Spanish-language long-form series called Hecho en USA (Made in America). "This is a huge community in the U.S., and we think that our readers are interested in learning more about them—and we're not just doing these stories for Latinos, but for everyone," said Cristina Silva, the *USA Today* national enterprise editor leading the series.[28]

Asian American Newspapers

The first Asian-language newspaper in the United States was *Golden Hill's News*, published in San Francisco in 1854. The Chinese weekly lasted only a few months, but it established a pattern of Asian American newspapers starting up to serve new immigrant populations.[29] In the 1980s, for example, hundreds of small papers emerged to serve immigrants from Pakistan, Laos, Cambodia, and China, helping readers both adjust to their new surroundings and retain ties to their traditional heritage. In addition to providing this type of support, these papers often cover major stories downplayed in the mainstream press. For example, in 2020 and 2021, the *Seattle Chinese Post*, the oldest Chinese-language newspaper in the Pacific Northwest, and the *Northwest Asian Weekly*, an English-language weekly serving the Asian community in the same region, reported on the rising violence against Asian Americans and Pacific Islanders in the United States and the reaction against anti-Asian rhetoric and hate crimes.[30]

There are several large Chinese-language newspapers currently being published in the United States. Founded in 1976, the *World Journal* is the largest U.S.-based Chinese-language paper. It is published daily in New York, where it is headquartered, with editions in several other cities across the United States and Canada.[31] Other major Chinese-language newspapers include the *China Press*, the U.S. edition of the Hong Kong–based *Sing Tao Daily*, and the *Epoch Times*, a once-small New York–based newspaper linked to the Chinese spiritual movement Falun Gong, which has grown in notoriety since 2016, when it began to use Facebook to promote right-wing disinformation and conspiracy theories.[32]

ASIAN NEWSPAPERS, such as the Hindi newspaper seen here, can help readers stay up to date on important events that are sometimes downplayed by traditional media.

Native American Newspapers

An activist Native American press has provided oppositional voices to mainstream American media since 1828, when the *Cherokee Phoenix* appeared in Georgia. The paper was burned down by white settlers six years later, but not before it covered President Andrew Jackson's defiance of treaties and the forced removal of the Cherokee from their homelands. Another prominent early paper was the

Cherokee Rose Bud, founded in 1848 by Cherokee women in the Oklahoma territory.

Native American newspapers have suffered a sharp decline in the past two decades, as Indigenous communities have dealt with even harsher economic circumstances than the rest of the United States. In 1998, there were a total of six hundred urban and reservation newspapers and bulletins. By 2018, there were just fifty-four urban and reservation newspapers and twenty-four newsletters. (The number of radio stations serving this population nearly doubled during this period, making it the strongest area of growth for Native American media.)[33] Two leading news organizations are *Indian Country Today*, owned by the Oneida nation in New York, and *Native News Online*, an independent newsroom based in Grand Rapids, Michigan.

Emmanuel Dunand/AFP/Getty Images

A major goal for today's Native American news media is to create more news organizations independent of tribal governments and their influence over reporting. According to a recent study, "Tribes own and mostly control 72 percent of all Native print and radio operations in the United States."[34] In addition, the need for Native American news organizations to counter poor representations in the mainstream press remains great. A 2020 study by the Native American Journalists Association found that "more than half of [the] national media coverage of Indigenous peoples in the U.S. [over an 18-month period] . . . relied on clichéd themes and stereotyping terminology," including "'poverty,' 'reservation,' 'drugs,' 'addiction,' 'alcohol' and 'drums.'"[35]

The Underground Press

The mid- to late 1960s saw an explosion of alternative newspapers. Labeled the **underground press** at the time, these papers questioned mainstream political policies and conventional values, often voicing radical opinions. Generally running on shoestring budgets, they were also erratic in meeting publication schedules. Springing up on college campuses and in major cities, underground papers were inspired by the writings of socialists and intellectuals from the 1930s and 1940s and by a new wave of thinkers and artists. Particularly inspirational were poets and writers (such as Allen Ginsberg, Jack Kerouac, LeRoi Jones, and Eldridge Cleaver) and "protest" musicians (including Bob Dylan, Pete Seeger, and Joan Baez). In criticizing social institutions, alternative papers questioned the official reports distributed by public relations agents, government spokespeople, and the conventional press.

During the 1960s, underground papers played a unique role in documenting social tension by including the voices of students, women, Black Americans, Native Americans, gay men and lesbians, and others whose opinions were often excluded from the mainstream press. The first and largest underground paper was the *Village Voice*, founded in Greenwich Village in 1955. From the very start and through its final print edition in 2017, the paper was distributed for free, surviving solely on advertising. Following its last print edition, the *Voice* continued online until 2018.[36] However, in 2021, new stories began to appear on the website, signaling the return of this storied publication.

Despite their irreverent tone, many underground papers turned a spotlight on racial and gender inequities and occasionally goaded mainstream journalism to examine social issues. Like the Black press, though, many early underground papers folded after the 1960s. Given their radical outlook, it was difficult for them to appeal to advertisers. Plus, the mainstream papers raided the alternatives, just as they'd done with the Black press, expanding their coverage of culture by hiring the underground's best writers. Still, almost one hundred papers are members of the Association of Alternative Newsmedia today.[37]

Newspaper Operations

Today, a weekly paper might employ only two or three people, while a small daily might have a newsroom of twelve to fifty journalists and a larger metro daily, several hundred. One of the largest reporting staffs is at the *New York Times*, an international news organization with about sixteen hundred journalists. The lack of diversity in U.S. newspaper newsrooms, which have long been staffed by white and mostly male journalists, remains a critical problem for newspapers and affects the tone and substance of their coverage.

Beyond the journalists themselves, other segments of newspaper operations include advertising (selling ad space), syndication and wire services (sharing or purchasing content), circulation (distributing the newspaper), and mechanical operations (assembling and printing the paper—although many newspapers have shut down their presses and outsourced this function to other newspapers). Most newspapers distinguish business operations from editorial or news functions. Journalists' and readers' praise or criticism usually rests on the quality of a paper's news and editorial components, but business and advertising concerns dictate whether papers will survive.

Newspaper Advertising

Most major daily papers prefer to devote one-half to two-thirds of their pages to advertisements. Newspapers carry everything from full-page spreads for department stores to shrinking classified ads, which consumers can still purchase for a few dollars to advertise used cars or old furniture (although many websites now do this for free). In most cases, ads are positioned in the paper first. The **newshole**—the leftover space not taken up by ads—accounts for the remaining content of daily newspapers, including front-page news. Despite the preference to run more ads rather than fewer, the number of ads in many newspapers has shrunk substantially in the past two decades. In fact, one analysis of a metro newspaper, the *Kansas City Star*, found that the typical 60:40 ratio (60% ads and 40% news content) in the printed pages of the newspaper is now closer to 10:90 (10% ads and 90% news).[38] This results in thinner print newspapers (fewer ads means fewer pages), higher subscription costs (to generate funds after advertising revenues decline), and cuts to newspaper staffs (because of lower revenue).

News and Editorial Responsibilities

On the news and editorial side of most larger papers, the chain of command begins with the publisher and owner at the top, then moves to the editor in chief and managing editor, who are in charge of the daily news-gathering and writing processes. Under the main editors, assistant editors have traditionally run different news divisions, including features, sports, photos, local news, state news, and wire service reports that contain major national and international news. These days, many editorial positions are being eliminated or condensed to the job of a single editor, whose chief responsibility is often ensuring that stories are first posted online (to give them the immediacy that radio and TV news have always had), then updated and prepared for the print edition. (For a look at how newspapers cover stories about their own employees unionizing in response to these types of labor conditions, see "Media Literacy and the Critical Process: Covering the Unionization of Newspaper Workers.")

Reporters work for editors. *General assignment reporters* handle all sorts of stories that might emerge—or "break"—in a given day. *Specialty reporters* are assigned to particular beats (police, courts, schools, local and national government) or topics (education, religion, health, environment, technology). At large dailies, *bureau reporters* file reports from other major cities, although recent consolidation and cutbacks have led to layoffs and the closing of bureaus outside a paper's city limits. Large daily papers feature columnists and critics who cover various aspects of culture, such as politics, books, television, movies, and food.

Newsroom Racial and Ethnic Diversity

In 1978, the American Society of Newspaper Editors (ASNE) set a goal of having "minority employment by the year 2000 equivalent to the percentage of minority persons within the national population"—a "fair and attainable goal," in its opinion.[39] Yet as the *Columbia Journalism Review* noted, "Newspapers have failed spectacularly at achieving that goal."[40] Even more than four decades later, the goal to build a diverse newspaper workforce has fallen short. People of color are about 40 percent of the U.S. population, but they make up only about 17 percent of print and online newsroom staff, and only 13 percent of newspaper management. Even in cities where there are more people of color than white people, white staff dominate the newsrooms of major newspapers. For example, the newsrooms of the *Wall Street Journal* and the *New York Times* are 81 percent white, while that of the *Washington Post* is 70 percent white.[41]

The need for newsroom diversity came to a head at the *Los Angeles Times* in 2020. The *Times* was bought by local owners in 2018, who hired more than one hundred journalists to build back its newsroom after years of cuts. Yet the newsroom of the *Times* was enormously unrepresentative of its local audience. By 2020, the paper employed 502 journalists; 61 percent of them were white, in a county where only 26 percent of the population is white. Latinos represented only 13 percent of the newsroom staff, though they make up nearly half of the county's population. Asian American

MEDIA LITERACY & THE CRITICAL PROCESS

Covering the Unionization of Newspaper Workers

With the economic challenges plaguing journalism over the last decade, the loss of jobs has affected what gets covered as news. But what about coverage of the changing news media business itself? Over the years, critics have claimed that business news pages tend to favor issues related to management over the roles of everyday employees, focusing on managers' promotions, for example, and downplaying news about the workers at the same businesses. So do local newspapers today cover the state of their own workers? One of the trends in U.S. newspapers in recent years has been reporters and other workers organizing into unions for collective bargaining in order to address the shortcomings of their labor conditions, such as a lack of diversity and job security.[1] Using information from the NewsGuild (the largest union of journalists in the United States and Canada), select a newspaper to see how effectively it covers its own workers and the unionization of the news media business in general.

1 DESCRIPTION

Visit the NewsGuild website (go to https://newsguild.org, or search for "NewsGuild") or Twitter account to find a newspaper with recent union activities, such as a unionization campaign, recognition of a new union by management, contract negotiations, or layoffs. Next, use LexisNexis or the newspaper's website (if possible) to track three to four weeks' worth of news in that newspaper to see if, and how well, it covers these union activities. If there is recent coverage, gauge the importance given to these stories (length might be one indicator) and whether these stories are positive or negative. Also check to see if competing news media in the same market (other newspapers, TV stations, radio stations) cover the union activities, and the quality of that coverage in comparison.

2 ANALYSIS

Look for patterns in the coverage. How many newspaper union stories were produced over the period? How many were positive? How many were negative? How many stories did you find in competing news media in the same market?

3 INTERPRETATION

What do some of the patterns mean? Did you find examples in which the newspaper's coverage of its workers' labor activities seems comprehensive and fair? Did the newspaper seem to favor its own business perspective? What does it mean if newspapers don't cover their own workers comprehensively and fairly? Could it make readers doubt how those newspapers cover workers at other businesses? Do you think news organizations are more likely to cover union activities at news organizations other than their own?

4 EVALUATION

Most newspapers have extensive business sections and often cover local markets, business leaders, and the ups and downs of the stock market. What should be the role of a newspaper in covering its own workers? Should a newspaper be transparent about its own operations—when it is profitable, when it is not profitable (and why), and the nature of compensation and job stability among its workforce? Why or why not? How would you evaluate this newspaper's reporting on its own workers?

5 ENGAGEMENT

E-mail the newspaper's editor to offer your findings, or submit them for publication in a letter to the editor. Note what the newspaper is doing well, and make a recommendation on how to improve its labor coverage, if necessary. You can also share your findings with the labor representative at the newspaper, and find out how the workers themselves feel about the way they are being covered by their own newspaper.

---

journalists at the *Times* matched the county's Asian American population at about 15 percent, but the newspaper had just 26 Black journalists (or 5.2% of the newsroom) in a county where 8 percent of the population is Black. As the *Times* reported about itself, staff members "openly chastised senior editors for allowing racial disparities to persist," leading to a "painful internal reckoning over glaring deficiencies and missteps regarding race and representation in its pages and its staff."[42] The executive editor later resigned, and the paper's owners vowed to diversify its reporting and editing staff and improve pay equity. Women and journalists of color at the *Times* won a pay equity settlement in late 2020 that affected nearly 240 former and current employees.[43]

✿ GRANDPA had To TAKE OUT the NAZIS. You have To WEAR A MASK and GET TAKEOUT.

POLITICAL CARTOONS are often syndicated features in newspapers that reflect the issues of the day.

Soledad O'Brien, a Black Latina journalist who has worked for NBC, CNN, and HBO, said that after the Black Lives Matter protests in 2020, journalists of color were having their moment by calling out institutional racism at news organizations across the country: "The thin ranks of people of color in American newsrooms have often meant us-and-them reporting, where everyone from architecture critics to real estate writers, from entertainment reporters to sports anchors, talk about the world as if the people listening or reading their work are exclusively white."[44]

Wire Services and Feature Syndication

Major daily papers might have dozens of local reporters and writers, but they still cannot cover the world or produce enough material to fill up the newshole each day. Since the founding of the Associated Press (AP) in 1846, newspapers have relied on wire services and syndicated feature services to supplement local coverage. A few major dailies, such as the *New York Times*, run their own wire services, selling their stories to other papers to reprint. Other agencies, such as the AP and United Press International (UPI), have hundreds of staffers stationed in major U.S. cities and world capitals, submitting stories and photos daily for distribution to newspapers across the country. Some U.S. papers also subscribe to foreign wire services, such as Agence France-Presse in Paris or Reuters in London.

Daily papers generally pay monthly fees for access to all wire stories. Although they use only a fraction of what is available over the wires, editors monitor wire services each day for important stories and ideas for local angles. Wire services have greatly expanded the reach and scope of news, as local editors depend on them when selecting statewide, national, or international reports for reprinting.

In addition, traditional **feature syndicates**, such as Creators, Andrews McMeel Syndication, and Tribune Content Agency, are businesses that contract with newspapers to provide work from the nation's best political writers, editorial cartoonists, comic-strip artists, and self-help columnists. These companies serve as brokers, distributing horoscopes and crossword puzzles as well as the political columns and comic strips that appeal to a wide audience. When a paper bids on and acquires the rights to a cartoonist or columnist, it signs exclusivity agreements with a syndicate to ensure that it is the only paper in the region to carry, say, Clarence Page, S. E. Cupp, Leonard Pitts, Connie Schultz, George Will, or cartoonist Mike Luckovich. Feature syndicates, like wire services, wield great influence in determining which writers and cartoonists gain national prominence.

Newspaper Ownership: Investment Companies Dominate the Industry

Edward Willis Scripps founded the first **newspaper chain**—a company that owns several papers throughout the country—in the 1890s. By the 1920s, there were about thirty chains in the United States, each owning an average of five papers. The emergence of chains paralleled the major business trend of the twentieth century: the movement toward oligopolies, in which a handful of corporations control each industry.

By 2001, the top ten chains controlled more than one-half of the nation's total daily newspaper circulation. Gannett, for example, the nation's largest chain, owned over eighty daily papers at that time (and hundreds of nondailies), ranging from small suburban papers to the *Cincinnati Enquirer*, the *Nashville Tennessean*, and *USA Today*.

But consolidation in newspaper chain ownership leveled off around 2005, when a decline in newspaper circulation and ad sales panicked investors, leading to drops in the stock value of newspapers. Many newspaper chains responded by significantly reducing their newsroom staffs and selling off individual papers. With the value of newspapers low, *hedge funds* and *private equity* firms—investment corporations that buy up other businesses to restructure them and increase profits—became the new newspaper barons of the twenty-first century, taking over traditional chains like Knight Ridder, Gannett, the Tribune Company, and McClatchy.

The new owners—investments firms like Alden Global Capital (doing business as Digital First Media and MediaNews Group), Chatham Asset Management, and New Media Investment Group (formerly GateHouse, now doing business as Gannett after it acquired that company in 2019)—have been derided by many journalists as "vulture capitalists" (including by the *Denver Post* editorial board, which lashed out at its newspaper's owner, Alden Global Capital, after its staff suffered severe cuts).[45] Other journalists, including one from a Gannett newspaper, charged that "the private equity formula of hollowing out local papers the better to extract windfalls is cynical and deliberate."[46] The main criticism of investment firm owners is that by deeply cutting staff to increase profits, they ultimately undermine the ability to create the basic product of journalism: news stories.

INVESTMENT FIRMS have bought up hundreds of newspapers in the past decade, making major cuts to newsroom staffs to extract more profits from their investment. This photo illustration shows the 142 members of the *Denver Post* staff in 2013, and the staff members remaining in 2018 after a series of cuts by owner Alden Global Capital.

Following more acquisitions in 2020, including Chatham Asset Management's purchase of the McClatchy newspaper chain (which was the second largest chain at the time), financial companies now control about 45 percent of the nation's newspaper circulation, including most of the country's large metro newspapers.[47] In some cities, investment firms have turned around and sold metro newspapers to what some call "benign billionaires," who are more interested in newspaper ownership for civic purposes than large profits.[48] But although that has worked in Minneapolis (the *Star Tribune*), Boston (the *Globe*), Los Angeles (the *Times*), and San Diego (*Union-Tribune*), it's not a viable solution for local ownership in most communities.

Even the last two remaining large newspaper chains, Tribune Publishing Company and Lee Enterprises, have not escaped the reach of the investment firms. By 2020, Alden Global Capital owned a 32 percent share of Tribune Publishing Company (with plans to increase its share) and a 7.1 percent share of Lee Enterprises. Meanwhile, another investment firm, the Retirement Systems of Alabama, was trying to shed its newspaper business, CNHI, a chain of more than ninety small newspapers (see the chapter-opening discussion). The remaining significant newspaper chains in the United States are private companies that own smaller dailies and weekly newspapers, including Adams Publishing Group, Ogden Newspapers, Paxton Media Group, Boone Newspapers, Community Media Group, and Landmark Media Enterprises.

Challenges and Changes Facing Newspapers Today

Publishers and journalists today face worrisome issues, such as the decline in newspaper readership and the shift of readers to digital platforms, which—although welcome—often doesn't generate the same subscription or ad revenue that the print platform generated. But new models for the twenty-first century point the way for the democratic function of newspapers to continue long past the life of the print format and the traditional advertising-based structure.

Declines in Readership, Newsrooms, and Newspapers

In the long view, the decline in daily newspaper readership actually began during the Great Depression with the rise of radio. Between 1931 and 1939, six hundred newspapers ceased operation. Another circulation crisis occurred from the late 1960s through the 1970s with the rise of network television viewing, greater competition from suburban weeklies, and metro newspapers' shift in focus from a

mass audience to a narrower, more upscale middle-class audience.[49] And with an increasing number of women working full-time outside the home, newspapers could no longer consistently count on one of their core readership groups. Yet despite these declines, the readership declines in the twenty-first century have been the most transformative. While many newspaper readers shifted from print to digital formats, many others just fell away from newspapers entirely, and many young readers never took up the habit in the first place. (For an alternative view, see "Global Village: The World's Biggest Newspapers.")

The unfortunate story for newspapers in the twenty-first century is that "between 2004 and 2019, total weekday circulation—including both dailies and weeklies—declined 45 percent, from more than 122 million to 68 million." These figures include digital readers, but digital readers are only counted in circulation figures if they are paying subscribers, and "newspapers have struggled with getting readers to pay even a small amount for online access."[50]

Historically, subscription revenue for newspapers has been relatively small. In the twentieth century, print newspapers generated most of their revenue—80 to 90 percent—through ad sales. By 2011, however, ad money was drying up; digital newspaper readers outnumbered print readers for the first time that year, but because digital ads were generating much less revenue than print, they could not cover the shortfall as the print model began to disappear. One reason for this is that by 2020, Google, Facebook, and Amazon had captured about 70 percent of digital advertising revenue.[51] That meant that a business that had bought ads in the local paper for decades might decide to advertise its products in a Google search, in a Facebook feed, or on Amazon Marketplace rather than spending money on a newspaper ad (digital or print). As more and more ad dollars have been redirected to digital megacorporations—billions of dollars, in total—the entire newspaper industry, including local papers, has suffered. Given the shortage of ad revenue, newspapers have relied more on subscription revenue, which now accounts for about half of a newspaper's earnings, but still falls short of what newspapers need to sustain themselves.

The decline in readership accompanies a decline in the number of journalists employed at newspapers. From 2008 to 2019, the number of employees in newspaper newsrooms declined by more than half, from about 71,000 to 35,000 workers.[52] Those drops in the number of journalists came as news organizations cut back circulation areas (i.e., newspapers like the *Atlanta Journal-Constitution* and *Wichita Eagle* pulled back from statewide or near-statewide circulation to cover only counties in their own metro region), as the loss of advertising revenue undermined the ability to afford a large newsroom staff, and as investment-driven owners cut personnel to further increase profits.

The loss of readers and journalists is linked to an inescapable fact: many newspapers in the United States have closed. Between 2005 and 2019, the United States lost a quarter of its newspapers (2,100), falling from about 8,800 newspapers at the beginning of 2005 to 6,700 newspapers at the end of 2019. Among the fallen newspapers were seventy dailies and more than two thousand weeklies/nondailies.[53] At the beginning of the chapter, we noted that these closures have resulted in *news deserts*—communities without a regular, reliable news source (see Figure 8.1)—while some of the newspapers that remain have suffered so much downsizing that they've become *ghost newspapers*, devoid of much actual local news. Because local newspapers generate most of the original reporting that circulates in a media environment, these losses are substantial. As summarized in a Pew Charitable Trusts story, "When a local newspaper dies, evidence shows civic engagement decreases, elected officials are less accountable, corruption is more pervasive and voter participation drops and becomes more polarized."[54]

FIGURE 8.1

WHERE HAVE NEWSPAPERS DISAPPEARED?

Penelope Muse Abernathy, UNC Center for Innovation and Sustainability in Local Media

WHERE HAVE NEWSPAPERS DISAPPEARED?

Since 2004, the U.S. has lost more than 2,100 newspapers.

● Weekly
● Daily

The World's Biggest Newspapers

While the United States has continued to experience a steep decline in newspaper readership and print advertising dollars, in many other nations—where Internet news is still emerging, and where the culture of reading a print paper is deeply ingrained—newspapers still reign supreme. For example, the World Association of News Publishers (WAN-IFRA) reported that in 2018, print circulation dropped by only 1 percent globally. Although newspapers in other countries are making the transition to digital, print still provides about 86 percent of overall publisher revenue. Even so, digital growth has helped to pull in new readers, and by 2018, as U.S. newspaper readership was plummeting, the paying news audience increased 0.5 percent worldwide, to 640 million per day.[1]

The following are some of the world's most widely circulated newspapers, which are concentrated in three countries: Japan, India, and China.

Japan

Yomiuri Shimbun, founded in 1874, is a national Japanese newspaper with both morning and evening editions. Its morning edition has a circulation of 7,954,126 (which, it likes to point out, is certified by the *Guinness Book of World Records* as the largest newspaper circulation in the world).[2] The newspaper has a readers' center with a staff of 40 who take questions, suggestions, and news tips from readers. *Yomiuri Shimbun* also publishes an English-language daily newspaper, the *Japan News*. The newspaper's holding company—the largest media conglomerate in Japan—also owns Japan's first baseball team, the Yomiuri Giants, as well as a symphony orchestra, the Nippon Television Network, and several commercial buildings in Tokyo. *Yomiuri Shimbun* is one of five national newspapers, which include *Asahi Shimbun*, *Mainichi Shimbun*, and *Chunichi Shimbun*.

Also in Japan, *Nikkei*, with a circulation of more than three million, is the world's largest financial newspaper (larger than the *Wall Street Journal* in the United States, with about 2.84 million in circulation, including digital). *Nikkei* also publishes the Nikkei Stock Index, which is the leading index of the Tokyo Stock Exchange. In 2015, it bought the London-based *Financial Times*, another leading daily business newspaper.

India

Dainik Bhaskar, established in Bhopal in 1958, is the third largest newspaper in the world (after *Yomiuri Shimbun* and *Asahi Shimbun*), and thus the largest newspaper in India, with sixty-five regional editions in Hindi, Marathi, and Gujatari languages.[3] It boasts coverage of more than two thousand cities across the country, and it released a mobile news app in 2017. The newspaper's holding company also owns India's largest radio network, which covers thirty cities in seven states, as well as a number of top websites, including the largest Bollywood site.

Dainik Jagran is the main rival for India's top newspaper. It was launched in 1942 and is headquartered in Kanpur. Other high-circulation newspapers in India include *Amar Ujala*, the *Times of India* (India's largest English-language newspaper, founded in 1838; the company owns several television channels as well), *Hindustan Dainik*, *Malayala Manorama* (established in 1890 and the leading newspaper in the Malayalam language), and *Rajasthan Patrika*.

China

Reference News, China's largest newspaper, translates and republishes news from other countries. Founded in 1931, it was originally developed as an insider publication for Chinese Communist Party officials and their families, but was permitted to circulate more broadly in the 1980s. It is published by the Xinhua News Agency, which has been the official state-run press agency of China since 1931.

The *People's Daily*, another widely circulated newspaper in China, has been the official newspaper of the Central Committee of the Chinese Communist Party since 1948. It offers editions in a number of other Asian and European languages, including English. Foreign policy experts read the *People's Daily* editorials to understand the statements and more nuanced positions of the Chinese government. The *People's Daily* also publishes *Global Times*, a nationalist tabloid with both Chinese and English editions. Other large newspapers include *Guangzhou Daily* and *Nanfang Daily*, just two of the many newspapers published by provincial or regional party committees. ●

Searching for Success in the Digital Era

Amid the downturn in the newspaper industry, a few national newspaper organizations have done comparatively well in the transition to a digital platform. The *New York Times* has been most notable. After committing to a digital strategy in 2011, when it first started charging for online content, the *Times* has slowly grown into the most successful digital newspaper in the United States. By mid-2020, the *Times'* digital revenue exceeded its print revenue for the first time, and by the end of 2020, it had surpassed more than seven million total paid subscribers, still attracting subscribers through some of the worst days of the coronavirus pandemic. The company's growth has been so substantial that it is likely to reach its goal of ten million subscribers by 2025. Yet the newspaper's 2020 announcement of digital success came with a sober reminder of a "worrying trend": "Digital readers were the only growth business for the *Times*. Every other unit fell."[55] The *Times* concluded that its online subscription business was its future, and that advertising "is likely to become a more narrow business for the *Times*, even as Facebook and Google continue to thrive in that area."[56]

Part of the *New York Times'* digital success has included its efforts to broaden its news platforms. Its overall digital subscriptions include subscribers to NYT Cooking (recipes) and NYT Games (crossword puzzles), which require separate subscriptions from the news content. The newspaper's *The Daily* podcast, an approximately half-hour dive into the reporting of a single news topic, averaged four million downloads per day in 2020, double the number from the previous year, and it encouraged listeners to subscribe to the *Times*. *The Weekly*, a similar investigative journalism program for television, debuted on FX and Hulu in 2019.

Like the *New York Times*, smaller newspapers have found that there is no single way to sustain a newspaper in the digital era, as each market is different. The *New York Times* has the luxury of growing its digital subscribers from a base of a national and international audience. However, nearly all other newspapers in the United States are geographically based to serve a region, metro area, or small town. One study of dailies, weeklies, and online news organizations found several dimensions that contributed to the resiliency of those kinds of U.S. newspapers:[57]

- **Local ownership.** Local owners are more vested in the community and treat the newspaper like a necessary asset.
- **Commitment to community.** This generally comes with local ownership and informs the editorial outlook of the news organization.
- **Diversified business.** Other affiliated revenue-generating businesses, such as a marketing firm, radio station, or TV station, can help to support the journalism of newspapers in economic downturns. But sustaining journalism must be central to the overall enterprise.
- **Health of the local economy.** For-profit newspapers have long depended on local banks, car dealerships, hospitals, and grocery stores as major advertisers. If they fail, the impact on a local newspaper is substantial.
- **Nonprofit status.** As an alternative to for-profits, many nonprofits have been better able to weather the ups and downs of the economy than their for-profit counterparts. Still, consistent nonprofit funding support is essential.
- **Little to no reduction of newsroom staff.** Staffing cutbacks—including furloughs and layoffs—reduce expenses in the short term but do long-term damage to the amount and quality of journalism, which is the primary reason someone subscribes to a newspaper in the first place.
- **No competing local media.** Newspapers are already subject to substantial advertising competition from digital powerhouses such as Google, Facebook, and Amazon. Given that challenge, it's easier to operate a newspaper without significant media competition at the local level, which can include other local newspapers, radio stations, and television stations that do local news.

The more of these factors that a local newspaper can have in place, the more likely it is that the paper's survival can be sustained.

New Models for Newspapers

As newspapers—and even many digital news organizations—suffered in the first decades of the twenty-first century, there were a number of ideas afloat for inventing new platforms for doing the work of newspaper journalism and for reinventing the very twentieth-century idea of newspapers as an ad-supported medium.

Nonprofit Journalism

One of the worst outcomes of staff cutbacks and fewer newspapers is the demise of watchdog journalism, especially at the state capital level. Although the federal government receives plenty of attention from all types of news media, it is newspapers that have traditionally filled the bulk of statehouse reporting corps. One nonprofit group responding to the resulting shortfall is States Newsroom, a North Carolina–based nonprofit established in 2017 that has funded twenty independent newsrooms in state capitals, including the digital *Arizona Mirror, Georgia Recorder, Michigan Advance, Ohio Capital Journal*, and *Pennsylvania Capital-Star*.[58] ProPublica, a nonprofit online investigative journalism organization established in 2007 (and the first online news organization to win a Pulitzer Prize), also announced that it would fund placing reporters in partner organizations to expand accountability journalism of state governments.[59]

One of the most ambitious projects to refill the nation's journalism ranks is Report for America (RFA). Launched by the GroundTruth Project—a journalism nonprofit organization—Report for America is modeled on the Peace Corps. News organizations that have gaps in their coverage can apply for an RFA corps member. Young journalists—many new college graduates—apply to serve those communities for a year, with an optional second year. The positions receive half funding from RFA, a 25 percent contribution from the local news organization, and 25 percent more from local donors. By 2020, RFA had 226 reporters placed in forty-six states, Washington, D.C., and Puerto Rico. Its goal is to place one thousand reporters by 2024.[60]

According to the Institute for Nonprofit News, there were more than 250 nonprofit news organizations in the United States by 2020, employing 3,500 people—2,300 of them journalists. The median staff size for nonprofit news organizations is 6.3 full-time employees. Foundations provide 48 percent of their support, with individual gifts supplying an additional 35 percent, and the rest coming from earned revenue or other charitable contributions.[61]

Journalism Goes "Vertical"

Another strategy for reinventing newspapers is the idea of creating **verticals**—niche-interest digital sites that bundle together related content (such as technology stories). Verticals can be helpful for selling targeted ads (for for-profit newspapers) and can bring attention to topic areas that have been overlooked by other news media.

USA Today, for example, has created opinion verticals on "Policing the USA" and "Race in America" as a way to aggregate all the relevant stories and opinion pieces on those topics.[62] The *New York Times* launched a "Parenting" vertical in 2019 with the idea of making it a stand-alone subscription product (like its NYT Cooking and NYT Games sites).[63] Not all verticals succeed. The *Boston Globe* shut down its Catholic news vertical "Crux" after two years in 2016. "We simply haven't been able to develop the financial model of big-ticket, Catholic-based advertisers that was envisioned," editors at the *Globe* said.[64]

A stand-alone nonprofit vertical called The 19th*, which was launched in Austin, Texas, in 2020 by several leading women journalists, specializes in news on "gender, politics, and policy." Its name refers to the Nineteenth Amendment, which gave U.S. women the right to vote in 1920, and the asterisk refers to the "unfinished business" that remains for women in U.S. politics.[65]

THE 19TH* is a nonprofit, nonpartisan digital journalism site established in 2020 and specializing in news on gender, politics, and policy. In its introduction to readers, the editors wrote, "The 19th Amendment remains unfinished business, a fact we acknowledge in our logo with an asterisk.... Our goal is to empower those we serve—particularly women, people of color and the LGBTQ+ community—with the information, resources and community they need to be equal participants in our democracy."

In this still image from California State Assembly video, Assemblywoman Buffy Wicks addresses lawmakers on a housing bill while holding her 1-month-old daughter, Elly. Wicks requested permission from legislative leaders to vote remotely or by proxy during the coronavirus pandemic but was denied. (CALIFORNIA STATE ASSEMBLY VIA AP)

AP Images/California State Assembly

Politics

Lawmaker who brought her newborn to the California statehouse joins other elected women challenging outdated notions around parenthood

Journalists Go Solo

By 2020, several notable journalists had opted to go solo by getting readers to subscribe directly to the reports and commentary on their personal newsletter sites. Platforms such as Substack, Medium, and Patreon have made it easier for journalists to become their own news organizations.

Solo journalists include Casey Newton (who left his job as technology editor at the *Verge* to create a Substack newsletter called *Platformer*), Lyz Lenz (who exited the *Cedar Rapids Gazette* to write a politics, culture, and feminism newsletter on Substack called *Men Yell at Me*), and labor journalist Mike Elk, who started the crowdfunded website *Payday Report* in 2016 after getting fired from *Politico*. Other notable solo journalists include Matthew Yglesias (formerly of *Vox*) and Glenn Greenwald (who left the *Intercept*).

Although going solo gives journalists the freedom to write whatever they wish, one writer warned that "very few 'Substack writers' will make 100% of their income from their newsletter," and instead will have to be entrepreneurial, mixing in work doing podcasts, speaking fees, book deals, consulting, and the like.[66]

"Replanting" Newspapers

The most large-scale new models for journalism seek to reinvent the very twentieth-century idea of newspapers as an ad-supported medium. Journalist Steve Waldman, one of the cofounders of Report for America, suggested that the 6,700 remaining newspapers in the United States could be "replanted." "Just as sickly plants can sometimes gain new life by being watered and repotted in healthier soil, could changing their ownership structures and sources of nourishment revive some dying newspapers?" he asked.[67]

This idea would require money to build a private nonprofit fund to enable newspapers to be transformed into community-based nonprofit papers. The idea would also require public policy changes to assist family-run newspapers in becoming nonprofits, in selling to a corporation with a specific mission for public benefit, or in selling to newspaper workers themselves, and also to encourage commercial owners to give newspapers back to their communities. According to Waldman, these commercial owners traditionally do one of two things with their newspapers: "milk them or close them." Instead, he says, "we want them to choose a third path: donate them."[68]

There are some local precedents for this idea. The *Tampa Bay Times*, the *Philadelphia Inquirer*, and the *Salt Lake City Tribune* have all been "replanted," leaving behind the for-profit model to become supported and owned by nonprofit foundations. Although they are still subject to the difficult economic environment for newspapers today, as nonprofits they can endure it better. Other organizations, such as Free Press and the NewsGuild, also support the idea of news organizations becoming nonprofits and providing tax credits to citizens to support more paid news subscriptions.[69]

PROPUBLICA began publishing in 2008 as a nonprofit investigative news site. Since its start, it has won six Pulitzer Prizes and many other awards, and its reports have resulted in a number of public policy reforms.

Newspapers and Democracy

Although there are other platforms for news media, newspapers hold a special position in the industry, as they account for about 50 percent of original news stories.[70] Thus, when newspapers die, the loss reverberates through the entire community and beyond. Studies have demonstrated that the loss of newspapers is associated with increased political polarization, lower voter turnout, and—without the watchdog press—higher local government expenses.[71]

Yet even before a newspaper closes, as citizens turn away from newspapers (including their websites and digital apps), the quality of the political news information people receive—information essential to being an engaged citizen—decreases. According to the Pew Research Center, about 25 percent of U.S. adults say they get political news primarily through a news website or app, 18 percent through social media, 16 percent through cable TV, 16 percent through local TV, 13 percent through network TV, 8 percent through radio, and 3 percent through print (with 1% giving no response). Yet not all news media are equal in terms of their association with "high political knowledge," according to Pew. Users of news websites/apps, radio, and print all rated between 41 and 45 percent in high political knowledge. Social media—the second-most common way people get their political news—was the second worst medium in fostering high political knowledge, at 17 percent. Only local TV scored lower, with only 10 percent of those who used it as their main source of political information considered to have high political knowledge.[72]

Although there are many readers who wax rhapsodic about the feel of a morning newspaper in their hands, a coffee and donut at their side, it is clear that in paper form, newspapers are on the endangered species list (note that only 3% of U.S. adults use it as their primary source of political news). But the more worrisome part of the future of newspapers isn't the loss of information on paper; it's the loss of the specific rigorous and wide-ranging culture and journalistic tradition that is practiced in newsrooms. Newspapers, the press, is the one medium that is mentioned by name in the First Amendment of the U.S. Constitution. Even in the imperfect democracy that was established more than 230 years ago, the founders considered a free press an essential partner in helping to make our democracy better.

Mark Edward Atkinson/Tracey Lee/Getty Images

READING A NEWSPAPER
About six out of ten Americans access newspapers each week, but most of them use a digital format. Slowly paging through a print newspaper is a tradition in decline; only about 16 percent of U.S. residents often get their news in the print format.

Chapter REVIEW 8

➔ QUIZ YOURSELF WITH LEARNINGCURVE

LearningCurve uses gamelike quizzing to help you review important concepts from this chapter. LearningCurve includes multiple-choice questions at different levels of difficulty, hints, feedback for both right and wrong answers, and links to e-book material for easy reference. Which answer choice would you select for the sample question below?

> Dayla has written a series of articles on the new honor code regulations on her campus. In her final article, she plans to writes about possibilities for the implementation of the code. Dayla's final article best exemplifies which of Walter Lippmann's press responsibilities?
>
> ◯ Suggest plans.
>
> ◯ Make a current record.
>
> ◯ Make a running analysis.
>
> ◯ Critique the plan.

➔ WATCH VIDEO ON KEY CONCEPTS

LaunchPad includes clips from movies, TV shows, online sources, and other media texts, in addition to interviews with media experts and newsmakers. In the videos for this chapter, you'll consider topics like the impact and importance of community newspapers.

➔ PRACTICE THE CRITICAL PROCESS

Use the Media Literacy Activity to walk through the critical process step-by-step and develop a critical perspective on what media means to you. In this chapter's activity, you'll compare and contrast the experience of reading a newspaper in print and online.

➔ LEARN KEY TERMS

Review the definitions for this chapter's key terms with LaunchPad's easy-to-use online flash cards. Try the cards in practice mode, or quiz yourself as a way to focus your study efforts. (Page numbers indicate where each term is highlighted within the chapter.)

news desert, 225
ghost newspapers, 227
partisan press, 228
penny papers, 228
human-interest stories, 229
wire services, 230
yellow journalism, 230
investigative journalism, 230

objective journalism, 233
inverted-pyramid style, 233
interpretive journalism, 234
literary journalism, 235
citizen journalism, 236
consensus-oriented
    journalism, 237

conflict-oriented
    journalism, 240
underground press, 243
newshole, 244
feature syndicates, 246
newspaper chain, 246
verticals, 251

REVIEW QUESTIONS

Use the questions below to revisit key themes and concepts from this chapter.

 Need more practice? Access the LearningCurve multiple-choice questions for this chapter, which are available on LaunchPad. All questions include feedback for correct and incorrect answers, hints, and links to e-book material.

Newspapers Today

1. What are ghost newspapers, and why are there more of them now than ever before?

How We Got Here: The Early Development of American Newspapers

2. What are the limitations of a press that serves only partisan interests? Why did the earliest papers appeal mainly to more privileged readers?
3. How did newspapers emerge as a mass medium during the penny press era? How did content changes make this happen?
4. What are the two main features of yellow journalism? How did Joseph Pulitzer and William Randolph Hearst contribute to newspaper history?

Competing Models of Modern Newspaper Journalism

5. Why did objective journalism develop? What are its characteristics? What are its strengths and limitations?
6. Why did interpretive forms of journalism develop in the modern era? What are the limits of objectivity?
7. How would you define *literary journalism*? Why did it emerge in such an intense way in the 1960s? How did literary journalism provide a critique of so-called objective news?
8. What is citizen journalism? What are its pros and cons?

The Business, Organization, and Ownership of Newspapers

9. What is the difference between consensus- and conflict-oriented journalism?
10. What role have ethnic, non-English-language, and oppositional newspapers played in the United States?
11. Define *wire service* and *feature syndication*.
12. Why did newspaper chains become an economic trend in the twentieth century?
13. What role do investment firms play as newspaper owners today?

Challenges and Changes Facing Newspapers Today

14. With traditional ownership in jeopardy today, what are some other possible business models for running a newspaper?
15. What are newspaper *verticals*, and what is their appeal to the newspaper industry?

Newspapers and Democracy

16. What is a newspaper's role in a democracy?
17. What makes newspaper journalism different from the journalism of other mass media?

QUESTIONING THE MEDIA

Use these critical-thinking questions to reflect on your own experiences with media and apply your understanding of media concepts.

 For an interactive experience, questions on topics like these are available for the iClicker student response system. Instructors can download the questions from LaunchPad and use them to poll the class.

1. What kinds of stories, topics, or issues are not being covered well by mainstream papers?
2. Why do you think people aren't reading U.S. daily newspapers as frequently as they once did?
3. Discuss whether investment firm owners are ultimately good or bad for the future of journalism.
4. What are *news deserts* and *ghost newspapers* and why should we care about them?
5. What are the roles of nonprofits in sustaining newspaper journalism?

Magazines in the Age of Specialization

9

The magazine industry website Folio got to the point quickly: "America can't get enough of Chip and Joanna."[1] Married couple Chip and Joanna Gaines became a household name with their show *Fixer Upper*, a popular home renovation program that aired on HGTV from 2013 to 2018. On each episode of *Fixer Upper*, Chip and Joanna take clients to visit three potential houses to buy and renovate. Once the client selects one of them, Chip leads the messy work of the renovation while Joanna handles the design and decorating details. The final reveal comes when a billboard-size image of the old building is pulled away to show the homeowners the house's dramatic transformation.

The key to the couple's TV success is their often funny on-screen chemistry. Chip is a jokester who likes to take sledgehammers to walls, and Joanna is the more organized taskmaster, with an eye for getting everything just right for the reveal. Viewers also sense their genuine dedication to their family (the couple has five children).

The Gaineses have translated their television sensation into an empire that goes well beyond cable programming. Chip and Joanna's Magnolia Homes and Magnolia Realty businesses have expanded to Magnolia Market—a renovated market, bakery, and garden that has turned the old industrial silo district of Waco, Texas, into a tourist attraction. Fans can also buy something from the couple's ever-expanding retail empire: Joanna's paint, textile, and furniture lines (including Magnolia Home sections inside Target department stores); an online Magnolia Market shop; and overnight stays at three historic homes in the Waco area that Chip and Joanna renovated and rent to guests for between $545 and $1,295 a night.

In addition to these endeavors, the couple and their brand have overseen another big debut: the creation of *Magnolia Journal*, which has had one of the most successful magazine launches in years. *Magnolia Journal* publishes four times a year and gives America even more Chip and Joanna. The magazine was launched by Meredith Corporation, which publishes titles based on other lifestyle

◀ HGTV's Joanna and Chip Gaines have expanded their media reach with a successful magazine brand, *Magnolia Journal*, and Magnolia Network for Discovery.

257

television celebrities, including *Martha Stewart Living* and *Rachael Ray in Season*. The company describes the magazine as "the first print extension of Joanna and Chip Gaines' powerhouse Magnolia brand. Inspiring readers to create their best homes, families and lives while making every moment count, the magazine covers entertaining, seasonally-driven celebrations, outdoor living, family, food, healthy lifestyle and more—all showcased through the Gaineses' signature rustic, back-to-our-roots aesthetic."[2]

With its base of millions of TV fans, the first issue of *Magnolia Journal* was so popular in October 2016 that Meredith had to go back to press after two weeks to increase the number of copies from 400,000 to 600,000. The company printed 750,000 copies to meet demand for the second issue, and in May 2017, Meredith announced that it would print over one million copies of the third issue: 700,000 just for the magazine's subscribers, and the rest for single-issue newsstand sales. Meredith reports that "the magazine has been a hit particularly with millennials."[3] By 2021, the magazine claimed a circulation of 1.2 million.[4]

One significant way in which the Gaineses built their television and magazine audience is through their social media activity. Joanna and Chip frequently post to Facebook, Twitter, Instagram, and Pinterest. These days, fans following them on social media can learn all about their latest project: the 2021 launch of Magnolia Network for Discovery, which includes new episodes of *Fixer Upper* premiering on streaming channel Discovery+. And, of course, there are plenty of posts throughout the year highlighting what's featured in the next new issue of *Magnolia Journal*.

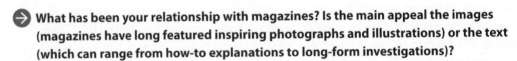 **What has been your relationship with magazines? Is the main appeal the images (magazines have long featured inspiring photographs and illustrations) or the text (which can range from how-to explanations to long-form investigations)?**

Magazines Today

From newer successes like *Magnolia Journal* to long-running titles like *Cosmopolitan* and *National Geographic*, magazines today are often thought of as brands—not unlike how *Magnolia Journal* represents one part of Chip and Joanna Gaines's ever-sprawling constellation of lifestyle media. Magazines today sometimes even exemplify the brands of specific companies or organizations, or take particular points of view on products or issues.

In some ways, magazines have always done this, championing certain political perspectives or lifestyles, or even the idea of a fair-minded democratic forum. Since the 1740s, magazines have played a key role in our social and cultural lives, becoming one of America's earliest forms of national mass media. They created some of the first spaces for discussing the broad issues of the age, including public education, the abolition of slavery, women's suffrage, literacy, and the Civil War.

In the nineteenth century, magazines became an educational forum for women, who were barred from higher education and from the nation's political life. At the turn of the twentieth century, magazines' probing reports paved the way for investigative journalism, while their use of engraving and photography provided a hint of the visual culture to come. Economically, magazines brought advertised products into households, hastening the rise of a consumer society.

Today, more than twenty thousand commercial, alternative, and noncommercial magazines are published in the United States annually. Like newspapers, radio, movies, and television, magazines reflect and construct portraits of American life. They are catalogs of daily events and experiences, but they also show us the latest products, fostering our consumer culture. We read magazines to learn something about our community, our nation, our world, and ourselves.

In this chapter, we will:

- Investigate the history of the magazine industry, highlighting the colonial and early American eras, the arrival of national magazines, and the development of photojournalism
- Focus on the age of muckraking and the rise of general-interest and consumer magazines in the modern American era
- Look at the decline of mass market magazines, the impact of TV and the Internet, and how numerous magazines have expanded their valuable brand identities across multiple media platforms
- Investigate the organization and economics of magazines and their function in a democracy

MAGAZINES increasingly use social media to extend their brands. In this Instagram post, Joanna Gaines promotes the summer 2020 issue of *Magnolia Journal*.

How We Got Here: The Early History of Magazines

The first magazines appeared in seventeenth-century France in the form of bookseller catalogs and notices that book publishers inserted in newspapers. In fact, the word *magazine* derives from the French term *magasin*, meaning "storehouse." The earliest magazines were storehouses of writing and reports taken mostly from newspapers. Today, the word **magazine** broadly refers to collections of articles, stories, and advertisements appearing in nondaily (such as weekly or monthly) periodicals that are usually published in the smaller tabloid format (with a glued or stapled binding on full-color glossy paper), rather than the larger broadsheet newspaper style.

Early British Magazines

The first political magazine, called the *Review*, appeared in London in 1704. Edited by political activist and novelist Daniel Defoe (author of *Robinson Crusoe*), the *Review* was printed sporadically until 1713. Like the *Nation*, the *National Review*, and the *Progressive* in the United States today, early European magazines were channels for political commentary and argument. These periodicals looked like newspapers of the time, but they appeared less frequently and were oriented toward broad domestic and political commentary rather than recent news.

Regularly published magazines or pamphlets, such as the *Tatler* and the *Spectator*, also appeared in England around this time. These publications offered poetry, politics, and philosophy for London's elite, and they served readerships of a few thousand. The first publication to use the term *magazine* was *Gentleman's Magazine*, which appeared in London in 1731 and consisted of reprinted articles from newspapers, books, and political pamphlets. Later, the magazine began publishing original work by such writers as Defoe, Samuel Johnson, and Alexander Pope.

Magazines in Colonial America

Without a substantial middle class, widespread literacy, or advanced printing technology, magazines developed slowly in colonial America. Like the partisan newspapers of the time, these

COLONIAL MAGAZINES
The first issue of Benjamin Franklin's *General Magazine and Historical Chronicle* appeared in January 1741. Although it lasted only six months, Franklin found success in other publications, like his annual *Poor Richard's Almanack*, which first appeared in 1732 and lasted twenty-five years.

magazines served politicians, the educated, and the merchant classes. Paid circulations were low—between one hundred and fifteen hundred copies. However, early magazines did serve the more widespread purpose of documenting a new nation coming to terms with issues of taxation, state versus federal power, treaties with Native Americans, public education, and the end of colonialism. George Washington, Alexander Hamilton, and John Hancock all wrote for early magazines, and Paul Revere worked as a magazine illustrator for a time.

The first colonial magazines appeared in Philadelphia in 1741, about fifty years after the first newspapers. Andrew Bradford's *American Magazine, or A Monthly View of the Political State of the British Colonies* debuted first. Three days later, Benjamin Franklin's *General Magazine and Historical Chronicle* appeared. Bradford's magazine lasted for only three issues, due to circulation and postal obstacles that Franklin, who had replaced Bradford as Philadelphia's postmaster, put in its way. But Franklin's rival magazine did not do much better, surviving for only six months.

Nonetheless, following the Philadelphia experiments, magazines began to emerge in the other colonies, beginning in Boston in the 1740s. The most successful magazines reprinted articles from leading London periodicals, keeping readers abreast of European events. These magazines included New York's *Independent Reflector* and the *Pennsylvania Magazine*, edited by activist Thomas Paine, which included political arguments that helped rally the colonies against British rule. By 1776, about a hundred colonial magazines had appeared and disappeared. Although historians generally consider them dull and uninspired, these magazines helped launch a new medium that caught on after the American Revolution.

U.S. Magazines and Specialization

After the revolution, the growth of the magazine industry in the newly independent United States remained slow. In a communication system where print was the only medium (newspapers, magazines, and books), the U.S. Post Office was the system of media distribution. The Postal Service Act of 1792 set low rates for newspapers, the young nation's preferred news medium. By contrast, delivery costs were set higher for magazines, and some postal carriers refused to carry magazines because of their weight. Only twelve magazines operated in 1800. By 1825, approximately one hundred magazines existed, although another five hundred or so had failed between 1800 and 1825. Nevertheless, during the first quarter of the nineteenth century, most communities had their own weekly magazines. Although these magazines did sell some advertising, they were usually in precarious financial straits due to their small circulations.

As the nineteenth century progressed, the idea of specialized magazines devoted to certain categories of readers developed. Many early magazines were overtly religious and boasted the largest readerships of the day. Literary magazines also emerged at this time. The *North American Review*, for instance, established the work of important writers, such as Ralph Waldo Emerson, Henry David Thoreau, and Mark Twain. In addition to religious and literary magazines, specialty magazines that addressed various professions, lifestyles, and topics also appeared, including the *American Farmer*, the *American Journal of Education*, the *American Law Journal*, *Medical Repository*, and the *American Journal of Science*. Such specialization spawned the modern trend of reaching readers who share a profession, a set of beliefs, cultural tastes, or a social identity, foreshadowing the affordances of technologies in the twentieth and twenty-first centuries that increasingly made the United States a *niche nation*.

General Interest, Women's, and Illustrated Magazines

In addition to the rise of specialized magazines and audiences, the nineteenth century saw the birth of the first **general-interest magazine**, which covered a wide variety of topics aimed at a broad national audience. In 1821, two young Philadelphia printers launched the *Saturday Evening Post*, which became the longest-running magazine in U.S. history. Like most magazines of the day, the early *Post* included a few original essays but "borrowed" many pieces from other sources. Eventually, however, the *Post* grew to incorporate news, poetry, essays, play reviews, and more. The *Post* published the writings of such prominent popular authors as Nathaniel Hawthorne and Harriet Beecher Stowe. Although the *Post* was a general-interest magazine, it was also the first major magazine to appeal directly to women, via its Lady's Friend column, which addressed women's issues.

With increases in literacy and public education, the development of faster printing technologies, and improvements in mail delivery (due to rail transportation), a market was created for more national magazines like the *Post*. Whereas in 1825 one hundred magazines struggled for survival, by 1850 nearly six hundred magazines were being published regularly. (Thousands of others lasted less than a year.)

Besides the move to national circulation, other important developments in the magazine industry were under way. In 1828, Sarah Josepha Hale started the first full magazine directed exclusively to a female audience: the *Ladies' Magazine*. In addition to carrying general-interest articles, the magazine advocated for women's education, work, and property rights. After nine years and marginal success, Hale merged her magazine with its main rival, *Godey's Lady's Book* (1830–1898), which she edited for the next forty years. By 1850, *Godey's*, known for its colorful fashion illustrations as well as its advocacy, achieved a circulation of forty thousand—at the time, the biggest distribution ever for a U.S. magazine. By 1860, circulation had swelled to 150,000, both a testament to its appeal to a growing middle class and a consequence of an 1852 statute that lowered magazine postal rates to the same as those for newspapers. Hale's magazine played a central role in educating working- and middle-class women, who were denied access to higher education throughout the nineteenth century.

Another major development in magazine publishing during the mid-nineteenth century was the arrival of illustration. Like the first newspapers, early magazines were totally dependent on the printed word. By the mid-1850s, drawings, engravings, woodcuts, and other forms of illustration had become a major feature of magazines. During the Civil War, many readers relied on *Harper's New Monthly Magazine* for its elaborate battlefield sketches. Publications like *Harper's* married visual language to the printed word, helping transform magazines into a mass medium. Bringing photographs into magazines took a bit longer. Mathew Brady and his staff of photographers, whose thirty-five hundred photos documented the Civil War, helped popularize photography by the 1860s. But it was not until the 1890s that magazines and newspapers possessed the technology to reproduce photos in print media.

COLOR ILLUSTRATIONS first became popular in the fashion sections of women's magazines in the mid-nineteenth century. The color for this fashion image from *Godey's Lady's Book* was added to the illustration by hand.

North Wind Picture Archives

The Arrival of Mass-Circulation National Magazines

In 1870, about twelve hundred magazines were produced in the United States. By 1890, that number had reached forty-five hundred, and by 1905, six thousand magazines existed (see Figure 9.1). Part of this surge in titles and readership was facilitated by the Postal Act of 1879, which further lowered postage rates for magazines, reducing distribution costs. Meanwhile, even faster presses and advances in mass-production printing, conveyor systems, and assembly lines reduced production costs and made large-circulation national magazines possible.[5]

The combination of reduced distribution and production costs enabled publishers to slash magazine prices. As prices dropped from thirty-five cents to fifteen cents and then to ten cents, the working class was gradually able to purchase national publications. By 1905, there were about twenty-five national magazines, available from coast to coast and serving millions of readers.[6] As jobs and the population began shifting from farms and small towns to urban areas, national magazines helped readers see themselves as part of broader **imagined communities**—that is, as members of a socially constructed group rather than as individuals with only local or regional identities. In addition, the dramatic growth of drugstores and dime stores, supermarkets, and department stores offered new venues and shelf space for selling consumer goods, including magazines.

As magazine circulation began to skyrocket, advertising revenue soared. The economics behind the rise of advertising was simple: A magazine publisher could dramatically expand circulation by dropping the price of an issue below the actual production cost for a single copy. The publisher recouped the loss through ad revenue, guaranteeing large readerships to advertisers who were willing to pay to reach more readers. The number of ad pages in national magazines proliferated. *Harper's*, for instance, devoted only seven pages to ads in the mid-1880s, nearly fifty pages in 1890, and more than ninety pages in 1900.[7]

Social Reform and the Muckrakers

The rise in magazine circulation coincided with rapid social change in America. While hundreds of thousands of Americans moved from the country to the city in search of industrial jobs, millions of new immigrants poured in. Thus, the nation that journalists had long written about had grown increasingly complex by the turn of the century. Many newspaper reporters became dissatisfied with the simplistic and conventional style of newspaper journalism and turned to magazines, where they were able to write at greater length and in greater depth about broader issues.

Printing the fiction and essays of the best writers of the day was one way to keep magazine circulation numbers high, but many magazines also engaged in aspects of *yellow journalism* by crusading for social reform on behalf of the public good. In the 1890s, for example, *Ladies' Home Journal* (*LHJ*) and its editor, Edward Bok, led the fight against unregulated patent medicines (unscientific or quack medicines that often contained nearly 50% alcohol), while other magazines took on poor living and working conditions, exposed unsanitary practices in various food industries, and unveiled government corruption.

In 1902, *McClure's Magazine* (1893–1933) touched off an investigative era in magazine reporting with a series of probing stories, including Ida Tarbell's "History of the Standard Oil Company," which took on John D. Rockefeller's oil monopoly,

FIGURE 9.1

THE GROWTH OF CONSUMER MAGAZINES PUBLISHED IN THE UNITED STATES

Data from: Association of Magazine Media, *Magazine Media Factbook,* www.magazine.org/common /Uploaded%20files/Factbook /MPA-2019-Factbook-ff1-lo.pdf. Historical data from John Tebbel and Mary Ellen Zuckerman, *The Magazine in America, 1741–1990* (New York: Oxford University Press, 1991); Theodore Peterson, *Magazines in the Twentieth Century* (Urbana: University of Illinois Press, 1964).

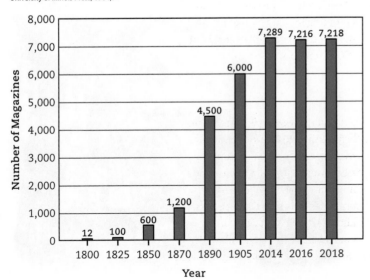

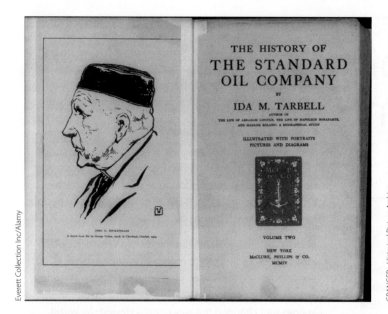

Everett Collection Inc/Alamy

GRANGER - Historical Picture Archive

IDA TARBELL (1857–1944), who is best known for her "History of the Standard Oil Company," once remarked on why she dedicated years of her life to investigating the company: "They had never played fair, and that ruined their greatness for me."[8] For Tarbell and other investigative journalists, or "muckrakers," exposing such corruption was a driving force behind their work.

and Lincoln Steffens's "Shame of the Cities," which tackled urban problems. In 1906, *Cosmopolitan* joined the fray with a series called "The Treason of the Senate," and *Collier's* magazine (1888–1957) developed "The Great American Fraud" series, focusing on patent medicines (whose ads accounted for 30% of the profits made by the American press by the 1890s). Much of this new reporting style was critical of American institutions. Angry with so much negative reporting, in 1906 President Theodore Roosevelt dubbed these investigative reporters **muckrakers** because they were willing to crawl through society's muck to uncover a story. "Muckraking" was a label that Roosevelt used with disdain, but it was worn with pride by reporters such as Ray Stannard Baker, Frank Norris, and Lincoln Steffens.

Influenced by Upton Sinclair's novel *The Jungle*—a fictional account of Chicago's meatpacking industry—and by the muckraking reports of *Collier's* and *LHJ*, in 1906 Congress passed the Meat Inspection Act and the Pure Food and Drug Act. Other reforms stemming from muckraking journalism and the politics of the era include antitrust laws for increased government oversight of business, a fair and progressive income tax, and direct election of U.S. senators.

The Era of General-Interest Magazines

The heyday of the muckraking era lasted into the mid-1910s, when America was drawn into World War I. After the war and through the 1950s, general-interest magazines were the most prominent publications, offering occasional investigative articles but also covering a wide variety of topics aimed at a broad national audience. A key aspect of these magazines was **photojournalism**—the use of photos to document the rhythms of daily life (see "Examining Ethics: The Evolution of Photojournalism"). High-quality photos gave general-interest magazines a visual advantage over radio, which was the most popular medium of the day. In 1920, about fifty-five magazines fit the general-interest category; by 1946, more than one hundred such magazines competed with radio networks for the national audience. The *Saturday Evening Post*, *Reader's Digest*, *Time*, and *Life* were among the most popular magazines of the era.

The Evolution of Photojournalism

BY CHRISTOPHER R. HARRIS

What we now recognize as photojournalism began with the assignment of photographer Roger Fenton, of the *Sunday Times of London*, to document the Crimean War in 1856. However, technical limitations did not allow direct reproduction of photodocumentary images in the publications of the day. Woodcut artists had to interpret the photographic images as black-and-white-toned woodblocks that could be reproduced by the presses of the period. Images interpreted by artists therefore lost the inherent qualities of photographic visual documentation: an on-site visual representation of facts for those who weren't present.

Woodcuts remained the basic method of press reproduction until 1880, when *New York Daily Graphic* photographer Stephen Horgan invented halftone reproduction using a dot-pattern screen. This screen enabled metallic plates to directly represent photographic images in the printing process; now periodicals could bring exciting visual reportage to their pages.

In the late-1890s, Jimmy Hare became the first photographer recognized as a photojournalist in the United States. Taken for *Collier's Weekly*, Hare's photoreportage on the sinking of the battleship *Maine* in 1898 near Havana, Cuba, established his reputation as a newsman traveling the world to bring back images of news events. Hare's images fed into growing popular support for Cuban independence from Spain and eventual U.S. involvement in the Spanish-American War.

In 1888, George Eastman brought photography to the working and middle classes when he introduced the first flexible-film camera from Kodak, his company in Rochester, New York. Gone were the bulky equipment and fragile photographic plates of the past. Now families and journalists could more easily and affordably document gatherings and events.

As photography became easier and more widespread, photojournalism began to take on an increasingly important social role. At the turn of the century, the documentary photography of Jacob Riis and Lewis Hine captured the harsh living and working conditions of the nation's many child laborers, including crowded ghettos and unsafe mills and factories. Reaction to these shockingly honest photographs resulted in public outcry and new laws against the exploitation of children. Photographs also brought the horrors of World War I to people far from the battlefields.

In 1923, visionaries Henry Luce and Briton Hadden published *Time*, the first modern photographic newsweekly; *Life* and *Fortune* soon followed. From coverage of the Roaring Twenties to the Great Depression, these magazines used images that changed the way people viewed the world.

Life, with its spacious 10-by-13-inch format and large photographs, became one of the most influential magazines in America, printing what are now classic images from World War II and the Korean War. Often, *Life* offered images that were unavailable anywhere else: Margaret Bourke-White's photographic proof of the unspeakably horrific concentration camps; W. Eugene Smith's gentle portraits of the humanitarian Albert Schweitzer in Africa; David Duncan's gritty images of the faces of U.S. troops fighting in Korea.

Television photojournalism made its quantum leap into the public mind as it documented the assassination of President Kennedy in 1963. In televised images that were broadcast and rebroadcast, the public witnessed the actual assassination and the confusing aftermath, including live coverage of both the murder of alleged assassin Lee Harvey Oswald and President Kennedy's funeral procession. Photojournalism also provided visual documentation of the turbulent 1960s, including aggressive photographic coverage of the Vietnam War—its protesters and supporters. Pulitzer Prize–winning photographer Eddie Adams shook the emotions of the American public with his photographs of a South Vietnamese general's summary execution of a suspected Vietcong terrorist. Closer to home, shocking images of the Civil Rights movement culminated in pictures of Birmingham police and police dogs attacking Civil Rights protesters.

In the 1970s, new computer technologies emerged and were embraced by print and television media worldwide. By the late 1980s, computers could transform images into digital form and easily manipulate them with sophisticated software programs. Today, a reporter can take a picture

MARGARET BOURKE-WHITE (1904–1971) was a photojournalist of many "firsts": first female photographer for *Life* magazine, first Western photographer allowed into the Soviet Union, first photographer to shoot the cover photo for *Life*, and first female war correspondent. Bourke-White (*right*) was well known for her photos of World War II—including concentration camps—but also for her documentation of the India-Pakistan partition, including a photo of Gandhi at his spinning wheel (*left*).

and within minutes send it to news offices in Tokyo, Berlin, and New York; moments later, the image can be used in a late-breaking TV story or sent directly to that organization's Twitter followers. Such digital technology has revolutionized photojournalism, perhaps even more than the advent of roll film did in the late nineteenth century. Today's photojournalists post entire interactive photo slideshows alongside stories, sometimes adding audio explaining their artistic and journalistic process. Their photographs live on through online news archives and through photojournalism blogs, such as the *Lens* of the *New York Times*, where photojournalists are able to gain recognition for their work and find new audiences.

However, there is a dark side to all this digital technology. Because of the absence of physical film, there is a loss of proof, or veracity, of the authenticity of images. Original film has qualities that make it easy to determine if it has been tampered with. Digital images, by contrast, can be easily altered, and such alteration can be very difficult to detect.

An egregious example of image-tampering involved the Ralph Lauren fashion model Filippa Hamilton. She appeared in a drastically Photoshopped advertisement that showed her hips as being thinner than her head—like a Bratz doll. The ad, published only in Japan, received

intense criticism when the picture went viral. The five-foot-ten, 120-pound model was subsequently dropped by the fashion label because, as Hamilton explained, "they said I was overweight and I couldn't fit in their clothes anymore."[1] In today's age of Photoshop, it is common practice to make thin female models look even thinner and make male models look unnaturally muscled. "Every picture has been worked on, some twenty, thirty rounds," Ken Harris, a fashion magazine photo-retoucher, said. "Going between the retoucher, the client, and the agency, . . . [photos] are retouched to death."[2] And since there is no disclaimer saying these images have been retouched, it can be hard for viewers to know the truth.

Photojournalists and news sources are confronted today with unprecedented concerns over truth-telling. In the past, trust in documentary photojournalism rested solely on the verifiability of images (what you see is what you get). This is no longer the case. Just as we must evaluate the words we read, now we must also take a more critical eye to the images we view.

Christopher R. Harris is a professor emeritus of the Department of Media Arts at Middle Tennessee State University. ●

Saturday Evening Post

Although it had been around since 1821, the *Saturday Evening Post* only achieved a modest circulation in the first seven decades of its life. But in 1897, Cyrus Curtis, who had already made *Ladies' Home Journal* the nation's top magazine, bought the *Post* and remade it into the first widely popular general-interest magazine. Curtis's strategy for reinvigorating the magazine included printing popular fiction and romanticizing American virtues through words and pictures (a *Post* tradition best depicted in the three-hundred-plus cover illustrations by Norman Rockwell). Curtis also featured articles that celebrated the business boom of the 1920s. This reversed the journalistic direction of the muckraking era, in which business corruption was often the focus. By the 1920s, the *Post* had reached two million in circulation, the first magazine to hit that mark, and its circulation grew to 6.3 million in 1960. The magazine stopped production in 1969 (as discussed later in the chapter), but today it has been rebooted by the nonprofit Saturday Evening Post Society and publishes six times a year.

Reader's Digest

The most widely circulated general-interest magazine during this period was *Reader's Digest*. Begun in a Greenwich Village basement in 1922 by Dewitt Wallace and Lila Acheson Wallace, *Reader's Digest* championed one of the earliest functions of magazines: printing condensed versions of selected articles from other magazines. With its inexpensive production costs, low price, and popular pocket-size format, the magazine's circulation climbed to over one million during the Great Depression, and by 1946, it was the nation's most popular magazine. By the mid-1980s, it was the most popular magazine in the world, with a circulation of twenty million in America and ten to twelve million abroad. But with a declining readership and two bankruptcies, its U.S. circulation base had dropped to three million by 2021, although the brand remained strong internationally, with editions in twenty-three countries.

Time

During the general-interest era, national newsmagazines such as *Time* were also major commercial successes. Established in 1923 by Henry Luce and Briton Hadden, *Time* developed a magazine brand of interpretive journalism, assigning reporter-researcher teams to cover stories, after which a rewrite editor would rework the article into a narrative with an interpretive point of view. *Time* had a circulation of 200,000 by 1930, increasing to more than 3 million by the mid-1960s. *Time*'s success encouraged prominent imitators, including *Newsweek* (established in 1933); *U.S. News & World Report* (1948); and, more recently, *The Week* (2001). By two decades into the 2000s, however, economic decline, competition from the web, and a shrinking number of readers and advertisers had taken their toll on the top newsweeklies. Today, *Time* and *The Week* still publish in print but focus more on their digital audiences, which number in the tens of millions. *Newsweek* ceased publication in 2012 but resumed printing in 2014 under new ownership. *U.S. News* switched to an all-digital format in 2010 (and is now most famous for its "America's Best Colleges" reports).

Life

Despite the commercial success of *Reader's Digest* and *Time* in the twentieth century, the magazines that truly symbolized the general-interest genre during this era were the oversized pictorial weeklies *Look* and *Life*. More than any other magazine of its day, *Life* developed an effective strategy for competing with popular radio by advancing photojournalism. Launched as a weekly by Henry Luce in 1936, *Life* appealed to the public's fascination with images (invigorated by the movie industry), radio journalism, and advertising and fashion photography. By the end of the 1930s, *Life* had a **pass-along readership**—the total number of people who come into contact with a single copy of a magazine—of more than seventeen million, rivaling the ratings of popular national radio programs.

    Life's first editor, Wilson Hicks—formerly a picture editor for the Associated Press—built a staff of renowned photographer-reporters who chronicled the world's ordinary and extraordinary events from the late 1930s through the 1960s. Among *Life*'s most famous photojournalists were Margaret Bourke-White, the first female war correspondent to fly combat missions during World War II, and Gordon Parks, who later became Hollywood's first Black director of major feature films. Today, *Life*'s photographic archive is hosted online by Google.

The Fall of General-Interest Magazines

The decline of weekly general-interest magazines, which had dominated the industry for thirty years, began in the 1950s. By 1957, both *Collier's* (founded in 1888) and *Woman's Home Companion* (founded in 1873) had folded. Each magazine had a national circulation of more than four million the year it died. No magazine with this kind of circulation had ever shut down before. Although some critics blamed poor management, both magazines were victims of changing consumer tastes, rising postal costs, falling ad revenues, and, perhaps most important, television, which began usurping the role of magazines as the preferred family medium. There would be even more notable general-interest magazines to fail in the ensuing decades.

Saturday Evening Post, *Look*, and *Life* Expire

Although *Reader's Digest* and women's magazines were not greatly affected by television, other general-interest magazines were. The *Saturday Evening Post* folded in 1969, *Look* in 1971, and *Life* in 1972. At the time, all three magazines were rated in the Top 10 in terms of paid circulation, and each had a readership that exceeded six million per issue. (A look at today's top-selling magazines—see Table 9.1—indicates just how large a readership this was.) Why did these magazines fold? First, to maintain these high circulation figures, their publishers were selling the magazines for far less than the cost of production.

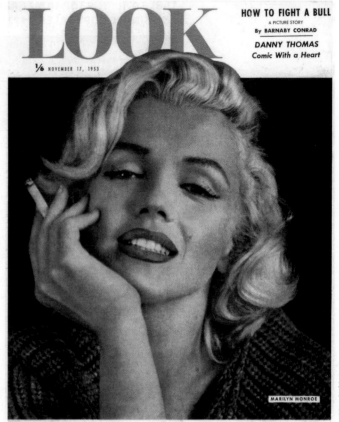

THE RISE AND FALL OF *LOOK* With large pages, beautiful photographs, and compelling stories on celebrities like Marilyn Monroe, *Look* entertained millions of readers from 1937 to 1971, emphasizing photojournalism to compete with radio. By the late 1960s, however, television had lured away national advertisers, postal rates had increased, and production costs had risen, forcing *Look* to fold despite a readership of more than eight million.

Second, the national advertising revenue pie that helped make up the cost differences for *Life*, *Look*, and the *Saturday Evening Post* now had to be shared with network television—and magazines' slices were getting smaller. *Life*'s high pass-along readership meant that it had a larger audience than many prime-time TV shows. But it cost more to have a single full-page ad in *Life* than it did to buy a minute of ad time during evening television. National advertisers were often forced to choose between the two, and in the late 1960s and early 1970s, television seemed a better buy to advertisers looking for the biggest audience.

Third, dramatic increases in postal rates had a particularly negative effect on oversized publications (those larger than the 8-by-10.5-inch standard). In the 1970s, postal rates increased by more than 400 percent for these magazines. *Post* and *Life* cut their circulations drastically to save money. The economic rationale here was that limiting the number of copies would reduce production and postal costs, enabling the magazines to lower their ad rates to compete with network television. Yet with their decreased circulation, these magazines became less attractive to advertisers trying to reach the largest general audience.

TABLE 9.1   THE TOP 10 MAGAZINES (RANKED BY PAID AND NONPAID U.S. CIRCULATION AND SINGLE-COPY SALES, 1972 VERSUS 2020)

1972		2020	
Rank/Publication	**Circulation**	**Rank/Publication**	**Circulation**
1 Reader's Digest	17,825,661	1 AARP The Magazine	23,608,221
2 TV Guide	16,410,858	2 AARP Bulletin	22,669,110
3 Woman's Day	8,191,731	3 Costco Connection	14,550,448
4 Better Homes and Gardens	7,996,050	4 American MainStreet	9,545,242
5 Family Circle	7,889,587	5 Better Homes and Gardens	7,649,079
6 McCall's	7,516,960	6 Game Informer	5,910,071*
7 National Geographic	7,260,179	7 Family Circle	4,035,811*
8 Ladies' Home Journal	7,014,251	8 Good Housekeeping	4,014,028
9 Playboy	6,400,573	9 People	3,502,833
10 Good Housekeeping	5,801,446	10 Reader's Digest	3,029,039

Data from: "Total Paid, Verified and Analyzed Non-paid Circulation," Alliance for Audited Media, June 30, 2020, https://abcas3
.auditedmedia.com/ecirc/magtitlesearch.asp.

* 2019 data; 2020 data not available. *Family Circle* magazine ended publication at the end of 2019.

The general-interest magazines that survived the competition for national ad dollars tended to be women's magazines, such as *Good Housekeeping, Better Homes and Gardens, Family Circle, Ladies' Home Journal*, and *Woman's Day*. These publications had smaller formats and depended primarily on supermarket sales rather than on expensive mail-delivered subscriptions (like *Life* and *Look*). However, the most popular magazines—including *Reader's Digest*—benefited not only from supermarket sales but also from their larger circulations (twice that of *Life*), their pocket size, and their small photo budgets. The failure of the *Saturday Evening Post, Look*, and *Life* as oversized general-audience weeklies ushered in a new era of specialization.

TV Guide Is Born and Points to the Future

While other magazines were just beginning to make sense of the impact of television on their readers, *TV Guide* appeared in 1953 with the novel idea of publishing TV program listings. Taking its cue from the pocket-size format of *Reader's Digest* and the supermarket sales strategy used by women's magazines, *TV Guide* forged a successful path as general-interest magazines declined by addressing a niche interest: the nation's growing fascination with television. The first issue sold a record 1.5 million copies in ten urban markets. Because many newspapers were not yet printing TV listings, by 1962 the magazine had become the first weekly to reach a circulation of eight million, with its seventy regional editions tailoring its listings to TV channels in specific areas of the country.

In 1988, media baron Rupert Murdoch acquired *TV Guide*'s parent company for $3 billion. Murdoch's News Corp. owned the new Fox network, and buying the then influential *TV Guide* ensured that the fledgling network would have its programs listed. In 2005, after years of declining circulation (TV schedules in local newspapers had increasingly undermined its regional editions), *TV Guide* became a full-size entertainment magazine, dropping its smaller digest format and its 140 regional editions. In 2008, *TV Guide*, once the most widely distributed magazine, was sold to a private venture capital firm for $1—less than the cost of a single issue. The TV Guide Network and TVGuide.com—both deemed the more valuable assets—were sold to the film company Lionsgate Entertainment for $255 million in 2009. But even with this decline, *TV Guide*'s decades of success had become a model for many other magazines seeking a market niche in the post-general-interest-magazine environment.

The Domination of Specialization

The general trend away from mass market publications and toward specialty magazines coincided with radio's move to specialized formats in the 1950s. With the rise of television, magazines ultimately reacted the same way that radio and movies did: They adapted. Radio developed formats for older and younger audiences, for rock fans and classical music fans. At the movies, filmmakers focused on more adult subject matter, which was off-limits to television's image as a family medium. And magazines traded their mass audience for smaller, discrete audiences that could be guaranteed to advertisers. This specialization continues today, as the magazine industry focuses on delivering its content online. At least seven of the nation's Top 25 magazines in circulation are now linked to membership in a specialized organization: *AARP The Magazine* and *AARP Bulletin* (for members of AARP), *Costco Connection* (for Costco warehouse club members), *Game Informer* (for a Pro membership card at GameStop stores), *AAA Living* (for members from a multistate region of the automobile organization AAA), *American Rifleman* (one of four magazines for members of the National Rifle Association), and *American Legion* (for members of the American Legion veterans organization). Linking a magazine subscription to organizational membership helps ensure audience loyalty to the magazine in the face of many competing media options.

Magazines are divided into two major groups depending on whether they include advertising: *commercial magazines*, which do include advertising, and *noncommercial magazines* and related periodicals, which do not. Commercial magazines are often further divided by advertiser type: consumer magazines (*In Style, Cosmopolitan*) carry a host of general consumer product ads; business or trade magazines (*Advertising Age, Progressive Grocer*) include ads for products and services for various occupational groups; and farm magazines (*Dairy Herd Management, Dakota Farmer*) contain ads for agricultural products and farming lifestyles. The noncommercial category includes everything from activist newsletters and scholarly journals to business newsletters created by companies for distribution to employees. Magazines such as *Ms., Consumer Reports,* and *Cook's Illustrated,* which rely solely on subscription and newsstand sales, accept no advertising and thus fit into the noncommercial periodical category.

In addition to grouping magazines by advertising approach, we can categorize popular consumer magazine styles by the demographic characteristics of their target audience—such as gender, age, or ethnic group—or by an audience interest area, such as entertainment, sports, literature, or tabloids.

Men's and Women's Magazines

One way the magazine industry competed with television was to reach niche audiences that were not being served by the new medium, creating magazines focused on more adult subject matter. *Playboy*, launched in 1953 by Hugh Hefner, was the first magazine to do this by undermining the conventional values of pre–World War II America and emphasizing previously taboo topics. Scraping together $7,000, Hefner published his first issue, which contained a nude calendar reprint of the actress Marilyn Monroe, along with male-focused articles that criticized alimony payments and gold-digging women. With the financial success of that first issue, which sold more than fifty thousand copies, Hefner was in business.

Playboy's circulation peaked in the 1960s at more than seven million, but it fell gradually throughout the 1970s as the magazine faced competition from imitators and video, as well as criticism for "packaging" and objectifying women for the enjoyment of men. *Playboy* and similar publications continue to publish, but newer men's magazines have shifted their focus to include health (*Men's Health*) and lifestyle (*Details* and *Maxim*).

Women's magazines had long demonstrated that gender-based publications were highly marketable, but during the era of specialization, the magazine industry sought the enormous market of magazine-reading women even more aggressively. *Better Homes and Gardens, Good Housekeeping, Ladies' Home Journal,* and *Woman's Day* focused on cultivating the image of women as homemakers and consumers. In the conservative 1950s and early 1960s, this formula proved to be enormously successful, but as the women's movement advanced in the late 1960s and into the 1970s, women's

magazines grew more contemporary and sophisticated, incorporating content related to feminism (such as in Gloria Steinem's *Ms.* magazine, which first appeared in 1972), women's sexuality (such as in *Cosmopolitan*, which became a young women's magazine in the 1960s [see "Global Village: Cosmopolitan Style Travels the World"]), and career and politics—topics previously geared primarily toward men. By 2020, only *Better Homes and Gardens* and *Good Housekeeping* remained in the Top 10 list of the highest-circulation magazines in the United States (see Table 9.1).

Entertainment, Sports, and Leisure Magazines

In the age of specialization, magazine executives have developed multiple magazines for fans of celebrities, soap operas, running, tennis, golf, hunting, quilting, antiquing, surfing, video games, and many other topics related to entertainment, sports, and leisure. Within those categories, magazines often specialize even further, targeting older or younger runners, men or women golfers, duck hunters or bird-watchers, and midwestern or southern antique collectors.

On the entertainment side, *People* has been a longtime best-seller. Launched in March 1974 by Time Inc. and featuring an abundance of celebrity profiles and human-interest stories, *People* turned a profit in just two years. That success is instructive, particularly because just two years earlier television had helped kill *Life* by draining away national ad dollars. Instead of using a bulky oversized format and relying on subscriptions, *People* chose to downsize—running short articles supported by plenty of photos—and generated most of its circulation revenue from newsstand and supermarket sales. *People* now ranks first in revenue from advertising and circulation sales, at more than $1.5 billion a year, and ranks ninth in the United States in terms of circulation (see Table 9.1). *People*'s success helps explain the host of magazines that try to emulate it, including *Us Weekly*, *Entertainment Weekly*, *In Touch Weekly*, *Star*, and *OK! People* has even spawned its own spin-offs, including *People en Español* and *People StyleWatch*—the latter a low-cost fashion magazine.

The most popular sports and leisure magazine is *Sports Illustrated*. Launched in 1954 by Time Inc., *Sports Illustrated* was initially aimed at well-educated middle-class men. In the years since, it has become the most successful general sports magazine in history, covering everything from major-league sports and mountain climbing to foxhunting and snorkeling. Although frequently criticized for its immensely profitable but exploitative yearly swimsuit edition, *Sports Illustrated* has also done major investigative pieces—for example, on racketeering in boxing and on land conservation. Its print circulation declined to 1.7 million in 2020 under new ownership, but it survived its main rival, *ESPN The Magazine*, which ended its run in 2019.

Founded in 1888, *National Geographic* promoted "humanized geography" and helped pioneer color photography in 1910. It was also the first publication to publish both undersea and aerial color photographs. In addition, many of *National Geographic*'s nature and culture specials on television, which began in 1965, rank among the most popular programs in the history of public television. *National Geographic*'s popularity grew slowly and steadily throughout the twentieth century, reaching one million in circulation in 1935 and ten million in the 1970s. In the late 1990s, its circulation of paid subscriptions slipped to under nine million. Other media ventures (for example, a cable channel and atlases) provided new revenue as circulation for the magazine continued to slide, falling to a little over two million in 2020. Despite its falling circulation, *National Geographic* is often recognized as one of the country's best magazines for its reporting and photojournalism. Today, *National Geographic* competes with other travel and geography magazines, such as *Discover*, *Smithsonian*, *Travel + Leisure*, and *Condé Nast Traveler*. The brand has also been extended to the National Geographic channel (launched in 2001 in the United States), National Geographic Films, and a museum in Washington, D.C.

SPECIALIZED MAGAZINES target a wide range of interests, from mainstream sports to hobbies such as model airplanes. Some of the more successful specialized magazines include *National Geographic* and *Sports Illustrated*.

Joshua Mitchell

Cosmopolitan Style Travels the World

In 1962, Helen Gurley Brown, one of the country's top advertising copywriters, wrote the best-selling book *Sex and the Single Girl*. After proposing a magazine modeled on the book's vision of strong, sexually liberated women, the Hearst Corporation hired her as editor in chief in 1965 to reinvent *Cosmopolitan*, at that time a women's illustrated literary monthly. The new *Cosmopolitan* helped spark a sexual revolution and was marketed to the "Cosmo Girl": a woman aged eighteen to thirty-four with an interest in love, sex, fashion, and her career.

Brown's vision of *Cosmo* continues today. It's the top women's fashion magazine in America, surpassing competitors like *Glamour*, *Marie Claire*, and *Vogue*. Beyond U.S. borders, *Cosmopolitan* is a global publication that publishes in more than eighty countries around the world, including Chile, Argentina, the U.K., Italy, Spain, France, Germany, the Netherlands, Russia, Malaysia, Singapore, Taiwan, Japan, Australia, South Africa, and the Middle East.

Although it looks much the same in every country, with a fashionable young woman on the cover and racy headlines ("Sofort Super Sex!" in Germany and "¡Ohhhh! Verdades y Mentiras sobre el Orgasmo" in Spain), some countries have to make adjustments for regional standards. For example, the editor of *Cosmopolitan Middle East*, which is mostly read in Dubai and Beirut, says, "We can't even publish the word 'sex,' . . . but [readers] are still thinking about sex and they're still wanting to meet a guy and they are still dating. We've just got to wrap it up and present it in a slightly more conservative fashion."[1] In Chile, the editor of *Cosmopolitan* says she covers contraception and women's reproductive health issues "a lot" because Chile has some of the most restrictive abortion laws in the world.[2] Oddly, news about one topic seems to transcend all cultural boundaries and be universally acceptable for *Cosmopolitan* international editions: the Kardashians.

The internationally known *Cosmopolitan* brand didn't always exist. In fact, *Cosmopolitan* had at least four format changes of much more limited international appeal before Helen Gurley Brown came along. The magazine was launched in 1886 as an illustrated monthly for the modern family (meaning it was targeted at married women), with articles on cooking, childcare, household decoration, and occasionally fashion, featuring illustrated images of women in the hats and high collars of the late-Victorian style.[3]

But the magazine was thin on content and almost folded. *Cosmopolitan* was saved in 1889, when journalist and entrepreneur John Brisben Walker gave it a second chance as an illustrated magazine of literature and insightful reporting. The magazine featured writers like Edith Wharton, Rudyard Kipling, and Theodore Dreiser, and it serialized entire books, including H. G. Wells's *The War of the Worlds*. And Walker, seeing the success of contemporary newspapers in New York, was not above stunt reporting. When Joseph Pulitzer's *New York World* sent reporter Nellie Bly to travel the world in less than eighty days in 1889 (challenging the premise of Jules Verne's 1873 novel, *Around the World in Eighty Days*), Walker sent reporter Elizabeth Bisland around the world in the opposite direction for a more literary travel account.[4] Walker's leadership turned *Cosmopolitan* into a respected magazine with increased circulation and a strong advertising base.

Walker sold *Cosmopolitan* at a profit to William Randolph Hearst (Pulitzer's main competitor) in 1905. Under Hearst, *Cosmopolitan* had its third rebirth—this time as a muckraking magazine. As magazine historians explain, Hearst was a U.S. representative who "had his eye on the presidency and planned to use his newspapers and the recently bought *Cosmopolitan* to stir up further discontent over the trusts and big business."[5] *Cosmopolitan's* first big muckraking series, David Graham Phillips's "The Treason of the Senate" in 1906, didn't help Hearst's political career, but it did boost the circulation of the magazine by 50 percent and was reprinted in Hearst newspapers for even more exposure.

By 1912, the progressive political movement that had given impetus to muckraking journalism was waning. *Cosmopolitan*, in its fourth incarnation, became a version of its former self—an illustrated literary monthly targeted to women, featuring short stories and serialized novels by such popular writers as Damon Runyon, Sinclair Lewis, and Faith Baldwin.

Cosmopolitan had great success as an upscale literary magazine, but by the early 1960s, the format had become outdated, and readership and advertising were on the decline. At this point, the magazine had its most radical makeover and became, under Brown's reign, the *Cosmo* we and young women around much of the world know today. ●

Magazines for the Ages

In the age of specialization, magazines have further delineated readers along ever-narrowing age lines, catering increasingly to very young readers and older readers, groups often ignored by mainstream television.

The first children's magazines appeared in New England in the late eighteenth century. Ever since, magazines such as *Youth's Companion*, *Scout Life* (the Boy Scouts' national publication since 1911, titled *Boys' Life* until 2021), *Highlights for Children*, and *Ranger Rick* have successfully targeted children of preschool and elementary-school age. The ad-free and subscription-only *Highlights for Children* topped the children's magazine category in 2020 with a circulation of about two million. In the popular arena, the leading female teen magazines have shown substantial growth; the top magazine for thirteen- to nineteen-year-olds is *Seventeen*, with a circulation of two million in 2020. (For a critical take on fashion magazines, see "Media Literacy and the Critical Process: Uncovering American Beauty.")

In targeting audiences by age, the most dramatic success has come from magazines aimed at readers over fifty, America's fastest-growing age segment. These publications have tried to meet the cultural interests of older Americans, who historically have not been prominently featured in mainstream consumer culture. AARP and its *AARP The Magazine* were founded in 1958 by retired California teacher Ethel Percy Andrus. Subscriptions to the bimonthly *AARP The Magazine* and the monthly *AARP Bulletin* come free when someone joins AARP and pays the modest membership fee ($16 in 2021). By the early 1980s, *AARP The Magazine*'s circulation was approaching seven million. However, with AARP signing up thirty thousand new members each week by the late 1980s, both *AARP The Magazine* and *AARP Bulletin* overtook *TV Guide* and *Reader's Digest* as the top circulated magazines. By 2020, both of these publications were reaching more than twenty-two million people a month, far surpassing the circulations of all other magazines (see Table 9.1). Article topics cover a range of lifestyle, travel, money, health, and entertainment issues, such as sex at age fifty-plus, secrets for spectacular vacations, and how poker can create a sharper mind.

IN SEPTEMBER 2020, *Vanity Fair* featured a cover story about Breonna Taylor, whose death by police gunfire in Louisville, Kentucky, earlier that year spurred months of protests. Taylor was a twenty-six-year-old medical technician at the time of her death. The cover image of Taylor was painted by Amy Sherald, the artist who painted the official portrait of former first lady Michelle Obama, which now hangs in the National Portrait Gallery.

THE GREAT FIRE: A Special Issue, Edited by TA-NEHISI COATES

VANITY FAIR

BREONNA TAYLOR
A BEAUTIFUL LIFE
Painting by
AMY SHERALD

Elite Magazines

Although long in existence, *elite magazines* grew in popularity during the age of specialization. Elite magazines are characterized by their combination of literature, criticism, humor, and journalism and by their appeal to highly educated audiences, often living in urban areas. Among the numerous elite publications that grew in stature during the twentieth century were the *Atlantic Monthly* (now the *Atlantic*), *Vanity Fair*, and *Harper's*.

However, the most widely circulated elite magazine is the *New Yorker*. Launched in 1925 by Harold Ross, the *New Yorker* became the first city magazine aimed at an upscale national audience. Over the years, the *New Yorker* has featured many of the twentieth century's most prominent biographers, writers, reporters, and humorists, including James Baldwin, Dorothy Parker, Lillian Ross, John Updike, E. B. White, and Garrison Keillor, as well as James Thurber's cartoons and Ogden Nash's poetry. By the mid-1960s, the *New Yorker*'s circulation hovered around 500,000; by 2020, its print circulation stood at 1.2 million. Along with the *New York Times*, the *New Yorker* in 2017 broke the story of dozens of women's charges of sexual harassment, assault, and rape against film producer Harvey Weinstein. The reports won Pulitzer Prizes for each publication and spurred a global #MeToo movement of women sharing their stories of workplace harassment, assault, and rape.

MEDIA LITERACY & THE CRITICAL PROCESS

Uncovering American Beauty

How do the United States' leading fashion magazines define beauty? One way to explore this question is by critically analyzing the covers of one of these magazines.

1 DESCRIPTION

Review a year's worth of recent covers from the same fashion magazine, such as *Allure, Cosmopolitan, Elle, Harper's Bazaar, InStyle, Marie Claire, Vanity Fair,* or *Vogue.* Covers of these magazines typically feature a head or body shot of a female model or celebrity, often dressed or posed provocatively and positioned against a solid-color background. (Some magazines, like *Vanity Fair,* occasionally feature male cover subjects. You can also review *GQ* or *Esquire* covers to analyze how men's beauty is defined.) Once you've looked at covers from a recent year, review a year's worth of covers from the same magazine from a decade earlier.

2 ANALYSIS

In comparing covers from a recent year with those from a decade earlier, what significant patterns do you notice? One thing to consider is race: There is likely a disproportionate number of white cover models, though you may also notice that things are improving somewhat in this regard, particularly in response to the #BlackLivesMatter movement. Be sure to consider potential issues of *tokenism* (are people of color featured only rarely?) and *colorism* (are the people of color featured more light-skinned than dark-skinned?).[1] In addition to race, there may be even more consistent patterns regarding body type.

3 INTERPRETATION

What do the patterns you've identified mean? For example, years ago, in an article commending *Elle* magazine for putting "plus-size comedic actress Melissa McCarthy" on its cover, the *Philadelphia Inquirer* called it a rare case of breaking the norm, asserting that "skinny and sexy sells magazines."[2] Is that still the norm? What seems to be the norm in terms of the cover models' race and shade of skin color? If you've chosen to review men on magazine covers, what do these covers say about male beauty? What does it mean when some people are excluded from being featured on the cover of a magazine?

4 EVALUATION

Magazines use cover models as aspirational objects for their readers—that is, as the people its readers would like to look like. Thus, these cover models can become the ideal image of beauty for the magazine's readers. Is this "ideal" of beauty expansive or limiting? Does it establish a certain body type, race, skin tone, or other feature as the standard of beauty?

5 ENGAGEMENT

Write a letter to the editor of the magazine you reviewed to share your findings and provide feedback—for example, to commend the magazine for progress it's made in representing more diverse people on its covers, or perhaps to critique a lack of progress made over the past decade. As an alternative, share your findings visually with the world via Instagram, a TikTok video, or another type of social media post.

Magazines Targeting Race, Ethnicity, and Sexuality

Magazines targeting readers of different demographics have existed since before the Civil War, including the African American antislavery magazines *Emancipator, Liberator,* and *Reformer.* One of the most influential early African American magazines, the *Crisis,* was founded by W. E. B. Du Bois in 1910 and is the official magazine of the NAACP (National Association for the Advancement of Colored People).

In the modern age, the major magazine publisher for African Americans was John H. Johnson. While working on a newsletter at a Chicago life insurance company, Johnson got the idea of starting a publication for Black readers. In 1942, Johnson launched *Negro Digest* on $500 borrowed against his mother's furniture. By 1945, the *Digest* had a circulation of more than 100,000, and its profits enabled Johnson and a small group of editors to start *Ebony,* a picture-text magazine modeled on *Life* but serving Black readers. The Johnson Publishing Company also successfully introduced *Jet,* a pocket-size supermarket magazine, in 1951. In 2016, Johnson Publishing Company sold *Ebony* and *Jet* (which had gone completely digital in 2014) to another company, which suspended *Ebony*'s

SELECTA is an upscale fashion magazine targeted at Hispanic women. The magazine is published in Spanish, although it maintains Twitter and Instagram feeds in English. Other popular magazines aimed at this audience include *Latina* and *People en Español*.

publication in 2019. Meanwhile, Johnson Publishing declared bankruptcy in 2019, but a group of nonprofit foundations bought its historic photo archive of four million prints and negatives, deemed "the most significant collection of photography depicting African-American life in the 20th century," for $30 million and planned to donate the archive to public institutions.[9] *Essence*, the first major magazine geared toward Black women, debuted in 1969. It remains in print and has a circulation of more than one million.

With increases in the Hispanic population and immigration, magazines appealing to Spanish-speaking readers have developed rapidly since the 1980s. In 1983, Armando de Armas's Hispanic Magazine Network began distributing Spanish-language versions of mainstream American magazines, including *Cosmopolitan en Español*; *Harper's Bazaar en Español*; and *Ring*, the prominent boxing magazine. These magazines target the most upwardly mobile segments of the growing American Hispanic population, which numbered more than sixty million—about 18.5 percent of the U.S. population—by 2020. Today, *People en Español* and *Vanidades* are among the top Hispanic magazines by ad revenue.

Although national magazines aimed at other groups were slow to arrive, there are magazines now that target virtually every race, culture, and ethnicity, including *Banana* (contemporary Asian culture), *Native Peoples*, and *Tikkun* (a progressive Jewish magazine).

Magazines also specialize in LGBTQ+ issues. *The Advocate*, founded in 1967 as a twelve-page newsletter, was the first major magazine to address issues of interest to gay men and lesbians. In the ensuing years, *The Advocate* has published some of the best journalism about antigay violence, policy issues affecting the LGBTQ+ community, and AIDS—topics often not well covered by the mainstream press. *Out* is the top LGBTQ+ style magazine. Both are owned by Pride Media, which also owns *Pride, Out Traveler, Chill*, and *Plus*. Other notable LGBTQ+ magazines include *Passport, GO, Curve*, and *FTM*.

Supermarket Tabloids

With headlines like "Obama Birth Certificate Is Fake!," "Hillary: 6 Months to Live" (published in 2015), and "Miracle After 67 Years of Blindness . . . Stevie Wonder Can See Again!," **supermarket tabloids** push the limits of both decency and credibility. Tabloid history can be traced to newspapers' use of graphics and pictorial layouts in the 1860s and 1870s, but the modern U.S. tabloid began with the founding of the *National Enquirer* by William Randolph Hearst in 1926. The *Enquirer* struggled until it was purchased in 1952 by Generoso Pope, who originally intended to use it to "fight for the rights of man" and "human decency and dignity."[10] In the interest of profit, however, Pope settled on the "gore formula" to transform the paper's anemic weekly circulation of seven thousand: "I noticed how auto accidents drew crowds and I decided that if it was blood that interested people, I'd give it to them."[11]

By the mid-1960s, the *Enquirer*'s circulation had jumped to over one million through the publication of bizarre human-interest stories, gruesome murder tales, violent accident accounts, unexplained-phenomena stories, and malicious celebrity gossip. By 1974, the magazine's weekly circulation had topped four million. Its popularity inspired the creation of other tabloids—like *Globe* (founded in 1954) and *Star* (founded by News Corp. in 1974)—and the adoption of a tabloid style by general-interest magazines such as *People* and *Us Weekly*. In 2019, after the *National Enquirer* was caught in its own scandal of paying to hide sexual misconduct allegations against Donald Trump, New York–based American Media, Inc., sold the magazine, along with *Globe* and the *National Examiner*, to the Hudson Group, a travel retailer and bookstore.

The Organization and Economics of Magazines

Given the great diversity in magazine content and ownership, it is hard to offer a common profile of a successful magazine. However, large or small, online or in print, most magazines deal with the same basic functions: production, content, ads, sales, and circulation. In this section, we discuss how magazines operate, the ownership structure behind major magazines, how smaller publications fulfill niche areas that even specialized magazines do not reach, and how magazines create specialized branded content to better connect with audiences and advertisers.

Inside Magazines: Creating Branded Content

The traditional magazine world of print publications featuring display advertisements has been the staple of the magazine industry for more than a century. Today, however, the magazine industry includes not just paper publications but brands that reach audiences in a variety of formats, such as print and digital editions, desktop/laptop computer platforms, the mobile web, and video. The relative popularity of these various platforms is shifting quickly, as Figure 9.2 illustrates. Mobile web and video content continue to expand their share of the magazine market, while print and digital editions and web (desktop/laptop) platforms are in decline. The good news for the magazine industry is that although the platforms used to reach its audience are changing rapidly, its total monthly brand audience—the number of people who experience magazines across all platforms—increased from 1.6 billion to 2 billion from 2014 to 2019. That means that magazine content remains popular, even if the methods used to deliver that content are different than they used to be.

Reconfigured as digital brands, magazines and their publishers offer multiple opportunities for advertisers to connect with audiences. For example, Condé Nast, a leading publisher, states, "We connect consumers to their passions, to the culture and ultimately to you, our marketing partners."[12] If that sounds somewhat vague, it's because new opportunities for magazine brand advertising are still being invented. So far, in addition to traditional display ads in its print and digital magazine versions, Condé Nast offers its advertising customers data about its digital readers, opportunities for **branded content** (specialized print, online, or video content produced and funded by individual advertisers), paid social media placements on platforms like Facebook and Instagram, custom video stories and features, and sponsored events. For example, *Condé Nast Traveler* worked with client Bank of America to develop "The Joy Index," a list of the ten most joyful places to visit. (In case you are wondering, Newfoundland, Canada; Shanghai, China; and Aarhus, Denmark topped the list.) The project created

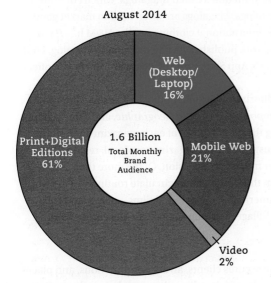

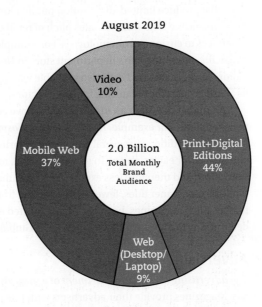

FIGURE 9.2

MAGAZINE BRAND AUDIENCE SHARE BY FORMAT, 2014–2019

Data from: "Magazine Media 360°: A Five-Year Review of Magazine Brand Vitality," December 5, 2019, www .magazine.org/MPA/Research/Five _Year_Review/Magazine/Research _Pages/MM360_Whitepaper.aspx.

content for the magazine and gave Bank of America a high-concept branded magazine feature to associate with its credit card service.[13]

By offering these varied advertising and sponsorship opportunities, magazine publishers (even nonprofits) hope to generate new revenue to replace declining print display ad earnings and subscription fees.

Departments and Duties

How are magazines created? Within most larger magazines, dedicated departments—including editorial and production, advertising and sales, and circulation and distribution—carry out various functions to get print and digital magazine content to readers on a regular basis.

Editorial and Production

The lifeblood of a magazine is its *editorial department*, which produces the magazine's content, excluding advertisements, and the *production staff* who design and compose print and digital editions and handle the technical side of other platforms, such as video. Like newspapers, most magazines have a chain of command that begins with a publisher and extends to the editor in chief, the managing editor, and a variety of subeditors. These subeditors oversee such functions as photography, illustrations, reporting and writing, copyediting, layout, and print and multimedia design. Magazine writers generally include contributing staff writers, who are specialists in certain fields, and freelance writers: non-staff professionals who are assigned to cover particular stories or a certain region of the country. Many magazines, especially those with small budgets, also rely on well-written unsolicited manuscripts to fill their pages. Most commercial magazines, however, reject more than 95 percent of unsolicited pieces.

Within editorial and production are newer sections that develop digital content built on the same magazine brand, including subscription newsletters, podcasts, and video. Video, in particular, has been a growth area for magazines, and it offers new ways to interact with audiences. For example, in 2020, *Vanity Fair*, one of the leading magazines to deploy video features, released a video interview of Billie Eilish for the fourth straight year in which she answered the same set of questions as in previous years— an amazing record of her thoughts on being a music superstar. One reader commented that the annual release of a new Eilish interview video "is basically a national holiday at this point."[14] *Allrecipes*, established in 1997 and owned by Meredith, and *Epicurious*, founded in 1995 by Condé Nast (and affiliated with Condé Nast brand *Bon Appétit* since 2015), are leading online food and recipe magazine sites. Both have flourished, as people increasingly search online for recipes and cooking videos.

Editorial and production staffs may also work on magazines that are produced exclusively for the consumers of a specific product or service. Many magazine publishers have custom publishing divisions that produce limited-distribution publications, sometimes called **magalogs**, which combine a glossy magazine style with the sales pitch of retail catalogs. Magalogs are often used to market goods or services to customers or employees. For example, international clothing retailer H&M has *H&M Magazine*, Bloomingdale's department store in New York publishes its *Bloomingdale's* magalog, Ford Motor Co. produces *My Ford* (also available via Apple's App Store), and Midwest grocer Hy-Vee offers its *Seasons* magazine, featuring food and recipes.

A number of American magazines also have editorial and production offices to serve international editions. For example, *Reader's Digest*, *Cosmopolitan*, *National Geographic*, and *Time* produce international editions in several languages, tailoring the familiar style of each magazine with stories and images for each country. In general, though, most American magazines are local, regional, or specialized and therefore less exportable than either movies or television. Of the more than seven thousand consumer titles, fewer than 2 percent from the United States circulate routinely in the world market. Such magazines, however—like exported American TV shows and films—play a key role in determining the look of global culture (see "Global Village: Cosmopolitan Style Travels the World").

Advertising and Sales

The *advertising and sales department* of a magazine secures clients, arranges promotions, and places ads. Magazines provide their advertisers with rate cards, which indicate how much they charge for a certain amount of advertising space on a printed page. A top-rated consumer magazine like *People* might charge more than $420,000 for a full-page color ad and about $135,400 for a one-third page black-and-white ad. However, in today's competitive world, most rate cards are not very meaningful: Almost all magazines offer 25 to 50 percent rate discounts to advertisers, particularly when they buy ads in multiple issues. Although fashion and general-interest magazines carry a higher percentage of ads than do political or literary magazines, the average print magazine contains about 40 percent ad copy and 60 percent editorial material, a figure that has remained fairly constant for the past decade.

As television advertising siphoned off national ad revenues in the 1950s, magazine publishers developed more creative ways to sell ads, including producing different editions of their print magazines to attract advertisers:

- **Regional editions** are national magazines whose content is tailored to the interests of different geographic areas. For example, *Sports Illustrated* often prints several regional versions of its College Football Preview and March Madness Preview editions, picturing regional stars on each of the covers.
- In **split-run editions**, the editorial content remains the same, but the magazine includes a few pages of ads purchased by local or regional companies. Most editions of *Time* and *Sports Illustrated*, for example, contain a number of pages reserved for regional ads.
- **Demographic editions**, meanwhile, are editions of magazines targeted to particular groups of consumers. In this case, market researchers identify subscribers primarily by occupation, class, and zip code. *Time* magazine, for example, developed special editions of its magazine to target high-income professional/managerial households. Demographic editions guarantee advertisers a particular magazine audience, one that enables them to pay lower rates for their ads because the ads will be run in only a limited number of copies of the magazine. The magazine can then compete with advertising in regional television or cable markets and in newspaper supplements.

Magazines today also sell digital ads and ads incorporated into their videos and podcasts, and profit from relationships that let other digital ad vendors sell advertisements alongside their digital content.

A few contemporary magazines, such as *Highlights for Children*, have decided not to carry ads and rely solely on subscriptions and newsstand sales. To protect the integrity of their various tests and product comparisons, *Consumer Reports* and *Cook's Illustrated* carry no advertising. To strengthen its editorial independence, *Ms.* magazine abandoned ads in 1990 after years of pressure from the food, cosmetics, and fashion industries to feature recipes and more complimentary stories.

Some advertisers and companies have canceled ads after a magazine featured an unflattering or critical article about a particular company or industry, an outcome that can put enormous pressure on editors not to offend advertisers.[15] In fact, the cozy relationships between some advertisers and magazines have led to a dramatic decline in investigative reporting, once central to popular magazines during the muckraking era.

Circulation and Distribution

Although the Internet was initially viewed as the death knell of print magazines, the industry now embraces it as a key way to distribute content. The Internet has become the place where print magazines like *Time* and *Entertainment Weekly* can extend their reach; where magazines like *FHM*, *PCWorld*, *Glamour*, and *O: The Oprah Magazine* can survive when their print version ends; and where online magazines like *Salon*, *Slate*, and *Wonderwall* can exist exclusively. Given the costs of paper, printing, and postage, and the flexibility of the web, mobile devices, and social media, magazines are increasingly being distributed across multiple digital formats. For example, while *Wired* has a print circulation of about 880,000 and estimates that its print edition reaches 3.6 million readers monthly, its digital version has 12.7 million unique users, its social media pages have 19.3 million followers, and its video views exceed 100 million.[16]

As Figure 9.3 demonstrates, certain magazines that might not have the highest annual print circulation figures are nevertheless thriving in the newer distribution platforms of web, mobile, and video. Other models for magazine distribution, such as the Apple News+ app, offer a subscription plan, with hundreds of weekly and monthly titles accessible for a monthly fee.

Print+Digital		Web		Mobile		Video	
MAGAZINE BRAND	**AUDIENCE**	**MAGAZINE BRAND**	**UNIQUE VISITORS**	**MAGAZINE BRAND**	**UNIQUE VISITORS**	**MAGAZINE BRAND**	**UNIQUE VISITORS**
1 AARP The Magazine	36,968	1 WebMD Magazine	11,682	1 People	50,543	1 Vogue	12,857
2 People	31,779	2 Allrecipes	10,134	2 Allrecipes	37,304	2 Vanity Fair	12,682
3 Better Homes and Gardens	29,465	3 The Atlantic	6,570	3 WebMD Magazine	31,393	3 GQ Gentlemen's Quarterly	10,259
4 National Geographic	28,106	4 Good Housekeeping	5,153	4 Good Housekeeping	29,333	4 Good Housekeeping	9,429
5 Sports Illustrated	17,014	5 People	5,109	5 The Atlantic	22,263	5 Architectural Digest	9,314
6 Good Housekeeping	16,827	6 Taste of Home	4,710	6 Taste of Home	20,465	6 Wired	7,986
7 Reader's Digest	16,533	7 AARP The Magazine	4,032	7 US Weekly	16,988	7 Bon Appetit	7,116
8 Time	15,185	8 Time	3,845	8 Cosmopolitan	16,886	8 Elle	5,373
9 Southern Living	14,740	9 Sports Illustrated	3,008	9 Women's Health	15,664	9 Allure	5,099
10 Food Network Magazine	13,108	10 New Yorker	2,969	10 Sports Illustrated	14,461	10 Allrecipes	4,745

FIGURE 9.3

THE MAGAZINE INDUSTRY'S LEADING BRANDS—MONTHLY AUDIENCE ACROSS PLATFORMS

Data from: Alliance for Audited Media, "Magazine Media 360°: Top 10 Magazine Brands," September 2020, https://f.hubspotusercontent10.net /hubfs/1932461/MM360/MM360 -Top10-Q3-Sept2020.pdf.

Of course, paper is still a leading format, and subscriptions account for more than 95 percent of print magazine distribution now that sales of more expensive single copies at retailers have declined. One tactic used by magazine *circulation departments* to increase subscription revenue is to encourage consumers to renew well in advance of their actual renewal dates. Magazines can thus invest and earn interest on early renewal money as a hedge against consumers who drop their subscriptions. Other strategies include **evergreen subscriptions**—those that automatically renew on a credit card account unless subscribers request that the automatic renewal be stopped—and *controlled circulations*, which provide readers with a magazine at no charge by targeting captive audiences, such as airline passengers or association members. These magazines' financial support comes solely from advertising or corporate sponsorship.

Major Magazine Chains

While there are many specialized magazines, only a few *magazine chains*, or large ownership groups, control the best-known brands in the magazine industry.

The Meredith Corporation, based in Des Moines, Iowa, specializes in women's lifestyle magazines, and it became the world's largest magazine publisher with its 2018 purchase of Time Inc. The $1.8 billion deal put the eponymously named *Time* magazine and other popular titles—*People, Sports Illustrated, Fortune, Golf Magazine, Travel + Leisure, Real Simple, Entertainment Weekly*—under the same corporate umbrella as Meredith's magazine titles, including *Better Homes and Gardens, Family Circle, Midwest Living*, and *Rachael Ray in Season*. Because Meredith's focus on women's lifestyle magazines did not mesh with all of its new Time Inc. magazines, it subsequently sold *Golf Magazine, Time, Sports Illustrated, Fortune*, and *Money*. Meredith's properties also include sixteen local television stations in the United States.

The Hearst Corporation, the leading magazine (and newspaper) chain in the early twentieth century, still remains a formidable publisher, with titles like *Cosmopolitan, Esquire, Elle, Car and Driver*, and *HGTV Magazine*. In 2018, Hearst bought Rodale, a family-owned magazine company that published health and wellness titles such as *Prevention, Runner's World*, and *Men's Health*. Hearst also owns interests in cable channels A&E, History, Lifetime, and ESPN; thirty local television stations; and a number of major newspapers.

Long a force in upscale consumer magazines, Condé Nast is a division of Advance Publications, which operates the Newhouse newspaper chain. The Condé Nast group controls several key magazines, including *Vanity Fair, GQ*, and *Vogue*.

Independently Owned Magazines

Although most well-known magazines are associated with the big ownership chains like Meredith or Hearst, a number of independently owned alternative magazines have long records of publication.

APPLE NEWS+, launched in 2019, is a subscription-based app featuring curated content from about three hundred magazines and newspapers. Available only on Apple and Mac devices for a monthly fee, the app provides full access to an array of digital magazines, including *People, National Geographic*, and *Sports Illustrated*, as well as featured articles and audio stories.

Alternative magazines have historically defined themselves in terms of politics—published by either the Left (the *Progressive*, *In These Times*, the *Nation*) or the Right (the *National Review*, *American Spectator*, the *Weekly Standard*). However, what constitutes an alternative magazine has broadened over time to include just about any publication considered outside the mainstream, ranging from environmental magazines to alternative lifestyle magazines to punk-zines—the magazine world's answer to punk rock. (**Zines**, pronounced "zeens," is a term used to describe self-published magazines.)

Occasionally, alternative magazines have become marginally mainstream. For example, during the conservative Reagan era in the 1980s, William F. Buckley's *National Review* saw its circulation swell to more than 100,000—enormous by alternative standards—only to grow to 170,390 in 2011 as a conservative counterpoint to President Barack Obama. However, the magazine didn't support Donald Trump as a Republican candidate, and it was ridiculed by him after he won the presidency. By 2021, the magazine's circulation had shrunk to 75,000.[17] On the Left, *Mother Jones* (named after labor organizer Mary Harris Jones), which champions muckraking and investigative journalism, had a circulation of about 185,000 in 2020 and about 8 million monthly readers across all platforms.[18]

Most alternative magazines, however, are content to swim outside the mainstream. These are the small magazines that typically include diverse political, cultural, religious, international, and environmental subject matter, such as *Against the Current*, *Y'all*, *Buddhadharma*, *Hot VWs*, *Jewish Currents*, *Small Farmer's Journal*, and *Humor Times*.

Magazines in a Democratic Society

Like other mass media, magazines are a major part of the cluttered media landscape. To stay competitive, the magazine industry has become fast-paced and high-risk, with frequent magazine launches and almost as many failures.

As an industry, magazine publishing—like advertising and public relations—has played a central role in transforming the United States from a producer society to a consumer society. Since the 1950s, though, individual magazines have not had the powerful national voice they once possessed, uniting different communities around important issues such as abolition and suffrage. Today, with so many specialized magazines appealing to distinct groups of consumers, magazines play a much-diminished role in creating a sense of national identity.

Contemporary commercial magazines provide essential information about politics, society, and culture, thus helping us think about ourselves as participants in a democracy. Unfortunately, however, these magazines have often identified their readers as consumers first and citizens second. With magazines dependent on advertising, and some of them being primarily *about* the advertising, controversial content sometimes has difficulty finding its way into magazine content. Increasingly, magazines are defining their readers merely as viewers of displayed products and purchasers of material goods. High-revenue magazines, especially those focusing on fashion, fitness, and lifestyle, can also shamelessly break down the firewall between the editorial and business departments. "Fluff" story copy serves as a promotional background for cosmetic, clothing, and gadget advertisements. Many digital magazines further break down that firewall, designing their pages so that a single click on a story or an image links readers to a page where they can purchase the item they clicked on.

At the same time, magazines have arguably had more freedom than other media to encourage and participate in democratic debates. More magazines circulate in the marketplace than do broadcast or cable television channels or yearly Hollywood releases, and good magazines can usually offer more analysis of and insight into society than other media outlets can. Moreover, many new magazines play an important role in uniting dispersed groups of readers, often giving underrepresented cultural groups, newly arrived immigrants, and other alternative groups a sense of membership in imagined communities, which sometimes coalesce into actual national political movements—for example, Black magazines that fostered the Civil Rights movement. In the midst of today's swirl of images, magazines and their advertisements certainly contribute to the commotion. But good magazines also maintain our connection to words, sustaining their vital role in our digital culture.

Chapter 9 REVIEW

LaunchPad
macmillan learning

Visit LaunchPad for *Media & Culture* at **launchpadworks.com** for additional learning tools:

➜ QUIZ YOURSELF WITH LEARNINGCURVE

LearningCurve uses gamelike quizzing to help you review important concepts from this chapter. LearningCurve includes multiple-choice questions at different levels of difficulty, hints, feedback for both right and wrong answers, and links to e-book material for easy reference. Which answer choice would you select for the sample question below?

> Martin works in the advertising department of a local car dealership. He buys ad space for the dealership in Sports Illustrated several times a year. What is this type of magazine edition called?
>
> ○ circulation
>
> ○ split-run
>
> ○ demographic
>
> ○ regional

➜ WATCH VIDEO ON KEY CONCEPTS

LaunchPad includes clips from movies, TV shows, online sources, and other media texts, in addition to interviews with media experts and newsmakers. In the videos for this chapter, you'll consider topics like magazine specialization and targeting audiences.

➜ PRACTICE THE CRITICAL PROCESS

Use the Media Literacy Activity to walk through the critical process step-by-step and develop a critical perspective on what media means to you. In this chapter's activity, you'll explore the underlying values that are communicated by different magazine articles.

➜ LEARN KEY TERMS

Review the definitions for this chapter's key terms with LaunchPad's easy-to-use online flash cards. Try the cards in practice mode, or quiz yourself as a way to focus your study efforts. (Page numbers indicate where each term is highlighted within the chapter.)

magazine, 259
general-interest magazine, 261
imagined communities, 262
muckrakers, 263
photojournalism, 263

pass-along readership, 266
supermarket tabloids, 274
branded content, 275
magalogs, 276
regional editions, 277

split-run editions, 277
demographic editions, 277
evergreen subscriptions, 278
zines, 279

REVIEW QUESTIONS

Use the questions below to revisit key themes and concepts from this chapter.

 Need more practice? Access the LearningCurve multiple-choice questions for this chapter, which are available on LaunchPad. All questions include feedback for correct and incorrect answers, hints, and links to e-book material.

Magazines Today

1. How did magazines contribute to the rise of a consumer society?
2. How have some magazines become significant brands in media empires?

How We Got Here: The Early History of Magazines

3. Why did magazines develop later than newspapers in the American colonies?
4. Why did most of the earliest magazines have so much trouble staying financially solvent?
5. How did magazines become national in scope?
6. What was the social impact of the most popular women's magazines in the nineteenth century?

The Arrival of Mass-Circulation National Magazines

7. What role did magazines play in social reform at the turn of the twentieth century?
8. When and why did general-interest magazines become so popular?
9. Why did some of the major general-interest magazines fail in the twentieth century?

The Domination of Specialization

10. Why do some magazines that are linked to membership in organizations have some of the highest circulations among magazines?
11. What triggered the move toward magazine specialization?
12. What are the most useful ways to categorize the magazine industry? Why?

The Organization and Economics of Magazines

13. What are the main digital platforms for magazines?
14. What is *branded content*, and why are some magazines developing this kind of content?
15. What are the differences between regional and demographic editions?
16. What are the major magazine chains, and what is their impact on the mass media industry in general?

Magazines in a Democratic Society

17. How do magazines serve a democratic society?
18. How does advertising affect what gets published in the editorial side of magazines?

QUESTIONING THE MEDIA

Use these critical-thinking questions to reflect on your own experiences with media and apply your understanding of media concepts.

 For an interactive experience, questions on topics like these are available for the iClicker student response system. Instructors can download the questions from LaunchPad and use them to poll the class.

1. What role did magazines play in America's political and social shift from being colonies of Great Britain to becoming an independent nation?
2. Why is the muckraking spirit—so important in popular magazines at the turn of the twentieth century—generally missing from magazines today?
3. If you were the marketing director at your favorite magazine, how would you increase your audience through the use of digital platforms?
4. Think of stories, ideas, and images (illustrations and photos) that do not appear in mainstream magazines. Why do you think this is so? (Compare your list with Project Censored, an annual list of the year's most underreported stories.)
5. Discuss whether your favorite magazines define you primarily as a consumer or as a citizen. Do you think magazines have a responsibility to educate their readers as both? What can they do to promote responsible citizenship?

Books and the Power of Print

10

Marvel comic books are all about the superheroes—Spider-Man, Black Widow, Thor, Falcon, Storm, the Hulk, Iron Man, Captain America, and their many powerful contemporaries. At the end of 2020, with the planet at risk from a deadly threat (which is just how most superhero stories begin), Marvel introduced a new comic book that, fittingly, featured some of the best and bravest heroes of the year: nurses.

Called *The Vitals: True Nurse Stories*, the comic book was a partnership between Allegheny Health Network (AHN), a hospital chain in western Pennsylvania, and Marvel, one of the world's leading comic book publishers (and a unit of Disney). Real-life accounts of AHN nurses, patients, and family members were turned into comic book stories by Marvel writers and artists. "If 2020 has cemented anything, it's that real heroes don't wear capes; they wear scrubs," Marvel said on the release of *The Vitals*.[1]

Comic books like *The Vitals* and graphic novels (comics' longer cousins) are a growing segment of the category known as *trade books*—the largest division of the book industry and the one that includes books aimed at general readers. Sales of comics and graphic novels reached a record $1.21 billion in North America in 2019.[2] Comic books and graphic novels now comprise 7.4 percent of the U.S. trade book market.

Overall, comics and graphic novels have made a big impact in the book industry and in popular culture, especially as the stories have been transformed into television and movie narratives. *The Witcher* (Netflix), *The Umbrella Academy* (Netflix), and *Watchmen* (HBO Max) are all critically acclaimed recent TV series adaptations of comics and graphic novels. Moreover, graphic novels and novelists have received some of the highest honors and cultural acclaim. Graphic novelists Alison Bechdel and Gene Luen Yang won MacArthur Fellowships in 2014 and 2016, respectively, and *March: Book 3*, the third installment in the memoir of Civil Rights leader and U.S. representative John Lewis, won a National Book Award in 2016, the first graphic novel to win the coveted honor.[3]

◀ Marvel Comics' *The Vitals: True Nurse Stories* celebrates the real stories of nurses in western Pennsylvania. Comics are one part of the book industry.

In spite of this acclaim and success, however, the *New York Times* made a decision in 2017 that dismayed those in the graphic novel business: the newspaper decided to drop its graphic novel best-seller lists, which had been in existence for eight years. The *Times*' lists of best-sellers are the most influential in the industry and can increase sales by communicating the most important books of the week. Starting with a single list of best-selling books in 1931, the *Times* now includes weekly lists for Combined Print & E-Book Fiction; Hardcover Fiction; Paperback Trade Fiction; similar categories for nonfiction; Advice, How-To & Miscellaneous; children's and young-adult categories; and monthly lists for categories including audio, business, and mass market titles.

In the wake of the *Times* dropping its best-seller lists for Hardcover Graphic Novel, Paperback Graphic Novel, and Manga (a graphic style originating in Japan), more than nine hundred people signed a letter in February 2018 asking the *Times*' publisher to "increase coverage of the comics medium and bring back the graphic novel bestseller list."[4]

The *New York Times* best-seller lists are important to the graphic novel industry, argued Susana Polo at *Polygon*. "Historically, comics in newspapers belonged to the gutter of journalistic practices," she explained, dating all the way back to the time of the Yellow Kid, the popular late nineteenth-century comic character over which Joseph Pulitzer and William Randolph Hearst battled for their New York newspapers.[5] The *New York Times* shunned comic strips, treating them as the stuff of the sensationalistic "yellow journalism" newspapers, so named after their fight over the Yellow Kid (see Chapter 8).

"The mere existence of The New York Times Graphic Novel bestseller lists was evidence of a sea change in the way its audience and its editors viewed sequential art (a fancy umbrella term for comics, comic strips, [and] cartoons)," Polo said.[6] Yet graphic novels and comics, one of the brightest segments of the book industry, were once again getting second-rate treatment from the *Times*.

And then came the best possible ending to the story. In 2019, the *New York Times* announced the return of a monthly Graphic Books and Manga best-seller list (combining fiction, nonfiction, children's, and adults' graphic novels, and manga). Why are they back? As the *Times* explained, "These lists are returning due to continued reader interest and market strength."[7] As *The Vitals: True Nurse Stories* illustrates, comics and graphic novels are not just a small niche about fictional superheroes; they are a format versatile enough to tell any kind of story, with pictures as powerful as the words.

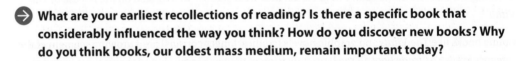 **What are your earliest recollections of reading? Is there a specific book that considerably influenced the way you think? How do you discover new books? Why do you think books, our oldest mass medium, remain important today?**

Books Today

The success of graphic novels and comics as a growing segment of the trade book market illustrates that the book industry, at more than five hundred years old, continues to evolve with readers. Its resiliency was also apparent in the 1950s and 1960s, when cultural forecasters worried that the popularity of television would spell the demise of a healthy book industry. Obviously, this did not happen. (Decades later, the Internet didn't kill off paper books either.) In 1950, more than 11,000 new book titles were introduced, and by 2018, publishers were producing nearly eighteen times that number—about 195,000 titles per year.[8] Despite the absorption of small publishing houses by big media corporations, thousands of publishers—mostly small independents—issue at least one title a year in the United States alone.

Our oldest mass medium is also still our most influential and diverse one. The portability and compactness of books—both print and electronic—make them the preferred medium in many situations (relaxing at the beach, resting in bed, traveling on buses or commuter trains), and books are still the main repository of history and everyday experience, passing along stories, knowledge, and wisdom from generation to generation.

In this chapter, we consider the long and significant relationship between books and culture.

We will:

- Trace the history of books, from Egyptian papyrus to downloadable e-books
- Examine the development of the printing press and how it transformed society, and investigate the rise of the book industry, from early publishers in Europe and colonial America to the development of publishing houses in the nineteenth and twentieth centuries
- Review the various types of books, and explore recent trends in the industry—including audio books, the convergence of books onto online platforms, and book digitization
- Consider the economic forces facing the book industry as a whole, from the decline of bookstore chains to the dominance of Amazon in our digital age
- Explore how books play a pivotal role in our culture by influencing everything from educational curricula to popular movies

Andy Cross/Getty Images

LITTLE FREE LIBRARIES are volunteer-based and work on the honor system: take a book, share a book. In 108 countries around the world, there are more than 100,000 registered Little Free Libraries. You can search for locations online—or find out how to start your own.

How We Got Here: The Early History of Books

Before books, or writing in general, oral cultures passed on information and values through the wisdom and memories of a community's elders or storytellers. Sometimes, however, these rich community traditions would be lost in the oral transmission. Print culture and the book gave future generations different and often more enduring records of authors' words.

Ever since the ancient Babylonians and Egyptians began experimenting with alphabets some five thousand years ago, people have found ways to preserve their written symbols. These first alphabets provided a fundamental element for the later development of books: a system for recording information. Initially, pictorial symbols and letters were drawn on wood strips or pressed with a stylus into clay tablets, then tied or stacked together. As early as 2400 BCE, the Egyptians wrote on **papyrus** (from which the word *paper* is derived), made from plant reeds found along the Nile River. They rolled these writings into scrolls, much as builders do today with blueprints. This method was adopted by the Greeks in 650 BCE and by the Romans (who imported papyrus from Egypt) in 300 BCE. Gradually, **parchment**—treated animal skin—replaced papyrus in Europe. Parchment was stronger, smoother, more durable, and less expensive because it did not have to be imported from Egypt.

At about the same time the Egyptians began using papyrus, the Babylonians recorded business transactions, government records, favorite stories, and local history on small tablets of clay. Around 1000 BCE, the Chinese began creating book-like objects, using strips of wood and bamboo tied together in bundles. Although the Chinese began making paper from cotton and linen around 105 CE, paper did not replace parchment in Europe until the thirteenth century because of its questionable durability.

The first protomodern book was probably produced in the fourth century by the Romans, who created the **codex**, a type of book made of sheets of parchment and sewn together along the edge, then bound with thin pieces of wood and covered with leather. Whereas scrolls had to be wound, unwound, and rewound, a codex could be opened to any page, and its configuration allowed writing on both sides of a page.

The Development of Manuscript Culture

The next step in the evolution of books occurred during the Middle Ages (400–1500 CE), when the Christian clergy strongly influenced what is known as **manuscript culture**, a period in which books were painstakingly lettered, decorated, and bound by hand. During this time, priests and monks advanced the art of bookmaking; in many ways, they may be considered the earliest professional editors. Known as *scribes*, they transcribed most of the existing philosophical tracts and religious texts of the period, especially versions of the Bible. Through tedious and painstaking work, scribes became the chief gatekeepers of recorded history and culture, promoting ideas they favored and censoring ideas that were out of line with contemporary Christian thought.

Many books from the Middle Ages were **illuminated manuscripts**. Often made for churches or wealthy clients, these books featured decorative, colorful designs and illustrations on each page. Their covers were made from leather, and some were embedded with precious gems or trimmed with gold and silver. During this period, scribes developed rules of punctuation, making distinctions between small and capital letters and placing space between words to make reading easier. (Older Roman writing used all capital letters, and the words ran together on a page, making reading a torturous experience.) Hundreds of illuminated manuscripts survive today in the rare-book collections of museums and libraries.

The Innovations of Block Printing and Movable Type

Whereas the work of the scribes in the Middle Ages led to advances in written language and the design of books, it did not lead to the mass proliferation of books, simply because each manuscript had to be painstakingly created one copy at a time. To make mechanically produced copies of pages, Chinese printers developed **block printing**—a technique in which sheets of paper were applied to blocks of inked wood, their raised surfaces depicting hand-carved letters and illustrations—as early as the third century. Although hand-carving each block, or "page," was time consuming, this printing breakthrough enabled multiple copies to be printed and then bound together. The oldest dated printed book still in existence is China's *Diamond Sutra* by Wang Chieh, from 868 CE. It consists of seven sheets pasted together and rolled up in a scroll. In 1295, explorer Marco Polo brought the block-printing technique from China to Europe, where block-printed books gradually became more popular as a literate middle-class populace began to emerge in large European cities.

THE SURVIVING COPY of the *Diamond Sutra* is a Sanskrit text translated into Chinese by Wang Chieh in 868 CE, discovered by a monk in 1900 among forty thousand scrolls sealed in a hidden cave in China. The full title of the sacred Buddhist text is *The Diamond That Cuts through Illusion*. The 17.5-foot scroll is housed at the British Library. The lower right side of the scroll notes the date and this message: "Reverently made for universal free distribution by Wang Jie on behalf of his two parents."

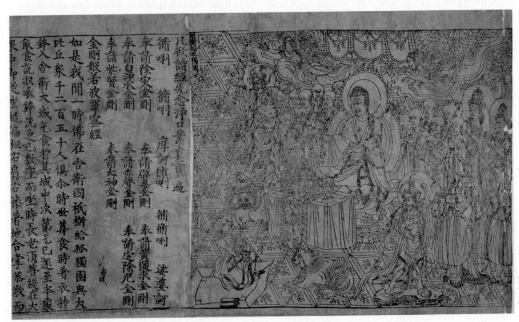

© British Library

The next step in printing was the radical development of **movable type**, first invented in China around the year 1000. Movable type featured individual characters made from reusable pieces of wood or metal, rather than entire hand-carved pages. Printers arranged the characters into various word combinations, greatly speeding up the time it took to create block pages. This process, also used in Korea as early as the thirteenth century, developed independently in Europe in the fifteenth century.

The Gutenberg Revolution: The Invention of the Printing Press

German inventor Johannes Gutenberg was responsible for the next great leap forward in printing. Between 1453 and 1456, Gutenberg used the principles of movable type to develop a mechanical **printing press**, which he adapted from the design of wine presses. Gutenberg's staff of printers produced the first "modern" books, including two hundred copies of a Latin Bible, twenty-one copies of which still exist. The Gutenberg Bible (as it's now known) required six presses, many printers, and several months to produce. It was printed on a fine calfskin-based parchment called **vellum**. The pages were hand-decorated, and the use of woodcuts made illustrations possible. Gutenberg and his printing assistants had not only found a way to make books a mass medium but formed the prototype for all mass production.

Printing presses spread rapidly across Europe in the late fifteenth and early sixteenth centuries. Chaucer's *Canterbury Tales* became the first English work to be printed in book form. Many early books were large, elaborate, and expensive, taking months to illustrate and publish; as such, they were typically purchased by aristocrats, royal families, religious leaders, and ruling politicians. Printers, however, gradually reduced the size of books and developed less expensive grades of paper, making books cheaper so that more people could afford them.

The social and cultural transformations ushered in by the spread of printing presses and books cannot be overestimated. As historian Elizabeth Eisenstein has noted, when people had the means and opportunity to learn for themselves by using maps, dictionaries, Bibles, and the writings of others, they could differentiate themselves as individuals; their social identities were no longer solely dependent on what their leaders told them or on the habits of their families, communities, or social class. The technology of printing presses permitted information and knowledge to spread outside local jurisdictions. Gradually, individuals had access to ideas far beyond their isolated experiences, and this permitted them to challenge the traditional wisdom and customs of their communities and leaders.[9]

The Birth of Publishing in the United States

In colonial America, English locksmith Stephen Daye set up a print shop in the late 1630s in Cambridge, Massachusetts. In 1640, Daye and his son Matthew printed the first colonial book, *The Whole Booke of Psalms* (known today as *The Bay Psalm Book*), marking the beginning of book publishing in the colonies. This collection of biblical psalms quickly sold out its first printing of 1,750 copies, even though fewer than 3,500 families lived in the colonies at the time. By the mid-1760s, all thirteen colonies had printing shops.

PULP FICTION The weekly paperback series *Tip Top Weekly*, published between 1896 and 1912, featured stories of the most popular dime novel hero of the day, the fictional Yale football star and heroic adventurer Frank Merriwell.

In 1744, Benjamin Franklin, who had worked in printing shops, imported Samuel Richardson's *Pamela; or, Virtue Rewarded* (1740) from Britain, the first novel reprinted and sold in colonial America. Both *Pamela* and Richardson's second novel, *Clarissa; or, The History of a Young Lady* (1747), connected with the newly emerging and literate middle classes—especially women, who were just beginning to gain a social identity as individuals apart from their fathers, husbands, and employers. Richardson's novels portray women in subordinate roles, but because they also depict women triumphing over tragedy, he is credited as one of the first popular writers to take the domestic life of women seriously.

By the early nineteenth century, the demand for books was growing. To meet this demand, it was necessary to reduce the cost of producing books. By the 1830s, machine-made paper replaced more expensive handmade varieties, cloth covers supplanted more expensive leather ones, and **paperback books** with cheaper paper covers (introduced from Europe) helped make books more accessible to the masses. Further reducing the cost of books, Erastus and Irwin Beadle introduced paperback **dime novels** (so called because they sold for five or ten cents) in 1860. Ann Stephens authored the first dime novel, *Malaeska: The Indian Wife of the White Hunter,* a reprint of a serialized story she had written in 1839 for *Ladies' Companion* magazine.[10] (A number of books in the nineteenth and twentieth centuries were serialized—printed in regular installments—in magazines and newspapers, foreshadowing later synergies between books, radio, TV, and film.) By 1870, dime novels had sold seven million copies. By 1885, one-third of all books published in the United States were popular paperbacks and dime novels, sometimes identified as **pulp fiction**—a reference to the cheap, machine-made pulp paper they were printed on.

In addition, the printing process became quicker and more mechanized. In the 1880s, the introduction of **linotype** machines enabled printers to save time by setting type mechanically using a typewriter-style keyboard, while the introduction of steam-powered and high-speed rotary presses permitted the production of more books at a lower cost. In the early 1900s, the development of **offset lithography** allowed books to be printed from photographic plates rather than from metal casts, greatly reducing the cost of color and illustrations and accelerating book production. With these developments, books disseminated further, preserving culture and knowledge and supporting a vibrant publishing industry.

Mass Publishing and the Book Industry

Throughout the nineteenth century, the rapid spread of knowledge and literacy as well as the Industrial Revolution spurred the emergence of the middle class. A demand for books promoted the development of the publishing industry, which capitalized on increased literacy and widespread compulsory education. Many early publishers were mostly interested in finding quality authors and publishing books of importance. But with the growth of advertising and the rise of a market economy in the latter half of the nineteenth century, publishing gradually became more competitive and more concerned with sales. With the emergence of a publishing industry, publishers organized their business into several categories of books, each with its own market segment.

The Formation of Publishing Houses

The U.S. book industry developed gradually in the nineteenth century with the formation of the early "prestigious" publishing houses: companies that tried to identify and produce the works of good writers.[11] Among the oldest American houses established at the time (all are now part of major media conglomerates) were J. B. Lippincott (1792); Harper & Bros. (1817), which became Harper & Row in 1962 and HarperCollins in 1990; Houghton Mifflin (1832); Little, Brown (1837); G. P. Putnam (1838); Scribner's (1842); E. P. Dutton (1852); Rand McNally (1856); and Macmillan (1869).

Between 1880 and 1920, the center of social and economic life shifted from rural farm production to an industrialized urban culture, and with the accompanying increase in literacy and leisure time and a wave of European immigrants interested in the English language and American culture, the demand for books grew. The burgeoning interest in reading caught the attention of entrepreneurs eager to profit by satisfying this demand. Thus, a savvy breed of publishing house, focused on marketing, was born. This new group of houses included Doubleday & McClure Company (1897), the McGraw-Hill Book Company (1909), Prentice-Hall (1913), Alfred A. Knopf (1915), Simon & Schuster (1924), and Random House (1925).

Despite the growth of the industry in the early twentieth century, book publishing sputtered from 1910 into the 1950s, as profits were adversely affected by the two world wars and the Great Depression. Radio and magazines fared better because they were generally less expensive and could more immediately cover topical issues during times of crisis. After World War II, however, book publishing bounced back, and the industry grew in five main market divisions defined by type of book.

Types of Books

The divisions of the U.S. book industry come from economic and structural categories developed by both publishers and trade organizations, such as the Association of American Publishers (AAP), the Book Industry Study Group (BISG), and the American Booksellers Association (ABA). The categories of book publishing that exist today include trade books (both adult and juvenile), professional books, textbooks (both elementary through high school, often called "el-hi," and college), religious books, and university press books. (For sales figures for the book types, see Figure 10.1.)

Trade Books

One of the most lucrative parts of the industry, **trade books** include hardbound and paperback books aimed at general readers and sold at commercial retail outlets. The industry distinguishes among the following subcategories:

- adult trade books (including hardbound and paperback fiction; current nonfiction and biographies; literary classics; books on hobbies, art, and travel; popular science, technology, and computer publications; self-help books; and cookbooks)
- juvenile books (ranging from preschool picture books to young-adult or young-reader books)
- comics and graphic novels

SCRIBNER'S—known more for its magazines in the late nineteenth century than for its books— became the most prestigious literary house of the 1920s and 1930s, publishing F. Scott Fitzgerald (*The Great Gatsby*, 1925) and Ernest Hemingway (*The Sun Also Rises*, 1926).

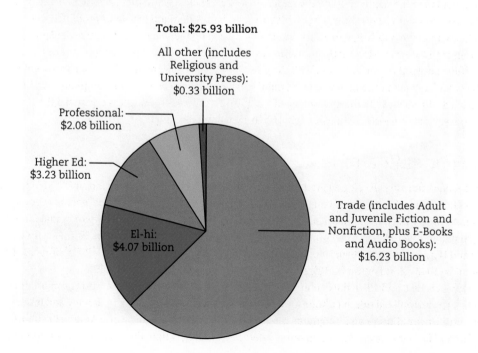

Total: $25.93 billion

All other (includes Religious and University Press): $0.33 billion

Professional: $2.08 billion

Higher Ed: $3.23 billion

El-hi: $4.07 billion

Trade (includes Adult and Juvenile Fiction and Nonfiction, plus E-Books and Audio Books): $16.23 billion

FIGURE 10.1

ESTIMATED U.S. BOOK REVENUE, 2019

Data from: Porter Anderson, "US Market Statistics: The AAP StatShot Annual Report for 2019," Publishing Perspectives, July 31, 2020, https://publishingperspectives .com/2020/07/united-states-market -statistics-aap-statshot-annual -report-for-2019.

Note: Figures are rounded to the nearest hundredth.

In the changing world of trade publishing, young-adult books and graphic novels have boosted the industry in recent years. The *Harry Potter* series alone created record-breaking first-press runs: 10.8 million for *Harry Potter and the Half-Blood Prince* (2005), and 12 million for the final book in the series, *Harry Potter and the Deathly Hallows* (2007).

Since the 1978 release of Will Eisner's *A Contract with God*, generally credited as the first graphic novel (which is longer than a comic and bound like a book in print form), interest in graphic novels has grown; in 2006, sales of graphic novels surpassed those of comic books. Given their strong stories and visual nature, many comics and graphic novels—including *X-Men*, *The Dark Knight*, *Watchmen*, and *Captain America*—have inspired movies (see the chapter-opening discussion). But graphic novels aren't only about warriors and superheroes; Raina Telgemeier's *Guts* and Jake Halpern and Michael Sloan's *Welcome to the New World* are both acclaimed graphic novels, but their characters are regular mortals in real settings.

Professional Books

The counterpart to professional trade magazines, **professional books** target various occupational groups and are not intended for the general consumer market. This area of publishing capitalizes on the growth of professional specialization that has characterized the job market. Traditionally, the industry has subdivided professional books into the areas of law, business, medicine, and technical-scientific works, with books in other professional areas accounting for a very small segment. These books are often bought by professional schools and university libraries; however, as a result of stagnating and declining library budgets, professional book sales have declined in recent years.

Textbooks

The most widely read secular book in U.S. history was *The Eclectic Reader*, an elementary-level reading textbook first written by William Holmes McGuffey, a Presbyterian minister and college professor. From 1836 to 1920, more than 100 million copies of this text were sold. Through stories, poems, and illustrations, *The Eclectic Reader* taught nineteenth-century schoolchildren to spell and read simultaneously—and to respect the nation's political and economic systems. Ever since the publication of the McGuffey reader (as it is often nicknamed), **textbooks** have served a nation

TEXTBOOKS Even though they may not be as much fun to read for some students as, say, graphic novels, textbooks and learning materials remain an important part of education today.

Ariel Skelley/DigitalVision/Getty Images

intent on improving literacy rates and public education. Elementary school textbooks found a solid market niche in the nineteenth century, while college textbooks boomed in the 1950s, when the GI Bill enabled hundreds of thousands of working- and middle-class men returning from World War II to attend college. The demand for textbooks further accelerated in the 1960s, as opportunities for women and people of color expanded. Textbooks are divided into elementary through high school (el-hi) texts, college texts, and vocational texts.

In about half of the states, local school districts determine which el-hi textbooks are appropriate for their students. In the other half of the states, including Texas and California—the two largest states—statewide adoption policies determine which texts can be used. If individual schools choose to use books other than those mandated, they are not reimbursed by the state for their purchases.

Unlike el-hi texts, which are subsidized by various states and school districts, college texts are paid for by individual students. For more than a decade, as tuition at most colleges and universities has been rising, the price of an academic year's worth of textbooks has been dropping due to the lower cost of digital textbooks, book rentals, and greater access to online retailers. Surveys indicate that each college student spent an average of $413 annually on required course materials in 2019–2020, down from $701 during the 2007–2008 academic year.[12] Increasingly, students have turned to online purchasing, through sites like Amazon, Barnes & Noble, eBay, eCampus.com, and textbooks.com, or renting digital or physical books through companies like Chegg or Campus Book Rentals. College bookstores have adapted by selling e-books and offering rentals as well, while book publishers have developed an inclusive access model, in which students receive deeply discounted digital textbooks as part of their course tuition.

Religious Books

The best-selling book of all time is the Bible, in all its diverse versions. Over the years, the success of Bible sales has created a large industry for religious books. After World War II, sales of religious books soared. Historians attribute the sales boom to economic growth and a nation seeking peace and security while facing the threat of "godless communism" and the Soviet Union.[13] By the 1960s, though, the scene had changed dramatically. The impact of the Civil Rights struggle, the Vietnam War, the sexual revolution, and the youth rebellion against authority led to declines in formal church membership. Not surprisingly, sales of some types of religious books dropped as well. To compete, many religious-book publishers extended their offerings to include serious secular titles on such topics as war and peace, race, poverty, gender, and civic responsibility.

Throughout this period of change, the publication of fundamentalist and evangelical literature remained steady. It then expanded rapidly during the 1980s, when the Republican Party began making political overtures to conservative groups and prominent TV evangelists. But more generally, inspirational books in this segment continue to thrive, including Lysa TerKeurst's bestseller *Forgiving What You Can't Forget: Discover How to Move On, Make Peace with Painful Memories, and Create a Life That's Beautiful Again* (2020).

University Press Books

The smallest division in the book industry is the nonprofit **university press**, which publishes scholarly works for small groups of readers interested in intellectually specialized areas, such as literary theory and criticism, art movements through history, and contemporary philosophy. Professors often try to secure book contracts from reputable university presses to increase their chances for *tenure*, a lifetime teaching contract. University presses range in size from very small, often producing fewer than a dozen titles a year, to the largest, Oxford University Press, which publishes more than six thousand titles a year.

University presses have not traditionally faced pressure to produce commercially viable books, preferring to encourage books about highly specialized topics by innovative thinkers. In fact, most university presses routinely lose money and are subsidized by their university. Even when they publish more commercially accessible titles, the lack of large marketing budgets can prevent such books from reaching mass audiences. While large commercial trade houses are often criticized for publishing only blockbuster books, university presses often suffer the opposite criticism—that they produce mostly obscure books that only a handful of scholars read. To offset costs and increase revenue, some presses are trying to form alliances with commercial houses to help promote and produce academic books that have a wider appeal.

Trends and Issues in Book Publishing

Ever since Harriet Beecher Stowe's abolitionist novel *Uncle Tom's Cabin* sold fifteen thousand copies in fifteen days back in 1852 (and three million total copies prior to the Civil War), many American publishers have sought out potential *best-sellers*, or blockbusters (just like in the movie business). While most authors are professional writers, the book industry also reaches out to famous media figures, who may pen a best-selling book (Barack Obama's *A Promised Land*, Mary L. Trump's *Too Much and Never Enough: How My Family Created the World's Most Dangerous Man*, Matthew McConaughey's *Greenlights*) or a commercial failure (Amy Schumer's *The Girl with the Lower Back Tattoo*). Other ways publishers attempt to ensure popular success include acquiring the rights to license popular film and television programs and releasing books in nonprint formats like e-books and audio books. Beyond the selling of books, issues affecting the industry include the preservation of older print books and the history of banned books and censorship.

Influences of Television and Film

There are two major facets in the relationship among books, television, and film: how TV and films can help sell books, and how books can serve as ideas for TV shows and movies. Through TV exposure, for example, books by or about talk-show hosts, actors, and politicians, such as Stephen Colbert, John Oliver, Matthew McConaughey, Barack and Michelle Obama, and Hillary Clinton, sell millions of copies—enormous numbers in a business in which selling 100,000 copies constitutes remarkable success. In national polls conducted from the 1980s through today, nearly 30 percent of respondents said they had read a book after seeing a story about it or a promotion on television.

In the 1990s and early 2000s, one of the most influential forces in promoting books on TV was afternoon talk-show host Oprah Winfrey. Winfrey began Oprah's Book Club as a monthly segment on her show in 1996 and became a major power broker in terms of selling books. Her selections became immediate best-sellers, generating tremendous excitement within the book industry. In 1996, for example, novelist Toni Morrison's nineteen-year-old book *Song of Solomon* became a paperback best-seller after Morrison appeared on the show. *The Oprah Winfrey Show* ended in 2011, but the book club was revived online in 2012, before returning to television in 2019 via the Apple TV+ program *Oprah's Book Club*. Actor Reese Witherspoon has also launched a book club—Reese's Book Club—for which she selects an adult and a YA (young adult) book each month that features a woman as the central character.

The film industry gets many of its story ideas from books (more than sixteen hundred feature-length movie adaptations in the United States since 1980), which results in enormous movie-rights revenues for the book industry and its authors.[14] Louisa May Alcott's *Little Women* (1868), Stephen King's *It* (1986), and Kevin Kwan's *Crazy Rich Asians* (2013) are just a few classic and recent books made into movies. Jenny Han's *To All the Boys I've Loved Before*

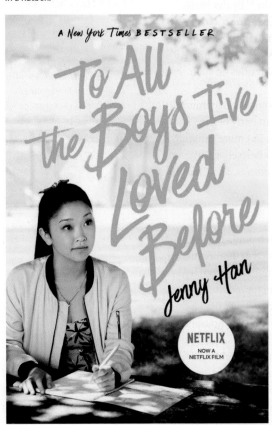

JENNY HAN'S *To All the Boys I've Loved Before* (2014), which was the first of a three-book young-adult series that was made into movies for Netflix, was based on the author's own experience of gaining closure by writing love letters to former boyfriends and then leaving them unsent, in a hatbox.

(2014) was also made into a Netflix movie in 2018; two sequels to the book became films for Netflix as well. Many of the most profitable movie successes for the book industry in recent years have emerged from fantasy works. J. K. Rowling's best-selling *Harry Potter* books became hugely popular movies, as did Peter Jackson's trilogy of J. R. R. Tolkien's enduringly popular *Lord of the Rings* (first published in the 1950s). In the cases of all these movies, the popularity of the films spurred renewed interest in the books.

Books have also inspired popular television programs, including *Game of Thrones* on HBO and *A Handmaid's Tale* on Hulu. Other recent adaptations of books to television include Julia Quinn's bestselling nine-book *Bridgerton* series (2000–2013), made into a popular Netflix series in 2020, and Walter Tevis's 1983 novel *The Queen's Gambit*, which also became an acclaimed miniseries for Netflix in 2020. In both of these cases, the television shows boosted sales of the original books.

E-Books

In 1971, Michael Hart, a student computer operator at the University of Illinois, typed up the text of the U.S. Declaration of Independence, and thus the idea of the **e-book**—a digital book read on a computer or a digital reading device—was born. Hart soon founded Project Gutenberg, which now offers more than forty thousand public domain books (older texts with expired copyrights) for free at www.gutenberg.org. However, the idea of *commercial e-books*—putting copyrighted books like current best-sellers in digital form—took a lot longer to gain traction.

In the 1990s, early portable reading devices were criticized for being too heavy, too expensive, or too difficult to read from, while their available e-book titles were scarce and had little cost advantage over full-price hardcover books. But in 2007, Amazon—the largest online bookseller—developed an e-reader (the Kindle) and an e-book store that seemed inspired by Apple's iPod and iTunes, which had already changed the music industry. The first Kindle had an easy-on-the-eyes electronic paper display, held more than two hundred books, and did something no other device before could do: wirelessly download e-books from Amazon's online bookstore. Moreover, most Kindle e-books sold for $9.99, less than half the price of most new hardcovers. This time, e-books caught on quickly.

In 2010, Apple introduced the iPad, a color touchscreen tablet that quickly outsold the Kindle. In 2011, Amazon responded by introducing the Kindle Fire, a color touchscreen tablet with web browsing, access to all the media on Amazon, and access to Amazon's Appstore. Today, Amazon's rivals in the e-book device business include Kobo, the most popular e-book device in Canada, which has had some success in the United States, and PocketBook, a popular e-book device in Europe. But the biggest rivals for e-readers are smartphones and tablets, which have undermined consumer demand for stand-alone e-reader devices.

By 2019, e-books accounted for 11.7 percent—$1.94 billion—of the U.S. trade book market. Yet revenue and unit sales of e-books declined by 30.8 percent from 2015 to 2019, suggesting a limit to reader demand for e-books and a resurgence in sales of printed books.[15]

Audio Books

Another major development in the book industry has been the merger of sound recording with publishing. *Audio books* generally feature actors or authors reading entire works or abridged versions of popular fiction and nonfiction trade books. Indispensable to many sightless readers and older readers whose vision is diminished, audio books are also popular among readers who do a lot of commuting or who want to listen to a book at home while doing something else, like exercising.

THE QUEEN'S GAMBIT is a 1983 novel by the late author Walter Tevis. The title refers to a series of opening moves in chess. The story follows the life of chess prodigy Beth Harmon and the struggles she faces, from her childhood in an orphanage to becoming an international chess champion. The limited television series was one of Netflix's most-watched programs of 2020.

The Queen's Gambit

By: Walter Tevis
Narrated by: Amy Landon
Length: 11 hrs and 50 mins
Categories: Literature & Fiction, Genre Fiction
★★★★★ 4.5 (4,865 ratings)

Audio books are readily available on the Internet for downloading to smartphones and other portable devices; in fact, between 2015 and 2019, the audio book format grew by almost 144 percent and now represents more than 8 percent of U.S. trade book revenue.[16] Amazon owns Audible, the largest provider of audio books.

Preserving and Digitizing Books

The preservation of older books—especially those from the nineteenth century printed on acid-based paper, which gradually deteriorates—is a vital issue in archiving that remains important to preservationists and book lovers alike. At the turn of the twentieth century, research initiated by libraries concerned with losing valuable older collections provided evidence that acid-based paper would eventually turn brittle and self-destruct. The paper industry did not respond, so in the 1970s, leading libraries began developing techniques to halt any further deterioration. Finally, by the early 1990s, motivated almost entirely by economics rather than by the cultural value of books, the paper industry began producing acid-free paper. Libraries and book conservationists, however, still focused their attention on older, at-risk books. Some institutions began photocopying original books onto acid-free paper and making the copies available to the public. Libraries then stored the originals, which were treated to halt further wear.

Another way to preserve books is through digital imaging. The most extensive digitization project is Google Books, which began in 2004 and features partnerships with the New York Public Library and about twenty major university research libraries—including Harvard, Michigan, Oxford, and Stanford—to scan millions of books and make them available online. The Authors Guild and the Association of American Publishers initially sued Google for digitizing copyrighted books without permission. Google argued that displaying only a limited portion of the books was legal under fair-use rules. After years of legal battles, a U.S. district court sided with Google's fair-use argument in 2013 and dismissed the lawsuit, a decision upheld in a 2015 appeal.

Another group, dissatisfied with Google Books' restriction of its scanned book content for use by other commercial search services, launched a nonprofit service in 2007. The Internet Archive's Open Library works with the Boston Public Library, several university libraries, Amazon, Microsoft, and Yahoo! to digitize millions of books with expired copyrights and make them freely available at OpenLibrary.org. In 2008, another group of universities formed the HathiTrust Digital Library to further archive and share digital collections. In 2010, these nonprofit archives joined other libraries to create the Digital Public Library of America.

Censorship and Banned Books

Over time, the wide circulation of books gave many ordinary people the same opportunities to learn that were once available to only the privileged few. However, as societies discovered the power associated with knowledge and the printed word, books were subjected to a variety of censors. Imposed by various rulers and groups intent on maintaining their authority, the censorship of books often prevented people from learning about the rituals and moral standards of other cultures. Political censors sought to banish "dangerous" books that promoted radical ideas or challenged conventional authority. In various parts of the world, some versions of the Bible, Karl Marx's *Das Kapital* (1867), *The Autobiography of Malcolm X* (1965), and Salman Rushdie's *The Satanic Verses* (1989) have all been banned at one time or another. In fact, one of the triumphs of the Internet is that it allows the digital passage of banned books into nations where printed versions have been outlawed. (For more on banned books, including one perspective on their value, see "Examining Ethics: Banned Books: Controversial Because They Are Real.")

Each year, the American Library Association (ALA) compiles a list of the most challenged books in the United States. Unlike an enforced ban, a **book challenge** is a formal request to have a book removed from a public or school library's collection. Common reasons for challenges include sexually explicit passages, offensive language, occult themes, violence, LGBTQ+ themes,

Banned Books: Controversial Because They Are Real

BY ELIZABETH KELSEY

Alcohol, drugs, sex, explicit content, foul language . . . these are just a few of the many reasons why books are challenged and banned in classrooms, libraries, and communities. As *The Perks of Being a Wallflower* by Stephen Chbosky and *Eleanor and Park* by Rainbow Rowell contain all these things, it's small wonder that they're among the most frequently banned books in the country, according to the American Library Association.

And why are they controversial?

Because their stories are *real*.

Perks, a 1999 novel, chronicles a year in the life of Charlie, an introverted first-year high school student who struggles to understand his place in the confusing world he sees around him. His friends expose him to sex, drugs, and alcohol, yes, but also to love, acceptance, and true friendship. They build him up as a person, as does his English teacher, Bill, who gives him extra assignments because he sees Charlie's potential—something no other teacher has ever expressed to him. *Perks* shows that while relationships with others can lead to pain and confusion, they also have the power to pick us up during difficult times. It highlights the need to participate in life, as Charlie notes in the book's thought-provoking final pages: "Sometimes, I think that the only perspective is to really be there, because it's okay to feel things. And be who you are about them."

Eleanor and Park is a 2013 novel set in 1980s Nebraska. It tells the story of the blossoming relationship between two sixteen-year-olds: Eleanor, an overweight redheaded girl with patched clothes and hair ribbons, and Park, a Korean American boy struggling with feelings of inadequacy in his family and society. Thrown together on the school bus when Park takes pity on Eleanor after she is bullied by classmates, awkward silences turn into a shared interest in comic books and mixtapes. Their initially platonic relationship escalates, filling a need that neither of them even knew they had. But faced with Eleanor's rocky domestic situation, the road is far from easy for the young lovers, and each stolen moment could be their last.

The book does depict domestic abuse, as Eleanor's stepfather is verbally and physically abusive. It also depicts extensive foul language, both from adults and from other teens but extremely rarely from Eleanor or Park. And this is the key point. The book does not glorify foul language or abuse; instead, its protagonists reject and condemn these things. While the other characters use foul language, Park uses headphones to shut it out, and Eleanor expresses how much she hates the language that her stepfather uses. Meanwhile, as Eleanor's mother's relationship with her stepfather declines, Eleanor's healthy relationship with Park grows. In fact, the book is able to build up healthy relationships and friendships *because* it illustrates the negative side of relationships as well. By depicting the ugly, Eleanor and Park's relationship becomes more beautiful by comparison, and the reader can appreciate its beauty and be inspired by it. "I don't like you, Park," Eleanor whispers at one point. "I think I live for you." Isn't that a relationship that we want to experience, that we want children to read about and know about?

Both books are not only socially important literary works but highly enjoyable reads, with an excellent mix of humor and heart. And, most importantly, both books create the sense that their characters are real people. Moreover, research has shown that reading about people whose lives are different from our own increases our sense of empathy.[1] Books can demonstrate what real can be—and what real should be allowed to be.

Real should be allowed to be redheaded teenage girls who worry about their families and struggle with intimacy. Real should be allowed to be high school boys who hurt their friends and partners because they're afraid and can't understand their own feelings. Real should be allowed to be messy, and students should be allowed to read it.

Elizabeth Kelsey is a graduate of the University of Northern Iowa, where she was a Presidential Scholar and a Spanish and TESOL major, with minors in digital journalism and Spanish-to-English translation. ●

A LIBRARY'S DISPLAY during the American Library Association's annual Banned Books Week. Examples of banned books featured here include the Bible and the children's story *A Wrinkle in Time* by Madeleine L'Engle (a graphic novel version).

promotion of a religious viewpoint, nudity, and racism. (The ALA defends the right of libraries to offer material with a wide range of views and does not support removing material on the basis of partisan or doctrinal disapproval.) Some of the most challenged books of the past decade include *The Absolutely True Diary of a Part-Time Indian* by Sherman Alexie, *Thirteen Reasons Why* by Jay Asher, the *Internet Girls* series by Lauren Myracle, and the *Captain Underpants* series by Dav Pilkey.

The Organization and Economics of the Book Industry

Compared with other media industries, book publishing has adapted to the digital turn and has been able to avoid huge declines in revenues. From the mid-1980s to 2019, total revenue rose from $9 billion to about $25.93 billion.[17] Within the industry, the concept of who or what constitutes a publisher varies widely. A publisher may be a large company that is a subsidiary of a global media conglomerate and occupies an entire office building, or a one-person home office operation that is run on a laptop computer.

Ownership Patterns

Like most mass media, commercial publishing is dominated by a handful of major corporations with ties to international media conglomerates. The two leading publishers, RELX Group and Thomson Reuters, are not names generally associated with popular best-selling books (see Table 10.1). Instead, both corporations specialize in informational publishing for specialized professions. RELX (based in the U.K., the Netherlands, and the United States) publishes leading scientific and medical journals and the LexisNexis legal and news database, whereas Thomson Reuters (based in Canada) operates the Westlaw online legal database and the Reuters international newswire service.

Mergers and consolidations have driven the book industry. For example, another of the world's largest publishers is Pearson, based in the U.K. Pearson established its book business in the 1920s and at one point owned the *Financial Times*, a 50 percent share in the *Economist* magazine, and nearly half of Penguin Random House (PRH)—the largest trade book publisher in the world. But in 2015, Pearson committed its focus to the el-hi, college, and professional publishing business, and is now deeply involved in publishing textbooks, digital materials, and assessment exams for the education market.

TABLE 10.1   WORLD'S FIFTEEN LARGEST BOOK PUBLISHERS (REVENUE IN MILLIONS OF DOLLARS), 2019

Rank 2019	Rank 2018	Publishing Group or Division	Parent Company	Parent Country	2019 Revenue (in $M)	2018 Revenue (in $M)
1	1	RELX Group	Reed Elsevier PLC & Reed Elsevier NV	U.K./U.S./NL	$5,636	$5,278
2	3	Thomson Reuters	The Woodbridge Co.	Canada	$5,277	$5,133
3	2	Pearson	Pearson PLC	U.K.	$5,084	$5,244
4	4	Bertelsmann	Bertelsmann AG	Germany	$4,156	$4,150
5	5	Wolters Kluwer	Wolters Kluwer	NL	$3,976	$3,757
6	6	Hachette Livre	Lagardère	France	$2,675	$2,576
7	7	Springer Nature	Springer Nature	Germany	$1,928	$1,897
8	8	Wiley	Wiley	U.S.	$1,800	$1,796
9	9	HarperCollins	News Corp.	U.S.	$1,754	$1,758
10	10	Scholastic	Scholastic	U.S.	$1,654	$1,628
11	11	Phoenix Publishing and Media	Phoenix Publishing and Media Co.	China	$1,634	$1,602
12	12	McGraw-Hill Education	Apollo Global Management	U.S.	$1,571	$1,597
13	13	Cengage Learning Holdings II	Apax and Omers Capital Partners	U.S./Canada	$1,442	$1,452
14	14	Holtzbrinck	Verlagsgruppe Georg von Holtzbrinck	Germany	$1,398	$1,419
15	15	Houghton Mifflin Harcourt	Houghton Mifflin Harcourt Co.	U.S./Cayman Islands	$1,319	$1,322

Data from: Jim Milliot, "RELX Group Remains the World's Top Publisher," *Publishers Weekly*, October 30, 2020, www.publishersweekly.com/pw/by-topic/industry-news/financial-reporting/article/84768-relx-group-remains-the-world-s-top-publisher.html.

In 1998, Germany's Bertelsmann shook up the book industry by adding Random House, which was then the largest U.S. book publisher, to its fold for $1.4 billion. Bertelsmann's book company subsidiaries include Ballantine Bantam Dell, Doubleday Broadway, Alfred A. Knopf, and the Random House Publishing Group. In 2013, Random House merged with Penguin Books (then owned by Pearson), creating Penguin Random House, which became the largest trade book publisher. Bertelsmann also owns large European magazine and television divisions.

Penguin Random House (now fully owned by Bertelsmann), Simon & Schuster (owned by ViacomCBS), Hachette (owned by French-based Lagardère), HarperCollins (owned by News Corp.), and Macmillan (owned by German-based Holtzbrinck) are currently the five largest trade book publishers in the United States. In 2020, Penguin Random House announced plans to buy Simon & Schuster for over $2 billion, which, if approved by regulators, would combine the first- and third-largest trade book publishers to create an enormous publishing entity, all owned by Bertelsmann. From a corporate viewpoint, executives have argued that large companies can financially support a number of smaller firms or *imprints* (publishing brand names) while allowing their editorial ideas to remain independent from the parent corporation. With thousands of independent presses competing with bigger corporations, book publishing continues to produce volumes on an enormous range of topics.

The Structure of Book Publishing

A small publishing house may be staffed with anywhere from a few people to as many as twenty people. Medium-size and large publishing houses employ hundreds of people. In the larger houses, divisions usually include acquisitions and development; copyediting, design, and production; marketing and sales; and administration and business. Similar to magazines, most publishing houses contract independent printers to produce their books.

Most publishers employ **acquisitions editors** to seek out and sign authors to contracts. For fiction, this might mean discovering talented writers through book agents or reading unsolicited manuscripts. For nonfiction, editors might examine manuscripts and letters of inquiry or match a known writer to a project (such as a celebrity biography). Acquisitions editors also handle **subsidiary rights** for an author—that is, selling the rights to a book for use in other media, such as the basis for a screenplay.

As part of their contracts, writers sometimes receive *advance money*, an early payment that is subtracted from royalties earned from book sales. Typically, an author's royalty (percentage of earnings for each sale) is between 5 and 15 percent of the print book price, with higher royalty rates for e-books. Amazon and book publishers have been experimenting with different price points for e-books—low enough to ensure good online sales but high enough to make publishing still profitable. Thus, while an author's royalty percentage may be much higher for e-books than for print, the author is receiving a larger percentage of what is usually a lower book price. New authors may receive little or no advance from a publisher, but commercially successful authors can receive millions. For example, author J. K. Rowling hauled in an estimated $7 million advance from Little, Brown & Company/Hachette for *The Casual Vacancy* (2012), her first novel after the *Harry Potter* series. Nationally recognized authors—including political leaders, sports figures, comedians, and movie stars—can also command large advances from publishers who are banking on the well-known person's commercial potential. One of the largest advances of this kind was the $65 million advance Barack and Michelle Obama jointly received from Penguin Random House for a book from each of them. Although the advance was record-breaking, Michelle Obama's *Becoming* became the best-selling book in the United States in 2018, and Barack Obama's *A Promised Land* was the best-selling book of 2020 (with a second volume of his presidential memoir planned for later release).

After a contract is signed, the acquisitions editor may turn the book over to a **developmental editor**, who provides the author with feedback, makes suggestions for improvements, and, in educational publishing, obtains advice from knowledgeable members of the academic community. If a book contains images, editors work with photo researchers to select photographs and pieces of art. Then the production staff enters the picture. While **copy editors** attend to specific problems in writing or length, production and **design managers** work on the look of the book, making decisions about type style, paper, cover design, and layout.

Simultaneously, plans are underway to market and sell the book. Decisions need to be made concerning the number of copies to print, ways to reach potential readers, and costs for promotion and advertising. For trade books and some scholarly books, publishing houses may send advance copies of a book to appropriate magazines and newspapers with the hope of receiving favorable reviews that can be used in promotional material. Prominent trade writers typically have book signings and travel the radio and TV talk-show circuit to promote their books. Unlike trade publishers, college textbook firms rarely sell directly to bookstores. Instead, they contact instructors through e-mail campaigns and sales representatives assigned to geographic regions.

To help create a best-seller, trade houses often distribute large illustrated cardboard bins, called *dumps*, to thousands of stores to display a book in bulk quantity. Like food merchants who buy eye-level shelf placement for their products in supermarkets, large trade houses buy shelf space to ensure prominent locations in bookstores. Similarly, publishers are required to pay for featured treatment on Amazon, Apple Books, and other digital stores. Publishers also buy ad space on the web and social media, on television and radio, in newspapers and magazines, and on buses and billboards—all in an effort to generate interest in a new book.

In recent years, the publishing industry has been subject to criticism about the lack of diversity in the industry, extending from the beginning of the process in acquisitions to the final marketing and promotion of a published book. A 2019 survey of diversity in the publishing business found that 76 percent of industry employees were white, 74 percent were cisgender women, 81 percent were heterosexual, and 89 percent were people without disabilities—all figures demonstrating little progress in creating a more diverse industry when compared to a similar 2015 survey.[18] The criticisms came to a head with the 2020 release of Jeanine Cummins's *American Dirt*, a story of a middle-class Mexican mother and son who flee a violent drug cartel and join other migrants in an attempt to reach safety in the United States. Nine publishing companies competed for the book, and Cummins won a seven-figure advance from Flatiron Books. Cummins's book was selected for Oprah's Book Club in January 2020, and it seemed destined to become a blockbuster. Yet several Latina and Latino writers charged that the book was filled with Mexican stereotypes and that Cummins, who had identified as a white writer, was out of touch with her subject matter. The publisher canceled the book tour, and Oprah invited Cummins on her Apple TV+ show to address the criticisms of three Latina writers.[19] According to *Publishers Weekly*, "Many members of the literary community, especially those who identify as Latinx, have argued that the book and the massive publicity push surrounding it is not only in poor taste but is proof that a predominantly white industry continues to preference white authors over authors of color."[20] For more on diversity and the structure of the book publishing industry, see "Media Literacy and the Critical Process: Publishing Gatekeepers: Who Gets to Write the Big Books?"

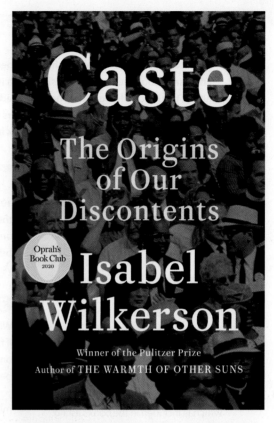

2020 WAS A YEAR FOR READING, and books relating to many of the year's struggles—those on racism, civil rights, and politics—led the nonfiction bestseller lists. Isabel Wilkerson's *Caste: The Origins of Our Discontents*, which examines the hierarchical and discriminatory caste social structure of the United States, India, and Nazi Germany, was one of the most acclaimed books of the year. Wilkerson, a Pulitzer Prize–winning journalist, published her first book, *The Warmth of Other Suns: The Epic Story of America's Great Migration* in 2010 with Random House. The book won the National Book Critics Circle Award and led to *Caste*, her second book with the same publisher.

MEDIA LITERACY & THE CRITICAL PROCESS

Publishing Gatekeepers: Who Gets to Write the Big Books?

As noted in this chapter, publishing has been subject to increasing criticism about the lack of diversity in the industry, which raises questions about the kinds of books that get acquired and promoted as potential blockbusters. How diverse are the authors of today's best-selling books, and does the diversity of those who work in the industry have an impact on who gets published?

① DESCRIPTION

Visit four current *New York Times* best-seller lists: Hardcover Fiction, Hardcover Nonfiction, Young Adult Hardcover, and Children's Picture Books. Research the Top 10 authors on each list, and then, as best as you are able, describe the race, ethnicity, and gender of each author. For example, you may be able to consult the author's biographical note on the book's website to see how the author self-identifies. Also note the publisher for each book, which is included in the book's listing.

② ANALYSIS

What patterns emerge? What is the race, ethnicity, and gender of the authors on these best-seller lists? Are there differences between authors on the fiction and nonfiction lists, or between authors of adult books and young-adult and children's books?

③ INTERPRETATION

Why do you think these patterns exist? If there are differences between authors on the fiction and nonfiction lists, or between authors of adult books and young-adult and children's books, what might be the reasons for this? For example, does the diversity of authors change with the age of the book's expected readers—that is, is there a generational expectation of more diversity from some age groups of readers than others?

④ EVALUATION

Overall, does the lack of diversity in the publishing industry seem to have a gatekeeping effect on the types of books and authors that publishers select and promote? Does the diversity of those who hold positions of influence in book publishing—and in other media industries—make a difference? Why or why not?

⑤ ENGAGEMENT

Go to the corporate website of one of the publishers of these best-selling books and read its diversity policy if one has been posted. E-mail the publisher to comment on the policy or, if appropriate, to ask why there isn't a publicly disclosed policy. Alternatively, write up the results of your findings on diversity in publishing, and either submit it to the arts editor of your campus newspaper or share it on social media.

Selling Books: Book Superstores and Independent Booksellers

Traditionally, the final part of the publishing process involves the business and order fulfillment stages—shipping books to thousands of commercial outlets and college bookstores. Warehouse inventories are monitored to ensure that enough physical copies of a book will be available to meet demand. Anticipating such demand, though, is tricky. No publisher wants to be caught short if a book becomes more popular than originally predicted or get stuck with books it cannot sell, as publishers must absorb the cost of returned books. Independent bookstores, which tend to order more carefully, return about 20 percent of books ordered; in contrast, mass merchandisers such as Walmart, Sam's Club, Target, and Costco, which routinely overstock popular titles, often return up to 40 percent. Returns this high can have a serious impact on a publisher's bottom line. For years, publishers have talked about doing away with the practice of allowing bookstores to return unsold books to the publisher for credit.

Today, about eleven thousand outlets sell books in the United States, including traditional bookstores, department stores, drugstores, used-book stores, and toy stores. Shopping-mall bookstores strengthened book sales beginning in the late 1960s, but it was the development of book superstores in the 1980s that truly reinvigorated the business. Following the success of a single Borders store established in Ann Arbor, Michigan, in 1971, a number of book chains began developing book superstores to cater to suburban areas as well as avid readers. A typical Barnes & Noble superstore stocks up to 200,000 titles. As superstores expanded, they began to sell music and coffee. Superstores like Borders and Barnes & Noble began to push out small, independent bookstores. In fact, between 1995 and 2000, 43 percent of independent bookstores in the United States closed.[21]

Yet the next wave of bookselling had already arrived. As Amazon (established in 1995) grew as an online bookseller (more on this in the next section), Borders lost more and more business, eventually closing its last brick-and-mortar store in 2011, making Barnes & Noble the last national bookstore retail chain in the United States. In 2019, Barnes & Noble was purchased by the British investment firm that also owned Waterstones, the largest U.K. bookseller. In 2021, Barnes & Noble remained the largest physical retailer of books in the United States, operating more than six hundred stores across all fifty states. Interestingly, Amazon invested in its own physical Amazon Books stores beginning in 2015. By 2021, Amazon Books had twenty-four U.S. locations, each with about five thousand titles—a fraction of the size of a Barnes & Noble superstore.

As superstores experienced difficulties, an amazing thing happened: independent bookstores came back. After a low point of just 1,651 independent bookstores in 2009, the number of stores

James Kirkikis/AGE Fotostock

INDEPENDENT BOOKSTORES City Lights Books in San Francisco is both an independent bookstore and an independent publisher, publishing nearly two hundred titles since launching in 1955, including poet Allen Ginsberg's revolutionary work *Howl*. Customers from around the world now come to browse through the landmark store's three floors and to see the place where beatniks like Ginsberg got their start. The iconic bookstore nearly closed in 2020 during the pandemic, but a GoFundMe campaign raised more than $500,000 to sustain it.

grew by 49 percent to more than 2,500 in 2019 (though the pandemic subsequently presented new challenges).[22] A professor at Harvard Business School says that the success of independent bookstores relates to the three Cs: community (responding to the distinct local needs of the place), curation (providing a thoughtfully selected inventory for customers), and convening (making the store a place for meetings, reading groups, events, and even parties).[23] For a look at the role of bookstores around the world, including independent bookstores, see "Global Village: Buenos Aires, the World's Bookstore Capital."

Selling Books Online

Beginning in the late 1990s, online booksellers created an entirely new book distribution system on the Internet. The strength of online sellers lies in their convenience and low prices, and especially their ability to offer backlist (older) titles and the works of less famous authors that retail stores aren't able to carry on their shelves. Online customers are also drawn to the interactive nature of these sites, which allow readers to post their own book reviews, read those of fellow customers, and receive book recommendations based on book searches and past purchases.

The trailblazer is Amazon, established in 1995 by then thirty-year-old Jeff Bezos, who left Wall Street to start a web-based business. Bezos realized that books were an untapped and ideal market for the Internet, with more than three million publications in print and plenty of distributors to fulfill orders. He moved to Seattle and started Amazon, so named because search engines like Yahoo! listed categories in alphabetical order, putting Amazon near the top of the list. In 1997, Barnes & Noble, the leading retail bookstore, launched its own online book site, BN.com. The site's success, however, remains dwarfed by that of Amazon.

Beyond selling books, Amazon's bigger objective was to transform the book industry itself, from one based on bound paper volumes to one based on digital files. The introduction of the Kindle e-reader in 2007 made Amazon the fastest book delivery system in the world. Instead of going to a bookstore or ordering from Amazon and waiting for the book to be delivered in a box, one could buy a digital version in a few seconds from the Amazon store. Amazon quickly grew to control about 90 percent of the e-book market, which it used as leverage to force book publishers to comply with its low prices or risk getting dropped from Amazon's bookstore (something that has happened to several independent book publishers who complained).[24] Amazon has done the same in print book sales, where it is also a major player.

Amazon's price slashing caused most of the major trade book publishing corporations to endorse Apple's *agency-model* pricing, in which the publishers set the book prices and the digital bookseller gets a 30 percent commission. When the U.S. Department of Justice ruled in 2013 that Apple and the major publishers had colluded to set book prices (thus denying consumers the lower prices that Amazon's deep discounts might offer), the booksellers responded that government investigators should be more concerned about Amazon and its growing power. Of particular concern to publishers is that Amazon has expanded into the domain of traditional publishers with the establishment of Amazon Publishing, which has grown rapidly since 2009. With a publishing arm that can sign authors to book contracts, distribution provided by the Amazon store, and millions of Kindle devices in the hands of readers, Amazon is a vertically integrated company and, traditional publishers fear, a too-powerful entity. Amazon ultimately agreed with the five major book publishers in 2014 and let them set e-book prices. At the same time, Amazon began undercutting the major publishers' e-books (often costing between $12.99 and $14.99) with e-books from independent publishers or from its own in-house publishing business, which typically cost between $2.99 and $5.99.

Buenos Aires, the World's Bookstore Capital

Buenos Aires, Argentina, has more bookstores per capita than any other city in the world—about 700, or roughly 25 bookstores per 100,000 people.[1] By comparison, Madrid has 16 bookstores per 100,000 people, Tokyo has 13, London has 10, Paris and New York each have 9, Amsterdam and Berlin each have 7, Los Angeles and Rio de Janeiro each have 5, Mumbai has 4, and Singapore has 3.

Buenos Aires also has some of the best bookstores in the world, including El Ateneo Gran Splendid. A former theater palace built in 1919, the ornate building was repurposed as a bookstore in 2007. The main floor and balconies are filled with bookshelves; the former theater boxes are now reading nooks; and the stage, framed by a crimson curtain, is a café. Juan Pablo Marciani, manager of El Ateneo, says books are a significant part of Argentinian culture: "Books represent us like the tango. We have a culture very rooted in print."[2] So how did Buenos Aires come to have such a bookish culture?

Javier Larrea/AGE Fotostock

Partly it's due to chance; the country's literary community grew and flourished with the influx of Spanish writers and publishers who fled to Argentina during the Spanish Civil War in the 1930s. It is also because of choice: Argentina doesn't charge sales tax on books. Heavy taxes on e-readers and tablets have kept e-book use low, too.[3] The local bookstore industry is also helped by the fact that Amazon.com doesn't have a retail website in Argentina. Even if Argentina did have Amazon, it's likely that the company wouldn't gain the dominant foothold it now has in the United States and the U.K. in digital and print book sales.

The main reason behind the country's thriving bookstore culture, though, is that Argentina has fixed book pricing (FBP). Several other countries have FBP as well, including Austria, France, Germany, Greece, Hungary, Italy, Japan, Lebanon, Mexico, the Netherlands, Norway, Portugal, South Korea, and Spain.[4] FBP rules among countries vary, but they generally require bookstores to limit price discounts during the first six months to two years following the release of a book. The rules usually apply to digital books as well. The effect is that all bookstores in a country sell the latest titles for roughly the same price, even digital books. Countries without FBP tend to have large book chains and Internet retailers, which can offer deep discounts on new releases and easily dominate the market.[5] In the United States, Amazon used the Internet to change the distribution rules and undercut the brick-and-mortar bookstores.

The effect of FBP can be seen when comparing the number of independent bookstores in various countries. The United Kingdom, which gave up FBP in the late 1990s, has lost one-third of its bookstores since 2005, and independents represent only about 4 percent of the bookseller market. In France, however (which has FBP), independent booksellers represent 22 percent of the market. France has about twenty-five hundred booksellers, more than the nineteen hundred booksellers in the United States, even though France is five times smaller in population.[6]

Catherine Blache of the French Publishers Association explains that in France, "booksellers compete not on price, therefore, but in terms of the variety of books they offer, their location and the quality of their customer service."[7] This is what the Buenos Aires bookselling market looks like, too. *Travel + Leisure* magazine recommends this special feature of life in Argentina's capital. "Buenos Aires is bursting with bookstores. . . . Enjoy the land that Amazon forgot."[8] ●

AMAZON'S WAREHOUSES go far beyond the stockrooms of a typical brick-and-mortar store, housing more than one hundred employees at each location, of which there are dozens across the United States and around the world. Though Amazon still uses these warehouses to support its massive fulfillment needs, it owns a lot of virtual businesses, too, including cloud storage, e-publishing, e-commerce sites like Zappos, and social media like Goodreads, which Amazon bought in 2013.

Sean Gallup/Getty Images

By 2018, Amazon still dominated the U.S. e-book market, with an 88.9 percent share of sales, while Apple iBooks accounted for 6.3 percent and all others held a share of 4.8 percent. Moreover, Amazon dominates in print books, selling 42 percent of U.S. print books in 2018.[25]

Self-Published Books

Because e-books make publishing and distribution costs low, **e-publishing** has enabled authors to sidestep traditional publishers. Internet-based publishing houses, such as Author Solutions (with imprint brands including Xlibris and iUniverse), Hillcrest Media, and Amazon's CreateSpace design and distribute books for a comparatively low price for aspiring authors who want to self-publish a title, which can be formatted for the Kindle or iPad. Although sales are typically low for such books, the low overhead costs allow higher royalty rates for the authors and lower retail prices for readers. Self-publishing opportunities also exist for audio books.

Sometimes self-published books make it to the best-seller lists. For example, British writer E. L. James's blockbuster erotic novel *Fifty Shades of Grey* was written as fan fiction and posted to a busy *Twilight* fan forum beginning in 2009, where thousands read and commented on it. In 2012, Vintage bought the rights to the *Fifty Shades* trilogy for more than $1 million.

Books and the Future of Democracy

In our digital age, the book-reading habits of children and adults have become a social concern. After all, books have played an important role not only in spreading the idea of democracy but also in connecting us to new ideas beyond our local experience. The impact of our oldest mass medium—the book—remains immense. Without the development of printing presses and books, the idea of democracy would be hard to imagine. From Harriet Beecher Stowe's *Uncle Tom's Cabin*, which helped bring an end to slavery in the 1860s, to Rachel Carson's *Silent Spring*, which led to reforms in the pesticide industry in the 1960s, to Michelle Alexander's *The New Jim Crow*, which delivered a scathing indictment of mass incarceration of people of color and supported a movement for criminal-justice reform in 2010, books have made a difference. They have told us things that we wanted—and needed—to know, and they have inspired us to action. And, despite what some people might believe, Americans are still reading books. A Pew Research Center study in 2019 found that 73 percent of Americans age eighteen and older had read at least one book in the past year.[26] Moreover, readers experienced books across several formats: print, e-books, and audio books. Among all Americans, the median number of books read per year is in the six-to-twelve range (that is, half of adults read more than six-to-twelve books, half read fewer).

That being said, a more worrisome trend is that 27 percent of adult Americans reported not reading or listening to any books in the previous twelve months—substantially more than back in 2011, when only 19 percent reported they hadn't read a book in the past year. Education and income are associated with reading (and not reading) books. Just 8 percent of college graduates reported that they had not read a book, compared to 44 percent of those with a high school education or less. Similarly, 14 percent of those in households with incomes of $75,000 or more reported not reading a book in the past year, compared to 36 percent of those in households with incomes of $30,000 or less. According to Pew, part of the problem may be technological: "Adults with lower levels of educational attainment are also among the least likely to own smartphones, a device that saw a substantial increase in usage for reading e-books."[27]

Although there remains substantial interest in books, many people are also concerned about their quality. Editors and executives often prefer to invest in commercially successful authors or those who have a built-in television, sports, or movie audience. In his book *The Death of Literature*, Alvin Kernan argues that serious literary work has been increasingly overwhelmed by the triumph of consumerism. People jump at craftily marketed celebrity biographies and popular fiction, he argues, but seldom read serious works. He contends that cultural standards have been undermined by marketing ploys that divert attention away from serious books and toward mass-produced works that are more easily consumed.[28]

Yet books and reading have survived the challenges of visual and digital culture. Developments such as e-books, audio books, graphic novels, and online booksellers have integrated aspects of print and digital culture into our daily lives. Most of these newer forms carry on the legacy of earlier books, transcending borders to provide personal stories, world history, and general knowledge to all who can read.

Since the early days of the printing press, books have helped us understand ideas and customs outside our own experiences. For democracy to work well, we must read. When we examine other cultures through books, we discover not only who we are and what we value but also who others are and what our common ties might be.

Chapter REVIEW 10

➔ QUIZ YOURSELF WITH LEARNINGCURVE

LearningCurve uses gamelike quizzing to help you review important concepts from this chapter. LearningCurve includes multiple-choice questions at different levels of difficulty, hints, feedback for both right and wrong answers, and links to e-book material for easy reference. Which answer choice would you select for the sample question below?

> While doing research, Robert visits an archive that holds a number of ancient Roman texts. Which item is Robert NOT likely to find in this archive?
>
> ○ a clay tablet
>
> ○ papyrus
>
> ○ a codex
>
> ○ parchment

➔ WATCH VIDEO ON KEY CONCEPTS

LaunchPad includes clips from movies, TV shows, online sources, and other media texts, in addition to interviews with media experts and newsmakers. In the videos for this chapter, you'll consider topics like the influence of Amazon and the relationship between books and movies.

➔ PRACTICE THE CRITICAL PROCESS

Use the Media Literacy Activity to walk through the critical process step-by-step and develop a critical perspective on what media means to you. In this chapter's activity, you'll reflect on the importance of books in people's lives.

➔ LEARN KEY TERMS

Review the definitions for this chapter's key terms with LaunchPad's easy-to-use online flash cards. Try the cards in practice mode, or quiz yourself as a way to focus your study efforts. (Page numbers indicate where each term is highlighted within the chapter.)

REVIEW QUESTIONS

Use the questions below to revisit key themes and concepts from this chapter.

 Need more practice? Access the LearningCurve multiple-choice questions for this chapter, which are available on LaunchPad. All questions include feedback for correct and incorrect answers, hints, and links to e-book material.

Chapter Opener and Books Today

1. What are some reasons that graphic novels and comics are a growing segment of the trade book division?

How We Got Here: The Early History of Books

2. What distinguishes the manuscript culture of the Middle Ages from the oral and print eras in communication?
3. Why was the printing press such an important and revolutionary invention?
4. Why were books particularly important to women readers during the early periods of American history?

Mass Publishing and the Book Industry

5. Why did publishing houses develop?
6. Why is the trade book division the most lucrative part of the book industry?
7. What factors have been causing a decline in the cost of an academic year's worth of college textbooks?

Trends and Issues in Book Publishing

8. What is the relationship between the book and the TV and movie industries?
9. Why did the Kindle succeed in the e-book market where other devices had failed?
10. What are the major issues in the debate over digitizing millions of books for web search engines?

The Organization and Economics of the Book Industry

11. What are the general divisions within a typical publishing house?
12. How have online booksellers affected bookstores and the publishing industry?
13. What are the concerns over Amazon's powerful role in determining book pricing and having its own publishing division?

Books and the Future of Democracy

14. Why are books an essential contribution to democratic life?

QUESTIONING THE MEDIA

Use these critical-thinking questions to reflect on your own experiences with media and apply your understanding of media concepts.

 For an interactive experience, questions on topics like these are available for the iClicker student response system. Instructors can download the questions from LaunchPad and use them to poll the class.

1. Now that so many books are available in digital formats, what advantages of the bound-book format are we sacrificing (if any)?
2. Given the digital nature of the book industry, if you were to self-publish a book, what marketing and distribution strategies would you use to help an audience find it?
3. Imagine that you are on a committee that oversees book choices for a high school library in your town. What policies do you think should guide the committee's selection of controversial books?
4. What is the cultural significance of a bound volume, particularly a religious holy book, such as the Bible or the Qur'an? If holy books are in digital form, does the format change their meaning?

Advertising and Commercial Culture

11

Hey, let's all live together in a big mansion in Southern California and create TikTok videos.

That's pretty much the idea behind Hype House, a collaboration that took up residence in a rented ($11,500 a month) Mediterranean-style mansion with palm trees and a swimming pool in late 2019 to churn out short videos for TikTok, one of the fastest-growing social media platforms.[1]

Millions of people have made TikTok videos, so what does it take to be part of the Hype House? According to TikTok star Chase "Lilhuddy" Hudson, who helps scout for new members, "You either have to be talented at something, or a weird funny mix, or extremely good looking."[2] (Being white seems to be a common attribute, too, at least among the founding members.)

"If you have all three, you're a TikTok god," said Alex Warren, another house member.[3]

If that's so, several TikTok gods were present at the creation of the Hype House.

Original Hype House member Addison Rae has 73 million TikTok followers, No. 2 on all of TikTok. She is beaten only by fellow dancer Charli D'Amelio, who was just fifteen when she made her first video in 2019 (and now is TikTok's No. 1 personality, with 105.2 million followers). D'Amelio was also an original Hype House member, along with her sister, Dixie (No. 7 on the list of TikTok stars, with 47.7 million followers), though they have since left the group.[4]

The founders of Hype House are Hudson (age seventeen when it was established) and Thomas Petrou (then twenty-one), who was a YouTube personality. Another founder, TikTokker Daisy Keech, left in March 2020 and filed a lawsuit in federal court, alleging that Hudson and Petrou cut their own sponsorship deals and pushed her out of the group. "The business has been completely stripped away from me," she told *Forbes* magazine.[5]

And that's the thing about Hype House and other content collaboratives—they're not about fun; they're businesses in which every member is an *influencer*, someone who is paid to promote products

◀ TikTok content collectives come and go, as agents try to package the most promising influencers. Girls in the Valley was a content (or "collab") house in Southern California that flamed out after several months.

309

on social media. For example, you will see plenty of TikTok videos with Hype House's resident influencers holding Bang energy drinks, one of their sponsors. Sometimes sponsors are musical artists, who hope that having an excerpt of one of their songs in an influencer's viral dance video will send it to the top of the streaming charts.

Once a social media personality has millions of followers, that person's attention can be worth a lot to advertisers. Different collaboration houses reflect individual styles and personalities that cater to different kinds of consumer goods: Hype House is "clean cut popular kids for whom life had always seemed easy"; Sway House is "edgy e-boys"; Girls in the Valley (which has since imploded) was "a sisterhood of Instagram baddies"; and, until coronavirus hit, Rihanna's Fenty Beauty House was a hub of Fenty Beauty product experimentation.[6]

Sponsorship money can be lucrative for top influencers, even for a single TikTok video. "You can make up to $30–$40,000 on a TikTok, if you are very big, yeah," Petrou said to *Entertainment Tonight*. But the money it costs advertisers to enlist Hype House—which has 20 million followers on TikTok, 6.6 million on Instagram, and 1.7 million on YouTube—can be worth it, Petrou argued. "If you want to reach 150 million people, and use 20 people to get there, in a very, very fast, like, manner, you can use us."[7]

Content collaboratives come and go as sponsorship opportunities emerge and dry up, but in 2020, Hype House carried on with some new influencers and a new LA house—a four-story complex twice the size of the original one. In December of that year, it was announced that the group signed a contract to create a Hype House collection of branded clothing and accessories, which would be available at retailers like Target and Kohl's. The company making the clothing and accessories praised the "immeasurable influence in the social sphere" of the Hype House creators.[8]

By 2021, there were an estimated 1 billion active monthly TikTok users worldwide, and about 100 million active monthly users in the United States. Sixty percent of U.S. TikTok users are Generation Z (born after 1996). As one independent social media advertising house in Utah noted, "If your brand's target audience includes anyone between the age[s of] 13 and 40, you should be on TikTok right now."[9] That demographic holds great interest for advertisers, and TikTok influencers (many of them Gen Zers themselves) are the leading contacts for access into that social sphere—at least for as long as they can maintain their audiences.

In which media format do you experience the most advertising? Would you rather pay more for media content without advertising, or would you prefer free content accompanied by advertising? How much advertising is too much for you?

Advertising Today

Today, advertisements are everywhere and in every media form. They take the human form of social media influencers (is this magical tea effective if my favorite social media personality says it is?). They show up in almost every Google search and get mixed in with user posts on Facebook, Instagram, Twitter, and most other social media sites. They clutter websites on the Internet and on the screens of our phones. They occupy up to half the space in most daily newspapers and consumer magazines. They fill our mailboxes and wallpaper the buses and subways we ride. Dotting the nation's highways, billboards promote fast-food and hotel chains, while neon signs announce the names of stores along major streets and strip malls. Ads are even found in the restrooms of malls, restaurants, and bars.

At local movie theaters, ads precede trailers for the latest Hollywood blockbusters. In the films themselves (and in TV shows, music videos, and digital games), **product placement**—the paid

YOU CAN'T STOP SPORT.

We may start from different places, but together we'll rise stronger.

appearance of particular goods in a narrative or scene—is a multimillion dollar business. Ads are part of a radio deejay's morning patter, get interspersed throughout most commercial podcasts, and routinely interrupt our favorite TV and cable programs. In addition, according to the Food Marketing Institute, the typical supermarket's shelves are filled with between thirty thousand and fifty thousand different brand-name packages, all functioning like miniature billboards. By some research estimates, Americans come into contact with five thousand or more advertisements and brand exposures each day (although we certainly cannot absorb all of them).[10]

Advertising comes in many forms, but in this chapter, we will concentrate on the more conspicuous advertisements that shape product images and brand-name identities. Because so much consumer advertising intrudes into daily life, ads are often viewed in a negative light. Although media executives agree that advertising is the foundation of a healthy media economy—far preferable to government-controlled media—audiences routinely complain about how many ads they are forced to endure, and they increasingly find ways to avoid them, like blocking pop-up ads with web browsers or subscribing to ad-free streaming video services. In response, market researchers routinely weigh consumers' tolerance—how many ads they are willing to tolerate to get "free" media content. Without consumer advertisements, however, many digital media industries would cease to function in their present forms. Advertising is the economic glue that holds most media industries together.

In this chapter, we will:

- Examine the historical development of advertising—an industry that helped transform numerous nations into consumer societies
- Look at the first U.S. ad agencies; early advertisements; the emergence of packaging, trademarks, and brand-name recognition; and the eventual shift to a more visually oriented culture
- Consider the direction and scope of contemporary advertising, including the influence of ad agencies and the dominance of digital advertising
- Outline the key persuasive techniques used in consumer advertising
- Investigate ads as a form of commercial speech, examine critical issues in advertising, and discuss the measures aimed at regulating advertising
- Look at political advertising and its impact on democracy

How We Got Here: The Development of American Advertising

Advertising has existed since 3000 BCE, when shop owners in ancient Babylon hung outdoor signs carved in stone and wood so that customers could spot their stores. Merchants in early Egyptian society hired town criers to walk through the streets, announcing the arrival of ships and listing the goods on board. Archaeologists searching Pompeii, the ancient Italian city destroyed when Mount

Vesuvius erupted in 79 CE, found advertising messages painted on walls. By 900 CE, many European cities featured town criers who not only called out the news of the day but also directed customers to various stores.

After the invention of the printing press in the mid-1400s, handbills, posters, and broadsides (long, newsprint-quality posters) became more common types of advertising. English booksellers printed brochures and bills announcing new publications as early as the 1470s, when posters advertising religious books were tacked onto church doors. In 1622, print ads imitating the oral style of criers appeared in the first English newspapers. Announcing land deals and ship cargoes, the first newspaper ads in colonial America ran in the *Boston News-Letter* in 1704.

To distinguish their approach from the commercialism of newspapers, early magazines refused to carry advertisements. By the mid-nineteenth century, though, most magazines contained ads, and most publishers started magazines hoping to earn advertising dollars. About 80 percent of these early advertisements covered three subjects: land sales, transportation announcements (stagecoach and ship schedules), and "runaways" (ads placed by farm and plantation owners seeking to re-enslave people who had escaped to freedom).

The First Advertising Agencies

In the early 1800s, little need existed for elaborate or extensive advertising, as few goods and services were available for sale. Before the Industrial Revolution, 90 percent of Americans lived in isolated areas and produced most of their own tools, clothes, and food. The minimal advertising that did exist usually featured local merchants selling goods and services in their own communities. In the United States, national advertising didn't begin in earnest until the 1850s, when railroads linking the East Coast to the Mississippi River valley began to transport consumer goods—as well as newspapers, handbills, and broadsides—across the country.

The first American advertising agencies were newspaper **space brokers**, individuals who purchased space in newspapers and sold it to various merchants. Newspapers, accustomed to a 25 percent nonpayment rate from advertisers, welcomed the space brokers, who paid up front. Brokers usually received discounts of 15 to 30 percent but sold the space to advertisers at the going rate. In 1841, Volney Palmer opened a prototype of the first ad agency in Boston; for a 25 percent commission from newspaper publishers, he sold space to advertisers.

Advertising in the 1800s

The first full-service modern ad agency, N. W. Ayer & Son, worked primarily for advertisers and product companies rather than for newspapers. Opening in 1869 in Philadelphia, the agency helped create, write, produce, and place ads in selected newspapers and magazines. The traditional payment structure at this time had the agency collecting a fee from its advertising client for each ad placed; the fee covered the price that each media outlet charged for placement of the ad plus a 15 percent commission for the agency. The more ads an agency placed, the larger the agency's revenue. Thus, agencies had little incentive to buy fewer ads on behalf of their clients. Nowadays, however, many advertising agencies work for a flat fee, and others are paid on a performance basis.

Trademarks and Packaging

During the mid-1800s, most manufacturers served retail store owners, who usually set their own prices by purchasing goods in large quantities. Manufacturers, however, came to realize that if their products were distinctive and associated with quality, customers would ask for them by name. This would allow manufacturers to dictate prices without worrying about being undersold by stores' generic products or bulk items. Advertising let manufacturers establish a special identity for their products, separate from those of their competitors.

Like many ads today, nineteenth-century advertisements often created the impression of significant differences among products when in fact very few differences actually existed. But when consumers began demanding certain products—either because of quality or because of advertising—manufacturers were able to raise the prices of their goods. With ads creating and maintaining brand-name recognition, retail stores had to stock the desired brands.

One of the first brand names, Smith Brothers, has been advertising cough drops since the early 1850s. Quaker Oats, the first cereal company to register a trademark, has used the image of William Penn, the Quaker who founded Pennsylvania in 1681, to project a company image of honesty, decency, and hard work since 1877. Other early and enduring brands include Campbell Soup, which came along in 1869; Levi Strauss overalls in 1873; Ivory Soap in 1879; and Eastman Kodak film in 1888. Many of these companies packaged their products in small quantities, thereby distinguishing them from the generic products sold in large barrels and bins.

Product differentiation associated with brand-name packaged goods represents the single biggest triumph of advertising. Studies suggest that although most ads are not very effective in the short run, over time they create demand by leading consumers to associate particular brands with quality. Not surprisingly, building or sustaining brand-name recognition is the focus of many product-marketing campaigns. But the costs that packaging and advertising add to products lead to many consumer complaints. The high price of many contemporary products is a direct result of advertising costs. For example, high-end jeans that cost $150 (or more) today are made from roughly the same inexpensive denim that has outfitted farmworkers since the 1800s. The difference now is that more than 90 percent of the jeans' cost goes toward advertising and profit.

Patent Medicines and Department Stores

By the end of the 1800s, two sources accounted for half of the revenue taken in by ad agencies: *patent medicines*—proprietary medicines sold with often-fraudulent claims about their curative powers—and department stores. Meanwhile, one-sixth of all print ads came from patent medicine and drug companies. Such ads ensured the financial survival of numerous magazines as "the role of the publisher changed from being a seller of a product to consumers to being a gatherer of consumers for the advertisers," according to Goodrum and Dalrymple in *Advertising in America*.[11] Bearing names like Lydia Pinkham's Vegetable Compound, Dr. Lin's Chinese Blood Pills, and William Radam's Microbe Killer, patent medicines were often made with water and 15 to 40 percent concentrations of ethyl alcohol. One patent medicine—Mrs. Winslow's Soothing Syrup—actually contained morphine. Powerful drugs in these medicines explain why people felt "better" after taking them; at the same time, they triggered lifelong addiction problems for many customers.

Many contemporary products, in fact, originated as medicines. Coca-Cola, for instance, was initially sold as a medicinal tonic and even contained traces of cocaine until 1903, when it was replaced with caffeine. Early Post and Kellogg's cereal ads promised to cure stomach and digestive problems. Many patent medicines made outrageous claims about what they could cure, leading to increased public cynicism. As a result, advertisers began to police their ranks and developed industry codes to restore consumer confidence. Partly to monitor patent medicine claims, Congress passed the Pure Food and Drug Act in 1906.

Along with patent medicine ads, department store ads were also becoming prominent in newspapers and magazines. By the early 1890s, more than 20 percent of ad space was devoted to department stores. At the time, these stores were often criticized for undermining small shops and businesses, where shopkeepers personally served customers. The more impersonal department stores allowed shoppers to browse and find brand-name goods themselves. Because these stores purchased merchandise in large quantities, they could generally sell the same products for less.

PATENT MEDICINES
Unregulated patent medicines, such as the ones represented in these ads, created a bonanza for nineteenth-century print media in search of advertising revenue. After several muckraking magazine reports about deceptive patent medicine claims, Congress passed the Pure Food and Drug Act in 1906.

Promoting Social Change and Dictating Values

As U.S. advertising became more pervasive, it contributed to major social changes in the twentieth century. First, it significantly influenced the transition from a producer-oriented to a consumer-oriented society. By stimulating demand for new products, advertising helped manufacturers create new markets and recover product start-up costs quickly. From farms to cities, advertising spread the word—first in newspapers and magazines and later on radio and television. Second, advertising promoted technological advances by arguing that new machines—such as vacuum cleaners, washing machines, and cars—could improve daily life. Third, advertising encouraged economic growth by increasing sales. To meet the demand generated by ads, manufacturers produced greater quantities, which reduced their costs per unit; however, they did not always pass these savings on to consumers.

Appealing to Female Consumers

By the early 1900s, advertisers and ad agencies came to believe that women, who constituted 70 to 80 percent of newspaper and magazine readers, controlled most household purchasing decisions. (This is still a fundamental principle of advertising today.) Ironically, more than 99 percent of the copywriters and ad executives at that time were men, based primarily in Chicago and New York. Copywriters emphasized stereotyped appeals to women, believing that simple ads with emotional and even irrational content worked best. Thus, early ad copy featured personal tales of "heroic" cleaning products and household appliances. The intention was to help female consumers feel good about defeating life's problems—an advertising strategy that endured throughout much of the twentieth century.

Dealing with Criticism

During the 1940s, the industry began to actively deflect criticism that advertising created needs that ordinary citizens never knew they had—for example, preventing body odor or removing female leg/armpit hair. Criticism of advertising grew as the industry appeared to be dictating values (e.g., about beauty standards or childhood necessities) as well as driving the economy. To promote a more positive image, during World War II the industry developed the War Advertising Council—a voluntary group of agencies and advertisers that organized war bond sales, blood donor drives, and the rationing of scarce goods.

The postwar extension of advertising's voluntary efforts became known as the Ad Council. This organization has earned praise over the years for its Smokey the Bear campaign ("Only you can prevent forest fires"); its crash test dummy spots for the Department of Transportation, which substantially increased seat belt use; and its extensive Coronavirus Response campaign in 2020–2021. Choosing a dozen worthy causes annually, the Ad Council continues to produce pro bono *public service announcements* (PSAs) on a wide range of topics, including suicide prevention, cancer screening, racial justice, texting and driving, and youth vaping.

Early Ad Regulation

The early 1900s saw the formation of several watchdog organizations. Partly to keep tabs on deceptive advertising, advocates in the business community in 1913 created the nonprofit Better Business Bureau, which now has more than one hundred branch offices in the United States. At the same time, advertisers wanted a formal service that would track newspaper readership, guarantee accurate audience measures, and ensure that papers would not overcharge ad agencies and their clients. As a result, publishers formed the Audit Bureau of Circulations (ABC) in 1914 (now known as the Alliance for Audited Media).

That same year, the government created the Federal Trade Commission (FTC), in part to help monitor advertising abuses. Thereafter, the industry urged self-regulatory measures to keep government interference at bay. For example, the American Association of Advertising Agencies (AAAA)—established in 1917—tried to minimize government oversight by urging ad agencies to refrain from making misleading product claims.

Finally, the advent of television dramatically altered advertising. With this new visual medium, ads increasingly intruded on daily life. Critics also discovered that some agencies were using **subliminal advertising**. This term, coined in the 1950s, refers to hidden or disguised print and visual messages that allegedly register in the subconscious and fool people into buying products. Noted examples of subliminal ads from that time include a "Drink Coca-Cola" ad embedded in a few frames of a movie and alleged hidden sexual activity drawn into liquor ads. Although research suggests that such ads are no more effective than regular ads, the National Association of Broadcasters banned the use of subliminal television ads in 1958.

The Influence of Visual Design

Until the 1960s, the shape and pitch of most U.S. ads were determined by a **slogan**, a phrase that attempts to sell a product by capturing its essence in words. With slogans such as "A Diamond Is Forever" (which De Beers introduced in 1948), the visual dimension of ads was merely a complement. Eventually, however, images asserted themselves, and visual style became dominant in U.S. advertising and ad agencies.

As a postmodern design phase developed in art and architecture in the late 1960s and early 1970s, a postmodern design era began to affect advertising as well. Part of this visual revolution was imported from non-U.S. schools of design; indeed, ad-rich magazines such as *Vogue* and *Vanity Fair*

PUBLIC SERVICE ANNOUNCEMENTS

The Ad Council has been creating public service announcements (PSAs) since 1942. Supported by contributions from individuals, corporations, and foundations, the council's PSAs are produced pro bono by ad agencies. This PSA is the result of the Ad Council's long-standing relationship with the nonprofit Keep America Beautiful campaign.

increasingly hired European designers as art directors. These directors tended to be less tied to U.S. word-driven print advertising because the European market required advertising that could transcend multiple linguistic boundaries. Thus, the image became paramount.

By the mid-1980s, the visual techniques of MTV, which initially modeled its style on advertising, influenced many ads and most agencies. MTV promoted a particular visual aesthetic—rapid edits, creative camera angles, compressed narratives, and staged performances. Video-style ads soon saturated television and featured such prominent performers as Paula Abdul, Ray Charles, Michael Jackson, Elton John, and Madonna. The popularity of MTV's visual style also started a trend in the 1980s to license hit songs for commercial tie-ins. By the twenty-first century, a wide range of short, polished musical performances and familiar songs—including Ateph Elidja's "Regular" (Apple iPhone), Club Yoko's "I Am" (Samsung Galaxy Note), and Patsy Cline's "Back in Baby's Arms" (Facebook)—were routinely being used in TV ads, essentially providing the model for the popular music that is used in TikTok videos today.

Most recently, the Internet, computers, and smartphones have had a significant impact on visual design in advertising. As the web became a mass medium in the 1990s, TV and print designs often mimicked the drop-down menu of computer interfaces. In the twenty-first century, visual design evolved in other ways, becoming more realistic and interactive as full-motion 3-D animation became a high-bandwidth multimedia standard. Design is also simpler, as ads and logos need to appear clearly on the small screens of smartphones, and more international, as agencies need to appeal to the global audiences of many companies by reflecting styles from around the world.

The Shape of Contemporary U.S. Advertising

Advertising has made a seismic shift in the last decade, from a world revolving around mega- and boutique agencies that influence consumer culture with carefully planned ad campaigns to Internet advertising that is data based and socially driven. Today, Google Adwords, social media influencers (and the products they promote), and predictive marketing using data gleaned from billions of Internet and mobile phone users are all defining a new era of advertising. In this section, we look at what agencies bring to the table, as well as the dominance of Internet ads.

Types of Advertising Agencies

About thirteen thousand ad agencies currently operate in the United States. In general, these agencies are classified as either **mega-agencies**—large ad firms that form when several agencies merge and that maintain regional offices worldwide—or small **boutique agencies**, which devote their talents to only a handful of select clients. During the coronavirus pandemic that hit the country in 2020, advertising suffered in some sectors (restaurants, entertainment, hotels, air travel) but thrived in others (home consumer goods, consumer electronics, pharmaceuticals, health care).

Mega-Agencies

Mega-agencies provide a full range of services, from advertising and public relations to their own in-house radio and TV production studios. In 2019, the top three global mega-agencies were WPP, Omnicom, and Publicis (see Figure 11.1). London-based WPP, founded in 1971 as Wire and Plastic Products, grew quickly in the 1980s as an ad firm by purchasing the largest U.S. ad firm at the time (J. Walter Thompson), one of the largest U.S. public relations agencies (Hill+Knowlton), and ad agency Ogilvy & Mather Worldwide. WPP's revenue in 2019 was $16.9 billion. The company today has about 107,000 employees worldwide and controls a number of leading U.S. advertising and PR agencies, including Young & Rubicam and Burson-Marsteller.

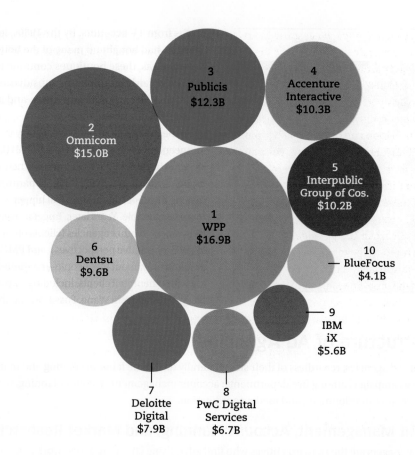

FIGURE 11.1

GLOBAL REVENUE FOR THE WORLD'S LARGEST AGENCIES (IN BILLIONS OF DOLLARS)

Data from: Ad Age, Marketing Fact Pack, 2021 edition, p. 19. Numbers rounded. 2019 worldwide revenue.

In 2021, New York–based Omnicom Group had about 70,000 employees operating in over one hundred countries; today, the company owns the global advertising firms BBDO Worldwide, DDB Worldwide, and TBWA Worldwide. Omnicom also owns three leading public relations agencies: FleishmanHillard, Ketchum, and Porter Novelli. Paris-based Publicis Groupe has a global reach through agencies like Leo Burnett Worldwide, the British agency Saatchi & Saatchi, and the public relations firm MSL Group. Publicis employees numbered more than 80,000 worldwide in 2021.

Mega-agencies like these traditionally dominate the world of advertising, but this dominance was disrupted in the 2010s by two occurrences. First, digital giants like Google and Facebook began capturing more of the advertising market, far eclipsing the largest mega-agencies in annual revenue. (More on those digital companies later in the chapter.) Second, four global business consulting firms—Accenture Interactive, Deloitte Digital, PwC Digital Services, and IBM iX—added digital and data-based advertising activities to the list of services they offered to clients, causing them to break into the Top 10 of the world's largest agencies. (See Figure 11.1.)

In addition, the coronavirus pandemic slammed the advertising industry beginning in 2020, causing some businesses to see sales increases, while others were forced to reduce their advertising budgets and struggled to stay afloat. The CEO of one New York–based ad firm said it was a time of reckoning for big advertising agencies, which can have big spending habits and expensive "ego gratification" headquarters. He speculated about "how a holding company CEO [could] justify to the CFO a $3 million lobby refurbishment during a time when those things don't matter to clients."[12]

Boutique Agencies

The visual revolutions in advertising during the 1960s elevated the standing of designers and graphic artists, who became closely identified with the look of particular ads. Breaking away from bigger agencies, many of these creative individuals formed small boutique agencies. Offering more personal services, the boutiques prospered, bolstered by innovative ad campaigns and increasing

STRAWBERRY FROG, an independent ad agency of about sixty people based in New York, focuses on transforming businesses through what it calls "movement marketing," an approach based on tapping into positive cultural values.

profits from TV accounts. By the 1980s, large agencies had bought up many of the boutiques. Nevertheless, these boutiques continue to operate as fairly autonomous subsidiaries within multinational corporate structures, and new boutiques continue to emerge.

One independent boutique agency in Minneapolis, Peterson Milla Hooks (PMH), made its name in 1999 with a boldly graphic national branding ad campaign for Target department stores, making the department store feel hipper and much more fashionable. Years later, Target moved most of its business to other agencies (clients often do this, to get new creative perspectives), and PMH—which employs only about fifty people—continued working with other retail clients, including Gap, Kohl's, Sleep Number, Vera Wang, Fossil, and Sephora.

The Structure of Ad Agencies

Traditional ad agencies, regardless of their size, generally divide the labor of creating and maintaining advertising campaigns among five departments: account management, account planning, market research, creative development, and media coordination.

Account Management, Account Planning, and Market Research

Account managers are the sales executives who find advertising clients and then work to maintain those client relationships—especially with the biggest-spending clients—for what they hope will be many years. Successful account managers often earn the highest salaries at ad agencies because they generate business for the firm and help their clients make more money through effective advertising.

Once client accounts have been established, the account planner's role is to develop an effective advertising strategy by combining the views of the client, the creative team, and consumers. Consumers' views are the most difficult to understand, so account planners coordinate **market research** to assess the behaviors and attitudes of consumers toward particular products long before any ads are created. Researchers, who might work in-house or be contracted by the agency, may study everything from possible names for a new product to the size of the copy for a print ad. Researchers also test new ideas and products with consumers to get feedback before developing final ad strategies. In addition, some researchers contract with outside polling firms to conduct regional and national studies of consumer preferences.

In 1978, the Stanford Research Institute (SRI), now called Strategic Business Insights (SBI), instituted its **Values and Lifestyles (VALS)** strategy. Using questionnaires, VALS researchers measure psychological factors and divide consumers into types. VALS research assumes that not every product suits every consumer and thus encourages advertisers to vary their sales slants to find market niches.

Over the years, the VALS system has been updated to reflect changes in consumer orientations. The most recent system classifies people by their primary consumer motivations: ideals, achievement, or self-expression. The ideals-oriented group, for instance, includes *thinkers*—those who "plan, research, and consider before they act."[13] VALS and similar research techniques ultimately provide advertisers with microscopic details about which consumers are most likely to buy which products.

Agencies and clients—particularly auto manufacturers—have relied heavily on VALS to determine the best placement for ads. VALS data suggest, for example, that consumers classified as *achievers* (who are "goal oriented" and "moderate") and *experiencers* (who "want everything") watch more sports and news programs; these groups also prefer luxury cars or sport-utility vehicles. *Thinkers*, on the other hand, favor TV dramas and documentaries and like the functionality of minivans or the gas efficiency of hybrids.

VALS researchers do not claim that most people fit neatly into one category. But many agencies believe that VALS research can give them an edge in markets where few differences in quality may actually exist among top-selling brands. Consumer groups, wary of such research, argue that this type of categorization reduces consumers to a handful of stereotypes, which is how they are then portrayed in advertisements. There are now VALS segmentations for other countries, including China, the Dominican Republic, Japan, Nigeria, the United Kingdom, and Venezuela.

Creative Development

Teams of writers and artists—many of whom regard ads as a commercial art form—make up the nerve center of the advertising business. The creative department outlines the rough sketches for print and online ads and then develops the words and graphics. For radio, the creative side prepares a working script, generating ideas for everything from choosing the narrator's voice to determining background sound effects. For television, the creative department develops a **storyboard**, a sort of blueprint or roughly drawn comic-strip version of the potential ad. For digital media, the creative team may develop websites, interactive tools and games, downloads, and **viral marketing**—short videos or other content that (marketers hope) will quickly gain widespread attention as users share it with friends online or by word of mouth.

Often the creative side of the business finds itself in conflict with the research side. In the 1960s, for example, agencies Doyle Dane Bernbach (DDB) and Ogilvy & Mather both downplayed research, instead championing the art of persuasion and what "felt right." Still, both the creative and the strategic sides of the business acknowledge that they cannot predict with any certainty which ads and which campaigns will succeed. Agencies say ads work best by slowly creating brand-name identities—that is, by associating certain products over time with quality and reliability in the minds of consumers.

Media Coordination: Planning and Placing Advertising

Ad agency media departments are staffed by **media planners** and **media buyers**. The planners analyze the effectiveness of various media channels, such as the Internet, television, radio, and outdoor advertising. The buyers negotiate rates and place ads with the specific media outlets that the planners determine to be best suited to reach the targeted audience and later measure the effectiveness of those ad placements. For instance, a company like Procter & Gamble, one of the world's leading advertisers, displays its more than three hundred major brands—most of them household products like Crest toothpaste and Pampers diapers—on TV shows and in magazines primarily targeted at women. To reach male viewers, media buyers encourage advertisers like beer companies to spend their ad budgets on cable and network sports programming, evening talk radio, or sports

CREATIVE ADVERTISING
The New York ad agency Doyle Dane Bernbach created a famous series of print and television ads for Volkswagen beginning in 1959 (*below left*) and helped usher in an era of creative advertising that combined a single-point sales emphasis with bold design, humor, and honesty. Arnold Worldwide, a Boston agency, continued the highly creative approach with its award-winning "Drivers wanted" campaign for the New Beetle (*below right*).

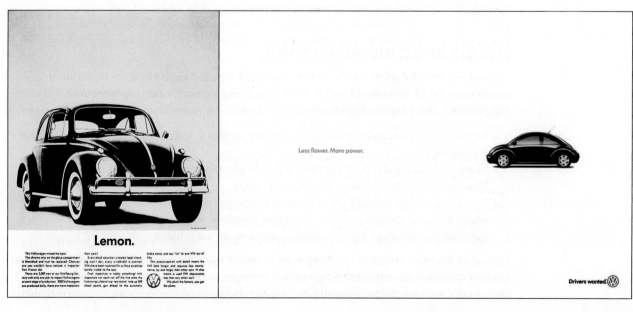

Lemon.

Less flower. More power.

Drivers wanted. ⓥ

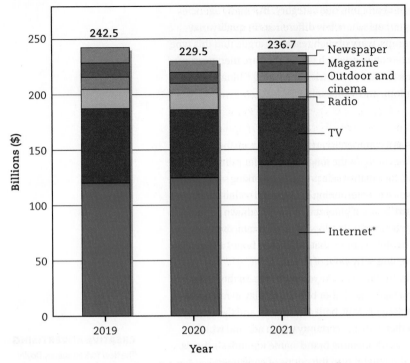

Note: Numbers are estimates and rounded; *includes online advertising for legacy media
(e.g., ads on newspaper websites)

FIGURE 11.2

WHERE WILL THE ADVERTISING DOLLARS GO?

Data from: Ad Age, *Marketing Fact Pack*, 2021 edition, p. 5

magazines. Ads can be even more precisely targeted in online searches and social media, which compile digital profiles on users for this purpose (more on this in a later section).

Advertisers often add incentive clauses to their contracts with agencies, raising the agency's compensation if sales goals are met and lowering it if goals are missed. Incentive clauses can sometimes encourage media planners and buyers to conduct repetitive **saturation advertising**, in which a variety of media are inundated with ads aimed at target audiences. The GEICO insurance campaign featuring a computer-animated gecko lizard is one of the most successful saturation campaigns in media history. The campaign started in 1998 (using an animated gecko to fill in for human actors during a Screen Actors Guild strike) and proved to be a popular device for helping consumers remember the similarly named insurance company. In later years, the gecko spoke with a working-class British Cockney accent to seem more memorable and relatable. GEICO was the most-advertised brand in the United States for the fourth consecutive year in 2020, and the gecko was the centerpiece of its advertising.[14]

The cost of advertising, especially on network television, increases each year. The Super Bowl remains the most expensive program for purchasing TV advertising, with thirty seconds of time costing an average of $5.5 million in 2021 on CBS—up from $4 million in 2014 on Fox. Running a thirty-second ad during a national prime-time TV show in 2020 cost anywhere from $10,500 (on the CW network's least expensive show) to more than $783,000 (on NBC's *Sunday Night Football*), depending on the popularity and ratings of the program. The prime-time average for a thirty-second TV spot was $99,678 in 2020, down from an all-time high of $129,600 in the prerecession year 2005.[15] (See Figure 11.2 for a look at how advertising dollars are spent by medium.)

Trends in Online Advertising

As noted earlier, digital giants Google and Facebook have captured a majority of the digital advertising market today (which includes web and mobile), far eclipsing the largest mega-agencies in annual advertising revenue. To get a sense of just how dominant the digital companies are, consider these numbers:

- WPP, the largest worldwide traditional ad agency, earned a total of $16.9 billion for its advertising work in 2019.
- That same year, Facebook earned $69.7 billion in advertising on the Facebook site, Instagram, Messenger, and affiliated websites and mobile apps.
- For its part, Google earned $134.8 billion on its advertising businesses in 2019—chiefly Google search ads, YouTube ads, and Google Network Members, which serve ads to apps and websites based on Internet users' demographics, location, or spending habits.

Because of this astounding success, Google and Facebook have often been called a *digital duopoly*, in that they controlled an estimated 60.7 percent of U.S. digital ad spending by 2020. Meanwhile, a third powerhouse, Amazon, has been growing its own digital advertising services; with Amazon included, the three companies controlled 69.3 percent of U.S. digital ad spending by 2020.[16] These digital companies, and to a lesser degree their more traditional agency rivals, have created a robust and thriving digital advertising market—one that outsells all other forms of advertising today.

Search and Mobile Bring Digital Dominance

When digital advertising first arrived in the mid-1990s, it was a modest business featuring *banner ads*, the print-like display ads that load across the top or side of a web page. Since that time, other formats have emerged, including video ads, sponsorships, and "rich media"—content like pop-up ads, pop-under ads, flash multimedia ads, and *interstitials*, which pop up in new screen windows as a user clicks to a new web page. Other forms of Internet advertising include classified ads and e-mail ads. (Unsolicited commercial e-mail—known as **spam**—constitutes 54 percent of all e-mails in the world's Internet traffic.)[17]

Paid search advertising soon became the dominant format of web advertising, which is how Google grew to be the world's top advertising company: It started selling sponsored links associated with its users' search terms. Back in 2004, digital ads accounted for just over 4 percent of global ad spending. By 2007, buoyed by the strength of Google's dominance in the search realm, that share had roughly doubled—to 8.2 percent. The Internet became the top advertising medium in 2017, and by 2021 it was projected to account for an estimated 57.5 percent of all ad spending in the United States.[18] In fact, although the coronavirus pandemic hurt the advertising industry, the Internet Advertising Bureau predicted that digital ad spending would increase by 6 percent in 2020, while all other media advertising would decline by 30 percent.[19]

Within digital advertising today, the biggest growth area by far is mobile ad sales—ads that appear mainly on smartphones. By 2021, mobile will account for more than 75 percent of all digital advertising (with the rest going to desktop web advertising).[20] The average time spent on mobile phones per day was one hour and thirty minutes back in 2012, but it jumped to three hours and six minutes per day in 2020 and is expected to climb to three hours and seventeen minutes by 2022. With more than a doubling of time spent on mobile phones in a decade, we should expect to see a continued emphasis on mobile advertising.[21]

The Rise of Digital Ad Agencies

The dominance of web and mobile advertising has influenced strategy at both ad agencies and digital companies. Because Internet advertising is the leading growth area, advertising mega-agencies have added digital media agencies and departments to develop and sell ads online. For example, WPP has Xaxis, Omnicom owns Proximity Worldwide, Publicis has Digitas and Publicis Sapient, and Interpublic operates R/GA. In addition, as noted earlier, global business consulting firms like Accenture and Deloitte now offer digital and data-based advertising services, which has made them among the most successful agencies in the advertising world.

Meanwhile, after realizing the potential of their online ad businesses, major web services like Google and Facebook began acquiring smaller Internet advertising agencies. Google bought DoubleClick, the biggest online ad server (which sends ads to appear on customers' web pages), in 2008, and it purchased AdMob, a mobile advertising company, in 2009. Today they are both part of Google's marketing platform, which sells ads to customers and offers data analytics. Similarly, Facebook bought Rel8tion, a mobile advertising company, in 2011, and Atlas Solutions, an online advertising company, in 2013.

DIGITAL ADVERTISING is growing rapidly. Social media sites like Facebook, Twitter, and Instagram gather a huge amount of information from their users every day, allowing advertisers to reach specific users by displaying ads for products related to their unique preferences and behaviors.

Online Marketers Target Individuals

Why are Internet ads so successful? Compared to ads in traditional media outlets, such as newspapers, magazines, radio, and television, Internet ads offer many advantages to advertisers. Perhaps the biggest advantage—and potentially the most disturbing part for citizens—is that marketers (digital ad agencies or digital firms like Google) can develop consumer profiles that direct targeted ads to specific website visitors.

Marketers collect information about each Internet user in a number of ways, including through cookies (see Chapter 2) and online surveys. Here's how it works: When we access free content on an app or a website like ESPN.com, we automatically give the site permission—through default browser settings or our explicit consent—to attach digital files called cookies to our browser, which then track our activities. Cookies can track every page we visit on a particular website, as well as the other websites we visit. If we fill out an ESPN survey to enter a contest or express our opinion about an athlete or a sports team, we are also creating data that ESPN can offer to its paying advertising clients. If we use our browser outside of ESPN.com—say, to search for topics like "camping equipment" or "Adidas shoes"—the cookies may collect that search history, too.

All this data, compiled by digital ad agencies or data-mining companies, largely without our knowledge, can be combined with our credit card/retail sales data (what we buy and where) and used to develop profiles about our tastes, interests, values, emotions, and purchasing power. This increasingly detailed data helps advertising companies target us with ads today and predict what we might want to purchase down the road. Called **predictive marketing**, this new marketing strategy is as much about creating a desire for future products and services as it is about getting people to buy current ones. "All of this data . . . is being fed into these systems that have almost no human supervision," says Sandy Parakilas, a former manager of Facebook's privacy issues and policy compliance, "and that are making better and better and better and better predictions about what we're gonna do . . . and who we are."[22]

Beyond cookies and surveys, the big tech companies—primarily Google, Facebook, and Apple—are poised to further consolidate control over our data profiles through a new authentication strategy called single sign-on (SSO). As users, when we sign in to a new website, we are often asked to use our Google or Facebook account name and password instead of creating new credentials. (Apple enables its customers to use Touch ID or Face ID to log in to affiliated sites.) This choice can simplify our lives, since we don't need to remember new logins and passwords for each site. But it also hands control over to these big tech companies, which can then gather data on our web preferences on each site. Google Chrome, the web's most popular browser, has begun to phase out cookies in favor of SSO. On the one hand, eliminating cookies gets rid of a major form of consumer tracking. On the other hand, SSO strategies further consolidate consumer information tracking among some of the world's most powerful digital companies.

All of the data compiled via strategies like cookies and SSO informs ad buying. Digital advertising agencies measure the effectiveness of Internet ads by tracking ad *impressions* (how often ads are seen) and *click-throughs* (how often ads are clicked on). This provides advertisers with specific data on the number of people who not only viewed the ad but also showed real interest by clicking on it. For advertisers, online ads are more beneficial because they can be precisely targeted and easily measured. For example, a business can use Google AdWords to create small ads that are linked to selected key words and geographic targeting (from global coverage to a small radius around a given location). AdWords tracks and graphs the performance of the ad's key words (through click-through and sales rates) and lets the advertiser update the campaign at any time. This kind of targeted advertising enables smaller companies—those with a $500 ad budget, for example—to place their ads in the same location as larger companies with multimillion-dollar ad budgets.

Smartphones in particular offer effective targeting of individuals, as they offer advertisers the opportunity to tailor ads according to a specific geographic location (e.g., a restaurant ad sent to someone in close proximity) or the user's demographic, since wireless providers already have that information.

Nikša Poleksić

SMARTWATCHES like the Galaxy Watch Active2 can track wearers' heart rhythms through ECG sensors, measure blood pressure, and even remind wearers to wash their hands periodically. They also connect to mobile phones.

Mobile phone companies have also developed unique applications for mobile advertising and searching. For example, iPhone's Siri can answer the question "Where can I get coffee?" with locations of coffee shops nearby. (Interestingly, Siri struggles with noncommercial requests.) Such apps are designed to help digital companies like Apple battle for a share of the mobile advertising market.

The next frontier in digital advertising may be wearables and IoT (Internet of Things) devices, which can be mined for even more personal data and may surrender much more information than an unsuspecting consumer might think. As one digital advertising CEO explained, "Internet-powered smart gadgets will synchronize with biorhythms and personal calendars. . . . Smartwatches, fitness trackers, sports watches and smart clothing will work as personal reminders and shopping advisors, closely synced to our lives."[23]

Advertising Dominates Social Media

Social media, such as Facebook, Twitter, and Instagram, provide a wealth of data for advertisers to mine. These sites and apps create an unprecedented public display of likes, dislikes, locations, and other personal information. And advertisers use such information to further refine their ability to send targeted ads that might interest users. Facebook and other sites (like Hulu) go even further by asking users if they dislike a particular ad. For example, choosing to hide a display ad on Facebook results in the question "Please tell us why you hid the ad," followed by the choices "repetitive," "too personal," "sensitive topic," "knows too much," "already purchased," and "irrelevant." All that information goes straight back to advertisers, so they can revise their advertising and try to engage you the next time.

Beyond allowing advertisers to target and monitor their ad campaigns, most social media platforms encourage advertisers to create their own online presence. For example, by 2021, Ben & Jerry's had a Facebook page with more than 8.8 million "likes." Despite appearances, profiles like these still constitute a form of advertising and serve to promote products to a growing online audience—for virtually no cost.

Companies and organizations also buy traditional paid advertisements on social media sites. A major objective of their *paid media* is to get *earned media*—to convince online consumers to promote products on their own. Imagine that the National Resources Defense Council buys an ad on Facebook that attracts your interest. That's a successful paid media ad for the council, but it's even more effective if it becomes earned media—that is, if you mark that you "like" it, which essentially gives the organization a personal endorsement. Knowing that you like the ad, your friends view it, too; as they pass it along, it gets more earned media and eventually becomes viral—an even greater advertising achievement. As the Nielsen rating service says about online earned media, "Study after study has shown that consumers trust their friends and peers more than anyone else when it comes to making a purchase decision."[24] Social media are helping advertisers use such personal endorsements to further their own products and marketing messages—basically, letting consumers do the work for them.

One controversy in online advertising is whether people have to disclose if they are being paid to promote a product. For example, influencers often review products or restaurants as part of their content. Some influencers with large followings have been paid (either directly or with gifts of free products or trips) to give positive reviews or promote products. When such instances came to light in 2008 and 2009, the influencers argued that they did not have to reveal that they were being compensated for posting their opinions. At the time, they were right. However, in 2009, the FTC released new guidelines that require influencers to disclose when an advertiser is compensating them to discuss a product.

In 2010, a similar controversy erupted when the FTC revealed that celebrities were being paid to tweet about their "favorite" products. In 2016, the FTC found Lord & Taylor in

IN 2021, BURGER KING announced a complete redesign of its image, with a revamped logo, new store exteriors and interiors, redesigned worker uniforms, and updated packaging to reflect a healthier, less complicated, and more plant-based food menu. The company's new approach and branding were prominently featured on social media, as shown in this post for the Impossible Whopper on Facebook. The Burger King Facebook page has more than 8.6 million "likes."

violation of the disclosure rules when the clothing company hired fifty influencers to wear the same designer dress within days of one another: "The dress not only sold out, but these sponsored posts reached 11.4 million Instagram users."[25] The bloggers did not disclose that their posts were sponsored, that the expensive dresses had been "gifted" to them, and that they earned an additional $1,000 and $4,000 each. The FTC fined Lord & Taylor an undisclosed amount.

A much bigger controversy and problem, especially for Facebook, involved a major data breach that undermined the trust of consumers using any social media platform. In late spring of 2018, news outlets reported that a firm known as Cambridge Analytica had collected private data from over fifty million Facebook users. The British company had then used the information in 2016 to try to influence voter behavior during the U.S. presidential election. In explaining the breach, TechAdvisory.org noted that Cambridge Analytica used the data "to create psychological profiles and invent better political drives to influence whom people would vote for. . . . There's no doubt that tens of thousands of users were manipulated into signing away their data without knowing it."[26] TechAdvisory.org recommends that consumers "remove third-party apps that use your Facebook account." While this is a good first step, issues of privacy and data breaches will continue to haunt digital companies and jeopardize their users because for some, the temptation to unethically or illegally influence millions of people is too great.

Persuasive Techniques in Contemporary Advertising

Ad agencies and product companies often argue that the main purpose of advertising is to inform consumers about available products in a straightforward way. Most consumer ads, however, merely create a mood or tell stories about products without revealing much else. A one-page magazine ad, a thirty-second TV spot, or a sponsored Facebook post gives consumers little information about how a product was made, how much it costs, or how it compares with similar brands. In managing space and time constraints, advertising agencies engage in a variety of persuasive techniques.

Conventional Persuasive Strategies

One of the most frequently used advertising approaches is the **famous-person testimonial**, in which a product is endorsed by a well-known person. Famous endorsers include Justin Timberlake for Bud Light, Taylor Swift for Diet Coke, and Beyoncé for Pepsi. Kim Kardashian revealed in a lawsuit that she makes $300,000 to $500,000 for an endorsement in a single Instagram post. She charges $5–$6 million for endorsement deals with multiple posts, plus a share of the company's profits.[27] Athletes earn some of the biggest endorsement contracts. For example, in 2015, Golden State Warrior guard Steph Curry made $12 million in endorsements for Under Armour, Degree, Kaiser Permanente, and Brita, among other companies. In 2016, JPMorgan Chase also added Curry (after buying the naming rights for Golden State's new arena). By 2020, Curry had accumulated $44 million in career endorsements, putting him sixth on the list of the world's highest-paid athletes. That same year, the two top endorsement earners were Roger Federer and Cristiano Ronaldo—at $100 million and $45 million, respectively, in career endorsement earnings.[28]

Another technique, the **plain-folks pitch**, associates a product with simplicity. Any ad that shows a "real" person (not a model or an actor) is using this technique. The Dove soap Campaign for Real Beauty, launched in 2004 by international consumer products corporation Unilever, famously used this approach by featuring women who were not flawless models (or retouched to appear so). The Subway restaurant chain plucked from obscurity Jared Fogle, an Indiana University student who claimed to have lost 245 pounds after a diet of Subway sandwiches, and hired him to serve as a plain-folks spokesperson for the fast-food company from 2000 to 2015. (In 2015, Fogle became a convicted sex offender after he pleaded guilty to having sex with minors and possessing child pornography. He is now serving a prison sentence until at least 2028, illustrating the potential downfall to companies that associate their brand with a real-life spokesperson.)

In contrast to the plain-folks pitch, the **snob-appeal approach** attempts to persuade consumers that using a product will maintain or elevate their social status. Advertisers selling jewelry, perfume, clothing, and luxury automobiles often use snob appeal. For example, the pricey bottled water brand Fiji ran ads in *Esquire* and other national magazines that said, "The label says Fiji because it's not bottled in Cleveland"—a jab intended to favorably compare the water bottled in the South Pacific to the drinking water of an industrial city in Ohio. (Fiji ended up withdrawing the ad after the Cleveland Water Department released test data showing that its water was purer than Fiji water.)

Another approach, the **bandwagon effect**, points out in exaggerated claims that *everyone* is using a particular product. Brands that refer to themselves as "America's favorite" or "the best" imply that consumers will be left behind if they ignore these products. A different technique, the **hidden-fear appeal**, plays on consumers' sense of insecurity. Deodorant, mouthwash, and shampoo ads frequently invoke anxiety, pointing out that only a specific product could relieve embarrassing personal hygiene problems and restore a person to social acceptability.

A final ad strategy, used more in local TV and radio campaigns than in national ones, has been labeled **irritation advertising**: creating product-name recognition by being annoying or obnoxious. Although both research and common sense suggest that irritating ads do not work very well, there have been exceptions. In the 1950s and 1960s, for instance, an aspirin company ran a TV ad illustrating a hammer pounding inside a person's brain. Critics and the product's own agency suggested that people bought the product, which sold well, to get relief from the ad as well as from their headaches. On the regional level, irritation ads are often used by appliance discount stores or local car dealers, who dress in outrageous costumes and yell at the camera.

ROMANCE IS ON

NEW REVLON ULTRA HD™ LIPSTICK

Revolutionary **wax-free gel technology** for a weightless feel and **true colour clarity** in one smooth swipe. 14 high-definition shades to love.

EMMA STONE WEARS HD GLADIOLUS #LoveIsOn

REVLON LOVE IS ON

"IS NOT A KISS THE VERY AUTOGRAPH OF LOVE?" HENRY FINCK, AUTHOR

The Advertising Archives

FAMOUS-PERSON TESTIMONIALS
Major stars used to be somewhat wary of appearing in ads (at least in the United States), but many brands now use celebrity endorsements. This Revlon campaign, for example, features actress Emma Stone.

The Association Principle

Historically, American car advertisements have shown automobiles in natural settings—on winding roads that cut through rugged mountain passes or across shimmering wheat fields—but rarely on congested city streets or in other urban settings where most driving actually occurs. Instead, the car—an example of advanced technology—merges seamlessly into the natural world.

This type of advertising exemplifies the **association principle**, a widely used persuasive technique that associates a product with a positive cultural value or image even if it has little connection to the product. For example, many ads displayed visual symbols of American patriotism in the wake of the 9/11 terrorist attacks in an attempt to associate products and companies with national pride. Media critic Leslie Savan noted that in trying "to convince us that there's an innate relationship between a brand name and an attitude," advertising may associate products with nationalism, happy families, success at school or work, natural scenery, freedom, or humor.[29]

Another popular use of the association principle is to claim that products are "real" and "natural"—possibly the most familiar adjectives associated with advertising. For example, Coke sells itself as "the real thing," and the cosmetics industry offers synthetic products that promise to make women look "natural." The adjectives *real* and *natural* saturate American ads yet almost always describe processed or synthetic goods. Green marketing has a similar problem, as it is associated with goods and services that aren't always environmentally friendly.

Philip Morris's Marlboro brand has used the association principle to completely transform its product image. In the 1920s, Marlboro began as a fashionable woman's cigarette. Back then, the company's ads equated smoking with a sense of freedom, attempting to appeal to women who had just won the right to vote. Marlboro, though, did poorly as a woman's product, and new campaigns in the 1950s and 1960s transformed the brand into a man's cigarette. Powerful images of active, rugged men dominated the ads. Often, Marlboro (in a long-running campaign from 1954 to 1999) associated its product with nature, displaying an image of a lone cowboy roping a calf, building a fence, or riding over a snow-covered landscape.

Interestingly, as a response to corporate mergers and public skepticism toward impersonal and large companies, a *disassociation corollary* also emerged in advertising to distance company associations that might be unfavorable. For example, in 1998, McDonald's made an investment in a small new Denver-based burrito chain, which enabled the chain—Chipotle Mexican Grill—to grow to hundreds of restaurants. By 2001, McDonald's held a majority stake in the company, but the McDonald's connection was never promoted in Chipotle marketing. In fact, Chipotle was phenomenally successful at a time when McDonald's wasn't, and the disassociation from McDonald's helped Chipotle maintain its reputation for being a different kind of fast-casual food chain. In 2006, McDonald's (which didn't get its wish to create very McDonald's-like drive-through windows and breakfast service at Chipotle) sold its entire stake in the chain; since then, the Chipotle company has grown five times larger.[30]

Advertising as Myth and Story

Another way to understand ads is to use **myth analysis**, which provides insight into how ads work at a general cultural level. Here, the term *myth* does not refer simply to an untrue story or an outright falsehood. Rather, myths (not unlike ancient Greek or Roman myths) are deeply held stories that circulate in a culture and help to define people, organizations, and social norms. Myths heighten the contrasts between a culture's values and those values' binary oppositions: who or what is good or bad, moral or immoral, and sacred or profane. According to myth analysis, we can understand advertising narratives in the same way—they have stories to tell and social conflicts to resolve. Three common mythical elements are found in many types of ads:

- Ads incorporate myths in mini-story form, featuring characters, settings, and plots.
- Most stories in ads involve conflicts, pitting one set of characters or social values against another.
- Such conflicts are negotiated or resolved by the end of the ad, usually by applying or purchasing a product. In advertising, the product and those who use it often emerge as the heroes of the story.

Even though the stories that ads tell are usually compressed into thirty seconds or onto a single page, they still include the traditional elements of narrative. But because of the short time or limited space available in an ad, they tap into powerful cultural myths to get the message across quickly. For instance, many SUV ads ask us to imagine ourselves driving out into the raw, untamed wilderness—a quiet, natural place that only, say, a Jeep can reach. The audience implicitly understands that the SUV can somehow, almost magically, take us out of our fast-paced, freeway-wrapped urban world, plagued with long commutes, traffic jams, and automobile exhaust. This implied conflict between the natural world and the manufactured world is apparently resolved by the image of the SUV in a natural setting. Although SUVs typically clog our urban and suburban highways, get low gas mileage, and create tons of air pollution particulates, the ads ignore these facts. Instead, they offer an alternative story about the wonders of nature, and the SUV amazingly becomes the vehicle that negotiates the conflict between city/suburban blight and the unspoiled wilderness.

Although most of us realize that ads create a fictional world, we often get caught up in their stories and myths. As media scholar Michael Schudson observed in his book *Advertising: The Uneasy Persuasion*, ads do not "make the mistake of *asking* for belief."[31] Instead, they are most effective when they operate like popular fiction, creating attitudes and reinforcing values while encouraging us to suspend our disbelief. Indeed, ads often work because the stories offer comfort about our deepest desires and conflicts—between men and women, nature and technology, tradition and change, the real and the artificial. Most contemporary consumer advertising does not provide much useful information about products. Rather, it tries to reassure us that through the use of familiar brand names, everyday tensions and problems can be managed (see "Media Literacy and the Critical Process: The Branded You").

MEDIA LITERACY & THE CRITICAL PROCESS

The Branded You

We live in a consumer culture—a specific kind of culture in which advertising messages promote a specific set of values. *Ad creep* has given us a media environment (and an overall environment) that makes those values seem natural. From commercial messages, we learn certain values: aspiration to a higher social status is good, more is better, happiness comes from buying things. Those values compete with other, often contradictory, values: happiness comes from personal relationships, from faith, from giving; minimalism is good; greed is bad.

How do we negotiate this clash of commercial and civic (or personal, family, or spiritual) values? One way to investigate these questions is to consider the following: To what extent are you influenced by brands?

1 DESCRIPTION

Take a look around your living space and list all the branded products you've purchased, including food, electronics, clothes, shoes, toiletries, and cleaning products. (If you are doing this activity in class, you can inventory all the clothing and personal items you have with you.)

2 ANALYSIS

Now organize your branded items into categories. For example, how many items of clothing are branded with athletic, university, or designer logos? How many of your brands are "premium" brands versus generic

brands (that is, a national/global brand versus an in-store brand or homemade or locally made item)? What do these brands suggest about your profile as a consumer?

3 INTERPRETATION

Why did you buy each particular product? Was it because you thought it was of superior quality? Because it was cheaper? Because your parents used this product (making it tried, trusted, and familiar)? Because it made you feel a certain way about yourself and you wanted to project that image to others? Have you ever purchased items without brands or removed logos once you

bought the product? Are you willing to pay more for a known brand, even when it seems to be exactly like a generic brand? Would you be embarrassed to be seen *not* using a premium brand? Why?

4 EVALUATION

As you become more conscious of our branded environment (and your participation in it), what is your assessment of U.S. consumer culture? Is there too much conspicuous branding? What is good and bad about the ubiquity of brand names in our culture? How does branding relate to the common American ethic of individualism? How does branding relate to your personal set of noncommercial values?

5 ENGAGEMENT

Read about action projects that confront commercialism—including Buy Nothing Day, Screen-Free Week, the Simplicity Collective, and Reverend Billy and the Church of Stop Shopping—and visit the Adbusters website. Then write a letter or an e-mail to a company about a product or an ad that you think is a problem, outlining your grievances. How does the company respond?

Product Placement

Product companies and ad agencies have become adept at *product placement*: strategically placing ads or buying space in movies, TV shows, comic books, video games, blogs, and music videos so that products appear as part of a story's set environment. For example, the film *Get Out* (2017) prominently featured Windows products, such as Bing, the Windows Phone, and the Surface Pro in multiple scenes; the TV show *The Biggest Loser*, which was sponsored by Subway, took contestants on field trips to the sandwich shop; and Instagram hired thirty "influencer dudes" to use Axe hair-styling products and create their own "Instagroom" videos.

Global product placement revenues continue to grow. In 2019, product placements in all media grew 14.5 percent, to a value of $20.57 billion.[32] Part of the growth has been in commercial-free streaming networks, which have captured a larger part of the TV viewing audience. Product placements have become the alternative way to get products in front of viewers on these networks with no commercials. For example, one of Netflix's top shows, *Stranger Things*, has been open to integrating products

PRODUCT PLACEMENT in movies and television is more prevalent than ever. On television, placement is often most visible in reality shows, while scripted series and films tend to be more subtle—but not all the time. In this scene from the Netflix series *Stranger Things*, Eggo waffles are prominently featured. Ironically, although it seemed like a clear product placement (and generated a boomlet of sales for Eggo), in this case the brand was used in the show solely for creative purposes. Later seasons of *Stranger Things* did include promotional deals for product placement.

© Netflix/Everett Collection, Inc

into its stories. The frozen waffle brand Eggo figured prominently in the first season as the character Eleven's favorite food (it actually wasn't a paid product placement but inspired paid placements in future seasons). In the second season, KFC played a significant role. By its third season in 2019, *Stranger Things* viewers saw product placements for Coke, Nike, Burger King, and Baskin Robbins.[33] Of course, with the third season of the show set in 1985, only products that existed at that time make sense for the story. Coke appears in at least three scenes during the season. The featured Coke product is actually New Coke, a formulation introduced in 1985 that was ultimately withdrawn in a marketing disaster for the soft drink company. In *Stranger Things*, though, people seem to like New Coke, as evidenced by one scene in which the main characters are in incredible danger but spend about a minute debating the merits of New Coke as character Lucas glugs down a can.[34] To accompany the product placement, Coke actually released a limited number of New Coke cans filled with the revived formulation of 1985-era New Coke.

But Netflix doesn't accept money for product placements. Instead, as *Fast Company* explained, Netflix is in the business of "copromotional" production placement—products are featured in the show in exchange for promotional support from the brand. Netflix has even hired former ad agency people to help with these agreements. "Even if money doesn't change hands, several brands featured in *Stranger Things* are spending big bucks to promote the show in their own marketing, resulting in a quid pro quo that ultimately helps Netflix attract more subscribers."[35] Companies like Coke benefit as well. According to one estimate, the *Stranger Things* copromotional arrangement was equivalent to Coke purchasing a $3.8 million ad, based on the screen time and number of viewers involved.

In the United States, the FTC requires social media influencers to disclose product placements by using a hashtag ("#ad" or "#sponsored") or by acknowledging the name of the sponsor, but it has no requirement for product placements to be disclosed in television programs. In contrast, Britain allows for some product placements in movies, entertainment, and sports programs but requires programs to alert viewers of such paid placements.[36]

Advertising and Concerns about Commercial Speech

In 1791, Congress passed and the states ratified the First Amendment to the U.S. Constitution, promising, among other guarantees, to "make no law . . . abridging the freedom of speech, or of the press." Over time, we have developed a shorthand label for the First Amendment, misnaming it the free-speech clause. The amendment ensures that citizens and journalists can generally say and write what they want, but it says nothing directly about **commercial speech**—any print or broadcast expression for which a fee is charged to organizations and individuals buying time or space in the mass media.

Whereas freedom of speech refers to the right to express thoughts, beliefs, and opinions in the abstract marketplace of ideas, commercial speech is about the right to circulate goods, services, and images in the concrete marketplace of products. For most of the history of mass media, only very wealthy citizens established political parties, and only multinational companies could routinely afford to purchase speech that reached millions. The Internet, however, has helped level that playing field. A cleverly edited mash-up or an entertaining speech—or apparently any video featuring cats captured on a smartphone—has the potential to go viral.

Although the media have not hesitated to carry product and service-selling advertisements and have embraced the concepts of influencers, infomercials, and home-shopping channels, they have also refused certain issue-based advertising that might upset their traditional advertisers. For example, whereas corporations experience little resistance when placing paid ads, many labor unions have had their print and broadcast ads rejected as "controversial."

Critical Issues in Advertising

In his 1957 book *The Hidden Persuaders*, Vance Packard expresses concern that advertising manipulates helpless consumers, attacks our dignity, and invades "the privacy of our minds."[37] According to this view, the advertising industry is all-powerful. Although consumers have historically been regarded as dupes by many critics, research reveals that the consumer mind is not as easy to predict as some advertisers once thought. One of the most disastrous campaigns of all time featured the now-famous "This is not your father's Oldsmobile" spots, which began running in 1989. Carmaker Oldsmobile (often called "Olds") and its ad agency, Leo Burnett, decided to market vehicles to a younger generation after sales started to decline. But the campaign backfired, apparently alienating older loyal customers (who may have felt abandoned by Olds and its catchy new slogan) and failing to lure younger buyers (who probably still had trouble getting past the name Olds). Sales fell further, and by 2005, GM had phased out its Oldsmobile division.[38]

As this example illustrates, most people are not easily persuaded by advertising. Over the years, studies have suggested that between 75 and 90 percent of new consumer products typically fail because they are not embraced by the buying public.[39] Despite public resistance to many new products, the ad industry has made contributions, including raising the American standard of living by promoting consumption (which in turn stimulated the economy) and financing most media industries. Yet serious concerns over the impact of advertising remain. Watchdog groups worry about the expansion of advertising's reach, and critics continue to condemn ads that stereotype or associate products with sex appeal, youth, and narrow definitions of beauty. Some of the most serious concerns involve children, teens, health, and representation.

Children and Advertising

The U.S. advertising industry often views children and teenagers as "consumer trainees," teaching them to desire advertised brands that they can ask their parents to buy and one day buy for themselves. For years, groups such as Action for Children's Television (ACT) worked to limit advertising aimed at children. In the 1980s, ACT fought particularly hard to curb program-length commercials: thirty-minute cartoon programs (such as *G.I. Joe*, *My Little Pony and Friends*, *The Care Bear Family*, and *He-Man and the Masters of the Universe*) developed for television syndication primarily to promote a line of toys. This commercial tradition continued with programs such as *Pokémon* and *SpongeBob SquarePants*.

In addition, parent groups have worried about the heavy promotion of sugar-coated cereals and fast-food products during children's programs. Pointing to European countries, where children's advertising is banned, these groups have pushed to minimize advertising directed at children. Congress, hesitant to limit commercial speech and faced with lobbying by the advertising industry, has responded weakly.

Because children and teenagers influence nearly $500 billion a year in family spending—on everything from snacks to cars—they are increasingly targeted by advertisers.[40] A Stanford University

SPEECH THAT REACHES MILLIONS Thirty-seven-year-old Nathan Apodaca, who works at a potato warehouse in Idaho, made one of 2020's most popular TikToks. Apodaca (@420doggface208 on TikTok) effortlessly skateboarding down a highway, drinking from a big jug of Ocean Spray Cran-Raspberry juice and lip-synching to Fleetwood Mac's 1977 song "Dreams" as he recorded himself with his outstretched arm, gave more than seventy-eight million viewers (in less than four months) one of the most chill videos of the year.

LaunchPad
macmillan learning
launchpadworks.com

Advertising and Effects on Children
Scholars and advertisers analyze the effects of advertising on children.

Discussion: In the video, some argue that using cute, kid-friendly imagery in ads can lead children to begin drinking; others dispute this claim. What do you think, and why?

study found that a single thirty-second TV ad can influence the brand choices of children as young as age two. The pull to use various methods of marketing to target children is becoming increasingly seductive as product placement and merchandising tie-ins become more prevalent.

For years, the battle over advertising in children's media was about television, but today, many children have shifted to their own personal digital screens. By 2020, according to one study, "nearly half (46%) of 2- to 4-year-olds and more than two-thirds (67%) of 5- to 8-year-olds ha[d] their own mobile device (tablet or smartphone)."[41] Children ages zero to eight spend an average of fifty-five minutes a day using mobile devices, with most time spent watching TV shows and videos and playing mobile games.[42] Advertisers have already found them there: A 2019 study of apps for children ages one to five revealed that 95 percent of them contained at least one form of advertising.[43]

Health and Advertising

One of the most enduring criticisms of advertising relates to its negative effects on people's health, as it can set cultural standards for body types that encourage eating disorders, and promote tobacco use and the abuse of alcohol and prescription drugs.

Eating Disorders.   Advertising has a powerful impact on the standards of beauty in our culture. A long-standing trend in advertising is the association of certain products with ultrathin female models, promoting a style of "attractiveness" that girls and women are invited to emulate. Even today, despite the popularity of fitness programs, most fashion models are much thinner than the average woman. Some forms of fashion and cosmetics advertising actually pander to individuals' insecurities and low self-esteem by promising the ideal body. Such advertising suggests standards of style and behavior that may not only be unattainable but also be harmful, leading to eating disorders such as anorexia and bulimia and an increase in cosmetic surgeries.

If advertising has been criticized for promoting skeleton-like beauty, it has also been blamed for the tripling of obesity rates in the United States since the 1960s, when only 14 percent of individuals were defined as obese; in 2018, 42.5 percent of U.S. individuals were defined as obese, according to the CDC. Corn syrup–laden soft drinks, fast food, and processed food are the staples of media ads and are major contributors to the nationwide weight problem. More troubling is that because an obese nation is good for business (creating a multibillion-dollar market for diet products, exercise equipment, and self-help books), media outlets see little reason to change current ad practices. The food and restaurant industry at first denied any connection between ads and the rise of U.S. obesity rates, instead blaming individuals who make bad choices. Increasingly, however, fast-food chains are offering healthier meals and listing calorie counts on various food items.

Tobacco.   One of the most sustained criticisms of advertising is its promotion of tobacco consumption. Opponents of tobacco advertising have become more vocal in the face of grim statistics: Each year, an estimated 480,000 Americans die from diseases related to nicotine addiction and poisoning. Tobacco ads disappeared from television in 1971 under pressure from Congress and the FCC. However, over the years, numerous ad campaigns have targeted teenage consumers of cigarettes. In 1988, for example, R. J. Reynolds updated its Joe Camel cartoon character, outfitting him with

MEDICAL ORGANIZATIONS such as the Ontario Medical Association seek to warn consumers about the dangers of obesity through powerful ads like the one seen here.

WARNING
TO STOP THE OBESITY CRISIS, GOVERNMENTS MUST APPLY THE LESSONS LEARNED FROM SUCCESSFUL ANTI-TOBACCO CAMPAIGNS

www.oma.org

Ontario Medical Association   @Ontariosdoctors

OMA

hipper clothes and sunglasses. Spending $75 million annually, the company put Joe on billboards and store posters and in sports stadiums and magazines. One study revealed that before 1988, fewer than 1 percent of teens under the age of eighteen smoked Camels. After the ad blitz, however, 33 percent of this age group preferred Camels.

In addition to young smokers, the tobacco industry has targeted other groups. In the 1960s, for instance, the advertising campaigns for Eve and Virginia Slims cigarettes (reminiscent of ads during the suffrage movement in the early 1900s) associated their products with women's liberation, equality, and slim fashion models. And in 1989, R. J. Reynolds introduced a cigarette called Uptown, targeting Black consumers. The ad campaign fizzled due to public protests by Black leaders and government officials. When these leaders pointed to the high concentration of cigarette billboards in Black neighborhoods and the high mortality rates among Black male smokers, the tobacco company withdrew the brand.

The government's position, which amounted to tolerating the tobacco industry with mild regulation, began to change in the mid-1990s, when new reports revealed that tobacco companies had known that nicotine was addictive as early as the 1950s and had withheld that information from the public. In 1998, the largest civil litigation settlement in U.S. history took place between forty-six U.S. states, five U.S. territories, the District of Columbia, and the major heads of the tobacco industry. The tobacco industry agreed to an unprecedented $206 billion settlement, which carried significant limits on the advertising and marketing of tobacco products.

The agreement's provisions banned cartoon characters in advertising, thus ending the use of the Joe Camel character; prohibited the industry from targeting young people in ads and marketing and from giving away free samples, tobacco-brand clothing, and other merchandise; and ended outdoor billboard and transit advertising. The agreement also banned tobacco company sponsorship of concerts and athletic events, and strictly limited its other corporate sponsorships. These provisions, however, do not apply to tobacco advertising abroad.

More recently, the battle against tobacco products has shifted to include vaping and nicotine delivery systems. By 2020, 3.6 million middle and high school students had used e-cigarettes. Health experts have blamed one company in particular, Juul Labs, for the fast rise of underage vaping, driven by sales of its mint and fruit flavor vaping devices. (A major stakeholder in Juul is Altria, a tobacco corporation.) The FDA has moved against vaping, first by regulating vaping under its tobacco rules in 2016 and then by ordering flavored vaping cartridges (the ones most preferred by minors) off the market in 2020. In addition, in 2019, federal law raised the minimum age for sale of tobacco products (including e-cigarettes) from eighteen to twenty-one.[44]

Alcohol.   Each year, about ninety-five thousand people in the United States die from alcohol-related or alcohol-induced diseases. In addition, another ten-thousand-plus die in car crashes involving drunk drivers. As a result, many of the same complaints regarding tobacco advertising are directed at alcohol ads. (See "Examining Ethics: Do Alcohol Ads Encourage Binge Drinking?") For example, one of the most popular beer ad campaigns of the late 1990s (featuring the Budweiser frogs croaking "Budweis-errrr") was accused of using cartoonlike animal characters to appeal to young viewers. In fact, the Budweiser ads would have been banned under the tough standards of the tobacco industry settlement, which prohibits the attribution of human characteristics to animals, plants, or other objects.

There is also a trend toward marketing high-end liquors to Black and Latino male populations. In one marketing campaign, Hennessy targeted young Black men with ads featuring hip-hop star Nas and sponsored events in Times Square and at the Governors Ball and Coachella music festivals. Hennessy also sponsored VIP parties with Latino deejays and hip-hop acts in Miami and Houston. In 2020, Hennessy signed a multiyear contract to be the Official Spirit of the NBA.

College students, too, have been heavily targeted by alcohol ads, particularly by the beer industry. Although colleges and universities have outlawed "beer bashes" hosted and supplied directly by major brewers, both Coors and Miller still employ student representatives to help create brand awareness. These students notify brewers of special events that might be sponsored by and linked to

Do Alcohol Ads Encourage Binge Drinking?

With clear evidence that cigarettes cause lung cancer, the tobacco industry in the early 1970s chose to pull all TV ads for cigarettes, in part to ward off the planned increase in public service ads that the government and nonprofit agencies were airing about the dangers of smoking. Similarly, ads for hard liquor (called "distilled spirits" by the industry) were not shown for decades in TV markets across the United States for fear of igniting anti-alcohol public service spots warning about alcoholism and heavy drinking, and countering TV commercials. Some ads for hard liquor have reappeared in recent years, but not all channels or shows will air them; often they appear on late-night or specialized programming.

Beer ads, however, have never been interrupted and remain ubiquitous, usually associating beer drinking with young people, sex appeal, and general good times. As such, the debates over alcohol ads continue, especially in light of the ritual of binge drinking that has bedeviled universities throughout the United States. The Centers for Disease Control and Prevention defines binge drinking as what

"typically happens when men consume 5 or more drinks or women consume 4 or more drinks in about 2 hours."[1]

According to the CDC, excessive alcohol use kills more than ninety-five thousand people in the United States each year, with almost half of these deaths due to binge drinking. Excessive drinking is also a massive drain to the economy, costing more than $249 billion each year. Most (77%) of these costs are due to binge drinking, which results in reduced workplace productivity, greater health-care expenses, and increased costs related to criminal justice and motor vehicle crashes.[2]

A Dartmouth University study, first released in 2015 and published in *JAMA Pediatrics*, demonstrated that "familiarity with and response to images of television alcohol marketing was associated with the subsequent onset of drinking."[3] The study surveyed more than twenty-five hundred people between the ages of fifteen and twenty-three, then reinterviewed fifteen hundred of them two years later.

In the reinterviews, 66 percent of high school students overall said they had tried alcohol, whereas only 21 percent said they had engaged in binge drinking. However, 29 percent of the fifteen- to seventeen-year-olds who had been exposed to alcohol ads reported binge drinking in the reinterviews. One coauthor of the study said, "It's very strong evidence that underage drinkers are not only exposed to the television advertising, but they also assimilate the messages. That process moves them forward in their drinking behavior."[4] Although the study argues that the efforts by beer and hard liquor advertisers to protect young people from the messages in their ads are ineffective, the Distilled Spirits Council disagrees, saying that the Dartmouth study was "driven by advocacy, not science."[5]

One ethical question raised by the 2015 study has to do with those who work in the ad business and the work they are asked to do. Many reputable ad agencies will ask new or potential employees if there are clients and products that they would not represent. In fact, some agencies specifically ask newly hired account executives if they would be willing to work for a tobacco or a liquor company—or, given what they know about childhood obesity and the low nutrition content in many fast foods, sugared cereals, and popular sodas, if they could represent these types of products.

It might be a useful exercise, then, to ask yourself, Are there products or companies I would not work for or represent in some capacity? Why or why not? Would I be willing to represent tobacco companies that wanted to place ads in magazines or a hard liquor company that wanted to advertise on TV? Why or why not? ●

BUD LIGHT attracted negative attention with a campaign about "turning no into yes," evoking the language of sexual assault.

a specific beer label. The images and slogans in alcohol ads often associate the products with power, romance, sexual prowess, or athletic skill. In reality, though, alcohol is a chemical depressant; it diminishes athletic ability and sexual performance, triggers addiction in roughly 10 percent of the U.S. population, and factors into many domestic abuse cases. In addition, at least half of the sexual assaults among college students involve alcohol consumption by the victim, perpetrator, or both.[45] One national study demonstrated "that young people who see more ads for alcoholic beverages tend to drink more."[46]

Prescription Drugs.   A final area of concern is the surge in prescription drug advertising. Spending on medical marketing in the United States increased from $17.7 billion to $29.9 billion between 1997 and 2016, with direct-to-consumer advertising for prescription drugs and health services responsible for most of the growth.[47] The FDA relaxed rules on marketing prescription drugs directly to consumers in 1997, setting off a huge rush of television pharmaceutical ads that has not abated. These ads have made household names of prescription drugs such as Nexium, Lipitor, Paxil, and Viagra. The ads are also very effective: One survey found that nearly one in three adults has talked to a doctor after seeing an ad for a prescription drug, and one in eight has received a prescription.[48]

The tremendous growth of prescription drug ads brings the potential for false and misleading claims, particularly because a brief TV advertisement cannot possibly communicate all the relevant cautionary information. By contrast, prescription pharmaceutical ads in print media are required by the FDA to include all the risks that appear in the drug's prescribing information; therefore, print ads are typically followed by pages of information in fine print, including warnings and possible side effects. More recently, direct-to-consumer prescription drug advertising has appeared in text messages and on Facebook. Pharmaceutical companies have also engaged in disease-awareness campaigns to build markets for their products. The United States and New Zealand are the only two nations to allow prescription drugs to be advertised directly to consumers.

Representation in Advertising

Because advertising is so omnipresent in society, its representations of people often establish cultural norms. Unfortunately, advertising's representations of people of color have long employed the worst stereotypes. Throughout the twentieth century, advertisers offended a broad range of people of color and differing ethnicities, with advertising campaigns like Aunt Jemima (1889–2021), Eskimo Pie (1921–2021), the Washington Redskins (1933–2020), Uncle Ben's (1946–2021), Miss Chiquita (since 1944), Mrs. Butterworth (since 1961), Frito Bandito (1967–1971), and Crazy Horse Malt Liquor (1992–2001).[49]

As noted earlier in the chapter, the vast majority of ad creators were men—white men—though a breakthrough came in 1971 when Tom Burrell and a business partner established what is now Burrell Communications. The Chicago ad agency would become a pioneer in marketing to Black audiences, with top clients including McDonald's, Coca-Cola, and Philip Morris. One of Burrell's famous insights is "Black people are not dark-skinned white people; we came on the American scene in a totally unique way: against our will and enslaved. And we know that those circumstances led to the formation of unique consumer attitudes and behavior patterns."[50] Burrell's success led to other ad agencies forming units dedicated to advertising for Black, Hispanic, and other audiences of color. By the end of the 1970s, advertising was far more diverse.[51]

Still, stereotypes persist. It took the protests of Black Lives Matter in 2020 for a number of corporations to retire longtime racist representations. For example, Quaker Oats (part of Pepsico) rebranded the Aunt Jemima pancake mix and syrup brands, which featured a Black woman in a stereotypical servile role, as Pearl Milling Company; Mars, Inc., announced it would rebrand the Uncle Ben's packaged rice brand (featuring a Black male servant) as Ben's Original; Dreyer's Ice Cream announced it would replace the Eskimo Pie name with Edy's Pies; and the Washington Redskins football team and Cleveland Indians baseball team announced they would abandon their team names. (See "Global Village: The Unfairness of Fairness Creams.")

The Unfairness of Fairness Creams

BY SHREYA SINGH

In the summer of 2020, Black Lives Matter protests rippled all over the world. In India, public figures—especially those in the entertainment industry—promoted #blacklivesmatter on social media, encouraging the movement's spread. However, activists soon noted their hypocrisy. These entertainers had also appeared in advertisements for Fair & Lovely, a popular skin lightening cream. They had silently endorsed colorism—discrimination based on shades of skin color—in India.

Why did an antiracism uprising spark conversations about colorism? Colorism in India, as well as in other previously colonized countries in South Asia and Africa, is believed to have internalized racist undertones. In these countries, centuries of colonial oppression by Britain and other European nations have bred a problematic rhetoric: That lighter skin is superior to and more desirable than darker skin tones, even within the same race. This notion of the superiority of light skin has seeped into popular culture through films, fashion, and advertisements and, in turn, influences the societal perception of beauty. Advertisements for skin lightening or fairness creams have recently come under fire for perpetuating colorist attitudes by equating beauty with fair skin and portraying people with dark skin as economically, socially, and professionally disadvantaged.

Fair & Lovely is the most widely used skin lightening product in India. It is manufactured by Hindustan Unilever Limited (HUL), the Indian subsidiary of Unilever, a multinational and multibillion-dollar company. Ever since the 1970s, when the first Fair & Lovely product hit the shelves, India has been one of Unilever's most coveted markets due to its vast population of about 1.3 billion people. According to a report by Bloomberg, Unilever made $500 million from Fair & Lovely product sales in India in 2019.[1] Although several other brands—including Garnier, Emami, Procter & Gamble, and Nivea—also manufacture fairness products, Unilever's Fair & Lovely has been at the core of the controversy due to its conspicuous name, massive market share, and overt promotion of colorism-based discrimination in advertisements.

The colorist rhetoric perpetuated by these advertisements is difficult to miss. For instance, Fair & Lovely ads routinely portray people with dark skin, mostly women, encountering difficulties in their careers, failing to find love or get married, and being treated as middle-class social outcasts. The narrative in such advertisements often takes a predictable turn when the woman discovers a skin-lightening product and her life changes for the better. The now "fair" woman suddenly gains confidence, becomes socially outgoing, and has better career prospects. These common themes and narratives have persisted throughout the years and appear in advertisements promoting other brands as well.

Skin-lightening products have ignited public criticism. In 2009, a women's advocacy group called Women of Worth launched a Dark Is Beautiful campaign with the

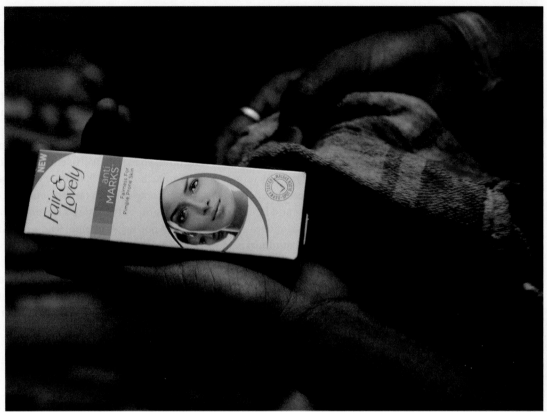

tagline "Stay unfair, stay beautiful." Since then, similar campaigns have condemned the toxicity of fairness products and challenged the narrative that people's worth can be measured by the fairness of their skin. In response, the Advertising Standards Council of India (ASCI) issued a set of guidelines in 2014 banning the depiction of darker skin as inferior. Advertisers then resorted to more subtle ways of endorsing the superiority of fair skin, and the underlying rhetoric remained unchanged.

In 2020, as racial justice advocates around the world took their demands to the streets in the form of demonstrations, protests, and silent marches, brands promoting color-based discrimination in India came into the spotlight. Activists demanded that international giants like Unilever and the public figures who endorse their products be held accountable for their role in perpetuating "beauty ideals borne out of racist standards."[2] Under this spotlight, Unilever and HUL rebranded their Fair & Lovely product as Glow & Lovely and dropped words such as "fairness," "whitening," and "lightening" from their product names and advertisements.

Reactions have been mixed. Some applaud Unilever for taking active body-positive and antiracist stances and are hopeful other brands will do the same. Others, though, have criticized Unilever for coming too late, doing too little, and just changing the branding and not the product itself.

Shreya Singh is from Kolkata, India, and received a master's degree from the Department of Communication and Media at the University of Northern Iowa. Her professional background is in advertising, and her research interests include postcolonial criticism and political rhetoric. ●

Often it is economic considerations that motivate companies to change or innovate advertising appeals. In the 1990s, groundbreaking campaigns featuring same-sex couples began to appear in the United States, as companies like IKEA and Subaru recognized this overlooked consumer niche.[52] Representation in advertising matters. A 2020 study by Procter & Gamble and GLAAD, the world's largest LGBTQ+ advocacy organization, found that "exposure to LGBTQ people in the media increases non-LGBTQ consumers' comfortability with LGBTQ people in their daily lives." Moreover, the study found that "non-LGBTQ consumers look favorably on companies that include LGBTQ people in advertising."[53]

Advertising's portrayals of women have changed over time as well. In family units, women had long been cast as the chief consumer for the household, so early advertisements for food, clothing, and home goods were typically targeted to them. As noted in Chapter 5, *soap operas* were daytime drama programs sponsored by soap product companies to target women at home listening to the radio in the 1930s, and later watching television. Advertising's stereotypical confinement of women to the household was reinforced by its portrayals of men, who—outside of their "masculine" spaces of the outdoors and work—were often portrayed as bumbling doofuses in matters of childcare and household chores. Ads in the twentieth century also commonly used women as ornamental sex objects to sell things to men, such as cars or beer. In the 1970s, advertisements began to portray women on more equal terms and as competent actors in all walks of life. Still, stereotypical portrayals of women in ads do persist, though they are much less frequent.

Watching Over Advertising

The Federal Trade Commission (FTC) plays a role in investigating deception in advertising. But a few nonprofit watchdog and advocacy organizations—Commercial Alert, as well as the Better Business Bureau and the National Consumers League—compensate in many ways for some of the shortcomings of the FTC and other government agencies in monitoring problems in advertising, including the excesses of commercialism and false and deceptive ads.

Excessive Commercialism

Since 1998, Commercial Alert—a nonprofit organization founded in part by longtime consumer advocate Ralph Nader and based in Portland, Oregon—has been working to "limit excessive commercialism in society" by informing the public about the ways that advertising has crept out of its "proper sphere." For example, Commercial Alert highlights the numerous deals for cross-promotion made between Hollywood studios and fast-food companies. These include DreamWorks' partnership with Hardee's for *TrollHunters: Tales of Arcadia* and Disney's partnership with McDonald's for family-friendly flicks like *The Incredibles 2* and *Ralph Breaks the Internet*.

These deals helped movie studios reach audiences that traditional advertising can't. As Jeffrey Godsick, president of consumer products at the former 21st Century Fox, once said, "We want to hit all the lifestyle points for consumers. Partners get us into places that are nonpurchasable (as media buys). McDonald's has access to tens of millions of people on a daily basis—that helps us penetrate the culture."[54]

With the help of only a few peer organizations, Commercial Alert is a lonely voice in checking the commercialization of U.S. culture. Its other activities have included challenges to specific marketing tactics, such as the potentially deceptive advertising practices to which the Instagram video app IGTV may expose users. In constantly questioning the role of advertising in democracy, the organization has aimed to strengthen noncommercial culture and limit the amount of corporate influence on publicly elected government bodies.

The FTC Takes On Puffery and Deception

Since the days when Lydia Pinkham's Vegetable Compound promised "a sure cure for all female weakness," false and misleading claims have haunted advertising. Over the years, the Federal Trade Commission, through its truth-in-advertising rules, has played an investigative role in substantiating the claims of various advertisers. A certain amount of *puffery*—ads featuring hyperbole and exaggeration—has usually been permitted, particularly when a product says it is "new and improved." However, ads become deceptive when they are likely to mislead reasonable consumers based on statements in the ad or because they omit information. Moreover, when a product claims to be "the best," "the greatest," or "preferred by four out of five doctors," FTC rules require scientific evidence to back up those claims.

Some famous examples of deceptive advertising have included a Campbell Soup TV ad in which marbles in the bottom of a soup bowl forced more bulky ingredients—and less water—to the surface. In another instance, a 1990 Volvo commercial featured a monster truck driving over a line of cars and crushing all but the Volvo; the company later admitted that the Volvo had been specially reinforced and the other cars' support columns had been weakened. A more subtle form of deception featured the Klondike Lite ice cream bar—"the 93 percent fat-free dessert with chocolate-flavored coating." The bars were indeed 93 percent fat-free—but only after the chocolate coating was removed.[55]

In 2020, the FTC filed a complaint in federal district court against a company that makes one of many quack "fit teas" that are marketed on social media. The FTC charged that the Teami 30 Day Detox Pack stated "without reliable scientific evidence" that it would "help consumers lose weight, and that its other teas fight cancer, clear clogged arteries, decrease migraines, treat and prevent flus, and treat colds."[56] The FTC also charged that ten well-known influencers did not adequately disclose that they were paid to promote the company's products on Instagram. "Companies need to back up health claims with credible science and ensure influencers prominently disclose that they're getting paid to promote a product," said Andrew Smith, director of the FTC's Bureau of Consumer Protection.[57]

When the FTC discovers deceptive ads, it usually requires advertisers to change them or remove them from circulation. The FTC can also impose monetary civil penalties for companies, and it occasionally requires an advertiser to run spots to correct the deceptive ads.

Kourtney Kardashian ✔ @kourtneykardash · May 16, 2016
Popeyes on a plane. #cheatday

♡ 123   ↻ 846   ♡ 5.2K

STEALTH ADS, like Kourtney Kardashian's Popeyes endorsement posted to her Twitter page, have been criticized for being deceptive and unethical, as consumers may not be aware that they are being subjected to an advertisement.

Advertising, Politics, and Democracy

Advertising as a profession came of age in the twentieth century, facilitating the shift of U.S. society from production-oriented small-town values to consumer-oriented urban lifestyles. With its ability to create consumers, advertising became the central economic support system for our mass media industries. Through its seemingly endless supply of pervasive and per-suasive strategies, advertising today saturates the cultural landscape. Products now blend in as props or even as "characters" in TV shows and movies. In addition, almost every national consumer product now has its own website to market itself to a global audience 365 days a year. With today's digital technology, ad images can even be made to appear in places where they don't really exist. For example, advertisements are now superimposed on the backstop wall behind the batter during televised baseball broadcasts. Viewers at home see the ads, but fans at the game do not.

Advertising's ubiquity, especially in the age of social media, raises serious questions about our privacy and the ease with which companies can gather data on our consumer habits. But an even more serious issue is the influence of ads on our lives as democratic citizens. With fewer and fewer large media conglomerates controlling advertising and commercial speech, what is the effect on free speech and political debate? In the future, how easy will it be to get heard in a marketplace where only a few large companies control access to that space?

Advertising's Role in Politics

Since the 1950s, political consultants have been imitating market-research and advertising techniques to sell their candidates, giving rise to **political advertising**, the use of ad techniques to promote a candidate's image and persuade the public to adopt a particular viewpoint. In the early days of tele-vision, politicians running for major office either bought or were offered half-hour blocks of time to discuss their views and the issues of the day. As advertising time became more valuable, however, local stations and the networks became reluctant to give away time in large chunks. Gradually, local TV stations began selling thirty-second spots to political campaigns, just as they sold time to product advertisers. Once in place, these stations would oppose any talk of reform to the political advertising system, as it had become quite profitable for them.

During the 1992 and 1996 presidential campaigns, third-party candidate Ross Perot restored the use of the half-hour time block when he ran political infomercials on cable and the networks. In 2008, Barack Obama also ran a half-hour infomercial; and in the 2012 presidential race, both major candidates and the various political organizations supporting them ran many online infomercials that were much longer than the standard thirty- to sixty-second TV spot. However, only very wealthy or well-funded candidates can afford such promotional strategies, and television does not usually provide free airtime to politicians (although some politicians, like Donald Trump, famously earned much more "free" time in the news than his rivals by frequently being outrageous). Questions about political ads continue to be asked: Can serious information on political issues be conveyed in thirty-second spots? Do repeated attack ads, which assault another candidate's character, so undermine citizens' confidence in the electoral process that they stop voting?[58] And how does a democratic society ensure that alternative political voices, which are not well financed or commercially viable, still receive a hearing?

In late October 2020, OpenSecrets.org estimated that the total cost of the 2020 federal elections—the presidential and congressional contests—would reach $14 billion, an unprecedented total and more than twice the amount spent just four years earlier.[59] Although small donors, whose donations are $200 or less, made a record 22.4 percent of contributions in the 2020 election, they were far outstripped by money coming from wealthy donors, political action committees, and rich politicians self-funding their own campaigns. The thirty-second political ad continued to be a

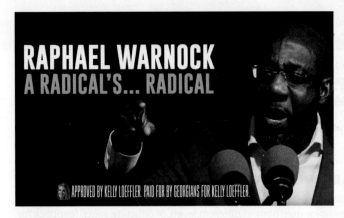

POLITICAL ADVERTISING People often complain about negative political advertising, but its continued use is evidence that it can be effective. In these competing ads from the 2020 Georgia Senate campaign that continued through a 2021 runoff election, opponents Raphael Warnock and Kelly Loeffler traded accusations.

campaign advertising staple in 2020, but political groups also spent more than $1 billion advertising on Google and Facebook. Like broadcasters, large digital corporations may be reluctant to support reforming the commercialized political system that sends so much money their way. Even so, because of public outcry about the proliferation of misleading ads on social media, Twitter banned political advertising in 2020, and Facebook banned political ads for a week before the 2020 election and for a period of time after the election, as it anticipated disputes over the results and potential interference in the 2021 Georgia Senate runoff election. Google and its subsidiary YouTube paused political ads for a time immediately after the election, too.

The Future of Advertising

Although commercialism—through packaging both products and politicians—has generated cultural feedback that is often critical of advertising's pervasiveness, the growth of the industry has not diminished. Ads continue to fascinate. Many consumers buy magazines or watch the Super Bowl just for the advertisements. Adolescents decorate their rooms with their favorite brands and identify with the images certain products convey. In 2021, advertising spending in the United States was projected to total $236.7 billion.[60]

A number of factors have made advertising's largely unchecked growth possible. Many Americans tolerate advertising as a necessary evil for maintaining the economy, and many others dismiss advertising as not believable and trivial. Thus, unwilling to admit its centrality to global culture, many citizens do not think advertising is significant enough to monitor or reform. Such attitudes have ensured advertising's pervasiveness and suggest the need to escalate our critical vigilance.

As individuals and as a society, we have developed an uneasy relationship with advertising. Favorite ads and commercial jingles remain part of our cultural world, yet we detest irritating and repetitive television commercials or invasive social media ads. We realize that without ads, many media industries would need to reinvent themselves. At the same time, we need to remain critical of what advertising has come to represent: the overemphasis on commercial acquisitions and images of material success, and the disparity between those who can afford to live comfortably in a commercialized society and those who cannot.

Chapter REVIEW 11

LAUNCHPAD FOR *MEDIA & CULTURE*

LaunchPad
macmillan learning

Visit LaunchPad for *Media & Culture* at **launchpadworks.com** for additional learning tools:

→ QUIZ YOURSELF WITH LEARNINGCURVE

LearningCurve uses gamelike quizzing to help you review important concepts from this chapter. LearningCurve includes multiple-choice questions at different levels of difficulty, hints, feedback for both right and wrong answers, and links to e-book material for easy reference. Which answer choice would you select for the sample question below?

> Jiffy, a company that produces baking products, bills its corn muffin mix as "America's Favorite." This is an example of which advertising approach?
>
> ○ irritation advertising
>
> ○ famous-person testimonial
>
> ○ bandwagon effect
>
> ○ hidden-fear appeal

→ WATCH VIDEO ON KEY CONCEPTS

LaunchPad includes clips from movies, TV shows, online sources, and other media texts, in addition to interviews with media experts and newsmakers. In the videos for this chapter, you'll consider topics like the effects of advertising on children and the impact of product placement in media.

→ PRACTICE THE CRITICAL PROCESS

Use the Media Literacy Activity to walk through the critical process step-by-step and develop a critical perspective on what media means to you. In this chapter's activity, you'll evaluate gender stereotyping in advertising.

→ LEARN KEY TERMS

Review the definitions for this chapter's key terms with LaunchPad's easy-to-use online flash cards. Try the cards in practice mode, or quiz yourself as a way to focus your study efforts. (Page numbers indicate where each term is highlighted within the chapter.)

product placement, 310
space brokers, 312
subliminal
  advertising, 315
slogan, 315
mega-agencies, 316
boutique agencies, 316
market research, 318
Values and Lifestyles
  (VALS), 318

storyboard, 319
viral marketing, 319
media planners, 319
media buyers, 319
saturation advertising, 320
spam, 321
predictive
  marketing, 322
famous-person
  testimonial, 324

plain-folks pitch, 324
snob-appeal approach,
  324
bandwagon effect, 325
hidden-fear appeal, 325
irritation advertising, 325
association principle, 325
myth analysis, 326
commercial speech, 328
political advertising, 338

REVIEW QUESTIONS

Use the questions below to revisit key themes and concepts from this chapter.

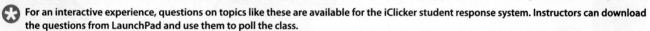

Need more practice? Access the LearningCurve multiple-choice questions for this chapter, which are available on LaunchPad. All questions include feedback for correct and incorrect answers, hints, and links to e-book material.

Chapter Opener and Advertising Today

1. How are social media influencers a form of advertising?

How We Got Here: The Development of American Advertising

2. Whom did the first ad agencies serve?
3. How did trademarks and packaging influence advertising?
4. Why did patent medicines and department stores figure so prominently in advertising in the late nineteenth century?
5. What role did advertising play in transforming America into a consumer society?
6. What influences did visual design exert on advertising?

The Shape of Contemporary U.S. Advertising

7. What are the differences between boutique agencies and mega-agencies?
8. What are the major divisions at most ad agencies? What is the function of each department?
9. What are the advantages of online and mobile advertising over traditional media, like newspapers and television?

Persuasive Techniques in Contemporary Advertising

10. How do the common persuasive techniques used in advertising work?
11. How does the association principle work, and why is it an effective way to analyze advertising?
12. What is the disassociation corollary?
13. What is product placement? Cite examples.

Advertising and Concerns about Commercial Speech

14. What is commercial speech?
15. What are four serious contemporary issues regarding health and advertising? Why is each issue controversial?
16. What is the difference between puffery and deception in advertising? How can the FTC regulate deceptive ads?

Advertising, Politics, and Democracy

17. What are some of the major issues involving political advertising?
18. What role does advertising play in a democratic society?

QUESTIONING THE MEDIA

Use these critical-thinking questions to reflect on your own experiences with media and apply your understanding of media concepts.

For an interactive experience, questions on topics like these are available for the iClicker student response system. Instructors can download the questions from LaunchPad and use them to poll the class.

1. What is your earliest recollection of watching a television commercial? Do you think the ad had a significant influence on you?
2. Why are so many people critical of advertising?
3. If you were (or are) a parent, what strategies would you use to explain an objectionable ad to your child or teenager? Provide an example.
4. Should advertising aimed at children be regulated? Support your response.
5. Should tobacco or alcohol advertising be prohibited? Why or why not? How would you deal with First Amendment issues regarding controversial ads?
6. Would you be in favor of regular commercial advertising on public television and radio as a means of financial support for these media? Explain your answer.
7. Is advertising at odds with the ideals of democracy? Why or why not?

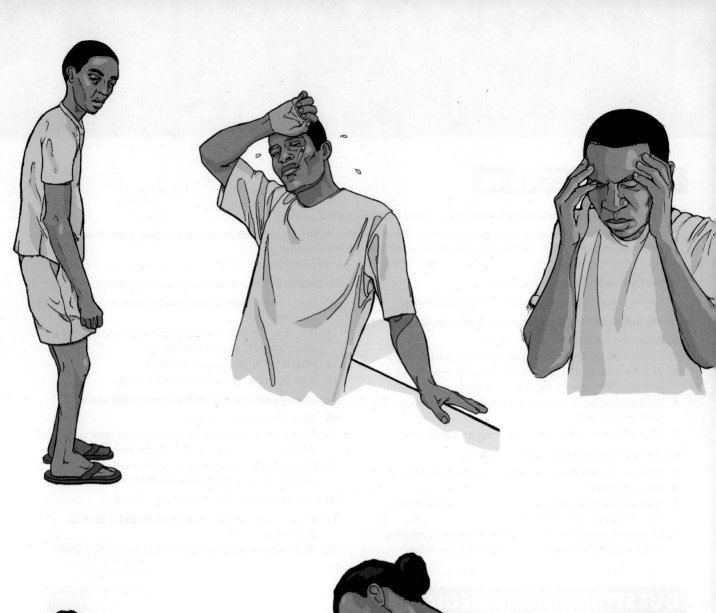

Public Relations and Framing the Message

12

It was a horrible disease—a virus—creating an unprecedented epidemic. Just a few months after the first cases emerged, health officials warned that the outbreak was moving so fast, it might soon be out of control.[1]

The COVID-19 pandemic, which spread around the world in 2020, began just this way, but this isn't the story of COVID—at least not yet. This story starts back in 2014 in West Africa, during the outbreak of another horrible pathogen: Ebola, a deadly and highly contagious disease, but one that was contained before it had a chance to spread across the globe.

The successful public relations (PR) work of global health organizations was an essential part of confronting the Ebola crisis. A group of public health professionals from the U.S. Centers for Disease Control and Prevention (CDC) were among those who played a significant role, helping to roll out a health communication effort—part of a PR specialty area called *health promotion*—designed to increase awareness and stop the spread.[2] The CDC deployed more than fifteen hundred staff members to the three affected countries in West Africa, while also working domestically to prepare the United States for a possible outbreak. Ultimately, the efforts of this and other teams helped to stop the disease's spread within West Africa and prevented a global pandemic—only fifteen deaths occurred in other countries.[3] But without a strong PR campaign promoting public health, this positive outcome may have turned out differently.

The strain of Ebola virus that struck West Africa in 2014 was the worst outbreak of the disease ever, and more than 11,300 people in Guinea, Liberia, and Sierra Leone died. The stakes facing public health officials were high, and the challenges were formidable: Different languages, cultures, religions, and governments within West Africa made it impossible to address everyone with a single communication campaign. There are an estimated thirty-one different languages and one hundred dialects in Liberia alone, with similar linguistic and cultural diversity in its two neighboring countries. Moreover, West African cultures have strong oral communication traditions and low literacy rates. Consequently, the

◀ The CDC developed these images to communicate symptoms of Ebola across West Africa.

CDC identified a "need for clear and literal illustrations that were also culturally appropriate."[4] One of the CDC tactics, then, was creating a set of illustrations showing the full range of symptoms for Ebola.

Health communication personnel also worked with the local media. For example, in Sierra Leone, health communication personnel from the CDC and other partner organizations trained about eighty radio, television, and print journalists on critical communication topics, including safe burial practices, early treatment, and addressing cultural beliefs that complicated containment.[5] The CDC also worked to deploy the activities that would later become a familiar part of the COVID-19 response: rapid diagnostic testing, contact tracing, training on methods for infection prevention and control (including quarantine), personal protective equipment for health-care workers, and early vaccinations for those at greatest risk.

Fast-forward to 2021—five years after the Ebola epidemic had been subdued in West Africa—and COVID-19 was raging across the United States, surpassing 550,000 deaths and more than 30 million cases. The United States had 4.25 percent of the world's population but about 20 percent of its COVID-19 deaths. Meanwhile, Guinea, Liberia, and Sierra Leone maintained relatively low levels of infection and deaths. These nations fought COVID-19 with lessons from Ebola—mask wearing, quarantines, and contact tracing. Conversely, the United States, which had been so involved in the region's successful Ebola health communication program, fared much worse during the pandemic.

During the Ebola outbreak, cultural beliefs in West Africa complicated the fight against the virus—for example, the idea that the disease was caused by witchcraft. In terms of combating COVID-19, the United States had its own unanticipated cultural beliefs to contend with: that the illness was caused by 5G mobile technology, that it could be cured by the antimalarial drug hydroxychloroquine or by drinking bleach, that vaccinations would come with microchips, and that mask wearing and social distancing were not public health necessities but rather political statements.

There were also problems with messaging in the United States. Health communication professionals need strong messaging guidelines to plan for health catastrophes, and the U.S. federal government actually has a 405-page manual titled *Crisis and Emergency Risk Communication*, completed in 2014.[6] But in the first year of the COVID-19 pandemic, the U.S. government didn't follow its own recommendations. Matthew Seeger, a crisis and risk communication professor, was one of the manual's coauthors. "For those of us in this field, [this COVID-19 response] is profoundly and deeply distressing," he said in March 2020. "We spent decades training people and investing in developing this competency. We know how to do this."[7]

A pandemic is a health crisis, but a corresponding communication crisis can also emerge without effective health promotion—a key part of a planned public relations effort to ensure the best possible communication in crises and emergencies.

 Do you think the U.S. health communication response to COVID-19 was adequate? What do you think went right? What went wrong? What other recent crisis events can you think of that had good public communication or that could have used better public relations planning?

Public Relations Today

The public health responses to Ebola and COVID-19 illustrate the complexity of communicating useful messages in ever-changing environments—environments where there can be conflicting information coming from a wide range of **stakeholders** (people who have an interest in what is happening). Even with the best public relations crisis and risk planning at an organization, there are no guarantees that things will go smoothly. Those at the top might decide they have a better way of doing things, and assumptions about the public—for example, expecting that everyone understands the basic process of science—can be flawed, requiring public relations professionals to quickly assess and adapt to new situations.

The Public Relations Society of America (PRSA), the field's leading professional organization, defines **public relations** as "a strategic communication process that builds mutually beneficial relationships between organizations and their publics." Thus, while public relations is sometimes practiced or portrayed as PR professionals communicating in only one direction—putting a certain kind of spin on a message to persuade people to accept a particular point of view—when it is done best, public relations is about two-way relationships, with communication and feedback between PR practitioners and their many audiences.

BRENDAN SMIALOWSKI/AFP/Getty Images

PR AND POLICY REFORM
The Keystone XL pipeline project, which would have carried oil from the tar sands of Alberta, Canada, to the U.S. Gulf of Mexico, was canceled in 2021 after years of strategic public relations work by environmental and Indigenous activists in the United States and Canada, taking on powerful oil companies.

Public relations covers a wide array of practices, such as shaping the public image of a politician or a celebrity, establishing or repairing communication between consumers and companies, and promoting government agencies and actions, especially during times of crisis. While public relations may sound very similar to advertising, which also seeks to persuade audiences, it is a different skill in a variety of ways. Advertising uses simple and fixed messages (e.g., "our appliance is the most efficient and affordable") that are transmitted directly to the public through the purchase of ads. Public relations involves more complex messages that may evolve over time (e.g., a political campaign that focuses on a series of small public listening sessions, or a long-term strategy designed to dispel unfavorable reports about a social media company violating users' privacy) and that may be transmitted to the public directly or indirectly, often through the news media. Of course, today's PR professionals also make frequent use of websites and social media tools, which enable their clients to interact with audiences firsthand, without having to go through more traditional news media gatekeepers.

The social and cultural impact of public relations has been immense. In its infancy, PR helped convince many American businesses of the value of nurturing the public, whose members became purchasers rather than producers of their own goods after the Industrial Revolution. PR sets the tone for the corporate image-building that characterizes our economic environment and for the battles of organizations taking sides on today's environmental, energy, and labor issues. Perhaps PR's most significant effect, however, has been on the political process, in which individuals and organizations—on both the Right and the Left—hire public relations staff to favorably shape their public images.

Critics of PR raise concerns about the communication inequalities that arise when some organizations have the ability to hire a huge PR firm or staff a PR department and others—such as nonprofits and smaller organizations—do not. Critics also worry about the impact that strategic communication can have on the nature of truth in a culture, if practitioners use their research to mold or manipulate the public in deceptive ways. But effective public relations can also be an equalizer, helping to generate public support for innovative ideas and policy reforms that challenge well-funded institutions and entrenched positions.

In this chapter, we will:

- Study the impact of public relations and the historical conditions that affected its development as a modern profession
- Look at nineteenth-century press agents and the role that railroad and utility companies played in developing corporate PR
- Consider the rise of modern PR, particularly the influences of former reporters Ivy Lee and the team of Edward Bernays and Doris Fleischman
- Explore the major practices and specialties of public relations
- Examine the reasons for the long-standing antagonism between journalists and members of the PR profession, and the social responsibilities of public relations in a democracy

How We Got Here: Early Developments in Public Relations

At the beginning of the twentieth century, the United States shifted to a consumer-oriented industrial society, which fostered the development of new products and services as people moved to cities to find work. During this transformation from farm to factory, advertising and PR emerged as professions. While advertising drew attention and customers to new products, PR began in part to help businesses fend off increased scrutiny from the muckraking journalists and emerging labor unions of the time.[8] But even before the formal beginning of the profession, there were those who developed ways to garner media exposure and publicity to promote their clients.

The First Publicists and Press Agents: P. T. Barnum and Buffalo Bill

The first PR practitioners were simply **press agents**: those in the early nineteenth century who sought to advance a client's image through media exposure, primarily via stunts staged for newspapers. For instance, people like Daniel Boone, who engineered various land-grab and real estate ventures, and Davy Crockett, who was involved in the massacre of Native Americans, employed press agents. Such individuals often wanted press agents to repair and reshape their reputations as cherished frontier legends or as respectable candidates for public office.

The most notorious press agent of the nineteenth century was Phineas Taylor (P. T.) Barnum, who used gross exaggeration, fraudulent stories, and staged events to secure newspaper coverage for his clients, his American Museum, and later his circus. Barnum's circus, dubbed "The Greatest Show on Earth," included General Tom Thumb (an entertainer with dwarfism), the Swedish soprano Jenny Lind, Jumbo the Elephant, and Joice Heth (who Barnum claimed was the 161-year-old nurse of George Washington but who was actually 80 when she died). These performers became some of the earliest nationally known celebrities because of Barnum's skill in using the media for promotion. Barnum understood that his audiences liked to be tricked. In newspapers and on handbills, he later often revealed the strategies behind his more elaborate hoaxes.

From 1883 to 1916, William F. Cody, who once killed buffalo for the railroads, promoted himself and his traveling show: "Buffalo Bill's Wild West and Congress of Rough Riders of the World." Cody's troupe—which featured Bedouins, Cossacks, and gauchos, as well as "cowboys and Indians"—re-created dramatic gunfights, the Civil War, and battles of the Old West. The show employed sharpshooter Annie Oakley and Lakota chief and healer Sitting Bull, whose legends were partially shaped by Cody's nine press agents. These agents were led by John Burke, who successfully promoted the show for its entire thirty-four-year run. Burke was one of the first press agents to use a wide variety of media channels to generate publicity: promotional newspaper stories, magazine articles and ads, dime novels, theater marquees, poster art, and early films. Burke and Buffalo Bill shaped many of the lasting myths about rugged American individualism and frontier expansion that were later adopted by books, radio programs, and Hollywood films depicting the American West. Along with Barnum, they were among the first to use **publicity**—a type of PR communication that uses various media messages to spread information about a person, a corporation, an issue, or a policy—to elevate entertainment culture to an international level.

Library of Congress, Prints & Photographs Division, [LC-DIG-ppmsca-54807]

EARLY PUBLIC RELATIONS Originally called "P. T. Barnum's Great Traveling Museum, Menagerie, Caravan, and Hippodrome," Barnum's circus merged with Bailey's circus in 1881 and again with the Ringling Bros. in 1919. Even with the ups and downs of the Ringling Bros. and Barnum & Bailey Circus over the decades, Barnum's original catchphrase, "The Greatest Show on Earth," endured through the circus's final performance on May 21, 2017.

Helping Big Business

As P. T. Barnum, Buffalo Bill, and John Burke demonstrated, the use of press agents brought with it enormous power to sway the public and to generate business. So it is not surprising that during the nineteenth century, America's largest industrial companies—particularly the railroads—also employed press agents to win favor in the court of public opinion.

The railroads began to use press agents to help them obtain federal funds. For example, Illinois Central was one of the first companies to use government *lobbyists* (people who try to influence the voting of lawmakers) to argue that railroad service between the North and the South was in the public interest and would ease tensions, unite the two regions, and prevent a war.

The railroad press agents successfully gained government support by developing some of the earliest publicity tactics. Their first strategy was simply to buy favorable news stories about rail travel from newspapers through direct bribes. Another practice was to engage in *deadheading*—giving reporters free rail passes with the tacit understanding that they would write glowing reports about rail travel. Eventually, wealthy railroads received the federal subsidies they wanted and increased their profits, while the American public shouldered much of the financial burden of rail expansion.

Having obtained construction subsidies, the larger rail companies turned their attention to bigger game—persuading the government to control rates and reduce competition, especially from smaller, more aggressive regional lines. Railroad lobbyists argued that federal support would lead to improved service and guaranteed quality because the government would be keeping a close watch. These lobbying efforts, accompanied by favorable publicity, led to passage of the Interstate Commerce Act in 1887, the first federal law to regulate private industry, which required railroads to publicize their shipping rates, banned special lower rates for certain freights or passengers, and established a commission to oversee enforcement of the law.[9] Historians have argued that, ironically, the PR campaign's success actually led to the decline of the railroads: Artificially maintained higher rates and burdensome government regulations forced smaller firms out of business and eventually drove many customers to other modes of transportation.

The Birth of Modern Public Relations

By the first decade of the twentieth century, reporters and muckraking journalists were investigating the promotional practices behind many companies. As an informed citizenry paid more attention, it became increasingly difficult for large firms to fool the press and mislead the public. With the rise of the middle class, increasing literacy rates and labor organization among the working classes, and the spread of information through print media, democratic ideals began to threaten the established order of business and politics—and the elite groups who managed them. Three pioneers of public relations—Ivy Lee and the duo of Edward Bernays and Doris Fleischman—emerged in this atmosphere to popularize an approach that emphasized shaping the interpretation of facts and "engineering consent."

Ivy Ledbetter Lee

Most nineteenth-century corporations and manufacturers cared little about public sentiment. By the early 1900s, though, executives had realized that their companies could sell more products if they were associated with positive public images and values. Into this public space stepped Ivy Ledbetter Lee, considered one of the founders of modern public relations. Lee understood that the public's attitude toward big corporations had changed. He counseled his corporate clients that honesty and directness were better PR devices than the deceptive practices of the nineteenth century, which had fostered suspicion and an anti-big-business sentiment.

A minister's son, an economics student at Princeton University, and a former reporter, Lee opened one of the first PR firms in the early 1900s with George Park. Lee quit the firm in 1906 to work for the Pennsylvania Railroad, which, following a rail accident, had hired him to help downplay

IVY LEE, a founding father of public relations (*above*), did more than just crisis work with large companies and business magnates. His PR clients also included transportation companies in New York City (*above right*) and aviator Charles Lindbergh.

unfavorable publicity. Lee's advice, however, was that Penn Railroad admit its mistake, vow to do better, and let newspapers in on the story. These suggestions ran counter to the then standard practice of hiring press agents to manipulate the media, yet Lee argued that a managed open relationship between business and the press would lead to a more favorable public image. In the end, Penn and subsequent clients, notably John D. Rockefeller, adopted Lee's successful strategies.

By the 1880s, Rockefeller controlled 90 percent of the nation's oil industry and suffered from periodic image problems, particularly after Ida Tarbell's powerful muckraking series about the ruthless business tactics practiced by Rockefeller and his Standard Oil Company appeared in *McClure's Magazine* in 1904. The Rockefeller and Standard Oil reputations reached a low point in April 1914, when tactics to stop union organizing erupted in tragedy at a coal company in Ludlow, Colorado. During a violent strike, fifty-three workers and their family members died, including thirteen women and children.

Lee was hired to contain the damaging publicity fallout. He immediately distributed a series of "fact" sheets to the press, telling the corporate side of the story and discrediting the tactics of the United Mine Workers, which had organized the strike. As he had done for Penn Railroad, Lee also brought in the press and staged photo opportunities. John D. Rockefeller Jr., who now ran the company, donned overalls and a miner's helmet and posed with the families of workers and union leaders. This was probably the first use of a PR campaign in a labor-versus-management dispute. Over the years, Lee completely transformed the wealthy family's image, urging the discreet Rockefellers to publicize their charitable work. To improve his image, the senior Rockefeller took to handing out dimes to children wherever he went—a strategic ritual that historians attribute to Lee.

Called "Poison Ivy" by corporate foes and critics within the press, Lee had a complex understanding of facts. For Lee, facts were elusive and malleable, begging to be forged and shaped. "Since crowds do not reason," he noted in 1917, "they can only be organized and stimulated through symbols and phrases."[10] In the Ludlow case, for instance, Lee noted that the women and children who died while retreating from the charging company-backed militia had overturned a stove, which caught fire and caused their deaths. One of his PR fact sheets implied that they had, in part, been victims of their own carelessness.

Edward Bernays and Doris Fleischman

The nephew of Sigmund Freud, former reporter Edward Bernays inherited the public relations mantle from Ivy Lee. Beginning in 1919, when he opened his own office, Bernays was the first person to apply the findings of psychology and sociology to public relations, referring to himself as a "public relations

Bettmann/Getty Images

The New York Public Library/Art Resource, NY

counselor" rather than a "publicity agent." Over the years, Bernays's client list included General Electric, the American Tobacco Company, General Motors, *Good Housekeeping* and *Time* magazines, Procter & Gamble, RCA, the government of India, the city of Vienna, and President Coolidge.

Bernays also worked for the Committee on Public Information during World War I, developing propaganda that supported America's entry into that conflict and promoting the image of President Woodrow Wilson as a peacemaker. Both efforts were among the first full-scale governmental attempts to mobilize public opinion. In addition, Bernays made key contributions to public relations education, teaching the first class called "public relations"—at New York University in 1923—and writing the field's first textbook, *Crystallizing Public Opinion*. For many years, his definition of PR was the standard: "Public relations is the attempt, by information, persuasion, and adjustment, to engineer public support for an activity, cause, movement, or institution."[11]

In the 1920s, Bernays was hired by the American Tobacco Company to develop a campaign to make smoking more publicly acceptable for women. Among other strategies, Bernays staged an event: placing women smokers in New York's 1929 Easter parade. He labeled cigarettes "torches of freedom" and encouraged women to smoke as a symbol of their newly acquired suffrage and independence from men. He also asked the women he placed in the parade to contact newspaper and newsreel companies in advance to announce their symbolic protest. The campaign received plenty of free publicity from newspapers and magazines. Within weeks of the parade, men-only smoking rooms in New York theaters began opening up to women.

Through much of his writing, Bernays suggested that emerging freedoms threatened the established hierarchical order. He thought it was important for experts and leaders to control the direction of American society: "The duty of the higher strata of society—the cultivated, the learned, the expert, the intellectual—is therefore clear. They must inject moral and spiritual motives into public opinion."[12] For the cultural elite to maintain order and control, they would have to win the consent of the larger public. As a result, Bernays described the shaping of public opinion through PR as the "engineering of consent." Like Ivy Lee, Bernays believed that public opinion was malleable and not always rational: In the hands of the right experts, leaders, and PR counselors, public opinion could be shaped into forms people could rally behind.[13] However, journalists like Walter Lippmann, who wrote the famous book *Public Opinion* in 1922, worried that PR professionals with hidden agendas, as opposed to journalists with professional detachment, held too much power over American public opinion.

Although Bernays often got more of the credit, his childhood friend Doris Fleischman was a pioneer in public relations work as well. Fleischman was a writer and assistant editor for the *New York Tribune* when Bernays hired her as a writer for his new firm in 1919, his first employee. When they married in 1922, she became an equal partner in the firm. Fleischman was a women's suffragist and an early feminist. She was a member of the Lucy Stone League, a women's rights organization that advocated for women to keep their name when they married. In fact, Fleischman was the first woman to receive a U.S. passport under her own last name in 1925. "I assume you will not wish me to travel under a false name," she wrote passport officials.[14]

Bernays's biographer wrote that Fleischman played a "central role in building the Bernays empire," taking part in the major decision making and writing press releases and stories for the Torches of Freedom campaign. Moreover, "she ghost-wrote scores of speeches and strategy papers that were delivered under her husband's name."[15] Fleischman was one of the first women to work in public relations, and she introduced PR to America's most powerful leaders through a pamphlet she edited called *Contact*. Because she opened up the profession to women from its inception, PR emerged as one of the few professions—apart from teaching and nursing—accessible to women who chose to work outside the home at that time. Today, more than 60 percent of PR professionals are women.

The Practice of Public Relations

Today, there are more than twelve thousand PR firms in the United States, plus thousands of additional PR departments within corporate, government, and nonprofit organizations.[16] Since the 1980s, the formal study of public relations has grown significantly at colleges and universities. By 2021, the Public Relations Student Society of America (PRSSA) had just under seven thousand members and over three hundred chapters in colleges and universities. As certified PR programs have expanded (often requiring courses in journalism), the profession has relied less and less on its traditional practice of recruiting journalists for its workforce. At the same time, new courses in professional ethics, issues management, and integrated marketing have expanded the responsibility of future practitioners. In this section, we discuss the differences between public relations agencies and in-house PR services and the various practices involved in performing PR.

The Business of Public Relations

As noted earlier, the Public Relations Society of America defines PR as "a strategic communication process that builds mutually beneficial relationships between organizations and their publics." To carry out this process, the PR industry uses two approaches. First, there are independent PR agencies whose sole job is to provide clients with PR services. Second, many organizations, which may or may not also hire independent PR firms, maintain their own in-house PR staffs to handle routine tasks, such as writing press releases, managing various media requests, staging special events, updating web and social media sites, and dealing with internal and external publics.

Many large PR firms are owned by, or are affiliated with, multinational communications holding companies, such as Omnicom, WPP, and Interpublic (see Table 12.1). However, the largest PR agency is Edelman, an independent firm started by Daniel J. Edelman in Chicago in 1952. Edelman was an innovator; his was the first public relations firm to bring clients and their products on television media tours, and one of the first American PR companies to do business in Asia. His son, Richard Edelman, now leads the company, which has represented corporate clients like KFC, Adobe, LinkedIn, Unilever, Samsung, and the Department of Defense for decades, and has about sixty global offices.

TABLE 12.1   THE TOP 5 PUBLIC RELATIONS FIRMS, 2019 (BY WORLDWIDE REVENUE, IN MILLIONS OF U.S. DOLLARS)

Rank	Agency	Parent Firm	Headquarters	Revenue
1	Edelman	Independent	Chicago	$892
2	Weber Shandwick	Interpublic	New York	$674
3	Burson Cohn & Wolfe	WPP	New York	$667
4	FleishmanHillard	Omnicom	St. Louis	$570
5	Ketchum	Omnicom	New York	$513

Data from: "Public Relations: Worldwide," *Advertising Age*, May 11, 2020, p. 14.

In contrast to external agencies, most PR work is done in-house at companies and organizations. Although America's largest companies typically retain external PR firms, almost every company involved in the manufacturing and service industries has an in-house PR department. Such departments are also a vital part of many professional organizations, such as the American Medical Association, the AFL-CIO, and the National Association of Broadcasters, as well as large nonprofit organizations, such as the American Cancer Society, the Arthritis Foundation, and most universities and colleges.

Performing Public Relations

Public relations, like advertising, pays careful attention to the needs of its clients—politicians, small businesses, industries, and nonprofit organizations—and to the perspectives of its targeted audiences: consumers and the general public, company employees, shareholders, media organizations, government agencies, and community and industry leaders. To do so, PR involves providing a multitude of services, including publicity, communication, public affairs, issues management, government relations, financial PR, community relations, industry relations, advertising, press agentry, promotion, media relations, social networking, and propaganda. This last service, **propaganda**, is communication strategically placed, either as advertising or as publicity, to gain public support for a special issue, program, or policy, such as a nation's war effort.

In addition, PR personnel (both PR technicians, who handle daily short-term activities, and PR managers, who counsel clients and manage activities over the long term) produce employee newsletters, manage client trade shows and conferences, appear on news programs, analyze complex issues and trends that may affect a client's future, and much more. Basic among these, however, are seven key activities: formulating a message through research, conveying the message through various channels (including Twitter and other social media accounts), building media relations to get stories placed in the news, planning and launching special events and media events, sustaining public support through community and consumer relations, maintaining client interests through government relations and lobbying, and managing communications during times of crisis.

WORLD WAR II was a time when the U.S. government used propaganda and other PR strategies to drum up support for the war. One of the more iconic posters at the time asked women to join the workforce.

MPI/Archive Photos/Getty Images

THE REAL COST⚠

Most vapes contain seriously addictive levels of nicotine.

Addiction isn't pretty.

MESSAGE FORMULATION One of the social media ads developed for the FDA's The Real Cost anti-tobacco campaign addresses vaping. This ad confronts the "cost-free" myth about vaping by emphasizing the risk for nicotine addiction and other health problems.

Research: Formulating the Message

One of the most essential practices in the PR profession is doing research. Just as advertising is driven today by consumer research, PR uses similar methods to begin to understand its various audiences. Because it has historically been difficult to determine why particular PR campaigns succeed or fail, research has become the key ingredient in PR forecasting. Like advertising, PR makes use of mail, telephone, and Internet surveys; focus group interviews; and social media analytics tools—such as Google Analytics, Klear, Keyhole, Sprout Social, and Twitter Analytics—to get a fix on an audience's perceptions of an issue, a policy, a program, or a client's image.

Research also helps PR professionals focus a campaign message. For example, after years of declining smoking rates, the Food and Drug Administration (FDA) was alarmed to discover an increase in youth smoking. The FDA conducted research with at-risk youths to determine what messages would be most effective among young people who were open to smoking or already experimenting with cigarettes. As a result, The Real Cost campaign launched nationwide in 2014 across a number of media platforms. The FDA even hired independent researchers to assess the effectiveness of the campaign among its target group. Over time, the campaign has evolved to keep up with tobacco-use trends. Most recently, The Real Cost campaign aims to communicate the dangers of vaping to the more than ten million young people ages twelve to seventeen "who have ever used e-cigarettes or are open to trying them."[17]

Conveying the Message

One of the chief day-to-day functions in public relations is creating and distributing PR messages for the news media or the public. There are several forms these messages can take, including press releases, VNRs, and various online options.

Press releases, or news releases, are announcements written in the style of news reports that present new information about an individual, a company, or an organization and pitch a story idea to the news media. In issuing press releases, PR agents hope that their client information will be picked up by the news media and transformed into news reports. Through press releases, PR firms manage the flow of information, controlling which media get what material in which order. (A PR agent may even reward a cooperative reporter by strategically releasing information.) News editors and broadcasters sort through hundreds of releases daily to determine which ones contain the most original ideas or are the most current. Most large media institutions rewrite and double-check the releases, but small market newspapers, radio, and TV stations often use them verbatim because of limited editorial resources. The more closely a press release resembles actual news copy, the more likely it is to be used, which is why many PR programs require students to take news writing classes and use the *AP Stylebook* (the style guide most commonly used by journalists). Press releases can be distributed via e-mail lists, by posting on the organization's website, and via Twitter. More than half of journalists follow Twitter to get news tips, so tweets can be effective in gaining news media coverage.

Since the introduction of portable video equipment in the 1970s, PR agencies and departments have also been issuing **video news releases (VNRs)**: thirty- to ninety-second visual press releases designed to mimic the style of a broadcast news report. Although networks and large TV news stations do not usually broadcast VNRs, news stations in small TV markets regularly use material from VNRs, which can be easily downloaded as a full news report, or in separate video clips that can be integrated into a report by the station. News stations have occasionally been criticized for using video footage from a VNR without acknowledging the source. In 2005, the FCC mandated that broadcast stations and cable operators must disclose the source of the VNRs they air.

The equivalent of VNRs for nonprofits are **public service announcements (PSAs)**: fifteen- to sixty-second audio or video reports that promote government programs, educational projects, volunteer agencies, or social reform. As part of their requirement to serve the public interest, broadcasters have been encouraged to carry free PSAs. Since the deregulation of broadcasting began in the 1980s, however, there has been less pressure and no minimum obligation for TV and radio stations to air PSAs.[18] When PSAs *do* run, they are frequently scheduled between midnight and 6:00 A.M., a less commercially valuable time slot.

The Internet is an essential avenue for distributing PR messages. Companies and organizations can upload and maintain their media kits online (including press releases, VNRs, images, executive bios, and organizational profiles), giving the traditional news media access to the information at any time. Social media have also transformed traditional PR communications, giving public relations professionals access to platforms (Twitter, Facebook, Instagram, YouTube) that have few or no formal gatekeepers. Now people can be "friends" and "followers" of companies and organizations. Executives, celebrities, and politicians can share observations and can seem more accessible and personable through social media posts. But social media's immediacy can also be a problem, especially for those who send messages into the public sphere without considering the ramifications. For PR professionals assessing an organization's "publics" today, anyone in the world can be a friend, a tough but honest critic, or a bad actor spewing disinformation.

Another issue with Internet communications is that sometimes these communications appear without complete disclosure, which is an unethical PR practice. For example, some PR firms have edited Wikipedia entries for their clients' benefit, a practice Wikipedia founder Jimmy Wales has repudiated as a conflict of interest. (See "Examining Ethics: Egg on The North Face—a Wikipedia Scandal.") As noted in Chapter 11, a growing number of companies also compensate influencers to subtly promote their products, unbeknownst to most audience members. In 2019, the Federal Trade Commission released rules for social media influencers to disclose their connections to companies.[19]

BUILDING RELATIONSHIPS The Internet enables organizations to build relationships with and convey messages to their audiences. With this tweet, the NBA uses social media to recognize Martin Luther King Jr. Day and connect King's values to its own and those of its fans.

Media Relations

PR managers specializing in media relations promote a client or an organization by securing publicity or favorable coverage in the news media. This is called **earned media**, and it means that the PR person has worked to earn the trust of a journalist to get the client's story reported in a news outlet. Earned media begins with building a relationship of trust with reporters and later involves sending out well-written press releases or interesting social media posts, holding newsworthy press conferences, or pitching stories to journalists that are not only helpful for the PR client but also beneficial for the news organization. This often requires an in-house PR person to speak on behalf of an organization or to direct reporters to experts who can provide information.

Media-relations specialists also perform damage control or crisis management when negative publicity occurs. Occasionally, in times of crisis—such as a scandal at a university or a safety recall by a car manufacturer—a PR spokesperson might be designated as the only source of information available to news media. Although journalists often resent being cut off from higher administrative levels and leaders, the institution or company in question often wants to manage the frame and delivery of information, which is easier with one contact person. In these situations, a game often develops between PR specialists and the media in which reporters attempt to circumvent the spokesperson and induce a knowledgeable insider to talk off the record, providing background details without being named directly as a source.

Egg on The North Face—a Wikipedia Scandal

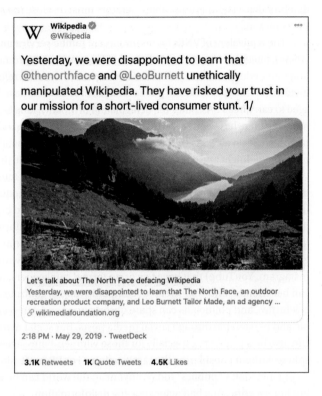

"**E**ver noticed that before going on a trip, everyone does a Google search? And most of the time, the first image [that comes up] is from Wikipedia?"

It was a rhetorical question, because the people who asked this question had already noticed this themselves, and proudly explained in a promotional video exactly how they used Wikipedia to game a Google search. In doing so, they did something that seemed brilliant but was completely unethical, and created a public relations problem for their client, the outdoor clothing and gear brand The North Face.

First, some background. Google, of course, is the world's dominant search engine. Wikipedia, a nonprofit online encyclopedia, is among the most visited websites in the United States and the world. Because Google's algorithmic search engine tends to list images from the most popular websites at the top of its returns, being able to place images on Wikipedia can be a powerful thing. Wikipedia's editors are volunteers but scrupulous, so if someone wants to add or alter a Wikipedia entry, the person shouldn't have a conflict of interest—that is, be working on behalf of a client or a close associate, or be working for pay.

This is where the ad agency Leo Burnett Tailor Made (headquartered in São Paulo, Brazil, and part of the communication holding company Publicis), working for client The North Face, made its ethical mistake. It decided to remove images of scenic outdoor destinations around the world and replace them with shots of people in The North Face gear in those same locations. As the agency boasted in its own video, "We photographed our brand in several adventurous places. Then we switched the Wikipedia photos for ours.

Simple as that." So, if you searched for places like Guarita State Park in Brazil, or the Cape Peninsula in South Africa, or the Storr in Scotland, you would have seen handsome young people who just happened to be wearing The North Face in the iconic location.

The Leo Burnett Tailor Made video gloated, "We hacked the results to reach one of the most difficult places: the top of the world's largest search engine. Paying absolutely nothing by just collaborating with Wikipedia." But there *was* a price to pay. Wikipedia charged that The North Face had "defaced" Wikipedia and lied about the organization's involvement in the campaign. In a tweet, The North Face responded, "We believe deeply in @Wikipedia's mission and apologize for engaging in activity inconsistent with those principles. Effective immediately, we have ended the campaign and moving forward, we'll commit to ensuring that our teams and vendors are better trained on the site policies."[1]

The Verge, which reported on the incident, concluded, "Maybe this article itself falls into the 'all publicity is good publicity' trap that North Face was expecting, but it's not really clear how taking advantage of an educational platform for free advertising does anything other than paint your own brand as greedy and disingenuous."[2] ●

PR agents who specialize in media relations also recommend the use of **paid media**—that is, advertising—to their clients when it seems appropriate. Unlike publicity, which is sometimes outside a PR agency's control, producing a paid advertisement may help focus a complex issue or a client's image. Earned media, however, carries the aura of legitimate news and thus has more credibility than advertising does.

Special Events and Media Events

Another public relations practice involves coordinating *special events* to raise the profile of corporate, organizational, or government clients. Typical special-events publicity often includes a corporate sponsor aligning itself with a cause or an organization that has positive stature among the general public. For example, John Hancock Financial has been the primary sponsor of the Boston Marathon since 1986 and funds the race's prize money. The company's corporate communications department also serves as the PR office for the race, operating the pressroom and creating the marathon's media guide and other press materials. Eighteen other sponsors—including Adidas, Gatorade, and JetBlue Airways—also pay to affiliate themselves with the Boston Marathon. At the local level, companies often sponsor a community parade or a charitable fundraising activity.

In contrast to a special event, a **media event** is any circumstance created for the sole purpose of gaining coverage in the media. Historian Daniel Boorstin coined a term—*pseudo-event*—to describe the contrived nature of these events in his influential 1962 book *The Image*. Typical media events include press conferences, TV and radio talk-show appearances, national awards shows, or any other staged activity aimed at drawing public attention and media coverage. In business, these events extend at least as far back as P. T. Barnum's publicity stunts, such as parading Jumbo the Elephant across the Brooklyn Bridge in the 1880s. In politics, Theodore Roosevelt's administration set up the first White House pressroom and held the first presidential press conferences in the early 1900s. These press conferences are media events that have become part of the expected governmental communication infrastructure, like the annual State of the Union address and the less formal annual media events at the White House, including the National Christmas Tree Lighting Ceremony, the National Thanksgiving Turkey Pardon, and the White House Easter Egg Roll.

Community and Consumer Relations

Another responsibility of PR is to sustain goodwill between an agency's clients and the public (see "Global Village: Public Relations and Bananas"). The public is often seen as two distinct audiences: communities and consumers. Companies have learned that sustaining close ties with their communities and neighbors not only enhances their image and attracts potential customers but also promotes the idea that the companies are good citizens. As a result, PR firms encourage companies to participate in community activities, such as hosting plant tours and open houses, making donations to national and local charities, and participating in town events like parades and festivals. More progressive companies may also get involved in unemployment and job-retraining programs or donate equipment and workers to community revitalization projects, such as Habitat for Humanity. One complicating factor is that as the United States has become more polarized, citizens are increasingly expecting companies to take a stand on certain political issues. As companies get entangled in these contentious topics, managing community relations has become more difficult.

Government Relations and Lobbying

While sustaining good relations with the public is a priority, so is maintaining connections with government agencies that have some say in how companies operate in a particular community, state, or nation. Both PR firms and the PR

COMMUNITY POP-UP EVENT FOR YOUR DOG
To subtly promote its SmartThings geo-tracker device, which can track luggage, keys, and (especially for this event) dogs, Samsung organized a daylong neighborhood pop-up community event in Greenwich Village in New York. The free event included an indoor pet park, gourmet dog treats, a caricature artist for dog portraits, and rescue dog adoptions.

Public Relations and Bananas

Doing public relations on behalf of bananas doesn't sound particularly necessary. After all, bananas are the number-one fresh fruit eaten in the United States, having long ago displaced apples in the top position. Yet the seemingly uncomplicated banana figures into the history of public relations, and not always in a good way.

In the early twentieth century, huge banana plantations were established in Colombia, Ecuador, Peru, Costa Rica, Guatemala, and Honduras. United Fruit (the predecessor of today's Chiquita Brands) was the dominant grower and importer of bananas and was particularly powerful in the small nations of Central America—in fact, too powerful. In 1951, Jacobo Árbenz, the new democratically elected president of Guatemala, proposed a number of reforms to raise the status of low-income agrarian Guatemalans. One of the reforms included redistributing idle, cultivable lands to peasants to lift them out of poverty. United Fruit owned some of those lands (which it had been given years earlier and on which it didn't pay property taxes). Unwilling to tolerate any limits on its control, United Fruit hired public relations pioneer Edward Bernays to work behind the scenes to build U.S. public opinion against the liberal Árbenz government, branding it as communist. In one of the worst moments for public relations and U.S. foreign policy, the CIA led a covert operation that deposed Guatemala's democratically elected administration in 1954 and installed a right-wing military dictator who was more to United Fruit's liking. Guatemala then endured decades of war while the CIA repeated similar covert interventions on behalf of U.S. business interests in several Latin American countries, giving rise to the term *banana republic*—a country in which a single dominant industry controls business and politics.

In another black eye for the banana industry, Dole and Del Monte, two of today's largest banana producers, were sued in 2012 by more than one thousand banana plantation workers for using a pesticide that had been banned in the United States in 1979. Bloomberg reported that the pesticide, dibromochloropropane (DBCP), "has been linked to sterility, miscarriages, birth defects, cancer, eye problems, skin disorders and kidney damage," and that workers said they had not been informed of the dangers or issued protective equipment.[1]

Now the good news for bananas and public relations: In 2001, Dole Food Company responded to increasing consumer interest by producing organic bananas for the first time. Although it still produces bananas that are not certified as organic, it is now the leading producer of organic bananas in the world. In 2007, Dole improved communication of its organic program by launching Doleorganic.com and labeling each bunch of organic bananas with a sticker that identifies the farm that produced them. The sticker reads "Visit the Farm at doleorganic .com" and includes the country of origin and a three-digit farm code. The website provides information about the banana farm in question; a Google map (viewers can zoom in on the satellite view to see the expanse of the farm); and photo albums containing shots of workers, plants, and facilities. The company says its Dole Organic site is evidence of the company's "corporate philosophy of adhering to the highest ethical conduct in all its business dealings, treatment of its employees, and social and environmental policies."[2] Considering the lack of transparency in the history of public relations for bananas, this is a good thing—for the countries where Dole does business, the company's workers, and its consumers. ●

Romeo Gacad/AFP/Getty Images

divisions within major corporations are especially interested in making sure that government regulation neither becomes burdensome nor reduces their control over their businesses.

Government PR specialists monitor new and existing legislation, create opportunities to ensure favorable publicity, and write press releases and direct-mail letters to persuade the public about the pros and cons of new regulations. In many industries, government relations has evolved into **lobbying**: the process of attempting to influence lawmakers to support and vote in favor of an organization's or industry's best interests. In seeking favorable legislation, some lobbyists contact government officials on a daily basis. In Washington, D.C., alone, there are more than eleven thousand registered lobbyists—and thousands more government-relations workers who aren't required to register under federal disclosure rules. Lobbying expenditures targeting the federal government were at $3.49 billion in 2020, more than twice the amount in 2000 (see Figure 12.1).[20]

Lobbying can often lead to ethical problems, as in the case of earmarks and astroturf lobbying. *Earmarks* are specific spending directives that are slipped into bills to accommodate the interests of lobbyists and are often the result of political favors or outright bribes. In 2006, lobbyist Jack Abramoff (dubbed "the Man Who Bought Washington" in *Time*) and several of his associates were convicted of corruption related to earmarks, leading to the resignation of leading House members and a decline in the use of earmarks.

Astroturf lobbying is phony grassroots public affairs campaigns engineered by public relations firms. PR firms deploy massive phone banks and computerized mailing lists to drum up support and create the impression that millions of citizens back their client's side of an issue. For instance, the Center for Consumer Freedom (CCF)—an organization that appears to serve the interests of consumers—is actually a creation of the Washington, D.C.–based PR firm Berman & Co. and is funded primarily by the restaurant, food, alcohol, and tobacco industries. According to SourceWatch, which tracks astroturf lobbying, "anyone who criticizes" tobacco and alcohol, as well as "processed food, fatty food, soda pop, pharmaceuticals, animal testing, overfishing, and pesticides . . . is likely to come under attack from CCF."[21]

Public relations firms do not always work for the interests of corporations, however. They also work for other clients, including consumer groups, labor unions, professional groups, religious

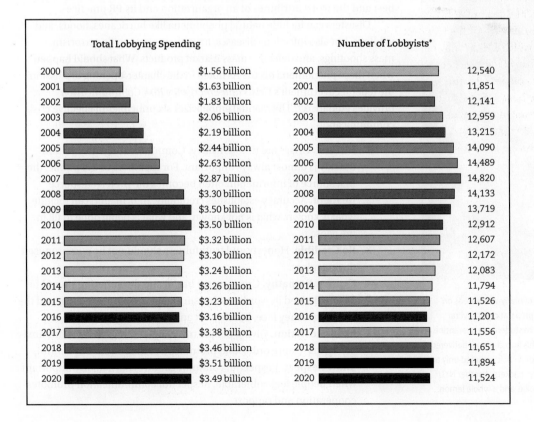

Total Lobbying Spending		Number of Lobbyists*	
2000	$1.56 billion	2000	12,540
2001	$1.63 billion	2001	11,851
2002	$1.83 billion	2002	12,141
2003	$2.06 billion	2003	12,959
2004	$2.19 billion	2004	13,215
2005	$2.44 billion	2005	14,090
2006	$2.63 billion	2006	14,489
2007	$2.87 billion	2007	14,820
2008	$3.30 billion	2008	14,133
2009	$3.50 billion	2009	13,719
2010	$3.50 billion	2010	12,912
2011	$3.32 billion	2011	12,607
2012	$3.30 billion	2012	12,172
2013	$3.24 billion	2013	12,083
2014	$3.26 billion	2014	11,794
2015	$3.23 billion	2015	11,526
2016	$3.16 billion	2016	11,201
2017	$3.38 billion	2017	11,556
2018	$3.46 billion	2018	11,651
2019	$3.51 billion	2019	11,894
2020	$3.49 billion	2020	11,524

FIGURE 12.1

TOTAL LOBBYING SPENDING AND NUMBER OF LOBBYISTS (2000–2020)

Data from: The Center for Responsive Politics based on data from the Senate Office of Public Records, updated January 23, 2021, www.opensecrets.org/federal-lobbying.

*the number of unique, registered lobbyists who have actively lobbied

organizations, and even foreign governments. For example, the New York–based PR firm BerlinRosen worked with the Service Employees International Union (SEIU) for several years on the Fight for $15 campaign, which seeks to increase the $7.25 federal hourly minimum wage, which has been in place since 2009, to $15. The campaign influenced several cities and states to raise their minimum wages to $15. The campaign got a big boost when a federal minimum wage increase to $15 was part of President Joe Biden's list of economic objectives in 2021.[22]

While lobbyists seek to influence government legislation, presidential administrations also use public relations—with varying degrees of success—to support their policies. From 2002 to 2008, the Bush administration's Defense Department operated a Pentagon Pundit program, secretly cultivating more than seventy retired military officers to appear on radio and television talk shows in order to shape public opinion about the Bush agenda. In 2008, the *New York Times* exposed the unethical program, and its story earned a Pulitzer Prize.[23] Barack Obama pledged to be more transparent on day one of his administration, but an Associated Press analysis during Obama's second term concluded that "the administration has made few meaningful improvements in the way it releases records."[24] And Donald Trump upended standard PR approaches altogether, routinely going around his communication team and tweeting messages on his own personal Twitter account, often disrupting efforts for carefully managed messaging. The Trump administration also dropped the practice of daily briefings and Q&A with the White House press corps, going for one full year without briefings until the White House Coronavirus Task Force began daily briefings in March 2020, then gradually ending these after less than two months.[25] In 2021, Joe Biden's White House brought back daily briefings with the press corps.

Public Relations during a Crisis

Since the Ludlow strike, one important duty of PR has been helping an organization handle a public crisis or tragedy, especially if the public assumes the organization is at fault. Disaster management may reveal the best and the worst attributes of an organization and its PR practices.

Disasters can include natural phenomena like hurricanes, floods, and wildfires, but also infectious diseases, hazardous waste spills, terrorism, mass shootings, and defective mass market products. What should happen when these phenomena occur? As noted in the chapter-opening discussion, the federal government's *Crisis and Emergency Risk Communication* provides guidance.[26] The manual emphasizes six principles of effective crisis communication:

1. **Be First.** Crises are time sensitive. Communicating information quickly is almost always important. For members of the public, the first source of information often becomes the preferred source.
2. **Be Right.** Accuracy establishes credibility. Information can include what is known, what is not known, and what is being done to fill in the gaps.
3. **Be Credible.** Honesty and truthfulness should not be compromised during crises.
4. **Express Empathy.** Crises create harm, and the suffering should be acknowledged in words. Addressing what people are feeling, and the challenges they face, builds trust and rapport.
5. **Promote Action.** Giving people meaningful things to do calms anxiety, helps restore order, and promotes a restored sense of control.
6. **Show Respect.** Respectful communication is particularly important when people feel vulnerable. Respectful communication promotes cooperation and rapport.

A BAD MEAL AT NYU was one of many viral videos on TikTok that documented the lack of preparation at several universities for delivering full meals to students in quarantine during the coronavirus pandemic. This boxed dinner delivered to an NYU student became bad PR for NYU: It included only an apple, a brownie, and salt and pepper packets. Another NYU student received a granola bar, a cookie, and a whole lemon.

The biggest global crisis thus far in the twenty-first century began in 2019, with the discovery of a virus named after that year: COVID-19. The coronavirus pandemic—the worst global pandemic in a century—created millions of crisis communication scenarios around the planet, as governments, businesses, hospitals, schools, churches, and other organizations were required to respond to the deadly virus. Some responded better than others. (See the chapter-opening discussion.)

A survey of PR professionals from around the world in late March 2020 (shortly after the World Health Organization declared COVID-19 a pandemic) identified New Zealand and Germany as the countries with the most impressive initial national communication response to COVID-19, while the United States was identified as having the worst. Global PR agency officials who were surveyed ranked the top three factors defining successful communication during the pandemic as "timeliness of communication," "communicating and updating policies," and "tone and language of messaging."[27]

As discussed at the beginning of the chapter, an effective public relations response is crucial to fighting a pandemic. According to two European health communication scholars, bad communication can be a virus itself. "Epidemics must be fought with the best biomedical and scientific tools and knowledge. . . . But we also have to deal with secondary, virtual epidemics that are taking place globally in news and social online media, where various correct and incorrect information, misconceptions, and rumors can be distributed without editorial control."[28]

Universities also had to respond to multiple stakeholders in light of the COVID-19 pandemic. According to university PR specialists, the two biggest crises to confront universities in spring 2020 were pivoting to remote instruction and deciding to send students home from campus. For remote instruction, the most effective universities quickly worked to get faculty up to speed on the best ways to deliver content. When it came to sending students home, the best university PR approaches paired the school closing announcements "with appeals for alumni to donate to student emergency funds to provide resources that would allow the institution to quickly buy additional technology or keep food banks open or ensure that they could . . . allow the students most at risk to remain on campus."[29] Of course, there were communication failures. Some universities quarantined students with COVID-19 in dirty isolation quarters ("I felt like a guinea pig," one University of Iowa student reported after her social media description of the "awful" experience went viral; the university apologized), while several others struggled to provide three sufficient meals a day to students quarantined in dorm rooms, prompting a raft of critical TikTok videos. (Regarding critiques about the quality and portions of food, a University of Georgia spokesperson reported that officials " 'are listening to [students'] feedback' and have implemented some changes.")[30]

Tensions between Public Relations and the Press

In 1932, Stanley Walker, an editor at the *New York Herald Tribune*, identified public relations agents as "mass-mind molders, fronts, mouthpieces, chiselers, moochers, and special assistants to the president."[31] Walker added that newspapers and PR firms would always remain enemies, even if PR professionals adopted a code of ethics (which they did in the 1950s) to "take them out of the red-light district of human relations."[32] Walker's tone captures the spirit of one of the most antagonistic—and mutually dependent—relationships in all of mass media.

Much of this antagonism, directed at public relations from the journalism profession, is cultural. Journalists have long considered themselves part of a public service profession, and some regard PR as having emerged as a profession created to distort the facts that reporters work hard to gather. Over time, reporters and editors developed the derogatory term **flack** to refer to a PR agent. The term—derived from the military word *flak*, meaning an antiaircraft artillery shell or a protective military jacket—symbolizes for journalists the protective barrier PR agents insert between their clients and the press. (Today, the *Associated Press Stylebook* defines *flack* simply as "slang for *press agent*.") Yet this antagonism belies journalism's dependence on public relations. Many editors, for instance, admit that more than half of their daily story ideas originate with PR people. In this section, we take a closer look at the relationship between journalism and public relations, which can be both adversarial and symbiotic.

launchpadworks.com

Give and Take: Public Relations and Journalism
This video debates the relationship between public relations and journalism.

Discussion: Are the similarities between public relations and journalism practices a good thing for the public? Why or why not?

Elements of Professional Friction

The relationship between journalism and PR is important and complex. Although journalism lays claim to independent traditions, the news media have become ever more reliant on public relations, which helps alert them to potential stories during a time in which there is an ever-increasing amount of information to sort through. Newspaper staff cutbacks, combined with television's need for local news events, have also expanded the news media's need for PR story ideas.

Another cause of tension is that PR firms often raid the ranks of reporting for new talent. Because most press releases are written to imitate news reports, the PR profession has always sought good writers who are well connected to sources and savvy about the news business. For instance, the fashion industry likes to hire former style or fashion news writers for its PR staff, and university information offices seek reporters who once covered higher education. But although reporters frequently move into PR, public relations practitioners seldom move into journalism; the money often isn't as good in journalism, and the news profession rarely accepts prodigal sons or daughters back into the fold once they have left reporting for public relations. Nevertheless, the professions remain codependent: PR needs journalists for publicity, and journalism needs PR for story ideas and access.

Undermining Facts and Blocking Access

Journalism's most prevalent criticism of public relations is that it sometimes works to counter the truths reporters seek to bring to the public. Modern public relations redefined and complicated the notion of what "facts" are. PR professionals demonstrated that the facts can be spun in a variety of ways, depending on what information is emphasized and what is downplayed. As Ivy Lee noted in 1925: "The effort to state an absolute fact is simply an attempt to achieve what is humanly impossible; all I can do is to give you my interpretation of the facts."[33] With practitioners like Lee showing the emerging PR profession how the truth could be interpreted, the journalist's role as a custodian of accurate information became much more difficult.

Journalists have also objected to PR professionals' blocking press access to key business leaders, political figures, and other newsworthy people. Before the prevalence of PR, reporters could talk to such leaders directly and obtain quotable information for their news stories. Now, however, journalists complain that PR agents often insert themselves between the press and the newsworthy, thus disrupting the journalistic tradition in which reporters would vie for interviews with top government and business leaders. Journalists further argue that PR agents are now able to manipulate reporters by giving exclusives only to journalists who are likely to cast a story in a favorable light or by cutting off a reporter's access to one of their newsworthy clients if that reporter has written unfavorably about the client in the past. (See "Media Literacy and the Critical Process: The Invisible Hand of PR.")

Promoting Publicity and Business as News

Another explanation for the professional friction between the press and PR involves simple economics. As Michael Schudson noted in his book *Discovering the News: A Social History of American Newspapers*, PR agents help companies "promote as news what otherwise would have been purchased in advertising."[34] Accordingly, Ivy Lee wrote to John D. Rockefeller after he gave money to Johns Hopkins University in 1914: "In view of the fact that this was not really news, and that the newspapers gave so much attention to it, it would seem that this was wholly due to the manner in which the material was 'dressed up' for newspaper consumption. It seems to suggest very considerable possibilities along this line."[35] News critics worry that this type of PR is taking media space and time away from those who do not have the financial resources or

THE INVISIBILITY OF PUBLIC RELATIONS is addressed in a series of six books by John Stauber and Sheldon Rampton. Both were staff members at the Center for Media and Democracy in Madison, Wisconsin. *Toxic Sludge Is Good for You!* is a critique of several public relations campaigns, including one to rebrand sewage sludge, which can contain an array of toxic substances, as "biosolids," which can be added to farm soil.

MEDIA LITERACY & THE CRITICAL PROCESS

The Invisible Hand of PR

John Stauber, founder of the Center for Media and Democracy and its publication *PRWatch*, which seeks to expose the activities of large PR firms, has described the PR industry as "a huge, invisible industry . . . that's really only available to wealthy individuals, large multinational corporations, politicians and government agencies."[1] How true is this? Is the PR industry so invisible?

1 DESCRIPTION

Test the so-called invisibility of the PR industry by seeing how often, and in what way, PR firms are discussed in the media. Using the Nexis database or a search engine, search U.S. news media coverage over the last six months for any mention of three prominent PR firms: Edelman, Weber Shandwick, and Burson Cohn & Wolfe.

2 ANALYSIS

What patterns emerge from the search? Possible patterns may have to do with personnel—that is, someone was hired, promoted, or fired. (These articles may be extremely brief, with only a quick mention of the firms.) Or these personnel-related articles may reveal connections between politicians or corporations and the PR industry. What about specific PR campaigns or articles that quote "experts" who work for Edelman, Weber Shandwick, or Burson Cohn & Wolfe?

3 INTERPRETATION

What do these patterns tell you about how the news media covers the PR industry? Was the coverage favorable? Was it critical or analytical? Did you learn anything about how the industry operates? Was the industry itself, its influencing strategies, and its wide reach across the globe visible in your search?

4 EVALUATION

PR firms—such as the three major firms in this search—have enormous power when it comes to influencing the public image of corporations, government bodies, and public policy initiatives in the United States and abroad. PR firms also have enormous influence over news content, yet the U.S. media do not always talk about this influence. Public relations firms aren't likely to reveal their power, but should journalism be more forthcoming about its role as a publicity vehicle for PR?

5 ENGAGEMENT

Visit the Center for Media and Democracy's website (PRWatch.org) and begin to learn about the unseen operations of the public relations industry. (You can also visit SpinWatch.org for similar critical analyses of PR in the United Kingdom.) Follow the Center for Media and Democracy's Twitter feed. Read some of the organization's books, join forum discussions, or attend a *PRWatch* event. Visit the organization's wiki site, SourceWatch (SourceWatch.org), and if you can, do some research of your own on PR and contribute an entry.

sophistication to become visible in the public eye. And there is another issue: If public relations can secure news publicity for clients, the added credibility of a journalistic context gives clients a status that the purchase of advertising cannot offer.

Another criticism is that PR firms with abundant resources clearly get more client coverage from the news media than do their lesser-known, or less well-funded, counterparts. For example, a business reporter at a large metro daily sometimes receives as many as a hundred press releases a day—far outnumbering the fraction of handouts generated by organized labor or grassroots organizations and putting these non-corporate organizations at a disadvantage. Workers and union leaders have long argued that the money that corporations allocate to PR leads to more favorable coverage for management positions in labor disputes. For example, standard news reports may feature subtle language choices, with "rational, coolheaded management making offers" and "hotheaded workers making demands." Walter Lippmann saw such differences in 1922 when he wrote, "If you study the way many a strike is reported in the press, you will find very often that [labor] issues are rarely in the headlines, barely in the leading paragraph, and sometimes not even mentioned anywhere."[36] This imbalance is particularly significant in that the great majority of workers are neither managers nor CEOs, and yet these workers receive little if any media coverage on a regular basis. Most newspapers now have business sections that focus on the work of various managers, but few have a labor, worker, or employee section.[37]

Shaping the Image of Public Relations

Dealing with a tainted past and journalism's hostility has led to the development of several image-enhancing strategies. In 1947, the PR industry formed its own professional organization, the PRSA (Public Relations Society of America). The PRSA is the largest organization devoted to the professional development of communications professionals. It is the recognized voice on ethics and professional standards, and it provides learning and networking opportunities for its members. It also offers operational support to the Universal Accreditation Board (UAB), a diverse group of educators and professionals who oversee the accreditation program for public relations. In addition to the PRSA, independent agencies devoted to uncovering shady or unethical public relations activities publish their findings in publications like *Strategies & Tactics*, *PRWeek*, and *PRWatch*. Ethical issues have become a major focus of the profession, with self-examination of these issues routinely appearing in public relations textbooks and professional newsletters (see Table 12.2). Still, the multibillion-dollar industry remains virtually invisible to the public, most of whom have never heard of even the largest agencies, like Edelman, Weber Shandwick, or Burson Cohn & Wolfe.

Over the years, as PR has subdivided itself into specialized areas, it has used more positive phrases—such as *institutional relations*, *corporate communications*, and *news and information services*—to describe what it does. Public relations' best press strategy, however, may be the limitations of the journalism profession itself. For most of the twentieth century, many reporters and editors clung to the ideal that journalism is, at its best, an objective institution that gathers information on behalf of the public. Reporters have only occasionally turned their pens, computers, and cameras on themselves to examine their own practices or their vulnerability to manipulation. Thus, by not challenging PR's more subtle strategies, many journalists have allowed PR professionals to interpret "facts" to their clients' advantage.

TABLE 12.2 PUBLIC RELATIONS SOCIETY OF AMERICA ETHICS CODE

In 2000, the PRSA approved a completely revised Code of Ethics, which included core principles, guidelines, and examples of improper conduct. Here is an excerpt from the code.

PRSA Member Statement of Professional Values
This statement presents the core values of PRSA members and, more broadly, of the public relations profession. These values provide the foundation for the Code of Ethics and set the industry standard for the professional practice of public relations. These values are the fundamental beliefs that guide our behaviors and decision-making process. We believe our professional values are vital to the integrity of the profession as a whole.
ADVOCACY
We serve the public interest by acting as responsible advocates for those we represent. We provide a voice in the marketplace of ideas, facts, and viewpoints to aid informed public debate.
HONESTY
We adhere to the highest standards of accuracy and truth in advancing the interests of those we represent and in communicating with the public.
EXPERTISE
We acquire and responsibly use specialized knowledge and experience. We advance the profession through continued professional development, research, and education. We build mutual understanding, credibility, and relationships among a wide array of institutions and audiences.
LOYALTY
We are faithful to those we represent, while honoring our obligation to serve the public interest.
FAIRNESS
We deal fairly with clients, employers, competitors, peers, vendors, the media, and the general public. We respect all opinions and support the right of free expression.

Data from: PRSA Code of Ethics at www.prsa.org.

Public Relations and Democracy

From the days of PR's formal origins in the early twentieth century, many people—especially journalists—have been skeptical of communications originating from public relations professionals. The bulk of the criticism leveled at public relations argues that the crush of information produced by PR professionals overwhelms traditional journalism. However, PR's most significant impact may be on the political process, especially when organizations hire public relations agencies to favorably shape or reshape a candidate's media image. In one example, former president Richard Nixon, who resigned from office in 1974 to avoid impeachment hearings regarding his role in the Watergate scandal, hired Hill+Knowlton to restore his postpresidency image. Through the firm's guidance, Nixon's writings—mostly on international politics—began appearing in Sunday op-ed pages. Nixon himself started showing up on television news programs like *Nightline* and spoke frequently before such groups as the American Newspaper Publishers Association and the Economic Club of New York. In 1984, after a media blitz by Nixon's PR handlers, the *New York Times* announced, "After a decade, Nixon is gaining favor," and *USA Today* trumpeted, "Richard Nixon is back." Before his death in 1994, Nixon, who never publicly apologized for his role in Watergate, saw a large portion of his public image shift from that of an arrogant, disgraced politician to that of a revered elder statesman.[38] Many media critics have charged that the press did not counterbalance this PR campaign and treated Nixon too reverently. In 2014, on the fortieth anniversary of the Watergate scandal, former CBS news anchor Dan Rather remembered Nixon's administration as a "criminal presidency" but added, "There has been an effort to change history, and in some ways it has been successful the last 40 years, saying well, it wasn't all that bad."[39]

Another critical area for public relations and democracy is how organizations integrate environmental claims into their public communications. In 1992, the Federal Trade Commission first issued its "Green Guides"—guidelines to ensure that environmental marketing practices don't run afoul of its prohibition against unfair or deceptive acts or practices, sometimes called **greenwashing**. As concern about climate change continues to grow, green marketing and public relations now extend into nearly every part of business and industry: product packaging (buzzwords include *recyclable, biodegradable, compostable, refillable, sustainable,* and *renewable*), buildings and textiles, renewable energy certificates, carbon offsets (funding projects to reduce greenhouse gas emissions in one place to offset carbon emissions produced elsewhere), labor conditions, and fair trade. Although there have been plenty of companies that make claims about providing green products and services, only some have infused environmentally sustainable practices throughout their corporate culture, and being able to tell the difference is essential to the public's understanding of environmental issues.

Though public relations often provides political information and story ideas, the PR profession bears only part of the responsibility for "spun" news; after all, it is the job of a PR agency to get favorable news coverage for the individual or group it represents. PR professionals police their ranks for unethical or irresponsible practices, but the news media should also monitor the public relations industry, as they do other government and business activities. Journalism itself also needs to institute changes that will make it less dependent on PR and more conscious of how its own practices play into the hands of spin strategies. A positive example of change on this front is that many major news organizations now offer regular critiques of the facts and falsehoods contained in political advertising and messaging. This media vigilance should be on behalf of citizens, who are entitled to robust, well-rounded debates on important social and political issues.

Like advertising and other forms of commercial speech, PR campaigns that result in free media exposure raise a number of questions regarding democracy and the expression of ideas. And like well-financed politicians, large companies and PR agencies can afford to invest in figuring out how to obtain favorable publicity. The question is not how to prevent that but how to ensure that other voices—less well financed and less commercial—also receive an adequate hearing. To that end, journalists need to become less willing conduits in the distribution of publicity. PR agencies, for their part, need to show clients that participating in the democratic process as responsible citizens can serve them well and enhance their image.

Chapter REVIEW 12

→ QUIZ YOURSELF WITH LEARNINGCURVE

LearningCurve uses gamelike quizzing to help you review important concepts from this chapter. LearningCurve includes multiple-choice questions at different levels of difficulty, hints, feedback for both right and wrong answers, and links to e-book material for easy reference. Which answer choice would you select for the sample question below?

How does public relations differ from advertising?

○ Public relations is purchased, but advertising is obtained without paying a fee.

○ Public relations uses simple, fixed messages, while advertising uses complex messages that may evolve over time.

○ Public relations messages may be transmitted to the public indirectly, while advertising messages are transmitted directly.

○ Public relations seeks to persuade audiences, but advertising does not.

→ WATCH VIDEO ON KEY CONCEPTS

LaunchPad includes clips from movies, TV shows, online sources, and other media texts, in addition to interviews with media experts and newsmakers. In the videos for this chapter, you'll consider topics like the relationship between PR practitioners and the press, plus other key issues in PR today.

→ PRACTICE THE CRITICAL PROCESS

Use the Media Literacy Activity to walk through the critical process step-by-step and develop a critical perspective on what media means to you. In this chapter's activity, you'll explore recent developments in the public relations industry.

→ LEARN KEY TERMS

Review the definitions for this chapter's key terms with LaunchPad's easy-to-use online flash cards. Try the cards in practice mode, or quiz yourself as a way to focus your study efforts. (Page numbers indicate where each term is highlighted within the chapter.)

stakeholders, 344
public relations, 345
press agents, 346
publicity, 346
propaganda, 351
press releases, 352

video news releases
 (VNRs), 352
public service
 announcements
 (PSAs), 353
earned media, 353

paid media, 355
media event, 355
lobbying, 357
astroturf lobbying, 357
flack, 359
greenwashing, 363

REVIEW QUESTIONS

Use the questions below to revisit key themes and concepts from this chapter.

 Need more practice? Access the LearningCurve multiple-choice questions for this chapter, which are available on LaunchPad. All questions include feedback for correct and incorrect answers, hints, and links to e-book material.

Public Relations Today

1. Who are *stakeholders* in a public relations campaign?

How We Got Here: Early Developments in Public Relations

2. What did people like P. T. Barnum and Buffalo Bill Cody contribute to the development of modern public relations?

3. How did railroads give the early forms of corporate public relations a bad name?

4. What contributions did Ivy Lee make toward the development of modern PR?

5. How did Edward Bernays and Doris Fleischman affect public relations?

The Practice of Public Relations

6. What are the two main business organizational structures for doing public relations?

7. What are press releases, and why are they important to reporters?

8. What is the difference between paid media and earned media?

9. Why have government relations and lobbying become increasingly important to the practice of PR?

Tensions between Public Relations and the Press

10. Explain the background of the antagonism between journalism and public relations.

11. How did PR change old relationships between journalists and their sources?

12. In what ways is conventional news like public relations?

Public Relations and Democracy

13. In what ways does the profession of public relations serve the rebranding of public officials and other famous people to obtain favorable publicity?

14. How do organizations integrate environmental claims into their public communications (greenwashing), and why?

15. How can unethical or irresponsible PR be prevented?

QUESTIONING THE MEDIA

Use these critical-thinking questions to reflect on your own experiences with media and apply your understanding of media concepts.

 For an interactive experience, questions on topics like these are available for the iClicker student response system. Instructors can download the questions from LaunchPad and use them to poll the class.

1. What do you think of when you hear the term *public relations*? What images come to mind? Where did these impressions come from?

2. What steps can reporters and editors take to monitor PR agents who manipulate the news media?

3. What are some key things an organization should do to respond effectively once a crisis hits?

Media Economics and the Global Marketplace

13

In December 2020, Christopher Nolan, the acclaimed director of films such as *Inception*, *Dunkirk*, and *Tenet*, was absolutely furious. In a statement released to the press, he fumed, "Some of our industry's biggest filmmakers and most important movie stars went to bed the night before thinking they were working for the greatest movie studio and woke up to find out they were working for the worst streaming service."[1]

What exactly was Nolan's problem? He had just found out that Warner Bros., one of the oldest and most venerable Hollywood movie studios, planned to make all seventeen of its 2021 films available on the HBO Max streaming service the same day they appeared in movie theaters. With this move, the studio was upending decades of industry norms and, some critics warned, threatening the very existence of movie theaters. Could theaters stay open if people could sidestep weekend crowds to watch the latest blockbuster on their home theater system?

The studio's announcement came amid the disruptions of the COVID-19 pandemic, which sank box-office ticket sales and led a number of Hollywood studios to release films on streaming services. Warner Bros.' announcement, however, suggested that the move might be more than just a temporary fix to deal with the crisis. As the *Hollywood Reporter* put it, "How do theaters survive this supposedly onetime, excused-by-the-pandemic move? Genies are hard to put back in the bottle—and no one believes Warners intended this to be temporary, anyway."[2]

The subsequent furor over Warner Bros.' plans has highlighted the complex nature of today's media industries, as well as their uncertain future. For one thing, it has revealed the changing relationship between movie studios and theater owners. Ann Sarnoff, chair and CEO of WarnerMedia Studios and Networks Group, claimed, "We agree with the fact that the theaters are very, very important. We want to keep the theaters in business."[3] Theater chain owners like AMC remained concerned.

◀ HBO Max, which launched in 2020 and is owned by media giant WarnerMedia, competes in the increasingly crowded online streaming marketplace alongside other big hitters, including Netflix, Disney+, and Amazon Prime. These services reflect a shifting landscape for the delivery of media content.

While theaters and studios had been in a mutually dependent relationship for decades, things were changing. The growing popularity of digital streaming services like Netflix and HBO Max made theater owners nervous—in the not-so-distant future, studios may not need to rely on theaters to show their movies.

The move also drew attention to the fact that Warner Bros. had just recently become part of a new media conglomerate. In 2018, the telecommunications giant AT&T purchased Time Warner and, along with it, Warner Bros. Studio and HBO. Time Warner was quickly restructured into WarnerMedia, as its new corporate owners looked to integrate the two massive corporations.

In May 2020, AT&T–WarnerMedia launched HBO Max, its answer to Netflix and Disney+ and the key to the company's future digital distribution plans. Many observers, including Nolan, believe WarnerMedia was using its slate of 2021 films as a *loss leader* to draw subscribers to HBO Max, meaning that the conglomerate was willing to make less money from those films as long as it would drive traffic to HBO Max.

While WarnerMedia and AT&T may have been happy with that trade-off, many of the producers, writers, actors, and directors involved in making the films—whose compensation packages might have included profit sharing from box-office ticket sales or who believe in the importance of the theatrical viewing experience—feared they would be the real losers. After all, they didn't have a financial stake in HBO Max.

For many Hollywood insiders, the decision confirmed that Warner Bros.' new corporate leaders did not understand the film business—especially the importance of building trust with the people who make movies. Nolan's angry response suggests they might have been right, as does that fact that, by May 2021, AT&T had already proposed spinning off WarnerMedia and merging it with Discovery, Inc. Only time will tell if WarnerMedia's move made sound business sense and how the role of theatrical exhibition will evolve. What is clear from the controversy it stirred up, however, is that the economics behind the media products we consume today are complex and volatile—with or without a pandemic.

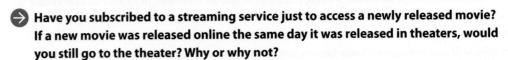 **Have you subscribed to a streaming service just to access a newly released movie? If a new movie was released online the same day it was released in theaters, would you still go to the theater? Why or why not?**

Media Economics Today

Over the last two decades, corporate takeovers and mergers have made our modern world distinct from that of earlier generations—at least in economic terms. What's at the heart of this brave new media world is a media landscape that has been forever altered by the emergence of the Internet; traditional media and telecommunications giants like Disney, NBC, and the *New York Times* have been joined and challenged by new digital giants like Amazon, Apple, Facebook, Google (Alphabet), and Microsoft. As WarnerMedia's controversial decision to stream its feature films demonstrates, media industries in our digital era are marked by unpredictable terrain. By altering how news, music, movies, TV, and advertisements are delivered, the Internet has forced most media businesses not only to rethink the content they provide but also to develop new business models and establish new partnerships to better navigate rapidly evolving economic conditions.

Economic forces significantly influence how media companies operate, and given the powerful cultural role the mass media play through the stories and information they circulate, the study of such economic conditions poses a number of complicated questions:

- What role should the government play in determining who owns the mass media and what kinds of media products are manufactured? Should it be a strong role, or should the government step back and let competition and market forces dictate what happens to mass media industries?
- Should citizen groups play a larger role in demanding that media organizations help maintain the quality of social and cultural life?
- Does the influence of American popular culture worldwide smother or encourage the growth of democracy and local cultures?
- Does the increasing concentration of economic power in the hands of several international corporations too severely restrict the number of voices in the media?

Answers to such questions can vary widely, generating debates that highlight the complex nature of media industries and differing views on corporate responsibility. Some people argue that because most media organizations are private, for-profit companies with a responsibility to maximize revenue and serve their shareholders, their fate should be left to the free market and their success measured in terms of efficiency and profitability. Others believe that because the products these media organizations make and sell are central to our culture, they have a greater responsibility to society than simply maximizing their own bottom line. They argue that people often want their culture to do certain things, such as reinforce specific moral standards, support national identity, reflect the diverse experience of all citizens, or promote democratic principles. As a result, critics have raised concerns about enormous media conglomerates, whose size and power might work against the public good by insulating corporations from public pressure and competition; reducing diversity in ownership and centralizing cultural influence in the hands of a few; and enabling corporations to exploit workers through exploitative labor practices.

Media scholars and others concerned about the negative impact of large companies can point to the fact that today's media industries are already very concentrated, as they are often structured in one of three ways: as a **monopoly**, an **oligopoly** (the most common structure), or a **limited competition** (typical of the radio and newspaper industries).[4] For a detailed explanation of these structures and how they constrain competition, refer to Figure 13.1.

In this chapter, as we examine the economic impact of these industry structures and related business strategies, we will:

- Explore the issues and tensions that are part of the current media economy
- Examine how changes in regulations, globalization, and the Internet fueled the rise of media conglomerates in the Information Age economy
- Examine major trends in today's media businesses, including major revenue streams, horizontal and vertical integration, and the importance of synergy and corporate flexibility
- Address social and political issues surrounding media economics, including struggles over antitrust laws, changing employment patterns, the media's role in building faith in free-market capitalism, and debates about cultural imperialism
- Consider the impact of media consolidation on democracy and on the diversity of the marketplace

Don Arnold/WireImage/Getty Images

OPRAH WINFREY has built a remarkable media empire in which she has frequently tried to blend profitability with social responsibility. From book publishing to filmmaking and television, where she got her start, Winfrey has established an expansive sphere of influence. After ending her talk show, she launched the cable TV network OWN in 2011, which, after a slow start, moved toward stability on the strength of scripted television programs. Discovery, Inc. is now the majority owner of OWN, but Winfrey remains its CEO.

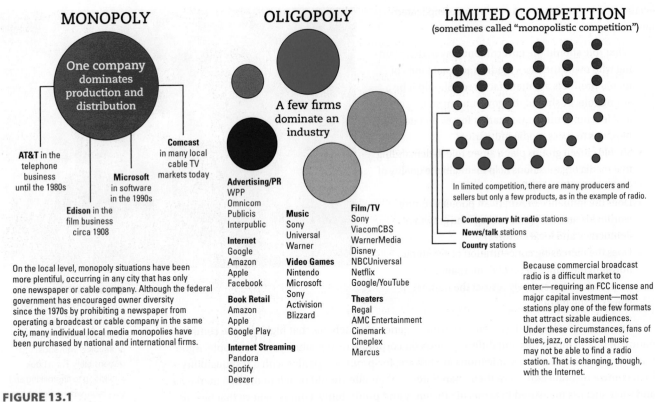

MONOPOLY

One company dominates production and distribution

AT&T in the telephone business until the 1980s

Edison in the film business circa 1908

Microsoft in software in the 1990s

Comcast in many local cable TV markets today

On the local level, monopoly situations have been more plentiful, occurring in any city that has only one newspaper or cable company. Although the federal government has encouraged owner diversity since the 1970s by prohibiting a newspaper from operating a broadcast or cable company in the same city, many individual local media monopolies have been purchased by national and international firms.

OLIGOPOLY

A few firms dominate an industry

Advertising/PR
WPP
Omnicom
Publicis
Interpublic

Internet
Google
Amazon
Apple
Facebook

Book Retail
Amazon
Apple
Google Play

Internet Streaming
Pandora
Spotify
Deezer

Music
Sony
Universal
Warner

Video Games
Nintendo
Microsoft
Sony
Activision
Blizzard

Film/TV
Sony
ViacomCBS
WarnerMedia
Disney
NBCUniversal
Netflix
Google/YouTube

Theaters
Regal
AMC Entertainment
Cinemark
Cineplex
Marcus

LIMITED COMPETITION
(sometimes called "monopolistic competition")

In limited competition, there are many producers and sellers but only a few products, as in the example of radio.

Contemporary hit radio stations
News/talk stations
Country stations

Because commercial broadcast radio is a difficult market to enter—requiring an FCC license and major capital investment—most stations play one of the few formats that attract sizable audiences. Under these circumstances, fans of blues, jazz, or classical music may not be able to find a radio station. That is changing, though, with the Internet.

FIGURE 13.1

MEDIA INDUSTRY STRUCTURES

How We Got Here: The Rise of Global Conglomerates

Today's massive media corporations were created as part of a large-scale shift that took place in the global economy. In the early twentieth century, the U.S. economy was driven by industrial manufacturing, mass production, and an intense rivalry between U.S.-based businesses and businesses from other nations. By the 1960s, however, the country began to transition to a postindustrial economy, as the Industrial Age slowly gave way to our current Information Age. Offices displaced factories as major work sites; centralized mass production often gave way to decentralized manufacturing that took place around the globe; many U.S. manufacturing jobs were replaced by lower-paid service work; and massive computer systems produced and processed data—a key product in the information economy. Mass media industries and their products both contributed to and were transformed by the evolving dynamics of this new economy.

During the Information Age, media industries have seen the emergence of massive corporations that operate on a global rather than just a national scale. The growth and power of such **conglomerates**—large corporations that develop through a series of mergers with and acquisitions of other companies—has been enabled and fueled by three factors: shifting regulations, globalization, and the Internet.

The Changing Role of Regulation

During the rise of industry in the nineteenth century, entrepreneurs such as John D. Rockefeller in oil and Andrew Carnegie in steel created monopolies in their respective industries. There was so little regulation of these newly powerful industries that the companies became notorious for their exploitative labor practices (including child labor), corrupt corporate conduct, and manipulation of the competitive landscape. Businesses that cooperated with one another to keep out competitors were called "trusts"—a term that became synonymous with corporations that used questionable business practices designed to drive out fair competition. Congress responded by passing three significant antitrust laws between 1890 and 1950 to increase competition among companies and prevent any one company from having too much control over a market:

- **1890—Sherman Antitrust Act**
  Outlawed monopoly practices and corporate trusts that often fixed prices to force competitors out of business.
- **1914—Clayton Antitrust Act**
  Prohibited manufacturers from selling only to dealers and contractors who agreed to reject the products of business rivals.
- **1950—The Celler-Kefauver Act**
  Limited any corporate mergers and joint ventures that reduced competition.

These laws remain in place today and are enforced by the Federal Trade Commission and the Antitrust Division of the Department of Justice.

The Move to Deregulation

Corporations chafed under the government's antitrust rules and other regulations. Using public relations tactics and aggressive lobbying campaigns, they worked to replace anticorporate sentiments that were prominent throughout the first half of the twentieth century (particularly in light of the Great Depression) with an alternative narrative that painted government regulation as bad for business and bad for America.[5] That narrative finally gained wide acceptance during the presidency of Ronald Regan (1981–1989), a period during which many controls on businesses were drastically weakened. This weakening, or *deregulation*, made it easier for corporations to take over or merge with companies in different sectors of the economy (producing highly diversified conglomerates) or companies in the same sector (producing oligopolies, as happened in air travel, finance, and media).[6]

One of the media sectors most visibly deregulated was broadcasting. In 1953, as television was expanding across the country, the FCC adopted the 7-7-7 Rule, which limited companies to owning no more than seven AM radio stations, seven FM radio stations, and seven television stations.[7] For more than thirty years, these ownership limitations helped ensure diversity in broadcast media ownership—and, with it, more diverse viewpoints. However, in the 1980s, the FCC began to weaken ownership limits. In 1984, the FCC expanded the rule to 12-12-12; it was increased to 18-18-12 in 1992, and then to 20-20-12 in 1994. Then, two years later, the Telecommunications Act of 1996 brought unprecedented deregulation to a broadcast industry that had been closely regulated for more than sixty years (see Figure 13.2). From 1996 onward, the following held true:

- A single company could now own an almost unlimited number of radio and TV stations.
- Telephone companies could now own TV and radio stations.
- Cable companies could now compete in the local telephone business.
- Cable companies could now freely raise rates.

Proponents of the Telecommunications Act of 1996 argued that these changes would spur the development of new competition, which would lower consumer prices. Instead, they spurred a series of mergers and acquisitions, resulting in ever-larger corporations that controlled cable, telephone, and broadband service—and still do today. Critics have argued that these changes actually resulted in increased prices. By 2015, for example, the average monthly cost of expanded basic cable service had grown to $69.03, a price increase almost triple the rate of inflation since 1995.[8]

Bettmann/Getty Images

ANTITRUST REGULATION
During the late nineteenth century, John D. Rockefeller Sr., considered the richest man in the world, controlled more than 90 percent of the U.S. oil refining business. But in 1911, antitrust regulations were used to bust up Rockefeller's powerful Standard Oil into more than thirty separate companies. He later hired PR guru Ivy Lee to refashion his negative image as a greedy corporate mogul.

FIGURE 13.2

U.S. BROADCAST OWNERSHIP DEREGULATION

From 1953 to 1984, the FCC enacted rules that prohibited a single company from owning more than seven AM radio stations, seven FM radio stations, and seven TV stations (called the 7-7-7 Rule):

7 AM
STATIONS

7 FM
STATIONS

7 TV
STATIONS

Also, a single person or company could own only one radio station per market. But ownership rules relaxed during the 1980s, and by 1994, the following was allowed:

20 AM
STATIONS

20 FM
STATIONS

12 TV
STATIONS

After the Telecommunications Act of 1996, several radio corporations quickly ballooned to include hundreds of stations. As a result, radio and television ownership became increasingly consolidated. Since 1996, one of the largest radio companies has been iHeartMedia (formerly Clear Channel Communications).

CLEAR CHANNEL COMMUNICATIONS GREW ASTRONOMICALLY AFTER THE TELECOMMUNICATIONS ACT OF 1996 WAS PASSED

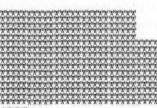

1972
1 FM STATION

1998
454 AM, FM, AND TV STATIONS

2005
1,200 AM, FM, AND TV STATIONS

Embracing Regulations That Boost Profits

While corporate lobbying efforts ended or eroded many regulations over the years, that's not the whole story. Media companies like Disney and Sony simultaneously fought to fortify other government regulations—most notably copyright, trademark, and patent laws. Newly strengthened regulations in these areas expanded the rights of owners of *intellectual property (IP)*—any product or invention that results from creativity, including movies, music, video games, and other outputs from media industries. For example, new regulations extended the window during which a product remains under copyright protection and increased penalties for violating copyright (see Chapter 16). These changes came at the expense of consumers, whose freedom to use the products covered by them became more limited. Corporations fought for stronger IP regulations because these protections were key to corporate profits in the emerging information economy, where leveraging creative content was becoming increasingly central to many business models. By increasing the value of copyrighted IP assets, strong IP regulations also helped fuel corporate mergers and acquisitions, like Disney's 2009 purchase of Marvel (which brought familiar characters like Captain America into Disney's fold) and Viacom's 2019 merger with CBS (which gave the conglomerate access to a library of 3,600 movies and 140,000 TV episodes).[9]

As these examples illustrate, although corporate lobbyists have painted government regulation as undue interference in a free market and bad for business for decades, the reality is more complicated. Instead of eliminating all regulations, corporations want to get rid of some regulations (those they don't like) and strengthen other regulations (those they do). A market free of any government regulation would be chaotic, which is something corporate leaders certainly don't want. As media citizens, we should understand that corporate calls for deregulation are actually calls for self-serving *reregulation*.

Globalization Expands Media Markets

Globalization is another factor that has influenced the strategies of media corporations in the information economy. While U.S. media companies distributed their products in international markets throughout the twentieth century, companies have become increasingly global in scale over the past fifty years thanks to large-scale changes in the global economy. After World War II, institutions like the World Trade Organization and trade agreements like the General Agreement on Tariffs and Trade established global systems for the flow of goods. These institutions also removed regulatory barriers that had protected national markets from international competitors, and fostered the rise of multinational corporations in various industries, including media and communication. These dynamics intensified in the 1990s, when the end of the Cold War and the opening of China further expanded the global market.

All these changes greatly transformed media industries. By the second half of the twentieth century, many U.S.-based media companies, finding it difficult to grow inside the highly saturated U.S. market, jumped at the chance to expand into less developed media markets around the world, where they could make additional revenue. (Roughly 80% of U.S. movies, for instance, fail to earn back their costs in U.S. theaters, so studios depend on foreign circulation and streaming revenue to cover losses.) This global expansion was helped along by the development of new technologies, such as the growth of direct broadcast satellite services during the 1990s, which made North American and European TV available around the world. Cable channels such as CNN and MTV quickly took their national acts to the international stage, and by the twenty-first century, CNN and MTV were available in more than two hundred countries.

Although the United States was a vital center in this emerging global media economy, other media hubs around the world also emerged, as countries like India, Nigeria, South Korea, and Brazil became major exporters of media content. Meanwhile, a number of venerable U.S. media companies were acquired by foreign conglomerates, which also served to increase globalization. In the 1980s, for example, Australian-based News Corp. purchased 20th Century Fox, and Japan-based electronics manufacturer Sony purchased CBS Records and Columbia Pictures.

The Internet and Digital Convergence

The Internet and the digital revolution have also shaped how media conglomerates have evolved in the information economy. For much of their history, media companies were part of discrete industries—that is, the newspaper business stood apart from book publishing, which was different from radio, which was different from the film industry. But the Internet changed all that. As printed newspapers, vinyl albums, analog TV signals, and celluloid films became digital and deliverable online, legacy (older) media companies were pushed to integrate across different parts of the media landscape. To avoid being siloed, or stuck, in just one media sector, companies repositioned themselves by merging with or acquiring other companies that could help them forge a profitable path through the rapidly evolving digital environment.

Meanwhile, as these older companies worked to reinvent themselves, an entirely new set of corporations appeared on the scene—Silicon Valley giants like Apple, Google, and Facebook, along with Amazon and Microsoft (see Figure 13.3). Each of these conglomerates came to dominate its corner of the Internet economy—hardware and software production, search, social media, and online retail—through aggressive mergers and acquisitions. In the early days of the digital economy, small start-ups popped up frequently. Often funded by venture capitalists, companies like Zappos, Instagram, YouTube, and Snapchat worked to find a niche market, connect with consumers, and get big fast. The successful start-ups then took one of two paths—either be acquired by a larger company (Google bought YouTube in 2006, Amazon bought Zappos in 2009, and Facebook bought Instagram in 2012) or go it alone and try to get even bigger (e.g., Snapchat) (see "Examining Ethics: Are the Big Digital Companies Too Big?").

Over time, the digital companies' business models drew them into greater collaboration and competition with legacy media companies like the *Washington Post*, Disney, and Time Warner, as everyone jostled to control the production, distribution, and exhibition of engaging content—the news, books, movies, TV shows, apps, and video games consumers wanted to access through digital devices. At first, the big tech companies relied entirely on other companies' content or on content created by their users, but as the market evolved, they began to see the advantages of controlling and creating their own content. Netflix, for example, started to produce its own TV series and movies and acquired StoryBots, a company focused on making education content for kids (see Figure 13.3 for more about Netflix). Meanwhile, Amazon established its own publishing divisions and produced its own television series. The push to control content will likely continue to produce new mergers and acquisitions in the future, as companies seek to supply programs for new streaming services and other endeavors.

The Internet, which has been heralded as a communication infrastructure that makes national barriers and national markets irrelevant, has also contributed to media globalization. Although certain countries have tightly controlled their digital borders, digital technologies and changing regulations have encouraged and enabled media companies to go global. By 2020, for example, Netflix was operating in over 190 countries after building strategic partnerships with scores of regional media and telecommunication companies. U.S.-based media companies are not alone in their focus on using the Internet for global expansion. A number of China-based corporations are challenging Silicon Valley and Hollywood for dominance around the world (see "Global Village: China's Dominant Media Corporations Rival America's").

NETFLIX IS A GLOBAL content delivery service operating in over 190 countries, and its offerings are tailored in varying ways to the local market. As this image from Netflix's service in Sweden illustrates, products imported from abroad—especially the United States—sit alongside locally or regionally produced content, like the Swedish film *En man som heter Ove* and the Norwegian TV comedy series *Dag*.

screenshot by Niclas Jonsson

Are the Big Digital Companies Too Big?

In the past few years, Americans have come to the realization that Apple, Google, Amazon, Microsoft, and Facebook structure much of our everyday lives. Can one imagine living without these five companies? But perhaps that is the problem. As reporter Dylan Byers says of the biggest tech companies, "It's their century. We're just living in it."[1]

With that realization comes reminders that all is not good with our big digital companies. Over the course of their relatively brief histories, all these companies have been under fire for a number of problems: deploying anticompetitive practices (Microsoft, Amazon, Facebook, and Google), recording users' private conversations on home digital assistants (Amazon), gobbling up too much of the advertising industry (Google and Facebook), being manipulated by trolls during U.S. elections (Google and Facebook), violating user privacy (Facebook), and slowing down mobile phone performance to spur sales (Apple).

But Amanda Lotz, a professor of media studies at Queensland University of Technology, warns us not to group these companies together like we do with Big Oil, Big Pharma, and Big Tobacco. "Because all of these companies provide services relating to computers, there is a tendency to lump them together, calling them 'Big Tech' or the 'Frightful Five.'" However, she continues, "lumping them together hides the fact they're very separate and distinct—not just as companies, but in terms of their business models and practices."[2]

Some analysts differentiate between the companies that make up the Big Five, suggesting that there may be meaningful differences in their sizes. For example, there is evidence that three of these companies have far too much dominance in their industries. The *Wall Street Journal* notes that "Facebook Inc., Google parent Alphabet Inc. and Amazon.com Inc. are enjoying profit margins, market dominance and clout that, according to economists and historians, suggest they're developing into a new category of monopolists."[3] Consider that Google and Facebook control over 60 percent of digital advertising in the United States, or that Amazon controls over 50 percent of U.S. online sales. Google's and Facebook's ability to corner the market on digital advertising has decimated the newspaper industry as it has moved online. Local advertisers that once supported local newspapers with their ads now often just place ads with one or both of these companies.

As dominant as they may be, however, giant companies' power can be surprisingly temporary. For example, Amazon's rise toppled big bookstore chains like Barnes & Noble, which had such enormous market power in the 1990s that they drove many small independent bookstores out of business. Amazon now sells everything and threatens brick-and-mortar grocery stores and department stores, including Walmart—a chain whose superstores had shattered many small-town business districts years before. If this cycle repeats, we would expect a new company to eventually emerge and challenge Amazon.

It is also important to remember that in the 1990s, it seemed that Microsoft had completely beaten its main rival, Apple, yet Apple came back. However, the U.S. Department of Justice's scrutiny of Microsoft during the 1990s limited the company's ability to engage in anticompetitive practices, which cleared a path for Apple and other competitors.[4] In 2020, the Department of Justice filed lawsuits against both Google and Facebook, asserting that the tech giants were engaged in anticompetitive business practices. Depending on the outcome of the cases, the future of the companies that impact our daily lives so greatly may be different—or not. Stay tuned. ●

FIGURE 13.3

RISE OF THE NEW DIGITAL MEDIA GIANTS

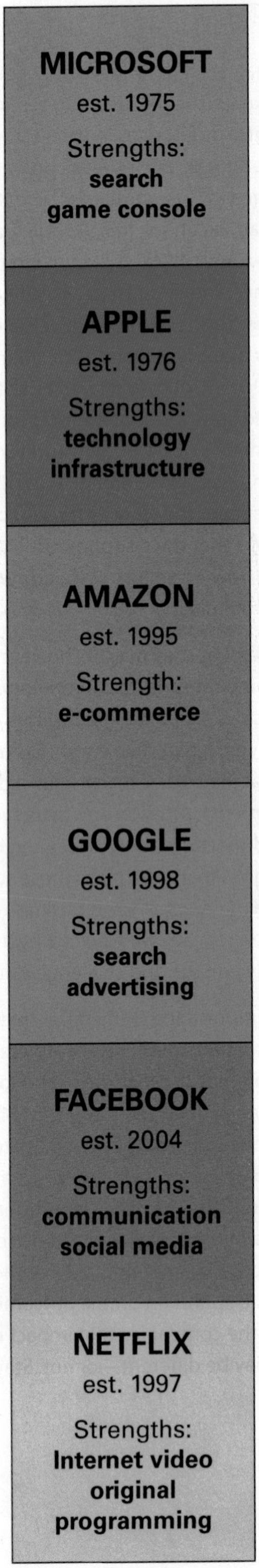

MICROSOFT

est. 1975

Strengths:
**search
game console**

APPLE

est. 1976

Strengths:
**technology
infrastructure**

AMAZON

est. 1995

Strength:
e-commerce

GOOGLE

est. 1998

Strengths:
**search
advertising**

FACEBOOK

est. 2004

Strengths:
**communication
social media**

NETFLIX

est. 1997

Strengths:
**Internet video
original
programming**

Microsoft, one of the wealthiest digital companies in the world, is making the transition from being the top software company (a business that is slowly in decline) to competing in the digital media world with its Bing search engine and devices like its successful Xbox game console and Surface tablet. Microsoft holds a small ownership share in Facebook and purchased the social media businesses LinkedIn ($26.2 billion) and GitHub ($7.5 billion) in 2016 and 2018, respectively. It has also entered the office collaboration market with its Microsoft Teams tool.

Apple's strength has been creating the technology and infrastructure to bring any media content to users' fingertips. When many traditional media companies didn't have the means to distribute online content easily, Apple developed the shiny devices (the iPod, iPhone, and iPad) and easy-to-use systems (the iTunes store) to do it, immediately transforming the media. Today, Apple has a hand in every media industry.

Amazon has grown into the largest e-commerce site in the world. In 2007, Amazon expanded beyond delivering physical products (e.g., bound books) to distributing digital products (e-books and downloadable music, movies, television shows, and more) on its digital devices (Kindle and Fire TV) and through its streaming service. Amazon, the top digital retailer, is also moving into physical locations, with Amazon bookstores and the purchase of the Whole Foods grocery chain.

Google's search advertising business is about 80 percent of its $181.69 billion annual revenue (2020). Google moved into the same digital media distribution business that Apple and Amazon offer via its Android phone operating system, Pixel phone, Chromebook, and Chromecast. The company used YouTube to enter the streaming video revolution; YouTube is now a platform for music videos, gaming, and, with YouTube TV, combined video-on-demand and live TV. Google's parent company is Alphabet. In 2020, the company's future grew less certain when the Justice Department filed an antitrust lawsuit that could break up the tech giant.

Facebook had more than 2.7 billion users worldwide in 2020, and the company continues to leverage those users (and the massive amounts of data they share about themselves) into advertising sales. With its 2014 purchase of the Oculus Rift virtual reality gaming headset ($2 billion), the company joined the four larger digital companies in having a device capable of accessing the Internet and digital media. Facebook introduced live streaming on its site in 2016 and gaming in 2018. Like Google, its future became more uncertain when the Justice Department filed an antitrust lawsuit against the company in 2020.

Netflix started out as a DVD-by-mail service in 1997 and began streaming in 2007. By 2021, it had over 204 million subscribers in over 190 countries, and it is the leading Internet television channel. Netflix does not market any hardware devices, but that may be an advantage in the fast-changing media business, where devices can quickly become obsolete. Netflix's popular original programming also gives it insurance against the high costs of buying programming from other media corporations and against competing streaming channels.

China's Dominant Media Corporations Rival America's

The 2020 rankings of the most valuable global brands show the dominance of digital media in our world. In 2020, seven of the Top 10 global brands were digital-media related (see figure). And although Amazon, Apple, Microsoft, Google, and Facebook are recognizable brands in the United States, two Chinese companies—Tencent and Alibaba—have been among the fastest-growing brands in the global Top 10.[1]

Both companies are leading forces in China's 770-million-plus Internet-user market and are stretching into other markets. Both are conglomerates like Amazon, spreading into a wide range of businesses. Founded in 1998, Tencent is currently one of China's largest web portals (at QQ.com); has the popular WeChat messaging, social media, and mobile payment apps (WeChat Pay, QQ Wallet); owns China's (and the world's) largest music streaming service and video game company in terms of revenue; and runs hundreds of other businesses in the areas of financial services, insurance, smartphones, and artificial intelligence.

Alibaba was started in 1999 as an e-commerce site, and it now challenges Amazon as the most popular destination for online transactions. It also runs Taobao Marketplace (bigger than eBay), the Fliggy travel platform, Alibaba Cloud, Alipay (an online payment business run by subsidiary Ant Financial), and music streaming and film production businesses.

By 2018, the two tech giants were vying with each other for dominance of the Chinese Internet industry in what *Fortune* magazine called "a clash of titans,"

battling it out "in an economy in which e-commerce is dominant in ways unthinkable in the U.S."[2] The *New York Times* mapped out their overlapping strategies: "They have both funded ventures that offer online education, make electric cars and rent out bicycles. For the giants, such initiatives represent new opportunities for people to use their digital wallets—Ant Financial's Alipay and Tencent's WeChat Pay—and new ways to collect data on consumer behavior."[3]

Until recently, Tencent and Alibaba had the support of the Chinese government, which placed few constraints on their rapid growth through mergers and acquisitions. In 2020, however, regulators fined both companies and announced new anti-monopoly rules. With little explanation from the country's leadership about their intentions, outside analysts debate whether they are "merely re-asserting their oversight power" or perhaps "have grown frustrated with the swagger of tech billionaires and want to teach them a lesson by breaking up their companies."[4] ●

TOP 10 GLOBAL BRANDS, 2020
Revenue in $billions
(Numbers have been rounded.)

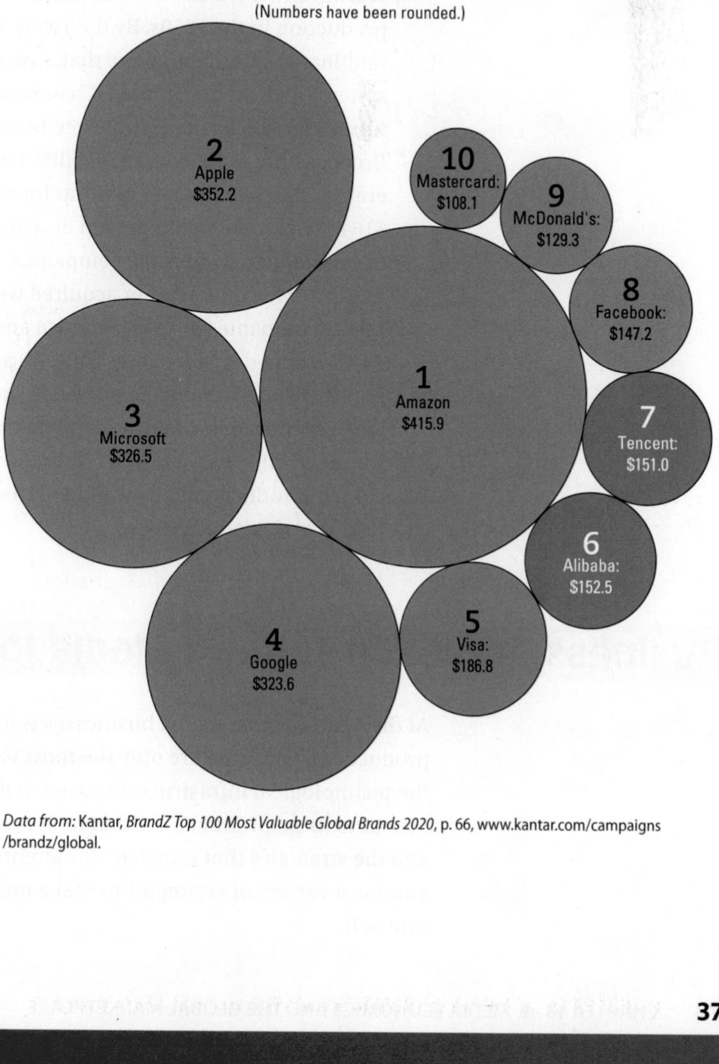

2 Apple $352.2
10 Mastercard: $108.1
9 McDonald's: $129.3
8 Facebook: $147.2
1 Amazon $415.9
3 Microsoft $326.5
7 Tencent: $151.0
6 Alibaba: $152.5
5 Visa: $186.8
4 Google $323.6

Data from: Kantar, *BrandZ Top 100 Most Valuable Global Brands 2020*, p. 66, www.kantar.com/campaigns/brandz/global.

Media Powerhouses and the Consolidation Frenzy

Changing regulations, globalization, and the Internet drove a conglomeration frenzy that began in earnest in the 1980s. In many instances, critics and regulators raised objections to the deals that created these enormous companies, warning that too much market power and cultural influence was being consolidated under the control of too few. However, though antitrust laws remained in place, they were applied infrequently and unevenly, especially in media industries. As a result, media competition was often sacrificed in favor of media consolidation.

Keeping track of the twists and turns of the many deals that have taken place can be dizzying, but a brief summary of two notable examples—NBC and Warner Bros.—illustrates how media companies that were once relatively autonomous have become part of global conglomerates.

- **NBC.** Founded in 1926 by the Radio Corporation of America, NBC was a dominant player in broadcast radio and TV for decades. In 1985, General Electric purchased RCA-NBC, making it part of the highly diversified electronics conglomerate. In 2004, GE merged its NBC subsidiary with France-based media conglomerate Vivendi Universal, forming NBCUniversal. In 2009, cable giant Comcast struck a deal to purchase a majority stake in NBCUniversal from GE and, in the process, produce the nation's largest media conglomerate. The proposal stirred up antitrust complaints from some consumer groups, and in 2010, Congress began hearings on whether uniting a major cable company and a major broadcasting network would decrease healthy competition and hurt consumers. In 2011, however, the FCC approved the deal. Today, Comcast's vast holdings include cable systems, cable channels, a mobile phone network, film studios, broadcast TV networks and stations, and amusement parks.

- **Warner Bros.** Warner Bros. began in 1923 as a Hollywood film studio and expanded into TV production in the 1950s. By the 1980s, it was a subsidiary of Warner Communications—a multinational conglomerate that also owned music labels, video game companies, cable systems and channels, and DC comics. In 1990, Warner Communications merged with Time Inc., a global magazine publisher, becoming Time Warner, which in turn bought Turner Broadcasting in 1995 in a $7.5 billion deal that made it the world's largest media conglomerate at the time. In 2001, dial-up Internet service provider AOL acquired Time Warner for $164 billion. The union proved disastrous when AOL's business model collapsed in the face of competition from cable companies' new high-speed broadband services. After selling off AOL in 2009, Time Warner acquired websites like Flixster and Rotten Tomatoes as well as media companies in Latin America and Europe. Then, in 2016, telecommunications giant AT&T reached a deal to buy Time Warner—a move that led the Justice Department to file an antitrust suit claiming the purchase would harm American consumers. In 2018, however, a judge approved the deal, and the new WarnerMedia—home of Warner Bros. and HBO Max— was born. In May 2021, AT&T announced plans to merge WarnerMedia with yet another major content producer and distributer—Discovery, Inc. and its slate of lifestyle channels like TLC, HGTV, and Food Network.

Business Trends in Today's Media Industries

At the heart of most media businesses is the production of engaging content. Yet the media products we consume are only the most visible part of complex business operations that include the technological infrastructures used to deliver content to consumers, the devices on which we consume these products, the marketing tactics that draw our attention to companies' offerings, and the strategies that translate our attention and desires into profits. Large global conglomerates employ a variety of strategies to make money, control the media landscape, and expand their role in it.

Making Money from Media Products

Media companies make money from media products in three main ways. First, companies generate revenue when consumers buy a book, video game, movie ticket, newspaper, or magazine—whether directly from them or through a retailer. This monetary transaction used to involve physical products bought at brick-and-mortar stores or by mail; now we often buy media products as digital files using a device media companies provide (e.g., an Apple iPhone or a Google Pixel).

Second, companies can make money by offering access to products via a subscription—a strategy used by the first newspapers and magazines. In contrast to per-item purchases, subscriptions promote habitual consumption and customer loyalty and offer companies predictable revenue streams. The subscription model has always played an important role in the business model of cable TV (especially for premium channels like HBO), but today it is also the dominant model for digital streaming services like Netflix and Peacock. Subscriptions are also reshaping the music business. Spotify, the Sweden-based music streaming service launched in 2008, offers customers access to sixty million songs and podcasts in exchange for a monthly fee.

Third, media companies can generate revenue through advertisements that support their products, such as those included on TV and radio shows, in newspapers and most magazines, and on many websites. While media products like TV shows or social media sites may seem free to us, we actually work for the "free" content by giving our time and attention to commercial sponsors. Advertisers pay more depending on how many of us are getting exposed to the ads and our potential buying power as an audience, which is often determined by the data companies collect about us (see Chapter 11). This is the main revenue structure for "free" over-the-air radio and TV broadcasting, websites, and social media platforms. Media companies similarly make money through product placement advertising in movies, television, and video games.

Media corporations can generate the most money when they combine aspects of these revenue-generating strategies, which happens when they get us to buy a media product or pay for a subscription (like a cable TV package or a newspaper or magazine subscription) *and* subsequently to be the target audience for advertising that comes with that product.

The Rise of Specialization

In most of today's media sectors, mass production (e.g., television programs targeted to mass audiences, or magazines designed to appeal to a broad cross section of the U.S. population) has been replaced by the cultivation of specialized niches. This development has been propelled by various interrelated economic factors. By the 1980s, for example, flexible production practices transformed factories, allowing manufacturers to produce a wider variety of consumer goods that could respond to or create more diverse consumer tastes. Meanwhile, computerized market research tools that could juggle numerous demographic and psychological variables pushed advertisers to sort consumers into ever narrower niche markets.

Media companies both responded and contributed to these changes. Looking for a way to compete with broadcast TV's domination of mass market advertising, magazines and later cable channels targeted narrower market segments, offering advertisers ways to reach niche markets. By the 1980s, network television followed suit, focusing less on a mass audience and more on the upscale eighteen-to-forty-nine demographic. By the 2000s, a wide range of media content was being produced and marketed with specific audiences in mind. Magazines such as *J-14* and *AARP The Magazine* flourished. Cable channels like Nickelodeon served the under-eighteen market, and the Hallmark Channel addressed female viewers over age fifty. Broadcast networks like CBS and NBC tilted their programs toward select demographics.

The interactive capabilities of Internet-based media pushed specialization even further. For example, Pinterest—the visual social media site where users "pin" images and videos to their "board"—creates a customized site that reflects each user's personal style on topics like home décor, apparel, food, crafts, and travel. To sign up for an account, users provide their name, e-mail address, and gender. During the sign-up process, there is also a box—prechecked by Pinterest—that gives users the option to "let Pinterest personalize your experience based on other sites you visit." Pinterest is just one example of **mass customization**—a term used to refer to marketing and production techniques that combine personalization with the cost benefits of efficient mass production.

Specialization has changed many aspects of how media products are produced and marketed. Although a few large global conglomerates control and profit from much of the media we consume, the dynamics of market specialization have helped open doors for more diverse producers and representations—to a degree. Media conglomerates' promotional rhetoric may suggest that they are focused on giving us what we want, but ultimately, these companies only address our interests in ways that are profitable for them. As a result, communities whose members can't pay or aren't appealing to enough advertisers will probably go unserved or underserved.

Horizontal Integration and Synergy

Beyond specialization, what really distinguishes the economics of media conglomerates today is horizontal integration and the opportunities for synergy it creates. **Horizontal integration** refers to the strategy in which a corporation owns companies involved in a wide array of media businesses. In other words, when a corporation owns a music label, a movie studio, a TV network, an amusement park, magazines, and websites, it is integrated horizontally, meaning across different sectors of the media landscape. Horizontal integration has been the goal of many of the mergers and acquisitions that produced today's massive conglomerates.

Being horizontally integrated is appealing to media corporations because it enables them to take advantage of *synergy*—opportunities to generate profits that come from interaction and cooperation among a conglomerate's cross-media subsidiaries. As one corporate executive put it, with synergy, "1 plus 1 equals 4."[10] When complementary companies combine, they can produce more revenue together than they could when operating separately. Synergistic benefits offer production efficiencies (e.g., a segment generated by a local sports reporter on an NBC-owned broadcast TV station can be repurposed on the NBC broadcast network's *Today* show, the cable channel NBCSN, and NBC Sports Radio—all owned by NBCUniversal). Even more importantly, they provide opportunities for cross-promotion, as the discussion of Disney later in the chapter will illustrate.

While horizontal integration can be the path to synergy windfalls, nothing is guaranteed. Many mergers and acquisitions fail to produce enough synergies to offset their costs.[11] (Consider,

IN JUNE 2018, AT&T completed a merger with Time Warner. (Pictured: The Warner Bros. studio lot in Burbank, California and AT&T's U-verse internet TV service.) According to Randall Stephenson, then chair and CEO of AT&T, "The content and creative talent at Warner Bros., HBO and Turner are first-rate. Combine all that with AT&T's strengths in direct-to-consumer distribution, and we offer customers a differentiated, high-quality, mobile-first entertainment experience."[12] The partnership was drawn into question in May 2021 when AT&T announced plans to spin off WarnerMedia into a joint venture with Discovery, Inc., a move that reflects the volatile dynamics of media mergers in the digital era.

for example, the disastrous AOL–Time Warner merger of 2001.) In other cases, internal conflicts can arise if one subsidiary feels it is being exploited for the benefit of another, like the controversy surrounding WarnerMedia examined in the chapter-opening discussion. The financial risks conglomerates take can also backfire. After the 2008 financial crisis, for example, many megamedia firms became *overleveraged*—that is, they could no longer pay what they owed on the loans they took out when they added companies to their empires. As a result, they are often forced to downsize (as Time Warner did when it sent AOL and Time Inc. adrift and as the New York Times Company did when it sold the *Boston Globe*) or they become vulnerable to being taken over themselves.

Vertical Integration

Media conglomeration can also result from companies seeking *vertical integration*, which occurs when a single company owns everything it needs to produce, distribute, and sell a certain product. As discussed in Chapter 7, during the 1930s and 1940s, Hollywood studios like Paramount and MGM were vertically integrated because they made their own movies, distributed their movies themselves, and owned theaters where audiences paid to see them. So rather than controlling some aspects of different media sectors (horizontal integration), they owned every aspect within one industry.

Being vertically integrated offers corporations numerous competitive advantages. In the studio system era, it helped the Big Five movie studios keep new competitors from entering the market—a fact that eventually led the Supreme Court to declare their practices anticompetitive and to make the studios vertically de-integrate by selling their theaters. For decades, media companies nervous about the appearance of monopolistic activities would insist that they were not pursuing vertical integration. In recent years, however, vertical integration has been making a comeback.

In today's digital streaming environment, where consumers increasingly access media content through broadband cable or wireless connections, companies that control that infrastructure—including Comcast, Spectrum, and AT&T—function somewhat like theater owners did in the 1940s. Given that, Comcast's acquisition of NBCUniversal in 2011 and AT&T's purchase of Time Warner in 2018 were considered moves toward vertical integration because they united companies that controlled the production and distribution of content (NBCUniversal and Time Warner) with companies that controlled the technology of delivery (Comcast and AT&T).

Such vertical integration, critics warn, could promote anticompetitive business practices. Comcast, for example, could decide to use its control over broadband service to promote NBCUniversal–related content and keep out content from competing companies. Writing about the Comcast–NBCUniversal merger, one industry analyst observed, "Vertical integration makes a ton of sense from a business perspective, but it makes life harder for start-ups that want to offer cable-alternative streaming services, which is probably the point."[13]

Disney: A Twenty-First Century Media Conglomerate

The Walt Disney Company is one of the most successful companies in leveraging its many properties to create synergies and adapt to a volatile media landscape. For example, in 2014, ABC broadcast the prime-time special *The Story of "Frozen": Making a Disney Animated Classic* to promote the Disney movie studio's enormous hit movie and soundtrack—and to hype ABC's *Once Upon a Time* series (which would soon feature a character from *Frozen*) along with Disney's next animated film, *Big Hero 6. Frozen* also tapped into a huge array of licensed merchandise and even *Frozen*-themed vacation trips by Disney's tour company and cruise line. Such promotional events and merchandise helped maintain interest in the story and characters for the *Frozen 2* sequel released in 2019. To fully understand the contemporary story of media economics, horizontal and vertical integration, and synergy, we need only examine the transformation of Disney from a struggling cartoon creator to one of the world's largest media conglomerates.

launchpadworks.com

Disney's Global Brand
Watch a clip from *Frozen*, one of Disney's biggest movies ever.

Discussion: What elements of *Frozen* might have contributed to its global popularity?

The Early Years

After Walt Disney's first cartoon company, Laugh-O-gram, went bankrupt in 1922, Disney moved to Hollywood and found his niche. He created Mickey Mouse (originally named Mortimer) for the first sound cartoons in the late 1920s, then developed the first animated feature-length film, *Snow White and the Seven Dwarfs* (1937), which won critical praise and became a box-office hit.

For much of the twentieth century, the Disney company set the standard for popular cartoons and children's culture. The *Silly Symphonies* series (1929–1939) established the studio's reputation for high-quality hand-drawn cartoons. Despite winning critical accolades, however, the studio barely broke even because cartoon projects took time—four years for *Snow White*—and commanded the company's attention. Operating on a much smaller scale than the Big Five, Disney remained a minor studio in the studio system era.

When the popularity of cartoon shorts in movie theaters declined, Disney expanded into other areas with its first nature documentary short, *Seal Island* (1949); its first live-action feature, *Treasure Island* (1950); and its first feature documentary, *The Living Desert* (1953).

In 1953, Disney started Buena Vista, a distribution company. This was the first step in making the studio into a major player. The company also began exploiting the power of its early cartoon features. *Snow White*, for example, was successfully rereleased in theaters to new generations of children before eventually going to videocassette and much later to DVD and streaming.

Disney was also among the first film studios to embrace television, launching a long-running prime-time show in 1954. Then, in 1955, Disneyland opened in Southern California. Eventually, Disney's theme parks would produce the bulk of the studio's revenues. (Walt Disney World in Orlando, Florida, began operation in 1971.)

Horizontal Integration, Vertical Integration, and Global Expansion

The death of Walt Disney in 1966 triggered a period of decline for the studio. But in 1984, a new management team, led by Michael Eisner, initiated a turnaround. The first Disney Store opened in 1987, offering the growing company a degree of vertical integration by enabling it to sell videotapes of its movies and related toys and clothing directly to consumers. The newly created Touchstone movie

▼ **A Brief Timeline of Disney**

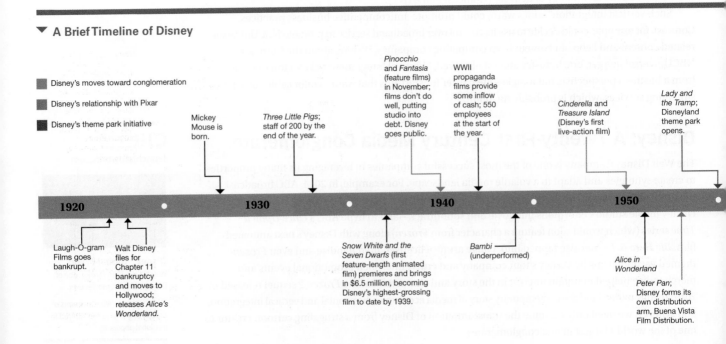

division reinvented the live-action cartoon with *Who Framed Roger Rabbit* (1988). A string of hand-drawn animated hits like *The Little Mermaid* (1989) and *The Lion King* (1994) followed. Disney also distributed computer-animated blockbusters from Pixar Animation Studios, including *Toy Story* (1995) and *The Incredibles* (2004); it later acquired Pixar outright and released movies like *Up* (2009), *Inside Out* (2015), and *Soul* (2020). Disney's in-house animation studio eventually got into the computer-animation business and achieved major success with *Frozen* (2013), *Moana* (2016), and *Ralph Breaks the Internet* (2018).

In the mid-1990s, Disney completed its evolution from a media company focused primarily on film to a horizontally and vertically integrated global media conglomerate. Through its purchase of Capital Cities/ABC Inc. in 1995, Disney became the owner of the cable sports channels ESPN and ESPN2, and later expanded the brand with ESPNEWS, ESPN Classic, and ESPNU channels; *ESPN The Magazine* (closed in 2019); ESPN Radio; and ESPN.com, beginning an era of sports monopolization.

In addition, it came to epitomize the synergistic possibilities of media consolidation (see Figure 13.4); for example, Disney can produce an animated feature for both theatrical release and streaming distribution. With its ABC network, it can promote Disney movies and television shows on programs like *Good Morning America*. A book version can be released through Disney's publishing arm, Disney Publishing Worldwide, and "the-making-of" versions and spinoff movie specials can appear on cable's Disney Channel or Freeform. Characters can become attractions at Disney's theme parks, which themselves have spawned Hollywood movies, such as the lucrative *Pirates of the Caribbean* franchise. In New York City, Disney renovated several Broadway theaters and launched versions of animated films like *The Lion King* and *Frozen* as musicals. And it sells DVDs and related merchandise at its string of Disney Stores.

Building on the international appeal of its cartoon features, Disney extended its global reach by opening Tokyo Disney Resort in 1983 and Disneyland Paris in 1991. Disney opened more venues in Asia, with Hong Kong Disneyland Resort in 2005 and Shanghai Disney Resort in 2016. Disney exemplifies the formula for becoming a "great media conglomerate" as defined in the book *Global Dreams*: "Companies able to use visuals to sell sound, movies to sell books, or software to sell hardware [will] become the winners in the new global commercial order."[14]

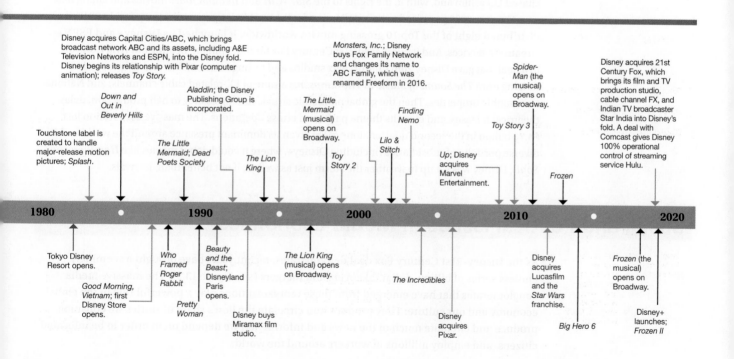

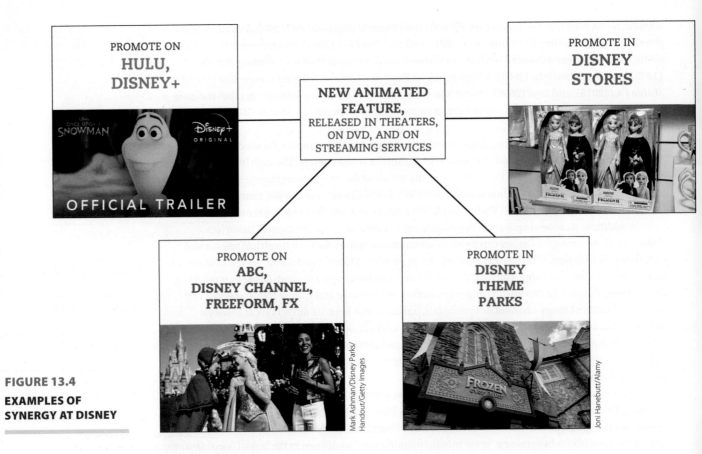

FIGURE 13.4

EXAMPLES OF SYNERGY AT DISNEY

Disney Today

By the early 2000s, Disney had grown into one of the world's biggest media conglomerates but looked to get even bigger. In 2006, new CEO Robert Iger merged Disney and Pixar. In 2009, Disney purchased Marvel Entertainment, bringing Iron Man and Captain America into the Disney family; in 2012, it purchased Lucasfilm and, with it, the rights to the *Star Wars* and *Indiana Jones* movies and characters.

The year 2020 promised to be a stellar one for the world's largest media company. In 2019, it distributed eight of the Top 10 grossing movies worldwide. It launched its Disney+ and ESPN+ streaming services. And it purchased 21st Century Fox for $71.3 billion, beating out rival Comcast in a deal that gave Disney 20th Century Fox studios and its huge trove of film and TV properties, ranging from *The Sound of Music* to *The Simpsons*; a suite of FX-related cable channels; and National Geographic properties. Then the global pandemic struck, forcing Disney to halt production, delay theatrical releases, and close its theme parks and cruise operations. The massive corporation lost $4.72 billion in the second quarter alone.[15] Yet given its dominant presence around the world and its diverse portfolio of subsidiaries (including Disney+, where it could debut movies like *Hamilton* and *Soul*), Disney was set up to weather the storm just as well as, if not better than, its rivals.

Social and Political Issues in Media Economics

As the Disney–21st Century Fox deal demonstrates, recent years have brought a seemingly endless series of billion-dollar takeovers and mergers (see Figure 13.5). The massive media conglomerates that have emerged from these reorganizations play a powerful role in the global economy and our culture: They produce and circulate the lion's share of stories we consume, produce and circulate much of the news and information we depend on in order to be informed citizens, and employ millions of workers around the world.

MERGER AND ACQUISITION	PRICE	OUTCOME
New York Times + *Boston Globe* (1993)	$1.1 billion	The *Times* sells the *Globe* to the owner of the Boston Red Sox for $70 million in 2013.
Disney + ABC (1995)	$19 billion	The television network and ESPN become huge profit centers for Disney.
Time Warner + Turner Broadcasting (1995)	$7.5 billion	Time Warner grows even bigger than the combined Disney and ABC and adds CNN, TBS, and other cable channels.
Tribune Media Company + Times Mirror Company (2000)	$8.3 billion	Biggest newspaper merger ever combines the *Chicago Tribune*, the *Los Angeles Times*, and several others. The company files for bankruptcy in 2008 and spins off its newspaper division in 2014 (like several other newspaper conglomerates).
AOL + Time Warner (2001)	$164 billion	Biggest media merger failure ever. Time Warner spins off AOL in 2009.
Google + YouTube (2006)	$1.65 billion	One of Google's best acquisitions; YouTube makes several billion dollars each year.
Sirius + XM (2008)	$13 billion	A merger of equals makes a bigger company but reduces the number of satellite radio companies to one.
Google + Motorola (2011)	$12.5 billion	Google sells the Motorola mobile phone business to Lenovo in 2014 for $2.91 billion.
Comcast + NBCUniversal (2011)	$28 billion	The biggest cable company becomes even bigger with a TV network, a movie studio, and more—though some of the media holdings are not at peak performance.
AT&T + T-Mobile (2011)	$39 billion	The Department of Justice opposes this merger, which would have reduced the number of cell phone carriers from four to three. AT&T loses, but consumers win with more choices and competitive pricing.
Universal Music Group + EMI (2012)	$1.3 billion	Universal becomes the biggest music company in the world, leaving only three major sound recording corporations.
Comcast + Time Warner Cable (2014)	$45.2 billion	A merger creating the most dominant wired broadband network in the United States is abandoned by Comcast in the face of regulatory opposition.
AT&T + DirecTV (2014)	$48.5 billion	The merger of the largest mobile phone service and the largest satellite television provider comes with a number of conditions imposed by the FCC to ensure that the company doesn't discriminate against other content providers.
Charter Cable + Time Warner Cable + Bright House Networks (2016)	$67.1 billion	The merger of three cable companies creates the second-largest cable and broadband firm, serving markets in forty states.
AT&T + Time Warner (2018)	$85 billion	AT&T rebrands Time Warner as WarnerMedia and restructures its vast entertainment and sports holdings.
Disney + 21st Century Fox (2019)	$71.3 billion	Disney gains control of 20th Century Fox studios and its vast archive of film and TV properties, a suite of FX-related cable channels, and National Geographic properties. Meanwhile, the Fox Broadcasting Company, Fox TV stations, and the Fox News cable channel become part of new Fox Corporation.

Given the scope of the control and influence these corporations have, critics argue that there has been far too little public debate surrounding the increasingly oligopolistic structure of global media. Consumer advocates and scholars urge us to think carefully about the varied ways in which media economics are entangled with social values and political power, specifically as they relate to struggles over antitrust laws, emerging labor practices, the power of free-market thinking, and cultural imperialism.

FIGURE 13.5

MAJOR MEDIA MERGERS AND ACQUISITIONS

The Impact of Media Ownership
Media critics and professionals debate the pros and cons of media conglomerates.

Discussion: This video argues that it is the drive for bottom-line profits that leads to conglomerates. What solution(s) might you suggest to make the media system work better?

Conflicts over Conglomeration

Antitrust laws are meant to ensure fair competition, and regulations are meant to promote diverse owners and political perspectives. These laws and regulations have been weakened or unevenly applied since the 1980s, enabling large-scale mergers and ownership concentration. Although there is an abundance of products in our vast media environment—thousands of daily and weekly newspapers, radio and television stations, cable channels, streaming services, magazines, and books— only a limited number of companies are in charge of those products.[16] Critics of media mergers like Ben Bagdikian fear that this represents a dangerous antidemocratic tendency, in which a handful of media moguls wield a disproportionate amount of economic control and political influence. It also leads to little diversity in media ownership. According to media journalist Kristal Brent Zook, "African American ownership remains particularly low, hovering at less than 1 percent of all television properties, and less than 2 percent of radio."[17] While corporations have enormous resources to advance their interests, consumer advocates and citizen groups have kept fighting to limit corporate supersizing.

Corporate Pressure

Scholars of *political economy*—a field of study that examines how laws and policies are influenced by economic factors—have demonstrated that corporations exert considerable pressure in shaping the laws intended to limit their power. Large media firms, for example, are among the most powerful lobbyists in Washington, D.C., and other political capitals. In their effort to get the public and lawmakers on their side, corporate leaders have tried—often quite successfully—to frame massive mergers as good for consumers.

If they expand horizontally rather than vertically (purchasing companies in different media-related fields rather than trying to control just one field), conglomerates can often evade monopoly charges—even as they grow large and powerful. During the 1990s, for example, as Disney grew into a media powerhouse that owned a TV network, cable channels, a movie studio, music labels, and theme parks, its CEO Michael Eisner defended the company's acquisitions, arguing that as long as large companies remain dedicated to quality, such mergers would benefit America.

The narrow interpretation of antitrust law that corporate leaders like Eisner advance (and many regulators seem to have bought into) underestimates the enormous market power of horizontally integrated conglomerates. Disney, for example, can leverage its various assets to gain incredible market visibility for its products while keeping costs down—a combo that makes it difficult for start-ups to break into the market at all. As a result, a few behemoth companies control most media production and distribution.

Eisner's defense raises other important questions: Who gets to decide what counts as a quality cultural product? If companies cannot make money on quality products, what happens? If Disney-owned ABC News, for example, cannot make a substantial profit, should Disney's managers cut back its national or international news staff? If they did, what would the effects of such layoffs be on the public mission of the news media and, consequently, on our political system? If this happens, how should citizens or the government respond? Media consolidation has limited the number of independent voices in the market and reduced the number of owners who might be able to innovate and challenge established economic powers. All of this has led to calls for a more aggressive application of antitrust laws.

Applying Antitrust Laws Today

While antitrust laws have been unevenly applied in recent decades, there have been instances when government regulators—often encouraged by citizen watchdog groups—have pushed back against proposed mergers. For example, when EchoStar (now the Dish Network) proposed to purchase DirecTV in 2001, a number of rural, consumer, and Latino organizations spoke out against the merger for several reasons. Latino organizations opposed the merger because in many U.S. markets, direct broadcast satellite (DBS) service offered the only Spanish-language television programming available at the time. The merger would have left the United States with just one major DBS company and created a virtual monopoly for EchoStar, which had fewer Spanish-language offerings than DirecTV. In 2002, the FCC declined to approve the merger, saying it would not serve the public interest.

In recent years, the government has also stepped in to regulate companies that it believes have already grown too powerful. This happened in 2020, when the Federal Trade Commission and the Justice Department filed separate antitrust lawsuits against Google and Facebook. According to the opening lines of the Justice Department's complaint against Google, "Two decades ago, Google became the darling of Silicon Valley as a scrappy startup with an innovative way to search the emerging internet. That Google is long gone. The Google of today is a monopoly gatekeeper for the internet, and one of the wealthiest companies on the planet."[18] In the case against Facebook, the government's lawyers cited the company's strategic efforts to eliminate competition in the social networking market by acquiring rivals Instagram and WhatsApp. According to the lawsuit, "Facebook's unlawfully maintained monopoly power gives it wide latitude to set the terms for how its users' private information is collected, used, and protected."[19] Both cases are expected to take years to decide, but how they are resolved will likely shape the organization of our media environment for decades.

The history of these cases highlights that the application of antitrust laws is always entangled in social and political contexts. The aura and excitement that surrounded Silicon Valley tech giants, for example, had helped insulate them from broad criticism for years. The lawsuits came only after years of lobbying by dogged citizen watchdog groups and after a series of national scandals related to the 2016 election revealed not only the companies' failure to protect user data but also the power of social media to influence democratic processes.

Employment Issues

Media corporations employ millions of workers around the world. That means their employment practices have reflected—and contributed to—the changes in income distribution and job security that occurred with the shift to an information economy.

One substantial change has been the number of workers who belong to labor unions, which has declined dramatically in recent years. Labor unions represented 34.8 percent of U.S. workers at their peak in 1954, when they saw success bargaining for middle-class wages. Over time, however, certain employers began to cut labor costs by busting unions. These de-unionization efforts were helped by the shift to an information economy, as many jobs—including those related to the manufacture of electronic devices like computers, TV sets, and smartphones—were exported to other countries with non-union workers and cheaper labor costs. Armed with the threat of outsourced jobs, corporations gained the upper hand over workers. According to the U.S. Department of Labor, union membership fell to 20.1 percent in 1983 and 10.3 percent in 2019.[20]

These changes, spurred on by changing regulations and a decline in worker protections, have altered social and political life in the Unites States, where inequality between the richest few and everyone else has been growing since the 1970s. This is apparent in the skyrocketing rate of executive compensation and the growing gulf between what executives earn and the typical pay of workers in corresponding industries. In 1965, the CEO-to-worker compensation ratio was 20:1, meaning that the typical CEO earned twenty times the salary of the typical worker in that industry. By 2019, the ratio was 320:1 (see Figure 13.6).[21] To earn a living in today's unstable gig economy, many people are forced to work two or three part-time positions for substandard pay. According to one January 2020 analysis, 44 percent of all workers in the United States earned "barely enough to live on," with median earnings of $10.22 per hour.[22] Such developments have influenced the political landscape, fueling populist movements on both the Left and the Right. Politicians as different as Donald Trump and Alexandria Ocasio-Cortez have tapped into the frustration many workers feel.

Global media conglomerates reflect these trends. While some media industry workers remain unionized (e.g., Hollywood writers, directors, and crews), many others struggle in highly precarious employment situations, working as temporary employees without any job stability or employer-provided health insurance. The cultural aura often associated with creative industries can exacerbate

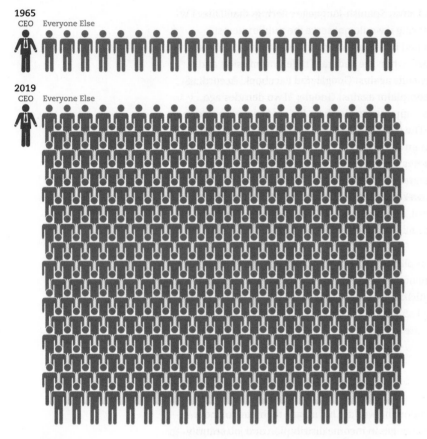

1965
CEO   Everyone Else

2019
CEO   Everyone Else

FIGURE 13.6

CEO-TO-WORKER WAGE GAP, 1965 AND 2019

Data from: Lawrence Mishel and Jori Kandra, "CEO Compensation Surged 14% in 2019 to $21.3 Million," Economic Policy Institute, August 18, 2020, www.epi.org/publication /ceo-compensation-surged-14 -in-2019-to-21-3-million-ceos-now -earn-320-times-as-much-as-a -typical-worker.

worker exploitation. For example, with a seemingly endless supply of creators and coders to make video games, the powerful gaming companies they work for have been able to keep salaries relatively low and working conditions stressful.

Beyond the people who produce media products, media industries also include a vast array of people who manufacture devices, install cable, keep amusement parks running, clean film studios, and work in scores of other jobs. In many cases, these workers' salaries reflect the general decline in wages that has happened across the U.S. economy. Meanwhile, the exorbitant salaries of media company CEOs mean that the media and communication industries have among the highest wage gaps. In 2019, Stephen Burke of NBCUniversal ($42.64 million), David Zaslav of Discovery, Inc. ($45.8 million), and Robert Iger of Disney ($47.5 million) were among the highest-paid CEOs of all publicly traded companies in the United States. To put the salary wage gap in perspective, it would take the average Disney employee 911 years to earn what Disney's Robert Iger made in 2019.[23]

Belief in the Free Market

Some media scholars have argued that media industries have played a unique role in advancing the shift to the information economy and justifying growing income inequality. As the source of much of the information we consume, media companies can shape what people pay attention to and provide frameworks for thinking about things—like free markets and consumer choice—in ways that advance their own interests.

Equating Free Markets with Democracy

In the 1920s and 1930s, when the Federal Communications Commission was making decisions about who should be awarded valuable broadcast licenses, commercial radio executives succeeded in portraying themselves as operating in the public interest while labeling their noncommercial radio counterparts in education, labor, or religion as mere voices of propaganda. In these early debates, corporate interests succeeded in misleadingly aligning the political ideas of democracy with the economic structures of capitalism.

As the United States waged a Cold War against communism in the 1950s and 1960s, it became increasingly difficult to criticize capitalism, which became a synonym for democracy in many circles. In this context, any criticism of capitalism was seen as an attack on the free marketplace. This, in turn, appeared to be a criticism of free speech, because the business community often sees its right to operate in a free marketplace as an extension of its right to buy commercial speech in the form of advertising. As longtime CBS chief William Paley told a group of educators in 1937, "He who attacks the fundamentals of the American system" of commercial broadcasting "attacks democracy itself."[24]

Broadcast historian Robert McChesney, discussing the rise of commercial radio during the 1930s, has noted that leaders like Paley "equated capitalism with the free and equal marketplace, the free and equal marketplace with democracy, and democracy with 'Americanism.'"[25] The collapse of the former Soviet Union and the new global world order it brought about in the 1990s is often portrayed as a triumph for democracy. As we now realize, however, it was primarily a victory for capitalism and free-market economies.

Consumer Choice versus Consumer Control

As many economists point out, capitalism is not structured democratically but arranged hierarchically, with powerful corporate leaders at the top and hourly wage workers at the bottom. But democracy, in principle, is built on a more horizontal model, in which all individuals have an equal opportunity to have their voices heard and their votes counted. In discussing free markets, economists distinguish between similar types of consumer power: *consumer control* over marketplace goods and freedom of *consumer choice*.[26] Most Americans and the citizens of other economically developed nations clearly have *consumer choice*: options among a range of media products. Yet consumers and even media employees have limited *consumer control*: power in deciding what kinds of media products get created and circulated.

Media companies work hard to obscure the distinction between these two types of power—to make consumers confuse the power to choose with the power to create. Interactive technologies and mass customization play into media conglomerates' hands on this point and help them co-opt what we might consider the innate human desire to play an active role in determining one's own culture. Netflix's website, for example, states: "Interactive titles are a fun new way to experience Netflix. In each title, you make choices for the characters, shaping the story as you go."[27] Of course, Netflix doesn't point out that all your available options are selected ahead of time by Netflix producers.

But despite conglomerates' outsized control of our media environment, there are places where more democratized production can take place. People who want to assert greater creative control might seek a forum on public access TV channels or on the fringes of media industries through independent newspapers, community radio stations, and online blogs and fan sites—places where citizens have a greater chance to control production and offer their opinions. Finally, social media platforms like Instagram and TikTok offer new options for consumer control in cultural production, though even in those instances, users operate within the constraints decided on by those platforms' corporate owners.

Ideology, Hegemony, and Media Stories

To understand people's belief in the free market and why our society hasn't participated in as much public discussion about wealth disparity, salary gaps, and corporate power as some people feel it should, it is helpful to understand the concepts of ideology and hegemony. **Ideology** refers to a way of thinking—a set of ideas and assumptions—that is shaped by a specific economic and social system. In turn, the ideas that make up that particular system (the economic status quo, for example) begin to seem natural and inevitable to the people living within that social system. This concept emerged to help explain human behavior in the face of societal inequality. In our society, some people and groups have more social power and economic resources than others, and their advantages often come at the expense of others. CEOs and corporate shareholders, for example, make more money when they keep average workers' salaries lower. Yet despite that fact, modern societies on the whole remain stable; the people who are disadvantaged by the status quo rarely revolt. Why not?

According to people like Italian philosopher and activist Antonio Gramsci (who wrote about these issues in the 1920s and 1930s), modern societies and economic systems remain relatively stable not because everyone is treated fairly or because those in power use the military or police to force people to obey, but because the ruling or dominant class persuades "the ruled to accept the system of beliefs of the ruling class and to share their social, cultural, and moral values."[28] In other words, social stability exists when the ideas that support the status quo are seen by everyone—including those disadvantaged by them—as simply common sense. Instead of seeing the status quo as the outcome of specific economic, political, and social forces that could be challenged and changed, people see it as simply the ways things are—as natural and inevitable. **Hegemony** is established when most of the public accepts or buys into a way of thinking about how the world works that favors the dominant class.

What are some examples of ideological assumptions that might support a certain economic status quo? If you believe that hard work always pays off, you may be more likely to assume that CEOs who earn millions of dollars deserve it. If you believe that men are from Mars and women are from Venus (as the title of a popular book claims), you might be more likely to believe that men should be paid more than women. If you believe that racial categories are biological, you may be more likely to believe that nothing will resolve the wealth gap that exists between Americans of different races.

The mass media play an important role in these processes, because the stories they tell and how they frame information can help to establish certain ways of thinking as natural and inevitable. Edward Bernays and Doris Fleischman, two of the founders of modern public relations (see Chapter 12), understood the power of the mass media as a tool to help their clients "engineer" public consent in their favor. According to Bernays and Fleischman, companies and rulers couldn't lead people—or get them to do what they wanted—until the people consented to what those companies or rulers were trying to do, whether it was accepting the idea of women smoking cigarettes or supporting going to war. For Bernays and Fleischman, the goal of a strategic public relations campaign was to convert their clients' goals into "common sense," then convince the public that those commonsense ideas were the "natural" way things worked.

For decades, corporate leaders and conservative thinkers have used a similar strategy, exploiting a long-standing American skepticism about government intervention in individuals' lives and working to convince people that government regulation of businesses limits freedom and should be minimized. But while Bernays and Fleischman and their clients may have strategically tried to engineer public consent for free-market policies, there doesn't have to be such a conscious and coordinated effort in order for certain ideas to be promoted through the media. Media professionals often simply tap into the dominant ideas circulating in society at the time and, in the process, reinforce them, thereby maintaining their hegemonic status.

For example, the American Dream—the powerful story (or *consensus narrative*) that America is a place where hard work pays off and all people can get ahead if they put their mind to it—has circulated for well over a century and provided inspiration for journalists and Hollywood's storytellers. It also helps explain why we might support economic plans and structures that may not be in our best interest. Because the American Dream has been woven into so much of our media and popular culture, many of us believe we have an equal chance of becoming rich and therefore successful and happy. So why would we do anything to disturb the economic structures that the dream is built on? In fact, in many versions of our American Dream story—from Hollywood films to political ads—the government often plays the role of villain, seeking to raise taxes or undermine rugged individualism and hard work. Pitted against the government in these stories, the protagonist is the little guy, at odds with

HGTV'S *SELF-MADE MANSIONS* (2020–) introduces viewers to entrepreneurs whose hard work allows them to search for their dream home. The show makes its investment in the American Dream ideology explicit. On the first episode, couple Jeremy and Jenn tour four grand homes. Jeremy admits that he's struggling with the feeling that he hasn't earned the right to buy a $1.7 million house. "You don't have to feel guilty," the host assures him. "It's not like you took anything away from anybody. This is what life is all about. . . . Like, you earned it. Awesome! That's really the American Dream, right?" Convinced, Jenn replies, "Totally."

burdensome regulations and bureaucratic oversight. However, many of these stories are produced and distributed by large media corporations and political leaders who rely on the rest of us to consent to the American Dream narrative in order to keep their privileged place in the status quo and reinforce this "commonsense" story as the way the world works. In this way, media products (and the powerful companies that control their production and circulation) can provide the commonsense narratives that keep the economic status quo relatively unchallenged and leave little room for alternatives.

Cultural Politics and Global Media

As we saw earlier in the chapter, the flow of media products became increasingly global during the twentieth century—a development that stirred up considerable debate about the consequences of this change. If media products help shape a society's culture, what happens when movies, music, and television shows produced by a media industry in one country become woven into the fabric of another country's popular culture? At first, these concerns were centered on U.S. culture's impact on the rest of the world. In more recent years, however, as the flow of global media has become more diverse and complicated, new questions about cultural influence are arising.

American Cultural Imperialism

For decades, U.S. media industries have been major exporters of media content. American music, movies, and TV shows often dominate international markets, shaping the cultures of other nations in a process known as **cultural imperialism**. (See "Media Literacy and the Critical Process: Cultural Imperialism and Movies" for more on the dominance of the American movie industry.) To the extent that U.S. popular culture reflects a particular brand of rebellious freedom and innovation, defenders of American global dominance have argued, its global influence can be a good thing; for example, it can nurture forms of political resistance and advance human rights in autocratic countries. Supporters also argue that a universal popular culture creates a global village and fosters communication across national boundaries.

Critics, on the other hand, argue that American cultural imperialism has a homogenizing effect, diluting the rich diversity of local traditions, values, and identities with a singular American-styled global culture. That culture is often thoroughly consumerist, materialist, and secular— values at odds with those in many parts of the world. And the affluent lifestyle represented in so many American media products can be jarringly disorienting for the two-thirds of the world's population for whom even middle-class economic stability is out of reach. Finally, the economics of global media production can make it difficult for local people to control their own media culture. American media companies have an economic advantage because they generally recoup most of their production costs in the United States before their programming is exported. That means they can dump big-budget TV shows and movies on foreign markets at bargain prices, undercutting and undermining local production companies trying to produce original content.

Complicating Global Cultural Influence

The dynamics of global media markets have always been more complicated than some accounts of cultural imperialism allow, and although U.S.-based conglomerates still dominate, global media markets are diversifying. First, research has shown that global audiences of media products never simply adopt American values wholesale. Instead, they interpret media texts through the lens of their own lives and adapt the values those texts carry to

CULTURAL IMPERIALISM Ever since Hollywood gained an edge in film production and distribution during World War I, U.S. movies have dominated the box office in Europe, in some years accounting for more than 80 percent of the revenues taken in by European theaters. Hollywood's reach has since extended throughout the world, including previously difficult markets such as China.

MEDIA LITERACY & THE CRITICAL PROCESS

Cultural Imperialism and Movies

In the 1920s, the U.S. film industry became the leader in the worldwide film business. The images and stories of American films are well known in nearly every corner of the world. But with major film production centers in places like India, China, Hong Kong, Japan, South Korea, Mexico, the United Kingdom, Germany, France, Russia, and Nigeria, to what extent do U.S. films dominate international markets today? Conversely, how often do international films get attention in the United States?

① DESCRIPTION

Using international box-office revenue listings (available with an Internet search), compare the recent weekly box-office rankings of the United States to those of five other countries. (Your sample could extend across several continents or focus on a specific region, like Southeast Asia.) Limit yourself to the top ten or fifteen films in the box-office ranking. Note where each film is produced (some films are joint productions of studios from two or more countries), and put your results in a table for comparison.

② ANALYSIS

What patterns emerged in each country's box-office rankings? What percentage of films came from the United States? What percentage of films were domestic productions in each country? What percentage of films came from countries other than the United States? In the United States, what percentage of films originated with studios from other countries?

③ INTERPRETATION

What do your discoveries mean? Can you make an argument for or against the existence of cultural imperialism by the United States? Are there film industries in other countries that dominate movie theaters in their region of the world? How would you critique the reverse of cultural imperialism, wherein films from other countries rarely break into the Top 10 box-office list? Does this happen in any countries you sampled?

④ EVALUATION

Given your interpretation, is cultural dominance by one country a good thing or a bad thing? Consider the potential advantages of creating a global village of shared popular culture versus the potential disadvantages of cultural imperialism. Also, is there any potential harm in a country's Top 10 box-office list being filled with domestic productions and rarely featuring international films?

⑤ ENGAGEMENT

Contact managers of your local movie theater (or executives at the headquarters of the chain that owns it) and ask them how they decide which films to screen. If they don't show many international films, ask them why. Be ready to provide them with a list of three to five international films released in the United States that haven't yet been screened in the theater. (Conduct an Internet search beforehand.)

the unique conditions of their own culture—a fact that makes the crudest predictions of a homogeneous, Americanized world unlikely to come true.[29]

In addition, the United States has never been the only media exporter. Brazil's telenovelas, Jamaica's reggae, and Japan's anime, for example, have been extremely popular around the world for decades. And in today's media landscape, new dynamics are helping producers in even more countries compete with U.S. products on the global market—including in the United States itself. Digital distribution, for example, has made it easier for American audiences (as well as others around the world) to discover K-pop from South Korea, Scandinavian police procedural TV series, and European League soccer. In this way, cultural influence now flows in multiple directions, not just outward from the United States.

Going forward, the appeal of the enormous Chinese media market will likely produce the biggest change to the political dynamics of global media. The controversy faced by the National Basketball Association in 2019, for example, may provide a hint of what the future holds. The NBA has been investing heavily in the Chinese market in hopes of building a profitable fan base there. When Daryl Morey, the Houston Rockets' general manager, tweeted his support for Hong Kong's pro-democracy protesters (who opposed the Chinese government's tightening grip on the province), the Chinese

Basketball Association quickly suspended cooperation with the Rockets, and Chinese media companies announced they would stop streaming Rockets games. In response, the NBA issued a statement in which they recognized that Morey's comments "deeply offended many of our friends and fans in China" and described the situation as "regrettable."[30] The response drew strong criticism from U.S. politicians across the political spectrum, who felt that the NBA had caved to Chinese censorship to preserve their economic interests. The incident not only illustrates the power of China's political influence on U.S. conglomerates doing business there but also suggests that this influence may end up affecting how these companies operate at home.

The Media Marketplace and Democracy

In the midst of today's major global transformations of economies, cultures, and societies, the best way to monitor the impact of transnational economies is through vigorous news attention and lively public discussion. This process, however, is being hampered. Beginning in the 1990s, news divisions, concerned about the bottom line, severely cut back the number of reporters assigned to cover international developments, eroding the reporting that a democratic citizenry needs in order to be informed and aware.

We live in a society in which often superficial consumer concerns, stock market quotes, and profit aspirations—rather than broader social issues—increasingly dominate the media agenda. In response, critics have posed some key questions: As consumers, do we care who owns the media as long as most of us have a broad selection of products? Do we care who owns the media as long as multiple voices *seem* to exist in the market?

The Effects of Media Consolidation on Democracy

In our current economy, global media conglomerates will continue to control more aspects of production, distribution, and exhibition. Of pressing concern is the impact of mergers on news operations, particularly the influence of large corporations on their news subsidiaries. These conglomerates have the capacity to determine what is covered in the national news, and they can divert funding from the news to other areas—for example, to promote their entertainment products.

Because of the growing consolidation of mass media, it has become increasingly difficult to sustain a public debate on economic issues. From a democratic perspective, the relationship of our mass media system to politics has been highly dysfunctional. Politicians in Washington, D.C., have regularly accepted millions of dollars in contributions from media conglomerates and their lobbying groups to finance their campaigns. Corporate donors were further strengthened by the Supreme Court's 2010 *Citizens United* case, which "ruled that the government may not ban political spending by corporations in candidate elections."[31] Writing for the 5–4 majority, Justice Anthony Kennedy stated, "If the First Amendment has any force, it prohibits Congress from fining or jailing citizens, or associations of citizens, for simply engaging in political speech." The ruling overturned two decades of precedents that had limited direct corporate spending on campaigns, including the Bipartisan Campaign Reform Act of 2002, which placed restrictions on buying TV and radio campaign ads.

As unfettered corporate political contributions count as "political speech," some corporations are experiencing backlash (or praise) once their customers discover their political positions. For example, in 2012, fast-food outlet Chick-fil-A's charitable foundation "was revealed to be funneling millions to groups that oppose gay marriage and, until recently, promoted gay 'cure' therapies," resulting in a firestorm of criticism from some but also a wave of support from others, the *Daily Beast* reported. In the same year, Amazon CEO Jeff Bezos and his then wife MacKenzie Scott donated $2.5 million of their own money to support a same-sex marriage referendum in Washington State, gaining both praise and criticism from some Amazon customers.[32]

Politicians have often turned to local television stations, spending record amounts during each election period to get their political ads on the air. In 2020, spending on the federal elections in the United States totaled a record $14 billion.[33] A large portion of that spending went to local broadcasters for commercials for congressional candidates and—in swing states like Florida, Michigan, and Wisconsin—for presidential candidates. But although local television stations have been happy to get part of the ever-increasing bounty of political ad money, the content of their news broadcasts is not substantial, particularly when it comes to covering politics.

In his 2020 book *Buying Reality: Political Ads, Money, and Local Television News*, University of Delaware professor Danilo Yanich noted that the vast majority of political ads on local television were focused on local political races. Yet "local television newscasts almost ignored them in their political coverage. Only 13 percent of the time devoted to political stories addressed local races." Instead, the stations "directed their attention overwhelmingly" to the presidential contest, leading Yanich to conclude that "the station owners' public interest obligation as license holders should press them to offer critical analyses of [the local political ad] content."[34] Thus, there is little news content to provide a counterpoint to all the allegations that might be hurled in the barrage of political ads.

The Media Reform Movement

The increasing consolidation of mainstream media power has prompted calls for *media reform*. John Nichols and Robert McChesney define media reform as "a catch-all phrase to describe the broad goals of a movement that says consolidated ownership of broadcast and cable media, chain ownership of newspapers, and telephone and cable-company colonization of the Internet pose a threat not just to the culture of the Republic but to democracy itself."[35] While our current era has spawned numerous grassroots organizations that challenge media to do a better job for the sake of democracy, there has not been much of an outcry from the general public about the kinds of concerns described by Nichols and McChesney. There is a reason for that. One key paradox of the Information Age is that for such economic discussions to be meaningful and democratic, they must be carried out in the popular media as well as in educational settings. Yet public debates and disclosures about the structure and ownership of the media are often not in the best economic interests of media owners.

THE PRESIDENT AND COFOUNDER of Free Press, a national nonpartisan organization dedicated to media reform, Robert McChesney is one of the foremost scholars of media economics in the United States. For ten years he hosted *Media Matters*, a radio call-in show in Central Illinois that discussed the relationship between politics and media. McChesney (*left*) published *People Get Ready: The Fight against a Jobless Economy and a Citizenless Democracy* (2016) with journalist and Free Press cofounder John Nichols (*right*).

JEFF CHIU/AP Images

Still, in some places, local groups, consumer movements, and political leaders are trying to address media issues that affect individual and community life. These media reform movements—coordinated by organizations such as Freepress.net—are generally united by shared concerns about the state of the media. The Internet has also made it possible for media reform groups to form globally, uniting around such issues as contesting censorship or monitoring the activities of multinational corporations. The push for media reform was also largely responsible for making more people aware of the stakes involved in debates over *net neutrality*, the idea that Internet service providers shouldn't censor or penalize particular websites and online services (see Chapter 2).

Given the exploitation of Facebook, Google, and other sites by trolls during recent elections and the revelation that millions of Facebook users had their data shared without their knowledge, perhaps we are more ready than ever to question some of the hierarchical and undemocratic arrangements of what McChesney, Nichols, and other reform critics call "Big Media." Even in the face of so many media mergers, the general public seems open to such examinations, as do an increasing number of politicians and policy makers. By better understanding media economics, we can play a more knowledgeable role in critiquing media organizations and evaluating their impact on democracy.

LAUNCHPAD FOR *MEDIA & CULTURE*

LaunchPad
macmillan learning

Visit LaunchPad for *Media & Culture* at **launchpadworks.com** for additional learning tools:

➜ QUIZ YOURSELF WITH LEARNINGCURVE

LearningCurve uses gamelike quizzing to help you review important concepts from this chapter. LearningCurve includes multiple-choice questions at different levels of difficulty, hints, feedback for both right and wrong answers, and links to e-book material for easy reference. Which answer choice would you select for the sample question below?

> Kevin has seen his cable bill increase every year, though the services and channels have remained the same. Comcast is the only cable provider in Kevin's region. Which of these led to Kevin's problem?
>
> ○ Great Depression
>
> ○ Federal Trade Commission
>
> ○ Clayton Antitrust Act of 1914
>
> ○ Telecommunications Act of 1996

➜ WATCH VIDEO ON KEY CONCEPTS

LaunchPad includes clips from movies, TV shows, online sources, and other media texts, in addition to interviews with media experts and newsmakers. In the videos for this chapter, you'll consider topics like the global popularity of Disney's *Frozen* and how ownership systems and profits shape media production.

➜ PRACTICE THE CRITICAL PROCESS

Use the Media Literacy Activity to walk through the critical process step-by-step and develop a critical perspective on what media means to you. In this chapter's activity, you'll research the latest developments in media ownership.

➜ LEARN KEY TERMS

Review the definitions for this chapter's key terms with LaunchPad's easy-to-use online flash cards. Try the cards in practice mode, or quiz yourself as a way to focus your study efforts. (Page numbers indicate where each term is highlighted within the chapter.)

monopoly, 369	conglomerates, 370	ideology, 389
oligopoly, 369	mass customization, 380	hegemony, 389
limited competition, 369	horizontal integration, 380	cultural imperialism, 391

REVIEW QUESTIONS

Use the questions below to revisit key themes and concepts from this chapter.

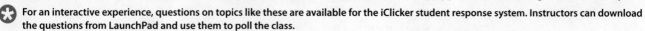 Need more practice? Access the LearningCurve multiple-choice questions for this chapter, which are available on LaunchPad. All questions include feedback for correct and incorrect answers, hints, and links to e-book material.

Media Economics Today

1. What questions and debates emerge as a result of the complex nature of media industries?
2. How are the three basic structures of mass media organizations—monopoly, oligopoly, and limited competition—different from one another?

How We Got Here: The Rise of Global Conglomerates

3. What role have regulatory changes, globalization, and the Internet played in the rise of conglomerates in the information economy?
4. While media conglomerates often fight hard to avoid government regulation, under what circumstances do they support increased regulation?

Business Trends in Today's Media Industries

5. What are the three main ways media companies generate revenue from media products?
6. What factors have helped fuel specialization in media markets, and what has the impact of this specialization been?
7. What is the difference between vertical and horizontal integration, and what benefits does each offer media companies?

8. Using Disney as an example, what role does synergy play in media mergers?

Social and Political Issues in Media Economics

9. How have corporations defended conglomeration? What questions do their efforts raise?
10. What employment trends have emerged in the information economy and media industries?
11. What is the difference between consumer choice and consumer control? How does each play out in everyday life?
12. What role do media companies play in shaping our ideas about the economic status quo?
13. What is cultural imperialism, and what does it have to do with the United States? What developments are complicating cultural imperialism and the dynamics of global media markets?

The Media Marketplace and Democracy

14. What do critics and activists fear most about the concentration of media ownership?
15. What are some promising signs regarding the relationship between media economics and democracy?

QUESTIONING THE MEDIA

Use these critical-thinking questions to reflect on your own experiences with media and apply your understanding of media concepts.

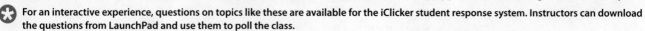 For an interactive experience, questions on topics like these are available for the iClicker student response system. Instructors can download the questions from LaunchPad and use them to poll the class.

1. Where do you come down in the debate over the complex role and responsibilities of media companies? Are they just businesses that should focus only on efficiency and profitability, or does their cultural impact mean that they have a wider social responsibility?
2. Why are consumers more likely to pay to access some digital content, like music and books, and less likely to pay for other content, like sports and news?

3. How does the concentration of media ownership limit the number of voices in the marketplace? Do we need rules limiting media ownership?
4. Is there such a thing as a global village? What does this concept mean to you?

The Culture of Journalism

VALUES, ETHICS, AND DEMOCRACY

14

It was written eye-level in big letters on the exterior of the door to the U.S. Capitol, so it was plain to see: "Murder the Media."

The writing on the door was one of the remnants of the January 6, 2021, insurrection in Washington, D.C., when a mob of thousands broke into the Capitol building in an attempt to disrupt the work of the House and Senate certifying Joe Biden's election as the next president of the United States. The insurrectionists wore red "Make America Great Again" hats (Donald Trump's slogan and brand), carried pro-Trump and confederate flags, and wore a mishmash of military gear. By the time they were finished, more than 140 people were injured and 5 were dead, including a member of the Capitol Police force.

At some point during the mob attack, two insurrectionists paused for a photo at the defaced door of the Capitol, smiling and giving a thumbs-up. One of the men wore a "Murder the Media" shirt and hat, according to court documents.

Inside, a few people from the mob accosted *New York Times* staff photographer Erin Schaff and threw her on the floor when they saw her press pass. "I started screaming for help as loudly as I could. No one came. People just watched. At this point, I thought I could be killed and no one would stop them. They ripped one of my cameras away from me, broke a lens on the other and ran away."[2]

Outside, extremists (some chanting "CNN sucks") smashed multiple video cameras and equipment set up by AP reporters, at which point others tried to burn it. Someone fashioned a noose out of a camera cable at the scene and hung it over a tree.

The anti-media sentiment that prompted these types of actions is not new. Presidents, politicians, and supporters of many parties and ideologies have voiced criticisms of the news media since the founding of the United States, resulting in sometimes antagonistic relationships between the people

Journalists were among those targeted during the January 6, 2021, assault on the U.S. Capitol. Upon seeing her press pass, rioters threw Erin Schaff of the *New York Times* to the floor, ripped away one of her cameras, and broke the other. They dragged John Minchillo—a photographer for the Associated Press who wore press identification—twenty feet and pushed him over a wall.[1] Rioters also smashed TV cameras (shown here) outside the building.

in power and those reporting on their activities. But until Donald Trump took office, no president had made disparagement of and violence against the media such a central part of his message, as Trump did in speeches, tweets, and direct confrontations with reporters. As the Committee to Protect Journalists noted, Trump "repeatedly called the press 'fake news,' 'the enemy of the people,' 'dishonest,' 'corrupt,' 'low life reporters,' 'bad people,' 'human scum' and 'some of the worst human beings you'll ever meet.'"[3] In 2017, Trump tweeted a manipulated video of himself (from footage originally taken at a wrestling event) tackling and punching a figure with a CNN logo for a face.[4]

Two weeks after the Capitol insurrection, Joe Biden became the new president. Yet even as Trump lost the presidency and was banned by his preferred social media platform, Twitter, for violating its user policies, an enormous news-related divide remained in the United States. A nationwide survey by Gallup (the polling company) and the Knight Foundation "found continued pessimism and further partisan entrenchment about how the news media delivers on its democratic mandate for factual, trustworthy information."[5] According to these organizations, fewer than half of the Americans polled in 2019—41 percent—said that "they have 'a great deal' or 'a fair amount' of confidence in the mass media to report the news 'fully, accurately and fairly.'"

Perhaps there is a way forward. Gallup and the Knight Foundation found that "Americans are more likely to trust local news than national news and to perceive less bias in local reporting than what they see nationally."[6] The good news is that local news media are not as bound up in the national culture wars. The bad news (as we noted in Chapter 8) is that key providers of local news—newspapers—are suffering immensely, leaving news deserts dotting the American landscape and giving citizens in those communities just one option: national news.

> **What are ways to rebuild citizens' confidence in the news media's ability to report fully, accurately, and fairly? Are citizens part of the problem? That is, do too many people now consume only news that reflects their political values?**

Journalism Today

Journalism—the work of reporters and editors—is an essential element of democracy. It is the only media business ("the press") that is specifically mentioned and protected by the U.S. Constitution, although journalism comes in many more forms these days than it did when that document was written. Today, journalism remains one of the cultural foundations of the United States, even as we simultaneously experience discouraging trends and exciting possibilities for telling news stories.

As noted in Chapter 8, newspapers are suffering, and the old commercial model of printing newspapers with a mix of news and advertisements is increasingly lacking in advertisers. Some newspapers, especially national dailies like the *New York Times*, the *Wall Street Journal*, and the *Washington Post*, have succeeded in growing their subscriber bases, though that supplementary source of revenue still doesn't cover the loss of advertising. The closure of small daily and metropolitan newspapers means investigative reporting suffers as well, since newspapers (compared to local radio and TV news) have been the main medium for that kind of accountability reporting.

The audience for national, cable, and local television news has dropped, but it still remains highly profitable. Local TV news, in particular, is viewed by people as being more trustworthy than national TV news, yet the TV news audience is mostly over the age of fifty. Younger people watch local TV news much less often and are much more familiar with the video news story style of digital news sites like Vice, Vox, NowThis, and Snapchat, which are characterized by somewhat longer stories with more animation, sound elements, and background information. (Interestingly, according to one study, subjects of all ages preferred the newer video reporting style of digital news sites to traditional local TV news reports.)[7]

The advent of the Internet has changed both the business model and the style of news. Daily newspapers and nightly newscasts have become 24/7 news operations and are competing for news customers with a wave of digital news start-ups on the same Internet platform. They are also competing with sites like Google, Facebook, and Twitter, which link to their stories (a fact the news industry likes) but have siphoned off billions in ad revenue that used to be placed with newspapers, magazines, radio, and television (a fact the news industry does not like, prompting it to engage in an increasingly visible battle with the social media giants over content and payment). The news media ecosystem has its pollution, too, in the form of opinion, propaganda, and conspiracy theory outlets masquerading as journalism.

In this chapter, we consider all these elements as we examine the changing news landscape and definitions of journalism. We will:

DIGITAL NEWS Millions of Americans get their news first through digital and social media sites like Google News. The digital sites and apps are not journalism organizations, but they aggregate and repost news media stories.

- Study the legacy of print-news conventions and rituals, including an emphasis on neutrality
- Investigate the impact of television and the Internet on news
- Consider the essential elements of news
- Explore the values underlying news and ethical problems confronting journalists
- Analyze the history of conservative media, which rose in opposition to the mainstream news media
- Consider contemporary developments in journalism and democracy—particularly the critique of objective-style journalism

How We Got Here: The Evolution of American Journalism

As noted in Chapter 8, journalism's beginning in North America was at a Boston newspaper, printed in 1690, which was banned after its first issue by the British colonial government. The creation of more newspapers and magazines followed, and the number of news publications expanded with the growing United States and in response to the guarantees provided by the First Amendment, adopted in 1791. With the advent of the telegraph (whose wire-laden poles often followed the line of train tracks) and the subsequent formation of the Associated Press in 1846, news could travel faster than any form of land transportation, and the news of the day was soon reported across the continental United States. Standards and values of what constitutes news developed during this long era of print news dominance.

The development of radio and then television made news even more omnipresent and immediate: a breaking news story could be communicated much faster than it would take for an "extra" edition to be written and printed by a newspaper. In part because of its very different format, television (which, unlike newspapers, entered the American scene first and foremost as an entertainment medium) made some of its own rules about what constitutes news.

The distribution and format of the news changed even more with the Internet, resolving the problems of space (which was now unlimited, compared to the structured pages of print) and time (also unlimited, compared to clock-bound radio and TV news reports). News could be more in-depth, more comprehensive, and available 24/7. Yet as the Internet broke these constraints on news delivery, the news lost many of its advertisers to sites like social media and Google, which promised easier and less expensive ways to reach consumers.

Let's take a closer look at these foundational periods in the evolution of journalism.

The Historical Foundations of Print Journalism

As noted in Chapter 8, with the beginnings of the penny press in the 1830s, there was a distinct shift in U.S. journalism away from the partisan press of the colonial era. The culture and business model of the penny press was more democratic: Instead of niche, partisan audiences, the target was mass audiences—a newspaper that would appeal to almost everyone. From that time and into the twentieth century, most newspapers would stake out the middle ground in order to best serve their financial interests.

Mass Audiences and Moderate Politics

During this period, newspapers weren't *neutral* as much as they were *centrist*—that is, moderate in their politics—and advertiser-friendly on the business side. William Greider, a former *Washington Post* editor, saw the tie between good business and balanced news: "If you [were] going to be a mass circulation journal, that mean[t] you [were] going to be talking simultaneously to lots of groups that [had] opposing views. So you [had] to modulate your voice and pretend to be talking to all of them."[8] Similarly, among the national newsmagazines, mainstream ones like *Time* or *Newsweek* appealed to a mass audience, and smaller niche periodicals appealed more directly to the political Left and Right.

An Ethic of Objectivity and Impartiality

By the 1920s, as mass newspaper publishers moved away from partisan politics, journalism was also becoming a more formal occupation. According to media historian Michael Schudson, journalists sought to "disaffiliate themselves" from the relatively new "public-relations specialists and propagandists who suddenly surrounded them."[9] The American Society of News Editors formed in 1922 and produced its first code of ethics, which included principles of sincerity, truthfulness, accuracy, and impartiality, and the statement that "news reports should be free from opinion or bias of any kind."[10] Those ideas became the touchstones for most of mainstream media in the twentieth century.

With this approach, newspapers could still take political positions on issues and candidate endorsements in their editorial pages, but ideally the news reporting operation would remain separate from that of the editorial board. Yet for everyday reporting, a big question remained: Exactly what did it mean for a journalist to be objective, impartial, and free from bias? As Schudson explained, "Most of the subjectivity in news is not idiosyncratic and personal but patterned and predictable. Journalists do not make decisions at random. . . . They depend on reliable shorthand, conventions, routines, habits, and assumptions about how, why and where to gather the news."[11]

What we think of as objective, impartial, or bias-free news emerged from the politically moderate mass-oriented news media and a set of reporting conventions created by these media (see "Reporting Rituals and the Legacy of Print Journalism" later in this chapter). These practices include the inverted-pyramid story form, the two-sides-to-every-story standard, the detached third-person writing voice, and the use of quotes from experts. Of course, while these conventions have become associated with the notion of impartiality, neutrality, and objectivity, they cannot guarantee that a story will be bias-free. As we note later in this chapter, there are several

Universal History Archive/Getty Images

BY THE 1920s, the conventions of objectivity and impartiality were solidly in place in U.S. journalism. One of the leading news organizations with this ethic was the *New York Times*, here reporting on Charles Lindbergh's successful transatlantic flight in 1927.

American values around which news stories are routinely framed, and the activity of deciding what is news (why tell this story but not that one?) is inherently a subjective process. The process operates more smoothly, though, because reporting conventions and routines suggest what stories should be covered, and long-held cultural values suggest *how* they should be covered.

In the second half of the twentieth century, TV news became the main news medium for most Americans. Most newspapers, folded into ever-larger newspaper chains, began to switch their business focus from a mass audience to a more upscale middle-class audience (considered to be more desirable to advertisers) and abandoned their working-class readers—many of whom later drifted to other, more specialized and politicized media, as we'll explain later in the chapter.

Stories Become Visual on TV News

In the 1950s, former radio reporter John Daly hosted the game show *What's My Line?* on CBS TV. When he began moonlighting as the evening news anchor on ABC, the network blurred the division between entertainment and information, foreshadowing what was to come. TV news could have high journalistic standards, but it had to entertain as well, and it was highly dependent on visuals—a limitation that still leads television news to under-cover some stories and overplay others.

In the early days, the most influential and respected television news program was CBS's *See It Now*. Coproduced by Fred Friendly and Edward R. Murrow, *See It Now* practiced a kind of TV journalism lodged somewhere between the neutral and the narrative traditions. Generally regarded as "the first and definitive" news documentary on American television, *See It Now* sought "to report in depth—to tell and show the American audience what was happening in the world using film as a narrative tool," according to A. William Bluem, author of *Documentary in American Television*.[12] Murrow worked as both the program's anchor and its main reporter, introducing the investigative model of journalism to television—a model that programs like *60 Minutes* and *Dateline* would later imitate.

While print journalists have long been expected to be detached, TV news has derived its credibility from live, on-the-spot reporting; believable imagery; and viewers' trust in the reporters and anchors (part of the reason former game-show hosts were among TV's early newscasters). In fact, from the early 1970s through the early 2000s, most annual polls indicated that the majority of viewers found TV news a more credible resource than print news. Viewers have tended to feel a personal regard for the local and national anchors who appear each evening on TV sets in their living rooms.

The visual nature of television news has created two characteristics unique to television: viewers often watch news based on what they think of the anchors, and the power of a story is often based on the strength of its visual images. As a result, TV news consultants tend to make recommendations to local stations and national networks about how to boost their ratings with visuals in mind.

EDWARD R. MURROW of CBS News, with microphone in hand, talks to a U.S. Marine in 1953 during the Korean War for his television news show *See It Now*. The company was holding a ridge on the Korean front.

Hulton Archive/Getty Images

Pretty Faces, Happy Talk, and the Crime Block: Packaging TV News

In the early 1970s, consultants advised the news director at a Milwaukee TV station that the evening anchor looked too old. The anchor, who showed a bit of gray, was replaced and went on to serve as the station's editorial director. He was thirty-two years old at the time. In the late 1970s, a reporter at the same station was fired because of a "weight problem," although that was not given as the official reason. Earlier that year, she had given birth to her first child. In 1983, Christine Craft, a former Kansas City television news anchor, was awarded $500,000 in damages in a sex discrimination suit against station KMBC (she eventually lost the monetary award when the station appealed). She had been fired because consultants believed she was too old, too unattractive, and not deferential enough to men.

Such stories are rampant in the annals of TV news, and they have helped create the stereotype of the half-witted but physically attractive news anchor, reinforced by images from popular culture—from Ron Burgundy in the *Anchorman* films to *The Morning Show* on Apple TV+. Although the situation has improved slightly today, national news consultants often continue to set the agenda for what stories local reporters should cover (lots of crime) as well as how they should look and sound (young, attractive, pleasant, and with no regional accent). News consultants have also advised stations to pair up older male and younger female anchors for their local TV news teams.

Another strategy favored by news consultants is *happy talk*: the ad-libbed or scripted banter that goes on among local news anchors, reporters, meteorologists, and sports reporters before and after news reports. Beginning in the 1970s, consultants started to recommend such chatter in order to create a more relaxed feeling on the news set and to foster the illusion of conversational intimacy with viewers. Some also believed that happy talk would counter much of that era's "bad news," which included coverage of riots and the Vietnam War. A strategy still used today, happy talk often appears forced and may create awkward transitions, especially when anchors must report on events that are sad or tragic.

In addition to pretty faces and happy talk, news consultants have advised news directors to invest in national prepackaged formats—such as Action News or Eyewitness News—which employ the same theme music and opening graphic visuals from market to market. Consultants have also suggested that stations lead their newscasts with *crime blocks*: groups of stories that recount the worst local criminal transgressions of the day. A cynical slogan soon developed in the industry: "If it bleeds, it leads." Depending on the local station, multiple studies continue to show that crime stories still dominate as the lead story on any typical evening newscast—far more than any other category of news.

Video News and Visual Evidence

Although photographs first made their way into newspapers in 1880 and newsreels became popular beginning in the early 1900s, nothing was as potent as video news images displayed on a television in one's own home. Over the past seventy years, television news has dramatized America's key events. Civil Rights activists, for instance, acknowledge that the movement benefited enormously

GEORGE FLOYD'S death at the hands of Minneapolis police in 2020, documented on video by a citizen journalist, became a worldwide story that inspired millions to join Black Lives Matter protests.

Christopher Mark Juhn/Anadolu Agency/Getty Images

from televised news that documented the experiences of Black Americans in the South in the 1960s. The news footage of southern police officers turning powerful water hoses on peaceful Civil Rights demonstrators, as well as the news images of "white only" and "colored only" signs in hotels and restaurants, created a context for understanding the disparity between Black and white in the 1950s and 1960s.

Other TV images are also embedded in the collective memory of many Americans, including the assassinations of John F. Kennedy and Martin Luther King Jr. in the 1960s, the turmoil of Watergate in the 1970s, the first space-shuttle disaster and the student uprisings in China in the 1980s, the Oklahoma City federal building bombing in the 1990s, the terrorist attacks on the Pentagon and World Trade Center in 2001, Hurricane Katrina in 2005, and the historic election of President Barack Obama in 2008. During these and other critical events, TV news has been a cultural reference point, often sparking national debates.

More recently, the Internet and *citizen journalism* (see Chapter 8) have helped TV news and other news organizations find the images necessary to tell major stories. Images of overflowing hospitals and morgues amid the COVID-19 pandemic in 2020; the death of George Floyd at the hands of Minneapolis police in 2020 and subsequent Black Lives Matter protests; and the insurrectionist attack on the U.S. Capitol in 2021 all became stories, or better-reported stories, because of video shot by citizens—some of whom, in the case of the Capitol attack, were participating in the event themselves.

Cable TV Pundits and Politics

The transformation of TV news by cable—with the arrival of CNN in 1980—led to dramatic changes in TV news delivery at the national level. Prior to cable news, most people tuned to their local and national news late in the afternoon or evening on a typical weekday, with each program lasting just thirty minutes. But since the 1980s (and supercharged with the rise of Internet news organizations in the 1990s), the 24/7 news cycle has meant that we can get TV news anytime, day or night, and the constant stream of new content has led to major changes in what is considered news.

Cable's Talking Heads

Because it is expensive to dispatch reporters to document stories or maintain foreign news bureaus to cover international issues, the much less expensive talking-head pundit has become a standard for cable news channels. Such a programming strategy requires few resources beyond the studio and a few guests. (Low-budget cable channel upstarts like Newsmax and OAN do little original reporting, just repeating their brand of political talking points.)

ANDERSON COOPER has been the primary anchor of *Anderson Cooper 360°* since 2003. Although the program is mainly taped and broadcast from its New York City studio and typically features reports of the day's main news stories, with added analyses from experts, Cooper is one of the few talking heads who reports live from the field fairly often for major news stories. Notably, he has done extensive coverage of the February 2011 uprisings in Egypt; the 2016 massacre at Pulse—a gay nightclub in Orlando—which claimed forty-nine lives; the aftermath of Hurricane Maria in Puerto Rico in 2017; and Hurricane Florence in the Carolinas in 2018.

Ted Soqui/Corbis/Getty Images

Today's main cable channels have built their evening programs along partisan lines and follow the model of journalism as opinion: Fox News goes far right with pundit stars like Tucker Carlson, Sean Hannity, and Laura Ingraham; MSNBC leans left with Rachel Maddow, Chris Hayes, and Lawrence O'Donnell; and CNN stakes out the relative middle with hosts like Jake Tapper, Anderson Cooper, and Don Lemon.

The Contemporary Journalist: Pundit or Reporter?
Journalists discuss whether the 24/7 news cycle encourages reporters to offer opinions more than facts.

Discussion: What might be the reasons why reporters should give opinions, and what might be the reasons why they shouldn't?

Journalism of Verification and Journalism of Assertion

On network television news and in most mainstream newspapers, the political tone remains centrist. But cable news, which has smaller audiences, takes a more niche approach on a 24/7 news cycle—a strategy that has been mimicked by some Internet-only news outlets. In today's fragmented media marketplace, going after niche audiences along political lines can be smart business, but it also changes the tone of some journalism organizations. In such an environment, we see the decline of the journalistic model that promoted fact-gathering and expertise and viewed objectivity as the ideal for news practice—what journalists Bill Kovach and Tom Rosenstiel call a "**journalism of verification**." Rising in its place is a new era of partisan news—a "**journalism of assertion**."[13] This transition is symbolized by the rise of the cable news pundit on Fox News and MSNBC who serves as a kind of "expert," in place of elements like verified facts, authentic documents, and actual experts. Like newspapers with separate editorial pages, some cable channels try to limit their opinion shows to only certain hours of the day, but others have let the journalism of assertion permeate their entire schedule.

Gen Z's high-speed rail meme dream, explained

Have you seen that high-speed rail map on Twitter? Gen Z is hoping President Biden has.

By Gabby Birenbaum | Mar 10, 2021, 8:00am EST

NEWS IN THE DIGITAL AGE Launched in 2014, Vox is a digital news organization whose goal is to "explain the news" using video, graphics, podcasts, and regular "explainer" features. As Vox says, "We live in a world of too much information and too little context. Too much noise and too little insight. That's where Vox's explainers come in."[14]

News Now: The Loss of Traditional Gatekeepers

As media like cable television and the Internet emerged in the 1980s and 1990s, they created many more openings for different kinds of news organizations to enter the scene. The small group of mainstream news media with centrist politics that had long set the news agenda for the nation—including the *New York Times*, the *Wall Street Journal*, the *Washington Post*, the major TV news networks (ABC, CBS, and NBC), and a few mainstream newsweekly magazines like *Time* and *Newsweek*—were no longer the sole gatekeepers. Newer news organizations like CNN operated by essentially the same mainstream rules, but upstarts like the *Huffington Post* (now *HuffPost*) on the left and the Drudge Report, Fox News, and Breitbart on the right defied the nonpartisan ethic of mainstream journalism. Meanwhile, news sites like BuzzFeed, Vox, and Vice tried to reinvent journalism by breaking stylistic conventions and targeting younger audiences.

Beyond these new news organizations, emerging digital giants and social media corporations in many ways have become the new gatekeepers of news, though they operate no news organizations themselves. Google has become the primary method for searching for news stories. Friends recommend news stories through Facebook and Twitter. Sites and apps like Yahoo!, Microsoft, Google, Snapchat, and Apple News+ curate news from other sources for readers, becoming the new "front page" for many. And YouTube, TikTok, and other social media can host user videos and stories that go viral, sometimes becoming citizen journalism that the mainstream news organizations use to set their agenda.

The net effect is that mainstream news organizations still continue their work and remain influential voices in setting the public agenda. But for better or worse, they aren't the only gatekeepers now. Public voices that have historically been marginalized have more methods for being heard, but so do propagandists and conspiracy theorists who attempt to poison public discourse.

The Essential Elements of News

Throughout most of the twentieth century, journalism in the United States sought to provide information that enabled citizens to make intelligent decisions. Today, this guiding principle faces serious threats. Why? First, we may just be producing too much information. Social critic Neil Postman recognized back in 1995 that with developments in media technology, society has developed an "information glut" that transforms news and information into "a form of garbage."[15] Postman believed that scientists, technicians, managers, and journalists merely pile up mountains of new data, which add to the problems and anxieties of everyday life. As a result, too much unchecked data—especially on the Internet—and too little thoughtful discussion emanate from too many channels of communication.

A second, related problem suggests that in the cacophony of information coming from media organizations, there is not always agreement on a basic set of shared facts. For example, in the November 2020 presidential election (which had one of the largest turnouts in decades and was also one of the most secure elections, according to the U.S. Department of Homeland Security), Joe Biden won the popular vote by more than seven million and the Electoral College by a vote of 306 to 232— roughly the same Electoral College result that his opponent, Donald Trump, had won with in 2016.[16] Yet despite those facts, Trump and partisan conservative media outlets (including Fox News) insisted the election had been stolen from Trump, ultimately leading a mob to attack the U.S. Capitol on January 6, 2021, in a failed attempt to stop Congress from certifying the vote (see the chapter-opening discussion). Even after that event, 73 percent of surveyed Republicans reported that they believed there was widespread voter fraud, while 95 percent of Democrats and 60 percent of independents said they did *not* believe there was widespread voter fraud.[17] As the Pew Center explained, Americans blame divisions like these on the fact that their "fellow citizens are too easily misled in the fractured social media and digital communications environment."[18]

This information-overloaded and fractured environment is where journalists work. Despite the distractions, most mainstream journalists today—like many who came before them in the twentieth century—seek to follow professional canons to serve citizens, verify facts, and remain independent monitors of power as they determine newsworthiness and remain cognizant of the cultural values that can influence how they frame their stories.

What Is News?

In a 1963 staff memo, NBC news president Reuven Frank outlined the narrative strategies integral to all news: "Every news story should . . . display the attributes of fiction, of drama. It should have structure and conflict, problem and denouement, rising and falling action, a beginning, a middle, and an end."[19] Despite Frank's candid insights, some journalists today are uncomfortable thinking of themselves as storytellers. Instead, they tend to describe themselves mainly as information-gatherers. But news does both things— gathering information and putting it in story form.

Of course, news doesn't tell just any kind of stories. Novels, television, and movies tell entertaining stories. Advertising tells sponsored stories. Public relations tells strategically purposeful stories. Propaganda tells intentionally misleading or harmful stories. **News** is defined here as the process of gathering information and making narrative reports—edited by journalists—that offer selected frames of reference; within those frames, news helps the public make sense of important events, political issues, cultural trends, prominent people, and unusual happenings in everyday life. What differentiates news stories from all other stories, then, is that they are reported, constructed, and edited by journalists, who follow certain professional standards and ideals.

In their classic book *The Elements of Journalism*, Bill Kovach and Tom Rosenstiel identify enduring core principles of journalism—elements that make journalism (the profession and the process that produces news) different from other kinds of communication. Their first five elements make clear the unique duties of journalism, as well as journalism's implicit bias: to help make public life go better.[20]

Mark Wilson/Getty Images

ONE ROLE of reporters is to serve as an independent monitor of power. Yamiche Alcindor, White House correspondent for *PBS NewsHour*, is a frequent questioner at White House press conferences.

- **"Journalism's first obligation is to the truth."** Journalists need to sort through the information and try to tell the most accurate version of the story possible (and follow up on that story to report new developments, if necessary). (See "Global Village: Authoritarians Use 'Fake News' Allegations as a Weapon" for a look at how some global leaders have questioned and countered this journalistic commitment to truth-telling.)
- **"Its first loyalty is to citizens."** Journalists are often confronted with competing loyalties to their news organization, their advertisers, and their sources, but they should consider citizens to be their ultimate boss. This means that they should do their reporting for *all* citizens, not just cater to a more affluent audience.
- **"Its essence is a discipline of verification."** In terms of reporting, journalists' practice insists on verification—confirming that what they've found is true. Verification, according to Kovach and Rosenstiel, is what the original concept of objectivity in journalism was meant to be—"a consistent method of testing information," not the unrealistic expectation that a human being can be neutral or free from all bias in reporting. As noted earlier, Kovach and Rosenstiel also make a distinction between the "journalism of assertion"—which often happens when TV pundits reinterpret parts of a news story for their own purposes—and the "journalism of verification," which is a painstaking process of getting the most reliable account of events.[21]
- **"Its practitioners must maintain an independence from those they cover."** Journalists need to disclose and avoid conflicts of interest (e.g., they shouldn't be reporting on a company in which they've invested or reviewing a theater production in which their friend is an actor).
- **"It must serve as an independent monitor of power."** According to an often-repeated 1902 saying from humorist Finley Peter Dunne, journalism should "comfort the afflicted and afflict the comfortable."[22] In the words of Kovach and Rosenstiel, that means journalism should give a voice to the voiceless and monitor the powerful (thereby fulfilling journalism's watchdog role).

Characteristics of News

Over time, a set of conventional criteria for determining **newsworthiness**—information most worthy of transformation into news stories—has evolved. Journalists are taught to select and develop news stories by relying on one or more of these criteria: timeliness, proximity, conflict, prominence, human interest, consequence, usefulness, novelty, and deviance.

Most issues and events that journalists select as news are *timely*, or *new*. Reporters, for example, cover speeches, meetings, crimes, and court cases that have just happened. In addition, most of these events have to occur close by, or in *proximity* to, readers and viewers.

Most news stories are narratives and thus contain a healthy dose of *conflict*—a key ingredient in narrative writing. In developing news narratives, reporters are encouraged to seek contentious quotes from those with opposing views. For example, stories on presidential elections almost always feature the most dramatic opposing Republican and Democratic positions. And many stories during the COVID-19 pandemic pitted the values of public health and safety against advocates of personal freedom and those with anti-government views.

Reader and viewer surveys indicate that most people identify more closely with other people than with abstract issues. Often this leads the news media to report stories that feature *prominent*, powerful, or influential people. Because these individuals often play a role in shaping the rules and values of a community, journalists have traditionally been responsible for keeping a watchful eye on them. Reporters also look for *human-interest* stories: extraordinary incidents that happen to "ordinary" people. In fact, reporters often relate a story about a complicated issue (such as

Authoritarians Use "Fake News" Allegations as a Weapon

In the Southeast Asian country of Myanmar, people of the Muslim Rohingya ethnic group have been persecuted, raped, and killed since 2016. In 2017, military forces and Buddhist extremists launched a genocidal campaign against the Rohingya, forcing more than 700,000 Rohingya to flee their towns and villages for Bangladesh and other countries in Southeast Asia.

But when a regional Myanmar state security minister was called to account for the genocide, the official responded, "There is no such thing as Rohingya. It is fake news."[1]

The use of the term "fake news," which was popularized by Donald Trump during his presidency, has "gone global," the *New York Times* reported, and "is now a cudgel for strongmen."[2] The phrase is used to discredit political opponents, restrict the freedom of news media, and jail perceived enemies. The *Times* reported, "Political leaders have invoked 'fake news' as justification for beating back media scrutiny" in countries including Angola, Somalia, Libya, Syria, Hungary, Poland, Turkey, Serbia, Russia, China, the Philippines, Thailand, Singapore, and Venezuela.[3]

Beyond accusing legitimate reports of being "fake news," authoritarian countries are using the phrase as a pretext for damaging the media and jailing journalists. According to the Freedom House—a nonprofit advocate of democracy and human rights—by 2018 "at least 17 countries [had] approved or proposed laws that would restrict online media in the name of fighting 'fake news' and online manipulation. Thirteen countries [had] prosecuted citizens for spreading allegedly false information."[4]

The idea of discrediting the press to suppress criticism has been around for more than a century. In Germany, *Lügenpresse* (lying press) was an insult aimed first at newspapers critical of the country's leaders during World War I

and then at newspapers critical of the rising Nazi regime in the following two decades. In recent years, the term has been resurrected by right-wing, xenophobic anti-immigrant groups in Germany.[5]

The list of dictators and autocrats who have attacked the news media with claims of "fake news" continues to grow:

- **China.** In 2017, "the Chinese state news agency denied a report that police had tortured Xie Yang, a human-rights activist, as 'essentially fake news.'"[6]
- **Hungary.** According to Bloomberg, "Hungarian authorities have started criminal probes against dozens of people for allegedly spreading fake news related to the pandemic, in a move that Prime Minister Viktor Orban's opponents say is meant to silence them." Authorities can jail people up to five years under the 2020 law.[7]
- **Russia.** The Russian Foreign Ministry now posts news stories it doesn't like (including those from U.S. news organizations) stamped with a giant red circle and the word "FAKE" in the middle.
- **Serbia.** After Kosovo—a former province of Serbia and now an independent state that Serbia does not recognize—announced its lifting of a tariff on imported Serbian goods, a Serbian government spokesperson denied it and said, "Don't get caught up in fake news."[8]
- **Syria.** President Bashar al-Assad rejected a report on the thousands who died in his military prisons. "You can forge anything these days. We are living in a fake-news era," he said.[9]
- **Venezuela.** President Nicolás Maduro accused the global media of bullying the country, noting, "This is what we call 'fake news' today, isn't it?"[10] ●

NEWSWORTHY

Employees of the *Washington Post* gather in the newsroom to celebrate its Pulitzer Prize winners in 2018. The newspaper won awards for its investigation of Russian interference in the 2016 election and its revelations of sexual misconduct allegations against candidate Roy Moore in the 2017 Senate election in Alabama. Both stories ranked high in several newsworthiness criteria.

unemployment, war, tax rates, health care, or homelessness) by illustrating its impact on an average person, family, or town.

Two other criteria for newsworthiness are *consequence* and *usefulness*. Many editors and reporters believe that some news must also be of consequence to a majority of readers or viewers. For example, stories about issues or events that affect a family's income or change a community's laws have consequence. Likewise, many people look for stories with a practical use: hints on buying a used car or choosing a college, strategies for training a pet or removing a stain.

Finally, news is often about the *novel* and the *deviant*. When events happen that are outside the routine of daily life, such as a seven-year-old girl attempting to pilot a plane across the country or an ex-celebrity involved in a drug deal, the news media are there. Reporters also cover events that appear to deviate from social norms, including murders, rapes, fatal car crashes, fires, and political scandals. For example, when a government report alleged that New York governor Andrew Cuomo undercounted the number of COVID-19 deaths in his state's nursing homes by several thousand, it represented the kind of deviant behavior that qualified as major news.

Values in American Journalism

Although newsworthiness criteria are a useful way to define which stories get covered, they do not reveal much about the cultural aspects of news. News is both a product and a process. It is a daily newsletter or push notification sharing breaking stories, as well as a set of subtle values and shifting rituals that have been adapted to historical and social circumstances, such as the partisan press values of the eighteenth century or the informational standards of the twentieth century.

As noted earlier, it remains a common cultural notion that journalists should be neutral and objective, but the very act of determining what is news—turning some events into reports and discarding many others—is a subjective process. Sociologist Herbert Gans, who studied the newsroom cultures of CBS, NBC, *Newsweek*, and *Time* in the 1970s, generalized that several basic "enduring values" have been shared by most American reporters and editors, which govern the stories they tell. The most prominent of these values, which persist to this day to varying degrees, are ethnocentrism, responsible capitalism, small-town pastoralism, and individualism.[23]

By **ethnocentrism**, Gans meant that in most news reporting, especially foreign coverage, reporters judge other countries and cultures on the basis of how "they live up to or imitate American practices and values." In identifying **responsible capitalism** as an underlying value, Gans contended that journalists sometimes naïvely assume that businesspeople compete with one another not primarily to maximize profits but "to create increased prosperity for all." Gans pointed out that although most reporters and editors condemn monopolies, "there is little implicit or explicit criticism of the oligopolistic nature of much of today's economy."[24]

Another value that Gans identified was the romanticization of **small-town pastoralism**: favoring the small over the large and the rural over the urban. Many journalists equate small-town life with innocence and harbor suspicions of cities, their governments, and urban experiences. Consequently, stories about rustic communities with crime or drug problems have often been framed as if the purity of country life had been contaminated by "mean" big-city values.[25]

Finally, **individualism**, according to Gans, remains the most prominent value underpinning daily journalism. Many idealistic reporters are attracted to this profession because it rewards the rugged tenacity needed to confront and expose corruption. Beyond this, individuals who overcome personal adversity are the subjects of many enterprising news stories.

Often, however, journalism that focuses on personal triumphs neglects to explain how large organizations and institutions work or fail. Many conventional reporters and editors are unwilling or unsure of how to tackle the problems raised by institutional decay. In addition, because they value their own individualism and are accustomed to working alone, many journalists dislike cooperating on team projects or participating in forums in which community members discuss their own interests and alternative definitions of news.[26]

Ethics and the News Media

Journalists frequently face ethical dilemmas. For example, when should reporters assert that a political figure lied, and what terminology should they use to do so—"lie" or "falsehood" or "baseless comment"? When should activists in the street be called "protesters," and when should they be called "rioters"? If violence is the standard for rioting, how much violence discredits all the other peaceful protesters? When should behavior, statements, or social practices be called "racist," and should terms like "racially charged" ever be used?[27] Editors need to weigh such decisions, since accuracy is paramount, but journalists also worry about charges of bias. (A follow-up question might be, bias against whom?)

What makes the predicament of these questions so tricky is that there are no easy-to-apply universal answers. Accurate and well-verified reporting is essential to finding the best version of the truth—"Stick to checking facts, rather than opinion," the *Associated Press Stylebook* recommends to reporters—but situations aren't always clear-cut. The *Stylebook* warns that "words can be true, false, exaggerated, a stretch, a selective use of data, partly or mostly true, etc.," so it is important to "use the most apt description that's supported by what the facts show."[28] Likewise, on the matter of identifying lies, the *Columbia Journalism Review* says that "calling a statement a 'deliberate falsehood' instead of a 'lie' isn't an attempt to excuse the behavior; it's an attempt to accurately, if perhaps overcautiously, describe what's going on."[29]

Figuring out how to select the words to convey the most accurate meaning is just one of the dilemmas journalists face. Other ethical quandaries can emerge in the everyday practice of reliable reporting. Professional codes of ethics can resolve some issues, but moral codes can also equip journalists with the needed structure to make ethical decisions.

Ethical Predicaments

What is the moral and social responsibility of journalists, not only for the stories they report but also for the actual events or issues they are shaping for millions of people? Wrestling with such media ethics involves determining the moral response to a situation through critical reasoning. Over decades, some of the most frequent ethical dilemmas encountered in newsrooms across the United States have involved intentional deception, privacy invasions, and conflicts of interest.

Deploying Deception

Ever since journalist Nellie Bly faked mental illness so she could report from inside an asylum in the 1880s, investigative journalists have used deception to get stories. Today, journalists continue to use disguises and assume false identities to gather information on social transgressions. Beyond legal considerations, though, a key ethical question comes into play: Does the end justify the means? For example, should a newspaper or TV newsmagazine use deceptive ploys to go undercover and expose a suspected fraudulent clinic that promises miracle cures at a high cost? Are news professionals justified in posing as clients desperate for a cure?

JOURNALIST NELLIE BLY dedicated her life to helping women and the poor as she laid the groundwork for what we know today as investigative journalism. Her undercover work—including time spent posing as a patient at the Women's Lunatic Asylum in New York City—exposed a need for reforms in the care of people with mental illnesses and economically marginalized members of society.

GRANGER - Historical Picture Archive

In terms of ethics, there are at least two major positions and multiple variations. At one end of the spectrum, *absolutist ethics* suggests that a moral society has laws and codes, including honesty, that everyone must live by. This means citizens, including members of the news media, should tell the truth at all times and in all cases. In other words, the ends (exposing a phony clinic) never justify the means (using deception to get the story). An editor who is an absolutist would cover this story by asking a reporter to find victims who have been ripped off by the clinic and then telling the story through their eyes. At the other end of the spectrum is *situational ethics*, which promotes ethical decisions on a case-by-case basis. If a greater public good could be served by using deceit, journalists and editors who believe in situational ethics would sanction deception as a practice.

Should journalists withhold information about their professional identity to get a quote or a story from an interview subject? Many sources and witnesses are reluctant to talk with journalists, especially about a sensitive subject that might jeopardize a job or hurt another person's reputation. Journalists know they can sometimes obtain information by posing as someone other than a journalist, such as a curious student or a concerned citizen.

Most newsrooms frown on such deception. In particular situations, though, such a practice might be condoned if reporters and their editors believed that the public needed the information. The ethics code adopted by the Society of Professional Journalists (SPJ) says to "avoid undercover or other surreptitious methods of gathering information unless traditional, open methods will not yield information vital to the public" (see Figure 14.1).

In one of the most noted cases of journalistic deception, courts ruled against ABC News after two producers misrepresented themselves and obtained jobs at the Food Lion grocery store chain in the Carolinas to document on video what the story alleged were unhealthy food practices in the meat department. In the suit, Food Lion did not dispute the report but charged that the journalists did not have permission to secretly videotape their store operations. In 1999, a U.S. court of appeals rejected the nearly $317,000 in punitive damages awarded to Food Lion in a lower court ruling but agreed that the journalists had trespassed and breached their duty of loyalty to their employer, upholding a nominal $2 award to the grocery chain.[30]

Invading Privacy

To achieve "the truth" or to "get the facts," journalists routinely straddle a line between "the public's right to know" and a person's right to privacy. In 2020, a new case involving these competing issues emerged in the Black Lives Matter protests that swept the nation. In cities across the country, some protesters called on photographers to not show their faces for fear of retaliation if images were posted on social media or the police used them for surveillance. The SPJ Code of Ethics urges reporters to "minimize harm." Yet news photographers have a legal right to photograph anything in public spaces. Legally, in the United States, "there is no such thing as being private in public," according to Frank LoMonte, the director of the University of Florida's Brechner Center for Freedom of Information. Akili Ramsess, executive director of the National Press Photographers Association, argues that protesters' faces are not only essential to the story but essential to the point of protests as well. "As photographers, we want the human connection. The whole purpose of demonstrations and civil disobedience is to put a human face on the issue and the best way to do that is to connect people to each other's humanity."[31]

Journalism's code of ethics says that journalists should "balance the public's need for information against potential harm or discomfort."[32] When these two ethical standards collide, should journalists err on the side of the public's right to know?

Conflict of Interest

Journalism's code of ethics also warns reporters and editors not to place themselves in positions that produce a **conflict of interest**—that is, any situation in which journalists may stand to benefit personally from stories they produce. "Refuse gifts, favors, fees, free travel and special treatment, and avoid political and other outside activities that may compromise integrity or impartiality, or may damage

Society of Professional Journalists

C⊙DE of ETHICS

PREAMBLE

Members of the Society of Professional Journalists believe that public enlightenment is the forerunner of justice and the foundation of democracy. Ethical journalism strives to ensure the free exchange of information that is accurate, fair and thorough. An ethical journalist acts with integrity.

The Society declares these four principles as the foundation of ethical journalism and encourages their use in its practice by all people in all media.

SEEK TRUTH AND REPORT IT

Ethical journalism should be accurate and fair. Journalists should be honest and courageous in gathering, reporting and interpreting information.

Journalists should:

▶ Take responsibility for the accuracy of their work. Verify information before releasing it. Use original sources whenever possible.

▶ Remember that neither speed nor format excuses inaccuracy.

▶ Provide context. Take special care not to misrepresent or oversimplify in promoting, previewing or summarizing a story.

▶ Gather, update and correct information throughout the life of a news story.

▶ Be cautious when making promises, but keep the promises they make.

▶ Identify sources clearly. The public is entitled to as much information as possible to judge the reliability and motivations of sources.

▶ Consider sources' motives before promising anonymity. Reserve anonymity for sources who may face danger, retribution or other harm, and have information that cannot be obtained elsewhere. Explain why anonymity was granted.

▶ Diligently seek subjects of news coverage to allow them to respond to criticism or allegations of wrongdoing.

▶ Avoid undercover or other surreptitious methods of gathering information unless traditional, open methods will not yield information vital to the public.

▶ Be vigilant and courageous about holding those with power accountable. Give voice to the voiceless.

▶ Support the open and civil exchange of views, even views they find repugnant.

▶ Recognize a special obligation to serve as watchdogs over public affairs and government. Seek to ensure that the public's business is conducted in the open, and that public records are open to all.

▶ Provide access to source material when it is relevant and appropriate.

▶ Boldly tell the story of the diversity and magnitude of the human experience. Seek sources whose voices we seldom hear.

▶ Avoid stereotyping. Journalists should examine the ways their values and experiences may shape their reporting.

▶ Label advocacy and commentary.

▶ Never deliberately distort facts or context, including visual information. Clearly label illustrations and re-enactments.

▶ Never plagiarize. Always attribute.

MINIMIZE HARM

Ethical journalism treats sources, subjects, colleagues and members of the public as human beings deserving of respect.

Journalists should:

▶ Balance the public's need for information against potential harm or discomfort. Pursuit of the news is not a license for arrogance or undue intrusiveness.

▶ Show compassion for those who may be affected by news coverage. Use heightened sensitivity when dealing with juveniles, victims of sex crimes, and sources or subjects who are inexperienced or unable to give consent. Consider cultural differences in approach and treatment.

▶ Recognize that legal access to information differs from an ethical justification to publish or broadcast.

▶ Realize that private people have a greater right to control information about themselves than public figures and others who seek power, influence or attention. Weigh the consequences of publishing or broadcasting personal information.

▶ Avoid pandering to lurid curiosity, even if others do.

▶ Balance a suspect's right to a fair trial with the public's right to know. Consider the implications of identifying criminal suspects before they face legal charges.

▶ Consider the long-term implications of the extended reach and permanence of publication. Provide updated and more complete information as appropriate.

ACT INDEPENDENTLY

The highest and primary obligation of ethical journalism is to serve the public.

Journalists should:

▶ Avoid conflicts of interest, real or perceived. Disclose unavoidable conflicts.

▶ Refuse gifts, favors, fees, free travel and special treatment, and avoid political and other outside activities that may compromise integrity or impartiality, or may damage credibility.

▶ Be wary of sources offering information for favors or money; do not pay for access to news. Identify content provided by outside sources, whether paid or not.

▶ Deny favored treatment to advertisers, donors or any other special interests, and resist internal and external pressure to influence coverage.

▶ Distinguish news from advertising and shun hybrids that blur the lines between the two. Prominently label sponsored content.

BE ACCOUNTABLE AND TRANSPARENT

Ethical journalism means taking responsibility for one's work and explaining one's decisions to the public.

Journalists should:

▶ Explain ethical choices and processes to audiences. Encourage a civil dialogue with the public about journalistic practices, coverage and news content.

▶ Respond quickly to questions about accuracy, clarity and fairness.

▶ Acknowledge mistakes and correct them promptly and prominently. Explain corrections and clarifications carefully and clearly.

▶ Expose unethical conduct in journalism, including within their organizations.

▶ Abide by the same high standards they expect of others.

The SPJ Code of Ethics is a statement of abiding principles supported by additional explanations and position papers (at spj.org) that address changing journalistic practices. It is not a set of rules, rather a guide that encourages all who engage in journalism to take responsibility for the information they provide, regardless of medium. The code should be read as a whole; individual principles should not be taken out of context. It is not, nor can it be under the First Amendment, legally enforceable.

FIGURE 14.1

SOCIETY OF PROFESSIONAL JOURNALISTS' CODE OF ETHICS

Courtesy Society of Professional Journalists (SPJ)

credibility."[33] Although small newspapers with limited resources and poorly paid reporters might accept such "freebies" as game tickets for their sportswriters and free meals for their restaurant critics, this practice does increase the likelihood of a conflict of interest that produces favorable or uncritical coverage.

On a broader level, ethical guidelines at many news outlets attempt to protect journalists from compromising positions. For instance, in most U.S. cities, journalists do not actively participate in politics or support social causes. Some journalists will not reveal their political affiliations, and some even decline to vote.

For these journalists, the rationale behind their decision is straightforward: Journalists should not place themselves in situations in which they might have to report on the misdeeds of an organization or a political party to which they belong. If a journalist has a tie to any group, and that group is later suspected of involvement in shady or criminal activity, the reporter's ability to report on that group would be compromised—along with the credibility of the news outlet for which the reporter works. Conversely, other journalists believe that not actively participating in politics or social causes means abandoning their civic obligations. They believe that fairness in their reporting, not total detachment from civic life, is their primary obligation.

Resolving Ethical Problems

When a journalist is criticized for ethical lapses or questionable reporting tactics, a typical response might be "I'm just doing my job" or "I was just getting the facts." Such explanations are troubling, though, because in responding this way, reporters are transferring personal responsibility for the story to a set of institutional rituals.

There are, of course, ethical alternatives to self-justifications such as "I'm just doing my job" that force journalists to think through complex issues. With the crush of deadlines and daily duties, most media professionals deal with ethical situations only on a case-by-case basis as issues arise. However, examining major ethical models and theories is a common strategy for addressing ethics on a general rather than a situational basis. The most well-known ethical standard, the Judeo-Christian command to "love your neighbor as yourself" (an idea that appears in most of the world's major religions), provides one foundation for constructing ethical guidelines. Although we cannot address all major moral codes here, a few key precepts can guide us:

- **Golden mean.** The Greek philosopher Aristotle offered an early ethical concept, the "golden mean"—a guideline for seeking balance between competing positions. For Aristotle, this was a desirable middle ground between extreme positions, usually with one regarded as deficient and the other as excessive. For example, Aristotle saw ambition as the balance between sloth and greed.
- **Categorical imperative.** Another ethical principle stems from the "categorical imperative," developed by German philosopher Immanuel Kant (1724–1804). This idea maintains that a society must adhere to moral codes that are universal and unconditional (that is, absolutist) and applicable in all situations at all times. With the categorical imperative, Kant asks us to choose the ethical rule for ourselves that we would want to have applied to everyone else. For example, do we really think it is acceptable to cheat on paying taxes? If everyone followed that rule, wouldn't society crumble under the weight of corruption? The First Amendment, which prevents Congress from abridging free speech and other rights, could be considered an example of an unconditional national law—although there is some speech, such as pornography, that federal law does not protect, bringing a measure of situational ethics to interpretations of the amendment.
- **Greatest good for the greatest number.** British philosophers Jeremy Bentham (1748–1832) and John Stuart Mill (1806–1873) promoted an ethical principle derived from "the greatest good for the greatest number," directing us "to distribute a good consequence to more people rather than to fewer, whenever we have a choice."[34]

Reporting Rituals and the Legacy of Print Journalism

Unfamiliar with being questioned themselves, many reporters are uncomfortable discussing their personal values or their strategies for getting stories. Nevertheless, a stock of rituals from the legacy of print journalism underlies the practice of mainstream news reporting today. These rituals include focusing on the present, relying on experts, balancing story conflict, and acting as adversaries toward leaders and institutions.

Focusing on the Present

In the 1840s, when the telegraph first enabled news to crisscross America instantly, modern journalism was born. To complement the new technical advances, editors called for a focus on the immediacy of the present. Modern front-page print journalism began to de-emphasize political analysis and historical context, accenting instead the new and the now.

Over time, the profession began drawing criticism for failing to offer historical, political, and social analyses. Journalism tends to reject "old news" for whatever new event or idea disrupts today's routines. For example, when several state governments moved to restrict mail-in voting in 2021, their stories were often reported in a vacuum, without much comparison to states like Colorado, which established a mail-in voting system in 2013—with bipartisan support—and has one of the highest voter registration and participation rates in the country.

Most news organizations still attend to the present in their stories, though the Internet has enabled them to include links to earlier reports, which provide some historical context for their audience. In keeping with this primary focus on the new and the now, journalists often emphasize two key elements: getting a good story and getting the story first. (To analyze current news stories, see "Media Literacy and the Critical Process: Telling Stories and Covering Disaster.")

Getting a Good Story

In 2018, the *New York Times* released a multipart podcast series called *Caliphate*, hosted by Rukmini Callimachi, the paper's star terrorism reporter. The podcast had a riveting real-life tale with an astonishing central character: a Canadian man who became radicalized and went to Syria to join ISIS, became one of their executioners, and then later escaped and returned to Canada. As Callimachi said in 2018, "It's an eye-opening account of his passage through the Islamic State."[35]

As it turned out, the man's story was a fraud, and by 2020, he was facing charges in Ontario for "perpetrating a terrorism hoax," NPR reported.[36] The *New York Times* reassigned Callimachi and returned its Peabody Award for the series. The newspaper's editors admitted that they were so taken by the unique story, they didn't diligently confirm the evidence. *New York Times* executive editor Dean Baquet said they fell for what they thought was a big story: "We were so in love with it that when we saw evidence that maybe he was a fabulist, when we saw evidence that he was making some of it up, we didn't listen hard enough."[37]

According to Don Hewitt, the late creator and longtime executive producer of *60 Minutes*, "There's a very simple formula if you're in Hollywood, Broadway, opera, publishing, broadcasting, newspapering. It's four very simple words—tell me a story."[38] But sometimes finding an unbelievable story can override concerns about accuracy—and the story may indeed turn out to be untrue.

Getting the Story First

In a discussion on public television about the press coverage of a fatal airline crash in Milwaukee in the 1980s, a news photographer was asked to talk about his role in covering the tragedy. Rather than discussing the poignant, heartbreaking aspects of witnessing the

THE *NEW YORK TIMES* multipart podcast series *Caliphate* conveyed a riveting story, but one of the central sources turned out to be a fraud.

MEDIA LITERACY & THE CRITICAL PROCESS

Telling Stories and Covering Disaster

Covering difficult stories—such as natural disasters like hurricanes or floods—may present challenges to journalists about how to frame their coverage. The opening sections, or leads, of news stories can vary depending on the source—print, broadcast, or online news—or even the editorial style of the news organization (e.g., some story leads are straightforward, whereas others are dramatic). And although journalists claim objectivity as a goal, it is unlikely that a professional in the storytelling business can approximate any sort of scientific objectivity. The best journalists can do is be fair, reporting and telling stories to their communities and nation by explaining the complicated and tragic experiences in words or pictures. To explore this type of coverage across various media outlets, compare versions of news stories about recent regional or national disasters.

① DESCRIPTION

Find print, online, and broadcast news versions of the same disaster story. Save copies of each story, and note the pictures chosen to tell the story.

② ANALYSIS

Find patterns in the coverage. How are the stories treated differently in print,

online, and on television? Are there similarities in the words chosen or the images used? What kinds of experiences are depicted? Who are the sources the reporters use to verify their information?

③ INTERPRETATION

What do these patterns suggest? Can you make any interpretations or arguments based on the kinds of disasters covered, the sources used, the areas covered, or the words and images chosen? How are the stories told in relation to their importance to the entire community or nation? How complex are the stories?

④ EVALUATION

Which stories are the strongest? Why? Which are the weakest? Why? Make a judgment on how well these disaster stories serve your interests as a citizen and the interests of the larger community or nation.

⑤ ENGAGEMENT

In an e-mail, share your findings with the relevant editors or TV news directors. Make suggestions for improved coverage, and cite strong stories that you admired. Share with the class how the editors and news directors responded.

aftermath of such an event, the excited photographer launched into a dramatic recounting of how he had slipped behind police barricades to snap the first grim photos, which later appeared in the *Milwaukee Journal*. As part of their socialization into the profession, reporters often learn not only to emotionally detach from a tragic event but also to evade authority figures in order to secure a story ahead of the competition.

The photographer's recollection points to the important role journalism plays in calling public attention to serious events and issues. Yet he also talked about the news-gathering process as a game that journalists play. It's now routine for local television stations, 24/7 cable news, newspapers, and online news organizations to run self-promotions about how they beat competitors to a story. In addition, during political elections, local television stations and networks project winners in particular races and often hype their projections when they are able to forecast results before the competition does (although on this front, they have become more cautious after incorrect projections in the 2016 presidential contest).

Journalistic *scoops* and exclusive stories attempt to portray reporters in a heroic light: They have won a race for facts, which they have gathered and presented ahead of their rivals. It is not always clear, though, how the public is better served by a journalist's claim to have gotten a story first. In some ways, 24/7 cable news, the Internet, and social media bloggers have intensified the race for getting a story first. With a fragmented audience and more media competing for news, mainstream news outlets can feel more pressure to lure an audience with "breaking news" stories—often through subscriptions to push notifications on mobile phones. Although readers and viewers might value the aggressiveness of reporters, the earliest reports are not necessarily better than, more accurate than, or as complete as stories written later, with more context and perspective.

This kind of scoop behavior, which has become rampant in the digital age, demonstrates pack or **herd journalism**, which occurs when reporters stake out a house; chase celebrities in packs; or follow a story in such herds that the entire profession comes under attack for invading people's privacy, exploiting their personal problems, or just plain getting the story wrong.

Relying on Experts

Another ritual of journalism—relying on outside sources—has made reporters heavily dependent on experts. Reporters, though often experts themselves in certain areas by virtue of having covered them over time, are not typically allowed to display their expertise overtly (although this seems to be shifting in some contexts, as we will address later in this section). Instead, they must seek outside authorities to give credibility to seemingly neutral reports. *What* daily reporters know is generally subordinate to *whom* they know.

The gap between those with expertise and those without it creates a need for reporters to serve as public mediators. With their access to experts, reporters transform specialized and insider knowledge into the everyday commonsense language of news stories. Reporters frequently use experts to create narrative conflict by pitting series of quotes against one another, and occasionally use experts to support a particular position. In addition, the use of experts enables journalists to distance themselves from daily experience; by doing so, they are able to attribute the responsibility for the events or issues reported in a story to those who are quoted.

To use experts, journalists must make direct contact with a source—by phone, text, e-mail, or video call, or in person. Journalists do not, however, heavily cite the work of other writers; that would violate reporters' desire not only to get a story first but to get it on their own. Interviews, rather than extensively researched interpretations, are the stuff of daily journalism.

The Lack of Diversity in Experts

A problem with using experts is that reporters on deadline often go back to the same list of familiar names in their phone or e-mail directories, and those sources have historically been predominantly male.[39] Gender and source representation numbers have not changed much, even recently. Adrienne LaFrance, a staff writer for the *Atlantic*, found this to be borne out in her own work in both 2013 and 2015: "Male dominance in global media is well documented, and has been for many decades. Both in newsrooms and in news articles, men are leaders—they make more money, get more bylines, spend more time on-camera, and are quoted far more often than women—by a ratio of about 3:1."[40] Other more recent research is consistent with these findings. A 2019 study found that men appeared more than twice as often as women in news photos on Facebook (a major source of news for many). Moreover, men appeared exclusively most often in photos for stories about the economy, sports, and immigration, whereas women appeared exclusively most often in photos for entertainment stories (but still less often than men).[41]

In addition, race seems to be a determinant in whether someone is consulted as an expert news source. A survey of media professionals in Minnesota by American Public Media found that 72 percent of the professionals contacted reported that people of color and Indigenous peoples are "rarely or never" consulted "as subject matter experts for stories *that are not explicitly about race and culture*."[42] Similarly, an NPR study found that "sources heard on the newsmagazines *Morning Edition* and *All Things Considered* were 70 percent male and 73 percent white, and about 40 percent came from Washington, D.C., California or New York combined." NPR reported that the problem is "media-wide" and prompted the creation of more diverse source banks and "discussion around newsrooms' unconscious biases."[43]

These numbers should not be surprising when we look at recent data on the number of women and people of color in print and broadcast newsrooms. A key 2019 Women's Media Center study reported that "women comprised 41.7 percent and people of color 22.6 percent of the overall

MEGAN TWOHEY, one of the *New York Times* reporters who reported a devastating account of the sexual abuse allegations against movie producer Harvey Weinstein in 2017, became a frequently interviewed source herself on the topic after winning a Pulitzer Prize and writing a book with reporting teammate Jodi Kantor titled *She Said: Breaking the Sexual Harassment Story That Helped Ignite a Movement.*

workforce in those responding newsrooms."[44] Similarly, as Nikole Hannah-Jones, racial injustice reporter for the *New York Times Magazine*, noted, "It's rare to see a woman of color in a mainstream newsroom in a very high management position, and, for those who are there, it's [often] been a struggle."[45] It is reasonable to conclude that if more women and people of color were in charge of newsrooms, more news sources would be women and people of color, and news would have a wider range of perspectives.

When Journalists Become the Experts

One interesting shift in the use of experts in recent years has been cable news, radio, and podcast programming that use journalists themselves as experts. Where the usual approach in some of these settings is the journalism of assertion, identified by its glib pundits, the discussion can be much more representative of the journalism of verification when reporters appear as guests to explain what they have uncovered. The *New York Times*, like other major news organizations, has rules for its journalists appearing in other media. "Staff members should avoid strident, theatrical forums that emphasize punditry and reckless opinion-mongering. Instead, we should offer thoughtful and retrospective analysis. Generally a staff member should not say anything on radio, television or the Internet that could not appear under his or her byline in The Times," the *Times* notes in its ethics handbook.[46] The *New York Times* showcases its own reporters, and often their audio recordings from the field, as the main content in episodes of *The Daily*, the newspaper's popular podcast.

Balancing Story Conflict

For most journalists, *balance* means presenting all sides of an issue without appearing to favor any one position. The quest for balance presents problems for journalists. On the one hand, time and space constraints do not always permit the presentation of *all* sides; in practice, this value has often been reduced to "telling *both* sides of a story." In recounting news stories as two-sided dramas, reporters often misrepresent the complexity of social issues. The abortion controversy, for example, is often treated as a story that pits two extreme positions (staunchly pro-life versus resolutely pro-choice) against each other. Yet people whose views fall somewhere between these positions are seldom represented (and studies show that this group actually represents the majority of Americans). In this manner, "balance" becomes a narrative device to generate story conflict, but it actually moves the story away from the most accurate representation.

On the other hand, although many journalists claim to be detached, they often stake out a moderate or middle-of-the-road position between the two sides represented in a story. In claiming neutrality and inviting readers to share their "detached" point of view, journalists offer a distant, third-person, all-knowing point of view (a narrative device that many novelists use as well), enhancing the impression of neutrality by making the reporter appear value-free (or valueless).

The claim for balanced stories, like the claim for neutrality, disguises the narrative aspects of journalists' work. After all, when choosing quotes for a story, reporters usually go with the most dramatic or conflict-oriented remarks that emerge from an interview, press conference, or public meeting. Choosing quotes sometimes has more to do with enhancing drama than with being fair, documenting an event, or establishing neutrality.

Acting as Adversaries

The value that many journalists take the most pride in is their adversarial relationship with the prominent leaders and major institutions they cover. The prime narrative frame for portraying this relationship is sometimes called a *gotcha story*, which refers to the moment when, through questioning, the reporter nabs "the bad guy," or wrongdoer.

This narrative strategy—part of the *tough questioning style* of some reporters—is frequently used in political reporting. Many journalists assume that leaders are hiding something and that the reporter's main job is to ferret out the truth through tenacious fact-gathering and gotcha questions. This stance locates the reporter on the citizens' side ("us") versus the political or business leaders' side ("them"). Many TV stations actually have an "on your side" reporter or team to advocate for viewers, like the 8 On Your Side team for KFLA News in Tampa Bay.

Critics of the tough questioning style of reporting argue that while it can reveal significant information, when overused it fosters a cynicism among journalists that actually harms the democratic process. Although journalists need to guard against becoming too cozy with their political sources, they sometimes go to the other extreme. By constantly searching for what politicians may be hiding, some reporters may miss other issues or other key stories. When journalists employ the gotcha model to cover news, being tough often becomes an end in itself. Thus, reporters believe they have done their job just by roughing up an interview subject. That is the moment the news becomes more performative than substantial, and not in service to citizens.

The Rise of Conservative Media and the Idea of "Fake News"

As we've seen throughout the chapter, during much of the twentieth century there was a general consensus that the American news media should be neutral in their reporting. Newspapers developed reputations as more conservative or more liberal based on the opinion pieces on their op-ed pages, but there was a shared belief across the political spectrum that democracy required news sources that adhered to standard journalistic practices in reporting—the journalistic practices we've just examined. This shared belief started to break down gradually, and now, in our current media environment, we face the twenty-first century predicament of American culture: a partisan divide and an inability for many citizens to even agree on the same set of facts.

As much as this partisan divide reflects twenty-first century attitudes toward media and journalism, it has its origins more than fifty years ago, when a group of conservative media figures, politicians, and activists reasoned that they could avoid mainstream media coverage that they considered unfair by building their own media system—a conservative media system—as a competing voice. The national mainstream media (that is, the traditional news gatekeepers) has largely understood its mission to be one that is not in service of a particular political party, organizing reporting around professional practices designed to limit political partisanship. But by design, the conservative news media that these early proponents envisioned, and ultimately realized, would be explicitly linked to the Republican Party and tied to conservative ideals.

One of the main objectives of this right-wing media—a controversial one, to be sure—was to work in counterpoint to the mainstream news media and label it as "liberal news" in support of the Democratic Party, thus damaging its perceived independence and credibility (the neutrality that was so highly valued by members of the mainstream press).

In the 1960s, according to media historian Nicole Hemmer, "conservatives began an active campaign against liberal bias, organized by groups like the Committee to Combat Bias in Broadcasting and Reed Irvine's Accuracy in Media."[47] There is a clear through line from that work, which framed mainstream media as "liberal," to the messages that emerged from Rush Limbaugh in the 1980s, Fox News in the 1990s, and Donald Trump in the 2010s. In 2016, for example, reporter Leslie Stahl of the CBS program *60 Minutes* asked then president-elect Trump why he kept "hammering" at the press. According to Stahl, Trump candidly replied, "You know why I do it? I do it to discredit you all and demean you all so that when you write negative stories about me no one will believe you."[48] Media scholars Kathleen Hall Jamieson and Joseph Cappella describe the results of such a strategy: "These conservative voices portray[ed] themselves as the reliable, trustworthy alternative to mainstream media, while at the same time attacking 'liberals' and dismissing or reframing information that undercut . . . conservative leaders or causes."[49]

Because of the significant impact conservative media has had on this country and the national conversation, considering its history and exploring its effects, as we will do here, can help us better understand both journalism and culture in the United States today—including the challenge of living in a society with competing media systems that focus on different approaches to news, different values, and even different truths.

And what of a comparable system on the left? Although there are certain liberal periodicals (such as the *Nation*, *Mother Jones*, *In These Times*, and *Jacobin*), blogs and sites like *HuffPost* and Democracy Now!, and some left-leaning opinion hosts on cable news channel MSNBC, there has not been a concerted effort by liberal activists, media figures, and politicians to create a separate media system on the left that would set itself in opposition to the mainstream media and rival the media ecosystem of the right. Although the media on the left have a liberal perspective, they remain largely committed to the journalism of verification.

The Origins of Conservative Media

Back in 1973, Nixon aide Patrick Buchanan (later White House communications director for Ronald Reagan) alluded to the idea of conservative cable television as an answer to what he called the "liberal bias" of the mainstream media. He wrote, "Dr. Milton Friedman, distinguished economist and disciple of Adam Smith, believes the long-range answer lies in more and competing channels of communication, to be achieved partly through the rapid expansion of cable television. But that is a long-range solution."[50] In just a few decades, as it turned out, conservatives like Buchanan would realize their dream for a powerful alternative media environment.

Early Foundations

As Nicole Hemmer explains in her book *Messengers of the Right: Conservative Media and the Transformation of American Politics*, the earliest activists in conservative media started in the 1940s and 1950s, establishing conservative magazines, publishing houses, and television and radio programs to spark a political movement.[51] But their foundational work remained mostly on the fringe.

A second wave of right-wing media network-building—one that began in the space of traditional media—got its start in 1976 with Rupert Murdoch's purchase of the *New York Post*, a newspaper that had been established in 1801 by Alexander Hamilton. As *Wall Street Journal* journalist and author Sarah Ellison noted, "Murdoch was unafraid to use his media outlets, particularly his tabloid newspapers, as instruments of influence."[52] As he had done in Britain and Australia before, Murdoch used the *Post* to boost his favored candidates.

Conservative media continued to develop in 1977 with the founding of CBN, the Christian Broadcasting Network. As former conservative media activist David Brock explained, the cable network and its program *The 700 Club* "moved beyond religious proselytizing to booking right-wing politicians and activists as talking heads."[53] CBN founder Pat Robertson became a 1988 Republican presidential candidate himself, and continued to wield political power with his program and the Christian Coalition (founded 1989), which still works to support conservative politicians today.

Conservative Radio Goes National

Beginning in 1987, right-wing media became much more powerful with nationwide conservative talk radio. From 1949 to 1987, the Federal Communications Commission had enforced a policy called the Fairness Doctrine, which allowed "a station to editorialize, provided it made air time available for the 'balanced presentation of all responsible viewpoints on particular issues.'"[54] But when President Ronald Reagan's FCC dropped the Fairness Doctrine, it changed the political tone of talk radio, swinging it from the middle to the right.[55]

Rush Limbaugh, talk radio's biggest star until his death in 2021, was the first conservative talk-radio host to go national in 1988. Author David Foster Wallace described Limbaugh as "a host of extraordinary, once-in-a-generation talent and charisma."[56] Limbaugh's constant political attacks on President Bill Clinton and his support of conservative causes on his national media platform helped Republicans gain control of the U.S. House of Representatives in 1994. Republicans acknowledged their debt to Limbaugh by naming him an "honorary member of their class."[57]

PRESIDENT TRUMP awarded talk-radio host Rush Limbaugh the Presidential Medal of Freedom at the 2020 State of the Union Address, thanking him for "the millions of people a day that you speak to and that you inspire." Limbaugh, who suffered from lung cancer at the time, died a year later.

Fox News Comes to TV

Perhaps the biggest institution in right-wing media is Fox News, the cable network founded by Rupert Murdoch in 1996 and run by Roger Ailes until his resignation in 2016. Ailes, who died in 2017, was a former Republican media operative who got his start with Richard Nixon's campaign. As NPR media correspondent David Folkenflik explained, "Above all, Ailes wanted Fox News to referee Republican Party politics."[58] He did this by putting former GOP politicians on the Fox payroll as commentators or show hosts. In announcing Ailes's resignation, Rupert Murdoch said, "Roger shared my vision of a great and independent television organization and executed it brilliantly over 20 great years. Fox News has given voice to those who were ignored by the traditional networks and has been one of the great commercial success stories of modern media."[59] Murdoch maintained control of Fox News after Ailes's departure, moving into Ailes's old office to personally run the network.[60] Even after Murdoch made a deal to sell 21st Century Fox to Disney for $52.4 billion in late 2017, Murdoch retained some properties, including Fox News.

Conservative Internet News Flourishes

Conservative activists have also used the Internet to effectively further the conservative message. One of these activists is Matt Drudge, who began the Drudge Report in 1994 and gained national notice a few years later, particularly with the leak of a journalist's story about Bill Clinton's affair with a White House intern. One of Drudge's assistants was Andrew Breitbart, who in 2007 began his own conservative Internet media sites that still feed the *echo chamber*—the idea that polarized media mutually reinforce the same political messaging.

Steve Bannon, a founding member of Breitbart's board, became executive chair after Andrew Breitbart's death in 2012. At Breitbart, Bannon was an early booster of Donald Trump's candidacy and joined his presidential campaign in August 2016 as chief executive. According to Jane Mayer in the *New Yorker*, Bannon was already an unofficial player in the Trump campaign: "A year before Bannon joined Trump's campaign staff, he described himself in [an] e-mail as Trump's de-facto 'campaign

BREITBART NEWS, founded in 2007, was called "the platform for the alt-right" in 2016 by Steve Bannon, then executive chair of the site.[69] Breitbart continues its alliance with political figures on the right and its mission to disparage those on the left.

manager,' because of the positive coverage that Breitbart was giving Trump," Mayer noted.[61] After Trump's election, Bannon served as chief strategist and senior counselor to President Trump until he was pushed out by chief of staff John F. Kelly in August 2017. Bannon didn't lose a step and immediately resumed his position as executive chair at Breitbart, which continues to function as a media outlet and conservative agitator.[62]

Stoking White Working-Class Grievances

Rush Limbaugh was the "first great promulgator of the Mainstream Media's Liberal Bias [MMLB] idea," David Foster Wallace argued, calling the strategy "brilliantly effective." As Wallace explained, "The MMLB concept functioned simultaneously as a standard around which Rush's audience could rally . . . and as a mechanism by which any criticism or refutation of conservative ideas could be dismissed."[63] Media historian Susan Douglas noted that Limbaugh tapped into "a sense among many Americans, and especially many men, that they were not being addressed or listened to by the mainstream media."[64] And there was an opportunity: Mainstream U.S. newspapers, especially as they became part of larger newspaper chains in the late 1960s and early 1970s, decided to stop focusing on a mass audience in their communities and instead appeal to a more upscale readership to please advertisers.[65] Conservative media wooed the abandoned working class—more specifically, the white working class.

Limbaugh articulated this idea in one of his many attacks: "The Media is now considered just another part of the arrogant, condescending, elite, and out-of-touch political structure which has ignored the people and their concerns and interests."[66] Thus, Limbaugh cultivated the notion that the mainstream media were part of the political structure that alienated white working-class Americans and held himself up as their chief advocate and ally.

Fox News tapped into the same grievances. Bill O'Reilly, once one of Fox News's biggest stars, described a typical complaint: "The no-spin truth is that the elite media think the degeneration of American popular culture is beneath them and not very important."[67] This "culture wars" element (grievances about the degeneration of various parts of American popular culture, which resonated strongly with members of the white working class) became crucial to Fox, Limbaugh, and others in the conservative media complex. Because of their fiscally conservative political position and that of their Republican sponsors and owners, they focused less on economic policy that directly addressed the problems of the working class—things like the merits of increasing the minimum wage, publicly funded health care, and full support for post–high school education for working-class people—and more on cultural wedge issues embraced by white conservative working-class voters, such as the "war" on Christmas, the importance of not taking a knee during the national anthem at football games, and the evils of immigration. Journalist Joshua Green reported that Andrew Breitbart used to say, "I want to change the cultural narrative." As Green explained, Breitbart "was less interested in trying to influence Washington directly than he was in going after the institutions (and the methods they employed, like 'political correctness') that he believed shaped this narrative."[68]

Allegations of "Fake News"

According to Jim Rutenberg, a *New York Times* writer, Donald Trump inherited a media ecosystem "tailor-made" for him.[70] Trump took his cues and style from conservative media, which had a built-in political base ready to activate. Trump also became part of the conservative media ecosystem himself, with his own printing press of sorts: his Twitter account, with tens of millions of followers.

With his Twitter account, Trump could communicate in short one-way blasts, taking down critics and praising his supporters, particularly those in the media. In a *New York Times* analysis of Trump's first 11,000 tweets as president (from January 20, 2017, to October 15, 2019), attacks on news organizations accounted for 1,308 tweets (nearly 12% of all tweets). Thirty-six of those tweets charged the news media as the "enemy of the people." But not all media organizations received such ire. In 758 tweets, Trump "praised or promoted Fox News and other conservative media," the *Times* reported.

After decades in which the conservative media—including Rush Limbaugh, Fox News, and Breitbart, among many others—discredited the mainstream media, it was an easy leap for Trump to use the term "fake news" in attempts to completely "delegitimize information or to delegitimize the positions of his opponents," according to Sam Woolley, director of propaganda research at the University of Texas at Austin's Center for Media Engagement. Trump used the term "fake news" about nine hundred times on Twitter during his four years as president.[71] (For more on fake news, including its origins, see "A Guide to Identifying Fake News" later in the text.) Although the content of Trump's tweets had long violated Twitter's user rules, it wasn't until two days after the January 6, 2021, insurrection at the Capitol—an act for which Trump was impeached a second time—that Trump was permanently suspended from Twitter. Yet the rest of the conservative media ecosystem carried on.

As we consider our current media environment, questions arise that deserve our consideration. Among them: How has the expansion of a conservative media ecosystem changed how Americans think of the mainstream media and how they approach politics? Should there be a comparable media system on the left, aligned with the Democratic Party? Would more diversity in news media ownership help to ensure more diversity in media coverage across a range of political perspectives?

TWITTER announced it was permanently suspending Donald Trump on January 8, 2021 (two days after the assault on Congress), "due to the risk of further incitement of violence."[72]

Robert Alexander/Archive Photos/Getty Images

Journalism's Role in Democracy

A journalism of verification is central to democracy: Both citizens and the media must have access to the information needed to make important decisions. As this chapter illustrates, however, this is a complicated idea. For example, mainstream media and conservative media—sharing space next to each other on cable television, on the Internet, and in social media—came up with entirely different ways of framing the largely peaceful activities of Black Lives Matter protesters in the summer of 2020, and framing the appropriate response to the COVID-19 pandemic. The basic principles of democracy require citizens and the media to question our leaders and government. But our split media system suggests very different ways to be a "patriotic" citizen.

Conventional journalists will fight ferociously for the principles that underpin journalism's basic tenets—freedom of the press, the obligation to question government, the public's right to know, and the belief that there are two sides to every story. These are mostly worthy ideals, but they do have limitations. For example, these tenets are coming under criticism, because taking a "neutral" stance does not acknowledge any moral or ethical duty for journalists to improve the quality of daily life. (See "Examining Ethics: Black Journalists Lead an 'Overdue Reckoning' on Objectivity.") Such conventional journalism values its news-gathering capabilities and the well-constructed news narrative but leaves the improvement of civic life to political groups, nonprofit organizations, business philanthropists, and individual citizens.

Black Journalists Lead an "Overdue Reckoning" on Objectivity

I t was a simple headline, but it got to the heart of recent criticisms of mainstream, objective-style journalism: "Journalists need to remember that not all news readers are white." The headline came from an article in *Nieman Reports*, the publication of the Nieman Foundation for Journalism at Harvard, which has a mission "to promote and elevate the standards of journalism."

The "standards of journalism" in this instance are journalism's long-lived notion of impartiality, neutrality, and objectivity. For Marc Lacey, national editor at the *New York Times* and the author of the article, it begins with whom the journalist imagines as the reader. "For too long, readers in the mainstream media have been presumed to be white," Lacey wrote. "I notice it regularly as I survey the news. White is the norm." Often, writers point out when someone is Black. Or Hispanic. Or Asian. But white subjects are rarely identified racially. Lacey says this kind of reporting comes from "a blind spot that lingers from the days when newsrooms were all white and readerships were presumed to be the same."[1]

As Black journalists covered the Black Lives Matter protests in Ferguson, Missouri, in 2014 and in Minneapolis and across the country in 2020, they brought to the forefront questions about the long-cherished values of impartiality, neutrality, and objectivity: Impartial from what perspective? Neutral from what center point? Objective by being detached from whose lives?

Wesley Lowery, a two-time Pulitzer Prize–winning journalist who works for CBS News (and formerly for the *Washington Post*), says that the concerns of Black journalists about the lack of diversity in newsrooms and how news is covered have been ignored for years. The protests of 2020 brought an "overdue reckoning" on the treatment of those journalists and how objectivity is defined. Mainstream journalism "considers objective truth to be decided almost exclusively by white reporters and their mostly white bosses. And those selective truths have been calibrated to avoid offending the sensibilities of white readers," Lowery says.[2]

Lowery advocates a rethinking of "neutral objectivity" in favor of "moral clarity." Instead of hiding facts in deferential euphemisms, journalists should speak the truth. In an example that emerges from the news about racial justice, Lowery says that instead of using terms like "officer-involved shooting," journalists should more clearly say what is simple and more accurate: "The police shot someone."[3] The problem, Lowery says, is that when judging whether a sentence, paragraph, or story will be objective to a reader, that reader "is invariably assumed to be white."[4] Journalism from a perspective of moral clarity should be able to call out racism and violations to human rights.

Ben Smith, the media columnist for the *New York Times*, says the business model of newspaper journalism in which readers pay for subscriptions, rather than having to appeal to "skittish advertisers" worried about offending customers, is enabling Lowery and others to push for new reporting standards. Smith says the shift "now feels irreversible."[5]

Lowery is leading the charge for change. "American view-from-nowhere, 'objectivity'-obsessed, both-sides journalism is a failed experiment," Lowery tweeted. "We need to fundamentally reset the norms of our field. . . . We need to rebuild our industry as one that operates from a place of moral clarity."[6] ●

Social Responsibility

Although reporters have traditionally thought of themselves first and foremost as observers and recorders, some journalists have acknowledged a social responsibility. Among them was James Agee in the 1930s. In his book *Let Us Now Praise Famous Men*, which was accompanied by the Depression-era photography of Walker Evans, Agee said that he regarded conventional journalism as dishonest, partly because the act of observing intruded on people and turned them into story characters that newspapers and magazines then exploited for profit.

Agee also worried that readers would retreat into the comfort of his writing—his narrative—instead of confronting what for many families was the horror of the Great Depression. For Agee, the question of responsibility extended not only to journalism and to himself but to the readers of his stories as well: "The reader is no less centrally involved than the authors and those of whom they tell."[73] Agee's self-conscious analysis provides insights into journalism's hidden agendas and the responsibility of all citizens to make public life better.

Making the Eagle Fly

Critics of journalism argue that when reporters are chiefly concerned with maintaining their antagonistic relationship to politics and are less willing to improve political discourse, news and democracy suffer. The late *Washington Post* columnist David Broder thought that national journalists like him—through rising salaries, prestige, and formal education—distanced themselves "from the people that [they] are writing for and have become much, much closer to people [they] are writing about."[74] Broder believed that journalists need to become activists, not for a particular party but for the political process and in the interest of reenergizing public life.

Yet mainstream journalism has to tackle an identity crisis created by decades of being discredited by a powerful and partisan competing media system. As noted in Chapter 8, there was a time in the early decades of the nation's history when a partisan press flourished, with some newspapers directly affiliating themselves with one of the many competing political parties. In addition, the early partisan press often slung stinging barbs at their opposition. But that is where the comparisons of the early partisan press and today's conservative media largely end.

Leading up to the 1800 election, James Callender—a Scottish immigrant and newspaper editor—wrote in his publication, "Take your choice, between Adams, war and beggary, and Jefferson, peace, and competency."[75] His favored candidate, Thomas Jefferson ended up winning the election, but not before the incumbent president John Adams jailed Callender for sedition for his disrespectful writing, which would be considered mild by today's standards. (Callender later turned on Jefferson and published rumors that the president had children with an enslaved woman, Sally Hemings.) At that time, politics and the partisan press were a rich man's game; only white male elites could vote, and there was no mass medium beyond print.

Now we live at a time when nearly all adults have the franchise to vote, and there are only two major political parties in the United States. When one news media system—the mainstream media—works largely in a realm of journalism of verification, and a competing media system—the conservative media—works largely in a realm of journalism of assertion, with an added political allegiance reminiscent of the old partisan press, we have arrived at a place where many Americans cannot agree on basic facts or a national consensus narrative going forward.

A functioning journalism system is essential to a functioning political system, across the ideological spectrum. The long-time journalist and political commentator Bill Moyers once said, "I've always thought the American eagle needed a left wing and a right wing . . . both would keep the great bird on course." But, Moyers said, when the left wing and the right wing aren't working in the same way, that bird won't fly.[76] We can ask ourselves: What should the news media do to help make our democracy function better? What is our responsibility, as citizens, and how can we encourage and support the news media to better serve us all?

Chapter 14 REVIEW

LAUNCHPAD FOR *MEDIA & CULTURE*

Visit LaunchPad for *Media & Culture* at launchpadworks.com for additional learning tools:

➜ QUIZ YOURSELF WITH LEARNINGCURVE

LearningCurve uses gamelike quizzing to help you review important concepts from this chapter. LearningCurve includes multiple-choice questions at different levels of difficulty, hints, feedback for both right and wrong answers, and links to e-book material for easy reference. Which answer choice would you select for the sample question below?

> Brooke is an editor for a weekly newspaper. At the weekly editorial meeting, she assigns a reporter to cover the story of a local family who started a 5K run to raise money for ALS research following the loss of their young son to the disease. Which criterion is Brooke most likely meeting as she assigns this story?
>
> ○ usefulness
>
> ○ prominence
>
> ○ human interest
>
> ○ deviance

➜ WATCH VIDEO ON KEY CONCEPTS

LaunchPad includes clips from movies, TV shows, online sources, and other media texts, in addition to interviews with media experts and newsmakers. In the videos for this chapter, you'll consider topics like facts versus opinions, objectivity in journalism, and journalistic ethics.

➜ PRACTICE THE CRITICAL PROCESS

Use the Media Literacy Activity to walk through the critical process step-by-step and develop a critical perspective on what media means to you. In this chapter's activity, you'll take a closer look at expert sources cited in reporting.

➜ LEARN KEY TERMS

Review the definitions for this chapter's key terms with LaunchPad's easy-to-use online flash cards. Try the cards in practice mode, or quiz yourself as a way to focus your study efforts. (Page numbers indicate where each term is highlighted within the chapter.)

journalism of verification, 406
journalism of assertion, 406
news, 407
newsworthiness, 408

ethnocentrism, 410
responsible capitalism, 410
small-town pastoralism, 410
individualism, 410

conflict of interest, 412
herd journalism, 417

REVIEW QUESTIONS

Use the questions below to revisit key themes and concepts from this chapter.

 **Need more practice?** Access the LearningCurve multiple-choice questions for this chapter, which are available on LaunchPad. All questions include feedback for correct and incorrect answers, hints, and links to e-book material.

Journalism Today

1. What do journalism organizations like and not like about having their stories linked in social media sites?

How We Got Here: The Evolution of American Journalism

2. How did the ethic of objectivity and impartiality develop?

3. How does television news derive its credibility, and how is that different from print news?

4. How have news consultants shaped the style and content of U.S. television news?

5. What are some of the visual images of television news that are embedded in American collective memory?

6. Why do many cable TV operations opt for more programs with pundits and political talk, rather than reporting?

7. What is the difference between the journalism of verification and the journalism of assertion?

8. How has news in the United States lost its traditional gatekeepers?

The Essential Elements of News

9. What are five essential elements of the news?

10. What are some of the important values in American journalism?

Ethics and the News Media

11. How do issues such as deception and privacy present ethical problems for journalists?

12. How can journalists avoid conflicts of interest?

13. What are the recommendations of Aristotle, Kant, and Bentham and Mill for resolving ethical dilemmas?

Reporting Rituals and the Legacy of Print Journalism

14. Why is getting a story first important to reporters?

15. Why have reporters become so dependent on experts?

The Rise of Conservative Media and the Idea of "Fake News"

16. Why did a conservative media system develop?

17. What is the reason for stoking white working-class grievances in conservative news?

18. What types of political leaders are charging that reports they don't like are "fake news"?

Journalism's Role in Democracy

19. Why can't Americans seem to agree on simple sets of facts?

QUESTIONING THE MEDIA

Use these critical-thinking questions to reflect on your own experiences with media and apply your understanding of media concepts.

 **For an interactive experience,** questions on topics like these are available for the iClicker student response system. Instructors can download the questions from LaunchPad and use them to poll the class.

1. What are your main criticisms of the state of news today? In your opinion, what are the news media doing well?

2. Is having both a mainstream news media and a competing conservative news media a sustainable information system for the United States?

3. Does the conservative media's approach count as journalism? Explain.

4. How would you go about changing journalism to make it more credible to a wider audience?

5. For a reporter, what are the dangers of both detachment from and involvement in public life?

6. How should news organizations ensure that there is more diversity in news sources?

Media Effects and Cultural Approaches to Research

15

Is it ever possible to depict suicide in a television show without also glamorizing it? That was the predicament for the creators of *13 Reasons Why*, a 2017 series on Netflix adapted by Brian Yorkey from the 2007 debut young-adult novel of the same name by Jay Asher.

The story, which spins out in thirteen episodes in its first Netflix season, follows teenager Clay Jensen (Dylan Minnette) as he returns home from school to find a mysterious box bearing his name on his porch. Inside he discovers cassette tapes recorded by Hannah Baker (Katherine Langford)—his classmate and crush—who tragically died from suicide two weeks earlier. On tape, Hannah explains that there are thirteen reasons why she decided to end her life. The reasons catalog betrayals, bullying, slut shaming, binge drinking, drunk driving, drug use, and rape. Hannah's suicide is depicted in the final episode. A reviewer for the *Guardian* concluded that the series was "too bleak to binge,"[1] but plenty of people did binge it—all the way through its concluding fourth season in 2020.

Like the book, the series prompted debate about a variety of issues—including bullying, depression, sexual consent, drug and alcohol abuse, and self-harm—as well as a number of research studies by scholars interested in exploring the meaning and influence of the show. From a *cultural studies perspective* (one of the two main approaches to conducting media research that we'll discuss in this chapter), one can look at the meaning of the show as a *text*—for example, by considering what the show says about sexual violence as a factor affecting young women's mental health. To do this, one study analyzed the show's themes of "toxic masculinity," "slut-shaming," and "the failure of adult systems to adequately respond to youth."[2] Other cultural studies approaches might look at how viewers responded to the show, or at the forces that brought this adaptation of the novel to Netflix, with musician and performer Selena Gomez as one of the show's producers.

◀ Hit Netflix original series *13 Reasons Why* raises important questions about teenage suicide. But does it go too far?

429

From a *media effects perspective* (the other approach to conducting media research that we'll discuss in this chapter, which focuses on how media affect individuals and society), there was concern that portrayals of suicide might glamorize suicide and induce copycats. In fact, the producers of the show later dropped the suicide scene from the first season's final episode in response to criticism.

Most of the social commentary about the show focused on potential media effects. For example, Mark Henick, a mental health advocate, argued that TV programs like *13 Reasons Why* can have several problematic features in their portrayal of suicide, including simplifying or romanticizing it, or portraying it as a viable option instead of seeking proper care for mental health issues. Graphic representations of suicide can also harm viewers, especially young and impressionable ones.[3]

In 2017, the *Atlantic* reported on a study in which public health researchers found that "Google queries about suicide rose by almost 20 percent in 19 days after the show came out, representing between 900,000 and 1.5 million more searches than usual regarding the subject."[4] The concern among many experts is that although the television series can increase awareness of the tragedy of suicide, it can also idealize it, sparking a *contagion effect*—a rise in the number of suicide attempts.

The show's creators defended the program and its story.[5] Nevertheless, about two months after the 2017 release of *13 Reasons Why* (which was already rated TV-MA), Netflix added stronger advisory warnings at the beginning of certain episodes and supplementary content to its companion 13ReasonsWhy.info website. The site contains videos addressing a number of the show's disturbing topics, such as sexual assault; a discussion guide; and links to mental health resources for help, including the Crisis Text Line and the National Suicide Prevention Lifeline.

LaunchPad
macmillan learning

launchpadworks.com

Suicide on TV
This clip from the first season of Netflix's *13 Reasons Why* introduces main character Clay Jensen.

Discussion: What messages about suicide do you think this clip sends to an audience?

Dan Reidenberg, psychologist and executive director of the national organization Suicide Awareness Voices of Education (SAVE), weighed in about the impact of the show: "Although it's created a conversation about suicide, it's not the right conversation." When Reidenberg was contacted by Netflix for guidance before the release of *13 Reasons Why,* he recommended that the company not release the show. That, of course, did not happen. In response, SAVE issued tips for discussing *13 Reasons Why* for people to use and share as a way to create the right conversation. The first point was this: "*13 Reasons Why* is a fictional story based on a widely known novel and is meant to be a cautionary tale."[6]

 How effective are programs like *13 Reasons Why* in bringing public attention to important social issues? Can you think of other media content—television shows, movies, music, digital games, or social media posts—that has fostered important discussions effectively?

Research on Media Today

Debates over television programs like *13 Reasons Why* are based on the idea that media have a powerful effect on individuals and society and are central to the ways we learn, work, and play. Especially since the advent of electronic media—including film, radio, and television—researchers have tried to understand the role of media in our lives. The Internet and social media have further expanded media research, especially when all of us can now be media creators as well as media users. Two major traditions of research have developed over the last century to investigate media's impact: media effects research and cultural studies research.

Media effects research attempts to understand, explain, and predict the effects of mass media on individuals and society and relies primarily on tools rooted in the scientific method. One of the first questions to animate this research was whether there is a connection between aggressive behavior and violence in the media, particularly in children and teens. In the late 1960s, government leaders—reacting to the social upheavals of that decade—set aside $1 million to examine this potential connection. Since that time, thousands of studies have told us what most teachers and parents believe instinctively: Violent scenes on television and in movies stimulate aggressive behavior in children and teens—especially young boys. This approach is based on the research traditions of psychology.

The other major area of media research is **cultural studies**, which developed later in the twentieth century as an alternative response to media effects research. This approach—which we first introduced in Chapter 1—tries to understand the complex relationship among media texts, the people who consume them, the institutions that produce them, the technologies used to create and distribute them, and the culture within which all the other factors exist. It uses a variety of research tools that emerged from sociology, history, political science, economics, and literary studies. Cultural studies scholars also examine how more powerful groups in society, including corporate and political elites, use media to circulate their messages and sustain a status quo that serves their interests.

In this chapter, we will:

- Examine the evolution of media research over time
- Focus on the two major strains of media research, investigating the strengths and limitations of each
- Conclude with a discussion of how media research interacts with democratic ideals

How We Got Here: Early Types of Media Research

In the early days of the United States, philosophical and historical writings tried to explain the nature of news and print media. For instance, the French political philosopher Alexis de Tocqueville, author of *Democracy in America*, noted differences between French and American newspapers in the early 1830s:

> In France the space allotted to commercial advertisements is very limited, and . . . the essential part of the journal is the discussion of the politics of the day. In America three quarters of the enormous sheet are filled with advertisements, and the remainder is frequently occupied by political intelligence or trivial anecdotes; it is only from time to time that one finds a corner devoted to passionate discussions like those which the journalists of France every day give to their readers.[7]

During most of the nineteenth century, media analysis was based on moral and political arguments, as demonstrated by the Tocqueville quote.[8]

More scientific approaches to mass media research did not begin to develop until the late 1920s and 1930s. In 1920, Walter Lippmann's *Liberty and the News* called on journalists to operate more like scientific researchers in gathering and analyzing factual material. Lippmann's next book, *Public Opinion* (1922), was the first to apply the principles of psychology to journalism. Described by media historian James Carey as "the founding book in American media studies,"[9] it led to an expanded understanding of the effects of the media, emphasizing data collection and numerical measurement. According to media historian Daniel Czitrom, by the 1930s "an aggressively empirical spirit, stressing new and increasingly sophisticated research techniques, characterized the study of modern communication in America."[10] Czitrom traces four trends between 1930 and 1960 that contributed to the rise of media effects research: propaganda analysis, public opinion research, social psychology studies, and marketing research.

Propaganda Analysis

After World War I, some media researchers began studying how governments used propaganda to advance the war effort. They found that during the war, governments routinely relied on propaganda divisions to spread "information" to the public. According to Czitrom, though propaganda was considered a positive force for mobilizing public opinion during the war, researchers after the war labeled propaganda negatively, calling it "partisan appeal based on half-truths and devious manipulation of communication channels."[11] Harold Lasswell's important 1927 study *Propaganda Technique in the World War* focused on propaganda in the media, defining it as "the control of opinion by significant symbols, . . . by stories, rumors, reports, pictures and other forms of social communication."[12] **Propaganda analysis** thus became a major early focus of mass media research.

Public Opinion Research

Researchers soon went beyond the study of war propaganda and began to focus on more general concerns about how the mass media filtered information and shaped public attitudes. In the face of growing media influence, Walter Lippmann distrusted the public's ability to function as knowledgeable citizens as well as journalism's ability to help the public separate truth from lies. In promoting the place of the expert in modern life, Lippmann celebrated the social scientist as part of a new expert class that could best make "unseen facts intelligible to those who have to make decisions."[13]

By the 1950s, *public opinion research*, conducted by social scientists, had become especially influential during political elections. On the upside, public opinion research on diverse populations provided insight into citizen behavior and social differences, especially during election periods or following major national events. On the downside, journalism became increasingly dependent on polls, particularly for political insight. Some critics argue that this heavy reliance on measured public opinion has adversely affected the active political involvement of American citizens. For example, many people do not vote because they have seen or read poll projections and have decided that their votes will not make a difference.

Another problem is the pervasive use of unreliable **pseudo-polls**, typically call-in, online, or person-in-the-street polls that the news media use to address a "question of the day." The National Council on Public Polls notes that "unscientific pseudo-polls are widespread and sometimes entertaining, but they never provide the kind of information that belongs in a serious report," and discourages news media from conducting them.[14]

Ethan Miller/Getty Images

PUBLIC OPINION RESEARCH Public opinion polls suggest that the American public's attitude toward same-sex marriage has evolved. Just weeks before the Supreme Court ruled same-sex marriage legal nationwide, a 2015 Pew Research poll reported that 57 percent of Americans were in favor of it—the same percentage of people who opposed it in a poll back in 2001. By 2020, 67 percent of Americans supported same-sex marriages.

Social Psychology Studies

While opinion polls measure public attitudes, *social psychology studies* measure the behavior and cognition of individuals. The most influential early social psychology studies, the Payne Fund Studies, encompassed a series of thirteen research projects conducted by social psychologists between 1929 and 1932. Named after the private philanthropic organization that funded the research, the Payne Fund Studies were a response to a growing national concern about the effects of motion pictures, which had become a popular pastime for young people in the 1920s. These studies, which were later used by some politicians to attack the movie industry, linked frequent movie attendance to juvenile delinquency, promiscuity, and other antisocial behaviors, arguing that movies took "emotional possession" of young filmgoers.[15]

In one of the Payne studies, for example, children and teenagers were wired with electrodes and galvanometers, mechanisms that detected any heightened response via the subject's skin. The researchers interpreted changes in the skin as evidence of emotional arousal. In retrospect, the findings hardly seem surprising: The youngest subjects in the group had the strongest reaction to violent or tragic movie scenes, whereas the teenage subjects reacted most strongly to scenes with romantic and sexual content. The researchers concluded that films could be dangerous for young children and might foster sexual promiscuity among teenagers. The conclusions of this and the other Payne studies contributed to the establishment of the Motion Picture Production Code, which tamed movie content from the 1930s through the 1950s (see Chapter 16). As forerunners of today's TV violence and aggression research, the Payne Fund Studies became the model for media effects research. (See Figure 15.1 for one example of a contemporary policy that developed out of this type of media research.)

Marketing Research

A fourth influential area of early media research, *marketing research*, developed when advertisers and product companies began conducting surveys on consumer buying habits in the 1920s. The

SOCIAL AND PSYCHOLOGICAL EFFECTS OF MEDIA Concerns about film violence are not new. The 1930 movie *Little Caesar* follows the career of gangster Rico Bandello (played by Edward G. Robinson, shown), who kills his way to the top of the crime establishment and gets the girl as well. The Motion Picture Production Code, which was established a few years after this movie's release, reined in sexual themes and profane language, set restrictions on film violence, and attempted to prevent audiences from sympathizing with bad guys like Rico.

FIGURE 15.1

TV PARENTAL GUIDELINES

The TV industry continues to study its self-imposed rating categories, promising to fine-tune them to ensure that the government keeps its distance. These standards are one example of a policy that was shaped in part by media research. Since the 1960s, research has attempted to demonstrate links between violent TV images and increased levels of aggression among children and adolescents.

Data from: TV Parental Guidelines Monitoring Board, accessed March 14, 2021, www.tvguidelines.org/resources/TheRatings.pdf.

The following categories apply to programs designed solely for children:

TV-Y All Children
This program is designed to be appropriate for all children. Whether animated or live-action, the themes and elements in this program are specifically designed for a very young audience, including children from ages 2–6. This program is not expected to frighten young children.

TV-Y7 Directed to Older Children
This program is designed for children age 7 and above. It may be more appropriate for children who have acquired the developmental skills needed to distinguish between make-believe and reality. Themes and elements in this program may include mild fantasy violence or comedic violence, or may frighten children under the age of 7. Therefore, parents may wish to consider the suitability of this program for their very young children.

TV-Y7-FV Directed to Older Children— Fantasy Violence
For those programs where fantasy violence may be more intense or more combative than other programs in this category, such programs will be designated **TV-Y7-FV**.

The following categories apply to programs designed for the entire audience:

TV-G General Audience
Most parents would find this program suitable for all ages. Although this rating does not signify a program designed specifically for children, most parents may let younger children watch this program unattended. It contains little or no violence, no strong language, and little or no sexual dialogue or situations.

TV-PG Parental Guidance Suggested
This program contains material that parents may find unsuitable for younger children. Many parents may want to watch it with their younger children. The theme itself may call for parental guidance and/or the program may contain one or more of the following: some suggestive dialogue (D), infrequent coarse language (L), some sexual situations (S), or moderate violence (V).

TV-14 Parents Strongly Cautioned
This program contains some material that many parents would find unsuitable for children under 14 years of age. Parents are strongly urged to exercise greater care in monitoring this program and are cautioned against letting children under the age of 14 watch unattended. This program may contain one or more of the following: intensely suggestive dialogue (D), strong coarse language (L), intense sexual situations (S), or intense violence (V).

TV-MA Mature Audiences Only
This program is specifically designed to be viewed by adults and therefore may be unsuitable for children under 17. This program may contain one or more of the following: crude indecent language (L), explicit sexual activity (S), or graphic violence (V).

---

emergence of commercial radio led to the first ratings system, which measured how many people were listening on a given night. By the 1930s, radio networks, advertisers, large stations, and advertising agencies all subscribed to ratings services. However, compared with print media, whose circulation departments kept careful track of customers' names and addresses, radio listeners were more difficult to trace. This problem precipitated the development of increasingly sophisticated marketing research methods to determine consumer preferences and media use, such as direct-mail diaries, television meters, phone surveys, telemarketing, and Internet tracking. In many instances, product companies paid consumers nominal amounts of money to take part in these studies.

Research on Media Effects

As concern about public opinion, propaganda, and the impact of media merged with the growth of journalism and mass communication departments in colleges and universities, media researchers looked more and more to behavioral science as the basis of their research. Between 1930 and 1970, as media historian Daniel Czitrom has noted, "Who says what to whom with what effect?" became the key question "defining the scope and problems of American communications research."[16] In addressing this question specifically, media effects researchers asked follow-up questions, such as, If children watch a lot of TV cartoons (stimulus or cause), will this repeated act influence their behavior toward their peers (response or effect)? For most of the twentieth century, media researchers and news reporters used different methods to answer similar sets of questions—who, what, when, and where—about our daily experiences (see "Media Literacy and the Critical Process: Wedding Media and the Meaning of the Perfect Wedding Day").

MEDIA LITERACY & THE CRITICAL PROCESS

Wedding Media and the Meaning of the Perfect Wedding Day

According to media researcher Erika Engstrom, the bridal industry in the United States generates $50 to $70 billion annually, with more than two million marriages a year.[1] Supporting that massive industry are books, magazines, websites, reality TV shows, and digital games (in addition to fictional accounts in movies and music) that promote the idea of what a "perfect" wedding should be. What values are wrapped up in these wedding narratives?

1 DESCRIPTION

Select three or four types of wedding media, then compare them. Possible choices include magazines such as *Brides*, *Bridal Guide*, and *Martha Stewart Weddings*; reality TV shows like *Say Yes to the Dress*, *Four Weddings*, and *Bride by Design*; websites like The Knot, Southern Bride, and WeddingWire; and games like *My Fantasy Wedding*, *Wedding Dash*, and *Imagine Wedding Designer*.

2 ANALYSIS

What patterns do you find in the wedding media? (Consider what isn't depicted as well as what is.) Are there limited ways in which femininity is defined? Do men have an equal role in the planning of wedding events? Are weddings depicted as something just for people who identify as heterosexual? Do the wedding media presume that weddings are first-time experiences for the couple getting married? What seem to be the standards in terms of consumption—the expense, size, and number of things to buy and rent to make a "perfect" day?

3 INTERPRETATION

What do the wedding media seem to say about what it is to be a woman or a man on one's wedding day? Do the wedding media always segment people into the female–male binary in their stories, or do they ever consider nonbinary genders? What do gender roles for the wedding itself suggest about appropriate gender roles for married life after the wedding? What do the wedding media infer about the appropriate level of consumption?

In other words, consider the role of wedding media in constructing *hegemony*. In their depiction of what makes a perfect wedding, do the stories attempt to get us to accept the dominant cultural values related to things like gender relations and consumerism?

4 EVALUATION

Come to a judgment about the wedding media analyzed. Are they good or bad regarding certain elements? Do they promote gender equality? Do they promote marriage equality (that is, LGBTQ+ marriage)? Do they offer alternatives for having a "perfect" day without buying into all the trappings of traditional weddings?

5 ENGAGEMENT

Talk to friends about what weddings are supposed to celebrate and whether an alternative conception of a wedding would be a better way to celebrate a union of two people. (In real life, if there is discomfort in talking about alternative ways to celebrate a wedding, that's probably the pressure of hegemony. Why is that pressure so strong?) Share your criticisms and ideas on wedding websites as well.

Early Theories of Media Effects

A major goal of scientific research is to develop theories or laws that can consistently explain or predict human behavior. The diverse ways in which people create media content and the varied impacts of the mass media, however, tend to defy predictable rules. Historical, economic, and political factors influence media industries, making it difficult to develop systematic theories that explain communication. Researchers developed a number of small theories, or models, that help explain individual behavior rather than the impact of the media on large populations. But before these small theories began to emerge in the 1970s, mass media research followed several other models. Developing between the 1930s and the 1970s, these major approaches included the hypodermic-needle model, the minimal-effects model, and the uses and gratifications model.

The Hypodermic-Needle Model

One of the earliest media theories attributed powerful effects to the mass media. A number of intellectuals and academics were fearful of the influence and popularity of film and radio in the 1920s and 1930s. Some social psychologists and sociologists who arrived in the United States after fleeing

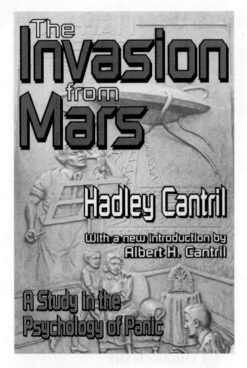

MEDIA EFFECTS?

In *The Invasion from Mars: A Study in the Psychology of Panic*, Hadley Cantril (1906–1969) argued against the hypodermic-needle model as an explanation for the panic that broke out after the *War of the Worlds* radio broadcast. A lifelong social researcher, Cantril also did a lot of work in public opinion research, even working with the government during World War II.

Germany and Nazism in the 1930s had watched Hitler use radio, film, and print media as propaganda tools. They worried that the popular media in America also had a strong hold on vulnerable audiences. The concept that powerful media affect weak audiences has been labeled the **hypodermic-needle model**, sometimes also called the *magic bullet theory* or the *direct-effects model*. It suggests that the media shoot their potent effects directly into unsuspecting victims.

One of the earliest challenges to this theory involved a study of Orson Welles's legendary October 30, 1938, radio broadcast of *War of the Worlds*, which presented H. G. Wells's Martian invasion novel in the form of a news report and frightened many listeners who didn't realize it was fictional (see Chapter 5). In a 1940 book-length study of the broadcast, *The Invasion from Mars: A Study in the Psychology of Panic*, radio researcher Hadley Cantril argued that contrary to expectations based on the hypodermic-needle model, not all listeners thought the radio program was a real news report. Instead, Cantril—after conducting personal interviews and a nationwide survey of listeners, and analyzing newspaper reports and listener mail to CBS Radio and the FCC—noted that although some did believe the radio report to be real (mostly those who missed the disclaimer at the beginning of the broadcast), the majority reacted due to the collective panic of others, not out of a gullible belief in anything transmitted through media. Although the hypodermic-needle model has been disproved over the years by social scientists, many people still attribute direct effects to the mass media, particularly in the case of children.

The Minimal-Effects Model

Cantril's research helped lay the groundwork for the **minimal-effects model**, or *limited model*. With the rise of empirical research techniques, social scientists began discovering and demonstrating that media alone cannot cause people to change their attitudes and behaviors. Based on tightly controlled experiments and surveys, researchers argued that people generally engage in **selective exposure** and **selective retention** with regard to the media. That is, people expose themselves to the media messages that are most familiar to them, and they retain the information that confirms the values and attitudes they already hold. Minimal-effects researchers have argued that in most cases, mass media reinforce existing behaviors and attitudes rather than change them. The findings from the first comprehensive study of children and television—by Wilbur Schramm, Jack Lyle, and Edwin Parker in the late 1950s—best capture the minimal-effects theory:

> For some children, under some conditions, some television is harmful. For other children under the same conditions, or for the same children under other conditions, it may be beneficial. For most children, under most conditions, most television is probably neither particularly harmful nor particularly beneficial.[17]

In addition, in important research published in *The Effects of Mass Communication* in 1960, Joseph Klapper found that the mass media influenced only those individuals who did not already hold strong views on an issue and had a greater impact on low-income and uneducated audiences. Solidifying the minimal-effects argument, Klapper concluded that strong media effects occur largely at an individual level and do not appear to have large-scale, measurable, and direct effects on society as a whole.[18]

The Uses and Gratifications Model

Like the minimal-effects model, the **uses and gratifications model** contested the notion of a passive media audience, instead considering people to be active agents in their media use. Under this model, researchers—usually using in-depth interviews to supplement survey questionnaires—studied the ways in which people used the media to satisfy various emotional or intellectual needs. Instead of asking, "What effects do the media have on us?" researchers asked, "Why do we use the media?"

Asking the *why* question enabled media researchers to develop inventories cataloging how people employed the media to fulfill their needs. For example, researchers noted that some individuals used the media to see authority figures elevated or toppled, to seek a sense of community and connectedness, to fulfill a need for drama and stories, and to confirm moral or spiritual values.[19]

Although the uses and gratifications model addressed the *functions* of the mass media for individuals, it did not address important questions related to the impact of the media on society. That meant that once researchers had accumulated substantial inventories of the uses and functions of media, they often did not move in new directions. Consequently, uses and gratifications did not become a dominant theory in media research. This model has received new attention in recent years, however, as researchers have begun to inventory some of the uses and gratifications of newer technologies like the Internet and social media—for example, the fact that these media can be used as tools for maintaining relationships with others.

Conducting Media Effects Research

Media effects research generally comes from either the private or the public sector, and each type has distinguishing features. *Private research*, sometimes called *proprietary research*, is generally conducted for a business, a corporation, or even a political campaign. It is usually applied research in the sense that the information it uncovers typically addresses some real-life problem or need, such as determining which campaign slogan would be most effective for a political candidate. *Public research*, in contrast, usually takes place in academic and government settings. It involves information that is often more *theoretical* than applied; it tries to clarify, explain, or predict the effects of mass media, such as the effects of social media use on civic and political participation.

Most media effects research today focuses on the effects of the media in such areas as learning, attitudes, aggression, and voting habits. This research employs the **scientific method**, a blueprint long used by scientists and scholars to study phenomena in systematic stages. The steps in the scientific method include the following:

1. Identifying the research problem
2. Reviewing existing research and theories related to the problem
3. Developing working *hypotheses*, or predictions, about what the study might find
4. Determining an appropriate method or research design
5. Collecting information or relevant data
6. Analyzing results to see if the hypotheses have been verified
7. Interpreting the implications of the study to determine whether they explain or predict the problem

The scientific method emphasizes *objectivity* (minimizing bias and judgments on the part of researchers), *reliability* (getting the same answers or outcomes from a study or measure during repeated testing), and *validity* (demonstrating that a study actually measures what it claims to measure).

In scientific studies, researchers pose one or more **hypotheses**: tentative general statements that predict the influence of an *independent variable* (the probable media cause) on a *dependent variable* (the thing being affected, such as an attitude or a behavior). For example, a researcher might hypothesize that frequent TV viewing among adolescents (independent variable) causes poor academic performance (dependent variable). Or another researcher might hypothesize that playing first-person shooter video games (independent variable) is associated with aggression in children (dependent variable).

J. R. Eyerman/The LIFE Picture Collection/Getty Images

USES AND GRATIFICATIONS

In 1952, audience members at the Paramount Theater in Hollywood donned 3-D glasses for the opening-night screening of *Bwana Devil*, the first full-length color 3-D film. The uses and gratifications model of research investigates the appeal of mass media, such as going out to the movies.

Broadly speaking, the methods for studying media effects on audiences have taken two forms—experiments and survey research. To supplement these approaches, researchers also use content analysis to count and document specific messages that circulate throughout mass media.

Experiments

Like all studies that use the scientific method, **experiments** in media research isolate some aspect of content; suggest a hypothesis; and manipulate variables to discover a particular medium's effect on attitude, emotion, or behavior. To test whether a hypothesis is true, researchers expose an *experimental group*—the group under study—to a selected media program or text. To ensure valid results, researchers also use a *control group*, which serves as a basis for comparison; this group is not exposed to the selected media content. Subjects are placed in either the experimental or the control group through **random assignment**, which simply means that each subject has an equal chance of being placed in either group.

For instance, to test the effects of violent films on preadolescent boys, a research study might take a group of ten-year-olds and randomly assign them to two groups. Researchers expose the experimental group to a violent action movie that the control group does not see. Later, both groups are exposed to a staged fight between two other boys so that the researchers can observe how each group responds to an actual physical confrontation. Researchers then determine whether or not there is a statistically measurable difference between the two groups' responses to the fight. For example, perhaps the control subjects try to break up the fight but the experimental subjects do not. Because the groups were randomly selected and the only measurable difference between them was the viewing of the movie, researchers may conclude that under these conditions, the violent film caused the difference in behavior (see the "Bobo doll" experiment photos under "Today's Leading Media Effects Theories").

When experiments carefully account for independent variables through random assignment, they generally work well to substantiate direct cause-effect hypotheses. Such research takes place both in laboratory settings and in field settings, where people can be observed using the media in their everyday environments. In field experiments, however, it is more difficult for researchers to control variables. In lab settings, researchers have more control, but other problems may occur. For example, when subjects are removed from the environments in which they regularly use media, they may act differently—often with fewer inhibitions—than they would in their everyday surroundings.

Experiments have other limitations as well. First, they are not generalizable to a larger population; they cannot tell us whether cause-effect results can be duplicated in settings or conditions that differ from those in the experiment. Second, most academic experiments today are performed on college students, who are convenient subjects for research but are not representative of the general public. Third, while most experiments are fairly good at predicting short-term media effects under controlled conditions, they do not predict how subjects will behave months or years later in the real world.

Survey Research

In the simplest terms, **survey research** is the collecting and measuring of data taken from a group of respondents. Using random sampling techniques aimed at producing a subject pool that is as representative of the population under investigation as possible (for example, all likely voters in Tennessee), this research method draws on much larger populations than those used in experimental studies. Surveys may be conducted through direct mail, personal interviews, telephone calls, e-mail, and websites, enabling survey researchers to accumulate large amounts of information by surveying diverse cross sections of people. These data help researchers examine demographic factors such as educational background, income level, race, ethnicity, gender, age, sexual orientation, and political affiliation, along with questions directly related to the survey topic.

Two other benefits of surveys are that the results can be generalizable to the larger society (if the sample is adequately large and representative), and that surveys can enable researchers to investigate populations in long-term studies. For example, survey research can measure subjects when they are ten, twenty, and thirty years old to track changes in how frequently they watch television and what kinds of programs they prefer at different ages. In addition, large government and academic survey databases are now widely available and contribute to the development of more long-range or **longitudinal studies**, which make it possible for social scientists to compare new studies with those conducted years earlier.

Like experiments, surveys have several drawbacks. First, survey investigators cannot account for all the variables that might affect media use; therefore, they cannot show cause-effect relationships. Survey research can, however, reveal **correlations**—or associations—between two variables. Second, the validity of survey questions is a chronic problem for survey practitioners. Surveys are only as good as the wording of their questions and the answer choices they present.

Content Analysis

Media researchers have also developed a social scientific method that allows them to analyze media messages. Known as **content analysis**, this method is a systematic approach to *coding* (or categorizing) and measuring media content in order to identify and quantify different types of media texts.

Although content analysis can be used to systematically code and measure any kind of media content, most studies have focused on television, film, and the Internet. Probably the most influential content analysis studies were conducted by George Gerbner and his colleagues at the University of Pennsylvania. Beginning in the late 1960s, they coded and counted acts of violence on network television. Combined with surveys, their annual "violence profiles" showed that heavy watchers of television, ranging from children to retired Americans, tend to overestimate the amount of violence that exists in the actual world.[20]

The limits of content analysis, however, have been well documented. First, this technique alone does not measure the effects of the messages on audiences; Gerbner needed to add surveys of TV viewers to his studies to suggest how violent media content may have affected them. Nor does it explain how those messages are presented. For example, a content analysis sponsored by the Kaiser Family Foundation examined more than eleven hundred television shows and found that 70 percent featured sexual content.[21] However, the study didn't explain how viewers interpreted the content or the context of the messages.

Second, problems of definition occur in content analysis. For instance, in the case of coding and counting acts of violence, how do researchers distinguish slapstick cartoon aggression from the violent murders or rapes in an evening police drama? Critics point out that such varied depictions may have diverse and subtle effects that are difficult to differentiate in content analysis.

Third, as content analysis grew to be a primary tool in media research, it sometimes pushed to the sidelines other ways of thinking about television and media content (see "Cultural Approaches to Media Research" later in this chapter). Broad questions concerning the media as a popular art form, as a measure of culture, as a democratic influence, or as a force for social control are difficult to address through strict measurement techniques. (In other words, it's difficult to tell whether social media environments like Facebook or Twitter are negatively affecting our culture or democracy by, say, counting how many posts containing disinformation users view each day.) Critics of content analysis, in fact, have objected to the kind of social science that reduces culture to acts of counting. Such criticism has addressed the tendency by some researchers to favor measurement accuracy over intellectual discipline and inquiry.[22]

Today's Leading Media Effects Theories

By the 1960s, the first departments of mass communication began graduating PhD-level researchers schooled in experiment and survey research techniques as well as content analysis. These researchers began documenting consistent patterns in mass communication and developing new theories. Five of the most influential theories that help explain media effects today are social learning theory, agenda-setting, the cultivation effect, the spiral of silence, and the third-person effect.

Social Learning Theory

Some of the most well-known studies that suggest a link between the mass media and behavior are the "Bobo doll" experiments, conducted on children by psychologist Albert Bandura and his colleagues at

LaunchPad
macmillan learning

launchpadworks.com

Media Effects Research
Experts discuss how media effects research informs media development.

Discussion: Why do you think the question of media's effects on children has continued to be such a big concern among researchers?

Courtesy of Albert Bandura

Stanford University in the 1960s. Bandura concluded that the experiments demonstrated a link between violent media programs, such as those on television, and aggressive behavior. Bandura developed **social learning theory** (later modified and renamed *social cognitive theory*) as a four-step process:

1. *Attention:* The subject must attend to the media and witness the aggressive behavior.
2. *Retention:* The subject must retain the memory for later retrieval.
3. *Motor reproduction:* The subject must be able to physically imitate the behavior.
4. *Motivation:* There must be a social reward or reinforcement to encourage modeling of the behavior.

Supporters of social learning theory often cite real-life imitations of media aggression as evidence of social learning theory at work. Yet critics note that many studies conclude just the opposite—that there is no link between media content and aggression. For example, millions of people have watched episodes of *How to Get Away with Murder* and *Breaking Bad* without subsequently exhibiting aggressive behavior. According to these critics, social learning theory simply makes television, film, and other media scapegoats for larger social problems relating to violence. Others suggest that experiencing media depictions of aggression can actually help viewers let off steam peacefully through a catharsis effect.

Agenda-Setting

A key phenomenon posited by contemporary media effects researchers is **agenda-setting**: the idea that when the mass media focus their attention on particular events or issues, they determine—that is, set the agenda for—the major topics of discussion for individuals and society. Essentially, agenda-setting researchers have argued that the mass media do not so much tell us what to think as *what to think about.* Traceable to Walter Lippmann's notion in the early 1920s that the media "create pictures in our heads," the first investigations into agenda-setting began in the 1970s.[23]

Over the years, agenda-setting research has demonstrated that the more stories the news media do on a particular subject, the more importance audiences attach to that subject. For instance, when the media seriously began to cover ecology issues after the first Earth Day in 1970, a much higher percentage of the population began listing the environment as a primary social concern in surveys. More recently, a 2020 report from the Pew Research Center found that "compared with a decade ago, more Americans today say protecting the environment and dealing with global climate change should be top priorities for the president and Congress." For example, in 2020, 64 percent of U.S. adults said "protecting the environment" should be a top government priority, compared to 40 percent in 2011. Moreover, 52 percent of U.S. adults said "dealing with global climate change" should be a top government priority, compared to 26 percent in 2011.[24] The survey data of people's attitudes toward climate change fit the agenda-setting theory because there was an increase in national media coverage of climate change in 2019, the period right before the survey was conducted.[25]

The Cultivation Effect

Another mass media phenomenon is the **cultivation effect**, an area of media effects research that has pushed researchers to focus not only on how media influence the way we feel or behave but also on how media affect the way we perceive reality. The cultivation effect suggests that the more time individuals spend viewing television and absorbing its viewpoints, the more likely their views of social reality will be "cultivated" by the images and portrayals they have seen.[26]

The major research in this area grew from the attempts of George Gerbner and his colleagues to make generalizations about the influence of televised violence. As a result of their efforts, the researchers concluded that although fewer than 1 percent of Americans are victims of violent crime in any single year, people who watch a lot of television tend to overestimate this percentage. Such exaggerated perceptions, Gerbner and his colleagues argued, are part of a "mean world" syndrome, in which viewers with heavy, long-term exposure to television violence are more likely to believe that the external world is a mean and dangerous place.

According to the cultivation effect, media messages interact in complicated ways with personal, social, political, and cultural factors; they are one of a number of important factors in determining individual behavior and defining social values. Some critics have charged that cultivation research has provided limited evidence to support its findings. In addition, some have argued that the cultivation effects recorded by Gerbner and his colleagues were so minimal as to be benign and that when compared side by side, the perceptions of heavy television viewers and nonviewers in terms of the "mean world" syndrome are virtually identical.

MEDIA COVERAGE OF TEXAS POWER OUTAGES

Days of extreme cold gripped Texas and other parts of the southern United States in February 2021, causing a regional disaster of extensive power outages and damage due to broken water pipes. Polling experts at Gallup said the widespread media coverage the power crisis attracted could put energy issues higher on the agenda of Americans.

The Spiral of Silence

Developed by German communication theorist Elisabeth Noelle-Neumann in the 1970s and 1980s, the **spiral of silence** theory links the mass media, social psychology, and the formation of public opinion. The theory proposes that those who believe that their views on controversial issues are in the minority will keep their views to themselves—that is, become silent—out of fear of social isolation, which diminishes or even silences alternative perspectives. The theory is based on social psychology studies, such as the classic conformity research studies of Solomon Asch in 1951. In Asch's study on the effects of group pressure, he demonstrated that a test subject is more likely to give clearly wrong answers to questions about line lengths if all other people in the room unanimously state the same incorrect answers. Noelle-Neumann argued that mass media, particularly television, can exacerbate this effect by communicating real or presumed majority opinions widely and quickly.

According to the theory, the mass media can help create a false, overstated majority; that is, a true majority of people holding a certain position can grow silent when they sense an opposing majority in the media. One criticism of the theory is that some people may fail to fall into a spiral of silence either because they don't monitor the media or because they mistakenly perceive that more people hold their position than really do. Noelle-Neumann also acknowledged that in many cases, "hard-core" nonconformists exist and remain vocal even in the face of social isolation and can ultimately prevail in changing public opinion.[27]

The Third-Person Effect

Identified in a 1983 study by W. Phillips Davison, the **third-person effect** theory suggests that people believe others are more affected by media messages than they are themselves.[28] In other words, it proposes the idea that people think they can escape the worst effects of media but must worry about

Our Masculinity Problem

There have been at least 187 mass shootings (defined as four or more killed) in the United States from August 1966 through April 2021, leaving 1,312 people dead—210 of which were children and teenagers. Half of these shootings have taken place since 2004, and the deadliest have occurred within the past few years.[1] Just some of those that made headlines include the Atlanta-area spa shootings in 2021 (8 killed); the Walmart shooting in El Paso, Texas, in 2019 (22 dead, 27 injured); the Marjory Stoneman Douglas High School shooting in Parkland, Florida, in 2018 (17 dead, 15 injured); the First Baptist Church shooting in Sutherland Springs, Texas, in 2017 (27 dead, 20 injured); the Route 91 Harvest festival shooting in Las Vegas in 2017 (59 dead, 851 injured); the Pulse nightclub shooting in Orlando in 2016 (49 dead, 53 injured); the Sandy Hook Elementary School shooting in Newtown, Connecticut, in 2012 (28 dead, 2 injured); the Century 16 movie theater shooting in Aurora, Colorado, in 2012 (12 dead, 58 injured); and the Virginia Tech shooting in 2007 (33 dead, 23 injured).

What are the reasons for these shootings? Our news media respond with a number of usual suspects: the easy availability of guns in the United States; influential movies, television shows, and video games; radicalization via the Internet; mental illness; bad parenting. But Jackson Katz—educator, author, and filmmaker (of *Tough Guise* and *Tough Guise 2*)—sees another major factor. The least-talked-about commonality in these shootings is the one so obvious most of us miss it: Nearly all the mass murderers are male (and usually white).

What would psychologists, pundits, and other talking heads be saying if women were responsible for nearly every mass shooting for more than three decades? "If a woman were the shooter," Katz says, "you can bet there would be all sorts of commentary about shifting cultural notions of femininity and how they might have contributed to her act."[2]

But women were involved in only five of the mass shootings; all the others had a man (or men) behind the trigger.

Archive Photos/Getty Images

those who are younger, less educated, more impressionable, or otherwise less capable of guarding against media influence. Under this theory, we might fear that other people will, for example, take Internet propaganda seriously, imitate violent movies, or get addicted to digital games, while dismissing the idea that any of those things could happen to us. This effect may explain why some people who have been duped by online conspiracy theories have such a hard time realizing it, since they think their eyes have been opened to the "real" truth and the rest of us are living in a false reality.

Evaluating Research on Media Effects

The mainstream models of media effects research have made valuable contributions to our understanding of mass media, submitting content and audiences to rigorous testing. This wealth of research exists

"Because men represent the dominant gender, their gender is rendered invisible in the discourse about violence," Katz says.[3] In fact, the dominance of masculinity is the norm in nearly all our mainstream mass media. Dramatic content is often about the performance of heroic, powerful masculinity (e.g., many action films, digital games, and sports). Similarly, humorous content often derives from calling into question the standards of masculinity (e.g., a man trying to cook, clean, or take care of a child). The same principles apply for the advertising that supports the content. How many automobile, beer, shaving cream, and food commercials peddle products that offer men a chance to maintain or regain their rightful masculinity?

Rachel Kalish and Michael Kimmel, sociologists at SUNY Stonybrook, critically analyzed the problem of mass shootings that usually end in suicide. They found that males and females have similar rates of suicide attempts in the general population. "Feeling aggrieved, wronged by the world—these are typical adolescent feelings, common to many boys and girls," they report.

The way these feelings play out in the general population, though, differs by gender. Female suicide behaviors are more likely to be self-harm as a cry for help. Male suicide behaviors, informed by social norms of masculinity, often result in a different outcome: "aggrieved entitlement." Kalish and Kimmel define this as "a gendered emotion, a fusion of that humiliating loss of manhood and the moral obligation and entitlement to get it back. And its gender is masculine."[4] So for men, it's self-harm, but first retaliatory harm against others. Retaliation, which is considered acceptable in lesser forms (think of all the cultural narratives in which the weak or aggrieved character finally gets his revenge), becomes horrifying when combined with the immediacy and lethal force of assault firearms. Aggrieved entitlement exactly fit the profile of Omar Mateen, the Orlando shooter with a record of misogyny and homophobia.[5]

There is some evidence that the gun industry understands the sense of masculine entitlement but uses that knowledge to sell guns, not to consider how guns might be misused. A marketing campaign begun in 2010 for the Bushmaster .223-caliber semiautomatic rifle showed an image of the rifle with the large tagline "Consider Your Man Card Reissued." The Bushmaster was the same civilian assault rifle used by the shooter who massacred twenty-eight people at the Newtown elementary school in 2012.

How do we find a way out of this cultural cycle? Katz says we need to make gender—especially male gender and an examination of how toxic masculinities are formed in our culture—central to our discussions about mass shootings. "It's not about guns or about mental illness. It's about gender," Katz says. "It's not a learned behavior; it's a taught behavior."[6] ●

launchpadworks.com

Masculinity on Screen
View a clip from a documentary featuring Jackson Katz.

Discussion: According to Katz, what is the relationship between media, masculinity, and violence? Do you agree with his assessment of this relationship?

partly because funding for studies on the effects of media on young people remains popular among politicians and has drawn ready government support since the 1960s. Media critic Richard Rhodes argues that media effects research is inconsistent and often flawed but continues to resonate with politicians and parents because it offers an easy-to-blame social cause for real-world violence.[29] (For more on real-world gun violence in the United States, see "Examining Ethics: Our Masculinity Problem.")

Funding restricts the scope of some media effects and survey research, particularly if government, business, or other administrative agendas do not align with researchers' interests. Other limits also exist, including the inability to address how media affect communities and social institutions. Because most media effects research operates best when examining media and individual behavior, fewer research studies explore media's impact on community and social life. Some research has begun to address these deficits and also to turn more attention to the increasing influence of media technology on international communication.

Cultural Approaches to Media Research

During the rise of media research in the twentieth century, approaches with a stronger historical and interpretive edge developed as well, often in direct opposition to the scientific models. In the late 1930s, some social scientists began to warn about the limits of "gathering data" and "charting trends," particularly when these kinds of research projects served only advertisers and media organizations and tended to be narrowly focused on individual behavior, ignoring questions like "Where are institutions taking us?" and "Where do we want them to take us?"[30]

In the United States in the 1960s, an important body of research—loosely labeled *cultural studies*—arose to challenge mainstream media effects theories. Since that time, cultural studies research has focused on how people make meaning, understand reality, and order experience by using cultural symbols that appear in the media. This research has attempted to make culture the centerpiece of media studies, focusing on how subtly mass communication shapes and is shaped by history, politics, and economics. Other cultural studies work examines the relationships between elite individuals and groups in government and politics, and how media play a role in sustaining the authority of elites and—occasionally—challenging their power (see "Global Village: International Media Research").

Early Developments in Cultural Studies Research

In Europe, media studies have always favored interpretive rather than scientific approaches; in other words, researchers there have approached the media as if they were literary or cultural critics rather than experimental or survey researchers. These approaches were built on the writings of political philosophers such as Karl Marx and Antonio Gramsci, who investigated how mass media support existing hierarchies in society. They examined how popular culture and sports distract people from redressing social injustices, and they addressed the subordinate status of particular social groups, something emerging media effects researchers were seldom doing.

In the United States, early criticism of media effects research came from the Frankfurt School, a group of European researchers who emigrated from Germany to America to escape Nazi persecution in the 1930s. Under the leadership of Max Horkheimer, T. W. Adorno, and Leo Lowenthal, this group pointed to at least three inadequacies of traditional scientific approaches to media research, arguing that they

- reduced large "cultural questions" to measurable and "verifiable categories,"
- depended on "an atmosphere of rigidly enforced neutrality,"
- refused to place "the phenomena of modern life" in a "historical and moral context."[31]

The researchers of the Frankfurt School did not completely reject the usefulness of measuring and counting data. They contended, however, that historical and cultural approaches were also necessary to focus critical attention on the long-range effects of the mass media on audiences.

Since the time of the Frankfurt School, criticisms of the media effects tradition and its methods have continued, with calls for more interpretive studies of the rituals of mass communication. Academics who have embraced a cultural approach to media research try to understand how media and culture are tied to patterns of communication in daily life.

Conducting Cultural Studies Research

As noted in Chapter 1, the cultural approach considers media texts, users, technologies, and industries within the larger context of culture. Cultural studies research also focuses on the investigation of daily experience, especially concerning issues of race, gender, class, and sexuality, and the unequal arrangements of power and status in contemporary society. Such research emphasizes how some social and cultural groups have been marginalized and ignored throughout history. Consequently, cultural studies have attempted to recover lost or silenced voices, particularly among Black, Native American, Asian and Asian American, Arab, Latino and Latina, Appalachian, LGBTQ+, immigrant, and women's cultures. The major analytical approaches in cultural studies research today are textual analysis, audience studies, and political economy studies.

Textual Analysis

In cultural studies research, **textual analysis** highlights the close reading and interpretation of cultural messages, including those found in books, movies, TV, and the Internet. It is the cultural studies alternative to media effects measurement methods like experiments, surveys, and content analysis. While media effects research approaches media messages with the tools of modern science—replicability, objectivity, and data—textual analysis looks at rituals, narratives, and meaning. One type of textual analysis is *framing research*, which looks at recurring media story structures, particularly in news stories. Media sociologist Todd Gitlin defines media frames as "persistent patterns of cognition, interpretation, and presentation, of selection, emphasis, and exclusion, by which symbol-handlers routinely organize discourse, whether verbal or visual."[32]

Although textual analysis has a long and rich history in film and literary studies, it became significant to media in 1974, when Horace Newcomb's *TV: The Most Popular Art* became the first serious academic book to analyze television shows. Newcomb studied why certain TV programs and formats became popular, especially comedies, westerns, mysteries, soap operas, news reports, and sports programs. Newcomb took television programs seriously, examining patterns in the most popular programs at the time, such as the *Beverly Hillbillies*, *Bewitched*, and *Dragnet*, which traditional researchers had usually snubbed or ignored. Trained as a literary scholar, Newcomb argued that content analysis and other social science approaches to popular media often ignored artistic traditions and social context. For Newcomb, "the task for the student of the popular arts is to find a technique through which many different qualities of the work—aesthetic, social, psychological—may be explored" and to discover "why certain formulas . . . are popular in American television."[33]

Before Newcomb's work, textual analysis generally focused only on "important" or highly regarded works of art—debates, films, poems, and books. But by the end of the 1970s, a new generation of

Lisa Wiltse/Corbis News/Getty Images

TEXTUAL ANALYSIS is the close reading and interpretation of cultural messages, including those found within popular culture. What might the popularity of programs about tot beauty pageants tell us about our culture?

International Media Research

Charlie Fong

LUWEI ROSE LUQIU, a renowned Chinese journalist who received her doctorate in media studies at Penn State University and is now a professor in Hong Kong, uses a microblog and social media to encourage conversation surrounding her journalistic work.

Outside the borders of the United States, the mass media can create markedly different types of content, generate diverse meanings among its audience, and be part of a distinctive media economy. Mass media researchers both inside and outside the United States often engage in international and comparative studies to illustrate the effects and meanings of media in a variety of contexts. Here are four examples of recent international media research:

China

How does political satire work in China, a country that is notorious for media censorship? For her article "The Cost of Humour: Political Satire on Social Media and Censorship in China" (2017), media researcher Luwei Rose Luqiu—then at Pennsylvania State University—interviewed many notable political satirists in China and conducted a content analysis of several satirical texts. In the study, Luqiu, a former journalist and blogger from Hong Kong, illustrates how the establishment of China's State Internet Information Office in 2011 increased censorship in the country. The government's constant threat of censorship on social media makes political satirists censor themselves, she explains, noting that

"the influence of those who continue to work is diminished because the government controls all Chinese social media platforms. However, political satire still has strong vitality thanks to collective action, such as the anonymous production, distribution and sharing of work on Chinese social media."[1]

media studies scholars, who had grown up on television and rock and roll, began to study less elite forms of culture, including fashion, tabloid magazines, rock music, soap operas, and professional wrestling, in an attempt to make sense of the most taken-for-granted aspects of everyday media culture. Today, cultural studies scholars could conduct a textual analysis of *Bridgerton*, a Cardi B song, *This American Life* podcasts, or TikTok videos. Often the study of what may seem to be minor elements of popular culture provides insight into broader meanings within our society. By shifting the focus to daily popular culture artifacts, cultural studies focuses scholarly attention not just on significant presidents, important religious leaders, prominent political speeches, or military battles but on the ways that more ordinary people organize experience and understand themselves and the social world.

Nigeria

Is Nigeria's film industry integrated into the larger global film entertainment industry? In "Global Nollywood: The Nigerian Movie Industry and Alternative Global Networks in Production and Distribution" (2012), Jade Miller—now an associate professor at Wilfrid Laurier University in Canada—investigated the global linkages of the Nigerian film industry ("Nollywood") by interviewing a variety of subjects, from Nigerian film producers to shop owners selling or renting Nollywood movies far from where they are produced. She found that Nollywood is international in scope but operates outside the dominant Hollywood-based film industry by working "via alternative global networks." This distinctive political economy "renders Nollywood as situated in an alternative media capital, central to alternative networks, while too informal to integrate into dominant networks, and it is from this position that we can best understand Nollywood's position in global media flows."[2]

Switzerland/France

How does news travel when it moves in multilingual newsrooms and across regions with different languages? Is anything lost in translation? Researcher Lucile Davier investigated these questions in "The Paradoxical Invisibility of Translation in the Highly Multilingual Context of News Agencies" (2014). Davier, of the University of Geneva, studied two European-based news agencies: Agence France-Presse (AFP) and Agence télégraphique suisse (ATS). Davier, who has worked as a professional translator herself, did fieldwork observations and interviews at the regional office of AFP in Geneva (in the French-speaking region of Switzerland) and at the head office of ATS in Bern (in Switzerland's German-speaking region). According to

Davier, the goal of the study was "to evaluate the potential consequences of a highly plurilingual production process on the one hand, and of the ostensible invisibility of multilingualism/translation on the other hand." Working on a deadline already presents challenges for journalists to remain accurate; adding in the presence of multiple languages in a workplace and the need to translate news into other languages increases the chance of inaccuracies. Davier found that "some of the interviewed journalists acknowledge the risks" of working in this fast-paced interlingual journalism environment, but that "the institutional denial of these possible biases may prevent news agencies from reducing them."[3]

Bahrain

What were the news media narrative frames in Bahrain during the Arab Spring protests there in 2011? Ahmed K. Al-Rawi, now an assistant professor at Simon Fraser University in British Columbia, Canada, analyzed this in the article "Sectarianism and the Arab Spring: Framing the Popular Protests in Bahrain" (2015). Al-Rawi noted that the anti-government protests in this Persian Gulf nation started in February 2011, and that allegiances were largely divided along the lines of the two main Muslim religious sects: the protesters were from Bahrain's Shiite majority and opposed the kingdom's monarchy, which is controlled by minority Sunnis. Al-Rawi explained that the monarchy in Bahrain framed the protests not as pleas for democracy but as the work of outside agitators—"an Iran-backed conspiracy against the Gulf in an attempt to spread Shiism and infiltrate into the region. This sectarian dimension became the dominant frame in order to discredit the cause of the mostly Shiite protestors who were asking for equal rights and job opportunities."[4] ●

Audience Studies

Cultural studies research that focuses on how people use and interpret cultural content is called **audience studies**, or *reader-response research*. Audience studies differ from textual analysis because the subject being researched is the audience for the text, not the text itself. For example, in one of the foundational audience studies, *Reading the Romance: Women, Patriarchy, and Popular Literature* (1984), Janice Radway studied a group of midwestern women who were fans of romance novels. Using her training in literary criticism and employing interviews and questionnaires, Radway investigated the meaning of romance novels to the women. She argued that reading romance novels functions as personal time for some women, whose complex family and work lives leave them very little time for themselves. The study also suggested that these particular romance-novel fans identified with the active, independent qualities of the romantic heroines they most admired. As a cultural study, Radway's work did not claim to be scientific, and her findings are not generalizable to all women. Rather, Radway was interested in investigating and interpreting the relationship between reading romantic fiction and living a conventional life.[34]

Radway's influential cultural research used a variety of interpretive methods, including literary analysis, interviews, and questionnaires. Most important, these studies helped define culture in broad terms—as being made up of both the *products* a society fashions and the *processes* that forge those products.

Political Economy Studies

A focus on the production of popular culture and the forces behind it is the topic of **political economy studies**, which specifically examine interconnections among economic interests, political power, and how that power is used. Among the major concerns of political economy studies is the increasing conglomeration of media ownership. This growing concentration of ownership means that the production of media content is being controlled by fewer and fewer organizations, investing those companies with more and more power. In addition, the domination of public discourse by for-profit corporations may mean that the bottom line for all public communication and popular culture is money, not democratic expression.

As noted earlier in this book, conglomerates like Google, Facebook, and Amazon prompt major concerns about excessive market power. But political economy researchers investigate the accumulation of power throughout the media and entertainment industries, not just in social media or e-commerce companies. For example, in 2020, a federal judge granted class action status to mixed martial artist (MMA) fighters so they could collectively sue Ultimate Fighting Championship (UFC) for using its monopoly power to keep pay "artificially low." The suit argued that the UFC bought up rival companies, enabling it to keep a larger share of its multibillion-dollar media and event profits and leaving a smaller share for the fighters.[35]

Political economy studies work best when combined with textual analysis and audience studies, which provide context for understanding the cultural content of a media product, its production process, and how the audience responds. For example, a major media corporation might, for commercial reasons, create a film and market it through a number of venues (political economy), but the film's meaning or popularity will make sense only within the historical and narrative contexts of the culture (textual analysis), and it may be interpreted by various audiences in ways both anticipated and unexpected (audience studies).

Cultural Studies' Theoretical Perspectives

Developed as an alternative to the predictive theories of social science research (e.g., if X happens, the result will be Y), cultural studies research on media is informed by more general perspectives about how the mass media interact with the world. Two foundational concepts in cultural studies research (in addition to ideology and hegemony, as explained in Chapter 13) are (1) the public sphere and (2) the idea of communication as culture.

The Public Sphere

The idea of the **public sphere**, defined as a space for critical public debate, was first advanced by German philosopher Jürgen Habermas in 1962.[36] Habermas, a professor of philosophy, studied late seventeenth-century and eighteenth-century England and France, and he found those societies to be increasingly influenced by free trade and the rise of the printing press. At that historical moment, an emerging middle class began to gather to discuss public life in coffeehouses, meeting halls, and pubs, and to debate the ideas of novels and other publications in literary salons and clubs. In doing so, this group (which did not yet include women, peasants, the working classes, and people of color) began to build a society beyond the control of aristocrats, royalty, and religious elites. The outcome of such critical public debate led to support for the right to assembly, free speech, and a free press.

Habermas's research is useful to cultural studies researchers when they consider how democratic societies and the mass media operate today. For Habermas, a democratic society should always work to create the most favorable communication situation possible—a public sphere. Basically, without an open communication system, there can be no democratically functioning society. This fundamental notion is the basis for some arguments on why an open, accessible mass media system is essential. However, Habermas warned that the mass media could also be an enemy of democracy; he cautioned modern societies to be aware of "the manipulative deployment of media power to procure mass loyalty, consumer demand, and 'compliance' with systematic imperatives" by those in power.[37]

GRANGER - Historical Picture Archive

Communication as Culture

As Habermas considered the relationship between communication and democracy, media historian James Carey considered the relationship between communication and culture. Carey rejected the linear model of communication—that is, that a message goes simply from sender to receiver. Carey argued that communication is more of a cultural ritual; he famously defined communication as "a symbolic process whereby reality is produced, maintained, repaired, and transformed."[38] Thus, communication creates our reality and maintains that reality in the stories we tell ourselves. For example, think about the novels and movies and the other stories, representations, and symbols that explicitly or tacitly supported discrimination against Black Americans in the United States before the Civil Rights movement. When events occur that question this reality (like protests and sit-ins in the 1950s and 1960s), communication may repair the culture with adjusted narratives or symbols, or it may completely transform the culture with new dominant symbols. Indeed, analysis of media culture in the 1960s and afterward (including books, movies, TV, music, and the Internet) suggests a U.S. culture undergoing repair and transformation. A similar symbolic process was set in motion with the social justice protests beginning in 2020 that ruptured the dominant-culture version of reality.

Carey's ritual view of communication leads cultural studies researchers to consider communication's symbolic process as culture itself. Everything that defines our culture—our language, food, clothing, architecture, media content, and the like—is a form of symbolic communication that signifies shared (but often still-contested) beliefs about culture at a point in historical time. From this viewpoint, then, cultural studies is tightly linked with communication studies.

Evaluating Cultural Studies Research

In opposition to media effects research, cultural studies research involves interpreting written, aural, and visual "texts" or artifacts as symbolic representations that contain cultural, historical, and political meaning. For example, the wave of shows like Netflix's *Black Mirror* and HBO's *Westworld* that appeared in the 2010s can be interpreted as a cultural response to concerns about digital technology and how it could be leading us in the direction of dystopia. Audiences were drawn to these dramas, whose technologies seemed similar enough to what was already possible that the effect was both entertaining and chilling. As James Carey put it, the cultural approach—unlike media effects research, which is grounded in the social sciences—"does not seek to explain human behavior, but to understand it. . . . It does not attempt to predict human behavior, but to diagnose human meanings."[40]

CULTURAL STUDIES researchers are interested in the production and meaning of a wide range of elements within communication culture, as well as audiences' responses to these. ABC's *Black-ish* seeks to challenge ideas about Black identity by satirically questioning stereotypes, biases, and myths.[39]

ABC/Photofest

In other words, a cultural approach does not provide explanations for laws that govern how mass media affect us. Rather, it offers interpretations of the stories, messages, and meanings that circulate throughout our culture.

One of the main strengths of cultural studies is the freedom it affords researchers to broadly interpret the impact of the mass media. Because cultural work is not bound by the precise control of variables, researchers can more easily examine the ties between media messages and the broader social, economic, and political world. For example, media effects research on politics has generally concentrated on election polls and voting patterns, while cultural research has broadened the discussion to examine class, gender, and cultural differences among voters and the various uses of power by individuals and institutions in authority. Following Horace Newcomb's work, cultural investigators have expanded the study of media content beyond "serious" works. These researchers have studied many popular forms, including music, movies, and prime-time television.

Just as media effects research has its limits, so does cultural studies research. Sometimes cultural studies have focused exclusively on the meanings of media programs or texts, ignoring their effect on audiences or the political economy of their production. Another criticism is that cultural studies is more interpretive, subjective, and political in comparison to media effects research. (Cultural studies scholars would likely respond that all research is political, as all research questions begin with a certain initial perspective or hypothesis about the relationship of media and culture.) Today, both media effects and cultural studies researchers have begun to look at the limitations of their work more closely, borrowing ideas from one another to better assess the complexity of the media's meaning and influence.

Media Research and Democracy

One charge frequently leveled at academic studies is that they fail to address the everyday problems of life, often seeming to have little practical application. The growth of mass media departments in colleges and universities has led to an increase in specialized jargon, which tends to alienate and exclude nonacademics. Although media research has built a growing knowledge base and dramatically advanced what we know about the impact of mass media on individuals and societies, the academic world has also built a barrier to that knowledge. That is, the larger public has often been excluded from access to the research process even though cultural research tends to identify with underrepresented groups. The scholarship is self-defeating if its complexity removes it from the daily experience of the groups it addresses. Researchers themselves have even found it difficult to speak to one another across domains because of discipline-specific language used to analyze and report findings. For example, understanding the elaborate statistical analyses used to document media effects requires special training.

In cultural research, the language used is often incomprehensible to students and to other audiences who use mass media, a fact that some academics have called attention to in controversial ways. For example, in 2018, scholars James Lindsay, Helen Pluckrose, and Peter Boghossian wrote twenty fake jargon-filled papers with outlandish premises and submitted them to academic journals in the humanities. By the time they announced their ruse, "seven of their articles had been accepted for publication by ostensibly serious peer-reviewed journals. Seven more were still going through various stages of the review process. Only six had been rejected," the *Atlantic* reported.[41] The experiment, which was praised by some scholars and derided by others as an act of "bad faith," was a more elaborate repeat of Alan Sokal's 1996 hoax in which his dense but meaningless article was published in the academic journal *Social Text*.[42] Like the later scholars, Sokal's point in submitting the article was to critique the use of trendy but impenetrable academic jargon.[43]

In addition, increasing specialization in the 1970s began isolating many researchers from life outside the university. Academics were locked away in their ivory towers, concerned with seemingly obscure matters to which the general public could not relate. Academics across many fields, however, began responding to this isolation and became increasingly active in community engagement and research with practical relevance to democracy and public life.

In recent years, public intellectuals have encouraged discussion about media in a digital world. Harvard law professor Lawrence Lessig has been a leading advocate of efforts to rewrite the nation's copyright laws to enable noncommercial "amateur culture" to flourish on the Internet. American University's Pat Aufderheide, longtime media critic for the alternative magazine *In These Times*, worked with independent filmmakers to develop the Documentary Filmmakers' Statement of Best Practices in Fair Use, which calls for documentary filmmakers to have reasonable access to copyrighted material for their work. In these examples, public intellectuals based on campuses help carry on the conversations of society and culture, actively circulating the most important new ideas of the day.

As ideas have moved outside academic circles, many of the lessons of media research have become mainstream. Findings from media effects research have long sensitized people to issues such as violence in television and movies. Cultural studies research has shaped societal conversations about representations of women, people of color, LGBTQ+ people, and the working class in media texts and in the industry. The criticisms have made their way into popular memes, the dialogues of YouTube media critics, and more formal industry appraisals, such as those that have condemned the lack of diversity among Academy Award nominees and among those holding positions of power in media industries. Political economy studies have highlighted the impact of oligopoly and monopoly ownership, leading to more scrutiny of the world's largest digital media corporations and inspiring entrepreneurs to build nonprofit and cooperative media organizations. In these moments, media research is very much a part of the necessary dialogue on the role of media in public life.

PUBLIC INTELLECTUALS

Melissa Harris-Perry is an author, a professor at Wake Forest University, and an editor-at-large at ELLE.com. From 2012 to 2016, she hosted an opinion show for MSNBC. Her most recent book is *Sister Citizen: Shame, Stereotypes, and Black Women in America*.

Chapter REVIEW 15

Visit LaunchPad for *Media & Culture* at **launchpadworks.com** for additional learning tools:

➜ QUIZ YOURSELF WITH LEARNINGCURVE

LearningCurve uses gamelike quizzing to help you review important concepts from this chapter. LearningCurve includes multiple-choice questions at different levels of difficulty, hints, feedback for both right and wrong answers, and links to e-book material for easy reference. Which answer choice would you select for the sample question below?

> Neve is researching the effects of news media on listeners and viewers, and she hopes to prove her hypothesis that most of the media audience is vulnerable to the strong, mesmerizing hold of mass media. Neve's theory is most closely aligned with which early theory of media effects?
>
> ○ the uses and gratifications model
>
> ○ the hypodermic-needle model
>
> ○ the limited model
>
> ○ the minimal-effects model

➜ WATCH VIDEO ON KEY CONCEPTS

LaunchPad includes clips from movies, TV shows, online sources, and other media texts, in addition to interviews with media experts and newsmakers. In the videos for this chapter, you'll consider topics like the impact of the TV show *13 Reasons Why* and the relationship between media, masculinity, and violence.

➜ PRACTICE THE CRITICAL PROCESS

Use the Media Literacy Activity to walk through the critical process step-by-step and develop a critical perspective on what media means to you. In this chapter's activity, you'll investigate instances where the media has been blamed for violence and other social problems.

➜ LEARN KEY TERMS

Review the definitions for this chapter's key terms with LaunchPad's easy-to-use online flash cards. Try the cards in practice mode, or quiz yourself as a way to focus your study efforts. (Page numbers indicate where each term is highlighted within the chapter.)

media effects research, 431
cultural studies, 431
propaganda analysis, 432
pseudo-polls, 432
hypodermic-needle model, 436
minimal-effects model, 436
selective exposure, 436
selective retention, 436
uses and gratifications model, 436

scientific method, 437
hypotheses, 437
experiments, 438
random assignment, 438
survey research, 438
longitudinal studies, 439
correlations, 439
content analysis, 439
social learning theory, 440
agenda-setting, 440

cultivation effect, 441
spiral of silence, 441
third-person effect, 441
textual analysis, 445
audience studies, 447
political economy studies, 448
public sphere, 448

REVIEW QUESTIONS

Use the questions below to revisit key themes and concepts from this chapter.

 Need more practice? Access the LearningCurve multiple-choice questions for this chapter, which are available on LaunchPad. All questions include feedback for correct and incorrect answers, hints, and links to e-book material.

Research on Media Today

1. What are the two major areas of media research?

How We Got Here: Early Types of Media Research

2. What were the earliest types of media studies, and what were their shortcomings as social scientific approaches?

3. What were the major influences that led to scientific media research?

Research on Media Effects

4. What are the differences between the hypodermic-needle model and the minimal-effects model in the history of media research?

5. What are the differences between experiments and surveys as media research strategies?

6. What is content analysis, and why is it significant?

7. What are the main ideas behind social learning theory, agenda-setting, the cultivation effect, the spiral of silence, and the third-person effect?

8. What are some strengths and limitations of modern media research?

Cultural Approaches to Media Research

9. Why did cultural studies develop in opposition to media effects research?

10. What are the features of cultural studies?

11. How is textual analysis different from content analysis?

12. What are some of the strengths and limitations of cultural studies research?

Media Research and Democracy

13. What is a major criticism about specialization in academic research at universities?

14. How have public intellectuals contributed to society's debates about the media?

QUESTIONING THE MEDIA

Use these critical-thinking questions to reflect on your own experiences with media and apply your understanding of media concepts.

For an interactive experience, questions on topics like these are available for the iClicker student response system. Instructors can download the questions from LaunchPad and use them to poll the class.

1. Think about instances in which the media have been blamed for a social problem. Could there be another, more accurate cause (an underlying variable) of that problem?

2. One charge leveled against a lot of media research—both the media effects and the cultural models—is that it has very little impact on our media institutions. Do you agree or disagree, and why?

3. Do you have a major concern about media in society that hasn't been, but should be, addressed by research? Explain your answer.

4. Can you think of a media issue on which researchers from different fields at a university could team up to study together? Explain.

Legal Controls and Freedom of Expression

16

There are two kinds of contests that occur with each election in the United States. The first and most important contest happens at the ballot box: people across the country vote in support of their preferred candidates for office. The second contest happens behind the scenes, and it's all about money.

In the United States, it's "one person, one vote" (as the saying goes), but when it comes to campaign contributions, it's more like "one person, as many dollars as that person can spend." Aspects of our present political system amount to a legal pay-to-play system in which the wealthiest among us can leverage indirect influence over elections (manipulating issues by buying lots of advertising) and more direct influence over legislation (manipulating politicians who desperately want money to pay for campaign advertising).[1]

There is plenty of evidence that a majority of Americans dislike this system. For example, a national survey in 2018 found that 77 percent of Americans agree that "there should be limits on the amount of money individuals and organizations can spend" on political campaigns and issues.[2] Yet at the same time, the oversized influence of wealthy contributors and businesses is protected by the First Amendment as a form of speech—even though the amendment says nothing explicitly about money. So how did we end up here?

Ironically, it began with Congress's intention to *control* the amount of money in elections. In 1974, emerging from the Watergate scandal (President Nixon's illegal tactics in the 1972 election), Congress amended federal election law to limit campaign contributions. But two years later, in *Buckley v. Valeo* (1976), the U.S. Supreme Court suggested for the first time that political contributions count as speech. The court argued that restrictions on campaign money would limit the number of ideas and issues expressed, the depth in which they would be discussed, and their audience reach. The court said campaign contributions count as speech "because virtually every means of communicating ideas in today's mass society requires the expenditure of money."

Over the ensuing years, Congress has tried to again rein in campaign finance with new laws, but federal courts, beholden to the idea that money equals speech, have always struck them down based on the First

◀ Protesters here are speaking out against the 2010 Supreme Court decision in the *Citizens United v. Federal Election Commission* case. This ruling protects corporations and labor unions, allowing them to spend unlimited amounts of money on TV and radio advertising during elections.

Amendment. This brings us to the current state of our national elections. The two main political parties and their supporters spent an estimated $14 billion on the 2020 election, more than the $6.5 billion spent in the 2016 election campaigns and the $6.3 billion spent in 2012.[3] The main explanation for why corporations and rich individuals can now spend extraordinary amounts on elections lies in another decision by the Supreme Court, *Citizens United v. Federal Election Commission* (2010). The five-to-four decision said that it was a violation of First Amendment free-speech rights for the federal government to limit spending for TV and radio advertising, usually done through organized Super PACs (political action committees), which are most often sponsored by corporate interests or super-rich donors.

While the Supreme Court decision ran counter to public opinion, many advocates on the political Right and some on the Left offered that the First Amendment means what it says: "Congress shall make no law . . . abridging the freedom of speech." Traditional First Amendment supporters like Gene Policinski of the First Amendment Center argue that the "good intentions" behind the idea of limiting campaign spending "don't justify ignoring a basic concept that the Supreme Court majority pointed out in its ruling: Nothing in the First Amendment provides for 'more or less' free-speech protection depending on who is speaking."[4]

An advantage in advertising spending is only one of many variables that help candidates win elections. Nevertheless, when it comes to buying expensive commercial speech and shaping the direction of a presidential campaign, those with limited financial means are at a clear disadvantage compared to those who have money. According to the Center for Responsive Politics, only 1.42 percent of the U.S. population gave more than $200 to "federal candidates, PACs, parties and outside groups" in the 2019–20 election cycle. Those donors (about 4.6 million out of the 330 million people in the United States) wielded great influence, accounting for 76 percent of all contributions.[5]

Given the *Citizens United* ruling, what can be done to give all citizens a voice in the campaign finance system and make each of us—in every income bracket—"patrons" of the political process?

 Is it fair to regulate campaign contributions just because wealthier people have more access to exercise the kind of "speech" that comes from money? Alternately, how might we rethink the system so that political campaigns are not dependent on expensive media advertising to disseminate their messages?

Free Expression Today

The cultural and political struggles over what constitutes free speech or free expression have defined American democracy. In 1989, when Supreme Court justice William Brennan Jr. was asked to comment on his favorite part of the Constitution, he replied, "The First Amendment, I expect. . . . Its enforcement gives us this society. The other provisions of the Constitution really only embellish it."[6] Of all the issues that involve the mass media and popular culture, none is more central—or explosive—than freedom of expression and the First Amendment. Our nation's historical development can often be traced to how much or how little we tolerated speech during particular periods.

The current era is as volatile a time as ever for free-speech issues, and the boundaries are uncertain with polarized politics and so many media platforms. The First Amendment guarantees that the federal government shall not abridge the freedom of speech or the press, as well as the freedom of religion, the right to assemble, and the right to petition the government. But not all citizens agree that the five freedoms protected in the First Amendment are essential to U.S. society. In a 2019 national survey, 29 percent of respondents, after having the text of the First Amendment read to them, agreed with the statement that "the First Amendment goes too far in the rights it guarantees."[7]

Moreover, many citizens are specifically unclear about their First Amendment rights regarding certain topics—especially social media. In the same national survey, 65 percent of respondents agreed that "social media companies violate users' First Amendment rights when they ban users based on the content of their posts." Companies actually do not violate First Amendment rights by banning someone, but this misconception illustrates an important point about the First Amendment (adopted more than 230 years ago) and our media environment today: The First Amendment protects us from *government* intrusion on speech and on the press, but media *corporations* (unimaginable as anything more than a printing press in 1791) can largely do whatever they like. When users sign up for a free social media account, they must consent to a lengthy agreement on terms of service and rules and policies. So, if users get banned based on the content of their posts, it is not due to a violation of First Amendment rights (that is, the government did not ban them); it is because they violated the social media company's policies to which they first agreed. If users violate Twitter's Hateful Conduct policy, for example, they can experience a range of consequences—the company can limit their tweet visibility, remove their tweet, place their account in read-only mode, or suspend them permanently.[8] But none of these consequences legally constitutes censorship.

The frustration for citizens today when it comes to online expression is that codes of conduct from digital media giants like Facebook, Google, Amazon, Apple, and Microsoft are inconsistent and erratically enforced. So, a great deal of misinformation gets distributed through their platforms, and only occasionally—often after complaints—do they strictly enforce their own rules on misinformation. Also, the companies are sometimes unsure if their rules should apply to everyone. For example, in March 2021, Twitter (a company founded back in 2006) asked the public to weigh in about "whether or not they believe world leaders should be subject to the same rules as others" on the site, and if "a world leader [does] violate a rule, what type of enforcement action is appropriate."[9] In addition, the two main political parties in Congress have been unable to agree on what responsibility the digital corporations should be required to take regarding misinformation on their platforms. Democrats have alleged that "big tech has failed to acknowledge the role they've played in fomenting and elevating blatantly false information to its online audiences," while Republicans have called such content moderation "censorship" and "repression."[10]

The responsibility of social media corporations for the content on their platforms is just one of the many ongoing debates over free expression, which also include copyright issues, obscenity and indecency, the right to privacy, the ability to show violent images in film and television, open access to the Internet, and the right of the press to publish government secrets.

In this chapter, we will:

- Examine free-expression issues, focusing on the implications of the First Amendment for a variety of media
- Investigate the models of expression, the origins of free expression, and the First Amendment
- Examine what constitutes censorship and how the First Amendment has been challenged and limited throughout U.S. history
- Focus on the impact of gag orders, shield laws, the use of cameras in the courtroom, and some of the clashes between the First Amendment and the Sixth Amendment

Aram Boghosian/Boston Globe/Getty Images

THIS MARINE, Elijah Collins, stands with a sign displaying the First Amendment as part of Occupy Boston in 2011. The nine-week protest, which included over one hundred tents occupying downtown Boston's financial district, decried income and wealth inequality in the United States. Shortly after this photo was taken, Boston police removed protesters and arrested forty-six people.

- Review the social and political pressures that gave rise to early censorship boards and the current film rating system
- Discuss First Amendment issues in broadcasting, considering why broadcasting has been treated differently from print media
- Explore the debates over free expression on the Internet, and how it presents new regulatory concerns beyond traditional applications of the First Amendment and FCC rules

How We Got Here: The Origins of Free Expression and a Free Press

When students from other cultures attend school in the United States, many are astounded by the number of books, news articles, editorials, cartoons, films, TV shows, and websites that make fun of U.S. presidents, the military, and the police. Throughout history, many countries' governments have jailed, or even killed, their citizens for such speech "violations." For instance, between 1992 and March 2021, fourteen hundred international journalists were killed in the line of duty, often because someone disagreed with what they wrote or reported. Almost nine hundred of those international journalists' deaths were outright murder.[11] In the United States, however, we have generally taken for granted our right to criticize and poke fun at the government and other authority figures, including via the press. Moreover, many of us are unaware of the ideas that underpin our freedoms and don't realize the extent to which our freedoms surpass those in most other countries.

JOURNALISTS IN MYANMAR After the February 2021 military coup in Myanmar, the government attacked the media, detaining at least eighteen journalists and censoring five news organizations by revoking their licenses, according to the Committee to Protect Journalists (CPJ). The military government also brought criminal charges against the independent newspaper *The Irrawaddy*, which publishes in Burmese and English, after it published a video critical of police. The CPJ called on the military government to allow journalists "to report freely and without fear of reprisal." In this photo, Police arrest *Myanmar Now* journalist Kay Zon Nwe in Yangon on February 27, 2021, as protesters demonstrate against the military coup.

YE AUNG THU/AFP/Getty Images

Freedom of Expression around the World

Beginning with the printing press in the 1400s, technological innovations expanded forms of mass communication and enabled messages to circulate more effectively beyond local areas. This increase in communication's reach made it a powerful political tool, prompting rulers and governments around the world to play a role in managing communication—a role that involved exercising varying degrees of control over the people and the press, such as laws licensing the press in England in the seventeenth and eighteenth centuries.

As their predecessors did, governments around the world today have varying approaches to the media. Some hold liberal views on the free expression of opinions, and others more strictly control speech and the media with the desire either to protect standards of truth and cultural morality or to shield their governments from criticism. (See Table 16.1 for the various models of expression.) In fact, in the 2020 ranking of 180 countries in the World Press Freedom Index by Reporters Without Borders, only 14 of the world's countries rated "good" in terms of press freedom (Norway was No. 1), and 33 rated "fairly good" (including the United States at No. 45). Of the rest, 63 rated "problematic," 47 rated "bad," and 23 rated "very bad" (with North Korea ranked last).[12]

TABLE 16.1   MODELS OF EXPRESSION

Models of Expression	
Since the mid-1950s, four conventional models for speech and media systems have been used as a good starting point from which to consider the widely differing ideas underlying free expression.[13] These models are distinguished by the levels of freedom permitted and by the attitudes of the ruling and political classes toward the freedoms granted to the average citizen.	
AUTHORITARIAN	• Developed in sixteenth-century England (at about the time the printing press first arrived), advocates of this model believe the general public needs guidance from an elite, educated ruling class. • Although media outlets are owned by private citizens, the government actively censors the media that critiques its actions, and supports the media that is sympathetic to its agenda and the agenda of the ruling class. • Today, authoritarian systems operate in many developing countries throughout Asia, Latin America, and Africa, where journalism often joins with government and business to foster economic growth, minimize political dissent, and promote social stability. Authoritarianism has also reemerged in European countries such as Hungary and Poland. In these societies, both reporters and citizens may be punished if they question leaders and the status quo too fiercely.
COMMUNIST or **STATE**	• Government "owns" the press and directly controls what it reports. Leaders believe the press should serve government goals. Ideas that challenge the basic premises of state authority are not tolerated. • A few countries still fit parts of this model, including China, Cuba, North Korea, and Turkmenistan.
SOCIAL RESPONSIBILITY	• Social responsibility characterizes the ideals of mainstream journalism in the United States. • The press is usually privately owned (although the government technically operates much of the broadcast media in most European democracies). • The press is free to function as a **Fourth Estate**—that is, an unofficial branch of government that monitors the legislative, judicial, and executive branches for abuses of power and provides information necessary for self-governance.
LIBERTARIAN	• A more radical extension of the social responsibility model is libertarianism, which encourages vigorous government criticism and supports the highest degree of individual and press freedoms. • Under this model, no restrictions are placed on the mass media or on individual speech. Libertarians tolerate the expression of everything, from publishing pornography to advocating anarchy. • In North America and Europe, many alternative newspapers, magazines, and websites operate on such a model.

Data from: Fred Siebert, Theodore Peterson, and Wilbur Schramm, *Four Theories of the Press* (Urbana: University of Illinois Press, 1956).

The First Amendment of the U.S. Constitution

To understand the development of free expression in the United States, we must first understand how the idea for a free press came about. In various European countries throughout the seventeenth century, in order to monitor—and punish, if necessary—the speech of editors and writers, governments controlled the circulation of ideas through the press by requiring printers to obtain licenses from them. However, in 1644, English poet John Milton, author of *Paradise Lost*, published his essay *Areopagitica*, which opposed government licenses for printers and defended a free press. Milton argued that all sorts of ideas, even false ones, should be allowed to circulate freely in a democratic society because eventually the truth would emerge. In 1695, England stopped licensing newspapers, and most of Europe followed.

Less than a hundred years later, the writers of the U.S. Constitution were ambivalent about the freedom of the press. It may come as a surprise to many Americans, but the Constitution as originally ratified in 1788 didn't include a guarantee of freedom of the press. Constitutional framer Alexander Hamilton thought it impractical to attempt to define "liberty of the press" and believed that whatever rights might be added to the Constitution, its security would ultimately depend on public opinion anyway. At that time, though, nine of the original thirteen states had charters defending the freedom of the press, and the states pushed to have federal guarantees of free speech and a free press approved at the first session of

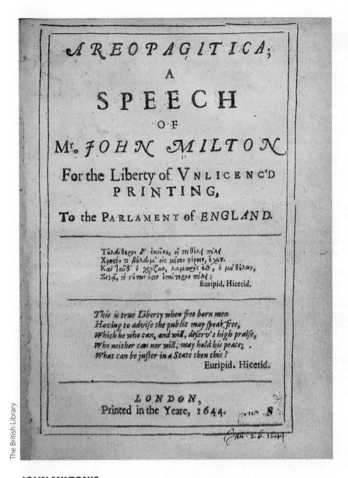

AREOPAGITICA;

A

SPEECH

OF

Mr. JOHN MILTON

For the Liberty of VNLICENC'D
PRINTING,

To the PARLAMENT of ENGLAND.

Euripid, Hicetid.

*This is true Liberty when free born men
Having to advise the public may speak free,
Which he who can, and will, deserv's high praise,
Who neither can nor will, may hold his peace;
What can be juster in a State then this?*

Euripid. Hicetid.

LONDON,
Printed in the Yeare, 1644.

JOHN MILTON'S

Areopagitica (1644) is one
of the most significant early
defenses of freedom of the
press.

the new Congress. The Bill of Rights, which contained the first ten amendments to the Constitution, was adopted in 1791.

The commitment to freedom of the press, however, was not resolute. In 1798, the Federalist Party, which controlled the presidency and Congress, passed the Sedition Act to silence opposition to an anticipated war against France. Led by President John Adams, the Federalists believed that defamatory articles by the opposition Democratic-Republican Party might stir up discontent against the government and undermine its authority. Over the next three years, twenty-five individuals were arrested and ten were convicted under the act, which was also used to prosecute anti-Federalist newspapers. After failing to curb opposition, the Sedition Act expired in 1801 during Thomas Jefferson's presidency. Jefferson, a Democratic-Republican who had challenged the act's constitutionality, pardoned all defendants convicted under it.[14] Ironically, the Sedition Act, the first major attempt to constrain the First Amendment, became the defining act in solidifying American support behind the notion of a free press. As journalism historian Michael Schudson explained, "Only in the wake of the Sedition Act did Americans boldly embrace a free press as a necessary bulwark of a liberal civil order."[15]

Jefferson's own views epitomize much of the conflicted current thinking about free expression in the media. Jefferson was an early champion of a free press, and in 1787, as the Constitution was being formed, famously said, "Were it left to me to decide whether we should have a government without newspapers, or newspapers without a government, I should not hesitate a moment to prefer the latter."[16] Nevertheless, as president, Jefferson had to withstand the vitriol and allegations of a partisan press. In 1807, near the end of his second term, Jefferson's idealism about the press had cooled, as he found newspapers to be filled with sensationalism and misinformation. "The man who never looks into a newspaper is better informed than he who reads them, inasmuch as he who knows nothing is nearer the truth than he whose mind is filled with falsehoods and errors," he remarked in a letter to a friend.[17]

The Limits to Freedom of Expression in the United States

Although the First Amendment says that "Congress shall make no law . . . abridging the freedom of speech, or of the press," there are in fact several laws or court decisions that put limits on the freedom of speech and the press. These include the ways in which censorship has been defined, and a number of unprotected forms of expression, including sedition, copyright infringement, libel, obscenity, and privacy violations. The First Amendment also collides with the Sixth Amendment when it comes to media coverage of some legal trials.

Censorship as Prior Restraint

In the United States, the First Amendment has theoretically prohibited censorship. Over time, Supreme Court decisions have defined censorship as instances of **prior restraint** by the government. This means that courts and governments cannot block any publication or speech before it actually occurs, on the principle that a law has not been broken until an illegal act has actually been committed. In 1931, for example, the Supreme Court determined in *Near v. Minnesota* that a Minneapolis newspaper could not

PRIOR RESTRAINT In 1971, Daniel Ellsberg surrendered to government prosecutors in Boston. Ellsberg was a former Pentagon researcher who turned against America's military policy in Vietnam and leaked information to the press. He was charged with unauthorized possession of top-secret federal documents. Later called the Pentagon Papers, the documents contained evidence on the military's bungled handling of the Vietnam War. In 1973, an exasperated federal judge dismissed the case when illegal government-sponsored wiretaps of Ellsberg's psychoanalyst came to light during the Watergate scandal.

be stopped from publishing "scandalous and defamatory" material about police and law officials who the paper felt were negligent in arresting and punishing local gangsters.[18] However, the Court left open the idea that the news media could be ordered to halt publication in exceptional cases. During a declared war, for instance, if a U.S. court judged that the publication of an article would threaten national security, such expression could be restrained prior to its printing. In fact, during World War I, the U.S. Navy seized all wireless radio transmitters to ensure control over critical information regarding weather conditions and troop movements that might inadvertently aid the enemy. In the 1970s, though, the Pentagon Papers decision and the *Progressive* magazine case tested important concepts underlying prior restraint.

The Pentagon Papers Case

In 1971, with the Vietnam War still in progress, Daniel Ellsberg, a former Defense Department employee, stole a copy of the forty-seven-volume report *History of U.S. Decision-Making Process on Vietnam Policy*. A thorough study of U.S. involvement in Vietnam since World War II, the report was classified by the government as top secret. Ellsberg and a friend leaked the study—nicknamed the Pentagon Papers—to the *New York Times* and the *Washington Post*. In June 1971, the *Times* began publishing articles based on the study. To block any further publication, the Nixon administration applied for and received a federal court injunction against the *Times*, arguing that the publication of these documents posed "a clear and present danger" to national security.

A lower U.S. district court supported the newspaper's right to publish, but the government's appeal put the case before the Supreme Court less than three weeks after the first article was published. In a six-to-three vote, the Court sided with the newspaper. Justice Hugo Black, in his majority opinion, attacked the government's attempt to suppress publication: "Both the history and language

MEDIA LITERACY & THE CRITICAL PROCESS

Who Knows the First Amendment?

Enacted in 1791, the First Amendment supports not just press and speech freedoms but also religious freedom and the right of people to protest and to "petition the government for a redress of grievances." It also says that "Congress shall make no law" abridging or prohibiting these five freedoms. To investigate some critics' charge that many citizens don't exactly know the protections offered in the First Amendment, conduct your own survey. Discuss with friends, family, or colleagues what they know or think about the First Amendment.

1 DESCRIPTION

Working alone or in small groups, find eight to ten people you know from two different age groups: (1) from your peers and friends or younger siblings, and (2) from your parents' or grandparents' generations. (Do not choose students from your class.) Interview your subjects individually, in person, by phone, or by FaceTime or Zoom, and ask them this question: "Would you approve of the following law if Congress were considering it?" Then offer the First Amendment, but don't tell them what it is. Ask them to respond to the following series of questions, adding any other questions that you think would be appropriate:

1. Do you agree or disagree with the freedoms? Explain.

2. Which do you support, and which, if any, do you think are excessive or provide too much freedom?

3. Do you recognize the law? (Note how many of your subjects identify it as the First Amendment to the U.S. Constitution and how many do not. Note the percentage from each age group.)

4. Optional: If you are willing to do so, please share your political leanings—Republican, Democrat, Independent, not sure, disaffected, apathetic, or other.

Record their answers.

2 ANALYSIS

What patterns emerge in the answers from the two groups? Are the answers similar or different? How? Note any differences in the answers based on gender, level of education, occupation, or other factors.

3 INTERPRETATION

What do these patterns mean? Are your interview subjects supportive or unsupportive of the First Amendment? What are their reasons?

4 EVALUATION

How do your interviewees judge the freedoms? In general, what did your interview subjects know about the First Amendment? What impresses you about your subjects' answers? Do you find anything alarming or troubling in their answers?

5 ENGAGEMENT

Research "free expression," and locate any national studies that are similar to this assignment. Then, check recent national surveys on attitudes toward the First Amendment—for example, the State of the First Amendment survey at www.freedomforuminstitute.org. Based on your research, educate others. Do a presentation at your college or university about the First Amendment.

---

of the First Amendment support the view that the press must be left free to publish news, whatever the source, without censorship, injunctions, or prior restraints."[19] (See "Media Literacy and the Critical Process: Who Knows the First Amendment?")

The *Progressive* Magazine Case

The issue of prior restraint for national security surfaced again in 1979 with an injunction being issued to the editors of the *Progressive*—a national left-wing magazine—to stop publication of an article titled "The H-Bomb Secret: How We Got It, Why We're Telling It." The dispute began when the editor of the magazine sent a draft to the Department of Energy to verify technical portions of the article. Believing that the article contained sensitive data that might damage U.S. efforts to halt the proliferation of nuclear weapons, the Department of Energy asked the magazine not to publish it. When the magazine said it would proceed anyway, the government sued the *Progressive* and asked a federal district court to block publication.

Judge Robert Warren sought to balance the *Progressive*'s First Amendment rights against the government's claim that the article would spread dangerous information and undermine national security. In an unprecedented action, Warren sided with the government, deciding that "a mistake in ruling against the United States could pave the way for thermonuclear annihilation for us all. In that event, our right to life is extinguished and the right to publish becomes moot."[20] During appeals and further litigation, several other publications, including the *Milwaukee Sentinel* and *Scientific American*, published their own articles related to the H-bomb, getting much of their information from publications already in circulation. None of these articles—including the one eventually published in the *Progressive* after the government dropped the case during an appeal— contained the precise technical details needed to actually design a nuclear weapon, nor did they provide information on where to obtain the sensitive ingredients.

Even though the *Progressive* article was eventually published, Warren's decision stands as the first time in American history that a prior-restraint order imposed in the name of national security actually stopped the initial publication of a controversial news report.

Unprotected Forms of Expression

Despite the First Amendment's defense of free speech, the federal government has made a number of laws that restrict forms of expression, especially concerning false or misleading advertising, expressions that intentionally threaten public safety, and certain speech that compromises war strategy and other issues of national security.

Beyond the federal government, state laws and local ordinances have on occasion curbed expression, and over the years the court system has determined that some kinds of expression do not merit protection under the Constitution, including seditious expression, copyright infringement, libel, obscenity, and privacy violations.

Seditious Expression

For more than a century after the Sedition Act of 1798, Congress passed no laws prohibiting dissenting opinion. But by the twentieth century, the sentiments of the Sedition Act reappeared in times of war. For instance, the Espionage Acts of 1917 and 1918, which were enforced during World Wars I and II, made it a federal crime to disrupt the nation's war effort, authorizing severe punishment for seditious statements.

In the landmark *Schenck v. United States* (1919) appeal case, the Supreme Court upheld the conviction of a Socialist Party leader, Charles T. Schenck, for distributing leaflets urging American men to protest the draft during World War I, in violation of the recently passed Espionage Act. In upholding the conviction, Justice Oliver Wendell Holmes's opinion included two of the more famous phrases in the First Amendment's legal history: "falsely shouting fire in a theatre," and speech that creates "a clear and present danger." He wrote:

> But the character of every act depends upon the circumstances in which it is done. The most stringent protection of free speech would not protect a man in falsely shouting fire in a theater and causing a panic.
>
> The question in every case is whether the words used are used in such circumstances and are of such a nature as to create a clear and present danger that they will bring about the substantive evils that Congress has a right to prevent.

In supporting Schenck's sentence—a ten-year prison term—Holmes noted that the Socialist leaflets were entitled to First Amendment protection, but only during times of peace. In creating the "clear and present danger" criterion for expression, the Supreme Court established the limits of the First Amendment.

Copyright Infringement

Appropriating a writer's or an artist's words or music without consent or payment is also a form of expression that is not protected as speech. A **copyright** legally protects the rights of authors and producers to their published or unpublished writing, music, lyrics, TV programs, movies, or graphic art designs. When Congress passed the first Copyright Act in 1790, it gave authors the right to control their

Richard Stonehouse/Camera Press/Redux

THE LIMITS OF COPYRIGHT The iconic album art for the Velvet Underground's 1967 debut—a banana print designed by artist Andy Warhol—has been a subject of controversy in recent years, as a copyright dispute between the Andy Warhol Foundation for the Visual Arts and the rock band has continued to flourish. The most recent disagreement occurred when the Warhol Foundation, which had previously accused the Velvet Underground of violating its claim to the print, announced plans to license the banana design for iPhone cases. Accusing the foundation of copyright violation, the band filed a copyright claim to the design, which a federal judge later dismissed.

published works for fourteen years, with the opportunity for a renewal for another fourteen years. At the end of the copyright period, the work would enter the **public domain**, which would give the public free access to the work. The idea was that a period of copyright control would give authors financial incentive to create original works, and that the public domain would give others incentive to create derivative works. Today, older works in the public domain nurture creativity and innovation by making it easier for people to use this material in new ways, such as in collages, mash-ups, and remakes.

Over the years, as artists lived longer and (more importantly) as corporate copyright owners became more common, copyright periods were extended by Congress. In 1976, Congress extended the copyright period to the life of the author plus fifty years, or seventy-five years for a corporate copyright owner. In 1998, as copyrights on such works as Disney's Mickey Mouse were approaching expiration, Congress again extended the copyright period for twenty additional years, for a total of ninety-five years—the eleventh time in forty years that the terms for copyright had been extended.[21] No additional extensions from Congress have followed, however, and works have again begun entering the public domain on the first day of each new year. January 1, 2021, for example, saw the release of works copyrighted in 1925. These included books like F. Scott Fitzgerald's *The Great Gatsby*, Ernest Hemingway's *In Our Time*, and Virginia's Woolf's *Mrs. Dalloway*; music like the songs "Sweet Georgia Brown," Irving Berlin's "Always," and works from Fats Waller, Jelly Roll Morton, Bessie Smith, and Gertrude "Ma" Rainey; and films like Harold Lloyd's *The Freshman* and Buster Keaton's *Go West.*[22]

Corporate owners have millions of dollars to gain by keeping their properties out of the public domain. Disney, a major lobbyist for the 1998 extension, would have lost its copyright to Mickey Mouse in 2004 but now continues to earn millions on its movies, T-shirts, and Mickey Mouse watches through 2024 (but not beyond that year, unless there is another last-minute extension). Warner/Chappell Music, which made up to $2 million a year in royalties from the popular "Happy Birthday to You" song, lost its copyright in a 2015 lawsuit in which a U.S. district court judge ruled that it didn't have a valid claim to the 120-year-old song.

Today, nearly every innovation in digital culture creates new questions about copyright law. For example, is a video mash-up that samples copyrighted sounds and images a copyright violation or a creative accomplishment protected under the concept of *fair use* (the same standard that enables students to legally quote attributed text from other works in their research papers)? Is it fair use for a blog to quote an entire newspaper article as long as it has a link and an attribution? Should news aggregators like Google News and Facebook pay something to financially strapped newspapers when they link to their articles? (In 2021, Google and Facebook did come to an agreement to pay News Corp. for its content in Australia, where News Corp. is a major publisher. The deal may serve as a model for the digital companies paying publishers in other countries.) One of the laws that tips the debates toward stricter enforcement of copyright is the Digital Millennium Copyright Act of 1998, which outlaws any action or technology that circumvents copyright protection systems. In other words, it may be illegal to merely create or distribute technology that enables someone to make illegal copies of digital content, such as a music file or a movie.

Libel

The biggest legal worry that haunts editors and publishers is the issue of libel, a form of expression that, unlike political expression, courts have not ruled to be protected as free speech under the First Amendment. **Libel** refers to defamation of character in written or broadcast form; libel is different from **slander**, which is spoken language that defames a person's character. Inherited from British common law, libel is generally defined as a false statement that holds a person up to public ridicule, contempt, or hatred, or injures a person's business or occupation. Examples of libelous statements include falsely accusing someone of professional dishonesty or incompetence (such as medical malpractice), falsely accusing a person of a crime (such as drug dealing), falsely stating that someone has a mental illness or engages in unacceptable behavior (such as public drunkenness), or falsely accusing a person of associating with a disreputable organization or cause (such as the Mafia or a neo-Nazi military group).

Determining Libel: Private Citizens vs. Public Officials.

Since 1964, the *New York Times v. Sullivan* case has served as the standard for libel law. The case stems from a 1960 full-page advertisement placed in the *New York Times* by the Committee to Defend Martin Luther King and the Struggle for Freedom in the South. Without naming names, the ad criticized the law enforcement tactics used in southern cities—including Montgomery, Alabama—to break up Civil Rights demonstrations. The ad condemned "southern violators of the Constitution" bent on destroying King and the movement. Taking exception, the city commissioner of Montgomery, L. B. Sullivan, sued the *Times* for libel, claiming the ad defamed him indirectly. Although Alabama civil courts awarded Sullivan $500,000, the newspaper's lawyers appealed to the Supreme Court, which unanimously reversed the ruling, holding that Alabama libel law violated the *Times'* First Amendment rights.[23]

As part of the *Sullivan* decision, the Supreme Court asked future civil courts to distinguish whether plaintiffs in libel cases are public officials or private individuals. Citizens with more "ordinary" jobs—such as city sanitation employees, undercover police informants, nurses, or unknown actors—are normally classified as private individuals. Private individuals have to prove

1. that the public statement about them was *false*,
2. that *damages* or actual injury occurred (such as the loss of a job, harm to reputation, public humiliation, or mental anguish), and
3. that the publisher or broadcaster was *negligent* in failing to determine the truthfulness of the statement.

There are two categories of public figures: (1) public celebrities (movie or sports stars) or people who "occupy positions of such pervasive power and influence that they are deemed public figures for all purposes" (presidents, senators, mayors), and (2) individuals who have thrown themselves—usually voluntarily but sometimes involuntarily—into the middle of "a significant public controversy," such as a lawyer defending a prominent client, an advocate for an antismoking ordinance, or a labor union activist.

Public officials have to prove falsehood, damages, and negligence like private individuals do, plus one additional factor: actual malice on the part of the news medium. **Actual malice** means that the reporter or editor knew that the statement was false and either printed (or broadcast) it anyway or acted with a reckless disregard for the truth. Because actual malice against a public official is hard to prove, it is difficult for public figures to win libel suits.

The *Sullivan* decision allowed news operations to aggressively pursue legitimate news stories without fear of continuous litigation. However, the mere threat of a libel suit still scares off many in the

Howard Sochurek/The LIFE Picture Collection/Getty Images

LIBEL AND THE MEDIA
A 1960 *New York Times* advertisement triggered one of the most influential and important libel cases in U.S. history by criticizing law enforcement tactics used against Martin Luther King (pictured above) and the Civil Rights movement. The behind-the-scenes machinations of King's later Alabama demonstrations are the subject of the film *Selma*.

AP Photo/Jacquelyn Martin

DOMINION VOTING SYSTEMS and Smartmatic USA brought multibillion dollar libel cases against several supporters of Donald Trump's campaign in 2021 for suggesting that their voting machines were rigged against Trump in the 2020 presidential election. The two companies are suing Fox News, as well as Trump lawyers Rudy Giuliani and Sidney Powell (both pictured here), for libel. Dominion additionally brought a defamation lawsuit against MyPillow CEO Mike Lindell, alleging that he was part of a "viral disinformation campaign" about Dominion.

news media. Plaintiffs may also belong to one of many vague classification categories, such as public high school teachers, police officers, and court-appointed attorneys. Individuals from these professions can end up being considered public or private citizens depending on a particular court's ruling.

Defenses against Libel Charges. Since the 1730s, the best defense against libel in American courts has been the truth. In most cases, if libel defendants can demonstrate that they printed or broadcast statements that were essentially true, such evidence usually bars plaintiffs from recovering any damages—even if their reputations were harmed.

In addition, there are other defenses against libel. Prosecutors, for example, who would otherwise be vulnerable to being accused of libel, are granted *absolute privilege* in a court of law so that they are not prevented from making accusatory statements toward defendants. The reporters who print or broadcast statements made in court are also protected against libel; they are granted conditional or **qualified privilege**, allowing them to report judicial or legislative proceedings even though the public statements being reported may be libelous.

Another defense against libel is the rule of **opinion and fair comment**. Generally, libel applies only to intentional misstatements of factual information rather than opinion, and therefore opinions are protected from libel. However, because the line between fact and opinion is often hazy, lawyers advise journalists to first set forth the facts on which a viewpoint is based and then state their opinion based on those facts. In other words, journalists should make it clear that a statement of opinion is a criticism and not an allegation of fact.

Libel laws also protect satire, comedy, and opinions expressed in reviews of books, plays, movies, and restaurants. Such laws may not, however, protect malicious statements in which plaintiffs can prove that defendants used their free-speech rights to mount a damaging personal attack.

Obscenity

For most of this nation's history, legislators have argued that **obscenity** does not constitute a legitimate form of expression protected by the First Amendment. The problem, however, is that little agreement has existed on how to define an obscene work. In the 1860s, a court could judge an entire book obscene if it contained a single passage believed capable of "corrupting" a person. In fact, throughout the nineteenth century, certain government authorities outside the courts—especially U.S. post office and customs officials—held the power to censor or destroy material they deemed obscene.

This began to change in the 1930s, during the trial involving the celebrated novel *Ulysses* by Irish writer James Joyce. Portions of *Ulysses* had been serialized in the early 1920s in an American magazine, *Little Review*, copies of which were later seized and burned by postal officials. The publishers of the magazine were fined $50 and nearly sent to prison. Because of the four-letter words contained in the novel and the book-burning and fining incidents, British and American publishing houses backed away from the book, and in 1928, the U.S. Customs Office officially banned *Ulysses* as an obscene work. Ultimately, however, Random House agreed to publish the work in the United States if it was declared "legal." Finally, in 1933, a U.S. judge ruled that an important literary work such as *Ulysses* was a legitimate, protected form of expression, even if portions of the book were deemed objectionable by segments of the population.

The current legal definition of obscenity derives from the 1973 *Miller v. California* case, which stated that to qualify as obscenity, the material must meet three criteria:

1. the average person, applying contemporary community standards, would find that the material as a whole appeals to prurient interest;
2. the material depicts or describes sexual conduct in a patently offensive way; and
3. the material, as a whole, lacks serious literary, artistic, political, or scientific value.

The *Miller* decision contained two important ideas. First, it acknowledged that different communities and regions of the country have different values and standards with which to judge obscenity. Second, it required that a work be judged *as a whole,* so that publishers could not use the loophole of inserting a political essay or literary poem into pornographic materials to demonstrate in court that their publications contained redeeming features.

Since the *Miller* decision, courts have granted great latitude to printed and visual obscenity. By the 1990s, major prosecutions had become rare—aimed mostly at child pornography—as the legal system accepted the concept that a free and democratic society must tolerate even repulsive kinds of speech. Most battles over obscenity are now online, where the global reach of the Internet has eclipsed the concept of community standards. A new complication in defining pornography has emerged with cases of "sexting," in which minors produce and send sexually graphic images of themselves via cell phones or the Internet (see "Examining Ethics: Is 'Sexting' Pornography?").

Privacy Violations

Whereas libel laws safeguard a person's character and reputation, the right to privacy protects an individual's peace of mind and personal feelings. In the simplest terms, the **right to privacy** addresses a person's right to be left alone, without the person's name, image, or daily activities becoming public property. Invasions of privacy occur in different situations, the most common of which are intrusion into someone's personal space via unauthorized tape recording, photographing, wiretapping, and the like; divulgence of personal records, such as health and phone records; disclosure of personal information, such as religion, sexual activities, or personal activities; and unauthorized appropriation of someone's image or name for advertising or other commercial purposes.

In general, the news media have been granted wide protections under the First Amendment to do their work. That means that public figures—like Britney Spears, Justin Bieber, Kanye West, the Kardashians, and ex–British royals Meghan and Harry—are subject to a great deal of media scrutiny and gossip in the United States. (With social media, many figures, like the Kardashians, preempt media coverage by sharing so much of themselves.) Public figures, however, have received some legal relief against privacy violations, as many local municipalities and states have passed "anti-paparazzi" laws that protect individuals from unwarranted scrutiny and surveillance of personal activities on private property or outside public forums.

In a recent test of the boundaries of privacy for public figures, a Florida jury in 2016 ordered gossip entertainment website *Gawker* to pay more than $140 million to Terry G. Bollea, better known as the former professional wrestler Hulk Hogan. In 2012, *Gawker* posted a brief excerpt of a grainy sex tape that showed Bollea having sex with his best friend's wife. *Gawker* argued that its actions were protected by the First Amendment and that Bollea was a public figure who had often talked about his sex life in

Is "Sexting" Pornography?

According to U.S. federal and state laws, when someone produces, transmits, or possesses images with graphic sexual depictions of minors, it is considered child pornography. Digital media have made the circulation of child pornography even more pervasive, according to a 2006 study on child pornography on the Internet. About one thousand people are arrested each year in the United States for child pornography, and according to a U.S. Department of Justice guide for police, they have few distinguishing characteristics other than being "likely to be white, male, and between the ages of 26 and 40."[1]

Now, a social practice has challenged the common wisdom of what can be considered obscenity and who can be considered child pornographers: What happens when the people who produce, transmit, and possess images with graphic sexual depictions of minors are minors themselves?

The practice in question is "sexting," the sending or receiving of sexual images via text message or the Internet. Sexting among minors occupies a gray area of obscenity law—yes, these are images of minors; but no, they don't fit the intent of child pornography laws, which are designed to stop the exploitation of children by adults.

While such messages are usually meant to be completely personal, technology makes it otherwise. "All control over the image is lost—it can be forwarded repeatedly all over the school, town, state, country and world," says Steven M. Dettelbach, U.S. attorney for the Northern District of Ohio.[2] And given the endless archives of the Internet, such images never really go away but can be accessed by anyone with enough skills to find them.

Research published in the journal *JAMA Pediatrics* in 2018 found that 14.8 percent of teenagers sent sexts, while 27.4 percent received them. Other sexting behaviors raise issues of privacy and bullying: 12 percent of teens forwarded a sext without consent, and 8.4 percent had a sext of them forwarded without consent.[3] A number of cases illustrate how young people engaging in sexting have gotten caught up in a legal system designed to punish pedophiles, including the following:

- In 2008, Florida resident Phillip Alpert, then eighteen, sent nude images of his sixteen-year-old girlfriend to friends after they got in an argument. He was convicted of child pornography and is required to be registered as a sex offender for the next twenty-five years.

- In 2009, three Pennsylvania girls took seminude pictures of themselves and sent the photos to three boys. All six minors were charged with child pornography. A judge later halted the charges in the interest of freedom of speech and parental rights.

- In Cañon City, Colorado, a 2015 texting scandal involving middle and high school students exchanging hundreds of nude photos resulted in student suspensions, a canceled high school football game, and a criminal investigation. Felony charges were a possibility, but the state district attorney ultimately decided not to bring charges against the students involved.

By 2021, twenty-six states (Arizona, Arkansas, Colorado, Connecticut, Florida, Georgia, Hawaii, Illinois, Indiana, Kansas, Louisiana, Nebraska, Nevada, New Jersey, New Mexico, New York, North Dakota, Oklahoma, Pennsylvania, Rhode Island, South Dakota, Texas, Utah, Vermont, Washington, and West Virginia) have responded to situations like these with sexting laws, so that teens involved in sexting generally face misdemeanor charges rather than being subject to the harsher felony laws against child pornography.[4] In the states without such laws, the charges are often at the discretion of prosecutors and courts and could be as harsh as a felony crime, with the accompanying fine, jail time, and permanent criminal record. How do you think sexting should be handled by the law? ●

launchpadworks.com

Sexting and Obscenity
This video clip discusses the legal and personal issues surrounding sexting and obscenity.

Discussion: What legal and social consequences did the teenager in this clip face after sharing nude images? Were those consequences fair? Unfair? Why?

HULK HOGAN'S successful lawsuit against the entertainment website *Gawker* set an important precedent for celebrities and other public figures, who are normally not protected against invasions of privacy.

media interviews. But the jury determined that Bollea's privacy had been violated, and it awarded him the huge sum for emotional distress, economic distress, and punitive damages. *Gawker* and Bollea reached a settlement of $31 million, but the litigation shattered *Gawker*, which declared bankruptcy and then sold itself (along with the company's other websites) to Univision for $135 million.

Bollea's victory in the *Gawker* decision was an important one, because the outcome may encourage future trials in which citizens or groups sue media organizations expressly to run them out of business, particularly when they are well-funded. (Bollea's case against *Gawker* received $10 million in support from billionaire Peter Theil.) For example, completely frivolous lawsuits have been filed by parties alleging libel or privacy violations, with an intent to "[impose] significant litigation costs and [chill] protected speech."[24] Such lawsuits, called SLAPPs (strategic lawsuits against public participation), are filed to intimidate media organizations with the prospect of big legal bills unless they back down from their reporting. By 2021, thirty states had anti-SLAPP laws in effect, enabling judges to quickly dismiss meritless retaliatory lawsuits before they damage a media organization or other defendants. Anti-SLAPP laws are considered to be pro–First Amendment. As the Reporters Committee for Freedom of the Press notes, "News organizations and individual journalists can use anti-SLAPP statutes to protect themselves from the financial threat of a groundless defamation case brought by a subject of an enterprise or investigative story."[25]

A number of laws also protect the privacy of regular citizens. For example, the Privacy Act of 1974 protects individuals' records from public disclosure unless individuals give written consent. The Electronic Communications Privacy Act of 1986 extended the law to computer-stored data and the Internet, although subsequent court decisions ruled that employees have no privacy rights in electronic communications conducted on their employer's equipment. The USA PATRIOT Act of 2001, however, weakened the earlier laws and gave the federal government more latitude in searching private citizens' records and intercepting electronic communications without a court order.

In early 2016, there was a brief but significant standoff between the FBI and Apple over the FBI getting access to an iPhone. The phone in question was recovered from one of the terrorists in the December 2015 attack in San Bernardino, California. A court ordered Apple to create a software key for the FBI to unlock the iPhone. Apple responded that writing such software would potentially make all iPhones subject to FBI scrutiny and could make them more susceptible to other hackers. A day before a court hearing on the matter, the FBI withdrew its case, saying that it had cracked the code to access the iPhone itself. Yet the question remains whether technology companies like Apple should provide customers (most of whom have the best intentions) with the most robust privacy possible or assist law enforcement agencies (which could include those from authoritarian nations) in gaining access to their products. In 2018, Apple announced it would close the software loophole that enabled the FBI to gain access to the iPhone.

First Amendment versus Sixth Amendment

Over the years, First Amendment protections of speech and the press have often clashed with the Sixth Amendment, which guarantees an accused individual in "all criminal prosecutions . . . the right to a speedy and public trial, by an impartial jury." In 1954, for example, the Sam Sheppard case garnered enormous nationwide publicity and became the inspiration for the TV show and film *The Fugitive*. Featuring lurid details about the murder of Sheppard's wife, the press editorialized in favor of Sheppard's quick arrest; some papers even pronounced him guilty. A prominent and wealthy osteopath, Sheppard was convicted of the murder, but twelve years later Sheppard's new lawyer, F. Lee Bailey, argued before the Supreme Court that his client had not received a fair trial because of prejudicial publicity in the press. The Court overturned the conviction and freed Sheppard.

Gag Orders and Shield Laws

A major criticism of recent criminal cases concerns the ways in which lawyers use the news media to comment publicly on cases that are pending or are in trial. After the Sheppard reversal in the 1960s, the Supreme Court introduced safeguards that judges could employ to ensure fair trials in heavily publicized cases. These included sequestering juries (meaning jurors are isolated from the public; the jury in the 2021 trial of the Minneapolis police officer accused of murdering George Floyd was sequestered during deliberations); moving cases to other jurisdictions; and placing restrictions, or **gag orders**, on lawyers and witnesses. In some countries, courts have issued gag orders to prohibit the press from releasing information or giving commentary that might prejudice jury selection or cause an unfair trial. In the United States, however, especially since a Supreme Court review in 1976, gag orders on the press have been struck down as a prior-restraint violation of the First Amendment.

In opposition to gag orders, **shield laws** have favored the First Amendment rights of reporters, protecting them from having to reveal their sources for controversial information used in news stories. The news media have argued that protecting the confidentiality of key sources maintains a reporter's credibility, protects a source from possible retaliation, and serves the public interest by providing information that citizens might not otherwise receive. In the 1960s, when the First Amendment rights of reporters clashed with Sixth Amendment fair-trial concerns, judges usually favored the Sixth Amendment arguments. In 1972, a New Jersey journalist became the first reporter jailed for contempt of court for refusing to identify sources in a probe of the Newark housing authority. But by 2021, forty-nine states (all except Wyoming) and the District of Columbia had adopted some type of shield law or established some shield law protection through legal precedent. However, there is still no federal shield law in the United States, leaving journalists exposed to subpoenas from federal prosecutors and courts.

Cameras in the Courtroom

The debates over limiting intrusive electronic broadcast equipment and photographers in U.S. courtrooms began with the sensationalized coverage of the Bruno Hauptmann trial in the mid-1930s.

AP Photo

Hauptmann was convicted and executed for the kidnap-murder of the nineteen-month-old son of Anne and Charles Lindbergh (the aviation hero who made the first solo flight across the Atlantic Ocean in 1927). During the trial, Hauptmann and his attorney complained that the circus atmosphere fueled by the presence of radio and flash cameras prejudiced the jury and turned the public against him.

After the trial, the American Bar Association amended its professional ethics code, Canon 35, stating that electronic equipment in the courtroom detracted "from the essential dignity of the proceedings." Calling for a ban on photographers and radio equipment, the association believed that if such elements were not banned, lawyers would begin playing to audiences and negatively alter the judicial process. For years after the Hauptmann trial, almost every state banned photographic, radio, and TV equipment from courtrooms.

As broadcast equipment became more portable and less obtrusive, however, and as television became the major news source for most Americans, courts gradually reevaluated their bans on broadcast equipment. In fact, in the early 1980s, the Supreme Court ruled that the presence of TV equipment did not make it impossible for a fair trial to occur, leaving it up to each state to implement its own system. The ruling opened the door for the debut of Court TV (now truTV) in 1991 and the televised O.J. Simpson trial of 1994 (the most publicized case in history). All states today allow television coverage of cases, although most states place certain restrictions on coverage of courtrooms, often leaving it up to the discretion of the presiding judge. While U.S. federal courts now allow limited TV coverage of their trials, the Supreme Court continues to ban TV from its proceedings; in 2000, however, the Court broke its anti-radio rule by permitting delayed radio broadcasts of the hearings on the Florida vote recount case that determined the winner of the 2000 presidential election. And during the COVID-19 pandemic in 2020, the Court permitted the public to listen to a live audio feed of its teleconference oral arguments via C-SPAN and other media.

As libel law and the growing acceptance of courtroom cameras indicate, the legal process has generally, though not always, tried to ensure that print and other news media are able to cover public issues broadly, without fear of reprisals.

Film and the First Amendment

When the First Amendment was ratified in 1791, even the most enlightened leaders of our nation could not have predicted the coming of visual media such as film and television. Consequently, new communication technologies have not always received the same kinds of protection under the First Amendment as those granted to speech or print media, including newspapers, magazines, and books. Movies, in existence since the late 1890s, earned legal speech protection only after a 1952 Supreme Court decision. Prior to that, social and political pressures led to both censorship and self-censorship in the movie industry.

Social and Political Pressures on the Movies

CENSORSHIP A native of Galveston, Texas, Jack Johnson (1878–1946) was the first Black heavyweight boxing champion, from 1908 to 1914. His stunning victory over white champion Jim Jeffries (who had earlier refused to fight Black boxers) in 1910 resulted in race riots across the country and led to a ban on the interstate transportation of boxing films. A 2005 Ken Burns documentary, *Unforgivable Blackness*, chronicles Johnson's life.

During the early part of the twentieth century, movies rose in popularity among European immigrants and others from modest socioeconomic groups. This, in turn, spurred the formation of censorship groups, which believed that the movies would undermine morality. During this time, according to media historian Douglas Gomery, criticism of movies converged on four areas: "the effects on children, the potential health problems, the negative influences on morals and manners, and the lack of a proper role for educational and religious institutions in the development of movies."[26]

Public pressure on movies came both from conservatives, who saw them as a potential threat to the authority of traditional institutions, and from progressives, who worried that children and adults were more attracted to movie houses than to social organizations and education centers. As a result, civic leaders publicly escalated their pressure, organizing local *review boards* that screened movies for their communities. In 1907, the Chicago City Council created an ordinance that gave the police authority to issue permits for the exhibition of movies. By 1920, more than ninety cities in the United States had some type of movie censorship board made up of vice squad officers, politicians, and citizens. By 1923, twenty-two states had established such boards.

Meanwhile, social pressure began to translate into law as politicians, wanting to please their constituencies, began to legislate against films. Support mounted for a federal censorship bill. When Jack Johnson won the heavyweight

Bettmann/Getty Images

championship in 1908, boxing films became the target of the first federal censorship law aimed at the motion-picture industry. In 1912, the government outlawed the transportation of boxing movies across state lines, preventing these films from going anywhere beyond their originating states and severely diminishing their market value. The laws against boxing films, however, had more to do with Johnson's race than with concern over violence in movies. The first Black heavyweight champion, he was perceived as a threat by many in the white community.[27]

The first Supreme Court decision regarding film's protection under the First Amendment was handed down in 1915 and went against the movie industry. In *Mutual v. Ohio*, the Mutual Film Company of Detroit sued the state of Ohio, whose review board had censored a number of the distributor's films. On appeal, the case arrived at the Supreme Court, which unanimously ruled that motion pictures were not a form of speech but "a business pure and simple" and, like a circus, merely a "spectacle" for entertainment with "a special capacity for evil." This ruling would stand as precedent for thirty-seven years, although a movement to create a national censorship board failed.

Early Self-Regulation in the Movie Industry

As the film industry expanded after World War I, the influence of public pressure and review boards began to affect movie studios and executives who wanted to ensure control over their economic well-being. In the early 1920s, a series of scandals rocked Hollywood: actress Mary Pickford's divorce and quick marriage to actor Douglas Fairbanks, director William Desmond Taylor's unsolved murder, and actor Wallace Reid's death from a drug overdose. But the most sensational scandal involved aspiring actress Virginia Rappe, who died a few days after a wild party in a San Francisco hotel hosted by popular silent-film comedian Fatty Arbuckle. After Rappe's death, the comedian was indicted for rape and manslaughter in a case that was sensationalized in the press. Although two hung juries could not reach a verdict, Arbuckle's career was ruined. Censorship boards across the country banned his films. And even though Arbuckle was acquitted at his third trial in 1922, the movie industry chose to send a signal about the kinds of values and lifestyles it would tolerate: Arbuckle was banned from acting in Hollywood. He later resurfaced to direct several films under the name Will B. Goode.

In response to the scandals, particularly the first Arbuckle trial, the movie industry formed the Motion Picture Producers and Distributors of America (MPPDA) and hired as its president Will Hays, a former Republican National Committee chair. Also known as the Hays Office, the MPPDA attempted to smooth out problems between the public and the industry. Hays banned promising actors or movie extras with even minor police records. He also developed an MPPDA public relations division, which stopped a national movement for a federal law censoring movies.

The Motion Picture Production Code

During the 1930s, the movie business faced a new round of challenges. First, various conservative and religious groups—including the influential Catholic Legion of Decency—increased their scrutiny of the industry. Second, deteriorating economic conditions during the Great Depression forced the industry to tighten self-regulation in order to maintain profits and keep harmful public pressure at bay. In 1927, the Hays Office had developed a list of "Don'ts and Be Carefuls" to steer producers and directors away from questionable sexual, moral, and social themes. Nevertheless, pressure for a more formal and sweeping code mounted. As a result, in the early 1930s the Hays Office established the Motion Picture Production Code to officially stamp Hollywood films with a moral seal of approval.

The Code laid out its mission in its first general principle: "No picture shall be produced which will lower the moral standards of those who see it. Hence the sympathy of the audience shall never be thrown to the side of crime, wrongdoing, evil or sin." The Code dictated how producers and directors should handle "methods of crime," "repellent subjects," and "sex hygiene." A section on profanity outlawed a long list of phrases and topics, including "toilet gags" and "traveling salesman and farmer's daughter jokes." Under "scenes of passion," the Code dictated that "excessive and lustful kissing, lustful embraces, suggestive postures and gestures are not to be shown," and it required that "passion should be treated in such manner as not to stimulate the lower and baser emotions." The section on religion revealed the influences of a Jesuit priest and a Catholic publisher, who helped write the Code: "No film or episode may throw ridicule on any religious faith," and "ministers of religion . . . should not be used as comic characters or as villains."[28] (See "Global Village: The Challenges of Film Censorship in China" for more about censoring content.)

Adopted by 95 percent of the industry, the Code influenced nearly every commercial movie made between the mid-1930s and the early 1950s. It also gave the industry a relative degree of freedom, enabling the major studios to remain independent of outside regulation. When television arrived, however, competition from the new family medium forced movie producers to explore more adult subjects.

The *Miracle* Case

In 1952, the Supreme Court heard the *Miracle* case—officially *Burstyn v. Wilson*—named after Roberto Rossellini's film *Il Miracolo* (*The Miracle*). The movie's distributor sued the head of the New York Film Licensing Board for banning the film. A few New York City religious and political leaders considered the 1948 Italian film sacrilegious and pressured the film board for the ban. In the film, an unmarried peasant girl is impregnated by a scheming vagrant who tells her that he is St. Joseph and she has conceived the baby Jesus. The importers of the film argued that censoring it constituted illegal prior restraint under the First Amendment. Because such an action could not be imposed on a print version of the same story, the film's distributor argued that the same freedom should apply to the film. The Supreme Court agreed, declaring movies "a significant medium for the communication of ideas." The decision granted films the same constitutional protections as those enjoyed by print media and other forms of speech. Even more important, the decision rendered most activities of film review boards unconstitutional because these boards had been engaged in prior restraint. Although a few local boards survived into the 1990s to handle complaints about obscenity, most of them had disbanded by the early 1970s.

The MPAA Rating System

The current voluntary movie rating system—the model for the advisory labels for music, television, and video games—developed in the late 1960s after discontent again mounted over movie content, spurred on by such films as 1965's *The Pawnbroker*, which contained brief female nudity, and 1966's *Who's Afraid of Virginia Woolf?*, which featured a level of profanity and sexual frankness that had not been seen before in a major studio film. In 1966, the movie industry hired Jack Valenti to run the MPAA (the Motion Picture Association of America, formerly the MPPDA), and in 1968 he established an industry board to rate movies. Eventually, G, PG, R, and X ratings emerged as guideposts for the suitability of films for various age groups. In 1984, prompted by the releases of *Gremlins* and *Indiana Jones and the Temple of Doom*, the MPAA added the PG-13 rating and sandwiched it between PG and R to distinguish slightly higher levels of violence or adult themes in movies that might otherwise qualify as PG-rated films.

The MPAA copyrighted all rating designations as trademarks except for the X rating, which was gradually appropriated as a promotional tool by the pornographic film industry. In fact, between 1972 and 1989, the MPAA stopped issuing the X rating. In 1990, however, based on protests from filmmakers over movies with adult sexual themes that they did not consider pornographic, the industry copyrighted the new NC-17 rating—no children age seventeen or under. In 1995, *Showgirls* became the first movie to intentionally seek an NC-17 to demonstrate that the rating was commercially viable. However, many theater chains refused to carry NC-17 movies, fearing economic sanctions and boycotts by their customers or religious groups. Many newspapers also refused to carry ads for NC-17 films. Panned by the critics, *Showgirls* flopped at the box office. Since then, the NC-17 rating has not

The Challenges of Film Censorship in China

With Netflix and Amazon Prime Video each available in about 200 countries around the globe, and YouTube Go in more than 130 countries, the flow of movie and television content across borders is becoming stronger. Yet these digital streaming services aren't available in China, the world's No. 2 box-office market. Getting international films onto movie screens in China remains a challenge.

As it works to expand its domestic film audience by building more movie theaters in smaller-tier cities throughout the country, China maintains a strong government censorship office. There are no movie ratings in China, so if a film gains the approval of the censors, it can be shown to any age group without restriction.

The lack of age guidelines in China can result in a unique audience mix in theaters, with children at adult-oriented films. As a writer for the *South China Morning Post* reported, "I've seen things you would never see in other countries: from impatient kids crying and then running around the aisles during a screening of Greek rape-revenge thriller *The Enemy Within* (2013) . . . to small boys egging on Jackie Chan and his co-stars as they dispose of caricatured Japanese villains."[1]

Before 2013, filmmakers had to submit screenplays for review. Now they are relieved of that requirement, but all films still need to be reviewed before screening. Hong Kong filmmaker Jevons Au listed some of the content limitations for filmmakers: "Crime stories cannot have too many details. Stories of corruption must end with the bad guy behind bars. No ghosts. No gay love stories. No religion. No nudity. No politics."[2]

In fact, the U.S.-produced *Ghostbusters* (2016) remake was banned in China because its guidelines reject movies that "promote cults or superstition." Sony's attempt to rename the movie "Super Power Dare Die Team" for the Chinese market did not move the censors to change their position.[3] Chinese censors also reject films that portray time travel. *Back to the Future* (1985) was banned because Chinese censors said it was a "disrespectful portrayal of history." Since 2011, all films and television shows with time-travel stories have been banned, eliminating screenings of such movies.[4]

Making films in China is a challenge as well, and the rules are reminiscent of the United States' Motion Picture Production Code, which lasted from the 1930s to the 1950s. The regulations are somewhat vague, so Chinese filmmakers learn to find ways to create their films as close to their original vision as possible without raising the ire of the censors.

"Does it mean compromises?" asked Hong Kong filmmaker Johnnie To. "Yes, very many. But the alternative is no movie in China."[5] ●

Zhang Yun/China News Service/Getty Images

THE CHINESE GOVERNMENT censors movie content but has no age guidelines in theaters.

proved commercially viable, and distributors avoid releasing films with the rating, preferring either to label such films "unrated" or to cut the films to earn an R rating, as happened with *Clerks* (1994), *Eyes Wide Shut* (1999), *Brüno* (2009), and *The Wolf of Wall Street* (2013). Today, there is mounting protest against the MPAA, which many argue is essentially a censorship board that limits the First Amendment rights of filmmakers.

Expression in the Media: Print, Broadcast, and Online

During the Cold War, a vigorous campaign led by Joseph McCarthy, an ultraconservative senator from Wisconsin, tried to rid both government and the media of so-called communist subversives who were allegedly challenging the American way of life. In 1950, a publication called *Red Channels: The Report of Communist Influence in Radio and Television* aimed "to show how the Communists have been able to carry out their plan of infiltration of the radio and television industry." (This report on radio and TV followed the congressional investigation of suspected communists in the film industry in 1947.) *Red Channels*, inspired by McCarthy and produced by a group of former FBI agents, named 151 performers, writers, and musicians who were "sympathetic" to communist or left-wing causes. Among those named were Leonard Bernstein, Will Geer, Dashiell Hammett, Lillian Hellman, Lena Horne, Burgess Meredith, Arthur Miller, Dorothy Parker, Pete Seeger, Irwin Shaw, and Orson Welles. For a time, all were banned from working in television and radio even though no one on the list was ever charged with a crime.[29]

Although the First Amendment protects an individual's right to hold controversial political views, it doesn't explicitly protect people from discrimination by private employers based on political affiliation. In this case, the network executives either sympathized with the anticommunist movement or feared losing ad revenue. At any rate, the networks did not stand up to the communist witch-hunters. In order to work, a banned or "suspected" performer required the support of the program's sponsor. Though *I Love Lucy*'s Lucille Ball, who in sympathy with her father once registered to vote as a communist in the 1930s, retained Philip Morris's sponsorship of her popular program, other performers were not as fortunate. Although no evidence was ever introduced to show how entertainment programs circulated communist propaganda, by the early 1950s the TV networks were asking actors and other workers to sign loyalty oaths denouncing communism—a low point for the First Amendment.

The communist witch-hunts demonstrated key differences between print and broadcast protections under the First Amendment. Whereas licenses for printers and publishers had been outlawed since the eighteenth century, commercial broadcasters themselves had asked the federal government to step in and regulate the airwaves in the late 1920s. At that time, the broadcasters had wanted the government to clear up technical problems, channel noise, noncommercial competition, and amateur interference. Ever since, most broadcasters have been trying to free themselves from the government intrusion they once demanded.

Bettmann/Getty Images

THE HOUSE UN-AMERICAN ACTIVITIES COMMITTEE attempted to expose performers, writers, and musicians as "communist subversives," banning them from working in Hollywood without any evidence of criminal wrongdoing. In 1947, movie stars like Humphrey Bogart, Evelyn Keyes, and Lauren Bacall, pictured here, visited Washington to protest the committee's methods.

The FCC Regulates Broadcasting

Drawing on the argument that limited broadcast signals constitute a scarce national resource, the Communications Act of 1934 mandated that radio broadcasters operate in "the public interest, convenience, or necessity." Since the 1980s, however, with cable and, later, DBS increasing channel capacity, station managers have lobbied to own their airwave assignments. Although the 1996 Telecommunications Act did not grant such ownership, stations continue to challenge the "public interest" statute. They argue that because the government is not allowed to dictate content in newspapers, it should not be allowed to control broadcasting via licenses or mandate any broadcast programming.

Two cases—*Red Lion Broadcasting Co. v. FCC* (1969) and *Miami Herald Publishing Co. v. Tornillo* (1974)—demonstrate the historic legal differences between broadcast and print. The *Red Lion* case began when WGCB, a small-town radio station in Red Lion, Pennsylvania, refused to give airtime to Fred Cook, author of a book that criticized Barry Goldwater, the Republican Party's presidential candidate in 1964. The Reverend Billy James Hargis, a conservative radio preacher and Goldwater fan, verbally attacked Cook on the air. Cook asked for response time from the two hundred stations that carried the Hargis attack. Most stations complied, granting Cook free reply time. But WGCB offered only to sell Cook time. He appealed to the FCC, which ordered the station to give Cook free time. The station refused, claiming that its First Amendment rights granted it control over its program content. On appeal, the Supreme Court sided with the FCC, deciding that whenever a broadcaster's rights conflict with the public interest, the public interest must prevail. In interpreting broadcasting as different from print, the Supreme Court upheld the 1934 Communications Act by reaffirming that broadcasters' responsibilities to program in the public interest may outweigh their right to program whatever they want.

In contrast, five years later, in *Miami Herald Publishing Co. v. Tornillo*, the Supreme Court sided with the newspaper. A political candidate, Pat Tornillo Jr., requested space to reply to an editorial opposing his candidacy. Previously, Florida had a right-to-reply law, which permitted a candidate to respond, in print, to editorial criticisms from newspapers. Counter to the *Red Lion* decision, the Court in this case struck down the Florida state law as unconstitutional. The Court argued that mandating that a newspaper give a candidate space to reply violated the paper's First Amendment right to control what it chose to publish. The two decisions demonstrate that the unlicensed print media receive protections under the First Amendment that have not always been available to licensed broadcast media.

Dirty Words, Indecent Speech, and Hefty Fines

In theory, communication law prevents the government from censoring broadcast content. Accordingly, the government may not interfere with programs or engage in prior restraint, although it may punish broadcasters for **indecency** or profanity after the fact. Earlier in this chapter we defined *obscenity*, which is pornographic material defined in a three-pronged test by the U.S. Supreme Court as not meriting protection under the First Amendment. *Indecency* or *profanity* is generally communication that does not rise to that level. According to the FCC, "indecent content portrays sexual or excretory organs or activities in a way that is patently offensive but does not meet the three-prong test for obscenity. Profane content includes 'grossly offensive' language that is considered a public nuisance."[30] In practice, complaints about indecency and profanity have usually involved brief depictions of nudity or profanity on the public airwaves (either radio or TV broadcasts). Streaming television, cable TV, satellite radio, and other subscription services don't operate on public airwaves, so rules against indecency and profanity don't apply to them.

The current precedent for regulating broadcast indecency stems from a complaint to the FCC in 1973. In the middle of the afternoon, WBAI—a nonprofit Pacifica network station in New York—aired George Carlin's famous comedy sketch about the seven dirty words that could not be uttered by broadcasters. A father, riding in a car with his fifteen-year-old son, heard the program and complained to the FCC, which sent WBAI a letter of reprimand. Although no fine was issued, the station appealed on principle and won its case in court. The FCC, however, appealed to the Supreme Court. Although no court had legally defined indecency (and still hasn't), the Supreme Court's unexpected ruling in

Everett Collection, Inc

FAMILY GUY has been the target of hundreds of thousands of indecency complaints, a majority of which have been filed by the Parents Television Council. The Federal Communications Commission evaluates shows based on occurrences of explicit language, violent content, or sexually obscene depictions. *Family Guy* has been at the center of moral controversy and criticism since its debut in 1999.

the 1978 *FCC v. Pacifica Foundation* case sided with the FCC and upheld the agency's authority to require broadcasters to air adult programming at times when children are not likely to be listening. The Court ruled that so-called indecent programming, though not in violation of federal obscenity laws, was a nuisance and could be restricted to late-evening hours. As a result, the FCC banned indecent programs from most stations between 6:00 A.M. and 10:00 P.M. In 1990, the FCC tried to ban such programs entirely. Although a federal court ruled this move unconstitutional, it still upheld the time restrictions intended to protect children.

This ruling provides the rationale for the indecency fines that the FCC has frequently leveled against programs and stations that have carried indecent programming during daytime and evening hours. While Howard Stern and his various bosses held the early record for racking up millions in FCC indecency fines in the 1990s—before Stern moved to unregulated satellite radio—the largest-ever fine was for $3.6 million, leveled in 2006 against 111 TV stations that broadcast a 2004 episode of the popular CBS program *Without a Trace* that depicted teenage characters taking part in a sexual orgy.

In 2006, after the FCC fined broadcasters for several instances of "fleeting expletives"—one-off, unplanned incidents of these words during live TV shows—the four major networks sued the FCC on the grounds that their First Amendment rights had been violated. In its fining flurry, the FCC was partly responding to organized campaigns aimed at Howard Stern's vulgarity and at the Janet Jackson exposed-breast incident during the 2004 Super Bowl halftime show. In 2006, Congress substantially increased the FCC's maximum allowable fine to $325,000 per incident of indecency—meaning that one fleeting expletive in a live entertainment, news, or sports program could cost millions of dollars in fines, as it is repeated on affiliate stations across the country. But in 2010, a federal appeals court rejected the FCC's policy against fleeting expletives, arguing that it was constitutionally vague and had a chilling effect on free speech "because broadcasters have no way of knowing what the FCC will find offensive."[31]

Political Broadcasts and Equal Opportunity

In addition to indecency rules, another law that the print media do not encounter is **Section 315** of the 1934 Communications Act, which mandates that during elections, broadcast, cable, and direct broadcast satellite stations must provide equal opportunity and response time for qualified political candidates. (References to newer technologies were added as updates to the original act.) In other words, if radio and TV stations or networks give or sell time to one candidate, they must give or sell the same opportunity to other candidates. Local stations and networks have fought this law for years, complaining that it has required them to give marginal third-party candidates with little hope of success equal airtime in political discussions. Radio and television stations and networks claim that because no similar rule applies to newspapers or magazines, the law violates their First Amendment right to control content. In fact, because of this rule, many stations avoid all political programming, ironically reversing the rule's original intention. The TV networks managed to get the law amended in 1959 to exempt newscasts, press conferences, and other events—such as political debates—that qualify as news. For instance, if a senator running for office appears in a news story, opposing candidates cannot invoke Section 315 and demand free time. The FCC has subsequently ruled that interview portions of programs like the *700 Club* and *TMZ* also count as news.

Due to Section 315, many stations from the late 1960s through the 1980s refused to air movies starring Ronald Reagan. Because his film appearances did not count as bona fide news stories, politicians opposing Reagan as a presidential candidate could demand free time in markets that ran old Reagan movies. For the same reason, TV stations in California banned the broadcast of Arnold Schwarzenegger movies in 2003, when he became a candidate for governor. And in November 2015, candidate Donald Trump's twelve minutes of screen time as the host of *SNL* opened the door for his competitors to demand equal screen time from NBC.

However, supporters of the equal opportunity law argue that it has provided forums for lesser-known candidates representing views counter to those of the Democratic and Republican parties, further noting that the other main way for alternative candidates to circulate their messages widely is to buy political ads, thus limiting serious outside contenders to wealthy candidates.

The Demise of the Fairness Doctrine

Considered an important corollary to Section 315, the **Fairness Doctrine** was to controversial issues what Section 315 is to political speech. Initiated in 1949, this FCC rule required stations to

1. air and engage in controversial-issue programs that affected their communities, and
2. provide competing points of view when offering such programming.

Over the years, broadcasters argued that mandating opposing views every time a program covered a controversial issue was a burden not required of the print media, and that it forced many of them to refrain from airing controversial issues. As a result, the Fairness Doctrine ended with little public debate in 1987 after a federal court ruled that it was merely a regulation rather than an extension of Section 315 law.

Since 1987, however, periodic support for reviving the Fairness Doctrine has surfaced. Its supporters argue that broadcasting is fundamentally different from—and more pervasive than—print media, requiring greater accountability to the public. Although many broadcasters disagree, supporters of fairness rules insist that as long as broadcasters are licensed as public trustees of the airwaves—unlike newspaper or magazine publishers—legal precedent permits the courts and the FCC to demand responsible content and behavior from radio and TV stations.

Communication Policy and the Internet

Many have looked to the Internet as the one true venue for unlimited free speech under the First Amendment because it is not regulated by the government, it is not subject to the Communications Act of 1934, and little has been done in regard to self-regulation. Its global expansion is comparable to that of the early days of broadcasting, when economic and technological growth outstripped law and regulation. At that time, noncommercial experiments by amateurs and engineering students provided a testing ground that commercial interests later exploited for profit. In much the same way, amateurs, students, and various interest groups have explored and extended the communication possibilities of the Internet, which were then transformed into money-making ventures by for-profit corporations.

Net Neutrality and the Rise of Corporate Power on the Internet

Initial debates about the Internet focused on First Amendment issues, such as pornography, obscenity, and how the government should control such speech. However, as the Internet developed (in a mostly unregulated market) and powerful companies came to dominate it, the bigger threat in the United States became big digital corporations. These corporations planned to charge some Internet services more money for better access and higher speeds, undermining the democratic principle that every service would receive equal access to the Internet.

In response, the FCC created net neutrality rules for wired (cable and DSL) broadband providers in late 2010, requiring that they provide the same level of access to all Internet services and content. But the FCC's net neutrality rules were rejected by federal courts twice. The courts argued that because the FCC had not defined the Internet as a utility, it couldn't regulate it in this manner. Telecommunication companies were pleased with the decision, as they don't want any rules governing how they distribute access to the Internet. However, many citizens and entrepreneurs opposed an unregulated system that would allow telecommunication companies to create fast lanes (for those who pay more) and slow lanes on the Internet. The debate generated a record number of comments to the FCC, the vast majority in favor of net neutrality.[32]

In February 2015, the FCC reclassified broadband Internet as a Title II utility and voted to approve net neutrality rules, which were upheld by the U.S. Court of Appeals in 2016, enabling the FCC to enforce open Internet standards on wired and mobile networks. Yet the battle over net neutrality continued, especially as presidential administrations changed. Donald Trump's appointee to chair the FCC moved to reconsider net neutrality, and in December 2017, the FCC voted to repeal the 2015 net neutrality policy on a 3–2 party-line vote, effective June 2018. Attorneys general from twenty-one states and the District of Columbia filed lawsuits to challenge the decision, and California disregarded the federal decision by adopting its own state-level net neutrality law. In 2021, Joe Biden's administration dropped the Trump administration's challenge to California's law, and Biden named his own chairperson of the FCC, who was an advocate of a national net neutrality policy.

Section 230: Should Corporations Be Responsible for Internet Content?

Although many digital corporations are happy to determine how Internet service gets provided, they do not want to be held responsible for the content that circulates on the Internet. As noted in Chapter 2, *Section 230* of the Communications Decency Act, approved in 1996, protects "interactive computer service" companies from civil liability for carrying or restricting material "that the provider or user considers to be obscene, lewd, lascivious, filthy, excessively violent, harassing, or otherwise objectionable, whether or not such material is constitutionally protected."[33] In practical terms, it means Facebook, Google, Twitter, Apple, and any other Internet content companies or providers are not responsible for any of the material on their platforms.

The legislative authors of Section 230 were hopeful for the Internet in 1996, writing that "The Internet and other interactive computer services offer a forum for a true diversity of political discourse, unique opportunities for cultural development, and myriad avenues for intellectual activity." Although the Internet still has all those things more than a quarter century later, there is also a good deal of content that those legislators never imagined: a flood of Russian propaganda, QAnon conspiracy theories, terrorist recruitment sites, and white supremacists live-streaming their own mass shootings. There are also many platforms, including Gab and Parler, that have a reputation for hosting extremist groups such as white supremacists, neo-Nazis, and conspiracy theorists.

Internet corporations are now at the forefront of determining what content belongs online and what does not. Responding to complaints, many mainstream social media platforms have stepped up enforcement of violations

FACEBOOK CEO Mark Zuckerberg testifies virtually before the House Judiciary Subcommittee in 2020.

Mandel Ngan/Pool/Getty Images

of their user agreements, resulting in such actions as placing warning messages over posts, taking down certain videos, removing apps from the Apple App Store or Google Play, and temporarily freezing or permanently suspending some accounts. Apple and TikTok already have strict policies regarding indecency and profanity, among other things. Other Internet media have been more liberal.

The question regarding Section 230 becomes, What's best for the common good? Democrats tend to want Section 230 altered or revoked, so that private companies take more responsibility and more proactively moderate objectionable material on their platforms. For example, in January 2020, then primary candidate Joe Biden said, "For [Facebook CEO Mark] Zuckerberg and other platforms . . . it should be revoked because it is not merely an internet company. It is propagating falsehoods they know to be false."[34] Many Republicans—concerned over actions such as former president Trump's permanent suspension from Twitter after tweets regarding the January 6, 2021, storming of the U.S. Capitol—aim for Section 230 to be altered or revoked but for different purposes: to punish platforms "that moderate in a biased or otherwise discriminatory way" or, in a more extreme proposal, to prevent companies from "removing content that they found 'objectionable.'"[35]

Section 230 was approved with the hopes that the Internet would flourish "to the benefit of all Americans," as the law states. But at this point, the question of what Internet corporations should do to benefit all Americans has no simple answer.

launchpadworks.com

Bloggers and Legal Rights
Legal and journalism scholars discuss the legal rights and responsibilities of bloggers.

Discussion: What are some of the advantages and disadvantages of audiences turning to blogs—rather than traditional sources—for news?

The First Amendment and Democracy

For most of our nation's history, citizens have counted on journalism to monitor abuses in government and business. During the muckraking period, writers like Upton Sinclair and Ida Tarbell made strong contributions by reporting on corporate expansion and the corruption that accompanied it. During the Watergate era in the 1970s, *Washington Post* journalists Bob Woodward and Carl Bernstein investigated treachery in the Nixon White House, eventually leading to the president's resignation. More recently, media investigations have given critical evidence to movements against sexual assault, police brutality, and political corruption. The First Amendment's power has been most effective when news media have engaged in their watchdog function over government and also investigated corporations and other institutions for bad conduct.

Although good journalism, and its support by the First Amendment, is still needed today, our current media system is a different kind of beast. The traditional gatekeepers of determining commonly held truths—the news media—are being challenged by an overwhelming amount of information. The most powerful gatekeepers are now Internet corporations, both big and small. In addition, as recent years have demonstrated, we all don't live in the same public sphere. Subscription streaming services, individual social media accounts, partisan cable news channels, and various corners of the Internet mean we are less likely to share the same consensus narratives and—more troubling—less likely to share the same basic truths.

The First Amendment, which is about preventing government censorship of speech and the media, may give us a way forward, particularly regarding the power that private corporations have over our communication. Free speech and a free press can be used to enable important societal discussions about the media environment, including how we think the media should best serve public life and free expression. As there are no clear gatekeepers now, all of us have the ability and the responsibility to become media watchdogs ourselves—watchdogs who learn from the past, care about the present, and are willing to map the media's future.

Chapter REVIEW 16

LAUNCHPAD FOR *MEDIA & CULTURE*

 **LaunchPad**
macmillan learning

Visit LaunchPad for *Media & Culture* at **launchpadworks.com** for additional learning tools:

⊙ QUIZ YOURSELF WITH LEARNINGCURVE

LearningCurve uses gamelike quizzing to help you review important concepts from this chapter. LearningCurve includes multiple-choice questions at different levels of difficulty, hints, feedback for both right and wrong answers, and links to e-book material for easy reference. Which answer choice would you select for the sample question below?

> Mary and Gary were a songwriting team in the 1920s. Their biggest hit, "Tallahassee Twirl," is now in the public domain. What does this mean?
>
> ○ Anyone can use the song "Tallahassee Twirl" without paying copyright royalties.
>
> ○ Mary's and Gary's heirs now own the copyright to "Tallahassee Twirl."
>
> ○ All earnings from the use of the song "Tallahassee Twirl" are applied to public works projects.
>
> ○ The copyright to "Tallahassee Twirl" is owned by the federal government.

⊙ WATCH VIDEO ON KEY CONCEPTS

LaunchPad includes clips from movies, TV shows, online sources, and other media texts, in addition to interviews with media experts and newsmakers. In the videos for this chapter, you'll consider topics like sexting and obscenity and the legal rights and responsibilities of bloggers.

⊙ PRACTICE THE CRITICAL PROCESS

Use the Media Literacy Activity to walk through the critical process step-by-step and develop a critical perspective on what media means to you. In this chapter's activity, you'll compare and contrast how media is regulated in various countries.

⊙ LEARN KEY TERMS

Review the definitions for this chapter's key terms with LaunchPad's easy-to-use online flash cards. Try the cards in practice mode, or quiz yourself as a way to focus your study efforts. (Page numbers indicate where each term is highlighted within the chapter.)

authoritarian model, 459
communist or state
  model, 459
social responsibility
  model, 459
Fourth Estate, 459
libertarian model, 459
prior restraint, 460

copyright, 463
public domain, 464
libel, 465
slander, 465
actual malice, 465
qualified privilege, 466
opinion and fair comment,
  466

obscenity, 466
right to privacy, 467
gag orders, 470
shield laws, 470
indecency, 477
Section 315, 478
Fairness Doctrine, 479

REVIEW QUESTIONS

Use the questions below to revisit key themes and concepts from this chapter.

Need more practice? Access the LearningCurve multiple-choice questions for this chapter, which are available on LaunchPad. All questions include feedback for correct and incorrect answers, hints, and links to e-book material.

Free Expression Today

1. How is the First Amendment's relationship to the government different from its relationship to media corporations?

How We Got Here: The Origins of Free Expression and a Free Press

2. Explain the various models of the news media that exist under different political systems.

3. What is the basic philosophical concept that underlies America's notion of free expression?

4. What was the Sedition Act of 1798, and what role did it have in the First Amendment?

The Limits to Freedom of Expression in the United States

5. How has censorship been defined historically?

6. What is the public domain, and why is it an important element in American culture?

7. Why is the case of *New York Times v. Sullivan* so significant in First Amendment history?

8. How has the Internet changed battles over what constitutes obscenity?

9. What issues are at stake when First Amendment and Sixth Amendment concerns clash?

Film and the First Amendment

10. Why were films not constitutionally protected as a form of speech until 1952?

11. Why did film review boards develop, and why did they eventually disband?

12. How did both the Motion Picture Production Code and the current movie rating system come into being?

Expression in the Media: Print, Broadcast, and Online

13. What's the difference between obscenity and indecency?

14. What is the significance of Section 315 of the Communications Act of 1934?

15. Why didn't broadcasters like the Fairness Doctrine?

16. Why is Section 230 so relevant to the debates over content moderation on the Internet?

The First Amendment and Democracy

17. How might the First Amendment give us a path forward in a media environment that challenges traditional gatekeepers with such an abundance of information?

QUESTIONING THE MEDIA

Use these critical-thinking questions to reflect on your own experiences with media and apply your understanding of media concepts.

For an interactive experience, questions on topics like these are available for the iClicker student response system. Instructors can download the questions from LaunchPad and use them to poll the class.

1. Have you ever had an experience in which you thought personal or public expression went too far and should be curbed? Explain. How might you remedy this situation?

2. If you owned a community newspaper and had to formulate a policy about what kinds of public comments could appear on your Internet site, would you choose to moderate comments in any way? If you chose to moderate (and eliminate) some comments, would you be acting as a censor in this situation? Why or why not?

3. Should the United States have a federal shield law to protect reporters?

4. What do you think of the current movie rating system? Should it be changed? Why or why not?

5. Should corporations, unions, and rich individuals be able to contribute any amount of money they want to support particular candidates and pay for TV ads? Why or why not?

A Guide to Identifying Fake News

Fake news, which has dominated the real news headlines in recent years, is unfortunately about as old as the United States itself, which is where it got its start—not surprisingly, with politics.

The presidential election of 1800 pitted two bitter rivals against each other: incumbent president John Adams of the Federalist Party (whose members included George Washington and Alexander Hamilton) and his challenger, Vice President Thomas Jefferson of the Democratic-Republican Party (whose members included James Madison and Aaron Burr).[1]

As we discussed in Chapter 8, many of America's earliest newspapers were part of either the *commercial press*, which catered to the merchant class, or the *partisan press*, which fervidly argued for the platform of the political party that subsidized the paper. Predictably, it was the partisan press that resorted to fake news—intentional *disinformation* in this case, which is false or misleading information spread knowingly by people with malicious intent (see Chapter 2). As the presidential campaign of 1800 grew more heated, some Federalist newspapers, allied with Adams, began to publish stories that Jefferson had died.[2] Eventually the truth came out in rival newspapers that he was alive and well—although it took time to spread the word, given the very slow pace of newspaper distribution by mail at the time—and Jefferson ultimately won the election.

Fast-forward to the 1830s, when another kind of fake news emerged. At that time, penny press papers (costing one cent) were a new form of journalism that targeted a mass readership, standing in contrast to the more expensive and narrowly focused six-cent partisan and commercial newspapers then in circulation.

The first notable penny press newspaper was the *New York Sun*, founded in 1833 by twenty-three-year-old printer Benjamin Day. Day's attempts to reach mass working-class audiences included such innovations as publishing human-interest stories that highlighted crime and scandal, and hiring newsboys to hawk newspapers on the street by shouting out headlines. His paper became New York's biggest, and he was soon contending with competing penny press start-ups intent on grabbing market share.

To help keep up with the demand for enticing news stories and outperform the competition, in 1835 Day hired writer Richard Adams Locke, who had arrived from England just a few years earlier. Locke had great success with his first story for the *Sun*, which focused on a long, lurid murder trial. Next, he wanted to write an even bigger story. Locke, who had a decent scientific acumen, was both fascinated and annoyed by some contemporary scientists who theorized that life and civilizations existed on the moon. He decided to satirize their way of thinking by writing a compelling account of life on the moon, concocting a made-up species of moon dwellers called "man-bats."

Beginning on August 25, 1835, the *Sun* ran a series of six articles describing a lunar Eden, complete with oceans, mountains, waterfalls, and cities featuring columned temples and coliseums.[3] Populating this lunar landscape were familiar and fantastic lunar beasts, including sheep, unicorns, beaver-like creatures on two legs, and four-legged giraffe-like creatures. The most sensational moon creatures, though, were the highest-order beings—flying families of *Vespertilio-homo*, Locke's pseudo-scientific Latin name for man-bats. The *Sun* described them this way: "They averaged four feet in height, were covered, except on the face, with short and glossy copper-colored hair, and had wings composed of a thin membrane, without hair, lying snugly upon their backs, from the top of their shoulders to the calves of their legs."[4]

Locke anonymously wrote the series but attributed the "Great Astronomical Discoveries" he described therein to a mix of real and fake elements. The discoveries were purportedly made by Sir John Herschel (a real astronomer of the time, whose father, also an astronomer, discovered Uranus) through his new hydro-oxygen telescope (which sounded somewhat believable but didn't exist) at the Cape of Good Hope in South Africa (a real place, where Herschel did indeed have an observatory), as reported in a supplement to the *Edinburgh Journal of Science* (a real journal but one that had ceased publication in 1832).[5]

LUNAR ANIMALS Within twenty-four hours of its first "man-bat" stories, the *New York Sun* began to sell thousands of copies of a lithograph titled *Lunar Animals and Other Objects, Discovered by Sir John Herschel in His Observatory at the Cape of Good Hope and Copied from Sketches in the "Edinburgh Journal of Science."*

Library of Congress, Prints & Photographs Division, [LC-DIG-pga-02667]

The *Sun*'s series, now known as "the Great Moon Hoax," consumed New Yorkers, and news of it spread to the Midwest. But after the series ended and suspicions grew that the story was fake, it drew sharp criticism from James Gordon Bennett, editor of the competing *New York Herald*, who wrote, "When that paper in order to get money out of a credulous public, seriously persists in averting its truth, it becomes highly improper, wicked, and in fact a species of impudent swindling."[6]

The *Sun*'s satire-turned-hoax emerged at a time when others were also playing on Americans' hopes, dreams, and gullibility. In that same year of 1835, a young P. T. Barnum created an exhibition featuring Joice Heth, an approximately 80-year-old Black woman who Barnum claimed was the 161-year-old nurse of George Washington. Thus began a long career for Barnum in exhibiting human beings he referred to as "curiosities" and bringing his masterful (and sometimes demeaning) promotional skills to traveling circuses and sideshows.

From Claypoole's Daily Advertiser, of July 3.

A gentleman who left Fredericktown last Friday, informs, that an account had been received there of the Death of THOMAS JEFFERSON, Esq. Vice-President of the United States, at his seat near Charlotteville. This report was corroborated last evening by a gentleman directly from Baltimore, who says that the same account had been received there from Winchester, and that it was generally believed.

DISINFORMATION, CIRCA 1800 The *Albany Gazette*, a Federalist newspaper (the party opposing Thomas Jefferson, who was a leader of the Democratic-Republican Party), reprinted this phony account of the death of Jefferson on July 7, 1800. Actually alive and well, Jefferson ended up winning the election.

The events of the 1800 election and the 1835 Great Moon Hoax are some of the earliest examples of fake news in the United States and demonstrate a tendency on the part of the media, politicians, and other performers to deceive the public—a tendency that has not abated as the country has grown older.

Defining Fake News

There are a number of definitions for fake news. At its most basic, fake news is news that is fraudulent. But in our complicated political world, it is important to distinguish the two main ways in which the term "fake news" is used. In the first sense, the term can be used as "an accusatory speech act," according to German philosophy professor Axel Gelfert. This is most commonly seen in cases where powerful politicians call something "fake news" with the hope of discrediting information with which they disagree. (See Chapter 14 for contemporary examples of fake news allegations in the United States and other countries.) In cases like these, the accusatory speech act is rarely followed by evidence showing that the information in question is actually fake.

The other way in which fake news is used is in reference to what Gelfert calls "a class of purportedly factual claims that are epistemically deficient."[7] In other words, it's something that pretends to be news (like the man-bat) but doesn't hold up to analysis. This is the type of fake news that we will primarily focus on here. Intent is important, and this flavor of fake news is purposely false (as opposed to news that is accidentally incorrect). Gelfert offers a useful definition: "Fake news is the deliberate presentation of (typically) false or misleading claims *as news*, where the claims are misleading *by design*."[8]

Types of Fake News

With technological advances today that go far beyond what existed at the time of the 1800 election and the 1835 Great Moon Hoax, fake news has grown in sophistication and speed.[9] Today, it is a phenomenon that spans five general categories of activity (see the figure provided). The five types of fake news constitute a continuum, from more helpful to democracy (satire) to more harmful to democracy (information anarchy).

| Satirists | Hoaxes and hucksters | Opinion entrepreneurs | Propagandists | Information anarchists |

More HELPFUL to democracy ←——————————————————→ More HARMFUL to democracy

Satirists

Satire has been around for centuries, but more recent work includes *Saturday Night Live*, *The Daily Show* (which Comedy Central once promoted as "America's Most Trusted Name in Fake News"), the *Onion*, *Full Frontal with Samantha Bee*, *Last Week Tonight with John Oliver*, and Seth Meyers's "A Closer Look" segments on his *Late Night* program.

Satire wears its "fake news" badge openly, using humor and detailed research to critique the news media and our political system. When done well, satire can be extremely effective as a critical voice, as *Time* magazine argued about Jon Stewart's time hosting *The Daily Show* from 1999 to 2015:

> So Stewart wasn't an actual news anchor. What his show did with comedy was a kind of journalism nonetheless, using satire and some thorough research of source material to analyze the news. . . . Any honest media critic knew that Stewart was doing the job better than the rest of us.[10]

The notable twentieth-century literary critic Northrup Frye pointed out satire's use of irony and sarcasm as a form of social criticism. He defined satire as "poetry assuming a special function of analysis, that is, of breaking up the lumber of stereotypes, fossilized beliefs, superstitious terrors, crank theories, pedantic dogmatisms, oppressive fashions, and all other things that impede the free movement of society."[11] He envisioned the satirist as a giant-killer of sorts, bravely speaking truth to power. Do you have a favorite satirist?

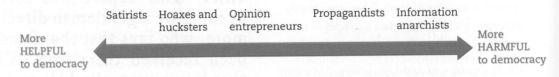

the ONION

HOME   LATEST   POLITICS   SPORTS   LOCAL   ENTERTAINMENT   THE TOPICAL   OGN

NEWS IN BRIEF

Biden Unveils $4 Trillion Bill For Dinosaur Statues, Giant Twine Balls To Restore Nation's Crumbling Highway Attractions

SATIRE Established in Madison, Wisconsin, in 1988 and now headquartered in Chicago, the *Onion* remains one of the country's leading satirical sites, regularly parodying political figures and the mainstream news.

Hoaxes and Hucksters

P. T. Barnum is America's most famous **huckster**—someone who aggressively promotes or sells products of questionable authenticity or value—and that spirit continues today with circus sideshows and Ripley's Believe It or Not! Odditoriums. Charles Ponzi and his fraudulent financial Ponzi scheme, the forged but convincing Hitler Diaries, King Tut's death curse for those who dare to disturb the ancient pharaoh's tomb, the Bigfoot film clip, and the Loch Ness Monster all fit into this category, too. Hoaxes and hucksters are often harmless (there is entertainment pleasure in having one's gullibility tested), but they can cause real harm, particularly with Ponzi scheme financial hoaxes or Internet phishing scams in which people are promised enormous amounts of money and end up losing everything. Can you come up with a recent example of a media hoax?

Opinion Entrepreneurs

Opinion entrepreneurs are media outlets—from websites and talk radio to newspapers and cable news—that seek to influence the news and public agenda, often with false or inaccurate stories. Researchers Peter Dreier and Christopher R. Martin first wrote of this kind of fake news in 2010, identifying opinion entrepreneurs as "non-elite individuals, businesses, and quasi-political organizations who, often by virtue of a web page or blog, work outside the traditional definitions of those who influence the news and public agenda. . . . Their influence is magnified by the fact that they work collaboratively, as part of a network, echoing the same message."[12] Breitbart.com and the Gateway Pundit, for example, have become well-known opinion entrepreneur websites.[13]

The most effective deployment of opinion entrepreneurs' messages has been through the fringes of the conservative media, which began to develop in the United States in the 1970s (see Chapter 14).[14] The structure of the conservative media mimics that of the mainstream media, but it functions in an entirely different way. Conservative media outlets have a clear political agenda to elect conservative candidates and support conservative issues, while the mainstream news media in the United States is generally moderate—more liberal or progressive in terms of certain social values and, because its owners are usually large media conglomerates, more conservative in regard to economic values. Mainstream media outlets aren't value-free—being moderate is a political position as well—but they generally remain committed to the journalism of verification. Opinion entrepreneurs covet the attention of mainstream news outlets and seek to get their opinions accepted as legitimate.

The fake news of opinion entrepreneurs usually starts as a report by one outlet or pundit, which then expands into an increasingly bigger issue as more opinion entrepreneurs join in and amplify that particular interpretation of a story. The work of opinion entrepreneurs is aided by social media sites such as Twitter, Facebook, and YouTube, where both plausible falsehoods and factual stories and events circulate. One of the most infamous instances of opinion entrepreneurialism is the "birther" story, which questioned former president Barack Obama's U.S. citizenship (and thus his qualifications to be president). It became a major news story only after the allegations of not-yet-president Donald Trump and several birther websites continued to gain steam by being shared and reshared on social media sites.[15] What other stories can you think of that were fueled by opinion entrepreneurs?

Propagandists

Propagandists are official state actors who spread coordinated partisan messages meant to propagate a point of view. Today North Korea, China, and Russia are the most easily identifiable propagandists, each with a secure hold on major national media outlets (either by owning them outright or by influencing them through coercion) and a sophisticated system of news and media that supports the goals of the regime both within the country and outside its borders.

WND

FRONT PAGE POLITICS U.S. WORLD

WND EXCLUSIVE

Sheriff's probe finds Obama birth certificate 'fake'

Years of forensics investigation confirmed 'birther' suspicions right all along

By Bob Unruh
Published December 15, 2016 at 6:09pm

WND (WorldNetDaily)—a fringe ultra-conservative site established in Washington, D.C., in 1997—often publishes unsupported conspiracy theories, including the "birther" conspiracy against former president Barack Obama.

Russia, for example, uses the news agency Sputnik and television channel RT to communicate its propaganda, along with "covert channels . . . that are almost always untraceable," the *New York Times* reported. The *Times* also noted:

> The flow of misleading and inaccurate stories is so strong that both NATO and the European Union have established special offices to identify and refute disinformation, particularly claims emanating from Russia. The Kremlin's clandestine methods have surfaced in the United States, too.[16]

Information Anarchists

Finally, fake news can be created via people we will call **information anarchists**. These are actors who want to stir the pot, make people angry with outrageous statements and allegations, and create doubt and mistrust (sometimes called **gaslighting**) in order to undermine the legitimacy of genuine news itself and create the perception that the truth might never be determined.

In recent presidential campaigns, Internet trolls—who often meet on Twitter, Reddit, or message boards like 4chan and 8kun—have taken delight in creating disruption.[17] One of the strangest cases of the information-anarchy type of fake news was the lie spread on 4chan and subversive websites alleging that a Washington, D.C., neighborhood pizzeria "was the home base of a child abuse ring led by Hillary Clinton and her campaign chief, John D. Podesta."[18] The fake news story prompted one North Carolina man to arrive at the restaurant with an assault rifle and fire shots during his own "investigation" into the crime.[19]

The disproved QAnon conspiracy theory, which posits that there is a global child sex-trafficking ring run by politically elite Satanic baby-eating pedophiles, is also the work of information anarchists. (The "Pizzagate" conspiracy theory regarding Clinton and Podesta is an early version of the QAnon conspiracy theory.) QAnon first emerged in October 2017 on 8chan (now relaunched as 8kun),

THE TERM "GASLIGHT" comes from British dramatist Patrick Hamilton's 1938 play *Gas Light* and the 1944 American film adaptation *Gaslight*, directed by George Cukor and starring Ingrid Bergman and Charles Boyer. In the film, Boyer plays Gregory Anton, a scheming husband who tries to convince his young wife, Paula (Bergman), that she is going mad. For example, each night the gaslights in their home dim while Gregory secretly hunts in the sealed attic for jewels he wants to steal, but Gregory convinces Paula that the dimming of the gaslights is all in her imagination.

FilmPublicityArchive/United Archives/Getty Images

where a supposed political insider named Q posted cryptic messages revealing "secrets" about these political elites. QAnon's persistent and vocal campaign labeling the mainstream media as "fake news" and "the enemy of the people"—because, according to the conspiracy theory, political elites run the media and are constantly trying to tear down Donald Trump, who is the only person who can stop them and their sex-trafficking ring—served to undermine faith in the press, and it became a vicious spiral that sucked in the theory's mixed-up followers: When all the mainstream media messages were discredited, only the confusingly cryptic messages of Q were believed to have any legitimacy.

The Critical Process: Identifying Fake News

With so many forms of fake news in circulation, it might be helpful to be reminded of what genuine, authentic journalism does. In their classic *The Elements of Journalism*, Bill Kovach and Tom Rosenstiel spell out several important factors of good journalism. (See Chapter 14 for more detail.) Perhaps the most important of these elements are the following: "Journalism's first obligation is to truth; its first loyalty is to citizens; its essence is a discipline of verification; its practitioners must maintain an independence from those they cover; and it must serve as an independent monitor of power."[20]

As Kovach and Rosenstiel also argue, safeguarding the enterprise of real journalism "is something for which both [citizens and journalists] bear some responsibility."[21] So, how do we, as critical consumers, decipher whether or not the news we're reading is real or fake? And when fake news is in play, how do we determine how harmful a particular piece of fake news might be to democracy?

As developed in Chapter 1, a media-literate perspective involves mastering five overlapping critical stages that build on one another:

- *Description:* paying close attention, taking notes, and researching the subject under study
- *Analysis:* discovering and focusing on significant patterns that emerge from the description stage
- *Interpretation:* asking and answering "What does that mean?" and "So what?" questions about one's findings
- *Evaluation:* arriving at a judgment about whether something is good, bad, or mediocre, which involves subordinating one's personal taste to the critical "bigger picture" resulting from the first three stages
- *Engagement:* taking some action that connects our critical perspective with our role as citizens and watchdogs who question our media institutions, adding our voice to the process of shaping the cultural environment

In the activity that follows, you will have the opportunity to examine several news stories covering a recent major news event and evaluate the validity of those stories.

Description

As a class, select a major news event to examine. It should be an event with plenty of media coverage from many different perspectives. Select five stories from a range of mainstream news organizations and lesser-known information sources, whether articles from websites or YouTube videos (instructors or students may gather materials). Read the articles and watch the videos to determine their themes.

Analysis

To conduct an analysis of the materials you have gathered, fill out the form "Is It Fake News?" for each story (select YES or NO in the right columns).[22] What similarities and differences between sources do you see as you analyze the coverage? What patterns present themselves?

Is It Fake News?

Select a story and use the questions below to analyze whether it might be fake news. For each question, choose YES or NO.

Story reviewed: _____

Fake Presentation (Credibility of the News Organization)			
1	Does the story provide reporter bylines and bios?	YES	NO
2	Does the site hosting the story include the news organization's address, telephone number, and e-mail address? (Tip: This information may be located in an "About" or a "Contact Us" link.)	YES	NO
3	Does the site provide a link or an e-mail address to report corrections or errors (possibly in an "About" or a "Contact Us" link)?	YES	NO

False or Misleading Claims (Validity of Claims)			
4	Does the story provide verification by doing original reporting that cites multiple sources?	YES	NO
5	Does the story use credible and verified quotes and information? (Be wary of stories that accept potentially untruthful or incorrect quotes or allegations from sources at face value and fail to question them.)	YES	NO
6	Does the story cite original sources? (Be wary of stories that rely heavily on secondary sources—including social media posts and other published media reports—as their main form of reporting.)	YES	NO
7	Does the story cite external sources beyond its own news organization? (Be wary of stories that rely on other stories from the same outlet.)	YES	NO
8	Does the story check out? That is, do other reliable news organizations verify the information it contains?	YES	NO

Misleading by Design (Ethics of Purpose)			
9	Does the site disclose the company's ownership and funding sources?	YES	NO
10	Does the news organization provide an accessible code of ethics or a standards and procedures document?	YES	NO
11	Does the story consider a range of valid opinions rather than just one position?	YES	NO
12	Does the story call out misleading information rather than give equal weight to truth and lies in the name of fairness?	YES	NO
13	Does the headline accurately represent what the story says rather than exaggerate?	YES	NO
14	Does the story avoid personal attacks on certain people or sources?	YES	NO
15	Does the story avoid using stereotypes to make its point?	YES	NO

Interpretation

In stage three of the critical process, we try to determine the meanings of the patterns that became evident during the analysis stage. Remember the definition of fake news that we are working with: "the deliberate presentation of (typically) false or misleading claims *as news*, where the claims are misleading *by design*."

Sometimes legitimate news organizations can make mistakes or have substandard documentation about their company and reporting processes. But the more you circle NO, the more concerned you should be that the story contains disinformation that could qualify as fake news. As your next step, consult multiple trusted news sources to find verified facts about the topic(s) discussed in the story. Do they match what you read about in the story itself?

In any stories that have major deficiencies, meaning they include a number of NO answers, determine the most common problems in terms of the following categories:

- Fake Presentation (credibility of the news organization)
- False or Misleading Claims (validity of claims)
- Misleading by Design (ethics of purpose)

What is the motivation of specific media organizations or websites in presenting or framing the stories the way they did? Are they trying to promote a certain political perspective? Trying to confuse the news audience? Something else?

JAMES O'KEEFE: OUR VIDEO IS 'HARD EVIDENCE' AND 'SMOKING GUN' OF VOTER FRAUD

SOURCES Does the story accept potentially untruthful or incorrect quotes or allegations from sources at face value and fail to question them? (Headline is from Breitbart, September 29, 2020.)

Top General Says There's Little Evidence To Corroborate New York Times Report That Russia Placed Bounties On American Soldiers

REPORTING Does the story rely heavily on secondary sources—including social media posts and other published media reports—as its main form of reporting? (Headline is from the Daily Caller, September 16, 2020; story includes no original reporting.)

Daily Kos: Rules of the Road

PURPOSE Does the story consider a range of valid opinions rather than just one position? (Site disclosure is from the Daily Kos.)

SITE PURPOSE
This is a site for Democrats.

AUTHORSHIP AND WORD CHOICE Does the story provide reporter bylines and bios? Does the story avoid personal attacks on certain people or sources? (Headline is from RedState, October 6, 2020.) Note that the identity of the author, referred to only as "Bonchie," is unclear; clicking on the name brings up a photo of Jerry Seinfeld.

Joe Biden Has an Idea So Dumb That it Will Make Your Brain Melt

By Bonchie | Oct 06, 2020 9:00 AM ET

Evaluation

In the evaluation stage, you take all the work you did in the first three stages and make informed judgments. Consider any fake news stories you've read and the possible motivation(s) behind each story. Consider the category of fake news each story falls into (satire, hoaxes and hucksters, opinion entrepreneurship, propaganda, or information anarchy). Based on the category, how harmful might these particular news stories be to democracy? Why?

Engagement

The fifth stage of the critical process encourages you to take action and use your voice for change. Using the knowledge you've gained during this activity, start identifying—perhaps as a class— the media outlets that support the best investigative journalism and news reporting. Combating harmful fake news begins with understanding the process of journalism, reading brilliant news stories (there are many talented journalists out there), and building a repertoire of trusted news sources. With this knowledge comes even more engagement; for example, you could help combat harmful fake news by writing letters to the editor of fake news sites describing the research you have compiled against their argument. You could also write letters to the editor of legitimate news sources thanking them for telling the whole story. Or you could spread awareness on social media, encouraging others to think critically about the type of news they're consuming. Reconsider the way you may have participated in sharing viral stories on social media in the past, and help steer others toward legitimate news.

 KEY TERMS

Review the definitions for this case study's key terms with LaunchPad's easy-to-use online flash cards. Try the cards in practice mode, or quiz yourself as a way to focus your study efforts. (Page numbers indicate where each term is highlighted within the case study.)

satire, 488
huckster, 488

opinion entrepreneurs, 489
propagandists, 489

information anarchists, 490
gaslighting, 490

Notes

1 Media, Culture, and Communication: A Critical Approach

1. John Sands, "Americans Are Losing Faith in an Objective Media. A New Gallup/Knight Study Explores Why," Knight Foundation, August 4, 2020, https://knightfoundation.org/articles/americans-are-losing-faith-in-an-objective-media-a-new-gallup-knight-study-explores-why.

2. Patrick B. O'Sullivan and Caleb T. Carr, "Masspersonal Communication: A Model Bridging the Mass-Interpersonal Divide," *New Media and Society* 20, no. 3 (March 2018): 1161–1180, https://doi.org/10.1177/1461444816686104.

3. Neil Postman, *Amusing Ourselves to Death: Public Discourse in the Age of Show Business* (New York: Penguin Books, 1985), 19.

4. James W. Carey, *Communication as Culture: Essays on Media and Society* (Boston: Unwin Hyman, 1989), 203.

5. Postman, *Amusing Ourselves to Death*, 65. See also Elizabeth Eisenstein, *The Printing Press as an Agent of Change*, 2 vols. (Cambridge: Cambridge University Press, 1979).

6. Douglas Blanks Hindman and Kenneth Weigand, "The Big Three's Prime-Time Decline: A Technological and Social Content," *Journal of Broadcasting and Electronic Media* 52 (March 2008): 119.

7. Henry Jenkins, *Confronting the Challenges of Participatory Culture: Media Education for the 21st Century*, The John D. and Catherine T. MacArthur Foundation Reports on Digital Media and Learning (Cambridge, MA: MIT Press, 2009).

8. W. J. T. Mitchell and Mark B. N. Hansen, introduction to *Critical Terms for Media Studies*, ed. W. J. T. Mitchell and Mark B. N. Hansen (Chicago: University of Chicago Press, 2010), xii.

9. "The Common Sense Census: Media Use by Tweens and Teens, 2019," Common Sense Media, 2019, www.commonsensemedia.org/research/the-common-sense-census-media-use-by-tweens-and-teens-2019.

10. This model reflects decades of cultural studies scholarship about the relationship between culture and the production, circulation, and reception of media. Related sources include Stuart Hall, "Encoding and Decoding in the Television Discourse," in *Essential Essays*, vol. 1, *Foundations of Cultural Studies*, ed. David Morley (Durham, NC: Duke University Press, 2019), 257–276; Julie D'Acci, "Cultural Studies, Television Studies, and the Crisis in the Humanities," in *Television after TV: Essays on a Medium in Transition*, ed. Lynn Spigel and Jan Olsson (Durham, NC: Duke University Press, 2004), 418–445; David Croteau and William Hoynes, *Media/Society: Technology, Industries, Content, and Users*, 6th ed. (Thousand Oaks, CA: Sage, 2019).

11. Ta-Nehisi Coates, "The Broad, Inclusive Canvas of Comics," *Atlantic*, February 3, 2015, www.theatlantic.com/entertainment/archive/2015/02/the-broad-inclusive-canvas-of-comics/385080.

12. Carey Purcell, "*Black Panther* Breaks Yet Another Record by Making Twitter History," *Forbes*, March 20, 2018, www.forbes.com/sites/careypurcell/2018/03/20/black-panther-breaks-yet-another-record-by-making-twitter-history/#6eb9d67b21e8.

13. Jerome Bruner, *Making Stories: Law, Literature, Life* (New York: Farrar, Straus & Giroux, 2002), 8.

14. See Carey, *Communication as Culture*.

15. For a historical discussion of culture, see Lawrence Levine, *Highbrow/Lowbrow: The Emergence of Cultural Hierarchy in America* (Cambridge, MA: Harvard University Press, 1988).

16. "Charlie Brooker: The Dark Side of Our Gadget Addiction," *Guardian*, December 1, 2011, www.theguardian.com/technology/2011/dec/01/charlie-brooker-dark-side-gadget-addiction-black-mirror.

▣ EXAMINING ETHICS Telling Stories about "Voices We Seldom Hear"

1. Christopher R. Martin, "People Ain't Gonna Come to Work If They Don't Feel Safe," *Working-Class Perspectives*, May 11, 2020, https://workingclassstudies.wordpress.com/2020/05/11/people-aint-gonna-come-to-work-if-they-dont-feel-safe.

2. Tony Thompson, "The Meatpacking Industry and COVID-19," LERA Industry Councils, May 21, 2020, https://lera.memberclicks.net/covid19-webinars.

3. Tim Jamison, "Watch Now: 1,031 Cases Reported at Tyson," *Waterloo Courier*, May 7, 2020, https://wcfcourier.com/news/local/watch-now-1-031-cases-reported-at-tyson/article_f9c48267-51ed-5b7d-9bb2-eac2648fd558.html#tracking-source=home-trending.

4. Ryan J. Foley, "Families of 3 Deceased Workers Sue Tyson over Iowa Outbreak," Associated Press, June 25, 2020, https://apnews.com/780cd323a100bee5acf2bb15c0ff6d1b.

5. COVID-19 Strike Wave Interactive Map, accessed May 15, 2020, https://paydayreport.com/covid-19-strike-wave-interactive-map.

6. Society of Professional Journalists, "SPJ Code of Ethics," September 16, 2014, www.spj.org/ethicscode.asp.

2 The Internet and Digital Media

1. Lev Grossman and Matt Vella, "iNeed?" *Time*, September 22, 2014, p. 44.

2. "Fear of Missing Out," J. Walter Thompson Intelligence, May 2011, quoted in Eric Barker, "This Is the Best Way to Overcome Fear of Missing Out," *Time*, accessed April 27, 2018, http://time.com/collection/guide-to-happiness/4358140/overcome-fomo.

3. Andrew K. Przybylski, Kou Murayama, Cody R. DeHaan, and Valerie Gladwell, "Motivational, Emotional, and Behavioral Correlates of Fear of Missing Out," *Computers in Human Behavior* 29 (2013): 1841–1848.

4. Melissa G. Hunt, Rachel Marx, Courtney Lipson, and Jordyn Young, "No More FOMO: Limiting Social Media Decreases Loneliness and Depression," *Journal of Social and Clinical Psychology* 37, no. 10 (2018): 763.

5. Ed Diener and Robert Biswas-Diener, *Happiness: Unlocking the Mysteries of Psychological Wealth* (Malden, MA: Wiley-Blackwell, 2008), 51.

6. "Internet Users Distribution in the World—2020 Q1," Internet World Stats, March 3, 2020, www.internetworldstats.com/stats.htm.

See also "Key Internet Statistics to Know in 2020 (Including Mobile)," Broadband Search, www.broadbandsearch.net/blog/internet-statistics#post-navigation-0.

7. Christo Petrov, "25+ Impressive Big Data Statistics for 2020," TechJury, updated September 10, 2020, https://techjury.net/blog/big-data-statistics/#gref.

8. Richard Harris, "More Data Will Be Created in 2017 Than the Previous 5,000 Years of Humanity," *App Developer Magazine*, December 23, 2016, https://appdevelopermagazine.com/more-data-will-be-created-in-2017-than-the-previous-5,000-years-of-humanity-/.

9. John Perry Barlow, "A Declaration of the Independence of Cyberspace," Electronic Frontier Foundation, February 8, 1996, www.eff.org/cyberspace-independence.

10. Barry Schwartz, "Google's Search Knows about over 130 Trillion Pages," November 14, 2016, https://searchengineland.com/googles-search-indexes-hits-130-trillion-pages-documents-263378.

11. "Search Engine Market Share Worldwide: Aug 2019–Aug 2020," StatCounter Global Stats, accessed September 2020, https://gs.statcounter.com/search-engine-market-share.

12. "47 U.S. Code § 230—Protection for Private Blocking and Screening of Offensive Material," Cornell Law School Legal Information Institute, www.law.cornell.edu/uscode/text/47/230.

13. Wikipedia, s.v. "Wikipedia," updated September 30, 2020, https://en.wikipedia.org/wiki/Wikipedia.

14. SimpleA, "What Is the Semantic Web?" November 20, 2018, https://simplea.com/Articles/what-is-the-semantic-web.

15. Alex Colon, "Samsung's Family Hub Fridge Could Be the First Truly Smart Appliance," *PC Magazine*, January 7, 2016, www.pcmag.com/article2/0,2817,2497578,00.asp.

16. Gillian Reagan, "The Evolution of Facebook's Mission Statement," *Observer*, July 13, 2009, https://observer.com/2009/07/the-evolution-of-facebooks-mission-statement.

17. "Herstory," Black Lives Matter, https://blacklivesmatter.com/herstory.

18. Brendan I. Koerner, "Why ISIS Is Winning the Social Media War," *Wired*, March 2016, www.wired.com/2016/03/isis-winning-social-media-war-heres-beat.

19. Meagan Flynn, "No One Who Watched New Zealand Shooter's Video Live Reported It to Facebook, Company Says," *Washington Post*, March 19, 2019, www.washingtonpost.com/nation/2019/03/19/new-zealand-mosque-shooters-facebook-live-stream-was-viewed-thousands-times-before-being-removed/#comments-wrapper.

20. Kate Conger, Mike Isaac, and Sheera Frenkel, "Twitter and Facebook Lock Trump's Accounts After Violence on Capitol Hill," *New York Times*, updated January 14, 2021, www.nytimes.com/2021/01/06/technology/capitol-twitter-facebook-trump.html.

21. Mary L. Gray, "Negotiating Identities/Queering Desires: Coming Out Online and the Remediation of the Coming-Out Story," *Journal of Computer-Mediated Communication* 14, no. 4 (July 2009): 1162–1189.

22. Kazutoshi Sasahara, Wen Chen, Hao Peng, Giovanni Luca Ciampaglia, Alessandro Flammini, and Filippo Menczer, "Social Influence and Unfollowing Accelerate the Emergence of Echo Chambers," *Journal of Computational Social Science*, September 11, 2020, https://doi.org/10.1007/s42001-020-00084-7.

23. Sasahara et al., "Social Influence and Unfollowing."

24. Lawrence Lessig, *Remix: Making Commerce Thrive in the Hybrid Economy* (London: Bloomsbury, 2008).

25. Isaac Stanley-Becker, "Fox News Removes Manipulated Images from Coverage of Seattle Protests," *Washington Post*, June 13, 2020, www.washingtonpost.com/politics/2020/06/13/fox-news-removes-manipulated-images-coverage-seattle-protests.

26. Team Counterpoint, "US Smartphone Market Share: By Quarter," August 17, 2020, Counterpoint, www.counterpointresearch.com/us-market-smartphone-share; Wayne Duggan, "Latest E-Commerce Market Share Numbers Highlight Amazon's Dominance," Yahoo! Finance, February 4, 2020, https://finance.yahoo.com/news/latest-e-commerce-market-share-185120510.html?guccounter=1; "Social Media Stats Worldwide: Aug 2019–Aug 2020," StatCounter Global Stats, accessed September 2020, https://gs.statcounter.com/social-media-stats; "Desktop Operating System Market Share Worldwide: Aug 2019–Aug 2020," StatCounter Global Stats, accessed September 2020, https://gs.statcounter.com/os-market-share/desktop/worldwide; "Search Engine Market Share Worldwide: Aug 2019–Aug 2020."

27. Liz Pelly, "The Problem with Muzak," *Baffler*, December 2017, https://thebaffler.com/salvos/the-problem-with-muzak-pelly.

28. Paul Mozur, "China Moves Towards 'Digital Totalitarian State' as Surveillance Technology Continues to Advance," *Independent*, December 18, 2019, www.independent.co.uk/news/world/asia/china-technology-surveillance-network-totalitarian-state-spy-a9251971.html.

29. Chris Buckley and Paul Mozur, "How China Uses High-Tech Surveillance to Subdue Minorities," *New York Times*, May 22, 2019, www.nytimes.com/2019/05/22/world/asia/china-surveillance-xinjiang.html.

30. Buckley and Mozur, "How China Uses High-Tech Surveillance."

31. Chris Anderson and Michael Wolff, "The Web Is Dead. Long Live the Internet," *Wired*, August 17, 2010, www.wired.com/magazine/2010/08/ff_webrip. See also Charles Arthur, "Walled Gardens Look Rosy for Facebook, Apple—and Would-Be Censors," *Guardian*, April 17, 2012, www.guardian.co.uk/technology/2012/apr/17/walled-gardens-facebook-apple-censors.

32. Arthur, "Walled Gardens."

33. Barry M. Leiner, Vinton G. Cerf, David D. Clark, Robert E. Kahn, Leonard Kleinrock, Daniel C. Lynch, Jon Postel, Larry G. Roberts, and Stephen Wolff, "Brief History of the Internet," Internet Society, 1997, www.internetsociety.org/internet/history-internet/brief-history-internet.

34. David Goldman, "Slow Comcast Speeds Were Costing Netflix Customers," CNN Business, August 29, 2014, https://money.cnn.com/2014/08/29/technology/netflix-comcast/index.html.

35. Rebecca R. Ruiz, "F.C.C. Sets Net Neutrality Rules," March 12, 2015, *New York Times*, www.nytimes.com/2015/03/13/technology/fcc-releases-net-neutrality-rules.html.

36. Ted Johnson, "State Attorneys General File Suit over FCC's Net Neutrality Repeal," *Variety*, January 16, 2018, http://variety.com/2018/biz/news/net-neutrality-fcc-state-attorneys-general-sue-1202665803.

37. "Montana Pushes Back on FCC Ruling to Enforce Net Neutrality," NPR, January 27, 2018, www.npr.org/2018/01/27/581343532/montana-pushes-back-on-fcc-ruling-to-enforce-net-neutrality.

38. "Banned Websites Awareness Day," American Library Association, accessed September 30, 2020, www.ala.org/aasl/advocacy/bwad.

39. Raymond Zhong and Kai Schultz, "With India's TikTok Ban, the World's Digital Walls Grow Higher," *New York Times*, updated July 11, 2020, www.nytimes.com/2020/06/30/technology/india-china-tiktok.html.

40. Brooke Auxier, Lee Rainie, Monica Anderson, Andrew Perrin, Madhu Kumar, and Erica Turner, "Americans and Privacy: Concerned, Confused, and Feeling Lack of Control over Their Personal Information," Pew Research Center, November 15, 2019, www.pewresearch.org/internet/2019/11/15/americans-and-privacy-concerned-confused-and-feeling-lack-of-control-over-their-personal-information.

41. See Federal Trade Commission, "Privacy Online: Fair Information Practices in the Electronic Marketplace," May 2000, www.ftc.gov/sites/default/files/documents/reports/

privacy-online-fair-information-practices-electronic-marketplace
-federal-trade-commission-report/privacy2000.pdf.

42. Elizabeth Warren, "Here's How We Can Break Up Big Tech,"
Medium, March 8, 2019, https://medium.com/@teamwarren
/heres-how-we-can-break-up-big-tech-9ad9e0da324c.

43. "Justice Department Sues Monopolist Google for Violating
Antitrust Laws," United States Department of Justice, October 20,
2020, www.justice.gov/opa/pr/justice-department-sues-monopolist
-google-violating-antitrust-laws.

44. Electronic Frontier Foundation, "About EFF," www.eff.org/about.

45. David Bollier, "Saving the Information Commons," Remarks to
American Library Association Convention, Atlanta, June 15, 2002,
www.ala.org/acrl/aboutacrl/directoryofleadership/committees
/copyright/piratesbollier.

46. Shannon Schumacher and Nicholas Kent, "8 Charts on Internet Use
around the World as Countries Grapple with COVID-19," Pew Research
Center, April 2, 2020, www.pewresearch.org/fact-tank/2020/04/02
/8-charts-on-internet-use-around-the-world-as-countries-grapple
-with-covid-19.

47. Monica Anderson and Madhumitha Kumar, "Digital Divide Persists
Even as Lower-Income Americans Make Gains in Tech Adoption,"
Pew Research Center, May 7, 2019, www.pewresearch.org/fact
-tank/2019/05/07/digital-divide-persists-even-as-lower-income
-americans-make-gains-in-tech-adoption.

48. Andrew Perrin, "Digital Gap between Rural and Nonrural America
Persists," Pew Research Center, May 31, 2019, www.pewresearch
.org/fact-tank/2019/05/31/digital-gap-between-rural-and-nonrural
-america-persists.

49. Anderson and Kumar, "Digital Divide Persists."

50. Toussaint Nothias, "Access Granted: Facebook's Free Basics in
Africa," Media, Culture, and Society 42, no. 3 (2020): 329–348, https//
doi.org/10.1177/0163443719890530.

51. Larry Madowo, "Is Facebook Undermining Democracy in Africa?"
BBC News, May 23, 2019, www.bbc.com/news/world-africa-48349671.

◼ EXAMINING ETHICS Algorithmic Bias

1. Joy Buolamwini, "Artificial Intelligence Has a Problem with Gender and
Racial Bias. Here's How to Solve It," Time, February 7, 2019, https://
time.com/5520558/artificial-intelligence-racial-gender-bias.

2. Buolamwini, "Artificial Intelligence."

3. Kashmir Hill, "Wrongly Accused by an Algorithm," New York Times,
June 24, 2020, www.nytimes.com/2020/06/24/technology/facial
-recognition-arrest.html.

4. Buolamwini, "Artificial Intelligence."

5. Jamie Condliffe, "The Week in Tech: Algorithmic Bias Is Bad.
Uncovering It Is Good," New York Times, November 15, 2019, www
.nytimes.com/2019/11/15/technology/algorithmic-ai-bias.html.

◼ GLOBAL VILLAGE Social Media Fraud and Elections

1. Alan Finlayson, "Facebook: The Election's Town Square," Guardian,
February 14, 2010, www.theguardian.com/commentisfree/2010/feb/14
/social-networking-facebook-general-election.

2. Omar L. Gallaga, "We've Reached a Turning Point for Facebook—and
Its Users," Dayton Daily News, December 3, 2016, D6.

3. Bill Priestap, "Statement Before the Senate Select Committee on
Intelligence," Washington, D.C., June 21, 2017, www.fbi.gov/news
/testimony/assessing-russian-activities-and-intentions-in-recent-elections.

4. Elizabeth Weise, "Russian Fake Accounts Showed Posts to
126 Million Facebook Users," USA Today, October 30, 2017, www
.usatoday.com/story/tech/2017/10/30/russian-fake-accounts
-showed-posts-126-million-facebook-users/815342001.

5. Mark Zuckerberg, Facebook, September 3, 2020, www.facebook.com
/zuck/posts/10112270823363411.

6. Dana Priest and Michael Birnbaum, "Europe Has Been Working to
Expose Russian Meddling for Years," Washington Post, June 25, 2017,
www.washingtonpost.com/world/europe/europe-has-been-working-to
-expose-russian-meddling-for-years/2017/06/25/e42dcece-4a09-11e7
-9669-250d0b15f83b_story.html.

◼ MEDIA LITERACY AND THE CRITICAL PROCESS
Harnessing Creativity by Stepping Away from Your Phone

1. Manoush Zomorodi, Bored and Brilliant: How Spacing Out Can Unlock
Your Most Productive and Creative Self (New York: St. Martin's Press,
2017), 128.

2. Zomorodi, Bored and Brilliant, 127.

3. Zomorodi, Bored and Brilliant, 129.

4. Zomorodi, Bored and Brilliant, 145.

5. Zomorodi, Bored and Brilliant, 164.

3 Digital Gaming and the Media Playground

1. See Megan Farokhmanesh, "First Game Tournament, 'Intergalactic
Spacewar Olympics,' Held 40 Years Ago," Polygon, October 20, 2012,
www.polygon.com/2012/10/20/3529662/first-game-tournament
-intergalactic-spacewar-olympics-held-40-years. See also Chris Baker,
"Stewart Brand Recalls First Spacewar Video Game Tournament,"
Rolling Stone, May 25, 2016, www.rollingstone.com/culture/news
/stewart-brand-recalls-first-spacewar-video-game-tournament-20160525.

2. Andrew Paradise, "The History behind a $5 Billion eSports Industry,"
TechCrunch, November 3, 2016, https://techcrunch.com/gallery
/the-history-behind-a-5-billion-esports-industry.

3. "U Unveils Roster for First Varsity eSports Team," University of Utah,
October 4, 2017, https://unews.utah.edu/university-of-utah-unveils
-roster-for-first-varsity-esports-team. See also Aaron Reiss, "What It's
Like to Be a Varsity Esports Player," Sports Illustrated, January 4, 2017,
www.si.com/tech-media/2017/01/04/varsity-esports-team-columbia
-college-league-legends.

4. Activate, Activate Tech & Media Outlook 2018, October 17, 2017,
www.slideshare.net/ActivateInc/activate-tech-media-outlook-2018.
See also Paradise, "History behind a $5 Billion eSports Industry."

5. Phuc Pham, "Facebook's Giant Step into eSports May Be a Look at Its
Future," Wired, January 18, 2018, www.wired.com/story/facebook-esl
-esports-streaming-partnership.

6. Irwin A. Kishner, "Esports Leagues Set to Level Up with Permanent
Franchises," Forbes, October 3, 2017, www.forbes.com/sites
/kurtbadenhausen/2017/10/03/esports-leagues-grow-up-with
-permanent-franchises/#3f3f9be121d6.

7. Phuc Pham, "Esports Zerg-Rush the Olympics—but Can They Become
Official Events?" Wired, February 9, 2018, www.wired.com/story
/esports-pyeongchang-olympics.

8. John Koetsier, "Is Gaming Recession-Proof? Record Revenue for
$160 Billion Industry with 2.7 Billion Players," Forbes, June 19, 2020,
www.forbes.com/sites/johnkoetsier/2020/06/19/is-gaming-recession
-proof-record-revenue-for-160-billion-industry-with-27-billion-players;
Kevin Webb, "The $120 Billion Gaming Industry Is Going through
More Change Than It Ever Has Before, and Everyone Is Trying to Cash
In," Business Insider, October 1, 2019, www.businessinsider.com
/video-game-industry-120-billion-future-innovation-2019-9. See also
Seth Schiesel, "Facebook to Introduce an App for Gaming," New York
Times, April 19, 2020, www.nytimes.com/2020/04/19/technology
/facebook-app-gaming.html.

9. Jeremy C. Owens and Emily Bary, "How the Pandemic Has Changed Tech in Its First 100 Days," MarketWatch, June 20, 2020, www.marketwatch.com/story/how-the-pandemic-has-changed-tech-in-its-first-100-days-2020-06-18.

10. Matthew Humphries, "Facebook Gaming Streaming App Launches," *PCMag*, April 20, 2020, www.pcmag.com/news/facebook-gaming-streaming-app-launches.

11. Schiesel, "Facebook to Introduce an App for Gaming."

12. Erkki Huhtamo, "Slots of Fun, Slots of Trouble: An Archaeology of Arcade Gaming," in *Handbook of Computer Game Studies*, ed. Joost Raessens and Jeffrey Goldstein (Cambridge, MA: MIT Press, 2005), 10.

13. Huhtamo, "Slots of Fun, Slots of Trouble," 9–10.

14. Seth Porges, "11 Things You Didn't Know about Pinball History," *Popular Mechanics*, September 1, 2009, www.popularmechanics.com/technology/g284/4328211-new.

15. Matthew Lasar, "*Spacewar!*, the First 2D Top-Down Shooter, Turns 50," *Ars Technica*, October 25, 2011, http://arstechnica.com/gaming/2011/10/spacewar-the-first-2d-top-down-shooter-turns-50.

16. Peter Ross Range, "The Space Age Pinball Machine," *New York Times*, September 15, 1974, www.nytimes.com/1974/09/15/archives/the-space-age-pinball-machine.html.

17. John Markoff, "Company News; Sony Starts a Division to Sell Game Machines," *New York Times*, May 19, 1994, www.nytimes.com/1994/05/19/business/company-news-sony-starts-a-division-to-sell-game-machines.html.

18. "PlayStation Network Monthly Active Users Reaches 103 Million," Sony Interactive Entertainment, January 6, 2020, www.playstation.com/en-us/corporate/press-releases/2020/playstation-network-monthly-active-users-reaches-103-million.

19. Chris Gaither, "Microsoft Explores a New Territory: Fun," *New York Times*, November 4, 2001, www.nytimes.com/2001/11/04/business/microsoft-explores-a-new-territory-fun.html.

20. Guinness World Records, "Hardware History II," *Guinness World Records Gamer's Edition 2008* (London: Guinness World Records, 2008), 27.

21. Sarah E. Needleman, "Mobile-Game Revenue to Surpass Console and PC, Study Says," *Wall Street Journal*, April 21, 2016, www.wsj.com/articles/mobile-game-revenue-to-surpass-console-and-pc-study-says-1461265949. See also Newzoo, *Global Games Market Report 2020*, April 2020, p. 14, https://newzoo.com/products/reports/global-games-market-report.

22. Keith Stuart, "Nintendo Game Boy—25 Facts for Its 25th Anniversary," *Guardian*, April 21, 2014, www.theguardian.com/technology/2014/apr/21/nintendo-game-boy-25-facts-for-its-25th-anniversary.

23. Entertainment Software Association, *2019 Essential Facts about the Computer and Video Game Industry*, May 2019, p. 9, www.theesa.com/wp-content/uploads/2019/05/ESA_Essential_facts_2019_final.pdf.

24. Sam Anderson, "Just One More Game . . . Angry Birds, Farmville and Other Hyperaddictive 'Stupid Games,'" *New York Times*, April 4, 2012, www.nytimes.com/2012/04/08/magazine/angry-birds-farmville-and-other-hyperaddictive-stupid-games.html.

25. "*Woozworld*," accessed May 31, 2020, www.woozworld.com.

26. Adrienne Shaw, *Gaming at the Edge: Sexuality and Gender at the Margins of Gamer Culture* (Minneapolis: University of Minnesota Press, 2014).

27. Pierre Lévy, *Collective Intelligence: Mankind's Emerging World in Cyberspace* (New York: Basic Books, 1997), xxviii.

28. Alex Wiltshire, "The Precarious Business of Living off Modding," *PCGamer*, August 13, 2018, www.pcgamer.com/the-precarious-business-of-living-off-modding.

29. Patrick Shanley, "Twitch Competitors Nibble at Overall Streaming Market Share but Still Fall Far Behind," *Hollywood Reporter*, December 19, 2019, www.hollywoodreporter.com/news/twitch-competitors-nibble-at-streaming-market-share-but-still-fall-far-behind-1263453. See also SuperData, *2019 Year in Review: Digital Games and Interactive Media*, January 2020, www.superdataresearch.com/reports/2019-year-in-review.

30. Lisa Respers France, "Drake and Ninja's *Fortnite* Battle Sets a New Twitch Record," CNN, March 15, 2020, www.cnn.com/2018/03/15/entertainment/drake-ninja-fortnite-twitch-battle/index.html.

31. "Industry Demographics," Fantasy Sports & Gaming Association, accessed June 2, 2020, https://thefsga.org/industry-demographics.

32. World Health Organization, "Gaming Disorder," *International Classification of Diseases for Mortality and Morbidity Statistics*, September 14, 2018, www.who.int/news-room/q-a-detail/gaming-disorder.

33. Cameren Rogers, "WHO Calls 'Gaming Disorder' Mental Health Condition," WebMD, June 20, 2018, www.webmd.com/mental-health/addiction/news/20180620/who-recognizes-gaming-disorder-as-a-condition.

34. Andrew Salmon, "Couple: Internet Gaming Addiction Led to Baby's Death," CNN, April 1, 2010, www.cnn.com/2010/WORLD/asiapcf/04/01/korea.parents.starved.baby/index.html.

35. "Rating Categories," Entertainment Software Rating Board," accessed June 6, 2020, www.esrb.org/ratings-guide; Entertainment Software Association, *2019 Essential Facts*, p. 20.

36. Anita Sarkeesian and Carolyn Petit, "Female Representation in Videogames Isn't Getting Any Better," *Wired*, June 14, 2019, www.wired.com/story/e3-2019-female-representation-videogames.

37. Patrick Markey and Charlotte N. Markey, "Vulnerability to Violent Video Games: A Review and Integration of Personality Research," *Review of General Psychology* 14, no. 2 (2010): 82–91.

38. Entertainment Software Association, *2019 Essential Facts*, pp. 14–19.

39. National Center for Women & Information Technology, "NCWIT Program Descriptions and Outcomes," April 15, 2020, www.ncwit.org/sites/default/files/file_type/ncwit_program_outcomes04152020web.pdf.

40. International Game Developers Association, *Developer Satisfaction Survey 2019 Summary Report*, November 20, 2019, p. 12, https://s3-us-east-2.amazonaws.com/igda-website/wp-content/uploads/2020/01/29093706/IGDA-DSS-2019_Summary-Report_Nov-20-2019.pdf.

41. Sarkeesian and Petit, "Female Representation in Videogames Isn't Getting Any Better."

42. International Game Developers Association, *Developer Satisfaction Survey*, p. 13.

43. Sandy Ong, "The Video Game Industry's Problem with Racial Diversity," *Newsweek*, October 13, 2016, www.newsweek.com/2016/10/21/video-games-race-black-protagonists-509328.html.

44. Keza MacDonald, "Not So White, Male and Straight: The Video Games Industry Is Changing," *Guardian*, February 19, 2020, www.theguardian.com/games/2020/feb/19/video-games-industry-diversity-women-people-of-colour.

45. "Avengers: Damage Control," The VOID, accessed June 7, 2020, www.thevoid.com/dimensions/marvel-avengers-vr/.

46. Koetsier, "Is Gaming Recession-Proof?"; Webb, "The $120 Billion Gaming Industry."

47. Ben Barrett, "*Overwatch* Just Reached 35 Million Players," *PCGamesN*, October 16, 2017, www.pcgamesn.com/overwatch/overwatch-sales-numbers.

48. Alex Pham, "*Star Wars: The Old Republic*—the Costliest Game of All Time?" *Los Angeles Times*, January 20, 2012, http://latimesblogs.latimes.com/entertainmentnewsbuzz/2012/01/star-wars-old-republic-cost.html.

49. "John Madden Net Worth," Celebrity Net Worth, accessed July 5, 2012, www.celebritynetworth.com/richest-athletes/nfl/john-madden-net-worth.

50. Samit Sarkar, "EA Maintains Exclusive Madden NFL License in Multiyear Renewal," Polygon, May 28, 2020, www.polygon.com/2020/5/28/21270647/madden-nfl-ea-renew-contract-license-football.

51. Joshua Brustein, "*Grand Theft Auto V* Is the Most Expensive Game Ever—and It's Almost Obsolete," *Bloomberg Businessweek*, September 18, 2013, www.businessweek.com/articles/2013-09-18/grand-theft-auto-v-is-the-most-expensive-game-ever-and-it-s-almost-obsolete? See also Superannuation, "How Much Does It Cost to Make a Big Video Game?" *Kotaku*, January 15, 2014, http://kotaku.com/how-much-does-it-cost-to-make-a-big-video-game-1501413649.

52. Erik Kain, "*Grand Theft Auto V* Crosses $1B in Sales, Biggest Entertainment Launch in History," *Forbes*, September 20, 2013, www.forbes.com/sites/erikkain/2013/09/20/grand-theft-auto-v-crosses-1b-in-sales-biggest-entertainment-launch-in-history.

53. GameStop Corp., "Fact Sheet," accessed February 17, 2018, http://news.gamestop.com/phoenix.zhtml?c=130125&p=factsheet.

54. Entertainment Software Association, *In-Game Advertising*, 2012, www.theesa.com/games-improving-what-matters/advertising.asp.

55. Activision Blizzard, *2019 Annual Report*, February 27, 2020, https://investor.activision.com/static-files/e610f6ff-cdf2-4f92-b373-4df046a590bb.

56. "Welcome to *Animal Crossing: New Horizons*," Nintendo, accessed June 7, 2020, www.animal-crossing.com/new-horizons.

57. "*Pandemic*," Asmodee Digital, accessed June 7, 2020, www.asmodee-digital.com/en/pandemic.

58. Matt Leacock, "No Single Player Can Win This Board Game. It's Called Pandemic." *New York Times*, March 25, 2020, www.nytimes.com/2020/03/25/opinion/pandemic-game-covid.html.

59. *Brown v. Entertainment Merchants Association*, 564 U.S. 786 (2011).

60. Ibid.

GLOBAL VILLAGE Phones in Hand, the World Finds Pokémon

1. Jon Fingas, "*Pokémon Go* Has Racked Up 1 Billion Downloads," Engadget, July 31, 2019, www.engadget.com/2019-07-31-pokemon-go-1-billion-downloads.html.

2. See "*Pokémon Go*: 10 Strangest PokéStop Locations," *Rolling Stone*, July 13, 2016, www.rollingstone.com/culture/pictures/pokemon-go-10-worst-pokestop-locations-20160713; Charles Pulliam-Moore, "Is There a *Pokémon Go* Gym in the Korean Demilitarized Zone?" Splinter, July 11, 2016, https://splinternews.com/is-there-a-pokemon-go-gym-in-the-korean-demilitarized-z-1793860155.

3. Paul Tassi, "Niantic and *Pokémon GO* Have Turned an Important Corner This Past Week," *Forbes*, November 24, 2017, www.forbes.com/sites/insertcoin/2017/11/24/niantic-and-pokemon-go-have-turned-an-important-corner-this-past-week/#646df3d472d1.

4. Matthew Gault, "Best Inventions 2019: Building New Worlds: *Minecraft Earth*," *Time*, December 2–December 9, 2019, https://time.com/collection/best-inventions-2019/5733093/minecraft-earth.

EXAMINING ETHICS The Gender Problem in Digital Games

1. Anita Sarkeesian, "It's Game Over for 'Gamers,'" *New York Times*, October 28, 2014, www.nytimes.com/2014/10/29/opinion/anita-sarkeesian-on-video-games-great-future.html.

2. Zachary Jason, "Game of Fear," *Boston Magazine*, April 28, 2015, www.bostonmagazine.com/news/article/2015/04/28/gamergate. (Note that Zoe Quinn now uses the pronouns *they/them*.)

3. KotakuInAction, "The Global Reach of Gamer-Gate?" May 12, 2015, www.reddit.com/r/KotakuInAction/comments/35ptw7/the_global_reach_of_gamergate.

4. Cory Marshall, "Dozens of Police, SWAT Respond to 'Swatting' Hoax in Southwest Portland," Katu.com, January 4, 2015, www.katu.com/news/local/Dozens-of-police-SWAT-respond-to-Swatting-hoax-in-Southwest-Portland--287467591.html.

5. Ross Miller, "Anita Sarkeesian to Create New Series Looking at Masculinity in Video Games," *Verge*, January 26, 2015, www.theverge.com/2015/1/26/7915385/new-feminist-frequency-series-on-masculinity-in-video-games. See also https://twitter.com/femfreq and www.youtube.com/user/feministfrequency.

6. Brianna Wu, "Can I Play, Too? Gender Equity in the Age of #Gamergate," University of Northern Iowa, March 31, 2015, www.youtube.com/watch?v=N-6fHFM_DdM.

7. Wu, "Can I Play, Too?"

8. Charlie Warzel, "How an Online Mob Created a Playbook for a Culture War," *New York Times*, August 15, 2019, www.nytimes.com/interactive/2019/08/15/opinion/what-is-gamergate.html.

◉ MEDIA LITERACY AND THE CRITICAL PROCESS
First-Person Shooter Games: Misogyny as Entertainment?

1. Seth Schiesel, "Way Down Deep in the Wild, Wild West," *New York Times*, May 16, 2010, www.nytimes.com/2010/05/17/arts/television/17dead.html.

4 Sound Recording and Popular Music

1. Andrew R. Chow, "Lil Nas X Talks 'Old Town Road' and the Billboard Controversy," *Time*, April 5, 2019, https://time.com/5561466/lil-nas-x-old-town-road-billboard.

2. Chow, "Lil Nas X Talks."

3. Chow, "Lil Nas X Talks."

4. Elias Leight, "Lil Nas X's 'Old Town Road' Was a Country Hit. Then Country Changed Its Mind," *Rolling Stone*, March 26, 2019, www.rollingstone.com/music/music-features/lil-nas-x-old-town-road-810844.

5. Leight, "Lil Nas X's 'Old Town Road.'"

6. Keith Caufield, "*Hollywood's Bleeding* Is Nielsen Music/MRC Data's Top Album of 2019, 'Old Town Road' Most-Streamed Song," *Billboard*, January 9, 2020, www.billboard.com/articles/business/chart-beat/8547651/post-malone-hollywoods-bleeding-nielsen-music-mrc-data-top-album-2019.

7. Sarah Perez, "Kids Now Spend Nearly as Much Time Watching TikTok as YouTube in US, UK and Spain," TechCrunch, June 4, 2020, https://techcrunch.com/2020/06/04/kids-now-spend-nearly-as-much-time-watching-tiktok-as-youtube-in-u-s-u-k-and-spain.

8. Elias Leight, "Surprising No One, TikTok Is Driving a Lot of New-Artist Growth," *Rolling Stone*, February 28, 2020, www.rollingstone.com/pro/news/chartmetric-breakthrough-artists-report-958401.

9. Mark Knowles, *The Wicked Waltz and Other Scandalous Dances: Outrage at Couple Dancing in the 19th and 20th Centuries* (McFarland: Jefferson, N.C., 2009).

10. Thomas Edison, quoted in Marshall McLuhan, *Understanding Media* (New York: McGraw-Hill, 1964), 276.

11. Shawn Fanning, quoted in Steven Levy, "The Noisy War over Napster," *Newsweek*, June 5, 2000, p. 46.

12. Ethan Smith, "LimeWire Found to Infringe Copyrights," *Wall Street Journal*, May 12, 2010, www.wsj.com/articles/SB1000142405274870424790457524057265442514.

13. Ben Sisario, "Spotify Hits 10 Million Subscribers, a Milestone," *New York Times*, May 21, 2014, www.nytimes.com/2014/05/22/business/media/spotify-hits-milestone-with-10-million-paid-subscribers.html.

14. IFPI, *IFPI Global Music Report 2016*, 2016, www.ifpi.org/downloads/GMR2016.pdf.

15. Stuart Dredge, "How Many Users Do Spotify, Apple Music and Other Big Music Streaming Services Have?" Music Ally, February 19, 2020, https://musically.com/2020/02/19/spotify-apple-how-many-users-big-music-streaming-services.

16. Republic Records Action Committee (@rractioncommittee), "Effective immediately, Republic Records will remove 'URBAN' from our verbiage," Instagram, June 5, 2020, www.instagram.com/p/CBD5FJZppti. See also Ethan Millman and Samantha Hissong, "How the Music Industry Can Keep Its Promises from Blackout Tuesday," *Rolling Stone*, June 9, 2020, www.rollingstone.com/pro/features/music-blackout-tuesday-week-later-1011478.

17. See Bruce Tucker, "'Tell Tchaikovsky the News': Postmodernism, Popular Culture and the Emergence of Rock 'n' Roll," *Black Music Research Journal* 9, no. 2 (Fall 1989): 280.

18. Robert Palmer, *Deep Blues: A Musical and Cultural History of the Mississippi Delta* (New York: Penguin, 1982), 15.

19. LeRoi Jones, *Blues People* (New York: Morrow Quill, 1963), 168.

20. Quoted in Dave Marsh and James Bernard, *The New Book of Rock Lists* (New York: Fireside, 1994), 15.

21. Mick Jagger, quoted in Jann S. Wenner, "Jagger Remembers," *Rolling Stone*, December 14, 1995, p. 66.

22. Little Richard, quoted in Charles White, *The Life and Times of Little Richard: The Quasar of Rock* (New York: Harmony Books, 1984), 65–66.

23. Tucker, "'Tell Tchaikovsky the News,'" 287.

24. Richard Harrington, "'A Wopbopaloobop'; and 'Alopbamboom,' as Little Richard Himself Would Be (and Was) First to Admit," *Washington Post*, November 12, 1984, final edition, sec. C1.

25. See Gerri Hershey, *Nowhere to Run: The Story of Soul Music* (New York: Penguin Books, 1984).

26. "2017 U.S. Music Year-End Report," Nielsen Music, January 3, 2018, www.nielsen.com/us/en/insights/reports/2018/2017-music-us-year-end-report.html.

27. Mark Mulligan, "Recorded Music Revenues Hit $21.5 Billion in 2019," MIDiA, March 5, 2020, www.midiaresearch.com/blog/recorded-music-revenues-hit-215-billion-in-2019.

28. Joshua Friedlander, "Year-End 2019 RIAA Music Revenues Report," February 2020, www.riaa.com/wp-content/uploads/2020/02/RIAA-2019-Year-End-Music-Industry-Revenue-Report.pdf.

29. Friedlander, "Year-End 2019 RIAA Music Revenues Report."

30. RIAA, "RIAA and GR&F Certification Audit Requirements RIAA Digital Single Award," February 2016, www.riaa.com/wp-content/uploads/2016/02/DIGITAL-SINGLE-AWARD-RIAA-AND-GRF-CERTIFICATION-AUDIT-REQUIREMENTS.pdf. See also RIAA, "RIAA and GR&F Certification Audit Requirements RIAA Album Award," June 2017, https://www.riaa.com/wp-content/uploads/2017/06/ALBUM-AWARD-RIAA-AND-GRF-CERTIFICATION-AUDIT-REQUIREMENTS.pdf.

31. Dmitry Pastukhov, "What Music Streaming Services Pay Per Stream (and Why It Actually Doesn't Matter)," *Soundcharts Blog*, June 26, 2019, https://soundcharts.com/blog/music-streaming-rates-payouts.

32. "2019–2020 Streaming Price Bible: YouTube Is STILL the #1 Problem to Solve," *The Trichordist* (blog), March 5, 2020, https://thetrichordist.com/2020/03/05/2019-2020-streaming-price-bible-youtube-is-still-the-1-problem-to-solve.

33. Value the Music (website), accessed October 21, 2020, https://valuethemusic.com.

34. Sean Hollister, "Spotify, the Leading Music Streaming App, Is Finally Profitable," Verge, February 6, 2019, www.theverge.com/2019/2/6/18214331/spotify-earnings-financial-announcement-profits-music-streaming-podcast.

35. Nat Hentoff, "Many Dreams Fueled Long Development of U.S. Music," *Milwaukee Journal*/United Press International, February 26, 1978, p. 2.

◘ **GLOBAL VILLAGE** Aya Nakamura: France's Global Pop Star

1. Robin Cannone, "Aya Nakamura, l'Europe est totalement gaga de la chanteuse française de 'Djadja,'" *Le Figaro*, October 11, 2018, www.lefigaro.fr/musique/2018/10/11/03006-20181011ARTFIG00167-aya-nakamura-l-europe-est-totalement-gaga-de-la-chanteuse-francaise-de-djadja.php.

2. Christelle Oyiri, "Aya Nakamura Is Flipping France's Rigid Rules, Beautifully," *Fader*, November 13, 2018, www.thefader.com/2018/11/13/aya-nakamura-is-flipping-frances-rigid-rules-beautifully.

3. "Aya Nakamura: French Rap Finding Its Global Voice," *IFPI Global Music Report 2019*, p. 27.

4. Oyiri, "Aya Nakamura."

◘ **EXAMINING ETHICS** YouTube: How One of the Richest Companies Shortchanges Music Artists

1. Tim Ingham, "It's Happened: The Major Labels Are Now Generating over $1M Every Hour from Streaming," Music Business Worldwide, February 25, 2020, www.musicbusinessworldwide.com/its-happened-the-major-labels-are-now-generating-over-1m-every-hour-from-streaming.

2. "YouTube Content ID," YouTube, September 28, 2010, www.youtube.com/watch?v=9g2U12SsRns.

3. Kit Smith, "57 Fascinating and Incredible YouTube Statistics," Brandwatch, February 21, 2020, www.brandwatch.com/blog/youtube-stats.

4. Alphabet Inc., "Form 10-K," February 3, 2020, https://abc.xyz/investor/static/pdf/20200204_alphabet_10K.pdf?cache=cdd6dbf.

5. "2019–2020 Streaming Price Bible: YouTube Is STILL the #1 Problem to Solve," *The Trichordist* (blog), March 5, 2020, https://thetrichordist.com/2020/03/05/2019-2020-streaming-price-bible-youtube-is-still-the-1-problem-to-solve.

6. "YouTube by the Numbers," YouTube for Press, YouTube, accessed July 11, 2020, www.youtube.com/about/press.

5 Popular Radio and the Origins of Broadcasting

1. "2020 Global Podcast Statistics, Demographics & Habits," PodcastHosting.com, last updated October 3, 2020, https://podcasthosting.org/podcast-statistics.

2. "Best of the Blogs: What Kind of Shows Dominate Podcasting?" Inside Radio, December 6, 2017, www.insideradio.com/free/best-of-the-blogs-what-kind-of-shows-dominate-podcasting/article_57b18af4-da5e-11e7-9d35-9bf13c627d23.html; "2018's Top Programming Trend Isn't a Format: It's Podcasting," Inside Radio, January 3, 2018, http://www.insideradio.com/2018-s-top-programming-trend-isn-t-a-format-it-s-podcasting/article_eee906ae-f057-11e7-a4e2-bb5440d19573.html.

3. Hannah Sayle, "*This American Life* Star Ira Glass on 3 Favorite Podcasts and 'The Power of Narrative,'" *Minneapolis Star Tribune*, May 9, 2019, www.startribune.com/this-american-life-star-ira-glass-on-3-favorite-podcasts-and-the-power-of-narrative/509700492.

4. Podtrac, "Top 20 Podcasts," June 2020, http://analytics.podtrac.com/podcast-rankings. See also Sarah Perez, "*NYT's The Daily* Now Reaches 2 Million Listeners per Day," TechCrunch, April 29, 2019, https://techcrunch.com/2019/04/29/nyts-the-daily-now-reaches-2-million-listeners-per-day.

5. Matthew Schneier, "The Voice of a Generation: Michael Barbaro Made the *New York Times* Podcast *The Daily* a Raging Success. Or Is It the Other Way Around?" Intelligencer, January 21, 2020, https://nymag.com/intelligencer/2020/01/michael-barbaro-the-daily-podcast-new-york-times.html.

6. American Public Media, "Programs: *The Daily*," www.apmdistribution.org/news/the-daily.

7. Jon Russell, "Spotify Says It Paid $340M to Buy Gimlet and Anchor," TechCrunch, February 14, 2019, https://techcrunch.com/2019/02/14/spotify-gimlet-anchor-340-million; Todd Spangler, "Spotify Is Paying Up to $196 Million in Cash to Acquire Bill Simmons' The Ringer," *Variety*, February 12, 2020, https://variety.com/2020/digital/news/spotify-acquires-the-ringer-196-million-cash-bill-simmons-1203502471.

8. Spangler, "Spotify Is Paying." See also Hank Green, "Spotify Wants to Become the Go-to for Podcasts. Creators and Audiences Should Worry," *Washington Post*, May 27, 2020, www.washingtonpost.com/opinions/spotify-wants-to-become-the-youtube-of-podcasts-it-would-be-terrible-for-the-industry/2020/05/27/394aec7c-a054-11ea-9590-1858a893bd59_story.html.

9. Daniel Ek, "Audio-First," For the Record, Spotify, February 6, 2019, https://newsroom.spotify.com/2019-02-06/audio-first.

10. Radio Advertising Bureau, "Radio Facts," Why Radio, accessed July 23, 2020, www.rab.com/whyradio.cfm#facts.

11. Captain Linwood S. Howeth, USN (Retired), *History of Communications-Electronics in the United States Navy* (Washington, D.C.: Government Printing Office, 1963), http://earlyradiohistory.us/1963hw.htm.

12. Margaret Cheney, *Tesla: Man out of Time* (New York: Touchstone, 2001).

13. William J. Broad, "Tesla, a Bizarre Genius, Regains Aura of Greatness," *New York Times*, August 28, 1984, http://query.nytimes.com/gst/fullpage.html?res=9400E4DD1038F93BA1575BC0A962948260&sec=health&spon=&partner=permalink&exprod=permalink.

14. Michael Pupin, "Objections Entered to Court's Decision," *New York Times*, June 10, 1934, p. E5.

15. Tom Lewis, *Empire of the Air: The Men Who Made Radio* (New York: HarperCollins, 1991), 73.

16. For a full discussion of early broadcast history and the formation of RCA, see Eric Barnouw, *Tube of Plenty* (New York: Oxford University Press, 1982); Susan Douglas, *Inventing American Broadcasting, 1899–1922* (Baltimore: Johns Hopkins University Press, 1987); Christopher Sterling and John Kitross, *Stay Tuned: A Concise History of American Broadcasting* (Belmont, Calif.: Wadsworth, 1990).

17. "How TV Licence Fees Compare around the World," News24, January 30, 2015, www.news24.com/News24/how-tv-licence-fees-compare-around-the-world-20150130; Helen Weeds, "Is the Television Licence Fee Fit for Purpose in the Digital Era?" *Economic Affairs* 36, no. 4 (2016): 2–20.

18. Paul Starr, *The Creation of Media: Political Origins of Modern Communications* (New York: Perseus, 2004), 341.

19. Robert W. McChesney, *Telecommunications, Mass Media, and Democracy: The Battle for Control of U.S. Broadcasting, 1928–1935* (New York: Oxford University Press, 1994).

20. Michele Hilmes, *Radio Voices: American Broadcasting, 1922–1952* (Minneapolis: University of Minnesota Press, 1997).

21. "*Amos 'n' Andy* Show," Museum of Broadcast Communications, https://museumtv.pastperfectonline.com/creator/EFD539AB-A602-445D-BE5C-992756161450.

22. Indiana University Archives of African American Music and Culture, "Golden Age of Black Radio—Part 1: The Early Years," 2016, https://artsandculture.google.com/exhibit/golden-age-of-black-radio-part-1-the-early-years-archives-of-african-american-music-and-culture/TgIyRMZB2RIOIQ?hl=en.

23. "Marshall Field Outlet Starts Air's First All-Negro Serial," *The Billboard*, March 15, 1947, p. 5.

24. Indiana University Archives, "Golden Age of Black Radio."

25. W. Joseph Campbell, *Getting It Wrong: Debunking the Greatest Myths in American Journalism*, 2nd ed. (Oakland: University of California Press, 2017), 27.

26. Pete Fornatale and Joshua E. Mills, *Radio in the Television Age* (New York: Harry N. Abrams, 1980).

27. Samantha Hissong, "Radio Is Quietly Scrubbing the Word 'Urban,' Sources Say," *Rolling Stone*, July 2, 2020, www.rollingstone.com/pro/news/iheartradio-mediabase-urban-radio-changes-1023730.

28. Ed Christman, "RIAA, Pandora, NARAS, NAB Square Off on Capitol Hill," *Billboard*, June 7, 2012, www.billboard.com/articles/business/1094327/riaa-pandora-naras-nab-square-off-on-capitol-hill.

29. "2020 Global Podcast Statistics, Demographics & Habits."

30. Radio Advertising Bureau, "Radio Facts."

31. Lance Venta, "BIA Reduces 2020 Ad Revenue Forecast by 10.7%," Radio Insight, June 25, 2020, https://radioinsight.com/headlines/189635/bia-reduces-2020-ad-revenue-forecast-by-10-7.

32. Venta, "BIA Reduces 2020 Ad Revenue Forecast by 10.7%."

33. Federal Communications Commission, "Broadcast Station Totals as of September 30, 2020," October 2, 2020, https://docs.fcc.gov/public/attachments/DOC-367270A1.pdf.

34. Glenn Peoples, "How 'Playola' Is Infiltrating Streaming Services: Pay for Play Is 'Definitely Happening,'" *Billboard*, August 19, 2015, www.billboard.com/articles/business/6670475/playola-promotion-streaming-services. See also Robert Cookson, "Spotify Bans 'Payola' on Playlists," *Financial Times*, August 20, 2015, www.ft.com/content/af1728ca-4740-11e5-af2f-4d6e0e5eda22.

35. Anne Steele, "FCC Asks Record Labels to Document Anti-Payola Policies," *Wall Street Journal*, January 22, 2020, www.wsj.com/articles/fcc-asks-record-labels-to-document-anti-payola-policies-11579737073. See also Elias Leight, "Want to Get on the Radio? Have $50,000?" *Rolling Stone*, August 6, 2019, www.rollingstone.com/pro/features/radio-stations-hit-pay-for-play-867825.

36. Peter DiCola, "False Premises, False Promises: A Quantitative History of Ownership Consolidation in the Radio Industry," Future of Music Coalition, December 13, 2006, http://futureofmusic.org/article/research/false-premises-false-promises.

37. Ben Sisario, "IHeartMedia, U.S.'s Largest Radio Broadcaster, Files for Bankruptcy," *New York Times*, March 15, 2018, www.nytimes.com/2018/03/15/business/media/iheartmedia-bankruptcy.html.

38. "About College Radio," Radio Survivor, June 10, 2020, www.radiosurvivor.com/learn-more/about-college-radio.

■ **EXAMINING ETHICS** How Did Talk Radio Become So One-Sided?

1. Nielsen, "Tops of 2019: Radio," December 18, 2019, www.nielsen.com/us/en/insights/article/2019/tops-of-2019-radio.

2. "State of the Media: Audio Today 2018, How America Listens," April 2018, https://www.nielsen.com/us/en/insights/report/2018/state-of-the-media-audio-today-2018/.

3. "The Top Talk Radio Audiences," *Talkers*, last updated October 2020, www.talkers.com/top-talk-audiences.

4. Robert L. Hilliard and Michael C. Keith, *Waves of Rancor: Tuning in the Radical Right* (Armonk, N.Y.: M.E. Sharpe, 1999), 15.

5. Robert D. Hershey Jr., "F.C.C. Votes Down Fairness Doctrine in a 4–0 Decision," *New York Times*, August 5, 1987, www.nytimes.com/1987/08/05/arts/fcc-votes-down-fairness-doctrine-in-a-4-0-decision.html.

6. Merrill Hartson, "President Vetoes Legislation to Make 'Fairness Doctrine' Permanent," Associated Press, June 20, 1987.

7. Hilliard and Keith, *Waves of Rancor*, 17.

8. Rush Limbaugh, *See, I Told You So* (New York: Pocket Star Books, 1993), 372.

9. Kathleen Hall Jamieson and Joseph N. Cappella, *Echo Chamber: Rush Limbaugh and the Conservative Media Establishment* (New York: Oxford University Press, 2008), 46.

10. "NPR Maintains Highest Ratings Ever," NPR, March 28, 2018, www.npr.org/about-npr/597590072/npr-maintains-highest-ratings-ever.

11. "Audio Today 2019."

◪ **GLOBAL VILLAGE** Radio Stories from Around the World

1. Matthew Weaver, "Local Radio Station Keeps Getting Hijacked by Song about Masturbation," *Guardian*, July 11, 2017, https://www.theguardian.com/tv-and-radio/2017/jul/11/local-radio-station-mansfield-hijacked-masturbation-winkers-song.

2. "Devastating South Sudan Coronavirus Deaths, 'A Tragedy That Can Be Prevented,' Security Council Hears," UN News, June 23, 2020, https://news.un.org/en/story/2020/06/1066952.

3. "Peacekeeping Radio Stations Provide COVID-19 Information to Vulnerable Communities in Conflict-Affected Countries," UN Peacekeeping, April 21, 2020, www.un.org/africarenewal/news/coronavirus/peacekeeping-radio-stations-provide-covid-19-information-vulnerable-communities-conflict.

4. Nicola Slawson, "Radio Monsoon Aims to Ensure Safety Reigns among Fishermen in South India," *Guardian*, April 24, 2017, www.theguardian.com/global-development/2017/apr/24/radio-monsoon-safety-fishermen-south-india-kerala.

5. "Explore the Government-Friendly Media Empire in Hungary," Center for Media, Data and Society, January 16, 2018, https://cmds.ceu.edu/article/2018-01-16/explore-government-friendly-media-empire-hungary.

6. "*Magyar Nemzet*, Lánchíd Rádió to Cease Operations Effective April 11th," *Budapest Beacon*, April 10, 2018, https://budapestbeacon.com/magyar-nemzet-lanchid-radio-to-cease-operations-effective-april-11th.

7. Marton Dunai, "Major Hungarian Opposition Newspaper to Close after Orban Victory," Reuters, April 10, 2018, www.reuters.com/article/us-hungary-election-media/major-hungarian-opposition-newspaper-to-close-after-orban-victory-idUSKBN1HH10S.

6   Television: From Broadcasting to Streaming

1. Todd Spangler, "*Tiger King* Nabbed over 34 Million U.S. Viewers in First 10 Days, Nielsen Says," *Variety*, April 8, 2020, https://variety.com/2020/digital/news/iger-king-nielsen-viewership-data-stranger-things-1234573602.

2. Michael Schneider, "100 Most Watched TV Shows of 2019–20: Winners and Losers," *Variety*, May 21, 2020, https://variety.com/2020/tv/news/most-popular-tv-shows-highest-rated-2019-2020-season-masked-singer-last-dance-1234612885.

3. Todd Spangler, "Netflix Reels in 10 Million Subscribers in Q2 as Coronavirus Tailwinds Continue," July 16, 2020, https://variety.com/2020/digital/news/netflix-q2-2020-10-million-subscribers-coronavirus-1234708562.

4. Nellie Andreeva, "*Evil* and *Unicorn* Headed to Netflix in One-Year Licensing Deal by CBS TV Studios Aimed at Finding New Audiences," Deadline, September 23, 2020, https://deadline.com/2020/09/evil-unicorn-netflix-in-one-year-licensing-deal-cbs-tv-studios-1234581581; Lesley Goldberg, "Shonda Rhimes' Netflix Deal Ups the Stakes in Hollywood's Battle for Ownership," *Hollywood Reporter*, August 16, 2017, www.hollywoodreporter.com/live-feed/shonda-rhimes-netflix-deal-ups-stakes-hollywoods-battle-ownership-1029850; Tatiana Siegel, "Obamas Settling into New Role as Netflix Producers," *Hollywood Reporter*, August 7, 2019, www.hollywoodreporter.com/news/obamas-settle-new-role-as-netflix-producers-1229621.

5. Chris Morris, "Netflix Will Spend over $17 Billion on Content in 2020," *Fortune*, January 16, 2020, https://fortune.com/2020/01/16/netflix-spending-content-2020-17-billion/#:~:text=A%20new%20forecast%20from%20BMO,competition%20continues%20to%20heat%20up.

6. Jill Disis, "Netflix Looks All Over the World to Feed a Growing Audience," CNN, February 27, 2018, http://money.cnn.com/2018/02/27/media/netflix-worldwide-content/index.html.

7. David Sims, "Why Netflix Is Releasing So Many New Shows in 2018," *Atlantic*, August 29, 2017, www.theatlantic.com/entertainment/archive/2017/08/now-dawns-the-age-of-peak-netflix/538263.

8. David Bauder, "Study Shows Explosive Growth in Time Spent Streaming TV," ABC News, February 12, 2020, https://abcnews.go.com/Lifestyle/wireStory/study-shows-explosive-growth-time-spent-streaming-tv-68940265.

9. Madhulika Sikka, "How Do Federal $$$ Get to Your Local Station?" PBS, February 22, 2018, www.pbs.org/publiceditor/blogs/pbs-public-editor/how-do-federal-get-to-your-local-station.

10. "Support," WFYI, accessed December 21, 2020, www.wfyi.org/federal-funding-for-public-broadcasting.

11. *United States v. Midwest Video Corp.*, 440 U.S. 689 (1979).

12. "Cable Television Coming of Age," *CQ Researcher,* December 27, 1985, https://library.cqpress.com/cqresearcher/document.php?id=cqresrre1985122700; Jill Goldsmith, "Traditional Pay TV in 'Death Spiral,' Analyst Says, with Penetration at 1995 Levels: Virtual MVPDs Also at Risk," May 8, 2020, https://deadline.com/2020/05/pay-tv-risk-penetration-falls-vmvpd-collapsed-1202929737; Alan Breznick, "US Pay-TV Penetration Falls from Peak," September 3, 2014, www.lightreading.com/video/video-services/us-pay-tv-penetration-falls-from-peak/d/d-id/710610.

13. Kristal Brent Zook, *Color by Fox: The Fox Network and the Revolution in Black Television* (New York: Oxford University Press, 1999), 4.

14. Ron Miller, "The 80s Were Big for TV," *Washington Post*, December 24, 1989, www.washingtonpost.com/archive/lifestyle/tv/1989/12/24/the-80s-were-big-for-tv/fce422b1-9857-4335-a1f6-ecb2461ac8c6.

15. "Broadband by the Numbers," NCTA, accessed December 21, 2020, www.ncta.com/broadband-by-the-numbers.

16. Sarah Whitten, "The Death of the DVD: Why Sales Dropped More Than 86% in 13 Years," CNBC, November 8, 2019, www.cnbc.com/2019/11/08/the-death-of-the-dvd-why-sales-dropped-more-than-86percent-in-13-years.html.

17. Cynthia Littleton, "*The Big Bang Theory* Finale Scores Huge Ratings, 18 Million Viewers," *Variety*, May 17, 2019, https://variety.com/2019/tv/news/the-big-bang-theory-finale-early-ratings-1203218476.

18. John Koblin, "Peak TV Hits a New Peak, with 532 Scripted Shows," *New York Times*, January 9, 2020, www.nytimes.com/2020/01/09/business/media/tv-shows-2020.html.

◪ **EXAMINING ETHICS** Is This Entertainment? TV Industry Reconsiders Police Shows

1. Dan Taberski, "Is the Show *Cops* Committing Crimes Itself?" *New York Times*, June 18, 2019, www.nytimes.com/2019/06/18/opinion/cops-podcast-investigation-abuse.html.

2. Alyssa Rosenberg, "Shut Down All Police Movies and TV Shows. Now." *Washington Post*, June 4, 2010, www.washingtonpost.com/opinions /2020/06/04/shut-down-all-police-movies-tv-shows-now.

3. Nellie Andreeva, "*Live PD* Canceled by A&E amid Ongoing Protests against Police Brutality," Deadline, June 10, 2020, https://deadline.com /2020/06/live-pd-canceled-ae-protests-against-police-brutality-george -floyd-1202956175.

◾ **GLOBAL VILLAGE** Telling and Selling Stories around the World

1. MPAA, "New Data: The American Film and Television Industry Continues to Drive Economic Growth in All 50 States," March 18, 2019, www.motionpictures.org/press/new-data-the-american-film-and -television-industry-continues-to-drive-economic-growth-in-all-50-states.

2. Quoted in Denise D. Bielby and C. Lee Harrington, *Global TV: Exporting Television and Culture in the World Market* (New York: New York University Press, 2008), 40.

3. "America's Cultural Role in the World Today," Access to International English, accessed May 31, 2018, access-internationalvg2.cappelendamm .no/c951212/artikkel/vis.html?tid=385685.

4. "America's Cultural Role."

7 Movies and the Power of Images

1. Eric Savitz, "Mickey, Meet Yoda: Disney to Buy Lucasfilm for $4.05 Billion," *Forbes*, October 30, 2012, www.forbes.com/sites/ericsavitz /2012/10/30/mickey-meet-yoda-disney-to-buy-lucasfilm-for-4-05 -billion/?sh=39f718dc2cd8.

2. Walt Disney Company, "Disney to Acquire Lucasfilm Ltd.," October 30, 2012, www.thewaltdisneycompany.com/disney-to-acquire-lucasfilm-ltd.

3. Dale Rutledge, "Disney Is Ready to Roll Out New *Star Wars* Sagas as One Story Ends," *Fortune*, January 14, 2020, https://fortune.com/2020 /01/14/star-wars-future-movies-television-video-games-theme-parks.

4. Charles Musser, *The Emergence of Cinema: The American Screen to 1907* (New York: Scribner's, 1991).

5. Douglas Gomery, *Shared Pleasures: A History of Movie Presentation in the United States* (Madison: University of Wisconsin Press, 1992), 18.

6. Douglas Gomery, *Movie History: A Survey* (Belmont, CA: Wadsworth, 1991), 167.

7. David Bordwell, Janet Staiger, and Kristin Thompson, *The Classical Hollywood Cinema: Film Style and Mode of Production to 1960* (New York: Columbia University Press), 1985.

8. Eric Barnouw, *Tube of Plenty: The Evolution of American Television*, rev. ed. (New York: Oxford University Press, 1982), 108–109.

9. See Douglas Gomery, "Who Killed Hollywood?" *Wilson Quarterly* (Summer 1991): 106–112.

10. Mike McPhate, "Hollywood's Inclusion Problem Extends beyond the Oscars, Study Says," *New York Times*, February 22, 2016, www.nytimes .com/2016/02/23/movies/hollywoods-inclusion-problem-extends -beyond-the-oscars-study-says.html; Cara Buckley, "More Women Than Ever Are Directing Major Films, Study Says," *New York Times*, January 2, 2020, www.nytimes.com/2020/01/02/movies/women-directors -hollywood.html.

11. "Our Story," Time's Up, accessed December 1, 2020, https://timesupnow .org/about/our-story.

12. "4 Indigenous-Made Films Premiering at the 2021 Sundance Film Festival," Sundance Institute, January 12, 2021, www.sundance.org/blogs /festival/indigenous-made-films-2021-festival.

13. See Gomery, *Shared Pleasures*, 171–180.

14. "Record Number of Films Produced," UNESCO, March 31, 2016, http://uis.unesco.org/en/news/record-number-films-produced; "Indian Feature Films Certified during the Year 2017," Film Federation of India, 2017, www.filmfed.org/IFF2017.html; Zheping Huang, "China's Top 10 Box Office Hits of All Time Include Four Domestic Films Released in 2017," Quartz, January 2, 2018, https://qz.com/1169192/chinas-all-time -top-10-box-office-list-has-four-domestic-films-released-in-2017-including -wolf-warrior-2.

15. A. O. Scott, "Tracing a 20-Year Odyssey across Hope and Despair," *New York Times*, October 5, 2004, www.nytimes.com/2004/10/05 /movies/tracing-a-20year-odyssey-across-hope-and-despair.html.

16. Motion Picture Association of America, *Theme Report 2019*, March 11, 2020, 18, www.motionpictures.org/wp-content/uploads/2020/03 /MPA-THEME-2019.pdf.

17. Motion Picture Association, 11.

18. Dave McNary, "U.S. Movie Ticket Sales Plunged 6% in 2017, Thanks to Lousy Summer," *Variety*, January 17, 2018, https://variety.com/2018 /film/box-office/u-s-movie-tickets-sold-2017-1202667483.

19. National Association of Theatre Owners, "Number of U.S. Movie Screens," accessed December 1, 2020, www.natoonline.org/data/us -movie-screens.

20. National Association of Theatre Owners, "Top 10 U.S. and Canadian Circuits," July 1, 2020, www.natoonline.org/data/top-10-circuits.

21. Motion Picture Association, *Theme Report 2019*, 11.

22. Motion Picture Association, 5.

23. Julianne Pepitone, "Americans Now Watch More Online Movies Than DVDs," CNN/Money, March 22, 2012, http://money.cnn.com /2012/03/22/technology/streaming-movie-sales/index.htm.

24. David Thorburn, "Television as an Aesthetic Medium," *Critical Studies in Mass Communication* 4 (June 1987): 168.

◾ **EXAMINING ETHICS** Breaking through Hollywood's Race Barrier

1. Gomery, *Shared Pleasures*, 155–170.

2. Scott Mendelson, "Box Office: Marvel's *Black Panther* Tops $1.1 Billion Worldwide," *Forbes*, March 14, 2018, www.forbes.com/sites/scottmendelson /2018/03/14/box-office-marvels-black-panther-tops-1-1-billion-worldwide /#6d47c4935ecb.

3. Jamil Smith, "The Revolutionary Power of *Black Panther*," *Time*, February 19, 2018, http://time.com/black-panther.

4. Sean Rameswaram, "Black Panther Is the Most Important Movie of 2018," *Today, Explained* (podcast), Vox, February 20, 2018, www.stitcher .com/podcast/stitcher/today-explained/e/53403504.

◾ **GLOBAL VILLAGE** Beyond Hollywood: Asian Cinema

1. Rob Cain, "Indian Movies Are Booming in America," *Forbes*, May 5, 2017, www.forbes.com/sites/robcain/2017/05/05/these-are-the-best-of -times-for-indian-movies-in-america/#17411a842b97; Box Office India, "Bahubali 2 Hits 1000 Crore Worldwide in Ten Days," May 8, 2017, www .boxofficeindia.com/report-details.php?articleid=2914.

2. Steve Rose, "*The Eight Hundred*: How China's Blockbusters Became a New Political Battleground," *Guardian*, September 18, 2020, www .theguardian.com/film/2020/sep/18/the-eight-hundred-how-chinas -blockbusters-became-a-new-political-battleground.

3. Douglas Parkes, "How Did the Hong Kong Film Industry Get So Big—and Why Did It Fall into Decline?" *South China Morning Post*, April 26, 2020, www.scmp.com/magazines/style/celebrity/article/3081457 /how-did-hong-kong-film-industry-get-so-big-and-why-did-it.

8 Newspapers: The Rise and Decline of Modern Journalism

1. Penelope Muse Abernathy, *News Deserts and Ghost Newspapers: Will Local News Survive?*, 2020, www.cislm.org/wp-content/uploads /2020/06/2020_News_Deserts_Report_Final-Version-to-Design -Hammer-6-23-1.pdf, 18.

2. Philip Napoli and Jessica Mahone, "Local Newspapers Are Suffering, but They're Still (By Far) the Most Significant Journalism Producers in Their Communities," Nieman Lab, September 9, 2019, www.niemanlab .org/2019/09/local-newspapers-are-suffering-but-theyre-still-by-far-the -most-significant-journalism-producers-in-their-communities.

3. Kristen Hare, "Here Are the Newsroom Layoffs, Furloughs and Closures Caused by the Coronavirus," Poynter, updated February 17, 2021, www.poynter.org/business-work/2020/here-are-the-newsroom-layoffs -furloughs-and-closures-caused-by-the-coronavirus.

4. "Gannett Announces First Quarter 2020 Results," press release, May 7, 2020, https://s1.q4cdn.com/307481213/files/doc_financials/2020/q1 /Q1-2020-Earnings-PR_vFINAL.pdf.

5. Penelope Muse Abernathy, *The Expanding News Desert*, 2018, www .cislm.org/wp-content/uploads/2018/10/The-Expanding-News-Desert -10_14-Web.pdf, 32.

6. CNHI, "Locations," accessed September 7, 2020, www.cnhi.com /locations.

7. Ken Doctor, "Newsonomics: The McClatchy Auction Ends Not with a Bang, but Only More Whimpers," Nieman Lab, July 13, 2020, www .niemanlab.org/2020/07/newsonomics-the-mcclatchy-auction-ends -not-with-a-bang-but-only-more-whimpers.

8. Doctor.

9. Douglas Burns, "*Centerville Iowegian*, Newspaper That Fought KKK in 1920s, Goes Down to Coronavirus in 2020," *Carroll Times Herald*, May 18, 2020, www.carrollspaper.com/opinion/centerville-iowegian -newspaper-that-fought-kkk-in-1920s-goes-down-to-coronavirus -in-2020/article_e5856890-9920-11ea-b44b-7fbd1fe848b7.html.

10. Abernathy, *News Deserts and Ghost Newspapers*, 8–9.

11. Joshua Benton, "The *Wall Street Journal* Joins the *New York Times* in the 2 Million Digital Subscriber Club," Nieman Lab, February 10, 2020, www.niemanlab.org/2020/02/the-wall-street-journal-joins-the -new-york-times-in-the-2-million-digital-subscriber-club.

12. Matthew Goodman, *The Sun and the Moon: The Remarkable True Account of Hoaxers, Showmen, Dueling Journalists, and Lunar Man-Bats in Nineteenth-Century New York* (New York: Basic Books, 2008).

13. William Randolph Hearst, quoted in Piers Brendon, *The Life and Death of the Press Barons* (New York: Atheneum, 1983), 134.

14. Michael Schudson, *Discovering the News: A Social History of American Newspapers* (New York: Basic Books, 1978), 23.

15. "The 2020 Pulitzer Prize Winner in Explanatory Reporting," The Pulitzer Prizes, 2020, www.pulitzer.org/winners/staff-washington-post.

16. See David T. Z. Mindich, "Edwin M. Stanton, the Inverted Pyramid, and Information Control," *Journalism Monographs* 140 (August 1993).

17. Curtis D. MacDougall, *The Press and Its Problems* (Dubuque: William C. Brown, 1964), 143, 189.

18. Walter Lippmann, *Liberty and the News* (New York: Harcourt, Brace and Howe, 1920), 92.

19. Tom Wolfe, quoted in Leonard W. Robinson, "The New Journalism: A Panel Discussion," in Ronald Weber, ed., *The Reporter as Artist: A Look at the New Journalism Controversy* (New York: Hastings House, 1974), 67. See also Tom Wolfe and E. E. Johnson, eds., *The New Journalism* (New York: Harper & Row, 1973).

20. Tom Wicker, *On Press* (New York: Viking, 1978), 3–5.

21. Richard Pérez-Peña and Timothy Williams, "Glare of Video Is Shifting Public's View of Police," *New York Times*, July 30, 2015, www.nytimes.com /2015/07/31/us/through-lens-of-video-a-transformed-view-of-police.html.

22. Your Voice Ohio, "Who We Are," accessed November 12, 2020, https://yourvoiceohio.org/who-we-are.

23. Abernathy, *News Deserts and Ghost Newspapers*, 43.

24. See Barbara K. Henritze, *Bibliographic Checklist of American Newspapers* (Baltimore: Clearfield, 2009).

25. Sara Atske, Michael Barthel, Galen Stocking, and Christine Tamir, "7 Facts about Black Americans and the News Media," Pew Research Center, August 7, 2019, www.pewresearch.org/fact-tank/2019/08/07/facts-about-black -americans-and-the-news-media. See also National Newspaper Publishers Association, "Current Members," https://nnpa.org/current-members.

26. Special thanks to Mary Lamonica and her students at New Mexico State University.

27. Craig Newmark Graduate School of Journalism at CUNY, "The State of the Latino News Media," June 2019, http://thelatinomediareport.journalism .cuny.edu/the-industry-at-a-glance.

28. Hanaa' Tameez, "With Hecho en USA, *USA Today* Wants to Tell Latinos' 'Everyday Stories about Navigating Life in America,'" Nieman Lab, January 30, 2020, www.niemanlab.org/2020/01/with-hecho-en-usa-usa-today-wants-to -tell-latinos-everyday-stories-about-navigating-life-in-america.

29. Kuei Chiu, "Asian Language Newspapers in the United States: History Revisited," Chinese American Librarians Association, November 12, 2008, www.cala-web.org/cala-ej/no-09/chiu.

30. "Not Your Model Minority March," *Northwest Asian Weekly*, April 22, 2021, http://nwasianweekly.com/2021/04/not-your-model-minority-march.

31. Chinese Advertising Agencies, "About *Chinese Daily News*," accessed September 3, 2014, www.chineseadvertisingagencies.com/mediaguide /Chinese-Daily-News.html.

32. Joshua Benton, "What Is the *Epoch Times*? A Vehicle for Pro-Trump Conspiracy Theories, and the Culmination of All That Facebook Has Encouraged," Nieman Lab, October 26, 2020, www.niemanlab.org/2020/10 /what-is-the-epoch-times-a-vehicle-for-pro-trump-conspiracy-theories -and-the-culmination-of-all-that-facebook-has-encouraged.

33. Jodi Rave, *American Indian Media Today: Tribes Maintain Majority Ownership as Independent Journalists Seek Growth*, Democracy Fund, November 2018, http://democracyfund.org/wp-content/uploads/2020/06/2018_DF _AmericanIndianMediaToday.pdf.

34. Rave, 13.

35. Native American Journalists Association, "2020 NAJA Media Spotlight Report," https://najanewsroom.com/2020-naja-media-spotlight-report.

36. "About the *Village Voice*," accessed June 4, 2018, www.villagevoice.com /about.

37. "About AAN," Association of Alternative Newsmedia, accessed November 14, 2020, https://aan.org/about.

38. Jim Fitzpatrick, "The Number of Newspaper Chains Is Shrinking, While Story Lengths—at Least at The *Star*—Are Exploding," *JimmyCsays* (blog), April 22, 2020, https://jimmycsays.com/2020/04/22/the-number -of-newspaper-chains-is-shrinking-while-story-lengths-at-least-at-the -star-are-exploding.

39. News Leaders Association, "ASNE Diversity History," accessed November 14, 2020, https://members.newsleaders.org/content.asp? contentid=57.

40. Gabriel Arana, "Decades of Failure," *Columbia Journalism Review*, Fall 2018, www.cjr.org/special_report/race-ethnicity-newsrooms-data.php.

41. Arana.

42. Meg James and Daniel Hernandez, "*L.A. Times* Faces Painful Reckoning over Race in Its Staff and Pages," *Los Angeles Times*, June 24, 2020, www.latimes.com/entertainment-arts/business/story/2020-06-24 /los-angeles-times-black-lives-matter-diversity.

43. Communication Workers of America, "Journalists of *L.A. Times* Win a Decisive Victory for Pay Equity," November 19, 2020, https://cwa-union.org/news/journalists-of-la-times-win-decisive-victory-for-pay-equity.

44. Soledad O'Brien, "Soledad O'Brien: A MeToo Moment for Journalists of Color," *New York Times*, July 4, 2020, www.nytimes.com/2020/07/04/opinion/soledad-obrien-racism-journalism.html.

45. "Editorial: As Vultures Circle, the *Denver Post* Must Be Saved," *Denver Post*, April 6, 2018, www.denverpost.com/2018/04/06/as-vultures-circle-the-denver-post-must-be-saved.

46. Robert Kuttner and Hildy Zenger, "Saving the Free Press from Private Equity," *American Prospect*, December 27, 2017, https://prospect.org/health/saving-free-press-private-equity.

47. Doctor, "Newsonomics."

48. Kuttner and Zenger, "Saving the Free Press."

49. Christopher R. Martin, *No Longer Newsworthy: How the Mainstream Media Abandoned the Working Class* (Ithaca, N.Y.: Cornell University Press, 2019).

50. Abernathy, *News Deserts and Ghost Newspapers*, 21.

51. Nicole Perrrin, "Facebook-Google Duopoly Won't Crack This Year," eMarketer, November 4, 2019, www.emarketer.com/content/facebook-google-duopoly-won-t-crack-this-year.

52. Abernathy, *News Deserts and Ghost Newspapers*, 22.

53. Abernathy, 11.

54. Erika Bolstad, "COVID-19 Is Crushing Newspapers, Worsening Hunger for Accurate Information," *Stateline* (blog), Pew Charitable Trusts, September 8, 2020, www.pewtrusts.org/en/research-and-analysis/blogs/stateline/2020/09/08/covid-19-is-crushing-newspapers-worsening-hunger-for-accurate-information.

55. Edmund Lee, "*New York Times* Hits 7 Million Subscribers as Digital Revenue Rises," *New York Times*, November 5, 2020, www.nytimes.com/2020/11/05/business/media/new-york-times-q3-2020-earnings-nyt.html.

56. Lee.

57. Christopher R. Martin, "What Makes Iowa Newspapers Resilient?" Center for Journalism & Liberty, September 25, 2020, www.journalismliberty.org/publications/what-makes-iowa-newspapers-resilient.

58. "About," States Newsroom, accessed November 14, 2020, https://statesnewsroom.com/about.

59. "ProPublica to Expand Local Reporting Network to Focus on State Governments," ProPublica, August 8, 2018, www.propublica.org/atpropublica/propublica-expanding-local-reporting-network-state-governments.

60. "About Us," Report for America, accessed November 14, 2020, www.reportforamerica.org/about-us.

61. Institute for Nonprofit News, "INN Index 2020: The State of Nonprofit News," June 15, 2020, https://inn.org/innindex.

62. See "Policing the USA," *USA Today*, www.usatoday.com/policing; "Race in America," *USA Today*, www.usatoday.com/opinion/race-in-america.

63. Laura Hazard Owen, "The *New York Times* Launches Its (Evidence-Driven, Non-judgy) Parenting Vertical, with an Eye toward Making It a Subscription Product," Nieman Lab, May 8, 2019, www.niemanlab.org/2019/05/the-new-york-times-launches-its-evidence-driven-non-judgy-parenting-vertical-with-an-eye-toward-making-it-a-subscription-product.

64. Laura Hazard Owen, "*The Boston Globe* Is Shutting Down Its Catholic Vertical Crux, Citing a Shortfall in Advertising," Nieman Lab, March 11, 2016, www.niemanlab.org/2016/03/the-boston-globe-is-shutting-down-its-catholic-vertical-crux-citing-a-shortfall-in-advertising.

65. "About," The 19th*, accessed November 14, 2020, https://19thnews.org/about.

66. Hunter Walk, "Why a Paid Newsletter Won't Be Enough Money for Most Writers (and That's Fine): The Multi-SKU Creator," *Medium*, November 23, 2020, https://hunterwalk.medium.com/why-a-paid-newsletter-wont-be-enough-money-for-most-writers-and-that-s-fine-the-multi-sku-f41daa074cdb.

67. Steve Waldman, "A Replanting Strategy: Saving Local Newspapers Squeezed by Hedge Funds," Center for Journalism & Liberty, September 18, 2020, www.journalismliberty.org/publications/replanting-strategy-saving-local-newspapers-squeezed-by-hedge-fund.

68. Waldman.

69. Craig Aaron and S. Derek Turner, "What a Journalism-Recovery Package Should Look Like during the COVID-19 Crisis," *Free Press*, May 2020, www.freepress.net/sites/default/files/2020-05/free_press_action_journalism_recovery_policies_final.pdf.

70. Jessica Mahone, Qun Wang, Philip Napoli, Matthew Weber, and Katie McCollough, *Who's Producing Local Journalism? Assessing Journalistic Output across Different Outlet Types*, News Measures Research Project, August 2019, https://dewitt.sanford.duke.edu/wp-content/uploads/2019/08/Whos-Producing-Local-Journalism_FINAL.pdf.

71. See Michael Barthel, Jesse Holcomb, Jessica Mahone, and Amy Mitchell, "Civic Engagement Strongly Tied to Local News Habits," Pew Research Center's Journalism Project, November 3, 2016, www.journalism.org/2016/11/03/civic-engagement-strongly-tied-to-local-news-habits; Kriston Capps, "The Hidden Costs of Losing Your City's Newspaper," Bloomberg, May 30, 2018, www.bloomberg.com/news/articles/2018-05-30/when-local-newspapers-close-city-financing-costs-rise; Chloe Reichel, "Political Polarization Increases after Local Newspapers Close," Journalist's Resource, November 19, 2018, journalistsresource.org/studies/politics/polarization/political-polarization-local-news-research.

72. Amy Mitchell, Mark Jurkowitz, J. Baxter Oliphant, and Elisa Shearer, "Americans Who Mainly Get Their News on Social Media Are Less Engaged, Less Knowledgeable," Pew Research Center, July 30, 2020, www.journalism.org/2020/07/30/americans-who-mainly-get-their-news-on-social-media-are-less-engaged-less-knowledgeable.

■ **EXAMINING ETHICS** The 1619 Project—Journalism Takes the Long View to Rethink History

1. Andrew Wind, "Waterloo Native Wins MacArthur Fellowship for Work on Inequality," *Waterloo Courier*, October 19, 2017, https://wcfcourier.com/news/local/education/waterloo-native-wins-macarthur-fellowship-for-work-on-inequality/article_6d8ab35a-5c6b-5d9c-95d8-af101f2d9a5d.html.

2. Alana Wise, "Trump Announces 'Patriotic Education' Commission, a Largely Political Move," NPR, September 17, 2020, www.npr.org/2020/09/17/914127266/trump-announces-patriotic-education-commission-a-largely-political-move. See also Sarah Ellison, "How the 1619 Project Took Over 2020," *Washington Post*, October 13, 2020, www.washingtonpost.com/lifestyle/style/1619-project-took-over-2020-inside-story/2020/10/13/af537092-00df-11eb-897d-3a6201d6643f_story.html.

■ **MEDIA LITERACY AND THE CRITICAL PROCESS** Covering the Unionization of Newspaper Workers

1. Steven Greenhouse, "Why Newsrooms Are Unionizing Now," Nieman Reports, Spring 2019, https://niemanreports.org/articles/why-newsrooms-are-unionizing-now.

■ **GLOBAL VILLAGE** The World's Biggest Newspapers

1. WAN-IFRA, "World Press Trends 2019: Facts and Figures," 2019, www.wptdatabase.org/world-press-trends-2019-facts-and-figures.

2. "Summary," *Yomiuri Shimbun*, https://info.yomiuri.co.jp/english/yomiuri.html.

3. Dainik Baskar Group, "Our History," 2020, www.dainikbhaskargroup.com/our-history.php.

9 Magazines in the Age of Specialization

1. Greg Dool, "Meredith's *Magnolia Journal* Ups the Ante Again," *Folio*, May 11, 2017, www.foliomag.com/merediths-magnolia-journal-ups-ante-industry-notes.

2. Meredith Corporation, "Meredith to Increase Distribution to 600,000 Copies for Premiere Issue of the *Magnolia Journal*," October 25, 2016, http://meredith.mediaroom.com/index.php?s=2311&item=137020.

3. Sara Guaglione, "*Magnolia Journal* Rakes in 700,000 Subscribers," MediaPost, May 10, 2017, www.mediapost.com/publications/article/300854/magnolia-journal-rakes-in-700000-subscribers.html.

4. Meredith Direct Media, "*Magnolia Journal*," accessed December 10, 2020, https://meredithdirectmedia.com/magazines/magnolia-journal.

5. See Theodore Peterson, *Magazines in the Twentieth Century* (Urbana: University of Illinois Press, 1964), 5.

6. See Richard Ohmann, *Selling Culture: Magazines, Markets, and Class at the Turn of the Century* (New York: Verso, 1996).

7. Peterson, *Magazines in the Twentieth Century*, 5.

8. "Ida Tarbell," American Experience, PBS, www.pbs.org/wgbh/americanexperience/features/rockefellers-tarbell.

9. Julie Bosman, "Ebony Photo Archives, Nearly Hidden, Will Be Made Public," *New York Times*, July 25, 2019, www.nytimes.com/2019/07/25/us/ebony-photographs-sale.html.

10. Generoso Pope, quoted in William H. Taft, *American Magazines for the 1980s* (New York: Hastings House, 1982), 226–227.

11. See S. Elizabeth Bird, *For Enquiring Minds: A Cultural Study of Supermarket Tabloids* (Knoxville: University of Tennessee Press, 1992), 24.

12. "Advertising," Condé Nast, accessed June 5, 2018, http://stag.condenast.com/advertising.

13. "The Joy Index—Bank of America," Condé Nast, 2020, https://cnx.condenast.com/work/the-joy-index.

14. FAKEMEME, "is basically a national holiday at this point," November 30, 2020, comment on "Billie Eilish: Same Interview, the Fourth Year," *Vanity Fair*, YouTube video, posted November 30, 2020, www.youtube.com/watch?v=hS2x1zl4rn0.

15. See Gloria Steinem, "Sex, Lies and Advertising," *Ms.*, July/August 1990, pp. 18–28.

16. "Wired U.S.," Condé Nast, accessed 2018, www.condenast.com/brands/wired#U.S.

17. "2021 Media Kit," *National Review*, www.nationalreview.com/wp-content/uploads/2020/11/2021-Media-Kit.pdf.

18. "Media Kit 2020," *Mother Jones*, https://assets.motherjones.com/advertising/2020/Mother_Jones_2020_MediaKit.pdf.

☐ EXAMINING ETHICS The Evolution of Photojournalism

1. Carrie Melago, "Ralph Lauren Model Filippa Hamilton: I Was Fired Because I Was Too Fat!" *New York Daily News*, October 14, 2009, www.nydailynews.com/life-style/fashion/ralph-lauren-model-filippa-hamilton-fired-fat-article-1.381093.

2. Ken Harris, quoted in Jesse Epstein, "Sex, Lies, and Photoshop," *New York Times*, March 8, 2009, http://video.nytimes.com/video/2009/03/09/opinion/1194838469575/sex-lies-and-photoshop.html.

☐ GLOBAL VILLAGE Cosmopolitan Style Travels the World

1. Tess Koman, "'We Can't Publish the Word 'Sex': What It's Like to Be the Editor of *Cosmo* Middle East," *Cosmopolitan*, October 20, 2015, www.cosmopolitan.com/career/news/a46331/cosmo-middle-east-brooke-sever-eic-interview.

2. Tess Koman, "What It's Like to Be the Editor of *Cosmo* Chile," *Cosmopolitan*, March 29, 2016, www.cosmopolitan.com/career/news/a55561/cosmo-chile-ignacia-uribe-eic-interview.

3. Sammye Johnson, "Promoting Easy Sex without the Intimacy: *Maxim* and *Cosmopolitan* Cover Lines and Cover Images," in *Critical Thinking about Sex, Love, and Romance in the Mass Media*, ed. Mary-Lou Galician and Debra L. Merskin (Mahwah, N.J.: Erlbaum, 2007), 55–74.

4. Karen S. H. Roggenkamp, "'Dignified Sensationalism': Elizabeth Bisland, *Cosmopolitan*, and Trips around the World," *American Periodicals: A Journal of History, Criticism, and Bibliography* 17, no. 1 (2007): 26–40.

5. John Tebbel and Mary Ellen Zuckerman, *The Magazine in America, 1741–1990* (New York: Oxford University Press, 1991), 116.

☐ MEDIA LITERACY AND THE CRITICAL PROCESS Uncovering American Beauty

1. Malaika Handa, "Colorism in High Fashion," *The Pudding*, April 2019, https://pudding.cool/2019/04/vogue.

2. Leah Kauffman, "*Elle* Covers Plus-Size Actress Melissa McCarthy on Cover," *Philadelphia Inquirer*, October 15, 2013, www.inquirer.com/philly/blogs/style/Elle-covers-up-plus-size-Melissa-McCarthy-on-cover.html.

10 Books and the Power of Print

1. "Marvel and AHN Team Up to Celebrate Nurses," Marvel, December 3, 2020, www.marvel.com/articles/comics/marvel-and-ahn-team-up-to-celebrate-nurses.

2. Calvin Reid, "2019 North American Comics Sales Rose 11%," *Publishers Weekly*, October 30, 2020, www.publishersweekly.com/pw/by-topic/industry-news/comics/article/84755-2019-north-american-comics-sales-grow-to-1-21-billion.html.

3. Michael Cavna, "Rep. John Lewis's National Book Award Win Is a Milestone Moment for Graphic Novels," *Washington Post*, November 17, 2016, www.washingtonpost.com/news/comic-riffs/wp/2016/11/17/rep-john-lewiss-national-book-award-win-is-a-milestone-moment-for-graphic-novels.

4. Calvin Reid, "Comics Industry Asks *NYT* to Restore Graphic Bestseller Lists," *Publishers Weekly*, February 7, 2018, www.publishersweekly.com/pw/by-topic/industry-news/comics/article/75998-comics-industry-asks-nyt-to-restore-graphic-bestseller-lists.html.

5. Susana Polo, "The *New York Times* Gave Up on Graphic Novels," *Polygon*, January 26, 2017, www.polygon.com/2017/1/26/14401436/new-york-times-graphic-novel-list.

6. Polo.

7. "The *New York Times* Updates and Expands Its Best-Sellers Lists," *New York Times*, September 26, 2019, www.nytco.com/press/the-new-york-times-updates-and-expands-its-best-sellers-lists.

8. *Library and Book Trade Almanac*, 64th ed., (Medford, NJ: Information Today, 2019).

9. Elizabeth Eisenstein, *The Printing Press as an Agent of Change* (Cambridge: Cambridge University Press, 1980).

10. See Quentin Reynolds, *The Fiction Factory: From Pulp Row to Quality Street* (New York: Street & Smith/Random House, 1955), 72–74.

11. For a comprehensive historical overview of the publishing industry and the rise of publishing houses, see John A. Tebbel, *A History of Book Publishing in the United States*, 4 vols. (New York: R. R. Bowker, 1972–1981).

12. OnCampus Research, "Think You Know Everything about the College Market? Think Again," National Association of College Stores, 2020, www.oncampusresearch.org.

13. See John P. Dessauer, *Book Publishing: What It Is, What It Does* (New York: R. R. Bowker, 1974), 48.

14. "Based on the Book," Mid-Continent Public Library, accessed June 6, 2018, www.mymcpl.org/books-movies-music/based-book.

15. Porter Anderson, "US Market Statistics: The AAP StatShot Annual Report for 2019," Publishing Perspectives, July 31, 2020, https://publishingperspectives.com/2020/07/united-states-market-statistics-aap-statshot-annual-report-for-2019.

16. Anderson.

17. Anderson.

18. John Maher, "New Lee and Low Survey Shows No Progress on Diversity in Publishing," *Publishers Weekly*, January 29, 2020, www.publishersweekly.com/pw/by-topic/industry-news/publisher-news/article/82284-new-lee-and-low-survey-shows-no-progress-on-diversity-in-publishing.html.

19. Concepción de León, "On *Oprah's Book Club, American Dirt* Author Faces Criticism," *New York Times*, March 6, 2020, www.nytimes.com/2020/03/06/books/american-dirt-oprah-book-club-apple-tv.html.

20. Maher, "New Lee and Low Survey."

21. Chloe Teasley, "Columbus Bookstores Have Independent Spirit," *ColumbusCEO*, May 7, 2018, www.columbusceo.com/business/20180507/columbus-bookstores-have-independent-spirit.

22. Pamela N. Danziger, "How Indie Bookstores Beat Amazon at the Bookselling Game: Lessons Here for Every Retailer," *Forbes*, February 12, 2020, www.forbes.com/sites/pamdanziger/2020/02/12/how-indie-bookstores-beat-amazon-at-the-bookselling-game-lessons-here-for-every-retailer/?sh=585ba7321a67.

23. Ryan Raffaelli, quoted in Carmen Nobel, "How Independent Bookstores Have Thrived in Spite of Amazon.com," *Working Knowledge*, November 20, 2017, https://hbswk.hbs.edu/item/why-independent-bookstores-haved-thrived-in-spite-of-amazon-com.

24. Steve Wasserman, "The Amazon Effect," *Nation*, May 29, 2012, www.thenation.com/article/168125/amazon-effect#.

25. Matt Day and Jackie Gu, "The Enormous Numbers behind Amazon's Market Reach," Bloomberg, March 27, 2019, www.bloomberg.com/graphics/2019-amazon-reach-across-markets.

26. Andrew Perrin, "Who Doesn't Read Books in America?" Pew Research Center, September 26, 2019, www.pewresearch.org/fact-tank/2019/09/26/who-doesnt-read-books-in-america.

27. Perrin.

28. Alvin Kernan, *The Death of Literature* (New Haven, Conn.: Yale University Press, 1990).

◾ EXAMINING ETHICS Banned Books: Controversial Because They Are Real

1. David Comer Kidd and Emanuele Castano, "Reading Literary Fiction Improves Theory of Mind," *Science*, October 18, 2013, 377–380, https://doi.org/10.1126/science.1239918.

◾ GLOBAL VILLAGE Buenos Aires, the World's Bookstore Capital

1. Debora Rey, "Bookstore Tourists Should Have One Destination on Their List: Buenos Aires," Associated Press, May 2, 2015, http://skift.com/2015/05/02/bookstore-tourists-should-have-one-destination-on-their-list-buenos-aires.

2. Rey.

3. Rey.

4. Porter Anderson, "Fixed Book Prices in Germany: Two New Studies Are Introduced in Berlin," *Publishing Perspectives*, November 8, 2019,

https://publishingperspectives.com/2019/11/fixed-book-prices-in-germany-two-new-studies-borsenverein-released-berlin.

5. Catherine Blache, "Why Fixed Book Price Is Essential for Real Competition," International Publishers Association, March 19, 2015, www.internationalpublishers.org/news/blog/entry/why-fixed-book-price-is-essential-for-real-competition.

6. International Publishers Association, *Global Fixed Book Price Report*, May 23, 2014, www.internationalpublishers.org/images/reports/2014/fixed-book-price-report-2014.pdf.

7. Blache, "Why Fixed Book Price Is Essential."

8. Matt Chesterton, "Top Bookstores in Buenos Aires," *Travel + Leisure*, August 2014, www.travelandleisure.com/local-experts/buenos-aires/top-bookstores-buenos-aires.

11 Advertising and Commercial Culture

1. "17 Questions with the TikTok Hype House," *Entertainment Tonight*, March 7, 2020, www.youtube.com/watch?v=ljJf5MfSpxw.

2. Taylor Lorenz, "Hype House and the Los Angeles TikTok Mansion Gold Rush," *New York Times*, January 3, 2020, www.nytimes.com/2020/01/03/style/hype-house-los-angeles-tik-tok.html.

3. Lorenz.

4. "The 20 Most Followed Accounts on TikTok," *Brandwatch Blog*, November 24, 2020, www.brandwatch.com/blog/most-followers-on-tiktok.

5. Abram Brown, "Founders Feud at Hype House Gets Nasty: An Armed Guard, a New Lawsuit—and a Breakaway Group of TikTok Stars," *Forbes*, March 26, 2020, www.forbes.com/sites/abrambrown/2020/03/26/founders-feud-at-hype-house-gets-nasty-an-armed-guard-a-new-lawsuit-and-a-breakaway-group-of-tiktok-stars.

6. Rebecca Jennings, "A TikTok House Divided: You're 19 Years Old. You Get Famous Overnight. You Move to LA. Now What?," Vox, October 1, 2020, www.vox.com/the-goods/21459677/tiktok-house-la-hype-sway-girls-in-the-valley.

7. "17 Questions with the TikTok Hype House."

8. "Hype House Teams with Jerry Leigh for Fashion Collection," License Global, December 2, 2020, www.licenseglobal.com/apparel/hype-house-teams-jerry-leigh-fashion-collection.

9. "TikTok Statistics—Updated January 2021," *WallaBlog*, January 1, 2021, https://wallaroomedia.com/blog/social-media/tiktok-statistics.

10. Sheree Johnson, "New Research Sheds Light on Daily Ad Exposures," SJ Insights, September 29, 2014, https://sjinsights.net/2014/09/29/new-research-sheds-light-on-daily-ad-exposures.

11. For a written and pictorial history of early advertising, see Charles Goodrum and Helen Dalrymple, *Advertising in America: The First 200 Years* (New York: Harry N. Abrams, 1990), 31.

12. Larissa Faw, "Strawberry Frog's Goodson: It's Time for Adland to Do a Big Rethink," StrawberryFrog, June 3, 2020, www.strawberryfrog.com/at-strawberryfrog-we-believe-that-the-most-effective-leadership-comes-through-activating-purpose-2-2.

13. "US Framework and VALS™ Types," Strategic Business Insights, accessed January 18, 2021, http://strategicbusinessinsights.com/vals/ustypes.shtml.

14. Ad Age, *Marketing Fact Pack 2021*, December 21, 2020, p. 8.

15. Jeanine Poggi, "What It Costs to Advertise in TV's Biggest Shows in 2020–2021 Season," *Ad Age*, October 30, 2020, https://adage.com/article/media/tvs-most-expensive-shows-advertisers-season/2281176.

16. Nicole Perrin, "Facebook-Google Duopoly Won't Crack This Year," eMarketer, November 4, 2019, www.emarketer.com/content/facebook-google-duopoly-won-t-crack-this-year.

17. "Global Spam Volume as Percentage of Total E-Mail Traffic from January 2014 to March 2020, by Month," Statista, June 24, 2020, www.statista.com/statistics/420391/spam-email-traffic-share.

18. Ad Age, *Marketing Fact Pack 2021*, p. 13.

19. Mike Vorhaus, "Digital Advertising Seeing Growth Despite Covid, While Traditional Media Suffers," *Forbes*, September 4, 2020, www.forbes.com/sites/mikevorhaus/2020/09/04/digital-advertising-seeing-growth-despite-covid-traditional-media-suffers/?sh=17a6c7b12f34.

20. Ad Age, *Marketing Fact Pack 2021*, p. 13.

21. Ad Age, "Time Spent Using Media," *Marketing Fact Pack 2016*, p. 21; Yoram Wurmser, "US Mobile Time Spent 2020," eMarketer, June 4, 2020, www.emarketer.com/content/us-mobile-time-spent-2020.

22. Jeff Orlowski, dir., *The Social Dilemma* (Exposure Labs, 2020, www.thesocialdilemma.com).

23. "Mobile and in-App Advertising Trends 2020," Business of Apps, January 27, 2020, www.businessofapps.com/ads/research/mobile-app-advertising-trends.

24. Jon Gibs and Sean Bruich, "NielsenFacebook Report the Value of Social Media Ad Impressions," April 20, 2010, www.nielsen.com/us/en/insights/news/2010/nielsenfacebook-ad-report.html.

25. See Alexandra Ilyashov, "That Dress You Saw All Over Instagram Last Year Might Be a *Really* Expensive Mistake," March 15, 2016, www.refinery29.com/2016/03/106064/lord-and-taylor-ootd-ads-campaign-ftc-settlement.

26. "The Facebook Data Breach Scandal Explained," May 1, 2018, www.techadvisory.org/2018/05/the-facebook-data-breach-scandal-explained.

27. Kate Taylor, "Kim Kardashian Revealed in a Lawsuit That She Demands up to Half a Million Dollars for a Single Instagram Post and Other Details about How Much She Charges for Endorsement Deals," Business Insider, May 9, 2020, www.businessinsider.com/how-much-kim-kardashian-charges-for-instagram-endorsement-deals-2019-5.

28. Kurt Badenhausen, "Highest Paid Athletes in the World," *Forbes*, May 21, 2020, www.forbes.com/athletes/#20eabc0c55ae.

29. Leslie Savan, "Op Ad: Sneakers and Nothingness," *Village Voice*, April 2, 1991, p. 43.

30. Hayley Peterson, "The Ridiculous Reason McDonald's Sold Chipotle and Missed Out on Billions of Dollars," Business Insider, May 22, 2015, www.businessinsider.com/the-ridiculous-reason-mcdonalds-sold-chipotle-2015-5.

31. Michael Schudson, *Advertising: The Uneasy Persuasion* (New York: Basic Books, 1984), 210.

32. PRWeb, "Global Product Placement Spend Up 14.5% to $20.6B in 2019, but COVID-19 Impact to End 10-Yr Growth Streak in 2020; Strong Rebound Seen in '21 on TV, Digital, Music Growth," news release, May 27, 2020, www.prweb.com/releases/global_product_placement_spend_up_14_5_to_20_6b_in_2019_but_covid_19_impact_to_end_10_yr_growth_streak_in_2020_strong_rebound_seen_in_21_on_tv_digital_music_growth/prweb17146134.htm.

33. Netflix, *Quarterly Report*, July 17, 2019, https://d18rn0p25nwr6d.cloudfront.net/CIK-0001065280/cf8e6005-a371-4989-aa7d-cc1bd4b4a871.pdf.

34. Palmer Haasch, "New Coke Is the Weirdest Pop Culture Throwback in *Stranger Things* 3," Polygon, July 6, 2019, www.polygon.com/2019/7/6/20683542/stranger-things-3-new-coke-1985-coca-cola-where-to-buy.

35. Jared Newman, "More Product Placements May Come to Netflix (but Don't Call Them Ads)," *Fast Company*, July 23, 2019, www.fastcompany.com/90380266/more-product-placements-may-come-to-netflix-but-dont-call-them-ads.

36. See "Product Placement on TV," Ofcom, www.ofcom.org.uk/tv-radio-and-on-demand/advice-for-consumers/television/product-placement-on-tv. See also "Why Are the Disclosure Rules for Influencers' Sponsored Content So Different Than They Are for TV Product Placements?" Fashion Law, July 3, 2019, www.thefashionlaw.com/why-are-the-disclosure-rules-for-influencers-sponsored-content-so-different-than-they-are-for-tv-product-placements; Federal Trade Commission, *Disclosures 101 for Social Media Influencers*, November 2019, www.ftc.gov/tips-advice/business-center/guidance/disclosures-101-social-media-influencers.

37. Vance Packard, *The Hidden Persuaders* (New York: Basic Books, 1957, 1978), 229.

38. See Eileen Dempsey, "Auld Lang Syne," *Columbus Dispatch*, December 28, 2000, p. 1G; John Reinan, "The End of the Good Old Days," *Minneapolis Star Tribune*, August 31, 2004, p. 1D.

39. See Schudson, *Advertising*, 36–43; Andrew Robertson, *The Lessons of Failure* (London: MacDonald, 1974).

40. Kim Campbell and Kent Davis-Packard, "How Ads Get Kids to Say, I Want It!" *Christian Science Monitor*, September 18, 2000, p. 1.

41. Victoria Rideout and Michael B. Robb, *The Common Sense Census: Media Use by Kids Age Zero to Eight* (San Francisco, CA: Common Sense Media, 2020), 4.

42. Rideout and Robb, 25.

43. Marisa Meyer, Victoria Adkins, Nalingna Yuan, Heidi Weeks, Yung-Ju Chang, Jenny Radesky, "Advertising in Young Children's Apps: A Content Analysis," *Journal of Developmental and Behavioral Pediatrics* 40, no. 1 (2019): 32–39, https://doi.org/10.1097/DBP.0000000000000622.

44. Nicole Wetsman, "FDA Announces New Crackdown on Flavored Vaping Products," The Verge, January 2, 2020, www.theverge.com/2020/1/2/21046933/fda-announces-new-crackdown-flavored-vaping-products-juul-trump-cigarette.

45. "Factors That Increase Sexual Assault Risk," National Institute of Justice, September 30, 2008, https://nij.ojp.gov/topics/articles/factors-increase-sexual-assault-risk.

46. Hilary Waldman, "Study Links Advertising, Youth Drinking," *Hartford Courant*, January 3, 2006, p. A1.

47. Lisa M. Schwartz and Steven Woloshin, "Medical Marketing in the United States, 1997–2016," *JAMA* 321, no. 1 (2019): 80–96. https://doi.org/10.1001/jama.2018.19320.

48. Alix Spiegel, "Selling Sickness: How Drug Ads Changed Health Care," National Public Radio, October 13, 2009, www.npr.org/templates/story/story.php?storyid=113675737.

49. Marguerite Ward and Melissa Wiley, "15 Racist Brands, Mascots, and Logos That Were Considered Just Another Part of American Life," Business Insider, July 13, 2020, www.businessinsider.com/15-racist-brand-mascots-and-logos-2014-6.

50. Chris Bodenner, "When Do Multicultural Ads Become Offensive? Your Thoughts," *Atlantic*, June 22, 2015, www.theatlantic.com/entertainment/archive/2015/06/advertising-race-1970s-stereotypes-offensive/395624.

51. Jason Chambers, *Madison Avenue and the Color Line; African Americans in the Advertising Industry* (Philadelphia: University of Pennsylvania Press, 2008).

52. "Case Study: IKEA," Marketing the Rainbow, 2016, https://marketingtherainbow.info/case%20studies/ikea.html; Alex Mayyasi and Priceonomics, "How Subarus Came to Be Seen as Cars for Lesbians," *Atlantic*, June 22, 2016, www.theatlantic.com/business/archive/2016/06/how-subarus-came-to-be-seen-as-cars-for-lesbians/488042.

53. "Procter & Gamble and GLAAD Study: Exposure to LGBTQ Representation in Media and Advertising Leads to Greater Acceptance of the LGBTQ Community," GLAAD, May 27, 2020, www.glaad.org/releases/procter-gamble-and-glaad-study-exposure-lgbtq-representation-media-and-advertising-leads.

54. Jeffrey Godsick, quoted in T. L. Stanley, "Hollywood Continues Its Fast-Food Binge," *Adweek*, June 6, 2009, www.adweek.com/news/advertising-branding/hollywood-continues-its-fast-food-binge-105907.

55. Douglas J. Wood, "Ad Issues to Watch for in '06," *Advertising Age*, December 19, 2005, p. 10.

56. "Tea Marketer Misled Consumers, Didn't Adequately Disclose Payments to Well-Known Influencers, FTC Alleges," Federal Trade Commission, March 6, 2020, www.ftc.gov/news-events/press-releases/2020/03/tea-marketer-misled-consumers-didnt-adequately-disclose-payments.

57. "Tea Marketer Misled Consumers."

58. See Stephen Ansolabehere and Shanto Iyengar, *Going Negative: How Attack Ads Shrink and Polarize the Electorate* (New York: Free Press, 1996).

59. "2020 Election to Cost $14 Billion, Blowing Away Spending Records," October 26, 2020, www.opensecrets.org/news/2020/10/cost-of-2020-election-14billion-update.

60. Ad Age, *Marketing Fact Pack 2021*, p. 5.

▣ EXAMINING ETHICS Do Alcohol Ads Encourage Binge Drinking?

1. "Binge Drinking," CDC, December 30, 2019, www.cdc.gov/alcohol/fact-sheets/binge-drinking.htm.

2. "Binge Drinking."

3. Amanda Stewart, "Alcohol Ads Have Heavy Impact on Underage Binge Drinking," *Design & Trend*, January 20, 2015. See also FoxNews.com, "Children's Health: TV Alcohol Ad Exposure Linked to Greater Chance of Underage Drinking," January 20, 2015, www.foxnews.com/health/tv-alcohol-ad-exposure-linked-to-greater-chance-of-underage-drinking.

4. Stewart, "Alcohol Ads Have Heavy Impact."

5. FoxNews.com, "Children's Health."

▣ GLOBAL VILLAGE The Unfairness of Fairness Creams

1. Nupur Acharya, "Unilever's Fair & Lovely Rebrand in India Will Be Costly: Jefferies," *Bloomberg*, June 26, 2020, www.bloomberg.com/news/articles/2020-06-26/unilever-india-unit-will-need-to-spend-big-to-rebrand-jefferies.

2. Mahita Gajanan, "Unilever Will Drop the Word 'Fair' from Its Skin-Lightening Creams. Experts Say It Does Not Combat Colorism," *Time*, June 26, 2020, https://time.com/5860313/unilever-fair-and-lovely-name-change-colorism.

12 Public Relations and Framing the Message

1. "WHO Director-General Dr Margaret Chan Speech on the Ebola Virus Disease Outbreak Delivered to the Presidents of Guinea, Liberia, Sierra Leone, and Cote d'Ivoire," ReliefWeb, August 1, 2014, https://reliefweb.int/report/guinea/who-director-general-dr-margaret-chan-speech-ebola-virus-disease-outbreak-delivered.

2. Sara R. Bedrosian, Cathy E. Young, Laura A. Smith, Joanne D. Cox, Craig Manning, Laura Pechta, Jana L. Telfer, et al., "Lessons of Risk Communication and Health Promotion—West Africa and United States," in "CDC's Response to the 2014–2016 Ebola Epidemic—West Africa and United States," supplement, *MMWR* 65, no. S3 (2016): 68–74, https://doi.org/10.15585/mmwr.su6503a10.

3. "CDC Releases Detailed History of the 2014–2016 Ebola Response in *MMWR*," CDC, July 7, 2016, www.cdc.gov/media/releases/2016/p0707-history-ebola-response.html.

4. Bedrosian et al., "Lessons of Risk Communication and Health Promotion," 69.

5. Bedrosian et al., "Lessons of Risk Communication and Health Promotion."

6. Barbara Reynolds and Matthew Seeger, *Crisis and Emergency Risk Communication, 2014 Edition* (Washington: U.S. Department of Health and Human Services, 2014).

7. Carolyn Y. Johnson and William Wan, "Trump Is Breaking Every Rule in the CDC's 450-Page Playbook for Health Crisis," *Washington Post*, March 14, 2020, www.washingtonpost.com/health/2020/03/14/cdc-manual-crisis-coronavirus-trump.

8. See Stuart Ewen, *PR! A Social History of Spin* (New York: Basic Books, 1996).

9. Marvin N. Olasky, "The Development of Corporate Public Relations, 1850–1930," *Journalism Monographs* 102 (April 1987): 14.

10. Ivy Lee, quoted in Anthony Fellow, *American Media History* (Boston: Cengage Learning, 2012), 202.

11. Edward Bernays, "The Theory and Practice of Public Relations: A Résumé," in E. L. Bernays, ed., *The Engineering of Consent* (Norman: University of Oklahoma Press, 1955), 3–25.

12. Edward Bernays, *Crystallizing Public Opinion* (New York: Horace Liveright, 1923), 217.

13. Michael Schudson, *Discovering the News: A Social History of American Newspapers* (New York: Basic Books, 1978), 136.

14. Susan Henry, *Anonymous in Their Own Names: Doris E. Fleischman, Ruth Hale, and Jane Grant* (Nashville: Vanderbilt University Press, 2012), 32.

15. Larry Tye, *The Father of Spin: Edward L. Bernays and the Birth of Public Relations* (New York: Picador, 1998), 124.

16. Glenn Gray, "Why Public Relations Agencies Are Evolving," *Forbes*, July 21, 2017, www.forbes.com/sites/forbescommunicationscouncil/2017/07/21/why-public-relations-agencies-are-evolving/#3b237c2317f4.

17. " 'The Real Cost' Youth E-Cigarette Prevention Campaign," Food and Drug Administration, January 13, 2021, www.fda.gov/tobacco-products/public-health-education/real-cost-campaign#1.

18. "Pt. 2: The Policy and Regulatory Landscape," Federal Communications Commission, June 9, 2011, www.fcc.gov/general/information-needs-communities.

19. Federal Trade Commission, *Disclosures 101 for Social Media Influencers*.

20. "Lobbying Data Summary," Center for Responsive Politics, updated January 23, 2021, www.opensecrets.org/federal-lobbying.

21. "Center for Consumer Freedom," SourceWatch, accessed August 20, 2014, www.sourcewatch.org/index.php?title=Center_for_Consumer_Freedom.

22. "Our Work: Labor Projects," BerlinRosen, accessed January 18, 2021, https://berlinrosen.com/labor/projects.

23. David Barstow, "Message Machine: Behind TV Analysis, Pentagon's Hidden Hand," *New York Times*, April 20, 2008, www.nytimes.com/2008/04/20/us/20generals.html.

24. Associated Press, "Open Government Study: Secrecy Up," *Politico*, March 16, 2014, www.politico.com/story/2014/03/open-government-study-secrecy-up-104715.html.

25. "One Full Year without a W.H. Press Secretary Briefing," CNN, March 8, 2020, www.cnn.com/videos/business/2020/03/08/one-full-year-without-a-w-h-press-secretary-briefing.cnn; "The White House Coronavirus Task Force Gives Its 1st Public Briefing in Months," *All Things Considered*, NPR, June 26, 2020, www.npr.org/2020/06/26/884039323/the-white-house-coronavirus-task-force-gives-its-1st-public-briefing-in-months.

26. Reynolds and Seeger, *Crisis and Emergency Risk Communication, 2014 Edition*.

27. "Covid-19 Comms: US Government Worst, New Zealand & Germany Best Say PR Pros," PRovoke, May 4, 2020, www.provokemedia.com/latest/article/covid-19-comms-us-government-worst-new-zealand-germany-best-say-pr-pros.

28. Joachim Allgaier and Anna Lydia Svalastog, "The Communication Aspects of the Ebola Virus Disease Outbreak in Western Africa—Do We Need to Counter One, Two, or Many Epidemics?" *Croatian Medical Journal* 56, no. 5 (2015): 496–499, https://doi.org/10.3325/cmj.2015.56.496.

29. EAB, "The Best (and Worst) COVID-19 Crisis Communications," April 28, 2020, https://eab.com/insights/podcast/strategy/best-worst-covid19-crisis-communications.

30. Rachel Schilke, " 'I Felt Like a Guinea Pig': Student's 'Awful' Quarantine Experience Prompts UI Apology," *Daily Iowan*, August 20, 2020, https://dailyiowan.com/2020/08/20/i-felt-like-a-guinea-pig-students-awful-quarantine-experience-prompts-university-of-iowa-apology; Tanya Chen, "College Students Are Sharing Fyre Fest–Like Meals in Quarantine While Some Say They're Not Even Getting Them," *Buzzfeed News*, August 20, 2020, www.buzzfeednews.com/article/tanyachen/nyu-uga-college-students-sharing-fyre-fest-quarantine-meals.

31. Stanley Walker, "Playing the Deep Bassoons," *Harper's*, February 1932, p. 365.

32. Walker, 370.

33. Ivy Lee, *Publicity* (New York: Industries, 1925), 21.

34. Schudson, *Discovering the News*, 136.

35. Ivy Lee, quoted in Ray Eldon Hiebert, *Courtier to the Crowd: The Story of Ivy Lee and the Development of Public Relations* (Ames: Iowa State University Press, 1966), 114.

36. Walter Lippmann, *Public Opinion* (New York: Free Press, 1922, 1949), 221.

37. Christopher R. Martin, *No Longer Newsworthy: How the Mainstream Media Abandoned the Working Class* (Ithaca, N.Y.: Cornell University Press, 2019).

38. See Alicia Mundy, "Is the Press Any Match for Powerhouse PR?," in *Impact of Mass Media*, ed. Ray Eldon Hiebert (White Plains, N.Y.: Longman, 1995), 179–188.

39. Dan Rather, interviewed in "Forty Years after Watergate: Carl Bernstein & Dan Rather with CNN's Candy Crowley," *State of the Union with Candy Crowley*, CNN, August 3, 2014, http://cnnpressroom.blogs.cnn.com/2014/08/03/forty-years-after-watergate-carl-bernstein-dan-rather-with-cnns-candy-crowley.

◼ **EXAMINING ETHICS** Egg on The North Face— a Wikipedia Scandal

1. The North Face, May 29, 2019, https://twitter.com/thenorthface/status/1133903040707059712.

2. Dami Lee, "North Face Tried to Scam Wikipedia to Get Its Products to the Top of Google Search," The Verge, May 29, 2019, www.theverge.com/2019/5/29/18644158/north-face-wikipedia-hack-leo-burnett-top-imagens.

◼ **GLOBAL VILLAGE** Public Relations and Bananas

1. Phil Milford, "Dole, Del Monte, Dow Chemical Sued over Banana Pesticide," Bloomberg, June 4, 2012, www.bloomberg.com/news/articles/2012-06-04/dole-del-monte-dow-chemical-sued-over-banana-pesticide.

2. "About Us," Dole Organic Program, accessed June 29, 2018, www.doleorganic.com/farms/index.php/about-us.

◼ **MEDIA LITERACY AND THE CRITICAL PROCESS** The Invisible Hand of PR

1. John Stauber, "Corporate PR: A Threat to Journalism?" *Background Briefing: Radio National*, March 30, 1997, www.abc.net.au/radionational/programs/backgroundbriefing/corporate-pr-a-threat-to-journalism/3563876.

13 Media Economics and the Global Marketplace

1. Ryan Lattanzio, "Christopher Nolan in 'Disbelief' over Warner Bros.' Max Shakeup: 'It's Very Messy,' " IndieWire, December 7, 2020, www.indiewire.com/2020/12/christopher-nolan-reacts-warner-bros-hbo-max-deal-1234603016.

2. Kim Masters, "Christopher Nolan Rips HBO Max as 'Worst Streaming Service,' Denounces Warner Bros.' Plan," *Hollywood Reporter*, December 7, 2020, www.hollywoodreporter.com/news/christopher-nolan-rips-hbo-max-as-worst-streaming-service-denounces-warner-bros-plan.

3. Aric Jenkins, "WarnerMedia Studios Chief on the Controversial Decision to Release New Movies on HBO Max," *Fortune*, December 19, 2020, https://fortune.com/2020/12/19/hbo-max-movies-warnermedia-studios-ann-sarnoff.

4. Douglas Gomery, "The Centrality of Media Economics," in *Defining Media Studies*, ed. Mark R. Levy and Michael Gurevitch (New York: Oxford University Press, 1994), 202.

5. Elizabeth Fones-Wolf, *Selling Free Enterprise: The Business Assault on Labor and Liberalism, 1945–60* (Urbana: University of Illinois Press, 1994).

6. David Harvey, *The Condition of Postmodernity: An Enquiry into the Origins of Cultural Change* (Oxford: Basil Blackwell, 1989), 171.

7. Alex S. Jones, "F.C.C. Raised Limits on Total Stations under One Owner," *New York Times*, July 27, 1984, www.nytimes.com/1984/07/27/business/fcc-raises-limit-on-total-stations-under-one-owner.html.

8. Federal Communications Commission, *Report on Cable Industry Prices*, October 12, 2016, https://docs.fcc.gov/public/attachments/DA-16-1166A1.pdf.

9. Jonathan Berr, "Here Is Everything You Need to Know about the Viacom-CBS Merger," *Forbes*, www.forbes.com/sites/jonathanberr/2019/11/26/here-is-everything-you-need-to-know-about-the-viacom-cbs-merger/?sh=71deb7cc147d.

10. Michael A. Hiltzik and James Flanigan, "Entertainment Merger Mania," *Los Angeles Times*, August 2, 1995, www.latimes.com/archives/la-xpm-1995-08-02-fi-30546-story.html.

11. Robert Sher, "Why Half of All M&A Deals Fail, and What You Can Do about It," *Forbes*, March 19, 2012, www.forbes.com/sites/forbesleadershipforum/2012/03/19/why-half-of-all-ma-deals-fail-and-what-you-can-do-about-it.

12. AT&T, "AT&T Completes Acquisition of Time Warner Inc," press release, June 15, 2018, https://about.att.com/story/att_completes_acquisition_of_time_warner_inc.html.

13. Chris Mills, "One Senator Makes a Good Case for Breaking Up Comcast-NBC," BGR, December 20, 2017, https://bgr.com/2017/12/20/comcast-nbc-pricing-unfair-doj-merger.

14. Richard J. Barnet and John Cavanagh, *Global Dreams: Imperial Corporations and the New World Order* (New York: Simon & Schuster, 1994), 131.

15. Dominic Rushe, "Walt Disney Sheds 28,000 Jobs at Theme Parks as Pandemic Bites," *Guardian*, September 29, 2020, www.theguardian.com/film/2020/sep/29/walt-disney-sheds-28000-jobs-theme-parks-pandemic.

16. Ben Bagdikian, *The Media Monopoly*, 6th ed. (Boston: Beacon Press, 2000), 222.

17. Kristal Brent Zook, "Blacks Own Just Ten U.S. Television Stations. Here's Why," *Washington Post*, August 17, 2015, www.washingtonpost.com/posteverything/wp/2015/08/17/blacks-own-just-10-u-s-television-stations-heres-why.

18. U.S. and Plaintiff States v. Google LLC, updated October 21, 2020, p. 3, www.justice.gov/atr/case-document/file/1329131/download.

See also "Justice Department Sues Monopolist Google for Violating Antitrust Laws," U.S. Department of Justice, updated October 21, 2020, www.justice.gov/opa/pr/justice-department-sues-monopolist-google-violating-antitrust-laws.

19. "U.S. Government and 48 State Attorneys General Files Lawsuit against Facebook," *Washington Post*, December 9, 2020, p. 7, www.washingtonpost.com/context/u-s-government-and-48-state-attorneys-general-files-lawsuit-against-facebook/5b97bd6f-8d7f-4ee2-b9ea-79a1fafc7661/?itid=lk_interstitial_manual_6. See also Tony Romm, "U.S., States Sue Facebook as an Illegal Monopoly, Setting Stage for Potential Breakup," *Washington Post*, December 9, 2020, www.washingtonpost.com/technology/2020/12/09/facebook-antitrust-lawsuit.

20. Bureau of Labor Statistics, "Union Members—2020," news release, January 22, 2021, www.bls.gov/news.release/pdf/union2.pdf.

21. Lawrence Mishel and Jori Kandra, "CEO Compensation Surged 14% in 2019 to $21.3 Million," Economic Policy Institute, August 18, 2020, www.epi.org/publication/ceo-compensation-surged-14-in-2019-to-21-3-million-ceos-now-earn-320-times-as-much-as-a-typical-worker.

22. Martha Ross and Nicole Bateman, "Low Unemployment Isn't Worth Much If the Jobs Barely Pay," Brookings, January 8, 2020, www.brookings.edu/blog/the-avenue/2020/01/08/low-unemployment-isnt-worth-much-if-the-jobs-barely-pay.

23. David Lieberman and Brent Lang, "What the Media's Most Powerful Executives Were Paid in 2019," *Variety*, May 20, 2020, https://variety.com/2020/biz/features/media-ceo-salaries-2019-bob-iger-rupert-murdoch-reed-hastings-1234611007.

24. William Paley, quoted in Robert W. McChesney, *Telecommunications, Mass Media, and Democracy: The Battle for Control of U.S. Broadcasting, 1928–1935* (New York: Oxford University Press, 1993), 251.

25. McChesney, *Telecommunications, Mass Media, and Democracy*, 264.

26. Edward Herman, "Democratic Media," *Z Papers*, January–March 1992, 23.

27. "Interactive TV Shows and Movies on Netflix," Help Center, Netflix, accessed December 19, 2020, https://help.netflix.com/en/node/62526.

28. Antonio Gramsci, *Selections from the Prison Notebooks* (New York: International Publishers, 1971), 12–13.

29. John Fiske, *Television Culture* (New York: Routledge, 1987); Elihu Katz and Tamar Leibes, "Mutual Aid in the Decoding of *Dallas*: Notes from a Cross-Cultural Study," in *Television in Transition*, ed. Phillip Drummond and Richard Paterson (London: British Film Institute, 1985), 187–198.

30. Ben DuBose, "Rockets GM Daryl Morey, NBA Issue Statements on China Incident," *USA Today*, October 6, 2019, https://rocketswire.usatoday.com/2019/10/06/rockets-gm-daryl-morey-nba-issue-statements-on-china-incident.

31. See Adam Liptak, "Justices, 5–4, Reject Corporate Spending Limit," *New York Times*, January 22, 2010, www.nytimes.com/2010/01/22/us/politics/22scotus.html.

32. David Sessions, "Chick-fil-A's Place in the Church of Fast Food," *Daily Beast*, July 29, 2012, www.thedailybeast.com/articles/2012/07/29/chick-fil-a-s-place-in-the-church-of-fast-food.html; Michael D. Shear, "Amazon's Founder Pledges $2.5 Million in Support of Same-Sex Marriage," *New York Times*, July 27, 2012, http://thecaucus.blogs.nytimes.com/2012/07/27/amazons-founder-pledges-2-5-million-in-support-of-same-sex-marriage.

33. Center for Responsive Politics, "2020 Election to Cost $14 Billion, Blowing Away Spending Records," October 28, 2020, www.opensecrets.org/news/2020/10/cost-of-2020-election-14billion-update.

34. Danilo Yanich, "Political Ads and Local TV News: Questions with Danilo Yanich," interview by Bill Arthur, Expanding News Desert, UNC Hussman School of Journalism and Media, December 28, 2020, www.usnewsdeserts.com/spotlight-research/political-ads-and-local-tv-news-questions-with-danilo-yanich.

35. John Nichols and Robert W. McChesney, "Who'll Unplug Big Media? Stay Tuned," *Nation*, May 29, 2008, www.thenation.com/article/wholl-unplug-big-media-stay-tuned.

◉ **EXAMINING ETHICS** Are the Big Digital Companies Too Big?

1. Dylan Byers, "Pacific," CNN Money, May 24, 2018, https://money.cnn.com/2018/05/24/technology/pacific-newsletter/index.html.

2. Amanda Lotz, " 'Big Tech' Isn't One Big Monopoly—It's 5 Companies All in Different Businesses," *Conversation*, March 23, 2018, https://theconversation.com/big-tech-isnt-one-big-monopoly-its-5-companies-all-in-different-businesses-92791.

3. Christopher Mims, "Tech's Titans Tiptoe toward Monopoly; Amazon, Facebook and Google May Be Repeating the History of Steel, Utility, Rail and Telegraph Empires Past—While Apple Appears Vulnerable," *Wall Street Journal*, May 31, 2018, www.wsj.com/articles/techs-titans-tiptoe-toward-monopoly-1527783845.

4. Charles Duhigg, "The Case against Google," *New York Times Magazine*, February 20, 2018, www.nytimes.com/2018/02/20/magazine/the-case-against-google.html.

◉ **GLOBAL VILLAGE** China's Dominant Media Corporations Rival America's

1. Kantar, *BrandZ Top 100 Most Valuable Global Brands 2020*, accessed December 19, 2020, www.kantar.com/campaigns/brandz/global.

2. Adam Lashinsky, "Alibaba v. Tencent: The Battle for Supremacy in China," *Fortune*, June 21, 2018, https://fortune.com/longform/alibaba-tencent-china-internet.

3. Raymond Zhong, "Worried about Big Tech? Chinese Giants Make America's Look Tame," *New York Times*, May 31, 2018, www.nytimes.com/2018/05/31/technology/china-tencent-alibaba.html.

4. "What's Behind China's Cracking Down on Tech Giants," Bloomberg News, updated January 22, 2021, www.bloomberg.com/news/articles/2020-11-13/dissecting-china-s-crackdown-on-its-internet-giants-quicktake.

14 The Culture of Journalism: Values, Ethics, and Democracy

1. Jordan Williams, "Journalist Accounts, Footage Suggest They Were Targeted in Capitol Riot," *The Hill*, January 8, 2021, https://thehill.com/homenews/media/533330-journalist-accounts-suggests-they-were-targeted-in-capitol-riot; *Independent*, "AP Photographer Violently Assaulted by Pro-Trump Mob," YouTube, January 8, 2021, www.youtube.com/watch?v=1Bs2HVS8fZc.

2. Nicholas Fando, Erin Schaff, and Emily Cochrane, " 'Senate Being Locked Down': Inside a Harrowing Day at the Capitol," *New York Times*, January 7, 2021, www.nytimes.com/2021/01/07/us/politics/capitol-lockdown.html.

3. Leonard Downie Jr., "The Trump Administration and the Media," Committee to Protect Journalists, April 16, 2020, https://cpj.org/reports/2020/04/trump-media-attacks-credibility-leaks.

4. Daniella Silva, "President Trump Tweets Wrestling Video of Himself Attacking 'CNN,'" NBC News, July 2, 2017, www.nbcnews.com/politics/donald-trump/president-trump-tweets-wwe-video-himself-attacking-cnn-n779031.

5. Gallup/Knight Foundation, *American Views 2020: Trust, Media and Democracy—a Deepening Divide*, November 9, 2020, https://knightfoundation.org/wp-content/uploads/2020/08/American-Views-2020-Trust-Media-and-Democracy.pdf.

6. Gallup/Knight Foundation.

7. Laura Hazard Owen, "How Can Local TV News Fix Its Young Person Problem? Maybe It Needs to Look More Like Vox," Nieman Lab, February 15, 2019, www.niemanlab.org/2019/02/how-can-local-tv-news-fix-its-young-person-problem-maybe-it-needs-to-look-more-like-vox.

8. William Greider, quoted in Mark Hertsgaard, *On Bended Knee: The Press and the Reagan Presidency* (New York: Farrar, Straus & Giroux, 1988), 78.

9. Michael Schudson, *The Sociology of News* (New York: W. W. Norton, 2003), 83.

10. Schudson, 82.

11. Schudson, 34.

12. A. William Bluem, *Documentary in American Television* (New York: Hastings House, 1965), 94.

13. Bill Kovach and Tom Rosenstiel, *The Elements of Journalism* (New York: Three Rivers Press, 2007), 78–112.

14. "Explainers," Vox, accessed March 12, 2021, www.vox.com/explainers.

15. Neil Postman, "Currents," *Utne Reader*, July/August 1995, p. 35.

16. Jen Kirby, "Trump's Own Officials Say 2020 Was America's Most Secure Election in History," Vox, November 13, 2020, www.vox.com/2020/11/13/21563825/2020-elections-most-secure-dhs-cisa-krebs.

17. "74% of Voters Say Democracy in the U.S. Is Under Threat, Quinnipiac University National Poll Finds; 52% Say President Trump Should Be Removed from Office," Quinnipiac Poll, January 11, 2021, https://poll.qu.edu/national/release-detail?ReleaseID=3686.

18. Michael Dimock, "How Americans View Trust, Facts, and Democracy Today," Pew Charitable Trust, February 19, 2020, www.pewtrusts.org/en/trust/archive/winter-2020/how-americans-view-trust-facts-and-democracy-today.

19. Reuven Frank, "Memorandum from a Television Newsman," reprinted as Appendix 2 in Bluem, *Documentary in American Television*, 276.

20. Kovach and Rosenstiel, *Elements of Journalism*.

21. Kovach and Rosenstiel.

22. David Sheddon, "Today in Media History: Mr. Dooley: 'The Job of the Newspaper Is to Comfort the Afflicted and Afflict the Comfortable,'" Poynter, October 7, 2014, www.poynter.org/reporting-editing/2014/today-in-media-history-mr-dooley-the-job-of-the-newspaper-is-to-comfort-the-afflicted-and-afflict-the-comfortable.

23. Herbert Gans, *Deciding What's News* (New York: Pantheon, 1979), 42.

24. Gans, 42, 46.

25. Gans, 48–50.

26. Gans, 50–51.

27. Pete Vernon, "Dancing around the Word 'Racist' in Coverage of Trump," *Columbia Journalism Review*, September 25, 2017, www.cjr.org/politics/trump-racism.php.

28. Associated Press, *The Associated Press Stylebook*, 55th ed. (New York: Basic Books, 2020), 198.

29. Pete Vernon, "Lie? Falsehood? What to Call the President's Words," *Columbia Journalism Review*, May 29, 2018, www.cjr.org/the_media_today/trump-lie-falsehood.php.

30. Reporters Committee for Freedom of the Press, "The Landmark Food Lion Case," Spring 2012, www.rcfp.org/journals/news-media-and-law-spring-2012/landmark-food-lion-case.

31. Eliana Miller and Nicole Asbury, "Photographers Are Being Called on to Stop Showing Protesters' Faces. Should They?" Poynter, June 4, 2020, www.poynter.org/ethics-trust/2020/should-journalists-show-protesters-faces.

32. Society of Professional Journalists, "SPJ Code of Ethics," September 6, 2014, www.spj.org/ethicscode.asp.

33. Society of Professional Journalists.

34. For reference and guidance on media ethics, see Clifford Christians, Mark Fackler, and Kim Rotzoll, *Media Ethics: Cases and Moral Reasoning*, 4th ed. (White Plains, N.Y.: Longman, 1995); Thomas H. Bivins, "A Worksheet for Ethics Instruction and Exercises in Reason," *Journalism Educator* 48, no. 2 (Summer 1993): 4–16.

35. David Folkenflik, "*New York Times* Retracts Core of Hit Podcast Series *Caliphate* on ISIS," NPR, December 18, 2020, www.npr.org/2020/12/18/944594193/new-york-times-retracts-hit-podcast-series-caliphate-on-isis-executioner.

36. Folkenflik.

37. Folkenflik.

38. Don Hewitt, interview by Richard Campbell, *60 Minutes*, CBS News, February 21, 1989.

39. See Ina Howard, "Power Sources: On Party, Gender, Race and Class, TV News Looks to the Most Powerful Groups," *Extra!*, Fairness and Accuracy in Reporting, May 1, 2002, https://fair.org/extra/power-sources. See also Sarah Marcharia and Marcus Burke, "Just 24% of News Sources Are Women. Here's Why That's a Problem," World Economic Forum, March 2, 2020, www.weforum.org/agenda/2020/03/women-representation-in-media.

40. See Adrienne LaFrance, "I Analyzed a Year of My Reporting for Gender Bias (Again)," *Atlantic*, February 17, 2016, www.theatlantic.com/technology/archive/2016/02/gender-diversity-journalism/463023.

41. Onyi Lam, Stefan Wojcik, Adam Hughes, and Brian Broderick, "Men Appear Twice as Often as Women in News Photos on Facebook," Pew Research Center, May 23, 2019, www.journalism.org/2019/05/23/men-appear-twice-as-often-as-women-in-news-photos-on-facebook.

42. Andi Egbert, "The Color of Expertise: Talking Heads and Talking Truth," APM Research Lab, April 23, 2019, www.apmresearchlab.org/blog/the-color-of-expertise.

43. Jeanine Santucci, "Source Diversity at NPR: Grassroots Initiatives Address Challenges," NPR, July 24, 2018, www.npr.org/sections/publiceditor/2018/07/24/631964116/source-diversity-at-npr-grassroots-initiatives-address-challenges.

44. Women's Media Center, *The Status of Women in the U.S. Media, 2019*, February 2019, 11, https://tools.womensmediacenter.com/page/-/WMCStatusofWomeninUSMedia2019.pdf.

45. Women's Media Center, *The Status of Women of Color in the U.S. News Media, 2018*, March 2018, 11, www.womensmediacenter.com/assets/site/reports/the-status-of-women-of-color-in-the-u-s-media-2018-full-report/Women-of-Color-Report-FINAL-WEB.pdf.

46. "Ethical Journalism: A Handbook of Values and Practices for the News and Editorial Departments," accessed March 3, 2021, *New York Times*, www.nytimes.com/editorial-standards/ethical-journalism.html.

47. Nicole Hemmer, "The Conservative War on Liberal Media Has a Long History," *Atlantic*, January 17, 2014, www.theatlantic.com/politics/archive/2014/01/the-conservative-war-on-liberal-media-has-a-long-history/283149.

48. Dan Mangan, "President Trump Told Lesley Stahl He Bashes Press 'to Demean You and Discredit You So . . . No One Will Believe' Negative Stories about Him," CNBC, May 22, 2018, www.cnbc.com/2018/05/22/trump-told-lesley-stahl-he-bashes-press-to-discredit-negative-stories.html.

49. Kathleen Hall Jamieson and Joseph N. Cappella, *Echo Chamber: Rush Limbaugh and the Conservative Media Establishment* (Oxford: Oxford University Press, 2008), x. See also David Brock, *The Republican Noise Machine* (New York: Crown, 2004).

50. Patrick J. Buchanan, *The New Majority: President Nixon at Mid-passage* (Philadelphia: Girard Bank, 1973), 21–22.

51. Nicole Hemmer, *Messengers of the Right: Conservative Media and the Transformation of American Politics* (Philadelphia: University of Pennsylvania Press, 2016).

52. Sarah Ellison, *War at the "Wall Street Journal": Inside the Struggle to Control an American Business Empire* (Boston: Houghton Mifflin Harcourt, 2010), xix.

53. David Brock, *The Republican Noise Machine* (New York: Crown, 2004), 190.

54. Robert L. Hilliard and Michael C. Keith, *Waves of Rancor: Tuning in the Radical Right* (Armonk, N.Y.: M.E. Sharpe, 1999), 15.

55. Hilliard and Keith, 17.

56. David Foster Wallace, "Host," *Atlantic*, April 2005, www.theatlantic.com/magazine/archive/2005/04/host/303812.

57. Jamieson and Cappella, *Echo Chamber*, 46.

58. David Folkenflik, "The Rise, Fall and Lasting Influence of Roger Ailes," NPR, July 23, 2016, www.npr.org/2016/07/23/487181154/the-rise-fall-and-lasting-influence-of-roger-ailes.

59. Marisa Guthrie, "Roger Ailes Resigns as Fox News Chief after Sexual Harassment Accusations," *Hollywood Reporter*, July 21, 2016, www.hollywoodreporter.com/news/roger-ailes-resigns-as-fox-913206.

60. Anousha Sakoui, "Rupert Murdoch's Grip on Power Put to the Test in Fox Upheaval," Bloomberg, May 22, 2017, www.bloomberg.com/news/articles/2017-05-22/rupert-murdoch-s-grip-on-power-put-to-the-test-in-fox-upheaval.

61. Jane Mayer, "The Reclusive Hedge-Fund Tycoon behind the Trump Presidency," *New Yorker*, March 27, 2017, www.newyorker.com/magazine/2017/03/27/the-reclusive-hedge-fund-tycoon-behind-the-trump-presidency.

62. Sarah Posner, "How Donald Trump's New Campaign Chief Created an Online Haven for White Nationalists," *Mother Jones*, August 22, 2016, www.motherjones.com/politics/2016/08/stephen-bannon-donald-trump-alt-right-breitbart-news.

63. Wallace, "Host."

64. Susan Douglas, "Letting the Boys Be Boys: Talk Radio, Male Hysteria, and Political Discourse in the 1980s," in *Radio Reader: Essays in the Cultural History of Radio*, ed. Michele Hilmes and Jason Loviglio (New York: Routledge, 2002), 487.

65. Christopher R. Martin, *No Longer Newsworthy: How the Mainstream Media Abandoned the Working Class* (Ithaca, N.Y.: Cornell University Press, 2019).

66. Rush Limbaugh, *The Way Things Ought to Be* (New York: Pocket Books, 1992), 268–269.

67. Bill O'Reilly, *Who's Looking Out for You?* (New York: Broadway Books, 2003), 78.

68. Joshua Green, *Devil's Bargain* (New York: Penguin Press, 2017), 86.

69. Posner, "How Donald Trump's New Campaign Chief Created an Online Haven for White Nationalists."

70. "The Legacy of Rush Limbaugh," *The Daily* (blog), *New York Times*, February 22, 2021, www.nytimes.com/2021/02/22/podcasts/the-daily/rush-limbaugh-conservatism-donald-trump.html.

71. Nate Rattner, "Trump's Election Lies Were among His Most Popular Tweets," CNBC, January 13, 2021, www.cnbc.com/2021/01/13/trump-tweets-legacy-of-lies-misinformation-distrust.html.

72. Twitter Inc., "Permanent Suspension of @realDonaldTrump," January 8, 2021, https://blog.twitter.com/en_us/topics/company/2020/suspension.html.

73. James Agee and Walker Evans, *Let Us Now Praise Famous Men* (Boston: Houghton Mifflin, 1960), xiv.

74. David Broder, quoted in "Squaring with the Reader: A Seminar on Journalism," *Kettering Review* (Winter 1992): 48.

75. Jill Lepore, "Party Time for a Young America," *New Yorker*, September 17, 2007, www.newyorker.com/magazine/2007/09/17/party-time.

76. Bill Moyers, "Bill Moyers' Speech to the National Conference for Media Reform," Free Press, May 16, 2005, www.beyondpesticides.org/assets/media/documents/documents/bhopalop_ed_cohen.pdf.

⊡ **GLOBAL VILLAGE** Authoritarians Use "Fake News" Allegations as a Weapon

1. Hannah Beech, "'No Such Thing as Rohingya': Myanmar Erases a History," *New York Times*, December 2, 2017, www.nytimes.com/2017/12/02/world/asia/myanmar-rohingya-denial-history.html.

2. Steven Erlanger, "'Fake News,' Trump's Obsession, Is Now a Cudgel for Strongmen," *New York Times*, December 12, 2017, www.nytimes.com/2017/12/12/world/europe/trump-fake-news-dictators.html.

3. Erlanger. See also Jason Schwartz, "Trump's 'Fake News' Mantra a Hit with Despots," Politico, December 8, 2017, www.politico.com/story/2017/12/08/trump-fake-news-despots-287129.

4. Freedom House, "The Rise of Digital Authoritarianism: Fake News, Data Collection and the Challenge to Democracy," press release, October 31, 2018, https://freedomhouse.org/article/rise-digital-authoritarianism-fake-news-data-collection-and-challenge-democracy.

5. Rick Noack, "The Ugly History of 'Lügenpresse,' a Nazi Slur Shouted at a Trump Rally," *Washington Post*, October 24, 2016, www.washingtonpost.com/news/worldviews/wp/2016/10/24/the-ugly-history-of-luegenpresse-a-nazi-slur-shouted-at-a-trump-rally/?utm_term=.d82a910e174b.

6. David Remnick, "Trump vs. the *Times*: Inside an Off-the-Record Meeting," *New Yorker*, July 30, 2018, www.newyorker.com/news/news-desk/trump-vs-the-times-inside-an-off-the-record-meeting.

7. Andras Gergely and Veronika Gulyas, "Orban Uses Crisis Powers for Detentions Under 'Fake News' Law," Bloomberg, May 13, 2020, www.bloomberg.com/news/articles/2020-05-13/orban-uses-crisis-powers-for-detentions-under-fake-news-law.

8. "Kosovo Lifts 100 Percent Tariff on Serbia; Belgrade Calls It 'Fake News,'" Radio Free Europe/Radio Liberty, April 1, 2020, www.rferl.org/a/kosovo-lifts-serbia-tariffs-belgrade/30521305.html.

9. Remnick, "Trump vs. the *Times*."

10. Remnick.

⊡ **EXAMINING ETHICS** Black Journalists Lead an "Overdue Reckoning" on Objectivity

1. Marc Lacey, "Journalists Need to Remember That Not All News Readers Are White," *Nieman Reports*, September 1, 2020, https://niemanreports.org/articles/journalists-need-to-remember-that-not-all-readers-are-white.

2. Wesley Lowery, "A Reckoning over Objectivity, Led by Black Journalists," *New York Times*, June 23, 2020, www.nytimes.com/2020/06/23/opinion/objectivity-black-journalists-coronavirus.html.

3. Lowery.

4. Lowery.

5. Ben Smith, "Inside the Revolts Erupting in America's Big Newsrooms," *New York Times*, June 7, 2020, www.nytimes.com/2020/06/07/business/media/new-york-times-washington-post-protests.html.

6. Wesley Lowery (@WesleyLowery), Twitter, June 3, 2020, 10:18 p.m., https://twitter.com/WesleyLowery/status/1268366363359354885.

15 Media Effects and Cultural Approaches to Research

1. Rebecca Nicholson, "*13 Reasons Why* Review—Sex, Drugs and Mixtapes in Netflix's High-School Horror Show," *Guardian*, March 31,

2017, www.theguardian.com/tv-and-radio/2017/mar/31/13-reasons-why-review-sex-drugs-and-mixtapes-in-netflix-high-school-horror-show.

2. Angelique Jenney and Deinera Exner-Cortens, "Toxic Masculinity and Mental Health in Young Women: An Analysis of *13 Reasons Why*," *Affilia* 33, no. 3 (2018), https://doi.org/10.1177/0886109918762492.

3. Mark Henick, "Why *13 Reasons Why* Is Dangerous," CNN, May 4, 2017, https://edition.cnn.com/2017/05/03/opinions/13-reasons-why-gets-it-wrong-henick-opinion/index.html.

4. Sophie Gilbert, "Did *13 Reasons Why* Spark a Suicide Contagion Effect?" *Atlantic*, August 1, 2017, www.theatlantic.com/entertainment/archive/2017/08/13-reasons-why-demonstrates-cultures-power/535518.

5. Joyce Chen, "Selena Gomez Defends *13 Reasons Why* as Honest Depiction of Teen Suicide," *Rolling Stone*, June 7, 2017, www.rollingstone.com/culture/news/selena-gomez-defends-13-reasons-why-as-honest-w486466.

6. "*Thirteen Reasons Why* Talking Points," Suicide Awareness Voices of Education, 2017, https://save.org/13-reasons-why.

7. Alexis de Tocqueville, *Democracy in America* (New York: Modern Library, 1835, 1840, 1945, 1981), 96–97.

8. Steve Fore, "Lost in Translation: The Social Uses of Mass Communications Research," *Afterimage*, no. 20 (April 1993): 10.

9. James Carey, *Communication as Culture: Essays on Media and Society* (Boston: Unwin Hyman, 1989), 75.

10. Daniel Czitrom, *Media and the American Mind: From Morse to McLuhan* (Chapel Hill: University of North Carolina Press, 1982), 122–125.

11. Czitrom, 123.

12. Harold Lasswell, *Propaganda Technique in the World War* (New York: Alfred A. Knopf, 1927), 9.

13. Walter Lippmann, *Public Opinion* (New York: Macmillan, 1922), 18.

14. Sheldon R. Gawiser and G. Evans Witt, *20 Questions a Journalist Should Ask about Poll Results*, 3rd ed., National Council on Public Polls, 2006, www.ncpp.org/?q=node/4.

15. See W. W. Charters, *Motion Pictures and Youth: A Summary* (New York: Macmillan, 1934); Garth Jowett, *Film: The Democratic Art* (Boston: Little, Brown, 1976), 220–229.

16. Czitrom, *Media and the American Mind*, 132. See also Harold Lasswell, "The Structure and Function of Communication in Society," in *The Communication of Ideas*, ed. Lyman Bryson (New York: Harper and Brothers, 1948), 37–51.

17. Wilbur Schramm, Jack Lyle, and Edwin Parker, *Television in the Lives of Our Children* (Stanford, Calif.: Stanford University Press, 1961), 1.

18. See Joseph Klapper, *The Effects of Mass Communication* (New York: Free Press, 1960).

19. For an early overview of uses and gratifications, see Jay Blumler and Elihu Katz, *The Uses of Mass Communication* (Beverly Hills, Calif.: Sage, 1974).

20. See George Gerbner, Larry Gross, Nancy Signorielli, Michael Morgan, and Marilyn Jackson-Beeck, "The Demonstration of Power: Violence Profile No. 10," *Journal of Communication* 29, no. 3 (1979): 177–196.

21. Kaiser Family Foundation, *Sex on TV 4* (Menlo Park, Calif.: Henry C. Kaiser Family Foundation, 2005).

22. Robert P. Snow, *Creating Media Culture* (Beverly Hills, Calif.: Sage, 1983), 47.

23. See Maxwell McCombs and Donald Shaw, "The Agenda-Setting Function of Mass Media," *Public Opinion Quarterly* 36, no. 2 (1972): 176–187.

24. Cary Funk and Brian Kennedy, "How Americans See Climate Change in 7 Charts," Pew Research Center, April 21, 2020, www.pewresearch.org/fact-tank/2020/04/21/how-americans-see-climate-change-and-the-environment-in-7-charts.

25. Michael Svoboda, "Media Coverage of Climate Change in 2019 Got Bigger—and Better," Yale Climate Connections, March 10, 2020, https://yaleclimateconnections.org/2020/03/media-coverage-of-climate-change-in-2019-got-bigger-and-better.

26. See Nancy Signorielli and Michael Morgan, *Cultivation Analysis: New Directions in Media Effects Research* (Newbury Park, Calif.: Sage, 1990).

27. John Gastil, *Political Communication and Deliberation* (Beverly Hills, Calif.: Sage, 2008), 60.

28. W. Phillips Davison, "The Third-Person Effect in Communication," *Public Opinion Quarterly* 47, no. 1 (1983): 1–15, https://doi.org/10.1086/268763.

29. Richard Rhodes, "The Media Violence Myth," American Booksellers Foundation for Free Expression, 2000.

30. Robert Lynd, *Knowledge for What? The Place of Social Science in American Culture* (Princeton, N.J.: Princeton University Press, 1939), 120.

31. Czitrom, *Media and the American Mind*, 143; Leo Lowenthal, "Historical Perspectives of Popular Culture," in *Mass Culture: The Popular Arts in America*, ed. Bernard Rosenberg and David White (Glencoe, Ill.: Free Press, 1957), 52.

32. Todd Gitlin, *The Whole World Is Watching* (Berkeley: University of California Press, 1980), 7.

33. Horace Newcomb, *TV: The Most Popular Art* (Garden City, N.Y.: Anchor Books, 1974), 19, 23.

34. See Janice Radway, *Reading the Romance: Women, Patriarchy, and Popular Literature* (Chapel Hill: University of North Carolina Press, 1984).

35. David Dayen, "It's Not a Big Tech Crackdown, It's an Anti-Monopoly Revolution," *American Prospect*, December 18, 2020, https://prospect.org/power/its-not-a-big-tech-crackdown-its-an-anti-monopoly-revolution.

36. Jürgen Habermas, *The Structural Transformation of the Public Sphere*, trans. Thomas Burger with Frederick Lawrence (Cambridge, Mass.: MIT Press, 1994).

37. Craig Calhoun, ed., *Habermas and the Public Sphere* (Cambridge, Mass.: MIT Press, 1994), 452.

38. James W. Carey, *Communication as Culture* (New York: Routledge, 1989), 23.

39. Abstract from Venise Berry, "*Blackish*: Deconstruction and the Changing Nature of Black Identity" (paper, Association for Education in Journalism and Mass Communication 2017 Conference, Chicago, Ill., August 10, 2018), www.aejmc.org/home/2017/06/esig-2017-abstracts.

40. James Carey, "Mass Communication Research and Cultural Studies: An American View," in *Mass Communication and Society*, ed. James Curran, Michael Gurevitch, and Janet Woollacott (London: Edward Arnold, 1977), 418, 421.

41. Yascha Mounk, "What an Audacious Hoax Reveals about Academia," *Atlantic*, October 5, 2018, www.theatlantic.com/ideas/archive/2018/10/new-sokal-hoax/572212.

42. Alexander C. Kafka, "'Sokal Squared': Is Huge Publishing Hoax 'Hilarious and Delightful' or an Ugly Example of Dishonesty and Bad Faith?," *Chronicle of Higher Education*, October 3, 2018, www.chronicle.com/article/Sokal-Squared-Is-Huge/244714.

43. Alan Sokal, quoted in Scott Janny, "Postmodern Gravity Deconstructed, Slyly," *New York Times*, May 18, 1996, p. 1. See also Editors of *Lingua Franca*, eds., *The Sokal Hoax: The Sham That Shook the Academy* (Lincoln, Neb.: Bison Press, 2000).

■ **MEDIA LITERACY AND THE CRITICAL PROCESS**
Wedding Media and the Meaning of the Perfect Wedding Day

1. Erika Engstrom, *The Bride Factory: Mass Media Portrayals of Women and Weddings* (New York: Peter Lang, 2012).

■ **EXAMINING ETHICS** Our Masculinity Problem

1. Bonnie Berkowitz and Chris Alcantara, "The Terrible Numbers That Grow with Each Mass Shooting," *Washington Post*, updated April 20, 2021, www.washingtonpost.com/graphics/2018/national/mass-shootings-in -america/.

2. Jackson Katz, "Memo to Media: Manhood, Not Guns or Mental Illness, Should Be Central in Newtown Shooting," *Huffington Post*, updated February 17, 2013, www.huffingtonpost.com/jackson-katz/men-gender-gun -violence_b_2308522.html.

3. Katz.

4. Rachel Kalish and Michael Kimmel, "Suicide by Mass Murder: Masculinity, Aggrieved Entitlement, and Rampage School Shootings," *Health Sociology Review* 19, no. 4 (2010): 451–464.

5. Dan Barry, Serge F. Kovaleski, Alan Blinder, and Mujib Mashal, "'Always Agitated. Always Mad': Omar Mateen, according to Those Who Knew Him," *New York Times*, June 18, 2016, www.nytimes.com/2016/06/19/us /omar-mateen-gunman-orlando-shooting.html.

6. Kelan Lyons, "Redefining 'Strong': Speaker Jackson Katz Discusses Rewiring Toxic Depictions of Masculinities in Media to Set Tone against Sexual Abuse," *Bryan (TX) Eagle*, October 5, 2017, https://theeagle.com /news/local/speaker-jackson-katz-discusses-rewiring-toxic-depictions -of-masculinities-in-media-to-set-tone-against/article_42a4d6ab-a5f8-5f06 -8501-6a2b58f1fb18.html.

■ **GLOBAL VILLAGE** International Media Research

1. Luwei Rose Luqiu, "The Cost of Humour: Political Satire on Social Media and Censorship in China," *Global Media and Communication* 13, no. 2 (2017): 123–138, https://doi.org/10.1177/1742766517704471.

2. Jade Miller, "Global Nollywood: The Nigerian Movie Industry and Alternative Global Networks in Production and Distribution," *Global Media and Communication* 8, no. 2 (2012): 117–133, https://doi.org /10.1177/1742766512444340.

3. Lucile Davier, "The Paradoxical Invisibility of Translation in the Highly Multilingual Context of News Agencies," *Global Media and Communication* 10, no. 1 (2014): 53–72, https://doi.org/10.1177 /1742766513513196.

4. Ahmed K. Al-Rawi, "Sectarianism and the Arab Spring: Framing the Popular Protests in Bahrain," *Global Media and Communication* 11, no. 1 (2015): 25–42, https://doi.org/10.1177/1742766515573550.

16 Legal Controls and Freedom of Expression

1. Allan J. Lichtman, "Who Rules America?" *The Hill*, August 12, 2014, http://thehill.com/blogs/pundits-blog/civil-rights/214857-who-rules -america.

2. Bradley Jones, "Most Americans Want to Limit Campaign Spending, Say Big Donors Have Greater Political Influence," Pew Research Center, May 8, 2018, www.pewresearch.org/fact-tank/2018/05/08/most-americans -want-to-limit-campaign-spending-say-big-donors-have-greater-political -influence.

3. "2020 Election to Cost $14 Billion, Blowing Away Spending Records," OpenSecrets.org, October 28, 2020, www.opensecrets.org/news/2020/10 /cost-of-2020-election-14billion-update.

4. Gene Policinski, "Amendment to Undo *Citizens United* Won't Do," First Amendment Center, September 21, 2011, www.firstamendmentcenter .org/amendment-to-undo-citizens-united-wont-do.

5. Center for Responsive Politics, "Donor Demographics," OpenSecrets. org, accessed March 20, 2021, www.opensecrets.org/elections-overview /donor-demographics?cycle=2020&display=A.

6. Nat Hentoff, *Living the Bill of Rights: How to Be an Authentic American* (New York: Harper, 1998).

7. Freedom Forum Institute, *The State of the First Amendment: 2019*, June 2019, www.freedomforuminstitute.org/wp-content/uploads/2019/06 /SOFAreport2019.pdf.

8. Twitter, "Our Range of Enforcement Options," accessed March 20, 2021, https://help.twitter.com/en/rules-and-policies/enforcement-options.

9. K. Bell, "Twitter Is 'Reviewing' Its Rules for World Leaders," Engadget, March 18, 2021, www.engadget.com/twitter-review-rules-for-world -leaders-213653810.html.

10. Lauren Feiner, "Facebook, Google and Twitter CEOs Will Make Another Appearance before Congress in March," CNBC, February 18, 2021, www.cnbc.com/2021/02/18/facebook-google-twitter-ceos-to-testify -before-congress-in-march.html; Lauren Feiner, "Facebook and Twitter Defend Election Safeguards and Moderation Practices before the Senate," CNBC, November 17, 2020, www.cnbc.com/2020/11/17/facebook -twitter-defend-election-moderation-practices-before-senate.html.

11. Committee to Protect Journalists, "1400 Journalists Killed," accessed March 28, 2021, https://cpj.org/data/killed.

12. Reporters Without Borders, 2020 World Press Freedom Index, accessed March 20, 2021, https://rsf.org/en/ranking.

13. Fred Siebert, Theodore Peterson, and Wilbur Schramm, *Four Theories of the Press* (Urbana: University of Illinois Press, 1956).

14. See Douglas M. Fraleigh and Joseph S. Tuman, *Freedom of Speech in the Marketplace of Ideas* (New York: St. Martin's Press, 1997), 71–73.

15. Michael Schudson, *The Good Citizen: A History of American Civic Life* (Cambridge, Mass.: Harvard University Press, 1998), 77.

16. Thomas Jefferson, "From Thomas Jefferson to Edward Carrington, 16 January 1787," National Archives, https://founders.archives.gov/documents /Jefferson/01-11-02-0047.

17. Thomas Jefferson, "Image 2 of Thomas Jefferson to John Norvell, June 11, 1807," Library of Congress, www.loc.gov/resource/mtj1.038_0592 _0594/?sp=2&st=text.

18. See Fraleigh and Tuman, *Freedom of Speech*, 125.

19. Hugo Black, quoted in "*New York Times Company v. U.S.*: 1971," in *Great American Trials: From Salem Witchcraft to Rodney King*, ed. Edward W. Knappman (Detroit: Visible Ink Press, 1994), 609.

20. Robert Warren, quoted in "*U.S. v. The Progressive*: 1979," in Knappman, *Great American Trials*, 684.

21. Lawrence Lessig, "Opening Plenary—Media at a Critical Juncture: Politics, Technology and Culture" (presentation, National Conference on Media Reform, Minneapolis, Minn., June 7, 2008).

22. Jennifer Jenkins, "Public Domain Day 2021," Center for the Study of Public Domain, 2021, https://web.law.duke.edu/cspd/publicdomainday /2021.

23. See Knappman, *Great American Trials*, 517–519.

24. Austin Vining and Sarah Matthews, "Introduction to Anti-SLAPP Laws," Reporters Committee for Freedom of the Press, accessed March 21, 2021, www.rcfp.org/introduction-anti-slapp-guide.

25. Reporters Committee for Freedom of the Press, "Understanding Anti-SLAPP Laws," accessed March 21, 2021, www.rcfp.org/resources /anti-slapp-laws/#recentantislappupdates.

26. Douglas Gomery, *Movie History: A Survey* (Belmont, Calif.: Wadsworth, 1991), 57.

27. See the excellent Vox video on this boxing match and its media coverage: "The Boxing Film That Was Banned around the World," Vox, February 21, 2021, www.youtube.com/watch?v=LmiBASu41-A.

28. See the appendix to Thomas Doherty, *Hollywood's Censor: Joseph I. Breen and the Production Code Administration* (New York: Columbia University Press, 2009), 351–363.

29. See Eric Barnouw, *Tube of Plenty: The Evolution of American Television*, rev. ed. (New York: Oxford University Press, 1982), 118–130.

30. FCC, "Obscene, Indecent and Profane Broadcasts," January 31, 2021, www.fcc.gov/consumers/guides/obscene-indecent-and-profane -broadcasts.

31. *Fox Television Stations, Inc., v. FCC*, 613 F.3d 317 (2d Cir. 2010).

32. Brooks Boliek, "Sorry, Ms. Jackson: FCC's New Record," *Politico*, September 10, 2014, www.politico.com/story/2014/09/fcc-net-neutrality -record-110818.html.

33. "47 U.S. Code § 230—Protection for Private Blocking and Screening of Offensive Material," Cornell Law School Legal Information Institute, www.law.cornell.edu/uscode/text/47/230.

34. Casey Newton, "Everything You Need to Know about Section 230," *The Verge*, December 29, 2020, www.theverge.com/21273768/section -230-explained-internet-speech-law-definition-guide-free-moderation.

35. Newton.

EXAMINING ETHICS Is "Sexting" Pornography?

1. Richard Wortley and Stephen Smallbone, *Child Pornography on the Internet*, U.S. Department of Justice, updated May 2012, https://popcenter .asu.edu/sites/default/files/child_pornography_on_the_internet.pdf.

2. Steven Dettelbach, quoted in Tracy Russo, "'Sexting' Town Hall Meeting Held in Cleveland," *Criminal Justice News* (blog), March 19, 2010, http://criminal-justice-online.blogspot.com/2010/03/sexting.html.

3. Sheri Madigan, Anh Ly, Christina L. Rash, Joris Van Ouytsel, and Jeff R. Temple, "Prevalence of Multiple Forms of Sexting Behavior among Youth: A Systematic Review and Meta-analysis," *JAMA Pediatrics* 172, no. 4 (2018): 327–335, https://doi.org/10.1001/jamapediatrics.2017.5314.

4. Cyberbullying Research Center, "Sexting Laws Across America," accessed March 21, 2021, https://cyberbullying.org/sexting-laws.

GLOBAL VILLAGE: The Challenges of Film Censorship in China

1. Clarence Tsui, "China's Film Censorship Paradox: Restricted Content, Unrestricted Access," *South China Morning Post*, March 8, 2017, www .scmp.com/magazines/post-magazine/arts-music/article/2076755 /chinas-film-censorship-paradox-restricted-content.

2. Ilaria Maria Sala, "'No Ghosts. No Gay Love Stories. No Nudity': Tales of Film-Making in China," *Guardian*, September 22, 2016, www.theguardian .com/film/2016/sep/22/tales-of-film-making-in-china-hollywood-hong -kong.

3. Patrick Brzeski, "*Ghostbusters* Denied Release in China," *Hollywood Reporter*, July 13, 2016, www.hollywoodreporter.com/news/ghostbusters -denied-release-china-910563.

4. Richard Hartley-Parkinson, "Great Scott! China Ban Films and TV Shows Featuring Time Travel (Just in Case Anyone Wants to Rewrite History)," *Daily Mail*, April 15, 2011, www.dailymail.co.uk/news/article-1376771 /Great-Scott-China-bans-time-travel-cinema-TV.html.

5. Sala, "'No Ghosts.'"

CASE STUDY: A Guide to Identifying Fake News

1. The approval of the Twelfth Amendment to the U.S. Constitution in 1804 enabled presidential and vice presidential candidates to run as a ticket, thus avoiding the awkward situation of Adams and Jefferson, who were elected president and vice president but were from opposing parties.

2. Nathan Connolly, Joanne Freeman, and Ed Ayers, "Fit to Print? A History of Fake News," *Backstory* (podcast), August 31, 2018, http:// backstoryradio.org/shows/fit-to-print.

3. David Uberti, "The Real History of Fake News," *Columbia Journalism Review*, December 15, 2016, www.cjr.org/special_report/fake_news _history.php.

4. Museum of Hoaxes, "The Great Moon Hoax of 1835/Day Four: Friday, August 28, 1835," http://hoaxes.org/text/display/the_great_moon_hoax_of _1835_text/P3.

5. Matthew Goodman, *The Sun and the Moon: The Remarkable True Account of Hoaxers, Showmen, Dueling Journalists, and Lunar Man-Bats in Nineteenth-Century New York* (New York: Basic Books, 2008).

6. James Gordon Bennett, quoted in Goodman, *Sun and the Moon*, 215.

7. Axel Gelfert, "Fake News: A Definition," *Informal Logic* 38, no. 1 (2018): 93.

8. Gelfert, 108.

9. Sapna Maheshwari, "How Fake News Goes Viral: A Case Study," *New York Times*, November 20, 2016, www.nytimes.com/2016/11/20 /business/media/how-fake-news-spreads.html.

10. James Poniewozik, "Jon Stewart, the Fake Newsman Who Made a Real Difference," *Time*, August 4, 2015, http://time.com/3704321/jon-stewart -daily-show-fake-news.

11. Northrup Frye, "The Nature of Satire," *University of Toronto Quarterly* 14, no. 1 (1944): 75–89.

12. Peter Dreier and Christopher R. Martin, "How ACORN Was Framed: Political Controversy and Media Agenda Setting," *Perspectives on Politics* 8, no. 3 (2010): 763.

13. Daniel Dale, "Fact Check: How a Group of Right-Wingers Spread a Lie That Black Lives Matter Stormed Iowa's Capitol," CNN, April 14, 2021, www.cnn.com/2021/04/13/politics/fact-check-iowa-capitol-black -lives-matter-stormed-lie/index.html.

14. See Kathleen Hall Jamieson and Joseph N. Cappella, *Echo Chamber: Rush Limbaugh and the Conservative Media Establishment* (Oxford: Oxford University Press, 2008); David Brock, *The Republican Noise Machine* (New York: Crown, 2004).

15. Vincent N. Pham, "Our Foreign President Barack Obama: The Racial Logics of Birther Discourses," *Journal of International and Intercultural Communication* 8, no. 2 (2015): 86–107.

16. David Sanger and Nick Corasaninti, "D.N.C. Says Russian Hackers Penetrated Its Files, Including Dossier on Donald Trump," *New York Times*, June 14, 2016, www.nytimes.com/2016/06/15/us/politics /russian-hackers-dnc-trump.html. See also Mark Scott and Melissa Eddy, "Europe Combats a New Foe of Political Stability: Fake News," *New York Times*, February 20, 2017, www.nytimes.com/2017/02/20/world/europe /europe-combats-a-new-foe-of-political-stability-fake-news.html.

17. See "The Revolution Starts at Noon," *This American Life* (podcast), WBEZ, January 17, 2017, www.thisamericanlife.org/radio-archives /episode/608/the-revolution-starts-at-noon; Jesse Singal, "How Internet Trolls Won the 2016 Presidential Election," *New York*, September 16, 2016, http://nymag.com/selectall/2016/09/how-internet-trolls-won -the-2016-presidential-election.html.

18. Cecelia Kang, "Fake News Onslaught Targets Pizzeria as Nest of Child-Trafficking," *New York Times*, November 21, 2016, www.nytimes .com/2016/11/21/technology/fact-check-this-pizzeria-is-not-a-child -trafficking-site.html.

19. Faiz Siddiqui and Susan Svrluga, "N.C. Man Told Police He Went to D.C. Pizzeria with Gun to Investigate Conspiracy Theory," *Washington Post*, December 5, 2016, www.washingtonpost.com/news/local/wp/2016/12/04/d-c-police-respond-to-report-of-a-man-with-a-gun-at-comet-ping-pong-restaurant/?utm_term=.2c36431b4bc6.

20. Bill Kovach and Tom Rosenstiel, *The Elements of Journalism: What Newspeople Should Know and the Public Should Expect* (New York: Three Rivers Press, 2007).

21. Kovach and Rosenstiel, 254.

22. We would like to acknowledge the excellent work of several researchers and projects that have identified many criteria for identifying fake news and misinformation, including the Trust Project, "The 8 Trust Indicators," https://thetrustproject.org/#indicators; Claire Wardle, "Fake News. It's Complicated," First Draft Footnotes, February 16, 2017, https://medium.com/1st-draft/fake-news-its-complicated-d0f773766c79; and the News Literacy Project, https://newslit.org.

Glossary

A&R (artist & repertoire) agents talent scouts of the music business who discover, develop, and sometimes manage artists.

access channels in cable television, a tier of nonbroadcast channels dedicated to local education, government, and the public.

acquisitions editors in the book industry, editors who seek out and sign authors to contracts.

actual malice in libel law, a reckless disregard for the truth, such as when a reporter or an editor knows that a statement is false and prints or airs it anyway.

addictive design principles developed from research into human behavior modification and used by social media companies to increase the time users spend on their platforms and to make checking for updates and messages a habit.

adult contemporary (AC) one of the oldest and most popular radio music formats, typically featuring a mix of news, talk, oldies, and soft rock.

affiliate in television, a TV station that, though independently owned, signs a contract to be part of a network and receives money to air the network's lineup; in exchange, the network reserves time slots, which it sells to national advertisers.

affordances the features or capabilities of a technology that help establish how we use it.

agenda-setting a media-research argument that says that when the mass media focus their attention on particular events or issues, they determine—that is, set the agenda for—the major topics of discussion for individuals and society.

album-oriented rock (AOR) the radio music format that features album cuts from mainstream rock bands.

AM (amplitude modulation) a type of radio and sound transmission that stresses the volume or height of radio waves.

A movies a term used during the studio system era to refer to highly anticipated films made with big budgets and famous stars.

analog a category of media products that encode information in ways that existed before binary code, such as records, which play songs using grooves carved into vinyl, or photographs, which are created when chemical processes convert images onto celluloid.

analog recording a recording that is made by capturing the fluctuations of sound waves and storing those signals in a record's grooves or a tape's continuous stream of magnetized particles—analogous to the actual sound.

analysis the second step in the critical process, it involves discovering and focusing on significant patterns that emerge from the description stage.

anthology dramas a popular form of early TV programming that brought live dramatic theater to television; influenced by stage plays, anthologies featured entirely different characters and settings each week.

arcades establishments that gather together multiple coin-operated games.

ARPAnet the network, designed by the U.S. Defense Department's Advanced Research Projects Agency (ARPA), that enabled researchers to pool computing power and established the cornerstone of what would become the Internet's infrastructure.

artists direct the fastest-growing segment of the sound recording industry; made up of artists who bypass all label representation, including indies, and sell their music directly to their audience, usually through streaming services.

association principle in advertising, a persuasive technique that associates a product with a cultural value or image, even if it has little connection to the actual product.

astroturf lobbying phony grassroots public affairs campaigns engineered by public relations firms.

audience studies cultural studies research that focuses on how people use and interpret cultural content. Also known as *reader-response research.*

audiotape lightweight magnetized strands of ribbon that make possible sound editing and multiple-track mixing; instrumentals or vocals can be recorded at one location and later mixed onto a master recording in another studio.

auteurs directors who have their own distinctive cinema style.

authoritarian model a model for journalism and speech in which the government actively censors the media that critiques its actions, and supports the media that is sympathetic to its agenda and the agenda of the ruling class.

avatar a graphic interactive "character" situated within the world of a game, such as *World of Warcraft.*

bandwagon effect an advertising strategy that uses exaggerated claims that *everyone* is using a particular product to encourage consumers to not be left behind.

basic cable channels in cable programming, a tier of channels composed of local broadcast signals, nonbroadcast access channels (for local government, education, and general public use), a few regional PBS stations, and a variety of cable channels downlinked from communication satellites.

battle royale games multiplayer digital games set in virtual environments in which players battle each other, but the number of players is limited as the objective is to be the last player (or team) surviving; a subset of MOBAs.

Big Five/Little Three from the 1920s through the late 1940s, the major movie studios that were vertically integrated and that dominated the industry; the Big Five were Paramount, MGM, Warner Brothers, Twentieth Century Fox, and RKO, and the Little Three were those studios that did not own many, or any, theaters: Columbia, Universal, and United Artists.

Big Three the broadcast networks—NBC, CBS, and ABC—that established an oligopoly in the television industry during the network era.

block booking an early tactic of movie studios to control exhibition, involving pressuring theater operators to agree to rent B movies in order to gain access to A movies.

blockbuster the type of big-budget special effects film that typically has a summer or holiday release date, heavy promotion, and lucrative merchandising tie-ins.

block printing a printing technique developed by early Chinese printers in which sheets of paper were applied to blocks of inked wood, whose raised surfaces depicted hand-carved letters and illustrations.

blues a type of music emerging from Black spirituals, ballads, and work songs from the rural South.

B movies marginal films with small budgets and unknown actors.

book challenge a formal request to have a book removed from a public or school library's collection.

boutique agencies in advertising, small regional ad agencies that devote their talents to only a handful of select clients.

branded content specialized print, online, or video content produced and funded by individual advertisers.

broadband data transmission over a fiber-optic cable—a signaling method that handles a wide range of frequencies and allows for fast download speeds.

broadcasting the transmission of radio waves or TV signals to a broad public audience.

browsers software applications that help users navigate the web, such as Chrome, Safari, and Firefox.

bulletin board services precursors to websites, where people with a shared interest in topics could post information and become part of a community.

byte the basic unit used to measure digital information.

cable franchise a mini-monopoly awarded by a city or town, usually for a fifteen-year period, to supply local cable service.

CATV (community antenna television) the first cable systems, which originated where mountains or tall buildings blocked TV signals; because of early technical and regulatory limits, CATV contained only twelve channels.

celluloid a transparent and pliable material that can hold a coating of chemicals sensitive to light.

chapter shows in television production, episodic series in which all story lines wrap up each week (for contrast, see **serial programs** and **episodic series**).

cinema verité French term for *truth film,* a documentary style that records fragments of everyday life unobtrusively; it often features a rough, grainy look and shaky, handheld camera work.

citizen journalism a grassroots movement wherein activist amateurs and concerned citizens, not professional journalists, use the Internet and social media to disseminate news and information; also known as *citizen media.*

classical Hollywood cinema a style of moviemaking that solidified during the 1930s and 1940s and was firmly rooted in the specific business practices of the studio system; its approach to narrative, reliance on genre, and adherence to specific production conventions have given Hollywood movies their flavor.

cloning in television production, it involves creating a new series by copying the key features of an innovative and popular program.

codex an early type of book made of sheets of parchment sewn together along the edge, then bound with thin pieces of wood and covered with leather.

collective intelligence the sharing of knowledge and ideas, particularly in the world of gaming.

commercial speech any print or broadcast expression for which a fee is charged to the organizations or individuals buying time or space in the mass media.

communication the creation and use of symbol systems that convey information and meaning (e.g., language, traffic lights, clothing, and photographs).

Communications Act of 1934 the far-reaching act that established the Federal Communications Commission (FCC) and the federal regulatory structure for U.S. broadcasting.

communist or state model a model for journalism and speech in which the government "owns" the press and directly controls what it reports; in this model, leaders believe the press should serve government goals, and ideas that challenge the basic premises of state authority are not tolerated.

compact discs (CDs) playback-only storage discs for music that incorporate pure and very precise digital techniques, thus eliminating noise during recording and editing sessions.

confirmation bias the tendency to socialize with people similar to ourselves as well as to favor information that conforms to our preexisting beliefs over information that challenges us.

conflict of interest considered unethical, any situation in which journalists may stand to benefit personally from stories they produce.

conflict-oriented journalism found in national and metropolitan daily newspapers, journalism in which front-page news is often defined primarily as events, issues, or experiences that deviate from social norms; journalists see their role as observers who monitor their city's institutions and problems.

conglomerates large corporations that develop through a series of mergers with and acquisitions of other companies.

consensus narratives stories that reflect certain values and assumptions about what the world is and should be like; for much of the twentieth century, the impact of the mass media helped establish these stories as well as a mainstream American culture and identity.

consensus-oriented journalism found in small communities, nondaily newspapers that promote social and economic harmony by providing community calendars and meeting notices and carrying articles on local schools, social events, town government, property crimes, and zoning issues.

consoles devices people use specifically to play video games.

contemporary hit radio (CHR) originally called *Top 40 radio*, this radio format encompasses everything from rap to pop-punk songs; it appeals to many teens and young adults.

content analysis in social science research, a systematic approach to categorizing and measuring media content in order to identify and quantify different types of media texts.

content delivery services (CDSs) companies whose business it is to gather and distribute TV content.

continuity editing in film production, the practice of producing and combining shots so that viewers experience a coherent flow of action as it plays out within and across scenes.

convergence the first definition refers to the technological merging of once distinct and incompatible formats into a single format, which can then be accessed through one device. The second definition refers to the trend of media companies merging together in order to better position themselves for a world in which all media can be digital.

cookies computer files that automatically collect and transfer information between a website and a user's browser, which makes revisiting a website easier or more personalized but also makes it possible for companies to track a user's browser history.

copy editors the people in magazine, newspaper, and book publishing who attend to specific problems in writing, such as style, content, and length.

copyright the legal right of authors and producers to own and control the use of their published or unpublished writing, music, lyrics, TV programs, movies, or graphic art designs.

cord cutting the act of consumers cutting the cord to expensive cable subscriptions in favor of a streaming subscription (and maybe free local television) in order to fulfill their viewing needs at a lower cost.

Corporation for Public Broadcasting (CPB) a private, nonprofit corporation created by Congress in 1967 to funnel federal funds to nonprofit radio and public television.

correlations observed associations between two variables.

cosplay a word coined by a Japanese writer that *combines* costume and *play*, cosplay refers to the performance of fans who dress in detailed costumes designed to accurately represent fictional characters of certain texts, such as comic books, novels, television shows, movies, and games.

country one of the strongest formats in the nation, this radio format includes such subdivisions as old-time, progressive, country rock, western swing, and country gospel.

cover music songs recorded or performed by someone other than the original writer or artist; in the 1950s, some white producers and artists capitalized on popular songs by Black artists by "covering" them.

critical process the process whereby a media-literate person or student studying the forms and practices of media communication employs the techniques of description, analysis, interpretation, evaluation, and engagement.

cultivation effect in media research, the idea that heavy television viewing leads individuals to perceive the world in ways that are consistent with television portrayals.

cultural imperialism the phenomenon of American music, movies, and TV shows dominating international markets and shaping the cultures of other nations.

cultural studies in media research, an approach that tries to understand the complex relationship between media texts, the people who consume them, the institutions that produce them, the technologies used to create and distribute them, and the culture within which all the other factors exist.

culture the forms and systems of expression that individuals, groups, and societies use to make sense of daily life, communicate with other people, and articulate their values.

data mining the gathering of data by online purveyors of content and merchandise that sometimes raises ethical issues.

deepfakes images or videos that use advanced digital editing technology to create fraudulent but convincing content.

demographic editions national magazines whose advertising is tailored to subscribers and readers according to occupation, class, and zip code.

description the first step in the critical process, it involves paying close attention, taking notes, and researching the subject under study.

designated market areas (DMAs) in television, market areas that TV stations are assigned to; there are 210 TV markets in the United States.

design managers publishing industry personnel who work on the look of a book, making decisions about type style, paper, cover design, and layout.

developmental editor in book publishing, the editor who provides authors with feedback, makes suggestions for improvements, and, in educational publishing, obtains advice from knowledgeable members of the academic community.

digital communication converts, or encodes, media content into combinations of ones and zeros (binary code) that are then reassembled, or *decoded*, when you play a video game on your console, view a picture on Instagram, download a textbook to your laptop, and more.

digital divide the growing contrast between the "information haves"—those who can afford to purchase a computer and pay for Internet services—and the "information have-nots"—those who may not be able to afford a computer or pay for Internet services.

digital recording a type of recording that translates sound waves into binary on-off pulses and stores that information as numerical codes. When a digital recording is played back, a microprocessor translates those numerical codes back into sounds and sends them to loudspeakers.

digital subchannel broadcast networks new networks that have emerged to provide content for local broadcast stations' multiplex channels.

dime novels sometimes identified as pulp fiction, these cheaply produced and low-priced novels were popular in the United States beginning in the 1860s.

direct broadcast satellite (DBS) a satellite-based service that for a monthly fee transmits TV programming directly to satellite dishes near or on customers' homes.

disinformation false or misleading information spread knowingly by people with malicious intent.

division of labor in the Hollywood studio system, an approach to filmmaking whereby studios broke up the tasks involved in making a movie into distinct jobs that helped studios maximize their resources and push a movie through production efficiently.

drive time in radio programming, the periods between 6 and 9 A.M. and between 4 and 7 P.M., when people are commuting to and from work or school; these periods constitute the largest listening audiences of the day.

earned media publicity or favorable coverage in the news media that a PR person works to secure for a client or an organization.

e-book a digital book read on a computer or a digital reading device.

electromagnetic waves invisible electronic impulses similar to visible light; electricity, magnetism, light, broadcast signals, and heat are part of such waves, which radiate in space at the speed of light, about 186,000 miles per second.

electronic publishing communication businesses, such as broadcasters or cable TV companies, that are entitled to choose what channels or content to carry.

engagement the fifth step in the critical process, it involves working to create a media world that best serves democracy by taking some action that connects our critical perspective with our role as citizens and watchdogs who question our media institutions, thereby adding our voice to the process of shaping the cultural environment.

episodic series in television, a narrative format in which the main characters remain the same from week to week; economic advantages include the fact that sets can be reused, actors sign long-term contracts, and script development is streamlined by placing a recurring cast of characters in new situations.

e-publishing Internet-based publishing houses that design and distribute books for comparatively low prices for authors who want to self-publish a title.

ethnocentrism an underlying value held by many U.S. journalists and citizens, it involves judging other countries and cultures according to how they live up to or imitate American practices and ideals.

evaluation the fourth step in the critical process, it involves arriving at a judgment about whether a cultural product is good, bad, or mediocre; this requires subordinating one's personal taste to the critical "bigger picture" resulting from the first three stages (description, analysis, and interpretation).

evergreen subscriptions magazine subscriptions that automatically renew on the subscriber's credit card.

experiments in regard to the mass media, research that isolates some aspect of content; suggests a hypothesis; and manipulates variables to discover a particular medium's effect on attitude, emotion, or behavior.

Fairness Doctrine repealed in 1987, this FCC rule required broadcast stations to both air and engage in controversial-issue programs that affected their communities and, when offering such programming, to provide competing points of view.

famous-person testimonial an advertising strategy that associates a product with the endorsement of a well-known person.

feature syndicates businesses that contract with newspapers to provide work from the nation's best political writers, editorial cartoonists, comic-strip artists, and self-help columnists; examples include Creators, Andrews McMeel Syndication, and Tribune Content Agency.

Federal Communications Commission (FCC) an independent U.S. government agency charged with regulating interstate and international communications by radio, telephone, television, wire, satellite, cable, and the Internet.

Federal Radio Commission (FRC) a body established in 1927 to oversee radio licenses and negotiate channel problems.

fiber-optic cables thin glass bundles of fiber capable of transmitting along cable wires thousands of messages converted to shooting pulses of light; these bundles of fiber can carry broadcast channels, telephone signals, and a variety of digital codes. They have dramatically increased the amount of data networks can handle and the speed by which data can be transmitted.

filter bubbles spaces where we are exposed only to ideas and opinions that match our existing beliefs.

first-run syndication in television, relates to any program originally produced for sale into syndication markets.

5G the next generation of mobile networking technology that promises download speeds on par with home broadband.

flack a derogatory term that, in journalism, is sometimes applied to a public relations agent.

FM (frequency modulation) a type of radio and sound transmission that offers static-free reception and greater fidelity and clarity than AM radio by accentuating the pitch, or distance, between radio waves.

folk music songs performed by untrained musicians and passed down mainly through oral traditions; it encompasses a wide range of music, from Appalachian fiddle tunes to the accordion-led zydeco of Louisiana.

folk-rock amplified folk music, often featuring politically overt lyrics; influenced by rock and roll.

format radio the concept of radio stations developing and playing specific styles (or formats) geared to listeners' age, race, or gender; in format radio, management, rather than deejays, controls programming choices.

Fourth Estate the notion that the press operates as an unofficial branch of government, monitoring the legislative, judicial, and executive branches for abuses of power.

franchise in television, programs created when producers leverage the name recognition of a popular television show to brand other series (e.g., the *Star Trek* franchise).

fringe time in television, the time slot either immediately before the evening's prime-time schedule (called *early fringe*) or immediately following the local evening news or the network's late-night talk shows (called *late fringe*).

gag orders legal restrictions prohibiting the press from releasing preliminary information that might prejudice jury selection or cause an unfair trial.

game developers in gaming, they write the code to design games.

gameplay the way in which a game's rules, rather than its visual or narrative style, structures how players interact with it.

game publishers the part of the gaming industry that releases games to the public.

gaslighting creating doubt and mistrust in order to undermine the legitimacy of news or information.

gatekeepers newspaper editors, network TV managers, record executives, and other media leaders who have traditionally decided which messages get circulated.

general-interest magazine type of magazine that addresses a wide variety of topics and is aimed at a broad national audience.

genre a narrative category in which conventions regarding similar characters, scenes, structures, and themes recur in combination.

ghost newspapers newspapers devoid of much actual local news that often carry only syndicated content from other newspapers.

GIFs animated images culled from popular culture that are often used in conversations over text or social media.

greenwashing unfair or deceptive acts or practices in environmental marketing.

grunge a subgenre of rock music that takes the spirit of punk and infuses it with more attention to melody.

guilds *or* **clans** in gaming, coordinated, organized team-like groups that can be either small and easygoing or large and demanding.

hegemony a condition that is established when most of the public accepts or buys into a way of thinking about how the world works that favors the dominant class.

herd journalism a situation in which reporters stake out a house; chase celebrities in packs; or follow a story in such herds that the entire profession comes under attack for invading people's privacy, exploiting their personal problems, or just plain getting the story wrong.

hidden-fear appeal an advertising strategy that plays on a sense of insecurity, trying to persuade consumers that only a specific product can offer relief.

high culture a symbolic expression that has come to be identified with "good taste" and higher education and supported by wealthy patrons; it is associated with fine art, which is available primarily in museums, theaters, and concert halls.

hip-hop a term for the Black urban culture that includes *rapping*, *cutting* (or *sampling*) by deejays, break dancing, street clothing, poetry slams, and graffiti art. Hip-hop music combines social politics, swagger, and confrontational lyrics that carry forward long-standing traditions from blues, R&B, soul, and rock and roll.

Hollywood Ten the nine screenwriters and one film director subpoenaed by the House Un-American Activities Committee (HUAC) who were sent to prison in the late 1940s for refusing to disclose their memberships or to identify communist sympathizers.

horizontal integration a corporate strategy that involves owning companies involved in a wide array of media businesses.

HTML (hypertext markup language) a language for displaying text, images, and other multimedia that allows users to link files to one another.

huckster someone who aggressively promotes or sells products of questionable authenticity or value.

human-interest stories news accounts that focus on the daily trials and tribulations of the human condition, often featuring ordinary individuals facing extraordinary challenges.

hypodermic-needle model an early model in mass communication research that attempted to explain media effects by suggesting that the media figuratively shoot their potent effects into unsuspecting victims; sometimes called the *magic bullet theory* or the *direct effects model*.

hypotheses in scientific studies, tentative general statements that predict the influence of an *independent variable* (the probable media cause) on a *dependent variable* (the thing being affected).

ideology a way of thinking—a set of ideas and assumptions—that is shaped by a specific economic and social system. These ideas, in turn, begin to seem natural and inevitable to the people living within that system.

illuminated manuscripts books from the Middle Ages that featured decorative, colorful designs and illustrations on each page.

imagined communities socially constructed groups that people identify with, rather than seeing themselves as individuals with only local or regional identities; created with the rise of national magazines.

indecency an issue related to appropriate broadcast content; the government may punish broadcasters for indecency or profanity after the fact, and over the years a handful of radio stations have had their licenses suspended or denied over indecent programming.

indies independent music and film production houses that work outside industry oligopolies; they often produce less mainstream music and film.

individualism an underlying value held by many U.S. journalists, it favors individual rights and responsibilities above group needs or institutional mandates.

information anarchists actors who want to stir the pot, make people angry with outrageous statements and allegations, and create doubt and mistrust (sometimes called *gaslighting*) in order to undermine the legitimacy of genuine news and create the perception that the truth might never be determined.

in-game advertisements integrated, often subtle advertisements—such as billboards, logos, or storefronts in a game—that can be either static or dynamic.

intellectual property (IP) any product or invention that results from creativity, including movies, music, video games, and other outputs from media industries.

Internet the vast network of fiber-optic lines, wireless connections, and satellite systems that links laptops, mobile phones, tablets, game consoles, smart TVs, and a growing array of smart devices to enormous data centers around the world.

Internet of Things technology that allows a growing array of devices—TVs, tablets, smartphones, cars, refrigerators, thermostats, traffic lights, and more—to communicate with each other and with the Internet.

Internet radio online radio stations that either stream simulcast versions of on-air radio broadcasts over the web or are created exclusively for the Internet.

Internet service providers (ISPs) companies that provide Internet access to homes and businesses for a fee.

interpretation the third step in the critical process, it involves asking and answering "What does that mean?" and "So what?" questions about one's findings.

interpretive journalism a type of journalism that involves analyzing and explaining key issues or events and placing them in a broader historical or social context.

inverted-pyramid style a style of journalism in which news reports begin with the most dramatic or newsworthy

information—answering *who*, *what*, *where*, and *when* (and less frequently *why* or *how*) questions at the top of the story—and then narrow down the information to presumably less significant details.

investigative journalism news reports that hunt out and expose corruption, particularly in business and government.

IP (Internet protocol) address the unique number that every device uses when communicating on the Internet.

irritation advertising an advertising strategy that tries to create product-name recognition by being annoying or obnoxious.

jazz an improvisational and mostly instrumental musical form that absorbs and integrates a diverse body of musical styles, including African rhythms, blues, and gospel.

journalism of assertion coined by journalists Bill Kovach and Tom Rosenstiel, a new era of partisan news, symbolized by the rise of the cable news pundit that serves as a kind of "expert," in place of elements like verified facts, authentic documents, and actual experts.

journalism of verification coined by journalists Bill Kovach and Tom Rosenstiel, a journalistic model that promotes fact-gathering and expertise and views objectivity as the ideal for news practice.

kinescope before the days of videotape, a device that worked by placing a film camera in front of a TV screen to capture a program as it aired; a 1950s technique for preserving television broadcasts.

kinetograph an early movie camera developed by Thomas Edison's assistant in the 1890s.

kinetoscope an early film projection system that served as a kind of peep show in which viewers looked through a hole and saw images moving on a tiny plate.

least objectionable programming (LOP) a strategy aimed at attracting as big a television audience as possible by not turning off any viewers.

libel in law, the defamation of character in written or broadcast form.

libertarian model a model for journalism and speech that encourages vigorous government criticism and supports the highest degree of freedom for individual speech and the press.

limited competition in media economics, a market with many producers and sellers but only a few differentiable products within a particular category; sometimes called *monopolistic competition*.

linear TV broadcast television's and cable's traditional approach to content delivery, in which a show airs at a specific time.

linotype a technology introduced in the nineteenth century that enabled printers to set type mechanically using a typewriter-style keyboard.

literary journalism news reports that adapt fictional storytelling techniques to nonfictional material and in-depth reporting; sometimes called *new journalism*.

Little Three See **Big Five/Little Three**.

lobbying in governmental public relations, the process of attempting to influence lawmakers to support and vote in favor of an organization's or industry's best interests.

longitudinal studies a term used for research studies that are conducted over long periods of time and often rely on large government and academic survey databases.

low culture a symbolic expression supposedly aligned with the "questionable" tastes of the masses, who enjoy the commercial "junk" circulated by the mass media, such as reality television, teen pop music, and violent video games.

low-power FM (LPFM) a new class of noncommercial radio stations approved by the FCC in 2000 to give voice to local groups lacking access to the public airwaves; each 100-watt station broadcasts to a small, community-based area.

magalogs limited-distribution publications that combine glossy magazine style with the sales pitch of retail catalogs; often used to market goods or services to customers or employees.

magazine a collection of articles, stories, and advertisements appearing in a nondaily periodical that is usually published in the smaller tabloid format.

magazine sponsorship in television, an advertising model developed in the late 1950s in which the networks managed the development of programs and sold thirty- or sixty-second ad spots to various sponsors (which appeared throughout the show, like ads in a magazine).

manuscript culture a period during the Middle Ages when priests and monks advanced the art of bookmaking, and books were painstakingly lettered, decorated, and bound by hand.

market research in advertising and public relations agencies, the department that uses social science techniques to assess the behaviors and attitudes of consumers toward particular products before any ads are created.

mass communication the process of designing cultural messages and stories and delivering them to increasingly large and diverse audiences through mass media channels like newspapers, magazines, movies, radio, and television.

mass customization a term used to refer to marketing and production techniques that combine personalization with the cost benefits of efficient mass production.

massively multiplayer online role-playing games (MMORPGs) role-playing games set in virtual fantasy worlds that require users to play through an avatar.

mass media the industries that produce and distribute songs, video games, movies, novels, news, Internet services, and other cultural products to a large number of people.

mass nation a society in which a large percentage of a diverse population goes to the same movies, listens to the same Top 40 hits, watches the same TV shows, and trusts the same evening news anchors.

masspersonal communication a method of communication that mixes and matches aspects of mass and interpersonal communication.

master recording royalties in sound recording, the percentage of the recording profits that the music label pays to the artist/band, as determined by the artist's/band's contract with the label.

mechanical royalties in sound recording, the royalties paid automatically to songwriters and publishers when music is sold or streamed, at a royalty rate of 9.1 cents per song.

media buyers in advertising, the individuals who negotiate rates and place ads with the specific media outlets that media planners determine are best suited to reach the targeted audience, then later measure the effectiveness of those ad placements.

media effects research a major tradition in mass communication research, it attempts to understand, explain, and predict the effects of mass media on individuals and society and relies primarily on tools rooted in the scientific method.

media environment media as the habitat in which we conduct almost every aspect of our daily lives.

media event in public relations, any circumstance created for the sole purpose of gaining coverage in the media.

media literacy an understanding of mass communication developed through the critical process—description, analysis, interpretation, evaluation, and engagement—that enables a person to become more engaged as a citizen and more discerning as a consumer of mass media products.

media planners in advertising, the individuals who analyze the effectiveness of various media channels.

mega-agencies in advertising, large ad firms that are formed when several agencies merge and that maintain worldwide regional offices; they provide both advertising and public relations services and operate in-house radio and TV production studios.

megaplexes movie theater facilities with fourteen or more screens.

microprocessors miniature circuits that process and store electronic signals.

mini-majors independent film studios that have modest market share.

minimal-effects model a mass communication research model based on tightly controlled experiments and surveys; it argues that people generally engage in selective exposure and selective retention with regard to the media; also known as the *limited model*.

misinformation false or misleading information spread by people who assume it's true.

modding the most advanced form of collective intelligence; slang for modifying game software or hardware.

modern era the term describing a historical era spanning from roughly the 1800s to the 1950s; coincided with the rise of mass communication industries, which were bound up with the era's faith in expertise, rationalism, and progress.

monopoly in media economics, an organizational structure that occurs when a single firm dominates production and distribution in a particular industry, either nationally or locally.

Morse code developed by American inventor Samuel Morse, a series of dots and dashes standing for letters of the alphabet sent via electrical impulses from a transmitter through a cable to a reception point.

movable type individual characters made from reusable pieces of wood or metal that printers arranged into various word combinations to speed up the time it took to create block pages; invented in China around the year 1000.

movie palaces during the studio system era, full-time single-screen movie theaters that offered a more hospitable moviegoing environment than nickelodeons, providing elegant décor usually reserved for high-society opera, ballet, symphony, and live theater.

MP3 an advanced type of audio compression that reduces file size, enabling audio to be uploaded to or downloaded from the Internet in a short amount of time.

muckrakers reporters who used a style of early-twentieth-century investigative journalism that emphasized a willingness to crawl around in society's muck to uncover a story.

multiplayer online battle arenas (MOBAs) virtual game environments in which players team up to destroy an opposing team's structure; can have hundreds of thousands of active players worldwide.

multiple-system operators (MSOs) large corporations that own numerous cable television systems.

multiplexing a process used by broadcast stations to transmit multiple digital channels via the single channel or frequency allocated to them by the FCC.

must-carry rules rules established by the FCC requiring all cable operators to include all local TV broadcasts on their system, thereby ensuring that local network affiliates, independent stations not carrying network programs, and public television channels would benefit from cable's clearer reception.

myth analysis a strategy for critiquing advertising that provides insights into how ads work on a cultural level; according to this strategy, ads are narratives with stories to tell and social conflicts to resolve.

narrative the structure underlying most media products, it includes two components: the story (what happens to whom) and the discourse (how the story is told).

narrative films movies that tell a story, with dramatic acting, clear storytelling, and engaging special effects.

narrowcasting any specialized electronic programming or media channel aimed at a target audience; occurred in television during the post-network era.

National Public Radio (NPR) noncommercial radio established in 1967 by the U.S. Congress to provide an alternative to commercial radio.

net neutrality the principle that all data being sent across the Internet should be treated the same—that is, it should have the same access to the network and travel across it at the same speed.

network a linked group of broadcast stations that share programming produced at a central location.

new Big Five the five major studios that currently rule the Hollywood commercial film business: Disney (which now owns the former 21st Century Fox), Warner Brothers, Universal, Sony Pictures, and Paramount.

news the process of gathering information and making narrative reports—edited by journalists—that offer select frames of reference and help the public make sense of important events, political issues, cultural trends, prominent people, and unusual happenings in everyday life.

news desert a term for counties across the United States that have no newspaper.

newshole the space left over in a newspaper for news content after all the ads are placed.

newspaper chain a large company that owns several papers throughout the country.

news/talk the fastest-growing radio format throughout much of the 1990s, dominated by news programs and talk shows.

newsworthiness information most worthy of transformation into news stories; the criteria that journalists use to select and develop news stories, including timeliness, proximity, conflict, prominence, human interest, consequence, usefulness, novelty, and deviance.

niche nation a society in which people navigate a more varied and complex media landscape. Rather than providing widely shared touchstones and consensus narratives, media technologies and media products sort us into narrow niches or subcultures, connecting us to some people but disconnecting us from others.

nickelodeons in the early 1900s, a form of movie theater whose name combines the admission price with the Greek word for "theater"; these makeshift movie theaters were often converted storefronts redecorated to mimic vaudeville theaters.

O & Os TV stations "owned and operated" by networks.

objective journalism a modern style of journalism that distinguishes factual reports from opinion columns; reporters strive to remain neutral toward the issue or event they cover, searching out competing points of view among the sources for a story.

obscenity expression that is not protected as speech if these three legal tests are all met: (1) the average person, applying contemporary community standards, would find that the material as a whole appeals to prurient interest; (2) the material depicts or describes sexual conduct in a patently offensive way; (3) the material, as a whole, lacks serious literary, artistic, political, or scientific value.

offset lithography a technology that enabled books to be printed from photographic plates rather than metal casts, reducing the cost of color and illustrations and accelerating book production.

oligopoly a business situation in which a few firms control most of an industry's production and distribution resources.

on-demand TV television programs that are made available to viewers to watch whenever they want.

online fantasy sports games in which players assemble teams and use actual sports results to determine scores in their online games. These games reach a mass audience, have a major social component, and take a managerial perspective on the game.

open-source software noncommercial software shared freely and developed collectively on the Internet.

opinion and fair comment a defense against libel that states that libel applies only to intentional misstatements of factual information rather than to statements of opinion.

opinion entrepreneurs media outlets—from websites and talk radio to newspapers and cable news—that seek to influence the news and public agenda, often with false or inaccurate stories.

opt-in policies policies that require websites to gain explicit permission from consumers before the sites can collect browsing history data.

option time a business tactic, now illegal, whereby a radio network in the 1920s and 1930s paid an affiliate station a set fee per hour for an option to control programming and advertising on that station.

over-the-top (OTT) a category of content delivery service that allows viewers to access TV content through an Internet connection without having to go through broadcast, cable, or satellite providers; examples include streaming sites like Hulu and Netflix.

Pacifica Foundation a radio broadcasting foundation established by radio reporter and World War II pacifist Lewis Kimball Hill to run experimental public stations.

paid media paid advertisements that PR agents use to help focus a complex issue or a client's image.

paperback books books made with relatively cheap paper covers, introduced in the United States (from Europe) in the 1830s.

papyrus one of the first substances to hold written language and symbols; produced from plant reeds found along the Nile River.

Paramount decision the 1948 U.S. Supreme Court decision that ended vertical integration in the film industry by forcing the studios to divest themselves of their theaters.

parchment treated animal skin that replaced papyrus as an early pre-paper substance on which to document written language.

participatory culture a culture in which it is relatively easy for people to create and share their own content and build connections with others that often reflect and deepen the dynamics of a niche nation.

partisan press an early dominant style of American journalism distinguished by opinion newspapers, which generally pushed the strategy of the particular political group that subsidized the paper.

pass-along readership the total number of people who come into contact with a single copy of a magazine.

payola the unethical (but not always illegal) practice of record promoters' paying deejays or radio programmers to play particular records.

peak TV an unprecedented surge in the production of original dramas, comedies, and limited series that occurred in the late 2010s, as competition for viewers' attention intensified among broadcast networks, cable channels, and streaming services.

penny arcade the first thoroughly modern indoor playground, filled with coin-operated games.

penny papers newspapers that, because of technological innovations in printing, were able to drop their price to one cent beginning in the 1830s, thereby making papers affordable to the working and emerging middle classes and enabling newspapers to become a genuine mass medium.

photojournalism the use of photos to document the rhythms of daily life.

pinball machine the most prominent mechanical game, in which players score points by manipulating the path of a metal ball on a slanted table sealed within a glass-covered case.

place shifting the practice of accessing stored media from different locations.

plain-folks pitch an advertising strategy that associates a product with simplicity.

podcasts spoken-word digital audio programs that can be downloaded or streamed.

political advertising the use of ad techniques to promote a candidate's image and persuade the public to adopt a particular viewpoint.

political economy studies an area of academic study that specifically examines interconnections among economic interests, political power, and how that power is used.

politics in the context of the cultural approach and our role as media citizens, refers to the process by which power, resources, status, and visibility get distributed in a society—usually unequally.

pop music popular music that appeals either to a wide cross section of the public or to sizable subcultures within the larger public.

populism a political approach that pits ordinary people against educated elites and the wealthy; nurtures a variety of movements on both the right and the left.

postmodern era the term describing a contemporary historical era spanning from roughly the 1950s to the present; a period marked by growing skepticism about expertise and the idea of progress.

predictive marketing a new marketing strategy that uses detailed data compiled by digital ad agencies or data-mining companies to target customers with ads today and predict what they might want to purchase down the road.

premium channels in cable programming, a tier of channels that subscribers can order at an additional monthly fee over their basic cable service; these provide access to recent Hollywood movies and original series, often without advertising.

press agents the earliest public relations practitioners, who sought to advance their clients' images through media exposure.

press releases in public relations, announcements written in the style of news reports that present new information about an individual, a company, or an organization, and pitch a story idea to the news media.

prime time in television programming, the hours between 8 and 11 P.M. (or 7 and 10 P.M. in the Midwest), when networks have traditionally drawn their largest audiences and charged their highest advertising rates.

printing press a fifteenth-century invention whose movable metallic type spawned modern mass communication by creating the first method for mass production; it not only reduced the size and cost of books—making them the first mass medium affordable to less affluent people—but provided the impetus for the Industrial Revolution, assembly-line production, modern capitalism, and the rise of consumer culture.

prior restraint the legal definition of censorship in the United States; it prohibits courts and governments from blocking any publication or speech before it actually occurs.

product differentiation offering products that are distinct from those of competitors and that target specific consumers.

product placement the paid appearance of particular goods in a narrative or scene in films, TV shows, music videos, and digital games.

product standardization offering generally uniform products as a way to make production efficient and reduce unpredictability for consumers.

professional books technical books that target various occupational groups and are not intended for the general consumer market.

progressive rock an alternative music format that developed as a backlash to the popularity of Top 40.

propaganda communication strategically placed, either as advertising or as publicity, to gain public support for a special issue, program, or policy.

propaganda analysis the study of propaganda's effectiveness in influencing and mobilizing public opinion.

propagandists official state actors who spread coordinated partisan messages meant to propagate a point of view.

protocols rules established by ARPA that allowed early supercomputers to join the ARPA network, or ARPAnet, and speak to one another.

pseudo-polls typically call-in, online, or person-in-the-street nonscientific polls that the news media use to address a "question of the day."

Public Broadcasting Act of 1967 the act by the U.S. Congress that established the Corporation for Public Broadcasting, which oversees the Public Broadcasting Service (PBS) and National Public Radio (NPR).

Public Broadcasting Service (PBS) noncommercial television established in 1967 by the U.S. Congress to provide an alternative to commercial television.

public domain the end of the copyright period for a work, at which point the public may begin to access it for free.

publicity in public relations, a type of communication that uses various media messages to spread information about a person, a corporation, an issue, or a policy.

public performance royalties in sound recording, the royalties paid out to songwriters and their publishers when music (live or recorded) is played publicly on radio; on television (e.g., a performance on *Saturday Night Live*); streaming; or in stores, bars, restaurants, and even stadiums and arenas.

public relations defined by the Public Relations Society of America (PRSA) as "a strategic communication process that builds mutually beneficial relationships between organizations and their publics."

public service announcements (PSAs) fifteen- to sixty-second audio or video reports, carried free by broadcasters, that promote government programs, educational projects, volunteer agencies, or social reform.

public sphere a space for critical public debate; the areas or arenas in social life—like coffeehouses and meeting halls—where people come together to engage in public debate and open communication.

PUGs in gaming, temporary teams usually assembled by matchmaking programs integrated into a game (short for "pick-up groups").

pulp fiction a term used to describe many late-nineteenth-century popular paperbacks and dime novels, which were constructed of cheap machine-made pulp paper.

punk rock rock music that challenges the orthodoxy and commercialism of the record business; it is characterized by simple chord structures, catchy melodies, and politically or socially challenging lyrics.

qualified privilege a legal right allowing journalists to report judicial or legislative proceedings even though the public statements being reported may be libelous.

quality audience the segment of television viewers advertisers pay the most money to reach.

Radio Act of 1912 act that attempted to address the problem of amateur radio operators cramming the airwaves.

Radio Act of 1927 in an attempt to restore order to the airwaves, an act that stated that licensees could operate on their assigned channels only as long as they operated to serve the "public interest, convenience, or necessity."

Radio Corporation of America (RCA) a company developed during World War I that was designed, with government approval, to pool radio patents; the formation of RCA gave the United States almost total control over the emerging mass medium of broadcasting.

radio waves a portion of the electromagnetic wave spectrum that was harnessed so that signals could be sent from a transmission point to a reception point.

random assignment a social science research method for assigning research subjects; it ensures that every subject has an equal chance of being placed in either the experimental group or the control group.

rating in TV audience measurement, a statistical estimate expressed as a percentage of TV households that watched a particular program (for contrast, see **share**).

read-only an early state of the web, in which websites were places people went to view information.

read-write when the web became interactive, a place where users could read information, contribute their own digital content, and directly engage with other users.

regional editions national magazines whose content is tailored to the interests of different geographic areas.

remix culture a term used to describe a society in which people are able to create and communicate by mixing, editing, combining, manipulating, or repurposing existing texts.

responsible capitalism an underlying value held by many U.S. journalists, it assumes that businesspeople should compete with one another not primarily to maximize profits but to create prosperity for all.

rhythm and blues (R&B) music that merges blues with big-band sounds.

right of fair use a legal doctrine that permits people to use copyrighted material without permission as long as the use does not compromise the material's value.

right to privacy addresses a person's right to be left alone, without the person's name, image, or daily activities becoming public property.

rockabilly music that mixes country (or hillbilly) influences, southern gospel, and Mississippi delta blues.

rock and roll music that merged the Black sounds of rhythm and blues, gospel, and Robert Johnson's screeching blues guitar with the white influences of country, folk, and pop vocals.

rotation in format radio programming, the practice of playing the top songs many times during the day.

satellite radio subscription-based radio service that delivers various radio formats nationally via satellite.

satire a technique that uses humor and detailed research to critique the news media and the political system.

saturation advertising the strategy of inundating a variety of media with ads aimed at target audiences.

scientific method a widely used research method that studies phenomena in systematic stages; it includes identifying a research problem, reviewing existing research, developing working hypotheses, determining an appropriate research design, collecting information, analyzing results to see if the hypotheses have been verified, and interpreting the implications of the study.

search engines sites or applications that offer an automated way to find content by allowing users to enter key words or queries to locate related web pages.

second-run syndication when older programs originally developed for and run on a specific broadcast network, cable channel, or streaming service are later made available to show on other content delivery services in the United States and overseas; commonly called *reruns*.

Section 230 part of the 1996 Communications Decency Act; it protects any company operating an "interactive computer service" from liability for anything published on their service by a third party.

Section 315 part of the 1934 Communications Act; it mandates that during elections, broadcast stations must provide equal opportunities and response time for qualified political candidates.

selective exposure the phenomenon whereby people expose themselves to the media messages that are most familiar to them.

selective retention the phenomenon whereby people retain information that confirms the values and attitudes they already hold.

Semantic Web version of the web in which web pages and databases are created in such a way that a computer—functioning as something akin to artificial intelligence—can examine the web's vast quantities of data and automatically provide useful solutions to people's needs.

serial programs episodic series in which story lines continue across episodes, such as soap operas (for contrast, see **chapter shows**).

share in TV audience measurement, a statistical estimate of the percentage of TV households who were watching TV who tuned in to a particular program (for contrast, see **rating**).

shield laws laws protecting the confidentiality of key interview subjects and reporters' rights not to reveal the sources of controversial information used in news stories.

single sponsorship in television and radio, a model of advertising popular during the 1930s and 1940s in which early programs were often developed, produced, and supported by a single sponsor.

slander in law, spoken language that defames a person's character.

slogan in advertising, a phrase that attempts to sell a product by capturing its essence in words.

small-town pastoralism an underlying value held by many U.S. journalists, it favors the small over the large and the rural over the urban.

snob-appeal approach an advertising strategy that attempts to persuade consumers that using a product will enable them to maintain or elevate their social status.

social learning theory a theory within media effects research that suggests a link between the mass media and behavior; later modified and renamed *social cognitive theory*.

social responsibility model a model for journalism and speech in which the press functions as a Fourth Estate, monitoring the three branches of government for abuses of power and providing information necessary for self-governance.

soul music that mixes gospel and blues with emotion and lyrics drawn from the American Black experience.

space brokers in the days before modern advertising, individuals who purchased space in newspapers and sold it to various merchants.

spam a computer term referring to unsolicited e-mail.

spin-off a type of television program in which the lead character was originally on another television series.

spiral of silence a theory that links the mass media, social psychology, and the formation of public opinion; the theory says that people who hold minority views on controversial issues tend to keep their views silent.

split-run editions editions of national magazines that include a few pages of ads purchased by local or regional companies.

stakeholders in public relations, people who have an interest in what is happening regarding a particular issue, event, or occurrence.

star system in the studio system, a method of cultivating and exploiting the allure of certain actors.

stereo the recording of two separate channels, or tracks, of sound, which enables more natural sound distribution.

storyboard in advertising, a blueprint or roughly drawn comic-strip version of a proposed advertisement.

studio system an arrangement in which five powerful movie studios took control of multiple aspects of the film industry, transforming an industry that had previously included many small competitive firms into an oligopoly during the 1920s.

subliminal advertising a 1950s term that refers to hidden or disguised print and visual messages that allegedly register in the subconscious and fool people into buying products.

subsidiary rights in the book industry, selling the rights to a book for use in other media, such as the basis for a screenplay.

supermarket tabloids newspapers that feature bizarre human-interest stories, gruesome murder tales, violent accident accounts, unexplained phenomena stories, and malicious celebrity gossip.

surveillance capitalism an increasingly important business model that involves making money by controlling the personal data of millions of users.

surveillance states societies in which governments conduct systematic mass surveillance on their populations.

survey research in social science research, a method of collecting and measuring data taken from a group of respondents.

synchronization royalties in sound recording, royalties earned when music is used in a film, TV show, or commercial.

syndication the process by which producers lease the right to air a program to local TV stations, cable channels, and streaming services.

synergy in media economics, opportunities to generate profits that come from interaction and cooperation among a conglomerate's different subsidiaries.

talkies movies with sound, beginning in 1927.

technological determinism a common but sometimes simplistic way of thinking that sees technology as an independent force that appears out of nowhere and changes everything.

Telecommunications Act of 1996 the sweeping update of telecommunications law that led to a wave of media consolidation.

telegraph invented in the 1840s, it sent electrical impulses through a cable from a transmitter to a reception point, transmitting Morse code.

tentpoles films that studios bank on being hits in order to offset losses on riskier films.

text anything that conveys meaning or communicates information; anything people "read" or interpret.

textbooks books made for the el-hi (elementary through high school) and college markets.

textual analysis in media research, a method for closely and critically examining and interpreting the meanings of culture, including architecture, fashion, books, movies, and TV programs.

theatrical exclusivity the window during which studios traditionally agree to make their films available only in theaters in order to drive ticket sales.

third-person effect the theory that people believe others are more affected by media messages than they are themselves.

time-shift the process whereby television viewers record programs to watch at their convenience.

Top 40 format the first radio format, in which stations played the forty most popular hits in a given week, as measured by record sales.

trade books hardbound and paperback books aimed at general readers and sold at commercial retail outlets; one of the most lucrative segments of the book industry.

transistors invented by Bell Laboratories in 1947, these small electrical devices receive and amplify radio signals, making portable radios possible.

underground press radical newspapers, run on shoestring budgets, that question mainstream political policies and conventional values; the term usually refers to a journalism movement of the 1960s.

universal access the notion that every citizen, regardless of income or location, should have the opportunity to use and benefit from a technology.

university press the segment of the book industry that publishes scholarly works for small groups of readers in intellectually specialized areas.

urban contemporary one of radio's more popular formats, targeting a wide variety of Black listeners, primarily in large cities.

uses and gratifications model a mass communication research model, usually employing in-depth interviews to supplement survey questionnaires, that argues that people use the media to satisfy various emotional desires or intellectual needs.

Values and Lifestyles (VALS) a market-research strategy that divides consumers into types and measures psychological factors, including how consumers think and feel about products and how they achieve (or do not achieve) the lifestyles to which they aspire.

vellum a handmade paper made from fine calfskin-based parchment; used to print the Gutenberg Bible.

vertical integration in media economics, the phenomenon of controlling a mass media industry at its three essential levels: production, distribution, and exhibition; the term is frequently used in reference to the film industry during the studio system era.

verticals in the newspaper industry, niche-interest digital sites that bundle together related content (such as technology stories).

video news releases (VNRs) in public relations, thirty- to ninety-second visual press releases designed to mimic the style of a broadcast news report.

video-on-demand (VOD) the term for any streaming video service in which viewers can select from that service's many offerings whenever they choose.

viral marketing short videos or other content that marketers hope will quickly gain widespread attention as users share it with friends online or by word of mouth.

vitascope a large-screen movie projection system developed by Thomas Edison.

voice-tracking a practice in which deejays digitally record their "breaks" between songs in advance so that they don't have to appear live for their shift.

walled gardens highly managed digital app environments that provide continuous streams of content directly to our feeds in an effort to discourage us from leaving to browse elsewhere; examples include Instagram, Facebook, and Pinterest.

wikis open and collaborative websites where people work together to edit and create content; the best-known example is Wikipedia.

Wireless Ship Act of 1910 legislation passed by Congress requiring all major U.S. seagoing ships carrying more than fifty passengers and traveling more than two hundred miles off the coast to be equipped with wireless equipment with a one-hundred-mile range.

wireless telegraphy the forerunner of radio, it is a form of voiceless point-to-point communication.

wireless telephony early experiments in wireless voice and music transmissions, which later developed into modern radio.

wire services commercial organizations, such as the Associated Press, that first relayed news stories and information around the country and the world using telegraph lines and later, using radio waves and digital transmissions.

World Wide Web a data-linking system for organizing and standardizing information on the Internet; "the web" enables computer-accessed information to associate with—or link to—other information, no matter where it is on the Internet.

yellow journalism a newspaper style or era that peaked in the 1890s, it emphasized sensational or overly dramatic stories and focused on early in-depth "detective" stories. Reporting during this period increasingly became a crusading force for common people, with the press assuming a watchdog role on their behalf.

zines a term used to describe self-published magazines.

Index

See media in action on LaunchPad

launchpadworks.com

Throughout this book, callouts direct you to **LaunchPad for *Media & Culture*,** where videos complement the material in the text. Below is a list of videos featured in the book and on LaunchPad, all sorted by chapter. For directions on how to access these videos online, please see the instructions on the next page.